Legend:
- s Block elements
- d Block elements
- p Block elements
- f Block elements

10	11	12	13	14	15	16	17	18
								4.002602
								2 He
			10.81 3 2.0 5 B	12.011 4, 2, −4 2.6 6 C	14.007 5, 4, 3, 2, −3 3.0 7 N	15.999 −2, −1 3.4 8 O	18.99840316 −1 4.0 9 F	20.1797 10 Ne
			26.9815385 3 1.6 13 Al	28.085 4, −4 1.9 14 Si	30.97376200 5, 3, −3 2.2 15 P	32.06 6, 4, 2, −2 2.6 16 S	35.45 7, 5, 3, 1, −1 3.2 17 Cl	39.948 18 Ar
58.6934 3, 2, 0 1.9 28 Ni	63.546 2, 1 1.9 29 Cu	65.38 2 1.7 30 Zn	69.723 3 1.8 31 Ga	72.630 4 2.0 32 Ge	74.921595 5, 3, −3 2.2 33 As	78.971 6, 4, −2 2.6 34 Se	79.904 7, 5, 3, 1, −1 3.0 35 Br	83.798 2 3.0 36 Kr
106.42 4, 2, 0 2.2 46 Pd	107.8682 2, 1 1.9 47 Ag	112.414 2 1.7 48 Cd	114.818 3 1.8 49 In	118.710 4, 2 2.0 50 Sn	121.760 5, 3, −3 2.1 51 Sb	127.60 6, 4, −2 2.1 52 Te	126.90447 7, 5, 1, −1 2.7 53 I	131.293 8, 6, 4, 2 2.6 54 Xe
195.084 4, 2, 0 2.3 78 Pt	196.966569 3, 1 2.5 79 Au	200.592 2, 1 2.0 80 Hg	204.38 3, 1 2.0 81 Tl	207.2 4, 2 2.3 82 Pb	208.98040 5, 3 2.0 83 Bi	208.9824* 6, 4, 2 2.0 84 Po	209.9871* 7, 5, 3, 1, −1 85 At	222.0176* 2 86 Rn
281.17* 110 Ds	282.17* 111 Rg	285.18* 112 Cn	285.18* 113 Uut	289.19* 114 Fl	289.19* 115 Uup	293.2* 116 Lv	294.21* 117 Uus	294.21* 118 Uuo

157.25 3 1.2 64 Gd	158.92535 4, 3 1.2 65 Tb	162.500 3 1.2 66 Dy	164.93033 3 1.2 67 Ho	167.259 3 1.2 68 Er	168.93422 3, 2 1.3 69 Tm	173.054 3, 2 70 Yb	174.9668 3 1.0 71 Lu

247.0704* 4, 3 96 Cm	247.0703* 4, 3 97 Bk	251.0796* 4, 3 98 Cf	252.083* 3 99 Es	257.0951* 3 100 Fm	258.0984* 3 101 Md	259.101* 3, 2 102 No	262.11* 3 103 Lr

ORGANC CHEMISTRY

About the Authors

K. PETER C. VOLLHARDT was born in Madr̶sed in Buenos Aires and Munich, studied at the University of Munich, got his P̶with Professor Peter Garratt at the University College, London, and was a pos̶oral fellow with Professor Bob Bergman (then) at the California Institute of Te̶ogy. He moved to Berkeley in 1974 when he began his efforts toward the dev̶ment of organocobalt reagents in organic synthesis, the preparation of theoretic̶interesting hydrocarbons, the assembly of novel transition metal arrays with p̶tial in catalysis, and the discovery of a parking space. Among other pleasant ex̶nces, he was a Studienstiftler, Adolf Windaus medalist, Humboldt Senior Scienti̶CS Organometallic Awardee, Otto Bayer Prize Awardee, A. C. Cope Scholar, Ja̶Society for the Promotion of Science Prize Holder, and recipient of the Medal̶ ̶e University Aix-Marseille and an Honorary Doctorate from The University o̶me Tor Vergata. He is the current Chief Editor of̶ ̶*lett*. Among his more than 350 publications, he tre̶es especially this textbook in organic chemistry, t̶lated into 13 languages. Peter is married to Marie̶é Sat, a French artist, and they have two children, ̶ma (b. 1994) and Julien (b. 1997), whose picture y̶can admire on p. 168.

NEIL E. SCHORE was b̶in Newark, New Jersey, in 1948. His education to̶him through the public school of the Bronx, New ̶k, and Ridgefield, New Jersey, after which he compl̶d a B.A. with honors in chemistry at the University o̶hennsylvania in 1969. Moving back to New York, ̶worked with the late Professor Nicholas J. Turro ̶Columbia University, studying photochemical and ̶ptophysical processes of organic compounds for his Ph.D. thesis. He first met Peter̶ollhardt when he and Peter were doing postdoctoral work in Professor Robert Berg̶n's laboratory at Cal Tech in the 1970s. Since joining the U.C. Davis faculty in 1976̶e has taught organic chemistry to more than 15,000 nonchemistry majors, winning s̶en teaching awards, publishing over 100 papers in various areas related to organic c̶mistry, and refereeing several hundred local youth soccer games. Neil is married̶t Carrie Erickson, a microbiologist at the U.C. Davis School of Veterinary Medic̶e. They have two children, Michael (b. 1981) and Stefanie (b. 1983), both of whom̶carried out experiments for this book.

ORGANIC CHEMISTRY
Structure and Function

SEVENTH EDITION

PETER VOLLHARDT
University of California at Berkeley

NEIL SCHORE
University of California at Davis

W.H. Freeman and Company
A Macmillan Higher Education Company

Publisher: Jessica Fiorillo
Acquisitions Editor: Bill Minick
Development Editor: Randi Blatt Rossignol
Marketing Manager: Debbie Clare
Media and Supplements Editor: Dave Quinn
Assistant Editor: Nick Ciani
Photo Editor: Robin Fadool
Photo Assistant: Eileen Liang
Photo Researcher: Dena Digilio Betz
Cover Designer: Blake Logan
Text Designer: Patrice Sheridan
Project Editing and Composition: Aptara®, Inc.
Illustrations: Network Graphics; Precision Graphics
Illustration Coordinator: Dennis Free at Aptara®, Inc.
Production Coordinator: Susan Wein
Printing and Binding: RR Donnelley

Library of Congress Control Number: 2013948560
ISBN-13: 978-1-4641-2027-5
ISBN-10: 1-4641-2027-7

Printed in the United States of America

First printing

W. H. Freeman and Company
41 Madison Avenue
New York, NY 10010
Houndmills, Basingstoke RG21 6XS, England

www.whfreeman.com

BRIEF CONTENTS

CONTENTS

PREFACE

A User's Guide to ORGANIC CHEMISTRY: Structure and Function

In this textbook, *Organic Chemistry: Structure and Function,* we present a logical framework for understanding contemporary organic chemistry. This framework emphasizes that the structure of an organic molecule determines how that molecule functions, be it with respect to its physical behavior or in a chemical reaction. In the seventh edition, we have strengthened the themes of understanding reactivity, mechanisms, and synthetic analysis to apply chemical concepts to realistic situations. We have incorporated new applications of organic chemistry in the life and material sciences. In particular, we have introduced some of the fundamentals of medicinal chemistry in over 70 new entries describing drug design, absorption, metabolism, mode of action, and medicinal terminology. We have expanded on improving students' ability to grasp concepts in a number of sections ("Keys to Success") and on their problem-solving skills by presenting step-by-step guides in Worked Examples. These and other innovations are illustrated in the following pages. *Organic Chemistry: Structure and Function* is offered in an online version to give students cost-effective access to all content from the text plus all student media resources. For more information, please visit our Web site at http://ebooks.bfwpub.com.

CONNECTING STRUCTURE AND FUNCTION

This textbook emphasizes that the structure of an organic molecule determines how that molecule functions. By understanding the connection between structure and function, we can learn to solve practical problems in organic chemistry.

Chapters 1 through 5 lay the foundation for making this connection. In particular, Chapter 1 shows how electronegativity is the basis for polar bond formation, setting the stage for an understanding of polar reactivity. Chapter 2 makes an early connection between acidity and electrophilicity, as well as their respective counterparts, basicity-nucleophilicity. Chapter 3 relates the structure of radicals to their relative stability and reactivity. Chapter 4 illustrates how ring size affects the properties of cyclic systems, and Chapter 5 provides an early introduction to stereochemistry. The structures of haloalkanes and how they determine haloalkane behavior in nucleophilic substitution and elimination reactions are the main topics of Chapters 6 and 7. Subsequent chapters present material on functional-group compounds according to the same scheme introduced for haloalkanes: nomenclature, structure, spectroscopy, preparations, reactions, and biological and other applications. The emphasis on structure and function allows us to discuss the mechanisms of all new important reactions concurrently, rather than scattered throughout the text. We believe this unified presentation of mechanisms benefits students by teaching them how to approach understanding reactions rather than memorizing them.

id shortening from emarkably, the *only* ifference is that the double bonds and ils are derivatives of ompounds containing and in Chapter 12, ties, generation, and

hapters, we learned wo major classes of -bonded functional on under appropriate this chapter we return re some additional ome. We shall then mine the reactions of r that they may be converted back into single-bonded sub-ion. Thus, we shall see how alkenes can serve as interme-/ersions. They are useful and economically valuable starting c fibers, construction materials, and many other industrially mple, addition reactions of many gaseous alkenes give oils as lass of compounds used to be called "olefins" (from *oleum* ed, "margarine" is a shortened version of the original name,

cis-9-Octadecenoic acid, also known as *oleic acid*, makes up more than 80% of natural olive oil extracted from the fruit of the European olive tree. It is acknowledged to be one of the most beneficial of all the food-derived fats and oils for human cardiovascular health. In contrast, the isomeric compound in which the double bond possesses trans instead of cis geometry has been found to have numerous adverse health effects.

C=C
Alkene double bond

UNDERSTANDING AND VISUALIZING REACTIONS AND THEIR MECHANISMS

The emphasis on structure (electronic and spatial) and function (in radical and ionic form) in the early chapters primes students for building a true grasp of reaction mechanisms, encouraging understanding over memorization.

Because visualizing chemical reactivity can be challenging for many students, we use many different graphical cues, animations, and models to help students envisage reactions and how they proceed mechanistically.

1. *Dissociation of a polar covalent bond into ions*

General case:
$$A\!-\!B \longrightarrow A^+ + :B^-$$

Movement of an electron pair converts the A–B covalent bond into a lone pair on atom B

The direction in which the pair of electrons moves depends on which of the two atoms is more electronegative. In the general case above, B is more electronegative than A, so B more readily accepts the electron pair to become negatively charged. Atom A becomes a cation.

Arrow points to Cl, the more electronegative atom

Chloride is released with an additional lone pair derived from the broken bond

Specific example (a):
$$H\!-\!\ddot{\underset{..}{C}l}: \longrightarrow H^+ + :\ddot{\underset{..}{C}l}:^-$$

Dissociation of the acid HCl to give a proton and chloride ion exemplifies this process: When breaking a polar covalent bond in this way, *draw the curved arrow starting at the center of the bond and ending at the more electronegative atom.*

Specific example (b):
$$H_3C\!-\!\underset{\underset{CH_3}{|}}{\overset{\overset{CH_3}{|}}{C}}\!-\!\ddot{\underset{..}{B}r}: \longrightarrow H_3C\!-\!\underset{\underset{CH_3}{|}}{\overset{\overset{CH_3}{|}}{C}}^+ + :\ddot{\underset{..}{B}r}:^-$$

In this example, dissociation features the breaking of a C–Br bond. You will note that its essential features are identical to those of example (a).

(Real Life 8-3).

In Summary Alkyllithium and alkylmagnesium reagents add to aldehydes and ketones to give alcohols in which the alkyl group of the organometallic reagent has formed a bond to the original carbonyl carbon.

8-9 KEYS TO SUCCESS: AN INTRODUCTION TO SYNTHETIC STRATEGY

The reactions introduced so far are part of the "vocabulary" of organic chemistry; unless we know the vocabulary, we cannot speak the language of organic chemistry. These reactions allow us to manipulate molecules and interconvert functional groups, so it is important to become familiar with these transformations—their types, the reagents used, the conditions under which they occur (especially when the conditions are crucial to the success of the process), and the limitations of each type.

This task may seem monumental, one that will require much memorization. But *it is made easier by an understanding of the reaction mechanisms.* We already know that reactivity can be predicted from a small number of factors, such as electronegativity, coulombic forces, and bond strengths. Let us see how organic chemists apply this understanding to devise useful synthetic strategies, that is, reaction sequences that allow the construction of a desired target in the minimum number of high-yielding steps.

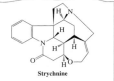

Strychnine

The total synthesis of the complex natural product strychnine (Section 25-8), containing seven fused rings and six stereocenters, has been steadily improved over a half-century of development of synthetic methods. The first synthesis, reported in 1954 by R. B. Woodward (Section 14-9), started from a simple indole derivative (Section 25-4) and required 28 synthetic steps to give the target in 0.00009% overall yield. A more recent synthesis (in 2011) took 12 steps and proceeded in 6% overall yield.

- ***NEW.*** Improved and expanded coverage of **electron-pushing arrows** in Sections 2-2 and 2-3. The use of electron-pushing arrows, introduced in these sections, is reinforced in Section 6-3 and numerous margin reminders in all subsequent chapters.

- ***NEW.*** Keys to Success sections teach and reinforce basic concepts and problem-solving techniques.

  - Chapter 2, Section 2-2: **KEYS TO SUCCESS: USING CURVED "ELECTRON-PUSHING" ARROWS TO DESCRIBE CHEMICAL REACTIONS**
  - Chapter 3, Section 3-6: **KEYS TO SUCCESS: USING THE "KNOWN" MECHANISM AS A MODEL FOR THE "UNKNOWN"**
  - Chapter 6, Section 6-9: **KEYS TO SUCCESS: CHOOSING AMONG MULTIPLE MECHANISTIC PATHWAYS**
  - Chapter 7, Section 7-8: **KEYS TO SUCCESS: SUBSTITUTION VERSUS ELIMINATION—STRUCTURE DETERMINES FUNCTION**
  - Chapter 8, Section 8-9: **KEYS TO SUCCESS: AN INTRODUCTION TO SYNTHETIC STRATEGY**
  - Chapter 18, Section 18-7: **COMPETITIVE REACTION PATHWAYS AND THE INTRAMOLECULAR ALDOL CONDENSATION**
  - Chapter 23, Section 23-1: **THE CLAISEN CONDENSATION WORKS BECAUSE HYDROGENS FLANKED BY TWO CARBONYL GROUPS ARE ACIDIC**

- **Interlude: A Summary of Organic Reaction Mechanisms,** following Chapter 14, summarizes the relatively few types of reaction mechanisms that drive the majority of organic reactions, thereby encouraging understanding over memorization.

Reaction

Mechanism

Model Building

- **Computer-generated ball-and-stick and space-filling models** help students recognize steric factors in many kinds of reactions. Icons in the page margins indicate where model building by students will be especially helpful for visualizing three-dimensional structures and dynamics.

- **Electrostatic potential maps** allow students to see how electron distributions affect the behavior of species in various interactions.

- **Icons** are employed to highlight the distinction between a **reaction** and its **mechanism.**

- **Model-building icons** encourage the student to build molecular models to illustrate the principle under discussion or to aid in the solution of a problem.

- **Reaction Summary Road Maps,** found at the ends of Chapters 8, 9, 11, 12, 13, 15, 17, 19, 20, and 21, provide one-page overviews of the reactivity of each major functional group. The **Preparation maps** indicate the possible origins of a functionality—that is, the precursor functional groups. The **Reaction maps** show what each functional group does. In both maps, reaction arrows are labeled with particular reagents and start from or end at specific reactants or products. Section numbers indicate where the transformation is discussed in the text.

STRONGER PEDAGOGY FOR SOLVING PROBLEMS

- *NEW.* **WHIP** **problem-solving strategy** is applied to Solved Exercises throughout the text.

  What does the problem ask?

  How to begin?

  Information needed?

  Proceed

  Beginning in Chapter 1, we introduce a novel and powerful approach to problem solving, the ***WHIP*** approach. We teach students how to recognize the fundamental types of questions they are likely to encounter, and explain the solution strategy in full detail.

- All in-chapter Solved Exercises begin with a **Strategy** section that emphasizes the reasoning students need to apply in attacking problems. The **Solution** arranges the steps logically and carefully, modeling good problem-solving skills.

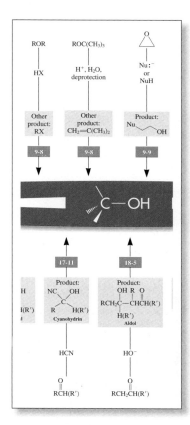

6-30. Analyzing Substrate Structures for S_N2 Reactivity

a. Which of the following compounds would be expected to react in an S_N2 manner at a reasonable rate with sodium azide, NaN_3, in ethanol? Which will not? Why not?

(iii)   (iv)   (v)   (vi)

SOLUTION

Let us apply the WHIP approach to break down the process of solving this problem.

*W*hat is the problem asking? This may be obvious—one merely has to identify which of the compounds shown reacts with azide in ethanol via an S_N2 process. However, there is a bit more to it, and the clue is the presence of the word "why" in the question. "How" and "why" questions invariably require a closer look at the situation, usually from a mechanistic perspective. It will be necessary to consider finer details of the S_N2 mechanism in light of the structures of each of the substrate molecules.

*H*ow to begin? Characterize each substrate in the context of the S_N2 process. Does it contain a viable leaving group? To what kind of carbon atom is the potential leaving group attached? Are other relevant structural features present?

*I*nformation needed? Does each of these six molecules contain a good leaving group? If necessary, look in Section 6-7 for guidance: To be a good leaving group, a species must be a weak base. Next, can you tell if the leaving group is attached to a primary, secondary, or tertiary carbon atom? See their definitions in Section 2-6. Anything else? Section 6-10 tells you what to look for: steric hindrance in the substrate that may obstruct the approach of the nucleophile.

*P*roceed. We identify first the molecules with good leaving groups. Referring to Table 6-4, we see that, as a general rule, only species that are the conjugate bases of strong acids (i.e., with pK_a values < 0) qualify. So, (i), (iv), and (vi) will not undergo S_N2 displacement. They lack good leaving groups: $^-NH_2$, ^-OH, and ^-CN are too strongly basic for this purpose (thus answering the "why not" for these three). Substrate (ii) contains a good leaving group, but the reaction site is a tertiary carbon and the S_N2 mechanism is sterically very unfavorable. That leaves substrates (iii) and (v), both of which are primary haloalkanes with minimal steric hindrance around the site of displacement. Both will transform readily by the S_N2 mechanism.

- **Try It Yourself Exercises.** Each in-chapter worked exercise is paired with a Try It Yourself problem that follows up on the concept being taught.

- **Caution statements** appear in many of the exercises, alerting students to potential pitfalls and how to avoid them.

A Wide Variety of Problem Types

Users and reviewers of past editions have often cited the end-of-chapter problems as a major strength of the book, both for the range of difficulty levels and the variety of practical applications. We highlight those end-of-chapter problems that are more difficult with a special icon:

- **Worked Examples: Integrating the Concepts** include worked-out, step-by-step solutions to problems involving several concepts from within chapters and from among several chapters. These solutions place particular emphasis on problem analysis, deductive reasoning, and logical conclusions.

- **Team Problems** encourage discussion and collaborative learning among students. They can be assigned as regular homework or as projects for groups of students to work on.

REAL CHEMISTRY BY PRACTICING CHEMISTS

An Emphasis on Practical Applications

Every chapter of this text features discussions of biological, medical, and industrial applications of organic chemistry, many of them new to this edition. In particular, as mentioned at the beginning, we have introduced some of the fundamentals of medicinal chemistry in over 70 new entries describing drug design, absorption, metabolism, mode of action, and medicinal terminology. Other topics range from advances in the development of "green," environmentally friendly methods in the chemical industry to new chemically based methods of disease diagnosis and treatment, and uses of transition metals and enzymes to catalyze reactions in pharmaceutical and medicinal chemistry. Some of these applications are found in the text discussion, others in the exercises and problems, and still others in the Real Life boxes. A new feature is margin entries called "Really?," which are meant to stimulate students' engagement by highlighting unusual and surprising aspects of the subject matter under discussion. A major application of organic chemistry, stressed throughout the text, is the synthesis of new products and materials. Many chapters contain specific syntheses of biological and medicinal importance.

NEW entries include:

Cubical Atoms by G. N. Lewis (Ch. 1, Really?, p. 14)
Elements in the Universe (Ch. 1, Really?, p. 31)
Stomach Acid, Peptic Ulcers, Pharmacology, and Organic Chemistry (Ch. 2, Real Life 2-1, p. 61)
Acidic and Basic Drugs (Ch. 2, p. 63)
The Longest Man-Made Linear Alkane (Ch. 2, Really?, p. 78)
Food Calories (Ch. 3, Really?, p. 123)
Conformational Drug Design (Ch. 4, p. 148)
Male Contraceptives (Ch. 4, Real Life 4-3, p. 157)
Ibuprofen Enantiomerization (Ch. 5, Really?, p. 180)
Fluorinated Pharmaceuticals (Ch. 6, Real Life 6-1, p. 213)
Halomethane Fumigants (Ch. 6, Really?, p. 216)
Solvation and Drug Activity (Ch. 6, p. 231)
An S_N2 Reaction at a Tertiary Carbon (Ch. 7, Really?, p. 269)
Alcohol Chain Length and Antimicrobial Activity (Ch. 8, p. 283)
Alcohol and Heartburn (Ch. 8, Really?, p. 284)
Don't Drink and Drive: The Breath Analyzer Test (Ch. 8, Real Life 8-2, p. 294)
Protecting-Group Strategy (Ch. 9, p. 350)
Oxacyclopropane: The Warhead of Drugs (Ch. 9, p. 356)
Scottish Whisky in Space (Ch. 9, Really?, p. 360)
Carbon has 15 Known Isotopes (Ch. 10, Really?, p. 411)
Structural Characterization of Natural and "Unnatural" Products (Ch. 10, Real Life 10-5, p. 419)
Various Forms of Radiation and Their Uses (Ch. 10, p. 425)
Bond Strength and Polarity Correlate with IR Absorptions (Ch. 11, p. 456)
IR Thermography (Ch. 11, Really?, p. 458)
L-DOPA and Parkinson's Disease (Ch. 12, p. 488)
Halohydroxylations in Nature (Ch. 12, p. 500)
Ethene is a Natural Plant and Fruit Hormone (Ch. 12, Really?, p. 522)
Carbon Allotropes: sp^3, sp^2, and sp (Ch. 13, p. 548)
Life is Under Kinetic Control (Ch. 14, Really?, p. 593)
Sunglasses on Demand (Ch. 14, p. 621)
The Sunburn Protection Factor (Ch. 15, Really?, p. 650)
Helicenes (Ch. 15, Really?, p. 660)
Sulfa Drugs: The First Antimicrobials (Ch. 15, p. 673)
Halogenated Drug Derivatives (Ch. 16, p. 700)
Sulfosalicylic Acid and Urine Testing (Ch. 16, Really?, p. 711)

Sunglasses on Demand

Self-darkening eyeglasses contain organic molecules that undergo thermally reversible photoisomerizations between two species that differ in their electronic spectra:

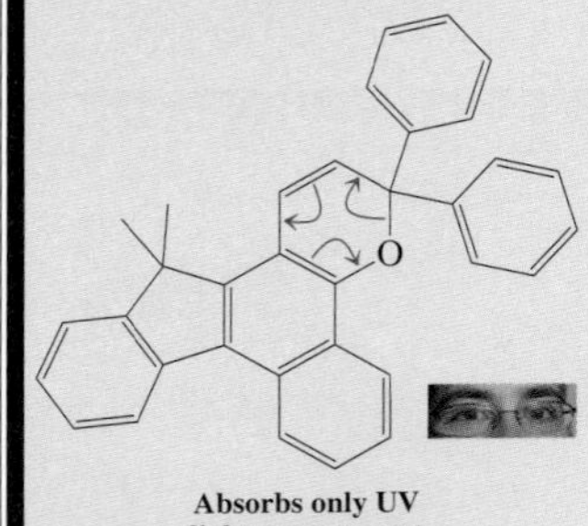

Absorbs only UV light: transparent

hv ↕ Δ

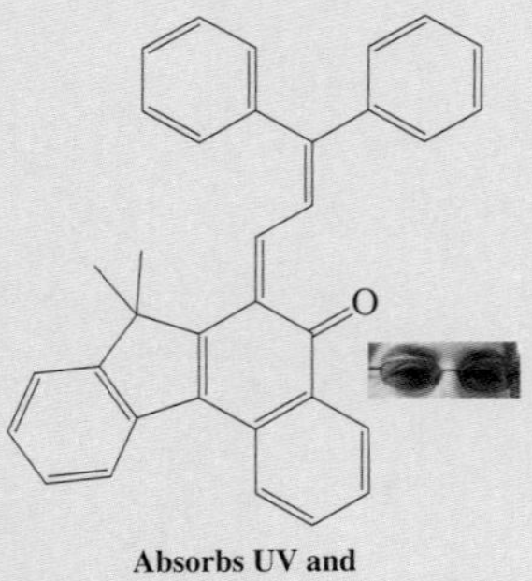

Absorbs UV and visible light

The top molecule is transparent in the visible range but absorbs the sun's UV rays to undergo electrocyclic ring opening to the bottom structure. The more extended conjugation in this isomer causes a shift of its λ_{max} to effect shading. In the dark, the system reverts thermally to its thermodynamically more stable state.

 Burnet moths use the glucose-bound cyanohydrin linamarin as an HCN reservoir for chemical defense. Enzymes catalyze the hydrolysis of the acetal unit to liberate acetone cyanohydrin, which then releases the toxic gas. Females seek out males with high levels of linamarin, which is passed on as a remarkable "nuptial gift" during their mating.

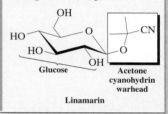

Glucose · Acetone cyanohydrin warhead · Linamarin

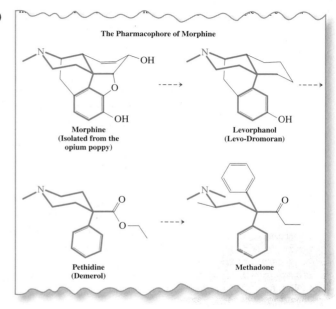

The Pharmacophore of Morphine

Morphine (Isolated from the opium poppy) → Levorphanol (Levo-Dromoran) →

Pethidine (Demerol) → Methadone

NEW AND UPDATED TOPICS

As with all new editions, each chapter has been carefully reviewed and revised.

NEW entries, updates, and improvements include:

How to obtain a Nobel Prize: peeling off graphene from graphite using Scotch tape.

Expanded and improved coverage of reactivity and selectivity (Ch. 3)
Updated coverage of the ozone layer (Ch. 3)
Updated presentation of diastereomeric relationships (Ch. 5)
New section: The S_N2 Reaction at a Glance (Ch. 6)
Improved section on retrosynthetic analysis (Ch. 8)
Improved presentation of π molecular orbital formation (Chs. 14 and 15)
New section: Nucleophilic trapping of carbocations is nonstereoselective (Ch. 12)
Expanded coverage of the stereochemistry of additions to alkenes (Ch. 12)
Revised section: Alkynes in Nature and Medicine (Ch. 13)
Updated coverage of carbon allotropes, including graphene (Ch. 15)
Expanded coverage of the reversibility of carbonyl reactions (Chs. 17 and 18)
New section: Enolate formation can be regioselective (Ch. 18)
Updated coverage of stereoselective aldol reactions in nature and in the laboratory: Organocatalysis (Ch. 18)
Expanded coverage of competitive pathways and reversibility in intramolecular aldol condensation reactions (Ch. 18)
Expanded coverage of soaps, unsaturated fatty acids, and bioplastics (Ch. 19)
New Road Map: Hydride Reductions (Ch. 20)
Updated and expanded coverage of physiologically active amines (Ch. 21)
Updated coverage of bisphenol A and resveratrol (Ch. 22)
Expanded and improved coverage of glutathione as an antioxidant (Ch. 22)
Revised coverage of the Claisen condensation (Ch. 23)
Updated "Top Ten" Drug List (Ch. 25)
Expanded coverage of nucleosides in medicine (Ch. 26)

SUPPLEMENTS

Student and Instructor Support

STUDENT ANCILLARY SUPPORT

We believe a student needs to interact with a concept several times in a variety of scenarios to obtain a practical understanding. With that in mind, W. H. Freeman has developed the most comprehensive student learning package available.

Printed Resources

Study Guide and Solutions Manual, **by Neil Schore, University of California, Davis**
ISBN: 1-4641-6225-5
Written by *Organic Chemistry* coauthor Neil Schore, this invaluable manual includes chapter introductions that highlight new materials, chapter outlines, detailed comments for each chapter section, a glossary, and solutions to the end-of-chapter problems, presented in a way that shows students how to reason their way to the answer.

Workbook for Organic Chemistry: Supplemental Problems and Solutions, **by Jerry Jenkins, Otterbein College**
ISBN: 1-4292-4758-4
Jerry Jenkins' extensive workbook provides approximately 80 problems per topic with full worked-out solutions. The perfect aid for students in need of more problem-solving practice, the *Workbook for Organic Chemistry* can be paired with any organic chemistry text on the

market. For instructors interested in online homework, W. H. Freeman has also placed these problems in WebAssign (see below).

Molecular Model Set
ISBN: 0-7167-4822-3
A modeling set offers a simple, practical way for students to see, manipulate, and investigate molecular behavior. Polyhedra mimic atoms, pegs serve as bonds, oval discs become orbitals. W. H. Freeman is proud to offer this inexpensive, best-of-its-kind kit containing everything you need to represent double and triple bonds, radicals, and long pairs of electrons—including more carbon pieces than are offered in other sets.

Free Media Resource

Student Companion Web Site
The *Organic Chemistry* Book Companion Web site, accessed at www.whfreeman.com/organic7e, provides a range of tools for problem solving and chemical explorations. They include, among others:

- Student self-quizzes
- An interactive periodic table of the elements
- Author lecture videos
- Animations
- Reaction and Nomenclature Exercises, which are drag-and-drop exercises designed for memorization
- Animated Mechanisms for reference and quizzing
- To access additional support, including the ChemCasts, Organic Flashcards, and ChemNews from *Scientific American*, students can upgrade their access through a direct subscription to the Premium component of the Web site.

Premium Media Resource

The *Organic Chemistry* Book Companion Web site, which can be accessed at www.whfreeman.com/organic7e, contains a wealth of Premium Student Resources. Students can unlock these resources with the click of a button, putting extensive concept and problem-solving support right at their fingertips. Some of the resources available are:

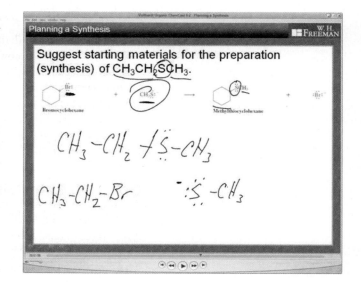

ChemCasts replicate the face-to-face experience of watching an instructor work a problem. Using a virtual whiteboard, the Organic ChemCast tutors show students the steps involved in solving key Worked Examples, while explaining the concepts along the way. The Worked Examples featured in the ChemCasts were chosen with the input of organic chemistry students.

ChemNews from *Scientific American* provides an up-to-the-minute streaming feed of organic chemistry-related new stories direct from *Scientific American* magazine. Stay on top of the latest happenings in chemistry, all in one easy place.

Spartan Student Discount

With purchase of this text, students can also purchase *Spartan Student* at a significant discount at www.wavefun.com/cart/spartaned.html using the code WHFOCHEM.

ELECTRONIC TEXTBOOK OPTIONS

For students interested in digital textbooks, W. H. Freeman offers *Organic Chemistry* in two easy-to-use formats.

The Multimedia-Enhanced e-Book

The multimedia-enhanced e-Book contains the complete text with a wealth of helpful functions. All student multimedia, including the ChemCasts, are linked directly from the e-Book pages. Students are thus able to access supporting resources when they need them—taking advantage of the "teachable moment" as students read. Customization functions include instructor and student notes, document linking, and editing capabilities.

The CourseSmart e-Textbook

The CourseSmart e-Textbook provides the full digital text, along with tools to take notes, search, and highlight passages. A free app allows access to CourseSmart e-Textbooks and Android and Apple devices, such as the iPad. They can also be downloaded to your computer and accessed without an Internet connection, removing any limitations for students when it comes to reading digital text. The CourseSmart e-Textbook can be purchased at www.coursesmart.com.

INSTRUCTOR ANCILLARY SUPPORT

Whether you're teaching the course for the first time or the hundredth time, the Instructor Resources that accompany *Organic Chemistry* should provide you with the resources you need to make the semester easy and efficient.

Electronic Instructor Resources

Instructors can access valuable teaching tools through www.whfreeman.com/organic7e. These password-protected resources are designed to enhance lecture presentations, and include all the illustrations from the textbook (in .jpg and PowerPoint format), Lecture PowerPoint slides, Clicker Questions, and more. Also available on the companion Web site are

- **New Molecular Modeling Problems**
  With this edition we now offer new molecular modeling problems for almost every chapter, which can be found on the text's companion Web site. The problems were written to be worked using the popular *Spartan Student* software. With purchase of this text, students can purchase *Spartan Student* at a significant discount from www.wavefun.com/cart/spartaned.html using the code WHFOCHEM. While the problems are written to be worked using *Spartan Student*, they can be completed using any electronic structure program that allows Hartree-Fock, density functional, and MP2 calculations.

ONLINE LEARNING ENVIRONMENTS

W. H. Freeman offers the widest variety of online homework options on the market.

WebAssign Premium

For instructors interested in online homework management, WebAssign Premium features a time-tested, secure online environment already used by millions of students worldwide. Featuring algorithmic problem generation and supported by a wealth of chemistry-specific learning tools, WebAssign Premium for *Organic Chemistry* presents instructors with a powerful assignment manager and study environment. WebAssign Premium provides the following resources:

- **Algorithmically generated problems:** Students receive homework problems containing unique values for computation, encouraging them to work out the problems on their own.
- **Complete access to the multimedia-enhanced e-Book**, from a live table of contents, as well as from relevant problem statements.

- Graded molecular drawing problems using the popular MarvinSketch application allow instructors to evaluate student understanding of molecular structure. The system evaluates virtually "drawn" molecular structures, returning a grade as well as helpful feedback for common errors.
- **Links to ChemCasts** are provided as hints and feedback to ensure a clearer understanding of the problems and the concepts they reinforce.

Sapling Learning

Sapling Learning provides highly effective interactive homework and instruction that improve student learning outcomes for the problem-solving disciplines. They offer an enjoyable teaching and effective learning experience that is distinctive in three important ways:

- **Ease of Use:** Sapling Learning's easy-to-use interface keeps students engaged in problem solving, not struggling with the software.
- **Targeted Instructional Content:** Sapling Learning increases student engagement and comprehension by delivering immediate feedback and targeted instructional content.
- **Unsurpassed Service and Support:** Sapling Learning makes teaching more enjoyable by providing a dedicated Masters- or Ph.D.-level colleague to service instructors' unique needs throughout the course, including content customization.

ACKNOWLEDGMENTS

We are grateful to the following professors who reviewed the manuscript for the seventh edition:

Marc Anderson, *San Francisco State University*
George Bandik, *University of Pittsburgh*
Anne Baranger, *University of California, Berkeley*
Kevin Bartlett, *Seattle Pacific University*
Scott Borella, *University of North Carolina—Charlotte*
Stefan Bossmann, *Kansas State University*
Alan Brown, *Florida Institute of Technology*
Paul Carlier, *Virginia Tech University*
Robert Carlson, *University of Kansas*
Toby Chapman, *University of Pittsburgh*
Robert Coleman, *Ohio State University*
William Collins, *Fort Lewis College*
Robert Corcoran, *University of Wyoming*
Stephen Dimagno, *University of Nebraska, Lincoln*
Rudi Fasan, *University of Rochester*
James Fletcher, *Creighton University*
Sara Fitzgerald, *Bridgewater College*
Joseph Fox, *University of Delaware*
Terrence Gavin, *Iona College*
Joshua Goodman, *University of Rochester*
Christopher Hadad, *Ohio State University*
Ronald Halterman, *University of Oklahoma*
Michelle Hamm, *University of Richmond*
Kimi Hatton, *George Mason University*
Sean Hightower, *University of North Dakota*
Shawn Hitchcock, *Illinois State University*
Stephen Hixson, *University of Massachusetts, Amherst*
Danielle Jacobs, *Rider University*
Ismail Kady, *East Tennessee State University*
Rizalia Klausmeyer, *Baylor University*

Krishna Kumar, *Tufts University*
Julie Larson, *Bemidji State University*
Carl Lovely, *University of Texas at Arlington*
Scott Lewis, *James Madison University*
Claudia Lucero, *California State University—Sacramento*
Sarah Luesse, *Southern Illinois University—Edwardsville*
John Macdonald, *Worcester Polytechnical Institute*
Lisa Ann McElwee-White, *University of Florida*
Linda Munchausen, *Southeastern Louisiana State University*
Richard Nagorski, *Illinois State University*
Liberty Pelter, *Purdue University—Calumet*
Jason Pontrello, *Brandeis University*
MaryAnn Robak, *University of California, Berkeley*
Joseph Rugutt, *Missouri State University—West Plains*
Kirk Schanze, *University of Florida*
Pauline Schwartz, *University of New Haven*
Trent Selby, *Mississippi College*
Gloria Silva, *Carnegie Mellon University*
Dennis Smith, *Clemson University*
Leslie Sommerville, *Fort Lewis College*
Jose Soria, *Emory University*
Michael Squillacote, *Auburn University*
Mark Steinmetz, *Marquette University*
Jennifer Swift, *Georgetown University*
James Thompson, *Alabama A&M University*
Carl Wagner, *Arizona State University*
James Wilson, *University of Miami*
Alexander Wurthmann, *University of Vermont*
Neal Zondlo, *University of Delaware*
Eugene Zubarev, *Rice University*

We are also grateful to the following professors who reviewed the manuscript for the sixth edition:

Michael Barbush, *Baker University*

Debbie J. Beard, *Mississippi State University*

Robert Boikess, *Rutgers University*

Cindy C. Browder, *Northern Arizona University*

Kevin M. Bucholtz, *Mercer University*

Kevin C. Cannon, *Penn State Abington*

J. Michael Chong, *University of Waterloo*

Jason Cross, *Temple University*

Alison Flynn, *Ottawa University*

Roberto R. Gil, *Carnegie Mellon University*

Sukwon Hong, *University of Florida*

Jeffrey Hugdahl, *Mercer University*

Colleen Kelley, *Pima Community College*

Vanessa McCaffrey, *Albion College*

Keith T. Mead, *Mississippi State University*

James A. Miranda, *Sacramento State University*

David A. Modarelli, *University of Akron*

Thomas W. Ott, *Oakland University*

Hasan Palandoken, *Western Kentucky University*

Gloria Silva, *Carnegie Mellon University*

Barry B. Snider, *Brandeis University*

David A. Spiegel, *Yale University*

Paul G. Williard, *Brown University*

Shmuel Zbaida, *Rutgers University*

Eugene Zubarev, *Rice University*

Peter Vollhardt thanks his colleagues at UC Berkeley, in particular Professors Anne Baranger, Bob Bergman, Carolyn Bertozzi, Ron Cohen, Matt Francis, John Hartwig, Darleane Hoffman, Tom Maimone, Richmond Sarpong, Rich Saykally, Andrew Streitwieser, and Dean Toste, for suggestions, updates, general discussions, and stimulus. He would also like to thank his administrative assistant, Bonnie Kirk, for helping with the logistics of producing and handling manuscript and galleys. Neil Schore thanks Dr. Melekeh Nasiri and Professor Mark Mascal for their ongoing comments and suggestions, and the numerous undergraduates at UC Davis who eagerly pointed out errors, omissions, and sections that could be improved or clarified. Our thanks go to the many people who helped with this edition. Jessica Fiorillo, acquisitions editor, and Randi Rossignol, development editor, at W. H. Freeman and Company, guided this edition from concept to completion. Dave Quinn, media editor, managed the media and supplements with great skill, and Nicholas Ciani, editorial assistant, helped coordinate our efforts. Also many thanks to Philip McCaffrey, managing editor, Blake Logan, our designer, and Susan Wein, production coordinator, for their fine work and attention to the smallest detail. Thanks also to Dennis Free at Aptara, for his unlimited patience.

CHAPTER 1

Structure and Bonding in Organic Molecules

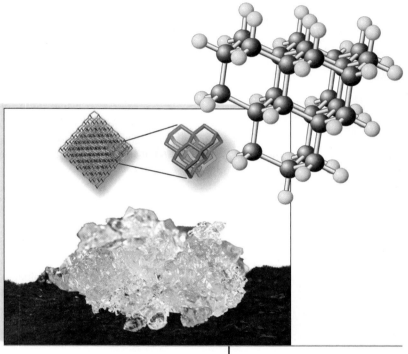

How do chemicals regulate your body? Why did your muscles ache this morning after last night's long jog? What is in the pill you took to get rid of that headache you got after studying all night? What happens to the gasoline you pour into the gas tank of your car? What is the molecular composition of the things you wear? What is the difference between a cotton shirt and one made of silk? What is the origin of the odor of garlic? You will find the answers to these questions, and many others that you may have asked yourself, in this book on organic chemistry.

Chemistry is the study of the structure of molecules and the rules that govern their interactions. As such, it interfaces closely with the fields of biology, physics, and mathematics. What, then, is organic chemistry? What distinguishes it from other chemical disciplines, such as physical, inorganic, or nuclear chemistry? A common definition provides a partial answer: *Organic chemistry is the chemistry of carbon and its compounds.* These compounds are called **organic molecules.**

Organic molecules constitute the chemical building blocks of life. Fats, sugars, proteins, and the nucleic acids are compounds in which the principal component is carbon. So are countless substances that we take for granted in everyday use. Virtually all the clothes that we wear are made of organic molecules—some of natural fibers, such as cotton and silk; others artificial, such as polyester. Toothbrushes, toothpaste, soaps, shampoos, deodorants, perfumes—all contain organic compounds, as do furniture, carpets, the plastic in light fixtures and cooking utensils, paintings, food, and countless other items. Consequently, organic chemical industries are among the largest in the world, including petroleum refining and processing, agrochemicals, plastics, pharmaceuticals, paints and coatings, and the food conglomerates.

Organic substances such as gasoline, medicines, pesticides, and polymers have improved the quality of our lives. Yet the uncontrolled disposal of organic chemicals has polluted the environment, causing deterioration of animal and plant life as well as injury and disease to humans. If we are to create useful molecules—*and* learn to control their effects—we need a knowledge of their properties and an understanding of their behavior. We must be able to apply the principles of organic chemistry.

Tetrahedral carbon, the essence of organic chemistry, exists as a lattice of six-membered rings in diamonds. In 2003, a family of molecules called *diamandoids* was isolated from petroleum. Diamandoids are subunits of diamond in which the excised pieces are capped off with hydrogen atoms. An example is the beautifully crystalline pentamantane (molecular model on top right and picture on the left; © 2004 *Chevron U.S.A. Inc. Courtesy of MolecularDiamond Technologies, ChevronTexaco Technology Ventures LLC),* which consists of five "cages" of the diamond lattice. The top right of the picture shows the carbon frame of pentamantane stripped of its hydrogens and its superposition on the lattice of diamond.

This chapter explains how the basic ideas of chemical structure and bonding apply to organic molecules. Most of it is a review of topics that you covered in your general chemistry courses, including molecular bonds, Lewis structures and resonance, atomic and molecular orbitals, and the geometry around bonded atoms.

Almost everything you see in this picture is made of organic chemicals.

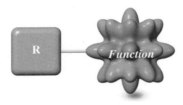

Carbon frame provides structure

Functional group imparts reactivity

H$_3$C—CH$_3$
Ethane

H$_2$C—C(H$_2$)—CH$_2$ ...

$$\begin{array}{c} \text{H}_2 \\ \text{C} \\ \text{H}_2\text{C} \quad \text{CH}_2 \\ \text{H}_2\text{C} \quad \text{CH}_2 \\ \text{C} \\ \text{H}_2 \end{array}$$
Cyclohexane

HC≡CH
Acetylene
(An alkyne)

H$_2$C=O
Formaldehyde
(An aldehyde)

$$\begin{array}{c} \text{O} \\ \| \\ \text{C} \\ \text{H}_3\text{C} \quad \text{CH}_3 \end{array}$$
Acetone
(A ketone)

H$_3$C—NH$_2$
Methylamine
(An amine)

1-1 | THE SCOPE OF ORGANIC CHEMISTRY: AN OVERVIEW

A goal of organic chemistry is to relate the structure of a molecule to the reactions that it can undergo. We can then study the steps by which each type of reaction takes place, and we can learn to create new molecules by applying those processes.

Thus, it makes sense to classify organic molecules according to the subunits and bonds that determine their chemical reactivity: These determinants are groups of atoms called **functional groups.** The study of the various functional groups and their respective reactions provides the structure of this book.

Functional groups determine the reactivity of organic molecules

We begin with the **alkanes,** composed of only carbon and hydrogen atoms ("hydrocarbons") connected by single bonds. They lack any functional groups and as such constitute the basic scaffold of organic molecules. As with each class of compounds, we present the systematic rules for naming alkanes, describe their structures, and examine their physical properties (Chapter 2). An example of an alkane is ethane. Its structural mobility is the starting point for a review of thermodynamics and kinetics. This review is then followed by a discussion of the strength of alkane bonds, which can be broken by heat, light, or chemical reagents. We illustrate these processes with the chlorination of alkanes (Chapter 3).

A Chlorination Reaction

$$CH_4 \ + \ Cl_2 \ \xrightarrow{\text{Energy}} \ CH_3\text{—}Cl \ + \ HCl$$

Next we look at cyclic alkanes (Chapter 4), which contain carbon atoms in a ring. This arrangement can lead to new properties and changes in reactivity. The recognition of a new type of isomerism in cycloalkanes bearing two or more substituents—either on the same side or on opposite sides of the ring plane—sets the stage for a general discussion of **stereoisomerism.** Stereoisomerism is exhibited by compounds with the same connectivity but differing in the relative positioning of their component atoms in space (Chapter 5).

We shall then study the haloalkanes, our first example of compounds containing a functional group—the carbon–halogen bond. The haloalkanes participate in two types of organic reactions: substitution and elimination (Chapters 6 and 7). In a **substitution** reaction, one halogen atom may be replaced by another; in an **elimination** process, adjacent atoms may be removed from a molecule to generate a double bond.

A Substitution Reaction

$$CH_3\text{—}Cl \ + \ K^+I^- \ \longrightarrow \ CH_3\text{—}I \ + \ K^+Cl^-$$

An Elimination Reaction

$$\begin{array}{c} \text{CH}_2\text{—CH}_2 \ + \ \text{K}^+ \ ^-\text{OH} \ \longrightarrow \ \text{H}_2\text{C}=\text{CH}_2 \ + \ \text{HOH} \ + \ \text{K}^+\text{I}^- \\ | \quad\quad | \\ \text{H} \quad\quad \text{I} \end{array}$$

Like the haloalkanes, each of the major classes of organic compounds is characterized by a particular functional group. For example, the carbon–carbon triple bond is the functional group of alkynes (Chapter 13); the smallest alkyne, acetylene, is the chemical burned in a welder's torch. A carbon–oxygen double bond is characteristic of aldehydes and ketones (Chapter 17); formaldehyde and acetone are major industrial commodities. The amines

(Chapter 21), which include drugs such as nasal decongestants and amphetamines, contain nitrogen in their functional group; methylamine is a starting material in many syntheses of medicinal compounds. We shall study the tools for identifying these molecular subunits, especially the various forms of spectroscopy (Chapters 10, 11, and 14). Organic chemists rely on an array of spectroscopic methods to elucidate the structures of unknown compounds. All of these methods depend on the absorption of electromagnetic radiation at specific wavelengths and the correlation of this information with structural features.

Subsequently, we shall encounter organic molecules that are especially crucial in biology and industry. Many of these, such as the carbohydrates (Chapter 24) and amino acids (Chapter 26), contain multiple functional groups. However, in *every* class of organic compounds, the principle remains the same: *The structure of the molecule determines the reactions that it can undergo.*

Synthesis is the making of new molecules

Carbon compounds are called "organic" because it was originally thought that they could be produced only from living organisms. In 1828, Friedrich Wöhler* proved this idea to be false when he converted the inorganic salt lead cyanate into urea, an organic product of protein metabolism in mammals (Real Life 1-1).

Wöhler's Synthesis of Urea

$$Pb(OCN)_2 \ + \ 2\,H_2O \ + \ 2\,NH_3 \ \longrightarrow \ 2\,H_2NCNH_2 \ + \ Pb(OH)_2$$

Lead cyanate   Water   Ammonia   Urea   Lead hydroxide

An organic molecular architect at work.

Synthesis, or the making of molecules, is a very important part of organic chemistry (Chapter 8). Since Wöhler's time, many millions of organic substances have been synthesized from simpler materials, both organic and inorganic.[†] These substances include many that also occur in nature, such as the penicillin antibiotics, as well as entirely new compounds. Some, such as cubane, have given chemists the opportunity to study special kinds of bonding and reactivity. Others, such as the artificial sweetener saccharin, have become a part of everyday life.

Typically, the goal of synthesis is to construct complex organic chemicals from simpler, more readily available ones. To be able to convert one molecule into another, chemists must know organic reactions. They must also know the physical factors that govern such processes, such as temperature, pressure, solvent, and molecular structure. This knowledge is equally valuable in analyzing reactions in living systems.

As we study the chemistry of each functional group, we shall develop the tools both for planning effective syntheses and for predicting the processes that take place in nature. But how? The answer lies in looking at reactions step by step.

Benzylpenicillin   **Cubane**   **Saccharin**

*Professor Friedrich Wöhler (1800–1882), University of Göttingen, Germany. In this and subsequent biographical notes, only the scientist's last known location of activity will be mentioned, even though much of his or her career may have been spent elsewhere.

[†]As of April 2012, the Chemical Abstracts Service had registered over 65 million chemical substances and more than 63 million genetic sequences.

REAL LIFE: NATURE 1-1 | Urea: From Urine to Wöhler's Synthesis to Industrial Fertilizer

Urination is the main process by which we excrete nitrogen from our bodies. Urine is produced by the kidneys and then stored in the bladder, which begins to contract when its volume exceeds about 200 mL. The average human excretes about 1.5 L of urine daily, and a major component is urea, about 20 g per liter. In an attempt to probe the origins of kidney stones, early (al)chemists, in the 18th century, attempted to isolate the components of urine by crystallization, but they were stymied by the cocrystallization with the also present sodium chloride. William Prout,* an English chemist and physician, is credited with the preparation of pure urea in 1817 and the determination of its accurate elemental analysis as CH_4N_2O. Prout was an avid proponent of the then revolutionary thinking that disease has a molecular basis and could be understood as such. This view clashed with that of the so-called vitalists, who believed that the functions of a living organism are controlled by a "vital principle" and cannot be explained by chemistry (or physics).

Into this argument entered Wöhler, an inorganic chemist, who attempted to make ammonium cyanate, $NH_4^+OCN^-$ (also CH_4N_2O), from lead cyanate and ammonia in 1828, but who obtained the same compound that Prout had characterized as urea. To one of his mentors, Wöhler wrote, "I can make urea without a kidney, or even a living creature." In his landmark paper, "On the Artificial Formation of Urea," he commented on his synthesis as a "remarkable fact, as it is an example of the artificial generation of an organic material from inorganic materials." He also alluded to the significance of the finding that a compound with an identical elemental composition as ammonium cyanate can have such completely different chemical properties, a forerunner to the recognition of isomeric compounds. Wöhler's synthesis of

urea forced his contemporary vitalists to accept the notion that simple organic compounds could be made in the laboratory. As you shall see in this book, over the ensuing decades, synthesis has yielded much more complex molecules than urea, some of them endowed with self-replicating and other "lifelike" properties, such that the boundaries between what is lifeless and what is alive are dwindling.

Apart from its function in the body, urea's high nitrogen content makes it an ideal fertilizer. It is also a raw material in the manufacture of plastics and glues, an ingredient of some toiletry products and fire extinguishers, and an alternative to rock salt for deicing roads. It is produced industrially from ammonia and carbon dioxide to the tune of 100 million tons per year worldwide.

The effect of nitrogen fertilizer on wheat growth: treated on the left; untreated on the right.

*Dr. William Prout (1785–1850), Royal College of Physicians, London.

Reactions are the vocabulary and mechanisms are the grammar of organic chemistry

When we introduce a chemical reaction, we will first show just the starting compounds, or **reactants** (also called **substrates**), and the **products.** In the chlorination process mentioned earlier, the substrates—methane, CH_4, and chlorine, Cl_2—may undergo a reaction to give chloromethane, CH_3Cl, and hydrogen chloride, HCl. We described the overall transformation as $CH_4 + Cl_2 \rightarrow CH_3Cl + HCl$. However, even a simple reaction such as this one may proceed through a complex sequence of steps. The reactants could have first formed one or more *unobserved* substances—call these X—that rapidly changed into the observed products. These underlying details of the reaction constitute the **reaction mechanism.** In our example, the mechanism consists of two major parts: $CH_4 + Cl_2 \rightarrow X$ followed by $X \rightarrow CH_3Cl + HCl$. Each part is crucial in determining whether the overall reaction will proceed.

Substances X in our chlorination reaction are examples of **reaction intermediates,** species formed on the pathway between reactants and products. We shall learn the mechanism of this chlorination process and the nature of the reaction intermediates in Chapter 3.

How can we determine reaction mechanisms? The strict answer to this question is, we cannot. All we can do is amass circumstantial evidence that is consistent with (or points to) a certain sequence of molecular events that connect starting materials and products ("the postulated mechanism"). To do so, we exploit the fact that organic molecules are no more than collections of bonded atoms. We can, therefore, study how, when, and how fast bonds break and form, in which way they do so in three dimensions, and how changes in substrate structure affect the outcome of reactions. Thus, although we cannot strictly prove a mechanism, we can certainly rule out many (or even all) reasonable alternatives and propose a most likely pathway.

In a way, the "learning" and "using" of organic chemistry is much like learning and using a language. You need the vocabulary (i.e., the reactions) to be able to use the right words, but you also need the grammar (i.e., the mechanisms) to be able to converse intelligently. Neither one on its own gives complete knowledge and understanding, but together they form a powerful means of communication, rationalization, and predictive analysis. To highlight the interplay between reaction and mechanism, icons are displayed in the margin at appropriate places throughout the text.

Before we begin our study of the principles of organic chemistry, let us review some of the elementary principles of bonding. We shall find these concepts useful in understanding and predicting the chemical reactivity and the physical properties of organic molecules.

● **Reaction**

● **Mechanism**

1-2 | COULOMB FORCES: A SIMPLIFIED VIEW OF BONDING

The bonds between atoms hold a molecule together. But what causes bonding? Two atoms form a bond only if their interaction is energetically favorable, that is, if energy—heat, for example—is released when the bond is formed. Conversely, breaking that bond requires the input of the same amount of energy.

The two main causes of the energy release associated with bonding are based on Coulomb's law of electric charge:

1. Opposite charges attract each other (electrons are attracted to protons).

2. Like charges repel each other (electrons spread out in space).

Bonds are made by simultaneous coulombic attraction and electron exchange

Each atom consists of a nucleus, containing electrically neutral particles, or neutrons, and positively charged protons. Surrounding the nucleus are negatively charged electrons, equal in number to the protons so that the net charge is zero. As two atoms approach each other, the positively charged nucleus of the first atom attracts the electrons of the second atom; similarly, the nucleus of the second atom attracts the electrons of the first atom. As a result, the nuclei are held together by the electrons located between them. This sort of bonding is described by **Coulomb's* law:** Opposite charges attract each other with a force inversely proportional to the square of the distance between the centers of the charges.

Charge separation is rectified by Coulomb's law, appropriately in the heart of Paris.

Coulomb's Law

$$\text{Attracting force} = \text{constant} \times \frac{(+)\,\text{charge} \times (-)\,\text{charge}}{\text{distance}^2}$$

This attractive force causes energy to be released as the neutral atoms are brought together. The resulting decrease in energy is called the **bond strength.**

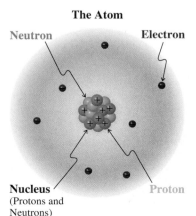

The Atom

Neutron   Electron

Nucleus   Proton
(Protons and
Neutrons)

---

*Lieutenant-Colonel Charles Augustin de Coulomb (1736–1806), Inspecteur Général of the University of Paris, France.

Figure 1-1 The changes in energy, *E*, that result when two atoms are brought into close proximity. At the separation defined as bond length, maximum bonding is achieved.

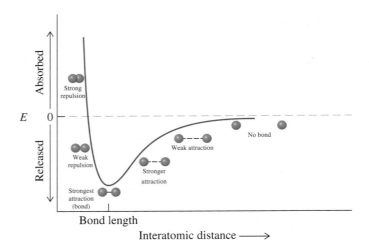

When the atoms reach a certain closeness, no more energy is released. The distance between the two nuclei at this point is called the **bond length** (Figure 1-1). Bringing the atoms closer together than this distance results in a sharp *increase* in energy. Why? As stated above, just as opposite charges attract, like charges repel. If the atoms are too close, the electron–electron and nuclear–nuclear repulsions become stronger than the attractive forces. When the nuclei are the appropriate bond length apart, the electrons are spread out around both nuclei, and attractive and repulsive forces balance for maximum bonding. The energy content of the two-atom system is then at a minimum, the most stable situation (Figure 1-2).

An alternative to this type of bonding results from the complete transfer of an electron from one atom to the other. The result is two charged *ions:* one positively charged, a *cation,* and one negatively charged, an *anion* (Figure 1-3). Again, the bonding is based on coulombic attraction, this time between two ions.

The coulombic bonding models of attracting and repelling charges shown in Figures 1-2 and 1-3 are highly simplified views of the interactions that take place in the bonding of atoms. Nevertheless, even these simple models explain many of the properties of organic molecules. In the sections to come, we shall examine increasingly more sophisticated views of bonding.

Figure 1-2 Covalent bonding. Attractive (solid-line) and repulsive (dashed-line) forces in the bonding between two atoms. The large spheres represent areas in space in which the electrons are found around the nucleus. The small circled plus sign denotes the nucleus.

Figure 1-3 Ionic bonding. An alternative mode of bonding results from the complete transfer of an electron from atom 1 to atom 2, thereby generating two ions whose opposite charges attract each other.

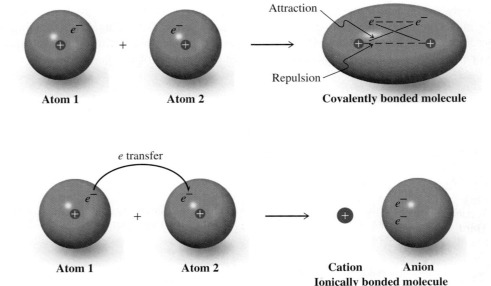

1-3 | IONIC AND COVALENT BONDS: THE OCTET RULE

We have seen that attraction between negatively and positively charged particles is a basis for bonding. How does this concept work in real molecules? Two extreme types of bonding explain the interactions between atoms in organic molecules:

1. A **covalent bond** is formed by the sharing of electrons (as shown in Figure 1-2).

2. An **ionic bond** is based on the electrostatic attraction of two ions with opposite charges (as shown in Figure 1-3).

We shall see that many atoms bind to carbon in a way that is intermediate between these extremes: Some ionic bonds have covalent character and some covalent bonds are partly ionic (polarized).

What are the factors that account for the two types of bonds? To answer this question, let us return to the atoms and their compositions. We start by looking at the periodic table and at how the electronic makeup of the elements changes as the atomic number increases.

The periodic table underlies the octet rule

The partial periodic table depicted in Table 1-1 includes those elements most widely found in organic molecules: carbon (C), hydrogen (H), oxygen (O), nitrogen (N), sulfur (S), chlorine (Cl), bromine (Br), and iodine (I). Certain reagents, indispensable for synthesis and commonly used, contain elements such as lithium (Li), magnesium (Mg), boron (B), and phosphorus (P). (If you are not familiar with these elements, refer to Table 1-1 or the periodic table on the inside cover.)

The elements in the periodic table are listed according to nuclear charge (number of protons), which equals the number of electrons. The nuclear charge increases by one with each element listed. The electrons occupy energy levels, or "shells," each with a fixed capacity. For example, the first shell has room for two electrons; the second, eight; and the third, 18. Helium, with two electrons in its shell, and the other noble gases, with eight electrons (called **octets**) in their outermost shells, are especially stable. These elements show very little chemical reactivity. All other elements (including carbon, see margin) lack octets in their outermost electron shells. *Atoms tend to form molecules in such a way as to reach an octet in the outer electron shell and attain a noble-gas configuration.* In the next two sections, we describe two extreme ways in which this goal may be accomplished: by the formation of pure ionic or pure covalent bonds.

Carbon Atom

Filled 1st shell

$C^{2,\,4}$

Unfilled 2nd shell: four valence electrons

Exercise 1-1

(a) Redraw Figure 1-1 for a weaker bond than the one depicted. (b) Write the elements in Table 1-1 from memory.

Table 1-1	Partial Periodic Table							Halogens	Noble gases
Period									
First	H^1								He^2
Second	$Li^{2,1}$	$Be^{2,2}$	$B^{2,3}$	$C^{2,4}$	$N^{2,5}$	$O^{2,6}$		$F^{2,7}$	$Ne^{2,8}$
Third	$Na^{2,8,1}$	$Mg^{2,8,2}$	$Al^{2,8,3}$	$Si^{2,8,4}$	$P^{2,8,5}$	$S^{2,8,6}$		$Cl^{2,8,7}$	$Ar^{2,8,8}$
Fourth	$K^{2,8,8,1}$							$Br^{2,8,18,7}$	$Kr^{2,8,18,8}$
Fifth								$I^{2,8,18,18,7}$	$Xe^{2,8,18,18,8}$

Note: The superscripts indicate the number of electrons in each principal shell of the atom.

In pure ionic bonds, electron octets are formed by transfer of electrons

Sodium (Na), a reactive metal, interacts with chlorine, a reactive gas, in a violent manner to produce a stable substance: sodium chloride. Similarly, sodium reacts with fluorine (F), bromine, or iodine to give the respective salts. Other alkali metals, such as lithium and potassium (K), undergo the same reactions. These transformations succeed because both reaction partners attain noble-gas character by the *transfer of outer-shell electrons,* called **valence electrons,** from the alkali metals on the left side of the periodic table to the halogens on the right.

Let us see how this works for the ionic bond in sodium chloride. Why is the interaction energetically favorable? First, it takes energy to remove an electron from an atom. This energy is the **ionization potential (IP)** of the atom. For sodium gas, the ionization energy amounts to 119 kcal mol^{-1}.* Conversely, energy may be released when an electron attaches itself to an atom. For chlorine, this energy, called its **electron affinity (EA),** is -83 kcal mol^{-1}. These two processes result in the transfer of an electron from sodium to chlorine. Together, they require a net energy *input* of $119 - 83 = 36$ kcal mol^{-1}.

$$[\text{Na}^{2,8,1} \xrightarrow{-1\,e} [\text{Na}^{2,8}]^+ \qquad \text{IP} = 119 \text{ kcal mol}^{-1} \ (498 \text{ kJ mol}^{-1})$$

$$\textbf{Sodium cation} \qquad \qquad \textbf{Energy input required}$$
$$\text{(Neon configuration)}$$

$$\text{Cl}^{2,8,7} \xrightarrow{+1\,e} [\text{Cl}^{2,8,8}]^- \qquad \text{EA} = -83 \text{ kcal mol}^{-1} \ (-347 \text{ kJ mol}^{-1})$$

$$\textbf{Chloride anion} \qquad \qquad \textbf{Energy released}$$
$$\text{(Argon configuration)}$$

$$\overline{\text{Na} + \text{Cl} \longrightarrow \text{Na}^+ + \text{Cl}^-} \qquad \overline{\text{Total} = 119 - 83 = 36 \text{ kcal mol}^{-1} \ (151 \text{ kJ mol}^{-1})}$$

Why, then, do the atoms readily form NaCl? The reason is their electrostatic attraction, which pulls them together in an ionic bond. At the most favorable interatomic distance [about 2.8 Å (angstroms) in the gas phase], this attraction releases (see Figure 1-1) about 120 kcal mol^{-1} (502 kJ mol^{-1}). This energy release is enough to make the reaction of sodium with chlorine energetically highly favorable [$+36 - 120 = -84$ kcal mol^{-1} $(-351$ kJ mol$^{-1})$].

Formation of Ionic Bonds by Electron Transfer

$$\text{Na}^{2,8,1} + \text{Cl}^{2,8,7} \longrightarrow [\text{Na}^{2,8}]^+ \, [\text{Cl}^{2,8,8}]^-, \text{ or NaCl } (-84 \text{ kcal mol}^{-1})$$

More than one electron may be donated (or accepted) to achieve noble-gas electronic configurations. Magnesium, for example, has two valence electrons. Donation to an appropriate acceptor produces the corresponding doubly charged cation, Mg^{2+}, with the electronic structure of neon. In this way, the ionic bonds of typical salts are formed.

A representation of how charge (re)distributes itself in molecules is given by electrostatic potential maps. These computer-generated maps not only show a form of the molecule's "electron cloud," they also use color to depict deviations from charge neutrality. Excess electron density—for example, a negative charge—is shown in colors shaded toward red; conversely, diminishing electron density—ultimately, a positive charge—is shown in colors shaded toward blue. Charge-neutral regions are indicated by green. The reaction of a sodium atom with a chlorine atom to produce Na^+Cl^- is pictured this way in the margin. In the product, Na^+ is blue, Cl^- is red.

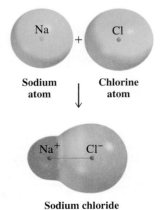

Na + Cl

Sodium   **Chlorine**
atom   **atom**

Na^+ Cl^-

Sodium chloride

---

*This book will cite energy values in the traditional units of kcal mol^{-1}, in which mol is the abbreviation for mole and a kilocalorie (kcal) is the energy required to raise the temperature of 1 kg (kilogram) of water by 1°C. In SI units, energy is expressed in joules (kg m^2 s^{-2}, or kilogram-meter^2 per second^2). The conversion factor is 1 kcal = 4184 J = 4.184 kJ (kilojoule), and we will list these values in parentheses in key places.

A more convenient way of depicting valence electrons is by means of dots around the symbol for the element. In this case, the letters represent the nucleus including all the electrons in the inner shells, together called the **core configuration.**

Valence Electrons as Electron Dots

Li·   ·Be   ·B·   ·C·   ·N·   :O·   :F·

Na·   ·Mg   ·Al·   ·Si·   ·P·   :S·   :Cl·

Electron-Dot Picture of Salts

$$Na· + ·\ddot{\underset{..}{Cl}}: \xrightarrow{1\,e\ transfer} Na^+ :\ddot{\underset{..}{Cl}}:^-$$

$$·Mg + 2·\ddot{\underset{..}{Cl}}: \xrightarrow{2\,e\ transfer} Mg^{2+} [:\ddot{\underset{..}{Cl}}:]_2^-$$

The hydrogen atom is unique because it may either lose an electron to become a bare nucleus, the **proton,** or accept an electron to form the **hydride ion,** [H, i.e., H:]$^-$, which possesses the helium configuration. Indeed, the hydrides of lithium, sodium, and potassium (Li^+H^-, Na^+H^-, and K^+H^-) are commonly used reagents.

H· $\xrightarrow{-1\,e}$ [H]$^+$ Bare nucleus   IP = 314 kcal mol^{-1} (1314 kJ mol^{-1})
        Proton

H· $\xrightarrow{+1\,e}$ [H:]$^-$ Helium configuration   EA = $-$18 kcal mol^{-1} ($-$75 kJ mol^{-1})
        Hydride ion

Exercise 1-2

Draw electron-dot pictures for ionic LiBr, Na_2O, BeF_2, $AlCl_3$, and MgS.

In covalent bonds, electrons are shared to achieve octet configurations

Formation of ionic bonds between two identical elements is difficult because the electron transfer is usually very unfavorable. For example, in H_2, formation of H^+H^- would require an energy input of nearly 300 kcal mol^{-1} (1255 kJ mol^{-1}). For the same reason, none of the halogens, F_2, Cl_2, Br_2, and I_2, has an ionic bond. The high IP of hydrogen also prevents the bonds in the hydrogen halides from being ionic. For elements nearer the center of the periodic table, the formation of ionic bonds is unfeasible, because it becomes more and more difficult to donate or accept enough electrons to attain the noble-gas configuration. Such is the case for carbon, which would have to shed four electrons to reach the helium electronic structure or add four electrons for a neon-like arrangement. The large amount of charge that would develop makes these processes very energetically unfavorable.

$$C^{4+} \xleftarrow{-4\,e} ·\overset{.}{\underset{.}{C}}· \xrightarrow{+4\,e} :\ddot{\underset{..}{C}}:^{4-}$$

Helium   **Very difficult**   **Neon**
configuration   **configuration**

Instead, **covalent bonding** takes place: The elements *share* electrons so that each atom attains a noble-gas configuration. Typical products of such sharing are H_2 and HCl. In HCl, the chlorine atom assumes an octet structure by sharing one of its valence electrons with that of hydrogen. Similarly, the chlorine molecule, Cl_2, is diatomic because both component atoms gain octets by sharing two electrons. Such bonds are called **covalent single bonds.**

Electron-Dot Picture of Covalent Single Bonds

$$H \cdot \; + \; \cdot H \longrightarrow H \colon H$$

$$H \cdot \; + \; \cdot \ddot{\underset{\cdot\cdot}{Cl}} \colon \longrightarrow H \colon \ddot{\underset{\cdot\cdot}{Cl}} \colon$$

Shared electron pairs

$$\colon \ddot{\underset{\cdot\cdot}{Cl}} \cdot \; + \; \cdot \ddot{\underset{\cdot\cdot}{Cl}} \colon \longrightarrow \colon \ddot{\underset{\cdot\cdot}{Cl}} \colon \ddot{\underset{\cdot\cdot}{Cl}} \colon$$

Because carbon has four valence electrons, it must acquire a share of four electrons to gain the neon configuration, as in methane. Nitrogen has five valence electrons and needs three to share, as in ammonia; and oxygen, with six valence electrons, requires only two to share, as in water.

$$\begin{array}{c} \quad\; H \\ H \colon \overset{\cdot\cdot}{\underset{\cdot\cdot}{C}} \colon H \\ \quad\; H \end{array} \qquad \begin{array}{c} \quad\;\; \cdot\cdot \\ H \colon \overset{}{\underset{\cdot\cdot}{N}} \colon H \\ \quad\; H \end{array} \qquad \begin{array}{c} \cdot\cdot \\ H \colon \overset{}{\underset{\cdot\cdot}{O}} \colon H \end{array}$$

Methane   **Ammonia**   **Water**

It is possible for one atom to supply both of the electrons required for covalent bonding. This occurs upon addition of a proton to ammonia, thereby forming $NH_4{}^+$, or to water, thereby forming H_3O^+.

$$\begin{array}{c} \;\; \cdot\cdot \\ H \colon \overset{}{\underset{\cdot\cdot}{N}} \colon \; + \; H^+ \longrightarrow \left[\begin{array}{c} \;\; H \\ \;\; \cdot\cdot \\ H \colon \overset{}{\underset{}{N}} \colon H \\ \;\; H \end{array} \right]^+ \qquad H \colon \overset{\cdot\cdot}{\underset{\cdot\cdot}{O}} \colon \; + \; H^+ \longrightarrow \left[\begin{array}{c} \cdot\cdot \\ H \colon \overset{}{\underset{}{O}} \colon H \\ H \end{array} \right]^+ \end{array}$$

Ammonium ion   **Hydronium ion**

Besides two-electron (**single**) bonds, atoms may form four-electron (**double**) and six-electron (**triple**) bonds to gain noble-gas configurations. Atoms that share more than one electron pair are found in ethene and ethyne.

$$\begin{array}{c} H \quad\;\;\; H \\ \;\; \overset{\cdot\cdot}{\underset{}{C}} \colon\colon \overset{\cdot\cdot}{\underset{}{C}} \\ H \quad\;\;\; H \end{array} \qquad H \colon C \colon\colon\colon C \colon H$$

Ethene   **Ethyne**
(Ethylene)*   **(Acetylene)***

The drawings above, with pairs of electron dots representing bonds, are also called **Lewis[†] structures.** We shall develop the general rules for formulating such structures in Section 1-4.

Exercise 1-3

Draw electron-dot structures for F_2, CF_4, CH_2Cl_2, PH_3, BrI, HO^-, H_2N^-, and H_3C^-. (Where applicable, the first element is at the center of the molecule.) Make sure that all atoms have noble-gas electron configurations.

In most organic bonds, the electrons are not shared equally: polar covalent bonds

The preceding two sections presented two extreme ways in which atoms attain noble-gas configurations by entering into bonding: pure ionic and pure covalent. In reality, most bonds are of a nature that lies between these two extremes: **polar covalent.** Thus, the ionic bonds in most salts have some covalent character; conversely, the covalent bonds to carbon have some ionic or polar character. Recall (Section 1-2) that both sharing of electrons *and*

---

*In labels of molecules, systematic names (introduced in Section 2-6) will be given first, followed in parentheses by so-called common names that are still used frequently.
[†]Professor Gilbert N. Lewis (1875–1946), University of California, Berkeley.

Table 1-2	Electronegativities of Selected Elements

Increasing electronegativity →

Increasing electronegativity ↑

			H 2.2			
Li 1.0	Be 1.6	B 2.0	C 2.6	N 3.0	O 3.4	F 4.0
Na 0.9	Mg 1.3	Al 1.6	Si 1.9	P 2.2	S 2.6	Cl 3.2
K 0.8						Br 3.0
						I 2.7

Note: Values established by L. Pauling and updated by A. L. Allred (see *Journal of Inorganic and Nuclear Chemistry*, 1961, *17*, 215).

coulombic attraction contribute to the stability of a bond. How polar are polar covalent bonds, and what is the direction of the polarity?

We can answer these questions by considering the periodic table and keeping in mind that the positive nuclear charge of the elements increases from left to right. Thus, the elements on the left of the periodic table are often called **electropositive,** electron donating, or "electron pushing," because their electrons are held by the nucleus less tightly than are those of elements to the right. These elements at the right of the periodic table are described as **electronegative,** electron accepting, or "electron pulling." Table 1-2 lists the relative electronegativities of some elements. On this scale, fluorine, the most electronegative of them all, is assigned the value 4.

Consideration of Table 1-2 readily explains why the most ionic (least covalent) bonds occur between elements at the two extremes (e.g., the alkali metal salts, such as sodium chloride). On the other hand, the purest covalent bonds are formed between atoms of equal electronegativity (i.e., identical elements, as in H_2, N_2, O_2, F_2, and so on) or in carbon–carbon bonds. However, most covalent bonds are between atoms of differing electronegativity, resulting in their **polarization.** The polarization of a bond is the consequence of a shift of the center of electron density in the bond toward the more electronegative atom. It is indicated in a very qualitative manner (using the Greek letter delta, δ) by designating a partial positive charge, δ^+, and partial negative charge, δ^-, to the respective less or more electronegative atom. The larger the difference in electronegativity, the bigger is the charge separation. As a rule of thumb, electronegativity differences of 0.3 to 2.0 units indicate polar covalent bonds; lesser values are typical of essentially "pure" covalent bonds, larger values of "pure" ionic ones.

The separation of opposite charges is called an electric **dipole,** symbolized by an arrow crossed at its tail and pointing from positive to negative. A polarized bond can impart polarity to a molecule as a whole, as in HF, ICl, and CH_3F.

Molecules Can Have Polar Bonds but No Net Polarization

Dipoles cancel

Carbon dioxide

Tetrachloromethane

Polar Bonds

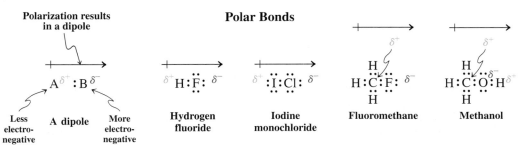

In symmetrical structures, the polarizations of the individual bonds may cancel, thus leading to molecules with no net polarization, such as CO_2 and CCl_4 (margin). To know whether a molecule is polar, we have to know its shape, because the net polarity is the vector sum of the bond dipoles. The electrostatic potential maps in the margin clearly illustrate the polarization in CO_2 and CCl_4, showing the respective carbon atoms shaded relatively blue, the attached, more electronegative atoms relatively red. Moreover, you can recognize how the shape of each molecule renders it nonpolar as a whole. There are two cautions in viewing electrostatic potential maps: (1) The scale on which the color differentials are rendered may vary. For example, a much more sensitive scale is used for the molecules in the margin of p. 11, in which the charges are only partial, than for NaCl on p. 8, in which the atoms assume full charges. Thus, it may be misleading to compare the electrostatic potential maps of one set of molecules with those of another, electronically very different group. Most organic structures shown in this book will be on a comparative scale, unless mentioned otherwise. (2) Because of the way in which the potential at each point is calculated, it will contain contributions from all nuclei and electrons in the vicinity. As a consequence, the color of the spatial regions around individual nuclei is not uniform.

Valence electron repulsion controls the shapes of molecules

Molecules adopt shapes in which electron repulsion (including both bonding and nonbonding electrons) is minimized. In diatomic species such as H_2 or LiH, there is only one bonding electron pair and one possible arrangement of the two atoms. However, beryllium fluoride, BeF_2, is a triatomic species. Will it be bent or linear? Electron repulsion is at a minimum in a **linear** structure, because the bonding and nonbonding electrons are placed as far from each other as possible, at 180°.* Linearity is also expected for other derivatives of beryllium, as well as of other elements in the same column of the periodic table.

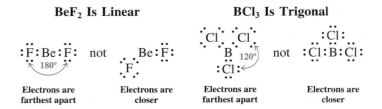

BeF₂ Is Linear   **BCl₃ Is Trigonal**

Electrons are farthest apart — Electrons are closer — Electrons are farthest apart — Electrons are closer

In boron trichloride, the three valence electrons of boron allow it to form covalent bonds with three chlorine atoms. Electron repulsion enforces a regular **trigonal** arrangement—that is, the three halogens are at the corners of an equilateral triangle, the center of which is occupied by boron, and the bonding (and nonbonding) electron pairs of the respective chlorine atoms are at maximum distance from each other, that is, 120°. Other derivatives of boron, and the analogous compounds with other elements in the same column of the periodic table, are again expected to adopt trigonal structures.

Applying this principle to carbon, we can see that methane, CH_4, has to be **tetrahedral.** Placing the four hydrogens at the vertices of a tetrahedron minimizes the electron repulsion of the corresponding bonding electron pairs.

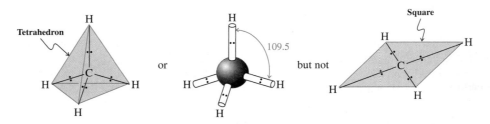

This method for determining molecular shape by minimizing electron repulsion is called the *valence-shell electron-pair repulsion* (*VSEPR*) method. Note that we often draw molecules such as BCl_3 and CH_4 as if they were flat and had 90° angles. *This depiction is for*

---

*This is true only in the gas phase. At room temperature, BeF_2 is a solid (it is used in nuclear reactors) that exists as a complex network of linked Be and F atoms, not as a distinct linear triatomic structure.

ease of drawing only. Do *not* confuse such two-dimensional drawings with the true three-dimensional molecular shapes (trigonal for BCl_3 and tetrahedral for CH_4).

Exercise 1-4

Show the bond polarization in H_2O, SCO, SO, IBr, CH_4, $CHCl_3$, CH_2Cl_2, and CH_3Cl by using dipole arrows to indicate separation of charge. (In the last four examples, place the carbon in the center of the molecule.)

Exercise 1-5

Ammonia, $:NH_3$, is not trigonal but pyramidal, with bond angles of 107.3°. Water, $H_2\overset{..}{\underset{..}{O}}$, is not linear but bent (104.5°). Why? (**Hint:** Consider the effect of the nonbonding electron pairs.)

In Summary There are two extreme types of bonding, ionic and covalent. Both derive favorable energetics from Coulomb forces and the attainment of noble-gas electronic structures. Most bonds are better described as something between the two types: the polar covalent (or covalent ionic) bonds. Polarity in bonds may give rise to polar molecules. The outcome depends on the shape of the molecule, which is determined in a simple manner by arrangement of its bonds and nonbonding electrons to minimize electron repulsion.

1-4 | ELECTRON-DOT MODEL OF BONDING: LEWIS STRUCTURES

Lewis structures are important for predicting geometry and polarity (hence reactivity) of organic compounds, and we shall use them for that purpose throughout this book. In this section, we provide rules for writing such structures correctly and for keeping track of valence electrons.

Lewis structures are drawn by following simple rules

The procedure for drawing correct electron-dot structures is straightforward, as long as the following rules are observed.

Rule 1. Draw the (given or desired) *molecular skeleton.* As an example, consider methane. The molecule has four hydrogen atoms bonded to one central carbon atom.

```
      H
    H C H      H H C H H
      H
    Correct       Incorrect
```

Rule 2. Count the number of available valence electrons. Add up all the valence electrons of the component atoms. Special care has to be taken with charged structures (anions or cations), in which case the appropriate number of electrons has to be added or subtracted to account for extra charges.

CH_4	4 H	4×1 electron $=$	4 electrons		HBr	1 H	1×1 electron $=$	1 electron
	1 C	1×4 electrons $=$	4 electrons			1 Br	1×7 electrons $=$	7 electrons
		Total	8 electrons				Total	8 electrons
H_3O^+	3 H	3×1 electron $=$	3 electrons		NH_2^-	2 H	2×1 electron $=$	2 electrons
	1 O	1×6 electrons $=$	6 electrons			1 N	1×5 electrons $=$	5 electrons
	Charge	$+1$	$= -1$ electron			Charge	-1	$= +1$ electron
		Total	8 electrons				Total	8 electrons

Rule 3. (The **octet rule**) *Depict all covalent bonds by two shared electrons, giving as many atoms as possible a surrounding electron octet, except for H, which requires a duet.* Make sure that the number of electrons used is *exactly* the number counted according to rule 2. Elements at the right in the periodic table may contain pairs of valence electrons not used for bonding, called **lone electron pairs** or just **lone pairs.**

Consider, for example, hydrogen bromide. The shared electron pair supplies the hydrogen atom with a duet, the bromine with an octet, because the bromine carries three lone electron pairs. Conversely, in methane, the four C–H bonds satisfy the requirement of the hydrogens and, at the same time, furnish the octet for carbon. Examples of correct and incorrect Lewis structures for HBr are shown below.

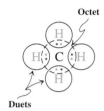

Nonbonding or lone electron pairs

Duet   Octet

Octet

Duets

| **Correct Lewis Structure** | **Incorrect Lewis Structures** |

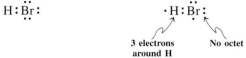   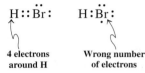

Frequently, the number of valence electrons is not sufficient to satisfy the octet rule only with single bonds. In this event, double bonds (two shared electron pairs) and even triple bonds (three shared pairs) are necessary to obtain octets. An example is the nitrogen molecule, N_2, which has ten valence electrons. An N–N single bond would leave both atoms with electron sextets, and a double bond provides only one nitrogen atom with an octet. It is the molecule with a triple bond that satisfies both atoms. You may find a simple procedure useful that gives you the total number of bonds needed in a molecule to give every atom an octet (or duet). Thus, after you have counted the supply of available electrons (rule 2), add up the total "electron demand," that is, two electrons for each hydrogen atom and eight for each other element atom. Then subtract supply from "demand" and divide by 2. For N_2, demand is 16 electrons, supply is 10, and hence the number of bonds is 3.

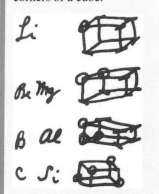
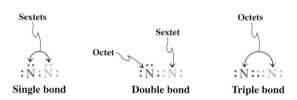

Sextets

Octet   Sextet   Octets

:N:N: :N::N: :N:::N:
Single bond   **Double bond**   **Triple bond**

Further examples of molecules with double and triple bonds are shown below.

Correct Lewis Structures

Ethene
(Ethylene)

Ethyne
(Acetylene)

Formaldehyde

In practice, another simple sequence may help. First, connect all mutually bonded atoms in your structure by single bonds (i.e., shared electron pairs); second, if there are any electrons left, distribute them as lone electron pairs to maximize the number of octets; and finally, if some of the atoms lack octet structures, change as many lone electron pairs into shared electron pairs as required to complete the octet shells (see also the Worked Examples 1-23 and 1-24).

Exercise 1-6

Draw Lewis structures for the following molecules: HI, $CH_3CH_2CH_3$, CH_3OH, HSSH, SiO_2 (OSiO), O_2, CS_2 (SCS).

Rule 4. Assign (formal) charges to atoms in the molecule. Each lone pair contributes two electrons to the valence electron count of an atom in a molecule, and each bonding (shared) pair contributes one. An atom is charged if this total is different from the outer-shell electron count in the free, nonbonded atom. Thus we have the formula

$$\text{Formal charge} = \left(\begin{array}{c} \text{number of outer-shell} \\ \text{electrons on the} \\ \text{free, neutral atom} \end{array} \right) - \left(\begin{array}{c} \text{number of unshared} \\ \text{electrons on the atom} \\ \text{in the molecule} \end{array} \right) - \frac{1}{2} \left(\begin{array}{c} \text{number of bonding} \\ \text{electrons surrounding the} \\ \text{atom in the molecule} \end{array} \right)$$

or simply

$$\textit{Formal charge} = \textit{number of valence electrons} - \textit{number of lone pair electrons}$$
$$- \frac{1}{2} \textit{number of bonding electrons}$$

The reason for the term *formal* is that, in molecules, charge is not localized on one atom but is distributed to varying degrees over its surroundings.

As an example, which atom bears the positive charge in the hydronium ion? Each hydrogen has a valence electron count of 1 from the shared pair in its bond to oxygen. Because this value is the same as the electron count in the free atom, the (formal) charge on each hydrogen is zero. The electron count on the oxygen in the hydronium ion is 2 (the lone pair) + 3 (half of 6 bonding electrons) = 5. This value is one short of the number of outer-shell electrons in the free atom, thus giving the oxygen a charge of +1. Hence the positive charge is assigned to oxygen.

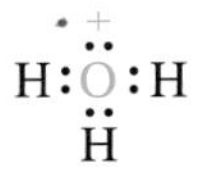

Another example is the nitrosyl cation, NO^+. The molecule bears a lone pair on nitrogen, in addition to the triple bond connecting the nitrogen to the oxygen atom. This gives nitrogen five valence electrons, a value that matches the count in the free atom; therefore the nitrogen atom has no charge. The same number of valence electrons (5) is found on oxygen. Because the free oxygen atom requires six valence electrons to be neutral, the oxygen in NO^+ possesses the +1 charge. Other examples are shown below.

Hydronium ion

Nitrosyl cation

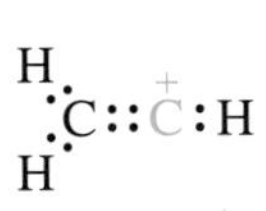

Ethenyl (vinyl) cation 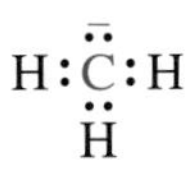 | Methyl anion 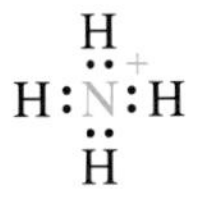 | Ammonium ion 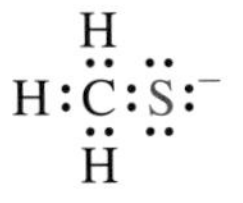 | Methanethiolate ion 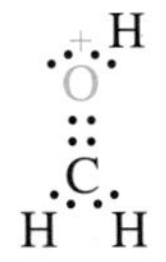 | Protonated formaldehyde

Charge-separated Lewis structures

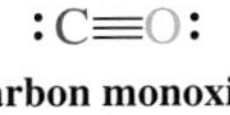

Carbon monoxide

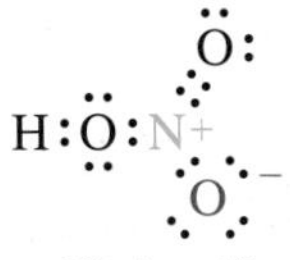

Nitric acid

Sometimes the octet rule leads to charges on atoms even in neutral molecules. The Lewis structure is then said to be **charge separated**. An example is carbon monoxide, CO. Some compounds containing nitrogen–oxygen bonds, such as nitric acid, HNO_3, also exhibit this property.

The octet rule does not always hold

The octet rule strictly holds only for the elements of the second row and then only if there is a sufficient number of valence electrons to satisfy it. Thus, there are three exceptions to be considered.

Exception 1. You will have noticed that all our examples of "correct" Lewis structures contain an even number of electrons; that is, all are distributed as bonding or lone pairs. This distribution is not possible in species having an odd number of electrons, such as nitrogen oxide (NO) and neutral methyl (methyl radical, $\cdot CH_3$; see Section 3-1).

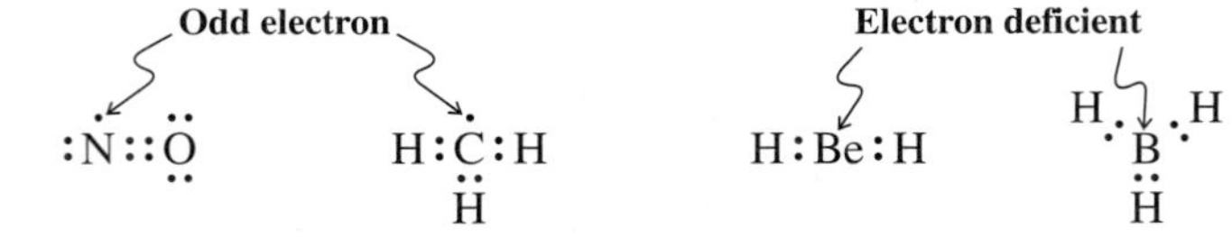

Nitrogen oxide **Methyl radical** **Beryllium hydride** **Borane**

Exception 2. Some compounds of the early second-row elements, such as BeH_2 and BH_3, have a deficiency of valence electrons.

Because compounds falling under exceptions 1 and 2 do not have octet configurations, they are unusually reactive and transform readily in reactions that lead to octet structures. For example, two molecules of $\cdot CH_3$ react with each other spontaneously to give ethane, $CH_3–CH_3$, and BH_3 reacts with hydride, H^-, to give borohydride, BH_4^-.

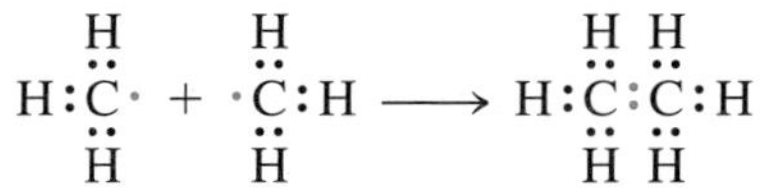

Ethane

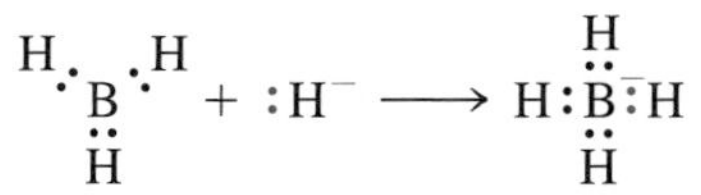

Borohydride

Exception 3. Beyond the second row, the simple Lewis model is not strictly applied, and elements may be surrounded by more than eight valence electrons, a feature referred to as **valence-shell expansion.** For example, phosphorus and sulfur (as relatives of nitrogen and oxygen) are trivalent and divalent, respectively, and we can readily formulate Lewis octet structures for their derivatives. But they also form stable compounds of higher valency, among them the familiar phosphoric and sulfuric acids. Some examples of octet and expanded-octet molecules containing these elements are shown below.

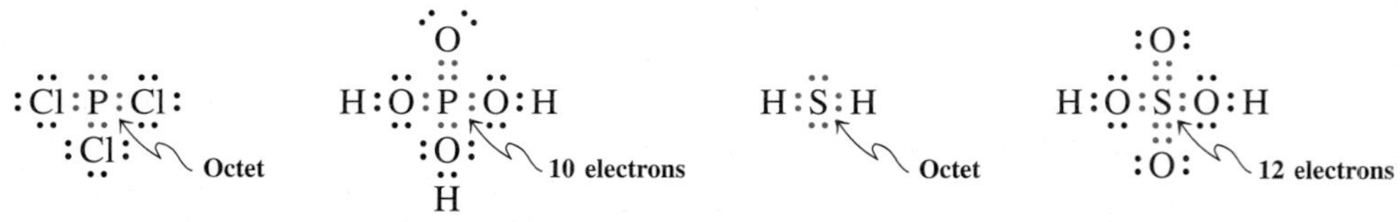

Phosphorous trichloride **Phosphoric acid** **Hydrogen sulfide** **Sulfuric acid**

An explanation for this apparent violation of the octet rule is found in a more sophisticated description of atomic structure by quantum mechanics (Section 1-6). However, you will notice that, even in these cases, you can construct dipolar forms in which the Lewis octet rule is preserved (see Section 1-5). Indeed, structural and computational data show that these formulations contribute strongly to the resonance picture of such molecules.

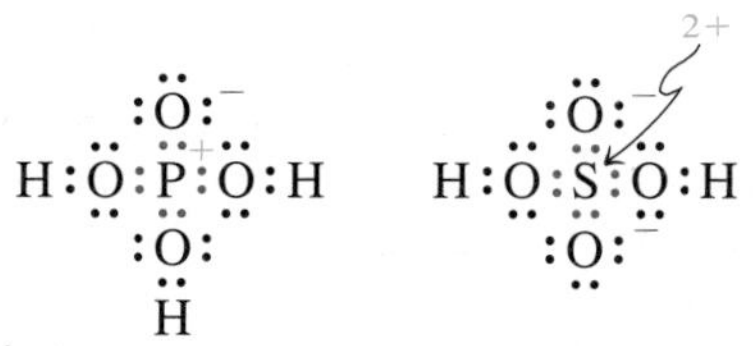

Covalent bonds can be depicted as straight lines

Electron-dot structures can be cumbersome, particularly for larger molecules. It is simpler to represent covalent single bonds by single straight lines; double bonds are represented by two lines and triple bonds by three. Lone electron pairs can either be shown as dots or simply omitted. The use of such notation was first suggested by the German chemist August Kekulé,* long before electrons were discovered; structures of this type are often called **Kekulé structures.**

Straight-Line Notation for the Covalent Bond

| Methane | Diatomic nitrogen | Ethene | Hydronium ion | Protonated formaldehyde |

Solved Exercise 1-7 | Working with the Concepts: Drawing Lewis Structures

Draw the Lewis structure of $HClO_2$ (HOClO), including the assignment of any charges to atoms.

Strategy
To solve such a problem, it is best to follow one by one the rules given in this section for drawing Lewis structures.

Solution
- *Rule 1:* The molecular skeleton is given as unbranched, as shown.
- *Rule 2:* Count the number of valence electrons:

$$H = 1, 2\ O = 12, Cl = 7, \text{total} = 20$$

- *Rule 3:* How many bonds (shared electron pairs) do we need? The supply of electrons is 20; the electron requirement is 2 for H and $3 \times 8 = 24$ electrons for the other three atoms, for a total of 26 electrons. Thus we need $(26 - 20)/2 = 3$ bonds.

   To distribute all valence electrons according to the octet rule, we first connect all atoms by two-electron bonds, H:O:Cl:O, using up 6 electrons. Second, we distribute the remaining 14 electrons to provide octets for all nonhydrogen atoms, (arbitrarily) starting at the left oxygen. This process requires in turn 4, 4, and 6 electrons, resulting in octet structures without needing additional electron sharing:

$$H:\overset{..}{\underset{..}{O}}:\overset{..}{\underset{..}{Cl}}:\overset{..}{\underset{..}{O}}:$$

- *Rule 4:* We determine any formal charges by noting any discrepancies between the "effective" valence electron count around each atom in the molecule we have found and its outer-shell count when isolated. For H in HOClO, the valence electron count is 1, which is the same as in the H atom, so it is neutral in the molecule. For the neighboring oxygen, the two values are again the same, 6. For Cl, the effective electron count is 6, but the neutral atom requires 7. Therefore, Cl bears a positive charge. For the terminal O, the electron counts are 7 (in the molecule) and 6 (neutral atom), giving it a negative charge. The final result is

$$H:\overset{..}{\underset{..}{O}}:\overset{+\ \ ..}{\underset{..}{Cl}}:\overset{-\ \ ..}{\underset{..}{O}}:$$

Exercise 1-8 | Try It Yourself

Draw Lewis structures of the following molecules, including the assignment of any charges to atoms (the order in which the atoms are attached is given in parentheses, when it may not be obvious from the formula as it is commonly written): SO, F_2O (FOF), BF_3NH_3 (F_3BNH_3), $CH_3OH_2^+$ ($H_3COH_2^+$), $Cl_2C{=}O$, CN^-, C_2^{2-}. (**Caution:** To draw Lewis structures correctly, it is essential that you know the number of valence electrons that belong to each atom. If you do not know this number, *look it up* before you begin. If a structure is charged, you must adjust the total number of valence electrons accordingly. For example, a species with a charge of -1 must have one electron more than the total number of valence electrons contributed by the constituent atoms.)

---

*Professor F. August Kekulé von Stradonitz (1829–1896), University of Bonn, Germany.

In Summary Lewis structures describe bonding by the use of electron dots or straight lines. Whenever possible, they are drawn so as to give hydrogen an electron duet and other atoms an electron octet. Charges are assigned to each atom by evaluating its electron count.

1-5 | RESONANCE FORMS

In organic chemistry, we also encounter molecules for which there are *several* correct Lewis structures.

The carbonate ion has several correct Lewis structures

Let us consider the carbonate ion, CO_3^{2-}. Following our rules, we can easily draw a Lewis structure (A) in which every atom is surrounded by an octet. The two negative charges are located on the bottom two oxygen atoms; the third oxygen is neutral, connected to the central carbon by a double bond and bearing two lone pairs. But why choose the bottom two oxygen atoms as the charge carriers? There is no reason at all—it is a completely arbitrary choice. We could equally well have drawn structure B or C to describe the carbonate ion. The three correct Lewis pictures are called **resonance forms.**

Resonance Forms of the Carbonate Ion

Red arrows ⌒ denote the movement of electron pairs: "electron pushing"

The individual resonance forms are connected by double-headed arrows and are placed within one set of square brackets. They have the characteristic property of being interconvertible by *electron-pair movement only,* indicated by red arrows, the nuclear positions in the molecule remaining *unchanged.* Note that, to turn A into B and then into C, we have to shift two electron pairs in each case. Such movement of electrons can be depicted by curved arrows, a procedure informally called "electron pushing."

The use of curved arrows to depict electron-pair movement is a useful technique that will prevent us from making the common mistake of changing the total number of electrons when we draw resonance forms. It is also advantageous in keeping track of electrons when formulating mechanisms (Sections 2-2 and 6-3).

But what is its true structure?

Does the carbonate ion have one uncharged oxygen atom bound to carbon through a double bond and two other oxygen atoms bound through a single bond each, both bearing a negative charge, as suggested by the Lewis structures? Or, to put it differently, are A, B, and C equilibrating isomers? *The answer is no.* If that were true, the carbon–oxygen bonds would be of different lengths, because double bonds are normally shorter than single bonds. But the carbonate ion is *perfectly symmetrical* and contains a trigonal central carbon, all C–O bonds being of equal length—between the length of a double and that of a single bond. The negative charge is evenly distributed over all three oxygens: It is said to be **delocalized,** in accord with the tendency of electrons to "spread out in space" (Section 1-2). In other words, none of the individual Lewis representations of this molecule is correct on its own. Rather, *the true structure is a composite of A, B, and C.*

Carbonate ion

The resulting picture is called a **resonance hybrid.** Because A, B, and C are equivalent (i.e., each is composed of the same number of atoms, bonds, and electron pairs), they contribute equally to the true structure of the molecule, but none of them by itself accurately represents it.

Because it minimizes coulombic repulsion, delocalization by resonance has a stabilizing effect: The carbonate ion is considerably more stable than would be expected for a doubly negatively charged organic molecule.

The word *resonance* may imply to you that the molecule vibrates or equilibrates from one form to another. This inference is incorrect. The molecule *never* looks like any of the individual resonance forms; it has only one structure, the resonance hybrid. Unlike substances in ordinary chemical equilibria, resonance forms are *not* real, although each makes a partial contribution to reality. The trigonal symmetry of carbonate is clearly evident in its electrostatic potential map shown on p. 18.

An alternative convention used to describe resonance hybrids such as carbonate is to represent the bonds as a combination of solid and dotted lines. The $\frac{2}{3}-$ sign here indicates that a partial charge ($\frac{2}{3}$ of a negative charge) resides on each oxygen atom (see margin). The equivalence of all three carbon–oxygen bonds and all three oxygens is clearly indicated by this convention. Other examples of resonance hybrids of octet Lewis structures are the acetate ion and the 2-propenyl (allyl) anion.

Acetate ion

2-Propenyl (allyl) anion

Resonance is also possible for nonoctet molecules. For example, the 2-propenyl (allyl) cation is stabilized by resonance.

Sextet

Sextet

2-Propenyl (allyl) cation

When drawing resonance forms, keep in mind that (1) pushing one electron pair toward one atom and away from another results in a movement of charge; (2) the relative positions of all the atoms stay unchanged—only electrons are moved; (3) equivalent resonance forms contribute equally to the resonance hybrid; and (4) the arrows connecting resonance forms are double headed ($\leftrightarrow$).

The recognition and formulation of resonance forms is important in predicting reactivity. For example, reaction of carbonate with acid can occur at any two of the three oxygens to give carbonic acid, H_2CO_3 (which is actually in equilibrium with CO_2 and H_2O).

Dotted-Line Notation of Carbonate as a Resonance Hybrid

Acetate ion

2-Propenyl anion

2-Propenyl cation

Carbonic acid

Acetic acid

Similarly, acetate ion is protonated at either oxygen to form acetic acid (see margin on p. 19). Analogously, the 2-propenyl anion is protonated at either terminus to furnish propene, and the corresponding cation reacts with hydroxide at either end to give the corresponding alcohol (see below).

You may find it easier to picture resonance by thinking about combining colors to produce a new one. For example, mixing yellow—one resonance form—and blue—a second resonance form—results in the color green: the resonance hybrid.

Enolate ion

Exercise 1-9

(a) Draw two resonance forms for nitrite ion, NO_2^-. What can you say about the geometry of this molecule (linear or bent)? (**Hint:** Consider the effect of electron repulsion exerted by the lone pair on nitrogen.) **(b)** The possibility of valence-shell expansion increases the number of feasible resonance forms, and it is often difficult to decide on one that is "best." One criterion that is used is whether the Lewis structure predicts bond lengths and bond angles with reasonable accuracy. Draw Lewis octet and valence-shell-expanded resonance forms for SO_2 (OSO). Considering the Lewis structure for SO (Exercise 1-8), its experimental bond length of 1.48 Å, and the measured S–O distance in SO_2 of 1.43 Å, which one of the various structures would you consider "best"?

Not all resonance forms are equivalent

The molecules described above all have equivalent resonance forms. However, many molecules are described by resonance forms that are not equivalent. An example is the enolate ion. The two resonance forms differ in the locations of both the double bond and the charge.

The Two Nonequivalent Resonance Forms of the Enolate Ion

Although both forms are contributors to the true structure of the ion, we shall see that one contributes more than the other. The question is, which one? If we extend our consideration of nonequivalent resonance forms to include those containing atoms without electron octets, the question becomes more general.

$$\left[\text{Octet} \longleftrightarrow \text{Nonoctet}\right] \text{ Resonance Forms}$$

Formaldehyde

Sulfuric acid

Such an extension requires that we relax our definitions of "correct" and "incorrect" Lewis structures and broadly regard *all* resonance forms as potential contributors to the true picture of a molecule. The task is then to recognize which resonance form is the most important one. In other words, which one is the **major resonance contributor?** Here are some guidelines.

Guideline 1. Structures with a maximum of octets are most important. In the enolate ion, all component atoms in either structure are surrounded by octets. Consider, however, the nitrosyl cation, NO^+: One resonance form has a positive charge on oxygen with electron octets around both atoms; the other form places the positive charge on nitrogen, thereby resulting in an electron sextet on this atom. Because of the octet rule, the second structure contributes less to the hybrid. Thus, the N–O linkage is closer to being a triple than a double bond, and more of the positive charge is on oxygen than on nitrogen. Similarly, the dipolar resonance form for formaldehyde (p. 20) generates an electron sextet around carbon, rendering it a minor contributor. The possibility of valence shell expansion for third-row elements (Section 1-4) makes the non-charge-separated picture of sulfuric acid with 12 electrons around sulfur a feasible resonance form, but the dipolar octet structure is better.

Guideline 2. Charges should be preferentially located on atoms with compatible electronegativity. Consider again the enolate ion. Which is the major contributing resonance form? Guideline 2 requires it to be the first, in which the negative charge resides on the more electronegative oxygen atom. Indeed, the electrostatic potential map shown in the margin on p. 20 confirms this expectation.

Looking again at NO^+, you might find guideline 2 confusing. The major resonance contributor to NO^+ has the positive charge on the more electronegative oxygen. In cases such as this, *the octet rule overrides the electronegativity criterion;* that is, guideline 1 takes precedence over guideline 2.

Guideline 3. Structures with less separation of opposite charges are more important resonance contributors than those with more charge separation. This rule is a simple consequence of Coulomb's law: Separating opposite charges requires energy; hence neutral structures are better than dipolar ones. An example is formic acid, shown below. However, the influence of the minor dipolar resonance form is evident in the electrostatic potential map in the margin: The electron density at the carbonyl oxygen is greater than that at its hydroxy counterpart.

Formic acid

In some cases, to draw octet Lewis structures, charge separation is necessary; that is, guideline 1 takes precedence over guideline 3. An example is carbon monoxide. Other examples are phosphoric and sulfuric acids, although valence-shell expansion allows the formulation of expanded octet structures (see also Section 1-4 and guideline 1).

When there are several charge-separated resonance forms that comply with the octet rule, the most favorable is the one in which the charge distribution best accommodates the relative electronegativities of the component atoms (guideline 2). In diazomethane, for example, nitrogen is more electronegative than carbon, thus allowing a clear choice between the two resonance contributors (see also the electrostatic potential map in the margin).

Diazomethane

Sextet

Nitrosyl cation

Major resonance contributor   Minor resonance contributor

Enolate ion

Major resonance contributor

Formic acid

Sextet

Minor   Major

Carbon monoxide

Diazomethane

Solved Exercise 1-10 | Working with the Concepts: Drawing Resonance Forms

Draw two all-octet resonance forms for nitrosyl chloride, ONCl. Which one is better?

Strategy

To formulate any octet structure, we follow the rules given in Section 1-4 for drawing Lewis structures. Once this task is completed, we can apply the procedures and guidelines from the present section to derive resonance forms and evaluate their relative contributions.

Solution

- *Rule 1:* The molecular skeleton is given as shown.
- *Rule 2:* Count the number of valence electrons:

$$N = 5, O = 6, Cl = 7, total = 18$$

- *Rule 3:* How many bonds (shared electron pairs) do we need? The supply of electrons is 18; and the electron requirement is $3 \times 8 = 24$ electrons for the three atoms. Thus we need $(24 - 18)/2 = 3$ bonds. Because there are only three atoms, there has to be a double bond.

    To distribute all valence electrons according to the octet rule, we first connect all atoms by two-electron bonds, O:N:Cl, using up 4 electrons. Second, we use 2 electrons for a double bond, arbitrarily added to the left portion to render O::N:Cl. Third, we distribute the remaining 12 electrons to provide octets for all atoms, (again arbitrarily) starting at the left with oxygen. This process requires in turn 4, 2, and 6 electrons, resulting in the octet structure Ö::N:C̈l:, which we shall label A.
- *Rule 4:* We determine any formal charges by noting any discrepancies between the "effective" valence electron count around each atom in A and its outer-shell count when it is isolated. For O, the two lone pairs and the double bond give a valence electron count of 6, as in the O atom; for N it is 5 and for Cl it is 7, again as found in the respective neutrals. Thus, there is no formal charge in A.
- We are now ready to formulate resonance forms of A by moving electron pairs. You should try to do so for all such electrons. You will soon find that only one type of electron movement furnishes an all-octet structure, namely, that shown on the left below, leading to B:

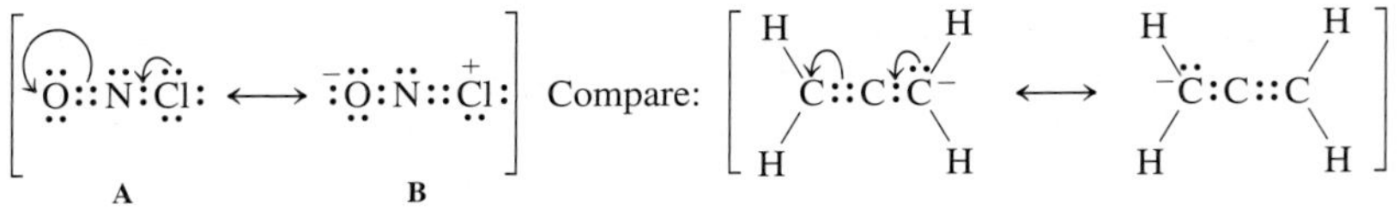

A   B
Nitrosyl chloride

Compare:

2-Propenyl (allyl) anion

This movement is similar to that in the 2-propenyl (allyl) anion (as shown on the right above) and related allylic resonance systems described in this section. Because we started with a charge-neutral formula, the electron movement to its resonance form generates charges: a positive one at the origin of the electron movement, a negative one at its terminus.
- Which one of the two resonance forms of ONCl is better? Inspection of the three guidelines provided in this section helps us to the answer: Guideline 3 tells us that less charge separation is better than more. Thus, charge-neutral A is a better descriptor of nitrosyl chloride than charge-separated B.

Exercise 1-11 | Try It Yourself

Draw resonance forms for the following two molecules. Indicate the more favorable resonance contributor in each case. **(a)** CNO⁻; **(b)** NO⁻.

To what extent can the assignment of a major resonance form help us predict reactivity? The answer is not simple, as it depends on the reagent, product stability, and other factors. For example, we shall see that the enolate ion can react at *either* oxygen *or* carbon with positively charged (or polarized) species (Section 18-1), even though, as shown earlier, more of the negative charge is at the oxygen. Another important example is the case of carbonyl compounds: Although the non-charge-separated form is dominant, the minor dipolar contributor, as shown earlier for formaldehyde, is the origin of the reactivity of the carbon–oxygen double bond, electron-rich species attacking the carbon, electron-poor ones the oxygen (Chapter 17).

In Summary Some molecules cannot be described accurately by one Lewis structure but exist as hybrids of several resonance forms. To find the most important resonance contributor, consider the octet rule, make sure that there is a minimum of charge separation, and place on the relatively more electronegative atoms as much negative and as little positive charge as possible.

1-6 | ATOMIC ORBITALS: A QUANTUM MECHANICAL DESCRIPTION OF ELECTRONS AROUND THE NUCLEUS

So far, we have considered bonds in terms of electron pairs arranged around the component atoms in such a way as to maximize noble-gas configurations (e.g., Lewis octets) and minimize electron repulsion. This approach is useful as a descriptive and predictive tool with regard to the number and location of electrons in molecules. However, it does not answer some simple questions that you may have asked yourself while dealing with this material. For example, why are some Lewis structures "incorrect" or, ultimately, why are noble gases relatively stable? Why are some bonds stronger than others, and how can we tell? What is so good about the two-electron bond, and what do multiple bonds look like? To get some answers, we start by learning more about how electrons are distributed around the nucleus, both spatially and energetically. The simplified treatment presented here has as its basis the theory of quantum mechanics developed independently in the 1920s by Heisenberg, Schrödinger, and Dirac.* In this theory, the movement of an electron around a nucleus is expressed in the form of equations that are very similar to those characteristic of waves. The solutions to these equations, called **atomic orbitals,** allow us to describe the probability of finding the electron in a certain region in space. The shapes of these regions depend on the energy of the electron.

The electron is described by wave equations

The classical description of the atom (Bohr[†] theory) assumed that electrons move on trajectories around the nucleus. The energy of each electron was thought to relate to the electron's distance from the nucleus. This view is intuitively appealing because it coincides with our physical understanding of classical mechanics. Yet it is incorrect for several reasons.

First, an electron moving in an orbit would give rise to the emission of electromagnetic radiation, which is characteristic of any moving charge. The resulting energy loss from the system would cause the electron to spiral toward the nucleus, a prediction that is completely at odds with reality.

Second, Bohr theory violates the Heisenberg uncertainty principle, because it defines the precise position and momentum of an electron at the same time.

Nucleus ⟨ ● Neutron / ● Proton

● Electron

Classical atom: electrons in "orbit" around the nucleus

---

*Professor Werner Heisenberg (1901–1976), University of Munich, Germany, Nobel Prize 1932 (physics); Professor Erwin Schrödinger (1887–1961), University of Dublin, Ireland, Nobel Prize 1933 (physics); Professor Paul Dirac (1902–1984), Florida State University, Tallahassee, Nobel Prize 1933 (physics).
[†]Professor Niels Bohr (1885–1962), University of Copenhagen, Denmark, Nobel Prize 1922 (physics).

Vibrating guitar strings

A better model is afforded by considering the wave nature of moving particles. According to de Broglie's* relation, a particle of mass m that moves with velocity v has a wavelength λ.

de Broglie Wavelength

$$\lambda = \frac{h}{mv}$$

in which h is Planck's[†] constant. As a result, an orbiting electron can be described by equations that are the same as those used in classical mechanics to define waves (Figure 1-4), similar to those exhibited in the vibrations of guitar strings. These "matter waves" have amplitudes with alternating positive and negative signs. Points at which the sign changes are called **nodes.** Waves that interact in phase reinforce each other, as shown in Figure 1-4B. Those out of phase interfere with each other to make smaller waves (and possibly even cancel each other), as shown in Figure 1-4C.

This theory of electron motion is called **quantum mechanics.** The equations developed in this theory, the **wave equations,** have a series of solutions called **wave functions,** usually described by the Greek letter psi, ψ. Their values around the nucleus are not directly identifiable with any observable property of the atom. However, *the squares ($\psi 2$) of their values at each point in space describe the probability of finding an electron at that point.* The physical realities of the atom make solutions attainable only for certain *specific energies.* The system is said to be **quantized,** similar to the fixed pitches of the six strings of a guitar.

Exercise 1-12

Draw a picture similar to Figure 1-4 of two waves overlapping such that their amplitudes cancel each other.

---

*Prince Louis-Victor de Broglie (1892 –1987), Nobel Prize 1929 (physics).
[†]Professor Max K. E. L. Planck (1858–1947), University of Berlin, Germany, Nobel Prize 1918 (physics).

Note: The $+$ and $-$ signs in Figure 1-4 refer to signs of the mathematical functions describing the wave amplitudes and have nothing to do with electrical charges.

Figure 1-4 (A) A wave. The signs of the amplitude are assigned arbitrarily. At points of zero amplitude, called nodes, the wave changes sign. (B) Waves with amplitudes of like sign (in phase) reinforce each other to make a larger wave. (C) Waves that are out of phase subtract from each other to make a smaller wave.

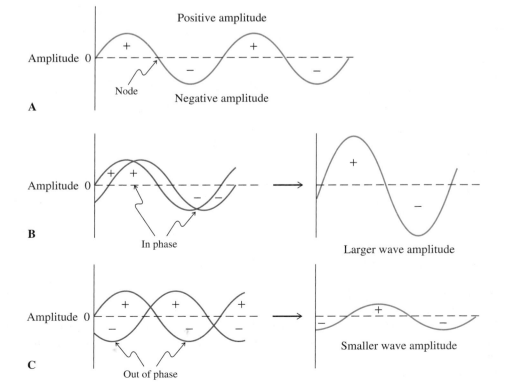

Atomic orbitals have characteristic shapes

Plots of wave functions in three dimensions typically have the appearance of spheres or dumb-bells with flattened or teardrop-shaped lobes. For simplicity, we may regard artistic renditions of atomic orbitals as indicating the regions in space in which the electron is likely to be found. Nodes separate portions of the wave function with opposite mathematical signs. The value of the wave function at a node is zero; therefore, the probability of finding electron density there is zero. Higher-energy wave functions have more nodes than do those of low energy.

Let us consider the shapes of the atomic orbitals for the simplest case, that of the hydrogen atom, consisting of a proton surrounded by an electron. The single lowest-energy solution of the wave equation is called the $1s$ orbital, the number 1 referring to the first (lowest) energy level. An orbital label also denotes the shape and number of nodes of the orbital. The $1s$ orbital is *spherically symmetric* (Figure 1-5) and has no nodes. This orbital can be represented pictorially as a sphere (Figure 1-5A) or simply as a circle (Figure 1-5B).

The next higher-energy wave function, the $2s$ orbital, also is unique and, again, spherical. The $2s$ orbital is larger than the $1s$ orbital; the higher-energy $2s$ electron is on the average farther from the positive nucleus. In addition, the $2s$ orbital has one node, a spherical surface of zero electron density separating regions of the wave function of opposite sign (Figure 1-6). As in the case of classical waves, the sign of the wave function on either side of the node is arbitrary, as long as it changes at the node. Remember that the sign of the wave function is not related to "where the electron is." As mentioned earlier, the probability of electron occupancy at any point of the orbital is given by the square of the value of the wave function. Moreover, the node does not constitute a barrier of any sort to the electron, which, in this description, is regarded not as a particle but as a wave.

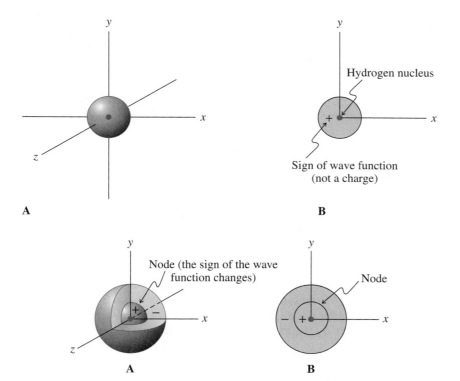

A

B

A

B

Animation

Note: Conceptually, you can relate the sequence of orbitals with various energies again to the strings of a guitar. The $1s$ orbital would correspond to the string with the lowest pitch (and frequency), the $2s$ orbital to the neighboring string. The next energetically higher three degenerate $2p$ levels would correspond to a guitar in which strings 3–5 are identical.

Figure 1-5 Representations of a $1s$ orbital. (A) The orbital is spherically symmetric in three dimensions. (B) A simplified two-dimensional view. The plus sign denotes the mathematical sign of the wave function and is *not a charge*.

Figure 1-6 Representations of a $2s$ orbital. Notice that it is larger than the $1s$ orbital and that a node is present. The + and − denote the signs of the wave function. (A) The orbital in three dimensions, with a section removed to allow visualization of the node. (B) The more conventional two-dimensional representation of the orbital.

After the $2s$ orbital, the wave equations for the electron around a hydrogen atom have three energetically equivalent solutions, the $2p_x$, $2p_y$, and $2p_z$ orbitals. Solutions of equal energy of this type are called **degenerate** (*degenus,* Latin, without genus or kind). As shown in Figure 1-7, p orbitals consist of two lobes that resemble a solid figure eight or dumbbell. A p orbital is characterized by its directionality in space. The orbital axis can be aligned with any one of the x, y, and z axes, hence the labels p_x, p_y, and p_z. The two lobes of opposite sign of each orbital are separated by a nodal plane through the atom's nucleus and perpendicular to the orbital axis.

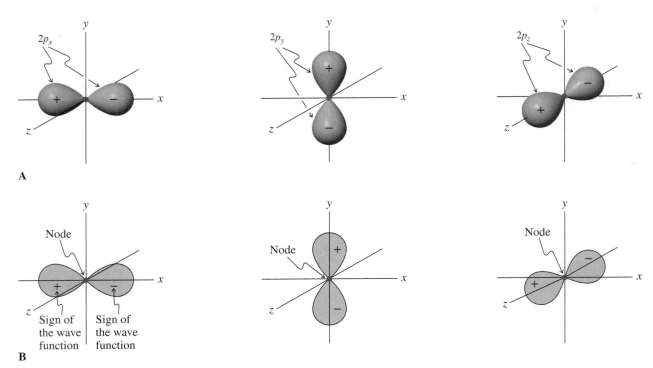

Figure 1-7 Representations of 2p orbitals (A) in three dimensions and (B) in two dimensions. Remember that the + and − signs refer to the wave functions and *not* to electrical charges. Lobes of opposite sign are separated by a nodal plane that is perpendicular to the axis of the orbital. For example, the p_x orbital is divided by a node in the yz plane.

Animation

The third set of solutions furnishes the 3s and 3p atomic orbitals. They are similar in shape to, but more diffuse than, their lower-energy counterparts and have two nodes. Still higher-energy orbitals (3d, 4s, 4p, etc.) are characterized by an increasing number of nodes and a variety of shapes. They are of much less importance in organic chemistry than are the lower orbitals. To a first approximation, the shapes and nodal properties of the atomic orbitals of other elements are very similar to those of hydrogen. Therefore, we may use s and p orbitals in a description of the electronic configurations of helium, lithium, and so forth.

The Aufbau principle assigns electrons to orbitals

Approximate relative energies of the atomic orbitals up to the 5s level are shown in Figure 1-8. With its help, we can give an electronic configuration to every atom in the periodic table. To do so, we follow three rules for assigning electrons to atomic orbitals:

1. Lower-energy orbitals are filled before those with higher energy.

2. No orbital may be occupied by more than two electrons, according to the **Pauli* exclusion principle.** Furthermore, these two electrons must differ in the orientation of their intrinsic angular momentum, their **spin.** There are two possible directions of the electron spin, usually depicted by vertical arrows pointing in opposite directions. An orbital is filled when it is occupied by two electrons of opposing spin, frequently referred to as **paired electrons.**

3. Degenerate orbitals, such as the p orbitals, are first occupied by one electron each, all of these electrons having the same spin. Subsequently, three more, each of opposite spin, are added to the first set. This assignment is based on **Hund's† rule.**

---

*Professor Wolfgang Pauli (1900–1958), Swiss Federal Institute of Technology (ETH) Zurich, Switzerland, Nobel Prize 1945 (physics).

†Professor Friedrich Hund (1896–1997), University of Göttingen, Germany.

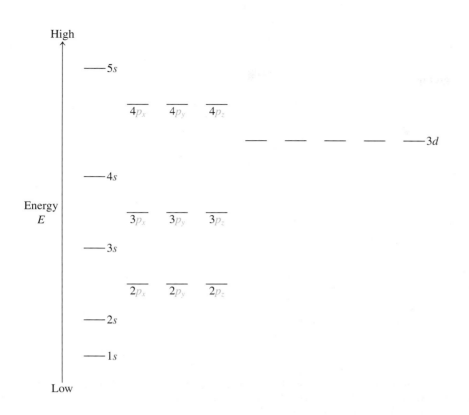

Figure 1-8 Approximate relative energies of atomic orbitals, corresponding roughly to the order in which they are filled in atoms. Orbitals of lowest energy are filled first; degenerate orbitals are filled according to Hund's rule.

With these rules in hand, the determination of electronic configuration becomes simple. Helium has two electrons in the $1s$ orbital and its electronic structure is abbreviated $(1s)^2$. Lithium $[(1s)^2(2s)^1]$ has one and beryllium $[(1s)^2(2s)^2]$ has two additional electrons in the $2s$ orbital. In boron $[(1s)^2(2s)^2(2p)^1]$, we begin the filling of the three degenerate $2p$ orbitals. This pattern continues with carbon and nitrogen, and then the addition of electrons of opposite spin for oxygen, fluorine, and neon fills all p levels. The electronic configurations of four of the elements are depicted in Figure 1-9. Atoms with completely

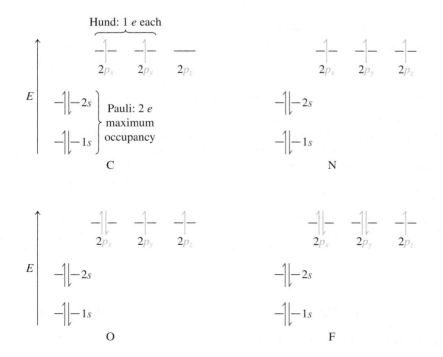

Figure 1-9 The most stable electronic configurations of carbon, $(1s)^2(2s)^2(2p)^2$; nitrogen, $(1s)^2(2s)^2(2p)^3$; oxygen, $(1s)^2(2s)^2(2p)^4$; and fluorine $(1s)^2(2s)^2(2p)^5$. Notice that the unpaired electron spins in the p orbitals are in accord with Hund's rule, and the paired electron spins in the filled $1s$ and $2s$ orbitals are in accord with the Pauli principle and Hund's rule. The order of filling the p orbitals has been arbitrarily chosen as p_x, p_y, and then p_z. Any other order would have been equally good.

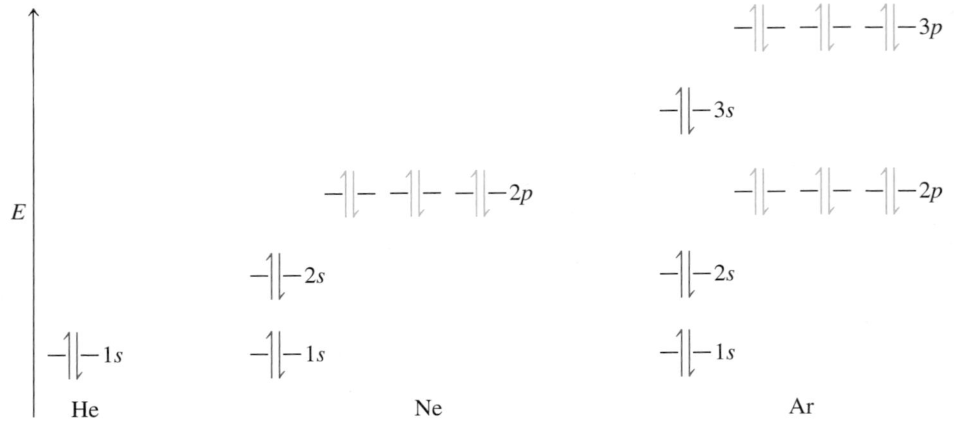

Figure 1-10 Closed-shell configurations of the noble gases helium, neon, and argon.

filled sets of atomic orbitals are said to have a **closed-shell configuration.** For example, helium, neon, and argon have this attribute (Figure 1-10). Carbon, in contrast, has an **open-shell configuration.**

The process of adding electrons one by one to the orbital sequence shown in Figure 1-8 is called the **Aufbau principle** (*Aufbau,* German, build up). It is easy to see that the Aufbau principle affords a rationale for the stability of the electron octet and duet. These numbers are required for closed-shell configurations. For helium, the closed-shell configuration is a $1s$ orbital filled with two electrons of opposite spin. In neon, the $2s$ and $2p$ orbitals are occupied by an additional eight electrons; in argon, the $3s$ and $3p$ levels accommodate eight more (Figure 1-10). The availability of $3d$ orbitals for the third-row elements provides an explanation for the phenomenon of valence-shell expansion (Section 1-4) and the loosening up of the strict application of the octet rule beyond neon.

Experimental verification and quantification of the relative ordering of the orbital energy levels depicted in Figures 1-8 through 1-10 can be obtained by measuring the ionization potentials of the corresponding electrons, that is, the energies required to remove these electrons from their respective orbitals. It takes more energy to do so from a $1s$ orbital than from a $2s$ orbital; similarly, ejecting an electron from a $2s$ level is more difficult than from its $2p$ counterpart, and so on. This makes intuitive sense: As we go from lower- to higher-lying orbitals, they become more diffuse and their associated electrons are located (on average) at increasing distances from the positively charged nucleus. Coulomb's law tells us that such electrons become increasingly less "held" by the nucleus.

Exercise 1-13

Using Figure 1-8, draw the electronic configurations of sulfur and phosphorus.

In Summary The motion of electrons around the nucleus is described by wave equations. Their solutions, atomic orbitals, can be symbolically represented as regions in space, with each point given a positive, negative, or zero (at the node) numerical value, the square of which represents the probability of finding the electron there. The Aufbau principle allows us to assign electronic configurations to all atoms.

1-7   MOLECULAR ORBITALS AND COVALENT BONDING

We shall now see how covalent bonds are constructed from the overlap of atomic orbitals.

The bond in the hydrogen molecule is formed by the overlap of 1s atomic orbitals

Let us begin by looking at the simplest case: the bond between the two hydrogen atoms in H_2. In a Lewis structure of the hydrogen molecule, we would write the bond as an electron pair shared by both atoms, giving each a helium configuration. How do we construct H_2 by

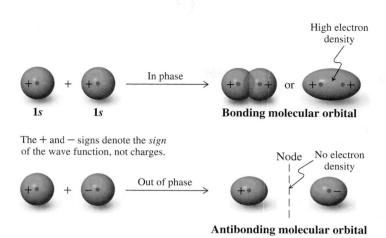

Figure 1-11 In-phase (bonding) and out-of-phase (antibonding) combinations of 1s atomic orbitals. Electrons in bonding molecular orbitals have a high probability of occupying the space *between* the atomic nuclei, as required for good bonding (compare Figure 1-2). The antibonding molecular orbital has a nodal plane where the probability of finding electrons is zero. Electrons in antibonding molecular orbitals are most likely to be found *outside* the space between the nuclei and therefore do not contribute to bonding.

using atomic orbitals? An answer to this question was developed by Pauling*: *Bonds are made by the in-phase overlap of atomic orbitals.* What does this mean? Recall that atomic orbitals are solutions of wave equations. Like waves, they may interact in a reinforcing way (Figure 1-4B) if the overlap is between areas of the wave function of the same sign, or *in phase.* They may also interact in a destructive way if the overlap is between areas of opposite sign, or *out of phase* (Figure 1-4C).

The in-phase overlap of the two 1s orbitals results in a new orbital of lower energy called a **bonding molecular orbital** (Figure 1-11). In the bonding combination, the wave function in the space between the nuclei is strongly reinforced. Thus, the probability of finding the electrons occupying this molecular orbital in that region is very high: a condition for bonding between the two atoms. This picture is strongly reminiscent of that shown in Figure 1-2. The use of two wave functions with *positive* signs for representing the in-phase combination of the two 1s orbitals in Figure 1-11 is arbitrary. Overlap between two *negative* orbitals would give identical results. In other words, it is overlap between *like* lobes that makes a bond, regardless of the sign of the wave function.

On the other hand, out-of-phase overlap between the same two atomic orbitals results in a destabilizing interaction and formation of an **antibonding molecular orbital.** In the antibonding molecular orbital, the amplitude of the wave function is canceled in the space between the two atoms, thereby giving rise to a node (Figure 1-11).

Thus, the net result of the interaction of the two 1s atomic orbitals of hydrogen is the generation of two molecular orbitals. One is bonding and lower in energy; the other is antibonding and higher in energy. Because the total number of electrons available to the system is only two, they are placed in the lower-energy molecular orbital: the two-electron bond. The result is a decrease in total energy, thereby making H_2 more stable than two free hydrogen atoms. This difference in energy levels corresponds to the strength of the H–H bond. The interaction can be depicted schematically in an energy diagram (Figure 1-12A).

It is now readily understandable why hydrogen exists as H_2, whereas helium is monatomic. The overlap of two filled atomic orbitals, as in He_2, with a total of four electrons, leads to bonding and antibonding orbitals, *both of which are filled* (Figure 1-12B). Therefore, making a He–He bond does not decrease the total energy.

The overlap of atomic orbitals gives rise to sigma and pi bonds

The formation of molecular orbitals by the overlap of atomic orbitals applies not only to the 1s orbitals of hydrogen but also to other atomic orbitals. In general, overlap of any *n* atomic orbitals gives rise to *n* molecular orbitals. For a simple two-electron bond, $n = 2$, and the two molecular orbitals are bonding and antibonding, respectively. The amount of energy by which the bonding level drops and the antibonding level is raised is called the

*Professor Linus Pauling (1901–1994), Stanford University, Nobel Prizes 1954 (chemistry) and 1963 (peace).

Figure 1-12 Schematic representation of the interaction of (A) two singly occupied atomic orbitals (as in H_2) and (B) two doubly occupied atomic orbitals (as in He_2) to give two molecular orbitals (MO). (Not drawn to scale.) Formation of an H–H bond is favorable because it stabilizes two electrons. Formation of an He–He bond stabilizes two electrons (in the bonding MO) but destabilizes two others (in the antibonding MO). Bonding between He and He thus results in no net stabilization. Therefore, helium is monatomic.

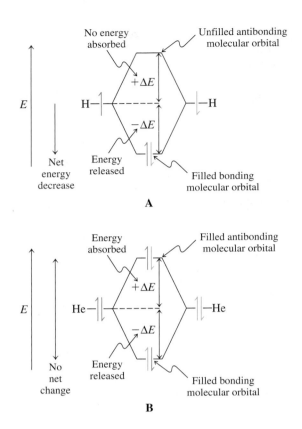

Maximizing Orbital Energy Splitting

- Same size of orbital
- Same energy of orbital
- Directionality in space

energy splitting. It indicates the strength of the bond being made and depends on a variety of factors. For example, *overlap is best between orbitals of similar size and energy.* Therefore, two $1s$ orbitals will interact with each other more effectively than a $1s$ and a $3s$.

Geometric factors also affect the degree of overlap. This consideration is important for orbitals with directionality in space, such as p orbitals. Such orbitals give rise to two types of bonds: one in which the atomic orbitals are aligned along the internuclear axis (parts A, B, C, and D in Figure 1-13) and the other in which they are perpendicular to it (part E). The first type is called a **sigma (σ) bond,** the second a **pi (π) bond.** All carbon–carbon single bonds are of the σ type; however, we shall find that double and triple bonds also have π components (Section 1-8).

Figure 1-13 Bonding between atomic orbitals. (A) $1s$ and $1s$ (e.g., H_2), (B) $1s$ and $2p$ (e.g., HF), (C) $2p$ and $2p$ (e.g., F_2), (D) $2p$ and $3p$ (e.g., FCl) aligned along internuclear axes, σ bonds; (E) $2p$ and $2p$ perpendicular to internuclear axis (e.g. $H_2C=CH_2$), a π bond. Note the arbitrary use of + and − signs to indicate in-phase interactions of the wave functions. In (D), also note the "figure 8 within a figure 8" dumbbell shape and more diffuse appearance of the $3p$ orbital when compared to its $2p$ counterpart.

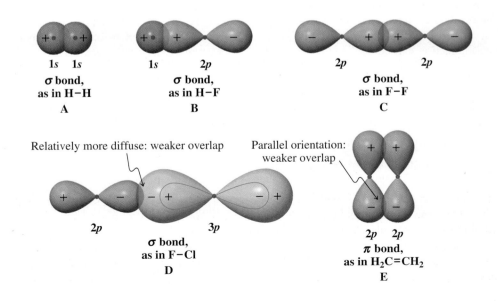

Solved Exercise 1-14	**Working with the Concepts: Orbital Splitting Diagrams**

Construct a molecular-orbital and energy-splitting diagram of the bonding in He_2^+. Is it favorable?

Strategy

To derive the molecular orbitals of the helium–helium bond, we first need to pick the appropriate atomic orbitals for overlap. The periodic table (Table 1-1) and the Aufbau principle (Figure 1-10) tell us that the orbital to use is $1s$. Therefore, a bond between two He atoms is made in the same manner as the bond between two H atoms (Figure 1-11)—by the overlap of two $1s$ atomic orbitals.

Solution

- In-phase interaction results in the lower-energy (relative to the starting $1s$ orbital) bonding molecular orbital. Out-of-phase interaction produces the higher-energy antibonding molecular orbital. The resulting energy diagram is essentially identical to those shown in Figure 1-12A and B, except that He_2^+ contains only three electrons.
- The Aufbau principle instructs us to "fill from the bottom up." Therefore two (bonding) electrons go into the lower energy level and one (antibonding) electron goes into the higher energy level.

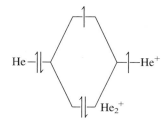

- The net effect is a favorable interaction (in contrast to neutral He_2, Figure 1-12B). Indeed, He_2^+ can be made in an electrical discharge of He^+ with He, showing that bond formation is favorable.

Exercise 1-15	**Try It Yourself**

Construct a molecular-orbital and energy-splitting diagram of the bonding in LiH. Is it favorable? (**Caution:** The overlapping orbital energies are not the same in this case. **Hint:** Consult Section 1-6, specifically the part describing the Aufbau principle. What are the electronic configurations of Li and H? The energy splitting between orbitals of unequal energies occurs such as to push the higher-lying level up, the lower-lying one down.)

In Summary We have come a long way in our description of bonding. First, we thought of bonds in terms of Coulomb forces, then in terms of covalency and shared electron pairs, and now we have a quantum mechanical picture. Bonds are a result of the overlap of atomic orbitals. The two bonding electrons are placed in the bonding molecular orbital. Because it is stabilized relative to the two initial atomic orbitals, energy is given off during bond formation. This decrease in energy represents the bond strength.

1-8 | HYBRID ORBITALS: BONDING IN COMPLEX MOLECULES

Let us now use quantum mechanics to construct bonding schemes for more complex molecules. How can we use atomic orbitals to build linear (as in BeH_2), trigonal (as in BH_3), and tetrahedral molecules (as in CH_4)? (See Lewis structures in margin).

Consider the molecule beryllium hydride, BeH_2. Beryllium has two electrons in the $1s$ orbital and two electrons in the $2s$ orbital. Without unpaired electrons, this arrangement does not appear to allow for bonding. However, it takes a relatively small amount of energy

H:Be:H

H· ·H
 ·B·
  ¨
  H

   H
H:C:H
 ¨
  H

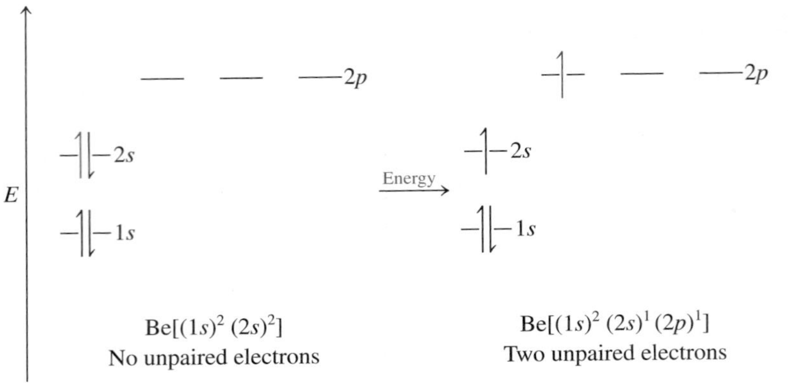

Figure 1-14 Promotion of an electron in beryllium to allow the use of both valence electrons in bonding.

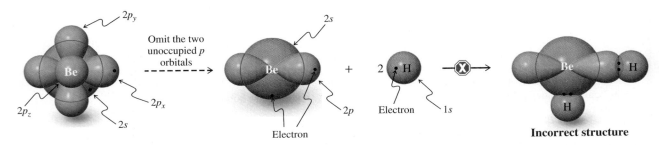

to promote one electron from the 2s orbital to one of the 2p levels (Figure 1-14)—energy that is readily compensated by eventual bond formation. Thus, in the $1s^22s^12p^1$ configuration, there are now two singly filled atomic orbitals available for bonding overlap.

One could propose bond formation in BeH_2 by overlap of the Be 2s orbital with the 1s orbital of one H and the Be 2p orbital with the second H (Figure 1-15). This scheme predicts two different bonds of unequal length, probably at an angle. However, electron repulsion predicts that compounds such as BeH_2 should have *linear* structures (Section 1-3). Experiments on related compounds confirm this prediction and also show that the bonds to beryllium are of *equal* length.*

Figure 1-15 Possible but incorrect bonding in BeH_2 by separate use of a 2s and a 2p orbital on beryllium. The 1s and the node in the 2s orbitals are not shown. Starting from the complete picture of the relevant orbitals around Be on the left, the two empty p orbitals are omitted subsequently for clarity. The dots indicate the valence electrons.

sp Hybrids produce linear structures

How can we explain this geometry in orbital terms? To answer this question, we use a quantum mechanical approach called **orbital hybridization.** Just as the mixing of atomic orbitals on different atoms forms molecular orbitals, the mixing of atomic orbitals on the same atom forms new **hybrid orbitals.**

When we mix the 2s and one of the 2p wave functions on beryllium, we obtain two new hybrids, called *sp* orbitals, made up of 50% s and 50% p character. This treatment rearranges the orbital lobes in space, as shown in Figure 1-16. The major parts of the orbitals, also called front lobes, point away from each other at an angle of 180°. There are two additional minor back lobes (one for each *sp* hybrid) with opposite sign. The remaining two p orbitals are left unchanged.

Overlap of the *sp* front lobes with two hydrogen 1s orbitals yields the bonds in BeH_2. The 180° angle that results from this hybridization scheme minimizes electron repulsion. The oversized front lobes of the hybrid orbitals also overlap better than do lobes of unhybridized orbitals; the result is energy reduction due to improved bonding.

---

*These predictions cannot be tested for BeH_2 itself, which exists as a complex network of Be and H atoms. However, both BeF_2 and $Be(CH_3)_2$ exist as individual molecules in the gas phase and possess the predicted structures.

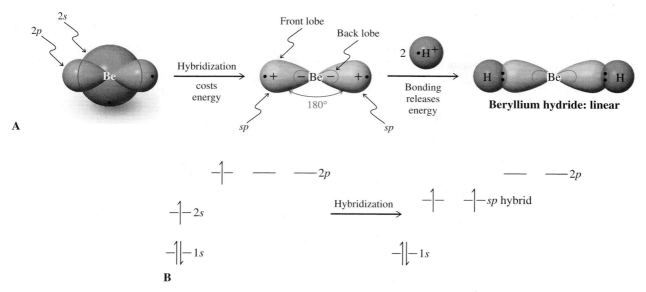

Figure 1-16 Hybridization in beryllium to create two *sp* hybrids. (A) The resulting bonding gives BeH$_2$ a linear structure. Both remaining *p* orbitals and the 1*s* orbital have been omitted for clarity. The sign of the wave function for the large *sp* lobes is opposite that for the small lobes. (B) The energy changes occurring on hybridization. The 2*s* orbital and one 2*p* orbital combine into two *sp* hybrids of intermediate energy. The 1*s* and remaining 2*p* energies remain the same.

Animation

Note that hybridization does not change the overall number of orbitals available for bonding. Hybridization of the four orbitals in beryllium gives a new set of four: two *sp* hybrids and two essentially unchanged 2*p* orbitals. We shall see shortly that carbon uses *sp* hybrids when it forms triple bonds.

*sp*² Hybrids create trigonal structures

Now let us consider an element in the periodic table with three valence electrons. What bonding scheme can be derived for borane, BH$_3$? Promotion of a 2*s* electron in boron to one of the 2*p* levels gives the three singly filled atomic orbitals (one 2*s*, two 2*p*) needed for forming three bonds. Mixing these atomic orbitals creates *three* new hybrid orbitals, which are designated *sp*² to indicate the component atomic orbitals: 67% *p* and 33% *s* (Figure 1-17). The third *p* orbital is left unchanged, so the total number of orbitals stays the same—namely, four.

Note: While beyond the scope of this discussion, a mathematical outcome of mixing any number of orbitals is the generation of an equal number of new (molecular or hybrid) orbitals. Hence, combination of one *s* and one *p* orbital produces two *sp* hybrid orbitals; one *s* and two *p* orbitals generate three *sp*² hybrid orbitals, and so on.

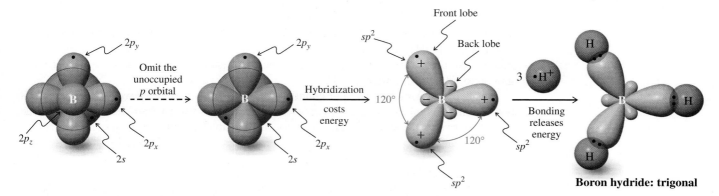

Figure 1-17 Hybridization in boron to create three *sp*² hybrids. The 1*s* and the node in the 2*s* orbitals are not shown. Starting from the complete picture of the relevant orbitals around B on the left, the empty *p* orbital is omitted subsequently for clarity. The resulting bonding gives BH$_3$ a trigonal planar structure. There are three front lobes of one sign and three back lobes of opposite sign. The remaining (omitted) *p* orbital is perpendicular to the molecular plane (the plane of the page; one lobe is above, the other below that plane). In analogy to Figure 1-16B, the energy diagram for the hybridized boron features three singly occupied, equal-energy *sp*² levels and one remaining empty 2*p* level, in addition to the filled 1*s* orbital.

Animation

The front lobes of the three sp^2 orbitals of boron overlap the respective $1s$ orbitals of the hydrogen atoms to give trigonal planar BH_3. Again, hybridization minimizes electron repulsion and improves overlap, conditions giving stronger bonds. The remaining unchanged p orbital is perpendicular to the plane incorporating the sp^2 hybrids. It is empty and does not enter significantly into bonding.

The molecule BH_3 is **isoelectronic** with the methyl cation, CH_3^+; that is, they have the same number of electrons. Bonding in CH_3^+ requires three sp^2 hybrid orbitals, and we shall see shortly that carbon uses sp^2 hybrids in double-bond formation.

sp^3 Hybridization explains the shape of tetrahedral carbon compounds

Consider the element whose bonding is of most interest to us: carbon. Its electronic configuration is $(1s)^2(2s)^2(2p)^2$, with two unpaired electrons residing in two $2p$ orbitals. Promotion of one electron from $2s$ to $2p$ results in four singly filled orbitals for bonding. We have learned that the arrangement in space of the four C–H bonds of methane that would minimize electron repulsion is tetrahedral (Section 1-3). To be able to achieve this geometry, the $2s$ orbital on carbon is hybridized with *all three* $2p$ orbitals to make *four* equivalent sp^3 orbitals with tetrahedral symmetry, each of 75% p and 25% s character and occupied by one electron. Overlap with four hydrogen $1s$ orbitals furnishes methane with four equal C–H bonds. The HCH bond angles are typical of a tetrahedron: 109.5° (Figure 1-18).

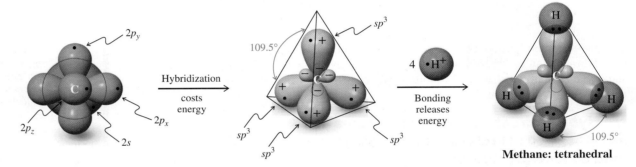

Animation

Figure 1-18 Hybridization in carbon to create four sp^3 hybrids. The resulting bonding gives CH_4 and other carbon compounds tetrahedral structures. The sp^3 hybrids contain small back lobes of sign opposite that of the front lobes. In analogy to Figure 1-16B, the energy diagram of sp^3-hybridized carbon contains four singly occupied, equal-energy sp^3 levels, in addition to the filled $1s$ orbital.

Any combination of atomic and hybrid orbitals may overlap to form bonds. For example, the four sp^3 orbitals of carbon can combine with four chlorine $3p$ orbitals, resulting in tetrachloromethane, CCl_4. Carbon–carbon bonds are generated by overlap of hybrid orbitals. In ethane, CH_3–CH_3 (Figure 1-19), this bond consists of two sp^3 hybrids, one from each of two CH_3 units. Any hydrogen atom in methane and ethane may be replaced by CH_3 or other groups to give new combinations.

In all of these molecules, and countless more, *carbon is approximately tetrahedral.* It is this ability of carbon to form chains of atoms bearing a variety of additional substituents that gives rise to the extraordinary diversity of organic chemistry.

Animation

Figure 1-19 Overlap of two sp^3 orbitals to form the carbon–carbon bond in ethane.

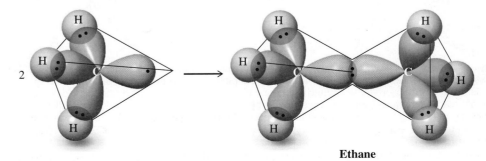

Ethane

Hybrid orbitals may contain lone electron pairs: ammonia and water

What sort of orbitals describe the bonding in ammonia and water (see Exercise 1-5)? Let us begin with ammonia. The electronic configuration of nitrogen, $(1s)^2(2s)^2(2p)^3$, explains why nitrogen is trivalent, three covalent bonds being needed for octet formation. We could use p orbitals for overlap, leaving the nonbonding electron pair in the $2s$ level. However, this arrangement does not minimize electron repulsion. The best solution is again sp^3 hybridization. Three of the sp^3 orbitals are used to bond to the hydrogen atoms, and the fourth contains the lone electron pair. The HNH bond angles (107.3°) in ammonia are almost tetrahedral (Figure 1-20). The effect of the lone electron pair is to reduce the bond angles in NH_3 from the ideal tetrahedral value of 109.5°. Because it is not shared, the lone pair is relatively close to the nitrogen. As a result, it exerts increased repulsion on the electrons in the bonds to hydrogen, thereby leading to the observed bond-angle compression.

In principle, the bonding in water could also be described by formal sp^3 hybridization on oxygen. However, the energetic cost is now too large (see Exercise 1-17). Nevertheless, for the sake of simplicity, we can extend the picture for ammonia to water as in Figure 1-20, with an HOH bond angle of 104.5°.

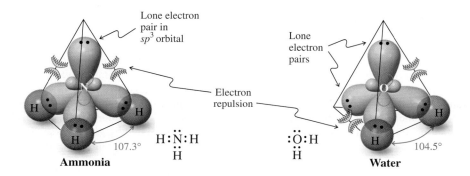

Figure 1-20 Bonding and electron repulsion in ammonia and water. The arcs indicate increased electron repulsion by the lone pairs located close to the central nucleus.

Pi bonds are present in ethene (ethylene) and ethyne (acetylene)

The double bond in alkenes, such as ethene (ethylene), and the triple bond in alkynes, such as ethyne (acetylene), are the result of the ability of the atomic orbitals of carbon to adopt sp^2 and sp hybridization, respectively. Thus, the σ bonds in ethene are derived entirely from carbon-based sp^2 hybrid orbitals: Csp^2–Csp^2 for the C–C bond, and Csp^2–$H1s$ for holding the four hydrogens (Figure 1-21). In contrast to BH_3, with an empty p orbital on boron, the leftover unhybridized p orbitals on the ethene carbons are occupied by one electron each, overlapping to form a π bond (recall Figure 1-13E). In ethyne, the σ frame is made up of bonds consisting of Csp hybrid orbitals. The arrangement leaves *two* singly occupied p orbitals on each carbon and allows the formation of two π bonds (Figure 1-21).

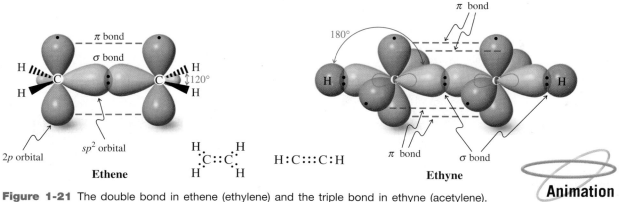

Figure 1-21 The double bond in ethene (ethylene) and the triple bond in ethyne (acetylene).

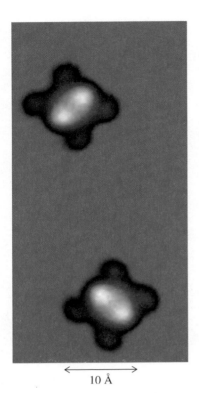

That the description of electrons as waves is not simply a mathematical construct but is "visibly real" can be demonstrated by using a device called a scanning tunneling microscope (STM). This instrument allows the mapping of electron distributions in molecules at the atomic level. The picture shows an orbital image of tetracyanoethene deposited on a Ag surface, taken at 7 K. *[Picture courtesy of Dr. Daniel Wegner, University of Münster, and Professor Michael F. Crommie, University of California at Berkeley.]*

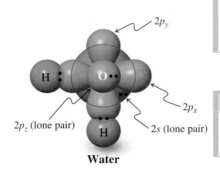

Water

Exercise 1-16

Draw a scheme for the hybridization and bonding in methyl cation, CH_3^+, and methyl anion, CH_3^-.

Solved Exercise 1-17   Working with the Concepts: Orbital Overlap in Water

Although it was convenient to describe water as containing an sp^3-hybridized oxygen atom, hybridization is unfavorable when compared to C in methane and N in ammonia. The reason is that the energy difference between the $2s$ and p orbitals in O is now so large that the energetic cost of hybridization can no longer be matched by the smaller number of bonds to H atoms (two, instead of four or three). Instead, the oxygen uses (essentially) unhybridized orbitals. Why is there a larger energy separation between the $2s$ and p orbitals in O? (**Hint:** As you proceed horizontally from C to F in the periodic table, the positive nuclear charge increases steadily. To review what effect this has on the energies of orbitals, consult the end of Section 1-6.)

Why should this phenomenon affect the process of hybridization adversely? Ponder these questions and then draw a picture of the orbitals in unhybridized O and their bonding in water. Use electron-repulsion arguments to explain the HOH bond angle of 104.5°.

Strategy

The intraatomic overlap of orbitals underlying hybridization is affected by the same features governing the quality of a bond made by interatomic overlap of orbitals: Overlap is best between orbitals of similar size and energy. Also, bonding is subject to the geometric constraints imposed by the molecule containing the hybridized atom (see also Problem 48). Proceeding along the series methane, ammonia, water, and hydrogen fluoride, the spherically symmetric $2s$ orbital is lowered more in energy than the corresponding p orbitals are by the increasing nuclear charge. One way to understand this trend is to picture the electrons in their respective orbitals: Those relatively closer to the nucleus ($2s$) are held increasingly more tightly than those farther away ($2p$). To the right of N in the periodic table, this discrepancy in energy makes hybridization of the orbitals difficult, because hybridization effectively moves the $2s$ electrons farther away from the nucleus. Thus the reduction in electron repulsion by hybridization is offset by the cost in Coulomb energy. However, we can still draw a reasonable overlap picture.

Solution

- We use the two singly occupied p orbitals on oxygen to overlap with the two respective hydrogen $1s$ orbitals (margin). In this picture, the two lone electron pairs reside in a p and the $2s$ orbital, respectively.
- Why, then, is the bond angle in water not 90°? Well, Coulomb's law (electron repulsion) still operates, whether we hybridize or not. Thus the two pairs of bonding electrons increase their distance by distorting bond angles to the observed value.

Exercise 1-18   Try It Yourself

Extrapolate the picture for water in the preceding exercise to the bonding in HF, which also uses unhybridized orbitals.

In Summary To minimize electron repulsion and maximize bonding in triatomic and larger molecules, we apply the concept of atomic-orbital hybridization to construct orbitals of appropriate shape. Combinations of s and p atomic orbitals create hybrids. Thus, a $2s$ and a $2p$ orbital mix to furnish two linear sp hybrids, the remaining two p orbitals being unchanged. Combination of the $2s$ with two p orbitals gives three sp^2 hybrids used in trigonal molecules. Finally, mixing the $2s$ with all three p orbitals results in the four sp^3 hybrids that produce the geometry around tetrahedral carbon.

1-9 | STRUCTURES AND FORMULAS OF ORGANIC MOLECULES

A good understanding of the nature of bonding allows us to learn how chemists determine the identity of organic molecules and depict their structures. Do not underestimate the importance of drawing structures. Sloppiness in drawing molecules has been the source of many errors in the literature and, perhaps of more immediate concern, in organic chemistry examinations.

To establish the identity of a molecule, we determine its structure

Organic chemists have many diverse techniques at their disposal with which to determine molecular structure. **Elemental analysis** reveals the **empirical formula,** which summarizes the kinds and ratios of the elements present. However, other procedures are usually needed to determine the molecular formula and to distinguish between structural alternatives. For example, the molecular formula C_2H_6O corresponds to *two* known substances: ethanol and methoxymethane (dimethyl ether). We can tell them apart on the basis of their physical properties—for example, their melting points, boiling points (b.p.'s), refractive indices, specific gravities, and so forth. Thus ethanol is a liquid (b.p. 78.5°C) commonly used as a laboratory and industrial solvent and present in alcoholic beverages. In contrast, methoxymethane is a gas (b.p. −23°C) used as a refrigerant. Their other physical and chemical properties differ as well. Molecules such as these, which have the same molecular formula but differ in the sequence (**connectivity**) in which the atoms are held together, are called **constitutional** or **structural isomers** (see also Real Life 1-1).

Exercise 1-19

Draw the two constitutional isomers with the molecular formula C_4H_{10}, showing all atoms and their bonds.

Two naturally occurring substances illustrate the biological consequences of such structural differences. Prostacyclin I_2 prevents blood inside the circulatory system from clotting. Thromboxane A_2, which is released when bleeding occurs, *induces* platelet aggregation, causing clots to form over wounds. Incredibly, these compounds are constitutional isomers (both have the molecular formula $C_{20}H_{32}O_5$) with only relatively minor connectivity differences. Indeed, they are so closely related that they are synthesized in the body from a common starting material (see Section 19-13 for details).

When a compound is isolated in nature or from a reaction, a chemist may attempt to identify it by matching its properties with those of known materials. Suppose, however, that the chemical under investigation is new. In this case, structural elucidation requires the use of other methods, most of which are various forms of spectroscopy. These methods will be dealt with and applied often in later chapters.

The most complete methods for structure determination are X-ray diffraction of single crystals and electron diffraction or microwave spectroscopy of gases. These techniques reveal the exact position of every atom, as if viewed under very powerful magnification. The structural details that emerge in this way for the two isomers ethanol and methoxymethane are depicted in the form of ball-and-stick models in Figure 1-22A and B.

Ethanol and Methoxymethane: Two Isomers

Ethanol
(b.p. 78.5°C)

Methoxymethane
(Dimethyl ether)
(b.p. −23°C)

Prostacyclin I_2

Thromboxane A_2

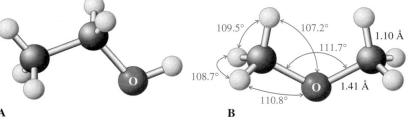

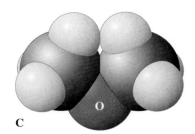

Figure 1-22 Three-dimensional representations of (A) ethanol and (B) methoxymethane, depicted by ball-and-stick molecular models. Bond lengths are given in angstrom units, bond angles in degrees. (C) Space-filling rendition of methoxymethane, taking into account the effective size of the electron "clouds" around the component nuclei.

Note the tetrahedral bonding around the carbon atoms and the bent arrangement of the bonds to oxygen, which is similar to that in water. A more accurate picture of the relative size of the component atoms in methoxymethane is given in Figure 1-22C, a space-filling model.

The visualization of organic molecules in three dimensions is essential for understanding their structures and, frequently, their reactivities. You may find it difficult to visualize the spatial arrangements of the atoms in even very simple systems. A good aid is a molecular model kit. You should acquire one and practice the assembly of organic structures. To encourage you in this practice and to indicate particularly good examples where building a molecular model can help you, the icon displayed in the margin will appear at the appropriate places in the text.

Model Building

Exercise 1-20

Use your molecular model kit to construct the two isomers with the molecular formula C_4H_{10}.

Several types of drawings are used to represent molecular structures

The representation of molecular structures was first addressed in Section 1-4, which outlined rules for drawing Lewis structures. We learned that bonding and nonbonding electrons are depicted as dots. A simplification is the straight-line notation (Kekulé structure), with lone pairs (if present) added again as dots. To simplify even further, chemists use **condensed formulas** in which most single bonds and lone pairs are omitted. The main carbon chain is written horizontally, the attached hydrogens usually to the right of the associated carbon atom. Other groups (the **substituents** on the main stem) are added through connecting vertical lines. Alternatively, a substituent may be placed on the same line as the main carbon chain, after the carbon to which it is attached and any hydrogens on that carbon.

The most economical notation of all is the **bond-line formula.** It portrays the carbon frame by zigzag straight lines, omitting all hydrogen atoms. Each unsubstituted terminus represents a methyl group, each apex a carbon atom, and all unspecified valences are assumed to be satisfied by single bonds to hydrogen.

Kekulé	**Condensed**	**Bond-Line Formulas**
	$CH_3CH_2CH_3$	
	$CH_3CHCH_2CH_2Br$ or $CH_3CHBrCH_2CH_2Br$	
	$CH_3\overset{O}{\overset{\|}{C}}CH=CH_2$ or $CH_3COCH=CH_2$	
	$HC\equiv CCH_2OH$	

Exercise 1-21

Draw condensed and bond-line formulas for each C_4H_{10} isomer.

Figure 1-22 calls attention to a problem: How can we draw the three-dimensional structures of organic molecules accurately, efficiently, and in accord with generally accepted conventions? For tetrahedral carbon, this problem is solved by the **hashed-wedged/solid-wedged line notation.** It uses a zigzag convention to depict the main carbon chain, now defined to lie *in the plane* of the page. Each apex (carbon atom) is then connected to two additional lines, one hashed wedged and one solid wedged, both pointing away from the chain. These represent the remaining two bonds to carbon; the hashed-wedged line corresponds to the bond that lies *below the plane* of the page, and the solid-wedged line to that lying *above that plane* (Figure 1-23). Substituents are placed at the appropriate termini. This convention is applied to molecules of all sizes, even methane (see Figure 1-23B–E). For simplicity, we will refer to the two types of bonds as "hashed" (instead of "hashed wedged") and "wedged" (instead of "solid wedged").

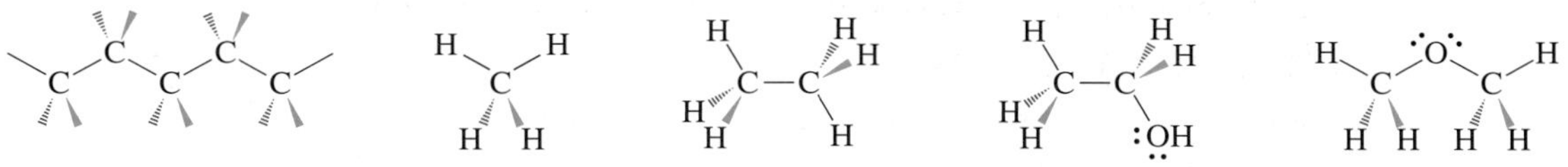

Figure 1-23 Hashed (red) and wedged (blue) line notation for (A) a carbon chain; (B) methane; (C) ethane; (D) ethanol; and (E) methoxymethane. Atoms attached by ordinary straight lines lie in the plane of the page. Groups at the ends of hashed lines lie below that plane; groups at the ends of wedges lie above it.

Exercise 1-22

(a) Draw hashed-wedged line formulas for each C_4H_{10} isomer. **(b)** Using the bond-line notation, redraw benzylpenicillin, cubane, and saccharin depicted on p. 3.

In Summary Determination of organic structures relies on the use of several experimental techniques, including elemental analysis and various forms of spectroscopy. Molecular models are useful aids for the visualization of the spatial arrangements of the atoms in structures. Condensed and bond-line notations are useful shorthand approaches to drawing two-dimensional representations of molecules, whereas hashed-wedged line formulas provide a means of depicting the atoms and bonds in three dimensions.

THE BIG PICTURE

What have we learned and where do we go from here?

Much of the material covered in this introductory chapter is probably familiar to you from introductory chemistry or even high school, perhaps in a different context. Here, the purpose was to recapitulate this knowledge as it pertains to the structure and reactivity of organic molecules. The fundamental take-home lessons for organic chemistry are these:

1. The importance of Coulomb's law (Section 1-2), as evidenced, for example, by atomic attraction (Section 1-3), relative electronegativity (Table 1-2), the electron repulsion model for the shapes of molecules (Section 1-3), and the choice of dominant resonance contributors (Section 1-5).

2. The tendency of electrons to spread out (delocalize), as manifested in resonance forms (Section 1-5) or bonding overlap (Section 1-7).

3. The correlation of the valence electron count (Sections 1-3 and 1-4) with the Aufbau principle (Section 1-6), and the associated stability of the elements in noble-gas–octet–closed-shell configurations obtained by bond formation (Sections 1-3, 1-4, and 1-7).

4. The characteristic shapes of atomic and molecular orbitals (Section 1-6), which provide a feeling for the location of the "reacting" electrons around the nuclei.

5. The overlap model for bonding (Section 1-7), which allows us to make judgments with respect to the energetics, directions, and overall feasibility of reactions.

With all this information in our grasp, we now have the tools to examine the structural and dynamic diversity of organic molecules, as well as their sites of reactivity.

WORKED EXAMPLES: INTEGRATING THE CONCEPTS

A General Strategy for Solving Problems in Organic Chemistry

Before attempting to solve a problem in organic chemistry (or any other subject, for that matter) it is essential to understand clearly what the question is asking. Read every question fully and carefully. The next stage is to decide logically how to start the problem-solving process. At this point, make an inventory of the information you will need as input for the solution. Find any necessary information that you do not already have. Lastly, organize a step-by-step protocol for devising the solution. Implement this protocol and do not skip steps.

This strategy is called the "WHIP" approach (as in, you can "whip" organic chemistry):

*W*hat is the problem asking?
*H*ow to begin?
*I*nformation needed?
*P*roceed step by step, without skipping any steps.

The following worked examples will illustrate the implementation of this approach.

1-23. Composing Lewis Structures: Octets

Sodium borohydride, $Na^+\ ^-BH_4$, is a reagent used in the conversion of aldehydes and ketones to alcohols (Section 8-6). It can be made by treating BH_3 with Na^+H^-:

$$BH_3\ +\ Na^+H^-\ \longrightarrow\ Na^+ \begin{bmatrix} H \\ H\ B\ H \\ H \end{bmatrix}^-$$

a. Draw the Lewis structure of $^-BH_4$.

SOLUTION

*W*hat the question is asking? To draw a Lewis structure.
*H*ow to begin? Use the rules of Section 1-4 as guides for the steps to follow.
*I*nformation needed? We need to know the molecular skeleton—the order in which the atoms are attached (*Rule 1*)—and the number of available valence electrons (*Rule 2*).
*P*roceed step-by-step:

Step 1. The molecular skeleton is indicated in the bracketed part of the above equation: a boron atom surrounded by four hydrogen atoms.

Step 2. What is the number of valence electrons? Answer (*Rule 2*):

4H	$= 4 \times 1$ electron	$= 4$ electrons
1 B		$= 3$ electrons
Charge $= -1$		$= 1$ electron
Total		8 electrons

Step 3. The octet rule (*Rule 3*), requiring 8 electrons to surround boron, plus the 2 necessary for each of the 4 hydrogens, gives us our total "electron demand" of 16. Our available supply from Step 2 above is 8. Subtracting supply from demand and dividing by 2 gives us the number of bonds required, namely 4. Placing two electrons each between boron and its four bonded hydrogens gives us the

necessary four bonds, consuming all the valence electrons at our disposal and completing the electron shell around every atom:

$$\text{H} : \overset{\displaystyle ..}{\underset{\displaystyle ..}{\text{H} : \text{B} : \text{H}}}$$
$$\text{H}$$

Step 4. We are not done yet! The species is charged, and we must assign this charge a formal location in order to finish with a complete, correct Lewis structure (*Rule 4*): Because each hydrogen has a valence electron count of one, namely that of the neutral atom, the –1 formal charge must reside on boron. This conclusion can be verified by counting the valence electrons around boron: it is four, half of the total of 8 electrons shared in bonds, and one more than the number associated with the free neutral atom. The correct Lewis structure is therefore

$$\text{H}$$
$$\text{H} : \overset{\displaystyle ..}{\underset{\displaystyle ..}{\text{B}}} : \text{H} \quad^-$$
$$\text{H}$$

b. What is the shape of the borohydride ion?

SOLUTION

*W*hat the question is asking is again straighforward. *H*ow to begin? We return to Section 1-3 for the necessary *I*nformation that electron repulsion controls the shape of simple molecules. Boron in borohydride is surrounded by four electron pairs, just as is carbon in methane. Reasoning by analogy we *P*roceed to the logical conclusion that, like methane, borohydride possesses a tetrahedral structure, consistent with sp^3 hybridization at boron (margin).

c. Draw an orbital picture of the attack of H:$^-$ on BH_3. What are the orbitals involved in overlap for bond formation?

SOLUTION

While the preceding two examples were relatively straightforward, we now encounter a problem that is more complicated. *W*hat the question is asking encompasses several parts.

First, we will need an orbital picture of the starting compounds. We will also need to identify the product and its orbital picture. Finally, we must try to determine how the orbitals we start with evolve into those at the finish. *H*ow to begin? Have we already encountered the *I*nformation we need? Yes we have. We can start by drawing the orbital pictures directly from Sections 1-6 and 1-8. For H:$^-$, we have a doubly occupied $1s$ orbital; for BH_3 we find a trigonal planar boron with three sp^2 orbitals forming three bonds to hydrogens and a remaining empty p orbital perpendicular to the molecular plane (Figure 1-17).

The process being considered is the reaction of H:$^-$ with BH_3, which we have seen both in Section 1-4 and again at the beginning of this problem, so we know the product: borohydride, BH_4^-. Which part of BH_3 is the hydride ion most likely to attack? The orbital picture provides the answer: The empty p orbital is an ideal target into which hydride can donate its electron pair to form a new bond.

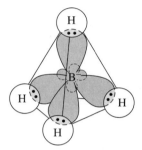

Borohydride ion

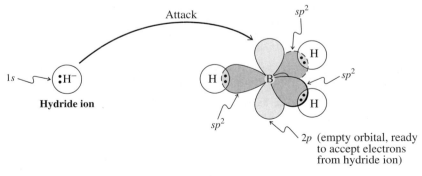

Boron hydride (borane)

Do the orbitals around boron remain unchanged during this process? To answer this question we need only refer to the structure of the product, borohydride, which we determined in part **b** of this problem. BH_4^- is tetrahedral, with sp^3 hybridization. You can readily imagine how the boron smoothly converts from being trigonal planar and sp^2 hybridized into an sp^3 hybridized tetrahedral structure. Thus the initial overlap of the hydride ion $1s$ orbital with the empty boron $2p$ orbital changes to the final overlap of the hydrogen $1s$ orbital with a boron sp^3 hybrid orbital. This process of rehybridization is quite common during the making and breaking of covalent bonds.

1-24. Composing Lewis Structures: Resonance Forms

Propyne can be deprotonated twice with very strong base (i.e., the base removes the two protons labeled by arrows) to give a dianion.

Two resonance forms of the anion can be constructed in which all three carbons have Lewis octets.

a. Draw both structures and indicate which is the more important one.

SOLUTION

Let us analyze the problem one step at a time:

Step 1. What structural information is embedded in the picture given for propyne dianion? Answer (Section 1-4, *Rule 1*): The picture shows the connectivity of the atoms—a chain of three carbons, one of the terminal atoms bearing two hydrogens.

Step 2. How many valence electrons are available? Answer (Section 1-4, *Rule 2*):

$$
\begin{array}{lll}
2\,\text{H} & = \quad 2 \times 1\ \text{electron} & = \quad 2\ \text{electrons} \\
3\,\text{C} & = \quad 3 \times 4\ \text{electrons} & = \ 12\ \text{electrons} \\
\text{Charge} = -2 & & = \quad 2\ \text{electrons} \\
\hline
\text{Total} & & \quad\ 16\ \text{electrons}
\end{array}
$$

Step 3. How do we get a Lewis octet structure for this ion? We can first determine the total number of bonds it will contain. Octets about the three carbons plus electron pairs for the two hydrogens require 28 electrons. Half of the difference between that and the 16 electrons available indicates that the structure must contain six bonds. Answer (Section 1-4, *Rule 3*): Using the connectivity given in the structure for propyne dianion, we can immediately dispose of eight of the available electrons in four bonds:

$$
\begin{array}{c}
\overset{\displaystyle\text{H}}{\underset{}{}} \\[-2pt]
\text{C:C:}\overset{\displaystyle\cdot\cdot}{\text{C}}\text{:H}
\end{array}
$$

Now, let us use the remaining eight electrons in the form of lone electron pairs to surround as many carbons as possible with octets. A good place to start is at the right, because that carbon requires only two electrons for this purpose. The center carbon needs two lone pairs and the carbon at the left has to make do (for the time being) with one additional pair of electrons:

$$
\begin{array}{c}
\overset{\displaystyle\text{H}}{} \\[-2pt]
\text{:C:}\overset{\displaystyle\cdot\cdot}{\underset{\displaystyle\cdot\cdot}{\text{C}}}\text{:}\overset{\displaystyle\cdot\cdot}{\text{C}}\text{:H}
\end{array}
$$

This structure leaves the carbon at the left with only four electrons. Thus, we have to change the two lone pairs at the center into two shared pairs, furnishing the following dot structure, which contains a total of six bonds, as predicted.

$$
\begin{array}{c}
\overset{\displaystyle\text{H}}{} \\[-2pt]
\text{:C:::C:}\overset{\displaystyle\cdot\cdot}{\underset{\displaystyle\cdot\cdot}{\text{C}}}\text{:H}
\end{array}
$$

Step 4. Now every atom has its duet or octet satisfied, but we still have to concern ourselves with charges. What are the charges on each atom? Answer (Section 1-4, *Rule 4*): Starting again at the right, we can quickly ascertain that the hydrogens are charge neutral. Each is attached to carbon through a shared electron pair, giving it an effective electron count of one, the same as in a free, neutral hydrogen atom. On the other hand, the carbon atom bears three shared electron pairs and one lone pair, thus having an effective electron count of five, one more electron than the number associated with the neutral nucleus. Hence one of the negative charges is located on the carbon at the right. The central carbon nucleus is surrounded by four shared electron pairs and is therefore neutral. Finally, the carbon at the left is attached to its neighbor by three shared pairs and has, in addition, two unbound electrons, giving it the other negative charge. The result is

$$\overset{\displaystyle \ \ \ \ \ \ \ \ \ \ \ \ H}{:\!\overset{-}{C}\!:::\!C\!:\!\overset{..}{\underset{..}{C}}\!:\!H}$$

Step 5. We can now address the question of resonance forms. Is it possible to move pairs of electrons in such a way as to generate another Lewis octet form? Answer (Section 1-5): Yes, by shifting the lone pair on the carbon at the right into a sharing position and at the same time moving one of the three shared pairs to the left into an unshared location:

$$\left[:\!\overset{-}{C}\!:::\!C\!:\!\overset{..}{\underset{..}{C}}\!:\!H \ \longleftrightarrow \ :\!\overset{2-}{C}\!::\!C\!::\!\overset{..}{\underset{..}{C}}\!:\!H \right]$$

The consequence of this movement is the transfer of the negative charge from the carbon at the right to the carbon at the left, which therefore becomes doubly negative.

Step 6. Which one of the two resonance pictures is more important? Answer (Section 1-5): Electron repulsion strongly disfavors the structure at the right, with its double negative charge on one carbon, and thus makes the structure at the left a more important resonance contributor.

A final point: You could have derived the first resonance structure much more quickly by considering the information given in the reaction scheme leading to the dianion. Thus, the bond-line formula of propyne represents its Lewis structure and the process of removing a proton each from the respective terminal carbons leaves these carbons with two lone electron pairs and the associated charges:

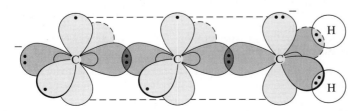

The important lesson to be learned from this final point is that, whenever you are confronted with a problem, you should take the time to complete an inventory (write it down) of all the information given explicitly or implicitly in stating the problem.

b. Propyne dianion adopts the following hybridization: $[CspCspCsp^2H_2]^{2-}$, in which the CH_2 terminus is sp^2 hybridized, unlike methyl anion, in which the carbon is sp^3 hybridized (Exercise 1-16). Provide an orbital drawing of the dianion to explain this preference in hybridization.

SOLUTION

You can construct an orbital picture simply by attaching one half (the CH_2 group) of the rendition of ethene in Figure 1-21 to that of ethyne without its hydrogens.

You can clearly see how the doubly filled p orbital of the CH_2 group enters into overlap with one of the π bonds of the alkyne unit, allowing the charge to delocalize as represented by the two resonance structures.

Important Concepts

1. Organic chemistry is the chemistry of **carbon** and its compounds.

2. **Coulomb's law** relates the attractive force between particles of opposite electrical charge to the distance between them.

3. **Ionic bonds** result from coulombic attraction of oppositely charged ions. These ions are formed by the complete transfer of electrons from one atom to another, typically to achieve a noble-gas configuration.

4. **Covalent bonds** result from electron sharing between two atoms. Electrons are shared to allow the atoms to attain noble-gas configurations.

5. **Bond length** is the average distance between two covalently bonded atoms. Bond formation releases energy; bond breaking requires energy.

6. **Polar bonds** are formed between atoms of differing electronegativity (a measure of an atom's ability to attract electrons).

7. The **shape of molecules** is strongly influenced by electron repulsion.

8. **Lewis structures** describe bonding by the use of valence electron dots. They are drawn so as to give hydrogen an electron duet and the other atoms electron octets (octet rule). Formal charge separation should be minimized but may be enforced by the **octet rule.**

9. When two or more Lewis structures differing only in the positions of the electrons are needed to describe a molecule, they are called **resonance forms.** None correctly describes the molecule, its true representation being an average **(hybrid)** of all its Lewis structures. If the resonance forms of a molecule are unequal, those which best satisfy the rules for writing Lewis structures and the electronegativity requirements of the atoms are more important.

10. The motion of electrons around the nucleus can be described by **wave equations.** The solutions to these equations are **atomic orbitals,** which roughly delineate regions in space in which there is a high probability of finding electrons.

11. An *s* **orbital** is spherical; a *p* **orbital** looks like two touching teardrops or a "spherical figure eight." The mathematical sign of the orbital at any point can be positive, negative, or zero (node). With increasing energy, the number of nodes increases. Each orbital can be occupied by a maximum of two electrons of opposite spin **(Pauli exclusion principle, Hund's rule).**

12. The process of adding electrons one by one to the atomic orbitals, starting with those of lowest energy, is called the **Aufbau principle.**

13. A **molecular orbital** is formed when two atomic orbitals overlap to generate a bond. Atomic orbitals of the same sign overlap to give a **bonding molecular orbital** of lower energy. Atomic orbitals of opposite sign give rise to an **antibonding molecular orbital** of higher energy and containing a node. The number of molecular orbitals equals the number of atomic orbitals from which they derive.

14. Bonds made by overlap along the internuclear axis are called σ **bonds;** those made by overlap of *p* orbitals perpendicular to the internuclear axis are called π **bonds.**

15. The mixing of orbitals on the same atom results in new **hybrid orbitals** of different shape. One *s* and one *p* orbital mix to give two **linear** *sp* **hybrids,** used, for example, in the bonding of BeH_2. One *s* and two *p* orbitals result in three **trigonal** sp^2 **hybrids,** used, for example, in BH_3. One *s* and three *p* orbitals furnish four **tetrahedral** sp^3 **hybrids,** used, for example, in CH_4. The orbitals that are not hybridized stay unchanged. Hybrid orbitals may overlap with each other. Overlapping sp^3 hybrid orbitals on different carbon atoms form the carbon–carbon bonds in ethane and other organic molecules. Hybrid orbitals may also be occupied by lone electron pairs, as in NH_3.

16. The composition (i.e., ratios of types of atoms) of organic molecules is revealed by **elemental analysis.** The **molecular formula** gives the number of atoms of each kind.

17. Molecules that have the same molecular formula but different connectivity order of their atoms are called **constitutional** or **structural isomers.** They have different properties.

18. **Condensed** and **bond-line formulas** are abbreviated representations of molecules. **Hashed-wedged line drawings** illustrate molecular structures in three dimensions.

Problems

25. Draw a Lewis structure for each of the following molecules and assign charges where appropriate. The order in which the atoms are connected is given in parentheses.

(a) ClF

(b) BrCN

(c) SOCl$_2$(ClSCl)

O

(d) CH$_3$NH$_2$

(e) CH$_3$OCH$_3$

(f) N$_2$H$_2$ (HNNH)

(g) CH$_2$CO

(h) HN$_3$ (HNNN)

(i) N$_2$O (NNO)

26. Using electronegativity values from Table 1-2 (in Section 1-3), identify polar covalent bonds in several of the structures in Problem 25 and label the atoms δ^+ and δ^-, as appropriate.

27. Draw a Lewis structure for each of the following species. Again, assign charges where appropriate.

(a) H$^-$

(b) CH$_3$$^-$

(c) CH$_3$$^+$

(d) CH$_3$

(e) CH$_3$NH$_3$$^+$

(f) CH$_3$O$^-$

(g) CH$_2$

(h) HC$_2$$^-$ (HCC)

(i) H$_2$O$_2$ (HOOH)

28. For each of the following species, add charges wherever required to give a complete, correct Lewis structure. All bonds and non-bonded valence electrons are shown.

(a) :O—C—H with H H on top and H H on bottom

(b) C=Ö$^{+1}$ with H on top-left, H on bottom-left, H on top-right, H on bottom-right

(c) H—C—H with H below, $^{-1}$

(d) H—N—Ö: with H on top, H below N, H on right, O above

(e) H—Ö, H—Ö, B=Ö, $^{-1}$

(f) H—Ö—N—Ö—H

29. (a) The structure of the bicarbonate (hydrogen carbonate) ion, HCO$_3$$^-$, is best described as a hybrid of several contributing resonance forms, two of which are shown here.

$$\left[\begin{array}{c} :O: \\ HO \quad O:^- \end{array} \longleftrightarrow \begin{array}{c} :Ö:^- \\ HO \quad O:^- \\ + \end{array} \right]$$

Bicarbonate is crucial for the control of body pH (for example, blood pH = 7.4). A more self-indulgent use is in baking soda, where it serves as a source of CO$_2$ gas, which gives bread and pastry their fluffy constituency.

(i) Draw at least one additional resonance form. (ii) Using curved "electron-pushing" arrows, show how these Lewis structures may

be interconverted by movement of electron pairs. (iii) Determine which form or forms will be the major contributor(s) to the real structure of bicarbonate, explaining your answer on the basis of the criteria in Section 1-5.

(b) Draw two resonance forms for formaldehyde oxime, H$_2$CNOH. As in parts (ii) and (iii) of (a), use curved arrows to interconvert the resonance forms and determine which form is the major contributor.

(c) Repeat the exercises in (b) for formaldehyde oximate ion, [H$_2$CNO]$^-$.

30. Several of the compounds in Problems 25 and 28 can have resonance forms. Identify these molecules and write an additional resonance Lewis structure for each. Use electron-pushing arrows to illustrate how the resonance forms for each species are derived from one another, and in each case indicate the major contributor to the resonance hybrid.

31. Draw two or three resonance forms for each of the following species. Indicate the major contributor or contributors to the hybrid in each case.

(a) OCN$^-$

(b) CH$_2$CHNH$^-$

O

(c) HCONH$_2$ (HCNH$_2$)

(d) O$_3$ (OOO)

(e) CH$_2$CHCH$_2$$^-$

(f) ClO$_2$$^-$ (OClO)

(g) HOCHNH$_2$$^+$

(h) CH$_3$CNO

32. Compare and contrast the Lewis structures of nitromethane, CH$_3$NO$_2$, and methyl nitrite, CH$_3$ONO. Write at least two resonance forms for each molecule. Based on your examination of the resonance forms, what can you say about the polarity and bond order of the two NO bonds in each substance? (Nitromethane is used as a solvent as well as building block in organic synthesis. The two embedded oxygens allow it to burn with less atmospheric oxygen, a quality exploited by drag racers, who add "nitro" to their fuel for extra power).

33. Write a Lewis structure for each substance. Within each group, compare (i) number of electrons, (ii) charges on atoms, if any, (iii) nature of all bonds, and (iv) geometry.

(a) chlorine atom, Cl, and chloride ion, Cl$^-$

(b) borane, BH$_3$, and phosphine, PH$_3$

(c) CF$_4$ and BrF$_4$$^-$ (C and Br are in the middle)

(d) nitrogen dioxide, NO$_2$, and nitrite ion, NO$_2$$^-$ (nitrogen is in the middle)

(e) NO$_2$, SO$_2$, and ClO$_2$ (N, S, and Cl are in the middle)

34. Use a molecular-orbital analysis to predict which species in each of the following pairs has the stronger bonding between atoms. (**Hint:** Refer to Figure 1-12.)

(a) H$_2$ or H$_2$$^+$

(b) He$_2$ or He$_2$$^+$

(c) O$_2$ or O$_2$$^+$

(d) N$_2$ or N$_2$$^+$

35. For each molecule below, predict the approximate geometry about each indicated atom. Give the hybridization that explains each geometry.

Br Br

(a) H$_2$C—CH$_2$

O

(b) H$_3$C—C—CH$_3$

(c)  $H_3C-O-CH=CH_2$

(d) 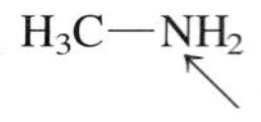 H_3C-NH_2

(e) 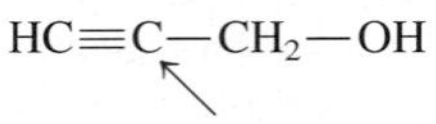 $HC\equiv C-CH_2-OH$

(f) 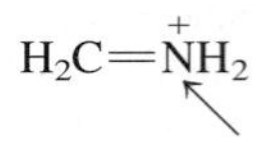 $H_2C=\overset{+}{N}H_2$

36. For each molecule in Problem 35, describe the orbitals that are used to form every bond to each of the indicated atoms (atomic s, p, hybrid sp, sp^2, or sp^3).

37. Draw and show the overlap of the orbitals involved in the bonds discussed in Problem 36.

38. Describe the hybridization of each carbon atom in each of the following structures. Base your answer on the geometry about the carbon atom.

(a) CH_3Cl

(b) CH_3OH

(c) $CH_3CH_2CH_3$

(d) $CH_2=CH_2$ (trigonal carbons)

(e) $HC\equiv CH$ (linear structure)

(f) 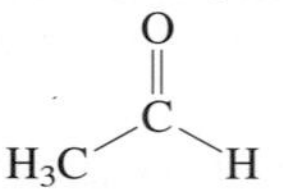 [structure: $H_3C-C(=O)-H$]

(g) 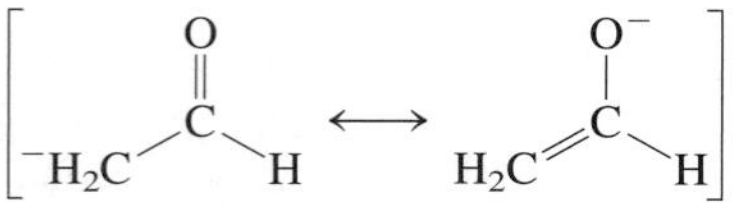 [resonance structures: $^-H_2C-C(=O)-H \longleftrightarrow H_2C=C(-O^-)-H$]

39. Depict the following condensed formulas in Kekulé (straight-line) notation. (See also Problem 42.)

(a) CH_3CN

(b) 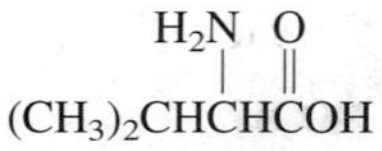 $(CH_3)_2CHCHCOH$ with H_2N and O substituents

(c) 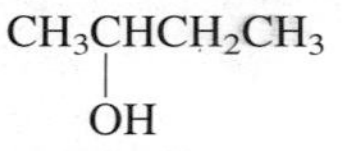 $CH_3CHCH_2CH_3$ with OH

(d) $CH_2BrCHBr_2$

(e) 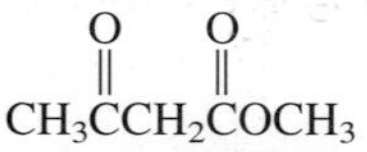 $CH_3CCH_2COCH_3$ with two O (double bonds)

(f) $HOCH_2CH_2OCH_2CH_2OH$

40. Convert the following bond-line formulas into Kekulé (straight-line) structures.

(a) 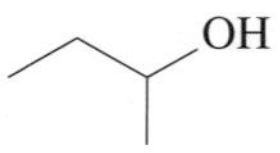 [bond-line structure with OH]

(b) 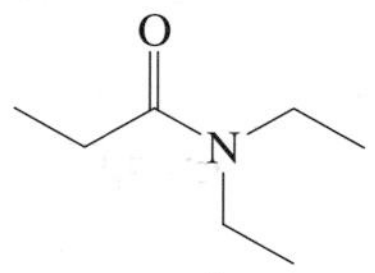 [bond-line structure with C=O and N]

(c) [bond-line structure with two Br]

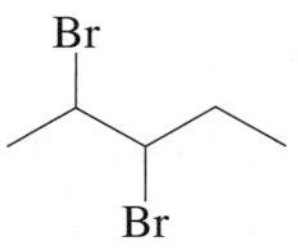

(d) 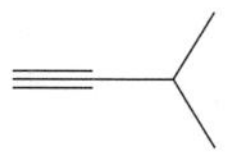 [bond-line structure with triple bond]

(e)  [bond-line structure with O and CN]

(f) 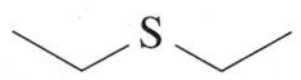 [bond-line structure with S]

41. Convert the following hashed-wedged line formulas into condensed formulas.

(a) 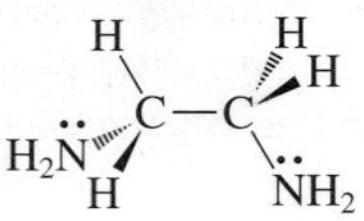 [hashed-wedged structure with H_2N, NH_2, H's]

(b) 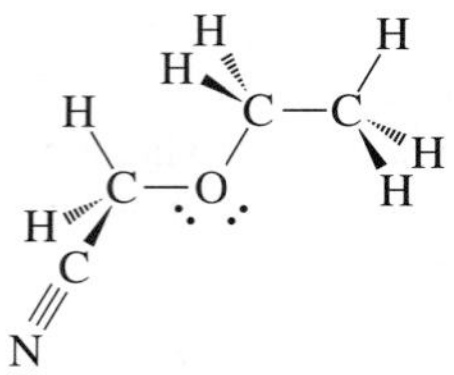 [hashed-wedged structure with C-O, CN, H's]

(c) [structure: $H-C$ with three Br]

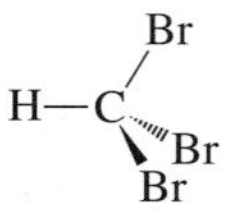

42. Depict the following Kekulé (straight-line) formulas in their condensed forms.

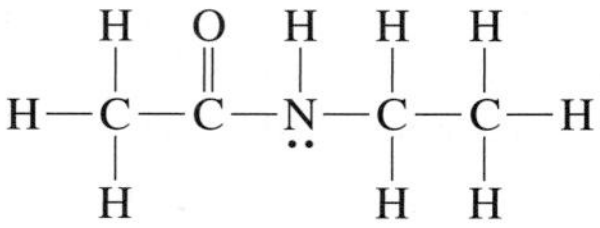

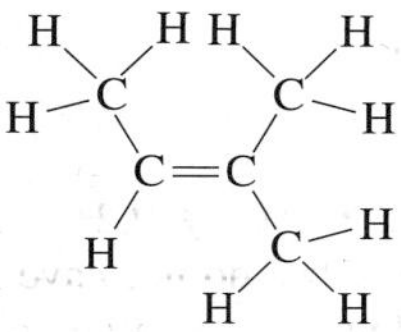

(a) 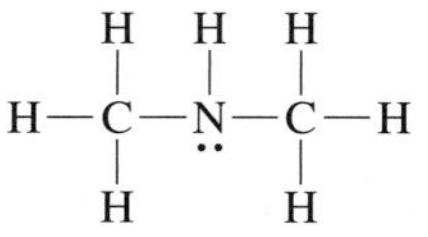 [structure: $H-C(H)(H)-N(H)-C(H)(H)-H$]

(b) [structure: $H-C(H)(H)-C(=O)-N(H)-C(H)(H)-C(H)(H)-H$]

(c) [structure: $H-S-C(H)(H)-C(H)(H)-C(H)(O-H)-C(H)(H)-H$]

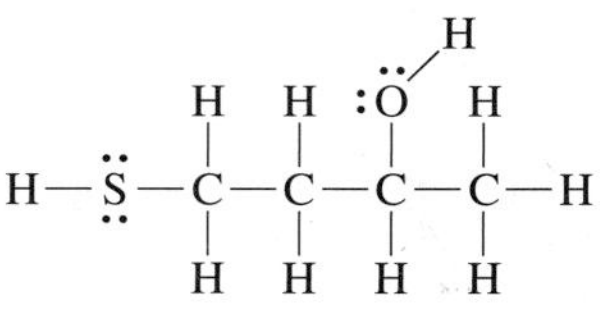

(d) 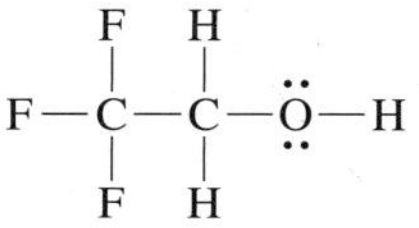 [structure: $F-C(F)(H)-C(F)(H)-\ddot{O}-H$]

(e) [structure: alkene with CH_3 and CH substituents]

(f) [structure: $H_2C=C(H)-C(=O)-C(H)(H)(H)$ type]

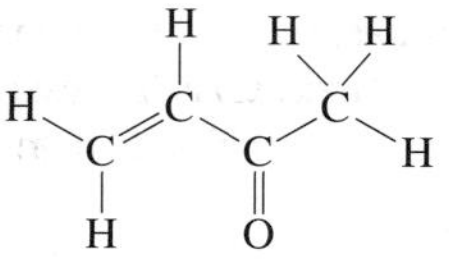

43. Redraw the structures depicted in Problems 39 and 42 using bond-line formulas.

44. Convert the following condensed formulas into hashed-wedged line structures.

(a) 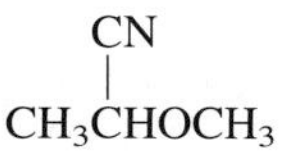 CH_3CHOCH_3 with CN

(b) $CHCl_3$

(c) $(CH_3)_2NH$

(d) 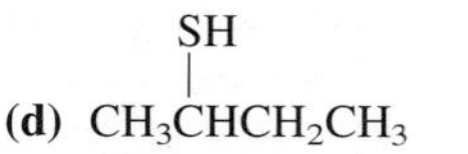 $CH_3CHCH_2CH_3$ with SH

45. Construct as many constitutional isomers of each molecular formula as you can for (a) C_5H_{12}; (b) C_3H_8O. Draw both condensed and bond-line formulas for each isomer.

46. Draw condensed formulas showing the multiple bonds, charges, and lone electron pairs (if any) for each molecule in the following pairs of constitutional isomers. (**Hint:** First make sure that you can draw a proper Lewis structure for each molecule.) Do any of these pairs consist of resonance forms?

(a) $HCCCH_3$ and H_2CCCH_2

(b) CH_3CN and CH_3NC

(c) CH_3CH and H_2CCHOH (with O above the first structure)

47. **CHALLENGE** Two resonance forms can be written for a bond between trivalent boron and an atom with a lone pair of electrons. **(a)** Formulate them for (i) $(CH_3)_2BN(CH_3)_2$; (ii) $(CH_3)_2BOCH_3$; (iii) $(CH_3)_2BF$. **(b)** Using the guidelines in Section 1-5, determine which form in each pair of resonance forms is more important. **(c)** How do the electronegativity differences between N, O, and F affect the relative importance of the resonance forms in each case? **(d)** Predict the hybridization of N in (i) and O in (ii).

48. **CHALLENGE** The unusual molecule [2.2.2]propellane is pictured below. On the basis of the given structural parameters, what hybridization scheme best describes the carbons marked by asterisks? (Make a model to help you visualize its shape.) What types of orbitals are used in the bond between them? Would you expect this bond to be stronger or weaker than an ordinary carbon–carbon single bond (which is usually 1.54 Å long)?

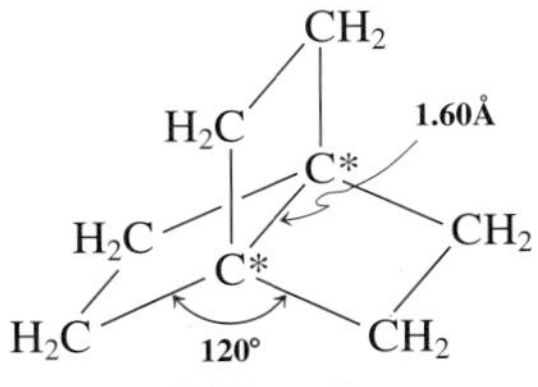

[2.2.2]Propellane

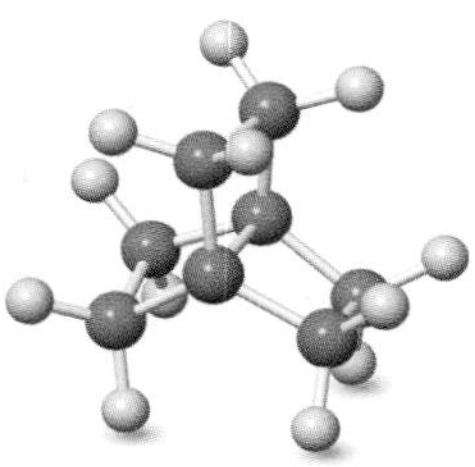

49. **CHALLENGE** **(a)** On the basis of the information in Problem 38, give the likely hybridization of the orbital that contains the unshared pair of electrons (responsible for the negative charge) in each of the following species: $CH_3CH_2^-$; $CH_2=CH^-$; $HC\equiv C^-$. **(b)** Electrons in sp, sp^2, and sp^3 orbitals do not have identical energies. Because the $2s$ orbital is lower in energy than a $2p$, the more s character a hybrid orbital has, the lower its energy will be. Therefore the sp^3 ($\frac{1}{4} s$ and $\frac{3}{4} p$ in character) is highest in energy, and the sp ($\frac{1}{2} s$, $\frac{1}{2} p$) lowest. Use this information to determine the relative abilities of the three anions in (a) to accommodate the negative charge. **(c)** The strength of an acid HA is related to the ability of its conjugate base A^- to accommodate negative charge. In other words, the ionization $HA \rightleftharpoons H^+ + A^-$ is favored for a more stable A^-. Although CH_3CH_3, $CH_2=CH_2$, and $HC\equiv CH$ are all weak acids, they are not equally so. On the basis of your answer to (b), rank them in order of acid strength.

50. A number of substances containing positively polarized carbon atoms have been labeled as "cancer suspect agents" (i.e., suspected carcinogens or cancer-inducing compounds). It has been suggested that the presence of such carbon atoms is responsible for the carcinogenic properties of these molecules. Assuming that the extent of polarization is proportional to carcinogenic potential, how would you rank the following compounds with regard to cancer-causing potency?

(a) CH_3Cl

(b) $(CH_3)_4Si$

(c) $ClCH_2OCH_2Cl$

(d) CH_3OCH_2Cl

(e) $(CH_3)_3C^+$

(*Note:* Polarization is only one of the many factors known to be related to carcinogenicity. Moreover, none of them shows the type of straightforward correlation implied in this question.)

51. Certain compounds, such as the one pictured below, show strong biological activity against cell types characteristic of prostate cancer. In this structure, locate an example of each of the following types of atoms or bonds: **(a)** a highly polarized covalent single bond; **(b)** a highly polarized covalent double bond; **(c)** a nearly nonpolar covalent bond; **(d)** an sp-hybridized carbon atom; **(e)** an sp^2-hybridized carbon atom; **(f)** an sp^3-hybridized carbon atom; **(g)** a bond between atoms of different hybridization; **(h)** the longest bond in the molecule; **(i)** the shortest bond in the molecule (excluding bonds to hydrogen).

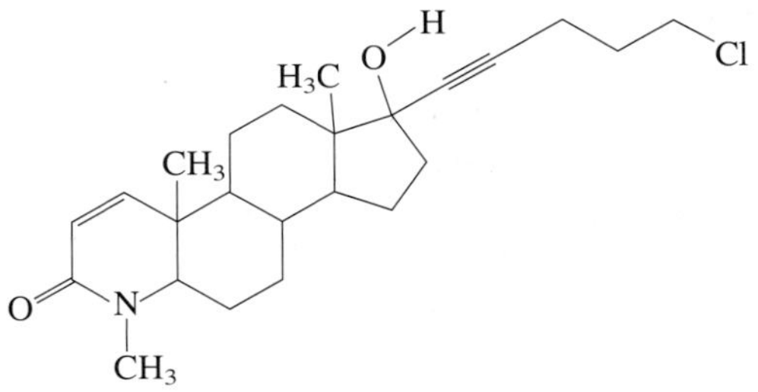

Team Problems

Team problems are meant to encourage discussion and collaborative learning. Try to solve team problems with a partner or small study group. Notice that the problems are divided into parts. Rather than tackling each part individually, discuss each section of the problem together. Try out the vocabulary that you learned in the chapter to question and convince yourselves that you are on the right track before you move to the next part. In general, the more you use the terms and apply the concepts presented in the text, the better you will become at correlating molecular structure and reactivity and thus visualizing bond breaking and bond making. You will begin to see the elegant patterns of organic chemistry and will not be a slave to memorization. The collaborative process used in partner or group study will force you to articulate your ideas. Talking out a solution with an "audience" instead of to yourself builds in checks and balances. Your teammates will not let you get away with, "Well, you know what I mean," because they probably do not. You become responsible to others as well as yourself. By learning from and teaching others, you solidify your own understanding.

52. Consider the following reaction:

$$CH_3CH_2CH_2\underset{O}{\overset{O}{C}}CH_3 + HCN \longrightarrow CH_3CH_2CH_2\underset{\underset{N}{C}}{\overset{H,\,O}{C}}CH_3$$

A   **B**

(a) Draw these condensed formulas as Lewis dot structures. Label the geometry and hybridization of the bold carbons in compounds A and B. Did the hybridization change in the course of the reaction?

(b) Draw the condensed formulas as bond-line structures.

(c) Examine the components of the reaction in light of bond polarity. Using the notation for partial charge separation, δ^+ and δ^-, indicate, on the bond-line structures, any polar bonds.

(d) This reaction is actually a two-step process: cyanide attack followed by protonation. Depict these processes by using the same "electron pair pushing" technique that we employed for resonance structures in Section 1-5, but now show the flow of electrons for the two steps. Clearly position the beginning (an electron pair) and the end (a positively polarized or charged nucleus) of your arrows.

Preprofessional Problems

Preprofessional problems are included to give you practice solving the type of problems found on exams required for entry into professional schools, such as the MCAT, DAT, chemistry GRE, and ACS exams, as well as on many undergraduate tests. Do these multiple-choice questions as you go through this course and then return to them before you take a professional school exam. These questions are to be answered "closed book"—that is, no periodic table, calculator, or the like.

53. A certain organic compound was found on combustion analysis to contain 84% carbon and 16% hydrogen (C = 12.0, H = 1.00). A molecular formula for the compound could be

(a) CH_4O (b) $C_6H_{14}O_2$ (c) C_7H_{16}
(d) C_6H_{10} (e) $C_{14}H_{22}$

54. The compound Br—Al—N—CH₂CH₃ (with Br, CH₃ above and Br, CH₃ below) has a formal charge of

(a) −1 on N   (b) +2 on N
(c) −1 on Al   (d) +1 on Br
(e) none of the above

55. The arrow in the structure points to a bond that is formed by

$CH_2{=}C$ with CH_3 and H

(a) overlap of an s orbital on H and an sp^2 orbital on C
(b) overlap of an s orbital on H and an sp orbital on C
(c) overlap of an s orbital on H and an sp^3 orbital on C
(d) none of the above

56. Which compound has bond angles nearest to 120°?

(a) O=C=S   (b) CHI_3
(c) $H_2C{=}O$ (d) H—C≡C—H
(e) CH_4

57. The pair of structures that are resonance hybrids is

(a) $H\overset{..}{O}{-}\overset{+}{C}HCH_3$ and $H\overset{+}{O}{=}CHCH_3$

(b) [square] and CH with CH₂, CH, CH₂

(c) $CH_3{-}\overset{:O:}{\overset{\|}{C}}{-}H$ and $CH_2{=}\overset{:\overset{..}{O}{-}H}{C}{-}H$

(d) $CH_3\overset{+}{C}H_2$ and $\overset{+}{C}H_2CH_3$

CHAPTER 2

Structure and Reactivity

Acids and Bases, Polar and
Nonpolar Molecules

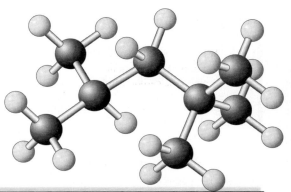

I n Chapter 1, we saw organic molecules containing several different types of bonds between various elements. Can we predict, on the basis of these structures, what kinds of chemical reactivity these substances will display? This chapter will begin to answer this question by showing how certain structural combinations of atoms in organic molecules, called functional groups, display characteristic and predictable behavior. We will see how the chemistry of acids and bases serves as a simple model for understanding the reactions of many functional groups, especially those containing polar bonds. We will pursue this analogy throughout the course, through the concepts of electrophiles and nucleophiles.

Most organic molecules contain a structural skeleton that carries the functional groups. This skeleton is a relatively nonpolar assembly consisting of carbon and hydrogen atoms connected by single bonds. The simplest class of organic compounds, the alkanes, lacks functional groups and is constructed entirely of singly bonded carbon and hydrogen. Therefore, alkanes serve as excellent models for the backbones of functionalized organic molecules. They are also useful compounds in their own right, as illustrated by the structure shown on this page, which is called 2,2,4-trimethylpentane and is an alkane in gasoline. By studying the alkanes we can prepare ourselves to better understand the properties of molecules containing functional groups. Therefore, in this chapter we will explore the names, physical properties, and structural characteristics of the members of the alkane family.

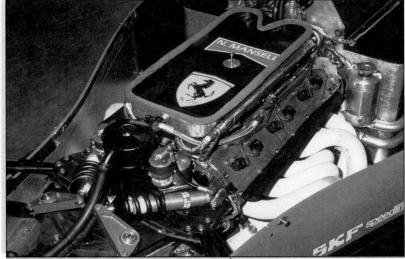

The branched alkane 2,2,4-trimethylpentane is an important component of gasoline and the standard on which the "octane rating" system for fuel efficiency is based. The car engine shown above requires high-octane fuel to achieve the performance for which it is famous.

Alkane single bond

2-1 | KINETICS AND THERMODYNAMICS OF SIMPLE CHEMICAL PROCESSES

The simplest chemical reactions may be described as equilibration between two distinct species. Such processes are governed by two fundamental considerations:

1. **Chemical thermodynamics,** which deals with the changes in energy that take place when processes such as chemical reactions occur. Thermodynamics controls the *extent* to which a reaction goes to completion.

2. **Chemical kinetics,** which concerns the velocity or rate at which the concentrations of reactants and products change. In other words, kinetics describes the *speed* at which a reaction goes to completion.

> What is a "favorable change in energy"? A favorable energy change is one in which the *energy content of a system decreases.* Energy tends to flow from systems of high energy content to those with lower energy content, just as a hot oven cools off when opened. See Problem 26.

These two principles are frequently related, but not necessarily so. Reactions that are thermodynamically very favorable often proceed faster than do less favorable ones. Conversely, some reactions are faster than others even though they result in a comparatively less stable product. A transformation that yields the most stable products is said to be under **thermodynamic control.** Its outcome is determined by the net favorable change in energy in going from starting materials to products. A reaction in which the product obtained is the one formed fastest is defined as being under **kinetic control.** Such a product may not be the thermodynamically most stable one. Let us put these statements on a more quantitative footing.

Equilibria are governed by the thermodynamics of chemical change

All chemical reactions are reversible, and reactants and products interconvert to various degrees. When the concentrations of reactants and products no longer change, the reaction is in a **state of equilibrium.** In many cases, equilibrium lies extensively (say, more than 99.9%) on the side of the products. When this occurs, the reaction is said to *go to completion.* (In such cases, the arrow indicating the reverse reaction is usually omitted and, for practical purposes, the reaction is considered to be irreversible.)

Equilibria are described by **equilibrium constants, K.** To find an equilibrium constant, divide the arithmetic product of the concentrations of the components on the right side of the reaction by that of the components on the left, all given in units of moles per liter (mol L^{-1}). A large value for K indicates that a reaction goes to completion; it is said to have a large **driving force.**

Reaction	Equilibrium Constant
$A \overset{K}{\rightleftharpoons} B$	$K = \dfrac{[B]}{[A]}$
$A + B \overset{K}{\rightleftharpoons} C + D$	$K = \dfrac{[C][D]}{[A][B]}$

If a reaction has gone to completion, a certain amount of energy has been released. The equilibrium constant can be related directly to the thermodynamic function called the **Gibbs* standard free energy change, $\Delta G°$.†** At equilibrium,

$$\Delta G° = -RT \ln K = -2.303\, RT \log K \text{ (in kcal mol}^{-1} \text{ or kJ mol}^{-1})$$

---

*Professor Josiah Willard Gibbs (1839–1903), Yale University, Connecticut.

† The symbol $\Delta G°$ refers to the free energy of a reaction with the molecules in their standard states (e.g., ideal molar solutions) after the reaction has reached equilibrium.

Table 2-1   Equilibria and Free Energy for A $\rightleftharpoons$ B: $K = $ [B]/[A]

		Percentage		$\Delta G°$	
	K	B	A	(kcal mol^{-1} at 25°C)	(kJ mol^{-1} at 25°C)
	0.01	0.99	99.0	+2.73	+11.42
	0.1	9.1	90.9	+1.36	+5.69
	0.33	25	75	+0.65	+2.72
	1	50	50	0	0
	2	67	33	−0.41	−1.72
	3	75	25	−0.65	−2.72
	4	80	20	−0.82	−3.43
	5	83	17	−0.95	−3.97
	10	90.9	9.1	−1.36	−5.69
	100	99.0	0.99	−2.73	−11.42
	1,000	99.9	0.1	−4.09	−17.11
	10,000	99.99	0.01	−5.46	−22.84

Increasing equilibrium constant (left arrow); More negative $\Delta G°$ (arrow).

in which R is the gas constant (1.986 cal K^{-1} mol^{-1} or 8.315 J K^{-1} mol^{-1}) and T is the absolute temperature in kelvin* (K). A negative $\Delta G°$ signifies a release of energy. The equation shows that a large value for K indicates a large favorable free energy change. At room temperature (25°C, 298 K), the preceding equation becomes

$$\Delta G° = -1.36 \log K \text{ (in kcal mol}^{-1})$$

This expression tells us that an equilibrium constant of 10 would have a $\Delta G°$ of −1.36 kcal mol^{-1} and, conversely, a K of 0.1 would have a $\Delta G° = +1.36$ kcal mol^{-1}. Because the relation is logarithmic, changing the $\Delta G°$ value affects the K value exponentially. When $K = 1$, starting materials and products are present in equal concentrations and $\Delta G°$ is zero (Table 2-1).

The free energy change is related to changes in bond strengths and the degree of energy dispersal in the system

The Gibbs standard free energy change is related to two other thermodynamic quantities: the change in **enthalpy, $\Delta H°$**, and the change in **entropy, $\Delta S°$**.

Gibbs Standard Free Energy Change

$$\Delta G° = \Delta H° - T\Delta S°$$

In this equation, T is again in kelvin and $\Delta H°$ in kcal mol^{-1} or kJ mol^{-1}, whereas $\Delta S°$ is in cal K^{-1} mol^{-1}, also called entropy units (e.u.), or J K^{-1} mol^{-1}.

The **enthalpy change** of a reaction, $\Delta H°$, is the heat absorbed or released at constant pressure during the course of the reaction. Enthalpy changes in chemical reactions relate mainly to differences between the strengths of the bonds in the products compared with those in the starting materials. Bond strengths are described quantitatively by bond-dissociation energies, $DH°$. The value of $\Delta H°$ for a reaction may be estimated by subtracting the sum of the $DH°$ values of the bonds formed from those of the bonds broken. Chapter 3 explores in detail bond-dissociation energies and their value in understanding chemical reactions.

*Temperature intervals in kelvin and degrees Celsius are identical. Temperature units are named after Lord Kelvin, Sir William Thomson (1824–1907), University of Glasgow, Scotland, and Anders Celsius (1701–1744), University of Uppsala, Sweden.

Enthalpy Change in a Reaction

$$\left(\begin{array}{c} \text{Sum of } DH^\circ \\ \text{of bonds broken} \end{array}\right) - \left(\begin{array}{c} \text{sum of } DH^\circ \\ \text{of bonds formed} \end{array}\right) = \Delta H^\circ$$

If the bonds formed are stronger than those broken, the value of ΔH° is negative and the reaction is defined as **exothermic** (releasing heat). In contrast, a positive ΔH° is characteristic of an **endothermic** (heat-absorbing) process. An example of an exothermic reaction is the combustion of methane, the main component of natural gas, to carbon dioxide and liquid water.

$$CH_4 + 2\,O_2 \longrightarrow CO_2 + 2\,H_2O_{liq} \qquad \Delta H^\circ = -213 \text{ kcal mol}^{-1}\ (-891 \text{ kJ mol}^{-1})$$
$$\textbf{Exothermic}$$

$$\Delta H^\circ = (\text{sum of } DH^\circ \text{ values of all the bonds in } CH_4 + 2\,O_2) - (\text{sum of } DH^\circ \text{ values}$$
$$\text{of all the bonds in } CO_2 + 2\,H_2O)$$

Opening a hot oven to permit the heat to disperse throughout a cooler room is entropically favorable: The total entropy of the system—oven + room—increases. The process distributes the heat from the smaller number of molecules inside the oven to the much larger number of molecules in the air and the surroundings of the room.

The exothermic nature of this reaction is due to the very strong bonds formed in the products. Many hydrocarbons release a lot of energy on combustion and are therefore valuable fuels.

If the enthalpy of a reaction depends strongly on changes in bond strength, what is the significance of ΔS°, the **entropy change?** You may be familiar with the concept that entropy is related to the order of a system: Increasing disorder correlates with an increase in the value of S°. However, the concept of "disorder" is not readily quantifiable and cannot be applied in a precise way to scientific situations. Instead, for chemical purposes, ΔS° is used to describe changes in energy dispersal. Thus, the value of S° increases with increasing *dispersal of energy content* among the constituents of a system. Because of the negative sign in front of the $T\Delta S^\circ$ term in the equation for ΔG°, a positive value for ΔS° makes a negative contribution to the free energy of the system. In other words, going from lesser to greater energy dispersal is thermodynamically favorable.

What is meant by energy dispersal in a chemical reaction? Consider a transformation in which the number of reacting molecules differs from the number of product molecules formed. For example, upon strong heating, 1-pentene undergoes cleavage into ethene and propene. This process is endothermic, primarily because a C–C bond is lost. It would not occur, were it not for entropy. Thus, two molecules are made from one, and this is associated with a relatively large positive ΔS°. After bond cleavage, the energy content of the system is distributed over a greater number of particles. At high temperatures, the $-T\Delta S^\circ$ term in our expression for ΔG° overrides the unfavorable enthalpy, making this a feasible reaction.

$$CH_3CH_2CH_2CH{=}CH_2 \longrightarrow CH_2{=}CH_2 + CH_3CH{=}CH_2 \qquad \Delta H^\circ = +22.4 \text{ kcal mol}^{-1}\ (+93.7 \text{ kJ mol}^{-1})$$

1-Pentene   **Ethene**   **Propene**   **Endothermic**
(Ethylene)

$$\Delta S^\circ = +33.3 \text{ cal K}^{-1} \text{ mol}^{-1} \text{ or e.u.}\ (+139.3 \text{ J K}^{-1} \text{ mol}^{-1})$$

Exercise 2-1

Calculate the ΔG° at 25°C for the preceding reaction. Is it thermodynamically feasible at 25°C? What is the effect of increasing T on ΔG°? What is the temperature at which the reaction becomes favorable? (**Caution:** ΔS° is in the units of *cal* K^{-1} mol^{-1}, whereas ΔH° is in *kcal* mol^{-1}. Don't forget that factor of 1000!)

In contrast, energy dispersal and entropy decrease when the number of product molecules is less than the number of molecules of starting materials. For example, the reaction of ethene (ethylene) with hydrogen chloride to give chloroethane is exothermic by -15.5 kcal mol^{-1}, but the entropy makes an unfavorable contribution to the $\Delta G°$; $\Delta S° = -31.3$ e.u.

$$CH_2{=}CH_2 + HCl \longrightarrow CH_3CH_2Cl$$

$$\Delta H° = -15.5 \text{ kcal mol}^{-1} (-64.9 \text{ kJ mol}^{-1})$$
$$\Delta S° = -31.3 \text{ e.u.} (-131.0 \text{ J K}^{-1} \text{ mol}^{-1})$$

Exercise 2-2

Calculate the $\Delta G°$ at 25°C for the preceding reaction. In your own words, explain why a reaction that combines two molecules into one should have a large negative entropy change.

In many organic reactions the change in entropy is small, and it often suffices to consider only the changes in bonding energy to estimate whether they are likely to occur or not. In those cases we will equate approximately $\Delta G°$ to $\Delta H°$. Exceptions are transformations in which the number of molecules on each side of the chemical equation differ (as shown in the examples above) or in which energy dispersal is greatly affected by profound structural changes, such as, for example, ring closures and ring openings (margin).

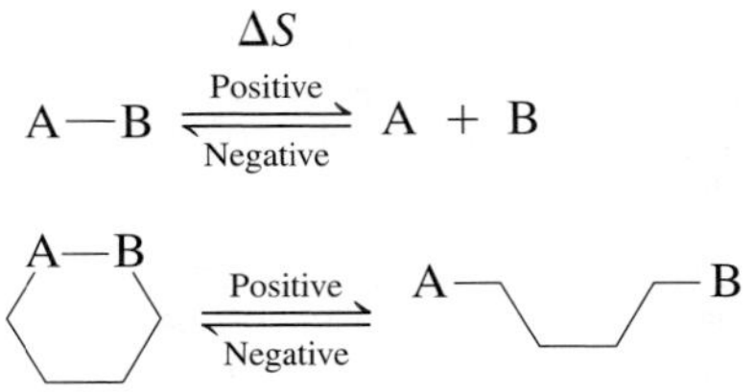

The rate of a chemical reaction depends on the activation energy

How fast is equilibrium established? The thermodynamic features of chemical reactions do not by themselves tell us anything about their rates. Consider the combustion of methane, mentioned earlier. This process releases 213 kcal mol^{-1} (-891 kJ mol^{-1}), a huge amount of energy, but we know that methane does not spontaneously ignite in air at room temperature. Why is this highly favorable combustion process so slow? The answer is that during the course of this reaction the potential energy of the system changes in the manner shown in Figure 2-1. This figure, which is an example of a **potential-energy diagram,** plots energy as a function of reaction progress. We measure reaction progress by the **reaction coordinate,** which describes the combined processes of bond breaking and bond formation that constitute the overall change from the structures of the starting compounds into those of the products. The energy first rises to a maximum, a point called the **transition state** (TS), before decreasing to the final value, which is the energy content of the product molecules. The energy of the transition state may be viewed as a barrier to be overcome in order for the reaction to take place. The energy input required to raise the energy of the starting compounds to that of the transition state is called the **activation energy,** E_a, of the

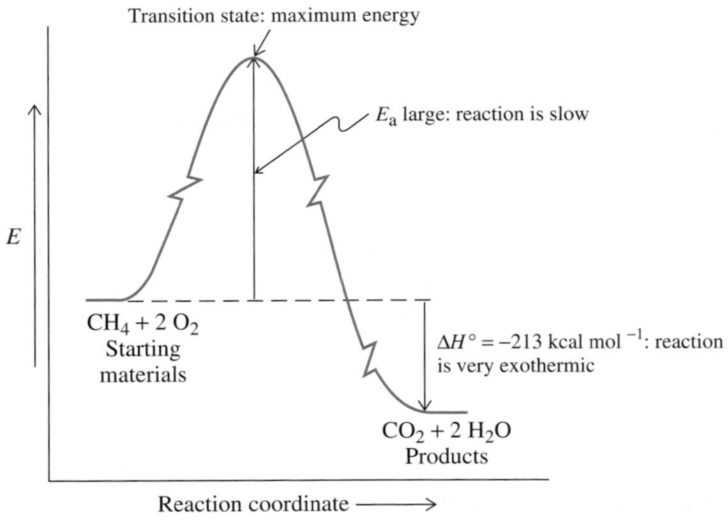

Organic Reactions with Significant Entropy Changes

$$\Delta S$$

$$A{-}B \underset{\text{Negative}}{\overset{\text{Positive}}{\rightleftarrows}} A + B$$

$$A{-}B \quad \underset{\text{Negative}}{\overset{\text{Positive}}{\rightleftarrows}} \quad A{\diagdown}{\diagup}{\diagdown}B$$

Transition state: maximum energy

E_a large: reaction is slow

E

CH$_4$ + 2 O$_2$
Starting
materials

$\Delta H° = -213$ kcal mol^{-1}: reaction is very exothermic

CO$_2$ + 2 H$_2$O
Products

Reaction coordinate ⟶

Figure 2-1 A (highly oversimplified) potential-energy diagram for the combustion reaction of methane. Despite its thermodynamic favorability, as shown by the large negative $\Delta H°$, the process is very slow because it has a high-energy transition state and a large activation energy. (In actuality, the process has many individual bond-breaking and bond-forming steps, and the applicable potential-energy diagram therefore has multiple maxima and minima.)

Sisyphus, of Greek mythology, was doomed forever to roll a boulder up a steep hill, only to see it roll back down just as he gets it to the top. His task is an extreme example of a process with a very large activation barrier. [*artpartner-images/Getty Images*]

reaction. The higher its value, the slower the process. The transition state for methane combustion is very high in energy, corresponding to a high E_a and a very low rate.

How can exothermic reactions have such high activation energies? As atoms move from their initial positions in the starting molecules, energy input is required for bond breaking to begin. At the transition state, where both partially broken old bonds and incompletely formed new ones are present, the overall loss of bonding reaches its greatest extent, and the energy content of the system is at its maximum. Beyond this point, continued strengthening of the new bonds releases energy until the atoms reach their final, fully bonded positions in the products.

Collisions supply the energy to get past the activation-energy barrier

Where do molecules get the energy to overcome the barrier to reaction? Molecules have *kinetic energy* as a result of their motion, but at room temperature the average kinetic energy is only about 0.6 kcal mol^{-1} (2.5 kJ mol^{-1}), far below many activation-energy barriers. To pick up enough energy, molecules must collide with each other or with the walls of the container. Each collision transfers energy between molecules.

A graph called a **Boltzmann* distribution curve** depicts the distribution of kinetic energy. Figure 2-2 shows that, although most molecules have only average speed at any given temperature, some molecules have kinetic energies that are much higher.

Figure 2-2 Boltzmann curves at two temperatures. At the higher temperature (green curve), there are more molecules of kinetic energy E than at the lower temperature (blue curve). Molecules with higher kinetic energy can more easily overcome the activation-energy barrier.

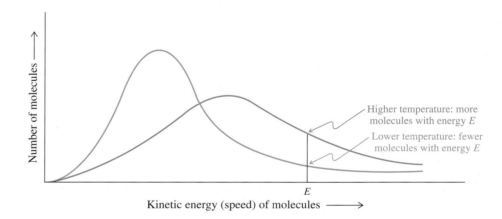

The shape of the Boltzmann curve depends on the temperature. At higher temperatures, as the average kinetic energy increases, the curve flattens and shifts toward higher energies. More molecules now have energy higher than is required by the transition state, so the speed of reaction increases. Conversely, at lower temperatures, the reaction rate decreases.

The concentration of reactants can affect reaction rates

Consider the addition of reagent A to reagent B to give product C:

$$A + B \longrightarrow C$$

In many transformations of this type, increasing the concentration of either reactant increases the rate of the reaction. In such cases, the transition-state structure incorporates both molecules A and B. The experimentally observed rate is expressed by

$$\text{Rate} = k[A][B] \text{ in units of mol L}^{-1}\text{ s}^{-1}$$

---

*Professor Ludwig Boltzmann (1844–1906), University of Vienna, Austria.

in which the proportionality constant, k, is also called the **rate constant** of the reaction. The rate constant equals the rate of the reaction at 1 molar concentrations of the two reactants, A and B. A reaction for which the rate depends on the concentrations of two molecules in this way is said to be a **second-order** reaction.

In some processes, the rate depends on the concentration of only one reactant, such as the hypothetical reaction

$$A \longrightarrow B$$
$$\text{Rate} = k[A] \text{ in units of mol } L^{-1} s^{-1}$$

A reaction of this type is said to be of **first order.**

Solved Exercise 2-3 | **Working with the Concepts: Using Rate Equations**

a. What is the reduction in the rate of a reaction that follows the first-order rate law, rate $= k[A]$, when half of the A has been consumed (i.e., after 50% conversion of the starting material)?

Strategy

We employ the WHIP approach, as we did in Chapter 1 for Worked Example 1-23.
*W*hat the question is asking seems straightforward enough, but it may be difficult to convert into a form that can be solved readily, especially if you have not encountered such a problem before.
*H*ow do you begin? The key is to realize that you need to *compare two rates,* the initial rate given above and the rate after the concentration of A has been halved. Can you come up with an equation to describe that new rate?
*I*nformation needed? For a first-order reaction, the rate equals the rate constant times the concentration of the starting material. After half of A has transformed, its new concentration is thus $0.5[A_0]$, where A_0 is the starting concentration of A.
So, *P*roceed:

Solution
- The rate after half of A has reacted is described by the equation $\text{rate}_{1/2} = k(0.5[A_0])$.
- The initial rate is described by the equation $\text{rate}_{initial} = k[A_0]$.
- We solve for $\text{rate}_{1/2}$ by substituting the second equation into the first:

$$\text{rate}_{1/2} = (0.5)\text{rate}_{initial}$$

The rate decreases to one-half its initial value.

b. Answer the same question for a second-order reaction, for which the rate law is rate $= k[A][B]$. Assume that the two starting materials A and B are initially present in equal amounts.

Strategy

As above, we must *compare two rates,* the initial rate given above and the rate after the concentrations of *both A and B* have been halved. Why? The initial amounts of A and B are equal, and according to our second-order equation, every A reacts with one B. So their concentrations drop by the same amount as the reaction proceeds. Can you come up with an equation to describe the new rate?

Solution
- The rate after half of both A and B have reacted is described by the equation $\text{rate}_{1/2} = k(0.5[A_0])(0.5[B_0])$, where $[A_0]$ and $[B_0]$ are the initial concentrations of A and B, respectively.
- The initial rate is described by the equation $\text{rate}_{initial} = k[A_0][B_0]$.
- We solve for $\text{rate}_{1/2}$ by substituting the second equation into the first:

$$\text{rate}_{1/2} = (0.5)(0.5)\text{rate}_{initial} = (0.25)\text{rate}_{initial}$$

The rate decreases to one-quarter its initial value.

Exercise 2-4 | Try It Yourself

The reaction described by the equation

$$CH_3Cl + NaOH \longrightarrow CH_3OH + NaCl$$

follows the second-order rate law, rate = $k[CH_3Cl][NaOH]$. When this reaction is carried out with starting concentrations $[CH_3Cl] = 0.2$ M and $[NaOH] = 1.0$ M, the measured rate is 1×10^{-4} mol L^{-1} s^{-1}. What is the rate after one-half of the CH_3Cl has been consumed? (**Caution:** The initial concentrations of the starting materials are *not* identical in this experiment. **Hint:** Determine how much of the NaOH has been consumed at this point and what its new concentration is, compared with its initial concentration.)

For Your Calibration

Approximate time to completion of a first order reaction at 20°C:

Ea	Reaction Time
10 kcal mol^{-1}	~10^{-5} seconds
15 kcal mol^{-1}	~0.1 seconds
20 kcal mol^{-1}	minutes
25 kcal mol^{-1}	days

The Arrhenius equation describes how temperature affects reaction rates

The kinetic energy of molecules increases when they are heated, which means that a larger fraction of them have sufficient energy to overcome the activation barrier E_a (Figure 2-2). A useful rule of thumb applies to many reactions: Raising the reaction temperature by 10 degrees (Celsius) causes the rate to increase by a factor of 2 to 3. The Swedish chemist Arrhenius* noticed the dependence of reaction rate k on temperature T. He found that his measured data conformed to the equation

Arrhenius Equation

$$k = Ae^{-E_a/RT} = A\left(\frac{1}{e^{E_a/RT}}\right)$$

The Arrhenius equation describes how rates of reactions with different activation energies vary with temperature. In this equation, R is again the gas constant and A is a factor with a value characteristic of a specific reaction. You can see readily that the larger the activation energy E_a, the slower the reaction. Conversely, the higher the temperature T, the faster the reaction. The A term can be imagined as the maximum rate constant that the reaction would have if every molecule had sufficient collisional energy to overcome the activation barrier. This will occur at very high temperature, when E_a/RT will be close to zero and $e^{-E_a/RT}$ approaches 1, thus rendering k nearly equal to A.

Increasing Rates of Chemical Reactions

Increasing temperature

Decreasing activation energy

Increasing concentration

Exercise 2-5

(a) Calculate $\Delta G°$ at 25°C for the reaction $CH_3CH_2Cl \rightarrow CH_2=CH_2 + HCl$ (the reverse of the reaction in Exercise 2-2). (b) Calculate $\Delta G°$ at 500°C for the same reaction. (**Hint:** Apply $\Delta G° = \Delta H° - T\Delta S°$ and do not forget to first convert degrees Celsius into kelvin.)

Exercise 2-6

For the reaction in Exercise 2-5, $A = 10^{14}$ and $E_a = 58.4$ kcal mol^{-1}. Using the Arrhenius equation, calculate k at 500°C for this reaction. $R = 1.986$ cal K^{-1} mol^{-1}. (**Caution:** Activation energies are given in the units of *kcal* mol^{-1}, while R is in *cal* K^{-1} mol^{-1}. Don't forget that factor of 1000!)

In Summary All reactions are described by equilibrating the concentrations of starting materials and products. On which side the equilibrium lies depends on the size of the equilibrium constant, which in turn is related to the Gibbs free energy changes, $\Delta G°$. An increase in the equilibrium constant by a factor of 10 is associated with a change in $\Delta G°$ of about -1.36 kcal mol^{-1} (-5.69 kJ mol^{-1}) at 25°C. The free energy change of a reaction is composed

*Professor Svante Arrhenius (1859–1927), Technical Institute of Stockholm, Sweden, Nobel Prize 1903 (chemistry), director of the Nobel Institute from 1905 until shortly before his death.

of changes in enthalpy, $\Delta H°$, and entropy, $\Delta S°$. Contributions to enthalpy changes stem mainly from variations in bond strengths; contributions to entropy changes arise from the relative dispersal of energy in starting materials and products. Whereas these terms define the position of an equilibrium, the rate at which it is established depends on the concentrations of starting materials, the activation barrier separating reactants and products, and the temperature. The relation among rate, E_a, and T is expressed by the Arrhenius equation.

2-2 | KEYS TO SUCCESS: USING CURVED "ELECTRON-PUSHING" ARROWS TO DESCRIBE CHEMICAL REACTIONS

As of April 2012, *Chemical Abstracts* had registered over 65 million chemical substances, all of which are formed by chemical reactions and all of which undergo chemical reactions. Clearly, memorizing even a tiny fraction of these transformations is not a practical approach to succeeding in a course in organic chemistry. Fortunately, reactions follow logical pathways defined by *reaction mechanisms,* and there are only a few dozen of those. Let us see how they can help us organize the task of learning organic chemistry.

Curved arrows show how starting materials convert to products

Bonds consist of electrons. Chemical change is defined as a process in which bonds are broken and/or formed. Therefore, *when chemistry takes place, electrons move.* It is the description of this electron movement that constitutes a reaction mechanism and is depicted by curved arrows.

A curved arrow (⌒) shows the "flow" of an electron pair from its point of origin, usually a lone pair or a covalent bond, to its destination. The "target" may be an atom that attracts the electrons by virtue of being relatively electronegative or electron deficient. Some examples follow.

1. *Dissociation of a polar covalent bond into ions*

General case:
$$A\overset{\frown}{}B \longrightarrow A^+ + :B^-$$

Movement of an electron pair converts the A–B covalent bond into a lone pair on atom B

The direction in which the pair of electrons moves depends on which of the two atoms is more electronegative. In the general case above, B is more electronegative than A, so B more readily accepts the electron pair to become negatively charged. Atom A becomes a cation.

Arrow points to Cl, the more electronegative atom

Chloride is released with an additional lone pair derived from the broken bond

Specific example (a):
$$H\overset{\frown}{}\ddot{\underset{\cdot\cdot}{C}l}: \longrightarrow H^+ + :\ddot{\underset{\cdot\cdot}{C}l}:^-$$

Dissociation of the acid HCl to give a proton and chloride ion exemplifies this process: When breaking a polar covalent bond in this way, *draw the curved arrow starting at the center of the bond and ending at the more electronegative atom.*

Specific example (b):
$$H_3C-\underset{\underset{CH_3}{|}}{\overset{\overset{CH_3}{|}}{C}}\overset{\frown}{}\ddot{\underset{\cdot\cdot}{B}r}: \longrightarrow H_3C-\underset{CH_3}{\overset{\overset{CH_3}{}}{C^+}} + :\ddot{\underset{\cdot\cdot}{B}r}:^-$$

In this example, dissociation features the breaking of a C–Br bond. You will note that its essential features are identical to those of example (a).

2. *Formation of a covalent bond from ions*

General case:

$$A^+ + :B^- \longrightarrow A—B$$

**The reverse of the previous process: A lone pair on
B moves toward A, becoming a new covalent bond between A and B**

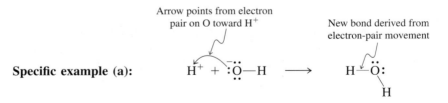

Arrow points from electron
pair on O toward H⁺

New bond derived from
electron-pair movement

Specific example (a):

$$H^+ + :\overset{..}{\underset{..}{O}}—H \longrightarrow H—\overset{..}{\underset{..}{O}}:$$
$$\hspace{7cm} |$$
$$\hspace{7cm} H$$

The acid-base reaction between hydrogen ion and hydroxide exemplifies this type of mechanism: When combining an anion with a cation, *draw the curved arrow starting at an electron pair on the anion and ending at the cation. NEVER start the curved arrow at the cation!* The arrow shows how *electrons* move, not atoms. Electrons move, and atoms follow.

Specific example (b):

$$H_3C—\overset{+}{\underset{\underset{CH_3}{|}}{\overset{\overset{CH_3}{|}}{C}}} \quad + :\overset{..}{\underset{..}{Br}}:^- \longrightarrow H_3C—\overset{\overset{CH_3}{|}}{\underset{\underset{CH_3}{|}}{C}}—\overset{..}{\underset{..}{Br}}:$$

This process is the reverse of example (b) of the preceding type 1 mechanisms of dissociation.

3. *Simultaneous making and breaking of two bonds: substitution reactions*

General case:

$$X:^- + \quad {}^{\delta+}A—B^{\delta-} \longrightarrow X—A + :B^-$$

**Movement of two electron pairs
results in substitution of one bond for another**

Electron pair of HCl bond is
"pushed" away from H and onto Cl

Chloride released with
additional lone pair

Specific example (a):

$$H—\overset{..}{\underset{..}{O}}:^- + H—\overset{..}{\underset{..}{Cl}}: \longrightarrow H—\overset{..}{\underset{..}{O}}: + :\overset{..}{\underset{..}{Cl}}:^-$$
$$\hspace{10cm} |$$
$$\hspace{10cm} H$$

New bond made from former
lone pair of hydroxide oxygen

Specific example (b):

$$H—\overset{..}{\underset{..}{O}}:^- + H—\overset{\overset{H}{|}}{\underset{\underset{H}{|}}{C}}—\overset{..}{\underset{..}{Cl}}: \longrightarrow :\overset{..}{\underset{..}{O}}—\overset{\overset{H}{|}}{\underset{\underset{H}{|}}{C}}—H + :\overset{..}{\underset{..}{Cl}}:^-$$

In both reactions, two electron pairs move, and one bond (to Cl) is broken while another (to O) is made. Example (a) shows hydroxide ion acting in its familiar role as a base, attacking and removing a proton from an acid. In example (b), an electron pair on hydroxide attacks a nonhydrogen atom, a carbon at the positive end of a polarized bond (Section 1-3). The carbon is said to be **electrophilic** (literally, "electron friendly"; *philos,* Greek, friend). In turn, the oxygen in hydroxide ion is described as **nucleophilic** ("nucleus friendly"). By convention, we use the term **nucleophile** to refer to a basic atom when it attacks an atom other than hydrogen.

When writing the mechanism for a substitution, *the head of the first arrow points to the tail of the second, and both arrows flow in sequence.* In the examples above, the two electron pairs move in the same direction, left to right. Think of the first electron pair pushing the second electron pair away. *NEVER point two arrows toward each other!*

4. *Reactions involving double (or triple) bonds: additions*

General case (a):

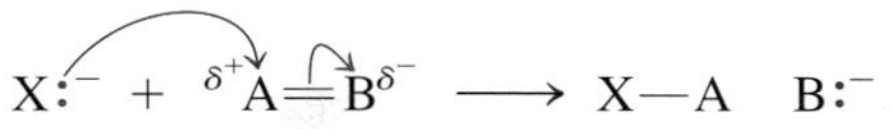

$$X\colon^- + \ ^{\delta +}A = B^{\delta -} \longrightarrow X - A \quad B\colon^-$$

**Movement of a lone electron pair toward a double bond results
in a new bond and changes the double bond to a single bond**

An atom with a lone pair may add to the δ^+ atom of a polarized multiple bond. As in the type 3 mechanism, the lone electron pair entering from the left "pushes" a second electron pair out to the right. This second pair is one of the two bonding electron pairs between A and B, so the original A=B double bond is reduced to an A–B single bond.

Specific example:

One of the two electron pairs of the
double bond is "pushed" onto O

New lone pair on O

New bond derived from former
lone pair on oxygen

One of the two bonds originally
between C and O remains

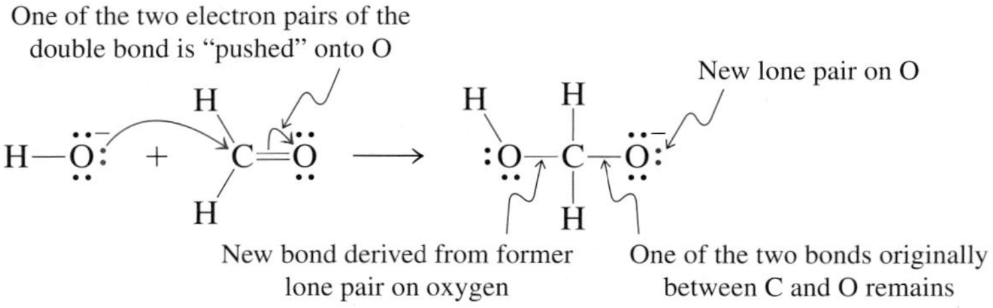

In this example, hydroxide behaves as a nucleophile and adds to the electrophilic carbon of the carbonyl function.

General case (b):

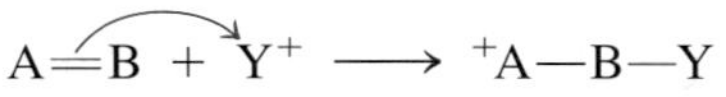

$$A = B + \ ^+Y \longrightarrow \ ^+A - B - Y$$

**Movement of one of the electron pairs of a double bond toward a
cation results in a new bond and changes the double bond into a single bond**

Specific example:

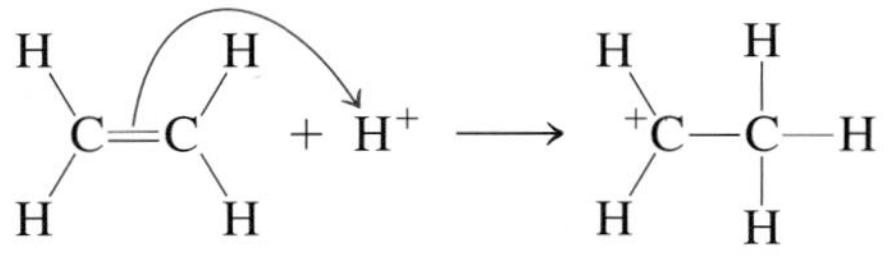

Movement of one of the electron pairs of a double bond toward a proton, also called "protonation" of a double bond, results in a carbocation, a species with a positively charged carbon atom. The proton is acting as an electrophile, attacking an electron pair in the double bond.

Exercise 2-7

From the categories above, identify the one to which each of the following reactions belongs. Draw appropriate curved arrows to show electron movement, and give the structure of the product. (**Hint:** First complete all Lewis structures by adding any missing lone pairs.)
(a) $CH_3O^- + H^+$; **(b)** $H^+ + CH_3CH = CHCH_3$; **(c)** $(CH_3)_2N^- + HCl$; **(d)** $CH_3O^- + H_2C = O$.

These examples are among the most common mechanistic transformation types that you will encounter in organic chemistry. There are several benefits of learning how to draw mechanisms using curved arrows. For starters, the technique allows you to keep track of all the electrons in your reacting species, because it *generates the correct Lewis structure of the product of a reaction automatically*. In addition, it provides a framework—the "grammar"—for tackling the formulation of possible modes of reactivity and therefore the writing of possible product structures.

In Summary Reaction mechanisms describe the electron motion that takes place when chemical bonds are formed or broken. Curved arrows are used to depict this electron motion. As you read on, try to associate each new reaction you encounter with one of the electron-movement patterns shown above and draw curved arrows that fit the pattern.

2-3 | ACIDS AND BASES

Brønsted and Lowry* have given us a simple definition of acids and bases: An **acid** is a proton donor and a **base** is a proton acceptor. Acidity and basicity are commonly measured in water. An acid donates protons to water, forming hydronium ions, whereas a base removes protons from water, forming hydroxide ions. Examples are the acid hydrogen chloride and the base ammonia. The electron flow in the case of the reaction of water with hydrogen chloride is indicated in the electrostatic potential maps below the equation. The red oxygen of water is protonated by the blue hydrogen in the acid to furnish as products the blue hydronium ion and red chloride ion.

| Water (Base) | Hydrogen chloride (Acid) | Hydronium ion (Conjugate acid of water) | Chloride ion (Conjugate base of HCl) |

| Ammonia (Base) | Water (Acid) | Ammonium ion (Conjugate acid of NH₃) | Hydroxide ion (Conjugate base of water) |

Acid and base strengths are measured by equilibrium constants

Water itself is neutral. It forms an equal number of hydronium and hydroxide ions by self-dissociation. The process is described by the equilibrium constant K_w, the self-ionization constant of water. At 25°C,

$$H_2O + H_2O \xrightleftharpoons{K_w} H_3O^+ + OH^- \qquad K_w = [H_3O^+][OH^-] = 10^{-14}\ mol^2L^{-2}$$

From the value for K_w, it follows that the concentration of H_3O^+ in pure water is $10^{-7}\ mol\ L^{-1}$.

---

*Professor Johannes Nicolaus Brønsted (1879–1947), University of Copenhagen, Denmark. Professor Thomas Martin Lowry (1874–1936), University of Cambridge, England.

REAL LIFE: MEDICINE 2-1 | Stomach Acid, Peptic Ulcers, Pharmacology, and Organic Chemistry

The human stomach produces, on average, 2 L of 0.02 M hydrochloric acid each day. The pH of "stomach juice" ranges between 1.0 and 2.5, falling as HCl production rises in response to the stimuli of tasting, smelling, or even looking at food. Stomach acid disrupts the natural folded shapes of protein molecules in food, exposing them to attack and breakdown by digestive enzymes.

You may wonder how the stomach protects itself from such strongly acidic conditions—after all, stomach tissue itself is constructed of protein molecules. The interior lining of the stomach is coated with a layer of cells called gastric mucosa, whose mucous secretions insulate the wall from acid attack. In response to the stimuli described above, certain cells just below the gastric mucosa release a signaling molecule called histamine that causes parietal cells in the lining to secrete HCl into the stomach. Some so-called acid-reducing medications such as ranitidine block histamine from reaching parietal cells, interrupting the signal that would produce stomach acid. Such products are useful in treating conditions such as hyperacidity, the secretion of unnecessarily large amounts of acid. Proton-pump inhibitors ("PPIs") such as omeprazole are the most powerful products on the market currently. They function by blocking directly the acid-producing engine (the "proton pump") in the parietal cells.

Peptic ulcers are sores in the stomach lining that expose it to acid attack. These sores result from infection by the bacterium *Helicobacter pylori*. *H. pylori* is susceptible to antibiotics such as amoxicillin, but the acid content of the stomach causes rapid destruction of the antibiotic. Thus the successful eradication of *H. pylori* infection and the cure of peptic ulcers relies on administration of the antibiotic together with a PPI. The inhibitor raises the pH of the stomach above 4, permitting the antibiotic to survive long enough to reach sites of infection deep in the gastric pits of the stomach lining. The development of this cure for peptic ulcers relied on close partnerships between organic chemists and pharmacologists. Chemists designed and synthesized potential drug candidate molecules, and pharmacologists enabled the optimization of the properties of those molecules through their studies of their biochemical and physiological properties. This partnership defines the field of pharmaceutical chemistry, a subset of the general area of chemical biology, the application of chemistry to solve biological problems.

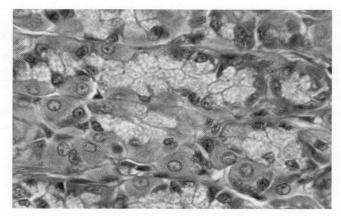

The parietal cells (orange) in the gastric glands of the stomach secrete hydrochloric acid upon activation by histamine.

The pH is defined as the negative logarithm of the value for $[H_3O^+]$.

$$pH = -\log[H_3O^+]$$

Thus, for pure water, the pH is +7. An aqueous solution with a pH lower than 7 is acidic; one with a pH higher than 7 is basic.

The acidity of a general acid, HA, is expressed by the following general equation, together with its associated equilibrium constant:

$$HA + H_2O \xrightleftharpoons{K} H_3O^+ + A^- \qquad K = \frac{[H_3O^+][A^-]}{[HA][H_2O]}$$

In dilute aqueous solution, $[H_2O]$ is constant at 55 mol L^{-1}, so this number may be incorporated into a new constant, the **acid dissociation constant, K_a.**

$$K_a = K[H_2O] = \frac{[H_3O^+][A^-]}{[HA]} \text{ mol } L^{-1}$$

Table 2-2	Relative Acidities of Common Compounds (25°C)		
Acid	K_a		pK_a
Hydrogen iodide, HI (strongest acid)	$\sim 1.0 \times 10^{10}$		-10.0
Hydrogen bromide, HBr	$\sim 1.0 \times 10^9$		-9.0
Hydrogen chloride, HCl	$\sim 1.0 \times 10^8$		-8.0
Sulfuric acid, H_2SO_4	$\sim 1.0 \times 10^3$		-3.0^a
Hydronium ion, H_3O^+	50		-1.7
Nitric acid, HNO_3	25		-1.4
Methanesulfonic acid, CH_3SO_3H	16	Increasing acidity	-1.2
Hydrogen fluoride, HF	6.3×10^{-4}		3.2
Acetic acid, CH_3COOH	2.0×10^{-5}		4.7
Hydrogen cyanide, HCN	6.3×10^{-10}		9.2
Ammonium ion, NH_4^+	5.7×10^{-10}		9.3
Methanethiol, CH_3SH	1.0×10^{-10}		10.0
Methanol, CH_3OH	3.2×10^{-16}		15.5
Water, H_2O	2.0×10^{-16}		15.7
Ethyne, $HC{\equiv}CH$	$\sim 1.0 \times 10^{-25}$		~ 25
Ammonia, NH_3	1.0×10^{-35}		35
Ethene, $H_2C{=}CH_2$	$\sim 1.0 \times 10^{-44}$		~ 44
Methane, CH_4 (weakest acid)	$\sim 1.0 \times 10^{-50}$		~ 50

Note: $K_a = [H_3O^+][A^-]/[HA] \text{ mol } L^{-1}$.

[a]First dissociation equilibrium

Like the concentration of H_3O^+ and its relation to pH, this measurement may be put on a logarithmic scale by the corresponding definition of pK_a.

$$pK_a = -\log K_a*$$

The pK_a is the pH at which the acid is 50% dissociated. An acid with a pK_a lower than 1 is defined as strong, one with a pK_a higher than 4 as weak. The acidities of several common acids are compiled in Table 2-2 and compared with those of compounds with higher pK_a values. Sulfuric acid and, with the exception of HF, the hydrogen halides, are very strong acids. Hydrogen cyanide, water, methanol, ammonia, and methane are decreasingly acidic, the last two being exceedingly weak.

The species A^- derived from acid HA is frequently referred to as its **conjugate base** (*conjugatus*, Latin, joined). Conversely, HA is the **conjugate acid** of base A^-. The strengths of two substances that are related as a conjugate acid-base pair are inversely related: The conjugate bases of strong acids are weak, as are the conjugate acids of strong bases. For example, HCl is a strong acid, because the equilibrium for its dissociation into H^+ and Cl^- is very favorable. The reverse process, reaction of Cl^- to combine with H^+, is unfavorable, therefore identifying Cl^- as a weak base.

$$H{-}\ddot{C}\ddot{l}: \; \rightleftharpoons \; H^+ + :\ddot{C}\ddot{l}:^- \qquad \textbf{Equilibrium lies on the right}$$

Strong acid   **Conjugate base is weak**

In contrast, dissociation of CH_3OH to produce CH_3O^- and H^+ is unfavorable; CH_3OH is a weak acid. The reverse, combination of CH_3O^- with H^+ is favorable; we therefore consider CH_3O^- to be a strong base.

*K_a carries the units of molarity, or mol L^{-1}, because it is the product of a dimensionless equilibrium constant K and the concentration $[H_2O]$, which equals 55 mol L^{-1}. However, the logarithm function can operate only on dimensionless numbers. Therefore, pK_a is properly defined as the negative log of the *numerical value* of K_a, which is K_a divided by the units of concentration. (For purposes of simplicity, we will omit the units of K_a in exercises and problems.)

$$H\!-\!\ddot{O}CH_3 \;\rightleftharpoons\; H^+ \;+\; {}^-\!:\!\ddot{O}CH_3 \qquad \textbf{Equilibrium lies on the left}$$

Weak acid **Conjugate base is strong**

For the sake of simplicity, we have used the symbol H^+ to depict the dissociating proton. In reality, the free proton does not exist in solution, but is always associated with the electron pair of another species present, typically the solvent. As we have seen, in water, H^+ is represented by the hydronium ion, H_3O^+. In methanol, it would be methoxonium ion $CH_3OH_2^+$, in methoxymethane (Figure 1-22B) it would be $(CH_3)_2OH^+$, and so on. We shall see later that in many organic reactions carried out under acidic conditions, there are several potential recipients of a dissociating proton, and it becomes cumbersome to show them all. In these cases, we will stick with the short notation H^+.

Exercise 2-8

Write the formula for the conjugate base of each of the following acids. **(a)** Sulfurous acid, H_2SO_3; **(b)** chloric acid, $HClO_3$; **(c)** hydrogen sulfide, H_2S; **(d)** dimethyloxonium, $(CH_3)_2OH^+$; **(e)** hydrogen sulfate, HSO_4^-.

Exercise 2-9

Write the formula for the conjugate acid of each of the following bases. **(a)** Dimethylamide, $(CH_3)_2N^-$; **(b)** sulfide, S^{2-}; **(c)** ammonia, NH_3; **(d)** acetone, $(CH_3)_2C\!=\!O$; **(e)** 2,2,2-trifluoroethoxide, $CF_3CH_2O^-$.

Exercise 2-10

Which is the stronger acid, nitrous (HNO_2, $pK_a = 3.3$) or phosphorous acid (H_3PO_3, $pK_a = 1.3$)? Calculate K_a for each.

We can estimate relative acid and base strengths from a molecule's structure

The relationship between structure and function is very evident in acid-base chemistry. Indeed, several specific structural features allow us to estimate, at least qualitatively, the relative strength of an acid HA. The guiding principle is as follows: The more stable the conjugate base—that is, the lower its base strength—the stronger will be the corresponding acid. Below are several important structural features that affect the weakness of the conjugate base A^-. We will refer to these effects as they become pertinent in subsequent sections.

1a. The increasing *electronegativity* of A as we proceed from left to right across a row in the periodic table. The more electronegative the atom to which the acidic proton is attached, the more polar the bond, and the more acidic the proton will be. For example, the increasing order of acidity in the series $H_4C < H_3N < H_2O < HF$ parallels the increasing electronegativity of A (Table 1-2).

Increasing electronegativity of A ⟶

$H_4C \quad H_3N \quad H_2O \quad HF$

Increasing acidity ⟶

Acidic and Basic Drugs

Most drugs, such as the analgesic aspirin and the decongestant ephedrine, are weak organic acids or bases. In the body they switch between the ionized and neutral forms, depending on pH. This ability is crucial to their potency: The neutral form diffuses more readily across nonpolar cell membranes to get to the target receptor site, but its ionized counterpart is more soluble in the aqueous blood plasma for distribution throughout the body.

Aspirin
(An acid)

Ephedrin
(A base)

1b. The electronegativity of an atom A depends on its *hybridization*. The greater the *s* character of the orbital on A involved in bonding with the acidic proton, the more electronegative A is, because it attracts the electrons in that orbital more strongly. Thus the acidity of a C–H bond increases as the hybridization of the carbon changes from sp^3 to sp^2 to sp:

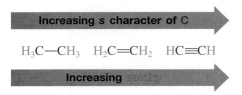

1c. The proximity of an atom A to another electronegative atom (or atoms) also stabilizes A^- and increases the acidity of the proton bonded to A. This property derives from the transmission of the electron-attracting power of the additional electronegative atom(s) through the bonds in the molecule and is termed an *inductive effect*.

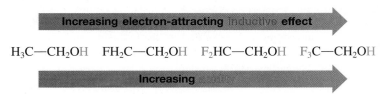

2. The increasing *size* of A as we proceed down a column in the periodic table. The acid strengths of the hydrogen halides increase in the order HF < HCl < HBr < HI. Dissociation to give H^+ and A^- is favored for larger A, because the overlap of its larger outer-shell orbital with the $1s$ hydrogen orbital is poor, weakening the H–A bond. In addition, the larger outer-shell orbitals permit the electrons to occupy a larger volume of space, thus reducing electron–electron repulsion in the resulting anion.*

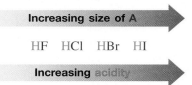

3. The *resonance* in A^- that allows delocalization of charge over several atoms. This effect is frequently enhanced by the presence of additional electronegative atoms in A^-. For example, acetic acid is more acidic than methanol. In both cases, an O–H bond dissociates into ions. Methoxide, the conjugate base of methanol, possesses a localized negative charge on oxygen. In contrast, the acetate ion has two resonance forms and is able to delocalize its charge onto a second oxygen atom. Thus in the acetate ion the negative charge is better accommodated (Section 1-5), stabilizing acetate and making it the weaker base.

Acetic Acid Is Stronger than Methanol Because Acetate Is Stabilized by Resonance

$$CH_3\ddot{O}{-}\overset{\frown}{H} + H_2\ddot{O} \rightleftharpoons CH_3{-}\ddot{O}{:}^- + H_3\ddot{O}^+$$

Weaker acid   **Stronger base**

$$\underset{\textbf{Stronger acid}}{CH_3\overset{\overset{\displaystyle :\ddot{O}:}{\|}}{C}{-}\ddot{O}{-}\overset{\frown}{H} + H_2\ddot{O}} \rightleftharpoons \left[\underset{\textbf{Weaker base}}{CH_3\overset{\overset{\displaystyle :O:}{\|}}{C}{\overset{\frown}{\ddot{O}}}{:}^- \longleftrightarrow CH_3\overset{\overset{\displaystyle :\ddot{O}:^-}{|}}{C}{=}\ddot{O}} \right] + H_3\ddot{O}^+$$

_____

*The bond-strength argument is often the only reason given for the acidity order in the hydrogen halides: HF possesses the strongest bond, HI the weakest. However, this correlation fails for the series H_4C, H_3N, H_2O, HF, in which the weakest acid, CH_4, also has the *weakest* H–A bond. As we will see in Chapter 3, bond strengths are only indirectly applicable to the process of dissociation of a bond into *ions*.

The effect of resonance is even more pronounced in sulfuric acid. The availability of *d* orbitals on sulfur enables us to write valence-shell-expanded Lewis structures containing as many as 12 electrons (Sections 1-4 and 1-5). Alternatively, charge-separated structures with one or two positive charges on sulfur can be used. Both representations indicate that the pK_a of H_2SO_4 should be low.

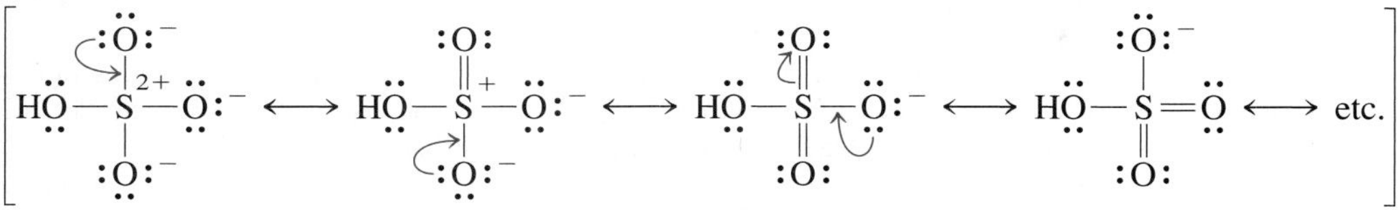

Hydrogen sulfate ion

As a rule, the acidity of HA increases to the right and down in the periodic table. Therefore, the basicity of A⁻ *decreases* in the same fashion.

The same molecule may act as an acid under one set of conditions and as a base under another. Water is the most familiar example of this behavior, but many other substances possess this capability as well. For instance, nitric acid acts as an acid in the presence of water but behaves as a base toward the more powerfully acidic H_2SO_4:

Nitric Acid Acting as an Acid

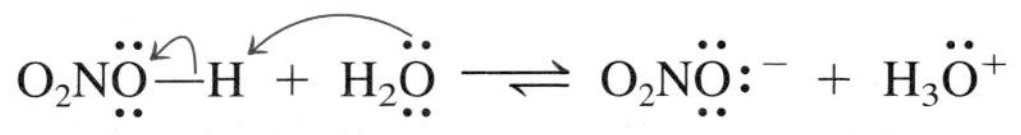

Nitric Acid Acting as a Base

Similarly, acetic acid protonates water, as shown earlier in this section, but is protonated by stronger acids such as HBr:

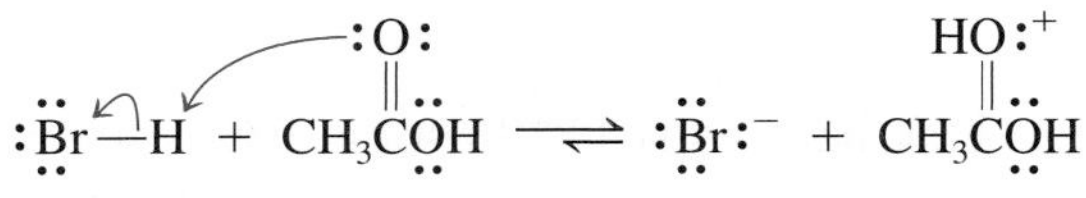

Exercise 2-11

Explain the site of protonation of acetic acid in the preceding equation. (**Hint:** Try placing the proton first on one, and then on the other of the two oxygen atoms in the molecule, and consider which of the two resulting structures is better stabilized by resonance.)

Solved Exercise 2-12 | Working with the Concepts: Determining the Stronger Acid

Which is the stronger acid, $CH_3\ddot{N}H_2$ or $CH_3\ddot{O}H$?

Let us employ the WHIP strategy to analyze and solve this problem.

*W*hat the question asks may seem straightforward: Determine which acid is stronger. However, it is not quite as simple. Each molecule contains two types of potentially acidic hydrogens, those attached to N and O, respectively, and those of the differing methyl groups. Therefore, we need to rephrase the question to be more specific: Which of the four types of hydrogens is the most acidic?

*H*ow to start? When asked to assess the strength of an acid, look at its conjugate base. *The most stable conjugate base (on the basis of atom electronegativity and size as well as stabilizing* inductive and resonance effects) *corresponds to the strongest acid.* So, draw all the possible conjugate bases by removing a proton (H$^+$) from the two respective locations in each molecule:

$$^-\!:\!CH_2\ddot{N}H_2 \qquad CH_3\ddot{N}H^- \qquad ^-\!:\!CH_2\ddot{O}H \qquad CH_3\ddot{O}\!:^-$$

Which is the most stable conjugate base? These four species have negative charges on three different atoms: C, N, and O. How do these atoms differ?

*I*nformation needed! They differ in their electronegativity: O > N > C (Table 1-2). The more electronegative an atom is, the more it attracts electrons and thus stabilizes the extra electron pair and negative charge in the corresponding conjugate base.

Finally, *P*roceed logically: (1) Oxygen is the most electronegative. So, with the extra electron pair and negative charge on O, $CH_3\ddot{O}\!:^-$ must be the most stable of the four conjugate bases above. (2) Given that conclusion, $CH_3\ddot{O}H$ has to be the stronger acid, and the hydrogen on the oxygen has to be the most acidic of its four hydrogen atoms.

Exercise 2-13 | **Try It Yourself**

a. On the basis of their respective pK_a values, determine which is the stronger acid, acetic acid (pK_a = 4.7) or benzoic acid (margin; pK_a = 4.2). By what factor do their acidities differ? (**Hint:** Convert the pK_a values into K_a values to make the comparison.)

b. On the basis of their respective structures, determine which is the stronger acid, acetic acid (CH_3COOH) or trichloroacetic acid (CCl_3COOH).

Benzoic acid

Lewis acids and bases interact by sharing an electron pair

A more generalized description of acid-base interaction in terms of electron sharing was introduced by Lewis. A **Lewis acid** is a species that contains an atom that is at least two electrons short of a closed outer shell. A **Lewis base** contains at least one lone pair of electrons. The symbol X denotes any halogen, while R represents an organic group (Section 2-4).

Lewis acids have unfilled valence shells

$$H^+$$

MgX$_2$, AlX$_3$, many transition metal halide salts

Lewis bases have available electron pairs

A Lewis base shares its lone pair with a Lewis acid to form a new covalent bond. A Lewis base–Lewis acid interaction may therefore be pictured by means of an arrow pointing in the direction that the electron pair moves—from the base to the acid. The Brønsted acid-base reaction between hydroxide ion and a proton is an example of a Lewis acid-base process as well.

Lewis Acid-Base Reactions

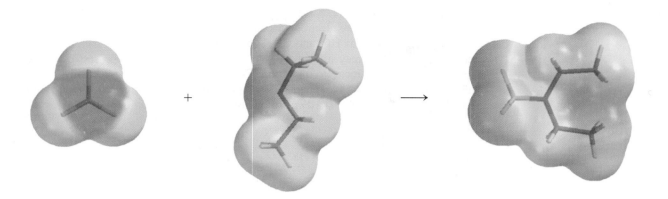

The reaction between boron trifluoride and ethoxyethane (diethyl ether) to give the Lewis acid-base reaction product is shown on the previous page, and also in the form of electrostatic potential maps (above). As electron density is transferred, the oxygen becomes more positive (blue) and the boron more negative (red).

As we have seen in Section 2-2, the dissociation of a Brønsted acid HA is just the reverse of the combination of the Lewis acid H^+ and the Lewis base A^-:

Dissociation of a Brønsted Acid

$$H \overset{\frown}{-} A \longrightarrow H^+ + :A^-$$

As a reminder (see Section 2-3), the use of the symbol for the free proton, H^+, in the above equation is for convenience, and we will encounter it in future reaction schemes and mechanisms. Nonetheless, bear in mind that *H^+ in solution is always associated with a Lewis basic species* such as a molecule of solvent.

Electrophiles and nucleophiles are similar to acids and bases

Many processes in organic chemistry exhibit characteristics of acid-base reactions. For example, heating an aqueous mixture of sodium hydroxide and chloromethane, CH_3Cl, produces methanol and sodium chloride. As noted in Section 2-2, this process involves the same kind of movement of two pairs of electrons as does the acid-base reaction between sodium hydroxide and HCl:

Reaction of Sodium Hydroxide and Chloromethane

$$Na^+ + H\overset{..}{\underset{..}{O}}:^- + CH_3-\overset{..}{\underset{..}{Cl}}: \xrightarrow{H_2O, \Delta} H\overset{..}{\underset{..}{O}}-CH_3 + :\overset{..}{\underset{..}{Cl}}:^- + Na^+$$

Methanol

> Note: The Na^+ ion is a nonparticipating "bystander"

Flow of Electrons Using Curved-Arrow Representation (Na^+ Omitted)

$$H\overset{..}{\underset{..}{O}}:^- + CH_3 \overset{\curvearrowright}{\underset{\curvearrowleft}{-}}\overset{..}{\underset{..}{Cl}}: \longrightarrow H\overset{..}{\underset{..}{O}}-CH_3 + :\overset{..}{\underset{..}{Cl}}:^-$$

Compare to a Brønsted acid-base reaction:

$$H\overset{..}{\underset{..}{O}}:^- + H \overset{\curvearrowright}{\underset{\curvearrowleft}{-}}\overset{..}{\underset{..}{Cl}}: \longrightarrow H\overset{..}{\underset{..}{O}}-H + :\overset{..}{\underset{..}{Cl}}:^-$$

Because the reaction between NaOH and CH_3Cl results in substitution of a nucleophile (hydroxide) for another atom or group in the starting organic molecule, it is called a **nucleophilic substitution.**

The terms *nucleophile* and *Lewis base* are synonymous. *All nucleophiles are Lewis bases.* Nucleophiles, often denoted by the abbreviation Nu, may be negatively charged, such as hydroxide, or neutral, such as water, but every nucleophile contains at least one unshared pair of electrons. All Lewis acids are electrophiles, as shown earlier in the examples of Lewis acid-base reactions. Species such as HCl and CH_3Cl have closed outer shells and therefore are not Lewis acids. However, they may still behave as electrophiles, because they possess polar bonds that impart electrophilic character to the H in HCl and the C in CH_3Cl, respectively.

Nucleophilic substitution is a general reaction of **haloalkanes,** organic compounds possessing carbon–halogen bonds. The two following equations are additional examples:

$$
\begin{array}{c}
\text{H} \\
| \\
CH_3\overset{|}{\underset{|}{C}}CH_2CH_3 \\
\overset{|}{\underset{\cdot\cdot}{:}Br:}
\end{array}
\; + \; :\ddot{I}:^- \longrightarrow
\begin{array}{c}
\text{H} \\
| \\
CH_3\overset{|}{\underset{|}{C}}CH_2CH_3 \\
\overset{|}{:I:}
\end{array}
\; + \; :\ddot{Br}:^-
$$

$$
CH_3CH_2\ddot{\underset{\cdot\cdot}{I}}: \; + \; :NH_3 \longrightarrow
\begin{array}{c}
\text{H} \\
|+ \\
CH_3CH_2\overset{|}{\underset{|}{N}}H \\
\text{H}
\end{array}
\; + \; :\ddot{I}:^-
$$

Solved Exercise 2-14 | Working with the Concepts: Using Curved Arrows

Using earlier examples in this section as models, add curved arrows to the first of the two reactions immediately above.

Strategy
In the organic substrate, identify the likely reactive bond and its polarization. Classify the other reacting species, and look for reactions in the text between similar types of substances.

Solution
- The C–Br bond is the site of reactivity in the substrate and is polarized (δ^+)C–Br(δ^-). Iodide is a Lewis base and therefore a potential nucleophile (electron-pair donor). Thus the situation resembles that found in the reaction between hydroxide and CH_3Cl just above and Example 3 at the beginning of the section.
- Follow the earlier patterns in order to add the appropriate arrows.

$$
:\ddot{I}:^- \; + \; H\overset{\delta+}{-}C\overset{\delta-}{-}\ddot{Br}: \longrightarrow I-C-H \; + \; :\ddot{Br}:^-
$$

Exercise 2-15 | Try It Yourself

Add curved arrows to the second of the reactions immediately above Exercise 2-14 in the text.

Although the haloalkanes in these examples contain different halogens and varied numbers and arrangements of carbon and hydrogen atoms, they behave very similarly toward nucleophiles. We conclude that it is *the presence of the carbon–halogen bond* that governs the chemical behavior of haloalkanes: The C–X bond is the structural feature that directs the chemical reactivity—*structure determines function*. The C–X bond constitutes the **functional group,** or the center of chemical reactivity, of the haloalkanes. In the next section, we shall introduce the major classes of organic compounds, identify their functional groups, and briefly preview their reactivity.

In Summary In Brønsted-Lowry terms, acids are proton donors and bases are proton acceptors. Acid-base interactions are governed by equilibria, which are quantitatively described by an acid dissociation constant K_a. Removal of a proton from an acid generates its conjugate base; attachment of a proton to a base forms its conjugate acid. Lewis bases donate an electron pair to form a covalent bond with Lewis acids, a process depicted by a curved arrow pointing from the lone pair of the base toward the acid. Electrophiles and nucleophiles are species in organic chemistry that interact very much like acids and bases. The carbon–halogen bond in the haloalkane is its functional group. It contains an electrophilic carbon atom, which reacts with nucleophiles in a process called nucleophilic substitution.

2-4   FUNCTIONAL GROUPS: CENTERS OF REACTIVITY

Many organic molecules consist predominantly of a backbone of carbons linked by single bonds, with only hydrogen atoms attached. However, they may also contain doubly or triply bonded carbons, as well as other elements. These atoms or groups of atoms tend to be sites of comparatively high chemical reactivity and are referred to as **functional groups** or **functionalities.** Such groups have characteristic properties, and *they control the reactivity of the molecule as a whole.*

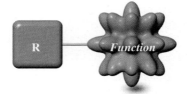

Carbon frame provides structure

Functional group imparts reactivity

Hydrocarbons are molecules that contain only hydrogen and carbon

We begin our study with hydrocarbons, which have the general empirical formula C_xH_y. Those containing only single bonds, such as methane, ethane, and propane, are called **alkanes.** Molecules such as cyclohexane, whose carbons form a ring, are called **cycloalkanes.** *Alkanes lack functional groups;* as a result, they are relatively nonpolar and unreactive. The properties and chemistry of the alkanes are described in this chapter and in Chapters 3 and 4.

Alkanes

CH_4 CH_3—CH_3 CH_3—CH_2—CH_3 CH_3—CH_2—CH_2—CH_3

Methane   Ethane   Propane   Butane

Double and triple bonds are the functional groups of **alkenes** and **alkynes,** respectively. Their properties and chemistry are the topics of Chapters 11 through 13.

Alkenes and Alkynes

CH_2=CH_2 $\overset{H}{\underset{CH_3}{}}$C=$CH_2$ $HC\equiv CH$ CH_3—$C\equiv CH$

Ethene (Ethylene)   Propene   Ethyne (Acetylene)   Propyne

A special hydrocarbon is **benzene,** C_6H_6, in which three double bonds are incorporated into a six-membered ring. Benzene and its derivatives are traditionally called **aromatic,** because some substituted benzenes do have a strong fragrance. Aromatic compounds, also called **arenes,** are discussed in Chapters 15, 16, 22, and 25.

Aromatic Compounds (Arenes)

Benzene

Methylbenzene (Toluene)

Cycloalkanes

Cyclopentane

Cyclohexane

Table 2-3	Common Functional Groups		
Compound class	**General structure**^{a}	**Functional group**	**Example**
Alkanes (Chapters 3, 4)	R—H	None	$CH_3CH_2CH_2CH_3$ **Butane**
Haloalkanes (Chapters 6, 7)	R—$\overset{..}{\underset{..}{X}}$: (X = F, Cl, Br, I)	—$\overset{..}{\underset{..}{X}}$:	CH_3CH_2—$\overset{..}{\underset{..}{Br}}$: **Bromoethane**
Alcohols (Chapters 8, 9)	R—$\overset{..}{\underset{..}{O}}$H	—$\overset{..}{\underset{..}{O}}$H	$\overset{\displaystyle H}{\overset{\displaystyle |}{(CH_3)_2C}}$—$\overset{..}{\underset{..}{O}}$H **2-Propanol (Isopropyl alcohol)**
Ethers (Chapter 9)	R—$\overset{..}{\underset{..}{O}}$—R′	—$\overset{..}{\underset{..}{O}}$—	CH_3CH_2—$\overset{..}{\underset{..}{O}}$—$CH_3$ **Methoxyethane (Ethyl methyl ether)**
Thiols (Chapter 9)	R—$\overset{..}{\underset{..}{S}}$H	—$\overset{..}{\underset{..}{S}}$H	CH_3CH_2—$\overset{..}{\underset{..}{S}}$H **Ethanethiol**
Alkenes (Chapters 11, 12)	$\overset{(H)R}{\underset{(H)R}{}}$C=C$\overset{R(H)}{\underset{R(H)}{}}$	$\diagdown$C=C$\diagup$	$\overset{CH_3}{\underset{CH_3}{}}$C=$CH_2$ **2-Methylpropene**
Alkynes (Chapter 13)	(H)R—C≡C—R(H)	—C≡C—	$CH_3C≡CCH_3$ **2-Butyne**
Aromatic compounds (Chapters 15, 16, 22)	(aromatic ring structure)	(aromatic ring structure)	**Methylbenzene (Toluene)**
Aldehydes (Chapters 17, 18)	$\overset{:O:}{\overset{\|}{R-C-H}}$	$\overset{:O:}{\overset{\|}{-C-H}}$	$\overset{:O:}{\overset{\|}{CH_3CH_2CH}}$ **Propanal**
Ketones (Chapters 17, 18)	$\overset{:O:}{\overset{\|}{R-C-R'}}$	$\overset{:O:}{\overset{\|}{-C-}}$	$\overset{:O:}{\overset{\|}{CH_3CH_2CCH_2CH_2CH_3}}$ **3-Hexanone**
Carboxylic acids (Chapters 19, 20)	$\overset{:O:}{\overset{\|}{R-C-\overset{..}{\underset{..}{O}}-H}}$	$\overset{:O:}{\overset{\|}{-C-\overset{..}{\underset{..}{O}}H}}$	$\overset{:O:}{\overset{\|}{CH_3CH_2C\overset{..}{\underset{..}{O}}H}}$ **Propanoic acid**
Anhydrides (Chapters 19, 20)	$\overset{:O:\quad:O:}{\overset{\|\quad\|}{R-C-\overset{..}{\underset{..}{O}}-C-R'(H)}}$	$\overset{:O:\quad:O:}{\overset{\|\quad\|}{-C-\overset{..}{\underset{..}{O}}-C-}}$	$\overset{:O::O:}{\overset{\|..\|}{CH_3CH_2COCCH_2CH_3}}$ **Propanoic anhydride**
Esters (Chapters 19, 20, 23)	$\overset{:O:}{\overset{\|}{(H)R-C-\overset{..}{\underset{..}{O}}-R'}}$	$\overset{:O:}{\overset{\|}{-C-\overset{..}{\underset{..}{O}}-}}$	$\overset{:O:}{\overset{\|}{CH_3CH_2C\overset{..}{\underset{..}{O}}CH_3}}$ **Methyl propanoate (Methyl propionate)**

^{a}The letter R denotes an alkyl group (see text). Different alkyl groups can be distinguished by adding primes to the letter R: R′, R″, and so forth.

Table 2-3 (continued)			
Compound class	General structure	Functional group	Example
Amides (Chapters 19, 20, 26)	$\overset{:O:}{\overset{\|}{R-C-\overset{..}{N}-R'(H)}} \atop R''(H)$	$\overset{:O:}{\overset{\|}{-C-\overset{..}{N}\big\langle}}$	$\overset{:O:}{\overset{\|}{CH_3CH_2CH_2\overset{..}{C}NH_2}}$ **Butanamide**
Nitriles (Chapter 20)	$R-C\equiv N:$	$-C\equiv N:$	$CH_3C\equiv N:$ **Ethanenitrile** **(Acetonitrile)**
Amines (Chapter 21)	$R-\overset{..}{N}-R'(H) \atop R''(H)$	$-\overset{..}{N}\big\langle$	$(CH_3)_3N:$ **N,N-Dimethylmethanamine** **(Trimethylamine)**

Many functional groups contain polar bonds

Polar bonds determine the behavior of many classes of molecules. Recall that polarity is due to a difference in the electronegativity of two atoms bound to each other (Section 1-3). We have already introduced the **haloalkanes,** which contain polar carbon–halogen bonds as their functional groups. In Chapters 6 and 7 we shall explore their chemistry in depth. Another functionality is the **hydroxy** group, –O–H, characteristic of **alcohols.** The characteristic functional unit of **ethers** is an oxygen bonded to two carbon atoms

$(-\overset{|}{\underset{|}{C}}-O-\overset{|}{\underset{|}{C}}-)$. The functional group in alcohols and in some ethers can be converted into

a large variety of other functionalities and are therefore important in synthetic transformations. This chemistry is the subject of Chapters 8 and 9.

Haloalkanes

$CH_3\overset{..}{\underset{..}{Cl}}:$ $CH_3CH_2\overset{..}{\underset{..}{Cl}}:$
Chloromethane   Chloroethane
(Methyl chloride)   (Ethyl chloride)
(Topical anesthetics)

Alcohols

$CH_3\overset{..}{\underset{..}{O}}H$ $CH_3CH_2\overset{..}{\underset{..}{O}}H$
Methanol   Ethanol

(Wood alcohol)   (Grain alcohol)

Ethers

$CH_3\overset{..}{\underset{..}{O}}CH_3$ $CH_3CH_2\overset{..}{\underset{..}{O}}CH_2CH_3$
Methoxymethane   Ethoxyethane
(Dimethyl ether)   (Diethyl ether)
(A refrigerant)   (An inhalation anesthetic)

The **carbonyl** functional group, C=O, is found in **aldehydes,** in **ketones,** and, in conjunction with an attached –OH, in the **carboxylic acids.** Aldehydes and ketones are discussed in Chapters 17 and 18, the carboxylic acids and their derivatives in Chapters 19 and 20.

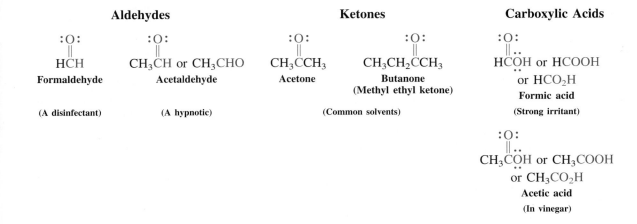

Aldehydes

$\overset{:O:}{\overset{\|}{HCH}}$ $\overset{:O:}{\overset{\|}{CH_3CH}}$ or CH_3CHO
Formaldehyde   **Acetaldehyde**

(A disinfectant)   (A hypnotic)

Ketones

$\overset{:O:}{\overset{\|}{CH_3CCH_3}}$ $\overset{:O:}{\overset{\|}{CH_3CH_2CCH_3}}$
Acetone   **Butanone** **(Methyl ethyl ketone)**

(Common solvents)

Carboxylic Acids

$\overset{:O:}{\overset{\|}{HC\overset{..}{O}H}}$ or $HCOOH$ or HCO_2H
Formic acid

(Strong irritant)

$\overset{:O:}{\overset{\|}{CH_3C\overset{..}{O}H}}$ or CH_3COOH or CH_3CO_2H
Acetic acid

(In vinegar)

Other elements give rise to further characteristic functional groups. For example, alkyl nitrogen compounds are **amines.** The replacement of oxygen in alcohols by sulfur furnishes **thiols.**

	Amines		**A Thiol**

$$CH_3\overset{..}{N}H_2 \qquad\qquad CH_3\overset{\overset{\displaystyle H}{|}}{\underset{..}{N}}CH_3 \text{ or } (CH_3)_2\overset{..}{N}H \qquad CH_3\overset{..}{\underset{..}{S}}H$$

Methanamine   **N-Methylmethanamine**   **Methanethiol**
(Methylamine)   **(Dimethylamine)**

(Used in tanning leather)   (Excreted after
we eat asparagus)

R represents a part of an alkane molecule

Table 2-3 depicts a selection of common functional groups, the class of compounds to which they give rise, a general structure, and an example. In the general structures, we commonly use the symbol **R** (for *radical* or *residue*) to represent an **alkyl group,** a molecular fragment derived by removal of one hydrogen atom from an alkane (Section 2-6). Therefore, a general formula for a haloalkane is R–X, in which R stands for any alkyl group and X for any halogen. Alcohols are similarly represented as R–O–H. In structures that contain multiple alkyl groups, we add a prime (′) or double prime (″) to R to distinguish groups that differ in structure from one another. Thus a general formula for an ether in which both alkyl groups are the same (a **symmetrical ether**) is R–O–R, whereas an ether with two dissimilar groups (an **unsymmetrical ether**) is represented by R–O–R′.

2-5 | STRAIGHT-CHAIN AND BRANCHED ALKANES

The functional groups in organic molecules are typically attached to a hydrocarbon scaffold constructed only with single bonds. Substances consisting entirely of single-bonded carbon and hydrogen atoms and lacking functional groups are called **alkanes.** They are classified into several types according to structure: the linear **straight-chain alkanes;** the **branched alkanes,** in which the carbon chain contains one or several branching points; and the cyclic alkanes, or **cycloalkanes,** which we shall cover in Chapter 4.

Model Building

A Straight-Chain Alkane	**A Branched Alkane**	**A Cycloalkane**

$$CH_3-CH_2-CH_2-CH_3 \qquad\qquad CH_3-\overset{\overset{\displaystyle CH_3}{|}}{\underset{\underset{\displaystyle CH_3}{|}}{C}}-H \qquad\qquad \begin{matrix} CH_2-CH_2 \\ | \qquad | \\ CH_2-CH_2 \end{matrix}$$

Butane, C$_4$H$_{10}$ **2-Methylpropane, C$_4$H$_{10}$**   **Cyclobutane, C$_4$H$_8$**
                                    (Isobutane)

Straight-chain alkanes form a homologous series

In the straight-chain alkanes, each carbon is bound to its two neighbors and to two hydrogen atoms. Exceptions are the two terminal carbon nuclei, which are bound to only one carbon atom and three hydrogen atoms. The straight-chain alkane series may be described by the general formula H–(CH$_2$)$_n$–H. Each member of this series differs from the next lower one by the addition of a methylene group, –CH$_2$–. Molecules that are related in this way are **homologs** of each other (*homos*, Greek, same as), and the series is a **homologous series.** Methane ($n = 1$) is the first member of the homologous series of the alkanes, ethane ($n = 2$) the second, and so forth.

Branched alkanes are constitutional isomers of straight-chain alkanes

Branched alkanes are derived from the straight-chain systems by removal of a hydrogen from a methylene (CH_2) group and replacement with an alkyl group. Both branched and straight-chain alkanes have the same general formula, C_nH_{2n+2}. The smallest branched alkane is 2-methylpropane. It has the same molecular formula as that of butane (C_4H_{10}) but different connectivity; the two compounds therefore form a pair of constitutional isomers (Section 1-9).

For the higher alkane homologs ($n > 4$), more than two isomers are possible. There are three pentanes, C_5H_{12}, as shown below. There are five hexanes, C_6H_{14}; nine heptanes, C_7H_{16}; and eighteen octanes, C_8H_{18}.

Table 2-4	Number of Possible Isomeric Alkanes, C_nH_{2n+2}
n	Isomers
1	1
2	1
3	1
4	2
5	3
6	5
7	9
8	18
9	35
10	75
15	4,347
20	366,319

The Isomeric Pentanes

$$CH_3-CH_2-CH_2-CH_2-CH_3$$

Pentane

$$CH_3-CH_2-\underset{\underset{CH_3}{|}}{\overset{\overset{CH_3}{|}}{CH}}$$

2-Methylbutane
(Isopentane)

$$CH_3-\underset{\underset{CH_3}{|}}{\overset{\overset{CH_3}{|}}{C}}-CH_3$$

2,2-Dimethylpropane
(Neopentane)

The number of possibilities in connecting n carbon atoms to each other and to $2n + 2$ surrounding hydrogen atoms increases dramatically with the size of n (Table 2-4).

Model Building

Exercise 2-16

(a) Draw the structures of the five isomeric hexanes. **(b)** Draw the structures of all the possible next higher and lower homologs of 2-methylbutane.

2-6 | NAMING THE ALKANES

The multiple ways of assembling carbon atoms and attaching various substituents accounts for the existence of the very large number of organic molecules. This diversity poses a problem: How can we systematically differentiate all these compounds by name? Is it possible, for example, to name all the C_6H_{14} isomers so that information on any of them (such as boiling points, melting points, reactions) might easily be found in the index of a handbook or in an online database? And is there a way to name a compound that we have never seen in such a way as to be able to draw its structure?

This problem of naming organic molecules has been with organic chemistry from its very beginning, but the initial method was far from systematic. Compounds have been named after their discoverers ("Nenitzescu's hydrocarbon"), after localities ("sydnones"), after their shapes ("cubane," "basketane"), and after their natural sources ("vanillin"). Many of these **common** or **trivial names** are still widely used. However, there now exists a precise system for naming the alkanes. **Systematic nomenclature,** in which the name of a compound describes its structure, was first introduced by a chemical congress in Geneva, Switzerland, in 1892. It has continually been revised since then, mostly by the International Union of Pure and Applied Chemistry (IUPAC). Table 2-5 gives the systematic names of the first 20 straight-chain alkanes. Their stems, mainly of Greek origin, reveal the number of carbon atoms in the chain. For example, the name heptadecane is composed of the Greek words *hepta*, seven, and *deka*, ten. The first four alkanes have special names that have been accepted as part of the systematic

Propane stored under pressure in liquefied form in canisters such as these is a common fuel for torches, lanterns, and outdoor cooking stoves.
[*Courtesy Bernzomtic, Columbus, OH.*]

Table 2-5	Names and Physical Properties of Straight-Chain Alkanes, C_nH_{2n+2}				
n	Name	Formula	Boiling point (°C)	Melting point (°C)	Density at 20°C (g mL^{-1})
1	Methane	CH_4	−161.7	−182.5	0.466 (at −164°C)
2	Ethane	CH_3CH_3	−88.6	−183.3	0.572 (at −100°C)
3	Propane	$CH_3CH_2CH_3$	−42.1	−187.7	0.5853 (at −45°C)
4	Butane	$CH_3CH_2CH_2CH_3$	−0.5	−138.3	0.5787
5	Pentane	$CH_3(CH_2)_3CH_3$	36.1	−129.8	0.6262
6	Hexane	$CH_3(CH_2)_4CH_3$	68.7	−95.3	0.6603
7	Heptane	$CH_3(CH_2)_5CH_3$	98.4	−90.6	0.6837
8	Octane	$CH_3(CH_2)_6CH_3$	125.7	−56.8	0.7026
9	Nonane	$CH_3(CH_2)_7CH_3$	150.8	−53.5	0.7177
10	Decane	$CH_3(CH_2)_8CH_3$	174.0	−29.7	0.7299
11	Undecane	$CH_3(CH_2)_9CH_3$	195.8	−25.6	0.7402
12	Dodecane	$CH_3(CH_2)_{10}CH_3$	216.3	−9.6	0.7487
13	Tridecane	$CH_3(CH_2)_{11}CH_3$	235.4	−5.5	0.7564
14	Tetradecane	$CH_3(CH_2)_{12}CH_3$	253.7	5.9	0.7628
15	Pentadecane	$CH_3(CH_2)_{13}CH_3$	270.6	10	0.7685
16	Hexadecane	$CH_3(CH_2)_{14}CH_3$	287	18.2	0.7733
17	Heptadecane	$CH_3(CH_2)_{15}CH_3$	301.8	22	0.7780
18	Octadecane	$CH_3(CH_2)_{16}CH_3$	316.1	28.2	0.7768
19	Nonadecane	$CH_3(CH_2)_{17}CH_3$	329.7	32.1	0.7855
20	Icosane	$CH_3(CH_2)_{18}CH_3$	343	36.8	0.7886

$$CH_3-\underset{\underset{H}{|}}{\overset{\overset{CH_3}{|}}{C}}-(CH_2)_n-CH_3$$

An isoalkane
(e.g., $n = 1$, isopentane)

$$CH_3-\underset{\underset{CH_3}{|}}{\overset{\overset{CH_3}{|}}{C}}-(CH_2)_n-H$$

A neoalkane
(e.g., $n = 2$, neohexane)

nomenclature but also all end in **-ane.** It is important to know these names, because they serve as the basis for naming a large fraction of all organic molecules. A few smaller branched alkanes have common names that still have widespread use. They make use of the prefixes **iso-** and **neo-** (margin), as in isobutane, isopentane, and neohexane.

Exercise 2-17

Draw the structures of isohexane and neopentane.

Alkyl groups

CH_3-

Methyl

CH_3-CH_2-

Ethyl

$CH_3-CH_2-CH_2-$

Propyl

As mentioned in Section 2-5, an **alkyl** group is formed by the removal of a hydrogen from an alkane. It is named by replacing the ending **-ane** in the corresponding alkane by **-yl,** as in methyl, ethyl, and propyl. Table 2-6 shows a few branched alkyl groups having common names. Note that some have the prefixes *sec-* (or *s-*), which stands for secondary, and *tert-* (or *t-*), for tertiary. These prefixes are used to classify sp^3-hybridized (tetrahedral) carbon atoms in organic molecules. A **primary** carbon is one attached directly to only one other carbon atom. For example, all carbon atoms at the ends of alkane chains are primary. The hydrogens attached to such carbons are designated primary hydrogens, and an alkyl group created by removing a primary hydrogen also is called primary. A **secondary** carbon is attached directly to two other carbon atoms, and a **tertiary** carbon to three others. Their hydrogens are labeled similarly. As shown in Table 2-6, removal of a secondary hydrogen results in a secondary alkyl group, and removal of a tertiary hydrogen in a tertiary alkyl group. Finally, a carbon bearing four alkyl groups is called **quaternary.**

Table 2-6	Branched Alkyl Groups			
Structure	Common name	Example of common name in use	Systematic name	Type of group

Structure	Common name	Example of common name in use	Systematic name	Type of group
CH_3—$\overset{\overset{\displaystyle CH_3}{|}}{\underset{\underset{\displaystyle H}{|}}{C}}$—	Isopropyl	CH_3—$\overset{\overset{\displaystyle CH_3}{|}}{\underset{\underset{\displaystyle H}{|}}{C}}$—Cl (Isopropyl chloride)	1-Methylethyl	Secondary
CH_3—$\overset{\overset{\displaystyle CH_3}{|}}{\underset{\underset{\displaystyle H}{|}}{C}}$—$CH_2$—	Isobutyl	CH_3—$\overset{\overset{\displaystyle CH_3}{|}}{\underset{\underset{\displaystyle H}{|}}{C}}$—$CH_3$ (Isobutane)	2-Methylpropyl	Primary
CH_3—CH_2—$\overset{\overset{\displaystyle CH_3}{|}}{\underset{\underset{\displaystyle H}{|}}{C}}$—	sec-Butyl	CH_3—CH_2—$\overset{\overset{\displaystyle CH_3}{|}}{\underset{\underset{\displaystyle H}{|}}{C}}$—$NH_2$ (sec-Butyl amine)	1-Methylpropyl	Secondary
CH_3—$\overset{\overset{\displaystyle CH_3}{|}}{\underset{\underset{\displaystyle CH_3}{|}}{C}}$—	tert-Butyl	CH_3—$\overset{\overset{\displaystyle CH_3}{|}}{\underset{\underset{\displaystyle CH_3}{|}}{C}}$—Br (tert-Butyl bromide)	1,1-Dimethylethyl	Tertiary
CH_3—$\overset{\overset{\displaystyle CH_3}{|}}{\underset{\underset{\displaystyle CH_3}{|}}{C}}$—$CH_2$—	Neopentyl	CH_3—$\overset{\overset{\displaystyle CH_3}{|}}{\underset{\underset{\displaystyle CH_3}{|}}{C}}$—$CH_2$—OH (Neopentyl alcohol)	2,2-Dimethylpropyl	Primary

Primary, Secondary, and Tertiary Carbons and Hydrogens

3-Methylpentane

The terms primary, secondary, tertiary, and quaternary are reserved for carbon atoms with exclusively single bonds. They are not applied to carbon atoms with double or triple bonds.

Exercise 2-18

Label the primary, secondary, and tertiary hydrogens in 2-methylpentane (isohexane).

The information in Table 2-5 enables us to name the first 20 straight-chain alkanes. How do we go about naming branched systems? A set of IUPAC rules makes this a relatively simple task, as long as they are followed carefully and in sequence.

IUPAC Rule 1. *Find the longest chain in the molecule and name it.* This task is not as easy as it seems. The problem is that, in the condensed formula, complex alkanes may be written in ways that mask the identity of the longest chain. Do *not* assume that it is always depicted horizontally! In the following examples, the longest chain, or **stem chain,** is clearly marked; the alkane stem gives the molecule its name. Groups other than hydrogen attached to the stem chain are called **substituents.**

The stem chain is shown in black in the examples in this section.

**A methyl-substituted butane
(A methylbutane)**

**An ethyl- and methyl-substituted decane
(An ethylmethyldecane)**

If a molecule has two or more chains of equal length, the chain with the largest number of substituents is the base stem chain.

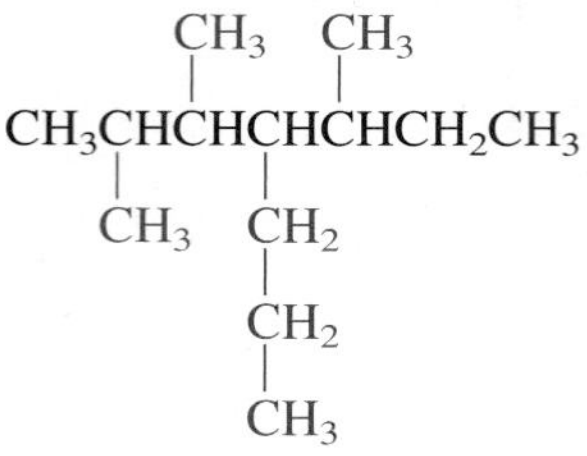

CH₃ CH₃	CH₃ CH₃

$$CH_3CHCHCH_2CHCH_2CH_3 \quad not \quad CH_3CHCHCHCHCH_2CH_3$$

4 substituents
A heptane
Correct stem chain

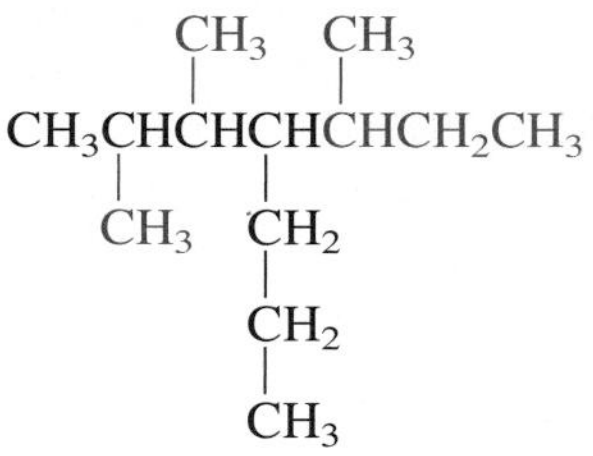

3 substituents
A heptane
Incorrect stem chain

Here are two more examples, drawn with the use of bond-line notation:

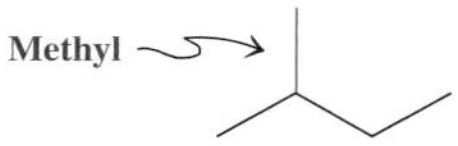

A methylbutane

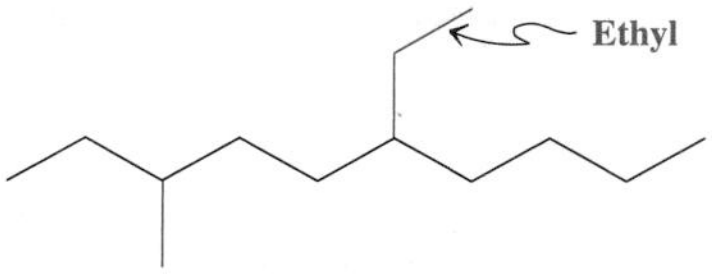

An ethylmethyldecane

IUPAC Rule 2. *Name all groups attached to the longest chain as alkyl substituents.* For straight-chain substituents, Table 2-5 can be used to derive the alkyl name. However, what if the substituent chain is branched? In this case, the same IUPAC rules apply to such complex substituents: First, find the longest chain in the substituent; next, name all *its* substituents.

IUPAC Rule 3. *Number the carbons of the longest chain beginning with the end that is closest to a substituent.*

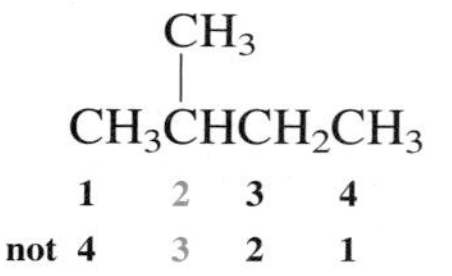

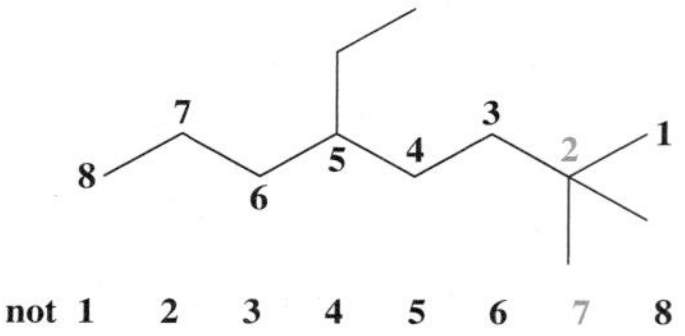

If there are two substituents at *equal* distance from the two ends of the chain, use the alphabet to decide how to number. The substituent to come first in alphabetical order is attached to the carbon with the lower number.

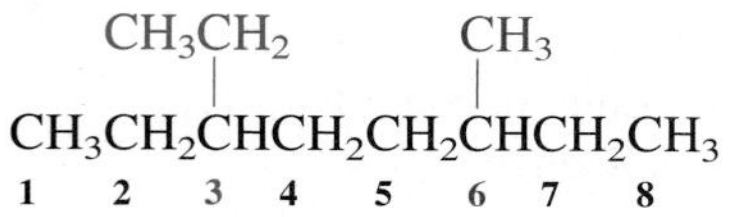

Ethyl before methyl

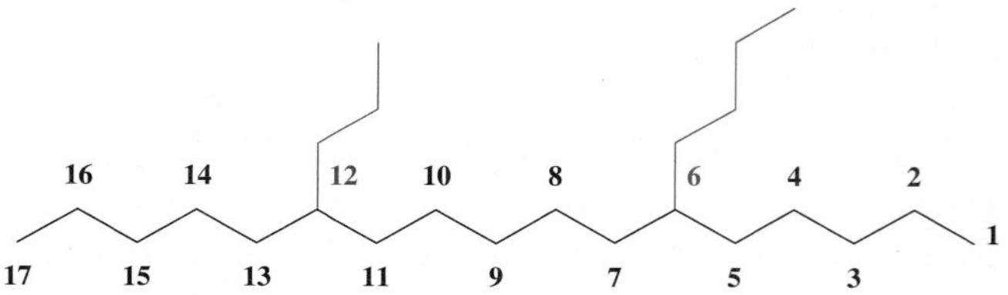

Butyl before propyl

What if there are three or more substituents? Then number the chain in the direction that gives the lower number at the *first difference* between the two possible numbering schemes. This procedure follows the **first point of difference principle.**

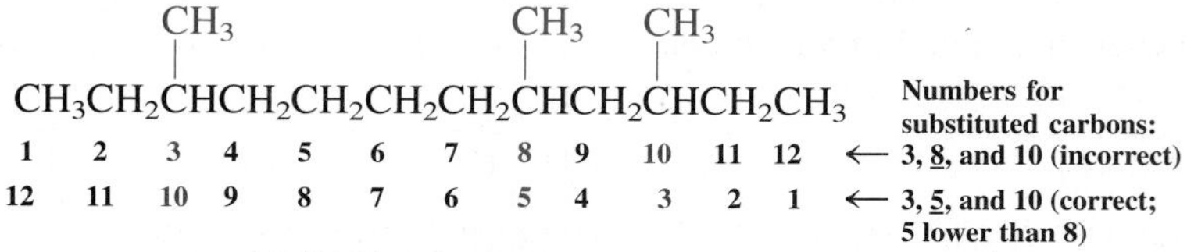

Numbers for substituted carbons:
← 3, **8**, and 10 (incorrect)
← 3, **5**, and 10 (correct; 5 lower than 8)

3,5,10-Trimethyldodecane

Substituent groups are numbered outward from the main chain, with C1 of the group being the carbon attached to the main stem.

IUPAC Rule 4. *Write the name of the alkane by first arranging all the substituents in alphabetical order (each preceded by the carbon number to which it is attached and a hyphen) and then adding the name of the stem.* Should a molecule contain more than one of a particular substituent, its name is preceded by the prefix di, tri, tetra, penta, and so forth. The positions of attachment to the stem are given collectively before the substituent name and are separated by commas. These prefixes, as well as *sec-* and *tert-*, are not considered in the alphabetical ordering, except when they are part of a complex substituent name.

5-Ethyl-2,2-dimethyloctane

("di" not counted in alphabetical ordering)

but

5-(1,1-Dimethylethyl)-3-ethyloctane

("di" counted: part of substituent name)

$$CH_3CHCH_2CH_3$$
with CH_3 substituent

2-Methylbutane

$$CH_3CHCHCH_3$$
with CH_3 and CH_3 substituents

2,3-Dimethylbutane

$$CH_3CHCH_2CH_2CHCH_2CCH_3$$
with CH_3, CH_3, CH_3CH_2, CH_3 substituents

4-Ethyl-2,2,7-trimethyloctane

4,5-Diethyl-3,6-dimethyldecane

$$CH_3CH_2CHCHCH_3$$
with CH_2CH_3 and CH_3 substituents

3-Ethyl-2-methylpentane

The five common group names in Table 2-6 are permitted by IUPAC: isopropyl, isobutyl, *sec*-butyl, *tert*-butyl, and neopentyl. These five are used universally in the course of normal communication between scientists, and it is necessary to know the structures to which they refer. Nonetheless, it is preferable to use systematic names, especially when searching for information about a chemical compound. The online databases containing such information are constructed to recognize the *systematic* names; therefore, use of a common name as input may not result in retrieval of a complete set of the information being sought.

The systematic name of a complex substituent should be enclosed in parentheses to avoid possible ambiguities. If a particular complex substituent is present more than once, a special set of prefixes is placed in front of the parenthesis: bis, tris, tetrakis, pentakis, and so on, for 2, 3, 4, 5, etc. In the chain of a complex substituent, the carbon numbered one (C1) is *always* the carbon atom directly attached to the stem chain.

4-(1-Ethylpropyl)-2,3,5-trimethylnonane

$$CH_3CH_2CH_2CHCH_2CH_2CH_3$$
with CH_3CH and CH_3 substituents

4-(1-Methylethyl)heptane
(4-Isopropylheptane)

5,8-Bis(1-methylethyl)-dodecane

Exercise 2-19

Write down the names of the preceding eight branched alkanes, close the book, and reconstruct their structures from those names.

To name haloalkanes, we treat the halogen as a substituent to the alkane framework. As usual, the longest (stem) chain is numbered so that the first substituent from either end receives the lowest number. Substituents are ordered alphabetically, and complex appendages are named according to the rules used for complex alkyl groups.

CH_3I

Iodomethane

$$CH_3\overset{\overset{\displaystyle CH_3}{|}}{\underset{\underset{\displaystyle CH_3}{|}}{C}}{-}Br$$

2-Bromo-2-methylpropane

$$FCH_2\overset{\overset{\displaystyle CH_3}{|}}{\underset{\underset{\displaystyle H}{|}}{C}}CH_3$$

1-Fluoro-2-methylpropane

6-(2-Chloro-2,3,3-trimethylbutyl)undecane

Common names are based on the older term *alkyl halide*. For example, the first three structures above have the common names methyl iodide, *tert*-butyl bromide, and isobutyl fluoride, respectively. Some chlorinated solvents have common names: for example, carbon tetrachloride, CCl_4; chloroform, $CHCl_3$; and methylene chloride, CH_2Cl_2.

Exercise 2-20

Draw the structure of 5-butyl-3-chloro-2,2,3-trimethyldecane.

Further instructions on nomenclature will be presented when new classes of compounds, such as the cycloalkanes, are introduced.

In Summary Four rules should be applied in sequence when naming a branched alkane: (1) Find the longest chain; (2) find the names of all the alkyl groups attached to the stem; (3) number the chain; (4) name the alkane, with substituent names in alphabetical order and preceded by numbers to indicate their locations. Haloalkanes are named in accord with the rules that apply to naming the alkanes, the halo substituent being treated the same as alkyl groups.

2-7 | STRUCTURAL AND PHYSICAL PROPERTIES OF ALKANES

The common structural feature of all alkanes is the carbon chain. This chain influences the physical properties of not only alkanes but also any organic molecules possessing such a backbone. This section will address the properties and physical appearance of such structures.

Alkanes exhibit regular molecular structures and properties

The structural features of the alkanes are remarkably regular. The carbon atoms are tetrahedral, with bond angles close to 109° and with regular C–H (≈ 1.10 Å) and C–C (≈ 1.54 Å) bond lengths. Alkane chains often adopt the zigzag patterns used in bond-line notation (Figure 2-3). To depict three-dimensional structures, we shall make use of the hashed-wedged line notation (see Figure 1-23). The main chain and a hydrogen at each end are drawn in the plane of the page (Figure 2-4).

Exercise 2-21

Draw zigzag hashed-wedged line structures for 2-methylbutane and 2,3-dimethylbutane.

The longest man-made linear alkane is $C_{390}H_{782}$, synthesized as a molecular model for polyethene (polyethylene). It crystallizes as an extended chain, but starts folding readily (picture) at its melting point of 132°C, in part due to attractive intramolecular London forces.

Figure 2-3 Ball-and-stick (top) and space-filling molecular models of hexane, showing the zigzag pattern of the carbon chain typical of the alkanes. [*Model sets courtesy of Maruzen Co., Ltd., Tokyo.*]

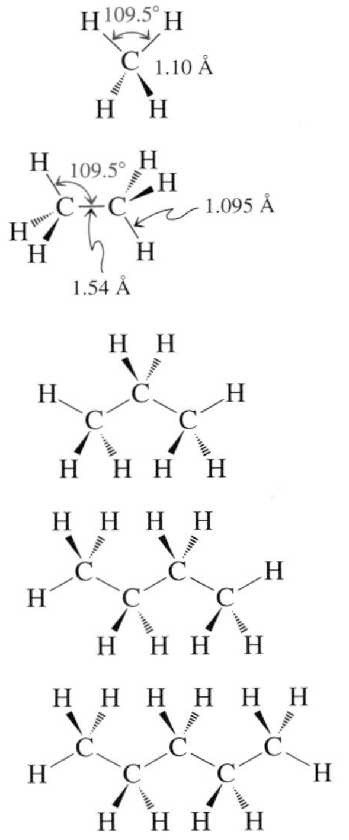

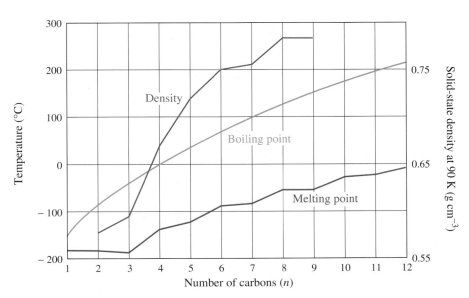

Figure 2-4 Hashed-wedged line structures of methane through pentane. Note the zigzag arrangement of the principal chain and two terminal hydrogens.

Figure 2-5 The physical constants of straight-chain alkanes. Their values increase with increasing size because London forces increase. Note that even-numbered systems have somewhat higher melting points than expected; these systems are more tightly packed in the solid state (notice their higher densities), thus allowing for stronger attractions between molecules.

The regularity in alkane structures suggests that their physical constants would follow predictable trends. Indeed, inspection of the data presented in Table 2-5 reveals regular incremental increases along the homologous series. For example, at room temperature (25°C), the lower homologs of the alkanes are gases or colorless liquids, the higher homologs are waxy solids. From pentane to pentadecane, each additional CH_2 group causes a 20–30°C increase in boiling point (Figure 2-5).

Attractive forces between molecules govern the physical properties of alkanes

Why are the physical properties of alkanes predictable? Such trends exist because of **intermolecular** or **van der Waals* forces.** Molecules exert several types of attractive forces on each other, causing them to aggregate into organized arrangements as solids and liquids. Most solid substances exist as highly ordered crystals. *Ionic* compounds, such as salts, are rigidly held in a crystal lattice, mainly by strong Coulomb forces. Nonionic but *polar* molecules, such as chloromethane (CH_3Cl), are attracted by weaker dipole–dipole interactions, also of coulombic origin (Sections 1-2 and 6-1). Finally, the *nonpolar* alkanes attract each other by **London[†] forces,** which are due to **electron correlation.** When one alkane molecule approaches another, repulsion of the electrons in one molecule by those in the other results in correlation of their movement. Electron motion causes temporary bond polarization in one molecule; correlated electron motion in the bonds of the other induces polarization in the opposite direction, resulting in attraction between the molecules. Figure 2-6 is a simple picture comparing ionic, dipolar, and London attractions.

---

*Professor Johannes D. van der Waals (1837–1923), University of Amsterdam, The Netherlands, Nobel Prize 1910 (physics).

[†]Professor Fritz London (1900–1954), Duke University, North Carolina. *Note:* In older references the term "van der Waals forces" referred exclusively to what we now call *London forces; van der Waals forces* now refers collectively to *all* intermolecular attractions.

Ion–Ion Interactions: Full Charges

Dipole–Dipole Interactions: Partial Charges

London Forces: Unsymmetrical Charge Distribution

δ^- δ^- δ^-
δ^+ δ^+ δ^-

A   B   C

Figure 2-6 (A) Coulombic attraction in an ionic compound: crystalline sodium acetate, the sodium salt of acetic acid. (B) Dipole–dipole interactions in solid chloromethane. The polar molecules arrange to allow for favorable coulombic attraction. (C) London forces in crystalline pentane. In this simplified picture, the electron clouds as a whole mutually interact to produce very small partial charges of opposite sign. The charge distributions in the two molecules change continually as the electrons continue to correlate their movements.

This surfboard is being waxed with paraffin to improve performance.

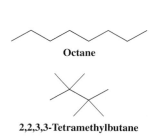

Octane

2,2,3,3-Tetramethylbutane

London forces are very weak. In contrast to Coulomb forces, which change with the square of the distance between charges, London forces fall off as the sixth power of the distance between molecules. There is also a limit to how close these forces can bring molecules together. At small distances, nucleus–nucleus and electron–electron repulsions outweigh these attractions.

How do these forces account for the physical constants of elements and compounds? The answer is that it takes energy, usually in the form of heat, to melt solids and boil liquids. For example, to cause melting, the attractive forces responsible for the crystalline state must be overcome. In an ionic compound, such as sodium acetate (Figure 2-6A), the strong interionic forces require a rather high temperature (324°C) for the compound to melt. In alkanes, melting points rise with increasing molecular size: Molecules with relatively large surface areas are subject to greater London attractions. However, these forces are still relatively weak, and even high-molecular-weight alkanes have rather low melting points. For example, a mixture of straight-chain alkanes from $C_{20}H_{42}$ to $C_{40}H_{82}$ constitutes paraffin wax, which melts below 64°C. Paraffin wax is distinct from normal wax, which is composed of long-chain carboxylic acid esters.

For a molecule to escape these same attractive forces in the liquid state and enter the gas phase, more heat has to be applied. When the vapor pressure of a liquid equals atmospheric pressure, boiling occurs. Boiling points of compounds are also relatively high if the intermolecular forces are relatively large. These effects lead to the smooth increase in boiling points seen in Figure 2-5.

Branched alkanes have smaller surface areas than do their straight-chain isomers. As a result, they are generally subject to smaller London attractions and are unable to pack as well in the crystalline state. The weaker attractions result in lower melting and boiling points. Branched molecules with highly compact shapes are exceptions. For example, 2,2,3,3-tetramethylbutane melts at +101°C because of highly favorable crystal packing (compare octane, m.p. −57°C). On the other hand, the greater surface area of octane compared with that of the more spherical 2,2,3,3-tetramethylbutane is clearly demonstrated in their boiling points (126°C and 106°C, respectively). Crystal packing differences also account for the slightly lower than expected melting points of odd-membered straight-chain alkanes relative to those of even-membered systems (Figure 2-5).

Bees pollinate flowers. We have all watched nature programs and been told by extremely authoritative-sounding narrators that "Instinct tells the bees which flower to pollinate . . . ," etc., etc. Instinct, schminstinct. *Sex* tells the bee which flower to pollinate. Female bees of the species *Andrena nigroaenea* produce a complex mixture of hydrocarbons whose fragrance attracts males of the same species. Such sex attractants, or pheromones (see Section 12-17), are ubiquitous in the animal kingdom and are typically quite species specific. The orchid *Ophrys sphegodes* relies on the male *Andrena* bee for pollination. Remarkably, in the orchid, the leaf wax has a composition almost identical to that of the *Andrena* pheromone mixture: The three major compounds in both the pheromone and in the wax are the straight-chain alkanes tricosane ($C_{23}H_{48}$), pentacosane ($C_{25}H_{52}$), and heptacosane ($C_{27}H_{56}$) in a 3:3:1 ratio. This is an example of what is termed "chemical mimicry," the use by one species of a chemical substance to elicit a desired, but not necessarily normal response from another species. The orchid is even more innovative than most plants, because its flower, whose shape and color already resemble those of the insect, also produces the pheromone-like mixture in high concentration. Thus the male bee is hopelessly attracted to this specific orchid by

what is termed by the discoverers of this phenomenon a case of "sexual swindle."

Over the past 25 years, numerous species-specific examples of such deception among plants have been discovered. The deception is so effective that reproductive success in the insect is compromised, with the potentially disastrous consequence to the plant of losing its population of insect pollinators. Fortunately, after being "swindled" once or twice, individual insects seem to recognize the deception and look for more "suitable" mates.

In Summary   Straight-chain alkanes have regular structures. Their melting points, boiling points, and densities increase with molecular size and surface area because of increasing attraction between molecules.

2-8 | ROTATION ABOUT SINGLE BONDS: CONFORMATIONS

We have considered how intermolecular forces can affect the physical properties of molecules. These forces act *between* molecules. In this section, we shall examine how the forces present *within* molecules (i.e., intramolecular forces) make some geometric arrangements of the atoms energetically more favorable than others. Later chapters will show how molecular geometry affects chemical reactivity.

Rotation interconverts the conformations of ethane

If we build a molecular model of ethane, we can see that the two methyl groups are readily rotated with respect to each other. The energy required to move the hydrogen atoms past each other, the *barrier to rotation*, is only 2.9 kcal mol^{-1} (12.1 kJ mol^{-1}). This value turns out to be so low that chemists speak of "free rotation" of the methyl groups. In general, *there is free rotation about all single bonds* at room temperature.

Figure 2-7 depicts the rotational movement in ethane by the use of hashed-wedged line structures (Section 1-9). There are two extreme ways of drawing ethane: the staggered conformation and the eclipsed one. If the **staggered conformation** is viewed along the C–C axis, each hydrogen atom on the first carbon is seen to be positioned perfectly between two hydrogen atoms on the second. The second extreme is derived from the first by a 60° turn of one of the methyl groups about the C–C bond. Now, if this **eclipsed conformation** is viewed along the C–C axis, all hydrogen atoms on the first carbon are directly opposite those on the second—that is, those on the first eclipse those on the second. A further 60° turn converts the eclipsed form into a new but equivalent staggered arrangement. Between

Model Building

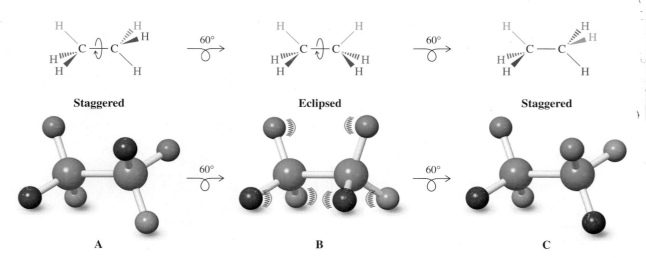

Staggered   **Eclipsed**   **Staggered**

A   B   C

Figure 2-7 Rotation in ethane: (A and C) staggered conformations; (B) eclipsed. There is virtually "free rotation" between conformers.

these two extremes, rotation of the methyl group results in numerous additional positions, referred to collectively as **skew conformations.**

The many forms of ethane (and, as we shall see, substituted analogs) created by such rotations are called **conformations** (also called **conformers**). All of them rapidly interconvert at room temperature. The study of their thermodynamic and kinetic behavior is **conformational analysis.**

Newman projections depict the conformations of ethane

A simple alternative to the hashed-wedged line structures for illustrating the conformers of ethane is the **Newman* projection.** We can arrive at a Newman projection from the hashed-wedged line picture by turning the molecule out of the plane of the page toward us and viewing it along the C–C axis (Figure 2-8A and B). In this notation, the front carbon

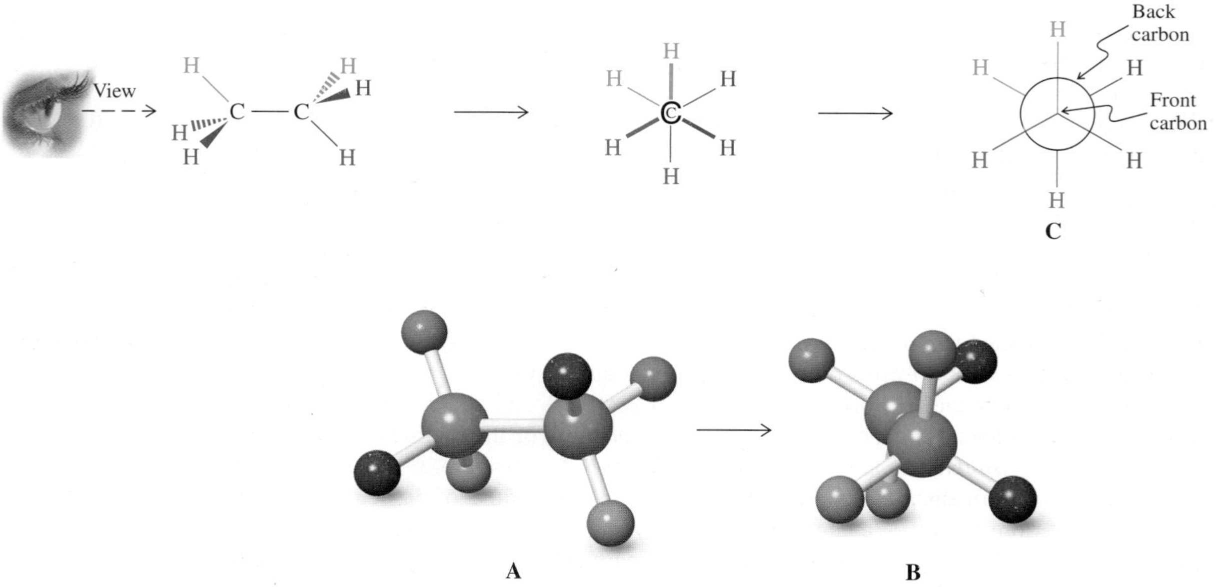

A   B

Figure 2-8 Representations of ethane. (A) Side-on views of the molecule. (B) End-on views of ethane, showing the carbon atoms directly in front of each other and the staggered positions of the hydrogens. (C) Newman projection of ethane derived from the view shown in (B). The "front" carbon is represented by the intersection of the bonds to its three attached hydrogens. The bonds from the remaining three hydrogens connect to the large circle, which represents the "back" carbon.

*Professor Melvin S. Newman (1908–1993), Ohio State University.

Figure 2-9 Newman projections and ball-and-stick models of staggered and eclipsed conformers of ethane. In these representations, the back carbon is rotated clockwise in increments of 60°.

Staggered   **Eclipsed**   **Staggered**

obscures the back carbon, but the bonds emerging from both are clearly seen. The front carbon is depicted as the point of juncture of the three bonds attached to it, one of them usually drawn vertically and pointing up. The back carbon is represented by a circle (Figure 2-8C). The bonds to this carbon project from the outer edge of the circle. The extreme conformational shapes of ethane are readily drawn in this way (Figure 2-9). To make the three rear hydrogen atoms more visible in eclipsed conformations, they are drawn somewhat rotated out of the perfectly eclipsing position.

The conformers of ethane have different potential energies

The conformers of ethane do not all have the same energy content. The staggered conformer is the most stable and lowest energy state of the molecule. As rotation about the C–C bond axis occurs, the potential energy rises as the structure moves away from the staggered geometry, through skewed shapes, finally reaching the eclipsed conformation. At the point of eclipsing, the molecule has its highest energy content, about 2.9 kcal mol^{-1} above the staggered conformation. The change in energy resulting from bond rotation from the staggered to the eclipsed conformation is called **rotational** or **torsional energy,** or **torsional strain.**

The origin of the torsional strain in ethane remains controversial. As rotation to the eclipsed geometry brings pairs of C–H bonds on the two carbons closer to each other, repulsion between the electrons in these bonds increases. Rotation also causes subtle changes in molecular orbital interactions, weakening the C–C bond in the eclipsed conformation. The relative importance of these effects has been a matter of debate for decades, with the most recent published theoretical research (2007) favoring electron repulsion as the major contributor to the rotational energy.

A potential-energy diagram (Section 2-1) can be used to picture the energy changes associated with bond rotation. In the diagram for rotation of ethane (Figure 2-10), the x axis denotes degrees of rotation, usually called **torsional angle.** Figure 2-10 sets 0° at the energy minimum of a staggered conformation, the most stable geometry of the ethane molecule.[*] Notice that the eclipsed conformer occurs at an energy maximum: Its lifetime is extremely short (less than 10^{-12} s), and it is in fact only a transition state between rapidly equilibrating staggered arrangements. The 2.9 kcal mol^{-1} energy difference between the staggered and eclipsed conformations therefore corresponds to the activation energy for the rotational process.

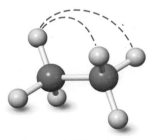

Staggered: More stable, because facing C–H bonds are at maximum distance (dashed lines)

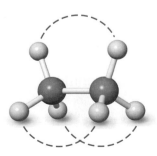

Eclipsed: Less stable, because facing C–H bonds are at minimum distance (dashed lines)

[*]Strictly speaking, the torsional (also called dihedral) angle in a chain of atoms A–B–C–D is defined as the angle between the planes containing A,B,C and B,C,D, respectively. Thus, in Figure 2-10 and subsequent figures in the next section, a torsional angle of 0° would correspond to one of the eclipsed conformations.

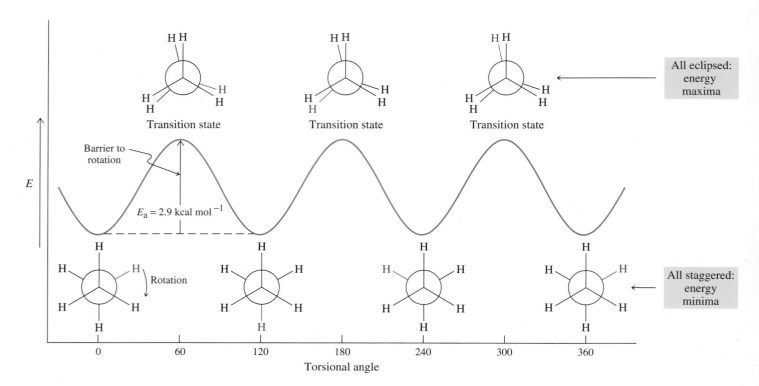

Figure 2-10 Potential-energy diagram of the rotational isomerism in ethane. Because the eclipsed conformations have the highest energy, they correspond to peaks in the diagram. These maxima may be viewed as transition states between the more stable staggered conformers. The activation energy (E_a) corresponds to the barrier to rotation.

All organic molecules with alkane-like backbones exhibit such rotational behavior. The sections that follow will illustrate these principles in more complex alkanes. Later chapters will show how the chemical reactivity of functionalized molecules can depend on their conformational characteristics.

In Summary Intramolecular forces control the arrangement of substituents on neighboring and bonded carbon atoms. In ethane, the relatively stable staggered conformations are inter-converted by rotation through higher-energy transition states in which the hydrogen atoms are eclipsed. Because the energy barrier to this motion is so small, rotation is extremely rapid at ordinary temperatures. A potential-energy diagram conveniently depicts the energetics of rotation about the C–C bond.

2-9 | ROTATION IN SUBSTITUTED ETHANES

How does the potential-energy diagram change when a substituent is added to ethane? Consider, for example, propane, whose structure is similar to that of ethane, except that a methyl group replaces one of ethane's hydrogen atoms.

Steric hindrance raises the energy barrier to rotation

Model Building

A potential-energy diagram for the rotation about a C–C bond in propane is shown in Figure 2-11. The Newman projections of propane differ from those of ethane only by the substituted methyl group. Again, the extreme conformations are staggered and eclipsed. However, the activation barrier separating the two is 3.2 kcal mol^{-1} (13.4 kJ mol^{-1}), slightly higher than that for ethane. This energy difference is due to unfavorable interference between the methyl substituent and the eclipsing hydrogen in the transition state, a phenomenon called **steric hindrance.** This effect arises from the fact that two atoms or groups of atoms cannot occupy the same region in space.

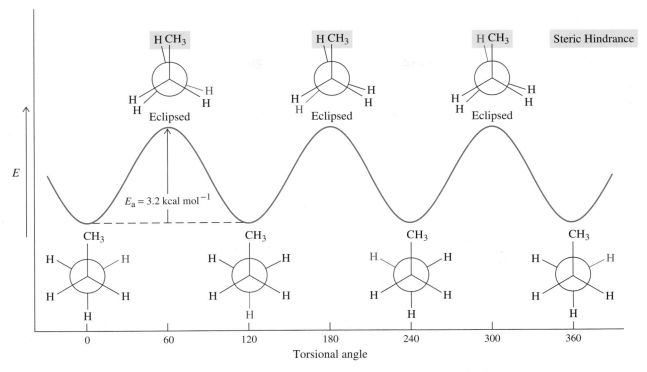

Figure 2-11 Potential-energy diagram of rotation about either C–C bond in propane. Steric hindrance increases the relative energy of the eclipsed form.

Steric hindrance in propane is actually worse than the E_a value for rotation indicates. Methyl substitution raises the energy not only of the eclipsed conformation, but also of the staggered (lowest-energy, or *ground* state) conformation, the staggered to a lesser extent because of less hindrance. However, because the activation energy is equal to the *difference* in energy between ground and transition states, the net result is only a small increase in E_a.

There can be more than one staggered and one eclipsed conformation: conformational analysis of butane

If we build a model and look at the rotation about the central C–C bond of butane, we find that there are more conformations than one staggered and one eclipsed (Figure 2-12). Consider the staggered conformer in which the two methyl groups are as far away from each other as possible. This arrangement, called *anti* (i.e., opposed), is the most stable because steric hindrance is minimized. Rotation of the rear carbon in the Newman projections in either direction (in Figure 2-12, the direction is clockwise) produces an eclipsed conformation with two CH_3–H interactions. This conformer is 3.6 kcal mol^{-1} (15.1 kJ mol^{-1}) higher in energy than the *anti* precursor. Further rotation furnishes a *new* staggered structure in which the two methyl groups are closer than they are in the *anti* conformation. To distinguish this conformer from the others, it is named *gauche* (*gauche*, French, in the sense of awkward, clumsy). As a consequence of steric hindrance, the *gauche* conformer is higher in energy than the *anti* conformer by about 0.9 kcal mol^{-1} (3.8 kJ mol^{-1}).

Further rotation (Figure 2-12) results in a *new* eclipsed arrangement in which the two methyl groups are superposed. Because the two bulkiest substituents eclipse in this conformer, it is energetically highest, 4.9 kcal mol^{-1} (20.5 kJ mol^{-1}) higher than the most stable *anti* structure. Further rotation produces another *gauche-*conformer. The activation energy for *gauche* ⇌ *gauche* interconversion is 4.0 kcal mol^{-1} (16.7 kJ mol^{-1}). A potential-energy diagram summarizes the energetics of the rotation (Figure 2-13). The most stable *anti* conformer is the most abundant in solution (about 72% at 25°C). Its less stable *gauche* counterpart is present in lower concentration (28%).

Model Building

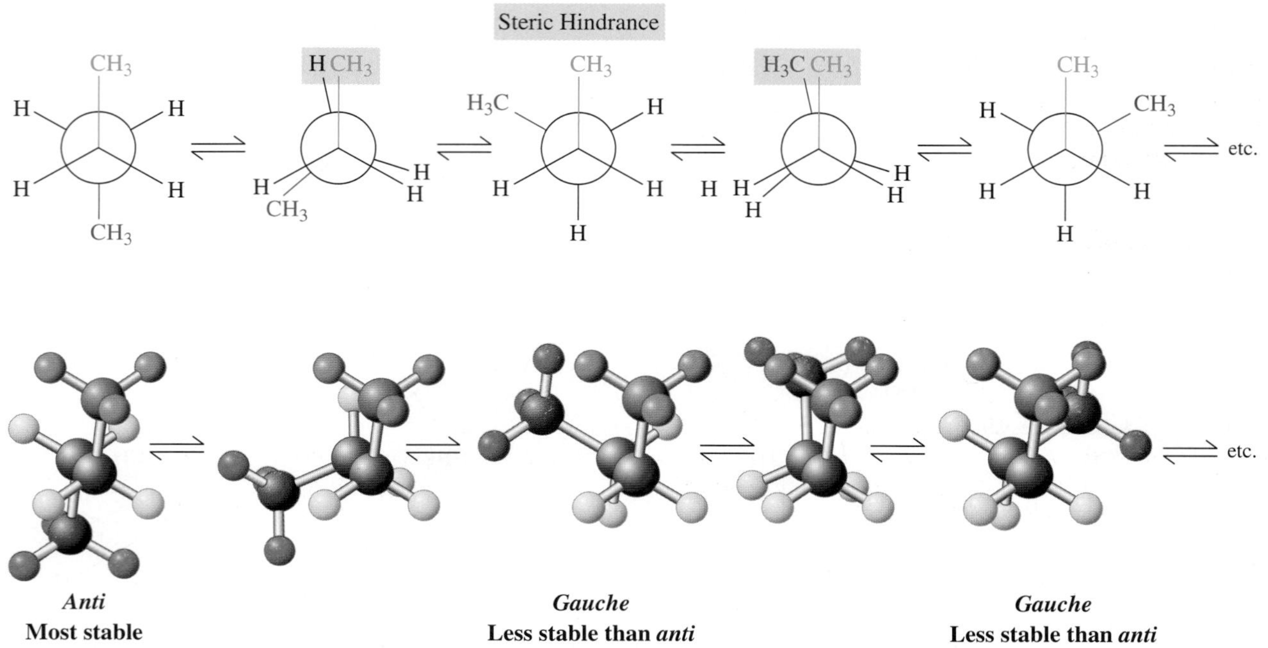

Steric Hindrance

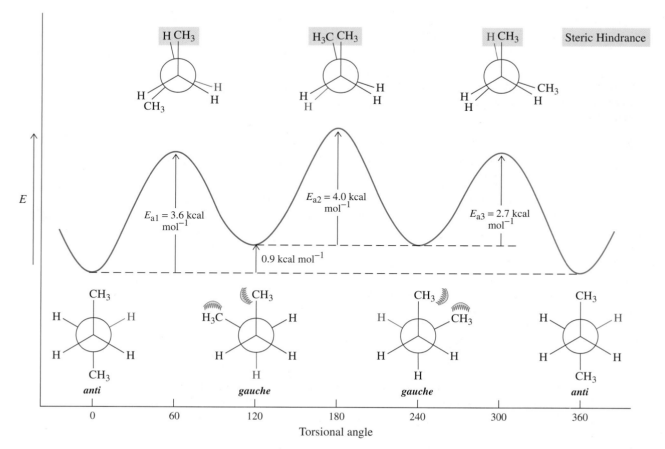

Anti
Most stable

Gauche
Less stable than *anti*

Gauche
Less stable than *anti*

Figure 2-12 Clockwise rotation of the rear carbon along the C2–C3 bond in a Newman projection (top) and a ball-and-stick model (bottom) of butane.

Steric Hindrance

$E_{a1} = 3.6$ kcal mol^{-1}

$E_{a2} = 4.0$ kcal mol^{-1}

$E_{a3} = 2.7$ kcal mol^{-1}

0.9 kcal mol^{-1}

anti   *gauche*   *gauche*   *anti*

0   60   120   180   240   300   360

Torsional angle

Figure 2-13 Potential-energy diagram of the rotation about the C2–C3 bond in butane. There are three processes: *anti → gauche* conversion with $E_{a1} = 3.6$ kcal mol^{-1}; *gauche → gauche* rotation with $E_{a2} = 4.0$ kcal mol^{-1}; and *gauche → anti* transformation with $E_{a3} = 2.7$ kcal mol^{-1}.

We can see from Figure 2-13 that knowing the difference in thermodynamic stability of two conformers (e.g., 0.9 kcal mol^{-1} between the *anti* and *gauche* isomers) and the activation energy for proceeding from the first to the second (e.g., 3.6 kcal mol^{-1}, 15.1 kJ mol^{-1}) allows us to estimate the activation barrier of the reverse reaction. In this case, E_a for the *gauche*-to-*anti* conversion is $3.6 - 0.9 = 2.7$ kcal mol^{-1} (11.3 kJ mol^{-1}).

Solved Exercise 2-22 | Working with the Concepts: Conformations

Draw a qualitative potential-energy diagram for rotation about the C3–C4 bond in 2-methylpentane. Show Newman projections for all conformations located at the maximum and minimum points on your graph. Describe similarities and differences with other molecules discussed in this section.

Strategy
We again turn to the WHIP approach to clarify our strategy.
What this question asks is to define the conformations about a *specific bond* of a *specific compound*. It is *not* asking you to draw conformations for a molecule you have already encountered, such as ethane or butane.
How to begin? Write the structure of the indicated molecule with all carbon–carbon bonds drawn out. Then mark the bond specified by the question (the one between carbon 3 and carbon 4) as shown below.

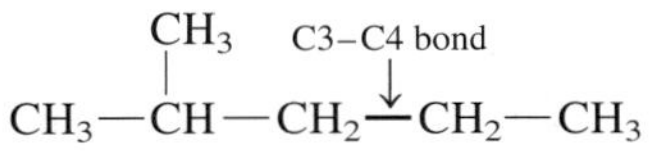

$$\text{CH}_3 \quad \text{C3–C4 bond}$$
$$\text{CH}_3\text{—CH—CH}_2\text{—CH}_2\text{—CH}_3$$

Next, identify the three atoms or groups attached to the carbons at either end of this bond. Carbon 3 contains two hydrogens and a 1-methylethyl (isopropyl) group, while carbon 4 contains two hydrogens and a methyl.
Information needed? Conformations are best represented in a specific way: Use the Newman projection stencil (margin) to construct one (any) conformation of the molecule, attaching the groups you have identified to their respective carbon.
Either one of the carbons 3 or 4 may be in front, but for the purposes of formulating a solution, let us choose carbon 3. Attach the two hydrogen atoms and the isopropyl group to the three bond lines that meet one another at 120° angles. Then add two hydrogens and a methyl group to the three lines emanating from the circle (which represents carbon 4).
With this initial conformation, you can **P**roceed.

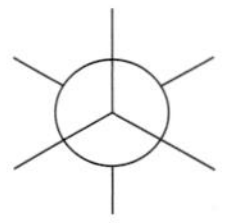

Stencil for a Newman projection

Solution
- This situation resembles rotation about the C2–C3 bond in butane (see Figures 2-12 and 2-13), and the Newman projections and potential-energy diagram are similar.

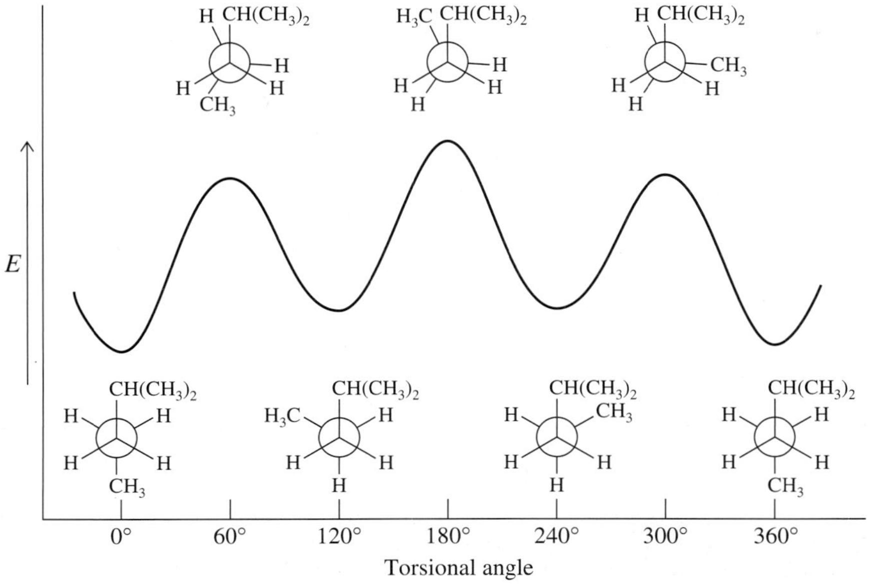

- The single difference is that one of the alkyl groups is an isopropyl substituent instead of CH_3. Because of its greater size, the energies of all its steric interactions will be larger, especially in conformations that bring the two alkyl groups close together. Therefore, the energy differences between the *anti* and all other conformations will increase, with the greatest increase at 180°.

Exercise 2-23 | **Try It Yourself**

Draw the expected potential-energy diagram for the rotation about the C2–C3 bond in 2,3-dimethylbutane. Include the Newman projections of each staggered and eclipsed conformation.

THE BIG PICTURE

The familiar chemistry of acids and bases provides a framework for understanding many of the most important reactions between organic molecules. Much of the chemistry we explore in the upcoming chapters expands on the concept that electrophiles and nucleophiles are mutually attractive species, analogous to acids and bases. By identifying polar sites in molecules, we can develop the ability to understand, and even to predict, what kinds of reactions these molecules will undergo.

The sites of reactivity in organic molecules are called functional groups. Functional groups serve to characterize the main classes of organic compounds. Beginning in Chapter 6 and continuing through to the end of the book, we will discuss these compound classes one by one, examining how their properties and reactivity arise from their structural characteristics.

Most organic molecules consist of hydrocarbon skeletons with functional groups attached. We began a study of these skeletons with a discussion of the names and structures of alkanes, hydrocarbons that lack functional groups. We presented concepts of molecular motion and their associated energy changes. These ideas will underlie the behavior of molecules of all types, and we will return to them frequently.

WORKED EXAMPLES: INTEGRATING THE CONCEPTS

2-24. Analyzing Structure and Function of Molecules

Consider the alkane shown in the margin.

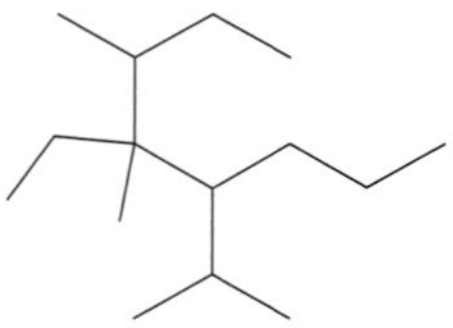

a. Name this molecule according to the IUPAC system.

SOLUTION

Step 1. Locate the main, or stem, chain, the longest one in the molecule (shown in black below). Do not be misled: The drawing of the stem chain can have almost any shape. The stem has eight carbons, so the base name is **octane.**

Step 2. Identify and name all substituents (shown in color): two **methyl** groups, an **ethyl** group, and a fourth, branched substituent. The branched substituent is named by first giving the number 1 (italicized in the illustration below) to the carbon that connects it to the main stem. By numbering away from the stem, we reach the number 2; therefore, the substituent is a derivative of the ethyl group (in green), onto which is attached a methyl group (red) at carbon 1. Thus, this substituent is called a **1-methylethyl** group.

Step 3. Number the stem chain, starting at the end closest to a carbon bearing a substituent. The numbering shown gives a methyl-substituted carbon the number 3. Numbering the opposite way would have C4 as the lowest numbered substituted carbon.

Step 4. Arrange the names of substituents alphabetically in the final name: *ethyl* comes first; then *methyl* comes before *methylethyl* (the "di" in *dimethyl,* denoting two methyl groups, is not considered in alphabetization because it is a multiplier of a substituent name and is therefore not considered part of the name). For more practice naming, see Problem 36.

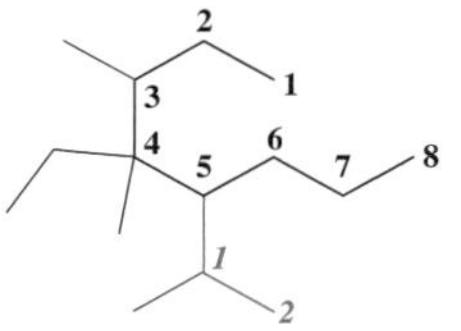

4-Ethyl-3,4-dimethyl-5-(1-methylethyl)octane

b. Draw structures to represent rotation about the C6–C7 bond. Correlate the structures that you draw with a qualitative potential energy diagram.

SOLUTION
Step 1. Identify the bond in question. Notice that much of the molecule can be treated simply as a large, complicated substituent on C6, the specific structure of which is unimportant. For the purpose of this question, this large substituent may be replaced by R. The "action" in this problem takes place between C6 and C7:

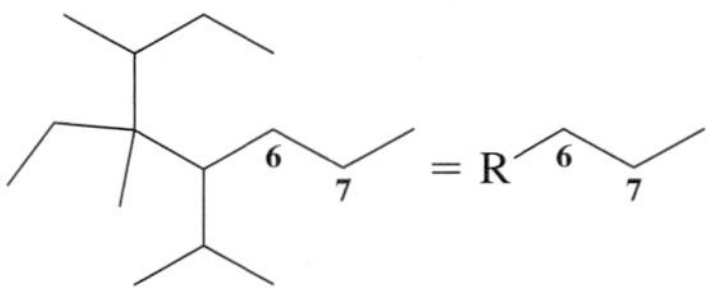

Step 2. Recognize that step 1 simplified the problem: Rotation about the C6–C7 bond will give results very similar to rotation about the C2–C3 bond in butane. The only difference is that a large R group has replaced one of the smaller methyl groups of butane.

Step 3. Draw conformations modeled after those of butane (Section 2-9) and superimpose them on an energy diagram similar to that in Figure 2-13. The only difference between this diagram and that for butane is that we do not know the exact heights of the energy maxima relative to the energy minima. However, we can expect them to be higher, qualitatively, because our R group is larger than a methyl group and thus can be expected to cause greater steric hindrance.

c. Two alcohols derived from this alkane are illustrated in the margin. Alcohols are categorized on the basis of the type of carbon atom that contains the –OH group (primary, secondary, or tertiary). Characterize the alcohols shown in the margin.

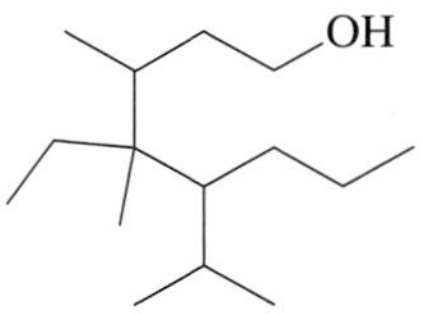

SOLUTION
In alcohol 1, the –OH group is located on a carbon atom that is directly attached to one other carbon, a primary carbon. Therefore, alcohol 1 is a primary alcohol. Similarly, the –OH group in alcohol 2 resides on a tertiary carbon (one attached to three other carbon atoms). It is a tertiary alcohol.

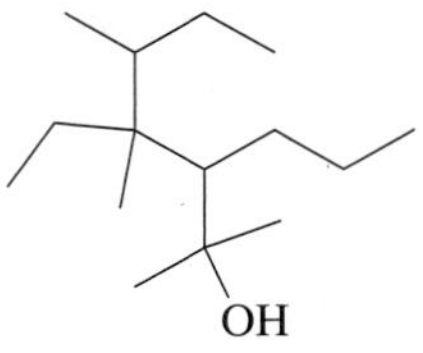

d. The –O–H bond in an alcohol is acidic to a similar degree to that in water. Primary alcohols have $K_a \approx 10^{-16}$; tertiary alcohols $K_a \approx 10^{-18}$. What are the approximate pK_a values for alcohols 1 and 2? Which is the stronger acid?

SOLUTION
The pK_a for alcohol 1 is approximately 16 ($-\log K_a$); that for alcohol 2 is about 18. Alcohol 1, with the lower pK_a value, is the stronger acid.

e. In which direction does the following equilibrium lie? Calculate K, the equilibrium constant, and $\Delta G°$, the free energy change, associated with the reaction as written in the left-to-right direction.

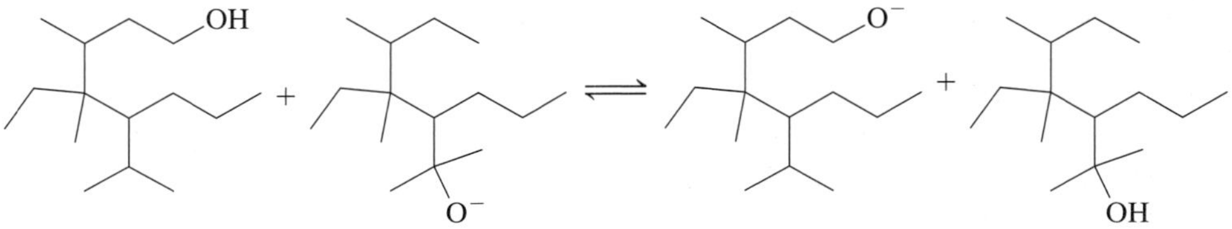

Alcohol 1

Alcohol 2

SOLUTION

The stronger acid (alcohol 1) is on the left; the weaker (alcohol 2) is on the right. Recall the relation between conjugate acids and bases: Stronger acids have weaker conjugate bases, and vice versa. Relatively speaking, therefore, we have

Alcohol 1	+	Conjugate base of alcohol 2	$\rightleftharpoons$	Conjugate base of alcohol 1	+	Alcohol 2
(Stronger acid)		**(Stronger base)**		**(Weaker base)**		**(Weaker acid)**

The equilibrium lies to the right, on the side of the *weaker* acid-base pair. Recall that $K > 1$ and $\Delta G° < 0$ for a reaction that is thermodynamically favorable as written from left to right; *use* this information to be sure to get the magnitude of K and the sign of $\Delta G°$ correct. The equilibrium constant, K, for the process is the ratio of the K_a values, $(10^{-16}/10^{-18}) = 10^2$ (not 10^{-2}). With reference to Table 2-1, a K value of 100 corresponds to a $\Delta G°$ of -2.73 kcal mol^{-1} (not $+2.73$). If the reaction were written in the opposite direction, with the equilibrium lying to the left, the correct values would be those in parentheses. For more practice with acids and bases, see Problem 28.

2-25. Dealing with Equilibria

a. Calculate the equilibrium concentrations of *gauche* and *anti* butane at 25°C using the data from Figure 2-13.

SOLUTION

We are given the relevant equations in Section 2-1. In particular, at 25°C the relationship between the Gibbs free energy and the equilibrium constant simplifies to $\Delta G° = -1.36 \log K$. The energy difference between the conformations is 0.9 kcal mol^{-1}. Substitution of this value into the equation gives $K = 0.219 = [gauche]/[anti]$. Conversion to percentages may be achieved by recognizing that % *gauche* = 100% × $[gauche]/([anti] + [gauche])$. Thus we find % *gauche* = 100% × $(0.219)/(1.0 + 0.219)$ = 18%, and therefore % *anti* = 82%. The problem is, we are given the answer on p. 85: At 25°C, butane consists of 28% *gauche* and 72% *anti*. What's gone wrong?

The error derives from the fact that the energy values given in Sections 2-8 and 2-9 are enthalpies, not free energies, and we have failed to include an *entropy* contribution to the free energy equation $\Delta G° = \Delta H° - T\Delta S°$. How do we correct this problem? We may look for equations to calculate $\Delta S°$, but let us use another, more intuitive approach. Examine Figure 2-13 again. Note that over a 360° rotation the butane molecule passes from its *anti* conformation through *two distinct gauche* conformations before returning to its original *anti* geometry. The entropy term arises from the availability of two *gauche* conformers equilibrating with a single *anti*. Thus we really have *three* species in equilibrium, not two. Is there a way to solve this problem without going through calculations to determine $\Delta S°$ and $\Delta G°$? The answer is yes, and it is not very difficult.

Returning to Figure 2-13, let us label the two gauche forms A and B to distinguish them from each other. When we calculated K in our original solution, what we actually determined was the value associated with equilibration between the *anti* and just *one* of the two *gauche* conformers, say *gauche*$_A$. Of course, the value for *gauche*$_B$ is identical, because the two *gauche* conformations are equal in energy. So, $K = [gauche_A]/[anti] = 0.219$, and $K = [gauche_B]/[anti] = 0.219$.

Finally, recognizing that total $[gauche] = [gauche_A] + [gauche_B]$, we therefore have % *gauche* = 100% × $([gauche_A] + [gauche_B])/([anti] + [gauche_A] + [gauche_B])$, giving us % *gauche* = 100% × $(0.219 + 0.219)/(1.0 + 0.219 + 0.219)$ = 30%, and therefore % *anti* = 70%, in much better agreement with the values given for the equilibrium percentages in the text.

b. Calculate the equilibrium concentrations of *gauche* and *anti* butane at 100°C.

SOLUTION

As above, we determine K from the enthalpy difference in Figure 2-13, using this value as to input for $\Delta G°$ in the more general equation $\Delta G° = -2.303 \, RT \log K$. We follow by correcting for the presence of two *gauche* conformations in the overall equilibrium. Don't forget that we must use degrees Kelvin, 373 K for T. Solving, we obtain $K = 0.297 = [gauche_A]/[anti] = [gauche_B]/[anti]$. Therefore, % *gauche* = 100% × $(0.297 + 0.297)/(1.0 + 0.297 + 0.297)$ = 37%, and therefore % *anti* = 63%.

The less stable conformations are more prevalent at higher temperatures, a straightforward consequence of the Boltzmann distribution, which states that more molecules have higher energies at higher temperatures (Section 2-1). For more practice with conformations, see Problems 44, 46, and 52.

Important Concepts

1. Chemical reactions can be described as equilibria controlled by **thermodynamic** and **kinetic** parameters. The change in the **Gibbs free energy,** $\Delta G°$, is related to the **equilibrium constant** by $\Delta G° = -RT \ln K = -1.36 \log K$ (at 25°C). The free energy has contributions from changes in **enthalpy,** $\Delta H°$, and **entropy,** $\Delta S°$: $\Delta G° = \Delta H° - T\Delta S°$. Changes in enthalpy are due mainly to differences between the strengths of the bonds made and those of the bonds broken. A reaction is **exothermic** when the former is larger than the latter. It is **endothermic** when there is a net loss in combined bond strengths. Changes in entropy are controlled by the relative degree of energy dispersal in starting materials compared with that in products. The greater the increase in energy dispersal, the larger a positive $\Delta S°$.

2. The rate of a chemical reaction depends mainly on the concentrations of starting material(s), the activation energy, and temperature. These correlations are expressed in the **Arrhenius equation:** rate constant $k = Ae^{-E_a/RT}$.

3. If the rate depends on the concentration of only one starting material, the reaction is said to be of **first order.** If the rate depends on the concentrations of two reagents, the reaction is of **second order.**

4. **Brønsted acids** are proton donors; **bases** are proton acceptors. Acid strength is measured by the **acid dissociation constant** K_a; $pK_a = -\log K_a$. Acids and their deprotonated forms have a **conjugate** relation. **Lewis acids** and **bases** are electron pair acceptors and donors, respectively.

5. Electron-deficient atoms attack electron rich atoms and are called **electrophiles.** Conversely, electron-rich atoms attack electron-poor atoms and are called **nucleophiles.** When a nucleophile, which may be either negatively charged or neutral, attacks an electrophile, it donates a lone electron pair to form a new bond with the electrophile.

6. An organic molecule may be viewed as being composed of a carbon skeleton with attached **functional groups.**

7. **Hydrocarbons** are made up of carbon and hydrogen only. Hydrocarbons possessing only single bonds are also called **alkanes.** They do not contain functional groups. An alkane may exist as a single continuous chain or it may be branched or cyclic. The empirical formula for the **straight-chain** and **branched alkanes** is C_nH_{2n+2}.

8. Molecules that differ only in the number of methylene groups, CH_2, in the chain are called **homologs** and are said to belong to a homologous series.

9. An sp^3 carbon attached directly to only one other carbon is labeled **primary.** A **secondary carbon** is attached to two and a **tertiary** to three other carbon atoms. The hydrogen atoms bound to such carbon atoms are likewise designated primary, secondary, or tertiary.

10. The **IUPAC rules** for naming saturated hydrocarbons are (a) find the longest continuous chain in the molecule and name it; (b) name all groups attached to the longest chain as alkyl substituents; (c) number the carbon atoms of the longest chain; (d) write the name of the alkane, citing all substituents as prefixes arranged in alphabetical order and preceded by numbers designating their positions.

11. Alkanes attract each other through weak **London forces,** polar molecules through stronger dipole–dipole interactions, and salts mainly through very strong ionic interactions.

12. Rotation about carbon–carbon single bonds is relatively easy and gives rise to **conformations** (conformers). Substituents on adjacent carbon atoms may be **staggered** or **eclipsed.** The eclipsed conformation is a transition state between staggered conformers. The energy required to reach the eclipsed state is called the activation energy for rotation. When both carbons bear alkyl or other groups, there may be additional conformers: Those in which the groups are in close proximity (60°) are ***gauche;*** those in which the groups are directly opposite (180°) each other are ***anti.*** Molecules tend to adopt conformations in which steric hindrance, as in *gauche* conformations, is minimized.

Problems

26. You have just baked a pizza and turned off your oven. When you open the door to cool the hot oven, what happens to the total enthalpy of the system "oven + room"? The total entropy of the system? How about the free energy? Is the process thermodynamically favorable? What can you say about the temperatures of the oven and of the kitchen after equilibrium has been reached?

27. The hydrocarbon propene (CH_3—CH=CH_2) can react in two different ways with bromine (Chapters 12 and 14).

(i) CH_3—CH=CH_2 + Br_2 $\longrightarrow$ CH_3—$\overset{\overset{\displaystyle Br}{|}}{CH}$—$\overset{\overset{\displaystyle Br}{|}}{CH_2}$

(ii) CH_3—CH=CH_2 + Br_2 $\longrightarrow$ $\overset{}{CH_2}$—CH=CH_2 + HBr
${|}$
$ Br$

(a) Using the bond strengths (kcal mol^{-1}) given in the table, calculate $\Delta H°$ for each of these reactions. **(b)** $\Delta S° \approx 0$ cal K^{-1} mol^{-1} for one of these reactions and -35 cal K^{-1} mol^{-1} for the other. Which reaction has which $\Delta S°$? Briefly explain your answer. **(c)** Calculate $\Delta G°$ for each reaction at 25°C and at 600°C. Are both of these reactions thermodynamically favorable at 25°C? At 600°C?

Bond	Average strength (kcal mol^{-1})
C—C	83
C=C	146
C—H	99
Br—Br	46
H—Br	87
C—Br	68

28. (i) Determine whether each species in the following equations is acting as a Brønsted acid or base, and label it. (ii) Indicate whether the equilibrium lies to the left or to the right. (iii) Estimate K for each equation if possible. (**Hint:** Use the data in Table 2-2.)

(a) H_2O + HCN $\rightleftharpoons$ H_3O^+ + CN^-
(b) CH_3O^- + NH_3 $\rightleftharpoons$ CH_3OH + NH_2^-
(c) HF + CH_3COO^- $\rightleftharpoons$ F^- + CH_3COOH
(d) CH_3^- + NH_3 $\rightleftharpoons$ CH_4 + NH_2^-
(e) H_3O^+ + Cl^- $\rightleftharpoons$ H_2O + HCl
(f) CH_3COOH + CH_3S^- $\rightleftharpoons$ CH_3COO^- + CH_3SH

29. Use curved arrows to show electron movement in each acid-base reaction in Problem 28.

30. Identify each of the following species as either a Lewis acid or a Lewis base, and write an equation illustrating a Lewis acid-base reaction for each one. Use curved arrows to depict electron-pair movement. Be sure that the product of each reaction is depicted by a complete, correct Lewis structure.

(a) CN^- **(b)** CH_3OH **(c)** $(CH_3)_2CH^+$
(d) $MgBr_2$ **(e)** CH_3BH_2 **(f)** CH_3S^-

31. For each example in Table 2-3, identify all polarized covalent bonds and label the appropriate atoms with partial positive or negative charges. (Do not consider carbon–hydrogen bonds.)

32. Characterize each of the following atoms as being either nucleophilic or electrophilic.

(a) Iodide ion, I^-
(b) Hydrogen ion, H^+
(c) Carbon in methyl cation, $^+CH_3$
(d) Sulfur in hydrogen sulfide, H_2S
(e) Aluminum in aluminum trichloride, $AlCl_3$
(f) Magnesium in magnesium oxide, MgO

33. Circle and identify by name each functional group in the compounds pictured.

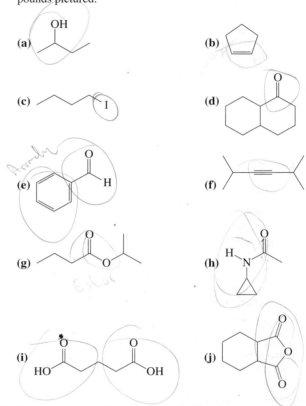

34. On the basis of electrostatics (Coulomb attraction), predict which atom in each of the following organic molecules might react with the indicated reagent. Write "no reaction" if none seems likely. (See Table 2-3 for the structures of the organic molecules.) **(a)** Bromoethane, with the oxygen of HO^-; **(b)** propanal, with the nitrogen of NH_3; **(c)** methoxyethane, with H^+; **(d)** 3-hexanone, with the carbon of CH_3^-; **(e)** ethanenitrile (acetonitrile), with the carbon of CH_3^+; **(f)** butane, with HO^-.

35. Use curved arrows to show the electron movement in each reaction in Problem 34.

36. Name the following molecules according to the IUPAC system of nomenclature.

(a) $CH_3CH_2CHCH_3$
$\overset{\displaystyle |}{CH}$
H_3CCH_3

(b) $CH_3\overset{\overset{\displaystyle CH_3}{|}}{CH}CH_2CH_2\overset{\overset{\displaystyle CH_3CHCH_2CH_3}{|}}{C}CH_2CH_2CH_2CH_3$
$\overset{\displaystyle |}{CH_3CHCH_3}$

(c) 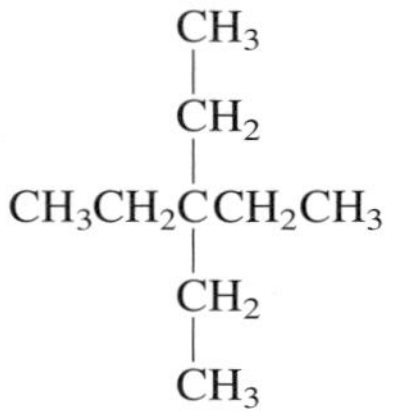
$$CH_3-CH_2-\overset{\overset{\displaystyle CH_3}{\overset{\displaystyle |}{CH_2}}}{\underset{\underset{\displaystyle CH_2}{\underset{\displaystyle |}{CH_3}}}{C}}-CH_2-CH_3$$

(d) 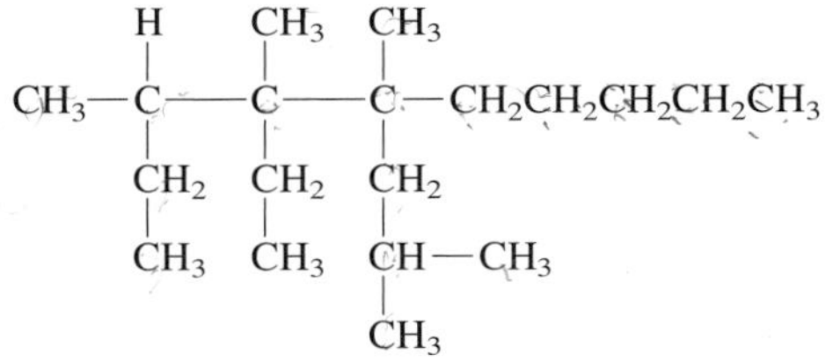
$$CH_3-\overset{\overset{\displaystyle H}{|}}{\underset{\underset{\displaystyle CH_2}{\underset{\displaystyle |}{CH_3}}}{C}}-\overset{\overset{\displaystyle CH_3}{|}}{\underset{\underset{\displaystyle CH_2}{\underset{\displaystyle |}{CH_3}}}{C}}-\overset{\overset{\displaystyle CH_3}{|}}{\underset{\underset{\displaystyle CH_2}{\underset{\displaystyle |}{\underset{\displaystyle CH-CH_3}{\underset{\displaystyle |}{CH_3}}}}}{C}}-CH_2CH_2CH_2CH_2CH_3$$

(e) $CH_3CH(CH_3)CH(CH_3)CH(CH_3)CH(CH_3)_2$

(f) 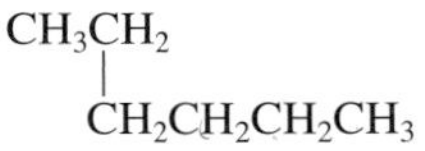
$$\overset{\overset{\displaystyle CH_3CH_2}{|}}{CH_2CH_2CH_2CH_3}$$

(g)  [structure]   (h) 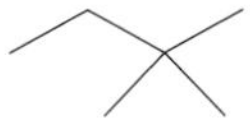 [structure]

(i) 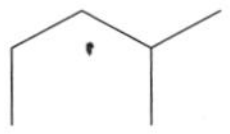 [structure]   (j) 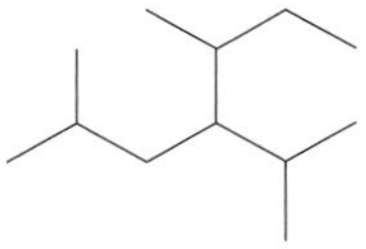 [structure]

37. Convert the following names into the corresponding molecular structures. After doing so, check to see if the name of each molecule as given here is in accord with the IUPAC system of nomenclature. If not, name the molecule correctly. **(a)** 2-methyl-3-propylpentane; **(b)** 5-(1,1-dimethylpropyl)nonane; **(c)** 2,3,4-trimethyl-4-butylheptane; **(d)** 4-*tert*-butyl-5-isopropyl-hexane; **(e)** 4-(2-ethylbutyl)decane; **(f)** 2,4,4-trimethylpentane; **(g)** 4-*sec*-butylheptane; **(h)** isoheptane; **(i)** neoheptane.

38. Draw the structures that correspond to the following names. Correct any names that are not in accord with the rules of systematic nomenclature.

(a) 4-Chloro-5-methylhexane
(b) 3-Methyl-3-propylpentane
(c) 1,1,1-Trifluoro-2-methylpropane
(d) 4-(3-Bromobutyl)nonane

39. Draw and name all possible isomers of C_7H_{16} (isomeric heptanes).

40. Identify the primary, secondary, and tertiary carbon atoms and the hydrogen atoms in each of the following molecules: **(a)** ethane; **(b)** pentane; **(c)** 2-methylbutane; **(d)** 3-ethyl-2,2,3,4-tetramethylpentane.

41. Identify each of the following alkyl groups as being primary, secondary, or tertiary, and give it a systematic IUPAC name.

(a) 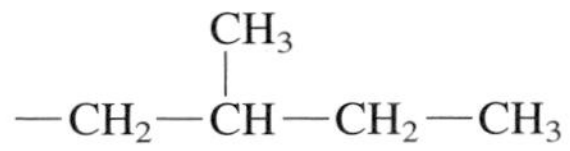
$$-CH_2-\overset{\overset{\displaystyle CH_3}{|}}{CH}-CH_2-CH_3$$

(b) 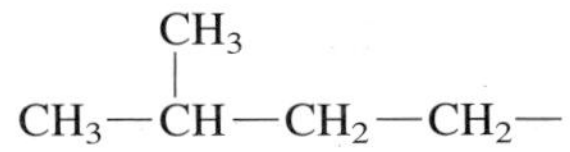
$$CH_3-\overset{\overset{\displaystyle CH_3}{|}}{CH}-CH_2-CH_2-$$

(c) 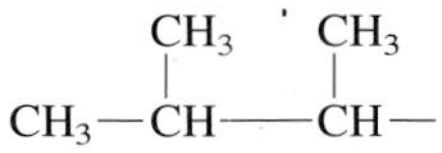
$$CH_3-\overset{\overset{\displaystyle CH_3}{|}}{CH}-\overset{\overset{\displaystyle CH_3}{|}}{CH}-$$

(d) 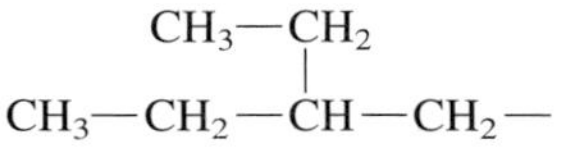
$$CH_3-CH_2-\overset{\overset{\displaystyle CH_3-CH_2}{|}}{CH}-CH_2-$$

(e) 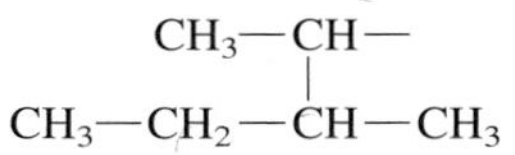
$$CH_3-CH_2-\overset{\overset{\displaystyle CH_3-CH-}{|}}{CH}-CH_3$$

(f) 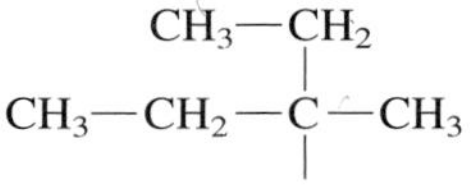
$$CH_3-CH_2-\overset{\overset{\displaystyle CH_3-CH_2}{|}}{\underset{\underset{\displaystyle CH_3}{|}}{C}}-CH_3$$

42. Does molecule **A** below contain a quaternary carbon atom? Does molecule **B**? For each, explain why or why not.

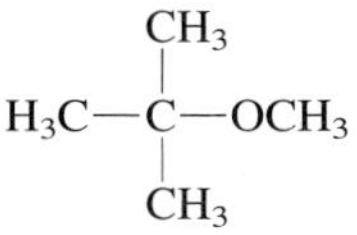

$$H_3C-\overset{\overset{\displaystyle CH_3}{|}}{\underset{\underset{\displaystyle CH_3}{|}}{C}}-OCH_3$$

A

$$H_3C-\overset{\overset{\displaystyle CH_3}{|}}{\underset{\underset{\displaystyle CH_3}{|}}{C}}-CH_3$$

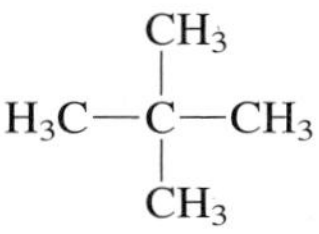

B

43. Rank the following molecules in order of increasing boiling point (without looking up the real values!): **(a)** 3-methylheptane; **(b)** octane; **(c)** 2,4-dimethylhexane; **(d)** 2,2,4-trimethylpentane.

44. Using Newman projections, draw each of the following molecules in its most stable conformation with respect to the bond indicated: **(a)** 2-methylbutane, C2–C3 bond; **(b)** 2,2-dimethylbutane, C2–C3 bond; **(c)** 2,2-dimethylpentane, C3–C4 bond; **(d)** 2,2,4-trimethylpentane, C3–C4 bond.

45. Based on the energy differences for the various conformations of ethane, propane, and butane in Figures 2-10, 2-11, and 2-13, determine the following:

(a) The energy associated with a single hydrogen–hydrogen eclipsing interaction
(b) The energy associated with a single methyl–hydrogen eclipsing interaction
(c) The energy associated with a single methyl–methyl eclipsing interaction
(d) The energy associated with a single methyl–methyl *gauche* interaction

46. At room temperature, 2-methylbutane exists primarily as two alternating conformations of rotation about the C2–C3 bond. About 90% of the molecules exist in the more favorable conformation and 10% in the less favorable one. **(a)** Calculate the free energy change ($\Delta G°$, more favorable conformation − less favorable conformation) between these conformations. **(b)** Draw a potential-energy diagram for rotation about the C2–C3 bond in 2-methylbutane. To the best of your ability, assign relative energy values to all the conformations on your diagram. **(c)** Draw Newman projections for all staggered and eclipsed conformers in (b) and indicate the two most favorable ones.

47. For each of the following naturally occurring compounds, identify the compound class(es) to which it belongs, and circle all functional groups.

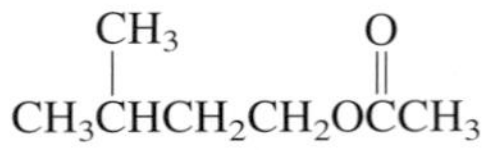

$$\underset{\substack{| \\ CH_3}}{CH_3CHCH_2CH_2O\overset{\displaystyle O}{\overset{\|}{C}}CH_3}$$

3-Methylbutyl acetate
(In banana oil)

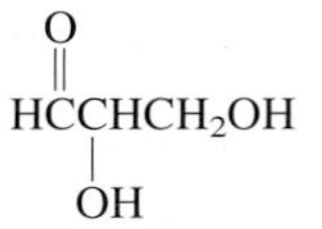

$$\underset{\substack{| \\ OH}}{H\overset{\displaystyle O}{\overset{\|}{C}}CHCH_2OH}$$

2,3-Dihydroxypropanal
(The simplest sugar)

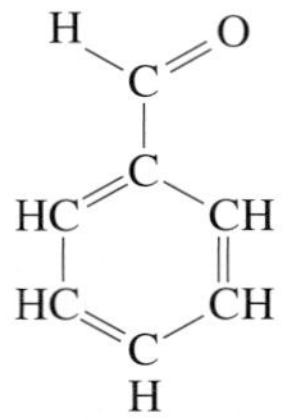

Benzaldehyde
(In fruit pits)

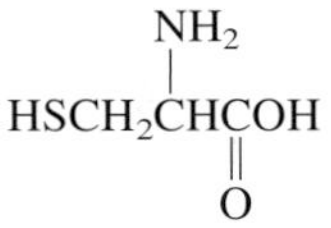

$$HSCH_2\underset{\substack{| \\ NH_2}}{CH}\overset{\displaystyle O}{\overset{\|}{C}}OH$$

Cysteine
(In proteins)

$$CH_3CH=CHC\equiv CC\equiv CCH=CHCH_2OH$$

Matricarianol
(From chamomile)

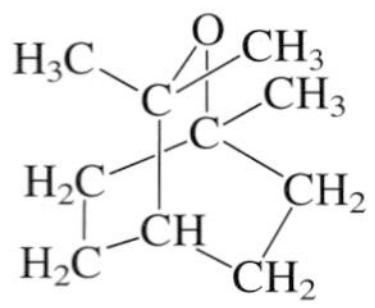

Cineole
(From eucalyptus)

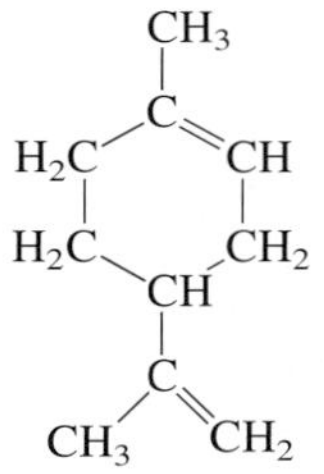

Limonene
(In lemons)

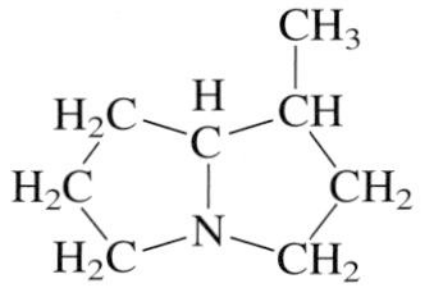

Heliotridane
(An alkaloid)

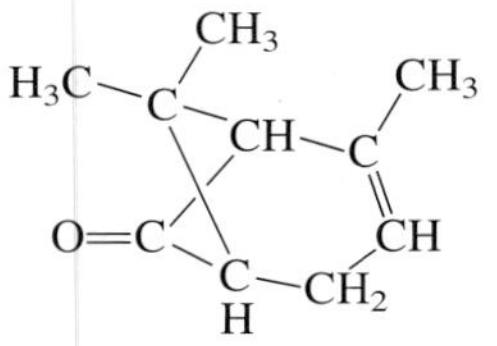

Chrysanthenone
(In chrysanthemums)

48. Give IUPAC names for all alkyl groups marked by dashed lines in each of the following biologically important compounds. Identify each group as a primary, secondary, or tertiary alkyl substituent.

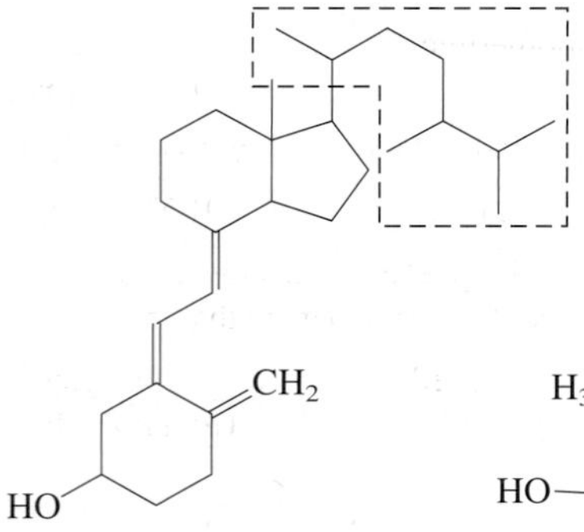

Vitamin D₄

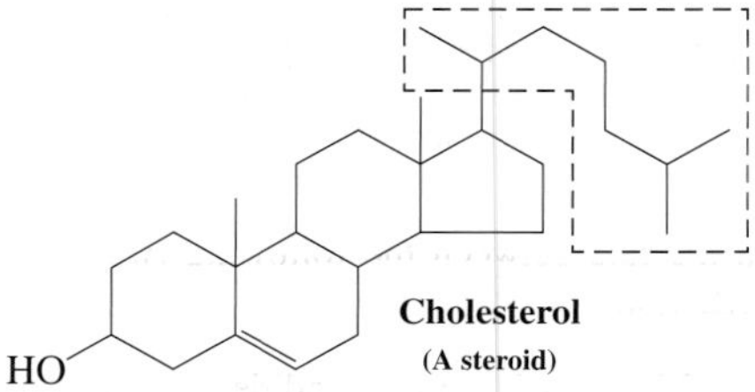

Cholesterol
(A steroid)

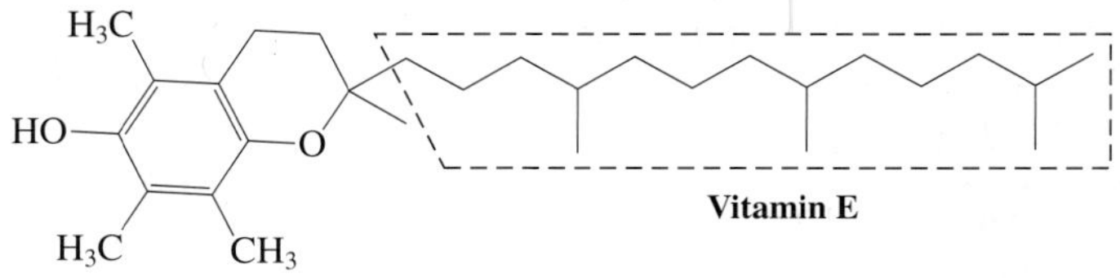

Vitamin E

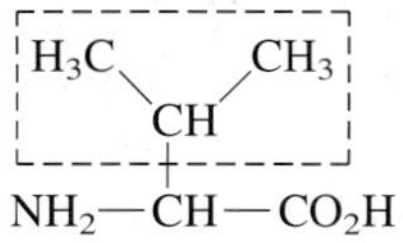

$$\underset{\substack{| \\ NH_2-CH-CO_2H}}{\underset{\substack{| \\ CH}}{H_3C\quad CH_3}}$$

Valine
(An amino acid)

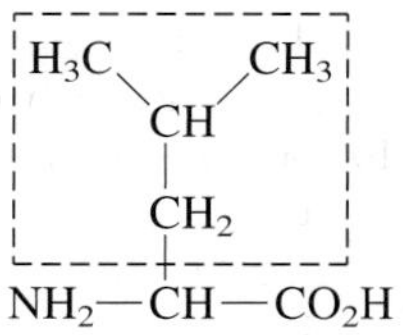

Leucine
(Another amino acid)

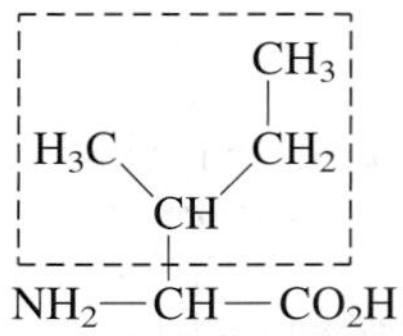

Isoleucine
(Still another amino acid)

49. **CHALLENGE** Using the Arrhenius equation, calculate the effect on k of increases in temperature of 10, 30, and 50 degrees (Celsius) for the following activation energies. Use 300 K (approximately room temperature) as your initial T value, and assume that A is a constant.

(a) $E_a = 15$ kcal mol^{-1}; **(b)** $E_a = 30$ kcal mol^{-1}; **(c)** $E_a = 45$ kcal mol^{-1}.

50. **CHALLENGE** The Arrhenius equation can be reformulated in a way that permits the experimental determination of activation energies. For this purpose, we take the natural logarithm of both sides and convert into the base 10 logarithm.

$$\ln k = \ln (Ae^{-E_a/RT}) = \ln A - E_a/RT \quad \text{becomes}$$

$$\log k = \log A - \frac{E_a}{2.3RT}$$

The rate constant k is measured at several temperatures T and a plot of $\log k$ versus $1/T$ is prepared, a straight line. What is the slope of this line? What is its intercept (i.e., the value of $\log k$ at $1/T = 0$)? How is E_a calculated?

51. Reexamine your answers to Problem 34. Rewrite each one in the form of a complete equation describing a Lewis acid-base process, showing the product and using curved arrows to depict electron-pair movement. [**Hint:** For (b) and (d), start with a Lewis structure that represents a second resonance form of the starting organic molecule.]

52. **CHALLENGE** The equation relating $\Delta G°$ to K contains a temperature term. Refer to your solution to Problem 46(a) to calculate the answers to the questions that follow. You will need to know that $\Delta S°$ for the formation of the more stable conformer of 2-methylbutane from the next most stable conformer is $+1.4$ cal K^{-1} mol^{-1}. **(a)** Calculate the enthalpy difference ($\Delta H°$) between these two conformers from the equation $\Delta G° = \Delta H° - T\Delta S°$. How well does this agree with the $\Delta H°$ calculated from the number of *gauche* interactions in each conformer? **(b)** Assuming that $\Delta H°$ and $\Delta S°$ do not change with temperature, calculate $\Delta G°$ between these two conformations at the following three temperatures: $-250°C$; $-100°C$; $+500°C$. **(c)** Calculate K for these two conformations at the same three temperatures.

Team Problem

53. Consider the difference in the rate between the following two second-order substitution reactions.

Reaction 1: The reaction of bromoethane and iodide ion to produce iodoethane and bromide ion is second order; that is, the rate of the reaction depends on the concentrations of both bromoethane and iodide ion:

$$\text{Rate} = k[CH_3CH_2Br][I^-] \text{ mol } L^{-1} s^{-1}$$

Reaction 2: The reaction of 1-bromo-2,2-dimethylpropane (neopentyl bromide) with iodide ion to produce neopentyl iodide and bromide ion is more than 10,000 times slower than the reaction of bromoethane with iodide ion.

$$\text{Rate} = k[\text{neopentyl bromide}][I^-] \text{ mol } L^{-1} s^{-1}$$

(a) Formulate each reaction by using bond-line structural drawings in your reaction scheme.
(b) Identify the reactive site of the starting haloalkane as primary, secondary, or tertiary.
(c) Discuss how the reaction might take place; that is, how the species would have to interact in order for the reaction to proceed. Remember that, because the reaction is second order, *both* reagents must be present in the transition state. Use your model

kit to help you visualize the trajectory of approach of the iodide ion toward the bromoalkane that enables the simultaneous iodide bond making and bromide bond breaking required by the second-order kinetics of these two reactions. Of all the possibilities, which one best explains the experimentally determined difference in rate between the reactions?
(d) Use hashed-wedged line structures to make a three-dimensional drawing of the trajectory on which you agree.

Preprofessional Problems

54. The compound 2-methylbutane has
  (a) no secondary H's
  (b) no tertiary H's
  (c) no primary H's
  (d) twice as many secondary H's as tertiary H's
  (e) twice as many primary H's as secondary H's

55.

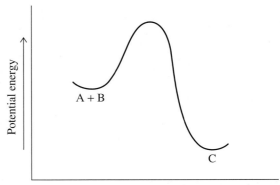

This energy profile diagram represents
  (a) an endothermic reaction   **(b)** an exothermic reaction
  (c) a fast reaction   **(d)** a termolecular reaction

56. In 4-(1-methylethyl)heptane, any H–C–C angle has the value
  (a) 120°   **(b)** 109.5°   **(c)** 180°   **(d)** 90°   **(e)** 360°

57. The structural representation shown below is a Newman projection of the conformer of butane that is
  (a) *gauche* eclipsed   **(b)** *anti* gauche
  (c) *anti* staggered   **(d)** *anti* eclipsed

58. Genipin is a Chinese herbal remedy that is effective against diabetes. To which of the compound classes below does genipin *not* belong?
  (a) alcohol   **(b)** alkene   **(c)** ester
  (d) ether   **(e)** ketone

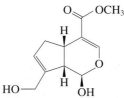

Genipin

CHAPTER 3

Reactions of Alkanes

Bond-Dissociation Energies, Radical Halogenation, and Relative Reactivity

Combustion of alkanes releases most of the energy that powers modern industrialized society. We saw in Chapter 2 that alkanes lack functional groups; that being the case, how does combustion occur? We will see in this chapter that alkanes are not very reactive, but that they do undergo several types of transformations. These processes, of which combustion is one example, do not involve acid-base chemistry. Instead, they are called **radical reactions.** Although we shall not explore radical reactions in great depth in this course, they play significant roles in biochemistry (such as aging and disease processes), the environment (destruction of the Earth's ozone layer), and industry (manufacture of synthetic fabrics and plastics).

Radical reactions begin with the breaking of a bond, or **bond dissociation.** We examine the energetics of this process and discuss the conditions under which it occurs. The majority of the chapter will deal with **halogenation,** a radical reaction in which a hydrogen atom in an alkane is replaced by halogen. The importance of halogenation lies in the fact that it introduces a reactive functional group, turning the alkane into a haloalkane, which is suitable for further chemical change. For each of these processes, we shall discuss the **mechanism** involved to explain in detail how the reaction occurs. We shall see that different alkanes, and indeed different bonds in the same alkane molecule, may react at different rates, and we shall see why this is so.

A carbon radical

Only a relatively limited number of mechanisms are needed to describe the very large number of reactions in organic chemistry. Mechanisms enable us to understand how and why reactions occur, and what products are likely to form in them. In this chapter we apply mechanistic concepts to explain the effects of halogen-containing chemicals on the stratospheric ozone layer. We conclude with a brief discussion of alkane combustion and show how that process serves as a useful source of thermodynamic information about organic molecules.

NASA's X-43A hypersonic research aircraft being dropped from under the wing of a B-52B Stratofortress on November 16, 2004. Most supersonic aircraft produce exhaust gases containing molecules such as nitric oxide (NO), whose radical reactions are destructive to the Earth's stratospheric ozone (O_3) layer. In the 1970s the United States abandoned plans to build a fleet of supersonic aircraft (SSTs, or supersonic transports) for just this reason. In contrast, the X-43A is hydrogen fueled, posing no risk to stratospheric ozone, and may represent the first step toward the development of environmentally acceptable high-speed flight. In 2008, Boeing flew successfully the first manned hydrogen-fuel-cell-powered aircraft, another aviation milestone.

3-1 | STRENGTH OF ALKANE BONDS: RADICALS

In Section 1-2 we explained how bonds are formed and that energy is released on bond formation. For example, bringing two hydrogen atoms into bonding distance produces 104 kcal mol^{-1} (435 kJ mol^{-1}) of heat (refer to Figures 1-1 and 1-12).

$$H\cdot + H\cdot \xrightarrow{\text{Bond making}} H\text{—}H \qquad \Delta H° = -104 \text{ kcal mol}^{-1} \; (-435 \text{ kJ mol}^{-1})$$
Released heat: exothermic

Consequently, breaking such a bond *requires* heat—in fact, the same amount of heat that was released when the bond was made. This energy is called **bond-dissociation energy, DH°,** and is a quantitative measure of the **bond strength.**

$$H\text{—}H \xrightarrow{\text{Bond breaking}} H\cdot + H\cdot \qquad \Delta H° = DH° = 104 \text{ kcal mol}^{-1} \; (435 \text{ kJ mol}^{-1})$$
Consumed heat: endothermic

Radicals are formed by homolytic cleavage

In our example, the bond breaks in such a way that the two bonding electrons divide equally between the two participating atoms or fragments. This process is called **homolytic cleavage** or **bond homolysis.** The separation of the two bonding electrons is denoted by two *single-barbed* or "fishhook" arrows that point from the bond to each of the atoms.

> A single-barbed arrow ⌒ shows the movement of a *single* electron.

Homolytic Cleavage: Bonding Electrons Separate

$$A\text{—}B \longrightarrow A\cdot + \cdot B$$
Radicals

The fragments that form have an unpaired electron, for example, H·, Cl·, CH$_3$·, and CH$_3$CH$_2$·. When these species are composed of more than one atom, they are called **radicals.** Because of the unpaired electron, radicals and free atoms are very reactive and usually cannot be isolated. However, radicals and free atoms are present in low concentration as unobserved *intermediates* in many reactions, such as the production of polymers (Chapter 12) and the oxidation of fats that leads to the spoilage of perishable foods (Chapter 22).

In Section 2-2 we introduced an alternative way of breaking a bond, in which the entire bonding electron pair is donated to one of the atoms. This process is **heterolytic cleavage** and results in the formation of **ions.**

:C̈l·

Chlorine atom

H
|
H—C—H
·

Methyl radical

H
|
H$_3$C—C·
|
H

Ethyl radical

Heterolytic Cleavage: Bonding Electrons Move as Pair

$$A\text{—}B \longrightarrow A^+ + :B^-$$
Ions

> A normal, double-barbed curved arrow ⌒ shows the movement of a *pair* of electrons.

Homolytic cleavage may be observed in nonpolar solvents or even in the gas phase. In contrast, heterolytic cleavage normally occurs in polar solvents, which are capable of stabilizing ions. Heterolytic cleavage is also restricted to situations where the electronegativies of atoms A and B and the groups attached to them stabilize positive and negative charges, respectively.

Dissociation energies, DH°, refer only to homolytic cleavages. They have characteristic values for the various bonds that can be formed between the elements. Table 3-1 lists dissociation energies of some common bonds. The larger the value for *DH°,* the stronger the corresponding bond. Note the relatively strong bonds to hydrogen, as in H–F and H–OH. However, even though these bonds have high *DH°* values, they readily undergo *heterolytic* cleavage in water to H$^+$ and F$^-$ or HO$^-$; *do not confuse homolytic with heterolytic processes.*

Table 3-1	Bond-Dissociation Energies of Various A–B Bonds in the Gas Phase [$DH°$ in kcal mol^{-1} (kJ mol^{-1})]						
	B in A–B						
A in A–B	**–H**	**–F**	**–Cl**	**–Br**	**–I**	**–OH**	**–NH$_2$**
H—	104 (435)	136 (569)	103 (431)	87 (364)	71 (297)	119 (498)	108 (452)
CH$_3$—	105 (439)	110 (460)	85 (356)	70 (293)	57 (238)	93 (389)	84 (352)
CH$_3$CH$_2$—	101 (423)	111 (464)	84 (352)	70 (293)	56 (234)	94 (393)	85 (356)
CH$_3$CH$_2$CH$_2$—	101 (423)	110 (460)	85 (356)	70 (293)	56 (234)	92 (385)	84 (352)
(CH$_3$)$_2$CH—	98.5 (412)	111 (464)	84 (352)	71 (297)	56 (234)	96 (402)	86 (360)
(CH$_3$)$_3$C—	96.5 (404)	110 (460)	85 (356)	71 (297)	55 (230)	96 (402)	85 (356)

Note: (a) $DH° = \Delta H°$ for the process A–B → A· + ·B. (b) These numbers are being revised continually because of improved methods for their measurement. (c) The trends observed for A–H bonds are significantly altered for polar A–B bonds because of dipolar contributions to $DH°$.

Bonds are strongest when made by overlapping orbitals that are closely matched in energy and size (Section 1-7). For example, the strength of the bonds between hydrogen and the halogens decreases in the order F > Cl > Br > I, because the *p* orbital of the halogen contributing to the bonding becomes larger and more diffuse along the series. Thus, the efficiency of its overlap with the relatively small 1*s* orbital on hydrogen diminishes. A similar trend holds for bonding between the halogens and carbon.

Increasing size of the halogen →

	CH$_3$–F	CH$_3$–Cl	CH$_3$–Br	CH$_3$–I	
$DH° =$	110	85	70	57	kcal mol^{-1}

Decreasing bond strength →

Solved Exercise 3-1 | Working with the Concepts: Understanding Bond Strengths

Compare the bond-dissociation energies of CH$_3$–F, CH$_3$–OH, and CH$_3$–NH$_2$. Why do the bonds get weaker along this series even though the orbitals participating in overlap become better matched in size and energy?

Strategy
What factors contribute to the strength of a bond? As mentioned above, the sizes and energies of the orbitals are very important. However, coulombic contributions may enhance covalent bond strength. Let's look at each factor separately to see if one outweighs the other in these three bonds.

Solution
- The better the match in size and energy of orbitals between two atoms in a bond, the better the bonding overlap will be (note Figure 1-2). However, the decrease in orbital size in the series C, N, O, F is slight, and therefore the overlap changes by only a very small amount going from C–F to C–O to C–N.
- In the progression from N to O to F, nuclear charge increases, giving rise to stronger attraction between the nucleus and the electrons. The increasing electronegativity in this series confirms this effect (Table 1-2). As the electronegativity of the element attached to carbon increases, so does its attraction for the shared electron pair in the covalent bond. Thus the polarity and the charge separation in the bond both increase, giving rise to a partial positive charge (δ^+) on carbon and a partial negative charge (δ^-) on the more electronegative atom.
- Coulombic attraction between these opposite charges supplements the bonding resulting from covalent overlap and makes the bond stronger. In this series of bonds, increasing coulombic attraction outweighs decreasing overlap going from N to O to F, giving the observed result.

> **Exercise 3-2** | Try It Yourself
>
> In the series C–C (in ethane, $H_3C–CH_3$), N–N (in hydrazine, $H_2N–NH_2$), O–O (in hydrogen peroxide, HO–OH), the bonds decrease in strength from 90 to 60 to 50 kcal mol^{-1}. Explain. (**Hint:** Lone pairs on adjacent atoms repel each other.)

The stability of radicals determines the C–H bond strengths

How strong are the C–H and C–C bonds in alkanes? The bond-dissociation energies of various alkane bonds are given in Table 3-2. Note that bond energies generally decrease with the progression from methane to primary, secondary, and tertiary carbon. For example, the C–H bond in methane is strong and has a high $DH°$ value of 105 kcal mol^{-1}. In ethane, this bond energy is less: $DH° = 101$ kcal mol^{-1}. The latter number is typical for primary C–H bonds, as can be seen for the bond in propane. The secondary C–H bond is even weaker, with a $DH°$ of 98.5 kcal mol^{-1}, and a tertiary carbon atom bound to hydrogen has a $DH°$ of only 96.5 kcal mol^{-1}.

C–H Bond Strength in Alkanes

$$CH_3–H \quad > \quad R_{primary}–H \quad > \quad R_{secondary}–H \quad > \quad R_{tertiary}–H$$

$$DH° = \quad 105 \qquad\qquad 101 \qquad\qquad 98.5 \qquad\qquad 96.5 \qquad\quad \text{kcal mol}^{-1}$$

→ Decreasing bond strength

A similar trend is seen for C–C bonds (Table 3-2). The extremes are the bonds in $H_3C–CH_3$ ($DH° = 90$ kcal mol^{-1}) and $(CH_3)_3C–CH_3$ ($DH° = 87$ kcal mol^{-1}).

Why do all of these dissociations exhibit different $DH°$ values? The radicals formed have different energies. For reasons we will address in the next section, radical stability *increases* along the series from primary to secondary to tertiary; consequently, the energy required to create them *decreases*.

Radical Stabilities

→ Increasing stability

$$·CH_3 \quad < \quad ·R_{primary} \quad < \quad ·R_{secondary} \quad < \quad ·R_{tertiary}$$

→ Decreasing $DH°$ of alkane R–H

Table 3-2	Bond-Dissociation Energies for Some Alkanes		
Compound	$DH°$ [kcal mol^{-1} (kJ mol^{-1})]	Compound	$DH°$ [kcal mol^{-1} (kJ mol^{-1})]
$CH_3{+}H$	105 (439)	$CH_3{+}CH_3$	90 (377)
$C_2H_5{+}H$	101 (423)	$C_2H_5{+}CH_3$	89 (372)
$C_3H_7{+}H$	101 (423)	$C_2H_5{+}C_2H_5$	88 (368)
$(CH_3)_2CHCH_2{+}H$	101 (423)	$(CH_3)_2CH{+}CH_3$	88 (368)
$(CH_3)_2CH{+}H$	98.5 (412)	$(CH_3)_3C{+}CH_3$	87 (364)
$(CH_3)_3C{+}H$	96.5 (404)	$(CH_3)_2CH{+}CH(CH_3)_2$	85.5 (358)
		$(CH_3)_3C{+}C(CH_3)_3$	78.5 (328)

Decreasing $DH°$ (left column); Decreasing $DH°$ (right column)

Note: See footnote for Table 3-1.

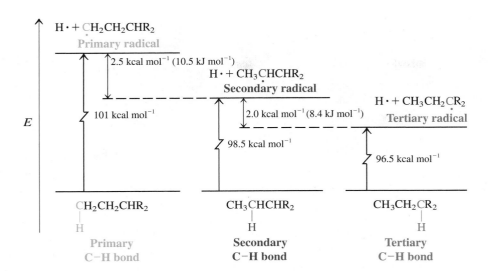

Figure 3-1 illustrates this finding in an energy diagram. We start (at the bottom) with an alkane containing primary, secondary, and tertiary C–H bonds. Primary bond dissociation is endothermic by $DH° = 101$ kcal mol^{-1}, that is, we are going *up* in energy by this amount to reach the primary radical. Secondary radical formation costs less, 98.5 kcal mol^{-1}. Thus, the secondary radical is more stable than the primary by 2.5 kcal mol^{-1}. To form the tertiary radical requires even less, 96.5 kcal mol^{-1}, and this radical is more stable than its secondary analog by 2.0 kcal mol^{-1} (or its primary analog by 4.5 kcal mol^{-1}).

Exercise 3-3

Which C–C bond would break first, the bond in ethane or that in 2,2-dimethylpropane?

In Summary Bond homolysis in alkanes yields radicals and free atoms. The heat required to do so is called the bond-dissociation energy, $DH°$. Its value is characteristic only for the bond between the two participating elements. Bond breaking that results in tertiary radicals demands less energy than that furnishing secondary radicals; in turn, secondary radicals are formed more readily than primary radicals. The methyl radical is the most difficult to obtain in this way.

3-2 | STRUCTURE OF ALKYL RADICALS: HYPERCONJUGATION

What is the reason for the ordering in stability of alkyl radicals? To answer this question, we need to inspect the alkyl radical structure more closely. Consider the structure of the methyl radical, formed by removal of a hydrogen atom from methane. Spectroscopic measurements have shown that the methyl radical, and probably other alkyl radicals, adopts a *nearly planar* configuration, best described by sp^2 hybridization (Figure 3-2). The unpaired electron occupies the remaining p orbital perpendicular to the molecular plane.

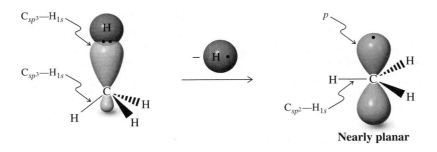

Figure 3-2 The hybridization change upon formation of a methyl radical from methane. The nearly planar arrangement is reminiscent of the hybridization in BH_3 (Figure 1-17).

Figure 3-1 The different energies needed to form radicals from an alkane $CH_3CH_2CHR_2$. Radical stability increases from primary to secondary to tertiary.

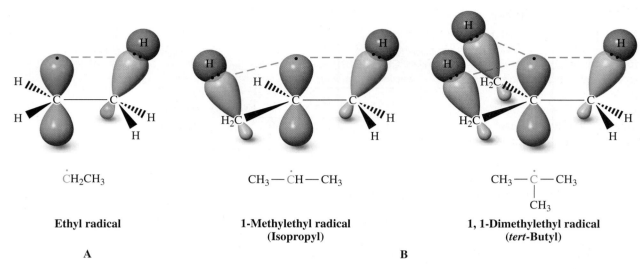

$\overset{\cdot}{C}H_2CH_3$

Ethyl radical

$CH_3 - \overset{\cdot}{C}H - CH_3$

**1-Methylethyl radical
(Isopropyl)**

$CH_3 - \overset{\cdot}{C} - CH_3$
           |
          CH_3

**1, 1-Dimethylethyl radical
(*tert*-Butyl)**

A   B

Figure 3-3 Hyperconjugation (green dashed lines) resulting from the donation of electrons in filled sp^3 hybrids to the partly filled p orbital in (A) ethyl and (B) 1-methylethyl and 1,1-dimethylethyl radicals. The resulting delocalization of electron density has a net stabilizing effect.

Model Building

Let us see how the planar structures of alkyl radicals help explain their relative stabilities. Figure 3-3A shows that there is a conformer in the ethyl radical in which a C–H bond of the CH_3 group is aligned with and overlaps one of the lobes of the singly occupied p orbital on the radical center. This arrangement allows the bonding pair of electrons in the σ orbital to delocalize into the partly empty p lobe, a phenomenon called **hyperconjugation.** The interaction between a filled orbital and a singly occupied orbital has a net stabilizing effect (recall Exercise 1-14). Both hyperconjugation and resonance (Section 1-5) are forms of electron delocalization. They are distinguished by type of orbital: Resonance normally refers to π-type overlap of p orbitals, whereas hyperconjugation incorporates overlap with the orbitals of σ bonds. *Radicals are stabilized by hyperconjugation.*

What if we replace the remaining hydrogens on the radical carbon with alkyl groups? Each additional alkyl group increases the hyperconjugation interactions further (Figure 3-3B). The order of stability of the radicals is a consequence of this effect. Notice in Figure 3-1 that the degree of stabilization arising from each hyperconjugative interaction is relatively small [2.0–2.5 kcal mol^{-1} (8.4–10.5 kJ mol^{-1})]: We shall see later that stabilization of radicals by resonance is considerably greater (Chapter 14). Another contribution to the relative stability of secondary and tertiary radicals is the greater relief of steric crowding between the substituent groups as the geometry changes from tetrahedral in the alkane to planar in the radical.

Even a cursory glance at the bond-dissociation energies between carbon and the more electronegative atoms in Table 3-1 suggests that hyperconjugation and radical stabilities alone do not provide a complete picture. For example, bonds between carbon and any of the halogens all display essentially identical $DH°$ regardless of the type of carbon. Several interpretations have been proposed to explain these observations. Polar effects are likely involved (as mentioned in the table footnote). In addition, the longer bonds between carbon and large atoms reduce the steric repulsion between atoms around that carbon, diminishing its influence on bond-dissociation energies.

3-3 | CONVERSION OF PETROLEUM: PYROLYSIS

Alkanes are produced naturally by the slow decomposition of animal and vegetable matter in the presence of water but in the absence of oxygen, a process lasting millions of years. The smaller alkanes—methane, ethane, propane, and butane—are present in natural gas, methane being by far its major component. Many liquid and solid alkanes are obtained from crude petroleum, but distillation alone does not meet the enormous demand for the lower-molecular-weight hydrocarbons needed for gasoline, kerosene, and other hydrocarbon-based fuels. Additional heating is required to break up longer-chain petroleum components into smaller molecules. How does this occur? Let us first look into the effect of strong heating on simple alkanes and then move on to petroleum.

High temperatures cause bond homolysis

When alkanes are heated to a high temperature, both C–H bonds and C–C bonds rupture, a process called **pyrolysis.** In the absence of oxygen, the resulting radicals can combine to form new higher- or lower-molecular-weight alkanes. Radicals can also remove hydrogen atoms from another alkane, a process called hydrogen abstraction, or from the carbon atom adjacent to another radical center to give alkenes, called radical disproportionation. Indeed, very complicated mixtures of alkanes and alkenes form in pyrolyses. Under special conditions, however, these transformations can be controlled to obtain a large proportion of hydrocarbons of a defined chain length.

Pyrolysis of Hexane

Examples of cleavage into radicals:

$$\overset{1}{CH_3}\overset{2}{CH_2}\overset{3}{CH_2}\overset{4}{CH_2}CH_2CH_3$$
Hexane

C1–C2 cleavage → $CH_3 \cdot + \cdot CH_2CH_2CH_2CH_2CH_3$

C2–C3 cleavage → $CH_3CH_2 \cdot + \cdot CH_2CH_2CH_2CH_3$

C3–C4 cleavage → $CH_3CH_2CH_2 \cdot + \cdot CH_2CH_2CH_3$

Example of radical combination:

$$CH_3CH_2CH_2CH_2CH_2 \cdot \quad \cdot CH_2CH_2CH_3 \longrightarrow CH_3CH_2CH_2CH_2CH_2CH_2CH_2CH_3$$
Octane

Example of hydrogen abstraction:

$$CH_3CH_2 \cdot \quad CH_3\overset{H}{\underset{|}{C}}HCH_3 \longrightarrow CH_3CH_2 + CH_3\dot{C}HCH_3$$
Ethane

Example of disproportionation:

$$CH_3CH_2CH_2 \cdot \quad \overset{H}{\underset{|}{C}}H_2-CH_2 \cdot \longrightarrow CH_3CH_2CH_2 + CH_2{=}CH_2$$
Propane   **Ethene**

Note how we used single-barbed (fishhook) arrows in these examples to show the formation of a new covalent bond by the combination of two electrons. In the hydrogen abstraction reactions, electrons from a bond being broken combine with unshared electrons to form new bonds.

Control of such processes frequently requires the use of special **catalysts,** such as crystalline sodium aluminosilicates, also called zeolites. For example, zeolite-catalyzed pyrolysis of dodecane yields a mixture in which hydrocarbons containing from three to six carbons predominate.

$$Dodecane \xrightarrow{\text{Zeolite, } 482°C,\ 2\ min} C_3 + C_4 + C_5 + C_6 + \text{other products}$$
$$\quad\quad\quad\quad\quad\quad\quad 17\% \quad 31\% \quad 23\% \quad 18\% \quad\quad\quad 11\%$$

The Function of a Catalyst

What is the function of the zeolite catalyst? It speeds up the pyrolysis, so that the process occurs at a lower temperature than otherwise would be the case. The catalyst also causes certain products to form preferentially. Such enhanced reaction *selectivity* is a feature frequently observed in catalyzed reactions. How does this happen?

In general, catalysts are additives that accelerate reactions. They enable new pathways through which reactants and products are interconverted, pathways that have lower activation energies (E_{cat}, than those available in the absence of the catalyst, E_a. In Figure 3-4, both uncatalyzed and catalyzed processes are shown in a simplified manner to consist of only a single step with one activation barrier. We shall see that most reactions involve more than one step. However, regardless of the number of steps, the catalyzed version of a reaction

Figure 3-4 Potential-energy diagram comparing catalyzed and uncatalyzed processes. Although each is shown as consisting of a single step, catalyzed reactions in particular typically proceed via multistep pathways.

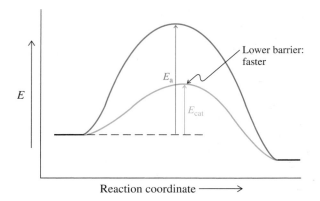

always has greatly reduced activation energies. While a catalyst is not consumed during the reaction it facilitates, it actively participates in it through the formation of intermediate reactive species from which it is ultimately regenerated. Therefore, only a small amount of a catalyst is needed to effect the conversion of a large amount of reactants. Catalysts modify the *kinetics* of reactions; that is, they change the rate at which equilibrium is established. However, catalysts do not affect the position of the equilibrium. The overall $\Delta H°$, $\Delta S°$, and hence $\Delta G°$ values for catalyzed and uncatalyzed processes are identical: A catalyst does *not* affect the overall *thermodynamics* of a reaction.

Many organic reactions occur at a useful rate only because of the presence of a catalyst. The catalyst may be an acid (a proton), a base (hydroxide), a metal surface or metal compound, or a complex organic molecule. In nature, *enzymes* usually fulfill this function. The degree of reaction acceleration induced by a catalyst can amount to many orders of magnitude. Enzyme-catalyzed reactions are known to take place 10^{19} times faster than the uncatalyzed processes. The use of catalysts allows many transformations to take place at lower temperatures and under much milder reaction conditions than would otherwise be possible.

Petroleum is an important source of alkanes

Breaking an alkane down into smaller fragments is also known as **cracking.** Such processes are important in the oil-refining industry for the production of gasoline and other liquid fuels from petroleum.

Petroleum, or crude oil, is believed to be the product of microbial degradation of living organic matter that existed several hundred million years ago. Crude oil, a dark viscous liquid, is primarily a mixture of several hundred different hydrocarbons, particularly straight-chain alkanes, some branched alkanes, and varying quantities of aromatic hydrocarbons. Distillation yields several fractions with a typical product distribution, as shown in Table 3-3. The composition of petroleum varies widely, depending on the origin of the oil.

Table 3-3	Product Distribution in a Typical Distillation of Crude Petroleum		
Amount (% of volume)	Boiling point (°C)	Carbon atoms	Products
1–2	<30	C_1–C_4	Natural gas, methane, propane, butane, liquefied petroleum gas (LPG)
15–30	30–200	C_4–C_{12}	Petroleum ether ($C_{5,6}$), ligroin (C_7), naphtha, straight-run gasoline[a]
5–20	200–300	C_{12}–C_{15}	Kerosene, heater oil
10–40	300–400	C_{15}–C_{25}	Gas oil, diesel fuel, lubricating oil, waxes, asphalt
8–69	>400 (Nonvolatiles)	>C_{25}	Residual oil, paraffin waxes, asphalt (tar)

[a]This refers to gasoline straight from petroleum, without having been treated in any way.

To increase the quantity of the much-needed gasoline fraction, the oils with higher boiling points are cracked by pyrolysis. Cracking the residual oil from crude petroleum distillation gives approximately 30% gas, 50% gasoline, and 20% higher-molecular-weight oils and a residue called petroleum coke.

REAL LIFE: SUSTAINABILITY 3-1 | Sustainability and the Needs of the 21st Century: "Green" Chemistry

Oil and natural gas supply much of the energy requirement of both the United States and most of the rest of the industrial world. In 2009, U.S. energy sources included gas (25%), oil (38%), coal (18%), nuclear (9%), hydroelectric and hydrothermal (3%), and other renewable (7%). After changing very little over more than a decade, reductions in percentage consumption of oil and coal are occurring, which have been offset by increased use of natural gas and renewable energy resources. Imported petroleum products, however, still make up a significant proportion of U.S. energy expenditure. These substances also constitute the raw materials of much of U.S. chemical industry, via their conversion into simpler hydrocarbons such as alkenes, the starting points for countless manufacturing processes. However, this petroleum-based economy is plagued by significant problems: It is energy intensive, suffers from the frequent necessity for toxic discharge, and generates waste in the form of by-products, solvents, and inorganic salts. It is also not sustainable in the future, because the earth's supply of petroleum is limited.

Chemists in academia and industry have responded by actively exploring alternative sources of raw materials. Less exploited fossil fuels such as methane are under investigation, as are renewable starting materials, typically derived from agricultural sources. The latter, consisting of wood, grain, plant parts and plant oils, and carbohydrates, are by far the most abundant. Plant growth consumes CO_2 via photosynthesis, a desirable feature in a time of concern over the increasing concentration of CO_2 in the atmosphere and its long-term effect on global climate. Conversion of these raw materials into useful products presents a significant challenge, however. Ideally, processes developed for these conversions should be both efficient and environmentally acceptable. What does this mean?

Over the past two decades the term *green chemistry* has been used increasingly to describe processes that meet a number of environmental requirements. The term was coined in 1994 by Dr. Paul T. Anastas of the U.S. Environmental Protection Agency (EPA) to denote chemical activities that strive to achieve the goals of environmental protection and sustainable development. Specifically, this means pollution prevention by reducing or eliminating the use or generation

The Alyeska Pipeline Marine Terminal, Valdez, Alaska. Alaska is second only to Texas in oil production in the United States.

of hazardous materials in design, manufacture, and application of chemical products, and switching from oil-based chemicals to those generated by nature. Some of the principles of green chemistry are (in abbreviated form)

1. It is better to prevent waste than to have to clean it up.

2. Synthetic methods should maximize the incorporation of all starting materials into the final products ("atom economy").

3. Reactions should use and generate substances that possess efficacy of function but little or no toxicity.

4. Energy requirements should be minimized by conducting reactions at ambient temperature and pressure.

5. Feedstocks should be renewable.

6. Catalytic processes are superior to stoichiometric ones.

A case of a green approach to the cracking of petroleum is the recent discovery of a catalytic process that converts linear alkanes into their higher and lower homologs with good selectivity. For example, when butane is passed over a Ta catalyst deposited on silica at 150°C, it undergoes metathesis (*metatithenai*, Greek, to transpose) to mainly propane and pentane:

$$H_3C-CH_2-CH_2-CH_3 \xrightarrow[150°C]{Ta \text{ on } SiO_2} H_3C-CH_2-CH_3 + H_3C-CH_2-CH_2-CH_3$$

Butane   **Propane**   **Pentane**

This process has no waste, complete atom economy, is nontoxic, occurs at a much lower temperature than conventional cracking, and is catalytic, fulfilling all the requirements of a green reaction. Methods such as these are forming a new paradigm for the practice of chemistry in the 21st century.

We have seen that alkanes undergo chemical transformations when subjected to pyrolysis, and that these processes include the formation of radical intermediates. Do alkanes participate in other reactions? In this section, we consider the effect of exposing an alkane (methane) to a halogen (chlorine). A **chlorination** reaction, in which radicals again play a key role, takes place, producing chloromethane and hydrogen chloride. We shall analyze each step in this transformation to establish the *mechanism* of the reaction.

Chlorine converts methane into chloromethane

When methane and chlorine gas are mixed in the dark at room temperature, no reaction occurs. The mixture must be heated to a temperature above 300°C (denoted by Δ) *or* irradiated with ultraviolet light (denoted by $h\nu$) before a reaction takes place. One of the two initial products is chloromethane, derived from methane in which a hydrogen atom is removed and replaced by chlorine. The other product of this transformation is hydrogen chloride. Further substitution leads to dichloromethane (methylene chloride), CH_2Cl_2; trichloromethane (chloroform), $CHCl_3$; and tetrachloromethane (carbon tetrachloride), CCl_4.

Reaction

Why should this reaction proceed? Consider its $\Delta H°$. Note that a C–H bond in methane ($DH° = 105$ kcal mol^{-1}) and a Cl–Cl bond ($DH° = 58$ kcal mol^{-1}) are broken, whereas the C–Cl bond of chloromethane ($DH° = 85$ kcal mol^{-1}) and an H–Cl linkage ($DH° = 103$ kcal mol^{-1}) are formed. The net result is the release of 25 kcal mol^{-1} in forming stronger bonds: The reaction is *exothermic* (heat releasing).

$$CH_3\!-\!H + :\!\overset{..}{\underset{..}{Cl}}\!-\!\overset{..}{\underset{..}{Cl}}: \xrightarrow{\Delta \text{ or } h\nu} CH_3\!-\!\overset{..}{\underset{..}{Cl}}: + H\!-\!\overset{..}{\underset{..}{Cl}}:$$

$$\begin{aligned}105 \qquad 58 \qquad\qquad 85 \qquad 103\end{aligned}$$

DH° (kcal mol^{-1})   **Chloromethane**

$$\begin{aligned}\Delta H° &= \text{energy input} - \text{energy output}\\ &= \Sigma DH° \text{ (bonds broken)} - \Sigma DH° \text{ (bonds formed)}\\ &= (105 + 58) - (85 + 103)\\ &= -25 \text{ kcal mol}^{-1} \ (-105 \text{ kJ mol}^{-1})\end{aligned}$$

Why, then, does the thermal chlorination of methane not occur at room temperature? The fact that a reaction is exothermic does not necessarily mean that it proceeds rapidly and spontaneously. Remember (Section 2-1) that the rate of a chemical transformation depends on its activation energy, which in this case is evidently high. Why is this so? What is the function of irradiation when the reaction does proceed at room temperature? Answering these questions requires an investigation of the mechanism of the reaction.

The mechanism explains the experimental conditions required for reaction

Mechanism

A **mechanism** is a detailed, step-by-step description of all the changes in bonding that occur in a reaction (Section 1-1). Even simple reactions may consist of several separate steps. The mechanism shows the sequence in which bonds are broken and formed, as well as the energy changes associated with each step. This information is of great value in both analyzing possible transformations of complex molecules and understanding the experimental conditions required for reactions to occur.

The mechanism for the chlorination of methane, in common with the mechanisms of most radical reactions, consists of three stages: initiation, propagation, and termination. Let us look at these stages and the experimental evidence for each of them in more detail.

Animation

ANIMATED MECHANISM:
Chlorination of methane

The chlorination of methane can be studied step by step

Experimental observation. Chlorination occurs when a mixture of CH_4 and Cl_2 is either heated to 300°C or irradiated with light, as mentioned earlier. Under such conditions, methane by itself is completely stable, but Cl_2 undergoes homolysis to two atoms of chlorine.
Interpretation. The first step in the mechanism of chlorination of methane is the heat- or light-induced homolytic cleavage of the Cl–Cl bond (which happens to be the weakest bond

in the starting mixture, with $DH° = 58$ kcal mol^{-1}). This event is required to start the chlorination process and is therefore called the **initiation** step. As implied by its name, the initiation step generates reactive species (in this case, chlorine atoms) that permit the subsequent steps in the overall reaction to take place.

Initiation: Homolytic cleavage of the Cl–Cl bond

$$:\ddot{C}l{-}\ddot{C}l: \xrightarrow{\Delta \text{ or } h\nu} 2 :\ddot{C}l\cdot \qquad \Delta H° = DH°(\text{Cl}_2)$$
$$= +58 \text{ kcal mol}^{-1} \ (+243 \text{ kJ mol}^{-1})$$

Chlorine atom

> **Note:** In this scheme and in those that follow, all radicals and free atoms are in green.

Experimental observation. Only a relatively small number of initiation events (e.g., photons of light) are necessary to enable a great many methane and chlorine molecules to undergo conversion into products.

Interpretation. After initiation has taken place, the subsequent steps in the mechanism are self-sustaining, or self-propagating; that is, they can occur many times without the addition of further chlorine atoms from the homolysis of Cl_2. Two **propagation** steps fulfill this requirement. In the first step, a chlorine atom attacks methane by abstracting a hydrogen atom. The products are hydrogen chloride and a methyl radical.

Propagation step 1: Abstraction of an H atom by $:\ddot{C}l\cdot$

$$:\ddot{C}l\cdot + \; H{-}\!\!\overset{\displaystyle H}{\underset{\displaystyle H}{C}}\!\!{-}H \longrightarrow :\ddot{C}l{-}H + \; \cdot\overset{\displaystyle H}{\underset{\displaystyle H}{C}}{-}H \qquad \Delta H° = DH°(\text{CH}_3\text{–H})$$
$$-\ DH°(\text{H–Cl})$$
$$= +2 \text{ kcal mol}^{-1} \ (+8 \text{ kJ mol}^{-1})$$

105   103

DH° (kcal mol^{-1})   **Methyl radical**

> **Careful!** The ionic reactions in Chapter 2 involved species such as chloride ion, $:\ddot{C}l:^-$. In contrast, the radical processes here feature radicals such as chlorine atoms, $:\ddot{C}l\cdot$, which are *neutral*. *Do not place a negative charge on the halogen atom in a radical reaction!*

The $\Delta H°$ for this transformation is positive; the process is *endothermic* (heat absorbing), and its equilibrium is slightly unfavorable. What is its activation energy, E_a? Is there enough heat to overcome this barrier? In this case, the answer is yes. An orbital description of the transition state (Section 2-1) of hydrogen removal from methane (Figure 3-5) reveals the details

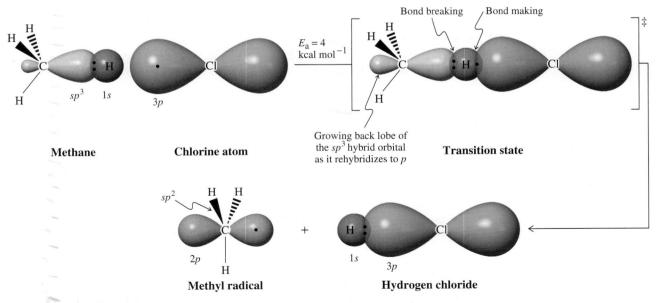

Figure 3-5 Approximate orbital description of the abstraction of a hydrogen atom by a chlorine atom to give a methyl radical and hydrogen chloride. Notice the rehybridization at carbon in the planar methyl radical. The additional three nonbonded electron pairs on chlorine have been omitted. The orbitals are not drawn to scale. The symbol ‡ identifies the transition state.

Figure 3-6 Potential-energy diagram of the reaction of methane with a chlorine atom. Partial bonds in the transition state are depicted by dotted lines. This process, propagation step 1 in the radical chain chlorination of methane, is slightly endothermic.

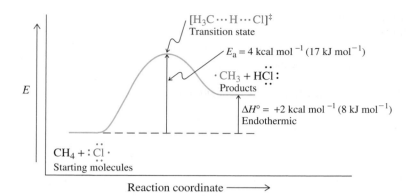

of the process. The reacting hydrogen is positioned between the carbon and the chlorine, partly bound to both: H–Cl bond formation has occurred to about the same extent as C–H bond breaking. The transition state, which is labeled by the symbol ‡, is located only about 4 kcal mol^{-1} above the starting materials. A potential-energy diagram describing this step is shown in Figure 3-6.

Propagation step 1 gives one of the products of the chlorination reaction: HCl. What about the organic product, CH$_3$Cl? Chloromethane is formed in the *second* propagation step. Here the methyl radical abstracts a chlorine atom from one of the starting Cl$_2$ molecules, thereby furnishing chloromethane *and a new chlorine atom*. The latter reenters propagation step 1 to react with a new molecule of methane. Thus, one propagation cycle is closed, and a new one begins, *without the need for another initiation step to take place*. Note how exothermic propagation step 2 is, −27 kcal mol^{-1}. It supplies the overall driving force for the reaction of methane with chlorine.

Propagation step 2: Abstraction of a Cl atom by ·CH$_3$

$$
\begin{array}{ccc}
\overset{\text{H}}{\underset{\text{H}}{\text{H}-\text{C}\cdot}} + \ :\ddot{\text{C}}\text{l}-\ddot{\text{C}}\text{l}: & \longrightarrow & \overset{\text{H}}{\underset{\text{H}}{\text{H}-\text{C}-\text{Cl}}} + \cdot\ddot{\text{C}}\text{l}:
\end{array}
\qquad
\begin{array}{l}
\Delta H^\circ = DH^\circ(\text{Cl}_2) - DH^\circ(\text{CH}_3-\text{Cl}) \\
 = -27 \text{ kcal mol}^{-1}\ (-113 \text{ kJ mol}^{-1})
\end{array}
$$

58   85

DH° (kcal mol^{-1})

The single-barbed "fishhook" arrows in these formulas show the movement of *single* electrons. The lone electron on the methyl carbon combines with one of the two electrons in the Cl–Cl bond, making a new C–Cl bond. Meanwhile, the second electron from the original Cl–Cl bond leaves with the departing free chlorine atom.

Because propagation step 2 is exothermic, the unfavorable equilibrium in the first propagation step is pushed toward the product side by the rapid depletion of its methyl radical product in the subsequent reaction.

$$
\text{CH}_4 + \text{Cl}\cdot \rightleftharpoons \underset{\substack{\text{Slightly}\\\text{unfavorable}}}{\text{CH}_3\cdot + \text{HCl}} \overset{\text{Cl}_2}{\underset{\substack{\text{Very favorable;}\\\text{"drives" first equilibrium}}}{\rightleftharpoons}} \text{CH}_3\text{Cl} + \text{Cl}\cdot + \text{HCl}
$$

The potential-energy diagram in Figure 3-7 illustrates this point by continuing the progress of the reaction begun in Figure 3-6. Propagation step 1 has the higher activation energy and is therefore slower than step 2. The diagram also shows that the overall ΔH° of the reaction is made up of the ΔH° values of the propagation steps: $+2 - 27 = -25$ kcal mol^{-1}. You can see that this is so by adding the equations for the two propagation steps.

$$
\begin{array}{lll}
 & & \Delta H^\circ \ [\text{kcal mol}^{-1}(\text{kJ mol}^{-1})] \\
:\ddot{\text{C}}\text{l}\cdot + \text{CH}_4 \longrightarrow \text{CH}_3\cdot + \text{H}\ddot{\text{C}}\text{l}: & & +2\ (+8) \\
\text{CH}_3\cdot + \ \text{Cl}_2 \longrightarrow \text{CH}_3\ddot{\text{C}}\text{l}: + :\ddot{\text{C}}\text{l}\cdot & & -27\ (-113) \\
\hline
\text{CH}_4 + \ \text{Cl}_2 \longrightarrow \text{CH}_3\ddot{\text{C}}\text{l}: + \text{H}\ddot{\text{C}}\text{l}: & & -25\ (-105)
\end{array}
$$

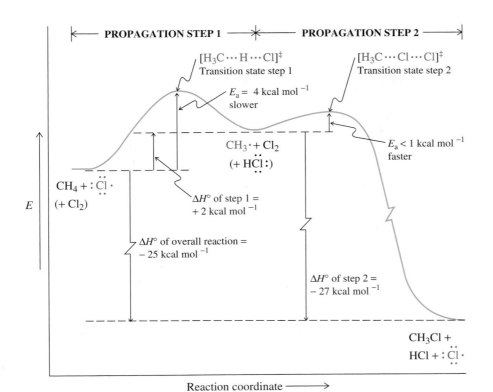

Figure 3-7 Complete potential-energy diagram for the formation of CH_3Cl from methane and chlorine. Propagation step 1 has the higher transition-state energy and is therefore slower. The $\Delta H°$ of the overall reaction $CH_4 + Cl_2 \rightarrow CH_3Cl + HCl$ amounts to -25 kcal mol^{-1} (-105 kJ mol^{-1}), the sum of the $\Delta H°$ values of the two propagation steps.

Experimental observation. Small amounts of *ethane* are identified among the products of chlorination of methane.

Interpretation. Radicals and free atoms are capable of undergoing direct covalent bonding with one another. In the methane chlorination process, three such combination processes are possible, one of which—the reaction of two methyl groups—furnishes ethane. The concentrations of radicals and free atoms in the reaction mixture are very low, however, and hence the chance of one radical or free atom finding another is small. Such combinations are therefore relatively infrequent. When such an event does take place, the propagation of the chains giving rise to the radicals or atoms is terminated. We thus describe these combination processes as **termination** steps.

Chain termination: Radical–radical combination

$$:\overset{..}{\underset{..}{Cl}}\cdot + \cdot\overset{..}{\underset{..}{Cl}}: \longrightarrow Cl-Cl$$

$$:\overset{..}{\underset{..}{Cl}}\cdot + \cdot CH_3 \longrightarrow CH_3-\overset{..}{\underset{..}{Cl}}:$$

$$CH_3\cdot + \cdot CH_3 \longrightarrow CH_3-CH_3$$

The mechanism for the chlorination of methane is an example of a **radical chain mechanism.**

A Radical Chain Mechanism

Initiation	Propagation steps	Chain termination
$X_2 \longrightarrow 2:\overset{..}{\underset{..}{X}}\cdot$	$:\overset{..}{\underset{..}{X}}\cdot + RH \longrightarrow R\cdot + H\overset{..}{\underset{..}{X}}:$	$:\overset{..}{\underset{..}{X}}\cdot + :\overset{..}{\underset{..}{X}}\cdot \longrightarrow X_2$
	$X_2 + R\cdot \longrightarrow R\overset{..}{\underset{..}{X}}: + :\overset{..}{\underset{..}{X}}\cdot$	$R\cdot + :\overset{..}{\underset{..}{X}}\cdot \longrightarrow RX$
		$R\cdot + R\cdot \longrightarrow R_2$

Consumed   Regenerated

Only a few halogen atoms are necessary for initiating the reaction, because the propagation steps are self-sufficient in $:\ddot{X}\cdot$. The first propagation step consumes a halogen atom, the second produces one. The newly generated halogen atom then reenters the propagation cycle in the first propagation step. In this way, a *radical chain* is set in motion that can drive the reaction for many thousands of cycles.

Solved Exercise 3-4 | Working with the Concepts: Radical Chain Mechanisms

Write a detailed mechanism for the light-initiated monochlorination of ethane, which furnishes chloroethane. Calculate $\Delta H°$ for each step.

Strategy

*W*hat the question asks is for you to work by analogy to the mechanism for the chlorination of methane given in this text section.

*H*ow do you begin? Find the overall equation for the reaction of chlorine with methane, but replace methane with ethane and make any other change that follows from that one. Then do the same for the initiation, propagation, and termination steps, again using the mechanism steps in the chlorination of methane as your model.

*I*nformation needed? Some bond energies will be the same and others will be different. Use the data in Tables 3-1 and 3-2, and apply the formula $\Delta H° = \Sigma DH°$ (bonds broken) $- \Sigma DH°$ (bonds formed) to the overall reaction as well as to each individual mechanistic step.

*P*roceed.

Solution

• The reaction equation is

$$CH_3CH_2{-}H + :\ddot{Cl}{-}\ddot{Cl}: \longrightarrow CH_3CH_2{-}\ddot{Cl}: + H{-}\ddot{Cl}:$$

$$\Delta H° = 101 + 58 - 84 - 103 = -28 \text{ kcal mol}^{-1}$$

The reaction is more exothermic than chlorination of methane, mainly because breaking the C–H bond in ethane requires less energy than breaking the C–H bond in methane.

• The initiation step in the mechanism is light-induced dissociation of Cl_2:

$$\text{Initiation} \qquad :\ddot{Cl}{-}\ddot{Cl}: \xrightarrow{h\nu} 2:\ddot{Cl}\cdot \qquad \Delta H° = +58 \text{ kcal mol}^{-1}$$

• Using the propagation steps for chlorination of methane as a model, write analogous steps for ethane. In step 1, a chlorine atom abstracts a hydrogen:

$$\text{Propagation 1} \qquad CH_3\ddot{C}H_2{-}H + \cdot\ddot{Cl}: \longrightarrow CH_3CH_2\cdot + H{-}\ddot{Cl}:$$

$$\Delta H° = 101 - 103 = -2 \text{ kcal mol}^{-1}$$

• In step 2, the ethyl radical formed in step 1 abstracts a chlorine atom from Cl_2:

$$\text{Propagation 2} \qquad CH_3CH_2\cdot + :\ddot{Cl}{-}\ddot{Cl}: \longrightarrow CH_3CH_2{-}\ddot{Cl}: + \cdot\ddot{Cl}:$$

$$\Delta H° = 58 - 84 = -26 \text{ kcal mol}^{-1}$$

Note that the sum of the $\Delta H°$ values for the two propagation steps gives us $\Delta H°$ for the overall reaction. This is because summing the species present in the two propagation steps cancels both ethyl radicals and chlorine atoms, leaving just the overall stoichiometry.

• Finally, we formulate the termination steps as follows:

$$\text{Termination} \qquad :\ddot{Cl}\cdot + \cdot\ddot{Cl}: \longrightarrow Cl_2 \qquad \Delta H° = -58 \text{ kcal mol}^{-1}$$

$$CH_3CH_2\cdot + \cdot\ddot{Cl}: \longrightarrow CH_3CH_2\ddot{Cl}: \qquad \Delta H° = -84 \text{ kcal mol}^{-1}$$

$$CH_3CH_2\cdot + \cdot CH_2CH_3 \longrightarrow CH_3CH_2CH_2CH_3 \qquad \Delta H° = -88 \text{ kcal mol}^{-1}$$

One of the practical problems in chlorinating methane is the control of product selectivity. As mentioned earlier, the reaction does not stop at the formation of chloromethane but continues to form di-, tri-, and tetrachloromethane by further substitution. A practical solution to this problem is the use of a large excess of methane in the reaction. Under such conditions, the reactive intermediate chlorine atom is at any given moment surrounded by many more methane molecules than product CH_3Cl. Thus, the chance of $Cl\cdot$ finding CH_3Cl to eventually make CH_2Cl_2 is greatly diminished, and product selectivity is achieved.

Exercise 3-5 | **Try It Yourself**

Write out the overall equation and the propagation steps of the mechanism for the chlorination of chloromethane to furnish dichloromethane, CH_2Cl_2. (**Caution:** Write out each step of the mechanism separately and completely. Be sure to include full Lewis structures of all species and all appropriate arrows to show electron movement.)

In Summary Chlorine transforms methane into chloromethane. The reaction proceeds through a mechanism in which heat or light causes a small number of Cl_2 molecules to undergo homolysis to chlorine atoms (initiation). The chlorine atoms induce and maintain a radical chain sequence consisting of two (propagation) steps: (1) hydrogen abstraction to generate the methyl radical and HCl and (2) conversion of $CH_3\cdot$ by Cl_2 into CH_3Cl and regenerated $Cl\cdot$. The chain is terminated by various combinations of radicals and free atoms. The heats of the individual steps are calculated by comparing the strengths of the bonds that are being broken with those of the bonds being formed.

3-5 | OTHER RADICAL HALOGENATIONS OF METHANE

Fluorine and bromine, but not iodine, also react with methane by similar radical mechanisms to furnish the corresponding halomethanes. The dissociation energies of X_2 (X = F, Br, I) are lower than that of Cl_2, thus ensuring ready initiation of the radical chain (Table 3-4).

Fluorine is most reactive, iodine least reactive

Let us compare the enthalpies of the first propagation step in the different halogenations of methane (Table 3-5). For fluorine, this step is exothermic by -31 kcal mol^{-1}. We have already seen that, for chlorine, the same step is slightly endothermic; for bromine, it is substantially so, and for iodine even more so. This trend has its origin in the decreasing bond strengths of the hydrogen halides in the progression from fluorine to iodine (Table 3-1). The strong hydrogen–fluorine bond is the cause of the high reactivity of fluorine atoms in hydrogen abstraction reactions. Fluorine is more reactive than chlorine, chlorine is more reactive than bromine, and the least reactive halogen atom is iodine.

The contrast between fluorine and iodine is illustrated by comparing potential-energy diagrams for their respective hydrogen abstractions from methane (Figure 3-8). The highly exothermic reaction of fluorine has a negligible activation barrier. Moreover, in its transition state, the fluorine atom is relatively far from the hydrogen that is being

Table 3-4	$DH°$ Values for the Elemental Halogens
Halogen	$DH°$ [kcal mol^{-1} (kJ mol^{-1})]
F_2	38 (159)
Cl_2	58 (243)
Br_2	46 (192)
I_2	36 (151)

Table 3-5	Enthalpies of the Propagation Steps in the Halogenation of Methane [kcal mol^{-1} (kJ mol^{-1})]			
Reaction	**F**	**Cl**	**Br**	**I**
$:\ddot{X}\cdot + CH_4 \longrightarrow \cdot CH_3 + H\ddot{X}:$	-31 (-130)	$+2$ ($+8$)	$+18$ ($+75$)	$+34$ ($+142$)
$\cdot CH_3 + X_2 \longrightarrow CH_3\ddot{X}: + :\ddot{X}\cdot$	-72 (-301)	-27 (-113)	-24 (-100)	-21 (-88)
$CH_4 + X_2 \longrightarrow CH_3\ddot{X}: + H\ddot{X}:$	-103 (-431)	-25 (-105)	-6 (-25)	$+13$ ($+54$)

Figure 3-8 Potential-energy diagrams: *(left)* the reaction of a fluorine atom with CH_4, an exothermic process with an early transition state; and *(right)* the reaction of an iodine atom with CH_4, an endothermic transformation with a late transition state. Both are thus in accord with the Hammond postulate.

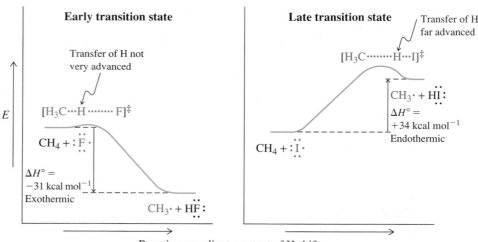

Early transition state   Late transition state

Reaction coordinate: extent of H shift

**Relative
Reactivities of X· in
Hydrogen Abstractions**
$F· > Cl· > Br· > I·$

Decreasing reactivity ▶

transferred, and the $H–CH_3$ distance is only slightly greater than that in CH_4 itself. Why should this be so? The $H–CH_3$ bond is about 30 kcal mol^{-1} (125 kJ mol^{-1}) weaker than that of $H–F$ (Table 3-1). Only a small shift of the H toward the F· is necessary for bonding between the two to overcome the bonding between hydrogen and carbon. If we view the reaction coordinate as a measure of the degree of hydrogen shift from C to F, the transition state is reached *early* and is much closer in appearance to the starting materials than to the products. ***Early transition states*** *are frequently characteristic of fast, exothermic processes.*

On the other hand, reaction of I· with CH_4 has a very high E_a [at least as large as its endothermicity, +34 kcal mol^{-1} (+142 kJ mol^{-1}); Table 3-5]. Thus, the transition state is not reached until the H–C bond is nearly completely broken and the H–I bond is almost fully formed. The transition state is said to occur *late* in the reaction pathway. It is substantially further along the reaction coordinate and is much closer in structure to the products of this process, CH_3· and HI. ***Late transition states*** *are frequently typical of relatively slow, endothermic reactions.* Together these rules concerning early and late transition states are known as the **Hammond* postulate.**

The second propagation step is exothermic

Let us now consider the second propagation step for each halogenation in Table 3-5. This process is exothermic for all the halogens. Again, the reaction is fastest and most exothermic for fluorine. The combined enthalpies of the two steps for the fluorination of methane result in a $\Delta H°$ of −103 kcal mol^{-1} (−431 kJ mol^{-1}). The formation of chloromethane is less exothermic and that of bromomethane even less so. In the latter case, the appreciably endothermic nature of the first step [$\Delta H° = +18$ kcal mol^{-1} (+75 kJ mol^{-1})] is barely overcome by the enthalpy of the second [$\Delta H° = −24$ kcal mol^{-1} (−100 kJ mol^{-1})], resulting in an energy change of only −6 kcal mol^{-1} (−25 kJ mol^{-1}) for the overall substitution. Finally, iodine does not react with methane to furnish methyl iodide and hydrogen iodide: The first step costs so much energy that the second step, although exothermic, cannot drive the reaction.

Exercise 3-6

When methane is allowed to react with an equimolar mixture of chlorine and bromine, only hydrogen abstraction by chlorine atoms is observed. Explain.

In Summary Fluorine, chlorine, and bromine react with methane to give halomethanes. All three reactions follow the radical chain mechanism described for chlorination. In these processes, the first propagation step is always the slower of the two. It becomes more exo-

---

*Professor George S. Hammond (1921–2005), Georgetown University, Washington, D.C.

thermic and its activation energy decreases in the progression from bromine to chlorine to fluorine. This trend explains the relative reactivity of the halogens, fluorine being the most reactive. Iodination of methane is endothermic and does not occur. Strongly exothermic reaction steps are often characterized by early transition states. Conversely, endothermic steps typically have late transition states.

3-6 | KEYS TO SUCCESS: USING THE "KNOWN" MECHANISM AS A MODEL FOR THE "UNKNOWN"

Section 3-4 presented the mechanism for the chlorination of methane in full, step-by-step detail. Section 3-5 discussed the reactions of methane with the other three halogens, but it did not actually display either the full equations for any of these reactions, nor did it illustrate any of the individual steps of their mechanisms. Why not? The reactions of methane with the four halogens all proceed through mechanisms that are *qualitatively identical* to one another. To write a mechanism for the reaction of methane with fluorine, bromine, or iodine you need only (1) copy the mechanism for reaction with chlorine and (2) replace every Cl in those equations with the symbol for the new halogen. The energy values will be different because the bond strengths are different, but the overall *appearance* of the mechanisms will be the same. Solved Exercise 3-4 already illustrated this strategy for the chlorination of ethane. Let us try some examples:

1. **Write the initiation step for fluorination of methane.**

   We use the light-induced dissociation of Cl_2 as our model:

   $$:\ddot{F}\overgroup{\frown}\ddot{F}: \xrightarrow{h\nu} 2 \cdot \ddot{F}:$$

2. **Write the second propagation step for bromination of methane.**

   Methyl radical + Cl_2 is our model:

   $$H_3C\cdot \; + \; :\ddot{Br}\overgroup{\frown}\ddot{Br}: \longrightarrow H_3C\text{---}\ddot{Br}: + \cdot\ddot{Br}:$$

   Don't forget: The Br that forms is a neutral atom, ***not a negatively charged bromide ion!***

3. **Write any termination step for iodination of methane.**

   Take your pick and just replace any Cl by I:

   $$H_3C\cdot \overgroup{\frown} \cdot \ddot{I}: \longrightarrow H_3C\text{---}I$$

In Summary Reasoning out complex problems by analogy, using models that have been presented previously, is an efficient method for both solving the problems and reinforcing your study of the patterns that govern the mechanisms of reactions.

3-7 | CHLORINATION OF HIGHER ALKANES: RELATIVE REACTIVITY AND SELECTIVITY

What happens in the radical halogenation of other alkanes? Will the different types of R–H bonds—namely, primary, secondary, and tertiary—react in the same way as those in methane? As we saw in Exercise 3-4, the monochlorination of ethane gives chloroethane as the product.

Reaction

Chlorination of Ethane

$$CH_3CH_3 + Cl_2 \xrightarrow{\Delta \text{ or } h\nu} CH_3CH_2Cl + HCl \qquad \Delta H° = -28 \text{ kcal mol}^{-1}$$
$$\text{Chloroethane} \qquad (-117 \text{ kJ mol}^{-1})$$

This reaction proceeds by a radical chain mechanism analogous to the one observed for methane. As in methane, all of the hydrogen atoms in ethane are indistinguishable from one another. Therefore, we observe only one monochlorination product, chloroethane, regardless of which hydrogen is initially abstracted by chlorine in the first propagation step.

Mechanism

Propagation Steps in the Mechanism of the Chlorination of Ethane

$$CH_3CH_3 + :\overset{..}{\underset{..}{Cl}}\cdot \longrightarrow CH_3CH_2\cdot + H\overset{..}{\underset{..}{Cl}}: \quad \Delta H° = -2 \text{ kcal mol}^{-1}$$
$$(-8 \text{ kJ mol}^{-1})$$

$$CH_3CH_2\cdot + Cl_2 \longrightarrow CH_3CH_2\overset{..}{\underset{..}{Cl}}: + :\overset{..}{\underset{..}{Cl}}\cdot \quad \Delta H° = -26 \text{ kcal mol}^{-1}$$
$$(-109 \text{ kJ mol}^{-1})$$

What can be expected for the next homolog, propane? The eight hydrogen atoms in propane fall into two groups: six primary hydrogens on carbons 1 and 3, and two secondary hydrogens on carbon 2. If chlorine atoms were to abstract and replace primary and secondary hydrogens at equal rates, we should expect to find a product mixture containing three times as much 1-chloropropane (75%) as 2-chloropropane (25%).

Expected Result from Chlorination of Propane if All Hydrogens React at the Same Rate

$$\text{Cl}_2 + CH_3CH_2CH_3 \xrightarrow{hv} CH_3CH_2CH_2Cl + \overset{\overset{\displaystyle Cl}{\displaystyle |}}{CH_3CHCH_3} + HCl$$

Propane	**1-Chloropropane**	**2-Chloropropane**
Six primary hydrogens (blue)		
Two secondary hydrogens (red)	Expected (statistical) ratio	
Ratio = 3:1	75% : 25%	

We call this outcome a **statistical product ratio** because it derives from the statistical fact that it is three times as likely for a chlorine atom to collide with a primary hydrogen in propane, of which there are six, as with a secondary hydrogen atom, of which there are only two. But is this the outcome that is observed? Actually, no.

Secondary C–H bonds are more reactive than primary ones

As we learned in Section 3-1, secondary C–H bonds are weaker ($DH° = 98.5$ kcal mol^{-1}) than primary ones ($DH° = 101$ kcal mol^{-1}). Abstraction of a secondary hydrogen is therefore more exothermic and proceeds with a smaller activation barrier (Figure 3-9). Thus secondary hydrogens *react faster* with chlorine than primary hydrogens. The result is that more 2-chloropropane forms than expected on the basis of the simple statistical ratio.

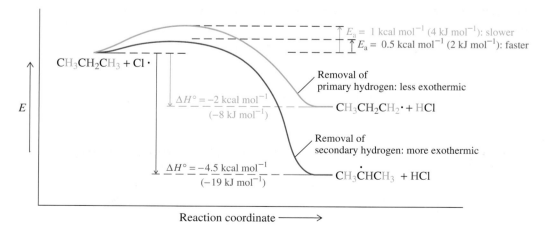

Figure 3-9 Hydrogen abstraction by a chlorine atom from the secondary carbon in propane is more exothermic and faster than that from the primary carbon.

Actual Experimental Result from Chlorination of Propane

$$Cl_2 + CH_3CH_2CH_3 \xrightarrow{h\nu} CH_3CH_2CH_2Cl + CH_3\overset{\overset{\displaystyle Cl}{|}}{C}HCH_3 + HCl$$

	43%		57%	
	1-Chloropropane		**2-Chloropropane**	
Expected (statistical) ratio	75%	:	25%	
C–H bond reactivity	Less (primary)		More (secondary)	
Experimental ratio	43%	:	57%	

Can we use this result to calculate the *relative reactivities of secondary and primary hydrogens* in the chlorination of propane? Yes: Since 6 primary hydrogens contribute to the 43% yield of 1-chloropropane, we can say that *one* primary hydrogen leads to 43% divided by 6 = about 7% of this product. Similarly, 2 secondary hydrogens give us a total of 57% yield of 2-chloropropane, so each one of them contributes 57% divided by 2, or about 28% yield to the total. Thus it follows that in the chlorination of propane, the relative reactivity of each secondary hydrogen compared to each primary hydrogen is 28/7 = 4. We say that chlorine is **selective** in preferring to remove secondary hydrogens versus primary hydrogens, and that its **selectivity** is 4:1.

Are *all* secondary hydrogens removed four times faster than all primary hydrogens in *all* radical chain reactions? The results above do hold in general for *chlorinations of alkanes* done under similar conditions (initiation by light at 25°C). At higher temperatures, however, collisions are more energetic, and the difference in ease of breaking of secondary *vs.* primary C–H bonds has less of an effect on the final product ratio. For example, at 600°C virtually every collision between a chlorine atom and *either* a secondary or a primary hydrogen in propane leads to reaction. Chlorination is therefore **unselective** at high temperatures, giving a statistical ratio of products. Another factor that we will encounter in subsequent sections is the effect on selectivity of changing the reacting species (to a different halogen, for example).

> Chlorine atoms remove secondary hydrogens from alkanes *four times faster* than they remove primary hydrogens.

Exercise 3-7

What do you expect the products of monochlorination of butane to be? In what ratio will they be formed at 25°C?

Tertiary C–H bonds are more reactive than either secondary or primary ones

Tertiary C–H bonds are even weaker ($DH° = 96.5$ kcal mol^{-1}) than either secondary ($DH° = 98.5$ kcal mol^{-1}) or primary ones ($DH° = 101$ kcal mol^{-1}). To determine the effect of this difference, let us consider the chlorination under light at 25°C of 2-methylpropane, a molecule containing one tertiary and nine primary hydrogens.

Chlorination of 2-Methylpropane: Statistical Expectation *vs.* Actual Experimental Result

$$Cl_2 + CH_3-\overset{\overset{\displaystyle CH_3}{|}}{\underset{\underset{\displaystyle CH_3}{|}}{C}}-H \xrightarrow{h\nu} ClCH_2-\overset{\overset{\displaystyle CH_3}{|}}{\underset{\underset{\displaystyle CH_3}{|}}{C}}-H + CH_3-\overset{\overset{\displaystyle CH_3}{|}}{\underset{\underset{\displaystyle CH_3}{|}}{C}}-Cl + HCl$$

		64%		36%	
2-Methylpropane		**1-Chloro-2-methylpropane**		**2-Chloro-2-methylpropane**	
Nine primary hydrogens (blue)					
One tertiary hydrogen (red)					
Ratio = 9:1					
Expected (statistical) ratio		90%	:	10%	
C–H bond reactivity		Less (primary)		More (tertiary)	
Experimental ratio		64%	:	36%	

As we did for propane, we use the experimental result to determine the relative reactivity at 25°C of a tertiary hydrogen relative to a primary one. Each of the nine primary hydrogens

> Chlorine atoms remove tertiary hydrogens from alkanes *five times faster* than they remove primary hydrogens.

contributes 64%/9, or about 7% to the final 1-chloro-2-methylpropane product formation. The single tertiary hydrogen is responsible for all 36% of the 2-chloro-2-methylpropane formed. Therefore, the tertiary hydrogen is 36/7, or about 5 times more reactive toward chlorination than one primary hydrogen.

Overall, at 25°C we can say that the relative reactivities of the three types of alkane C–H bonds in chlorination are approximately

$$\text{Tertiary} : \text{secondary} : \text{primary} = 5 : 4 : 1$$

← **Increase in reactivity of R-H**

The result agrees qualitatively quite well with the relative reactivity expected from consideration of bond strength: The tertiary C–H bond is weaker than the secondary, and the secondary in turn is weaker than the primary.

Solved Exercise 3-8 | Working with the Concepts: Determining Product Ratios from Selectivity Data

Consider the monochlorination of 2-methylbutane. How many different products do you expect? Estimate their yields.

Strategy

The first step is to identify all nonequivalent groups of hydrogens in the starting alkane and count how many hydrogens are in each group. This will tell you how many different products to expect from the reaction. To calculate the *relative* yield of each product, multiply the number of hydrogens in the *starting alkane* that give rise to that product by the relative reactivity corresponding to that type of hydrogen (primary, secondary, or tertiary). To find the absolute percentage yield, normalize to 100% by dividing each relative yield by the sum of the yields of all the products.

Solution

- 2-Methylbutane contains nine primary hydrogens, two secondary hydrogens, and one tertiary hydrogen. However, the nine primary hydrogens are not all equivalent; that is, they are not all indistinguishable from one another. Instead, we can discern two distinct groups of primary hydrogens. How do we know these groups are distinct from each other? Chlorination of any one of the six hydrogens in group A gives 1-chloro-2-methylbutane, whereas reaction of any one of the three in group B gives 1-chloro-3-methylbutane. These products are constitutional isomers of each other and have different names—this is the clue that tells us that they arise from replacement of hydrogens in distinct, nonequivalent groups. Thus, we can obtain four structurally different products instead of three:

Chlorination of 2-Methylbutane

1-Chloro-2-methylbutane (Chlorination at A) | **1-Chloro-3-methylbutane** (Chlorination at B) — Substitution at primary carbons

2-Chloro-3-methylbutane (Chlorination at C) — Substitution at the secondary carbon

2-Chloro-2-methylbutane (Chlorination at D) — Substitution at the tertiary carbon

- Carrying out the calculations as described in the **strategy** above gives us the following table:

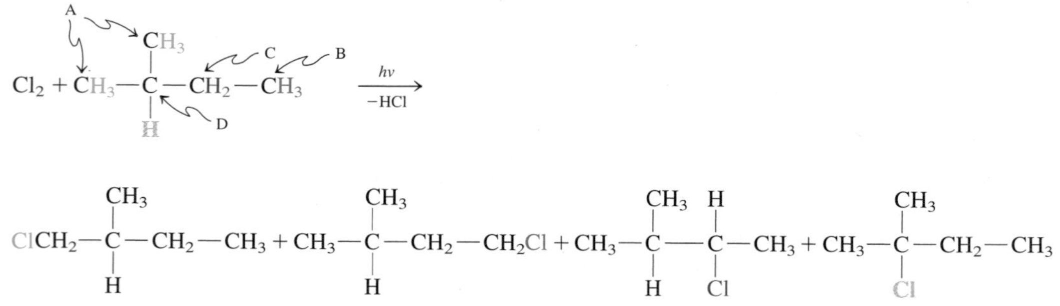

Product	Relative yield	Absolute yield
1-Chloro-2-methylbutane (A, six primary)	$6 \times 1 = 6$	$6/22 = 0.27 = 27\%$
1-Chloro-3-methylbutane (B, three primary)	$3 \times 1 = 3$	$3/22 = 0.14 = 14\%$
2-Chloro-3-methylbutane (C, two secondary)	$2 \times 4 = 8$	$8/22 = 0.36 = 36\%$
2-Chloro-2-methylbutane (D, one tertiary)	$1 \times 5 = 5$	$5/22 = 0.23 = 23\%$
Sum of relative yields of all four products	22	

In Summary The relative reactivity of primary, secondary, and tertiary hydrogens follows the trend expected on the basis of their relative C–H bond strengths. Relative reactivity ratios can be calculated by factoring out statistical considerations. These ratios are temperature dependent, with greater selectivity at lower temperatures.

3-8 | SELECTIVITY IN RADICAL HALOGENATION WITH FLUORINE AND BROMINE

How selectively do halogens other than chlorine halogenate the alkanes? Table 3-5 and Figure 3-8 show that fluorine is the most reactive halogen: Hydrogen abstraction is highly exothermic and has negligible activation energy. Conversely, bromine is much less reactive, because the same step has a large positive $\Delta H°$ and a high activation barrier. Does this difference affect their selectivity in halogenation of alkanes?

To answer this question, consider the reactions of fluorine and bromine with 2-methylpropane. Single fluorination at 25°C furnishes two possible products, in a ratio very close to that expected statistically.

Fluorination of 2-Methylpropane

$$F_2 + (CH_3)_3CH \xrightarrow{h\nu} FCH_2 - \underset{\underset{CH_3}{|}}{\overset{\overset{CH_3}{|}}{C}} - H \quad + \quad (CH_3)_3CF \quad + \quad HF$$

	86%		14%
	1-Fluoro-2-methylpropane (Isobutyl fluoride)		**2-Fluoro-2-methylpropane** (*tert*-Butyl fluoride)
Expected (statistical) ratio	90%	:	10%
C–H bond reactivity	Less (primary)		More (tertiary)
Experimental ratio	86%	:	14%

Fluorine thus displays very little selectivity. Why? Because the transition states for the two competing processes are reached very early, their energies and structures are similar to each other, as well as similar to those of the starting material (Figure 3-10).

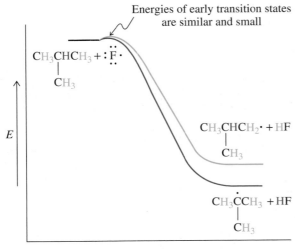

Reaction coordinate ⟶

Figure 3-10 Potential-energy diagram for the abstraction of a primary or a tertiary hydrogen by a fluorine atom from 2-methyl-propane. The energies of the respective early transition states are almost the same and barely higher than that of starting material (i.e., both E_a values are close to zero), resulting in little selectivity.

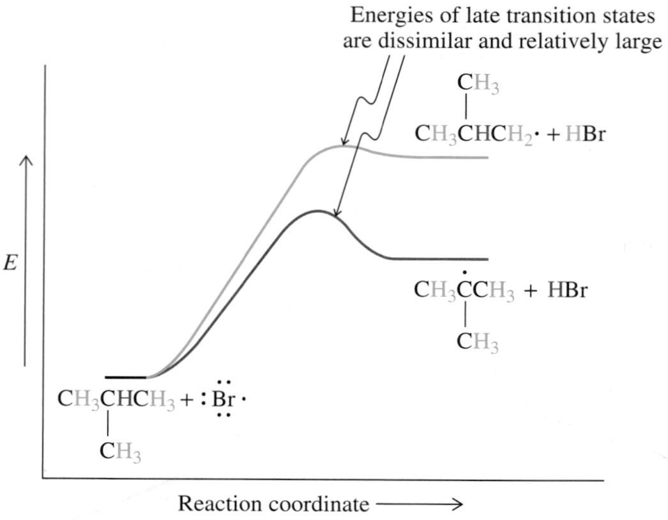

Figure 3-11 Potential-energy diagram for the abstraction of a primary or a tertiary hydrogen of 2-methylpropane by a bromine atom. The two late transition states are dissimilar in energy, indicative of the energy difference between the resulting primary and tertiary radicals, respectively, leading with greater selectivity to the products.

Energies of late transition states are dissimilar and relatively large

$$CH_3CHCH_2 \cdot + HBr$$ (with CH_3 substituent)

$$CH_3\overset{\cdot}{C}CH_3 + HBr$$ (with CH_3 substituent)

$$CH_3CHCH_3 + \overset{..}{:}\overset{.}{\underset{..}{Br}}\cdot$$ (with CH_3 substituent)

E

Reaction coordinate ⟶

Conversely, *bromination of the same compound is highly selective,* giving the tertiary bromide almost exclusively. Hydrogen abstractions by bromine have *late* transition states in which extensive C–H bond breaking and H–Br bond making have occurred. Thus, their respective structures and energies resemble those of the corresponding radical products. As a result, the activation barriers for the reaction of bromine with primary and tertiary hydrogens, respectively, will differ by almost as much as the difference in stability between primary and tertiary radicals (Figure 3-11), a difference leading to the observed high selectivity (more than 1700:1).

> Bromine atoms remove secondary hydrogens *80 times faster* and tertiary hydrogens *1700 times faster* than they remove primary hydrogens.

Bromination of 2-Methylpropane

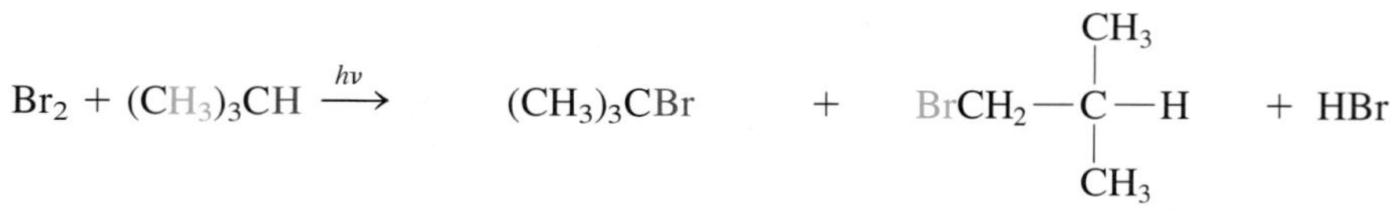

$$Br_2 + (CH_3)_3CH \xrightarrow{h\nu} (CH_3)_3CBr + BrCH_2-\overset{\overset{\displaystyle CH_3}{|}}{\underset{\underset{\displaystyle CH_3}{|}}{C}}-H + HBr$$

	>99%		<1%
	2-Bromo-2-methylpropane (*tert*-Butyl bromide)		1-Bromo-2-methylpropane (Isobutyl bromide)
Expected (statistical) ratio	10%	:	90%
C–H bond reactivity	More (tertiary)		Less (primary)
Experimental ratio	99.94%	:	0.06%

In Summary Increased reactivity goes hand in hand with reduced selectivity in radical substitution reactions. Fluorine and chlorine, the more reactive halogens, discriminate between the various types of C–H bonds much less than does the less reactive bromine (Table 3-6).

Table 3-6	Relative Reactivities of the Four Types of Alkane C–H Bonds in Halogenations		
C–H bond	F· (25°C, gas)	Cl· (25°C, gas)	Br· (150°C, gas)
CH_3–H	0.5	0.004	0.002
RCH_2–H[a]	1	1	1
R_2CH–H	1.2	4	80
R_3C–H	1.4	5	1700

[a]For each halogen, reactivities with four types of alkane C–H bonds are normalized to the reactivity of the primary C–H bond.

3-9 | SYNTHETIC RADICAL HALOGENATION

Halogenation converts nonfunctional alkanes into functionalized haloalkanes, which (as we shall soon see) are useful starting materials for a variety of subsequent transformations. Thus, devising successful and cost-effective halogenations is of practical value. To do so, we must take into account safety, convenience, selectivity, efficiency, and cost of starting materials and reagents—considerations of green chemistry (see Real Life 3-1).

Fluorinations are unattractive, because fluorine is relatively expensive and corrosive. Even worse, its reactions are often violently uncontrollable. Radical iodinations, at the other extreme, fail because of unfavorable thermodynamics.

Chlorinations are important, particularly in industry, simply because chlorine is inexpensive. (It is prepared by electrolysis of hydrogen chloride, HCl.) The drawback to chlorination is low selectivity, so the process results in mixtures of isomers that are difficult to separate. To circumvent the problem, an alkane that contains a single type of hydrogen can be used as a substrate, thus giving (at least initially) a single product. Cyclopentane is one such alkane. We depict it using the bond-line notation (Section 1-9).

Chlorination of a Molecule with Only One Type of Hydrogen

Cyclopentane   Chlorocyclopentane
(Large excess)

To minimize production of compounds with more than one chlorine atom, chlorine is used as the limiting reagent (Section 3-4). Even then, multiple substitution can complicate the reaction. Conveniently, the more highly chlorinated products have higher boiling points and can be separated by distillation.

| Solved Exercise 3-10 | Working with the Concepts: Evaluating Synthetic Utility |

Would you consider the monochlorination of methylcyclopentane (margin) to be a synthetically useful reaction?

Strategy
Synthetically useful reactions are ones that produce a single product with high selectivity and in good yield. Does this one? The starting compound contains 12 hydrogen atoms. Consider the product of replacement of each of these hydrogens by chlorine. Are they all the same, or are they structurally different? If more than one can form, we must estimate their relative amounts.

Solution
The molecule's 12 hydrogens divide into three primary (on the CH_3 group), one tertiary (on C1), and eight secondary (on C2–C5). In addition, the eight secondary hydrogens are divided into two groups, four on C2 and C5, and four on C3 and C4. Thus several isomeric products must result from monochlorination. The only way this reaction can still be synthetically useful is if one of these isomers forms in much higher yield than all of the others.

CH_3

Methylcyclopentane

Recalling the relative reactivities of hydrogens toward chlorination (tertiary = 5, secondary = 4, and primary = 1), it is evident that all of the above products will form in significant amounts. By multiplying the number of hydrogens in each group in the starting structure by the relative reactivity corresponding to its type, we find that the actual ratios for A, B, C, and D will be 5 (1 × 5) : 3 (3 × 1) : 16 (4 × 4) : 16 (4 × 4). This process will *not* be synthetically useful!

REAL LIFE: MEDICINE 3-2 | Chlorination, Chloral, and DDT: The Quest to Eradicate Malaria

1,1,1-Trichloro-2,2-bis(4-chlorophenyl)ethane
(DDT, for "dichlorodiphenyltrichloroethane")

Chlorination of ethanol to produce trichloroacetaldehyde, CCl_3CHO, was first described in 1832. Its hydrated form is called chloral and is a powerful hypnotic with the nickname "knockout drops." Chloral is a key reagent in the synthesis of DDT, which was first prepared in 1874 and was demonstrated by Paul Mueller* in 1939 to be a powerful insecticide. The use of DDT in the suppression of insect-borne diseases was perhaps best described in a 1970 report by the U.S. National Academy of Sciences: "To only a few chemicals does man owe as great a debt as to DDT. . . . In little more than two decades, DDT has prevented 500 million human deaths, due to malaria, that otherwise would have been inevitable." DDT effectively kills the *Anopheles* mosquito, the main carrier of the malaria parasite. Malaria afflicts hundreds of millions of people worldwide and claims over 2 million lives each year, mostly children under the age of five.

Although its toxicity to mammals is low, DDT is rather resistant to biodegradation. Its accumulation in the food chain makes it a hazard to birds and fish; in particular, DDT interferes with proper eggshell development in many species. DDT was banned by the U.S. Environmental Protection Agency in 1972. However, because of its remarkable efficacy in controlling malaria, 12 countries in which the disease is a major health problem are still using DDT, albeit in a highly controlled manner.

Normal peregrine falcon egg on the left and DDT poisoned egg on the right.

*Dr. Paul Mueller (1899–1965), J. R. Geigy Co., Basel, Switzerland, Nobel Prize 1948 (physiology or medicine).

Exercise 3-11 | Try It Yourself

Would you expect either monochlorination or monobromination of 2,3-dimethylbutane to be a synthetically useful process?

Because bromination is selective (and bromine is a liquid), it is frequently the method of choice for halogenating an alkane on a relatively small scale in the research laboratory. Reaction occurs at the more substituted carbon, even in statistically unfavorable situations. Typical solvents are chlorinated methanes (CCl_4, $CHCl_3$, CH_2Cl_2), which are comparatively unreactive with bromine.

In Summary Even though it is more expensive, bromine is the reagent of choice for selective radical halogenations. Chlorinations furnish product mixtures, a problem that can be minimized by choosing alkanes with only one type of hydrogen and treating them with a deficiency of chlorine.

3-10 | SYNTHETIC CHLORINE COMPOUNDS AND THE STRATOSPHERIC OZONE LAYER

We have seen that bond homolysis can be caused by heat and light. Such chemical events can occur on a grand scale in nature and may have serious environmental consequences. This section explores an example of radical chemistry that has had a significant effect on our lives and will continue to do so for at least the next 50 years.

The ozone layer shields the Earth from high-energy ultraviolet light

Earth's atmosphere consists of several distinct layers. The lowest layer, the troposphere, which extends to about 15 km in altitude, is the region where weather occurs. Above it, extending to about 50 km, is the stratosphere. Its density is too low to sustain life, but it is the home of the **ozone layer,** which is critical to life on Earth. Ozone (O_3) forms in the stratosphere when high-energy solar radiation splits O_2 into oxygen atoms, which may react with additional molecules of O_2. The ozone thus generated absorbs ultraviolet (UV) light in the 200- to 300-nm wavelength range. Such radiation is capable of destroying the complex biomolecules necessary for living systems. Ozone serves as a natural atmospheric filter, blocking up to 99% of this light from reaching the surface of the Earth and thus protecting life from damage.

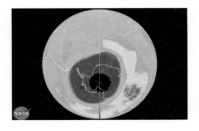

A color-enhanced view of the upper atmosphere above Antarctica shows that region in which the concentration of ozone has dropped below 35% of normal (violet and gray areas). The ozone hole has shown considerable variation in size and shape, reaching its greatest extent—12 million square miles (30 million square km)—in both 2003 and 2006. In 2011 the hole peaked at 10 million square miles (25 million square km), still twice the size of Antarctica and large enough to pose a hazard to people in South America.

Formation of Ozone in the Stratosphere and Its Absorption of Dangerous UV Light

Ozone formation: (1) $O_2 \xrightarrow{h\nu} \cdot \ddot{\underset{..}{O}} \cdot$ followed by (2) $O_2 + \cdot \ddot{\underset{..}{O}} \cdot \longrightarrow O_3$
Ozone

Absorption of dangerous UV light by ozone: $O_3 \xrightarrow{h\nu} \cdot \ddot{\underset{..}{O}} \cdot + O_2$
Ozone

CFCs release chlorine atoms upon UV irradiation

Chlorofluorocarbons (CFCs), also known as **freons,** are halogenated alkanes containing fluorine and chlorine. Until recently, CFCs were the most widely used synthetic organic compounds in modern society. Their ability to absorb large quantities of heat upon vaporization made them popular refrigerants. Remarkably, virtually every country of the world agreed in 1987 to take CFCs completely out of use. Why? This event dates its origins to the late 1960s and early 1970s, when chemists Johnston, Crutzen, Rowland, and Molina* uncovered radical mechanisms that could convert CFCs into reactive species capable of destroying ozone in the Earth's stratosphere (Figure 3-12).

Upon irradiation by UV light from the sun, the C–Cl bonds in CFC molecules undergo homolysis, giving rise to chlorine atoms.

Common CFCs

$CFCl_3$
CFC-11 (Freon 11)

CF_2Cl_2
CFC-12 (Freon 12)

CF_3Cl
CFC-13 (Freon 13)

$CFCl_2CF_2Cl$
CFC-113 (Freon 113)

Initiation step: Freon 13 is dissociated by sunlight

$$F_3C \!-\! \ddot{\underset{..}{C}}l\!: \xrightarrow{h\nu} F_3C \cdot + :\!\ddot{\underset{..}{C}}l \cdot$$

Chlorine atoms, in turn, react efficiently with ozone in a radical chain sequence.

*Professor Harold S. Johnston (1920–2012), University of California at Berkeley; Professor Paul Crutzen (b. 1933), Max Planck Institute, Mainz, Germany, Nobel Prize 1995 (chemistry); Professor F. Sherwood Rowland (1927–2012), University of California at Irvine, Nobel Prize 1995 (chemistry); Professor Mario Molina (b. 1943), Massachusetts Institute of Technology, Nobel Prize 1995 (chemistry).

Figure 3-12 Ozone destroying chemicals emanate from Earth to reach the stratosphere.

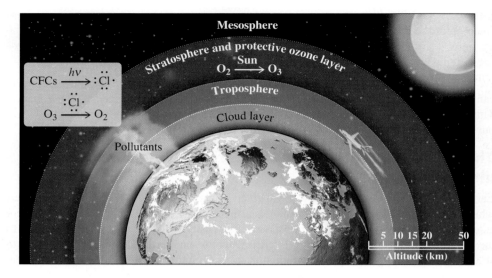

Propagation steps: Ozone is decomposed by a radical chain reaction

$$:\ddot{\underset{..}{C}}l\cdot + O_3 \longrightarrow :\ddot{\underset{..}{C}}l{-}\ddot{\underset{..}{O}}\cdot + O_2$$

$$:\ddot{\underset{..}{C}}l{-}\ddot{\underset{..}{O}}\cdot + O \longrightarrow O_2 + :\ddot{\underset{..}{C}}l\cdot$$

Sunburned whales are appearing off the coast of Mexico. The culprit? The hole in the ozone layer. Whales are particularly vulnerable to the sun's damage in part because they need to spend extended periods of time on the ocean's surface to breathe, socialize, and feed their young. The worry is that this skin damage will lead to skin cancer.

CFC Substitutes

CH$_2$FCF$_3$

HFC-134a

CHClF$_2$

HCFC-22

CHCl$_2$CF$_3$

HCFC-123

CH$_3$CCl$_2$F

HCFC-141b

CH$_3$CClF$_2$

HCFC-142b

The net result of these two steps is the conversion of a molecule of ozone and an oxygen atom into two molecules of ordinary oxygen. As in the other radical chain processes that we have examined in this chapter, however, the reactive species consumed in one propagation step ($:\ddot{\underset{..}{C}}l\cdot$) is regenerated in the other. As a consequence, a small concentration of chlorine atoms is capable of destroying many molecules of ozone. Does such a process actually occur in the atmosphere?

Total stratospheric ozone has decreased dramatically

Since measurements of atmospheric composition were first made, significant decreases in stratospheric ozone have been recorded. Seasonal variations over the poles have been the most noticeable. In early spring in the Southern Hemisphere, ozone levels over the polar regions drop dramatically, resulting in an "ozone hole" of enormous size containing less than 15% of the normal ozone concentration. In the past decade, the extent of this region has often exceeded 10 million square miles—2.5 times the area of Europe—which has put residents of southern South America at risk of exposure to dangerous levels of UV radiation at wavelengths most associated with eye damage and skin cancer. Evidence of a northern springtime "ozone hole" has been accumulating recently as well, affecting potentially hundreds of millions of people. Average worldwide ozone reduction peaked at about 6% in the 1990s but had lessened to about 4% by 2010; each reduction of stratospheric ozone density of 1% is estimated to give rise to a 2–3% rise in skin cancers. Stratospheric ozone levels are expected to return to those of the early 1980s by about mid-century.

Systematic studies carried out over the past 25 years have left no doubt that chlorine atoms derived from human-made substances such as CFCs are largely responsible. In the extreme cold of the polar stratosphere in late winter and early spring, clouds containing nitrogen oxides form that enhance the ozone-destroying effects of chlorine. Satellite measurements of ClO correlate directly with ozone depletion values. Furthermore, the observation of stratospheric HF, which has no other atmospheric source besides the light-induced breakdown of CFCs in the presence of hydrocarbons, strongly supports these conclusions.

The world continues to search for CFC substitutes

The Montreal Protocol on Substances That Deplete the Ozone Layer was signed in 1987 and called for a 50% reduction in output of CFCs by 1998. Increasingly alarming measurements of ozone depletion led to amendments that ultimately set a final full phaseout date of

December 31, 1995. Since that time, first hydrochlorofluorocarbons (HCFCs) and later hydro-fluorocarbons (HFCs) have been developed as CFC replacements. The HCFCs are destroyed by light at lower altitudes than CFCs, and thus pose lesser threats to the ozone layer. They are scheduled to be phased out of use by 2030. HFC-134a has become a popular CFC substitute in refrigerators and air conditioners in the United States. HFCs are not ozone depletors but are potent greenhouse gases and thus are not long-term solutions as CFC replacements.

3-11 | COMBUSTION AND THE RELATIVE STABILITIES OF ALKANES

Let us review what we have learned in this chapter so far. We started by defining bond strength as the energy required to cleave a molecule homolytically. Some typical values were then presented in Tables 3-1 and 3-2 and explained through a discussion of relative radical stabilities, a major factor being the varying extent of hyperconjugation. We then used this information to calculate the $\Delta H°$ values of the steps making up the mechanism of radical halogenation, a discussion leading to an understanding of reactivity and selectivity. It is clear that knowing bond-dissociation energies is a great aid in the thermochemical analysis of organic transformations, an idea that we shall explore on numerous occasions later on. How are these numbers found experimentally?

Chemists determine bond strengths by first establishing the relative energy contents of entire molecules, or their relative positions along the energy axis in our potential-energy diagrams. The reaction chosen for this purpose is complete oxidation (literally, "burning"), or **combustion,** a process common to almost all organic structures, in which all carbon atoms are converted into CO_2 (gas) and all of the hydrogens into H_2O (liquid).

Both products in the combustion of alkanes have a very low energy content, and hence their formation is associated with a large negative $\Delta H°$, released as heat.

$$2\ C_nH_{2n+2} + (3n + 1)\ O_2 \longrightarrow 2n\ CO_2 + (2n + 2)\ H_2O + \text{heat of combustion}$$

The heat released in the burning of a molecule is called its **heat of combustion, $\Delta H°_{comb}$.** Many heats of combustion have been measured with high precision, thus allowing comparisons of the relative energy content of the alkanes (Table 3-7) and other compounds.

Big Mac
kcal 490

Really ?

Our bodies use the caloric content of food for energy production by (stepwise) oxidation to give ultimately CO_2 and H_2O, much like combustion. Do the dietary calorie values correspond? The answer is yes, but not quite. First, a "food calorie," as recorded on a supermarket label, actually means a kilocalorie, a mislabeling by a factor of 1000. Moreover, these numbers are given as per gram, or ounce, or volume, adding to the confusion. Second, the "heat of food" is less than that of the corresponding heat of combustion, because not all of what we eat is fully metabolized. Part of it may be excreted either untouched, such as ethanol in breath and in urine, or in partly oxidized form, such as urea from protein. Other materials are difficult to digest and pass through our body with little change—for example, alkanes. For calibration, the approximate percentages of metabolizable energies of basic nutrients are protein 70%, fat 95%, and carbohydrate 97%.

Table 3-7	Heats of Combustion [kcal mol^{-1} (kJ mol^{-1}), Normalized to 25°C] of Various Organic Compounds	
Compound (state)	**Name**	**$\Delta H°_{comb}$**
CH_4 (gas)	Methane	$-212.8\ (-890.4)$
C_2H_6 (gas)	Ethane	$-372.8\ (-1559.8)$
$CH_3CH_2CH_3$ (gas)	Propane	$-530.6\ (-2220.0)$
$CH_3(CH_2)_2CH_3$ (gas)	Butane	$-687.4\ (-2876.1)$
$(CH_3)_3CH$ (gas)	2-Methylpropane	$-685.4\ (-2867.7)$
$CH_3(CH_2)_3CH_3$ (gas)	Pentane	$-845.2\ (-3536.3)$
$CH_3(CH_2)_3CH_3$ (liquid)	Pentane	$-838.8\ (-3509.5)$
$CH_3(CH_2)_4CH_3$ (gas)	Hexane	$-1002.5\ (-4194.5)$
$CH_3(CH_2)_4CH_3$ (liquid)	Hexane	$-995.0\ (-4163.1)$
(liquid)	Cyclohexane	$-936.9\ (-3920.0)$
CH_3CH_2OH (gas)	Ethanol	$-336.4\ (-1407.5)$
CH_3CH_2OH (liquid)	Ethanol	$-326.7\ (-1366.9)$
$C_{12}H_{22}O_{11}$ (solid)	Cane sugar (sucrose)	$-1348.2\ (-5640.9)$

Note: Combustion products are CO_2 (gas) and H_2O (liquid).

Figure 3-13 Butane has a higher energy content than does 2-methylpropane, as measured by the release of energy on combustion. Butane is therefore thermodynamically less stable than its isomer.

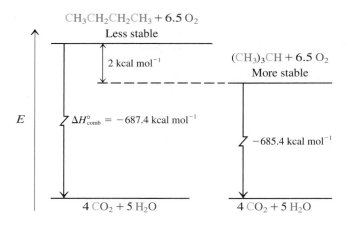

Such comparisons have to take into account the physical state of the compound undergoing combustion (gas, liquid, solid). For example, the difference between the heats of combustion of liquid and gaseous ethanol corresponds to its heat of vaporization, $\Delta H^{\circ}_{vap} = 9.7$ kcal mol^{-1} (40.6 kJ mol^{-1}).

It is not surprising that the ΔH°_{comb} of alkanes increases with chain length, simply because there is more carbon and hydrogen to burn along the homologous series. Conversely, isomeric alkanes contain the same number of carbons and hydrogens, and one might expect that their respective combustions would be equally exothermic. However, that is not the case.

A comparison of the heats of combustion of isomeric alkanes reveals that their values are usually *not* the same. Consider butane and 2-methylpropane. The combustion of butane has a ΔH°_{comb} of -687.4 kcal mol^{-1}, whereas its isomer releases $\Delta H^{\circ}_{comb} = -685.4$ kcal mol^{-1}, 2 kcal mol^{-1} less (Table 3-7). This finding shows that 2-methylpropane has a *smaller* energy content than does butane, because combustion yielding the identical kind and number of products produces less energy (Figure 3-13). Butane is said to be *thermodynamically less stable* than its isomer. Exercise 3-12 reveals the origin of this energy difference.

> Molecules with high energy content are thermodynamically less stable than molecules with low energy content.

Exercise 3-12

The hypothetical thermal conversion of butane into 2-methylpropane should have a $\Delta H^{\circ} = -2.0$ kcal mol^{-1}. What value do you obtain by using the bond-dissociation data in Table 3-2? (Use $DH^{\circ} = 89$ kcal mol^{-1} for the methyl–propyl bond in butane.)

In Summary The heats of combustion values of alkanes and other organic molecules give quantitative estimates of their energy content and, therefore, their relative stabilities.

THE BIG PICTURE

Alkanes lack functional groups, so they do not undergo the kinds of electrophile–nucleophile reactions typical of functionalized molecules. In fact, alkanes are pretty unreactive. However, under appropriate conditions, they undergo homolytic bond cleavage to form radicals, which are reactive species containing odd numbers of electrons. This is another situation in which the *structure* of a class of compounds determines their *function*. Unlike heterolytic processes, which normally proceed via movement of *pairs* of electrons to form or break bonds, homolytic chemistry utilizes the splitting of covalent bonds to give unpaired *single* electrons, as well as their combination to give new bonds.

In organic chemistry, radical reactions are not encountered as frequently as those of polar functional groups. However, radicals play prominent roles in biological, environmental, and industrial chemistry.

The halogenation of alkanes, a radical process in which hydrogen is replaced by halogen, forms the haloalkane functional group. Examination of halogenation allows us to learn about several features common to most transformations, including the way information about a reaction mechanism may be obtained from experimental observations, the relationship between thermodynamics and kinetics, and notions of reactivity and selectivity. The products of halogenation, the haloalkanes, are the starting compounds for a wide variety of reactions, as we will see in Chapters 6 through 9.

Before we examine other classes of compounds and their properties, we need to learn more about the structures and, in particular, the geometric shapes of organic molecules. In Chapter 4 we discuss compounds that contain atoms in rings and in Chapter 5 we study additional forms of isomerism. The ideas we introduce are a necessary background as we begin a systematic study in the chapters that follow of polar reactions of haloalkanes and alcohols.

WORKED EXAMPLES: INTEGRATING THE CONCEPTS

3-13. Mechanism and Enthalpy of a Radical Reaction

Iodomethane reacts with hydrogen iodide under free radical conditions (hv) to give methane and iodine. The overall equation of the reaction is

$$CH_3I + HI \xrightarrow{hv} CH_4 + I_2$$

a. Explain how this process occurs.

SOLUTION

*W*hat is this problem asking? Your key to answering this question lies in the word "how." Reactions occur via mechanisms, described using electron-pushing curved arrows. This problem is really about writing a mechanism for the process: The *mechanism* is the explanation.

*H*ow to begin? The problem states that this reaction occurs "under free radical conditions." We know that radical reactions consist of three stages: initiation, propagation, and termination. Therefore, propose an **initiation** step, followed by likely **propagation** and **termination** steps.

*I*nformation? Tables 3-1 and 3-4 will give us the needed bond strength data. As in Solved Exercise 3-4, we may rely on the mechanisms in the text as models for our full solution.

*P*roceed.

Step 1. Begin by proposing a probable **initiation** step. Recall—from Section 3-4, for example—that initiation steps of radical reactions include cleavage of the *weakest* bond in the starting compounds. According to Tables 3-1 and 3-4, that is the carbon–iodine bond in CH_3I, with $DH° = 57$ kcal mol^{-1}. Therefore

Initiation step

$$H_3C\overset{\frown}{}\ddot{\underset{\cdot\cdot}{I}}: \xrightarrow{hv} H_3C\cdot + \cdot\ddot{\underset{\cdot\cdot}{I}}:$$

Step 2. Again following the model in Section 3-4, propose a **propagation** step in which one of the species produced in the initiation step reacts with one of the molecules shown in the overall equation of the reaction. Try to design the step so that one of its products corresponds to a molecule formed in the overall transformation and the other is a species that can give rise to a second propagation step. The possibilities are

(i) $H_3C\cdot + H\overset{\frown}{}\ddot{\underset{\cdot\cdot}{I}}: \longrightarrow CH_4 + \cdot\ddot{\underset{\cdot\cdot}{I}}:$

(iii) $:\ddot{I}\cdot + :\ddot{I}\overset{\frown}{}H \longrightarrow I_2 + \cdot H$

(ii) $H_3C\cdot + H\overset{\frown}{}CH_2I \longrightarrow CH_4 + \cdot CH_2I$

(iv) $:\ddot{I}\cdot + :\ddot{I}\overset{\frown}{}CH_3 \longrightarrow I_2 + \cdot CH_3$

Propagation steps (i) and (ii) both convert methyl radical into methane by removing a hydrogen atom from HI and CH_3I, respectively. Processes (iii) and (iv) show the removal of an iodine atom by another

> "How" and "why" are organic chemistry code words. When you see them in a question about a reaction, they nearly always mean "Explain by writing a mechanism."

iodine atom from either HI or CH_3I, giving I_2. All four propagation steps turn a molecule of start-ing material in the overall equation of the reaction into a molecule of product. How do we choose the correct ones? *Look at the radical products of each hypothetical propagation step. The two correct equations are those in which the product radical of one supplies the starting radical for the other.* Propagation step (i) consumes a methyl radical and produces an iodine atom. Step (iv) consumes iodine and produces methyl. Therefore, steps (i) and (iv) are the correct steps of a propagation cycle.

Propagation steps

(i) $H_3C\cdot \ + \ H—\ddot{\underset{..}{I}}: \ \longrightarrow \ CH_4 \ + \ \cdot\ddot{\underset{..}{I}}:$ (iv) $:\ddot{\underset{..}{I}}\cdot \ + \ :\ddot{\underset{..}{I}}—CH_3 \ \longrightarrow \ I_2 \ + \ \cdot CH_3$

Step 3. Finally, combination of *any pair of radicals* to give a single molecule constitutes a legitimate **termination** step. There are three:

Termination steps: $:\ddot{\underset{..}{I}}\cdot \ + \ \cdot\ddot{\underset{..}{I}}: \ \longrightarrow \ I_2$ $H_3C\cdot \ + \ \cdot CH_3 \ \longrightarrow \ H_3C—CH_3$ (ethane)

$H_3C\cdot \ + \ \cdot\ddot{\underset{..}{I}}: \ \longrightarrow \ H_3C—I$

b. Calculate the enthalpy changes, $\Delta H°$, associated with the overall reaction and all of the mechanistic steps. Use Tables 3-1, 3-2, and 3-4, as appropriate.

SOLUTION

Breaking a bond requires energy *input,* forming a bond gives rise to energy *output,* and $\Delta H° = $ (energy in) − (energy out). For the overall reaction, we have the following bond strength values to consider:

$$CH_3—I + H—I \overset{hv}{\longrightarrow} CH_3—H + I—I$$
$$DH°: \quad 57 \qquad 71 \qquad\qquad 105 \qquad 36$$

The answer is $\Delta H° = (57 + 71) − (105 + 36) = −13$ kcal mol^{-1} (see also Table 3-5).

For the mechanistic steps, the same principle applies. With one exception, the same four $DH°$ values just shown are all that you need, because they correspond to the only four bonds that are either made or broken in any of the steps in the mechanism.

Initiation step: $\Delta H° = DH°$ (CH_3–I) $= +57$ kcal mol^{-1}

Propagation step (i): $\Delta H° = DH°$ (H–I) $− DH°$ (CH_3–H) $= −34$ kcal mol^{-1}

Propagation step (iv): $\Delta H° = DH°$ (CH_3–I) $− DH°$ (I–I) $= +21$ kcal mol^{-1}

Notice that the sum of the $\Delta H°$ values for the two propagation steps equals $\Delta H°$ for the overall reac-tion. *This is always true.*

Termination steps: $\Delta H° = −DH°$ for the bond formed; $−36$ kcal mol^{-1} for I_2, $−57$ kcal mol^{-1} for CH_3I, and $−90$ kcal mol^{-1} for the C–C bond in ethane.

3-14. Combining Mechanisms and Bond Strength Data to Predict Products

Consider the process described in Exercise 3-6, the reaction between methane and equimolar amounts of Cl_2 and Br_2. Analyze the full process mechanistically and predict what products you expect to form.

SOLUTION

*W*hat is the problem asking? Two things: You are to write out the mechanism steps that are involved in the reaction system and decide what organic product or products you expect to form.

*H*ow to begin? Begin with the propagation steps, *because they describe the formation of the products.*

*I*nformation needed? Refer back to Exercise 3-6 (and its solution in Appendix A): The problem relates to radical halogenation of methane, which is depicted in Sections 3-4 and 3-5. Enthalpy data from Tables 3-1 and 3-3 and from the text are also likely to be useful.

*P*roceed, step by step:

Initiation

Both Cl_2 and Br_2 undergo dissociation to atoms under the influence of heat or light. So, both chlorine and bromine atoms are present.

Propagation step 1

Although both chlorine and bromine atoms are *capable* of reacting with methane, we note that the reaction of chlorine proceeds with $\Delta H° = +2$ kcal mol^{-1} and $E_a = 4$ kcal mol^{-1} (Section 3-4), whereas the reaction of bromine has $\Delta H° = +18$ kcal mol^{-1} and $E_a \approx 19\text{-}20$ kcal mol^{-1} (Section 3-5). The large difference in E_a means that chlorine will abstract a hydrogen atom from methane very much faster than will bromine. Thus for all practical purposes the only first propagation step we need to consider is

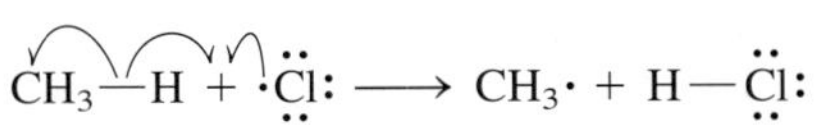

$$CH_3\text{—}H + \cdot\ddot{C}l\colon \longrightarrow CH_3\cdot + H\text{—}\ddot{C}l\colon$$

Does this mean that only CH_3Cl will form as a final product? If you jumped to this conclusion, you failed to "*P*roceed logically, step by step. — Do *not* skip any steps!" You also got the problem wrong. Why? Ask yourself the following question: Does propagation step 1 include the formation of the final product? No! The products of this step are HCl and methyl radical, not CH_3Cl. At this point we are not yet ready to answer the question of the final product. We *must* look at propagation step 2 first. And here, things get interesting.

Propagation step 2

Propagation step 1 forms methyl radicals. Propagation step 2 is the reaction of methyl radical with the halogen X_2 (X_2 = either Cl_2 or Br_2) to give a halogen atom and the final product, CH_3X (X = either Cl or Br). What have we learned in this chapter about the reactions of methyl radical with these halogens? From Sections 3-4 and 3-5 we find

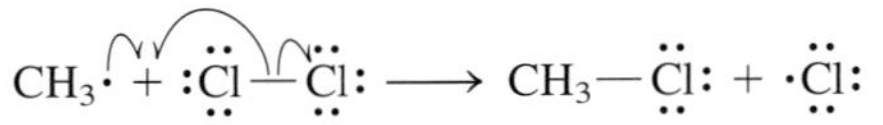

$$CH_3\cdot + \colon\ddot{C}l\text{—}\ddot{C}l\colon \longrightarrow CH_3\text{—}\ddot{C}l\colon + \cdot\ddot{C}l\colon \qquad \Delta H° = -27 \text{ kcal mol}^{-1}$$
$$E_a \approx 0 \text{ kcal mol}^{-1}$$

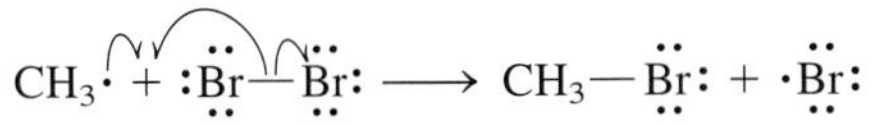

$$CH_3\cdot + \colon\ddot{B}r\text{—}\ddot{B}r\colon \longrightarrow CH_3\text{—}\ddot{B}r\colon + \cdot\ddot{B}r\colon \qquad \Delta H° = -24 \text{ kcal mol}^{-1}$$
$$E_a \approx 0 \text{ kcal mol}^{-1}$$

Look carefully! Methyl radicals, once formed, have the option to attack *either* Cl_2 or Br_2 in reactions that are almost equally exothermic and, more importantly, because of their very low activation energies, *almost equally fast!* It is here, in propagation step 2, where the choice of forming either CH_3Cl or CH_3Br as the final product is made. We find that this propagation step proceeds at about the same rate regardless of whether methyl radical reacts with bromine or chlorine. Thus, we arrive at the observed experimental result, namely that CH_3Cl or CH_3Br are formed in *almost equal amounts*.

This quite counterintuitive outcome would be impossible to predict (or understand, after the experimental fact) without analyzing the mechanistic details as we have done. Notice that both CH_3Cl and CH_3Br are derived from methyl radicals that arose from the reaction of methane with *only* chlorine atoms in propagation step 1. For more practice with radical reactions, see Problems 24, 39, and 40.

Important Concepts

1. The $\Delta H°$ of **bond homolysis** is defined as the **bond-dissociation energy,** $DH°$. Bond homolysis gives radicals or free atoms.

2. The C–H bond strengths in the alkanes decrease in the order

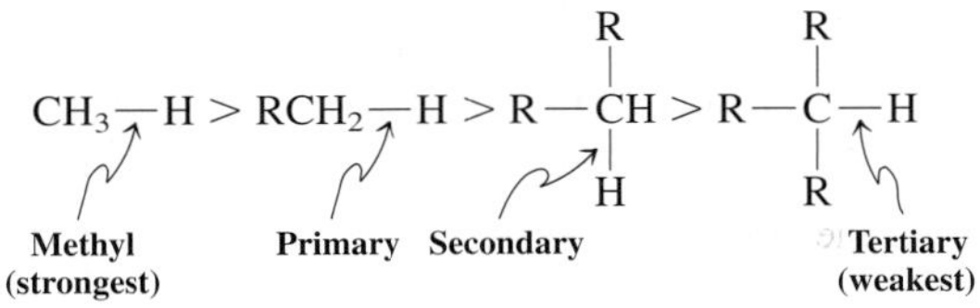

$$CH_3\text{—}H > RCH_2\text{—}H > R\text{—}\underset{\underset{\displaystyle H}{|}}{\overset{\overset{\displaystyle R}{|}}{C}}H > R\text{—}\underset{\underset{\displaystyle R}{|}}{\overset{\overset{\displaystyle R}{|}}{C}}\text{—}H$$

Methyl   **Primary** **Secondary**   **Tertiary**
(strongest)   **(weakest)**

because the order of stability of the corresponding alkyl radicals is

$$CH_3\cdot \; < \; RCH_2\cdot \; < \; R\overset{R}{\underset{}{-}}\overset{\cdot}{C}H\cdot \; < \; R\overset{R}{\underset{R}{-}}\overset{}{C}\cdot$$

Methyl   **Primary**   **Secondary**   **Tertiary**
(least stable)   **(most stable)**

This is the order of increasing stabilization due to **hyperconjugation.**

3. **Catalysts** speed up the establishment of an equilibrium between starting materials and products.

4. Alkanes react with halogens (except iodine) by a **radical chain mechanism** to give haloalkanes. The mechanism consists of **initiation** to create a halogen atom, two **propagation steps,** and various **termination steps.**

5. In the first propagation step, the slower of the two, a hydrogen atom is abstracted from the alkane chain, a reaction resulting in an alkyl radical and HX. Hence, **reactivity** increases from I_2 to F_2. **Selectivity** decreases along the same series, as well as with increasing temperature.

6. The **Hammond postulate** states that fast, exothermic reactions are typically characterized by **early transition states,** which are similar in structure to the starting materials. On the other hand, slow, endothermic processes usually have **late** (product-like) **transition states.**

7. The $\Delta H°$ for a reaction may be calculated from the $DH°$ values of the bonds affected in the process as follows:

$$\Delta H° = \Sigma \, DH°_{\text{bonds broken}} - \Sigma \, DH°_{\text{bonds formed}}$$

8. The $\Delta H°$ for a radical halogenation process equals the sum of the $\Delta H°$ values for the propagation steps.

9. The relative reactivities of the various types of alkane C–H bonds in halogenations can be estimated by factoring out statistical contributions. They are roughly constant under identical conditions and follow the order

      Primary   Secondary   Tertiary
CH_4 <   CH   <   CH   <   CH

The reactivity differences between these types of CH bonds are greatest for bromination, making it the most *selective* radical halogenation process. Chlorination is much less selective, and fluorination shows very little selectivity.

10. The $\Delta H°$ of the combustion of an alkane is called the **heat of combustion, $\Delta H°_{\text{comb}}$.** The heats of combustion of isomeric compounds provide an experimental measure of their relative stabilities.

Problems

15. Label the primary, secondary, and tertiary hydrogens in each of the following compounds.

   (a) $CH_3CH_2CH_2CH_3$ **(b)** $CH_3CH_2CH_2CH_2CH_3$

   (c)    **(d)**

16. Within each of the following sets of alkyl radicals, name each radical; identify each as either primary, secondary, or tertiary; rank in order of decreasing stability; and sketch an orbital picture of the most stable radical, showing the hyperconjugative interaction(s).

   (a) $CH_3CH_2\overset{\cdot}{C}HCH_3$ and $CH_3CH_2CH_2CH_2\cdot$

   (b) $(CH_3CH_2)_2CHCH_2\cdot$ and $(CH_3CH_2)_2\overset{\cdot}{C}CH_3$

   (c) $(CH_3)_2CH\overset{\cdot}{C}HCH_3, (CH_3)_2\overset{\cdot}{C}CH_2CH_3$, and $(CH_3)_2CHCH_2CH_2\cdot$

17. Write as many products as you can think of that might result from the pyrolytic cracking of propane. Assume that the only initial process is C–C bond cleavage.

18. Answer the question posed in Problem 17 for **(a)** butane and **(b)** 2-methylpropane. Use the data in Table 3-2 to determine the bond most likely to cleave homolytically, and use that bond cleavage as your first step.

19. Calculate $\Delta H°$ values for the following reactions.

   (a) $H_2 + F_2 \rightarrow$ 2 HF; **(b)** $H_2 + Cl_2 \rightarrow$ 2 HCl;
   (c) $H_2 + Br_2 \rightarrow$ 2 HBr; **(d)** $H_2 + I_2 \rightarrow$ 2 HI;
   (e) $(CH_3)_3CH + F_2 \rightarrow (CH_3)_3CF + HF$;
   (f) $(CH_3)_3CH + Cl_2 \rightarrow (CH_3)_3CCl + HCl$;
   (g) $(CH_3)_3CH + Br_2 \rightarrow (CH_3)_3CBr + HBr$;
   (h) $(CH_3)_3CH + I_2 \rightarrow (CH_3)_3CI + HI$.

20. For each compound in Problem 15, determine how many constitutional isomers can form upon monohalogenation.

(**Hint:** Identify all groups of hydrogens that reside in distinct structural environments in each molecule.)

21. (a) Using the information given in Sections 3-6 and 3-7, write the products of the radical monochlorination of (i) pentane and (ii) 3-methylpentane. **(b)** For each, estimate the ratio of the isomeric monochlorination products that would form at 25°C. **(c)** Using the bond strength data from Table 3-1, determine the $\Delta H°$ values of the propagation steps for the chlorination of 3-methylpentane at C3. What is the overall $\Delta H°$ value for this reaction?

22. Write in full the mechanism for monobromination of methane. Be sure to include initiation, propagation, and termination steps.

23. Sketch potential-energy/reaction-coordinate diagrams for the two propagation steps of the monobromination of methane (Problem 22).

24. Write a mechanism for the radical bromination of the hydrocarbon benzene, C_6H_6 (for structure, see Section 2-4). Use propagation steps similar to those in the halogenation of alkanes, as presented in Sections 3-4 through 3-6. Calculate $\Delta H°$ values for each step and for the reaction as a whole. How does this reaction compare thermodynamically with the bromination of other hydrocarbons? Data: $DH° (C_6H_5–H) = 112$ kcal mol^{-1}; $DH° (C_6H_5–Br) = 81$ kcal mol^{-1}. Note the **Caution** in Exercise 3-5.

25. Sketch potential-energy/reaction-coordinate diagrams for the two propagation steps of the monobromination of benzene (Problem 24).

26. Identify each of the diagrams you drew in Problem 25 as showing an early or a late transition state.

27. Write the major organic product(s), if any, of each of the following reactions.

(a) $CH_3CH_3 + I_2 \xrightarrow{\Delta}$

(b) $CH_3CH_2CH_3 + F_2 \longrightarrow$

(c) 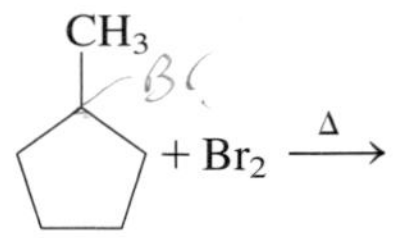 [cyclopentane with CH₃ and BC] $+ Br_2 \xrightarrow{\Delta}$

(d) 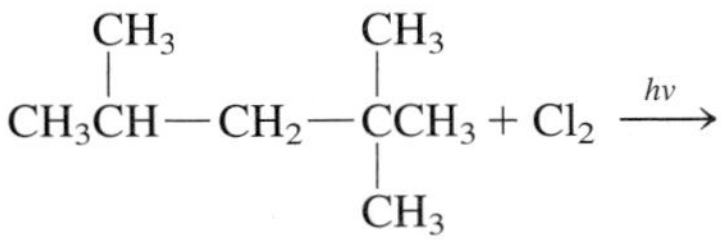 $CH_3\overset{CH_3}{\underset{}{CH}}-CH_2-\overset{CH_3}{\underset{CH_3}{C}}CH_3 + Cl_2 \xrightarrow{hv}$

(e) 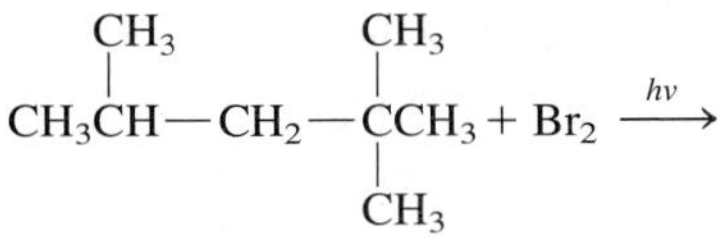 $CH_3\overset{CH_3}{\underset{}{CH}}-CH_2-\overset{CH_3}{\underset{CH_3}{C}}CH_3 + Br_2 \xrightarrow{hv}$

28. Calculate product ratios in each of the reactions in Problem 27. Use relative reactivity data for F_2 and Cl_2 at 25°C and for Br_2 at 150°C (Table 3-6).

29. Which, if any, of the reactions in Problem 27 give the major product with reasonable selectivity (i.e., are useful "synthetic methods")?

30. (a) What would be the major organic product of monobromination of pentane at 125°C? **(b)** Draw Newman projections of all possible staggered conformations arising from rotation about the C2–C3 bond for this product molecule. **(c)** Draw a qualitative

graph of potential energy versus torsional angle for C2–C3 rotation in this molecule. (**Note:** A bromine atom is considerably smaller, sterically, than is a methyl group.)

31. (a) Sketch a potential energy/reaction coordinate graph showing the two propagation steps for the monobromination of pentane to give the major product (Problem 30). Use $DH°$ information from this chapter (Tables 3-1, 3-2, and 3-4, as appropriate). **(b)** Indicate the locations of the transition states and whether each is early or late. **(c)** Sketch a similar graph for reaction of pentane with I_2. How does it differ from the graph for bromination?

32. At room temperature, 1,2-dibromoethane exists as an equilibrium mixture in which 89% of the molecules are in an *anti* conformation and 11% are *gauche*. The comparable ratio for butane under the same circumstances is 72% *anti* and 28% *gauche*. Suggest an explanation for the difference, bearing in mind that Br is sterically *smaller* than CH_3 (see Problem 30). (**Hint:** Consider the polarity of a C–Br bond and consequent electrostatic effects.)

33. Write balanced equations for the combustion of each of the following substances (molecular formulas may be obtained from Table 3-7): **(a)** methane; **(b)** propane; **(c)** cyclohexane; **(d)** ethanol; **(e)** sucrose.

34. Propanal ($CH_3CH_2\overset{O}{\overset{||}{C}}H$) and acetone ($CH_3\overset{O}{\overset{||}{C}}CH_3$) are isomers with the formula C_3H_6O. The heat of combustion of propanal is -434.1 kcal mol^{-1}, that of acetone -427.9 kcal mol^{-1}. **(a)** Write a balanced equation for the combustion of either compound. **(b)** What is the energy difference between propanal and acetone? Which has the lower energy content? **(c)** Which substance is more thermodynamically stable, propanal or acetone? (**Hint:** Draw a diagram similar to that in Figure 3-13.)

35. Sulfuryl chloride (SO_2Cl_2, see below for structure) is a liquid reagent that may be used for chlorinations of alkanes as a substitute for gaseous elemental chlorine. Propose a mechanism for chlorination of CH_4 using sulfuryl chloride. (**Hint:** Follow the usual model for a radical chain process, substituting SO_2Cl_2 for Cl_2 where appropriate.)

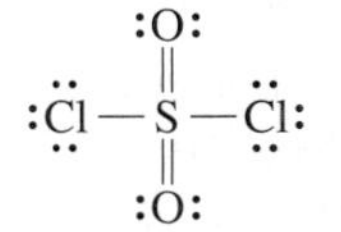

Sulfuryl chloride
(b.p. 69°C)

36. Use the Arrhenius equation (Section 2-1) to estimate the ratio of the rate constants k for the reactions of a C–H bond in methane with a chlorine atom and with a bromine atom at 25°C. Assume that the A values for the two processes are equal, and use $E_a = 19$ kcal mol^{-1} for the reaction between Br· and CH_4.

37. **CHALLENGE** When an alkane with different types of C–H bonds, such as propane, reacts with an equimolar mixture of Br_2 and Cl_2, the selectivity in the formation of the brominated products is much worse than that observed when reaction is carried out with Br_2 alone. (In fact, it is very similar to the selectivity for *chlorination*.) Explain.

38. Bromination of 1-bromopropane gives the following results:

$$CH_3CH_2CH_2Br \xrightarrow{Br_2,\ 200°C}$$
$$CH_3CH_2CHBr_2 + CH_3CHBrCH_2Br + BrCH_2CH_2CH_2Br$$
$$90\% \qquad\qquad 8.5\% \qquad\qquad 1.5\%$$

Calculate the relative reactivities of the hydrogens on each of the three carbons toward bromine atoms. Compare these results with those from a simple alkane such as propane, and suggest explanations for any differences.

39. A hypothetical alternative mechanism for the halogenation of methane has the following propagation steps.

(i) $X\cdot + CH_4 \longrightarrow CH_3X + H\cdot$
(ii) $H\cdot + X_2 \longrightarrow HX + X\cdot$

(a) Using $DH°$ values from appropriate tables, calculate $\Delta H°$ for these steps for any one of the halogens. (b) Compare your $\Delta H°$ values with those for the accepted mechanism (Table 3-5). Do you expect this alternative mechanism to compete successfully with the accepted one? (**Hint:** Consider activation energies.)

40. **CHALLENGE** The addition of certain materials called radical inhibitors to halogenation reactions causes the reactions to come to virtually a complete stop. An example is the inhibition by I_2 of the chlorination of methane. Explain how this inhibition might come about. (**Hint:** Calculate $\Delta H°$ values for possible reactions of the various species present in the system with I_2, and evaluate the possible further reactivity of the products of these I_2 reactions.)

41. Typical hydrocarbon fuels (e.g., 2,2,4-trimethylpentane, a common component of gasoline) have very similar heats of combustion when calculated in kilocalories *per gram*. (a) Calculate heats of combustion per gram for several representative hydrocarbons in Table 3-7. (b) Make the same calculation for ethanol (Table 3-7). (c) In evaluating the feasibility of "gasohol" (90% gasoline and 10% ethanol) as a motor fuel, it has been estimated that an automobile running on pure ethanol would get approximately 40% fewer miles per gallon than would an identical automobile running on standard gasoline. Is this estimate consistent with the results in (a) and (b)? What can you say in general about the fuel capabilities of oxygen-containing molecules relative to hydrocarbons?

42. Two simple organic molecules that have been used as fuel additives are methanol (CH_3OH) and 2-methoxy-2-methylpropane [*tert*-butyl methyl ether, $(CH_3)_3COCH_3$]. The $\Delta H°_{comb}$ values for these compounds in the gas phase are -182.6 kcal mol^{-1} for methanol and -809.7 kcal mol^{-1} for 2-methoxy-2-methylpropane. (a) Write balanced equations for the complete combustion of each of these molecules to CO_2 and H_2O. (b) Using Table 3-7, compare the $\Delta H°_{comb}$ values for these compounds with those for alkanes with similar molecular weights.

43. **CHALLENGE** Figure 3-9 compares the reactions of $Cl\cdot$ with the primary and secondary hydrogens of propane. (a) Draw a similar diagram comparing the reactions of $Br\cdot$ with the primary and secondary hydrogens of propane. (**Hint:** First obtain the necessary $DH°$ values from Table 3-1 and calculate $\Delta H°$ for both the primary and the secondary hydrogen abstraction reactions. Other data: $E_a = 15$ kcal mol^{-1} for $Br\cdot$ reacting with a primary C–H bond and $E_a = 13$ kcal mol^{-1} for $Br\cdot$ reacting with a secondary C–H bond.) (b) Which among the transition

states of these reactions would you call "early," and which "late"? (c) Judging from the locations of the transition states of these reactions along the reaction coordinate, should they show greater or lesser radical character than do the corresponding transition states for chlorination (Figure 3-9)? (d) Is your answer to (c) consistent with the selectivity differences between $Cl\cdot$ reacting with propane and $Br\cdot$ reacting with propane? Explain.

44. Two of the propagation steps in the $Cl\cdot/O_3$ system consume ozone and oxygen atoms (which are necessary for the production of ozone), respectively (Section 3-10).

$$Cl + O_3 \longrightarrow ClO + O_2$$
$$ClO + O \longrightarrow Cl + O_2$$

Calculate $\Delta H°$ for each of these propagation steps. Use the following data: $DH°$ for $ClO = 64$ kcal mol^{-1}; $DH°$ for $O_2 = 120$ kcal mol^{-1}; $DH°$ for an O–O$_2$ bond in $O_3 = 26$ kcal mol^{-1}. Write the overall equation described by the combination of these steps and calculate its $\Delta H°$. Comment on the thermodynamic favorability of the process.

Team Problem

45. (a) Provide an IUPAC name for each of the isomers that you drew in Exercise 2-16 (a). (b) For each isomer that you drew and named here, give all the free radical monochlorination and monobromination products that are structurally isomeric. (c) Referring to Table 3-6, discuss which starting alkane and which halogen will yield the least number of isomeric products.

Preprofessional Problems

46. The reaction $CH_4 + Cl_2 \longrightarrow CH_3Cl + HCl$ is an example of

(a) neutralization   (b) an acidic reaction
(c) an isomerization   (d) an ionic reaction
(e) a radical chain reaction

47.
$$CH_2Cl$$
$$|$$
$$CH_2-CHCH_3$$
$$|$$
$$CH_3CH_2CH_2CHCH_2CH_2CH_2CH_3$$

The sum of all the digits that appear in the (IUPAC) name for this compound is which of the following?
(a) Five   (b) Six   (c) Seven
(d) Eight   (e) Nine

48. In a competition reaction, equimolar amounts of the four alkanes shown were allowed to react with a limited amount of Cl_2 at 300°C. Which one of these alkanes would be depleted most from the mixture?
(a) Pentane   (b) 2-Methylpropane
(c) Butane   (d) Propane

49. The reaction of CH_4 with Cl_2 to yield CH_3Cl and HCl is well known. On the basis of the values in the short table below, the enthalpy $\Delta H°$ (kcal mol^{-1}) of this reaction is
(a) $+135$ (b) -135 (c) $+25$ (d) -25

Bond-Dissociation Energies $DH°$ (kcal mol^{-1})

H–Cl	103	Cl–Cl	58
H_3C–Cl	85	H_3C–H	105

CHAPTER 4

Cycloalkanes

W hen you hear or read the word *steroids,* two things probably come to mind immediately: athletes who "take steroids" illegally to develop their muscles, and "the pill" used for birth control. But what do you know about steroids aside from this general association? What is their structure and function? How does one steroid differ from another? Where are they found in nature?

An example of a naturally occurring steroid is diosgenin, obtained from root extracts of the Mexican yam and used as a starting material for the synthesis of several commercial steroids. Most striking is the number of rings in the compound.

Diosgenin

Steroids have had a major beneficial effect on human well-being, as medicines and in the control of fertility. However, abuse of steroids as performance-enhancing drugs in competitive athletics has surfaced occasionally. Thus, the sports world was shaken when illicit use of the designer steroid tetrahydrogestrinone (THG) was discovered in 2003—"designed" to avoid detection in doping tests.

Hydrocarbons containing single-bonded carbon atoms arranged in rings are known as **cyclic alkanes, carbocycles** (in contrast to heterocycles, Chapter 25), or **cycloalkanes.** The majority of organic compounds occurring in nature contain rings. Indeed, so many fundamental biological

The root of the Mexican yam.

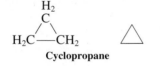

A carbocycle

functions depend on the chemistry of ring-containing compounds that life as we know it could not exist in their absence.

This chapter deals with the names, physical properties, structural features, and conformational characteristics of the cycloalkanes. Because of their cyclic nature, members of this class of compounds can exhibit new types of strain, such as ring strain and transannular interactions. We end with the biochemical significance of selected carbocycles and their derivatives, including steroids.

4-1 | NAMES AND PHYSICAL PROPERTIES OF CYCLOALKANES

Cycloalkanes have their own names under IUPAC rules, and their properties are generally different from those of their noncyclic (also called acyclic) analogs with the same number of carbons. We will see that they also exhibit a kind of isomerism unique to cyclic molecules.

The names of the cycloalkanes follow IUPAC rules

We can construct a molecular model of a cycloalkane by removing a hydrogen atom from each terminal carbon of a model of a straight-chain alkane and allowing these carbons to form a bond. The empirical formula of a cycloalkane is C_nH_{2n} or $(CH_2)_n$. The system for naming members of this class of compounds is straightforward: Alkane names are preceded by the prefix **cyclo-.** Three members in the homologous series—starting with the smallest, cyclopropane—are shown in the margin, written both in condensed form and in bond-line notation.

$$H_2C\overset{\overset{\displaystyle H_2}{C}}{\frown}CH_2$$

Cyclopropane

$$\begin{array}{c} H_2C-CH_2 \\ | \qquad | \\ H_2C-CH_2 \end{array}$$

Cyclobutane

$$\begin{array}{c} \overset{\displaystyle H_2}{C} \\ H_2C \qquad CH_2 \\ H_2C \qquad CH_2 \\ \overset{\displaystyle C}{H_2} \end{array}$$

Cyclohexane

> ### Exercise 4-1
>
> Make molecular models of cyclopropane through cyclododecane. Compare the relative conformational flexibility of each ring with that of others within the series and with that of the corresponding straight-chain alkanes.

Naming a substituted cyclic alkane requires numbering the individual ring carbons only if more than one substituent is attached to the ring. In monosubstituted systems, the carbon of attachment is defined as carbon 1 of the ring. For polysubstituted compounds, take care to provide the lowest possible numbering sequence. When two such sequences are possible, the alphabetical order of the substituent names takes precedence. Radicals derived from cycloalkanes by abstraction of a hydrogen atom are **cycloalkyl radicals.** Substituted cycloalkanes are therefore sometimes named as cycloalkyl derivatives. In general, the smaller unit is treated as a substituent to the larger one—for example, methylcyclopropane (not cyclopropylmethane) and cyclobutylcyclohexane (not cyclohexylcyclobutane).

Methylcyclopropane

(No numbering)

1-Ethyl-1-methylcyclobutane

(Alphabetical)

1-Chloro-2-methyl-4-propylcyclopentane

(Alphabetical; not 2-chloro-1-methyl-4-propylcyclopentane)

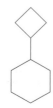

Cyclobutylcyclohexane

(Smaller ring is substituent)

Disubstituted cycloalkanes can be stereoisomers

Inspection of molecular models of disubstituted cycloalkanes in which the two substituents are located on different carbons shows that *two isomers are possible* in each case. In one isomer, the two substituents are positioned on the *same* face, or side, of the ring; in the other isomer, they are on *opposite* faces. Substituents on the same face are called **cis** (*cis*, Latin, on the same side); those on opposite faces, **trans** (*trans*, Latin, across).

We can use hashed-wedged line structures to depict the three-dimensional arrangement of substituted cycloalkanes. The positions of any remaining hydrogens are not always shown.

Model Building

Stereoisomers of 1,2-Dimethylcyclopropane

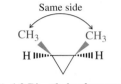

cis-1,2-Dimethylcyclopropane

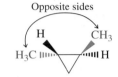

trans-1,2-Dimethylcyclopropane

A Constitutional Isomer of 1,2-Dimethylcyclopropane

Methylcyclobutane

Stereoisomers of 1-Bromo-2-chlorocyclobutane

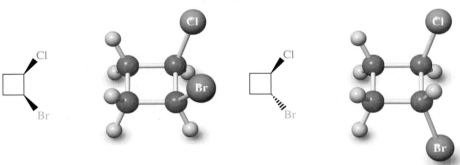

cis-1-Bromo-2-chlorocyclobutane   *trans*-1-Bromo-2-chlorocyclobutane

A Constitutional Isomer of 1-Bromo-2-chlorocyclobutane

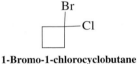

1-Bromo-1-chlorocyclobutane

Conformational Isomers of Butane

anti-Butane

Rotation

gauche-Butane

Cis and trans isomers are **stereoisomers**—compounds that have identical connectivities (i.e., their atoms are attached in the same sequence) but differ in the arrangement of their atoms in space. They are distinct from constitutional or structural isomers (Sections 1-9 and 2-5), which are compounds with differing connectivities among atoms. Conformations (Sections 2-8 and 2-9) also are stereoisomers by this definition. However, unlike cis and trans isomers, which can be interconverted only by *breaking* bonds (try it on your models), conformers are readily equilibrated by *rotation* about bonds. Stereoisomerism will be discussed in more detail in Chapter 5.

Because of the possibility of structural and cis-trans isomerism, a variety of structural possibilities exist in substituted cycloalkanes. For example, there are eight isomeric bromomethylcyclohexanes (three of which are shown below), all with different and distinct physical and chemical properties.

Two hydrogens are understood to be present

One H is understood to be present

BrCH₂

(Bromomethyl)-cyclohexane

1-Bromo-1-methyl-cyclohexane

cis-1-Bromo-2-methylcyclohexane

Name the compound shown in the margin according to the IUPAC rules.

Strategy

We first need to establish whether to name this molecule as a straight-chain alkane or a cycloalkane. We then use the IUPAC rules in Section 2-6 in conjunction with the new rules for naming cycloalkanes to arrive at the correct name.

Solution

- The longest straight-chain piece in the structure is ethyl, with a two-carbon stem, whereas the rings contain three and eight carbons, respectively. Therefore, ethyl should be named as a substituent.
- Cyclooctane is larger than cyclopropane: The molecule should be named as a substituted cyclooctane.
- In numbering the eight-membered ring, we notice that the substitution pattern is symmetrical, that is, 1,5-. Which carbon is assigned number 1 and which number 5 is decided on the basis of IUPAC rule 3: We number the ring in the direction that gives the lower number at the *first difference* between the two possible numbering schemes. The choices are 1,1,5 versus 1,5,5. The first of these is proper (1 < 5), and therefore the carbon with the two substituents (the ethyl groups) is assigned the number 1.
- The substituent names are ethyl, specifically diethyl, and cyclopropyl, specifically dimethylcyclopropyl.
- We need to specify the positions and stereochemistry of the methyl groups on cyclopropyl. For the first, the point of ring attachment is defined as "1," therefore we are dealing with a 2,3-dimethylcyclopropyl substituent. For the second, the two methyl groups are cis.
- We can now name the alkane, placing the substituents in alphabetical order (IUPAC rule 4). In this respect, recall that the prefix "di" in "diethyl" is *not* counted, because it is merely multiplying the substituent whose name is "ethyl." In contrast, the "di" in "dimethylcyclopropyl" is part of that complex substituent's name, and therefore *is* counted in the alphabetization. Thus, dimethylcyclopropyl comes before diethyl (alphabetized as "ethyl"). The result is 5-(*cis*-2,3-dimethylcyclopropyl)-1,1-diethylcyclooctane.

Preceding Exercise 4-2, we showed the structures and names of three isomeric bromomethylcyclohexanes. Do the same for the other five isomers.

The properties of the cycloalkanes differ from those of their straight-chain analogs

The physical properties of a few cycloalkanes are recorded in Table 4-1. Note that, compared with the corresponding straight-chain alkanes (Table 2-5), the cycloalkanes have higher boiling and melting points as well as higher densities. These differences are due in large part to increased London interactions of the relatively more rigid and more symmetric cyclic systems. In comparing lower cycloalkanes possessing an odd number of carbons with those having an even number, we find a pronounced alternation in their melting points. This phenomenon has been ascribed to differences in crystal packing forces between the two series.

In Summary Names of the cycloalkanes are derived in a straightforward manner from those of the straight-chain alkanes. In addition, the position of a single substituent is defined to be C1. Disubstituted cycloalkanes can give rise to cis and trans isomers, depending on the relative spatial orientation of the substituents. Physical properties parallel those of the straight-chain alkanes, but individual values for boiling and melting points and for densities are higher for the cyclic compounds of equal carbon number.

Table 4-1	Physical Properties of Various Cycloalkanes		
Cycloalkane	Melting point (°C)	Boiling point (°C)	Density at 20°C (g mL^{-1})
Cyclopropane (C_3H_6)	-127.6	-32.7	0.617^b
Cyclobutane (C_4H_8)	-50.0	-12.5	0.720
Cyclopentane (C_5H_{10})	-93.9	49.3	0.7457
Cyclohexane (C_6H_{12})	6.6	80.7	0.7785
Cycloheptane (C_7H_{14})	-12.0	118.5	0.8098
Cyclooctane (C_8H_{16})	14.3	148.5	0.8349
Cyclododecane ($C_{12}H_{24}$)	64	160 (100 torr)	0.861
Cyclopentadecane ($C_{15}H_{30}$)	66	110 (0.1 torr^a)	0.860

^{a}Sublimation point.
^{b}At 25°C.

4-2 | RING STRAIN AND THE STRUCTURE OF CYCLOALKANES

The molecular models made for Exercise 4-1 reveal obvious differences between cyclopropane, cyclobutane, cyclopentane, and so forth, and the corresponding straight-chain alkanes. One notable feature of the first two members in the series is how difficult it is to close the ring without breaking the plastic tubes used to represent bonds. This problem is called **ring strain.** The reason for it lies in the tetrahedral carbon model. The C–C–C bond angles in, for example, cyclopropane (60°) and cyclobutane (90°) differ considerably from the tetrahedral value of 109.5°. As the ring size increases, strain diminishes. Thus, cyclohexane can be assembled without distortion or strain.

Does this observation tell us anything about the relative stability of the cycloalkanes—for example, as measured by their heats of combustion, $\Delta H°_{comb}$? How does strain affect structure and function? This section and Section 4-3 address these questions.

The heats of combustion of the cycloalkanes reveal the presence of ring strain

Section 3-11 introduced one measure of the stability of a molecule: its heat content. We also learned that the heat content of an alkane can be estimated by measuring its heat of combustion, $\Delta H°_{comb}$ (Table 3-7). To find out whether there is something special about the stability of cycloalkanes, we could compare their heats of combustion with those of the analogous straight-chain alkanes. Such a direct comparison is flawed, however, because the empirical formula of cycloalkanes, C_nH_{2n}, differs from that of normal alkanes, C_nH_{2n+2}, by two hydrogens (see margin). To solve this problem, we take a roundabout approach, based on the recognition that we can rewrite the formula for cycloalkanes as $(CH_2)_n$. Thus, if we had an experimental number for the contribution of a "strain-free" CH_2 fragment to the $\Delta H°_{comb}$ of straight-chain alkanes, then the corresponding $\Delta H°_{comb}$ of a cycloalkane should simply be multiples of this number. If it is not, it might signal the presence of strain.

The $\Delta H°_{comb}$ of a Strain-Free Cycloalkane Should Be Multiples of $\Delta H°_{comb}(CH_2)$

$$\Delta H°_{comb}(C_nH_{2n}) = n \times \Delta H°_{comb}(CH_2)$$

How do we obtain a $\Delta H°_{comb}$ value of CH_2? Let us turn to Table 3-7 and the combustion data for the straight-chain alkanes. We note that $\Delta H°_{comb}$ increases by about the same amount with each successive member of the homologous series: about 157 kcal mol^{-1} for each additional CH_2 moiety.

Model Building

Combustion of Cyclohexane Versus Hexane

+ 9 O_2

↓

$6 CO_2$ + 6 H_2O

+ 10 O_2

↓

$6 CO_2$ + 7 H_2O

$\Delta H^{\circ}_{\text{comb}}$ Values for the Series of Straight-Chain Alkanes

CH₃CH₂CH₃ (gas)	−530.6	Increment = −156.8
CH₃CH₂CH₂CH₃ (gas)	−687.4	Increment = −157.8 kcal mol⁻¹
CH₃(CH₂)₃CH₃ (gas)	−845.2	Increment = −157.3
CH₃(CH₂)₄CH₃ (gas)	−1002.5	

When averaged over a large number of alkanes, this value approaches 157.4 kcal mol⁻¹ (658.6 kJ mol⁻¹), our requisite value for $\Delta H^{\circ}_{\text{comb}}(\text{CH}_2)$!

Armed with this number, we can calculate the expected $\Delta H^{\circ}_{\text{comb}}$ of the cycloalkanes, $(\text{CH}_2)_n$, namely, $-(n \times 157.4)$ kcal mol⁻¹. For example, for cyclopropane, $n = 3$, hence its $\Delta H^{\circ}_{\text{comb}}$ should be −472.2 kcal mol⁻¹; for cyclobutane, it should be −629.6 kcal mol⁻¹; and so on (Table 4-2, column 2). However, when we measure the actual heats of combustion of these molecules, they turn out to be generally *larger in magnitude* (Table 4-2, column 3). Thus, for cyclopropane, the experimental value is −499.8 kcal mol⁻¹, a discrepancy between the expected and observed values of 27.6 kcal mol⁻¹. Therefore, cyclopropane is more energetic than expected for a strainless molecule. The extra energy is attributed to a property of cyclopropane of which we are already aware because of the model we built: *ring strain*. The strain per CH₂ group in this molecule is 9.2 kcal mol⁻¹.

Cyclopropane gas (from tanks, as shown above) was used in medicine until the early 1950s as a general anesthetic. It was administered by inhalation as a mixture with oxygen—an explosive concoction, in part due to the release of ring strain upon combustion!

Ring Strain in Cyclopropane

Calculated for strainless molecule:

$$\Delta H^{\circ}_{\text{comb}} = -(3 \times 157.4) = -472.2 \text{ kcal mol}^{-1}$$

Measured: $\Delta H^{\circ}_{\text{comb}} = -499.8$ kcal mol⁻¹

Strain: $499.8 - 472.2 = 27.6$ kcal mol⁻¹

A similar calculation for cyclobutane (Table 4-2) reveals a ring strain of 26.3 kcal mol⁻¹, or about 6.6 kcal mol⁻¹ per CH₂ group. In cyclopentane, this effect is much smaller, the total strain amounting to only 6.5 kcal mol⁻¹, and cyclohexane is virtually strain free. However, succeeding members of the series again show considerable strain until we reach very large rings (see Section 4-5). Because of these trends, organic chemists have loosely defined four groups of cycloalkanes:

1. *Small rings* (cyclopropane, cyclobutane)

2. *Common rings* (cyclopentane, cyclohexane, cycloheptane)

3. *Medium rings* (from 8- to 12-membered)

4. *Large rings* (13-membered and larger)

Table 4-2	Calculated and Experimental Heats of Combustion in kcal mol⁻¹ (kJ mol⁻¹) of Various Cycloalkanes				
Ring size (C_n)	$\Delta H^{\circ}_{\text{comb}}$ (calculated)		$\Delta H^{\circ}_{\text{comb}}$ (experimental)		Total strain	Strain per CH₂ group
3	−472.2	(−1976)	−499.8	(−2091)	27.6 (115)	9.2 (38)
4	−629.6	(−2634)	−655.9	(−2744)	26.3 (110)	6.6 (28)
5	−787.0	(−3293)	−793.5	(−3320)	6.5 (27)	1.3 (5.4)
6	−944.4	(−3951)	−944.5	(−3952)	0.1 (0.4)	0.0 (0.0)
7	−1101.8	(−4610)	−1108.2	(−4637)	6.4 (27)	0.9 (3.8)
8	−1259.2	(−5268)	−1269.2	(−5310)	10.0 (42)	1.3 (5.4)
9	−1416.6	(−5927)	−1429.5	(−5981)	12.9 (54)	1.4 (5.9)
10	−1574.0	(−6586)	−1586.0	(−6636)	14.0 (59)	1.4 (5.9)
11	−1731.4	(−7244)	−1742.4	(−7290)	11.0 (46)	1.1 (4.6)
12	−1888.8	(−7903)	−1891.2	(−7913)	2.4 (10)	0.2 (0.8)
14	−2203.6	(−9220)	−2203.6	(−9220)	0.0 (0.0)	0.0 (0.0)

Note: The calculated numbers are based on the value of −157.4 kcal mol⁻¹ (−658.6 kJ mol⁻¹) for a CH₂ group.

What kinds of effects contribute to the ring strain in cycloalkanes? We answer this question by exploring the detailed structures of several of these compounds.

Strain affects the structures and conformational function of the smaller cycloalkanes

As we have just seen, the smallest cycloalkane, *cyclopropane,* is much less stable than expected for three methylene groups. Why should this be? The reason is twofold: torsional strain and bond-angle strain.

The structure of the cyclopropane molecule is represented in Figure 4-1. We notice first that all methylene hydrogens are eclipsed, much like the hydrogens in the eclipsed conformation of ethane (Section 2-8). We know that the energy of the eclipsed form of ethane is higher than that of the more stable staggered conformation because of **eclipsing (torsional) strain.** This effect is also present in cyclopropane. Moreover, the carbon skeleton in cyclopropane is by necessity flat and quite rigid, and bond rotation that might relieve eclipsing strain is very difficult.

Second, we notice that cyclopropane has C–C–C bond angles of 60°, a significant deviation from the "natural" tetrahedral bond angle of 109.5°. How is it possible for three supposedly tetrahedral carbon atoms to maintain a bonding relation at such highly distorted angles? The problem is perhaps best illustrated in Figure 4-2, in which the bonding in the strain-free "open cyclopropane," the trimethylene diradical ·CH$_2$CH$_2$CH$_2$·, is compared with that in the closed form. We can see that the two ends of the trimethylene diradical cannot "reach" far enough to close the ring without "bending" the two C–C bonds already present. However, if all three C–C bonds in cyclopropane adopt a bent configuration (interorbital angle 104°, see Figure 4-2B), overlap is sufficient for bond formation. The energy needed to distort the tetrahedral carbons enough to close the ring is called **bond-angle strain.** The ring strain in cyclopropane is derived from a combination of eclipsing and bond-angle contributions.

As a consequence of its structure, cyclopropane has relatively weak C–C bonds [$DH° = 65$ kcal mol^{-1} (272 kJ mol^{-1})]. This value is low [recall that the C–C strength in ethane is 90 kcal mol^{-1} (377 kJ mol^{-1})] because breaking the bond opens the ring and relieves ring strain. For example, reaction with hydrogen in the presence of a palladium catalyst opens the ring to give propane.

$$\triangle \;+\; H_2 \;\xrightarrow{\text{Pd catalyst}}\; CH_3CH_2CH_3 \qquad \Delta H° = -37.6 \text{ kcal mol}^{-1}\,(-157 \text{ kJ mol}^{-1})$$
Propane

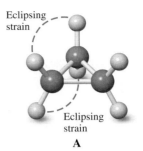

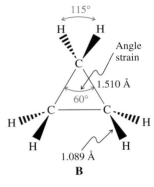

Figure 4-1 Cyclopropane: (A) molecular model; (B) bond lengths and angles.

Model Building

$DH° = 65$ kcal mol^{-1}

$DH° = 90$ kcal mol^{-1}

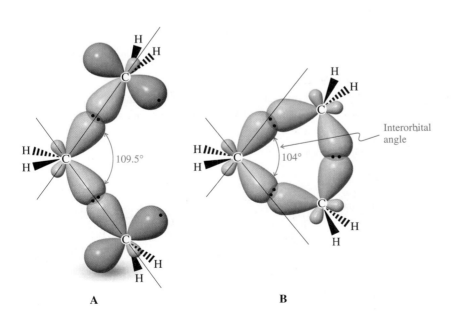

A   B

Figure 4-2 Orbital picture of (A) the trimethylene diradical and (B) the bent bonds in cyclopropane. Only the hybrid orbitals forming C–C bonds are shown. Note the interorbital angle of 104° in cyclopropane.

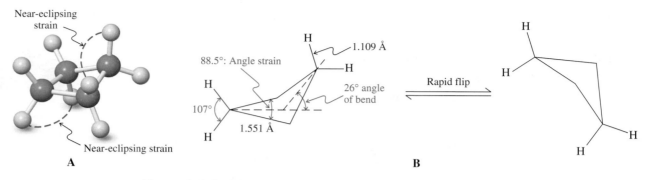

Figure 4-3 Cyclobutane: (A) molecular model; (B) bond lengths and angles. The nonplanar molecule "flips" rapidly from one conformation to another.

Exercise 4-4

trans-1,2-Dimethylcyclopropane is more stable than *cis*-1,2-dimethylcyclopropane. Why? Draw a picture to illustrate your answer. Which isomer liberates more heat on combustion?

What about higher cycloalkanes? The structure of *cyclobutane* (Figure 4-3) reveals that this molecule is not planar but puckered, with an approximate bending angle of 26°. The nonplanar structure of the ring, however, is not very rigid. The molecule "flips" rapidly from one puckered conformation to the other. Construction of a molecular model shows why distorting the four-membered ring from planarity is favorable: It partly relieves the strain introduced by the eight eclipsing hydrogens. Moreover, bond-angle strain is considerably reduced relative to that in cyclopropane, although maximum overlap is, again, only possible with the use of bent bonds. The C–C bond strength in cyclobutane also is low [about 63 kcal mol⁻¹ (264 kJ mol⁻¹)] because of the release of ring strain on ring opening and the consequences of relatively poor overlap in bent bonds. Cyclobutane is less reactive than cyclopropane but undergoes similar ring-opening processes.

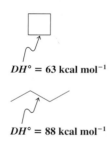

$DH° = 63$ kcal mol⁻¹

$DH° = 88$ kcal mol⁻¹

$DH° = 81$ kcal mol⁻¹

$$ \square \ + \ H_2 \ \xrightarrow{\text{Pd catalyst}} \ CH_3CH_2CH_2CH_3 $$
Butane

Cyclopentane might be expected to be planar because the angles in a regular pentagon are 108°, close to tetrahedral. However, such a planar arrangement would have *ten* H–H eclipsing interactions. The puckering of the ring reduces this effect, as indicated by the **envelope** structure of the molecule (Figure 4-4). Although puckering relieves eclipsing, it also increases bond-angle strain. The envelope conformation is a compromise in which the energy of the system is minimized.

Overall, cyclopentane has relatively little ring strain, and its C–C bond strength [$DH° = 81$ kcal mol⁻¹ (338 kJ mol⁻¹)] approaches that in acyclic alkanes (Table 3-2). As a consequence, it does not show the unusual reactivity of three- or four-membered rings.

Figure 4-4 Cyclopentane: (A) molecular model of the half-chair conformation; (B) bond lengths and angles. The molecule is flexible, with little strain.

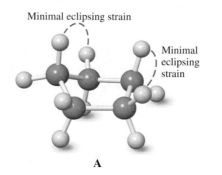

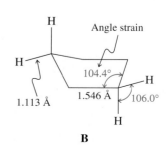

Solved Exercise 4-5 | Working with the Concepts: Estimating Strain

The heat of reaction of hydrogen with the exotic-looking hydrocarbon bicyclo[2.1.0]pentane (A), a strained bicyclic alkane (see Section 4-6), to give cyclopentane has been measured to be -56 kcal mol^{-1}. This reaction is considerably more exothermic than that for cyclopropane on p. 137 (-37.6 kcal mol^{-1}), indicating relatively more strain. How would you estimate the strain energy in A? (**Hint:** To tackle this problem, it is useful to review Section 2-1 and consult Tables 3-1, 3-2, and 4-2.)

$$\Delta H^\circ = -56 \text{ kcal mol}^{-1}$$

Strategy 1

A possibly quickest way to estimate the strain in A is to check Table 4-2 and simply add the strains of the component rings cyclopropane (27.6 kcal mol^{-1}) and cyclobutane (26.3 kcal mol^{-1}): 53.9 kcal mol^{-1}. (**Caution:** This approach ignores the likely effect that the two rings will have on their respective strains when sharing a bond. You can verify this effect by building a model of cyclobutane and then converting it to a model of A. The four-membered ring flattens completely, and the bond angles of the cyclopropane CH$_2$ bridge are considerably more distorted than those to the corresponding two hydrogens in cyclobutane.)

Strategy 2

Another way to approach this problem is to estimate the strain in the shared bond in A and equate this value to the overall strain in the molecule. To do so, we need to determine the corresponding bond strength and compare it to that of a presumed unstrained model, for example, the central bond in 2,3-dimethylbutane, $DH^\circ = 85.5$ kcal mol^{-1} (Table 3-2). How do we do this? We can use the heat of the reaction given in the problem and apply the equation given in Section 2-1, in which the enthalpy change of a reaction was related to the changes in bond strengths.

Solution

- Rewrite the reaction of A with hydrogen and label the relevant bonds with the available data (in kcal mol^{-1}) from Tables 3-1 and 3-2.

$$\Delta H^\circ = -56 \text{ kcal mol}^{-1}$$

- Apply the equation from Section 2-1:

$$\Delta H^\circ = \Sigma(\text{strengths of bonds broken}) - \Sigma(\text{strengths of bonds made})$$
$$-56 = (104 + x) - 197$$
$$x = 37 \text{ kcal mol}^{-1}$$

- This is a very weak bond indeed! Compared to the central C–C bond in 2,3-dimethylbutane (85.5 kcal mol^{-1}), its strain is 48.5 kcal mol^{-1}.
- Does this number reflect the strain in A completely? (**Caution:** Not quite, because the product cyclopentane has some residual strain of 6.5 kcal mol^{-1} that is not released in the reaction of A.) Hence, a reasonable estimate of the ring strain in A is $48.5 + 6.5 = 55$ kcal mol^{-1}, which is pretty close to our "quick solution," the sum of the ring strains of the component rings (53.9 kcal mol^{-1}). Gratifyingly, it is also close to the experimental value, based on the heat of combustion: 57.3 kcal mol^{-1}.

Exercise 4-6 | Try It Yourself

The strain energy in the hydrocarbon A shown in the margin is 50.7 kcal mol^{-1}. Estimate its heat of reaction with hydrogen to give cyclohexane.

A

Model Building

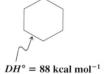

$DH° = 88$ kcal mol^{-1}

Chair

Boat

A chair and a boat. Do you see them in cyclohexane?

4-3 | CYCLOHEXANE: A STRAIN-FREE CYCLOALKANE

The cyclohexane ring is one of the most common and important structural units in organic chemistry. Its substituted derivatives exist in many natural products (see Section 4-7), and an understanding of its conformational mobility is an important aspect of organic chemistry. Table 4-2 reveals that, within experimental error, cyclohexane is unusual in that it is free of bond-angle or eclipsing strain. Why?

The chair conformation of cyclohexane is strain free

A hypothetical planar cyclohexane would suffer from 12 H–H eclipsing interactions and sixfold bond-angle strain (a regular hexagon requires 120° bond angles). However, one conformation of cyclohexane, obtained by moving carbons 1 and 4 out of planarity in opposite directions, is in fact strain free (Figure 4-5). This structure is called the **chair conformation** of cyclohexane (because it resembles a chair), in which eclipsing is completely absent, and the bond angles are very nearly tetrahedral. As seen in Table 4-2, the calculated $\Delta H°_{comb}$ of cyclohexane (-944.4 kcal mol^{-1}) based on a strain-free $(CH_2)_6$ model is very close to the experimentally determined value (-944.5 kcal mol^{-1}). Indeed, the C–C bond strength, $DH° = 88$ kcal mol^{-1} (368 kJ mol^{-1}), is normal (Table 3-2).

Looking at the molecular model of cyclohexane enables us to recognize the conformational stability of the molecule. If we view it along (any) one C–C bond, we can see the staggered arrangement of all substituent groups along it. We can visualize this arrangement by drawing a Newman projection of that view (Figure 4-6). Because of its lack of strain, cyclohexane is as inert as a straight-chain alkane.

Cyclohexane also has several less stable conformations

Cyclohexane also adopts other, less stable conformations. One is the **boat form,** in which carbons 1 and 4 are out of the plane in the *same* direction (Figure 4-7). The boat is less stable than the chair form by 6.9 kcal mol^{-1}. One reason for this difference is the eclipsing of eight hydrogen atoms at the base of the boat. Another is steric hindrance (Section 2-9) due to the close

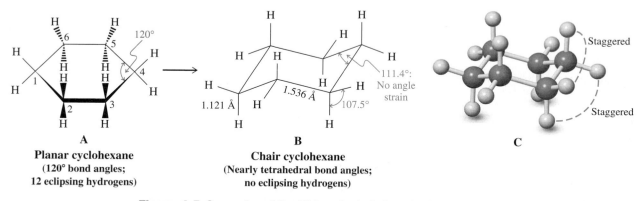

A
Planar cyclohexane
(120° bond angles;
12 eclipsing hydrogens)

B
Chair cyclohexane
(Nearly tetrahedral bond angles;
no eclipsing hydrogens)

C

Figure 4-5 Conversion of the (A) hypothetical planar cyclohexane into the (B) chair conformation, showing bond lengths and angles; (C) molecular model. The chair conformation is strain free.

Figure 4-6 View along one of the C–C bonds in the chair conformation of cyclohexane. Note the staggered arrangement of all substituents.

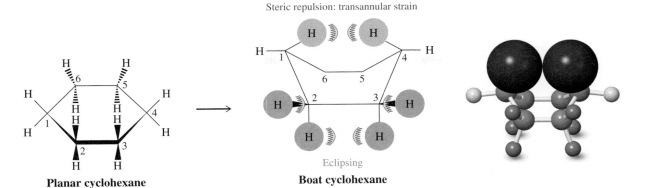

Planar cyclohexane   **Boat cyclohexane**

Figure 4-7 Conversion of the hypothetical planar cyclohexane into the boat form. In the boat form, the hydrogens on carbons 2, 3, 5, and 6 are eclipsed, thereby giving rise to torsional strain. The "inside" hydrogens on carbons 1 and 4 interfere with each other sterically in a transannular interaction. The space-filling size of these two hydrogens, reflecting the actual size of their respective electron clouds, is depicted in the ball-and-stick model on the right.

proximity of the two inside hydrogens in the boat framework. The distance between these two hydrogens is only 1.83 Å, small enough to create an energy of repulsion of about 3 kcal mol^{-1} (13 kJ mol^{-1}). This effect is an example of **transannular strain,** that is, strain resulting from steric crowding of two groups across a ring (*trans,* Latin, across; *anulus,* Latin, ring).

Boat cyclohexane is fairly flexible. If one of the C–C bonds is twisted relative to another, this form can be somewhat stabilized by partial removal of the transannular interaction. The new conformation obtained is called the **twist-boat** (or **skew-boat**) **conformation** of cyclohexane (Figure 4-8). The stabilization relative to the boat form amounts to about 1.4 kcal mol^{-1}. As shown in Figure 4-8, two twist-boat forms are possible. They interconvert rapidly, with the boat conformer acting as a *transition state* (verify this with your model). Thus, the boat cyclohexane is not a normally isolable species, the twist-boat form is present in very small amounts, and the chair form is the major conformer (Figure 4-9). The activation barrier separating the most stable chair from the boat forms is 10.8 kcal mol^{-1}. We shall see that the equilibration depicted in Figure 4-9 has important structural consequences with respect to the positions of substituents on the cyclohexane ring.

Cyclohexane has axial and equatorial hydrogen atoms

The chair-conformation model of cyclohexane reveals that the molecule has two types of hydrogens. Six carbon–hydrogen bonds are nearly parallel to the principal molecular axis (Figure 4-10) and hence are referred to as **axial;** the other six are nearly perpendicular to the axis and close to the equatorial plane and are therefore called **equatorial.***

Being able to draw cyclohexane chair conformations will help you learn the chemistry of six-membered rings. Several rules are useful.

How to Draw Chair Cyclohexanes

1. Draw the chair so as to place the C2 and C3 atoms below and slightly to the right of C5 and C6, with apex 1 pointing downward on the left and apex 4 pointing upward on the right. Ideally, bonds straight across the ring (namely, bonds 1–6 and 3–4; 2–3 and 5–6; 1–2 and 4–5) should appear parallel to one another.

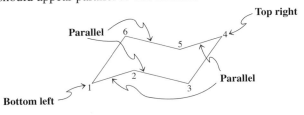

Boat

Twist (skew) boat

Figure 4-8 Twist-boat to twist-boat flipping of cyclohexane proceeds through the boat conformation.

Animation

ANIMATION: Fig. 4-9, cyclohexane potential energy diagram

---

*An equatorial plane is defined as being perpendicular to the axis of rotation of a rotating body and equidistant from its poles, such as the equator of the planet Earth.

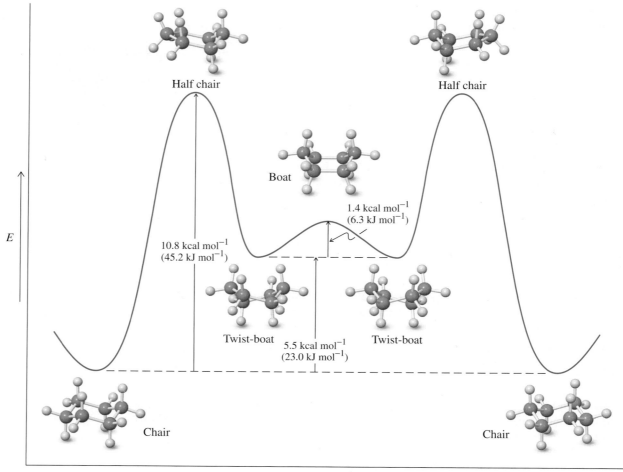

E

Reaction coordinate to conformational interconversion of cyclohexane ⟶

Figure 4-9 Potential-energy diagram for the chair–chair interconversion of cyclohexane through the twist-boat and boat forms. In the progression from left to right, the chair is converted into a twist boat (by the twisting of one of the C–C bonds) with an activation barrier of 10.8 kcal mol^{-1}. The transition state structure is called a half chair. The twist-boat form flips (as shown in Figure 4-8) through the boat conformer as a transition state (1.4 kcal mol^{-1} higher in energy) into another twist-boat structure, which relaxes back into the (ring-flipped) chair cyclohexane. Use your molecular models to visualize these changes.

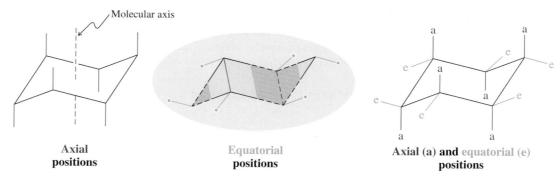

Figure 4-10 The axial and equatorial positions of hydrogens in the chair form of cyclohexane. The blue shading represents the equatorial plane encompassing the (blue) equatorial hydrogens. The yellow and green shaded areas are located above and below that plane, respectively.

2. Add all the axial bonds as vertical lines, pointing downward at C1, C3, and C5 and upward at C2, C4, and C6. In other words, the axial bonds alternate up-down around the ring.

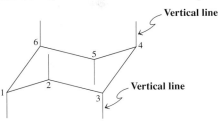

3. Draw the two equatorial bonds at C1 and C4 at a slight angle from horizontal, pointing upward at C1 and downward at C4, parallel to the bond between C2 and C3 (or between C5 and C6).

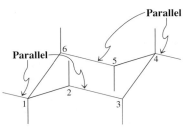

4. This rule is the most difficult to follow: Add the remaining equatorial bonds at C2, C3, C5, and C6 by aligning them *parallel* to the C–C bond "once removed," as shown below.

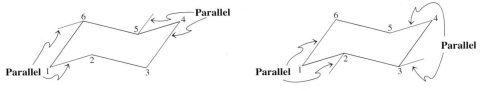

Exercise 4-7

(a) Draw Newman projections of the carbon–carbon bonds in cyclopropane, cyclobutane, cyclopentane, and cyclohexane in their most stable conformations. Use the models that you prepared for Exercise 4-1 to assist you and refer to Figure 4-6. What can you say about the (approximate) torsion angles between adjacent C–H bonds in each? (b) Draw the following molecules in their chair conformation. Place the ring atom at the top of the flat stencils shown below so that it appears at the top right of the chair rendition.

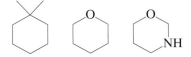

Conformational flipping interconverts axial and equatorial hydrogens

What happens to the identity of the equatorial and axial hydrogens when we let chair cyclohexane equilibrate with its boat forms? You can follow the progress of conformational interconversion in Figure 4-9 with the help of molecular models. Starting with the chair structure on the left, you simply "flip" the CH$_2$ group farthest to the left (C1 in the preceding section) upward through the equatorial plane to generate the boat conformers. If you now return the molecule to the chair form not by a reversal of the movement but by the

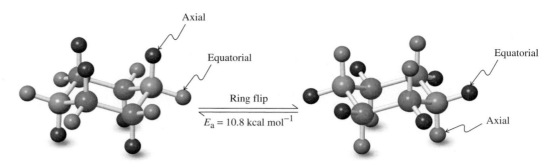

Figure 4-11 Chair–chair interconversion ("ring flipping") in cyclohexane. In the process, which is rapid at room temperature, a (green) carbon at one end of the molecule moves up while its counterpart (also green) at the other end moves down. All groups originally in axial positions (red in the structure at the left) become equatorial, and those that start in equatorial positions (blue) become axial.

Animation

ANIMATION: Fig. 4-11, cyclohexane ring flip

equally probable alternative—namely, the flipping downward of the opposite CH_2 group (C4)—the original sets of axial and equatorial positions have traded places. In other words, cyclohexane undergoes chair–chair interconversions ("flipping") in which all axial hydrogens in one chair become equatorial in the other and vice versa (Figure 4-11). The activation energy for this process is 10.8 kcal mol^{-1} (Figure 4-9). As suggested in Sections 2-8 and 2-9, this value is so low that, at room temperature, the two chair forms interconvert rapidly (approximately 200,000 times per second).

The two chair forms shown in Figure 4-11 are, except for the color coding, identical. We can lift this degeneracy by introducing substituents: Now the chair with a substituent in the equatorial position is different from its conformer, in which the substituent is axial. The preference for one orientation over the other strongly affects the stereochemistry and reactivity of cyclohexanes. We will describe the consequences of such substitution in the next section.

In Summary The discrepancy between calculated and measured heats of combustion in the cycloalkanes can be largely attributed to three forms of strain: bond angle (deformation of tetrahedral carbon), eclipsing (torsional), and transannular (across the ring). Because of strain, the small cycloalkanes are chemically reactive, undergoing ring-opening reactions. Cyclohexane is strain free. It has a lowest-energy chair as well as additional higher-energy conformations, particularly the boat and twist-boat structures. Chair–chair interconversion is rapid at room temperature; it is a process in which equatorial and axial hydrogen atoms interchange their positions.

4-4 | SUBSTITUTED CYCLOHEXANES

We can now apply our knowledge of conformational analysis to substituted cyclohexanes. Let us look at the simplest alkylcyclohexane, methylcyclohexane.

Axial and equatorial methylcyclohexanes are not equivalent in energy

In methylcyclohexane, the methyl group occupies either an equatorial or an axial position. Are the two forms equivalent? Clearly not. In the equatorial conformer, the methyl group extends into space away from the remainder of the molecule. In contrast, in the axial conformer, the methyl substituent is close to the other two axial hydrogens on the same side of the molecule. The distance to these hydrogens is small enough (about 2.7 Å) to result in steric repulsion, another example of transannular strain. Because this effect is due to axial substituents on carbon atoms that have a 1,3-relation (in the drawing, 1,3 and 1,3′), it is called a **1,3-diaxial interaction.** This interaction is the same as that resulting in the *gauche* confor-

Combustion of hydrocarbons (Sections 3-11 and 4-2) typically starts with the abstraction of an H atom by O_2. Methylcyclohexane burns particularly well, because of the presence of the relatively weak tertiary C–H bond (Section 3-1), and is therefore used as an additive to jet fuel.

mation of butane (Section 2-9). Thus, the axial methyl group is *gauche* to two of the ring carbons (C3 and C3′); when it is in the equatorial position, it is *anti* to the same nuclei.

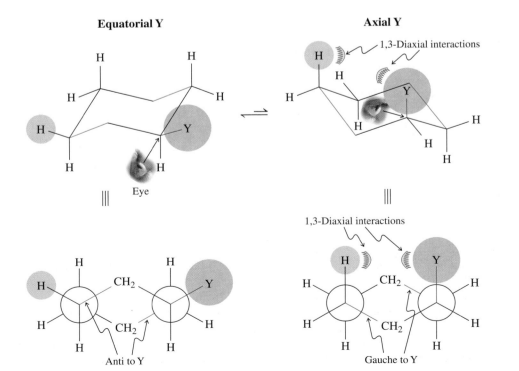

No 1,3-Diaxial interactions
More stable

1,3-Diaxial interactions
Less stable

Ratio = 95:5

Model Building

The two forms of chair methylcyclohexane are in equilibrium. *The equatorial conformer is more stable* by 1.7 kcal mol^{-1} (7.1 kJ mol^{-1}) and is favored by a ratio of 95:5 at 25°C (Section 2-1). The activation energy for chair–chair interconversion is similar to that in cyclohexane itself [about 11 kcal mol^{-1} (46 kJ mol^{-1})], and equilibrium between the two conformers is established rapidly at room temperature.

The unfavorable 1,3-diaxial interactions to which an axial substituent is exposed are readily seen in Newman projections of the ring C–C bond bearing that substituent. In contrast with that in the axial form (*gauche* to two ring bonds), the substituent in the equatorial conformer (*anti* to the two ring bonds) is away from the axial hydrogens (Figure 4-12).

Exercise 4-8

Calculate K for equatorial versus axial methylcyclohexane from the $\Delta G°$ value of 1.7 kcal mol^{-1}. Use the expression $\Delta G°$ (in kcal mol^{-1}) = $-1.36 \log K$. (**Hint:** If $\log K = x$, then $K = 10^x$.) How well does your result agree with the 95:5 conformer ratio stated in the text?

Equatorial Y

Axial Y

1,3-Diaxial interactions

Eye

1,3-Diaxial interactions

Anti to Y

Gauche to Y

Figure 4-12 A Newman projection of a substituted cyclohexane. The conformation with an axial Y substituent is less stable because of 1,3-diaxial interactions, only one of which is shown (see the "eye" for the choice of the Newman projection). Axial Y is *gauche* to the ring bonds shown in green; equatorial Y is *anti*.

Table 4-3	Change in Free Energy on Flipping from the Cyclohexane Conformer with the Indicated Substituent Equatorial to the Conformer with the Substituent Axial		

Substituent	$\Delta G°$ [kcal mol^{-1} (kJ mol^{-1})]	Substituent	$\Delta G°$ [kcal mol^{-1} (kJ mol^{-1})]
H	0 (0)	F	0.25 (1.05)
CH$_3$	1.70 (7.11)	Cl	0.52 (2.18)
CH$_3$CH$_2$	1.75 (7.32)	Br	0.55 (2.30)
(CH$_3$)$_2$CH	2.20 (9.20)	I	0.46 (1.92)
(CH$_3$)$_3$C	≈ 5 (21)		
O‖HOC	1.41 (5.90)	HO	0.94 (3.93)
		CH$_3$O	0.75 (3.14)
		H$_2$N	1.4 (5.9)
O‖CH$_3$OC	1.29 (5.40)		

Note: In all examples, the more stable conformer is the one in which the substituent is equatorial.

(arrow: Increasing size)   (arrow: Increasing $\Delta G°$)

The energy differences between the axial and the equatorial forms of many monosubstituted cyclohexanes have been measured; several are given in Table 4-3. In many cases (but not all), particularly for alkyl substituents, the energy difference between the two forms increases with the size of the substituent, a direct consequence of increasing unfavorable 1,3-diaxial interactions. This effect is particularly pronounced in (1,1-dimethylethyl)cyclohexane (*tert*-butylcyclohexane). The energy difference here is so large (about 5 kcal mol^{-1}) that very little (about 0.01%) of the axial conformer is present at equilibrium.

Solved Exercise 4-9	Working with the Concepts: Building Models to Visualize Sterics

The $\Delta G°$ value for the equatorial to axial flip of cyclohexylcyclohexane is the same as that for (1-methylethyl)cyclohexane, 2.20 kcal mol^{-1}. Is this reasonable? Explain.

Strategy
When trying to deal with conformational problems, a good strategy is to build molecular models.

Solution
- Molecular models reveal that cyclohexylcyclohexane may be viewed as a cyclic analog of (1-methylethyl)cyclohexane, in which the two methyl groups have been connected by a (CH$_2$)$_3$ chain (Section 4-1).
- Both structures are floppy and have several conformers, but you will find that the seemingly more bulky (chair) cyclohexyl substituent in (cyclohexyl)cyclohexane can rotate away from the cyclohexane core in such a way as to avoid 1,3-diaxial contact beyond that encountered by its (1-methylethyl) counterpart. Thus, the equivalence of the free energy changes is reasonable.

A

B

Exercise 4-10	Try It Yourself

The isomeric hydrocarbons A and B (see margin) both exhibit preferred conformations in which the methyl groups are equatorial, yet B is more stable than A by 2.3 kcal mol^{-1}. What is the origin of this difference? (**Caution:** Are A and B ring flip isomers? **Hint:** Build models and look at the nature of the conformations of the methyl substituted cyclohexane ring.)

Model Building

Substituents compete for equatorial positions

To predict the more stable conformer of a more highly substituted cyclohexane, the cumulative effect of placing substituents either axially or equatorially must be considered, in addition to their potential mutual 1,3-diaxial or 1,2-*gauche* (Section 2-9) interactions. For many cases, we can ignore the last two and simply apply the values of Table 4-3 for a prediction.

Let us look at some isomers of dimethylcyclohexane to illustrate this point. In 1,1-dimethylcyclohexane, one methyl group is always equatorial and the other axial. The two chair forms are identical, and hence their energies are equal.

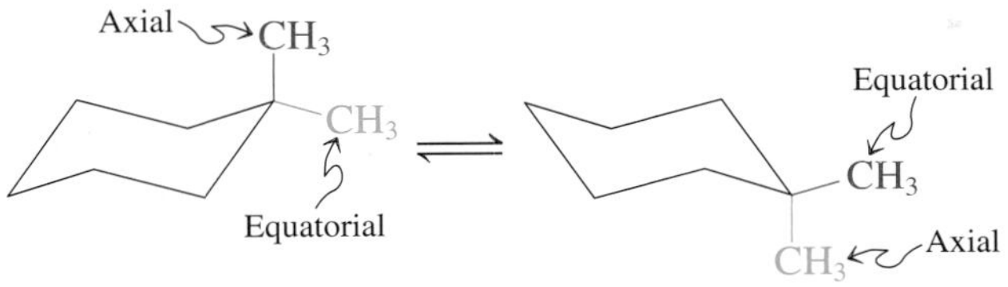

One CH₃ axial
One CH₃ equatorial

One CH₃ axial
One CH₃ equatorial

1,1-Dimethylcyclohexane
(Conformations equal in energy, equally stable)

Similarly, in *cis*-1,4-dimethylcyclohexane, both chairs have one axial and one equatorial substituent and are of equal energy.

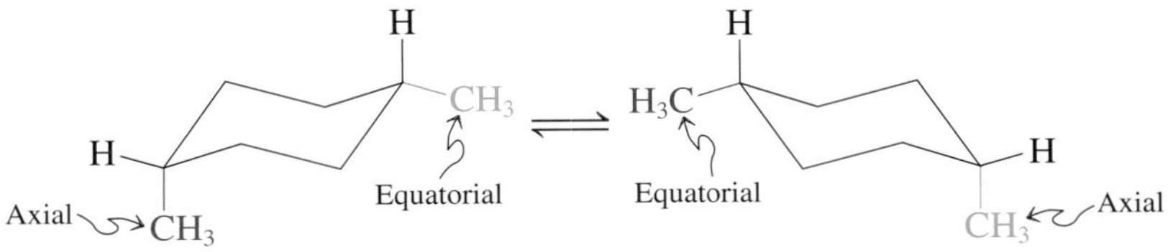

One axial, one equatorial

One axial, one equatorial

***cis*-1,4-Dimethylcyclohexane**

> The bonds to both methyl groups point downward; they are cis (i.e., on the same face of the ring) *regardless* of conformation.

On the other hand, the trans isomer can exist in two different chair conformations: one having two axial methyl groups (diaxial) and the other having two equatorial groups (diequatorial).

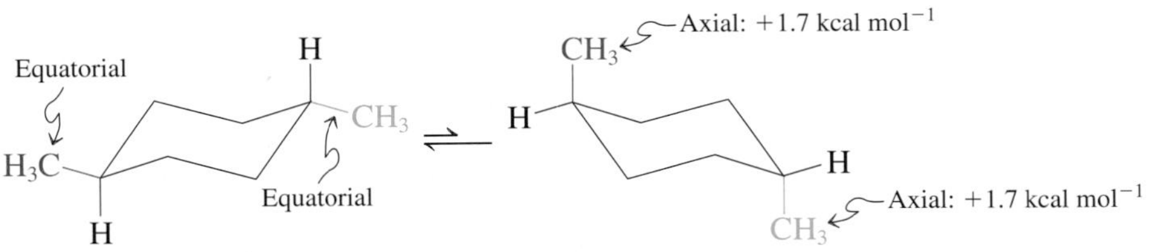

Diequatorial methyls
More stable

Diaxial methyls
Less stable: +3.4 kcal mol⁻¹ (14.2 kJ mol⁻¹)

***trans*-1,4-Dimethylcyclohexane**

> The bond to one methyl group points downward, the other upward. They are trans (i.e., on opposite faces of the ring) *regardless* of conformation.

Experimentally, the diequatorial form is preferred over the diaxial form by 3.4 kcal mol⁻¹, exactly twice the $\Delta G°$ value for monomethylcyclohexane. Indeed, this additive behavior of the data given in Table 4-3 applies to many other substituted cyclohexanes. For example, the $\Delta G°$ (diaxial ⇌ diequatorial) for *trans*-1-fluoro-4-methylcyclohexane is -1.95 kcal mol⁻¹ $[-(1.70$ kcal mol⁻¹ for CH_3 plus 0.25 kcal mol⁻¹ for F)]. Conversely, in *cis*-1-fluoro-4-methylcyclohexane, the two groups compete for the equatorial positions and the corresponding $\Delta G° = -1.45$ kcal mol⁻¹ $[-(1.70$ kcal mol⁻¹ minus 0.25 kcal mol⁻¹)], with the larger methyl winning out over the smaller fluorine.

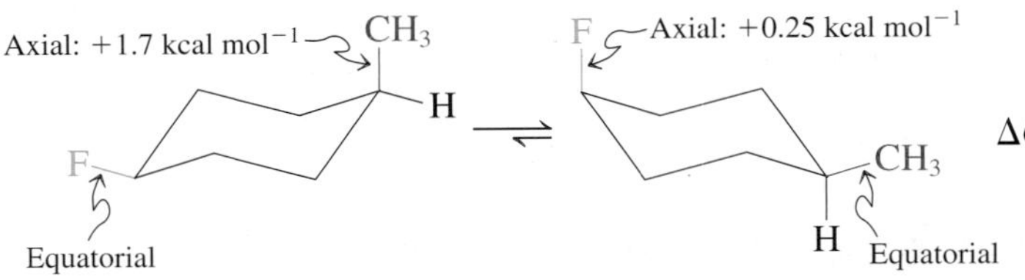

$\Delta G° = -1.45$ kcal mol⁻¹ (-6.07 kJ mol⁻¹)

Large group axial
Small group equatorial
Less stable

Small group axial
Large group equatorial
More stable

***cis*-1-Fluoro-4-methylcyclohexane**

Conformational Drug Design

Fentanyl is an extensively used, potent opioid analgesic.

Fentanyl
100 times as potent as morphine

It has been proposed that the drug binds to its opioid receptor site in the form of the less stable ring-flip conformer containing an axial amino substituent. Indeed, strategic introduction of an adjacent cis- (but not trans! Make models!) methyl group (which prefers the equatorial position) places the nitrogen axial, causing a dramatic increase in activity.

***cis*-Methylfentanyl**
2600 times as potent as morphine

Exercise 4-11

Calculate $\Delta G°$ for the equilibrium between the two chair conformers of **(a)** 1-ethyl-1-methylcyclohexane; **(b)** *cis*-1-ethyl-4-methylcyclohexane; **(c)** *trans*-1-ethyl-4-methylcyclohexane.

Exercise 4-12

Draw both chair conformations for each of the following isomers: **(a)** *cis*-1,2-dimethylcyclohexane; **(b)** *trans*-1,2-dimethylcyclohexane; **(c)** *cis*-1,3-dimethylcyclohexane; **(d)** *trans*-1,3-dimethylcyclohexane. Which of these isomers always have equal numbers of axial and equatorial substituents? Which exist as equilibrium mixtures of diaxial and diequatorial forms?

Solved Exercise 4-13 | Working with the Concepts: Finding the Most Stable Cycloalkane

(a) Rank the following four isomeric hydrocarbons in order of decreasing stability: (a) *cis*-1-ethyl-2-propylcyclopropane; (b) 1-cyclopropylpentane; (c) ethylcyclohexane; (d) *cis*-1,4-dimethylcyclohexane.

What does the problem ask? To establish an approximate order of energy content for the four structures, and convert that to relative stability.

How to start? Drawing the four structures carefully is essential in order to recognize what features are present that may stabilize or destabilize each one by changing its energy content. Notice that this problem includes a nomenclature component: Each name must be converted to a structure. Finally, to make the comparison fair, we draw the most stable possible conformation of each one. Making models can be very helpful for visualizing the molecules in three dimensions.

Information: The lower the energy content of a structure, the more stable it is. Steric crowding raises energy content and decreases stability (Section 2-9). So: A staggered conformation is more stable than an eclipsed one; *anti* is more stable than *gauche*; equatorial is more stable than axial (Section 4-4). In each case the more stable option has less steric crowding. Are there other destabilizing effects? Small rings are subject to bond-angle strain (Section 4-2). We will need data in order to evaluate the quantitative importance of each of these issues.

Proceed: Draw all four structures. Use the guidance of Section 4-3 to draw good chair cyclohexanes. As indicated above, identify each structural feature that affects stability. How important is each one? Find the quantitative information: Bond-angle strain in cyclopropanes is the largest destabilizing factor (27.6 kcal mol^{-1}; Section 4-2). Then we must add to that the enforced eclipsing in (b): at least as bad as the 4.9 kcal mol^{-1} in the highest-energy conformation of butane (Figure 2-13). Cyclohexanes do not suffer from bond-angle strain, but (d) cannot avoid one axial substituent (1.7 kcal mol^{-1}, Table 4-3). So, we have

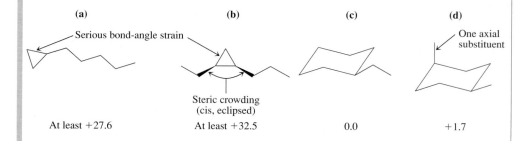

(a)	**(b)**	**(c)**	**(d)**
Serious bond-angle strain	Steric crowding (cis, eclipsed)		One axial substituent
At least +27.6	At least +32.5	0.0	+1.7

Approximate relative energies in kcal mol^{-1} are given below each structure. With these data in hand, apply the logic: Higher energy content = less stable. The stability order is therefore (c) most stable, then (d), then (a), and finally (b) least stable. (If needed, you could even give approximate energy differences by adding the energy-content increments for each compound.)

(b) Although the substituent values in Table 4-3 are additive and may be used to indicate the position of the equilibrium between two substituted cyclohexane conformers, the observed $\Delta G°$ values can be perturbed by additional 1,3-diaxial or 1,2-*gauche* interactions between groups. For example, like *trans*-1,4-dimethylcyclohexane, its isomer *cis*-1,3-dimethylcyclohexane exists in a diequatorial–diaxial equilibrium and hence should exhibit the same $\Delta G°$ value of 3.4 kcal mol^{-1}. However, the measured value is larger (5.4 kcal mol^{-1}). Explain. (**Hint:** For *cis*-1,3-dimethylcyclohexane, look closely at all 1,3-diaxial interactions and compare them with those of diaxial *trans*-1,4-dimethylcyclohexane.)

Strategy
Again, a good strategy for addressing conformational issues is to build models. First construct a model for *cis*-1,3-dimethylcyclohexane and perform the ring flip from diequatorial to diaxial. What is different in this system compared with methylcyclohexane "taken twice"?

Solution
- In the diequatorial conformer, the two methyl groups are each located in an area of space that is the same as that in methylcyclohexane.
- Is this true also for the diaxial conformer? The answer is no. The $\Delta G° = 1.7$ kcal mol^{-1} for ring flip in axial methylcyclohexane arises from two CH$_3$/H-1,3-diaxial interactions [one CH$_3$ with two hydrogens, as in the picture for methylcyclohexane shown on p. 145; see also Figure 4-12], 0.85 kcal mol^{-1} each. In *cis*-1,3-dimethylcyclohexane, the diaxial conformation suffers also from two CH$_3$/H interactions (two CH$_3$ groups with a single H) for the same total of 1.7 kcal mol^{-1}. The extra strain originates from the *proximity of the two axial methyls*, amounting to 3.7 kcal mol^{-1}.

- Therefore, the diaxial conformation of *cis*-1,3-dimethylcyclohexane is less stable than the diequatorial form by more than 3.4 kcal mol^{-1}.

Exercise 4-14 | **Try It Yourself**

Like *cis*-1,3-dimethylcyclohexane in Exercise 4-13, its isomer *trans*-1,2-dimethylcyclohexane exists in a diequatorial–diaxial equilibrium and hence would be expected to exhibit the same $\Delta G°$ value as *trans*-1,4-dimethylcyclohexane, namely, 3.4 kcal mol^{-1}. However, the measured value is smaller (2.5 kcal mol^{-1}). Explain. (**Hint:** Take into consideration the proximity of the two methyl groups; see *gauche–anti* butane, Section 2-9.)

In Summary The conformational analysis of cyclohexane enables us to predict the relative stability of its various conformers and even to approximate the energy differences between two chair conformations. Bulky substituents, particularly a 1,1-dimethylethyl group, tend to shift the chair–chair equilibrium toward the side in which the large substituent is equatorial.

The largest man-made cycloalkane is $C_{288}H_{576}$. Because of attractive London forces (Section 2-7), the molecule curls up into a spherical shape.

4-5 | LARGER CYCLOALKANES

Do similar relations hold for the larger cycloalkanes? Table 4-2 shows that cycloalkanes with rings larger than that of cyclohexane also have more strain. This strain is due to a combination of bond-angle distortion, partial eclipsing of hydrogens, and transannular steric repulsions. It is not possible for medium-sized rings to relieve all of these strain-producing interactions in a

Model Building

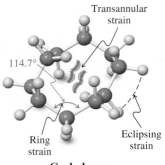

Transannular strain

114.7°

Ring strain

Eclipsing strain

Cyclodecane

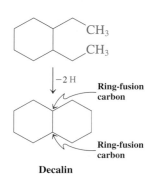

CH_3
CH_3

$\downarrow$ −2 H

Ring-fusion carbon

Ring-fusion carbon

Decalin

Model Building

single conformation. Instead, a compromise solution is found in which the molecule equilibrates among several geometries that are very close in energy. One such conformation of cyclodecane, which has a strain energy of 14 kcal mol⁻¹ (59 kJ mol⁻¹), is shown in the margin.

Essentially strain-free conformations are attainable only for large-sized cycloalkanes, such as cyclotetradecane (Table 4-2). In such rings, the carbon chain adopts a structure very similar to that of the straight-chain alkanes (Section 2-7), having staggered hydrogens and an all-*anti* configuration. However, even in these systems, the attachment of substituents usually introduces various amounts of strain. Most cyclic molecules described in this book are not strain free.

4-6 | POLYCYCLIC ALKANES

The cycloalkanes discussed so far contain only one ring and therefore may be referred to as monocyclic alkanes. In more complex structures—the bi-, tri-, tetra-, and higher poly-cyclic hydrocarbons—two or more rings share carbon atoms. Many of these compounds exist in nature with various alkyl or functional groups attached. Let us look at some of the wide variety of possible structures.

Polycyclic alkanes may contain fused or bridged rings

Molecular models of polycyclic alkanes can be readily constructed by linking the carbon atoms of two alkyl substituents in a monocyclic alkane. For example, if you remove two hydrogen atoms from the methyl groups in 1,2-diethylcyclohexane and link the resulting two CH_2 groups, the result is a new molecule with the common name decalin. In decalin, two cyclohexanes share two adjacent carbon atoms, and the two rings are said to be **fused.** Compounds constructed in this way are called **fused bicyclic** ring systems and the shared carbon atoms are called the **ring-fusion carbons.** Groups attached to ring-fusion carbons are called **ring-fusion substituents.**

When we treat a molecular model of *cis*-1,3-dimethylcyclopentane in the same way, we obtain another carbon skeleton, that of norbornane. Norbornane is an example of a **bridged bicyclic** ring system. In bridged bicyclic systems, two nonadjacent carbon atoms, the **bridgehead** carbons, belong to both rings.

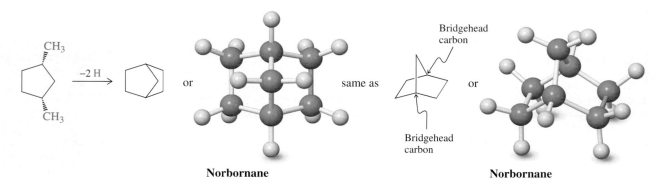

CH_3

−2 H →

or

same as

Bridgehead carbon

or

Bridgehead carbon

CH_3

Norbornane

Norbornane

If we think of one of the rings as a substituent on the other, we can identify stereochemical relations at ring fusions. In particular, bicyclic ring systems can be cis or trans fused. The stereochemistry of the ring fusion is most easily determined by inspecting the ring-fusion substituents. For example, the ring-fusion hydrogens of *trans*-decalin are trans with respect to each other, whereas those of *cis*-decalin have a cis relation (Figure 4-13).

Exercise 4-15

Construct molecular models of both *cis*- and *trans*-decalin. What can you say about their conformational mobility?

Figure 4-13 Conventional drawings and chair conformations of *trans*- and *cis*-decalin. The trans isomer contains only equatorial carbon–carbon bonds at the ring fusion, whereas the cis isomer possesses two equatorial C–C bonds (green) and two axial C–C linkages (red), one with respect to each ring.

Equatorial C–C bonds

trans-Decalin

Axial C–C bonds   Equatorial C–C bonds

cis-Decalin

Strain = 66.5 kcal mol⁻¹
Bicyclobutane

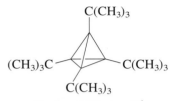

Strain = 129 kcal mol⁻¹
Tetrakis(1,1-dimethylethyl)-tetrahedrane

Do hydrocarbons have strain limits?

Seeking the limits of strain in hydrocarbon bonds is a fascinating area of research that has resulted in the synthesis of many exotic molecules. What is surprising is how much bond-angle distortion a carbon atom is able to tolerate. A case in point in the bicyclic series is bicyclobutane, whose strain energy is 66.5 kcal mol⁻¹ (278 kJ mol⁻¹), making it remarkable that the molecule exists at all. Yet it can be isolated and stored.

A series of strained compounds attracting the attention of synthetic chemists possess a carbon framework geometrically equivalent to the Platonic solids: the *tetrahedron* (tetrahedrane), the *hexahedron* (cubane), and the pentagonal *dodecahedron* (dodecahedrane; see margin). In these polyhedra, all faces are composed of equally sized rings—namely, cyclopropane, cyclobutane, and cyclopentane, respectively. The hexahedron was synthesized first in 1964, a C_8H_8 hydrocarbon shaped like a cube and accordingly named cubane. The experimental strain energy [166 kcal mol⁻¹ (695 kJ mol⁻¹)] is more than the total strain of six cyclobutanes. Although tetrahedrane itself is unknown, a tetra(1,1-dimethylethyl) derivative was synthesized in 1978. Despite the measured strain (from ΔH°_{comb}) of 129 kcal mol⁻¹ (540 kJ mol⁻¹), the compound is stable and has a melting point of 135°C. The synthesis of dodecahedrane was achieved in 1982. It required 23 synthetic operations, starting from a simple cyclopentane derivative. The last step gave 1.5 mg of pure compound. Although small, this amount was sufficient to permit complete characterization of the molecule. Its melting point at 430°C is extraordinarily high for a C_{20} hydrocarbon and is indicative of the symmetry of the compound. For comparison, icosane, also with 20 carbons, melts at 36.8°C (Table 2-5). As you might expect on the basis of its component five-membered rings, the strain in dodecahedrane is "only" 61 kcal mol⁻¹ (255 kJ mol⁻¹), much less than that of its lower homologs.

Tetrahedrane (C₄H₄)

Strain = 166 kcal mol⁻¹
Cubane (C₈H₈)

In Summary Carbon atoms in bicyclic compounds are shared by rings in either fused or bridged arrangements. A great deal of strain may be tolerated by carbon in its bonds, particularly to other carbon atoms. This capability has allowed the preparation of molecules in which carbon is severely deformed from its tetrahedral shape.

Strain = 61 kcal mol⁻¹
Dodecahedrane (C₂₀H₂₀)

4-7 CARBOCYCLIC PRODUCTS IN NATURE

Let us now take a brief look at the variety of cyclic molecules created in nature. **Natural products** are organic compounds produced by living organisms. Some of these compounds, such as methane, are extremely simple; others have great structural complexity. Scientists have attempted to classify the multitude of natural products in various ways. Generally, four schemes are followed, in which these products are classified according to (1) chemical structure, (2) physiological activity, (3) organism or plant specificity (taxonomy), and (4) biochemical origin.

The brief glimpse into the world of polycyclic hydrocarbons in Section 4-6 provides further examples of the diversity of the carbon frame in organic chemistry (see Table 2-4 for the number of possible isomeric acyclic alkanes). Whenever possible, such molecules adopt structures in which cyclohexane rings exist in the chair configuration, for example, *trans*-decalin. We constructed this molecule by fusing one cyclohexane ring to another at the equatorial positions. Another way of building all-chair cyclohexane polycycles is via the axial positions. Thus, if we add three axial CH_2 groups to cyclohexane and connect them by a single CH unit, we arrive at an all-chair tetracyclic cage, $C_{10}H_{16}$, called adamantane (*adamantinos,* Greek, diamond, read on why). Build a model and you will recognize its simple symmetry: All four faces are identical.

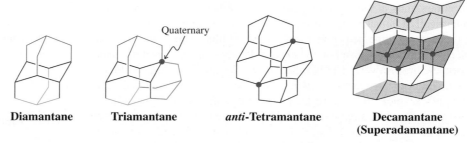

Cyclohexane   Adamantane

Adamantane is the most stable of the family of $C_{10}H_{16}$ isomers. Its tight shape allows it to pack unusually well in the solid, as reflected in the high melting point of 270°C (see Section 2-7). For comparison, its acyclic counterpart decane melts at −29.7°C (Table 2-5). Adamantane was discovered in crude oil in 1933, can be readily synthesized, and is now commercially available in kilogram quantities.

If we apply the capping process described above to three axial positions of any of the rings in adamantane, we get diamantane, a polycycle composed of seven chair cyclohexanes and endowed with six equivalent faces. Continued capping gives triamantane, the first in this series to contain a quaternary carbon. From triamantane, we can construct three tetramantanes in this way, only one of which is shown,

Quaternary

Diamantane   **Triamantane**   *anti*-**Tetramantane**   **Decamantane (Superadamantane)**

bearing two quaternary carbons. The number of possible isomers increases rapidly for the higher "oligomantanes"; there are 7 pentamantanes, 24 hexamantanes, and 88 heptamantanes. Decamantane, $C_{35}H_{36}$, is a unique isomer dubbed "superadamantane" because of its compact symmetry. It exhibits a quaternary carbon surrounded by four other quaternary carbons. Its structure illustrates how one can envisage building a three-dimensional array of layers of linked chair cyclohexanes (highlighted in color) made up of pure carbon. Such a polymer is known: It is the crystalline form of carbon named diamond (see also Real Life 15-1)!

Because the oligomantanes represent "microdiamonds" in which all peripheral carbons are saturated by hydrogen (so-called hydrogen-terminated diamond), they have been given the name *diamandoids*. The higher diamandoids have only been known since 2003 (see Chapter 1 Opening), when a group at Chevron in California identified members of the series up to undecamantane in high-boiling petroleum distillates. Interest in these molecules stems from not only their relationship to diamond and their potential to act as seed crystals in industrial diamond production, but also from their stability and inertness. Thus, they provide biologically compatible scaffolds for new drugs, cosmetics, and polymeric materials.

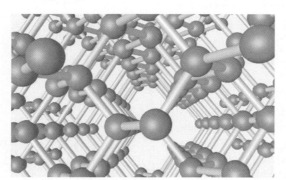

Inside the carbon lattice of diamond.

The Golden Jubilee, the world's biggest diamond.

Organic chemists are interested in natural products for many reasons. Many of these compounds are powerful drugs, others function as coloring or flavoring agents, and yet others are important raw materials. A study of animal secretions furnishes information concerning the ways in which animals use chemicals to mark trails, defend themselves from predators, and attract the opposite sex. Investigations of the biochemical pathways by which an organism metabolizes and otherwise transforms a compound are sources of insight into the detailed workings of the organism's bodily functions. Two classes of natural products, terpenes and steroids, have received particularly close attention from organic chemists.

A Peruvian Matsés Indian collecting frogs to extract "sapo" frog poison used for medicinal and ritual purposes.

Terpenes are constructed in plants from isoprene units

Most of you have smelled the strong odor emanating from freshly crushed plant leaves or orange peels. This odor is due to the liberation of a mixture of volatile compounds called **terpenes,** usually containing 10, 15, or 20 carbon atoms. Well over 30,000 terpenes are found in nature. They are used as food flavorings (the extracts from cloves and mint), as perfumes (roses, lavender, sandalwood), and as solvents (turpentine).

Terpenes are synthesized in the plant by the linkage of at least two molecular units, each containing five carbon atoms. The structure of these units is like that of 2-methyl-1,3-butadiene (isoprene), and so they are referred to as **isoprene units.** Depending on how many isoprene units are incorporated into the structure, terpenes are classified as mono- (C_{10}), sesqui- (C_{15}), or diterpenes (C_{20}). (The isoprene building units are shown in color in the examples given here.)

2-Methyl-1,3-butadiene
(Isoprene)

Isoprene unit in terpenes
(Some contain double bonds)

Chrysanthemic acid is a monocyclic terpene containing a three-membered ring. Its esters are found in the flower heads of pyrethrum (*Chrysanthemum cinerariaefolium*) and are naturally occurring insecticides. A cyclobutane is present in grandisol, the sex-attracting chemical used by the male boll weevil (*Anthonomus grandis*).

trans-**Chrysanthemic acid (R = H)**
trans-**Chrysanthemic esters (R ≠ H)**

Grandisol

Many ceremonial objects are made of pleasant-smelling sandalwood. The black bull shown is the focal point of a Balinese funeral.

Menthol (peppermint oil) is an example of a substituted cyclohexane natural product, whereas camphor (from the camphor tree) and β-cadinene (from juniper and cedar) are simple bicyclic terpenes, the first a norbornane system, the second a decalin derivative.

Menthol

Camphor

β-Cadinene

Taxol (paclitaxel) is a complex, functionalized diterpene isolated from the bark of the Pacific yew tree, *Taxus brevifolia,* in 1962 as part of a National Cancer Institute program in search of natural products exhibiting anticancer activity.

The Pacific yew tree: the source of taxol.

Taxol

Taxol proved to be perhaps the most interesting of more than 100,000 compounds extracted from more than 35,000 plant species and is one of the most effective antitumor drugs on the market. Because roughly six trees must be sacrificed to treat one patient, there has been much effort to improve efficacy and availability and to increase yields. Most of this work has been carried out by synthetic organic chemists, leading to the first two total syntheses of taxol in 1994. (A "total" synthesis is one that makes the target starting from simple, preferably commercial, compounds containing five carbons or less). In addition, chemists found ways to convert more abundant natural products of related structure to taxol itself, a procedure described as "semisynthesis." As a result, the taxol used in the treatment of cancer is derived from a compound found in the needles of the common English yew, a much more readily available and nonsacrificial source.

Exercise 4-16

Draw the more stable chair conformation of menthol.

Exercise 4-17

The structures of two terpenes utilized by insects in defense (see Section 12-17) are shown in the margin. Classify them as mono-, sesqui-, or diterpenes. Identify the isoprene units in each.

Exercise 4-18

After reviewing Section 2-4, specify the functional groups present in the terpenes shown in Section 4-7.

Really? Steroids are ubiquitous but not necessarily abundant in nature. In order to obtain the first samples of pure, crystalline epiandrosterone (in 1931), German chemists Adolf Butenandt (Nobel prize in 1939) and Kurt Tscherning had to distill over 17,000 L of male urine to separate 50 mg of material.

Steroids are tetracyclic natural products with powerful physiological activities

Steroids are abundant in nature, and many derivatives have physiological activity. They function frequently as **hormones,** which are regulators of biochemical processes. In the human body, they control sexual development and fertility, in addition to other functions. Because of this feature, many steroids, often the products of laboratory synthesis, are used in the treatment of cancer, arthritis, allergies, and in birth control.

In the steroids, three cyclohexane rings are fused in such a way as to form an angle, also called **angular.** The ring junctions are usually trans, as in *trans*-decalin. The fourth ring is a cyclopentane; its addition gives the typical tetracyclic structure. The four rings are labeled A, B, C, D, and the carbons are numbered according to a scheme specific to steroids. Many steroids have methyl groups attached to C10 and C13 and oxygen at C3 and C17. In addition, longer side chains may be found at C17. The trans fusion of the rings allows for a least-strained all-chair configuration in which the methyl groups and hydrogen atoms at the ring junctions occupy axial positions. These features are illustrated below with epiandrosterone, a steroid found in normal human urine.

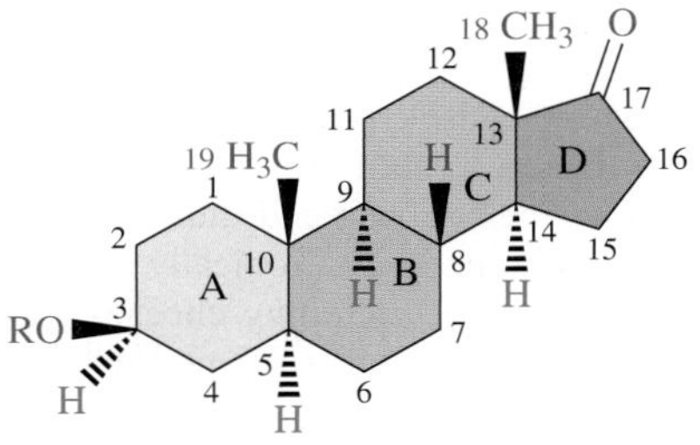

**Steroid nucleus
(R = H, Epiandrosterone)**

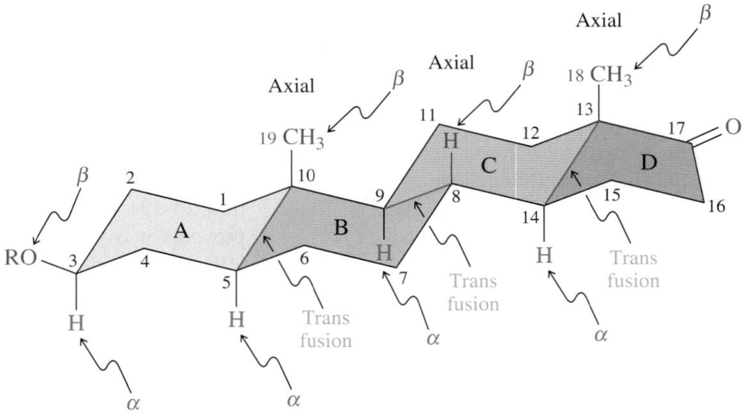

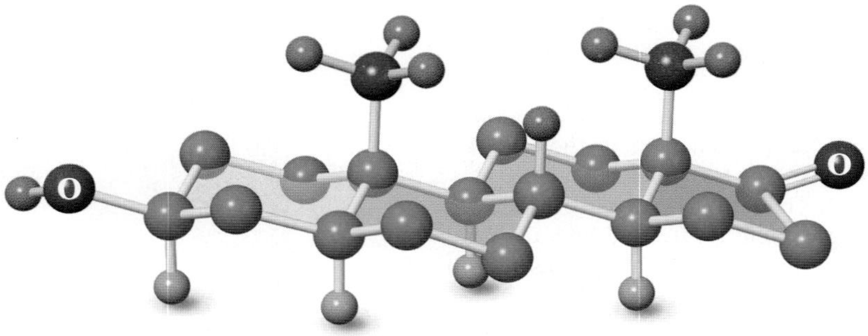

Epiandrosterone

Groups attached above the plane of the steroid molecule as written are β substituents, whereas those below are referred to as α. Thus, the structure of epiandrosterone has a 3β-OH, 5α-H, 10β-CH$_3$, and so forth.

Among the most abundant steroids, cholesterol is present in almost all human and animal tissue (Real Life 4-2). Bile acids are produced in the liver as part of a fluid delivered to the duodenum to aid in the emulsification, digestion, and absorption of fats. An example is cholic acid. Cortisone, used extensively in the treatment of rheumatoid inflammations, is one of the adrenocortical hormones produced by the outer part (cortex) of the adrenal glands. These hormones participate in regulating the electrolyte and water balance in the body, as well as in protein and carbohydrate metabolism.

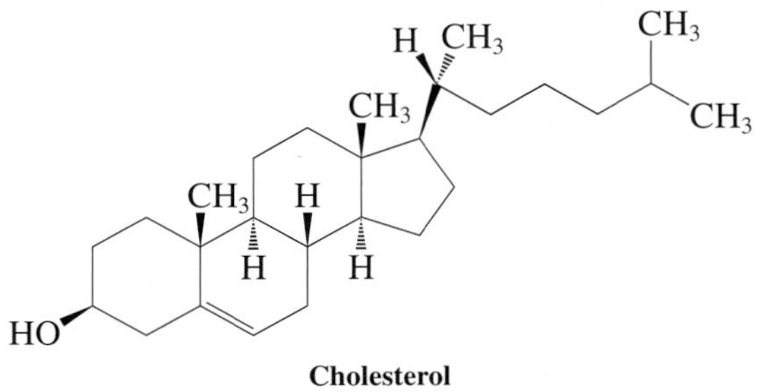

Cholesterol

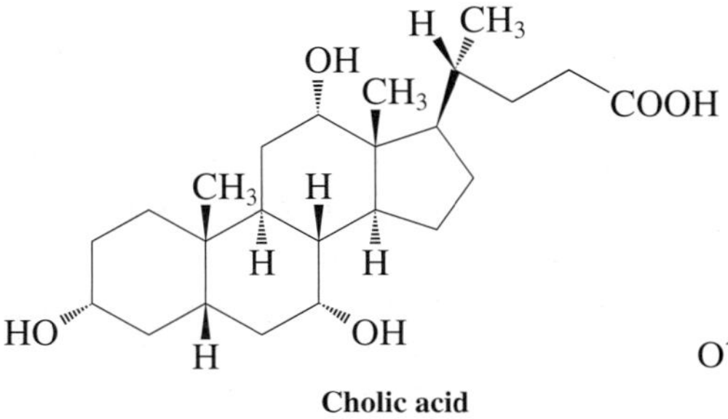

Cholic acid

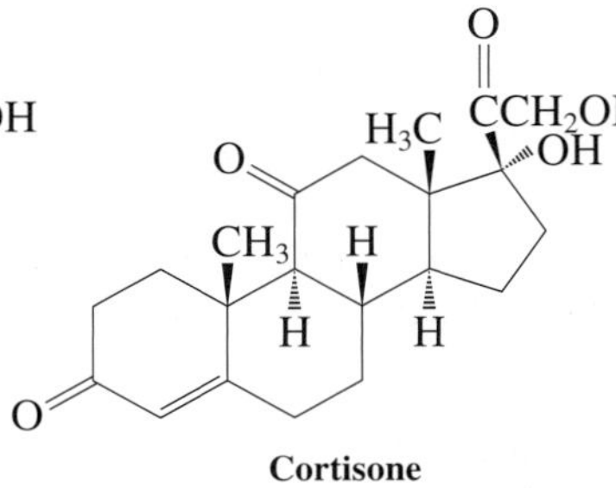

Cortisone

The sex hormones are divided into three groups: (1) the male sex hormones, or *androgens;* (2) the female sex hormones, or *estrogens;* and (3) the pregnancy hormones, or *progestins.*

REAL LIFE: MEDICINE 4-2   Cholesterol: How Is It Bad and How Bad Is It?

"Too much cholesterol!" How often have you heard this admonition as you were about to dig into your favorite three-egg breakfast or chocolate-fudge dessert? The reason for the warning is that high levels of cholesterol have been implicated in atherosclerosis and heart disease. Atherosclerosis is the buildup of plaques that can narrow or even block your blood vessels. In the heart, such an event can cause a heart attack. Plaques can break and travel through your bloodstream, causing havoc elsewhere. For example, a blocked blood vessel in the brain can lead to a stroke.

About 20% of the population of the United States has cholesterol levels that exceed the recommended total. Typical adults have about 150 g in their body and for a good reason—cholesterol is vital for the running of the body. It is an essential building block of cell membranes, especially of the nervous system, the brain, and the spinal cord. It is also a key chemical intermediate in the biological production of other steroids, especially the sex and the corticoid hormones, including cortisone. We need cholesterol to produce bile acids, which in turn are key chemicals that help digest the fats we consume, and we need it to form vitamin D, which enables us to utilize calcium in bone construction.

You have probably heard much talk about "good" and "bad" cholesterol. These adjectives refer to the proteins that attach themselves to the molecule to make so-called lipoproteins, which are water soluble and hence allow cholesterol to be transported in the blood (for the general structure of proteins, see Section 26-4). There are two types of these aggregates, low-density lipoprotein (LDL) and high-density lipoprotein (HDL). LDL carries cholesterol from the liver to other parts of the body, wherever it is needed. HDL acts as a cholesterol sweep and delivers it to the liver for conversion into bile acids. If this careful equilibrium is disturbed such that too much LDL is present, that is "bad," because the excess is deposited in the form of the dangerous plaques mentioned earlier. Interestingly, a study in 2012 found that raising HDL levels has no beneficial effect with respect to the risk of heart disease. Hence, the term "good" for HDL is a misnomer.

The opening warning notwithstanding, very little cholesterol comes from our food. Rather, our body, especially our liver, makes about a gram a day, which is four times as much as even a high-cholesterol diet would supply. Why is there so much fuss, then? It has to do with that 15 to 20%

from our food intake. In people with moderately elevated cholesterol levels, watching their diet may make all the difference. This is where fats come in (see also Section 20-5). To digest fats, we need bile acids. Excessive fat intake stimulates bile acid production, in turn increasing cholesterol synthesis and therefore cholesterol blood levels. Therefore, a balanced low-fat, cholesterol-free diet helps in maintaining the recommended level of cholesterol: 200 mg per 100 mL of blood.

When such a diet is not enough, drugs come to the rescue. One type of drug, such as hydroxypropyl methylcellulose (HPMC; for the structure of cellulose, see Section 24-12), binds cholesterol in the stomach and thus prevents it from being absorbed by the body. Ironically, HPMC is used as a thickener in foods, including cheesecake and desserts! Another type of drug turns off production of cholesterol directly in the liver and has been used with spectacular success in the last decade. Examples are atorvastatin (Lipitor) and rosuvastatin (Crestor), both "top-10" prescription drugs with combined sales of close to $20 billion (see Table 25-1).

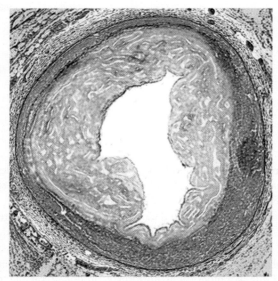

Cross section of a human coronary artery of the heart blocked by cholesterol plaques (yellow).

Testosterone is the principal male sex hormone. Produced by the testes, it is responsible for male (masculine) characteristics (deep voice, facial hair, general physical constitution). Synthetic testosterone analogs are used in medicine to promote muscle and tissue growth (anabolic steroids; *ana-*, Greek, up—i.e., "anabolic," the opposite of "metabolic"), for example, in patients with muscular atrophy. Unfortunately, such steroids are also abused and consumed illegally, most commonly by "body builders" and athletes, even though the health risks are numerous, including liver cancer, coronary heart disease, and sterility. Estradiol is the principal female sex hormone. It was first isolated by extraction of 4 tons of sow ovaries, yielding only a few milligrams of pure steroid. Estradiol is responsible for the development of the secondary female characteristics and participates in the control of the menstrual cycle.

An example of a progestin is progesterone, responsible for preparing the uterus for implantation of the fertilized egg.

Testosterone

Estradiol

Progesterone

The structural similarity of the steroid hormones is remarkable, considering their widely divergent activity. Steroids are the active ingredients of "the pill," functioning as an antifertility agent for the control of the female menstrual cycle and ovulation. It is estimated that over 100 million women throughout the world take "the pill" as the primary form of contraception (Real Life: Medical 4-3). Despite these advances, almost half of the pregnancies in the United States are unintended, more than 3 million per year.

In Summary There is great variety in the structure and function of naturally occurring organic products, as manifested in the terpenes and the steroids. Natural products will be frequently introduced in subsequent chapters to illustrate the presence and chemistry of a functional group, to demonstrate synthetic strategy or the use of reagents, to picture three-dimensional relations, and to exemplify medicinal applications. Several classes of natural products will be discussed more extensively: fats (Sections 19-13 and 20-5), carbohydrates (Chapter 24), alkaloids (Section 25-8), and amino and nucleic acids (Chapter 26).

REAL LIFE: MEDICINE 4-3 | Controlling Fertility: From "the Pill" to RU-486 to Male Contraceptives

The menstrual cycle is controlled by three protein hormones from the pituitary gland. The follicle-stimulating hormone (FSH) induces the growth of the egg, and the luteinizing hormone (LH) induces its release from the ovaries and the formation of an ovarian tissue called the corpus luteum. The third pituitary hormone (luteotropic hormone, also called luteotropin or prolactin), stimulates the corpus luteum and maintains its function.

As the cycle begins and egg growth is initiated, the tissue around the egg secretes increasing quantities of estrogens. When a certain concentration of estrogen in the bloodstream has been reached, the production of FSH is turned off. The egg is released at this stage in response to LH. At the time of ovulation, LH also triggers the formation of the corpus luteum, which in turn begins to secrete increasing amounts of progesterone. This last hormone suppresses any further ovulation by turning off the production of LH. If the egg is not fertilized, the corpus luteum regresses and the ovum and the endometrium (uterine lining) are expelled (men-

A fertilized human egg before cleavage (*zygote*), approximately 100 μm in size.

struation). Pregnancy, on the other hand, leads to increased production of estrogens and progesterone to prevent pituitary hormone secretion and thus renewed ovulation.

The birth control pill consists of a mixture of synthetic potent estrogen and progesterone derivatives (more potent than the natural hormones), which, when taken throughout most of the menstrual cycle, prevent both development of the ovum and ovulation by turning off production of both FSH and LH. The female body is essentially being tricked into believing that it is pregnant. Some of the commercial pills contain a combination of norethindrone and ethynylestradiol. Other preparations consist of similar analogs with minor structural variations.

Norethindrone, R = CH₃
Levonorgestrel, R = CH₃CH₂

Ethynylestradiol

RU-486 (mifepristone)

Gestrinone   **Tetrahydrogestrinone (THG)**

Postcoital contraceptives, such as Preven or Plan B, contain levonorgestrel and/or ethynylestradiol. RU-486 (mifepristone) is a synthetic steroid that binds to the progesterone receptors in the woman's uterus, thus blocking the action of progesterone. Used in combination with a prostaglandin (Real Life 11-1) that induces uterine contractions, RU-486 effects abortion when administered in early pregnancy. The drug has been available in Europe since 1988, and, after much discussion and testing, the Food and Drug Administration approved it for the U.S. market in 2000.

You will have noticed the presence of the C≡C triple bond attached to C17 in these synthetic compounds, an alteration that makes the drugs particularly potent. Such triple bonds can be hydrogenated ("saturated") with hydrogen in the presence of a catalyst (Section 13-6). It is suspected that it was this simple alteration of the commercial drug gestrinone that led to the dopant tetrahydrogestrinone (THG; see Chapter Opening).

What about male contraceptives? For a variety of reasons, including the simple fact that the "pill" works, but also cultural biases and the formidable challenges associated with developing any drug, research has been lagging. Nevertheless, there has been interest since ancient times in exploring ingredients of herbs, seeds, plants, and fruits as male antifertility agents. Modern pharmaceutical strategies are varied, including targeting testosterone production or activity. For example, the steroid trestolone significantly reduces sperm count by inhibiting the release of testosterone and other hormones that are vital for spermatogenesis. Importantly, the so-induced sterility is reversible. A number of nonsteroidal structures show promise by affecting a male's reproductive system in other ways. Thus, nifedipine alters the metabolism of sperm so as to sabotage its ability to fertilize an egg, whereas phenoxybenzamine blocks ejaculation. None of these compounds has made it to market yet.

Trestolone   **Nifedipine**   **Phenoxybenzamine**

THE BIG PICTURE

In this chapter, we have expanded our knowledge of organic structure and function to include cyclic and polycyclic frameworks. In particular, we have seen again how much three-dimensionality matters in explaining and predicting the behavior of organic molecules. Their structures are ultimately based on the spatial distribution of bonding electrons surrounding the nuclear components (Sections 1-3 and 1-6 to 1-9) and, from a more mechanical perspective, on the conformational flexibility inherent in C–C bonds (Sections 2-8 and 2-9).

One consequence of three-dimensionality is the existence of stereoisomers, namely cis and trans isomers of substituted cycloalkanes (Section 4-1). We shall see in Chapter 5 that the phenomenon of stereoisomerism is more general and occurs in acyclic molecules as well. These concepts influence such diverse areas as relative reactivities and biological effectiveness. Because of its fundamental importance and its powerful utility in biological applications, stereochemistry constitutes a recurring theme through the remainder of this book.

WORKED EXAMPLES: INTEGRATING THE CONCEPTS

4-19 Practicing the Drawing of Cyclohexane Stereoisomers

a. 1,2,3,4,5,6-Hexachlorocyclohexane exists as a number of cis-trans isomers. Using the flat cyclohexane stencil and hashed-wedged lines, draw all of them.

SOLUTION

Before starting to solve this part by random trial and error, consider a more systematic approach. We start with the simplest case of all chlorines positioned cis to one another (using all wedged lines) and then look at the various permutations obtained by placing a progressively increasing number of the substituents trans (hashed lines). We can stop at the stage of three chlorines "up" and three chlorines "down" because "two up and four down" is the same as "four up and two down," and so on. The first two cases, "six up" (A) and "five up and one down" (B), are unique because only one structural possibility exists for each:

A   B   C   D   E

Now we look at the "four up and two down" isomers. There are three differing ways of placing two chlorines "down": Their positioning along the six-membered ring can be only 1,2 (C), 1,3 (D), or 1,4 (E), because the options 1,5 and 1,6 are the same as 1,3 and 1,2, respectively. Finally, a similar line of thought leads to the realization that placing three chlorines "down" can be done only in three unique ways: 1,2,3 (F), 1,2,4 (G), and 1,3,5 (H):

F   G   H

b. The so-called γ-isomer (margin) is an insecticide (lindane, Gammexane, Kwell) for the treatment of food crops, livestock, and pets. It is best known in its formulation as a shampoo for the eradication of lice in humans. Because of its toxicity, its use was severely restricted in 2009. Draw the two chair conformations of this compound. Which is more stable?

γ-Hexachlorocyclohexane

SOLUTION

Note that the γ-isomer corresponds to our structure E in (a). To help in converting the flat cyclohexane stencil structures into their chair renditions, look at Figure 4-11 and note the alternating relation of the members of the two sets of hydrogens. As we proceed around the ring, we can see that neighboring members (i.e., those in a "1,2"-relation) of either set (i.e., axial or equatorial neighbors) always have a trans disposition. On the other hand, the 1,3-relation is always cis, whereas that describable as 1,4 is again trans. Conversely, in a hashed-wedged line structure, when two neighboring substituents (i.e., 1,2-relation) are cis, one of the substituents will be axial and the other equatorial in either of the two chair conformations of the molecule. When they are trans, one chair picture will show them diequatorial, the other diaxial. This assignment alternates in 1,3-related substituents: cis results in a diaxial-diequatorial pair of chair conformers, trans in an axial-equatorial or equatorial-axial pair, and so on. These relations are summarized in Table 4-4.

Table 4-4	Relation of Cis-Trans Stereochemistry in Substituted Cyclohexanes to Equatorial-Axial Positions in the Two Chair Forms	
cis-1,2	Axial-equatorial	Equatorial-axial
trans-1,2	Axial-axial	Equatorial-equatorial
cis-1,3	Axial-axial	Equatorial-equatorial
trans-1,3	Axial-equatorial	Equatorial-axial
cis-1,4	Axial-equatorial	Equatorial-axial
trans-1,4	Axial-axial	Equatorial-equatorial

The two chair forms of γ-hexachlorocyclohexane thus look as shown here:

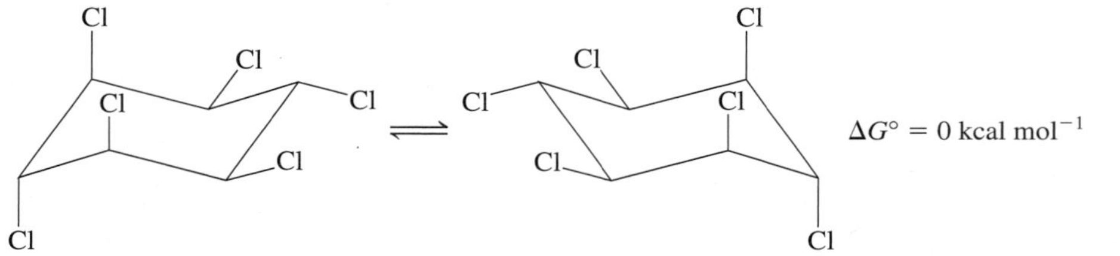

$\Delta G° = 0 \text{ kcal mol}^{-1}$

Either structure has three equatorial and three axial chlorine substituents; hence they are equal in energy.

c. For which isomer do you expect the energy difference between the two cyclohexane chair forms to be largest? Estimate the $\Delta G°$ value.

SOLUTION

The biggest $\Delta G°$ for chair–chair flip is that between an all-equatorial and all-axial hexachlorocyclohexane. Application of Table 4-4 and inspection of Figure 4-11 reveal that this relation holds only for the all-trans isomer. Table 4-3 gives the $\Delta G°$ (equatorial-axial) for Cl as 0.52 kcal mol^{-1}; hence, for our example, $\Delta G° = 6 \times 0.52 = 3.12$ kcal mol^{-1}.

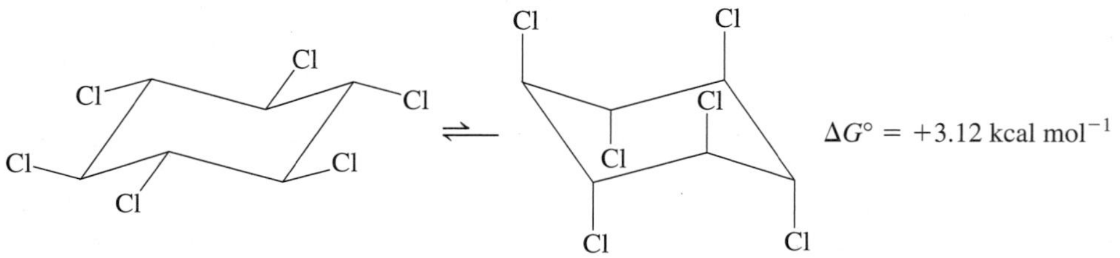

$\Delta G° = +3.12 \text{ kcal mol}^{-1}$

all-*trans*-Hexachlorocyclohexane

Note that this value is only an estimate. For example, it ignores the six Cl–Cl *gauche* interactions in the all-equatorial form, which would reduce the energy difference between the two conformers. However, it also disregards the six 1,3-diaxial interactions in the all-axial chair, which would counteract this effect. For additional examples try Problems 30–33.

4-20 Formulating a Mechanism

a. The strained tetracyclic alkane A is thought to isomerize thermally to the cyclic alkene B. Suggest a possible mechanism.

SOLUTION

When dealing with mechanistic problems, it is advantageous to carefully assimilate all of the given information, not unlike the situation encountered by an experimentalist who may have observed such a reaction and is faced with the same challenge. First, compound A is said to be strained. How so? *Answer* (Sections 4-2 and 4-6): It is a polycyclic alkane containing three linked four-membered rings. Second, the A-to-B conversion is an isomerization. We can ascertain quickly the truth of this statement by establishing that both compounds have the same molecular formula, $C_{12}H_{18}$. Third, the process is "purely" thermal: There are no other reagents that could attack A, add to it, or subtract from it. Fourth, what topological (connectivity) changes have occurred? *Answer:* A tetracycle is unraveled to a monocycle; in other words, the three cyclobutane rings are opened, leaving a cyclododecatriene. Fifth, what are the qualitative bond changes? *Answer:* Three single bonds are lost, three double bonds are gained.

Armed with the results of this analysis, we can now write a mechanism: We need to break the three C–C bonds by pyrolysis (Section 3-3). The result is seemingly a very exotic hexaradical; however, closer inspection shows that the six radical centers divide into three respective pairs of electrons on adjacent carbons: the three double bonds. Thus, a possible mechanism of this process requires only one step.

b. Regardless of the mechanism, is the isomerization thermodynamically feasible? Estimate the approximate $\Delta H°$, using 65 kcal mol^{-1} for the strength of the π bond (Section 11-2).

SOLUTION

First, let us be clear about what the question asks. It does *not* ask us whether it is kinetically feasible to break the three cyclobutane bonds, either as shown in part (a) or by some other sequence. Instead, the issue is where the equilibrium A ⇌ B lies: Is it in the forward direction (Section 2-1)? We have learned how to get an estimate of the answer: by calculating $\Delta H° =$ (sum of strengths of bonds broken) − (sum of strengths of bonds made) (Sections 2-1 and 3-4). The second term in this equation is easy: We are making three π bonds, worth 195 kcal mol^{-1}. How do we deal with the three σ bonds that are broken? Table 3-2 provides an estimate for the strength of a $C_{sec}–C_{sec}$ bond: 85.5 kcal mol^{-1}. In our case, this bond is weakened by ring strain (Section 4-2, Table 4-2): $85.5 − 26.3 = 59.2$ kcal mol^{-1}. Multiplied by 3, this provides the first term of our $\Delta H°$ equation, 177.6 kcal mol^{-1}. Thus, the $\Delta H°$ (A ⇌ B) ≈ −17.4 kcal mol^{-1}. Conclusion: Indeed, this reaction should go! Note how the strain inherent in structure A affects its function in the thermolysis. Without strain, the reaction would be endothermic.

Important Concepts

1. **Cycloalkane** nomenclature is derived from that of the straight-chain alkanes.

2. All but the 1,1-disubstituted cycloalkanes exist as two isomers: If both substituents are on the same face of the molecule, they are **cis;** if they are on opposite faces, they are **trans.** Cis and trans isomers are **stereoisomers**—compounds that have identical connectivities but differ in the arrangement of their atoms in space.

3. Some cycloalkanes are **strained.** Distortion of the bonds about tetrahedral carbon introduces **bond-angle strain. Eclipsing (torsional) strain** results from the inability of a structure to adopt staggered conformations about C–C bonds. Steric repulsion between atoms across a ring leads to **transannular strain.**

4. Bond-angle strain in the small cycloalkanes is largely accommodated by the formation of bent bonds.

5. Bond-angle, eclipsing, and other strain in the cycloalkanes larger than cyclopropane (which is by necessity flat) can be accommodated by deviations from planarity.

6. Ring strain in the small cycloalkanes gives rise to reactions that result in opening of the ring.

7. Deviations from planarity lead to conformationally mobile structures, such as **chair, boat,** and **twist-boat** cyclohexane. Chair cyclohexane is almost strain free.

8. Chair cyclohexane contains two types of hydrogens: **axial** and **equatorial.** These interconvert rapidly at room temperature by a conformational **chair–chair ("flip") interconversion,** with an activation energy of 10.8 kcal mol^{-1} (45.2 kJ mol^{-1}).

9. In monosubstituted cyclohexanes, the $\Delta G°$ of equilibration between the two chair conformations is substituent dependent. Axial substituents are exposed to **1,3-diaxial interactions.**

10. In more highly substituted cyclohexanes, substituent effects are often **additive,** the bulkiest substituents being the most likely to be equatorial.

11. Completely strain-free cycloalkanes are those that can readily adopt an all-*anti* conformation and lack transannular interactions.

12. **Bicyclic** ring systems may be **fused or bridged.** Fusion can be cis or trans.

13. Natural products are generally classified according to structure, physiological activity, taxonomy, and biochemical origin. Examples of the last class are the **terpenes,** of the first the **steroids.**

14. Terpenes are made up of **isoprene** units of five carbons.

15. Steroids contain three angularly fused cyclohexanes (A, B, C rings) attached to the cyclopentane D ring. Beta substituents are above the molecular plane, alpha substituents below.

16. An important class of steroids are the **sex hormones,** which have a number of physiological functions, including the control of fertility.

Problems

21. Write as many structures as you can that have the formula C_5H_{10} and contain one ring. Name them.

22. Write as many structures as you can that have the formula C_6H_{12} and contain one ring. Name them.

23. Name the following molecules according to the IUPAC nomenclature system.

(a)

(b) H$_3$C   CH(CH$_3$)$_2$

(c)   Cl   Cl

(d)   CH$_3$

(e)   Br   Br

(f)   Br   Br

24. Draw structural representations of each of the following molecules. Give a systematic name for any compound whose name is not in accord with IUPAC nomenclature: **(a)** isobutylcyclopentane; **(b)** cyclopropylcyclobutane; **(c)** cyclohexylethane; **(d)** (1-ethylethyl)cyclohexane; **(e)** (2-chloropropyl)cyclopentane; **(f)** *tert*-butylcycloheptane.

25. Draw structural representations of each of the following molecules: **(a)** *trans*-1-chloro-2-ethylcyclopropane; **(b)** *cis*-1-bromo-2-chlorocyclopentane; **(c)** 2-chloro-1,1-diethylcyclopropane; **(d)** *trans*-2-bromo-3-chloro-1,1-diethylcyclopropane; **(e)** *cis*-1,3-dichloro-2,2-dimethylcyclobutane; **(f)** *cis*-2-chloro-1,1-difluoro-3-methylcyclopentane.

26. The kinetic data for the radical chain chlorination of several cycloalkanes (see the table below) illustrate that the C–H bonds of cyclopropane and, to a lesser extent, cyclobutane are somewhat abnormal. **(a)** What do these data tell you about the strength of the cyclopropane C–H bond and the stability of the cyclopropyl radical? **(b)** Suggest a reason for the stability characteristics of the cyclopropyl radical. (**Hint:** Consider bond-angle strain in the radical relative to cyclopropane itself.)

Reactivity per Hydrogen Toward Cl·	
Cycloalkane	Reactivity
Cyclopentane	0.9
Cyclobutane	0.7
Cyclopropane	0.1

Note: Relative to cyclohexane = 1.0; at 68°C, *hv*, CCl$_4$ solvent.

27. Write out a mechanism for the radical monobromination of cyclohexane, showing initiation, propagation, and termination steps. Draw the product in its most stable conformation.

28. Use the data in Tables 3-2 and 4-2 to estimate the $DH°$ value for a C–C bond in (**a**) cyclopropane; (**b**) cyclobutane; (**c**) cyclopentane; and (**d**) cyclohexane.

29. Draw each of the following substituted cyclobutanes in its two interconverting "puckered" conformations (Figure 4-3). When the two conformations differ in energy, identify the more stable shape and indicate the form(s) of strain that raise the relative energy of the less stable one. (**Hint:** Puckered cyclobutane has axial and equatorial positions similar to those in chair cyclohexane.)

(**a**) Methylcyclobutane

(**b**) *cis*-1,2-Dimethylcyclobutane

(**c**) *trans*-1,2-Dimethylcyclobutane

(**d**) *cis*-1,3-Dimethylcyclobutane

(**e**) *trans*-1,3-Dimethylcyclobutane

Which is more stable: *cis*- or *trans*-1,2-dimethylcyclobutane; *cis*- or *trans*-1,3-dimethylcyclobutane?

30. For each of the following cyclohexane derivatives, indicate (i) whether the molecule is a cis or trans isomer and (ii) whether it is in its most stable conformation. If your answer to (ii) is no, flip the ring and draw its most stable conformation.

(**a**) Cl / CH₃

(**b**) OCH₃ / NH₂

(**c**) HO / CH(CH₃)₂

(**d**) CH₃ / C(=O)OCH₃

(**e**) CH₂CH₃ / I

(**f**) Br / NH₂

(**g**) H₃C / OCH₃

(**h**) F / HO

(**i**) O=C–OH / Br

(**j**) Cl / CH(CH₃)₂

31. Using the data in Table 4-3, calculate the $\Delta G°$ for ring flip to the other conformation of the molecules depicted in Problem 30. Make sure that the sign (i.e., positive or negative) of your values is correct.

32. Draw the most stable conformation for each of the following substituted cyclohexanes; then, in each case, flip the ring and redraw the molecule in the higher energy chair conformation: (**a**) cyclohexanol; (**b**) *trans*-3-methylcyclohexanol (see structures below); (**c**) *cis*-1-methyl-3-(1-methylethyl)cyclohexane; (**d**) *trans*-1-ethyl-3-methoxycyclohexane (see structure below); (**e**) *trans*-1-chloro-4-(1,1-dimethylethyl)cyclohexane.

OH
Cyclohexanol

OH / ''CH₃
***trans*-3-Methylcyclohexanol**

CH₂CH₃ / ''OCH₃
***trans*-1-Ethyl-3-methoxycyclohexane**

33. For each molecule in Problem 32, estimate the energy difference between the most stable and next best conformation. Calculate the approximate ratio of the two at 300 K.

34. Sketch a potential-energy diagram (similar to that in Figure 4-9) for methylcyclohexane showing the two possible chair conformations at the left and right ends of the reaction coordinate for conformational interconversion.

35. Draw all the possible all-chair conformers of cyclohexylcyclohexane.

36. What is the most stable of the four *boat* conformations of methylcyclohexane, and why?

37. The most stable conformation of *trans*-1,3-bis(1,1-dimethylethyl)cyclohexane is not a chair. What conformation would you predict for this molecule? Explain.

38. The bicyclic hydrocarbon formed by the fusion of a cyclohexane ring with a cyclopentane ring is known as hexahydroindane (below). Using the drawings of *trans*- and *cis*-decalin for reference (Figure 4-13), draw the structures of *trans*- and *cis*-hexahydroindane, showing each ring in its most stable conformation.

Hexahydroindane

39. On viewing the drawings of *cis*- and *trans*-decalin in Figure 4-13, which do you think is the more stable isomer? Estimate the energy difference between the two isomers.

40. Several tricyclic compounds exist in nature with a cyclopropane ring fused to a *cis*-decalin structure, as shown in the molecule tricyclo[5.4.0¹,³.0¹,⁷]undecane (below). In various countries, some of these substances have a history of use as folk medicines for purposes such as contraception. Make a model of this compound. How does the cyclopropane ring affect the conformations of the two cyclohexane rings? The cyclohexane rings in *cis*-decalin itself are capable of (simultaneous) chair–chair interconversion (recall Exercise 4-13). Is the same true in tricyclo[5.4.0¹,³.0¹,⁷]undecane?

Tricyclo[5.4.0¹,³.0¹,⁷]undecane

41. The naturally occurring sugar glucose (Chapter 24) exists in the two isomeric cyclic forms shown below. These are called α and β, respectively, and they are in equilibrium by means of chemical processes that are introduced in Chapter 17. Glucose is the fuel of all living cells. Human blood contains about 0.1% glucose. It is made in plants from CO_2 and H_2O, by using our planet's ultimate fuel source, the Sun, in a process called photosynthesis. The "by-product," O_2, is equally important for the maintenance of complex life.

α form of glucose

β form of glucose

(a) Which of the two forms is more stable?

(b) At equilibrium the two forms are present in a ratio of approximately 64:36. Calculate the free energy difference that corresponds to this equilibrium ratio. How closely does the value you obtained correlate with the data in Table 4-3?

42. Identify each of the following molecules as a monoterpene, a sesquiterpene, or a diterpene (all names are common).

(a) **Geraniol**

(b) **Eremanthin**

(c) **Eudesmol**

(d) **Ipomeamarone**

(e) **Genipin**

(f) **Castoramine**

(g) **Cantharidin**

(h) **Vitamin A**

43. Circle and identify by name each functional group in the structures pictured in Problem 42.

44. Find the 2-methyl-1,3-butadiene (isoprene) units in each of the naturally occurring organic molecules pictured in Problem 42.

45. Circle and identify by name all the functional groups in any three of the steroids illustrated in Section 4-7. Label any polarized bonds with partial positive and negative charges (δ^+ and δ^-).

46. Several additional examples of naturally occurring molecules with strained ring structures are shown here.

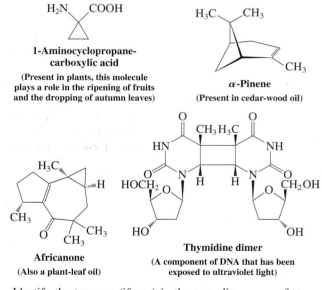

1-Aminocyclopropane-carboxylic acid

(Present in plants, this molecule plays a role in the ripening of fruits and the dropping of autumn leaves)

α-Pinene

(Present in cedar-wood oil)

Africanone

(Also a plant-leaf oil)

Thymidine dimer

(A component of DNA that has been exposed to ultraviolet light)

Identify the terpenes (if any) in the preceding group of structures. Find the 2-methyl-1,3-butadiene units in each structure and classify the latter as a mono-, sesqui-, or diterpene.

47. **CHALLENGE** If cyclobutane were flat, it would have exactly 90° C–C–C bond angles and could conceivably use pure p

orbitals in its C–C bonds. What would be a possible hybridization for the carbon atoms of the molecule that would allow all the C–H bonds to be equivalent? Exactly where would the hydrogens on each carbon be located? What are the real structural features of the cyclobutane molecule that contradict this hypothesis?

48. Compare the structure of cyclodecane in an all-chair conformation with that of *trans*-decalin. Explain why all-chair cyclodecane is highly strained, and yet *trans*-decalin is nearly strain free. Make models.

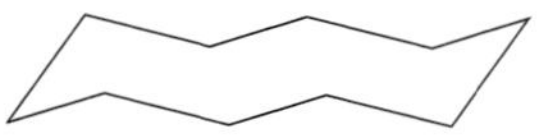

All-chair cyclodecane 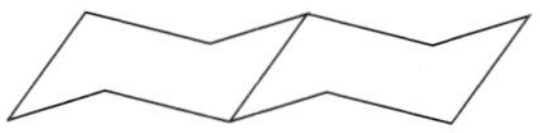   ***trans*-Decalin**

49. Fusidic acid is a steroid-like microbial product that is an extremely potent antibiotic with a broad spectrum of biological activity. Its molecular shape is most unusual and has supplied important clues to researchers investigating the methods by which steroids are synthesized in nature.

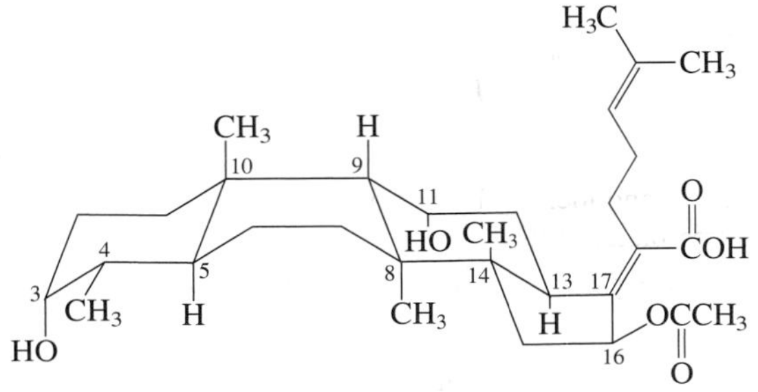

Fusidic acid

(a) Locate all the rings in fusidic acid and describe their conformations.

(b) Identify all ring fusions in the molecule as having either cis or trans geometry.

(c) Identify all groups attached to the rings as being either α- or β-substituents.

(d) Describe in detail how this molecule differs from the typical steroid in structure and stereochemistry. (As an aid to answering these questions, the carbon atoms of the framework of the molecule have been numbered.)

50. The enzymatic oxidation of alkanes to produce alcohols is a simplified version of the reactions that produce the adrenocortical steroid hormones. In the biosynthesis of corticosterone from progesterone (Section 4-7), two such oxidations take place successively (*a, b*). It is thought that the monooxygenase enzymes act as complex oxygen-atom donors in these reactions. A suggested mechanism, as applied to cyclohexane, consists of the two steps shown below the biosynthesis.

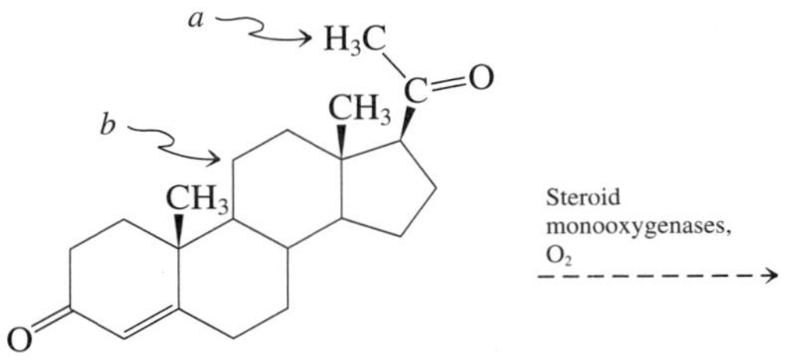

Progesterone

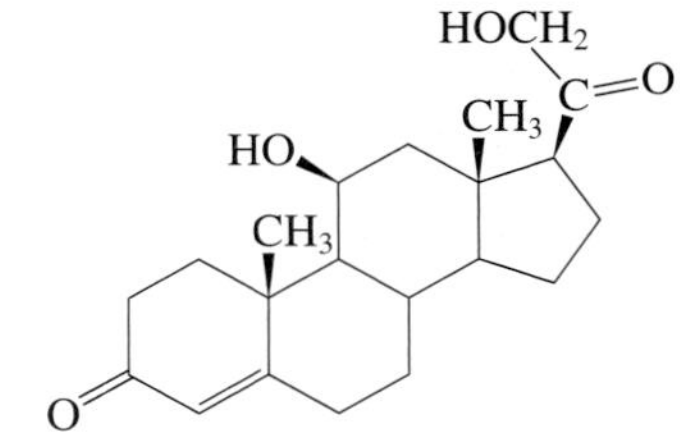

Corticosterone

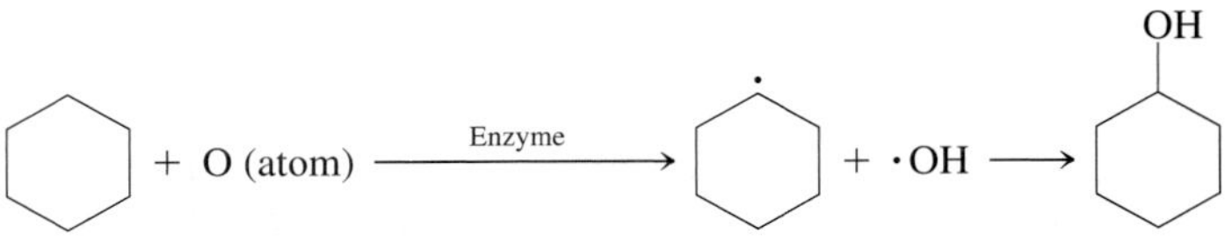

Calculate $\Delta H°$ for each step and for the overall oxidation reaction of cyclohexane. Use the following $DH°$ values: cyclohexane C–H bond, 98.5 kcal mol^{-1}; bond in O–H radical, 102.5 kcal mol^{-1}; cyclohexanol C–O bond, 96 kcal mol^{-1}.

51. **CHALLENGE** Iodobenzene dichloride, formed by the reaction of iodobenzene and chlorine, is a reagent for the chlorination of alkane C–H bonds. As predicted by the valence electron-pair repulsion method (VSEPR; Section 1-3), iodobenzene dichloride adopts a "T-shaped" geometry, in this way keeping the electrons at maximum distance from each other.

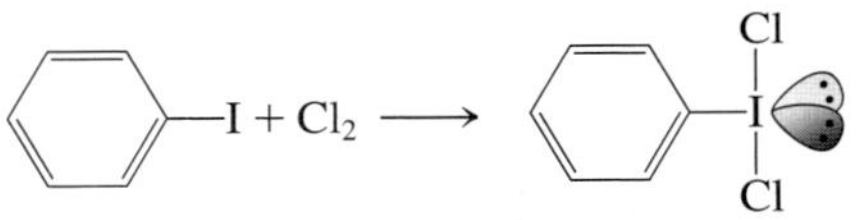

Iodobenzene dichloride

(a) Propose a radical chain mechanism for the chlorination of a typical alkane RH by iodobenzene dichloride. To get you started, the overall equation for the reaction is given below, as is the initiation step.

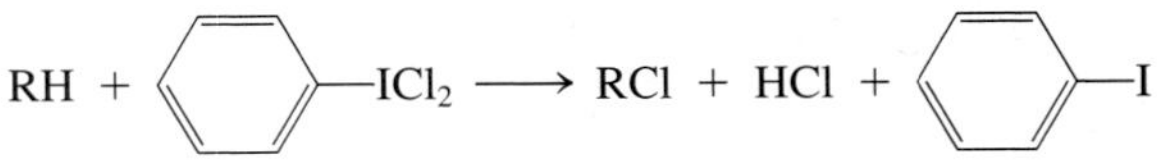

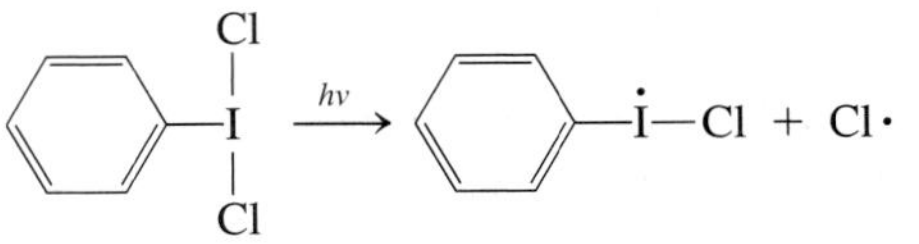

(b) Radical chlorination of typical steroids by iodobenzene dichloride gives, predominantly, three isomeric monochlorination products:

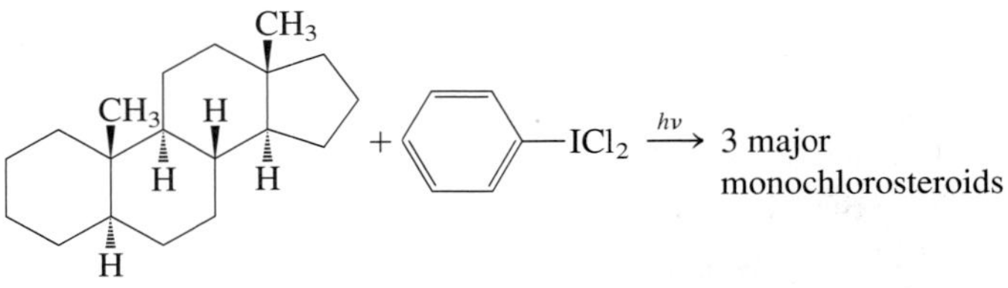

On the basis of both reactivity (tertiary, secondary, primary) considerations and steric effects (which might hinder the approach of a reagent toward a C–H bond that might otherwise be reactive), predict the three major sites of chlorination in the steroid molecule. Either make a model or carefully analyze the drawings of the steroid nucleus in Section 4-7.

52. **CHALLENGE** As Problem 50 indicates, the enzymatic reactions that introduce functional groups into the steroid nucleus in nature are highly selective, unlike the laboratory chlorination described in Problem 51. However, by means of a clever adaptation of this reaction, it is possible to partly mimic nature's selectivity in the laboratory. Two such examples are illustrated at the right.

(a)

(b)

Propose reasonable explanations for the results of these two reactions. Make a model of the product of the addition of Cl$_2$ to each iodocompound (compare Problem 51) to help in analyzing each system.

Team Problem

53. Consider the following compounds:

A   B

Conformational analysis reveals that, though compound A exists in a chair conformation, compound B does not.
(a) Make a model of A. Draw chair conformations and label the substituents as equatorial or axial. Circle the most stable conformation. (Note that the carbonyl carbon is sp^2 hybridized and therefore the attached oxygen is neither equatorial nor axial. Do not let that lead you astray.)
(b) Make a model of B. Consider both transannular and *gauche* interactions in your analysis of its two chair forms. Discuss the steric problems of these conformations in comparison with those of A. Illustrate the key points of your discussion with Newman projections. Suggest a less sterically encumbered conformation for B.

Preprofessional Problems

54. Which of the following cycloalkanes has the greatest ring strain?

(a) Cyclopropane
(b) Cyclobutane
(c) Cyclohexane
(d) Cycloheptane

55. The molecule below has **(a)** one axial chlorine and one sp^2 carbon, **(b)** one axial chlorine and two sp^2 carbons, **(c)** one equatorial chlorine and one sp^2 carbon, or **(d)** one equatorial chlorine and two sp^2 carbons.

56. In the structure below:

(a) The D is equatorial
(b) The methyls are both equatorial
(c) The Cl is axial
(d) The deuterium is axial

57. Which of the following structures has the smallest heat of combustion?

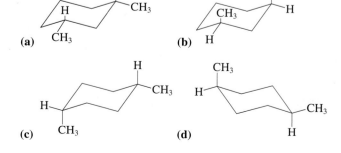

CHAPTER 5

Stereoisomers

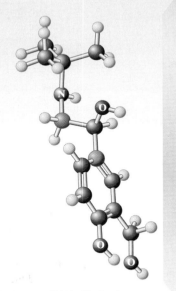
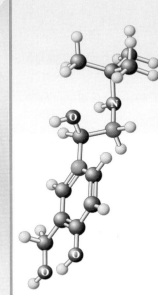

S-(+)-Albuterol   R-(−)-Albuterol

H ave you ever looked at yourself in the mirror in the morning and exclaimed: "That can't be me!" Well, you were right. What you see, your mirror image, is not identical with you: You and your mirror image are *nonsuperimposable*. You can demonstrate this fact by trying to shake hands with your mirror counterpart: As you reach out with your right hand, your mirror image will offer you its left hand! We shall see that many molecules have this property—namely, that image and mirror image are nonsuperimposable and therefore not identical. How do we classify such structures? Do their functions differ, and if so, how?

Because they have the same molecular formula, these molecules are isomers, but of a different kind from those encountered so far. The preceding chapters dealt with two kinds of isomerism: constitutional (also called structural) and stereo (Figure 5-1). **Constitutional isomerism** describes compounds that have identical molecular formulas but differ in the order in which the individual atoms are connected (Sections 1-9 and 2-5).

The physiological effects of the image and mirror image of the leading bronchodilator albuterol are dramatically different. The *R* image increases the bronchial airway diameter, whereas the *S* mirror image cancels this effect and is a suspected inflammatory agent.

Constitutional Isomers

C_4H_{10} $CH_3CH_2CH_2CH_3$

Butane

$$H_3C-\underset{\underset{CH_3}{|}}{\overset{\overset{CH_3}{|}}{CH}}$$

2-Methylpropane

C_2H_6O CH_3CH_2OH CH_3OCH_3
Ethanol   **Methoxymethane**
(Dimethyl ether)

Figure 5-1 Relations among isomers of various types.

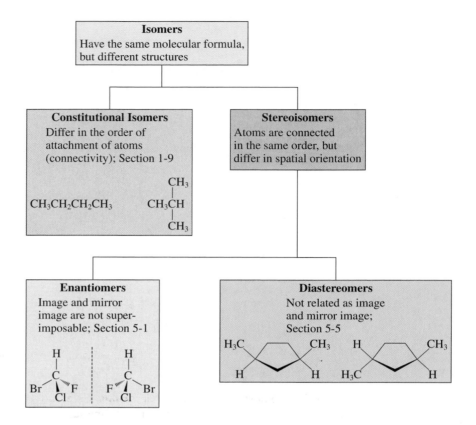

Isomers
Have the same molecular formula, but different structures

Constitutional Isomers
Differ in the order of attachment of atoms (connectivity); Section 1-9

$CH_3CH_2CH_2CH_3$ CH_3CH with CH_3 above and CH_3 below

Stereoisomers
Atoms are connected in the same order, but differ in spatial orientation

Enantiomers
Image and mirror image are not superimposable; Section 5-1

Diastereomers
Not related as image and mirror image; Section 5-5

Julien and Paloma Sat-Vollhardt and their mirror image.

Model Building

Stereoisomerism describes isomers whose atoms are connected in the same order but differ in their spatial arrangement. Examples of stereoisomers include the relatively stable and isolable cis-trans isomers and the rapidly equilibrating (and usually not isolable) conformational ones (Sections 2-5 through 2-8 and 4-1).

Stereoisomers

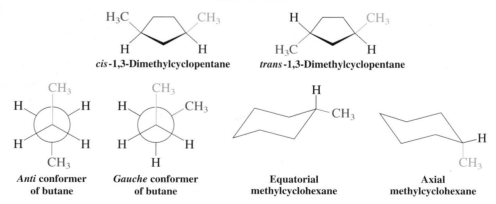

cis-1,3-Dimethylcyclopentane   *trans*-1,3-Dimethylcyclopentane

Anti conformer of butane   *Gauche* conformer of butane   Equatorial methylcyclohexane   Axial methylcyclohexane

---

Exercise 5-1

Are cyclopropylcyclopentane and cyclobutylcyclobutane isomers?

Exercise 5-2

Draw additional (conformational) stereoisomers of methylcyclohexane. (**Hint:** Use molecular models in conjunction with Figure 4-8.)

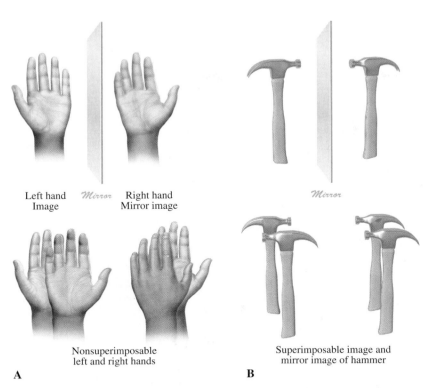

Figure 5-2 (A) Left and right hands as models for mirror-image stereoisomerism. (B) Image and mirror image of a hammer are superimposable.

Left hand   Right hand
Image   *Mirror*   Mirror image

Mirror

Nonsuperimposable   Superimposable image and
left and right hands   mirror image of hammer

A   **B**

This chapter introduces another type of stereoisomerism, **mirror-image stereoisomerism.** Molecules in this class are said to possess "handedness," referring to the fact that your left hand is not superimposable on your right hand, yet one hand can be viewed as the mirror image of the other (Figure 5-2A). In this way, hands are different from objects that are superimposable with their mirror images, such as a hammer (Figure 5-2B). The property of handedness in molecules is very important in nature, because most biologically relevant compounds are either "left-" or "right-handed." As such, they react differently with each other, much as shaking your friend's right hand is very different from shaking his or her left hand. A summary of isomeric relations is depicted in Figure 5-1.

5-1 | CHIRAL MOLECULES

How can a molecule exist as two nonsuperimposable mirror images? Consider the radical bromination of butane. This reaction proceeds mainly at one of the secondary carbons to furnish 2-bromobutane. A molecular model of the starting material *seems* to show that either of the two hydrogens on that carbon may be replaced to give only one form of 2-bromobutane (Figure 5-3). Is this really true, however?

Reaction

Figure 5-3 Replacement of one of the secondary hydrogens in butane results in two stereoisomeric forms of 2-bromobutane.

Chiral molecules cannot be superimposed on their mirror images

Look more closely at the 2-bromobutanes obtained by replacing either of the methylene hydrogens with bromine. In fact, the two structures are nonsuperimposable and therefore *not identical* (see the following page). The two molecules are related as object and mirror image, and to convert one into the other would require the breaking of bonds. A molecule that is not superimposable on its mirror image is said to be **chiral.** Each isomer of the image–mirror image pair is called an **enantiomer** (*enantios,* Greek, opposite). In our example of the bromination of butane, a 1:1 mixture of enantiomers is formed.

Model Building

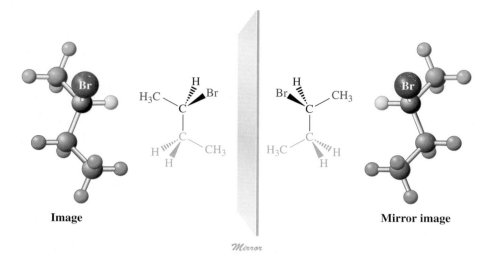

Image　　　　　**Mirror image**

Mirror

Not superimposable with image

Image from above in same position

Mirror image from above rotated to match up the ethyl (blue) and methyl (green) parts

The two enantiomers of 2-bromobutane are nonsuperimposable

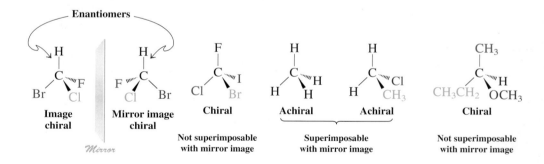

Enantiomers

Image chiral　　**Mirror image chiral**

Mirror

Chiral

Not superimposable with mirror image

Achiral　　**Achiral**

Superimposable with mirror image

Chiral

Not superimposable with mirror image

Stereocenter

Image　　**Mirror image**

Mirror

(C* = a stereocenter based on asymmetric carbon)

In contrast with chiral molecules, such as 2-bromobutane, compounds having structures that *are* superimposable on their mirror images are **achiral.** Examples of chiral and achiral molecules are shown above. The first two chiral structures depicted are enantiomers of each other.

All the chiral examples contain an atom that is connected to four *different* substituent groups. Such a nucleus is called an **asymmetric atom** (e.g., asymmetric carbon) or a **stereocenter.** Centers of this type are sometimes denoted by an asterisk. *Molecules with one stereocenter are always chiral.* (We shall see in Section 5-6 that structures incorporating more than one such center need *not* be chiral.)

Exercise 5-3

Among the natural products shown in Section 4-7, which are chiral and which are achiral? Give the number of stereocenters in each case.

The symmetry in molecules helps to distinguish chiral structures from achiral ones

The word *chiral* is derived from the Greek *cheir,* meaning "hand" or "handedness." Human hands have the mirror-image relation that is typical of enantiomers (see Figure 5-2A). Among the many other objects that are chiral are shoes, ears, screws, and spiral staircases. On the other hand, there are many achiral objects, such as balls, ordinary water glasses, hammers (Figure 5-2B), and nails.

Many chiral objects, such as spiral staircases, do not have stereocenters. The same is true for many chiral molecules. *Remember that the only criterion for chirality is the nonsuperimposable nature of object and mirror image.* In this chapter, we shall confine our discussion to molecules that are chiral as a result of the presence of stereocenters. But how do we determine whether a molecule is chiral or not? As you have undoubtedly already noticed, it is not

---

REAL LIFE: NATURE 5-1 | **Chiral Substances in Nature**

As we saw in the introduction to this chapter, the human body is chiral. Indeed, handedness pervades our macroscopic natural world. Moreover, there is usually a prevalence of a chiral image compared to its mirror image. For example, most of us are right handed, the heart is left and the liver is right, the bindweed plant winds around a support in a left-handed helix, the honeysuckle prefers the opposite, right-handed shells dominate (on both sides of the Equator), and so on.

Houses of snails: the ratio of right-handed (on the left) to left-handed is 20,000:1.

This preferential handedness is also present in the nanoscopic world of molecules; in fact, it is the presence of chiral molecular building blocks that often imparts macroscopic chirality. Thus, many chiral organic compounds exist in nature as only one enantiomer, although some are present as both. Specific handedness is associated with a specific biological function, dictated by the presence of chiral receptor

sites in the body (see Real Life 5-5). For example, natural alanine is an abundant amino acid that is found in only one form. Lactic acid, however, is present in blood and muscle fluid as one enantiomer but in sour milk and some fruits and plants as a mixture of the two.

**2-Aminopropanoic acid
(Alanine)**   **2-Hydroxypropanoic acid
(Lactic acid)**

Another example is carvone [2-methyl-5-(1-methylethenyl)-2-cyclohexenone], which contains a stereocenter in a six-membered ring. This carbon atom may be thought of as bearing four different groups, if we consider the ring itself to be two separate and different substituents. They are different because, starting from the stereocenter, the clockwise sequence of atoms differs from the counterclockwise sequence. Carvone is found in nature in either enantiomeric form, each having a very characteristic odor: One enantiomer gives caraway its characteristic smell, whereas the other is responsible for the flavor of spearmint.

In this book, you will encounter numerous examples of chiral molecules from nature that exist as only one enantiomer.

Caraway fruits ("seeds")

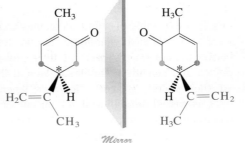

Mirror

Spearmint

Carvone enantiomers

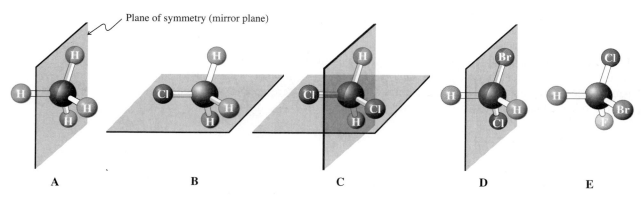

Plane of symmetry (mirror plane)

A   B   C   D   E

Figure 5-4 Examples of planes of symmetry: (A) methane has six planes of symmetry (incorporating the six edges of the tetrahedral frame), only one of which is shown; (B) chloromethane has three such planes, only one of which is shown; (C) dichloromethane has only two; (D) bromochloromethane only one; and (E) bromochlorofluoromethane has none. Chiral molecules cannot have a plane of symmetry.

Model Building

always easy to tell. A foolproof way is to construct molecular models of the molecule and its mirror image and look for superimposability. However, this procedure is very time consuming. A simpler method is to look for symmetry in the molecule under investigation.

For most organic molecules, we have to consider only one test for chirality: the presence or absence of a plane of symmetry. A **plane of symmetry (mirror plane)** is one that bisects the molecule so that the part of the structure lying on one side of the plane mirrors the part on the other side. For example, methane has six planes of symmetry, chloromethane has three, dichloromethane two, bromochloromethane one, and bromochlorofluoromethane none (Figure 5-4).

How do we use this idea to distinguish a chiral molecule from an achiral one? *Chiral molecules cannot have a plane of symmetry.* For example, the first four methanes in Figure 5-4 are clearly achiral because of the presence of a mirror plane. You will be able to classify most molecules in this book as chiral or achiral simply by identifying the presence or absence of a plane of symmetry.

Exercise 5-4

Draw pictures of the following common achiral objects, indicating the plane of symmetry in each: a ball, an ordinary water glass, a hammer, a chair, a suitcase, a toothbrush.

Exercise 5-5

Write the structures of all dimethylcyclobutanes. Specify those that are chiral. Show the mirror planes in those that are not.

In Summary A chiral molecule exists in either of two stereoisomeric forms called enantiomers, which are related as object and nonsuperimposable mirror image. Most chiral organic molecules contain stereocenters, although chiral structures that lack such centers do exist. A molecule that contains a plane of symmetry is achiral.

Enantiomers of 2-Bromobutane

Identical physical properties, except:

(+) Enantiomer   (−) Enantiomer

↓   ↓

Rotates plane of light *clockwise:* **dextrorotatory**

Rotates plane of light *counterclockwise:* effect **levorotatory**

5-2 | OPTICAL ACTIVITY

Our first examples of chiral molecules were the two enantiomers of 2-bromobutane. If we were to isolate each enantiomer in pure form, we would find that we could not distinguish between them on the basis of their physical properties, such as boiling points, melting points, and densities. This result should not surprise us: Their bonds are identical and so are their energy contents. However, when a special kind of light, called plane-polarized light, is passed through a sample of one of the enantiomers, the plane of polarization of the incoming light is *rotated* in one direction (either clockwise or counterclockwise). When the same experiment is repeated with the other enantiomer, the plane of the polarized light is rotated by exactly the same amount *but in the opposite direction.*

An enantiomer that rotates the plane of light in a clockwise sense as the viewer faces the light source is **dextrorotatory** (*dexter,* Latin, right), and the compound is (arbitrarily) referred to as the (+) enantiomer. Consequently, the other enantiomer, which will effect counterclockwise rotation, is **levorotatory** (*laevus,* Latin, left) and called the (−) enantiomer.

This special interaction with light is called **optical activity,** and enantiomers are frequently called **optical isomers.**

Optical rotation is measured with a polarimeter

What is plane-polarized light, and how is its rotation measured? Ordinary light can be thought of as bundles of electromagnetic waves that oscillate simultaneously in all planes perpendicular to the direction of the light beam. When such light is passed through a material called a polarizer, all but one of these light waves are "filtered" away, and the resulting beam oscillates in only one plane: **plane-polarized light** (Figure 5-5).

Figure 5-5 Measuring the optical rotation of the (−) enantiomer of 2-bromobutane with a polarimeter.

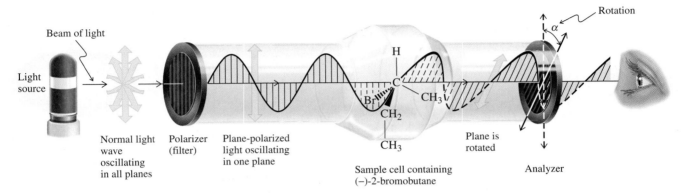

When light travels through a molecule, the electrons around the nuclei and in the various bonds interact with the electric field of the light beam. If a beam of plane-polarized light is passed through a chiral substance, the electric field interacts differently with, say, the "left" and "right" halves of the molecule. This interaction results in a rotation of the plane of polarization, called **optical rotation;** the sample giving rise to it is referred to as **optically active.**

Optical rotations are measured by using a **polarimeter** (Figure 5-5). In this instrument, light is first plane polarized and subsequently traverses a cell containing the sample. The angle of rotation of the plane of polarization is measured by aligning another polarizer—called the analyzer—so as to maximize the transmittance of the light beam to the eye of the observer. The measured rotation (in degrees) is the **observed optical rotation,** α, of the sample. Its value depends on the concentration and structure of the optically active molecule, the length of the sample cell, the wavelength of the light, the solvent, and the temperature. To avoid ambiguities, chemists have agreed on a standard value of the **specific rotation,** $[\alpha]$, for each compound. This quantity (which is solvent dependent) is defined as

Note: The presence of chiral molecules in a substance does not mean that the sample will *necessarily* exhibit optical activity. The material must contain *an excess of one enantiomer of at least one chiral compound over the other* for observation of optical activity to be possible.

Specific Rotation*

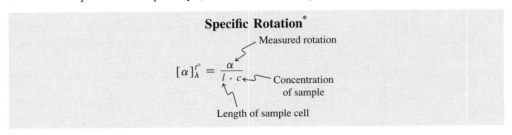

$$[\alpha]_{\lambda}^{t^{\circ}} = \frac{\alpha}{l \cdot c}$$

where $[\alpha]$ = specific rotation
  t = temperature in degrees Celsius
  λ = wavelength of incident light; for a sodium vapor lamp, which is commonly used for this purpose, the yellow D emission line (usually indicated simply by D) has λ = 589 nm.
  α = observed optical rotation in degrees
  l = length of sample container in decimeters; its value is frequently 1 (i.e., 10 cm).
  c = concentration (grams per milliliter of solution)

*The dimensions of $[\alpha]$ are deg cm^2 g^{-1}, the units (for l = 1) 10^{-1} deg cm^2 g^{-1}. (Remember: 1 mL = 1 cm^3.) Because of their awkward appearance, it is common practice to give $[\alpha]$ without units, in contrast with the observed rotation α (degrees). Moreover, for practical reasons of solubility, some reference works list c in grams per 100 mL, in which case the observed rotation has to be multiplied by 100.

Table 5-1 **Specific Rotations of Various Chiral Compounds $[\alpha]_D^{25°C}$**

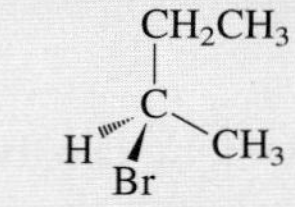

(−)-2-Bromobutane   −23.1

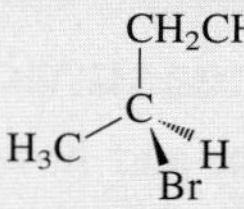

(+)-2-Bromobutane   +23.1

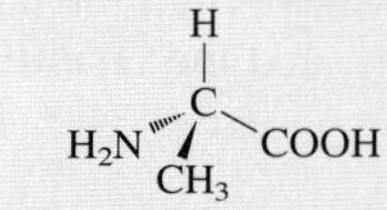

(+)-2-Aminopropanoic acid [(+)-Alanine]   +8.5

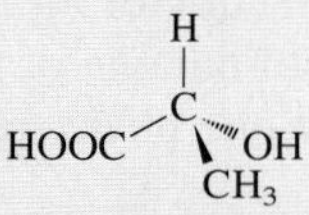

(−)-2-Hydroxypropanoic acid [(−)-Lactic acid]   −3.8

Note: Pure liquid for the haloalkane; in aqueous solution for the acids.

Exercise 5-6

A solution of 0.1 g mL^{-1} of common table sugar (the naturally occurring form of sucrose) in water in a 10-cm cell exhibits a clockwise optical rotation of 6.65°. Calculate $[\alpha]$. Does this information tell you $[\alpha]$ for the enantiomer of natural sucrose?

The specific rotation of an optically active molecule is a physical constant characteristic of that molecule, just like its melting point, boiling point, and density. Four specific rotations are recorded in Table 5-1.

Optical rotation indicates enantiomeric composition

As mentioned, enantiomers rotate plane-polarized light by equal amounts but in opposite directions. Thus, in 2-bromobutane the (−) enantiomer rotates this plane counterclockwise by 23.1°, its mirror image (+)-2-bromobutane clockwise by 23.1°. It follows that a 1:1 mixture of (+) and (−) enantiomers shows no rotation and is therefore optically inactive. Such a mixture is called a **racemic mixture.** If one enantiomer equilibrates with its mirror image, it is said to undergo **racemization.** For example, amino acids such as (+)-alanine (Table 5-1) have been found to undergo very slow racemization in fossil deposits, resulting in reduced optical activity.

The optical activity of a sample of a chiral molecule is directly proportional to the ratio of the two enantiomers. It is at a maximum when only one enantiomer is present, and the sample is optically pure. It is zero when the two enantiomers are present in equal amounts, and the sample is racemic and optically inactive. In practice, one often encounters mixtures in which one enantiomer is in excess of the other. The **enantiomer excess (*ee*)** tells us by how much:

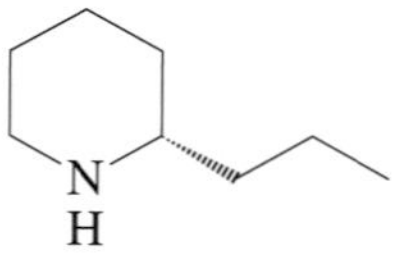

Enantiomer excess (*ee*) = % of major enantiomer − % of minor enantiomer

Since a racemate constitutes a 1:1 mixture of the two (*ee* = 0), the *ee* is a measure of how much one enantiomer is present in excess of racemate. The *ee* can be obtained from the % optical rotation of such a mixture relative to that of the pure enantiomer, also called **optical purity:**

Optical Purity and Enantiomer Excess

$$\text{Enantiomer excess } (ee) = \text{optical purity} = \frac{[\alpha]_{\text{Mixture}}}{[\alpha]_{\text{Pure enantiomer}}} \times 100\%$$

(+)-Coniine

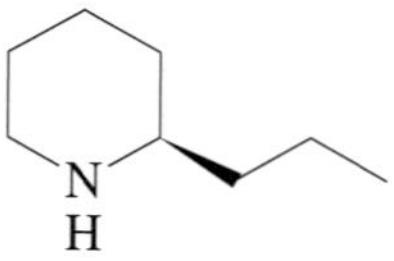

(−)-Coniine

Coniine is the lethal ingredient in poison hemlock and the means by which Socrates was put to death in 399 BC. It exists in the plant as a near-racemic mixture, with the (more poisonous) (+) enantiomer predominating.

Solved Exercise 5-7 | Working with the Concepts: *Ee* and Optical Purity

A solution of (+)-alanine from a fossil exhibits a value of $[\alpha] = +4.25$. What is its *ee* and optical purity? What is the actual enantiomer composition of the sample, and how is the measured optical rotation derived from it?

Strategy
We need to look up what the specific rotation of pure (+)-alanine is and then use the preceding equation to get the answers.

Solution
- Table 5-1 gives us the specific rotation of pure (+)-alanine: +8.5.
- Our equation tells us: Enantiomer excess (*ee*) = optical purity = (4.25/8.5) × 100% = 50%.
- This means that 50% of the sample is pure (+) isomer and the other 50% is racemic. Because the racemic portion consists of equal amounts of (+) and (−) enantiomers, the actual composition of the sample is 75% (+)- and 25% (−)-alanine.
- The 25% (−) enantiomer cancels the rotation of a corresponding amount of (+) enantiomer. This mixture is therefore 75% − 25% = 50% optically pure, and the observed optical rotation is one-half that of the pure dextrorotatory enantiomer.

Exercise 5-8 | Try It Yourself

What is the optical rotation of a sample of (+)-2-bromobutane that is 75% optically pure? What percentages of (+) and (−) enantiomers are present in this sample? Answer the same questions for samples of 50% and 25% optical purity.

In Summary Two enantiomers can be distinguished by their optical activity, that is, their interaction with plane-polarized light as measured in a polarimeter. One enantiomer always rotates such light clockwise (dextrorotatory), the other counterclockwise (levorotatory) by the same amount. The specific rotation, $[\alpha]$, is a physical constant possible only for chiral molecules. The interconversion of enantiomers leads to racemization and the disappearance of optical activity.

5-3 | ABSOLUTE CONFIGURATION: *R,S* SEQUENCE RULES

How do we establish the structure of one pure enantiomer of a chiral compound? And, once we know the answer, is there a way to name it unambiguously and distinguish it from its mirror image?

X-ray diffraction can establish the absolute configuration

Virtually all the physical characteristics of one enantiomer are identical with those of its mirror image, except for the sign of optical rotation. Is there a correlation between the sign of optical rotation and the actual spatial arrangement of the substituent groups, the **absolute configuration?** Is it possible to determine the structure of an enantiomer by measuring its $[\alpha]$ value? The answer to both questions is, unfortunately, no. *There is no straightforward correlation between the sign of rotation and the structure of the particular enantiomer.* For example, conversion of lactic acid (Table 5-1) into its sodium salt changes the sign (and degree) of rotation, even though the absolute configuration at the stereocenter is unchanged (see margin).

$[\alpha]_D^{25°C} = -3.8$

(−)-Lactic acid
(Levorotatory)

NaOH, H_2O

$[\alpha]_D^{25°C} = +13.5$

(+)-Sodium lactate
(Dextrorotatory)

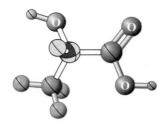

Structure of (+)-lactic acid as deduced from X-ray diffraction analysis.

If the sign of rotation does not tell us anything about structure, how do we know which enantiomer of a chiral molecule is which? Or, to put it differently, how do we know that the levorotatory enantiomer of 2-bromobutane has the structure indicated in Table 5-1 (and therefore the dextrorotatory enantiomer the mirror-image configuration)? The answer is that such information can be obtained only through single-crystal X-ray diffraction analysis (Section 1-9 and the picture in the margin). This does not mean that every chiral compound must be submitted to X-ray analysis to ascertain its structure. Absolute configuration can also be established by chemical correlation with a molecule whose own structure has been proved by this method. For example, knowing the stereocenter in (−)-lactic acid by X-ray analysis also provides the absolute configuration of the (+)-sodium salt (i.e., the same).

Stereocenters are labeled *R* or *S*

To name enantiomers unambiguously, we need a system that allows us to indicate the handedness in the molecule, a sort of "left-hand" versus "right-hand" nomenclature. Such a system was developed by three chemists, R. S. Cahn, C. Ingold, and V. Prelog.*

Let us see how the handedness around an asymmetric carbon atom is labeled. The first step is to rank all four substituents in the order of decreasing priority, the rules of which will be described shortly. Substituent *a* has the highest priority, *b* the second highest, *c* the third, and *d* the lowest. Next, we position the molecule (mentally, on paper, or by using a molecular model set) so that the lowest-priority substituent is placed as far away from us as possible (Figure 5-6). This process results in two (and only two) possible arrangements of the remaining substituents. If the progression from *a* to *b* to *c* is counterclockwise, the configuration at the stereocenter is named *S* (*sinister,* Latin, left). Conversely, if the progression is clockwise, the center is *R* (*rectus,* Latin, right). The symbol *R* or *S* is added as a prefix in parentheses to the name of the chiral compound, as in (*R*)-2-bromobutane and (*S*)-2-bromobutane. A racemic mixture can be designated *R,S*, if necessary, as in (*R,S*)-bromochlorofluoromethane. The sign of the rotation of plane-polarized light may be added if it is known, as in (*S*)-(+)-2-bromobutane and (*R*)-(−)-2-bromobutane. It is important to remember, however, that the symbols *R* and *S* are *not* necessarily correlated with the sign of α.

Priority Color Scheme

$a > b > c > d$

Figure 5-6 Assignment of *R* or *S* configuration at a tetrahedral stereocenter. The group of lowest priority is placed as far away from the observer as possible. In many of the structural drawings in this chapter, the color scheme shown here is used to indicate the priority of substituents—in decreasing order, red > blue > green > black.

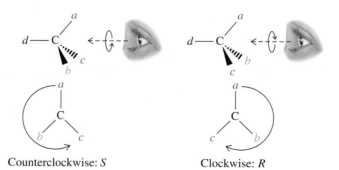

Counterclockwise: *S*   Clockwise: *R*

Sequence rules assign priorities to substituents

Before applying the *R,S* nomenclature to a stereocenter, we must first assign priorities by using sequence rules.

*Dr. Robert S. Cahn (1899–1981), Fellow of the Royal Institute of Chemistry, London; Professor Christopher Ingold (1893–1970), University College, London; Professor Vladimir Prelog (1906–1998), Swiss Federal Institute of Technology (ETH), Zürich, Nobel Prize 1975 (chemistry).

Rule 1. We look first at the atoms attached directly to the stereocenter. A substituent atom of higher atomic number takes precedence over one of lower atomic number. Consequently, the substituent of lowest priority is hydrogen. In regard to isotopes, the atom of higher atomic mass receives higher priority.

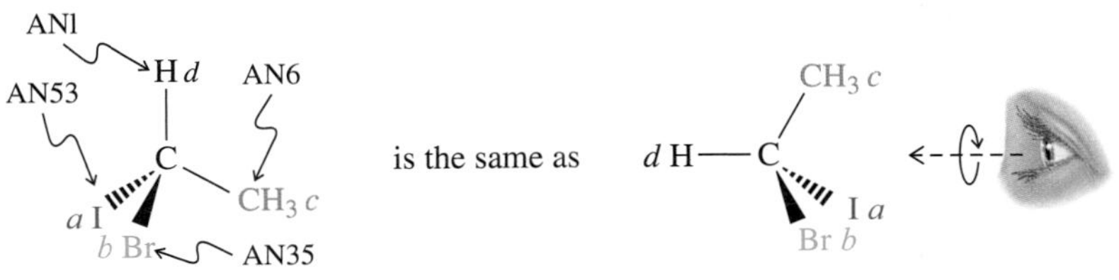

AN = atomic number

(*R*)-1-Bromo-1-iodoethane

Rule 2. What if two substituents have the same rank when we consider the atoms directly attached to the stereocenter? In such a case, we proceed along the two respective substituent chains until we reach a point of difference.

For example, an ethyl substituent takes priority over methyl. Why? At the point of attachment to the stereocenter, each substituent has a carbon nucleus, equal in priority. Farther from that center, however, methyl has only hydrogen atoms, but ethyl has a carbon atom (higher in priority).

> **Note:** The substituents are attached through the dangling bond to the stereocenter in the molecule under discussion.

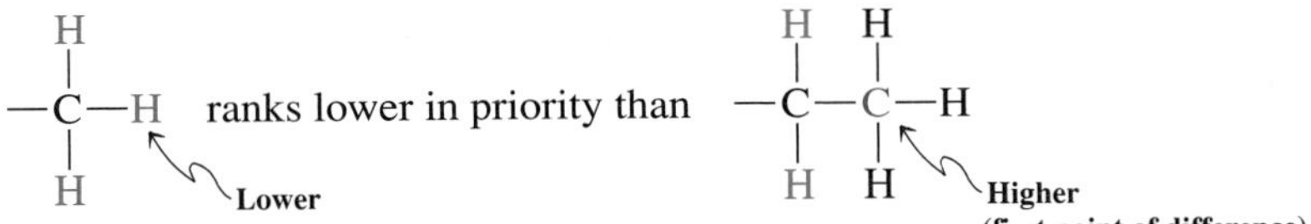

However, 1-methylethyl takes precedence over ethyl because, at the first carbon, ethyl bears only one other carbon substituent, but 1-methylethyl bears two. Similarly, 2-methylpropyl takes priority over butyl but ranks lower than 1,1-dimethylethyl.

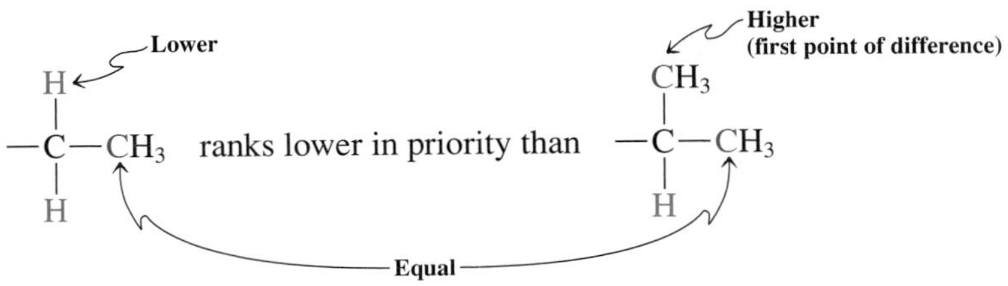

Ranking in —C$_4$H$_9$

—CH$_2$CH$_2$CH$_2$CH$_3$

Butyl

ranks lower in priority than

CH$_3$
|
—CH$_2$CCH$_3$
|
H

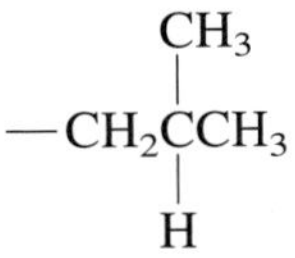

2-Methylpropyl

ranks lower in priority than

CH$_3$
|
—CCH$_3$
|
CH$_3$

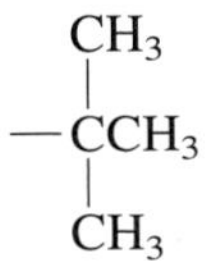

1,1-Dimethylethyl
(***tert*-butyl**)

We must remember that the decision on priority is made at the *first* point of difference along otherwise similar substituent chains. When that point has been reached, the constitution of the remainder of the chain is irrelevant.

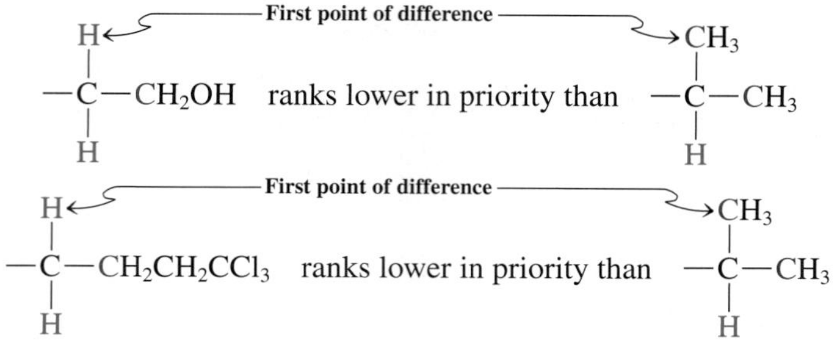

When we reach a point along a substituent chain at which it branches, we choose the branch that is higher in priority. When two substituents have similar branches, we rank the elements in those branches until we reach a point of difference.

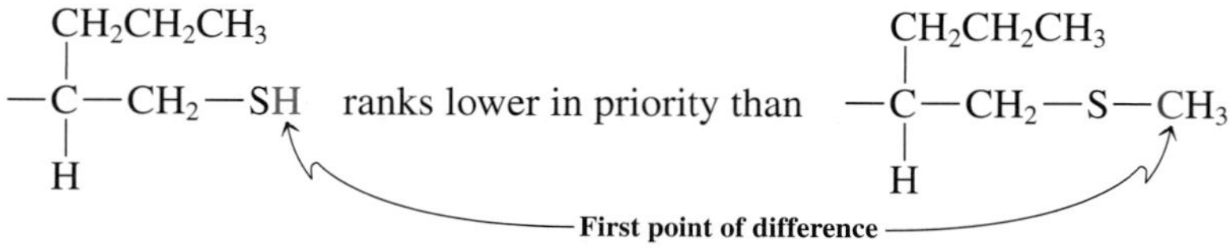

Two examples are shown below.

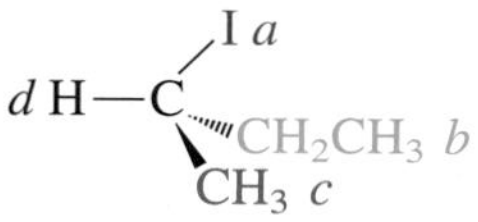

(R)-2-Iodobutane 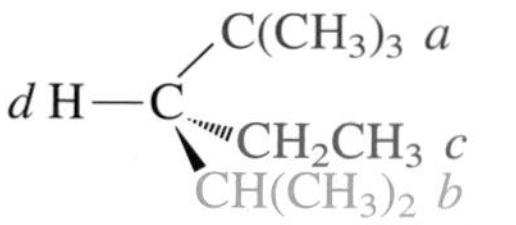   (S)-3-Ethyl-2,2,4-trimethylpentane

Rule 3. Double and triple bonds are treated as if they were single, and the atoms in them are duplicated or triplicated at each end by the respective atoms at the other end of the multiple bond.

The red atoms shown in the groups on the right side of the display are not really there. They are added only for the purpose of assigning a relative priority to each of the corresponding groups to their left.

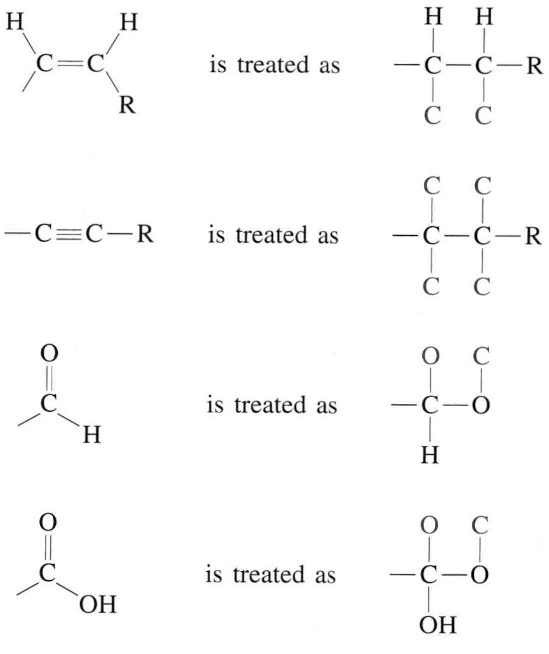

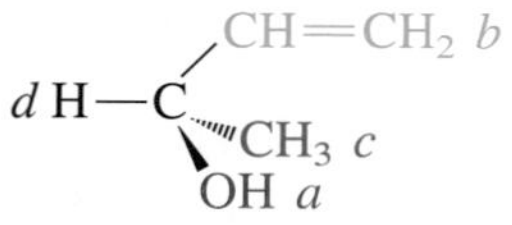

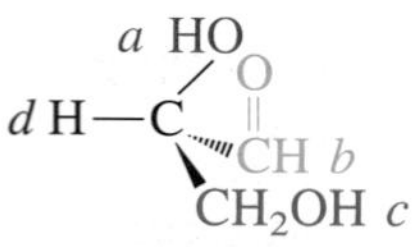

Examples are shown in the margin and in the structures of *S*- and *R*-albuterol that appear in the Chapter Opening.

Exercise 5-9

Draw the structures of the following substituents and within each group rank them in order of decreasing priority. **(a)** Methyl, bromomethyl, trichloromethyl, ethyl; **(b)** 2-methyl-propyl (isobutyl), 1-methylethyl (isopropyl), cyclohexyl; **(c)** butyl, 1-methylpropyl (*sec*-butyl), 2-methylpropyl (isobutyl), 1,1-dimethylethyl (*tert*-butyl); **(d)** ethyl, 1-chloroethyl, 1-bromoethyl, 2-bromoethyl.

Solved Exercise 5-10 | Working with the Concepts: Assigning *R* and *S*

Assign the absolute configuration of ($-$)-2-bromobutane, depicted in Table 5-1.

Strategy
To establish a molecule's absolute configuration, we cannot rely on whether it is levorotatory or dextrorotatory. Instead, we focus on the stereocenter and arrange the molecule in space in such a way that the substituent with lowest priority is facing away from us. Therefore, the first task is to assign priorities and the second step is to arrange the molecule in space as required.

Solution
Let us look at ($-$)-2-bromobutane, reproduced below (labeled **A**) as it appears in Table 5-1.

- According to the Cahn–Ingold–Prelog rules, Br is *a*, CH_2CH_3 is *b*, CH_3 is *c*, and H is *d*, as depicted in structure **B**.
- Arranging the molecule in space can be difficult at first, but becomes easier with practice. A safe approach is described in Figure 5-6, in which you move the tetrahedral frame so as to place the C-*d* bond in the plane of the page pointing to the left and imagine looking down this axis from the right. For structure **B**, this procedure means rotating the carbon atom to reach structure **C** (with the C-*d* bond in the plane of the page) and then rotating the molecule clockwise to reach **D**. Looking at **D** from the right as shown reveals the absolute configuration: *R*.
- As you do more of these assignments, you will become increasingly adept at seeing the molecule in three dimensions and viewing it so that the trio *a*, *b*, *c* is pointing toward you and group *d* is pointing away from you.

Exercise 5-11 | Try It Yourself

Assign the absolute configuration of the remaining three molecules depicted in Table 5-1.

Exercise 5-12

Draw one enantiomer of your choice (specify which, *R* or *S*) of 2-chlorobutane, 2-chloro-2-fluorobutane, and (HC≡C)(CH$_2$=CH)C(Br)(CH$_3$). (**Hint:** Do not set out to try to draw a specific enantiomer. Rather, draw one at random, but in a view that allows easy determination of the stereocenter as being either *R* or *S*.)

To assign correctly the stereostructure of stereoisomers, we must develop a fair amount of three-dimensional "vision," or "stereoperception." In most of the structures that we have used to illustrate the priority rules, the lowest-priority substituent is located at the left of the carbon center and in the plane of the page, and the remainder of the substituents at the right, the upper-right group also being positioned in this plane. However, we already know that this is not the only way of drawing hashed-wedged line structures; others are equally correct. Consider some of the structural drawings of (*S*)-2-bromobutane (on the next page). These are simply different views of the same molecule.

Model Building

Four (of Many) Ways of Depicting (S)-2-Bromobutane

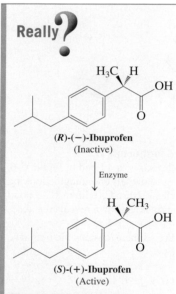

(R)-(−)-Ibuprofen
(Inactive)

Enzyme

(S)-(+)-Ibuprofen
(Active)

The painkiller ibuprofen, commonly available under the names Advil and Motrin, is sold as a racemate. Even though the active ingredient is the (S)-(+) enantiomer ("dexibuprofen"), the racemate is almost as active as the (S) form. As luck (for the discoverers) would have it (see also Problem 64), an enzyme in the body, α-methylacyl-CoA racemase, converts the inactive (R) to the (S) isomer to the tune of 63%, thus adding value to the racemic mixture.

The Fischer Stencil

is shown as

In Summary The sign of optical rotation cannot be used to establish the absolute configuration of a stereoisomer. Instead, X-ray diffraction (or chemical correlations) must be used. We can express the absolute configuration of the chiral molecule as *R* or *S* by applying the sequence rules, which allow us to rank all substituents in order of decreasing priority. Turning the structures so as to place the lowest-priority group at the back causes the remaining substituents to be arranged in clockwise (*R*) or counterclockwise (*S*) fashion.

5-4   FISCHER PROJECTIONS

A **Fischer** (Real Life 5-2) **projection** is a simplified way of depicting tetrahedral carbon atoms and their substituents in two dimensions. With this method, the molecule is drawn in the form of a cross, the central carbon being at the point of intersection. The horizontal lines signify bonds directed *toward* the viewer; the vertical lines are pointing *away*. Hashed-wedged line structures have to be arranged in this way to facilitate their conversion into Fischer projections.

**Conversion of the Hashed-Wedged Line Structures
of 2-Bromobutane into Fischer Projections (of the Stereocenter)**

Hashed-wedged line structure	Fischer projection	Hashed-wedged line structure	Fischer projection
(R)-2-Bromobutane		(S)-2-Bromobutane	

Notice that just as there are several ways of depicting a molecule in the hashed-wedged line notation, there are several correct Fischer projections of the same stereocenter.

Two Additional Projections of (R)-2-Bromobutane

A simple mental procedure that will allow you to safely convert any hashed-wedged line structure into a Fischer projection is to picture yourself at the molecular level and grasp any two substituents of the hashed-wedged line rendition with your hands while facing the central carbon, as shown on the next page. In this cartoon, these substituents are labeled arbitrarily *a* and *c*. If you then imagine descending on the page while holding the molecule,

REAL LIFE: HISTORY 5-2 | Absolute Configuration: A Historical Note

Before the X-ray diffraction technique was developed, the absolute configurations of chiral molecules were unknown. Amusingly, the first assignment of a three-dimensional structure to a chiral molecule was a *guess* made more than a century ago by the sugar chemist Emil Fischer.* He did so in a paper in 1891 to simplify the depiction of the complicated stereochemical relationships of grape sugar (glucose) to other sugars, a topic that we will visit in Chapter 24. His choice (by chemical correlation) assigned the naturally occurring dextrorotatory enantiomer of 2,3-dihydroxypropanal (glyceraldehyde) the three-dimensional structure labeled D-glyceraldehyde. The label "D" does not refer to the sign of rotation of plane-polarized light but to the spatial arrangement of the substituent groups.

Professor Emil Fischer in his laboratory.

---

*Professor Emil Fischer (1852–1919), University of Berlin, Germany, Nobel Prize 1902 (chemistry).
†Professor Johannes M. Bijvoet (1892–1980), University of Utrecht, The Netherlands.

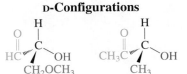

$[\alpha]_D^{25°C} = +8.7$

D-(+)-2,3-Dihydroxypropanal
[D-(+)-Glyceraldehyde]

$[\alpha]_D^{25°C} = -8.7$

L-(−)-2,3-Dihydroxypropanal
[L-(−)-Glyceraldehyde]

The other isomer was called L-glyceraldehyde. All chiral compounds that could be converted into D-(+)-glyceraldehyde by reactions that did not affect the configuration at the stereocenter were assigned the D configuration, and their mirror images the L.

D-Configurations

L-Configurations

In 1951, the Dutch crystallographer Johannes Bijvoet† established the absolute configurations of these compounds by the X-ray diffraction analysis of a salt of tartaric acid, which had been correlated by chemical means with Fischer's sugars and, in turn, glyceraldehyde. As Bijvoet states in his paper (abbreviated), "The result is that Emil Fischer's convention . . . appears to answer the reality." Lucky guess!

D,L nomenclature is still used for sugars (Chapter 24) and amino acids (Chapter 26).

---

the two remaining substituents (positioned behind the carbon center) will submerge below it, and your left and right hands will position the two horizontal, wedged-line groups in the proper orientation. This procedure places the two remaining hashed-line groups vertically (juxtaposing your head and feet, respectively).

Model Building

A Simple Mental Exercise:
Conversion of Hashed-Wedged Line Structures into Fischer Projections

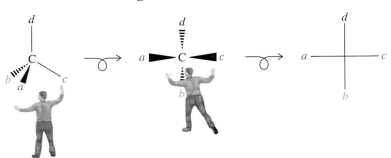

Having achieved this conversion, you can change one Fischer projection into another of the same molecule by using certain manipulations: rotations and substituent switches. However, we shall see next that care has to be taken so you don't inadvertently interconvert *R* and *S* configurations.

Rotating a Fischer projection may or may not change the absolute configuration

What happens when we rotate a Fischer projection in the plane of the page by 90°? Does the result depict the spatial arrangement of the original molecule? The definition of a Fischer projection—horizontal bonds are pointed above, vertical ones below the plane of the page—tells us that the answer is clearly no, because this rotation has *switched* the relative spatial disposition of the two sets: The result is a picture of the enantiomer. On the other hand, rotation by 180° is fine, because horizontal and vertical lines have not been interchanged: The resulting drawing represents the same enantiomer.

Exercise 5-13

Draw Fischer projections for all the molecules in Exercises 5-10 through 5-12.

Exchanging substituents in a Fischer projection also changes the absolute configuration

As is the case for hashed-wedged line structures, there are several Fischer projections of the same enantiomer, a situation that may lead to confusion. How can we quickly ascertain whether two Fischer projections are depicting the same enantiomer or two mirror images? We have to find a sure way to convert one Fischer projection into another in a manner that either leaves the absolute configuration unchanged or converts it into its opposite. It turns out that this task can be achieved by simply making substituent groups trade places. As we can readily verify by using molecular models, any *single* such exchange turns one enantiomer into its mirror image. Two such exchanges (we may select different substituents every time) produce the original absolute configuration. As shown on the next page, this operation merely results in a different view of the same molecule.

$$
\begin{array}{c}
\text{CH}_3 \\
\text{Cl} \!-\!\!\!\!\!\!\!\!\!\!\!\!-\! \text{Br} \\
\text{CH}_2\text{CH}_3 \\
\textbf{S}
\end{array}
\xrightarrow{\text{CH}_3 \rightleftharpoons \text{Br}}
\begin{array}{c}
\text{Br} \\
\text{Cl} \!-\!\!\!\!\!\!\!\!\!\!\!\!-\! \text{CH}_3 \\
\text{CH}_2\text{CH}_3 \\
\textbf{R}
\end{array}
\xrightarrow{\text{CH}_3\text{CH}_2 \rightleftharpoons \text{Br}}
\begin{array}{c}
\text{CH}_2\text{CH}_3 \\
\text{Cl} \!-\!\!\!\!\!\!\!\!\!\!\!\!-\! \text{CH}_3 \\
\text{Br} \\
\textbf{S}
\end{array}
$$

S   changes into   R   changes back into   S

(The double arrow denotes two groups trading places)

We now have a simple way of establishing whether two different Fischer projections depict the same or opposite configurations. If the conversion of one structure into another takes an even number of exchanges, the structures are identical. If it requires an odd number of such moves, the structures are mirror images of each other.

Consider, for example, the two Fischer projections A and B. Do they represent molecules having the same configuration? The answer is found quickly. We convert A into B by two exchanges; so A equals B.

$$
\begin{array}{c}
\text{Cl} \\
\text{H} \!-\!\!\!\!\!\!\!\!\!\!\!\!-\! \text{CH}_3 \\
\text{CH}_2\text{CH}_3 \\
\textbf{A}
\end{array}
\xrightarrow{\text{H} \rightleftharpoons \text{Cl}}
\begin{array}{c}
\text{H} \\
\text{Cl} \!-\!\!\!\!\!\!\!\!\!\!\!\!-\! \text{CH}_3 \\
\text{CH}_2\text{CH}_3
\end{array}
\xrightarrow{\text{Cl} \rightleftharpoons \text{CH}_3}
\begin{array}{c}
\text{H} \\
\text{H}_3\text{C} \!-\!\!\!\!\!\!\!\!\!\!\!\!-\! \text{Cl} \\
\text{CH}_2\text{CH}_3 \\
\textbf{B}
\end{array}
$$

Exercise 5-14

Draw the hashed-wedged line structures corresponding to Fischer projections A and B, above. Is it possible to transform A into B by means of a rotation about a single bond? If so, identify the bond and the degree of rotation required. Use models if necessary.

Fischer projections tell us the absolute configuration

Fischer projections allow us to assign absolute configurations without having to visualize the three-dimensional arrangement of the atoms. For this purpose, we

First draw the molecule as any Fischer projection.

Next, we rank all the substituents in accord with the sequence rules.

Finally, if necessary, we place group d on top by a double exchange.

With d at the top, the three groups of priority a, b, and c can adopt only two arrangements: either clockwise or counterclockwise. The first corresponds unambiguously to the R, the second to the S configuration.

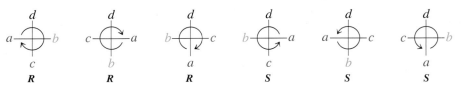

Solved Exercise 5-15	Working with the Concepts: Assigning *R* and *S* Using Fischer Projections

What is the absolute configuration of

$$
\begin{array}{c}
Br \\
| \\
H \text{——} D \\
| \\
CH_3
\end{array}
$$

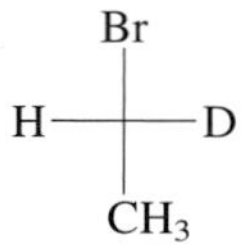

Strategy

This molecule, 1-bromo-1-deuterioethane, is chiral because of the presence of a carbon atom bearing four different substituents, one of which happens to be deuterium, an isotope of hydrogen. The first task is to assign priorities, and the second step is to arrange the Fischer projection such as to place the lowest-priority substituent on top.

Solution

- According to the Cahn–Ingold–Prelog rules, Br is *a*, CH₃ is *b*, D is *c*, and H is *d*.
- To simplify matters, we replace the substituents with their stereochemical priority designators:

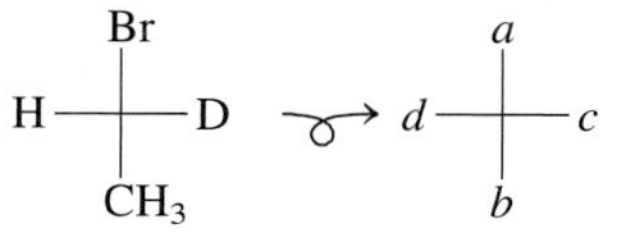

$$
\begin{array}{c}
Br \\
| \\
H \text{——} D \\
| \\
CH_3
\end{array}
\quad \rightarrow \quad
\begin{array}{c}
a \\
| \\
d \text{——} c \\
| \\
b
\end{array}
$$

- We next apply the (any) "double-exchange" protocol (which leaves the absolute configuration intact) so as to place *d* on top. In the approach chosen below, *d* trades places with *a*, and *b* with *c*. This arranges the remainder in a clockwise fashion: *R*.

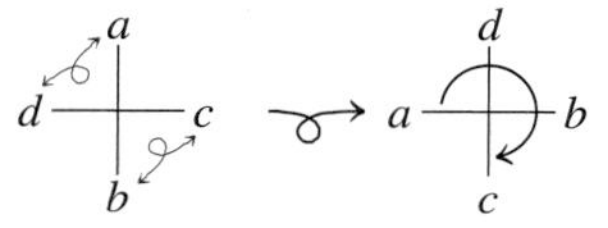

$$
\begin{array}{c}
a \\
| \\
d \text{——} c \\
| \\
b
\end{array}
\quad \rightarrow \quad
\begin{array}{c}
d \\
a \text{——} b \\
c \\
R
\end{array}
$$

- We could have done other double exchanges, as long as *d* winds up on top. Try, for example, switching *d/a* and subsequently *a/c*, or the sequence *d/c*, *a/d*, and confirm that you always get the same answer!

Exercise 5-16	Try It Yourself

Assign the absolute configuration of the three molecules depicted below.

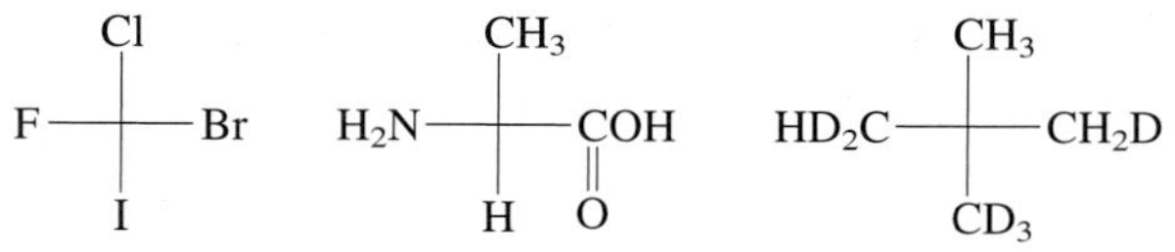

$$
\begin{array}{c}
Cl \\
| \\
F \text{——} Br \\
| \\
I
\end{array}
\qquad
\begin{array}{c}
CH_3 \\
| \\
H_2N \text{——} COH \\
| \quad\; \| \\
H \quad\; O
\end{array}
\qquad
\begin{array}{c}
CH_3 \\
| \\
HD_2C \text{——} CH_2D \\
| \\
CD_3
\end{array}
$$

"Chirally deuterated neopentane" (made in 2007)

Exercise 5-17

Convert the Fischer projections in Exercises 5-15 and 5-16 into hashed-wedged line formulas and determine their absolute configurations by using the procedure described in Section 5-3. When the lowest-priority group is at the top in a Fischer projection, is it in front of the plane of the page or behind it? Does this explain why the procedure outlined on p. 183 for determination of configuration from Fischer projections succeeds?

In Summary A Fischer projection is a convenient way of drawing chiral molecules. We can rotate such projections in the plane by 180° (retains absolute configuration) but not by 90° (changes absolute configuration). Switching substituents reverses absolute configuration, if done an odd number of times, but leaves it intact when the number of such exchanges is even. By placing the substituent of lowest priority on top, we can readily assign the absolute configuration.

5-5   MOLECULES INCORPORATING SEVERAL STEREOCENTERS: DIASTEREOMERS

Many molecules contain several stereocenters. Because the configuration about each center can be R or S, several possible structures emerge, all of which are isomeric.

Two stereocenters can give four stereoisomers: chlorination of 2-bromobutane at C3

In Section 5-1 we described how a carbon-based stereocenter can be created by the radical halogenation of butane. Let us now consider the chlorination of racemic 2-bromobutane to give (among other products) 2-bromo-3-chlorobutane. The introduction of a chlorine atom at C3 produces a new stereocenter in the molecule. This center may have either the R or the S configuration. This reaction can be conveniently shown using Fischer projections. For this purpose, we draw the stem as a vertical line and the stereocenters as horizontal lines. How many stereoisomers are possible for 2-bromo-3-chlorobutane? There are four, as can be seen by completing a simple exercise in permutation. Each stereocenter can be either R or S, and, hence, the possible combinations are RR, RS, SR, and SS. You can readily establish the existence of four stereoisomers by recognizing that each halogen substituent can be placed either to the right or to the left of the stem, respectively, for a total of four combinations (see below, margin of p. 186, and Figure 5-7).

Model Building

$$CH_3 \overset{*}{-} \underset{\underset{Br}{|}}{C} - CH_2 - CH_3$$

One stereocenter

$Cl_2, h\nu \downarrow -HCl$

$$CH_3 - \underset{\underset{Br}{|}}{\overset{\overset{H}{|}}{C}} \overset{**}{-} \underset{\underset{H}{|}}{\overset{\overset{Cl}{|}}{C}} - CH_3$$

Two stereocenters
2-Bromo-3-chlorobutane

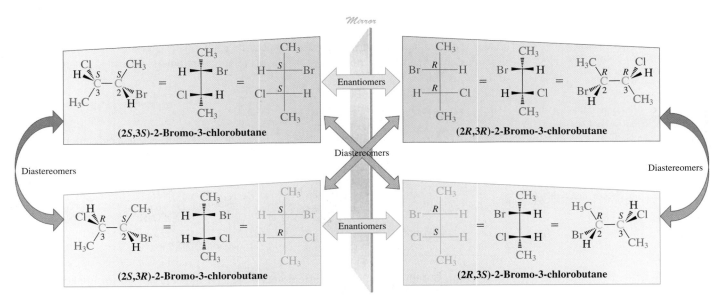

Four stereoisomers of 2-bromo-3-chlorobutane

Racemic 3-bromobutane

Figure 5-7 The four stereoisomers of 2-bromo-3-chlorobutane. Each molecule is the enantiomer of one of the other three (its mirror image) and is at the same time a diastereomer of each of the remaining two. For example, the 2R,3R isomer is the enantiomer of the 2S,3S compound and is a diastereomer of both the 2S,3R and the 2R,3S structures. Notice that two structures are enantiomers only when they possess the opposite configuration at *every* stereocenter.

Chlorinating racemic 2-bromobutane at C3 is stereochemically equivalent to pairing two different pairs ("racemates") of shoes: four combinations.

Isoleucine is an important amino acid that cannot be made by the body and therefore needs to be an ingredient of our diet. Relatively high amounts of isoleucine are found in eggs, chicken, lamb, cheese, and fish.

SOLUTION: The center under scrutiny is *S*.

Figure 5-8 Assigning the absolute configuration at C3 in 2-bromo-3-chlorobutane. We consider the group containing the stereocenter C2 merely as one of the four substituents. Priorities (also noted in color) are assigned in the usual way (Cl > CHBrCH$_3$ > CH$_3$ > H), giving rise to the representation shown in the center. Two exchanges place the substituent of lowest priority (hydrogen) at the top of the Fischer projection to facilitate assignment.

Because all horizontal lines in Fischer projections signify bonds directed toward the viewer, the Fischer stencil represents an *eclipsed* conformation and therefore does not depict the molecule in its most stable form, which is *anti*. This is illustrated below for (2*S*,3*S*)-2-bromo-3-chlorobutane (see also Figure 5-7).

(2*S*,3*S*)-2-Bromo-3-chlorobutane: From Eclipsed Fischer Projection to *anti* Conformation

Fischer projection   *anti*-**Hashed-wedged line structure**   **Newman projection**

To make stereochemical assignments, one treats each stereocenter *separately*, and the group containing the other stereocenter is regarded as a simple substituent (Figure 5-8).

By looking closely at the structures of the four stereoisomers (Figure 5-7), we see that there are two related pairs of compounds: an *R,R/S,S* pair and an *R,S/S,R* pair. The members of each individual pair are mirror images of each other and therefore enantiomers. Conversely, each member of one pair is not a mirror image of either member of the other pair; hence, they are not enantiomeric with respect to each other. Stereoisomers that are not related as object and mirror image, and therefore are not enantiomers, are called **diastereomers** (*dia*, Greek, across).

Exercise 5-18

The two amino acids isoleucine and alloisoleucine are depicted below in staggered conformations. Convert both into Fischer projections. (Keep in mind that Fischer projections are views of molecules *in eclipsed conformations*.) Are these two compounds enantiomers or diastereomers?

Isoleucine   **Alloisoleucine**

In contrast with enantiomers, diastereomers, because they are *not* mirror images of each other, are molecules with *different physical and chemical properties* (see, e.g., Real Life 5-3). Their steric interactions and energies differ. They can be separated by fractional distillation, crystallization, or chromatography. They have different melting and boiling points (see margin of the next page) and different densities, just as constitutional isomers do. In addition, they have different specific rotations.

REAL LIFE: NATURE 5-3 | Stereoisomers of Tartaric Acid

$$HO \quad COOH$$
$$H \cdots C - C \cdots OH$$
$$HOOC \quad H$$

(+)-Tartaric acid

$[\alpha]_D^{20°C} = +12.0$
m.p. 168–170°C
Density (g mL⁻¹) $d = 1.7598$

$$H \quad COOH$$
$$HO \cdots C - C \cdots H$$
$$HOOC \quad OH$$

(−)-Tartaric acid

$[\alpha]_D^{20°C} = -12.0$
m.p. 168–170°C
$d = 1.7598$

$$HO \quad COOH$$
$$H \cdots C - C \cdots H$$
$$HOOC \quad OH$$

***meso*-Tartaric acid**

$[\alpha]_D^{20°C} = 0$
m.p. 146–148°C
$d = 1.666$

Tartaric acid (systematic name 2,3-dihydroxy-butanedioic acid) is a naturally occurring dicarboxylic acid containing two stereocenters with identical substitution patterns. Therefore it exists as a pair of enantiomers (which have identical physical properties but which rotate plane-polarized light in opposite directions) and an achiral meso compound (with different physical and chemical properties from those of the chiral diastereomers).

The dextrorotatory enantiomer of tartaric acid is widely distributed in nature. It is present in many fruits, and its monopotassium salt is found as a deposit during the fermentation of grape juice. Pure levorotatory tartaric acid is rare, as is the meso isomer.

Tartaric acid is of historical significance because it was the first chiral molecule whose racemate was separated into the two enantiomers. This happened in 1848, long before it was recognized that carbon could be tetrahedral in organic molecules. By 1848, natural tartaric acid had been shown to be dextrorotatory and the racemate had been isolated from grapes. [In fact, the words "race-mate" and "racemic" are derived from an old common name for this form of tartaric acid, racemic acid (*racemus*, Latin, cluster of grapes)]. The French chemist Louis

Pasteur's polarimeter

Pasteur* obtained a sample of the sodium ammonium salt of this acid and noticed that there were two types of crystals: One set was the mirror image of the second. In other words, the crystals were chiral.

By manually separating the two sets, dissolving them in water, and measuring their optical rotation, Pasteur found one of the crystalline forms to be the pure salt of (+)-tartaric acid and the other to be the levo-rotatory form. Remarkably, the chirality of the individual molecules in this rare case had given rise to the macroscopic property of chirality in the crystal. He concluded from his observation that the molecules themselves must be chiral. These findings and others led in 1874 to the first proposal, by van't Hoff and Le Bel[†] independently, that saturated carbon has a tetra-hedral bonding arrangement and is not, for example, square planar. (Why is the idea of a planar carbon incompatible with that of a stereocenter?)

*Professor Louis Pasteur (1822–1895), Sorbonne, Paris.
†Professor Jacobus H. van't Hoff (1852–1911), University of Amsterdam, Nobel Prize 1901 (chemistry); Dr. J. A. Le Bel (1847–1930), Ph.D., Sorbonne, Paris.

Exercise 5-19

What are the stereochemical relations (identical, enantiomers, diastereomers) of the following four molecules? Assign absolute configurations at each stereocenter.

$$\begin{array}{c} CH_3 \\ H - \!\!\!-\!\!\!- F \\ H - \!\!\!-\!\!\!- CH_2CH_3 \\ CH_3 \end{array}$$
1

H F
(with H CH₃)
2

H F
(with H CH₃)
3

F H
(with H CH₃)
4

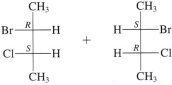

2-Bromo-3-chlorobutane

R,S/S,R-Racemate:
b.p. 31–33°C (at 16 torr)

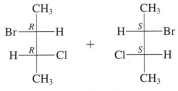

2-Bromo-3-chlorobutane

R,R/S,S-Racemate:
b.p. 38–38.5°C (at 16 torr)

Cis and trans isomers are cyclic diastereomers

It is instructive to compare the stereoisomers of 2-bromo-3-chlorobutane with those of a cyclic analog, 1-bromo-2-chlorocyclobutane (Figure 5-9). In both cases, there are four stereoisomers: *R,R*, *S,S*, *R,S*, and *S,R*. In the cyclic compound, however, the stereoisomeric relation of the first pair to the second is easily recognized: One pair has cis stereochemistry, the other trans. Cis and trans isomers (Section 4-1) in cycloalkanes are in fact diastereomers.

Model Building

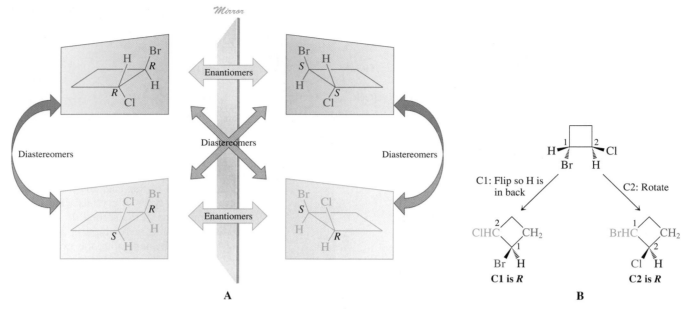

Figure 5-9 (A) The diastereomeric relation of *cis*- and *trans*-1-bromo-2-chlorocyclobutane. (B) Stereochemical assignment of the *R,R* stereoisomer. Recall that the color scheme indicates the priority order of the groups around each stereocenter: red > blue > green > black.

More than two stereocenters means still more stereoisomers

What structural variety do we expect for a compound having three stereocenters? We may again approach this problem by permuting the various possibilities. If we label the three centers consecutively as either *R* or *S*, the following sequence emerges:

RRR   RRS   RSR   SRR   RSS   SRS   SSR   SSS

Powers of 2	
n	2^n
0	1
1	2
2	4
3	8
4	16
10	1,024
20	1,048,576
30	1,073,741,824

a total of eight stereoisomers. They can be arranged to reveal a division into four enantiomer pairs of diastereomers.

Image	*RRR*	*RRS*	*RSS*	*SRS*
Mirror image	*SSS*	*SSR*	*SRR*	*RSR*

Generally, *a compound with n stereocenters can have a maximum of 2^n stereoisomers.* Therefore, a compound having three such centers gives rise to a maximum of eight stereoisomers; one having four produces sixteen; one having five, thirty-two; and so forth. The structural possibilities are quite staggering for larger systems (margin).

Exercise 5-20

Draw all the stereoisomers of 2-bromo-3-chloro-4-fluoropentane.

In Summary The presence of more than one stereocenter in a molecule gives rise to diastereomers. These are stereoisomers that are not related to each other as object and mirror image. Whereas enantiomers have opposite configurations at every respective stereocenter, two diastereomers do not. A molecule with n stereocenters may exist in as many as 2^n stereoisomers. In cyclic compounds, cis and trans isomers are diastereomers.

Model Building

5-6 | MESO COMPOUNDS

We saw that the molecule 2-bromo-3-chlorobutane contains two distinct stereocenters, each with a *different* halogen substituent. How many stereoisomers are to be expected if both centers are identically substituted?

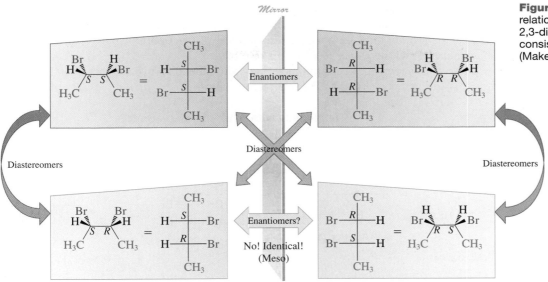

Figure 5-10 The stereochemical relations of the stereoisomers of 2,3-dibromobutane. The lower pair consists of identical structures. (Make a model.)

Two identically substituted stereocenters give rise to only three stereoisomers

Consider, for example, 2,3-dibromobutane, which can be obtained by the radical bromination of 2-bromobutane. As we did for 2-bromo-3-chlorobutane, we have to consider four structures, resulting from the various permutations in *R* and *S* configurations (Figure 5-10).

$$CH_3\overset{*}{C}CH_2CH_3 \xrightarrow[-HBr]{Br_2,\ h\nu} H_3C\overset{*}{-}\overset{|}{C}-\overset{*}{C}-CH_3$$

One stereocenter   **Two stereocenters**

2,3-Dibromobutane

The first pair of stereoisomers, with *R,R* and *S,S* configurations, is clearly recognizable as a pair of enantiomers. However, a close look at the second pair reveals that (*S,R*) and mirror image (*R,S*) are superimposable and therefore identical. Thus, the *S,R* diastereomer of 2,3-dibromobutane is achiral and not optically active, even though it contains two stereocenters. The identity of the two structures can be readily confirmed by using molecular models.

A compound that contains two (or, as we shall see, even more than two) stereocenters but is superimposable with its mirror image is a **meso compound** (*mesos,* Greek, middle). A characteristic feature of a meso compound is the *presence of an internal mirror plane,* which divides the molecule such that one half is the mirror image of the other half. For example, in 2,3-dibromobutane, the *2R* center is the reflection of the *3S* center. This arrangement is best seen in an eclipsed hashed-wedged line structure (Figure 5-11). The presence

Brominating racemic 2-bromobutane at C3 is stereochemically equivalent to pairing two identical pairs ("racemates") of shoes: only three distinct combinations.

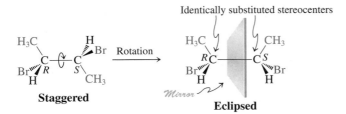

Figure 5-11 *meso*-2,3-Dibromobutane contains an internal mirror plane when rotated into the eclipsed conformation shown. A molecule with more than one stereocenter is meso and achiral as long as it contains a mirror plane in any readily accessible conformation. Meso compounds possess identically substituted stereocenters.

of a mirror plane in *any* energetically accessible conformation of a molecule (Sections 2-8 and 2-9) is sufficient to make it achiral (Section 5-1). As a consequence, 2,3-dibromobutane exists in the form of three stereoisomers only: a pair of (necessarily chiral) enantiomers and an achiral meso diastereomer (see also margin of p. 189).

Meso diastereomers can exist in molecules with more than two stereocenters. Examples are 2,3,4-tribromopentane and 2,3,4,5-tetrabromohexane.

Model Building

Meso Compounds with Multiple Stereocenters

Mirror plane

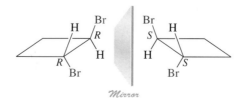

Chiral or meso?

Note: The presence of both *R* and *S* stereocenters in a structure does not mean that the compound is *automatically* meso. For it to be meso, its stereocenters must have identical sets of substituents, and the molecule as a whole must have an internal mirror plane of symmetry.

Exercise 5-21

Draw all the stereoisomers of 2,4-dibromo-3-chloropentane.

Cyclic compounds may also be meso

It is again instructive to compare the stereochemical situation in 2,3-dibromobutane with that in an analogous cyclic molecule: 1,2-dibromocyclobutane. We can see that *trans*-1,2-dibromocyclobutane exists as two enantiomers (*R,R* and *S,S*) and may therefore be optically active. The cis isomer, however, has an internal mirror plane and is meso, achiral, and optically inactive (Figure 5-12).

Notice that we have drawn the ring in a planar shape in order to illustrate the mirror symmetry, although we know from Chapter 4 that cycloalkanes with four or more carbons in the ring are not flat. Is this justifiable? Generally yes, because such compounds, like their acyclic analogs, possess a variety of conformations that are readily accessible at room temperature (Sections 4-2 through 4-4 and Section 5-1). At least one of these conformations will contain the necessary mirror plane to render achiral any cis-disubstituted cycloalkane with identically constituted stereocenters. For simplicity, cyclic compounds may usually be treated *as if they were planar* for the purpose of identifying a mirror plane.

Exercise 5-22

Draw each of the following compounds, representing the ring as planar. Which ones are chiral? Which are meso? Indicate the location of the mirror plane in each meso compound. (**a**) *cis*-1,2-Dichlorocyclopentane; (**b**) its trans isomer; (**c**) *cis*-1,3-dichlorocyclopentane; (**d**) its trans isomer; (**e**) *cis*-1,2-dichlorocyclohexane; (**f**) its trans isomer; (**g**) *cis*-1,3-dichlorocyclohexane; (**h**) its trans isomer.

Figure 5-12 The trans isomer of 1,2-dibromocyclobutane is chiral; the cis isomer is a meso compound and optically inactive.

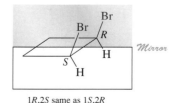

Enantiomers of
chiral diastereomer
trans-**1,2-Dibromocyclobutane**

1*R*,2*S* same as 1*S*,2*R*

Meso diastereomer
cis-**1,2-Dibromocyclobutane**

Exercise 5-23

For each meso compound in Exercise 5-22, draw the conformation that contains the mirror plane. Refer to Sections 4-2 through 4-4 to identify energetically accessible conformations of these ring systems.

In Summary Meso compounds are diastereomers containing a molecular plane of symmetry. They are therefore superimposable on their mirror images and achiral. Molecules with two or more identically substituted stereocenters may exist as meso stereoisomers.

5-7 | STEREOCHEMISTRY IN CHEMICAL REACTIONS

We have seen that a chemical reaction, such as the halogenation of an alkane, can introduce chirality into a molecule. How exactly does this occur? To find the answer, we need to look more closely at the conversion of achiral butane into chiral 2-bromobutane, which gives racemic material. Once we have done so, we shall be able to understand the halogenation of 2-bromobutane and the effect of the chiral environment of a stereocenter already present in a molecule on the stereochemistry of the reaction.

Mechanism

The radical mechanism explains why the bromination of butane results in a racemate

The radical bromination of butane at C2 creates a chiral molecule (Figure 5-3). This happens because one of the methylene hydrogens is replaced by a new group, furnishing a stereocenter—a carbon atom with four different substituents.

In the first step of the mechanism for radical halogenation (Sections 3-4 and 3-7), one of these two hydrogens is abstracted by the attacking bromine atom. It does not matter which of the two is removed: This step does not generate a stereocenter. It furnishes a planar, sp^2-hybridized, and therefore achiral radical. The radical center has two equivalent reaction sites—the two lobes of the p orbital (Figure 5-13)—that are equally susceptible to attack by bromine in the second step. We can see that the two transition states resulting in the respective enantiomers of 2-bromobutane are mirror images of each other. They are enantiomeric and therefore energetically equivalent (see also margin). The rates of formation of R and S products are hence equal, and a racemate is formed. In general, *the formation of chiral compounds* (e.g., 2-bromobutane) *from achiral reactants* (e.g., butane and bromine) *yields racemates. Or, optically inactive starting materials furnish optically inactive products.**

The bicycle above was rendered starkly chiral as a result of being hit by a car from the right. Had the same accident occurred from the left side, it would have generated the mirror image below.

The presence of a stereocenter affects the outcome of the reaction: chlorination of (S)-2-bromobutane

Now we understand why the halogenation of an achiral molecule gives a racemic halide. What products can we expect from the halogenation of a chiral, enantiomerically pure molecule? Or, to put it differently, how does the presence of a stereocenter in the structure affect its function in a reaction?

For example, consider the radical chlorination of the S enantiomer of 2-bromobutane. In this case, the chlorine atom has several options for attack: the two terminal methyl groups, the single hydrogen at C2, and the two hydrogens on C3. Let us examine each of these reaction paths.

---

*We shall see later that it *is* possible to generate optically active products from optically inactive starting materials if we use an optically active reagent or catalyst (see, for example, Real Life 5-4).

Figure 5-13 The creation of racemic 2-bromobutane from butane by radical bromination at C2. Abstraction of either methylene hydrogen by bromine gives an achiral radical. Reaction of Br_2 with this radical is equally likely at either the top or the bottom face, a condition leading to a racemic mixture of products.

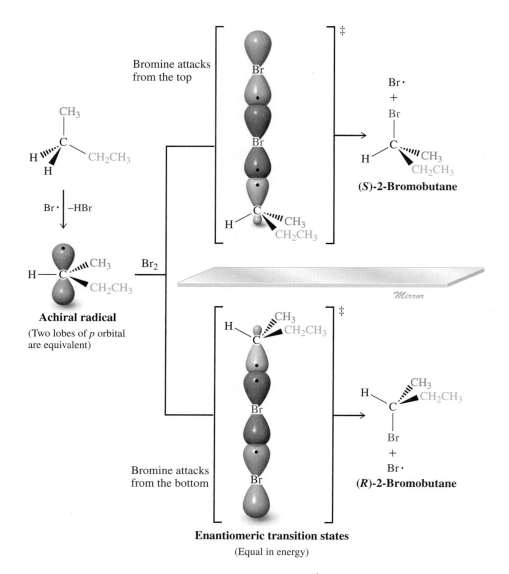

Bromine attacks from the top

(S)-2-Bromobutane

$Br \cdot$ | $-HBr$

Achiral radical

(Two lobes of *p* orbital are equivalent)

Br_2

Mirror

Bromine attacks from the bottom

(R)-2-Bromobutane

Enantiomeric transition states

(Equal in energy)

Chlorination of (S)-2-Bromobutane at Either C1 or C4

Remember the use of color to denote group priorities:
a Highest—red
b Second highest—blue
c Third highest—green
d Lowest—black

Group priorities have switched

$Cl_2, h\nu$ at C1 $-HCl$

$Cl_2, h\nu$ at C4 $-HCl$

Optically active **2R**

Optically active **2S**

Optically active **3S**

Chlorination of either terminal methyl group is straightforward, proceeding at C1 to give 2-bromo-1-chlorobutane or at C4 to give 3-bromo-1-chlorobutane. In the latter, the original C4 has now become C1, to maintain the lowest possible substituent numbering. *Both of these chlorination products are optically active because the original stereocenter is left intact.* Note, however, that conversion of the C1 methyl into a chloromethyl unit changes the sequence of priorities around C2. Thus, although the stereocenter itself does not participate in the reaction, its designated configuration changes from S to R.

What about halogenation at C2, the stereocenter? The product from chlorination at C2 of (S)-2-bromobutane is 2-bromo-2-chlorobutane. Even though the substitution pattern at the stereocenter has changed, the molecule remains chiral. However, an attempt to measure the [α] value for the product would reveal the absence of optical activity: *Halogenation at the stereocenter leads to a racemic mixture.* How can this be explained? For the answer, we must look again at the structure of the radical formed in the course of the reaction mechanism.

A racemate forms in this case because hydrogen abstraction from C2 furnishes a planar, sp^2-hybridized, achiral radical.

Note: Radicals lack brains. They *do not remember* which enantiomer they came from *when the hydrogen removed renders a stereocenter planar and achiral.* Thus, in this particular experiment, it makes *no difference* whether the original starting material was *S*, *R*, or any mixture of the two: The product must always be racemic, because it derives from an achiral radical.

Chlorination of (S)-2-Bromobutane at C2

Optically active

2*S*

Achiral

Optically inactive
(A racemate)

50% 2*S*   50% 2*R*

REAL LIFE: MEDICINE 5-4 | **Chiral Drugs—Racemic or Enantiomerically Pure?**

Until the early 1990s, most synthetic chiral medicines were prepared as racemic mixtures and sold as such. The reasons were mainly of a practical nature. Reactions that convert an achiral into a chiral molecule ordinarily produce racemates (Section 5-7). In addition, often both enantiomers have comparable physiological activity or one of them (the "wrong" one) is inactive; therefore resolution was deemed unnecessary. Finally, resolution of racemates on a large scale is expensive and adds substantially to the cost of drug development.

However, in a number of cases, one of the enantiomers of a drug has been found to act as a blocker of the biological receptor site, thus diminishing the activity of the other enantiomer. Worse, one of the enantiomers may have a completely different, and sometimes toxic, spectrum of activity. This phenomenon is quite general, since nature's receptor sites are handed (Real Life 5-5). Several examples of different bioactive behavior between enantiomers are shown below.

Bitter   **Sweet**

Asparagine
(Amino acid; see Chapter 26)

In 1806, asparagine was the first amino acid to be isolated from nature, from the juice of asparagus.

Active   **Inactive**

Dichlorprop
(Herbicide)

Antagonist   **Bronchodilator**

Albuterol
(see also Chapter Opening)

Weed control.

REAL LIFE: MEDICINE 5-4 | (Continued)

Because of these findings, the U.S. Food and Drug Administration (FDA) revised its guidelines for the commercialization of chiral drugs, making it more advantageous for companies to produce single enantiomers of medicinal products. The logistics of testing pure enantiomers are simpler, the biological efficiencies of the drugs are higher, and, potentially, the lifetime of a patent on a successful drug can be further extended by switching sales from a racemate to the active enantiomer ("chiral switch"). The result has been a flurry of research activities designed to improve resolution of racemates or, even better, to develop methods of *enantioselective synthesis*. The essence of this approach is that used by nature in enzyme-catalyzed reactions (see the oxidation of dopamine in Section 5-7): An achiral starting material is converted into the chiral product in the presence of an enantiopure environment, often a chiral catalyst. Because enantiomeric transition states (Figure 5-13) become diastereomeric in such an environment (Figure 5-14; note that, in this case, the chiral "environment" of the reacting carbon is provided by the neighboring stereocenter), high stereoselectivity can be

achieved. Such selectivity is a nice example of adherence to some of the principles of green chemistry: avoiding generation of 50% "waste" in the form of the wrong enantiomer, the associated cumbersome separation (Section 5-8), atom economy, and catalysis. As shown below, such methods have been applied to the syntheses of drugs such as the antiarthritic and analgesic naproxen and the antihypertensive propranolol in high enantiomeric purity.

To give you an idea of the importance of this emerging technology, the worldwide market for chiral drugs is approaching $300 billion per year; 80% of small-molecule drugs approved currently by the FDA are chiral (see also Table 25-1 for top-selling chiral drugs); and the Nobel prize in chemistry 2001 was given to three researchers who made ground-breaking discoveries in the field of enantioselective catalysis.*

*Dr. William S. Knowles (1917–2012), Monsanto Company, St. Louis, Missouri; Professor Ryoji Noyori (b. 1938), Nagoya University, Japan; Professor K. Barry Sharpless (b. 1941), The Scripps Research Institute, La Jolla, California.

Chlorination can occur from either side through enantiomeric transition states of equal energy, as in the bromination of butane (Figure 5-13), producing (S)- and (R)-2-bromo-2-chlorobutane at equal rates and in equal amounts. The reaction is an example of a transformation in which an optically active compound leads to an optically inactive product (a racemate).

Exercise 5-24

What halogenations of (S)-2-bromobutane, other than the ones described above, would furnish optically inactive products?

Model Building

The chlorination of (S)-2-bromobutane at C3 does not affect the existing chiral center. However, *the formation of a second stereocenter gives rise to diastereomers*. Specifically, attachment of chlorine to the left side of C3 in the drawing gives (2S,3S)-2-bromo-3-chlorobutane, whereas attachment to the right side gives its 2S,3R diastereomer.

REAL LIFE: MEDICINE 5-5 | Why Is Nature "Handed"?

In this chapter, we have seen that many of the organic molecules in nature are chiral. More important, most natural compounds in living organisms not only are chiral, but also are present in only one enantiomeric form. An example of an entire class of such compounds consists of the *amino acids,* which are the component units of *polypeptides.* The large polypeptides in nature are called *proteins* or, when they catalyze biotransformations, *enzymes.*

Absolute Configuration of Natural Amino Acids and Polypeptides

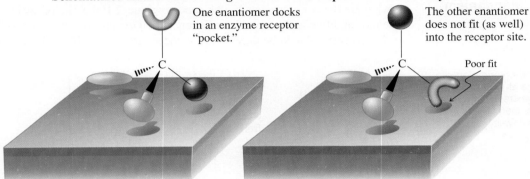

Schematized Enantiomer Recognition in the Receptor Site of an Enzyme

One enantiomer docks in an enzyme receptor "pocket."

The other enantiomer does not fit (as well) into the receptor site.

Poor fit

Being made up of smaller chiral pieces, enzymes arrange themselves into bigger conglomerates that also are chiral and show handedness. Thus, much as a right hand will readily distinguish another right hand from a left hand, enzymes (and other biomolecules) have "pockets" that, by virtue of their stereochemically defined features, are capable of recognizing and processing only one of the enantiomers in a racemate. The differences in physiological activity of the

In the quest to explore space and find signs for extraterrestial life, the Cassini mission was launched in 1997 to reach Saturn in 2004. The spacecraft has been orbiting this planet ever since, exploring its surroundings, including its moons. In the photo shown, Earth is an invisible spec in the top left quadrant.

two enantiomers of a chiral drug are based on this recognition (Real Life 5-4). A good analogy is that of a chiral key fitting only its image (not mirror image) lock. The chiral environment provided by these structures is also able to effect highly enantioselective conversions of achiral starting materials into enantiopure, chiral products. In this way, how nature preserves and proliferates its own built-in chirality can be readily understood (at least in principle).

What is more difficult to understand is how the enantiomeric homogeneity of nature arose in the first place; in other words, why was only one stereochemical configuration of the amino acids chosen but not the other? Trying to understand this mystery has fascinated many scientists, because it is very likely linked to the evolution of life as we know it. Speculation ranges from the invocation of a chance separation of enantiomers ("spontaneous resolution") to the postulate of the operation of a chiral physical force, such as handed radiation (as observed during the decay of radioactive elements or in so-called circularly polarized light). Another hypothesis suggests that enantiomeric excess (and perhaps life itself) was simply imported from another planet, with meteorites as carriers (thus really begging the question). A lot of effort has been expended in trying to detect nonracemic amino acids in meteor (and other planetary) samples, so far without success.

Chlorination of (S)-2-Bromobutane at C3

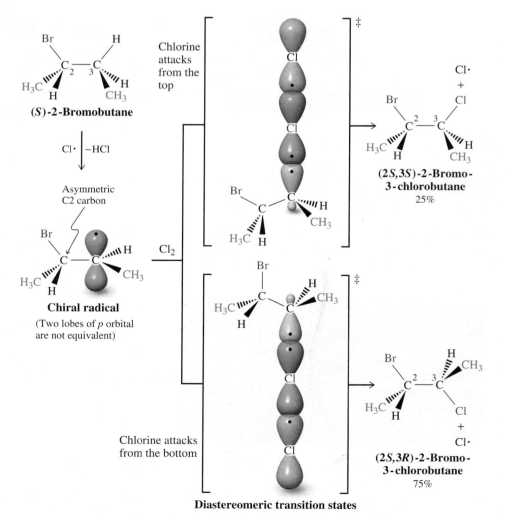

The chlorination at C2 results in a 1:1 mixture of enantiomers. Does the reaction at C3 also give an equimolar mixture of diastereomers? The answer is no. This finding is readily explained on inspection of the two transition states leading to the product (Figure 5-14). Abstraction of either one of the hydrogens results in a radical center at C3. In contrast with the radical formed in the chlorination at C2, however, the two faces of this radical are *not* mirror images of each other, because the radical retains the asymmetry of the original molecule as a result of the presence of the stereocenter at C2. Thus, the two sides of the *p* orbital are not equivalent.

What are the consequences of this nonequivalency? If the rate of attack at the two faces of the radical differ, as one would predict on steric grounds, then the rates of formation of the two diastereomers should be different, as is indeed found: (2S,3R)-2-Bromo-3-chlorobutane is

Mechanism

Figure 5-14 The chlorination of (S)-2-bromobutane at C3 produces the two diastereomers of 2-bromo-3-chlorobutane in unequal amounts as a result of the chirality at C2.

Diastereomeric transition states

(Different in energy)

preferred over the 2S,3S isomer by a factor of 3 (see Figure 5-14). The two transition states leading to products are not mirror images of each other and are not superimposable: They are diastereomeric. They therefore have different energies and represent different pathways.

The chlorination of racemic 2-bromobutane gives racemates

In the preceding discussion, we have used the enantiomerically pure starting material (S)-2-bromobutane to illustrate the stereochemical consequences of further halogenation. This choice was arbitrary with respect to absolute configuration; we might have equally well picked the R enantiomer. The result will be the same, except that all optically active products, namely, those of chlorination at C1, C3, and C4, will exhibit the opposite configurations to those shown: (2S)-2-bromo-1-chlorobutane, (2R,3R)- and (2R,3S)-2-bromo-3-chlorobutane (in a 1:3 ratio), and (3R)-3-bromo-1-chlorobutane, respectively. Attack at C2 will again provide a racemate, because the stereochemistry is lost at the stereocenter during the process. What about carrying out this reaction with racemic 2-bromobutane?

Solved Exercise 5-25	**Working with the Concepts: Writing All the Products of the Halogenation of a Chiral Compound**

Formulate all the products of the monobromination of (R)-1-bromo-1-deuterioethane A (see also Exercise 5-15), and specify whether they are chiral and optically active or not optically active. Remember that D is an isotope of H and will react qualitatively in the same way.

Strategy
Let's first list the possible sites for attack on **A** by Br·. They are the H on C1, the D on C1, and the three methyl hydrogens of C2. Now we can inspect the outcome of each abstraction and see how the ensuing radical forms product.

Solution
• Attack at H of C1

This process generates an achiral radical; that fact alone ensures that the product would be at least racemic, hence not optically active. However, in the case of bromination, this point is irrelevant because the product has lost its stereocenter anyway, since C1 now has two bromine atoms.
• Attack at D of C1

The situation here is similar: An achiral radical is produced, but its lack of stereochemistry is irrelevant because the product 1,1-dibromoethane is achiral.
• Attack at H of C2

Attack at C2 leaves the stereocenter untouched. Thus, the intermediate radical is chiral and so is the product 1,2-dibromo-1-deuterioethane (R), which is therefore optically active.

Remember WHIP

*W*hat

*H*ow

*I*nformation

*P*roceed

Exercise 5-26 | Try It Yourself

Write the structures of the products of monobromination of (S)-2-bromopentane at each carbon atom. Name the products and specify whether they are chiral or achiral, whether they will be formed in equal or unequal amounts, and which will be in optically active form.

Recalling the rule that states that "optically inactive starting materials furnish optically inactive products," we expect racemates for all products. Thus, attack at C1, C2, or C4 will result in racemic 2-bromo-1-chlorobutane, 2-bromo-2-chlorobutane, and 3-bromo-1-chlorobutane, respectively. Importantly, attack at C3 will still give two compounds, even though racemic, namely, the 2S,3S/2R,3R (25%) and 2S,3R/2R,3S diastereomers (75%) of 2-bromo-3-chlorobutane.

What are the conventions of writing chemical equations when racemates are involved? Unless specifically indicated by the *R/S* notation, the sign of optical rotation, or some surrounding text, it is assumed that all ingredients in a reaction are racemic. To avoid the clutter of writing both enantiomers in such cases, only one is shown, the equimolar presence of the other being tacitly assumed. The chlorination of racemic 2-bromobutane at C3 is then written as follows:

75%   25%

Exercise 5-27 |

Write the products of the monochlorination of bromocyclohexane at C2. (**Caution:** Is the starting material chiral?)

Stereoselectivity is the preference for one stereoisomer

A reaction that leads to the predominant (or exclusive) formation of one of several possible stereoisomeric products is **stereoselective.** For example, the chlorination of (S)-2-bromobutane at C3 is stereoselective, as a result of the chirality of the radical intermediate. The corresponding chlorination at C2, however, is not stereoselective: The intermediate is achiral and a racemate is formed.

How much stereoselectivity is possible? The answer depends very much on substrate, reagents, the particular reaction in question, and conditions. In the laboratory, chemists use enantiomerically pure reagents or catalysts to convert achiral compounds into one enantiomer of product (enantioselectivity; see Real Life 5-4). In nature, enzymes perform this job (see Real Life 5-5). In all cases, it is the handedness of that reagent, catalyst, or enzyme that is responsible for introducing the stereocenter compatible with their own chirality. An example from nature is the enzyme-catalyzed oxidation of dopamine to (−)-norepinephrine, discussed in detail in Problem 65 at the end of the chapter. The chiral reaction environment created by the enzyme gives rise to 100% stereoselectivity in favor of the enantiomer shown. The situation is very similar to shaping flexible achiral objects with your hands. For example, clasping a piece of modeling clay with your left hand furnishes a shape that is the mirror image of that made with your right hand.

Dopamine plays a major role in reward-driven learning. The pleasure you will experience when you receive your graduation diploma is due to an increase of the level of dopamine in your brain.

Dopamine   **(−)-Norepinephrine**

In Summary Chemical reactions, as exemplified by radical halogenation, can be stereoselective or not. Starting from achiral materials, such as butane, a racemic (nonstereoselective) product is formed by halogenation at C2. The two hydrogens at the methylene carbons of butane are equally susceptible to substitution, the halogenation step in the mechanism of radical bromination proceeding through an achiral intermediate and two enantiomeric transition states of equal energy. Similarly, starting from chiral and enantiomerically pure 2-bromobutane, chlorination of the stereocenter also gives a racemic product. However, stereoselectivity is possible in the formation of a new stereocenter, because the chiral environment retained by the molecule results in two unequal modes of attack on the intermediate radical. The two transition states have a diastereomeric relation, a condition that leads to the formation of products at unequal rates.

5-8 | RESOLUTION: SEPARATION OF ENANTIOMERS

As we know, the generation of a chiral structure from an achiral starting material furnishes a racemic mixture. How, then, can pure enantiomers of a chiral compound be obtained?

One possible approach is to start with the racemate and separate one enantiomer from the other. This process is called the **resolution** of enantiomers. Some enantiomers, such as those of tartaric acid, crystallize into mirror-image shapes, which can be manually separated (as done by Pasteur; see Real Life 5-3). However, this process is time consuming, not economical for anything but minute-scale separations, and applicable only in rare cases.

A better strategy for resolution is based on the different physical properties of diastereomers. Suppose we can find a reaction that converts a racemate into a mixture of diastereomers. All the R forms of the original enantiomer mixture should then be separable from the corresponding S forms by fractional crystallization, distillation, or chromatography of the diastereomers. How can such a process be developed? The trick is to add an enantiomerically pure reagent that will attach itself to the components of the racemic mixture. For example, we can imagine reaction of a racemate, $X_{R,S}$ (in which X_R and X_S are the two enantiomers), with an optically pure compound Y_S (the choice of the S configuration is arbitrary; the pure R mirror image would work just as well). The reaction produces two optically active diastereomers, $X_R Y_S$ and $X_S Y_S$, separable by standard techniques (Figure 5-15). Now the bond between X and Y in each of the separated and purified diastereomers is broken, liberating X_R and X_S in their enantiomerically pure states. In addition, the optically active agent Y_S may be recovered and reused in further resolutions (see also margin).

What we need, then, is a readily available, enantiomerically pure compound, Y, that can be attached to the molecule to be resolved in an easily reversible chemical reaction. In fact, nature has provided us with a large number of pure optically active molecules that can be used. An example is (+)-2,3-dihydroxybutanedioic acid [(+)-(R,R)-tartaric acid]. A popular reaction employed in the resolution of enantiomers is salt formation between acids and bases. For example, (+)-tartaric acid functions as an effective resolving agent of racemic amines. Figure 5-16 shows how this works for 3-butyn-2-amine. The racemate is first treated with (+)-tartaric acid to form two diastereomeric tartrate salts. The salt incorporating the R-amine crystallizes on standing and can be filtered away from the solution, which contains the more soluble salt of the S-amine. Treatment of the (+) salt with aqueous base liberates the free amine, (+)-(R)-3-butyn-2-amine. Similar treatment of the solution gives the (−)-S enantiomer (evidently slightly less pure: Note the slightly lower optical rotation). This process is just one of many ways in which the formation of diastereomers can be used in the resolution of racemates.

A very convenient way of separating enantiomers without the necessity of isolating diastereomers is by so-called **chiral chromatography** (Figure 5-17). The principle is the same as that illustrated in Figure 5-16, except that the optically active auxiliary [such as (+)-tartaric acid or any other suitable cheap optically active compound] is immobilized on a solid support (such as silica gel, SiO_2, or aluminum oxide, Al_2O_3). This material is then used to fill a column, and a solution of the racemate is allowed to pass through it. The individual enantiomers will reversibly bind to the chiral support to different extents (because

Resolution is similar to sorting your shoes (in the dark) with one foot (let us say the right) as a chiral auxiliary. The combination right foot–right shoe is completely different from that of right foot–left shoe, as the young lady in the photo discovers.

Figure 5-15 Flowchart for the separation (resolution) of two enantiomers. The procedure is based on conversion into separable diastereomers by means of reaction with an optically pure reagent.

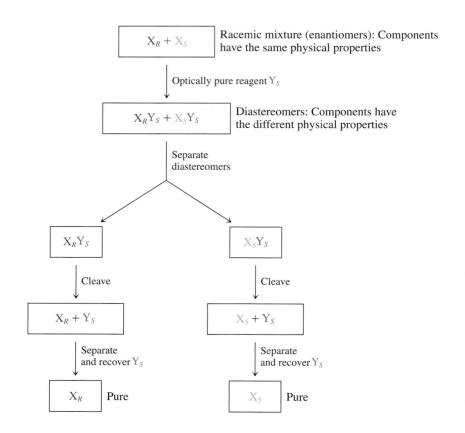

Figure 5-16 Resolution of 3-butyn-2-amine with (+)-2,3-dihydroxybutanedioic [(+)-tartaric] acid. It is purely accidental that the [α] values for the two diastereomeric tartrate salts are similar in magnitude and of opposite sign.

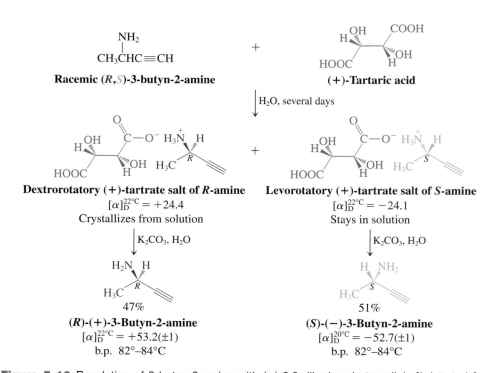

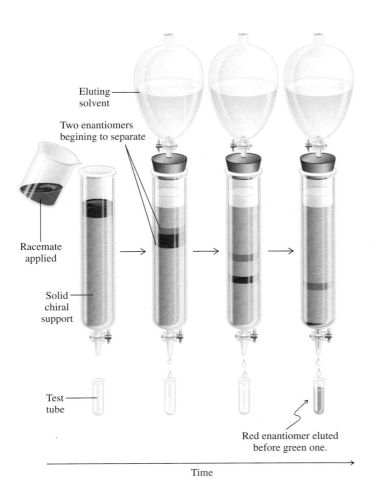

Figure 5-17 Resolution of a racemate on a chiral column. The sample is applied to the top of a column filled with an enantiopure chiral support. One enantiomer (green) interacts more strongly with the support than the other (red) and is relatively slow to pass through the column. Therefore, the red enantiomer is eluted before its green mirror image. Commercial columns often use the glucose polymer cellulose (Section 24-12) as the chiral stationary phase.

Eluting solvent

Two enantiomers begining to separate

Racemate applied

Solid chiral support

Test tube

Red enantiomer eluted before green one.

Time

this interaction is diastereomeric) and therefore be held on the column for different lengths of time (retention time). Therefore, one enantiomer will elute from the column before the other, enabling separation.

THE BIG PICTURE

The end of this chapter marks a milestone in your learning of organic chemistry. From here on, we will add very few *fundamentally* new ideas to our understanding of molecular structure. Rather, we will build on the basic principles outlined so far. You may find it useful to refer back to Chapters 1–5 as often as needed as a reminder of the several complementary viewpoints from which we have examined molecular structure:

1. Molecules consist of nuclei surrounded by electrons. The rules that govern their interactions are Coulomb's law, Lewis structures, and orbitals (Chapter 1).

2. The hydrocarbon skeletons of organic molecules, as exemplified by the alkanes, can be modeled as chains of linked carbon atoms that rotate, flip, have spatial requirements, and can be distorted away from ideal bond angles (Chapters 2 and 4).

3. The diversity of structures in organic chemistry is due to the ability of carbon atoms to assemble chains in linear, branched, cyclic, or polycyclic forms, all of which can bear multiple substituents and functional groups. Because of carbon's tetrahedral bonding, organic molecules assume a variety of three-dimensional shapes, with variations that have important consequences in their physical properties and reactivity.

4. When tetrahedral carbon bears four different substituents, it constitutes a stereocenter, giving rise to enantiomers—image and nonsuperimposable mirror image. Two or more such stereocenters generate new stereoisomers, differing only in the spatial arrangement of their component groups.

5. Organic molecules can be attacked by external reagents. Attack often occurs at nuclei attached to the remainder of the molecule by relatively weak bonds, or at polar functions, and is subject to steric and electronic constraints.

Most of the rest of this book examines in turn the classes of organic compounds characterized by their functional group. We focus primarily on the reactions these compounds undergo, emphasizing how the mechanism of each reaction is affected by the details of molecular structure. There are only a limited number of different reaction mechanisms; understanding them and the conditions that favor one over the other is the key to understanding organic chemistry.

WORKED EXAMPLES: INTEGRATING THE CONCEPTS

5-28. Figuring out All Possible Products of a Radical Halogenation

Selectivity in chemical reactions is a primary goal of the synthetic chemist. We have learned how such selectivity may be achieved, at least to some extent, in radical halogenations: in Sections 3-7 and 3-8, with respect to the type of hydrogen to be replaced (e.g., primary versus secondary versus tertiary) and, in Section 5-7, with respect to stereochemistry. You will have recognized that, because of the reactivity of radicals and the planarity of the carbon-centered radical intermediates, radical halogenations often lack selectivity. Thus, any synthetic plan considering them must take into account all possible outcomes of a proposed conversion. For example, looking again at our picture of the generalized steroid nucleus (Section 4-7), you can see that there are many types of hydrogens, including tertiary ones (shown explicitly), all of which are, in principle, susceptible to abstraction by a halogen atom.

Because the steroids are important biological molecules, their selective functionalization has been the focus of attention for many researchers. By developing carefully controlled conditions with special halogenating agents, chemists have been able to restrict attack not only to the tertiary centers, but also selectively to C5, C9, or C14 (shown in red; see also Problems 48–50 of Chapter 4). The following problem illustrates the kind of analyses that they undertook with a less complex cyclohexane fragment of the steroid nucleus.

How many products are there of the radical monobromination of (S)-2-bromo-1,1-dimethylcyclohexane at C2 and C6? Draw the structure of the starting material, name the resulting dibromodimethylcyclohexanes, label them as chiral or achiral, specify whether they are formed in equal or unequal amounts, and state whether they are optically active or not.

SOLUTION

What: We begin by drawing the structure of our starting material, first ignoring stereochemistry (A).

How: We then designate the priority sequence (B) according to the rules in Section 5-3. We now have a choice of two enantiomeric arrangements (C and D), and the task is to orient the molecule in our minds in such a way as to place the substituent of lowest priority (H atom) as far away as possible. To assist in this mental exercise, picture yourself at the molecular scale (shrunk by a factor of 10^{10}) and stand on the stereocenter in question with the C–H bond pointing away from you. The three remaining substituents will now surround you either in clockwise (*R*) or counterclockwise (*S*) fashion: D is the correct structure of the *S* enantiomer (margin).

Now we are ready to introduce bromine at either C2 or C6. *I*nformation: It is important here to remember the mechanism of free-radical halogenation: The crucial intermediate is a radical center—in our case, at either C2 (E) or C6 (F)—that can be attacked by halogen from either side of the *p* orbital (Section 3-4).

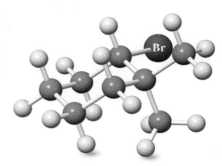

(S)-2-Bromo-1,1-dimethylcyclohexane

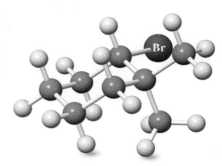

| E | F | G | H | I |

*P*roceed: In E, the molecule is symmetrical, and the rate of attack from the top is equal to that from the bottom. If the halogenation were to be executed by using F_2 or Cl_2, C_1 would remain a stereocenter, the *R* and *S* enantiomers having been formed in equal amounts (racemate; Section 5-7, Figure 5-13). However, in our case, bromination at C2 removes the asymmetry of this carbon: Compound G is achiral and hence not optically active.

Turning to F, the situation is different. Here, the presence of the unchanged original stereocenter (C2 in D) makes the two faces of the intermediate radical center unequal. Two diastereomers (H and I) are formed at unequal rates and therefore in unequal amounts (Section 5-7, Figure 5-14). In H, *cis*-2,6-dibromo-1,1-dimethylcyclohexane, the second bromine is attached in such a way as to introduce a mirror plane into the molecule: H is a meso compound, achiral, and hence not optically active (Section 5-6). Another way of describing what has happened is that the chirality of C2 in D—namely, *S*—is canceled out by the introduction of its "mirror image" at C6—namely, *R*. The two stereoisomers are indistinguishable because (2*S*,6*R*)-H is the same as (2*R*,6*S*)-H. (You can verify that statement by simply rotating compound H about the dashed line representing the mirror plane.)

On the other hand, compound I, (2*S*,6*S*)-2,6-dibromo-1,1-dimethylcyclohexane, contains no mirror plane: The molecule is chiral, enantiomerically pure, and, hence, optically active. In other words, the reaction leaves the stereochemical integrity and identity of C2 intact, generating only one enantiomer of the product, which is nonsuperimposable with its mirror image, the (2*R*,6*R*) diastereomer (Section 5-5).

5-29. The Stereochemical Consequences of Hydrogenating Limonene

We shall learn in Sections 11-5 and 12-2 that double bonds in alkenes can be hydrogenated by using hydrogen gas and specific metal catalysts (Section 3-3), leading to the corresponding alkanes:

The two enantiomers of limonene (shown in the margin) smell quite differently. The *S*-isomer is present in the cones of spruce trees and has a turpentine-like odor; the *R*-isomer gives oranges their characteristic fragrance. *R*-Limonene is a by-product of the juice industry and is the major constituent of citrus peel oil. Every year, more than 110 million pounds of the oil are made in the United States alone. Draw the respective products of the hydrogenation of both double bonds in (*R*)- and (*S*)-limonene. Are these products isomers, identical, chiral, achiral, or optically active/inactive?

SOLUTION

First draw the respective products of double hydrogenation of (*R*)- and (*S*)-limonene. As indicated above, the two hydrogens can be added from either the top or the bottom of the π bond (see Figure 1-21). This is of no consequence for hydrogenation of the substituent but is important for the outcome of the ring hydrogenation: One mode leads to a trans, the other to a cis disubstituted cyclohexane. Thus, we obtain two stereoisomers from each enantiomer. How do the respective pairs of

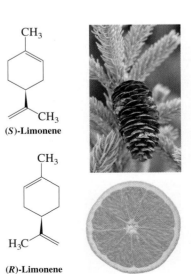

(S)-Limonene

(R)-Limonene

products from the (R) and (S) starting materials relate to each other? It is apparent that the two trans and the two cis isomers are superimposable: They are identical. In other words, the two enantiomers of limonene give an identical mixture of stereoisomers. Are they chiral? The answer is no: The resulting 1,4-disubstituted cyclohexanes contain a mirror plane. Consequently, hydrogenation of limonene causes symmetrization of the molecule, rendering it achiral. Therefore, the products are optically inactive.

Important Concepts

1. Isomers have the same molecular formula but are different compounds. Constitutional (structural) isomers differ in the order in which the individual atoms are connected. Stereoisomers have the same connectivity but differ in the three-dimensional arrangement of the atoms. **Mirror-image stereoisomers** are related to each other as object and mirror image.

2. An object that is not superimposable on its mirror image is **chiral.**

3. A carbon atom bearing four different substituents (**asymmetric carbon**) is an example of a **stereocenter.**

4. Two stereoisomers that are related to each other as image–nonsuperimposable mirror image are called **enantiomers.**

5. A compound containing one stereocenter is chiral and exists as a pair of enantiomers. A 1:1 mixture of enantiomers is a **racemate (racemic mixture).**

6. Chiral molecules cannot have a plane of symmetry (mirror plane). If a molecule has a **mirror plane,** then it is **achiral.**

7. **Diastereomers** are stereoisomers that are not related to each other as object to mirror image. Cis and trans isomers of cyclic compounds are examples of diastereomers.

8. Two stereocenters in a molecule result in as many as four stereoisomers—two diastereomerically related pairs of enantiomers. The maximum number of stereoisomers that a compound with n stereocenters can have is 2^n. This number is reduced when equivalently substituted stereocenters give rise to a plane of symmetry. A molecule containing stereocenters *and* a mirror plane is identical with its mirror image (achiral) and is called a **meso compound.** The presence of a mirror plane in any energetically accessible conformation of a molecule is sufficient to make it achiral.

9. Most of the physical properties of enantiomers are the same. A major exception is their interaction with **plane-polarized light:** One enantiomer will rotate the polarization plane clockwise (**dextrorotatory**), the other counterclockwise (**levorotatory**). This phenomenon is called **optical activity.** The extent of the rotation is measured in degrees and is expressed by the **specific rotation, [α].** Racemates and meso compounds show zero rotation. The **enantiomer excess** or **optical purity** of an unequal mixture of enantiomers is given by

$$\text{enantiomer excess } (ee) = \text{optical purity} = \left(\frac{[\alpha]_{\text{observed}}}{[\alpha]} \right) \times 100\%$$

10. The "handedness" of a stereocenter (its absolute configuration) is revealed by X-ray diffraction and can be assigned as R or S by using the **sequence rules** of Cahn, Ingold, and Prelog.

11. **Fischer projections** provide stencils for the quick drawing of molecules with stereocenters.

12. Chirality can be introduced into an achiral compound by radical halogenation. When the transition states are enantiomeric (related as object and mirror image), the result is a racemate because the faces of the planar radical react at equal rates.

13. Radical halogenation of a chiral molecule containing one stereocenter will give a racemate if the reaction takes place at the stereocenter. When reaction elsewhere leads to two diastereomers, they will be formed in unequal amounts.

14. The preference for the formation of one stereoisomer, when several are possible, is called **stereoselectivity.**

15. The separation of enantiomers is called **resolution.** It is achieved by the reaction of the racemate with the pure enantiomer of a chiral compound to yield separable diastereomers. Chemical removal of the chiral reagent frees both enantiomers of the original racemate. Another way of separating enantiomers is by **chiral chromatography** on an optically active support.

Problems

30. Classify each of the following common objects as being either chiral or achiral. Assume in each case that the object is in its simplest form, without decoration or printed labels. (**a**) A ladder; (**b**) a door; (**c**) an electric fan; (**d**) a refrigerator; (**e**) Earth; (**f**) a baseball; (**g**) a baseball bat; (**h**) a baseball glove; (**i**) a flat sheet of paper; (**j**) a fork; (**k**) a spoon; (**l**) a knife.

31. Each part of this problem lists two objects or sets of objects. As precisely as you can, describe the relation between the two sets, using the terminology of this chapter; that is, specify whether they are identical, enantiomeric, or diastereomeric. (**a**) An American toy car compared with a British toy car (same color and design but steering wheels on opposite sides); (**b**) two left shoes compared with two right shoes (same color, size, and style); (**c**) a pair of skates compared with two left skates (same color, size, and style); (**d**) a right glove on top of a left glove (palm to palm) compared with a left glove on top of a right glove (palm to palm; same color, size, and style).

32. For each pair of the following molecules, indicate whether its members are identical, structural isomers, conformers, or stereoisomers. How would you describe the relation between conformations when they are maintained at a temperature too low to permit them to interconvert?

(a) $CH_3CH_2CH_2CH$ with CH_3 and CH_3 and $CH_3CH_2CHCH_2CH_3$ with CH_3

(b)

(c) $ClCH_2CH_2$ ... OH and CH_3CH with Cl ... OH

(d)

(e) $CH_3CCH_2CH_2CH_3$ with Cl and Br and $CH_3CHCH_2CHCH_3$ with Br and Cl

(f)

(g)

and

(h)

and

33. Which of the following compounds are chiral? (**Hint:** Look for stereocenters.)

(**a**) 2-Methylheptane
(**b**) 3-Methylheptane
(**c**) 4-Methylheptane
(**d**) 1,1-Dibromopropane
(**e**) 1,2-Dibromopropane
(**f**) 1,3-Dibromopropane
(**g**) Ethene, $H_2C=CH_2$
(**h**) Ethyne, $HC\equiv CH$

(**i**) Benzene, (Note: Like ethene, benzene contains all sp^2-hybridized carbons and is therefore planar.)

(**j**) Epinephrine,

(**k**) Vanillin,

(**l**) Citric acid,

(**m**) Ascorbic acid,

(**n**) *p*-Menthane-1,8-diol (terpin hydrate),

(o) Petridine (Demerol),

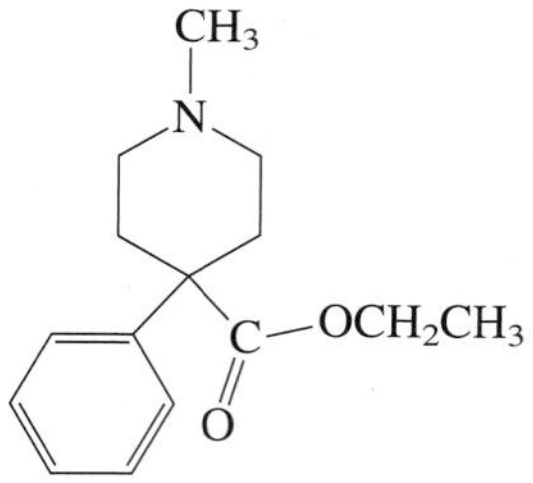

34. Each of the following molecules has the molecular formula $C_5H_{12}O$ (check for yourself). Which ones are chiral?

(a)

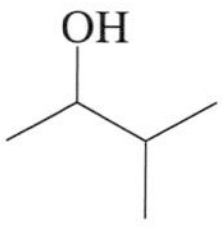

(b)

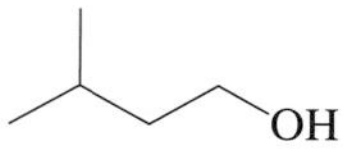

(c)

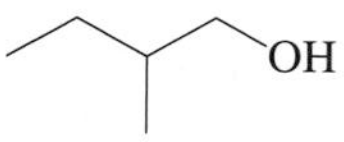

(d)

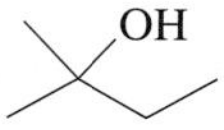

(e)

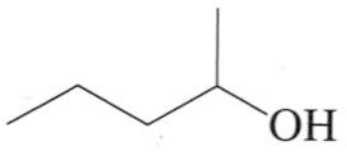

(f)

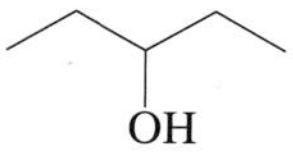

35. Draw either one of the enantiomers for each chiral molecule in Problem 34 and label its stereocenter as *R* or *S*.

36. Which of the following cyclohexane derivatives are chiral? For the purpose of determining the chirality of a cyclic compound, the ring may generally be treated as if it were planar.

(a)

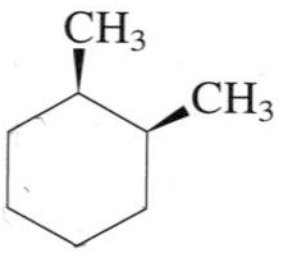

(b)

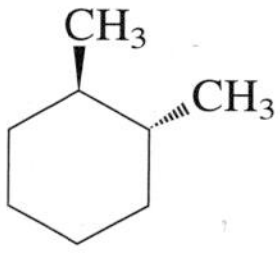

(c)

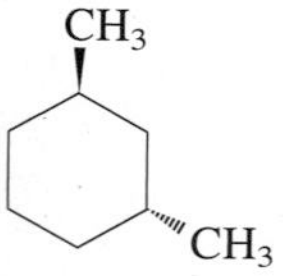

(d)

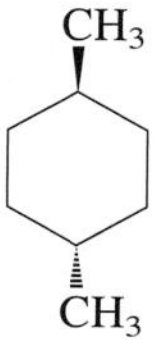

37. Label every stereocenter in the molecules in Problem 36 as *R* or *S*.

38. Circle each chiral molecule. Put a star (*) next to each chiral carbon and label it as *R* or *S*.

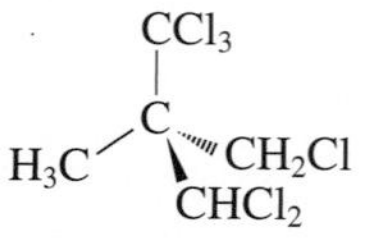

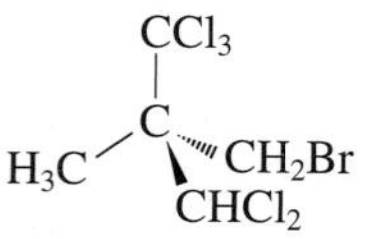

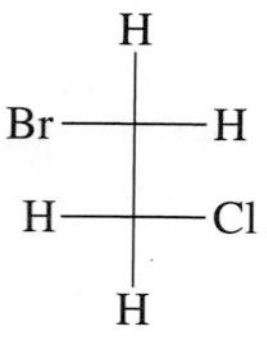

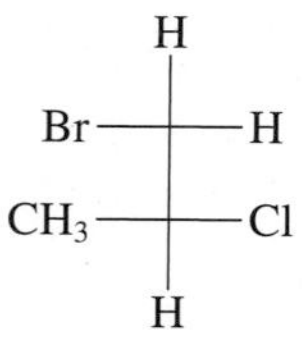

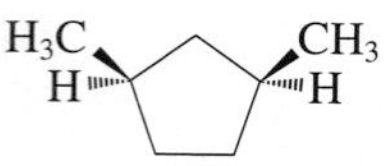

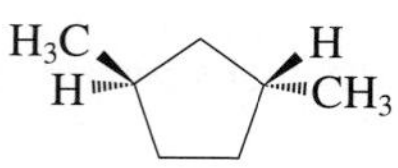

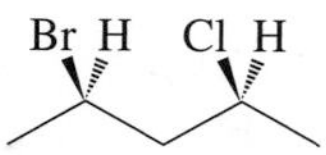

39. For each pair of structures shown, indicate whether the two species are constitutional isomers, enantiomers, diastereomers of one another, or identical molecules.

(a)

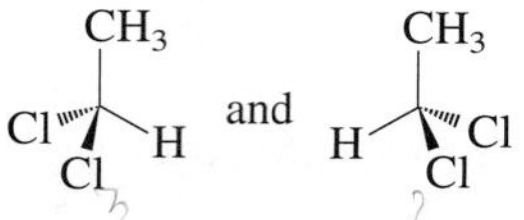

(b)

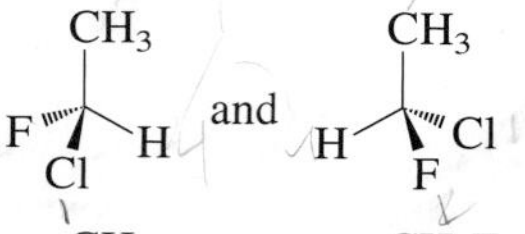

(c)

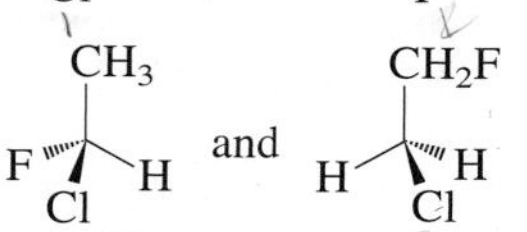

(d)

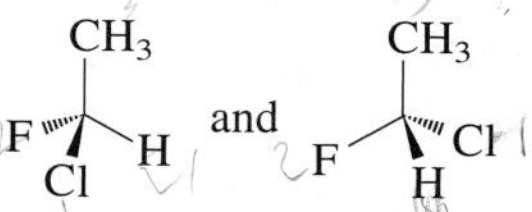

(e)

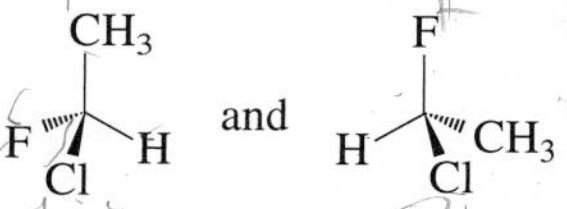

(f)

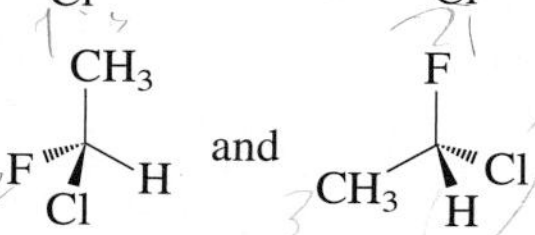

(g)

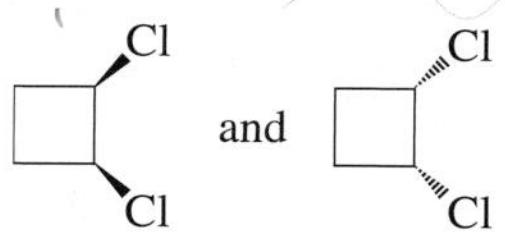

(h)

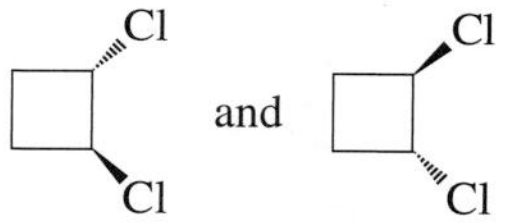

(i)

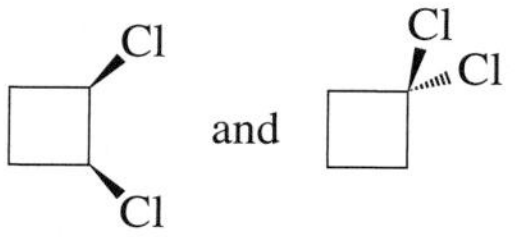

(j)

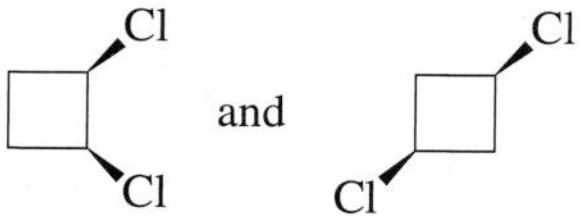

(k)

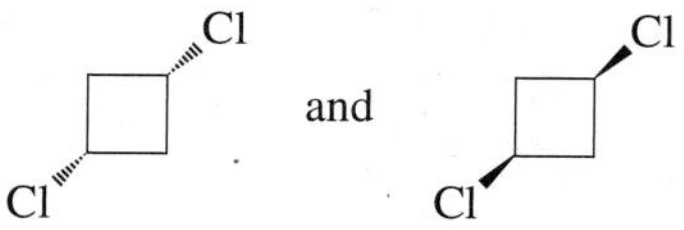

(l)

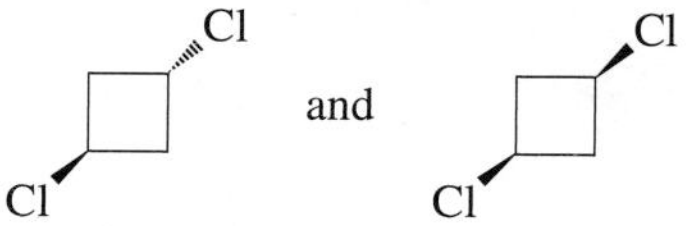

(m)

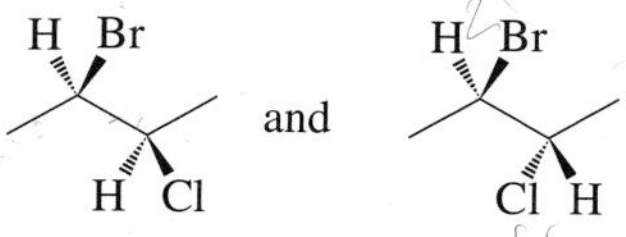

(n)

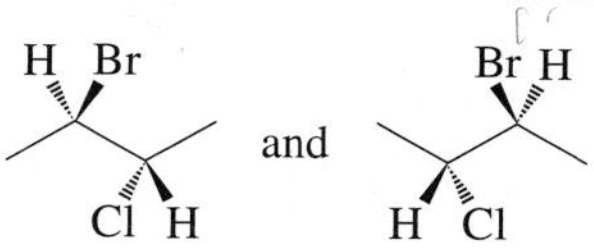

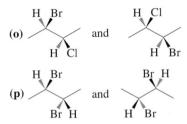

(o) and

(p) and

40. For each of the following formulas, identify every structural isomer containing one or more stereocenters, give the number of stereoisomers for each, and draw and fully name at least one of the stereoisomers in each case.

(a) C_7H_{16} (b) C_8H_{18} (c) C_5H_{10}, with one ring

41. Assign the appropriate designation of configuration (R or S) to the stereocenter in each of the following enantiomers. (**Hint:** Regarding cyclic structures containing stereocenters, treat the ring as if it were two separate substituents that happen to be attached to each other at the far end of the molecule—look for the first point of difference, just as you would for acyclic structures.)

(a)

(b)

(c)

(d)

(e)

(f)

(g)

(h)

(i)

(j)

42. Mark the stereocenters in each of the chiral molecules in Problem 33. Draw any single stereoisomer of each of these molecules, and assign the appropriate designation (R or S) to each stereocenter.

43. The two enantiomers of carvone [systematic name: 2-methyl-5-(1-methylethenyl)-2-cyclohexenone; Real Life 5-1] are drawn below. Which is R and which is S?

CH₃

CH₃

O

O

H₂C=C H

H₂C=C H

CH₃

CH₃

(+)-Carvone
(In caraway seeds)

(−)-Carvone
(In spearmint)

44. Draw structural representations of each of the following molecules. Be sure that your structure clearly shows the configuration at the stereocenter. (**Hint:** You may find it useful to first draw the enantiomer whose configuration is easiest for you to determine and then, if necessary, modify your structure to fit the one requested in the problem.) (a) (R)-2-chloropentane; (b) (S)-2-methyl-3-bromohexane; (c) (S)-1,3-dichlorobutane; (d) (R)-2-chloro-1,1,1-trifluoro-3-methylbutane.

45. Draw structural representations of each of the following molecules. Be sure that your structure clearly shows the configuration at each stereocenter. (a) (R)-3-bromo-3-methylhexane; (b) (3R,5S)-3,5-dimethylheptane; (c) (2R,3S)-2-bromo-3-methylpentane; (d) (S)-1,1,2-trimethylcyclopropane; (e) (1S,2S)-1-chloro-1-trifluoromethyl-2-methylcyclobutane; (f) (1R,2R,3S)-1,2-dichloro-3-ethylcyclohexane.

46. Draw and name all possible stereoisomers of $(CH_3)_2CHCHBrCHClCH_3$.

47. CHALLENGE For each of the following questions, assume that all measurements are made in 10-cm polarimeter sample containers. (a) A 10-mL solution of 0.4 g of optically active 2-butanol in water displays an optical rotation of −0.56°. What is its specific rotation? (b) The specific rotation of sucrose (common sugar) is +66.4. What would be the observed optical rotation of such a solution containing 3 g of sucrose? (c) A solution of pure (S)-2-bromobutane in ethanol is found to have an observed $\alpha = 57.3°$. If $[\alpha]$ for (S)-2-bromobutane is 23.1, what is the concentration of the solution?

48. Natural epinephrine (adrenaline), $[\alpha]_D^{25°C} = -50$, is used to treat cardiac arrest and sudden severe allergic reactions. Its enantiomer is medically worthless and is, in fact, toxic. You, a pharmacist, are given a solution said to contain 1 g of epinephrine in 20 mL of liquid, but the optical purity is not specified. You place it in a polarimeter (10-cm tube) and get a reading of −2.5°. What is the optical purity of the sample? Is it safe to use medicinally?

49. Sodium hydrogen (S)-glutamate [(S)-monosodium glutamate], $[\alpha]_D^{25°C} = +24$, is the active flavor enhancer known as MSG. The condensed formula of MSG is shown below. (a) Draw the structure of the S enantiomer of MSG. (b) If a commercial sample of MSG were found to have a $[\alpha]_D^{25°C} = +8$, what would be its optical purity? What would be the percentages of the S and R enantiomers in the mixture? (c) Answer the same questions for a sample with $[\alpha]_D^{25°C} = +16$.

$$\underset{O}{\overset{NH_2}{HOCCHCH_2CH_2CO^-Na^+}}$$

50. The molecule below is menthol, with the stereochemistry omitted (see Section 4-7). (a) Identify all stereocenters in menthol. (b) How many stereoisomers exist for the menthol structure? (c) Draw all the stereoisomers of menthol, and identify all pairs of enantiomers.

OH

Menthol

51. **CHALLENGE**   Natural (−)-menthol, the volatile oil primarily responsible for the flavor and aroma of peppermint, is the 1*R*,2*S*,5*R*-stereoisomer. **(a)** Identify (−)-menthol from the structures you drew for Problem 50, part (b). **(b)** Another of the naturally occurring diastereomers of menthol is (+)-isomenthol, the 1*S*,2*R*,5*R*-stereoisomer. Identify (+)-isomenthol among your structures. **(c)** A third is (+)-neomenthol, the 1*S*,2*S*,5*R*-compound. Find (+)-neomenthol among your structures. **(d)** Based on your understanding of the conformations of substituted cyclohexanes (Section 4-4), what is the stability order (from most stable to least) for the three diastereomers, menthol, isomenthol, and neomenthol?

52. Of the stereoisomers described in the two problems above, (−)-menthol ($[\alpha]_D = -51$) and (+)-neomenthol ($[\alpha]_D = +21$) are the major constituents in mentha oil, their main natural source. The menthol–neomenthol mixture in a natural sample of mentha oil exhibits ($[\alpha]_D = -33$). What are the percentages of menthol and neomenthol in this oil?

53. For each of the following pairs of structures, indicate whether the two compounds are identical or enantiomers of each other.

(a)

$$\begin{array}{c}CH_2CH_3\\ |\\ C\\ Br \diagup \ \ \diagdown CH_2CH_3\\ Cl \end{array} \quad \text{and} \quad \begin{array}{c}CH_2CH_3\\ |\\ C\\ CH_3CH_2 \diagup \ \ \diagdown Br\\ Cl \end{array}$$

(b)

$$\begin{array}{c}CH_3\\ H \!-\!\!\!-\!\!\!- Cl\\ Br\end{array} \quad \text{and} \quad \begin{array}{c}CH_3\\ C\\ Cl \diagup \ \ \diagdown Br\\ H\end{array}$$

(c)

$$\begin{array}{c}CH_3\\ Cl \!-\!\!\!-\!\!\!- CF_3\\ OCH_3\end{array} \quad \text{and} \quad \begin{array}{c}OCH_3\\ F_3C \!-\!\!\!-\!\!\!- CH_3\\ Cl\end{array}$$

(d)

$$\begin{array}{c}H\\ H_2N \!-\! C \!\!\rightarrow\!\! CO_2H\\ CH(CH_3)_2\end{array} \quad \text{and} \quad \begin{array}{c}NH_2\\ H \!-\!\!\!-\!\!\!- CH(CH_3)_2\\ CO_2H\end{array}$$

54. Determine the *R* or *S* designation for each stereocenter in the structures in Problem 53.

55. The compound pictured below is a sugar called (−)-arabinose. Its specific rotation is −105. **(a)** Draw the enantiomer of (−)-arabinose. **(b)** Does (−)-arabinose have any other enantiomers? **(c)** Draw a diastereomer of (−)-arabinose. **(d)** Does (−)-arabinose have any other diastereomers? **(e)** If possible, predict the specific rotation of the structure that you drew for (a). **(f)** If possible, predict the specific rotation of the structure that you drew for (c). **(g)** Does (−)-arabinose have any optically inactive diastereomers? If it does, draw one.

$$\begin{array}{c}H\diagdown \ _{\diagup O}\\ C\\ HO \!-\!\!\!-\!\!\!- H\\ H \!-\!\!\!-\!\!\!- OH\\ H \!-\!\!\!-\!\!\!- OH\\ CH_2OH\end{array}$$

(−)-Arabinose

(+)-Arabinose, the enantiomer of the sugar shown above, is sold as a low-calorie sweetener.

56. Write the complete IUPAC name of the following enantiomer (do not forget stereochemical designations).

$$\begin{array}{c}CH_2CH_3\\ |\\ C\\ H \diagup \ \ \diagdown CH_2CH_2Cl\\ Cl\end{array} \quad C_5H_{10}Cl_2$$

Reaction of this compound with 1 mol of Cl_2 in the presence of light produces several isomers of the formula $C_5H_9Cl_3$. For each part of this problem, give the following information: How many stereoisomers are formed? If more than one is formed, are they generated in equal or unequal amounts? Designate every stereocenter in each stereoisomer as *R* or *S*.

(a) Chlorination at C3   **(b)** Chlorination at C4
(c) Chlorination at C5

57. Monochlorination of methylcyclopentane can result in several products. Give the same information as that requested in Problem 56 for the monochlorination of methylcyclopentane at C1, C2, and C3.

58. Draw all possible products of the chlorination of (*S*)-2-bromo-1,1-dimethylcyclobutane. Specify whether they are chiral or achiral, whether they are formed in equal or unequal amounts, and which are optically active when formed.

59. Illustrate how to resolve racemic 1-phenylethanamine (shown below), using the method of reversible conversion into diastereomers.

$$\begin{array}{c}H\diagdown \ _{\diagup NH_2}\\ C\\ \diagup \ \ \diagdown CH_3\end{array}$$

1-Phenylethanamine

60. Draw a flowchart that diagrams a method for the resolution of racemic 2-hydroxypropanoic acid (lactic acid, Table 5-1), using (*S*)-1-phenylethanamine.

61. How many different stereoisomeric products are formed in the monobromination of **(a)** racemic *trans*-1,2-dimethylcyclohexane and **(b)** pure (*R*,*R*)-1,2-dimethylcyclohexane? **(c)** For your answers to (a) and (b), indicate whether you expect equal or unequal amounts of the various products to be formed. Indicate to what extent products can be separated on the basis of having different physical properties (e.g., solubility, boiling point).

62. **CHALLENGE**   Make a model of *cis*-1,2-dimethylcyclohexane in its most stable conformation. If the molecule were rigidly locked into this conformation, would it be chiral? (Test your answer by making a model of the mirror image and checking for superimposability.)

Flip the ring of the model. What is the stereoisomeric relation between the original conformation and the conformation after flipping the ring? How do the results that you have obtained in this problem relate to your answer to Problem 36, part (a)?

63. Morphinane is the parent substance of the broad class of chiral molecules known as the morphine alkaloids. Interestingly, the (+) and (−) enantiomers of the compounds in this family have rather different physiological properties. The (−) compounds, such as morphine, are "narcotic analgesics" (painkillers), whereas the (+) compounds are "antitussives" (ingredients in cough syrup). Dextromethorphan is one of the simplest and most common of the latter.

Morphinane   **Dextromethorphan**

(a) Locate and identify all the stereocenters in dextromethorphan. (b) Draw the enantiomer of dextromethorphan. (c) As best you can (it is not easy), assign R and S configurations to all the stereocenters in dextromethorphan.

64. We will learn in Chapter 18 that hydrogens on the carbon atom adjacent to the carbonyl functional group (C=O) are acidic. The compound (S)-3-methyl-2-pentanone (below) loses its optical activity when it is dissolved in a solution containing a catalytic amount of base. Explain.

(S)-3-Methyl-2-pentanone

65. The enzymatic introduction of a functional group into a biologically important molecule is not only specific with regard to the location at which the reaction occurs in the molecule (see Chapter 4, Problem 50), but also usually specific in the stereochemistry obtained. The biosynthesis of epinephrine first requires that a hydroxy group be introduced specifically to produce (−)-norepinephrine from the achiral substrate dopamine. (The completion of the synthesis of epinephrine will be presented in Problem 71 of Chapter 9.) Only the (−) enantiomer is functional in the appropriate physiological manner, so the synthesis must be highly stereoselective.

Dopamine

(−)-Norepinephrine

(a) Is the configuration of (−)-norepinephrine R or S? (b) In the absence of an enzyme, would the transition states of a radical oxidation leading to (−)- and (+)-norepinephrine be of equal or unequal energy? What term describes the relation between these transition states? (c) In your own words, describe how the enzyme must affect the energy of these transition states to favor production of the (−) enantiomer. Does the enzyme have to be chiral or can it be achiral?

Team Problem

66. Studies have shown that one stereoisomeric form of compound A is an effective agent against certain types of neurodegenerative disorders. Recognize that structure A contains a decalin-type system, as illustrated in structure B, and that the nitrogen can be treated just like a carbon.

A   B

(a) Use your model kits to analyze the ring juncture. Make models of the cis as well as the trans ring juncture of structure B. You should have four different models. Identify the stereochemical relation between them as diastereomeric or enantiomeric. Draw the isomers and assign the R or S configuration to the stereocenters at the ring fusion.

(b) Although the trans ring juncture is the energetically more favorable one, the compound with cis ring juncture is the stereoisomer of structure A that shows biological activity. Make models of structure A that have the cis ring juncture exclusively. Set the stereochemistry of C3 as shown in structure A and vary the center at C6 in relation to that at C3. Again, there are four different models. Draw them and convince yourselves that none of them are enantiomers by assigning the R or S configuration to all four of the stereocenters in each of the compounds.

(c) The stereoisomer of compound A that shows the greatest biological activity has a cis ring fusion with substituents at C3 and C6 that are both equatorial. Which of the stereoisomers that you drew encompasses these constraints? Identify it by recording the absolute configuration at C3, C4a, C6, and C8a.

Preprofessional Problems

67. Which compound will *not* exhibit optical activity? (Note that these are all Fischer projections.)

68. The enantiomer of H—$\overset{\text{Cl}}{\underset{\underset{S}{\text{CH}_3}}{|}}$—CH$_2CH_3$

(a) is CH$_3$CH$_2$—$\overset{\text{Cl}}{\underset{\underset{R}{\text{CH}_3}}{|}}$—H

(b) can exist only at low temperatures

(c) is nonisomeric

(d) is incapable of existence

69. The molecule that is of the *R* configuration according to the Cahn–Ingold–Prelog convention is (remember these are Fischer projections):

(a) H$_3$C—$\overset{\text{H}}{\underset{\text{CH}_3}{|}}$—CH$_2$Cl

(b) H$_3$C—$\overset{\text{H}}{\underset{\text{CH}_2\text{Br}}{|}}$—CH$_2$Cl

(c) H$_3$C—$\overset{\text{CH}_2\text{Br}}{\underset{\text{H}}{|}}$—CH$_2$Cl

(d) H$_3$C—$\overset{\text{H}}{\underset{\text{CH}_2\text{Br}}{|}}$—CH$_2$F

(e) H$_3$C—$\overset{\text{CH}_2\text{Br}}{\underset{\text{CH}_2\text{Cl}}{|}}$—CH$_2$Br

70. Which compound is *not* a meso compound?

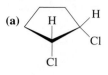

(a)

(b)

(c)

(d)

(e)

CHAPTER 6

Properties and Reactions of Haloalkanes

Bimolecular Nucleophilic Substitution

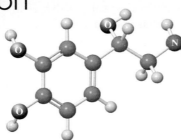

rganic chemistry provides us with innumerable ways to convert one substance into another. The products of these transformations are literally all around us. Recall from Chapter 2, however, that functional groups are the centers of reactivity in organic molecules; before we can make practical use of organic chemistry, we must develop our ability to work with these functional groups. In Chapter 3 we examined halogenation of alkanes, a process by which the carbon–halogen group is introduced into an initially unfunctionalized structure. Where do we go from here?

In this chapter we turn to the chemistry of the products of halogenation, the haloalkanes. We shall see how the polarized carbon–halogen bond governs the reactivity of these substances and how it can be converted into other functional groups. On the basis of the kinetics observed for a common reaction of haloalkanes, we introduce a new mechanism and learn the effects of different solvents on its progress. We shall review principles that govern the general mechanistic behavior of molecules with polar functional groups. Finally, we shall begin to apply these principles and see the role they play in many conversions of halogenated organic compounds into other substances, such as amino acids—the building blocks of proteins.

A haloalkane

6-1 | PHYSICAL PROPERTIES OF HALOALKANES

The physical properties of the haloalkanes are quite distinct from those of the corresponding alkanes. To understand these differences, we must consider the size of the halogen substituent and the polarity of the carbon–halogen bond. Let us see how these factors affect bond strength, bond length, molecular polarity, and boiling point.

The bond strength of C–X decreases as the size of X increases

The C–X bond-dissociation energies in the halomethanes, CH_3X, decrease along the series F, Cl, Br, I. At the same time, the C–X bond lengths increase (Table 6-1). The bond between carbon and halogen is made up mainly by the overlapping of an sp^3 hybrid orbital on carbon

In the body, the nitrogen of the amino group in noradrenaline attacks the methyl group in S-adenosylmethionine by nucleophilic substitution to give adrenaline. Adrenaline is a "fight-or-flight" hormone, released into the bloodstream during stress and emergencies, and is responsible for the "rush" felt during thrilling experiences.

Table 6-1	C–X Bond Lengths and Bond Strengths in CH₃X	
Halo-methane	Bond length (Å)	Bond strength [kcal mol⁻¹ (kJ mol⁻¹)]
CH₃F	1.385	110 (460)
CH₃Cl	1.784	85 (356)
CH₃Br	1.929	70 (293)
CH₃I	2.139	57 (238)

Increasing bond length → Decreasing bond strength

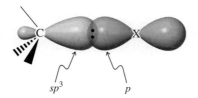

Figure 6-1 Bond between an alkyl carbon and a halogen. The size of the *p* orbital is substantially larger than that shown for X = Cl, Br, or I.

Chloromethane

Dipole–Dipole Attraction

with a *p* orbital on the halogen (Figure 6-1). In the progression from fluorine to iodine in the periodic table, the size of the halogen increases, and the electron cloud around the halogen atom becomes more diffuse. As a consequence of this growing mismatch in size between the halogen *p* orbital and the relatively compact orbital on carbon, bonding overlap diminishes along the series, leading to both longer and weaker C–X bonds. This phenomenon is general: *Short bonds are stronger than longer bonds.*

The C–X bond is polarized

The leading characteristic of the haloalkanes is their polar C–X bond. Recall from Section 1-3 that halogens are more electronegative than carbon. Thus, the electron density along the C–X bond is displaced in the direction of X, giving the halogen a partial negative charge (δ^-) and the carbon a partial positive charge (δ^+). This polarization can be seen in the electrostatic potential map of chloromethane shown in the margin. The chlorine atom is electron rich (red), whereas the region around the carbon atom is electron poor (blue). How does this bond polarization govern the chemical behavior of the haloalkanes? As we saw in Chapter 2, the electrophilic δ^+ carbon atom is subject to attack by anions and other electron-rich, nucleophilic species. Cations and other electron-deficient species, however, attack the δ^- halogen.

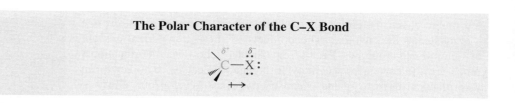

The Polar Character of the C–X Bond

Haloalkanes have higher boiling points than the corresponding alkanes

The polarity of the C–X bond affects the physical properties of the haloalkanes in predictable ways. Their boiling points are generally higher than those of the corresponding alkanes (Table 6-2), largely because of coulombic attraction between the δ^+ and δ^- ends of C–X bond dipoles in the liquid state (*dipole–dipole interaction;* Figure 2-6).

Boiling points also rise with increasing size of X, the result of greater London interactions (Section 2-7). Recall that London forces arise from mutual correlation of electrons among molecules (Figure 2-6). This effect is strongest when the outer electrons are not held very tightly around the nucleus, as in the larger atoms. To measure it, we define the **polarizability** of an atom or group as the degree to which its electron cloud is deformed under the influence of an external electric field. The more polarizable an atom or group, the stronger its London interactions, and the higher will be the boiling point.

Table 6-2	Boiling Points of Haloalkanes (R–X)				
		Boiling Point (°C)			
R	X =	H	F	Cl	Br	I
CH₃		−161.7	−78.4	−24.2	3.6	42.4
CH₃CH₂		−88.6	−37.7	12.3	38.4	72.3
CH₃(CH₂)₂		−42.1	−2.5	46.6	71.0	102.5
CH₃(CH₂)₃		−0.5	32.5	78.4	101.6	130.5
CH₃(CH₂)₄		36.1	62.8	107.8	129.6	157.0
CH₃(CH₂)₇		125.7	142.0	182.0	200.3	225.5

REAL LIFE: MEDICINE 6-1 | Fluorinated Pharmaceuticals

In Chapter 3 we noted the significant role of organofluorine compounds in atmospheric science, specifically with respect to ozone depletion. Substances containing carbon–fluorine bonds have also had a major impact in medicine. The C–F bond is the shortest and strongest of all the carbon–halogen bonds (Table 6-1); indeed, it is only slightly longer but much harder to break than a C–H bond. It is very polar, $\overset{\delta^+}{C}-\overset{\delta^-}{F}$, and has lone pairs that can participate in hydrogen bonding with δ^+ hydrogens. As a result, replacement of a C–H bond with C–F in a potential medicinal compound can dramatically influence its pharmaceutically relevant biochemical properties, including effectiveness and tendency to produce side effects. Similarly, fluorine substitution modifies the physical characteristics of the substance such as water solubility and ability to pass through cell membranes, affecting how it can be introduced into the body. Finally, the strong C–F linkage resists metabolic breakdown, allowing the drug to survive intact in the body longer, enhancing its pharmaceutical effect. Given these features, it is perhaps not surprising that as many as 20% of pharmaceuticals on the market today, including several of the most widely used, incorporate one or more C–F bonds. Examples include the proton-pump inhibitor (Real Life 2-1) lansoprazole (Prevacid), the cholesterol-lowering drug atorvastatin (Lipitor,) the antiasthmatic fluticasone propionate (Flonase), and the anesthetics halothane (Fluothane) and sevoflurane (Sojourn).

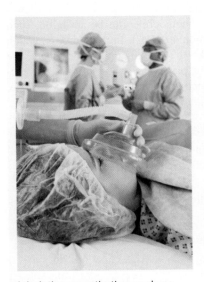

Inhalation anesthetics such as halothane, $CF_3CHBrCl$, or the newer sevoflurane, $(CF_3)_2CHOCH_2F$, derive their biological activity from the polar nature of their C–X bonds.

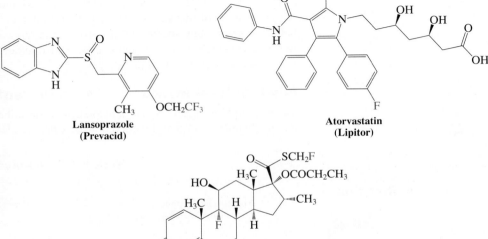

**Lansoprazole
(Prevacid)**

**Atorvastatin
(Lipitor)**

**Fluticasone propionate
(Flonase)**

Applications and Hazards of Haloalkanes: "Greener" Alternatives

The properties of haloalkanes have made this class of compounds a rich source of commercially useful substances. For example, fully halogenated liquid bromomethanes, such as $CBrF_3$ and $CBrClF_2$ ("Halons"), are extremely effective fire retardants. Heat-induced cleavage of the weak C–Br bond releases bromine atoms, which suppress combustion by inhibiting the free-radical chain reactions occurring in flames (see Problem 40 of Chapter 3). Like Freon refrigerants, however, bromoalkanes are ozone depleting (Section 3-10) and have been banned for all uses except fire-suppression systems in aircraft engines. Phosphorus tribromide, PBr_3, a non-ozone-depleting liquid with a high weight percent of bromine, is a promising replacement. In 2006, a PBr_3-based fire-suppression cartridge system (under the trade name PhostrEx) was approved by both the U.S. Environmental Protection Agency (EPA) and the U.S. Federal Aviation Administration (FAA). It is now in commercial use in the Eclipse 500 jet aircraft.

The polarity of the carbon–halogen bond makes haloalkanes useful for applications such as dry cleaning of clothing and degreasing of mechanical and electronic components. Alternatives

An Eclipse 500 jet taking off.

for these purposes include fluorinated solvents such as 1,1,1,2,2,3,4,5,5,9-decafluoropentane ($CF_3CF_2CHFCHFCF_3$), a DuPont product that does not decompose to release ozone-destroying halogen atoms because the C–F bond is strong. This solvent is safe, stable, usable for a wide variety of industrial functions, and may be readily recovered and recycled. Problem 50 on p. 245 introduces yet another class of "green" solvents—ionic liquids—that are revolutionizing industrial chemistry.

In Summary The halogen orbitals become increasingly diffuse along the series F, Cl, Br, I. Hence, (1) the C–X bond strength decreases; (2) the C–X bond becomes longer; (3) for the same R, the boiling points increase; (4) the polarizability of X becomes greater; and (5) London interactions increase. We shall see next that these interrelated effects also play an important role in the reactions of haloalkanes.

6-2 | NUCLEOPHILIC SUBSTITUTION

Haloalkanes contain an electrophilic carbon atom, which may react with nucleophiles—substances that contain an unshared electron pair. The nucleophile can be an anion, such as hydroxide ($^-$:ÖH), or a neutral species, such as ammonia (:NH_3). In this process, which we call **nucleophilic substitution,** the reagent attacks the haloalkane and replaces the halide. A great many species are transformed in this way, particularly in solution. The reaction occurs widely in nature and can be controlled effectively even on an industrial scale. Let us see how it works in detail.

Nucleophiles attack electrophilic centers

The nucleophilic substitution of a haloalkane is described by either of two general equations. Recall (Section 2-2) that the curved arrows denote electron-pair movement.

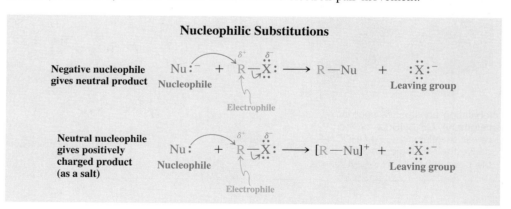

In the first example, a negatively charged nucleophile reacts with a haloalkane to yield a neutral substitution product. In the second example, an uncharged Nu produces a positively charged product, which, together with the counterion, constitutes a salt. In both cases, the group displaced is the halide ion, :Ẍ:$^-$, which is called the **leaving group.** We shall see later that there are leaving groups other than :Ẍ:$^-$. Specific examples of these two types of nucleophilic substitution are shown in Table 6-3. As will be the case in many equations and mechanisms that follow, nucleophiles, electrophiles, and leaving groups are shown here in red, blue, and green, respectively. The general term **substrate** (*substratus,* Latin, to have been subjected) is applied to the organic starting material—in this case, the haloalkane—that is the target of attack by a nucleophile.

Nucleophilic substitution exhibits considerable diversity

Nucleophilic substitution changes the functional group in a molecule. A great many nucleophiles are available to participate in this process; therefore, a wide variety of new molecules is accessible through substitution. Note that Table 6-3 depicts only methyl, primary, and secondary halides. In Chapter 7 we shall see that tertiary substrates behave differently toward these

Color code
Nucleophiles: red
Electrophiles: blue
Leaving groups: green

Table 6-3	The Diversity of Nucleophilic Substitution			
Reaction number	Substrate	Nucleophile	Product	Leaving group
1.	$\overset{..}{CH_3Cl}\!:$ **Chloromethane**	$+$ $\overset{..}{HO}\!:^-$	$\longrightarrow$ $CH_3\overset{..}{O}H$ **Methanol**	$+$ $:\overset{..}{\underset{..}{Cl}}\!:^-$
2.	$CH_3CH_2\overset{..}{I}\!:$ **Iodoethane**	$+$ $CH_3\overset{..}{O}\!:^-$	$\longrightarrow$ $CH_3CH_2\overset{..}{O}CH_3$ **Methoxyethane**	$+$ $:\overset{..}{\underset{..}{I}}\!:^-$
3.	$\overset{\displaystyle H}{\underset{\displaystyle :\overset{..}{\underset{..}{Br}}:}{\underset{|}{\overset{|}{CH_3CCH_2CH_3}}}$ **2-Bromobutane**	$+$ $:\overset{..}{\underset{..}{I}}\!:^-$	$\longrightarrow$ $\overset{\displaystyle H}{\underset{\displaystyle :\overset{..}{\underset{..}{I}}:}{\underset{|}{\overset{|}{CH_3CCH_2CH_3}}}$ **2-Iodobutane**	$+$ $:\overset{..}{\underset{..}{Br}}\!:^-$
4.	$\overset{\displaystyle H}{\underset{\displaystyle CH_3}{\underset{|}{\overset{|}{CH_3CCH_2\overset{..}{I}\!:}}}$ **1-Iodo-2-methylpropane**	$+$ $:N{\equiv}C:^-$	$\longrightarrow$ $\overset{\displaystyle H}{\underset{\displaystyle CH_3}{\underset{|}{\overset{|}{CH_3CCH_2C{\equiv}N:}}}$ **3-Methylbutanenitrile**	$+$ $:\overset{..}{\underset{..}{I}}\!:^-$
5.	Br on cyclohexane **Bromocyclohexane**	$+$ $CH_3\overset{..}{S}\!:^-$	$\longrightarrow$ SCH_3 on cyclohexane **Methylthiocyclohexane**	$+$ $:\overset{..}{\underset{..}{Br}}\!:^-$
6.	$CH_3CH_2\overset{..}{I}\!:$ **Indoethane**	$+$ $:NH_3$	$\longrightarrow$ $CH_3CH_2\overset{+}{N}H_3$ **Ethylammonium iodide**	$+$ $:\overset{..}{\underset{..}{I}}\!:^-$
7.	$CH_3\overset{..}{Br}\!:$ **Bromomethane**	$+$ $:P(CH_3)_3$	$\longrightarrow$ $\overset{\displaystyle CH_3}{\underset{\displaystyle CH_3}{\underset{|}{\overset{|+}{CH_3PCH_3}}}$ **Tetramethylphosphonium bromide**	$+$ $:\overset{..}{\underset{..}{Br}}\!:^-$

Note: Remember that nucleophiles are red, electrophiles are blue, and leaving groups are green. Anionic nucleophiles give neutral products (Reactions 1–5). Neutral nucleophiles give salts as products (Reactions 6 and 7).

nucleophiles and that secondary halides may sometimes give other products as well. Methyl and primary haloalkanes give the "cleanest" substitutions, relatively free of side products.

Let us inspect these transformations in greater detail. In reaction 1, a hydroxide ion, typically derived from sodium or potassium hydroxide, displaces chloride from chloromethane to give methanol. This substitution is a general synthetic method for converting a methyl or primary haloalkane into an alcohol.

A variation of this transformation is reaction 2. Methoxide ion reacts with iodoethane to give methoxyethane, an example of the synthesis of an ether (Section 9-6).

In reactions 1 and 2, the species attacking the haloalkane is an anionic oxygen nucleophile. Reaction 3 shows that a halide ion may function not only as a leaving group, but also as a nucleophile.

Reaction 4 depicts a carbon nucleophile, cyanide (often supplied as sodium cyanide, $Na^+{}^-CN$), and leads to the formation of a new carbon–carbon bond, an important means of modifying molecular structure.

Really? Halomethanes are popular soil fumigants that kill agricultural pests by methylating nucleophilic N- and S-containing biomolecules through substitution reactions. (The picture shows the fumigation of a strawberry field with bromomethane and subsequent coverage with plastic.) However, the chemical is also toxic to humans, and, in addition, it is an ozone depletor, so its continued use is controversial.

Reaction 5 shows the sulfur analog of reaction 2, demonstrating that nucleophiles in the same column of the periodic table react similarly to give analogous products. This conclusion is also borne out by reactions 6 and 7. However, the nucleophiles in these two reactions are *neutral*, and the expulsion of the negatively charged leaving group results in a cationic species, an ammonium or phosphonium salt, respectively.

All of the nucleophiles shown in Table 6-3 are quite reactive, but not all for the same reasons. Some are reactive because they are strongly basic (HO^-, CH_3O^-). Others are weak bases (I^-) whose nucleophilicity derives from other characteristics. Notice that, in each example, the leaving group is a halide ion. Halides are unusual in that they may serve as leaving groups as well as nucleophiles (therefore making reaction 3 reversible). However, the same is *not* true of some of the other nucleophiles in Table 6-3 (in particular, the strong bases); the equilibria of their reactions lie strongly in the direction shown. These topics are addressed in Sections 6-7 and 6-8, as are factors that affect the reversibility of displacement reactions. First, however, we shall examine the mechanism of nucleophilic substitution.

Exercise 6-1

What are the substitution products of the reaction of 1-bromobutane with (**a**) $:\ddot{I}:^-$; (**b**) $CH_3CH_2\ddot{O}:^-$; (**c**) N_3^-; (**d**) $:As(CH_3)_3$; (**e**) $(CH_3)_2\ddot{S}e$?

Solved Exercise 6-2 | Working with the Concepts: Planning a Synthesis

Suggest starting materials for the preparation (synthesis) of $CH_3CH_2SCH_3$.

Strategy

*W*hat: The question does not specify a method for preparing this molecule, but it makes sense to use our newest reaction, nucleophilic substitution.

*H*ow: A powerful method for designing synthetic preparations involves working *backward* from the structure of the target molecule and is called **retrosynthetic analysis.** We demonstrate the idea here and will return to it in Section 8-9. Begin by rephrasing the question as "What substances must react by nucleophilic substitution to give the desired product?" Write the structure in full, so as to see clearly all the bonds it contains, and identify one that might be formed in the course of nucleophilic substitution.

Solution

*I*nformation: Reaction 5 in Table 6-3 gives us a model for a process that forms a sulfur compound with two C–S bonds.

*P*roceed in the same way, even though the problem does not tell us which halide leaving group to displace by the sulfur nucleophile.

- We may choose any one that will work, namely, chloride, bromide, or iodide:

Delete C–S bond

$$H_3C-\overset{..}{\underset{..}{S}}\diagdown_{CH_2}\diagup^{CH_3} \Longrightarrow H_3C-\overset{..}{\underset{..}{S}}:^- \quad Br-CH_2-CH_3$$

Attach any suitable halide leaving group

This type of arrow means "derives from"

- We finish by writing the preparation in the *forward* direction, the way we would actually carry it out:

$$CH_3\overset{..}{\underset{..}{S}}:^- + CH_3CH_2\overset{..}{\underset{..}{Br}}: \longrightarrow CH_3\overset{..}{\underset{..}{S}}CH_2CH_3 + :\overset{..}{\underset{..}{Br}}:^-$$

- Notice that we could have just as easily conducted our reverse analysis by deleting the bond between the sulfur and the methyl carbon, rather than the ethyl. That would give us a second, equally correct method of preparation:

$$CH_3\overset{..}{\underset{..}{I}}: + {}^-:\overset{..}{\underset{..}{S}}CH_2CH_3 \longrightarrow CH_3\overset{..}{\underset{..}{S}}CH_2CH_3 + :\overset{..}{\underset{..}{I}}:^-$$

As before, the choice of halide leaving group is immaterial.

Exercise 6-3 | Try It Yourself

Suggest starting materials for the preparation of $(CH_3)_4N^+I^-$. (**Hint:** Look for a reaction in Table 6-3 that gives a similar product.)

In Summary Nucleophilic substitution is a fairly general reaction for primary and secondary haloalkanes. The halide functions as the leaving group, and several types of nucleophilic atoms enter into the process.

6-3 | REACTION MECHANISMS INVOLVING POLAR FUNCTIONAL GROUPS: USING "ELECTRON-PUSHING" ARROWS

In our consideration in Chapter 3 of radical halogenation, we found that a knowledge of its mechanism was helpful in explaining the experimental characteristics of the process. The same is true for nucleophilic substitution and, indeed, virtually every chemical process that we encounter. Nucleophilic substitution is an example of a polar reaction: It includes charged species and polarized bonds. Recall (Chapter 2) that an understanding of electrostatics is essential if we are to comprehend how such processes take place. Opposite charges attract—nucleophiles are attracted to electrophiles—and this principle provides us with a basis for understanding the mechanisms of polar organic reactions. In this section, we expand the concept of *electron flow* and review the conventional methods for illustrating polar reaction mechanisms by *moving electrons* from electron-rich to electron-poor sites.

Curved arrows depict the movement of electrons

As we learned in Section 2-3, acid-base processes require electron movement. Let us briefly reexamine the Brønsted-Lowry process in which the acid HCl donates a proton to a molecule of water in aqueous solution:

Depiction of a Brønsted-Lowry Acid-Base Reaction by Using Curved Arrows

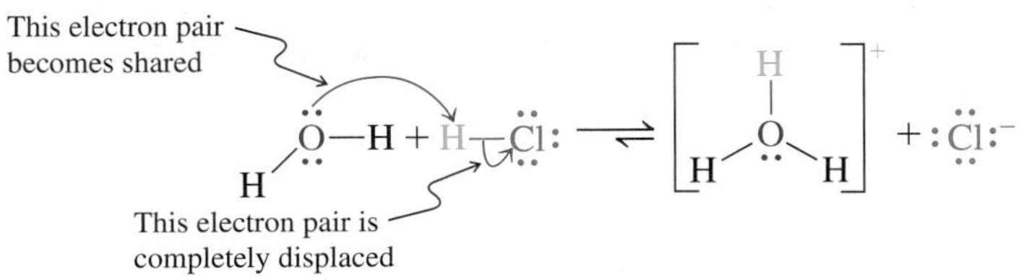

Notice that the arrow starting at the lone pair on oxygen and ending at the hydrogen of HCl does *not* imply that the lone electron pair departs from oxygen completely; it just becomes a *shared* pair between that oxygen atom and the atom to which the arrow points. In contrast, however, the arrow beginning at the H–Cl bond and pointing toward the chlorine atom *does* signify heterolytic cleavage of the bond; that electron pair becomes separated from hydrogen and ends up entirely on the chloride ion.

Exercise 6-4

Use curved arrows to depict the flow of electrons in each of the following acid-base reactions. (**a**) Hydrogen ion + hydroxide ion; (**b**) fluoride ion + boron trifluoride, BF_3; (**c**) ammonia + hydrogen chloride; (**d**) hydrogen sulfide, H_2S, + sodium methoxide, $NaOCH_3$; (**e**) dimethyloxonium ion, $(CH_3)_2OH^+$, + water; (**f**) the self-ionization of water to give hydronium ion and hydroxide ion.

Curved "electron-pushing" arrows are the means by which we describe mechanisms in organic chemistry. We have already noted the close parallels between acid-base reactions and reactions between organic electrophiles and nucleophiles (Section 2-3). Curved arrows show how nucleophilic substitution occurs when a lone pair of electrons on a nucleophile is transformed into a new bond with an electrophilic carbon, "pushing" a bonding pair of electrons away from that carbon onto a leaving group. However, nucleophilic substitution is just one of many kinds of processes for which electron-pushing arrows are used to depict the mechanisms of electrophile–nucleophile interactions. The following examples reprise several of the reaction types introduced in Chapter 2.

Curved-Arrow Representations of Several Common Types of Mechanisms

$H-\ddot{O}:^- + -\overset{\mid}{\underset{\mid}{C}}-\ddot{\underset{\cdot\cdot}{C}l}:$	$\xrightarrow{\text{Nucleophilic substitution}}$ $-\overset{\mid}{\underset{\mid}{C}}-\ddot{O}H + :\ddot{\underset{\cdot\cdot}{C}l}:^-$	Compare with Brønsted acid-base reaction
$-\overset{\mid}{\underset{\mid}{C}}-\ddot{\underset{\cdot\cdot}{C}l}:$	$\xrightarrow{\text{Dissociation}}$ $-\overset{+}{C}\diagdown + :\ddot{\underset{\cdot\cdot}{C}l}:^-$	Reverse of Lewis acid– Lewis base reaction
$H-\ddot{O}:^- + \diagup\hspace{-0.3em}C=\ddot{O}$	$\xrightarrow{\text{Nucleophilic addition}}$ $-\overset{\overset{\displaystyle H\ddot{O}:}{\mid}}{\underset{\mid}{C}}-\ddot{O}:^-$	Only one of the two bonds between C and O is cleaved
$\diagup\hspace{-0.3em}\overset{}{C}=\overset{}{C}\hspace{-0.3em}\diagdown + H^+$	$\xrightarrow{\text{Electrophilic addition}}$ $\diagup\hspace{-0.3em}\overset{+}{C}-\overset{\overset{\displaystyle H}{\mid}}{\underset{\mid}{C}}-$	Carbon–carbon double bond acting as a Lewis base

In every case the curved arrows start from either a lone electron pair on an atom or the center of a bond. Curved arrows *never* start at electron-deficient atoms, such as H^+ (last equation): The movement of a proton is depicted by an arrow pointing *from an electron source* (lone pair or bond) *toward the proton*. Although this may seem counterintuitive at first, it is a very important aspect of the curved-arrow formalism. Curved arrows represent movement of electrons, not atoms.

The first and third examples illustrate a characteristic property of electron movement: If an electron pair moves toward an atom, that atom must have a "place to put that electron pair," so to speak. In nucleophilic substitution, the carbon atom in a haloalkane has a filled outer shell; another electron pair cannot be added without displacement of the electron pair bonding carbon to halogen. The two electron pairs can be viewed as "flowing" in a synchronous manner: As one pair arrives at the closed-shell atom, the other departs, thereby preventing violation of the octet rule at carbon. When you depict electron movement with curved arrows, *it is absolutely essential to keep in mind the rules for drawing Lewis structures.* Correct use of electron-pushing arrows helps in drawing such structures, because all electrons are moved to their proper destinations.

There are other types of processes, but, surprisingly, *not that many.* One of the most powerful consequences of studying organic chemistry from a mechanistic point of view is the way in which this approach highlights similarities between types of polar reactions even if the specific atoms and bonds are not the same.

Exercise 6-5

Identify the electrophilic and nucleophilic sites in the four mechanisms shown earlier as curved-arrow representations.

Exercise 6-6

Write out in detail the equations for the reactions in Exercise 6-2, using curved arrows to denote the movement of electron pairs.

Exercise 6-7

Rewrite each reaction of Table 6-3, adding curved arrows to indicate the flow of electrons.

Exercise 6-8

Propose a curved-arrow depiction of the flow of electrons in the following processes, which will be considered in detail in this chapter and in Chapter 7.

(a) $-\overset{\scriptstyle /}{\underset{\scriptstyle \backslash}{C}}{}^{+} + Cl^{-} \longrightarrow -\overset{\scriptstyle |}{\underset{\scriptstyle |}{C}}-Cl$ (b) $HO^{-} + \overset{\scriptstyle H}{\underset{\scriptstyle /}{\overset{\scriptstyle |}{\underset{\scriptstyle |}{C}}}}{}^{+}_{}\overset{\scriptstyle |}{\underset{\scriptstyle |}{C}}- \longrightarrow H_2O + \overset{\scriptstyle \backslash}{\underset{\scriptstyle /}{C}}{=}\overset{\scriptstyle /}{\underset{\scriptstyle \backslash}{C}}$

In Summary Curved arrows depict movement of electron pairs in reaction mechanisms. Electrons move from nucleophilic, or Lewis basic, atoms toward electrophilic, or Lewis acidic, sites. If a pair of electrons approaches an atom already containing a closed shell, a pair of electrons must depart from that atom so as not to exceed the maximum capacity of its valence orbitals.

6-4 A CLOSER LOOK AT THE NUCLEOPHILIC SUBSTITUTION MECHANISM: KINETICS

Many questions can be raised at this stage. What are the kinetics of nucleophilic substitution, and how does this information help us determine the underlying mechanism? What happens with optically active haloalkanes? Can we predict relative rates of substitution? These questions will be addressed in the remainder of this chapter.

When a mixture of chloromethane and sodium hydroxide in water is heated (denoted by the uppercase Greek letter *delta*, Δ, at the right of the arrow in the equation in the margin), a high yield of two compounds—methanol and sodium chloride—is the result. This outcome, however, does not tell us anything about *how* starting materials are converted into products. What experimental methods are available for answering this question?

One of the most powerful techniques employed by chemists is the measurement of the *kinetics* of the reaction (Section 2-1). By comparing the rate of product formation beginning with several different concentrations of the starting materials, we can establish the rate equation, or **rate law,** for a chemical process. Let us see what this experiment tells us about the reaction of chloromethane with sodium hydroxide.

Mechanism

$CH_3Cl + NaOH$

$\downarrow H_2O, \Delta$

$CH_3OH + NaCl$

"Δ" means that the reaction mixture is heated.

The reaction of chloromethane with sodium hydroxide is bimolecular

We can monitor rates by measuring either the disappearance of one of the reactants or the appearance of one of the products. When we apply this method to the reaction of chloromethane with sodium hydroxide, we find that the rate depends on the initial concentrations of *both* of the reagents. For example, doubling the concentration of hydroxide doubles the rate at which the reaction proceeds. Likewise, at a fixed hydroxide concentration, doubling the concentration of chloromethane has the same effect. Doubling the concentrations of both increases the rate by a factor of 4. These results are consistent with a *second-order* process (Section 2-1), which is governed by the following rate equation:

$$Rate = k[CH_3Cl][HO^-] \text{ mol } L^{-1} s^{-1}$$

All the examples given in Table 6-3 exhibit such second-order kinetics: Their rates are directly proportional to the concentrations of both substrate and nucleophile.

Solved Exercise 6-9 | Working with the Concepts: Concentrations and Rates

When a solution containing 0.01 M sodium azide ($Na^+N_3^-$) and 0.01 M iodomethane in methanol at 0°C is monitored kinetically, the results reveal that iodide ion is produced at a rate of 3.0×10^{-10} mol L^{-1} s^{-1}. Write the formula of the organic product of this reaction and calculate its rate constant k. What would be the rate of appearance of I^- for an initial concentration of reactants of $[NaN_3] = 0.02$ M and $[CH_3I] = 0.01$ M?

Strategy
Write the formula describing the reaction by finding a closely analogous example in Table 6-3. Then determine k by solving the rate equation using the given information.

Solution
• Reaction 1 in Table 6-3 is the model. The nucleophile is azide rather than hydroxide, and the substrate is iodomethane rather than chloromethane. Thus,

$$CH_3I + Na^+N_3^- \longrightarrow CH_3N_3 + Na^+I^-$$

• The rate of appearance of I^- is the same as the rate of appearance of the organic product and the rate of disappearance of both starting materials. Solve the equation for k:

$$3.0 \times 10^{-10} \text{ mol } L^{-1} \text{ s}^{-1} = k(10^{-2} \text{ mol } L^{-1})(10^{-2} \text{ mol } L^{-1})$$
$$k = 3.0 \times 10^{-6} \text{ L mol}^{-1} \text{ s}^{-1}$$

• Now use k to solve for the new rate given the changed set of initial concentrations.

$$\text{New rate} = (3.0 \times 10^{-6} \text{ L mol}^{-1} \text{ s}^{-1})(2 \times 10^{-2} \text{ mol } L^{-1})(10^{-2} \text{ mol } L^{-1})$$
$$= 6.0 \times 10^{-10} \text{ mol } L^{-1} \text{ s}^{-1}$$

(**Hint:** As a shortcut in problems like this, simply multiply the original rate by the factor by which any concentration has changed. **Caution:** Consider only changes in concentration of substances *that appear in the rate equation.*)

Exercise 6-10 | Try It Yourself

What is the rate of appearance of I^- in the reaction in Exercise 6-9 for the following initial concentrations of reactants? **(a)** $[NaN_3] = 0.03$ M and $[CH_3I] = 0.01$ M; **(b)** $[NaN_3] = 0.02$ M and $[CH_3I] = 0.02$ M; **(c)** $[NaN_3] = 0.03$ M and $[CH_3I] = 0.03$ M.

What kind of mechanism is consistent with a second-order rate law? The simplest is one in which the two reactants interact in a single step. We call such a process **bimolecular,** and the general term applied to substitution reactions of this type is **bimolecular nucleophilic substitution,** abbreviated as S_N2 (S stands for substitution, N for nucleophilic, and 2 for bimolecular).

Bimolecular nucleophilic substitution is a concerted, one-step process

Bimolecular nucleophilic substitution is a one-step transformation: The nucleophile attacks the haloalkane, with simultaneous expulsion of the leaving group. Bond making takes place *at the same time* as bond breaking. Because the two events occur "in concert," we call this process a **concerted** reaction.

We can envisage two stereochemically distinct alternatives for such concerted displacements. The nucleophile could approach the substrate from the same side as the leaving

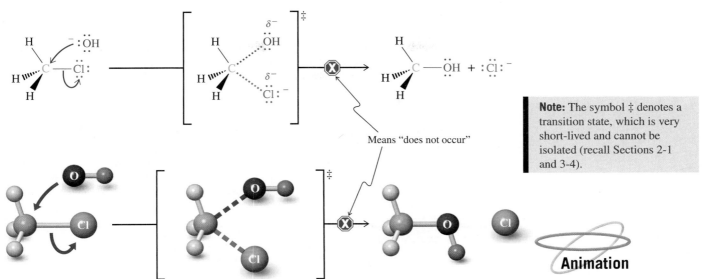

Note: The symbol ‡ denotes a transition state, which is very short-lived and cannot be isolated (recall Sections 2-1 and 3-4).

Means "does not occur"

Figure 6-2 Hypothetical frontside nucleophilic substitution (does not occur). The (hypothetical) transition state is enclosed in brackets and labeled with the ‡ symbol.

Animation

ANIMATED MECHANISM:
Nucleophilic substitution (S_N2)

This classic billiards shot is a model for backside displacement in the S_N2 reaction. When the white ball hits the orange one, the orange ball does not move, but the red ball is driven away.

group, one group exchanging for the other. This pathway is called **frontside displacement** (Figure 6-2). As we shall see in the next section, it does not occur. The second possibility is a **backside displacement,** in which the nucleophile approaches carbon from the side opposite the leaving group (Figure 6-3). In both equations, an electron pair from the negatively charged hydroxide oxygen moves toward carbon, creating the C–O bond, while that of the C–Cl linkage shifts onto chlorine, thereby expelling the latter as $:\ddot{C}l:^-$. In either of the two respective transition states, the negative charge is distributed over both the oxygen and the chlorine atoms.

Note that the formation of the transition state is *not* a separate step; the transition state only describes the geometric arrangement of the reacting species as they pass through the *maximum energy point* of a *single-step* process (Section 2-1).

Figure 6-3 Backside nucleophilic substitution. Attack is from the side *opposite* the leaving group. The concerted nature of bond making (to OH) and bond breaking (from Cl) is indicated by the dotted lines, which signify the partial bonding of both to carbon in the transition state.

Exercise 6-11

Draw representations of the hypothetical frontside and backside displacement mechanisms for the S_N2 reaction of sodium iodide with 2-bromobutane (Table 6-3). Use arrows like those shown in Figures 6-2 and 6-3 to represent electron-pair movement.

In Summary The reaction of chloromethane with hydroxide to give methanol and chloride, as well as the related transformations of a variety of nucleophiles with haloalkanes, are examples of the bimolecular process known as the S_N2 reaction. Two single-step mechanisms—frontside attack and backside attack—may be envisioned for the reaction. Both are concerted processes, consistent with the second-order kinetics obtained experimentally. Can we distinguish between the two? To answer this question, we return to a topic that we have considered in detail: stereochemistry.

Mechanism

Model Building

Animation

ANIMATED MECHANISM:
Nucleophilic substitution (S_N2)

6-5 | FRONTSIDE OR BACKSIDE ATTACK? STEREOCHEMISTRY OF THE S_N2 REACTION

When we compare the structural drawings in Figures 6-2 and 6-3 with respect to the arrangement of their component atoms in space, we note immediately that in the first conversion the three hydrogens stay put and to the left of the carbon, whereas in the second they have "moved" to the right. In fact, the two methanol pictures are related as object and mirror image. In this example, the two are superimposable and therefore indistinguishable—properties of an achiral molecule. The situation is entirely different for a chiral haloalkane in which the electrophilic carbon is a stereocenter.

The S_N2 reaction is stereospecific

Consider the reaction of (S)-2-bromobutane with iodide ion. Frontside displacement should give rise to 2-iodobutane with the *same* (S) configuration as that of the substrate; backside displacement should furnish a product with the *opposite* configuration.

What is actually observed? It is found that (S)-2-bromobutane gives (R)-2-iodobutane on treatment with iodide: *This and all other S_N2 reactions proceed with **inversion of configuration.*** A process whose mechanism requires that each stereoisomer of the starting material transform into a specific stereoisomer of product is described as **stereospecific.** The S_N2 reaction is therefore stereospecific, proceeding by a backside displacement mechanism to give inversion of configuration at the site of the reaction.

In the three equations that follow, the progress of the reaction of (S)-2-bromobutane with iodide ion is shown using conventional drawings, molecular models, and electrostatic potential maps. You can see that in the transition state, the negative charge on the nucleophile has spread partly onto the leaving group. As the reaction comes to completion, the leaving group evolves to a fully charged anion. In the electrostatic potential map of the transition state, this process is reflected in the attenuated red color around the two halogen nuclei, compared with the full red visible in the starting and ending halide ions. Note that in the reaction schemes preceding the electrostatic potential renditions, we use the mechanistic color scheme of green for the leaving group, and not red. In other respects, the colors match, as you would expect.

Stereochemistry of the Backside Displacement Mechanism for S_N2 Reactions

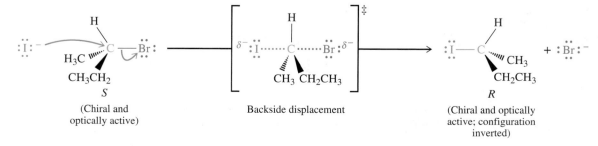

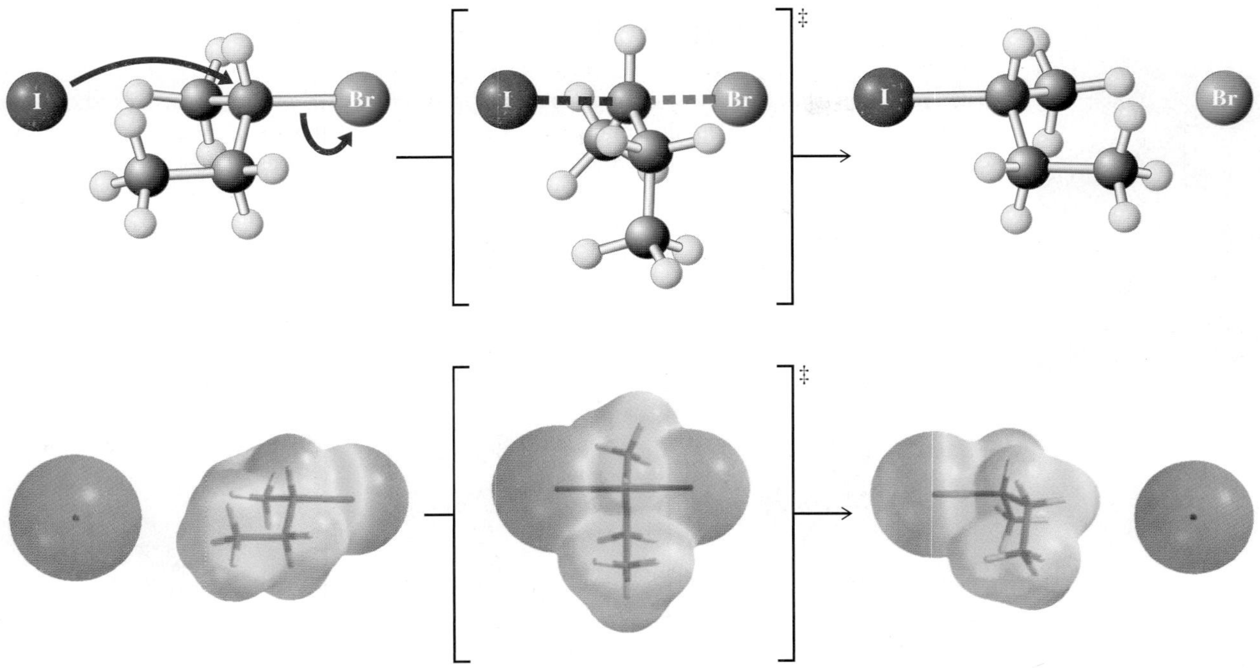

The stereochemistry of displacement at a primary carbon is more difficult to observe directly, because a primary carbon atom is attached to two hydrogens in addition to the leaving group: It is not a stereocenter. This obstacle may be overcome by replacing one of the two hydrogen atoms by deuterium, the hydrogen isotope with mass = 2. The result is a stereocenter at the primary carbon and a chiral molecule. This strategy has been employed to confirm that S_N2 displacement at a primary carbon atom does indeed occur with inversion of configuration, as the example below illustrates.

Stereochemistry of S_N2 Displacement at a Primary Carbon Atom

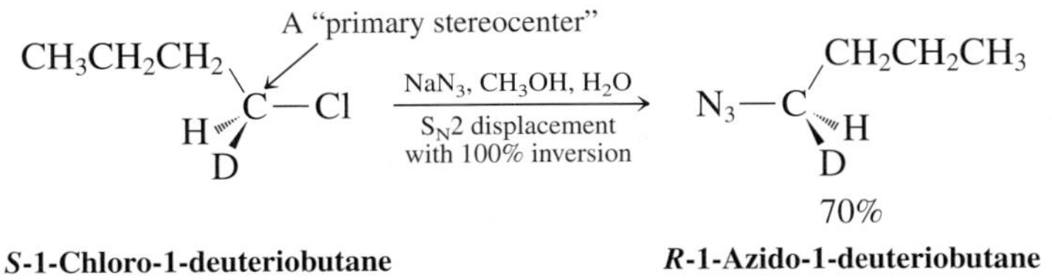

S-1-Chloro-1-deuteriobutane

(Chiral and optically active)

R-1-Azido-1-deuteriobutane

(Chiral and optically active; configuration inverted)

The nucleophile, azide ion (N_3^-), gives rise to stereospecific backside displacement of chloride, giving the azidoalkane product with the inverted configuration at the chiral carbon.

Exercise 6-12

Write the products of the following S_N2 reactions: **(a)** (R)-3-chloroheptane + $Na^+\ {}^-SH$; **(b)** (S)-2-bromooctane + $N(CH_3)_3$; **(c)** (3R,4R)-4-iodo-3-methyloctane + $K^+\ {}^-SeCH_3$.

Exercise 6-13

Write the structures of the products of the S_N2 reactions of cyanide ion with **(a)** meso-2,4-dibromopentane (double S_N2 reaction); **(b)** trans-1-iodo-4-methylcyclohexane.

The transition state of the S$_N$2 reaction can be described in an orbital picture

The transition state for the S$_N$2 reaction can be described in orbital terms, as shown in Figure 6-4. As the nucleophile approaches the back lobe of the sp^3 hybrid orbital used by carbon to bind the halogen atom, the rest of the molecule becomes planar at the transition

Former President George W. Bush experiences inversion of configuration.

Figure 6-4 Orbital description of backside attack in the S$_N$2 reaction. The process is reminiscent of the inversion of an umbrella exposed to gusty winds.

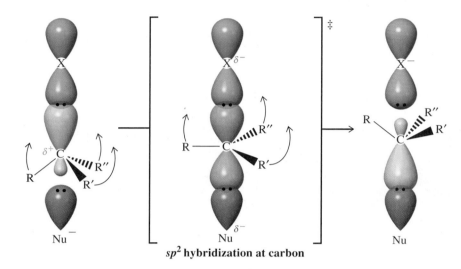

sp^2 hybridization at carbon

state by changing the hybridization at carbon to sp^2. As the reaction proceeds to products, the inversion motion is completed and the carbon returns to the tetrahedral sp^3 configuration. A depiction of the course of the reaction using a potential energy–reaction coordinate diagram is shown in Figure 6-5.

Animation

ANIMATED MECHANISM:
Nucleophilic substitution (S$_N$2)

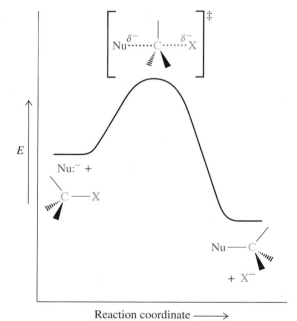

Figure 6-5 Potential energy diagram for an S$_N$2 reaction. The process takes place in a single step, with a single transition state.

6-6 | CONSEQUENCES OF INVERSION IN S$_N$2 REACTIONS

What are the consequences of the inversion of stereochemistry in the S$_N$2 reaction? Because the reaction is stereospecific, we can design ways to use displacement reactions to synthesize a desired stereoisomer.

We can synthesize a specific enantiomer by using S$_N$2 reactions

Consider the conversion of 2-bromooctane into 2-octanethiol in its reaction with hydrogen sulfide ion, HS$^-$. If we were to start with optically pure R bromide, we would obtain only S thiol and none of its R enantiomer.

Inversion of Configuration of an Optically Pure Compound by S$_N$2 Reaction

$$HS:^- \ + \quad \overset{H}{\underset{CH_3}{CH_3(CH_2)_4CH_2}}C-Br: \longrightarrow HS-\overset{H}{\underset{CH_3}{C}}CH_2(CH_2)_4CH_3 \ + \ :Br:^-$$

(*R*)-2-Bromooctane
([α] = −34.6)

(*S*)-2-Octanethiol
([α] = +36.4)

Color code for priorities (see Section 5-3)
Highest: red
Second highest: blue
Third highest: green
Lowest: black

But what if we wanted to convert (*R*)-2-bromooctane into the *R* thiol? One technique uses a sequence of *two* S$_N$2 reactions, each resulting in inversion of configuration at the stereocenter. For example, an S$_N$2 reaction with iodide would first generate (*S*)-2-iodooctane. We would then use this haloalkane with an inverted configuration as the substrate in a second displacement, now with HS$^-$ ion, to furnish the *R* thiol. This double inversion sequence of two S$_N$2 processes gives us the result we desire, a net **retention of configuration.**

Model Building

Using Double Inversion to Give Net Retention of Configuration

(*R*)-2-Bromooctane
([α] = −34.6)
→ First inversion of configuration →
(*S*)-2-Iodooctane
([α] = +46.3)
→ Second inversion of configuration →
(*R*)-2-Octanethiol
([α] = −36.4)

Exercise 6-14

As we saw for carvone (Problem 43 of Chapter 5), enantiomers can sometimes be distinguished by odor and flavor. 3-Octanol and some of its derivatives are examples: The dextrorotatory compounds are found in natural peppermint oil, whereas their (−) counterparts contribute to the essence of lavender. Show how you would synthesize optically pure samples of each enantiomer of 3-octyl acetate, starting with (*S*)-3-iodooctane. (The conversion of acetates into alcohols will be shown in Section 8-5.)

$$\overset{O}{\underset{|}{\overset{\|}{OCCH_3}}}$$
CH$_3$CH$_2$CHCH$_2$CH$_2$CH$_2$CH$_3$
3-Octyl acetate

Solved Exercise 6-15 | **Working with the Concepts: Stereochemical Consequences of S$_N$2 Displacement**

Treatment of (*S*)-2-iodooctane with NaI in solution causes the optical activity of the starting organic compound to disappear. Explain.

Strategy
What: If you write out the equation for this reaction, you will notice something unusual about it: This S$_N$2 reaction uses iodide as the nucleophile as well as the leaving group. Therefore, *iodide displaces iodide*. This is the key insight to approaching the problem.

Solution

*H*ow: The optical activity of (*S*)-2-iodooctane originates from the fact that it is chiral and a single enantiomer. Its structure appears in the text on the previous page. The stereocenter is C2, the carbon bearing the iodine atom. (*S*)-2-Iodooctane is a secondary haloalkane and, as we have seen in several examples in this chapter, it may undergo S$_N$2 reaction, which proceeds by backside displacement and inversion at the site of reactivity.

*I*nformation: As noted earlier, I$^-$ is *both* a good nucleophile and a good leaving group.

*P*roceed: Because it functions in both roles in this reaction, the transformation occurs rapidly. Each time displacement occurs, the stereocenter undergoes stereochemical inversion. Because the process is fast, it takes place multiple times for every substrate molecule, inverting the stereochemistry each time. Ultimately, this leads to an equilibrium (i.e., racemic) mixture of (*R*) and (*S*) stereoisomers of the starting (and ending) compound.

Exercise 6-16 | Try It Yourself

Amino acids are the building blocks of peptides and proteins in nature. They may be prepared in the laboratory by S$_N$2 displacement of the halogen in 2-halocarboxylic acids using ammonia as the nucleophile, as illustrated by the conversion of 2-bromopropanoic acid into alanine.

$$\underset{\substack{\textbf{2-Bromopropanoic} \\ \textbf{acid}}}{\overset{\overset{\displaystyle Br}{|}}{CH_3CHCOOH}} \xrightarrow[-HBr]{\substack{NH_3,\ H_2O, \\ 25°C,\ 4\ days}} \underset{\textbf{Alanine}}{\overset{\overset{\displaystyle {}^+NH_3}{|}}{CH_3CHCOO^-}}$$

The stereocenter in alanine, like that in most naturally occuring amino acids, has the *S* configuration. Draw both a clear stereochemical structure for (*S*)-alanine and one for the enantiomer of 2-bromopropanoic acid that would be required to produce (*S*)-alanine according to the equation above.

> Among the amino acids that are made by the body ("nonessential" amino acids), (*S*)-alanine is the most abundant, contributing about 8% to the amino acid sequence of proteins.

In substrates bearing more than one stereocenter, inversion takes place *only* at the carbons that undergo reaction with the incoming nucleophile. Note that the reaction of (2*S*,4*R*)-2-bromo-4-chloropentane with excess cyanide ion results in a meso product. This outcome is particularly readily recognized using Fischer projections.

Model Building

S$_N$2 Reactions of Molecules with Two Stereocenters

Reactive stereocenters (both possess good leaving groups)

2*S*,4*R* + $^-$CN Excess $\xrightarrow{\text{Ethanol (Solvent)}}$ 2*R*,4*S*: Meso + Br$^-$ + Cl$^-$

Reactive stereocenter / Inert stereocenter (no leaving group)

2*S*,3*R* + ·I$^-$ $\xrightarrow{\text{Acetone (Solvent)}}$ 2*R*,3*R* + Br$^-$

In these equations, ethanol and acetone, respectively, are the *solvents* for the indicated transformations. These solvents are polar (Section 1-3) and particularly good at dissolving salts. We shall come back to the influence of the nature of the solvent on the S_N2 reaction in Section 6-8. In the second example, notice that the reaction taking place at C2 has no effect on the stereocenter at C3.

Exercise 6-17

As an aid in the prediction of stereochemistry, organic chemists often use the guideline that "diastereomers produce diastereomers." Replace the starting compound in each of the two preceding examples with one of its diastereomers, and write the product of S_N2 displacement with the nucleophile shown. Are the resulting structures in accord with this "rule"?

Similarly, nucleophilic substitution of a substituted halocycloalkane may change the stereochemical relation between the substituents. For example, in the disubstituted cyclohexane below, the stereochemistry changes from cis to trans.

Model Building

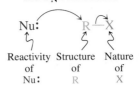

cis-**1-Bromo-**
3-methylcyclohexane   *trans*-**1-Iodo-**
                                  3-methylcyclohexane

In Summary Inversion of configuration in the S_N2 reaction has distinct stereochemical consequences. Optically active substrates give optically active products, unless the nucleophile and the leaving group are the same or meso compounds are formed. In cyclic systems, cis and trans stereochemical relations may be interconverted.

6-7 STRUCTURE AND S_N2 REACTIVITY: THE LEAVING GROUP

The relative facility of S_N2 displacements depends on several factors, including the nature of the leaving group, the reactivity of the nucleophile (which is affected by the choice of reaction solvent), and the structure of the alkyl portion of the substrate. We employ kinetics as our tool to evaluate the degree to which changes in each of these *structural* features affect their *function* in the S_N2 reaction. We begin by examining the leaving group. Subsequent sections will address the nucleophile and the substrate.

Leaving-group ability is a measure of the ease of its displacement

As a general rule, nucleophilic substitution occurs only when the group being displaced, X, is readily able to depart, taking with it the electron pair of the C–X bond. Are there structural features that might allow us to predict, at least qualitatively, whether a leaving group is "good" or "bad"? Not surprisingly, the relative rate at which it can be displaced, its **leaving-group ability,** can be correlated with its capacity to accommodate a negative charge. Remember that a certain amount of negative charge is transferred to the leaving group in the transition state of the reaction (Figure 6-4).

For the halogens, leaving-group ability increases along the series from fluorine to iodine. Thus, iodide is regarded as a "good" leaving group; fluoride, however, is so "poor" that S_N2 reactions of fluoroalkanes are rarely observed.

Some Variables Affecting the S_N2 Reaction

Nu:⁻   R⌒X

Reactivity   Structure   Nature
of   of   of
Nu:⁻   R   X

Leaving-Group Ability

$$I^- > Br^- > Cl^- > F^-$$
Best   **Worst**

⬅   **Increasing**

Exercise 6-18

Predict the product of the reaction of 1-chloro-6-iodohexane with one equivalent of sodium methyl-selenide ($Na^{+\,-}SeCH_3$).

Halides are not the only groups that can be displaced by nucleophiles in S_N2 reactions. Other examples of good leaving groups are sulfur derivatives of the type $ROSO_3^-$ and RSO_3^-, such as methyl sulfate ion, $CH_3OSO_3^-$, and various sulfonate ions. Alkyl sulfate and sulfonate leaving groups are used so often that trivial names, such as mesylate, triflate, and tosylate, have found their way into the chemical literature.

Sulfate and Sulfonate Leaving Groups

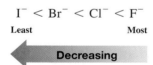

| Methyl sulfate ion | Methanesulfonate ion (Mesylate ion) | Trifluoromethanesulfonate ion (Triflate ion) | 4-Methylbenzenesulfonate ion (*p*-Toluenesulfonate ion, tosylate ion) |

Weak bases are good leaving groups

Basicity

$$I^- < Br^- < Cl^- < F^-$$

Least   Most

⬅ **Decreasing**

Is there some characteristic property that distinguishes good leaving groups from poor ones? Yes: *Leaving-group ability is inversely related to base strength.* Weak bases are best able to accommodate negative charge and are the best leaving groups. Among the halides, iodide is the weakest base and therefore the best leaving group in the series. Sulfates and sulfonates are weak bases as well.

Is there a way to recognize weak bases readily? The weaker X^- is as a base, the stronger is its conjugate acid HX. Therefore, *good leaving groups are the conjugate bases of strong acids.* This rule applies to the four halides: HF is the weakest of the conjugate acids, HCl is stronger, and HBr and HI are stronger still. Table 6-4 lists a number of acids, in order of descending strength, and their pK_a values. Their conjugate bases are listed alongside in order of increasing strength and, therefore, decreasing leaving-group ability.

Table 6-4	Base Strengths and Leaving Groups						
Conjugate acid		**Leaving group**		**Conjugate acid**		**Leaving group**	
Strong	pK_a	*Good*		*Weak*	pK_a	*Poor*	
HI (strongest)	−10.0	I^- (best)		HF	3.2	F^-	
HBr	−9.0	Br^-		CH_3CO_2H	4.7	$CH_3CO_2^-$	
HCl	−8.0	Cl^-		HCN	9.2	NC^-	
H_2SO_4	−3.0	HSO_4^-		CH_3SH	10.0	CH_3S^-	
H_3O^+	−1.7	H_2O		CH_3OH	15.5	CH_3O^-	
CH_3SO_3H	−1.2	$CH_3SO_3^-$		H_2O	15.7	HO^-	
				NH_3	35	H_2N^-	
				H_2 (weakest)	38	H^- (worst)	

(Stronger acid → / Better leaving group →)

Predict the relative acidities within each of the following groups. Review Section 2-3 if necessary. **(a)** H$_2$S, H$_2$Se; **(b)** PH$_3$, H$_2$S; **(c)** HClO$_3$, HClO$_2$; **(d)** HBr, H$_2$Se; **(e)** NH$_4$$^+$, H$_3O^+$. Within each of the groups, identify the conjugate bases and predict their relative leaving-group abilities.

Predict the relative basicities within each of the following groups. **(a)** $^-$OH, $^-$SH; **(b)** $^-$PH$_2$, $^-$SH; **(c)** I$^-$, $^-$SeH; **(d)** HOSO$_2$$^-$, HOSO$_3$$^-$. Predict the relative acidities of the conjugate acids within each group.

In Summary The leaving-group ability of a substituent correlates with the strength of its conjugate acid. Both depend on the ability of the leaving group to accommodate negative charge. In addition to the halides Cl$^-$, Br$^-$, and I$^-$, sulfates and sulfonates (such as methane- and 4-methylbenzenesulfonates) are good leaving groups. Good leaving groups are weak bases, the conjugate bases of strong acids. We shall return in Section 9-4 to uses of sulfates and sulfonates as leaving groups in synthesis.

6-8 STRUCTURE AND S$_N$2 REACTIVITY: THE NUCLEOPHILE

Now that we have looked at the effect of the leaving group, let us turn to a consideration of nucleophiles. How can we predict their relative nucleophilic strength, their **nucleophilicity?** We shall see that nucleophilicity depends on a variety of factors: charge, basicity, solvent, polarizability, and the nature of substituents. To grasp the relative importance of these effects, let us analyze the outcome of a series of comparative experiments.

Increasing negative charge increases nucleophilicity

If the same nucleophilic atom is used, does charge play a role in the reactivity of a given nucleophile as determined by the rate of its S$_N$2 reaction? The following experiments answer this question.

Experiment 1

$$CH_3\ddot{\underset{\cdot\cdot}{Cl}}: + H\ddot{\underset{\cdot\cdot}{O}}:^- \longrightarrow CH_3\ddot{\underset{\cdot\cdot}{O}}H + :\ddot{\underset{\cdot\cdot}{Cl}}:^- \quad \textbf{Fast}$$

$$CH_3\ddot{\underset{\cdot\cdot}{Cl}}: + H_2\ddot{\underset{\cdot\cdot}{O}} \longrightarrow CH_3\ddot{\underset{\cdot\cdot}{O}}H_2{}^+ + :\ddot{\underset{\cdot\cdot}{Cl}}:^- \quad \textbf{Very slow}$$

Experiment 2

$$CH_3\ddot{\underset{\cdot\cdot}{Cl}}: + H_2\ddot{N}:^- \longrightarrow CH_3\ddot{N}H_2 + :\ddot{\underset{\cdot\cdot}{Cl}}:^- \quad \textbf{Very fast}$$

$$CH_3\ddot{\underset{\cdot\cdot}{Cl}}: + H_3N: \longrightarrow CH_3\ddot{N}H_3{}^+ + :\ddot{\underset{\cdot\cdot}{Cl}}:^- \quad \textbf{Slower}$$

Conclusion. Of a pair of nucleophiles containing the same reactive atom, the species with a negative charge is the more powerful nucleophile. Or, of a base and its conjugate acid, the base is always more nucleophilic. This finding is intuitively very reasonable. Because nucleophilic attack is characterized by the formation of a bond with an electrophilic carbon center, the more negative the attacking species, the faster the reaction should be.

Predict which member in each of the following pairs is a better nucleophile. **(a)** HS$^-$ or H$_2$S; **(b)** CH$_3$SH or CH$_3$S$^-$; **(c)** CH$_3$NH$^-$ or CH$_3$NH$_2$; **(d)** HSe$^-$ or H$_2$Se.

Nucleophilicity decreases to the right in the periodic table

Experiments 1 and 2 compared pairs of nucleophiles containing the same nucleophilic element (e.g., oxygen in H_2O versus HO^- and nitrogen in H_3N versus H_2N^-). What about nucleophiles of similar structure but with different nucleophilic atoms? Let us examine the elements along one row of the periodic table.

Experiment 3

$$CH_3CH_2\ddot{B}r: + H_3N: \longrightarrow CH_3CH_2NH_3^+ + :\ddot{B}r:^- \qquad \textbf{Fast}$$

$$CH_3CH_2\ddot{B}r: + H_2\ddot{O} \longrightarrow CH_3CH_2\ddot{O}H_2^+ + :\ddot{B}r:^- \qquad \textbf{Very slow}$$

Experiment 4

$$CH_3CH_2\ddot{B}r: + H_2\ddot{N}:^- \longrightarrow CH_3CH_2\ddot{N}H_2 + :\ddot{B}r:^- \qquad \textbf{Very fast}$$

$$CH_3CH_2\ddot{B}r: + H\ddot{O}:^- \longrightarrow CH_3CH_2\ddot{O}H + :\ddot{B}r:^- \qquad \textbf{Slower}$$

Conclusion. Nucleophilicity again appears to correlate with basicity: The more basic species is the more reactive nucleophile. Therefore, in the progression from the left to the right of the periodic table, nucleophilicity decreases. The approximate order of reactivity for nucleophiles in the first row is

◄ Increasing basicity

$$H_2N^- > HO^- > NH_3 > F^- > H_2O$$

◄ Increasing nucleophilicity

Observations using other nucleophiles demonstrate that the trends revealed in Experiments 1–4 are generally applicable to all of the nonmetallic elements (groups 15–17) of the periodic table. Increasing negative charge (Experiments 1 and 2) usually has a greater effect than moving one group to the left (Experiments 3 and 4). Thus, in the order of reactivity shown above, both HO^- and NH_3 are more nucleophilic than water, but HO^- is more nucleophilic than NH_3.

Exercise 6-22

In each of the following pairs of molecules, predict which is the more nucleophilic. **(a)** Cl^- or CH_3S^-; **(b)** $P(CH_3)_3$ or $S(CH_3)_2$; **(c)** $CH_3CH_2Se^-$ or Br^-; **(d)** H_2O or HF.

Should basicity and nucleophilicity be correlated?

The parallels between nucleophilicity and basicity first described in Section 2-3 make sense: Strong bases typically make good nucleophiles. However, a fundamental difference between the two properties is based on how they are measured. Basicity is a *thermodynamic* property, measured by an equilibrium constant:

Recall from Section 2-3 that the species we refer to as bases or nucleophiles are the same: The distinction is in their mode of action. When attacking a proton, they are called a *base* (often shown as A^- or B:); when attacking any other nucleus, for example, carbon, they are called a *nucleophile* (often shown as Nu^- or Nu:).

$$A^- + H_2O \underset{}{\overset{K}{\rightleftharpoons}} AH + HO^- \qquad K = \text{equilibrium constant}$$

In contrast, nucleophilicity is a *kinetic* phenomenon, quantified by comparing rates of reactions:

$$Nu^- + R{-}X \xrightarrow{k} Nu{-}R + X^- \qquad k = \text{rate constant}$$

Despite these inherent differences, we have observed good correlation between basicity and nucleophilicity in the cases of charged versus neutral nucleophiles along a row of the periodic table. What happens if we look at nucleophiles in a column of the periodic table? We shall find that the situation changes, because now solvent plays a role.

Solvation impedes nucleophilicity

If it is a general rule that nucleophilicity correlates with basicity, then the elements considered from top to bottom of a column of the periodic table should show decreasing nucleophilic power. Recall (Section 2-3) that basicity decreases in an analogous fashion. To test this prediction, let us consider another series of experiments. In the equations below, we have explicitly added the solvent methanol to the reaction scheme, because, as we shall see, consideration of the solvent will be important in understanding the outcome of these experiments.

Experiment 5

$$CH_3CH_2CH_2\ddot{O}SCH_3 \ + \ :\overset{..}{\underset{..}{C}}l:^- \ \xrightarrow[\text{(Solvent)}]{CH_3OH} \ CH_3CH_2CH_2\overset{..}{\underset{..}{C}}l: \ + \ ^-O_3SCH_3 \qquad \textbf{Slow}$$

$$CH_3CH_2CH_2\ddot{O}SCH_3 \ + \ :\overset{..}{\underset{..}{B}}r:^- \ \xrightarrow[\text{(Solvent)}]{CH_3OH} \ CH_3CH_2CH_2\overset{..}{\underset{..}{B}}r: \ + \ ^-O_3SCH_3 \qquad \textbf{Faster}$$

$$CH_3CH_2CH_2\ddot{O}SCH_3 \ + \ :\overset{..}{\underset{..}{I}}:^- \ \xrightarrow[\text{(Solvent)}]{CH_3OH} \ CH_3CH_2CH_2\overset{..}{\underset{..}{I}}: \ + \ ^-O_3SCH_3 \qquad \textbf{Fastest}$$

Experiment 6

$$CH_3CH_2CH_2\overset{..}{\underset{..}{B}}r: \ + \ CH_3\overset{..}{\underset{..}{O}}:^- \ \xrightarrow[\text{(Solvent)}]{CH_3OH} \ CH_3CH_2CH_2\ddot{O}CH_3 \ + \ :\overset{..}{\underset{..}{B}}r:^- \qquad \textbf{Not very fast}$$

$$CH_3CH_2CH_2\overset{..}{\underset{..}{B}}r: \ + \ CH_3\overset{..}{\underset{..}{S}}:^- \ \xrightarrow[\text{(Solvent)}]{CH_3OH} \ CH_3CH_2CH_2\ddot{S}CH_3 \ + \ :\overset{..}{\underset{..}{B}}r:^- \qquad \textbf{Very fast}$$

Conclusion. Surprisingly, nucleophilicity *increases* in the progression down the periodic table, a trend *directly opposing* that expected from the basicity of the nucleophiles tested. For example, in the series of halides, iodide is the fastest, although it is the weakest base.

◀ Increasing basicity

$$F^- \ < \ Cl^- \ < \ Br^- \ < \ I^-$$

Increasing nucleophilicity in CH$_3$OH ▶

Moving one column to the left in the periodic table, sulfide nucleophiles are more reactive than the analogous oxide systems, and, as other experiments have shown, their selenium counterparts are even more reactive. Thus, this column exhibits the same trend as that observed for the halides. The phenomenon is general for other columns in the periodic table.

How can these trends be explained? An important consideration is the interaction of the solvent methanol with the anionic nucleophile. We have largely ignored the solvent in our discussion of organic reactions so far, in particular, radical halogenations (Chapter 3), in which they play an insignificant role. Nucleophilic substitution features polar starting materials and a polar mechanism, and the nature of the solvent becomes more important. Let us see how the solvent can become involved.

Solvation and Drug Activity

In the design of a drug, medicinal chemists try to optimize the three-dimensional fit to the drug target's receptor site (see, e.g., Real Life 5-5). However, equally important to efficacy are the *aqueous solvation energies* of both partners in this interaction. Optimal binding is offset by unfavorable desolvation of water molecules. Desolvation also occurs during the crossing of membranes during the journey of the drug from (usually) the stomach to its desired destination, and therefore affects the bioavailability of a compound.

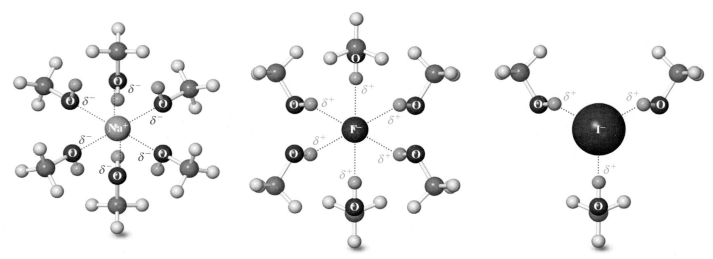

Animation

ANIMATED MECHANISM:
Nucleophilic substitution (S_N2)

Table 6-5	Polar Aprotic Solvents

$:\!O\!:$
$\|$
CH_3CCH_3
Acetone

$CH_3C\!\equiv\!N\!:$
**Ethanenitrile
(Acetonitrile)**

$:\!O\!:$
$\| \;..$
$HCN(CH_3)_2$
**N,N-Dimethylformamide
(DMF)**

$:\!O\!:$
$\|$
$CH_3\overset{..}{S}CH_3$
**Dimethyl sulfoxide
(DMSO)**

$:\!O\!:$
$\|$
P
$(CH_3)_2\overset{..}{N} \; | \; \overset{..}{N}(CH_3)_2$
$\overset{..}{N}\!:$
$(CH_3)_2$
**Hexamethylphosphoric
triamide
(HMPA)**

$\overset{..}{O}\!:$
$+ \;/$
CH_3N
$\overset{..}{\underset{..}{O}}\!:^-$
Nitromethane

Figure 6-6 (A) Solvation of Na^+ by ion–dipole interactions with methanol. (B) Approximate representation of the relatively dense solvation of the small F^- ion by hydrogen bonds to methanol. (C) Approximate representation of the comparatively diminished solvation of the large I^- ion by hydrogen bonds to methanol. The tighter solvent shell around F^- reduces its ability to participate in nucleophilic substitution reactions.

When a solid dissolves, the intermolecular forces that held it together (Section 2-7; Figure 2-6) are replaced by intermolecular forces between molecules and solvent. Such molecules, especially the ions derived from the starting salts of many S_N2 reactions, are said to be **solvated.** Salts dissolve well in alcohols and water, because these solvents contain highly polarized $^{\delta+}H\!-\!O^{\delta-}$ bonds that act by ion–dipole interactions. Thus, cations are solvated by the negatively polarized oxygens (Figure 6-6A), anions by the positively polarized hydrogens (Figure 6-6B and C). This solvation of anions is particularly strong, because the small size of the hydrogen nucleus makes the $\delta+$ charge relatively dense. We shall study these interactions, called **hydrogen bonds,** more closely in Chapter 8. Solvents capable of hydrogen bonding are also called **protic,** in contrast to **aprotic** solvents, such as acetone, which will be discussed later.

Returning to the problem of our experimental results: What accounts for the increasing nucleophilicity of negatively charged nucleophiles from the top to the bottom of a column of the periodic table? The answer is that *solvation weakens the nucleophile* by forming a shell of solvent molecules around the nucleophile and thus impeding its ability to attack an electrophile. As we move down the periodic table, such as from F^- to I^-, the solvated ion becomes larger and its charge more diffuse. As a result, solvation is diminished along the series and nucleophilicity increases. Figures 6-6B and C depict this effect for F^- and I^-. The smaller fluoride ion is much more heavily solvated than the larger iodide. Is this true in other solvents as well?

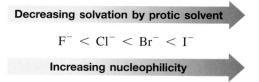

Decreasing solvation by protic solvent →

$$F^- \;<\; Cl^- \;<\; Br^- \;<\; I^-$$

Increasing nucleophilicity →

Aprotic solvents: the effect of solvation is diminished

Other solvents that are useful in S_N2 reactions are highly polar but aprotic. Several common examples are shown in Table 6-5; all lack protons capable of hydrogen bonding but do exhibit polarized bonds. Nitromethane even exists as a charge-separated species.

Polar, aprotic solvents also dissolve salts by ion–dipole interactions, albeit not as well as protic solvents. Because they cannot form hydrogen bonds, they solvate anionic nucleophiles relatively weakly. The consequences are twofold. First, compared to protic solvents, the *reactivity of the nucleophile is raised,* sometimes dramatically. For example, bromomethane reacts with

Animation

ANIMATED MECHANISM:
Nucleophilic substitution (S$_N$2)

Table 6-6	Relative Rates of S$_N$2 Reactions of Iodomethane with Chloride Ion in Various Solvents $CH_3I + Cl^- \xrightarrow[k_{rel}]{Solvent} CH_3Cl + I^-$		
Solvent			**Relative rate**
Formula	**Name**	**Classification**	**(k_{rel})**
CH$_3$OH	Methanol	Protic	1
HCONH$_2$	Formamide	Protic	12.5
HCONHCH$_3$	*N*-Methylformamide	Protic	45.3
HCON(CH$_3$)$_2$	*N,N*-Dimethylformamide	Aprotic	1,200,000

potassium iodide 500 times faster in acetone than in methanol. Table 6-6 compares the rates of S$_N$2 reactions of iodomethane with chloride in three protic solvents—methanol, formamide, and *N*-methylformamide—and one *aprotic* solvent, *N,N*-dimethylformamide (DMF). (Formamide and *N*-methylformamide can form hydrogen bonds by virtue of their polarized N–H linkages.) The rate of reaction in DMF is more than a million times greater than it is in methanol.

The second consequence of comparatively weaker solvation of anions by aprotic solvents is that the nucleophilicity trend observed in protic solvents inverts. Thus, while the reactivity of all anions increases, that of the smaller ones *increases more* than that of the others. For many nucleophiles, including the halide series, base strength overrides solvation: Back to our original expectation!

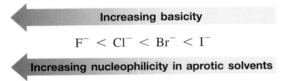

Increasing basicity

$$F^- < Cl^- < Br^- < I^-$$

Increasing nucleophilicity in aprotic solvents

Increasing polarizability improves nucleophilic power

The solvation effects just described should be very pronounced only for charged nucleophiles. Nevertheless, the degree of nucleophilicity increases down the periodic table, even for *uncharged nucleophiles,* for which solvent effects should be much less strong, for example, H$_2$Se > H$_2$S > H$_2$O, and PH$_3$ > NH$_3$. Therefore, there must be an additional factor that comes into play.

This factor is the polarizability of the nucleophile (Section 6-1). Larger elements have larger, more diffuse, and more polarizable electron clouds. These electron clouds allow for more effective orbital overlap in the S$_N$2 transition state (Figure 6-7). The result is a lower transition-state energy and faster nucleophilic substitution.

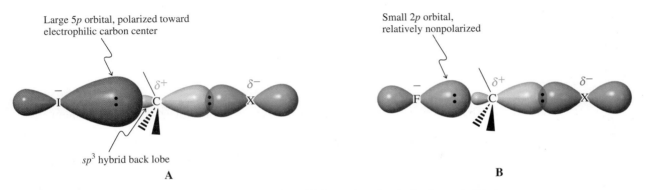

Figure 6-7 Comparison of I$^-$ and F$^-$ in the S$_N$2 reaction. (A) In protic solvents the larger iodide is a better nucleophile, in part because its polarizable 5p orbital is distorted toward the electrophilic carbon atom. (B) The tight, less polarizable 2p orbital on fluoride does not interact as effectively with the electrophilic carbon at a point along the reaction coordinate comparable to the one for (A).

Exercise 6-23

Which species is more nucleophilic: **(a)** CH_3SH or CH_3SeH; **(b)** $(CH_3)_2NH$ or $(CH_3)_2PH$?

Sterically hindered nucleophiles are poorer reagents

We have seen that the bulk of the surrounding solvent may adversely affect the power of a nucleophile, another example of steric hindrance (Section 2-9). Such hindrance may also be built into the nucleophile itself in the form of bulky substituents. The effect on the rate of reaction can be seen in Experiment 7.

Experiment 7

$$CH_3\ddot{I}: + CH_3\ddot{O}:^- \longrightarrow CH_3\ddot{O}CH_3 + :\ddot{I}:^- \quad \textbf{Fast}$$

Model Building

$$CH_3\ddot{I}: + CH_3\underset{\underset{CH_3}{|}}{\overset{\overset{CH_3}{|}}{C}}\ddot{O}:^- \longrightarrow CH_3\ddot{O}\underset{\underset{CH_3}{|}}{\overset{\overset{CH_3}{|}}{C}}CH_3 + :\ddot{I}:^- \quad \textbf{Slower}$$

Conclusion. Sterically bulky nucleophiles react more slowly.

Exercise 6-24

Which of the two nucleophiles in the following pairs will react more rapidly with bromomethane?

(a) CH_3S^- or $CH_3\underset{\underset{CH_3}{|}}{\overset{\overset{CH_3}{|}}{C}}HS^-$ **(b)** $(CH_3)_2NH$ or $(CH_3\underset{\underset{CH_3}{|}}{\overset{\overset{CH_3}{|}}{C}}H)_2NH$

Table 6-7	Relative Rates of Reaction of Various Nucleophiles with Iodomethane in Methanol (Protic Solvent)
Nucleophile	**Relative rate**
CH_3OH	1
NO_3^-	~32
F^-	500
$CH_3\overset{\overset{O}{\|\|}}{C}O^-$	20,000
Cl^-	23,500
$(CH_3CH_2)_2S$	219,000
NH_3	316,000
CH_3SCH_3	347,000
N_3^-	603,000
Br^-	617,000
CH_3O^-	1,950,000
CH_3SeCH_3	2,090,000
CN^-	5,010,000
$(CH_3CH_2)_3As$	7,940,000
I^-	26,300,000
HS^-	100,000,000

Increasing nucleophilicity (downward arrow)

Nucleophilic substitutions may be reversible

The halide ions Cl^-, Br^-, and I^- are both good nucleophiles and good leaving groups. Therefore, their S_N2 reactions are reversible. For example, in acetone, the reactions between lithium chloride and primary bromo- and iodoalkanes form an equilibrium that lies on the side of the chloroalkane products:

$$CH_3CH_2CH_2CH_2I + LiCl \overset{Acetone}{\rightleftharpoons} CH_3CH_2CH_2CH_2Cl + LiI$$

This result correlates with the relative stabilities of the product and starting material, which favor the chloroalkane. However, this equilibrium may be driven in the reverse direction by a simple "trick": Whereas all of the lithium halides are soluble in acetone, solubility of the *sodium* halides decreases dramatically in the order $NaI > NaBr > NaCl$, the last being virtually insoluble in this solvent. Indeed, the reaction between NaI and a primary or secondary chloroalkane in acetone is *completely* driven to the side of the iodoalkane (the reverse of the reaction just shown) by the precipitation of NaCl:

$$CH_3CH_2CH_2CH_2Cl + NaI \overset{Acetone}{\rightleftharpoons} CH_3CH_2CH_2CH_2I + NaCl\downarrow$$

Insoluble in acetone

The direction of the equilibrium in reaction 3 of Table 6-3 may be manipulated in exactly the same way. However, when the nucleophile in an S_N2 reaction is a strong base (e.g., HO^- or CH_3O^-; see Table 6-4), it will be incapable of acting as a leaving group. In such cases, K_{eq} will be very large and displacement will essentially be an irreversible process (Table 6-3, reactions 1 and 2).

In Summary Nucleophilicity is controlled by a number of factors. Increased negative charge and progression from right to left and down (protic solvent) or up (aprotic solvent) the periodic table generally increase nucleophilic power. Table 6-7 compares the reactivity of a range of nucleophiles relative to that of the very weakly nucleophilic methanol (arbitrarily set at 1). We can confirm the validity of the conclusions of this section by inspecting the various entries. The use of aprotic solvents improves nucleophilicity, especially of smaller anions, by eliminating hydrogen bonding.

6-9 | KEYS TO SUCCESS: CHOOSING AMONG MULTIPLE MECHANISTIC PATHWAYS

As indicated in Section 2-2, a reaction mechanism moves all electrons to their logical destinations, generating automatically the structures of the products of the reaction. Thus chemists frequently solve problems by "thinking mechanistically"—applying a reasonable mechanism to reacting species so as to predict a product. But what does one do if multiple pathways are possible, and each leads to a different product? This situation is not at all uncommon and will likely be troublesome at first, but it may be mastered, with practice, within the WHIP problem-solving protocol. Here are some suggestions:

When determining **W**hat the problem is asking, make note of *every* piece of information provided in the statement of the problem. Not all of them may be useful or even relevant, but if you have noted them all, you may find an important clue to the solution.

Mechanisms can be complex. **H**ow you begin will likely determine whether you will reach the correct solution. Sometimes the steps in a possible sequence will be readily reversible, so the pathway may not lead to any stable structure. In this case you will simply have to start over and try to find a new trail that points to a more productive direction. Alternatively, two sequences may both end in reasonable compounds, but one may be much faster than the other and provide the predominant or exclusive product. Or, your arrow pushing may generate a molecule that is energetically too unfavorable to be feasible (e.g., too strained, electron deficient, and the like). Your **I**nformation base is the critical component of evaluating these situations. In particular:

1. When you write the structures that your proposed mechanism generates, can you recognize any that are not "chemically reasonable"? Examples might be those that violate the octet rule, have bonding patterns completely unlike any you have previously encountered, or are unlikely to form for other reasons. The ability to recognize high-energy, unstable species in the course of writing a mechanistic proposal will quickly tell you that you are not on the right track. For example, consider mixing the two species shown:

$$CH_4 + :\overset{..}{I}\overset{..}{:}^- \longrightarrow \ ?$$

Will they transform to new species?

Rather than trying to guess the answer, we can propose a mechanism and see where it leads. Based on the content of this chapter, let us think in terms of an S_N2 displacement:

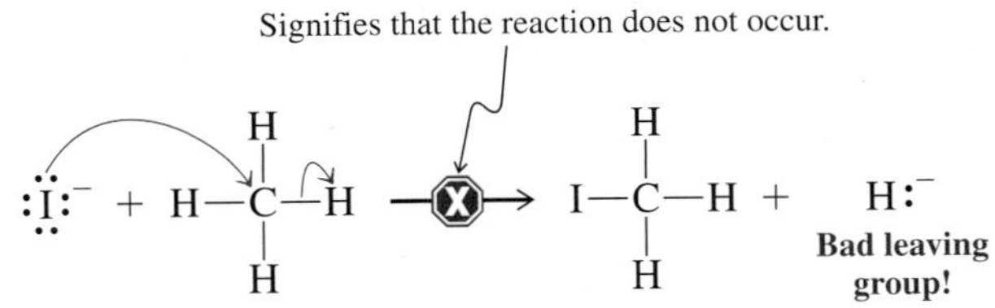

Signifies that the reaction does not occur.

Bad leaving group!

Completing the arrow pushing as shown above reveals that one product is a molecule of iodomethane, but as the leaving group we find hydride ion, which we notice from Table 6-4 is a strong base and therefore *a bad leaving group*. The conclusion is that, even though we can write a formally correct mechanism, the reaction cannot take place because one of the products, hydride ion, is too high in energy—too unstable—to be a leaving group in a nucleophilic displacement reaction. In this case there are no alternatives, and the proper answer is that no reaction occurs.

Caution! *Make sure* that the substrate in a proposed nucleophilic substitution reaction contains a good leaving group. If it does not, *reaction is unlikely to take place!*

2. Have you considered *all* the forms of reactivity that your starting materials may exhibit? This section points out that many nucleophiles are also strong bases. Exercise 6-25 illustrates a scenario in which this fact plays an important role.

Solved Exercise 6-25 | **Working with the Concepts: Suggesting a Reaction Product by Mechanistic Reasoning**

Treatment of 4-chloro-1-butanol, $:\ddot{C}lCH_2CH_2CH_2CH_2\ddot{O}H$, with NaOH in DMF solvent leads to rapid formation of a compound with the molecular formula C_4H_8O. Propose a structure for this product and suggest a mechanism for its formation.

Strategy

"Think mechanistically": Consider the reaction pathways available. If the first pathway you choose doesn't work, try to refine the problem—what is the change that takes place in the molecule, and how could this change come about?

Solution

• The most obvious mechanism to try is an S_N2 trajectory for the substrate and hydroxide:

$$HO\ddot{:}^- + H\ddot{O}CH_2CH_2CH_2{-}CH_2{-}\ddot{C}l: \longrightarrow H\ddot{O}CH_2CH_2CH_2CH_2\ddot{O}H + :\ddot{C}l:^-$$

Unfortunately, the product in this equation cannot be correct, because its molecular formula is $C_4H_{10}O_2$, not C_4H_8O.

• Let's consider another approach. The substrate has the molecular formula C_4H_9OCl. Therefore, the change in its conversion to C_4H_8O is loss of one hydrogen and the chlorine—that is, a molecule of the strong acid HCl. How could we effect this change?

• Hydroxide is a base as well as a nucleophile; therefore, a reasonable alternative to the (incorrect) S_N2 reaction above is an acid-base reaction with the most acidic hydrogen in the substrate:

$$HO\ddot{:}^- + H{-}\ddot{O}CH_2CH_2CH_2CH_2\ddot{C}l: \longrightarrow {}^-:\ddot{O}CH_2CH_2CH_2CH_2\ddot{C}l: + H_2\ddot{O}$$

The formula of the organic product of this transformation is $C_4H_8OCl^-$, only a chloride ion away from the correct product. How can we induce chloride ion to leave without adding any external species? By displacement using the nucleophilic negatively charged oxygen at the *opposite end* of the molecule, forming a ring:

$$:\ddot{O}{-}CH_2CH_2CH_2{-}CH_2{-}\ddot{C}l: \longrightarrow \underset{H_2C{-}CH_2}{\overset{\overset{\ddot{O}:}{H_2C\qquad CH_2}}{}} + :\ddot{C}l:^-$$

Indeed, such intramolecular S_N2 reactions are widely used for the synthesis of cyclic compounds.

You may ask why this reaction proceeds as it does. There are two main reasons. First, Brønsted-Lowry acid-base reactions—proton transfers from one basic atom to another—are generally faster than other processes. So removal of a proton from the hydroxy group of the substrate by hydroxide ion (second equation) is faster than that same hydroxide displacing chloride in an S_N2 process (first equation). Second, reactions that form five- and six-membered rings are generally favored, both kinetically and thermodynamically, over mechanistically analogous processes between two separate molecules. Thus, the internal displacement of chloride by alkoxide in the last equation is preferred over the previously mentioned S_N2 option. In this example, the internal displacement produces two species (the cyclic product and chloride ion) from one, giving rise to an increase in energy dispersal and therefore a favorable entropy change.

Exercise 6-26 | **Try It Yourself**

Gentle warming of a solution of 5-chloro-1-pentanamine, $Cl(CH_2)_5NH_2$, in ethoxyethane (diethyl ether, $CH_3CH_2OCH_2CH_3$) solvent causes precipitation of a white solid. This solid is found to be a salt. Suggest a structure for this compound and explain its formation.

In Summary Determining whether and how a reaction may occur requires reasoning that takes into account the possible mechanism(s). The structural requirements of a reaction mechanism must be met by the substrates and reagents; otherwise, they will not transform. Furthermore, when multiple pathways are possible, you should try to find information that allows discrimination among them (such as their relative rates or energetic favorability).

6-10 | STRUCTURE AND S$_N$2 REACTIVITY: THE SUBSTRATE

Finally, does the structure of the alkyl portion of the substrate, particularly in the vicinity of the atom bearing the leaving group, affect the rate of nucleophilic attack? Once again, we can get a sense of comparative reactivities by looking at relative rates of reaction. Let us examine the kinetic data that have been obtained.

Branching at the reacting carbon decreases the rate of the S$_N$2 reaction

What happens if we successively replace each of the hydrogens in a halomethane with a methyl group? Will this affect the rate of its S$_N$2 reactions? In other words, what are the relative bimolecular nucleophilic reactivities of methyl, primary, secondary, and tertiary halides? Kinetic experiments show that reactivities decrease rapidly in the order shown in Table 6-8.

We can find an explanation by comparing the transition states for these four substitutions. Figure 6-8A shows this structure for the reaction of chloromethane with hydroxide ion. The carbon is surrounded by the incoming nucleophile, the outgoing leaving group, and three substituents (all hydrogen in this case). Although the presence of these five groups increases the crowding about the carbon relative to that in the starting halomethane, the hydrogens do not give rise to serious steric interactions with the nucleophile because of their small size. However, replacement of one hydrogen by a methyl group, as in a haloethane, creates substantial steric repulsion with the incoming nucleophile, thereby raising the transition-state energy (Figure 6-8B). This effect significantly retards nucleophilic attack. If we continue to replace hydrogen atoms with methyl groups, we find that steric hindrance to nucleophilic attack increases dramatically. The two methyl groups in the secondary substrate severely shield the backside of the carbon attached to the leaving group; the rate of

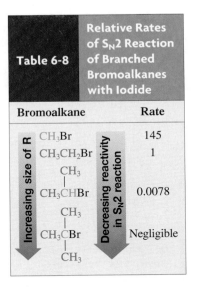

Table 6-8	Relative Rates of S$_N$2 Reaction of Branched Bromoalkanes with Iodide	
Bromoalkane		**Rate**
CH$_3$Br		145
CH$_3$CH$_2$Br		1
CH$_3$CHBr with CH$_3$		0.0078
CH$_3$CBr with CH$_3$ and CH$_3$		Negligible

(Increasing size of R ↓ ; Decreasing reactivity in S$_N$2 reaction ↓)

Model Building

Figure 6-8 Transition states for S$_N$2 reactions of hydroxide ion with
(A) chloromethane,
(B) chloroethane,
(C) 2-chloropropane, and
(D) 2-chloro-2-methylpropane.

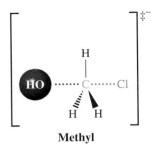
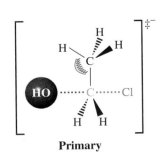
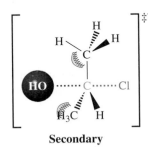

Methyl

Primary

Secondary
(Slow reaction: hydrogens on two methyl groups interfere)

Tertiary
(Negligible S$_N$2 reaction; too much steric hindrance)

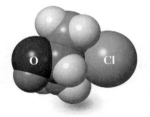

A

B

C

D

reaction diminishes considerably (Figure 6-8C and Table 6-8). Finally, in the tertiary substrate, in which a third methyl group is present, access to the backside of the halide-bearing carbon is nearly blocked (Figure 6-8D); the transition state for S_N2 substitution is high in energy, and displacement of a tertiary haloalkane by this mechanism is rarely observed. To summarize, as we successively replace the hydrogens of a halomethane by methyl groups (or alkyl groups in general), S_N2 reactivity decreases in the following order:

Relative S_N2 Displacement Reactivity of Haloalkanes

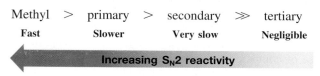

Methyl	>	primary	>	secondary	≫	tertiary
Fast		Slower		Very slow		Negligible

⬅ **Increasing S_N2 reactivity**

> Do cyclic molecules pose problems for you? Focus on the site of reaction: In the structures for Exercise 27(a), the bromine-bearing carbons are secondary and tertiary, respectively.

Exercise 6-27

Predict the relative rates of the S_N2 reaction of cyanide with these pairs of substrates.

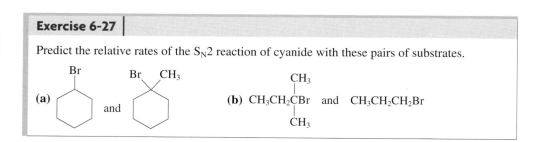

(a) [cyclohexane with Br] and [cyclohexane with Br and CH₃]

(b) CH_3CH_2CBr (with CH₃ above and CH₃ below)   and $CH_3CH_2CH_2Br$

Now that we have seen the effect of major structural changes on substrate reactivity in the S_N2 process, we are in a position to evaluate the effects of more subtle structural modifications. In all cases, we shall find that steric hindrance to attack at the backside of the reacting carbon is the most important consideration.

Lengthening the chain by one or two carbons reduces S_N2 reactivity

As we have seen, the replacement of one hydrogen atom in a halomethane by a methyl group (Figure 6-8B) causes significant steric hindrance and reduction of the rate of S_N2 reaction. Chloroethane is about two orders of magnitude less reactive than chloromethane in S_N2 displacements. Will elongation of the chain of the primary alkyl substrate by the addition of methylene (CH_2) groups further reduce S_N2 reactivity? Kinetic experiments reveal that 1-chloropropane reacts about half as fast as chloroethane with nucleophiles such as I^-.

Does this trend continue as the chain gets longer? The answer is *no*: Higher haloalkanes, such as 1-chlorobutane and 1-chloropentane, react at about the same rate as does 1-chloropropane.

Model Building

Again, an examination of the transition states to backside displacement provides an explanation for these observations. In Figures 6-9A and 6-9B, one of the hydrogens on the methyl carbon of chloroethane is partially obstructing the path of attack of the incoming nucleophile. The 1-halopropanes have an additional methyl group near the reacting carbon center. If reaction occurs from the most stable *anti* conformer of the substrate, the incoming nucleophile faces severe steric hindrance (Figure 6-9C). However, rotation to a *gauche* conformation before attack gives an S_N2 transition state similar to that derived from a haloethane (Figure 6-9D). The propyl substrate exhibits only a small decrease in reactivity relative to the ethyl, the decrease resulting from the energy input needed to attain a *gauche* conformation. Further chain elongation has no effect, because the added carbon atoms do not increase steric hindrance around the reacting carbon in the transition state.

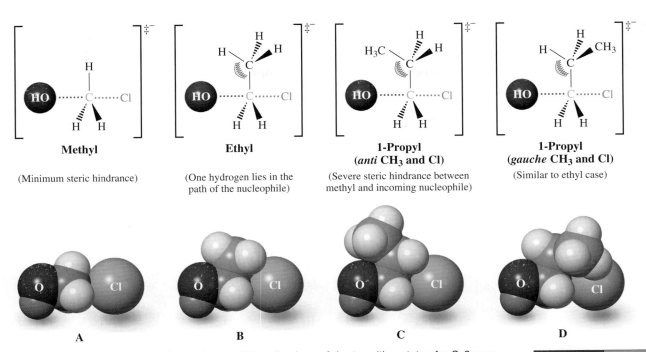

Methyl

(Minimum steric hindrance)

Ethyl

(One hydrogen lies in the path of the nucleophile)

1-Propyl
(*anti* CH$_3$ and Cl)

(Severe steric hindrance between methyl and incoming nucleophile)

1-Propyl
(*gauche* CH$_3$ and Cl)

(Similar to ethyl case)

A   B   C   D

Figure 6-9 Hashed-wedged line and space-filling drawings of the transition states for S$_N$2 reactions of hydroxide ion with (A) chloromethane; (B) chloroethane; and (C and D) two conformers of 1-chloropropane: (C) *anti* and (D) *gauche*. Steric interference is illustrated strikingly in the space-filling drawings. Partial charges have been omitted for clarity. (See Figure 6-3.)

Branching next to the reacting carbon also retards substitution

What about multiple substitution at the position *next to* the electrophilic carbon? Let us compare the reactivities of bromoethane and its derivatives (Table 6-9). A dramatic decrease in rate is seen on further substitution: 1-Bromo-2-methylpropane is ~25 times less reactive toward iodide than is 1-bromopropane, and 1-bromo-2,2-dimethylpropane is virtually inert. Branching at positions farther from the site of reaction has a much smaller effect.

We know that rotation into a *gauche* conformation is necessary to permit nucleophilic attack on a 1-halopropane (Figure 6-10A). We can use the same picture to understand the data in Table 6-9. For a 1-halo-2-methylpropane, the only conformation that permits the nucleophile to approach the backside of the reacting carbon experiences *two gauche* methyl–halide interactions, a considerably worse situation (Figure 6-10B). With the addition of a third methyl group, as in a 1-halo-2,2-dimethylpropane, commonly known as a *neopentyl* halide, backside attack is blocked almost completely (Figure 6-10C).

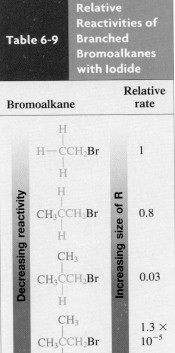

	Relative Reactivities of Branched Bromoalkanes with Iodide
Table 6-9	

Bromoalkane	Relative rate
H—CCH$_2$Br	1
CH$_3$CCH$_2$Br	0.8
CH$_3$CCH$_2$Br	0.03
CH$_3$CCH$_2$Br	1.3 × 10^{-5}

Decreasing reactivity →   Increasing size of R →

Exercise 6-28

Predict the order of reactivity in the S$_N$2 reaction of

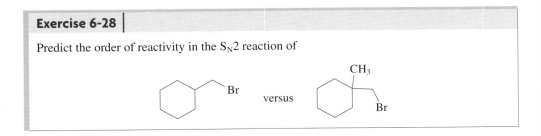

versus

In Summary The structure of the alkyl part of a haloalkane can have a pronounced effect on the rate of nucleophilic attack. Simple chain elongation beyond three carbons has little effect on the rate of the S$_N$2 reaction. However, increased branching leads to strong steric hindrance and rate retardation.

As in Exercise 6-27, don't be bewildered by the cyclic nature of these substrates. Focus on the reaction site (primary) and its neighborhood (branched). Which of the compounds in Table 6-9 are the respective closest relatives?

Model Building

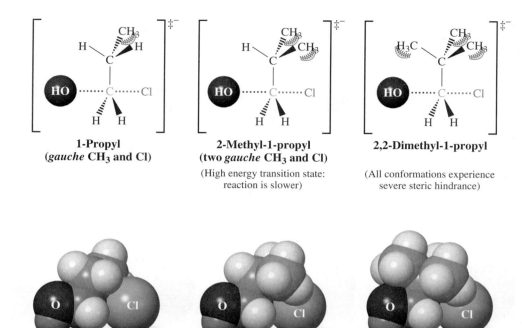

1-Propyl
(*gauche* CH₃ and Cl)

2-Methyl-1-propyl
(two *gauche* CH₃ and Cl)

(High energy transition state:
reaction is slower)

2,2-Dimethyl-1-propyl

(All conformations experience
severe steric hindrance)

A   **B**   **C**

Figure 6-10 Hashed-wedged line and space-filling renditions of the transition states for S_N2 reactions of hydroxide ion with (A) 1-chloropropane, (B) 1-chloro-2-methylpropane, and (C) 1-chloro-2,2-dimethylpropane. Increasing steric hindrance from a second *gauche* interaction reduces the rate of reaction in (B). S_N2 reactivity in (C) is eliminated almost entirely because a methyl group prevents backside attack by the nucleophile in all accessible conformations of the substrate. (See also Figures 6-8 and 6-9.)

6-11 | THE S_N2 REACTION AT A GLANCE

Figure 6-11 summarizes the factors that affect the energy of the transition state and therefore the rate of the S_N2 reaction:

Nucleophilicity   Increases to the left (more basic Nu) and down (more polarizable Nu) the periodic table.

Solvation   Impedes nucleophilicity by forming a solvent shell around Nu, particularly with protic solvents and for charged, small Nu^-. Solvation is much attenuated with aprotic solvents.

Steric hindrance   Slows the reaction through substituents at and adjacent to the reacting center.

Leaving-group ability   Increases with decreasing basicity of L.

Figure 6-11 Factors that influence the transition state of the S_N2 reaction: nucleophilicity, solvation, steric hindrance, and leaving-group ability.

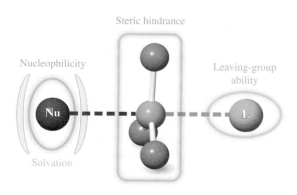

Steric hindrance

Nucleophilicity

Leaving-group ability

Nu

L

Solvation

THE BIG PICTURE

In the first five chapters, we examined the most basic ideas of organic chemistry. We applied principles of general chemistry, such as orbital theory, thermodynamics, and kinetics, to organic molecules. We also introduced topics of particular relevance to organic chemistry, such as isomerism and stereochemistry. The properties and reactions of alkanes gave us an opportunity to observe many of these basic principles in action.

The products of halogenation of alkanes, the haloalkanes, are functionalized molecules. Their carbon–halogen bond is a typical polar functional group and reacts by bimolecular nucleophilic substitution, or the S_N2 mechanism. This process, involving backside displacement, is one of the fundamental mechanisms of organic chemistry. Its characteristics include sensitivity to the identity of the nucleophile, substrate structure, leaving group, and solvent. Close attention to these details enables us to understand *why* the reaction occurs and what makes it easier or more difficult—in short, how changes in *structure* affect *function*. With this understanding comes the ability to predict: to *extrapolate by analogy* from examples of reactions we have seen to reactions we have not seen. Rational prediction is an essential component of scientific reasoning, which we illustrate often throughout this book, beginning in the very next chapter. There, we shall expand our study of haloalkanes and examine additional processes they can undergo. Our continued focus will be the effect that structural changes have on chemical behavior.

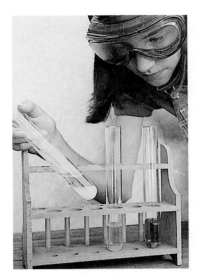

Young Stefanie Schore (now an accomplished violinist) providing a visual demonstration of relative S_N2 reactivity. The three test tubes contain, from left to right, solutions of 1-bromobutane, 2-bromopropane, and 2-bromo-2-methylpropane in acetone, respectively. Addition of a few drops of NaI solution to each causes immediate formation of NaBr (white precipitate) from the primary bromoalkane (*left*), slow NaBr precipitation only after warming from the secondary substrate (*center*), and no NaBr formation at all from the tertiary halide even after extended heating (*right*).

SOLVED EXERCISES: INTEGRATING THE CONCEPTS

6-29. Effects of Changing Reaction Conditions for S_N2 Processes

a. Write a mechanism and final product for the reaction between sodium ethoxide, $NaOCH_2CH_3$, and bromoethane, CH_3CH_2Br, in ethanol solvent, CH_3CH_2OH.

SOLUTION

The mechanism is backside attack in which the nucleophilic atom of the reagent attacks the atom of the substrate that contains the leaving group (Section 6-5). We begin by identifying each of these components. The nucleophilic atom is the negatively charged oxygen atom in ethoxide ion, CH_3CH_2–$\mathbf{O}^-$. Attack occurs at the carbon attached to bromine in the substrate molecule, CH_3–CH_2Br:

$$CH_3CH_2\ddot{\underset{\cdot\cdot}{O}}{:}^- \quad \overset{H_3C}{\underset{\underset{H}{H}}{\longrightarrow}}C\text{—}\ddot{\underset{\cdot\cdot}{Br}}{:} \longrightarrow CH_3CH_2\ddot{\underset{\cdot\cdot}{O}}\text{—}CH_2CH_3 + {:}\ddot{\underset{\cdot\cdot}{Br}}{:}^-$$

The products are bromide ion and ethoxyethane, $CH_3CH_2\ddot{O}CH_2CH_3$, an ether.

b. How would the preceding reaction be affected by each of the following changes?
   1. Replace bromoethane with fluoroethane.
   2. Replace bromoethane with bromomethane.
   3. Replace sodium ethoxide with sodium ethanethiolate, $NaSCH_2CH_3$.
   4. Replace ethanol with dimethylformamide (DMF).

SOLUTION

   1. Table 6-4 tells us that fluoride is a stronger base than bromide and, therefore, a poorer leaving group. The reaction would still take place but would be very much slower. (The actual rate decrease is of the order of 10^4.)

   2. The carbon containing the leaving group in bromomethane is less sterically hindered than that in bromoethane, so the rate of reaction would increase (Section 6-10). The product of the reaction would be $CH_3OCH_2CH_3$, methoxyethane.

   3. Both ethoxide and ethanethiolate are negatively charged. Oxygen in ethoxide is more basic than sulfur in ethanethiolate (Table 6-4), but the sulfur atom in ethanethiolate is larger, more polarizable, and less tightly solvated in the hydrogen-bonding ethanol solvent (compare Figure 6-6). We know that strong bases are good nucleophiles, but base strength is outweighed by the increased

polarizability and reduced solvation of the larger atoms within the same column of the periodic table (Section 6-8). Ethanethiolate reacts hundreds of times faster, giving as a product $CH_3CH_2SCH_2CH_3$, an example of a sulfide (Section 9-10).

4. Conversion from a protic, hydrogen-bonding solvent to a polar, aprotic one accelerates the reaction enormously by reducing solvation of the negatively charged oxygen atom (compare Table 6-6). See Problem 57 for a similar exercise.

6-30. Analyzing Substrate Structures for S_N2 Reactivity

a. Which of the following compounds would be expected to react in an S_N2 manner at a reasonable rate with sodium azide, NaN_3, in ethanol? Which will not? Why not?

(i)   (ii)   (iii)   (iv)   (v)   (vi)

SOLUTION

Let us apply the WHIP approach to break down the process of solving this problem.

*W*hat is the problem asking? This may be obvious—one merely has to identify which of the compounds shown reacts with azide in ethanol via an S_N2 process. However, there is a bit more to it, and the clue is the presence of the word "why" in the question. "How" and "why" questions invariably require a closer look at the situation, usually from a mechanistic perspective. It will be necessary to consider finer details of the S_N2 mechanism in light of the structures of each of the substrate molecules.

*H*ow to begin? Characterize each substrate in the context of the S_N2 process. Does it contain a viable leaving group? To what kind of carbon atom is the potential leaving group attached? Are other relevant structural features present?

*I*nformation needed? Does each of these six molecules contain a good leaving group? If necessary, look in Section 6-7 for guidance: To be a good leaving group, a species must be a weak base. Next, can you tell if the leaving group is attached to a primary, secondary, or tertiary carbon atom? See their definitions in Section 2-6. Anything else? Section 6-10 tells you what to look for: steric hindrance in the substrate that may obstruct the approach of the nucleophile.

*P*roceed. We identify first the molecules with good leaving groups. Referring to Table 6-4, we see that, as a general rule, only species that are the conjugate bases of strong acids (i.e., with pK_a values < 0) qualify. So, (i), (iv), and (vi) will not undergo S_N2 displacement. They lack good leaving groups: $^-NH_2$, ^-OH, and ^-CN are too strongly basic for this purpose (thus answering the "why not" for these three). Substrate (ii) contains a good leaving group, but the reaction site is a tertiary carbon and the S_N2 mechanism is sterically very unfavorable. That leaves substrates (iii) and (v), both of which are primary haloalkanes with minimal steric hindrance around the site of displacement. Both will transform readily by the S_N2 mechanism.

b. Compare the rates at which the S_N2 reactions of substrates (iii) and (v) with azide will proceed.

SOLUTION

Following the WHIP protocol, we look for any differences between the two substrates that might be significant in the context of the S_N2 mechanism. Both are comparable in steric bulk with respect to backside displacement: Both possess branching only at the remote γ-carbon, sterically without adverse consequences. That leaves as the only reasonable deciding factor the identity of the leaving group itself. Bromide is better than chloride in this respect (HBr is a stronger acid than HCl—Table 6-4) and is more readily displaced. Therefore we expect the rate of reaction of (iii) to be greater. Problem 56 requires similar reasoning.

c. Name substrates (iii) and (v) according to the IUPAC system.

SOLUTION

Review Sections 2-6 and 4-1 if necessary.

(iii) 1-Bromo-3-methylbutane   (v) (2-Chloroethyl)cyclopentane

Important Concepts

1. A **haloalkane,** commonly termed an alkyl halide, consists of an alkyl group and a halogen.

2. The physical properties of the haloalkanes are strongly affected by the polarization of the C–X bond and the polarizability of X.

3. Reagents bearing lone electron pairs are called **nucleophilic** when they attack positively polarized centers (other than protons). The latter are called **electrophilic.** When such a reaction leads to displacement of a substituent, it is a **nucleophilic substitution.** The group being displaced by the nucleophile is the **leaving group.**

4. The kinetics of the reaction of nucleophiles with primary (and most secondary) haloalkanes are second order, indicative of a **bimolecular** mechanism. This process is called **bimolecular nucleophilic substitution (S$_N$2 reaction).** It is a **concerted reaction,** one in which bonds are simultaneously broken and formed. Curved arrows are typically used to depict the flow of electrons as the reaction proceeds.

5. The S$_N$2 reaction is **stereospecific** and proceeds by **backside displacement,** thereby producing **inversion of configuration** at the reacting center.

6. An orbital description of the **S$_N$2 transition state** includes an sp^2-hybridized carbon center, partial bond making between the nucleophile and the electrophilic carbon, and simultaneous partial bond breaking between that carbon and the leaving group. Both the nucleophile and the leaving group bear partial charges.

7. **Leaving-group ability,** a measure of the ease of displacement, is roughly proportional to the strength of the conjugate acid. Especially good leaving groups are weak bases such as chloride, bromide, iodide, and the sulfonates.

8. **Nucleophilicity** increases (a) with negative charge, (b) for elements farther to the left and down the periodic table, and (c) in polar aprotic solvents.

9. **Polar aprotic solvents** accelerate S$_N$2 reactions because the nucleophiles are well separated from their counterions but are not tightly solvated.

10. **Branching** at the reacting carbon or at the carbon next to it in the substrate leads to steric hindrance in the S$_N$2 transition state and decreases the rate of bimolecular substitution.

Problems

31. Name the following molecules according to the IUPAC system.

(a) CH_3CH_2Cl

(b) $BrCH_2CH_2Br$

(c) $CH_3CH_2CHCH_2F$
         $|$
      CH_2CH_3

(d) $(CH_3)_3CCH_2I$

(e) 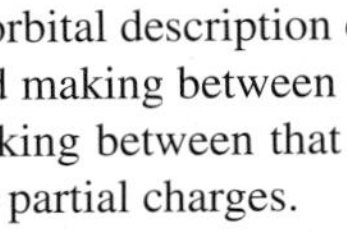 [cyclohexyl]—CCl_3

(f) $CHBr_3$

32. Draw structures for each of the following molecules: (a) 3-ethyl-2-iodopentane; (b) 3-bromo-1,1-dichlorobutane; (c) *cis*-1-(bromomethyl)-2-(2-chloroethyl)cyclobutane; (d) (trichloromethyl)cyclopropane; (e) 1,2,3-trichloro-2-methylpropane.

33. Draw and name all possible structural isomers having the formula C_3H_6BrCl.

34. Draw and name all structurally isomeric compounds having the formula $C_5H_{11}Br$.

35. For each structural isomer in Problems 33 and 34, identify all stereocenters and give the total number of stereoisomers that can exist for the structure.

36. For each reaction in Table 6-3, identify the nucleophile, its nucleophilic atom (draw its Lewis structure first), the electrophilic atom in the substrate, and the leaving group.

37. A second Lewis structure can be drawn for one of the nucleophiles in Problem 36. (a) Identify it and draw its alternate structure

(which is simply a second resonance form). (b) Is there a second nucleophilic atom in the nucleophile? If so, rewrite the reaction of Problem 36, using the new nucleophilic atom, and write a correct Lewis structure for the product.

38. For each reaction shown here, identify the nucleophile, its nucleophilic atom, the electrophilic atom in the substrate molecule, and the leaving group. Write the organic product of the reaction.

(a) $CH_3I + NaNH_2 \rightarrow$

(b) 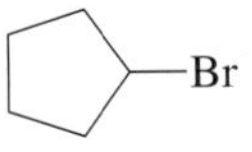 [cyclopentyl]—$Br + NaSH \rightarrow$

(c) 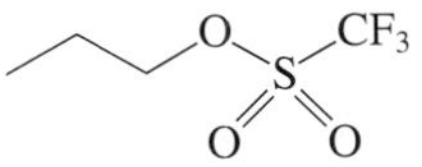 [structure: $\sim\!\!O\!-\!S(=O)(=O)\!-\!CF_3$] $+ NaI \rightarrow$

(d) 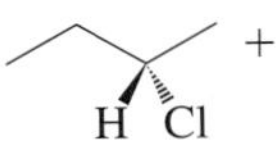 [structure with H and Cl] $+ NaN_3 \rightarrow$

(e) $CH_3Cl +$ [structure with N—CH_3] $\rightarrow$

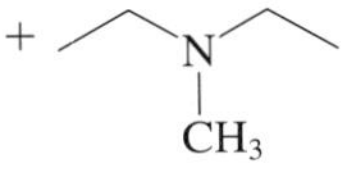

(f) 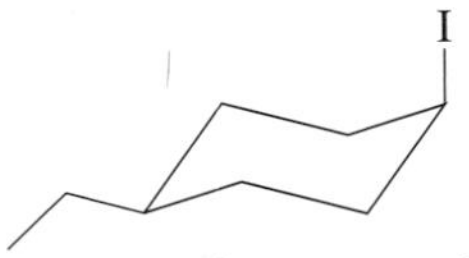 [cyclohexane structure with I] $+ KSeCN \rightarrow$

39. For each reaction presented in Problem 38, write out the mechanism using the curved-arrow notation.

40. A solution containing 0.1 M CH_3Cl and 0.1 M KSCN in DMF reacts to give CH_3SCN and KCl with an initial rate of 2×10^{-8} mol L^{-1} s^{-1}. **(a)** What is the rate constant for this reaction? **(b)** Calculate the initial reaction rate for each of the following sets of reactant concentrations: (i) $[CH_3Cl]$ = 0.2 M, [KSCN] = 0.1 M; (ii) $[CH_3Cl]$ = 0.2 M, [KSCN] = 0.3 M; (iii) $[CH_3Cl]$ = 0.4 M, [KSCN] = 0.4 M.

41. Write the product of each of the following bimolecular substitutions. The solvent is indicated above the reaction arrow.

(a) $CH_3CH_2CH_2Br + Na^+I^-$ $\xrightarrow{\text{Acetone}}$

(b) $(CH_3)_2CHCH_2I + Na^{+-}CN$ $\xrightarrow{\text{DMSO}}$

(c) $CH_3I + Na^{+-}OCH(CH_3)_2$ $\xrightarrow{(CH_3)_2CHOH}$

(d) $CH_3CH_2Br + Na^{+-}SCH_2CH_3$ $\xrightarrow{CH_3OH}$

(e) [cyclopentyl]$-CH_2Cl + CH_3CH_2SeCH_2CH_3$ $\xrightarrow{\text{Acetone}}$

(f) $(CH_3)_2CHOSO_2CH_3 + N(CH_3)_3$ $\xrightarrow{(CH_3CH_2)_2O}$

42. Determine the *R/S* designations for both starting materials and products in the following S_N2 reactions. Which of the products are optically active?

(a) $CH_3\!-\!\overset{\text{H}}{\underset{\text{CH}_2\text{CH}_3}{\overset{|}{\underset{|}{C}}}}\!-\!Cl$ + Br^-

(b) $H_3C\overset{\text{Cl}}{\underset{\text{H}}{\cdots}}\cdots\overset{\text{H}}{\underset{\text{Br}}{\cdots}}CH_3$ + 2 I^-

(c) [cyclohexane with Cl and HO] + $^-OCCH_3$ (with O)

(d) [cyclohexane with Cl and HO] + $^-OCCH_3$ (with O)

43. For each reaction presented in Problems 41 and 42, write out the mechanism using curved-arrow notation.

44. List the product(s) of the reaction of 1-bromopropane with each of the following reagents. Write "no reaction" where appropriate. (**Hint:** Carefully evaluate the nucleophilic potential of each reagent.)

(a) H_2O **(b)** H_2SO_4 **(c)** KOH   **(d)** CsI
(e) NaCN   **(f)** HCl   **(g)** $(CH_3)_2S$ **(h)** NH_3
(i) Cl_2 **(j)** KF

45. Formulate the potential product of each of the following reactions. As you did in Problem 44, write "no reaction" where appropriate. (**Hint:** Identify the expected leaving group in each of the substrates and evaluate its ability to undergo displacement.)

(a) $CH_3CH_2CH_2CH_2Br + K^{+-}OH$ $\xrightarrow{CH_3CH_2OH}$

(b) $CH_3CH_2I + K^+Cl^-$ $\xrightarrow{\text{DMF}}$

(c) [phenyl]$-CH_2Cl + Li^{+-}OCH_2CH_3$ $\xrightarrow{CH_3CH_2OH}$

(d) $(CH_3)_2CHCH_2Br + Cs^+I^-$ $\xrightarrow{CH_3OH}$

(e) $CH_3CH_2CH_2Cl + K^{+-}SCN$ $\xrightarrow{CH_3CH_2OH}$

(f) $CH_3CH_2F + Li^+Cl^-$ $\xrightarrow{CH_3OH}$

(g) $CH_3CH_2CH_2OH + K^+I^-$ $\xrightarrow{\text{DMSO}}$

(h) $CH_3I + Na^{+-}SCH_3$ $\xrightarrow{CH_3OH}$

(i) $CH_3CH_2OCH_2CH_3 + Na^{+-}OH$ $\xrightarrow{H_2O}$

(j) $CH_3CH_2I + K^{+-}O\overset{\text{O}}{\overset{\|}{C}}CH_3$ $\xrightarrow{\text{DMSO}}$

46. Show how each of the following transformations might be achieved.

(a) (R)-$CH_3\overset{OSO_2CH_3}{\underset{}{\overset{|}{C}}}HCH_2CH_3$ $\longrightarrow$ (S)-$CH_3\overset{N_3}{\underset{}{\overset{|}{C}}}HCH_2CH_3$

(b) $\begin{array}{c}CH_3\\H\!-\!\!-\!Br\\CH_3O\!-\!\!-\!H\\CH_3\end{array}$ $\longrightarrow$ $\begin{array}{c}CH_3\\H\!-\!\!-\!CN\\CH_3O\!-\!\!-\!H\\CH_3\end{array}$

(c) [bicyclic with H, H, Br] $\longrightarrow$ [bicyclic with H, H, SCH_3]

(d) [piperidine $N-CH_3$] $\longrightarrow$ [piperidinium $N^+(CH_3)(H_3C)$]

47. Rank the members of each of the following groups of species in the order of basicity, nucleophilicity, and leaving-group ability. Briefly explain your answers. **(a)** H_2O, HO^-, $CH_3CO_2^-$; **(b)** Br^-, Cl^-, F^-, I^-; **(c)** $^-NH_2$, NH_3, $^-PH_2$; **(d)** ^-OCN, ^-SCN; **(e)** F^-, HO^-, $^-SCH_3$; **(f)** H_2O, H_2S, NH_3.

48. Give the product(s) of each of the following reactions. Write "no reaction" as your answer, if appropriate.

(a) $CH_3CH_2CH_2CH_3 + Na^+Cl^-$ $\xrightarrow{CH_3OH}$

(b) $CH_3CH_2Cl + Na^{+-}OCH_3$ $\xrightarrow{CH_3OH}$

(c) [Newman projection with Br, H_3C, H, H_3C, H_3C, H, H] + Na^+I^- $\xrightarrow{\text{Acetone}}$

(d) $H\cdots\overset{Cl}{\underset{CH_3CH_2 \quad CH_3}{\overset{|}{C}}}$ + $Na^{+-}SCH_3$ $\xrightarrow{\text{Acetone}}$

(e) $CH_3\overset{OH}{\underset{}{\overset{|}{C}}}HCH_3 + Na^{+-}CN$ $\longrightarrow$

(f) $CH_3\overset{OSO_2CH_3}{\underset{}{\overset{|}{C}}}HCH_3$ + HCN $\xrightarrow{CH_3CH_2OH}$

(g) $CH_3\overset{OSO_2CH_3}{\underset{}{\overset{|}{C}}}HCH_3$ + $Na^{+-}CN$ $\xrightarrow{CH_3CH_2OH}$

(h) H_3C-[benzene]-$\overset{O}{\underset{O}{\overset{\|}{\underset{\|}{S}}}}OCH_2CH_2\overset{CH_3}{\underset{CH_3}{\overset{|}{C}}}H$ + $K^{+-}SCN$ $\xrightarrow{CH_3OH}$

(i) $CH_3CH_2NH_2 + Na^+Br^- \xrightarrow{DMSO}$

(j) $CH_3I + Na^{+-}NH_2 \xrightarrow{NH_3}$

49. For each reaction presented in Problem 48 that actually proceeds to a product, write out the mechanism using the curved-arrow notation.

50. The substance 1-butyl-3-methylimidazolium (BMIM) hexafluorophosphate (below) is a liquid at room temperature, even though it is a salt composed of positive and negative ions. BMIM and other ionic liquids constitute a new class of solvents for organic reactions, because they are capable of dissolving both organic and inorganic substances. More important, they are relatively benign environmentally, or "green," because they can be easily separated from reaction products and reused virtually indefinitely. Therefore they do not constitute a waste-disposal problem, unlike conventional solvents. **(a)** How would you characterize BMIM as a solvent? Polar or nonpolar? Protic or aprotic? **(b)** How would changing the solvent from ethanol to BMIM affect the rate of the nucleophilic substitution reaction between sodium cyanide and 1-chloropentane?

$CH_3CH_2CH_2CH_2$

PF_6^-

1-Butyl-3-methylimidazolium
(BMIM) hexafluorophosphate

51. (2S,3S)-3-Hydroxyleucine is an amino acid (Chapter 26) that is a key component in the structures of many "depsipeptide" medicinal agents, such as sanjoinine (below). **(a)** Find the part of the sanjoinine molecule that is derived structurally from (2S,3S)-3-hydroxyleucine. **(b)** Although many depsipeptide antibiotics occur in nature, the quantities available are too small to be useful pharmaceutically; thus these molecules must be synthesized. (2S,3S)-3-Hydroxyleucine, which is also not available in quantity from nature, must be synthesized as well. Possible starting materials are the four diastereomers of 2-bromo-3-hydroxy-4-methylpentanoic acid (below). Draw structural formulas for each of these diastereomers and identify which of the four should be the best starting material for a preparation of (2S,3S)-3-hydroxyleucine.

(2S,3S)-3-Hydroxyleucine

2-Bromo-3-hydroxy-4-methyl-
pentanoic acid

Sanjoinine

52. **CHALLENGE** Iodoalkanes are readily prepared from the corresponding chloro compounds by S_N2 reaction with sodium iodide in acetone. This particular procedure is especially useful because the inorganic by-product, sodium chloride, is insoluble in acetone; its precipitation drives the equilibrium in the desired direction. Thus, it is not necessary to use excess NaI, and the

process goes to completion in a very short time. Because of its great convenience, this method is named after its developer (the Finkelstein reaction). In an attempt to synthesize optically pure (R)-2-iodoheptane, a student prepared a solution of (S)-2-chloroheptane in acetone. In order to ensure success, he added excess sodium iodide and allowed the mixture to stir over the weekend. His yield of 2-iodoheptane was high, but, to his dismay, he found that his product was racemic. Explain.

53. **CHALLENGE** Using the information in Chapters 3 and 6, propose the best possible synthesis of each of the following compounds with propane as your organic starting material and any other reagents needed. [**Hint:** On the basis of the information in Sections 3-7 and 3-8, you should not expect to find very good answers for (a), (c), and (e). One general approach is best, however.]

(a) 1-Chloropropane   **(b)** 2-Chloropropane
(c) 1-Bromopropane   **(d)** 2-Bromopropane
(e) 1-Iodopropane   **(f)** 2-Iodopropane

54. Propose two syntheses of *trans*-1-methyl-2-(methylthio)cyclohexane (below), beginning with the starting compound **(a)** *cis*-1-chloro-2-methylcyclohexane; **(b)** *trans*-1-chloro-2-methylcyclohexane.

55. In each pair of molecules that follows, indicate the member of the pair that would be better suited in its indicated function for an S_N2 reaction.

(a) Nucleophile: NH_3, PH_3

(b) Substrate:

(c) Solvent:

(d) Leaving group: CH_3OH, CH_3SH

56. Rank each of the following sets of molecules in order of increasing S_N2 reactivity.

(a) CH_3CH_2Br, CH_3Br, $(CH_3)_2CHBr$

(b) $(CH_3)_2CHCH_2CH_2Cl$, $(CH_3)_2CHCH_2Cl$, $(CH_3)_2CHCl$

(c) CH_3CH_2Cl, CH_3CH_2I, —Cl

(d) $(CH_3CH_2)_2CHCH_2Br$, $CH_3CH_2CH_2CHBr$, $(CH_3)_2CHCH_2Br$
         |
        CH_3

57. Predict the effect of the changes given below on the rate of the reaction

$CH_3Cl + {}^-OCH_3 \xrightarrow{CH_3OH} CH_3OCH_3 + Cl^-.$

(a) Change substrate from CH_3Cl to CH_3I; **(b)** change nucleophile from CH_3O^- to CH_3S^-; **(c)** change substrate from CH_3Cl to $(CH_3)_2CHCl$; **(d)** change solvent from CH_3OH to $(CH_3)_2SO$.

58. The following table presents rate data for the reactions of CH_3I with three different nucleophiles in two different solvents. What is the significance of these results regarding relative reactivity of nucleophiles under different conditions?

Nucleophile	k_{rel}, CH$_3$OH	k_{rel}, DMF
Cl$^-$	1	1.2×10^6
Br$^-$	20	6×10^5
NCSe$^-$	4000	6×10^5

59. Explain the outcome of the following transformations mechanistically.

(a) HSCH$_2$CH$_2$Br + NaOH $\xrightarrow{\text{CH}_3\text{CH}_2\text{OH}}$ [triangle with S]

(b) BrCH$_2$CH$_2$CH$_2$CH$_2$CH$_2$Br + NaOH $\xrightarrow{\text{DMF}}$ [hexagon with O]

Excess

(c) BrCH$_2$CH$_2$CH$_2$CH$_2$CH$_2$Br + NH$_3$ $\xrightarrow{\text{CH}_3\text{CH}_2\text{OH}}$ [hexagon with N–H]

Excess

60. **CHALLENGE** S$_N$2 reactions of halocyclopropane and halocyclobutane substrates are very much slower than those of analogous acyclic secondary haloalkanes. Suggest an explanation for this finding. (**Hint:** Consider the effect of bond-angle strain on the energy of the transition state; see Figure 6-4.)

61. Nucleophilic attack on halocyclohexanes is also somewhat retarded compared with that on acyclic secondary haloalkanes, even though in this case bond-angle strain is *not* an important factor. Explain. (**Hint:** Make a model, and refer to Chapter 4 and Section 6-10.)

Team Problem

62. Compounds A through H are isomeric bromoalkanes with the molecular formula C$_5$H$_{11}$Br. With your team, draw all eight constitutional isomers. Indicate any stereocenter(s), but do not label it (them) as *R* or *S* until you have completed your analysis. Using the data in the following column, assign structures to A through H. Divide the problem into equal parts to share the effort of finding a solution. Reconvene and discuss your

analysis. At this point, you should indicate the stereochemistry with wedged and hashed lines as appropriate.

- Treatment of compounds A through G with NaCN in DMF followed second-order kinetics and showed the following relative rates:

$$A \cong B > C > D \cong E > F >> G$$

- Compound H does not undergo the S$_N$2 reaction under the preceding conditions.
- Compounds C, D, and F were found to be optically active, each having *S* absolute configuration at the stereocenter. Substitution reactions of D and F with NaCN in DMF proceeded with inversion of configuration, while treatment of C in the same way proceeded with retention of configuration.

Preprofessional Problems

63. The S$_N$2 reaction mechanism best applies to
(a) cyclopropane and H$_2$
(b) 1-chlorobutane and aqueous NaOH
(c) KOH and NaOH
(d) ethane and H$_2$O

64. The reaction CH$_3$Cl + OH$^-$ $\longrightarrow$ CH$_3$OH + Cl$^-$ is first order in both chloromethane and hydroxide. Given the rate constant $k = 3.5 \times 10^{-3}$ mol L^{-1} s^{-1}, what is the observed rate at the following concentrations?

[CH$_3$Cl] = 0.50 mol L^{-1} [OH$^-$] = 0.015 mol L^{-1}

(a) 2.6×10^{-5} mol L^{-1} s^{-1} **(b)** 2.6×10^{-6} mol L^{-1} s^{-1}
(c) 2.6×10^{-3} mol L^{-1} s^{-1} **(d)** 1.75×10^{-3} mol L^{-1} s^{-1}
(e) 1.75×10^{-5} mol L^{-1} s^{-1}

65. Which ion is the strongest nucleophile in aqueous solution?
(a) F$^-$ **(b)** Cl$^-$ **(c)** Br$^-$ **(d)** I$^-$
(e) all of these are equally strong

66. Only one of the following processes will occur measurably at room temperature. Which one?
(a) :F̈—C̈l:
(b) :N≡C:⁻ ↷ CH$_3$—Ï:
(c) :N≡N: ↷ CH$_3$—Ï:
(d) :Ö=Ö: ↷ CH$_2$=CH$_2$

CHAPTER 7

Further Reactions of Haloalkanes

Unimolecular Substitution and Pathways of Elimination

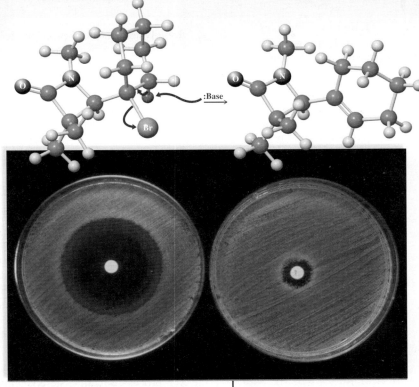

We have learned that the S_N2 displacement process is an important reaction pathway for haloalkanes. But is it the only mechanism for displacement available? Or are there other, fundamentally different types of transformations that haloalkanes undergo? In this chapter, we shall see that haloalkanes can indeed follow reaction pathways other than S_N2 displacement, especially if the haloalkanes are tertiary or secondary. In fact, bimolecular substitution is only one of *four* possible modes of reaction. The other three modes are unimolecular substitution and two types of elimination processes. The elimination processes give rise to double bonds through loss of HX and serve as our introduction into the preparation of multiply bonded organic compounds.

7-1 | SOLVOLYSIS OF TERTIARY AND SECONDARY HALOALKANES

We have learned that the rate of the S_N2 reaction diminishes drastically when the reacting center changes from primary to secondary to tertiary. These observations, however, pertain only to *bimolecular* substitution. Secondary and tertiary halides do undergo substitution, but by another mechanism. In fact, these substrates transform readily, even in the presence of weak nucleophiles, to give substitution products.

For example, when 2-bromo-2-methylpropane (*tert*-butyl bromide) is mixed with water, it is rapidly converted into 2-methyl-2-propanol (*tert*-butyl alcohol) and hydrogen bromide. Water is the nucleophile here, even though it is poor in this capacity. Such a transformation, in which a substrate undergoes substitution by *solvent* molecules, is called **solvolysis,** such as methanolysis, ethanolysis, and so on. When the solvent is water, the term **hydrolysis** is applied.

An Example of Solvolysis: Hydrolysis

Poor nucleophile
yet fast reaction!

$$CH_3CBr: + H-\overset{..}{\underset{..}{O}}H \underset{\text{Relatively fast}}{\rightleftharpoons} CH_3C\overset{..}{\underset{..}{O}}H + H\overset{..}{\underset{..}{Br}}:$$

2-Bromo-2-methylpropane
(*tert*-Butyl bromide)

2-Methyl-2-propanol
(*tert*-Butyl alcohol)

Medicinal chemists use many reactions to explore structure-activity relationships in physiologically active compounds. Above, the bromocyclohexyl substituent to a β-lactam is converted to a cyclohexenyl group by elimination of HBr. β-Lactams are four-membered ring amides featured in the structure of many antibiotics, such as penicillin and cephalosporin, and their modification is essential in combating drug resistance. The photo shows Petri dish cultures of two strains of Staphylococcus aureus bacteria (opaque and grey), an organism that causes boils, abscesses, and urinary tract infections. At left, one bacterial strain shows sensitivity to penicillin (white pellet) as indicated by the clear zone of inhibited growth around it. At right, a second strain of bacteria shows resistance to the drug and its growth is not inhibited.

Reaction

**Methyl and Primary
Haloalkanes:
Unreactive in Solvolysis**

CH_3Br
CH_3CH_2Br
$CH_3CH_2CH_2Br$
**Essentially no reaction with H_2O
at room temperature**

2-Bromopropane is hydrolyzed similarly, albeit much more slowly, whereas 1-bromopropane, bromoethane, and bromomethane are *unchanged* under these conditions.

Hydrolysis of a Secondary Haloalkane

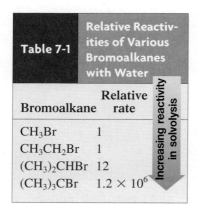

2-Bromopropane **(Isopropyl bromide)**	**2-Propanol** **(Isopropyl alcohol)**

Solvolysis also takes place in alcohol solvents.

Solvolysis of 2-Chloro-2-methylpropane in Methanol

Solvolysis—the
solvent is also
the nucleophile

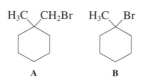

2-Chloro- **2-methylpropane**	**2-Methoxy-** **2-methylpropane**

Table 7-1 | **Relative Reactivities of Various Bromoalkanes with Water**

Bromoalkane	Relative rate
CH_3Br	1
CH_3CH_2Br	1
$(CH_3)_2CHBr$	12
$(CH_3)_3CBr$	1.2×10^6

Increasing reactivity in solvolysis

The relative rates of reaction of 2-bromopropane and 2-bromo-2-methylpropane with water to give the corresponding alcohols are shown in Table 7-1 and are compared with the corresponding rates of hydrolysis of their unbranched counterparts. Although the process gives the products expected from an S_N2 reaction, the order of reactivity is *reversed* from that found under typical S_N2 conditions. Thus, primary halides are very slow in their reactions with water, secondary halides are more reactive, and tertiary halides are about *1 million times* as fast as primary ones.

These observations suggest that the mechanism of solvolysis of secondary and, especially, tertiary haloalkanes *must be different* from that of bimolecular substitution. To understand the details of this transformation, we shall use the same methods that we used to study the S_N2 process: kinetics, stereochemistry, and the effect of substrate structure and solvent on reaction rates.

H_3C CH_2Br H_3C Br

        A   **B**

Exercise 7-1

Whereas compound A (shown in the margin) is completely stable in ethanol, B is rapidly converted into another compound. Explain.

7-2 | UNIMOLECULAR NUCLEOPHILIC SUBSTITUTION

In this section, we shall learn about a new pathway for nucleophilic substitution. Recall that the S_N2 reaction

• Has second-order kinetics

• Generates products stereospecifically with inversion of configuration

• Is fastest with halomethanes and successively slower with primary and secondary halides

• Takes place only extremely slowly with tertiary substrates, if at all

In contrast, solvolyses

- Follow a *first-order* rate law
- Are *not* stereospecific
- Are characterized by the *opposite* order of reactivity

Let us see how these findings can be accommodated mechanistically.

Solvolysis follows first-order kinetics

In Chapter 6, the kinetics of reaction between halomethanes and nucleophiles revealed a bimolecular transition state: The rate of the S_N2 reaction is proportional to the concentration of *both* ingredients. Similar studies have been carried out by varying the concentrations of 2-bromo-2-methylpropane and water in formic acid (a polar solvent of very low nucleophilicity) and measuring the rates of solvolysis. The results of these experiments show that *the rate of hydrolysis of the bromide is proportional to the concentration of only the starting halide, **not** the water.*

$$\text{Rate} = k[(CH_3)_3CBr] \text{ mol L}^{-1}\text{ s}^{-1}$$

What does this observation mean? First, it is clear that the haloalkane has to undergo some transformation on its own before anything else takes place. Second, because the final product contains a hydroxy group, water (or, in general, any nucleophile) must enter the reaction, but at a later stage and not in a way that will affect the rate law. The only way to explain this behavior is to postulate that any steps that follow the initial reaction of the halide are relatively fast. In other words, *the observed rate is that of the slowest step in the sequence:* the **rate-determining step.** It follows that only those species taking part in the transition state of this step enter into the rate expression: in this case, only the starting haloalkane.

In analogy, think of the rate-determining step as a bottleneck. Imagine a water hose with several attached clamps restricting the flow (Figure 7-1). We can see that the rate at which the water will spew out of the end is controlled by the narrowest constriction. If we were to reverse the direction of flow (to model the reversibility of a reaction), again the rate of flow would be controlled by this point. Such is the case in transformations consisting of more than one step—for example, solvolysis. What, then, are the steps in our example?

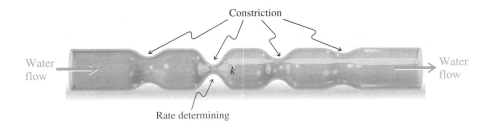

Constriction

Water flow

k

Water flow

Rate determining

Figure 7-1 The rate *k* at which water flows through this hose is controlled by the narrowest constriction.

The mechanism of solvolysis includes carbocation formation

The hydrolysis of 2-bromo-2-methylpropane is said to proceed by **unimolecular nucleophilic substitution,** abbreviated S_N1. The number 1 indicates that only one molecule, the haloalkane, participates in the rate-determining step: The rate of the reaction does *not* depend on the concentration of the nucleophile. The mechanism consists of three steps.

Mechanism

Animation

ANIMATED MECHANISM:
Nucleophilic substitution
(S_N1) of $(CH_3)_3CBr$ with HOH

Heterolytic cleavage of the carbon–halogen bond separates opposite charges, making this step slow and rate determining.

The terms "trap" or "capture" describe the fast attack on a reactive intermediate by another chemical species.

Step 1. The rate-determining step is the dissociation of the haloalkane to an alkyl cation and bromide, a process we have already seen in Section 2-2.

Dissociation of Halide to Form a Carbocation

$$CH_3\overset{\overset{\displaystyle CH_3}{|}}{\underset{\underset{\displaystyle CH_3}{|}}{C}}\!\!-\!\!\ddot{\ddot{Br}}: \underset{\text{Rate determining}}{\rightleftharpoons} CH_3\overset{\overset{\displaystyle CH_3}{|}}{\underset{\underset{\displaystyle CH_3}{|}}{C}}{}^{+} + :\ddot{\ddot{Br}}:^{-}$$

1,1-Dimethylethyl cation
(*tert*-Butyl cation)

This conversion is an example of heterolytic cleavage. The hydrocarbon product contains a positively charged central carbon atom attached to three other groups and bearing only an electron sextet. Such a structure is called a **carbocation.**

Step 2. The 1,1-dimethylethyl (*tert*-butyl) cation formed in step 1 is a powerful electrophile that is immediately trapped by the surrounding water. This process can be viewed as a nucleophilic attack by the solvent on the electron-deficient carbon.

Nucleophilic Attack by Water

$$CH_3\overset{\overset{\displaystyle CH_3}{|}}{\underset{\underset{\displaystyle CH_3}{|}}{C}}{}^{+} + :\overset{\displaystyle H}{\underset{\displaystyle H}{\ddot{O}}} \overset{\text{Fast}}{\rightleftharpoons} CH_3\overset{\overset{\displaystyle CH_3}{|}}{\underset{\underset{\displaystyle CH_3}{|}}{C}}\!\!-\!\!\overset{\displaystyle H}{\underset{\displaystyle H}{\overset{+}{\ddot{O}}}}:$$

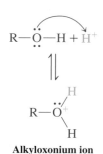

$$R\!-\!\ddot{O}\!-\!H + \overset{+}{H}$$

$$\updownarrow$$

$$R\!-\!\overset{\overset{\displaystyle H}{\diagup}}{\ddot{O}}{}^{+}_{\diagdown H}$$

Alkyloxonium ion

An alkyloxonium ion

The resulting species is an example of an **alkyloxonium ion,** the conjugate acid of an alcohol—in this case 2-methyl-2-propanol, the eventual product of the sequence.

Step 3. Like the hydronium ion, H_3O^+, the first member of the series of oxonium ions,* all alkyloxonium ions are strong acids. They are therefore readily deprotonated by the water in the reaction medium to furnish the final alcohol.

---

*Indeed, IUPAC recommends the use of the name oxonium ion instead of hydronium ion for H_3O^+.

Deprotonation

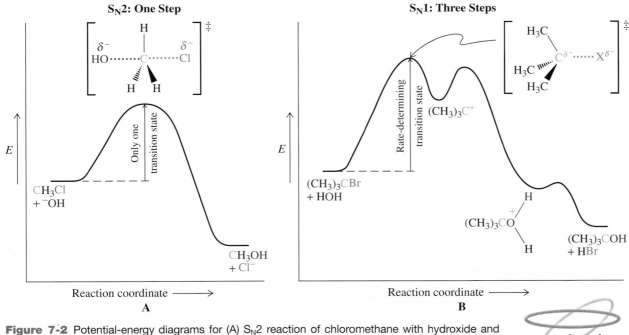

$$CH_3C^+ \!-\! \overset{\cdot\cdot}{O}\!: \;+\; \overset{\cdot\cdot}{O}H_2 \;\overset{Fast}{\rightleftharpoons}\; CH_3\overset{\cdot\cdot}{C}OH \;+\; H\overset{+}{\overset{\cdot\cdot}{O}}H_2$$

(with CH₃ groups above and below the central carbon, H on the oxygen)

Alkyloxonium ion
(Strongly acidic)

2-Methyl-2-propanol

Figure 7-2 compares the potential-energy diagrams for the S_N2 reaction of chloromethane with hydroxide ion and the S_N1 reaction of 2-bromo-2-methylpropane with water. The S_N1 diagram exhibits three transition states, one for each step in the mechanism. The first has the highest energy—and thus is rate determining—because it requires the separation of opposite charges.

Exercise 7-2

Using the bond-strength data in Table 3-1, calculate the $\Delta H°$ for the hydrolysis of 2-bromo-2-methylpropane to 2-methyl-2-propanol and hydrogen bromide. [**Caution:** Don't get confused by the ionic mechanism of these reactions. The $\Delta H°$ is a measure of the thermicity of the overall transformation, regardless of mechanism (Section 2-1).]

S_N2: One Step   **S_N1: Three Steps**

$$\left[\begin{array}{c} H \\ \overset{\delta^-}{HO}\cdots\cdots\overset{|}{\underset{\underset{H\quad H}{}}{C}}\cdots\cdots\overset{\delta^-}{Cl} \end{array}\right]^{\ddagger}$$

$$\left[\begin{array}{c} H_3C \\ H_3C^{\text{w}}\,C^{\delta^+}\cdots\cdots X^{\delta^-} \\ H_3C \end{array}\right]^{\ddagger}$$

E (vertical axis) — Only one transition state

$CH_3Cl + {}^-OH$

$CH_3OH + Cl^-$

Reaction coordinate →
A

E (vertical axis) — Rate-determining transition state

$(CH_3)_3C^+$

$(CH_3)_3CBr + HOH$

$(CH_3)_3C\overset{+}{O}\big\langle{}^{H}_{H}$

$(CH_3)_3COH + HBr$

Reaction coordinate →
B

Figure 7-2 Potential-energy diagrams for (A) S_N2 reaction of chloromethane with hydroxide and (B) S_N1 hydrolysis of 2-bromo-2-methylpropane. Whereas the S_N2 process takes place in a single step, the S_N1 mechanism consists of three distinct events: rate-determining dissociation of the haloalkane into a halide ion and a carbocation, nucleophilic attack by water on the carbocation to give an alkyloxonium ion, and proton loss to furnish the final product. *Note:* For clarity, inorganic species have been omitted from the intermediate stages of (B).

Animation

ANIMATED MECHANISM: Nucleophilic substitution (S_N1) of ((CH₃)₃CBr with HOH

All three steps of the mechanism of solvolysis are reversible. The overall equilibrium can be driven in either direction by the suitable choice of reaction conditions. Thus, a large excess of nucleophilic solvent ensures complete solvolysis. In Chapter 9, we shall see how this reaction can be reversed to permit the synthesis of tertiary haloalkanes from alcohols.

In Summary The kinetics of haloalkane solvolysis lead us to a three-step mechanism. The crucial, rate-determining step is the initial dissociation of a leaving group from the starting material to form a carbocation. Because only the substrate molecule participates in the rate-limiting step, this process is called *unimolecular nucleophilic substitution*, S_N1. Let us now see what other experimental observations can tell us about the S_N1 mechanism.

7-3 | STEREOCHEMICAL CONSEQUENCES OF S_N1 REACTIONS

Model Building

Animation

ANIMATED MECHANISM:
Nucleophilic substitution
(S_N1) of $(CH_3)_3CBr$ with HOH

The proposed mechanism of unimolecular nucleophilic substitution has predictable stereochemical consequences because of the structure of the intermediate carbocation. To minimize electron repulsion, the positively charged carbon assumes trigonal planar geometry, the result of sp^2 hybridization (Sections 1-3 and 1-8). Such an intermediate is therefore achiral (make a model). Hence, starting with an optically active secondary or tertiary haloalkane in which the stereocenter bears the departing halogen, under conditions favorable for S_N1 reaction we should obtain racemic products (Figure 7-3). This result is, in fact, observed in many solvolyses. In general, the formation of racemic products from optically active substrates is strong evidence for the intermediate being a symmetrical, achiral species, such as a carbocation.

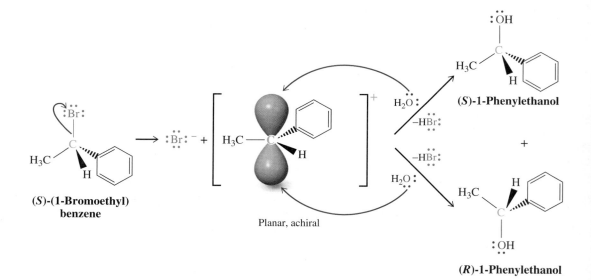

Figure 7-3 The mechanism of hydrolysis of (S)-(1-bromoethyl)benzene explains the stereochemistry of the reaction. Initial ionization furnishes a planar, achiral carbocation. This ion, when trapped with water, yields racemic alcohol.

Exercise 7-3

(R)-3-Bromo-3-methylhexane loses its optical activity when dissolved in nitromethane, a highly polar but nonnucleophilic solvent. Explain by a detailed mechanism. (**Caution:** When writing mechanisms, use "arrow pushing" to depict electron flow; write out every step separately; formulate complete structures, including charges and relevant electron pairs; and draw explicit reaction arrows to connect starting materials or intermediates with their respective products. Don't use shortcuts, and don't be sloppy!)

Solved Exercise 7-4	Working with the Concepts: Stereochemical Consequences of S$_N$1 Displacement

Gentle warming of (2R,3R)-2-iodo-3-methylhexane in methanol gives two stereoisomeric methyl ethers. How are they related to each other? Explain mechanistically.

Strategy

The substrate is secondary; therefore, substitution can proceed by either the S$_N$1 or the S$_N$2 mechanism. Let's consider the reaction conditions to see which is more likely and what its consequences will be.

Solution

- The reaction takes place in methanol, CH$_3$OH, a poor nucleophile (disfavoring S$_N$2) but a very polar, protic solvent, well suited for dissociation of secondary and tertiary haloalkanes into ions (favoring S$_N$1).
- Dissociation of the excellent leaving group I$^-$ from C2 gives a trigonal planar carbocation. Methanol may attack from either face (compare mechanism steps 1 and 2 in Section 7-2), giving two stereoisomeric oxonium ions. The positively charged oxygen makes the attached hydrogen very acidic; after proton loss, two stereoisomeric ethers result (mechanism step 3 in Section 7-2; see also Figure 7-3). We have here another example of solvolysis (specifically, methanolysis), because the nucleophile is the solvent (methanol).

Caution! When describing the S$_N$1 mechanism, *avoid the following two very common errors:* (1) Do *not* dissociate CH$_3$OH to give methoxide (CH$_3$O$^-$) and a proton before bonding with the cationic carbon. Methanol is a *weak acid* whose dissociation is thermodynamically not favored. (2) Do *not* dissociate CH$_3$OH to give a methyl cation and hydroxide ion. Although the presence of the OH functional group in alcohols may remind you of the formulas of inorganic hydroxides, alcohols are *not* sources of hydroxide ion.

- The two stereoisomeric ether products are diastereomers; 2S,3R and 2R,3R. At the reaction site, C2, both R and S configurations result from the two possible pathways of methanol attack, a and b. At C3 a stereocenter where no reaction occurs, the original R configuration remains unchanged.

Exercise 7-5	Try It Yourself

Hydrolysis of molecule A (shown in the margin) gives two alcohols. Explain.

A

7-4 | EFFECTS OF SOLVENT, LEAVING GROUP, AND NUCLEOPHILE ON UNIMOLECULAR SUBSTITUTION

As in S$_N$2 reactions (Section 6-10), varying the solvent, the leaving group, and the nucleophile greatly affects unimolecular substitution.

Mechanism

Polar solvents accelerate the S_N1 reaction

Heterolytic cleavage of the C–X bond in the rate-determining step of the S_N1 reaction entails a transition-state structure that is highly polarized (Figure 7-4), leading eventually to two fully charged ions. In contrast, in a typical S_N2 transition state, charges are not created; rather, they are dispersed (see Figure 6-4).

Because of this polar transition state, the rate of an S_N1 reaction increases as solvent polarity is increased. The effect is particularly striking when the solvent is changed from aprotic to protic. For example, hydrolysis of 2-bromo-2-methylpropane is much faster in pure water than in a 9:1 mixture of acetone and water. The protic solvent accelerates the S_N1 reaction because it stabilizes the transition state shown in Figure 7-4 by hydrogen bonding with the leaving group. Remember that, in contrast, the S_N2 reaction is accelerated in polar *aprotic* solvents, mainly because of a solvent effect on the reactivity of the *nucleophile*.

Effect of Solvent on the Rate of an S_N1 Reaction

			Relative rate
$(CH_3)_3CBr$	$\xrightarrow[\text{More polar solvent}]{100\% \ H_2O}$	$(CH_3)_3COH \ + \ HBr$	400,000
$(CH_3)_3CBr$	$\xrightarrow[\text{Less polar solvent}]{90\% \ \text{acetone, } 10\% \ H_2O}$	$(CH_3)_3COH \ + \ HBr$	1

The solvent nitromethane, CH_3NO_2 (see Table 6-5), is both highly polar and virtually nonnucleophilic. Therefore, it is useful in studies of S_N1 reactions with nucleophiles other than solvent molecules.

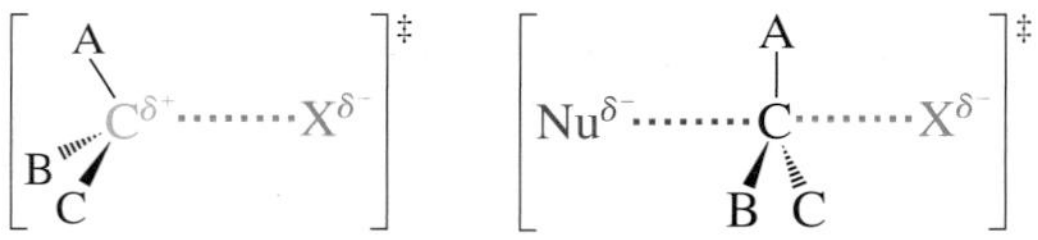

Figure 7-4 The respective transition states for the S_N1 and S_N2 reactions explain why the S_N1 process is strongly accelerated by polar solvents. Heterolytic cleavage entails charge separation, a process aided by polar solvation.

S_N1
Opposite charges
are separated

S_N2
Negative charge
is dispersed

The S_N1 reaction speeds up with better leaving groups

Because the leaving group departs in the rate-determining step of the S_N1 reaction, it is not surprising that the rate of the reaction increases as the leaving-group ability of the departing group improves. Thus, tertiary iodoalkanes more readily undergo solvolysis than do the corresponding bromides, and bromides are, in turn, more reactive than chlorides. Sulfonates are particularly prone to departure.

Relative Rate of Solvolysis of RX (R = Tertiary Alkyl)

$$X = -OSO_2R' > -I > -Br > -Cl$$

◄ Increasing rate

The strength of the nucleophile affects the product distribution but not the reaction rate

Does changing the nucleophile affect the rate of S_N1 reaction? The answer is no. Recall that, in the S_N2 process, the rate of reaction increases significantly as the nucleophilicity of the attacking species improves. However, because the rate-determining step of unimolecular substitution does *not* include the nucleophile, changing its structure (or concentration) does *not* alter the rate of disappearance of the haloalkane. Nevertheless, when two or more nucleophiles compete for capture of the intermediate carbocation, their relative strengths and concentrations may greatly affect the *product distribution*.

For example, hydrolysis of a solution of 2-chloro-2-methylpropane gives the expected 2-methyl-2-propanol (*tert*-butyl alcohol), with a rate constant k_1. Quite a different result is obtained when the same experiment is carried out in the presence of the soluble salt calcium formate: 1,1-Dimethylethyl formate (*tert*-butyl formate) replaces the alcohol as the product, but the reaction still proceeds at exactly the *same* rate k_1. In this case, formate ion, a better nucleophile than water, wins out in competition for bonding to the intermediate carbocation. The rate of disappearance of the starting material is determined by k_1 (regardless of the product eventually formed), but the relative yields of the *products* depend on the relative reactivities and concentrations of the competing nucleophiles.

Competing Nucleophiles in the S$_N$1 Reaction

$$(CH_3)_3COH + HCl$$
2-Methyl-2-propanol

$(CH_3)_3CCl$

$+$

HOH $\xrightarrow[\text{Rate determining}]{k_1}$ $(CH_3)_3C^+ + Cl^-$

$+$

½ Ca(OCH)$_2$
Calcium formate

Water and formate ion compete for cation

$$(CH_3)_3COCH + ½ CaCl_2$$
1,1-Dimethylethyl formate
(*tert*-Butyl formate)

Solved Exercise 7-6 | **Working with the Concepts: Nucleophile Competition**

A solution of 1,1-dimethylethyl (*tert*-butyl) methanesulfonate in polar aprotic solvent containing equal amounts of sodium fluoride and sodium bromide produces 75% 2-fluoro-2-methylpropane and only 25% 2-bromo-2-methylpropane. Explain. (**Hint:** Refer to Section 6-8 and Problem 58 of Chapter 6 for information regarding relative nucleophilic strengths of the halide ions in aprotic solvents.)

Note: The word "Explain" in a problem usually means that you need to *think about the mechanism* in order to find a solution.

Strategy
The substrate is tertiary; therefore, substitution can proceed readily only via the S$_N$1 mechanism. Equal amounts of the two nucleophiles are present, but the two substitution products do not form in equivalent yields. The explanation must lie in differences in nucleophile strength and rate of carbocation trapping under the reaction conditions.

Solution
• Hydrogen bonding is absent in polar aprotic solvents, so nucleophilicity is determined by polarizability and basicity.
• Bromide, the larger ion, is more polarizable, but fluoride is the stronger base (see either Table 2-2 or 6-4). Which effect wins? They are closely balanced, but the table in Problem 58 of Chapter 6 provides the answer: In DMF (Table 6-5), Cl$^-$, the stronger base, is about twice as nucleophilic as Br$^-$. Fluoride is even more basic; consequently, it wins out over other halide ions in attacking the carbocation intermediate.

Exercise 7-7 | **Try It Yourself**

Predict the major substitution product that would result from mixing 2-bromo-2-methylpropane with concentrated aqueous ammonia. (**Caution:** Although ammonia in water forms ammonium hydroxide according to the equation $NH_3 + H_2O \rightleftharpoons NH_4^+ {}^-OH$, the K_{eq} for this process is very small. Thus the concentration of hydroxide is quite low.)

In Summary We have seen further evidence supporting the S_N1 mechanism for the reaction of tertiary (and secondary) haloalkanes with certain nucleophiles. The stereochemistry of the process, the effects of the solvent and the leaving-group ability on the rate, and the absence of such effects when the strength of the nucleophile is varied, are consistent with the unimolecular route.

7-5   EFFECT OF THE ALKYL GROUP ON THE S_N1 REACTION: CARBOCATION STABILITY

What is so special about tertiary haloalkanes that they undergo conversion by the S_N1 pathway, whereas primary systems follow S_N2? How do secondary haloalkanes fit into this scheme? Somehow, the degree of substitution at the reacting carbon must control the pathway followed in the reaction of haloalkanes (and related derivatives) with nucleophiles. We shall see that only secondary and tertiary systems can form carbocations. For this reason, tertiary halides, *whose steric bulk inhibits them from undergoing S_N2 reactions,* substitute almost exclusively by the S_N1 mechanism, primary haloalkanes only by S_N2, and secondary haloalkanes by either route, depending on conditions.

Carbocation stability increases from primary to secondary to tertiary

We have learned that primary haloalkanes undergo *only* bimolecular nucleophilic substitution. In contrast, secondary systems often transform through carbocation intermediates and tertiary systems virtually always do. The reasons for this difference are twofold. First, steric hindrance increases along the series, thereby slowing down S_N2. Second, increasing alkyl substitution stabilizes carbocation centers. Only secondary and tertiary cations are energetically feasible under the conditions of the S_N1 reaction.

Relative Stability of Carbocations

$$CH_3CH_2CH_2\overset{+}{C}H_2 \ < \ CH_3CH_2\overset{+}{C}HCH_3 \ < \ (CH_3)_3\overset{+}{C}$$

Primary   <   Secondary   <   Tertiary

→ Increasing carbocation stability

Now we can see why tertiary haloalkanes undergo solvolysis so readily. Because tertiary carbocations are more stable than their less substituted relatives, they form more easily. But what is the reason for this order of stability?

Hyperconjugation stabilizes positive charge

Note that the order of carbocation stability parallels that of the corresponding radicals. Both trends have their roots in the same phenomenon: *hyperconjugation.* Recall from Section 3-2 that hyperconjugation is the result of overlap of a *p* orbital with a neighboring bonding molecular orbital, such as that of a C–H or a C–C bond. In a radical, the *p* orbital is singly filled; in a carbocation, it is empty. In both cases, the alkyl group donates electron density to the electron-deficient center and thus stabilizes it. Figure 7-5 compares the orbital pictures of the methyl and 1,1-dimethylethyl (*tert*-butyl) cations and depicts the electrostatic potential maps of the methyl, ethyl, 1-methylethyl (isopropyl), and 1,1-dimethylethyl cations. Figure 7-6 shows the structure of the tertiary butyl system, which is stabilized enough to be isolated and characterized by X-ray diffraction measurements.

Secondary systems undergo both S_N1 and S_N2 reactions

As you will have gathered from the preceding discussion, secondary haloalkanes exhibit the most varied substitution behavior. *Both* S_N2 and S_N1 reactions are possible: Steric hindrance slows but does not preclude bimolecular nucleophilic attack. At the same time, unimolecular dissociation becomes competitive because of the relative stability of secondary

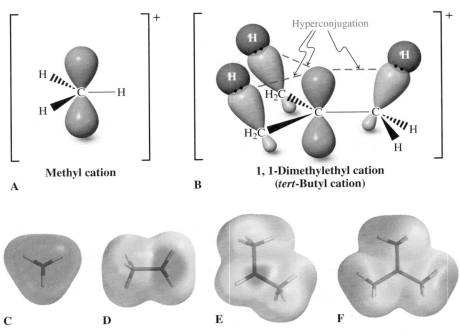

Methyl cation

A

1, 1-Dimethylethyl cation
(*tert*-Butyl cation)

B

Hyperconjugation

C   D   E   F

Figure 7-5 (A) The partial orbital picture of the methyl cation reveals why it cannot be stabilized by hyperconjugation. (B) In contrast, the 1,1-dimethylethyl (*tert*-butyl) cation benefits from three hyperconjugative interactions. The electrostatic potential maps of the (C) methyl, (D) ethyl, (E) 1-methylethyl (isopropyl), and (F) 1,1-dimethylethyl (*tert*-butyl) cations show how the initially strongly electron-deficient (blue) central carbon increasingly loses its blue color along the series because of increasing hyperconjugation.

carbocations. The pathway chosen depends on the reaction conditions: the solvent, the leaving group, and the nucleophile.

If we use a substrate carrying a very good leaving group, a poor nucleophile, and a polar protic solvent (S_N1 conditions), then *unimolecular* substitution is favored. If we employ a high concentration of a good nucleophile, a polar aprotic solvent, and a haloalkane bearing a reasonable leaving group (S_N2 conditions), then *bimolecular* substitution predominates. Table 7-2 summarizes our observations regarding the reactivity of haloalkanes toward nucleophiles.

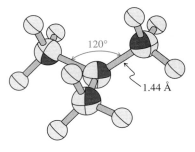

Figure 7-6 X-ray crystal structure determination for the 1,1-dimethylethyl (*tert*-butyl) cation. The four carbons lie in a plane with 120° C–C–C bond angles, consistent with sp^2 hybridization at the central carbon. The C–C bond length is 1.44 Å, shorter than normal single bonds (1.54 Å), a consequence of hyperconjugative overlap.

Substitution of a Secondary Substrate Under S_N1 Conditions

$$H_3C \quad :O: \qquad\qquad H_3C$$
$$C\!-\!OSCF_3 \xrightarrow{H\ddot{O}} \quad C\!-\!\ddot{O}H \;+\; CF_3SO_3H$$
$$H_3C\!\!\!\!\! \diagdown\! H \quad :O: \qquad\qquad H_3C\!\!\!\!\! \diagdown\! H$$

Substitution of a Secondary Haloalkane Under S_N2 Conditions

$$H_3C \qquad\qquad\qquad\qquad CH_3$$
$$C\!-\!\ddot{B}r: \;+\; CH_3\ddot{S}:^- \xrightarrow{\text{Acetone}} CH_3\ddot{S}\!-\!C\!\!\diagup\! CH_3 \;+\; :\ddot{B}r:^-$$
$$H_3C\!\!\!\!\! \diagdown\! H \qquad\qquad\qquad\qquad H$$

Model Building

Table 7-2	Reactivity of R–X in Nucleophilic Substitutions: R–X + Nu⁻ ⟶ R–Nu + X⁻	
R	**S_N1**	**S_N2**
CH_3	Not observed in solution (methyl cation too high in energy)	Frequent; fast with good nucleophiles and good leaving groups
Primary	Not observed in solution (primary carbocations too high in energy)[a]	Frequent; fast with good nucleophiles and good leaving groups, slow when branching at C2 is present in R
Secondary	Relatively slow; best with good leaving groups in polar protic solvents	Relatively slow; best with high concentrations of good nucleophiles in polar aprotic solvents
Tertiary	Frequent; particularly fast in polar, protic solvents and with good leaving groups	Extremely slow

(S_N1 better ↓ on left side; S_N2 better ↑ on right side)

[a]Exceptions are resonance-stabilized carbocations; see Chapter 14.

A visual demonstration of relative S_N1 reactivity. The three test tubes contain, from left to right, solutions of 1-bromobutane, 2-bromopropane, and 2-bromo-2-methylpropane in ethanol, respectively. Addition of a few drops of $AgNO_3$ solution to each causes immediate formation of a heavy AgBr precipitate from the *tert*-bromoalkane (*right*), less AgBr precipitation from the secondary substrate (*center*), and no AgBr formation from the primary halide (*left*).

Solved Exercise 7-8 | Working with the Concepts: Secondary Haloalkanes

Explain the following results.

(a) ![reaction a] + CN^- →(Acetone)→ product
   Cl H   H CN
   R   S

(b) ![reaction b] + CH_3OH → product
   I H   OCH_3
   R   R + S

What is requested is an *explanation*: your cue that the answer lies in the consideration of *mechanisms*.

How to begin? Look at the species involved in each reaction, and evaluate how their properties may determine what mechanistic pathway is followed.

Information needed? Consult Table 7-2. Both reactions have secondary substrates with good leaving groups, giving the options of substitution by either an S_N1 or an S_N2 pathway. To establish which one dominates in each case, we need to consider the nucleophiles and solvents.

Proceed. Use the text examples that immediately precede this exercise for guidance.

• In reaction (a), cyanide ion, CN^-, is a good nucleophile (Table 6-7), and acetone is a polar aprotic solvent, a combination that favors the S_N2 mechanism. Backside attack occurs, leading to inversion at the site of displacement (see Figure 6-4).

• In reaction (b), methanol, CH_3OH, is both the solvent and the nucleophile. As in Exercise 7-4, we have the conditions for solvolysis via the S_N1 mechanism, leading to both enantiomeric ether products.

Exercise 7-9 | Try It Yourself

Would treatment of (R)-2-chlorobutane with aqueous ammonia be a good synthetic method for the preparation of (R)-2-butanamine, (R)-$CH_3CH_2CH(NH_2)CH_3$? Why or why not? Can you think of a better route?

The first five sections of this chapter have given us the background for understanding how the S_N1 mechanism takes place and what factors favor its occurrence. It is useful to keep in mind that two conditions must be satisfied before dissociation of a carbon–halogen bond into ions can occur: The carbon atom must be secondary or tertiary so that the carbocation has sufficient thermodynamic stability to form, *and* the reaction must take place in a polar solvent capable of interacting with and stabilizing *both* positive and negative ions. Nevertheless, carbocations are common intermediates that will appear in the reactions of many of the compound classes that we will study in the chapters to come.

Which is "greener": S_N1 or S_N2?

The contrast between the stereochemical outcomes of the S_N1 and S_N2 mechanisms directly affects the comparative utility of the two processes in synthesis. The S_N2 process is *stereospecific*: Reaction of a single stereoisomeric substrate gives a single stereoisomeric product (Section 6-6). In contrast, virtually all reactions that proceed via the S_N1 mechanism at a stereocenter give mixtures of stereoisomers. And it gets worse: The chemistry of carbocations, the intermediates in all S_N1 reactions, is complex. As we shall see in Chapter 9, these species are prone to rearrangements, frequently resulting in complicated collections of products. In addition, carbocations undergo another important transformation to be described next: *loss of a proton* to furnish a double bond.

In the final analysis, S_N1 reactions, unlike S_N2 processes, are of limited use in synthesis because they fail the first two criteria of "green" reactions (see Real Life 3-1): They are poor in atom efficiency and wasteful overall, because they tend to lead to mixtures of stereoisomeric substitution products as well as other organic compounds. S_N2 is "greener."

Unusually Stereoselective S_N1 Displacement in Anticancer Drug Synthesis

S_N1 displacements normally give mixtures of stereoisomeric products. The high-energy carbocation intermediate reacts with the first nucleophilic species it encounters, regardless of which lobe of the carbocation p orbital the nucleophile approaches. The example shown below is a *very* unusual exception: A secondary haloalkane, a good leaving group (bromide), a highly polar, protic solvent but a poor nucleophile (water)—ideal circumstances for an S_N1 reaction—and displacement of bromide by water occurs with over 90% *retention* of configuration!

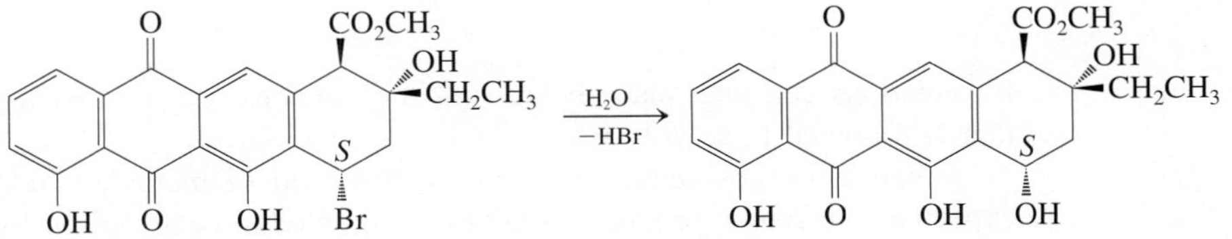

The structure of the relevant carbocation is shown at the right. Approach of a nucleophile toward the top lobe of its p orbital is partly blocked by the ethyl group two carbon atoms over and to a lesser extent by the ester function one carbon farther away (green). In addition, the hydroxy group on the bottom face of the ring "directs" nucleophilic addition of a water molecule from below by hydrogen bonding, as shown.

This stereochemical result is of critical importance because the product, named aklavinone, is a component in a powerful anticancer drug called aclacinomycin A. This compound belongs to a class of chemotherapeutic agents called anthracyclines, whose clinical utility is compromised by

their toxicity; aclacinomycin is less cardiotoxic than other anthracyclines and thus has been under careful study by medical researchers for over two decades.

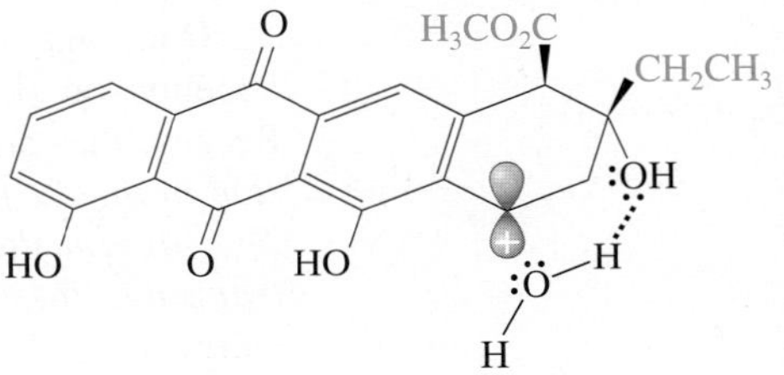

In Summary Tertiary haloalkanes are reactive in the presence of nucleophiles even though they are too sterically hindered to undergo S_N2 reactions readily: The tertiary carbocation is readily formed because it is stabilized by hyperconjugation. Subsequent trapping by a nucleophile, such as a solvent (solvolysis), results in the product of nucleophilic substitution. Primary haloalkanes do not react in this manner: The primary cation is too highly energetic (unstable) to be formed in solution. The primary substrate follows the S_N2 route. Secondary systems are converted into substitution products through either pathway, depending on the nature of the leaving group, the solvent, and the nucleophile.

7-6 | UNIMOLECULAR ELIMINATION: E1

We know that carbocations are readily attacked by nucleophiles at the positively charged carbon. However, this is not their only mode of reaction. A competing alternative is deprotonation by the nucleophile acting as a base, furnishing a new class of compounds, the alkenes. This process is possible because the proton neighboring the positive charge is unusually acidic.

Competition Between Nucleophilic and Basic Attack on a Carbocation

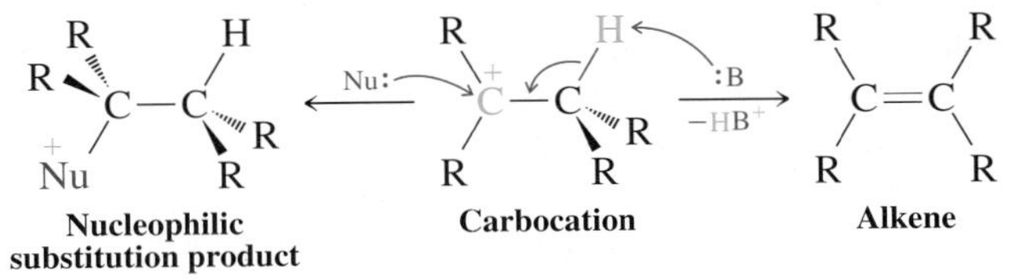

Starting from a haloalkane, the overall transformation constitutes the removal of HX with the simultaneous generation of a double bond. The general term for such a process is **elimination,** abbreviated **E.**

● **Reaction**

Elimination

Eliminations can take place by several mechanisms. Let us first establish the mechanism that is followed in solvolysis.

When 2-bromo-2-methylpropane is dissolved in methanol, it disappears rapidly. As expected, the major product, 2-methoxy-2-methylpropane, arises by solvolysis. However, there is also a significant amount of another compound, 2-methylpropene, the product of *elimination* of HBr from the original substrate. Thus, in competition with the S_N1 process, which leads to displacement of the leaving group, another mechanism transforms the tertiary halide, giving rise to the alkene. What is this mechanism? Is it related to the S_N1 reaction?

Once again we turn to a kinetic analysis and find that the rate of alkene formation depends on the concentration of *only* the starting halide; the reaction is first order. Because they are unimolecular, eliminations of this type are labeled **E1.** *The rate-determining step in the E1 process is the same as that in S_N1 reactions: dissociation to a carbocation. This intermediate then has a second pathway at its disposal along with nucleophilic trapping: loss of a proton from a carbon adjacent to the one bearing the positive charge.*

Competition Between E1 and S_N1 in the Methanolysis of 2-Bromo-2-methylpropane

Model Building

How exactly is the proton lost? Figure 7-7 on the next page depicts this process with orbitals. Although we often show protons that evolve in chemical processes by using the notation H^+, "free" protons do not participate under the conditions of ordinary organic reactions. A Lewis base typically removes the proton (Section 2-3). In aqueous solution, water plays this role, giving H_3O^+; here, the proton is carried off by CH_3OH as $CH_3OH_2^+$, an alkyloxonium ion. The carbon left behind rehybridizes from sp^3 to sp^2. As the C–H bond breaks, its electrons shift to overlap in a π fashion with the vacant p orbital at the neighboring cationic center. The result is a hydrocarbon containing a double bond: an alkene. The complete mechanism is shown on the following page.

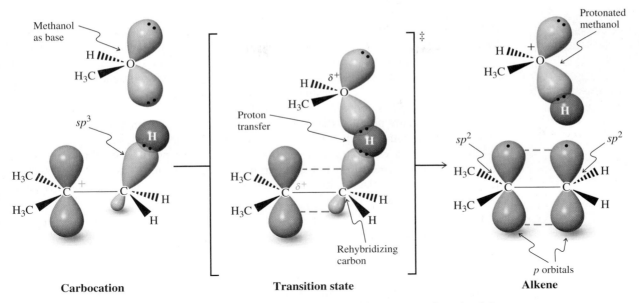

Carbocation   **Transition state**   **Alkene**

Figure 7-7 The alkene-forming step in unimolecular elimination (E1): deprotonation of a 1,1-dimethylethyl (*tert*-butyl) cation by the solvent methanol. In an orbital description of proton abstraction, an electron pair on the oxygen atom in the solvent attacks a hydrogen on a carbon adjacent to that bearing the positive charge. The proton is transferred, leaving an electron pair behind. As the carbon rehybridizes from sp^3 to sp^2, these electrons redistribute over the two p orbitals of the new double bond.

The E1 Reaction Mechanism

Any hydrogen positioned on *any carbon next to the center bearing the leaving group* can participate in the E1 reaction. The 1,1-dimethylethyl (*tert*-butyl) cation has nine such hydrogens, each of which is equally reactive. In this case, the product is the same regardless of the identity of the proton lost. In other cases, more than one product may be obtained. These pathways will be discussed in more detail in Chapter 11.

The E1 Reaction Can Give Product Mixtures

S_N1 product

E1 products

*This notation indicates that the elements of the acid have been removed from the starting material. In reality, the proton ends up protonating a Lewis base. This symbolism will be used occasionally in other elimination reactions in this book.

Table 7-3	Ratio of S_N1 to E1 Products in the Hydrolyses of 2-Halo-2-methyl-propanes at 25°C	
X in $(CH_3)_3CX$		Ratio $S_N1:E1$
Cl		95:5
Br		95:5
I		96:4

The identity of the leaving group should have no effect on the ratio of substitution to elimination, because the carbocation formed is the same in each case. This is indeed observed (Table 7-3). The product ratio may be affected by the addition of base, but at low base concentration this effect is usually small. Recall that strong bases are usually strong nucleophiles as well (Section 6-8), so addition of a base will generally not greatly favor deprotonation of the carbocation at the expense of nucleophilic attack, and the ratio of E1 to S_N1 products remains approximately constant. Indeed, elimination by the E1 mechanism is usually no more than a *minor side reaction* accompanying S_N1 substitution. Is there a way to make elimination the major outcome, to give alkenes as major products? Yes: By using *high* concentrations of strong base, the proportion of elimination rises dramatically. This effect does not arise from a change in the E1:S_N1 ratio, however. Instead, a new mechanistic pathway for elimination is responsible. The next section describes this process.

Exercise 7-10

When 2-bromo-2-methylpropane is dissolved in aqueous ethanol at 25°C, a mixture of $(CH_3)_3COCH_2CH_3$ (30%), $(CH_3)_3COH$ (60%), and $(CH_3)_2C=CH_2$ (10%) is obtained. Explain.

In Summary Carbocations formed in solvolysis reactions are not only trapped by nucleophiles to give S_N1 products but also deprotonated in an elimination (E1) reaction. In this process, the nucleophile (usually the solvent) acts as a base.

7-7 | BIMOLECULAR ELIMINATION: E2

In addition to S_N2, S_N1, and E1 reactions, there is a fourth pathway by which haloalkanes may react with nucleophiles *that are also strong bases*: elimination by a *bimolecular* mechanism. This method is the one employed when alkene formation is the desired outcome.

Strong bases effect bimolecular elimination

The preceding section taught us that unimolecular elimination may compete with substitution. However, a dramatic change of the kinetics is observed at higher concentrations of strong base. The rate of alkene formation becomes proportional to the concentrations of both the starting halide *and* the base: The kinetics of elimination are now second order and the process is called **bimolecular elimination**, abbreviated **E2**.

Reaction

Kinetics of the E2 Reaction of 2-Chloro-2-methylpropane

$$(CH_3)_3CCl + Na^{+-}OH \xrightarrow{k} CH_2=C(CH_3)_2 + NaCl + H_2O$$

$$\text{Rate} = k[(CH_3)_3CCl][^-OH] \text{ mol L}^{-1}\text{ s}^{-1}$$

What causes this change in mechanism? Strong bases (such as hydroxide, HO^-, and alkoxides, RO^-) can attack haloalkanes before carbocation formation. The target is a hydrogen on a carbon atom *next to* the one carrying the leaving group. This reaction pathway is not restricted to tertiary halides, although in secondary and primary systems it must compete with the S_N2 process. Section 7-8 will describe the conditions under which S_N2 or E2 reaction predominates with these substrates.

Exercise 7-11

What products do you expect from the reaction of bromocyclohexane with hydroxide ion?

Exercise 7-12

Give the products (if any) of the E2 reaction of the following substrates: CH_3CH_2I; CH_3I; $(CH_3)_3CCl$; $(CH_3)_3CCH_2I$.

E2 reactions proceed in one step

The bimolecular elimination mechanism consists of a *single step*. The bonding changes that occur in its transition state are shown here with electron-pushing arrows; in Figure 7-8, they are shown with orbitals. Three changes take place:

1. Deprotonation by the base

2. Departure of the leaving group

3. Rehybridization of the reacting carbon centers from sp^3 to sp^2 to furnish the two p orbitals of the emerging double bond

Mechanism

Model Building

The E2 Reaction Mechanism

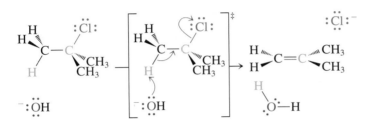

All three changes take place *simultaneously*: The E2 is a one-step, *concerted* process.

Notice that the E1 (Figure 7-7) and E2 mechanisms are very similar, differing only in the sequence of events. In the bimolecular reaction, proton abstraction and leaving-group departure are simultaneous, as depicted in the transition state above (for a Newman projection, see margin on the next page). In the E1 process, the halide leaves first, followed by an attack by

Animation

ANIMATED MECHANISM: Elimination (E2) reaction of 2-chloro-2-methylpropane

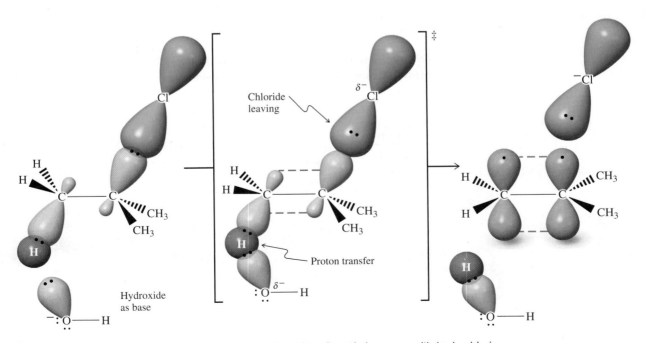

Figure 7-8 Orbital description of the E2 reaction of 2-chloro-2-methylpropane with hydroxide ion.

Newman Projection of E2 Transition State

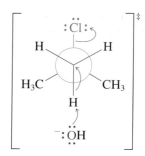

Relative Reactivity in the E2 Reaction

RI > RBr > RCl > RF

Increasing

the base. A good way of thinking about the difference is to imagine that the strong base participating in the E2 reaction is more aggressive. It does not wait for the tertiary or secondary halide to dissociate but attacks the substrate directly.

Experiments elucidate the detailed structure of the E2 transition state

What is the experimental evidence in support of a one-step process with a transition state like that depicted in Figure 7-8? There are three pieces of relevant information. First, the second-order rate law requires that both the haloalkane and the base take part in the rate-determining step. Second, better leaving groups result in faster eliminations. This observation implies that the bond to the leaving group is partially broken in the transition state.

Exercise 7-13

Explain the result in the reaction shown below.

$$Cl-\text{\large ⬡-⬡}-I \xrightarrow{CH_3O^-} Cl-\text{\large ⬡-⬡}$$

The third observation not only strongly suggests that both the C–H and the C–X bonds are broken in the transition state, it also describes their relative orientation in space when this event takes place. Figure 7-8 illustrates a characteristic feature of the E2 reaction: its stereochemistry. The substrate is pictured as reacting in a conformation that places the breaking C–H and C–X bonds in an *anti* relation. How can we establish the structure of the transition state with such precision? For this purpose, we can use the principles of conformation and stereochemistry. Treatment of *cis*-1-bromo-4-(1,1-dimethylethyl)cyclohexane with strong base leads to rapid bimolecular elimination to the corresponding alkene. In contrast, under the same conditions, the trans isomer reacts only very slowly. Why? When we examine the most stable chair conformation of the cis compound, we find that two hydrogens are located *anti* to the axial bromine substituent. This geometry is very similar to that required by the E2 transition state, and consequently elimination is easy. Conversely, the trans system has no C–H bonds aligned *anti* to the equatorial leaving group (make a model). E2 elimination in this case would require either ring-flip to a diaxial conformer (see Section 4-4) or removal of a hydrogen *gauche* to the bromine, both of which are energetically costly. The latter would be an example of an elimination proceeding through an unfavorable *syn* transition state (*syn,* Greek, together). We shall return to E2 elimination in Chapter 11.

Model Building

Anti Elimination Occurs Readily for *cis-* but Not for *trans*-1-Bromo-4-(1,1-dimethylethyl)cyclohexane

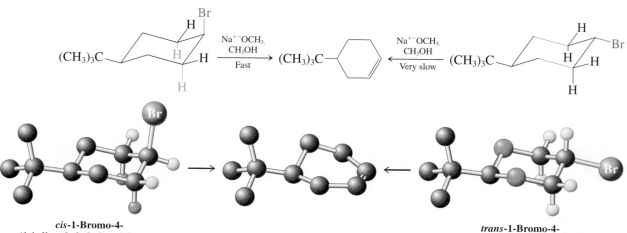

cis-1-Bromo-4-(1,1-dimethylethyl)cyclohexane
(**Two** *anti* hydrogens)

trans-1-Bromo-4-(1,1-dimethylethyl)cyclohexane
(**No** *anti* hydrogens; only *anti* ring carbons)

Solved Exercise 7-14 | **Working with the Concepts: Elimination Rates and Mechanisms**

The rate of elimination of *cis*-1-bromo-4-(1,1-dimethylethyl)cyclohexane is proportional to the concentration of both substrate and base, but that of the trans isomer is proportional *only* to the concentration of the substrate. Explain.

Remember: "Explain" = think about mechanism.

Strategy
In this problem we are given rate information that we need to explain. Throughout Chapters 6 and 7 we have seen how reaction kinetics can help define mechanism. Apply the lessons: Consider the kinetic order of each reaction and the consequences of the corresponding mechanism.

Solution
- Because the base-promoted elimination of *cis*-1-bromo-4-(1,1-dimethylethyl)cyclohexane occurs with a rate proportional to *both* substrate *and* base, it must be following the E2 pathway.
- The E2 reaction strongly favors an *anti* orientation of the leaving group with respect to the proton being removed. The graphic (at the bottom of p. 264, left) shows that this orientation is already present in the most stable chair conformation of the substrate molecule. Thus the E2 mechanism readily occurs, with the base removing an H (blue) and initiating simultaneous expulsion of Br (green).
- In contrast, in the trans isomer (graphic, right) the best conformation places the Br equatorial; the C–Br bond is *anti* to two C–C bonds of the cyclohexane ring (blue). No hydrogen on either adjacent carbon is *anti* with respect to the Br. E2 cannot readily occur from this conformation.
- For the trans isomer to undergo easy E2 reaction, it would first have to undergo ring-flip in order for the Br to become axial. The energy of this ring-flip is prohibitively unfavorable, however, because the result would be a very high-energy conformation with the bulky tertiary butyl group also in an axial position (see Table 4-3).
- Indeed, the rate of elimination from the trans isomer is proportional only to the substrate concentration and *not* that of the base, indicating that the mechanism is E1, not E2.
- Considering the possible options (below), the kinetic experiments tell us that the preferred pathway is initial dissociation of the leaving group, leading to unimolecular elimination. Flipping the chair to give the proper conformation for E2 does not compete.

Caution: In an elimination reaction, the hydrogen removed by base is one attached to a carbon atom *adjacent* to the carbon bearing the leaving group. *Do not remove an H from the same carbon that contains the leaving group!*

Conformation required for *anti* elimination: too high in energy

Product forms via E1 pathway instead

Exercise 7-15 | **Try It Yourself**

The isomer of 1,2,3,4,5,6-hexachlorocyclohexane shown in the margin undergoes E2 elimination 7000 times *more slowly* than any of its stereoisomers. Explain.

In Summary Strong bases react with haloalkanes not only by substitution, but also by elimination. The kinetics of these reactions are second order, an observation that points to a bimolecular mechanism. An *anti* transition state is preferred, in which the base abstracts a proton at the same time as the leaving group departs.

7-8 | KEYS TO SUCCESS: SUBSTITUTION VERSUS ELIMINATION—STRUCTURE DETERMINES FUNCTION

The multiple reaction pathways—S_N2, S_N1, E2, and E1—that haloalkanes may follow in the presence of nucleophiles may seem confusing. Given the many parameters that affect the relative importance of these transformations, are there some simple guidelines that might allow us to predict, at least roughly, what the outcome of any particular reaction will be? The answer is a cautious yes. This section will explain how consideration of *base strength* and *steric bulk* of the reacting species can help us decide whether substitution or elimination will predominate. We shall see that variation of these parameters may even permit control of the reaction pathway.

Weakly basic nucleophiles give substitution

Good nucleophiles that are weaker bases than hydroxide give good yields of S_N2 products with primary and secondary halides and of S_N1 products with tertiary substrates. Examples include I^-, Br^-, RS^-, N_3^-, $RCOO^-$, and PR_3. Thus, 2-bromopropane reacts with both iodide and acetate ions cleanly through the S_N2 pathway, with virtually no competing elimination.

$$\underset{\overset{|}{H}}{\overset{\overset{CH_3}{|}}{CH_3CBr}} + Na^+ I^- \xrightarrow{\text{Acetone}} \underset{\overset{|}{H}}{\overset{\overset{CH_3}{|}}{CH_3CI}} + Na^+ Br^-$$

$$\underset{\overset{|}{H}}{\overset{\overset{CH_3}{|}}{CH_3CBr}} + \overset{\overset{O}{\|}}{CH_3CO^-} Na^+ \xrightarrow{\text{Acetone}} \underset{\overset{|}{H}\;\overset{\|}{O}}{\overset{\overset{CH_3}{|}}{CH_3COCCH_3}} + Na^+ Br^-$$
$$100\%$$

Weak nucleophiles such as water and alcohols react at appreciable rates only with secondary and tertiary halides, substrates capable of following the S_N1 pathway. Unimolecular elimination is usually only a minor side reaction.

$$\underset{CH_3CH_2CHCH_2CH_3}{\overset{\overset{Br}{|}}{}} \xrightarrow{H_2O,\ CH_3OH,\ 80°C} \underset{\underset{85\%}{CH_3CH_2CHCH_2CH_3}}{\overset{OH}{\overset{|}{}}} + \underset{15\%}{CH_3CH=CHCH_2CH_3}$$

Strongly basic nucleophiles give more elimination as steric bulk increases

We have seen (Section 7-7) that strong bases may give rise to elimination through the E2 pathway. Is there some straightforward way to predict how much elimination will occur in competition with substitution in any particular situation? Yes, but other factors need to be considered. Let us examine the reactions of sodium ethoxide, a strong base, with several halides, measuring the relative amounts of ether and alkene produced in each case.

Reactions of simple primary halides with strongly basic nucleophiles give mostly S_N2 products. As steric bulk is increased around the carbon bearing the leaving group, substitution is retarded relative to elimination because an attack at carbon is subject to more steric hindrance than is an attack on hydrogen. Thus, branched primary substrates give about equal amounts of S_N2 and E2 reaction, whereas E2 is the major outcome with secondary substrates.

The S_N2 mechanism is disfavored for tertiary halides. S_N1 and E1 pathways compete under neutral or weakly basic conditions. However, high concentrations of strong base give exclusively E2 reaction.

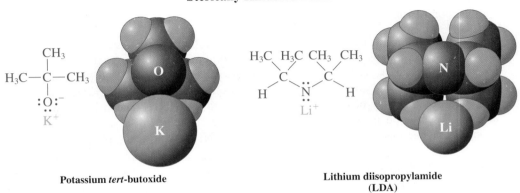

$$CH_3CH_2CH_2Br \xrightarrow[-HBr]{CH_3CH_2O^-Na^+, \, CH_3CH_2OH} CH_3CH_2CH_2OCH_2CH_3 \; + \; \overset{H_3C}{\underset{H}{\diagdown}}C=C\overset{H}{\underset{H}{\diagup}}$$

Strong base — Solvent

Ether (Substitution product) **Alkene (Elimination product)**

Primary 91% 9%

$$CH_3\overset{CH_3}{\underset{H}{\underset{|}{C}}}CH_2Br \xrightarrow[-HBr]{CH_3CH_2O^-Na^+, \, CH_3CH_2OH} CH_3\overset{CH_3}{\underset{H}{\underset{|}{C}}}CH_2OCH_2CH_3 \; + \; \overset{H_3C}{\underset{H_3C}{\diagdown}}C=C\overset{H}{\underset{H}{\diagup}}$$

Branched primary 40% 60%

$$CH_3\overset{CH_3}{\underset{H}{\underset{|}{C}}}Br \xrightarrow[-HBr]{CH_3CH_2O^-Na^+, \, CH_3CH_2OH} CH_3\overset{CH_3}{\underset{H}{\underset{|}{C}}}OCH_2CH_3 \; + \; \overset{H_3C}{\underset{H}{\diagdown}}C=C\overset{H}{\underset{H}{\diagup}}$$

Secondary 13% 87%

(left vertical arrow: Increasing steric bulk; right vertical arrow: Increasing proportion of E2 product)

Sterically hindered basic nucleophiles favor elimination

We have seen that primary haloalkanes react by substitution with good nucleophiles, including strong bases. The situation changes when the steric bulk of the nucleophile hinders attack at the electrophilic carbon. In this case, elimination may predominate, even with primary systems, through deprotonation at the less hindered periphery of the molecule.

$$CH_3CH_2CH_2CH_2Br \xrightarrow[-HBr]{(CH_3)_3CO^-K^+, \, (CH_3)_3COH} CH_3CH_2CH=CH_2 \; + \; CH_3CH_2CH_2CH_2OC(CH_3)_3$$

Strong, sterically hindered base — Solvent

85% 15%

Two examples of sterically hindered bases that are frequently employed in eliminations are potassium *tert*-butoxide and lithium diisopropylamide (LDA). The first contains a tertiary alkyl group on oxygen; the second has two secondary alkyl groups on nitrogen. For use in such reactions, these bases are frequently dissolved in their conjugate acids, 2-methyl-2-propanol (*tert*-butyl alcohol; $pK_a = 18$) and 1-methyl-*N*-(1-methylethyl)ethanamine (diisopropylamine; $pK_a = 36$), respectively.

Sterically Hindered Bases

Potassium *tert*-butoxide

Lithium diisopropylamide (LDA)

In Summary We have identified three principal factors that affect the competition between substitution and elimination: basicity of the nucleophile, steric hindrance in the haloalkane, and steric bulk around the nucleophilic (basic) atom.

Factor 1. Base strength of the nucleophile

Weak Bases	Strong Bases
H_2O,* ROH,* PR_3, halides, RS^-, N_3^-, NC^-, $RCOO^-$	HO^-, RO^-, H_2N^-, R_2N^-
Substitution more likely	Likelihood of elimination increased

Factor 2. Steric hindrance around the reacting carbon

Sterically Unhindered	Sterically Hindered
Primary haloalkanes	Branched primary, secondary, tertiary haloalkanes
Substitution more likely	Likelihood of elimination increased

Factor 3. Steric hindrance in the nucleophile (strong base)

Sterically Unhindered	Sterically Hindered
HO^-, CH_3O^-, $CH_3CH_2O^-$, H_2N^-	$(CH_3)_3CO^-$, $[(CH_3)_2CH]_2N^-$
Substitution may occur	Elimination strongly favored

For simple predictive purposes, we assume that each of these factors is of equal importance in determining the ratio of elimination to substitution. Thus, "the majority rules." This method of analysis is quite reliable. Verify that it applies to the examples of this section and the summary section that follows.

Exercise 7-16

Which nucleophile in each of the following pairs will give a higher elimination:substitution product ratio in reaction with 1-bromo-2-methylpropane?

(a) $N(CH_3)_3$, $P(CH_3)_3$ (b) H_2N^-, $(CH_3\overset{\displaystyle CH_3}{\underset{|}{CH}})_2N^-$ (c) I^-, Cl^-

---

Exercise 7-17

In all cases where substitution and elimination compete, higher reaction temperatures lead to greater proportions of elimination products. Thus, the amount of elimination accompanying hydrolysis of 2-bromo-2-methylpropane doubles as the temperature is raised from 25 to 65°C, and that from reaction of 2-bromopropane with ethoxide rises from 80% at 25°C to nearly 100% at 55°C. Explain.

7-9 | SUMMARY OF REACTIVITY OF HALOALKANES

Table 7-4 summarizes the substitution and elimination chemistry of primary, secondary, and tertiary haloalkanes. Each entry presents the major mechanism(s) observed for a given combination of substrate and nucleophile type.

Primary Haloalkanes. Unhindered primary alkyl substrates always react in a bimolecular way and almost always give predominantly substitution products, except when sterically hindered strong bases, such as potassium *tert*-butoxide, are employed. In these cases, the S_N2 pathway is slowed down sufficiently for steric reasons to allow the E2 mechanism to take over. Another way of reducing substitution is to introduce branching. However, even in these cases, good nucleophiles still furnish predominantly substitution products. Only strong bases, such as alkoxides, RO^-, or amides, R_2N^-, tend to react by elimination.

Exercise 7-18

Write the structure of the major organic product of the reaction of 1-bromopropane with (a) NaCN in acetone; (b) NaOCH$_3$ in CH$_3$OH; (c) $(CH_3)_3COK$ in $(CH_3)_3COH$.

---

*Reacts only with S_N1 substrates; no reaction with simple primary or methyl halides.

Table 7-4   Likely Mechanisms by Which Haloalkanes React with Nucleophiles (Bases)

Type of haloalkane	Poor nucleophile (e.g., H_2O)	Weakly basic, good nucleophile (e.g., I^-)	Strongly basic, unhindered nucleophile (e.g., CH_3O^-)	Strongly basic, hindered nucleophile (e.g., $(CH_3)_3CO^-$)
		Type of nucleophile (base)		
Methyl	No reaction	S_N2	S_N2	S_N2
Primary				
Unhindered	No reaction	S_N2	S_N2	E2
Branched	No reaction	S_N2	E2	E2
Secondary	Slow S_N1, E1	S_N2	E2	E2
Tertiary	S_N1, E1	S_N1, E1	E2	E2

Exercise 7-19

Write the structure of the major organic product of the reaction of 1-bromo-2-methylpropane with (a) NaI in acetone; (b) $NaOCH_2CH_3$ in CH_3CH_2OH.

Primary (and methyl) haloalkanes react so exceedingly slowly with poor nucleophiles that for practical purposes we consider the combination to give "no reaction."

Secondary Haloalkanes. Secondary alkyl systems undergo, depending on conditions, both eliminations and substitutions by either possible pathway: uni- or bimolecular. Good nucleophiles favor S_N2, strong bases result in E2, and weakly nucleophilic polar media give mainly S_N1 and E1.

Exercise 7-20

Write the structure of the major organic product of the reaction of 2-bromopropane with (a) CH_3CH_2OH; (b) $NaSCH_3$ in CH_3CH_2OH; (c) $NaOCH_2CH_3$ in CH_3CH_2OH.

Tertiary Haloalkanes. Tertiary systems eliminate (E2) with concentrated strong base and are substituted in nonbasic media (S_N1). Bimolecular substitution is almost never observed, but elimination by E1 accompanies S_N1.

Exercise 7-21

Write the structure of the major organic product of the reaction of (a) 2-bromo-2-methylbutane with water in acetone; (b) 3-chloro-3-ethylpentane with $NaOCH_3$ in CH_3OH.

Exercise 7-22

Predict which reaction in each of the following pairs will have a higher E2:E1 product ratio and explain why.

(a) $CH_3CH_2\overset{\underset{|}{CH_3}}{C}HBr \xrightarrow{CH_3OH}$? $CH_3CH_2\overset{\underset{|}{CH_3}}{C}HBr \xrightarrow{CH_3O^-Na^+, CH_3OH}$?

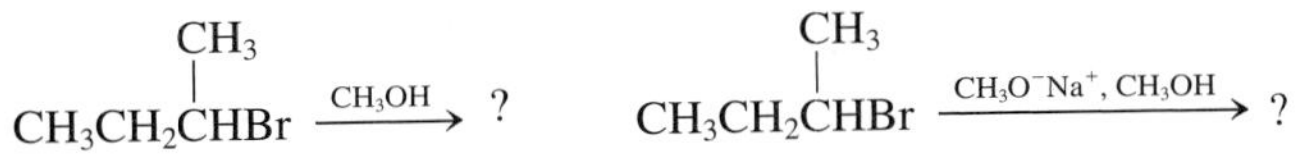

(b)

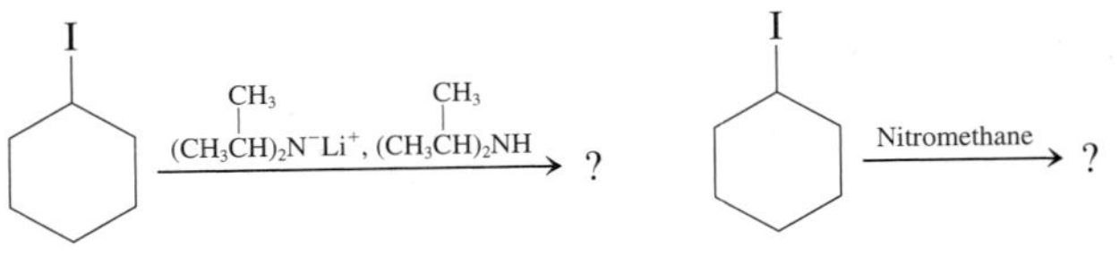

Really? Never say "never" ... really!

The oxonium ion shown below, with its positive oxygen attached to three tertiary carbons, is stable and immune to solvolysis! Evidently, its tightly constrained structure impedes unimolecular C–O bond cleavage that would give a carbocation. Basic nucleophiles give an E2 product. Equally surprisingly, azide, N_3^-, attacks the tertiary center by S_N2, the only such case reported in the literature.

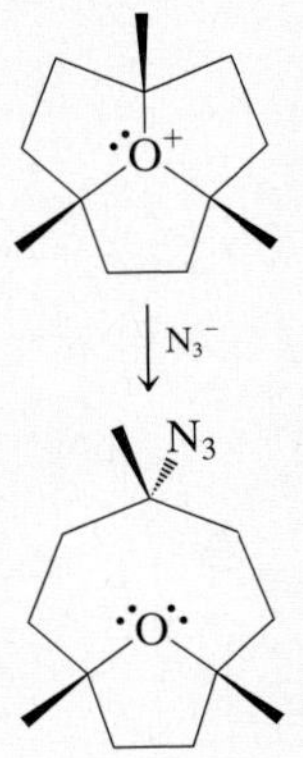

THE BIG PICTURE

We have completed our study of haloalkanes by describing three new reaction pathways—S_N1, E1, and E2—that are available to this class of compounds, in addition to the S_N2 process. We see now that *details* can make a big difference! Small changes in substrate structure can completely alter the course followed by the reaction of a haloalkane with a nucleophile. Similarly, correctly identifying a reaction product may require that we be able to assign both nucleophile and base strength to a reagent.

The final two sections in this chapter showed how the interplay of these factors affects the outcome of a reaction. However, learning to solve problems of this type correctly requires the use of accurate information. Therefore, we make the following **study suggestion:** *Practice* solving problems first with your textbook and your class notes *open,* so that you can find the correct, applicable information as you are trying to devise a solution. Remember, much of the problem solving you will be doing throughout this course will require extrapolation from examples presented to you in various contexts. If you are comfortable studying in small groups, propose reaction problems to one another, and analyze their components in order to determine what process or processes are likely to occur.

Where do we go from here? We next examine compounds containing polar carbon–oxygen single bonds, comparing them with polar carbon–halogen bonds. The alcohols will be first; they add the feature of a polar O–H bond as well. Over the next two chapters we shall see that some of the conversions of these and related compounds follow very similar mechanistic pathways, as do the reactions of haloalkanes. Such parallels between the structures of functional groups and the mechanisms of their reactions provide a useful framework for comprehending the extensive body of information in organic chemistry.

WORKED EXAMPLES: INTEGRATING THE CONCEPTS

7-23. Choosing Between Substitution and Elimination

Consider the reaction shown below. Will it proceed by substitution or by elimination? What factors determine the most likely mechanism? Write the expected product and the mechanism by which it forms.

> **Remember WHIP**
> *What*
> *How*
> *Information*
> *Proceed*

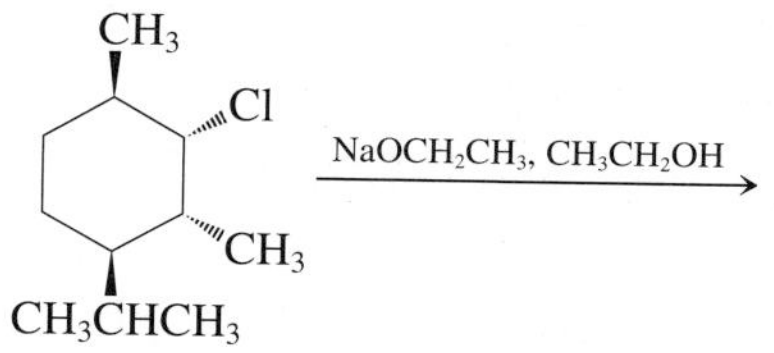

SOLUTION

The fact that the compound is cyclic does not change how we approach the problem. The substrate is secondary because the leaving group is attached to a secondary carbon. The nucleophile (ethoxide ion, derived from $NaOCH_2CH_3$) is a good one, but, like hydroxide and methoxide (see Table 6-4), it is also a strong base. Based on the evaluation criteria in Sections 7-8 and 7-9, the combination of a secondary substrate and a strongly basic nucleophile will favor elimination by the E2 pathway (Table 7-4). This mechanism has specific requirements with regard to the relative geometric arrangement of the leaving group and the hydrogen being removed by the base: They must possess an *anti* conformation with respect to the carbon–carbon bond between them (Section 7-7). In order to prepare a reasonable structural representation of the molecule, we must make use of what we learned in Chapter 4 (Section 4-4) about conformations of substituted cyclohexanes. The best available shape is the chair shown below, in which two of the substituents, including the bulky 1-methylethyl (isopropyl) group, are equatorial. Note that the leaving group is axial, and one of the neighboring carbons possesses an axial hydrogen, positioned in the required orientation *anti* to the chlorine:

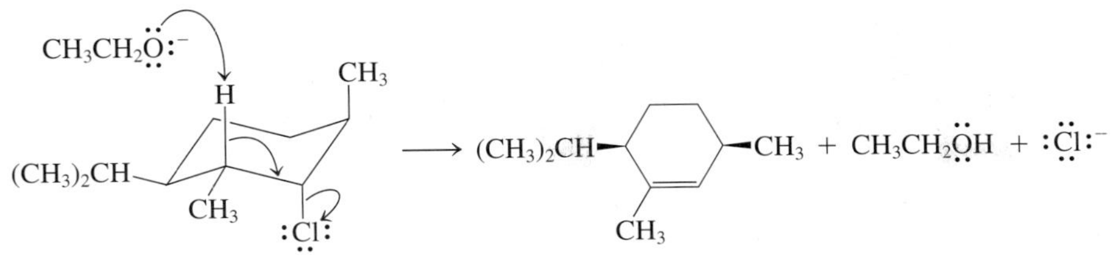

We draw curved arrows appropriate to the one-step E2 mechanism, arriving at the product structure shown, a cyclic alkene, with the double bond located in the indicated position. The methyl group on the alkene carbon moves into the plane of the double bond (recall from Section 1-8 that alkene carbons exhibit trigonal planar geometry—sp^2 hybridization). Notice that the remaining two alkyl groups, which started cis, remain cis with respect to each other, because no chemical change occurred to the ring carbons on which they were located.

Avoid the very common error of removing a hydrogen from the wrong carbon atom in an E2 process. In the correct mechanism the hydrogen comes off a carbon atom *adjacent* to that bearing the leaving group; *never* from the same carbon.

7-24. Reactivity of a Tertiary Substrate as a Function of Solvent and Nucleophile

a. 2-Bromo-2-methylpropane (*tert*-butyl bromide) reacts readily in nitromethane with chloride and iodide ions.

1. Write the structures of the substitution products, and write the complete mechanism by which one of them is formed.

2. Assume equal concentrations of all reactants, and predict the relative rates of these two reactions.

3. Which reaction will give more elimination? Write its mechanism.

SOLUTION

1. We begin by analyzing the participating species and then recognizing just what kind of reaction is likely to take place. The substrate is a haloalkane with a good leaving group attached to a *tertiary* carbon. According to Table 7-4, displacement by the S_N2 mechanism is not a probable option, but S_N1, E1, and E2 processes are possibilities. Chloride and iodide are good nucleophiles and weak bases, suggesting that substitution should predominate to give $(CH_3)_3CCl$ and $(CH_3)_3CI$ as the products, respectively (Section 7-8). The mechanism (S_N1) is as shown in Section 7-2 except that, subsequent to initial ionization of the C–Br bond to give the carbocation, halide ion attacks at carbon to give the final product directly. The very polar nitromethane is a good solvent for S_N1 reactions (Section 7-4).

2. We learned in Section 7-4 that different nucleophilic power has no effect on the rates of unimolecular processes. The rates should be (and, experimentally, are) identical.

3. This part requires a bit more thought. According to Table 7-4 and Section 7-8, elimination by the E1 pathway always accompanies S_N1 displacement. However, increasing the base strength of the nucleophile may "turn on" the E2 mechanism, increasing the proportion of elimination product. Referring to Tables 6-4 and 6-7, we see that chloride is more basic (and less nucleophilic) than iodide. More elimination is indeed observed with chloride than with iodide. The mechanism is as shown in Figure 7-7, with chloride acting as the base to remove a proton from the carbocation.

b. The table in the margin presents data for the reactions that take place when the chloro compound shown here is dissolved in acetone containing varying quantities of water and sodium azide, NaN_3:

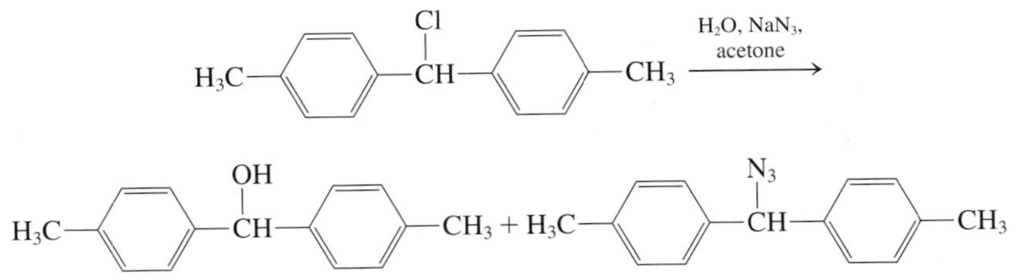

H_2O %	[N_3^-]	RN_3 %	k_{rel}
10	0 M	0	1
10	0.05 M	60	1.5
15	0.05 M	60	7
20	0.05 M	60	22
50	0.05 M	60	*
50	0.10 M	75	*
50	0.20 M	85	*
50	0.50 M	95	*

*Too fast to measure.

In the table, $H_2O\%$ is the percentage of water by volume in the solvent, $[N_3^-]$ is the initial concentration of sodium azide, $RN_3\%$ is the percentage of organic azide in the product mixture (the remainder is the alcohol), and k_{rel} is the relative rate constant for the reaction, derived from the rate at which the starting material is consumed. The initial concentration of substrate is 0.04 M in all experiments. Answer the following questions.

1. Describe and explain the effects of changing the percentage of H_2O on the rate of the reaction and on the product distribution.

2. Do the same for the effects of changing $[N_3^-]$. Additional information: The reaction rates shown are the same when other ions—for example, Br^- or I^-—are used instead of azide.

SOLUTION

1. We begin by examining the data in the table, specifically rows 2–5, which compare reactions in the presence of different amounts of water at constant azide concentration. The rate of substitution increases rapidly as the proportion of water goes up, but the ratio of the two products stays the same: 60% azide and 40% alcohol. These two results suggest that the only effect of increasing the amount of water is to make the solvent environment more polar, thereby speeding up the initial ionization of the substrate. Even when the proportion of water is only 10%, it is present in great excess and is trapping carbocations as fast as it can, relative to the rate at which azide ion reacts with the same intermediates (Section 7-2).

2. We note from rows 1 and 2 in the table that the reaction rate rises by about 50% when NaN_3 is added. Without further information, we might assume that this effect is a consequence of the occurrence of the S_N2 mechanism. Were that to be the case, however, other anions should affect the rate differently. However, we were told that bromide and iodide, far more powerful nucleophiles, affect the measured rate in exactly the same way as does azide. We can explain this observation only by assuming that displacement is entirely by the S_N1 mechanism, and added ions affect the rate only by increasing the polarity of the solution and speeding up ionization (Section 7-4).

In rows 5–8 of the table, we note that increasing the amount of azide ion increases the amount of azide-containing product that is formed. At the higher concentrations, azide, a better nucleophile than water, is better able to compete for reaction with the carbocation intermediate.

New Reactions

1. Bimolecular Substitution—S_N2 (Sections 6-2 through 6-11, 7-5)

Primary and secondary substrates only

Direct backside displacement with 100% inversion of configuration

2. Unimolecular Substitution—S_N1 (Sections 7-1 through 7-5)

Secondary and tertiary substrates only

Through carbocation: Chiral systems are racemized

3. Unimolecular Elimination—E1 (Section 7-6)

Secondary and tertiary substrates only

Through carbocation

4. Bimolecular Elimination—E2 (Section 7-7)

$$CH_3CH_2CH_2I \xrightarrow{:B^-} CH_3CH=CH_2 + BH + I^-$$

Simultaneous elimination of leaving group and neighboring proton

Important Concepts

1. Secondary haloalkanes undergo slow and tertiary haloalkanes undergo fast **unimolecular substitution** in polar media. When the solvent serves as the nucleophile, the process is called **solvolysis.**

2. The slowest, or rate-determining, step in unimolecular substitution is dissociation of the C–X bond to form a **carbocation** intermediate. Added strong nucleophile changes the product but not the reaction rate.

3. Carbocations are stabilized by **hyperconjugation:** Tertiary are the most stable, followed by secondary. Primary and methyl cations are too unstable to form in solution.

4. **Racemization** often results when unimolecular substitution takes place at a chiral carbon.

5. **Unimolecular elimination** to form an alkene accompanies substitution in secondary and tertiary systems.

6. High concentrations of strong base may bring about **bimolecular elimination.** Expulsion of the leaving group accompanies removal of a hydrogen from the neighboring carbon by the base. The stereo chemistry indicates an *anti* conformational arrangement of the hydrogen and the leaving group.

7. Substitution is favored by unhindered substrates and small, less basic nucleophiles.

8. Elimination is favored by hindered substrates and bulky, more basic nucleophiles.

Problems

25. What is the major substitution product of each of the following solvolysis reactions?

(a) $CH_3\underset{\underset{CH_3}{|}}{\overset{\overset{CH_3}{|}}{C}}Br \xrightarrow{CH_3CH_2OH}$

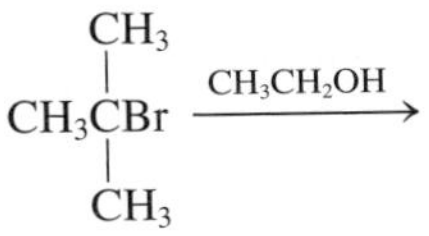

(b) $(CH_3)_2\underset{\underset{}{}}{\overset{\overset{Br}{|}}{C}}CH_2CH_3 \xrightarrow{CF_3CH_2OH}$

(c) $\xrightarrow{CH_3OH}$

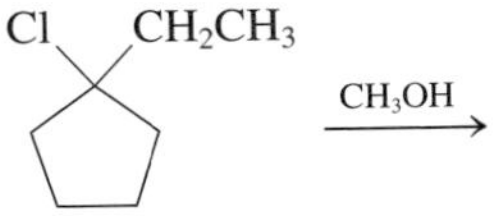

(d) $\xrightarrow{HCOH}$

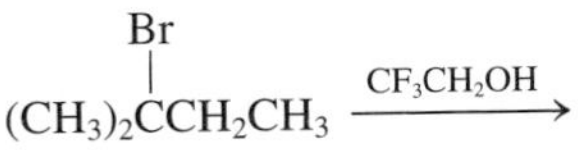

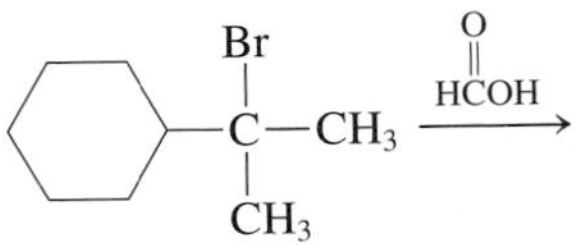

(e) $CH_3\underset{\underset{CH_3}{|}}{\overset{\overset{CH_3}{|}}{C}}Cl \xrightarrow{D_2O}$

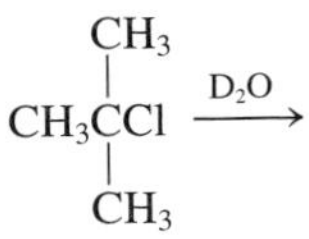

(f) $CH_3\underset{\underset{CH_3}{|}}{\overset{\overset{CH_3}{|}}{C}}Cl \xrightarrow{OD}$

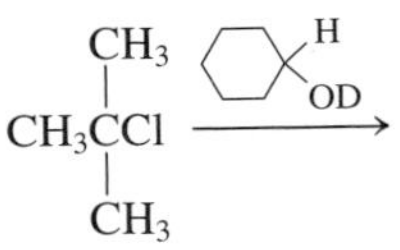

26. For each reaction presented in Problem 25, write out the complete, step-by-step mechanism using curved-arrow notation. Be sure to show each individual step of each mechanism *separately,* and show the complete structures of the products of that step before going on to the next.

27. Write the two major substitution products of the reaction shown below. (a) Write a mechanism to explain the formation of each of them. (b) Monitoring the reaction mixture reveals that an *isomer* of the starting material is generated as an intermediate. Draw its structure and explain how it is formed.

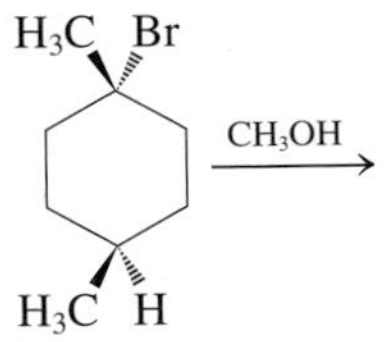

28. Give the two major substitution products of the following reaction.

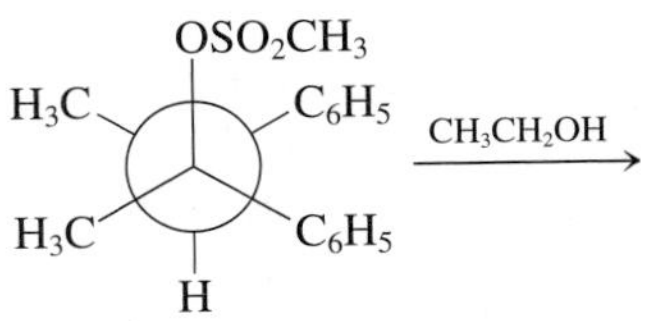

29. How would each reaction in Problem 25 be affected by the addition of each of the following substances to the solvolysis mixture?

(a) H_2O (b) KI

(c) NaN_3 (d) $CH_3CH_2OCH_2CH_3$ (**Hint:** Low polarity.)

30. Rank the following carbocations in decreasing order of stability.

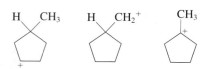

31. Rank the compounds in each of the following groups in order of decreasing rate of solvolysis in aqueous acetone.

(a) $CH_3CHCH_2CH_2Cl$ $CH_3CHCHCH_3$ $CH_3CCH_2CH_3$
(with CH$_3$ and Cl substituents as shown)

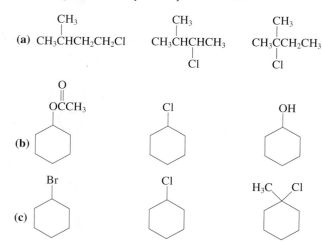

(b)

(c)

32. Give the products of the following substitution reactions. Indicate whether they arise through the S_N1 or the S_N2 process. Formulate the detailed mechanisms of their generation.

(a) $(CH_3)_2CHOSO_2CF_3 \xrightarrow{CH_3CH_2OH}$

(b) (cyclopentane with CH$_3$ and Br) $\xrightarrow{\text{Excess } CH_3SH, CH_3OH}$

(c) $CH_3CH_2CH_2CH_2Br \xrightarrow{(C_6H_5)_3P, DMSO}$

(d) $CH_3CH_2CHClCH_2CH_3 \xrightarrow{NaI, acetone}$

33. Give the product of each of the following substitution reactions. Which of these transformations should proceed faster in a polar, aprotic solvent (such as acetone or DMSO) than in a polar, protic solvent (such as water or CH_3OH)? Explain your answer on the basis of the mechanism that you expect to be operating in each case.

(a) $CH_3CH_2CH_2Br + Na^+ \ ^-CN \longrightarrow$
(b) $(CH_3)_2CHCH_2I + Na^+ N_3^- \longrightarrow$
(c) $(CH_3)_3CBr + HSCH_2CH_3 \longrightarrow$
(d) $(CH_3)_2CHOSO_2CH_3 + HOCH(CH_3)_2 \longrightarrow$

34. Propose a synthesis of (R)-$CH_3CHN_3CH_2CH_3$, starting from (R)-2-chlorobutane.

35. Two substitution reactions of (S)-2-bromobutane are shown here. Show their stereochemical outcomes.

(S)-$CH_3CH_2CHBrCH_3 \xrightarrow{HCOH}$

(S)-$CH_3CH_2CHBrCH_3 \xrightarrow{HCO^-Na^+, DMSO}$

36. Propose a stereocontrolled synthesis of *cis*-1-acetoxy-3-methyl-cyclopentane (below), starting from *trans*-1-chloro-3-methylcyclopentane.

cis-**1-Acetoxy-
3-methylcyclopentane**

37. The two seemingly similar reactions shown below differ in their outcomes.

$CH_3CH_2CH_2CH_2Br \xrightarrow{NaOH, CH_3CH_2OH} CH_3CH_2CH_2CH_2OH$

$CH_3CH_2CH_2CH_2Br \xrightarrow{NaSH, CH_3CH_2OH} CH_3CH_2CH_2CH_2SH$

The first proceeds in high yield. The yield of the product in the second, however, is diminished by the formation of $(CH_3CH_2CH_2CH_2)_2S$ in substantial quantities. Discuss the formation of this by-product mechanistically, and explain why it occurs in the second case but not in the first.

38. Write all possible E1 products of each reaction in Problem 25.

39. Formulate the complete step-by-step mechanisms for all the E1 processes that you identified in Problem 38.

40. Write the products of the following elimination reactions. Specify the predominant mechanism (E1 or E2) and formulate it in detail.

(a) $(CH_3CH_2)_3CBr \xrightarrow{NaNH_2, NH_3}$

(b) $CH_3CH_2CH_2CH_2Cl \xrightarrow{KOC(CH_3)_3, (CH_3)_3COH}$

(c) (diphenyl CH with Br) $\xrightarrow{\text{Excess KOH, } CH_3CH_2OH}$

(d) (cyclohexane with Cl and CH$_3$) $\xrightarrow{NaOCH_3, CH_3OH}$

41. From the list of reagents (a)–(f) below, choose all those that are most likely to give primarily (i) S_N2 reaction with primary RX; (ii) E2 reaction with primary RX; (iii) S_N2 reaction with secondary RX; (iv) E2 reaction with secondary RX.

(a) $NaSCH_3$ in CH_3OH
(b) $(CH_3)_2CHOLi$ in $(CH_3)_2CHOH$
(c) $NaNH_2$ in liquid NH_3
(d) KCN in DMSO

(e) (piperidine anion with Li$^+$) in (piperidine)

(f) $CH_3CH_2CH_2CONa$ in DMF

42. Predict the major product(s) that should form from reaction between 1-bromobutane and each of the following substances. By which reaction mechanism is each formed—S_N1, S_N2, E1, or E2? If it appears that a reaction will either not take place or be exceedingly slow, write "no reaction." Assume that each reagent is present in large excess. The solvent for each reaction is given.

(a) KCl in DMF
(b) KI in DMF
(c) KCl in CH_3NO_2
(d) NH_3 in CH_3CH_2OH
(e) $NaOCH_2CH_3$ in CH_3CH_2OH
(f) CH_3CH_2OH
(g) $KOC(CH_3)_3$ in $(CH_3)_3COH$
(h) $(CH_3)_3P$ in CH_3OH
(i) CH_3CO_2H

43. Predict the major product(s) and mechanism(s) for reaction between 2-bromobutane (*sec*-butyl bromide) and each of the reagents in Problem 42.

44. Predict the major product(s) and mechanism(s) for reaction between 2-bromo-2-methylpropane (*tert*-butyl bromide) and each of the reagents in Problem 42.

45. Three reactions of 2-chloro-2-methylpropane are shown here. **(a)** Write the major product of each transformation. **(b)** Compare the rates of the three reactions. Assume identical solution polarities and reactant concentrations. Explain mechanistically.

$$(CH_3)_3CCl \xrightarrow{H_2S,\ CH_3OH}$$

$$(CH_3)_3CCl \xrightarrow{CH_3\overset{O}{\overset{\|}{C}}O^-K^+,\ CH_3OH}$$

$$(CH_3)_3CCl \xrightarrow{CH_3O^-K^+,\ CH_3OH}$$

46. Give the major product(s) of the following reactions. Indicate which of the following mechanism(s) is in operation: S_N1, S_N2, E1, or E2. If no reaction takes place, write "no reaction."

(a) (cyclopentane with CH_2Cl and H) $\xrightarrow[(CH_3)_3COH]{KOC(CH_3)_3,}$

(b) $CH_3\overset{F}{\overset{|}{C}}HCH_2CH_3 \xrightarrow[acetone]{KBr,}$

(c) $H_3C\overset{CH_2CH_3}{\underset{H}{-\overset{|}{\underset{|}{C}}-}}Br \xrightarrow{H_2O}$

(d) (cyclohexane with I) $\xrightarrow{NaNH_2,\ liquid\ NH_3}$

(e) $(CH_3)_2CHCH_2CH_2CH_2Br \xrightarrow{NaOCH_2CH_3,\ CH_3CH_2OH}$

(f) $H_3C\overset{Br}{\underset{CH_2CH_3}{\overset{|}{\underset{|}{\cdots C\cdots}}}}CH_2CH_2CH_3 \xrightarrow{NaI,\ nitromethane}$

(g) (cyclopentane with OH and H) $\xrightarrow[CH_3CH_2OH]{KOH,}$

(h) $Cl-$(cyclohexane)$-CH_2CH_2CH_2Br \xrightarrow[CH_3OH]{Excess\ KCN,}$

(i) OSO_2-(benzene ring)$-CH_3 \xrightarrow{NaSH,\ CH_3CH_2OH}$
(R)-$CH_3CH_2CHCH_3$

(j) (cyclohexane with I and CH_2CH_3) $\xrightarrow{CH_3OH}$

(k) $(CH_3)_3CC\overset{Br}{\overset{|}{H}}CH_3 \xrightarrow{KOH,\ CH_3CH_2OH}$

(l) $CH_3CH_2Cl \xrightarrow{CH_3\overset{O}{\overset{\|}{C}}OH}$

47. Fill in the blanks in the following table with the major product(s) of the reaction of each haloalkane with the reagents shown.

Haloalkane	Reagent			
	H_2O	$NaSeCH_3$	$NaOCH_3$	$KOC(CH_3)_3$
CH_3Cl	___	___	___	___
$CH_3CH_2CH_2Cl$	___	___	___	___
$(CH_3)_2CHCl$	___	___	___	___
$(CH_3)_3CCl$	___	___	___	___

48. Indicate the major mechanism(s) (simply specify S_N2, S_N1, E2, or E1) required for the formation of each product that you wrote in Problem 47.

49. For each of the following reactions, indicate whether the reaction would work well, poorly, or not at all. Formulate alternative products, if appropriate.

(a) $CH_3CH_2\overset{}{\underset{Br}{\overset{|}{C}H}}CH_3 \xrightarrow{NaOH,\ acetone} CH_3CH_2\overset{}{\underset{OH}{\overset{|}{C}H}}CH_3$

(b) $CH_3\overset{H_3C}{\overset{|}{C}}HCH_2Cl \xrightarrow{CH_3OH} CH_3\overset{H_3C}{\overset{|}{C}}HCH_2OCH_3$

(c) (cyclohexane with H, Cl) $\xrightarrow{HCN,\ CH_3OH}$ (cyclohexane with H, CN)

(d) $CH_3-\overset{CH_3}{\underset{CH_3SO_2O}{\overset{|}{\underset{|}{C}}}}-CH_2CH_2CH_2CH_2OH \xrightarrow{Nitromethane}$ H_3C-(tetrahydropyran ring with O)$-H_3C$

(e) (cyclopentane with H_3C, CH_2I) $\xrightarrow{NaSCH_3,\ CH_3OH}$ (cyclopentane with H_3C, CH_2SCH_3)

(f) $CH_3CH_2CH_2Br \xrightarrow{NaN_3,\ CH_3OH} CH_3CH_2CH_2N_3$

(g) $(CH_3)_3CCl \xrightarrow{NaI,\ nitromethane} (CH_3)_3CI$

(h) $(CH_3CH_2)_2O \xrightarrow{CH_3I} (CH_3CH_2)_2\overset{+}{O}CH_3 + I^-$

(i) $CH_3I \xrightarrow{CH_3OH} CH_3OCH_3$

(j) $(CH_3CH_2)_3COCH_3 \xrightarrow{NaBr,\ CH_3OH} (CH_3CH_2)_3CBr$

(k) $CH_3\overset{CH_3}{\overset{|}{C}}HCH_2CH_2Cl \xrightarrow{NaOCH_2CH_3,\ CH_3CH_2OH} CH_3\overset{CH_3}{\overset{|}{C}}HCH=CH_2$

(l) $CH_3CH_2CH_2CH_2Cl \xrightarrow{NaOCH_2CH_3,\ CH_3CH_2OH} CH_3CH_2CH=CH_2$

50. Propose syntheses of the following molecules from the indicated starting materials. Make use of any other reagents or solvents that you need. In some cases, there may be no alternative but to employ a reaction that results in a mixture of products. If so, use reagents and conditions that will maximize the yield of the desired material (compare Problem 53 of Chapter 6).

(a) $CH_3CH_2CHICH_3$, from butane

(b) $CH_3CH_2CH_2CH_2I$, from butane

(c) $(CH_3)_3COCH_3$, from methane and 2-methylpropane

(d) Cyclohexene, from cyclohexane

(e) Cyclohexanol, from cyclohexane

(f) ⟨S...S ring⟩, from 1,3-dibromopropane

51. **CHALLENGE** [(1-Bromo-1-methyl)ethyl]benzene, shown below, undergoes solvolysis in a unimolecular, strictly first-order process. The reaction rate for [RBr] = 0.1 M RBr in 9:1 acetone:water is measured to be 2×10^{-4} mol $L^{-1}s^{-1}$. **(a)** Calculate the rate constant k from these data. What is the product of this reaction? **(b)** In the presence of 0.1 M LiCl, the rate is found to increase to 4×10^{-4} mol $L^{-1}s^{-1}$, although the reaction still remains strictly first order. Calculate the new rate constant k_{LiCl} and suggest an explanation. **(c)** When 0.1 M LiBr is present instead of LiCl, the measured rate *drops* to 1.6×10^{-4} mol $L^{-1}s^{-1}$. Explain this observation, and write the appropriate chemical equations to describe the reactions.

$$RBr = \text{⟨phenyl⟩}-\underset{\underset{CH_3}{|}}{\overset{\overset{CH_3}{|}}{C}}-Br$$

52. In this chapter we have encountered many examples of S_N1 solvolysis reactions, all of which proceed according to the following scheme:

$$R\overset{\curvearrowright}{-}X \xrightarrow{\text{Rate}_1 = k_1[RX]} X^- + R^+ \xrightarrow{\text{Rate}_2 = k_2[R^+][Nu\text{:}]} R\overset{+}{-}\ddot{O}H_2$$

Loss of a proton gives the final product. Although there is considerable evidence for the intermediacy of a carbocation, it is not directly observed normally because its combination with a nucleophile is so rapid. Recently, examples of S_N1 solvolyses have been found that give rise to very unusual observations. One example is

$$CH_3O-\text{⟨phenyl⟩}-\underset{\underset{H}{|}}{\overset{\overset{Cl}{|}}{C}}-\text{⟨phenyl⟩}-OCH_3 \xrightarrow{CF_3CH_2OH}$$

$$CH_3O-\text{⟨phenyl⟩}-\underset{\underset{H}{|}}{\overset{\overset{CH_2CF_3}{\overset{|}{O}}}{C}}-\text{⟨phenyl⟩}-OCH_3$$

Upon mixing the colorless substrate and solvent, a reddish-orange color is observed immediately, signaling the formation of an intermediate carbocation. This color fades over a period of about a minute, and analysis of the solution reveals the presence of the final product in 100% yield. **(a)** There are two reasons for the buildup of a detectable concentration of carbocation in this case. One is that the carbocation derived from dissociation of this particular substrate is unusually stable (for reasons we will explore in Chapter 22). The other is that the solvent (2,2,2-trifluoroethanol) is an

unusually poor nucleophile, even compared with ordinary alcohols such as ethanol. Suggest an explanation for the poor nucleophilicity of the solvent. **(b)** What can you say about the relative rates of the two steps (rate$_1$ and rate$_2$), and how do they compare to those in the usual S_N1 reaction mechanism? **(c)** How might increasing carbocation stability and decreasing solvent nucleophilicity affect the relative magnitudes of rate$_1$ and rate$_2$ in an S_N1 process? **(d)** Write the complete mechanism for the reaction above.

53. Match each of the following transformations to the correct reaction profile shown here, and draw the structures of the species present at all points on the energy curves marked by capital letters.

(i)

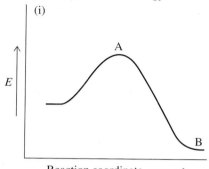

(ii)

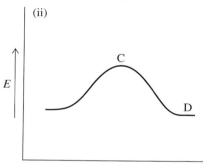

(iii)

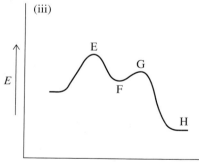

(iv)

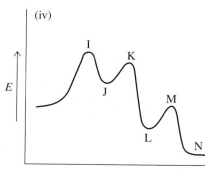

(a) $(CH_3)_3CCl + (C_6H_5)_3P \longrightarrow$
(b) $(CH_3)_2CHI + KBr \longrightarrow$
(c) $(CH_3)_3CBr + HOCH_2CH_3 \longrightarrow$
(d) $CH_3CH_2Br + NaOCH_2CH_3 \longrightarrow$

54. Formulate the structure of the most likely product of the following reaction of 4-chloro-4-methyl-1-pentanol in neutral polar solution.

$$(CH_3)_2\overset{\overset{\displaystyle Cl}{|}}{C}CH_2CH_2CH_2OH \longrightarrow HCl + C_6H_{12}O$$

In strongly *basic* solution, the starting material again converts into a molecule with the molecular formula $C_6H_{12}O$, but with a completely different structure. What is it? Explain the difference between the two results.

55. The following reaction can proceed through both E1 and E2 mechanisms.

$$C_6H_5CH_2\overset{\overset{\displaystyle CH_3}{|}}{\underset{\underset{\displaystyle CH_3}{|}}{C}}Cl \xrightarrow{\text{NaOCH}_3, \text{ CH}_3\text{OH}} C_6H_5CH=C(CH_3)_2 + C_6H_5CH_2\overset{\overset{\displaystyle CH_3}{|}}{C}=CH_2$$

The E1 rate constant $k_{E1} = 1.4 \times 10^{-4}\,s^{-1}$ and the E2 rate constant $k_{E2} = 1.9 \times 10^{-4}\,L\,mol^{-1}\,s^{-1}$; 0.02 M haloalkane. **(a)** What is the predominant elimination mechanism with 0.5 M $NaOCH_3$? **(b)** What is the predominant elimination mechanism with 2.0 M $NaOCH_3$? **(c)** At what concentration of base does exactly 50% of the starting material react by an E1 route and 50% by an E2 pathway?

56. The compound below is an example of a methyl ester. Methyl esters react with lithium iodide to give lithium carboxylate salts. The solvent in this example is pyridine (below).

Pyridine

$$\xrightarrow[\text{pyridine}]{\text{Li}^+\text{I}^-} \quad + \quad CH_3I$$

Suggest several experiments that would allow you to determine the likely mechanism of this process.

57. **CHALLENGE** Ethers containing the 1,1-dimethylethyl (*tert*-butyl) group are readily cleaved with dilute, strong acid, as shown in the example below.

$$\xrightarrow{CF_3CO_2H,\ H_2O}$$

$$\text{—OH} + \overset{H_3C}{\underset{H_3C}{>}}C=CH_2$$

Suggest a plausible mechanism for this process. What role might the strong acid play?

58. Give the mechanism and major product for the reaction of a secondary haloalkane in a polar aprotic solvent with the following nucleophiles. The pK_a value of the conjugate acid of the nucleophile is given in parentheses.

(a) N_3^- (4.6) (b) H_2N^- (35)
(c) NH_3 (9.5) (d) HSe^- (3.7)
(e) F^- (3.2) (f) $C_6H_5O^-$ (9.9)
(g) PH_3 (−12) (h) NH_2OH (6.0)
(i) NCS^- (−0.7)

59. Cortisone is an important steroidal anti-inflammatory agent. Cortisone can be synthesized efficiently from the alkene shown here.

Alkene   **Cortisone**

Of the following three chlorinated compounds, two give reasonable yields of the alkene shown above by E2 elimination with base, but one does not. Which one does not work well, and why? What does it give during attempted E2 elimination? (**Hint:** Consider the geometry of each system.)

A   **B**

C

60. **CHALLENGE** The chemistry of derivatives of *trans*-decalin is of interest because this ring system is part of the structure of steroids. Make models of the brominated systems (i and ii) to help you answer the following questions.

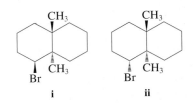

i   ii

(a) One of the molecules undergoes E2 reaction with $NaOCH_2CH_3$ in CH_3CH_2OH considerably faster than does the other. Which molecule is which? Explain. (b) The following deuterated analogs of systems i and ii react with base to give the products shown.

i-deuterated

$\xrightarrow{NaOCH_2CH_3, \ CH_3CH_2OH}$

(All D retained)

ii-deuterated

$\xrightarrow{NaOCH_2CH_3, \ CH_3CH_2OH}$

(All D lost)

Specify whether *anti* or *syn* eliminations have taken place. Draw the conformations that the molecules must adopt for elimination to occur. Does your answer to (b) help you in solving (a)?

Team Problem

61. Consider the general substitution-elimination reactions of the bromoalkanes.

$$R\text{—}Br \xrightarrow{Nu/base} R\text{—}Nu + alkene$$

How do the reaction mechanisms and product formation differ when the structure of the substrate and reaction conditions change? To begin to unravel the nuances of bimolecular and unimolecular substitution and elimination reactions, focus on the treatment of bromoalkanes A through D under conditions (a) through (e). Divide the problem evenly among yourselves so that each of you tackles the questions of reaction mechanism(s) and qualitative distribution of product(s), if any. Reconvene to discuss your conclusions and come to a consensus. When you are explaining a reaction mechanism to the rest of the team, use curved arrows to show the flow of electrons. Label the stereochemistry of starting materials and products as R or S, as appropriate.

A   B   C   D

(a) NaN_3, DMF   (b) LDA, DMF

(c) NaOH, DMF   (d) $CH_3\overset{O}{\overset{\|}{C}}O^-Na^+$, $CH_3\overset{O}{\overset{\|}{C}}OH$

(e) CH_3OH

Preprofessional Problems

62. Which of the following haloalkanes will undergo hydrolysis most rapidly?

(a) $(CH_3)_3CF$ (b) $(CH_3)_3CCl$

(c) $(CH_3)_3CBr$ (d) $(CH_3)_3CI$

63. The reaction

$$(CH_3)_3CCl \xrightarrow{CH_3O^-} \underset{H_3C}{\overset{H_3C}{>}}C{=}CH_2$$

is an example of which of the following processes?

(a) E1   (b) E2   (c) S_N1 (d) S_N2

64. In this transformation,

$$A \xrightarrow{H_2O, \ acetone} CH_3CH_2\underset{OH}{\overset{}{C}}(CH_3)_2$$

what is the best structure for A?

(a) $BrCH_2CH_2CH(CH_3)_2$

(b) $CH_3CH_2\underset{CH_3}{\overset{CH_3}{C}}Br$

(c) $CH_3CH_2\underset{CH_2Br}{\overset{CH_3}{C}H}$

(d) $CH_3CHCH(CH_3)_2$ with Br below

65. Which of the following isomeric carbocations is the most stable?

(a)   (b)

(c)   (d)

66. Which reaction intermediate is involved in the following reaction?

$$2\text{-Methylbutane} \xrightarrow{Br_2, \ h\nu} \underset{\text{(not the major product)}}{2\text{-bromo-3-methylbutane}}$$

(a) A secondary radical   (b) A tertiary radical

(c) A secondary carbocation   (d) A tertiary carbocation

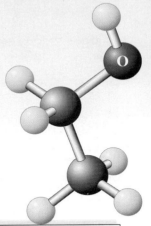

CHAPTER 8 | Hydroxy Functional Group: Alcohols

Properties, Preparation, and Strategy of Synthesis

What is your first thought when you hear the word "alcohol"? Undoubtedly, whether pleasant or not, it is connected in some way to ethanol, which is the alcohol in alcoholic beverages. The euphoric effects of (limited) ethanol consumption have been known and purposely used for thousands of years. This is perhaps not surprising, because ethanol is naturally generated by the fermentation of carbohydrates. For example, the addition of yeast to an aqueous sugar solution leads to the evolution of CO_2 and the formation of ethanol.

$$C_6H_{12}O_6 \xrightarrow{\text{Yeast enzymes}} 2\ CH_3CH_2OH + 2\ CO_2$$

Sugar   **Ethanol**

Fermentation is currently employed on a large scale to supply the vast quantities of ethanol needed as a renewable "green" fuel source, so-called bioethanol or, as a 10% additive to gasoline, gasohol. A variety of feed stocks, such as sugar cane, corn, switchgrass, and straw, can be converted quite efficiently to ethanol in this way (see Real Life 3-1). Worldwide production in 2012 is estimated at 40 billion liters.

Ethanol is a member of a large family of compounds called **alcohols.** This chapter introduces you to some of their chemistry. From Chapter 2, we know that alcohols have carbon backbones bearing the substituent OH, the **hydroxy**

The fermentation of the juice from crushed grapes produces ethanol in wine, in this picture from Bulgaria, made the traditional way.

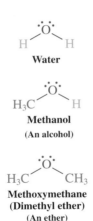

Water

Methanol
(An alcohol)

Methoxymethane
(Dimethyl ether)
(An ether)

Ground corn being loaded onto a barge on the Mississippi River after it has been used for ethanol production. This product is known as distiller's grain and constitutes commercial feed for cattle.

functional group. Alcohols may be viewed as derivatives of water in which one hydrogen has been replaced by an alkyl group. Replacement of the second hydrogen gives an **ether** (Chapter 9). The hydroxy function is readily converted to other functional groups, such as the C=C double bond in alkenes (Chapters 7, 9, and 11) or the C=O bond in aldehydes and ketones (this chapter and Chapter 17).

Alcohols are abundant in nature and varied in structure (see, e.g., Section 4-7). Simple alcohols are used as solvents; others aid in the synthesis of more complex molecules. They are a good example of how functional groups shape the structure and function of organic compounds.

Our discussion begins with the naming of alcohols, followed by a brief description of their structures and other physical properties, particularly in comparison with those of the alkanes and haloalkanes. Finally, we examine the preparation of alcohols, which will introduce our first study of strategies for efficiently synthesizing new organic compounds.

8-1 | NAMING THE ALCOHOLS

Like other compounds, alcohols may have both systematic and common names. Systematic nomenclature treats alcohols as derivatives of alkanes. The ending *-e* of the alkane is replaced by **-ol.** Thus, an alkane is converted into an **alkanol.** For example, the simplest alcohol is derived from methane: methanol. Ethanol stems from ethane, propanol from propane, and so on. In more complicated, branched systems, the name of the alcohol is based on the longest chain *containing the OH substituent*—not necessarily the longest chain in the molecule.

A methylheptanol   A methylpropyloctanol

To locate positions along the chain, number each carbon atom beginning from the end closest to the OH group. The names of other substituents along the chain can then be added to the alkanol stem as prefixes. Complex alkyl appendages are named according to the IUPAC rules for hydrocarbons (Section 2-6), and enantiomers according to the *R,S* rules (Section 5-4). When there is more than one hydroxy substituent along the alkane stem, the molecule is called a diol, triol, and so on.

$$\overset{3}{C}H_3\overset{2}{C}H_2\overset{1}{C}H_2OH$$

1-Propanol

$$\overset{1}{C}H_3\overset{2}{\underset{\underset{OH}{|}}{C}}\overset{H}{\underset{|}{C}}\overset{3}{H}_2\overset{4}{C}H_2\overset{5}{C}H_3$$

2-Pentanol

(3*R*)-2,2,5-Trimethyl-3-hexanol

1,4-Butanediol

Note: In *acyclic* alkanols, the OH-bearing carbon receives the number 1 only when it is located at the *end* of the stem chain.

Cyclic alcohols are called **cycloalkanols.** Here the carbon carrying the functional group automatically receives the number 1.

Cyclohexanol   1-Ethylcyclopentanol   *cis*-3-Chlorocyclobutanol

(−)-2-Hydroxypropanoic acid
[(−)-Lactic acid]

When named as a substituent, the OH group is called *hydroxy.* This occurs when a functional group taking higher precedence, such as in hydroxycarboxylic acids, is present (see margin structure). Like haloalkanes, alcohols can be classified as primary, secondary, or tertiary.

$$RCH_2OH$$
A primary alcohol

$$
\begin{array}{c}
OH \\
| \\
RCR' \\
| \\
H
\end{array}
$$
A secondary alcohol

$$
\begin{array}{c}
OH \\
| \\
RCR' \\
| \\
R''
\end{array}
$$
A tertiary alcohol

Exercise 8-1

Draw the structures of the following alcohols: **(a)** (*S*)-3-methyl-3-hexanol; **(b)** *trans*-2-bromocyclopentanol; **(c)** 2,2-dimethyl-1-propanol (neopentyl alcohol).

Exercise 8-2

Name the following compounds.

$$
\begin{array}{c}
CH_3 \quad OH \\
| \qquad | \\
\textbf{(a)} \;\; CH_3CHCH_2CHCH_3
\end{array}
$$

(b) CH_3CH_2 — (cyclohexane with OH)

(c) (structure with Br, Cl, and OH)

In common nomenclature, the name of the alkyl group is followed by the word *alcohol,* written separately. Common names are found in the older literature; although it is best not to use them, we should be able to recognize them.

$$CH_3OH$$
Methyl alcohol

$$
\begin{array}{c}
CH_3 \\
| \\
CH_3CH \\
| \\
OH
\end{array}
$$
Isopropyl alcohol

$$
\begin{array}{c}
CH_3 \\
| \\
CH_3COH \\
| \\
CH_3
\end{array}
$$
***tert*-Butyl alcohol**

In Summary Alcohols can be named as alkanols (IUPAC) or alkyl alcohols. In IUPAC nomenclature, the name is derived from the chain bearing the hydroxy group, whose position is given the lowest possible number.

8-2 STRUCTURAL AND PHYSICAL PROPERTIES OF ALCOHOLS

The hydroxy functional group strongly shapes the physical characteristics of the alcohols. It affects their molecular structure and allows them to enter into hydrogen bonding. As a result, it raises their boiling points and increases their solubilities in water.

The structure of alcohols resembles that of water

Figure 8-1 shows how closely the structure of methanol resembles those of water and of methoxymethane (dimethyl ether). In all three, the bond angles reflect the effect of electron repulsion and increasing steric bulk of the substituents on the central oxygen. Although it is not strictly correct (see Exercise 1-17), you can think of the oxygen as sp^3 hybridized, as in ammonia and methane (Section 1-8), with nearly tetrahedral bond angles around the heteroatom. The two lone electron pairs are then placed into two nonbonding sp^3 hybrid orbitals.

The O–H bond is considerably shorter than the C–H bond, in part because of the high electronegativity of oxygen relative to that of carbon. Remember that electronegativity (Table 1-2) determines how tightly nuclei hold all their surrounding electrons, including the bonding electrons. Consistent with this bond shortening is the order of bond strengths: $DH^\circ_{O-H} = 104$ kcal mol^{-1} (435 kJ mol^{-1}); $DH^\circ_{C-H} = 98$ kcal mol^{-1} (410 kJ mol^{-1}).

The electronegativity of oxygen causes an unsymmetrical distribution of charge in alcohols. This effect polarizes the O–H bond so that the hydrogen has a partial positive charge and gives rise to a molecular dipole (Section 1-3), similar to that observed for water.

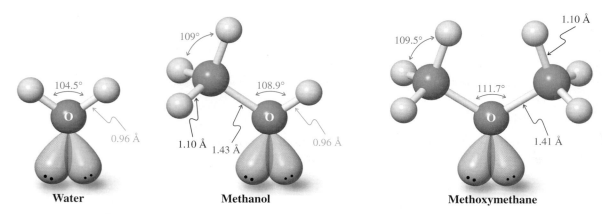

Figure 8-1 The similarity in structure of water, methanol, and methoxymethane.

The consequences of polarization are seen clearly in the electrostatic potential maps of water and methanol.

Bond and Molecular Dipoles of Water and Methanol

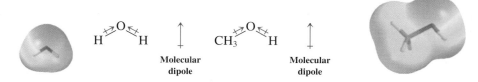

Water clusters in the gas phase have been analyzed by physical chemists. This picture (courtesy of Professor Richard Saykally) shows the lowest-energy structure of the water hexamer against a background of the structure of ice. The yellow lines depict hydrogen bonds.

Hydrogen bonding raises the boiling points and water solubilities of alcohols

In Section 6-1 we invoked the polarity of the haloalkanes to explain why their boiling points are higher than those of the corresponding nonpolar alkanes. The polarity of alcohols is similar to that of the haloalkanes. Does this mean that the boiling points of haloalkanes and alcohols correspond? Inspection of Table 8-1 shows that they do not: Alcohols have unusually high boiling points, much higher than those of comparable alkanes and haloalkanes.

The explanation lies in hydrogen bonding. Hydrogen bonds may form between the oxygen atoms of one alcohol molecule and the hydroxy hydrogen atoms of another. Alcohols build up an extensive network of these interactions (Figure 8-2). Although hydrogen bonds are longer and much weaker $DH° ≈ 5-6$ kcal mol^{-1} ($21-25$ kJ mol^{-1}) than the covalent O–H linkage ($DH° = 104$ kcal mol^{-1}), so many of them form that their combined strength makes it difficult for molecules to escape the liquid. The result is a higher boiling point.

Table 8-1	Physical Properties of Alcohols and Selected Analogous Haloalkanes and Alkanes				
Compound	IUPAC name	Common name	Melting point (°C)	Boiling point (°C)	Solubility in H$_2$O at 23°C
CH$_3$OH	Methanol	Methyl alcohol	−97.8	65.0	Infinite
CH$_3$Cl	Chloromethane	Methyl chloride	−97.7	−24.2	0.74 g/100 mL
CH$_4$	Methane		−182.5	−161.7	3.5 mL (gas)/100 mL
CH$_3$CH$_2$OH	Ethanol	Ethyl alcohol	−114.7	78.5	Infinite
CH$_3$CH$_2$Cl	Chloroethane	Ethyl chloride	−136.4	12.3	0.447 g/100 mL
CH$_3$CH$_3$	Ethane		−183.3	−88.6	4.7 mL (gas)/100 mL
CH$_3$CH$_2$CH$_2$OH	1-Propanol	Propyl alcohol	−126.5	97.4	Infinite
CH$_3$CH$_2$CH$_3$	Propane		−187.7	−42.1	6.5 mL (gas)/100 mL
CH$_3$CH$_2$CH$_2$CH$_2$OH	1-Butanol	Butyl alcohol	−89.5	117.3	8.0 g/100 mL
CH$_3$(CH$_2$)$_4$OH	1-Pentanol	Pentyl alcohol	−79	138	2.2 g/100 mL

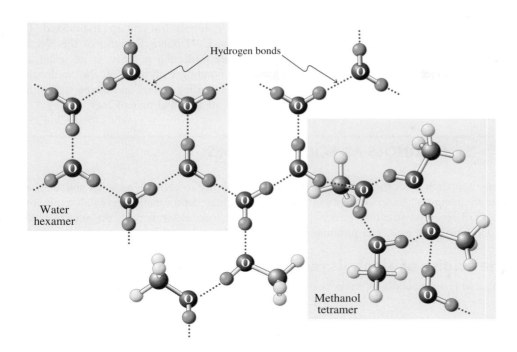

Hydrogen bonds

Water hexamer

Methanol tetramer

Figure 8-2 Hydrogen bonding in an aqueous solution of methanol. The molecules form a complex three-dimensional array, and only one layer is depicted here. Pure water (as, for example, in ice) tends to arrange itself in cyclic hexamer units (screen at the top left); pure small alcohols prefer a cyclic tetramer structure (screen at the bottom right).

The effect is even more pronounced in water, which has two hydrogens available for hydrogen bonding (see Figure 8-2). This phenomenon explains why water, with a molecular weight of only 18, has a boiling point of 100°C. Without this property, water would be a gas at ordinary temperatures. Considering the importance of water in all living organisms, imagine how the absence of liquid water would have affected the development of life on our planet.

Hydrogen bonding in water and alcohols is responsible for another property: Many alcohols are appreciably water soluble (Table 8-1). This behavior contrasts with that of the nonpolar alkanes, which are poorly solvated by this medium. Because of their characteristic insolubility in water, alkanes are said to be **hydrophobic** (*hydro,* Greek, water; *phobos,* Greek, fear). So are most alkyl chains. The hydrophobic effect has its origin in two phenomena. First, dissolution of alkyl chains in water requires breaking up the hydrogen-bonded network of the solvent. Second, the alkyl moieties can self-aggregate by London forces (Section 2-7).

In contrast to the hydrophobic behavior of alkyl groups, the OH group and other polar substituents, such as COOH and NH₂, are **hydrophilic:** They enhance water solubility.

As the values in Table 8-1 show, the larger the alkyl (hydrophobic) part of an alcohol, the lower its solubility in water. At the same time, the alkyl group increases the solubility of the alcohol in nonpolar solvents (Figure 8-3). The "water-like" structure of the lower alcohols, particularly methanol and ethanol, makes them excellent solvents for polar compounds and even salts. It is not surprising, then, that alcohols are popular *protic solvents* in the S_N2 reaction (Section 6-8).

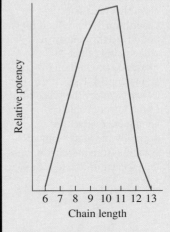

Methanol   1-Pentanol

Figure 8-3 The hydrophobic and hydrophilic parts of methanol and 1-pentanol (space-filling models). The polar functional group dominates the physical properties of methanol: The molecule is completely soluble in water but only partially so in hexane. Conversely, the increased size of the hydrophobic part in 1-pentanol leads to infinite solubility in hexane but reduced solubility in water (Table 8-1).

In Summary The oxygen in alcohols (and ethers) is tetrahedral and sp^3 hybridized. The covalent O–H bond is shorter and stronger than the C–H bond. Because of the electronegativity of the oxygen, alcohols exhibit appreciable molecular polarity, as do water and ethers. The hydroxy hydrogen enters into hydrogen bonding with other alcohol molecules. These properties lead to a substantial increase in the boiling points and in the solubilities of alcohols in polar solvents relative to those of the alkanes and haloalkanes.

8-3 | ALCOHOLS AS ACIDS AND BASES

Many applications of the alcohols depend on their ability to act both as acids and as bases. (See the review of these concepts in Section 2-3.) Thus, deprotonation gives alkoxide ions. We shall see how structure affects pK_a values. The lone electron pairs on oxygen render alcohols basic as well, and protonation results in alkyloxonium ions.

The acidity of alcohols resembles that of water

The acidity of alcohols in water is expressed by the equilibrium constant K.

$$\text{R}\ddot{\text{O}}\text{—H} + \text{H}_2\ddot{\text{O}} \xrightleftharpoons{K} \text{R}\ddot{\text{O}}:^- + \text{H}_3\text{O}:^+$$

Alkoxide ion

Making use of the constant concentration of water (55 mol L^{-1}; Section 2-3), we derive a new equilibrium constant K_a.

$$K_a = K[\text{H}_2\text{O}] = \frac{[\text{H}_3\text{O}^+][\text{RO}^-]}{[\text{ROH}]} \text{ mol L}^{-1}, \text{ and } pK_a = -\log K_a$$

Table 8-2 lists the pK_a values of several alcohols. A comparison of these values with those given in Table 2-2 for mineral and other strong acids shows that alcohols, like water, are fairly weak acids. Their acidity is far greater, however, than that of alkanes and haloalkanes.

Why are alcohols acidic, whereas alkanes and haloalkanes are not? The answer lies in the relatively strong electronegativity of the oxygen to which the proton is attached, which stabilizes the negative charge of the alkoxide ion.

To drive the equilibrium between alcohol and alkoxide to the side of the conjugate base, it is necessary to use a base *stronger* than the alkoxide formed (i.e., a base derived from a conjugate acid *weaker* than the alcohol; see also Section 9-1). An example is the reaction of sodium amide, NaNH$_2$, with methanol to furnish sodium methoxide and ammonia.

$$\text{CH}_3\ddot{\text{O}}\text{—H} + \text{Na}^+ \; ^-:\ddot{\text{N}}\text{H}_2 \xrightleftharpoons{K} \text{CH}_3\ddot{\text{O}}:^- \text{Na}^+ + :\text{NH}_3$$

$pK_a = 15.5$ **Sodium amide**   **Sodium methoxide** $pK_a = 35$

This equilibrium lies well to the right ($K \approx 10^{35-15.5} = 10^{19.5}$), because methanol is a much stronger acid than is ammonia, or, conversely, because amide is a much stronger base than is methoxide.

Table 8-2	pK_a Values of Alcohols in Water		
Compound	pK_a	Compound	pK_a
HOH	15.7	ClCH$_2$CH$_2$OH	14.3
CH$_3$OH	15.5	CF$_3$CH$_2$OH	12.4
CH$_3$CH$_2$OH	15.9	CF$_3$CH$_2$CH$_2$OH	14.6
(CH$_3$)$_2$CHOH	17.1	CF$_3$CH$_2$CH$_2$CH$_2$OH	15.4
(CH$_3$)$_3$COH	18		

It is sometimes sufficient to generate alkoxides in less than stoichiometric equilibrium concentrations. For this purpose, we may add an alkali metal hydroxide to the alcohol.

$$CH_3CH_2\overset{..}{\underset{..}{O}}\!-\!H + Na^+ \, ^-\!:\overset{..}{O}H \overset{K}{\rightleftharpoons} CH_3CH_2\overset{..}{\underset{..}{O}}:^- Na^+ + H_2\overset{..}{\underset{..}{O}}$$

$pK_a = 15.9$ $pK_a = 15.7$

With this base present, approximately half of the alcohol exists as the alkoxide, if we assume equimolar concentrations of starting materials. If the alcohol is the solvent (i.e., present in large excess), however, essentially all of the base exists in the form of the alkoxide.

Solved Exercise 8-3 | Working with the Concepts: Estimating Acid-Base Equilibria

You want to prepare potassium methoxide by treatment of methanol with KCN. Will this procedure work?

Strategy
What: We need to visualize the desired reaction by writing it down.
How: We then add to the equation the pK_a values of the acids on each side (consult Table 2-2 or 6-4, and Table 8-2).
Information: If the pK_a of the (conjugate) acid on the right is more than 2 units larger than that of methanol on the left, the equilibrium will lie >99% to the right ($K > 100$).
Proceed:

Solution
• The equilibrium reaction and the associated pK_a values are

$$CH_3OH + K^+ CN^- \rightleftharpoons CH_3O^- K^+ + HCN$$

$pK_a = 15.5$ $pK_a = 9.2$

• The pK_a of HCN is 6.3 units smaller than that of methanol; it is a much stronger acid.
• The equilibrium will lie to the left; $K = 10^{-6.3}$. This approach to preparing potassium methoxide will not work.

Exercise 8-4 | Try It Yourself

Which of the following bases are strong enough to cause essentially complete deprotonation of methanol? The pK_a of the conjugate acid is given in parentheses.
(a) $CH_3CH_2CH_2CH_2Li$ (50); **(b)** CH_3CO_2Na (4.7); **(c)** $LiN[CH(CH_3)_2]_2$ (LDA, 36);
(d) KH (38); **(e)** CH_3SNa (10).

Steric disruption and inductive effects control the acidity of alcohols

Table 8-2 shows an almost million-fold variation in the acidity of the alcohols. A closer look at the first column reveals that the acidity decreases (pK_a increases) from methanol to primary, secondary, and finally tertiary systems.

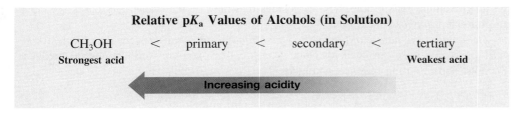

Relative pK_a Values of Alcohols (in Solution)

CH₃OH   <   primary   <   secondary   <   tertiary
Strongest acid   **Weakest acid**

Increasing acidity

This ordering has been ascribed to steric disruption of solvation and to hydrogen bonding in the alkoxide (Figure 8-4). Because solvation and hydrogen bonding stabilize the negative charge on oxygen, interference with these processes leads to an increase in pK_a.

The second column in Table 8-2 reveals another contribution to the pK_a of alcohols: The presence of halogens increases acidity. Recall that the carbon of the C–X bond is positively

Figure 8-4 The smaller methoxide ion is better solvated than is the larger tertiary butoxide ion. (S = solvent molecules).

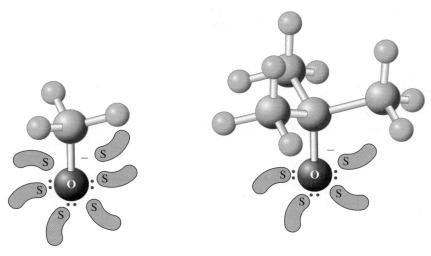

Inductive Effect of the Chlorine in 2-Chloroethoxide

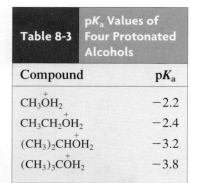

$$\overset{\leftrightarrow}{Cl}-\overset{\leftrightarrow}{CH_2}-\overset{\leftrightarrow}{CH_2}-\overset{\cdots}{\underset{\cdots}{O}}\!:^-$$

⬅ **Increasing inductive effect**

polarized as a result of the high electronegativity of X (Sections 1-3 and 6-1). Electron withdrawal by the halogen also causes atoms farther away to be slightly positively charged. This phenomenon of transmission of charge, both negative and positive, through the σ bonds in a chain of atoms is called an **inductive effect.** Here it stabilizes the negative charge on the alkoxide oxygen by electrostatic attraction. The inductive effect in alcohols increases with the number of electronegative groups but decreases with distance from the oxygen.

Exercise 8-5

Rank the following alcohols in order of increasing acidity.

OH   OH   OH   OH

(structures of cyclohexanol, 4-chlorocyclohexanol, 2-chlorocyclohexanol, 3-chlorocyclohexanol)

Exercise 8-6

Which side of the following equilibrium reaction is favored (assume equimolar concentrations of starting materials)?

$$(CH_3)_3CO^- + CH_3OH \rightleftharpoons (CH_3)_3COH + CH_3O^-$$

The lone electron pairs on oxygen make alcohols weakly basic

Alcohols may also be basic, although weakly so. Very strong acids are required to protonate the OH group, as indicated by the low pK_a values (strong acidity) of their conjugate acids, the alkyloxonium ions (Table 8-3). Molecules that may be both acids and bases are called **amphoteric** (*ampho,* Greek, both).

The amphoteric nature of the hydroxy functional group characterizes the chemical reactivity of alcohols. In strong acids, they exist as alkyloxonium ions, in neutral media as alcohols, and in strong bases as alkoxides.

Table 8-3	pK_a Values of Four Protonated Alcohols
Compound	**pK_a**
$CH_3\overset{+}{O}H_2$	−2.2
$CH_3CH_2\overset{+}{O}H_2$	−2.4
$(CH_3)_2CH\overset{+}{O}H_2$	−3.2
$(CH_3)_3C\overset{+}{O}H_2$	−3.8

Alcohols Are Amphoteric

$$R-\overset{\overset{\displaystyle H}{|}}{\underset{\underset{\displaystyle H}{|}}{\overset{+}{O}}}: \quad \underset{\text{Mild base}}{\overset{\text{Strong acid}}{\rightleftharpoons}} \quad R\overset{\cdots}{\underset{\cdots}{O}}H \quad \underset{\text{Mild acid}}{\overset{\text{Strong base}}{\rightleftharpoons}} \quad R\overset{\cdots}{\underset{\cdots}{O}}:^-$$

Alkyloxonium ion   **Alcohol**   **Alkoxide ion**

In Summary Alcohols are amphoteric. They are acidic by virtue of the electronegativity of the oxygen and are converted into alkoxides by strong bases. In solution, the steric bulk of branching inhibits solvation of the alkoxide, thereby raising the pK_a of the corresponding alcohol. Electron-withdrawing substituents close to the functional group lower the pK_a by the inductive effect. Alcohols are also weakly basic and can be protonated by strong acids to furnish alkyloxonium ions.

8-4 | INDUSTRIAL SOURCES OF ALCOHOLS: CARBON MONOXIDE AND ETHENE

Let us now turn to the preparation of alcohols. We start in this section with methods of special importance in industry. Subsequent sections address procedures that are used more generally in the synthetic laboratory to introduce the hydroxy functional group into a wide range of organic molecules.

Methanol is made on a multibillion-pound scale from a pressurized mixture of CO and H_2 called **synthesis gas.** The reaction involves a catalyst consisting of copper, zinc oxide, and chromium(III) oxide.

$$CO + 2\,H_2 \xrightarrow{\text{Cu–ZnO–Cr}_2\text{O}_3,\ 250°C,\ 50\text{–}100\ \text{atm}} CH_3OH$$

Changing the catalyst to rhodium or ruthenium leads to 1,2-ethanediol (ethylene glycol), an important industrial chemical that is the principal component of automobile antifreeze.

$$2\,CO + 3\,H_2 \xrightarrow{\text{Rh or Ru, pressure, heat}} \begin{array}{c} CH_2\text{—}CH_2 \\ | \quad\ \ | \\ OH \quad OH \end{array}$$

1,2-Ethanediol
(Ethylene glycol)

Other reactions that would permit the selective formation of a given alcohol from synthesis gas are the focus of much current research, because synthesis gas is readily available by the gasification of coal or other biomass in the presence of water, or by the partial oxidation of methane.

$$\text{Coal} \xrightarrow{\text{Air, H}_2\text{O, }\Delta} x\,CO + y\,H_2 \xleftarrow{\text{Air}} CH_4$$

The world's smallest direct methanol fuel cell.

The methanol produced from synthesis gas costs only about $1 per gallon and, because its heat content is high [182.5 kcal mol^{-1} (763.6 kJ mol^{-1}); compare values in Table 3-7], it has become the cornerstone of a methanol-based fuel strategy. A key development has been the methanol fuel cell, in which the electrons released upon the conversion of methanol to CO_2 are used to power cars and electronic devices, such as cell phones, laptop computers, and small power-generating units for domestic use.

Ethanol is prepared in large quantities by fermentation of sugars (see Chapter Opening) or by the phosphoric acid-catalyzed hydration of ethene (ethylene). The hydration (and other addition reactions) of alkenes are considered in detail in Chapter 12.

$$CH_2\text{=}CH_2 + HOH \xrightarrow{\text{H}_3\text{PO}_4,\ 300°C} \begin{array}{c} CH_2\text{—}CH_2 \\ | \qquad | \\ H \qquad OH \end{array}$$

In Summary The industrial preparation of methanol and 1,2-ethanediol proceeds by reduction of carbon monoxide with hydrogen. Ethanol is prepared by fermentation or the acid-catalyzed hydration of ethene (ethylene).

8-5 | SYNTHESIS OF ALCOHOLS BY NUCLEOPHILIC SUBSTITUTION

On a smaller than industrial scale, we can prepare alcohols from a wide variety of starting materials. For example, conversions of haloalkanes into alcohols by S_N2 and S_N1 processes featuring hydroxide and water, respectively, as nucleophiles were described in Chapters 6 and 7. These methods are not as widely used as one might think, however, because the required halides are often accessible only from the corresponding alcohols (Chapter 9). They also suffer from the usual drawbacks of nucleophilic substitution: Bimolecular elimination can be a major side reaction of hindered systems, and tertiary halides form carbocations that may undergo E1 reactions. Some of these drawbacks are overcome by the use of polar, aprotic solvents (Table 6-5).

Alcohols by Nucleophilic Substitution

92%

50%   21%

90%   10%

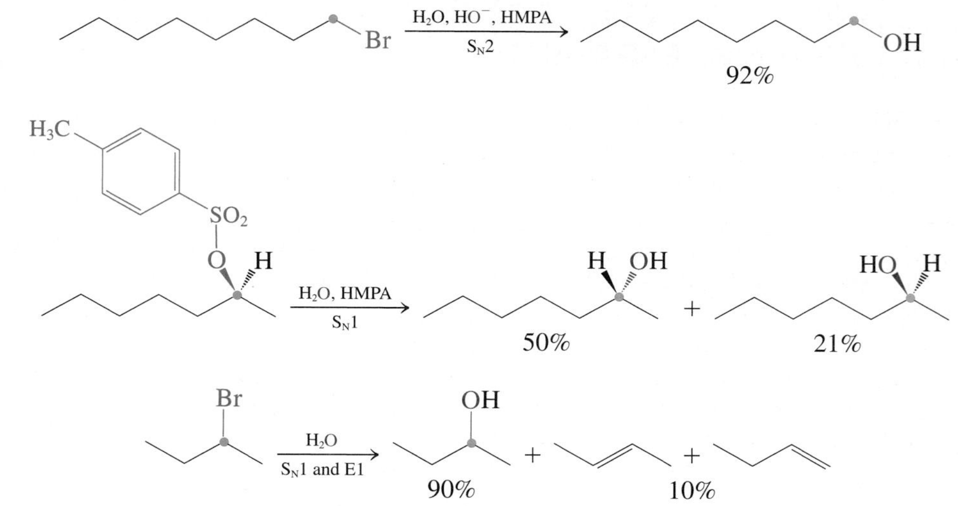

Exercise 8-7

Show how you might convert the following haloalkanes into alcohols:
(a) Bromoethane; (b) chlorocyclohexane; (c) 3-chloro-3-methylpentane.

A way around the problem of elimination in S_N2 reactions of oxygen nucleophiles with secondary or sterically encumbered, branched primary substrates is the use of less basic functional equivalents of water, such as acetate (Section 6-8). The resulting alkyl acetate (an ester) can then be converted into the desired alcohol by aqueous hydroxide. We shall consider this reaction, known as *ester hydrolysis,* in Chapter 20.

Alcohols from Haloalkanes by Acetate Substitution—Hydrolysis

Step 1. Acetate formation (S_N2 reaction)

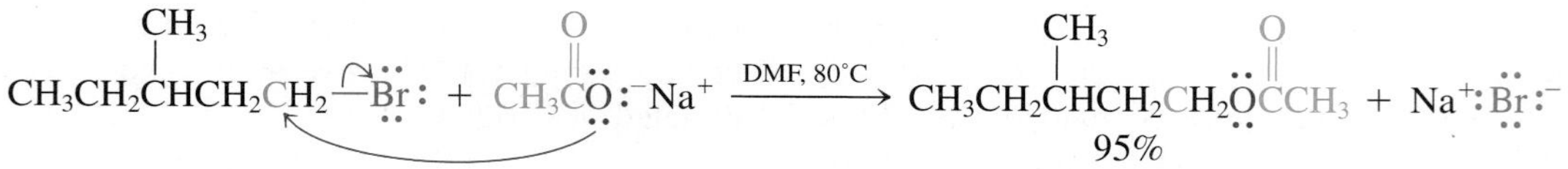

95%

1-Bromo-3-methylpentane   **3-Methylpentyl acetate**
                                      (An ester)

Step 2. Conversion into the alcohol (ester hydrolysis)

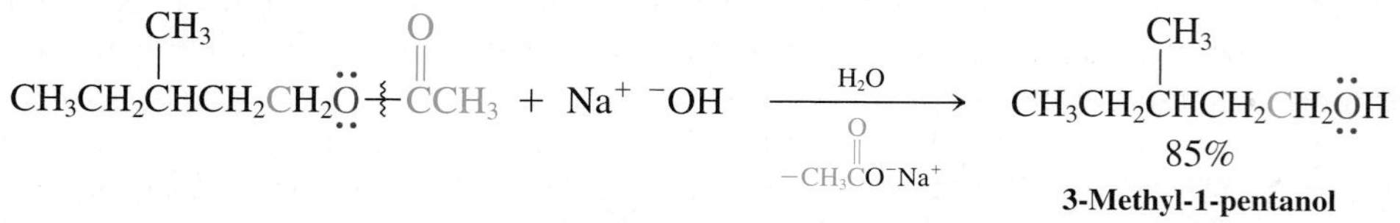

85%

3-Methyl-1-pentanol

In Summary Alcohols may be prepared from haloalkanes by nucleophilic substitution, provided the haloalkane is readily available and side reactions such as elimination do not interfere.

8-6 | SYNTHESIS OF ALCOHOLS: OXIDATION–REDUCTION RELATION BETWEEN ALCOHOLS AND CARBONYL COMPOUNDS

This section describes an important synthesis of alcohols: reduction of aldehydes and ketones. Later, we shall see that aldehydes and ketones may be converted into alcohols by addition of organometallic reagents, with resulting formation of a new carbon–carbon bond. Because of this versatility of aldehydes and ketones in synthesis, we shall also illustrate their preparation by oxidation of alcohols.

Oxidation and reduction have special meanings in organic chemistry

We can readily recognize inorganic oxidation and reduction processes as the loss and gain of electrons, respectively (margin). With organic compounds, it is often less clear whether electrons are being gained or lost in a reaction. Hence, organic chemists find it more useful to define oxidation and reduction in other terms. A process that adds electronegative atoms such as halogen or oxygen to, or removes hydrogen from, a molecule constitutes an **oxidation.** Conversely, the removal of halogen or oxygen or the addition of hydrogen is defined as **reduction.** You can readily visualize this definition in the step-by-step oxidation of methane, CH_4, to carbon dioxide, CO_2.

A zinc metal strip reduces a blue Cu^{2+} salt solution. While the Zn dissolves as Zn^{2+}, Cu metal forms a black precipitate: $Cu^{2+} + Zn \rightarrow Cu + Zn^{2+}$.

Step-by-Step Oxidation of CH_4 to CO_2

$$CH_4 \xrightarrow{+O} CH_3OH \xrightarrow{-2H} H_2C=O \xrightarrow{+O} \overset{\displaystyle O}{\overset{\displaystyle \|}{HCOH}} \xrightarrow{-2H} O=C=O$$

This definition of an oxidation-reduction relation allows us to relate alcohols to aldehydes and ketones. Addition of two hydrogen atoms to the double bond of a carbonyl group constitutes reduction to the corresponding alcohol. Aldehydes give primary alcohols; ketones give secondary alcohols. The reverse process, removal of hydrogen to furnish carbonyl compounds, is an example of oxidation. Together, these processes are referred to as **redox reactions.**

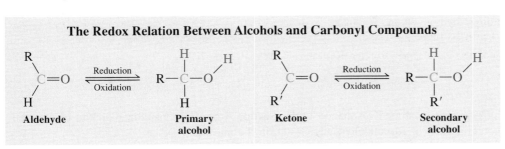

The Redox Relation Between Alcohols and Carbonyl Compounds

Aldehyde — Primary alcohol — Ketone — Secondary alcohol

How are such processes carried out in the laboratory? The remainder of this section introduces the most common methods used to effect the reduction of carbonyl compounds and the oxidation of alcohols.

Alcohols can form by hydride reduction of the carbonyl group

Conceptually, the easiest way to reduce a carbonyl group would be to add hydrogen, H–H, across the carbon–oxygen double bond directly. Although this can be done, it requires high pressures and special catalysts. A more convenient way is a polar process, in which hydride ion, H:⁻, and a proton, H⁺, are delivered to the double bond, either simultaneously or sequentially. The net result is the same, because H:⁻ + H⁺ = H–H. How does this work in practice?

The electrons in the carbonyl group are not distributed evenly between the two component atoms. Because oxygen is more electronegative than carbon, the carbon of a carbonyl group is electrophilic and the oxygen is nucleophilic. This polarization can be represented

REAL LIFE: MEDICINE 8-1 | Oxidation and Reduction in the Body

In biological systems, alcohols are metabolized by oxidation to carbonyl compounds. For example, ethanol is converted into acetaldehyde by the cationic oxidizing agent *nicotinamide adenine dinucleotide* (abbreviated as NAD$^+$; see Real Life 25-2). The process is catalyzed by the enzyme alcohol dehydrogenase. (This enzyme also catalyzes the reverse process, reduction of aldehydes and ketones to alcohols; see Problems 58 and 59 at the end of this chapter.) When the two enantiomers of 1-deuterioethanol are subjected to the enzyme, the biochemical oxidation is found to be stereospecific, NAD$^+$ removing only the hydrogen marked by the arrow in the first reaction below from C1 of the alcohol (Real Life 25-2).

Other alcohols are similarly oxidized biochemically. The relatively high toxicity of methanol ("wood alcohol") is due largely to its oxidation to formaldehyde, which interferes specifically with a system responsible for the transfer of one-carbon fragments between nucleophilic sites in biomolecules.

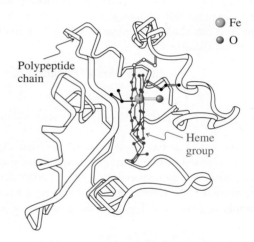

(*S*)-1-Deuterioethanol   (*R*)-1-Deuterioethanol

The capability of alcohols to undergo enzymatic oxidation makes them important relay stations in metabolism. One of the functions of the metabolic degradation of the food we eat is its controlled "burning" (i.e., combustion; see Section 3-11) to release the heat and chemical energy required to run our bodies. Another function is the selective introduction of functional groups, especially hydroxy groups, into unfunctionalized parts of molecules—in other words, alkyl substituents. This process is called hydroxylation. The *cytochrome* proteins are crucial biomolecules that help to accomplish this task. These molecules are present in almost all living cells and emerged about 1.5 billion years ago, before the development of plants and animals as separate species. Cytochrome P-450 (see Section 22-9) uses O$_2$ to accomplish the direct hydroxylation of organic molecules. In the liver, this process serves to

● Fe
● O

Polypeptide chain

Heme group

Cytochrome model

by a charge-separated resonance form (Section 1-5). It is visualized in the electrostatic potential map of formaldehyde, H$_2$C=O, in the margin.

Formaldehyde

Polar Character of the Carbonyl Function

Electrophilic   Nucleophilic

Therefore, it should be possible to add hydride to carbon and proton to oxygen, provided that suitable reagents containing nucleophilic hydrogen are available. Such reagents are sodium borohydride, Na$^+$ $^-$BH$_4$, and lithium aluminum hydride, Li$^+$ $^-$AlH$_4$. The anions in these species are electronically and structurally similar to methane (see Problem 23 of Chapter 1), but because boron and aluminum are to the left of carbon in the periodic table (Table 1-1), the anions are negatively charged. Therefore, the hydrogens are "hydridic" and capable of attacking the carbonyl carbon by transferring with their bonding electron pair to detoxify substances that are foreign to the body (xenobi-otic), many of which are the medicines that we take. Often, the primary effect of hydroxylation is simply to impart greater water solubility, thereby accelerating the excretion of a drug and thus preventing its accumulation to toxic levels.

Selective hydroxylation is important in steroid synthesis (Section 4-7). For example, progesterone is converted by triple hydroxylation at C17, C21, and C11 into cortisol. Not only does the protein pick specific positions as targets for introducing functional groups with complete stereoselectivity,

it also controls the sequence in which these reactions take place. You can get an inkling of the origin of this selectivity when you inspect the cytochrome model shown on the opposite page.

The active site is an Fe atom tightly held by a strongly bound heme group (see Section 26-8) embedded in the cloak of a polypeptide (protein) chain. The Fe center binds O_2 to generate an Fe–O_2 species, which is then reduced to H_2O and Fe=O. This oxide reacts as a radical (Section 3-4) with the R–H unit as shown, producing an Fe–OH intermediate in the presence of R·. The carbon-based radical then abstracts OH to furnish the alcohol.

Progesterone

Cytochrome P-450, O_2

Cortisol

$$Fe^{3+} \xrightarrow{+e, O_2} Fe^{3+}\!-\!O_2^- \xrightarrow{+e} Fe^{3+}\!-\!O_2^{2-}$$

$$\xrightarrow[-H_2O]{2\,H^+} \left[Fe^{3+}\!=\!O \longleftrightarrow \cdot Fe^{4+}\!-\!O\cdot \right]$$

$$\xrightarrow{RH} \left[Fe^{3+}\!-\!OH + R\cdot \right] \longrightarrow Fe^{3+} + ROH$$

The steric and electronic environment provided by the polypeptide mantle allows substrates, such as progesterone, to approach the active iron site only in very specific orientations, leading to preferential oxidation at only certain positions, such as C17, C21, and C11.

generate an alkoxide ion. You can visualize this transformation by pushing electrons starting from the B–H bond and ending at the carbonyl oxygen (see margin). In a separate (or simultaneous) process the alkoxide oxygen is protonated, either by solvent (alcohol in the case of $NaBH_4$), or by aqueous work-up (for $LiAlH_4$).

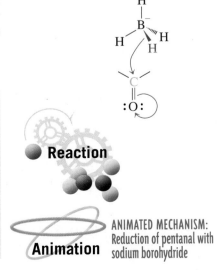

Reaction

ANIMATED MECHANISM: Reduction of pentanal with sodium borohydride

Animation

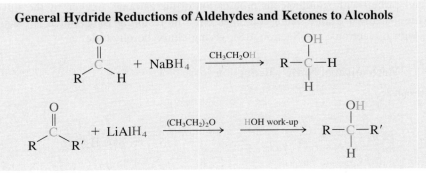

General Hydride Reductions of Aldehydes and Ketones to Alcohols

Note: The reduction of cyclobutanone introduces a short convention to describe several step sequences. In step 1, the starting material is reacted with $LiAlH_4$ in ethoxyethane (diethyl ether). In step 2, the product of this transformation is treated with aqueous acid. It is important to understand and use this convention correctly. For example, mixing the reagents of 1 and 2 will cause violent hydrolysis of $LiAlH_4$.

Examples of Hydride Reductions of Aldehydes and Ketones to Alcohols

$$CH_3CH_2CH_2CH_2\overset{O}{\overset{\|}{C}}H \xrightarrow[CH_3CH_2OH]{NaBH_4,} CH_3CH_2CH_2CH_2\overset{OH}{\underset{H}{\overset{|}{C}H}}$$

Pentanal   85%

1-Pentanol

$$\text{Cyclobutanone} \xrightarrow[\substack{2.\ H^+,\ H_2O}]{\substack{1.\ LiAlH_4,\\ (CH_3CH_2)_2O^*}} \text{Cyclobutanol}$$

90%

Cyclobutanone   Cyclobutanol

Exercise 8-8

Formulate all of the expected products of $NaBH_4$ reduction of the following compounds. (**Hint:** Remember the possibility of stereoisomerism.)

(a) $CH_3\overset{O}{\overset{\|}{C}}CH_2CH_2CH_3$ (b) $CH_3CH_2\overset{O}{\overset{\|}{C}}CH_2CH_3$ (c) $CH_3CH_2\overset{O}{\overset{\|}{C}}\overset{CH_3}{\underset{H}{\overset{|}{C}}}CH_2CH_3$

Exercise 8-9

Because of electronic repulsion, nucleophilic attack on the carbonyl function does not occur perpendicular (90° angle) to the π bond, but at an angle (107°) away from the negatively polarized oxygen. Consequently, the nucleophile approaches the target carbon in relatively close proximity to its substituents. For this reason, hydride reductions can be stereoselective, with the delivery of hydrogen from the less hindered side of the substrate molecule. Predict the likely stereochemical outcome of the treatment of compound A with $NaBH_4$. (**Hint:** Draw the chair form of **A**.)

A

Why not use the simpler reagents LiH or NaH (Section 1-3) for such reductions? The reason is the reduced basicity of hydride in the form of BH_4^- and AlH_4^-, as well as the higher solubility of the B and Al reagents in organic solvents. For example, free hydride ion is a powerful base that is instantly protonated by protic solvents [see Exercise 8-4(d)], but attachment to boron in BH_4^- moderates its reactivity considerably, thus allowing $NaBH_4$ to be used in solvents such as ethanol. In this medium, the reagent donates hydride to the carbonyl carbon with simultaneous protonation of the carbonyl oxygen by the solvent. The ethoxide generated from ethanol combines with the resulting BH_3 (which is electron deficient, with 6 electrons; see Section 1-8), giving ethoxyborohydride.

Mechanism

Mechanism of $NaBH_4$ Reduction

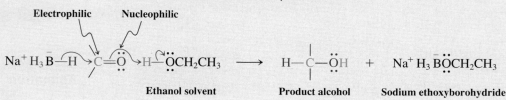

Ethanol solvent   Product alcohol   Sodium ethoxyborohydride

Ethoxyborohydride, in turn, may attack three more carbonyl substrates until all the hydride atoms of the original reagent have been used up. As a result, one equivalent of borohydride is capable of reducing *four* equivalents of aldehyde or ketone to alcohol. In the end, the boron reagent has been converted into tetraethoxyborate, $^-B(OCH_2CH_3)_4$.

Lithium aluminum hydride is more reactive than sodium borohydride (and therefore less selective; see Section 8-7 and later chapters). Because Al is less electronegative (more electropositive) than B (Table 1-2), the hydrogens in $^-AlH_4$ are less strongly bound to the metal and more negatively polarized. They are thus much more basic (as well as nucleophilic) and are attacked vigorously by water and alcohols to give hydrogen gas. Reductions utilizing lithium aluminum hydride are therefore carried out in aprotic solvents, such as ethoxyethane (diethyl ether).

Reaction of Lithium Aluminum Hydride with Protic Solvents

$$LiAlH_4 \ + \ 4\,CH_3\ddot{O}H \ \xrightarrow{\text{Fast}} \ LiAl(\ddot{O}CH_3)_4 \ + \ 4\,H-H\uparrow$$

Addition of lithium aluminum hydride to an aldehyde or ketone furnishes initially an alkoxyaluminum hydride, which continues to deliver hydride to three more carbonyl groups, in this way reducing a total of four equivalents of aldehyde or ketone. Work-up with water consumes excess reagent, hydrolyzes the tetraalkoxyaluminate to aluminum hydroxide, $Al(OH)_3$, and releases the product alcohol.

Mechanism

Animation

ANIMATED MECHANISM:
Reduction of cyclobutanone with lithium aluminum hydride

Mechanism of Lithium Aluminum Hydride Reduction

Exercise 8-10

Formulate reductions that would give rise to the following alcohols: **(a)** 1-decanol; **(b)** 4-methyl-2-pentanol; **(c)** cyclopentylmethanol; **(d)** 1,4-cyclohexanediol.

Alcohol synthesis by reduction can be reversed: chromium reagents

We have just learned how to make alcohols from aldehydes and ketones by reduction with hydride reagents. The reverse process is also possible: Alcohols may be oxidized to produce aldehydes and ketones. A useful reagent for this purpose is a transition metal in a high oxidation state: chromium(VI). In this form, chromium has a yellow-orange color. Upon exposure to an alcohol, the Cr(VI) species is reduced to the diagnostic deep green Cr(III) (see Real Life 8-2). The reagent is usually supplied as a dichromate salt ($K_2Cr_2O_7$ or $Na_2Cr_2O_7$) or as CrO_3. Oxidation of secondary alcohols to ketones is often carried out in aqueous acid, in which all of the chromium reagents generate varying amounts of chromic acid, H_2CrO_4, depending on pH.

$$CrO_3 + H_2O$$

$$\Big\updownarrow {\scriptstyle pH>6}$$

$$CrO_4{}^{2-}$$

$$\Big\updownarrow {\scriptstyle pH=2-6}$$

$$HCrO_4{}^- + Cr_2O_7{}^{2-}$$

$$\Big\updownarrow {\scriptstyle pH<1}$$

$$H_2CrO_4$$

REAL LIFE: MEDICINE 8-2 | Don't Drink and Drive: The Breath Analyzer Test

Most drunk driving tests rely on the oxidation of ethanol in the breath of potentially intoxicated drivers. They work because of the diffusion of blood alcohol through the lungs into the breath, with a measured distribution ratio of roughly 2100:1 (i.e., 2100 mL of breath contains as much ethanol as 1 mL of blood). An older method was based on the chemistry described in this section and measured the color change from Cr(VI) (orange) to Cr(III) (green). The suspect was asked to blow into a tube containing $K_2Cr_2O_7$ and H_2SO_4 supported on powdered silica gel (SiO_2) for a duration of 10–20 s; any alcohol present was revealed by the progressive color change from orange to green along the tube.

Modern variants are more sophisticated and accurate. They include the use of mini–gas chromatographs, infrared spectrometers (Section 11-8), and the–currently most popular–electrochemical analyzers. The latter contain a fuel cell that generates electrical current when supplied with ethanol. Ethanol is oxidized at the anode of this electrochemical device all the way to acetic acid, while oxygen is reduced to water at the cathode. The rate of the electron flow (the current) is proportional to the amount of alcohol in the sample and is shown on a display.

Current (converted to blood alcohol content)

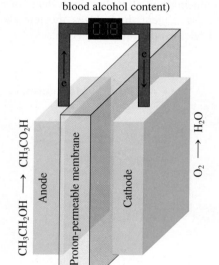

Anode

$$CH_3CH_2OH + H_2O \rightarrow CH_3CO_2H + 4\,H^+ + 4\,e$$

Cathode

$$O_2 + 4\,H^+ + 4\,e \rightarrow 2\,H_2O$$

Overall

$$CH_3CH_2OH + O_2 \rightarrow CH_3CO_2H + H_2O$$

Some people claim that a breath analyzer can be tricked into producing a "false negative" by (among others) smoking, chewing coffee beans, eating garlic, or ingesting chlorophyll preparations beforehand: These *claims* are *false*. (For the physiological effects of ethanol, see Section 9-11.)

Reaction

Oxidation of a Secondary Alcohol to a Ketone with Aqueous Cr(VI)

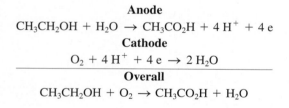

$$\xrightarrow{Na_2Cr_2O_7,\ H_2SO_4,\ H_2O}$$

96%

In water, primary alcohols tend to *overoxidize* to carboxylic acids, as shown for 1-propanol. The reason is that aldehydes in water are in equilibrium with the corresponding diols, derived by addition of water. One of the hydroxy functions of the diol then reacts further with the chromium reagent to the carboxylic acid. We shall discuss the hydration of aldehydes and ketones in Chapter 17.

$$CH_3CH_2CH_2OH \xrightarrow[\text{H}_2\text{SO}_4,\ \text{H}_2\text{O}]{\text{K}_2\text{Cr}_2\text{O}_7,} CH_3CH_2\overset{\overset{\displaystyle O}{\|}}{C}H \underset{}{\rightleftharpoons} \overset{\text{H}^+,\ \text{H}_2\text{O}}{} CH_3CH_2\overset{\overset{\displaystyle OH}{|}}{\underset{\underset{\displaystyle OH}{|}}{C}}H \xrightarrow{\text{Overoxidation}} CH_3CH_2\overset{\overset{\displaystyle O}{\|}}{C}OH$$

Propanal   1,1-Propanediol   Propanoic acid

In the absence of water, however, aldehydes are not susceptible to overoxidation. Therefore, a water-free form of Cr(VI) has been developed by reaction of CrO_3 with HCl, followed by the addition of the organic base pyridine. The result is the oxidizing agent **pyridinium chlorochromate,** abbreviated as $pyH^+\ CrO_3Cl^-$ or just **PCC** (margin), in which the hydrophobic cation portion of the salt imparts organic solvent solubility. It gives excellent yields of aldehydes upon exposure to primary alcohols in dichloromethane solvent.

PCC Oxidation of a Primary Alcohol to an Aldehyde

$$CH_3(CH_2)_8CH_2OH \xrightarrow{pyH^+\ CrO_3Cl^-,\ CH_2Cl_2} CH_3(CH_2)_8\overset{\overset{\displaystyle O}{\|}}{C}H$$
92%

PCC oxidation conditions are often also used with secondary alcohols, because the relatively nonacidic reaction conditions minimize side reactions (e.g., carbocation formation; Sections 7-2, 7-3, and 9-3) and often give better yields than does the aqueous chromate method. Tertiary alcohols are unreactive toward oxidation by Cr(VI) because they do not carry hydrogens next to the OH function and therefore cannot readily form a carbon–oxygen double bond.

Pyridine (a base) + HCl

↓ Pyridinium salt formation

Pyridinium salt with Cl^-

↓ CrO_3

Pyridinium chlorochromate (PCC or pyH⁺CrO₃Cl⁻)

Exercise 8-11

Formulate the product(s) of each of the following steps. What can you say about stereochemistry?

(a)

$$\xrightarrow[\text{H}_2\text{SO}_4,\ \text{H}_2\text{O}]{\text{Na}_2\text{Cr}_2\text{O}_7,} \xrightarrow{\text{NaBH}_4}$$

cis-**4-Methylcyclohexanol**

(b) $\xrightarrow[\text{2. H}_2\text{O work-up}]{\text{1. Excess LiAlH}_4}$ $C_7H_{16}O_2$ (two alcohols)

(c) Optically active $\xrightarrow{\text{pyH}^+ \text{CrO}_3\text{Cl}^-}$ Optically inactive

---

Chromic esters are intermediates in alcohol oxidation

What is the mechanism of the chromium(VI) oxidation of alcohols? The first step is formation of an intermediate called a **chromic ester;** the oxidation state of chromium stays unchanged in this process.

Chromic Ester Formation from an Alcohol

$$RCH_2\ddot{O}H + H\ddot{O}\overset{\overset{\displaystyle :O:}{\|}}{\underset{\underset{\displaystyle :O:}{\|}}{Cr(VI)}}\ddot{O}H \rightleftharpoons RCH_2\ddot{O}\overset{\overset{\displaystyle :O:}{\|}}{\underset{\underset{\displaystyle :O:}{\|}}{Cr(VI)}}\ddot{O}H + H_2\ddot{O}$$

Chromic acid   Chromic ester

Mechanism

The next step in alcohol oxidation is equivalent to an E2 reaction. Here water (or pyridine, in the case of PCC) acts as a mild base, removing the proton next to the alcohol oxygen. This proton is made unusually acidic by the electron-withdrawing power of the Cr(VI) (remember that it wants to become reduced!). $HCrO_3$ is an exceptionally good leaving group, because the donation of an electron pair to chromium changes its oxidation state by two units, yielding Cr(IV).

Aldehyde Formation from a Chromic Ester: an E2 Reaction

$$R-\overset{H}{\underset{H}{\overset{|}{C}}}-\overset{:\ddot{O}\quad\ddot{O}H}{\underset{}{O:}}\overset{Cr(VI)}{\underset{}{O:}} \longrightarrow \overset{H}{\underset{R}{C}}=\ddot{O} + H_3O:^+ + {}^-O_3Cr(IV)H$$

$H_2\ddot{O}$

In contrast with the kinds of E2 reactions considered so far, this elimination furnishes a carbon–oxygen instead of a carbon–carbon double bond. The Cr(IV) species formed undergoes a redox reaction with itself to Cr(III) and Cr(V); the latter may function as an oxidizing agent independently. Eventually, all Cr(VI) is reduced to Cr(III).

Exercise 8-12

Formulate a synthesis of each of the following carbonyl compounds from the corresponding alcohol.

(a) $CH_3CH_2\overset{O}{\overset{||}{C}}CH(CH_3)_2$
(b) [structure with H and CHO]
(c) CH_3CH_2 [structure with O and CH_3]

In Summary Reductions of aldehydes and ketones by hydride reagents constitute general syntheses of primary and secondary alcohols, respectively. The reverse reactions, oxidations of primary alcohols to aldehydes and secondary alcohols to ketones, are achieved with chromium(VI) reagents. Use of pyridinium chlorochromate (PCC) prevents overoxidation of primary alcohols to carboxylic acids.

8-7 ORGANOMETALLIC REAGENTS: SOURCES OF NUCLEOPHILIC CARBON FOR ALCOHOL SYNTHESIS

The reduction of aldehydes and ketones with hydride reagents is a useful way of synthesizing alcohols. This approach would be even more powerful if, instead of hydride, we could use a source of *nucleophilic carbon*. Attack by a carbon nucleophile on a carbonyl group would give an alcohol and simultaneously form a carbon–carbon bond. This kind of reaction—adding carbon atoms to a molecule—is of fundamental practical importance for synthesizing new compounds from simpler reactants.

To achieve such transformations, we need to find a way of making carbon-based nucleophiles, $R:^-$. This section describes how to reach this goal. Metals, particularly lithium and magnesium, act on haloalkanes to generate new compounds, called **organometallic reagents,** in which a carbon atom of an organic group is bound to a metal. These species are strong bases and good nucleophiles and as such are extremely useful in organic syntheses.

Alkyllithium and alkylmagnesium reagents are prepared from haloalkanes

Organometallic compounds of lithium and magnesium are most conveniently prepared by direct reaction of a haloalkane with the metal suspended in ethoxyethane (diethyl ether) or oxacyclopentane (tetrahydrofuran, THF). The reactivity of the haloalkanes increases in the

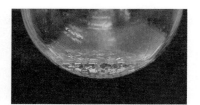

Sequence of events during the preparation of a Grignard reagent. From top to bottom: magnesium chips submerged in ether; beginning of Grignard reagent formation, after addition of the organic halide; reaction mixture showing increasing dissolution of magnesium; the final reagent solution, ready for further transformation.

order Cl < Br < I; the relatively unreactive fluorides are not normally used as starting materials in these reactions. Organomagnesium compounds, RMgX, are also called **Grignard reagents,** named after their discoverer, F. A. Victor Grignard.*

Alkyllithium Synthesis

$$CH_3Br + 2 \ Li \xrightarrow{\text{(CH}_3\text{CH}_2)_2\text{O, } 0°-10°\text{C}} CH_3Li + LiBr$$

Methyl-
lithium

Alkylmagnesium (Grignard) Synthesis

1-Methylethyl-
magnesium **iodide**

Alkyllithium compounds and Grignard reagents are rarely isolated; they are formed in solution and used immediately in the desired reaction. Sensitive to air and moisture, they must be prepared and handled under rigorously air- and water-free conditions. Simple examples, such as methyllithium, methylmagnesium bromide, butyllithium, and others, are commercially available.

The formulas RLi and RMgX oversimplify the true structures of these reagents. Thus, as written, the metal ions are highly electron deficient. To make up the desired electron octet, they function as Lewis acids (Section 2-3) and attach themselves to the Lewis basic solvent molecules. For example, alkylmagnesium halides are stabilized by bonding to two ether molecules. The solvent is said to be **coordinated** to the metal. This coordination is rarely shown in equations, but it is crucial for the formation of the Grignard species.

Grignard Reagents Are Coordinated to Solvent

$$R{-}X \ + \ Mg \xrightarrow{\text{(CH}_3\text{CH}_2)_2\text{O}}$$

Methyllithium

The alkylmetal bond is strongly polar

Alkyllithium and alkylmagnesium reagents have strongly polarized carbon–metal bonds; the strongly electropositive metal (Table 1-2) is the positive end of the dipole, as shown in the margin for CH_3Li and CH_3MgCl. The degree of polarization is sometimes referred to as "percentage of ionic bond character." The carbon–lithium bond, for example, has about 40% ionic character and the carbon–magnesium bond 35%. Such systems react chemically as if they contained a negatively charged carbon. To symbolize this behavior, we can show the carbon–metal bond with a resonance form that places the full negative charge on the carbon atom: a **carbanion.** Carbanions, R^-, are related to alkyl radicals, $R\cdot$ (Section 3-2), and carbocations, R^+ (Section 7-5), by successive removal of one electron. Because of charge repulsion, the carbon in carbanions assumes sp^3 hybridization and a tetrahedral structure (Exercise 1-16).

**Methylmagnesium
chloride**

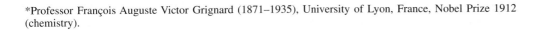

*Professor François Auguste Victor Grignard (1871–1935), University of Lyon, France, Nobel Prize 1912 (chemistry).

**Carbon–Metal Bond
in Alkyllithium and Alkylmagnesium Compounds**

Carbanion

$$\left[\begin{array}{c} \overset{|}{\underset{|}{C}} \overset{\delta^-}{}\overset{\delta^+}{} M \end{array} \longleftrightarrow \begin{array}{c} \overset{|}{\underset{|}{C}} :^- M^+ \end{array} \right]$$

Polarized   **Charge separated**
M = metal

The preparation of alkylmetals from haloalkanes illustrates an important principle in synthetic organic chemistry: **reverse polarization.** In a haloalkane, the presence of the electronegative halogen turns the carbon into an electrophilic center. Upon treatment with a metal, the $C^{\delta+}-X^{\delta-}$ unit is converted into $C^{\delta-}-M^{\delta+}$. In other words, the direction of polarization is reversed. Reaction with a metal (metallation) has turned an electrophilic carbon into a nucleophilic center.

The alkyl group in alkylmetals is strongly basic

Carbanions are very strong bases. In fact, alkylmetals are much more basic than are amides or alkoxides, because carbon is considerably less electronegative than either nitrogen or oxygen (Table 1-2) and much less capable of supporting a negative charge. Recall (Table 2-2, Section 2-3) that alkanes are *extremely* weak acids: The pK_a of methane is estimated to be 50. It is not surprising, therefore, that carbanions are such strong bases: They are, after all, the *conjugate bases of alkanes.* Their basicity makes organometallic reagents moisture sensitive and incompatible with OH or similarly acidic functional groups. Therefore, it is impossible to make organolithium or Grignard species from haloalcohols or halocarboxylic acids. On the other hand, such alkylmetals can be used as efficient bases to turn alcohols into their corresponding alkoxides (see Section 8-3). The by-product is an alkane. The outcome of this transformation is predictable on purely electrostatic grounds.

Alkoxide Formation with Methyllithium

$$(CH_3)_3C\overset{\delta^-}{O}-\overset{\delta^+}{H} + \overset{\delta^+}{Li}-\overset{\delta^-}{CH_3} \longrightarrow (CH_3)_3CO^- \ Li^+ + H-CH_3$$

2-Methyl-2-propanol   **Methyllithium**   **Lithium *tert*-butoxide**   **Methane**
(*tert*-Butylalcohol) $pK_a \simeq 50$
$pK_a = 18$

Similarly, organometals hydrolyze with water—often violently—to produce a metal hydroxide and alkane.

Hydrolysis of an Organometallic Reagent

$$\underset{\substack{\text{3-Methylpentylmagnesium} \\ \text{bromide}}}{CH_3CH_2\overset{\overset{\displaystyle CH_3}{|}}{C}HCH_2CH_2MgBr} + HOH \longrightarrow \underset{\substack{100\% \\ \text{3-Methylpentane}}}{CH_3CH_2\overset{\overset{\displaystyle CH_3}{|}}{C}HCH_2CH_2H} + BrMgOH$$

The sequence Grignard (or alkyllithium) formation, also called metallation, followed by hydrolysis converts a haloalkane into an alkane. A more direct way of achieving the same goal is the reaction of a haloalkane with the powerful hydride donor lithium aluminum hydride, an S_N2 displacement of halide by H^-. The less reactive $NaBH_4$ is incapable of performing this substitution.

$$CH_3(CH_2)_7CH_2 \!-\! Br \xrightarrow[-LiBr]{LiAlH_4, \ (CH_3CH_2)_2O} CH_3(CH_2)_7CH_2 \!-\! H$$

1-Bromononane   **Nonane**

Another useful application of metallation–hydrolysis is the introduction of hydrogen isotopes, such as deuterium, into a molecule by exposure of the organometallic compound to labeled water (see margin).

Introduction of Deuterium by Reaction of an Organometallic Reagent with D_2O

$$(CH_3)_3CCl \xrightarrow[2.\ D_2O]{1.\ Mg} (CH_3)_3CD$$

Solved Exercise 8-13	Working with the Concepts: Making a Deuterated Hydrocarbon

Show how you would prepare monodeuteriocyclohexane from cyclohexane.

Strategy

What: You are asked to replace one of the hydrogens in your starting material by deuterium.

How: The best way to approach a solution to this problem is to work backward, that is, ask the question: What do I know about making a deuterated alkane?

Information: The answer is given in the preceding paragraphs: You have learned two ways to convert a *haloalkane* into a deuterated alkane. The two reagents employed are $LiAlD_4$ or Mg, followed by D_2O. This problem requires one of these reagents and a halocyclohexane. How can you make a halocyclohexane from cyclohexane? The answer is in Chapter 3: radical halogenation.

Proceed: Putting everything together, a possible solution scheme is

Note: You may have noticed that we are beginning to apply our knowledge of organic chemistry to more complex problems. This process is not unlike learning a language: Each reaction may be viewed as part of the vocabulary, and now we are learning to form sentences. The "sentence" to be written here leads from cyclohexane to singly deuterated cyclohexane. We shall see in Section 8-9 that for our sentence to make sense, it is easiest to work backward from the product.

Exercise 8-14	Try It Yourself

You have a small amount of precious CD_3OH, but what you really need is completely deuterated CD_3OD. How can you make it?

In Summary Haloalkanes can be converted into organometallic compounds of lithium or magnesium (Grignard reagents) by reaction with the respective metals in ether solvents. In these compounds, the alkyl group is negatively polarized, a charge distribution opposite that found in the haloalkane. Although the alkyl–metal bond is to a large extent covalent, the carbon attached to the metal behaves as a strongly basic carbanion, exemplified by its ready protonation.

8-8 ORGANOMETALLIC REAGENTS IN THE SYNTHESIS OF ALCOHOLS

Among the most useful applications of organometallic reagents of magnesium and lithium are those in which the negatively polarized alkyl group reacts as a nucleophile. Like the hydrides, these reagents can attack the carbonyl group of an aldehyde or ketone to produce an alcohol (upon aqueous work-up). The difference is that a new carbon–carbon bond is formed in the process.

Reaction

**Alcohol Syntheses from
Aldehydes, Ketones, and Organometallics**

M = Li or MgX

Following the flow of electrons can help us understand the reaction. In the first step, the nucleophilic alkyl group in the organometallic compound attacks the carbonyl carbon. As an electron pair from the alkyl group shifts to generate the new carbon–carbon linkage, it "pushes" two electrons from the double bond onto the oxygen, thus producing a metal alkoxide. The addition of a dilute aqueous acid furnishes the alcohol by hydrolyzing the metal–oxygen bond, another example of aqueous work-up.

The reaction of organometallic compounds with *formaldehyde* results in *primary alcohols*. In the electrostatic potential maps of the example below, the electron-rich (orange-red) carbon of the butylmagnesium bromide is seen to attack the electron-poor (blue) carbon of formaldehyde to give 1-pentanol.

**Formation of a Primary Alcohol from
a Grignard Reagent and Formaldehyde**

$$CH_3CH_2CH_2CH_2-MgBr \ + \ \overset{H}{\underset{H}{C}}=O \ \xrightarrow{(CH_3CH_2)_2O} \ CH_3CH_2CH_2CH_2-\overset{H}{\underset{H}{C}}-\ddot{O}:^- \ ^+MgBr$$

**Butylmagnesium
bromide**   **Formaldehyde**

$$\downarrow H^+, H_2O$$

$$CH_3CH_2CH_2CH_2-\overset{H}{\underset{H}{C}}OH \ + \ HOMgBr$$

93%
1-Pentanol

However, *aldehydes* other than formaldehyde convert into *secondary alcohols*.

**Formation of a Secondary Alcohol from
a Grignard Reagent and an Aldehyde**

$$CH_3CH_2CH_2CH_2-MgBr \ + \ CH_3\overset{O}{\overset{\|}{C}}H \ \xrightarrow[2.\ H^+,\ H_2O]{1.\ (CH_3CH_2)_2O} \ CH_3CH_2CH_2CH_2-\overset{CH_3}{\underset{H}{C}}OH$$

**Butylmagnesium
bromide**   **Acetaldehyde**

78%
2-Hexanol

Ketones furnish *tertiary alcohols*.

Formation of a Tertiary Alcohol from a Grignard Reagent and a Ketone

$$CH_3CH_2CH_2CH_2-MgBr + CH_3\overset{O}{\underset{\|}{C}}CH_3 \xrightarrow[\text{2. H}^+, \text{H}_2\text{O}]{\text{1. THF}} CH_3CH_2CH_2CH_2-\overset{\overset{\displaystyle CH_3}{|}}{\underset{\underset{\displaystyle CH_3}{|}}{C}}OH$$

Butylmagnesium bromide   **Acetone**   95%
                                                         2-Methyl-2-hexanol

---

Exercise 8-15

Write a synthetic scheme for the conversion of 2-bromopropane, $(CH_3)_2CHBr$, into 2-methyl-1-propanol, $(CH_3)_2CHCH_2OH$.

Exercise 8-16

Propose efficient syntheses of the following products from starting materials containing no more than four carbons.

(a) $CH_3(CH_2)_4OH$

(b) $CH_3CH_2CH_2\overset{\overset{\displaystyle OH}{|}}{C}HCH_2CH_2CH_3$

(c) cyclobutane with $C(CH_3)_3$ and $-OH$ substituents

(d) $CH_3CH_2CH_2\overset{\overset{\displaystyle OH}{|}}{\underset{\underset{\displaystyle CH_3}{|}}{C}}CH_2CH_3$

Although the nucleophilic addition of alkyllithium and Grignard reagents to the carbonyl group provides us with a powerful C–C bond-forming transformation, nucleophilic attack is too slow on haloalkanes and related electrophiles, such as those encountered in Section 6-7. This kinetic problem is what enables us to make the organometallic reagents described in Section 8-7: The product alkylmetal does not attack the haloalkane from which it is made (Real Life 8-3).

In Summary Alkyllithium and alkylmagnesium reagents add to aldehydes and ketones to give alcohols in which the alkyl group of the organometallic reagent has formed a bond to the original carbonyl carbon.

8-9 | KEYS TO SUCCESS: AN INTRODUCTION TO SYNTHETIC STRATEGY

The reactions introduced so far are part of the "vocabulary" of organic chemistry; unless we know the vocabulary, we cannot speak the language of organic chemistry. These reactions allow us to manipulate molecules and interconvert functional groups, so it is important to become familiar with these transformations—their types, the reagents used, the conditions under which they occur (especially when the conditions are crucial to the success of the process), and the limitations of each type.

This task may seem monumental, one that will require much memorization. But *it is made easier by an understanding of the reaction mechanisms.* We already know that reactivity can be predicted from a small number of factors, such as electronegativity, coulombic forces, and bond strengths. Let us see how organic chemists apply this understanding to devise useful synthetic strategies, that is, reaction sequences that allow the construction of a desired target in the minimum number of high-yielding steps.

Strychnine

The total synthesis of the complex natural product strychnine (Section 25-8), containing seven fused rings and six stereocenters, has been steadily improved over a half-century of development of synthetic methods. The first synthesis, reported in 1954 by R. B. Woodward (Section 14-9), started from a simple indole derivative (Section 25-4) and required 28 synthetic steps to give the target in 0.00006% overall yield. A more recent synthesis (in 2011) took 12 steps and proceeded in 6% overall yield.

REAL LIFE: CHEMISTRY 8-3 | What Magnesium Does Not Do, Copper Can: Alkylation of Organometallics

The general coupling reaction of a haloalkane, containing a positively polarized carbon, with an alkylmetal, containing a negatively polarized carbon, is quite exothermic.

$$\overset{\delta^+}{R}-\overset{\delta^-}{X} + \overset{\delta^-}{R'}-\overset{\delta^+}{M} \longrightarrow R-R' + MX$$

Yet, in the case of Li and Mg, such couplings are either too slow at room temperature or lead to product mixtures on heating. As this process constitutes one of the most fundamental C–C bond-forming reactions, it is not surprising that synthetic chemists have devoted considerable effort to the solution of the problem, an effort that is ongoing. A solution is provided by copper salts as catalysts. Catalysts enable reactions to proceed faster, through lower-energy transition states and mechanisms (Section 3-3).

$$\text{Br} \diagup\diagdown O \diagdown + CH_3(CH_2)_5CH_2MgCl \xrightarrow[-\ MgBrCl]{5\% \ CuI} CH_3(CH_2)_8OCH_2CH_3$$
$$82\%$$

The method has been applied on an industrial scale in the manufacture of muscalure, the sex attractant of the housefly. In conjunction with a toxic ingredient, it is used for pest control, particularly in poultry, swine, beef and dairy cattle facilities, and stables (see also Section 12-17).

$$CH_3(CH_2)_7 \underset{H \quad\quad H}{\diagup\diagdown} (CH_2)_7CH_2Br + CH_3(CH_2)_3CH_2MgBr \xrightarrow[-\ MgBr_2]{CuI} CH_3(CH_2)_7 \underset{H \quad\quad H}{\diagup\diagdown} (CH_2)_{12}CH_3$$
$$80\%$$

Muscalure

The mechanism of these reactions proceeds through organocopper species, also called *cuprates* (see Section 18-10), which can be generated and used stoichiometrically, for example, from alkyllithium reagents.

Fatal attraction to muscalure spells doom to the common housefly.

Let us begin with a few examples in which we predict reactivity on mechanistic grounds. Then we shall turn to synthesis—the making of molecules. How do chemists develop new synthetic methods, and how can we make a "target" molecule as efficiently as possible? The two topics are closely related. The second, known as **total synthesis,** usually requires a series of reactions. In studying these tasks, therefore, we will also be reviewing much of the reaction chemistry that we have considered so far.

Mechanisms help in predicting the outcome of a reaction

First, recall how we predict the outcome of a reaction. What are the factors that let a particular mechanism go forward? Here are three examples.

How to Predict the Outcome of a Reaction on Mechanistic Grounds

Example 1. What happens when you add I⁻ to $FCH_2CH_2CH_2Br$?

$$ICH_2CH_2CH_2Br \xleftarrow{\overset{I^-}{\otimes}} FCH_2CH_2CH_2Br \xrightarrow{I^-} FCH_2CH_2CH_2I$$
$$\textbf{Not formed}$$

Explanation. Bromide is a better leaving group than fluoride.

$$2\ CH_3CH_2CH_2CH_2Li + CuI \longrightarrow (CH_3CH_2CH_2CH_2)_2CuLi + LiI$$

Lithium dibutylcuprate

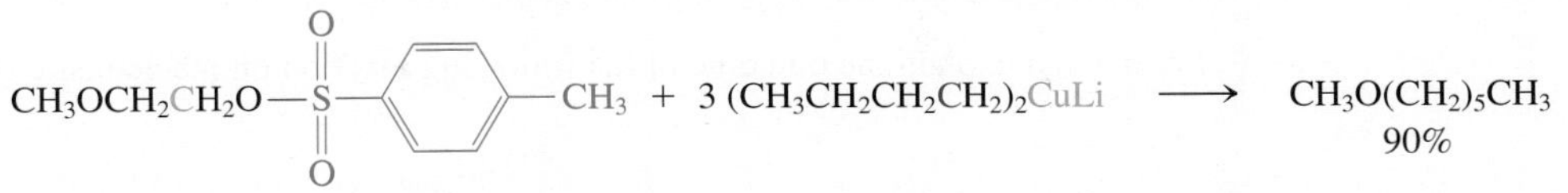

$$CH_3OCH_2CH_2O - S(=O)_2 - C_6H_4 - CH_3 + 3\ (CH_3CH_2CH_2CH_2)_2CuLi \longrightarrow CH_3O(CH_2)_5CH_3$$
90%

More recently, a number of variations on this theme have been explored, employing M = Zn, Sn, Al, and others, in the presence of catalysts based on Ni, Pd, Fe, and Rh, just to name a few. The aim is to improve not only efficiency, but also functional group tolerance. For example, unlike alkyllithium and Grignard reagents, the corresponding Zn compounds do not attack carbonyl functions.

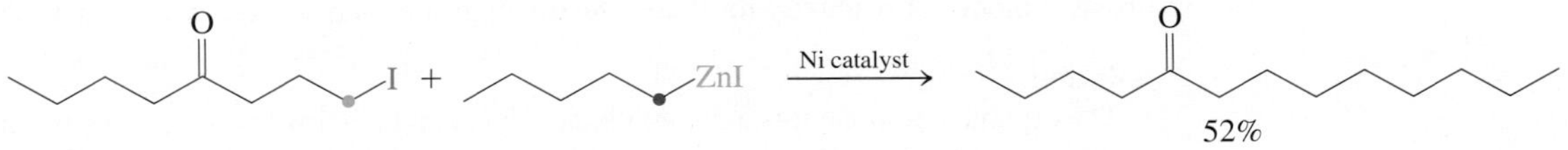

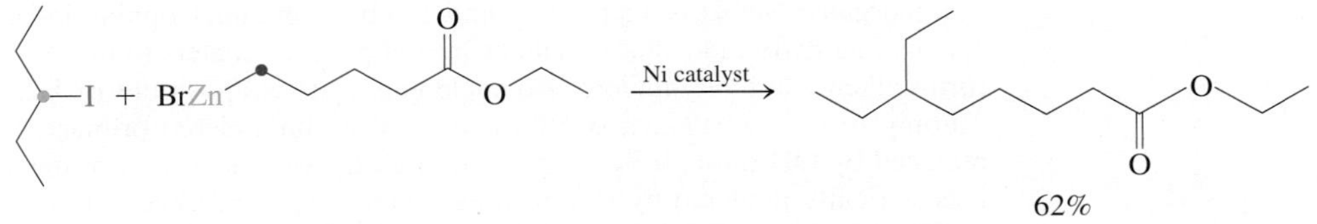

52%

62%

In these cases, the mechanism is not direct nucleophilic substitution, but rather assembly of the two fragments of R and R′ around the catalyst, as schematized in a simplified manner below.

$$R - X \xrightarrow{+\ Ni} R - Ni - X \xrightarrow[-\ ZnX_2]{R'ZnX} R - Ni - R' \xrightarrow[-\ Ni]{} R - R'$$

The development of transition metal-catalyzed C–C bond-forming reactions has seen an explosive growth during the past decade, and widely used methods employing metals in the coupling to alkenes and alkynes will be discussed in Real Life 12-4 and 13-1, and in Sections 13-9, 18-10, and 20-2.

Example 2. How does a Grignard reagent add to a carbonyl group?

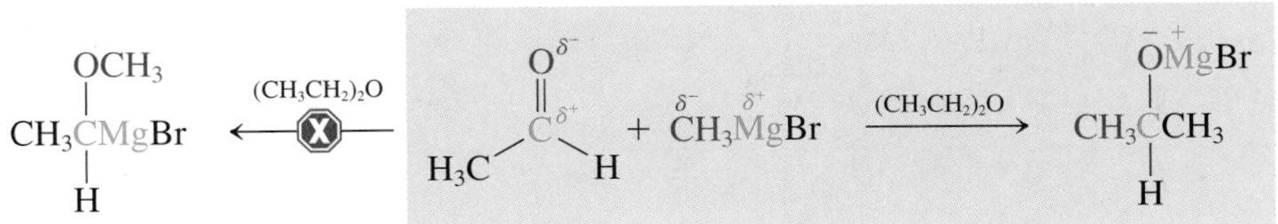

Explanation. The positively polarized carbonyl carbon forms a bond to the negatively polarized alkyl group of the organometallic reagent.

Example 3. What is the product of the radical bromination of methylcyclohexane?

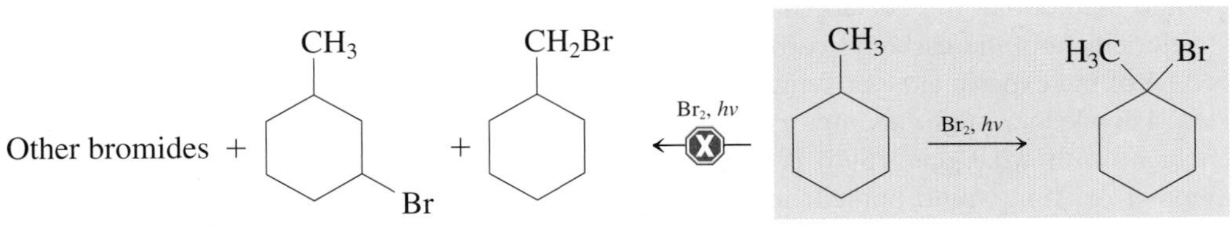

Explanation. The tertiary C–H bond is weaker than a primary or secondary C–H bond, and Br_2 is quite selective in radical halogenations.

Solved Exercise 8-17	**Working with the Concepts: How to Apply Mechanistic Knowledge to Predict the Outcome of a Reaction**

Predict and explain the outcome of the following reaction on mechanistic grounds.

$$\underset{\text{CH}_2\text{Cl}}{\text{ClCH}_2\text{CH}_2\text{CH}_2\overset{|}{\text{C}}(\text{CH}_3)_2} + \text{NaOH} \xrightarrow{\text{H}_2\text{O}}$$

Strategy
The first step is to identify the functional sites in the two starting materials. Then you can list the possible modes of reactivity for these functional groups and sort out which ones best apply.

Solution
- The organic component is a dihaloalkane. Hence, it contains two reaction sites that might be subject to the chemistry described in Chapters 6 and 7: S_N2, S_N1, E2, and E1.
- The inorganic NaOH is a strong unhindered base and nucleophile. Inspection of Table 7-4 reveals that hydroxide attacks haloalkanes at primary centers to make alcohols by S_N2, but forms alkenes at more hindered (to nucleophilic attack) positions by E2.
- Turning to the haloalkane, one Cl resides at an unhindered primary center; it should be replaced by OH through S_N2. The second Cl is also bound to a primary carbon; however, it is sterically hindered by β branching. Such steric hindrance retards nucleophilic attack, resulting in favorable E2, but only in cases in which a β hydrogen is available for deprotonation. In the present case, the carbon is neopentyl-like and E2 is not possible. Therefore, no reaction occurs at this center. Consequently, the product is

Exercise 8-18	**Try It Yourself**

Predict and explain the outcome of the following reactions on mechanistic grounds.

(a) $\underset{\text{Br}}{\text{ClCH}_2\text{CH}_2\text{CH}_2\overset{|}{\text{C}}(\text{CH}_3)_2} + \text{CH}_3\text{CH}_2\text{OH} \longrightarrow$

(b) $\underset{\text{OH}}{\text{HOCH}_2\text{CH}_2\text{CH}_2\overset{|}{\text{C}}(\text{CH}_3)_2} \xrightarrow{\text{PCC, CH}_2\text{Cl}_2}$

New reactions lead to new synthetic methods

New reactions are found by design or by accident. For example, consider how two different students might discover the reactivity of a Grignard reagent with a ketone to give an alcohol. The first student, knowing about electronegativity and the electronic makeup of ketones, would predict that the nucleophilic alkyl group of the Grignard species should attach itself to the electrophilic carbonyl carbon. This student would be pleased by the successful outcome of the experiment, verifying chemical principles in practice. The second student, with less knowledge, might attempt to dilute a particularly concentrated solution of a Grignard reagent with what one might conceive to be a perfectly good polar solvent: acetone. A violent reaction would immediately reveal that this notion is incorrect, and further investigation would uncover the powerful potential of the reagent in alcohol synthesis.

When a reaction has been discovered, it is important to show its scope and its limitations. For this purpose, many different substrates are tested, side products (if any) noted, new functional groups subjected to the reaction conditions, and mechanistic studies carried out. Should these investigations prove the new reaction to be generally applicable, it is added as a new synthetic method to the organic chemist's arsenal.

Because a reaction leads to a very specific change in a molecule, it is frequently useful to emphasize the general nature of this "molecular alteration." A simple example is the addition of a Grignard or alkyllithium reagent to formaldehyde. What is the structural change in this transformation? A one-carbon unit is added to an alkyl group. The method is valuable because it allows a straightforward one-carbon extension, also called a *homologation*.

Even though our synthetic vocabulary at this stage is relatively limited, we already have quite a number of molecular alterations at our disposal. For example, bromoalkanes are excellent starting points for numerous transformations.

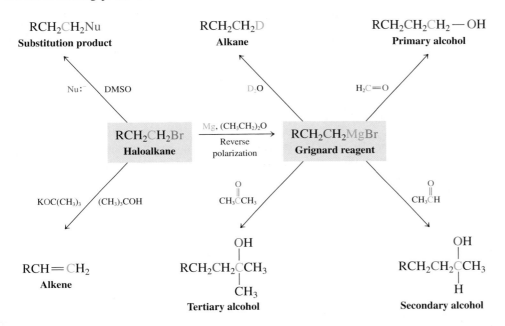

Each one of the products in the scheme can enter into further transformations of its own, thereby leading to more complicated products.

When we ask, "What good is a reaction? What sort of structures can we make by applying it?" we address a problem of *synthetic methodology*. Let us ask a different question. Suppose that we want to prepare a specific target molecule. How would we go about devising an efficient route to it? How do we find suitable starting materials? The problem with which we are dealing now is *total synthesis*.

Organic chemists want to make complex molecules for specific purposes. For example, certain compounds might have valuable medicinal properties but are not readily available from natural sources. Biochemists need a particular isotopically labeled molecule to trace metabolic pathways. Physical organic chemists frequently design novel structures to study. There are many reasons for the total synthesis of organic molecules.

Whatever the final target, a successful synthesis is characterized by brevity and high overall yield. The starting materials should be readily available, preferably commercially, and inexpensive. The principles of "green" chemistry need to be addressed (see Real Life 3-1), minimizing safety and environmental concerns, such as potentially dangerous reaction conditions and ingredients and the production of toxic waste.

Retrosynthetic analysis simplifies synthesis problems

Many compounds that are commercially available and inexpensive are also small, containing six or fewer carbon atoms. Therefore, the most frequent task facing the synthetic planner is that of building up a larger, complicated molecule from smaller, simple fragments. The

---

Homologation

R—M
Alkyl group
+
H_2C=O
One-carbon unit
↓
R—CH_2—OH

A "Green" Reduction

H_2, 17 atm,
chiral Ru
catalyst

94%
(*R*)-1,2-Propanediol

Hydrogen is the "greenest" reducing agent. Large-scale reductions of carbonyl compounds in industry are preferably carried out by catalytic hydrogenation (even though pressure is needed), in this case using a chiral catalyst to give only one enantiomer of the product.

Retrosynthesis Tree

Complex target compound

↓ Think of any reaction that will turn the precursor into the target

Less complex (smaller) precursor 1

↓ Think of any reaction that will turn compound 2 into precursor 1

Even smaller compound 2

⇓

Continue as far as necessary to reach a given starting material

---

best approach to the preparation of the target is to work its synthesis *backward* on paper, an approach called **retrosynthetic analysis*** (*retro*, Latin, backward), as schematized in the margin. In this analysis, strategic carbon–carbon bonds *in the target* are "broken" at points where bond formation seems possible. This way of thinking backward may seem strange to you at first, because you are accustomed to learning reactions in a forward way—for example, "A plus B *gives* C." Retrosynthesis requires that you think of this process in the reverse manner—for example, "C is *derived* from A plus B" (recall Exercise 6-2, p. 216).

Why retrosynthesis? The answer is that, in any "building" of a complex framework from simple building blocks, the number of possibilities of adding pieces increases drastically when going forward and includes myriad "dead-end" options. In contrast, in working backward, complexity decreases and unworkable solutions are minimized. A simple analogy is a jigsaw puzzle (margin below): It is clearly easier to dismantle step by step than it is to assemble. For

example, consider the following task: Design a synthesis of 3-hexanone $CH_3CH_2\overset{\overset{\displaystyle O}{\|}}{C}CH_2CH_2CH_3$, beginning with any organic molecules containing no more than three carbons. We employ the WHIP approach to break down the problem into its components.

*W*hat does the problem ask? Actually, several things. The product is a ketone, so you have to make that functional group. It has six carbons, and you are only allowed starting molecules with three, so you have to construct (at least) one carbon–carbon bond. You also have to decide what starting materials to use, and what reactions to do with them. That's quite a lot!

*H*ow to start (1): You need to make a ketone. Do you know any reactions that generate ketones?

*I*nformation (1): So far, only one—oxidation of a secondary alcohol with a chromium (VI) reagent (Section 8-6).

*P*roceed (1): Retrosynthetically, we say that the ketone *derives from* a secondary alcohol (below, left). An actual reaction that provides the ketone is shown below on the right. It must be the final step in your answer:

Retrosynthetic Analysis

"Retrosynthetic arrow" means "derives from"

Synthetic Step (Forward Direction)

$\xrightarrow[]{Na_2Cr_2O_7,\ H_2SO_4}$

Make sure you include the actual reagents above the reaction arrow

A jigsaw puzzle: Taking it apart is easier than putting it together.

The double-shafted arrow—a "retrosynthetic arrow"—indicates a so-called **strategic disconnection.** We recognize that the bond "broken" in this analysis above on the left, the second bond between C and O, is one that we can construct by using a transformation that we know, the oxidation of the alcohol 3-hexanol. That process, depicted above on the right, will be the final step in our answer.

*W*hat does the problem ask next? You are now faced with a new synthesis problem: making the starting material for *this* reaction (3-hexanol). It is a secondary alcohol.

*H*ow to start (2): Ask yourself, do you know any ways to make secondary alcohols?

*I*nformation (2): You know two ways, reduction of a ketone with a hydride reagent (Section 8-6) and addition of a Grignard reagent to an aldehyde (Section 8-8).

*P*roceed (2): Check these two options for feasibility. Reduction of a ketone is not a useful method here, because the ketone we would need to reduce to make 3-hexanol is the one we are trying to make: 3-hexanone. We would just be going around in circles. Turning to the second route, consider combinations of Grignard reagents and aldehydes that will give us 3-hexanol. Thinking retrosynthetically, we find that, there are two combinations, because

---

*Pioneered by Professor Elias J. Corey (b. 1928). Harvard University, Nobel Prize 1990 (chemistry).

we can make either the carbon–carbon bond to the left of the alcohol carbon (bond a), or the one to the right (bond b):

Retrosynthetic Analysis of 3-Hexanol: Two Options

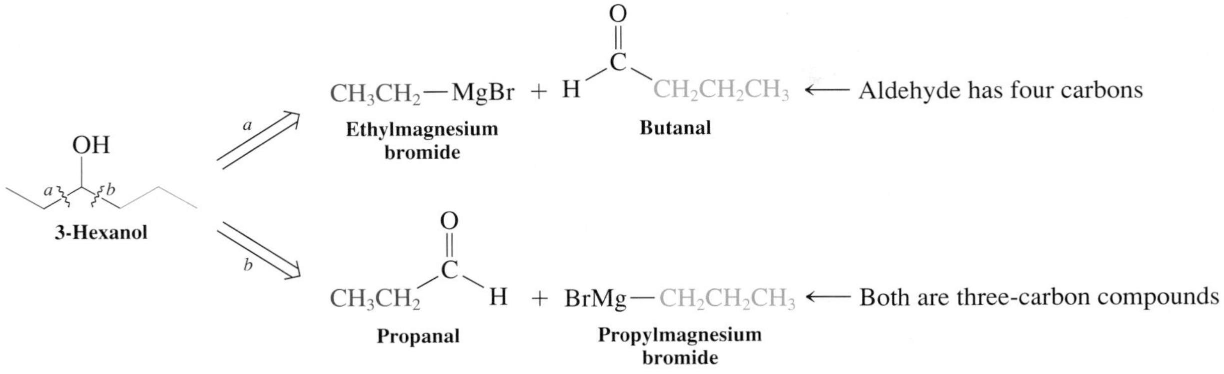

CH_3CH_2—MgBr + H $\overset{\overset{O}{\|}}{C}$ $CH_2CH_2CH_3$ ⟵ Aldehyde has four carbons

Ethylmagnesium bromide   **Butanal**

CH_3CH_2 $\overset{\overset{O}{\|}}{C}$ H + BrMg—$CH_2CH_2CH_3$ ⟵ Both are three-carbon compounds

Propanal   **Propylmagnesium bromide**

Both of these retrosynthetic disconnections depict reasonable strategies: We can make bond a, but one of the starting materials is a four-carbon aldehyde. Because we are allowed to use only starting materials with three carbons or less, we would have to build that aldehyde from smaller molecules. On the other hand, we can construct bond b from two three-carbon components of the Grignard reaction. The required Grignard reagent is generated from a haloalkane, such as 1-bromopropane, and magnesium (Section 8-7). The final synthetic scheme can be presented now as follows:

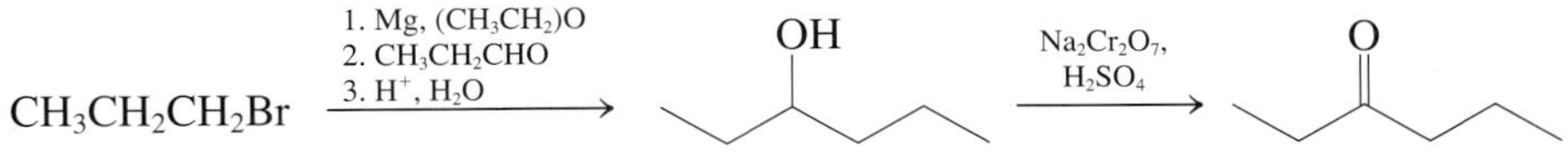

$CH_3CH_2CH_2Br$ $\xrightarrow{\substack{1.\ Mg,\ (CH_3CH_2)O \\ 2.\ CH_3CH_2CHO \\ 3.\ H^+,\ H_2O}}$ [OH structure] $\xrightarrow{\substack{Na_2Cr_2O_7, \\ H_2SO_4}}$ [O structure]

As mentioned, the two approaches shown above for the preparation of 3-hexanol via Grignard chemistry are both reasonable. In general, however, retrosynthetic disconnections should be made to provide molecular pieces that are as equally sized as possible. Therefore, our final approach combining two three-carbon molecules is the superior one, even if we had not had the constraints specified in the problem.

In a similar vein, an alternative retrosynthesis of 3-hexanol using an S_N2 reaction (Section 8-5) is a generally inferior solution as well, because it does not *simplify* the target structure:

An Inferior Retrosynthetic Plan for the Preparation of 3-Hexanol

$\overset{OH}{CH_3CH_2CH_2CHCH_2CH_3}$ $\Longrightarrow$ $NaOCCH_3$ + $\overset{Br}{CH_3CH_2CH_2CHCH_2CH_3}$

As you gain familiarity with the reactions at your disposal, practice finding the most effective ways to use them in the synthesis of more complex target molecules from less complex starting materials!

Retrosynthetic analysis aids in alcohol construction

Let us apply retrosynthetic analysis to the preparation of a tertiary alcohol, 4-ethyl-4-nonanol. Because of their steric encumbrance and hydrophobic nature, this alcohol and its homologs have important industrial applications as cosolvents and additives in certain polymerization processes (Section 12-14). There are two steps to follow at each stage of the process. First, we identify all possible strategic disconnections, "breaking" all bonds that can be formed by reactions that we know. Second, we evaluate the relative merits of these disconnections, seeking the one that best simplifies the target structure. The strategic bonds in 4-ethyl-4-nonanol are those around the functional group. There are three disconnections leading to simpler precursors. Path *a* cleaves the ethyl group from C4, suggesting as the starting materials for its construction ethylmagnesium bromide and 4-nonanone. Cleavage *b* is an alternative possibility leading to a propyl Grignard reagent and 3-octanone as precursors. Finally, disconnection *c* reveals a third synthesis route derived from the addition of pentylmagnesium bromide to 3-hexanone.

Partial Retrosynthetic Analysis of the Synthesis of 4-Ethyl-4-nonanol

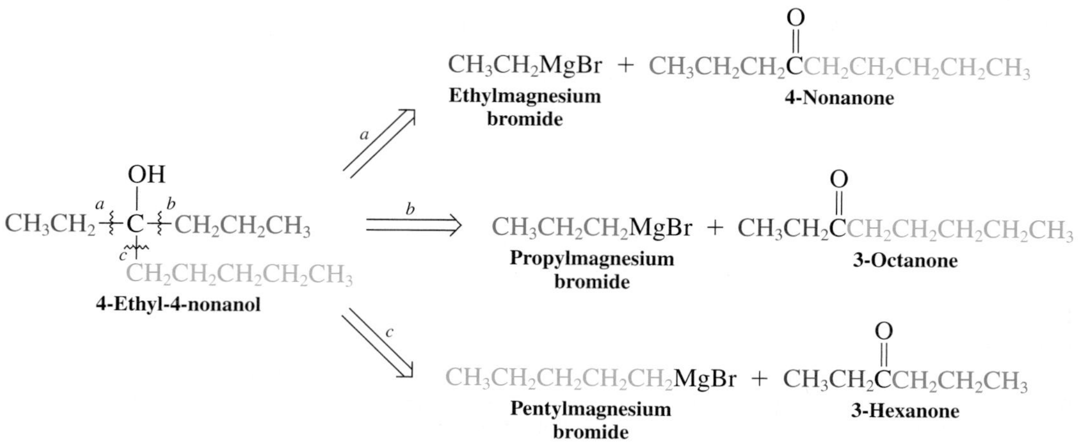

Evaluation reveals that pathway *c* is best: The necessary building blocks are almost equal in size, containing five and six carbons; thus, this disconnection provides the greatest simplification in structure.

Exercise 8-19

Apply retrosynthetic analysis to 4-ethyl-4-nonanol, disconnecting the carbon–*oxygen* bond. Does this lead to an efficient synthesis? Explain.

Can we pursue either of the fragments arising from disconnection by pathway *c* to even simpler starting materials? Yes; recall (Section 8-6) that ketones are obtained from the oxidation of secondary alcohols by Cr(VI) reagents. We may therefore envision preparation of 3-hexanone from the corresponding alcohol, 3-hexanol.

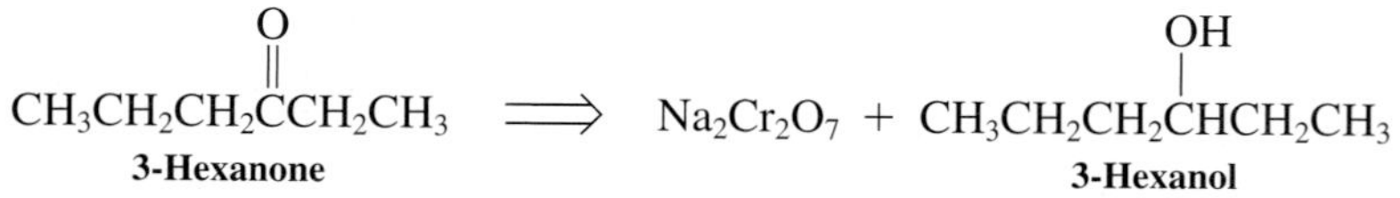

Because we earlier identified an efficient disconnection of 3-hexanol into two three-carbon fragments, we are now in a position to present our complete synthetic scheme:

Synthesis of 4-Ethyl-4-nonanol

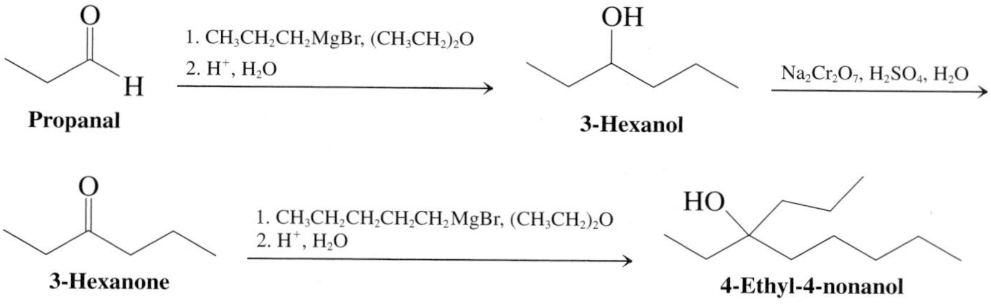

This example illustrates a very powerful general sequence for the construction of complex alcohols: first, Grignard or organolithium addition to an aldehyde to give a secondary alcohol; then oxidation to a ketone; and finally, addition of another organometallic reagent to give a tertiary alcohol.

Utility of Alcohol Oxidations in Synthesis

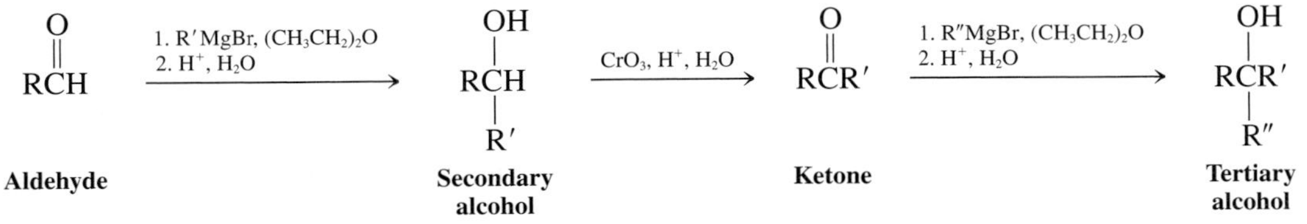

| Aldehyde | Secondary alcohol | Ketone | Tertiary alcohol |

Solved Exercise 8-20 | **Working with the Concepts: Starting with the Target and Going Backward**

Write a retrosynthetic analysis of 3-cyclobutyl-3-heptanol from starting materials containing four carbons or less.

Strategy
We apply the steps discussed for retrosynthetic analysis: Identify all possible strategic disconnections, then evaluate the relative merits of these disconnections. In this case, our evaluation has to take into account the restriction that starting materials contain four carbons or less.

Solution
- Applying what we have learned so far, we can dissect the product retrosynthetically in three possible ways: *a*, *b*, and *c*.

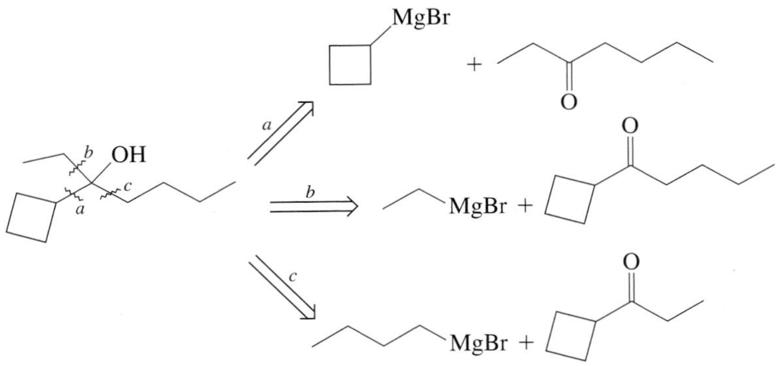

- All of them break down the target into smaller fragments, but none of these schemes provides pieces of the stipulated size: four carbons or less. Thus, the resulting ketones contain seven or nine carbon atoms, respectively, requiring their independent synthesis from correspondingly smaller molecules.
- Because the ketone arising from pathway *b* is too large to be put together directly from two four-carbon pieces, disconnections *a* and *c* appear best to pursue in our analysis. Since we do not know (yet) how to make ketones directly by C–C bond-forming reactions, we "modify" them retrosynthetically by writing the corresponding alcohols, which we know how to disassemble by further retrosynthesis (and we know how to oxidize in the forward sense, Section 8-6). The alcohol structures can then be subjected to additional potential C–C disconnections (as indicated by the wavy lines):

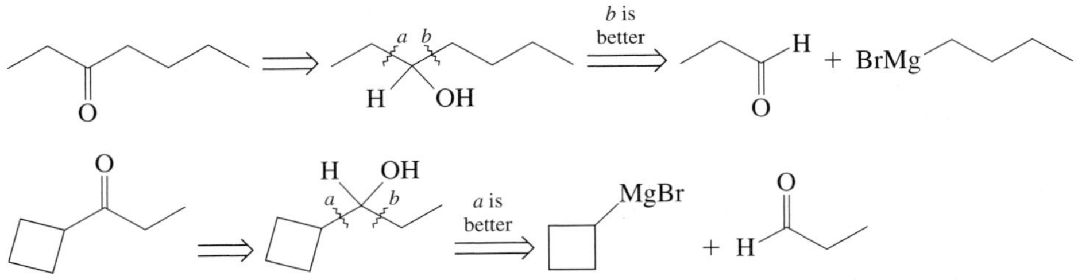

Remember WHIP

*W*hat
*H*ow
*I*nformation
*P*roceed

In both cases, only one of those disconnections provides the required four- and three-carbon fragments.
- We have two perfectly reasonable solutions to our problem: one that introduces the cyclobutyl portion early and one that does so late. Is one preferable to the other? One may argue yes, namely, the second pathway. The strained ring is sensitive and subject to side reactions, so bringing it in late in the synthesis is advantageous.

Exercise 8-21 | Try It Yourself

Show how you would prepare 2-methyl-2-propanol from methane as the only organic starting material.

Watch out for pitfalls in planning syntheses

There are several considerations to keep in mind when practicing synthetic chemistry that will help to avoid designing unsuccessful or low-yielding approaches to a target molecule.

First, try to minimize the total number of transformations required to convert the initial starting material into the desired product.

This point is so important that, in some cases, it is worthwhile to accept a low-yield step if it allows a significant shortening of the synthetic sequence. For example (with the assumption that all starting materials are of comparable cost), a seven-step synthesis in which each step has an 85% yield is inferior to a four-step synthesis with three yields at 95% and one at 45%. The overall efficiency in the first sequence comes to $(0.85 \times 0.85 \times 0.85 \times 0.85 \times 0.85 \times 0.85 \times 0.85) \times 100 = 32\%$, whereas the second synthesis, in addition to being three steps shorter, gives $(0.95 \times 0.95 \times 0.95 \times 0.45) \times 100 = 39\%$.

In these examples, all steps take place consecutively, a procedure called **linear synthesis.** In general, it is better to approach complex targets through two or more concurrent routes, as long as the overall number of steps is the same, a strategy described as **convergent synthesis.** Although a simple overall-yield calculation is not possible for a convergent strategy, you can readily convince yourself of its increased efficiency by comparing the actual *amounts* of starting materials required by the two approaches to make the same amount of product. In the following example, 10 g of a product H is prepared first in three steps (50% each) by a linear sequence A → B → C → H and second by a convergent one starting from D and F, respectively, through E and G. If we assume (for the sake of simplicity) that the molecular weights of these compounds are all the same, the first preparation requires 80 g of starting materials and the second only (a combined) 40 g.

$$A \xrightarrow{50\%} B \xrightarrow{50\%} C \xrightarrow{50\%} H$$

80 g   40 g   20 g   10 g

Linear synthesis of H

$$D \xrightarrow{50\%} E$$
20 g   10 g
$$\searrow \xrightarrow{50\%} H$$
10 g
$$F \xrightarrow{50\%} G \nearrow$$
20 g   10 g

Convergent synthesis of H

Second, do not use reagents whose molecules have functional groups that would interfere with the desired reaction.

For example, treating a hydroxyaldehyde with a Grignard reagent leads to an acid-base reaction that destroys the organometallic reagent, and not to carbon–carbon bond formation.

A possible solution to this problem would be to add two equivalents of Grignard reagent: *one* to react with the acidic hydrogen as shown, the *other* to achieve the desired addition to the carbonyl group. Another solution is to "protect" the hydroxy function in the form of an ether. We will encounter this strategy in Section 9-8.

Do not try to make a Grignard reagent from a bromoketone. Such a reagent is not stable and will, as soon as it is formed, decompose by reacting with its own carbonyl group (in the same or another molecule). We will learn how to protect a carbonyl function in Section 17-8.

2,2-Disubstituted Hindered Haloalkanes

> Third, take into account any mechanistic and structural constraints affecting the reactions under consideration.

For example, radical brominations are more selective than chlorinations. Keep in mind the structural limitations on nucleophilic reactions, and do not forget the lack of reactivity of the 2,2-dimethyl-1-halopropanes. Although sometimes difficult to recognize, many haloalkanes have such hindered structures and are similarly unreactive. Nevertheless, such systems do form organometallic reagents and may be further functionalized in this manner. For example, treatment of the Grignard reagent made from 1-bromo-2,2-dimethylpropane with formaldehyde leads to the corresponding alcohol.

$$(CH_3)_3CCH_2Br \xrightarrow[\substack{3.\ H^+,\ H_2O}]{\substack{1.\ Mg \\ 2.\ CH_2=O}} (CH_3)_3CCH_2CH_2OH$$
1-Bromo-2,2-dimethylpropane   **3,3-Dimethyl-1-butanol**

Tertiary haloalkanes, if incorporated into a more complex framework, also are sometimes difficult to recognize. Remember that tertiary halides do not undergo S_N2 reactions but eliminate in the presence of bases.

Expertise in synthesis, as in many other aspects of organic chemistry, develops largely from practice. Planning the synthesis of complex molecules requires a review of the reactions and mechanisms covered in earlier sections. The knowledge thus acquired can then be applied to the solution of synthetic problems.

THE BIG PICTURE

Where do we stand now, and where do we go from here? With Chapter 8, we have begun to discuss a second important functional class of compounds after the haloalkanes, namely, the alcohols. We used the haloalkanes to introduce two major mechanistic pathways: radical reactions (in order to make haloalkanes; Chapter 3) and ionic pathways (to demonstrate their reactivity in substitution and elimination, Chapters 6 and 7). In contrast, we have used the alcohols to introduce new types of reactions: oxidations, reductions, and organometallic additions to aldehydes and ketones. This discussion has allowed us to examine the idea of how to plan a synthesis. It requires us to start with a given product and work backward (retrosynthetic analysis) to determine what reactants and what kind of reaction conditions we need to obtain this product. As we proceed through the course and build up our knowledge of families of compounds and their chemistry, we shall keep returning to synthetic strategies as a way of classifying and applying this information.

This chapter and the next follow the format we shall use to present all future functional groups: (a) how to name them; (b) a description of their structures and general properties; (c) how to synthesize them; and (d) what kinds of reactions they undergo and how we can apply these reactions in various ways.

WORKED EXAMPLES: INTEGRATING THE CONCEPTS

8-22. Practicing the Retrosynthesis of a Complex Alcohol

Tertiary alcohols are important additives in some industrial processes utilizing Lewis acidic metal compounds (Section 2-3) as catalysts. The alcohol provides the metal with a sterically protecting and hydrophobic environment (see Figure 8-3; see also Real Life 8-1), which ensures solubility in organic solvents, longer lifetimes, and selectivity in substrate activation. Preparation of these tertiary alcohols typically follows the synthetic principles outlined in Section 8-9.

Starting from cyclohexane and using any other building blocks containing four or fewer carbons, in addition to any necessary reagents, formulate a synthesis of tertiary alcohol A.

A

SOLUTION

Before we start a random trial-and-error approach to solving this problem, it is better to take an inventory of what is given. First, we are given cyclohexane, and we note that this unit shows up as a substituent in tertiary alcohol A. Second, a total of seven additional carbons appears in the product, so our synthesis will require some additional stitching together of smaller fragments because we cannot use compounds containing more than four carbons. Third, target A is a tertiary alcohol, which should be amenable to the retrosynthetic analysis introduced in Section 8-9 (M = metal):

Approach *a* is clearly the one of choice because it breaks up tertiary alcohol A into evenly sized fragments B and C.

Having chosen route *a* as the most suitable way to find direct precursors to A, we proceed to work backward further: What are the appropriate precursors to B and C, respectively? Compound B is readily traced by retrosynthesis to our starting hydrocarbon, cyclohexane: The precursor to the organometallic compound B must be a halocyclohexane, which in turn can be made from cyclohexane by free-radical halogenation:

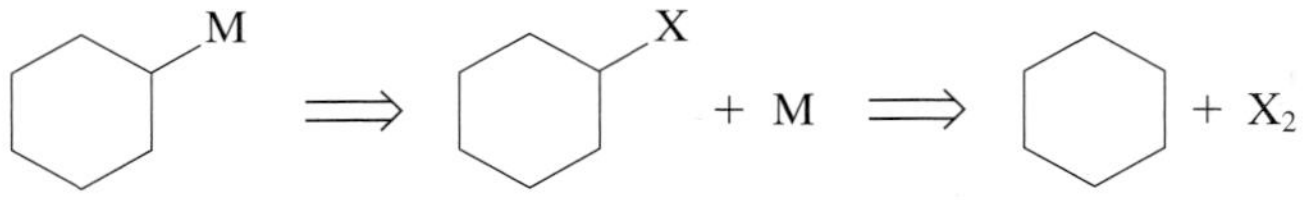

Ketone C must be broken into two smaller components. The best breakdown would be a "four + three" carbon combination: It is the most evenly sized solution, and it suggests the use of cyclobutyl intermediates. Because the only C–C disconnection that we know at this stage is that of alcohols, the first retrosynthetic step from C is its precursor alcohol (plus a chromium oxidant). Further retrosynthesis then provides the required starting pieces D and E.

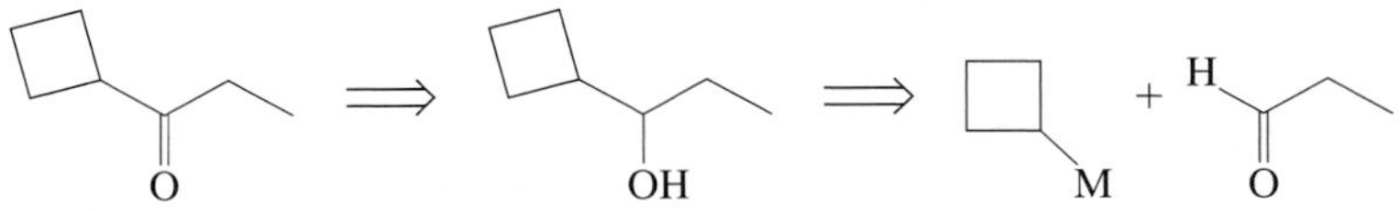

Now we can write the detailed synthetic scheme in a forward mode, with cyclohexane and pieces D and E as our starting materials:

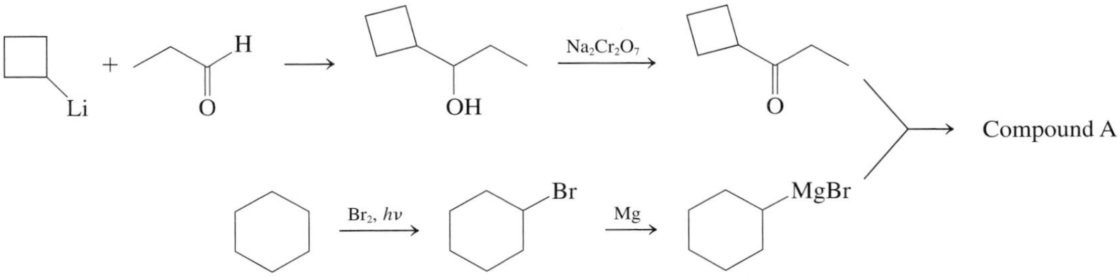

A final note: In this and subsequent synthetic exercises in this book, retrosynthetic analysis requires that you have command of all reactions not only in a forward fashion (i.e., starting material + reagent → product) but also in reverse (i.e., product ← starting material + reagent). The two sets address two different questions. The first asks: What are all the possible products that I can make from my starting material in the presence of all the reagents that I know? The second asks: What are all the conceivable starting materials that, with the appropriate reagents, will lead to my product? The two types of schematic summaries of reactions that you will see at the end of this and subsequent chapters emphasize this point.

8-23. A Flashback to General Chemistry: Balancing Equations

In this chapter we introduced redox reactions interconverting alcohols with aldehydes and ketones. The reagents employed were Cr(VI) (in the form of chromates, for example $Na_2Cr_2O_7$) and H$^-$ (in the form of $NaBH_4$ and $LiAlH_4$). Organic chemists usually don't worry about the inorganic products of such processes, because they are routinely discarded. However, for the purposes of electron bookkeeping (and in experimental recipes), it is useful (essential) to write balanced equations, showing how much of any given starting material is "going into" the reaction and how much of any possible product is "coming out." Most of you have dealt with the problem of balancing equations in introductory chemistry, but usually only for redox exchanges between metals. Can you balance the following general oxidation of a primary alcohol to an aldehyde?

$$RCH_2OH \ + \ H_2SO_4 \ + \ Na_2Cr_2O_7 \ \longrightarrow \ R\overset{\overset{\displaystyle O}{\|}}{C}H \ + \ Cr_2(SO_4)_3 \ + \ Na_2SO_4 \ + \ H_2O$$

SOLUTION

It is best to think of this transformation as two separate reactions taking place simultaneously: (1) the oxidation of the alcohol, (2) the reduction of the Cr(VI) species to Cr(III). These two parts are called **half-reactions.** In water, which is the common solvent for this and similar redox processes, we balance the half-reactions by

a. treating any hydrogen atoms consumed or produced as H_3O^+ (or, for simplicity, H^+),
b. treating any oxygen atoms consumed or produced as H_2O (in acidic solutions) or ^-OH (in basic solution),
c. adding electrons explicitly to the side deficient in negative charge.

Let us apply these guidelines to equation (1), which can be thought of as the removal of two hydrogens from the alcohol. These hydrogens are written as protons on the product side (rule a), charge-balanced by two added electrons (rule c):

$$\text{RCH}_2\text{OH} \longrightarrow \overset{\displaystyle \overset{\text{O}}{\|}}{\text{RCH}} + 2\,\text{H}^+ + 2\,e \tag{1}$$

Turning to the half-reaction of the chromium species, we know that $Cr_2O_7{}^{2-}$ is changed to (two) Cr^{3+} ions:

$$\text{Cr}_2\text{O}_7{}^{2-} \longrightarrow 2\,\text{Cr}^{3+}$$

We note that seven oxygen atoms need to appear on the right side; rule (b) stipulates seven molecules of H_2O:

$$\text{Cr}_2\text{O}_7{}^{2-} \longrightarrow 2\,\text{Cr}^{3+} + 7\,\text{H}_2\text{O}$$

This change requires that 14 hydrogens be added to the left side; rule (a) says 14 H^+ atoms:

$$14\,\text{H}^+ + \text{Cr}_2\text{O}_7{}^{2-} \longrightarrow 2\,\text{Cr}^{3+} + 7\,\text{H}_2\text{O}$$

Is the charge balanced? No, rule (c) tells us to add six electrons to the left side, resulting in balanced equation (2).

$$14\,\text{H}^+ + \text{Cr}_2\text{O}_7{}^{2-} + 6\,e \longrightarrow 2\,\text{Cr}^{3+} + 7\,\text{H}_2\text{O} \tag{2}$$

Inspection of the two half-reactions shows that, as written, (1) generates $2e$ (an oxidation, Section 8-6) and (2) consumes $6e$ (a reduction, Section 8-6). Since a chemical equation does not show electrons, we need to balance the production and consumption of electrons. We do this by simply multiplying (1) by 3:

$$3\,\text{RCH}_2\text{OH} \longrightarrow 3\,\overset{\displaystyle \overset{\text{O}}{\|}}{\text{RCH}} + 6\,\text{H}^+ + 6\,e \tag{3}$$

We now combine the two balanced half-reactions by adding them to each other, that is, (3) + (2), a procedure that cancels out the electrons to give equation (4).

$$3\,\text{RCH}_2\text{OH} \longrightarrow 3\,\overset{\displaystyle \overset{\text{O}}{\|}}{\text{RCH}} + 6\,\text{H}^+ + 6\,e \tag{3}$$

$$14\,\text{H}^+ + \text{Cr}_2\text{O}_7{}^{2-} + 6\,e \longrightarrow 2\,\text{Cr}^{3+} + 7\,\text{H}_2\text{O} \tag{2}$$

$$3\,\text{RCH}_2\text{OH} + 14\,\text{H}^+ + \text{Cr}_2\text{O}_7{}^{2-} \longrightarrow 3\,\overset{\displaystyle \overset{\text{O}}{\|}}{\text{RCH}} + 6\,\text{H}^+ + 2\,\text{Cr}^{3+} + 7\,\text{H}_2\text{O} \tag{4}$$

In this form, (4) contains H^+ on both sides. We can simplify it by removing the "excess" H^+ to furnish equation (5).

$$3\,\text{RCH}_2\text{OH} + 8\,\text{H}^+ + \text{Cr}_2\text{O}_7{}^{2-} \longrightarrow 3\,\overset{\displaystyle \overset{\text{O}}{\|}}{\text{RCH}} + 2\,\text{Cr}^{3+} + 7\,\text{H}_2\text{O} \tag{5}$$

Finally, we add nonreacting "spectator ions" of this reaction to show the proper stoichiometry in equation (6).

$$3\ RCH_2OH\ +\ 4\ H_2SO_4\ +\ Na_2Cr_2O_7\ \longrightarrow\ 3\ R\overset{\displaystyle O}{\overset{\|}{C}}H\ +\ Cr_2(SO_4)_3\ +\ Na_2SO_4\ +\ 7\ H_2O \quad (6)$$

The balanced equation (6) very nicely reveals why the reaction is run in acidic medium: H_2SO_4 is consumed. It also highlights the oxidizing power of dichromate: One mole is sufficient to effect the oxidation of three moles of alcohol.

New Reactions

1. Acid-Base Properties of Alcohols (Section 8-3)

$$R-\overset{\overset{\displaystyle H}{|}}{\underset{\underset{\displaystyle H}{|}}{\overset{+}{O}}}H \quad \underset{H^+}{\overset{}{\rightleftharpoons}} \quad ROH \quad \overset{Base\ :B^-}{\rightleftharpoons} \quad RO^- \ + \ BH$$

Alkyloxonium ion   **Alcohol**   **Alkoxide**

Acidity: RO–H≈HO–H>H₂N–H>H₃C–H
Basicity: RO⁻≈HO⁻<H₂N⁻<H₃C⁻

Laboratory Preparation of Alcohols

2. Nucleophilic Displacement of Halides and Other Leaving Groups by Hydroxide (Section 8-5)

$$RCH_2X\ +\ HO^-\ \xrightarrow[S_N2]{H_2O}\ RCH_2OH\ +\ X^-$$

X = halide, sulfonate
Primary, secondary (tertiary undergoes elimination)

$$R\overset{\underset{\underset{\displaystyle R'}{|}}{}}{C}HX\ +\ CH_3\overset{\overset{\displaystyle O}{\|}}{C}O^-\ \xrightarrow{S_N2}\ R\overset{\underset{\underset{\displaystyle R'}{|}}{}}{C}HO\overset{\overset{\displaystyle O}{\|}}{C}CH_3\ \xrightarrow[\text{Ester hydrolysis}]{HO^-}\ R\overset{\underset{\underset{\displaystyle R'}{|}}{}}{C}HOH$$

$$R'\overset{\overset{\displaystyle R}{|}}{\underset{\underset{\displaystyle R''}{|}}{C}}X\ \xrightarrow[S_N1]{H_2O,\ acetone}\ R'\overset{\overset{\displaystyle R}{|}}{\underset{\underset{\displaystyle R''}{|}}{C}}OH$$

Best method for tertiary

3. Reduction of Aldehydes and Ketones by Hydrides (Section 8-6)

$$R\overset{\overset{\displaystyle O}{\|}}{C}H\ \xrightarrow{NaBH_4,\ CH_3CH_2OH}\ RCH_2OH \qquad R\overset{\overset{\displaystyle O}{\|}}{C}R'\ \xrightarrow{NaBH_4,\ CH_3CH_2OH}\ R\overset{\overset{\displaystyle OH}{|}}{\underset{\underset{\displaystyle H}{|}}{C}}R'$$

$$R\overset{\overset{\displaystyle O}{\|}}{C}H\ \xrightarrow[2.\ H^+,\ H_2O]{1.\ LiAlH_4,\ (CH_3CH_2)_2O}\ RCH_2OH \qquad R\overset{\overset{\displaystyle O}{\|}}{C}R'\ \xrightarrow[2.\ H^+,\ H_2O]{1.\ LiAlH_4,\ (CH_3CH_2)_2O}\ R\overset{\overset{\displaystyle OH}{|}}{\underset{\underset{\displaystyle H}{|}}{C}}R'$$

Aldehyde   **Primary**   **Ketone**   **Secondary**
                         alcohol   **alcohol**

Oxidation of Alcohols

4. Chromium Reagents (Section 8-6)

$$RCH_2OH\ \xrightarrow{PCC,\ CH_2Cl_2}\ R\overset{\overset{\displaystyle O}{\|}}{C}H \qquad R\overset{\overset{\displaystyle OH}{|}}{C}HR'\ \xrightarrow{Na_2Cr_2O_7,\ H_2SO_4}\ R\overset{\overset{\displaystyle O}{\|}}{C}R'$$

Primary alcohol   **Aldehyde** **Secondary alcohol**   **Ketone**

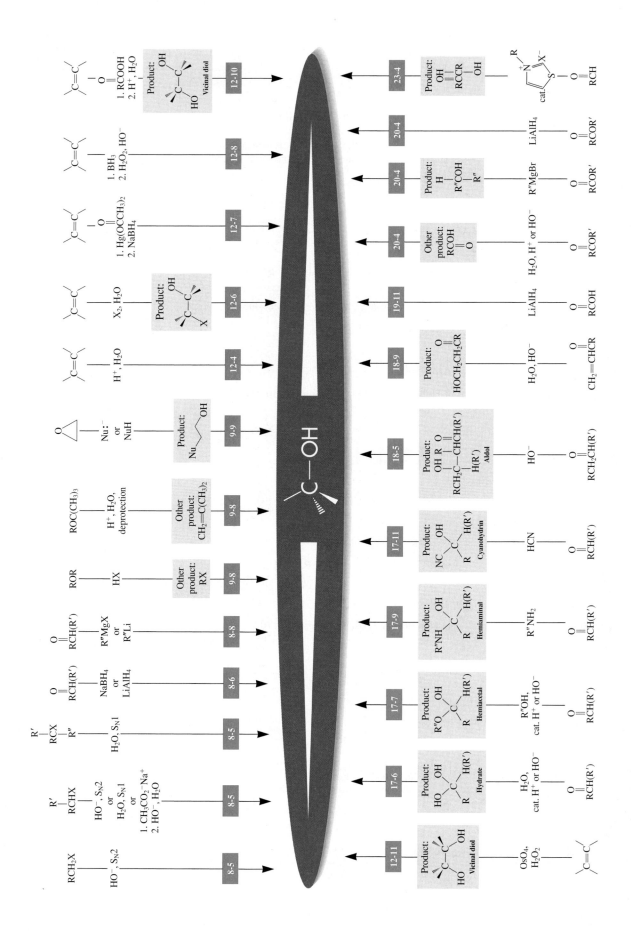

Preparation of Alcohols

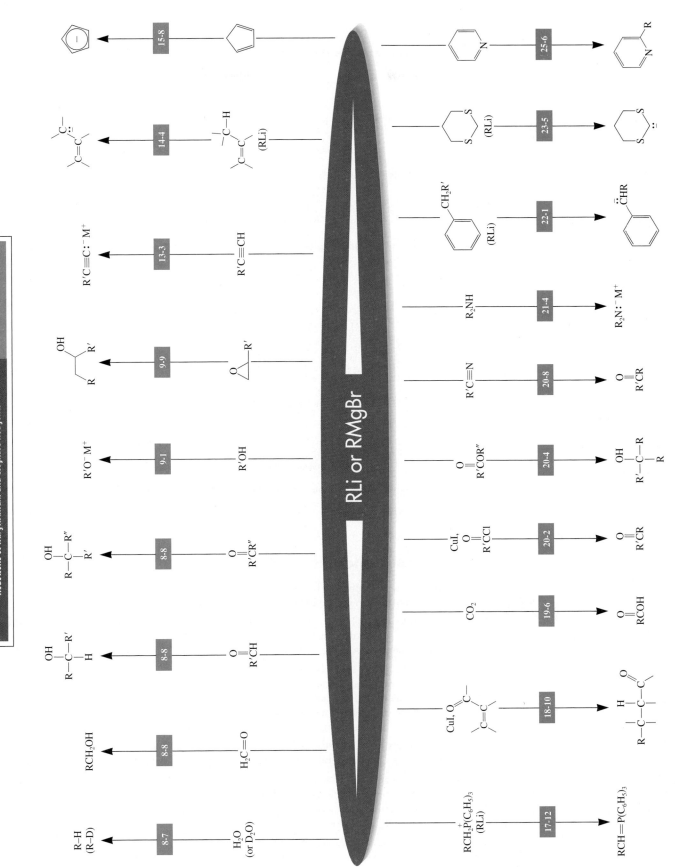

Reactions of Alkyllithium and Grignard Reagents

section number	

RLi or RMgBr

15-8

14-4

13-3

9-9

9-1

8-8

8-8

8-8

8-7

25-6

23-5 (RLi)

22-1 (RLi)

21-4

20-8

20-4

20-2 CuI

19-6

18-10 CuI

17-12 (RLi)

R'C≡C:⁻ M⁺

R'O⁻ M⁺

RCH₂OH

R–H (R–D)

R₂N:⁻ M⁺

RCOH

R'OH

R'C≡CH

H₂C=O

H₂O (or D₂O)

R₂NH

R'C≡N

CO₂

RCH₂P(C₆H₅)₃⁺ (RLi)

RCH=P(C₆H₅)₃

317

Organometallic Reagents

5. Reaction of Metals with Haloalkanes (Section 8-7)

$$RX \ + \ Li \ \xrightarrow{\ (CH_3CH_2)_2O\ } \ RLi$$

Alkyllithium reagent

$$RX \ + \ Mg \ \xrightarrow{\ (CH_3CH_2)_2O\ } \ RMgX$$

Grignard reagent

R cannot contain acidic groups such as O–H or electrophilic groups such as C=O.

6. Hydrolysis (Section 8-7)

$$RLi \quad or \quad RMgX \ + \ H_2O \ \longrightarrow \ RH$$

$$RLi \quad or \quad RMgX \ + \ D_2O \ \longrightarrow \ RD$$

7. Reaction of Organometallic Compounds to Aldehydes and Ketones (Section 8-8)

$$RLi \quad or \quad RMgX \ + \ CH_2{=}O \ \longrightarrow \ RCH_2OH$$

                                  Formaldehyde   **Primary alcohol**

$$
\begin{array}{ccc}
 & \overset{\displaystyle O}{\overset{\displaystyle \|}{}} & \overset{\displaystyle OH}{\overset{\displaystyle |}{}} \\
RLi \quad or \quad RMgX \ + \ R'CH & \longrightarrow & RCR' \\
 & & \overset{|}{H}
\end{array}
$$

                         Aldehyde   **Secondary alcohol**

$$
\begin{array}{ccc}
 & \overset{\displaystyle O}{\overset{\displaystyle \|}{}} & \overset{\displaystyle OH}{\overset{\displaystyle |}{}} \\
RLi \quad or \quad RMgX \ + \ R'CR'' & \longrightarrow & RCR' \\
 & & \overset{|}{R''}
\end{array}
$$

                         Ketone   **Tertiary alcohol**

Aldehyde or ketone cannot contain other groups that react with
organometallic reagents such as O–H or other C=O groups.

8. Alkanes from Haloalkanes and Lithium Aluminum Hydride (Section 8-7)

$$RX + LiAlH_4 \ \xrightarrow{\ (CH_3CH_2)_2O\ } \ RH$$

Important Concepts

1. Alcohols are **alkanols** in IUPAC nomenclature. The stem containing the functional group gives the alcohol its name. Alkyl and halo substituents are added as prefixes.

2. Like water, alcohols have a **polarized** and short O–H bond. The hydroxy group is **hydrophilic** and enters into **hydrogen bonding.** Consequently, alcohols have unusually high boiling points and, in many cases, appreciable water solubility. The alkyl part of the molecule is **hydrophobic.**

3. Again like water, alcohols are **amphoteric:** They are both acidic and basic. Complete deprotonation to an **alkoxide** takes place with bases whose conjugate acids are considerably weaker than the alcohol. Protonation gives an **alkyloxonium ion.** In solution, the order of acidity is primary > secondary > tertiary alcohol. Electron-withdrawing substituents increase the acidity (and reduce the basicity).

4. The conversion of the electrophilic alkyl group in a haloalkane, $C^{\delta+}{-}X^{\delta-}$, into its nucleophilic analog in an **organometallic compound,** $C^{\delta-}{-}M^{\delta+}$, is an example of **reverse polarization.**

5. The carbon atom in the **carbonyl group,** C=O, of an aldehyde or a ketone is electrophilic and therefore subject to attack by nucleophiles, such as hydride in **hydride reagents** or alkyl in organometallic compounds. Subsequent to aqueous work-up, the products of such transformations are alcohols.

6. The **oxidation** of alcohols to aldehydes and ketones by chromium(VI) reagents opens up important synthetic possibilities based on further reactions with organometallic reagents.

7. **Retrosynthetic analysis** aids in planning the synthesis of complex organic molecules by identifying strategic bonds that may be constructed in an efficient sequence of reactions.

Problems

24. Name the following alcohols according to the IUPAC nomenclature system. Indicate stereochemistry (if any) and label the hydroxy groups as primary, secondary, or tertiary.

(a) CH$_3$CH$_2$CHCH$_3$ with OH

(b) CH$_3$CHCH$_2$CHCH$_2$CH$_3$ with Br and OH

(c) HOCH$_2$CH(CH$_2$CH$_2$CH$_3$)$_2$

(d) structure with CH$_2$Cl, H, C, H$_3$C, OH

(e) cyclobutane with CH$_2$CH$_3$ and OH

(f) cyclodecane ring with OH and Br

(g) C(CH$_2$OH)$_4$

(h) Fischer projection: CH$_2$OH / H—OH / H—OH / CH$_2$OH

(i) cyclopentane with OH and CH$_2$CH$_2$OH

(j) Fischer projection: CH$_2$OH / H$_3$C—Cl / CH$_2$CH$_3$

25. Draw the structures of the following alcohols. **(a)** 2-(Trimethylsilyl)ethanol; **(b)** 1-methylcyclopropanol; **(c)** 3-(1-methylethyl)-2-hexanol; **(d)** (*R*)-2-pentanol; **(e)** 3,3-dibromocyclohexanol.

26. Rank each group of compounds in order of increasing boiling point. **(a)** Cyclohexane, cyclohexanol, chlorocyclohexane; **(b)** 2,3-dimethyl-2-pentanol, 2-methyl-2-hexanol, 2-heptanol.

27. Explain the order of water solubilities for the compounds in each of the following groups. **(a)** Ethanol > chloroethane > ethane; **(b)** methanol > ethanol > 1-propanol.

28. 1,2-Ethanediol exists to a much greater extent in the *gauche* conformation than does 1,2-dichloroethane. Explain. Would you expect the *gauche* : *anti* conformational ratio of 2-chloroethanol to be similar to that of 1,2-dichloroethane or more like that of 1,2-ethanediol?

29. The most stable conformation of *trans*-1,2-cyclohexanediol is the chair in which both hydroxy groups are equatorial. **(a)** Draw the structure or, better yet, make a model of the compound in this conformation. **(b)** Reaction of this diol with the chlorosilane R$_3$SiCl, R = (CH$_3$)$_2$CH (isopropyl), gives the corresponding disilyl ether shown below. Remarkably, this transformation causes the chair to flip, giving a conformation where both silyl ether groups are in *axial* positions. Explain this observation by means of either structural drawings or models.

cyclohexane with OSiR$_3$ and OSiR$_3$

30. Rank the compounds in each group in order of decreasing acidity.

(a) CH$_3$CHClCH$_2$OH, CH$_3$CHBrCH$_2$OH, BrCH$_2$CH$_2$CH$_2$OH

(b) CH$_3$CCl$_2$CH$_2$OH, CCl$_3$CH$_2$OH, (CH$_3$)$_2$CClCH$_2$OH

(c) (CH$_3$)$_2$CHOH, (CF$_3$)$_2$CHOH, (CCl$_3$)$_2$CHOH

31. Write an appropriate equation to show how each of the following alcohols acts as, first, a base, and, second, an acid in solution. How do the base and acid strengths of each compare with those of methanol? **(a)** (CH$_3$)$_2$CHOH; **(b)** CH$_3$CHFCH$_2$OH; **(c)** CCl$_3$CH$_2$OH.

32. Given the pK_a values of -2.2 for CH$_3$OH$_2^+$ and 15.5 for CH$_3$OH, calculate the pH at which **(a)** methanol will contain exactly equal amounts of CH$_3$OH$_2^+$ and CH$_3$O$^-$; **(b)** 50% CH$_3$OH and 50% CH$_3$OH$_2^+$ will be present; **(c)** 50% CH$_3$OH and 50% CH$_3$O$^-$ will be present.

33. Do you expect hyperconjugation to be important in the stabilization of alkyloxonium ions (e.g., ROH$_2^+$, R$_2$OH$^+$)? Explain your answer.

34. Evaluate each of the following possible alcohol syntheses as being good (the desired alcohol is the major or only product), not so good (the desired alcohol is a minor product), or worthless. (**Hint:** Refer to Section 7-9 if necessary.)

(a) CH$_3$CH$_2$CH$_2$CH$_2$Cl $\xrightarrow{\text{H}_2\text{O, CH}_3\text{CCH}_3}$ CH$_3$CH$_2$CH$_2$CH$_2$OH

(b) CH$_3$OSO$_2$—⟨benzene⟩—CH$_3$ $\xrightarrow{\text{HO}^-, \text{H}_2\text{O}, \Delta}$ CH$_3$OH

(c) cyclohexane-I $\xrightarrow{\text{HO}^-, \text{H}_2\text{O}, \Delta}$ cyclohexane-OH

(d) CH$_3$CHCH$_2$CH$_2$CH$_3$ (I) $\xrightarrow{\text{H}_2\text{O}, \Delta}$ CH$_3$CHCH$_2$CH$_2$CH$_3$ (OH)

(e) CH$_3$CHCH$_3$ (CN) $\xrightarrow{\text{HO}^-, \text{H}_2\text{O}, \Delta}$ CH$_3$CHCH$_3$ (OH)

(f) CH$_3$OCH$_3$ $\xrightarrow{\text{HO}^-, \text{H}_2\text{O}, \Delta}$ CH$_3$OH

(g) cyclopentane with H$_3$C, Br $\xrightarrow{\text{H}_2\text{O}}$ cyclopentane with H$_3$C, OH

(h) CH$_3$CHCH$_2$Cl (CH$_3$) $\xrightarrow{\text{HO}^-, \text{H}_2\text{O}, \Delta}$ CH$_3$CHCH$_2$OH (CH$_3$)

35. For every process in Problem 34 that gives the designated product in poor yield, suggest a superior method if possible.

36. Give the major product(s) of each of the following reactions. Aqueous work-up steps (when necessary) have been omitted.

(a) CH$_3$CH=CHCH$_3$ $\xrightarrow[\textbf{(Hint: See Section 8-4.)}]{\text{H}_3\text{PO}_4, \text{H}_2\text{O}, \Delta}$

(b) CH$_3$CCH$_2$CH$_2$CCH$_3$ (two C=O) $\xrightarrow[\text{2. H}^+, \text{H}_2\text{O}]{\text{1. LiAlH}_4, (\text{CH}_3\text{CH}_2)_2\text{O}}$

(c) cyclohexane-CHO $\xrightarrow{\text{NaBH}_4, \text{CH}_3\text{CH}_2\text{OH}}$

(d) 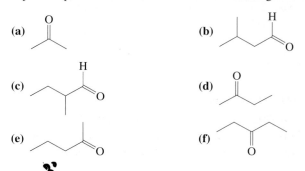 Br — LiAlH₄, (CH₃CH₂)₂O →

(e) CH₃, (CH₃)₂CH — NaBH₄, CH₃CH₂OH →

(f) H, H — NaBH₄, CH₃CH₂OH →

37. What is the direction of the following equilibrium? (**Hint:** The pK_a for H_2 is about 38.)

$$H^- + H_2O \rightleftharpoons H_2 + HO^-$$

38. Formulate the product of each of the following reactions. The solvent in each case is $(CH_3CH_2)_2O$.

(a) CH_3CH (O) — 1. LiAlD₄ / 2. H⁺, H₂O →

(b) CH_3CH (O) — 1. LiAlH₄ / 2. D⁺, D₂O →

(c) CH_3CH_2I — LiAlD₄ →

39. Write out a mechanism for every reaction depicted in Problem 38.

40. Give the major product(s) of each of the following reactions [after work-up with aqueous acid in (d), (f), and (h)].

(a) $CH_3(CH_2)_5CHCH_3$ (Cl) — Mg, (CH₃CH₂)₂O →

(b) Product of (a) — D₂O →

(c) Br — Li, (CH₃CH₂)₂O →

(d) Product of (c) + (cyclopentanone) →

(e) $CH_3CH_2CH_2Cl$ + Mg — (CH₃CH₂)₂O →

(f) Product of (e) + (phenyl)–CCH₃ (O) →

(g) (cyclobutyl)Br + 2 Li — (CH₃CH₂)₂O →

(h) 2 mol product of (g) + 1 mol $CH_3CCH_2CH_2CCH_3$ (O, O) →

41. The common practice of washing laboratory glassware with acetone can lead to unintended consequences. For example, a student plans to carry out the preparation of methylmagnesium iodide,

CH₃MgI, which he will add to benzaldehyde, C_6H_5CHO. What compound is he intending to synthesize after aqueous work-up? Using his freshly washed glassware, he carries out the procedure and finds that he has produced an unexpected tertiary alcohol as a product. What substance did he make? How did it form?

42. Which of the following halogenated compounds can be used successfully to prepare a Grignard reagent for alcohol synthesis by subsequent reaction with an aldehyde or ketone? Which ones cannot and why?

(a) H_3C CH_3 —Br

(b) H OH —Cl

(c) I ·H —OCH₃

(d) H, Cl, O

(e) H, Br —H

(**Hint:** See Problem 49 of Chapter 1.)

43. Give the major product(s) of each of the following reactions (after aqueous work-up). The solvent in each case is ethoxyethane (diethyl ether).

(a) (cyclopropyl)—MgBr + HCH (O) →

(b) CH_3CHCH_2MgCl (CH₃) + CH_3CH (O) →

(c) $C_6H_5CH_2Li$ + C_6H_5CH (O) →

(d) CH_3CHCH_3 (MgBr) + (cyclohexanone, O) →

(e) (cyclopentyl H MgCl) + (H, O aldehyde) →

44. For each reaction presented in Problem 43, write out the complete, step-by-step mechanism using curved-arrow notation. Include the aqueous acid work-up.

45. Write the structures of the products of reaction of ethylmagnesium bromide, CH_3CH_2MgBr, with each of the following carbonyl compounds. Identify any reaction that gives more than one stereoisomeric product, and indicate whether you would expect the products to form in identical or in differing amounts.

(a) (O)

(b) (H, O)

(c) (H, O)

(d) (O)

(e) (O)

(f) (O)

(g)

(h)

(i)

(j)

46. Give the expected major product of each of the following reactions. PCC is the abbreviation for pyridinium chlorochromate (Section 8-6).

Carboxylic acid

(a) $CH_3CH_2CH_2OH$ $\xrightarrow{Na_2Cr_2O_7, H_2SO_4, H_2O}$

(b) $(CH_3)_2CHCH_2OH$ $\xrightarrow{PCC, CH_2Cl_2}$

(c) $\xrightarrow{Na_2Cr_2O_7, H_2SO_4, H_2O}$

(d) $\xrightarrow{PCC, CH_2Cl_2}$

(e) $\xrightarrow{PCC, CH_2Cl_2}$

47. Write out a mechanism for every reaction depicted in Problem 46.

48. Give the expected major product of each of the following reaction *sequences*. PCC refers to pyridinium chlorochromate.

(a) $(CH_3)_2CHOH$ $\xrightarrow{\begin{array}{l}1.\ CrO_3, H_2SO_4, H_2O\\2.\ CH_3CH_2MgBr, (CH_3CH_2)_2O\\3.\ H^+, H_2O\end{array}}$

(b) $CH_3CH_2CH_2CH_2Cl$ $\xrightarrow{\begin{array}{l}1.\ ^-OH, H_2O\\2.\ PCC, CH_2Cl_2\\3.\ \text{(cyclopentyl)}-Li, (CH_3CH_2)_2O\\4.\ H^+, H_2O\end{array}}$

(c) Product of (b) $\xrightarrow{\begin{array}{l}1.\ CrO_3, H_2SO_4, H_2O\\2.\ LiAlD_4, (CH_3CH_2)_2O\\3.\ H^+, H_2O\end{array}}$

49. **CHALLENGE** Unlike Grignard and organolithium reagents, organometallic compounds of the most electropositive metals (Na, K, etc.) react rapidly with haloalkanes. As a result, attempts to convert RX into RNa or RK by reaction with the corresponding metal lead to alkanes by a reaction called *Wurtz coupling*.

$$2\ RX + 2\ Na \longrightarrow R{-}R + 2\ NaX$$

which is the result of

$$R{-}X + 2\ Na \longrightarrow R{-}Na + NaX$$

followed rapidly by

$$R{-}Na + R{-}X \longrightarrow R{-}R + NaX$$

When it was still in use, the Wurtz coupling reaction was employed mainly for the preparation of alkanes by the coupling of two identical alkyl groups (e.g., equation 1 below). Suggest a reason why Wurtz coupling might not be a useful method for coupling two *different* alkyl groups (equation 2).

$$2\ CH_3CH_2CH_2Cl + 2\ Na \longrightarrow$$
$$CH_3CH_2CH_2CH_2CH_2CH_3 + 2\ NaCl \quad (1)$$

$$CH_3CH_2Cl + CH_3CH_2CH_2Cl + 2\ Na \longrightarrow$$
$$CH_3CH_2CH_2CH_2CH_3 + 2\ NaCl \quad (2)$$

50. The reaction of two equivalents of Mg with 1,4-dibromobutane produces compound A. The reaction of A with two equivalents of CH_3CHO (acetaldehyde), followed by work-up with dilute aqueous acid, produces compound B, having the formula $C_8H_{18}O_2$. What are the structures of A and B?

51. Suggest the best synthetic route to each of the following simple alcohols, using in each case a simple alkane as your initial starting molecule. What are some disadvantages of beginning syntheses with alkanes?

(a) Methanol	**(b)** Ethanol	**(c)** 1-Propanol
(d) 2-Propanol	**(e)** 1-Butanol	**(f)** 2-Butanol
(g) 2-Methyl-2-propanol		

52. For each alcohol in Problem 51, suggest (if possible) a synthetic route that starts with, first, an aldehyde and, second, a ketone.

53. Outline the best method for preparing each of the following compounds from an appropriate alcohol.

(a)

(b) $CH_3CH_2CH_2CH_2COOH$

(c)

(d) CH_3CHCCH_3 with CH_3 and O

(e) CH_3CH (with $=O$)

54. Suggest three different syntheses of 2-methyl-2-hexanol. Each route should utilize one of the following starting materials. Then use any number of steps and any other reagents needed.

(a)

(b)

(c)

55. Devise three different syntheses of 3-octanol starting with **(a)** a ketone; **(b)** an aldehyde; **(c)** an aldehyde different from that employed in (b).

56. Fill in the missing reagent(s) needed to convert each molecule into the next one pictured in the synthetic scheme below. If a transformation requires more than one step, number the reagents for the individual steps sequentially.

57. Waxes are naturally occurring esters (alkyl alkanoates) containing long, straight alkyl chains. Whale oil contains the wax 1-hexadecyl hexadecanoate, as shown below. How would you synthesize this wax, using an S_N2 reaction?

$$CH_3(CH_2)_{14}\overset{\overset{\displaystyle O}{\|}}{C}O(CH_2)_{15}CH_3$$

1-Hexadecyl hexadecanoate

58. The reduced form of the coenzyme nicotinamide adenine dinucleotide (NAD$^+$, Real Life 8-1) is abbreviated NADH. In the presence of a variety of enzyme catalysts, it acts as a biological hydride donor, capable of reducing aldehydes and ketones to alcohols, according to the general formula

$$\overset{\overset{\displaystyle O}{\|}}{R C R} + NADH + H^+ \xrightarrow{\text{Enzyme}} \overset{\overset{\displaystyle OH}{|}}{R C H R} + NAD^+$$

The COOH functional group of carboxylic acids is not reduced. Write the products of the NADH reduction of each of the molecules below.

(a) $CH_3\overset{\overset{\displaystyle O}{\|}}{C}H + NADH \xrightarrow{\text{Alcohol dehydrogenase}}$

(b) $CH_3\overset{\overset{\displaystyle O}{\|}}{C}\overset{\overset{\displaystyle O}{\|}}{C}OH + NADH \xrightarrow{\text{Lactate dehydrogenase}}$

2-Oxopropanoic acid   **Lactic acid**
(Pyruvic acid)

(c) $HO\overset{\overset{\displaystyle O}{\|}}{C}CH_2\overset{\overset{\displaystyle O}{\|}}{C}\overset{\overset{\displaystyle O}{\|}}{C}OH + NADH \xrightarrow{\text{Malate dehydrogenase}}$

2-Oxobutanedioic acid   **Malic acid**
(Oxaloacetic acid)

59. Reductions by NADH (Problem 58) are stereospecific, with the stereochemistry of the product controlled by an enzyme (see Real Life 8-1). The common forms of lactate and malate dehydrogenases produce exclusively the *S* stereoisomers of lactic and malic acids, respectively. Draw these stereoisomers.

60. **CHALLENGE** Chemically modified steroids have become increasingly important in medicine. Give the possible product(s) of the following reactions. In each case, identify the major stereoisomer formed on the basis of delivery of the attacking reagent from the less hindered side of the substrate molecule. (**Hint:** Make models and refer to Section 4-7.)

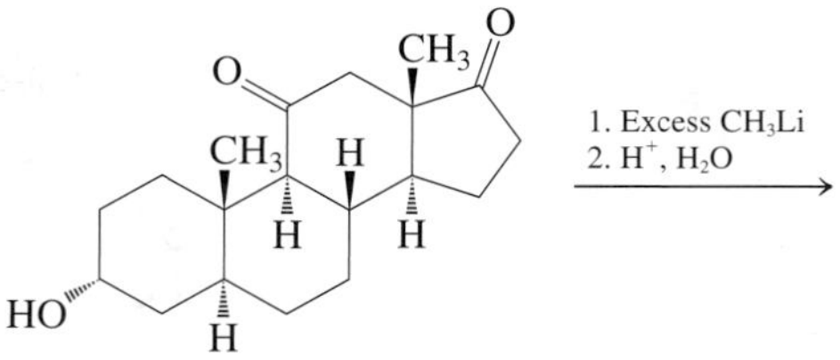

61. **CHALLENGE** Why do the two reactions shown in Problem 60 both require the use of excess CH$_3$MgI and CH$_3$Li, respectively? How many equivalents of the organometallic reagents are needed in each case? What are the products of the reaction at each functionalized site in each molecule?

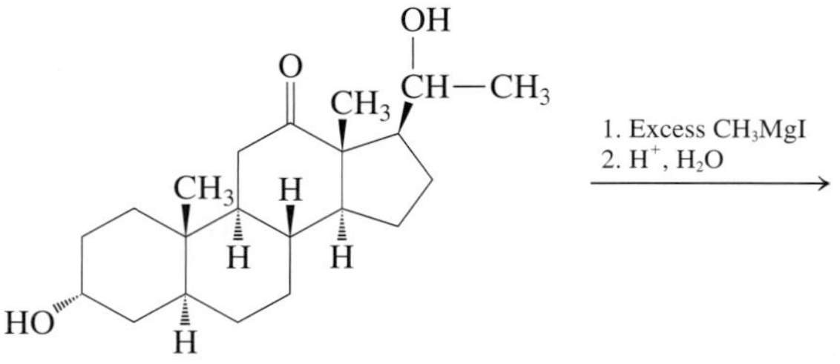

Team Problem

62. Your team has been asked to devise a synthesis of the tertiary alcohol 2-cyclohexyl-2-butanol, A, a molecule that imparts a lily-of-the-valley "fresh" fragrance to perfumes. Your laboratory is well stocked with the usual organic and inorganic reagents and solvents. An inventory check reveals that there are many appropriate bromoalkanes and alcohols on hand. As a group, analyze alcohol A retrosynthetically and propose all possible strategic disconnections. Check the inventory to see if a particular route is feasible in terms of available starting materials. Then divide the proposed routes evenly among yourselves to evaluate the merits or pitfalls of these strategies. Write a detailed synthetic plan based on your chosen retrosynthesis for the synthesis of 2-cyclohexyl-2-butanol. Reconvene to defend or reject these plans. Finally, take into consideration the prices of your starting materials. Which one of your routes to A is the cheapest?

Target Molecule		Inventory (Price)

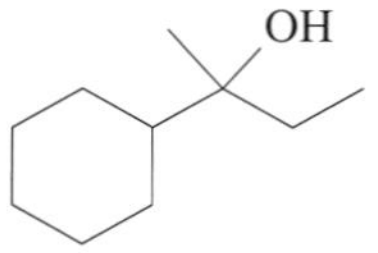

2-Cyclohexyl-2-butanol

A

2-Bromobutane ($79/500 g)
Bromocyclohexane ($94/kg)
Bromoethane ($55/kg)
Bromomethane ($811/kg)
2-Butanol ($46/kg)

Cyclohexanol ($38/kg)
1-Cyclohexylethanol ($79/5 g)
Cyclohexylmethanol ($45/25 g)
(Bromomethyl)cyclohexane ($183/100 g)

Preprofessional Problems

63. A compound known to contain only C, H, and O gives the following upon microanalysis (atomic weights: C = 12.0, H = 1.00, O = 16.0): 52.1% C, 13.1% H. It is found to have a boiling point of 78°C. Its structure is

(a) CH_3OCH_3 (b) CH_3CH_2OH
(c) $HOCH_2CH_2CH_2CH_2OH$ (d) $HOCH_2CH_2CH_2OH$
(e) none of these

64. The compound whose structure is $(CH_3)_2CHCH_2CHCH_2CH_3$ is best named (IUPAC):
                                                          $\overset{\displaystyle |}{OH}$

(a) 2-methyl-4-hexanol
(b) 5-methyl-3-hexanol
(c) 1,4,4-trimethyl-2-butanol
(d) 1-isopropyl-2-hexanol

65. In this transformation, what is the best structure for "A"?

$$A \xrightarrow[\text{2. H}^+, \text{H}_2\text{O (work-up)}]{\text{1. LiAlH}_4, \text{ dry ether}}$$

(structure: cyclopentane ring with H and OH on one carbon, and two CH₂CH₃ groups on another carbon)

(a) (cyclopentanone)

(b) (cyclopentanone with CH₂CH₃ and CH₂CH₃ substituents)

(c) (cyclopentanone with CH₂CH₃ and CH₂CH₃ substituents)

(d) (cyclopentenone with CH₂CH₃ and CH₂CH₃ substituents)

66. Ester hydrolysis is best illustrated by

(a) $CH_3OCCH_3 \xrightarrow{H^+, H_2O} CH_3OCH_3 + CO$ (with C=O on the second carbon)

(b) $CH_3OCCH_3 \xrightarrow{H^+, H_2O} CH_3OH + HOCCH_3$ (with C=O groups)

(c) $CH_3OCH_2OH \xrightarrow{H_2O} CH_3OH + H-\overset{\text{O}}{\underset{}{C}}-H$

(d) $CH_3OH + CH_3CO_2H \xrightarrow{H_2O} CH_3OCCH_3$ (with C=O)

CHAPTER 9

Further Reactions of Alcohols and the Chemistry of Ethers

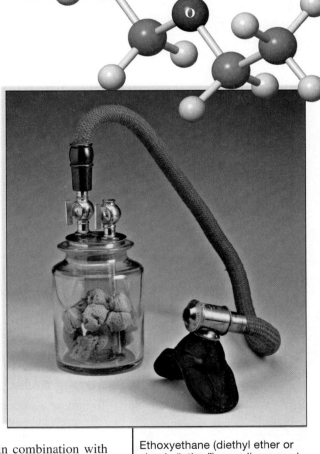

o you remember the fizzing that occurred when you (or your teacher) dropped a pellet of sodium into water? The violent reaction that you observed was due to the conversion of the metal and water into NaOH and H_2 gas. Alcohols, which can be regarded as "alkylated water" (Section 8-2), undergo the same reaction, albeit less vigorously, to give NaOR and H_2. This chapter will describe this and further transformations of the hydroxy substituent.

Figure 9-1 depicts a variety of reaction modes available to alcohols. Usually at least one of the four bonds marked *a*, *b*, *c*, or *d* is cleaved. In Chapter 8 we learned that oxidation to aldehydes and ketones breaks bonds *a* and *d*. We found that the use of this reaction in combination with additions of organometallic reagents provides us with the means of preparing alcohols of considerable structural diversity.

To further explore the reactions of alcohols, we start by reviewing their acidic and basic properties. Deprotonation at bond *a* furnishes alkoxides (Section 8-3), which are valuable both as strong bases and as nucleophiles (Section 7-8). Strong acids transform alcohols into alkyloxonium ions (Section 8-3), converting OH (a poor leaving group) into H_2O (a good leaving group). Subsequently, bond *b* may break, thereby leading to substitution; or elimination can take place by cleavage of bonds *b* and *c*. We shall see that the carbocation intermediates arising from acid treatment of secondary and tertiary alcohols have a varied chemistry.

An introduction to the preparation of esters and their applications in synthesis is followed by the chemistry of ethers and sulfur compounds. Alcohols, ethers, and their sulfur analogs occur widely in nature and have numerous applications in industry and medicine.

Ethoxyethane (diethyl ether or simply "ether") was discovered as an anesthetic in 1846, prompting a newspaper headline exclaiming "We have conquered pain." The photo shows an ether inhaler from that period. It consists of a glass vessel that contains ether-soaked sponges, connected to a facemask by tubing. Apart from its flammability, ether is fairly safe, but has the undesirable postanesthetic effects of headaches and nausea. It has been superseded by a host of more effective pain-killing drugs, but is still popular in some developing nations.

Figure 9-1 Four typical reaction modes of alcohols. In each mode, one or more of the four bonds marked *a–d* are cleaved (wavy line denotes bond cleavage): (*a*) deprotonation by base; (*b*) protonation by acid followed by uni- or bimolecular substitution; (*b, c*) elimination; and (*a, d*) oxidation.

Sodium reacts with water vigorously, releasing hydrogen gas.

9-1 REACTIONS OF ALCOHOLS WITH BASE: PREPARATION OF ALKOXIDES

As described in Section 8-3, alcohols can be both acids and bases. In this section, we shall review the methods by which the hydroxy group of alcohols is deprotonated to furnish their conjugate bases, the alkoxides.

Strong bases are needed to deprotonate alcohols completely

To remove a proton from the OH group of an alcohol (Figure 9-1, cleavage of bond *a*), we must use a base stronger than the alkoxide. Examples include lithium diisopropylamide (Section 7-8), butyllithium (Section 8-7), and alkali metal hydrides (Section 8-6, Exercise 8-4), such as potassium hydride, KH. Such hydrides are particularly useful because the only by-product of the reaction is hydrogen gas.

Three Ways of Making Methoxide from Methanol

$$CH_3OH \;+\; \underset{\substack{\text{Lithium diisopropylamide}}}{Li^{+-}NCH(CH_3)_2 \atop \overset{CH(CH_3)_2}{|}} \;\overset{K=10^{24.5}}{\rightleftharpoons}\; CH_3O^-Li^+ \;+\; \underset{pK_a=40}{HNCH(CH_3)_2 \atop \overset{CH(CH_3)_2}{|}}$$

$pK_a = 15.5$

$$\underset{pK_a=15.5}{CH_3OH} \;+\; \underset{\text{Butyllithium}}{CH_3CH_2CH_2CH_2Li} \;\overset{K=10^{34.5}}{\rightleftharpoons}\; CH_3O^-Li^+ \;+\; \underset{pK_a=50}{CH_3CH_2CH_2CH_2H}$$

$$\underset{pK_a=15.5}{CH_3OH} \;+\; \underset{\substack{\text{Potassium}\\ \text{hydride}}}{K^+H^-} \;\overset{K=10^{22.5}}{\rightleftharpoons}\; CH_3O^-K^+ \;+\; \underset{pK_a=38}{H—H}$$

Exercise 9-1

We shall encounter later many reactions that require catalytic base, for example, catalytic sodium methoxide in methanol. Assume that you would like to make such a solution containing 10 mmol of $NaOCH_3$ in 1 liter of CH_3OH. Would it be all right simply to add 10 mmol of NaOH to the solvent? (**Caution:** Simply comparing pK_a values (Table 2-2) is not enough. **Hint:** See Section 2-3.)

Alkali metals also deprotonate alcohols—but by reduction of H$^+$

Another common way of obtaining alkoxides is by the reaction of alcohols with alkali metals, such as lithium. Such metals reduce water—in some cases, violently—to yield alkali metal hydroxides and hydrogen gas. When the more reactive metals (sodium, potassium, and cesium) are exposed to water in air, the hydrogen generated can ignite spontaneously or even detonate.

$$2\,H\!-\!OH \;+\; 2\,M\,(Li,\,Na,\,K,\,Cs) \;\longrightarrow\; 2\,M^{+\,-}OH \;+\; H_2$$

The alkali metals act similarly on the alcohols to give alkoxides, but the transformation is less vigorous. Here are two examples.

Alkoxides from Alcohols and Alkali Metals

$$2\,CH_3CH_2OH \;+\; 2\,Na \;\longrightarrow\; 2\,CH_3CH_2O^-Na^+ \;+\; H_2$$
$$2\,(CH_3)_3COH \;+\; 2\,K \;\longrightarrow\; 2\,(CH_3)_3CO^-K^+ \;+\; H_2$$

The reactivity of the alcohols used in this process decreases with increasing substitution, methanol being most reactive and tertiary alcohols least reactive.

Relative Reactivity of ROH with Alkali Metals

$$R = CH_3 > primary > secondary > tertiary$$

Decreasing reactivity →

2-Methyl-2-propanol reacts so slowly that it can be used to safely destroy potassium residues in the laboratory.

What are alkoxides good for? We have already seen that they can be useful reagents in organic synthesis. For example, the reaction of hindered alkoxides with haloalkanes gives elimination.

$$CH_3CH_2CH_2CH_2Br \xrightarrow[\text{E2}]{(CH_3)_3CO^-K^+,\,(CH_3)_3COH} CH_3CH_2CH\!=\!CH_2 \;+\; (CH_3)_3COH \;+\; K^+Br^-$$

Less branched alkoxides attack primary haloalkanes by the S_N2 reaction to give ethers. This method is described in Section 9-6.

In Summary A strong base will convert an alcohol into an alkoxide by an acid-base reaction. The stronger the base, the more the equilibrium is displaced to the alkoxide side. Alkali metals react with alcohols by reduction to generate hydrogen gas and an alkoxide. This process is retarded by steric hindrance.

9-2 | REACTIONS OF ALCOHOLS WITH STRONG ACIDS: ALKYLOXONIUM IONS IN SUBSTITUTION AND ELIMINATION REACTIONS OF ALCOHOLS

We have seen that heterolytic cleavage of the O–H bond in alcohols is readily achieved with strong bases. Can we break the C–O linkage (bond *b*, Figure 9-1) as easily? Yes, but now we need acid. Recall (Section 2-3) that water has a high pK_a (15.7): It is a weak acid. Consequently, hydroxide, its conjugate base, is an exceedingly poor leaving group. *For alcohols to undergo substitution or elimination reactions, the OH must first be converted into a better leaving group.*

Haloalkanes from primary alcohols and HX: Water can be a leaving group in S_N2 reactions

The simplest way of turning the hydroxy substituent in alcohols into a good leaving group is to protonate the oxygen to form an alkyloxonium ion. Recall (Section 8-3) that this process ties up one of the oxygen lone pairs by forming a bond to a proton. The positive charge therefore resides on the oxygen atom. *Protonation changes OH from a bad leaving group into neutral water, a good leaving group.*

This reaction is reversible and, under normal conditions, the equilibrium lies on the side of unprotonated alcohol. However, this is immaterial if a nucleophile is present in the mixture that is capable of trapping the oxonium species. For example, alkyloxonium ions derived from primary alcohols are subject to such nucleophilic attack. Thus, the butyloxonium ion resulting from the treatment of 1-butanol with concentrated HBr undergoes displacement by bromide to form 1-bromobutane. The originally nucleophilic (red) oxygen is protonated by the electrophilic proton (blue) to give the alkyloxonium ion, containing an electrophilic (blue) carbon and H_2O as a leaving group (green). In the subsequent S_N2 reaction, bromide acts as a nucleophile.

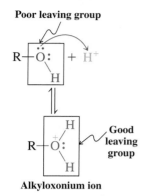

Poor leaving group

Good leaving group

Alkyloxonium ion

Primary Bromoalkane Synthesis from an Alcohol

$$CH_3CH_2CH_2CH_2\ddot{O}H \ + \ H\ddot{B}r\colon \ \rightleftharpoons \ CH_3CH_2CH_2CH_2\overset{+}{O}H_2 \ + \ \colon\!\ddot{B}r\colon^- \ \longrightarrow$$

$$CH_3CH_2CH_2CH_2\ddot{B}r\colon \ + \ H_2\ddot{O}$$

The reaction of primary alcohols with concentrated HI proceeds in similar fashion to furnish primary iodoalkanes (margin). On the other hand, concentrated HCl is sluggish in turning primary alcohols into the corresponding chloroalkanes, because chloride ion is a relatively weak nucleophile under protic conditions (Section 6-8). Therefore, this conversion, while possible, is usually carried out by other reagents (Section 9-4). In general, the acid-catalyzed S_N2 reaction of *primary* alcohols with HBr or HI is a good way of preparing simple *primary* haloalkanes. What about secondary and tertiary alcohols?

Iodoalkane Synthesis

$$HO(CH_2)_6OH \ + \ 2\ HI$$
1,6-Hexanediol

$$\downarrow$$

$$I(CH_2)_6I \ + \ 2\ H_2O$$
85%
1,6-Diiodo-hexane

Secondary and tertiary alcohols and HX: Water can be a leaving group to form carbocations in S_N1 and E1 reactions

Alkyloxonium ions derived from secondary and tertiary alcohols, in contrast with their primary counterparts, lose water with increasing ease to give the corresponding carbocations. The reason for this difference in behavior is the difference in carbocation stability (Section 7-5). Primary carbocations are too high in energy to be accessible under ordinary laboratory conditions, whereas secondary and tertiary carbocations are generated with increasing ease. Thus, primary alkyloxonium ions undergo only S_N2 reactions, whereas their secondary and tertiary relatives enter into S_N1 and E1 processes. When good nucleophiles are present, we observe S_N1 products.

Reactivity of Oxonium Ions

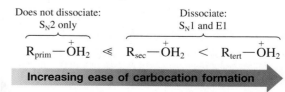

Does not dissociate:
S_N2 only

Dissociate:
S_N1 and E1

$$R_{prim}\!-\!\overset{+}{O}H_2 \ \ll \ R_{sec}\!-\!\overset{+}{O}H_2 \ < \ R_{tert}\!-\!\overset{+}{O}H_2$$

Increasing ease of carbocation formation

Synthetically, this fact is exploited in the preparation of *tertiary* haloalkanes from *tertiary* alcohols in the presence of excess concentrated aqueous hydrogen halide. The product forms in minutes at room temperature. The mechanism is precisely the reverse of that of solvolysis (Section 7-2).

Conversion of 2-Methyl-2-propanol into 2-Bromo-2-methylpropane

$$(CH_3)_3COH + HBr \rightleftharpoons (CH_3)_3CBr + H_2O$$

Excess

Mechanism of the S$_N$1 Reaction of Tertiary Alcohols with Hydrogen Halides

The reason for the success of this process is that relatively low temperatures are sufficient to generate the tertiary carbocation, thus largely preventing competing E1 reactions (Section 7-6). Indeed, at higher temperatures (or in the absence of good nucleophiles), elimination becomes dominant.* This explains why protonated *secondary* alcohols show the most complex behavior in the presence of HX, following S$_N$2, S$_N$1, and E1 pathways: They are relatively hindered, compared with their primary counterparts (thus reduced S$_N$2 reactivity), and relatively slow in forming carbocations, compared with their tertiary counterparts (e.g., retarded S$_N$1 reactivity); review Section 7-9.

The E1 reaction, here called **dehydration** because it results in the loss of a molecule of water (Figure 9-1, breaking bonds *b* and *c*; see also Sections 9-3 and 9-7), is one of the methods for the synthesis of alkenes (Section 11-7). Rather than the "nucleophilic" acids HBr and HI, so called because the conjugate base is a good nucleophile, "nonnucleophilic" acids, such as H_3PO_4 or H_2SO_4, are employed.

ANIMATED MECHANISM:
Alcohol dehydration

Alcohol Dehydration by the E1 Mechanism

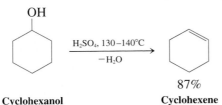

87%
Cyclohexanol   Cyclohexene

Mechanism of the Dehydration of Cyclohexanol

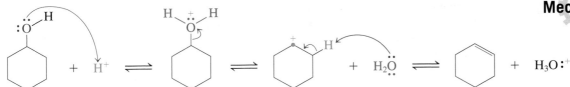

In the example, the conjugate base of the acid is the poor nucleophile HSO_4^-, and proton loss from the intermediate carbocation is observed. Dehydration of tertiary alcohols is even easier, often occurring at slightly above room temperature.

Exercise 9-2

Write the structure of the product that you expect from the reaction of 4-methyl-1-pentanol with concentrated aqueous HI. Give the mechanism of its formation.

---

*The preference for elimination at high temperatures has its origin in its relatively large positive entropy (Section 2-1), as two molecules (alkene + H_2O) result from one (alcohol), and the $\Delta S°$ term in the expression $\Delta G° = \Delta H° - T\Delta S°$ increases with temperature (see also Exercise 7-17).

Exercise 9-3

Write the structure of the products that you expect from the reaction of 1-methylcyclohexanol with **(a)** concentrated HCl and **(b)** concentrated H_2SO_4. Compare and contrast the mechanisms of the two processes. (**Hint:** Compare the relative nucleophilicity of Cl^- and HSO_4^-. **Caution:** When writing mechanisms, use "arrow pushing" to depict electron flow; write out every step separately; formulate complete structures, including charges and relevant electron pairs; and draw explicit reaction arrows to connect starting materials or intermediates with their respective products. Don't use shortcuts, and don't be sloppy!)

In Summary Treatment of alcohols with strong acid leads to protonation to give alkyloxonium ions, which, when primary, undergo S_N2 reactions in the presence of good nucleophiles. Alkyloxonium ions from secondary or tertiary alcohols convert into carbocations, which furnish products of substitution and elimination (dehydration).

$$
\overset{..}{R\overset{..}{O}H} \xrightarrow{H^+} R\overset{H}{\underset{H}{\overset{|}{\pm}}}\overset{..}{O}:
\begin{cases}
\xrightarrow[R = \text{prim}]{X^-, S_N2} RX + H_2\overset{..}{\underset{..}{O}} \\
\xrightarrow[R = \text{sec, tert}]{-H_2O} R^+
\begin{cases}
\xrightarrow{X^-, S_N1} RX \\
\xrightarrow{-H^+, E1} \text{Alkene}
\end{cases}
\end{cases}
$$

9-3 CARBOCATION REARRANGEMENTS

When alcohols are transformed into carbocations, the carbocations themselves are subject to rearrangements. The two rearrangements, known as hydride shifts and alkyl shifts, can occur with most types of carbocations. The rearranged molecules can then undergo further S_N1 or E1 reactions. The result is likely to be a complex mixture, unless we can establish a thermodynamic driving force toward one specific product.

Hydride shifts give new S_N1 products

Treatment of 2-propanol with concentrated hydrogen bromide gives 2-bromopropane, as expected. However, exposure of the more highly substituted secondary alcohol 3-methyl-2-butanol to the same reaction conditions produces a surprising result. The expected S_N1 product, 2-bromo-3-methylbutane, is only a minor component of the reaction mixture; the major product is 2-bromo-2-methylbutane.

Reaction

Normal S_N1 Reaction of an Alcohol (No Rearrangement)

$$
\underset{CH_3\overset{OH}{\overset{|}{C}}HCH_3}{} + HBr
$$

$$\downarrow 0°C$$

$$
\underset{CH_3\overset{Br}{\overset{|}{C}}HCH_3}{} + H—OH
$$

Animation

Hydride Shift in the S_N1 Reaction of an Alcohol with HBr

$$
\underset{\overset{|}{H_3C}\ \overset{|}{H}}{\overset{H\quad OH}{CH_3\overset{|}{C}—\overset{|}{C}CH_3}} \xrightarrow{HBr, 0°C} \underset{\overset{|}{H_3C}\ \overset{|}{H}}{\overset{H\quad Br}{CH_3\overset{|}{C}—\overset{|}{C}CH_3}} + \underset{\overset{|}{H_3C}\ \overset{|}{H}}{\overset{Br\quad H}{CH_3\overset{|}{C}—\overset{|}{C}CH_3}} + H—OH
$$

3-Methyl-2-butanol   **Minor product** 2-Bromo-3-methylbutane (Normal product)   **Major product** 2-Bromo-2-methylbutane (Rearranged product)

What is the mechanism of this transformation? The answer is that *carbocations can undergo rearrangement* by **hydride shifts,** in which the hydrogen (yellow) moves *with both electrons* from its original position to the neighboring carbon atom. Initially, protonation of the alcohol followed by loss of water gives the expected secondary carbocation. A shift of the tertiary hydrogen to the electron-deficient neighbor then generates a tertiary cation, *which is more stable*. This species is finally trapped by bromide ion to give the rearranged S_N1 product.

Mechanism of Carbocation Rearrangement

Mechanism

Step 1. Protonation   **Step 2.** Loss of water

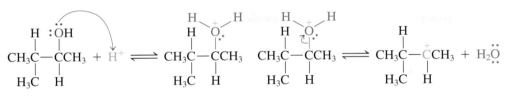

Step 3. Hydride shift   **Step 4.** Trapping by bromide

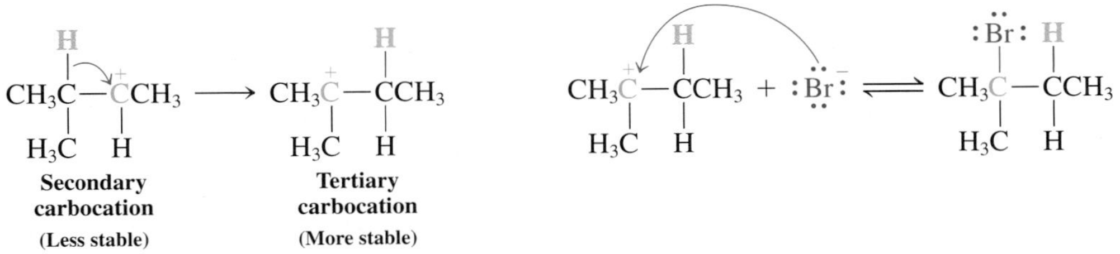

Secondary carbocation (Less stable) → Tertiary carbocation (More stable)

The details of the transition state of the observed hydride shift are shown schematically in Figure 9-2. A simple rule to remember when executing hydride shifts in carbocations is that *the hydrogen and the positive charge formally exchange places* between the two neighboring carbon atoms participating in the reaction.

Hydride shifts of carbocations are generally very fast—faster than S_N1 or E1 reactions. In part, this rapidity is due to hyperconjugation, which weakens the C–H bond (Section 7-5 and Figure 9-2B). They are particularly favored when the new carbocation is more stable than the original one, as in the example depicted in Figure 9-2.

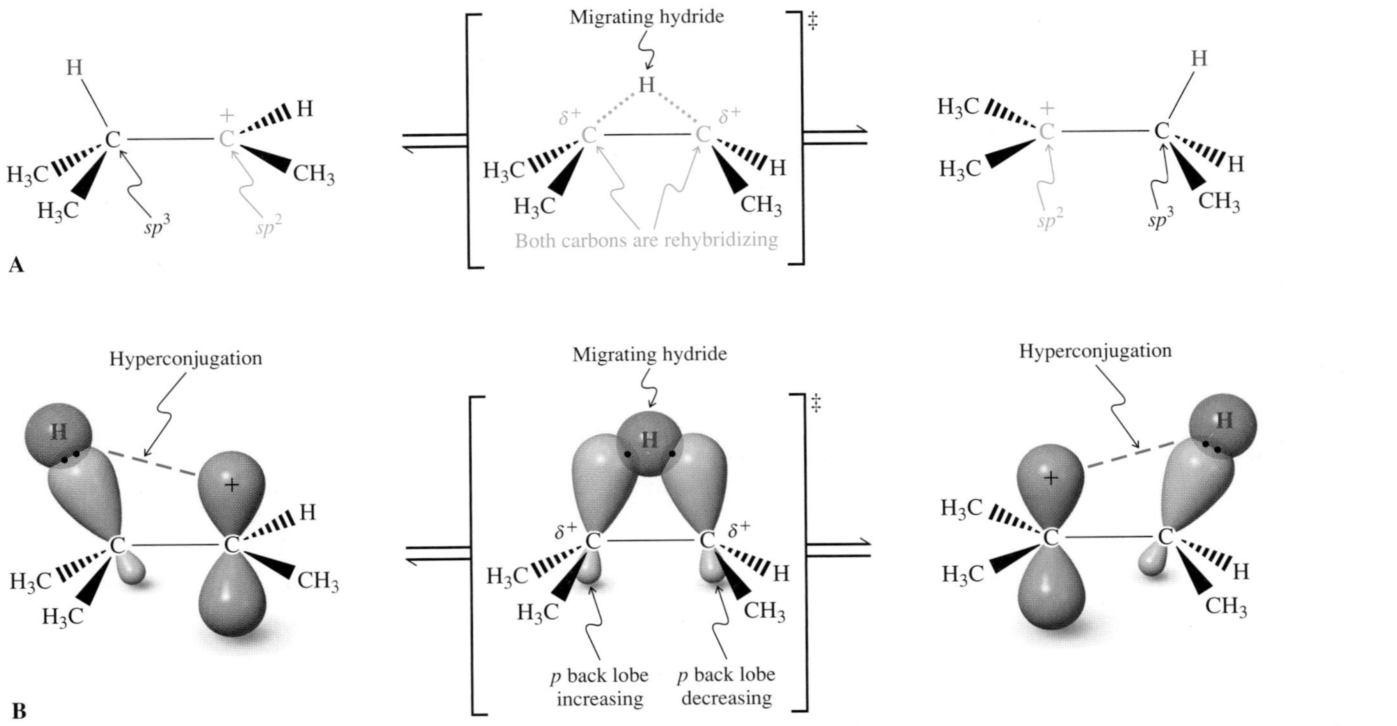

Note: Color is used to indicate the electrophilic (blue), nucleophilic (red), and leaving-group (green) character of the reacting centers. Therefore, a color may "switch" from one group or atom to another as the reaction proceeds.

Migrating hydride

Both carbons are rehybridizing

A

Hyperconjugation   Migrating hydride   Hyperconjugation

p back lobe increasing   *p* back lobe decreasing

B

Figure 9-2 The rearrangement of a carbocation by a hydride shift: (A) dotted-line notation; (B) orbital picture. Note that the migrating hydrogen and the positive charge exchange places. In addition, you can see how hyperconjugation weakens the C–H bond by effecting some electron transfer into the empty neighboring π orbital.

Solved Exercise 9-4 | Working with the Concepts: Formulating a Carbocation Rearrangement

2-Methylcyclohexanol, on treatment with HBr, gives 1-bromo-1-methylcyclohexane. Explain by a mechanism.

Strategy

Let us dissect the problem along the lines of taking an inventory.

- What topological/connectivity changes are taking place? *Answer:* The six-membered ring stays intact, but the functionality has migrated from a secondary to a tertiary carbon.
- What is the change in molecular formula? *Answer:* We start with $C_7H_{14}O$ and wind up with $C_7H_{13}Br$. The net change is that OH is being replaced by Br; no carbons are added or subtracted from the starting material.
- What thoughts come to mind when considering the functional group and the reagent to which it is exposed? *Answer:* We have an alcohol that is treated with HBr, a strong acid. The hydroxy function will be protonated, turning it into a good leaving group, and hence we have to consider substitution and elimination as possible reaction pathways.

Solution

- The answer to point 1 (rearrangement) and consideration of point 3 (alcohol and acid) strongly implicate an acid-catalyzed rearrangement via a carbocation.
- To arrive at the required carbocation, we need to protonate the alcohol and dissociate the leaving group. The result is a secondary carbocation.

Secondary carbocation

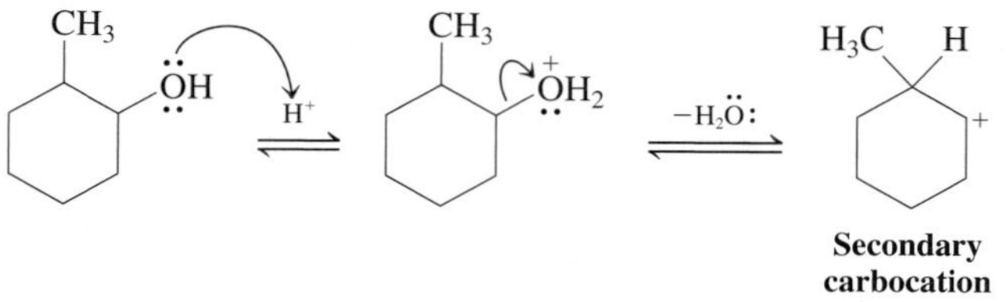

- We know that the product contains the functional substituent Br at the neighboring tertiary center. Thus, to transfer functionality, we have to invoke a hydride shift.
- The ensuing tertiary carbocation is then trapped by bromide to give the product.

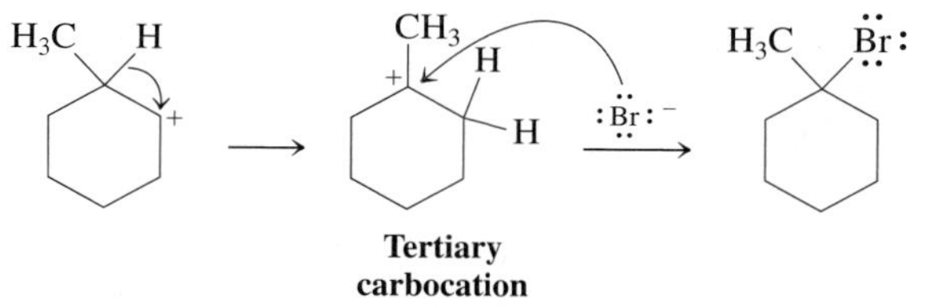

Tertiary carbocation

Exercise 9-5 | Try It Yourself

Predict the major product from the following reactions.

(a) 2-Methyl-3-pentanol + H_2SO_4, CH_3OH solvent   **(b)**

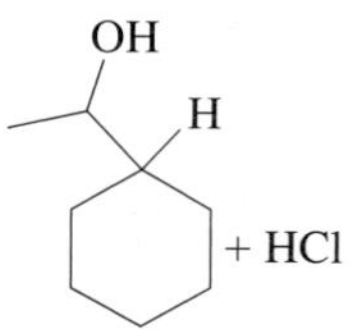

Primary carbocations are too unstable to be formed by rearrangement. However, carbocations of comparable stability—for example, the pairs secondary–secondary or tertiary–tertiary—equilibrate readily. In this case, any added nucleophile will trap all carbocations present, furnishing mixtures of products.

$$CH_3\underset{\underset{\displaystyle H}{|}}{\overset{\overset{\displaystyle OH}{|}}{C}}CH_2CH_2CH_3 \xrightarrow{\text{HBr, 0°C}} CH_3\underset{\underset{\displaystyle H}{|}}{\overset{\overset{\displaystyle Br}{|}}{C}}CH_2CH_2CH_3 \ + \ CH_3CH_2\underset{\underset{\displaystyle H}{|}}{\overset{\overset{\displaystyle Br}{|}}{C}}CH_2CH_3$$

Carbocation rearrangements take place regardless of the nature of the precursor to the carbocation: alcohols (this section), haloalkanes (Chapter 7), and alkyl sulfonates (Section 6-7). For example, solvolysis of 2-bromo-3-ethyl-2-methylpentane in ethanol (ethanolysis) gives the two possible tertiary ethers.

Rearrangement in Solvolysis of a Haloalkane

$$\underset{\underset{\displaystyle \begin{matrix}\text{2-Bromo-3-ethyl-}\\\text{2-methylpentane}\end{matrix}}{}}{CH_3\underset{\underset{\displaystyle H_3C}{|}}{\overset{\overset{\displaystyle Br}{|}}{C}}-\underset{\underset{\displaystyle CH_2CH_3}{|}}{\overset{\overset{\displaystyle H}{|}}{C}}CH_2CH_3} \xrightarrow{CH_3CH_2OH} \underset{\underset{\displaystyle \begin{matrix}\text{2-Ethoxy-3-ethyl-}\\\text{2-methylpentane}\\\text{(Normal product)}\end{matrix}}{}}{CH_3\underset{\underset{\displaystyle H_3C}{|}}{\overset{\overset{\displaystyle CH_3CH_2O}{|}}{C}}-\underset{\underset{\displaystyle CH_2CH_3}{|}}{\overset{\overset{\displaystyle H}{|}}{C}}CH_2CH_3} \ + \ \underset{\underset{\displaystyle \begin{matrix}\text{3-Ethoxy-3-ethyl-}\\\text{2-methylpentane}\\\text{(Rearranged product)}\end{matrix}}{}}{CH_3\underset{\underset{\displaystyle H_3C}{|}}{\overset{\overset{\displaystyle H}{|}}{C}}-\underset{\underset{\displaystyle CH_2CH_3}{|}}{\overset{\overset{\displaystyle OCH_2CH_3}{|}}{C}}CH_2CH_3} \ + \ H-Br$$

Exercise 9-6

Give a mechanism for the preceding reaction. Then predict the outcome of the reaction of 2-chloro-4-methylpentane with methanol. (**Hint:** Try *two successive* hydride shifts to the most stable carbocation.)

Carbocation rearrangements also give new E1 products

How does the rearrangement of intermediates affect the outcome of reactions under conditions that favor elimination? At elevated temperatures and in relatively nonnucleophilic media, rearranged carbocations yield alkenes by the E1 mechanism (Sections 7-6 and 9-2). For example, treatment of 2-methyl-2-pentanol with aqueous sulfuric acid at 80°C gives the same major alkene product as that formed when the starting material is 4-methyl-2-pentanol. The conversion of the latter alcohol includes a hydride shift of the initial carbocation followed by deprotonation.

Rearrangement in E1 Elimination

$$\underset{\underset{\displaystyle \text{2-Methyl-2-pentanol}}{}}{CH_3\underset{\underset{\displaystyle CH_3}{|}}{\overset{\overset{\displaystyle OH}{|}}{C}}-CH_2CH_2CH_3} \xrightarrow[-H_2O]{H_2SO_4,\ 80°C} \underset{\underset{\displaystyle \begin{matrix}\text{Major product}\\\text{2-Methyl-2-pentene}\end{matrix}}{}}{\underset{\displaystyle H_3C}{}\underset{\displaystyle H_3C}{}C=C\underset{\displaystyle H}{}\underset{\displaystyle CH_2CH_3}{}} \xleftarrow[\underset{\displaystyle \text{With rearrangement}}{-H_2O}]{H_2SO_4,\ 80°C} \underset{\underset{\displaystyle \text{4-Methyl-2-pentanol}}{}}{CH_3\underset{\underset{\displaystyle CH_3}{|}}{\overset{\overset{\displaystyle H}{|}}{C}}-CH_2\underset{\underset{\displaystyle H}{|}}{\overset{\overset{\displaystyle OH}{|}}{C}}CH_3}$$

> **Note:** Carbocations can always give S_N1 and E1 products. Their relative ratios depend on the structure of the cation as well as the nucleophilicity of potential trapping agents and the temperature.

Exercise 9-7

(**a**) Give mechanisms for the preceding E1 reactions. (**b**) Treatment of 4-methylcyclohexanol with hot acid gives 1-methylcyclohexene. Explain by a mechanism. (**Hint:** Consider several sequential H shifts.)

Other carbocation rearrangements are due to alkyl shifts

Carbocations, particularly when lacking suitable (secondary and tertiary) hydrogens next to the positively charged carbon, can undergo another mode of rearrangement, known as **alkyl group migration** or **alkyl shift.**

Reaction

Rearrangement by Alkyl Shift in S$_N$1 Reaction

$$H_3C \quad CH_3$$
$$CH_3C-COH \xrightarrow[-HOH]{HBr} CH_3C-CCH_3$$
$$H_3C \quad H \qquad\qquad\qquad H_3C \quad H$$

Br CH$_3$

94%

3,3-Dimethyl-2-butanol   **2-Bromo-2,3-dimethylbutane**

As in the hydride shift, the migrating group takes its electron pair with it to form a bond to the neighboring positively charged carbon. *The moving alkyl group and the positive charge formally exchange places.*

Mechanism of Alkyl Shift

Mechanism

$$H_3C \quad CH_3 \qquad H_3C \qquad\qquad CH_3 \qquad\qquad :Br: CH_3$$
$$CH_3C-COH + H^+ \underset{+H_2O}{\overset{-H_2O}{\rightleftharpoons}} CH_3C-CCH_3 \xrightarrow{CH_3\ shift} CH_3C-CCH_3 \underset{-:Br:^-}{\overset{+:Br:^-}{\rightleftharpoons}} CH_3C-CCH_3$$
$$H_3C \quad H \qquad H_3C \quad H \qquad\qquad H_3C \quad H \qquad\qquad H_3C \quad H$$

The rates of alkyl and hydride shifts are comparable when leading to carbocations of similar stability. However, either type of migration is faster when furnishing tertiary carbocations relative to those ending in their secondary counterparts. This explains why, in the preceding discussion of hydride shifts, alkyl group rearrangement was not observed: The less substituted cation would have been the result. Exceptions to this observation are found only if there are other compelling reasons for the preference of alkyl migration, such as electronic stabilization or steric relief (see Worked Example 26 and Problem 61).

Solved Exercise 9-8 | **Working with the Concepts: Formulating a More Complex Carbocation Rearrangement**

Molecules of the type A have been shown to dehydrate to B. Formulate a mechanism.

$$\xrightarrow[-H_2O]{H_2SO_4}$$

A   B

Remember WHIP

*W*hat
*H*ow
*I*nformation
*P*roceed

Strategy

First we need to look carefully at the function(s) in A and B, as well as the reagent: We recognize the presence of a tertiary alcohol that is being treated with acid. These conditions are suggestive of the formation of a carbocation. Moreover, the net reaction is dehydration, indicating an E1 process. Second, it is important to inspect the carbon skeleton in A and compare it with that in B: We can see that the methyl group has migrated to its neighboring carbon. Conclusion: We are dealing with a carbocation rearrangement featuring a methyl shift.

Solution

• The first steps are protonation followed by loss of water to generate the carbocation.

$$+ H^+ \longrightarrow \xrightarrow{-H_2O}$$

- This cation has, in principle, various options. For example, it could be trapped by HSO_4^- as a nucleophile (S_N1). However, HSO_4^- is very weak in this respect (it is a poor base; Section 6-8). Moreover, even if it did form a bond to the cationic center, this would be readily reversible (HSO_4^- is an excellent leaving group; Section 6-7). It could also lose a proton in two possible ways. This is also reversible and not observed. Instead, this cation provides an opportunity to move the methyl group to the neighboring position, as in B, by rearrangement.
- Finally, deprotonation by HSO_4^- (written as "proton loss") provides the product.

Exercise 9-9 | Try It Yourself

At 100°C, 3,3-dimethyl-2-butanol gives three products of E1 reaction: one derived from the carbocation present prior to rearrangement, the other two from that formed after an alkyl shift has taken place. Give the structures of these elimination products.

Primary alcohols may undergo rearrangement

Treatment of a primary alcohol with HBr or HI normally produces the corresponding haloalkane through S_N2 reaction of the alkyloxonium ion (Section 9-2). However, it is possible in some cases to observe alkyl and hydride shifts to primary carbons bearing leaving groups, even though primary carbocations are not formed in solution. For example, treating 2,2-dimethyl-1-propanol (neopentyl alcohol) with strong acid causes rearrangement, despite the fact that a primary carbocation cannot be an intermediate.

Rearrangement in a Primary Substrate

$$CH_3\underset{\underset{CH_3}{|}}{\overset{\overset{CH_3}{|}}{C}}CH_2OH \xrightarrow[-H-OH]{HBr, \Delta} CH_3\underset{\underset{CH_3}{|}}{\overset{\overset{Br}{|}}{C}}CH_2CH_3$$

2,2-Dimethyl-1-propanol
(Neopentyl alcohol)   **2-Bromo-2-methylbutane**

Reaction

In this case, after protonation to form the alkyloxonium ion, steric hindrance interferes with direct displacement by bromide (Section 6-10). Instead, water leaves *at the same time* as a methyl group migrates from the neighboring carbon, thus bypassing formation of a primary carbocation.

Nucleophile: red
Electrophile: blue
Leaving group: green

Mechanism of Concerted Alkyl Shift

Mechanism

Rearrangements of primary substrates are relatively difficult processes, usually requiring elevated temperatures and long reaction times.

In Summary Another mode of reactivity of carbocations, in addition to regular S$_N$1 and E1 processes, is rearrangement by hydride or alkyl shifts. In such rearrangements, the migrating group delivers its bonding electron pair to a positively charged carbon neighbor, exchanging places with the charge. Rearrangement may lead to a more stable cation—as in the conversion of a secondary cation into a tertiary one. Primary alcohols also can undergo rearrangement, but they do so by concerted pathways and not through the intermediacy of primary cations.

9-4 | ESTERS FROM ALCOHOLS AND HALOALKANE SYNTHESIS

Reaction of alcohols with carboxylic acids converts them to **organic esters,** also called **carboxylates** or **alkanoates** (Table 2-3). They are formally derived from the carboxylic acids by replacement of the hydroxy group with alkoxy. One can formulate a corresponding set of **inorganic esters** derived from inorganic acids, such as those based on phosphorus and sulfur in various oxidation states. In such inorganic esters, the attachment of the

Organic and Inorganic Esters

heteroatom turns the normally poor leaving group OH in alcohols into a good leaving group (highlighted in the green boxes), which can be used in the synthesis of haloalkanes (see also Section 9-2). We have already mentioned the good leaving-group ability of sulfate and sulfonate groups in S$_N$2 reactions (Section 6-7). Here, we shall see how specific phosphorus and sulfur reagents accomplish this task:

$$R-OH \xrightarrow{\text{Reagent}} R-L \xrightarrow{X^-} R-X$$

Alcohol / Inorganic ester / Haloalkane

Alcohols react with carboxylic acids to give organic esters

Alcohols react with carboxylic acids in the presence of catalytic amounts of a strong inorganic acid, such as H$_2$SO$_4$ or HCl, to give organic esters and water, a process called **esterification.** Starting materials and products in this transformation form an equilibrium that can be shifted in either direction. The formation and reactions of organic esters will be presented in detail in Chapters 19 and 20.

Esterification

$$CH_3COH + CH_3CH_2OH \xrightleftharpoons{H^+} CH_3COCH_2CH_3 + HOH$$

Acetic acid / Ethanol solvent / Ethyl acetate

Haloalkanes can be made from alcohols through inorganic esters

Because of the difficulties and complications that can be encountered in the acid-catalyzed conversions of alcohols into haloalkanes (Section 9-2), several alternatives have been developed. These methods rely on a number of inorganic reagents that are capable of changing the hydroxy function into a good leaving group under milder conditions. Thus, primary and secondary alcohols react with phosphorus tribromide, a readily available commercial compound, to give bromoalkanes and phosphorous acid. This method constitutes a general way of making bromoalkanes from alcohols. All three bromine atoms are transferred from phosphorus to alkyl groups.

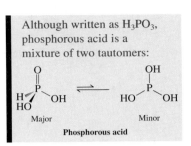

Although written as H_3PO_3, phosphorous acid is a mixture of two tautomers:

Phosphorous acid

Bromoalkane Synthesis by Using PBr₃

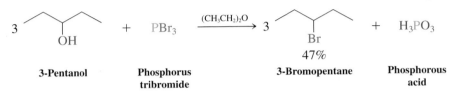

| 3-Pentanol | Phosphorus tribromide | 3-Bromopentane | Phosphorous acid |

Reaction

What is the mechanism of action of PBr_3? In the first step, the alcohol attacks the phosphorus reagent as a nucleophile to form a protonated inorganic ester, a derivative of phosphorous acid.

Mechanism

Step 1

Next, $HOPBr_2$, a good leaving group, is displaced (S_N2) by the bromide generated in step 1, producing the haloalkane.

Step 2

This method of haloalkane synthesis is especially efficient because $HOPBr_2$ continues to react successively with two more molecules of alcohol, converting them into haloalkane as well.

$$2\ RCH_2\ddot{O}H\ +\ H\ddot{O}PBr_2\ \longrightarrow\longrightarrow\ 2\ RCH_2\ddot{B}r:\ +\ H_3PO_3$$

What if, instead of a bromoalkane, we want the corresponding iodoalkane? The required phosphorus triiodide, PI_3, is best generated in the reaction mixture in which it will be used, because it is a reactive species. We do this by adding red elemental phosphorus and elemental iodine to the alcohol (margin). The reagent is consumed as soon as it is formed.

$$CH_3(CH_2)_{14}CH_2OH$$
$$\downarrow P, I_2, \Delta$$
$$CH_3(CH_2)_{14}CH_2I$$
$$85\%$$
$$+$$
$$H_3PO_3$$

A chlorinating agent commonly used to convert alcohols into chloroalkanes is thionyl chloride, $SOCl_2$. Simply warming an alcohol in its presence results in the evolution of SO_2 and HCl and the formation of the chloroalkane.

Chloroalkane Synthesis with SOCl₂

$$CH_3CH_2CH_2OH\ +\ SOCl_2\ \longrightarrow\ CH_3CH_2CH_2Cl\ +\ O{=}S{=}O\ +\ HCl$$
$$91\%$$

Reaction

Mechanistically, the alcohol RCH_2OH again first forms an inorganic ester, RCH_2O_2SCl.

Mechanism

Step 1

$$RCH_2-\overset{..}{\underset{..}{O}}H \ + \ :\overset{..}{\underset{..}{Cl}}\overset{\overset{\displaystyle :O:}{\|}}{\underset{}{S}}\overset{..}{\underset{..}{Cl}}: \longrightarrow \ RCH_2-\overset{..}{\underset{..}{O}}:\overset{\overset{\displaystyle \cdot\overset{..}{O}\cdot}{\|}}{\underset{}{S}}-\overset{..}{\underset{..}{Cl}}: \ + \ H^+ \ + \ :\overset{..}{\underset{..}{Cl}}:^-$$

The chloride ion created in this process then acts as a nucleophile and attacks the ester, an S_N2 reaction yielding one molecule each of SO_2 and HCl.

Step 2

$$H^+ \ + \ :\overset{..}{\underset{..}{Cl}}:^- \ + \ \overset{}{\underset{R}{CH_2}}-\overset{..}{\underset{..}{O}}:\overset{\overset{\displaystyle \cdot\overset{..}{O}\cdot}{\|}}{\underset{}{S}}\overset{..}{\underset{..}{Cl}}: \longrightarrow \ :\overset{..}{\underset{..}{Cl}}-CH_2R \ + \ :\overset{..}{O}=S=\overset{..}{O}: \ + \ H\overset{..}{\underset{..}{Cl}}:$$

(CH₃CH₂)₃N :
N,N-Diethylethanamine
(Triethylamine)
+
HCl
↓
$(CH_3CH_2)_3\overset{+}{N}H \ Cl^-$

The reaction works better in the presence of an amine, which neutralizes the hydrogen chloride generated. One such reagent is *N,N*-diethylethanamine (triethylamine), which forms the corresponding ammonium chloride under these conditions.

Alkyl sulfonates are versatile substrates for substitution reactions

The inorganic esters in the reactions of $SOCl_2$ are special examples of leaving groups derived from sulfur-based acids. They are related to the sulfonates (Section 6-7). Alkyl sulfonates contain excellent leaving groups and can be readily prepared from the corresponding sulfonyl chlorides and an alcohol. A mild base such as pyridine or a tertiary amine is often added to remove the HCl formed.

Synthesis of an Alkyl Sulfonate

$$\underset{\textbf{2-Methyl-1-propanol}}{\overset{\overset{\displaystyle CH_3}{|}}{CH_3CHCH_2OH}} \ + \ \underset{\substack{\textbf{Methanesulfonyl}\\\textbf{chloride}\\\textbf{(Mesyl chloride)}}}{\overset{\overset{\displaystyle O}{\|}}{\underset{\underset{\displaystyle O}{\|}}{CH_3SCl}}} \ + \ \underset{\textbf{Pyridine}}{\overset{\displaystyle \bigcirc\!\!\!\!N}{:}} \ \longrightarrow \ \underset{\substack{\textbf{2-Methylpropyl}\\\textbf{methanesulfonate}\\\textbf{(2-Methylpropyl mesylate)}}}{\overset{\overset{\displaystyle CH_3 \quad O}{| \quad \|}}{\underset{\underset{\displaystyle O}{\|}}{CH_3CHCH_2OSCH_3}}} \ + \ \underset{\substack{\textbf{Pyridinium}\\\textbf{hydrochloride}}}{\overset{\displaystyle \bigcirc\!\!\!\!\overset{+}{N}Cl^-}{\underset{\displaystyle H}{|}}}$$

Unlike the inorganic esters derived from phosphorus tribromide and thionyl chloride, alkyl sulfonates are often crystalline solids that can be isolated and purified before further reaction. They then can be used in reactions with a variety of nucleophiles to give the corresponding products of nucleophilic substitution.

> **Sulfonate Intermediates in Nucleophilic Displacement of the Hydroxy Group in Alcohols**
>
> $$R-OH \ \longrightarrow \ R-\overset{\overset{\displaystyle O}{\|}}{\underset{\underset{\displaystyle O}{\|}}{OSR'}} \ \longrightarrow \ R-Nu$$

The displacement of sulfonate groups by halide ions (Cl^-, Br^-, I^-) readily yields the corresponding haloalkanes, particularly with primary and secondary systems, in which S_N2 reactivity is good. In addition, however, alkyl sulfonates allow replacement of the hydroxy group by *any* good nucleophile: They are not limited to halides alone, as is the case with hydrogen, phosphorus, and thionyl halides.

Substitution Reactions of Alkyl Sulfonates

$CH_3CH_2CH_2{-}\ddot{O}{-}\overset{\displaystyle :O:}{\underset{\displaystyle :O:}{\overset{\|}{\underset{\|}{S}}}}{-}CH_3$ + $:\ddot{I}:^-$ $\longrightarrow$ $CH_3CH_2CH_2{-}\ddot{I}:$ + $^-:\ddot{O}{-}\overset{\displaystyle :O:}{\underset{\displaystyle :O:}{\overset{\|}{\underset{\|}{S}}}}{-}CH_3$

90%

$CH_3{-}\overset{\displaystyle H}{\underset{\displaystyle CH_3}{C}}{-}\ddot{O}{-}\overset{\displaystyle :O:}{\underset{\displaystyle :O:}{\overset{\|}{\underset{\|}{S}}}}{-}\text{⟨⟩}{-}CH_3$ + $CH_3CH_2\ddot{S}:^-$ $\longrightarrow$ $CH_3\overset{\displaystyle}{\underset{\displaystyle CH_3}{CH}}{-}\ddot{S}CH_2CH_3$ +

85%

$O{=}\overset{\displaystyle :\ddot{O}:^-}{\underset{\displaystyle}{S}}{=}O$ on a ring with CH_3

Exercise 9-10 | Try It Yourself

What is the product of the reaction sequence shown in the margin?

(margin structure: cyclohexane ring with OH and CH₃ substituents)
1. CH_3SO_2Cl
2. NaI

Exercise 9-11 | Try It Yourself

Supply reagents with which you would prepare the following haloalkanes from the corresponding alcohols.

(a) $I(CH_2)_6I$ **(b)** $(CH_3CH_2)_3CCl$ **(c)** (branched structure with Br)

In Summary Alcohols react with carboxylic acids by loss of water to furnish organic esters. They also react with inorganic halides, such as PBr_3, $SOCl_2$, and RSO_2Cl, by loss of HX to produce inorganic esters. These inorganic esters contain good leaving groups in nucleophilic substitutions that are, for example, displaced by halide ions to give the corresponding haloalkanes.

9-5 | NAMES AND PHYSICAL PROPERTIES OF ETHERS

Ethers may be thought of as derivatives of alcohols in which the hydroxy proton has been replaced by an alkyl group. We now introduce this class of compounds systematically. This section gives the rules for naming ethers and describes some of their physical properties.

In the IUPAC system, ethers are alkoxyalkanes

The IUPAC system for naming **ethers** treats them as alkanes that bear an alkoxy substituent— that is, as alkoxyalkanes. The smaller substituent is considered part of the alkoxy group and the larger chain defines the stem.

In common names, the names of the two alkyl groups are followed by the word *ether*. Hence, CH_3OCH_3 is dimethyl ether, $CH_3OCH_2CH_3$ is ethyl methyl ether, and so forth.

$CH_3\ddot{O}CH_2CH_3$

Methoxyethane

$CH_3CH_2\overset{\displaystyle}{\underset{\displaystyle CH_3}{\ddot{O}\overset{\displaystyle CH_3}{\underset{\displaystyle}{C}}}}CH_3$

2-Ethoxy-2-methylpropane

cis-1-Ethoxy-2-
methoxy**cyclopentane**

Ethers are generally fairly unreactive (except for strained cyclic derivatives; see Section 9-9) and are therefore frequently used as solvents in organic reactions. Some of these ether solvents are cyclic; they may even contain several ether functions. All have common names.

Ether Solvents and Their Names

$CH_3CH_2OCH_2CH_3$
Ethoxyethane
(Diethyl ether)

1,4-Dioxacyclohexane
(1,4-Dioxane)

$CH_3OCH_2CH_2OCH_3$
1,2-Dimethoxyethane
(Glycol dimethyl ether,
glyme)

Oxacyclopentane
(Tetrahydrofuran,
THF)

Cyclic ethers are members of a class of cycloalkanes in which one or more carbons have been replaced by a *heteroatom*—in this case, oxygen. (A **heteroatom** is defined as any atom except carbon and hydrogen.) Cyclic compounds of this type, called **heterocycles,** are discussed more fully in Chapter 25.

The simplest system for naming cyclic ethers is based on the **oxacycloalkane** stem, in which the prefix *oxa* indicates the replacement of carbon by oxygen in the ring. Thus, three-membered cyclic ethers are oxacyclopropanes (other names used are oxiranes, epoxides, and ethylene oxides), four-membered systems are oxacyclobutanes, and the next two higher homologs are oxacyclopentanes (tetrahydrofurans) and oxacyclohexanes (tetrahydropyrans). The compounds are numbered by starting at the oxygen and proceeding around the ring.

The absence of hydrogen bonding affects the physical properties of ethers

The molecular formula of simple alkoxyalkanes is $C_nH_{2n+2}O$, identical with that of the alkanols. However, because of the absence of hydrogen bonding, the boiling points of ethers are much lower than those of the corresponding isomeric alcohols (Table 9-1). The two smallest members of the series are water miscible, but ethers become less water soluble as the hydrocarbon residues increase in size. For example, methoxymethane is completely water soluble, whereas ethoxyethane forms only an approximately 10% aqueous solution.

Polyethers solvate metal ions: crown ethers and ionophores

Model Building

Cyclic polyethers that contain multiple ether functional groups based on the 1,2-ethanediol unit are called **crown ethers,** so named because the molecules adopt a crown-like conformation in the crystalline state and, presumably, in solution. The polyether 18-crown-6 is shown in Figure 9-3. The number 18 refers to the total number of atoms in the ring, and 6 to the number of oxygens. The electrostatic potential map on the right of Figure 9-3 shows

Table 9-1	Boiling Points of Ethers and the Isomeric 1-Alkanols			
Ether	**Name**	**Boiling point (°C)**	**1-Alkanol**	**Boiling point (°C)**
CH_3OCH_3	Methoxymethane (Dimethyl ether)	−23.0	CH_3CH_2OH	78.5
$CH_3OCH_2CH_3$	Methoxyethane (Ethyl methyl ether)	10.8	$CH_3CH_2CH_2OH$	82.4
$CH_3CH_2OCH_2CH_3$	Ethoxyethane (Diethyl ether)	34.5	$CH_3(CH_2)_3OH$	117.3
$(CH_3CH_2CH_2CH_2)_2O$	1-Butoxybutane (Dibutyl ether)	142	$CH_3(CH_2)_7OH$	194.5

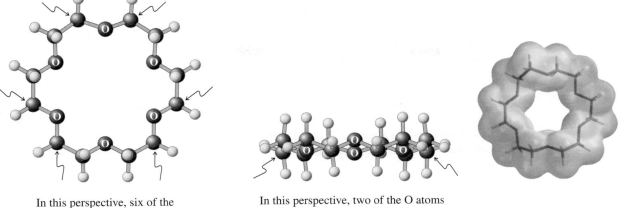

In this perspective, six of the H atoms are hidden by the attached carbons (marked by arrows).

In this perspective, two of the O atoms are masked by the attached carbons (marked by arrows).

Figure 9-3 The crown-like structural arrangement of 18-crown-6.

that the inside of the ring is quite electron rich due to the electron pairs on the ether oxygens, which renders them Lewis basic. These lone pairs give cyclic polyethers striking solvating power, in which several such oxygen atoms may surround and bind metal cations. Crown ethers are therefore capable of solubilizing ordinary salts in organic solvents. For example, potassium permanganate ($KMnO_4$), a deep-violet solid, is completely insoluble in benzene but readily dissolves when 18-crown-6 is added. This solution is useful because it allows oxidations with potassium permanganate to be carried out in organic solvents. Dissolution is possible by effective solvation of the metal ion by the six crown oxygens.

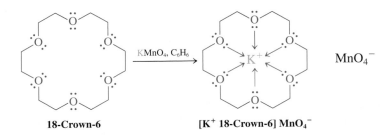

18-Crown-6   **[K⁺ 18-Crown-6] MnO₄⁻**

Space-filling model of the cation [K⁺ 18-crown-6].

The size of the "cavity" in the crown ether can be tailored to allow for the selective binding of only certain cations—namely, those whose ionic radius is best accommodated by the polyether. This concept has been extended successfully into three dimensions by the synthesis of polycyclic ethers, also called **cryptands** (*kryptos*, Greek, hidden), which are highly selective in alkali and other metal binding (Figure 9-4). The significance of these

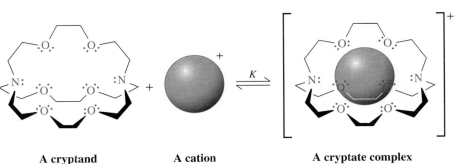

A cryptand   **A cation**   **A cryptate complex**

Figure 9-4 The binding of a cation by a polycyclic ether (cryptand) to form a complex (cryptate). The system shown selectively binds the potassium ion, with a binding constant of $K = 10^{10}$. The order of selectivity is K⁺ > Rb⁺ > Na⁺ > Cs⁺ > Li⁺. The binding constant for lithium is about 100. Thus, the total range within the series of alkali metals spans eight orders of magnitude.

compounds was recognized by the award of the Nobel Prize in chemistry in 1987, shared by Cram, Lehn, and Pedersen.*

Crown ethers and cryptands are often called **ion transport agents** and are part of the general class of **ionophores** (*-phoros,* Greek, bearing, hence "ion bearing"), compounds that organize themselves around cations by coordination. In general, the result of this interaction is that the polar hydrophilic nature of the ion is masked by a hydrophobic shell, hence making the ion much more soluble in nonpolar solvents. In nature, ionophores can transport ions through hydrophobic cell membranes. The ion balance between the inside and outside of the cell is carefully regulated to ensure cell survival, and therefore any undue disruption causes cell destruction. This property is put to medicinal use in fighting invading organisms with polyether antibiotics. However, because ion transport affects nerve transmission, some naturally occurring ionophores are also deadly neurotoxins (see below and the cover of this book).

Monensin
(Antibiotic from a *Streptomyces* strain)

Tetrodotoxin
(The nerve toxin in puffer fish or fugu)

Brevetoxin B
(Marine neurotoxin produced by algae associated with red tide)

Occurrence of red tide on a beach near Queensland, New Zealand. Red tides are vast phytoplankton blooms that turn the sea red, brown, or green. Algae that proliferate during such events can produce polyether neurotoxins, like brevetoxin B, which have been responsible for massive fish kills and poisonings of humans who ate affected seafood. Brevetoxin B binds to sodium channels on nerve and muscle cell membranes, causing cell death.

In Summary Acyclic ethers can be named as alkoxyalkanes or as alkyl ethers. Their cyclic counterparts are called oxacycloalkanes. Ethers have lower boiling points than do alcohols of comparable size, because they cannot hydrogen bond to each other. The Lewis basicity of the oxygen lone pairs, particularly in polyethers, allows efficient metal ion complexation.

9-6 | WILLIAMSON ETHER SYNTHESIS

Alkoxides are excellent nucleophiles. This section describes their use in the most common method for the preparation of ethers.

Ethers are prepared by S_N2 reactions

The simplest way to synthesize an ether is to have an alkoxide react with a primary halo-alkane or a sulfonate ester under typical S_N2 conditions (Chapter 6). This approach is known as the **Williamson†** ether synthesis. The alcohol from which the alkoxide is derived can

*Professor Donald J. Cram (1919–2001), University of California at Los Angeles; Professor Jean-Marie Lehn (b. 1939), University of Strasbourg and Collège de France, Paris; Dr. Charles J. Pedersen (1904–1989), E. I. du Pont de Nemours & Company, Wilmington, Delaware.

†Professor Alexander W. Williamson (1824–1904), University College, London.

be used as the solvent (if inexpensive), but other polar molecules, such as dimethyl sulfoxide (DMSO) or hexamethylphosphoric triamide (HMPA), are often better (Table 6-5).

Williamson Ether Syntheses

$$CH_3CH_2CH_2CH_2OH, 14\ h$$
$$\text{or DMSO, }9.5\ h$$
$$-Na^+Cl^-$$

60% (butanol solvent)
95% (DMSO solvent)
1-Butoxybutane

REAL LIFE: NATURE 9-1 | Chemiluminescence of 1,2-Dioxacyclobutanes

A 2-bromohydroperoxide   **3,3,4,4-Tetramethyl-1,2-dioxacyclobutane**   **Acetone**
                               (A 1,2-dioxetane)

$$\xrightarrow{\Delta} 2\ CH_3CCH_3 + h\nu$$

An intramolecular Williamson-type reaction of a special kind is that in which a 2-bromohydroperoxide is the reactant. The product is a 1,2-dioxacyclobutane (1,2-dioxetane). This compound is unusual because it decomposes to the corresponding carbonyl compounds with emission of light (*chemiluminescence*). Dioxacyclobutanes seem to be responsible for the *bioluminescence* of certain species in nature. Terrestrial organisms, such as the firefly, the glowworm, and certain click beetles, are well-known light emitters. However, most bioluminescent species live in the ocean; they range from microscopic bacteria and plankton to fish. The emitted light serves many purposes and seems to be important in courtship and communication, sex differentiation, finding prey, and hiding from or scaring off predators.

An example of a chemiluminescent molecule in nature is firefly luciferin. The base oxidation of this molecule furnishes a dioxacyclobutanone intermediate that decomposes in a manner analogous to that of 3,3,4,4-tetramethyl-1,2-dioxacyclobutane to give a complex heterocycle, carbon dioxide, and emitted light.

Bioluminescence is extraordinarily efficient. For example, the firefly converts about 40% of the energy of the underlying chemical process into visible light. For your calibration, a normal light bulb is only 10% efficient, most of the (electric) energy being emitted as heat.

Male and female fireflies glowing in concert.

Firefly luciferin

$$\xrightarrow{\text{Base, O}_2}$$

1,2-Dioxacyclobutanone intermediate

$$+\ CO_2 + h\nu$$

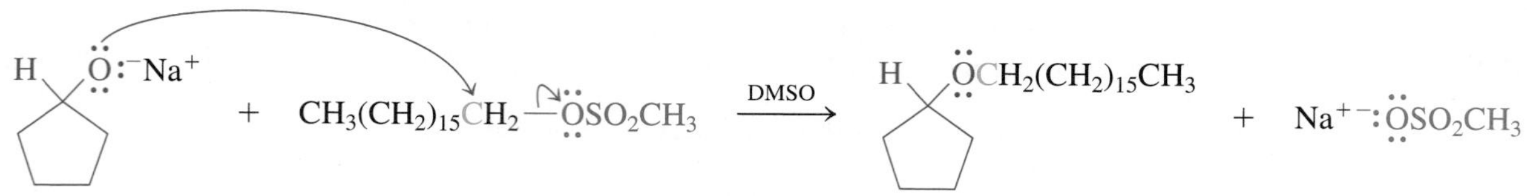

91%
Cyclopentoxyheptadecane

Nucleophile: red
Electrophile: blue
Leaving group: green

Because alkoxides are strong bases, their use in ether synthesis is restricted to primary unhindered alkylating agents; otherwise, a significant amount of E2 product is formed (Section 7-8).

Exercise 9-12

The following ethers can, in principle, be made by more than one Williamson ether synthesis. Consider the relative merits of your approaches. **(a)** 1-Ethoxybutane; **(b)** 2-methoxypentane; **(c)** propoxycyclohexane; **(d)** 1,4-diethoxybutane. [**Cautions:** Alkoxides are strong bases. What is wrong with 4-bromo-1-butanol as a starting material for **(d)**?]

Cyclic ethers can be prepared by intramolecular Williamson synthesis

The Williamson ether synthesis is also applicable to the preparation of cyclic ethers, starting from haloalcohols. Figure 9-5 depicts the reaction of hydroxide ion with a bromoalcohol. The black curved lines denote the chain of carbon atoms linking the functional groups. The mechanism consists of initial formation of a bromoalkoxide by fast proton transfer to the base, followed by ring closure to furnish the cyclic ether. The ring closure is an example of an intramolecular displacement. Cyclic ether formation is usually much faster than the side reaction shown in Figure 9-5, namely intermolecular displacement of bromide by hydroxide to give a diol. The reason is entropy (Section 2-1). In the intramolecular reaction, the two reacting centers are in the same molecule; in the transition state, one molecule of bromoalkoxide turns into two molecules of products, the ether and the leaving group. In the intermolecular reaction, the alkoxide and the electrophile have to be brought together at an entropic cost to the transition state, and, overall, the number of molecules stays the same. In those cases in which the intermolecular S_N2 reaction competes with its intramolecular counterpart, the intermolecular process can be effectively suppressed by using high dilution conditions, which drastically reduce the rate of bimolecular processes (Section 2-1).

Intramolecular Williamson synthesis allows the preparation of cyclic ethers of various sizes, including small rings.

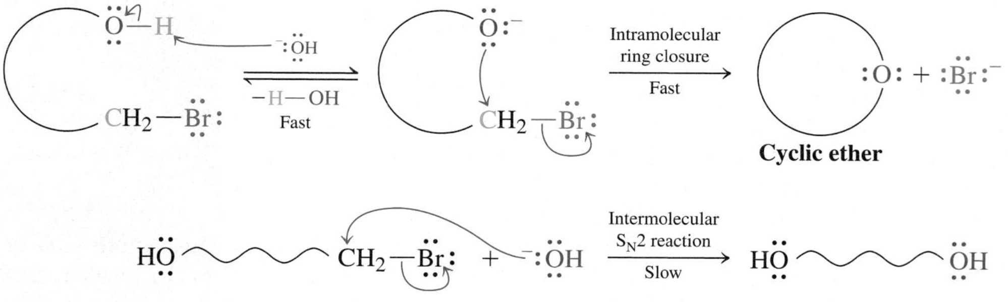

Reminder on Entropy

Positive (favorable) with increasing disorder or (more rigorously) increasing dispersal of the energy content of the system.

Figure 9-5 The mechanism of cyclic ether synthesis from a bromoalcohol and hydroxide ion (*upper reactions*). A competing but slower side reaction, direct displacement of bromide by hydroxide, is also shown (*lower reaction*). The curved lines denote a chain of carbon atoms.

Cyclic Ether Synthesis

$$\overset{\cdot\cdot}{H\overset{\cdot\cdot}{O}}CH_2CH_2\overset{\cdot\cdot}{\overset{\cdot\cdot}{Br}}: \;+\; H\overset{\cdot\cdot}{\overset{\cdot\cdot}{O}}:^- \;\longrightarrow\;$$

Oxacyclopropane
(Oxirane, ethylene oxide)

$$+ \;:\overset{\cdot\cdot}{\overset{\cdot\cdot}{Br}}:^- \;+\; H\overset{\cdot\cdot}{\overset{\cdot\cdot}{O}}H$$

$$H\overset{\cdot\cdot}{\overset{\cdot\cdot}{O}}(CH_2)_4CH_2\overset{\cdot\cdot}{\overset{\cdot\cdot}{Br}}: \;+\; H\overset{\cdot\cdot}{\overset{\cdot\cdot}{O}}:^- \;\longrightarrow\;$$

Oxacyclohexane
(Tetrahydropyran)

$$+ \;:\overset{\cdot\cdot}{\overset{\cdot\cdot}{Br}}:^- \;+\; H\overset{\cdot\cdot}{\overset{\cdot\cdot}{O}}H$$

The blue dot indicates the carbon atom in the product that corresponds to the site of ring closure in the starting material.

Exercise 9-13

The product of the reaction of 5-bromo-3,3-dimethyl-1-pentanol with hydroxide is a cyclic ether. Suggest a mechanism for its formation.

Ring size controls the speed of cyclic ether formation

A comparison of the relative rates of cyclic ether formation reveals a surprising fact: Three-membered rings form quickly, about as fast as five-membered rings. Six-membered ring systems, four-membered rings, and the larger oxacycloalkanes are generated more slowly.

Relative Rates of Cyclic Ether Formation

$$k_3 \geq k_5 > k_6 > k_4 \geq k_7 > k_8$$

k_n = reaction rate, n = ring size

What effects are at work here? Since we are concerned with rates, we need to compare the structures and hence energies of the transition states of the intramolecular Williamson ether synthesis. We shall find that the answer is composed of both *enthalpic* and *entropic contributions* (Section 2-1). Recall that enthalpy reflects changes not only in bond strengths during a reaction, but also in strain (Section 4-2). Entropy, on the other hand, is related to changes in the extent of order (or energy dispersal) in the system. What are the differences between the various transition states for ring formation with respect to these quantities?

As we proceed from larger ring to three-membered ring closures, the most obvious enthalpy effect is ring strain. If it were dominant, the most strained rings should be formed at the lowest rate, even though the full effect of strain may not be felt in the structure of the respective transition states. However, this is not observed, and we need to look for other factors that complicate this simple analysis. One such factor is entropy.

To understand how entropy comes into play, put yourself into the position of the nucleophilic, negatively charged oxygen in search of the backside of the electrophilic carbon bearing the leaving group. Clearly, the closer you are to the target, the easier it is to find it. If the target is remote, the intervening chain has to be arranged or "ordered" in such a way as to bring it closer to you. This is more difficult when the chain gets longer.

In molecular terms, to reach the transition state for ring formation, the opposite ends of the molecule must approach each other. In the ensuing conformation, rotation about the intervening bonds becomes restricted, and energy dispersal in the molecule is reduced: The entropy change is unfavorable (negative). This effect is most severe for long chains, making the formation of medium-sized and larger rings relatively most difficult. In addition, their rates of formation suffer from eclipsing, *gauche,* and transannular strain (Section 4-5). In contrast, cyclization of shorter chains requires less restriction of bond rotation and therefore a smaller unfavorable reduction in entropy: Three- to six-membered rings are generated relatively quickly. In fact, on entropy grounds alone, the relative rates for ring closure should be $k_3 > k_4 > k_5 > k_6$. Superposition of ring strain effects then leads to the observed trends listed above (see also margin structures).

Model Building

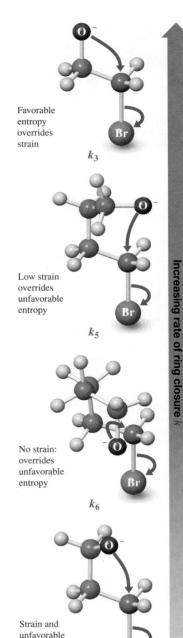

Favorable entropy overrides strain
k_3

Low strain overrides unfavorable entropy
k_5

No strain: overrides unfavorable entropy
k_6

Strain and unfavorable entropy team up
k_4

Increasing rate of ring closure k

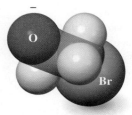

The oxygen in 2-bromoethoxide is "pushing" into the electrophilic carbon.

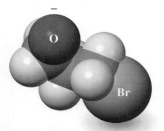

The oxygen in 3-bromopropoxide is at a comfortable distance from the electrophilic carbon.

Recent studies have shown that entropy alone cannot explain why three-membered rings are generated fastest. A second enthalpic phenomenon, which has been called the *proximity effect,* operates, especially in 2-haloalkoxides. To understand it, you need to remember that all S_N2 reactions suffer to a varying degree from steric hindrance of the nucleophile (Section 6-10). In 2-haloalkoxides (and related three-membered ring precursors), the nucleophilic part of the molecule is so close to the electrophilic carbon that part of the strain of the transition state is already present in the ground state. In other words, the molecule is activated along the reaction coordinate of the "normal" (unstrained) substitution process. This rate-accelerating proximity effect is dramatically reduced in four-membered ring synthesis (and so is the entropy advantage), while the ring strain is still large; therefore, ring closure to oxacyclobutanes is comparatively slow (see also space-filling models in margin). As we shall find, the general conclusions reached in this section apply also to other ring closing transformations that we shall encounter in later chapters.

The intramolecular Williamson synthesis is stereospecific

The Williamson ether synthesis proceeds with inversion of configuration at the carbon bearing the leaving group, in accord with expectations based on an S_N2 mechanism. The attacking nucleophile approaches the electrophilic carbon from the opposite side of the leaving group. Only one conformation of the haloalkoxide can undergo efficient substitution. For example, oxacyclopropane formation requires an *anti* arrangement of the nucleophile and the leaving group. The alternative two *gauche* conformations cannot give the product (Figure 9-6).

Gauche: unsuitable for S_N2 **Gauche:** unsuitable for S_N2 **Anti:** correct conformation for S_N2

Figure 9-6 Only the *anti* conformation of a 2-bromoalkoxide allows for oxacyclopropane formation. The two *gauche* conformers cannot undergo intramolecular backside attack at the bromine-bearing carbon.

Model Building

| **Solved Exercise 9-14** | Working with the Concepts: Stereochemistry of the Intramolecular Williamson Ether Synthesis |

(1*R*,2*R*)-2-Bromocyclopentanol reacts rapidly with sodium hydroxide to yield an optically inactive product. In contrast, the (1*S*,2*R*) isomer is much less reactive. Explain.

What you are being asked to do is to explain a result. So, you need to consider mechanisms. **H**ow to begin? Draw the structures of the two isomeric substrates (even better, build models!), so you can visualize their differences:

(1R,2R)-2-Bromocyclopentanol
The nucleophilic oxygen and the leaving group are trans *(anti)*

(1S,2R)-2-Bromocyclopentanol
Here the nucleophile and leaving group are cis *(syn)*

Information needed? Given the mechanistic options available to these compounds, how might their stereochemical difference affect their reactivity? **P**roceed.

• Figure 9-6 showed us that base treatment of 2-haloalkanols deprotonates the OH group. The resulting alkoxide then displaces the neighboring halide to produce an oxacyclopropane. Unlike the situation shown in Figure 9-6, however, our two 2-bromocyclopentanols contain a ring that restricts the conformational freedom around the bond connecting the two functional groups. In the (1R,2R) isomer, the alkoxide oxygen and the halogen are arranged trans to each other and hence in perfect alignment for intramolecular backside displacement of bromide by alkoxide. The resulting oxacyclopropane has a mirror plane and is meso and achiral.

Meso

- In contrast, the (1S,2R) isomer contains the same functions on the same side of the five-membered ring. To proceed to (the same) product, a frontside displacement is needed, which is much more difficult. Thus, the alkoxide derived from the 1R,2R-diastereomer reacts faster than that generated from the 1S,2R-isomer.

Exercise 9-15 | **Try It Yourself**

Bromoalcohol A transforms rapidly in the presence of sodium hydroxide to give the corresponding oxacyclopropane, whereas its diastereomer B does not. Why? [**Caution:** Unlike the previous problem, both substrates are *trans*-bromoalcohols. **Hint:** Draw the most stable cyclohexane chair conformers of both isomers (Section 4-4) and picture the respective transition states for the intramolecular Williamson ether synthesis.]

A   B

In Summary Ethers are prepared by the Williamson synthesis, an S_N2 reaction of an alkoxide with a haloalkane. This reaction works best with primary halides or sulfonates that do not undergo ready elimination. Cyclic ethers are formed by the intramolecular version of this method. The relative rates of ring closure in this case are highest for three- and five-membered rings.

9-7 | SYNTHESIS OF ETHERS: ALCOHOLS AND MINERAL ACIDS

An even simpler, albeit less selective, route to ethers is the reaction of a strong inorganic acid (e.g., H_2SO_4) with an alcohol. Protonation of the OH group in one alcohol generates water as a leaving group. Nucleophilic displacement of this leaving group by a second alcohol then results in the corresponding alkoxyalkane.

Alcohols give ethers by both S_N2 and S_N1 mechanisms

We have learned that treating primary alcohols with HBr or HI furnishes the corresponding haloalkanes through intermediate alkyloxonium ions (Section 9-2). However, when strong nonnucleophilic acids—such as sulfuric acid—are used at elevated temperatures, the main products are ethers.

Symmetrical Ether Synthesis from a Primary Alcohol with Strong Acid

$$2\ CH_3CH_2OH \xrightarrow[\substack{\text{Relatively high}\\\text{temperature}}]{H_2SO_4,\ 130°C} CH_3CH_2OCH_2CH_3\ +\ HOH$$

Reaction

Mechanism

In this reaction, the strongest nucleophile present in solution is the unprotonated starting alcohol. As soon as one alcohol molecule has been protonated, nucleophilic attack begins, the ultimate products being an ether and water.

Mechanism of Ether Synthesis from Primary Alcohols: Protonation and S_N2

$$CH_3CH_2\ddot{O}H + CH_3CH_2\ddot{O}H \underset{-H^+}{\overset{+H^+}{\rightleftharpoons}} CH_3CH_2-\overset{+}{\underset{H}{\overset{H}{\ddot{O}}}} \overset{S_N2}{\rightleftharpoons} \underset{CH_3CH_2}{\overset{CH_3CH_2}{\ddot{O}}}{-}\overset{+}{H} + H_2\ddot{O} \underset{+H^+}{\overset{-H^+}{\rightleftharpoons}} \underset{CH_3CH_2}{\overset{CH_3CH_2}{\ddot{O}}} + H_3\overset{+}{O}:$$

with $CH_3CH_2\ddot{O}H$ attacking.

E2 Mechanism for the Acid-Catalyzed Dehydration of 1-Propanol

(mechanism diagram showing 1-propanol → propene + H_2O)

S_N1 Mechanism for the Acid-Catalyzed Ether Formation from 2-Propanol

(mechanism diagram)

Only symmetric ethers can be prepared by this method.

At even higher temperatures (see footnote on p. 329), elimination of water to generate an alkene is observed. This reaction proceeds by an E2 mechanism (Sections 7-7 and 11-7), in which the neutral alcohol serves as the base that attacks the alkyloxonium ion (margin).

Alkene Synthesis from a Primary Alcohol and Strong Acid at Elevated Temperature: E2

$$\underset{\textbf{1-Propanol}}{CH_3\underset{|}{\overset{H}{C}}HCH_2OH} \xrightarrow[\text{Relatively high temperature}]{H_2SO_4, 180°C} \underset{\textbf{Propene}}{CH_3CH{=}CH_2} + HOH$$

Secondary and tertiary ethers can also be made by acid treatment of secondary and tertiary alcohols. However, in these cases, a carbocation is formed initially and is then trapped by an alcohol (S_N1), as described in Section 9-2.

Symmetrical Ether Synthesis from a Secondary Alcohol

$$2\ \underset{\textbf{2-Propanol}}{CH_3\underset{H}{\overset{OH}{C}}CH_3} \xrightarrow[\text{Relatively low temperature}]{H_2SO_4, 40°C} \underset{\substack{75\% \\ \textbf{2-(1-Methylethoxy)propane} \\ \textbf{(Diisopropyl ether)}}}{(CH_3)_2CHOCH(CH_3)_2} + HOH + H^+$$

The major side reaction follows the E1 pathway (Sections 9-2, 9-3, and 11-7), which, again, becomes dominant at higher temperatures.

It is harder to synthesize ethers containing two different alkyl groups, because mixing two alcohols in the presence of an acid usually results in mixtures of all three possible products. However, mixed ethers containing one tertiary and one primary or secondary alkyl substituent can be prepared in good yield in the presence of dilute acid. Under these conditions, the much more rapidly formed tertiary carbocation is trapped by the other alcohol.

Synthesis of a Mixed Ether from a Tertiary Alcohol

$$\underset{\textbf{}}{CH_3\underset{CH_3}{\overset{CH_3}{C}}OH} + CH_3CH_2OH\ (\text{excess}) \xrightarrow[-HOH]{15\%\ \text{aqueous}\ H_2SO_4,\ 40°C} \underset{\substack{95\% \\ \textbf{2-Ethoxy-2-methylpropane}}}{CH_3\underset{CH_3}{\overset{CH_3}{C}}OCH_2CH_3}$$

Exercise 9-16

Write mechanisms for the following two reactions. **(a)** 1,4-Butanediol + H⁺ → oxacyclopentane (tetrahydrofuran); **(b)** 5-methyl-1,5-hexanediol + H⁺ → 2,2-dimethyloxacyclohexane (2,2-dimethyltetrahydropyran).

Ethers also form by alcoholysis

As we know, tertiary and secondary ethers may also form by the alcoholysis of the corresponding haloalkanes or alkyl sulfonates (Section 7-1). The starting material is simply dissolved in an alcohol until the S_N1 process is complete (see margin).

Exercise 9-17

You now know several ways of constructing an ether from an alcohol and a haloalkane. Which approach would you choose for the preparation of **(a)** 2-methyl-2-(1-methylethoxy)butane; **(b)** 1-methoxy-2,2-dimethylpropane? [**Hint:** The product for (a) is a tertiary ether, that for (b) is a neopentyl ether.]

In Summary Ethers can be prepared by treatment of alcohols with acid through S_N2 and S_N1 pathways, with alkyloxonium ions or carbocations as intermediates, and by alcoholysis of secondary or tertiary haloalkanes or alkyl sulfonates.

9-8 | REACTIONS OF ETHERS

As mentioned earlier, ethers are normally rather inert. They do, however, react slowly with oxygen by radical mechanisms to form hydroperoxides and peroxides. Because peroxides can decompose explosively, extreme care should be taken with samples of ethers that have been exposed to air for several days.

Peroxides from Ethers

$$2\ \text{ROCH} \quad + \quad O_2 \quad \longrightarrow \quad 2\ \text{ROC—O—OH} \quad \longrightarrow \quad \text{ROC—O—O—COR}$$

An ether
hydroperoxide   An ether peroxide

A more useful reaction is cleavage by strong acid. The oxygen in ethers, like that in alcohols, may be protonated to generate alkyloxonium ions. The subsequent reactivity of these ions depends on the alkyl substituents. With *primary* groups and strong nucleophilic acids such as HBr, S_N2 displacement takes place.

Primary Ether Cleavage with HBr

$$CH_3CH_2OCH_2CH_3 \xrightarrow{\text{HBr}} CH_3CH_2Br \ + \ CH_3CH_2OH$$

Ethoxyethane   Bromoethane   Ethanol

Mechanism of Primary Ether Cleavage: S_N2

$$CH_3CH_2\ddot{O}CH_2CH_3 \rightleftharpoons CH_3\overset{+}{C}H_2{-}\overset{H}{\underset{CH_2CH_3}{\overset{..}{\overset{+}{O}}}} \rightleftharpoons CH_3CH_2Br \ + \ H\ddot{O}CH_2CH_3$$

Alkyloxonium ion

The alcohol formed as the second product may in turn be attacked by additional HBr to give more of the bromoalkane.

H₃C   Cl

1-Chloro-1-methyl-cyclohexane

| CH₃CH₂OH

H₃C   OCH₂CH₃

+ H⁺ + Cl⁻

86%
1-Ethoxy-1-methylcyclohexane

Reaction

Mechanism

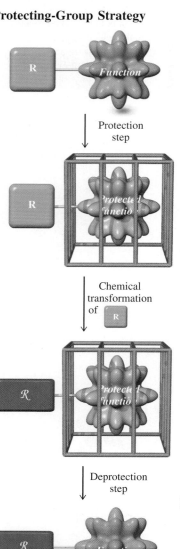

**Oxacyclohexane
(Tetrahydropyran)**

Exercise 9-18

Treatment of methoxymethane with hot concentrated HI gives two equivalents of iodomethane. Suggest a mechanism.

Exercise 9-19

Reaction of oxacyclohexane (tetrahydropyran; shown in the margin) with hot concentrated HI gives 1,5-diiodopentane. Give a mechanism for this reaction.

Oxonium ions derived from *secondary* ethers can transform by either S_N2 or S_N1 (E1) reactions, depending on the system and conditions (Section 7-9 and Tables 7-2 and 7-4). For example, 2-ethoxypropane is protonated by aqueous HI and then converted into 2-propanol and iodoethane by selective attack by iodide at the less hindered primary center.

Protecting-Group Strategy

Primary-Secondary Ether Cleavage with HI: S_N2 at Primary Center

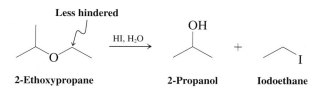

2-Ethoxypropane   **2-Propanol**   **Iodoethane**

Tertiary butyl ethers function to protect alcohols

Ethers containing *tertiary* alkyl groups transform even in dilute acid to give intermediate tertiary carbocations, which are either trapped by S_N1 processes, when good nucleophiles are present, or deprotonated in their absence:

Primary-Tertiary Ether Cleavage with Dilute Acid: S_N1 and E1 at Tertiary Center

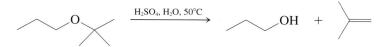

Because tertiary ethers are made under similarly mild conditions from alcohols (Section 9-7), they act as **protecting groups** for the hydroxy function. A protecting group renders a specific functionality in a molecule unreactive with respect to reagents and conditions that would normally transform it. Such protection allows chemistry to be carried out elsewhere in a molecule without interference. Subsequently, the original function is restored (deprotection). A protecting group has to be reversibly installed, readily and in high yield. Such is the case with tertiary ethers, in which the original alcohol is protected from base, organometallic reagents, oxidants, and reductants. Another method of alcohol protection is esterification (Section 9-4; Real Life 9-2).

Protection of Alcohols as Tertiary Butyl Ethers

$$ROH \xrightarrow[-H_2O]{(CH_3)_3COH,\ H^+} ROC(CH_3)_3 \xrightarrow{\substack{\text{Carry out reactions on R by}\\\text{using Grignard reagents,}\\\text{oxidizing agents, etc.}}} R'OC(CH_3)_3 \xrightarrow{H^+,\ H_2O} R'OH$$

Protection step   **Protected alcohol**   **R changed into R′**   **Deprotection**

Using protecting groups is a common procedure in organic synthesis, which enables chemists to carry out many transformations that would otherwise be impossible. We shall see other protecting strategies in conjunction with other functional groups later in the course.

REAL LIFE: MEDICINE 9-2 | Protecting Groups in the Synthesis of Testosterone

Protecting groups, as described in the text, are essential parts of many organic syntheses. An example is the synthesis of the sex hormone testosterone (Section 4-7) from a cholesterol-derived starting material. Natural sources of steroid hormones are far too limited to meet the needs of medicine and research; these molecules must be synthesized. In our case, the hydroxy function at C3 and carbonyl function at C17 of the starting material have to "trade places" to furnish the desired testosterone precursor. In other words, selective reduction of the carbonyl group at C17 and oxidation of the hydroxy group at C3 are required. In the scheme that was executed, you will note that all the "action" takes place at the centers of reactivity C3 and C17, the remainder of the seemingly complex steroid molecule simply providing a scaffold.

Thus, protection by formation of the 1,1-dimethylethyl (*tert*-butyl) ether at C3 is followed by reduction at C17. A second protection step at C17 is esterification (Section 9-4). Esters are stable in dilute acid, which hydrolyzes tertiary ethers. This strategy allows the hydroxy group at C3 to be freed and oxidized to a carbonyl while that at C17 remains protected. Exposure to strong acid finally converts the product of the sequence shown here into testosterone.

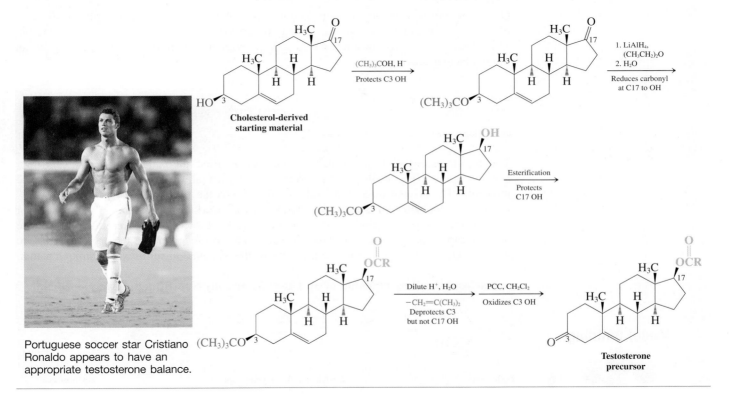

Portuguese soccer star Cristiano Ronaldo appears to have an appropriate testosterone balance.

Exercise 9-20

Show how you would achieve the following interconversions (the dashed arrow indicates that several steps are required). (**Hint:** You need to protect the OH function in each case.)

(a) $BrCH_2CH_2CH_2OH \dashrightarrow DCH_2CH_2CH_2OH$

(b)

In Summary Ethers are cleaved by (strong) acids. Protonation of an ether containing methyl or primary alkyl groups gives an alkyloxonium ion that is subject to S_N2 attack by nucleophiles. Carbocation formation follows protonation when secondary and tertiary groups are present, leading to S_N1 and E1 products. The hydroxy group of alcohols can be protected in the form of a *tert*-butyl ether.

9-9 | REACTIONS OF OXACYCLOPROPANES

Although ordinary ethers are relatively inert, the strained structure of the oxacyclopropanes makes possible nucleophilic ring-opening reactions. This section presents details of these processes.

Nucleophilic ring opening of oxacyclopropanes by S_N2 is regioselective and stereospecific

Oxacyclopropane is subject to bimolecular ring opening by anionic nucleophiles. Because of the symmetry of the substrate, substitution occurs to the same extent at either carbon. The reaction proceeds by nucleophilic attack, with the ether oxygen functioning as an intramolecular leaving group.

$$CH_2 - CH_2 \quad + \quad CH_3\ddot{S}\text{:}^- \quad \longrightarrow \quad {}^-\text{:}\ddot{O}CH_2CH_2 - \ddot{S}CH_3 \xrightarrow[-H\ddot{O}\text{:}^-]{H-\ddot{O}H} \quad H\ddot{O}CH_2CH_2\ddot{S}CH_3$$

Oxacyclopropane   **2-Methylthioethanol**

This S_N2 transformation is unusual for two reasons. First, alkoxides are usually very poor leaving groups. Second, the leaving group does not actually "leave"; it stays bound to the molecule. The driving force is the release of strain as the ring opens.

What is the situation with unsymmetric systems? Consider, for example, the reaction of 2,2-dimethyloxacyclopropane with methoxide. There are *two* possible reaction sites: at the primary carbon (*a*), to give 1-methoxy-2-methyl-2-propanol, and at the tertiary carbon (*b*), to yield 2-methoxy-2-methyl-1-propanol. Evidently, this system transforms solely through path *a*.

Nucleophilic Ring Opening of an Unsymmetrically Substituted Oxacyclopropane

Model Building

1-Methoxy-2-methyl-2-propanol $\xleftarrow[a]{CH_3O^-, CH_3OH}$ 2,2-Dimethyloxacyclopropane $\xrightarrow[b]{CH_3O^-, CH_3OH}$ ⊗   2-Methoxy-2-methyl-1-propanol (Not formed)

Is this result surprising? No, because, as we know, if there is more than one possibility, S_N2 attack will be at the *less* substituted carbon center (Section 6-10). This selectivity in the nucleophilic opening of substituted oxacyclopropanes is referred to as **regioselectivity**, because, of two possible and similar "regions," the nucleophile attacks only one.

In addition, when the ring opens at a stereocenter, inversion is observed. Thus, we find that the rules of nucleophilic substitution developed for simple alkyl derivatives also apply to strained cyclic ethers.

Hydride and organometallic reagents convert strained ethers into alcohols

The highly reactive lithium aluminum hydride is able to open the rings of oxacyclopropanes, a reaction leading to alcohols. Ordinary ethers, lacking the ring strain of oxacyclopropanes, do not react with $LiAlH_4$. The reaction also proceeds by the S_N2 mechanism. Thus, in unsymmetric systems, the hydride attacks the less substituted side; when the reacting carbon constitutes a stereocenter, inversion is observed.

Ring Opening of an Oxacyclopropane by Lithium Aluminum Hydride

$$\underset{\text{Less hindered}}{\overset{\displaystyle \overset{O}{\overset{\displaystyle C \!-\! C}{\underset{H \;\; H}{}}}}{}} \quad \xrightarrow{\begin{array}{l}\text{1. LiAlH}_4, \text{(CH}_3\text{CH}_2)_2\text{O} \\ \text{2. H}^+, \text{H}_2\text{O}\end{array}} \quad$$

Inversion on Oxacyclopropane Opening

$$\xrightarrow{\begin{array}{l}\text{1. LiAlD}_4 \\ \text{2. H}^+, \text{H}_2\text{O}\end{array}}$$

99.4%

D and OH are trans, not cis

Model Building

| **Working with the Concepts: A Retrosynthetic Analysis of an Oxacyclopropane**

Applying the principles of retrosynthetic analysis, as described in Section 8-9, which oxacyclopropane would be the best precursor to racemic 3-hexanol after treatment with LiAlH$_4$, followed by acidic aqueous work-up?

Strategy

The first thing to do is write the structure of 3-hexanol. Then look to see how many pathways lead to this structure from oxacyclopropanes and examine each path for feasibility.

Solution

- We recognize two possible retrosynthetic paths to 3-hexanol from oxacyclopropanes: removal of an anti H:$^-$ with simultaneous ring closure either to the "left" side or the "right" side. In our drawing, these two pathways are indicated by *a* and *b*, respectively.

| 3-Hexanol | Via *a* | Via *b* |

- Now that we have drawn the two possible precursors to 3-hexanol, let us see which one is better at making the desired product when it is reacted with LiAlH$_4$. (**Caution:** Remember that both carbons of an oxacyclopropane are electrophilic, so attack by hydride can occur in two possible ways.)
- Inspection of the precursor derived from retrosynthetic path *a* shows that it is unsymmetrical. Because both ring carbons are equally hindered, ring opening by hydride will give the two isomers, 2- and 3-hexanol.
- On the other hand, retrosynthetic path *b* furnishes a symmetric oxacyclopropane in which the regiochemistry of hydride opening is immaterial. Hence this precursor is best.

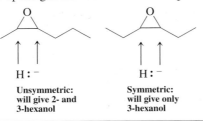

Unsymmetric:
will give 2- and
3-hexanol

Symmetric:
will give only
3-hexanol

Exercise 9-22 | Try It Yourself

(2R)-Butanol can be made by LiAlH$_4$ reduction of an oxacyclopropane. Which one?

In contrast with haloalkanes (Section 8-8), oxacyclopropanes are sufficiently reactive electrophiles to be attacked by organometallic compounds. Thus, Grignard reagents and alkyllithium compounds undergo 2-hydroxyethylation by ether ring opening, following the S$_N$2 mechanism. This reaction constitutes a two-carbon homologation of an alkyl chain, as opposed to the one-carbon homologation of alkyl organometallic reagents by formaldehyde (Sections 8-8 and 8-9).

Oxacyclopropane Ring Opening by a Grignard Reagent: 2-Hydroxyethylation

$$H_2C-CH_2 \quad + \quad CH_3CH_2CH_2CH_2MgBr \quad \xrightarrow[\text{2. H}^+\text{, H}_2\text{O}]{\text{1. THF}} \quad CH_3CH_2CH_2CH_2CH_2CH_2OH$$

$$62\%$$

Oxacyclopropane   Butylmagnesium bromide   1-Hexanol: "2-hydroxyethylated butyl"

Exercise 9-23 |

Propose an efficient synthesis of 3,3-dimethyl-1-butanol from starting materials containing no more than four carbons. (**Hint:** Consider the product retrosynthetically as a 2-hydroxyethylated *tert*-butyl.)

REAL LIFE: CHEMISTRY 9-3 | Hydrolytic Kinetic Resolution of Oxacyclopropanes

As we pointed out in Real Life 5-5, nature is "handed" and shows great, if not exclusive, preference for reactions with only one of two enantiomers of a chiral compound. This preference is of particular significance in drug development, because usually only one enantiomer of a chiral drug is effective (Real Life 5-4). Therefore, the preparation of single enantiomers is an important "green" challenge for the synthetic chemist (Real Life 3-1). The classic way of meeting this challenge has been the resolution of racemates via the (readily reversible) reaction with an optically pure compound generating diastereomers that can be separated by chromatography or fractional crystallization (Section 5-8). The approach is equivalent to using a collection of right hands to separate a collection of pairs of shoes. Once all hands are "on," the resulting collection is divided into two groups, namely, the combinations right hand/right shoe and right hand/left shoe. These groups are not related by mirror symmetry, hence they are diastereomeric.

In our example, the most obvious feature is that members of the respective groups are shaped completely differently and could be separated by an achiral device, such as a sieve. From the resulting two piles, the right hands would be recovered and recycled, leaving the right shoes separated from the left ones.

A much better machine would be a fishing device, in which the hook would be shaped like a right (or left) foot. This machine would pull out only right shoes from the pile, allowing their selective tagging, for example by attaching a weight. Left and right shoes could then be separated by an achiral device on the basis of their differing weights. Such a process at the molecular level is called *catalytic kinetic resolution*. An example is the hydrolysis of methyloxacyclopropane with basic water. Normally, starting with a racemate, the result is racemic 1,2-propanediol. This is to be expected, as the two respective transition states for the reactions of R and S starting ether are enantiomeric (Section 5-7).

$$H_3C \overset{O}{\triangle} R \quad + \quad H_3C \overset{O}{\triangle} S \quad \xrightarrow{\text{H}_2\text{O, HO}^-} \quad \overset{OH}{\underset{R}{\diagup}}\text{—OH} \quad + \quad \overset{OH}{\underset{S}{\diagup}}\text{—OH}$$

Racemic   Racemic
methyloxacyclopropane   1,2-propanediol

However, in the presence of an enantiomer of a chiral cobalt catalyst (the "right foot" in our device above), water attacks the R form of starting material much more rapidly than the S counterpart, converting it selectively to (2R)-1,2-

propanediol (our "selective tagging"), leaving behind pure (S)-oxacyclopropane. The reason is the chiral nature of the catalyst, rendering the two respective transition states of the reaction diastereomeric. Thus, they are of unequal energy,

Acids catalyze oxacyclopropane ring opening

Ring opening of oxacyclopropanes is also catalyzed by acids. The reaction in this case proceeds through initial cyclic alkyloxonium ion formation followed by ring opening as a result of nucleophilic attack.

Acid-Catalyzed Ring Opening of Oxacyclopropane

Mechanism of Acid-Catalyzed Ring Opening

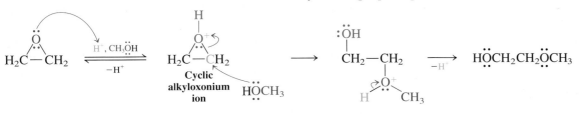

The anionic nucleophilic opening of oxacyclopropanes just discussed is regioselective and stereospecific. What about acid-catalyzed ring opening—is that also regioselective and stereospecific? Yes, but the details are different. Thus, acid-catalyzed methanolysis of 2,2-dimethyloxacyclopropane proceeds by exclusive ring opening at the *more* hindered carbon.

allowing one enantiomer of oxacyclopropane to hydrolyze faster than the other.

The structure of the catalyst enantiomer is shown below. You can see the chiral environment around the metal, provided by the substituted cyclohexane scaffold. The cobalt attacks selectively the lone pairs on the (*R*)-oxacyclopropane substrate as a Lewis acid (Section 2-3), thus facilitating ring opening by water.

Cobalt catalyst

Using the mirror image of the catalyst shown gives the complementary results to our example: Only (*S*)-methyloxacyclopropane is attacked to furnish (*S*)-diol, leaving behind unreacted *R* starting material. Such highly functionalized small chiral building blocks are of great value in the synthesis of medicines and other fine chemicals and therefore in great demand by synthetic chemists. As a result, the above kinetic resolution has been refined to require less than 1 kg of catalyst to make 1 ton of product and is being used by Daiso Co. in Japan on a 50-ton/year scale.

Reaction

**Protonated
2,2-dimethyloxacyclopropane**

<hr>

**Oxacyclopropane:
The Warhead of Drugs**

Fosfomycin

Cysteine

Enzyme

The antibiotic fosfomycin works by interfering with bacterial cell-wall synthesis through oxacyclopropane ring opening. Thus, the enzyme crucial for wall construction is deactivated by reaction of the SH group of one of its cysteine amino acids (for structure, see Problem 45 of Chapter 2) with the strained ether function.

<hr>

Acid-Catalyzed Ring Opening of 2,2-Dimethyloxacyclopropane

Nucleophile attacks
this carbon

H_2C—C
CH$_3$
CH$_3$

2,2-Dimethyloxacyclopropane

$\xrightarrow{H_2SO_4,\ CH_3OH}$

HO   CH$_3$
CH$_2$—C—CH$_3$
OCH$_3$

2-Methoxy-2-methyl-1-propanol

Why is the more hindered position attacked? Protonation at the oxygen of the ether generates a reactive intermediate alkyloxonium ion with substantially polarized oxygen–carbon bonds. This polarization places partial positive charges on the ring carbons. Because alkyl groups act as electron donors (Section 7-5), more positive charge is located on the tertiary than on the primary carbon. You can see this difference in the electrostatic potential map in the margin, in which the molecule is viewed from the perspective of the attacking nucleophile. The tertiary carbon at the bottom is more positive (blue) than the primary neighbor above (green). The proton in the back is strongly blue. The color-energy scale of this map was changed to make this subtle gradation in shading visible.

**Mechanism of Acid-Catalyzed Ring Opening
of 2,2-Dimethyloxacyclopropane by Methanol**

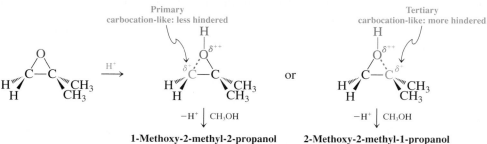

1-Methoxy-2-methyl-2-propanol
(Not formed)

2-Methoxy-2-methyl-1-propanol

This uneven charge distribution counteracts steric hindrance: Methanol is attracted by coulombic forces more to the tertiary than to the primary center. Although the result is clear-cut in this example, it is less so in cases in which the two carbons are not quite as different. For example, mixtures of isomeric products are formed by acid-catalyzed ring opening of 2-methyloxacyclopropane.

Why do we not simply write the isomeric free carbocations as intermediates in the acid-catalyzed ring openings? The reason is that the cyclic oxonium ion has an octet structure, whereas the carbocation isomer has a carbon with an electron sextet. Indeed, experimentally, inversion is observed when reaction takes place at a stereocenter. Like the reaction of oxacyclopropanes with anionic nucleophiles, the acid-catalyzed process includes backside displacement—in this case, on a highly polarized cyclic alkyloxonium ion.

<hr>

Exercise 9-24

Predict the major product of ring opening of 2,2-dimethyloxacyclopropane on treatment with each of the following reagents. (**a**) LiAlH$_4$, then H$^+$, H$_2$O; (**b**) CH$_3$CH$_2$CH$_2$MgBr, then H$^+$, H$_2$O; (**c**) CH$_3$SNa in CH$_3$OH; (**d**) dilute HCl in CH$_3$CH$_2$OH; (**e**) concentrated aqueous HBr.

<hr>

In Summary Although ordinary ethers are relatively inert, the ring in oxacyclopropanes can be opened both regioselectively and stereospecifically. For anionic nucleophiles, the usual rules of bimolecular nucleophilic substitution hold: Attack is at the less hindered carbon center, which undergoes inversion. Acid catalysis, however, changes the regioselectivity (but not the stereospecificity): Attack is at the more hindered center. Hydride and organometallic reagents behave like other anionic nucleophiles, furnishing alcohols by an S$_N$2 pathway.

9-10 | SULFUR ANALOGS OF ALCOHOLS AND ETHERS

Sulfur is located directly below oxygen in the periodic table, and therefore one might expect the sulfur analogs of alcohols and ethers to behave in a rather similar manner. In this section, we shall see whether this assumption is true.

The sulfur analogs of alcohols and ethers are thiols and sulfides

The sulfur analogs of alcohols, R–SH, are called **thiols** in the IUPAC system (*theion*, Greek, brimstone—an older name for sulfur). The ending *thiol* is added to the alkane stem to yield the alkanethiol name. The SH group is referred to as **mercapto,** from the Latin *mercurium,* mercury, and *captare,* to capture, because of its ability to precipitate mercury (and other heavy metal) ions. Its location is indicated by numbering the longest chain, as in alkanol nomenclature. The mercapto functional group has lower precedence than hydroxy.

$CH_3\ddot{S}H$
Methanethiol

$CH_3CH_2\overset{\overset{\displaystyle CH_3}{|}}{\underset{4\quad3\quad2\quad1}{C}}HCH_2\ddot{S}H$
2-Methyl-1-butanethiol

$CH_3CH_2\overset{\overset{\displaystyle :\ddot{S}H}{|}}{C}HCH_2CH_3$
3-Pentanethiol

(cyclohexane ring with $:\ddot{S}H$)
Cyclohexanethiol

$HSCH_2CH_2\ddot{O}H$
2-Mercaptoethanol

The sulfur analogs of ethers (common name, thioethers) are called **sulfides,** as in alkyl ether nomenclature. The RS group is named **alkylthio,** the RS⁻ group **alkanethiolate.**

$CH_3\ddot{S}CH_2CH_3$
Ethyl methyl sulfide

$CH_3\overset{\overset{\displaystyle CH_3}{|}}{\underset{\underset{\displaystyle CH_3}{|}}{C}}\ddot{S}(CH_2)_6CH_3$
1,1-Dimethylethyl heptyl sulfide

$CH_3\ddot{S}CH_2CH_2\ddot{O}H$
2-Methylthioethanol

$CH_3\ddot{S}{:}^-$
Methanethiolate ion

Thiols are less hydrogen bonded and more acidic than alcohols

Sulfur, because of its large size, its diffuse orbitals, and the relatively nonpolarized S–H bond (Table 1-2), does not enter into hydrogen bonding very efficiently. Thus, the boiling points of thiols are not as abnormally high as those of alcohols; rather, their volatilities lie close to those of the analogous haloalkanes (Table 9-2).

Partly because of the relatively weak S–H bond, thiols are also more acidic than water, with pK_a values ranging from 9 to 12. They can therefore be more readily deprotonated by hydroxide and alkoxide ions.

Acidity of Thiols

$$R\ddot{S}H \;+\; H\ddot{O}{:}^- \;\rightleftharpoons\; R\ddot{S}{:}^- \;+\; HOH$$

$pK_a = 9–12$ $pK_a = 15.7$
More acidic   **Less acidic**

Thiols and sulfides react much like alcohols and ethers

Many reactions of thiols and sulfides resemble those of their oxygen analogs. The sulfur in these compounds is even more nucleophilic and much less basic than the oxygen in alcohols and ethers. Therefore, thiols and sulfides are readily made through nucleophilic attack by RS⁻ or HS⁻ on haloalkanes, with little competing elimination. A large excess of the HS⁻ is used in the preparation of thiols to ensure that the product does not react with the starting halide to give the dialkyl sulfide.

$$CH_3\overset{\overset{\displaystyle CH_3}{|}}{C}HBr + Na^{+-}SH \xrightarrow{CH_3CH_2OH} CH_3\overset{\overset{\displaystyle CH_3}{|}}{C}HSH + Na^+Br^-$$

Excess   **2-Propanethiol**

Table 9-2	Comparison of the Boiling Points of Thiols, Haloalkanes, and Alcohols
Compound	**Boiling point (°C)**
CH_3SH	6.2
CH_3Br	3.6
CH_3Cl	−24.2
CH_3OH	65.0
CH_3CH_2SH	37
CH_3CH_2Br	38.4
CH_3CH_2Cl	12.3
CH_3CH_2OH	78.5

Sulfides are prepared in an analogous way by alkylation of thiols in the presence of base, such as hydroxide. The base generates the alkanethiolate, which reacts with the haloalkane by an S_N2 process. Because of the strong nucleophilicity of thiolates, there is no competition from hydroxide in this displacement.

Sulfides by Alkylation of Thiols

$$RSH \ + \ R'Br \ \xrightarrow{\text{NaOH}} \ RSR' \ + \ NaBr \ + \ H_2O$$

The nucleophilicity of sulfur also explains the ability of sulfides to attack haloalkanes to furnish **sulfonium ions.**

95%
Trimethylsulfonium iodide

Sulfonium salts are subject to nucleophilic attack at carbon, the sulfide functioning as the leaving group (see also Chapter 6 Opening).

$$HO:^- \ + \ CH_3\overset{+}{S}(CH_3)_2 \ \longrightarrow \ HOCH_3 \ + \ S(CH_3)_2$$

Exercise 9-25

Sulfide A (below) is a powerful poison known as "mustard gas," a devastating chemical warfare agent used in World War I and again in the eight-year war between Iraq and Iran in the 1980s. The specter of chemical and biological weapons loomed again during the Persian Gulf war of 1990–1991, and a medical condition known as "Gulf war syndrome" has been, at times, ascribed to the suspected exposure of ground troops to chemical and perhaps biological agents during the campaign. The Geneva protocol of 1925 explicitly bans the use of chemical and biological weapons. The 1982 Biological and 1993 (ratified by the United States in 1997) Chemical Weapons Conventions ban possession of such materials, but there is great concern about compliance and enforcement. One of the problems is the relative ease with which such toxic chemicals can be produced, as highlighted in this problem. **(a)** Propose a synthesis of A starting with oxacyclopropane. (**Hint:** Your retrosynthetic analysis should proceed through the diol precursor to A.) **(b)** Its mechanism of action is believed to include sulfonium salt B, which is thought to react with nucleophiles in the body. How is compound B formed, and how would it react with nucleophiles?

$$ClCH_2CH_2SCH_2CH_2Cl$$
A

$$ClCH_2CH_2\overset{+}{S}\underset{CH_2}{\overset{CH_2}{\big<}} \ Cl^-$$
B

Valence-shell expansion of sulfur accounts for the special reactivity of thiols and sulfides

As a third-row element with d orbitals, sulfur's valence shell can expand to accommodate more electrons than are allowed by the octet rule (Section 1-4). We have already seen that, in some of its compounds, sulfur is surrounded by 10 or even 12 valence electrons, and this capacity enables sulfur compounds to undergo reactions inaccessible to the corresponding oxygen analogs. For example, oxidation of thiols with strong oxidizing agents, such as hydrogen peroxide or potassium permanganate, gives the corresponding sulfonic acids. In this way, methanethiol is converted into methanesulfonic acid. Sulfonic acids react with PCl_5 to give sulfonyl chlorides, which are used in sulfonate synthesis, as discussed in Section 9-4.

More careful oxidation of thiols, by the use of iodine, results in the formation of **disulfides,** the sulfur analogs of peroxides (Section 9-8). Disulfides are readily reduced back to thiols by mild reducing agents, such as aqueous sodium borohydride.

Soldiers wearing chemical protective gear.

CH_3SH
Methanethiol

$\downarrow$ $KMnO_4$

$$CH_3\overset{O}{\underset{O}{\overset{\|}{\underset{\|}{S}}}OH}$$

Methane-sulfonic acid

The Thiol–Disulfide Redox Reaction

Oxidation:

$$2\ CH_3CH_2CH_2S{-}H\ +\ I_2\ \longrightarrow\ CH_3CH_2CH_2S{-}SCH_2CH_2CH_3\ +\ 2\ HI$$

1-Propanethiol   **Dipropyl disulfide**

Reduction:

$$CH_3CH_2CH_2S{-}SCH_2CH_2CH_3\ +\ NaBH_4\ \xrightarrow{\ H_2O\ }\ 2\ CH_3CH_2CH_2SH$$

Disulfide formation by oxidation of thiols and its reverse are important biological processes, although nature uses much milder reagents and conditions than those depicted above. Many proteins and peptides contain free SH groups that form bridging disulfide linkages. Nature exploits this mechanism to link amino acid chains. By thus helping to control the shape of enzymes in three dimensions, the mechanism makes biocatalysis far more efficient and selective.

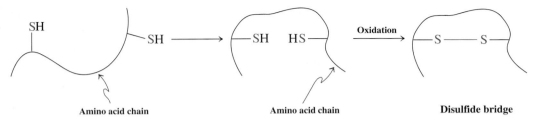

Amino acid chain   **Amino acid chain**   **Disulfide bridge**

Sulfides are readily oxidized to **sulfones,** a transformation proceeding through a **sulfoxide** intermediate. For example, oxidation of dimethyl sulfide first gives dimethyl sulfoxide (DMSO), which subsequently furnishes dimethyl sulfone. Dimethyl sulfoxide has already been mentioned as a highly polar nonprotic solvent of great use in organic chemistry, particularly in nucleophilic substitutions (see Section 6-8 and Table 6-5).

In Summary The naming of thiols and sulfides is related to the system used for alcohols and ethers. Thiols are more volatile, more acidic, and more nucleophilic than alcohols. Thiols and sulfides can be oxidized, thiols to disulfides or sulfonic acids and sulfides to sulfoxides and sulfones.

$$CH_3\overset{..}{\underset{..}{S}}CH_3$$

Dimethyl sulfide

$$\downarrow H_2O_2$$

$$CH_3\overset{\overset{\textstyle :O:}{\|}}{\underset{..}{S}}CH_3$$

Dimethyl sulfoxide (DMSO)

$$\downarrow H_2O_2$$

$$CH_3\overset{\overset{\textstyle :O:}{\|}}{\underset{\underset{\textstyle :O:}{\|}}{S}}CH_3$$

Dimethyl sulfone

9-11 | PHYSIOLOGICAL PROPERTIES AND USES OF ALCOHOLS AND ETHERS

Since we live in an oxidizing atmosphere, it is not surprising that oxygen is found in abundance in the chemicals of nature. Many of them are alcohols and ethers with varied biological functions, exploited by medicinal chemists in drug synthesis. Industrial chemists produce alcohols and ethers on a large scale, to be used as solvents and synthetic intermediates. This section gives a glimpse of the versatile uses of these compound classes.

Methanol, made in large amounts by the catalytic hydrogenation of carbon monoxide (Section 8-4), is sold as a solvent for paint and other materials, as a fuel for camp stoves and soldering torches, and as a synthetic intermediate. It is highly poisonous—ingestion or chronic exposure may lead to blindness. Death from ingestion of as little as 30 mL has been reported. It is sometimes added to commercial ethanol to render it unfit for consumption (denatured alcohol). The toxicity of methanol is thought to be due to metabolic oxidation to formaldehyde, $CH_2{=}O$, which interferes with the physiochemical processes of vision. Further oxidation to formic acid, HCOOH, causes acidosis, an unusual lowering of the blood pH. This condition disrupts oxygen transport in the blood and leads eventually to coma.

Methanol has been studied as a possible precursor of gasoline. For example, certain zeolite catalysts (Section 3-3) allow the conversion of methanol into a mixture of hydrocarbons,

Denatured ethanol contains the much more poisonous methanol.

ranging in length from four-carbon chains to ten-carbon ones, with a composition that, on distillation, yields largely gasoline (see Table 3-3).

$$n\,CH_3OH \xrightarrow{\text{Zeolite, }340°\text{–}375°C} \underset{67\%}{C_nH_{2n+2}} + \underset{6\%}{C_nH_{2n}} + \underset{27\%}{\text{aromatics}}$$

In 2011, a Scottish whisky distillery sent unmatured spirit and charred oak in two separate samples into space in a Russian cargo ship. Once in orbit, the contents were mixed and left to mature over two years under zero gravity conditions. In collaboration with space scientists, the experiment hopes to learn much about how flavors develop, with applications to food and perfume chemistry. For scotch fans, it may lead to the discovery of new chemical building blocks adding to the tasting spectrum.

Ethanol—diluted by various amounts of flavored water—is an alcoholic beverage. It is classified pharmacologically as a general depressant, because it induces a nonselective, reversible depression of the central nervous system. Approximately 95% of the alcohol ingested is metabolized in the body (usually in the liver) to products that are transformed eventually into carbon dioxide and water. Although it is high in calories, ethanol has little nutritional value.

The rate of metabolism of most drugs increases with their concentration in the liver, but this is not true for alcohol, which degrades linearly with time. An adult metabolizes about 10 mL of pure ethanol per hour, roughly the ethanol content of a cocktail, a shot of spirits, or a can of beer. As few as two or three drinks—depending on a person's weight, the ethanol content of the drink, and the speed with which it is consumed—can produce a level of alcohol in the blood that is more than 0.08%, a concentration at or above the legal limit beyond which the operation of a motorized vehicle is prohibited in much of the United States.

Ethanol is poisonous. Its lethal concentration in the bloodstream has been estimated at 0.4%. Its effects include progressive euphoria, disinhibition, disorientation, and decreased judgment (drunkenness), followed by general anesthesia, coma, and death. It dilates the blood vessels, producing a "warm flush," but it actually decreases body temperature. Although long-term ingestion of moderate amounts (the equivalent of about two beers a day) does not appear to be harmful, larger amounts can be the cause of a variety of physical and psychological symptoms, usually described by the general term *alcoholism*. These symptoms include hallucinations, psychomotor agitation, liver diseases, dementia, gastritis, and addiction.

Ethanol destined for human consumption is prepared by fermentation of sugars and starch (rice, potatoes, corn, wheat, flowers, fruit, etc.; Chapter 24). Fermentation is catalyzed by enzymes in a multistep sequence that converts carbohydrates into ethanol and carbon dioxide.

Alcohol disease causes a normal liver (on the left) to become fatty (middle), and eventually cirrhotic (right).

$$\underset{\textbf{Starch}}{(C_6H_{10}O_5)_n} \xrightarrow{\text{Enzymes}} \underset{\textbf{Glucose}}{C_6H_{12}O_6} \xrightarrow{\text{Enzymes}} \underset{\textbf{Ethanol}}{2\,CH_3CH_2OH} + 2\,CO_2$$

Interest in ethanol production from such "green" sources ("bioethanol") has surged because of its potential as a gasoline additive ("gasohol") or even substitute (see Chapter 8 Opening). For example, in Brazil, which has the world's largest biofuels program, ethanol provides about 50% of the country's automotive fuel needs. While not quite as calorific as the hydrocarbon mixture in gasoline (see Table 3-7), ethanol burns more efficiently and cleanly:

$$CH_3CH_2OH + 3\,O_2 \xrightarrow{\text{Combustion}} 2\,CO_2 + 3\,H_2O \quad -326.7\ \text{kcal mol}^{-1}$$

Therefore, a biofuels economy, in essence, converts glucose to carbon dioxide and water:

$$\underset{\textbf{Glucose}}{C_6H_{12}O_6} + 6\,O_2 \longrightarrow 6\,CO_2 + 6\,H_2O + \text{heat (car mileage)}$$

This process is sustainable because green plants use sunlight for the reverse process, called *photosynthesis* (Real Life 24-1):

$$6\,CO_2 + 6\,H_2O \xrightarrow[\text{Photosynthesis}]{\text{Sunlight}} \underset{\textbf{Glucose}}{C_6H_{12}O_6} + 6\,O_2$$

Thus, simply put, the idea is to power our cars with sunlight in a roundabout way. In practice, however, switching to biofuels is not without problems, primarily because of the sheer size of our fuel needs. For example, in 2011, the United States committed 40% of its corn crop for the preparation of 7 billion gallons of ethanol. Compare this to our *daily* consumption of 370 million gallons of gasoline, and you realize the scope of this undertaking! The potential negative impact of committing large agricultural areas of our planet to biofuel production on the environment, the economy, and the supply (hence price) of food is a matter of concern. On a more fundamental level, some critics argue that, at least for certain crops, the complete process of growing, harvesting, fermenting, and then distributing biofuels consumes more energy than it returns, dealing a fatal blow to the viability of such efforts.

Commercial alcohol that is not intended as a beverage is made industrially by hydration of ethene (Section 8-4). It is used, for example, as a solvent in perfumes, varnishes, and shellacs and as a synthetic intermediate, as demonstrated in earlier equations.

1,2-Ethanediol (ethylene glycol) is prepared by oxidation of ethene to oxacyclopropane, followed by hydrolysis, in quantities exceeding 20 million tons globally. Its low melting point ($-11.5°C$), its high boiling point ($198°C$), and its complete miscibility with water make it a useful antifreeze. Its toxicity is similar to that of other simple alcohols.

Ethylene glycol is effective in the deicing of airplanes.

$$CH_2{=}CH_2 \xrightarrow{\text{Oxidation}} \underset{\substack{\text{Oxacyclopropane} \\ \text{(Ethylene oxide)}}}{\triangle} \xrightarrow{H_2O} \underset{\substack{\text{1,2-Ethanediol} \\ \text{(Ethylene glycol)}}}{HOCH_2CH_2OH}$$

Ethene
(Ethylene)

1,2,3-Propanetriol (glycerol, glycerine), $HOCH_2CHOHCH_2OH$, is a viscous greasy substance, soluble in water, and nontoxic. It is obtained by alkaline hydrolysis of triglycerides, the major component of fatty tissue. The sodium and potassium salts of the long-alkyl-chain acids produced from fats ("fatty acids," Chapter 19) are sold as soaps.

$$\underset{\substack{\text{Triglyceride} \\ \text{("Fat")}}}{\begin{array}{c} O \\ \| \\ CH_2OCR \\ | \\ O \\ \| \\ HCOCR \\ | \\ O \\ \| \\ CH_2OCR \end{array}} \xrightarrow{H_2O,\ NaOH} \underset{\substack{\text{1,2,3-Propanetriol} \\ \text{(Glycerol, glycerine)}}}{\begin{array}{c} CH_2OH \\ | \\ HCOH \\ | \\ CH_2OH \end{array}} + \underset{\text{Soap}}{\overset{O}{\underset{}{RC\overset{\|}{O}^-Na^+}}}$$

R = long alkyl chain

Phosphoric esters of 1,2,3-propanetriols (phosphoglycerides; Section 20-4) are primary components of cell membranes.

1,2,3-Propanetriol is present in lotions and other cosmetics, as well as in medicinal preparations. Treatment with nitric acid gives a trinitrate ester known as *nitroglycerine,* used medicinally in the treatment of the symptoms of angina, especially the chest pain caused by insufficient flow of blood to the heart. The drug relaxes the blood vessels, thus increasing blood flow. A totally different application of nitroglycerin is as an extremely powerful explosive. The explosive potential of this substance results from its shock-induced, highly exothermic decomposition to gaseous products (N_2, CO_2, H_2O gas, O_2), raising temperatures to more than $3000°C$ and creating pressures higher than 2000 atmospheres in a fraction of a second (see also Real Life 16-1).

Ethoxyethane (diethyl ether) was at one time employed as a general anesthetic (see Chapter Opening). It produces unconsciousness by depressing central nervous system activity. Because of adverse effects such as irritation of the respiratory tract and extreme nausea, its use has been discontinued, and 1-methoxypropane (methyl propyl ether, "neothyl") and other compounds have replaced it in such applications. Ethoxyethane and other ethers are explosive when they are mixed with air.

Oxacyclopropane (oxirane, ethylene oxide) is a large-volume industrial chemical intermediate and a fumigating agent for seeds and grains. In nature, oxacyclopropane derivatives control insect metamorphosis (see Real Life 12-1) and are formed in the course of

$$\begin{array}{c} CH_2OH \\ | \\ HCOH \\ | \\ CH_2OH \end{array} + 3\ HONO_2$$

$$\downarrow$$

$$\underset{\text{Nitroglycerine}}{\begin{array}{c} CH_2ONO_2 \\ | \\ HCONO_2 \\ | \\ CH_2ONO_2 \end{array}} + 3\ H_2O$$

enzyme-catalyzed oxidation of aromatic hydrocarbons, often leading to highly carcinogenic (cancer-causing) products (see Section 16-7).

Many *natural products,* some of which are quite active physiologically, contain alcohol and ether groups. For example, morphine is a powerful analgesic. Its synthetic acetate derivative, heroin, is a widely abused street drug. Tetrahydrocannabinol is the main active ingredient in marijuana *(cannabis),* whose mood-altering effects have been known for thousands of years. Efforts are ongoing in the United States and worldwide to legalize marijuana for medical purposes, on the basis of the finding that it provides relief to patients afflicted with cancer, AIDS, multiple sclerosis, and other diseases from the effects of nausea, pain, and loss of appetite.

The opium poppy is the source of morphine, the active ingredient in opium.

Morphine
(R = H)
Heroin
$\left(\begin{array}{c} R = CCH_3 \\ \parallel \\ O \end{array}\right)$

Tetrahydrocannabinol

The lower-molecular-weight thiols and sulfides are most notorious for their foul smell. *Ethanethiol* is detectable by its odor even when it is diluted in 50 million parts of air. The major volatile components of the skunk's defensive spray are 3-methyl-1-butanethiol, *trans-2-butene-1-thiol,* and *trans-2-butenyl methyl disulfide.* The all-too-familiar human "BO" (body odor) emanating from sweaty armpits was analyzed by chemists in the perfume industry in 2004. The major chemical culprit is 3-mercapto-3-methylhexan-1-ol, specifically the obnoxious *S*-enantiomer. It is excreted admixed with 25% of its mirror image, which, curiously, has a fruity odor.

$$CH_3$$
$$|$$
$$CH_3CHCH_2CH_2SH$$

3-Methyl-1-butanethiol

trans-2-Butene-1-thiol

trans-2-Butenyl methyl disulfide

3-Mercapto-3-methylhexan-1-ol

(R,S)-2-(4-Methyl-3-cyclohexenyl)-2-propanethiol

Strangely enough, when they are highly diluted, sulfur compounds can have a rather pleasant odor. For example, the smell of freshly chopped onions or garlic is due to the presence of low-molecular-weight thiols and sulfides (Real Life 9-4). Dimethyl sulfide is a component of the aroma of black tea. The compound 2-(4-methyl-3-cyclohexenyl)-2-propanethiol (margin) is responsible for the unique taste of grapefruit, in which it is present in concentrations below the parts-per-billion (ppb; i.e., 1-in-10^9) range. It can be tasted at even lower concentrations, at a dilution of 10^{-4} ppb. In other words, you can notice the presence of 1 mg of this compound when it is dissolved in 10 million liters of water!

Many beneficial drugs contain sulfur in their molecular framework. Particularly well known are the *sulfonamides,* or *sulfa drugs,* powerful antibacterial agents (Section 15-10):

Sulfadiazine
(An antibacterial drug)

Diaminodiphenylsulfone
(An antileprotic drug)

REAL LIFE: MEDICINE 9-4 | Garlic and Sulfur

What a culinary delight it is to augment your meal with the flavorful components of the genus *Allium*: garlic, onion, leeks, chives, scallions, and shallots! The odorants in all of these foods are based on the same element: sulfur. What is surprising is that, in many cases, the compounds giving rise to the desirable odor are not actually present in the intact plants but are biosynthesized upon crushing, frying, or boiling the "starting material." For example, a clove of garlic does not itself smell, and uncut onions are neither flavorful nor bring tears to your eyes.

With regard to garlic, crushing the clove releases so-called allinase enzymes that convert sulfoxide precursors into intermediate sulfenic acids. These acids subsequently dimerize with the loss of water to the flavorants, such as allicin. Garlic generates a host of other compounds in this way, all containing the functional groups of sulfides, RSR′,

Component of intact garlic

A sulfenic acid → Allicin (A flavorant)

sulfoxides, RSR′, and disulfides, RSSR′. Interestingly, some of these compounds are medicinally active. For example, allicin is a powerful antibacterial agent. Before modern antibiotics became available, garlic preparations were used in the treatment of typhus, cholera, dysentery, and tuberculosis.

It is likely that the garlic plant uses these compounds as chemical warfare agents against invading organisms. In China, a significant reduction in gastric cancer risk has been noted to parallel the consumption of garlic. Garlic is alleged to prevent and fight the common cold, it inhibits blood platelet aggregation, and it may help to regulate blood sugar levels. Several studies suggesting that it lowers cholesterol levels were refuted by a National Institutes of Health–funded clinical trial in 2007.

Among the most notable "negative" effects of garlic is bad breath, originating from the lungs by way of the blood and not, as you might have thought, from garlic traces in your mouth. Indeed, ingested garlic can persist in your urine for 3–4 days. Allicin is readily absorbed through the skin (a property it shares with dimethyl sulfoxide). Thus, it has been claimed that rubbing garlic on the foot soon leads to the taste of garlic in the mouth, a claim confirmed by one of the authors of this text.

The flavor of garlic, leeks, and onions is due to the extrusion of volatile sulfur compounds upon cutting.

In Summary Alcohols and ethers have various uses, both as chemical raw materials and as medicinal agents. Many of their derivatives can be found in nature; others are readily synthesized.

THE BIG PICTURE

We have now completed our coverage of alcohols, the second major class of functional-group compounds in this text. This does not mean that we are done with them, however. On the contrary, alcohols show up again in every one of the following chapters, often in conjunction with new functional-group substituents.

In this chapter, we also examined the last complication that may be encountered in the reactions of RX with nucleophiles/bases—carbocation rearrangements. We shall see later that skeletal rearrangements are also possible by other mechanisms, but acid catalysis is by far the most important.

Before we go on to present other classes of organic compounds, with other functional groups, we shall examine some of the key analytical techniques used by organic chemists to determine molecular structure. You should now have enough experience with the fundamentals of structure and function to appreciate that subtle differences in structures can lead to distinct changes in the electronic environment of molecules. By analyzing the interaction of molecules with various forms of electromagnetic radiation, chemists can infer a wealth of information about molecular structure. This forms the basis for the fundamental analytical tool available to us: spectroscopy.

WORKED EXAMPLES: INTEGRATING THE CONCEPTS

9-26. Working with Carbocation Rearrangements

Treatment of alcohol A with acidic methanethiol gives sulfide B. Explain by a mechanism.

SOLUTION

This is an example of a mechanistic, rather than a synthetic, problem. In other words, to solve it, we cannot add any reagents as in a multistep synthetic sequence; what we see is what we have to work with. Let us take an inventory of what we have:

1. The (tertiary) alcohol function disappears and the (secondary) thioether unit (from CH_3SH) is introduced.

2. The four-membered ring turns into cyclopentyl.

3. The molecular formula of compound A, $C_7H_{14}O$, turns into that of compound B, $C_8H_{16}S$.

Focusing on the alkyl part attached to the respective functional group, we can rewrite these changes as C_7H_{13}–OH → C_7H_{13}–SCH_3.

4. The reaction medium contains catalytic acid in the presence of a tertiary alcohol.

What can we conclude from this information? We have a carbocation rearrangement (Section 9-3) in which the strained cyclobutane ring undergoes expansion to a substituted cyclopentane. The initial carbocation must derive from a protonation–water loss sequence (Section 9-2), as applied to A, and product B must be formed through S_N1 capture of the rearranged cation by CH_3SH.

We can now begin to formulate a step-by-step description of these thoughts.

Step 1. The hydroxy group is protonated and leaves as H_2O.

Step 2. The tertiary carbocation undergoes ring expansion by alkyl shift (the migrating carbon is represented by a dot).

Step 3. The new carbocation is trapped by the relatively (compared with water) nucleophilic sulfur of CH_3SH, followed by proton loss to give product B (Section 9-10).

Visually, the most difficult step to follow in this sequence is step 2, because it consists of a fairly extreme topological change: The migrating carbon "drags" with it the appended chain, which is part of the ring. A good way of eliminating confusion is to label the "action pieces" in your molecule, as was done in the scheme for step 2, and to keep in mind the "bare bones" of an alkyl (or H) shift. Thus, only three key atoms take part: the cationic center, which will receive the migrating group; the neighboring carbon, which will become charged; and the migrating atom. A simple aid in remembering the basic feature of a carbocation rearrangement is the slogan: "The charge and the migrating center trade places."

Finally, note that rearrangement step 2 converts a tertiary into a secondary carbocation. The driving force is the release of ring strain in going from a four- (26.3 kcal mol^{-1} strain) to a five-membered ring (6.5 kcal mol^{-1} strain; Section 4-2). In our particular case, the secondary carbocation is trapped by the highly nucleophilic sulfur of the thiol before it rearranges further by methyl migration to the tertiary counterpart, a potential source of other products, which are not observed. Problems 36 and 61 provide further practice with related mechanisms.

9-27. Applying Stereospecific Oxacyclopropane Ring Openings

Write a synthetic scheme that will convert enantiomer A to enantiomer B efficiently.

SOLUTION

This is an example of a synthetic, rather than mechanistic, problem. In other words, to solve it, we need to use reagents, specify conditions, and use as many steps as we see fit to get us from starting material to product. Let us take an inventory:

1. The strained oxacyclopropane ring has been opened in the product, which is another ether.

2. The molecular formula of A, C_4H_8O, is altered to that in B, $C_{11}H_{22}O$. The added increment is therefore C_7H_{14}. An obvious component of this increment is the cyclopentyl group, C_5H_9. Subtraction leaves C_2H_5. This looks like ethyl; however, comparison of the substituents in A (two methyls) with the substituents in B suggests that the extra two carbons are due to the introduction of two methyls (in addition to the added cyclopentyl group).

3. We have two stereocenters (both S) in starting material and only one (S) in the product. Remember that R,S-nomenclature does not correlate necessarily with changes in absolute configuration, only with changes in the priority sequence of substituents at a stereocenter (Section 5-3). How does the remaining stereocenter in B then relate to those in A? We get the answer by rewriting B in a conformation B′ that is visually clearer with respect to the stereochemical arrangement around the stereocenter, with the CH$_3$ and H connected to the chain by a wedged and hashed line, respectively, as in A. This reveals that the ether oxygen in A has been replaced by a cyclopentyl group with inversion.

What hints do we derive from this analysis? It is clear that the stereocenter in B can be derived from A by nucleophilic ring opening with a cyclopentyl organometallic:

However, that would provide an alcohol C, not an ether, and the oxygen-bearing carbon misses the extra methyl group.

Let us work retrosynthetically. Ether disconnection by retro-Williamson synthesis (Section 9-6) gives the tertiary alcohol D. How do we unravel D to C? *Answer:* We go back to Section 8-9, recognizing that complex alcohols can be made from simpler ones by organometallic additions to carbonyl compounds:

The solution therefore looks as follows:

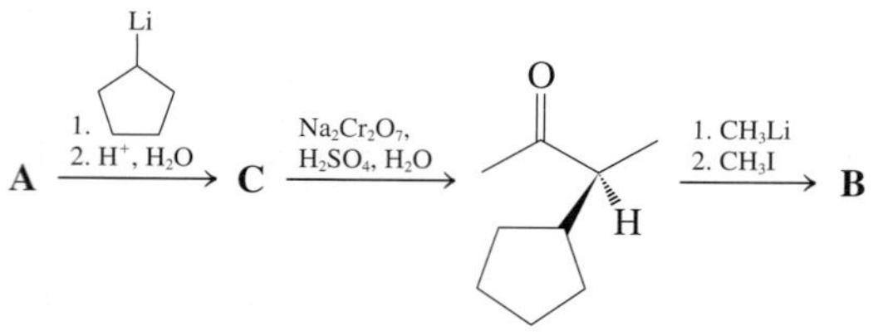

$$A \xrightarrow[\text{2. H}^+, \text{H}_2\text{O}]{\text{1.}} C \xrightarrow[\text{H}_2\text{SO}_4, \text{H}_2\text{O}]{\text{Na}_2\text{Cr}_2\text{O}_7,} \xrightarrow[\text{2. CH}_3\text{I}]{\text{1. CH}_3\text{Li}} B$$

Problem 55 contains similar synthesis exercises.

New Reactions

1. Alkoxides from Alcohols (Sections 8-3, 9-1)

Using strong bases

$$ROH \underset{}{\overset{\text{Strong base}}{\rightleftharpoons}} RO^-$$

Examples of strong bases: $Li^{+-}N[CH(CH_3)_2]_2$; $CH_3CH_2CH_2CH_2Li$; K^+H^-

Using alkali metals

$$ROH + M \longrightarrow RO^-M^+ + \tfrac{1}{2}H_2$$

M = Li, Na, K

Haloalkanes from Alcohols

2. Using Hydrogen Halides (Sections 8-3, 9-2, 9-3)

Primary $ROH \xrightarrow{\text{Conc. HX}} RX$ Secondary or tertiary $ROH \xrightarrow{\text{Conc. HX}} RX$

X = Br or I (S_N2 mechanism)   X = Cl, Br, or I (S_N1 mechanism)

3. Using Phosphorus Reagents (Section 9-4)

$$3\ ROH + PBr_3 \longrightarrow 3\ RBr + H_3PO_3$$
$$6\ ROH + 2\ P + 3\ I_2 \longrightarrow 6\ RI + 2\ H_3PO_3$$

S_N2 mechanism with primary and secondary ROH
Less likelihood of carbocation rearrangements than with HX

4. Using Sulfur Reagents (Section 9-4)

$$ROH + SOCl_2 \xrightarrow{\text{N(CH}_2\text{CH}_3)_3} RCl + SO_2 + (CH_3CH_2)_3\overset{+}{N}H\ Cl^-$$

$$ROH + R'SO_2Cl \longrightarrow \underset{\substack{\text{Alkyl} \\ \text{sulfonate}}}{ROSO_2R'} \xrightarrow{\text{Nu}^-, \text{DMSO}} RNu + R'SO_3^-$$

Carbocation Rearrangements in Alcohols

5. Carbocation Rearrangements by Alkyl and Hydride Shifts (Section 9-3)

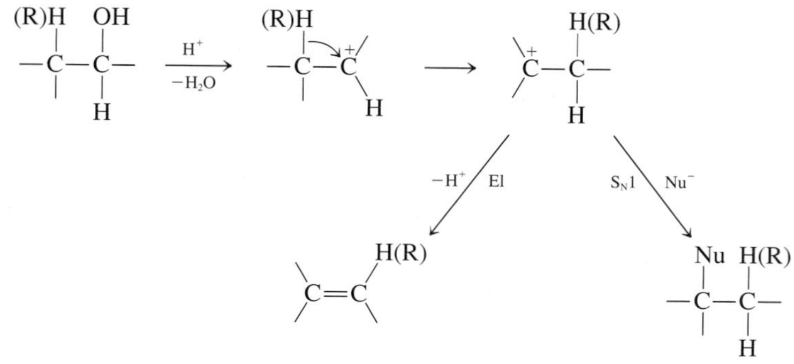

6. Concerted Alkyl Shifts from Primary Alcohols (Section 9-3)

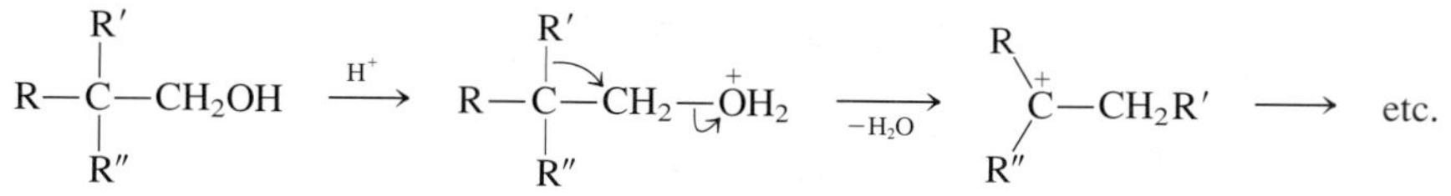

Elimination Reactions of Alcohols

7. Dehydration with Strong Nonnucleophilic Acid (Sections 9-2, 9-3, 9-7, 11-5)

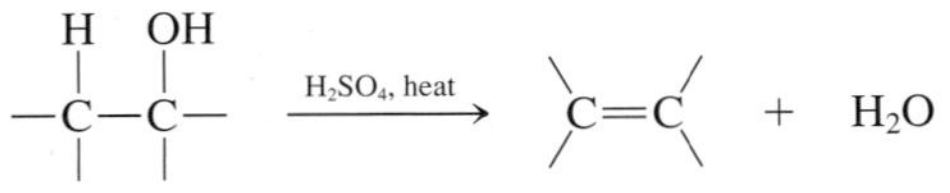

$$-\underset{|}{\overset{H}{C}}-\underset{|}{\overset{OH}{C}}- \xrightarrow{H_2SO_4,\ heat} \underset{/}{\overset{\backslash}{C}}=\underset{\backslash}{\overset{/}{C}} \ + \ H_2O$$

Carbocation rearrangements may occur

Temperature required
Primary ROH: $170°-180°C$ (E2 mechanism)
Secondary ROH: $100°-140°C$ (usually E1)
Tertiary ROH: $25°-80°C$ (E1 mechanism)

Preparation of Ethers

8. Williamson Synthesis (Section 9-6)

$$ROH \xrightarrow{NaH,\ DMSO} RO^- \ Na^+ \xrightarrow[S_N2]{R'X,\ DMSO} ROR'$$

R' must be methyl or primary
ROH can be primary or secondary (tertiary alkoxides usually give E2 products, unless R' = methyl)
Ease of intramolecular version forming cyclic ethers: $k_3 \geq k_5 > k_6 > k_4 \geq k_7 > k_8$
(k_n = reaction rate, n = ring size)

9. Mineral Acid Method (Section 9-7)
Primary alcohols:

$$RCH_2OH \xrightarrow{H^+,\ low\ temperature} RCH_2\overset{+}{O}H_2 \xrightarrow[-H_2O]{RCH_2OH,\ 130°-140°C} RCH_2OCH_2R$$

Secondary alcohols:

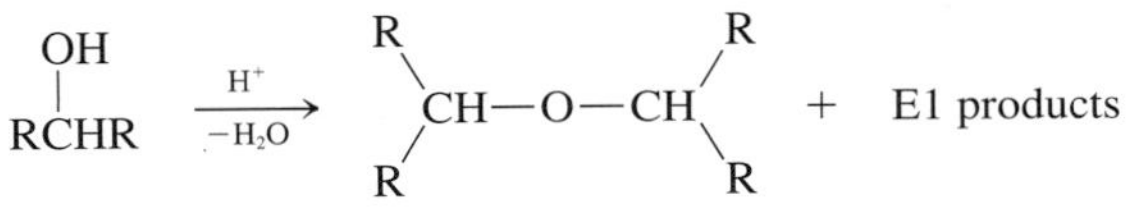

$$\underset{RCHR}{\overset{OH}{|}} \xrightarrow[-H_2O]{H^+} \underset{R}{\overset{R}{\underset{|}{\overset{|}{CH}}}}-O-\underset{R}{\overset{R}{\underset{|}{\overset{|}{CH}}}} \ + \ E1\ products$$

Tertiary alcohols:

$$R_3COH \ + \ R'OH \xrightarrow[S_N1,\ -\ H_2O]{NaHSO_4,\ H_2O} R_3C-OR' \ + \ E1\ products$$

R' = (mainly) primary

Reactions of Ethers

10. Cleavage by Hydrogen Halides (Section 9-8)

$$ROR \xrightarrow{Conc.\ HX} RX \ + \ ROH \xrightarrow{Conc.\ HX} 2\ RX$$

X = Br or I
Primary R: S_N2 mechanism
Secondary R: S_N1 or S_N2 mechanism
Tertiary R: S_N1 mechanism

11. Nucleophilic Opening of Oxacyclopropanes (Sections 9-9, 25-2)
Anionic nucleophiles:

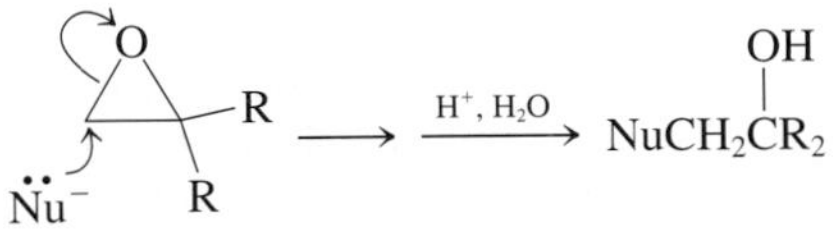

$$\xrightarrow{} \xrightarrow{H^+,\ H_2O} Nu CH_2 \underset{OH}{\overset{OH}{\underset{|}{C}}R_2}$$

Examples of Nu⁻: HO⁻, RO⁻, RS⁻

Acid-catalyzed opening:

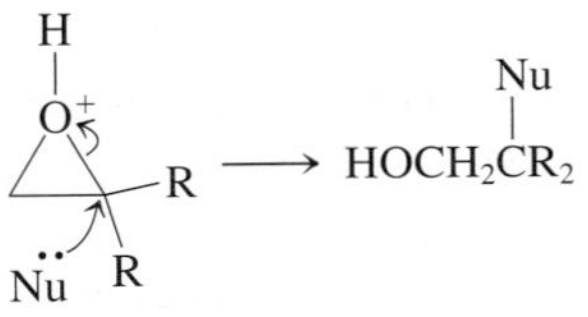

$$\xrightarrow{} HOCH_2\underset{}{\overset{Nu}{\underset{|}{C}}R_2}$$

Examples of Nu: H_2O, ROH, halide

12. Nucleophilic Opening of Oxacyclopropane by Lithium Aluminum Hydride (Section 9-9)

$$\text{H}_2\text{C}\overset{\text{O}}{-}\text{CH}_2 \xrightarrow[\text{2. H}^+,\text{ H}_2\text{O}]{\text{1. LiAlH}_4,\text{ (CH}_3\text{CH}_2)_2\text{O}} \text{CH}_3\text{CH}_2\text{OH}$$

13. Nucleophilic Opening of Oxacyclopropane by Organometallic Compounds (Section 9-9)

$$\text{RLi}\quad \text{or}\quad \text{RMgX} + \text{H}_2\text{C}\overset{\text{O}}{-}\text{CH}_2 \xrightarrow{\text{THF}} \xrightarrow{\text{H}^+,\text{ H}_2\text{O}} \text{RCH}_2\text{CH}_2\text{OH}$$

Sulfur Compounds

14. Preparation of Thiols and Sulfides (Section 9-10)

$$\text{RX} + \text{HS}^- \longrightarrow \text{RSH}$$
$$\qquad\quad \text{Excess} \qquad\quad \text{Thiol}$$

$$\text{RSH} + \text{R}'\text{X} \xrightarrow{\text{Base}} \text{RSR}'$$
$$\qquad\qquad\qquad\qquad \text{Alkyl sulfide}$$

15. Acidity of Thiols (Section 9-10)

$$\text{RSH} + \text{HO}^- \rightleftharpoons \text{RS}^- + \text{H}_2\text{O} \qquad pK_a(\text{RSH}) = 9\text{–}12$$
$$\text{Acidity of RSH} > \text{H}_2\text{O} \sim \text{ROH}$$

16. Nucleophilicity of Sulfides (Section 9-10)

$$\text{R}_2\overset{..}{\underset{..}{\text{S}}} + \text{R}'\text{X} \longrightarrow \text{R}_2\overset{+}{\text{S}}\text{R}'\ \text{X}^-$$
$$\qquad\qquad\qquad\qquad \text{Sulfonium salt}$$

17. Oxidation of Thiols (Section 9-10)

$$\text{RSH} \xrightarrow{\text{KMnO}_4\text{ or H}_2\text{O}_2} \text{RSO}_3\text{H} \qquad\qquad \text{RSH} \underset{\text{NaBH}_4}{\overset{\text{I}_2}{\rightleftharpoons}} \text{RS}-\text{SR}$$
$$\qquad\qquad\quad \text{Alkanesulfonic acid} \qquad\qquad\qquad\qquad\qquad \text{Dialkyl disulfide}$$

18. Oxidation of Sulfides (Section 9-10)

$$\overset{..}{\underset{..}{\text{R}\text{S}\text{R}'}} \xrightarrow{\text{H}_2\text{O}_2} \text{R}\overset{\text{O}}{\overset{\|}{\underset{..}{\text{S}}}}\text{R}' \xrightarrow{\text{H}_2\text{O}_2} \text{R}\overset{\text{O}}{\underset{\text{O}}{\overset{\|}{\underset{\|}{\text{S}}}}}\text{R}'$$
$$\qquad\qquad\qquad \text{Dialkyl sulfoxide} \qquad \text{Dialkyl sulfone}$$

Important Concepts

1. The reactivity of ROH with alkali metals to give **alkoxides** and hydrogen follows the order R = CH$_3$ > primary > secondary > tertiary.

2. In the presence of acid and a nucleophilic counterion, primary alcohols undergo S$_N$2 reactions. Secondary and tertiary alcohols tend to form **carbocations** in the presence of acid, capable of **E1** and S$_N$1 product formation, before and after **rearrangement.**

3. **Carbocation rearrangements** take place by **hydride** and **alkyl** group **shifts.** They usually result in interconversion of secondary carbocations or conversion of a secondary into a tertiary carbocation. Primary **alkyloxonium** ions can rearrange by a concerted process consisting of loss of water and simultaneous hydride or alkyl shift to give secondary or tertiary carbocations.

4. Synthesis of primary and secondary haloalkanes can be achieved with less risk of rearrangement by methods using **inorganic esters.**

5. Ethers are prepared by either the **Williamson ether synthesis** or by reaction of alcohols with strong nonnucleophilic acids. The first method is best when S$_N$2 reactivity is high. In the latter case, elimination (dehydration) is a competing process at higher temperatures.

6. **Crown ethers** and **cryptands** are examples of **ionophores,** polyethers that coordinate around metal ions, thus rendering them soluble in hydrophobic media.

7. Whereas nucleophilic **ring opening of oxacyclopropanes** by anions is at the less substituted ring carbon according to the rules of the S_N2 reaction, acid-catalyzed opening favors the more substituted carbon, because of charge control of nucleophilic attack.

8. Sulfur has more diffuse orbitals than does oxygen. In **thiols,** the S–H bond is less polarized than the O–H bond in alcohols, thus leading to **diminished hydrogen bonding.** Because the S–H bond is also weaker than the O–H bond, the **acidity** of thiols is **greater** than that of alcohols.

9. **Note on color use:** Throughout the main parts of the text, beginning in Chapter 6, reacting species in mechanisms and most examples of new transformations are color coded **red** for **nucleophiles, blue** for **electrophiles,** and **green** for **leaving groups.** Color coding is *not* used in exercises, summaries of new reactions, or chapter-end problems.

Problems

28. On which side of the equation do you expect each of the following equilibria to lie (left or right)?

(a) $(CH_3)_3COH + K^+ \ ^-OH \rightleftharpoons (CH_3)_3CO^-K^+ + H_2O$

(b) $CH_3OH + NH_3 \rightleftharpoons CH_3O^- + NH_4^+$ ($pK_a = 9.2$)

(c) $CH_3CH_2OH +$ $\rightleftharpoons$

$CH_3CH_2O^-Li^+ +$ ($pK_a = 40$)

(d) $NH_3 (pK_a = 35) + Na^+H^- \rightleftharpoons$

$Na^+ \ ^-NH_2 + H_2 (pK_a \sim 38)$

29. Give the expected major product of each of the following reactions.

(a) $CH_3CH_2CH_2OH \xrightarrow{Conc. HI}$

(b) $(CH_3)_2CHCH_2CH_2OH \xrightarrow{Conc. HBr}$

(c) $\xrightarrow{Conc. HI}$

(d) $(CH_3CH_2)_3COH \xrightarrow{Conc. HCl}$

30. For each reaction in Problem 29, write out a detailed step-by-step mechanism.

31. For each of the following alcohols, write the structure of the alkyloxonium ion produced after protonation by strong acid; if the alkyloxonium ion is capable of losing water readily, write the structure of the resulting carbocation; if the carbocation obtained is likely to be susceptible to rearrangement, write the structures of all new carbocations that might be reasonably expected to form.

(a) $CH_3CH_2CH_2OH$

(b) $CH_3\underset{\underset{}{\overset{\overset{OH}{|}}{}}CHCH_3$

(c) $^-CH_3CH_2CH_2CH_2OH$

(d) $(CH_3)_2CHCH_2OH$

(e) $(CH_3)_3CCH_2CH_2OH$

(f)

(g) $(CH_3)_3C\underset{}{\overset{}{}}OH$

(h)

32. Write all products of the reaction of each of the alcohols in Problem 31 with concentrated H_2SO_4 under elimination conditions.

33. Write all sensible products of the reaction of each of the alcohols in Problem 31 with concentrated aqueous HBr.

34. Give detailed mechanisms and final products for the reaction of 3-methyl-2-pentanol with each of the reagents that follow.

(a) NaH   (b) Concentrated HBr   (c) PBr_3

(d) $SOCl_2$ (e) Concentrated H_2SO_4 at 130°C

(f) Dilute H_2SO_4 in $(CH_3)_3COH$

35. Primary alcohols are often converted into bromides by reaction with NaBr in H_2SO_4. Explain how this transformation works and why it might be considered a superior method to that using concentrated aqueous HBr.

$$CH_3CH_2CH_2CH_2OH \xrightarrow{NaBr, H_2SO_4} CH_3CH_2CH_2CH_2Br$$

36. What are the most likely product(s) of each of the following reactions?

a) $\xrightarrow{CH_3CH_2OH, H_2SO_4}$

(b) $CH_3\underset{\underset{CH_3}{|}}{\overset{\overset{CH_3}{|}}{C}}CH_2OH \xrightarrow{Conc. HI}$

(c) $\xrightarrow{Conc. H_2SO_4, 180°C}$

(d) $CH_3\underset{\underset{CH_3}{|}}{\overset{\overset{CH_3 \ \ \ I}{| \ \ \ |}}{C}}CHCH_3 \xrightarrow{H_2O}$

37. Give the expected main product of the reaction of each of the alcohols in Problem 31 with PBr_3. Compare the results with those of Problem 33.

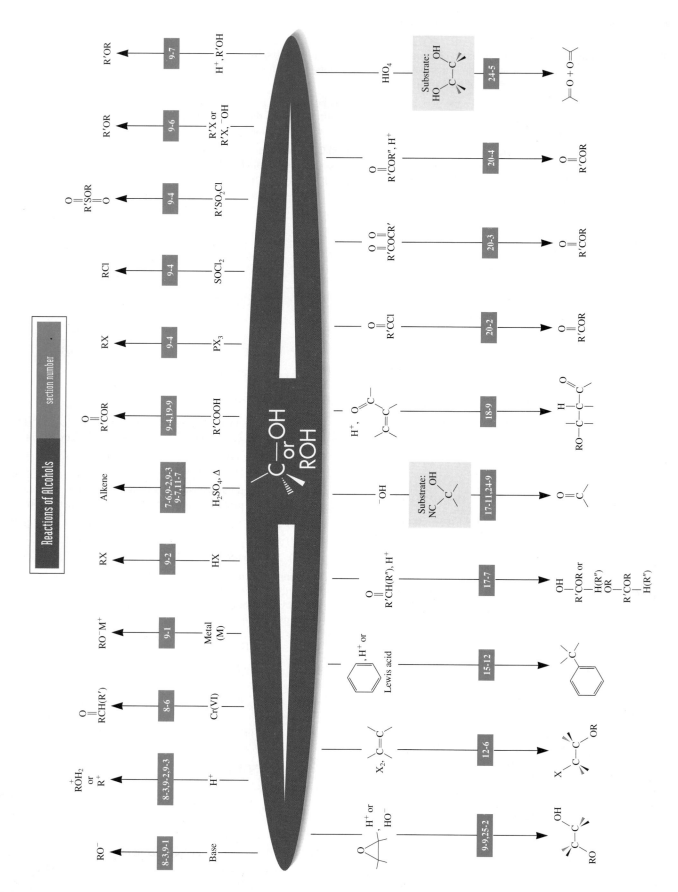

Reactions of Alcohols

38. Give the expected product(s) of the reaction of 1-pentanol with each of the following reagents.

(a) $K^+ \, ^-OC(CH_3)_3$　　(b) Sodium metal
(c) CH_3Li　　(d) Concentrated HI
(e) Concentrated HCl　　(f) FSO_3H
(g) Concentrated H_2SO_4 at 130°C　　(h) Concentrated H_2SO_4 at 180°C
(i) CH_3SO_2Cl, $(CH_3CH_2)_3N$　　(j) PBr_3
(k) $SOCl_2$　　(l) $K_2Cr_2O_7 + H_2SO_4 + H_2O$
(m) PCC, CH_2Cl_2　　(n) $(CH_3)_3COH + H_2SO_4$ (as catalyst)

39. Give the expected product(s) of the reaction of *trans*-3-methylcyclopentanol with each of the reagents in Problem 38.

40. Suggest a good synthetic method for preparing each of the following haloalkanes from the corresponding alcohols.

(a) $CH_3CH_2CH_2Cl$　　(b) $CH_3CH_2CHCH_2Br$ with CH_3 substituent

(c) 　　(d) $CH_3CHCH(CH_3)_2$ with I substituent

41. Name each of the following molecules according to IUPAC.

(a) $(CH_3)_2CHOCH_2CH_3$　　(b) $CH_3OCH_2CH_2OH$
(c) cyclopentyl–O–cyclopentyl
(d) $(ClCH_2CH_2)_2O$
(e) H_3C, OCH_3 cyclopentane
(f) CH_3O–cyclohexane–OCH_3
(g) CH_3OCH_2Cl

42. Explain why the boiling points of ethers are lower than those of the isomeric alcohols. Would you expect the relative water solubilities to differ in a similar way?

43. Suggest the best syntheses for each of the following ethers. Use alcohols or haloalkanes or both as your starting materials.

(a), (b), (c), (d), (e), (f) [structures]

44. Write the expected major product(s) of each of the following attempted ether syntheses.

(a) $CH_3CH_2CH_2Cl + CH_3CH_2CHCH_2CH_3$ (with O^-) $\xrightarrow{DMSO}$

(b) $CH_3CH_2CH_2O^- + CH_3CH_2CHCH_2CH_3$ (with Cl) $\xrightarrow{HMPA}$

(c) $+ CH_3I \xrightarrow{DMSO}$

(d) $(CH_3)_2CHO^- + (CH_3)_2CHCH_2CH_2Br \xrightarrow{(CH_3)_2CHOH}$

(e) [structures] $\xrightarrow{Cyclohexanol}$

(f) [structure] $+ CH_3CH_2I \xrightarrow{DMSO}$

45. For each reaction in Problem 44, write out a detailed step-by-step mechanism.

46. For each synthesis proposed in Problem 44 that is not likely to give a good yield of ether product, suggest an alternative synthesis beginning with suitable alcohols or haloalkanes that will give a superior result. (**Hint:** See Problem 25 of Chapter 7.)

47. (a) What would be the product of reaction of *trans*-2-bromocyclooctanol (below) with NaOH? (b) Compare the effect of entropy on the transition state of this reaction with its effect in the reactions presented in Figure 9-6 and in Exercise 9-14.

[structure] *trans*-**2-Bromocyclooctanol**

48. Propose efficient syntheses for each of the following ethers, using haloalkanes or alcohols as starting materials.

(a) $CH_3CH_2CHOCH_2CH_3$ (with CH_3)　　(b) [structure] $OCH_2CH_2CH_2CH_3$
(c) [structure]　　(d) [structure]

49. Give the major product(s) of each of the following reactions.

(a) $CH_3CH_2OCH_2CH_2CH_3 \xrightarrow{Excess\ conc.\ HI}$
(b) $CH_3OCH(CH_3)_2 \xrightarrow{Excess\ conc.\ HBr}$
(c) $CH_3OCH_2CH_2OCH_3 \xrightarrow{Excess\ conc.\ HI}$
(d) [structure] $\xrightarrow{Excess\ conc.\ HBr}$
(e) [structure] $\xrightarrow{Excess\ conc.\ HBr}$
(f) [structure] $\xrightarrow{Excess\ conc.\ HBr}$

50. Give the expected major product of reaction of 2,2-dimethyloxacyclopropane with each of the following reagents.

(a) Dilute H_2SO_4 in CH_3OH (b) $Na^+ \ ^-OCH_3$ in CH_3OH
(c) Dilute, aqueous HBr   (d) Concentrated HBr
(e) CH_3MgI, then H^+, H_2O (f) C_6H_5Li, then H^+, H_2O

51. Propose a synthesis of ![cyclohexane with CH2CH2CH2OH and OH groups] $CH_2CH_2CH_2OH$ / OH beginning

with cyclohexanone, ![cyclohexanone] =O, and 3-bromopropanol.

[**Hint:** Beware of a possible pitfall in planning this synthesis (recall Section 8-9).]

52. Cleavage of tertiary butyl ethers requires the use of an aqueous acid (Problem 57 of Chapter 7, and Section 9-8). Why do strong bases not cleave ethers (other than oxacyclopropanes)?

53. Provide an IUPAC name for each of the structures pictured below.

(a) ![oxirane with ethyl] (b) ![oxetane with methyls]

(c) CH_3O ![furan ring] CH_2Cl (d) ![pyran ring] —OH

(e) ![oxepane ring with O] (f) ![seven-membered ring with two O]

54. Give the major product(s) of each of the following reactions. (**Hint:** The strained oxacyclobutanes react like oxacyclopropanes.)

(a) ![oxacyclopropane] $\xrightarrow{Na^+ \ ^-NH_2, \ NH_3}$

(b) H ![oxacyclopropane with CH3] CH_3 $\xrightarrow{Na^+ \ ^-SCH_2CH_3, \ CH_3CH_2OH}$

(c) ![oxacyclobutane] $\xrightarrow{\text{Excess conc. HBr}}$

(d) ![oxacyclobutane with CH3] CH_3 / CH_3 $\xrightarrow{\text{Dilute HCl in } CH_3OH}$

(e) ![oxacyclobutane with CH3] CH_3 / CH_3 $\xrightarrow{Na^+ \ ^-OCH_3 \text{ in } CH_3OH}$

(f) ![oxacyclopropane with CH3] CH_3 / CH_3 $\xrightarrow[2. \ H^+, \ H_2O]{1. \ LiAlD_4, \ (CH_3CH_2)_2O}$

(g) ![oxacyclopropane with CH3] CH_3 / CH_3 $\xrightarrow[2. \ H^+, \ H_2O]{1. \ (CH_3)_2CHMgCl, \ (CH_3CH_2)_2O}$

(h) ![oxacyclopropane] $\xrightarrow[2. \ H^+, \ H_2O]{1. \ \text{(cyclopentyl)}-Li, \ (CH_3CH_2)_2O}$

55. For each alcohol in Problem 51 of Chapter 8, suggest a synthetic route that starts with an oxacyclopropane (if possible).

56. Give the major product(s) expected from each of the reactions shown below. Watch stereochemistry (see model of starting material below).

(a) H ![oxacyclopropane with CH3, CH3, H] CH_3 / CH_3 H $\xrightarrow{\text{Dilute } H_2SO_4 \text{ in } CH_3CH_2OH}$

(b) H ![oxacyclopropane with CH3, CH3, H] CH_3 / CH_3 H $\xrightarrow[2. \ H^+, \ H_2O]{1. \ LiAlD_4, \ (CH_3CH_2)_2O}$

57. Name each of the following compounds according to IUPAC.

(a) ![cyclopropyl] $-CH_2SH$ (b) $CH_3CH_2\overset{\overset{\displaystyle CH_3}{|}}{C}HSCH_3$

(c) $CH_3CH_2CH_2SO_3H$ (d) CF_3SO_2Cl

58. In each of the following pairs of compounds, indicate which is the stronger acid and which is the stronger base. (a) CH_3SH, CH_3OH; (b) HS^-, HO^-; (c) H_3S^+, H_2S.

59. Give reasonable products for each of the following reactions.

(a) $ClCH_2CH_2CH_2CH_2Cl \xrightarrow{\text{One equivalent } Na_2S}$

(b) ![cyclohexane with Br and CH3] $\xrightarrow{KSH}$ CH_3 / Br

(c) ![bicyclic epoxide with H, H] H / H $\xrightarrow{KSH}$

(d) $CH_3CH_2\overset{\overset{\displaystyle CH_3CH_2}{|}}{\underset{\underset{\displaystyle CH_3CH_2}{|}}{C}}Br \xrightarrow{CH_3SH}$

(e) $CH_3\overset{\underset{\displaystyle SH}{|}}{C}HCH_3 \xrightarrow{I_2}$

(f) ![six-membered ring with O and S] O S $\xrightarrow{\text{Excess } H_2O_2}$

60. Give the structures of compounds A, B, and C (with stereochemistry) from the information in the following scheme. (**Hint:** A is acyclic.) To what compound class does the product belong?

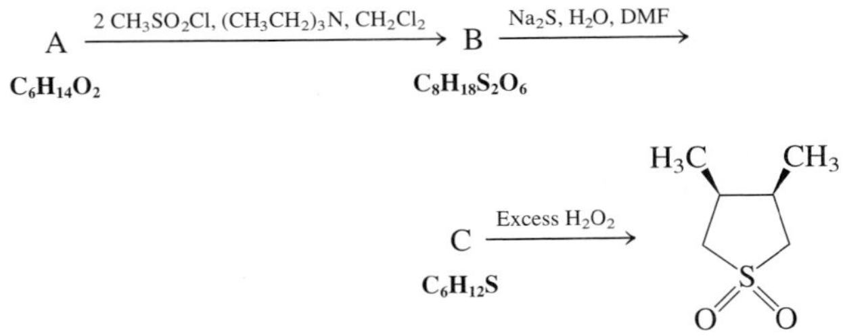

$$A \xrightarrow{\text{2 CH}_3\text{SO}_2\text{Cl, (CH}_3\text{CH}_2)_3\text{N, CH}_2\text{Cl}_2} B \xrightarrow{\text{Na}_2\text{S, H}_2\text{O, DMF}}$$

$C_6H_{14}O_2 \qquad\qquad C_8H_{18}S_2O_6$

$$C \xrightarrow{\text{Excess H}_2\text{O}_2}$$

$C_6H_{12}S$

61. In an attempt to make 1-chloro-1-cyclobutylpentane, the following reaction sequence was employed. The actual product isolated, however, was not the desired molecule but an isomer of it. Suggest a structure for the product and give a mechanistic explanation for its formation. (**Hint:** See Worked Example 9-26.)

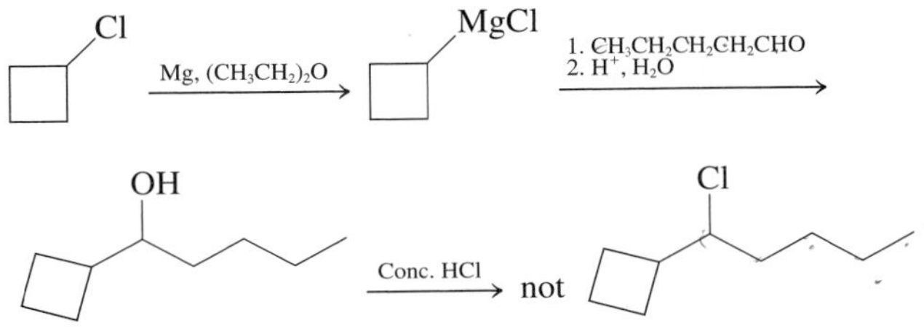

62. Suggest better methods for the final step in Problem 61.

63. **CHALLENGE** In an early study of the stereochemistry of nucleophilic displacements, optically pure (*R*)-1-deuterio-1-pentanol was treated with 4-methylphenylsulfonyl (tosyl) chloride to make the corresponding tosylate. The tosylate was then treated with excess ammonia to convert it to 1-deuterio-1-pentanamine:

(*R*)-CH$_3$CH$_2$CH$_2$CH$_2$CHDOH
(*R*)-1-Deuterio-1-pentanol

$$\xrightarrow{\text{Excess NH}_3} \text{CH}_3\text{CH}_2\text{CH}_2\text{CH}_2\text{CHDNH}_2$$
1-Deuterio-1-pentanamine

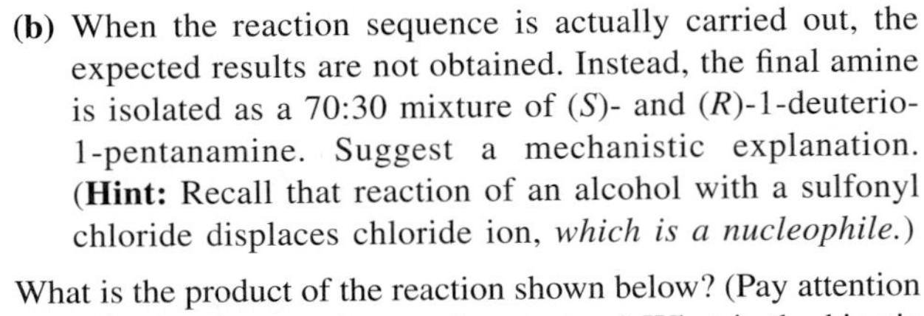

(a) Describe the stereochemistry that you expect to observe at C1 of both the intermediate tosylate and the final amine.

(b) When the reaction sequence is actually carried out, the expected results are not obtained. Instead, the final amine is isolated as a 70:30 mixture of (*S*)- and (*R*)-1-deuterio-1-pentanamine. Suggest a mechanistic explanation. (**Hint:** Recall that reaction of an alcohol with a sulfonyl chloride displaces chloride ion, *which is a nucleophile.*)

64. What is the product of the reaction shown below? (Pay attention to stereochemistry at the reacting centers.) What is the kinetic order of this reaction?

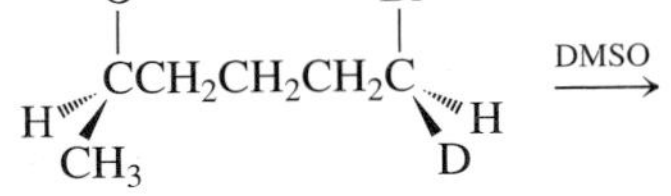

65. **CHALLENGE** Propose syntheses of the following molecules, choosing reasonable starting materials on the basis of the principles of synthetic strategy introduced in preceding chapters, particularly in Section 8-9. Suggested positions for carbon–carbon bond formation are indicated by wavy lines.

(a)

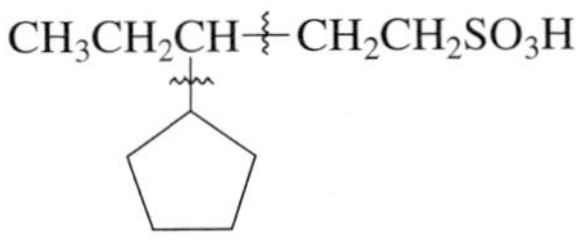

CH$_3$CH$_2$CH$\frac{}{}$CH$_2$CH$_2$SO$_3$H

(b) CH$_3$CH$_2$CH$_2$$\frac{}{}C\frac{}{}$CHO

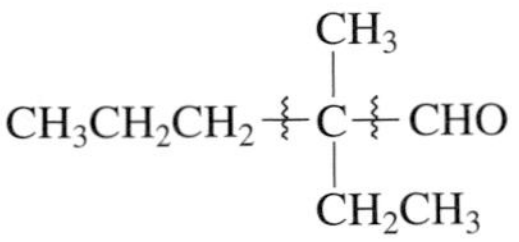

66. Give efficient syntheses of each of the following compounds, beginning with the indicated starting material.

(a) *trans*-1-Bromo-2-methylcyclopentane, from *cis*-2-methylcyclopentanol

(b)

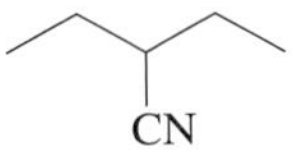

from 3-pentanol

(c) 3-Chloro-3-methylhexane, from 3-methyl-2-hexanol

(d)

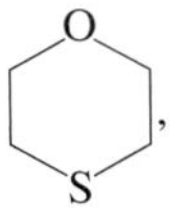

, from 2-bromoethanol (two equivalents)

67. Compare the following methods of alkene synthesis from a general primary alcohol. State the advantages and disadvantages of each one.

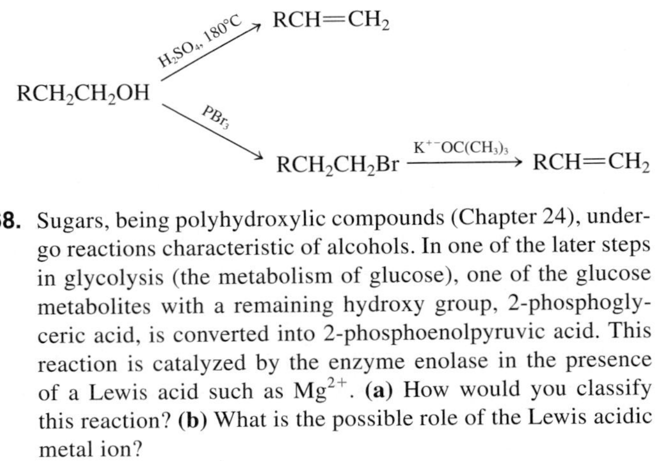

68. Sugars, being polyhydroxylic compounds (Chapter 24), undergo reactions characteristic of alcohols. In one of the later steps in glycolysis (the metabolism of glucose), one of the glucose metabolites with a remaining hydroxy group, 2-phosphoglyceric acid, is converted into 2-phosphoenolpyruvic acid. This reaction is catalyzed by the enzyme enolase in the presence of a Lewis acid such as Mg^{2+}. (a) How would you classify this reaction? (b) What is the possible role of the Lewis acidic metal ion?

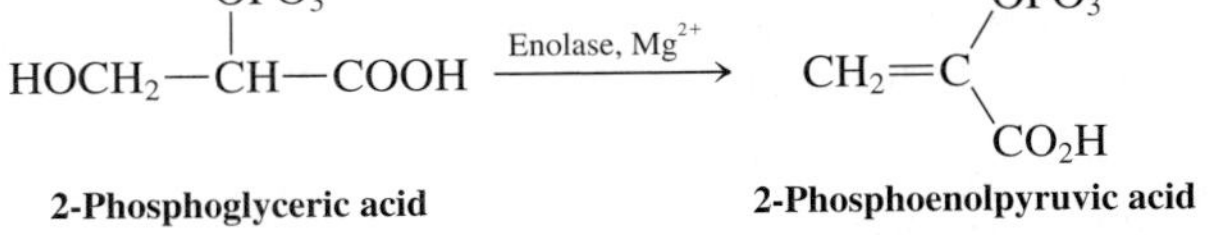

2-Phosphoglyceric acid   2-Phosphoenolpyruvic acid

69. The formidable-looking molecule 5-methyltetrahydrofolic acid (abbreviated 5-methyl-FH$_4$) is the product of sequences of biological reactions that convert carbon atoms from a variety of simple molecules, such as formic acid and the amino acid histidine, into methyl groups.

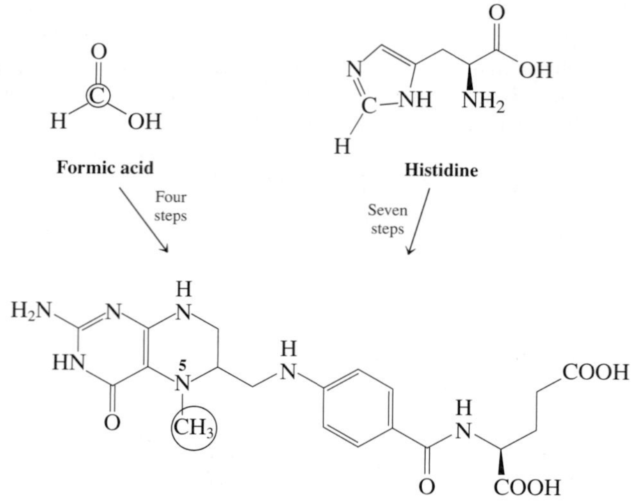

Formic acid   **Histidine**

Four steps / Seven steps

5-Methyltetrahydrofolic acid
(5-Methyl-FH₄)

The simplest synthesis of 5-methyltetrahydrofolic acid is from tetrahydrofolic acid (FH₄) and trimethylsulfonium ion, a reaction carried out by microorganisms in the soil.

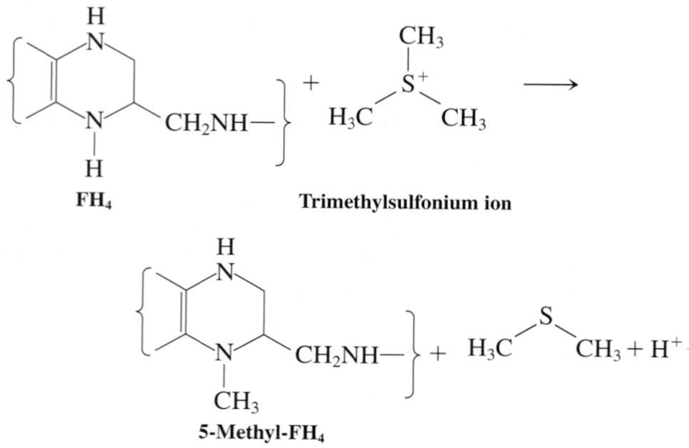

FH₄   **Trimethylsulfonium ion**

5-Methyl-FH₄

(a) Can this reaction be reasonably assumed to proceed through a nucleophilic substitution mechanism? Write the mechanism, using the "electron-pushing" arrow notation. (b) Identify the nucleophile, the nucleophilic and electrophilic atoms participating in the reaction, and the leaving group. (c) On the basis of the concepts presented in Sections 6-7, 6-8, 9-2, and 9-9, are all the groups that you identified in (b) behaving in a reasonable way in this reaction? Does it help to know that species such as H_3S^+ are very strong acids (e.g., pK_a of $CH_3SH_2^+$ is -7)?

70. CHALLENGE   The role of 5-methyl-FH₄ (Problem 69) in biology is to serve as a donor of methyl groups to small molecules. The synthesis of the amino acid methionine from homocysteine is perhaps the best-known example.

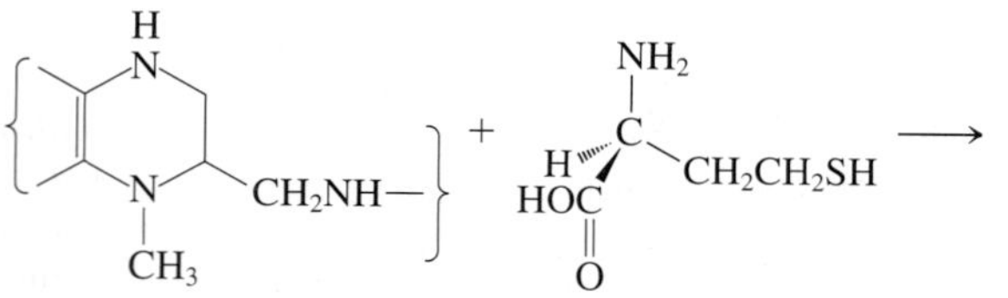

5-Methyl-FH₄   **Homocysteine**

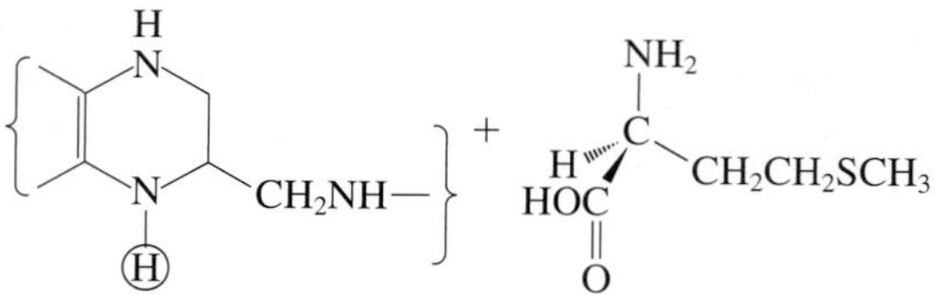

FH₄   **Methionine**

For this challenge, answer the same questions that were posed in Problem 69. The pK_a of the circled hydrogen in FH₄ is 5. Does this pose a difficulty with any feature of your mechanism? In fact, methyl transfer reactions of 5-methyl-FH₄ require a proton source. Review the material in Section 9-2, especially the subsection titled "Haloalkanes from primary alcohols and HX." Then suggest a useful role for a proton in the reaction illustrated here.

71. Epinephrine (adrenalin; see also Chapter 6 Opening) is produced in your body in a two-step process that accomplishes the transfer of a methyl group from methionine (Problem 70) to norepinephrine (see reactions 1 and 2 below). (a) Explain in detail what is going on mechanistically in these two reactions, and analyze the role played by the molecule of ATP. (b) Would you expect methionine to react directly with norepinephrine? Explain. (c) Propose a laboratory synthesis of epinephrine from norepinephrine.

Reaction 1

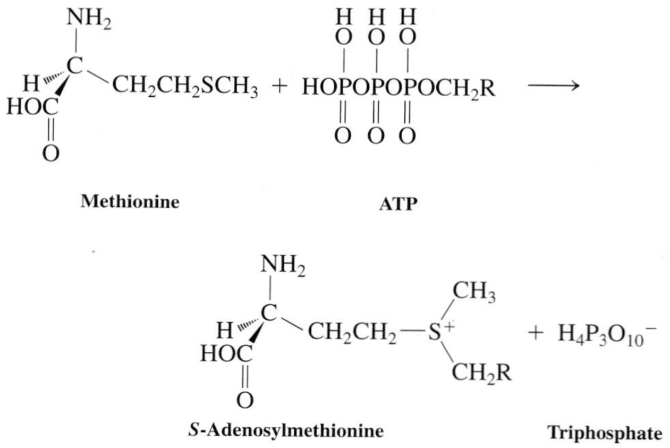

Methionine   **ATP**

S-Adenosylmethionine   **Triphosphate**

Reaction 2

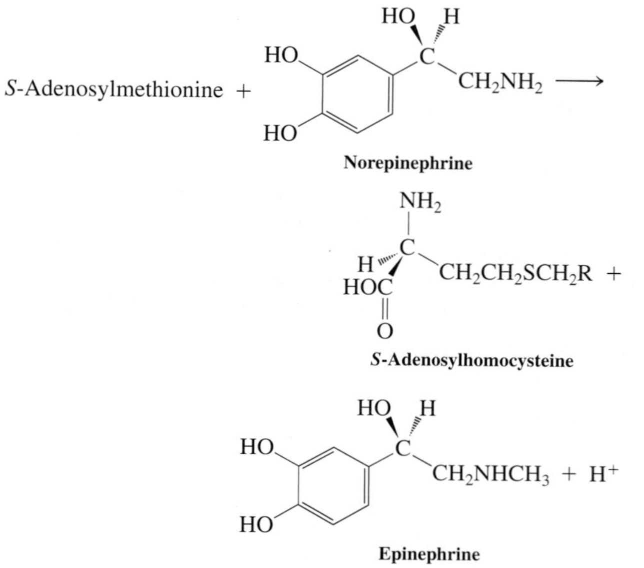

S-Adenosylmethionine +

Norepinephrine

S-Adenosylhomocysteine

Epinephrine

72. **(a)** Only the trans isomer of 2-bromocyclohexanol can react with sodium hydroxide to form an oxacyclopropane-containing product. Explain the lack of reactivity of the cis isomer. [**Hint:** Draw the available conformations of both the cis and trans isomers around the C1–C2 bonds (compare Figure 4-12). Use models if necessary.] **(b)** The synthesis of some oxacyclopropane-containing steroids has been achieved by use of a two-step procedure starting with steroidal bromoketones. Suggest suitable reagents for accomplishing a conversion such as the following one.

(c) Do any of the steps in your proposed sequence have specific stereochemical requirements for the success of the oxacyclopropane-forming step?

73. Freshly cut garlic contains allicin (below), a compound responsible for the true garlic odor (see Real Life 9-4). Propose a short synthesis of allicin, starting with 3-chloropropene.

Allicin
(A flavorant)

(a) Using the curved-arrow formalism (Section 6-3), show the flow of electrons in the attack of the base on the various cyclohexane conformers. Reconvene and present your mechanisms to your teammates, justifying the structural assignments of A–D. Find an explanation for the qualitative rate differences and the divergent course of the reactions of A and B versus C and D.

(b) When compounds A–D are exposed to conditions favoring bromide dissociation in the presence of Ag^+ salts (to accelerate heterolysis with formation of insoluble AgBr), A, C, and D give the same products as those obtained on treatment with base. Discuss the mechanism as a group.

(c) Curiously, compound B traverses another pathway under the conditions described in (b); that is, rearrangement to the aldehyde E. Discuss a possible mechanism for this ring contraction. (**Hint:** Keep in mind the principles outlined in Section 9-3. The mechanism proceeds through a hydroxycation. What is the driving force for its formation?)

Team Problem

74. There are four diastereomers (A–D, below) of (4S)-2-bromo-4-phenylcyclohexanol. As a team, formulate their structures and draw each diastereomer in the most stable chair conformation (see Table 4-3; the $\Delta G°$ value for axial versus equatorial C_6H_5 is 2.9 kcal mol^{-1}). Divide your team into equal groups to consider the outcome of the reaction of each isomer with base ($^-$OH).

Preprofessional Problems

75. The compound below

(C₇H₁₄O)

is best named (IUPAC)

(a) 3,5-dimethylcyclopentyl ether
(b) 3,5-dimethylcyclopentane-oxo
(c) cis-3,5-dimethyloxacyclohexane
(d) trans-3,5-dimethyloxacyclohexane

76. The first step in the detailed mechanism for the dehydration of 1-propanol with concentrated H_2SO_4 would be

(a) loss of OH^-

(b) formation of a sulfate ester

(c) protonation of the alcohol

(d) loss of H^+ by the alcohol

(e) elimination of H_2O by the alcohol

77. Identify the nucleophile in the following reaction:

$$RX + H_2O \longrightarrow ROH + H^+X^-$$

(a) X^- (b) H^+ (c) H_2O (d) ROH   (e) RX

78. Which is the method of choice for preparing the ether $(CH_3CH_2)_3COCH_3$?

(a) $CH_3Br + (CH_3CH_2)_3CO^-K^+$

(b) $(CH_3CH_2)_3COH + CH_3MgBr$

(c) $(CH_3CH_2)_3CMgBr + CH_3OH$

(d) $(CH_3CH_2)_3CBr + CH_3O^-K^+$

CHAPTER 10 | Using Nuclear Magnetic Resonance Spectroscopy to Deduce Structure

A researcher prepares to insert a sample into the probe of one of the most powerful NMR spectrometers in the world. Its superconducting magnet is nearly 15 ft tall, weighs 40 tons, and produces a magnetic field of over 21 tesla—400,000 times stronger than that of Earth.

We have seen that one of the main goals of studying organic chemistry is to appreciate how details of molecular structure affect the ways in which molecules function in reactions—whether those reactions take place in industrial settings, in laboratory syntheses, or within the body. But how do we know the detailed structures of molecules? How can we identify new products or be sure that we have isolated the product we want from a reaction mixture? Wouldn't it be nice if we had some technique that could allow us to recognize the presence of certain nuclei in a molecule, count their relative numbers, describe the nature of their electronic environment, and tell us how they are connected to other atoms?

We do have such a technique, known as nuclear magnetic resonance (NMR) spectroscopy. Not only does this method enable us to identify the structure of an organic molecule, it can be applied to the imaging of whole body organs, a variation known as magnetic resonance imaging (MRI). Just as NMR spectroscopy has become one of the most powerful tools for the organic chemist, MRI has become one of the most powerful applications in medical diagnosis.

We begin with a brief consideration of how classical physical measurements and chemical tests can help in determining a compound's structure. Then we shall discuss how spectroscopy works, how we can interpret its results, and what information we can obtain from the latest advances in spectroscopic instruments and techniques.

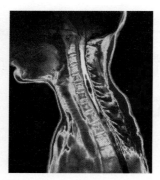

MRI of the neck area of a patient afflicted with cervicodorsal syringomyelia. This condition is characterized by the presence of fluid-filled cavities in the spinal cord substance.

10-1   PHYSICAL AND CHEMICAL TESTS

Let us imagine that you have run a reaction yielding an unidentified compound. To study our sample, we must first purify it—by chromatography, distillation, or recrystallization. We can then compare its melting point, boiling point, and other physical properties with data for known compounds. Even when our measurements match values in the literature (or appropriate handbooks), we cannot be reasonably certain of the identity and structure of our molecule. Moreover, many substances made in the laboratory are new: No published data are available. We need ways to determine their structures *for the first time*.

Elemental analysis will reveal the sample's gross chemical composition. Chemical tests of the compound can then help us to identify its functional groups. For example, we saw in Section 1-9 that we can distinguish between methoxymethane and ethanol on the basis of their physical properties. Section 9-1 indicated how we could do so also on the basis of differing reactivity, for example, in the presence of sodium (ethanol will form sodium ethoxide and hydrogen; methoxymethane is inert).

The problem becomes considerably more difficult for larger molecules, which vary far more in structure. What if a reaction gave us an alcohol of molecular formula $C_7H_{16}O$? A test with sodium metal would reveal a hydroxy functional group—but not an unambiguous structure. In fact, there are many possibilities, only three of which are shown here.

Three Structural Possibilities for an Alcohol $C_7H_{16}O$

$$CH_3(CH_2)_5CH_2OH \qquad \underset{\underset{CH_3}{|}}{\overset{\overset{CH_3}{|}}{CH_3CCH_2CH_2CH_2OH}} \qquad \underset{\underset{CH_3}{|}}{\overset{\overset{CH_2CH_3}{|}}{CH_3CCH_2CH_2OH}}$$

Exercise 10-1

Write the structures of several secondary and tertiary alcohols having the molecular formula $C_7H_{16}O$.

To differentiate among these alternatives, a modern organic chemist makes use of another tool: spectroscopy.

10-2   DEFINING SPECTROSCOPY

Spectroscopy is a technique for analyzing the structure of molecules, usually based on differences in how they absorb electromagnetic radiation. Although there are many types of spectroscopy, four are used most often in organic chemistry: (1) nuclear magnetic resonance (NMR) spectroscopy; (2) infrared (IR) spectroscopy; (3) ultraviolet (UV) spectroscopy; and (based on a different principle) (4) mass spectrometry (MS). The first, **NMR spectroscopy,** probes the structure in the vicinity of individual nuclei, particularly hydrogens and carbons, and provides the most detailed information regarding the atomic connectivity of a molecule.

We begin with a simple overview of spectroscopy as it relates to NMR, IR, and UV. Then we describe how a spectrometer works. Finally, we consider the principles and applications of NMR spectroscopy in more detail. We shall return to the other major forms of spectroscopy in Chapters 11 and 14.

Molecules undergo distinctive excitations

Electromagnetic radiation can be described in the form of waves (or particles; Section 1-6). A wave is defined by its wavelength λ (see margin) or by its frequency ν. The two are related by the expressions

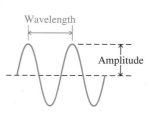

$$\nu = \frac{c}{\lambda} \qquad \text{or} \qquad \lambda\nu = c$$

where c is the speed of the radiation, "the speed of light," 3×10^{10} cm s^{-1}. Frequency is reported in cycles per second (cps) or hertz (Hz, after the German physicist R. H. Hertz). Spectroscopy works because molecules absorb electromagnetic radiation in discrete "packets" of energy, or **quanta.** Absorption occurs only when radiation supplying exactly the right packet reaches the compound under investigation. If the frequency of the incident radiation is ν, the packet has energy $\Delta E = h\nu$ (Figure 10-1).

The absorbed energy causes electronic or mechanical "motion" in the molecule, a process called **excitation.** This motion also is quantized and, because a molecule can undergo many different kinds of excitation, each kind of motion requires its own distinctive energy. X-rays, for example, which are a form of high-energy radiation, can promote electrons in atoms from inner shells to outer ones; this change, called an **electronic transition,** requires energy higher than 300 kcal mol^{-1}. Ultraviolet radiation and visible light, in contrast, excite valence-shell electrons, typically from a filled bonding molecular orbital to an unfilled antibonding one (see Figure 1-12A); here the energy needed ranges from 40 to 300 kcal mol^{-1}. We perceive the visible range of electromagnetic radiation as colors. Infrared radiation causes vibrational excitation of a compound's bonds ($\Delta E = $ 2–10 kcal mol^{-1}), whereas quanta of microwave radiation cause bond rotations to occur ($\Delta E = \sim 10^{-4}$ kcal mol^{-1}). Finally, radio waves can produce changes in the alignment of nuclear magnetism in a magnetic field ($\Delta E = \sim 10^{-6}$ kcal mol^{-1}); in the next section we shall see how this phenomenon is the basis of nuclear magnetic resonance spectroscopy.

Figure 10-2 depicts the various forms of radiation, the energy (ΔE) related to each form, the corresponding wavelengths, and frequencies. Note that frequency can also be given in units of wavenumbers, defined as $\tilde{\nu} = 1/\lambda$, the number of waves per centimeter and an energy measure used in infrared spectroscopy. Wavelengths, λ, in nanometers (nm) are employed in UV and visible spectroscopy.

We shall come back to Figure 10-2 repeatedly, as we discuss the various kinds of spectroscopy. For the moment, just remember that the energy of radiation increases with increasing frequency (ν) or wavenumber ($\tilde{\nu}$) but decreasing wavelength (λ) (see margin).

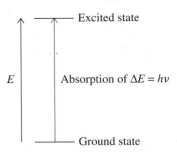

Figure 10-1 Absorption of energy occurs when incident radiation has exactly the right frequency ν so that its energy $h\nu$ equals the energy difference ΔE between the ground state and the excited state of a molecule [ν, frequency of absorbed radiation; h (Planck's constant) = 6.626×10^{-34} J s].

Decreasing wavelength λ

Increasing frequency ν

Increasing energy E

Exercise 10-2

What type of radiation (in wavelengths, λ) would be minimally required to initiate the radical chlorination of methane? [**Hint:** The initiation step requires the breaking of the Cl–Cl bond (see Section 3-4).]

A spectrometer records the absorption of radiation

As illustrated in Figure 10-1, the absorption of quanta of radiation by a molecule brings about transitions from its (normal) ground state to (a variety of) excited states. Spectroscopy is a procedure by which these absorptions can be mapped by instruments called **spectrometers.**

Figure 10-3 shows the principle of a spectrometer. It contains a source of electromagnetic radiation with a frequency in the region of interest, such as the visible, infrared, or radio. The apparatus is designed so that radiation of a specific wavelength range (NMR, IR, UV, etc.) passes through the sample. In the traditional continuous-wave (CW) spectrometer, the frequency of this radiation is changed continuously, and its intensity is measured at a detector and recorded on calibrated paper. In the absence of absorption, the sweep of radiation appears as a straight line, the **baseline.** Whenever the sample absorbs electromagnetic radiation, however, the resulting change in intensity at the detector registers as a **peak,** or deviation from the baseline. The resulting pattern is the **spectrum** (Latin, appearance, apparition) of the sample.

New-generation spectrometers use a different and much faster recording technique, in which a pulse of electromagnetic radiation that covers the entire frequency range under scrutiny (NMR, IR, UV) is used to obtain the whole spectrum instantly. Moreover, rather than simple absorption, as in traditional CW instruments, the decay of the absorption event with time is recorded, a procedure that requires a more elaborate computer analysis, called **Fourier transform (FT),** after the French mathematician Joseph Fourier (1768–1830). Apart from the speed of this technique, multiple pulse accumulation of the same spectrum allows for much higher sensitivity, which is of great value when only small amounts of sample are available.

Figure 10-2 The spectrum of electromagnetic radiation. The top line is an energy scale, in units of kilocalories per mole (kilojoules per mole in parentheses), increasing from right to left. The next line contains the corresponding wavenumbers, $\tilde{\nu}$, in units of reciprocal centimeters. The types of radiation associated with the principal types of spectroscopy and the transitions induced by each type are given in the middle. A wavelength scale is at the bottom (λ, in units of nanometers, 1 nm = 10^{-9} m; micrometers, 1 μm = 10^{-6} m; millimeters, mm; and meters, m). ΔE (kcal mol^{-1}) = 28,600/λ (nm).

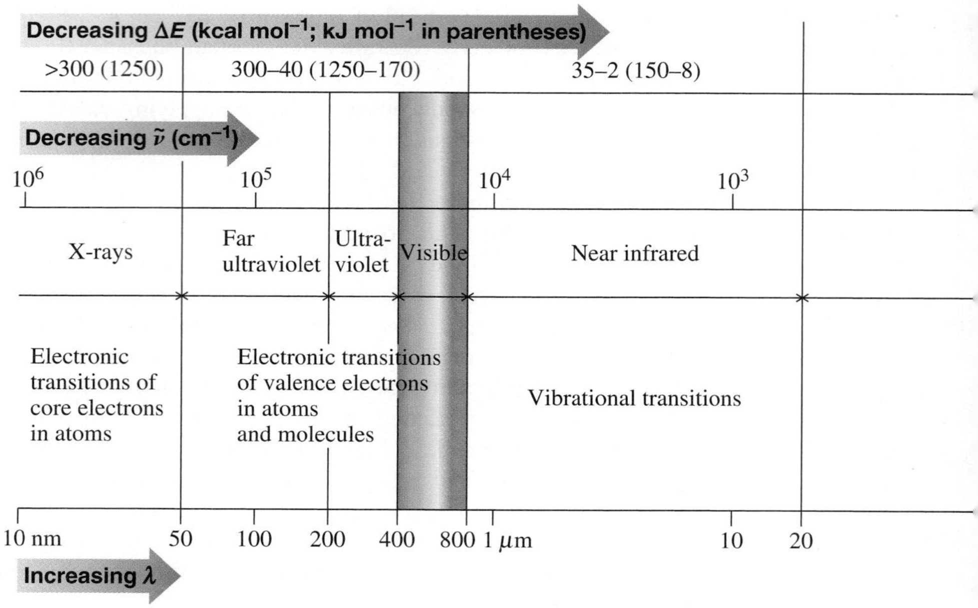

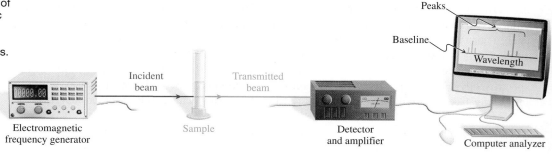

Decreasing ΔE (kcal mol^{-1}; kJ mol^{-1} in parentheses)

| >300 (1250) | 300–40 (1250–170) | 35–2 (150–8) |

Decreasing $\tilde{\nu}$ (cm^{-1})

10^6 10^5 10^4 10^3

| X-rays | Far ultraviolet | Ultra-violet | Visible | Near infrared |

| Electronic transitions of core electrons in atoms | Electronic transitions of valence electrons in atoms and molecules | Vibrational transitions |

10 nm   50   100   200   400   800   1 μm   10   20

Increasing λ

Figure 10-3 General diagram of a spectrometer. Electromagnetic radiation passes through and interacts with a sample by absorption at certain frequencies. The incident beam is altered to a transmitted beam and the changes detected, amplified, and processed by computer to generate a spectrum.

Peaks

Baseline

Wavelength

Incident beam

Transmitted beam

Electromagnetic frequency generator

Sample

Detector and amplifier

Computer analyzer

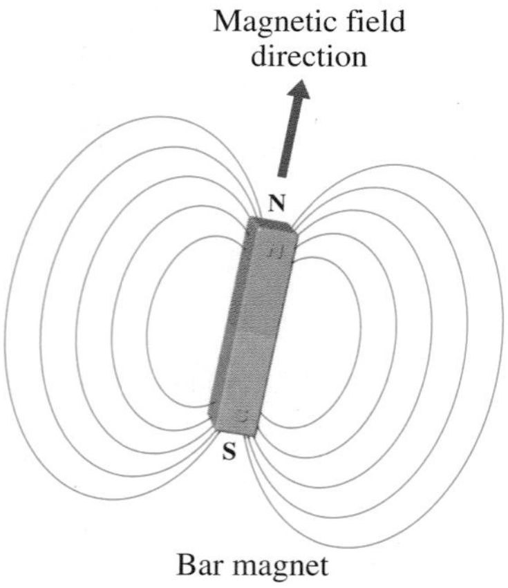

Magnetic field direction

N

S

Bar magnet

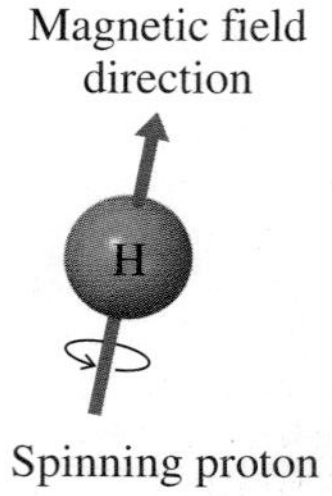

Magnetic field direction

H

Spinning proton

The Spinning Proton Creates a Magnetic Field

In Summary Molecules absorb electromagnetic radiation in discrete quanta of incident energy measurable by spectroscopy. Spectrometers scan samples with radiation of varying wavelength as it passes through a sample of the compound under investigation, resulting in a plot of the absorptions occurring at certain energies: the spectrum.

10-3 | HYDROGEN NUCLEAR MAGNETIC RESONANCE

Nuclear magnetic resonance spectroscopy requires low-energy radiation in the radio-frequency (RF) range. This section presents the principles behind this technique.

Nuclear spins can be excited by the absorption of radio waves

Many atomic nuclei can be thought of as spinning around an axis and are therefore said to have a **nuclear spin.** One of those nuclei is hydrogen, written as ^{1}H (the hydrogen isotope of mass 1) to differentiate it from other isotopes [deuterium (^{2}H), tritium (^{3}H)]. Let us consider the simplest form of hydrogen, the proton. Because the proton is positively charged, its spinning motion creates a magnetic field (as does any moving charged particle). The net result is that a proton may be viewed as a tiny cylindrical (bar) magnet floating freely in solution or in space (see margin). When the proton is exposed to an *external* magnetic field of strength H_0, it may have one of two orientations: It may be aligned either with H_0, an energetically favorable choice, or (unlike a normal bar magnet) against H_0, an orientation that is higher in energy. The two possibilities are designated the α and β **spin states,** respectively (Figure 10-4).

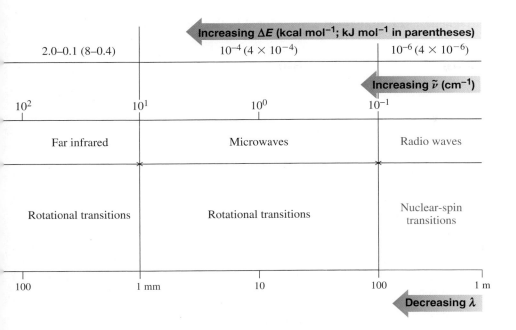

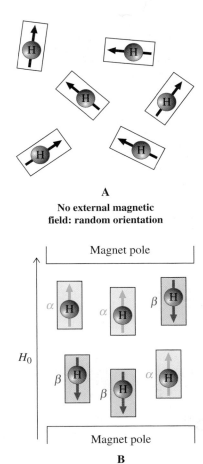

A

No external magnetic field: random orientation

B

In an external magnetic field: alignment with (α) or against (β) H_0

Figure 10-4 (A) Single protons (H) act as tiny magnets. (B) In a magnetic field, H_0, the nuclear spins align about equally with (α) or against (β) the field.

These two energetically different states afford the necessary condition for spectroscopy. Irradiation of the sample at just the right frequency to bridge the difference in energy between the α and β states produces **resonance,** an absorption of energy, ΔE, as an α proton "flips" to the β spin state. This phenomenon is illustrated for a pair of protons in Figure 10-5. After excitation, the nuclei relax and return to their original states by a variety of pathways (which will not be discussed here). At resonance, therefore, there is continuous excitation and relaxation.

As you might expect, increasing the magnetic field strength H_0 makes the α → β spin flip more difficult. Indeed, the difference in energy ΔE between the two spin states is directly proportional to H_0. Consequently, since $\Delta E = h\nu$, the resonance frequency is also proportional to the magnetic field strength. You can see this relationship in commercial spectrometers, for which the size of the magnet field is given in units of tesla (T)* and the corresponding hydrogen resonance frequency in megahertz (MHz).

---

*Nikola Tesla (1856–1943), American inventor (of Serbian origin), physicist, and mechanical and electrical engineer.

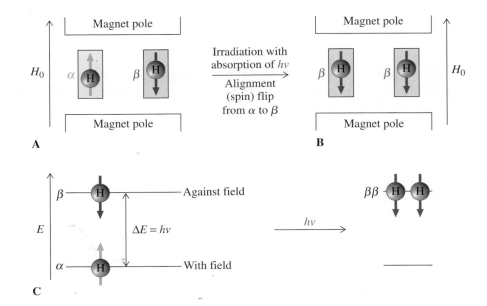

A

Irradiation with absorption of $h\nu$
→
Alignment (spin) flip from α to β

B

β — Against field

E

$\Delta E = h\nu$

$h\nu$ →

α — With field

C

ββ

Figure 10-5 (A) A simpler rendition of Figure 10-4B: In an external magnetic field, protons align about equally with (lower energy α) and against (higher energy β) the field, differing in energy by ΔE, as shown in part C. (B) Irradiation with energy of the right frequency ν causes absorption, "flipping" the nuclear spin of a proton in the α state to the β state (also called resonance). (C) An energy diagram, showing a proton gaining the energy $\Delta E = h\nu$ and undergoing "spin flip" from α to β. When looking at this figure, remember that the two nuclei depicted are surrounded by the bulk sample in which there is only a slight excess of α over β states. The process of absorption serves to bring this ratio closer to 1:1.

Magnetic Field Strength Is Proportional to Resonance Frequency

Increasing H_0 →

Magnetic field strength H_0 (tesla):	2.11	4.23	7.05	11.8	14.1	21.1	T
Hydrogen resonance frequency ν (megahertz):	90	180	300	500	600	900	MHz

Increasing ν →

To give you a feeling for the size of these magnets, the maximum intensity of Earth's magnetic field anywhere on its surface is about 0.00007 T.

How much energy must be expended for the spin of a proton to flip from α to β? Because $\Delta E_{\beta-\alpha} = h\nu$, we can calculate how much. The amount is very small, that is, $\Delta E_{\beta-\alpha}$ at 300 MHz is of the order of 3×10^{-5} kcal mol^{-1} (1.5×10^{-4} kJ mol^{-1}). Equilibration between the two states is fast, and typically only slightly more than half of all proton nuclei in a magnetic field will adopt the α state, the remainder having a β spin. At resonance, this difference is diminished, as α spins flip to β spins, but the nearly equal proportionality is not perturbed much.

Many nuclei undergo magnetic resonance

Hydrogen is not the only nucleus capable of magnetic resonance. Table 10-1 lists a number of nuclei that are responsive to NMR and are of importance in organic chemistry, as well as several that lack NMR activity. In general, nuclei composed of an odd number of protons, such as ^1H (and its isotopes), ^{14}N, ^{19}F, and ^{31}P, or an odd number of neutrons, such as ^{13}C, exhibit magnetic behavior. On the other hand, when *both* the proton and the neutron counts are even, as in ^{12}C or ^{16}O, the nuclei are nonmagnetic.

Upon exposure to equal magnetic fields, *different NMR-active nuclei will resonate at different values of ν*. For example, if we were to scan a hypothetical spectrum of a sample of chlorofluoromethane, CH_2ClF, in a 7.05-T magnet, we would observe six absorptions corresponding to the six NMR-active nuclei in the sample: the highly abundant ^1H, ^{19}F, ^{35}Cl, and ^{37}Cl, and the much less plentiful ^{13}C (1.11%) and ^2H (0.015%), as shown in Figure 10-6.

High-resolution NMR spectroscopy can differentiate nuclei of the same element

Consider now the NMR spectrum of chloro(methoxy)methane (chloromethyl methyl ether), $ClCH_2OCH_3$. A sweep at 7.05 T from 0 to 300 MHz would give one peak for each element present (Figure 10-7A). Just as a microscope allows you to magnify a detail of the macroscopic world, we can "peek" into any of these signals and expand them to reveal much

Table 10-1	NMR Activity and Natural Abundance of Selected Nuclei				
Nucleus	NMR activity	Natural abundance (%)	Nucleus	NMR activity	Natural abundance (%)
^1H	Active	99.985	^{16}O	Inactive	99.759
^2H (D)	Active	0.015	^{17}O	Active	0.037
^3H (T)	Active	0	^{18}O	Inactive	0.204
^{12}C	Inactive	98.89	^{19}F	Active	100
^{13}C	Active	1.11	^{31}P	Active	100
^{14}N	Active	99.63	^{35}Cl	Active	75.53
^{15}N	Active	0.37	^{37}Cl	Active	24.47

Abbreviations: D, deuterium; T, tritium.

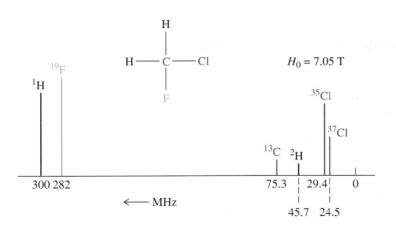

$$H_0 = 7.05 \text{ T}$$

^{19}F

^1H

^{35}Cl

^{37}Cl

^{13}C ^2H

300 282   75.3   29.4   0

⟵ MHz

45.7 24.5

Figure 10-6 A hypothetical NMR spectrum of CH$_2$ClF at 7.05 T. Because each NMR-active nucleus resonates at a characteristic frequency, six lines are observed. We show them here with similar heights for simplicity, even though the natural abundance of the isotopes ^{2}H and ^{13}C is much less than that of the others. The spectrum, as shown, could not be scanned in a single experiment, because most NMR instruments are tuned to only one nucleus, such as ^{1}H, at any given time.

REAL LIFE: SPECTROSCOPY 10-1 | Recording an NMR Spectrum

To take an NMR spectrum, the sample to be studied (a few milligrams) is usually dissolved in a solvent (0.3–0.5 mL), preferably one that does not contain any atoms that themselves absorb in the NMR range under investigation. Typical solvents are deuterated, such as trichlorodeuteriomethane (deuteriochloroform), CDCl$_3$; hexadeuterioacetone, CD$_3$COCD$_3$; hexadeuteriobenzene, C$_6$D$_6$; and octadeuteriooxacyclopentane (octadeuteriotetrahydrofuran), C$_4$D$_8$O. The effect of replacing hydrogen by deuterium is to remove any solvent peaks from the proton spectrum. Note that the resonance frequency of deuterium lies in a completely different spectral region than that of hydrogen (^{1}H) (Figure 10-6). The solution is transferred

into an NMR sample container (a cylindrical glass tube), which is inserted into a superconducting magnet (left photograph). To make sure that all molecules in the sample are rapidly averaged with respect to their position in the magnetic field, the NMR tube is rapidly spun by an air jet inside a radio-frequency coil (see diagram). The sample is irradiated with a pulse of RF across the entire spectral region, the ensuing response is registered by the detector unit, and the decay of the spectral signals with time is Fourier transformed by the computer to give the final spectrum. The right photograph depicts a student at a workstation, analyzing her NMR data.

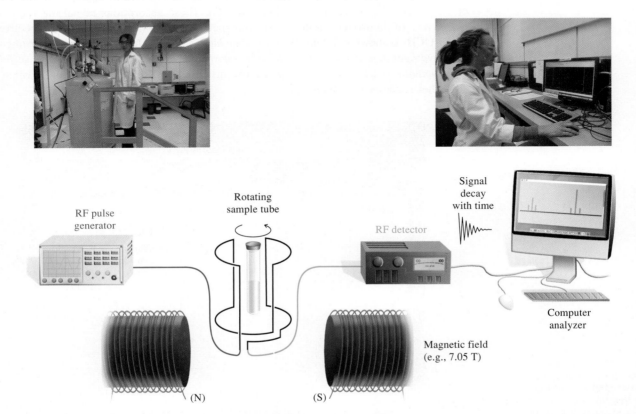

Figure 10-7 High resolution can reveal additional peaks in an NMR spectrum. (A) At low resolution, the spectrum of $ClCH_2OCH_3$ at 7.05 T shows six peaks for the six NMR active isotopes present in the molecule. (B) At high resolution, the hydrogen spectrum shows two peaks for the two sets of hydrogens (one shown in blue in the structure, the other in red). Note that the high-resolution sweep covers only 0.001% of that at low resolution. (C) The high-resolution ^{13}C spectrum (see Section 10-9) shows peaks for the two different carbon atoms in the molecule.

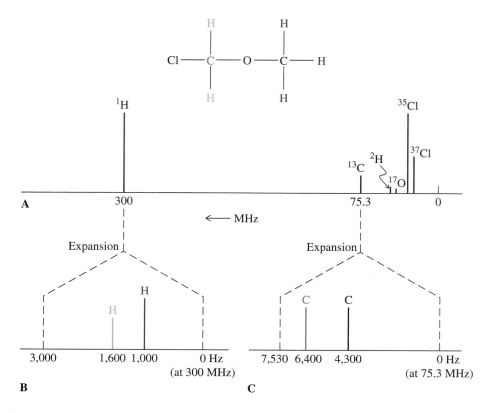

more. Thus, using the technique called **high-resolution NMR spectroscopy,** we can study the hydrogen resonance from 300,000,000 to 300,003,000 Hz. We find that what appeared to be only one peak in that region actually consists of two peaks that were not resolved at first (Figure 10-7B). Similarly, the high-resolution ^{13}C spectrum measured in the vicinity of 75.3 MHz shows two peaks (Figure 10-7C). These absorptions reveal the presence of *two* types of hydrogens and carbons, respectively. An actual ^1H NMR spectrum of $ClCH_2OCH_3$ is shown in Figure 10-8. *Because high-resolution NMR spectroscopy distinguishes both hydrogen and carbon atoms in different structural environments, it is a powerful tool for elucidating structures.* The organic chemist uses NMR spectroscopy more often than any other spectroscopic technique.

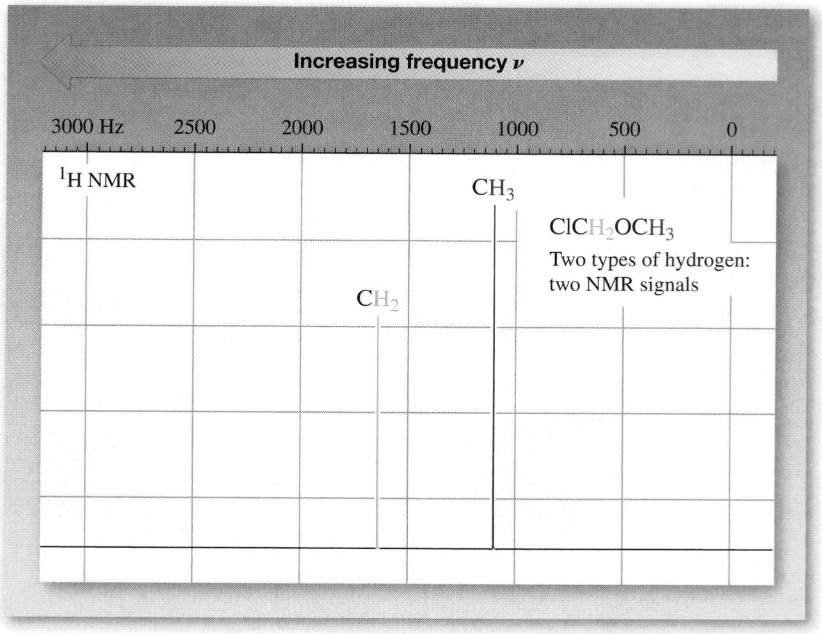

Figure 10-8 300-MHz ^1H NMR spectrum of chloro(methoxy)methane. Because the frequency range of interest starts at 300 MHz, we set this frequency to be 0 Hz at the right-hand side of the recording paper, for simplicity.

In Summary Certain nuclei, such as ^1H and ^{13}C, can be viewed as tiny atomic magnets that, when exposed to a magnetic field, can align with (α) or against the field (β). These two states are of unequal energy, a condition exploited in nuclear magnetic resonance spectroscopy. At resonance, radio-frequency radiation is absorbed by the nucleus to effect α-to-β transitions (excitation). Nuclei in the β state relax to the α state by giving off energy (in the form of a miniscule amount of heat). The resonance frequency, which is characteristic of the nucleus and its environment, is proportional to the strength of the external magnetic field.

10-4 USING NMR SPECTRA TO ANALYZE MOLECULAR STRUCTURE: THE PROTON CHEMICAL SHIFT

Why do the two different groups of hydrogens in chloro(methoxy)methane give rise to distinct NMR peaks? How does the molecular structure affect the position of an NMR signal? This section will answer these questions.

The position of an NMR absorption, also called the **chemical shift,** depends on the electron density around the hydrogen. This density, in turn, is controlled by the structural environment of the observed nucleus. Therefore, *the NMR chemical shifts of the hydrogens in a molecule are important clues for determining its molecular structure*. At the same time, the structure of a molecule determines how it "functions" in an NMR experiment.

The position of an NMR signal depends on the electronic environment of the nucleus

The high-resolution ^1H NMR spectrum of chloro(methoxy)methane depicted in Figure 10-8 reveals that the two kinds of hydrogens give rise to two separate resonance absorptions. What is the origin of this effect? It is the differing electronic environments of the respective hydrogen nuclei. A free proton is essentially unperturbed by electrons. Organic molecules, however, contain covalently bonded hydrogen nuclei, *not* free protons, and the electrons in these bonds affect nuclear magnetic resonance absorptions.*

Bound hydrogens are surrounded by orbitals whose electron density varies depending on the polarity of the bond, the hybridization of the attached atom, and the presence of electron-donating or -withdrawing groups. When a nucleus surrounded by electrons is exposed to a magnetic field of strength H_0, these electrons move in such a way as to generate a small **local magnetic field, h_{local},** *opposing* H_0. As a consequence, the total field strength near the hydrogen nucleus is *reduced,* and the nucleus is thus said to be **shielded** from H_0 by its electron cloud (Figure 10-9). The degree of shielding depends on the amount

---

*In discussions of NMR, the terms *proton* and *hydrogen* are frequently (albeit incorrectly) interchanged. "Proton NMR" and "protons in molecules" are used even in reference to covalently bound hydrogen.

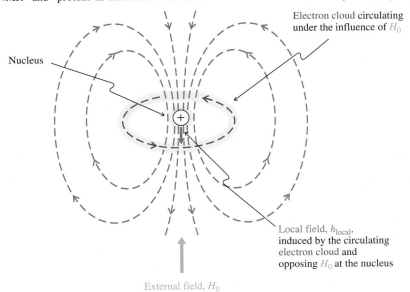

Nucleus

Electron cloud circulating under the influence of H_0

Local field, h_{local}, induced by the circulating electron cloud and opposing H_0 at the nucleus

External field, H_0

Figure 10-9 The external field, H_0, causes movement of the bonding electrons around a hydrogen nucleus. This current, in turn, generates a local magnetic field opposing H_0. [You may recognize this as Lenz's law, named for Russian physicist Heinrich Friedrich Emil Lenz (1804–1865). Note that the direction of electron movement is opposite that of the corresponding electric current, which is defined as flowing from anode (+) to cathode (−)].

Figure 10-10 Effect of shielding on a covalently bound hydrogen. The bare nucleus, H^+, which has no shielding bonding electrons, is least *shielded;* in other words, its signal occurs at the left of the spectrum, or at *low field*. A hydrogen attached to, for example, carbon is *shielded* by the surrounding bonding electrons, and thus its signal appears farther to the right or at *high field*.

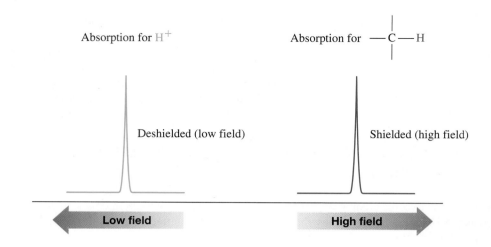

of electron density surrounding the nucleus. Adding electrons increases shielding; their removal causes **deshielding.**

What is the effect of shielding on the relative position of an NMR absorption? As a result of the way in which NMR spectra are displayed, *shielding* causes a displacement of the absorption peak to the *right* in the spectrum. *Deshielding* shifts such a peak to the *left* (Figure 10-10). Rather than using the terms *right* and *left,* chemists are accustomed to terminology that is derived from the old way (pre-FT) in which NMR spectra were recorded. Here, to compensate for shielding, the external field strength H_0 is increased to achieve resonance (remember that H_0 is proportional to ν), and therefore the peak is said to be at **high field,** or shifted **upfield** (to the right). Conversely, deshielding causes signals to appear at **low field** (to the left).

Because each chemically distinct hydrogen has a unique electronic environment, it gives rise to a characteristic resonance. Moreover, *chemically equivalent hydrogens show peaks at the same position.* Chemically equivalent hydrogens are those related by symmetry, such as the hydrogens of methyl groups, the methylene hydrogens in butane, or all the hydrogens in cycloalkanes (see, however, Section 10-5).

^{1}H NMR: Different Types of Hydrogens Give Rise to Different Signals

One type of H: one signal	Two types of H: two signals	Three types of H: three signals	Three types of H: three signals

We describe in more detail some tests for chemical equivalency in the next section, but for simple molecules such equivalence is obvious. An example is the NMR spectrum of 2,2-dimethyl-1-propanol in Figure 10-11: There are three absorptions—one (most shielded) for the nine equivalent methyl hydrogens of the tertiary butyl group, another for the OH, and a third (most deshielded) for the CH_2 hydrogens.

The chemical shift describes the position of an NMR peak

How are spectral data reported? As noted earlier, most hydrogen absorptions in 300-MHz ^{1}H NMR fall within a range of 3000 Hz. Rather than record the exact frequency of each resonance, we measure it relative to an internal standard, the compound tetramethylsilane, $(CH_3)_4Si$. Its 12 equivalent hydrogens are shielded relative to those in most organic molecules, resulting in a resonance line conveniently removed from the usual spectral range. The position of the NMR absorptions of a compound under investigation can then be measured (in hertz) relative to the internal standard. In this way, the signals of, for example,

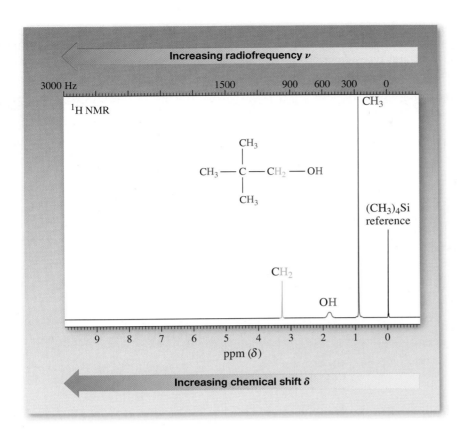

Figure 10-11 300-MHz ^{1}H NMR spectrum of 2,2-dimethyl-1-propanol (containing a little tetramethylsilane as an internal standard) in deuterated chloroform, CDCl$_3$. Three peaks are observed for the three sets of different hydrogens. (The scale at the bottom indicates the chemical shift in δ, the distance from tetramethylsilane, to be defined in the next subsection.)

2,2-dimethyl-1-propanol (Figure 10-11) are reported as being located 266, 541, and 978 Hz downfield from (CH$_3$)$_4$Si.

A problem with these numbers, however, is that *they vary with the strength of the applied magnetic field*. Because field strength and resonance frequency are directly proportional, doubling or tripling the field strength will double or triple the distance (in hertz) of the observed peaks relative to (CH$_3$)$_4$Si. To make it easier to compare reported literature spectra, we standardize the measured frequency by dividing the distance to (CH$_3$)$_4$Si (in hertz) by the frequency of the spectrometer. This procedure yields a *field-independent* number, the **chemical shift δ.**

The Chemical Shift

$$\delta = \frac{\text{distance of peak from (CH}_3)_4\text{Si in hertz}}{\text{spectrometer frequency in megahertz}} \text{ ppm}$$

The chemical shift is reported in units of parts per million (ppm). For (CH$_3$)$_4$Si, δ is defined as 0.00. The NMR spectrum of 2,2-dimethyl-1-propanol in Figure 10-11 would then be reported in the following format: ^{1}H NMR (300 MHz, CDCl$_3$) δ = 0.89, 1.80, 3.26 ppm.

Exercise 10-3

With a 90-MHz NMR instrument, the three signals for 2,2-dimethyl-1-propanol are recorded at 80, 162, and 293 Hz, respectively, downfield from (CH$_3$)$_4$Si. Calculate the δ values and compare them with those obtained at 300 MHz.

Functional groups cause characteristic chemical shifts

The reason that NMR is such a valuable analytical tool is that it can identify certain types of hydrogens in a molecule. Each type has a characteristic chemical shift depending on its structural environment. The hydrogen chemical shifts typical of standard organic structural units are listed in Table 10-2. It is important to be familiar with the chemical shift ranges for the structural types described so far: alkanes, haloalkanes, ethers, alcohols, aldehydes, and ketones. Others will be discussed in more detail in subsequent chapters.

Table 10-2 Typical Hydrogen Chemical Shifts in Organic Molecules

Type of hydrogen[a]	Chemical shift δ in ppm	
Primary alkyl, RCH_3	0.8–1.0	Alkane and alkane-like hydrogens
Secondary alkyl, RCH_2R'	1.2–1.4	
Tertiary alkyl, R_3CH	1.4–1.7	
Allylic (next to a double bond), $R_2C{=}C\!\!\begin{smallmatrix}CH_3\\R'\end{smallmatrix}$	1.6–1.9	Hydrogens adjacent to unsaturated functional groups
Benzylic (next to a benzene ring), $ArCH_2R$	2.2–2.5	
Ketone, $RCCH_3$ ‖ O	2.1–2.6	
Alkyne, $RC{\equiv}CH$	1.7–3.1	
Chloroalkane, RCH_2Cl	3.6–3.8	Hydrogens adjacent to electronegative atoms
Bromoalkane, RCH_2Br	3.4–3.6	
Iodoalkane, RCH_2I	3.1–3.3	
Ether, RCH_2OR'	3.3–3.9	
Alcohol, RCH_2OH	3.3–4.0	
Terminal alkene, $R_2C{=}CH_2$	4.6–5.0	Alkene hydrogens
Internal alkene, $R_2C{=}CH$ | R'	5.2–5.7	
Aromatic, ArH	6.0–9.5	
Aldehyde, RCH ‖ O	9.5–9.9	
Alcoholic hydroxy, ROH	0.5–5.0	(variable)
Thiol, RSH	0.5–5.0	(variable)
Amine, RNH_2	0.5–5.0	(variable)

[a] R, R', alkyl groups; Ar, aromatic group (not argon).

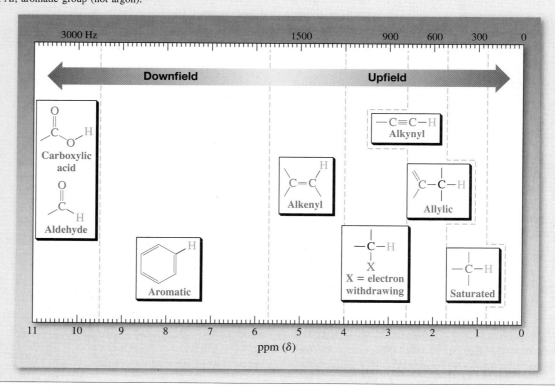

Table 10-3	The Deshielding Effect of Electronegative Atoms	
CH_3X	Electronegativity of X (from Table 1-2)	Chemical shift δ (ppm) of CH_3 group
CH_3F	4.0	4.26
CH_3OH	3.4	3.40
CH_3Cl	3.2	3.05
CH_3Br	3.0	2.68
CH_3I	2.7	2.16
CH_3H	2.2	0.23

(Increasing electronegativity →) (Increasing chemical shift →)

Note that the absorptions of the alkane hydrogens occur at relatively high field ($\delta = 0.8–1.7$ ppm). A hydrogen close to an electron-withdrawing group or atom (such as a halogen or oxygen) is shifted to relatively lower field: Such substituents *deshield* their neighbors. Table 10-3 shows how adjacent heteroatoms affect the chemical shifts of a methyl group. The more electronegative the atom, the more deshielded are the methyl hydrogens relative to methane. Several such substituents exert a cumulative effect, as seen in the series of the three chlorinated methanes shown in the margin. The deshielding influence of electron-withdrawing groups diminishes rapidly with distance. This "tapering off" is seen in the electrostatic potential map of 1-bromopropane in the margin. The area around the bromine-bearing carbon is relatively electron deficient (blue). Progressing farther along the propyl chain reveals shading that tends first toward green, then orange-yellow, signifying increasing electron density.

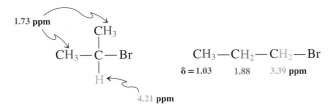

1.73 ppm

CH_3

CH_3—$\underset{\underset{H}{|}}{\overset{\overset{}{|}}{C}}$—Br

4.21 ppm

CH_3—CH_2—CH_2—Br
$\delta = 1.03$ 1.88   3.39 ppm

Cumulative Deshielding in Chloromethanes

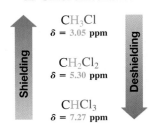

CH_3Cl
$\delta = 3.05$ ppm

CH_2Cl_2
$\delta = 5.30$ ppm

$CHCl_3$
$\delta = 7.27$ ppm

(Shielding ↑)   (Deshielding ↓)

1-Bromopropane

Exercise 10-4

Explain the assignment of the ^1H NMR signals of chloro(methoxy)methane (see color scheme, Figure 10-8). (**Hint:** Consider the number of electronegative neighbors to each type of hydrogen.)

Exercise 10-5

Provide the expected δ values in the ^1H NMR spectra of the following compounds by consulting Table 10-2.

(a) $CH_3CH_2OCH_2CH_3$

(b) H₃C⟍   ⟋CH₃

(c) H_3C—CHO

(d) H—≡—⟍⟋—OH

NMR can be used to detect the undesirable presence of acetic acid in unopened bottles of aged wine. Here, Dr. April Weekly and Prof. Matthew Augustine from the University of California at Davis are getting ready to take the NMR spectrum of a 1959 bottle of Bordeaux. (*Photo by N. Schore.*)

As noted in Table 10-2, hydroxy, mercapto, and amino hydrogens absorb over a range of frequencies. In spectra of such samples, the absorption peak of the proton attached to the heteroatom may be relatively broad. This variability of chemical shift is due to hydrogen bonding and proton exchange, and depends on temperature, concentration, and the presence of water (i.e., moisture). In simple terms, these effects alter the electronic environment of the hydrogen nuclei. When such a broad line is observed, it usually indicates the presence of OH, SH, or NH_2 (NHR) groups (see Figure 10-11).

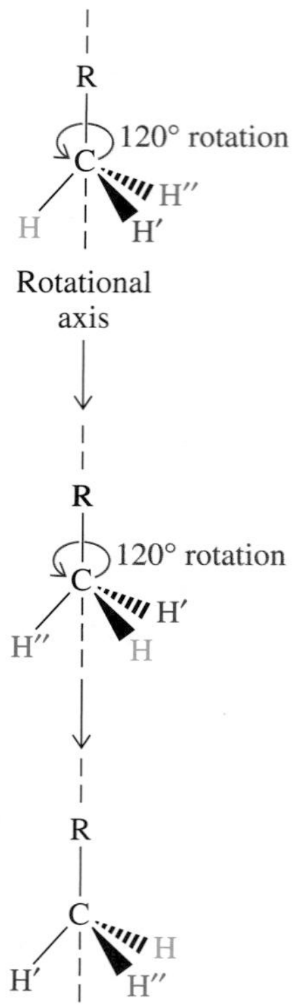

Figure 10-12 Rotation of a methyl group as a test of symmetry.

In Summary The various hydrogen atoms present in an organic molecule can be recognized by their characteristic NMR peaks at certain chemical shifts, δ. An electron-poor environment is deshielded and leads to low-field (high-δ) absorptions, whereas an electron-rich environment results in shielded or high-field peaks. The chemical shift δ is measured in parts per million by dividing the difference in hertz between the measured resonance and that of the internal standard, tetramethylsilane, $(CH_3)_4Si$, by the spectrometer frequency in megahertz. The NMR spectra for the OH groups of alcohols, the SH groups of thiols, and the NH_2 (NHR) groups of amines exhibit characteristically broad peaks with concentration- and moisture-dependent δ values.

10-5 | TESTS FOR CHEMICAL EQUIVALENCE

In the NMR spectra presented so far, two or more hydrogens occupying positions that are chemically equivalent give rise to only *one* NMR absorption. It can be said, in general, that *chemically equivalent protons have the same chemical shift*. However, we shall see that it is not always easy to identify chemically equivalent nuclei. We shall resort to the symmetry operations presented in Chapter 5 to help us decide on the expected NMR spectrum of a specific compound.

Molecular symmetry helps establish chemical equivalence

To establish chemical equivalence, we have to recognize the symmetry of molecules and their substituent groups. As we know, one form of symmetry is the presence of a mirror plane (Section 5-1, Figure 5-4). Another is rotational equivalence. For example, Figure 10-12 demonstrates how two successive 120° rotations of a methyl group allow each hydrogen to occupy the position of either of the other two without effecting any structural change. Thus, in a rapidly rotating methyl group, all hydrogens are equivalent and should have the same chemical shift. We shall see shortly that this is indeed the case.

Application of the principles of rotational or mirror symmetry or both allows the assignment of equivalent nuclei in other compounds (Figure 10-13).

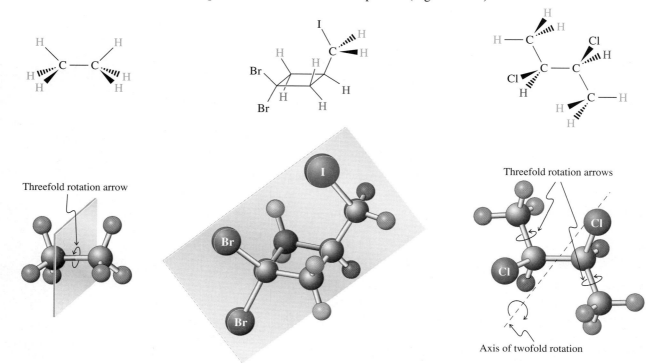

Figure 10-13 The recognition of rotational and mirror symmetry in organic molecules allows the identification of chemical-shift–equivalent hydrogens. The different colors distinguish among nuclei giving rise to separate absorptions with distinct chemical shifts.

Solved Exercise 10-6 | **Working with the Concepts: How to Establish the Presence of Equivalent Hydrogens**

How many ^1H NMR absorptions would you expect for $CH_3OCH_2CH_2OCH_2CH_2OCH_3$?

Strategy
The best way to approach this kind of problem is to make a model or draw the structure in detail, showing all hydrogens in place.

Then you need to identify any mirror planes or rotational axes that render groups of hydrogens equivalent.

Solution
- We can recognize the vertical *mirror plane 1,* which coincides with the molecular plane and makes all wedged hydrogens equivalent to their hashed neighbors.
- Next, we can see a second *mirror plane 2,* through the central oxygen lying perpendicular to the molecular plane, which makes the left half of the molecule the same as the right half.
- Finally, the methyl hydrogens are equivalent by *rotation.*
- There are no symmetry operations that would turn one methylene group into its neighbor, and the methyl groups are obviously distinct from the methylenes.

We would therefore anticipate three proton resonance signals arising from the three types of hydrogens distinguished by color, below.

Exercise 10-7 | **Try It Yourself**

How many ^1H NMR absorptions would you expect for **(a)** 2,2,3,3-tetramethylbutane; **(b)** oxacyclopropane?

Conformational interconversion may result in equivalence on the NMR time scale

Let us look more closely at two more examples, chloroethane and cyclohexane. Chloroethane should have two NMR peaks because it has the two sets of equivalent hydrogens; cyclohexane has 12 chemically equivalent hydrogen nuclei and is expected to show only one absorption. However, are these expectations really justified? Consider the possible conformations of these two molecules (Figure 10-14).

Begin with chloroethane. The most stable conformation is the staggered arrangement, in which one of the methyl hydrogens (H_{b_3} in the first Newman projection in Figure 10-14)

Model Building

A

Rotation of the methyl group, located in the back

B

Ring flip

Figure 10-14 (A) Newman projections of chloroethane. In the initial conformation, H_{b_1} and H_{b_2} are located *gauche* and H_{b_3} *anti* to the chlorine, and therefore are not in the same environments. However, fast rotation moves each type of H through all positions, thus averaging all the methyl hydrogens on the NMR time scale. (B) In any given conformation of cyclohexane, the axial hydrogens are different from the equatorial ones. However, conformational flip is rapid on the NMR time scale, equilibrating axial and equatorial hydrogens, so only one average signal is observed. Colors are used here to distinguish among environments and thus to indicate distinct chemical shifts.

CH_3CH_2Cl
b   a

is located *anti* with respect to the chlorine atom. We expect this particular nucleus to have a chemical shift different from the two *gauche* hydrogens (H_{b_1} and H_{b_2}). In fact, however, the NMR spectrometer cannot resolve that difference, because the fast rotation of the methyl group averages the signals for H_b. This rotation is said to be "fast on the NMR time scale." The resulting absorption appears at an average δ of the two signals expected for H_b.

In theory, it should be possible to slow the rotation in chloroethane by cooling the sample. In practice, "freezing" the rotation is very difficult to do, because the activation barrier to rotation is only a few kilocalories per mole. We would have to cool the sample to about −180°C, at which point most solvents would solidify—and ordinary NMR spectroscopy would not be possible.

A similar situation is encountered for cyclohexane. Here, fast conformational isomerism causes the axial hydrogens to be in equilibrium with the equatorial ones on the NMR time scale (Figure 10-14B); so, at room temperature, the NMR spectrum shows only one sharp line at δ = 1.36 ppm. However, in contrast with that for chloroethane, the process is slow enough at −90°C that, instead of a single absorption, two are observed: one for the six axial hydrogens at δ = 1.12 ppm, the other for the six equatorial hydrogens at δ = 1.60 ppm. The conformational isomerization in cyclohexane is frozen on the NMR time scale at this temperature because the activation barrier to ring flip is much higher [E_a = 10.8 kcal mol^{-1} (45.2 kJ mol^{-1}); Section 4-3] than the barrier to rotation in chloroethane.

In general, the lifetime of a molecule in an equilibrium must be of the order of about a second to allow its resolution by NMR. If the molecule is substantially shorter lived, an *average* spectrum is obtained. Organic chemists use such temperature-dependent NMR spectra to measure the rates of chemical processes and thus their activation parameters (Section 2-1). As a simple analogy, you can relate the phenomenon of NMR time scale to that of vision. If you consider your eyes a "spectrometer," you can "resolve" events only if they occur below a certain speed. Try moving one hand back and forth in front of you. At a rate of once per second, your view is sharp. Now try it at five times per second: Your hand will appear as an averaged blur (see also margin).

This ceiling fan is beginning to rotate fast enough to look blurred.

Exercise 10-8

How many signals would you expect in the ^{1}H NMR spectrum of bromocyclohexane? (**Caution:** Even considering fast ring flip, do the hydrogens located cis to the bromine ever become equivalent to those located trans? Build a model!)

Solved Exercise 10-9 | **Working with the Concepts: ¹H NMR Spectra of Two Stereoisomers**

How many signals would you expect in the ¹H NMR spectra of *cis-* and *trans-*1,2-dichloro-cyclobutane?

Strategy

As always with stereochemical problems, it is useful to build models. Draw both molecules and their stereochemistry using the hashed-wedged line notation. Then search for symmetry elements: mirror planes and rotational axes.

Solution

• Looking at the cis isomer, you will recognize that the molecule contains a mirror plane, bisecting the bonds C1–C2 and C3–C4, the left half (as we look at it in the first structure below) being mirrored by the right half, making it a meso stereoisomer (Section 5-6). This renders the corresponding pairs of hydrogens equivalent (as indicated in color). Note that the green hydrogens, located cis to the chlorine substituents, cannot be equivalent to the blue ones, which are located trans (see also Exercise 10-8). Therefore, we would expect three signals in the ¹H NMR spectrum of this isomer.

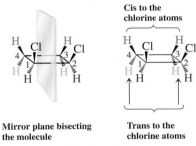

Mirror plane bisecting
the molecule

*cis-*1,2-Dichlorocyclobutane

• Turning to the trans isomer, there is no mirror plane, making the molecule chiral (Section 5-6), but there is an axis of rotation. This symmetry property divides the molecule again into three pairs of equivalent hydrogens (indicated in color), but now apportioned differently at C3 and C4: Unlike the cis isomer, the like-colored hydrogens have a trans relationship. The green hydrogens are unique, because they are positioned cis to the nearest-neighbor chlorine atom, trans to the remote one. In contrast, the blue hydrogens are located trans to the nearest chlorine and cis to the other. Thus, this compound, too, will give rise to three signals, but, because it is stereoisomeric to the cis isomer, with differing chemical shifts. On the other hand, its mirror image will be identical in its NMR spectral properties (Section 5-2).

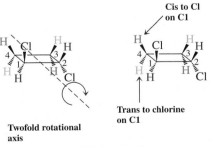

Twofold rotational
axis

*trans-*1,2-Dichlorocyclobutane

Exercise 10-10 | **Try It Yourself**

Revisit Section 5-5: How many signals would you expect in the ¹H NMR spectra of the two diastereomers of 2-bromo-3-chlorobutane? (**Caution:** Is there any symmetry in these molecules?)

REAL LIFE: MEDICINE 10-2 | Magnetic Resonance Imaging (MRI) in Medicine

After NMR spectroscopy was introduced to organic chemistry in the late 1960s, it did not take long for physicists and chemists to ask whether the technique could be used in medical diagnosis. In particular, if a spectrum is viewed as a sort of image of a molecule, why not image sections of the human (or animal) body? The answer emerged between the early 1970s and mid-1980s and relied not on the usual information embedded in chemical shifts, integration, and spin–spin splitting, but on a different phenomenon: *proton relaxation times*. Thus, the rate at which a hydrogen that has been induced to undergo the $\alpha \rightarrow \beta$ spin flip "relaxes" back to the α state is not constant but depends on the environment. Relaxation times can range from milliseconds to seconds and affect the line shapes of the corresponding signal. In the body, the hydrogens of water attached to the surface of biological molecules have been found to relax faster than those in the free fluid. In addition, there are slight differences depending on the nature of the tissue or structure to which the water is bound. For example, water in some cancerous tumors has a shorter relaxation time than that in healthy cells. These differences can be used to image the inside of the human body by *magnetic resonance imaging,* or *MRI*. In this application, a patient's entire body is placed within the poles of a large electromagnet, and proton NMR spectra are

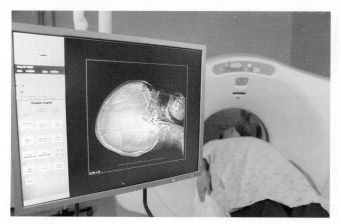

MRI brain scan.

collected and computer processed to give a series of cross-sectional plots of signal intensity. These cross-sectional plots are combined to produce a three-dimensional image of tissue proton density, as shown in the photo on the left.

Because most of the signal is due to water, variations from normal water-density patterns may be detected and used in diagnosis. Improvements during the last decade have shortened the time needed for analysis from minutes to seconds or less, permitting direct viewing of essentially every part of the body and monitoring changes in their environment instantaneously. Blood flow, kidney secretion, chemical imbalance, vascular condition, pancreatic abnormalities, cardiac function, and many other conditions of medical importance are now readily made visible. The Nobel prize in medicine was awarded in 2003 for the discovery of the use of MRI.* MRI is particularly helpful in detecting abnormalities that are not readily found by CAT (computerized axial tomography[†]) scans and conventional X-rays. Unlike other imaging methods, the technique is *noninvasive,* requiring neither ionizing radiation nor the injection of radioactive substances for visualization.

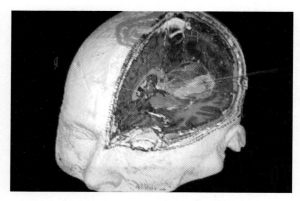

3D virtual reality map of a brain tumor (green), derived by combining multiple MRI scans. This image was used for computer-aided surgery; the red line to the right is the surgical point of attack.

*Professor Paul C. Lauterbur (1929–2007), University of Illinois at Urbana-Champaign, Illinois; Professor Sir Peter Mansfield (b. 1933), University of Nottingham, UK.

[†]Tomography is a method for taking pictures of a specific plane of an object.

In Summary The properties of symmetry, particularly mirror images and rotations, help to establish the chemical-shift equivalence or nonequivalence of the hydrogens in organic molecules. Those structures that undergo rapid conformational changes on the NMR time scale show only averaged spectra at room temperature. In some cases, these processes may be "frozen" at low temperatures to allow distinct absorptions to be observed.

10-6 | INTEGRATION

So far we have looked only at the *position* of NMR peaks. We shall see in this section that another useful feature of NMR spectroscopy is its ability to measure the relative integrated *intensity* of a signal, which is proportional to the relative number of nuclei giving rise to that absorption.

Integration reveals the relative number of hydrogens responsible for an NMR peak

The more hydrogens of one kind there are in a molecule, the more intense is the corresponding NMR absorption relative to the other signals. By measuring the area under a peak (the "integrated area") and comparing it with the corresponding peak areas of other signals, we can estimate the nuclear ratios quantitatively. For example, in the spectrum of 2,2-dimethyl-1-propanol (Figure 10-15A), three signals are observed, with relative areas of 9:2:1.

These numbers are obtained by computer and can be plotted on top of the regular spectrum by choosing the **integration** mode. In this mode, at the onset of an absorption peak, the recorder pen moves vertically upward a distance proportional to the area under the peak. It then again moves horizontally until the next peak is reached, and so forth. A ruler can be used to measure the distance by which the horizontal line is displaced at every

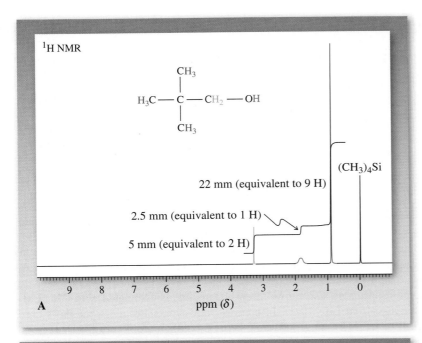

Figure 10-15 Integrated 300-MHz ^{1}H NMR spectra of (A) 2,2-dimethyl-1-propanol and (B) 1,2-dimethoxyethane, in $CDCl_3$ with added $(CH_3)_4Si$. In (A), the integrated areas measured by a ruler are 5:2.5:22 (in mm). Normalization through division by the smallest number gives a peak ratio of 2:1:9. Note that the integration gives only *ratios*, not absolute values for the number of hydrogens present in the sample. Thus, in (B), the integrated peak ratio is ~3:2, yet the compound contains hydrogens in a ratio of 6:4 (see also margin below).

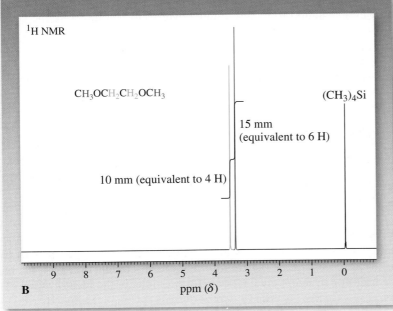

Other Molecules with NMR Integration Ratios of 3:2

CH_3OCH_2Cl

peak. *The relative sizes of these displacements furnish the **ratio** of hydrogens giving rise to the various signals.* Figure 10-15 depicts the ^{1}H NMR spectra of 2,2-dimethyl-1-propanol and 1,2-dimethoxyethane, including plots of the integration.

Such integrated plots are useful when the spectrum is unusually complex, either because the molecule has many types of hydrogens or because the sample is impure or a mixture, and visual inspection becomes advantageous. Normally, the computer provides digital readouts of integrated peak intensity automatically. Therefore, in subsequent spectra, these values will be given above the corresponding signals in numerical form.

Chemical shifts and peak integration can be used to determine structure

Consider the three products obtained in the monochlorination of 1-chloropropane, $CH_3CH_2CH_2Cl$. All have the same molecular formula $C_3H_6Cl_2$ and very similar physical properties (such as boiling points).

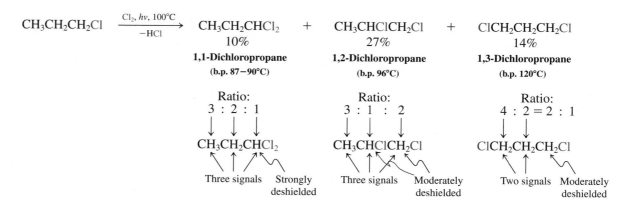

NMR spectroscopy clearly distinguishes all three isomers. 1,1-Dichloropropane contains three types of nonequivalent hydrogens, a situation giving rise to three NMR signals in the ratio of 3:2:1. The single hydrogen absorbs at relatively low field ($\delta = 5.93$ ppm) because of the cumulative deshielding effect of the two halogen atoms; the others absorb at relatively high field ($\delta = 1.01$ and 2.34 ppm).

1,2-Dichloropropane also shows three sets of signals associated with CH_3, CH_2, and CH groups (see also Real Life 10-3). In contrast, however, their chemical shifts are quite different: Each of the last two groups now bears a halogen atom and gives rise to a low-field signal as a result ($\delta = 3.68$ ppm for the CH_2, and $\delta = 4.17$ ppm for the CH). Only one signal, shown by integration to represent three hydrogens and therefore the CH_3 group, is at relatively high field ($\delta = 1.70$ ppm).

Finally, 1,3-dichloropropane shows only two peaks ($\delta = 3.71$ and 2.25 ppm) in a relative ratio of 2:1, a pattern clearly distinct from those of the other two isomers. By this means, the structures of the three products are readily assigned by a simple measurement.

Exercise 10-11

Chlorination of chlorocyclopropane gives three compounds of molecular formula $C_3H_4Cl_2$. Draw their structures and describe how you would differentiate them by ^{1}H NMR. (**Hint:** Look for symmetry. Use the deshielding effect of chlorine and integration.)

In Summary The NMR spectrometer in the integration mode records the relative areas under the various peaks, values that represent the relative numbers of hydrogens giving rise to these absorptions. This information, in conjunction with chemical shift data, can be used for structure elucidation—for example, in the identification of isomeric compounds.

| 10-7 | SPIN–SPIN SPLITTING: THE EFFECT OF NONEQUIVALENT NEIGHBORING HYDROGENS |

The high-resolution NMR spectra presented so far have rather simple line patterns—single sharp peaks, also called **singlets.** The compounds giving rise to these spectra have one feature in common: In each compound, nonequivalent hydrogens are separated by at least one carbon or oxygen atom. These examples were chosen for good reason, because neighboring hydrogen nuclei can complicate the spectrum as the result of a phenomenon called **spin–spin splitting** or **spin–spin coupling.**

Figure 10-16 shows that the NMR spectrum of 1,1-dichloro-2,2-diethoxyethane has four absorptions, characteristic of four sets of hydrogens (H_a–H_d). Instead of single peaks, they adopt more complex patterns called **multiplets:** two two-peak absorptions, or **doublets** (blue and green); one of four peaks, a **quartet** (black); and one of three peaks, a **triplet** (red). The detailed appearance of these multiplets depends on the number and kind of hydrogen atoms directly adjacent to the nuclei giving rise to the absorption.

In conjunction with chemical shifts and integration, spin–spin splitting frequently helps us arrive at a complete structure for an unknown compound. How can we interpret this information?

One neighbor splits the signal of a resonating nucleus into a doublet

Let us first consider the two doublets of relative integration 1, assigned to the two single hydrogens H_a and H_b. The splitting of these peaks is explained by the behavior of nuclei in an external magnetic field: They are like tiny magnets aligned with (α) or against (β) the field. The energy difference between the two states is minuscule (see Section 10-3), and at room temperature their populations are nearly equal. In the case under consideration, this means that there are two magnetic types of H_a—approximately half next to an H_b in the α state, the other half with a neighboring H_b in its β state. Conversely, H_b has two types of neighboring H_a—half of them in the α, half in the β state. H_a and H_b recognize that they are magnetic neighbors by communicating through the three bonds between them. What are the consequences of this phenomenon in the NMR spectrum?

A proton of type H_a that has as its neighbor an H_b aligned *with* the field is exposed to a total magnetic field that is strengthened by the addition of that due to the α spin of H_b. To achieve resonance for this type of H_a, a smaller external field strength is required than

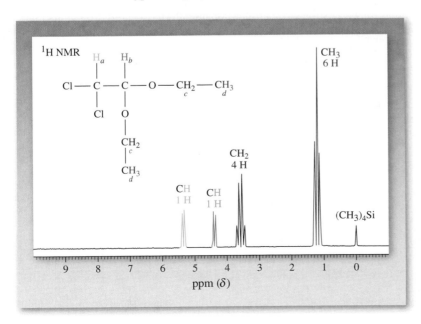

Figure 10-16 Spin–spin splitting in the 90-MHz ^1H NMR spectrum of 1,1-dichloro-2,2-diethoxyethane. The splitting patterns include two doublets, one triplet, and one quartet for the four types of protons. These multiplets reveal the effect of adjacent hydrogens. *Note:* The relative assignments of H_a and H_b are not obvious (see Table 10-3) and can be made only on consideration of additional data.

Figure 10-17 The effect of a hydrogen nucleus on the chemical shift of its neighbor is an example of spin–spin splitting. Two peaks are generated, because the hydrogen under observation has two types of neighbors. (A) When the neighboring nucleus H_b is in its α state, it adds a local field, h_{local}, to H_0, thereby resulting in a downfield shift of the H_a peak. (B) When the neighboring nucleus is in its β state, its local field opposes the external one, and as a result the H_a peak is shifted to higher field. (C) The observed peak pattern is a doublet.

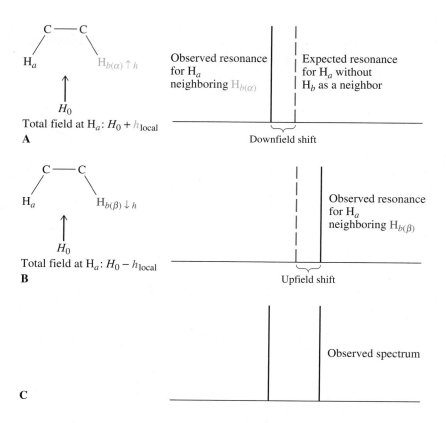

that necessary for H_a in the absence of a perturbing neighbor. A peak at lower field than that expected is observed (Figure 10-17A). However, this absorption is due to only half the H_a protons. The other half have H_b in its β state as a neighbor. Because H_b in its β state is aligned *against* the external field, the strength of the local field around H_a in this case is *diminished*. To achieve resonance, the external field H_0 has to be increased; an upfield shift is observed (Figure 10-17B). The resulting spectrum is a doublet (Figure 10-17C).

Because the local contribution of H_b to H_0, whether positive or negative, is of the same magnitude, the downfield shift of the hypothetical signal equals the upfield shift. The single absorption expected for a neighbor-free H_a is said to be *split* by H_b into a doublet. Integration of each peak of this doublet shows a 50% contribution of each hydrogen. The chemical shift of H_a is reported as the center of the doublet (Figure 10-18).

The signal for H_b is subject to similar considerations. This hydrogen also has two types of hydrogens as neighbors—$H_{a(\alpha)}$ and $H_{a(\beta)}$. Consequently its absorption lines appear in the form of a doublet. So, in NMR jargon, H_b is split by H_a. The amount of this mutual splitting is equal; that is, the distance (in hertz) between the individual peaks making up

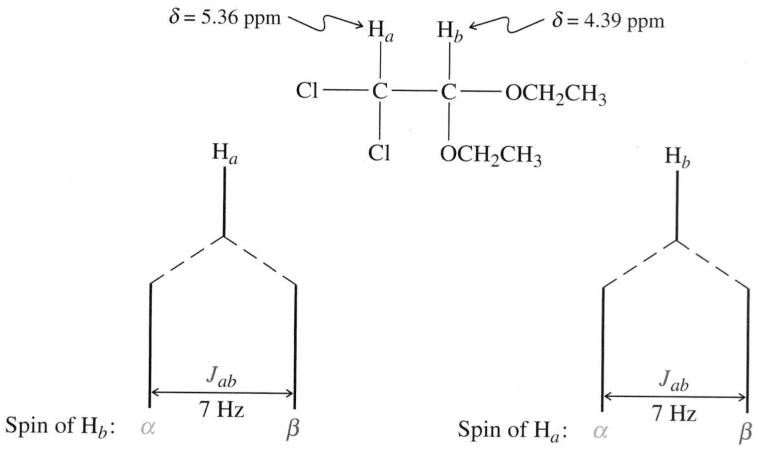

Figure 10-18 Spin–spin splitting between H_a and H_b in 1,1-dichloro-2,2-diethoxyethane. The coupling constant J_{ab} is the same for both doublets. The chemical shift is reported as the center of the doublet in the following format: $\delta_{H_a} = 5.36$ ppm (d, $J = 7$ Hz, 1 H), $\delta_{H_b} = 4.39$ ppm [d, $J = 7$ Hz, 1 H, in which "d" stands for the splitting pattern (doublet) and the last entry refers to the integrated value of the absorption].

each doublet is identical. This distance is termed the **coupling constant, *J*.** In our example, $J_{ab} = 7$ Hz (Figure 10-18). Because the coupling constant is related only to magnetic field contributions by neighboring nuclei communicating through intervening bonds, it is *independent of the external field strength.* Coupling constants *remain unchanged* regardless of the field strength of the NMR instrument being used.

Spin–spin splitting is generally observed only between hydrogens that are immediate neighbors, bound either to the same carbon atom [**geminal coupling** (*geminus*, Latin, twin)] or to two adjacent carbons [**vicinal coupling** (*vicinus*, Latin, neighboring)]. Hydrogen nuclei separated by more than two carbon atoms are usually too far apart to exhibit appreciable coupling. Moreover, *equivalent nuclei do not exhibit mutual spin–spin splitting.* For example, the NMR spectrum of ethane, CH_3–CH_3, consists of a *single line* at $\delta = 0.85$ ppm, similar to that of cyclohexane, C_6H_{12}, which exhibits a singlet at $\delta = 1.36$ ppm (at room temperature; see Section 10-5). Another example is 1,2-dimethoxyethane (Figure 10-15B), which gives rise to two singlets in NMR—one for the methyl groups and one for the chemical-shift–equivalent central methylene hydrogens. Splitting is observed only between nuclei with *different* chemical shifts.

Local-field contributions from more than one hydrogen are additive

How do we handle nuclei with two or more neighboring hydrogens? It turns out that we must consider the effect of each neighbor separately. Let us return to the spectrum of 1,1-dichloro-2,2-diethoxyethane shown in Figure 10-16. In addition to the two doublets assignable to H_a and H_b, this spectrum exhibits a triplet due to the methyl protons H_d and a quartet assignable to the methylene hydrogens H_c. Because these two nonequivalent sets of nuclei are next to each other, vicinal coupling is observed as expected. However, compared with the peak patterns for H_a and H_b, those for H_c and H_d are considerably more complicated. They can be understood by expanding on the explanation used for the mutual coupling of H_a and H_b.

Consider first the triplet whose chemical shift and integrated value allow it to be assigned to the hydrogens H_d of the two methyl groups. Instead of one peak, we observe three, in the approximate ratio $1:2:1$. The splittings must be due to coupling to the adjacent methylene groups—but how?

The three equivalent methyl hydrogens in each ethoxy group have two equivalent methylene hydrogens as their neighbors, and each of these methylene hydrogens may adopt the α or β spin orientation. Thus, each H_d may "see" its two H_c neighbors as an $\alpha\alpha$, $\alpha\beta$, $\beta\alpha$, or $\beta\beta$ combination (Figure 10-19). Those methyl hydrogens that are adjacent to the first possibility, $H_{c(\alpha\alpha)}$, are exposed to a twice-strengthened local field and give rise to a lower-field absorption. In the $\alpha\beta$ or $\beta\alpha$ combination, one of the H_c nuclei is aligned with the external field and the other is opposed to it. The net result is no net local-field contribution at H_d. In these cases, a spectral peak should appear at a chemical shift identical with the one expected if there were no coupling between H_c and H_d. Moreover, because *two* equivalent combinations of neighboring H_cs ($H_{c(\alpha\beta)}$ and $H_{c(\beta\alpha)}$) contribute to this signal (instead of only one, as did $H_{c(\alpha\alpha)}$ to the first peak), its approximate height should be double that of the first peak, as observed. Finally, H_d may have the $H_{c(\beta\beta)}$ combination as its neighbor. In this case, the local field subtracts from the external field

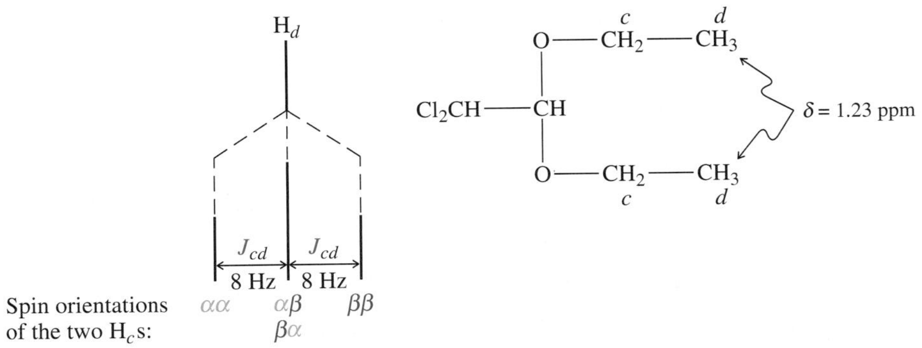

Coupling Between Close-Lying Hydrogens

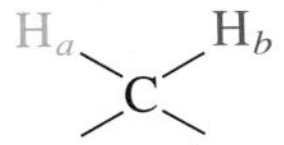

J_{ab}, geminal coupling, variable 0–18 Hz

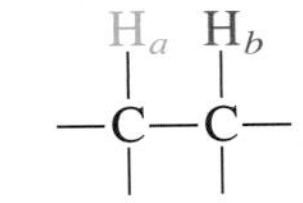

J_{ab}, vicinal coupling, typically 6–8 Hz

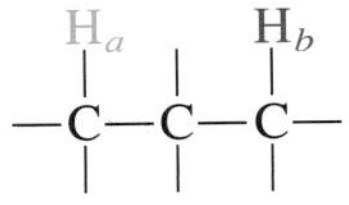

J_{ab}, 1,3-coupling, usually negligible

Spin orientations of the two H_cs:

Figure 10-19 Nucleus H_d is represented by a three-peak NMR pattern because of the presence of three magnetically nonequivalent neighbor combinations: $H_{c(\alpha\alpha)}$, $H_{c(\alpha\beta}$ and $_{\beta\alpha)}$, and $H_{c(\beta\beta)}$. The chemical shift of the absorption is reported as that of the center line of the triplet: $\delta_{H_d} = 1.23$ ppm (t, $J = 8$ Hz, 6 H, in which "t" stands for triplet).

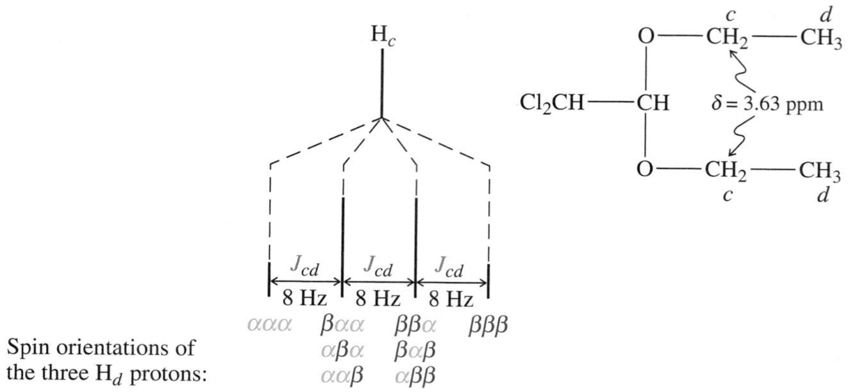

Figure 10-20 Splitting of H_c into a quartet by the various spin combinations of H_d. The chemical shift of the quartet is reported as its midpoint: δ_{H_c} = 3.63 ppm (q, J = 8 Hz, 4 H, in which "q" stands for quartet).

and an upfield peak of relative intensity 1 is produced. The resulting pattern for H_d is a *1:2:1 triplet* with a total integration corresponding to six hydrogens (because there are two methyl groups). The coupling constant J_{cd}, measured as the distance between each pair of adjacent peaks, is 8 Hz.

The quartet observed in Figure 10-16 for H_c can be analyzed in the same manner (Figure 10-20). This nucleus is exposed to four different types of H_d proton combinations as neighbors: one in which all protons are aligned with the field ($H_{d(\alpha\alpha\alpha)}$); three equivalent arrangements in which one H_d is opposed to the external field and the other two are aligned with it ($H_{d(\beta\alpha\alpha, \alpha\beta\alpha, \alpha\alpha\beta)}$); another set of three equivalent arrangements in which only one proton remains aligned with the field ($H_{d(\beta\beta\alpha, \beta\alpha\beta, \alpha\beta\beta)}$); and a final possibility in which all H_ds oppose the external magnetic field ($H_{d(\beta\beta\beta)}$). The resulting spectrum is predicted—and observed—to consist of a *1:3:3:1 quartet* (integrated intensity 4). The coupling constant J_{cd} is identical with that measured in the triplet for H_d (8 Hz).

In many cases, spin–spin splitting is given by the $N + 1$ rule

We can summarize our analysis so far by a set of simple rules:

1. Equivalent nuclei located adjacent to one neighboring hydrogen resonate as a *doublet*.

2. Equivalent nuclei located adjacent to two hydrogens of a second set of equivalent nuclei resonate as a *triplet*.

3. Equivalent nuclei located adjacent to a set of three equivalent hydrogens resonate as a *quartet*.

Table 10-4 shows the expected splitting patterns for nuclei adjacent to N equivalent neighbors. The NMR signals of these nuclei *split into $N + 1$ peaks,* a result known as the **$N + 1$ rule.** Their relative ratio is given by a mathematical mnemonic device called

Note: The splitting pattern of a hydrogen NMR signal gives you the number of *neighboring* hydrogens. It provides no information about the absorbing hydrogen itself.

Table 10-4	NMR Splittings of a Set of Hydrogens with *N* Equivalent Neighbors and Their Integrated Ratios (Pascal's Triangle)		
Equivalent neighboring (*N*) hydrogens	Number of peaks (*N* + 1)	Name for peak pattern (abbreviation)	Integrated ratios of individual peaks
0	1	Singlet (s)	1
1	2	Doublet (d)	1 : 1
2	3	Triplet (t)	1 : 2 : 1
3	4	Quartet (q)	1 : 3 : 3 : 1
4	5	Quintet (quin)	1 : 4 : 6 : 4 : 1
5	6	Sextet (sex)	1 : 5 : 10 : 10 : 5 : 1
6	7	Septet (sep)	1 : 6 : 15 : 20 : 15 : 6 : 1

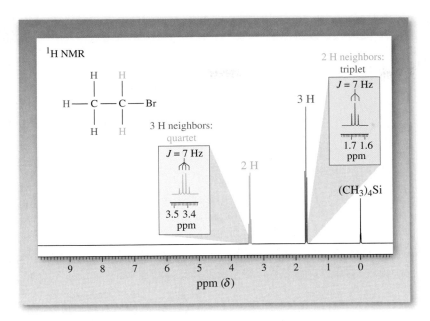

Figure 10-21 The 300-MHz ^{1}H NMR spectrum of bromoethane illustrates the $N + 1$ rule. The methylene group, which has three equivalent neighbors, appears as a quartet at $\delta = 3.43$ ppm, $J = 7$ Hz. The methyl hydrogens, which have two equivalent neighbors, absorb as a triplet at $\delta = 1.67$ ppm, $J = 7$ Hz.

Pascal's* triangle. Each number in this triangle is the sum of the two numbers closest to it in the line above. The splitting patterns of two common alkyl groups, ethyl and 1-methylethyl (isopropyl), are shown in Figures 10-21 and 10-22, respectively. In both spectra, the relative intensities of the individual peaks of the respective multiplets conform (roughly) to those predicted by Pascal's triangle. As a result, the outermost lines of the septet for the central hydrogen in 2-iodopropane (Figure 10-22) are barely visible and readily missed. This problem is general for hydrogen signals split by couplings to multiple neighbors, and you need to be careful in their interpretation. The task is aided by integration of the multiplets, which reveals the relative number of associated hydrogens.

It is important to remember that neighboring nuclei split one another *mutually*. In other words, the observation of one split absorption necessitates the presence of another split signal in the spectrum. Moreover, the coupling constants for these patterns must be the same. Some frequently encountered multiplets and the corresponding structural units are shown in Table 10-5.

*Blaise Pascal (1623–1662), French mathematician, physicist, and religious philosopher.

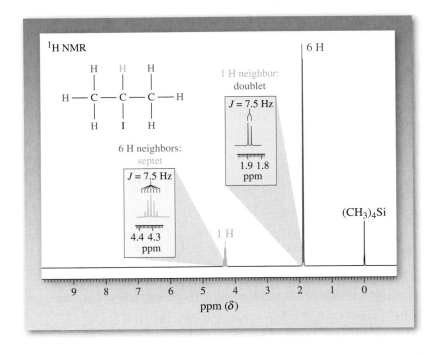

Figure 10-22 300-MHz ^{1}H NMR spectrum of 2-iodopropane: $\delta = 4.31$ (sep, $J = 7.5$ Hz, 1 H), 1.88 (d, $J = 7.5$ Hz, 6 H) ppm. The six equivalent nuclei on the two methyl groups give rise to a septet for the tertiary hydrogen ($N + 1$ rule).

Table 10-5	Frequently Observed Spin–Spin Splittings in Common Alkyl Groups		

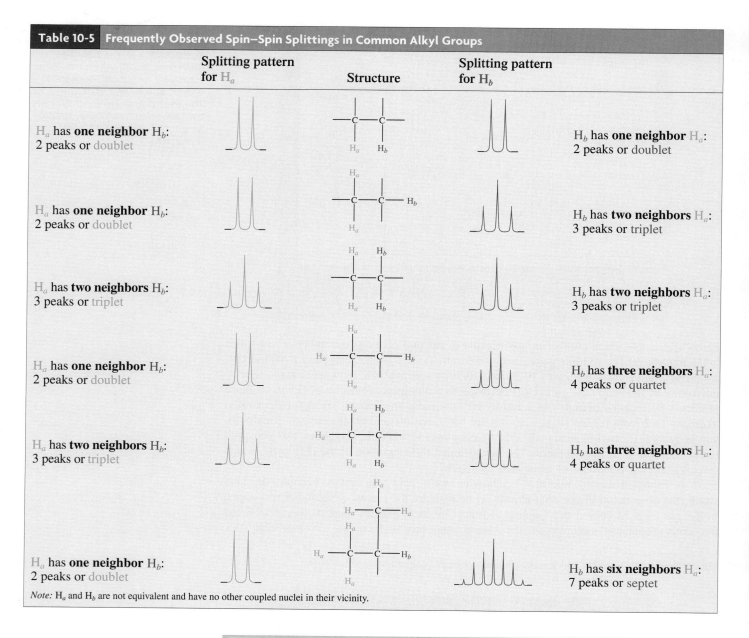

Splitting pattern for H_a	Structure	Splitting pattern for H_b

H_a has **one neighbor** H_b: 2 peaks or doublet — H_b has **one neighbor** H_a: 2 peaks or doublet

H_a has **one neighbor** H_b: 2 peaks or doublet — H_b has **two neighbors** H_a: 3 peaks or triplet

H_a has **two neighbors** H_b: 3 peaks or triplet — H_b has **two neighbors** H_a: 3 peaks or triplet

H_a has **one neighbor** H_b: 2 peaks or doublet — H_b has **three neighbors** H_a: 4 peaks or quartet

H_a has **two neighbors** H_b: 3 peaks or triplet — H_b has **three neighbors** H_a: 4 peaks or quartet

H_a has **one neighbor** H_b: 2 peaks or doublet — H_b has **six neighbors** H_a: 7 peaks or septet

Note: H_a and H_b are not equivalent and have no other coupled nuclei in their vicinity.

Exercise 10-12

Predict the NMR spectra of (**a**) ethoxyethane (diethyl ether); (**b**) 1,3-dibromopropane; (**c**) 2-methyl-2-butanol; (**d**) 1,1,2-trichloroethane. Specify approximate chemical shifts, relative abundance (integration), and multiplicities.

Solved Exercise 10-13 | Working with the Concepts: Using Chemical Shift, Integration, and Spin–Spin Splitting in Structure Elucidation

There are several isomeric alcohols and ethers of molecular formula $C_5H_{12}O$ (see, for example, 2,2-dimethyl-1-propanol; Figures 10-11 and 10-15; and Problem 47). Two of these, A and B, exhibit the following ^1H NMR spectra.

A: $\delta = 1.19$ (s, 9 H), 3.21 (s, 3 H) ppm
B: $\delta = 0.93$ (t, 3 H), 1.20 (t, 3 H), 1.60 (sextet, 2 H), 3.37 (t, 2 H), 3.47 (q, 2 H) ppm

Deduce the structures of both isomers.

Strategy

What you are asked to do is to come up with two structures that match both the given composition and the respective spectroscopic data. Your answers must be *completely* consistent with all the given information. Agreement between the structures you suggest and just some of the data is not sufficient.

How to begin? You may be tempted to solve this problem by drawing all isomers of $C_5H_{12}O$ and then trying to match each structure with the given NMR spectra. This approach is very time consuming. Instead, it is better to extract as much information as possible from the spectra before formulating trial structures. In particular, often we can immediately associate certain patterns with certain pieces (substructures) of the molecule (Table 10-5). Once identified, we can subtract the atoms in the substructures from $C_5H_{12}O$, leaving smaller fragments to consider. Finally, piece together your fragments to give the answer most consistent with all the information.

Information needed? Refer to material in the chapter, especially Tables 10-2 and 10-5, and Sections 10-5 and 10-7.

Proceed.

Solution

Let us start with A.
- The presence of two singlets is an indication of extensive symmetry (Section 10-5).
- The absence of a peak integrating for 1 H excludes the presence of an OH function; hence, the molecule is an ether.
- A singlet integrating for 3 H is a strong indication of the presence of a CH_3 substituent. The singlet's chemical shift suggests that it is attached to the ether oxygen (Table 10-2).
- Subtracting CH_3O from $C_5H_{12}O$ leaves C_4H_9, which must be the source of the other singlet at higher field, in the alkane region (Table 10-2).
- A singlet integrating for 9 H is a strong indication for the presence of three equivalent CH_3 substituents: The solution is *tert*-butyl, $C(CH_3)_3$. Combining these pieces provides the answer for A: $CH_3OC(CH_3)_3$. (Familiar? See the Chapter Opening).

Turning to B, we can follow the same kind of analysis.
- This molecule exhibits five signals, all of which are split. Again, there is no single hydrogen peak, ruling out the presence of a hydroxy group.
- We note that two of the δ values are relatively large, identifying the two pieces attached to oxygen: the triplet for 2 H at $\delta = 3.37$ ppm and the quartet for 2 H at $\delta = 3.47$ ppm. This indicates an unsymmetrical substructure, $X–CH_2OCH_2–Y$.
- We can guess the nature of the neighbors X and Y from the coupling patterns of the CH_2 groups: One must be CH_3 to cause the appearance of a quartet; the other must be another CH_2 fragment to generate a triplet. Thus, you could consider strongly $CH_3CH_2OCH_2CH_2$—as a substructure.
- Subtracting this fragment from $C_5H_{12}O$ leaves only CH_3 as a final piece; hence a possible solution is $CH_3CH_2OCH_2CH_2CH_3$. Does this structure fit the other data?
- Looking to higher field, we have two triplets, each arising from three equivalent hydrogens. This is a "dead giveaway" for two separate CH_3 groups, each attached to a CH_2 neighbor—in other words, two ethyl moieties. Finally, a sextet integrating for 2 H indicates the presence of a CH_2 group with five hydrogen neighbors. Both of these pieces of information are consistent with the proposed solution, leaving $CH_3CH_2OCH_2CH_2CH_3$, as the only viable structure.

Exercise 10-14 | Try It Yourself

Another isomer of $C_5H_{12}O$ exhibits the following ^1H NMR spectrum:

$\delta = 0.92$ (t, 3 H), 1.20 (s, 6 H), 1.49 (q, 2 H), 1.85 (br s, 1 H) ppm

What is its structure? (**Hint:** The singlet at 1.85 ppm is broad.)

In Summary Spin–spin splitting occurs between vicinal and geminal nonequivalent hydrogens. Usually, N equivalent neighbors will split the absorption of the observed hydrogen into $N + 1$ peaks, their relative intensities being in accord with Pascal's triangle. The common alkyl groups give rise to characteristic NMR patterns.

10-8 | SPIN–SPIN SPLITTING: SOME COMPLICATIONS

The rules governing the appearance of split peaks outlined in Section 10-7 are somewhat idealized. Thus, when there is a relatively small difference in δ between two absorptions, more complicated patterns (complex multiplets) are observed that are not interpretable without the use of computers. Moreover, the $N + 1$ rule may not be applicable in a direct way if two or more types of neighboring hydrogens are coupled to the resonating nucleus with fairly different coupling constants. Finally, the hydroxy proton may appear as a singlet (see Figure 10-11) even if vicinal hydrogens are present. Let us look in turn at each of these complications.

Close-lying peak patterns may give rise to non-first-order spectra

A careful look at the spectra shown in Figures 10-16, 10-21, and 10-22 reveals that the relative intensities of the splitting patterns do not conform to the idealized peak ratios expected from consideration of Pascal's triangle: The patterns are not completely symmetric, but skewed. Specifically, the two multiplets of two mutually coupled hydrogens are skewed toward each other such that the intensity of the lines facing each other is slightly larger than expected. The exact intensity ratios dictated by Pascal's triangle and the $N + 1$ rule are observed only when the difference between the resonance frequencies of coupled protons is much larger than their coupling constant: $\Delta\nu \gg J$. Under these circumstances, the spectrum is said to be **first order.*** However, when these differences become smaller, the expected peak pattern is subject to increasing distortion.

In extreme cases, the simple rules devised in Section 10-7 do not apply any more, the resonance absorptions assume more complex shapes, and the spectra are said to be **non-first order.** Although such spectra can be simulated with the help of computers, this treatment is beyond the scope of the present discussion.

Particularly striking examples of non-first-order spectra are those of compounds that contain alkyl chains. Figure 10-23 shows an NMR spectrum of octane, which is not first order because all nonequivalent hydrogens (there are four types) have very similar chemical shifts. All the methylenes absorb as one broad multiplet. In addition, there is a highly distorted triplet for the terminal methyl groups.

Non-first-order spectra arise when $\Delta\nu \approx J$, so it should be possible to "improve" the appearance of a multiplet by measuring a spectrum at higher field, because the resonance frequency is proportional to the external field strength, whereas the coupling constant J is independent of field (Section 10-7). Thus, at increasingly higher field, the individual multiplets become more and more separated (resolved), and the non-first-order effect of a close-lying absorption gradually disappears. The effect is similar to that of a magnifying glass in the observation of ordinary objects. The eye has a relatively low resolving power and cannot discern the fine structure of a sample that becomes evident only on appropriate magnification.

Greater field strength has a dramatic effect on the spectrum of 2-chloro-1-(2-chloroethoxy)ethane (Figure 10-24). In this compound, the deshielding effect of the oxygen

---

*This expression derives from the term *first-order theory,* that is, one that takes into account only the most important variables and terms of a system.

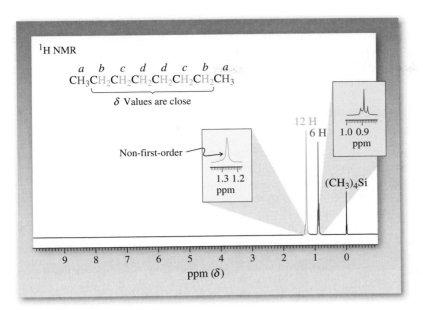

Figure 10-23 300-MHz ^{1}H NMR spectrum of octane. Compounds that contain alkyl chains often display such non-first-order patterns.

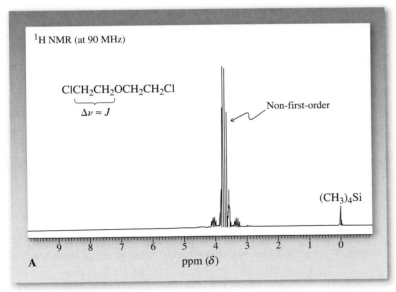

A

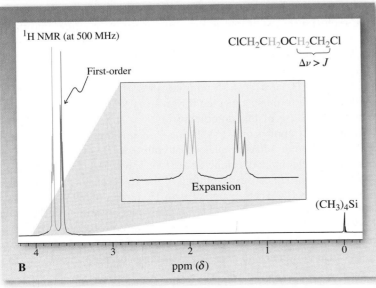

B

Figure 10-24 The effect of increased field strength on a non-first-order NMR spectrum: 2-chloro-1-(2-chloroethoxy)ethane at (A) 90 MHz; (B) 500 MHz. At high field strength, the complex multiplet observed at 90 MHz is simplified into two slightly distorted triplets, as might be expected for two mutually coupled CH_2 groups.

is about equal to that of the chlorine substituent. As a consequence, the two sets of methylene hydrogens give rise to very close-lying peak patterns. At 90 MHz, the resulting absorption has a symmetric shape but is very complicated, exhibiting more than 32 peaks of various intensities. However, recording the NMR spectrum with a 500-MHz spectrometer (Figure 10-24B) produces a first-order pattern.

Coupling to nonequivalent neighbors may modify the simple N + 1 rule

When hydrogens are coupled to two sets of nonequivalent neighbors, complicated splitting patterns may result. The spectrum of 1,1,2-trichloropropane illustrates this point (Figure 10-25). In this compound, the hydrogen at C2 is located between a methyl and a CHCl$_2$ group, and it is coupled to the hydrogens of each group independently and with different coupling constants.

> **Sequential N + 1 Rule**
>
> When there are two types of neighboring hydrogens, numbering N and n, with two differing J values, J_1 and J_2, respectively, the simple $N + 1$ rule becomes:
>
> Number of peaks = $(N_{J_1} + 1) \times (n_{J_2} + 1)$

Figure 10-25 300-MHz ^{1}H NMR spectrum of 1,1,2-trichloropropane. Nucleus H$_b$ gives rise to a quartet of doublets at $\delta = 4.35$ ppm: eight peaks.

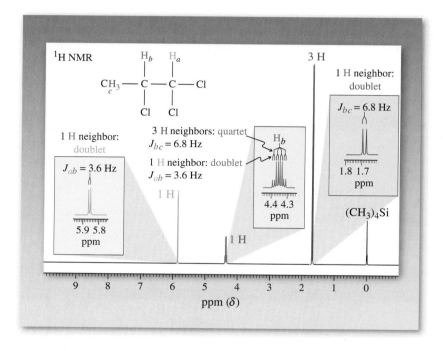

Let us analyze the spectrum in detail. We first notice two doublets, one at low field ($\delta = 5.86$ ppm, $J = 3.6$ Hz, 1 H) and one at high field ($\delta = 1.69$ ppm, $J = 6.8$ Hz, 3 H). The low-field absorption is assignable to the hydrogen at C1 (H$_a$), adjacent to two deshielding halogens; and the methyl hydrogens (H$_c$) resonate as expected at highest field. In accord with the $N + 1$ rule, each signal is split into a doublet because of coupling with the hydrogen at C2 (H$_b$). The resonance of H$_b$, however, is quite different in appearance from what we expect. The nucleus giving rise to this absorption has a total of four hydrogens as its neighbors: H$_a$ and three H$_c$. Application of the $N + 1$ rule suggests that a quintet should be observed. However, the signal for H$_b$ at $\delta = 4.35$ ppm consists of *eight* lines, with relative intensities that do not conform to those expected for ordinary splitting patterns (see Tables 10-4 and 10-5). What is the cause of this complexity?

The $N + 1$ rule strictly applies only to splitting by *equivalent* neighbors. In this molecule, we have two sets of different adjacent nuclei that couple to H$_b$ *with different coupling constants*. The effect of these couplings can be understood, however, if we apply the $N + 1$ rule sequentially. The methyl group causes a splitting of the H$_b$ resonance into a quartet, with the relatively larger $J_{bc} = 6.8$ Hz. Then, coupling with H$_a$ further splits *each peak* in this quartet into a doublet, with the smaller $J_{ab} = 3.6$ Hz, both splits resulting in the observed eight-line pattern (Figure 10-26). The hydrogen at C2 is said to be split into a quartet of doublets.

The hydrogens on C2 of 1-bromopropane also couple to two nonequivalent sets of neighbors. In this case, however, the resulting splitting pattern appears to conform with the

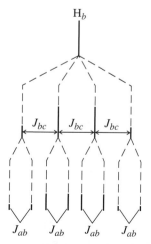

Figure 10-26 The splitting pattern for H$_b$ in 1,1,2-trichloropropane follows the sequential $N + 1$ rule. Each of the four lines arising from coupling to the methyl group is further split into a doublet by the hydrogen on C1.

REAL LIFE: SPECTROSCOPY 10-3 | The Nonequivalence of Diastereotopic Hydrogens

You have probably assumed that all methylene (CH$_2$) groups contain equivalent hydrogens, which give rise to only one signal in the NMR spectrum. Such is indeed the case when there is a symmetry element present that renders them equivalent, such as a mirror plane or axis of rotation. For example, the methylene hydrogens in butane or cyclohexane have this property. However, this symmetry is readily removed by substitution (Chapters 3 and 5), a change that has profound effects not only with respect to stereoisomerism (Chapter 5), but also to NMR spectroscopy.

Consider, for example, turning cyclohexane into bromocyclohexane, such as by radical bromination. Not only does this alteration make C1, C2, C3, and C4 different, but also all of the hydrogens cis to the bromine substituent have become different from those trans. In other words, the CH$_2$ groups exhibit nonequivalent geminal hydrogens. These hydrogens have separate chemical shifts and show mutual geminal (as well as vicinal) coupling. The resulting 300-MHz spectrum is quite complex, as shown on the right. Only the C1 hydrogen at $\delta = 4.17$ ppm, deshielded by the neighboring bromine, is readily assigned.

Methylene hydrogens that are not equivalent are called **diastereotopic.** This word is based on the stereochemical outcome of replacing either of these hydrogens by a substituent: diastereomers. For example, in the case of bromocyclohexane, substituting the wedged (red) hydrogen at C2 results in a cis product. The same alteration involving the hashed (blue) hydrogen generates the trans isomer. You can verify the same outcome of this procedure when applied to C3 and C4, respectively.

You might think that it is the rigidity of the cyclic frame that imparts this property to diastereotopic hydrogens. However, this is not true, as you can realize by reviewing Section 5-5. The chlorination of 2-bromobutane at C3 resulted in two diastereomers; therefore, the methylene group in 2-bromobutane also contains diastereotopic protons. We can recognize and generalize the origin of this effect: The presence of the stereocenter precludes a mirror plane through the CH$_2$ carbon, and rotation is ineffective in making the two hydrogens equivalent. To illustrate the presence of diastereotopic hydrogens in such chiral, acyclic molecules, we return to one of the three monochlorination products of 1-chloropropane, discussed in Section 10-6, namely, 1,2-dichloropropane. The 300-MHz ^{1}H NMR spectrum shown reveals *four* signals (instead of the normally expected three), two of which are due to the diastereotopic hydrogens at C1. Specifically, these hydrogens appear as doublets of doublets at $\delta = 3.58$ and 3.76 ppm, because they coupled to each other with $J_{\text{geminal}} = 10.8$ Hz, and then with individual coupling constants to the hydrogen at C2 ($J_{\text{vicinal}} = 4.7$ and 9.1 Hz, respectively).

Frequently, diastereotopic hydrogens have such close chemical shifts that their nonequivalence is not visible in an NMR spectrum. For example, in the chiral 2-bromohexane, all three methylene groups contain diastereotopic hydrogens. In this compound, only the two methylenes closest to the stereocenter reveal their presence. The third one is too far removed for the asymmetry of the stereocenter to have a measurable effect.

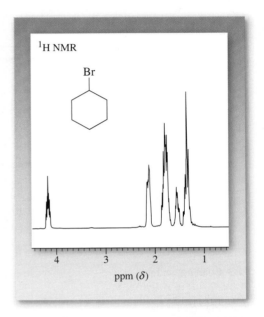

Bromocyclohexane

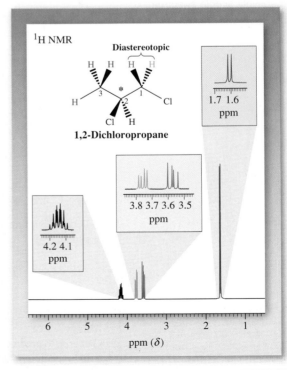

1,2-Dichloropropane

Figure 10-27 300-MHz ^{1}H NMR spectrum of 1-bromopropane.

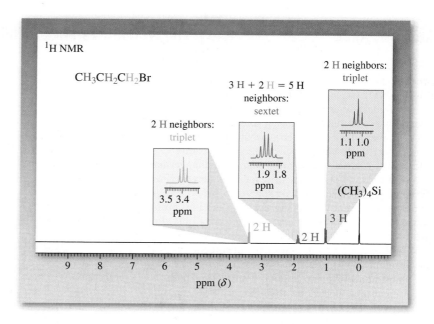

$N + 1$ rule, and a (slightly distorted) sextet is observed (Figure 10-27). The reason is that the coupling constants to the two different groups are very similar, about 6–7 Hz. Although an analysis similar to that given earlier for 1,1,2-trichloropropane would lead us to predict as many as 12 lines in this signal (a quartet of triplets), the nearly equal coupling constants cause many of the lines to overlap, thus simplifying the pattern (Figure 10-28). The hydrogens in many simple alkyl derivatives display similar coupling constants and, therefore, spectra that are in accord with the $N + 1$ rule.

Figure 10-28 Splitting pattern expected for H$_b$ in a propyl derivative when $J_{ab} \approx J_{bc}$. Several of the peaks coincide, giving rise to a deceptively simple spectrum: a sextet.

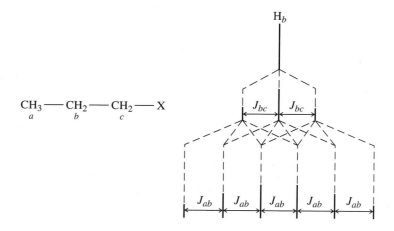

Solved Exercise 10-15 | Working with the Concepts: Applying the $N + 1$ Rule

Predict the coupling patterns for the boldface hydrogen in the structure shown, first according to the simple $N + 1$ rule and then according to the sequential $N + 1$ rule.

$$H_3C-\underset{\underset{\textbf{H}}{|}}{\overset{\overset{CH_3}{|}}{C}}-CH_2-OH$$

Strategy

In general, to predict splitting patterns of specific hydrogens, you need to be clear about the identity of their "neighborhood." It is therefore often useful to build a model and to draw the structure out fully, showing all hydrogens. In this way, you can recognize symmetry and establish the different types of hydrogen neighbors and their relative abundance.

Solution

- Applying the simple $N + 1$ rule is easy: All we have to do is count the number of neighbors to the bold hydrogen to determine N. In this case, we have two CH_3 units and one CH_2 unit, adding up to 8 H. Therefore, the bold hydrogen should appear as a nonet or nine lines, their relative intensities following Pascal's triangle (Table 10-4): $1:8:28:56:70:56:28:8:1$. In practice, you don't have to figure out these exact numbers, and it suffices to look for a symmetrical pattern around the most intense center line, ascending from the left and descending to the right. You would need to amplify the spectrum to make sure that you don't miss the outermost lines.

    However, a symmetrical nonet is what is expected only if the coupling constants to the methyl and methylenes hydrogens, respectively, are the same.

- The sequential $N + 1$ rule applies if the J values to unequal neighbors differ. In that case, we first determine the splitting pattern expected if there was only one type of neighbor present. For example, if we pick the CH_3 groups, it would be 6 H, giving rise to a septet. We then apply the additional splitting caused by another neighbor, in our case the CH_2 group, and split each line of the original septet into a triplet. The result is a septet of triplets, 21 lines. This sequence is arbitrary and could have been reversed, resulting in a triplet of septets, the same 21 lines. In practice, the coupling pattern with the larger J value is quoted first, followed by the others, in order of decreasing size. You may be curious what the actual experimental splitting pattern looks like: It is a nonet, with $J = 6.5$ Hz.

Exercise 10-16 | Try It Yourself

Predict the coupling patterns for the boldface hydrogens in the structures below, first according to the simple $N + 1$ rule and then according to the sequential $N + 1$ rule.

(a) $BrCH_2CH_2CH_2Cl$

(b) $CH_3CHCHCl_2$
       |
      OCH_3

(c) $Cl_2CHCHCHCH_3$
     | |
   CH_3S Br

(d)

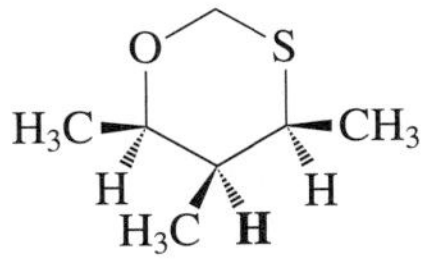

Exercise 10-17 |

In Section 10-6 you learned how to distinguish among the three monochlorination products of 1-chloropropane—1,1-, 1,2-, and 1,3-dichloropropane—by using just chemical-shift data and chemical integration. Would you also expect to tell them apart on the basis of their coupling patterns?

Fast proton exchange decouples hydroxy hydrogens

With our knowledge of vicinal coupling, let us now return to the NMR of alcohols. We note in the NMR spectrum of 2,2-dimethyl-1-propanol (Figure 10-11) that the OH absorption appears as a single peak, devoid of any splitting. This is curious, because the hydrogen is adjacent to two others, which should cause its appearance as a triplet. The CH_2 hydrogens that show up as a singlet should in turn appear as a doublet with the same coupling constant. Why, then, do we not observe spin–spin splitting? The reason is that the weakly acidic OH hydrogens are rapidly transferring, both between alcohol molecules and to traces of water, on the NMR time scale at room temperature. As a consequence of this process, the NMR spectrometer sees only an average signal for the OH hydrogen. No coupling is visible, because the binding time of the proton to the oxygen is too short (about 10^{-5} s). It follows that the CH_2 nuclei are similarly uncoupled, a condition resulting in the observed singlet.

Rapid Proton Exchange Between Alcohols with Various CH_2–α,β Spin Combinations Averages δ_{OH}

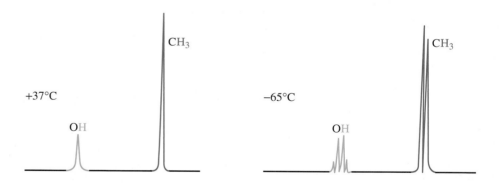

Rapid α-Proton with β-Proton Exchange Averages δ_{CH_2}

Absorptions of this type are said to be **decoupled** by **fast proton exchange.** The exchange may be slowed by removal of traces of water or acid or by cooling. In these cases, the OH bond retains its integrity long enough (more than 1 s) for coupling to be observed on the NMR time scale. An example is shown in Figure 10-29 for methanol. At 37°C, two singlets are observed, corresponding to the two types of hydrogens, both devoid of spin–spin splitting. However, at −65°C, the expected coupling pattern is detectable: a quartet and a doublet.

Figure 10-29 Temperature dependence of spin–spin splitting in methanol. The singlets at 37°C illustrate the effect of fast proton exchange in alcohols. (*After H. Günther, NMR-Spektroskopie, Georg Thieme Verlag, Stuttgart, 1973.*)

Rapid magnetic exchange "self-decouples" chlorine, bromine, and iodine nuclei

All halogen nuclei are magnetic. Therefore, the ^1H NMR spectra of haloalkanes would be expected to exhibit spin–spin splitting owing to the presence of these nuclei (in addition to the normal H–H coupling). In practice, only fluorine exhibits this effect, in much the same way as the proton does but with much larger J values. Thus, for example, the ^1H NMR spectrum of CH_3F exhibits a doublet with $J = 81$ Hz. Because fluoroorganic compounds are a relatively specialised area, we shall not deal with their NMR spectroscopy any further.

Turning to the other halides, inspection of the spectra of the haloalkanes depicted in Figures 10-16, 10-21, 10-22, 10-24, 10-25, and 10-27 (fortunately) reveals the *absence* of any visible spin–spin splitting by these nuclei. The reason for this observation lies in their relatively fast internal magnetic equilibration on the NMR time scale, precluding their recognition by the adjacent hydrogens as having differing alignments with respect to the external magnetic field H_0. They "self-decouple," in contrast with the "exchange decoupling" exhibited by the hydroxy protons.

In Summary The peak patterns in many NMR spectra are not first order, because the differences between the chemical shifts of nonequivalent hydrogens are close to the values of the corresponding coupling constants. Use of higher-field NMR instruments may improve the appearance of such spectra. Coupling to nonequivalent hydrogen neighbors occurs separately, with different coupling constants. In some cases, they are sufficiently dissimilar to allow for an analysis of the multiplets. In many simple alkyl derivatives, they are sufficiently similar ($J = 6$–7 Hz) that the spectra observed are simplified to those predicted in accordance with the $N + 1$ rule. Vicinal coupling through the oxygen in alcohols is frequently not observed because of decoupling by fast proton exchange.

10-9 | CARBON-13 NUCLEAR MAGNETIC RESONANCE

Proton nuclear magnetic resonance is a powerful method for determining organic structures because most organic compounds contain hydrogens. Of even greater potential utility is NMR spectroscopy of carbon. After all, by definition, *all* organic compounds contain this element. In combination with ^{1}H NMR, it has become the most important analytical tool in the hands of the organic chemist. We shall see in this section that ^{13}C NMR spectra are much simpler than ^{1}H NMR spectra, because we can avoid the complications of spin–spin splitting.

Carbon NMR utilizes an isotope in low natural abundance: ^{13}C

Carbon NMR is possible. However, there is a complication: The most abundant isotope of carbon, carbon-12, is not active in NMR. Fortunately, another isotope, carbon-13, is present in nature at a level of about 1.11%. Its behavior in the presence of a magnetic field is the same as that of hydrogen. One might therefore expect it to give spectra very similar to those observed in ^{1}H NMR spectroscopy. This expectation turns out to be only partly correct, because of a couple of important (and very useful) differences between the two types of NMR techniques.

Carbon-13 NMR (^{13}C NMR) spectra used to be much more difficult to record than hydrogen spectra, not only because of the low natural abundance of the nucleus under observation, but also because of the much weaker magnetic resonance of ^{13}C. Thus, under comparable conditions, ^{13}C signals are about 1/6000 as strong as those for hydrogen. FT NMR (Sections 10-3 and 10-4) is particularly valuable here, because multiple radio-frequency pulsing allows the accumulation of much stronger signals than would normally be possible.

One advantage of the low abundance of ^{13}C is the absence of carbon–carbon coupling. Just like hydrogens, two adjacent carbons, if magnetically nonequivalent (as they are in, e.g., bromoethane), split each other. In practice, however, such splitting is not observed. Why? Because coupling can occur only if two ^{13}C isotopes come to lie next to each other. With the abundance of ^{13}C in the molecule at 1.11%, this event has a very low probability (roughly 1% of 1%; i.e., 1 in 10,000). Most ^{13}C nuclei are surrounded by only ^{12}C nuclei, which, having no spin, do not give rise to spin–spin splitting. This feature simplifies ^{13}C NMR spectra appreciably, reducing the problem of their analysis to a determination of the coupling patterns resulting from any attached hydrogens.

Figure 10-30 depicts the ^{13}C NMR spectrum of bromoethane (for the ^{1}H NMR spectrum, see Figure 10-21). The chemical shift δ is defined as in ^{1}H NMR and is determined relative to an internal standard, normally the carbon absorption in $(CH_3)_4Si$. The chemical-shift range of carbon is much larger than that of hydrogen. For most organic compounds, it covers a distance of about 200 ppm, in contrast with the relatively narrow spectral "window" (10 ppm) of hydrogen. Figure 10-30 reveals the relative complexity of the absorptions caused by extensive ^{13}C–H spin–spin splittings. Not surprisingly, directly bound hydrogens are most strongly coupled ($\sim$125–200 Hz). Coupling tapers off, however, with increasing distance from the ^{13}C nucleus under observation, such that the two-bond coupling constant $J_{^{13}C-C-H}$ is in the range of only 0.7–6.0 Hz.

Really? Carbon has 15 (!) known isotopes, from ^{8}C to ^{23}C, only three of which occur on Earth: the stable ^{12}C and ^{13}C; and the radioisotope ^{14}C, with a half-life of 5700 years. ^{14}C is generated in trace quantities in the upper atmosphere by bombardment of ^{14}N with high-energy cosmic rays and is the basis for "carbon dating" in archeology. The least stable isotope is ^{8}C, with a half-life of 2×10^{-21} s.

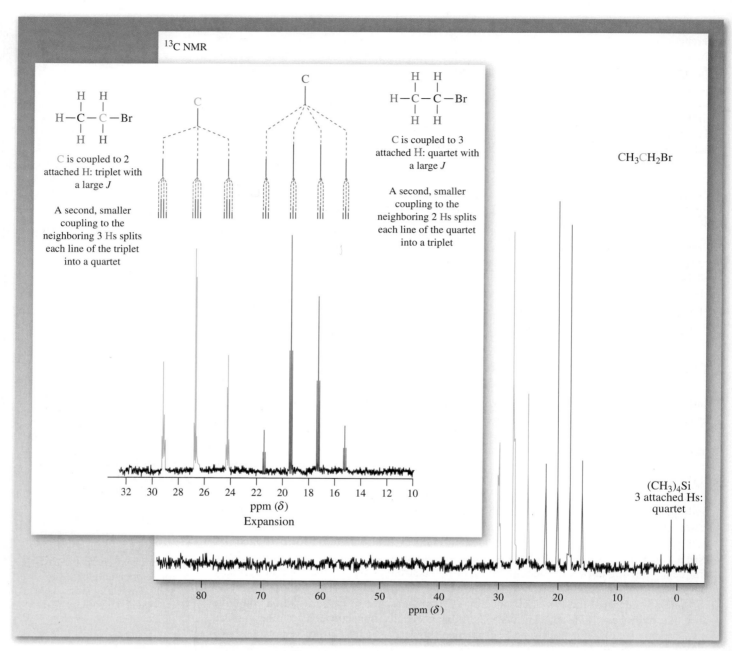

Figure 10-30 ^{13}C NMR spectrum of bromoethane, showing the complexity of ^{13}C–H coupling. There is an upfield quartet ($\delta = 18.3$ ppm, $J = 126$ Hz) and a downfield triplet ($\delta = 26.6$ ppm, $J = 151$ Hz) resonance for the two carbon atoms. Note the large chemical-shift range. Tetramethylsilane, defined to be located at $\delta = 0$ ppm as in ^{1}H NMR, absorbs as a quartet ($J = 118$ Hz) because of coupling of each carbon to three equivalent hydrogens. The inset shows a part of the spectrum expanded horizontally to reveal the fine splitting of each of the main peaks that is due to coupling of each ^{13}C to protons on the neighboring carbon. Thus, each line of the upfield quartet (red) is split again into a triplet with $J = 3$ Hz, and each line of the downfield triplet (blue) is split again into a quartet with $J = 5$ Hz.

You may wonder why it is that we observe hydrogen couplings in ^{13}C NMR spectroscopy, yet we do not notice the converse, namely, carbon couplings, in ^{1}H NMR. The answer lies in the low natural abundance of the NMR-active ^{13}C isotope and the high natural abundance of the ^{1}H nucleus. Thus, we could not detect ^{13}C coupling in our proton spectra, because 99% of the attached carbons were ^{12}C. On the other hand, the corresponding carbon spectra reveal ^{1}H coupling, because 99.9% of our sample contains this isotope of hydrogen (Table 10-1).

Exercise 10-18

Predict the ^{13}C NMR spectral pattern of 1-bromopropane (for the ^1H NMR spectrum, see Figure 10-27). (**Hint:** Use the sequential $N + 1$ rule.)

Hydrogen decoupling gives single lines

A technique that completely removes ^{13}C–H coupling is called **broad-band hydrogen (or proton) decoupling.** This method employs a strong, broad radio-frequency signal that covers the resonance frequencies of all the hydrogens and is applied at the same time as the ^{13}C spectrum is recorded. For example, in a magnetic field of 7.05 T, carbon-13 resonates at 75.3 MHz, hydrogen at 300 MHz (Figure 10-7). To obtain a proton-decoupled carbon spectrum at this field strength, we irradiate the sample at both frequencies. The first radio-frequency signal produces carbon magnetic resonance. Simultaneous exposure to the second signal causes all the hydrogens to undergo rapid $\alpha \rightleftharpoons \beta$ spin flips, fast enough to average their local magnetic field contributions. The net result is the absence of coupling. Use of this technique simplifies the ^{13}C NMR spectrum of bromoethane to two single lines, as shown in Figure 10-31.

The power of proton decoupling becomes evident when spectra of relatively complex molecules are recorded. *Every magnetically distinct carbon gives only one single peak in the ^{13}C NMR spectrum.* Consider, for example, a hydrocarbon such as methylcyclohexane. Analysis by ^1H NMR is made very difficult by the close-lying chemical shifts of the eight different types of hydrogens. However, a proton-decoupled ^{13}C spectrum shows only five peaks, clearly depicting the presence of the five different types of carbons and revealing the twofold symmetry in the structure (Figure 10-32). These spectra also exhibit a limitation in ^{13}C NMR spectroscopy: Integration is not usually straightforward. As a consequence of the broad-band decoupling, peak intensities no longer correspond to numbers of nuclei.

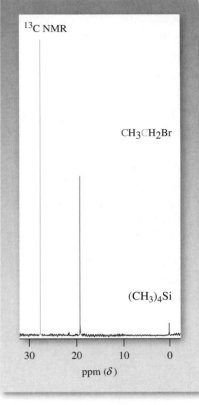

Figure 10-31 This 62.8-MHz ^{13}C NMR spectrum of bromoethane was recorded with broad-band decoupling at 250 MHz. All lines simplify to singlets, including the absorption for $(CH_3)_4Si$.

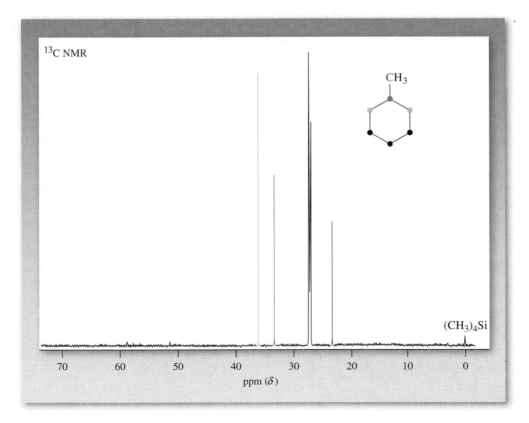

Figure 10-32 ^{13}C NMR spectrum of methylcyclohexane with hydrogen decoupling. Each of the five magnetically different types of carbon in this compound gives rise to a distinct peak: $\delta = 23.1$, 26.7, 26.8, 33.1, and 35.8 ppm.

Table 10-6　Typical ^{13}C NMR Chemical Shifts

Type of carbon	Chemical shift δ (ppm)
Primary alkyl, RCH_3	5–20
Secondary alkyl, RCH_2R'	20–30
Tertiary alkyl, R_3CH	30–50
Quaternary alkyl, R_4C	30–45
Allylic, $R_2C{=}CCH_2R'$	20–40
$\quad\quad\quad\quad\;\;\mid$	
$\quad\quad\quad\quad\;\;R''$	
Chloroalkane, RCH_2Cl	25–50
Bromoalkane, RCH_2Br	20–40
Ether or alcohol, RCH_2OR' or RCH_2OH	50–90
Carboxylic acids, $RCOOH$	170–180
Aldehyde or ketone, $R\overset{\overset{\displaystyle O}{\|}}{C}H$ or $R\overset{\overset{\displaystyle O}{\|}}{C}R'$	190–210
Alkene, aromatic, $R_2C{=}CR_2$	100–160
Alkyne, $RC{\equiv}CR$	65–95

Table 10-6 shows that carbon, like hydrogen (Table 10-2), has characteristic chemical shifts depending on its structural environment. As in ^{1}H NMR, electron-withdrawing groups cause deshielding, and the chemical shifts go up in the order primary < secondary < tertiary carbon. Apart from the diagnostic usefulness of such δ values, knowing the number of different carbon atoms in the molecule can aid structural identification. Consider, for example, the difference between methylcyclohexane and its isomers with the molecular formula C_7H_{14}. Many of them differ in the number of nonequivalent carbons and therefore give distinctly different carbon spectra. Notice how much the (lack of) symmetry in a molecule affects the complexity of the carbon spectrum.

Number of ^{13}C Peaks in Some C_7H_{14} Isomers

Four peaks　　　Four peaks　　　Three peaks　　　One peak

Exercise 10-19

How many peaks would you expect in the proton-decoupled ^{13}C NMR spectra of the following compounds? (**Hint:** Look for symmetry.)

(a) 2,2-Dimethyl-1-propanol   (b)   (c)

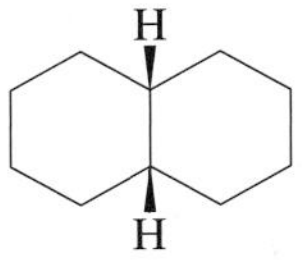

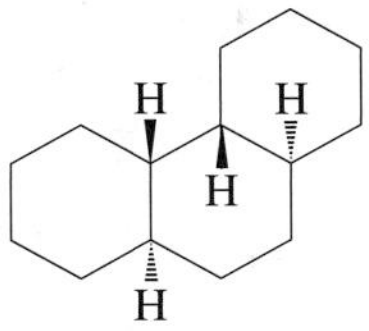

(d) 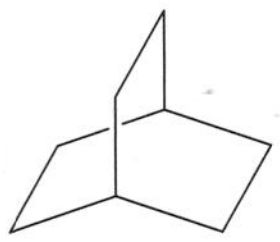   (e) *cis*-1,4-Dimethylcyclohexane, at 20°C and at −60°C. (**Hint:** Review Sections 4-4 and 10-5.)

Solved Exercise 10-20 | Working with the Concepts: Differentiating Isomers by ^{13}C NMR

In Exercise 2-16, part (a), you formulated the structures of the five possible isomers of hexane, C_6H_{14}. One of them shows three peaks in the ^{13}C NMR spectrum at $\delta = 13.7, 22.7,$ and 31.7 ppm. Deduce its structure.

Strategy
We need to write down all possible isomers and see to what extent symmetry (or the lack thereof) influences the number of peaks expected for each (using the labels *a*, *b*, *c*, and *d*.)

Solution

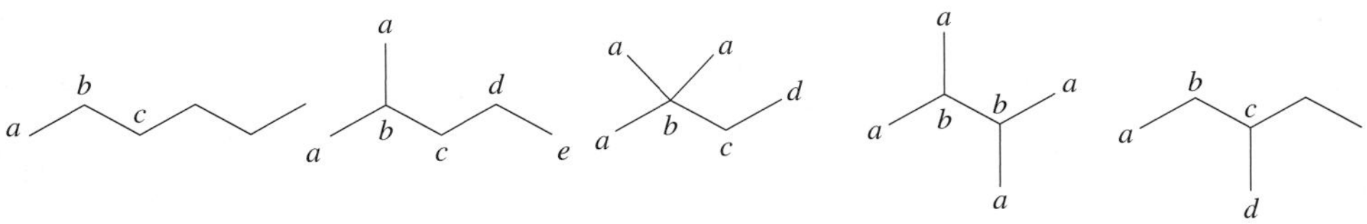

    Hexane   2-Methylpentane   2,2-Dimethylbutane   2,3-Dimethylbutane   3-Methylpentane

• Only one isomer has only three different carbons: hexane.

Exercise 10-21 | Try It Yourself

A researcher finds two unlabeled bottles in the stock room. She knows that one of them contains the sugar D-ribose, the other D-arabinose (both shown as Fischer projections below), but she does not know which bottle contains which sugar. Armed with a supply of $Na^+{}^-BH_4$ and equipped with an NMR spectrometer, how could she tell?

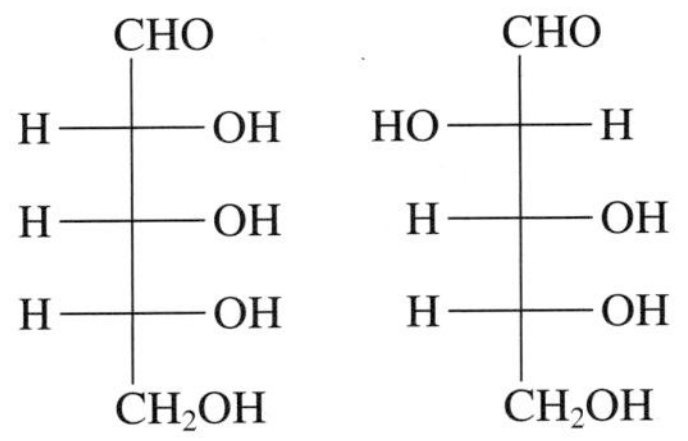

D-Ribose   D-Arabinose

Advances in FT NMR greatly aid structure elucidation: DEPT ^{13}C and 2D-NMR

The FT technique for the measurement of NMR spectra is extremely versatile, allowing data to be collected and presented in a variety of ways, each providing information about the structure of molecules. Most recent advances are due to the development of sophisticated time-dependent pulse sequences, including the application of *two-dimensional NMR, or 2D NMR* (Real Life 10-4). With these methods, it is now possible to establish coupling (and therefore bonding) between close-lying hydrogens *(homonuclear correlation)* or connected carbon and hydrogen atoms *(heteronuclear correlation)*. Thus, ^{1}H and ^{13}C NMR allow the determination of molecular connectivity by measuring the magnetic effect of neighboring atoms on one another along a carbon chain.

An example of such a pulse sequence, now routine in the research laboratory, is the **DEPT ^{13}C NMR** spectrum **(distortionless enhanced polarization transfer),** which tells you what type of carbon gives rise to a specific signal in the normal ^{13}C spectrum: CH_3, CH_2, CH, or $C_{quaternary}$. It avoids the complications arising from a proton-coupled ^{13}C NMR spectrum (see Figure 10-30), particularly overlapping multiplets of close-lying carbon signals. The DEPT method consists of a combination of spectra run with differing pulse sequences: the normal broad-band decoupled spectrum and a set of spectra that reveals *signals only of carbons bound to three (CH$_3$), two (CH$_2$), and one hydrogen (CH),* respectively. Figure 10-33 depicts a series of such spectra for limonene (see also Problem 29 of Chapter 5).

The first is the normal proton-decoupled spectrum (Figure 10-33A), which exhibits the expected number of lines (10) and groups them into the six alkyl and four alkenyl carbon signals at high and at low field, respectively. The remaining spectra specifically identify the three types of possible carbons bearing hydrogens, CH_3 (red; Figure 10-33B),

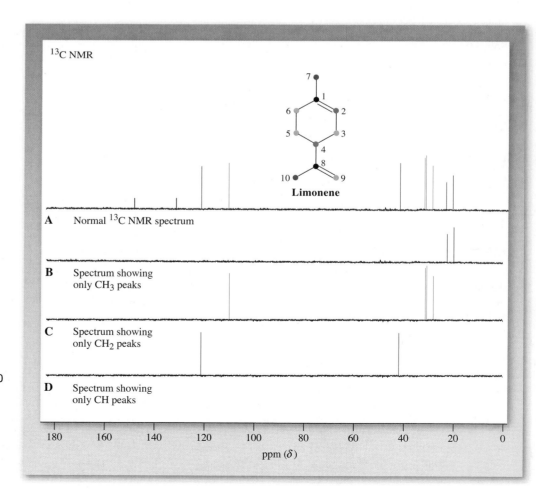

Figure 10-33 The DEPT ^{13}C NMR method applied to limonene: (A) broad-band decoupled spectrum revealing the six alkyl carbon signals at high field (20–40 ppm) and four alkenyl carbon signals at low field (108–150 ppm; see Table 10-6); (B) spectrum showing only the two CH$_3$ signals for C7 and C10 (red); (C) spectrum showing only the four CH$_2$ signals for C3, C5, C6, and C9 (blue); (D) spectrum showing only the two CH signals for C2 and C4 (green). The additional lines in (A) are assigned to the quaternary carbons C1 and C8 (black).

REAL LIFE: SPECTROSCOPY 10-4 | How to Determine Atom Connectivity in NMR

We have learned that NMR is of great help in proving the structure of molecules. It tells us what types of nuclei are present and how many of them are in a compound. It can also reveal connectivity by observing spin coupling through bonds. This last phenomenon can be pictured in a special type of plot, which is obtained by the application of sophisticated pulse techniques and the associated computer analysis: *two-dimensional (2D) NMR*. In this method, two spectra of the molecule are plotted on the horizontal and vertical axes. Mutually coupled signals are indicated as "blotches," or correlation peaks, on the *x–y* graph. Because this plot correlates nuclei that are in close proximity (and therefore exhibit spin–spin splitting), this type of spectroscopy is also called *correlation spectroscopy* or *COSY*.

Let us pick an example we already know: 1-bromopropane (see Figure 10-27). The COSY spectrum is shown below.

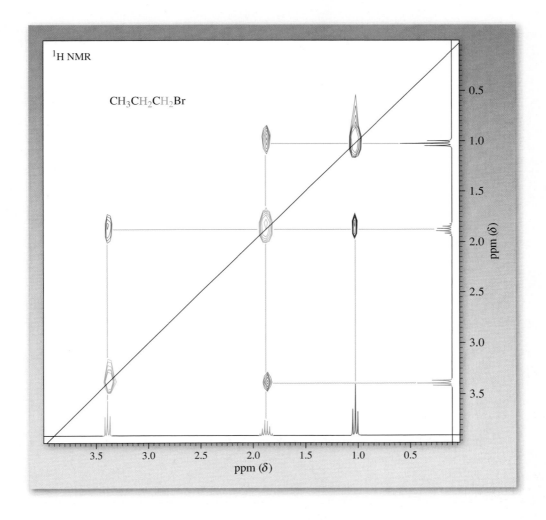

The two spectra along the *x* and *y* axes are identical to that in Figure 10-27. The marks on the diagonal line correlate the same peaks in each spectrum (indicated by the colored "blotches") and can be ignored. It is the off-diagonal correlations (indicated by the orange lines) that prove the connectivity of the hydrogens in the molecule. Thus, the (red) methyl triplet at $\delta = 1.05$ ppm has a cross peak (indicated by the black "blotches") with the (green) methylene sextet at $\delta = 1.88$ ppm, establishing their relation as neighbors. Similarly, the (blue) peaks for the methylene hydrogens, deshielded by the bromine to appear at $\delta = 3.40$ ppm, also correlate with the center signal at $\delta = 1.88$ ppm. Finally, the center absorption shows cross peaks with the signals of both of its neighbors, thus proving the structure.

REAL LIFE: SPECTROSCOPY 10-4 | **(Continued)**

Of course, the example of 1-bromopropane is overly simple and was chosen only to demonstrate the workings of this technique. Think of 1-bromohexane instead. Here, the rapid assignment of the individual methylene groups is made possible by COSY spectroscopy. Starting from either of the two signals that are readily recognized—namely, those due to the methyl (a unique high-field triplet) or the CH_2Br protons (a unique low-field triplet)—we can "walk along" the correlated signals from one methylene neighbor

to the next until the complete connectivity of the structure is evident.

Connectivity is disclosed even more powerfully by *heteronuclear correlation spectroscopy (HETCOR)*. Here, an ^1H NMR spectrum is juxtaposed with a ^{13}C spectrum. This allows the mapping of a molecule along its entire C–H framework by revealing the identity of C–H bonded fragments. The principle is illustrated again with 1-bromopropane in the spectrum shown below.

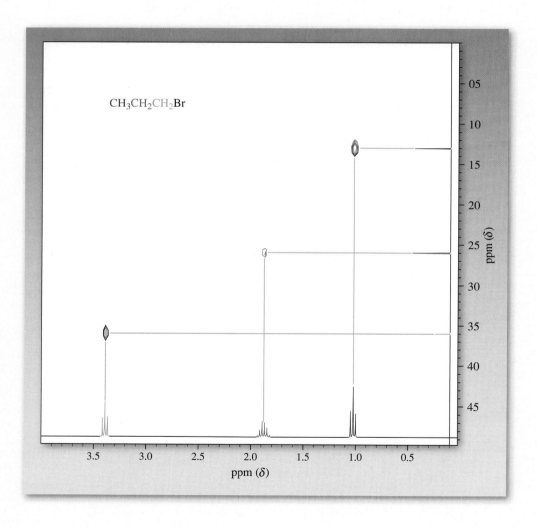

The ^{13}C NMR spectrum is plotted on the y axis and exhibits three peaks. On the basis of what we know about carbon NMR, we can readily assign the signal at $\delta = 13.0$ ppm to the methyl carbon, that at $\delta = 26.2$ ppm to the methylene, and that at $\delta = 36.0$ ppm to the bromine-bearing carbon. But even without this knowledge, these assignments are immediately evident from the cross-correlations of the

respective hydrogen signals (which we have already assigned) with those of their bound carbon counterparts. Thus, the methyl hydrogen triplet correlates with the carbon signal at $\delta = 13.0$ ppm, the sextet for the methylenes has a cross peak with the carbon signal at $\delta = 26.2$ ppm, and the downfield proton triplet is connected to the carbon with a signal at $\delta = 36.0$ ppm.

Structural Characterization of Natural and "Unnatural" Products: An Antioxidant from Grape Seeds and a Fake Drug in Herbal Medicines

Viniferone

The extracts of grape seeds are being touted in the prevention of a range of maladies, including heart disease, cancer, and psoriasis.

The world of plants is a rich source of useful pharmacological, therapeutic, and chemoprotective substances. The 15-carbon compound viniferone (a sesquiterpene; see Section 4-7) was isolated from grape seeds in 2004. It is one of a variety of substances called grape seed proanthocyanidins, which are highly active against radicals (Chapter 3) and oxidative stress (see Section 22-9). Only 40 mg of the compound was obtained from 10.5 kg of grape seeds; therefore, elemental analysis, chemical tests, and any other procedures that would destroy the tiny amount of available material could not be used in structural elucidation. Instead, a combination of spectroscopic techniques (NMR, IR, MS, and UV) was employed, leading to the structure shown here, which was later confirmed by X-ray crystallography.

The characterization of viniferone relied in part on ^1H and ^{13}C NMR data. Proton chemical shifts provided evidence for key pieces of the structure. Thus, there were three signals in the range reserved for internal alkene and aromatic hydrogens (shown in orange in the structural drawings) between $\delta = 5.9$ and 6.2 ppm (consult Table 10-2 for typical proton chemical shifts). Another three signals were observed at $\delta > 3.8$ ppm, reflecting the presence of three protons (red) attached to oxygen-bearing carbons. Finally, there were four absorptions with relatively low δ values between 2.5 and 3.1 ppm, assignable to the two pairs of (green) diastereotopic hydrogens (see Real Life 10-3).

^{13}C NMR was equally helpful (consult Table 10-6 for typical carbon chemical shifts), as it clearly revealed the two C=O carbons at $\delta = 171.0$ and 173.4 ppm and indicated the presence of eight alkene and benzene carbons ($\delta > 95$ ppm). The three tetrahedral carbons attached to oxygen showed up as peaks between $\delta = 67$ and 81 ppm, and the two remaining tetrahedral carbon signals occurred at $\delta = 28.9$ and 37.4 ppm. All of these were identified further by establishing the number of attached hydrogens by techniques equivalent to DEPT NMR.

Spin–spin splitting and correlated (COSY) proton spectra (Real Life 10-4) confirmed the structural assignment. For example, focusing on the six-membered ether ring, the proton on C2 ($\delta = 4.61$ ppm) showed a doublet pattern due to coupling to its neighbor on C3 ($\delta = 3.90$ ppm). Similarly, the two diastereotopic protons at C4 ($\delta = 2.53$ and 3.02 ppm) gave rise to a doublet of doublets each, because of mutual coupling and independent coupling of each to the proton on C3. Not surprisingly, the proton on C3 produced an unresolved multiplet. Additional aspects of the structural determination of viniferone will be presented in Chapter 14 (Real Life 14-4).

Turning to the "unnatural," the world of synthetic drugs abounds with illicit compounds, sometimes dangerous, often devoid of any useful activity. A case in point is herbal dietary supplements (HDSs), a globally expanding market with annual sales of close to $100 billion per year. Their success is based largely on the common belief that natural products are safer than synthetic ones (see also Real Life 25-4), even though these concoctions have not undergone efficacy and toxicity trials, and their ease of acquisition without a prescription, typically from websites. Ironically, to boost claims, some manufacturers are illegally adding prescription drugs to the mix, defeating the whole purpose of HDSs in the first place. An example is herbal preparations as alternatives to the treatment of erectile dysfunction, tackled clinically with success by drugs such as sildenafil (Viagra, Chapter 25 Opening and Problem 29 of Chapter 25) and the close analog vardenafil (Levitra). In several products, these supplements have been found to be laced with the actual pharmaceuticals, a matter of serious concern for drug regulators and, of course,

REAL LIFE: MEDICINE 10-5 | **(Continued)**

Sildenafil (R = CH₃)
Vardenafil (R = CH₃CH₂)

"Acetylvardenafil"

users. Such addends are readily identified by spectroscopic techniques, including NMR spectroscopy. For example, vardenafil shows three aromatic signals between $\delta = 7.4$ and 8.0 ppm (orange) and another three absorptions at $\delta = 2.9$–4.2 ppm for the 12 hydrogens next to the relatively electron-withdrawing nitrogen and oxygen atoms (red). The remaining alkyl segments exhibit peaks at higher field, most notably a methyl singlet at $\delta = 2.5$ ppm and three methyl triplets in the range $\delta = 0.9$–1.4 ppm (green). The ^{13}C NMR spectrum reveals the expected 21 signals, including six distinct CH_2 carbons (by DEPT NMR), four of which (attached to N and O) ranging from $\delta = 43$ to 51 ppm, the remaining two at $\delta = 20$ and 27 ppm, and four distinct CH_3 peaks. The other resonances are at lower field, including one C=O carbon at $\delta = 155$ ppm. Drug enforcement agencies record these (and other) spectra routinely and match up the data electronically with those deposited in data banks. To avoid such detection in drug screens, perpetrators have resorted to using counterfeit analogs of the real thing. For example, in 2011, an "all-natural lifestyle enhancement supplement" was found to contain a new (and untested) compound labeled "acetylvardenafil." Spectroscopic detective work relied in part on NMR analysis. Thus, while the spectra were very similar to those of authentic vardenafil, there were additional peaks. Thus, in the proton spectrum, a singlet at

$\delta = 3.78$ ppm revealed an extra CH_2 group, while in the carbon spectrum, in addition to the signal arising from the CH_2 fragment, there was an additional carbonyl absorption at $\delta = 194.7$ ppm. Cleverly, but not sufficiently so to bypass the scrutiny of modern analytical techniques, the designer chemists had replaced the SO_2 linker by that of an acetyl!

"Natural Viagra" and other herbal remedies offered at a Moroccan market, where French is spoken.

CH₂ (blue; Figure 10-33C), and CH (green; Figure 10-33D). The quaternary carbon signals (black) don't show up in the last three experiments and are located by subtracting all of the lines in Figure 10-33B–D from the complete spectrum in Figure 10-33A.

For the remainder of the text, whenever the depiction or description of a ^{13}C NMR spectrum includes spectral assignments to CH_3, CH_2, CH, or $C_{quaternary}$ carbons, they are based on a DEPT experiment.

We can apply ^{13}C NMR spectroscopy to the problem of the monochlorination of 1-chloropropane

In Section 10-6 we learned how we could use ^{1}H NMR chemical shifts and integration to distinguish among the three isomers of dichloropropane arising from the chlorination of 1-chloropropane. Exercise 10-12 addressed the use of spin–spin splitting patterns as a complementary means of solving this problem. How does ^{13}C NMR fare in this task? Our prediction is straightforward: Both 1,1- and 1,2-dichloropropane should exhibit three carbon signals each, but spaced significantly differently because the 1,1-isomer has the two electron-withdrawing chlorine atoms located on the same carbon (hence no chlorines on the remaining two), whereas 1,2-dichloropropane bears one chlorine each on C1 and C2. In contrast, 1,3-dichloropropane would be clearly distinct from the other two isomers because of its symmetry: Only two lines should be observed. The experimental data are shown in the margin and confirm our expectations. The specific assignments of the two deshielded chlorine-bearing carbons in 1,2-dichloropropane (as indicated in the margin) can be made on the basis of the DEPT technique: The signal at 49.5 ppm appears as a CH_2, and that at 55.8 ppm as a CH in the DEPT-90 experiment.

You can see from this example how ^1H NMR and ^{13}C NMR spectroscopy complement each other. ^1H NMR spectra provide an estimate of the electronic environment (i.e., electron rich versus electron poor) of a hydrogen nucleus under observation (δ), a measure of its relative abundance (integration), and an indication of how many neighbors (and their number of types) it has (spin–spin splitting). Proton-decoupled ^{13}C NMR provides the total number of chemically distinct carbons, their electronic environment (δ), and, in the DEPT mode, even the quantity of their attached hydrogens. Application of both techniques to the solution of a structural problem is not unlike the methods used to solve a crossword puzzle. The horizontal entries (such as the data provided by ^1H NMR spectroscopy) have to fit the vertical ones (i.e., the corresponding ^{13}C NMR information) to provide the correct answer.

Three Signals

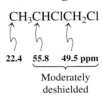

10.1   34.9   73.2 ppm

Strongly
deshielded

Three Signals

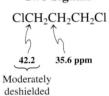

22.4   55.8   49.5 ppm

Moderately
deshielded

Two Signals

$ClCH_2CH_2CH_2Cl$

42.2   35.6 ppm

Moderately
deshielded

Exercise 10-22

Are bicyclic compounds A and B shown below readily distinguished by their proton-decoupled ^{13}C NMR spectra? Would DEPT spectra be of use in solving this problem?

A   B

In Summary ^{13}C NMR requires FT techniques because of the low natural abundance of the carbon-13 isotope and its intrinsically lower sensitivity in this experiment. ^{13}C–^{13}C coupling is not observed, because the scarcity of the isotope in the sample renders the likelihood of neighboring ^{13}C nuclei negligible. ^{13}C–H coupling can be measured but is usually removed by broad-band proton decoupling, providing single lines for each distinct carbon atom in the molecule under investigation. The ^{13}C NMR chemical-shift range is large, about 200 ppm for organic structures. ^{13}C NMR spectra cannot usually be integrated, but the DEPT experiment allows the identification of each signal as arising from CH_3, CH_2, CH, or $C_{quaternary}$ units, respectively.

THE BIG PICTURE

In your study of organic chemistry so far, there may have been moments during which you wondered, how do chemists know this? How do they establish the veracity of a structure? How do they follow the kinetics of disappearance of a molecule? How is an equilibrium constant determined? When do they know that a reaction is over? The introduction of NMR spectroscopy provides a first glimpse of the variety of practical tools available to answer these questions. Other forms of spectroscopy can yield other important information about molecules. We shall introduce them in conjunction with the functional groups for which they are particularly diagnostic. Spectroscopy, especially NMR spectroscopy, is the key to identifying the different classes of organic compounds and is an important part of each subsequent chapter in the book. You may find it helpful to review the material in this chapter as you examine new spectra later on.

WORKED EXAMPLES: INTEGRATING THE CONCEPTS

10-23. Using NMR Spectra to Assign Structures of Unknowns

A researcher executed the following reaction sequence in a preparation of (S)-2-chlorobutane:

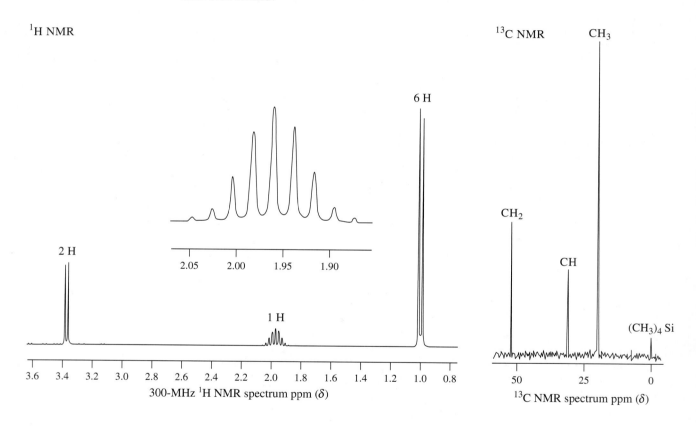

(R)-2-Methyloxacyclopropane

(S)-2-Chlorobutane

Careful preparative gas chromatography of the reaction product (b.p. 68.2°C) allowed the separation of a very small amount of another compound, C_4H_9Cl (b.p. 68.5°C), which was optically inactive and exhibited the NMR spectra depicted below. What is the structure of this compound and how could it have been formed?

SOLUTION

Once more, let us apply the WHIP approach to break down the process of solving this problem.

*W*hat is the problem asking? We are first to deduce the structure of a minor product of the given reaction sequence. Subsequently, we are asked to explain "how" this compound forms. As mentioned earlier, "how" and "why" questions normally require some mechanistic insight. We'll take things in order.

*H*ow to begin? The given molecular composition of the unknown compound, C_4H_9Cl, tells us that it is an isomer of the observed major product. We have both its ¹H and ¹³C NMR spectra available for analysis. We also know the reagents that convert the starting material to the major product. Which pieces of information are the most useful ones to help us get started? We could try to come up with alternative results of the two reactions in the sequence without examining the spectroscopic data at all, but, realistically, sooner or later it will be necessary to use the spectra to confirm any hypothetical structure we propose. It makes much more sense to start with them and at least find out whether they give us an unambiguous answer.

*I*nformation needed? Sections 10-4 through 10-7 reveal the information contained in an ^{1}H NMR spectrum; Section 10-9 covers ^{13}C NMR. The carbon spectrum is simpler. As a general rule of strategy, try to extract the most information you can from the sources that are the *simplest* to analyze. Start with the ^{13}C NMR spectrum; then proceed to the ^{1}H NMR spectrum.

*P*roceed. The formula of the unknown is C_4H_9Cl. It has four carbon atoms, but the ^{13}C NMR spectrum (assignments by DEPT) displays only *three* lines. Therefore, one of these three lines must arise from two equivalent carbon atoms. The line farthest downfield, at $\delta = 51$ (CH$_2$) ppm, is the most deshielded and therefore most likely to be the unique carbon bearing the chlorine atom (Table 10-6). This reasoning suggests the presence of substructure –CH$_2$Cl. The other two lines are in the alkyl region, at about $\delta = 20$ (CH$_3$) and 31 (CH) ppm. One of these two must be due to two equivalent carbon nuclei. We can use the molecular formula and our knowledge of substructure –CH$_2$Cl to identify which one. Thus, if the signal (for CH) at $\delta = 31$ ppm were due to two identical carbons, then we would arrive at a molecular formula of CH$_3$ + 2 CH + CH$_2$Cl = C$_4$H$_7$Cl, not a match for that given. On the other hand, if the signal at $\delta = 20$ (CH$_3$) ppm corresponded to two carbons, then we have 2 CH$_3$ + CH + CH$_2$Cl = C$_4$H$_9$Cl, the correct formula. Therefore, the unknown contains two equivalent CH$_3$ groups. We could connect these fragments to give us an immediate likely solution, but let us exercise a little patience and see what the ^{1}H NMR spectrum has to tell us.

This spectrum reveals three types of hydrogens at about $\delta = 1.0$, 1.9, and 3.4 ppm. Following the same logic as that applied in the above assignment of the most deshielded carbon atom, the most deshielded hydrogens must be those attached to the chlorine-bearing carbon (Section 10-4). The integrated values for the three hydrogen signals are 6, 1, and 2, respectively, totaling to the nine hydrogens present (Section 10-6). Finally, there is spin–spin splitting. Both the highest and lowest field signals are doublets, with almost identical J values. This means that each of these sets of hydrogens (6 + 2, or 8 in all) has a single neighbor (the ninth H). That hydrogen shows up in the middle as a nine-line pattern, as expected by the $N + 1$ rule ($N = 8$; Section 10-7).

Now let us combine this information in a structural assignment. As with most puzzles, one can arrive at the answer in several ways. In NMR spectral problems, it is often best to start with the formulation of partial structures, as dictated by the ^{1}H NMR spectrum, and use the other information as corroborating evidence. Thus, the high-field doublet integrating for 6 H indicates a (CH$_3$)$_2$CH–substructure. Similarly, the low-field counterpart points to –CH$_2$CH–. Combining the two provides –CH$_2$CH(CH$_3$)$_2$ and, adding the Cl atom, the solution: the achiral (hence optically inactive) 1-chloro-2-methylpropane ClCH$_2$CH(CH$_3$)$_2$. This assignment is confirmed by the ^{13}C NMR spectrum, the highest field line being due to the presence of the two equivalent methyl carbons. The center line is due to the tertiary, the most deshielded absorption due to the chlorine-bearing carbon (Table 10-6). You can confirm your solution in another manner, taking advantage of the relatively small molecular formula; thus, there are only four possible chlorobutane isomers: CH$_3$CH$_2$CH$_2$CH$_2$Cl, CH$_3$CHCH$_2$CH$_3$ (our major

$\qquad\qquad\qquad\qquad\qquad\qquad\qquad\qquad\qquad\qquad\qquad\qquad$ |
$\qquad\qquad\qquad\qquad\qquad\qquad\qquad\qquad\qquad\qquad\qquad\qquad$ Cl

product), (CH$_3$)$_2$CHCH$_2$Cl (the minor product), and (CH$_3$)$_3$CCl. They differ drastically in their ^{1}H and ^{13}C NMR spectra with respect to number of signals, chemical shifts, integration, and multiplicities. (Verify.)

The second aspect of this problem is mechanistic. How do we get 1-chloro-2-methylpropane from the preceding reaction sequence? The answer presents itself on retrosynthetic analysis, using the reagents given in our initial scheme:

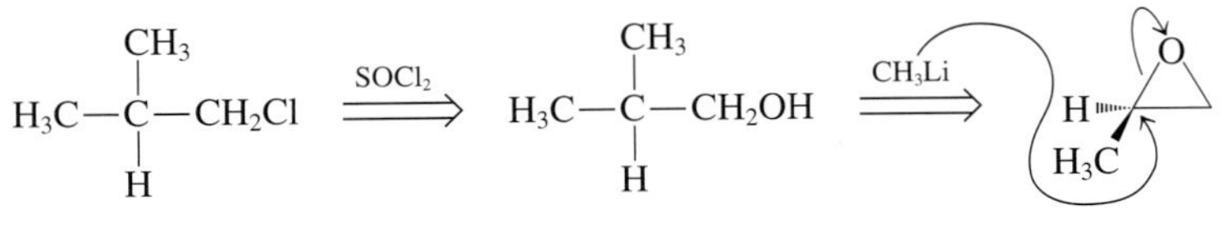

The only way to obtain a product with two CH$_3$ groups attached to the same carbon is to have the methyl group of CH$_3$Li add to the carbon of the oxacyclopropane that already contains one CH$_3$. Therefore, the observed minor product is the result of nucleophilic ring opening of the starting compound by attack at the more hindered position, usually neglected because it is less favored.

10-24. Using NMR Spectra to Uncover Rearrangements

a. A graduate student took the ^{1}H and ^{13}C NMR spectra of optically pure (1R,2R)-*trans*-1-bromo-2-methylcyclohexane (A) in deuterated nitromethane (CD$_3$NO$_2$) as a solvent (Table 6-5) and recorded the following values: ^{1}H NMR spectrum: $\delta = 1.06$ (d, 3 H), 1.42 (m, 6 H), 1.90 (m, 2 H), 2.02 (m, 1 H), 3.37 (m, 1 H) ppm; ^{13}C NMR (DEPT): $\delta = 16.0$ (CH$_3$), 23.6 (CH$_2$), 23.9 (CH$_2$), 30.2 (CH$_2$), 33.1 (CH$_2$), 35.0 (CH), 43.2 (CH) ppm. Assign these spectra as best you can with the help of Tables 10-2 and 10-6, respectively.

^{1}H NMR Information

Chemical shift

Integration

Spin–spin splitting

^{13}C NMR Information

Chemical shift

DEPT

SOLUTION

¹H NMR spectrum. All of the signals, with the exception of the highest field doublet, appear as complex multiplets. This is not surprising, considering that each hydrogen (except for the three CH_3 nuclei) is magnetically distinct, that there are multiple vicinal and geminal couplings all around the ring, and that the δ values (except for those in the immediate vicinity of the bromine substituent) are close. The exception is the methyl group appearing at highest field, as expected, and as a doublet due to coupling to the neighboring tertiary hydrogen. To complete our assignments, we have to rely more than usual on chemical shifts and integration.

What is the expected effect of bromine on its vicinity? *Answer:* Br is deshielding (Section 10-4), mostly its direct neighbors, the effect tapering off with distance. Indeed, the δ values divide into two groups. One set has higher ($\delta = 3.37, 2.02, 1.90$ ppm), the other lower values ($\delta = 1.42, 1.06$ ppm). The most deshielded single hydrogen at $\delta = 3.37$ ppm is readily assignable to that at C1 next to bromine. The next most deshielded positions are its neighbors at C2 and C6. Since the signal at $\delta = 2.02$ ppm integrates for only 1 H, it must be the single tertiary hydrogen at C2; the peak at $\delta = 1.90$ ppm can then be assigned to the CH_2 group at C6. This choice is also consistent with the general appearance of secondary hydrogen signals at higher field than those of tertiary hydrogens (Table 10-2). The remaining six hydrogens at $\delta = 1.42$ ppm are not resolved, and all absorb at around the same place.

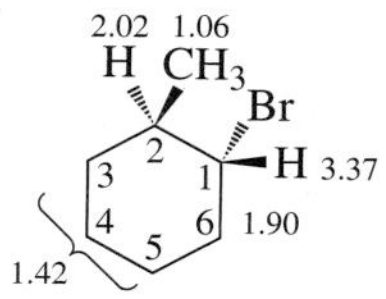

A

¹H NMR assignments (ppm)

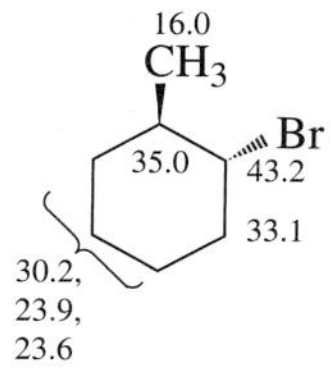

A

¹³C assignments (ppm)

¹³C NMR spectrum. Using the DEPT correlations in conjunction with the deshielding effect of bromine allows a ready assignment to the extent shown above.

b. To the student's surprise, the optical activity of the sample decreased with time. Concurrently, the NMR spectrum of A diminished in intensity and new peaks appeared of an optically inactive isomer B. ¹H NMR: $\delta = 1.44$ (m, 6 H), 1.86 (m, 4 H), 1.89 (s, 3 H) ppm; ¹³C NMR: $\delta = 20.8$ (CH_2), 26.7 (CH_2), 28.5 (CH_3), 37.6 (C_{quat}), 41.5 (CH_2) ppm. Following the disappearance of A revealed a rate $= k[A]$. What is B, and how is it formed?

SOLUTION

Let us analyze the spectra of B, particularly in comparison to those of A. In the ¹H NMR spectrum, we note that the number of signals has decreased from five to only three. Moreover, the low-field peak at $\delta = 3.37$ ppm for the hydrogen next to bromine in A has disappeared and the methyl doublet is now a singlet, moved to lower field relative to its original position. In the ¹³C NMR spectrum, we recognize a similar simplification (hence symmetrization) from seven to five peaks. Moreover, the tertiary carbons (CH) have disappeared, there are only three types of CH_2, and a quaternary carbon has appeared. The relatively deshielded CH_3 carbon is also evident.

Conclusions. The bromine atom must now be located at the same position as the CH_3 substituent, as in the achiral 1-bromo-1-methylcyclohexane:

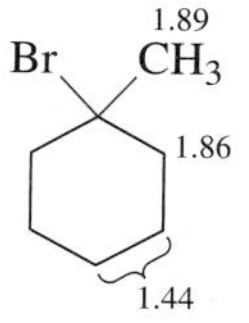

B

¹H NMR assignments (ppm)

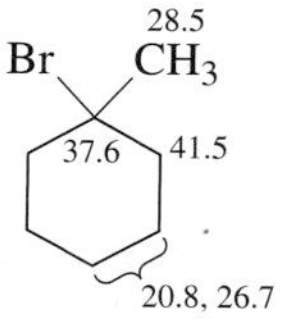

B

¹³C NMR assignments (ppm)

How could that rearrangement have happened? *Answer:* We have a case of an S_N1 reaction (Section 7-2) proceeding through a rearranging carbocation (Section 9-3):

The mechanism explains all of the observations: The loss of optical activity, the first-order disappearance of A, and the formation of B.

Important Concepts

1. **NMR** is the most important spectroscopic tool in the elucidation of the **structures** of organic molecules.

2. **Spectroscopy** is possible because molecules exist in various energetic forms, those at lower energy being convertible into states of higher energy by absorption of discrete quanta of **electromagnetic radiation.**

3. NMR is possible because certain nuclei, especially ^1H and ^{13}C, when exposed to a strong magnetic field, align **with** it (α) or **against** it (β). The α-to-β transition can be effected by radio-frequency radiation, leading to **resonance** and a spectrum with characteristic absorptions. The higher the external field strength, the higher the resonance frequency. For example, a magnetic field of 7.05 T causes hydrogen to absorb at 300 MHz, a magnetic field of 14.1 T causes it to absorb at 600 MHz.

4. **High-resolution NMR** allows for the differentiation of hydrogen and carbon nuclei in different chemical environments. Their characteristic positions in the spectrum are measured as the **chemical shift, δ,** in ppm from an internal standard, tetramethylsilane.

5. The chemical shift is highly dependent on the presence (causing **shielding**) or absence (causing **deshielding**) of electron density. Shielding results in relatively high-field peaks [to the right, toward $(CH_3)_4Si$], deshielding in low-field ones. Therefore, electron-donor substituents shield, and electron-withdrawing components deshield. The protons on the heteroatoms of alcohols, thiols, and amines show variable chemical shifts and often appear as broad peaks because of hydrogen bonding and exchange.

6. Chemically equivalent hydrogens and carbons have the same chemical shift. Equivalence is best established by the application of **symmetry** operations, such as those using **mirror planes** and **rotations.**

7. The number of hydrogens giving rise to a peak is measured by **integration.**

8. The number of hydrogen neighbors of a nucleus is given by the **spin–spin splitting** pattern of its NMR resonance, following the **$N + 1$ rule.** Equivalent hydrogens show no mutual spin–spin splitting.

9. When the chemical-shift difference between coupled hydrogens is comparable to their coupling constant, **non-first-order spectra** with complicated patterns are observed.

10. When the constants for coupling to nonequivalent types of neighboring hydrogens are different, the **$N + 1$ rule** is applied **sequentially.**

11. **Carbon NMR** utilizes the low-abundance ^{13}C isotope. Carbon–carbon coupling is not observed in ordinary ^{13}C spectra. Carbon–hydrogen coupling can be removed by proton decoupling, thereby simplifying most ^{13}C spectra to a collection of single peaks.

12. **DEPT** ^{13}C NMR allows the assignment of absorptions to CH_3, CH_2, CH, and quaternary carbons, respectively.

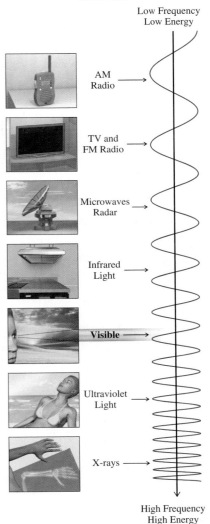

Various Forms of Radiation and Their Uses

Low Frequency
Low Energy

AM Radio

TV and FM Radio

Microwaves Radar

Infrared Light

Visible

Ultraviolet Light

X-rays

High Frequency
High Energy

Problems

25. Where on the chart presented in Figure 10-2 would the following be located: AM radio waves ($\nu \sim 1$ MHz = 1000 kHz = 10^6 Hz = 10^6 s^{-1}, or cycles s^{-1}); FM broadcast frequencies ($\nu \sim 100$ MHz = 10^8 s^{-1})?

26. Convert each of the following quantities into the specified units. **(a)** 1050 cm^{-1} into λ, in μm; **(b)** 510 nm (green light) into ν, in s^{-1} (cycles s^{-1}, or hertz); **(c)** 6.15 μm into $\tilde{\nu}$, in cm^{-1}; **(d)** 2250 cm^{-1} into ν, in s^{-1} (Hz).

27. Convert each of the following quantities into energies, in kcal mol^{-1}. **(a)** A bond rotation of 750 wavenumbers (cm^{-1}); **(b)** a bond vibration of 2900 wavenumbers (cm^{-1}); **(c)** an electronic transition of 350 nm (ultraviolet light, capable of sunburn); **(d)** the broadcast frequency of the audio signal of TV channel 6 (87.25 MHz; before the advent of digital TV in 2009); **(e)** a "hard" X-ray with a 0.07-nm wavelength.

28. Calculate to three significant figures the amount of energy absorbed by a hydrogen when it undergoes an α-to-β spin flip in the field of **(a)** a 2.11-T magnet ($\nu = 90$ MHz); **(b)** an 11.75-T magnet ($\nu = 500$ MHz).

29. For each of the following changes, indicate whether it corresponds to moving to the right or to the left in an NMR spectrum. **(a)** Increasing radio frequency (at constant magnetic field strength); **(b)** increasing magnetic field strength (at constant radio frequency; moving "upfield"; Section 10-4); **(c)** increasing chemical shift; **(d)** increased shielding.

30. Sketch a hypothetical low-resolution NMR spectrum showing the positions of the resonance peaks for all magnetic nuclei for each of the following molecules. Assume an external magnetic field of 2.11 T. How would the spectra change if the magnetic field were 8.46 T?

(a) CFCl$_3$ (Freon 11; see Section 3-10)

(b) CH$_3$CFCl$_2$ (HCFC-141b; see Section 3-10)

(c) CF$_3$—$\overset{\displaystyle \text{Cl}}{\underset{\displaystyle \text{Br}}{\text{C}}}$—H   (Halothane; see Real Life 6-1)

31. If the NMR spectra of the molecules in Problem 30 were recorded by using high resolution for each nucleus, what differences would be observed?

32. The ^{1}H NMR spectrum of CH$_3$COCH$_2$C(CH$_3$)$_3$, 4,4-dimethyl-2-pentanone, taken at 300 MHz, shows signals at the following positions: 307, 617, and 683 Hz downfield from tetramethylsilane. **(a)** What are the chemical shifts (δ) of these signals? **(b)** What would their positions be in hertz, relative to tetramethylsilane, if the spectrum were recorded at 90 MHz? At 500 MHz? **(c)** Assign each signal to a set of hydrogens in the molecule.

33. Order the ^{1}H NMR signals of the following compounds by chemical-shift position (lowest to highest). Which one is the most upfield? The most downfield?

(a) H$_3$C—CH$_3$

(b) H$_2$C=CH$_2$

(c) H$_3$C—O—CH$_3$

(d) $\overset{\displaystyle \text{O}}{\overset{\displaystyle \|}{\text{H}_3\text{C}-\text{C}-\text{CH}_3}}$

(e) (benzene ring structure)

(f) H$_3$C—$\overset{\displaystyle \text{CH}_3}{\underset{}{\text{N}}}$—CH$_3$

34. Which hydrogens in the following molecules exhibit the more downfield signal relative to (CH$_3$)$_4$Si in the NMR experiment? Explain.

(a) (CH$_3$)$_2$O or (CH$_3$)$_3$N

(b) CH$_3$$\overset{\displaystyle \text{O}}{\overset{\displaystyle \|}{\text{C}}}OCH_3$
  ↑ or ↑

(c) CH$_3$CH$_2$CH$_2$OH
  ↑ or ↑

(d) (CH$_3$)$_2$S or (CH$_3$)$_2$S=O

35. How many signals would be present in the ^{1}H NMR spectrum of each of the cyclopropane derivatives shown below? Consider carefully the geometric environments around each hydrogen.

(a) (cyclopropane with Br) **(b)** (cyclopropane with Br, Br) **(c)** (cyclopropane with Br, Br)

(d) (cyclopropane with Br) **(e)** (cyclopropane with Br, Br)

36. How many signals would be present in the ^{1}H NMR spectrum of each of the following molecules? What would the *approximate* chemical shift be for each of these signals? Ignore spin–spin splitting.

(a) CH$_3$CH$_2$CH$_2$CH$_3$

(b) CH$_3$$\underset{\displaystyle \text{Br}}{\text{CHCH}_3}$

(c) HOCH$_2$$\overset{\displaystyle \text{CH}_3}{\underset{\displaystyle \text{CH}_3}{\text{C}}}$Cl

(d) CH$_3$$\underset{\displaystyle \text{CH}_3}{\text{CHCH}_2\text{CH}_3}$

(e) CH$_3$$\overset{\displaystyle \text{CH}_3}{\underset{\displaystyle \text{CH}_3}{\text{C}}}NH_2$

(f) CH$_3$CH$_2$CH(CH$_2$CH$_3$)$_2$

(g) CH$_3$OCH$_2$CH$_2$CH$_3$

(h) $\overset{\displaystyle \text{H}_2\text{C}-\text{CH}_2}{\underset{\displaystyle \text{H}_2\text{C}-\text{C}}{}}$ (with O)

(i) CH$_3$CH$_2$—$\overset{\displaystyle \text{O}}{\overset{\displaystyle \|}{\text{C}}}$—H

(j) CH$_3$$\overset{\displaystyle \text{CH}_3\text{O}}{\text{CH}}$—$\overset{\displaystyle \text{CH}_3}{\underset{\displaystyle \text{CH}_3}{\text{C}}}$—CH$_3$

37. For each compound in each of the following groups of isomers, indicate the number of signals in the ^{1}H NMR spectrum, the *approximate* chemical shift of each signal, and the integration ratios for the signals. Ignore spin–spin splitting. Indicate whether all the isomers in each group can be distinguished from one another by these three pieces of information alone.

(a) CH$_3$$\overset{\displaystyle \text{CH}_3}{\underset{\displaystyle \text{Br}}{\text{C}}}CH_2CH_3$, BrCH$_2$$\overset{\displaystyle \text{CH}_3}{\text{CHCH}_2\text{CH}_3}$, CH$_3$$\overset{\displaystyle \text{CH}_3}{\text{CHCH}_2\text{CH}_2\text{Br}}$

(b) ClCH$_2$CH$_2$CH$_2$CH$_2$OH, CH$_3$$\overset{\displaystyle \text{CH}_2\text{Cl}}{\text{CHCH}_2\text{OH}}$, CH$_3$$\overset{\displaystyle \text{CH}_3}{\underset{\displaystyle \text{Cl}}{\text{C}}}CH_2$OH

(c) ClCH$_2$$\overset{\displaystyle \text{CH}_3}{\underset{\displaystyle \text{Br}}{\text{C}}}$—CHCH$_3$ (with CH$_3$), ClCH$_2$$\overset{\displaystyle \text{CH}_3}{\text{CH}}$—$\overset{\displaystyle \text{CH}_3}{\underset{\displaystyle \text{Br}}{\text{C}}}CH_3$,

ClCH$_2$$\overset{\displaystyle \text{CH}_3}{\underset{\displaystyle \text{CH}_3}{\text{C}}}$—CHCH$_3$ (with Br), ClCH$_2$CHCCH$_3$ (with CH$_3$, Br CH$_3$)

38. ^{1}H NMR spectra for two haloalkanes are shown below. Propose structures for these compounds that are consistent with the spectra. **(a)** $C_5H_{11}Cl$, spectrum A; **(b)** $C_4H_8Br_2$, spectrum B.

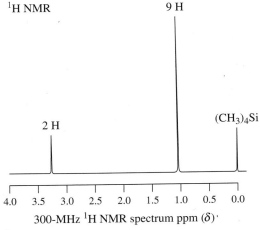

^{1}H NMR

9 H

2 H

(CH$_3$)$_4$Si

4.0   3.5   3.0   2.5   2.0   1.5   1.0   0.5   0.0

300-MHz ^{1}H NMR spectrum ppm (δ)

A

^{1}H NMR

6 H

2 H

(CH$_3$)$_4$Si

4.0   3.5   3.0   2.5   2.0   1.5   1.0   0.5   0.0

300-MHz ^{1}H NMR spectrum ppm (δ)

B

39. The following ^{1}H NMR signals are for three molecules with ether functional groups. All the signals are singlets (single, sharp peaks). Propose structures for these compounds. **(a)** $C_3H_8O_2$, $\delta = 3.3$ and 4.4 ppm (ratio 3:1); **(b)** $C_4H_{10}O_3$, $\delta = 3.3$ and 4.9 ppm (ratio 9:1); **(c)** $C_5H_{12}O_2$, $\delta = 1.2$ and 3.1 ppm (ratio 1:1). Compare and contrast these spectra with that of 1,2-dimethoxyethane (Figure 10-15B).

40. **(a)** The ^{1}H NMR spectrum of a ketone with the molecular formula $C_6H_{12}O$ has $\delta = 1.2$ and 2.1 ppm (ratio 3:1). Propose a structure for this molecule. **(b)** Each of two isomeric molecules related to the ketone in part (a) has the molecular formula $C_6H_{12}O_2$. Their ^{1}H NMR spectra are described as follows: isomer 1, $\delta = 1.5$ and 2.0 ppm (ratio 3:1); isomer 2, $\delta = 1.2$ and 3.6 ppm (ratio 3:1). All signals in these spectra are singlets. Propose structures for these compounds. To what compound class do they belong?

41. List the four important features of ^{1}H NMR and the information you can derive from them. (**Hint:** See Sections 10-4 through 10-7.)

42. Describe in which ways the ^{1}H NMR spectra of the compounds below would be similar and how they would differ. Address each of the four issues you listed in Problem 41. To which compound class does each of these compounds belong?

$$CH_3CH_2\overset{\displaystyle O}{\overset{\|}{-C}}-O-CH_3 \qquad CH_3-\overset{\displaystyle O}{\overset{\|}{C}}-O-CH_2CH_3$$

$$CH_3CH_2-\overset{\displaystyle O}{\overset{\|}{C}}-CH_3 \qquad CH_3CH_2CH_2-\overset{\displaystyle O}{\overset{\|}{C}}-H$$

43. Below are shown three $C_4H_8Cl_2$ isomers on the left and three sets of ^{1}H NMR data that one would expect on application of the simple $N + 1$ rule on the right. Match the structures to the proper spectral data. (**Hint:** You may find it helpful to sketch the spectra on a piece of scratch paper.)

(a) CH$_3$CH$_2$CHCH$_2$ (with Cl Cl above)   **(i)** $\delta = 1.5$ (d, 6 H) and 4.1 (q, 2 H) ppm

(b) CH$_3$CHCHCH$_3$ (with Cl Cl above)   **(ii)** $\delta = 1.6$ (d, 3 H), 2.1 (q, 2 H), 3.6 (t, 2 H), and 4.2 (sex, 1 H) ppm

(c) CH$_3$CHCH$_2$CH$_2$ (with Cl Cl above)   **(iii)** $\delta = 1.0$ (t, 3 H), 1.9 (quin, 2 H), 3.6 (d, 2 H), and 3.9 (quin, 1 H) ppm

44. Predict the spin–spin splitting that you would expect to observe in the NMR spectra of each compound in Problem 36. (**Reminder:** Hydrogens attached to oxygen and nitrogen do not normally exhibit spin–spin splitting.)

Remember

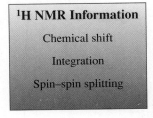

^{1}H NMR Information

Chemical shift

Integration

Spin–spin splitting

45. Predict the spin–spin splitting that you would expect to observe in the NMR spectra of each compound in Problem 37.

46. The ^{1}H NMR chemical shifts are given for each of the following compounds. As best you can, assign each signal to the proper group of hydrogens in the molecule and sketch a spectrum for each compound, incorporating spin–spin splitting whenever appropriate. **(a)** Cl$_2$CHCH$_2$Cl, $\delta = 4.0$ and 5.8 ppm; **(b)** CH$_3$CHBrCH$_2$CH$_3$, $\delta = 1.0, 1.7, 1.8,$ and 4.1 ppm; **(c)** CH$_3$CH$_2$CH$_2$COOCH$_3$, $\delta = 1.0, 1.7, 2.3,$ and 3.6 ppm; **(d)** ClCH$_2$CHOHCH$_3$, $\delta = 1.2, 3.0, 3.4,$ and 3.9 ppm.

47. ^{1}H NMR spectra C through F (see the following) correspond to four isomeric alcohols with the molecular formula $C_5H_{12}O$. Try to assign their structures.

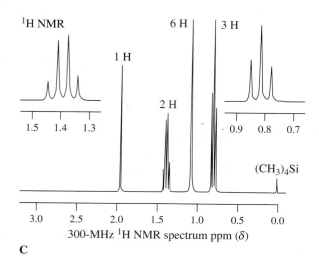

C

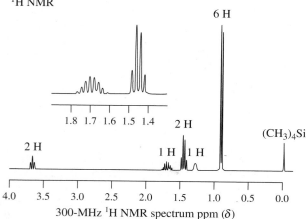

D

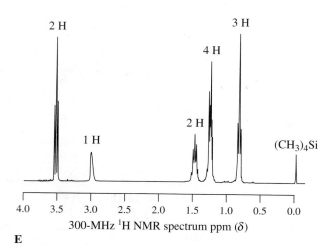

E

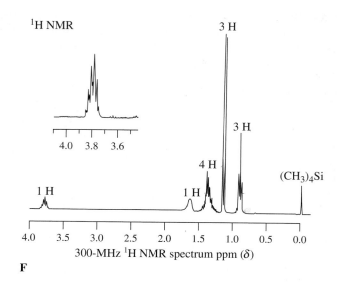

F

48. Sketch ^{1}H NMR spectra for the following compounds. Estimate chemical shifts (see Section 10-4) and show the proper multiplets for peaks that exhibit spin–spin coupling.
(a) $CH_3CH_2OCH_2Br$;   (b) $CH_3OCH_2CH_2Br$;
(c) $CH_3CH_2CH_2OCH_2CH_2CH_3$; (d) $CH_3CH(OCH_3)_2$.

49. A hydrocarbon with the formula C_6H_{14} gives rise to ^{1}H NMR spectrum G (below). What is its structure? This molecule has a structural feature similar to that of another compound whose spectrum is illustrated in this chapter. What molecule is that? Explain the similarities and differences in the spectra of the two.

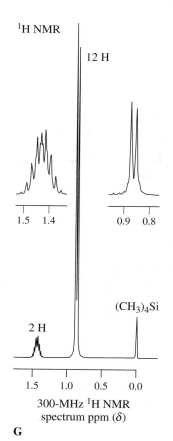

G

50. Treatment of the alcohol corresponding to NMR spectrum D in Problem 47 with hot concentrated HBr yields a substance with the formula $C_5H_{11}Br$. Its ^1H NMR spectrum exhibits signals at $\delta = 1.0$ (t, 3 H), 1.2 (s, 6 H), and 1.6 (q, 2 H) ppm. Explain. (**Hint:** See NMR spectrum C in Problem 42.)

51. The ^1H NMR spectrum of 1-chloropentane is shown at 60 MHz (spectrum H) and 500 MHz (spectrum I). Explain the differences in appearance of the two spectra, and assign the signals to specific hydrogens in the molecule.

^1H NMR

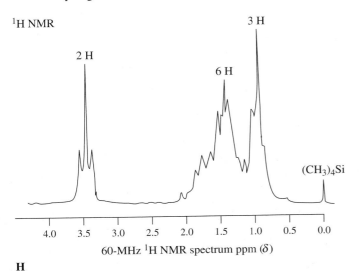

60-MHz ^1H NMR spectrum ppm (δ)

H

^1H NMR

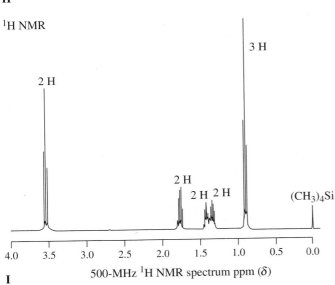

500-MHz ^1H NMR spectrum ppm (δ)

I

52. Describe the spin–spin splitting patterns that you expect for each signal in the ^1H NMR spectra of the five bromocyclopropane derivatives illustrated in Problem 35. Note that in these compounds the geminal coupling constants (i.e., between nonequivalent hydrogens on the same carbon atom—Section 10-7) and trans vicinal coupling constants are smaller (ca. 5 Hz) than the cis vicinal coupling constants (ca. 8 Hz).

53. Can the three isomeric pentanes be distinguished unambiguously from their broad-band proton-decoupled ^{13}C NMR spectra *alone?* Can the five isomeric hexanes be distinguished in this way?

54. Predict the ^{13}C NMR spectra of the compounds in Problem 36, with and without proton decoupling.

Remember

^{13}C NMR Information

Chemical shift

DEPT

55. Rework Problem 37 as it pertains to ^{13}C NMR spectroscopy.

56. How would the DEPT ^{13}C spectra of the compounds discussed in Problems 35 and 37 differ in appearance from the ordinary ^{13}C spectra?

57. From each group of three molecules, choose the one whose structure is most consistent with the proton-decoupled ^{13}C NMR data. Explain your choices. (**a**) $CH_3(CH_2)_4CH_3$, $(CH_3)_3CCH_2CH_3$, $(CH_3)_2CHCH(CH_3)_2$; $\delta = 19.5$ and 33.9 ppm. (**b**) 1-Chlorobutane, 1-chloropentane, 3-chloropentane; $\delta = 13.2$, 20.0, 34.6, and 44.6 ppm. (**c**) Cyclopentanone, cycloheptanone, cyclononanone; $\delta = 24.0$, 30.0, 43.5, and 214.9 ppm. (**d**) $ClCH_2CHClCH_2Cl$, $CH_3CCl_2CH_2Cl$, $CH_2{=}CHCH_2Cl$; $\delta = 45.1$, 118.3, and 133.8 ppm. (**Hint:** Consult Table 10-6.)

58. Propose a reasonable structure for each of the following molecules on the basis of the given molecular formula and of the ^1H and proton-decoupled ^{13}C NMR data. (**a**) $C_7H_{16}O$, spectra J and K (below; * = CH_2 by DEPT); (**b**) $C_7H_{16}O_2$, spectra L and M (next page; the assignments in M were made by DEPT).

^1H NMR

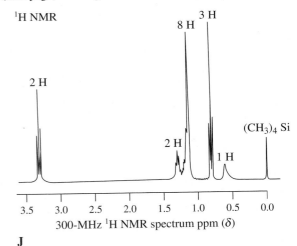

300-MHz ^1H NMR spectrum ppm (δ)

J

^{13}C NMR

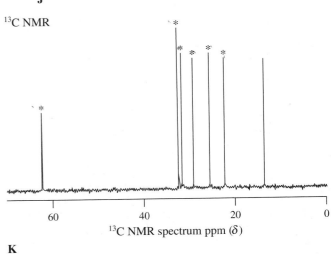

^{13}C NMR spectrum ppm (δ)

K

¹H NMR

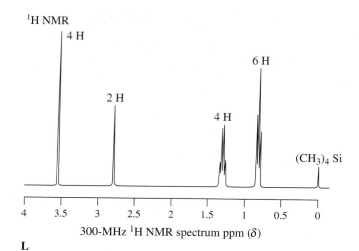

300-MHz ¹H NMR spectrum ppm (δ)

L

¹³C NMR

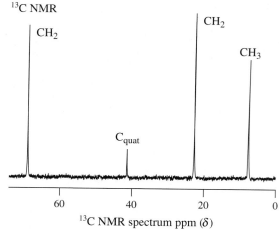

¹³C NMR spectrum ppm (δ)

M

59. **CHALLENGE** The ¹H NMR spectrum of cholesteryl benzoate (see Section 4-7) is shown as spectrum N. Although complex, it contains a number of distinguishing features. Analyze the absorptions marked by integrated values. The inset is an expansion of the signal at δ = 4.85 ppm and exhibits an approximately first-order splitting pattern. How would you describe this pattern? (**Hint:** The peak patterns at δ = 2.5, 4.85, and 5.4 ppm are simplified by the occurrence of chemical shift and/or coupling constant equivalencies.)

Cholesteryl Benzoate and LCDs

Cholesteryl benzoate was the first material for which liquid crystal properties were discovered: an ordered state of matter between that of a liquid fluid and a solid crystal. When exposed to an electric field, the initial order is disturbed, changing the transparency to incident light. This response is rapid and forms the basis for the use of liquid crystals in displays ("LCDs").

¹H NMR

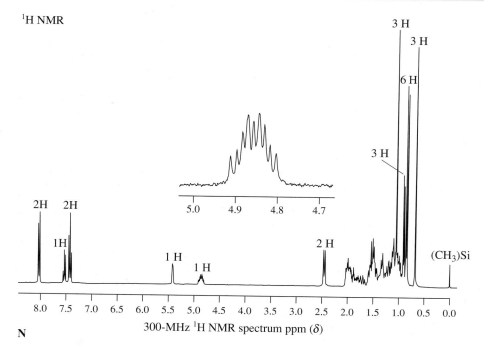

300-MHz ¹H NMR spectrum ppm (δ)

N

Cholesteryl benzoate

60. CHALLENGE   The terpene α-terpineol has the molecular formula $C_{10}H_{18}O$ and is a constituent of pine oil. As the *-ol* ending in the name suggests, it is an alcohol. Use its ^1H NMR spectrum (spectrum O, below) to deduce as much as you can about the structure of α-terpineol. [**Hints:** (1) α-Terpineol has the same 1-methyl-

4-(1-methylethyl)cyclohexane framework found in a number of other terpenes (e.g., carvone, Problem 43 of Chapter 5). (2) In your analysis of spectrum O, concentrate on the most obvious features (peaks at δ = 1.1, 1.6, and 5.4) and use chemical shifts, integrations, and the splitting of the δ 5.4 signal (inset) to help you.]

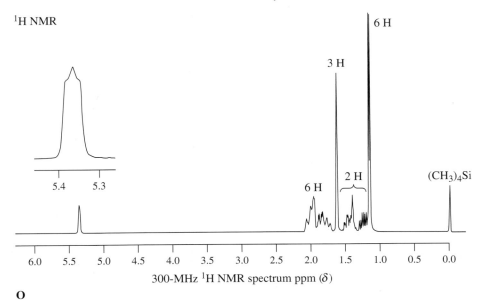

^1H NMR

300-MHz ^1H NMR spectrum ppm (δ)

O

61. CHALLENGE   Study of the solvolysis of derivatives of menthol [5-methyl-2-(1-methylethyl)cyclohexanol] has greatly enhanced our understanding of these types of reactions. Heating the isomer of the 4-methylbenzenesulfonate ester shown below in 2,2,2-trifluoro-ethanol (a highly ionizing solvent of low nucleophilicity) leads to two products with the molecular formula $C_{10}H_{18}$. **(a)** The major product displays 10 different signals in its ^{13}C NMR spectrum. Two of them are at relatively low field, about δ = 120 and 145 ppm, respectively. The ^1H NMR spectrum exhibits a multiplet near δ = 5 ppm (1 H); all other signals are upfield of δ = 3 ppm. Identify this compound. **(b)** The minor product gives only seven ^{13}C signals. Again, two are at low field (δ ≈ 125 and 140 ppm), but, in contrast with the ^1H NMR data on the major isomer, there are no signals at lower field than δ = 3 ppm. Identify this product and explain its formation mechanistically. **(c)** When the solvolysis is carried out starting with the ester labeled with deuterium at C2, the ^1H spectrum of the resulting major product isomer in part (a) reveals a significant reduction of the intensity of the signal at δ = 5 ppm, a result indicating the *partial* incorporation of deuterium at the position associated with this peak. How might this result be explained? [**Hint:** The answer lies in the mechanism of formation of the minor product in part (b).]

CH₃

2

OSO₂——⟨⟩——CH₃

(D)H   CH(CH₃)₂

CF₃CH₂OH, Δ ——→

two $C_{10}H_{18}$ products

Team Problem

62. Your team is faced with a puzzle. Four isomeric compounds, A–D, with the molecular formula C_4H_9BrO react with KOH to produce E–G with the molecular formula C_4H_8O. Molecules A and B yield compounds E and F, respectively. The NMR spectra of compounds C and D are identical, and both furnish the same product, G. Although some of the starting materials are optically active, none of the products is. Moreover, each of E, F, and G displays only two ^1H NMR signals of varying chemical shifts, none of them located between δ 4.6 and 5.7 ppm. Both the respective resonances of E and G are complex, whereas F exhibits two singlets. Proton-decoupled ^{13}C NMR spectra for E and G show only two peaks, whereas F exhibits three. Using this spectral information, work together to determine which isomers of C_4H_9BrO will yield the respective isomers of C_4H_8O. When you have matched reactant and product, divide the task of predicting the proton and carbon NMR spectra of E, F, and G among yourselves. Estimate the ^1H and ^{13}C chemical shifts for all, and predict the respective DEPT spectra.

Preprofessional Problems

63. The molecule $(CH_3)_4Si$ (tetramethylsilane) is used as an internal standard in ^1H NMR spectroscopy. One of the following properties makes it especially useful. Which one?

(a) Highly paramagnetic   (b) Highly colored
(c) Highly volatile   (d) Highly nucleophilic

64. One of the following compounds will show a doublet as part of its ^1H NMR spectrum. Which one?

(a) CH_4 (b) $ClCH(CH_3)_2$

(c) $CH_3CH_2CH_3$

(d) 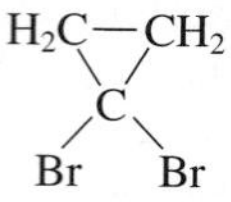 $H_2C\!-\!CH_2$ with C below bearing Br and Br

65. In the ^1H NMR spectrum of 1-fluorobutane, the most deshielded hydrogens are those bound to

(a) C_4 (b) C_3 (c) C_2 (d) C_1

66. One of the following compounds will have one peak in its ^1H NMR spectrum and two peaks in its ^{13}C NMR spectrum. Which one?

(a) 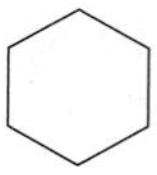 [hexagon]

(b) 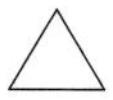 [triangle]

(c) $CH_3\!-\!CH_3$

(d) 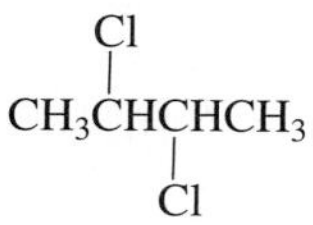 $CH_3CHCHCH_3$ with Cl above first CH and Cl below second CH

(e) 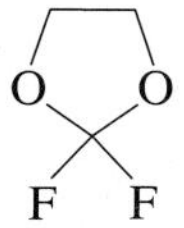 [five-membered ring with O—O at top and C bearing F F at bottom]

CHAPTER 11

Alkenes: Infrared Spectroscopy and Mass Spectrometry

What differentiates solid shortening from liquid cooking oil? Remarkably, the *only* significant structural difference is that the liquid contains carbon–carbon double bonds and the solid does not. Cooking oils are derivatives of **alkenes,** the simplest organic compounds containing multiple bonds. In this chapter and in Chapter 12, we shall investigate the properties, generation, and reactivity of alkenes.

In the preceding several chapters, we learned that haloalkanes and alcohols, two major classes of compounds containing single-bonded functional groups, may undergo elimination under appropriate conditions to form alkenes. In this chapter we return to these processes and explore some additional features that affect their outcome. We shall then proceed in Chapter 12 to examine the reactions of alkenes, and we shall discover that they may be converted back into single-bonded substances by the process of addition. Thus, we shall see how alkenes can serve as intermediaries in many synthetic conversions. They are useful and economically valuable starting materials for plastics, synthetic fibers, construction materials, and many other industrially important substances. For example, addition reactions of many gaseous alkenes give oils as products, which is why this class of compounds used to be called "olefins" (from *oleum facere,* Latin, to make oil). Indeed, "margarine" is a shortened version of the original name,

cis-9-Octadecenoic acid, also known as *oleic acid,* makes up more than 80% of natural olive oil extracted from the fruit of the European olive tree. It is acknowledged to be one of the most beneficial of all the food-derived fats and oils for human cardiovascular health. In contrast, the isomeric compound in which the double bond possesses trans instead of cis geometry has been found to have numerous adverse health effects.

$$C = C$$

Alkene double bond

oleomargarine, for this product.* Because alkenes can undergo addition reactions, they are described as **unsaturated** compounds. In contrast, alkanes, which possess the maximum number of single bonds and thus are inert with respect to addition, are referred to as **saturated.**

We begin with the names and physical properties of the alkenes and show how we evaluate the relative stability of their isomers. A review of elimination reactions allows us to further our discussion of alkene preparation.

We also introduce two additional methods for determining molecular structure: a second type of spectroscopy—infrared (IR) spectroscopy—and a technique for determining the elemental composition of a molecule—mass spectrometry (MS). These methods complement NMR by ascertaining directly the presence or absence of functional groups and their characteristic bonds (O—H, C=C, etc.), as well as their arrangement in the overall structure.

11-1 | NAMING THE ALKENES

A carbon–carbon double bond is the characteristic functional group of the alkenes. Their general formula is C_nH_{2n}, the same as that for the cycloalkanes.

Like other organic compounds, some alkenes are still known by common names, in which the *-ane* ending of the corresponding alkane is replaced by **-ylene.** Substituent names are added as prefixes.

Common Names of Typical Alkenes

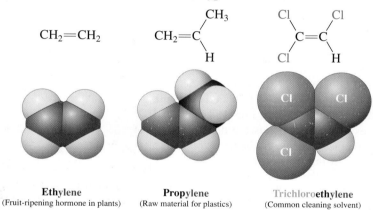

$$CH_2{=}CH_2 \qquad CH_2{=}C\begin{smallmatrix} CH_3 \\ \\ H \end{smallmatrix} \qquad \begin{smallmatrix} Cl \\ \\ Cl \end{smallmatrix}C{=}C\begin{smallmatrix} Cl \\ \\ H \end{smallmatrix}$$

Ethylene
(Fruit-ripening hormone in plants)

Propylene
(Raw material for plastics)

Trichloroethylene
(Common cleaning solvent)

In IUPAC nomenclature, the simpler ending **-ene** is used instead of -ylene, as in ethene and propene. More complicated systems require adaptations and extensions of the rules for naming alkanes (Section 2-6).

Rule 1. To name the stem, find the longest chain that *includes both* carbons making up the double bond. The molecule may have longer carbon chains, but ignore them.

$$CH_2{=}CHCHCH_2CH_3 \qquad CH_2{=}CHCH(CH_2)_4CH_3 \qquad CH_3CH_2CH_2CH_2C{=}CCH_2CH_2CH_2CH_3$$
with CH_3 above, $CH_2CH_2CH_3$ above, and H_3C, CH_2CH_3 above

A methylpentene

A propyloctene
(Not a hexene or a nonane derivative)

An ethylmethyldecene
(Not a pentene or a heptene or an octene derivative)

Rule 2. Indicate the location of the double bond in the main chain by number, starting at the end *closer* to the double bond. (Cycloalkenes do not require the numerical prefix, but the carbons making up the double bond are assigned the numbers 1 and 2, unless another group takes precedence; see rule 6.) Alkenes that have the same molecular formula but differ in the

*The name margarine itself originates indirectly from the Greek, *margaron,* pearl, and directly from margaric acid, the common name given to one of the constituent fatty acids of margarine, heptadecanoic acid, because of the shiny, "pearly" crystals it forms.

location of the double bond (such as 1-butene and 2-butene) are constitutional isomers that are also called **double-bond isomers.** A 1-alkene is also referred to as a **terminal alkene;** the others are called **internal.** Note that alkenes are easily depicted in line notation.

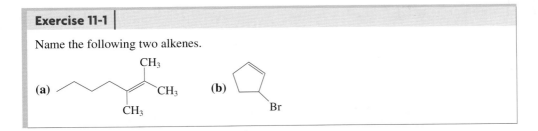

$$\overset{1}{C}H_2=\overset{2}{C}H\overset{3}{C}H_2\overset{4}{C}H_3$$

1-Butene
(A terminal alkene;
not 3-butene)

$$\overset{1}{C}H_3\overset{2}{C}H=\overset{3}{C}H\overset{4}{C}H_3$$

2-Butene
(An internal alkene
and a double-bond
isomer of 1-butene)

2-Pentene
(Not 3-pentene)

Cyclohexene

Rule 3. Add substituents and their positions to the alkene name as prefixes. If the alkene stem is symmetric, begin from the end that gives the first substituent along the chain the lowest possible number.

$$\overset{1}{C}H_2=\overset{2}{C}H\overset{3}{C}H\overset{4}{C}H_2\overset{5}{C}H_3$$
with CH₃ on C3

3-Methyl-1-pentene

3-Methylcyclohexene
(Not 6-methylcyclohexene)

$$\overset{1}{C}H_3\overset{2}{C}H\overset{3}{C}H=\overset{4}{C}H\overset{5}{C}H_2\overset{6}{C}H_3$$
with CH₃ on C2

2-Methyl-3-hexene
(Not 5-methyl-3-hexene)

Exercise 11-1

Name the following two alkenes.

(a) [structure with CH₃, CH₃, CH₃ groups]   (b) [cyclopentene structure with Br]

Rule 4. Identify any stereoisomers. In a 1,2-disubstituted ethene, the two substituents may be on the same side of the molecule or on opposite sides. The first stereochemical arrangement is called cis and the second trans, in analogy to the cis-trans names of the disubstituted cycloalkanes (Section 4-1). Two alkenes of the same molecular formula differing only in their stereochemistry are called **geometric** or **cis-trans isomers** and are examples of diastereomers: stereoisomers that are not mirror images of each other.

Same side of
double bond
cis

Opposite sides of
double bond
trans

cis-**2-Butene**   *trans*-**2-Butene**   *cis*-**4-Chloro-2-pentene**

Exercise 11-2

Name the following three alkenes.

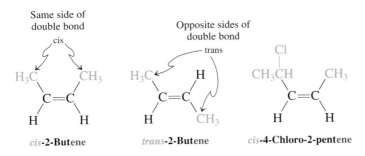

(a)   (b)   (c) Br

In the smaller substituted cycloalkenes, the double bond can exist only in the cis configuration. The trans arrangement is prohibitively strained (as building a model reveals). However, in larger cycloalkenes, trans isomers are stable.

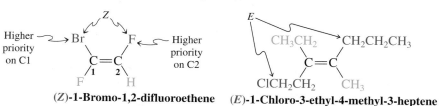

3-Fluoro-1-methylcyclopentene   **1-Ethyl-2,4-dimethylcyclohexene**   *trans*-**Cyclodecene**

(In both cases, only the cis isomer is stable)

Rule 5. The labels *cis* and *trans* cannot be applied when there are three or four different substituents attached to the double-bond carbons. An alternative system for naming such alkenes has been adopted by IUPAC: the **E,Z system.** In this convention, the sequence rules devised for establishing priority in *R,S* names (Section 5-3) are applied separately to the two groups on each double-bonded carbon. When the two groups of higher priority are on opposite sides, the molecule is of the *E* configuration (E from *entgegen,* German, opposite). When the two substituents of higher priority appear on the same side, the molecule is a *Z* isomer (Z from *zusammen,* German, together).

(Z)-1-Bromo-1,2-difluoroethene   (E)-1-Chloro-3-ethyl-4-methyl-3-heptene

---

Exercise 11-3

Name the following three alkenes.

(a)   (b)   (c)

Rule 6. With the alkenes, we are introducing a new functional group after the alcohols. This rule addresses the problem that arises when we have both functions in a compound: Do we call it an alkene or an alcohol? The answer is that we give the hydroxy functional group precedence over the double bond. Therefore, we name alcohols containing double bonds **alkenols,** and the stem *incorporating both functions* is numbered to give the carbon bearing the OH group the lowest possible assignment. Note that the last *e* in alkene is dropped in the naming of alkenols.

$\overset{3}{C}H_2 = \overset{2}{C}H\overset{1}{C}H_2OH$

2-Propen-1-ol

(Not 1-propen-3-ol)

(Z)-5-Chloro-3-ethyl-4-hexen-2-ol

(The two stereocenters are unspecified)

3-Cyclohexen-1-ol

(Not 1-cyclohexen-4-ol)

Exercise 11-4

Draw the structures of the following molecules: **(a)** *trans*-3-penten-1-ol; **(b)** 2-cyclohexen-1-ol.

Rule 7. Substituents containing a double bond are named **alkenyl;** for example, ethenyl (common name, vinyl), 2-propenyl (allyl), and *cis*-1-propenyl.

$$CH_2{=}CH{-} \qquad CH_2{=}CH{-}CH_2{-}$$

Ethenyl   **2-Propenyl**   *cis*-**1-Propenyl**
(Vinyl)   **(Allyl)**

As usual, the numbering of a substituent chain begins at the point of attachment to the basic stem. The example in the margin constitutes an alkenol. However, we cannot incorporate both functional groups into the stem name. Therefore, the double bond is part of the substituent to the cyclooctanol.

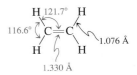

trans-**3-(4-Pentenyl)cyclooctanol**

Exercise 11-5

(a) Draw the structure of *trans*-2-ethenylcyclopropanol. **(b)** Name the structure shown in the margin.

11-2 | STRUCTURE AND BONDING IN ETHENE: THE PI BOND

The carbon–carbon double bond in alkenes has special electronic and structural features. This section reviews the hybridization of the carbon atoms in this functional group, the nature of its two bonds (σ and π), and their relative strengths. We consider ethene, the simplest of the alkenes.

The double bond consists of sigma and pi components

Ethene is planar, with two trigonal carbon atoms and bond angles close to 120° (Figure 11-1). Therefore, both carbon atoms are best described as being sp^2 hybridized (Section 1-8; Figure 1-21). Two sp^2 hybrids on each carbon atom overlap with hydrogen 1s orbitals to form the four C–H σ bonds. The remaining sp^2 orbitals, one on each carbon, overlap with each other to form the carbon–carbon σ bond. Each carbon also possesses a 2p orbital; these are aligned parallel to each other and are close enough to overlap, forming the π **bond** (Figure 11-2A). The electron density in a π bond is distributed over both carbons *above and below* the molecular plane, as indicated in Figure 11-2B.

Figure 11-1 Molecular structure of ethene.

The pi bond in ethene is relatively weak

How much do the σ and π bonds each contribute to the total double-bond strength? We know from Section 1-7 that bonds are made by overlap of orbitals and that their relative strengths depend on the effectiveness of this overlap. Therefore, we can expect overlap in a σ bond to be considerably better than that in a π bond, because the sp^2 orbitals lie

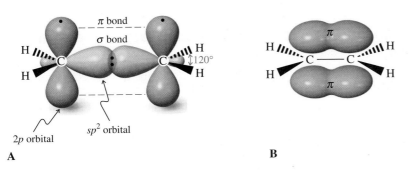

π bond

σ bond

$2p$ orbital sp^2 orbital

A   **B**

Figure 11-2 An orbital picture of the double bond in ethene. The σ carbon–carbon bond is made by sp^2–sp^2 overlap. The pair of p orbitals perpendicular to the ethene molecular plane overlap to form the additional π bond. For clarity, this overlap is indicated in (A) by the dashed green lines; the orbital lobes are shown artificially separated. Another way of presenting the π bond is depicted in (B), in which the "π-electron cloud" is above and below the molecular plane.

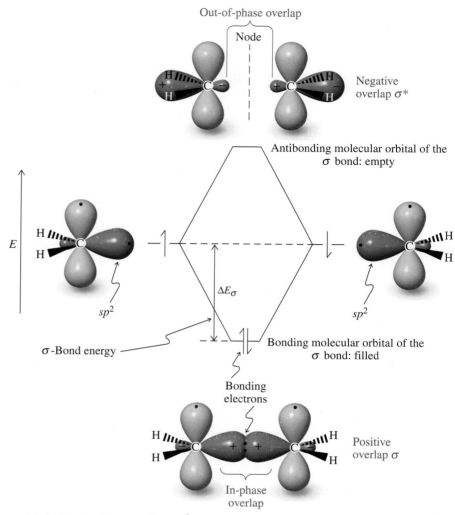

Recall: Mixing (any) two orbitals generates two new molecular orbitals.

Figure 11-3 Overlap between two sp^2 hybrid orbitals (containing one electron each; shown in red) determines the relative strength of the σ bond of ethene. In-phase interaction between regions of the wave function having the *same* sign reinforces bonding (compare in-phase overlap of waves, Figure 1-4B) and creates a *bonding molecular orbital*. [Recall: These signs do *not* refer to charges; the + designations are chosen arbitrarily (see Figure 1-11).] Both electrons end up occupying this orbital and have a high probability of being located near the internuclear axis. The orbital stabilization energy, ΔE_σ, corresponds to the σ-bond strength. The out-of-phase interaction, between regions of *opposite* sign (compare Figure 1-4C), results in an unfilled *antibonding molecular orbital* (designated σ^*) with a node.

along the internuclear axis (Figure 11-2). This situation is illustrated in energy-level-interaction diagrams (Figures 11-3 and 11-4) analogous to those used to describe the bonding in the hydrogen molecule (Figures 1-11 and 1-12). Figure 11-5 combines (for comparison) our predictions of the relative energies of the two pairs of molecular orbitals (σ and π) that make up the double bond in ethene.

Thermal isomerization allows us to measure the strength of the pi bond

How do these predictions of the π-bond strength compare with experimental values? We can measure the energy required to interconvert the cis form of a substituted alkene—say, 1,2-dideuterioethene—with its trans isomer. In this process, called **thermal isomerization,** the two p orbitals making up the π bond are rotated 180°. At the midpoint of this rotation—90°—the π (but not the σ) bond has been broken (Figure 11-6). Thus, the activation energy for the reaction can be roughly equated with the π energy of the double bond.

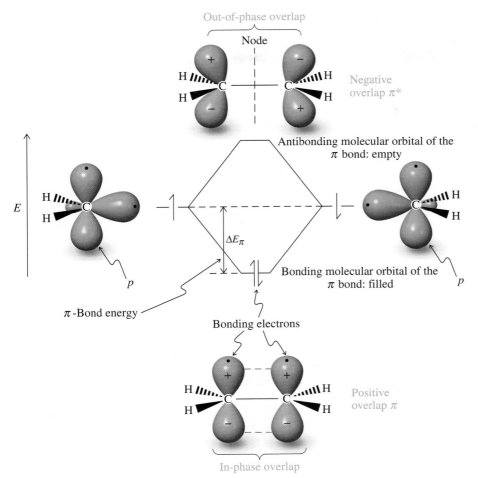

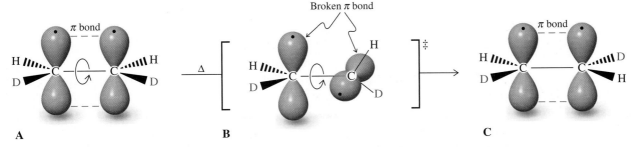

Figure 11-4 Compare this picture of the formation of the π bond in ethene with Figure 11-3. In-phase interaction between two parallel *p* orbitals (containing one electron each; shown in blue) results in positive overlap and a filled bonding π orbital. The representation of this orbital indicates the probability of finding the electrons between the carbons above and below the molecular plane. Because π overlap is less effective than σ, the stabilization energy, ΔE_π, is smaller than ΔE_σ. The π bond is therefore weaker than the σ bond. The out-of-phase interaction results in the antibonding molecular orbital π*.

Figure 11-5 Energy ordering of the molecular orbitals making up the double bond. The four electrons occupy only bonding orbitals.

Thermal isomerization occurs, but only at high temperatures (>400°C). Its activation energy is 65 kcal mol^{-1} (272 kJ mol^{-1}), a value we may assign to the strength of the π bond. Below 300°C, double bonds are configurationally stable; that is, cis stays cis and trans stays trans. The strength of the double bond in ethene—the energy required for dissociation into two CH$_2$ fragments—is 173 kcal mol^{-1} (724 kJ mol^{-1}). Consequently the C–C σ bond amounts

Figure 11-6 Thermal isomerization of *cis*-dideuterioethene to the trans isomer requires breaking the π bond. The reaction proceeds from starting material (A) through rotation around the C–C bond until it reaches the point of highest energy, the transition state (B). At this stage, the two *p* orbitals used to construct the π bond are perpendicular to each other. Further rotation in the same direction results in a product in which the two deuterium atoms are trans (C).

Remember:
The symbol ‡ in Figure 11-6 denotes a transition state.

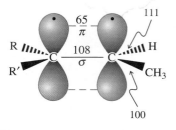

Figure 11-7 Approximate bond strengths in an alkene (in kcal mol^{-1}). Note the relative weakness of the π bond.

to about 108 kcal mol^{-1} (452 kJ mol^{-1}, Figure 11-7). Note that the other σ bonds to the alkenyl carbon are also stronger than the analogous bonds in alkanes (Table 3-2). This effect is due largely to improved bonding overlap involving the relatively compact sp^2 orbitals. As one consequence, the strongly bound alkenyl hydrogens are *not* abstracted in radical reactions. Instead, additions to the weaker π bond characterize the reactivity of alkenes (Chapter 12).

In Summary The characteristic hybridization scheme for the double bond of an alkene accounts for its physical and electronic features. Alkenes contain a planar double bond, incorporating trigonal carbon atoms. Their hybridization explains the strong σ and weaker π bonds, stable cis and trans isomers, and the strength of the alkenyl-substituent bond. Alkenes are prone to undergo addition reactions, in which the weaker π bond but not the C–C σ bond is broken.

11-3 | PHYSICAL PROPERTIES OF ALKENES

The carbon–carbon double bond alters many of the physical properties of alkenes relative to those of alkanes. Exceptions are boiling points, which are very similar, primarily a result of similar London forces (Figure 2-6C). Like their alkane counterparts, ethene, propene, and the butenes are gases at room temperature. Melting points, however, depend in part on the packing of molecules in the crystal lattice, a function of molecular shape. The double bond in cis-disubstituted alkenes imposes a U-shaped bend in the molecule that disrupts packing and reduces the melting point, usually below that of either the corresponding alkane or isomeric trans alkene (Table 11-1). A cis double bond is responsible for the sub-room-temperature melting point of vegetable oil.

Depending on their structure, alkenes may exhibit weak dipolar character. Why? Bonds between alkyl groups and an alkenyl carbon are polarized in the direction of the sp^2-hybridized atom, because the degree of s character in an sp^2 hybrid orbital is greater than that in an sp^3. Electrons in orbitals with increased s character are held closer to the nucleus than those in orbitals containing more p character. This effect makes the sp^2 carbon relatively electron withdrawing (although much less so than strongly electronegative atoms such as O and Cl) and creates a weak dipole along the substituent–alkenyl carbon bond. To put it differently, *alkyl substituents are inductive electron donors to the π bond.*

In cis-disubstituted alkenes, the two individual dipoles combine to give a net molecular dipole. These dipoles are opposed in trans-disubstituted alkenes and tend to cancel each other. The more polar cis-disubstituted alkenes often have slightly higher boiling points than their trans counterparts. The boiling-point difference is greater when the individual bond dipoles are larger, as in the case of the two 1,2-dichloroethene isomers, whose highly electronegative chlorine atoms also cause the direction of the bond dipoles to be reversed.

Table 11-1	Comparison of Melting Points of Alkenes and Alkanes	
Compound		**Melting point (°C)**
Butane		−138
trans-2-Butene		−106
cis-2-Butene		−139
Pentane		−130
trans-2-Pentene		−135
cis-2-Pentene		−180
Hexane		−95
trans-2-Hexene		−133
cis-2-Hexene		−141
trans-3-Hexene		−115
cis-3-Hexene		−138

Polarization in Alkenes: Alkyl Groups are Inductive Electron Donors

Net dipole

H$_3$C   CH$_3$
   C=C
H   H
b.p. 4°C

H$_3$C   H
   C=C
H   CH$_3$
b.p. 1°C
No net dipole

Net dipole

Cl   Cl
   C=C
H   H
b.p. 60°C

Cl   H
   C=C
H   Cl
b.p. 48°C
No net dipole

Another consequence of the electron-attracting character of the sp^2 carbon is the increased acidity of the alkenyl hydrogen. Whereas ethane has an approximate pK_a of 50, ethene is somewhat more acidic with a pK_a of 44. Even so, ethene is a very weak acid compared with other compounds, such as the carboxylic acids or alcohols.

Acidity of the Ethenyl Hydrogen

$$CH_3—CH_2—H \underset{K \approx 10^{-50}}{\rightleftharpoons} CH_3—\ddot{C}H_2^- + H^+$$

Ethyl anion

Relatively more acidic

$$CH_2=C\underset{H}{\overset{H}{\diagup}} \underset{K \approx 10^{-44}}{\rightleftharpoons} CH_2=\ddot{C}H^- + H^+$$

Ethenyl (vinyl) anion

Exercise 11-6

Ethenyllithium (vinyllithium) is not generally prepared by direct deprotonation of ethene but rather from chloroethene (vinyl chloride) by metallation (Section 8-7).

$$CH_2=CHCl + 2\ Li \xrightarrow{(CH_3CH_2)_2O} \underset{60\%}{CH_2=CHLi} + LiCl$$

Upon treatment of ethenyllithium with acetone followed by aqueous work-up, a colorless liquid is obtained in 74% yield. Propose a structure.

In Summary The presence of the double bond does not greatly affect the boiling points of alkenes, compared with alkanes, but cis-disubstituted alkenes usually have lower melting points than their trans isomers, because the cis compounds pack less well in the solid state. Alkenyl hydrogens are more acidic than those in alkanes because of the electron-withdrawing character of the sp^2-hybridized alkenyl carbon.

11-4 | NUCLEAR MAGNETIC RESONANCE OF ALKENES

The double bond exerts characteristic effects on the ^{1}H and ^{13}C chemical shifts of alkenes (see Tables 10-2 and 10-6). Let's see how to make use of this information in structural assignments.

The pi electrons exert a deshielding effect on alkenyl hydrogens

Figure 11-8 shows the ^{1}H NMR spectrum of *trans*-2,2,5,5-tetramethyl-3-hexene. Only two signals are observed, one for the 18 equivalent methyl hydrogens and one for the 2 alkenyl protons. The absorptions appear as singlets because the methyl hydrogens are too far away from the alkenyl hydrogens to produce detectable coupling. The low-field resonance of the alkenyl hydrogens (δ = 5.30 ppm) is typical of hydrogen atoms bound to alkenyl carbons. Terminal alkenyl hydrogens (RR'C=CH$_2$) resonate at δ = 4.6–5.0 ppm, their internal counterparts (RCH=CHR') at δ = 5.2–5.7 ppm.

Why is deshielding so pronounced for alkenyl hydrogens? Although the electron-withdrawing character of the sp^2-hybridized carbon is partly responsible, another phenomenon is more important: *the movement of the electrons in the π bond*. When subjected to an external magnetic field perpendicular to the double-bond axis, these electrons enter into a circular motion. This motion induces a local magnetic field that *reinforces* the external

Figure 11-8 300-MHz
^{1}H NMR spectrum of *trans*-
2,2,5,5-tetramethyl-3-hexene,
illustrating the deshielding effect
of the π bond in alkenes. It reveals
two sharp singlets for two sets of
hydrogens: the 18 methyl hydrogens at $\delta = 0.97$ ppm and 2 highly
deshielded alkenyl protons at
$\delta = 5.30$ ppm.

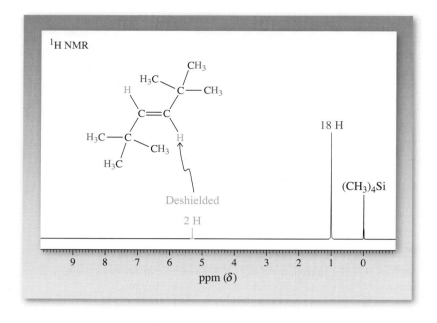

Figure 11-9 Movement of electrons in the π bond causes pronounced deshielding of alkenyl
hydrogens. An external field, H_0,
induces a circular motion of the
π electrons (shown in red) above
and below the plane of a double
bond. This motion in turn induces
a local magnetic field (shown in
green) that opposes H_0 at the
center of the double bond but
reinforces it in the regions occupied by the alkenyl hydrogens.

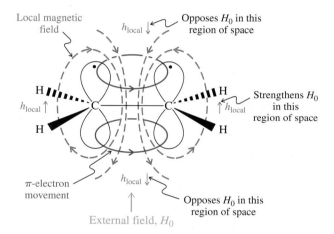

field at the edge of the double bond (Figure 11-9). As a consequence, the alkenyl hydrogens
are strongly deshielded (Section 10-4).

Exercise 11-7

The hydrogens on methyl groups attached to alkenyl carbons resonate at about $\delta = 1.6$ ppm (see
Table 10-2). Explain the deshielding of these hydrogens relative to hydrogens on methyl groups
in alkanes. (**Hint:** Try to apply the principles in Figure 11-9.)

Cis coupling through a double bond is different from trans coupling

When a double bond is not substituted symmetrically, the alkenyl hydrogens are nonequivalent, a situation leading to observable spin–spin coupling such as that shown in the spectra of *cis*- and *trans*-3-chloropropenoic acid (Figure 11-10). Note that the coupling constant

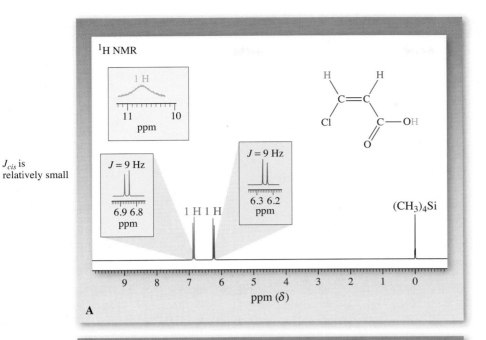

J_{cis} is
relatively small

A

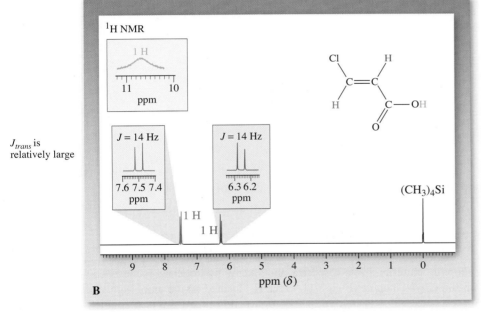

J_{trans} is
relatively large

B

Figure 11-10 300-MHz
^{1}H NMR spectra of (A) *cis*-3-
chloropropenoic acid and (B) the
corresponding trans isomer. The
two alkenyl hydrogens are nonequivalent and coupled. The
broad carboxylic acid proton
($-CO_2H$) resonates at $\delta =$
10.80 ppm and is shown in
the inset.

for the hydrogens situated cis ($J = 9$ Hz) is different from that for the hydrogens arranged trans ($J = 14$ Hz). Table 11-2 gives the magnitudes of the various possible couplings around a double bond. Although the range of J_{cis} overlaps that of J_{trans}, within a set of isomers J_{cis} is always smaller than J_{trans}. In this way cis and trans isomers can be readily distinguished.

Coupling between hydrogens on adjacent carbon atoms, such as J_{cis} and J_{trans}, is called **vicinal.** Coupling between nonequivalent hydrogens on the same carbon atom is referred to as **geminal.** In alkenes, geminal coupling is usually small (Table 11-2). Coupling to neighboring alkyl hydrogens (**allylic,** see Section 11-1) and across the double bond (**1,4-** or **long-range**) also is possible, sometimes giving rise to complicated spectral patterns. Thus,

Table 11-2	Coupling Constants Around a Double Bond		
		J (Hz)	
Type of coupling	**Name**	**Range**	**Typical**
(cis C=C structure)	Vicinal, cis	6–14	10
(trans C=C structure)	Vicinal, trans	11–18	16
(geminal C=C structure)	Geminal	0–3	2
(C=C–C–H structure)	None	4–10	6
(allylic structure)	Allylic, (1,3)-cis or -trans	0.5–3.0	2
(1,4 structure)	(1,4)- or long-range	0.0–1.6	1

the simple rule devised for saturated systems, discounting coupling between hydrogens farther than two intervening atoms apart, does not hold for alkenes.

Further coupling leads to more complex spectra

The "doublet of doublet" splitting patterns for the alkene hydrogens illustrate the sequential $N + 1$ rule (Section 10-8). One hydrogen split by one neighbor by J_1 and another neighbor by J_2 will display $(N_{J_1} + 1) \times (N_{J_2} + 1) = 2 \times 2 = 4$ peaks.

The spectra of 3,3-dimethyl-1-butene and 1-pentene illustrate the potential complexity of the coupling patterns. In both spectra, the alkenyl hydrogens appear as complex multiplets. In 3,3-dimethyl-1-butene (Figure 11-11A), H_a, located on the more highly substituted carbon atom, resonates at lower field ($\delta = 5.86$ ppm) and in the form of a doublet of doublets with two relatively large coupling constants (trans $J_{ab} = 18$ Hz, cis $J_{ac} = 10.5$ Hz). Hydrogens H_b and H_c also absorb as a doublet of doublets each because of their respective coupling to H_a and their small mutual coupling (geminal $J_{bc} = 1.5$ Hz). In the spectrum of 1-pentene (Figure 11-11B), additional coupling due to the attached alkyl group (see Table 11-2) creates a relatively complex pattern for the alkenyl hydrogens, although the two sets (terminal and internal) are clearly differentiated. In addition, the electron-withdrawing effect of the sp^2 carbon and the movement of the π electrons (Figure 11-9) cause a slight deshielding of the directly attached (allylic) CH_2 group. The magnitude of the coupling between these hydrogens and the neighboring alkenyl hydrogen is about the same (6–7 Hz) as the coupling with the two CH_2 hydrogens on the other side. As a result, the multiplet for this allylic CH_2 group appears as a quartet (with additional long-range couplings to the terminal alkenyl hydrogens), in accordance with the simple $N + 1$ rule: $N = (2 \text{ H from } CH_2) + \left(1 \text{ H from } =C \right) = 3.$

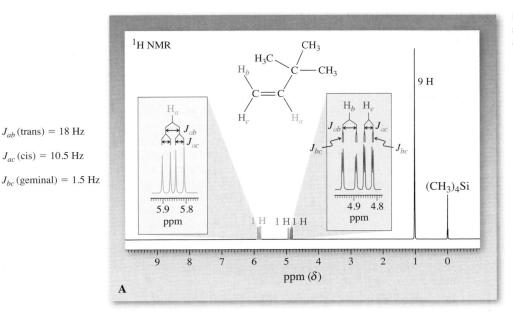

¹H NMR

J_{ab} (trans) = 18 Hz

J_{ac} (cis) = 10.5 Hz

J_{bc} (geminal) = 1.5 Hz

Figure 11-11 300-MHz ¹H NMR spectra of (A) 3,3-dimethyl-1-butene and (B) 1-pentene.

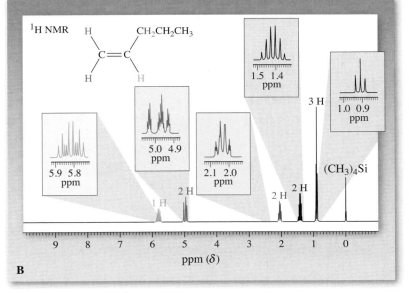

Alkenyl hydrogens show non-first-order multiplets

The sequential $N + 1$ rule nicely explains the splitting of the signal at δ 5.82 ppm in Figure 11-11B (for the blue H). That hydrogen has four neighbors, one that splits it by 16 Hz (J_{trans}), another by 10 Hz (J_{cis}), and the remaining two (on the CH$_2$ group) by 8 Hz (J_{vicinal}). The result is $(N_{J_{\text{trans}}} + 1) \times (N_{J_{\text{cis}}} + 1) \times (N_{J_{\text{vic}}} + 1) = 2 \times 2 \times 3 = 12$ peaks, two pairs of which overlap, leaving 10 that can be readily seen.

Solved Exercise 11-8	**Working with the Concepts: Interpreting NMR Spectra of Alkenes**

Ethyl 2-butenoate (ethyl crotonate), $CH_3CH{=}CHCO_2CH_2CH_3$, in CCl$_4$ has the following ¹H NMR spectrum: δ = 1.24 (t, J = 7 Hz, 3 H), 1.88 (dd, J = 6.8, 1.7 Hz, 3 H), 4.13 (q, J = 7 Hz, 2 H), 5.81 (dq, J = 16, 1.7 Hz, 1 H), and 6.95 (dq, J = 16, 6.8 Hz, 1 H) ppm; dd denotes a doublet of doublets, dq a doublet of quartets. Assign the various hydrogens and indicate whether the double bond is substituted cis or trans (consult Table 11-2).

Strategy

You are given both the compound's structure and the NMR data. Therefore, this problem consists of deciphering the information. The ¹H NMR spectrum shows five signals, one for each of the five

Remember WHIP

*W*hat
*H*ow
*I*nformation
*P*roceed

distinct hydrogen environments in the molecule. Each signal has an integration value telling us the number of hydrogens giving rise to that signal, and that is where we start. Then we analyze the location and splitting (if any) of each signal in turn.

Solution

• Beginning with the highest-field (smallest δ ppm value) signal, at $\delta = 1.24$ ppm, it integrates as 3 H; therefore, it must arise from one of the two CH_3 groups in the molecule. Looking further, we see that it is split into a triplet, three lines. According to the $N + 1$ rule for spin–spin splitting, this tells us that the CH_3 group giving rise to this signal is adjacent to a carbon bearing two hydrogens, the CH_2 group in the structure (2 neighboring hydrogens + 1 = 3 lines). We can find the signal for this CH_2 at $\delta = 4.13$; as expected, it is a quartet (3 neighboring methyl hydrogens + 1 = 4 lines). By a process of elimination, the CH_3 group on the alkene carbon must be responsible for the signal at $\delta = 1.88$. This signal is split as a *doublet of doublets,* indicative of splitting to *two distinct* single hydrogens: the two different alkene hydrogens.

• We now turn to these final two hydrogens, which give signals at $\delta = 5.81$ and 6.95 ppm, respectively. Both are *doublets of quartets*—in other words, eight-line patterns that appear as pairs of quartets. What does this imply? Each is evidently split into a quartet by the CH_3 on the alkene carbon ($3 + 1 = 4$). Also, because the two alkene hydrogens are neighbors and in dissimilar environments, they split each other, causing the doublet splitting of each pattern (1 neighbor + 1 = 2 lines). Which is which? The alkene hydrogen at $\delta = 6.95$ ppm shows the larger quartet splitting, 6.8 Hz, suggesting that it is the one adjacent to the CH_3 group. The other alkene hydrogen, at $\delta = 5.81$ ppm, shows a much smaller quartet splitting of 1.7 Hz, consistent with its increased distance to the methyl. Notice that the order in which the information is presented allows us to assign each coupling constant to each splitting: given "dq, $J = 16, 6.8$ Hz," we infer that the first splitting (d) corresponds to the first J value (16 Hz), and the second (q) to the second J (6.8 Hz).

• Finally, with $J = 16$ Hz for the mutual splitting of the alkene hydrogens, we conclude from the data in Table 11-2 that they are trans to each other. In problems such as this, it is often instructive to reproduce all the information pictorially, as shown below. Notice that this representation is especially useful in revealing the mutual couplings by their identical J values, for example, 6.8 Hz for both the methyl hydrogens and the alkenyl hydrogen on the left. Identify the others on your own.

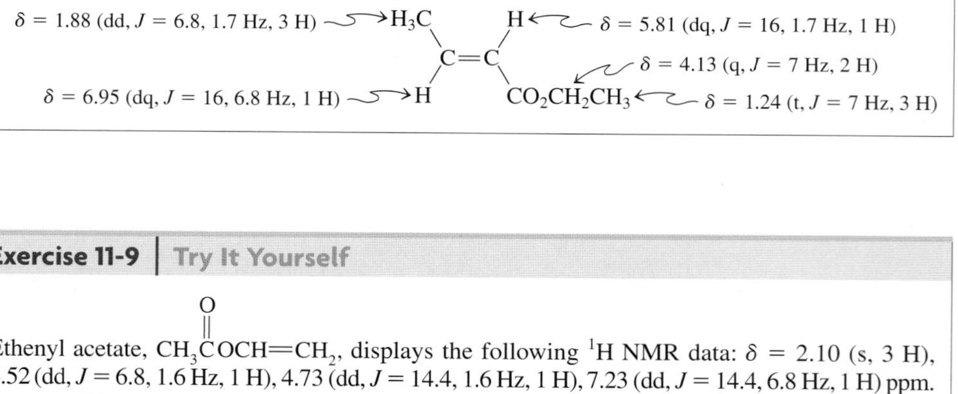

$\delta = 1.88$ (dd, $J = 6.8, 1.7$ Hz, 3 H) H_3C H $\delta = 5.81$ (dq, $J = 16, 1.7$ Hz, 1 H)

$\delta = 6.95$ (dq, $J = 16, 6.8$ Hz, 1 H) H $CO_2CH_2CH_3$ $\delta = 4.13$ (q, $J = 7$ Hz, 2 H) $\delta = 1.24$ (t, $J = 7$ Hz, 3 H)

Table 11-3 Comparison of ^{13}C NMR Absorptions of Alkenes with the Corresponding Alkane Carbon Chemical Shifts (in ppm)

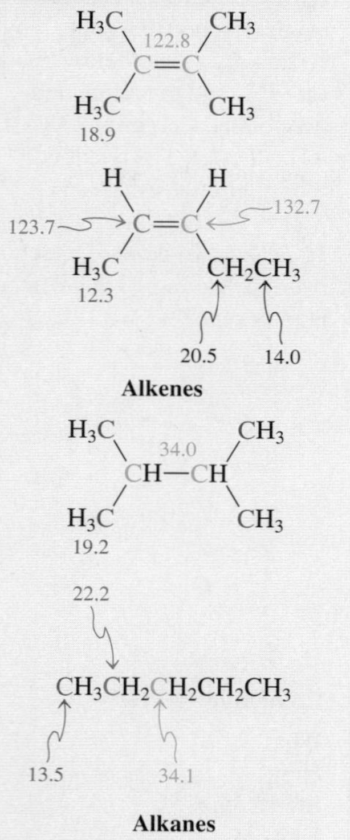

Alkenes

Alkanes

Exercise 11-9 | **Try It Yourself**

Ethenyl acetate, $CH_3\overset{\text{O}}{\underset{\|}{C}}OCH\!=\!CH_2$, displays the following ^{1}H NMR data: $\delta = 2.10$ (s, 3 H), 4.52 (dd, $J = 6.8, 1.6$ Hz, 1 H), 4.73 (dd, $J = 14.4, 1.6$ Hz, 1 H), 7.23 (dd, $J = 14.4, 6.8$ Hz, 1 H) ppm. Interpret this spectrum.

Alkenyl carbons are deshielded in ^{13}C NMR

The carbon NMR absorptions of the alkenes also are highly revealing. Relative to alkanes, the corresponding alkenyl carbons (with similar substituents) absorb at about 100 ppm lower field (see Table 10-6). Two examples are shown in Table 11-3, in which the carbon chemical shifts of an alkene are compared with those of its saturated counterpart. Recall that, in broad-band decoupled ^{13}C NMR spectroscopy, all magnetically unique carbons absorb as sharp single lines (Section 10-9). It is therefore very easy to determine the presence of sp^2 carbons by this method.

| REAL LIFE: MEDICINE 11-1 | **NMR of Complex Molecules: The Powerfully Regulating Prostaglandins** |

NMR analysis is widely used to determine the structures of complex molecules containing multiple functional groups. The first three compounds shown are members of the prostaglandin (PG) family of naturally occurring, biologically active substances. The ^{1}H NMR spectra of these PGs reveal some aspects of their structures, but overall they are quite complicated, with many overlapping signals. In contrast, ^{13}C NMR permits rapid distinction between PG derivatives, merely by counting the peaks in three chemical-shift ranges. For example, PGE$_2$ is readily distinguished by the presence of two signals near $\delta = 70$ ppm for two hydroxy carbons, four alkene ^{13}C resonances between $\delta = 125$ and 140 ppm, and two carbonyls above $\delta = 170$ ppm.

Prostaglandins are extremely potent hormone-like substances with many biological functions, including muscle stimulation, inhibition of platelet aggregation, lowering of blood pressure, enhancement of inflammatory reactions, and induction of labor in childbirth. Indeed, the anti-inflammatory properties of aspirin (see Real Life 22-2) are due to its ability to suppress prostaglandin biosynthesis. An undesirable side effect of aspirin use is gastric ulceration, because some prostaglandins function to protect the stomach lining. The synthetic prostaglandin-like substance misoprostol exhibits a similar protective effect and is frequently administered together with aspirin or other anti-inflammatory agents to prevent ulcer formation.

In Summary Hydrogen NMR is highly effective in establishing the presence of double bonds in organic molecules. Alkenyl hydrogens and carbons are strongly deshielded. The order of coupling is $J_{gem} < J_{cis} < J_{trans}$. Coupling constants for allylic substituents display characteristic values as well. ^{13}C NMR identifies alkenyl carbons by their unusually low-field chemical shifts, compared with those of alkane carbons.

11-5 CATALYTIC HYDROGENATION OF ALKENES: RELATIVE STABILITY OF DOUBLE BONDS

When an alkene and hydrogen gas are mixed in the presence of catalysts such as palladium or platinum, two hydrogen atoms add to the double bond to give the saturated alkane (see Section 12-2). This reaction, which is called **hydrogenation,** is very exothermic. The heat released, called **heat of hydrogenation,** is typically about -30 kcal mol^{-1} (-125 kJ mol^{-1}) per double bond.

Hydrogenation of an Alkene

$$\text{C=C} + \text{H—H} \xrightarrow{\text{Pd or Pt}} \text{—C—C—}$$

$\Delta H° \approx -30$ kcal mol^{-1}

Heats of hydrogenation can be measured so accurately that they may be used to determine the relative energy contents, and therefore the thermodynamic stabilities, of alkenes. Let us see how this is done.

The heat of hydrogenation is a measure of stability

In Section 3-11 we presented one method for determining relative stability: measuring heat of combustion. The less stable the molecule, the greater its energy content, and the more energy is released in this process. A very similar connection can be established using heats of hydrogenation.

For example, what are the relative stabilities of the three isomers 1-butene, *cis*-2-butene, and *trans*-2-butene? Hydrogenation of each isomer leads to the same product: butane. If their respective energy contents are equal, their heats of hydrogenation should also be equal; however, as the reactions in Figure 11-12 illustrate, they are not. The most heat is evolved by hydrogenation of the terminal double bond, the next most exothermic reaction is that with *cis*-2-butene, and finally the trans isomer gives off the least heat. Therefore, the thermodynamic stability of the butenes increases in the order 1-butene < *cis*-2-butene < *trans*-2-butene (Figure 11-12).

Highly substituted alkenes are most stable; trans isomers are more stable than cis isomers

The results of the preceding hydrogenation reactions may be generalized: The relative stability of the alkenes increases with increasing substitution, and trans isomers are usually more stable than their cis counterparts. The first trend is due in part to hyperconjugation. Just as the stability of a radical increases with increasing alkyl substitution (Section 3-2), the p orbitals of a π bond can be stabilized by alkyl substituents.

The fat molecules in butter (and hard margarines) are highly saturated, whereas those in vegetable oils have a high proportion of cis-alkene functions. Partial hydrogenation of these oils yields soft (tub) margarine.

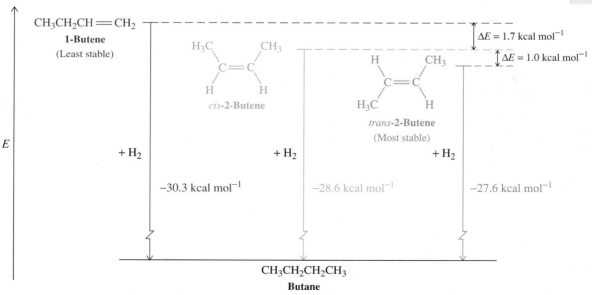

Figure 11-12 The relative energy contents of the butene isomers, as measured by their heats of hydrogenation, tell us their relative stabilities. The diagram is not drawn to scale.

The second finding is easily understood by looking at molecular models. In cis-disubstituted alkenes, the substituent groups frequently crowd each other. This steric interference is energetically unfavorable and absent in the corresponding trans isomers (Figure 11-13).

Relative Stabilities of the Alkenes

$$CH_2{=}CH_2 \quad < \quad RCH{=}CH_2 \quad < \quad \underset{\text{(cis)}}{\overset{R}{\underset{H}{\diagup}}{C}{=}{C}\overset{R}{\underset{H}{\diagdown}}} \quad < \quad \underset{\text{(trans)}}{\overset{H}{\underset{R}{\diagup}}{C}{=}{C}\overset{R}{\underset{H}{\diagdown}}} \quad < \quad \overset{R}{\underset{R}{\diagup}}{C}{=}{C}\overset{R}{\underset{H}{\diagdown}} \quad < \quad \overset{R}{\underset{R}{\diagup}}{C}{=}{C}\overset{R}{\underset{R}{\diagdown}}$$

Least stable ————— Increasing alkene stability ————→ **Most stable**

Decreasing heat of hydrogenation ————→

Exercise 11-10

Rank the following alkenes in order of stability of the double bond to hydrogenation (order of $\Delta H°$ of hydrogenation): 2,3-dimethyl-2-butene, *cis*-3-hexene, *trans*-4-octene, and 1-hexene.

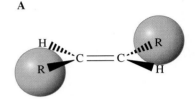

Cycloalkenes are exceptions to the generalization that trans alkenes are more stable than their cis isomers. In the medium-ring and smaller members of this class of compounds (Section 4-2), the trans isomers are much more strained (Section 11-1). The smallest isolated simple trans cycloalkene is *trans*-cyclooctene. It is 9.2 kcal mol^{-1} (38.5 kJ mol^{-1}) less stable than the cis isomer and has a highly twisted structure.

Figure 11-13 (A) Steric congestion in cis-disubstituted alkenes and (B) its absence in trans alkenes explain the greater stability of the trans isomers.

Exercise 11-11

Alkene A hydrogenates to compound B with an estimated release of 65 kcal mol^{-1}, more than double the values of the hydrogenations shown in Figure 11-12. Explain.

$\Delta H° = -65$ kcal mol^{-1}

A   B

Model Building

Molecular Model of
***trans*-Cyclooctene**

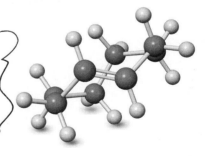

In Summary The relative energies of isomeric alkenes can be estimated by measuring their heats of hydrogenation. The more energetic alkene has a higher $\Delta H°$ of hydrogenation. Stability increases with increasing substitution because of hyperconjugation. Trans alkenes are more stable than their cis isomers because of steric hindrance. Exceptions are the small- and medium-ring cycloalkenes, in which cis substitution is more stable than trans substitution because of ring strain.

11-6 | PREPARATION OF ALKENES FROM HALOALKANES AND ALKYL SULFONATES: BIMOLECULAR ELIMINATION REVISITED

With the physical aspects of alkene structure and stability as a background, let us now return to the various ways in which alkenes can be made. The most general approach is by *elimination,* in which two adjacent groups on a carbon framework are removed. The E2 reaction (Section 7-7) is the most common laboratory source of alkenes. Another method of alkene synthesis, the dehydration of alcohols, is described in Section 11-7.

General Elimination

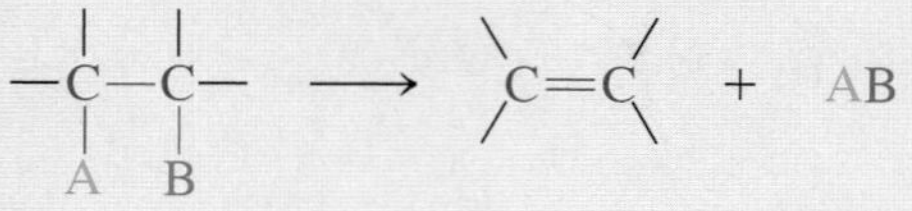

$$-\overset{|}{\underset{A}{C}}-\overset{|}{\underset{B}{C}}- \longrightarrow \overset{}{\underset{}{C}}=\overset{}{\underset{}{C} } + \quad AB$$

Regioselectivity in E2 reactions depends on the base

In Chapter 7 we discussed how haloalkanes (or alkyl sulfonates) in the presence of strong base can undergo elimination of the elements of HX with simultaneous formation of a carbon–carbon double bond. With many substrates, removal of a hydrogen can take place from more than one carbon atom in a molecule, giving rise to constitutional (double-bond) isomers. In such cases, can we control which hydrogen is removed—that is, the *regio-selectivity* of the reaction (Section 9-9)? The answer is yes, to a limited extent. A simple example is the elimination of hydrogen bromide from 2-bromo-2-methylbutane. Reaction with sodium ethoxide in hot ethanol furnishes mainly 2-methyl-2-butene, but also some 2-methyl-1-butene.

E2 Reaction of 2-Bromo-2-methylbutane with Ethoxide

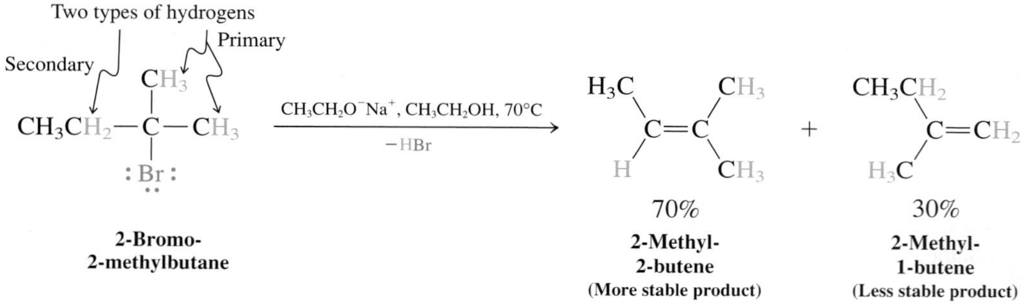

● **Reaction**

Reminder: "–HBr" under the reaction arrow indicates the species removed from the starting material in the elimination reaction.

Two types of hydrogens

Secondary | Primary

$$CH_3CH_2\!-\!\underset{\underset{\displaystyle :\ddot{Br}:}{|}}{\overset{\displaystyle CH_3}{\underset{\displaystyle |}{C}}}\!-\!CH_3 \xrightarrow[{-HBr}]{CH_3CH_2O^-Na^+,\ CH_3CH_2OH,\ 70°C}$$

2-Bromo-
2-methylbutane

$$\underset{H}{\overset{H_3C}{>}}C=C\underset{CH_3}{\overset{CH_3}{<}} \quad + \quad \underset{H_3C}{\overset{CH_3CH_2}{>}}C=CH_2$$

70%   30%

2-Methyl-
2-butene
(More stable product)

2-Methyl-
1-butene
(Less stable product)

Mechanism

In our example, the major product contains a trisubstituted double bond, so it is thermodynamically more stable than the minor product. Indeed, many eliminations are regioselective in this way, with the more stable product predominating. This result can be explained by analysis of the transition state of the reaction (Figure 11-14). Elimination of HBr proceeds

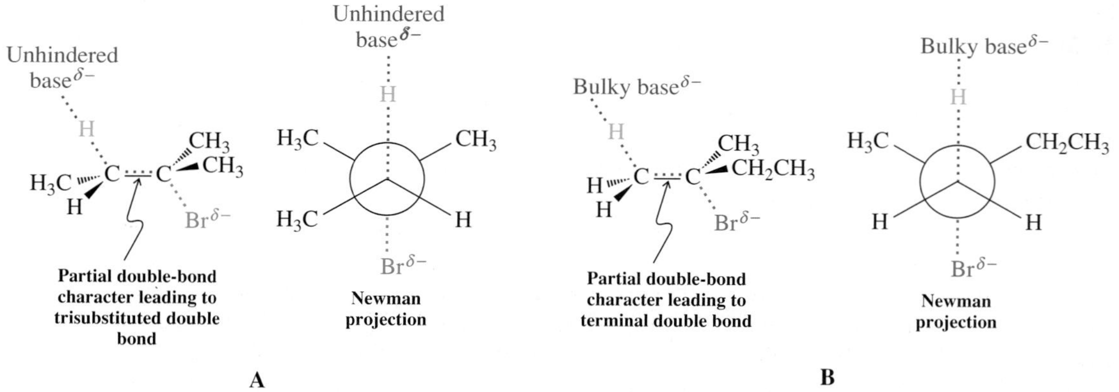

Unhindered
base$^{\delta-}$

Unhindered
base$^{\delta-}$

Bulky base$^{\delta-}$

Bulky base$^{\delta-}$

Partial double-bond
character leading to
trisubstituted double
bond

Newman
projection

Partial double-bond
character leading to
terminal double bond

Newman
projection

A

B

Figure 11-14 The two transition states leading to products in the dehydrobromination of 2-bromo-2-methylbutane. With unhindered base ($CH_3CH_2O^-Na^+$), transition state A is preferred over transition state B because there are more substituents around the partial double bond (Saytzev rule). With hindered base [$(CH_3)_3CO^-K^+$], transition state B is preferred over transition state A because there is less steric hindrance to abstraction of the primary hydrogens (Hofmann rule).

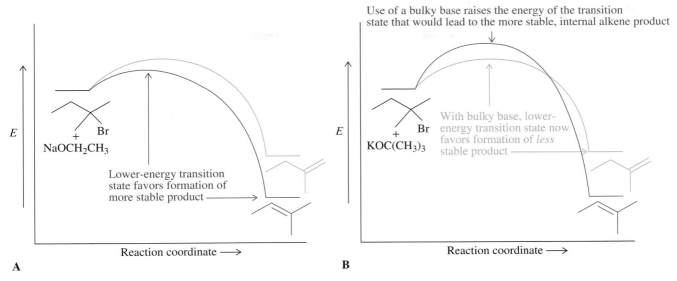

Figure 11-15 Potential energy diagrams for E2 reactions of 2-bromo-2-methylbutane with (A) sodium ethoxide (Saytzev rule) and (B) potassium *tert*-butoxide (Hofmann rule).

through attack by the base on one of the neighboring hydrogens situated *anti* to the leaving group. In the transition state, there is partial C–H bond rupture, partial C–C double-bond formation, and partial cleavage at C–Br (compare Figure 7-8). The transition state leading to 2-methyl-2-butene is slightly more stabilized than that generating 2-methyl-1-butene (Figure 11-15A). The more stable product is formed faster because *the structure of the transition state of the reaction resembles that of the products.* Elimination reactions of this type that lead to the more highly substituted alkene are said to follow the **Saytzev* rule:** The double bond forms preferentially between the carbon that contained the leaving group and *the most highly substituted adjacent carbon* that bears a hydrogen.

A different product distribution is obtained when a more hindered base is used; more of the thermodynamically *less* favored *terminal* alkene is generated.

**E2 Reaction of 2-Bromo-2-methylbutane with *tert*-Butoxide,
a Hindered Base**

Reaction

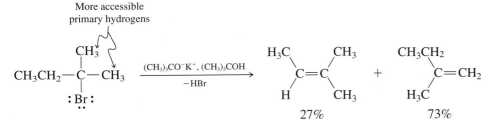

To see why the terminal alkene is now favored, we again examine the transition state. Removal of a secondary hydrogen (from C3 in the starting bromide) is sterically more difficult than abstracting one of the more exposed methyl hydrogens. When the bulky base *tert*-butoxide is used, the energy of the transition state leading to the more stable product is increased by steric interference relative to that leading to the less substituted isomer; thus, the less substituted isomer becomes the major product (Figure 11-15B). An E2 reaction that generates the thermodynamically less favored isomer is said to follow the **Hofmann rule,**[†] named after the chemist who investigated a series of eliminations that proceeded with this particular mode of regioselectivity (Section 21-8).

---

*Alexander M. Saytzev (also spelled Zaitsev or Saytzeff; 1841–1910), Russian chemist.
[†]Professor August Wilhelm von Hofmann (1818–1892), University of Berlin.

Solved Exercise 11-12 | **Working with the Concepts: Regioselectivity in Elimination**

When the following reaction is carried out with *tert*-butoxide in 2-methyl-2-propanol (*tert*-butyl alcohol), two products, A and B, are formed in the ratio 23 : 77. When ethoxide in ethanol is used, this ratio changes to 82 : 18. What are products A and B, and how do you explain the difference in their ratios in the two experiments?

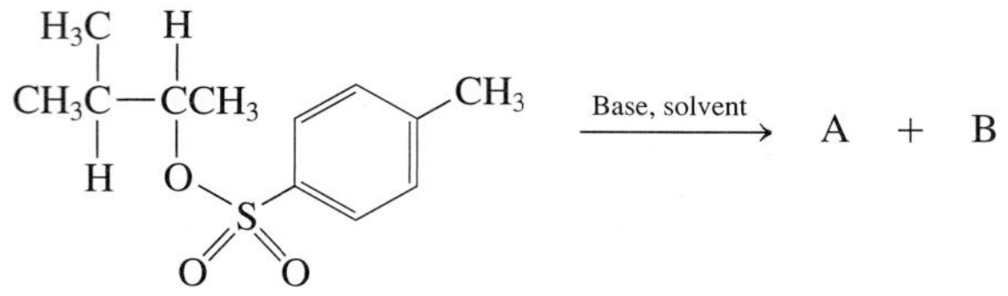

Strategy

The difference between the two experiments is the use of a hindered base (*tert*-butoxide) in one and an unhindered base (ethoxide) in the other. Following the ideas presented in the text, we should expect A and B to be regioisomeric products of E2 reactions, arising from the removal of hydrogens from different carbon atoms adjacent to the one bearing the sulfonate leaving group.

Solution

- The bulky *tert*-butoxide is more likely to abstract a hydrogen atom from the less hindered adjacent (methyl) carbon atom, giving mostly B, the Hofmann-rule product:

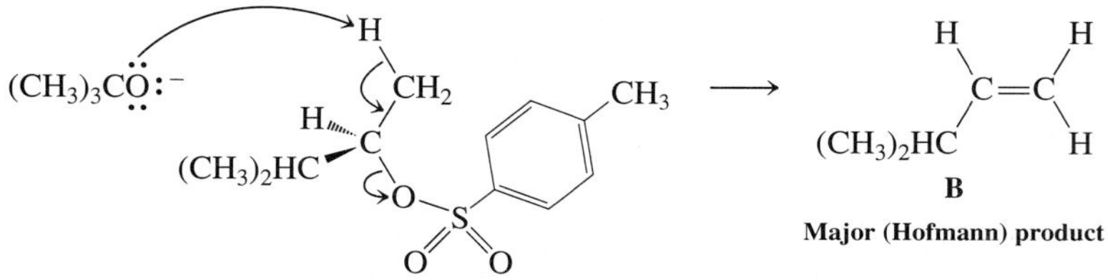

B

Major (Hofmann) product

- On the other hand, ethoxide preferentially removes a hydrogen from the tertiary carbon atom on the other side, because the result is the more stable trisubstituted alkene A, the Saytzev product:

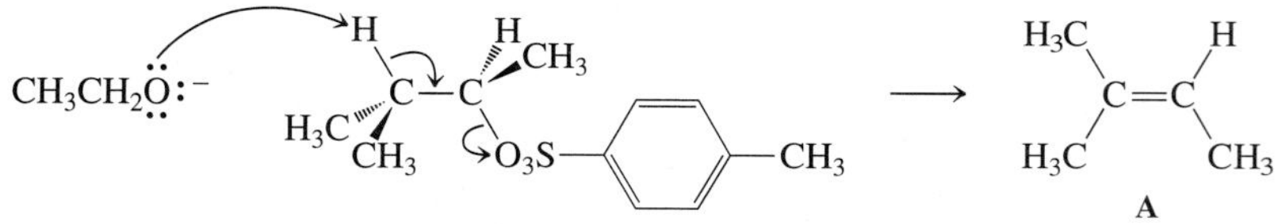

A

Major (Saytzev) product
(More stable)

Exercise 11-13 | **Try It Yourself**

(a) E2 reaction of 2-bromo-2,3-dimethylbutane, $(CH_3)_2CBrCH(CH_3)_2$, gives two products, A and B, in a 79 : 21 ratio using ethoxide in ethanol but a 27 : 73 ratio with *tert*-butoxide in 2-methyl-2-propanol. What are A and B? **(b)** Use of $(CH_3CH_2)_3CO^-$ as the base gives an 8 : 92 ratio of A and B. Explain.

E2 reactions often favor trans over cis

Depending on the structure of the alkyl substrate, the E2 reaction can lead to cis-trans alkene mixtures, in some cases with selectivity. For example, treatment of 2-bromopentane with sodium ethoxide furnishes 51% *trans*- and only 18% *cis*-2-pentene, the remainder of the product being the terminal regioisomer. The outcome of this and related reactions appears to be controlled again to some extent by the relative thermodynamic stabilities of the products, the more stable trans double bond being formed preferentially.

Stereoselective Dehydrobromination of 2-Bromopentane

$$CH_3CH_2CH_2\overset{\overset{\displaystyle CH_3}{|}}{\underset{\underset{\displaystyle H}{|}}{\overset{..}{\underset{..}{C}}Br}}: \quad \xrightarrow[\substack{-HBr}]{\substack{CH_3CH_2O^-Na^+, \\ CH_3CH_2OH}} \quad \underset{51\%}{\overset{\displaystyle CH_3CH_2}{\underset{\displaystyle H}{}}C=C\overset{\displaystyle H}{\underset{\displaystyle CH_3}{}}} \quad + \quad \underset{18\%}{\overset{\displaystyle CH_3CH_2}{\underset{\displaystyle H}{}}C=C\overset{\displaystyle CH_3}{\underset{\displaystyle H}{}}} \quad + \quad \underset{31\%}{CH_3CH_2CH_2CH=CH_2}$$

Unfortunately from a synthetic viewpoint, complete trans selectivity is rare in E2 reactions. Chapter 13 deals with alternative methods for the preparation of stereochemically pure cis and trans alkenes.

Some E2 processes are stereospecific

Recall (Section 7-7) that the preferred transition state of elimination places the proton to be removed and the leaving group *anti* with respect to each other. Thus, before an E2 reaction takes place, bond rotation to an *anti* conformation occurs. This fact has additional consequences when reaction may lead to Z or E stereoisomers. For example, the E2 reaction

Model Building

Stereospecificity in the E2 Reaction of 2-Bromo-3-methylpentane

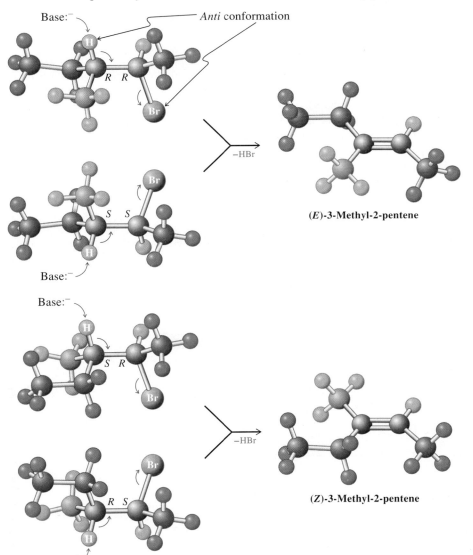

(E)-3-Methyl-2-pentene

(Z)-3-Methyl-2-pentene

> Stereospecificity in the E2 process arises from the *anti* conformation of the substrate, as specified by the reaction mechanism.

of the two diastereomers of 2-bromo-3-methylpentane to give 3-methyl-2-pentene is stereo-specific. Both the (R,R) and the (S,S) isomer yield *exclusively* the (E) isomer of the alkene. Conversely, the (R,S) and (S,R) diastereomers give only the (Z) alkene. (Build models!)

As shown in the three-dimensional structures on the previous page, *anti* elimination of HBr dictates the eventual configuration around the double bond. The reaction is stereospecific: One diastereomer (and its mirror image) produces only one stereoisomeric alkene, the other diastereomer furnishing the opposite configuration.

Exercise 11-14

Which diastereomer of 2-bromo-3-deuteriobutane gives (E)-2-deuterio-2-butene, and which diastereomer gives the Z isomer?

In Summary Alkenes are most generally made by E2 reactions. Usually, the thermodynamically more stable internal alkenes are formed faster than the terminal isomers (Saytzev rule). Bulky bases may favor the formation of the products with thermodynamically less stable (e.g., terminal) double bonds (Hofmann rule). Elimination may be stereoselective, producing greater quantities of trans isomers than their cis counterparts from racemic starting materials. It also may be stereospecific, certain haloalkane diastereomers furnishing only one of the two possible stereoisomeric alkenes.

11-7 PREPARATION OF ALKENES BY DEHYDRATION OF ALCOHOLS

We have seen that treatment of alcohols with mineral acid at elevated temperatures results in alkene formation by loss of water, a process called **dehydration,** which proceeds by E1 or E2 pathways (Chapters 7 and 9). This section reviews this chemistry, now from the perspective of the alkene product. The usual way in which to dehydrate an alcohol is to heat it in the presence of sulfuric or phosphoric acid at relatively high temperatures (120–170°C).

Acid-Mediated Dehydration of Alcohols

$$-\underset{\underset{H}{|}}{\overset{|}{C}}-\underset{\underset{:OH}{|}}{\overset{|}{C}}- \quad \xrightarrow{\text{Acid, } \Delta} \quad \underset{/}{\overset{\backslash}{C}}=\underset{\backslash}{\overset{/}{C}} \quad + \quad \ddot{H}\ddot{O}H$$

The ease of elimination of water from alcohols increases with increasing substitution of the hydroxy-bearing carbon.

Relative Reactivity of Alcohols (RÖH) in Dehydration Reactions

R = primary < secondary < tertiary

Increasing ease of dehydration →

$$CH_3CH_2\ddot{O}H \quad \xrightarrow[-HOH]{\text{Conc. } H_2SO_4, \, 170°C} \quad CH_2=CH_2$$

Primary alcohol

$$CH_3\underset{\underset{H}{|}}{\overset{\overset{\ddot{H\ddot{O}}:}{|}}{C}}-\underset{\underset{H}{|}}{\overset{H}{C}}CH_3 \quad \xrightarrow[-HOH]{50\% \, H_2SO_4, \, 100°C} \quad CH_3CH=CHCH_3 \quad + \quad CH_2=CHCH_2CH_3$$

Secondary alcohol 　　　　　　　　　　　　　　　　　　　　80%　　　　　　　　　　　Trace

$$(CH_3)_3C\ddot{O}H \quad \xrightarrow[-HOH]{\text{Dilute } H_2SO_4, \, 50°C} \quad H_2C=\underset{\underset{CH_3}{|}}{\overset{\overset{CH_3}{|}}{C}}$$

Tertiary alcohol 　　　　　　　　　　　　　　　　　　　　　　　　　　　　　100%

Increasing ease of dehydration ↓

Secondary and tertiary alcohols dehydrate by the unimolecular elimination pathway (E1), discussed in Sections 7-6 and 9-2. Protonation of the weakly basic hydroxy oxygen forms an alkyloxonium ion, now containing water as a good potential leaving group. Loss of H_2O supplies the respective secondary or tertiary carbocations and deprotonation furnishes the alkene. The reaction is subject to all the side reactions of which carbocations are capable, particularly hydrogen and alkyl shifts (Section 9-3).

Dehydration with Rearrangement

The mechanism of this reaction begins with the initial protonation of oxygen. Then, attack by hydrogen sulfate ion or another alcohol molecule effects bimolecular elimination of a proton from one carbon atom and a water molecule from the other.

Exercise 11-15

Referring to Sections 7-6 and 9-3, write a mechanism for the preceding reaction. (**Caution:** As emphasized repeatedly, when writing mechanisms, use "arrow pushing" to depict electron flow; write out every step separately; formulate complete structures, including charges and relevant electron pairs; and draw explicit reaction arrows to connect starting materials or intermediates with their respective products. Don't use shortcuts, and don't be sloppy!)

Typically, the thermodynamically most stable alkene or alkene mixture results from unimolecular dehydration in the presence of acid. Thus, whenever possible, the most highly substituted system is generated; if there is a choice, trans-substituted alkenes predominate over the cis isomers. For example, acid-catalyzed dehydration of 2-butanol furnishes the equilibrium mixture of butenes, consisting of 74% *trans*-2-butene, 23% of the cis isomer, and only 3% 1-butene.

Treatment of primary alcohols with mineral acids at elevated temperatures also leads to alkenes; for example, ethanol gives ethene and 1-propanol yields propene (Section 9-7).

$$CH_3CH_2CH_2OH \xrightarrow{\text{Conc. } H_2SO_4, 180°C} CH_3CH=CH_2$$

The mechanism of this reaction begins with the initial protonation of oxygen. Then, attack by hydrogen sulfate ion or another alcohol molecule effects bimolecular elimination of a proton from one carbon atom and a water molecule from the other.

Exercise 11-16

(a) Propose a mechanism for the formation of propene from 1-propanol upon treatment with hot concentrated H_2SO_4. (b) Propene is also formed when propoxypropane (dipropyl ether) is subjected to the same conditions. Explain.

$$CH_3CH_2CH_2OCH_2CH_2CH_3 \xrightarrow{\text{Conc. } H_2SO_4, 180°C} 2\ CH_3CH=CH_2 + H_2O$$

In Summary Alkenes can be made by dehydration of alcohols. Secondary and tertiary systems proceed through carbocation intermediates, whereas primary alcohols can undergo E2 reactions from the intermediate alkyloxonium ions. All systems are subject to rearrangement and thus frequently give mixtures.

11-8 | INFRARED SPECTROSCOPY

The remaining sections in this chapter deal with two additional methods for the determination of the structures of organic compounds: **infrared (IR) spectroscopy** and **mass spectrometry (MS).** IR spectroscopy is a very useful tool because it is capable of detecting the characteristic bonds of many functional groups through their absorption of infrared light. IR spectroscopy measures the vibrational excitation of atoms around the bonds that connect them. The positions of the absorption lines associated with this excitation depend on the types of functional groups present, and the IR spectrum as a whole displays a pattern unique for each individual substance.

Absorption of infrared light causes molecular vibrations

At energies slightly lower than those of visible radiation, light causes **vibrational excitation** of the bonds in a molecule. This part of the electromagnetic spectrum is the infrared region (see Figure 10-2). The intermediate range, or **middle infrared,** is most useful to the organic chemist. IR absorption bands are described by either the wavelength, λ, of the absorbed light in micrometers (10^{-6} m; $\lambda \approx 2.5-16.7$ μm; see Figure 10-2) or its reciprocal value, called wavenumber, $\tilde{\nu}$ (in units of cm^{-1}; $\tilde{\nu} = 1/\lambda$). Thus, a typical infrared spectrum ranges from $\tilde{\nu} = 600$ to 4000 cm^{-1}, and the energy changes associated with absorption of this radiation range from 1 to 10 kcal mol^{-1} (4 to 42 kJ mol^{-1}).

Figure 10-3 described the general principles of a spectrometer, which apply also to an infrared instrument. Modern systems use sophisticated rapid-scan techniques and are linked with computers. This equipment allows for data storage, spectra manipulation, and computer library searches, so that unknown compounds can be matched with stored spectra.

We can envision vibrational excitation simply by thinking of two atoms, A and B, linked by a *flexible* bond. Picture the atoms as two masses connected by a bond that stretches and compresses at a certain frequency, ν, like a spring (Figure 11-16). In this picture, the frequency of the vibrations between two atoms depends both on the strength of the bond between them and on their atomic weights. In fact, it is governed by Hooke's* law, just like the motion of a spring.

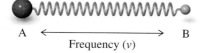

A ⟵ Frequency (ν) ⟶ B

Figure 11-16 Two unequal masses on an oscillating ("vibrating") spring: a model for vibrational excitation of a bond.

Hooke's Law and Vibrational Excitation

$$\tilde{\nu} = k\sqrt{f\frac{(m_1 + m_2)}{m_1 m_2}}$$

$\tilde{\nu}$ = vibrational frequency in wavenumbers (cm^{-1})
k = constant
f = force constant, indicating the strength of the spring (bond)
m_1, m_2 = masses of attached atoms

This equation might lead us to expect every individual bond in a molecule to show one specific absorption band in the infrared spectrum. However, in practice, an interpretation of the entire infrared spectrum is considerably more complex and beyond the needs of the organic chemist. This is because molecules that absorb infrared light undergo not only bond stretching, but also various bending motions (Figure 11-17), as well as combinations of the two. The bending vibrations are mostly of weaker intensity, they overlap with other absorptions, and they may show complicated patterns. In addition, for a bond to absorb infrared light, its vibrational motion must cause a change in the molecular dipole. Therefore, vibrations of polar bonds give strong infrared absorption bands, whereas absorptions associated with nonpolar bonds may be weak or entirely absent. The practicing organic chemist can find good use for IR spectroscopy for two reasons: The vibrational bands of many functional

---

Stronger bonds give higher-frequency ($\tilde{\nu}$) IR absorptions:
$\tilde{\nu}_{O-H} > \tilde{\nu}_{N-H} > \tilde{\nu}_{C-H}$, and
$\tilde{\nu}_{C\equiv C} > \tilde{\nu}_{C=C} > \tilde{\nu}_{C-C}$

More polar bonds give more intense (I) IR absorptions:
$I_{O-H} > I_{N-H}, I_{C=O} > I_{C=C}$, and
$I_{C\equiv N} > I_{C\equiv C}$

---

*Professor Robert Hooke (1635–1703), physicist at Gresham College, London.

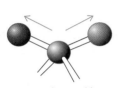
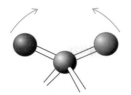

Symmetric stretching vibration (both outside atoms move away from or toward the center)

Symmetric bending vibration in a plane (scissoring)

Symmetric bending vibration out of a plane (twisting)

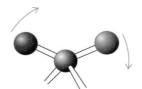
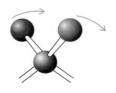

Asymmetric stretching vibration (as one atom moves toward the center, the other moves away)

Asymmetric bending vibration in a plane (rocking)

Asymmetric bending vibration out of a plane (wagging)

Figure 11-17 Various vibrational modes around tetrahedral carbon. The motions are labeled symmetric and asymmetric stretching or bending, scissoring, rocking, twisting, and wagging.

groups appear at characteristic frequencies; furthermore, the entire infrared spectrum of any given compound is unique in its pattern and can be distinguished from that of any other substance.

Functional groups have typical infrared absorptions

Table 11-4 lists the characteristic stretching wavenumber values for the bonds (shown in red) in some common structural units. Notice how most absorb in the region above 1500 cm^{-1}. We shall show the IR spectra typical of new functional groups when we introduce each of the corresponding compound classes in subsequent chapters.

Figures 11-18 and 11-19 show the IR spectra of pentane and hexane. Above 1500 cm^{-1}, we see only the C–H stretching absorptions typical of alkanes, in the range from 2840 to 3000 cm^{-1}; no functional groups are present, and the two spectra are very similar in this region. However, below 1500 cm^{-1}, the spectra differ, as the scans at higher sensitivity show. This is the **fingerprint region,** in which absorptions due to C–C bond stretching and C–C and C–H bending motions overlap to give complicated patterns. Bands at approximately 1460, 1380, and 730 cm^{-1} are common to all saturated hydrocarbons.

Figure 11-20 shows the IR spectrum of 1-hexene. A characteristic feature of alkenes when compared with alkanes is the stronger C_{sp2}–H bond, which should therefore have a band at higher energy in the IR spectrum. Indeed, as Figure 11-20 shows, there is a sharp spike at 3080 cm^{-1}, due to this stretching mode, at a slightly higher wavenumber than the remainder of the C–H stretching absorptions. A useful rule of thumb is that C_{sp3}–H bonds give rise to peaks below 3000 cm^{-1}, whereas C_{sp2}–H bonds absorb above 3000 cm^{-1}. According to Table 11-4, the C=C stretching band should appear between about 1620 and 1680 cm^{-1}. Figure 11-20 shows a sharp band at 1640 cm^{-1} assigned to this vibration. The other major peaks are the result of bending motions. For example, the two signals at 915 and 995 cm^{-1} are typical of a terminal alkene.

Two other strong bending modes may be used as a diagnostic tool for the substitution pattern in alkenes. One mode results in a single band at 890 cm^{-1} and is characteristic of 1,1-dialkylethenes; the other gives a sharp band at 970 cm^{-1} and is produced by the C_{sp3}–H bending mode of a trans double bond. The C=C stretching absorption of internal alkenes is usually less intense than that in terminal alkenes, because vibration of an internal C=C

Warm objects release their energy as heat through the emission of infrared light. This radiation can be imaged by infrared thermography, allowing the taking of pictures in the absence of visible light ("night vision"). Medical applications include the monitoring of healing processes and the detection of disease through fever screening. This technique is also used for the early diagnosis of (breast) cancer, because the "hyperactivity" of precancerous tissue results in relatively elevated local temperatures.

Thermographic photograph of a passenger in the airport terminal of Sofia, Bulgaria, aimed at detecting dangerous infections, such as Severe Acute Respiratory Syndrome (SARS).

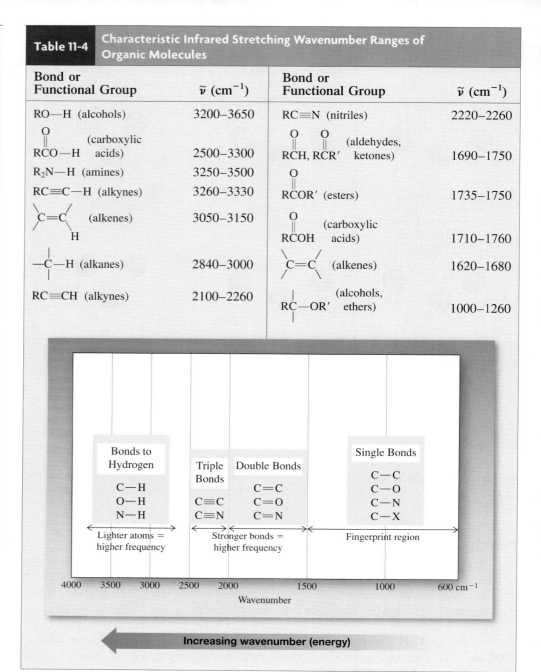

Table 11-4	Characteristic Infrared Stretching Wavenumber Ranges of Organic Molecules		
Bond or Functional Group	$\tilde{\nu}$ (cm⁻¹)	Bond or Functional Group	$\tilde{\nu}$ (cm⁻¹)
RO—H (alcohols)	3200–3650	RC≡N (nitriles)	2220–2260
RCO—H (carboxylic acids)	2500–3300	RCH, RCR′ (aldehydes, ketones)	1690–1750
R₂N—H (amines)	3250–3500	RCOR′ (esters)	1735–1750
RC≡C—H (alkynes)	3260–3330		
C=C (alkenes) / H	3050–3150	RCOH (carboxylic acids)	1710–1760
—C—H (alkanes)	2840–3000	C=C (alkenes)	1620–1680
RC≡CH (alkynes)	2100–2260	RC—OR′ (alcohols, ethers)	1000–1260

Bonds to Hydrogen
C—H
O—H
N—H
← Lighter atoms = higher frequency →

Triple Bonds
C≡C
C≡N

Double Bonds
C=C
C=O
C=N
← Stronger bonds = higher frequency →

Single Bonds
C—C
C—O
C—N
C—X
← Fingerprint region →

4000 3500 3000 2500 2000 1500 1000 600 cm⁻¹
Wavenumber

← Increasing wavenumber (energy)

Figure 11-18 IR spectrum of pentane. Note the format: Wavenumber is plotted (decreasing from left to right) against percentage of transmittance. A 100% transmittance means *no* absorption; therefore "peaks" in an IR spectrum point *downward*. The spectrum shows absorbances at $\tilde{\nu}_{C-H\ stretch}$ = 2960, 2930, and 2870 cm⁻¹; $\tilde{\nu}_{C-H\ bend}$ = 1460, 1380, and 730 cm⁻¹. The region between 600 and 1300 cm⁻¹ is also shown recorded at higher sensitivity (in red), revealing details of the pattern in the fingerprint region.

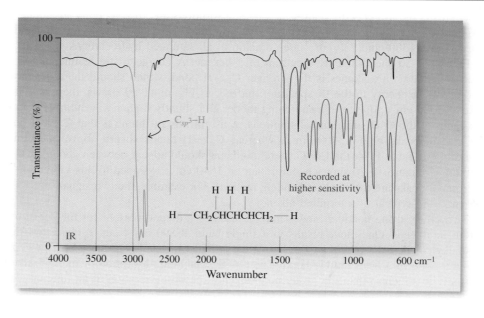

C_{sp^3}–H

H H H
H—CH₂CHCHCHCH₂—H

Recorded at higher sensitivity

IR

100
Transmittance (%)
0
4000 3500 3000 2500 2000 1500 1000 600 cm⁻¹
Wavenumber

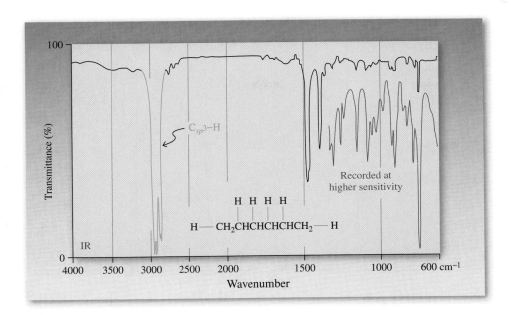

Figure 11-19 IR spectrum of hexane. Comparison with that of pentane (Figure 11-18) shows that the location and appearance of the major bands are very similar, but the two fingerprint regions show significant differences at higher recorder sensitivity (in red).

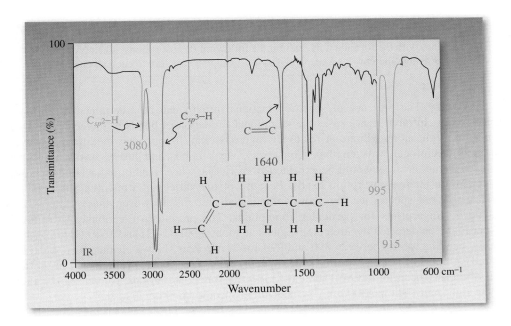

Figure 11-20 IR spectrum of 1-hexene: $\tilde{\nu}_{C_{sp^2}-H\ stretch} = 3080\ cm^{-1}$; $\tilde{\nu}_{C=C\ stretch} = 1640\ cm^{-1}$; $\tilde{\nu}_{C_{sp^2}-H\ bend} = 995$ and $915\ cm^{-1}$.

bond causes less change in the molecular dipole. In highly symmetric molecules, such as *trans*-3-hexene, the band for the C=C vibration may be too weak to be observed readily. However, the alkenyl C–H bend in *trans*-3-hexene still gives a very strong, sharp absorption at $970\ cm^{-1}$. In conjunction with NMR (Section 11-4), the presence or absence of such bands allows for fairly reliable structural assignments of specifically substituted double bonds.

The O–H stretching absorption is the most characteristic band in the IR spectra of alcohols (Chapters 8 and 9), appearing as a readily recognizable intense, broad absorption over a wide range ($3200–3650\ cm^{-1}$, Figure 11-21). The broadness of this band is due to hydrogen bonding to other alcohol molecules or to water. The C–O bond gives rise to a sharp peak at around $1100\ cm^{-1}$. In contrast, the C–X bonds of haloalkanes (Chapters 6 and 7) possess IR stretching frequencies at energies too low ($<800\ cm^{-1}$) to be generally useful for characterization.

Drinking, Driving, and IR Spectroscopy

The strong IR bands of ethanol at 3360 and $1050\ cm^{-1}$ are recorded by the Intoxilyzer instrument to detect alcohol in your breath.

Figure 11-21 IR spectrum of cyclohexanol: $\tilde{\nu}_{O-H\ stretch}$ = 3345 cm^{-1}; $\tilde{\nu}_{C-O}$ = 1070 cm^{-1}; Note the broad and very intense peak arising from the polar O–H bond.

Infrared Bending Frequencies for Alkenes

R   H
 \   /
  C = C
 /   \
H   H
915, 995 cm^{-1}

R   H
 \   /
  C = C
 /   \
R   H
890 cm^{-1}

R   H
 \   /
  C = C
 /   \
H   R
970 cm^{-1}

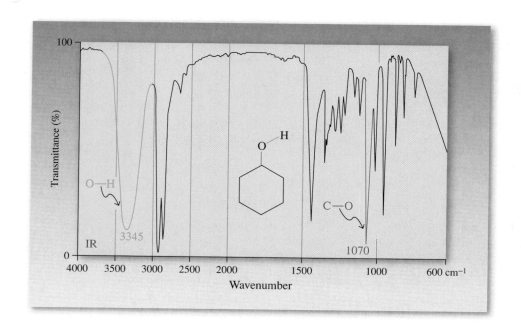

Exercise 11-17

Three alkenes with the formula C_4H_8 exhibit the following IR absorptions: alkene A, 964 cm^{-1}; alkene B, 908 and 986 cm^{-1}; alkene C, 890 cm^{-1}. Assign a structure to each alkene.

In Summary The presence of specific functional groups can be ascertained by infrared spectroscopy. Infrared light causes the vibrational excitation of bonds in molecules. Strong bonds and light atoms vibrate at relatively high stretching frequencies measured in wavenumbers (reciprocal wavelengths). Conversely, weak bonds and heavy atoms absorb at lower wavenumbers, as would be expected from Hooke's law. Highly polar bonds tend to exhibit more intense absorption bands. Because of the variety of stretching and bending modes, infrared spectra usually show complicated patterns. However, these patterns are diagnostic fingerprints for particular compounds. The presence of variously substituted alkenes may be detected by stretching signals at about 3080 (C–H) and 1640 (C=C) cm^{-1} and by bending modes between 890 and 990 cm^{-1}. Alcohols show a characteristic band for the OH group in the range from 3200 to 3650 cm^{-1}. As a general rule, bands in the left half of the IR spectrum (above 1500 cm^{-1}) identify functional groups, whereas absorptions in the right half (below 1500 cm^{-1}) characterize specific compounds.

11-9 | MEASURING THE MOLECULAR MASS OF ORGANIC COMPOUNDS: MASS SPECTROMETRY

In the examples and problems dealing with structure determinations of organic compounds, we have so far been given the molecular formula of the "unknown." How is this information obtained? Elemental analysis (Section 1-9) gives us an *empirical* formula, which tells us the *ratios* of the different elements in a molecule. However, empirical and molecular formulas are not necessarily identical. For example, elemental analysis of cyclohexane merely reveals the presence of carbon and hydrogen atoms in a 1:2 ratio; it does not tell us that the molecule contains six carbons and twelve hydrogens.

To determine molecular masses, the chemist turns to another important physical technique used to characterize organic molecules: **mass spectrometry.** This section begins with a description of the apparatus used and the physical principle on which it is based.

Subsequently, we shall consider processes by which molecules fragment under the conditions required for molecular mass measurement, giving rise to characteristic recorded patterns called **mass spectra.** Mass spectra can help chemists distinguish between constitutional isomers and identify the presence of many functions such as hydroxy and alkenyl groups.

The mass spectrometer distinguishes ions by mass

Mass spectrometry is not a form of spectroscopy in the conventional sense, because no radiation is absorbed (Section 10-2). A sample of an organic compound is introduced into an inlet chamber (Figure 11-22, upper left). It is vaporized, and a small quantity is allowed to leak into the source chamber of the spectrometer. Here the neutral molecules (M) pass through a beam of high-energy [usually 70 eV, or about 1600 kcal mol^{-1} (6700 kJ mol^{-1})] electrons. Upon electron impact, some of the molecules lose an electron to form the corresponding radical cation, M$^{+\cdot}$, called the **parent** or **molecular ion.** Most organic molecules undergo only a single ionization.

Ionization of a Molecule on Electron Impact

$$\text{M} \quad + \quad e\,(70\,\text{eV}) \quad \longrightarrow \quad \text{M}^{+\cdot} \quad + \quad 2\,e$$

Neutral molecule   **Ionizing beam**   **Radical cation (Molecular ion)**

As charged particles, the molecular ions are next accelerated to high velocity by an electric field. (Molecules that are not ionized remain in the source chamber to be pumped away.) The accelerated M$^{+\cdot}$ ions are then subjected to a magnetic field, which deflects them from a linear to a circular path. The curvature of this path is a function of the strength of the magnetic field. Much as in an NMR spectrometer (Section 10-3), the strength of the magnetic field can be varied and therefore adjusted to give the exact degree of curvature to the path of the ions necessary to direct them through the collector slit to the collector, where they are detected and counted. Because lighter species are deflected more than heavier ones, the strength of the field necessary to direct the ions through the slit to the

Molecular Masses of Organic Molecules

$$CH_4$$
m/z = 16

$$CH_3OH$$
m/z = 32

$$\underset{\underset{CH_3COCH_3}{}}{\overset{\displaystyle O}{\overset{\|}{}}}$$

CH$_3$COCH$_3$
m/z = 74

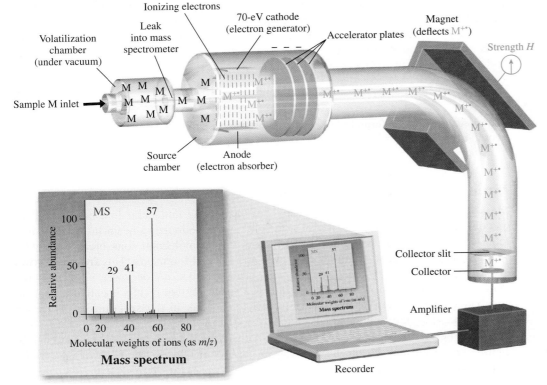

Figure 11-22 Diagram of a mass spectrometer.

collector is a function of the mass of $M^{+\bullet}$ and therefore of the original molecule M. Thus, at a given magnetic field strength, *only ions of a specific mass can pass through the collector slit.* All others will collide with the internal walls of the instrument. Finally, the arrival of ions at the collector is translated electronically into a signal and recorded on a chart. The chart plots mass-to-charge ratio, m/z (on the abscissa), versus peak height (on the ordinate); peak height measures the relative number of ions with a given m/z ratio. Because only singly charged species are normally formed, $z = 1$, and m/z equals the mass of the ion being detected.

Exercise 11-18

Three unknown compounds containing only C, H, and O give rise to the following molecular weights. Draw as many reasonable structures as you can. (a) $m/z = 46$; (b) $m/z = 30$; (c) $m/z = 56$.

High-resolution mass spectrometry reveals molecular formulas

Consider substances with the following molecular formulas: C_7H_{14}, $C_6H_{10}O$, $C_5H_6O_2$, and $C_5H_{10}N_2$. All possess the same **integral mass;** that is, to the nearest integer, all four would be expected to exhibit a molecular ion at $m/z = 98$. However, the atomic weights of the elements are composites of the masses of their naturally occurring isotopes, *which are not integers.* Thus, if we use the atomic masses for the most abundant isotopes of C, H, O, and N (Table 11-5) to calculate the **exact mass** corresponding to each of the aforementioned molecular formulas, we see significant differences.

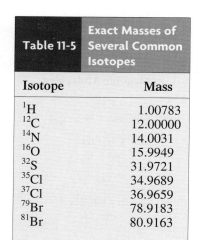

Table 11-5	Exact Masses of Several Common Isotopes
Isotope	**Mass**
^1H	1.00783
^{12}C	12.00000
^{14}N	14.0031
^{16}O	15.9949
^{32}S	31.9721
^{35}Cl	34.9689
^{37}Cl	36.9659
^{79}Br	78.9183
^{81}Br	80.9163

Exact Masses of Four Compounds with $m/z = 98$

C_7H_{14}	$C_6H_{10}O$	$C_5H_6O_2$	$C_5H_{10}N_2$
98.1096	98.0732	98.0368	98.0845

Can we use mass spectrometry to differentiate these species? Yes. Modern **high-resolution mass spectrometers** are capable of distinguishing among ions that differ in mass by as little as a few thousandths of a mass unit. We can therefore measure the exact mass of any molecular ion. By comparing this experimentally determined value with that calculated for each species possessing the same integral mass, we can assign a molecular formula to the unknown ion. High-resolution mass spectrometry is the most widely used method for determining the molecular formulas of unknowns.

Exercise 11-19

Choose the molecular formula that matches the exact mass. (a) $m/z = 112.0888$, C_8H_{16}, $C_7H_{12}O$, or $C_6H_8O_2$; (b) $m/z = 86.1096$, C_6H_{14}, $C_4H_6O_2$, or $C_4H_{10}N_2$.

Molecular ions undergo fragmentation

Mass spectrometry gives information not only about the molecular ion, but also about its component structural parts. Because the energy of the ionizing beam far exceeds that required to break typical organic bonds, some of the molecular ions break apart into virtually all possible combinations of neutral and ionized fragments. This **fragmentation** gives rise to a number of additional mass-spectral peaks, *all of lower mass* than the molecular ion from which they are derived. The spectrum that results is called the **mass-spectral fragmentation pattern.** The most intense peak in the spectrum is called the **base peak.** Its relative intensity is defined to be 100, and the intensities of all other peaks are described as a percentage of the intensity of the base peak. The base peak in a mass spectrum may be the molecular ion peak or it may be the peak of one of the fragment ions.

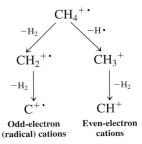

Figure 11-23 Mass spectrum of methane. At the left is the spectrum actually recorded; at the right is the tabulated form, the largest peak (base peak) being defined as 100%. For methane, the base peak at $m/z = 16$ is due to the parent ion. Fragmentation gives rise to peaks of lower mass.

Tabulated Spectrum

m/z	Relative Abundance (%)	Molecular or Fragment ion
17	1.1	$(M + 1)^{+\cdot}$
16	100.0 (base peak)	$M^{+\cdot}$ (parent ion)
15	85.0	$(M - 1)^{+}$
14	9.2	$(M - 2)^{+\cdot}$
13	3.9	$(M - 3)^{+}$
12	1.0	$(M - 4)^{+\cdot}$

For example, the mass spectrum of methane contains, in addition to the molecular ion peak, lines for CH_3^+, $CH_2^{+\cdot}$, CH^+, and $C^{+\cdot}$ (Figure 11-23). These lines are formed by the processes shown in the margin. The relative abundances of these species, as indicated by the heights of the peaks, give a useful indication of the relative ease of their formation. It can be seen that the first C–H bond is cleaved readily, the $m/z = 15$ peak reaching 85% of the abundance of the molecular ion, which in this case is the base peak. The breaking of additional C–H bonds is increasingly difficult, and the corresponding ions have lower relative abundance. Section 11-10 considers the process of fragmentation in more detail and reveals how fragmentation patterns may be used as an aid to molecular structure determination.

Fragmentation of Methane in the Mass Spectrometer

$$CH_4^{+\cdot}$$

$-H_2 \swarrow \qquad \searrow -H\cdot$

$$CH_2^{+\cdot} \qquad\qquad CH_3^{+}$$

$\downarrow -H_2 \qquad\qquad \downarrow -H_2$

$$C^{+\cdot} \qquad\qquad CH^{+}$$

Odd-electron (radical) cations   **Even-electron cations**

REAL LIFE: MEDICINE 11-2 | Detecting Performance-Enhancing Drugs Using Mass Spectrometry

Recent and rather notorious cases of using illicit substances to enhance athletic performance have highlighted the technology that has made the detection of these substances possible (see Chapter 4 Opening). An instrument called a gas chromatograph (GC) separates a test sample into its individual components, each of which is analyzed by high-resolution mass spectrometry. The success of this method in detecting "cheating" rests in its exquisite sensitivity and extremely high quantitative precision.

In cases of suspected administration of anabolic steroids such as testosterone (Section 4-7), two approaches are used. The first compares the ratio of testosterone (T) to its stereoisomer, epitestosterone (E; identical to T except that the hydroxy group on the five-membered ring is "down" instead of "up"). E and T occur naturally in humans in roughly equal amounts, but E, unlike T, does not enhance performance. Taking synthetic T alters the T:E ratio and is easily detected. Therefore, some athletes have taken synthetic T and E together, to maintain a T:E ratio within normal limits. Mass spectrometry identifies these situations, because of a quirk of biology: Synthetic steroids are made from plant-derived precursors, which have a slightly lower ^{13}C content (relative to ^{12}C) than do steroids biosynthesized naturally in the human body. The differences are quite small (parts per thousand)

but readily detectable: After the steroids are separated from the samples by GC, they undergo combustion to CO_2, and the mass spectrometer measures the ratio of $^{13}CO_2$ to $^{12}CO_2$. A $^{13}C:{}^{12}C$ ratio that departs significantly from that found in normal human steroids and that approaches that found in plant-derived synthetics is considered to be strong evidence of "doping."

American cyclist Lance Armstrong leaving the doping test station during the 2001 Tour de France.

Mass spectra reveal the presence of isotopes

An unusual feature in the mass spectrum of methane is a small (1.1%) peak at $m/z = 17$; it is designated $(M + 1)^{+\cdot}$. How is it possible to have an ion present that has an extra mass unit? The answer lies in the fact that carbon is not isotopically pure. About 1.1% of natural carbon is the ^{13}C isotope (see Table 10-1), giving rise to the additional peak. In the mass spectrum of ethane, the height of the $(M + 1)^{+\cdot}$ peak, at $m/z = 31$, is about 2.2% that of the parent ion. The reason for this finding is statistical. The chance of finding a ^{13}C atom in a compound containing two carbons is double that expected of a one-carbon molecule. For a three-carbon moiety, it would be threefold, and so on.

Other elements, too, have naturally occurring higher isotopes: Hydrogen (deuterium, ^{2}H, about 0.015% abundance), nitrogen (0.366% ^{15}N), and oxygen (0.038% ^{17}O, 0.200% ^{18}O) are examples. These isotopes also contribute to the intensity of peaks at masses higher than $M^{+\cdot}$, but less so than ^{13}C.

Fluorine and iodine are isotopically pure. However, chlorine (75.53% ^{35}Cl; 24.47% ^{37}Cl) and bromine (50.54% ^{79}Br; 49.46% ^{81}Br) each exist as a mixture of two isotopes and give rise to readily identifiable isotopic patterns. For example, the mass spectrum of 1-bromopropane (Figure 11-24) shows two peaks of nearly equal intensity at $m/z = 122$ and 124. Why? The isotopic composition of the molecule is a nearly 1:1 mixture of $CH_3CH_2CH_2{}^{79}Br$ and $CH_3CH_2CH_2{}^{81}Br$. Similarly, the spectra of monochloroalkanes exhibit ions two mass units apart in a 3:1 intensity ratio, because of the presence of about 75% $R^{35}Cl$ and 25% $R^{37}Cl$. Peak patterns such as these are useful in revealing the presence of chlorine or bromine.

Exercise 11-20

What peak pattern do you expect for the molecular ion of dibromomethane?

Exercise 11-21

Nonradical compounds containing C, H, and O have even molecular weights, those containing C, H, O, and an odd number of N atoms have odd molecular weights, but those with an even number of N atoms are even again. Explain.

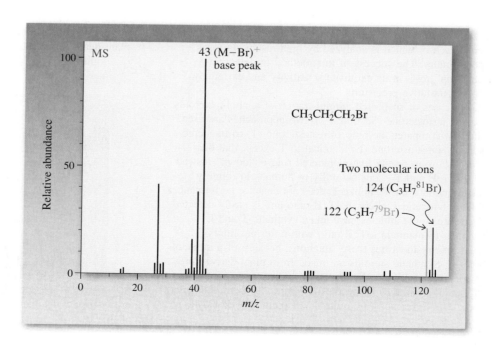

Figure 11-24 Mass spectrum of 1-bromopropane. Note the nearly equal heights of the peaks at $m/z = 122$ and 124, owing to the almost equal abundances of the two bromine isotopes.

In Summary Molecules can be ionized by an electron beam at 70 eV to give radical cations that are accelerated by an electric field and then separated by the different deflections that they undergo in a magnetic field. In a mass spectrometer, this effect is used to measure the molecular masses of molecules. At high resolution, the mass of the molecular ion permits the determination of the molecular formula. The molecular ion is usually accompanied by less massive fragments and isotopic "satellites," owing to the presence of less abundant isotopes. In some cases, such as with Cl and Br, more than one isotope may be present in substantial quantities.

11-10 | FRAGMENTATION PATTERNS OF ORGANIC MOLECULES

Upon electron impact, molecules dissociate by breaking weaker bonds first and then stronger ones. Because the original molecular ion is positively charged, its dissociation typically gives one neutral fragment and one that is cationic. Normally the cationic fragment that forms bears its charge at the center that is most capable of stabilizing a positive charge. This section illustrates how the preference to break weaker bonds and to form more stable carbocationic fragments combines to make mass spectrometry a powerful tool for molecular structure determination.

Fragmentation is more likely at a highly substituted center

The mass spectra of the isomeric hydrocarbons pentane, 2-methylbutane, and 2,2-dimethylpropane (Figures 11-25, 11-26, and 11-27) reveal the relative ease of the several possible C–C bond-dissociation processes. In each case, the molecular ion produces a relatively weak peak; otherwise, the spectra of the three compounds are very different.

Pentane fragments by C–C bond-breaking in four possible ways, each of which gives a carbocation and a radical. Only the positively charged cation is observed in the mass spectrum (see margin); the radical, being neutral, is "invisible." For example, one process breaks the C1–C2 bond to give a methyl cation and a butyl radical. The peak at $m/z = 15$ for CH_3^+ in the mass spectrum (Figure 11-25) is very weak, consistent with the instability of this carbocation (Section 7-5). Similarly, the peak at $m/z = 57$ is weak, because it derives from fragmentation into a butyl cation and a methyl radical, and although the primary butyl cation is more stable than CH_3^+, the methyl radical is a high-energy species. Thus this mode of fragmentation is not favored. The favored bond cleavages give peaks

Fragment Ions from Pentane

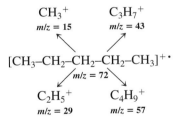

CH_3^+ $C_3H_7^+$
$m/z = 15$ $m/z = 43$

$[CH_3–CH_2–CH_2–CH_2–CH_3]^{+\cdot}$
$m/z = 72$

$C_2H_5^+$ $C_4H_9^+$
$m/z = 29$ $m/z = 57$

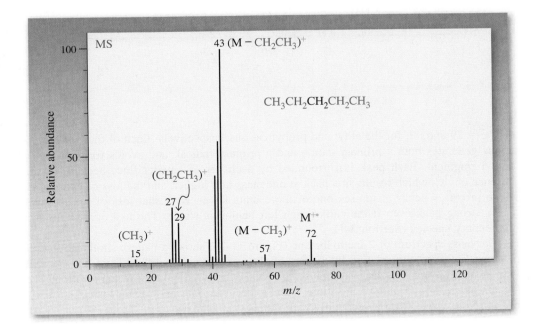

Figure 11-25 Mass spectrum of pentane, revealing that all C–C bonds in the chain have been ruptured.

Figure 11-26 Mass spectrum of 2-methylbutane. The peaks at $m/z = 43$ and 57 result from preferred fragmentation around C2 to give secondary carbocations.

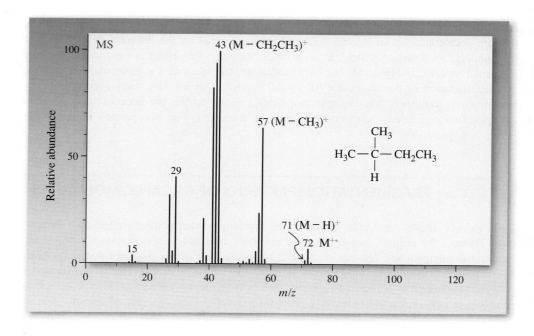

Figure 11-27 Mass spectrum of 2,2-dimethylpropane. Only a very weak molecular ion peak is seen, because the fragmentation to give a tertiary cation is favored.

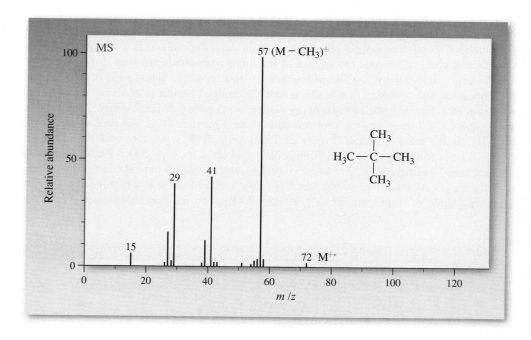

at $m/z = 29$ and 43, for the ethyl and propyl cations, respectively. Each of these fragmentations generates both a primary cation and a primary radical, and avoids formation of a methyl fragment. Each peak is surrounded by a cluster of smaller lines because of the presence of ^{13}C, which results in a peak at one mass unit higher, and the loss of hydrogens, which produces peaks at masses one or more units lower. Note that loss of H· does not give a strong peak even if the carbocation left behind is stable: The hydrogen atom is a high-energy species (Section 3-1).

The mass spectrum of 2-methylbutane (Figure 11-26) shows a pattern similar to that of pentane; however, the relative intensities of the various peaks differ. Thus, the peak at $m/z = 71$ $(M - 1)^+$ is larger, because loss of H· from C2 gives a tertiary cation. The signals at $m/z = 43$ and 57 are more intense, because both arise from loss of an alkyl radical from C2 to form a secondary carbocation.

Cleavage of C–C bonds to give more stable carbocations is the main fragmentation mode seen for alkanes in the mass spectrometer. C–H bonds, which are stronger (Table 3-2), cleave less readily.

Preferred Fragmentation of 2-Methylbutane

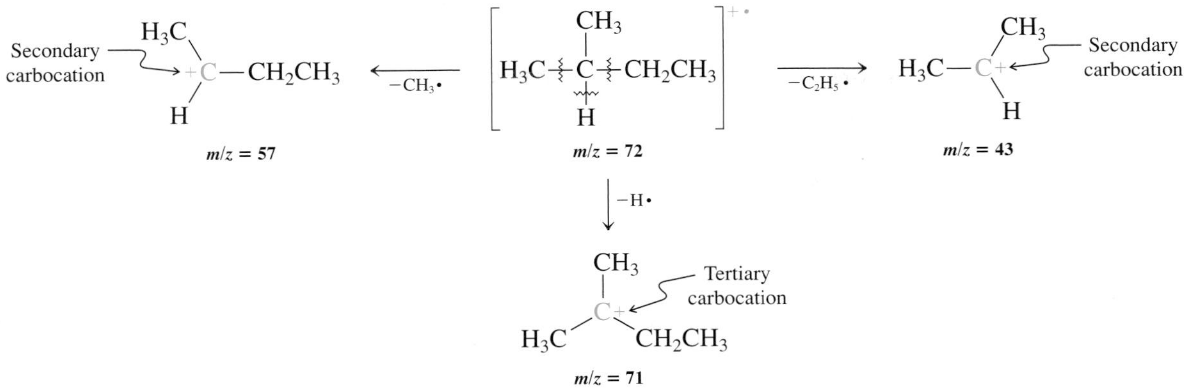

The preference for fragmentation at a highly substituted center is even more pronounced in the mass spectrum of 2,2-dimethylpropane (Figure 11-27). Here, loss of a methyl radical from the molecular ion produces the 1,1-dimethylethyl (*tert*-butyl) cation as the base peak at $m/z = 57$. This fragmentation is so easy that the molecular ion is barely visible. The spectrum also reveals peaks at $m/z = 41$ and 29, the result of complex structural reorganizations, such as the carbocation rearrangements considered in Section 9-3.

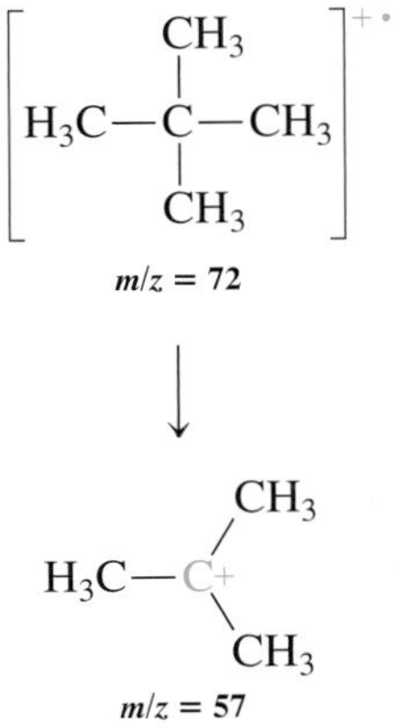

Fragmentations also help to identify functional groups

Particularly easy fragmentation of relatively weak bonds is also seen in the mass spectra of the haloalkanes. The fragment ion $(M - X)^+$ is frequently the base peak in these spectra. A similar phenomenon is observed in the mass spectra of alcohols, which eliminate water to give a large $(M - H_2O)^{+\bullet}$ peak 18 mass units below the parent ion (Figure 11-28). The bonds to the C–OH group also readily dissociate in a process called **α cleavage**, leading to resonance-stabilized hydroxycarbocations:

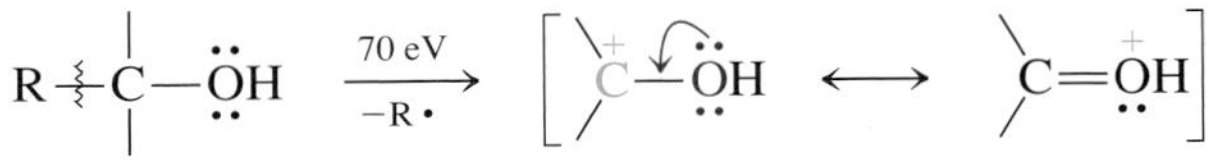

The strong peak at $m/z = 31$ in the mass spectrum of 1-butanol is due to the hydroxymethyl cation, $^+CH_2OH$, which arises from α cleavage.

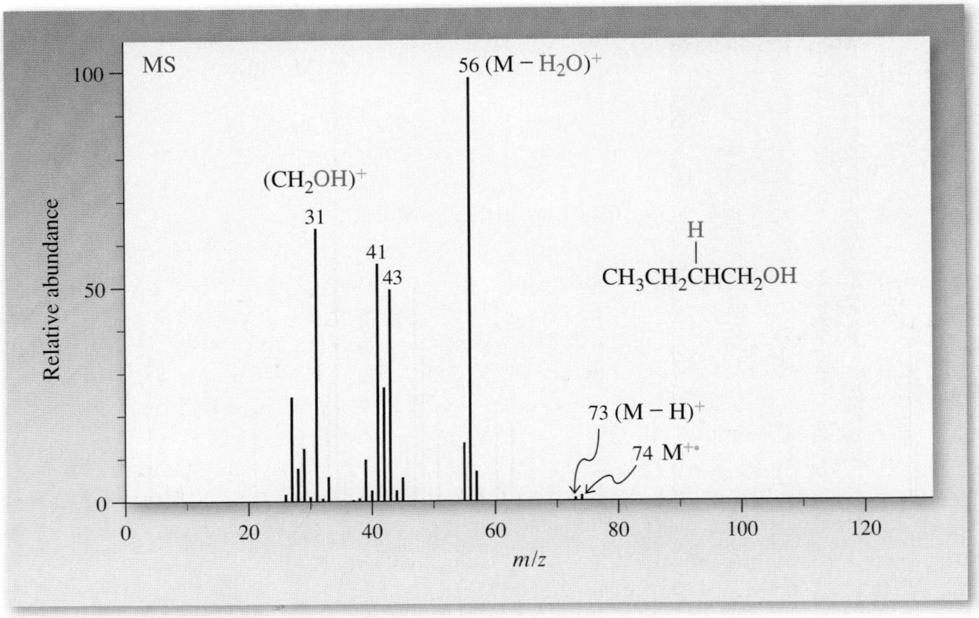

Figure 11-28 Mass spectrum of 1-butanol. The parent ion, at $m/z = 74$, gives rise to a small peak because of ready loss of water to give the ion at $m/z = 56$. Other fragment ions due to α cleavage are propyl ($m/z = 43$), 2-propenyl (allyl) ($m/z = 41$), and hydroxymethyl ($m/z = 31$).

Alcohol Fragmentation by Dehydration and α Cleavage

$$\begin{bmatrix} HO & H \\ | & | \\ R-C-CHR' \\ | \\ H \end{bmatrix}^{+\cdot} \longrightarrow [RCH=CHR']^{+\cdot} + H_2O$$

$$\textbf{M}^{+\cdot} \qquad\qquad (\textbf{M} - \textbf{18})^{+\cdot}$$

> Fragmentation in the mass spectrometer favors formation of more stable cations. Thus fragmentation of alcohols gives resonance-stabilized hydroxycarbocations.

$$-R\cdot \swarrow \qquad -H\cdot \downarrow \qquad -R'CH_2\cdot \searrow$$

$$\begin{bmatrix} HO \\ | \\ C \\ / \ \backslash \\ H \quad CH_2R' \end{bmatrix}^{+} \qquad \begin{bmatrix} HO \\ | \\ C \\ / \ \backslash \\ R \quad CH_2R' \end{bmatrix}^{+} \qquad \begin{bmatrix} HO \\ | \\ C \\ / \ \backslash \\ R \quad H \end{bmatrix}^{+}$$

Exercise 11-22

Try to predict the appearance of the mass spectrum of 3-methyl-3-heptanol.

Alkenes fragment to give resonance-stabilized cations

The fragmentation patterns of alkenes also reflect a tendency to break weaker bonds and to form more stable cationic species. The bonds one atom removed from the alkene function—the so-called *allylic* bonds—are relatively easily broken, because the result is a resonance-stabilized carbocation. For example, the mass spectra of terminal straight-chain alkenes such as 1-butene reveal formation of the 2-propenyl (allyl) cation, at $m/z = 41$, the base peak in the spectrum (Figure 11-29A).

Figure 11-29 Mass spectra of (A) 1-butene, showing a peak at $m/z = 41$ from cleavage to give the resonance-stabilized 2-propenyl (allyl) cation; (B) 2-hexene, showing similar cleavage between C4 and C5 to give the 2-butenyl cation, with $m/z = 55$.

Allylic bond

$$\begin{bmatrix} CH_2=CH-CH_2 \ \vdots\ CH_3 \end{bmatrix}^{+\cdot} \xrightarrow{-CH_3\cdot} \begin{bmatrix} CH_2=CH-\overset{+}{CH_2} \\ \updownarrow \\ \overset{+}{CH_2}-CH=CH_2 \end{bmatrix}$$

$$m/z = 56$$

2-Propenyl (allyl) cation
$m/z = 41$

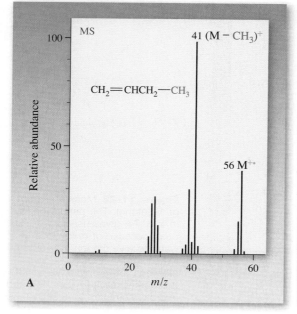

A

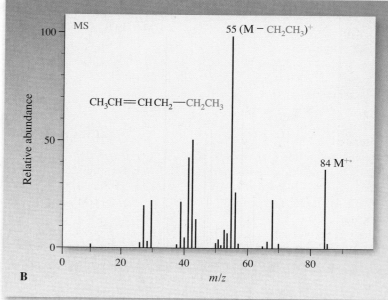

B

Branched and internal alkenes fragment similarly at allylic bonds. Figure 11-29B shows the mass spectrum of 2-hexene, in which the base peak at $m/z = 55$ corresponds to formation of the resonance-stabilized 2-butenyl cation.

$$\left[CH_3-CH=CH-CH_2\!\!\not|\!\!-CH_2-CH_3\right]^{+\,\bullet} \xrightarrow{-C_2H_5\,\bullet} \left[\begin{array}{c} CH_3-CH=CH\overset{\frown}{\text{—}}\overset{+}{C}H_2 \\ \updownarrow \\ CH_3-\overset{+}{C}H-CH=CH_2 \end{array}\right]$$

$m/z = 84$

2-Butenyl cation
$m/z = 55$

Exercise 11-23

The mass spectrum of 4-methyl-2-hexene shows peaks at $m/z = 69$, 83, and 98. Explain the origin of each.

In Summary Fragmentation patterns can be interpreted for structural elucidation. For example, the radical cations of alkanes cleave to form the most stable positively charged fragments, haloalkanes fragment by rupture of the carbon–halogen bond, alcohols readily dehydrate and undergo α cleavage, and alkenes break an allylic bond to form a resonance-stabilized carbocation.

11-11 | DEGREE OF UNSATURATION: ANOTHER AID TO IDENTIFYING MOLECULAR STRUCTURE

NMR and IR spectroscopy and mass spectrometry are important tools for determining the structure of an unknown. However, concealed in the molecular formula of every compound is an additional piece of information that can make the job easier. Consider a saturated, acyclic alkane: Its molecular formula is C_nH_{2n+2}. In contrast, an acyclic alkene containing one double bond contains two hydrogens less: C_nH_{2n}; it is called unsaturated. Cycloalkanes also have the general formula C_nH_{2n}. You can see that hydrocarbons containing several double bonds and/or rings will deviate from the "saturated" formula C_nH_{2n+2} by the appropriate number of fewer hydrogens. The **degree of unsaturation** is defined as the *sum of the numbers of rings and π bonds* present in the molecule. Table 11-6 illustrates the relation between molecular formula, structure, and degree of unsaturation for several hydrocarbons.

As Table 11-6 shows, each increase in the degree of unsaturation corresponds to a decrease of *two* hydrogens in the molecular formula. Therefore, starting with the general formula for acyclic alkanes (saturated; degree of unsaturation = 0), C_nH_{2n+2} (Section 2-5), the degree of unsaturation may be determined for any hydrocarbon merely by comparing the actual number of hydrogens present with the number required for the molecule to be saturated, namely, $2n + 2$, where n = the number of carbon atoms present. For example, what is the degree of unsaturation in a hydrocarbon of the formula C_5H_8? A *saturated* compound with five carbons has the formula C_5H_{12} (C_nH_{2n+2}, with $n = 5$). Because C_5H_8 is four hydrogens short of being saturated, the degree of unsaturation is $4/2 = 2$. All molecules with this formula contain a combination of rings and π bonds adding up to two.

The presence of heteroatoms may affect the calculation. Let us compare the molecular formulas of several saturated compounds: ethane, C_2H_6, and ethanol, C_2H_6O, have the same number of hydrogen atoms; chloroethane, C_2H_5Cl, has one less; ethanamine, C_2H_7N, one more. The number of hydrogens required for saturation is reduced by the presence of halogen, increased when nitrogen is present, and unaffected by oxygen. We can generalize

Some C_5H_8 Hydrocarbons

(Two π bonds)

(One π bond + one ring)

(Two rings)

Table 11-6	Degree of Unsaturation as a Key to Structure	
Formula	Representative Structures	Degree of Unsaturation
C_6H_{14}		0
C_6H_{12}	(one π bond) ; (one ring)	1
C_6H_{10}	(two π bonds) ; (one π bond + one ring) (two rings)	2
C_6H_8	(three π bonds) ; (two π bonds + one ring) (one π bond + two rings)	3

the procedure for determination of the degree of unsaturation from a molecular formula as follows.

Step 1. Determine from the number of carbons (n_C), halogens (n_X), and nitrogens (n_N) in the molecular formula the number of hydrogens required for the molecule to be saturated, H_{sat}.

$$H_{sat} = 2n_C + 2 - n_X + n_N \qquad \text{(Oxygen and sulfur are disregarded.)}$$

Step 2. Compare H_{sat} with the actual number of hydrogens in the molecular formula, H_{actual}, to determine the degree of unsaturation.

$$\text{Degree of unsaturation} = \frac{H_{sat} - H_{actual}}{2}$$

Alternatively, these steps may be combined into one formula:

$$\text{Degree of unsaturation} = \frac{2n_C + 2 + n_N - n_H - n_X}{2}$$

Exercise 11-24

Calculate the degree of unsaturation indicated by each of the following molecular formulas.
(a) C_5H_{10}; **(b)** $C_9H_{12}O$; **(c)** C_8H_7ClO; **(d)** $C_8H_{15}N$; **(e)** $C_4H_8Br_2$.

Solved Exercise 11-25 | **Working with the Concepts: Degree of Unsaturation in Structure Determination**

Spectroscopic data for three compounds with the molecular formula C_5H_8 are given below; m denotes a complex multiplet. Assign a structure to each compound. (**Hint:** One is acyclic; the others each contain one ring.) **(a)** IR: 910, 1000, 1650, 3100 cm^{-1}; ^{1}H NMR: $\delta = 2.79$ (t, $J = 8$ Hz), 4.8–6.2 (m) ppm, integrated intensity ratio of the signals = 1:3. **(b)** IR: 900, 995, 1650, 3050 cm^{-1}; ^{1}H NMR: $\delta = 0.5$–1.5 (m), 4.8–6.0 (m) ppm, integrated intensity ratio of the signals = 5:3. **(c)** IR 1611, 3065 cm^{-1}; ^{1}H NMR: $\delta = 1.5$–2.5 (m), 5.7 (m) ppm, integrated intensity ratio of the signals = 3:1. Is there more than one possibility?

Strategy

Find the degree of unsaturation in order to set limits on the numbers of π bonds and rings. Use IR data to determine the presence or absence of π bonds. Then continue by using NMR data to define a reasonable structure possibility.

Solution

- The molecular formula C_5H_8 corresponds to a degree of unsaturation = 2. Therefore, all molecules must contain a combination of π bonds and rings adding up to two. Taking each in turn:

  (a) Four bands are listed in the IR spectrum; those at 1650 and 3100 cm^{-1} correspond unambiguously to alkene C=C and alkenyl C–H stretching motions. The bands at 910 and 1000 cm^{-1} are strongly suggestive of the presence of the terminal –CH=CH$_2$ group (Section 11-8). The NMR spectrum displays two signals in an intensity ratio of 1 : 3. Because the molecule contains eight hydrogens total, this information implies two groups of hydrogens, one containing two and the other six. The two-hydrogen signal, showing a triplet at δ = 2.79, would appear to be due to a –CH$_2$– unit, coupled to two neighboring hydrogens on the surrounding carbon atoms. The signal for six hydrogens is in the range δ = 4.8−6.2, typical of alkenyl hydrogens and must therefore correspond to two –CH=CH$_2$ groups. Combining these fragments, we obtain the structure CH$_2$=CH–CH$_2$–CH=CH$_2$.

  (b) The IR data are essentially identical to those in (a), implying the presence of a –CH=CH$_2$ unit. The NMR spectrum shows a five-hydrogen signal at high field and a three-hydrogen signal in the alkenyl region. The three-hydrogen signal is consistent with –CH=CH$_2$, leaving us with a C_3H_5 fragment with NMR signals in the alkane region. Therefore, the molecule must contain a ring as the second degree of unsaturation, leaving ▷—CH=CH$_2$ as the only answer.

  (c) The IR data are limited to C=C and alkenyl C–H bands, without additional information from absorptions below 1000 cm^{-1}. We again rely most heavily on the NMR spectrum, which again shows two signals, with the one at higher field three times as intense as the one at lower field. The molecule must contain six alkyl and two alkenyl hydrogens. There is no way to construct such a molecule with the correct formula and more than one double bond, so a ring must be present as well. The most straightforward option is

  ⬠ , with ▷—CH$_2$—CH$_3$ as a second possibility. The latter is less likely because its NMR would show a clear triplet for the –CH$_3$ group, not a feature noted in the data.

Exercise 11-26 | Try It Yourself

Propose a structure for another compound with the molecular formula C_5H_8. Its IR spectrum, however, shows no absorption whatsoever in the region between 1600 and 2500 cm^{-1}.

In Summary The degree of unsaturation is equal to the sum of the numbers of rings and π bonds in a molecule. Calculation of this parameter makes solving structure problems from spectroscopic data easier.

THE BIG PICTURE

In this chapter, we looked at alkenes, a compound class characterized by the carbon–carbon double bond. In Chapters 7 and 9, we learned that alkenes are prepared synthetically by elimination reactions of haloalkanes and alcohols. In this chapter, we examined these reactions in more depth. We saw that the structure of the base determines what products will form in E2 elimination from haloalkanes. Similarly, the structure of an alcohol undergoing acid-catalyzed dehydration determines what mechanism takes place and how easily it occurs.

We also examined three additional methods that organic chemists use for determining molecular structure: calculation of the degree of unsaturation from a molecular formula, infrared (IR) spectroscopy, and mass spectrometry (MS). Mass spectrometry is the most important method for deriving the molecular formula; the degree of unsaturation tells us the total number of rings and π bonds in a molecule; IR spectra help identify the presence of the characteristic bonds of many functional groups. Combined with NMR spectra, this information is normally enough to determine a molecule's structure. In later chapters, we shall cover ultraviolet (UV) spectroscopy, which provides information on multiple π bonds.

In the next chapter, we turn to the reactions of alkenes, to see how their unsaturated structure defines their chemical behavior and the compounds into which they may be converted.

WORKED EXAMPLES: INTEGRATING THE CONCEPTS

11-27. Using Mechanistic Details to Predict a Reaction Product Correctly

Write the major product(s) of the following reaction:

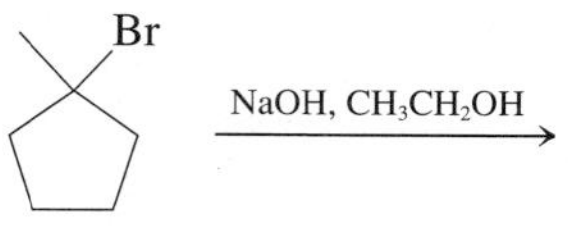

What the question asks seems straightforward: Predict the product of a reaction of the type we have seen several times, going back to Chapters 6 and 7. Although this problem does not ask *explicitly* "how" or "why"—which are our clues to thinking about mechanisms—you know that haloalkanes may transform by a variety of pathways under the given conditions. Hence, mechanistic thinking is of the essence.

How to start? Identify the starting organic molecule and the reagent by the compound categories to which they belong. See if that leads you to a solution.

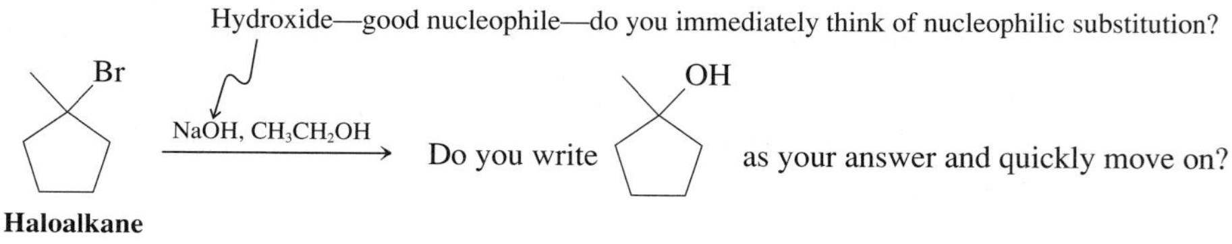

Slow down! Take a second look: The substrate is *tertiary*, and ⁻OH is a *strong base* as well.

Information: Table 7-4 tells you **how** a reaction between a tertiary substrate and a strong base proceeds: through elimination, by the E2 mechanism. But you are still not finished. More than one isomer can be formed. Section 11-6 covers the regiochemistry of elimination: Unhindered bases (such as hydroxide) give the more stable alkene (Saytzev rule). The identification of the major product relies on **I**nformation regarding relative alkene stabilities that were presented in Section 11-5: The alkene on the right is trisubstituted and is therefore favored over its disubstituted isomer on the left.

Now, finally, you can **P**roceed:

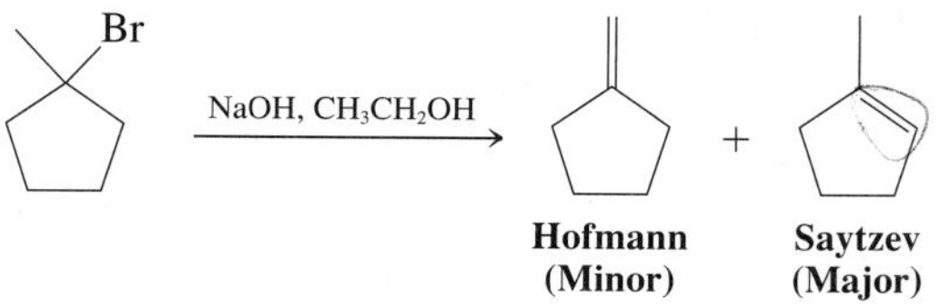

11-28. Spectroscopy as an Aid to Product Identification

Acid-catalyzed dehydration of 2-methyl-2-pentanol (dilute H_2SO_4, 50°C) gives one major product and one minor product. Elemental analysis reveals that both contain only carbon and hydrogen in a 1:2 atom ratio, and high-resolution MS gives a mass of 84.0940 for the molecular ions of both compounds. Spectroscopic data for these substances are as follows:

1. Major product: IR: 1660 and 3080 cm^{-1}; ^{1}H NMR: $\delta = 0.91$ (t, $J = 7$ Hz, 3 H), 1.60 (s, 3 H), 1.70 (s, 3 H), 1.98 (quin, $J = 7$ Hz, 2 H), and 5.08 (t, $J = 7$ Hz, 1 H) ppm.

2. Minor product: IR: 1640 and 3090 cm^{-1}; ^{1}H NMR: $\delta = 0.92$ (t, $J = 7$ Hz, 3 H), 1.40 (sex, $J = 7$ Hz, 2 H), 1.74 (s, 3 H), 2.02 (t, $J = 7$ Hz, 2 H), and 4.78 (s, 2 H) ppm.

Deduce the structures of the products, suggest mechanisms for their formation, and discuss why the major product forms in the greater amount.

SOLUTION

First, we write the structure of the starting material (alcohol nomenclature, Section 8-1) and what we know about the reaction:

$$\overset{\displaystyle :\ddot{O}H}{\diagdown\diagup}\diagdown\diagup \qquad \xrightarrow{\text{Dilute } H_2SO_4,\ 50°C}$$

2-Methyl-2-pentanol

This is an acid-catalyzed reaction of a tertiary alcohol (Section 11-7). Even though we may know enough to be able to make a sensible prediction, let us proceed by interpretation of the spectra first, and see if the answer that we get is consistent with our expectations.

Both compounds have the same molecular formula—they are isomers. The elemental analysis gives the empirical formula of CH_2. The exact mass of CH_2 is $12.000 + 2(1.00783) = 14.01566$. Thus, the MS data tell us that the molecular formula is $6(CH_2) = C_6H_{12}$, because $84.0940/14.01566 = 6$.

For the major product, peaks at 1660 and 3080 cm^{-1} in the IR spectrum are in the ranges for the alkene C=C and C–H bond-stretching frequencies (1620–1680 and 3050–3150 cm^{-1}, Table 11-3). Armed with this information, we turn to the NMR spectrum and immediately look for signals in the region characteristic of alkene hydrogens, $\delta = 4.6$–5.7 ppm (Table 10-2 and Section 11-4). Indeed, we find one at $\delta = 5.08$ ppm. The information following the chemical shift position (t, $J = 7$ Hz, 1 H) tells us that the signal is split into a triplet with a coupling constant of 7 Hz and has a relative integrated intensity corresponding to one hydrogen. The spectrum also shows three signals with integrations of 3 H each; it is usually reasonable to assume that simple signals with intensities of 3 between $\delta = 0$–4 ppm indicate the presence of methyl groups, unless information to the contrary appears. Two of the methyl groups are singlets, and the other appears as a triplet. Finally, a signal integrating for 2 H at $\delta = 1.98$ ppm is split into five lines. Assuming that a CH_2 group gives rise to the latter signal, we have the following fragments to consider in piecing together a structure: CH=C, 3 CH_3, and CH_2. The sum of atoms in these fragments is C_6H_{12}, in agreement with the formula given for the product and giving us confidence that we are on the right track. There is only a limited number of ways in which to connect these pieces into a proper structure:

$$\underset{\text{CH}_3—\text{CH}=\overset{\displaystyle \overset{\text{CH}_3}{|}}{\text{C}}—\text{CH}_2—\text{CH}_3}{} \text{ (ignoring stereochemistry)} \qquad \text{and} \qquad \text{CH}_3—\text{CH}_2—\text{CH}=\overset{\displaystyle \overset{\text{CH}_3}{|}}{\text{C}}—\text{CH}_3$$

We can either use our chemical knowledge or, again, turn to the spectroscopy to determine which is correct. With the use of the splitting patterns in the NMR spectrum, a quick decision may be made. Using the $N + 1$ rule (Section 10-7), we see that, under ideal conditions, an NMR signal will be split by N neighboring hydrogens into $N + 1$ lines. In the first structure, the alkenyl hydrogen is neighbor to a CH_3 group and should therefore appear with $3 + 1 = 4$ lines, a quartet. The actual spectrum shows this signal as a triplet. Furthermore, this structure contains three methyl groups, but they have 0, 1, and 2 neighboring hydrogens, respectively, and should appear as one singlet, one doublet, and one triplet—again, in disagreement with the actual spectrum. In contrast, the second structure fits: the alkenyl hydrogen has two neighbors, consistent with the observed triplet splitting, and two of the three CH_3 groups are on an alkene carbon lacking a neighboring hydrogen and should be singlets. Using our chemical knowledge, we note that this correct structure has the same carbon connectivity as that of the starting material, whereas the incorrect one would have required a rearrangement.

Turning to the minor product, we use the same logic: Again, the IR spectrum shows an alkene C=C stretch (at 1640 cm^{-1}) and an alkenyl C–H stretch at 3090 cm^{-1}. The NMR spectrum contains a singlet at $\delta = 4.78$, integrating for 2 H: two alkenyl hydrogens. It also shows two methyl signals, as well as two others integrating 2 H for each. Therefore, we have 2 CH_3, 2 CH_2, and 2 alkenyl H, adding again (with the two alkenyl carbons) to C_6H_{12}. Even though many possible combinations can be devised at this point, we can make use of NMR splitting information to close in rapidly on the answer. One of the CH_3 groups is a singlet, meaning that it must be attached to a carbon that lacks hydrogens. If we look at the preceding array of fragments, this latter carbon can only be an alkenyl carbon. Thus, we have CH_3–**C**=C, in which the bold carbon is *not* attached to hydrogen. Therefore, by process of elimination, both alkenyl hydrogens must be attached to the *other* alkenyl carbon: CH_3–C=**CH₂**. The remaining fragments can be attached to this piece in only one way, giving the final structure:

$$\underset{\text{CH}_3—\overset{\displaystyle \overset{\text{CH}_2—\text{CH}_2}{|}}{\text{C}}=\text{CH}_2}{}$$

Thus, we can complete the equation stated at the beginning of the problem as follows:

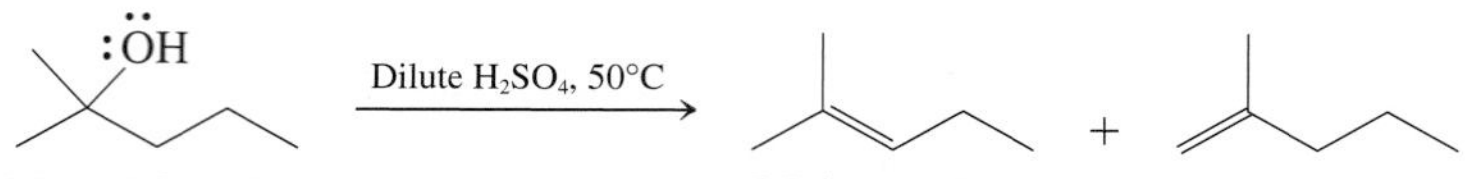

2-Methyl-2-pentanol →(Dilute H₂SO₄, 50°C) Major product +

Is this result in accord with our chemical expectations? Let us consider the mechanism (Section 11-7). Dehydration of a secondary or a tertiary alcohol under acidic conditions begins with protonation of the oxygen atom, giving a good potential leaving group (water). Departure of the leaving group results in a carbocation, and removal of a proton from a neighboring carbon atom (most likely by a second molecule of alcohol acting as a Lewis base) gives the alkene, overall an E1 process:

> **Reminder**
>
> Free H⁺ does not exist in solution, but is attached to any available electron pair, such as the oxygen of water (or the HSO₄ anion, not shown) in the adjacent scheme.

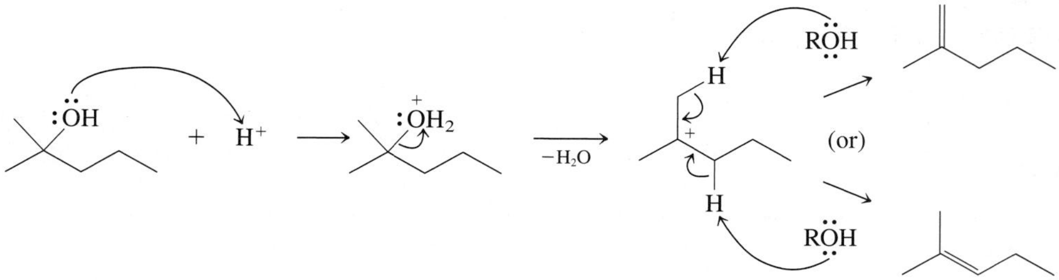

The major product is the one with the more substituted double bond, the thermodynamically more stable alkene (Sections 11-5 and 11-7), as is typical in E1 dehydrations.

New Reactions

1. Hydrogenation of Alkenes (Section 11-5)

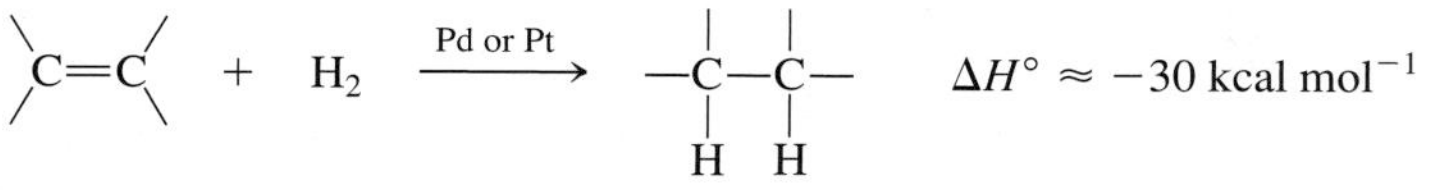

$$\Delta H° \approx -30 \text{ kcal mol}^{-1}$$

Order of stability of the double bond

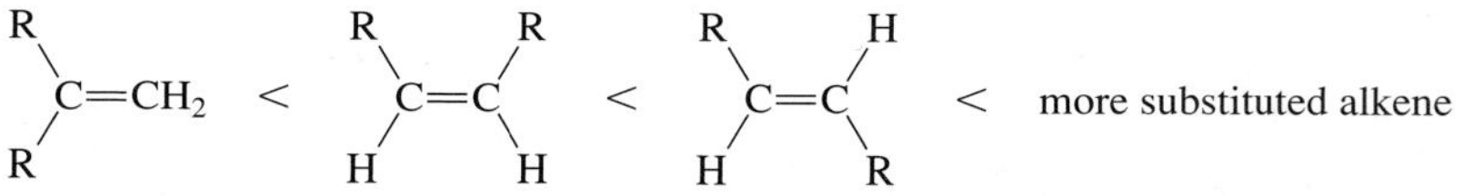

< more substituted alkene

Preparation of Alkenes

2. From Haloalkanes, E2 with Unhindered Base (Section 11-6)

Saytzev rule

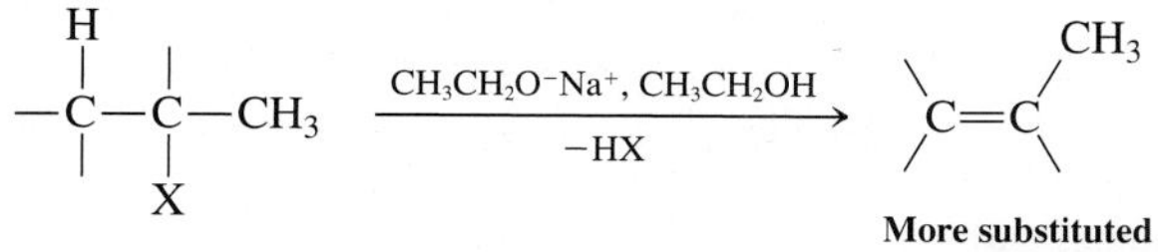

More substituted (more stable) alkene

3. From Haloalkanes, E2 with Sterically Hindered Base (Section 11-6)

Hofmann rule

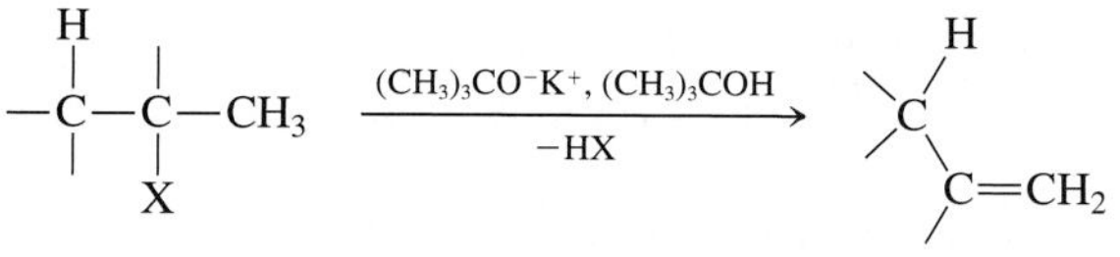

Less substituted (less stable) alkene

4. Stereochemistry of E2 Reaction (Section 11-6)

Anti elimination

5. Dehydration of Alcohols (Section 11-7)

Most stable alkene is major product
Primary: E2 mechanism
Secondary, tertiary: E1 mechanism
Carbocations may rearrange

Order of reactivity: **primary < secondary < tertiary**

Important Concepts

1. **Alkenes** are **unsaturated** molecules. Their IUPAC names are derived from alkanes, the longest chain incorporating the double bond serving as the stem. **Double-bond isomers** include **terminal, internal, cis,** and **trans** arrangements. Tri- and tetrasubstituted alkenes are named according to the ***E,Z* system,** in which the *R,S* priority rules apply.

2. The double bond is composed of a σ bond and a π bond. The σ bond is obtained by overlap of the two sp^2 hybrid lobes on carbon, the π bond by interaction of the two remaining p orbitals. The **π bond** is weaker (~ 65 kcal mol^{-1}) than its σ counterpart (~ 108 kcal mol^{-1}) but strong enough to allow for the existence of stable cis and trans isomers.

3. The functional group in the alkenes is flat, **sp^2 hybridization** being responsible for the possibility of creating dipoles and for the relatively high acidity of the alkenyl hydrogen.

4. Alkenyl hydrogens and carbons appear at **low field** in ^{1}H NMR ($\delta = 4.6$–5.7 ppm) and ^{13}C NMR ($\delta = 100$–140 ppm) experiments, respectively. J_{trans} is larger than J_{cis}, $J_{geminal}$ is very small, and $J_{allylic}$ is variable but small.

5. The relative stability of isomeric alkenes can be established by comparing **heats of hydrogenation.** It decreases with decreasing substitution; trans isomers are more stable than cis.

6. Elimination of haloalkanes (and other alkyl derivatives) may follow the **Saytzev rule** (nonbulky base, internal alkene formation) or the **Hofmann rule** (bulky base, terminal alkene formation). Trans alkenes as products predominate over cis alkenes. Elimination is **stereospecific,** as dictated by the *anti* transition state.

7. **Dehydration** of alcohols in the presence of strong acid usually leads to a mixture of products, with the most stable alkene being the major constituent.

8. **Infrared spectroscopy** measures **vibrational excitation.** The energy of the incident radiation ranges from about 1 to 10 kcal mol^{-1} ($\lambda \approx 2.5 - 16.7$ μm; $\tilde{\nu} \approx 600$–4000 cm^{-1}). Characteristic peaks are observed for certain functional groups, a consequence of stretching, bending, and other modes of vibration, and their combination. Moreover, each molecule exhibits a characteristic infrared spectral pattern in the **fingerprint** region below 1500 cm^{-1}.

9. Alkanes show IR bands characteristic of C–H bonds in the range from 2840 to 3000 cm^{-1}. The C=C stretching absorption for alkenes is in the range from 1620 to 1680 cm^{-1}, that for the alkenyl C–H bond is about 3100 cm^{-1}. Bending modes sometimes give useful peaks below 1500 cm^{-1}. Alcohols are usually characterized by a broad peak for the O–H stretch between 3200 and 3650 cm^{-1}.

10. **Mass spectrometry** is a technique for ionizing molecules and separating the resulting ions magnetically by molecular mass. Because the ionizing beam has high energy, the ionized molecules also **fragment** into smaller particles, all of which are separated and recorded as the **mass spectrum** of a compound. **High-resolution mass spectral data** allow determination of molecular formulas from **exact mass** values. The presence of certain elements (such as Cl, Br) can be detected by their isotopic patterns. The presence of fragment-ion signals in mass spectra can be used to deduce the structure of a molecule.

11. **Degree of unsaturation** (number of rings + number of π bonds) is calculated from the molecular formula by using the equation

$$\text{Degree of unsaturation} = \frac{H_{sat} - H_{actual}}{2}$$

where $H_{sat} = 2n_C + 2 - n_X + n_N$ (disregard oxygen and sulfur).

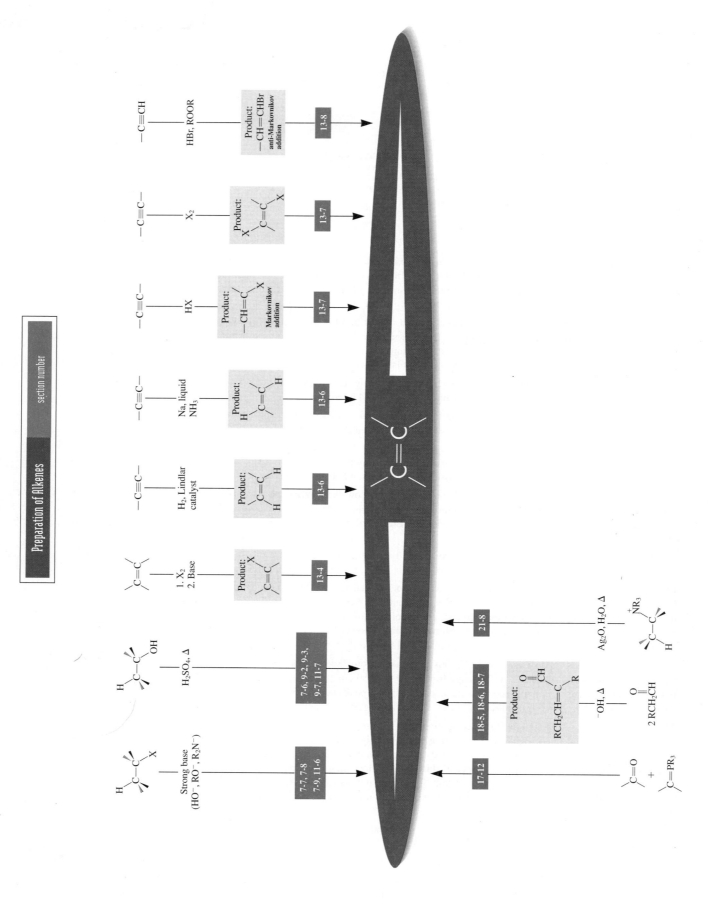

Preparation of Alkenes

section number

476

Problems

29. Draw the structures of the molecules with the following names.

   (a) 4,4-Dichloro-*trans*-2-octene

   (b) (*Z*)-4-bromo-2-iodo-2-pentene

   (c) 5-Methyl-*cis*-3-hexen-1-ol

   (d) (*R*)-1,3-Dichlorocycloheptene

   (e) (*E*)-3-Methoxy-2-methyl-2-buten-1-ol

30. Name each of the following molecules in accord with the IUPAC system of nomenclature.

 (a)

 (b)

(c)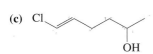

(d)

(e) HO CF₃

(f) Cl Cl

(g) CH₃O OCH₃

(h)

 (i)

31. Name each of the compounds below. Use cis/trans and/or *E/Z* designations, if appropriate, to designate stereochemistry.

   (a)
$$\underset{H_3C}{\overset{Cl}{}}C=C\underset{CH_3}{\overset{H}{}}$$

   (b)
$$\underset{H_3C}{\overset{H}{}}C=C\underset{H}{\overset{CH_2Cl}{}}$$

   (c)
$$\underset{H_3C}{\overset{H_3C}{}}C=C\underset{Cl}{\overset{H}{}}$$

32. Of each pair of the following compounds, which one should have the higher dipole moment? The higher boiling point? **(a)** *cis*- or *trans*-1,2-Difluoroethene; **(b)** *Z*- or *E*-1,2-difluoropropene; **(c)** *Z*- or *E*-2,3-difluoro-2-butene

33. Draw the structures of each of the following compounds, rank them in order of acidity, and circle the most acidic hydrogen(s) in each: cyclopentane, cyclopentanol, cyclopentene, 3-cyclopenten-1-ol.

34. Assign structures to the following molecules on the basis of the indicated ¹H NMR spectra A–E. Consider stereochemistry, where applicable.

   (a) C₄H₇Cl, NMR spectrum A;

   (b) C₅H₈O₂, NMR spectrum B;

   (c) C₄H₈O, NMR spectrum C;

   (d) another C₄H₈O, NMR spectrum D (next page);

   (e) C₃H₄Cl₂, NMR spectrum E (next page).

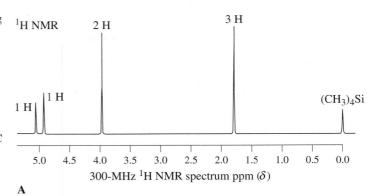

A

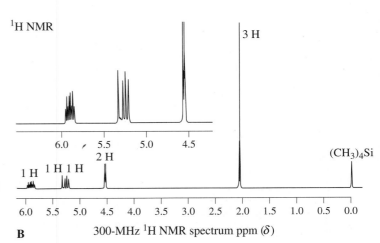

B

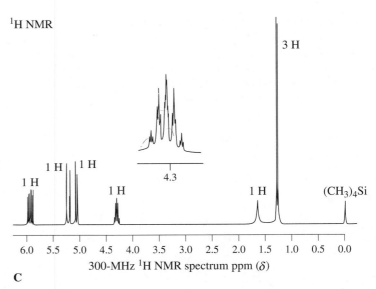

C

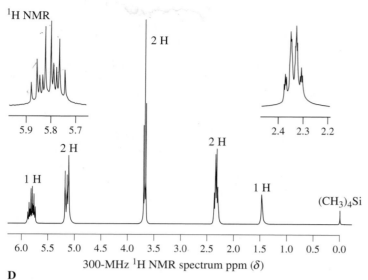

¹H NMR

300-MHz ¹H NMR spectrum ppm (δ)

D

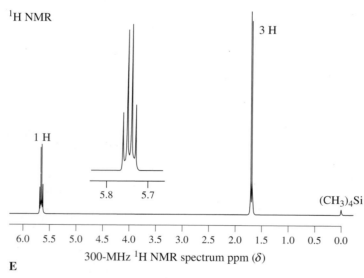

¹H NMR

300-MHz ¹H NMR spectrum ppm (δ)

E

35. Explain the splitting patterns in ¹H NMR spectrum D in detail. The insets are fivefold expansions.

36. For each of the pairs of alkenes below, indicate whether measurements of polarity alone would be sufficient to distinguish the compounds from one another. Where possible, predict which compound would be more polar.

37. Place the alkenes in each group in order of increasing stability of the double bond and increasing heat of hydrogenation.

38. Write the structures of as many simple alkenes as you can that, upon catalytic hydrogenation with H_2 over Pt, will give as the product **(a)** 2-methylbutane; **(b)** 2,3-dimethylbutane; **(c)** 3,3-dimethylpentane; **(d)** 1,1,4-trimethylcyclohexane. In each case in which you have identified more than one alkene as an answer, rank the alkenes in order of stability.

39. The reaction between 2-bromobutane and sodium ethoxide in ethanol gives rise to three E2 products. What are they? Predict their relative amounts.

40. What key structural feature distinguishes haloalkanes that give more than one stereoisomer on E2 elimination (e.g., 2-bromobutane, Problem 39) from those that give only a single isomer exclusively (e.g., 2-bromo-3-methylpentane, Section 11-6)?

41. Write the most likely major product(s) of each of the following haloalkanes with sodium ethoxide in ethanol or potassium *tert*-butoxide in 2-methyl-2-propanol (*tert*-butyl alcohol). **(a)** Chloromethane; **(b)** 1-bromopentane; **(c)** 2-bromopentane; **(d)** 1-chloro-1-methylcyclohexane; **(e)** (1-bromoethyl)-cyclopentane; **(f)** (2R,3R)-2-chloro-3-ethylhexane; **(g)** (2R,3S)-2-chloro-3-ethylhexane; **(h)** (2S,3R)-2-chloro-3-ethylhexane.

42. Referring to the data in Worked Example 27, predict how the rate of E2 reaction between 1-bromopropane and sodium ethoxide in ethanol would compare with those of the three substrates discussed in that problem under the same reaction conditions.

43. Draw Newman projections of the four stereoisomers of 2-bromo-3-methylpentane in the conformation required for E2 elimination. (See the structures labeled "Stereospecificity in the E2 Reaction of 2-Bromo-3-methylpentane" on p. 453.) Are the reactive conformations also the most stable conformations? Explain.

44. Referring to the answer to Problem 38 of Chapter 7, predict (qualitatively) the relative amounts of isomeric alkenes that are formed in the elimination reactions shown.

45. Referring to the answers to Problem 30 of Chapter 9, predict (qualitatively) the relative yields of all the alkenes formed in each reaction.

46. Compare and contrast the major products of dehydrohalogenation of 2-chloro-4-methylpentane with **(a)** sodium ethoxide in ethanol and **(b)** potassium *tert*-butoxide in 2-methyl-2-propanol (*tert*-butyl alcohol). Write the mechanism of each process. Next consider the reaction of 4-methyl-2-pentanol with concentrated H_2SO_4 at 130°C, and compare its product(s) and the mechanism of its (their) formation with those from the dehydrohalogenations in (a) and (b). (**Hint:** The dehydration gives as its major product a molecule that is not observed in the dehydrohalogenations.)

47. Referring to Problem 59 of Chapter 7, write the structure of the alkene that you would expect to be formed as the major product from E2 elimination of each of the chlorinated steroids shown.

48. 1-Methylcyclohexene is more stable than methylenecyclohexane (A, below), but methylenecyclopropane (B) is more stable than 1-methylcyclopropene. Explain.

CH₂   CH₂

A   B

49. Give the products of bimolecular elimination from each of the following isomeric halogenated compounds.

(a)   Br H
           | |
   C₆H₅—C—C—C₆H₅
           | |
           H CH₃

(b)   Br H
           | |
   C₆H₅—C—C—C₆H₅
           | |
           H₃C H

One of these compounds undergoes elimination 50 times faster than the other. Which compound is it? Why? (**Hint:** See Problem 41.)

50. Explain in detail the differences between the mechanisms giving rise to the following two experimental results.

CH₃
    Cl   —Na⁺ ⁻OCH₂CH₃, CH₃CH₂OH→
CH(CH₃)₂

CH₃

CH(CH₃)₂
100%

CH₃
    Cl   —Na⁺ ⁻OCH₂CH₃, CH₃CH₂OH→
CH(CH₃)₂

CH₃   CH₃

CH(CH₃)₂   +   CH(CH₃)₂
25%   75%

51. The molecular formulas and ¹³C NMR data (in ppm) for several compounds are given here. The type of carbon, as revealed from DEPT spectra, is specified in each case. Deduce a structure for each compound. **(a)** C_4H_6: 30.2 (CH_2), 136.0 (CH); **(b)** C_4H_6O: 18.2 (CH_3), 134.9 (CH), 153.7 (CH), 193.4 (CH); **(c)** C_4H_8: 13.6 (CH_3), 25.8 (CH_2), 112.1 (CH_2), 139.0 (CH); **(d)** $C_5H_{10}O$: 17.6 (CH_3), 25.4 (CH_3), 58.8 (CH_2), 125.7 (CH), 133.7 ($C_{quaternary}$); **(e)** C_5H_8: 15.8 (CH_2), 31.1 (CH_2), 103.9 (CH_2), 149.2 ($C_{quaternary}$); **(f)** C_7H_{10}: 25.2 (CH_2), 41.9 (CH), 48.5 (CH_2), 135.2 (CH).

(**Hint:** This one is difficult. The molecule has one double bond. How many rings must it have?)

52. Data from both ordinary and DEPT ¹³C NMR spectra for several compounds with the formula C_5H_{10} are given here. Deduce a structure for each compound. **(a)** 25.3 (CH_2); **(b)** 13.3 (CH_3), 17.1 (CH_3), 25.5 (CH_3), 118.7 (CH), 131.7 ($C_{quaternary}$); **(c)** 12.0 (CH_3), 13.8 (CH_3), 20.3 (CH_2), 122.8 (CH), 132.4 (CH).

53. From the Hooke's law equation, would you expect the C–X bonds of common haloalkanes (X = Cl, Br, I) to have IR bands at higher or lower wavenumbers than are typical for bonds between carbon and lighter elements (e.g., oxygen)?

54. Convert each of the following IR frequencies into micrometers.
(a) 1720 cm⁻¹ (C=O)   **(b)** 1650 cm⁻¹ (C=C)
(c) 3300 cm⁻¹ (O–H)   **(d)** 890 cm⁻¹ (alkene bend)
(e) 1100 cm⁻¹ (C–O)   **(f)** 2260 cm⁻¹ (C≡N)

55. Match each of the following structures with the IR data that correspond best. Abbreviations: w, weak; m, medium; s, strong; br, broad. **(a)** 905 (s), 995 (m), 1040 (m), 1640 (m), 2850–2980 (s), 3090 (m), 3400 (s, br) cm⁻¹; **(b)** 2840 (s), 2930 (s) cm⁻¹; **(c)** 1665 (m), 2890–2990 (s), 3030 (m) cm⁻¹; **(d)** 1040 (m), 2810–2930 (s), 3300 (s, br) cm⁻¹.

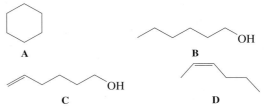

A   B OH

C OH   D

56. You have just entered the chemistry stockroom to look for several isomeric bromopentanes. There are three bottles on the shelf marked $C_5H_{11}Br$, but their labels have fallen off. The NMR machine is broken, so you devise the following experiment in an attempt to determine which isomer is in which bottle: You first treat a sample of the contents in each bottle with NaOH in aqueous ethanol, and then you determine the IR spectrum of each product or product mixture. Here are the results:

(i) $C_5H_{11}Br$ isomer in bottle A $\xrightarrow{NaOH}$ IR bands at 1660, 2850–3020, and 3350 cm⁻¹

(ii) $C_5H_{11}Br$ isomer in bottle B $\xrightarrow{NaOH}$ IR bands at 1670 and 2850–3020 cm⁻¹

(iii) $C_5H_{11}Br$ isomer in bottle C $\xrightarrow{NaOH}$ IR bands at 2850–2960 and 3350 cm⁻¹

(a) What do the data tell you about each product or product mixture?

(b) Suggest possible structures for the contents of each bottle.

57. An organic compound exhibits IR spectrum F. From the group of structures below, choose the one that matches the spectrum best.

OH

O

O

H   OCH₃

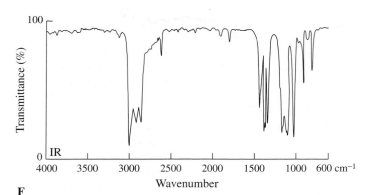

F

58. The three compounds hexane, 2-methylpentane, and 3-methyl-pentane correspond to the three mass spectra shown below. Match each compound with the spectrum that best fits its structure on the basis of the fragmentation pattern.

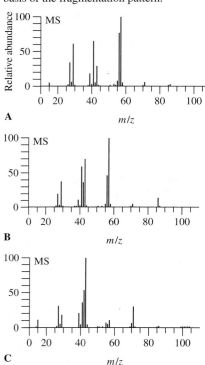

A

B

C

59. Assign as many peaks as you can in the mass spectrum of 1-bromopropane (Figure 11-24).

60. The following table lists selected mass-spectral data for three isomeric alcohols with the formula $C_5H_{12}O$. On the basis of the peak positions and intensities, suggest structures for each of the three isomers. A dash means that the peak is very weak or absent entirely.

Relative Peak Intensities

m/z	Isomer A	Isomer B	Isomer C
88 M$^+$	—	—	—
87 (M − 1)$^+$	2	2	—
73 (M − 15)$^+$	—	7	55
70 (M − 18)$^+$	38	3	3
59 (M − 29)$^+$	—	—	100
55 (M − 15 − 18)$^+$	60	17	33
45 (M − 43)$^+$	5	100	10
42 (M − 18 − 28)$^+$	100	4	6

61. Determine the molecular formulas corresponding to each of the following structures. For each structure, calculate the number of degrees of unsaturation from the molecular formula and evaluate whether your calculations agree with the structures shown.

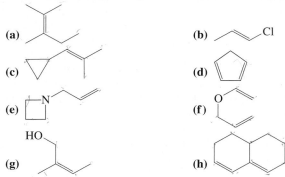

62. Calculate the degree of unsaturation that corresponds to each of the following molecular formulas. (a) C_7H_{12}; (b) $C_8H_7NO_2$; (c) C_6Cl_6; (d) $C_{10}H_{22}O_{11}$; (e) $C_6H_{10}S$; (f) $C_{18}H_{28}O_2$.

63. A hydrocarbon with an exact molecular mass of 96.0940 exhibits the following spectroscopic data: ^1H NMR: $\delta = 1.3$ (m, 2 H), 1.7 (m, 4 H), 2.2 (m, 4 H), and 4.8 (quin, $J = 3$ Hz, 2 H) ppm; ^{13}C NMR: $\delta = 26.8, 28.7, 35.7, 106.9,$ and 149.7 ppm. The IR spectrum is shown below (spectrum G). Hydrogenation furnishes a product with an exact molecular mass of 98.1096. Suggest a structure for the compound consistent with these data.

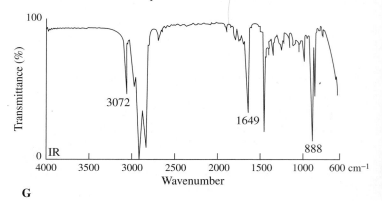

G

64. The isolation of a novel form of molecular carbon, C_{60}, was reported in 1990. The substance has the shape of a soccer ball of carbon atoms and possesses the nickname "buckyball" (you don't want to know the IUPAC name). Hydrogenation produces a hydrocarbon with the molecular formula $C_{60}H_{36}$. What is the degree of unsaturation in C_{60}? In $C_{60}H_{36}$? Does the hydrogenation result place limits on the numbers of π bonds and rings in "buckyball"? (More on C_{60} is found in Real Life 15-1.)

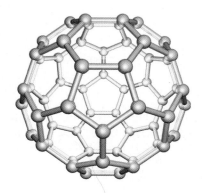

65. **CHALLENGE**   You have just been named president of the famous perfume company, Scents "R" Us. Searching for a hot new item to market, you run across a bottle labeled only $C_{10}H_{20}O$, which contains a liquid with a wonderfully sweet rose aroma. You want more, so you set out to elucidate its structure. Do so from the following data. (i) ^1H NMR: clear signals at $\delta = 0.94$ (d, $J = 7$ Hz, 3 H), 1.63 (s, 3 H), 1.71 (s, 3 H), 3.68 (t, $J = 7$ Hz, 2 H), 5.10 (t, $J = 6$ Hz, 1 H) ppm; the other 8 H have overlapping absorptions in the range $\delta = 1.3$–2.2 ppm. (ii) ^{13}C NMR (^1H decoupled): $\delta = 60.7, 125.0, 130.9$ ppm; seven other signals are upfield of $\delta = 40$ ppm. (iii) IR: $\tilde{\nu} = 1640$ and 3350 cm^{-1}. (iv) Oxidation with buffered PCC (Section 8-6) gives a compound with the molecular formula $C_{10}H_{18}O$. Its spectra show the following changes compared with the starting material: ^1H NMR: signal at $\delta = 3.68$ ppm is gone, but a new signal is seen at $\delta = 9.64$ ppm; ^{13}C NMR: signal at $\delta = 60.7$ ppm is gone, replaced by one at $\delta = 202.1$ ppm; IR: loss of signal at $\tilde{\nu} = 3350$ cm^{-1}; new peak at $\tilde{\nu} = 1728$ cm^{-1}. (v) Hydrogenation gives $C_{10}H_{22}O$, indentical with that formed on hydrogenation of the natural product geraniol (see below).

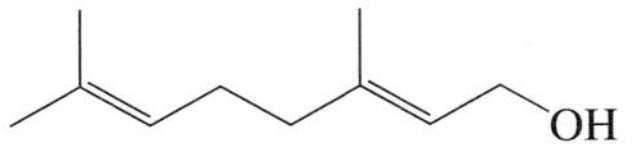

Geraniol

66. Using the information in Table 11-4, match up each set of the following IR signals with one of these naturally occurring compounds: camphor; menthol; chrysanthemic ester; epiandrosterone. You can find the structures of the natural products in Section 4-7. **(a)** 3355 cm^{-1}; **(b)** 1630, 1725, 3030 cm^{-1}; **(c)** 1730, 3410 cm^{-1}; **(d)** 1738 cm^{-1}.

67. **CHALLENGE**   Identify compounds A, B, and C from the following information and explain the chemistry that is taking place. Reaction of the alcohol shown below with 4-methylbenzenesulfonyl chloride in pyridine produced A ($C_{15}H_{20}O_3S$). Reaction of A with lithium diisopropylamide (LDA, Section 7-8) produces a single product, B (C_8H_{12}), which displays in its ^1H NMR a two-proton multiplet at about $\delta = 5.6$ ppm. If, however, compound A is treated with NaI before the reaction with LDA, two products are formed: B and an isomer, C, whose NMR shows a multiplet at $\delta = 5.2$ ppm that integrates as only one proton.

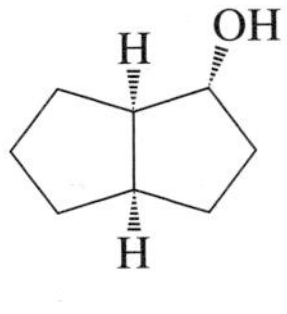

68. **CHALLENGE**   The *citric acid cycle* is a series of biological reactions that plays a central role in cell metabolism. The cycle includes dehydration reactions of both malic and citric acids, yielding fumaric and aconitic acids, respectively (all common names). Both proceed strictly by enzyme-catalyzed *anti* elimination mechanisms.

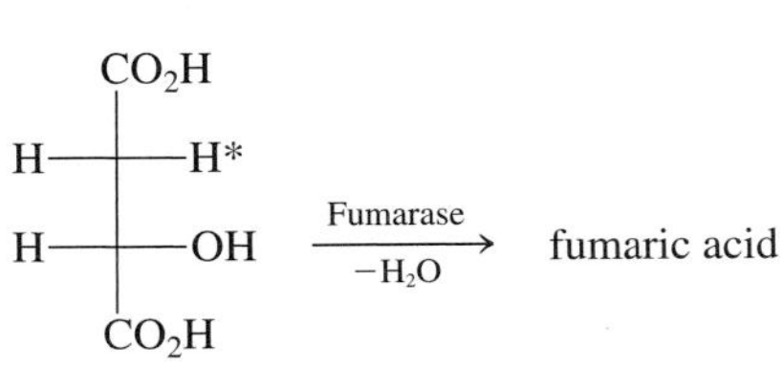

Malic acid

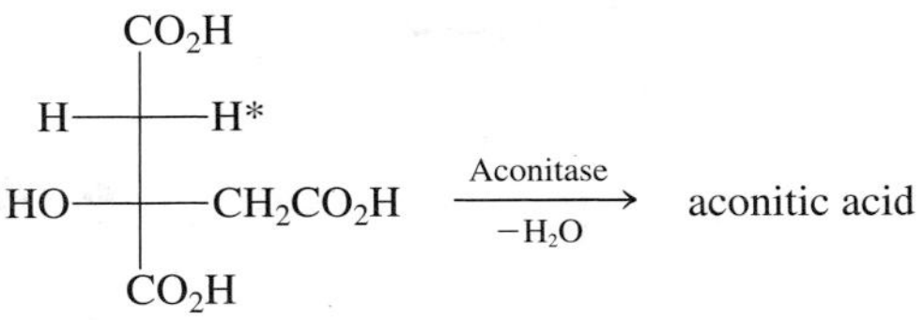

Citric acid

(a) In each dehydration, only the hydrogen identified by an asterisk is removed, together with the OH group on the carbon below. Write the structures for fumaric and aconitic acids as they are formed in these reactions. Make sure that the stereochemistry of each product is clearly indicated. **(b)** Specify the stereochemistry of each of these products, using either cis-trans or *E,Z* notation, as appropriate. **(c)** Isocitric acid (shown below) also is dehydrated by aconitase. How many stereoisomers can exist for isocitric acid? Remembering that this reaction proceeds through *anti* elimination, write the structure of a stereoisomer of isocitric acid that will give on dehydration the same isomer of aconitic acid that is formed from citric acid. Label the chiral carbons in this isomer of isocitric acid, using *R,S* notation.

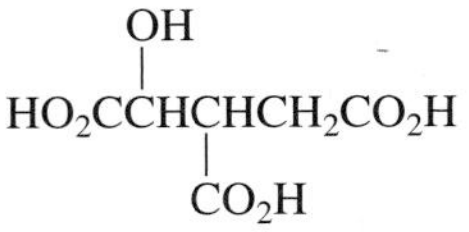

Isocitric acid

Team Problem

69. The following data indicate that the dehydration of certain amino acid derivatives is stereospecific.

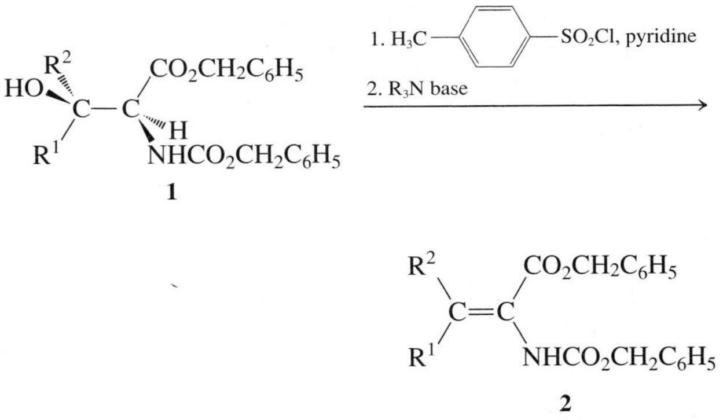

	R^1	R^2
a	CH_3	H
b	H	CH_3
c	$CH(CH_3)_2$	H
d	H	$CH(CH_3)_2$

Divide the task of analyzing these data among yourselves to determine the nature of the stereocontrolled eliminations. Assign the absolute configuration (*R,S*) to compounds **1a**–**1d** and the *E,Z* configuration to compounds **2a**–**2d**. Draw a Newman projection of the active conformation of each starting compound (**1a**–**1d**). As a team, apply your understanding of this information to determine the absolute configuration of the unassigned stereocenter (marked by an asterisk) in compound **3**, which was dehydrated to afford compound **4**, an intermediate in an approach to the synthesis of compound **5**, an antitumor agent.

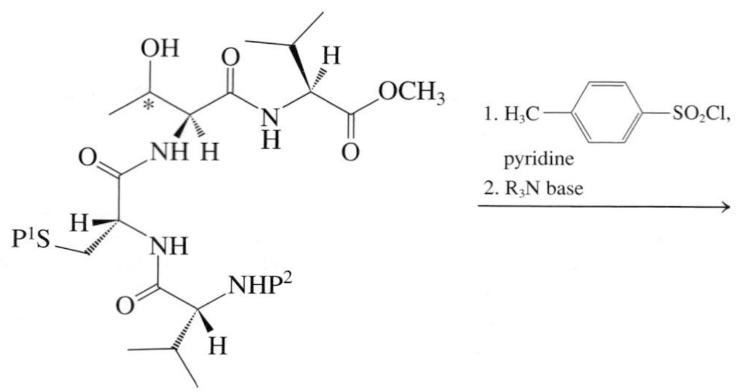

3

(P¹ and P² are protecting groups)

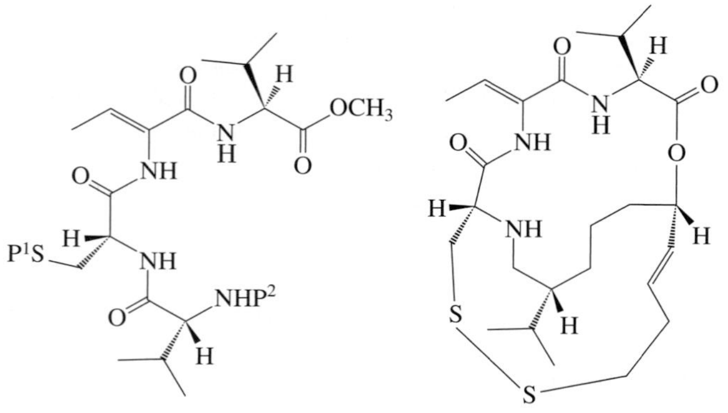

1. H_3C—⟨ ⟩—SO_2Cl,

pyridine

2. R_3N base

4

5

Preprofessional Problems

70. What is the empirical formula of compound A (see below)?

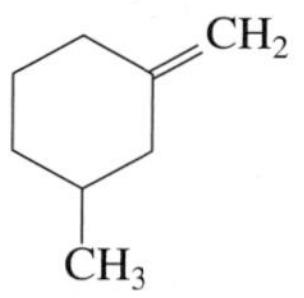

A

(a) C_8H_{14}; (b) C_8H_{16}; (c) C_8H_{12}; (d) C_4H_7

71. What is the degree of unsaturation in cyclobutane?

(a) Zero; (b) one; (c) two; (d) three

72. What is the IUPAC name for compound B (see below)?

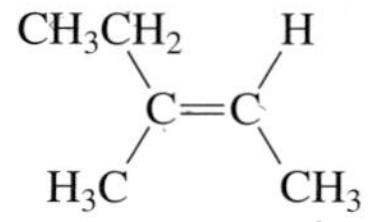

B

(a) (E)-2-Methyl-3-pentene; (b) (E)-3-methyl-2-pentene;
(c) (Z)-2-methyl-3-pentene; (d) (Z)-3-methyl-2-pentene

73. Which of the following molecules would have the lowest heat of hydrogenation?

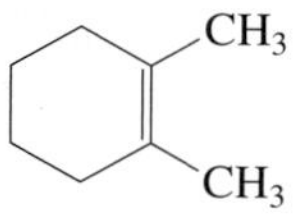

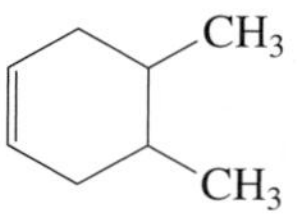

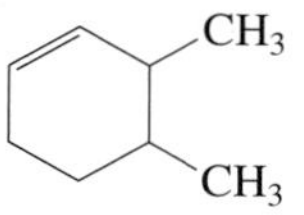

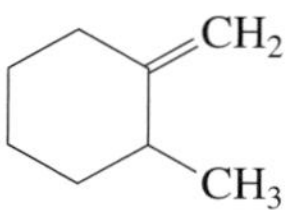

74. A certain hydrocarbon containing eight carbons was found to have two degrees of unsaturation but no absorption bands in the IR spectrum at 1640 cm^{-1}. The best structure for this compound is

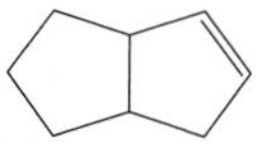

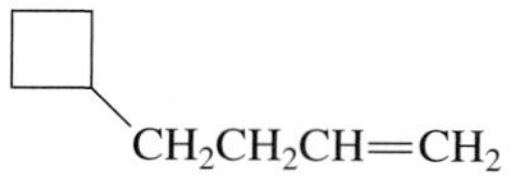

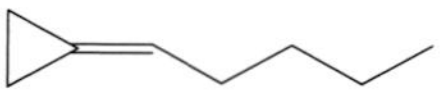

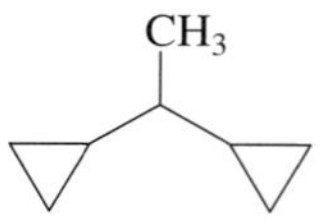

CHAPTER 12 | Reactions of Alkenes

Take a look around your room. Can you imagine how different it would appear if every polymer-derived material (including everything made of plastic) were to be removed? Polymers have had an enormous effect indeed on modern society. The chemistry of alkenes underlies our ability to produce polymeric materials of diverse structure, strength, elasticity, and function. In the later sections of this chapter, we shall investigate the processes that give rise to such substances. They are, however, only a subset of the varied types of transformations that alkenes undergo.

Additions constitute the largest group of alkene reactions and lead to saturated products. Through addition, we can take advantage of the fact that the alkene functional group bridges *two* carbons, and we can elaborate on the molecular structure at either or both of these carbons. Fortunately for us, most additions to the π bond are exothermic: They are almost certain to take place if a mechanistic pathway is available.

Beyond our simple ability to add to the double bond, other features further enhance the usefulness and versatility of addition reactions. Many alkenes possess defined stereochemistry (*E* and *Z*) and, as we shall see in our discussions, *many of their addition reactions proceed in a stereochemically defined manner.* By combining these facts with the realization that additions to unsymmetric alkenes may also take place regioselectively, we shall find that we have the ability to exert a large measure of control over the course of these reactions and, consequently, the structures of the products that are formed. This control has been exquisitely refined in applications toward the synthesis of enantiomerically pure pharmaceuticals (see Real Life 5-4 and 12-2, and Section 12-2).

We begin with a discussion of hydrogenation, focusing on the details of catalytic activation. Then we turn to the largest class of addition processes, those in which electrophiles such as protons, halogens, and metal ions are added to the alkene. Other additions that will contribute further to our synthetic repertoire include hydroboration, several oxidations (which can lead to complete rupture of the double bond if desired), and radical reactions. Each of these transformations takes us in a different direction; the Reaction Summary Road Map at the end of the chapter provides an overview of the interconversions leading to and from this versatile compound class.

Modern electronic devices (for example, the iPhone5 shown in the photo) rely on batteries that can be recharged thousands of times. Polymerization of 1,1-difluoroethene gives a high-performance membrane (polyvinylidene fluoride) that allows charge to flow between the individual cells of lithium-ion batteries but protects the battery from internal short-circuits and catastrophic failure. Li-ion polymer batteries offer significant advantages in weight and energy capacity over earlier Li-ion and nickel-based designs. They are seeing increased use in consumer electronics such as cellular phones and laptop computers and are under development for applications in hybrid vehicles.

12-1 | WHY ADDITION REACTIONS PROCEED: THERMODYNAMIC FEASIBILITY

The carbon–carbon π bond is relatively weak, and the chemistry of the alkenes is governed largely by its reactions. The most common transformation is **addition** of a reagent A–B to give a saturated compound. In the process, the A–B bond is broken, and A and B form single bonds to carbon. Thus, the *thermodynamic feasibility* of this process depends on the strength of the π bond, the dissociation energy DH°_{A-B}, and the strengths of the newly formed bonds of A and B.

Addition to the Alkene Double Bond

$$\text{C=C} \quad + \quad A-B \quad \xrightarrow{\Delta H^\circ = ?} \quad -\overset{A}{\underset{|}{C}}-\overset{B}{\underset{|}{C}}-$$

Recall that we can *estimate* the ΔH° of such reactions by subtracting the combined strength of the bonds made from that of the bonds broken (Section 3-4):

$$\Delta H^\circ = (DH^\circ_{\pi \text{ bond}} + DH^\circ_{A-B}) - (DH^\circ_{C-A} + DH^\circ_{C-B})$$

in which C stands for carbon.

Table 12-1 gives the DH° values [obtained by using the data from Tables 3-1 and 3-4 and by equating the strength of the π bond to 65 kcal mol^{-1} (272 kJ mol^{-1})] and the estimated ΔH° values for various additions to ethene. In all the examples, the combined strength of the bonds formed exceeds, sometimes significantly, that of the bonds broken. Therefore, thermodynamically, *additions to alkenes should proceed to products with release of energy.*

Table 12-1 Estimated ΔH° (all values in kcal mol^{-1}) for Additions to Ethene[a]					
$CH_2{=}CH_2$	+	A—B	$\longrightarrow$	H$-\overset{A}{\underset{H}{C}}-\overset{B}{\underset{H}{C}}-$H	
$DH^\circ_{\pi \text{ bond}}$		DH°_{A-B}		DH°_{A-C} $\quad$ DH°_{B-C}	$\sim\Delta H^\circ$
			Hydrogenation		
$CH_2{=}CH_2$ 65	+	H—H 104	$\longrightarrow$	$\overset{H}{\underset{}{CH_2}}-\overset{H}{\underset{}{CH_2}}$ 101 $\quad$ 101	-33
			Bromination		
$CH_2{=}CH_2$ 65	+	:Br—Br: 46	$\longrightarrow$	H$-\overset{:Br:}{\underset{H}{C}}-\overset{:Br:}{\underset{H}{C}}-$H 70 $\quad$ 70	-29
			Hydrochlorination		
$CH_2{=}CH_2$ 65	+	H—Cl: 103	$\longrightarrow$	H$-\overset{H}{\underset{H}{C}}-\overset{:Cl:}{\underset{H}{C}}-$H 101 $\quad$ 84	-17
			Hydration		
$CH_2{=}CH_2$ 65	+	H—OH 119	$\longrightarrow$	H$-\overset{H}{\underset{H}{C}}-\overset{:OH}{\underset{H}{C}}-$H 101 $\quad$ 94	-11

[a]These values are only estimates: They do not take into account the changes in C–C and C–H σ-bond strengths that accompany changes in hybridization.

Exercise 12-1

Calculate the $\Delta H°$ for the addition of H_2O_2 to ethene to give 1,2-ethanediol (ethylene glycol) $[DH°_{HO-OH} = 49 \text{ kcal mol}^{-1} \text{ (205 kJ mol}^{-1})].$

12-2 | CATALYTIC HYDROGENATION

The simplest reaction of the double bond is its saturation with hydrogen. As discussed in Section 11-5, this reaction allows us to estimate the relative stability of substituted alkenes from their heats of hydrogenation. The process requires a catalyst, which may be either heterogeneous or homogeneous—that is, either insoluble or soluble in the reaction medium.

Hydrogenation takes place on the surface of a heterogeneous catalyst

The hydrogenation of an alkene to an alkane, although exothermic, does not take place even at elevated temperatures. Ethene and hydrogen can be heated in the gas phase to 200°C for prolonged periods without any measurable change. However, as soon as a catalyst is added, hydrogenation proceeds at a steady rate even at room temperature. The catalysts frequently are insoluble materials such as palladium (e.g., dispersed on carbon, Pd–C), platinum (Adams's* catalyst, PtO_2, which is converted into colloidal platinum metal in the presence of hydrogen), and nickel (finely dispersed, as in a preparation called Raney[†] nickel, Ra–Ni).

The major function of the catalyst is the activation of hydrogen to generate metal-bound hydrogen on the catalyst surface (Figure 12-1). Without the metal, thermal cleavage of the strong H–H bond is energetically prohibitive. Solvents commonly used in such hydrogenations include methanol, ethanol, acetic acid, and ethyl acetate.

*Professor Roger Adams (1889–1971), University of Illinois at Urbana–Champaign.
[†]Dr. Murray Raney (1885–1966), Raney Catalyst Company, South Pittsburg, Tennessee.

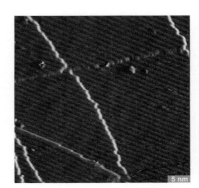

Scanning tunneling microscope (STM) image of a catalytic Pt surface. STM provides pictures at atomic resolution (the bar at the bottom right indicates 5 nm = 50 Å). You can see the highly ordered patterns of parallel arrays of Pt atoms in brown. The yellow "crosslines" indicate steps on the surface, on which much of the catalytic activity takes place. *(Courtesy Professor Gabor A. Somorjai and Dr. Feng Tao, University of California at Berkeley.)*

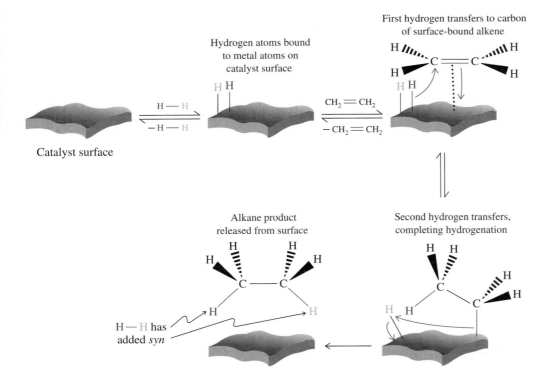

Figure 12-1 Mechanism of catalytic hydrogenation of ethene to produce ethane. The hydrogens bind to the catalyst surface and are delivered to the carbons of the surface-adsorbed alkene.

Solved Exercise 12-2 | **Working with the Concepts: Cis-trans Isomerization Under Alkene Hydrogenation Conditions**

During the catalytic hydrogenation of the cis double bond in oleic acid, a naturally occurring fatty acid, the formation of some of the trans isomer is observed. Explain.

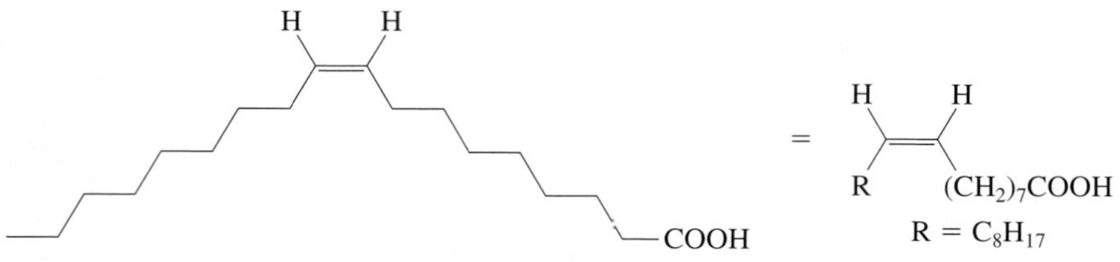

Oleic acid

$R = C_8H_{17}$

Strategy

What are we trying to accomplish? We need to find a way for an isomerized alkene to emerge from a catalytic hydrogenation reaction mixture.

How to begin? *Take the hint:* The question asks you to *explain* a chemical result. That is your cue that an examination of the *mechanism* is in order.

Information needed? The mechanism is already presented to you on the previous page.

Proceed, step by step.

Examine the mechanism in Figure 12-1. Look for a pathway that could (1) allow rotation about the bond between the originally double-bonded carbons followed by (2) regeneration of a double bond with a trans configuration.

Solution

• The mechanism possesses two key features that are applicable to this problem. First, each of the first three steps is *reversible.* Second, the two hydrogen atoms add to the carbons of the original double bond *one at a time.* Let us see how we can make use of these characteristics to define an isomerization pathway.

• Write out the mechanism, beginning with the binding of the alkene to the catalyst surface. (**Caution:** Do *not* combine steps! Each step in *any* mechanism should be written separately. Otherwise, you may miss a critical intermediate.) The mechanism, through the addition of the first hydrogen, goes as follows:

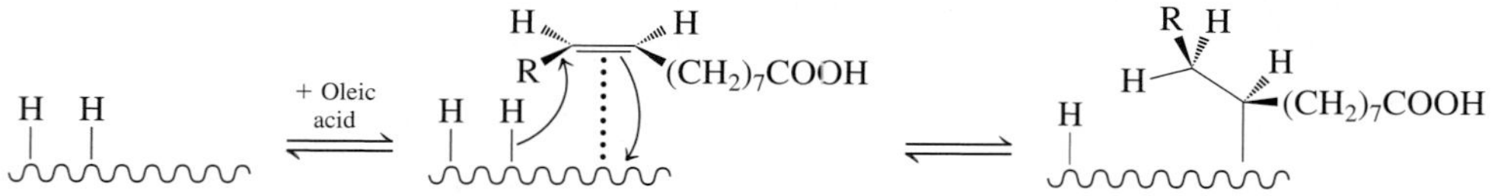

Single-hydrogen transfer intermediate

• In accordance with the general description in Figure 12-1, we know what happens next: The second hydrogen transfers over, and the hydrogenated product [stearic acid, $CH_3(CH_2)_{16}COOH$] is released. But notice that this final step is irreversible. Once it occurs, there is no way back to any alkene, cis or trans. Therefore, the cis-trans isomerization pathway must involve only the species we have written in the partial mechanism above.

• Look closely at the single-hydrogen transfer intermediate. It has a *single bond* between the two former alkene carbons that is conformationally flexible (Section 2-9). Carrying out a 120° rotation and *reversing* the hydrogen transfer that led to this species gives us an alkene with a trans double bond:

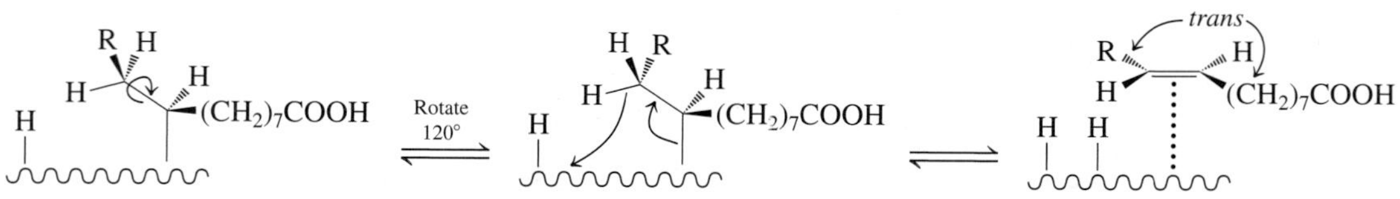

• Release of this alkene from the catalyst surface completes the isomerization. This transformation is exactly what occurs during the partial hydrogenation of vegetable oils, which is the commercial process for the formation of margarine and other partially saturated fats. The trans double bonds in these products—the so-called **trans fatty acids**—give rise to various adverse health effects that are described in more detail in Real Life 19-3.

**Two Topologies of
Alkene Addition**

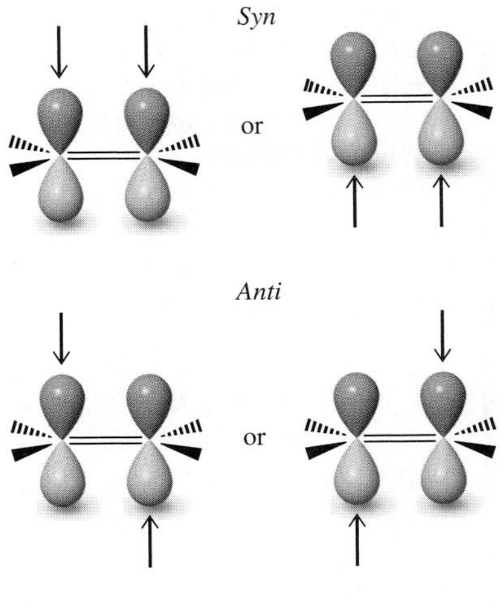

Syn

or

Anti

or

Exercise 12-3 | Try It Yourself

During the course of the catalytic hydrogenation of 3-methyl-1-butene, some 2-methyl-2-butene is observed. Explain.

Hydrogenation is stereospecific

An important feature of addition reactions to alkenes is their potential stereochemistry. Thus, any of the reagents A–B depicted in Table 12-1 can, in principle, add in two topologically selective ways: from one side of the double bond, called *syn* **addition,** or from opposite sides, termed *anti* **addition** (see margin). Alternatively, there may be no selectivity, and both modes may be observed. The outcome is crucially dependent on the mechanism of the reaction.

In the case of catalytic hydrogenation, inspection of Figure 12-1 reveals that the delivery of the two hydrogens occurs *syn*; we find that it is *stereospecific* (Section 6-5). For example, 1-ethyl-2-methylcyclohexene is hydrogenated over platinum to give specifically *cis*-1-ethyl-2-methylcyclohexane. Addition of hydrogen can be from above or from below the molecular plane with equal probability. Therefore, each stereocenter is generated as both image and mirror image, and the product is racemic.

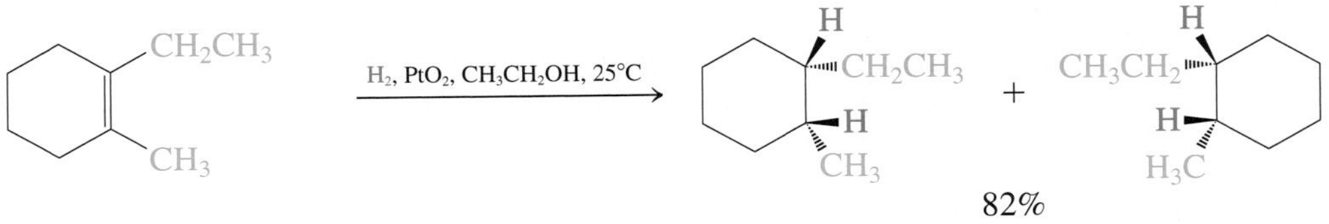

1-Ethyl-2-methylcyclohexene

H₂, PtO₂, CH₃CH₂OH, 25°C

82%

***cis*-1-Ethyl-2-methylcyclohexane
(Racemic)**

In the above scheme, we have shown both enantiomers of product explicitly. However, recall that in reactions in which racemic or achiral starting materials furnish racemic chiral products, we avoid the clutter or writing both enantiomers and show only one (the choice is arbitrary); the equimolar presence of the other is tacitly assumed (Section 5-7).

Chiral catalysts permit enantioselective hydrogenation

When steric hindrance inhibits hydrogenation on one face of a double bond, addition will take place exclusively to the *less hindered* face. This principle has been used to develop enantioselective or so-called *asymmetric hydrogenation*. The process employs homogeneous (soluble) catalysts, consisting of a metal, such as rhodium, and an *enantiopure chiral phosphine ligand,* which binds to the metal. A typical example is the Rh complex of the diphosphine (*R,R*)-DIPAMP (margin). After coordination of the alkene double bond and a molecule of H₂ to rhodium, hydrogenation occurs via *syn* addition, just as in the case of insoluble metal catalysts.

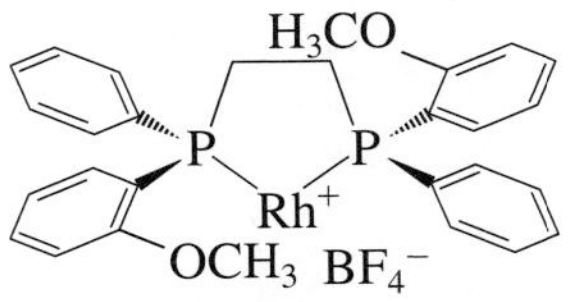

Rh-(*R,R*)-DIPAMP⁺ BF₄⁻

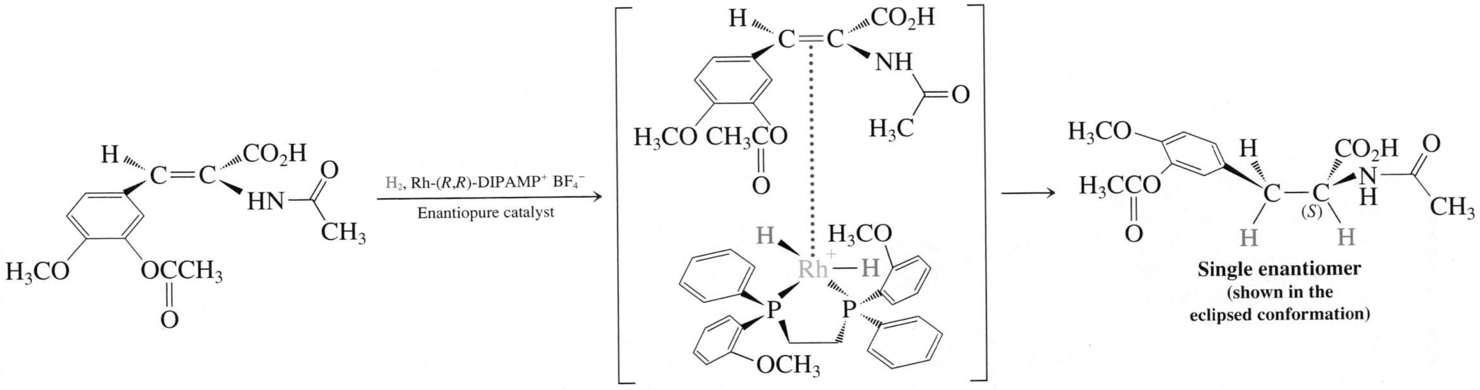

H₂, Rh-(*R,R*)-DIPAMP⁺ BF₄⁻
Enantiopure catalyst

**Hydrogen is transferred
from Rh to only one face
of the double bond**

**Single enantiomer
(shown in the
eclipsed conformation)**

Dopamine

L-DOPA

(shown in the
eclipsed conformation)

However, the asymmetrically arranged bulky groups in the chiral ligand prevent addition of hydrogen to one of the faces of the double bond, resulting in the stereoselective formation of only one of the two possible enantiomers of the hydrogenation product (see also Real Life 5-4 and 9-3).

This approach has proven to be a powerful method for synthesis of enantiomerically pure compounds of pharmaceutical significance. The industrial synthesis of L-DOPA (margin), the anti-Parkinson's disease agent, employs in its key step the asymmetric hydrogenation of the alkene shown to give exclusively the necessary S stereoisomer of the reduction product.

Exercise 12-4

Catalytic hydrogenation of (S)-2,3-dimethyl-1-pentene gives only one optically active product. Show the product and explain the result. [**Hint:** Does addition of H_2 either (1) create a new stereocenter or (2) affect any of the bonds around the stereocenter already present?]

In Summary Hydrogenation of the double bond in alkenes requires a catalyst. This transformation occurs stereospecifically by *syn* addition and, when there is a choice, from the least hindered side of the molecule. This principle underlies the development of enantioselective hydrogenation using chiral catalysts.

12-3   BASIC AND NUCLEOPHILIC CHARACTER OF THE PI BOND: ELECTROPHILIC ADDITION OF HYDROGEN HALIDES

As noted earlier, the π electrons of a double bond are not as strongly bound as those of a σ bond. This π-electron cloud, located above and below the molecular plane of the alkene, is polarizable and capable of acting as a base or nucleophile, just like the lone electron pairs of typical Lewis bases (Section 2-3). The relatively high electron density of the double bond in 2,3-dimethylbutene is indicated (red) in its electrostatic potential map in the margin. In the upcoming sections, we shall discuss reactions between the alkene π bond as a nucleophile and a variety of electrophiles. As in hydrogenation, the final outcome of such reactions is addition. However, there are several different mechanisms for these transformations, collectively called *electrophilic additions*. We shall see that they may or may not be regioselective and stereospecific. We begin with the simplest electrophile, the proton.

Electrophilic attack by protons gives carbocations

The proton of a strong acid may add to a double bond (acting as a base) to yield a carbocation (Section 2-2). This process is the reverse of the deprotonation step in the E1 reaction and has the same transition state (Figure 7-7). When a good nucleophile is present, particularly at low temperatures, the carbocation is intercepted to give the **electrophilic addition** product. For example, treatment of alkenes with hydrogen halides leads to the corresponding haloalkanes. The electrostatic potential map version of the general scheme shows the flow of electron density that occurs during this process.

Mechanism of Electrophilic Addition of HX to Alkenes

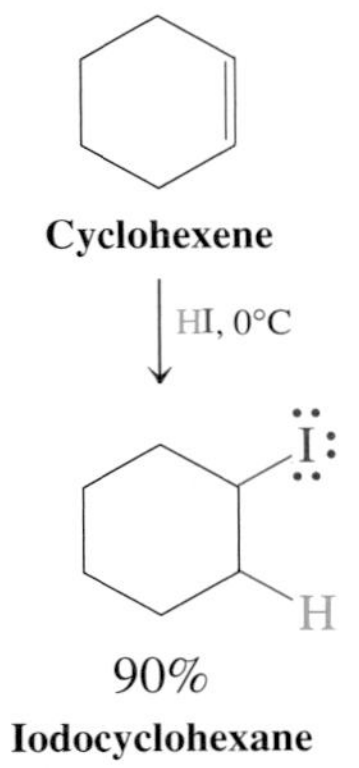

In the first step, an electron pair moves from the π bond (orange-red in the image at the left) toward the electrophilic (purple) proton to form a new σ bond. The map of the carbocation product of this step (center) shows that electron deficiency is now centered on the cationic carbon. Next, addition of the negative (and, therefore, red) halide ion is shown. In the resulting haloalkane product at the right, the polarity of the new C–X bond is reflected by the orange-red color for the strongly δ^- halogen atom, and the range of colors from greenish-blue to purple for the rest of the structure, indicating the distribution of δ^+ character among the remaining atoms.

In a typical experiment, the gaseous hydrogen halide, which may be HCl, HBr, or HI, is bubbled through pure or dissolved alkene. Alternatively, HX can be added in a solvent, such as acetic acid. Aqueous work-up furnishes the haloalkane in high yield (margin).

Cyclohexene

HI, 0°C

90%
Iodocyclohexane

Nucleophilic trapping of carbocations is nonstereoselective

The intervention of carbocations in the electrophilic additions of hydrogen halides to alkenes has important stereochemical consequences. Consider, for example, the hydrochlorination of *cis*- and *trans*-2-butene. Both give the same product, 2-chlorobutane, as a racemic mixture. We can understand this outcome by formulating the mechanism.

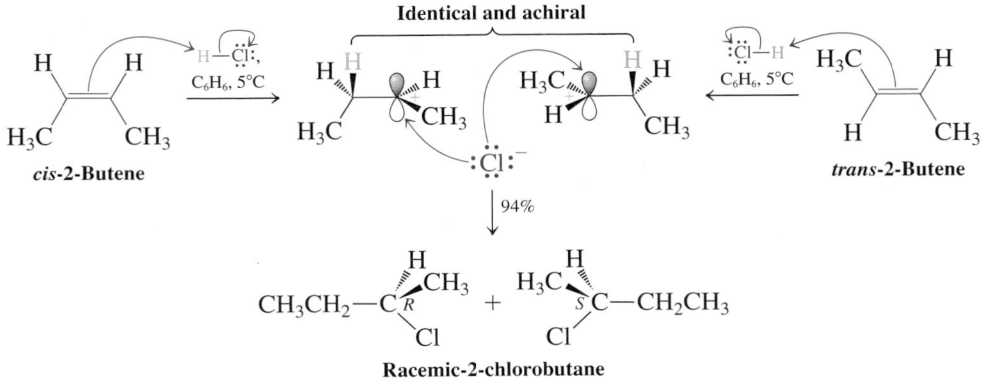

Identical and achiral

cis-2-Butene   *trans*-2-Butene

94%

Racemic-2-chlorobutane

You can see that initial protonation of either isomer removes the stereochemical information embedded in the double bond (cis versus trans) by creating an achiral CH_2 group next to an achiral sp^2 carbon to furnish the *same* intermediate cation. Chloride can intercept this species by attack from either the top or the bottom via enantiomeric transition states (Figure 5-13) to give racemic 2-chlorobutane.

What if both steps, protonation and nucleophilic trapping, engender two adjacent stereocenters? In this case two diastereomers are produced, both in racemic form. A case in point is the hydrochlorination of 1,2-dimethylcyclohexene, as shown below. In the first step, protonation occurs equally likely from either face of the double bond (again through enantiomeric transition states; Figure 5-13). Therefore, the ensuing cation, while chiral, is racemic (only one enantiomer is shown). In the second step, chloride can attack the cationic carbon, again from either the top or the bottom side. Now the transition states are diastereomeric (Figure 5-14) giving the two products in unequal amounts, but both as racemates.

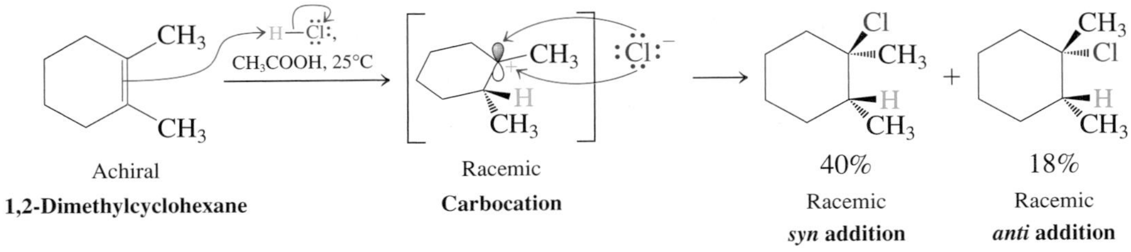

Achiral	**Racemic**	40%	18%
1,2-Dimethylcyclohexane	**Carbocation**	Racemic	Racemic
		syn addition	*anti* addition

You will note that topologically, the outcome is one of concurrent *syn* and *anti* addition, unlike the case of stereospecific *syn* hydrogenation in the preceding section.

Exercise 12-5

Write out two step-by-step mechanisms for the addition of HI to cyclohexene shown above (p. 489, margin). In the first, use a free proton as the electrophile. In the second, use undissociated HI in the electrophilic addition step. Make sure to include all necessary curved arrows to depict electron-pair movement.

The Markovnikov rule predicts regioselectivity in electrophilic additions

So far we have only looked at hydrohalogenations of symmetric alkenes, in which the sense of addition, H⁺ to one, X⁻ to the other carbon, was immaterial. What about unsymmetric alkenes? Will there be regioselectivity? To answer these questions, let us consider the reaction of propene with hydrogen chloride. Two products are possible: 2-chloropropane and 1-chloropropane. However, the only product observed is 2-chloropropane.

Reaction

Animation

ANIMATED MECHANISM: Electrophilic addition of HCl to propene

Regioselective Electrophilic Addition to Propene

$$CH_3CH{=}CH_2 \xrightarrow{H\ddot{C}l:} CH_3CHCH_2 \quad \text{but no} \quad CH_3CHCH_2$$

Less substituted carbon: proton attaches here   **2-Chloropropane** ($:\ddot{C}l:H$)   **1-Chloropropane** ($H:\ddot{C}l:$)

Similarly, reaction of 2-methylpropene with hydrogen bromide gives only 2-bromo-2-methylpropane, and 1-methylcyclohexene combines with HI to furnish only 1-iodo-1-methylcyclohexane.

Two Further Examples of Regioselective Additions

We can see from these examples that, if the carbon atoms participating in the double bond are not equally substituted, *the proton from the hydrogen halide attaches itself to the less substituted carbon.* As a consequence, the halogen ends up at the more substituted carbon. This phenomenon, referred to as the **Markovnikov* rule,** can be explained by what we know about the mechanism of electrophile additions of protons to alkenes. *The key is the relative stability of the resulting carbocation intermediates.*

Consider the addition of HCl to propene. The regiochemistry of the reaction is determined in the first step, in which the proton attacks the π system to give an intermediate carbocation. Carbocation generation is rate determining; once it occurs, reaction with chloride proceeds quickly. Let us look at the crucial first step in more detail. The proton may attack either of the two carbon atoms of the double bond. Addition to the internal carbon leads to the primary propyl cation.

Protonation of Propene at C2—More Substituted Carbon (Does Not Occur)

Transition state 1   **Primary carbocation** (Not observed) $CH_3CH_2CH_2{}^+$

Partial positive charge on primary carbon (unfavorable)

*Professor Vladimir V. Markovnikov (1838–1904), University of Moscow, formulated his rule in 1869.

In contrast, protonation at the terminal carbon results in the formation of the secondary 1-methylethyl (isopropyl) cation.

Protonation of Propene at C1—Less Substituted Carbon

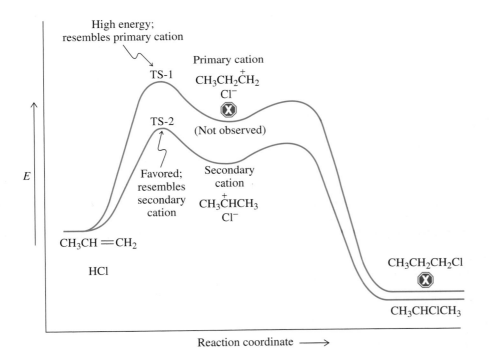

Mechanism

As we know, a primary carbocation is too unstable to be a reasonable reaction intermediate in solution. In contrast, secondary cations form relatively readily. In addition, notice that the transition states for the two possible modes of addition show positive charge building up on primary and secondary carbons, respectively. *Thus, the energies, and stabilities, of the transition states will reflect the relative energies of the cations to which they lead.* The energy of the transition state (and, therefore, the activation energy) leading to the secondary cation is lower, meaning that this cation forms much faster. Figure 12-2 is a potential-energy diagram showing the two competing pathways. We see that these are late transition states, whose energies closely resemble those of the product cations.

Figure 12-2 Potential-energy diagram for the two possible modes of HCl addition to propene. Transition state 1 (TS-1), which leads to the higher-energy primary propyl cation, is less favored than transition state 2 (TS-2), which gives the 1-methylethyl (isopropyl) cation.

On the basis of this analysis, we can rephrase the empirical Markovnikov rule: HX adds to unsymmetric alkenes in such a way that *the initial protonation gives the more stable carbocation.* For alkenes that are similarly substituted at both sp^2 carbons, product mixtures are to be expected, because carbocations of comparable stability are formed. By analogy with other carbocation reactions (e.g., S_N1, Section 7-3), when addition to an achiral alkene generates a chiral product, this product is obtained as a racemic mixture.

Exercise 12-6

Predict the outcome of the addition of HBr to (**a**) 1-hexene; (**b**) *trans*-2-pentene; (**c**) 2-methyl-2-butene; (**d**) 4-methylcyclohexene. How many isomers can be formed in each case?

> **Exercise 12-7**
>
> Draw a potential-energy diagram for the reaction in (c) of Exercise 12-6.

Electrophilic protonation may be followed by carbocation rearrangements

Remember: The migrating group and the positive charge trade places in a carbocation rearrangement.

In the absence of a good nucleophile, carbocation rearrangements may occur following addition of an electrophile to the alkene double bond (Section 9-3). Rearrangements are favored in electrophilic additions of acids whose conjugate bases are poor nucleophiles. An example is trifluoroacetic acid, CF_3CO_2H. Its trifluoroacetate counterion is much less nucleophilic than are halide ions. Thus, addition of trifluoroacetic acid to 3-methyl-1-butene gives only about 43% of the normal product of Markovnikov addition. The major product results from a hydride shift that converts the initial secondary cation into a more stable tertiary cation before the trifluoroacetate can attach.

Rearrangement Accompanying Addition of Trifluoroacetic Acid to 3-Methyl-1-butene

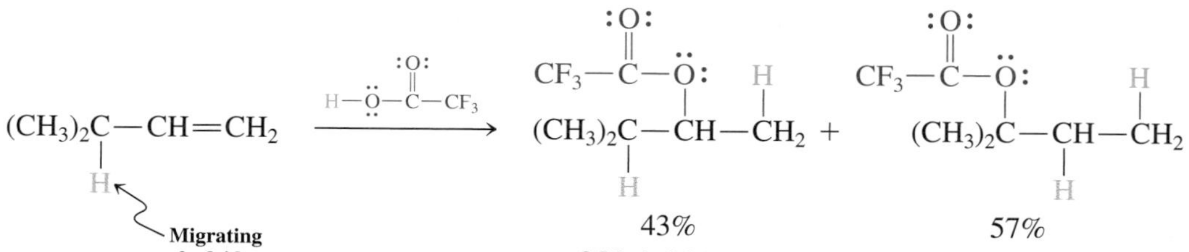

43%	57%
3-Methyl-2-butyl trifluoroacetate	2-Methyl-2-butyl trifluoroacetate
Product of normal Markovnikov addition	Product resulting from carbocation rearrangement

> **Exercise 12-8**
>
> Write a detailed step-by-step mechanism for the reaction depicted above. Refer to Section 9-3 if necessary.

The extent of carbocation rearrangement is difficult to predict: It depends on the alkene structure, the solvent, the strength and concentration of the nucleophile, and the temperature. In general, rearrangements are favored under strongly acidic, nucleophile-deficient conditions.

In Summary Additions of hydrogen halides to alkenes are electrophilic reactions that begin with protonation of the double bond to give a carbocation. Trapping of the carbocation by halide ion gives the final product. The Markovnikov rule predicts the regioselectivity of hydrohalogenation to haloalkanes. As in any carbocation reaction, rearrangements may occur if good nucleophiles are absent.

12-4 | ALCOHOL SYNTHESIS BY ELECTROPHILIC HYDRATION: THERMODYNAMIC CONTROL

So far, we have seen attack on the double bond by a proton, followed by nucleophilic attachment of its counterion to the intermediate carbocation. Can other nucleophiles participate? Upon exposure of an alkene to an *aqueous* solution of sulfuric acid, which has a poorly nucleophilic counterion, *water* acts as the nucleophile to trap the carbocation formed by initial protonation. Overall, the elements of water add to the double bond, an **electrophilic hydration.** The addition follows the Markovnikov rule in that H^+ adds to the less substituted carbon and the OH group ends up at the more substituted one. Because water is a poor nucleophile, carbocation rearrangements can intervene in the hydration process.

This addition process is the reverse of the acid-induced elimination of water from alcohols (dehydration, Section 11-7). Its mechanism is the same in reverse, as illustrated in the hydration of 2-methylpropene, a reaction of industrial importance leading to 2-methyl-2-propanol (*tert*-butyl alcohol).

Electrophilic Hydration

$$H_3C\diagdown C=CH_2 \xrightarrow{\text{50\% H\ddot{O}H, H}_2\text{SO}_4} H_3C-\underset{\underset{:\ddot{O}H}{|}}{\overset{\overset{CH_3}{|}}{C}}-CH_2-H$$

2-Methylpropene   92%
                          2-Methyl-2-propanol

Reaction

Mechanism of the Hydration of 2-Methylpropene

$$H_3C\diagdown C=CH_2 \underset{-H^+}{\overset{+H^+}{\rightleftharpoons}} \underset{CH_3}{\overset{H_3C\diagup \diagdown CH_3}{C^+}} \underset{-HOH}{\overset{+H\ddot{O}H}{\rightleftharpoons}} CH_3\overset{CH_3}{\underset{CH_3}{C}}-\overset{H}{\underset{H}{\ddot{O}}} \underset{+H^+}{\overset{-H^+}{\rightleftharpoons}} CH_3\overset{CH_3}{\underset{CH_3}{C}}-\ddot{O}H$$

Protonation   Nucleophilic   Deprotonation
                     attack by H$_2$Ö

Mechanism

Alkene hydration and alcohol dehydration are equilibrium processes

In the mechanism of alkene hydration, *all the steps are reversible.* The proton acts only as a catalyst and is not consumed in the overall reaction. Indeed, without the acid, hydration would not occur; alkenes are stable in neutral water. The presence of acid, however, establishes an equilibrium between alkene and alcohol. This equilibrium can be driven toward the alcohol by using low reaction temperatures and a large excess of water. Conversely, we have seen (Section 11-7) that treating the alcohol with *concentrated* acid favors dehydration, especially at elevated temperatures.

$$\text{Alcohol} \underset{\text{H}_2\text{SO}_4,\ \text{excess H}_2\text{O, low temperature}}{\overset{\text{Conc. H}_2\text{SO}_4,\ \text{high temperature}}{\rightleftharpoons}} \text{alkene} + \text{H}_2\text{O}$$

Hydration–Dehydration Equilibrium

$$RCH=CH_2 + H_2\ddot{O}$$

$$\Updownarrow \text{ Catalytic H}^+$$

$$\underset{:\ddot{O}H}{\overset{}{RCHCH_3}}$$

Exercise 12-9

Treatment of 2-methylpropene with catalytic deuterated sulfuric acid (D$_2$SO$_4$) in D$_2$O gives (CD$_3$)$_3$COD. Explain by a mechanism.

The reversibility of alkene protonation leads to alkene equilibration

In Section 11-7 we explained that the acid-catalyzed dehydration of alcohols gives mixtures of alkenes in which the more stable isomers predominate. Equilibrating carbocation rearrangements, followed by E1, are partly responsible for these results. However, more important for the attainment of thermodynamic equilibration is that the E1 process is *reversible:* As we discussed above, alkenes can be protonated by acid to form carbocations.

Let us illustrate this feature of reversible protonation by envisaging what can happen when 2-butanol is dehydrated under acidic conditions. To avoid complications arising from S$_N$1, we will employ sulfuric acid, which possesses a nonnucleophilic counterion. The first thing to happen is protonation of the hydroxy group. Then, loss of water from the protonated alcohol gives the corresponding secondary carbocation. This species can undergo E1 in three different ways, to give the three observed products: 1-butene, *cis*-2-butene, and *trans*-2-butene. The *initial* ratio of these isomers is controlled by the relative energies of the transition states leading to them: It is under kinetic control (Section 2-1). However, under the strongly acidic conditions, a proton can re-add to the double bond of any of the isomers. In the case of the two 2-butenes, this process engenders the corresponding secondary carbocations. In the case of 1-butene, and as noted earlier, regioselective Markovnikov addition leads to the same ion. Because this cation may again lose a proton to give any of the same three alkene isomers, the net effect is *interconversion* of the isomers to an *equilibrium mixture,* in which the thermodynamically most stable isomer is the major component. This system is therefore an example of a reaction that is under *thermodynamic control* (Section 2-1).

Thermodynamic Control in the Acid-Catalyzed Dehydration of 2-Butanol

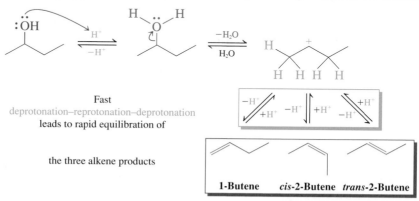

Fast
deprotonation–reprotonation–deprotonation
leads to rapid equilibration of

the three alkene products

1-Butene   *cis*-2-Butene   *trans*-2-Butene

By this procedure, less stable alkenes may be catalytically converted into their more stable isomers (see below and margin).

Acid-Catalyzed Equilibration of Alkenes

$(CH_3)_3C$⎯⎯⎯⎯⎯$C(CH_3)_3$
        $C=C$ → Catalytic H^+ →
H⎯⎯⎯⎯⎯⎯⎯H
Cis

$(CH_3)_3C$⎯⎯⎯⎯⎯H
        $C=C$
H⎯⎯⎯⎯⎯⎯$C(CH_3)_3$
Trans

Exercise 12-10

Write a mechanism for this rearrangement. What is the driving force for the reaction?

H_3C ⟍ (vinyl cyclohexane with methyl) → H^+ → H_3C ⟍ (cyclohexene with CH_2CH_3)

In Summary The carbocation formed by addition of a proton to an alkene may be trapped by water to give an alcohol, the reverse of alkene synthesis by alcohol dehydration. Reversible protonation equilibrates alkenes in the presence of acid, thereby forming a thermodynamically controlled mixture of isomers.

12-5 | ELECTROPHILIC ADDITION OF HALOGENS TO ALKENES

Halogens, although they do not appear to contain electrophilic atoms, add to the double bond of alkenes, giving vicinal dihalides. These compounds have uses as solvents for dry cleaning and degreasing and as antiknock additives for gasoline.

Margin figures:

H ⟍ ⟋ H
    C Disubstituted

Terminal
(less stable)

↓ Catalytic H^+

H
$H—C—H$ Trisubstituted

Internal
(more stable)

Halogen addition works best for chlorine and bromine. The reaction of fluorine is too violent to be generally employed, and iodine addition is not normally favorable thermodynamically.

Halogenation of Alkenes

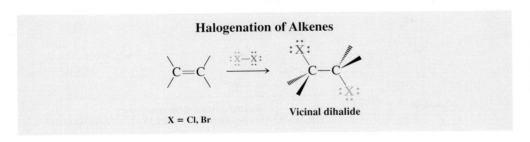

X = Cl, Br

Vicinal dihalide

Addition of bromine to an alkene results in almost immediate loss of the red-brown color of Br_2 as the reaction takes place.

Exercise 12-11

Calculate (as in Table 12-1) the $\Delta H°$ values for the addition of F_2 and I_2 to ethene. (For $DH°_{X_2}$, see Section 3-5.)

Bromine addition is particularly easy to observe because bromine solutions immediately change from red to colorless when exposed to an alkene. This phenomenon is sometimes used as a color test for unsaturation.

Halogenations are best carried out at or below room temperature in inert halogenated solvents such as tetrachloromethane (carbon tetrachloride).

Electrophilic Halogen Addition of Br$_2$ to 1-Hexene

$$CH_3(CH_2)_3CH{=}CH_2 \xrightarrow{\ :\!Br{-}Br\!:,\ CCl_4\ } CH_3(CH_2)_3CHCH_2\ddot{B}r\!:$$

$$\underset{:\ddot{B}r:}{\big|}$$

1-Hexene

90%

1,2-Dibromohexane

Reminder: Two Topologies of Alkene Addition

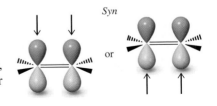

Halogen additions to double bonds may seem to be similar to hydrogenations. However, their mechanism is quite different, as revealed by the stereochemistry of bromination; similar arguments hold for the other halogens.

Bromination takes place through *anti* addition

What is the stereochemistry of bromination? Do the two bromine atoms add from the same side of the double bond (*syn,* as in catalytic hydrogenation) or from opposite sides (see margin)? Let us examine the bromination of cyclohexene. Addition on the same side should give *cis*-1,2-dibromocyclohexane; the alternative would result in *trans*-1,2-dibromocyclohexane. The second pathway is borne out by experiment—only **anti addition** is observed. Because *anti* addition to the two reacting carbon atoms can take place with equal probability in two possible ways—in either case, from both above and below the π bond—the product is racemic.

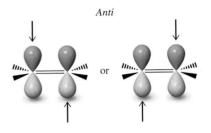

Anti Bromination of Cyclohexene

83%

Racemic *trans*-1,2-dibromocyclohexane

Reaction

With acyclic alkenes, the reaction is also cleanly stereospecific. For example, *cis*-2-butene is brominated to furnish a racemic mixture of (2*R*,3*R*)- and (2*S*,3*S*)-2,3-dibromobutane; *trans*-2-butene results in the meso diastereomer.

Reaction

Model Building

Stereospecific 2-Butene Bromination

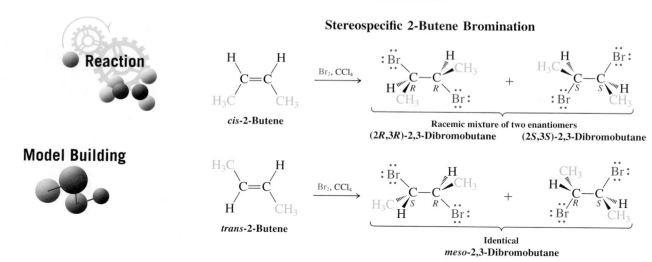

cis-2-Butene $\xrightarrow{\text{Br}_2,\ \text{CCl}_4}$

Racemic mixture of two enantiomers
(2*R*,3*R*)-2,3-Dibromobutane   (2*S*,3*S*)-2,3-Dibromobutane

trans-2-Butene $\xrightarrow{\text{Br}_2,\ \text{CCl}_4}$

Identical
meso-2,3-Dibromobutane

Cyclic bromonium ions explain the stereochemistry

Electron Sextet and Octet Isomers of Cyclic Onium Ions

Sextet   Octet

Sextet   Octet

How does bromine attack the electron-rich double bond even though it does not appear to contain an electrophilic center? The answer lies in the polarizability of the Br–Br bond, which is prone to heterolytic cleavage upon reaction with a nucleophile. The π-electron cloud of the alkene is nucleophilic and attacks one end of the bromine molecule, with simultaneous displacement of the second bromine atom as bromide ion in an S_N2-like process.

What is the product of this process? We might expect a carbocation, in analogy with the proton additions discussed in Sections 12-3 and 12-4. However, a more stable structure is that of a cyclic **bromonium ion,** in which the bromine bridges both carbon atoms of the original double bond to form a three-membered ring (Figure 12-3). As in the case of protonated oxacyclopropanes (Section 9-9), this isomer contains an electronic octet configuration, avoiding the electron-deficient carbocation formulation (margin). The structure of this ion is rigid, and it may be attacked by bromide ion only on the side opposite the bridging bromine atom. The leaving group is the bridging bromine. The three-membered ring is thus opened stereospecifically *anti*. Hence, cyclohexene furnishes only *trans*-1,2-dibromocyclohexane, and the 2-butene isomers result in the two respective diastereomers of 2,3-dibromobutane.

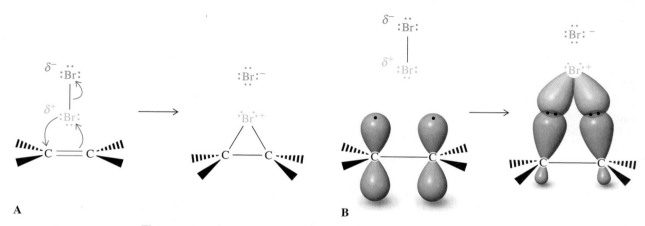

Figure 12-3 (A) Electron-pushing picture of cyclic bromonium ion formation. The alkene (red) acts as a nucleophile to displace bromide ion (green) from bromine. The molecular bromine behaves as if it were strongly polarized, one atom as a bromine cation, the other as an anion. (B) Orbital picture of bromonium ion formation.

Since we start with achiral starting materials, all products (if chiral) are racemates (Section 5-7). If the first step of addition of bromine to the double bond had given a carbocation, the bromide ion released in this process could have attacked the positively charged carbon atom from either side, resulting in both *syn* and *anti* addition products, which is not observed.

Formation and Nucleophilic Opening of a Cyclic Bromonium Ion

Mechanism

Animation

ANIMATED MECHANISM: Stereospecific bromination of z-butenes

Exercise 12-12

Draw the intermediate in the bromination of cyclohexene, using the following conformational picture. Show why the product is racemic. What can you say about the initial conformation of the product?

Conformational flip in cyclohexene

The halogenation of alkenes should not be confused with halogenation of *alkanes* (Sections 3-4 through 3-9). The addition to alkenes follows a mechanism in which a nucleophile (the alkene π bond) interacts with an electrophilic species (such as molecular Cl_2 or Br_2) via movement of electron pairs. *Alkane* halogenation is a radical process that requires an initiation step to generate halogen atoms. This step typically needs heat, light, or a radical initiator (such as a peroxide) and proceeds via a mechanism involving movement of single electrons.

In Summary Halogens add as electrophiles to alkenes, producing vicinal dihalides. The reaction begins with the formation of a bridged halonium ion. This intermediate is opened stereospecifically by the halide ion displaced in the initial step to give overall *anti* addition to the double bond. In subsequent sections, we shall see that other stereochemical outcomes are possible, depending on the electrophile.

12-6 | THE GENERALITY OF ELECTROPHILIC ADDITION

Halogens are just one of many electrophile–nucleophile combinations that add to alkene double bonds. In this section we begin a survey of some of the most important of these processes, beginning with addition of halogens (the electrophile source) in the presence of water (the nucleophile). The products, 2-haloalcohols, commonly known as halohydrins, are widely used in a number of industrial and synthetic applications. In particular, they are important intermediates for the synthesis of oxacyclopropanes (epoxides, Section 9-9).

The bromonium ion can be trapped by other nucleophiles

The creation of a bromonium ion in alkene brominations suggests that, in the presence of other nucleophiles, competition might be observed in the trapping of the intermediate. For example, bromination of cyclopentene in water as solvent gives the vicinal bromoalcohol (common name, bromohydrin). In this case, the bromonium ion is attacked by water, which is present in large excess. The net transformation is the *anti* addition of Br and OH to the double bond. The other product formed is HBr. The corresponding chloroalcohols (chlorohydrins) can be made from chlorine in water through a chloronium ion intermediate.

Bromoalcohol (Bromohydrin) Synthesis

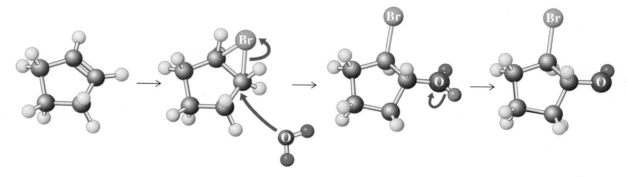

Cyclopentene

trans-2-
Bromocyclopentanol

Exercise 12-13

Write the expected product from the reaction of **(a)** *trans*-2-butene and **(b)** *cis*-2-pentene with aqueous chlorine. Show the stereochemistry clearly.

Vicinal Haloether Synthesis

Vicinal haloalcohols undergo intramolecular ring closure in the presence of base to give oxacyclopropanes (Section 9-9) and are therefore useful intermediates in organic synthesis.

Oxacyclopropane Formation from an Alkene Through the Haloalcohol

73%   70%

If alcohol is used as a solvent instead of water in these halogenations, the corresponding vicinal haloethers are produced, as shown in the margin.

76%

trans-1-Bromo-2-methoxy-cyclohexane

Halonium ion opening can be regioselective

In contrast with addition of two identical halogens, mixed additions to double bonds can pose regiochemical problems. Is the addition of Br and OH (or OR) to an unsymmetric double bond selective? The answer is yes. For example, 2-methylpropene is converted by aqueous bromine into only 1-bromo-2-methyl-2-propanol; none of the alternative regioisomer, 2-bromo-2-methyl-1-propanol, is formed.

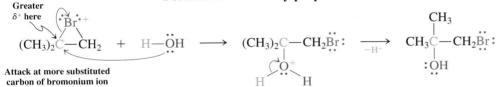

$$\underset{\substack{\text{1-Bromo-2-methyl-} \\ \text{2-propanol}}}{\overset{82\%}{\underset{\overset{|}{CH_3}}{CH_3\overset{\overset{\ddot{O}H}{|}}{\underset{}{C}}CH_2\ddot{B}r}}} \quad \text{but no} \quad \text{—⊗→} \quad \underset{\substack{\text{2-Bromo-2-methyl-} \\ \text{1-propanol}}}{\overset{\overset{\ddot{B}r}{|}}{\underset{\overset{|}{CH_3}}{CH_3\underset{}{C}CH_2\ddot{O}H}}}$$

The electrophilic halogen in the product always becomes linked to the less substituted carbon of the original double bond. The subsequently added nucleophile attaches to the more highly substituted center.

How can this be explained? The situation is very similar to the acid-catalyzed nucleophilic ring opening of oxacyclopropanes (Section 9-9), in which the intermediate contains a protonated oxygen in the three-membered ring. In both reactions, *the nucleophile attacks the more highly substituted carbon of the ring, because this carbon is more positively polarized than the other.*

Regioselective Opening of the Bromonium Ion Formed from 2-Methylpropene

Greater δ⁺ here

Attack at more substituted carbon of bromonium ion

$$(CH_3)_2\overset{\overset{\overset{+}{\ddot{B}r}}{\curvearrowleft}}{\underset{}{C}}—CH_2 \; + \; H—\ddot{O}H \; \longrightarrow \; (CH_3)_2\overset{}{\underset{\underset{H \quad H}{\overset{+}{O}}}{C}}—CH_2\ddot{B}r \; \xrightarrow{-H^+} \; CH_3\overset{\overset{CH_3}{|}}{\underset{\underset{:\ddot{O}H}{|}}{C}}—CH_2\ddot{B}r$$

Mechanism

A simple rule of thumb is that electrophilic additions of unsymmetric reagents of this type add in a Markovnikov-like fashion, with the electrophilic unit emerging at the less substituted carbon of the double bond. Mixtures are formed only when the two carbons are not sufficiently differentiated [see Exercise 12-14, part (b)].

Exercise 12-14

What are the product(s) of the following reactions?

(a) $CH_3CH{=}CH_2 \xrightarrow{Cl_2,\ CH_3OH}$ **(b)**   [structure: cyclohexene with H₃C substituent] $\xrightarrow{Br_2,\ H_2O}$

Solved Exercise 12-15 | Working with the Concepts: Mechanism of Electrophilic Addition to Alkenes

Write a mechanism for the reaction shown in Exercise 12-14, part (a).

Strategy
This problem is quite similar to the transformation shown in the margin of the preceding page. We have simply replaced bromine with chlorine.

Solution
• Initial attack of the alkene π bond on a molecule of Cl_2 gives rise to a cyclic *chloronium* ion:

$$CH_3{-}\underset{}{\overset{\overset{:\ddot{Cl}{-}\ddot{Cl}:}{\curvearrowleft}}{CH}}{\overset{}{=}}CH_2 \; \longrightarrow \; CH_3{-}\overset{\overset{+}{\ddot{Cl}}}{\underset{}{CH}}{-}CH_2 \; + \; :\ddot{Cl}:^-$$

Halohydroxylations in Nature

Nature exploits the chemistry described in this section with the help of enzymes. For example, a peroxidase from the fungus *Caldariomyces fumago* manages to bromohydroxylate the perfumery alcohol citronellol to the diastereomeric bromodiols following Markovnikov's rule. Of course, the enzyme does not use the corrosive bromine to achieve this transformation, but rather NaBr in the presence of H_2O_2 as an oxidizing agent.

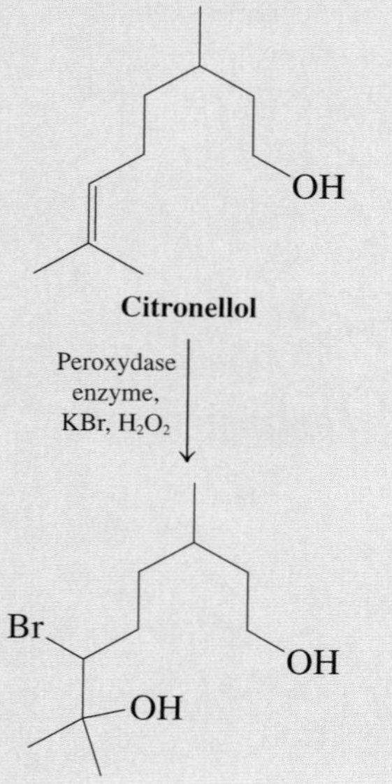

Citronellol

Peroxydase enzyme, KBr, H_2O_2

• This species is not symmetric: In particular, of the two carbons bonded to the positive chlorine, the internal (secondary) carbon is more positively polarized. Attack of the nucleophilic solvent methanol occurs preferentially at this center. Loss of a proton from the resulting oxonium ion completes the reaction:

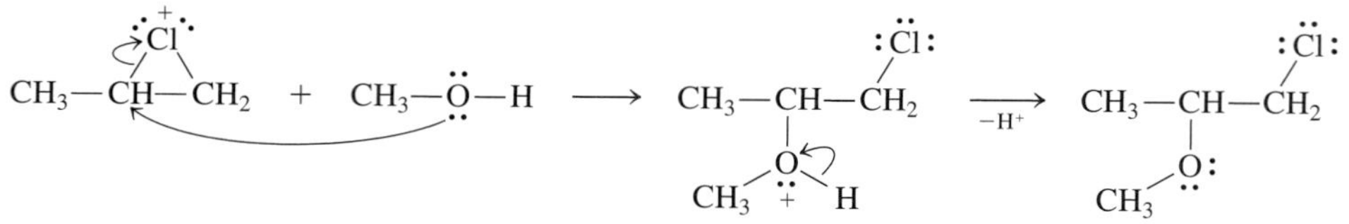

Exercise 12-16 | **Try It Yourself**

Write a mechanism for the reaction shown in Exercise 12-14, part (b). [**Caution:** The presence of the methyl group on the cyclohexene ring has a significant stereochemical effect (compared with the other examples in the section)! **Hint:** The initial addition of halogen to the alkene π bond can occur either from the same face of the ring that contains the methyl group or from the opposite face. How many isomers do you expect to result from this addition?]

Exercise 12-17 |

What would be a good alkene precursor for a racemic mixture of (2R,3R)- and (2S,3S)-2-bromo-3-methoxypentane? What other isomeric products might you expect to find from the reactions that you propose?

In general, alkenes can undergo stereo- and regiospecific addition reactions with reagents of the type A–B, in which the A–B bond is polarized such that A acts as the electrophile A^+ and B as the nucleophile B^-. Table 12-2 shows how such reagents add to 2-methylpropene.

Table 12-2 Reagents A–B That Add to Alkenes by Electrophilic Attack

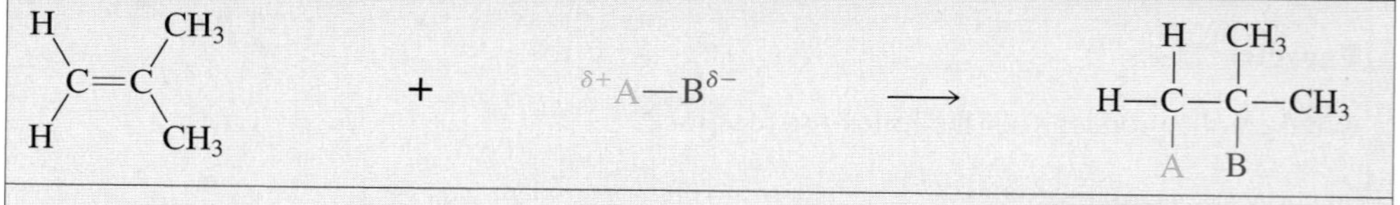

Name	Structure	Addition product to 2-methylpropene
Bromine chloride	:Br̈—C̈l:	:B̈rCH₂C(CH₃)₂ \ :C̈l:
Cyanogen bromide	:Br̈—CN:	:B̈rCH₂C(CH₃)₂ \ CN:
Iodine chloride	:Ï—C̈l:	:ÏCH₂C(CH₃)₂ \ :C̈l:
Sulfenyl chlorides	RS̈—C̈l:	RS̈CH₂C(CH₃)₂ \ :C̈l:
Mercuric salts	XHg—Xᵃ, HÖH	XHgCH₂C(CH₃)₂ \ :ÖH

ᵃX here denotes acetate.

In Summary Halonium ions are subject to stereospecific and regioselective ring opening in a manner that is mechanistically very similar to the nucleophilic opening of protonated oxacyclopropanes. Halonium ions can be trapped by halide ions, water, or alcohols to give vicinal dihaloalkanes, haloalcohols, or haloethers, respectively. The principle of electrophilic additions can be applied to any reagent A–B containing a polarized or polarizable bond.

12-7 OXYMERCURATION–DEMERCURATION: A SPECIAL ELECTROPHILIC ADDITION

The last example in Table 12-2 is an electrophilic addition of a mercuric salt to an alkene. The reaction is called **mercuration,** and the resulting compound is an alkylmercury derivative, from which the mercury can be removed in a subsequent step. One particularly useful reaction sequence is **oxymercuration–demercuration,** in which mercuric acetate acts as the reagent. In the first step (oxymercuration), treatment of an alkene with this species in the presence of water leads to the corresponding addition product.

Oxymercuration

 **Reaction**

Mercuric acetate   Alkylmercuric acetate

In the subsequent demercuration, the mercury-containing substituent is replaced by hydrogen through treatment with sodium borohydride in base. The net result is hydration of the double bond to give an alcohol.

Demercuration

1-Methylcyclopentanol

Oxymercuration is *anti* stereospecific and regioselective. This outcome implies a mechanism similar to that for the electrophilic addition reactions discussed so far. The mercury reagent initially dissociates to give an acetate ion and a cationic mercury species. The latter acts as a Lewis acid and attacks the Lewis basic alkene double bond, furnishing a mercurinium ion, with a structure similar to that of a cyclic bromonium ion. The water that is present attacks the more substituted carbon (Markovnikov rule regioselectivity) to give an alkylmercuric acetate intermediate. Replacement of mercury by hydrogen (demercuration) is achieved by sodium borohydride reduction through a complex and only incompletely understood mechanism. It is not stereospecific.

The alcohol obtained after demercuration is the same as the product of Markovnikov hydration (Section 12-4) of the starting material. However, oxymercuration–demercuration is a valuable alternative to acid-catalyzed hydration, because no carbocation is involved; therefore *oxymercuration–demercuration is not susceptible to the rearrangements that commonly occur under acidic conditions* (Section 12-3). Its use is limited by the expense of the mercury reagent and its toxicity, which requires careful removal of mercury from the product and safe disposal.

REAL LIFE: MEDICINE 12-1 | Juvenile Hormone Analogs in the Battle Against Insect-Borne Diseases

Juvenile hormone (JH) is a substance that controls metamorphosis in insects. It is produced by the male wild silk moth, *Hyalophora cecropia L.,* and its presence delays the maturation of insect larvae until the appropriate

developmental stage is reached. Exposure to JH causes insect metamorphosis to stall at the pupa stage: Mosquitoes exposed to JH fail to develop into biting, egg-laying adults. JH therefore has the potential for use in controlling mosquito-borne diseases such as malaria, yellow fever, and West Nile virus. This potential is limited by the instability of JH and the difficulty with which it can be isolated naturally or prepared synthetically. As a result, analogs have been sought that are more stable, suitably bioactive, and easy to prepare.

The synthetic compound methoprene possesses all of these desirable features. Its synthesis utilizes several reactions we have studied, including oxymercuration–demercuration to give the tertiary methyl ether, ester hydrolysis (Section 8-5), and PCC oxidation of a primary alcohol to produce an aldehyde (Section 8-6).

Whereas earlier attempts to prepare JH analogs gave compounds with very poor activity, methoprene is up to 1000 times more bioactive than JH toward a variety of pests. It is effective against fleas, mosquitoes, and fire ants and is now marketed under a number of trade names. Methoprene may be used indoors to eliminate flea infestations, greatly reducing the need for conventional insecticides. Although the product does not kill insects that have already reached adulthood, the eggs they lay after exposure do not develop into adults. Spreading methoprene as granules in areas where mosquitoes may breed prevents their survival past the pupal stage. Methoprene possesses relatively low toxicity toward vertebrates and, unlike chlorinated insecticides such as DDT (Real Life 3-2), it does not persist in the environment. Although it is stable enough to be effective for weeks to months after application, it is degraded over time by sunlight into innocuous smaller molecules. Methoprene and several other JH analogs have therefore become important new tools for pest management.

The pupa of an *Anopheles* mosquito is shown breathing through two tubes that penetrate the surface of its watery environment. *Anopheles* is a species that carries the malaria parasite. Juvenile hormone stops mosquito development at the pupa stage.

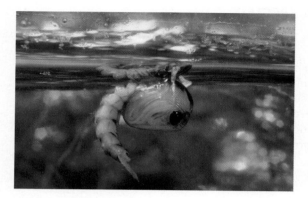

Juvenile hormone

Methoprene (JH analog)

Mechanism of Oxymercuration–Demercuration

Step 1. Dissociation

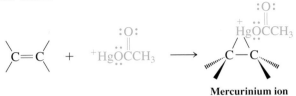

Step 2. Electrophilic attack

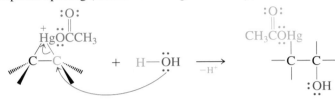

Mercurinium ion

Step 3. Nucleophilic opening (Markovnikov regioselectivity)

Alkylmercuric acetate

Step 4. Reduction

When the oxymercuration of an alkene is executed in an alcohol solvent, demercuration gives an ether, as shown in the margin.

Mechanism

Ether Synthesis by Oxymercuration– Demercuration

1-Hexene

1. $Hg(OCCH_3)_2$, CH_3OH
2. $NaBH_4$, NaOH, H_2O

65%
2-Methoxyhexane

Solved Exercise 12-18 | **Working with the Concepts: Addressing a Difficult Mechanism Problem**

Explain the result shown below.

1. $Hg(OCCH_3)_2$
2. $NaBH_4$, NaOH, H_2O

HOCH₂ CH₂OH

H_2C CH₂OH
42%

Reminder: "Explain" = look at the mechanism.

Strategy
Start by addressing the apparent dissimilarity between the structure of the substrate and that of the product. You can clarify the pathway to a solution by numbering the carbon atoms in the substrate and then identifying the corresponding atoms in the product. Then you can begin the process of *thinking mechanistically*—following the steps of the process as we understand them to see where they lead.

Solution
• Begin with mercuration of the double bond:

$Hg(O_2CCH_3)_2$ $CH_3CO_2Hg^+$

HOCH₂ CH₂OH
  7   8

- In all previous examples, the next step involved a molecule of solvent—either water or alcohol—which provided a nucleophilic oxygen atom to add to one carbon of the mercurinium ion and open the ring. In this case, however, we can see that an oxygen atom *already present in the substrate* (at C7) becomes attached to one of the original double-bond carbons (C2) in the product. This process is one of *intramolecular* bond formation. After removal of the mercury by $NaBH_4$, we arrive at the final product:

Exercise 12-19 | Try It Yourself

The reaction below proceeds to a cyclic product that is an isomer of the starting material. Suggest a structure for it. (**Hint:** Think mechanistically. Begin with the appropriate electrophilic attack. Then identify and utilize a nucleophilic atom *already present in the substrate* in order to complete the addition process. **Caution:** There is a regioselectivity issue to address. Use the examples presented in the chapter section to guide you.)

$$\text{1. Hg(O}_2\text{CCH}_3)_2$$
$$\text{2. NaBH}_4,$$
$$\text{NaOH, H}_2\text{O}$$

In Summary Oxymercuration–demercuration is a synthetically useful method for converting alkenes regioselectively (following the Markovnikov rule) into alcohols or ethers. Carbocations are not involved; therefore, rearrangements do not occur.

12-8 | HYDROBORATION–OXIDATION: A STEREOSPECIFIC ANTI-MARKOVNIKOV HYDRATION

So far in this chapter, we have seen two ways to add the elements of water to alkenes to give alcohols. This section presents a third method, which complements the other two synthetically by giving in a different regiochemical outcome. The process involves a reaction that lies mechanistically between hydrogenation and electrophilic addition: hydroboration of double bonds. The alkylboranes that result can be oxidized to alcohols.

The boron–hydrogen bond adds across double bonds

Borane, BH_3, adds to double bonds without catalytic activation, a reaction called **hydroboration** by its discoverer, H. C. Brown.*

Hydroboration of Alkenes

Borane   An alkylborane   A trialkylborane

---

*Professor Herbert C. Brown (1912–2004), Purdue University, West Lafayette, Indiana, Nobel Prize 1979 (chemistry).

Borane (which by itself exists as a dimer, B_2H_6) is commercially available in ether and tetrahydrofuran (THF). In these solutions, borane exists as a Lewis acid-base complex with the ether oxygen (see Sections 2-3 and 9-5), an aggregate that allows the boron to have an electron octet (for the molecular-orbital picture of BH_3, see Figure 1-17).

How does the B–H unit add to the π bond? The π bond is electron rich and borane is electron poor. Therefore, it is reasonable to formulate an initial Lewis acid-base complex similar to that of a bromonium ion (Figure 12-3), requiring the participation of the empty p orbital on BH_3. This shifts electron density from the alkene to boron. Subsequently, one of the hydrogens is transferred by means of a four-center transition state to one of the alkene carbons, while the boron shifts to the other. The stereochemistry of the addition is *syn*. All three B–H bonds are reactive in this way. The boron in the product alkylborane is again electron deficient. The electrostatic potential maps of the general scheme shown below (on a scale that maximizes the desired color changes) show how the boron in borane starts out as an electron-deficient species (blue), becomes more electron rich (red) in the complex, and then loses the electron density gained as it proceeds to product (blue).

Borane–THF complex

Mechanism of Hydroboration

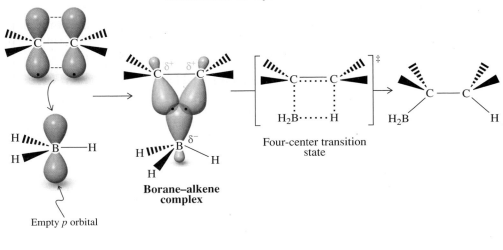

Four-center transition state

Borane–alkene complex

Empty p orbital

Mechanism

Animation

ANIMATED MECHANISM:
Hydroboration–oxidation

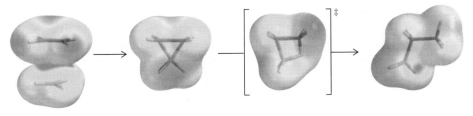

Hydroboration is not only stereospecific (*syn* addition), it is also regioselective. Unlike the electrophilic additions described previously, steric more than electronic factors primarily control the regioselectivity: The boron binds to the less hindered (less substituted) carbon. The reactions of the trialkylboranes resulting from these hydroborations are of special interest to us, as the next section will explain.

The oxidation of alkylboranes gives alcohols

Trialkylboranes can be oxidized with basic aqueous hydrogen peroxide to furnish alcohols in which the hydroxy function has replaced the boron atom. The net result of the two-step sequence, **hydroboration–oxidation,** is the addition of the elements of water to a double bond. In contrast with the hydrations described in Sections 12-4 and 12-7, however, those using borane proceed with the opposite regioselectivity: In this sequence, the OH group ends up at the *less* substituted carbon, an example of **anti-Markovnikov addition.**

Regioselectivity of Hydroboration

$$3\ RCH{=}CH_2\ +\ BH_3$$

Less hindered carbon

$$RCH_2CH_2\underset{\underset{CH_2CH_2R}{|}}{B}CH_2CH_2R$$

Each of the three B–H bonds in BH_3 adds to a molecule of alkene, resulting in a trialkylborane product.

Reaction

Hydroboration–Oxidation Sequence

$$3\ RCH{=}CHR \xrightarrow{BH_3,\ THF} (RCH_2CHR)_3B \xrightarrow{H_2O_2,\ NaOH,\ H_2O} 3\ RCH_2\overset{R}{\underset{}{\overset{|}{C}}}H\ddot{O}H$$

$$(CH_3)_2CHCH_2CH{=}CH_2 \xrightarrow[2.\ H_2O_2,\ NaOH,\ H_2O]{1.\ BH_3,\ THF} (CH_3)_2CHCH_2CH_2CH_2\ddot{O}H$$

4-Methyl-1-pentene   80%
                                          4-Methyl-1-pentanol

In the mechanism of alkylborane oxidation, the nucleophilic hydroperoxide ion attacks the electron-poor boron atom. The resulting species undergoes a rearrangement in which an alkyl group migrates with its electron pair—and with *retention* of configuration—to the neighboring oxygen atom, thus expelling a hydroxide ion in the process. Although hydroxide as a leaving group is rare (Section 6-7), it is facilitated here by the presence of the adjacent negative charge on boron. We have encountered a very similar transformation in the acid-catalyzed rearrangement of 2,2-dimethyl-1-propanol (neopentyl alcohol; Section 9-3), in which water is made to leave by the migrating methyl, as shown in the margin.

Mechanism

Mechanism of Alkylborane Oxidation

$$-\overset{/}{B}\overset{\curvearrowleft}{\underset{R}{|}} + \ \ :\overset{..}{\underset{..}{O}}{-}\overset{..}{O}H \longrightarrow -\overset{|}{\underset{R}{B}}{\overset{\curvearrowright}{=}}\overset{..}{O}{-}\overset{..}{O}H \xrightarrow{-H\overset{..}{O}:^-} -\overset{/}{B}\overset{..}{\underset{..}{O}}{-}R$$

Hydroperoxide ion

Two Similar Rearrangement Steps

Alkylborane hydroperoxide rearrangement

Protonated 2,2-dimethyl-1-propanol rearrangement

$$\overset{CH_3}{\underset{CH_3}{H_3C{-}\overset{|}{\underset{|}{C}}{\overset{\curvearrowright}{-}}CH_2{-}\overset{\curvearrowleft}{\overset{..}{O}}H_2}}$$

This process is repeated until all three alkyl groups have migrated to oxygen atoms, finally forming a triakyl borate $(RO)_3B$. This inorganic ester is then hydrolyzed by base to the alcohol and sodium borate.

$$(RO)_3B\ +\ 3\ NaOH \xrightarrow{H_2O} Na_3BO_3\ +\ 3\ ROH$$

Because borane additions to double bonds and subsequent oxidation are so selective, this sequence allows the stereospecific and regioselective synthesis of alcohols from alkenes. The anti-Markovnikov regioselectivity of the hydroboration–oxidation sequence complements that of acid-catalyzed hydration and oxymercuration–demercuration. In addition, hydroboration, like oxymercuration, occurs without the participation of carbocations; therefore, *rearrangements are not observed.*

A Stereospecific and Regioselective Alcohol Synthesis by Hydroboration–Oxidation

1-Methylcyclopentene $\xrightarrow{BH_3,\ THF}$ **Product of *syn* addition of B–H across C=C bond.** $\xrightarrow{H_2O_2,\ NaOH,\ H_2O}$ 86%
trans-2-Methylcyclopentanol

Note cis relationship between H and OH

Exercise 12-20

Give the products of hydroboration–oxidation of (**a**) propene and (**b**) (*E*)-3-methyl-2-pentene. Show the stereochemistry clearly.

In Summary Hydroboration–oxidation constitutes another method for hydrating alkenes. The initial addition is *syn* and regioselective, the boron shifting to the less hindered carbon. Oxidation of alkyl boranes with basic hydrogen peroxide gives anti-Markovnikov alcohols with retention of configuration of the alkyl group.

12-9 DIAZOMETHANE, CARBENES, AND CYCLOPROPANE SYNTHESIS

Cyclopropanes make interesting synthetic targets. Their highly strained structures (Section 4-2) are fascinating to study, as are the functional contributions they make to a variety of naturally occurring biological compounds (Section 4-7). Cyclopropanes may be quite readily prepared by the addition of reactive species called **carbenes** to the double bond of alkenes. Carbenes have the general structure $R_2C:$, in which the central carbon atom possesses an electron sextet. Although neutral, carbenes are electron deficient and act as electrophiles toward alkenes.

Diazomethane forms methylene, which converts alkenes into cyclopropanes

The unusual substance **diazomethane,** CH_2N_2, is a yellow, highly toxic, and explosive gas. It decomposes on exposure to light, heat, or copper metal by loss of N_2. The result is the highly reactive species **methylene,** $H_2C:$, the simplest carbene.

$$H_2\overset{..}{\overset{-}{C}}\!\!-\!\!\overset{+}{N}\!\!\equiv\!\!N: \xrightarrow{\text{hv or Δ or Cu}} H_2C: \;+\; :N\!\!\equiv\!\!N:$$

Diazomethane   **Methylene**

When methylene is generated in the presence of compounds containing double bonds, addition takes place to furnish cyclopropanes. The process, another Lewis acid-base addition, is usually stereospecific, with retention of the original configuration of the double bond.

Methylene Additions to Double Bonds

40%
Bicyclo[4.1.0]heptane

70%
cis-**Diethylcyclopropane**

Exercise 12-21

Diazomethane is the simplest member of the class of compounds called *diazoalkanes* or *diazo compounds,* $R_2C\!\!=\!\!N_2$. When diazo compound A is irradiated in heptane solution at $-78°C$, it gives a hydrocarbon, C_4H_6, exhibiting three signals in ^1H NMR and two signals in ^{13}C NMR spectroscopy, all in the aliphatic region. Suggest a structure for this molecule.

$$CH_2\!\!=\!\!CHCH_2CH\!\!=\!\!\overset{+}{N}\!\!=\!\!\overset{..}{N}\!:^-$$

A

Halogenated carbenes and carbenoids also give cyclopropanes

Cyclopropanes may also be synthesized from halogenated carbenes, which are prepared from halomethanes. For example, treatment of trichloromethane (chloroform) with a strong base causes an unusual elimination reaction in which both the proton and the leaving group are removed from the same carbon. The product is dichlorocarbene, which gives cyclopropanes when generated in the presence of alkenes.

Dichlorocarbene from Chloroform and Its Trapping by Cyclohexene

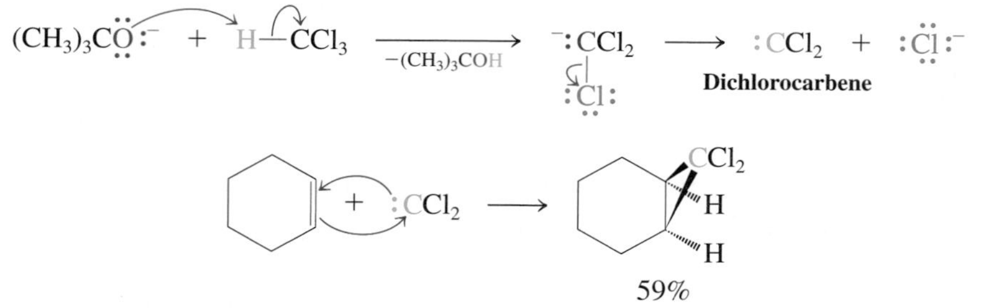

$(CH_3)_3C\ddot{O}:^- \quad + \quad H-CCl_3 \xrightarrow[-(CH_3)_3COH]{} \quad :CCl_2 \longrightarrow \quad :CCl_2 \quad + \quad :\ddot{C}l:^-$

:CCl₂ with Cl — **Dichlorocarbene**

59%

In another route to cyclopropanes, diiodomethane is treated with zinc powder (usually activated with copper) to generate ICH_2ZnI, called the **Simmons-Smith* reagent.** This species is an example of a **carbenoid,** or carbene-like substance, because, like carbenes, it also converts alkenes into cyclopropanes stereospecifically. Use of the Simmons-Smith reagent in cyclopropane synthesis avoids the hazards associated with diazomethane preparation.

Simmons-Smith Reagent in Cyclopropane Synthesis

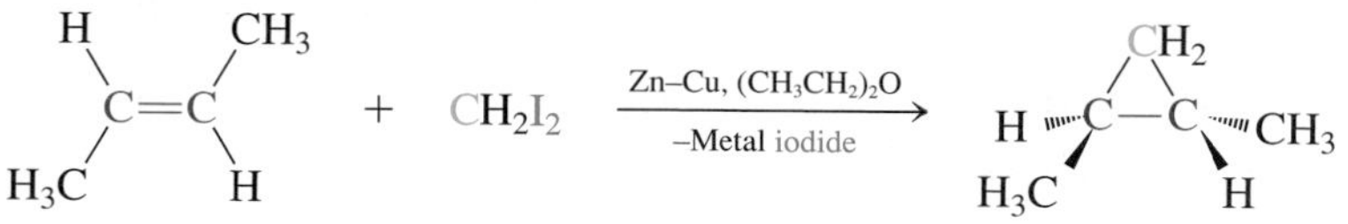

An impressive example of the use of the Simmons-Smith reagent in the construction of natural products is the highly unusual, potent antifungal agent FR-900848, obtained in 1990 from a fermentation broth of *Streptoverticillium fervens* and first synthesized in 1996. Its most noteworthy feature is the fatty acid residue, which contains five cyclopropanes, four of which are contiguous and all of which were made by Simmons-Smith cyclopropanations.

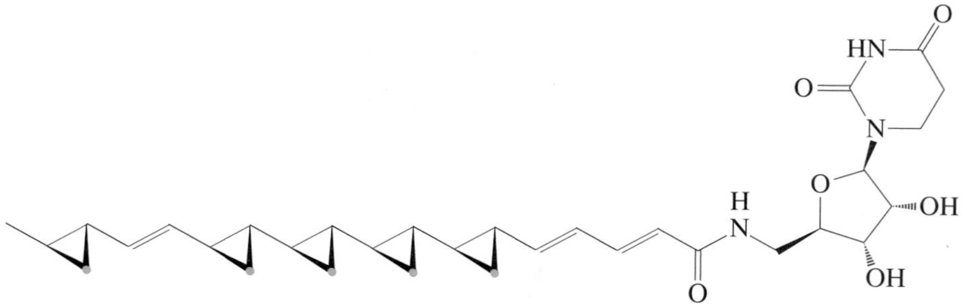

FR-900848

In Summary Diazomethane is a useful synthetic intermediate as a methylene source for forming cyclopropanes from alkenes. Halogenated carbenes, which are formed by dehydrohalogenation of halomethanes, and the Simmons-Smith reagent, a carbenoid arising from the reaction of diiodomethane with zinc, also convert alkenes into cyclopropanes. Additions of carbenes to alkenes differ from other addition processes because a *single carbon atom* becomes bonded to *both* alkene carbons.

12-10 | OXACYCLOPROPANE (EPOXIDE) SYNTHESIS: EPOXIDATION BY PEROXYCARBOXYLIC ACIDS

This section describes how an electrophilic oxidizing agent is capable of introducing a single *oxygen* atom to connect to both carbons of a double bond. This produces oxacyclopropanes, which may, in turn, be converted into vicinal *anti* diols. Sections 12-11 and 12-12

---

*Dr. Howard E. Simmons (1929–1997) and Dr. Ronald D. Smith (b. 1930), both with E. I. du Pont de Nemours and Company, Wilmington, Delaware.

will show methods for the attachment of oxygen atoms to *each* alkene carbon to give vicinal *syn* diols by partial double-bond cleavage or carbonyl compounds by complete double-bond cleavage.

Peroxycarboxylic acids deliver oxygen atoms to double bonds

The OH group in peroxycarboxylic acids, $\overset{\overset{O}{\|}}{R\text{COOH}}$, contains an *electrophilic* oxygen. These compounds react with alkenes by adding this oxygen to the double bond to form oxacyclopropanes. The other product of the reaction is a carboxylic acid. The transformation is of value because, as we know, oxacyclopropanes are versatile synthetic intermediates (Section 9-9). It proceeds at room temperature in an inert solvent, such as chloroform, dichloromethane, or benzene. This reaction is commonly referred to as **epoxidation,** a term derived from *epoxide,* one of the older common names for oxacyclopropanes. A popular peroxycarboxylic acid for use in the research laboratory is *meta*-chloroperoxybenzoic acid (MCPBA). For large-scale and industrial purposes, however, the somewhat shock-sensitive (explosive) MCPBA has been replaced by magnesium monoperoxyphthalate (MMPP).

Oxacyclopropane Formation: Epoxidation of a Double Bond

The transfer of oxygen is stereospecifically *syn,* the stereochemistry of the starting alkene being retained in the product. For example, *trans*-2-butene gives *trans*-2,3-dimethyloxacyclopropane; conversely, the *cis*-2-butene yields *cis*-2,3-dimethyloxacyclopropane.

trans-2-Butene   meta-Chloroperoxybenzoic acid   trans-2,3-Dimethyloxacyclopropane
(MCPBA)

85%

What is the mechanism of this oxidation? It is related but not quite identical to electrophilic halogenation (Section 12-5). In epoxidation, we can write a cyclic transition state in which the electrophilic oxygen is added to the π bond at the same time as the peroxycarboxylic acid proton is transferred to its own carbonyl group, releasing a molecule of carboxylic acid, which is a good leaving group. The two new C–O bonds in the oxacyclopropane product are formally derived from the electron pairs of the alkene π bond and of the cleaved O–H linkage.

Mechanism of Oxacyclopropane Formation

Peroxycarboxylic Acids

Electrophilic oxygen

A peroxycarboxylic acid

Peroxyethanoic
(peracetic) acid

meta-Chloro-
peroxybenzoic acid
(MCPBA)

● **Reaction**

Magnesium
monoperoxyphthalate
(MMPP)

Mechanism

Animation

ANIMATED MECHANISM:
Oxacyclopropanation

Exercise 12-22

Outline a short synthesis of *trans*-2-methylcyclohexanol from cyclohexene. (**Hint:** Review the reactions of oxacyclopropanes in Section 9-9.)

In accord with the electrophilic mechanism, the reactivity of alkenes toward peroxycarboxylic acids increases with alkyl substitution (see Section 11-3), allowing for selective oxidations. For example,

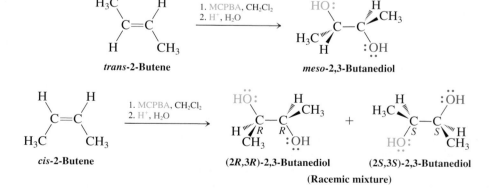

Epoxidation only at more electron-rich double bond

CH₃COOH (1 equivalent), CHCl₃, 10°C

Disubstituted   Monosubstituted

86%

Hydrolysis of oxacyclopropanes furnishes the products of *anti* dihydroxylation of an alkene

Treatment of oxacyclopropanes with water in the presence of catalytic acid or base leads to ring opening to the corresponding vicinal diols. These reactions follow the mechanisms described in Section 9-9: The nucleophile (water or hydroxide) attacks the side opposite the oxygen in the three-membered ring, so the net result of the oxidation–hydrolysis sequence constitutes an ***anti* dihydroxylation** of an alkene. In this way, *trans*-2-butene gives *meso*-2,3-butanediol, whereas *cis*-2-butene furnishes the racemic mixture of the 2*R*,3*R* and 2*S*,3*S* enantiomers.

Vicinal *Anti* Dihydroxylation of Alkenes

1. MCPBA, CH₂Cl₂
2. H⁺, H₂O

Synthesis of Isomers of 2,3-Butanediol

Model Building

1. MCPBA, CH₂Cl₂
2. H⁺, H₂O

***trans*-2-Butene**   ***meso*-2,3-Butanediol**

1. MCPBA, CH₂Cl₂
2. H⁺, H₂O

***cis*-2-Butene**   **(2*R*,3*R*)-2,3-Butanediol**   **(2*S*,3*S*)-2,3-Butanediol**

(Racemic mixture)

Exercise 12-23

Give the products obtained by treating the following alkenes with MCPBA and then aqueous acid: (**a**) 1-hexene; (**b**) cyclohexene; (**c**) *cis*-2-pentene; (**d**) *trans*-2-pentene.

In Summary Peroxycarboxylic acids supply oxygen atoms to convert alkenes into oxacyclopropanes (epoxidation). Oxidation–hydrolysis reactions with peroxycarboxylic acids furnish vicinal diols in a stereospecifically *anti* manner.

Osmium tetroxide

12-11 | VICINAL *SYN* DIHYDROXYLATION WITH OSMIUM TETROXIDE

Osmium tetroxide reacts with alkenes in a two-step process to give the corresponding vicinal diols in a stereospecifically *syn* manner. The process therefore complements the epoxidation–hydrolysis sequence described in the previous section, which proceeds with *anti* selectivity.

Vicinal *Syn* Dihydroxylation with Osmium Tetroxide

$$C=C \xrightarrow[\text{2. } H_2S]{\text{1. } OsO_4, \text{ THF, } 25°C} HO-C-C-OH$$

Reaction

The process leads initially to an isolable cyclic ester, which is reductively hydrolyzed with H_2S or bisulfite, $NaHSO_3$. For example,

90%

What is the mechanism of this transformation? The initial reaction of the π bond with osmium tetroxide constitutes a concerted (Section 6-4) addition in which three electron pairs move simultaneously to give a cyclic ester containing Os(VI). This process can be viewed as an electrophilic attack on the alkene: Two electrons flow from the alkene onto the metal, which is reduced [Os(VIII) → Os(VI)]. For steric reasons, the product can form only in a way that introduces the two oxygen atoms on the *same* face of the double bond—*syn*. This intermediate is usually not isolated but converted upon reductive work-up into the free diol.

Mechanism of the Osmium Tetroxide Oxidation of Alkenes

Mechanism

Because OsO_4 is expensive and highly toxic, a commonly used modification calls for the use of only catalytic quantities of the osmium reagent and stoichiometric amounts of another oxidizing agent such as H_2O_2, which serves to reoxidize reduced osmium.

An older reagent for vicinal *syn* dihydroxylation of alkenes is potassium permanganate, $KMnO_4$. Although this reagent functions in a manner that is mechanistically similar to OsO_4, it is less useful for synthesizing diols because of a tendency to give poorer yields owing to overoxidation. Potassium permanganate solutions, which are deep purple, are useful as a color test for alkenes, however: Upon reaction, the purple reagent is immediately converted into the brown precipitate of its reduction product, MnO_2.

Potassium Permanganate Test for Alkene Double Bonds

$$C=C + \underset{\text{Dark purple}}{KMnO_4} \longrightarrow HO-C-C-OH + \underset{\substack{\text{Brown}\\\text{precipitate}}}{MnO_2}$$

REAL LIFE: MEDICINE 12-2 | Synthesis of Antitumor Drugs: Sharpless Enantioselective Oxacyclopropanation (Epoxidation) and Dihydroxylation

In the decade of the 1990s, a significant shift occurred in the way in which new pharmaceuticals were synthesized. Before this time, most of the available methods for the preparation of chiral molecules in enantiomerically pure form were impractical on an industrial scale. Thus, typically, racemic mixtures were generated, although in many cases only one of the enantiomers in such mixtures possessed the desired activity. However, fundamental conceptual advances in catalysis changed this situation. Some of the most useful examples are a series of highly enantioselective oxidation reactions of double bonds developed by K. B. Sharpless (see Real Life 5-4).

The first such process is a variant of the oxacyclopropanation reaction discussed in Section 12-10, as applied specifically to 2-propenyl (allylic) alcohols. However, instead of a peroxycarboxylic acid, the reagent is *tert*-butyl hydroperoxide in the presence of titanium (IV) isopropoxide ("Sharpless epoxidation"), the function of the chiral auxiliary being assumed by tartaric acid diethyl ester (Real Life 5-3). The naturally occurring (+)-[2R,3R]-diethyl tartrate and its nonnatural (−)-(2S,3S) mirror image are both commercial products. One delivers oxygen to one face of the double bond, the other to the opposite face, as shown below, giving either enantiomer of the oxacyclopropane product with high enantiomer excess (Section 5-2).

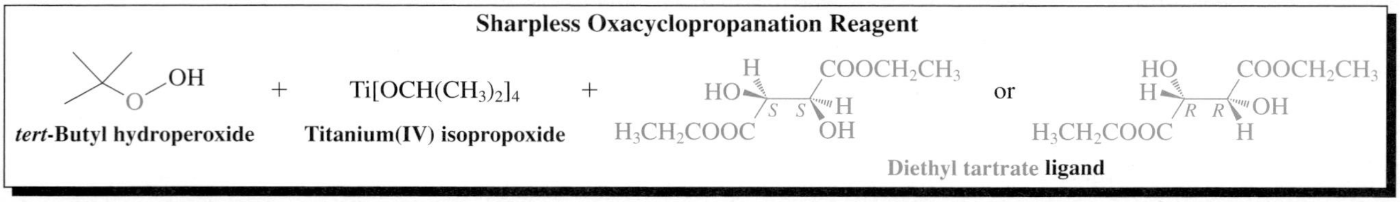

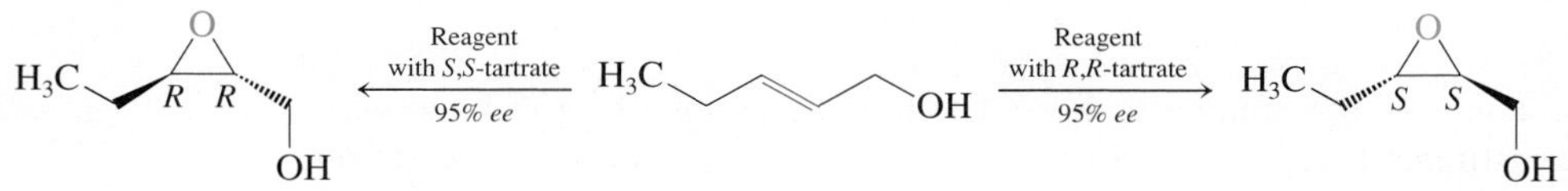

The role of the chiral coordinated ligand is to provide a pocket into which the substrate can enter in only one spatial orientation (see also Real Life 9-3 and Section 12-2). In this respect, it bears the characteristics of many enzymes, biological catalysts that function in essentially the same way (see Real Life 5-5 and Chapter 26). In the absence of the chiral ligand, a racemic mixture forms.

The Sharpless enantioselective oxacyclopropanation has been exploited in the synthesis of many chiral, enantiomerically pure building blocks for the construction of important drugs, such as the powerful antitumor agent aclacinomycin A (Real Life 7-1).

Sharpless applied the same principle of using a central metal that can hold a chiral directing group proximal to an alkene substrate in an enantioselective version of the OsO_4-catalyzed alkene dihydroxylation (Section 12-11). Here, the essence of the chiral auxiliary is an amine derived from the family of natural alkaloids called the *cinchona* (Section 25-8). One of these amines is dihydroquinine, which is added in the linked dimeric form shown on the right. Instead of H_2O_2 as the stoichiometric oxidant (Section 12-11), Fe^{3+} [as $K_3Fe(CN)_6$] is employed. This method has been applied by E. J. Corey (Section 8-9) to the enantioselective synthesis of ovalicin, a member of a class of fungal-derived natural products called antiangiogenesis agents: They inhibit the growth of new blood vessels and therefore have the capability to cut off the blood supply to solid tumors. The key to Corey's synthetic route is the enantioselective

syn dihydroxylation of the achiral cyclohexene derivative shown on the next page.

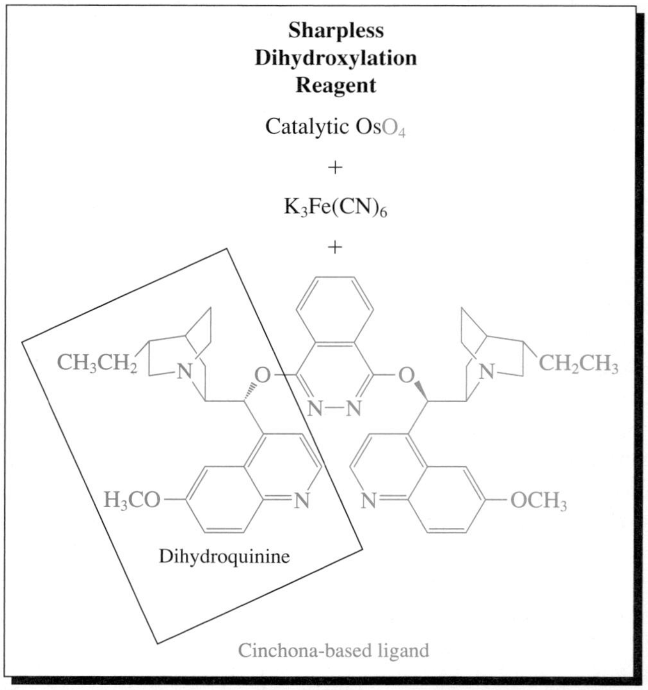

REAL LIFE: MEDICINE 12-2 | (Continued)

A recently identified synthetic derivative of ovalicin, called TNP-470, possesses powerful antiangiogenesis activity, is chemically stable, nontoxic, noninflammatory, and potentially amenable to oral administration. Several preliminary studies revealed the effectiveness of TNP-470 against tumors in animals. By early 2004, TNP-470 had been entered in numerous human clinical trials to determine its applicability toward the treatment of cancers of the breast, brain, cervix, liver, and prostate, as well as AIDS-related Kaposi sarcoma, lymphoma, and leukemia. Unfortunately, as is often the case in drug testing, clinical trials were suspended, in this case because of neurological side effects. These problems were solved by attaching polymeric side chains at the chloride end of the molecule to prevent it from crossing the blood–brain barrier. The resulting drug, Lodamin, is nontoxic, and can be taken orally. It is hoped that Lodamin will pass upcoming clinical trials and become available to patients in the future.

Key Step Toward Ovalicin

(−)-Ovalicin

TNP-470

Exercise 12-24

The stereochemical consequences of the vicinal *syn* dihydroxylation of alkenes are complementary to those of vicinal *anti* dihydroxylation. Show the products (indicate stereochemistry) of the vicinal *syn* dihydroxylation of *cis*- and *trans*-2-butene.

In Summary Osmium tetroxide, either stoichiometrically or catalytically together with a second oxidizing agent, converts alkenes into *syn*-1,2-diols. A similar reaction of purple potassium permanganate is accompanied by decolorization, a result that makes it a useful test for the presence of double bonds.

12-12 | OXIDATIVE CLEAVAGE: OZONOLYSIS

Although oxidation of alkenes with osmium tetroxide breaks only the π bond, other reagents may rupture the σ bond as well. The most general and mildest method of oxidatively cleaving alkenes is through the reaction with ozone, **ozonolysis.** The products are carbonyl compounds.

Ozone, O_3, is produced in the laboratory in an instrument called an *ozonator,* in which an arc discharge generates 3–4% ozone in a dry oxygen stream. The gas mixture is passed through a solution of the alkene in methanol or dichloromethane. The first isolable intermediate is a species called an **ozonide,** which is reduced directly in a subsequent step by exposure to zinc in acetic acid or by reaction with dimethyl sulfide. The net result of the ozonolysis–reduction sequence is the cleavage of the molecule at the carbon–carbon double bond; oxygen becomes attached to each of the carbons that had originally been doubly bonded.

Ozone is a blue gas that condenses to a dark blue, highly unstable liquid. Ozone is a powerful bacteriocide. As a result, ozonators are used for disinfecting water in pools and spas.

Reaction

Ozonolysis Reaction of Alkenes

$$\text{C=C} \xrightarrow{O_3} \text{Ozonide} \xrightarrow[\text{- [O]}]{\text{Reduction}} \text{C=O} + \text{O=C}$$

Ozonide   **Carbonyl products**

$$\underset{\text{(Z)-3-Methyl-2-pentene}}{\text{CH}_3\text{CH}_2\text{C=CCH}_3} \xrightarrow[\text{2. Zn, CH}_3\text{COOH}]{\text{1. O}_3\text{, CH}_2\text{Cl}_2} \underset{\underset{\text{2-Butanone}}{90\%}}{\text{CH}_3\text{CH}_2\overset{\text{O}}{\overset{\|}{\text{C}}}\text{CH}_3} + \underset{\text{Acetaldehyde}}{\text{CH}_3\overset{\text{O}}{\overset{\|}{\text{CH}}}}$$

The mechanism of ozonolysis proceeds through initial electrophilic addition of ozone to the double bond, a transformation that yields the so-called **molozonide.** In this reaction, as in several others already presented, six electrons move in concerted fashion in a cyclic transition state. The molozonide is unstable and breaks apart into a carbonyl fragment and a carbonyl oxide fragment through another cyclic six-electron rearrangement. Recombination of the two fragments as shown yields the ozonide.

Mechanism of Ozonolysis

Mechanism

Animation

ANIMATED MECHANISM: Ozonolysis

Step 1. Molozonide formation and cleavage

A molozonide   **A carbonyl oxide**

Step 2. Ozonide formation and reduction

Ozonide

Exercise 12-25

An unknown hydrocarbon of the molecular formula $C_{12}H_{20}$ exhibited an ^1H NMR spectrum with a complex multiplet of signals between 1.0 and 2.2 ppm. Ozonolysis of this compound gave two equivalents of cyclohexanone, whose structure is shown in the margin. What is the structure of the unknown?

Exercise 12-26

Give the products of the following reactions.

(a) $\xrightarrow[\text{2. (CH}_3)_2\text{S}]{\text{1. O}_3\text{, CH}_2\text{Cl}_2}$

(b) $\xrightarrow[\text{2. Zn, CH}_3\text{COH}]{\text{1. O}_3\text{, CH}_2\text{Cl}_2}$

(c) $\xrightarrow[\text{2. (CH}_3)_2\text{S}]{\text{1. O}_3\text{, CH}_2\text{Cl}_2}$

Solved Exercise 12-27	Working with the Concepts: Deducing the Structure of an Ozonolysis Substrate

What is the structure of the following starting material?

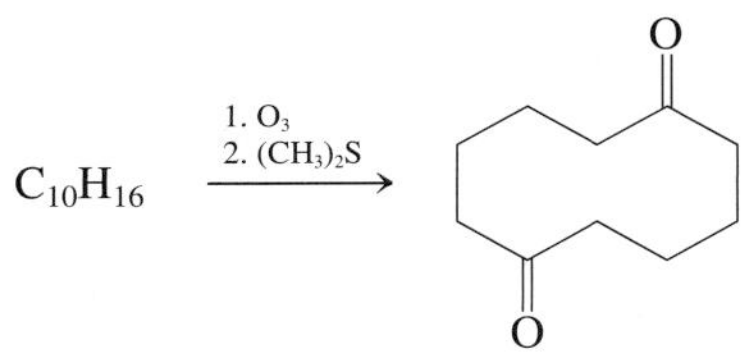

$C_{10}H_{16}$ $\xrightarrow{\substack{1.\ O_3 \\ 2.\ (CH_3)_2S}}$

Strategy

Begin by *counting atoms:* How does the molecular formula of the product compare to that of the starting material? Then *consider the reaction:* What kind of reaction is it and what transformation does it achieve? Putting this information together should enable you to reconstruct the starting material.

Solution

• The molecular formula of the product is $C_{10}H_{16}O_2$, which is identical to the starting material plus two oxygen atoms. This information simplifies the problem: How were these oxygens introduced? Can we imagine undoing that process in order to identify the original structure?

• The reaction is ozonolysis, which achieves the overall transformation

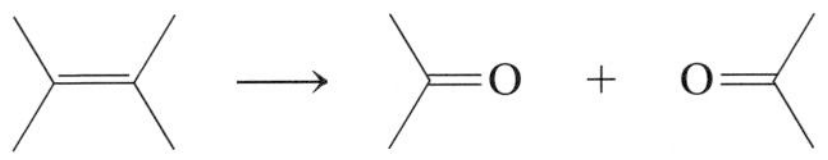

that is, the addition of two oxygen atoms to the original starting species, which is exactly the change observed in the problem.

• Reconstructing the starting molecule therefore requires nothing more than excising the two oxygen atoms and connecting the two carbonyl carbons by a double bond:

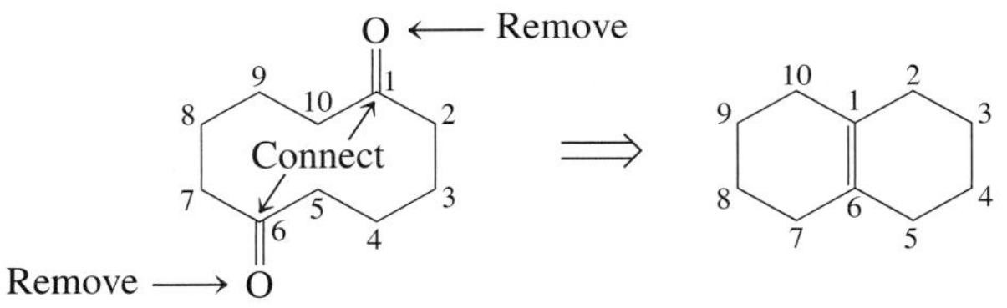

On first sight, this seems easier said than done, because of the way in which we have written the dicarbonyl product. However, if we number the carbons as shown, and recall that carbon–carbon single bonds give rise to flexible molecules with multiple conformations, we find that connecting these two atoms is not so difficult.

Exercise 12-28	Try It Yourself

Suggest a structure for a substance that, upon ozonolysis followed by treatment with $(CH_3)_2S$, gives as the sole product $CH_3COCH_2CH_2CH_2CH_2CHO$. (**Hint:** Begin by writing out a bond-line formula of this product so that you can clearly see its structure and number its carbon atoms.)

In Summary Ozonolysis followed by reduction yields aldehydes and ketones. Mechanistically, the reactions presented in Sections 12-10 through 12-12 are related in as much as the attack of an electrophilic oxidizing agent leads to rupture of the π bond. Unlike the reaction sequences studied in Sections 12-10 and 12-11, however, ozonolysis causes cleavage of both the π and the σ bonds.

12-13 | RADICAL ADDITIONS: ANTI-MARKOVNIKOV PRODUCT FORMATION

In this section, all radicals and single atoms are shown in green, as in Chapter 3.

Radicals, lacking a closed outer shell of electrons, are capable of reacting with double bonds. However, a radical requires only one electron for bond formation, unlike the electrophiles presented in this chapter so far, which consume both electrons of the π bond upon addition. The product of radical addition to an alkene is an alkyl radical, and the final products exhibit anti-Markovnikov regiochemistry, similar to the products of hydroboration–oxidation (Section 12-8).

Reaction

Markovnikov Addition of HBr

$CH_3CH_2CH{=}CH_2$

(Freshly distilled)

↓ HBr
24 h

:Ör:
|
$CH_3CH_2CHCH_2H$
90%

Markovnikov product
(By *ionic* mechanism)

Anti-Markovnikov Addition of HBr

$CH_3CH_2CH{=}CH_2$

(Exposed to oxygen)

↓ HBr
4 h

H
|
$CH_3CH_2CHCH_2\ddot{B}r$:
65%

Anti-Markovnikov product
(By *radical* mechanism)

Hydrogen bromide can add to alkenes in anti-Markovnikov fashion: a change in mechanism

When freshly distilled 1-butene is exposed to hydrogen bromide, clean Markovnikov addition to give 2-bromobutane is observed. This result is in accord with the ionic mechanism for electrophilic addition of HBr discussed in Section 12-3. Curiously, the same reaction, when carried out with a sample of 1-butene that has been exposed to air, proceeds much more quickly and gives an entirely different result. In this case, we isolate 1-bromobutane, formed by anti-Markovnikov addition.

This change caused considerable confusion in the early days of alkene chemistry, because one researcher would obtain only one hydrobromination product, whereas another would obtain a different product or mixtures from a seemingly identical reaction. The mystery was solved by Kharasch* in the 1930s, when it was discovered that the culprits responsible for anti-Markovnikov additions were radicals formed from peroxides, ROOR, in alkene samples that had been stored in the presence of air. In practice, to effect anti-Markovnikov hydrobromination, radical initiators, such as peroxides, are added deliberately to the reaction mixture.

The mechanism of the addition reaction under these conditions is not an ionic sequence; rather, it is a *much faster* **radical chain sequence.** The reason is that the activation energies of the component steps of radical reactions are very small, as we observed earlier during the discussion of the radical halogenation of alkanes (Section 3-4). Consequently, in the presence of radicals, anti-Markovnikov hydrobromination simply outpaces the regular addition pathway. The initiation steps are

1. the homolytic cleavage of the weak RO–OR bond [$DH° \approx 39$ kcal mol^{-1} (163 kJ mol^{-1})], and
2. reaction of the resulting alkoxy radical with hydrogen bromide.

The driving force for the second (exothermic) step is the formation of the strong O–H bond. The bromine atom generated in this step initiates chain propagation by attacking the double bond. One of the π electrons combines with the unpaired electron on the bromine atom to form the carbon–bromine bond. The other π electron remains on carbon, giving rise to a radical.

The halogen atom's attack is *regioselective,* creating the relatively more stable secondary radical rather than the primary one. This result is reminiscent of the ionic additions of hydrogen bromide (Section 12-3), except that the roles of the proton and bromine are reversed. In the ionic mechanism, a proton attacks first to generate the more stable carbocation, which is then trapped by bromide ion. *In the radical mechanism, a bromine atom is the attacking species,* creating the more stable radical center. The alkyl radical subsequently reacts with HBr by abstracting a hydrogen and regenerating the chain-carrying bromine atom. Both propagation steps are exothermic, and the reaction proceeds rapidly. As usual, termination is by radical combination or by some other removal of the chain carriers (Section 3-4).

*Professor Morris S. Kharasch (1895–1957), University of Chicago.

Mechanism of Radical Hydrobromination

Initiation steps

$$RO\!-\!OR \xrightarrow{\Delta} 2\ RO\cdot \qquad \Delta H° \approx +39\ kcal\ mol^{-1}$$
$$(+163\ kJ\ mol^{-1})$$

$$RO\cdot\ +\ H\!:\!Br\!: \xrightarrow{\Delta} RO H\ +\ :Br\cdot \qquad \Delta H° \approx -17\ kcal\ mol^{-1}$$
$$(-71\ kJ\ mol^{-1})$$

Mechanism

Animation

ANIMATED MECHANISM: Radical hydrobromination of 1-butene

Propagation steps

$$\underset{CH_3CH_2}{\overset{H}{C}}\!=\!CH_2\ +\ \cdot Br\!: \longrightarrow CH_3CH_2\dot{C}H\!-\!CH_2Br\!: \qquad \Delta H° \approx -5\ kcal\ mol^{-1}$$

Secondary radical $\qquad (-21\ kJ\ mol^{-1})$

$$CH_3CH_2\dot{C}HCH_2Br\ +\ H\!:\!Br\!: \longrightarrow CH_3CH_2\overset{H}{\underset{|}{C}}HCH_2Br\!:\ +\ :Br\cdot \qquad \Delta H° \approx -11.5\ kcal\ mol^{-1}$$
$$(-48\ kJ\ mol^{-1})$$

Are radical additions general?

Hydrogen chloride and hydrogen iodide do not give anti-Markovnikov addition products with alkenes; in both cases, one of the propagating steps is endothermic and consequently so slow that the chain reaction terminates. As a result, HBr is the *only* hydrogen halide that adds to an alkene under radical conditions to give anti-Markovnikov products. Additions of HCl and HI proceed only by ionic mechanisms to give normal Markovnikov products regardless of the presence or absence of radicals. Other reagents such as thiols, however, do undergo radical additions to alkenes.

Radical Addition of a Thiol to an Alkene

$$CH_3CH\!=\!CH_2\ +\ CH_3CH_2\ddot{S}H \xrightarrow{ROOR} CH_3\overset{}{\underset{\underset{H}{|}}{C}}HCH_2\ddot{S}CH_2CH_3$$

Ethanethiol $\qquad\qquad\qquad$ **Ethyl propyl sulfide**

(CH$_3$)$_3$C$\diagdown$O$\diagup$O$\diagdown$C(CH$_3$)$_3$

Bis(1,1-dimethylethyl) peroxide
(Di-*tert*-butyl peroxide)

In this example, the initiating alkoxy radical abstracts a hydrogen from sulfur to yield CH$_3$CH$_2\ddot{S}\cdot$, which then attacks the double bond. Bis(1,1-dimethylethyl) peroxide (di-*tert*-butyl peroxide) and dibenzoyl peroxide are commercially available initiators for such radical addition reactions.

H$_5$C$_6$—C(=O)—O—O—C(=O)—C$_6$H$_5$

Dibenzoyl peroxide

Exercise 12-29

Ultraviolet irradiation of a mixture of 1-octene and diphenylphosphine, (C$_6$H$_5$)$_2$PH, furnishes 1-(diphenylphosphino)octane by radical addition. Write a plausible mechanism for this reaction.

$$(C_6H_5)_2PH + H_2C\!=\!CH(CH_2)_5CH_3 \xrightarrow{h\nu} (C_6H_5)_2P\!-\!CH_2\!-\!CH_2(CH_2)_5CH_3$$

As we saw in the case of hydroboration (Section 12-8), anti-Markovnikov additions are synthetically useful because their products complement those obtained from ionic additions. The ability to control regiochemistry is an important feature in the development of new synthetic methods.

In Summary Radical initiators alter the mechanism of the addition of HBr to alkenes from ionic to radical chain. The consequence of this change is anti-Markovnikov regioselectivity. Other species, most notably thiols, but not HCl or HI, are capable of undergoing similar reactions.

12-14 | DIMERIZATION, OLIGOMERIZATION, AND POLYMERIZATION OF ALKENES

Is it possible for alkenes to react with one another? Indeed it is, but only in the presence of an appropriate catalyst—for example, an acid, a radical, a base, or a transition metal. In this reaction the unsaturated centers of the alkene monomer (*monos,* Greek, single; *meros,* Greek, part) are linked to form dimers, trimers, **oligomers** (*oligos,* Greek, few, small), and ultimately **polymers** (*polymeres,* Greek, of many parts), substances of great industrial importance.

Polymerization

Monomers   Polymer

Carbocations attack pi bonds

Treatment of 2-methylpropene with hot aqueous sulfuric acid gives two dimers: 2,4,4-trimethyl-1-pentene and 2,4,4,-trimethyl-2-pentene. This transformation is possible because 2-methylpropene can be protonated under the reaction conditions to furnish the 1,1-dimethylethyl (*tert*-butyl) cation. This species can attack the electron-rich double bond of 2-methylpropene with formation of a new carbon–carbon bond. Electrophilic addition proceeds according to the Markovnikov rule to generate the more stable carbocation. Subsequent deprotonation from either of two adjacent carbons furnishes a mixture of the two observed products.

Reaction

Dimerization of 2-Methylpropene

$$CH_2\!=\!C(CH_3)_2 \quad + \quad CH_2\!=\!C(CH_3)_2 \quad \xrightarrow{\text{H}^+} \quad CH_3\overset{\underset{\displaystyle CH_3}{|}}{\underset{\underset{\displaystyle CH_3}{|}}{C}}CH_2\overset{\underset{\displaystyle }{}}{C}\!=\!CH_2 \quad + \quad CH_3\overset{\underset{\displaystyle CH_3}{|}}{\underset{\underset{\displaystyle CH_3}{|}}{C}}CH\!=\!C(CH_3)_2$$

2,4,4-Trimethyl-1-pentene   2,4,4-Trimethyl-2-pentene

Mechanism

Mechanism of Dimerization of 2-Methylpropene

From deprotonation *a*   From deprotonation *b*

Repeated attack can lead to oligomerization and polymerization

The two dimers of 2-methylpropene tend to react further with the starting alkene. For example, when 2-methylpropene is treated with strong acid under more concentrated conditions, trimers, tetramers, pentamers, and so forth, are formed by repeated electrophilic attack of intermediate carbocations on the double bond. This process, which leads to alkane chains of intermediate length, is called **oligomerization.**

Oligomerization of the 2-Methylpropene Dimers

$$CH_3\overset{\overset{\displaystyle CH_3}{|}}{\underset{\underset{\displaystyle CH_3}{|}}{C}}CH=C(CH_3)_2 \quad + \quad CH_3\overset{\overset{\displaystyle CH_3}{|}}{\underset{\underset{\displaystyle CH_3}{|}}{C}}CH_2\overset{\overset{\displaystyle CH_2}{\|}}{C}{\underset{\underset{\displaystyle CH_3}{}}{}} \xrightarrow{H^+} CH_3\overset{\overset{\displaystyle CH_3}{|}}{\underset{\underset{\displaystyle CH_3}{|}}{C}}CH_2\overset{+}{C}\overset{\overset{\displaystyle CH_3}{|}}{\underset{\underset{\displaystyle CH_3}{|}}{}} \quad + \quad CH_2=\overset{\overset{\displaystyle CH_3}{|}}{\underset{\underset{\displaystyle CH_3}{|}}{C}} \longrightarrow$$

$$CH_3\overset{\overset{\displaystyle CH_3\ CH_3}{|\ \ \ |}}{\underset{\underset{\displaystyle CH_3\ CH_3}{|\ \ \ |}}{C}}CH_2\overset{}{C}CH_2\overset{+}{C}\overset{\overset{\displaystyle CH_3}{|}}{\underset{\underset{\displaystyle CH_3}{|}}{}} \quad + \quad CH_2=\overset{\overset{\displaystyle CH_3}{|}}{\underset{\underset{\displaystyle CH_3}{|}}{C}} \longrightarrow CH_3\overset{\overset{\displaystyle CH_3\ CH_3\ CH_3}{|\ \ \ |\ \ \ |}}{\underset{\underset{\displaystyle CH_3\ CH_3\ CH_3}{|\ \ \ |\ \ \ |}}{C}}CH_2CCH_2CCH_2\overset{+}{C}\overset{\overset{\displaystyle CH_3}{|}}{\underset{\underset{\displaystyle CH_3}{|}}{}} \quad etc.$$

Oligomerization continues to give polymers containing many subunits. To control the temperature of these highly exothermic reactions, to minimize E_1 (see footnote on p. 329), and to maximize polymer length, industrial processes are executed with extensive cooling.

Polymerization of 2-Methylpropene

$$n\ CH_2=C(CH_3)_2 \xrightarrow{H^+,\ -100°C} H-(CH_2-\overset{\overset{\displaystyle CH_3}{|}}{\underset{\underset{\displaystyle CH_3}{|}}{C}})_{n-1}-CH_2\overset{\overset{\displaystyle CH_3}{|}}{C}=CH$$

Poly(2-methylpropene)
(Polyisobutylene)

In Summary Catalytic acid causes alkene–alkene additions to occur, a process that forms dimers, trimers, oligomers containing several components, and finally polymers, which are composed of a great many alkene subunits.

12-15 | SYNTHESIS OF POLYMERS

Many alkenes are suitable monomers for polymerization. Polymerization is exceedingly important in the chemical industry, because many polymers have desirable properties, such as durability, inertness to many chemicals, elasticity, transparency, and electrical and thermal resistance.

Although the production of polymers has contributed to pollution—many of them are not biodegradable—they have varied uses as synthetic fibers, films, pipes, coatings, and molded articles. Polymers are also being used increasingly as coatings for medical implants. Names such as polyethylene, poly(vinyl chloride) (PVC), Teflon, polystyrene, Orlon, and Plexiglas (Table 12-3) have become household words.

Acid-catalyzed polymerizations, such as that described for poly(2-methylpropene), are carried out with H_2SO_4, HF, and BF_3 as the initiators. Because they proceed through carbocation intermediates, they are also called *cationic polymerizations.* Other mechanisms of polymerizations are *radical, anionic,* and *metal catalyzed.*

Radical polymerizations lead to commercially useful materials

An example of **radical polymerization** is that of ethene in the presence of an organic peroxide at high pressures and temperatures. The reaction proceeds by a mechanism that, in its initial stages, resembles that of the radical addition to alkenes (Section 12-13). The peroxide initiators cleave into alkoxy radicals, which begin polymerization by addition to the double bond of ethene. The alkyl radical thus created attacks the double bond of another

This spectacular dress designed by Spanish born designer Paco Rabanne would not have been possible without synthetic polymers.

Table 12-3 Common Polymers and Their Monomers

Monomer	Structure	Polymer (common name)	Structure	Uses
Ethene	$H_2C=CH_2$	Polyethylene	$-(CH_2CH_2)_n-$	Food storage bags, containers
Chloroethene (vinyl chloride)	$H_2C=CHCl$	Poly(vinyl chloride) (PVC)	$-(CH_2CH)_n-$ $\|$ Cl	Pipes, vinyl fabrics
Tetrafluoroethene	$F_2C=CF_2$	Teflon	$-(CF_2CF_2)_n-$	Nonstick cookware
Ethenylbenzene (styrene)	(C$_6$H$_5$)CH=CH$_2$	Polystyrene	$-(CH_2CH)_n-$ (C$_6$H$_5$)	Foam packing material
Propenenitrile (acrylonitrile)	$H_2C=C$ with H and $C\equiv N$	Orlon	$-(CH_2CH)_n-$ $\|$ CN	Clothing, synthetic fabrics
Methyl 2-methyl-propenoate (methyl methacrylate)	$H_2C=C$ with CH_3 and $COCH_3$ (C=O)	Plexiglas	$-(CH_2C)_n-$ with CH_3 and CO_2CH_3	Impact-resistant paneling
2-Methylpropene (isobutylene)	$H_2C=C$ with CH_3 and CH_3	Elastol	$-(CH_2C)_n-$ with CH_3 and CH_3	Oil-spill clean-up

Although plastic waste has become a significant disposal problem, some polymers provide environmental benefits through their ability to absorb many times their weight in organic pollutants. For example, Elastol has been used in the clean-up of oil spills.

This penguin is the victim of our oil-based economy: oil spill off the coast of South Africa.

Mechanism

ethene molecule, furnishing another radical center, and so on. Termination of the polymerization can be by dimerization, disproportionation of the radical, or other radical-trapping reactions (Section 3-4).

Mechanism of Radical Polymerization of Ethene

Initiation steps

$$RO-OR \longrightarrow RO\cdot$$

$$RO\cdot + CH_2=CH_2 \longrightarrow ROCH_2-\dot{C}H_2$$

Propagation steps

$$R\ddot{O}CH_2CH_2\cdot + CH_2=CH_2 \longrightarrow R\ddot{O}CH_2CH_2CH_2CH_2\cdot$$

$$R\ddot{O}CH_2CH_2CH_2CH_2\cdot \xrightarrow{(n-1)\ CH_2=CH_2} R\ddot{O}-(CH_2CH_2)_n-CH_2CH_2\cdot$$

Polyethene (polyethylene) produced in this way does not have the expected linear structure. *Branching* occurs by abstraction of a hydrogen along the growing chain by another radical center followed by chain growth originating from the new radical. The average molecular weight of polyethene is almost 1 million.

Polychloroethene [poly(vinyl chloride), PVC] is made by similar radical polymerization. Interestingly, the reaction is regioselective. The peroxide initiator and the intermediate chain radicals add only to the unsubstituted end of the monomer, because the radical center formed next to chlorine is relatively stable. Thus, PVC has a very regular *head-to-tail structure* of molecular weight in excess of 1.5 million. Although PVC itself is fairly hard and brittle, it can be softened by addition of carboxylic acid esters (Section 20-4), called **plasticizers** (*plastikos,* Greek, to form). The resulting elastic material is used in "vinyl leather," plastic covers, and garden hoses.

$$CH_2=CHCl \xrightarrow{ROOR} -(CH_2CH)_n-$$
$$\overset{|}{Cl}$$

Polychloroethene
[Poly(vinyl chloride)]

Exposure to chloroethene (vinyl chloride) has been linked to the incidence of a rare form of liver cancer (angiocarcinoma). The Occupational Safety and Health Administration (OSHA) has set limits to human exposure of less than an average of 1 ppm per 8-h working day per worker.

An iron compound, $FeSO_4$, in the presence of hydrogen peroxide promotes the radical polymerization of propenenitrile (acrylonitrile). **Polypropenenitrile** (poly-acrylonitrile), $-(CH_2CHCN)_n-$, also known as Orlon, is used to make fibers. Similar polymerizations of other monomers furnish Teflon and Plexiglas.

Exercise 12-30

Prior to 2005, Saran Wrap was made by radical polymerization of 1,1-dichloroethene and chloroethene together. Propose a structure. Note: This is a "copolymerization," in which the two monomers alternate in the final polymer.

Anionic polymerizations require initiation by bases

Anionic polymerizations are initiated by strong bases such as alkyllithiums, amides, alkoxides, and hydroxide. For example, methyl 2-cyanopropenoate (methyl α-cyanoacrylate) polymerizes rapidly in the presence of even small traces of hydroxide. When spread between two surfaces, it forms a tough, solid film that cements the surfaces together. For this reason, commercial preparations of this monomer are marketed as Super Glue.

What accounts for this ease of polymerization? When the base attacks the methylene group of α-cyanoacrylate, it generates a carbanion whose negative charge is located next to the nitrile and ester groups, both of which are strongly electron withdrawing. The anion is stabilized because the nitrogen and oxygen atoms polarize their multiple bonds in the sense $^{\delta+}C\equiv N^{\delta-}$ and $^{\delta+}C\equiv O^{\delta-}$ and because the charge can be delocalized by resonance.

Anionic Polymerization of Super Glue (Methyl α-Cyanoacrylate)

**Methyl
2-Cyanopropenoate**
(α-**Cyanoacrylate, Super Glue**)

**Radical
Polymerization
of Ethene**

$$n\ CH_2=CH_2$$

ROOR,
1000 atm,
>100°C

$$-(CH_2-CH_2)_n-$$

**Polyethene
(Polyethylene)**

(**Plus branched
isomers**)

**Branching in
Polyethene
(Polyethylene)**

Metal-catalyzed polymerizations produce highly regular chains

An important **metal-catalyzed polymerization** is that initiated by Ziegler-Natta* catalysts. They are typically made from titanium tetrachloride and a trialkylaluminum, such as triethylaluminum, $Al(CH_2CH_3)_3$. The system polymerizes alkenes, particularly ethene, at relatively low pressures with remarkable ease and efficiency.

Although we shall not consider the mechanism here, two features of Ziegler-Natta polymerization are the regularity with which substituted alkane chains are constructed from substituted alkenes, such as propene, and the high linearity of the chains. The polymers that result possess higher density and much greater strength than those obtained from radical polymerization. An example of this contrast is found in the properties of polyethene (polyethylene) prepared by the two methods. The chain branching that occurs during radical polymerization of ethene results in a flexible, transparent material *(low-density polyethylene)* used for food storage bags, whereas the Ziegler-Natta method produces a tough, chemically resistant plastic *(high-density polyethylene)* that may be molded into containers.

In Summary Alkenes are subject to attack by carbocations, radicals, anions, and transition metals to give polymers. In principle, any alkene can function as a monomer. The intermediates are usually formed according to the rules that govern the stability of charges and radical centers.

12-16 | ETHENE: AN IMPORTANT INDUSTRIAL FEEDSTOCK

Ethene (ethylene) can serve as a case study for the significance of alkenes in industrial chemistry. This monomer is the basis for the production of polyethene (polyethylene), millions of tons of which are manufactured in the United States annually. The major source of ethene is the pyrolysis of petroleum, or hydrocarbons derived from natural gas, such as ethane, propane, other alkanes, and cycloalkanes (Section 3-3).

Apart from its direct use as a monomer, ethene is the starting material for many other industrial chemicals. For example, acetaldehyde is obtained in the reaction of ethene with water in the presence of a palladium(II) catalyst, air, and $CuCl_2$. The product formed initially, ethenol (vinyl alcohol), is unstable and spontaneously rearranges to the aldehyde (see Chapters 13 and 18). The catalytic conversion of ethene into acetaldehyde is also known as the *Wacker*[†] *process.*

The Wacker Process

$$CH_2{=}CH_2 \xrightarrow{\text{H}_2\text{O, O}_2, \text{ catalytic PdCl}_2, \text{ CuCl}_2} CH_2{=}CH\ddot{O}H \longrightarrow \underset{\substack{\text{Acetaldehyde}}}{\overset{\displaystyle :\overset{\displaystyle \|}{O}:}{CH_3CH}}$$

Ethenol
(Vinyl alcohol)
(Unstable)

Chloroethene (vinyl chloride) is made from ethene by a chlorination–dehydrochlorination sequence in which addition of Cl_2 produces 1,2-dichloroethane. This compound is converted into the desired product by elimination of HCl.

Chloroethene (Vinyl Chloride) Synthesis

$$CH_2{=}CH_2 \xrightarrow{\text{Cl}_2} \underset{\substack{\text{1,2-Dichloroethane}}}{\overset{\displaystyle CH_2{-}CH_2}{\underset{\displaystyle :\ddot{Cl}: \quad :\ddot{Cl}:}{\big|\qquad\big|}}} \xrightarrow[-\text{HCl}]{\Delta} \underset{\substack{\text{Chloroethene} \\ \text{(Vinyl chloride)}}}{CH_2{=}CH\ddot{Cl}:}$$

Really? Ethene is a natural plant and fruit hormone, involved in the regulation of the maturing of fruit, opening of flowers, shedding of leaves, and chemical defense. The gas is used at the parts-per-million level in commercial ripening chambers. Ancient cultures gashed fruits (which induces ethene production) or exposed them to burning incense (incomplete combustion generates ethene) to hasten maturation. The compound emanates from all parts of higher plants, including leaves, stems, and roots. You can test this at home: If the bananas you bought are too green, place them overnight in a paper bag with an apple or tomato. They will be yellow in the morning.

Five stages of banana ripening.

---

*Professor Karl Ziegler (1898–1973), Max Planck Institute for Coal Research, Mülheim, Germany, Nobel Prize 1963 (chemistry); Professor Giulio Natta (1903–1979), Polytechnic Institute of Milan, Nobel Prize 1963 (chemistry).

[†]Dr. Alexander Wacker (1846–1922), Wacker Chemical Company, Munich, Germany.

Oxidation of ethene with oxygen in the presence of silver furnishes oxacyclopropane (ethylene oxide), the hydrolysis of which gives 1,2-ethanediol (ethylene glycol) (Section 9-11). Hydration of ethene gives ethanol (Section 9-11).

$$CH_2=CH_2 \xrightarrow{O_2,\ catalytic\ Ag} H_2C\overset{\overset{\ddot{O}}{\diagup\!\diagdown}}{-}CH_2 \xrightarrow{H^+,\ H_2O} \overset{:\ddot{O}H\ \ :\ddot{O}H}{\underset{CH_2-CH_2}{|\ \ \ \ \ |}}$$

Oxacyclopropane   **1,2-Ethanediol**
(Ethylene oxide)   **(Ethylene glycol)**

In Summary Ethene is a valuable source of various industrial raw materials, including ethanol, 1,2-ethanediol (ethylene glycol), and several important monomers for the polymer industry.

12-17 | ALKENES IN NATURE: INSECT PHEROMONES

Many natural products contain π bonds; several were mentioned in Sections 4-7 and 9-11. This section describes a specific group of naturally occurring alkenes, the **insect pheromones** (*pherein*, Greek, to bear; *hormon*, Greek, to stimulate).

Insect Pheromones

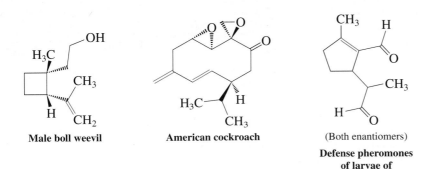

European vine moth   **Japanese beetle**

Male boll weevil   **American cockroach**   (Both enantiomers)

Defense pheromones of larvae of chrysomelid beetle

This pheromone trap removes the males of the pea moth *Cydia nigricana* from circulation.

Bombykol

Pheromones are chemical substances used for communication within a living species. There are sex, trail, alarm, and defense pheromones, to mention a few. Many insect phero- mones are simple alkenes; they are isolated by extraction of certain parts of the insect and separation of the resulting product mixture by chromatographic techniques. Often only minute quantities of the bioactive compound can be obtained, in which case the synthetic organic chemist can play a very important role in the design and execution of total syntheses. Inter- estingly, the specific activity of a pheromone frequently depends on the configuration around the double bond (e.g., *E* or *Z*), as well as on the absolute configuration of any chiral centers present (*R*, *S*) *and* the composition of isomer mixtures. For example, the sex attractant for the male silkworm moth, 10-*trans*-12-*cis*-hexadecadien-1-ol (known as bombykol, margin), is 10 billion times more active in eliciting a response than the 10-*cis*-12-*trans* isomer, and 10 *trillion* times more active than the trans, trans compound.

REAL LIFE: MEDICINE 12-3 | Alkene Metathesis Transposes the Termini of Two Alkenes: Construction of Rings

One of the most startling examples of metal-catalyzed chemistry is alkene metathesis, a reaction in which two alkenes exchange their double-bonded carbon atoms, as shown in general form here.

$$
\begin{array}{c}
W_2C{=}CX_2 \\
+ \\
Y_2C{=}CZ_2
\end{array}
\xrightleftharpoons{\text{Catalyst}}
\begin{array}{c}
CW_2 \\
\| \\
CY_2
\end{array}
+
\begin{array}{c}
CX_2 \\
\| \\
CZ_2
\end{array}
$$

The equilibrium in this reversible process may be driven by removal of one of the four components (Le Chatelier's principle). This idea has been employed to form medium- and large-sized rings that are otherwise very difficult to construct, because of unfavorable strain and entropy factors. The example below illustrates ring closure of an acyclic starting compound with terminal double bonds. The products are a cyclic alkene and ethene, which, being a gas, evolves rapidly out of the reaction mixture, driving the equilibrium toward product.

$$
\left(
\begin{array}{c}
C{=}CH_2 \\
H \\
H \\
C{=}CH_2
\end{array}
\right.
\xrightleftharpoons{\text{Catalyst}}
\left(
\begin{array}{c}
H \\
C \\
\| \\
C \\
H
\end{array}
\right.
+
\begin{array}{c}
CH_2 \\
\| \\
CH_2
\end{array}
$$

A recent, extraordinary application of methathesis is the synthesis of ciguatoxin by Hirama.* Ciguatoxin is produced by marine microorganisms associated with algae and is accumulated by some 400 species of warm-water reef fish.

*Professor Masahiro Hirama (b. 1948), Tohoku University, Japan.

Ciguatoxin is 100 times more poisonous than brevetoxin, the "red tide" toxin (Section 9-5), and is responsible for more human poisoning from seafood consumption than any other substance: More than 20,000 people become sick in this way each year, developing gastrointestinal, cardiovascular, and neurological abnormalities that can lead to paralysis, coma, and death.

The tiny amounts of ciguatoxin in fish are far too small to have any effect on the flavor or odor of the food. Thus, a supply of this material is needed to develop sensitive detection methods. Ciguatoxin contains 13 ether rings and 30 stereocenters, making it a formidable synthetic target. Its synthesis was an amazing 12-year effort, culminating in the linkage of two polycyclic molecules containing five and seven rings, respectively, and closure of the final nine-membered "F" ring using alkene metathesis. The catalyst, developed by

The coral grouper can turn lethal when ciguatoxin accumulates in its tissue.

As is the case for the juvenile hormones (Real Life 12-1), pheromone research affords an important opportunity for achieving pest control. Minute quantities of sex pheromones can be used per acre of land to confuse male insects about the location of their female partners. These pheromones can thus serve as lures in traps to remove insects effectively without spraying crops with large amounts of other chemicals. It is clear that organic chemists in collaboration with insect biologists will make important contributions in this area in the years to come.

THE BIG PICTURE

We saw in Chapter 11 that in the double bonds of alkenes the π bond is weaker than the σ bond. Thus, the general process of addition, in which π bonds are replaced by σ bonds, is generally energetically favorable. We also found that although the C=C bond is nonpolar, it has a higher concentration of electrons than does a single bond. These two features are the basis of the reactivity of the alkene function. Thus, the alkene double bond acts as a nucleophile, and many of its reactions begin with attack by an electrophile on the π electrons.

In this chapter, we found that alkene chemistry subdivides into several categories, including reactions with simple electrophiles, such as the proton, to give carbocations, followed by

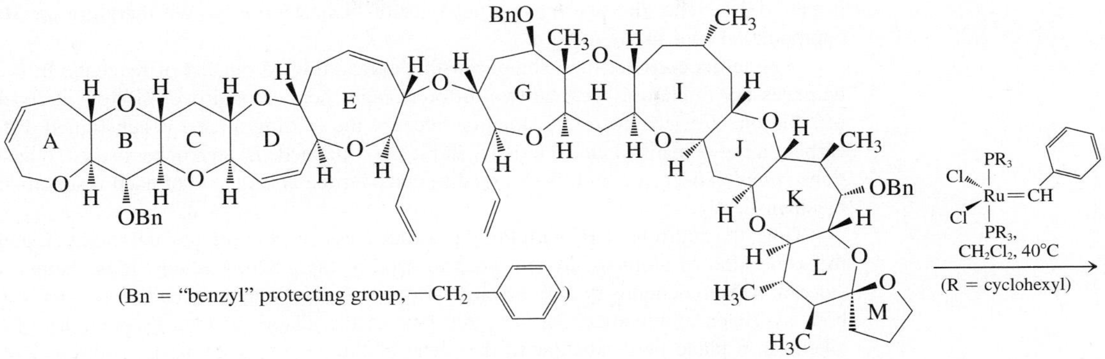

(Bn = "benzyl" protecting group, —CH₂—)

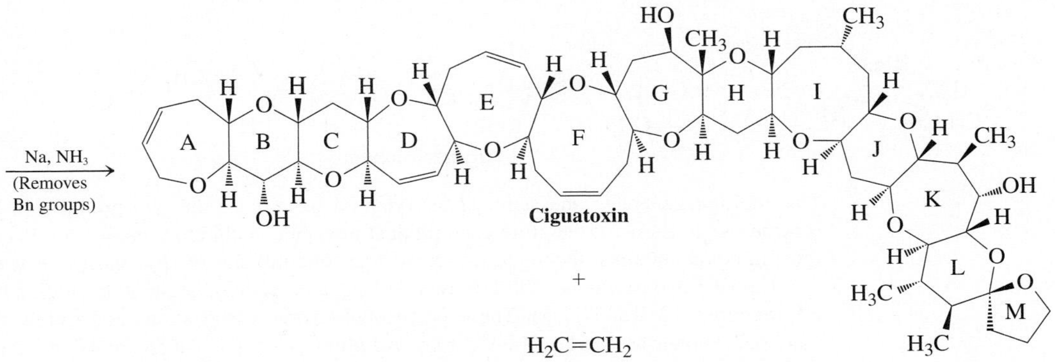

Na, NH₃
(Removes
Bn groups)

Ciguatoxin

+

H₂C=CH₂

Grubbs,* contains a ruthenium atom linked by a double bond to carbon, an example of a so-called metal carbene complex. Metal carbenes were proposed (by Chauvin)[†] and confirmed as intermediates in alkene metathesis decades ago. Subsequently,

Grubbs and Schrock[‡] prepared stable carbenes of Ru and Mo, respectively, that catalyze the process in a well-controlled manner. Alkene metathesis is now one of the most reliable and widely used methods for medium- and large-ring construction.

---

*Professor Robert H. Grubbs (b. 1942), California Institute of Technology, Nobel Prize 2005 (chemistry).
[†]Professor Yves Chauvin (b. 1930), Institut Français du Pétrole, Rueil-Malmaison, France, Nobel Prize 2005 (chemistry).
[‡]Professor Richard R. Schrock (b. 1945), Massachusetts Institute of Technology, Nobel Prize 2005 (chemistry).

---

attachment of a nucleophile to give the final addition product. These processes are very much like the familiar acid-base reactivity we discussed in Chapter 2. Alkenes also undergo reactions that proceed through more complex pathways, often involving ring formation. Issues of regio- and stereochemistry apply, depending on the specific mechanism, but the underlying theme of nucleophile–electrophile interaction is almost always present.

Many of the same ideas in alkene chemistry also apply to the carbon–carbon triple bond. In Chapter 13, we shall examine the behavior of alkynes and see that much of their chemistry is a direct extension of alkene reactivity to a system with two π bonds.

WORKED EXAMPLES: INTEGRATING THE CONCEPTS

12.31. Reviewing Reactions of Alkenes

Compare and contrast the addition reactions of each of the following reagents with (*E*)-3-methyl-3-hexene: H_2 (catalyzed by PtO_2), HBr, dilute aqueous H_2SO_4, Br_2 in CCl_4, mercuric acetate in H_2O, and B_2H_6 in THF. Consider regiochemistry and stereochemistry. Which of these reactions can be used to synthesize alcohols? In what respects do the resulting alcohols differ?

SOLUTION

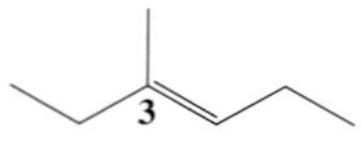

First, we need to identify the starting material's structure. Recall (Section 11-1) that the designation "*E*" describes the stereoisomer in which the two highest-priority groups (according to Cahn-Ingold-Prelog guidelines, Section 5-3) are on opposite sides of the double bond (i.e., trans to each other). In 3-methyl-3-hexene, the two ethyl groups are of highest priority. We therefore are starting with the compound shown in the margin.

Now let us consider the behavior of this alkene toward our list of reagents. In each case, it may be necessary to choose between two different regiochemical and two different stereochemical modes of addition. We recognize this situation because the starting alkene is substituted differently at each carbon, a regiochemical characteristic, and it has a defined (*E*) stereochemistry. To answer such questions *completely* correctly, it is essential to consider the *mechanism* of each reaction—that is, to *think* mechanistically.

Thus, the addition of H_2 with PtO_2 as a catalyst is an example of catalytic hydrogenation. Because the same kind of atom (hydrogen) adds to each of the alkene carbons, regiochemistry is not a consideration. Stereochemistry may be one, however. Catalytic hydrogenation is a *syn* addition, in which both hydrogen atoms attach to the same face of the alkene π bond (Section 12-2). If we view the alkene in a plane perpendicular to the plane of the page (Figure 12-1), addition will be on the top face 50% of the time and on the bottom face 50% of the time:

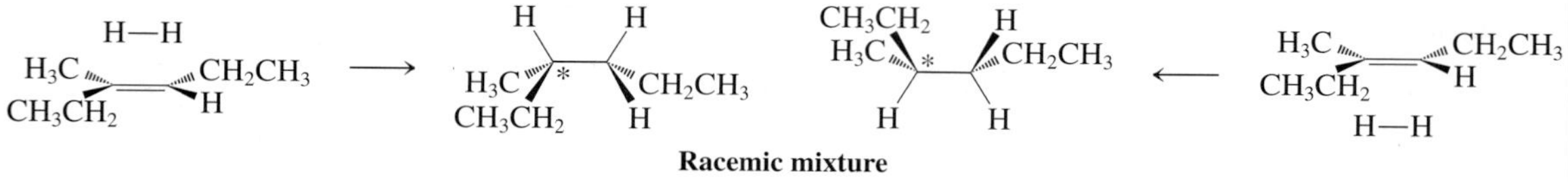

Racemic mixture

The addition generates one stereocenter (marked in each of the two products in the center of the scheme by an asterisk); therefore each product molecule is chiral (Section 5-1). Because the products form in equal amounts, the consequence is a racemic mixture of (*R*)- and (*S*)-3-methylhexane.

The next two reactions, with HBr and with aqueous H_2SO_4, begin with addition of the electrophile H^+ (Sections 12-3 and 12-4). These processes generate carbocations and follow the regiochemical guideline known as Markovnikov's rule: Addition is by attachment of H^+ to the less substituted alkenyl carbon to give the more stable carbocation. The carbocation is trapped by any available nucleophile—Br^- in the case of HBr, and H_2O in the case of aqueous H_2SO_4. Both steps proceed without stereoselectivity and, because the carbocation formed is already tertiary, rearrangement to a more stable carbocation is not possible. So we have the following result:

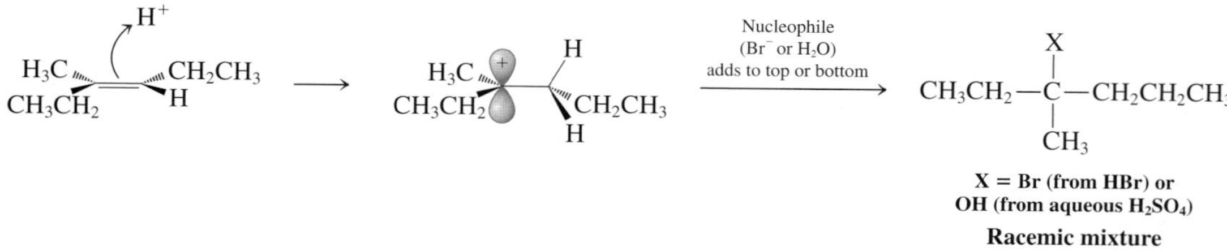

X = Br (from HBr) or
OH (from aqueous H_2SO_4)
Racemic mixture

The next two are examples of additions of electrophiles that form bridged cationic intermediates: a cyclic bromonium ion in the first case (Section 12-5) and a cyclic mercurinium ion in the second (Section 12-7). Additions therefore proceed stereospecifically *anti,* because the cyclic ion can be attacked only from the direction opposite the location of the bridging electrophile (Figure 12-3). With Br_2, identical atoms add to both alkene carbons, so regiochemistry is not a consideration. In oxymercuration, the nucleophile is water, and it will add to the most substituted alkene carbon, because the latter is tertiary and possesses the greatest partial positive charge. Stereochemistry must be considered because addition of the electrophile occurs with equal probability from the top and from the bottom and stereocenters are created. So we have

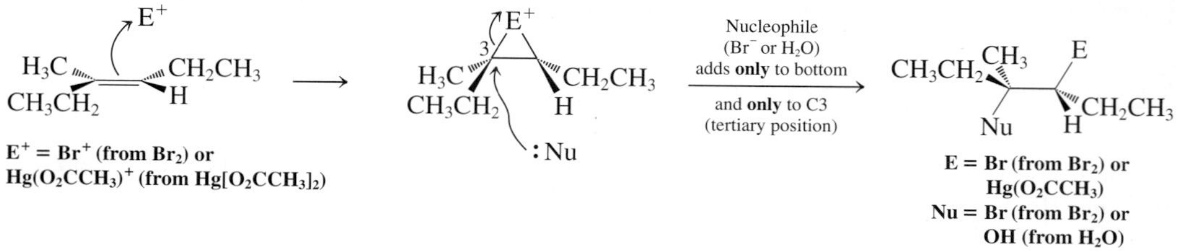

E^+ = Br^+ (from Br_2) or
$Hg(O_2CCH_3)^+$ (from $Hg[O_2CCH_3]_2$)

Nucleophile
(Br^- or H_2O)
adds **only** to bottom
and **only** to C3
(tertiary position)

E = Br (from Br_2) or
$Hg(O_2CCH_3)$
Nu = Br (from Br_2) or
OH (from H_2O)

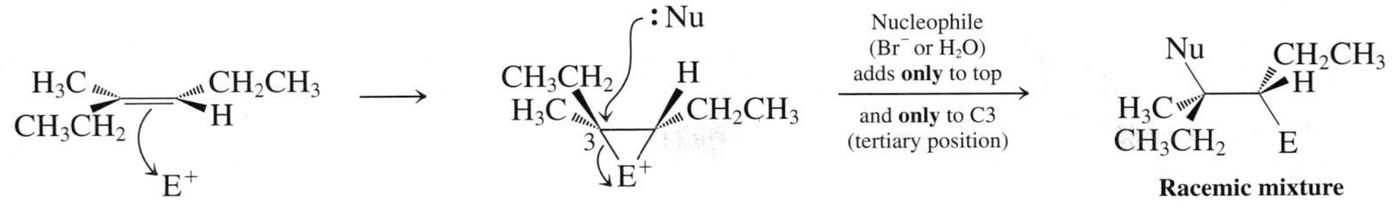

Equal amounts—a racemic mixture—of the two chiral, enantiomerically related products are formed.

Finally, we come to hydroboration. Again, both regio- and stereoselectivity must be considered. As in hydrogenation, the stereochemistry is *syn;* unlike the preceding electrophilic additions, the regiochemistry is anti-Markovnikov: Boron attaches to the *less* substituted alkene carbon:

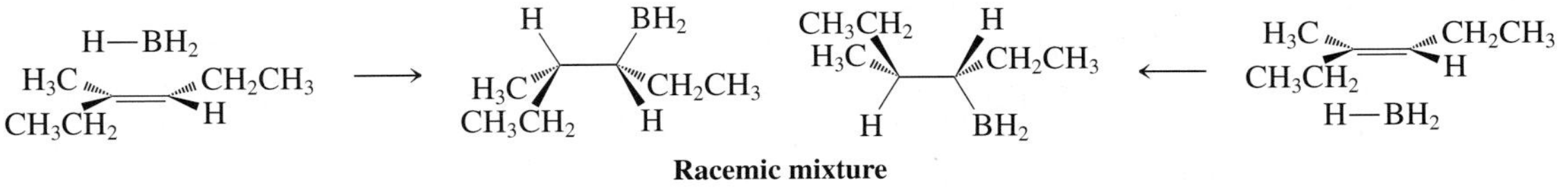

For simplicity, addition of only one of the B–H bonds is shown. The picture strongly resembles that of hydrogenation, but, because of the unsymmetric nature of the reagent, both alkene carbons are now transformed into stereocenters.

Three of the six reactions are well suited for the synthesis of alcohols: acid-catalyzed hydration, oxymercuration (after reduction of the C–Hg bond by $NaBH_4$), and hydroboration (by oxidation of the C–B bond by H_2O_2). Compare the alcohol from hydration (shown earlier) with that from demercuration of the oxymercuration product:

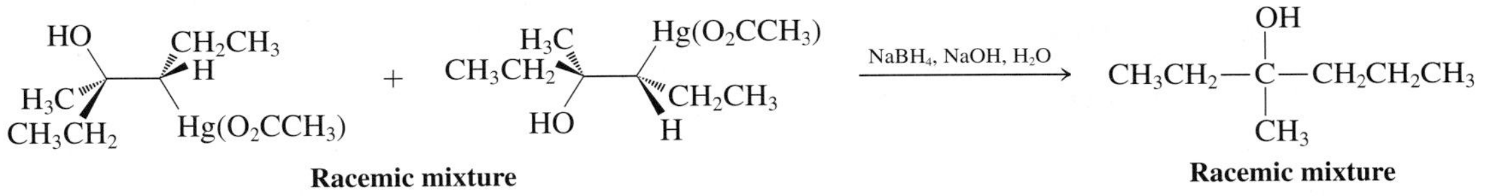

They are the same. Had a rearrangement taken place during hydration, this outcome might not have been the case (Section 9-3).

Oxidation of the hydroboration product gives a different, regioisomeric alcohol having two stereocenters, also as a racemic mixture (the boron atoms are shown without the additional alkyl groups):

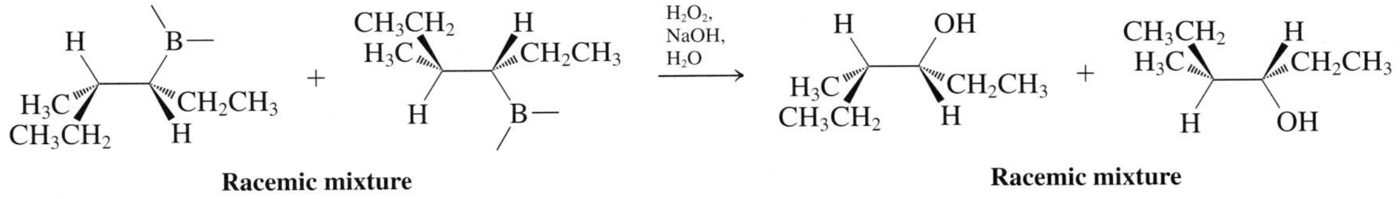

12.32. Using Spectroscopy to Determine the Product of a Sequence of Reactions

Thujene, which occurs in numerous plant oils, possesses the molecular formula $C_{10}H_{16}$ and is a monoterpene (Section 4-7). Several chemical and spectroscopic characteristics of thujene follow. What does each of these pieces of information tell you about the structure of thujene? (i) Thujene reacts instantly with one equivalent of $KMnO_4$ in aqueous solution to discharge the purple color of permanganate and form a brown precipitate. Additional $KMnO_4$ is not decolorized. (ii) Hydroboration–oxidation of thujene forms a compound $C_{10}H_{18}O$, called *thujyl alcohol,* whose ^1H NMR spectrum shows a 1 H signal at $\delta = 3.40$ ppm. (iii) Oxymercuration–demercuration of thujene forms a different alcohol $C_{10}H_{18}O$, whose ^1H NMR spectrum shows no signals downfield of $\delta = 3$ ppm. (iv) Ozonolysis of thujene gives

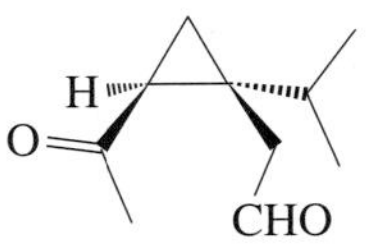

SOLUTION

Beginning with the molecular formula, we may ascertain that thujene possesses three degrees of unsaturation (Section 11-11): $[(2 \times 10 + 2) - 16]/2 = 6/2 = 3$. (i) The fact that thujene reacts with only one equivalent of $KMnO_4$ implies that only one of these unsaturations is a π bond (Section 12-11); the other two must be rings. (ii) The NMR spectrum of thujyl alcohol is consistent only with a secondary alcohol:

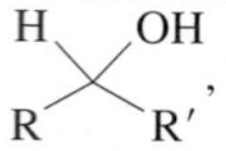

$$\underset{R \quad R'}{\overset{H \quad OH}{\diagdown / \diagup}}$$

, because the signal in the region between $\delta = 3$ and 4 integrates for only one hydrogen. (iii) In contrast, the oxymercuration product lacks an NMR signal in this region, meaning that this product cannot have a hydrogen on the HO-bearing carbon: In other words, it must be a tertiary alcohol. Taken together, these three pieces of information tell us that thujene has two rings and a trisubstituted double bond,

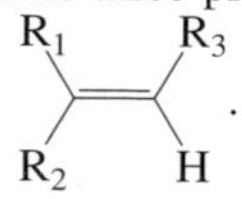

$$\underset{R_2 \quad H}{\overset{R_1 \quad R_3}{\diagdown / \diagup}}$$

. The result of ozonolysis completes the picture, because it gives a product with 10 carbons. Therefore, this product is simply thujene with its alkene function cleaved into two separate C=O units, one being an aldehyde and the other a ketone. Reversing this process, we have

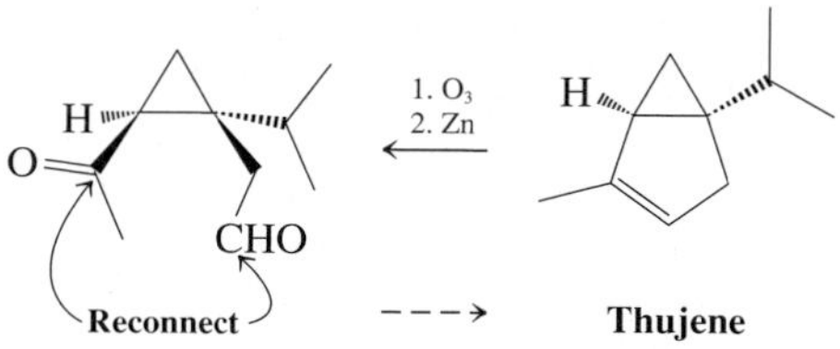

New Reactions

1. General Addition to Alkenes (Section 12-1)

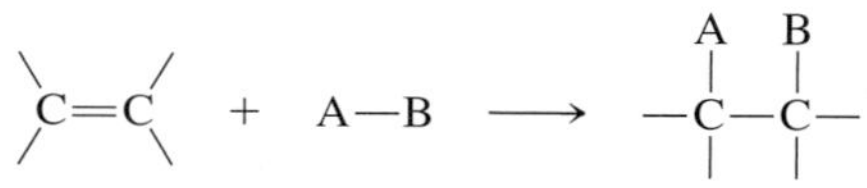

2. Hydrogenation (Section 12-2)

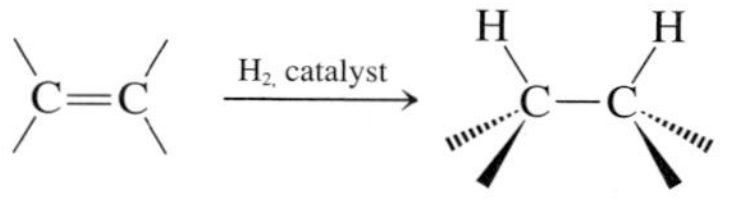

Syn addition

Typical catalysts: PtO_2, Pd–C, Ra–Ni

Electrophilic Additions

3. Hydrohalogenation (Section 12-3)

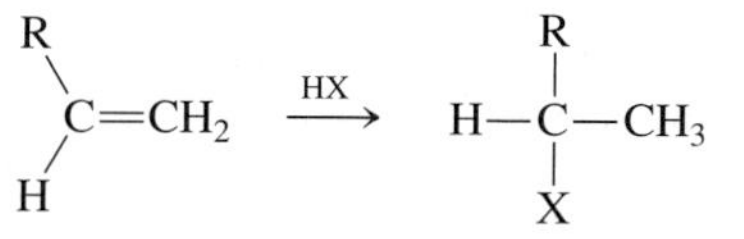

Regiospecific
(Markovnikov rule)

Through more stable carbocation

4. Hydration (Section 12-4)

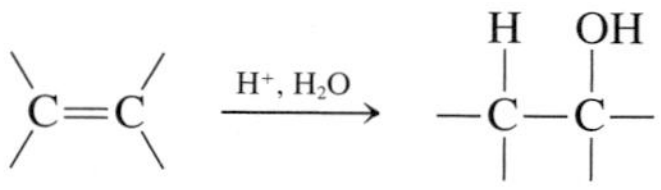

Through more stable carbocation

5. Halogenation (Section 12-5)

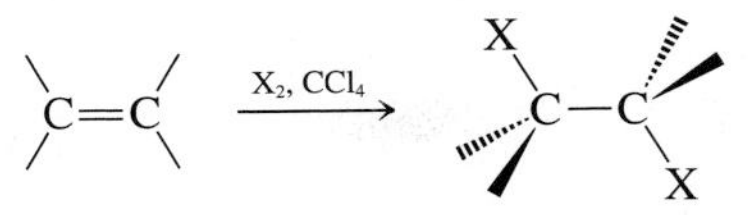

Stereospecific (*anti*)

$X_2 = Cl_2$ or Br_2, but not I_2

6. Vicinal Haloalcohol Synthesis (Section 12-6)

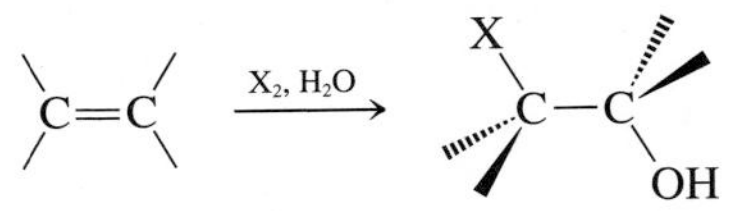

OH attaches to more substituted carbon

7. Vicinal Haloether Synthesis (Section 12-6)

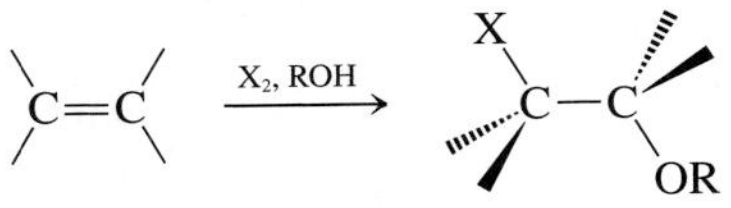

OR attaches to more substituted carbon

8. General Electrophilic Additions (Section 12-6, Table 12-2)

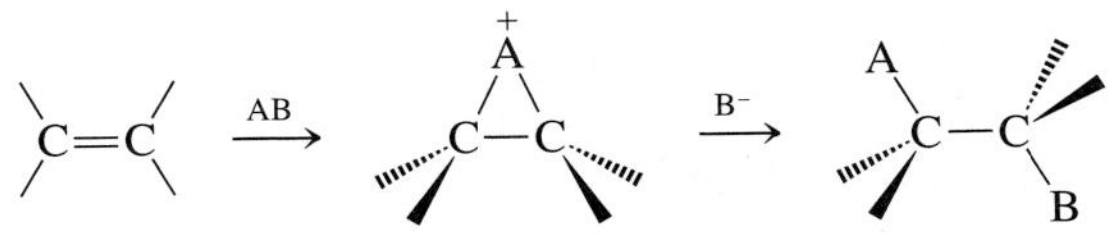

A = electropositive, B = electronegative
B attaches to more substituted carbon

9. Oxymercuration–Demercuration (Section 12-7)

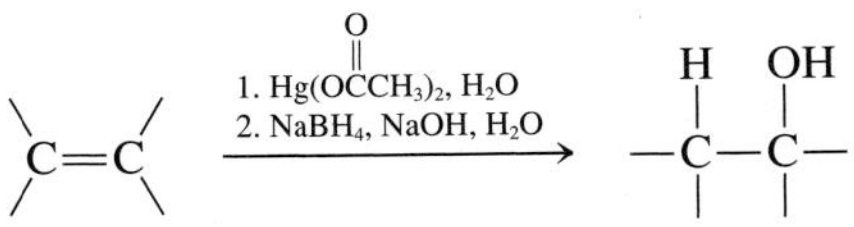

Initial addition is *anti*, through mercurinium ion 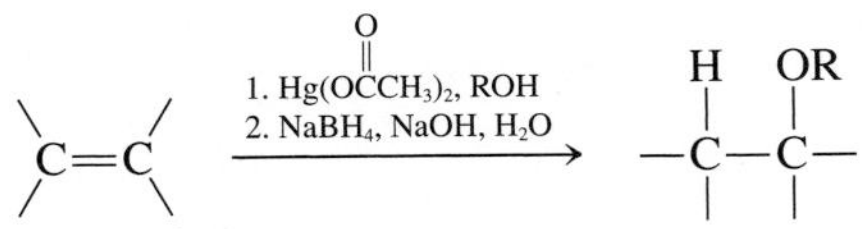   **OH or OR attaches to more substituted carbon**

10. Hydroboration (Section 12-8)

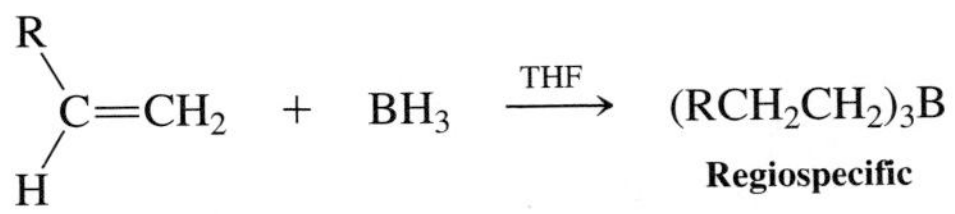

B attaches to less substituted carbon

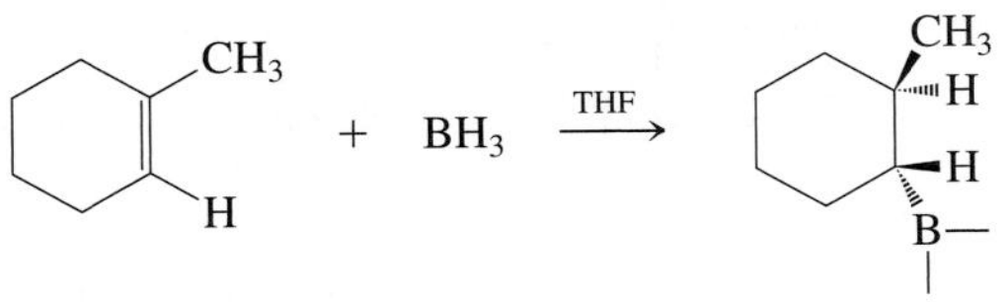

Regiospecific

Stereospecific (*syn*)
and anti-Markovnikov

11. Hydroboration–Oxidation (Section 12-8)

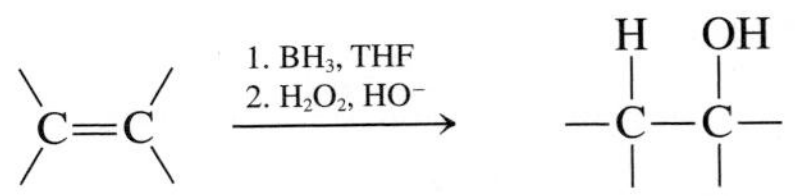

Stereospecific (*syn*)
and anti-Markovnikov

OH attaches to less substituted carbon

12. Carbene Addition for Cyclopropane Synthesis (Section 12-9)

Using diazomethane:

$$R\text{—}CH=CH\text{—}R' \ + \ CH_2N_2 \ \xrightarrow{\ h\nu \text{ or } \Delta \text{ or Cu}\ } \ \triangle \overset{}{\underset{R \quad R'}{}}$$

Stereospecific

Other sources of carbenes or carbenoids:

$$CHCl_3 \ \xrightarrow{\text{Base}} \ :CCl_2 \qquad CH_2I_2 \ \xrightarrow{\text{Zn–Cu}} \ ICH_2ZnI$$

Oxidation

13. Oxacyclopropane Formation (Section 12-10)

$$C=C \ \xrightarrow[\text{RCOOH, CH}_2\text{Cl}_2]{} \ \underset{\triangle}{C\text{—}C} \ + \ RCOH$$

Stereospecific (*syn*)

14. Vicinal *Anti* Dihydroxylation (Section 12-10)

$$C=C \ \xrightarrow[\substack{1.\ \text{RCOOH, CH}_2\text{Cl}_2 \\ 2.\ \text{H}^+,\ \text{H}_2\text{O}}]{} \ HO\text{—}C\text{—}C\text{—}OH \ + \ RCOH$$

15. Vicinal *Syn* Dihydroxylation (Section 12-11)

$$C=C \ \xrightarrow[]{1.\ \text{OsO}_4,\ 2.\ \text{H}_2\text{S};\ \text{or catalytic OsO}_4,\ \text{H}_2\text{O}_2} \ HO\text{—}C\text{—}C\text{—}OH$$

Through cyclic intermediates

16. Ozonolysis (Section 12-12)

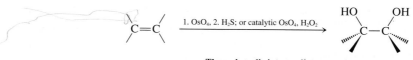

$$C=C \ \xrightarrow[\substack{1.\ \text{O}_3,\ \text{CH}_3\text{OH} \\ 2.\ (\text{CH}_3)_2\text{S};\ \text{or Zn, CH}_3\text{COH}}]{} \ C=O \ + \ O=C$$

Through molozonide and ozonide intermediates

Radical Additions

17. Radical Hydrobromination (Section 12-13)

$$C=CH_2 \ \xrightarrow[]{\text{HBr, ROOR}} \ \underset{H}{\overset{H \quad Br}{\text{—}C\text{—}C\text{—}H}}$$

Anti-Markovnikov
Does not occur with HCl or HI

18. Other Radical Additions (Section 12-13)

$$C=C \ \xrightarrow[]{\text{RSH, ROOR}} \ \overset{H \quad SR}{\text{—}C\text{—}C\text{—}}$$

Anti-Markovnikov

Monomers and Polymers

19. Dimerization, Oligomerization, and Polymerization (Sections 12-14 and 12-15)

$$n \quad \diagdown C{=}C \diagup \quad \xrightarrow{\text{H}^+ \text{ or RO}\cdot \text{ or B}^-} \quad -(\overset{|}{\underset{|}{C}}-\overset{|}{\underset{|}{C}})_n-$$

Important Concepts

1. The reactivity of the double bond manifests itself in exothermic **addition** reactions leading to **saturated** products.

2. The **hydrogenation** of alkenes is immeasurably slow unless a **catalyst** capable of splitting the strong H–H bond is used. Possible catalysts are palladium on carbon, platinum (as PtO_2), and Raney nickel. Addition of hydrogen is subject to steric control, the least hindered face of the least substituted double bond frequently being attacked preferentially.

3. As a Lewis base, the π bond is subject to attack by acid and **electrophiles,** such as H^+, X_2, and Hg^{2+}. If the initial intermediate is a free **carbocation,** the more highly substituted carbocation is formed. Alternatively, a **cyclic onium ion** is generated subject to nucleophilic ring opening at the more substituted carbon. Carbocation formation leads to control of regiochemistry (**Markovnikov rule**); onium ion formation leads to control of both regio- and stereochemistry.

4. Mechanistically, **hydroboration** lies between hydrogenation and electrophilic addition. The first step is π complexation to the electron-deficient boron, whereas the second is a concerted transfer of the hydrogen to carbon. **Hydroboration–oxidation** results in the **anti-Markovnikov hydration** of alkenes.

5. **Carbenes** and carbenoids are useful for the synthesis of **cyclopropanes** from alkenes.

6. **Peroxycarboxylic acids** may be thought of as containing an electrophilic oxygen atom, transferable to alkenes to give **oxacyclopropanes.** The process is often called *epoxidation*.

7. Osmium tetroxide acts as an electrophilic oxidant of alkenes; in the course of the reaction, the oxidation state of the metal is reduced by two units. Addition takes place in a concerted *syn* manner through cyclic six-electron transition states to give vicinal diols.

8. **Ozonolysis** followed by reduction yields carbonyl compounds derived by cleavage of the double bond.

9. In **radical chain additions** to alkenes, the chain carrier adds to the π bond to create the more highly substituted radical. This method allows for the anti-Markovnikov hydrobromination of alkenes, as well as the addition of thiols and some halomethanes.

10. Alkenes react with themselves through initiation by charged species, radicals, or some transition metals to give **polymers.** The initial attack at the double bond yields a reactive intermediate that perpetuates carbon–carbon bond formation.

Problems

33. With the help of the $DH°$ values given in Tables 3-1 and 3-4, calculate the $\Delta H°$ values for addition of each of the following molecules to ethene, using 65 kcal mol^{-1} for the carbon–carbon π bond strength.

(a) Cl_2

(b) IF ($DH° = 67$ kcal mol^{-1})

(c) IBr ($DH° = 43$ kcal mol^{-1})

(d) HF

(e) HI

(f) HO–Cl ($DH° = 60$ kcal mol^{-1})

(g) Br–CN ($DH° = 83$ kcal mol^{-1}; $DH°$ for C_{sp3}–CN $= 124$ kcal mol^{-1})

(h) CH_3S–H ($DH° = 88$ kcal mol^{-1}; $DH°$ for C_{sp3}–S $= 60$ kcal mol^{-1})

34. The bicyclic alkene car-3-ene, a constituent of turpentine, undergoes catalytic hydrogenation to give only one of the two possible stereoisomeric products. The product has the common name *cis*-carane, indicating that the methyl group and the cyclopropane ring are on the same face of the cyclohexane ring. Suggest an explanation for this stereochemical outcome.

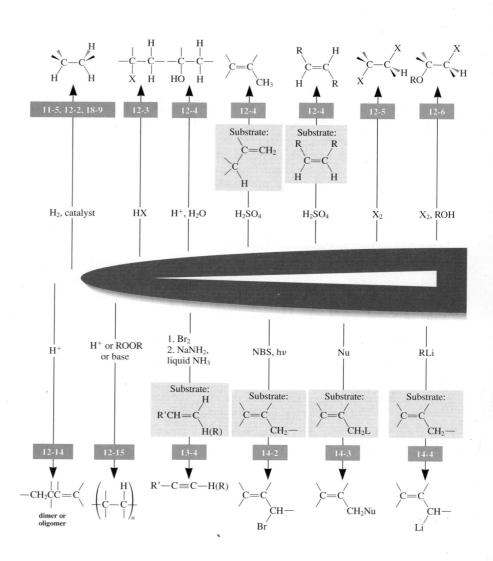

35. Give the expected major product of catalytic hydrogenation of each of the following alkenes. Clearly show and explain the stereochemistry of the resulting molecules.

(a)   **(b)**   **(c)**

36. Would you expect the catalytic hydrogenation of a small-ring cyclic alkene such as cyclobutene to be more or less exothermic than that of cyclohexene? (**Hint:** Which has more bond-angle strain, cyclobutene or cyclobutane?)

37. Give the expected major product from the reaction of each alkene with (i) peroxide-free HBr and (ii) HBr in the presence of peroxides.

(**a**) 1-Hexene; (**b**) 2-methyl-1-pentene; (**c**) 2-methyl-2-pentene; (**d**) (Z)-3-hexene; (**e**) cyclohexene.

38. Give the product of addition to Br_2 to each alkene in Problem 37. Pay attention to stereochemistry.

39. What alcohol would be obtained from treatment of each alkene in Problem 37 with aqueous sulfuric acid? Would any

section number

of these alkenes give a different product upon oxymercuration–demercuration? Upon hydroboration–oxidation?

40. Give the reagents and conditions necessary for each of the following transformations and comment on the thermodynamics of each (a) cyclohexanol → cyclohexene; (b) cyclohexene → cyclohexanol; (c) chlorocyclopentane → cyclopentene; (d) cyclopentene → chlorocyclopentane.

41. Problem 51 of Chapter 6 presented a strategy for the synthesis of the amino acid (2S,3S)-3-hydroxyleucine, requiring as the starting material a specific stereoisomer of 2-bromo-3-hydroxy-4-methylpentanoic acid. Addition of bromine and water to the methyl ester of 4-methyl-2-pentenoic acid (below) gives the corresponding ester of 2-bromo-3-hydroxy-4-methylpentanoic acid. (a) Which stereoisomer of the unsaturated ester, cis or trans, is needed for this addition to give the necessary stereoisomer of the product? (b) Can this strategy give the bromoalcohol as a single enantiomer or not? Explain, mechanistically.

$$(CH_3)_2CHCH{=}CHCO_2CH_3$$

**4-Methyl-2-pentenoic acid
methyl ester**

42. Formulate the product(s) that you would expect from each of the following reactions. Show stereochemistry clearly.

(a) 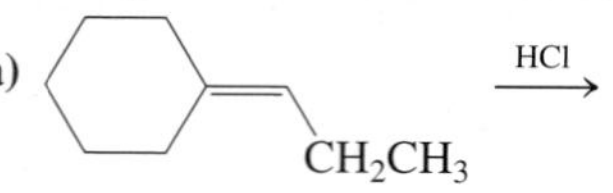 [cyclohexane with =CH-CH₂CH₃ substituent] $\xrightarrow{HCl}$

(b) *trans*-3-Heptene $\xrightarrow{Cl_2}$

(c) 1-Ethylcyclohexene $\xrightarrow{Br_2,H_2O}$

(d) Product of (c) $\xrightarrow{NaOH, H_2O}$

(e) 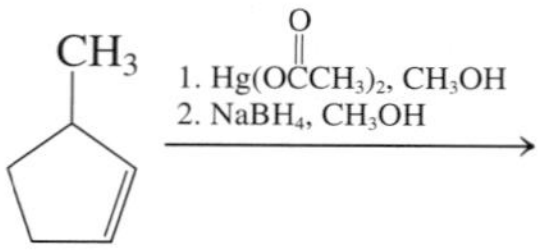 [methylcyclopentene structure] $\xrightarrow[\text{2. NaBH}_4,\text{ CH}_3\text{OH}]{\text{1. Hg(OCCH}_3)_2,\text{ CH}_3\text{OH}}$

(f) *cis*-2-Butene $\xrightarrow{Br_2,\text{ excess Na}^+N_3^-}$

(g) 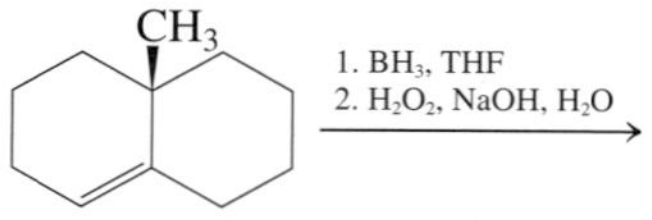 [decalin structure with CH₃] $\xrightarrow[\text{2. H}_2\text{O}_2,\text{ NaOH, H}_2\text{O}]{\text{1. BH}_3,\text{ THF}}$

43. Show how you would synthesize each of the following molecules from an alkene of appropriate structure (your choice).

(a) 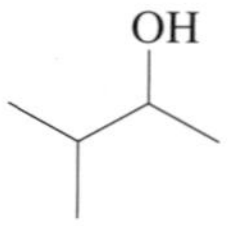 [structure with OH]

(b) 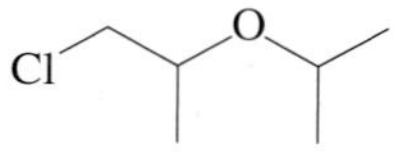 Cl [structure with O]

(c) 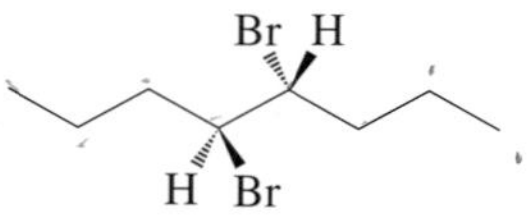 [structure with Br H / H Br] (*meso–4R,5S–isomer*)

(d) 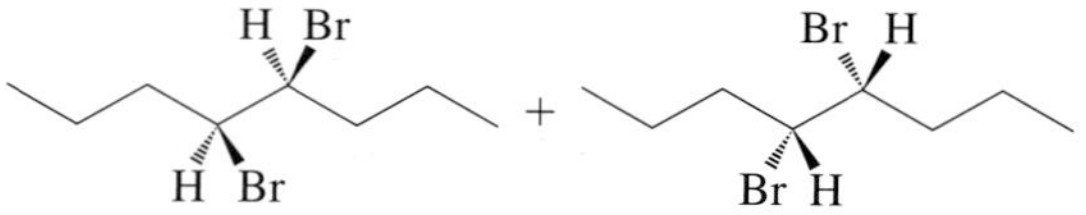 [structure H Br / H Br] + [structure Br H / Br H] (Racemate of 4R,5R and 4S,5S isomers)

(e) 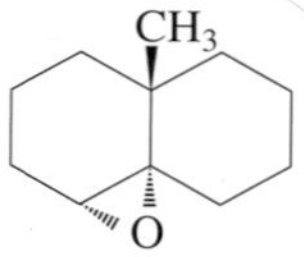 [decalin with CH₃ and O]

(f) 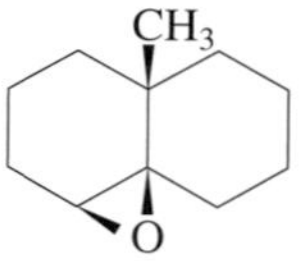 [decalin with CH₃ and O] (More challenging. **Hint:** See Section 12-6.)

44. Propose efficient methods for accomplishing each of the following transformations. Most will require more than one step.

(a) 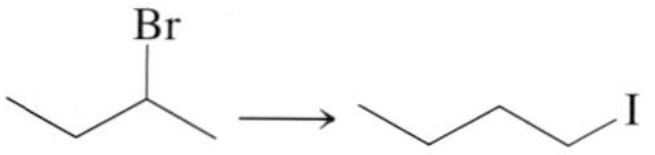 [structure with Br] ⟶ [structure with I]

(b) 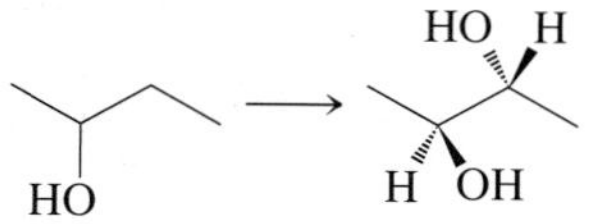 [structure with HO] ⟶ [structure with HO H / H OH] (*meso–2R,3S–isomer*)

(c) 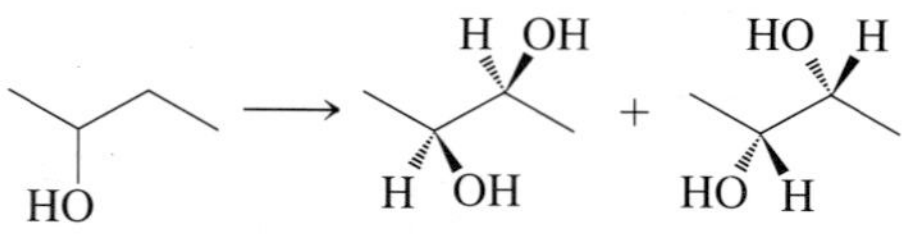 [structure with HO] ⟶ [structure H OH / H OH] + [structure HO H / HO H] (Racemate of 2R,3R and 2S,3S isomers)

(d) 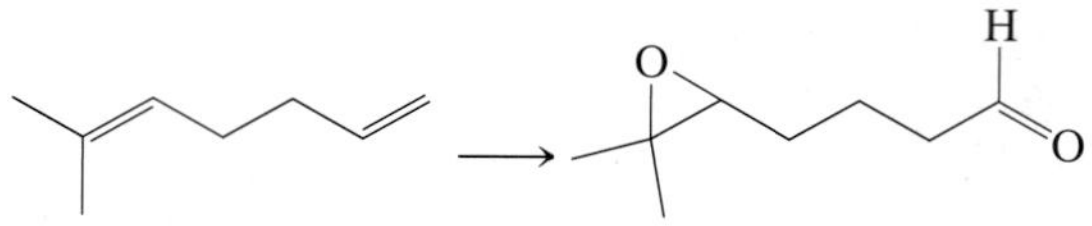 [structure] ⟶ [epoxy aldehyde structure]

45. Reaction review. Without consulting the Reaction Road Map on pp. 532–533, suggest a reagent to convert a general alkene,

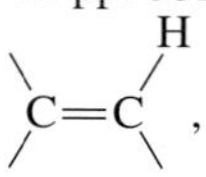

$$\text{C}=\text{C} \quad \text{(with H)}$$

, into each of the following types of compounds.

(a) 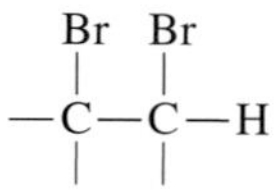 —C—C—H (Br, Br)

(b) 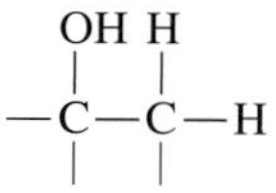 —C—C—H (OH, H) (Markovnikov product)

(c) 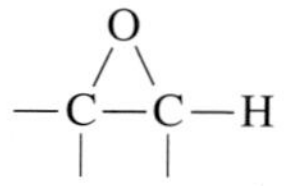 —C—C—H (epoxide, O)

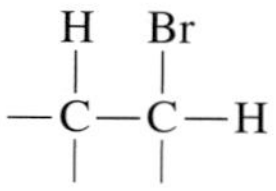

(d) 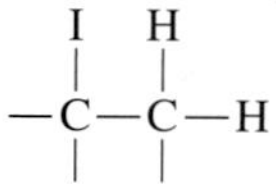 —C—C—H (I, H)

(e) 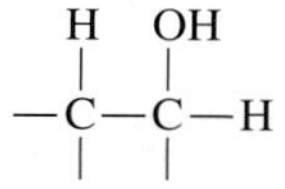 —C—C—H (H, OH) (anti-Markovnikov product)

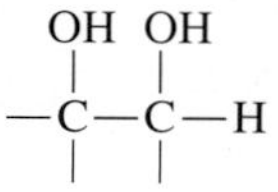

(f) 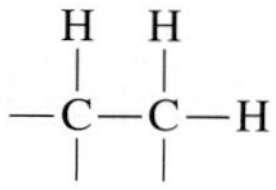 —C—C—H (H, H)

(g) —C—C—H (OH, OH)

(h) —C—C—H (H, Br) (anti-Markovnikov product)

(i) 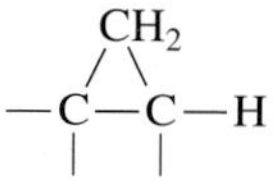 —C—C—H (CH₂)

(j) 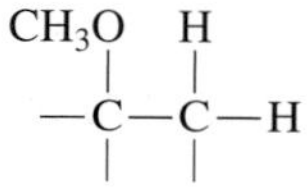 —C—C—H (CH₃O, H)

(k) 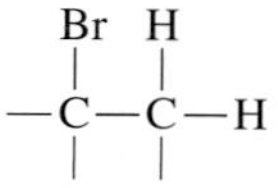 —C—C—H (Br, H) (Markovnikov product)

(l) 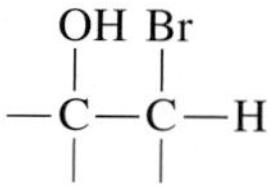 —C—C—H (OH, Br)

(m) 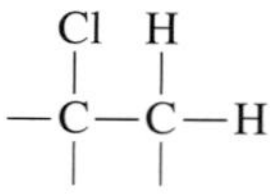 —C—C—H (Cl, H)

(n) 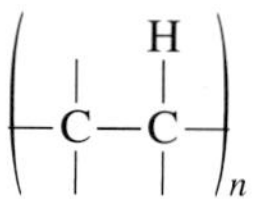 $\left(\!-\text{C}-\text{C}-\!\right)_n$ (H) (polymer)

(o) 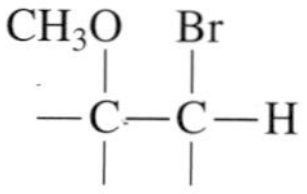 —C—C—H (CH₃O, Br)

(p) 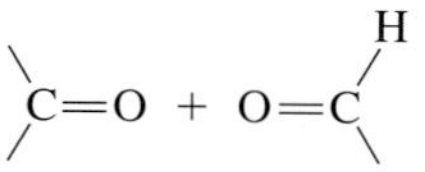 C=O + O=C (H)

(q) 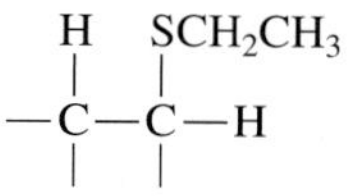 —C—C—H (H, SCH₂CH₃) (anti-Markovnikov product)

46. Give the expected product of reaction of 2-methyl-1-pentene with each of the following reagents.

(a) H₂, PtO₂, CH₃CH₂OH **(b)** D₂, Pd–C, CH₃CH₂OH
(c) BH₃, THF then NaOH + H₂O₂ **(d)** HCl
(e) HBr **(f)** HBr + peroxides

(g) HI + peroxides

(h) H_2SO_4 + H_2O

(i) Cl_2

(j) ICl

(k) Br_2 + CH_3CH_2OH

(l) CH_3SH + peroxides

(m) MCPBA, CH_2Cl_2

(n) OsO_4, then H_2S

(o) O_3, then Zn + $CH_3\overset{\overset{\displaystyle O}{\|}}{C}OH$

(p) $Hg(O\overset{\overset{\displaystyle O}{\|}}{C}CH_3)_2$ + H_2O, then $NaBH_4$

(q) Catalytic H_2SO_4 + heat

47. What are the products of reaction of (E)-3-methyl-3-hexene with each of the reagents in Problem 46?

48. Write the expected products of reaction of 1-ethylcyclopentene with each of the reagents in Problem 46.

49. Write out detailed step-by-step mechanisms for the reactions in parts **(c)**, **(e)**, **(f)**, **(h)**, **(j)**, **(k)**, **(m)**, **(n)**, **(o)**, and **(p)** of Problem 46.

50. What alkene monomer gives the polymer shown below?

$$\left(\begin{array}{c} CH_3\ H \\ | \ \ | \\ -C-C- \\ | \ \ | \\ H \ \ H \end{array}\right)_n$$

51. Give the expected major product from reaction of 3-methyl-1-butene with each of the following reagents. Explain any differences in the products mechanistically.

(a) 50% aqueous H_2SO_4;

(b) $Hg(O\overset{\overset{\displaystyle O}{\|}}{C}CH_3)_2$ in H_2O, followed by $NaBH_4$;

(c) BH_3 in THF, followed by NaOH and H_2O_2.

52. Answer the question posed in Problem 51 for cyclohexylethene.

53. Give the expected major product of reaction of magnesium monoperoxyphthalate (MMPP) with each alkene. In each case, also give the structure of the material formed upon hydrolysis in aqueous acid of the initial product.

(a) 1-Hexene;   **(b)** (Z)-3-ethyl-2-hexene;

(c) (E)-3-ethyl-2-hexene;   **(d)** (E)-3-hexene;

(e) 1,2-dimethylcyclohexene.

54. Give the expected major product of reaction of OsO_4, followed by H_2S, with each alkene in Problem 53.

55. Give the expected major product of reaction of CH_3SH in the presence of peroxides with each alkene in Problem 53.

56. Propose a mechanism for the peroxide-initiated reaction of CH_3SH with 1-hexene.

57. Write the expected products of each of the following reactions.

(a) (E)-2-Pentene + $CHCl_3$ $\xrightarrow{KOC(CH_3)_3, \ (CH_3)_3COH}$

(b) 1-Methylcyclohexene + CH_2I_2 $\xrightarrow{Zn-Cu, \ (CH_3CH_2)_2O}$

(c) Propene + CH_2N_2 $\xrightarrow{Cu, \ \Delta}$

(d) (Z)-1,2-Diphenylethene + $CHBr_3$ $\xrightarrow{KOC(CH_3)_3, \ (CH_3)_3COH}$

(e) (E)-1,3-Pentadiene + 2 CH_2I_2 $\xrightarrow{Zn-Cu, \ (CH_3CH_2)_2O}$

(f) $CH_2{=}CHCH_2CH_2CH_2CHN_2$ $\xrightarrow{hv}$

58. ^{1}H NMR spectrum A corresponds to a molecule with the formula C_3H_5Cl. The compound shows significant IR bands at 730 (see Problem 53 of Chapter 11), 930, 980, 1630, and 3090 cm^{-1}. **(a)** Deduce the structure of the molecule. **(b)** Assign each NMR signal to a hydrogen or group of hydrogens. **(c)** The "doublet" at $\delta = 4.05$ ppm has $J = 6$ Hz. Is this in accord with your assignment in (b)? **(d)** This "doublet," upon fivefold expansion, becomes a doublet of triplets (inset, spectrum A), with $J \approx$ Hz for the triplet splittings. What is the origin of this triplet splitting? Is it reasonable in light of your assignment in (b)?

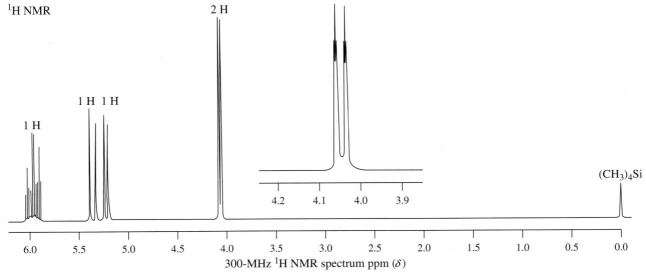

^{1}H NMR

2 H

1 H 1 H

1 H

(CH$_3$)$_4$Si

4.2   4.1   4.0   3.9

6.0   5.5   5.0   4.5   4.0   3.5   3.0   2.5   2.0   1.5   1.0   0.5   0.0

300-MHz ^{1}H NMR spectrum ppm (δ)

A

59. Reaction of C_3H_5Cl (Problem 58, spectrum A) with Cl_2 in H_2O gives rise to two products, both $C_3H_6Cl_2O$, whose spectra are shown in B and C. Reaction of either of these products with KOH yields the same molecule C_3H_5ClO (spectrum D, below). The insets show expansions of some of the multiplets. The IR spectrum reveals bands at 720 and 1260 cm^{-1} and the absence of signals between 1600 and 1800 cm^{-1} and between 3200 and 3700 cm^{-1}. **(a)** Deduce the structures of the compounds giving rise to spectra B, C, and D. **(b)** Why does reaction of the starting chloride compound with Cl_2 in H_2O give two isomeric products? **(c)** Write mechanisms for the formation of the product C_3H_5ClO from both isomers of $C_3H_6Cl_2O$.

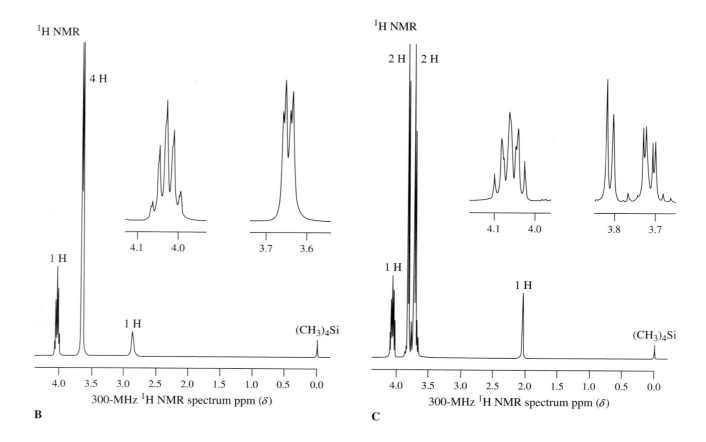

B

C

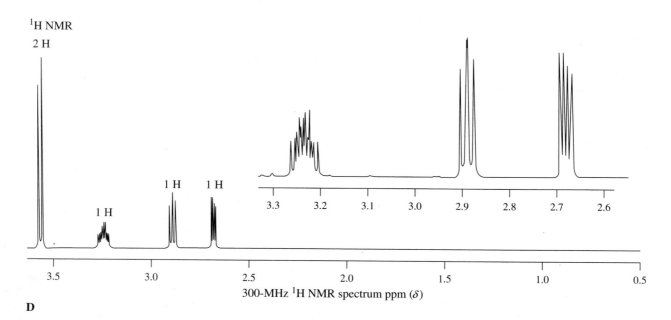

D

60. ¹H NMR spectrum E corresponds to a molecule with the formula C_4H_8O. Its IR spectrum has important bands at 945, 1015, 1665, 3095, and 3360 cm^{-1}. **(a)** Determine the structure of the unknown. **(b)** Assign each NMR and IR signal. **(c)** Explain the splitting patterns for the signals at $\delta = 1.3$, 4.3, and 5.9 ppm (see inset for 10-fold expansion).

¹H NMR

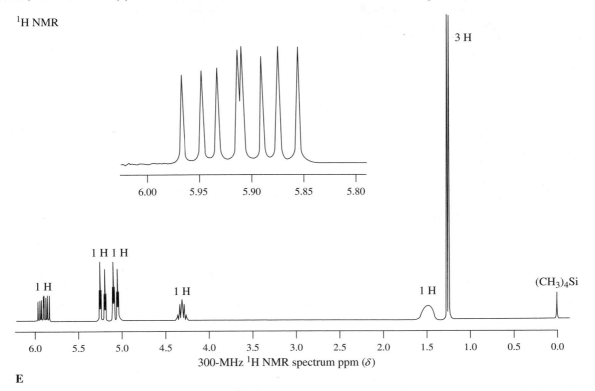

E

61. Reaction of the compound corresponding to spectrum E with $SOCl_2$ produces a chloroalkane, C_4H_7Cl, whose NMR spectrum is almost identical with spectrum E, except that the broad signal at $\delta = 1.5$ ppm is absent. Its IR spectrum shows bands at 700 (Problem 53 of Chapter 11), 925, 985, 1640, and 3090 cm^{-1}. Treatment with H_2 over PtO_2 results in C_4H_9Cl (spectrum F, below). Its IR spectrum reveals the absence of all the bands quoted for its precursor, except for the signal at 700 cm^{-1}. Identify these two molecules.

¹H NMR

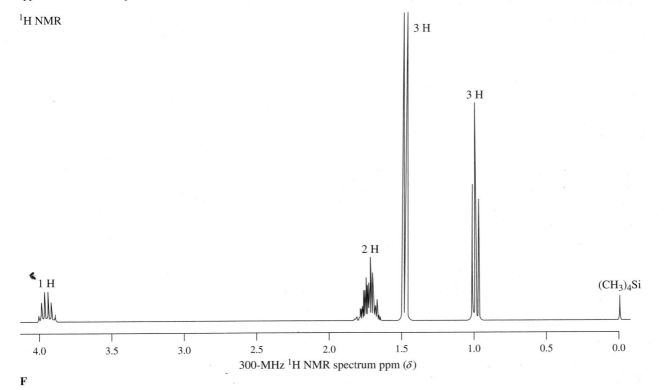

F

62. The mass spectra of both of the compounds described in Problem 61 show two molecular ion peaks, two mass units apart, in an intensity ratio of about 3 : 1. Explain.

63. Give the structure of an alkene that will give the following carbonyl compounds upon ozonolysis followed by reduction with $(CH_3)_2S$.

(a) CH_3CHO only

(b) CH_3CHO and CH_3CH_2CHO

(c) $(CH_3)_2C=O$ and $H_2C=O$

(d) $CH_3CH_2\overset{\overset{\displaystyle O}{\|}}{C}CH_3$ and CH_3CHO

(e) Cyclopentanone and CH_3CH_2CHO

64. **CHALLENGE** Plan syntheses of each of the following compounds, utilizing retrosynthetic-analysis techniques. Starting compounds are given in parentheses. However, other simple alkanes or alkenes also may be used, as long as you include at least one carbon–carbon bond-forming step in each synthesis.

(a) $CH_3CH_2\overset{\overset{\displaystyle O}{\|}}{C}\overset{\overset{\displaystyle }{\underset{\underset{\displaystyle CH_3}{|}}{C}}}HCH_3$ (propene)

(b) $CH_3CH_2CH_2\overset{\underset{\underset{\displaystyle Cl}{|}}{C}}HCH_2CH_2CH_3$ (propene, again)

(c) 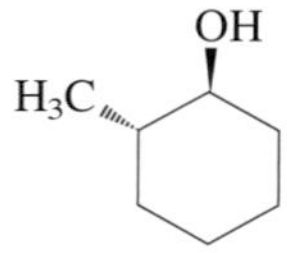 (cyclohexene)

65. Show how you would convert cyclopentane into each of the following molecules.

(a) *cis*-1,2-Dideuteriocyclopentane

(b) *trans*-1,2-Dideuteriocyclopentane

(c)

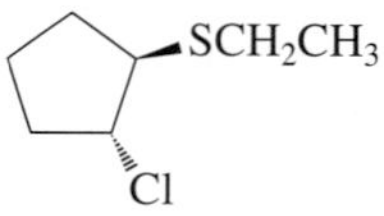

(d)

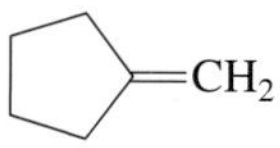

(e)

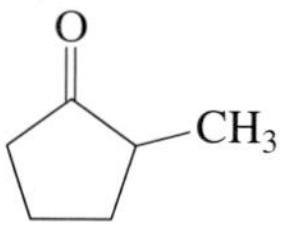

(f) 1,2-Dimethylcyclopentene

(g) *trans*-1,2-Dimethyl-1,2-cyclopentanediol

66. Give the expected major product(s) of each of the following reactions.

(a) $CH_3OCH_2CH_2CH=CH_2$ $\xrightarrow{\begin{array}{l}\text{1. Hg(OCCH}_3)_2,\\ \text{CH}_3\text{OH}\\ \text{2. NaBH}_4,\\ \text{CH}_3\text{OH}\end{array}}$

(b) $H_2C=\overset{\overset{\displaystyle CH_3}{|}}{\underset{\underset{\displaystyle CH_2OH}{|}}{C}}$ $\xrightarrow{\begin{array}{l}\text{1. CH}_3\text{COOH, CH}_2\text{Cl}_2\\ \text{2. H}^+,\text{ H}_2\text{O}\end{array}}$

(c) 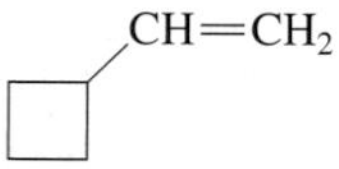 $\xrightarrow{\text{Conc. HI}}$

(d) 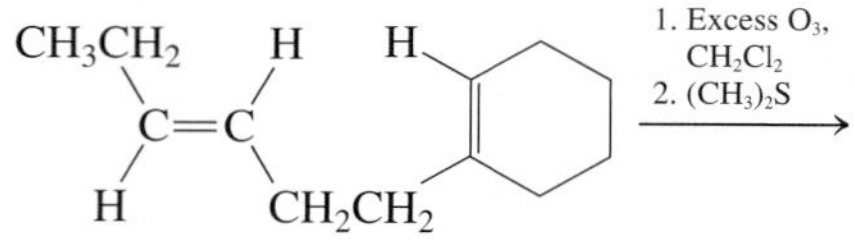 $\xrightarrow{\begin{array}{l}\text{1. Excess O}_3,\\ \text{CH}_2\text{Cl}_2\\ \text{2. (CH}_3)_2\text{S}\end{array}}$

(e) 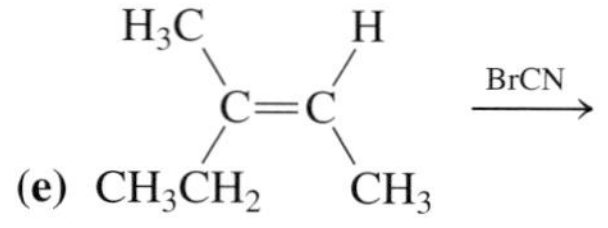 $\xrightarrow{\text{BrCN}}$

(f) 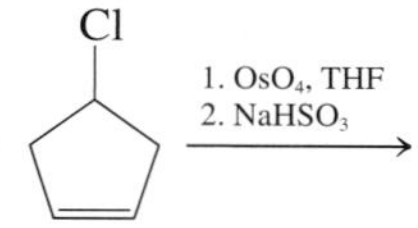 $\xrightarrow{\begin{array}{l}\text{1. OsO}_4, \text{THF}\\ \text{2. NaHSO}_3\end{array}}$

(g) $CH_3CH=CH_2$ $\xrightarrow{\text{Catalytic HF}}$

(h) $CH_2=CHNO_2$ $\xrightarrow{\text{Catalytic KOH}}$
(**Hint:** Draw Lewis structures for the NO_2 group.)

67. (*E*)-5-Hepten-1-ol reacts with the following reagents to give products with the indicated formulas. Determine their structures and explain their formation by detailed mechanisms. (a) HCl, $C_7H_{14}O$ (no Cl!); (b) Cl_2, $C_7H_{13}ClO$ (IR: 740 cm^{-1}; nothing between 1600 and 1800 cm^{-1} and between 3200 and 3700 cm^{-1}).

68. When a cis alkene is mixed with a small amount of I_2 in the presence of heat or light, it isomerizes to some trans alkene. Propose a detailed mechanism to account for this observation.

69. Treatment of α-terpineol (Problem 60 of Chapter 10) with aqueous mercuric acetate followed by sodium borohydride reduction leads predominantly to an isomer of the starting compound $(C_{10}H_{18}O)$ instead of a hydration product. This isomer is the chief component in oil of eucalyptus and, appropriately enough, is called eucalyptol. It is popularly used as a flavoring for otherwise foul-tasting medicines because of its pleasant spicy taste and aroma. Deduce a structure for eucalyptol on the basis of sensible mechanistic chemistry and the following proton-decoupled ^{13}C NMR data. (**Hint:** The IR spectrum shows nothing between 1600 and 1800 cm^{-1} or between 3200 and 3700 cm^{-1}!)

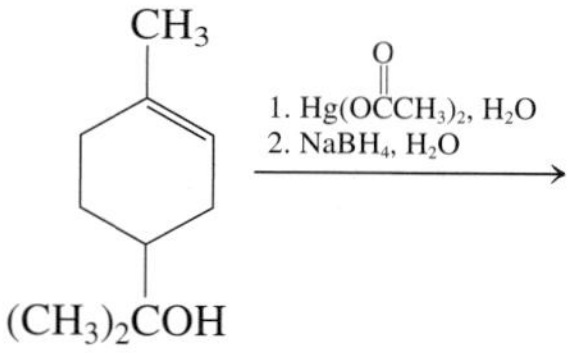

$\xrightarrow{\begin{array}{l}\text{1. Hg(OCCH}_3)_2, \text{H}_2\text{O}\\ \text{2. NaBH}_4, \text{H}_2\text{O}\end{array}}$ eucalyptol, $(C_{10}H_{18}O)$

^{13}C NMR: $\delta = 22.8, 27.5, 28.8, 31.5, 32.9, 69.6,$ and 73.5 ppm

α-Terpineol

70. Both borane and MCPBA react highly selectively with molecules, such as limonene, that contain double bonds in very different environments. Predict the products of reaction of limonene with (**a**) one equivalent of BH_3 in THF, followed by basic aqueous H_2O_2, and (**b**) one equivalent of MCPBA in CH_2Cl_2. Explain your answers.

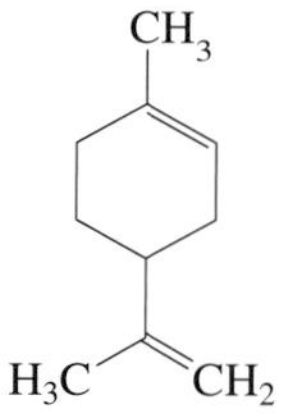

Limonene

71. Oil of marjoram contains a pleasant, lemon-scented substance, $C_{10}H_{16}$ (compound G). Upon ozonolysis, G forms two products. One of them, H, has the formula $C_8H_{14}O_2$ and can be independently synthesized in the following way.

$$\text{structure} \xrightarrow[\text{2. } H_2C\!-\!CHCH_3]{\text{1. Mg, } (CH_3CH_2)_2O} C_8H_{16}O \xrightarrow[]{\text{PCC, } CH_2Cl_2}$$
I

$$C_8H_{14}O \xrightarrow[\text{3. PCC, } CH_2Cl_2]{\begin{array}{l}\text{1. } BH_3, \text{THF}\\ \text{2. } H_2O_2, \text{NaOH}\end{array}} \textbf{H}$$
J

From this information, propose reasonable structures for compounds G through J.

72. **CHALLENGE** Humulene and α-caryophyllene alcohol are terpene constituents of carnation extracts. The former is converted into the latter by acid-catalyzed hydration in one step. Write a mechanism. (**Hint:** The mechanism includes cation-induced double-bond isomerization, cyclizations, and rearrangements that may involve hydrogen and alkyl-group migrations. Two of the intermediates in the mechanistic sequence are shown; five carbon atoms are identified by starts to help you track their positions through the process.)

Humulene α-**Caryophyllene alcohol**

via   and

73. Predict the product(s) of ozonation of humulene (Problem 72), followed by reduction with zinc in acetic acid. If you had not known the structure of humulene ahead of time, would the identities of these ozonolysis products have enabled you to determine it unambiguously?

74. **CHALLENGE** Caryophyllene ($C_{15}H_{24}$) is an unusual sesquiterpene familiar to you as a major cause of the odor of cloves. Determine its structure from the following information. (**Caution:** The structure is totally different from that of α-caryophyllene alcohol in Problem 72.)

Reaction 1

$$\text{Caryophyllene} \xrightarrow{H_2, \text{Pd–C}} C_{15}H_{28}$$

Reaction 2

$$\text{Caryophyllene} \xrightarrow[\text{2. Zn, } CH_3COH]{\text{1. } O_3, CH_2Cl_2} \text{structure} + CH_2\!=\!O$$

Reaction 3

$$\text{Caryophyllene} \xrightarrow[\text{2. } H_2O_2, \text{NaOH, } H_2O]{\text{1. One equivalent of } BH_3, \text{THF}}$$

$$C_{15}H_{26}O \xrightarrow[\text{2. Zn, } CH_3COH]{\text{1. } O_3, CH_2Cl_2} \text{structure}$$

An isomer, isocaryophyllene, gives the same products as caryophyllene upon hydrogenation and ozonolysis. Hydroboration–oxidation of isocaryophyllene gives a $C_{15}H_{26}O$ product isomeric to the one shown in reaction 3; however, ozonolysis converts this compound into the same final product shown. In what way do caryophyllene and its isomer differ?

75. Beginning with methylcyclohexane, propose a synthesis of the cyclohexane derivative shown below. Use a retrosynthetic approach to keep your synthesis short and efficient, and employ reaction(s) that will furnish the regiochemistry and the relative stereochemistry given in the target structure.

Team Problem

76. The selectivity of hydroboration increases with increasing bulkiness of the borane reagent.

(a) For example, 1-pentene is selectively hydroborated in the presence of *cis*- and *trans*-2-pentene when treated with bis(1,2-dimethylpropyl)borane (disiamylborane) or with 9-borabicyclo[3.3.1]-nonane, 9-BBN. Divide the task of formulating the structure of the starting alkene used in preparing both of these bulky borane reagents among yourselves. Make models to visualize the features of these reagents that direct the structural selectivity.

$$[(CH_3)_2CHCH]_2BH$$
with CH_3 above the CH

Bis(1,2-dimethylpropyl)borane
(Disiamylborane)

9-BBN

(b) In an enantioselective approach to making secondary alcohols, two equivalents of one enantiomer of α-pinene are treated with BH_3. The resulting borane reagent is treated with *cis*-2-butene followed by basic hydrogen peroxide to yield optically active 2-butanol.

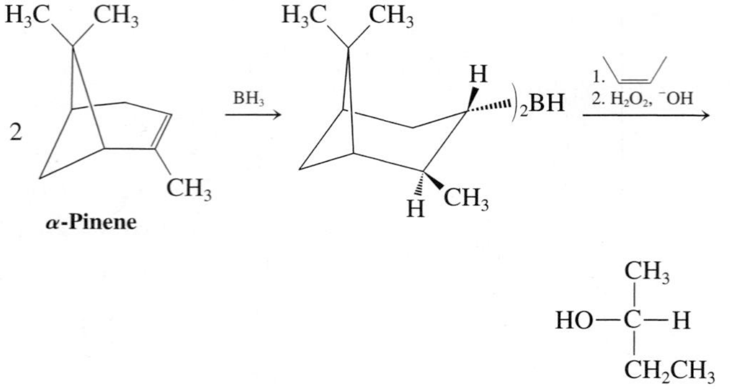

α-Pinene

$$HO-\underset{\underset{CH_2CH_3}{|}}{\overset{\overset{CH_3}{|}}{C}}-H$$

Optically active

Share your model kits to make a model of α-pinene and the resulting borane reagent. Discuss what is directing the enantioselectivity of this hydroboration–oxidation reaction. What products besides 2-butanol result from the oxidation step?

Preprofessional Problems

77. A chiral compound, C_5H_8, upon simple catalytic hydrogenation, yields an achiral compound, C_5H_{10}. What is the best name for the former? **(a)** 1-Methylcyclobutene; **(b)** 3-methylcyclobutene; **(c)** 1,2-dimethylcyclopropene; **(d)** cyclopentene.

78. A chemist reacted 300 g of 1-butene with excess Br_2 (in CCl_4) at 25°C. He isolated 418 g of 1,2-dibromobutane. What is the percent yield? (Atomic weights: C = 12.0, H = 1.00, Br = 80.0.) **(a)** 26; **(b)** 36; **(c)** 46; **(d)** 56; **(e)** 66.

79. *trans*-3-Hexene and *cis*-3-hexene differ in one of the following ways. Which one? **(a)** Products of hydrogenation; **(b)** products of ozonolysis; **(c)** products of Br_2 addition in CCl_4; **(d)** products of hydroboration–oxidation; **(e)** products of combustion.

80. Which reaction intermediate is believed to be part of the following reaction?

$$RCH=CH_2 \xrightarrow{\text{HBr, ROOR}} RCH_2CH_2Br$$

(a) Radical; **(b)** carbocation; **(c)** oxacyclopropane; **(d)** bromonium ion.

81. When 1-pentene is treated with mercuric acetate, followed by sodium borohydride, which of the following compounds is the resulting product? **(a)** 1-Pentyne; **(b)** pentane; **(c)** 1-pentanol; **(d)** 2-pentanol.

CHAPTER 13

Alkynes

The Carbon–Carbon Triple Bond

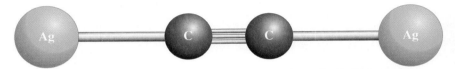

Alkynes are hydrocarbons that contain carbon–carbon triple bonds. It should not come as a surprise that their characteristics resemble the properties and behavior of alkenes, their double-bonded cousins. In this chapter we shall see that, like alkenes, alkynes find numerous uses in a variety of modern settings. For example, the polymer derived from the parent compound, ethyne (HC≡CH), can be fashioned into electrically conductive sheets usable in lightweight, all-polymer batteries. Ethyne is also a substance with a relatively high energy content, a property that is exploited in oxyacetylene torches. A variety of alkynes, both naturally occurring and synthetic, have found use in medicine for their antibacterial, antiparasitic, and antifungal activities.

Alkyne triple bond

Because the —C≡C— functional group contains two π linkages (which are mutually perpendicular; recall Figure 1-21), its reactivity is much like that of the double bond. For example, like alkenes, alkynes are electron rich and subject to attack by electrophiles. Many of the alkenes that serve as monomers for the production of polymeric fabrics, elastics, and plastics are prepared by electrophilic addition reactions to ethyne and other alkynes. Alkynes can be made by elimination reactions similar to those used to generate alkenes, and they are likewise more stable when the multiple bond is internal rather than terminal. A further, and useful, feature is that the alkynyl hydrogen is much more acidic than its alkenyl or alkyl counterpart, a property that permits easy deprotonation by strong bases. The resulting alkynyl anions are valuable nucleophilic reagents in synthesis.

We begin with discussions of the naming, structural characteristics, and spectroscopy of the alkynes. Subsequent sections introduce methods for the synthesis of compounds in this class and the typical reactions they undergo. We end with an overview of the extensive industrial uses and physiological characteristics of alkynes.

Because of their high degree of unsaturation, alkynes are highly energetic and can decompose explosively. These properties manifest themselves particularly in metal salts of ethyne. For example, silver acetylide, made by bubbling ethyne gas through an aqueous silver nitrate solution, is extremely sensitive to impact, friction, heat, and static electricity, all of which cause violent explosions. These are unusual in as much as no gases are produced, only a cloud of solid Ag and carbon particles.

Common Names for Alkynes

HC≡CH

Acetylene

CH₃C≡CCH₃

Dimethylacetylene

CH₃CH₂CH₂C≡CH

Propylacetylene

13-1 | NAMING THE ALKYNES

A carbon–carbon triple bond is the functional group characteristic of the **alkynes.** The general formula for the alkynes is C_nH_{2n-2}, the same as that for the cycloalkenes. The common names for many alkynes are still in use, including *acetylene,* the common name of the smallest alkyne, C_2H_2. Other alkynes are treated as its derivatives—for example, the alkylacetylenes.

The IUPAC rules for naming alkenes (Section 11-1) also apply to alkynes, the ending **-yne** replacing *-ene.* A number indicates the position of the triple bond in the main chain.

$$HC≡CH \qquad CH_3C≡CCH_3 \qquad \overset{1}{C}H_3\overset{2}{C}≡\overset{3}{C}\overset{4}{C}\overset{Br}{H}\overset{5}{C}H_2\overset{6}{C}H_3 \qquad \overset{4}{C}H_3\overset{3}{C}\overset{CH_3}{\underset{CH_3}{C}}\overset{2}{C}≡\overset{1}{C}H$$

Ethyne   **2-Butyne**   **4-Bromo-2-hexyne**   **3,3-Dimethyl-1-butyne**
                                                (An internal alkyne)   (A terminal alkyne)

Alkynes having the general structure RC≡CH are **terminal,** whereas those with the structure of RC≡CR′ are **internal.**

Substituents bearing a triple bond are **alkynyl** groups. Thus, the substituent –C≡CH is named **ethynyl;** its homolog –CH₂C≡CH is **2-propynyl** (propargyl). Like alkanes and alkenes, alkynes can be depicted in straight-line notation.

trans-**1,2-Diethynylcyclohexane**   **2-Propynylcyclobutane**   **2-Propyn-1-ol**
                                          (Propargylcyclobutane)   (Propargyl alcohol)

$$\overset{1}{C}H_2\overset{2}{C}≡\overset{3}{C}H \qquad HC\overset{3}{≡}\overset{2}{C}\overset{1}{C}H_2OH$$

In IUPAC nomenclature, a hydrocarbon containing both double and triple bonds is called an **alkenyne.** The chain is numbered starting from the end closest to either of the functional groups. When a double bond and a triple bond are at equidistant positions from either terminus, the *double* bond is given the lower number. Alkynes incorporating the hydroxy function are named **alkynols.** Note the omission of the final *e* of -ene in -enyne and of -yne in -ynol. The OH group takes precedence over both double and triple bonds in the numbering of a chain.

$$\overset{6}{C}H_3\overset{5}{C}H_2\overset{4}{C}H=\overset{3}{C}H\overset{2}{C}≡\overset{1}{C}H \qquad \overset{1}{C}H_2=\overset{2}{C}H\overset{3}{C}H_2\overset{4}{C}≡\overset{5}{C}H$$

3-Hexen-1-yne   **1-Penten-4-yne**   **5-Hexyn-2-ol**
(Not 3-hexen-5-yne)   (Not 4-penten-1-yne)   (Not 1-hexyn-5-ol)

Exercise 13-1

Give the IUPAC names for **(a)** all the alkynes of composition C_6H_{10};

(b)

$$\begin{array}{c} H_3C \\ \\ H \end{array} C-C≡CH$$
$$CH=CH_2$$

(c) all butynols. Remember to include and designate stereoisomers.

13-2 | PROPERTIES AND BONDING IN THE ALKYNES

The nature of the triple bond helps explain the physical and chemical properties of the alkynes. In molecular-orbital terms, we shall see that the carbons are *sp* hybridized, and the four singly filled *p* orbitals form two perpendicular π bonds.

Alkynes are relatively nonpolar

Alkynes have boiling points very similar to those of the corresponding alkenes and alkanes. Ethyne is unusual in that it has no boiling point at atmospheric pressure; rather, it sublimes at $-84°C$. Propyne (b.p. $-23.2°C$) and 1-butyne (b.p. $8.1°C$) are gases, whereas 2-butyne is barely a liquid (b.p. $27°C$) at room temperature. The medium-sized alkynes are distillable liquids.

Ethyne is linear and has strong, short bonds

In ethyne, the two carbons are sp hybridized (Figure 13-1A). One of the hybrid orbitals on each carbon overlaps with hydrogen, and a σ bond between the two carbon atoms results from mutual overlap of the remaining sp hybrids. The two perpendicular p orbitals on each carbon contain one electron each. These two sets overlap to form two perpendicular π bonds (Figure 13-1B). Because π bonds are diffuse, the distribution of electrons in the triple bond resembles a cylindrical cloud (Figure 13-1C). As a consequence of hybridization and the two π interactions, the strength of the triple bond is about 229 kcal mol^{-1}, considerably stronger than either the carbon–carbon double or single bonds (margin). As with alkenes, however, the alkyne π bonds are much weaker than the σ component of the triple bond, a feature that gives rise to much of its chemical reactivity. The C–H bond-dissociation energy of terminal alkynes is also substantial: 131 kcal mol^{-1} (548 kJ mol^{-1}).

Dissociation Energies of C–C Bonds

HC≡CH

$DH° = 229$ kcal mol^{-1}
(958 kJ mol^{-1})

$H_2C=CH_2$

$DH° = 173$ kcal mol^{-1}
(724 kJ mol^{-1})

H_3C-CH_3

$DH° = 90$ kcal mol^{-1}
(377 kJ mol^{-1})

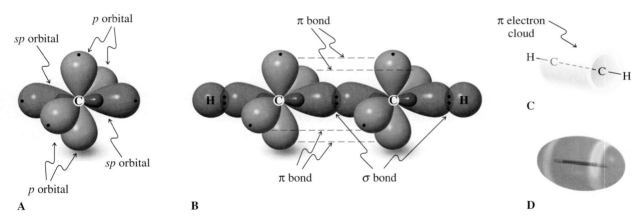

Figure 13-1 (A) Orbital picture of sp-hybridized carbon, showing the two perpendicular p orbitals. (B) The triple bond in ethyne: The orbitals of two sp-hybridized CH fragments overlap to create a σ bond and two π bonds. (C) The two π bonds produce a cylindrical electron cloud around the molecular axis of ethyne. (D) The electrostatic potential map reveals the (red) belt of high electron density around the central part of the molecular axis.

Because both carbon atoms in ethyne are sp hybridized, its structure is linear (Figure 13-2). The carbon–carbon bond length is 1.20 Å, shorter than that of a double bond (1.33 Å, Figure 11-1). The carbon–hydrogen bond also is short, again because of the relatively large degree of s character in the sp hybrids used for bonding to hydrogen. The electrons in these orbitals (and in the bonds that they form by overlapping with other orbitals) reside relatively close to the nucleus and produce shorter (and stronger) bonds.

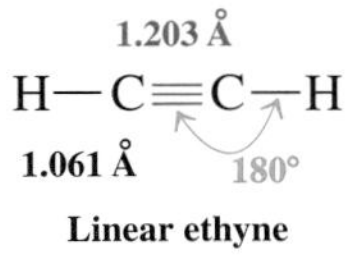

1.203 Å
H—C≡C—H
1.061 Å   180°
Linear ethyne

Figure 13-2 Molecular structure of ethyne.

Alkynes are high-energy compounds

The alkyne triple bond is characterized by a concentration of four π electrons in a relatively small volume of space. The resulting electron–electron repulsion contributes to the relative weakness of the two π bonds and to a very high energy content of the alkyne molecule itself. Because of this property, alkynes often react with the release of considerable amounts of energy and must be handled with care: They polymerize very easily and are prone to explosive decomposition. Ethyne can be shipped in pressurized cylinders that contain acetone and porous fillers such as pumice as stabilizers.

The high energy content of ethyne is reflected in its heat of combustion of 311 kcal mol^{-1}. As shown in the equation for ethyne combustion, this energy is distributed among only three

The high temperatures required for welding are attained by combustion of ethyne (acetylene).

product molecules, one of water and two of CO_2, causing each to be heated to extremely high temperatures (>2500°C), sufficient for use in welding torches.

Combustion of Ethyne

$$HC\equiv CH \ + \ 2.5\,O_2 \ \longrightarrow \ 2\,CO_2 \ + \ H_2O \qquad \Delta H° = -311 \text{ kcal mol}^{-1}$$
$$(-1301 \text{ kJ mol}^{-1})$$

As we found in our discussion of alkene stabilities (Section 11-5), heats of hydrogenation also provide convenient measures of the relative stabilities of alkyne isomers. In the presence of catalytic amounts of platinum or palladium on charcoal, the two isomers of butyne hydrogenate by addition of two molar equivalents of H_2 to produce butane. Just as we discovered in the case of alkenes, hydrogenation of the internal alkyne isomer releases less energy, allowing us to conclude that 2-butyne is the more stable of the two. Hyperconjugation is the reason for the greater relative stability of internal compared with terminal alkynes.

$$CH_3CH_2C\equiv CH \ + \ 2\,H_2 \ \xrightarrow{\text{Catalyst}} \ CH_3CH_2CH_2CH_3 \qquad \Delta H° = -69.9 \text{ kcal mol}^{-1}$$
$$(-292.5 \text{ kJ mol}^{-1})$$

$$CH_3C\equiv CCH_3 \ + \ 2\,H_2 \ \xrightarrow{\text{Catalyst}} \ CH_3CH_2CH_2CH_3 \qquad \Delta H° = -65.1 \text{ kcal mol}^{-1}$$
$$(-272.4 \text{ kJ mol}^{-1})$$

Exercise 13-2

Are the heats of hydrogenation of the butynes consistent with the notion that alkynes are high-energy compounds? Explain. (**Hint:** Compare these values with the heats of hydrogenation of alkene double bonds.)

Terminal alkynes are remarkably acidic

Relative Stabilities of the Alkynes

$$RC\equiv CH \ < \ RC\equiv CR'$$

More stable

Deprotonation of 1-Alkynes

$$RC\equiv C-H \ + \ :\bar{B}$$
$$\downarrow$$
$$RC\equiv C:^- \ + \ HB$$

In Section 2-3 you learned that the strength of an acid, H–A, increases with increasing electronegativity, or electron-attracting capability, of atom A. Is the electronegativity of an atom the same in all structural environments? The answer is no: Electronegativity varies with hybridization. Electrons in s orbitals are more strongly attracted to an atomic nucleus than are electrons in p orbitals. As a consequence, an atom with hybrid orbitals high in s character (e.g., sp, with 50% s and 50% p character) will be slightly more electronegative than the same atom with hybrid orbitals with less s character (sp^3, 25% s and 75% p character). This effect is indicated below in the electrostatic potential maps of ethane, ethene, and ethyne. The increasingly positive polarization of the hydrogen atoms is reflected in their increasingly blue shadings, whereas the carbon atoms become more electron rich (red) along the series. The relatively high s character in the carbon hybrid orbitals of terminal alkynes makes them more acidic than alkanes and alkenes. The pK_a of ethyne, for example, is 25, remarkably low compared with that of ethene and ethane.

Relative Acidities of Alkanes, Alkenes, and Alkynes

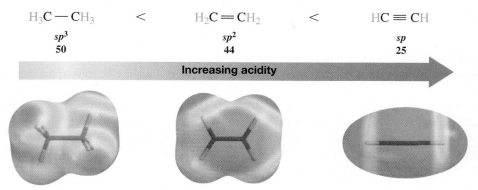

	H_3C-CH_3	<	$H_2C=CH_2$	<	$HC\equiv CH$
Hybridization:	sp^3		sp^2		sp
pK_a:	50		44		25

Increasing acidity →

This property is useful, because strong bases such as sodium amide in liquid ammonia, alkyllithiums, and Grignard reagents can deprotonate terminal alkynes to the corresponding **alkynyl anions.** These species react as bases and nucleophiles, much like other carbanions (Section 13-5).

Deprotonation of a Terminal Alkyne

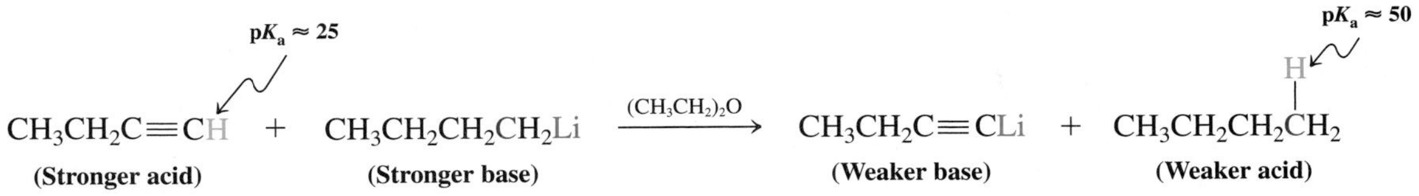

$$\text{CH}_3\text{CH}_2\text{C} \equiv \overset{pK_a \approx 25}{\text{CH}} \; + \; \text{CH}_3\text{CH}_2\text{CH}_2\text{CH}_2\text{Li} \xrightarrow{\text{(CH}_3\text{CH}_2)_2\text{O}} \text{CH}_3\text{CH}_2\text{C} \equiv \text{CLi} \; + \; \overset{pK_a \approx 50}{\text{CH}_3\text{CH}_2\text{CH}_2\overset{H}{\text{CH}_2}}$$

(Stronger acid)   (Stronger base)   (Weaker base)   (Weaker acid)

Solved Exercise 13-3 | Working with the Concepts: Deprotonation of Alkynes

What is the equilibrium constant, K_{eq}, for the acid-base reaction shown above? Does its value explain why the reaction is written with only a forward arrow, suggesting that it is "irreversible"?

Strategy
Recall how pK_a values relate to acid dissociation constants. Use this information to determine the value for K_{eq}.

Solution
- The pK_a is the negative logarithm of the acid dissociation constant. Dissociation of the alkyne therefore has a $K_a \approx 10^{-25}$, very unfavorable, at least in comparison with the more familiar acids. However, butyllithium is the conjugate base of butane, which has a $K_a \approx 10^{-50}$. As an acid, butane is 25 orders of magnitude weaker than is the terminal alkyne. Thus, butyllithium is that much stronger a base compared with the alkynyl anion.
- The K_{eq} for the reaction is found by dividing the K_a for the acid on the left by the K_a for the acid on the right: $10^{-25}/10^{-50} = 10^{25}$. The reaction is *very* favorable in the forward direction, so much so that for all practical purposes it may be considered to be irreversible. (**Caution:** Use common sense to avoid major errors in solving acid-base problems, such as deciding that the equilibrium lies the wrong direction. Use this **Hint:** The favored direction for an acid-base reaction converts the stronger acid/stronger base pair into the weaker acid/weaker base pair.)

Exercise 13-4 | Try It Yourself

Strong bases other than those mentioned here for the deprotonation of alkynes were introduced earlier. Two examples are potassium *tert*-butoxide and lithium diisopropylamide (LDA). Would either (or both) of these compounds be suitable for making ethynyl anion from ethyne? Explain, in terms of their pK_a values.

In Summary The characteristic hybridization scheme for the triple bond of an alkyne controls its physical and electronic features. It is responsible for strong bonds, the linear structure, and the relatively acidic alkynyl hydrogen. In addition, alkynes are highly energetic compounds. Internal isomers are more stable than terminal ones, as shown by the relative heats of hydrogenation.

13-3 | SPECTROSCOPY OF THE ALKYNES

Alkenyl hydrogens (and carbons) are deshielded and give rise to relatively low-field NMR signals compared with those in saturated alkanes (Section 11-4). In contrast, alkynyl hydrogens have chemical shifts at relatively high field, much closer to those in alkanes. Similarly, the *sp*-hybridized carbons absorb in a range between that recorded for alkenes and alkanes. Alkynes, especially terminal ones, are also readily identified by IR spectroscopy. Finally, mass spectrometry can be a useful tool for identification and structure elucidation of alkynes.

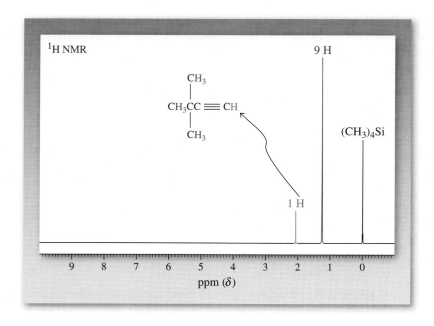

The NMR absorptions of alkyne hydrogens show a characteristic shielding

Unlike alkenyl hydrogens, which are deshielded and give ^{1}H NMR signals at δ = 4.6–5.7 ppm, protons bound to sp-hybridized carbon atoms are found at δ = 1.7–3.1 ppm (Table 10-2). For example, in the NMR spectrum of 3,3-dimethyl-1-butyne, the alkynyl hydrogen resonates at δ = 2.06 ppm (Figure 13-3).

Why is the terminal alkyne hydrogen so shielded? Like the π electrons of an alkene, those in the triple bond enter into a circular motion when an alkyne is subjected to an external magnetic field (Figure 13-4). However, the cylindrical distribution of these electrons (Figure 13-1C) now allows the major direction of this motion to be perpendicular to

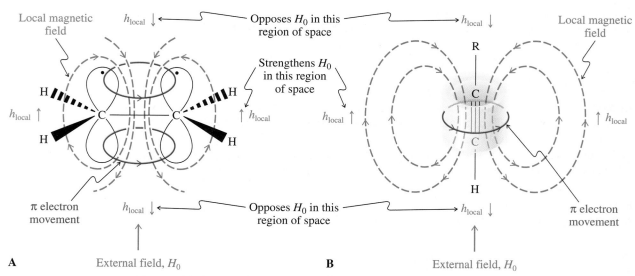

Figure 13-4 Electron circulation in the presence of an external magnetic field generates local magnetic fields that cause the characteristic chemical shifts of alkenyl and alkynyl hydrogens. (A) Alkenyl hydrogens are located in a region of space where h_{local} reinforces H_0. Therefore, these protons are relatively deshielded. (B) Electron circulation in an alkyne generates a local field that *opposes* H_0 in the vicinity of the alkynyl hydrogen, thus causing shielding.

that in alkenes and to generate a local magnetic field that *opposes* H_0 in the vicinity of the alkyne hydrogen. The result is a strong *shielding* effect that cancels the deshielding tendency of the electron-withdrawing *sp*-hybridized carbon and gives rise to a relatively high-field chemical shift.

The triple bond transmits spin–spin coupling

The alkyne functional group transmits coupling so well that the terminal hydrogen is split by the hydrogens across the triple bond, even though it is separated from them by three carbons. This result is an example of long-range coupling. The coupling constants are small and range from about 2 to 4 Hz. Figure 13-5 shows the NMR spectrum of 1-pentyne. The alkynyl hydrogen signal at $\delta = 1.94$ ppm is a triplet ($J = 2.5$ Hz) because of coupling to the two equivalent hydrogens at C3, which appear at $\delta = 2.16$ ppm. The latter, in turn, give rise to a doublet of triplets, representing coupling to the two hydrogens at C4 ($J = 6$ Hz) as well as that at C1 ($J = 2.5$ Hz).

Long-Range Coupling in Alkynes

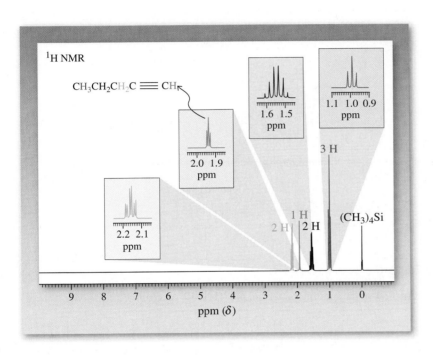

Figure 13-5 300-MHz ^{1}H NMR spectrum of 1-pentyne showing coupling between the alkynyl (green) and propargylic (blue) hydrogens.

Predict the first-order splitting pattern in the ^{1}H NMR spectrum of 3-methyl-1-butyne.

Strategy
First, write out the structure. Then identify groups of hydrogens within coupling distance of each other, both neighboring and long range. Finally, use information regarding approximate values of coupling constants (and the $N + 1$ rule) to generate expected splitting patterns.

Solution
• The structure of the molecule is

$$\underset{3 \quad\quad 2 \quad 1}{CH_3 - CH - C\equiv CH}$$
$$\overset{\displaystyle CH_3}{\big|}$$

• The two methyl groups are equivalent and give one signal that is split into a doublet by the single hydrogen atom at C3 ($N + 1 = 2$ lines). The coupling constant (J value) for this splitting is the typical 6–8 Hz found in saturated systems (Section 10-7).

- The alkynyl —C≡CH hydrogen at C1 experiences long-range coupling to the same H at C3, appearing also as a doublet, but *J* is smaller, about 3 Hz.
- Finally, the signal for the hydrogen at C3 displays a more complex pattern. The 6–8-Hz splitting by the six hydrogens of the methyl groups gives a septet (*N* + 1 = 7 lines). Each line of this septet is further split by the additional 3-Hz coupling to the alkynyl H. As the actual spectrum below shows, the outermost lines of this signal, a doublet of septets, are so small that they are barely visible (see Tables 10-4 and 10-5). (**Caution:** When interpreting ¹H NMR spectra, be aware of the very low intensity of the outer lines in highly split signals. In fact, it is prudent to assume that such signals may consist of more lines than are readily visible.)

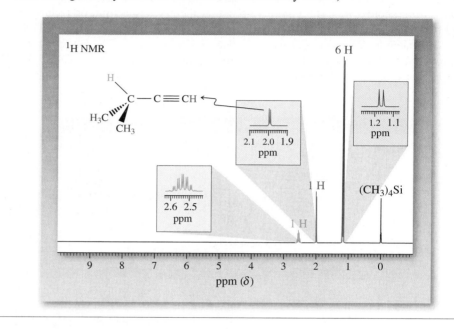

> Pure structured carbon exists in nature predominantly in the form of the three-dimensional diamond (all *sp*³) and two-dimensional graphite (all *sp*²; see also Real Life 15-1). The one-dimensional version of a polymeric acetylene chain (all *sp*) has remained elusive, but synthetic chemists have come close by making well-defined oligomers containing up to 44 contiguous *sp*-hybridized carbons! The ¹³C NMR spectrum shows a range of peaks centered around 63.7 ppm, extrapolated to be the likely chemical shift of the infinite polymer.

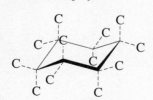

Diamond: *sp*³

Graphite: *sp*²

Carbyne: *sp*

Exercise 13-6 | Try It Yourself

Predict the first-order splitting pattern in the ¹H NMR spectrum of 2-pentyne.

The ¹³C NMR chemical shifts of alkyne carbons are distinct from those of the alkanes and alkenes

Carbon-13 NMR spectroscopy also is useful in deducing the structure of alkynes. For example, the triple-bonded carbons in alkyl-substituted alkynes resonate in the range of δ = 65–95 ppm, quite separate from the chemical shifts of analogous alkane (δ = 5–45 ppm) and alkene (δ = 100–150 ppm) carbon atoms (Table 10-6).

Typical Alkyne ¹³C NMR Chemical Shifts

HC≡CH   HC≡CCH₂CH₂CH₂CH₃   CH₃CH₂C≡CCH₂CH₃

δ = 71.9   68.6   84.0   18.6   31.1   22.4   14.1   81.1   15.6   13.2 ppm

Terminal alkynes give rise to two characteristic infrared absorptions

Infrared spectroscopy is helpful in identifying terminal alkynes. Characteristic stretching bands appear for the alkynyl hydrogen at 3260–3330 cm⁻¹ and for the C≡C triple bond at 2100–2260 cm⁻¹. There is also a diagnostic $\tilde{\nu}_{C_{sp}-H}$ bending absorption at 640 cm⁻¹

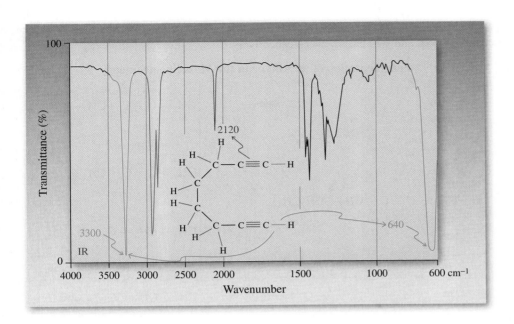

Figure 13-6 IR spectrum of 1,7-octadiyne:
$\tilde{\nu}_{C_{sp}-H \text{ stretch}} = 3300 \text{ cm}^{-1}$;
$\tilde{\nu}_{C \equiv C \text{ stretch}} = 2120 \text{ cm}^{-1}$;
$\tilde{\nu}_{C_{sp}-H \text{ bend}} = 640 \text{ cm}^{-1}$.

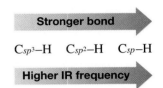

Stronger bond ⟶

C_{sp^3}–H C_{sp^2}–H C_{sp}–H

Higher IR frequency ⟶

(Figure 13-6). Such data are especially useful when ^{1}H NMR spectra are complex and difficult to interpret. However, the band for the C≡C stretching vibration in internal alkynes is often weak, like that for internal alkenes (Section 11-8), thus reducing the value of IR spectroscopy for characterizing these systems.

Mass spectral fragmentation of alkynes gives resonance-stabilized cations

The mass spectra of alkynes, like those of alkenes, frequently show prominent molecular ions. Thus high-resolution measurements can reveal the molecular formula and therefore the presence of two degrees of unsaturation derived from the presence of the triple bond.

In addition, fragmentation at the carbon once removed from the triple bond is observed, giving resonance-stabilized cations. For example, the mass spectrum of 3-heptyne (Figure 13-7) shows an intense molecular ion peak at $m/z = 96$ and loss of both methyl

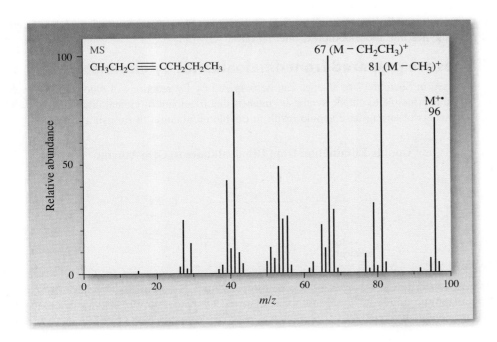

Figure 13-7 Mass spectrum of 3-heptyne, showing $M^{+\bullet}$ at $m/z = 96$ and important fragments at $m/z = 67$ and 81 arising from cleavage of the C1–C2 and C5–C6 bonds.

(cleavage *a*) and ethyl (cleavage *b*) fragments to give two different stabilized cations, with $m/z = 81$ and 67 (base peak), respectively:

Fragmentation of an Alkyne in the Mass Spectrometer

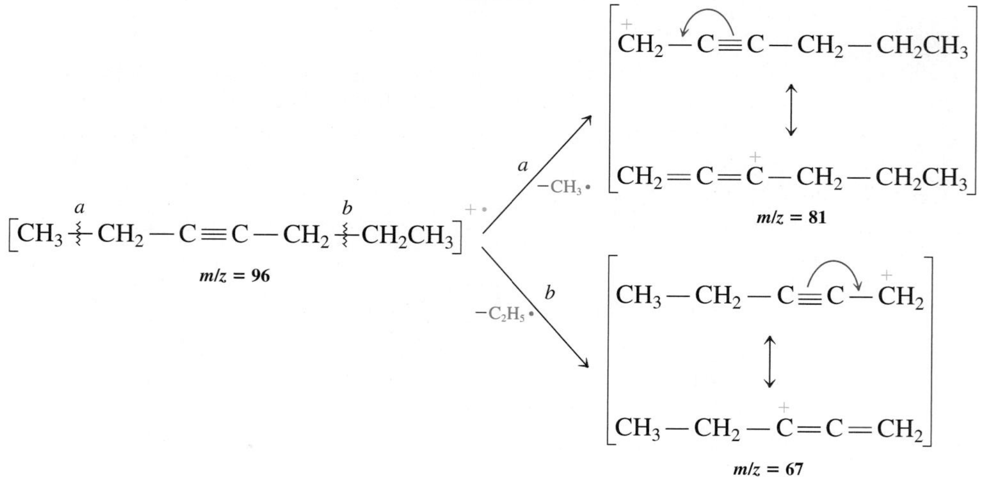

Unfortunately, under the high energy conditions of the mass spectrometry experiment, migration of the triple bond can occur. Thus this fragmentation is not typically very useful for identifying the location of the triple bond in a longer-chain alkyne.

In Summary The cylindrical π cloud around the carbon–carbon triple bond induces local magnetic fields that lead to NMR chemical shifts for alkynyl hydrogens at higher fields than those of alkenyl protons. Long-range coupling is observed through the C≡C linkage. Infrared spectroscopy provides a useful complement to NMR data, displaying characteristic bands for the C≡C and ≡C–H bonds of terminal alkynes. In the mass spectrometer, alkynes fragment to give resonance-stabilized cations.

13-4 PREPARATION OF ALKYNES BY DOUBLE ELIMINATION

The two basic methods used to prepare alkynes are double elimination from 1,2-dihaloalkanes and alkylation of alkynyl anions. This section deals with the first method, which provides a synthetic route to alkynes from alkenes; Section 13-5 addresses the second, which converts terminal alkynes into more complex, internal ones.

Alkynes are prepared from dihaloalkanes by elimination

As discussed in Section 11-6, alkenes can be prepared by E2 reactions of haloalkanes. Application of this principle to alkyne synthesis suggests that treatment of vicinal dihaloalkanes with two equivalents of strong base should result in double elimination to furnish a triple bond.

Double Elimination from Dihaloalkanes to Give Alkynes

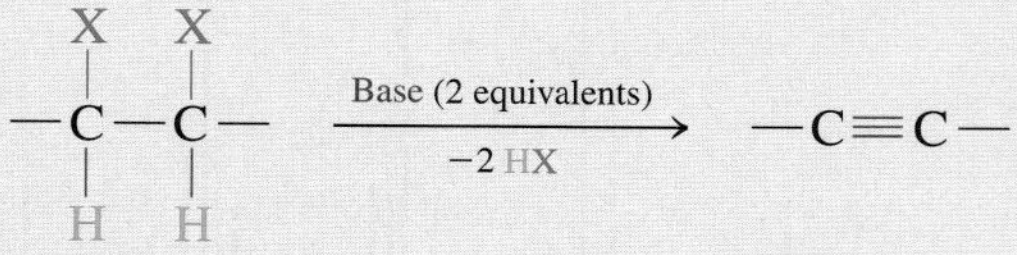

Indeed, addition of 1,2-dibromohexane (prepared by addition of Br_2 to 1-hexene, Section 12-5) to sodium amide in liquid ammonia followed by evaporation of solvent and aqueous work-up gives 1-hexyne.

Example of Double Dehydrohalogenation to Give an Alkyne

$$\underset{\underset{Br}{|}}{CH_3CH_2CH_2CH_2CH}-CH_2Br \xrightarrow[\underset{-2\ HBr}{}]{\begin{array}{l}1.\ 3\ NaNH_2,\ liquid\ NH_3\\2.\ H_2O\end{array}} CH_3CH_2CH_2CH_2C\equiv CH$$

Three equivalents of $NaNH_2$ are necessary in the preparation of a terminal alkyne because, as this alkyne forms, its acidic terminal hydrogen (Section 13-2) immediately protonates an equivalent amount of base. Eliminations in liquid ammonia are usually carried out at its boiling point, $-33°C$.

Because vicinal dihaloalkanes are readily available from alkenes by halogenation, this sequence, called **halogenation–double dehydrohalogenation,** is a ready means of converting alkenes into the corresponding alkynes.

**A Halogenation–Double Dehydrohalogenation
Used in Alkyne Synthesis**

1. Br₂, CCl₄
2. NaNH₂, liquid NH₃
3. H₂O

| 1,5-Hexadiene | 53% 1,5-Hexadiyne |

Exercise 13-7

Illustrate the use of halogenation–double dehydrohalogenation in the synthesis of the alkynes **(a)** 2-pentyne; **(b)** 1-octyne; **(c)** 2-methyl-3-hexyne.

Haloalkenes are intermediates in alkyne synthesis by elimination

Dehydrohalogenation of dihaloalkanes proceeds through the intermediacy of haloalkenes, also called **alkenyl halides.** Although mixtures of *E*- and *Z*-haloalkenes are in principle possible, with diastereomerically pure vicinal dihaloalkanes only one product is formed because elimination proceeds stereospecifically *anti* (Section 11-6).

Newman projection

Anti elimination | 1 equivalent NaOCH₃, CH₃OH

An alkenyl halide

$$\underset{H}{\overset{X}{\diagdown}}C=C\diagup + BH + X^-$$

Exercise 13-8

Give the structure of the bromoalkene intermediate in the bromination–dehydrobromination of *cis*-2-butene to 2-butyne. Do the same for the trans isomer. (**Caution:** There is stereochemistry involved in both steps. **Hint:** Refer to Section 12-5 for useful information, and use models.)

The stereochemistry of the intermediate haloalkene is of no consequence when the sequence is used for alkyne synthesis. Both *E*- and *Z*-haloalkenes eliminate with base to give the same alkyne.

In Summary Alkynes are made from vicinal dihaloalkanes by double elimination. Alkenyl halides are intermediates, being formed stereospecifically in the first elimination.

13-5 | PREPARATION OF ALKYNES FROM ALKYNYL ANIONS

Alkynes can also be prepared from other alkynes. The reaction of terminal alkynyl anions with alkylating agents, such as primary haloalkanes, oxacyclopropanes, aldehydes, or ketones, results in carbon–carbon bond formation. As we know (Section 13-2), such anions are readily prepared from terminal alkynes by deprotonation with strong bases (mostly alkyllithium reagents, sodium amide in liquid ammonia, or Grignard reagents). Alkylation with methyl or primary haloalkanes is typically done in liquid ammonia or in ether solvents. The process is unusual, because ordinary alkyl organometallic compounds are unreactive in the presence of haloalkanes. Alkynyl anions are an exception, however.

Alkylation of an Alkynyl Anion

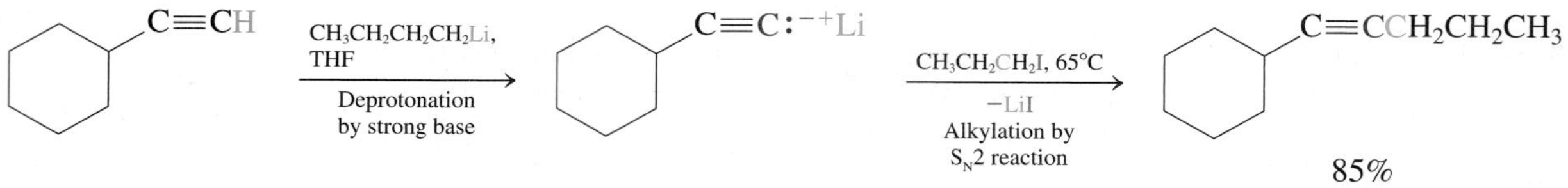

85%
1-Pentynylcyclohexane

Caution! These alkylations follow the S$_N$2 mechanism. Therefore, only methyl and primary haloalkanes are suitable alkylating agents (substrates).

Attempted alkylation of alkynyl anions with secondary and tertiary halides leads to E2 products because of the strongly basic character of the nucleophile (recall Section 7-8). Ethyne itself may be alkylated in a series of steps through the selective formation of the monoanion to give mono- and dialkyl derivatives.

Alkynyl anions react with other carbon electrophiles such as oxacyclopropanes and carbonyl compounds in the same manner as do other organometallic reagents (Sections 8-8 and 9-9).

Reactions of Alkynyl Anions

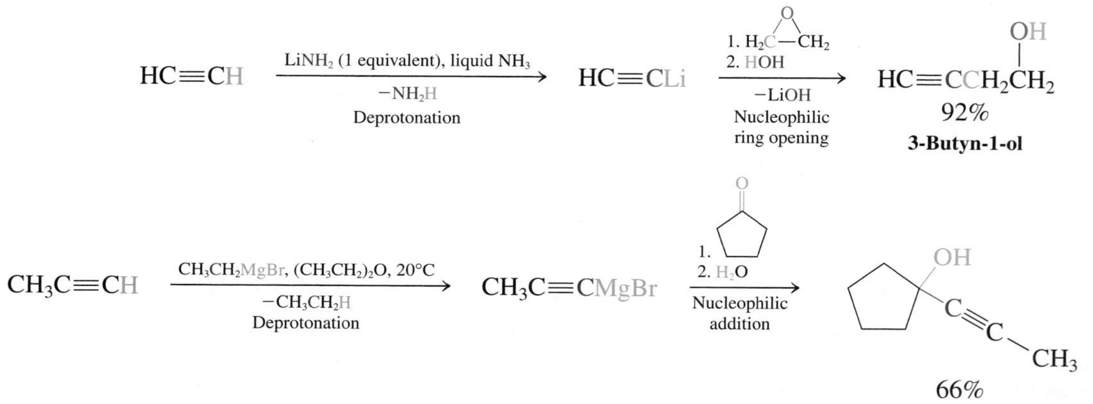

92%
3-Butyn-1-ol

66%
1-(1-Propynyl)cyclopentanol

Exercise 13-9

Suggest efficient and short syntheses of these two compounds. (**Hint:** Review Section 8-9.)

(a) 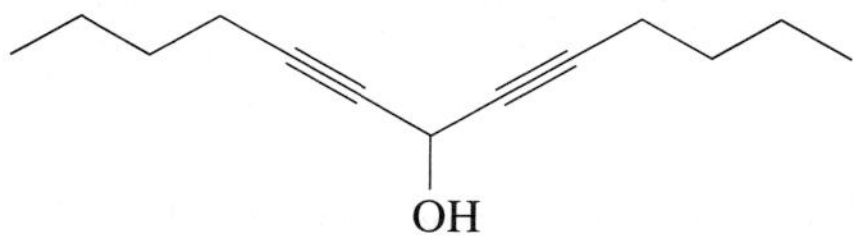 from

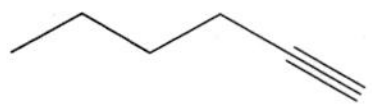

(b) 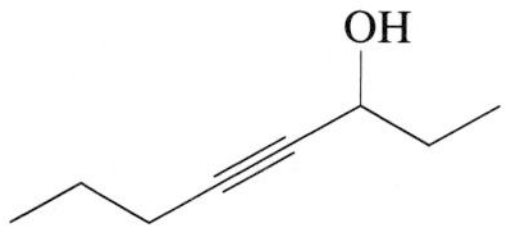 from ethyne

---

Exercise 13-10

3-Butyn-2-ol is an important raw material in the pharmaceutical industry. It is the starting point for the synthesis of a variety of medicinally valuable alkaloids (Section 25-8), steroids (Section 4-7), and prostaglandins (Real Life 11-1 and Section 19-13), as well as vitamins E (Section 22-9) and K. Propose a short synthesis of 3-butyn-2-ol by using the techniques outlined in this section.

In Summary Alkynes can be prepared from other alkynes by alkylation with primary haloalkanes, oxacyclopropanes, or carbonyl compounds. Ethyne itself can be alkylated in a series of steps.

13-6 | REDUCTION OF ALKYNES: THE RELATIVE REACTIVITY OF THE TWO Pi BONDS

Now we turn from the preparation of alkynes to the characteristic reactions of the triple bond. In many respects, alkynes are like alkenes, except for the availability of *two π* bonds. Thus, alkynes can undergo additions, such as hydrogenation and electrophilic attacks.

Addition of Reagents A–B to Alkynes

$$R-C \equiv C-R \xrightarrow{A-B} \underset{A}{\overset{R}{C}} = \underset{B}{\overset{R}{C}} \quad or \quad \underset{A}{\overset{R}{C}} = \underset{R}{\overset{B}{C}} \xrightarrow{A-B} A-\underset{A}{\overset{R}{\underset{|}{C}}}-\underset{B}{\overset{R}{\underset{|}{C}}}-B \quad or \quad A-\underset{B}{\overset{R}{\underset{|}{C}}}-\underset{A}{\overset{R}{\underset{|}{C}}}-B$$

In this section we introduce two new hydrogen addition reactions: step-by-step hydrogenation and dissolving-metal reduction by sodium to give cis and trans alkenes, respectively.

Cis alkenes can be synthesized by catalytic hydrogenation

Alkynes can be hydrogenated under the same conditions used to hydrogenate alkenes. Typically, platinum or palladium on charcoal is suspended in a solution containing the alkyne and the mixture is exposed to a hydrogen atmosphere. Under these conditions, the triple bond is saturated completely.

Complete Hydrogenation of Alkynes

$$CH_3CH_2CH_2C \equiv CCH_2CH_3 \xrightarrow{H_2,\ Pt} CH_3CH_2CH_2CH_2CH_2CH_2CH_3$$

3-Heptyne   100%
                                        Heptane

Hydrogenation is a stepwise process that may be stopped at the intermediate alkene stage by the use of modified catalysts, such as the **Lindlar* catalyst.** This catalyst is palladium that has been precipitated on calcium carbonate and treated with lead acetate and quinoline. The surface of the metal rearranges to a less active configuration than that of palladium on carbon so that only the first π bond of the alkyne is hydrogenated. As with catalytic hydrogenation of alkenes (Section 12-2), the addition of H₂ is a *syn* process (see margin). As a result, this method affords a stereoselective synthesis of cis alkenes from alkynes.

Hydrogenation with Lindlar Catalyst

3-Heptyne $\xrightarrow[\text{\textit{syn} Addition of H}_2]{H_2,\ \text{Lindlar catalyst, }25°C}$ *cis*-3-Heptene

3-Heptyne   100%
                             cis-**3-Heptene**

Two Topologies of Addition to One of the π Bonds of an Alkyne

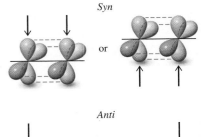

Syn

Anti

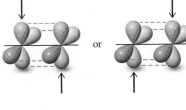

Lindlar Catalyst
5% Pd–CaCO₃,

$$Pb(O\overset{\overset{\displaystyle O}{\|}}{C}CH_3)_2,$$

Quinoline

---

*Dr. Herbert H. M. Lindlar (b. 1909), Hoffman–La Roche Ltd., Basel.

Exercise 13-11

Write the structure of the product expected from the following reaction.

$$\text{H}_2, \text{Lindlar catalyst, } 25°\text{C} \longrightarrow$$

Exercise 13-12

The perfume industry makes considerable use of naturally occurring substances such as those obtained from rose and jasmine extracts. In many cases, the quantities of fragrant oils available by natural product isolation are so small that it is necessary to synthesize them. Examples are the olfactory components of violets, which include *trans*-2-*cis*-6-nonadien-1-ol and the corresponding aldehyde. An intermediate in their large-scale synthesis is *cis*-3-hexen-1-ol, whose industrial preparation is described as "a closely guarded secret." Using the methods in this and the preceding sections, propose a synthesis from 1-butyne.

Some perfumes have star quality (from left to right): Jean Paul Gaultier MaDame Perfume, Paris Hilton Fairy Dust, Armani Prive Oranger Alhambra, and Jeanne Lanvin.

With a method for the construction of cis alkenes at our disposal, we might ask: Can we modify the reduction of alkynes to give only trans alkenes? The answer is yes, with a different reducing agent and through a different mechanism.

Sequential one-electron reductions of alkynes produce trans alkenes

When we use *sodium metal* dissolved in liquid ammonia (**dissolving-metal reduction**) as the reagent for the reduction of alkynes, we obtain trans alkenes as the products. For example, 3-heptyne is reduced to *trans*-3-heptene in this way. Unlike sodium amide in liquid ammonia, which functions as a strong base, elemental sodium in liquid ammonia acts as a powerful electron donor (i.e., a reducing agent).

Reaction

Dissolving-Metal Reduction of an Alkyne

$$\begin{array}{c}\text{1. Na, liquid NH}_3\\ \text{2. H}_2\text{O}\end{array}\longrightarrow$$

86%

3-Heptyne   *trans*-**3-Heptene**

In the first step of the mechanism of this reduction, the π framework of the triple bond accepts one electron to give a radical anion. This anion is protonated by the ammonia solvent (step 2) to give an alkenyl radical, which is further reduced (step 3) by accepting another electron to give an alkenyl anion. This species is again protonated (step 4) to give the product alkene, which is stable to further reduction. The trans stereochemistry of the final alkene is set in the first two steps of the mechanism, which give rise preferentially to the less sterically hindered trans alkenyl radical. Under the reaction conditions (liquid NH_3, $-33°C$), the second one-electron transfer takes place faster than cis-trans equilibration of the radical. This type of reduction typically provides >98% stereochemically pure trans alkene.

Mechanism of the Reduction of Alkynes by Sodium in Liquid Ammonia

Step 1. One-electron transfer

R groups adopt trans-like geometry to minimize steric repulsion

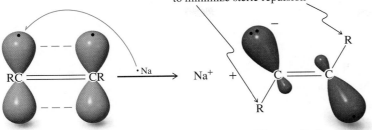

Mechanism

Alkyne radical anion

A

Step 2. First protonation

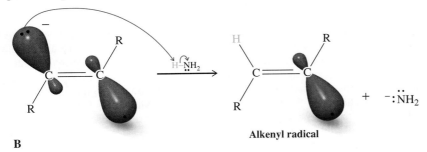

Alkenyl radical

B

Step 3. Second one-electron transfer

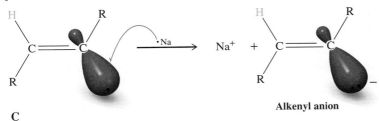

Alkenyl anion

C

Step 4. Second protonation

Trans alkene

D

The equation below illustrates the application of dissolving-metal reduction in the synthesis of the sex pheromone of the spruce budworm, which is the most destructive pest to the spruce and fir forests of North America. The pheromone "lure" is employed at hundreds of sites in the United States and Canada as part of an integrated pest-management strategy (Section 12-17). The key reaction is reduction of 11-tetradecyn-1-ol to the corresponding trans alkenol. Subsequent oxidation to the aldehyde completes the synthesis.

The spruce budworm, a serious pest.

Primary alcohol

Oxidized by PCC to aldehyde

$$HO(CH_2)_{10}C{\equiv}CCH_2CH_3 \xrightarrow[\text{Reduction}]{\text{Na, liquid NH}_3}$$

$$\underset{H}{HO(CH_2)_{10}}\overset{H}{\underset{CH_2CH_3}{C{=}C}} \xrightarrow[\text{(Section 8-6)}]{\text{PCC, CH}_2\text{Cl}_2 \\ \text{Oxidation}} \underset{O}{\overset{\parallel}{HC(CH_2)_9}}\overset{H}{\underset{CH_2CH_3}{C{=}C}}$$

11-Tetradecyn-1-ol

***trans*-11-Tetradecen-1-ol**

Sex pheromone of the spruce budworm

Solved Exercise 13-13 | Working with the Concepts: Selectivity in Reduction

When 1,7-undecadiyne (11 carbons) was treated with a mixture of sodium *and* sodium amide in liquid ammonia, only the internal bond was reduced to give *trans*-7-undecen-1-yne. Explain. (**Hint:** What reaction takes place between sodium amide and a terminal alkyne? Note that the pK_a of NH$_3$ is 35.)

Strategy
First, write out the equation for the reaction. Then consider the substrate's functionality in the context of the reaction conditions.

Solution
- The equation is

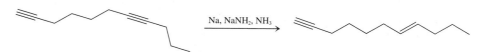

- The conditions are strongly reducing (Na), but also strongly *basic* (NaNH$_2$). We learned earlier in the chapter that the pK_a of a terminal alkynyl hydrogen is about 25. Sodium amide, which is the conjugate base of the exceedingly weak acid ammonia, readily deprotonates the terminal alkyne, giving an alkynyl anion, RC≡C:$^-$.
- The dissolving-metal reduction process requires electron transfer to the triple bond. However, the negative charge on the deprotonated *terminal* alkyne repels any attempt to introduce additional electrons, rendering that particular triple bond immune to reduction. Therefore, only the internal triple bond is reduced, producing a trans alkene.

Exercise 13-14 | Try It Yourself

What should be the result of the treatment of 2,7-undecadiyne with a mixture of excess sodium and sodium amide in liquid ammonia? Explain any differences between this outcome and that in Exercise 13-13.

In Summary Alkynes are very similar in reactivity to alkenes, except that they have two π bonds, both of which may be saturated by addition reactions. Hydrogenation of the first π bond, which gives cis alkenes, is best achieved by using the Lindlar catalyst. Alkynes are converted into trans alkenes by treatment with sodium in liquid ammonia, a process that includes two successive one-electron reductions.

13-7 | ELECTROPHILIC ADDITION REACTIONS OF ALKYNES

As a center of high electron density, the triple bond is readily attacked by electrophiles. This section describes the results of three such processes: addition of hydrogen halides, reaction with halogens, and hydration. The hydration is catalyzed by mercury(II) ions. As is the case in electrophilic additions to unsymmetrical alkenes (Section 12-3), the Markovnikov rule is followed in transformations of terminal alkynes: The electrophile adds to the terminal (less substituted) carbon atom.

Addition of hydrogen halides forms haloalkenes and geminal dihaloalkanes

The addition of hydrogen bromide to 2-butyne yields (*Z*)-2-bromobutene. The mechanism is analogous to that of hydrogen halide addition to an alkene (Section 12-3).

Addition of a Hydrogen Halide to an Internal Alkyne

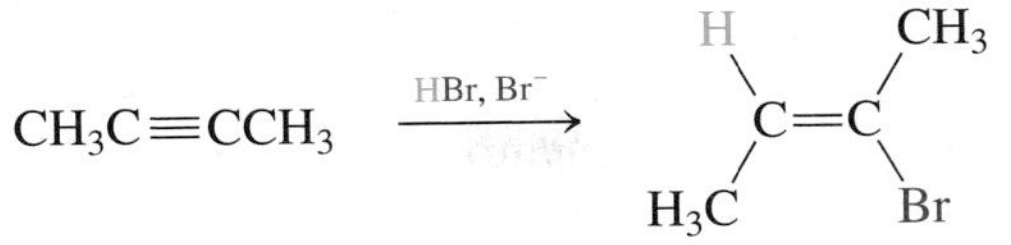

$$CH_3C{\equiv}CCH_3 \xrightarrow{\text{HBr, Br}^-}$$

H, CH₃ / C=C / H₃C, Br

60%

(Z)-2-Bromobutene

The stereochemistry of this type of addition is typically *anti,* particularly when excess halide ion is used. A second molecule of hydrogen bromide may also add, with regioselectivity that follows Markovnikov's rule, giving the product with both bromine atoms bound to the same carbon, a **geminal** dihaloalkane.

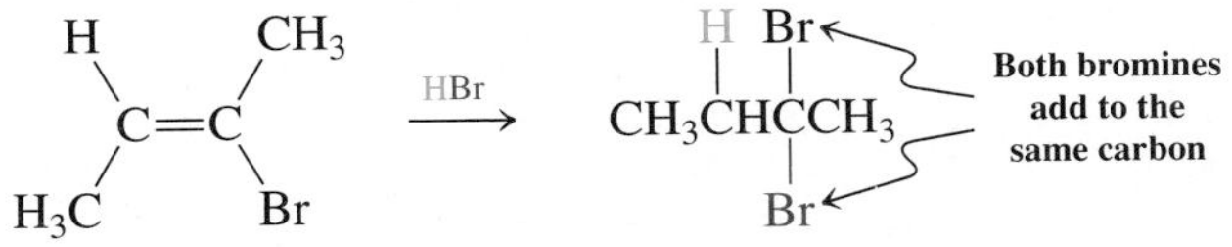

H, CH₃ / C=C / H₃C, Br $\xrightarrow{\text{HBr}}$ CH_3CHCCH_3 (with H, Br on one carbon and Br below)

Both bromines add to the same carbon

90%

2,2-Dibromobutane

The addition of hydrogen halides to terminal alkynes also proceeds in accord with the Markovnikov rule.

Addition to a Terminal Alkyne

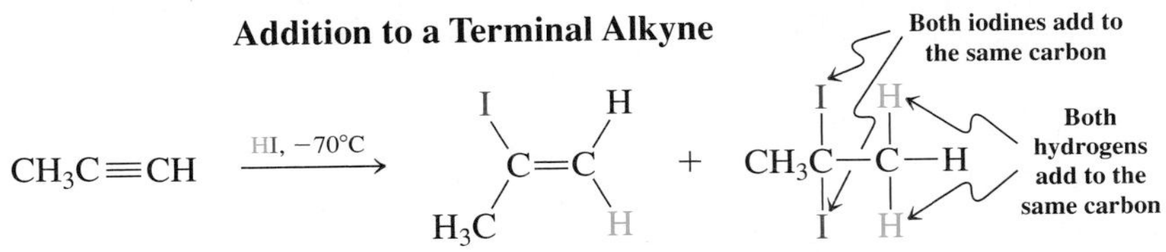

$$CH_3C{\equiv}CH \xrightarrow{\text{HI, } -70°C}$$

I, H / C=C / H₃C, H   + $CH_3C{-}C{-}H$ (with I, I and H, H)

Both iodines add to the same carbon

Both hydrogens add to the same carbon

35%   65%

It is usually difficult to limit such reactions to addition of a single molecule of HX.

Exercise 13-15

Write a step-by-step mechanism for the addition of HBr twice to 2-butyne to give 2,2-dibromobutane. Show clearly the structure of the intermediate formed in each step.

Halogenation also takes place once or twice

Electrophilic addition of halogen to alkynes proceeds through the intermediacy of isolable vicinal dihaloalkenes, the products of a single *anti* addition. Reaction with additional halogen gives tetrahaloalkanes. For example, halogenation of 3-hexyne gives the expected (*E*)-dihaloalkene and the tetrahaloalkane.

Double Halogenation of an Alkyne

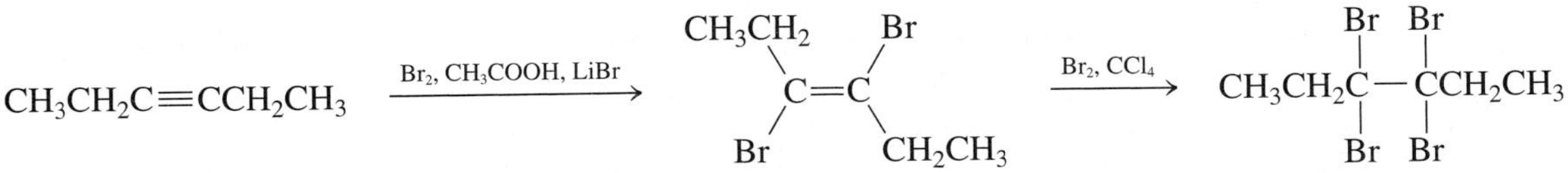

$$CH_3CH_2C{\equiv}CCH_2CH_3 \xrightarrow{\text{Br}_2,\ CH_3COOH,\ LiBr}$$

CH₃CH₂, Br / C=C / Br, CH₂CH₃ $\xrightarrow{\text{Br}_2,\ CCl_4}$ $CH_3CH_2C{-}CCH_2CH_3$ (with Br, Br and Br, Br)

3-Hexyne   99%   95%

          (*E*)-3,4-Dibromo-3-hexene   **3,3,4,4-Tetrabromohexane**

Exercise 13-16

Give the products of addition of one and two molecules of Cl_2 to 1-butyne.

Mercuric ion-catalyzed hydration of alkynes furnishes ketones

In a process analogous to the hydration of alkenes, water can be added to alkynes in a Markovnikov sense to give alcohols—in this case **enols,** in which the hydroxy group is attached to a double-bond carbon. As mentioned in Section 12-16, enols spontaneously rearrange to the isomeric carbonyl compounds. This process, called **tautomerism,** interconverts two isomers by simultaneous proton and double-bond shifts. The enol is said to **tautomerize** to the carbonyl compound, and the two species are called **tautomers** (*tauto,* Greek, the same; *meros,* Greek, part). We shall look at tautomerism more closely in Chapter 18 when we investigate the behavior of carbonyl compounds. Hydration followed by tautomerism converts alkynes into ketones. The reaction is catalyzed by Hg(II) ions.

Hydration of Alkynes

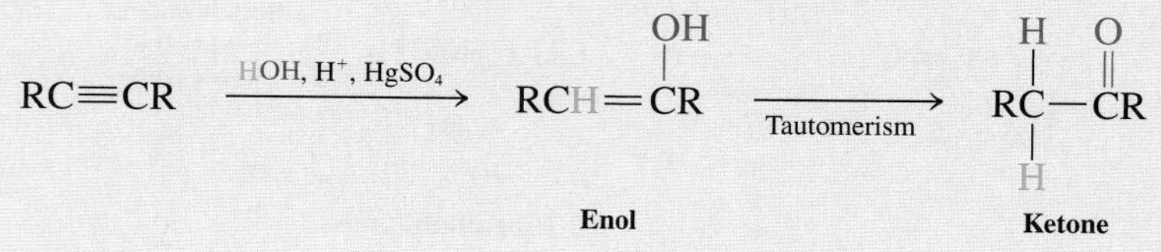

$$RC\equiv CR \xrightarrow{\text{HOH, H}^+, \text{HgSO}_4} RCH=CR(OH) \xrightarrow{\text{Tautomerism}} RCH_2-C(=O)R$$

Enol   Ketone

Hydration follows Markovnikov's rule: Terminal alkynes give methyl ketones.

Hydration of a Terminal Alkyne

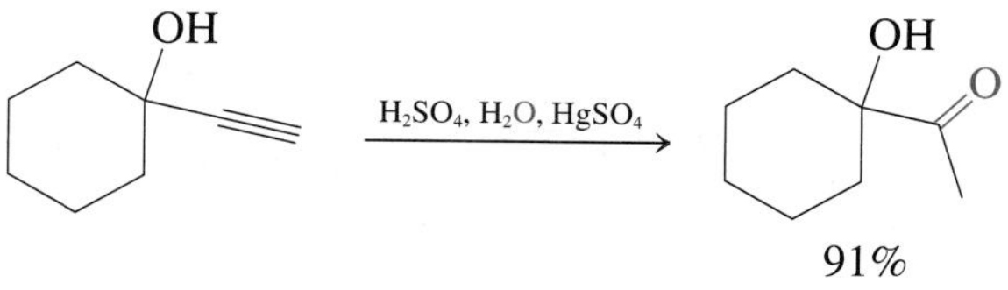

91%

Exercise 13-17

Draw the structure of the enol intermediate in the reaction above.

Symmetric internal alkynes give a single carbonyl compound; unsymmetric systems lead to a mixture of ketones.

Hydration of Internal Alkynes

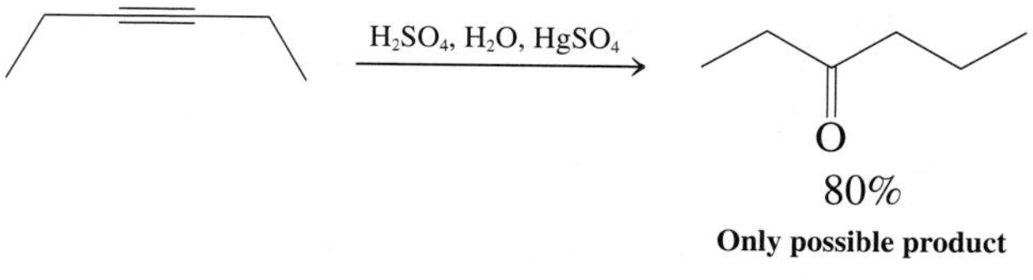

80%
Only possible product

Example of Hydration of an Internal Alkyne That Gives a Mixture of Two Ketones

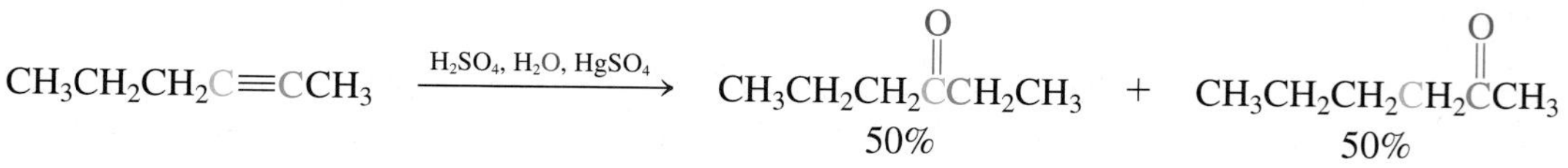

$$CH_3CH_2CH_2C\equiv CCH_3 \xrightarrow{\text{H}_2\text{SO}_4, \text{H}_2\text{O}, \text{HgSO}_4} CH_3CH_2CH_2\overset{O}{\overset{\|}{C}}CH_2CH_3 + CH_3CH_2CH_2CH_2\overset{O}{\overset{\|}{C}}CH_3$$

50%   50%

Exercise 13-18

Give the products of mercuric ion-catalyzed hydration of **(a)** ethyne; **(b)** propyne; **(c)** 1-butyne; **(d)** 2-butyne; **(e)** 2-methyl-3-hexyne.

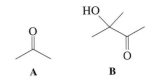

A   **B**

Solved Exercise 13-19 | Working with the Concepts: Using Alkynes in Synthesis

Propose a synthetic scheme that will convert compound A into B (see margin). [**Hint:** Consider a route that proceeds through the alkynyl alcohol $(CH_3)_2CC\equiv CH$.]

Strategy
The hint reveals to us a possible retrosynthetic analysis of the problem:

Solution
- Adding the anion to acetone gives the necessary intermediate alcohol:

$$HC\equiv CH \xrightarrow[\text{liquid } NH_3]{LiNH_2 \text{ (1 equivalent),}} HC\equiv CLi \xrightarrow[\text{2. } H_2O]{1.} HO$$

- Finally, hydration of the terminal alkyne function, as illustrated for the cyclohexyl derivative shown on the previous page, completes the synthesis:

$$\xrightarrow{H_2SO_4, H_2O, HgSO_4}$$

Let us consider what we have learned so far in this chapter that can be helpful. This section has shown how alkynes can be converted to ketones by mercury ion-catalyzed hydration. Section 13-5 introduced a new strategy for the formation of carbon–carbon bonds through the use of alkynyl anions. Beginning with the three-carbon ketone A (acetone), our first task is to add a two-carbon alkynyl unit. Referring to Section 13-5, we can use any of several methods to convert ethyne into the corresponding anion.

Exercise 13-20 | Try It Yourself

Propose a synthesis of *trans*-3-hexene starting with 1-butyne.

In Summary Alkynes can react with electrophiles such as hydrogen halides and halogens either once or twice. Terminal alkynes transform in accord with the Markovnikov rule. Mercuric ion-catalyzed hydration furnishes enols, which convert into ketones by a process called tautomerism.

13-8 | ANTI-MARKOVNIKOV ADDITIONS TO TRIPLE BONDS

Just as methods exist to permit anti-Markovnikov additions to double bonds (Sections 12-8 and 12-13), similar techniques allow additions to terminal alkynes to be carried out in an anti-Markovnikov manner.

Radical addition of HBr gives 1-bromoalkenes

As with alkenes, hydrogen bromide can add to triple bonds by a radical mechanism in an anti-Markovnikov fashion if light or other radical initiators are present. Both *syn* and *anti* additions are observed.

$$CH_3(CH_2)_3C\equiv CH \xrightarrow{\text{HBr, ROOR}} CH_3(CH_2)_3CH=CHBr$$

74%

1-Hexyne   *cis*- and *trans*-**1-Bromo-1-hexene**

Aldehydes result from hydroboration–oxidation of terminal alkynes

Terminal alkynes are hydroborated in a regioselective, anti-Markovnikov fashion, the boron attacking the less hindered carbon. However, with borane itself, this reaction leads ultimately to sequential hydroboration of both π bonds. To stop at the alkenylborane stage, bulky borane reagents, such as dicyclohexylborane, are used.

Hydroboration of a Terminal Alkyne

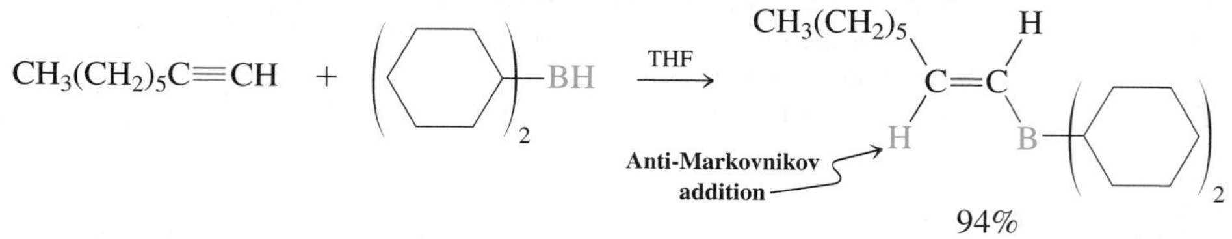

1-Octyne   **Dicyclohexylborane**   **Dicyclohexyl (*E*-1-octenyl) borane**

Exercise 13-21

Dicyclohexylborane is made by a hydroboration reaction. What are the starting materials for its preparation?

Like alkylboranes (Section 12-8), alkenylboranes can be oxidized to the corresponding alcohols—in this case, to terminal enols that spontaneously rearrange to aldehydes.

Hydroboration–Oxidation of a Terminal Alkyne

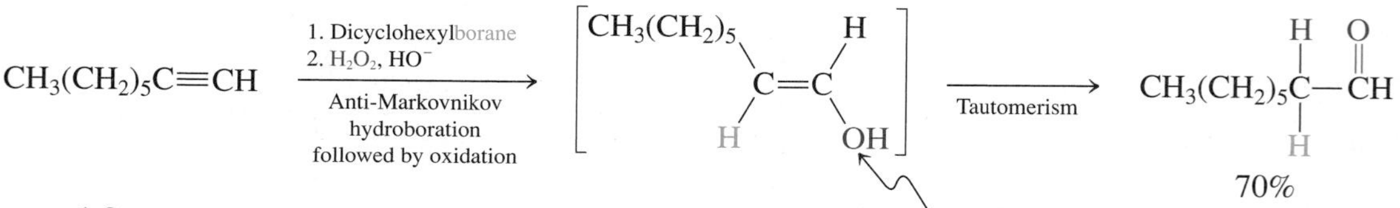

1-Octyne   **Enol**   OH on less substituted carbon   70%   **Octanal**

Exercise 13-22

Give the products of hydroboration–oxidation of **(a)** ethyne; **(b)** 1-propyne; **(c)** 1-butyne.

Exercise 13-23

Outline a synthesis of the following molecule from 3,3-dimethyl-1-butyne.

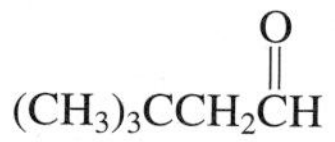

$$(CH_3)_3CCH_2\overset{\displaystyle O}{\overset{\|}{C}}H$$

In Summary HBr in the presence of peroxides undergoes anti-Markovnikov addition to terminal alkynes to give 1-bromoalkenes. Hydroboration–oxidation with bulky boranes furnishes intermediate enols that tautomerize to the final product aldehydes.

13-9 | CHEMISTRY OF ALKENYL HALIDES

We have encountered haloalkenes—alkenyl halides—as intermediates in both the preparation of alkynes by dehydrohalogenation and also the addition to alkynes of hydrogen halides. Alkenyl halides have become increasingly important as synthetic intermediates in recent years as a result of developments in organometallic chemistry. These systems do not, however, follow the mechanisms familiar to us from our survey of the haloalkanes (Chapters 6 and 7). This section discusses their reactivity.

Alkenyl halides do not undergo S$_N$2 or S$_N$1 reactions

Unlike haloalkanes, alkenyl halides are relatively unreactive toward nucleophiles. Although we have seen that, with strong bases, alkenyl halides undergo elimination reactions to give alkynes, they do not react with weak bases and relatively nonbasic nucleophiles, such as iodide. Similarly, S$_N$1 reactions do not normally take place, because the intermediate alkenyl cations are species of high energy.

Alkenyl halides, however, can react through the intermediate formation of alkenyl organometallics (see Exercise 11-6). These species allow access to a variety of specifically substituted alkenes.

Alkenyl Organometallics in Synthesis

Metal catalysts couple alkenyl halides to alkenes in the Heck reaction

In the presence of soluble complexes of metals such as Ni and Pd, alkenyl halides undergo carbon–carbon bond formation with alkenes to produce dienes. In this process, called the **Heck* reaction,** a molecule of hydrogen halide is liberated.

The Heck Reaction

*Professor Richard F. Heck (b. 1931), University of Delaware, Nobel Prize 2010 (chemistry).

| **Metal-Catalyzed Stille, Suzuki, and Sonogashira Coupling Reactions**

Three additional processes, the Stille, Suzuki, and Sonogashira* reactions, further broaden the scope of transition metal-catalyzed bond-forming processes. All utilize catalytic palladium or nickel; the differences lie in the nature and functionality of the substrates commonly employed.

In the Stille coupling, Pd catalyzes the direct linkage between alkenyl halides and alkenyltin compounds:

Stille Coupling Reaction

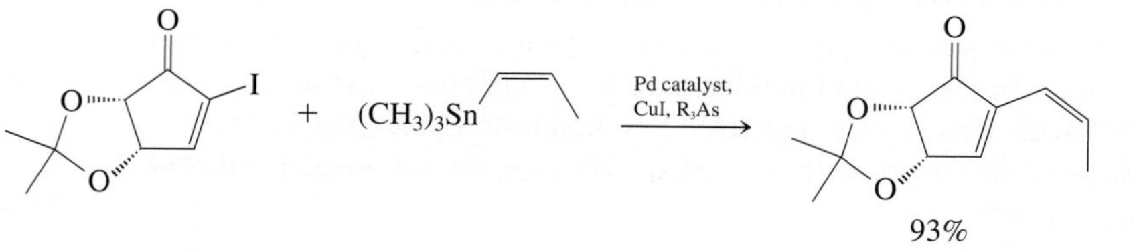

93%

Copper(I) iodide and an arsenic-derived ligand, R₃As, facilitate this very efficient process. The product shown was converted into a close relative of a microbially derived natural product that inhibits a factor associated with immune and inflammation responses. This factor also affects HIV activation and cell-death processes that are disrupted in cancer.

The Suzuki reaction replaces tin with boron and provides a different spectrum of utility. In particular,

Suzuki Coupling Reaction

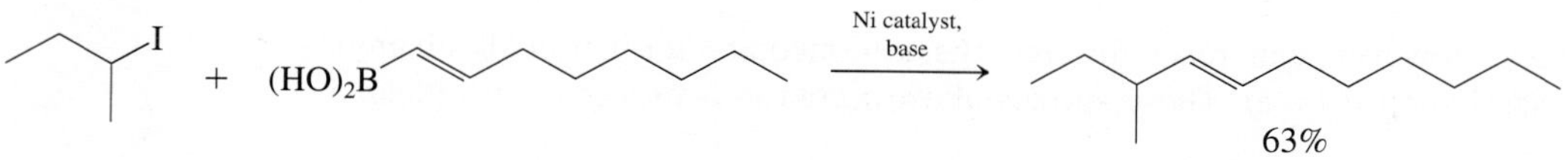

63%

*Professor John K. Stille (1930–1990), Colorado State University; Professor Akira Suzuki (b. 1930), Kurashiki University, Japan, Nobel Prize 2010 (chemistry); Professor Kenkichi Sonogashira (b. 1931), Osaka City University, Japan.

In common with other transition metal-catalyzed cross-couplings (see Real Life 8-3), assembly of the fragments around the catalyst precedes carbon–carbon bond formation. A simplified mechanism for the Heck reaction begins with attack of the metal on the alkenyl halide to give an alkenylmetal halide (1). The alkene then complexes with the metal (2), and inserts itself into the carbon–metal bond, forming the new carbon–carbon linkage (3). Finally, elimination of HX in an E2-like manner gives the diene product and frees the metal catalyst (4).

Mechanism of the Heck Reaction

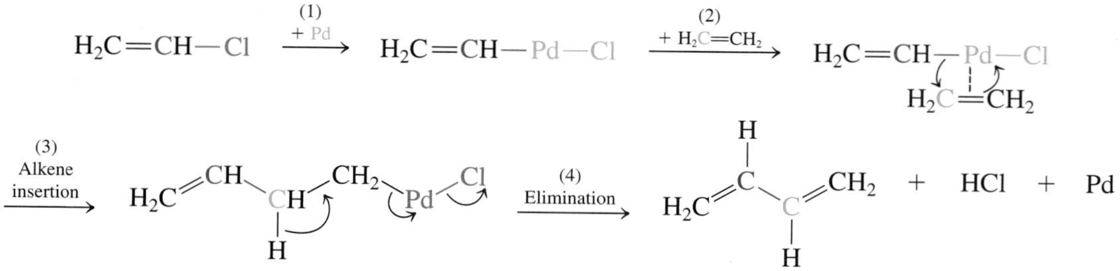

The growing popularity of the Heck reaction arises from both its versatility and its efficiency. In particular, it requires only a very small amount of catalyst compared with the quantity of the substrates; typically, 1% of palladium acetate in the presence of a phosphine ligand (R₃P) is sufficient.

Suzuki coupling succeeds with primary and secondary haloalkanes, which are poor Stille substrates. In the example below, Ni gives better results than does Pd.

The boron-containing substrate (a **boronic acid**) is efficiently prepared by hydroboration of a terminal alkyne with a special reagent, catechol borane:

Preparation of an Alkenylboronic Acid

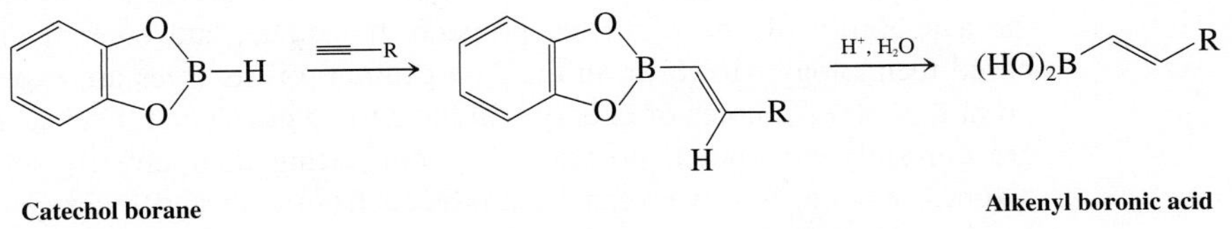

Catechol borane   Alkenyl boronic acid

Boronic acids are prepared commercially in very large quantities, and the Suzuki coupling has become a major industrial process. Boronic acids are stable and easier to handle than organotin compounds, which are toxic and must be handled with great care.

Finally, the Sonogashira reaction has a niche of its own as a preferred method for linking alkenyl and alkynyl

moieties. As in the Stille process, Pd, CuI, and ligands derived from nitrogen-group elements are employed. However, there is no need for tin; terminal alkynes react directly. The added base removes the HI by-product.

Sonogashira Coupling Reaction

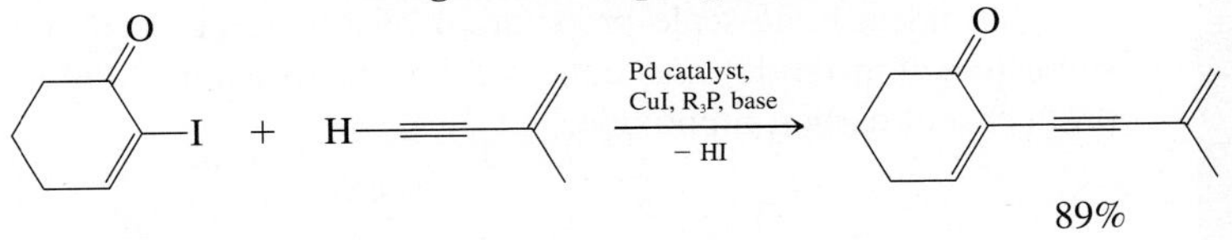

89%

Examples of Heck Reactions

72%

67%

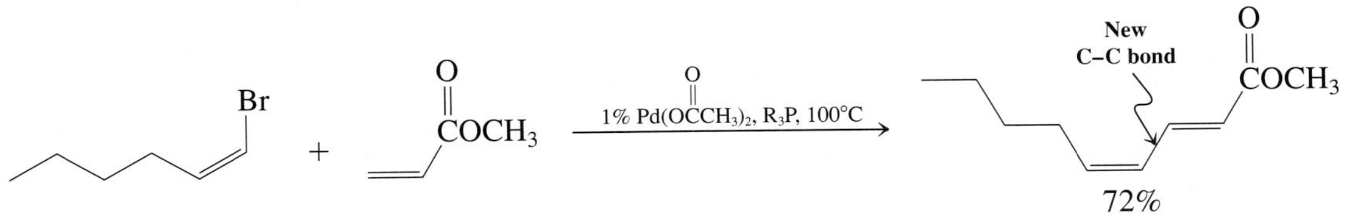

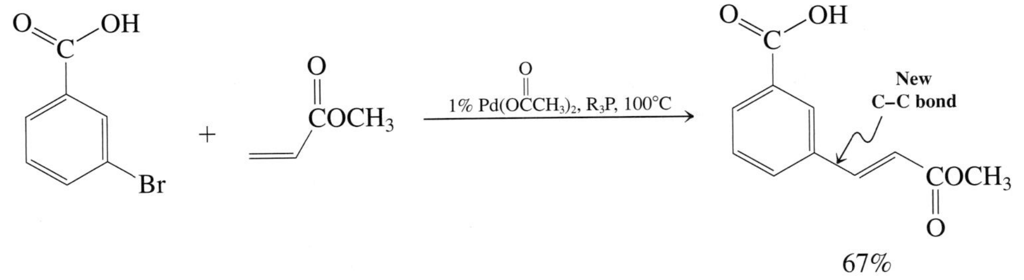

Exercise 13-24

Write out a detailed step-by-step mechanism for the first of the two examples of Heck reactions above.

In Summary Alkenyl halides are unreactive in nucleophilic substitutions. However, they can participate in carbon–carbon bond-forming reactions after conversion to alkenyllithium or alkenyl Grignard reagents, or in the presence of transition-metal catalysts such as Ni and Pd.

13-10 | ETHYNE AS AN INDUSTRIAL STARTING MATERIAL

Ethyne was once one of the four or five major starting materials in the chemical industry for two reasons: Addition reactions to one of the π bonds produce useful alkene monomers (Section 12-15), and it has a high heat content. Its industrial use has declined because of the availability of cheap ethene, propene, butadiene, and other hydrocarbons through oil-based technology. However, in the 21st century, oil reserves are expected to dwindle to the point that other sources of energy will have to be developed. One such source is coal. There are currently no known processes for converting coal directly into the aforementioned alkenes; ethyne, however, can be produced from coal and hydrogen or from coke (a coal residue obtained after removal of volatiles) and limestone through the formation of calcium carbide. Consequently, it may once again become an important industrial raw material.

Production of ethyne from coal requires high temperatures

The high energy content of ethyne requires the use of production methods that are costly in energy. One process for making ethyne from coal uses hydrogen in an arc reactor at temperatures as high as several thousand degrees Celsius.

Vivid demonstration of the combustion of ethyne, generated by the addition of water to calcium carbide.

$$\text{Coal} \quad + \quad \text{H}_2 \quad \xrightarrow{\Delta} \quad \underset{\textbf{33\% conversion}}{\text{HC}\equiv\text{CH}} \quad + \quad \text{nonvolatile salts}$$

The oldest large-scale preparation of ethyne proceeds through calcium carbide. Limestone (calcium oxide) and coke are heated to about 2000°C, which results in the desired product and carbon monoxide.

$$\underset{\textbf{Coke}}{3\,\text{C}} \quad + \quad \underset{\textbf{Lime}}{\text{CaO}} \quad \xrightarrow{2000°\text{C}} \quad \underset{\textbf{Calcium carbide}}{\text{CaC}_2} \quad + \quad \text{CO}$$

The calcium carbide is then treated with water at ambient temperatures to give ethyne and calcium hydroxide.

$$\text{CaC}_2 \quad + \quad 2\,\text{H}_2\text{O} \quad \longrightarrow \quad \text{HC}\equiv\text{CH} \quad + \quad \text{Ca(OH)}_2$$

Ethyne is a source of valuable monomers for industry

Ethyne chemistry underwent important commercial development in the 1930s and 1940s in the laboratories of Badische Anilin and Sodafabriken (BASF) in Ludwigshafen, Germany. Ethyne under pressure was brought into reaction with carbon monoxide, carbonyl compounds, alcohols, and acids in the presence of catalysts to give a multitude of valuable raw materials to be used in further transformations. For example, nickel carbonyl catalyzes the addition of carbon monoxide and water to ethyne to give propenoic (acrylic) acid. Similar exposure to alcohols or amines instead of water results in the corresponding acid derivatives. All of these products are valuable monomers (see Section 12-15).

Industrial Chemistry of Ethyne

$$\text{HC}\equiv\text{CH} \quad + \quad \text{CO} \quad + \quad \text{H}_2\text{O} \quad \xrightarrow{\text{Ni(CO)}_4,\ 100\ \text{atm},\ >250°\text{C}} \quad \underset{\substack{\textbf{Propenoic acid}\\ \textbf{(Acrylic acid)}}}{\overset{\text{H}}{\underset{\text{H}}{}}\text{C}=\text{CHCOOH}}$$

Polymerization of propenoic (acrylic) acid and its derivatives produces materials of considerable utility. The polymeric esters (**polyacrylates**) are tough, resilient, and flexible polymers that have replaced natural rubber (see Section 14-10) in many applications. Poly(ethyl acrylate) is used for O-rings, valve seals, and related purposes in automobiles. Other polyacrylates are found in biomedical and dental appliances, such as dentures.

The addition of formaldehyde to ethyne is achieved with high efficiency by using copper acetylide as a catalyst.

$$HC\equiv CH \quad + \quad CH_2=O \xrightarrow{Cu_2C_2-SiO_2,\ 125°C,\ 5\ atm} HC\equiv CCH_2OH \quad \text{or} \quad HOCH_2C\equiv CCH_2OH$$

2-Propyn-1-ol
(Propargyl alcohol)

2-Butyne-1,4-diol

The resulting alcohols are useful synthetic intermediates. For example, 2-butyne-1,4-diol is a precursor for the production of oxacyclopentane (tetrahydrofuran, one of the solvents most frequently employed for Grignard and organolithium reagents) by hydrogenation, followed by acid-catalyzed dehydration.

Oxacyclopentane (Tetrahydrofuran) Synthesis

$$HOCH_2C\equiv CCH_2OH \xrightarrow{Catalyst,\ H_2} HO(CH_2)_4OH \xrightarrow[\ -H_2O\ ]{\substack{H_3PO_4,\ pH\ 2,\\ 260-280°C,\ 90-100\ atm}}$$

99%
Oxacyclopentane
(Tetrahydrofuran, THF)

Several technical processes have been developed in which reagents $^{\delta+}A-B^{\delta-}$ in the presence of a catalyst add to the triple bond. For example, the catalyzed addition of hydrogen chloride gives chloroethene (vinyl chloride), and addition of hydrogen cyanide produces propenenitrile (acrylonitrile).

Addition Reactions of Ethyne

$$HC\equiv CH \quad + \quad HCl \xrightarrow{Hg^{2+},\ 100-200°C} \overset{H}{\underset{H}{C}}{=}CHCl$$

Chloroethene
(Vinyl chloride)

$$HC\equiv CH \quad + \quad HCN \xrightarrow{Cu^+,\ NH_4Cl,\ 70-90°C,\ 1.3\ atm} \overset{H}{\underset{H}{C}}{=}CHCN$$

80–90%
Propenenitrile
(Acrylonitrile)

Poly(vinyl chloride) is used extensively in the construction industry (pipes, "vinyl" siding, window and door frames), for the insulation on electrical cables, in clothing (such as in Goth and Punk fashion), for medical devices, and the manufacture of ordinary garden hoses.

In 2009, the world produced 1.9 million tons of **acrylic fibers,** polymers containing at least 85% propenenitrile (acrylonitrile). Their applications include clothing (Orlon), carpets, and insulation. Copolymers of acrylonitrile and 10–15% vinyl chloride have fire-retardant properties and are used in children's sleepwear.

In Summary Ethyne was once, and may again be in the future, a valuable industrial feed stock because of its ability to react with a large number of substrates to yield useful monomers and other compounds having functional groups. It can be made from coal and H_2 at high temperatures, or it can be prepared from calcium carbide by hydrolysis. Some of the industrial reactions that it undergoes are carbonylation, addition of formaldehyde, and addition reactions with HX.

13-11 | ALKYNES IN NATURE AND IN MEDICINE

Although alkynes are comparatively scarce in nature, they have been isolated from a wide variety of plant species, higher fungi, and marine invertebrates. Well over a thousand such compounds are now known, and many show interesting physiological activity. The first natural alkyne to be isolated, in 1826, was the antitumor dehydromatricaria ester, from the chamomile flower. Capillin, an oil extracted from chrysanthemums, has fungicidal activity, in addition to inhibiting cell proliferation in colon, pancreatic, and lung tumors. Reactive enediyne (—C≡C—CH=CH—C≡C—) and trisulfide (RSSSR) groups characterize a new class of extremely potent antibiotic–antitumor agents discovered in the late 1980s, such as calicheamicin and esperamicin.

trans-Dehydromatricaria ester
(Antitumor agent)

Capillin
(Antifungal and carcinostatic)

Calicheamicin (X = H)
Esperamicin (X = OR′)
R and R′ = sugars (Chapter 24)

Enediyne group

Turning to the subject of many an adventure story, ichthyothereol is the active ingredient of a poisonous substance used in arrowheads by the Indians of the Lower Amazon River Basin. It causes convulsions in mammals. And histrionicotoxin is isolated from the skin of "poison arrow frog," a highly colorful species of the genus _Dendrobates_. The frog secretes this compound and similar ones as defensive venoms and mucosal-tissue irritants against both mammals and reptiles.

"Poison arrow" frog.

Ichthyothereol
(Convulsant)

Histrionicotoxin

Many drugs have been modified by synthesis to contain alkyne substituents, with the aim to increase bioavailability and activity, and at the same time reduce potential toxicity. For example, ethynyl estrogens, such as 17-ethynylestradiol, are considerably more potent birth-control agents than are the naturally occurring hormones (see Section 4-7).

17-Ethynylestradiol

BIIB021
(Tumor inhibitor)

EC144
(Second-generation drug)

The reasons for the beneficial effect of this alteration are not clear, but one contributor is the compactness of the hydrophobic alkyne unit and its shape of a "rigid rod" analog of ethyl or similar-sized groups. A current case study that illustrates this point features an anticancer agent in clinical trials, BIIB021. In initial screening experiments, the drug was found to bind strongly to a test protein called Heat Shock Protein 90 (Hsp90). Its name relates to its function to protect cells when they are stressed by sudden temperature rises. This protein also controls the folding of other proteins, many of which are involved in carcinogenesis, so blocking it with pharmaceuticals might be beneficial in the treatment of cancer (as turned out to be correct in this instance). The X-ray picture of the binding site (margin) reveals a smooth fit of the molecule to its surroundings, but a narrow region of empty space above the five-membered ring nitrogen (usually filled by solvent molecules). Consequently, a second-generation drug was designed, EC144, modified to include an alkyne group pointed in exactly the direction of the "hole." This drug turns out to be superior in its binding to Hsp90 and also in its tumor-inhibitory properties in mice.

In Summary The alkyne unit is present in a number of physiologically active natural and synthetic compounds.

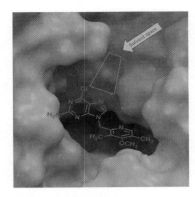

Binding of BIIB021 to Hsp90 (shown as the cloud-like polypeptide surface). There is an unoccupied region of space above the five-membered ring nitrogen.

THE BIG PICTURE

As we said at the end of the previous chapter, much of what we have encountered in our examination of the alkynes represents an extension of what we learned regarding alkenes. Addition reactions take place under very similar conditions, obeying the same rules of regio- and stereochemistry. Reagents such as the hydrogen halides and the halogens may add once or twice. Addition of the elements of water to one of the π bonds takes us in a new direction, however: The resulting alkenol (or enol, for short) undergoes rearrangement (tautomerism) to an aldehyde or a ketone. Finally, terminal alkynes display a form of reactivity that does not normally appear in the alkenes (or alkanes, for that matter): The $-C{\equiv}C-H$ hydrogen is unusually acidic. Its deprotonation gives rise to nucleophilic anions capable of forming new carbon–carbon bonds by reaction with a variety of functional groups possessing electrophilic carbon atoms.

In the next chapter we shall examine compounds containing multiple double bonds, including some made by the Heck process, a new organometallic reaction that we have just encountered. The same principles that have appeared repeatedly in Chapters 11–13 will continue to underlie the behavior of the systems that we cover next.

WORKED EXAMPLES: INTEGRATING THE CONCEPTS

13-25. Designing an Alkyne-Based Synthesis

Propose an efficient synthesis of 2,7-dimethyl-4-octanone, using organic building blocks containing no more than four carbon atoms.

2,7-Dimethyl-4-octanone

SOLUTION
We begin by analyzing the problem retrosynthetically (Section 8-9). What methods do we know for the preparation of ketones? We can oxidize alcohols (Section 8-6). Is this a productive line of analysis? If we look at the corresponding alcohol precursor, we can imagine synthesizing it by addition of an appropriate organometallic reagent to an aldehyde to form either bond *a* or bond *b* (Section 8-9):

Target molecule   **Alcohol precursor**

Let us count carbon atoms in the fragments necessary for each of these synthetic pathways. To make bond *a*, we need to add a four-carbon organometallic to a six-carbon aldehyde. The bond *b* alternative would employ 2 five-carbon building blocks. Remember the restriction that only four-carbon starting materials are allowed. From this point of view, neither of the preceding options is overly attractive. We shall shortly look again at route *a* but not route *b*. Do you see why? The latter would require initial construction of 2 five-carbon pieces, whereas the former needs formation of only 1 six-carbon unit from fragments containing four carbons or fewer.

Now let us consider a second, fundamentally different ketone synthesis—hydration of an alkyne (Section 13-7). Either of two precursors, 2,7-dimethyl-3-octyne and 2,7-dimethyl-4-octyne, will lead to the target molecule. As shown here, however, only the latter, *symmetric* alkyne undergoes hydration to give just one ketone, regardless of the initial direction of addition.

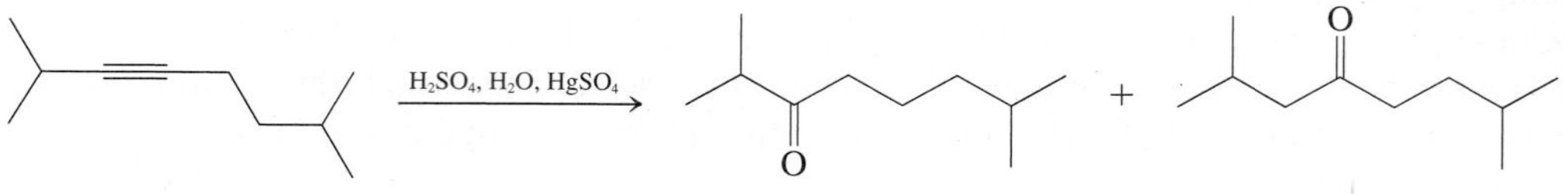

2,7-Dimethyl-3-octyne   H₂SO₄, H₂O, HgSO₄

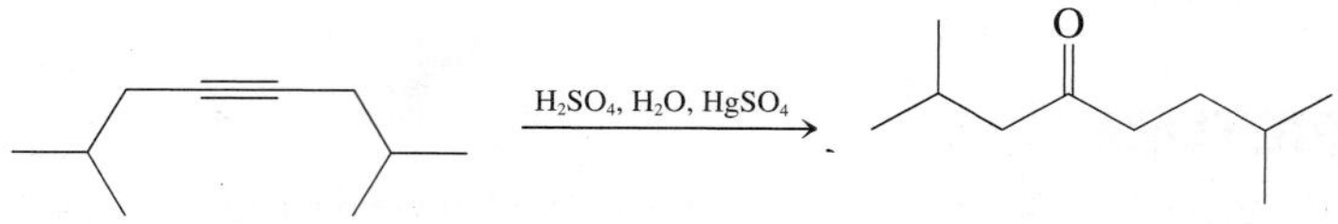

2,7-Dimethyl-4-octyne   H₂SO₄, H₂O, HgSO₄

With 2,7-dimethyl-4-octyne appearing to be the precursor of choice, we continue by investigating its synthesis from building blocks of four carbons or fewer. The alkylation of terminal alkynes (Section 13-5) affords us a method of bond formation that divides the molecule into three suitable fragments, shown in the following analysis:

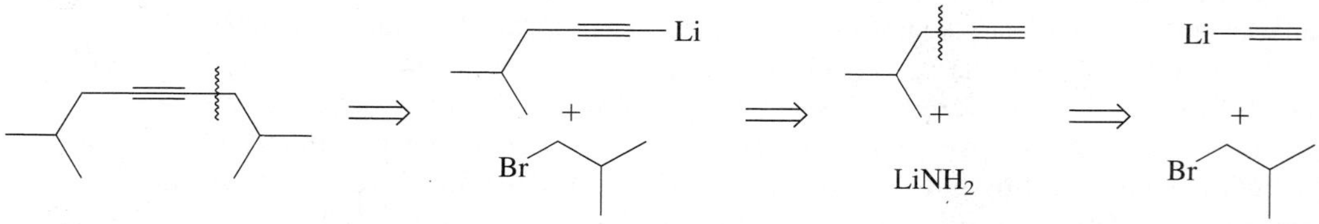

The synthesis follows directly:

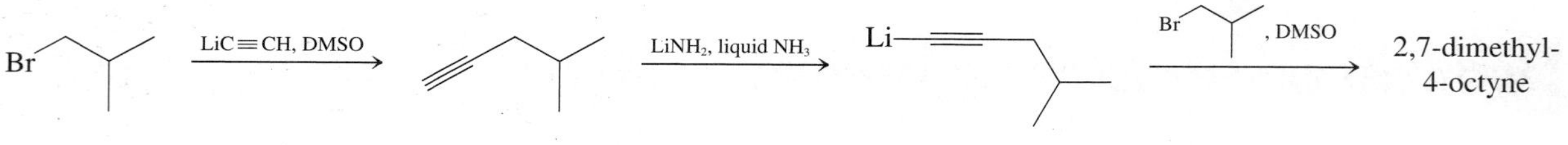

LiC≡CH, DMSO   LiNH₂, liquid NH₃   Br⟍, DMSO   2,7-dimethyl-4-octyne

Although this three-step synthesis is the most efficient answer, a related approach derives from our earlier consideration of ketone synthesis with the use of an alcohol. It, too, proceeds through an alkyne. Construction of bond *a* of the target molecule, shown earlier, requires addition of an organometallic reagent to a six-carbon aldehyde, which, in turn, may be produced through hydroboration–oxidation (Section 13-8) of the terminal alkyne shown in the preceding scheme.

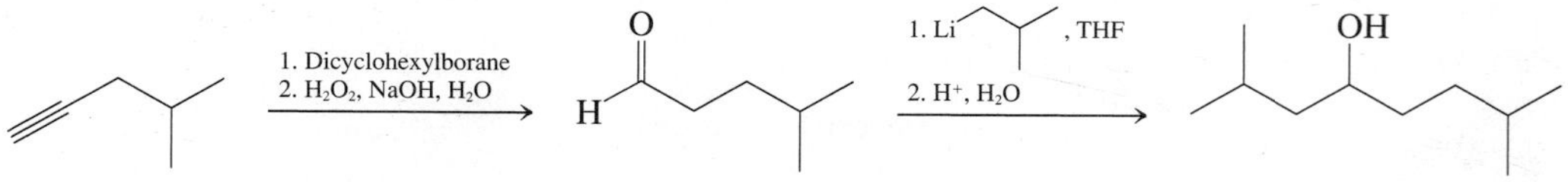

1. Dicyclohexylborane
2. H₂O₂, NaOH, H₂O

1. Li⟍, THF
2. H⁺, H₂O

Oxidation of this alcohol by using a Cr(VI) reagent (Section 8-6) completes a synthesis that is just slightly longer than the optimal one described first.

13-26. Predicting the Product of a New Reaction by Thinking Mechanistically

Predict the product you would expect from the treatment of a terminal alkyne with bromine in water solvent, i.e.,

$$CH_3CH_2C\equiv CH \xrightarrow{Br_2, H_2O}$$

SOLUTION

Consider the problem mechanistically. Bromine adds to π bonds to form a cyclic bromonium ion, which is subject to ring opening by any available nucleophile. In the similar reaction of alkenes (Sections 12-5 and 12-6), nucleophilic attack is directed to the more substituted alkene carbon, namely, the carbon atom bearing the larger partial positive charge. Following that analogy in the case at hand and using water as the nucleophile, we may postulate the following as reasonable mechanistic steps:

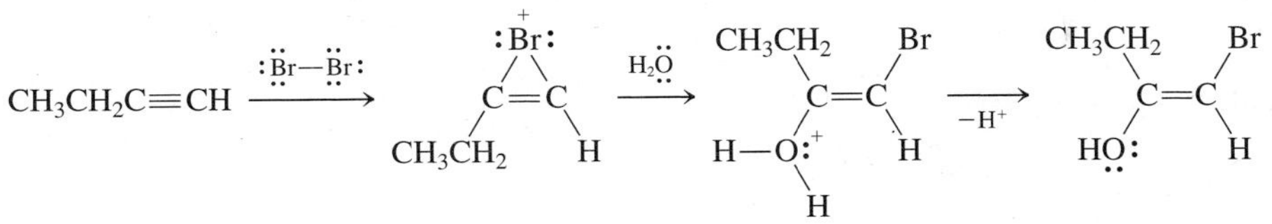

The product of the sequence is an enol, which, as we have seen (Section 13-8), is unstable and rapidly

tautomerizes into a carbonyl compound. In this case, the ultimate product is $CH_3CH_2-\overset{\overset{\displaystyle :O:}{||}}{C}-CH_2Br$, a bromoketone.

New Reactions

1. Acidity of 1-Alkynes (Section 13-2)

$$RC\equiv CH + :B^- \rightleftharpoons RC\equiv C:^- + BH$$

$pK_a \approx 25$

Base (B): $NaNH_2$–liquid NH_3; RLi–$(CH_3CH_2)_2O$; $RMgX$–THF

Preparation of Alkynes

2. Double Elimination from Dihaloalkanes (Section 13-4)

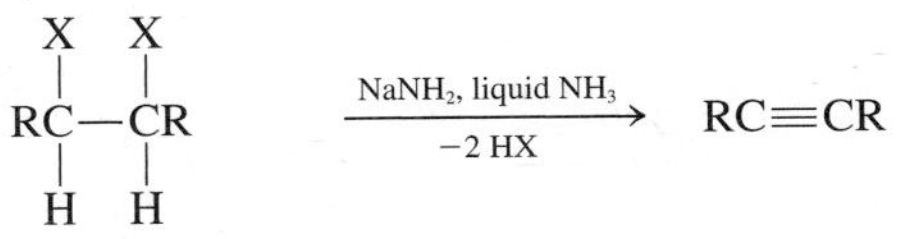

Vicinal dihaloalkane

3. From Alkenes by Halogenation–Dehydrohalogenation (Section 13-4)

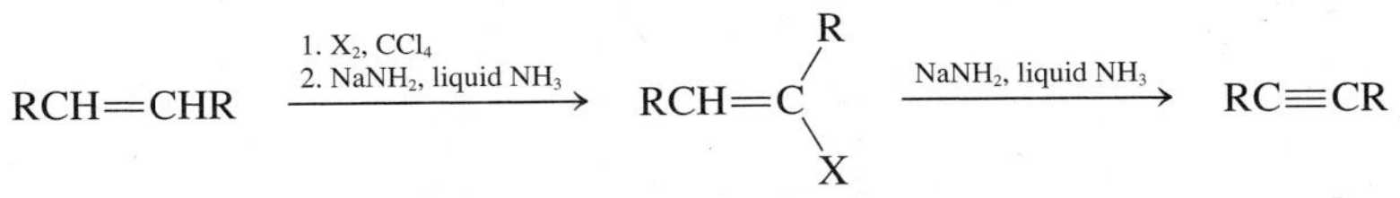

Alkenyl halide intermediate

Conversion of Alkynes into Other Alkynes

4. Alkylation of Alkynyl Anions (Section 13-5)

$$RC\equiv CH \xrightarrow[\text{2. R'X}]{\text{1. }NaNH_2,\text{ liquid }NH_3} RC\equiv CR'$$

S_N2 reaction: R' must be primary

5. Alkylation with Oxacyclopropane (Section 13-5)

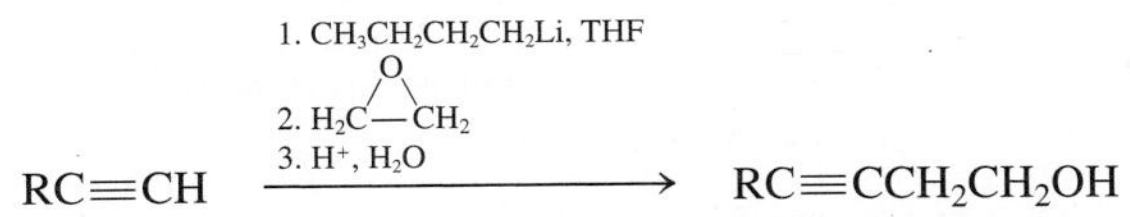

$$RC\equiv CH \xrightarrow[\substack{\text{2. } H_2C-CH_2 \\ \text{3. } H^+, H_2O}]{\text{1. }CH_3CH_2CH_2CH_2Li,\text{ THF}} RC\equiv CCH_2CH_2OH$$

Attack takes place at less substituted carbon in unsymmetric oxacyclopropanes

6. Alkylation with Carbonyl Compounds (Section 13-5)

$$RC{\equiv}CH \xrightarrow[\substack{2.\ R'CR'' \\ 3.\ H^+, H_2O}]{\substack{1.\ CH_3CH_2CH_2CH_2Li,\ THF \\ \overset{O}{\parallel}}} RC{\equiv}C\underset{\underset{R''}{\overset{|}{|}}}{\overset{\overset{OH}{|}}{C}}R'$$

Reactions of Alkynes

7. Hydrogenation (Section 13-6)

$$RC{\equiv}CR \xrightarrow{\text{Catalyst, } H_2} RCH_2CH_2R \qquad \Delta H° \approx -70 \text{ kcal mol}^{-1}$$

Catalysts: Pt, Pd–C

$$RC{\equiv}CR \xrightarrow{H_2,\ \text{Lindlar catalyst}} \underset{\substack{\text{Cis alkene}}}{\overset{\substack{H \qquad H}}{\underset{R \qquad R}{C{=}C}}} \qquad \Delta H° \approx -40 \text{ kcal mol}^{-1}$$

8. Reduction with Sodium in Liquid Ammonia (Section 13-6)

$$RC{\equiv}CR \xrightarrow[\substack{2.\ H^+, H_2O}]{\substack{1.\ \text{Na, liquid } NH_3}} \underset{\substack{\text{Trans alkene}}}{\overset{\substack{H \qquad R}}{\underset{R \qquad H}{C{=}C}}}$$

9. Electrophilic (and Markovnikov) Additions: Hydrohalogenation, Halogenation, and Hydration (Section 13-7)

$$RC{\equiv}CR \xrightarrow{HX} RCH{=}CXR \xrightarrow{HX} \underset{\substack{\text{Geminal dihaloalkane}}}{RCH_2CX_2R}$$

$$RC{\equiv}CH \xrightarrow{2\ HX} RCX_2CH_3$$

$$RC{\equiv}CR \xrightarrow{Br_2,\ Br^-} \underset{\substack{\text{Mainly trans}}}{\overset{\substack{R \qquad Br}}{\underset{Br \qquad R}{C{=}C}}} \xrightarrow{Br_2} RCBr_2CBr_2R$$

$$RC{\equiv}CR \xrightarrow{Hg^{2+},\ H_2O} RCH_2\overset{\overset{O}{\parallel}}{C}R$$

10. Radical Addition of Hydrogen Bromide (Section 13-8)

$$RC{\equiv}CH \xrightarrow{HBr,\ ROOR} \underset{\substack{\textbf{Anti-Markovnikov}}}{RCH{=}CHBr}$$

Br attaches to less substituted carbon

11. Hydroboration (Section 13-8)

$$RC{\equiv}CH \xrightarrow{R'_2BH,\ THF} \underset{\substack{H \qquad BR'_2}}{\overset{\substack{R \qquad H}}{C{=}C}}$$

**Anti-Markovnikov and
stereospecific (*syn*) addition**

B attaches to less substituted carbon

Dicyclohexylborane (R′ = ⬡ **—)**

12. Oxidation of Alkenylboranes (Section 13-8)

13. Alkenyl Organometallics (Section 13-9)

14. Heck Reaction (Section 13-9)

Important Concepts

1. The rules for **naming alkynes** are essentially the same as those formulated for alkenes. Molecules with both double and triple bonds are called **alkenynes,** the double bond receiving the lower number if both are at equivalent positions. Hydroxy groups are given precedence in numbering alkynyl alcohols (**alkynols).**

2. The **electronic structure** of the triple bond reveals two π bonds, perpendicular to each other, and a σ bond, formed by two overlapping sp hybrid orbitals. The strength of the triple bond is about 229 kcal mol^{-1}; that of the alkynyl C—H bond is 131 kcal mol^{-1}. Triple bonds form **linear structures** with respect to other attached atoms, with short C—C (1.20 Å) and C—H (1.06 Å) bonds.

3. The high s character at C1 of a terminal alkyne makes the bound hydrogen relatively **acidic** (p$K_a \approx 25$).

4. The **chemical shift** of the alkynyl hydrogen is low ($\delta = 1.7$–3.1 ppm) compared with that of alkenyl hydrogens because of the shielding effect of an induced electron current around the molecular axis caused by the external magnetic field. The triple bond allows for long-range coupling. **IR spectroscopy** indicates the presence of the C≡C and ≡C—H bonds in terminal alkynes through bands at 2100–2260 cm^{-1} and 3260–3330 cm^{-1}, respectively.

5. The **elimination** reaction with vicinal dihaloalkanes proceeds regioselectively and stereospecifically to give alkenyl halides.

6. Selective *syn* **dihydrogenation** of alkynes is possible with Lindlar catalyst, the surface of which is less active than palladium on carbon and therefore not capable of hydrogenating alkenes. Selective *anti* **hydrogenation** is possible with sodium metal dissolved in liquid ammonia because simple alkenes cannot be reduced by one-electron transfer. The stereochemistry is set by the greater stability of a trans disubstituted alkenyl radical intermediate.

7. Alkynes generally undergo the same addition reactions as alkenes; these reactions may take place twice in succession. Hydration of alkynes is unusual. It requires an Hg(II) catalyst, and the initial product, an **enol,** rearranges to a ketone by **tautomerism.**

8. To stop the **hydroboration** of terminal alkynes at the alkenylboron intermediate stage, modified dialkylboranes—particularly dicyclohexylborane—are used. Oxidation of the resulting alkenylboranes produces enols that tautomerize to aldehydes.

9. The **Heck reaction** links alkenes to alkenyl halides in a metal-catalyzed process.

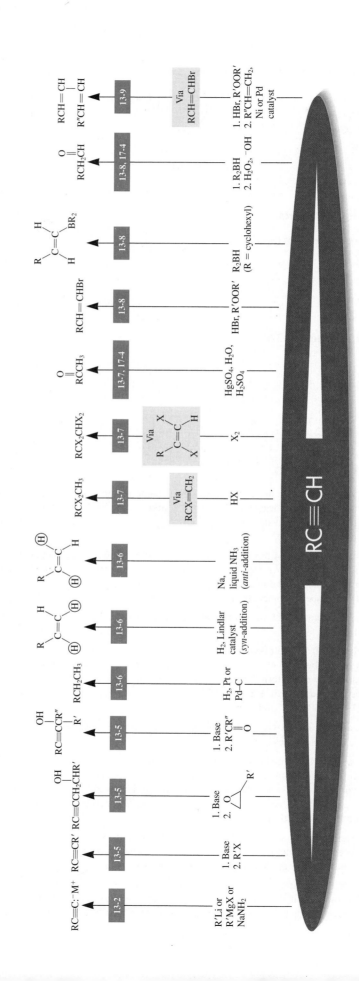

Reactions of Alkynes

$RC \equiv CH$

13-2 $\longrightarrow$ $RC \equiv C:^- M^+$
R'Li or R'MgX or NaNH$_2$

13-5 $\longrightarrow$ $RC \equiv CR'$
1. Base
2. R'X

13-5 $\longrightarrow$ $RC \equiv CCH_2CHR'$ $\overset{OH}{|}$
1. Base
2. $\triangle$O$\diagdown$R'

13-5 $\longrightarrow$ $RC \equiv CCR''$ $\overset{OH}{\underset{R'}{|}}$
1. Base
2. R'CR'' $\overset{O}{\|}$

13-6 $\longrightarrow$ RCH_2CH_3
H$_2$, Pt or Pd–C

13-6 $\longrightarrow$ R, H / C=C \ ⒽH
H$_2$, Lindlar catalyst (*syn*-addition)

13-6 $\longrightarrow$ R, Ⓗ / C=C \ HⒽ
Na, liquid NH$_3$ (*anti*-addition)

13-7 $\longrightarrow$ RCX_2CH_3
Via $RCX=CH_2$
HX

13-7 $\longrightarrow$ RCX_2CHX_2
Via R,X / C=C \ X CHX,H (X X)
X$_2$

13-7, 17-4 $\longrightarrow$ $RCCH_3$ $\overset{O}{\|}$
HgSO$_4$, H$_2$O, H$_2$SO$_4$

13-8 $\longrightarrow$ $RCH=CHBr$
HBr, R'OOR'

13-8 $\longrightarrow$ R, H / C=C \ H BR$_2$
R$_2$BH (R = cyclohexyl)

13-8, 17-4 $\longrightarrow$ RCH_2CH $\overset{O}{\|}$
1. R$_2$BH
2. H$_2$O$_2$, $^-$OH

13-9 $\longrightarrow$ RCH=CH— R''CH=CH
Via RCH=CHBr
1. HBr, R'OOR'
2. R''CH=CH$_2$ Ni or Pd catalyst

section number

Problems

27. Draw the structures of the molecules with the following names.
  (a) 1-Chloro-1-butyne
  (b) (Z)-4-Bromo-3-methyl-3-penten-1-yne
  (c) 4-Hexyn-1-ol

28. Name each of the compounds below, using the IUPAC system of nomenclature.

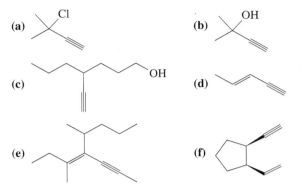

(a)   **(b)**
(c)   **(d)**
(e)   **(f)**

29. Compare C–H bond strengths in ethane, ethene, and ethyne. Reconcile these data with hybridization, bond polarity, and acidity of the hydrogen.

30. Compare the C2–C3 bonds in propane, propene, and propyne. Should they be any different with respect to either bond length or bond strength? If so, how should they vary?

31. Predict the order of acid strengths in the following series of cationic species: $CH_3CH_2NH_3^+$, $CH_3CH=NH_2^+$, $CH_3C\equiv NH^+$. [**Hint:** Look for an analogy among hydrocarbons (Section 13-2).]

32. The heats of combustion for three compounds with the molecular formula C_5H_8 are as follows: cyclopentene, $\Delta H_{comb} = -1027$ kcal mol^{-1}; 1,4-pentadiene, $\Delta H_{comb} = -1042$ kcal mol^{-1}; and

1-pentyne, $\Delta H_{comb} = -1052$ kcal mol^{-1}. Explain in terms of relative stability and bond strengths.

33. Rank in order of decreasing stability.
  (a) 1-Heptyne and 3-heptyne

  (b) ⬡—C≡CCH₃,   ⬡—CH₂C≡CH,

  and   ⬡ (cyclooctyne)

  (**Hint:** Make a model of the third structure. Is there anything unusual about its triple bond?)

34. Deduce structures for each of the following. **(a)** Molecular formula C_6H_{10}; NMR spectrum A (below); no strong IR bands between 2100 and 2300 or 3250 and 3350 cm^{-1}. **(b)** Molecular formula C_7H_{12}; NMR spectrum B (next page); IR bands at about 2120 and 3330 cm^{-1}. **(c)** The percentage composition is 71.41% carbon and 9.59% hydrogen (the remainder is O), and the exact molecular mass is 84.0584; NMR and IR spectra C (next page). The inset in NMR spectrum C provides better resolution of the signals between 1.6 and 2.4 ppm.

35. The IR spectrum of 1,8-nonadiyne displays a strong, sharp band at 3300 cm^{-1}. What is the origin of this absorption? Treatment of 1,8-nonadiyne with NaNH₂, then with D₂O, leads to the incorporation of two deuterium atoms, leaving the molecule unchanged otherwise. The IR spectrum reveals that the peak at 3300 cm^{-1} has disappeared, but a new one is present at 2580 cm^{-1}. **(a)** What is the product of this reaction? **(b)** What new bond is responsible for the IR absorption at 2580 cm^{-1}? **(c)** Using Hooke's law, calculate the approximate expected position of this new band from

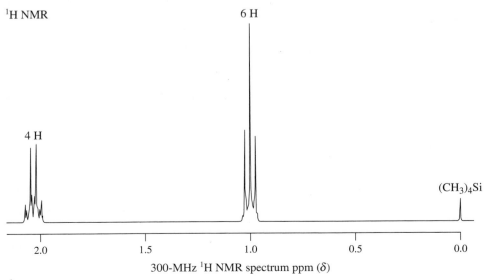

^{1}H NMR

6 H

4 H

(CH₃)₄Si

2.0   1.5   1.0   0.5   0.0

300-MHz ^{1}H NMR spectrum ppm (δ)

A

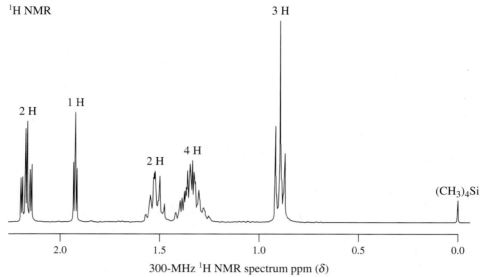

¹H NMR

2 H   1 H   2 H   4 H   3 H   (CH₃)₄Si

2.0   1.5   1.0   0.5   0.0

300-MHz ¹H NMR spectrum ppm (δ)

B

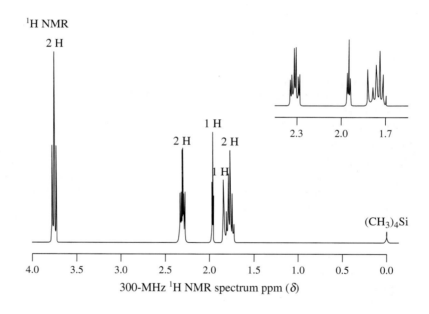

¹H NMR

2 H   2 H   1 H   2 H   1 H   (CH₃)₄Si

2.3   2.0   1.7

4.0   3.5   3.0   2.5   2.0   1.5   1.0   0.5   0.0

300-MHz ¹H NMR spectrum ppm (δ)

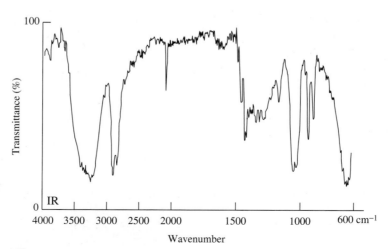

100

Transmittance (%)

0

IR

4000   3500   3000   2500   2000   1500   1000   600 cm⁻¹

Wavenumber

C

the structure of the original molecule and its IR spectrum. Assume that *k* and *f* have not changed.

36. Write the expected product(s) of each of the following reactions.

(a) CH₃CH₂CHCHCH₂Cl $\xrightarrow{\text{3 NaNH}_2,\ \text{liquid NH}_3}$
with CH₃ and Cl substituents

(b) CH₃OCH₂CH₂CH₂CHCHCH₃ $\xrightarrow{\text{2 NaNH}_2,\ \text{liquid NH}_3}$
with Br Br substituents

(c) *meso*-CH₃CHCH₂CHCHCH₂CHCH₃ $\xrightarrow[\text{CH}_3\text{OH}]{\text{NaOCH}_3\ (1\ \text{equivalent}),}$
with CH₃ Cl Cl CH₃ substituents

(d) (4R,5R)-CH₃CHCH₂CHCHCH₂CHCH₃ $\xrightarrow[\text{CH}_3\text{OH}]{\text{NaOCH}_3\ (1\ \text{equivalent}),}$
with CH₃ Cl Cl CH₃ substituents

37. (a) Write the expected product of the reaction of 3-octyne with Na in liquid NH₃. **(b)** When the same reaction is carried out with cyclooctyne [Problem 33, part (b)], the product is *cis*-cyclooctene, not *trans*-cyclooctene. Explain, mechanistically.

38. Write the expected major product of reaction of 1-propynyllithium, CH₃C≡C⁻Li⁺, with each of the following molecules in THF.

(a) CH₃CH₂Br

(b) [structure with Cl]

(c) Cyclohexanone

(d) [cyclopentane with CHO group]

(e) CH₃CH—CH₂ (epoxide, O)

(f) [bicyclic ketone with CH₃ and H]

39. Write the mechanism and final product for the reaction of 1-propynyllithium with *trans*-2,3-dimethyloxacyclopropane.

40. Which of the following methods is best suited as a high-yield synthesis of 2-methyl-3-hexyne, [structure] ?

(a) [structure] $\xrightarrow{\text{H}_2,\ \text{Lindlar's catalyst}}$

(b) [structure with Cl Cl] $\xrightarrow{\text{NaNH}_2,\ \text{liquid NH}_3}$

(c) [structure] $\xrightarrow[\text{2. NaNH}_2,\ \text{liquid NH}_3]{\text{1. Cl}_2,\ \text{CCl}_4}$

(d) [structure]—Li + [structure]Br

(e) [structure]—Li + [structure]Br

41. Propose reasonable syntheses of each of the following alkynes, using the principles of retrosynthetic analysis. Each alkyne func-

tional group in your synthetic target should come from a *separate* molecule, which may be any two-carbon compound (e.g., ethyne, ethene, ethanal).

(a) [structure]

(b) [structure with OH]

(c) [structure with OH]

(d) (CH₃)₃CC≡CH [Be careful! What is wrong with (CH₃)₃CCl + ⁻:C≡CH?]

42. Draw the structure of (*R*)-4-deuterio-2-hexyne. Propose a suitable retro-S$_N$2 precursor of this compound.

43. Reaction review. Without consulting the Reaction Road Map on p. 572, suggest reagents to convert a general alkyne RC≡CH into each of the following types of compounds.

(a) [structure] R C=C Br (Br and H)

(b) R—C—C—H (Br Br / Br Br)

(c) R—C—C—H (O, H) (Markovnikov product)

(d) R—C—C—H (I H / I H)

(e) R—C≡C:⁻M⁺

(f) R—C≡C—C—R″ (OH, R′)

(g) R—C≡C—R′

(h) R—C≡C—CH₂—CH₂OH

(i) R—C—C—H (H H / H H)

(j) [structure] R C=C H (H and H)

(k) R—C—C—H (H O) (anti-Markovnikov product)

44. Give the expected major product of the reaction of propyne with each of the following reagents.
(a) D₂, Pd–CaCO₃, Pb(O₂CCH₃)₂, quinoline;
(b) Na, ND₃;
(c) 1 equivalent HI;
(d) 2 equivalents HI;
(e) 1 equivalent Br₂;
(f) 1 equivalent ICl;
(g) 2 equivalents ICl;
(h) H₂O, HgSO₄, H₂SO₄;
(i) dicyclohexylborane, then NaOH, H₂O₂.

45. What are the products of the reactions of dicyclohexylethyne with the reagents in Problem 44?

46. Write the structures of the initially formed enol tautomers in the reactions of propyne and dicyclohexylethyne with dicyclohexylborane followed by NaOH and H₂O₂ (Problems 44, part i, and 45, part i).

47. Give the products of the reactions of your first two answers to Problem 45 with each of the following reagents.

(a) H_2, Pd–C, CH_3CH_2OH; (b) Br_2, CCl_4;

(c) BH_3, THF, then NaOH, H_2O_2;

(d) MCPBA, CH_2Cl_2; (e) OsO_4, then H_2S.

48. Propose several syntheses of *cis*-3-heptene, beginning with each of the following molecules. Note in each case whether your proposed route gives the desired compound as a major or minor final product.

(a) 3-Chloroheptane; (b) 4-chloroheptane;

(c) 3,4-dichloroheptane; (d) 3-heptanol;

(e) 4-heptanol; (f) *trans*-3-heptene;

(g) 3-heptyne.

49. Propose reasonable syntheses of each of the following molecules, using an alkyne at least once in each synthesis.

(a) [structure with Br, Cl] (b) [structure with I, I]

(c) meso-2,3-Dibromobutane

(d) Racemic mixture of (2R,3R)-and (2S,3S)-2,3-dibromobutane

(e) [structure: Br—C(CH₃)—Cl, H—C(CH₃)—Cl] (f) [ketone structure]

(g) [diol structure with OH, HO] (h) [cyclopentane aldehyde structure]

(i) [cyclohexene vinyl structure]

50. Show how the Heck reaction might be employed to synthesize each of the following molecules.

(a) [structure with $COCH_3$] (b) [stilbene structure]

51. Propose a reasonable structure for calcium carbide, CaC_2, on the basis of its chemical reactivity (Section 13-10). What might be a more systematic name for it?

52. Propose *two different* syntheses of linalool, a terpene found in cinnamon, sassafras, and orange flower oils. Start with the eight-carbon ketone shown here and use ethyne as your source of the necessary additional two carbons in both syntheses.

[ketone structure]

[linalool structure with OH]

Linalool

53. The synthesis of chamaecynone, the volatile oil of the Benihi tree, requires the conversion of a chloroalcohol into an alkynyl ketone. Propose a synthetic strategy to accomplish this task.

[structure HO...Cl] → ? [structure with O and C≡CH]

Eventually → [structure with O and C≡CH]

Chamaecynone

54. CHALLENGE Synthesis of the sesquiterpene bergamotene, a trace component of cannabis oil, proceeds from the alcohol shown here. Suggest a sequence to complete the synthesis.

[HOH_2C...CH_3 structure] → ? [structure with CH_3]

Bergamotene

55. CHALLENGE An unknown molecule displays ^1H NMR and IR spectra D (next page). Reaction with H_2 in the presence of the Lindlar catalyst gives a compound that, after ozonolysis and treatment with Zn in aqueous acid, gives rise to one equivalent of CH_3CCH and two of HCH. What was the structure of the original molecule?

56. CHALLENGE Formulate a plausible mechanism for the hydration of ethyne in the presence of mercuric chloride. (**Hint:** Review the hydration of alkenes catalyzed by mercuric ion, Section 12-7.)

57. A synthesis of the sesquiterpene farnesol requires the conversion of a dichloro compound into an alkynol, as shown below. Suggest a way of achieving this transformation. (**Hint:** Devise a conversion of the starting compound into a terminal alkyne.)

[structure with Cl, Cl] → ? [structure with $C≡C—CH_2OH$]

Eventually → [farnesol structure with OH]

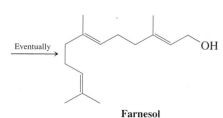

Farnesol

¹H NMR

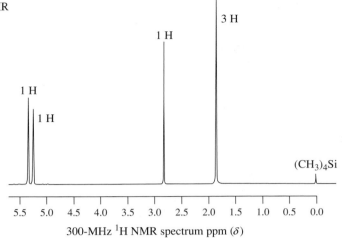

D

IR

Team Problem

58. Your team is studying the problem of an intramolecular ring closure of enediyne systems important in the total synthesis of dynemicin A, which exhibits potent antitumor activity.

One research group tried the following approaches to effect this process. Unfortunately, all were unsuccessful. Divide the schemes among yourselves and assign structures to compounds A through D. (Note: R′ and R″ are protecting groups.)

Dynemicin A

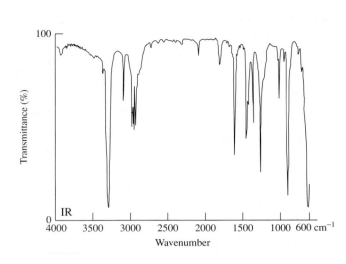

3.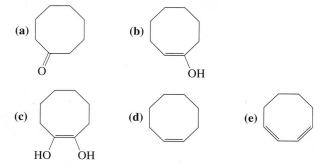

A successful model study (shown here) provided an alternative strategy toward the completion of the total synthesis.

Discuss the advantages of this approach and apply it to the appropriate compound in approaches 1 through 3.

Preprofessional Problems

59. The compound whose structure is H–C≡C(CH$_2$)$_3$Cl is best named (IUPAC)

(a) 4-chloro-1-pentyne;
(b) 5-chloropent-1-yne;
(c) 4-pentyne-1-chloroyne;
(d) 1-chloropent-4-yne.

60. A nucleophile made by deprotonation of propyne is

(a) $^-$:CH$_2$CH$_3$;
(b) $^-$:HC=CH$_2$;
(c) $^-$:C≡CH;
(d) $^-$:C≡CCH$_3$;
(e) $^-$:HC=CHCH$_3$.

61. When cyclooctyne is treated with dilute, aqueous sulfuric acid and HgSO$_4$, a new compound results. It is best represented as

(a)   (b)   (c)   (d)   (e)

62. From the choices shown on the next page, pick the one that best describes the structure of compound A.

$$A \xrightarrow[270°C, 100\ atm]{H_3PO_4,\ pH\ 2,}$$

(a) HOCH$_2$CH(CH$_2$)$_2$OH
 |
 CH$_3$

(b) HOCH$_2$CHCH$_2$OH
 |
 CH$_3$

(c) HC≡CCHCH$_2$OH
 |
 CH$_3$

(d) HC≡CCH$_2$CHCH$_2$OH
 |
 CH$_3$

63. From the choices shown below, pick the one that best describes the structure of compound A.

$$CH_3\underset{\underset{Br}{|}}{\overset{\overset{Br}{|}}{C}}-\underset{\underset{Br}{|}}{\overset{\overset{Br}{|}}{C}}CH_2OH \xleftarrow{Br_2\,(2\ equivalents)} A$$

(A primary alcohol;
%C = 68.6,
%H = 8.6, and
%O = 22.9)

$$\xrightarrow{H_2\,(2\ equivalents),\ Raney\ Ni} \text{1-butanol}$$

(a) CH$_2$=CHCH$_2$CH$_2$OH (b) ▷—CH$_2$OH
(c) CH$_3$C≡CCH$_2$OH (d) CH$_3$CH=CH—CH=CHOH

(e) OH

CHAPTER 14

Delocalized Pi Systems

Investigation by Ultraviolet and Visible Spectroscopy

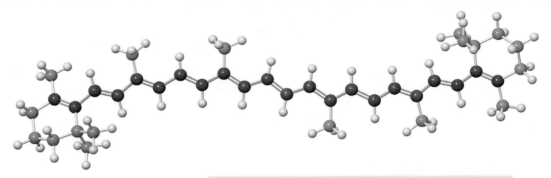

We live in a universe of color. Our ability to perceive and distinguish thousands of hues and shades of color is tied to the ability of molecules to absorb different frequencies of visible light. In turn, this molecular property is frequently a consequence of the presence of multiple π bonds.

In the preceding three chapters we introduced the topic of compounds containing carbon–carbon π bonds, the products of overlap between two adjacent parallel *p* orbitals. We found that addition reactions to these chemically versatile systems provided entries both to relatively simple products, of use in synthesis, and to more complex products, including polymers—substances that have affected modern society enormously. In this chapter we expand further on all these themes by studying compounds in which *three or more* parallel *p* orbitals participate in π-type overlap. The electrons in such orbitals are therefore shared by three or more atomic centers and are said to be **delocalized.**

β-Carotene is a pigment that is important for photosynthesis. It is part of the general family of carotenoids, produced in nature to the tune of 100 million tons per year and responsible for the orange color of carrots and many other fruits and vegetables. Its color is due to the presence of 11 contiguous double bonds, which cause the absorption of visible light.

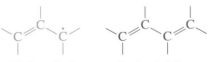

Allylic radical   **Conjugated diene**

Our discussion begins with the 2-propenyl system—also called allyl—containing three interacting *p* orbitals. We then proceed to compounds that contain *several* double bonds: dienes and higher analogs. These compounds give rise to some of the most widely used polymers in the modern world, found in everything from automobile tires to the plastic cases around the computers used to prepare the manuscript for this textbook.

The special situation of alternating double and single bonds gives rise to **conjugated** dienes, trienes, and so forth, possessing more extended delocalization of their π electrons. These substances illustrate new modes of reactivity, including thermal and photochemical cycloadditions and ring closures, which are among the most powerful methods for the synthesis of cyclic compounds such as steroidal pharmaceuticals. These processes exemplify a fundamentally new class of mechanisms: *pericyclic* reactions, which is the last type of mechanism we shall consider in this book. Following this chapter, we take the opportunity in an Interlude to review all the major types of organic processes covered in this course. We conclude the chapter itself with a discussion of light absorption by molecules with delocalized π systems. This process forms the basis of ultraviolet and visible spectroscopy.

14-1 | OVERLAP OF THREE ADJACENT p ORBITALS: ELECTRON DELOCALIZATION IN THE 2-PROPENYL (ALLYL) SYSTEM

What is the effect of a neighboring double bond on the reactivity of a carbon center? Three key observations answer this question.

Observation 1. The primary carbon–hydrogen bond in propene is relatively weak, only 87 kcal mol^{-1}.

Dissociation Energies of Various C–H Bonds

$CH_2\!\!=\!\!CHCH_2\!\!-\!\!H$
$DH° = 87$ kcal mol^{-1} (364 kJ mol^{-1})

$(CH_3)_3C\!\!-\!\!H$
$DH° = 96.5$ kcal mol^{-1} (404 kJ mol^{-1})

$(CH_3)_2CH\!\!-\!\!H$
$DH° = 98.5$ kcal mol^{-1} (412 kJ mol^{-1})

$CH_3CH_2\!\!-\!\!H$
$DH° = 101$ kcal mol^{-1} (423 kJ mol^{-1})

Decreasing bond strength

Propene → 2-Propenyl radical + H· $DH° = 87$ kcal mol^{-1}

A comparison with the values found for other hydrocarbons (see margin) shows that it is even weaker than a tertiary C–H bond. *Evidently, the 2-propenyl radical enjoys some type of special stability.*

Observation 2. In contrast with saturated primary haloalkanes, 3-chloropropene dissociates relatively fast under S_N1 (solvolysis) conditions and undergoes rapid unimolecular substitution through a carbocation intermediate.

3-Chloropropene → 2-Propenyl cation → 3-Methoxypropene (S_N1 product)

This finding clearly contradicts our expectations (recall Section 7-5). *It appears that the cation derived from 3-chloropropene is somehow more stable than other primary carbocations.* By how much? The ease of formation of the 2-propenyl cation in solvolysis reactions has been found to be roughly equal to that of a secondary carbocation.

Observation 3. The pK_a of propene is about 40.

2-Propenyl anion

Thus, propene is considerably more acidic than propane (p$K_a \approx 50$), and *the formation of the propenyl anion by deprotonation appears to be unusually favored.*

How can we explain these three observations?

Delocalization stabilizes 2-propenyl (allyl) intermediates

Each of the preceding three processes generates a reactive carbon center—a radical, a carbocation, or a carbanion, respectively—that is adjacent to the π framework of a double bond. This arrangement seems to impart special stability. Why? The reason is electron delocalization: Each species may be described by a pair of equivalent contributing resonance forms (Section 1-5). These three-carbon intermediates have been given the name **allyl** (followed by the appropriate term: radical, cation, or anion). The activated carbon is called **allylic.**

Resonance Representation of Delocalization in the 2-Propenyl (Allyl) System

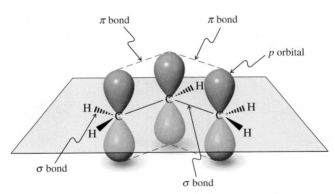

The stabilization of the 2-propenyl (allyl) system by resonance can also be described in terms of molecular orbitals. Each of the three carbons is sp^2 hybridized and bears a p orbital perpendicular to the molecular plane (Figure 14-1). *Make a model: The structure is symmetric, with equal C–C bond lengths.*

> Remember that resonance forms are *not* isomers but partial molecular representations. The true structure (the resonance hybrid) is derived by their superposition, better represented by the dotted-line drawings at the right of the classical picture.

Model Building

The 2-propenyl (allyl) pi system is represented by three molecular orbitals

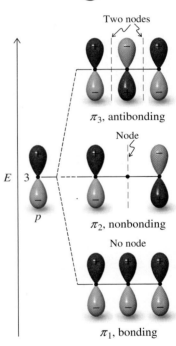

Figure 14-2 The three π molecular orbitals of 2-propenyl (allyl), obtained by combining three adjacent atomic p orbitals. Note the drop in energy of the bonding molecular orbital π_1, reflecting the increased stability of the system. Molecular orbital π_2 remains at the same level as the starting p orbitals and is therefore called a nonbonding molecular orbital.

Figure 14-1 goes here (description of p orbitals overlap)

Figure 14-1 The three p orbitals in the 2-propenyl (allyl) group overlap, giving a symmetric structure with delocalized electrons. The σ framework is shown as black lines.

Ignoring the σ framework, we can combine the three p orbitals mathematically to give three π molecular orbitals. This process is analogous to mixing two atomic orbitals to give two molecular orbitals describing a π bond (Figures 11-2 and 11-4), except that there is now a third atomic orbital. Of the three resulting molecular orbitals, shown in Figure 14-2, one (π_1) is bonding and has no nodes, one (π_2) is *nonbonding* (in other words, it has the same energy as a noninteracting p orbital) and has one node, and one (π_3) is antibonding, with two nodes. We can use the Aufbau principle to fill up the π molecular orbitals in Figure 14-2 with the appropriate number of electrons for the 2-propenyl cation, radical, and anion as shown in Figure 14-3. The cation, with a total of two electrons, contains only one filled orbital, π_1.

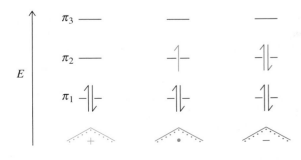

Figure 14-3 The Aufbau principle is used to fill up the π molecular orbitals of 2-propenyl (allyl) cation, radical, and anion. In each case, the total energy of the π electrons is lower than that of three noninteracting p orbitals. Partial cation, radical, or anion character is present at the end carbons in these systems, a result of the location of the lobes in the π_2 molecular orbital.

> **Recall:** Mixing (any) three orbitals (here three p orbitals) generates three new molecular orbitals.

For the radical and the anion, we place one or two additional electrons, respectively, into the nonbonding orbital, π_2. The total π-electron energy of each system is lower (more favorable) than that expected from three noninteracting p orbitals—because π_1 is greatly stabilized and filled with two electrons in all cases, whereas the antibonding level, π_3, stays empty throughout.

The resonance formulations for the three 2-propenyl species indicate that it is mainly the two *terminal* carbons that accommodate the charges in the ions or the odd electron in the radical. The molecular-orbital picture is consistent with this view: The three structures differ only in the number of electrons present in molecular orbital π_2, which possesses a node passing through the central carbon; therefore, very little of the electron excess or deficiency will show up at this position. The electrostatic potential maps of the three 2-propenyl systems show their delocalized nature. (The cation and anion have been rendered at an attenuated scale to tone down the otherwise overwhelming intensity of color). To some extent, especially in the cation and anion, you can also discern the relatively greater charge density at the termini. Remember that these maps take into account all electrons in all orbitals, σ and π.

**Partial Electron Density Distribution in the
2-Propenyl (Allyl) System**

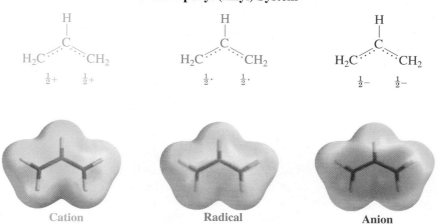

| Cation | Radical | Anion |

In Summary Allylic radicals, cations, and anions are unusually stable. In Lewis terms, this stabilization is readily explained by electron delocalization. In a molecular-orbital description, the three interacting p orbitals form three new molecular orbitals: One is considerably lower in energy than the p level, another one stays the same, and a third is higher in energy. Because only the first two are populated with electrons, the total π energy of the system is lowered.

14-2 RADICAL ALLYLIC HALOGENATION

A consequence of delocalization is that resonance-stabilized allylic intermediates can readily participate in reactions of unsaturated molecules. For example, although halogens can add to alkenes to give the corresponding vicinal dihalides (Section 12-5) by an ionic mechanism, the course of this reaction is changed with added radical initiators (or on irradiation) and with the halogen present only in low concentrations. These conditions slow the ionic addition pathway sufficiently to allow a faster radical chain mechanism to take over, leading to **radical allylic substitution.***

---

*A complete explanation for this change requires a detailed kinetic analysis that is beyond the scope of this book. Suffice it to say that at low bromine concentrations the competing addition processes are reversible, and allylic substitution wins out.

Radical Allylic Halogenation

$$CH_2\!\!=\!\!CHCH_3 \xrightarrow{\text{X}_2 \text{ (low conc.), ROOR or } h\nu} CH_2\!\!=\!\!CHCH_2X \ + \ HX$$

A reagent that is frequently used in allylic brominations in the laboratory is *N*-bromobutanimide (*N*-bromosuccinimide, NBS) suspended in tetrachloromethane. This species is nearly insoluble in CCl$_4$ and is a steady source of very small amounts of bromine formed by reaction with trace impurities of HBr (see margin).

NBS as a Source of Bromine

Reaction

N-Bromobutanimide
(*N*-Bromosuccinimide, NBS)   Butanimide

For example, NBS converts cyclohexene to 3-bromocyclohexene.

$$h\nu, \text{CCl}_4$$

85%
3-Bromocyclohexene

Mechanism of Br$_2$ Generation from NBS

Protonation of oxygen (see also Exercise 2–11)

Proton shift (tautomerism; Section 13-7)

The bromine reacts with the alkene by a radical chain mechanism (Section 3-4). The process is initiated by light or by traces of radical initiators that cause dissociation of Br$_2$ into bromine atoms. Propagation of the chain involves abstraction of a weakly bound allylic hydrogen by Br·.

Mechanism of Allylic Bromination

Initiation step

$$:\!\ddot{\text{Br}}\!-\!\ddot{\text{Br}}\!: \xrightarrow{h\nu} 2 \ :\!\ddot{\text{Br}}\cdot$$

Mechanism

Propagation steps

H Abstraction

$DH° = 87$ kcal mol^{-1}

$DH° = 87$ kcal mol^{-1}

Allylic Radical

Animation

ANIMATED MECHANISM:
Radical allylic halogenation

The resonance-stabilized radical may then react with Br$_2$ at either end of the allylic system to furnish an allylic bromide and regenerate Br·, which continues the chain. Thus, alkenes that form unsymmetric allylic radicals can give mixtures of products upon treatment with NBS. For example,

$$CH_2=CH(CH_2)_5CH_3 \xrightarrow[-HBr]{NBS, ROOR} \left[CH_2=CH\overset{\cdot}{C}H(CH_2)_4CH_3 \longleftrightarrow \cdot CH_2CH=CH(CH_2)_4CH_3\right]$$

1-Octene

$$\xrightarrow[\text{(from NBS)}]{Br_2} \Big\downarrow -Br\cdot \qquad\qquad \xrightarrow[\text{(from NBS)}]{Br_2} \Big\downarrow -Br\cdot$$

$$\overset{Br}{\underset{|}{CH_2=CHCH(CH_2)_4CH_3}} \quad + \quad BrCH_2CH=CH(CH_2)_4CH_3$$

28%   72%

3-Bromo-1-octene   **1-Bromo-2-octene**
                                                      (Mixture of *cis* and *trans*)

Exercise 14-1

Ignoring stereochemistry, give all the isomeric monobromoheptenes that are possible from NBS treatment of *trans*-2-heptene.

Allylic chlorinations are important in industry because chlorine is relatively cheap. For example, 3-chloropropene (allyl chloride) is made commercially by the gas-phase chlorination of propene at 400°C. It is a building block for the synthesis of epoxy resin and many other useful substances.

$$CH_3CH=CH_2 \quad + \quad Cl_2 \xrightarrow{400°C} ClCH_2CH=CH_2 \quad + \quad HCl$$

3-Chloropropene
(Allyl chloride)

Exercise 14-2

Predict the outcome of the allylic monochlorination of the following substrates.

(a) Cyclohexene   **(b)**    **(c)** 1-Methylcyclohexene

---

The biochemical degradation of unsaturated molecules frequently involves radical abstraction of allylic hydrogens by oxygen-containing species. Such processes will be discussed in Section 22-9.

In Summary Under radical conditions, alkenes containing allylic hydrogens enter into allylic halogenation. A particularly good reagent for allylic bromination is *N*-bromobutanimide (*N*-bromosuccinimide, NBS).

14-3   NUCLEOPHILIC SUBSTITUTION OF ALLYLIC HALIDES: S$_N$1 AND S$_N$2

As our example of 3-chloropropene in Section 14-1 shows, allylic halides dissociate readily to produce allylic cations. These can be trapped at either end by nucleophiles in S$_N$1 reactions. Allylic halides also readily undergo S$_N$2 transformations.

Allylic halides undergo S$_N$1 reactions

The ready dissociation of allylic halides has important chemical consequences. Different allylic halides may give identical products upon solvolysis if they dissociate to the same allylic cation. For example, the hydrolysis of either 1-chloro-2-butene or 3-chloro-1-butene results in the same alcohol mixture. The reason is that the same allylic cation is the intermediate.

Hydrolysis of Isomeric Allylic Chlorides

$$CH_3CH\!=\!CHCH_2Cl \xrightarrow[-Cl^-]{Dissociation} \left[\begin{array}{c} \overset{4}{C}H_3\overset{3}{C}H\!=\!\overset{2}{C}H\overset{1}{C}H_2{}^+ \\ \updownarrow \\ \overset{4}{C}H_3\overset{+}{\underset{4}{C}}H\overset{3}{C}H\!=\!\overset{2}{C}H_2 \end{array}\right] \xleftarrow[-Cl^-]{Dissociation} \overset{\displaystyle Cl}{\overset{|}{CH_3CHCH\!=\!CH_2}}$$

1-Chloro-2-butene   **Allylic cation**   **3-Chloro-1-butene**

$$\xrightarrow[\text{trapping}]{\text{Nucleophilic}} \Big\downarrow \text{HOH}$$

$$CH_3CH\!=\!CHCH_2OH \;+\; \overset{\displaystyle OH}{\overset{|}{CH_3CHCH\!=\!CH_2}} \;+\; H^+$$

2-Buten-1-ol   **3-Buten-2-ol**

Reaction

Mechanism

Reminder

Free H^+ does not exist in solution, but is attached to any available electron pair, such as the oxygen of hydroxy or the chloride ion in the adjacent scheme.

Exercise 14-3

Hydrolysis of (*R*)-3-chloro-1-butene gives racemic 3-buten-2-ol in addition to 2-buten-1-ol. Explain. (**Hint:** Review Section 7-3.)

Solved Exercise 14-4 | **Working with the Concepts: Allylic Alcohols and Acid**

Treatment of 3-buten-2-ol with cold hydrogen bromide gives 1-bromo-2-butene and 3-bromo-1-butene. Explain by a mechanism.

Strategy

First we convert the text problem into a balanced equation. Then we examine the structure of 3-buten-2-ol, identifying its functional groups. Subsequently we consider the reaction conditions and decide how they will affect these functional groups, either individually or together.

Remember WHIP

*W*hat
*H*ow
*I*nformation
*P*roceed

Solution

$$\overset{\displaystyle OH}{\overset{|}{CH_3CHCH\!=\!CH_2}} \underset{}{\overset{HBr}{\rightleftharpoons}} \overset{\displaystyle Br}{\overset{|}{CH_3CHCH\!=\!CH_2}} + CH_3CH\!=\!CHCH_2Br + H_2O$$

- 3-Buten-2-ol is a secondary *and allylic* alcohol.
- Recall (Section 9-2) that alcohols are protonated in the presence of strong acids and that, depending on structure, the resulting oxonium ions can react by S_N2 or S_N1 pathways. In the present case, clearly an S_N1 reaction is indicated.
- The resonance-stabilized allylic cation is trapped by bromide at either of the two allylic termini, producing the observed products.

$$\overset{\displaystyle OH}{\overset{|}{CH_3CHCH\!=\!CH_2}} \underset{}{\overset{HBr}{\rightleftharpoons}} \left[\begin{array}{c} CH_3\overset{+}{C}HCH\!=\!CH_2 \\ \updownarrow \\ CH_3CH\!=\!CHCH_2{}^+ \end{array}\right] + H_2O + Br^- \longrightarrow$$

$$\overset{\displaystyle Br}{\overset{|}{CH_3CHCH\!=\!CH_2}} + CH_3CH\!=\!CHCH_2Br + H_2O$$

Exercise 14-5 | Try It Yourself

Write a mechanism for the following transformation:

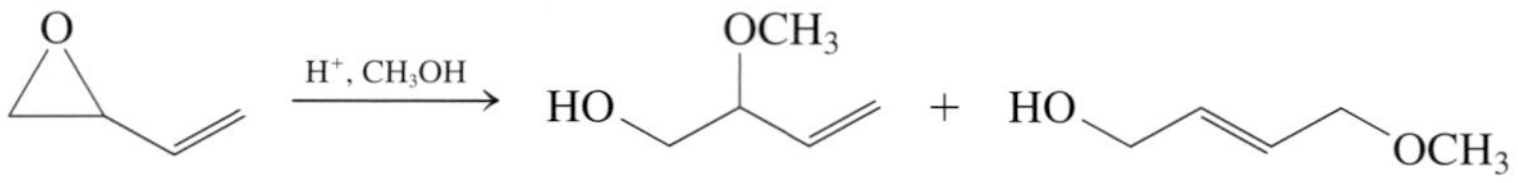

Allylic halides can also undergo S$_N$2 reactions

S$_N$2 reactions of allylic halides with good nucleophiles (Section 6-8) are faster than those of the corresponding saturated haloalkanes. Two factors contribute to this acceleration. One is that the allylic carbon is attached to a relatively electron-withdrawing sp^2 hybridized carbon (as opposed to sp^3; Section 13-2), making it more electrophilic. The second is that overlap between the double bond and the p orbital in the transition state of the S$_N$2 displacement (see Figure 6-4) is stabilizing, resulting in a relatively low activation barrier.

The S$_N$2 Reactions of 3-Chloro-1-propene and 1-Chloropropane

	Relative rate

$$CH_2{=}CHCH_2Cl \ + \ I^- \ \xrightarrow{\text{Acetone, 50°C}} \ CH_2{=}CHCH_2I \ + \ Cl^- \qquad 73$$

$$CH_3CH_2CH_2Cl \ + \ I^- \ \xrightarrow{\text{Acetone, 50°C}} \ CH_3CH_2CH_2I \ + \ Cl^- \qquad 1$$

Exercise 14-6 |

The solvolysis of 3-chloro-3-methyl-1-butene in acetic acid at 25°C gives initially a mixture containing mostly the structurally isomeric chloride with some of the acetate. After a longer period of time, no allylic chloride remains and the acetate is the only product present. Explain this result.

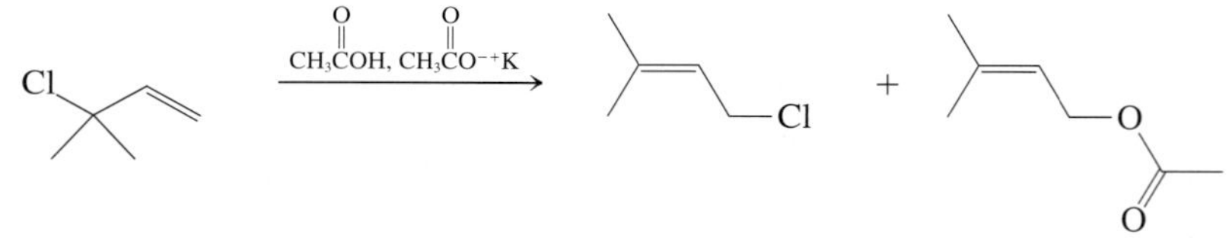

| 3-Chloro-3-methyl-1-butene | 1-Chloro-3-methyl-2-butene | 3-Methyl-2-butenyl acetate |

In Summary Allylic halides undergo both S$_N$1 and S$_N$2 reactions. The allylic cation intermediate of the S$_N$1 reaction may be trapped by nucleophiles at either terminus, leading to product mixtures in the case of unsymmetrical systems. With good nucleophiles, allylic halides undergo S$_N$2 reaction more rapidly than the corresponding saturated substrates.

14-4 | ALLYLIC ORGANOMETALLIC REAGENTS: USEFUL THREE-CARBON NUCLEOPHILES

Propene is appreciably more acidic than propane because of the relative stability of the conjugated carbanion that results from deprotonation (Section 14-1). Therefore, allylic lithium reagents can be made from propene derivatives by proton abstraction by an alkyllithium. The process is facilitated by *N,N,N',N'*-tetramethylethane-1,2-diamine (tetramethylethylenediamine, TMEDA), a good solvating agent.

Allylic Deprotonation

$$CH_3CH_2CH_2CH_2Li \quad + \quad H_2C=C\begin{array}{c}CH_3\\\\CH_3\end{array} \xrightarrow{(CH_3)_2NCH_2CH_2N(CH_3)_2 \text{ (TMEDA)}} H_2C=C\begin{array}{c}CH_3\\\\CH_2Li\end{array} \quad + \quad CH_3CH_2CH_2CH_2-H$$

Another way of producing an allylic organometallic is Grignard formation. For example,

$$CH_2=CHCH_2Br \xrightarrow{\text{Mg, THF, 0°C}} CH_2=CHCH_2MgBr$$

3-Bromo-1-propene   **2-Propenylmagnesium bromide**

Like their alkyl counterparts (Section 8-8), allylic lithium and Grignard reagents can react as nucleophiles. This is useful, because the embedded double bond allows for further functional-group transformations (Chapter 12).

Exercise 14-7

Show how to accomplish the following conversion in as few steps as possible. (**Hint:** Allylic organometallic compounds react with ketones in the same manner as ordinary organometallic reagents.)

In Summary Alkenes tend to be deprotonated at the allylic position, a reaction resulting in the corresponding delocalized anions. Allylic lithium or Grignard reagents can be made from the corresponding halides. Like their alkyl analogs, allyl organometallics function as nucleophiles.

14-5 TWO NEIGHBORING DOUBLE BONDS: CONJUGATED DIENES

Now that we have caught a glimpse of the consequences of delocalization over three atoms, it should be interesting to learn what happens if we go one step further. Let us consider the addition of a fourth *p* orbital, which results in two double bonds separated by one single bond: a **conjugated diene** (*conjugatio*, Latin, union). In these compounds, delocalization again results in stabilization, as measured by heats of hydrogenation. Extended π overlap is also revealed in their molecular and electronic structures and in their chemistry.

Hydrocarbons with two double bonds are named dienes

Conjugated dienes have to be contrasted with their **nonconjugated** isomers, in which the two double bonds are separated by saturated carbons, and the **allenes** (or **cumulated** dienes), in which the π bonds share a single *sp* hybridized carbon and are perpendicular to each other (Figure 14-4). You can see the contrasting π-electron distribution in conjugated and nonconjugated dienes in the electrostatic potential maps in the margin. In 1,3-butadiene, the regions of π-electron density (red) overlap, while in 1,4-pentadiene they are separated by a methylene group devoid of π electrons. In 1,2-propadiene (allene), these electrons are close, but are located in orthogonal (perpendicular) regions of space.

The names of conjugated and nonconjugated dienes are derived from those of the alkenes in a straightforward manner. The longest chain incorporating both double bonds is

The Simplest Conjugated and Nonconjugated Dienes

$$CH_2=CH-CH=CH_2$$

1,3-Butadiene (Conjugated)

$$CH_2=CHCH_2-CH=CH_2$$

1,4-Pentadiene (Nonconjugated)

$$CH_2=C=CH_2$$

1,2-Propadiene (Allene, nonconjugated)

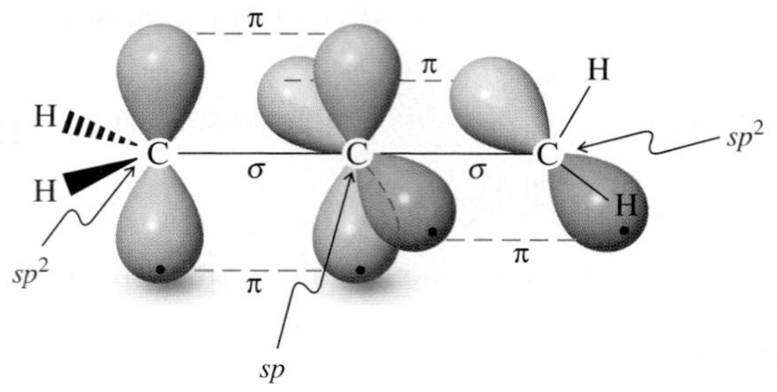

Figure 14-4 The two π bonds of an allene share a single carbon and are perpendicular to each other.

$$CH_2 = C = CH_2$$

found and then numbered to indicate the positions of the functional and substituent groups. If necessary, cis-trans or E,Z prefixes indicate the geometry around the double bonds. Cyclic dienes are named accordingly.

trans-1,3-Pentadiene

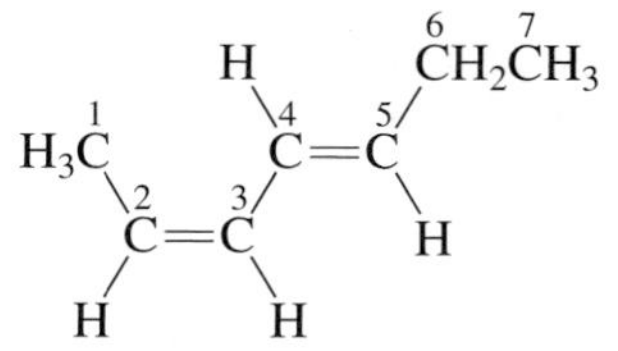

cis-2-*trans*-4-Heptadiene

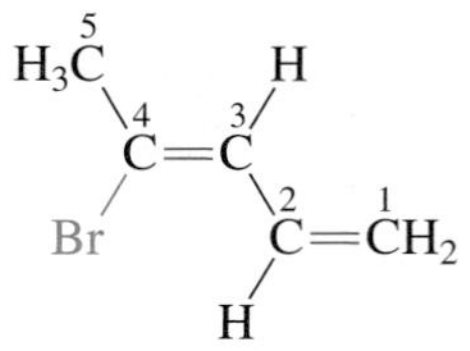

(**Z**)-4-Bromo-1,3-pentadiene

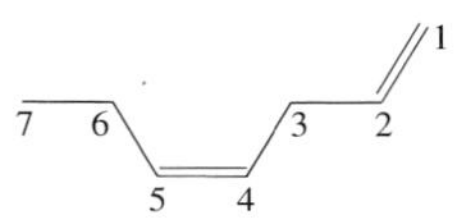

cis-1,4-Heptadiene
(A nonconjugated diene)

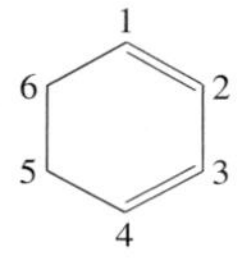

1,3-Cyclohexadiene

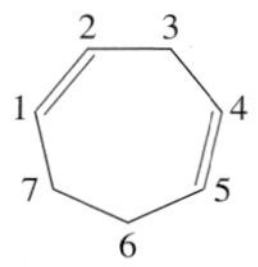

1,4-Cycloheptadiene
(A nonconjugated cyclic diene)

Exercise 14-8

Suggest names or draw structures, as appropriate, for the following compounds.

(**a**)

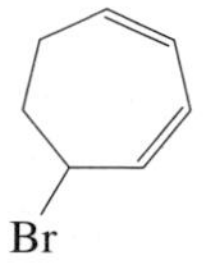

(**b**)

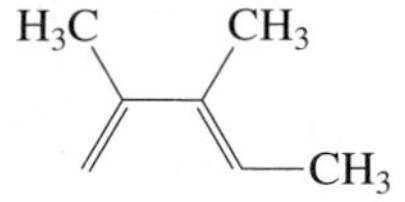

(**c**) *cis*-3,6-Dimethyl-1,4-cyclohexadiene   (**d**) *cis,cis*-1,4-Dibromo-1,3-butadiene

---

Conjugated dienes are more stable than nonconjugated dienes

The preceding sections noted that delocalization of electrons makes the allylic system especially stable. Does a conjugated diene have the same property? If so, that stability should be manifest in its heat of hydrogenation. We know that the heat of hydrogenation of a terminal alkene is about -30 kcal mol^{-1} (see Section 11-5). A compound containing two *noninteracting* (i.e., separated by one or more saturated carbon atoms) terminal double bonds should exhibit a heat of hydrogenation roughly twice this value, about -60 kcal mol^{-1}. Indeed, catalytic hydrogenation of either 1,5-hexadiene or 1,4-pentadiene releases just about that amount of energy.

Heat of Hydrogenation of Nonconjugated Alkenes

$$CH_3CH_2CH{=}CH_2 \ + \ H_2 \ \xrightarrow{Pt} \ CH_3CH_2CH_2CH_3 \qquad \Delta H° = -30.3 \text{ kcal mol}^{-1} \ (-127 \text{ kJ mol}^{-1})$$

$$CH_2{=}CHCH_2CH_2CH{=}CH_2 \ + \ 2 \text{ H}_2 \ \xrightarrow{Pt} \ CH_3(CH_2)_4CH_3 \qquad \Delta H° = -60.5 \text{ kcal mol}^{-1} \ (-253 \text{ kJ mol}^{-1})$$

$$CH_2{=}CHCH_2CH{=}CH_2 \ + \ 2 \text{ H}_2 \ \xrightarrow{Pt} \ CH_3(CH_2)_3CH_3 \qquad \Delta H° = -60.8 \text{ kcal mol}^{-1} \ (-254 \text{ kJ mol}^{-1})$$

When the same experiment is carried out with the conjugated diene, 1,3-butadiene, *less* energy is produced.

Heat of Hydrogenation of 1,3-Butadiene

$$CH_2{=}CH{-}CH{=}CH_2 \ + \ 2 \text{ H}_2 \ \xrightarrow{Pt} \ CH_3CH_2CH_2CH_3 \qquad \Delta H° = -57.1 \text{ kcal mol}^{-1} \qquad (-239 \text{ kJ mol}^{-1})$$

The difference, about 3.5 kcal mol^{-1} (15 kJ mol^{-1}), is due to a stabilizing interaction between the two double bonds, as illustrated in Figure 14-5.

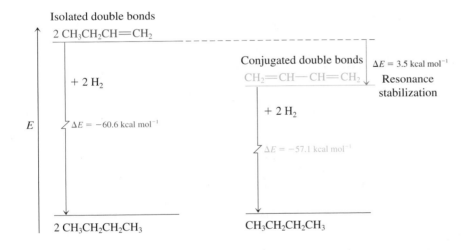

Figure 14-5 The difference between the heats of hydrogenation of two molecules of 1-butene (a terminal monoalkene) and one molecule of 1,3-butadiene (a doubly terminal conjugated diene) reveals their relative stabilities. The value, about 3.5 kcal mol^{-1}, is a measure of the stabilization of 1,3-butadiene due to conjugation.

Exercise 14-9

The heat of hydrogenation of *trans*-1,3-pentadiene is −54.2 kcal mol^{-1}, 6.6 kcal mol^{-1} less than that of 1,4-pentadiene, even lower than that expected from the stabilizing energy of 1,3-butadiene. Explain. (**Hint:** See Section 11-5.)

Conjugation in 1,3-butadiene results from overlap of the pi bonds

How do the two double bonds in 1,3-butadiene interact? The answer lies in the alignment of their π systems, an arrangement that permits the p orbitals on C2 and C3 to overlap (Figure 14-6A). The π interaction that results is weak but nevertheless amounts to a few kilocalories per mole because the π electrons are delocalized over the system of four p orbitals.

Besides adding stability to the diene, this π interaction also raises the barrier to rotation about the single bond to about 4 kcal mol^{-1} (17 kJ mol^{-1}). An inspection of models shows that the molecule can adopt two possible extreme coplanar conformations. In one, designated **s-cis**, the two π bonds lie on the same side of the C2–C3 axis; in the other, called **s-trans**, the π bonds are on opposite sides (Figure 14-6B). The prefix *s* refers to the fact that the bridge between C2 and C3 constitutes a *single* bond. The s-cis form is almost

Model Building

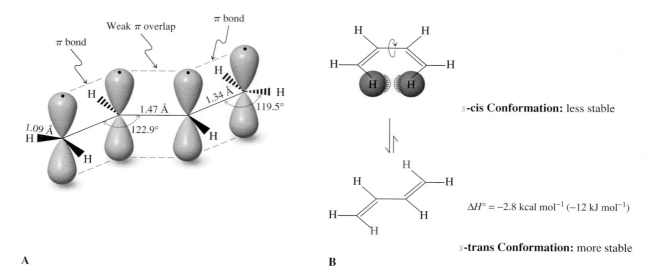

A

B

s-cis Conformation: less stable

$\Delta H° = -2.8$ kcal mol^{-1} (-12 kJ mol^{-1})

s-trans Conformation: more stable

Figure 14-6 (A) The structure of 1,3-butadiene. The central bond is shorter than that in an alkane (1.54 Å for the central C–C bond in butane). The *p* orbitals aligned perpendicularly to the molecular plane form a contiguous interacting array. (B) 1,3-Butadiene can exist in two planar conformations. The *s*-cis form shows steric hindrance because of the proximity of the two "inside" hydrogens highlighted in red.

3 kcal mol^{-1} (12.5 kJ mol^{-1}) less stable than the *s*-trans conformation because of the steric interference between the two hydrogens on the inside of the diene unit.*

Exercise 14-10

The dissociation energy of the central C–H bond in 1,4-pentadiene is only 77 kcal mol^{-1}. Explain. (**Hint:** See Sections 14-1 and 14-2 and draw the product of H atom abstraction.)

The π-electronic structure of 1,3-butadiene may be described by constructing four molecular orbitals out of the four *p* atomic orbitals (Figure 14-7).

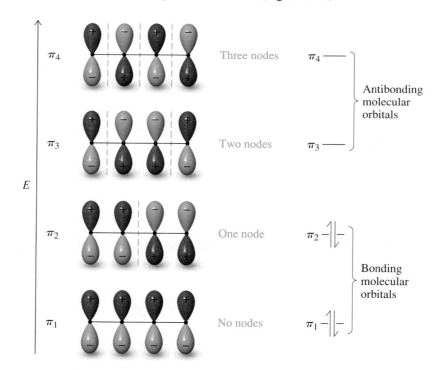

Recall: Mixing four *p* orbitals generates four new molecular orbitals.

π_4 Three nodes π_4 ———

Antibonding molecular orbitals

π_3 Two nodes π_3 ———

π_2 One node π_2

π_1 No nodes π_1

Bonding molecular orbitals

Figure 14-7 A π-molecular-orbital description of 1,3-butadiene. Its four electrons are placed into the two lowest π (bonding) orbitals, π_1 and π_2.

*The *s*-cis conformation is very close in energy to a nonplanar conformation in which the two double bonds are *gauche* (Section 2-9). Whether the *s*-cis or the *gauche* conformation is more stable remains a subject of controversy.

In Summary Dienes are named according to the rules formulated for ordinary alkenes. Conjugated dienes are more stable than dienes containing two isolated double bonds, as measured by their heats of hydrogenation. Conjugation is manifested in the molecular structure of 1,3-butadiene, which has a relatively short central carbon–carbon bond with a small barrier to rotation of more than 6 kcal mol^{-1} (25 kJ mol^{-1}). The two conformers, *s-trans* and *s-cis*, differ in energy by about 3 kcal mol^{-1} (12.5 kJ mol^{-1}). The molecular-orbital picture of the π system in 1,3-butadiene shows two bonding and two antibonding orbitals. The four electrons are placed in the first two bonding levels.

14-6 | ELECTROPHILIC ATTACK ON CONJUGATED DIENES: KINETIC AND THERMODYNAMIC CONTROL

Reaction

Does the structure of conjugated dienes affect their reactivity? Although more stable thermodynamically than dienes with isolated double bonds, conjugated dienes are actually *more reactive* kinetically in the presence of electrophiles and other reagents. 1,3-Butadiene, for example, readily adds 1 mol of cold hydrogen bromide. Two isomeric addition products are formed: 3-bromo-1-butene and 1-bromo-2-butene.

$$CH_2=CH-CH=CH_2 \ + \ HBr \ \xrightarrow{0°C} \ HCH_2-\overset{\overset{\displaystyle Br}{|}}{CH}-CH=CH_2 \ + \ HCH_2-CH=CH-\overset{\overset{\displaystyle Br}{|}}{CH_2}$$

70%   30%

3-Bromo-1-butene   **1-Bromo-2-butene**

The generation of the first product is readily understood in terms of ordinary alkene chemistry. It is the result of a Markovnikov addition to one of the double bonds. But what about the second product?

The presence of 1-bromo-2-butene is explained by the reaction mechanism. Initial protonation at C1 gives the thermodynamically most favored allylic cation.

Mechanism

Protonation of 1,3-Butadiene

$$\overset{+}{CH_2}-CH-CH=CH_2 \longleftarrow \overset{H^+}{\underset{\substack{\text{Attack} \\ \text{at C2}}}{\boxed{\times}}} \overset{1}{CH_2}=\overset{2}{CH}-\overset{3}{CH}=\overset{4}{CH_2} \underset{\substack{\text{Attack} \\ \text{at C1}}}{\xrightarrow{H^+}}$$

Primary nondelocalized cation not formed

$$\left[\overset{\overset{\displaystyle H}{|}}{CH_2}-\overset{+}{CH}-CH=CH_2 \longleftrightarrow \overset{\overset{\displaystyle H}{|}}{CH_2}-CH=CH-\overset{+}{CH_2} \right] \text{ is the same as}$$

Delocalized allylic cation formed exclusively

This cation can be trapped by bromide in two possible ways to form the two observed products: At the terminal carbon, it yields 1-bromo-2-butene; and, at the internal carbon, it furnishes 3-bromo-1-butene. The 1-bromo-2-butene is said to result from *1,4-addition* of hydrogen bromide to butadiene, because reaction has taken place at C1 and C4 of the original diene. The other product arises by the normal 1,2-addition.

Animation

ANIMATED MECHANISM: Addition of HBr to 1,3-butadiene

Nucleophilic Trapping of the Allylic Cation Formed on Protonation of 1,3-Butadiene

$$CH_3\overset{\overset{\displaystyle }{}}{C}HCH=CH_2 \ \underset{\substack{\text{Attack at} \\ \text{internal carbon}}}{\xleftarrow{Br^-}} \ \overset{H}{\underset{CH_3CH \ \ \ \ \overset{+}{} \ \ \ CH_2}{C}} \ \underset{\substack{\text{Attack at} \\ \text{terminal carbon}}}{\xrightarrow{Br^-}} \ CH_3CH=CHCH_2Br$$

Br

1,2-Addition   **1,4-Addition**

Many electrophilic additions to dienes give rise to product mixtures by both modes of addition, because of the intermediacy of allylic cations. For example, the bromination of 1,3-butadiene proceeds through the (bromomethyl) allylic cation, instead of the cyclic bromonium ion encountered for normal alkenes (Section 12-5; see also Exercise 14-13).

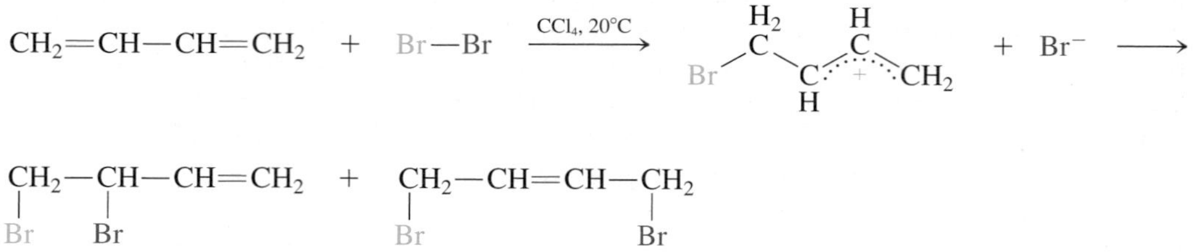

$$CH_2=CH-CH=CH_2 \ + \ Br-Br \ \xrightarrow{CCl_4, \ 20°C} \ \underset{Br}{} \overset{H_2 \quad H}{\underset{H}{C}} \ + \ Br^- \longrightarrow$$

$$\underset{\underset{Br}{|} \quad \underset{Br}{|}}{CH_2-CH-CH=CH_2} \ + \ \underset{\underset{Br}{|} \qquad \qquad \underset{Br}{|}}{CH_2-CH=CH-CH_2}$$

54%   46%

3,4-Dibromo-1-butene   **1,4-Dibromo-2-butene**

Conjugated dienes also function as monomers in polymerizations induced by electrophiles, radicals, and other initiators (see Sections 12-14 and 12-15) and, as such, will be discussed in Section 14-10.

Exercise 14-11

Conjugated dienes can be made by the methods applied to the preparation of ordinary alkenes. Propose syntheses of **(a)** 2,3-dimethylbutadiene from 2,3-dimethyl-1,4-butanediol; **(b)** 1,3-cyclohexadiene from cyclohexane.

Exercise 14-12

Write the products of 1,2-addition and 1,4-addition of **(a)** HBr and **(b)** DBr to 1,3-cyclohexadiene. (**Caution:** What is unusual about the products of 1,2- and 1,4-addition of HX to unsubstituted cyclic 1,3-dienes?)

Changing product ratios: kinetic and thermodynamic control

When the hydrobromination of 1,3-butadiene is carried out at 40°C rather than 0°C, a curious result is obtained: Instead of the original 70:30 mixture of 1,2- and 1,4-adducts, we now observe a 15:85 ratio of the corresponding bromobutenes:

Hydrobromination of 1,3-Butadiene at 0°C: Kinetic Control

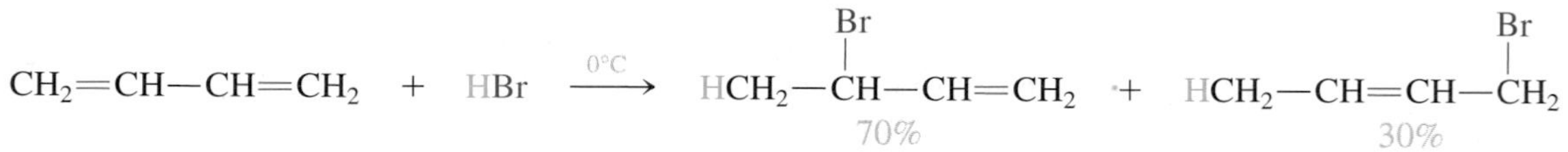

$$CH_2=CH-CH=CH_2 \ + \ HBr \ \xrightarrow{0°C} \ \underset{}{HCH_2}-\overset{\overset{Br}{|}}{CH}-CH=CH_2 \ + \ HCH_2-CH=CH-\overset{\overset{Br}{|}}{CH_2}$$

70%   30%

3-Bromo-1-butene   **1-Bromo-2-butene**

More   Less

Hydrobromination of 1,3-Butadiene at 40°C: Thermodynamic Control

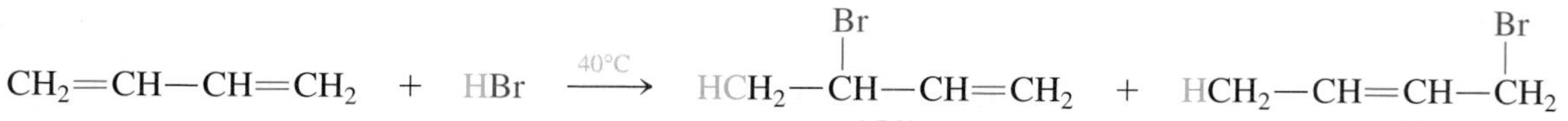

$$CH_2=CH-CH=CH_2 \ + \ HBr \ \xrightarrow{40°C} \ HCH_2-\overset{\overset{Br}{|}}{CH}-CH=CH_2 \ + \ HCH_2-CH=CH-\overset{\overset{Br}{|}}{CH_2}$$

15%   85%

Less   More

This ratio of bromides is obtained also by simply heating the initial 70:30 mixture or, more strikingly, either one of the pure isomers. How can this be explained?

To understand these results, we have to go back to what we discussed in Section 2-1 regarding the kinetics and thermodynamics, or the rates and equilibria, that govern the outcome of reactions. From our observations, it is clear that at the higher temperature the two isomers are in equilibrium, their ratio reflecting their relative thermodynamic stability:

1-bromo-2-butene (with its internal double bond; Section 11-5) is a bit more stable than 3-bromo-1-butene. In general, a reaction whose ratio of products reflects their thermodynamic stability is said to be under **thermodynamic control.** Such is the case in the hydrobromination of 1,3-butadiene at 40°C.

What is happening at 0°C? At that temperature, the two isomers are not interconverting and, hence, thermodynamic control is not attained. What, then, is the origin of the nonthermodynamic ratio? The answer lies in the relative rates at which the two products are generated from the intermediate allylic cation (drawn as the central starting point of the scheme below): 3-Bromo-1-butene, although thermodynamically less stable, is formed faster than 1-bromo-2-butene. In general, a reaction whose ratio of products reflects their relative rates of formation (hence, the relative heights of the respective activation barriers) is said to be under **kinetic control.** Such is the case in the hydrobromination of 1,3-butadiene at 0°C.

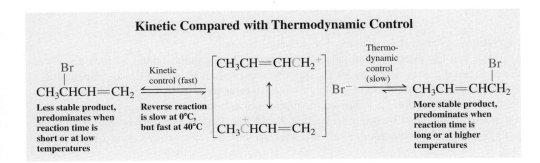

Kinetic Compared with Thermodynamic Control

A potential energy diagram of our reaction (Figure 14-8) illustrates the lower activation barrier (and larger rate k_1) associated with the formation of the less stable product and the higher barrier (and lower rate k_2) leading to its more stable counterpart. The crucial feature is the reversibility (rate k_{-1}) of the kinetic step of trapping the intermediate cation. At 0°C, the less stable, kinetic 3-bromo-1-butene product predominates, because reversal of its formation is relatively slow. At 40°C, this product enters into rapid equilibrium with its cationic precursor and, eventually, with the thermodynamically more stable 1-bromo-2-butene.

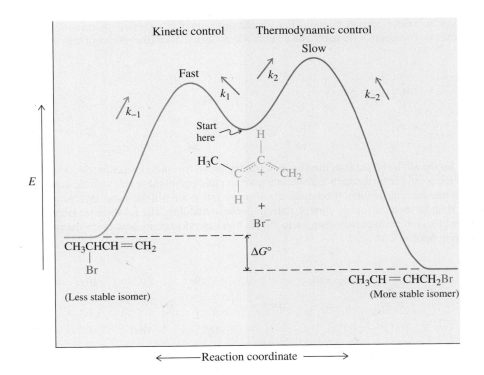

Really? Life itself is under kinetic control. Unlike the "static" chemical systems discussed here, this control is dynamic, in that energy input is required to maintain the continuance of replication, metabolism, supply of heat, etc. You can picture this situation as that of a running engine, which will appear stable until you turn off the power. Similarly, life (as allegorized by the tropical fish tank in the photo) ends when the supply of oxygen or nutrients is cut off and the body is allowed to reach "thermodynamic equilibrium" by decay.

Figure 14-8 Kinetic control (to the left) as compared to thermodynamic control (to the right) in the reaction of 1-methyl-2-propenyl cation with bromide ion (middle).

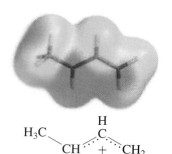

1-Methyl-2-propenyl cation

Why does the less stable product form with a lower activation barrier? Attack of the nucleophile (bromide, in this case) is faster at C3, the more substituted carbon, for a combination of reasons. When HBr protonates the terminal carbon atom of the diene, the freed bromide ion is initially relatively near to the adjacent carbon (C3) of the newly formed allylic cation. In addition, this cation is asymmetric; its positive charge is *unequally* distributed between carbons 1 and 3. More positive charge resides at the secondary carbon C3, as is reflected in the electrostatic-potential map shown in the margin. This position has less electron density (more blue) than the unsubstituted terminus. Thus, attack at C3 is favored kinetically by its greater partial positive charge and by its proximity to the newly released nucleophile.

Solved Exercise 14-13 | Working with the Concepts: Kinetic Versus Thermodynamic Control

Bromination of 1,3-butadiene (see p. 592) at 60°C gives 3,4-dibromo-1-butene and 1,4-dibromo-2-butene in a 10:90 ratio, whereas at −15°C this ratio is 60:40. Explain.

*W*hat you must do is explain two observations. As always, look to mechanisms for a rationale.

*H*ow to begin? Formulate a mechanism that connects starting materials with products. Once you have done so, look closely at the individual steps and intermediates to see if there is a reason for the difference in product ratios at the two temperatures.

*I*nformation needed? We are dealing here with the bromination of a double bond, so refer back to Sections 12-5 and 12-6 for pertinent mechanistic information. Specifically, Section 12-5 showed the initial formation of a bromonium ion A (see structure below), which is then subject to nucleophilic attack by bromide, leading to *anti* addition. Section 12-6 revealed that when the double bond is substituted by alkyl groups, the bromonium ion is distorted, as in B, to reflect carbocation-like character on the more substituted carbon.

*P*roceed. In the case of 1,3-butadiene, the substituent is ethenyl (vinyl). The bromonium ion exists in an open form in order to permit delocalization of the positive charge by resonance. As a general rule, the possibility of stabilization by resonance is the dominant factor in determining the structure of carbocations. Therefore, the relevant intermediate is allylic cation C.

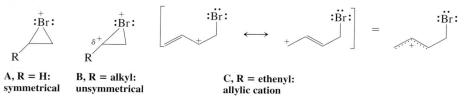

A, R = H: B, R = alkyl: C, R = ethenyl:
symmetrical   unsymmetrical   allylic cation

- Cation C can now be trapped by bromide ion at either its terminal or internal carbon. For reasons explained above, attack by bromide on the internal carbon appears slightly faster (kinetic control), giving the initial low temperature ratio of 60:40 of 3,4-dibromo-1-butene and *trans*-1,4-dibromo-2-butene. At −15°C, these two compounds are stable and do not undergo dissociation (the reverse of their formation).

Kinetic control:

60 : 40

- Warming to 60°C provides enough energy for the allylic bromide to function as a leaving group (by an S_N1 reaction; Section 14-3) and thus to regenerate the allylic cation. At this temperature, the products and the cation are in rapid equilibrium, although the concentration of the higher-energy cation is very small. While the relative rates of bromide attack on the cation are still the same, this factor becomes irrelevant, because the product distribution is under thermodynamic control, reflecting relative stability: The 1,4-dibromo isomer is more stable than the 3,4-dibromo isomer, and the observed product ratio changes to 10:90. You can calculate by how much the two isomers differ in energy by applying the equation from Section 2-1.

Thermodynamic control:

60 : 40   10 : 90

Exercise 14-14 | Try It Yourself

In the reaction shown below, which product is the result of 1,2- and 1,4-addition, respectively, and which is the kinetic and thermodynamic product, respectively?

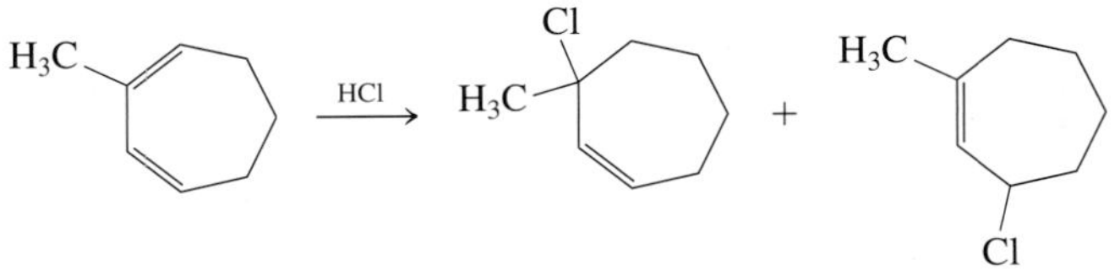

In Summary Conjugated dienes are electron rich and are attacked by electrophiles to give intermediate allylic cations on the way to 1,2- and 1,4-addition products. These reactions may be subject to kinetic control at relatively low temperatures. At relatively higher temperatures, the kinetic product ratios may change to thermodynamic product ratios, when such product formation is reversible.

14-7 | DELOCALIZATION AMONG MORE THAN TWO Pi BONDS: EXTENDED CONJUGATION AND BENZENE

What happens when more than two conjugated double bonds are present in a molecule? Does its reactivity increase? And what happens if the molecule is cyclic? Does it react in the same way as its linear analog? In this section we begin to answer these questions.

Extended pi systems are thermodynamically stable but kinetically reactive

When more than two double bonds are in conjugation, the molecule is called an **extended π system.** An example is 1,3,5-hexatriene, the next higher double-bond homolog of 1,3-butadiene. This compound is quite reactive and polymerizes readily, particularly in the presence of electrophiles. Despite its reactivity as a delocalized π system, it is also relatively stable thermodynamically.

The increased reactivity of this extended π system is due to the low activation barriers for electrophilic additions, which proceed through highly delocalized carbocations. For example, the bromination of 1,3,5-hexatriene produces a substituted pentadienyl cation intermediate that can be described by three resonance structures.

Bromination of 1,3,5-Hexatriene

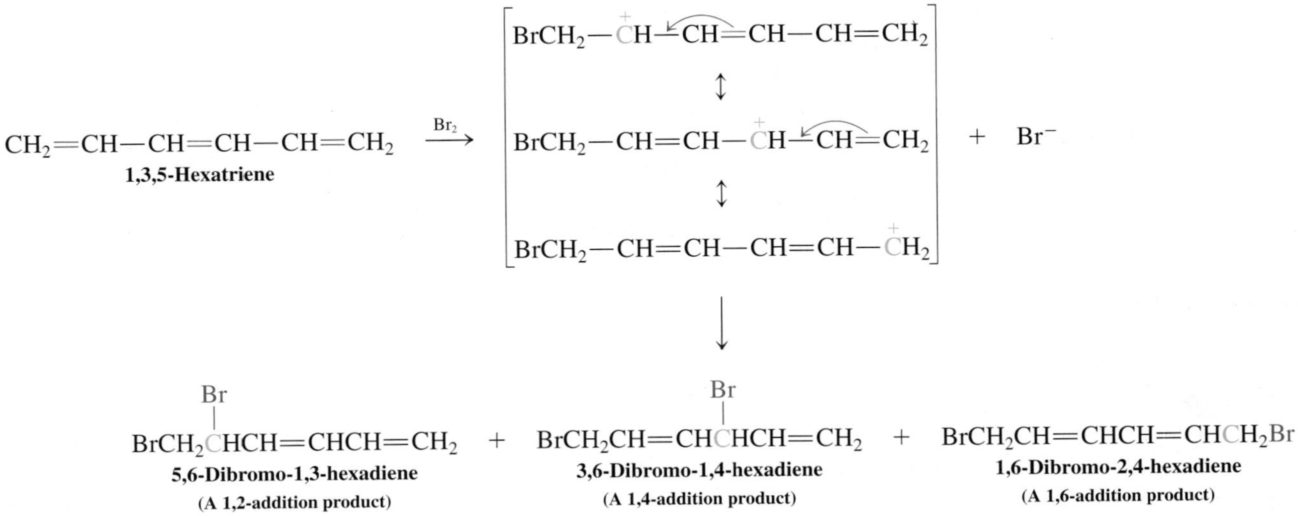

The final mixture is the result of 1,2-, 1,4-, and 1,6-additions, the last product being the most favored thermodynamically, because it retains an internal conjugated diene system.

Exercise 14-15

Upon treatment with two equivalents of bromine, 1,3,5-hexatriene has been reported to give moderate amounts of 1,2,5,6-tetrabromo-3-hexene. Write a mechanism for the formation of this product.

Some highly extended π systems are found in nature. An example is β-carotene, the orange coloring agent in carrots (see Chapter Opening), and its biological degradation product, vitamin A (retinol, below; see Real Life 18-2). Compounds of this type can be very reactive because there are many potential sites for attack by reagents that add to double bonds. In contrast, some cyclic conjugated systems may be considerably less reactive, depending on the number of π electrons (see Chapter 15). The most striking example of this effect is benzene, the cyclic analog of 1,3,5-hexatriene.

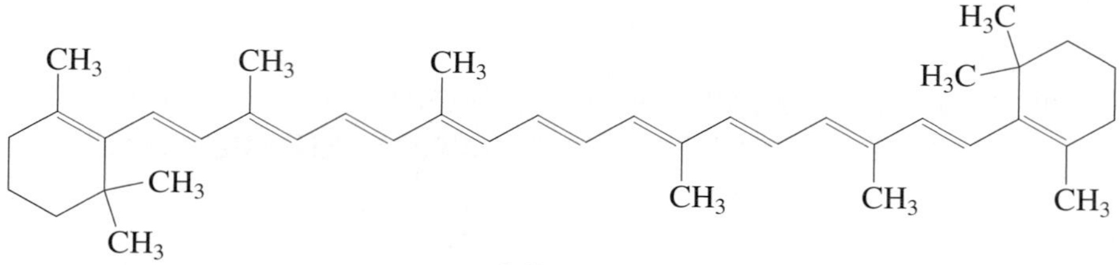

β-Carotene

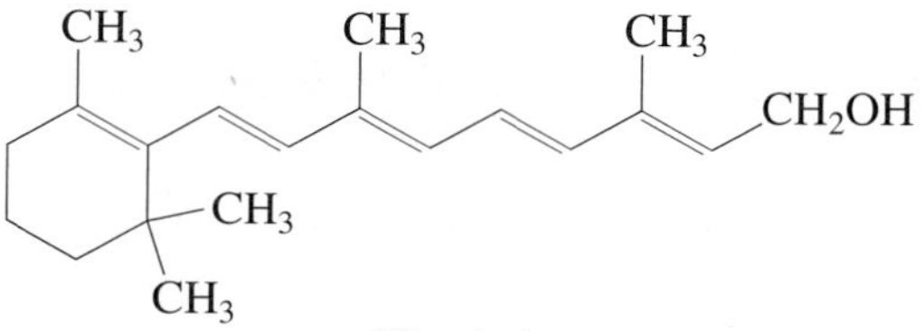

Vitamin A
(Retinol)

Benzene and Its Resonance Structures

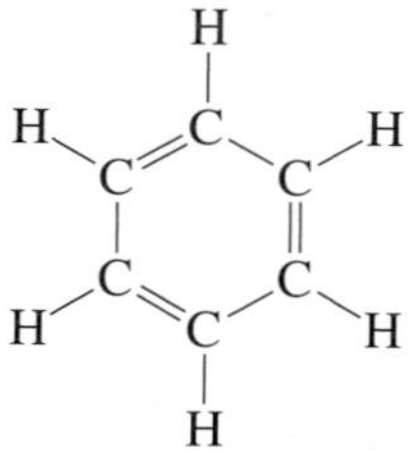

Benzene

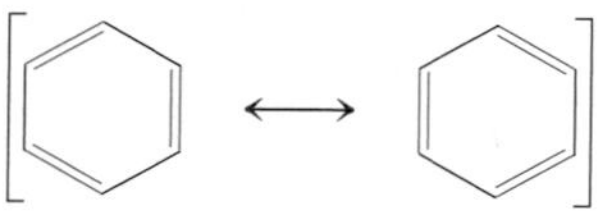

Model Building

Benzene, a conjugated cyclic triene, is unusually stable

Cyclic conjugated systems are special cases. The most common examples are the cyclic triene C_6H_6, better known as benzene, and its derivatives (Chapters 15, 16, and 22). In contrast to hexatriene, benzene is unusually stable, both thermodynamically and kinetically, because of its special electronic makeup (see Chapter 15). That benzene is unusual can be seen by drawing its resonance forms: There are two *equally* contributing Lewis structures. Benzene does not readily undergo addition reactions typical of unsaturated systems, such as catalytic hydrogenation, hydration, halogenation, and oxidation. In fact, because of its low reactivity, benzene can be used as a solvent in organic reactions.

Benzene Is Unusually Unreactive

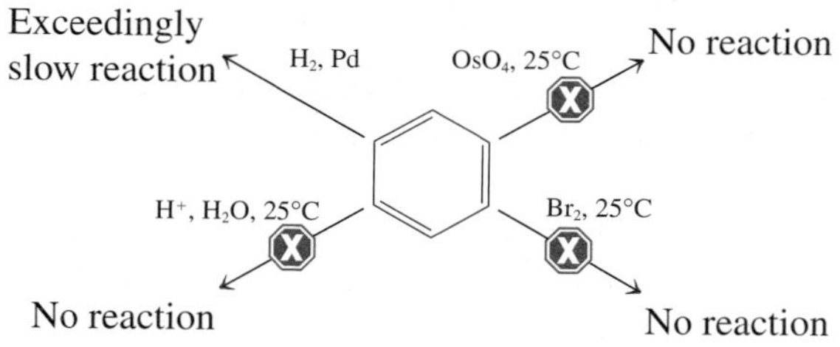

In the chapters that follow, we shall see that the unusual lack of reactivity of benzene is related to the number of π electrons present in its cyclically conjugated array—namely, six. The next section introduces a reaction that is made possible only because its transition state benefits from six-electron cyclic overlap.

In Summary Acyclic extended conjugated systems show increasing thermodynamic stability but also kinetic reactivity because of the many sites open to attack by reagents and the ease of formation of delocalized intermediates. In contrast, the cyclohexatriene benzene is unusually stable and unreactive.

14-8 | A SPECIAL TRANSFORMATION OF CONJUGATED DIENES: DIELS-ALDER CYCLOADDITION

Conjugated double bonds participate in more than just the reactions typical of the alkenes, such as electrophilic addition. This section describes a process in which conjugated dienes and alkenes combine to give substituted cyclohexenes. In this transformation, known as Diels-Alder cycloaddition, the atoms at the ends of the diene add to the alkene double bond, thereby closing a ring. The new bonds form simultaneously and stereospecifically.

The cycloaddition of dienes to alkenes gives cyclohexenes

When a mixture of 1,3-butadiene and ethene is heated in the gas phase, a remarkable reaction takes place in which cyclohexene is formed by the simultaneous generation of two new carbon–carbon bonds. This is the simplest example of the **Diels-Alder* reaction,** in which a conjugated diene adds to an alkene to yield cyclohexene derivatives. The Diels-Alder reaction is in turn a special case of the more general class of **cycloaddition reactions** between π systems, the products of which are called **cycloadducts.** In the Diels-Alder reaction, an assembly of four conjugated atoms containing four π electrons reacts with a double bond containing two π electrons. For that reason, the reaction is also referred to as a *[4 + 2]cycloaddition.* The four-carbon component is simply called *diene,* the alkene is labeled **dienophile,** "diene loving."

Reaction

Diels-Alder Cycloaddition of Ethene and 1,3-Butadiene

$$
\underset{\substack{\textbf{1,3-Butadiene}\\ \textbf{(Four }\pi\textbf{ electrons)}}}{\overset{\text{HC}}{\underset{\text{HC}}{\diagdown}}\!\!\begin{array}{c}\text{CH}_2\\ \\ \text{CH}_2\end{array}} \;+\; \underset{\substack{\textbf{Ethene}\\ \textbf{(Two }\pi\textbf{ electrons)}}}{\overset{\text{CH}_2}{\underset{\text{CH}_2}{\|}}} \;\xrightarrow[\text{[4 + 2] Cycloaddition}]{200°C}\; \underset{\substack{\textbf{Cyclohexene}\\ \textbf{(A cycloadduct)}}}{\begin{array}{c}\text{H}_2\\ \text{C}\\ \text{HC}\diagup\;\diagdown\text{CH}_2\\ \| \qquad |\\ \text{HC}\diagdown\;\diagup\text{CH}_2\\ \text{C}\\ \text{H}_2\\ \mathbf{20\%}\end{array}}
$$

What makes a good Diels-Alder reaction? Reactivity of the diene and dienophile

The prototype reaction of butadiene and ethene actually does not work very well and gives only low yields of cyclohexene. It is much better to use an *electron-poor alkene* with an *electron-rich diene.* Substitution of the alkene with electron-attracting groups and of the diene with electron-donating groups therefore creates excellent reaction partners (see margins on this and the next page).

The trifluoromethyl group, for example, is inductively (Section 8-3) electron attracting owing to its highly electronegative fluorine atoms. The presence of such a substituent enhances the Diels-Alder reactivity of an alkene. Conversely, alkyl groups are electron

$$\underset{\text{H}}{\overset{\text{F}_3\text{C}}{\diagdown}}\text{C}=\text{CH}_2$$

3,3,3-Trifluoro-1-propene
(An electron-poor alkene)

*Professor Otto P. H. Diels (1876–1954), University of Kiel, Germany, Nobel Prize 1950 (chemistry); Professor Kurt Alder (1902–1958), University of Köln, Germany, Nobel Prize 1950 (chemistry).

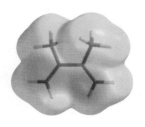

H₃C   CH₃
    C—C
H₂C   CH₂

2,3-Dimethyl-1,3-butadiene
(An electron-rich diene)

donating by induction (Section 11-3) and hyperconjugation (Sections 7-5 and 11-5); their presence increases electron density and is beneficial to dienes in the Diels-Alder reaction. The electrostatic potential maps in the margins on this and the preceding page illustrate these effects. The double bond with the trifluoromethyl substituent has less electron density (yellow) than the double bonds bearing methyl groups (red).

Other alkenes have substituents that interact with double bonds by resonance. For example, carbonyl-containing groups and nitriles are good electron acceptors by this effect. Double bonds bearing such substituents are electron poor because of the contribution of resonance forms that place a positive charge on an alkene carbon atom.

Groups That Are Electron Withdrawing by Resonance

Some examples of the trend in reactivity of dienophiles and dienes are

Dienophile

Diene

Increasing reactivity

Exercise 14-16

Classify each of the following alkenes as electron poor or electron rich, relative to ethene. Explain your assignments.

(a) $H_2C\!=\!CHCH_2CH_3$ (b)   (c)   (d)

Exercise 14-17

The double bond in nitroethene, $H_2C\!=\!CHNO_2$, is electron poor, and that in methoxyethene, $H_2C\!=\!CHOCH_3$, is electron rich. Explain, using resonance structures.

Examples of reaction partners that undergo efficient Diels-Alder cycloaddition are 2,3-dimethyl-1,3-butadiene and propenal (acrolein).

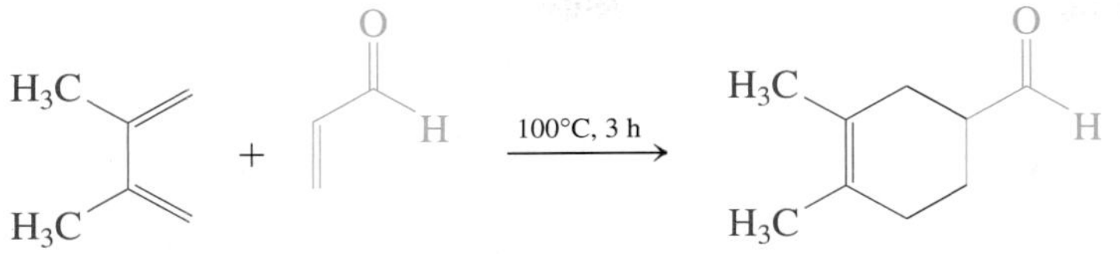

| **2,3-Dimethyl-1,3-butadiene** | **Propenal (Acrolein)** | 100°C, 3 h | **90%** **Diels-Alder cycloadduct** |

The carbon–carbon double bond in the cycloadduct is electron rich and sterically hindered. Thus, it does not react further with additional diene.

The parent 1,3-butadiene, without additional substituents, is electron rich enough to undergo cycloadditions with electron-poor alkenes.

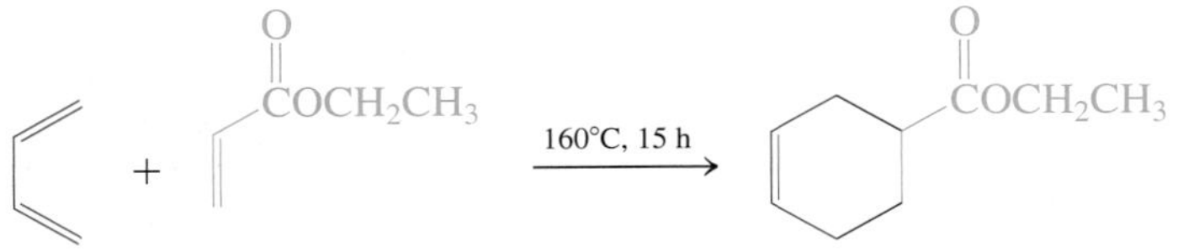

Ethyl propenoate (Ethyl acrylate)   160°C, 15 h   **94%**

Many typical dienes and dienophiles have common names, owing to their widespread use in synthesis (Table 14-1).

Table 14-1	Typical Dienes and Dienophiles in the Diels-Alder Reaction

Dienes

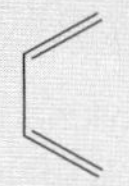

| **1,3-Butadiene** | **2,3-Dimethyl-1,3-butadiene** | **trans,trans-2,4-Hexadiene** | **1,3-Cyclopentadiene** | **1,3-Cyclohexadiene** |

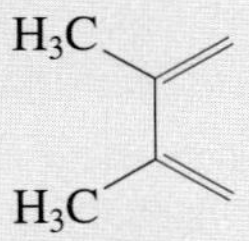

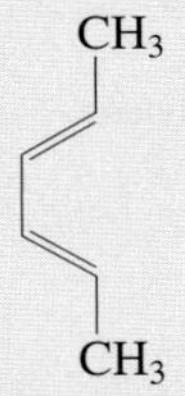

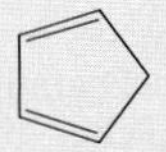

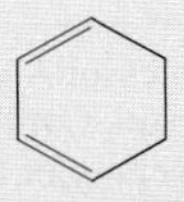

Dienophiles

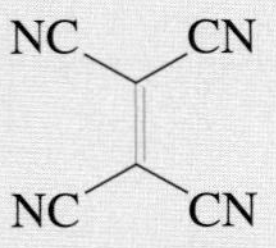

| **Tetracyanoethene** | **cis-1,2-Dicyanoethene** | **Dimethyl cis-2-butenedioate (Dimethyl maleate)** | **Dimethyl trans-2-butenedioate (Dimethyl fumarate)** |

| **2-Butenedioic anhydride (Maleic anhydride)** | **Dimethyl butynedioate (Dimethyl acetylene-dicarboxylate)** | **Propenal (Acrolein)** | **Methyl propenoate (Methyl acrylate)** |

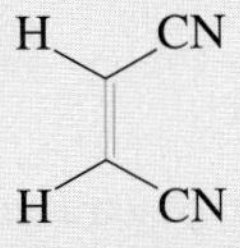

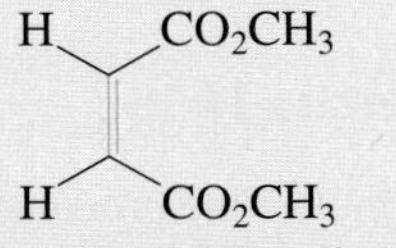

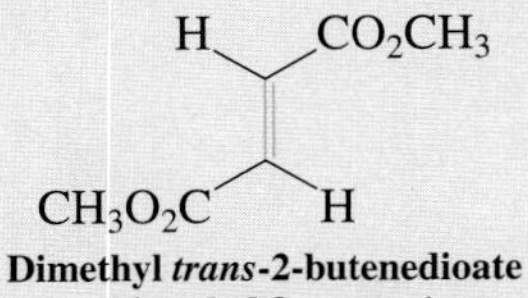

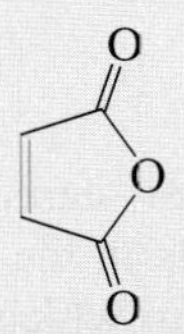

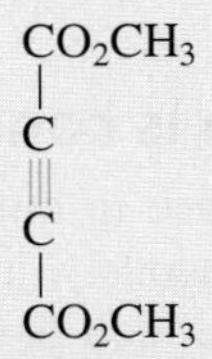

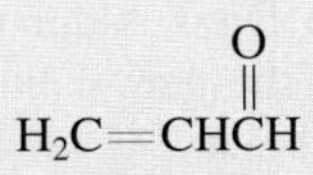

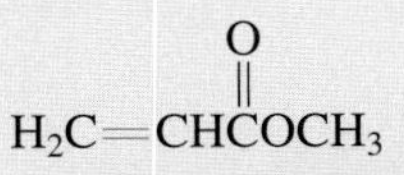

REAL LIFE: MATERIALS 14-1 | Organic Polyenes Conduct Electricity

Can you imagine replacing all the copper wire in our electrical power lines and appliances with an organic polymer? A giant step toward achieving this goal was made in the late 1970s by Heeger, MacDiarmid, and Shirakawa,* for which they received the Nobel Prize in 2000. They synthesized a polymeric form of ethyne (acetylene) that conducts electricity as metals do. This discovery caused a fundamental change in how organic polymers ("plastics") were viewed. Indeed, normal plastics are used to insulate and protect us from electrical currents.

What is so special about polyethyne (polyacetylene)? For a material to be conductive, it has to have electrons that are free to move and sustain a current, instead of being localized, as in most organic compounds. In this chapter, we have seen how such delocalization is attained by linking sp^2 hybridized carbon atoms in a growing chain: conjugated polyenes. We have also learned how a positive charge, a single electron, or a negative charge can "spread out" along the π network, not

unlike a molecular wire. Polyacetylene has such a polymeric structure, but the electrons are still too rigid to move with the facility required for conductivity. To achieve this goal, the electronic frame is "activated" by either removing electrons (oxidation) or adding them (reduction), a transformation called *doping*. The electron hole (positive charge) or electron pair (negative charge) delocalize over the polyenic structure in much the same way as that shown for extended allylic chains in Section 14-6. In the original breakthrough experiment, polyacetylene, made from acetylene by transition metal-catalyzed polymerization (see Section 12-15), was doped with iodine, resulting in a spectacular 10-million-fold increase in conductivity. Later refinement improved this figure to 10^{11}, essentially organic copper!

The black, shiny, flexible foil of polyacetylene (polyethyne) made by polymerization of gaseous ethyne.

trans-Polyacetylene → (−1e, Oxidation (doping)) → Conducting polyacetylene

*Professor Alan J. Heeger (b. 1936), University of California at Santa Barbara, California; Professor Alan G. MacDiarmid (1927–2007), University of Pennsylvania, Philadelphia, Pennsylvania; Professor Hideki Shirakawa (b. 1936), University of Tsukuba, Japan.

1,2-Dimethylenecyclohexane

Exercise 14-18

Formulate the products of [4 + 2]cycloaddition of tetracyanoethene with
(a) 1,3-butadiene; **(b)** cyclopentadiene; **(c)** 1,2-dimethylenecyclohexane (see margin).

The Diels-Alder reaction is concerted

The Diels-Alder reaction takes place in one step. Both new carbon–carbon single bonds and the new π bond form simultaneously, just as the three π bonds in the starting materials break. As mentioned earlier (Section 6-4), one-step reactions, in which bond breaking happens at the same time as bond making, are *concerted*. The concerted nature of this transformation can be depicted in either of two ways: by a dotted circle, representing the six delocalized π electrons, or by electron-pushing arrows. Just as six-electron cyclic overlap

Because of its sensitivity to air and moisture, poly-acetylene is difficult to use in practical applications. However, the idea of using extended π systems to impart organic conductivity can be exploited with a range of materials, all of which have proven utility. Many of these contain especially stabilized cyclic 6-π-electron units, such as benzene (Section 15-2), pyrrole, and thiophene (Section 25-4).

Organic Conductors and Applications

Poly(*p*-phenylene vinylene)

(Electroluminescent displays, such as in mobile phones)

Polythiophene

(Field-effect transistors, such as in supermarket checkouts; antistatic, such as in photographic film)

Polyaniline

(Conductor; electromagnetic shielding of electronic circuits; antistatic, such as in carpets)

Polypyrrole

(Electrolyte; screen coating; sensing devices)

Apart from these applications in electronics, conducting polymers can be made to "light up" when excited by an electric field, a phenomenon called *electroluminescence* that has gained enormous utility in the form of organic light-emitting diodes (OLEDs). Simply put, such organic materials can be viewed as organic light "bulbs." They are relatively light and flexible and encompass a broad spectrum of colors. Because organic polymers are, in principle, readily processed into any shape or form, they provide novel, flexible displays, such as books, luminous cloth, and wall decorations. The future, indeed, looks bright!

These bright colors are thanks to OLEDs.

stabilizes benzene (Section 14-7), the Diels-Alder process benefits from the presence of such an array in its transition state.

Two Pictures of the Transition State of the Diels-Alder Reaction

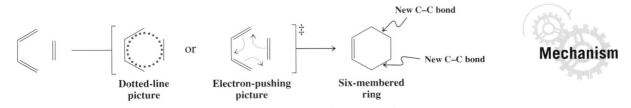

| Dotted-line picture | or | Electron-pushing picture | | Six-membered ring |

New C–C bond

New C–C bond

Mechanism

An orbital representation (Figure 14-9) clearly shows bond formation by overlap of the *p* orbitals of the dienophile with the terminal *p* orbitals of the diene. While these four carbons rehybridize to sp^3, the remaining two internal diene *p* orbitals give rise to the new π bond.

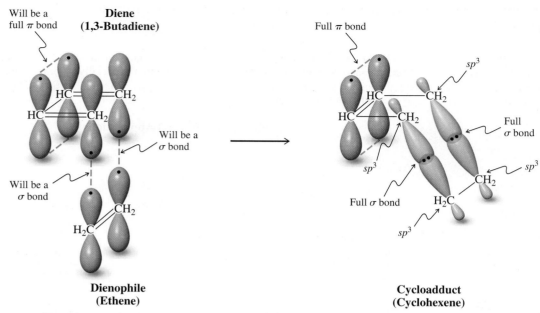

Figure 14-9 Orbital representation of the Diels-Alder reaction between 1,3-butadiene and ethene. The two *p* orbitals at C1 and C4 of 1,3-butadiene and the two *p* orbitals of ethene interact, as the reacting carbons rehybridize to sp^3 to maximize overlap in the two resulting new single bonds. At the same time, π overlap between the two *p* orbitals on C2 and C3 of the diene increases to create a full double bond.

The mechanism of the Diels-Alder reaction requires that both ends of the diene point in the same direction to be able to reach the dienophile carbons simultaneously. This means that the diene has to adopt the energetically slightly less favorable *s*-cis conformation, relative to the more stable *s*-trans form (Figure 14-6).

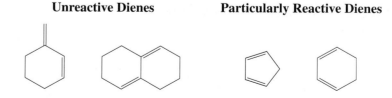

This necessity affects the rates of the cycloaddition: When the *s*-cis form is particularly hindered or impossible, the reaction slows down or does not occur. Conversely, when the diene is constrained to *s*-cis, the transformation is accelerated.

Unreactive Dienes   **Particularly Reactive Dienes**

The Diels-Alder reaction is stereospecific

As a consequence of the concerted mechanism, the Diels-Alder reaction is *stereospecific*. For example, reaction of 1,3-butadiene with dimethyl *cis*-2-butenedioate (dimethyl maleate, a cis alkene) gives dimethyl *cis*-4-cyclohexene-1,2-dicarboxylate. *The stereochemistry at the original double bond of the dienophile is retained in the product.* In the complementary reaction, dimethyl *trans*-2-butenedioate (dimethyl fumarate, a trans alkene) gives the trans adduct.

In the Diels-Alder Reaction, the Stereochemistry of the Dienophile Is Retained

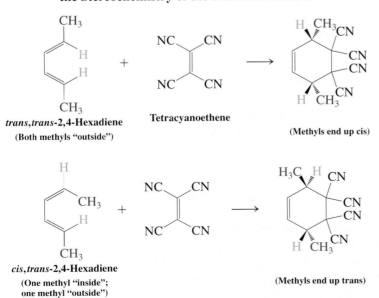

Model Building

Dimethyl *cis*-2-butenedioate
(Dimethyl maleate)
(Cis starting material)

150–160°C, 20 h

Dimethyl *cis*-4-cyclohexene-1,2-dicarboxylate
(Cis product)

68%

Dimethyl *trans*-2-butenedioate
(Dimethyl fumarate)
(Trans starting material)

200–205°C, 3.5 h

Dimethyl *trans*-4-cyclohexene-1,2-dicarboxylate
(Trans product)

95%

Similarly, *the stereochemistry of the diene also is retained.* Note that the cycloadducts depicted here contain stereocenters and may be either meso or chiral. However, since we begin with achiral starting materials, the ensuing products, when chiral, are formed as racemates via two equal-energy transition states (see, for example, Sections 5-7 and 12-5). In other words, the stereospecificity exhibited by the Diels-Alder process refers to relative, not absolute, stereochemistry. As always, we are depicting only one enantiomer of a chiral (but racemic) product, as in the case of the cycloadditions of dimethyl *trans*-2-butenedioate (above) and *cis,trans*-2,4-hexadiene (below).

In the Diels-Alder Reaction, the Stereochemistry of the Diene Is Retained

Model Building

trans,trans-2,4-Hexadiene
(Both methyls "outside")

Tetracyanoethene

(Methyls end up cis)

cis,trans-2,4-Hexadiene
(One methyl "inside";
one methyl "outside")

(Methyls end up trans)

Solved Exercise 14-19 | Working with the Concepts: Diels-Alder Reactions

Draw the product of the following Diels-Alder cycloaddition:

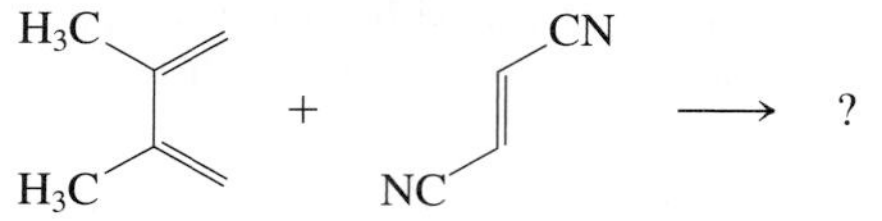

Strategy

In dealing with problems that feature Diels-Alder reactions, it is useful to remind yourself of the spatial approach of diene and dienophile, ideally using molecular models. Once you feel comfortable with visualizing how the reactants align in the transition state, you can practice drawing it.

Solution

• Using Figure 14-9 as a guide, draw, in perspective, the two components of the reaction before cycloaddition.
• Complete your drawing by moving the six participating electrons appropriately to generate the two new σ bonds and the new π bond.
• If your drawing looks distorted, straighten it out to a regular cyclohexene ring, adding the substituents, including their relative stereochemistry.

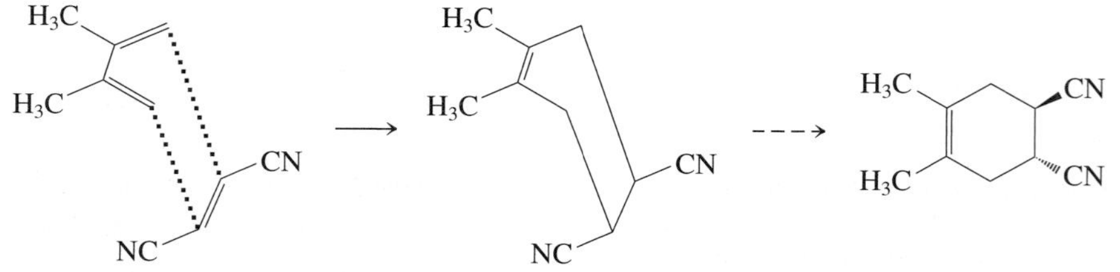

Exercise 14-20 | Try It Yourself

Add structures of the missing starting materials to the following Diels-Alder reaction schemes.

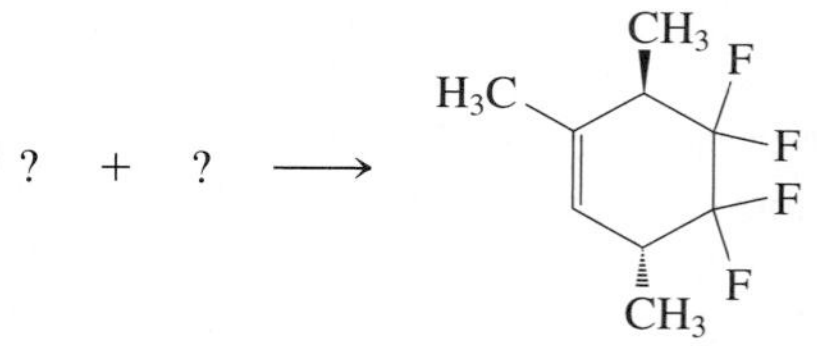

Exercise 14-21 |

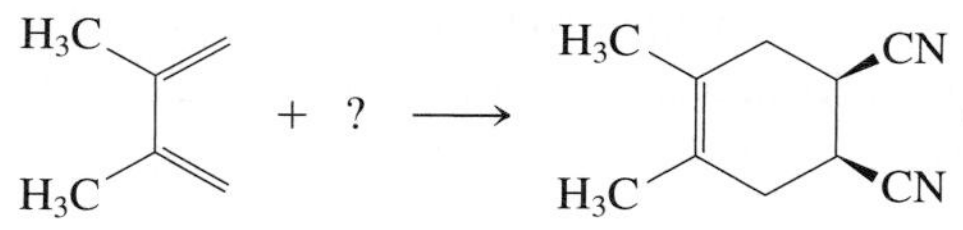

cis,trans-2,4-Hexadiene reacts very sluggishly in [4 + 2]cycloadditions; the trans,trans isomer does so much more rapidly. Explain. [**Hint:** The Diels-Alder reaction requires the *s*-cis arrangement of the diene (Figure 14-9, and see Figure 14-6).]

Diels-Alder cycloadditions follow the endo rule

The Diels-Alder reaction is highly stereocontrolled, not only with respect to the substitution pattern of the original double bonds, but also with respect to the orientation of the starting materials relative to each other. Consider the reaction of 1,3-cyclopentadiene with dimethyl *cis*-2-butenedioate. Two products are conceivable, one in which the two ester substituents on the bicyclic frame are on the same side (cis) as the methylene bridge, the other in which they are on the side opposite (trans) from the bridge. The first is called the **exo adduct,** the second the **endo adduct** (*exo,* Greek, outside; *endo,* Greek, within). The terms refer to the position of groups in bridged systems. Exo substituents are placed cis with respect to the shorter bridge; endo substituents are positioned trans to this bridge. In general, in an exo addition, the substituents on the dienophile point away from the diene. Conversely, in an endo addition they point toward the diene.

Exo and Endo Cycloadditions to Cyclopentadiene

Hydrogens point toward the diene

CO_2CH_3

CO_2CH_3

Substituents point away from the diene

Exo addition →

Shorter bridge (methylene)

CO_2CH_3 ← Exo groups

CO_2CH_3

Substituents point toward the diene

H_3CO_2C

H_3CO_2C

Endo addition →

Hydrogens point away from the diene

CO_2CH_3

CO_2CH_3 ← Endo groups

The Diels-Alder reaction usually proceeds with *endo selectivity,* that is, the product in which the activating electron-withdrawing group of the dienophile is located in the endo position is formed faster than the alternative exo isomer. This occurs even though the exo product is often more stable than its endo counterpart. This observation is referred to as the **endo rule.** The preference for endo cycloaddition has its origin in a variety of steric and electronic influences on the transition state of the reaction. Although the endo transition state is only slightly lower in energy, this is sufficient to control the outcome of most Diels-Alder reactions we shall encounter. Mixtures may ensue in the case of highly substituted systems or when several different activating substituents are present.

Model Building

The Endo Rule

+

$COCH_3$

O

$\xrightarrow{80°C}$

endo

CH_3O

O

H

H

$\xrightarrow{\text{Endo addition}}$

H

$COCH_3$

O

91%

Methyl propenoate

Endo product

REAL LIFE: SUSTAINABILITY 14-2 | The Diels-Alder Reaction is "Green"

In the Diels-Alder reaction, both starting materials are consumed to give a new product without generating any additional materials. Such transformations are called "atom economical," because all of the atoms of starting materials are incorporated in the product. Atom-economical reactions are a key ingredient of green chemistry (see Real Life 3-1), and the Diels-Alder reaction adheres to several of its principles. Thus, it generates no (or little) waste, incorporates all starting materials in the product(s), and retains functionality. It normally does not require protecting groups; it can be run with neat reagents, thus avoiding solvents; and it is often efficient enough to allow crystallization or distillation of pure product, thus avoiding column chromatography. Many Diels-Alder reactions need some heating to proceed at reasonable rates, but this problem can be overcome by the application of catalysts that allow room-temperature transformations. For example, Lewis acids (Section 2-3) accelerate cycloadditions greatly. A case in which this effect has been quantified is shown below. Such catalysis also affects exo/endo ratios and can result in enantioselectivity when optically active catalysts are employed (see, for example, Real Life 5-4 and 9-3, and Section 12-2). The effect of the Lewis acid is to activate the dienophile by complexation to the carbonyl oxygen lone pairs, rendering the carbonyl group even more electron withdrawing.

Water, the greenest solvent

Sometimes solvents are unavoidable, as in intramolecular reactions (Exercise 14-24), when high dilution is important (Section 9-6). The green solvent of choice is, naturally, water. In fact, as shown in the example, water is not only feasible as a solvent, it can speed up Diels-Alder reactions on its own, in addition to improving stereoselectivity, especially when combined with Lewis acid catalysis. This effect of water has been ascribed to hydrogen bonding and hydrophobic effects (Section 8-2) on the transition state.

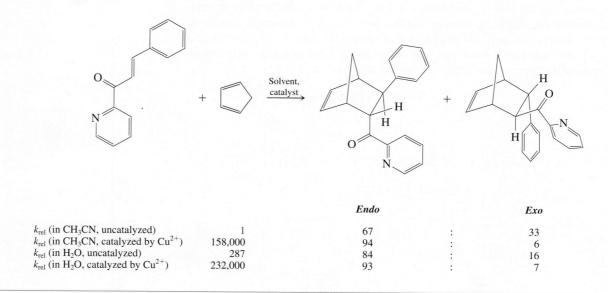

		Endo		*Exo*
k_{rel} (in CH_3CN, uncatalyzed)	1	67	:	33
k_{rel} (in CH_3CN, catalyzed by Cu^{2+})	158,000	94	:	6
k_{rel} (in H_2O, uncatalyzed)	287	84	:	16
k_{rel} (in H_2O, catalyzed by Cu^{2+})	232,000	93	:	7

Animation

ANIMATED MECHANISM:
Diels-Alder cycloaddition
(endo rule)

The relative stereochemistry of the product of a general Diels-Alder reaction that follows the endo rule is illustrated on top of the next page. As an aid to keeping track of the substituents and where they are going, we can use the general labels "o" (for *o*utside, as in outside the half-circle marked by the diene carbon sequence) and "i" (for *i*nside) for the two possible stereochemical orientations of groups attached to the end of the diene. We then label the substituents on the dienophile with respect to their orientation in the transition state of the reaction as either endo or exo. The structure of the expected product with all substituents in place is shown on the right of the equation. You can see that "o" is always cis to "endo." This scheme allows you to get the structure of your product quickly, without the need for stereodrawings. However, it does not replace a thorough understanding of the principles that led to it!

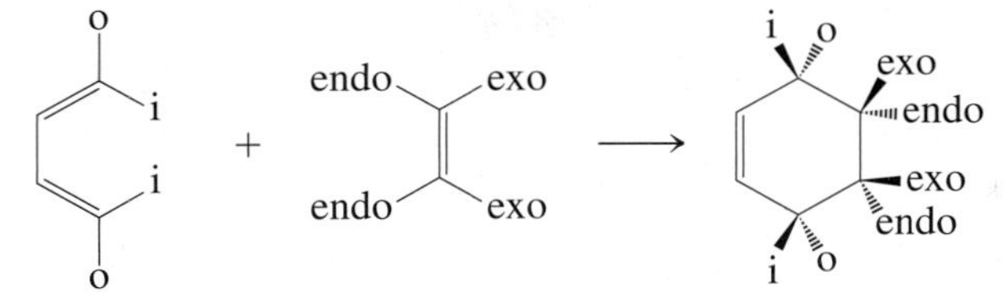

o = "outside"
i = "inside"

Solved Exercise 14-22 | Working with the Concepts: The Endo Rule

Draw the product of the reaction of *trans,trans*-2,4-hexadiene with methyl propenoate (show the stereochemistry clearly).

Strategy
First, we draw the structures of the two reagents, *trans,trans*-2,4-hexadiene and methyl propenoate. Next, to get the stereochemistry of the product right, we need to line up the reagents as in the transition state picture in Figure 14-9, one above the other. The ester function of the dienophile is, according to the endo rule, expected to be positioned endo.

Solution
• Following the prescribed steps gives us

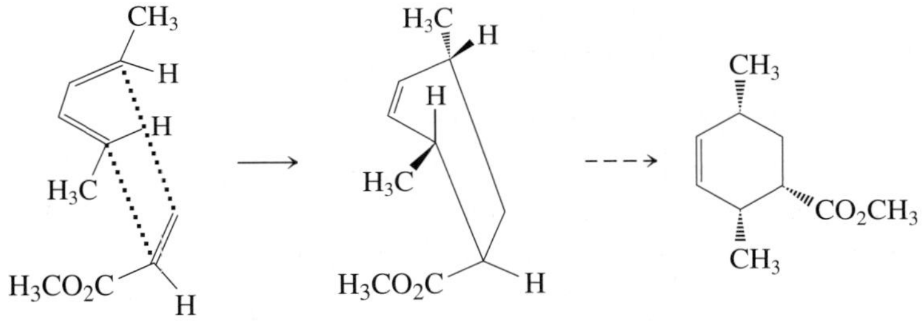

• We can check the result, using our general schematic formalism on top of this page. For this purpose, we label all substituents in the reagents: The two methyl groups in the diene are located "outside" and therefore are designated "o." The ester function in the dienophile is labeled endo. Plugging in these labeled groups into our generalized product structure confirms our solution.

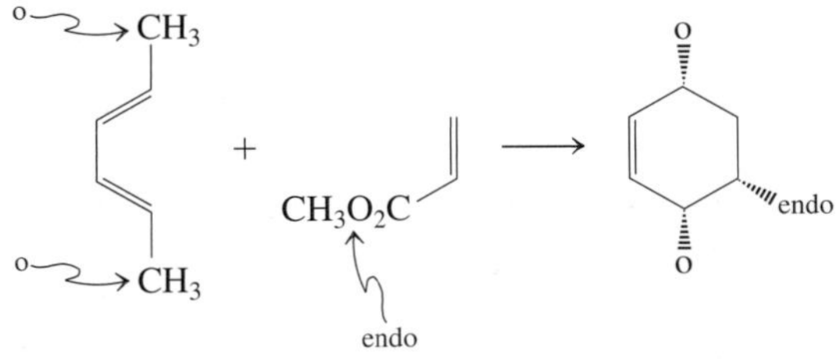

Exercise 14-23 | Try It Yourself

Predict the products of the following reactions (show the stereochemistry clearly): **(a)** *trans*-1,3-pentadiene with 2-butenedioic anhydride (maleic anhydride); **(b)** 1,3-cyclopentadiene with dimethyl *trans*-2-butenedioate (dimethyl fumarate).

Exercise 14-24

The Diels-Alder reaction can also occur in an intramolecular fashion. Draw the two transition states leading to products in the following reaction.

$$\xrightarrow[75\%]{180°C, 5 \text{ h}}$$

65 : 35

In Summary The Diels-Alder reaction is a concerted cycloaddition that proceeds best between electron-rich 1,3-dienes and electron-poor dienophiles to furnish cyclohexenes. It is stereospecific with respect to the stereochemistry of the double bonds and with respect to the arrangements of the substituents on diene and dienophile: It follows the endo rule.

14-9 ELECTROCYCLIC REACTIONS

The Diels-Alder reaction couples the ends of two separate π systems. Can rings be formed by the linkage of the termini of a *single* conjugated di-, tri-, or polyene? Yes, and this section will describe the conditions under which such ring closures (and their reverse), called **electrocyclic reactions,** take place. Cycloadditions and electrocyclic reactions belong to a class of transformations called **pericyclic** (*peri,* Greek, around), because they exhibit transition states with a cyclic array of nuclei and electrons.

Electrocyclic transformations are driven by heat or light

Let us consider first the conversion of 1,3-butadiene into cyclobutene. This process is endothermic because of ring strain. Indeed, the reverse reaction, ring *opening* of cyclobutene, occurs readily upon heating. However, ring *closure of cis*-1,3,5-hexatriene to 1,3-cyclohexadiene is exothermic and takes place thermally. Is it possible to drive these transformations in the thermally disfavored directions?

We know that in a thermal reaction this is a difficult task, because equilibrium is governed by thermodynamics (Section 2-1). However, the problem can be surmounted in some cases by using light, so-called **photochemical reactions.** In these, absorption of a photon by the starting material excites the molecule into a higher energy state. We have seen how such absorptions form the basis for spectroscopy (Section 10-2; see also Section 14-11). Molecules can relax from such excited states to furnish thermodynamically less stable products than starting material(s). We shall not deal with the details of photochemistry in this text, but we do note that it allows electrocyclic reaction equilibria to be driven in the energetically unfavorable direction. Therefore, irradiation of 1,3-cyclohexadiene with light of appropriate frequency will cause its conversion to its triene isomer. Similarly, irradiation of 1,3-butadiene effects ring closure to cyclobutene.

Photochemical reactions are used increasingly in "green" technology. This reactor on the roof of Complutense University of Madrid is used for water disinfection. A polymer-supported dye absorbs sunlight to convert oxygen to a more reactive state ("singlet oxygen"), which destroys harmful bacteria in water (picture courtesy of Professor Guillermo Orellana, UCM).

Electrocyclic Reactions

cis-**1,3,5-Hexatriene** $\underset{h\nu}{\overset{\Delta}{\rightleftharpoons}}$ **1,3-Cyclohexadiene**

$\Delta H° = -14.5 \text{ kcal mol}^{-1} (-60.7 \text{ kJ mol}^{-1})$
Ring closure to six-membered ring is exothermic

Cyclobutene $\underset{h\nu}{\overset{\Delta}{\rightleftharpoons}}$ **1,3-Butadiene**

$\Delta H° = -9.7 \text{ kcal mol}^{-1} (-40.6 \text{ kJ mol}^{-1})$
Ring opening of four-membered ring is exothermic

Exercise 14-25

Give the products obtained on heating the following compounds.

(a) [structure]   (b) [structure]

Electrocyclic reactions are concerted and stereospecific

Like the Diels-Alder cycloaddition, electrocyclic reactions are concerted and stereospecific. Thus, the thermal isomerization of *cis*-3,4-dimethylcyclobutene gives only *cis,trans*-2,4-hexadiene.

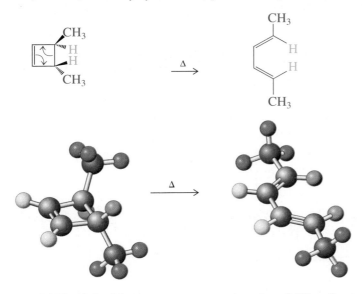

cis-**3,4-Dimethylcyclobutene**   *cis,trans*-**2,4-Hexadiene**

Heating its isomer, *trans*-3,4-dimethylcyclobutene, provides only *trans,trans*-2,4-hexadiene.

trans-**3,4-Dimethylcyclobutene**   *trans,trans*-**2,4-Hexadiene**

Model Building

Figure 14-10 takes a closer look at these processes. As the bond between carbons C3 and C4 in the cyclobutene is broken, these carbon atoms must rehybridize from sp^3 to sp^2

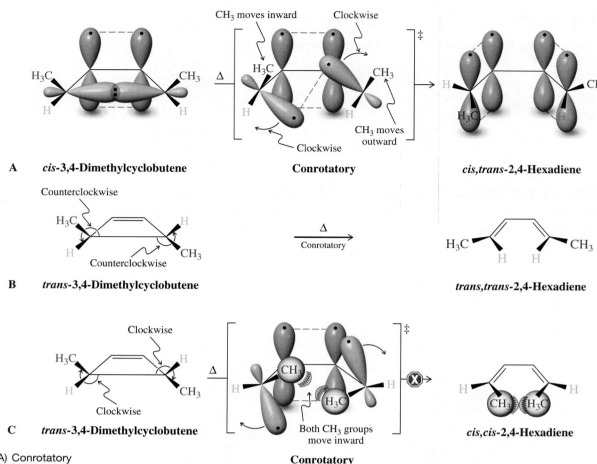

A   *cis*-3,4-Dimethylcyclobutene   **Conrotatory**   *cis,trans*-2,4-Hexadiene

B   *trans*-3,4-Dimethylcyclobutene   *trans,trans*-2,4-Hexadiene

C   *trans*-3,4-Dimethylcyclobutene   *cis,cis*-2,4-Hexadiene

Conrotatory

Figure 14-10 (A) Conrotatory ring opening of *cis*-3,4-dimethylcyclobutene. Both reacting carbons rotate clockwise. The *sp³* hybrid lobes in the ring change into *p* orbitals, the carbons becoming *sp²* hybridized. Overlap of these *p* orbitals with those already present in the cyclobutene starting material creates the two double bonds of the cis,trans diene. (B) Similar conrotatory opening of *trans*-3,4-dimethylcyclobutene in a counterclockwise fashion proceeds to the trans,trans diene. (C) The alternative clockwise conrotatory opening of *trans*-3,4-dimethylcyclobutene does not occur because of steric encumbrance in the transition state.

and rotate to permit overlap between the emerging *p* orbitals and those originally present. In such thermal cyclobutene ring openings, the carbon atoms are found to rotate *in the same direction,* either both clockwise or both counterclockwise. This mode of reaction is called a **conrotatory** process. In the case of *cis*-3,4-dimethylcyclobutene, both the clockwise and counterclockwise courses result in the same product, *cis,trans*-2,4-hexadiene. However, for *trans*-3,4-dimethylcyclobutene, two products are possible. The counterclockwise mode leads to the observed *trans,trans*-2,4-hexadiene. Rotation in the opposite direction would form the corresponding cis,cis isomer but is sterically encumbered and not observed.

Fascinatingly, the photochemical closure (**photocyclization**) of butadiene to cyclobutene proceeds with stereochemistry exactly *opposite* that observed in the thermal opening. In this case, the products arise by rotation of the two reacting carbons in opposite directions. In other words, if one rotates clockwise, the other does so counterclockwise. This mode of movement is called **disrotatory** (Figure 14-11).

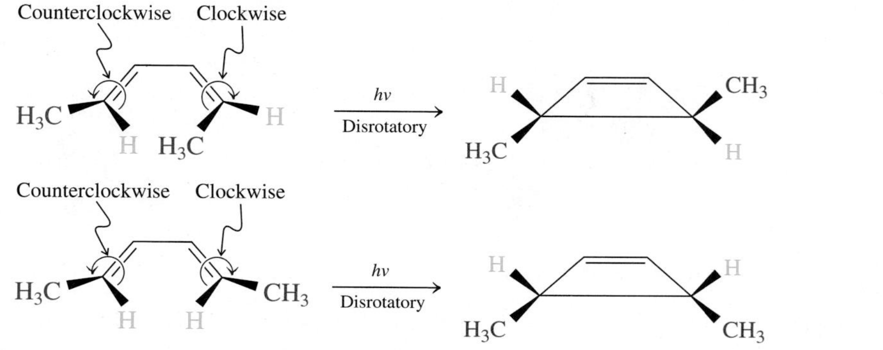

Figure 14-11 Disrotatory photochemical ring closure of *cis,trans*- and *trans,trans*-2,4-hexadiene. In the disrotatory mode, one carbon rotates clockwise, the other counterclockwise.

Can these observations be generalized? Let us look at the stereochemistry of the *cis*-1,3,5-hexatriene–cyclohexadiene interconversion. Surprisingly, the six-membered ring is formed thermally by the disrotatory mode, as can be shown by using derivatives. For example, heated *trans,cis,trans*-2,4,6-octatriene gives *cis*-5,6-dimethyl-1,3-cyclohexadiene, and *cis,cis,trans*-2,4,6-octatriene converts into *trans*-5,6-dimethyl-1,3-cyclohexadiene, both disrotatory closures.

Stereochemistry of the Thermal 1,3,5-Hexatriene Ring Closure

Model Building

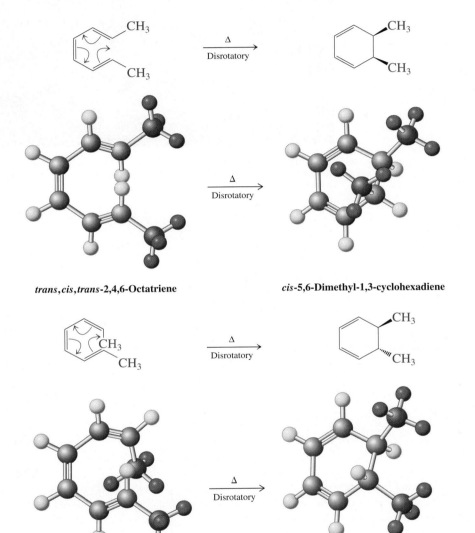

trans,cis,trans-2,4,6-Octatriene

cis-5,6-Dimethyl-1,3-cyclohexadiene

cis,cis,trans-2,4,6-Octatriene

trans-5,6-Dimethyl-1,3-cyclohexadiene

In contrast, the corresponding photochemical reactions occur in conrotatory fashion.

Stereochemistry of the Photochemical 1,3,5-Hexatriene Ring Closure

Model Building

REAL LIFE: MEDICINE 14-3 | **An Electrocyclization Cascade in Nature: Immunosuppressants from Streptomyces Cultures**

Streptomyces is a group of bacteria found predominantly in soil and in decaying vegetation. You may have noticed them in action by the "earthy" odor of compost heaps. These bacteria are prolific in producing medicinally useful, biologically active compounds and have therefore been investigated thoroughly by organic chemists. Among the components of some *Streptomyces* cultures are polyenes, including the conjugated tetraene spectinabilin. In 2001, the promising potent immunosuppressants called SNF 4435 C and D (an acronym referring to the biological screening method) were discovered in the same cultures. Chemists noticed that these compounds were isomers of spectinabilin and speculated that they were formed by a cascade of two electrocyclizations.

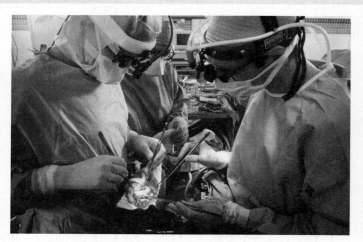

Surgeons prepare to insert a new heart into a transplant patient. Immunosuppressants will be essential to avoid rejection by the immune system of the recipient.

Spectinabilin

SNF 4435 C

SNF 4435 D

Spectinabilin is an all-trans tetraene and therefore incapable of undergoing ring closure. It was found, however, that sunlight induced cis-trans isomerization of the two internal double bonds furnishing the cis,cis isomer that is perfectly set up to enter into an 8π conrotatory cyclization. Because

there is a stereocenter (*), there are two modes of conrotatory movement (clockwise and counterclockwise), giving two isomeric cyclooctatrienes. These compounds then undergo subsequent 6π disrotatory closure to the two SNF isomers.

Formation of SNF 4435 Isomers from Spectinabilin

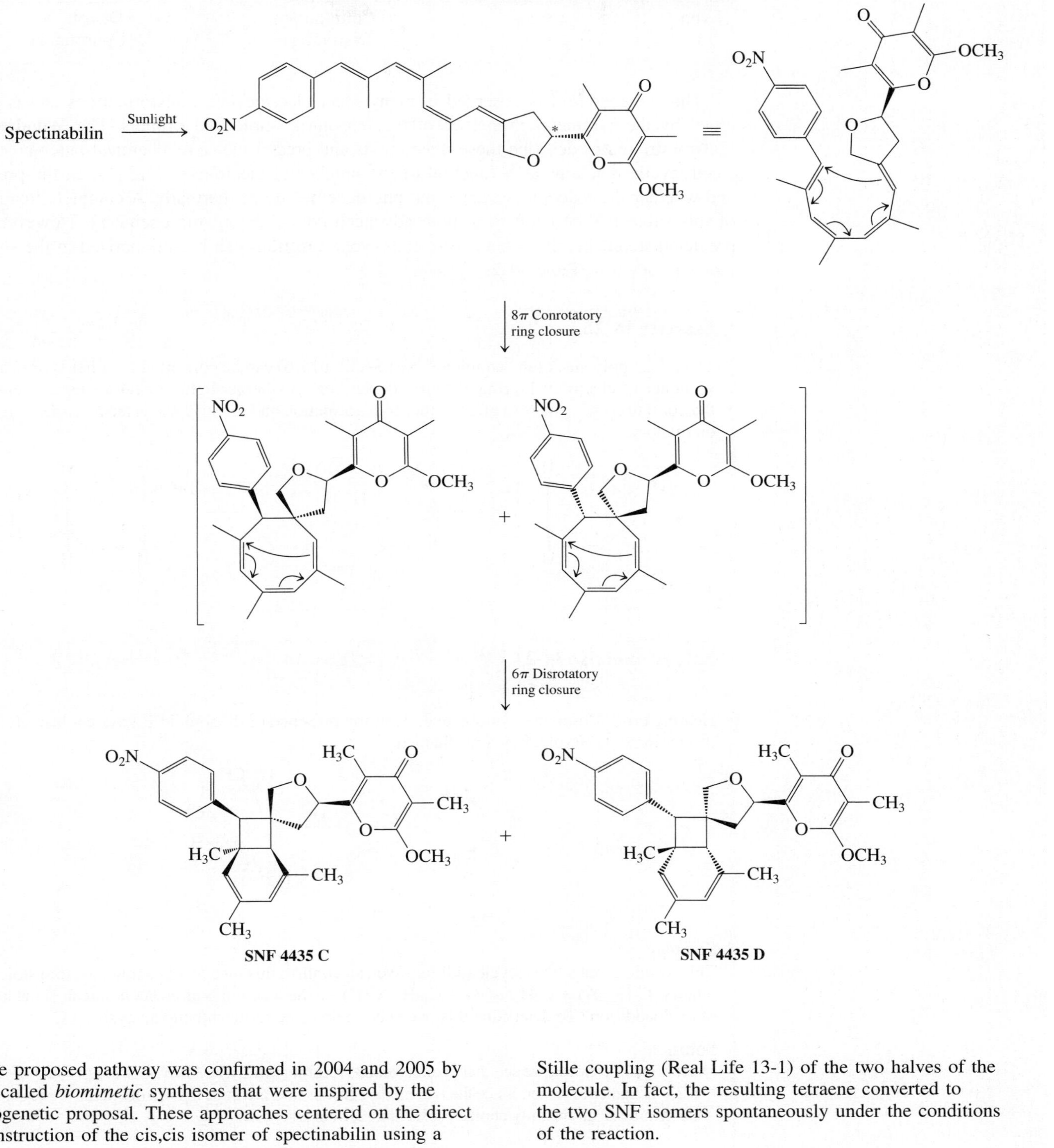

Spectinabilin $\xrightarrow{\text{Sunlight}}$

8π Conrotatory
ring closure

6π Disrotatory
ring closure

SNF 4435 C

SNF 4435 D

The proposed pathway was confirmed in 2004 and 2005 by so-called *biomimetic* syntheses that were inspired by the biogenetic proposal. These approaches centered on the direct construction of the cis,cis isomer of spectinabilin using a Stille coupling (Real Life 13-1) of the two halves of the molecule. In fact, the resulting tetraene converted to the two SNF isomers spontaneously under the conditions of the reaction.

Table 14-2	Stereochemical Course of Electrocyclic Reactions (Woodward-Hoffmann Rules)	
Number of pairs of participating electrons	**Thermal process**	**Photochemical process**
Even	Conrotatory	Disrotatory
Odd	Disrotatory	Conrotatory

This stereocontrol is observed in many other electrocyclic transformations and is governed by the symmetry properties of the relevant π molecular orbitals. The **Woodward-Hoffmann* rules** describe these interactions and predict the stereochemical outcome of all electrocyclic reactions as a function of the number of electrons taking part in the process and whether the reaction is carried out photochemically or thermally. A complete treatment of this subject is best left to a more advanced course in organic chemistry. However, the predicted stereochemical course of electrocyclic reactions can be summarized in the simple manner shown in Table 14-2.

Exercise 14-26

The cyclic polyene A (an "annulene"; see Section 15-6) can be converted to either B or C by a sequence of electrocyclic ring closures, depending on whether light or heat are used. Identify the conditions necessary to effect either transformation and identify each step as either con- or disrotatory.

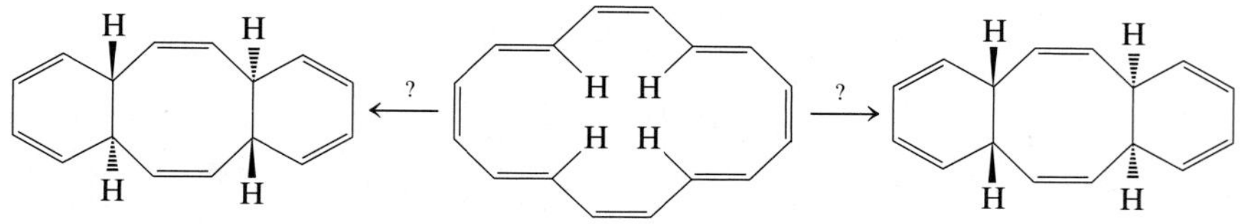

B   A   C

Solved Exercise 14-27 | Working with the Concepts: An Electrocyclic Reaction with a Twist

Heating *cis*-3,4-dimethylcyclobutene, A, in the presence of dienophile B gave exclusively the diastereomer C. Explain by a mechanism.

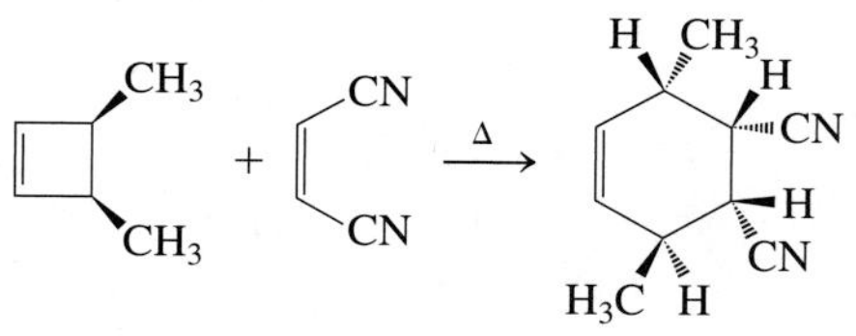

A   B   C

Strategy

This reaction looks like a cycloaddition. We can confirm this idea by checking the atom stoichiometry: $C_6H_{10}(A) + C_4H_2N_2(B) = C_{10}H_{12}N_2(C)$, so the reaction is atom economical. What kind of cycloaddition? To determine this, we need to do some retrosynthetic analysis on C.

Solution

• Working backward, we see that cyclohexene C resembles a Diels-Alder addition product of B to a 2,4-hexadiene isomer. Since the two methyl groups in C are trans to each other, the diene cannot be symmetrical: the only choice is *cis,trans*-2,4-hexadiene D:

---

*Professor Robert B. Woodward (1917–1979), Harvard University, Cambridge, Massachusetts, Nobel Prize 1965 (chemistry); Professor Roald Hoffmann (b. 1937), Cornell University, Ithaca, New York, Nobel Prize 1981 (chemistry).

- D must be derived from isomer A by thermal, conrotatory electrocyclic ring opening.
- Is the stereochemistry of the cycloaddition of B to D exo or endo? Inspection of the relative positioning of the substituents at the four contiguous stereocenters shows it to be endo, as in E.

Exercise 14-28 | Try It Yourself

Irradiation of ergosterol gives provitamin D_2, a precursor of vitamin D_2 (a deficiency of which causes softening of the bones, especially in children). Is the ring opening conrotatory or disrotatory? (**Caution:** The product is written in a more stable conformation than that obtained upon ring opening.)

Ergosterol   Provitamin D_2 Vitamin D_2

In Summary Conjugated dienes and hexatrienes are capable of (reversible) electrocyclic ring closures to cyclobutenes and 1,3-cyclohexadienes, respectively. The diene–cyclobutene system prefers thermal conrotatory and photochemical disrotatory modes. The triene–cyclohexadiene system reacts in the opposite way, proceeding through thermal disrotatory and photochemical conrotatory rearrangements. The stereochemistry of such electrocyclic reactions is governed by the Woodward-Hoffmann rules.

14-10 | POLYMERIZATION OF CONJUGATED DIENES: RUBBER

Like simple alkenes (Section 12-15), conjugated dienes can be polymerized. The elasticity of the resulting materials has led to their use as synthetic rubbers. The biochemical pathway to natural rubber features an activated form of the five-carbon unit 2-methyl-1,3-butadiene (isoprene, see Section 4-7), which is an important building block in nature.

1,3-Butadiene can form cross-linked polymers

When 1,3-butadiene is polymerized at C1 and C2, it yields a polyethenylethene (polyvinylethylene).

1,2-Polymerization of 1,3-Butadiene

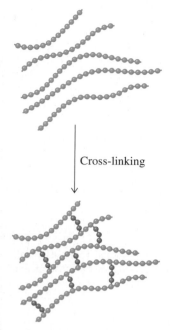

Cross-linking

Figure 14-12 Cross-linking underlies the elasticity of poly-butadiene chains in rubber.

Elasticity in action: This bouncing ball was photographed using colored strobe lights, flashing 50 times per second, giving about 75 separate images. The sequence shows that the ball's trajectory is a parabola. The ball moves fastest near the ground and slowest at the top of an arc. The horizontal speed is approximately constant, and only the vertical speed is changing (due to gravity). The height of successive bounces is less as the ball loses energy by hitting the ground.

Alternatively, polymerization at C1 and C4 gives either *trans*-polybutadiene, *cis*-polybutadiene, or a mixed polymer.

1,4-Polymerization of 1,3-Butadiene

$$n\ CH_2{=}CH{-}CH{=}CH_2 \xrightarrow{\text{Initiator}} {-}(CH_2{-}CH{=}CH{-}CH_2)_n{-}$$

cis- or *trans*-**Polybutadiene**

Butadiene polymerization is unique in that the product itself may be unsaturated. The double bonds in this initial polymer may be linked by further treatment with added chemicals, such as radical initiators, or by radiation. In this way, **cross-linked polymers** arise (Figure 14-12), in which individual chains have been connected into a more rigid framework. Cross-linking generally increases the density and hardness of such materials. It also greatly affects a property characteristic of butadiene polymers: **elasticity.** The individual chains in *most* polymers can be moved past each other, a process allowing for their molding and shaping. In cross-linked systems, however, such deformations are rapidly reversible: The chain snaps back to its original shape (more or less). Such elasticity is characteristic of rubbers.

Synthetic rubbers are derived from poly-1,3-dienes

Polymerization of 2-methyl-1,3-butadiene (isoprene, Section 4-7) by a Ziegler-Natta catalyst (Section 12-15) results in a synthetic rubber *(polyisoprene)* of almost 100% Z configuration. Similarly, 2-chloro-1,3-butadiene furnishes an elastic, heat- and oxygen-resistant polymer called neoprene, with trans chain double bonds. Several million tons of synthetic rubber are produced in the United States every year.

$$n\ H_2C{=}\overset{\overset{\textstyle CH_3}{|}}{C}{-}CH{=}CH_2 \xrightarrow{\text{TiCl}_4,\ \text{AlR}_3}$$

2-Methyl-1,3-butadiene

(Z)-Polyisoprene

$$n\ H_2C{=}\overset{\overset{\textstyle Cl}{|}}{C}{-}CH{=}CH_2 \xrightarrow{\text{TiCl}_4,\ \text{AlR}_3}$$

2-Chloro-1,3-butadiene

Neoprene

Natural *Hevea* rubber is a 1,4-polymerized (Z)-poly(2-methyl-1,3-butadiene), similar in structure to polyisoprene. To increase its elasticity, it is treated with hot elemental sulfur in a process called **vulcanization** (*Vulcanus*, Latin, the Roman god of fire), which creates sulfur cross-links. This reaction was discovered by Goodyear[*] in 1839. One of the earliest and most successful uses of the product, Vulcanite, was the manufacture of dentures that could be molded for good fit. Before the 1860s, false teeth were embedded in animal bone, ivory, or metal. The swollen appearance of George Washington's lips (visible on U.S. currency) is attributed to his ill-fitting ivory dentures. Nowadays, dentures are made from acrylics (Section 13-10). Rubber remains an indispensable component of many commercial products, including tires (the major use), shoes, rainwear, and other clothing incorporating elastic fibers.

Copolymers in which the double bonds of 1,3-butadiene undergo polymerization with the double bonds of other alkenes have assumed increasing importance in recent years. By varying the proportions of the different monomers in the polymerization mix, the properties of the final product may be "tuned" over a considerable range. One such substance is a three-component copolymer of propenenitrile, 1,3-butadiene, and ethenylbenzene, known as ABS (for *a*crylonitrile/*b*utadiene/*s*tyrene copolymer). The diene imparts the rubber-like property of flexibility, whereas the nitrile hardens the polymer. The result is a highly versatile

---

[*]Charles Goodyear (1800–1860), American inventor, Washington, D.C.

material that may be fashioned into sheets or molded into virtually any shape. Its strength and ability to tolerate deformation and stress have allowed its utilization in everything from clock mechanisms to camera and computer cases to automobile bodies and bumpers.

Polyisoprene is the basis of natural rubber

How is rubber made in nature? Plants construct the polyisoprene framework of natural rubber by using as a building block 3-methyl-3-butenyl pyrophosphate (isopentenyl pyrophosphate). This molecule is an ester of pyrophosphoric acid and 3-methyl-3-buten-1-ol. An enzyme equilibrates a small amount of this material with the 2-butenyl isomer, an allylic pyrophosphate.

Biosynthesis of the Two Isomers of 3-Methylbutenyl Pyrophosphate

Latex, the precursor of natural rubber, is drained from the bark of *Hevea brasiliensis.*

A wheel hub cover made of ABS copolymer.

Although the subsequent processes are enzymatically controlled, they can be formulated simply in terms of familiar mechanisms (OPP = pyrophosphate).

Reaction

Mechanism

Mechanism of Natural Rubber Synthesis

Step 1. Ionization to stabilized (allylic) cation

Step 2. Electrophilic attack

Step 3. Proton loss

Geranyl pyrophosphate

Step 4. Second oligomerization

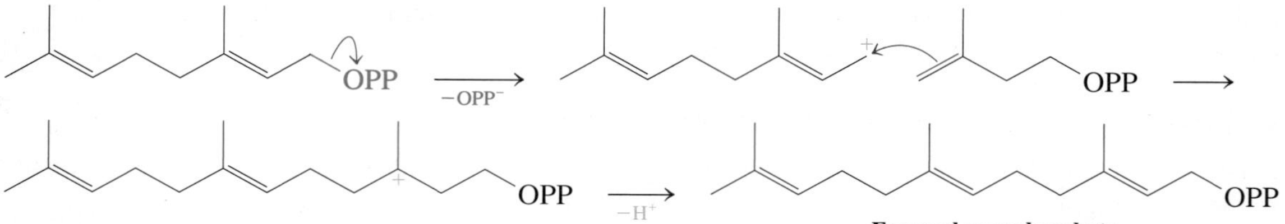

Farnesyl pyrophosphate

In the first step, ionization of the allylic pyrophosphate gives an allylic cation. Attack by a molecule of 3-methyl-3-butenyl pyrophosphate, followed by proton loss, yields a dimer called geranyl pyrophosphate. Repetition of this process leads to natural rubber.

Many natural products are composed of 2-methyl-1,3-butadiene (isoprene) units

Many natural products are derived from 3-methyl-3-butenyl pyrophosphate, including the terpenes first discussed in Section 4-7. Indeed, the structures of terpenes can be dissected into five-carbon units connected as in 2-methyl-1,3-butadiene. Their structural diversity can be attributed to the multiple ways in which 3-methyl-3-butenyl pyrophosphate can couple. The monoterpene geraniol and the sesquiterpene farnesol, two of the most widely distributed substances in the plant kingdom, form by hydrolysis of their corresponding pyrophosphates.

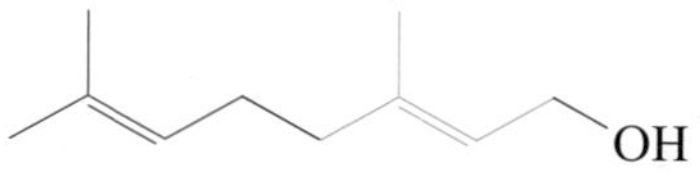

Geraniol 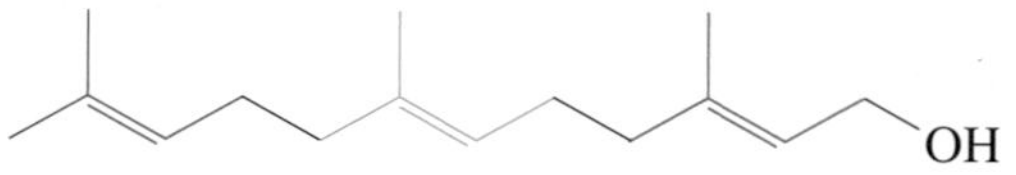   **Farnesol**

Coupling of two molecules of farnesyl pyrophosphate leads to squalene, a biosynthetic precursor of the steroid nucleus (Section 4-7).

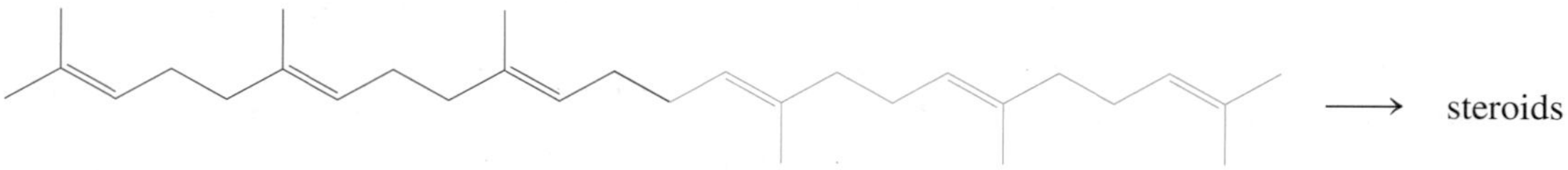

Squalene $\longrightarrow$ steroids

Bicyclic substances, such as camphor, a chemical used in mothballs, nasal sprays, and muscle rubs, are built up from geranyl pyrophosphate by enzymatically controlled electrophilic carbon–carbon bond formations.

Camphor Biosynthesis from Geranyl Pyrophosphate

Geranyl pyrophosphate

is the same as

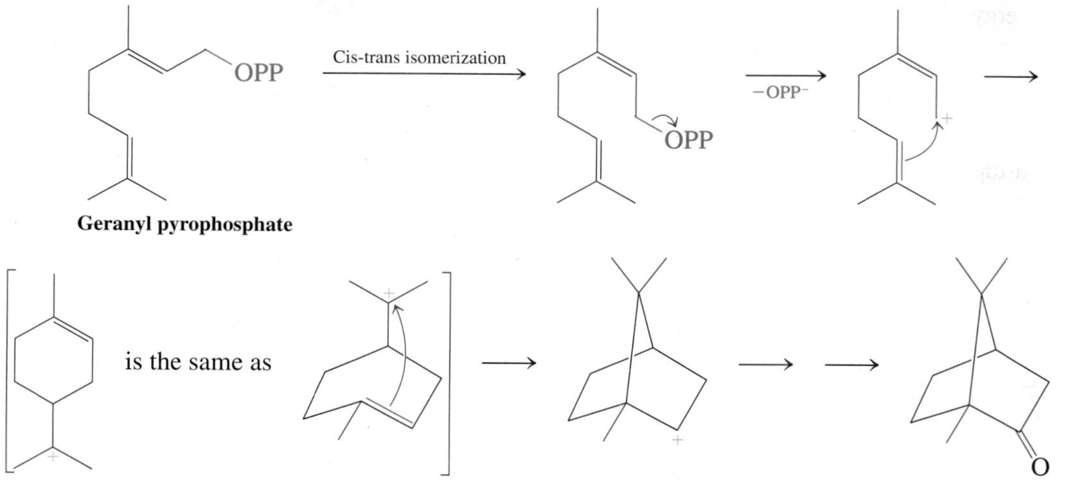

Camphor

Other higher terpenes are constructed by similar cyclization reactions.

In Summary 1,3-Butadiene polymerizes in a 1,2 or 1,4 manner to give polybutadienes with various amounts of cross-linking and therefore variable elasticity. Synthetic rubber can be made from 2-methyl-1,3-butadiene and contains varied numbers of *E* and *Z* double bonds. Natural rubber is constructed by isomerization of 3-methyl-3-butenyl pyrophosphate to the 2-butenyl system, ionization, and electrophilic (step-by-step) polymerization. Similar mechanisms account for the incorporation of 2-methyl-1,3-butadiene (isoprene) units into the polycyclic structure of terpenes.

14-11 | ELECTRONIC SPECTRA: ULTRAVIOLET AND VISIBLE SPECTROSCOPY

In Section 10-2, we explained that organic molecules may absorb radiation at various wavelengths. Spectroscopy is possible because absorption is restricted to quanta of defined energies, hv, to effect specific excitations with energy change ΔE.

$$\Delta E = hv = \frac{hc}{\lambda} \qquad (c = \text{velocity of light})$$

Figure 10-2 divided the range of electromagnetic radiation into various subsections, from high-energy X-rays to low-energy radiowaves. Among these subsections, one stood out by its representation in colors, the visible spectrum. Indeed, this is the only region of the electromagnetic spectrum that the human body, with our eyes as the "spectrometer," is able to resolve. The effect of other forms of radiation on us is less well defined: X-rays and ultraviolet light (sunburn!) are destructive, infrared radiation is perceived as heat, and micro- and radio waves are undetectable.

To understand how colors come about, we have to go back to an experiment by Isaac Newton, who showed that white light, when passed though a prism, can be dispersed into the full color spectrum, much like a rainbow, for which water droplets fulfil the function of the prism. Thus, what we perceive as white light is actually the effect of "broad-band irradiation" (to borrow a term from NMR spectroscopy; Section 10-9) on the light receptors in our retina (see Real Life 18-2). Consequently, we see an object (or compound) as colored when it absorbs light from part of the visible spectrum, reflecting the remainder. For example, when an object absorbs blue light, we see it as orange; when it absorbs green light, we see it as purple. Orange absorption turns the object blue and purple absorption turns it green. In organic chemistry, such colored compounds are most frequently those containing an array of conjugated double bonds. The excitation of electrons in these bonds happens to match the energy of visible light (and, as it turns out, also much of the ultraviolet spectrum), giving them their color. Recall the orange color of β-carotene (Section 14-7) in the Chapter Opening, and you are certainly familiar with the effects of indigo (margin) on blue jeans.

In this section, we shall quantify the color of organic compounds by spectroscopy in the wavelength range from 400 to 800 nm, called **visible spectroscopy** (see Figure 10-2 and margin on top of the next page). We also examine the range from 200 to 400 nm, called **ultraviolet spectroscopy.** Because these two wavelength ranges are so close to each other, they are typically measured simultaneously with the same spectrometer. Both techniques are particularly useful for investigating the electronic structures of unsaturated molecules and for measuring the extent of their conjugation.

A UV-visible spectrometer is constructed according to the general scheme in Figure 10-3. As in NMR, samples are usually dissolved in solvents that do not absorb in the spectral region under scrutiny. Examples are ethanol, methanol, and cyclohexane, none of which has absorption bands above 200 nm. The events triggered by electromagnetic radiation at UV and visible wavelengths lead to the excitation of electrons from filled bonding (and sometimes nonbonding) molecular orbitals to unfilled antibonding molecular orbitals. These changes in electron energies are recorded as **electronic spectra.** As in NMR, FT instruments have greatly improved sensitivity and ease of spectral acquisition.

Sunlight is split into its component colors (the visible spectrum) by raindrops to give a rainbow.

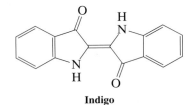

Indigo

The blue in blue jeans is due to the presence of indigo.

The Visible Spectrum

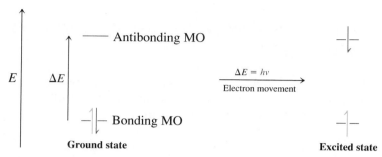

Color	Wavelength
violet	380–450 nm
blue	450–495 nm
green	495–570 nm
yellow	570–590 nm
orange	590–620 nm
red	620–750 nm

Ultraviolet and visible light give rise to electronic excitations

Consider the bonds in an average molecule: We can safely assume that, except for lone pairs, all electrons occupy bonding molecular orbitals. The compound is said to be in its **ground electronic state.** Electronic spectroscopy is possible because ultraviolet radiation and visible light have sufficient energy to transfer many such electrons to antibonding orbitals, thereby creating an **excited electronic state** (Figure 14-13). Dissipation of the absorbed energy may occur in the form of a chemical reaction (cf. Section 14-9), as emission of light (fluorescence, phosphorescence), or simply as emission of heat.

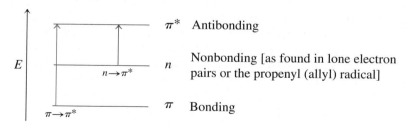

Figure 14-13 Electronic excitation by transfer of an electron from a bonding to an antibonding orbital converts a molecule from its ground electronic state into an excited state.

Organic σ bonds have a large gap between bonding and antibonding orbitals. To excite electrons in such bonds requires wavelengths much below the practical range (<200 nm). As a result, the technique has found its major uses in the study of π systems, in which filled and unfilled orbitals are much closer in energy. Excitation of such electrons gives rise to **$\pi \rightarrow \pi^*$ transitions.** Nonbonding (n) electrons are even more readily promoted through **$n \rightarrow \pi^*$ transitions** (Figure 14-14). Because the number of π molecular orbitals is equal to that of the component p orbitals, the simple picture presented in Figure 14-14 is rapidly complicated by extending conjugation: The number of possible transitions skyrockets, and with it the complexity of the spectra.

$$E$$

π^* Antibonding

n Nonbonding [as found in lone electron pairs or the propenyl (allyl) radical]

$n \rightarrow \pi^*$

π Bonding

$\pi \rightarrow \pi^*$

Figure 14-14 Electronic transitions in a simple π system. The wavelength of radiation required to cause them is revealed as a peak in the ultraviolet or visible spectrum.

A typical UV spectrum is that of 2-methyl-1,3-butadiene (isoprene), shown in Figure 14-15. The position of a peak is defined by the wavelength at its maximum, a λ_{max} value (in nanometers). Its intensity is reflected in the **molar extinction coefficient** or **molar absorptivity, ϵ,** that is characteristic of the molecule. The value of ϵ is calculated by dividing the measured peak height (absorbance, A) by the molar concentration, C, of the sample (assuming a standard cell length of 1 cm).

$$\epsilon = \frac{A}{C}$$

The value of ϵ can range from less than a hundred to several hundred thousand. It provides a good estimate of the efficiency of light absorption. Electronic spectral absorption bands are frequently broad, as in Figure 14-15, not the sharp lines typical of many NMR spectra.

The orange-red β-carotene and deep blue azulene differ in their π-electronic structure.

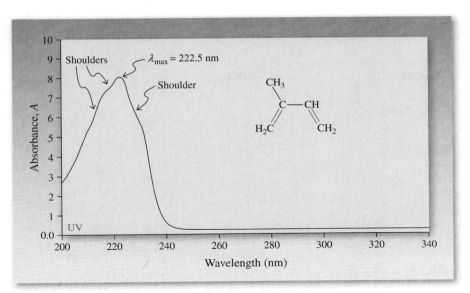

Figure 14-15 Ultraviolet spectrum of 2-methyl-1,3-butadiene in methanol, λ_{max} = 222.5 nm (ϵ = 10,800). The indentations at the sides of the main peak are called shoulders.

Electronic spectra tell us the extent of delocalization

Electronic spectra often indicate the size and degree of delocalization in an extended π system. The more double bonds there are in conjugation, the longer is the wavelength for the lowest-energy excitation (and the more peaks will appear in the spectrum). For example, ethene absorbs at λ_{max} = 171 nm, and an unconjugated diene, such as 1,4-pentadiene, absorbs at λ_{max} = 178 nm. A conjugated diene, such as 1,3-butadiene, absorbs at much lower energy (λ_{max} = 217 nm). Further extension of the conjugated system leads to corresponding incremental increases in the λ_{max} values, as shown in Table 14-3. The hyperconjugation of alkyl groups and the improved π overlap in rigid, planar cyclic systems both seem to contribute. Beyond 400 nm (in the visible range), molecules become colored: first yellow, then orange, red, violet, and finally blue-green. For example, the contiguous array of 11 double bonds in β-carotene (Section 14-7) is responsible for its characteristic intense orange appearance (λ_{max} = 480 nm).

Why should larger conjugated π systems have more readily accessible and lower-energy excited states? The answer is pictured in Figure 14-16. As the overlapping *p*-orbital array

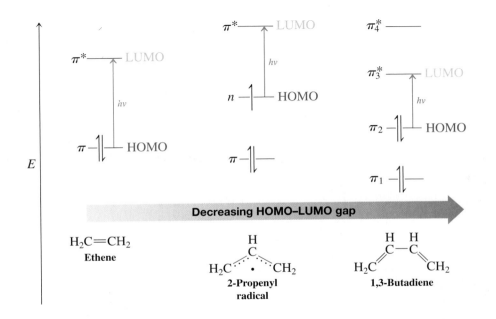

Decreasing HOMO–LUMO gap

Sunglasses on Demand

Self-darkening eyeglasses contain organic molecules that undergo thermally reversible photoisomerizations between two species that differ in their electronic spectra:

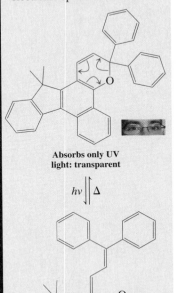

Absorbs only UV light: transparent

$h\nu \Big\Downarrow \Delta$

Absorbs UV and visible light

The top molecule is transparent in the visible range but absorbs the sun's UV rays to undergo electrocyclic ring opening to the bottom structure. The more extended conjugation in this isomer causes a shift of its λ_{max} to effect shading. In the dark, the system reverts thermally to its thermodynamically more stable state.

Figure 14-16 The energy gap between the highest occupied and lowest unoccupied molecular orbitals (abbreviated HOMO and LUMO, respectively) decreases along the series ethene, the 2-propenyl (allyl) radical, and butadiene. Excitation therefore requires less energy and is observed at longer wavelengths.

Table 14-3 λ_{max} **Values for the Lowest Energy Transitions in Ethene and Conjugated Pi Systems**

Alkene structure	Name	λ_{max} (nm)	ϵ
	Ethene	171	15,500
	1,4-Pentadiene	178	Not measured
	1,3-Butadiene	217	21,000
	2-Methyl-1,3-butadiene	222.5	10,800
	trans-1,3,5-Hexatriene	268	36,300
	trans,trans-1,3,5,7-Octatetraene	330	Not measured
	2,5-Dimethyl-2,4-hexadiene	241.5	13,100
	1,3-Cyclopentadiene	239	4,200
	1,3-Cyclohexadiene	259	10,000
	A steroid diene	282	Not measured
	A steroid triene	324	Not measured
	A steroid tetraene	355	Not measured
(For structure, see Section 14-7.)	β-Carotene (vitamin A precursor)	497 (orange appearance)	133,000
	Azulene, a cyclic conjugated hydrocarbon	696 (blue-violet appearance)	150

REAL LIFE: SPECTROSCOPY 14-4 | The Contributions of IR, MS, and UV to the Characterization of Viniferone

^{1}H and ^{13}C NMR data enabled the elucidation of much of the structure of the grape-seed-derived antioxidant viniferone (Real Life 10-5), shown again below. In particular, the presence of two carbonyl carbons and a total of eight alkene or benzene carbons was revealed. Also, a six-membered ether ring was strongly indicated. Sorting out the rest of the structure was greatly aided by complementary information derived from mass, IR, and UV measurements.

Although ^{13}C NMR gave the number of carbon atoms, high-resolution mass spectrometry revealed the entire molecular formula as $C_{15}H_{14}O_8$ and thus the presence of nine degrees of unsaturation (15 carbons $\times$ 2 = 30; 30 + 2 = 32; 32 − 14 hydrogens = 18; 18/2 = 9). IR spectroscopy is very characteristic for five-membered rings such as the one in viniferone: Bands at 1700, 1760, and 1790 cm^{-1} for various individual and combined vibrational modes of the C=O groups, and a very broad, strong absorption between 2500 and 3300 cm^{-1} for the carboxylic acid O—H also appear in the IR spectra of the simpler model compound shown below.

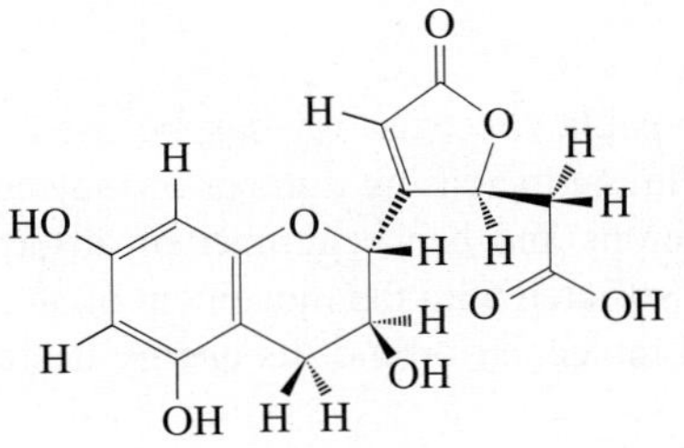

Viniferone 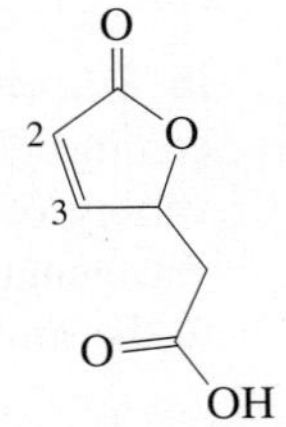   Model compound

The site of the linkage between this ring and the rest of the structure of viniferone was apparent from the ^{1}H NMR spectrum. In the simpler model, the alkene hydrogens appear at δ = 6.12 (H2) and 7.55 (H3) ppm; viniferone shows only one signal at δ = 6.19. The absence of any signal in the vicinity of δ = 7.5–7.6 ppm implies that the point of connection is C3.

Finally, UV spectroscopy helps to identify the oxygen-substituted benzene ring by the presence of absorptions in the vicinity of 275 nm for the cyclic conjugated moiety. This assignment also completes the identification of the nine degrees of unsaturation: the two C=O groups, the C=C double bond in the five-membered ring, the three double bonds in the benzene ring, and the three rings themselves.

gets longer, the energy gap between filled and unfilled orbitals gets smaller, and more bonding and antibonding orbitals are available to give rise to additional electronic excitations.

Finally, conjugation in cyclopolyenes is governed by a separate set of rules, to be introduced in the next two chapters. Just compare the electronic spectra of benzene, which is colorless (Figure 15-6), with that of azulene, which is deep blue (Figure 14-17), and then compare both with the data in Table 14-3.

Exercise 14-29

Arrange the following compounds in the order of their increasing λ_{max} values.

(a) 1,3,5-Cycloheptatriene   (b) 1,5-Hexadiene   (c) 1,3-Cyclohexadiene

(d) 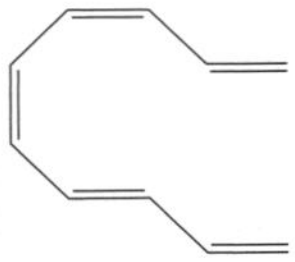   (e) Polyacetylene   (f)

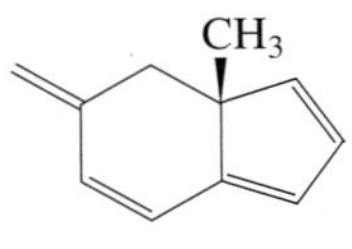

Figure 14-17 UV–visible spectrum of azulene in cyclohexane. The absorbance is plotted as log ε to compress the scale. The horizontal axis, representing wavelength, also is nonlinear.

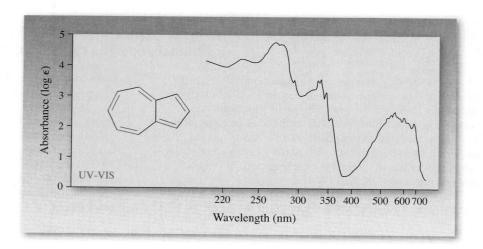

In Summary UV and visible spectroscopy can be used to detect electronic excitations in conjugated molecules. With an increasing number of molecular orbitals, there is an increasing variety of possible transitions and hence number of absorption bands. The band of longest wavelength is typically associated with the movement of an electron from the highest occupied to the lowest unoccupied molecular orbital. Its energy decreases with increasing conjugation.

THE BIG PICTURE

We have progressed from a single electron in a single p orbital, as in radicals (Sections 3-1 and 3-2), to a pair of electrons in neighboring p orbitals, as in double bonds (Chapter 12), to two pairs of electrons in neighboring p orbitals, as in triple bonds (Chapter 13), to the present chapter, which extends this theme. We now know that an infinite number of p orbitals can line up in carbon sequences that exhibit the phenomenon of conjugation.

Conjugation delocalizes electrons and therefore charges, as in a molecular wire. It can also be viewed as a way by which one end of a molecule may "communicate" with the other chemically. For example, protonation of a conjugated polyene has a profound effect on the molecule along the entire chain of sp^2-hybridized carbons. Even more impressive are the pericyclic reactions, following the last new kind of mechanism that we shall encounter in this book. For example, in the Diels-Alder cycloaddition, the respective termini of the diene and the dienophile add to each other in concert, bond formation and bond breaking occurring at the same time. Similar communication is evident in spectral data, particularly in UV–visible spectroscopy.

Where does this lead us? Returning to the analogy of a conjugated polyene as a molecular wire, we can ask a simple question: What happens if we touch the ends of that wire to complete a closed circuit? A lot, as the next two chapters will reveal.

WORKED EXAMPLES: INTEGRATING THE CONCEPTS

14-30. Exploring the Reactivity of Conjugated Systems

a. Propose a reasonable mechanism for the transformation of *trans,trans*-2,4-hexadien-1-ol (sorbyl alcohol) into *trans*-5-ethoxy-1,3-hexadiene.

$$\text{Sorbyl alcohol} \xrightarrow{\text{H}^+,\ \text{CH}_3\text{CH}_2\text{OH}} \textit{trans}\text{-5-Ethoxy-1,3-hexadiene}$$

Sorbyl alcohol ⟶ **trans-5-Ethoxy-1,3-hexadiene** (OCH_2CH_3)

SOLUTION

This process takes place in acidic ethanol solution (otherwise known as wine). We begin by inspecting the structures of the starting and final molecules to find that (1) an alcohol has become an ether and (2) the double bonds have moved. Let us consider relevant information that we have seen concerning these functional groups. The conversion of two alcohols into an ether in the presence of acid was introduced in Section 9-7. Protonation of one alcohol molecule gives an alkyloxonium ion, whose leaving group (water) may be replaced in either an S_N2 or an S_N1 fashion by a molecule of a second alcohol:

$$CH_3CH_2\overset{+}{O}H_2 \quad + \quad CH_3CH_2OH \quad \xrightarrow{S_N2} \quad CH_3CH_2OCH_2CH_3$$

$$\underset{\underset{CH_3}{|}}{\overset{\overset{CH_3}{|}}{CH_3\overset{+}{C}OH_2}} \quad + \quad CH_3CH_2OH \quad \xrightarrow{S_N1} \quad \underset{\underset{CH_3}{|}}{\overset{\overset{CH_3}{|}}{CH_3COCH_2CH_3}}$$

Can either of these processes be adapted to the present question? Sorbyl alcohol is an allylic alcohol, and we have just seen (Section 14-3) that allylic *halides* readily undergo both S_N2 and S_N1 displacement reactions. A good deal of Chapter 9 dealt with comparing and contrasting the behavior of haloalkanes with that of alcohols: Protonation of the OH group of an alcohol opens the way to both substitution and elimination chemistry. We should therefore expect that allylic alcohols will be similarly reactive. We need next to consider whether the S_N2 or the S_N1 mechanism is more appropriate.

The fact that the double bonds have moved in the course of the reaction is a helpful piece of information. Referring again to Section 14-3 and also 14-6, we note that the allylic cation that results from dissociation of a leaving group in the first step of an S_N1 reaction is delocalized, and nucleophiles can therefore attach at more than one site. Let us explore this line of reasoning by examining the carbocation derived from protonation and loss of water from sorbyl alcohol:

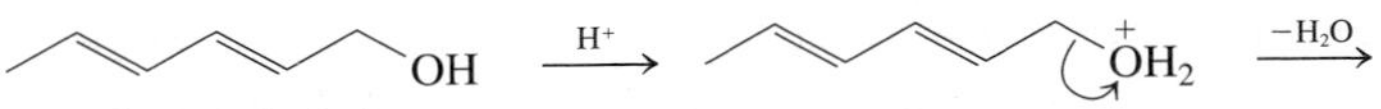

Sorbyl alcohol

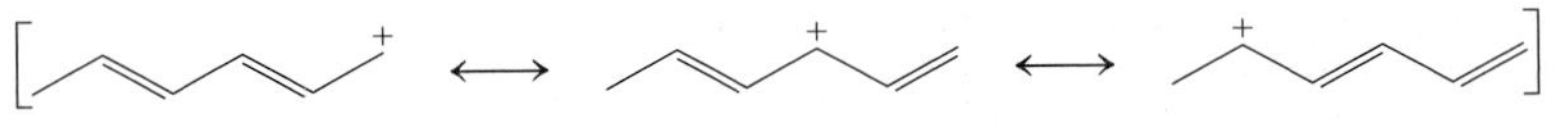

Like an allylic cation, this carbocation is delocalized. However, the initial system is a more extended conjugated one, and the resulting cation is a hybrid of three, rather than two, contributing resonance forms. As in electrophilic addition to a conjugated triene (Section 14-7), an incoming nucleophile has the option of attaching to any of three positions. In this particular case, attachment of ethanol to a secondary carbon, where a relatively high fraction of the positive charge resides (Section 14-6), is the predominant (kinetic) result.

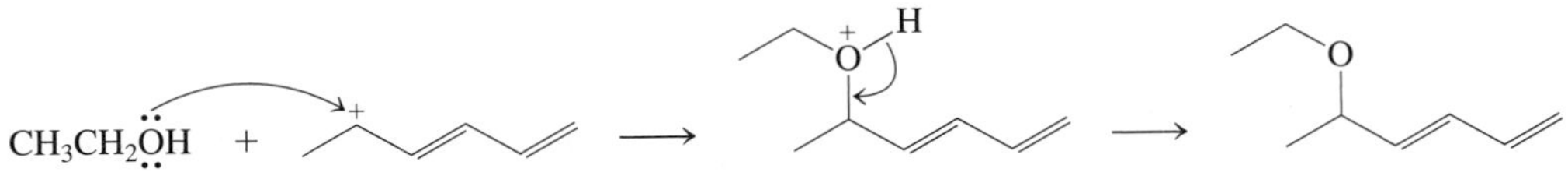

b. Esters of the food preservative sorbic acid (*trans,trans*-2,4-hexadienoic acid) undergo Diels-Alder reactions. Predict the major product to arise from the heating of ethyl sorbate with 2-butenedioic anhydride (maleic anhydride, see structure, Table 14-1). Consider carefully all stereochemical aspects of the process.

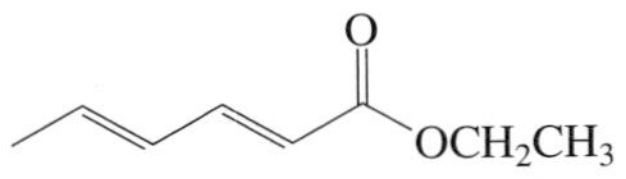

**Ethyl *trans,trans*-2,4-hexadienoate
(Ethyl sorbate)**

SOLUTION

The Diels-Alder reaction is a cycloaddition between a diene, in this case the ethyl sorbate, and a dienophile, usually an electron-poor alkene (Section 14-8). We should first have a look at the two reaction partners to visualize the new bonding connections that this process will effect. To do so, we need to rotate about the single bond between carbon 3 and carbon 4 of ethyl sorbate to place its double bonds in the necessary conformation. We do this carefully, so as not to mistakenly change the stereochemistry of the double bonds: They both start out trans and must be that way after we have finished. Next, we connect the ends of the diene system to the carbon atoms of the dienophile's alkene function (dotted lines below), giving us the connectivity of the product, but without stereochemistry specified:

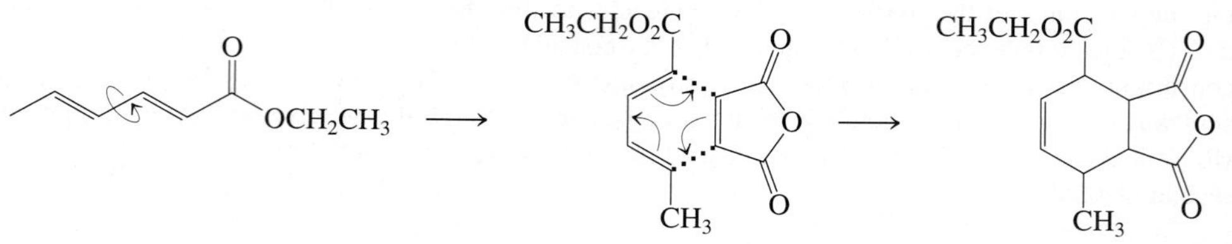

To complete the problem, we must finally take into account two details of the Diels-Alder reaction: (1) Stereochemical relations in the components are retained in the course of the reaction; and (2) unsaturated substituents on the double bond of the dienophile prefer to locate themselves underneath the diene system (in endo positions) in the course of the cycloaddition. Making use of illustrations similar to those in the text, we can visualize the process as follows:

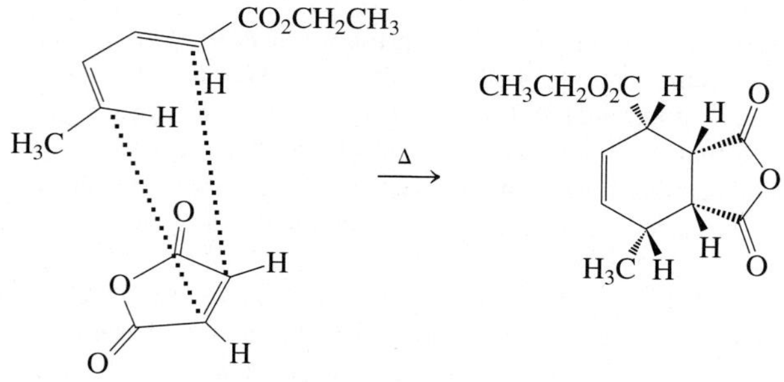

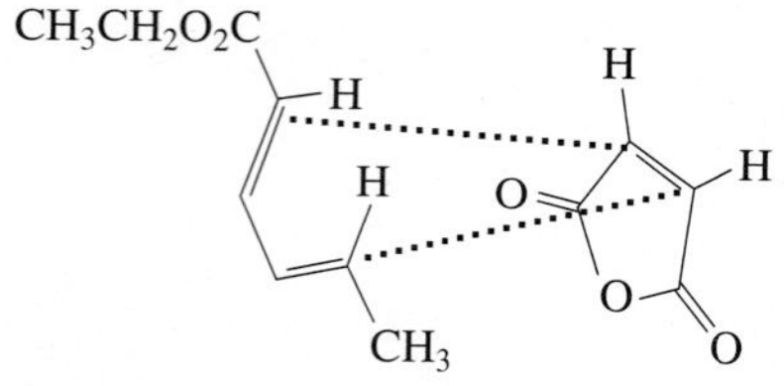

It takes some careful examination to see how the spatial arrangement of the groups in the sketch of the transition state translates into the final positions in the product. As you work on this aspect of the problem, look in particular at each carbon taking part in the formation of the two new single bonds [dotted lines in transition state (left side of reaction)]. Examine the positions of the substituents on these carbons relative to the two new bonds. A view that may be useful for this purpose is one in which the whole picture is rotated by 90° (see margin). It is reasonably clear from this perspective that the four hydrogen atoms will end up all cis to each other in the newly formed cyclohexene ring. Try Problems 56 and 68 for more practice with mechanisms.

14-31. Retrosynthesis Featuring a Diels-Alder Reaction

Because of its superb stereoselectivity, the Diels-Alder reaction is often used as a key step in the construction of *acyclic* building blocks containing several defined stereocenters. The strategy is to retrosynthetically convert the target into a cyclohexane derivative that is accessible by [4 + 2] cycloaddition. Suggest a synthesis of (racemic) compound A from starting materials containing four carbons or less.

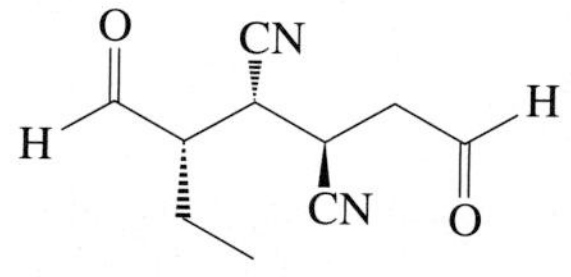

A

SOLUTION

On first sight, this problem seems impossible. The trick is not to focus on the immediate solution, but on the process of retrosynthetic analysis. Thus, the first task is to find a precursor to A that is an appropriately substituted cyclohexene. We can worry about how to make it (by a Diels-Alder reaction) later.

We note that A is a substituted hexane containing aldehyde functions at the two termini. Is there a retrosynthetic step that will link these two functional carbons to make a six-membered ring? The answer is in Section 12-12: (reverse) ozonolysis, a process that will connect the two carbonyl carbons through a C—C double bond. To draw the resulting cyclohexene B with the correct stereochemistry, it helps to first redraw A in an angular fashion, by rotation around the indicated bonds, before removing the two oxygens and closing the ring.

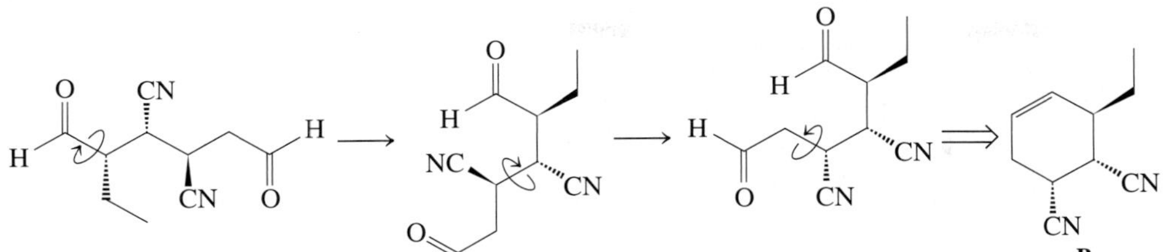

This, like any, cyclohexene can be dismantled by a retro-endo Diels-Alder process (Section 14-8), providing the two components C and D of the forward reaction in the required stereochemical form.

Dienophile D is readily available (Table 14-1). *cis*-1,3-Hexadiene, C, will have to be made, and there are many ways by which this task can be accomplished. The crucial feature of

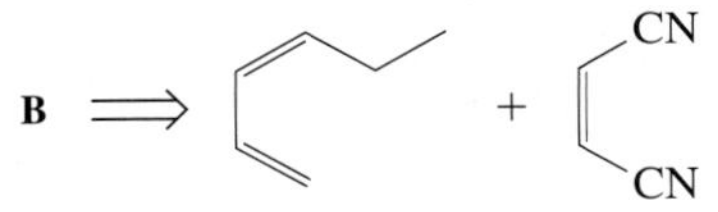

its construction is the stereochemistry of the double bond. How do we synthesize cis alkenes? The answer is in Section 13-6: the Lindlar-catalyzed hydrogenation of alkynes. Hence, a good precursor of C is enyne E, which can be derived from F and iodoethane (Section 13-5).

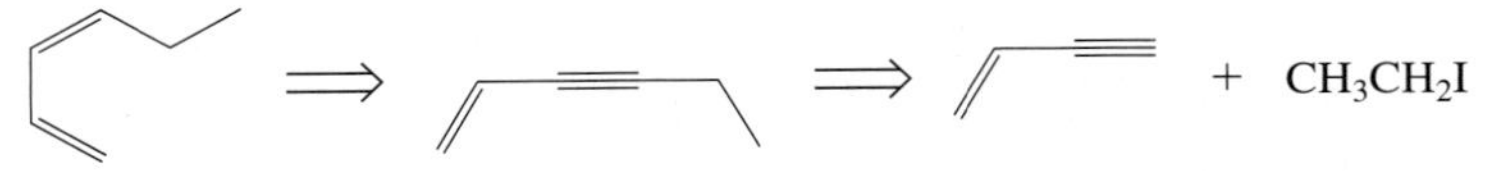

But-1-en-3-yne ("vinylacetylene"), F, is made industrially by the CuCl-catalyzed dimerization of ethyne. How would you make it? We learned that alkenes can be converted to alkynes by halogenation-double dehydrohalogenation (Section 13-4). This approach suggests 1,3-butadiene as a starting material. We have seen, however, that halogenation of 1,3-butadiene, such as bromination, is not as straightforward as that of a monoalkene: 1,2- and 1,4-addition occurs (Section 14-6). Is that a problem? *Answer:* No. The 1,2-dihalobutene will eliminate normally, first to 2-bromobutadiene (via deprotonation at the more acidic allylic position) and then to F:

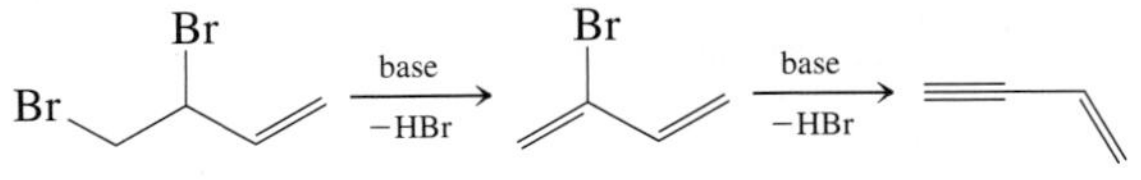

The 1,4-dihalobutene will also eliminate, but in a manner that uses the conjugative ability of the double bond, producing 1-bromo-1,3-butadiene as an intermediate, which then goes on to product.

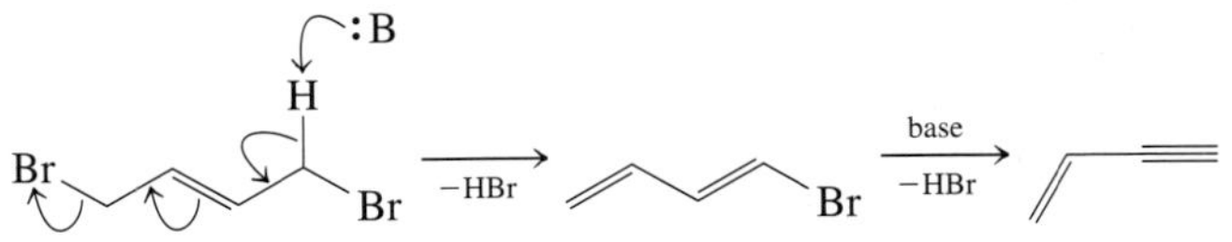

On the basis of this analysis, write a forward scheme for the synthesis of A, starting with 1,3-butadiene, bromoethane, and D as the sources of carbon.

New Reactions

1. Radical Allylic Halogenations (Section 14-2)

$$RCH_2CH=CH_2 \xrightarrow[\text{$DH°$ of allylic C–H bond} \approx 87 \text{ kcal mol}^{-1}]{NBS, CCl_4, h\nu} RCHCH=CH_2 \ (Br) \ + \ RCH=CHCH_2Br$$

2. **S$_N$2 Reactivity of Allylic Halides (Section 14-3)**

$$CH_2{=}CHCH_2X \ + \ Nu{:}^- \ \xrightarrow{\text{Acetone}} \ CH_2{=}CHCH_2Nu \ + \ X^-$$

Faster than ordinary primary halides

3. **Allylic Grignard Reagents (Section 14-4)**

$$CH_2{=}CHCH_2Br \ \xrightarrow{\text{Mg, (CH}_3\text{CH}_2)_2\text{O}} \ CH_2{=}CHCH_2MgBr$$

Can be used in additions to carbonyl compounds

4. **Allyllithium Reagents (Section 14-4)**

$$RCH_2CH{=}CH_2 \ \xrightarrow{\text{CH}_3\text{CH}_2\text{CH}_2\text{CH}_2\text{Li, TMEDA}} \ R\overset{..}{\overset{-}{C}}HCH{=}CH_2 \ Li^+$$

pK_a of allylic C–H bond ≈ 40

5. **Hydrogenation of Conjugated Dienes (Section 14-5)**

$$CH_2{=}CH{-}CH{=}CH_2 \ \xrightarrow{\text{H}_2,\ \text{Pd–C, CH}_3\text{CH}_2\text{OH}} \ CH_3CH_2CH_2CH_3 \qquad \Delta H° = -57.1 \text{ kcal mol}^{-1}$$

but compare

$$CH_2{=}CH{-}CH_2{-}CH{=}CH_2 \ \xrightarrow{\text{H}_2,\ \text{Pd–C, CH}_3\text{CH}_2\text{OH}} \ CH_3(CH_2)_3CH_3 \qquad \Delta H° = -60.8 \text{ kcal mol}^{-1}$$

6. **Electrophilic Reactions of 1,3-Dienes: 1,2- and 1,4-Addition (Section 14-6)**

$$CH_2{=}CH{-}CH{=}CH_2 \ \xrightarrow{\text{HX}} \ CH_2{=}CHCH(X)CH_3 \ + \ XCH_2CH{=}CHCH_3$$

$$CH_2{=}CH{-}CH{=}CH_2 \ \xrightarrow{\text{X}_2} \ CH_2{=}CHCH(X)CH_2X \ + \ XCH_2CH{=}CHCH_2X$$

7. **Thermodynamic Compared with Kinetic Control in S$_N$1 Reactions of Allylic Derivatives (Section 14-6)**

$$CH_3CH{=}CHCH_2X \ \overset{\text{Slow}}{\longleftarrow} \ CH_3CH{=}CH\overset{+}{C}H_2 \ + \ X^- \ \overset{\text{Fast}}{\rightleftharpoons} \ CH_3CH(X)CH{=}CH_2$$

More stable product   **Stability of primary allylic cation ≈ ordinary secondary cation**   **Less stable product**

(Formation reversible at higher temperature)

8. **Diels-Alder Reaction (Concerted and Stereospecific, Endo Rule) (Section 14-8)**

A = electron acceptor
Requires *s*-cis diene; better with electron-poor dienophile

9. **Electrocyclic Reactions (Section 14-9)**

10. Polymerization of 1,3-Dienes (Section 14-10)

1,2-Polymerization

$$2n \ CH_2=CH-CH=CH_2 \xrightarrow{\text{Initiator}} $$

$$-(CH-CH_2-CH-CH_2)_n-$$

(with pendant groups $\overset{CH_2}{\underset{\parallel}{CH}}$ on each CH)

1,4-Polymerization

$$n \ CH_2=CH-CH=CH_2 \xrightarrow{\text{Initiator}} -(CH_2-CH=CH-CH_2)_n-$$
$$\text{Cis or trans}$$

11. 3-Methyl-3-butenyl Pyrophosphate as a Biochemical Building Block (Section 14-10)

$$CH_2=\overset{CH_3}{\underset{\mid}{C}}-CH_2CH_2OPP \underset{\text{Enzyme}}{\rightleftharpoons} (CH_3)_2C=CHCH_2OPP \longrightarrow (CH_3)_2C=CHCH_2{}^+ \ + \ {}^-OPP$$

3-Methyl-3-butenyl
pyrophosphate

Allylic cation

Pyrophosphate
ion

C–C bond formation

Important Concepts

1. The 2-propenyl **(allyl)** system is stabilized by **resonance.** Its molecular-orbital description shows the presence of three π molecular levels: one bonding, one nonbonding, and one antibonding. Its structure is symmetric, any charges or odd electrons being equally distributed between the two end carbons.

2. The chemistry of the 2-propenyl (allyl) cation is subject to both **thermodynamic and kinetic control.** Nucleophilic trapping may occur more rapidly at an internal carbon that bears relatively more positive charge, giving the thermodynamically less stable product. The kinetic product may rearrange to its thermodynamic isomer by dissociation followed by eventual thermodynamic trapping.

3. The stability of allylic radicals allows **radical halogenations** of alkenes at the allylic position.

4. The S_N2 reaction of allylic halides is accelerated by orbital overlap in the transition state.

5. The special stability of allylic anions allows **allylic deprotonation** by a strong base, such as butyllithium–TMEDA.

6. 1,3-Dienes reveal the effects of **conjugation** by their relative stability (compared to nonconjugated systems) and a relatively short internal bond (1.47 Å).

7. Electrophilic attack on 1,3-dienes leads to the preferential formation of allylic cations.

8. **Extended conjugated systems** are reactive because they have many sites for attack and the resulting intermediates are stabilized by resonance.

9. Benzene has special stability because of **cyclic delocalization.**

10. The **Diels-Alder reaction** is a concerted stereospecific **cycloaddition reaction** of an *s*-cis diene to a dienophile; it leads to cyclohexene derivatives. It follows the **endo rule.**

11. Conjugated dienes and trienes equilibrate with their respective cyclic isomers by concerted and stereospecific **electrocyclic reactions.**

12. **Polymerization** of 1,3-dienes results in 1,2- or 1,4-additions to give polymers that are capable of further **cross-linking.** Synthetic rubbers can be synthesized in this way. Natural rubber is made

by electrophilic carbon–carbon bond formation involving biosynthetic five-carbon cations derived from 3-methyl-3-butenyl pyrophosphate.

13. Ultraviolet and visible spectroscopy gives a way of estimating the extent of conjugation in a molecule. Peaks in **electronic spectra** are usually broad and are reported as λ_{max} (nm). Their relative intensities are given by the **molar absorptivity** (extinction coefficient) ϵ.

Problems

32. Draw all resonance forms and a representation of the appropriate resonance hybrid for each of the following species.

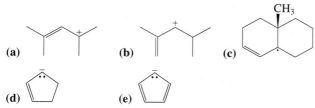

(a)   **(b)**   **(c)**

(d)   **(e)**

33. For each species in Problem 32, indicate the resonance form that is the major contributor to the resonance hybrid. Explain your choices.

34. Illustrate by means of appropriate structures (including all relevant resonance forms) the initial species formed by **(a)** breaking the weakest C–H bond in 1-butene; **(b)** treating 4-methylcyclohexene with a powerful base (e.g., butyllithium–TMEDA); **(c)** heating a solution of 3-chloro-1-methylcyclopentene in aqueous ethanol.

35. Rank primary, secondary, tertiary, and allylic radicals in order of decreasing stability. Do the same for the corresponding carbocations. Do the results indicate something about the relative ability of hyperconjugation and resonance to stabilize radical and cationic centers?

36. Give the major product(s) of each of the following reactions.

(a) [structure with OH] Conc. HBr

(b) [cyclohexene structure with CH₃ and Cl] H₂O

(c) [cyclopentane with Br and CH=CH₂] CH₃CH₂OH

(d) [cyclohexane with CH₃ and CH₂I] CH₃COH (O)

(e) [cyclohexane with CH₃ and CH₂I] KSCH₃, DMSO

(f) Cl—[chain]—OH CH₃NO₂, Δ

37. Formulate detailed mechanisms for the reactions in Problem 36, parts (a), (c), (e), and (f).

38. Rank primary, secondary, tertiary, and (primary) allylic chlorides in approximate order of **(a)** decreasing S_N1 reactivity; **(b)** decreasing S_N2 reactivity.

39. Rank the following six molecules in approximate order of decreasing S_N1 reactivity and decreasing S_N2 reactivity.

(a) [structure with Cl]   **(b)** [structure with Cl]

(c) [structure with Cl]   **(d)** [structure with Cl]

(e) [structure with Cl]   **(f)** [structure with Cl]

40. How would you expect the S_N2 reactivities of simple saturated primary, secondary, and tertiary chloroalkanes to compare with the S_N2 reactivities of the compounds in Problem 39? Make the same comparison for S_N1 reactivities.

41. Give the major product(s) of each of the following reactions.

(a) [cyclohexene with H₃C, I, and H] H₂O

(b) [cyclohexene with CH₃ and H] NBS, CCl₄, ROOR

(c) (S)-CH₃CH₂CHCH=CH₂ (with CH₃) NBS, CCl₄, ROOR

(d) [alkene: CH₃CH₂ and CH₂CH₃ with C=C and H, H] CH₃CH₂CH₂CH₂Li, TMEDA

(e) Product of (d) 1. CH₃CH, THF (O); 2. H⁺, H₂O

(f) $(CH_3)_2C=CH$ [C with H, Br, H₃C] KSCH₃, DMSO

42. Write a detailed step-by-step mechanism to show how each of the products arises from the reaction in Problem 41, part (a).

43. The following reaction sequence gives rise to two isomeric products. What are they? Explain the mechanism of their formation.

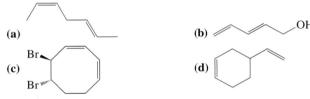

1. Mg
2. D₂O

44. Starting with cyclohexene, propose a reasonable synthesis of the cyclohexene derivative shown below.

45. Give a systematic name to each of the following molecules.

(a)

(b)

(c)

(d)

46. Compare the allylic bromination reactions of 1,3-pentadiene and 1,4-pentadiene. Which should be faster? Which is more energetically favorable? How do the product mixtures compare?

$$CH_2=CH-CH=CH-CH_3 \xrightarrow{NBS, ROOR, CCl_4}$$

$$CH_2=CH-CH_2-CH=CH_2 \xrightarrow{NBS, ROOR, CCl_4}$$

47. We learned in Section 14-6 that electrophilic additions to conjugated dienes at low reaction temperatures give kinetic product ratios. Furthermore, these kinetic mixtures may change to mixtures with thermodynamic product ratios when the temperature is raised. Do you think that cooling the thermodynamic product mixture back to the original low reaction temperature will change it back to the original kinetic ratio? Why or why not?

48. Compare the addition of H⁺ to 1,3-pentadiene and 1,4-pentadiene (see Problem 46). Draw the structures of the products. Draw a qualitative reaction profile showing both dienes and both proton addition products on the same graph. Which diene adds the proton faster? Which one gives the more stable product?

49. What products would you expect from the electrophilic addition of each of the following reagents to 1,3-cycloheptadiene? **(a)** HI; **(b)** Br₂ in H₂O; **(c)** IN₃; **(d)** H₂SO₄ in CH₃CH₂OH. (**Hint:** For b and c, see Exercise 14-13, specifically drawings A–C).

50. Give the products of the reaction of *trans*-1,3-pentadiene with each of the reagents in Problem 49.

51. What are the products of reaction of 2-methyl-1,3-pentadiene with each of the reagents in Problem 49?

52. Write a detailed step-by-step mechanism for the formation of each of the products in Problem 51.

53. Give the products expected from reaction of deuterium iodide (DI) with **(a)** 1,3-cycloheptadiene; **(b)** *trans*-1,3-pentadiene; **(c)** 2-methyl-1,3-pentadiene. In what way does the observable result of reaction of DI differ from that of reaction of HI with these same substrates [compare with parts (a) of Problems 49–51]?

54. Arrange the following carbocations in order of decreasing stability. Draw all possible resonance forms for each of them.

(a) $CH_2=CH-\overset{+}{C}H_2$ **(b)** $CH_2=\overset{+}{C}H$

(c) $CH_3\overset{+}{C}H_2$ **(d)** $CH_3-CH=CH-\overset{+}{C}H-CH_3$

(e) $CH_2=CH-CH=CH-\overset{+}{C}H_2$

55. Sketch the molecular orbitals for the pentadienyl system in order of ascending energy (see Figures 14-2 and 14-7). Indicate how many electrons are present, and in which orbitals, for **(a)** the radical; **(b)** the cation; **(c)** the anion (see Figures 14-3 and 14-7). Draw all reasonable resonance forms for any one of these three species.

56. Dienes may be prepared by elimination reactions of substituted allylic compounds. For example,

$$H_3C-\underset{\underset{CH_3}{|}}{C}=CH-CH_2OH \xrightarrow{\text{Catalytic } H_2SO_4, \Delta} H_2C=\underset{\underset{CH_3}{|}}{C}-CH=CH_2$$

$$H_3C-\underset{\underset{CH_3}{|}}{C}=CH-CH_2Cl \xrightarrow{\text{LDA, THF}} H_2C=\underset{\underset{CH_3}{|}}{C}-CH=CH_2$$

Propose detailed mechanisms for each of these 2-methyl-1,3-butadiene (isoprene) syntheses.

57. Give the structures of all possible products of the acid-catalyzed dehydration of vitamin A (Section 14-7).

58. Propose a synthesis of each of the following molecules by Diels-Alder reactions.

(a)

(b)

(c)

59. The haloconduritols are members of a class of compounds called glycosidase inhibitors. These substances possess an array of intriguing biological functions from antidiabetic and antifungal to activity against HIV virus and cancer metastasis. Stereoisomeric mixtures of bromoconduritol (below) are commonly used in studies of these properties. One recent synthesis of these compounds proceeds via the bicyclic ethers A and B. **(a)** Identify the starting materials for a one-step preparation of both of these ethers via a Diels-Alder reaction. **(b)** Which of the starting molecules in your answer is the diene, and which is the dienophile? **(c)** This Diels-Alder reaction gives an 80:20 mixture of B and A. Explain.

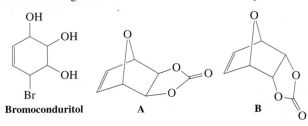

Bromoconduritol   **A**   **B**

60. Give the product(s) of each of the following reactions.

(a) 3-Chloro-1-propene (allyl chloride) + NaOCH₃

(b) *cis*-2-Butene + NBS, peroxide (ROOR)

(c) 3-Bromocyclopentene + LDA

(d) *trans,trans*-2,4-Hexadiene + HCl

(e) *trans,trans*-2,4-Hexadiene + Br₂, H₂O

(f) 1,3-Cyclohexadiene + methyl propenoate (methyl acrylate)

(g) 1,2-Dimethylenecyclohexane + methyl propenoate (methyl acrylate)

61. **CHALLENGE** Propose an efficient synthesis of the cyclohexenol below, beginning exclusively with acyclic starting materials and employing sound retrosynthetic analysis strategy. [**Hint:** A Diels-Alder reaction may be useful, but take note of structural features in dienes and dienophiles that permit Diels-Alder reactions to work well (Section 14-8).]

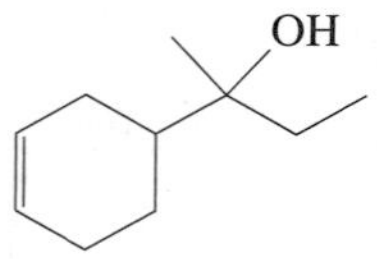

62. Dimethyl azodicarboxylate (see below) takes part in the Diels-Alder reaction as a dienophile. Write the structure of the product of cycloaddition of this molecule with each of the following dienes. (a) 1,3-Butadiene; (b) *trans,trans*-2,4-hexadiene; (c) 5,5-dimethoxycyclopentadiene; (d) 1,2-dimethylenecyclohexane. Ignore the stereochemistry at nitrogen in the products (amines undergo rapid inversion, as we shall see in Section 21-2).

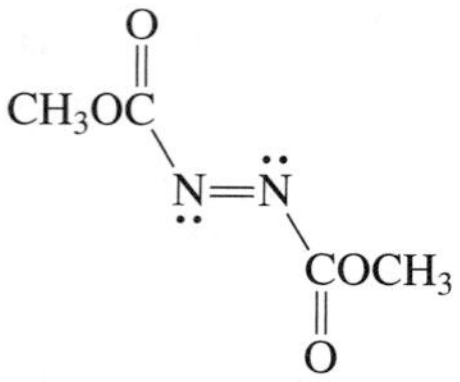

Dimethyl azodicarboxylate

63. Bicyclic diene A reacts readily with appropriate alkenes by the Diels-Alder reaction, whereas diene B is totally unreactive. Explain.

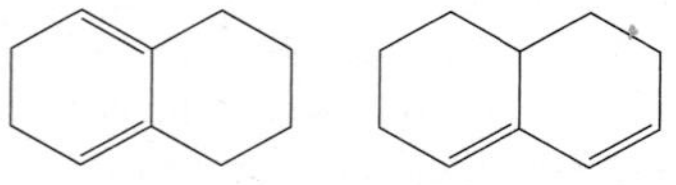

A   **B**

64. Formulate the expected product of each of the following conversions.

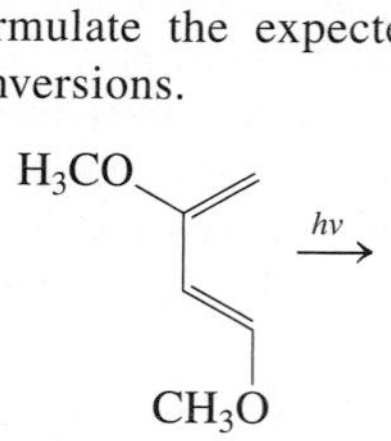

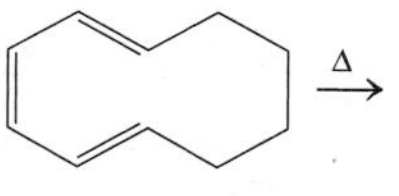

65. Which of the reactions shown below will occur under the influence of heat? Light?

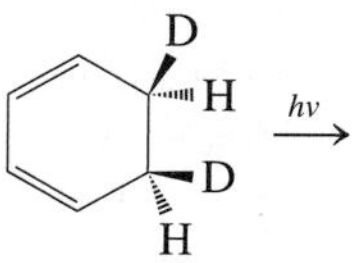

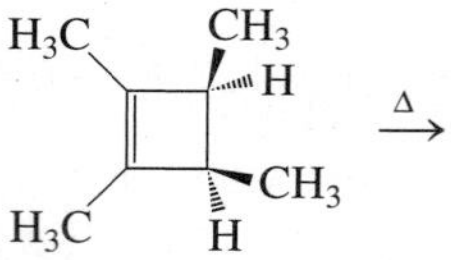

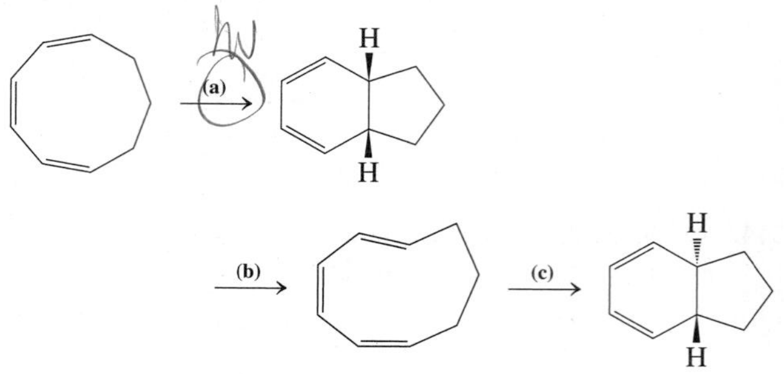

66. Explain the following reaction sequence. (**Hint:** Remember the Heck reaction, Section 13-9).

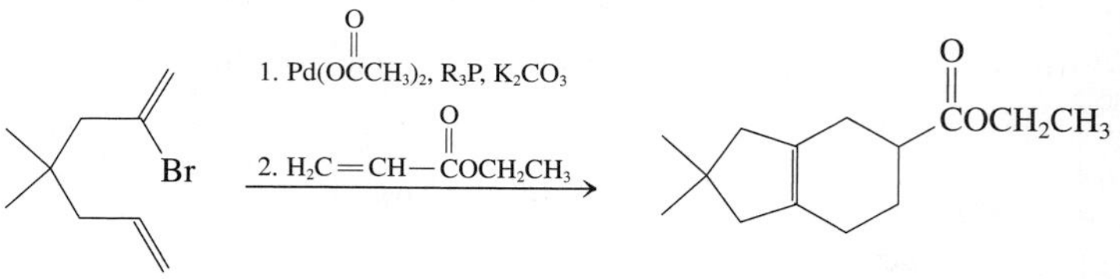

67. Give abbreviated structures of each of the following compounds: (a) (*E*)-1,4-poly-2-methyl-1,3-butadiene [(*E*)-1,4-polyisoprene]; (b) 1,2-poly-2-methyl-1,3-butadiene (1,2-polyisoprene); (c) 3,4-poly-2-methyl-1,3-butadiene (3,4-polyisoprene); (d) copolymer of 1,3-butadiene and ethenylbenzene (styrene, C₆H₅CH=CH₂, SBR, used in automobile tires); (e) copolymer of 1,3-butadiene and propenenitrile (acrylonitrile, CH₂=CHCN, latex); (f) copolymer of 2-methyl-1,3-butadiene (isoprene) and 2-methylpropene (butyl rubber, for inner tubes).

68. The structure of the terpene limonene is shown below (see also Exercise 5-29). Identify the two 2-methyl-1,3-butadiene (isoprene) units in limonene. (a) Treatment of isoprene with catalytic amounts of acid leads to a variety of oligomeric products, one of which is limonene. Devise a detailed mechanism for the acid-catalyzed conversion of two molecules of isoprene into limonene. Take care to use sensible intermediates in each step. (b) Two molecules of isoprene may also be converted into limonene by a completely different mechanism, which takes place in the strict absence of catalysts of any kind. Describe this mechanism. What is the name of the reaction?

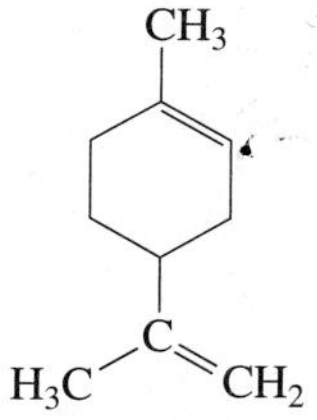

Limonene

69. **CHALLENGE** The carbocation derived from geranyl pyrophosphate (Section 14-10) is the biosynthetic precursor of not only camphor, but also limonene (Problem 68) and α-pinene (Problem 46 of Chapter 4). Formulate mechanisms for the formation of the latter two compounds.

70. What is the longest-wavelength electronic transition in each of the following species? Use molecular-orbital designations, such as *n*→π*, π₁→π₂, in your answer. (**Hint:** Construct a molecular-orbital energy diagram, like that in Figure 14-16, for each.) (**a**) 2-Propenyl (allyl) cation; (**b**) 2-propenyl (allyl) radical; (**c**) formaldehyde, $H_2C=O$; (**d**) N_2; (**e**) pentadienyl anion (Problem 55); (**f**) 1,3,5-hexatriene.

71. Ethanol, methanol, and cyclohexane are commonly used solvents for UV spectroscopy because they do not absorb radiation of wavelength longer than 200 nm. Why not?

72. The ultraviolet spectrum of a 2×10^{-4} M solution of 3-penten-2-one exhibits a $\pi \rightarrow \pi^*$ absorption at 224 nm with $A = 1.95$ and an $n \rightarrow \pi^*$ band at 314 nm with $A = 0.008$. Calculate the molar absorptivities (extinction coefficients) for these bands.

73. In a published synthetic procedure, acetone is treated with ethenyl (vinyl) magnesium bromide, and the reaction mixture is then neutralized with strong aqueous acid. The product exhibits the ^1H NMR spectrum shown below. What is its structure? When the reaction mixture is (improperly) allowed to remain in contact with aqueous acid for too long, an additional new compound is observed. Its ^1H NMR spectrum has peaks at $\delta = 1.70$ (s, 3 H), 1.79 (s, 3 H), 2.25 (broad s, 1 H), 4.10 (d, $J = 8$ Hz, 2 H), and 5.45 (t, $J = 8$ Hz, 1 H) ppm. What is the structure of the second product, and how did it get there?

^1H NMR

6 H

5.9   5.8   5.1   5.0   4.9   4.8

1 H   1 H   1 H $(CH_3)_4Si$
       1 H

6.0 5.5 5.0 4.5 4.0 3.5 3.0 2.5 2.0 1.5 1.0 0.5 0.0

300-MHz ^1H NMR spectrum ppm (δ)

74. **CHALLENGE**   Farnesol is a molecule that makes flowers smell good (lilacs, for instance). Treatment with hot concentrated H_2SO_4 converts farnesol first into bisabolene and finally into cadinene, a compound of the volatile oils of junipers and cedars. Propose detailed mechanisms for these conversions.

Farnesol

Bisabolene   Cadinene

75. The ratio of 1,2- to 1,4-addition of Br_2 to 1,3-butadiene (Section 14-6) is temperature dependent. Identify the kinetic and thermodynamic products, and explain your choices.

76. Diels-Alder cycloaddition of 1,3-butadiene with the cyclic dienophile shown below takes place at only one of the two carbon–carbon double bonds in the latter to give a single product. Give its structure and explain your answer. Watch stereochemistry.

This transformation was the initial step in the total synthesis of cholesterol (Section 4-7), completed by R. B. Woodward (see Section 14-9) in 1951. This achievement, monumental for its time, revolutionized synthetic organic chemistry.

Team Problem

77. As a team, consider the following historic preparation of a tris(1,1-dimethylethyl) derivative of Dewar benzene, B, by the photochemical isomerisation of 1,2,4-tris(1,1-dimethylethyl) benzene by van Tamelen and Pappas (1962). B does not revert to A via either a thermal or a photochemical electrocyclic mechanism. Formulate a mechanism for the conversion of A to B and explain the kinetic robustness of B with respect to the regeneration of A.

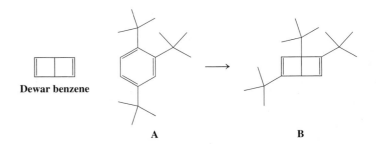

Dewar benzene

A   B

Preprofessional Problems

78. How many nodes are present in the LUMO (lowest unoccupied molecular orbital) of 1,3-butadiene?

(**a**) Zero; (**b**) one; (**c**) two; (**d**) three; (**e**) four

79. Arrange the following three chlorides in decreasing order of S_N1 reactivity.

$CH_3CH_2CH_2Cl$ $H_2C=CHCHCH_3$ $CH_3CH_2CHCH_3$
                                        | |
                                        Cl   Cl

A   B   C

(**a**) A > B > C; (**b**) B > C > A; (**c**) B > A > C; (**d**) C > B > A.

80. When cyclopentadiene is treated with tetracyanoethene, a new product results. Its most likely structure is

(a)

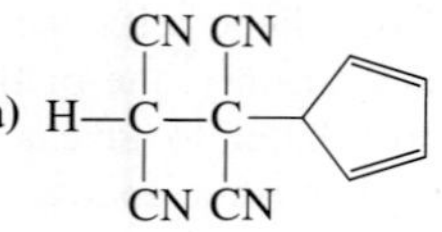

(b)

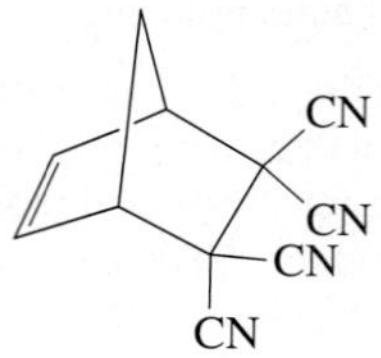

(c)

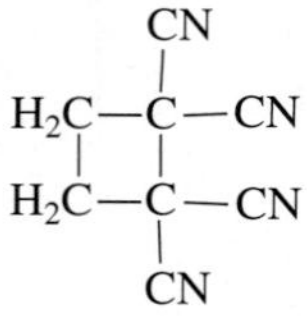

(d)

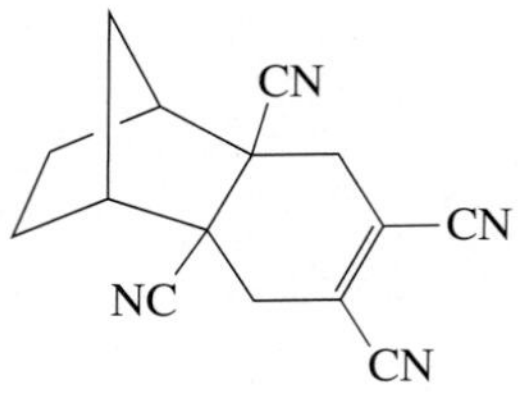

81. Which common analytical method will most clearly and rapidly distinguish A from B?

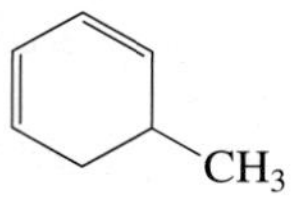

A 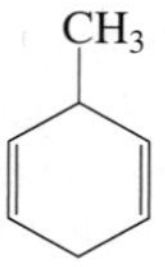   B

(a) IR spectroscopy; (b) UV spectroscopy;
(c) combustion analysis; (d) visible spectroscopy.

Interlude

A SUMMARY OF ORGANIC REACTION MECHANISMS

Although we are only just past the halfway point in our survey of organic chemistry, with the completion of Chapter 14 we have in fact now seen examples of each of the three major classes of organic transformations: radical, polar, and pericyclic processes. This section summarizes all of the individual mechanism types that we have so far encountered in each of these reaction classes.

Radical reactions follow chain mechanisms

Radical reactions begin with the generation of a reactive odd-electron intermediate by means of an initiation step and convert starting materials into products through a chain of propagation steps. We have seen both **radical substitutions** (Chapter 3) and **radical additions** (Chapter 12). Substitution is capable of introducing a functional group into a previously unfunctionalized molecule; radical addition is an example of functional-group interconversion. These individual subcategories are summarized in Table 1.

Polar reactions constitute the largest class of organic transformations

Interactions of polarized, or charged, species lead to the greatest variety in organic chemistry and to the largest number of reaction mechanism types: the typical chemistry of the organic functional groups. Two mechanisms each for **substitution** and **elimination** were first presented in Chapters 6 and 7. Both unimolecular and bimolecular pathways were found to be possible for each mechanism, depending on the structure of the substrate and, in some cases, the reaction conditions. With the introduction of functional groups containing π bonds, we encountered polar **addition reactions** in two distinct forms: nucleophilic in Chapter 8 and electrophilic in Chapter 12. These processes are summarized in Table 2.

Pericyclic reactions lack intermediates

The final class of reactions is those characterized by cyclic transition states in which there is continuous overlap of a cyclic array of orbitals. These processes take place in a single step, without the intervention of any intermediate species. They may combine multiple components to introduce new rings, as in the Diels-Alder and other **cycloaddition reactions,** or they may take the form of ring-opening and ring-closing processes—the **electrocyclic reactions.** Examples are presented in Table 3.

Table 1	Types of Radical Reactions

1. Radical Substitution

Mechanism: Radical chain   (Section 3-4)

INITIATION

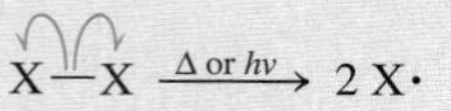

$$X \!-\! X \xrightarrow{\Delta \text{ or } h\nu} 2\,X\cdot$$

PROPAGATION

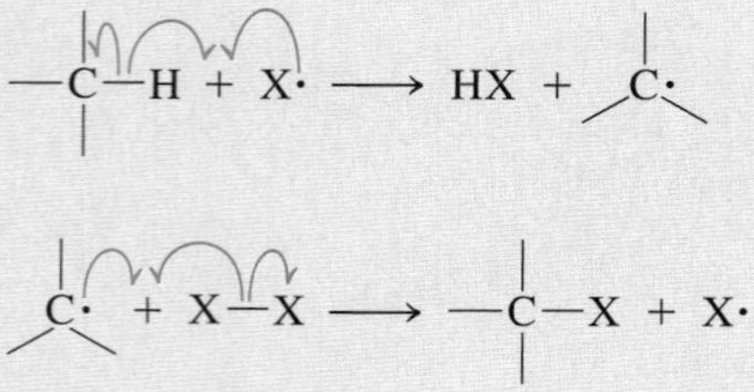

$$-\overset{|}{\underset{|}{C}}\!-\!H + X\cdot \longrightarrow HX + \overset{|}{C}\cdot$$

$$\overset{|}{C}\cdot + X \!-\! X \longrightarrow -\overset{|}{\underset{|}{C}}\!-\!X + X\cdot$$

TERMINATION

Examples: $\overset{|}{C}\cdot + X\cdot \longrightarrow -\overset{|}{\underset{|}{C}}\!-\!X$

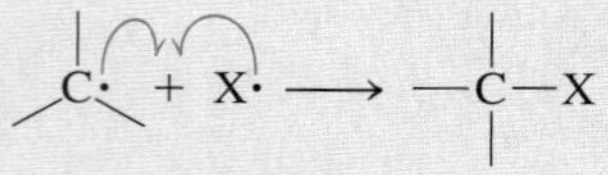

(Alkanes) $RH + X_2 \xrightarrow{h\nu} RX + HX$ (Sections 3-4 through 3-9)

(Allylic systems) $CH_2\!=\!CHCH_3 + X_2 \xrightarrow{h\nu} CH_2\!=\!CHCH_2X + HX$ (Sections 14-2 and 22-9)

2. Radical Addition

Mechanism: Radical chain   (Section 12-13)

Examples:

(Alkenes) $RCH\!=\!CH_2 + HBr \xrightarrow{\text{Peroxides}} RCH_2CH_2Br$ (Sections 12-13 and 12-15)

**Anti-Markovnikov
product**

(Alkynes) $RC\!\equiv\!CH + HBr \xrightarrow{\text{Peroxides}} RCH\!=\!CHBr$ (Section 13-8)

Table 2	Types of Polar Reactions

1. Bimolecular Nucleophilic Substitution

Mechanism: Concerted backside displacement (S_N2) (Sections 6-2, 6-4, and 6-5)

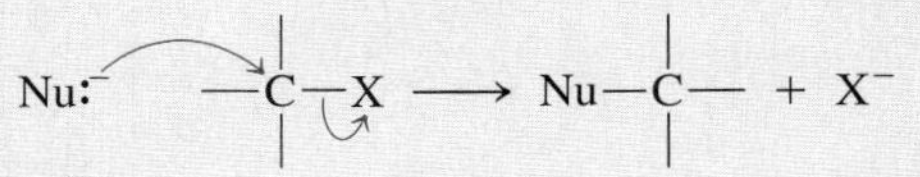

$$Nu:^- \quad \overset{|}{\underset{|}{C}}-X \longrightarrow Nu-\overset{|}{\underset{|}{C}}- \ + \ X^-$$

Example:

$$HO^- + CH_3Cl \longrightarrow CH_3OH + Cl^-$$
(Sections 6-2 through 6-9)

**100% Inversion
at a stereocenter**

2. Unimolecular Nucleophilic Substitution

Mechanism: Carbocation formation–nucleophilic attack (S_N1; usually accompanied by E1) (Section 7-2)

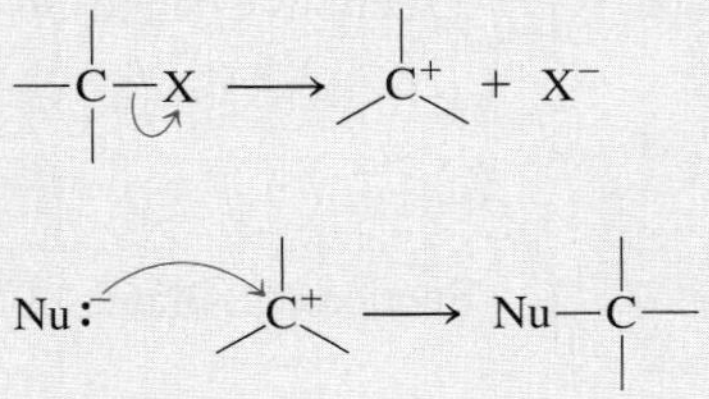

$$-\overset{|}{\underset{|}{C}}-X \longrightarrow \overset{+}{C} + X^-$$

$$Nu:^- \quad \overset{+}{C} \longrightarrow Nu-\overset{|}{\underset{|}{C}}-$$

Example:

$$H_2O + (CH_3)_3CCl \longrightarrow (CH_3)_3COH + HCl$$
(Sections 7-2 through 7-5)

**Racemization
at a stereocenter**

3. Bimolecular Elimination

Mechanism: Concerted deprotonation–π-bond formation–expulsion of leaving group (E2) (Section 7-7)

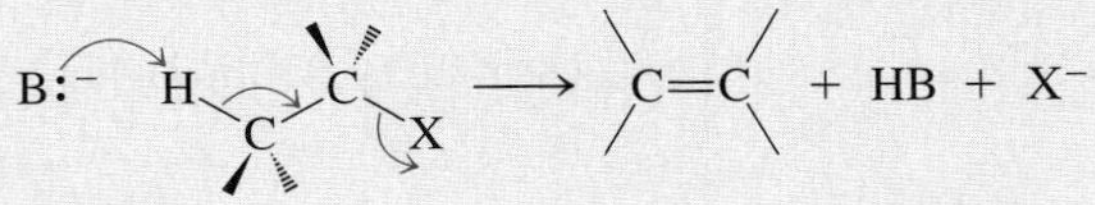

$$B:^- \quad H \underset{C}{\overset{C}{\diagdown}} X \longrightarrow C=C + HB + X^-$$

Example:

$$CH_3CH_2O^- + CH_3CHClCH_3 \longrightarrow CH_3CH_2OH + CH_3CH=CH_2 + Cl^-$$
(Sections 7-7 and 11-6)

Anti transition state preferred

4. Unimolecular Elimination

Mechanism: Carbocation formation–deprotonation and π-bond formation (E1) accompanies S_N1 (Section 7-6)

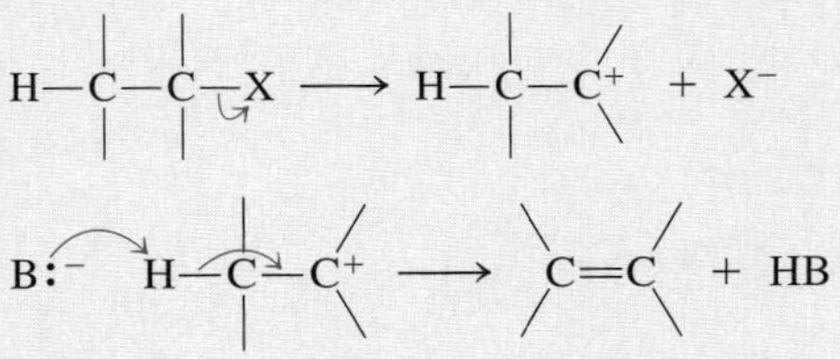

$$H-\overset{|}{\underset{|}{C}}-\overset{|}{\underset{|}{C}}-X \longrightarrow H-\overset{|}{\underset{|}{C}}-\overset{+}{C} + X^-$$

$$B:^- \quad H-\overset{|}{C}-\overset{+}{C} \longrightarrow C=C + HB$$

Example:

$$(CH_3)_3CCl \xrightarrow{H_2O} (CH_3)_2C=CH_2 + HCl$$
(Sections 7-6 and 11-7)

Table 2	Types of Polar Reactions (continued)

5. Nucleophilic Addition

Mechanism: Nucleophilic addition–protonation   (Sections 8-6 and 8-8)

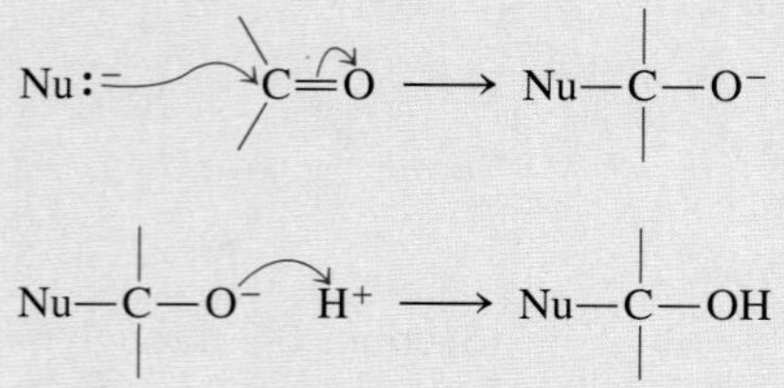

$$Nu:^- \ \ \diagup C{=}O \longrightarrow Nu{-}C{-}O^-$$

$$Nu{-}C{-}O^- \ \ H^+ \longrightarrow Nu{-}C{-}OH$$

Examples:

(Hydride reagents) $NaBH_4 + (CH_3)_2C{=}O \longrightarrow (CH_3)_2CHOH$ (Section 8-6)

(Organometallic reagents) $RMgX + (CH_3)_2C{=}O \longrightarrow R{-}\underset{CH_3}{\overset{CH_3}{C}}{-}OH$ (Sections 8-8 and 14-4)

6. Electrophilic Addition

Mechanism: Electrophilic addition–nucleophilic attack   (Sections 12-3 and 12-5)

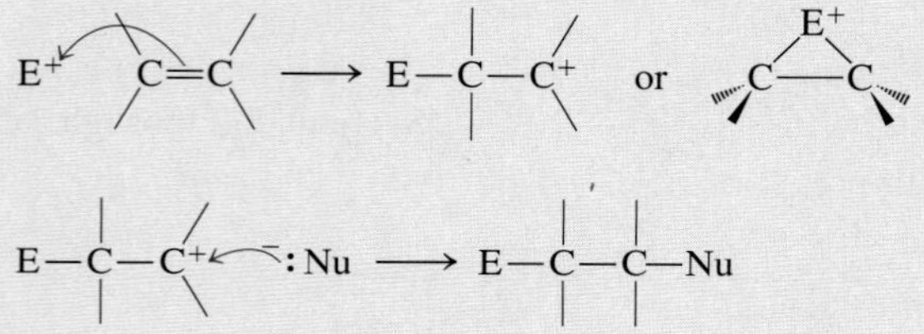

$$E^+ \ \ C{=}C \longrightarrow E{-}C{-}C^+ \quad or \quad \overset{E^+}{\underset{}{C{-}C}}$$

$$E{-}C{-}C^+ \ \ :Nu \longrightarrow E{-}C{-}C{-}Nu$$

Example:

(Alkenes) $RCH{=}CH_2 + HBr \longrightarrow R\underset{}{\overset{Br}{C}}HCH_3$ (Sections 12-3 through 12-7)

Markovnikov product

Table 3	Types of Pericyclic Reactions

1. Cycloaddition

Mechanism: Concerted, through cyclic array of electrons   (Section 14-8)

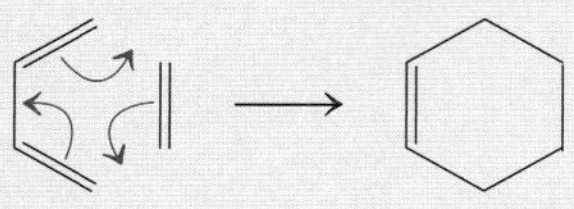

Example:

(Diels-Alder reaction)

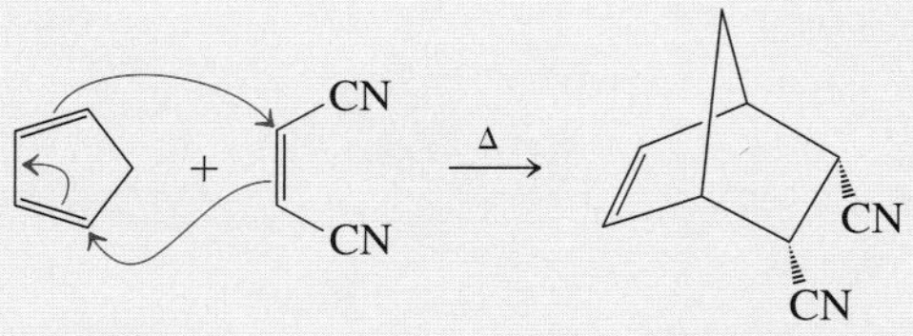

**Stereospecific,
endo product preferred**

2.   Electrocyclic Reactions

Mechanism: Concerted, through cyclic array of electrons   (Section 14-9)

Examples:

(Cyclobutene → butadiene)

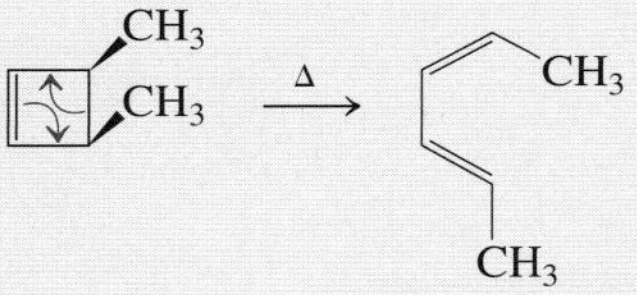

Conrotatory thermally

(Hexatriene → cyclohexadiene)

(Section 14-9)

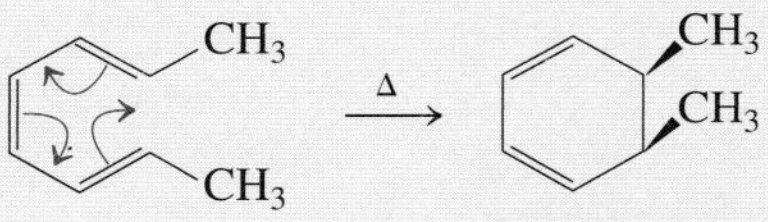

Disrotatory thermally

CHAPTER 15 | Benzene and Aromaticity
Electrophilic Aromatic Substitution

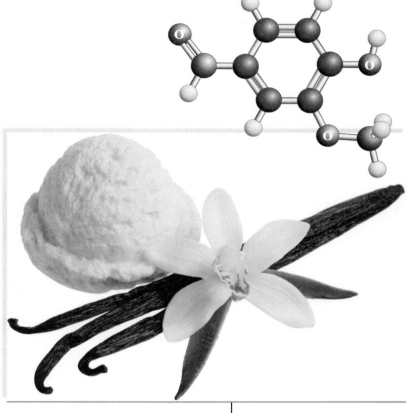

During the early nineteenth century, the oil used to illuminate the streets of London and other cities came from rendered whale fat (called "blubber"). Eager to determine its components, in 1825 the English scientist Michael Faraday* heated whale oil to obtain a colorless liquid (b.p. 80.1 °C, m.p. 5.5 °C) that had the empirical formula CH. This compound posed a problem for the theory that carbon had to have four valences to other atoms, and it was of particular interest because of its unusual stability and chemical inertness. The molecule was named **benzene** and was eventually shown to have the molecular formula C_6H_6. Annual production in the United States alone is 7 million liters per year.

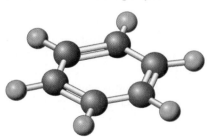

Benzene

The structure of vanillin is that of a substituted benzene ring, highlighted in red here and in benzene on the left. Vanillin is the essential component of oil of vanilla, extracted from the fermented bean pods of the vanilla orchid (*Vanilla fragrans*). It is the world's second most expensive spice (after saffron), and global demand is estimated at 2000 tons per year.

Benzene has four degrees of unsaturation (see Table 11-6), satisfied by a 1,3,5-cyclohexatriene structure (one ring, three double bonds) first proposed by Kekulé[†] and Loschmidt,[‡] yet it does not exhibit the kind of reactivity expected for a conjugated triene. In other words, its structure does not seem to correlate with its observed function. However, benzene is not completely unreactive. For example, it reacts with bromine, albeit only in the

*Professor Michael Faraday (1791–1867), Royal Institute of Chemistry, London.
†Professor F. August Kekulé; see Section 1-4.
‡Professor Josef Loschmidt (1821–1895), University of Vienna, Austria.

presence of catalytic amounts of a Lewis acid, such as iron tribromide, $FeBr_3$ (Section 15-9). Surprisingly, the result is not addition but substitution to furnish bromobenzene.

Benzene + Br—Br $\xrightarrow{FeBr_3}$ **Bromobenzene** (Substitution product) + HBr

Addition product not formed

The formation of only one monobromination product is nicely consistent with the sixfold symmetry of the structure of benzene. Further reaction with bromine introduces a second halogen atom to give the three isomeric 1,2-, 1,3-, and 1,4-dibromobenzenes.

$\xrightarrow[-HBr]{Br_2, FeBr_3}$

1,2-Dibromobenzene (the same as 1,6-dibromobenzene) + **1,3-Dibromobenzene** + **1,4-Dibromobenzene**

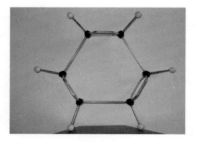

Kekulé's perception of the structure of benzene.

Historically, the observation of only one 1,2-dibromobenzene product posed another puzzle. If the molecule has a cyclohexatriene topology with alternating single and double bonds, two isomers should have been formed: 1,2- and 1,6-dibromobenzene, in which the two substituents are connected by double or single carbon–carbon bonds, respectively. Kekulé solved this puzzle brilliantly by boldly proposing that benzene should be viewed as a set of two rapidly equilibrating (he used the word "oscillating") cyclohexatriene *isomers,* which would render 1,2- and 1,6-dibromobenzene indistinguishable. We now know that this notion was not quite right. In terms of modern electronic theory, benzene is a single compound, best described by two equivalent cyclohexatrienic *resonance* forms (Section 14-7).

Why does the cyclic array of π electrons in benzene impart unusual stability? How can we quantify this observation? Considering the diagnostic power of NMR (Section 11-4) and electronic spectroscopy (Section 14-11) in establishing the presence of delocalized π systems, can we expect characteristic spectral signals for benzene? In this chapter we shall answer these questions. We look first at the system of naming substituted benzenes, second at the electronic and molecular structure of the parent molecule, and third at the evidence for its unusual stabilizing energy, the **aromaticity** of benzene. This aromaticity and the special structure of benzene affect its spectral properties and reactivity. We shall see what happens when two or more benzene rings are fused to give more extended π systems. Similarly, we shall compare the properties of benzene with those of other cyclic conjugated polyenes. Finally, we shall study the special mechanism by which substituents are introduced into the benzene ring: **electrophilic aromatic substitution.**

is the same as

15-1 | NAMING THE BENZENES

Benzene and its derivatives were originally called **aromatic compounds** because many of them have a strong aroma. Benzene, even though its odor is not particularly pleasant, is viewed as the "parent" aromatic molecule. Whenever the symbol for the benzene ring with its three double bonds is written, it should be understood to represent only one of a pair of contributing resonance forms. Alternatively, the ring is sometimes drawn as a regular hexagon with an inscribed circle.

Many monosubstituted benzenes are named by adding a substituent prefix to the word "benzene."

Fluorobenzene   **Nitrobenzene**   **(1,1-Dimethylethyl)benzene**
(*tert*-Butylbenzene)

There are three possible arrangements of disubstituted benzenes. These arrangements are designated by the prefixes **1,2-** (**ortho,** or **o-**) for adjacent substituents, **1,3** (**meta-,** or ***m-***) for 1,3-disubstitution, and **1,4** (**para-,** or ***p-***) for 1,4-disubstitution. The substituents are listed in alphabetical order.

1,2-Dichlorobenzene   **1-Bromo-3-nitrobenzene**   **1-Ethyl-4-(1-methylethyl)benzene**
(*o*-Dichlorobenzene)   **(*m*-Bromonitrobenzene)**   **(*p*-Ethylisopropylbenzene)**

To name tri- and more highly substituted derivatives, the six carbons of the ring are numbered so as to give the substituents the lowest set of numbers, as in cyclohexane nomenclature.

1-Bromo-2,3-dimethylbenzene   **1,2,4-Trinitrobenzene**   **1-Ethenyl-3-ethyl-5-ethynylbenzene**

The following benzene derivatives will be encountered in this book.

Methylbenzene   **1,2-Dimethylbenzene**   **1,3,5-Trimethylbenzene**   **Ethenylbenzene**   **Methoxybenzene**
(Toluene)   **(*o*-Xylene)**   **(Mesitylene)**   **(Styrene)**   **(Anisole)**

(Common industrial and laboratory solvents)   **(Used in polymer manufacture)**   **(Used in perfume)**

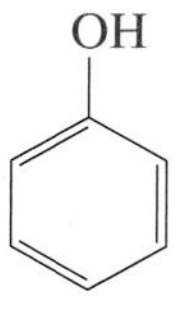

Benzenol
(Phenol)
(An antiseptic and anesthetic used in sore throat sprays)

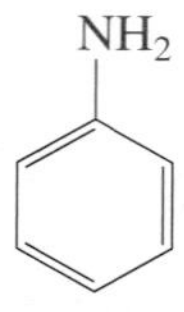

Benzenamine
(Aniline)
(Used in dye manufacture)

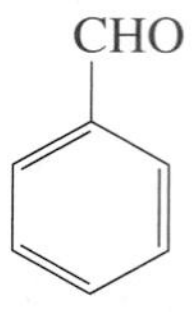

Benzenecarbaldehyde
(Benzaldehyde)
(Artificial almond flavoring)

1-Phenylethanone
(Acetophenone)
(A hypnotic drug)

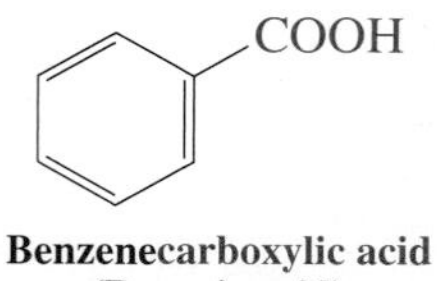

Benzenecarboxylic acid
(Benzoic acid)
(A food preservative)

We shall employ IUPAC nomenclature for all but three systems. In accord with the indexing preferences of *Chemical Abstracts,* the common names phenol, benzaldehyde, and benzoic acid will be used in place of their systematic counterparts.

Ring-substituted derivatives of such compounds are named by numbering the ring positions or by using the prefixes *o-*, *m-*, and *p-*. The substituent that gives the compound its base name is placed at carbon 1.

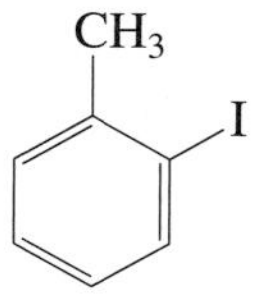

1-Iodo-2-methylbenzene
(*o*-Iodotoluene)

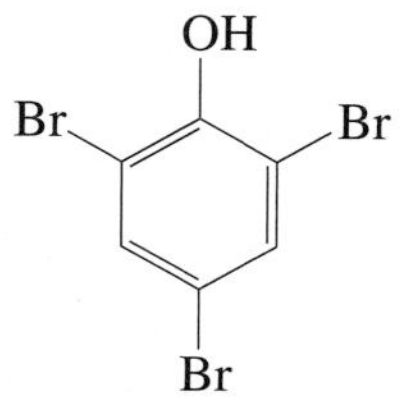

2,4,6-Tribromophenol

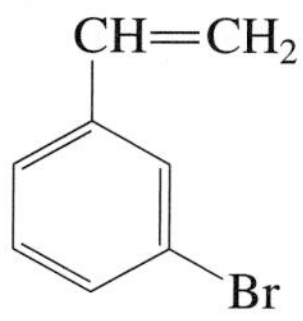

1-Bromo-3-ethenylbenzene
(*m*-Bromostyrene)

A number of the common names for aromatic compounds refer to their fragrance and natural sources. Several of them have been accepted by IUPAC. As before, a consistent logical naming of these compounds will be adhered to as much as possible, with common names mentioned in parentheses.

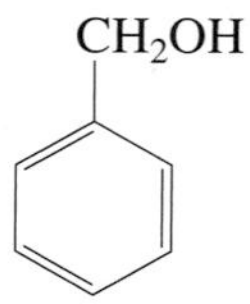

Aromatic Flavoring Agents

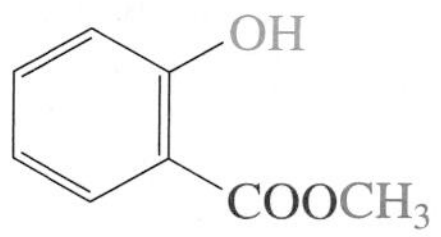

Phenylmethanol
(Benzyl alcohol)

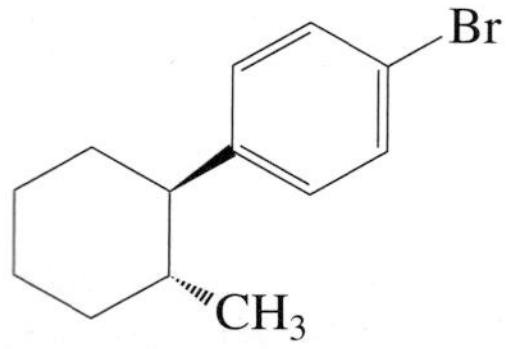

Methyl 2-hydroxybenzoate
(Methyl salicylate, oil of wintergreen flavor)

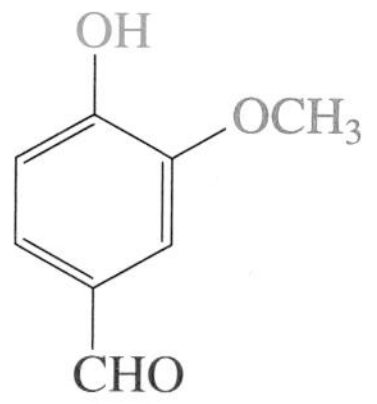

4-Hydroxy-3-methoxybenzaldehyde
(Vanillin, vanilla flavor)

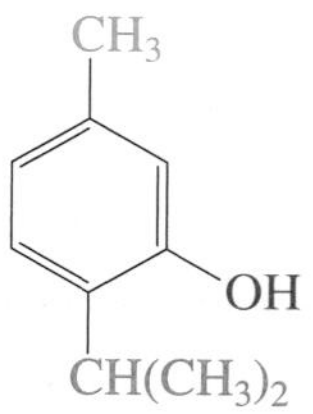

5-Methyl-2-(1-methylethyl)phenol
(Thymol, thyme flavor)

trans-1-(4-Bromophenyl)-2-methylcyclohexane

The generic term for substituted benzenes is **arene.** An arene as a substituent is referred to as an **aryl group,** abbreviated **Ar.** The parent aryl substituent is **phenyl,** C_6H_5. The $C_6H_5CH_2-$ group, which is related to the 2-propenyl (allyl) substituent (Sections 14-1 and 22-1), is called **phenylmethyl (benzyl)** (see examples in margin).

Exercise 15-1

Write systematic and common names of the following substituted benzenes.

(a) Cl ... NO$_2$ **(b)** CH$_3$ ... D **(c)** OH ... NO$_2$ ... NO$_2$

Exercise 15-2

Draw the structures of **(a)** (1-methylbutyl)benzene; **(b)** 1-ethenyl-4-nitrobenzene (*p*-nitrostyrene); **(c)** 2-methyl-1,3,5-trinitrobenzene (2,4,6-trinitrotoluene—the explosive TNT; see also Real Life 16-1).

Exercise 15-3

The following names are incorrect. Write the correct form. **(a)** 3,5-Dichlorobenzene; **(b)** *o*-aminophenyl fluoride; **(c)** *p*-fluorobromobenzene.

In Summary Simple monosubstituted benzenes are named by placing the substituent name before the word "benzene." For more highly substituted systems, 1,2-, 1,3-, and 1,4- (or ortho-, meta-, and para-) prefixes indicate the positions of disubstitution. Alternatively, the ring is numbered, and substituents labeled with these numbers are named in alphabetical order. Many simple substituted benzenes have common names.

15-2 STRUCTURE AND RESONANCE ENERGY OF BENZENE: A FIRST LOOK AT AROMATICITY

Benzene is unusually unreactive: At room temperature, benzene is inert to acids, H$_2$, Br$_2$, and KMnO$_4$—reagents that readily add to conjugated alkenes (Section 14-6). The reason for this poor reactivity is that the cyclic six-electron arrangement imparts a special stability in the form of a large resonance energy (Section 14-7). We shall first review the evidence for the structure of benzene and then estimate its resonance energy by comparing its heat of hydrogenation with those of model systems that lack cyclic conjugation, such as 1,3-cyclohexadiene.

The benzene ring contains six equally overlapping *p* orbitals

If benzene were a conjugated triene, a "cyclohexatriene," we should expect the C–C bond lengths to alternate between single and double bonds. In fact, experimentally, the benzene molecule is a completely symmetrical hexagon (Figure 15-1), with equal C–C bond lengths of 1.39 Å, between the values found for the single bond (1.47 Å) and the double bond (1.34 Å) in 1,3-butadiene (Figure 14-6).

Figure 15-2 shows the electronic structure of the benzene ring. All carbons are sp^2 hybridized, and each *p* orbital overlaps to an equal extent with its two neighbors. The resulting delocalized electrons form circular π clouds above and below the ring. The

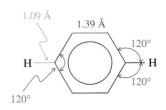

Figure 15-1 The molecular structure of benzene. All six C–C bonds are equal; all bond angles are 120°.

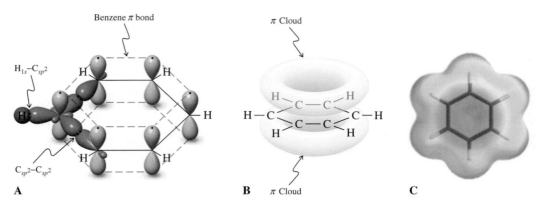

Figure 15-2 Orbital picture of the bonding in benzene. (A) The σ framework is depicted as straight lines except for the bonding to one carbon, in which the p orbital and the sp^2 hybrids are shown explicitly. (B) The six overlapping p orbitals in benzene form a π-electron cloud located above and below the molecular plane. (C) The electrostatic potential map of benzene shows the relative electron richness of the ring and the even distribution of electron density over the six carbon atoms.

symmetrical structure of benzene is a consequence of the interplay between the σ and π electrons in the molecule. The symmetrical σ frame acts in conjunction with the delocalized π frame to enforce the regular hexagon.

Benzene is especially stable: heats of hydrogenation

A way to establish the relative stability of a series of alkenes is to measure their heats of hydrogenation (Sections 11-5 and 14-5). We may carry out a similar experiment with benzene, relating its heat of hydrogenation to those of 1,3-cyclohexadiene and cyclohexene. These molecules are conveniently compared because hydrogenation changes all three into cyclohexane.

The hydrogenation of cyclohexene is exothermic by -28.6 kcal mol^{-1}, a value expected for the hydrogenation of a cis double bond (Section 11-5). The heat of hydrogenation of 1,3-cyclohexadiene ($\Delta H^\circ = -54.9$ kcal mol^{-1}) is slightly less than double that of cyclohexene because of the resonance stabilization in a conjugated diene (Section 14-5); the energy of that stabilization is $(2 \times 28.6) - 54.9 = 2.3$ kcal mol^{-1} (9.6 kJ mol^{-1}).

$$\text{(cyclohexene)} \quad + \quad H_2 \quad \xrightarrow{\text{Catalytic Pt}} \quad \text{(cyclohexane)} \qquad \Delta H^\circ = -28.6 \text{ kcal mol}^{-1}\,(-120 \text{ kJ mol}^{-1})$$

$$\text{(1,3-cyclohexadiene)} \quad + \quad 2\,H_2 \quad \xrightarrow{\text{Catalytic Pt}} \quad \text{(cyclohexane)} \qquad \Delta H^\circ = -54.9 \text{ kcal mol}^{-1}\,(-230 \text{ kJ mol}^{-1})$$

Armed with these numbers, we can calculate the expected heat of hydrogenation of benzene, as though it were simply composed of three double bonds like that of cyclohexene, but with the extra resonance stabilization of conjugation as in 1,3-cyclohexadiene.

$$\text{(benzene)} \quad + \quad 3\,H_2 \quad \xrightarrow{\text{Catalyst}} \quad \text{(cyclohexane)} \qquad \Delta H^\circ = ?$$

ΔH° = 3 (ΔH° of hydrogenation of ⬡) + 3 (resonance correction in ⬡)

= $(3 \times -28.6) + (3 \times 2.3)$ kcal mol^{-1}

= $-85.8 + 6.9$ kcal mol^{-1}

= -78.9 kcal mol^{-1} (-330 kJ mol^{-1})

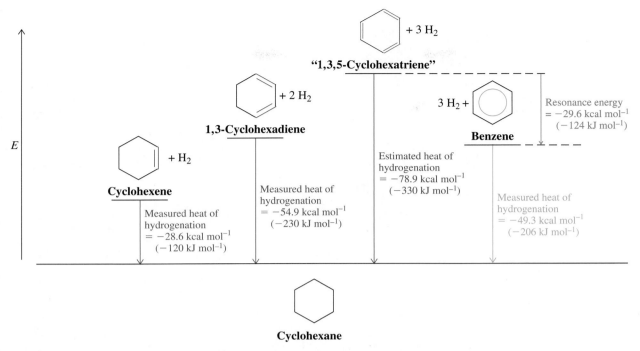

Figure 15-3 Heats of hydrogenation provide a measure of benzene's unusual stability. Experimental values for cyclohexene and 1,3-cyclohexadiene allow us to estimate the heat of hydrogenation for the hypothetical "1,3,5-cyclohexatriene." Comparison with the experimental $\Delta H°$ for benzene gives a value of approximately 29.6 kcal mol^{-1} for the aromatic resonance energy.

Now let us look at the experimental data. Although benzene is hydrogenated only with difficulty (Section 14-7), special catalysts carry out this reaction, so the heat of hydrogenation of benzene can be measured: $\Delta H° = -49.3$ kcal mol^{-1}, much less than the -78.9 kcal mol^{-1} predicted.

Figure 15-3 summarizes these results. It is immediately apparent that benzene is *much* more stable than a cyclic triene containing alternating single and double bonds. The difference is the **resonance energy** of benzene, about 30 kcal mol^{-1} (126 kJ mol^{-1}). Other terms used to describe this quantity are *delocalization energy, aromatic stabilization,* or simply the **aromaticity** of benzene. The original meaning of the word *aromatic* has changed with time, now referring to a thermodynamic property rather than to odor.

In Summary The structure of benzene is a regular hexagon made up of six sp^2-hybridized carbons. The C–C bond length is between those of a single and a double bond. The electrons occupying the p orbitals form a π cloud above and below the plane of the ring. The structure of benzene can be represented by two equally contributing cyclohexatriene resonance forms. Hydrogenation to cyclohexane releases about 30 kcal mol^{-1} less energy than is expected on the basis of nonaromatic models. This difference is the resonance energy of benzene.

15-3 | PI MOLECULAR ORBITALS OF BENZENE

We have just examined the atomic orbital picture of benzene. Now let us look at the molecular orbital picture, comparing the six π molecular orbitals of benzene with those of 1,3,5-hexatriene, the open-chain analog. Both sets are the result of the contiguous overlap of six p orbitals, yet the cyclic system differs considerably from the acyclic one. A comparison of the energies of the bonding orbitals in these two compounds shows that cyclic conjugation of three double bonds is better than acyclic conjugation.

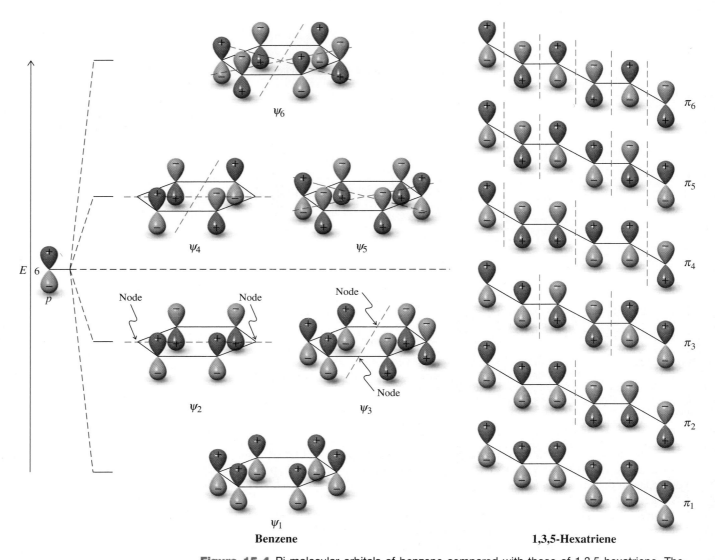

Recall: Mixing six *p* orbitals generates six new molecular orbitals.

Figure 15-4 Pi molecular orbitals of benzene compared with those of 1,3,5-hexatriene. The orbitals are shown at equal size for simplicity. Favorable overlap (bonding) takes place between orbital lobes of equal sign. A sign change is indicated by a nodal plane (dashed line). As the number of such planes increases, so does the energy of the orbitals. Note that benzene has two sets of degenerate (equal energy) orbitals, the lower energy set occupied (ψ_2, ψ_3), the other not (ψ_4, ψ_5), as shown in Figure 15-5.

Cyclic overlap modifies the energy of benzene's molecular orbitals

Figure 15-4 compares the π molecular orbitals of benzene with those of 1,3,5-hexatriene. The acyclic triene follows a pattern similar to that of 1,3-butadiene (Figure 14-7), but with two more molecular orbitals: The orbitals all have different energies, the number of nodes increasing in the procession from π_1 to π_6. The picture for benzene is different in all respects: different orbital energies, two sets of degenerate (equal energy) orbitals, and completely different nodal patterns.

Is the cyclic π system more stable than the acyclic one? To answer this question, we have to compare the combined energies of the three filled bonding orbitals in both. Figure 15-5 shows the answer: *The cyclic π system is stabilized relative to the acyclic one.* On going from 1,3,5-hexatriene to benzene, two of the bonding orbitals (π_1 and π_3) are lowered in energy and one is raised; the effect of the one raised in energy is more than offset by the two that are lowered.

Inspection of the signs of the wave functions at the terminal carbons of 1,3,5-hexatriene (Figure 15-4) explains why the orbital energies change in this way: Linkage of C1 and C6 causes *p* orbital overlap that is in phase for π_1 and π_3 but out of phase for π_2.

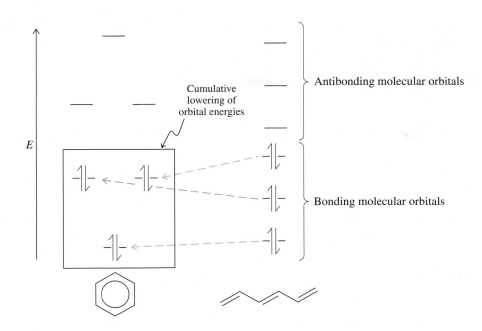

Figure 15-5 Energy levels of the π molecular orbitals in benzene and 1,3,5-hexatriene. In both, the six π electrons fill the three bonding molecular orbitals. In benzene, two of them are lower in energy and one is higher than the corresponding orbitals in 1,3,5-hexatriene. Overall, energy is reduced and stability increased in going from the acyclic to the cyclic system.

Some reactions have aromatic transition states

The relative stability of six cyclically overlapping orbitals accounts in a simple way for several reactions that proceed readily by what has seemed to be a complicated, concerted movement of three electron pairs: the Diels-Alder reaction (Section 14-8), osmium tetroxide addition to alkenes (Section 12-11), and the first step of ozonolysis (Section 12-12). In all three processes, a transition state occurs with cyclic overlap of six electrons in π orbitals (or orbitals with π character). This electronic arrangement is similar to that in benzene and is energetically more favored than the alternative, sequential bond breaking and bond making. Such transition states are called aromatic.

Aromatic Transition States

Diels-Alder reaction   **Osmium tetroxide addition**   **Ozonolysis**

Exercise 15-4

If benzene were a cyclohexatriene, 1,2-dichloro- and 1,2,4-trichlorobenzene should each exist as two isomers. Draw them.

Exercise 15-5

The thermal ring opening of cyclobutene to 1,3-butadiene is exothermic by about 10 kcal mol^{-1} (Section 14-9). Conversely, the same reaction for benzocyclobutene, A, to compound B (shown in the margin) is *endothermic* by the same amount. Explain.

A

B

$\Delta H° \sim +10$ kcal mol^{-1}

In Summary Two of the filled π molecular orbitals of benzene are lower in energy than are those in 1,3,5-hexatriene. Benzene is therefore stabilized by considerably more resonance energy than is its acyclic analog. A similar orbital structure also stabilizes aromatic transition states.

Really ? The potency of sunscreen lotions is given by the sunburn protection factor (SPF), which is a relative measure of the amount of solar UV radiation that is necessary to cause sunburn. As the SPF value increases, sunburn protection increases. However, its value does not correlate with exposure time; that is, doubling the strength of your lotion does not mean that you can lie in the sun safely twice as long. The reason is that the sun's intensity varies greatly with the time of day (peak at midday), geographic location (greater at lower latitudes), and the presence of clouds. Other factors that affect your risk of damage are skin type and the amount and frequency with which the sunscreen is applied.

15-4 | SPECTRAL CHARACTERISTICS OF THE BENZENE RING

The electronic arrangement in benzene and its derivatives gives rise to characteristic ultraviolet spectra. The hexagonal structure is also manifest in specific infrared bands. What is most striking, cyclic delocalization causes induced ring currents in NMR spectroscopy, resulting in unusual deshielding of hydrogens attached to the aromatic ring. Moreover, the different coupling constants of 1,2- (ortho), 1,3- (meta), and 1,4- (para) hydrogens in substituted benzenes indicate the substitution pattern.

The UV–visible spectrum of benzene reveals its electronic structure

Cyclic delocalization in benzene gives rise to a characteristic arrangement of energy levels for its molecular orbitals (Figure 15-4). In particular, the energetic gap between bonding and antibonding orbitals is relatively large (Figure 15-5). Is this manifested in its electronic spectrum? As shown in Section 14-11, the answer is yes: Compared with the spectra of the acyclic trienes, smaller λ_{max} values are expected for benzene and its derivatives. You can verify this effect in Figure 15-6: The highest-wavelength absorption for benzene occurs at 261 nm, closer to that of 1,3-cyclohexadiene (259 nm, Table 14-3) than to that of 1,3,5-hexatriene (268 nm).

The ultraviolet and visible spectra of aromatic compounds vary with the introduction of substituents; this phenomenon has been exploited in the tailored synthesis of dyes (Section 22-11). Simple substituted benzenes absorb between 250 and 290 nm. For example, the water-soluble 4-aminobenzoic acid (*p*-aminobenzoic acid, or PABA, see margin), $4\text{-}H_2N\text{-}C_6H_4\text{-}CO_2H$, has a λ_{max} at 289 nm, with a rather high extinction coefficient of 18,600. Because of this property, it is used in sunscreens, in which it filters out the most dangerous ultraviolet radiation emanating from the sun in this wavelength region. Some individuals cannot tolerate PABA because of an allergic skin reaction, and many recent sun creams use other compounds for protection from the sun ("PABA-free").

COOH

NH₂

4-Aminobenzoic acid
(***p*-Aminobenzoic acid, PABA**)

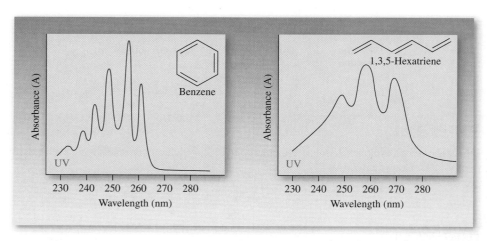

Figure 15-6 The distinctive ultraviolet spectra of benzene: $\lambda_{max}(\epsilon) = 234(30), 238(50), 243(100), 249(190), 255(220), 261(150)$ nm; and 1,3,5-hexatriene: $\lambda_{max}(\epsilon) = 247(33,900), 258(43,700), 268(36,300)$ nm. The extinction coefficients ϵ of the absorptions of 1,3,5-hexatriene are very much larger than those of benzene; therefore the spectrum at the right was taken at lower concentration.

The infrared spectrum reveals substitution patterns in benzene derivatives

The infrared spectra of benzene and its derivatives show characteristic bands in three regions. The first is at 3030 cm^{-1} for the phenyl–hydrogen stretching mode. The second ranges from 1500 to 2000 cm^{-1} and includes aromatic ring C–C stretching vibrations. Finally, a useful set of bands due to C–H out-of-plane bending motions is found between 650 and 1000 cm^{-1}.

Typical Infrared C–H Out-of-Plane Bending Vibrations for Substituted Benzenes (cm^{-1})

690–710	735–770	690–710	790–840
730–770		750–810	

Their precise location indicates the specific substitution pattern. For example, 1,2-dimethylbenzene (*o*-xylene) has this band at 738 cm^{-1}, the 1,4 isomer at 793 cm^{-1}; the 1,3 isomer (Figure 15-7) shows two absorptions in this range, at 690 and 765 cm^{-1}.

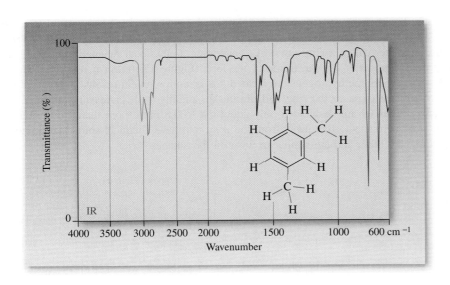

Figure 15-7 The infrared spectrum of 1,3-dimethylbenzene (*m*-xylene). There are two C–H stretching absorptions, one due to the aromatic bonds (3030 cm^{-1}), the other to saturated C–H bonds (2920 cm^{-1}). The two bands at 690 and 765 cm^{-1} are typical of 1,3-disubstituted benzenes.

The mass spectrum of benzene indicates stability

Figure 15-8 depicts the mass spectrum of benzene. Noticeable is the lack of any significant fragmentation, attesting to the unusual stability of the cyclic six-electron framework (Section 15-2). The (M + 1)$^{+\bullet}$ peak has a relative height of 6.8%, as expected for the relative abundance of ^{13}C in a molecule made up of six carbon atoms.

The NMR spectra of benzene derivatives show the effects of an electronic ring current

^{1}H NMR is a powerful spectroscopic technique for the identification of benzene and its derivatives. The cyclic delocalization of the aromatic ring gives rise to unusual deshielding, which causes the ring hydrogens to resonate at very low field ($\delta \approx 6.5$–8.5 ppm), even lower than the already rather deshielded alkenyl hydrogens ($\delta \approx 4.6$–5.7 ppm, see Section 11-4).

Chemical Shifts of Allylic and Benzylic Hydrogens

$CH_2{=}CH{-}CH_3$

Allylic:
$\delta = 1.68$ ppm

Benzylic:
$\delta = 2.35$ ppm

Figure 15-8 The mass spectrum of benzene reveals very little fragmentation.

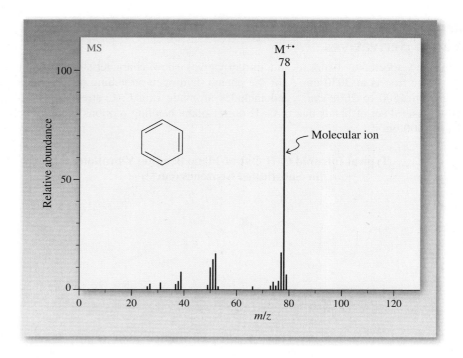

The ^{1}H NMR spectrum of benzene, for example, exhibits a sharp singlet for the six equivalent hydrogens at $\delta = 7.27$ ppm. How can this strong deshielding be explained? In a simplified picture, the cyclic π system with its delocalized electrons may be compared to a loop of conducting metal. When such a loop is exposed to a perpendicular magnetic field (H_0), the electrons circulate (called a **ring current**), generating a new local magnetic field (h_{local}). This induced field opposes H_0 on the inside of the loop (Figure 15-9), but it reinforces H_0 on the outside—just where the hydrogens are located. Such reinforcement results in deshielding. The effect is strongest close to the ring and diminishes rapidly with increasing distance from it. Thus, benzylic nuclei are deshielded only about 0.4 to 0.8 ppm more than their allylic counterparts, and hydrogens farther away from the π system have chemical shifts that do not differ much from each other and are similar to those in the alkanes.

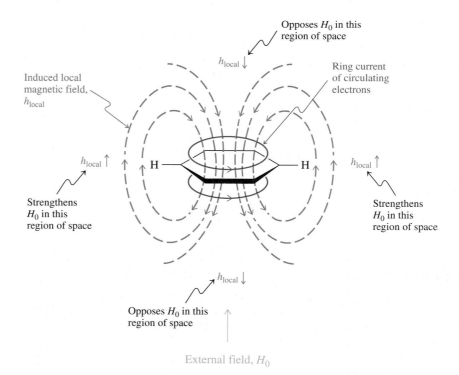

Figure 15-9 The π electrons of benzene may be compared to those in a loop of conducting metal. Exposure of this loop of electrons to an external magnetic field H_0 causes them to circulate. This "ring current" generates a local field, reinforcing H_0 on the outside of the ring. Thus, the hydrogens resonate at lower field.

Whereas benzene exhibits a sharp singlet in its NMR spectrum, substituted derivatives may have more complicated patterns. For example, introduction of one substituent renders the hydrogens positioned ortho, meta, and para nonequivalent and subject to mutual coupling. An example is the NMR spectrum of bromobenzene, in which the signal for the ortho hydrogens is shifted slightly downfield relative to that of benzene. Moreover, all the nonequivalent protons are coupled to one another, thus giving rise to a complicated spectral pattern (Figure 15-10).

Figure 15-11 shows the NMR spectrum of 4-(N,N-dimethylamino)benzaldehyde. The large chemical shift difference between the two sets of hydrogens on the ring results in a near-first-order pattern of two doublets. The observed coupling constant is 9 Hz, a typical splitting between ortho protons.

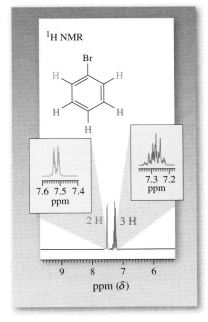

Figure 15-10 A part of the 300-MHz ^{1}H NMR spectrum of bromobenzene, a non-first-order spectrum.

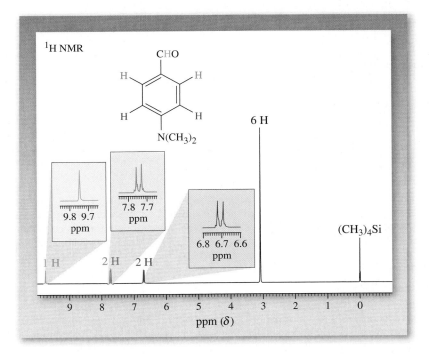

Figure 15-11 300-MHz ^{1}H NMR spectrum of 4-(N,N-dimethylamino) benzaldehyde (p-dimethylamino-benzaldehyde). In addition to the two aromatic doublets (J = 9 Hz), there are singlets for methyl (δ = 3.09 ppm) and the aldehydic hydrogens (δ = 9.75 ppm).

A first-order spectrum revealing all three types of coupling is shown in Figure 15-12. 1-Methoxy-2,4-dinitrobenzene (2,4-dinitroanisole) bears three ring hydrogens with different chemical shifts and distinct splittings. The hydrogen ortho to the methoxy group appears at

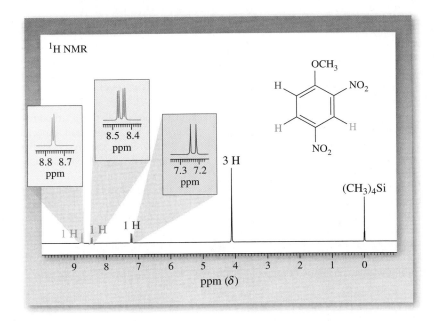

Figure 15-12 300-MHz ^{1}H NMR spectrum of 1-methoxy-2,4-dinitrobenzene (2,4-dinitroanisole).

Aromatic Coupling Constants

J_{ortho} = 6–9.5 Hz
J_{meta} = 1.2–3.1 Hz
J_{para} = 0.2–1.5 Hz

^{13}C NMR Data of Two Substituted Benzenes (ppm)

Remember WHIP

What

How

Information

Proceed

δ = 7.23 ppm as a doublet with a 9-Hz ortho coupling. The hydrogen flanked by the two nitro groups (at δ = 8.76 ppm) also appears as a doublet, with a small (3-Hz) meta coupling. Finally, we find the remaining ring hydrogen absorbing at δ = 8.45 ppm as a doublet of doublets because of simultaneous coupling to the two other ring protons. Para coupling between the hydrogens at C3 and C6 is too small (<1 Hz) to be resolved; it is evident as a slight broadening of the resonances of these protons.

In contrast with ^{1}H NMR, ^{13}C NMR chemical shifts in benzene derivatives are dominated by hybridization and substituent effects. Because the induced ring current flows directly above and below the aromatic carbons (Figure 15-9), they are less affected by it. Moreover, the relatively large ^{13}C chemical shift range, about 200 ppm, makes ring current contributions (only a few ppm) less noticeable. Therefore, benzene carbons exhibit chemical shifts similar to those in alkenes, between 120 and 135 ppm when unsubstituted (see margin). Benzene itself exhibits a single line at δ = 128.7 ppm.

Exercise 15-6

Can the three isomeric trimethylbenzenes be distinguished solely on the basis of the number of peaks in their proton-decoupled ^{13}C NMR spectra? Explain.

Solved Exercise 15-7 | Working with the Concepts: Using Spectral Data to Assign the Structure of a Substituted Benzene

A hydrocarbon shows a molecular-ion peak in the mass spectrum at m/z 134. The associated [M + 1] ion at m/z 135 has about 10% of the intensity of the molecular ion. The largest peak in the spectrum is a fragment ion at m/z 119. The other spectral data for this compound are as follows: ^{1}H NMR (90 MHz): δ = 1.22 ppm (d, J = 7.0 Hz, 6 H), 2.28 (s, 3 H), 2.82 (septet, J = 7.0 Hz, 1 H), and 7.02 (broad s, 4 H) ppm; ^{13}C NMR: δ = 21.3, 24.2, 38.9, 126.6, 128.6, 134.8, and 145.7 ppm; IR: $\tilde{\nu}$ = 3030, 2970, 2880, 1515, 1465, and 813 cm^{-1}; UV: $\lambda_{max}(\epsilon)$ = 265(450) nm. What is its structure?

Strategy

The spectra supply a lot of information, so your first problem is *Where do I start?* Typically, researchers look at the most important data first—the mass and the ^{1}H NMR spectrum—and then use the other spectra as corroborating evidence. The following solution employs only one of several strategies that you could apply. It is often not necessary to dwell in such depth on the information embedded in the individual spectra, and you may find it more useful first to identify the most important pieces, such as the molecular ion in the mass spectrum, the spectral regions and/or splitting patterns in the ^{1}H NMR spectrum, the number of ^{13}C signals, and a characteristic IR frequency or UV absorption. Whenever you are ready to formulate a (sub)structure, you should do so, and keep validating (or discarding) it as you analyze additional data. Trial and error is the key!

Solution

- The mass spectrum is already quite informative. There are not many hydrocarbons with a molecular ion of m/z 134 for which a sensible molecular formula can be derived. For example, the maximum number of carbons has to be less than C_{12} (mass 144, too large). If it were C_{11}, the molecular formula would have to be $C_{11}H_2$, ruled out by a quick look at the ^{1}H NMR spectrum, which shows more than two hydrogens in the molecule. (Besides, how many structures can you formulate with the composition $C_{11}H_2$?) The next choice, $C_{10}H_{14}$, looks good, especially since going down once again in the number of carbons to C_9 would require the presence of 26 hydrogens, an impossible proposition, because the maximum number of hydrogens is dictated by the generic formula C_nH_{2n+2} (Section 2-5).
- $C_{10}H_{14}$ means four degrees of unsaturation (Section 11-11), suggesting the presence of a combination of rings, double bonds, and triple bonds. The last option is less likely because of the absence of a band at ~2200 cm^{-1} in the IR spectrum (Section 13-3). This leaves us with double bonds and/or rings.
- A check of the ^{1}H NMR spectrum does not show signals for normal double bonds, but instead an aromatic peak. A benzene ring has the postulated four degrees of unsaturation. Hence we are dealing with an aromatic compound.

- Looking at the fragmentation pattern in the mass spectrum, the mass ion at m/z 119 points to the loss of a methyl group, which must therefore be present in our molecule.

  Turning to the ^{1}H NMR spectrum in more detail and remembering that we have 14 hydrogens in our structure, we can now probe how these hydrogens are distributed. The ^{1}H NMR spectrum reveals 4 H in the aromatic region, all very close in chemical shift, giving rise to a broad singlet. The remainder of the hydrogen signals appear to be due to alkyl groups: One of them is clearly a methyl substituent at $\delta = 2.28$ ppm; the others are part of a more complex array, containing two sets of mutually coupled nuclei. Closer inspection shows them to be a septet (indicating six equivalent neighbors) and a doublet (indicating a single hydrogen as a neighbor), pointing to the presence of a 1-methylethyl substituent (Table 10-5). The relatively large δ values for the single methyl and the tertiary hydrogen of the 1-methylethyl groups suggest that these nuclei are close to the aromatic ring (Table 10-2).

- Moving on to the ^{13}C NMR spectrum, the preceding analysis requires the presence of (at least) three sp^3-hybridized carbon peaks (Table 10-6). Indeed, there are three signals at $\delta = 21.3$, 24.2, and 38.9 ppm, and no additional peaks in this region. Instead, we see four resonances in the aromatic range. Since a benzene ring has six carbons, there must be some symmetry to the molecule.

- The IR spectrum reveals the presence of $C_{aromatic}$–H units ($\tilde{\nu} = 3030$ cm^{-1}), and the signal at $\tilde{\nu} = 813$ cm^{-1} indicates a para disubstituted benzene.

- Finally, the electronic spectrum shows the presence of a conjugated system, obviously a benzene ring.

- Putting all of this information together suggests a benzene ring with two substituents, a methyl and a 1-methylethyl group. The ^{13}C NMR spectrum rules out ortho or meta disubstitution, leaving only 1-methyl-4-(1-methylethyl)benzene as the solution (see margin).

1-Methyl-4-(1-methylethyl)benzene
(Common name cymene; constituent of a number of volatile oils from plants, such as cumin and thyme)

Exercise 15-8 | Try It Yourself

Deduce the structure of a compound $C_{16}H_{16}Br_2$ that exhibits the following spectral data: UV: λ_{max} (log ϵ) = 226sh(4.33), 235sh(4.55), 248(4.78), 261(4.43), 268(4.44), 277(4.36) nm (sh = shoulder); IR: $\tilde{\nu}$ = 614, 892, 1060, 1362, 1456, 2233, 2933, 2964, 3030 cm^{-1}; ^{1}H NMR (300 MHz): δ = 1.06 (t, J = 7.4 Hz, 6 H), 1.64 (sextet, J = 7.0, 4 H), 2.41 (t, J = 7.0 Hz, 4 H), 7.45 (s, 1 H), 7.75 (s, 1 H) ppm; ^{13}C NMR: δ = 13.51, 21.54, 21.89, 78.26, 96.68, 124.4, 125.2, 135.2, 136.8 ppm. (**Caution:** Look closely at the NMR spectra with reference to the molecular formula. Moreover, the IR spectrum contains important information. **Hint:** There is symmetry in the molecule.)

In Summary Benzene and its derivatives can be recognized and structurally characterized by their spectral data. Electronic absorptions take place between 250 and 290 nm. The infrared vibrational bands are found at 3030 cm^{-1} ($C_{aromatic}$–H), from 1500 to 2000 cm^{-1} (C–C), and from 650 to 1000 cm^{-1} (C–H out-of-plane bending). Most informative is NMR, with low-field resonances for the aromatic hydrogens and carbons. Coupling is largest between the ortho hydrogens, smaller in their meta and para counterparts.

15-5 | POLYCYCLIC AROMATIC HYDROCARBONS

When several benzene rings are fused together to give more extended π systems, the molecules are called **polycyclic benzenoid** or **polycyclic aromatic hydrocarbons (PAHs)**. In these structures, two or more benzene rings share two or more carbon atoms. Do these compounds also enjoy the special stability of benzene? The next two sections will show that they largely do.

There is no simple system for naming these structures, so we shall use their common names. The fusion of one benzene ring to another results in a compound called naphthalene. Further fusion can occur in a linear manner to give anthracene, tetracene, pentacene, and so on,

REAL LIFE: MATERIALS 15-1 | Compounds Made of Pure Carbon: Graphite, Graphene, Diamond, and Fullerenes

Elements can exist in several forms, called allotropes, depending on conditions and modes of synthesis. Thus, elemental carbon can arrange in more than 40 configurations, most of them amorphous (i.e., noncrystalline), such as coke (Sections 3-3 and 13-10), soot, carbon black (as used in printing ink), and activated carbon (as used in air and water filters). You probably know best two crystalline modifications of carbon: graphite and diamond. Graphite, the most stable carbon allotrope, is a completely fused polycyclic benzenoid π system, consisting of layers arranged in an open honeycomb pattern and 3.35 Å apart. The fully delocalized nature of these sheets (all carbons are sp^2 hybridized) gives rise to their black color and conductive capability. Graphite's lubricating property is the result of the ready mutual sliding of its component planes, greatly aided by air molecules (or other vapors) trapped between the layers. The "lead" of pencils is graphitic carbon, and the black pencil marks left on a sheet of paper consist of rubbed-off layers of the element.

In the colorless diamond, the carbon atoms (all sp^3 hybridized) form an insulating network of cross-linked cyclohexane chair conformers (see also Real Life 4-1). Diamond is the densest and hardest (least deformable) material known. It is also less stable than graphite, by 0.45 kcal/g C atom, and transforms into graphite at high temperatures or when subjected to high-energy radiation, a little-appreciated fact in the jewelry business.

A spectacular discovery was made in 1985 by Curl, Kroto, and Smalley* (for which they received the Nobel Prize in 1996): buckminsterfullerene, C_{60}, a new, spherical allotrope of carbon in the shape of a soccer ball. They found that laser evaporation of graphite generated a variety of carbon clusters in the gas phase, the most abundant of which contained 60 carbon atoms. The best way of assembling such a cluster while satisfying the tetravalency of carbon is to formally "roll up" 20 fused benzene rings and to connect the dangling valencies in such a way as to generate 12 pentagons: a so-called truncated icosahedron with 60 equivalent vertices—the shape of a soccer ball. The molecule was named after Buckminster Fuller[†] because its shape is reminiscent of the "geodesic domes" designed by him. It is soluble in organic solvents, and the ^{13}C NMR spectrum shows a single line at $\delta = 142.7$ ppm, in the expected range (Sections 15-4 and 15-6). Because of its curvature, the constituent benzene rings in C_{60} are strained and the energy content relative to graphite is 10.16 kcal/g C atom. This strain is manifested in a rich chemistry, including electrophilic, nucleophilic, radical, and concerted addition reactions (Chapter 14). The enormous interest spurred by the discovery of C_{60} rapidly led to a number of exciting developments, such as the design of multikilogram synthetic methods (commercial material sells for less than $1 per gram); the isolation of many other larger carbon clusters, dubbed "fullerenes," such as the rugby-ball–shaped C_{70}; chiral systems (e.g., as in C_{84}); isomeric forms; fullerenes encaging host atoms, such as He and metal nuclei ("endohedral fullerenes"); the synthesis of conducting salts (e.g., Cs_3C_{60}, which becomes superconducting at 40 K); and medical applications (e.g., C_{60} inhibits the HIV virus). Moreover, reexamination of the older literature and newer studies have revealed that C_{60} and other fullerenes are produced simply upon incomplete combustion of organic matter under certain conditions or by varied heat treatments of soot and therefore have probably been "natural products" on our planet since early in its formation.

From a materials point of view, perhaps most useful has been the synthesis of graphitic tubules, so-called nanotubes, based on the fullerene motif. Nanotubes are even

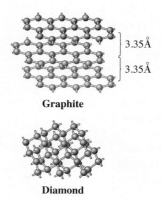

Graphite

Diamond

Buckminsterfullerene C_{60} C_{70} Chiral C_{84}

*Professor Robert F. Curl (b. 1933), Rice University, Houston, Texas; Professor Harold W. Kroto (b. 1939), University of Sussex, England; Professor Richard E. Smalley (1943–2005), Rice University, Houston, Texas.

[†]Richard Buckminster Fuller (1895–1983), American architect, inventor, and philosopher.

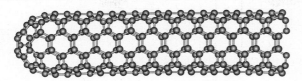

Carbon nanotube

harder than diamond, yet elastic, and show unusual magnetic and electrical (metallic) properties. There is the real prospect that nanotubes may replace the computer chip as we currently know it in the manufacture of a new generation of faster and smaller computers (see also Real Life 14-1). Nanotubes also function as a molecular "packaging material" for other structures, such as metal catalysts and even biomolecules. Thus, carbon in the fullerene modification has taken center stage in the new field of nanotechnology, aimed at the construction of devices at the molecular level.

Coming full circle in the breathtakingly fast evolution of carbon research, another profound advance was made in 2004 by Geim and Novoselov,* when they discovered that single sheets of graphite, called graphene, could be literally peeled off bulk graphite by using Scotch tape! The bond length in graphene is 1.42 Å. It absorbs 2.3% of incident light and is therefore visible to the naked eye. When damaged, it self-repairs upon exposure to hydrocarbons, a testament to the extraordinary aromaticity of the polybenzenoid array. Apart from its potential applications in optoelectronics, graphene has remarkable mechanical properties: It is 200 times stronger than steel, in addition to being lighter, harder, more flexible, and readily and completely recyclable. Stay tuned for the next breakthrough. . . .

An example of a geodesic dome (of the type whose design was pioneered by Buckminster Fuller) forms part of the entrance to EPCOT Center, Disney World, Florida. It is 180 feet high with a diameter of 165 feet.

How to obtain a Nobel Prize: peeling off graphene from graphite using Scotch tape.

*Professor Andre K. Geim (b. 1958), University of Manchester, England, Nobel Prize 2010 (physics); Professor Kostya S. Novoselov, (b. 1974), University of Manchester, England, Nobel Prize 2010 (physics).

a series called the **acenes. Angular fusion** ("annulation") results in phenanthrene, which can be further annulated to a variety of other benzenoid polycycles.

Naphthalene

Anthracene

Tetracene
(Naphthacene)

Phenanthrene

Model Building

Each structure has its own numbering system around the periphery. A quaternary carbon is given the number of the preceding carbon in the sequence followed by the letters *a*, *b*, and so on, depending on how close it is to that carbon.

Exercise 15-9

Name the following compounds or draw their structures.

(a) 2,6-Dimethylnaphthalene **(b)** 1-Bromo-6-nitrophenanthrene **(c)** 9,10-Diphenylanthracene

(d)

(e)

The ferocious Formosan termite first appeared in the United States in the 1960s. It produces naphthalene as a poison for chemical warfare.

Naphthalene is aromatic: a look at spectra

In contrast with benzene, which is a liquid, naphthalene is a colorless crystalline material with a melting point of 80°C. It is probably best known as a moth repellent and insecticide, although in these capacities it has been partly replaced by chlorinated compounds such as 1,4-dichlorobenzene (*p*-dichlorobenzene).

The spectral properties of naphthalene strongly suggest that it shares benzene's delocalized electronic structure and thermodynamic stability. The ultraviolet and NMR spectra are particularly revealing. The ultraviolet spectrum of naphthalene (Figure 15-13) shows a pattern typical of an extended conjugated system, with peaks at wavelengths as long as 320 nm. On the basis of this observation, we conclude that the electrons are delocalized more extensively than in benzene (Section 15-2 and Figure 15-6). Thus, the added four π electrons enter into efficient overlap with those of the attached benzene ring. In fact, it is possible to draw several resonance forms.

Resonance Forms of Naphthalene

Alternatively, the continuous overlap of the 10 *p* orbitals and the fairly even distribution of electron density can be shown as in Figure 15-14.

According to these representations, the structure of naphthalene should be symmetric, with planar and almost hexagonal benzene rings and two perpendicular mirror planes bisecting the

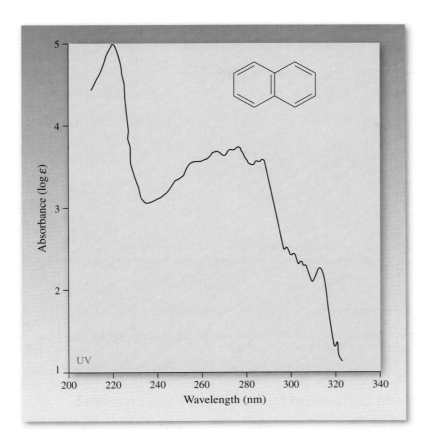

Figure 15-13 Extended π conjugation in naphthalene is manifest in its UV spectrum (measured in 95% ethanol). The complexity and location of the absorptions are typical of extended π systems.

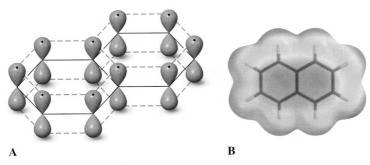

A   **B**

Figure 15-14 (A) Orbital picture of naphthalene, showing its extended overlap of p orbitals. (B) Electrostatic potential map, revealing its electron density distribution over the 10 carbon atoms.

molecule. X-ray crystallographic measurements confirm this prediction (Figure 15-15). The C–C bonds deviate only slightly in length from those in benzene (1.39 Å), and they are clearly different from pure single (1.54 Å) and double bonds (1.33 Å).

Further evidence of aromaticity is found in the ^{1}H NMR spectrum of naphthalene (Figure 15-16). Two symmetric multiplets can be observed at $\delta = 7.49$ and 7.86 ppm, characteristic of aromatic hydrogens deshielded by the ring-current effect of the π-electron loop (see Section 15-4, Figure 15-9). Coupling in the naphthalene nucleus is very similar to that in substituted benzenes: $J_{ortho} = 7.5$ Hz, $J_{meta} = 1.4$ Hz, and $J_{para} = 0.7$ Hz. The ^{13}C NMR spectrum shows three lines with chemical shifts that are in the range of other benzene derivatives (see margin). Thus, on the basis of structural and spectral criteria, naphthalene is aromatic.

Figure 15-15 The molecular structure of naphthalene. The bond angles within the rings are 120°.

1.42 Å
1.37 Å
1.39 Å   1.40 Å
121°

^{13}C NMR Data of Naphthalene (ppm)

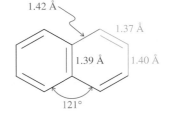

134.4   128.5   126.5

Exercise 15-10

A substituted naphthalene, $C_{10}H_8O_2$, gave the following spectral data: ^{1}H NMR: $\delta = 5.20$ (broad s, 2 H), 6.92 (dd, $J = 7.5$ Hz and 1.4 Hz, 2 H), 7.00 (d, $J = 1.4$ Hz, 2 H), and 7.60 (d, $J = 7.5$ Hz, 2 H) ppm; ^{13}C NMR: $\delta = 107.5$, 115.3, 123.0, 129.3, 136.8, and 155.8 ppm; IR: $\tilde{\nu} = 3300$ cm^{-1} (broad). What is its structure? [**Hints:** Review the values for $J_{ortho,\,meta,\,para}$ in benzene (Section 15-4). The number of NMR peaks is less than maximum.]

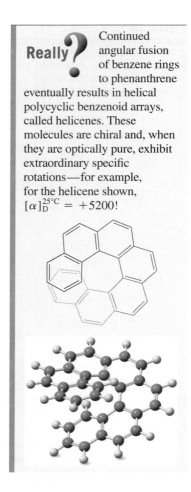

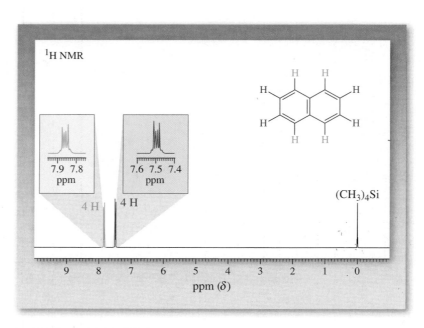

Figure 15-16 The 300-MHz ^{1}H NMR spectrum of naphthalene reveals the characteristic deshielding due to a π-electronic ring current.

Most fused benzenoid hydrocarbons are aromatic

These aromatic properties of naphthalene hold for most of the other polycyclic benzenoid hydrocarbons. It appears that the cyclic delocalization in the individual benzene rings is not significantly perturbed by the fact that they have to share at least one π bond. Linear and angular fusion of a third benzene ring onto naphthalene result in the systems anthracene and phenanthrene. Although isomeric and seemingly very similar, they have different thermodynamic stabilities: Anthracene is about 6 kcal mol^{-1} (25 kJ mol^{-1}) less stable than phenanthrene, even though both are aromatic. Enumeration of the various resonance forms of the molecules explains why. Anthracene has only four, and only two contain two fully aromatic benzene rings (red in the structures shown here). Phenanthrene has five, three of which incorporate two aromatic benzenes, one even three benzenes.

Resonance in Anthracene

Resonance in Phenanthrene

Exercise 15-11

Draw all the possible resonance forms of tetracene (naphthacene, Section 15-5). What is the maximum number of completely aromatic benzene rings in these structures?

In Summary The physical properties of naphthalene are typical of an aromatic system. Its UV spectrum reveals extensive delocalization of all π electrons, its molecular structure shows bond lengths and bond angles very similar to those in benzene, and its ^{1}H NMR spectrum reveals deshielded ring hydrogens indicative of an aromatic ring current. Other polycyclic benzenoid hydrocarbons have similar properties and are considered aromatic.

15-6 OTHER CYCLIC POLYENES: HÜCKEL'S RULE

The special stability and reactivity associated with cyclic delocalization is not unique to benzene and polycyclic benzenoids. Thus, we shall see that other cyclic conjugated polyenes can be aromatic, but only if they contain $(4n + 2)$ π electrons ($n = 0, 1, 2, 3, \ldots$). In contrast, $4n$ π circuits may be *destabilized* by conjugation, or are **antiaromatic.** This pattern is known as **Hückel's* rule.** Nonplanar systems in which cyclic overlap is disrupted sufficiently to impart alkene-like properties are classified as **nonaromatic.** Let us look at some members of this series, starting with 1,3-cyclobutadiene.

1,3-Cyclobutadiene, the smallest cyclic polyene, is antiaromatic

1,3-Cyclobutadiene, a $4n$ π system ($n = 1$), is an air-sensitive and extremely reactive molecule in comparison to its analogs 1,3-butadiene and cyclobutene. Not only does the molecule have none of the attributes of an aromatic molecule like benzene, it is actually destabilized through π overlap by more than 35 kcal mol^{-1} (146 kJ mol^{-1}) and therefore is antiaromatic. As a consequence, its structure is rectangular, and the two diene forms represent *isomers,* equilibrating through a symmetrical transition state, rather than *resonance forms.*

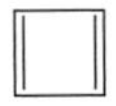

Cyclobutadiene: antiaromatic

1,3-Cyclobutadiene Is Unsymmetrical

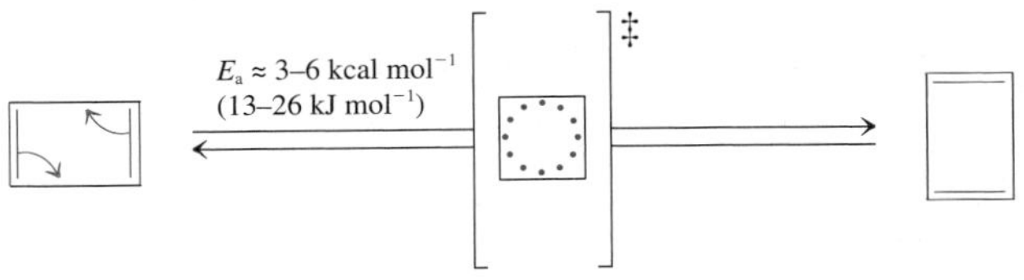

$E_a \approx 3\text{–}6$ kcal mol^{-1}
(13–26 kJ mol^{-1})

Transition state

Free 1,3-cyclobutadiene can be prepared and observed only at very low temperatures. The reactivity of cyclobutadiene can be seen in its rapid Diels-Alder reactions, in which it can act as both diene (shown in red) and dienophile (blue).

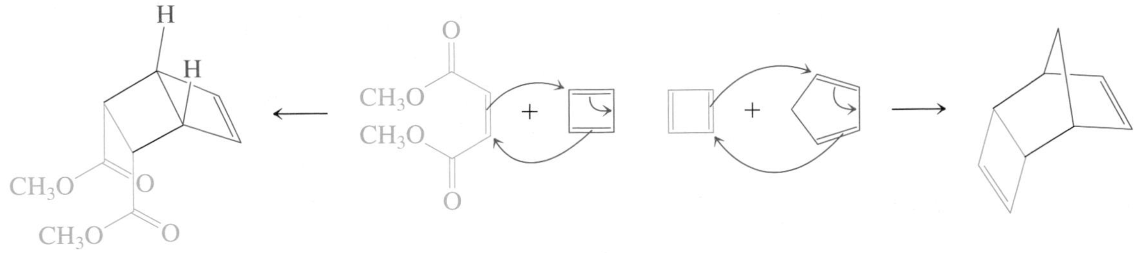

Model Building

---

*Professor Erich Hückel (1896–1984), University of Marburg, Germany.

Exercise 15-12

1,3-Cyclobutadiene dimerizes at temperatures as low as −200°C to give the two products shown. Explain mechanistically.

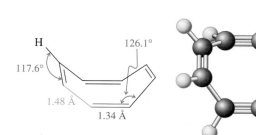

Substituted cyclobutadienes are less reactive because of steric protection, particularly if the substituents are bulky; they have been used to probe the spectroscopic features of the cyclic system of four π electrons. For example, in the ¹H NMR spectrum of 1,2,3-tris(1,1-dimethylethyl)cyclobutadiene (1,2,3-tri-*tert*-butylcyclobutadiene), the ring hydrogen resonates at δ = 5.38 ppm, at much higher field than expected for an aromatic system. This and other properties of cyclobutadiene show that it is quite unlike benzene.

1,3,5,7-Cyclooctatetraene is nonplanar and nonaromatic

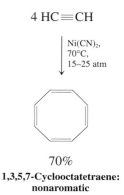

4 HC≡CH

Ni(CN)₂,
70°C,
15–25 atm

70%
**1,3,5,7-Cyclooctatetraene:
nonaromatic**

Let us now examine the properties of the next *higher* cyclic polyene analog of benzene, 1,3,5,7-cyclooctatetraene, another $4n$ π cycle (n = 2). Is it antiaromatic, like 1,3-cyclobutadiene? First prepared in 1911 by Willstätter,* this substance is now readily available from a special reaction, the nickel-catalyzed cyclotetramerization of ethyne. It is a yellow liquid (b.p. 152 °C) that is stable if kept cold but that polymerizes when heated. It is oxidized by air, catalytically hydrogenated to cyclooctane, and subject to electrophilic additions and to cycloaddition reactions. This chemical reactivity is diagnostic of a normal polyene (Section 14-7).

Spectral and structural data confirm the ordinary alkene nature of cyclooctatetraene. Thus, the ¹H NMR spectrum shows a sharp singlet at δ = 5.68 ppm, typical of an alkene. The molecular-structure determination reveals that cyclooctatetraene is actually nonplanar and tub-shaped (Figure 15-17). The double bonds are nearly orthogonal (perpendicular) and not conjugated. Conclusion: The molecule is *nonaromatic*.

Exercise 15-13

On the basis of the molecular structure of 1,3,5,7-cyclooctatetraene, would you describe its double bonds as conjugated (i.e., does it exhibit extended π overlap)? Would it be correct to draw two resonance forms for this molecule, as we do for benzene? (**Hint:** Build molecular models of the two forms of 1,2-dimethylcyclooctatetraene.)

Figure 15-17 The molecular structure of 1,3,5,7-cyclooctatetraene. Note the alternating single and double bonds of this nonplanar, nonaromatic molecule.

H   126.1°
117.6°
1.48 Å
1.34 Å

*Professor Richard Willstätter (1872–1942), Technical University, Munich, Germany, Nobel Prize 1915 (chemistry).

Exercise 15-14

Cyclooctatetraene A exists in equilibrium with less than 0.05% of a bicychlic isomer B, which is trapped by Diels-Alder cycloaddition to 2-butenedioic anhydride (maleic anhydride, Table 14-1) to give compound C.

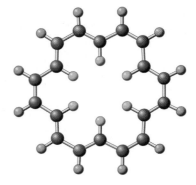

A   C

What is isomer B? Show a mechanism for the A → B → C interconversion. (**Hint:** Work backward from C to B and review Section 14-9.)

Only cyclic conjugated polyenes containing (4n + 2) pi electrons are aromatic

Unlike cyclobutadiene and cyclooctatetraene, certain higher cyclic conjugated polyenes are aromatic. All of them have two properties in common: They contain $(4n + 2)$ π electrons, and they are sufficiently planar to allow for delocalization.

The first such system was prepared in 1956 by Sondheimer*; it was 1,3,5,7,9,11,13,15,17-cyclooctadecanonaene, containing 18 π electrons $(4n + 2, n = 4)$. To avoid the use of such cumbersome names, Sondheimer introduced a simpler system for naming cyclic conjugated polyenes. He named completely conjugated monocyclic hydrocarbons $(CH)_N$ as **[N]annulenes,** in which N denotes the ring size. Thus, cyclobutadiene would be called [4]annulene; benzene, [6]annulene; cyclooctatetraene, [8]annulene. The first almost unstrained aromatic system in the series after benzene is [18]annulene.

[18]Annulene
(1,3,5,7,9,11,13,15,17-Cyclooctadecanonaene)

Exercise 15-15

The three isomers of [10]annulene shown here have been prepared and exhibit nonaromatic behavior. Why? (**Hint:** Build models!)

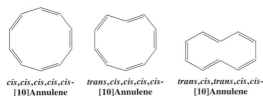

cis,cis,cis,cis,cis-
[10]Annulene

trans,cis,cis,cis,cis-
[10]Annulene

trans,cis,trans,cis,cis-
[10]Annulene

*Professor Franz Sondheimer (1926–1981), University College, London.

[18]Annulene: aromatic

[18]Annulene is fairly planar and shows little alternation of the single and double bonds. The extent of the delocalization of its π electrons is pictured in the electrostatic potential map in the margin. Like benzene, it can be described by a set of two equivalent resonance forms. In accord with its aromatic character, the molecule is relatively stable and undergoes electrophilic aromatic substitution. It also exhibits a benzene-like ring-current effect in its ^{1}H NMR spectrum (see Problems 64 and 65).

Since the preparation of [18]annulene, many other annulenes have been made. As long as they are (nearly) planar and delocalized, those with $(4n + 2)$ π electrons, such as benzene and [18]annulene, are aromatic, whereas those with $4n$ π electrons, such as cyclobutadiene and [16]annulene, are antiaromatic. When cyclic delocalization is prohibited by angle or steric strain, such as in cyclooctatetraene or [10]annulene (Exercise 15-15), the systems are nonaromatic. Of course, cyclic polyenes in which there is no contiguous array of p orbitals are not annulenes and therefore also nonaromatic.

The Annulenes and Other Cyclic Polyenes

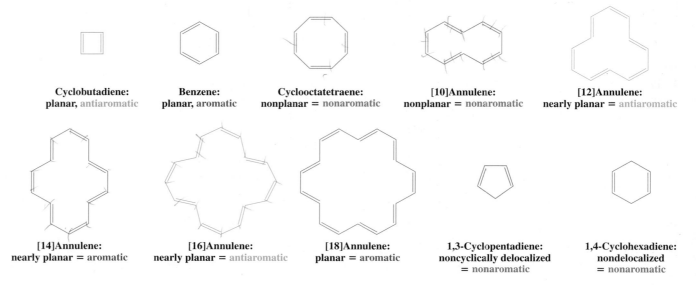

| Cyclobutadiene: planar, antiaromatic | Benzene: planar, aromatic | Cyclooctatetraene: nonplanar = nonaromatic | [10]Annulene: nonplanar = nonaromatic | [12]Annulene: nearly planar = antiaromatic |

| [14]Annulene: nearly planar = aromatic | [16]Annulene: nearly planar = antiaromatic | [18]Annulene: planar = aromatic | 1,3-Cyclopentadiene: noncyclically delocalized = nonaromatic | 1,4-Cyclohexadiene: nondelocalized = nonaromatic |

The alternating behavior of the annulenes between aromatic and antiaromatic had been predicted earlier by the theoretical chemist Hückel, who formulated this $(4n + 2)$ rule in 1931. Hückel's rule expresses the regular molecular-orbital patterns calculated for planar, cyclic conjugated polyenes. The p orbitals mix to give an equal number of π molecular orbitals, as shown in Figure 15-18. For example, the four p orbitals of cyclobutadiene result

Figure 15-18 (A) Hückel's $4n + 2$ rule is based on the regular pattern of the π molecular orbitals in cyclic conjugated polyenes. They are equally spaced and degenerate, except for the highest and lowest ones, which are unique. (B) Molecular-orbital levels in 1,3-cyclobutadiene. Four π electrons are not enough to result in a closed shell (in other words, doubly filled molecular orbitals), so the molecule is not aromatic. (C) The six π electrons in benzene produce a closed-shell configuration, so benzene is aromatic.

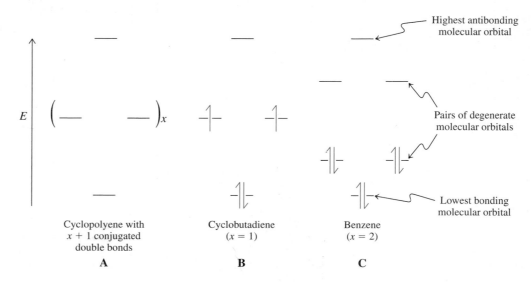

in four molecular orbitals, the six p orbitals of benzene in six molecular orbitals, and so on. These orbitals occur in symmetrically disposed degenerate pairs, except for the lowest bonding and highest antibonding ones, which are unique. Thus, cyclobutadiene has one such degenerate pair of molecular orbitals, benzene has two, and so on. A closed-shell system is possible only if all bonding molecular orbitals are occupied (see Section 1-7), that is, only if there are $(4n + 2)\pi$ electrons. On the other hand, $4n\ \pi$ cycles always contain a pair of singly occupied orbitals, an unfavorable electronic arrangement.

Exercise 15-16

On the basis of Hückel's rule, label the following molecules as aromatic or antiaromatic. **(a)** [30]Annulene; **(b)** [20]annulene; **(c)** *trans*-15,16-dihydropyrene; **(d)** the deep blue (see Figure 14-17) azulene; **(e)** S-indacene.

trans-**15,16-Dihydropyrene**   **Azulene**   **S-indacene**

In Summary Cyclic conjugated polyenes are aromatic if their π electron count is $4n + 2$. This number corresponds to a completely filled set of bonding molecular orbitals. Conversely, $4n\ \pi$ systems have open-shell, antiaromatic structures that are unstable, are reactive, and lack aromatic ring-current effects in ^{1}H NMR. Finally, when steric constraints impose nonplanarity, cyclic polyenes behave as nonaromatic alkenes.

15-7 | HÜCKEL'S RULE AND CHARGED MOLECULES

Hückel's rule also applies to charged molecules, as long as cyclic delocalization can occur. Their aromaticity is reflected in relative thermodynamic and kinetic stability, the observation of ring currents in the NMR experiment, and the absence of bond alternation in crystal structures. This section shows how charged aromatic systems can be prepared.

The cyclopentadienyl anion and the cycloheptatrienyl cation are aromatic

1,3-Cyclopentadiene is unusually acidic [$pK_a \approx 16$; comparable to alcohols (Section 8-3)] because the cyclopentadienyl anion resulting from deprotonation contains a delocalized, aromatic system of six π electrons. The negative charge is equally distributed over all five carbon atoms. For comparison, the pK_a of propene is 40. An electrostatic potential map of the molecule is shown in the margin, on a scale that attenuates the otherwise overwhelming effect of the negative charge.

Cyclopentadienyl anion is aromatic: six π electrons cyclically delocalized over *five* carbon atoms.

Aromatic Cyclopentadienyl Anion

$pK_a = 16$

Cycloheptatrienyl cation is aromatic: six π electrons cyclically delocalized over *seven* carbon atoms.

In contrast, the cyclopentadienyl cation, a system of four π electrons, can be produced only at low temperature and is extremely reactive.

When 1,3,5-cycloheptatriene is treated with bromine, a stable salt is formed, cycloheptatrienyl bromide. In this molecule, the organic cation contains six delocalized π electrons, and the positive charge is equally distributed over seven carbons (as shown in the electrostatic potential map in the margin). Even though it is a carbocation, the system is remarkably unreactive, as is expected for an aromatic system. In contrast, the cycloheptatrienyl anion is antiaromatic, as indicated by the much lower acidity of cycloheptatriene ($pK_a = 39$) compared with that of cyclopentadiene.

Aromatic Cycloheptatrienyl Cation

Exercise 15-17

Draw an orbital picture of **(a)** the cyclopentadienyl anion and **(b)** the cycloheptatrienyl cation (consult Figure 15-2).

Exercise 15-18

The rate of solvolysis of compound A in 2,2,2-trifluoroethanol at 25°C exceeds that of compound B by a factor of 10^{14}. Explain.

A   **B**

Exercise 15-19

On the basis of Hückel's rule, label the following molecules aromatic or antiaromatic.

(a) Cyclopropenyl cation; **(b)** cyclononatetraenyl anion; **(c)** cycloundecapentaenyl anion.

Nonaromatic cyclic polyenes can form aromatic dianions and dications

Cyclic systems of $4n$ π electrons can be converted into their aromatic counterparts by two-electron oxidations and reductions. For example, cyclooctatetraene is reduced by alkali metals to the corresponding aromatic dianion. This species is planar, contains fully delocalized electrons, and is relatively stable. It also exhibits an aromatic ring current in ^{1}H NMR.

Nonaromatic Cyclooctatetraene Forms an Aromatic Dianion

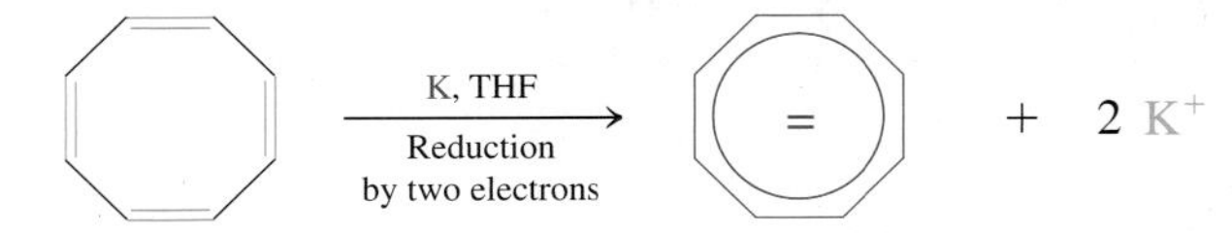

Eight π electrons,
nonaromatic

K, THF
Reduction
by two electrons

Ten π electrons,
aromatic

$+ \; 2 \; K^+$

Similarly, [16]annulene can be either reduced to its dianion or oxidized to its dication, both products being aromatic. On formation of the dication, the configuration of the molecule changes.

Aromatic [16]Annulene Dication and Dianion from Antiaromatic [16]Annulene

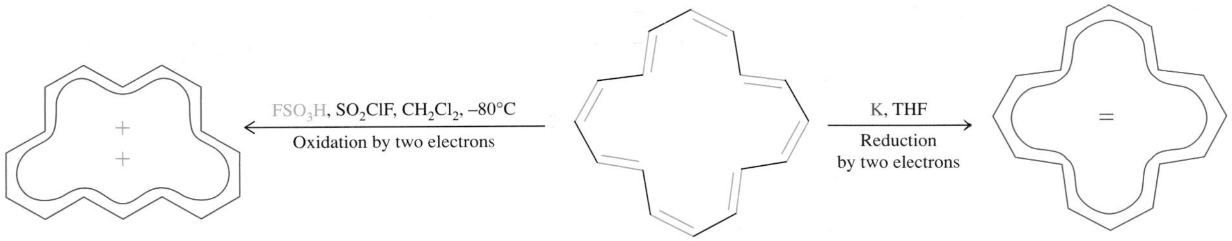

Fourteen π electrons,
aromatic

FSO₃H, SO₂ClF, CH₂Cl₂, −80°C
Oxidation by two electrons

[16]Annulene
Sixteen π electrons,
antiaromatic

K, THF
Reduction
by two electrons

Eighteen π electrons,
aromatic

Solved Exercise 15-20 | Working with the Concepts: Recognizing Aromaticity in Charged Molecules

Azulene [see Exercise 15-16, part (d)] is readily attacked by electrophiles at C1, by nucleophiles at C4. Explain.

Strategy
We need to formulate first the various resonance forms of the species resulting from these two modes of attack. Inspection of these structures might then provide the answer.

Solution
- Attack by electrophiles at C1:

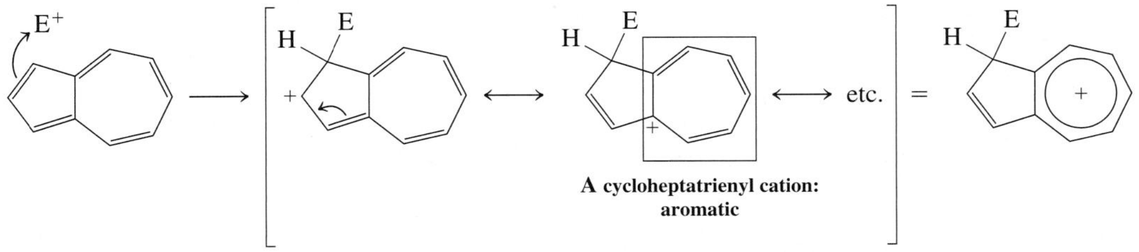

A cycloheptatrienyl cation:
aromatic

Attack by E^+ at C1 generates a fused, aromatic cycloheptatrienyl cation framework.
- Attack by nucleophiles at C4:

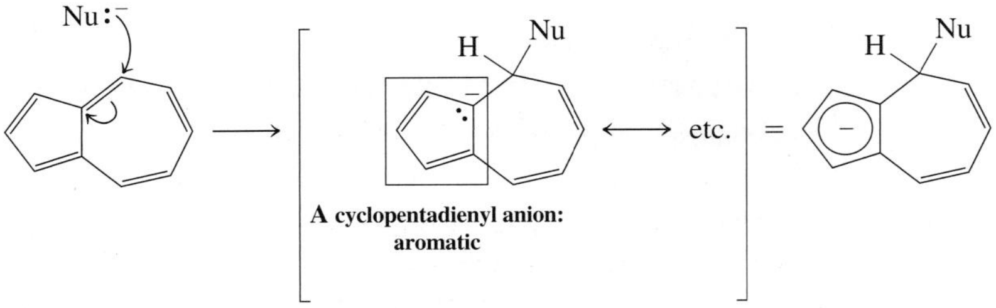

A cyclopentadienyl anion:
aromatic

Attack by Nu^- at C4 generates a fused, aromatic cyclopentadienyl anion framework.

Exercise 15-21 | Try It Yourself

The triene A can be readily deprotonated twice to give the stable dianion B. However, the neutral analog of B, the tetraene C (pentalene), is extremely unstable. Explain.

A   B   C

In Summary Charged species may be aromatic, provided they exhibit cyclic delocalization and obey the $4n + 2$ rule.

15-8 SYNTHESIS OF BENZENE DERIVATIVES: ELECTROPHILIC AROMATIC SUBSTITUTION

In this section we begin to explore the reactivity of benzene, the prototypical aromatic compound. The aromatic stability of benzene makes it relatively unreactive, despite the presence of three formal double bonds. As a result, its chemical transformations require forcing conditions and proceed through new pathways. Not surprisingly, however, most of the chemistry of benzene features attack by electrophiles. We shall see in Section 22-4 that attack by nucleophiles is rare but possible, provided that a suitable leaving group is present.

Benzene undergoes substitution reactions with electrophiles

Benzene is attacked by electrophiles, but, in contrast to the corresponding reactions of alkenes, this reaction results in *substitution* of hydrogens—**electrophilic aromatic substitution**—*not addition* to the ring.

Reaction

Electrophilic Aromatic Substitution

$$+ \quad E^+X^- \quad \longrightarrow \quad + \quad H^+X^-$$

Electrophile

Mechanism

Under the conditions employed for these processes, nonaromatic conjugated polyenes would rapidly polymerize. However, the stability of the benzene ring allows it to survive. Let us begin with the general mechanism of electrophilic aromatic substitution.

Electrophilic aromatic substitution in benzene proceeds by addition of the electrophile followed by proton loss

The mechanism of electrophilic aromatic substitution has two steps. First, the electrophile E^+ attacks the benzene nucleus, much as it would attack an ordinary double bond. The resonance stabilized cationic intermediate thus formed then loses a proton to regenerate the aromatic ring. Note two important points in the formulation of this general mechanism. First, always show the hydrogen at the site of the initial electrophilic attack. Second, the positive charge in the resulting cation is indicated by three resonance forms and is located ortho and para to the carbon that has been attacked, a result of the rules for drawing resonance forms (Sections 1-5 and 14-1).

Mechanism of Electrophilic Aromatic Substitution

Step 1. Electrophilic attack

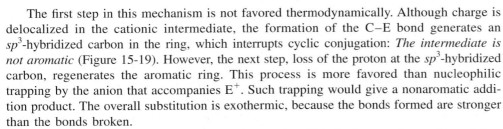

Step 2. Proton loss

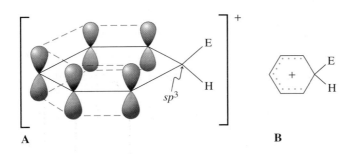

The first step in this mechanism is not favored thermodynamically. Although charge is delocalized in the cationic intermediate, the formation of the C–E bond generates an sp^3-hybridized carbon in the ring, which interrupts cyclic conjugation: *The intermediate is not aromatic* (Figure 15-19). However, the next step, loss of the proton at the sp^3-hybridized carbon, regenerates the aromatic ring. This process is more favored than nucleophilic trapping by the anion that accompanies E^+. Such trapping would give a nonaromatic addition product. The overall substitution is exothermic, because the bonds formed are stronger than the bonds broken.

Figure 15-20 depicts a potential energy diagram in which the first step is rate determining, a kinetic finding that applies to most electrophiles that we shall encounter. The subsequent

Figure 15-19 (A) Orbital picture of the cationic intermediate resulting from attack by an electrophile on the benzene ring. Aromaticity is lost because cyclic conjugation is interrupted by the sp^3-hybridized carbon. The four electrons in the π system are not shown. (B) Dotted-line notation to indicate delocalized nature of the charge in the cation.

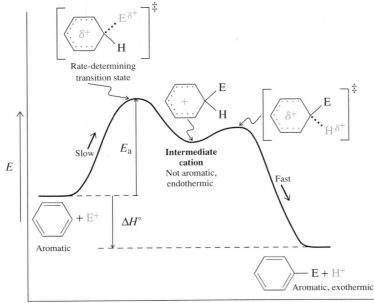

Reminder

Free H^+ does not exist in solution but is attached to any available electron pair. In electrophilic aromatic substitution, it may be those located on the counterion to E^+, other additives, or solvent.

Figure 15-20 Potential-energy diagram describing the course of the reaction of benzene with an electrophile. The first transition state is rate determining. Proton loss is relatively fast. The overall rate of the reaction is controlled by E_a; the amount of exothermic energy released is given by $\Delta H°$.

loss of proton is much faster than initial electrophilic attack because it leads to the aromatic product in an exothermic step, which furnishes the driving force for the overall sequence.

The following sections look more closely at the most common reagents employed in this transformation and the details of the mechanism.

Exercise 15-22

We learned in Section 11-5 that the $\Delta H°$ of the hydrogenation of *cis*-2-butene is -28.6 kcal mol^{-1}. Taking this system as a model for a double bond in benzene, estimate the corresponding $\Delta H°$ of the hydrogenation of benzene to 1,3-cyclohexadiene. What is the difference between these two double bonds? (**Hint:** Consult Figure 15-3.)

In Summary The general mechanism of electrophilic aromatic substitution begins with electrophilic attack by E^+ to give an intermediate, charge-delocalized but nonaromatic cation in a rate-determining step. Subsequent fast proton loss regenerates the (now substituted) aromatic ring.

15-9 | HALOGENATION OF BENZENE: THE NEED FOR A CATALYST

An example of electrophilic aromatic substitution is halogenation. Benzene is normally unreactive in the presence of halogens, because halogens are not electrophilic enough to disrupt its aromaticity. However, the halogen may be activated by Lewis acidic catalysts, such as ferric halides (FeX$_3$) or aluminum halides (AlX$_3$), to become a much more powerful electrophile.

How does this activation work? Lewis acids have the ability to accept electron pairs. When a halogen such as bromine is exposed to FeBr$_3$, the two molecules combine in a Lewis acid-base reaction.

Activation of Bromine by the Lewis Acid FeBr$_3$

$$:\overset{..}{\underset{..}{Br}}-\overset{..}{\underset{..}{Br}}: \rightharpoonup FeBr_3 \longrightarrow \left[:\overset{..}{\underset{..}{Br}}\overset{+}{\underset{..}{\overset{..}{Br}}}-\bar{F}eBr_3 \longleftrightarrow \overset{..}{\underset{..}{Br}}=\overset{+}{\underset{..}{Br}}-\bar{F}eBr_3 \right]$$

In this complex, the Br–Br bond is polarized, thereby imparting electrophilic character to the bromine atoms. Electrophilic attack on benzene is at the terminal bromine, allowing the other bromine atom to depart with the good leaving group FeBr$_4^-$. In terms of electron flow, you can also view this process as a nucleophilic substitution of [Br$_2$FeBr$_3$] by the benzene double bond, not unlike an S$_N$2 reaction (see also Figure 15-21).

Electrophilic Attack on Benzene by Activated Bromine

The FeBr$_4^-$ formed in this step now functions as a base, abstracting a proton from the cyclohexadienyl cation intermediate. This transformation not only furnishes the two products of the reaction, bromobenzene and hydrogen bromide, but also regenerates the FeBr$_3$ catalyst.

Bromobenzene Formation

Sidebar (left margin)

Bromination of Benzene

Br$_2$, FeBr$_3$

Br

Bromobenzene

Reaction

Mechanism

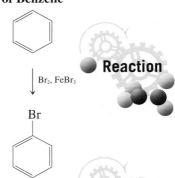

Figure 15-21 X-ray structure of the carbocation arising from attack of bromine on hexamethylbenzene. This cation can be isolated because it is unusually stabilized by the methyl substituents and because it lacks a proton on the sp^3-hybridized carbon, thus disabling bromoarene formation.

A quick calculation confirms that electrophilic bromination of benzene is exothermic. A phenyl–hydrogen bond (approximately 112 kcal mol^{-1}, Table 15-1) and a bromine molecule (46 kcal mol^{-1}) are lost in the process. Counterbalancing this loss is the formation of a phenyl–bromine bond ($DH° = 81$ kcal mol^{-1}) and an H–Br bond ($DH° = 87.5$ kcal mol^{-1}). Thus, the overall reaction is exothermic by $158 - 168.5 = -10.5$ kcal mol^{-1} (43.9 kJ mol^{-1}).

As in the radical halogenation of alkanes (Section 3-8), the exothermic nature of aromatic halogenation decreases down the periodic table. Fluorination is so exothermic that direct reaction of fluorine with benzene is explosive. Chlorination, on the other hand, is controllable but requires the presence of an activating catalyst, such as aluminum chloride or ferric chloride. The mechanism of this reaction is identical with that of bromination. Finally, electrophilic iodination with iodine is endothermic and thus not normally possible. Much like the radical halogenation of alkanes, electrophilic chlorination and bromination of benzene (and substituted benzenes, Chapter 16) introduces functionality that can be utilized in further reactions, in particular C–C bond formations through organometallic reagents (see Problem 54, Section 13-9, and Real Life 13-1).

Table 15-1		Dissociation Energies [$DH°$] of Bonds A–B [kcal mol^{-1} (kJ mol^{-1})]
A	**B**	**$DH°$**
F	F	38 (159)
Cl	Cl	58 (243)
Br	Br	46 (192)
I	I	36 (151)
F	C$_6$H$_5$	126 (527)
Cl	C$_6$H$_5$	96 (402)
Br	C$_6$H$_5$	81 (339)
I	C$_6$H$_5$	65 (272)
C$_6$H$_5$	H	112 (469)
F	H	135.8 (568)
Cl	H	103.2 (432)
Br	H	87.5 (366)
I	H	71.3 (298)

Exercise 15-23

When benzene is dissolved in D$_2$SO$_4$, its ^{1}H NMR absorption at $\delta = 7.27$ ppm disappears and a new compound is formed having a molecular weight of 84. What is it? Propose a mechanism for its formation. (**Caution:** In all mechanisms of electrophilic aromatic substitution, always draw the H atom at the site of electrophilic attack.)

Exercise 15-24

Professor G. Olah* and his colleagues exposed benzene to the especially strong acid system HF–SbF$_5$ in an NMR tube and observed a new ^{1}H NMR spectrum with absorptions at $\delta = 5.69$ (2 H), 8.22 (2 H), 9.42 (1 H), and 9.58 (2 H) ppm. Propose a structure for this species.

In Summary The halogenation of benzene becomes more exothermic as we proceed from I$_2$ (endothermic) to F$_2$ (exothermic and explosive). Chlorinations and brominations are achieved with the help of Lewis acid catalysts that polarize the X–X bond and activate the halogen by increasing its electrophilic power.

15-10 | NITRATION AND SULFONATION OF BENZENE

In two other typical electrophilic substitutions of benzene, the electrophiles are the nitronium ion (NO$_2^+$), leading to nitrobenzene, and sulfur trioxide (SO$_3$), giving benzenesulfonic acid.

Benzene is subject to electrophilic attack by the nitronium ion

To bring about nitration of the benzene ring at moderate temperatures, it is not sufficient just to treat benzene with concentrated nitric acid. Because the nitrogen in HNO$_3$ has little electrophilic power, it must somehow be activated. Addition of concentrated sulfuric acid serves this purpose by protonating the nitric acid. Loss of water then yields the **nitronium**

Reaction

Nitration of Benzene

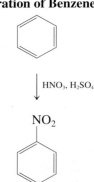

HNO$_3$, H$_2$SO$_4$

NO$_2$

Nitrobenzene

---

*Professor George A. Olah (b. 1927), University of Southern California, Los Angeles, Nobel Prize 1994 (chemistry).

ion, NO_2^+, a strong electrophile, with much of its positive charge residing on nitrogen, as shown in the electrostatic potential map.

Activation of Nitric Acid by Sulfuric Acid

Nitric acid

Nitronium ion

The nitronium ion, with its positively charged nitrogen, then attacks benzene.

Mechanism

Mechanism of Aromatic Nitration

Nitrobenzene

Aromatic nitration is the best way to introduce nitrogen-containing substituents into the benzene ring. The nitro group functions as a directing group in further substitutions (Chapter 16) and as a masked amino function (Section 16-5), as unraveled in benzenamines (anilines; Section 22-10).

Sulfonation is reversible

Concentrated sulfuric acid does not sulfonate benzene at room temperature. However, a more reactive form, called fuming sulfuric acid, permits electrophilic attack by SO_3. Commercial fuming sulfuric acid is made by adding about 8% of sulfur trioxide, SO_3, to the concentrated acid. Because of the strong electron-withdrawing effect of the three oxygens, the sulfur in SO_3 is electrophilic enough to attack benzene directly. Subsequent proton transfer results in the sulfonated product, benzenesulfonic acid.

Mechanism

Mechanism of Aromatic Sulfonation

Benzenesulfonic acid

Aromatic sulfonation is readily reversible. The reaction of sulfur trioxide with water to give sulfuric acid is so exothermic that heating benzenesulfonic acid in dilute aqueous acid completely reverses sulfonation.

Hydration of SO₃

Reverse Sulfonation: Hydrolysis

$$\text{C}_6\text{H}_5\text{—SO}_3\text{H} \xrightarrow{\text{H}_2\text{O, catalytic H}_2\text{SO}_4, \ 100°\text{C}} \text{C}_6\text{H}_5\text{—H} + \text{HOSO}_3\text{H}$$

The reversibility of sulfonation may be used to control further aromatic substitution processes. The ring carbon containing the substituent is blocked from attack, and electrophiles are directed to other positions. Thus, the sulfonic acid group can be introduced to serve as a *directing blocking group* and then removed by reverse sulfonation. Synthetic applications of this strategy will be discussed in Section 16-5.

Benzenesulfonic acids have important uses

Sulfonation of substituted benzenes is used in the synthesis of detergents. Thus, long-chain branched alkylbenzenes are sulfonated to the corresponding sulfonic acids, then converted into their sodium salts. Because such detergents are not readily biodegradable, they have been replaced by more environmentally acceptable alternatives. We shall examine this class of compounds in Chapter 19.

Aromatic Detergent Synthesis

$$\text{R—C}_6\text{H}_4 \xrightarrow{\text{SO}_3, \text{H}_2\text{SO}_4} \text{R—C}_6\text{H}_4\text{—SO}_3\text{H} \xrightarrow[-\text{H}_2\text{O}]{\text{NaOH}} \text{R—C}_6\text{H}_4\text{—SO}_3{}^-\text{Na}^+$$

R = branched alkyl group

Another application of sulfonation is to the manufacture of dyes, because the sulfonic acid group imparts water solubility (Chapter 22).

Sulfonyl chlorides, the acid chlorides of sulfonic acids (see Section 9-4), are usually prepared by reaction of the sodium salt of the acid with PCl_5 or $SOCl_2$.

Preparation of Benzenesulfonyl Chloride

$$\text{C}_6\text{H}_5\text{—SO}_2\text{—O}^-\text{Na}^+ + \text{PCl}_5 \longrightarrow \text{C}_6\text{H}_5\text{—SO}_2\text{—Cl} + \text{POCl}_3 + \text{Na}^+\text{Cl}^-$$

Sulfonyl chlorides are frequently employed in synthesis. For example, recall that the hydroxy group of an alcohol may be turned into a good leaving group by conversion of the alcohol into the 4-methylbenzenesulfonate (*p*-toluenesulfonate, tosylate; Sections 6-7 and 9-4).

Sulfonyl chlorides are important precursors of **sulfonamides,** many of which are chemotherapeutic agents, such as the sulfa drugs discovered in 1932 (Section 9-11). Sulfonamides are derived from the reaction of a sulfonyl chloride with an amine. Sulfa drugs specifically contain the 4-aminobenzenesulfonamide (sulfanilamide) function. Their mode of action is based on their structural similarity with the central fragment of folic acid. The sulfanilamide interferes with the bacterial enzymes that help to synthesize folic acid (Real Life 25-3), thereby depriving them of an essential nutrient and thus causing bacterial

---

Reaction

Sulfonation of Benzene

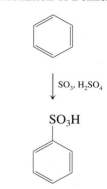

$$\downarrow \text{SO}_3, \text{H}_2\text{SO}_4$$

Benzenesulfonic acid

Animation

ANIMATED MECHANISM: Electrophilic aromatic sulfonation of benzene

---

Sulfa Drugs: The First Antimicrobials

The discovery of the sulfa drugs had a huge impact on medicine because they constituted the first antibiotics (preceding the advent of penicillin) available to the general public. They saved the lives of tens of thousands of patients, including soldiers during World War II. The photo shows Navy Corpsman Lavern L. Hamer, left, of Jordan, MN, pouring a sulfa drug on the wound of a Marine in the battle for the Marshall Islands.

cell death. Humans acquire the acid (vitamin B$_9$) through their diet and are not affected by the drug.

Sulfa Drugs

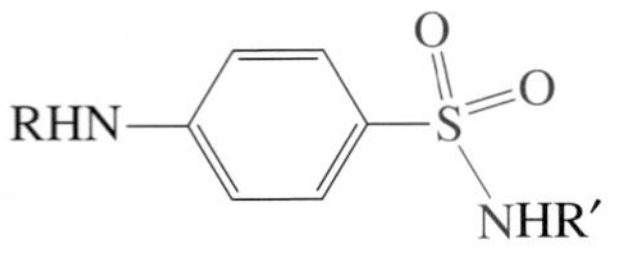

General structure

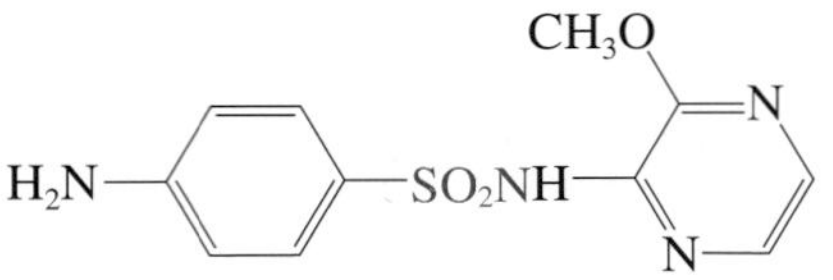

**Sulfalene
(Kelfizina)**
(Antiparasitic)

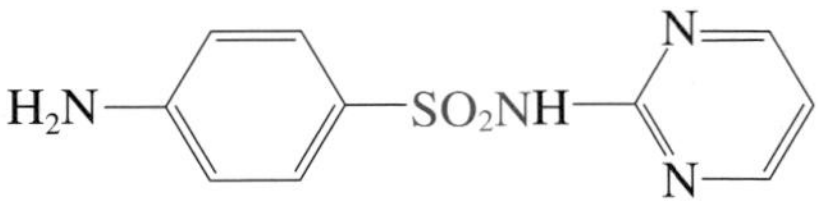

**Sulfadiazine
(Antimalarial)**

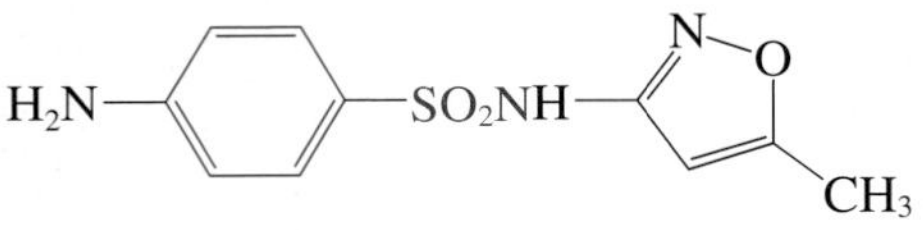

**Sulfamethoxazole
(Gantanol)**
(Antibacterial, used to treat urinary infections)

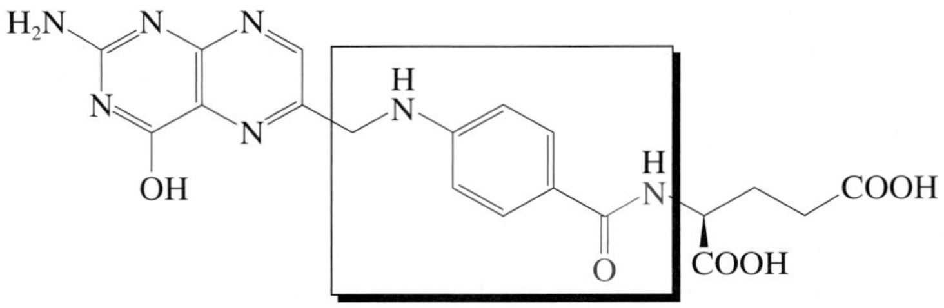

Folic acid

About 15,000 sulfa derivatives have been synthesized and screened for antibacterial activity; some have become new drugs. With the advent of newer generations of antibiotics, the medicinal use of sulfa drugs has greatly diminished, but their discovery was a milestone in the systematic development of medicinal chemistry.

Exercise 15-25

Formulate mechanisms for **(a)** the reverse of sulfonation; **(b)** the hydration of SO$_3$.

In Summary Nitration of benzene requires the generation of the nitronium ion, NO$_2^+$, which functions as the active electrophile. The nitronium ion is formed by the loss of water from protonated nitric acid. Sulfonation is achieved with fuming sulfuric acid, in which sulfur trioxide, SO$_3$, is the electrophile. Sulfonation is reversed by hot aqueous acid. Benzenesulfonic acids are used in the preparation of detergents, dyes, compounds containing leaving groups, and sulfa drugs.

15-11 FRIEDEL-CRAFTS ALKYLATION

None of the electrophilic substitutions mentioned so far has led to carbon–carbon bond formation, one of the primary challenges in organic chemistry. In principle, such reactions could be carried out with benzene in the presence of a sufficiently electrophilic carbon-based electrophile. This section introduces the first of two such transformations, the **Friedel-Crafts* reactions.** The secret to the success of both processes is the use of a Lewis acid, usually aluminum chloride. In the presence of this reagent, haloalkanes attack benzene to form alkylbenzenes.

---

*Professor Charles Friedel (1832–1899), Sorbonne, Paris; Professor James M. Crafts (1839–1917), Massachusetts Institute of Technology, Cambridge, MA.

In 1877, Friedel and Crafts discovered that a haloalkane reacts with benzene in the presence of an aluminum halide. The resulting products are the alkylbenzene and hydrogen halide. This reaction, which can be carried out in the presence of other Lewis acid catalysts, is called the **Friedel-Crafts alkylation** of benzene.

Friedel-Crafts Alkylation

+ RX $\xrightarrow{\text{AlX}_3}$ + HX

● **Reaction**

The reactivity of the haloalkane increases with the polarity of the C–X bond in the order RI < RBr < RCl < RF. Typical Lewis acids are BF_3, $SbCl_5$, $FeCl_3$, $AlCl_3$, and $AlBr_3$.

Friedel-Crafts Alkylation of Benzene with Chloroethane

CH_3CH_2Cl + [benzene with H] $\xrightarrow{\text{AlCl}_3,\ 25°C}$ [benzene with CH_2CH_3] + HCl

28%
Ethylbenzene

With primary halides, the reaction begins with coordination of the Lewis acid to the halogen of the haloalkane, much as in the activation of halogens in electrophilic halogenation. This coordination places a partial positive charge on the carbon bearing the halogen, rendering it more electrophilic. Attack on the benzene ring is followed by proton loss in the usual manner, giving the observed product.

Mechanism of Friedel-Crafts Alkylation with Primary Haloalkanes

Step 1. Haloalkane activation

$$RCH_2-\overset{..}{\underset{..}{X}}\!: \;\; +\; AlX_3 \;\rightleftharpoons\; R\overset{\delta+}{C}H_2:\overset{+}{\underset{..}{X}}\!:\bar{A}lX_3$$

Mechanism

Step 2. Electrophilic attack

[benzene ring] + $H_2\overset{\delta+}{C}$ with R, $\overset{+}{\underset{..}{X}}$—$\bar{A}lX_3$ $\longrightarrow$ [cyclohexadienyl cation with CH$_2$R and H] + AlX_4^-

Recall: Section 9-2

$RCH_2-\overset{..}{\underset{..}{O}}H \quad H^+$

↓

$RCH_2-\overset{+}{\underset{..}{O}}H_2$

$\quad$:Nu ↓

RCH_2-Nu
+ H_2O

Step 3. Proton loss

[cyclohexadienyl cation]—CH_2R + $:\overset{..}{\underset{..}{X}}\!-\bar{A}lX_3$ $\longrightarrow$ [benzene]—CH_2R + HX + AlX_3

With secondary and tertiary halides, free carbocations are usually formed as intermediates; these species attack the benzene ring in the same way as the cation NO_2^+.

Exercise 15-26

Write a mechanism for the formation of (1,1-dimethylethyl)benzene (*tert*-butylbenzene) from 2-chloro-2-methylpropane (*tert*-butyl chloride), benzene, and catalytic $AlCl_3$.

Intramolecular Friedel-Crafts alkylations can be used to fuse a new ring onto the benzene nucleus.

An Intramolecular Friedel-Crafts Alkylation

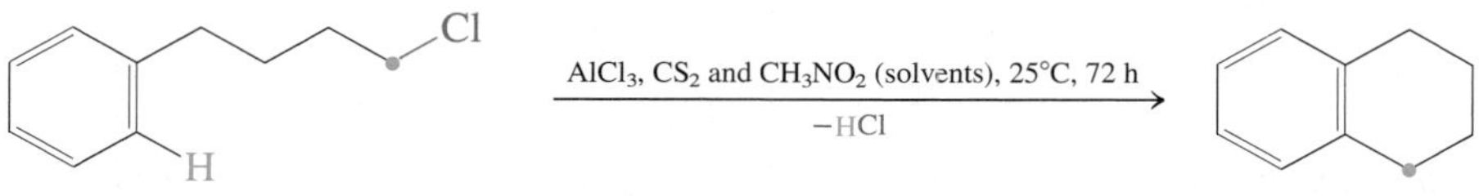
Cl
H
AlCl3, CS2 and CH3NO2 (solvents), 25°C, 72 h
−HCl

31%
Tetralin
(Common name)

Friedel-Crafts alkylations can be carried out with any starting material that functions as a precursor to a carbocation, such as an alcohol or alkene (Sections 9-2 and 12-3).

Friedel-Crafts Alkylations Using Other Carbocation Precursors

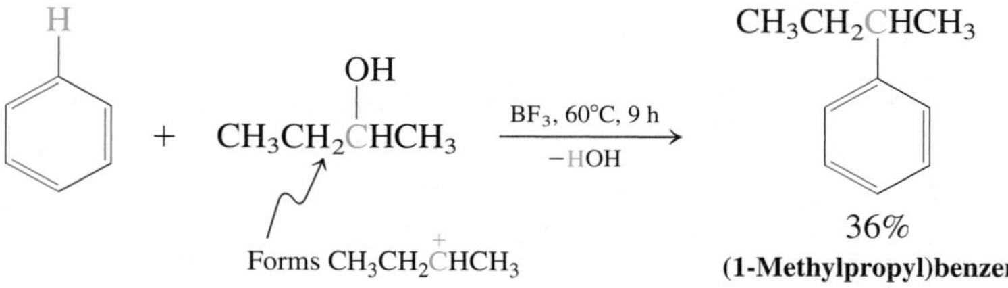
H
OH
CH3CH2CHCH3
BF3, 60°C, 9 h
−HOH
CH3CH2CHCH3
Forms CH3CH2CHCH3

36%
(1-Methylpropyl)benzene

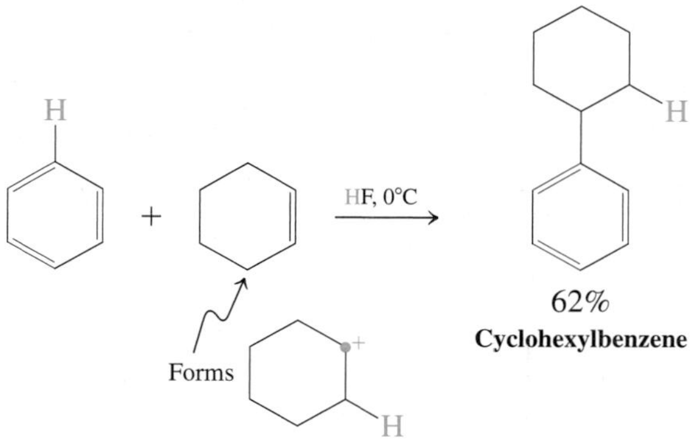
H
HF, 0°C
H
Forms
H

62%
Cyclohexylbenzene

Unreactive Halides in the Friedel-Crafts Alkylation

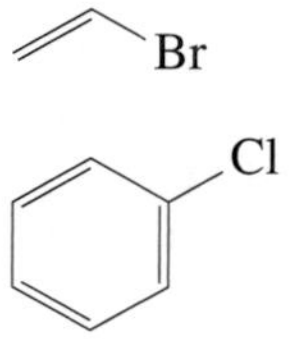
Br
Cl

Friedel-Crafts vinylations or arylations employing haloalkenes or haloarenes do not work (margin). The reason is that the corresponding cations are energetically inaccessible (Sections 13-9 and 22-10).

Exercise 15-27

In 2010, more than 3.4 million tons of (1-methylethyl)benzene (isopropylbenzene or cumene), an important industrial intermediate in the manufacture of phenol (Section 22-4), was synthesized in the United States from propene and benzene in the presence of phosphoric acid. Write a mechanism for its formation in this reaction.

Solved Exercise 15-28	Working with the Concepts: Reversible Friedel-Crafts Alkylations

Heating any of the three isomeric dimethylbenzenes with HF in the presence of BF_3 to 80°C leads to the equilibrium mixture shown below. Formulate a mechanism for these isomerizations, starting with 1,2-dimethylbenzene and simply using H^+ to represent the acid. Why is the equilibrium concentration of the 1,2-isomer the lowest?

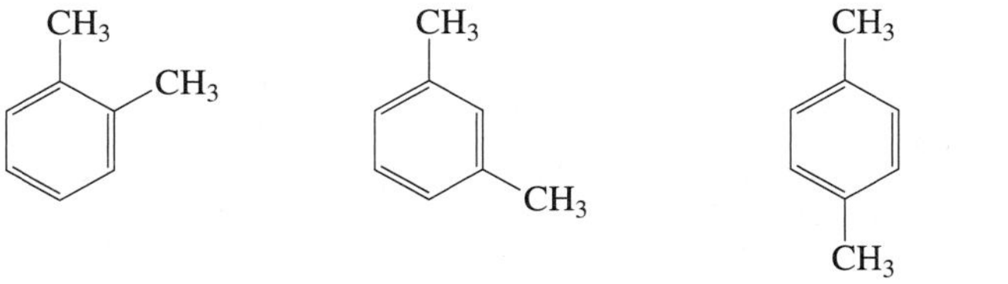

18%   58%   24%
1,2-Dimethylbenzene   **1,3-Dimethylbenzene**   **1,4-Dimethylbenzene**
(*o*-Xylene)   (*m*-Xylene)   (*p*-Xylene)

Strategy
Topologically, these isomerizations are reminiscent of alkyl shifts, which we studied earlier in connection with carbocation rearrangements (Section 9-3). To generate a carbocation from a benzene derivative, we protonate with acid (Section 15-8, Exercises 15-23 and 15-24). Protonation can occur anywhere on the ring and is reversible.

Solution
• There are three possible distinct delocalized carbocations (write them), but the one of interest results from protonation at the methyl-bearing carbon, shown below as its resonance forms.

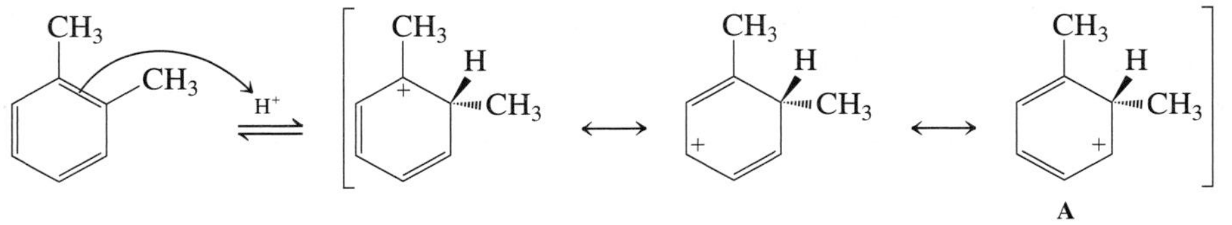

• Inspection of the resonance contributors reveals that one of them, form A, looks exactly like the type of carbocation drawn in Section 9-3 for H and alkyl shifts.
• Execution of a methyl shift from A gives a new carbocation, which can deprotonate to give 1,3-dimethylbenzene.

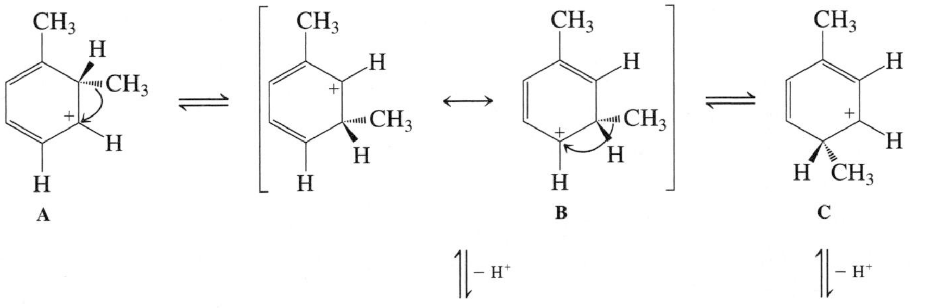

1,3-Dimethylbenzene   1,4-Dimethylbenzene

Alternatively, another methyl shift (best visualized from resonance form B) gives C, which then loses a proton to give the 1,4-isomer. Therefore, the mechanism is indeed a sequence of simple alkyl shifts occurring in protonated benzene.
• Why is 1,2-dimethylbenzene the minor component of the mixture? You have probably guessed the answer: steric hindrance, like that seen in cis alkenes (Figure 11-13).

Exercise 15-29 | Try It Yourself

Intramolecular Friedel-Crafts alkylation of A provided the expected B, but also 10% of C, suggesting the operation of a competing pathway to the normal mechanism. Propose such a pathway.

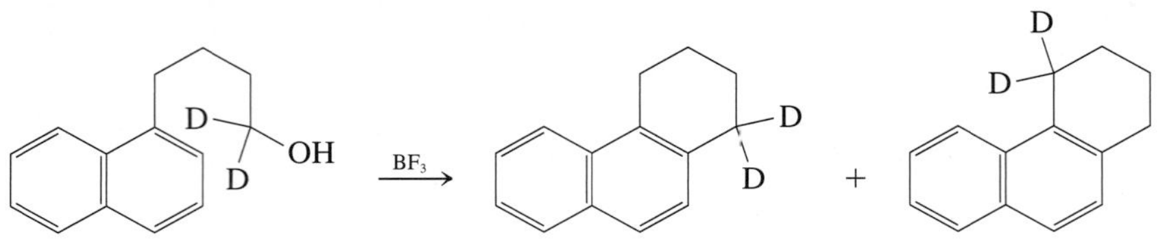

In Summary The Friedel-Crafts alkylation produces carbocations (or their equivalents) capable of electrophilic aromatic substitution by formation of aryl–carbon bonds. Haloalkanes, alkenes, and alcohols can be used to achieve aromatic alkylation in the presence of a Lewis or mineral acid.

15-12 | LIMITATIONS OF FRIEDEL-CRAFTS ALKYLATIONS

The alkylation of benzenes under Friedel-Crafts conditions is accompanied by two important side reactions: One is *polyalkylation;* the other, *carbocation rearrangement.* Both cause the yield of the desired products to diminish and lead to mixtures that may be difficult to separate.

Consider first polyalkylation. Benzene reacts with 2-bromopropane in the presence of $FeBr_3$ as a catalyst to give products of both single and double substitution. The yields are low because of the formation of many by-products.

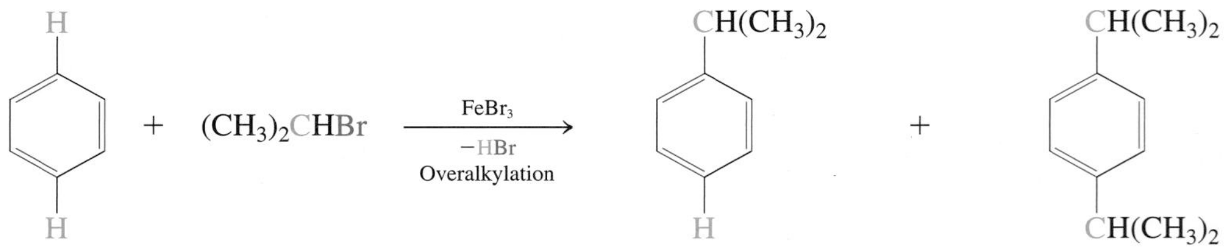

25%
**(1-Methylethyl)benzene
(Isopropylbenzene)**

15%
**1,4-Bis(1-methylethyl)benzene
(*p*-Diisopropylbenzene)**

The electrophilic aromatic substitutions that we studied in Sections 15-9 and 15-10 can be stopped at the monosubstitution stage. Why do Friedel-Crafts alkylations have the problem of multiple electrophilic substitution? It is because the substituents differ in electronic structure (a subject discussed in more detail in Chapter 16). Bromination, nitration, and sulfonation introduce an electron-withdrawing group into the benzene ring, which renders the product *less* susceptible than the starting material to electrophilic attack. In contrast, an alkylated benzene is more electron rich than unsubstituted benzene and thus *more* susceptible to electrophilic attack.

Exercise 15-30 |

Treatment of benzene with chloromethane in the presence of aluminum chloride results in a complex mixture of tri-, tetra-, and pentamethylbenzenes. One of the components in this mixture crystallizes out selectively: m.p. = 80°C; molecular formula = $C_{10}H_{14}$; ^1H NMR: δ = 2.27 (s, 12 H) and 7.15 (s, 2 H) ppm; ^{13}C NMR: δ = 19.2, 131.2, and 133.8 ppm. Draw a structure for this product.

The second side reaction in aromatic alkylation is skeletal rearrangement (Section 9-3). For example, the attempted propylation of benzene with 1-bromopropane and $AlCl_3$ produces (1-methylethyl)benzene.

$$\text{(benzene, H)} + CH_3CH_2CH_2Br \xrightarrow[\substack{-HBr \\ \text{Rearranged} \\ \text{alkyl group}}]{AlCl_3} \text{(benzene, } CH(CH_3)_2)$$

Reaction

The starting haloalkane rearranges by a hydride shift to the thermodynamically favored 1-methylethyl (isopropyl) cation in the presence of the Lewis acid.

Mechanism

Rearrangement of 1-Bromopropane to 1-Methylethyl (Isopropyl) Cation

$$CH_3\overset{\overset{\displaystyle H}{|}}{CH}-CH_2-Br + AlCl_3 \longrightarrow CH_3\overset{+}{C}HCH_3 + Br\overline{A}lCl_3$$

1-Methylethyl (isopropyl) cation

Solved Exercise 15-31	**Working with the Concepts: Rearrangements in Friedel-Crafts Alkylations**

Attempted alkylation of benzene with 1-chlorobutane in the presence of $AlCl_3$ gave not only the expected butylbenzene, but also, as a major product, (1-methylpropyl)benzene. Write a mechanism for this reaction.

Strategy

First write an equation for the described transformation:

$$\text{(benzene)} + \text{(butyl-Cl)} \xrightarrow[-HCl]{AlCl_3} \text{(butylbenzene)} + \text{((1-methylpropyl)benzene)}$$

Butylbenzene   **(1-Methylpropyl)-benzene**

Consider the mechanism for each product separately.

Solution

• The first product is derived readily by a normal Friedel-Crafts alkylation:

$$\text{(benzene)} + \overset{+}{C}l-\overline{A}lCl_3 \longrightarrow \text{(arenium ion, } \overset{+}{H}) + \cdot\cdot\overline{C}l-\overline{A}lCl_3$$

$$\longrightarrow \text{(butylbenzene)} + HCl + AlCl_3$$

- The second product contains a rearranged butyl group, the result of electrophilic aromatic substitution by a secondary butyl cation. The required rearrangement is brought about by a Lewis acid-catalyzed hydride shift:

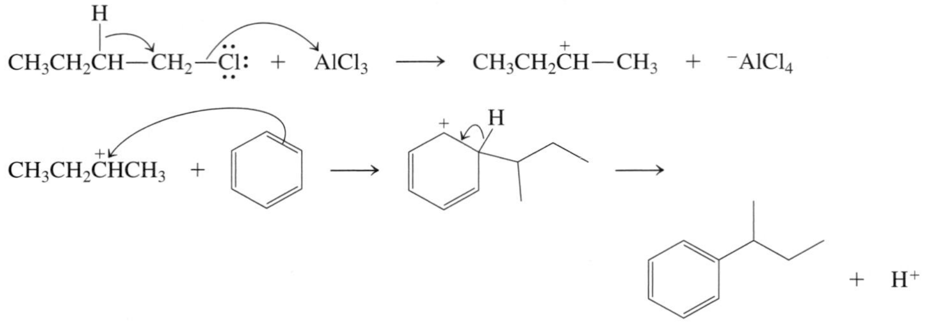

$$CH_3CH_2\overset{\overset{\displaystyle H}{|}}{CH}-CH_2\!-\!\ddot{\underset{\displaystyle ..}{Cl}}\!: \; + \; AlCl_3 \longrightarrow CH_3CH_2\overset{+}{CH}-CH_3 \; + \; {}^-AlCl_4$$

$$CH_3CH_2\overset{+}{C}HCH_3 \; + \; \text{(benzene)} \longrightarrow \text{(arenium ion)} \longrightarrow \text{(sec-butylbenzene)} \; + \; H^+$$

Exercise 15-32 | **Try It Yourself**

Write a mechanism for the following reaction:

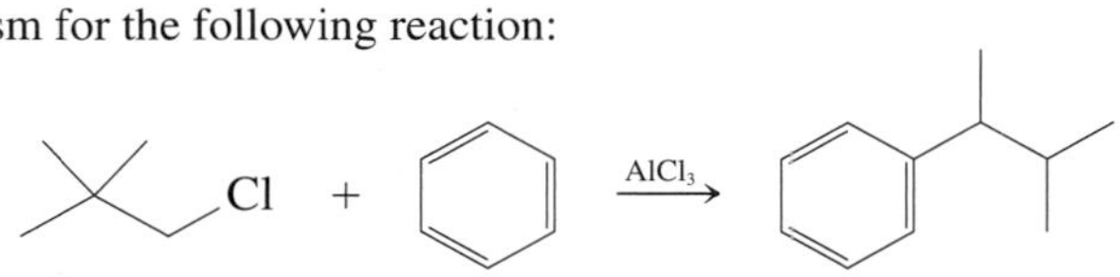

$$\text{(neopentyl chloride)}\;Cl \; + \; \text{(benzene)} \xrightarrow{AlCl_3} \text{(product)}$$

Because of these limitations, Friedel-Crafts alkylations are rarely used in synthetic chemistry. Can we improve this process? A more useful reaction would require an electrophilic carbon species that could not rearrange and that would, moreover, deactivate the ring to prevent further substitution. There is such a species—an acylium cation—and it is used in the second Friedel-Crafts reaction, the topic of the next section.

In Summary Friedel-Crafts alkylation suffers from overalkylation and skeletal rearrangements by both hydrogen and alkyl shifts.

15-13 | FRIEDEL-CRAFTS ACYLATION (ALKANOYLATION)

The second electrophilic aromatic substitution that forms carbon–carbon bonds is **Friedel-Crafts alkanoylation** (as in butanoylation, pentanoylation, and so on). A more popular alternative systematic name for this process is **acylation,** which is the term we shall use. IUPAC retains the common naming of formyl for $H\overset{\overset{\displaystyle O}{\|}}{C}-$ and acetyl for $CH_3\overset{\overset{\displaystyle O}{\|}}{C}-$ (Section 17-1), hence the corresponding acylations are called formylation and acetylation. Acylation proceeds through the intermediacy of **acylium cations,** with the general structure $RC\!\equiv\!O\!:^+$. This section describes how these ions readily attack benzene to form ketones.

Friedel-Crafts Acylation

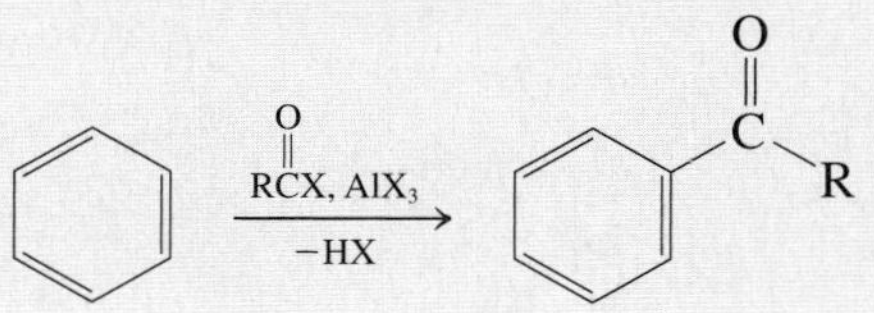

$$\text{(benzene)} \xrightarrow[\text{$-$HX}]{RCX,\, AlX_3} \text{(phenyl ketone)}$$

Friedel-Crafts acylation employs acyl chlorides

Benzene reacts with acyl halides in the presence of an aluminum halide to give 1-phenylalkanones (phenyl ketones). An example is the preparation of 1-phenylethanone (acetophenone) from benzene and acetyl chloride, by using aluminum chloride as the Lewis acid.

Friedel-Crafts Acylation of Benzene with Acetyl Chloride

$$\text{benzene-H} + CH_3\overset{O}{\underset{\|}{C}}Cl \xrightarrow[\substack{2.\ H_2O,\ H^+ \\ \text{(work-up)}}]{1.\ AlCl_3} \text{phenyl-}\overset{O}{\underset{\|}{C}}\text{-}CH_3 + HCl$$

61%
1-Phenylethanone
(Acetophenone)

Acyl chlorides are reactive derivatives of carboxylic acids. They are readily formed from carboxylic acids by reaction with thionyl chloride, $SOCl_2$. (We shall explore this process in detail in Chapter 19.)

Preparation of an Acyl Chloride

$$R\overset{O}{\underset{\|}{C}}OH + SOCl_2 \longrightarrow R\overset{O}{\underset{\|}{C}}Cl + SO_2 + HCl$$

Even though it is unstable and reactive, the formyl cation, $H–C\equiv O^+$, is a fundamental small organic molecule that is (relatively) abundant in such diverse environments as hot flames and cold interstellar space. It has been detected in the gas surrounding the comet Hale-Bopp, a spectacular visitor to Earth's skies in 1997.

Acyl halides react with Lewis acids to produce acylium ions

The key reactive intermediates in Friedel-Crafts acylations are acylium cations. These species can be formed by the reaction of acyl halides with aluminum chloride. The Lewis acid initially coordinates to the carbonyl oxygen because of resonance (see Exercise 2-11). This complex is in equilibrium with an isomer in which the aluminum chloride is bound to the halogen. Dissociation then produces the acylium ion, which is stabilized by resonance and, unlike alkyl cations, is not prone to rearrangements. As shown in the electrostatic potential map of the acetyl cation in the margin, most of the positive charge (blue) resides on the carbonyl carbon.

Acylium Ions from Acyl Halides

Acetyl cation
[Although the resonance form with a positive charge on oxygen is the major contributor to the structure of this ion, the presence of a lone pair on oxygen causes it to appear less positive (green instead of blue) in the electrostatic potential map].

$$R\overset{:O:}{\underset{:X:}{C}} + AlCl_3 \rightleftharpoons \left[R\overset{\overset{+}{:O}\text{—}\bar{A}lCl_3}{\underset{:X:}{C}} \longleftrightarrow R\overset{\overset{..}{:O}\text{—}\bar{A}lCl_3}{\underset{:X:^+}{C}} \right] \rightleftharpoons$$

$$R\overset{:O:}{\underset{\overset{+}{:X:}\ \bar{A}lCl_3}{C}} \rightleftharpoons \left[R\text{—}C\equiv\overset{+}{O}: \longleftrightarrow R\text{—}\overset{+}{C}=\overset{..}{O}: \right] + :\overset{..}{X}\text{—}\bar{A}lCl_3$$

Acylium ion

Sometimes carboxylic anhydrides are used in acylation in place of the halides. These molecules react with Lewis acids in a similar way.

Acylium Ions from Carboxylic Anhydrides

$$R\overset{:O:\ :O:}{\underset{:O:}{C\quad C}}R + AlCl_3 \rightleftharpoons \left[R\overset{\overset{+}{:O}\text{—}\bar{A}lCl_3\ :O:}{\underset{:O:}{C\quad C}}R \longleftrightarrow R\overset{..\ \bar{A}lCl_3\ :O:}{\underset{\overset{+}{O}}{C\quad C}}R \right] \rightleftharpoons$$

$$R\overset{:O:\ :O:}{\underset{\overset{..\ +}{O}}{C\quad C}}R \rightleftharpoons \left[R\text{—}C\equiv\overset{+}{O}: \longleftrightarrow R\text{—}\overset{+}{C}=\overset{..}{O}: \right] + R\overset{:O:}{\underset{\overset{..}{O}\text{—}\bar{A}lCl_3}{C}}$$
$$\bar{A}lCl_3$$

Acylium ions undergo electrophilic aromatic substitution

The acylium ion is sufficiently electrophilic to attack benzene by the usual aromatic substitution mechanism.

Electrophilic Acylation

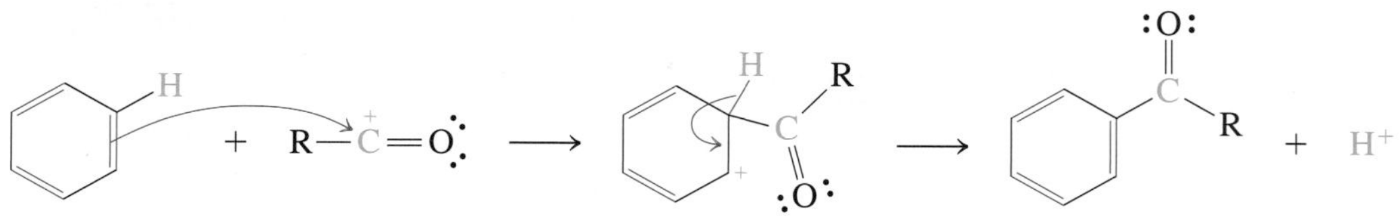

Mechanism

Because the newly introduced acyl substituent is electron withdrawing (see Sections 14-8 and 16-1), it deactivates the ring and protects it from further substitution. Therefore polyacylation does not occur, whereas polyalkylation does (Section 15-12). The effect is accentuated by the formation of a strong complex between the aluminum chloride catalyst and the carbonyl function of the product ketone.

Lewis Acid Complexation with 1-Phenylalkanones

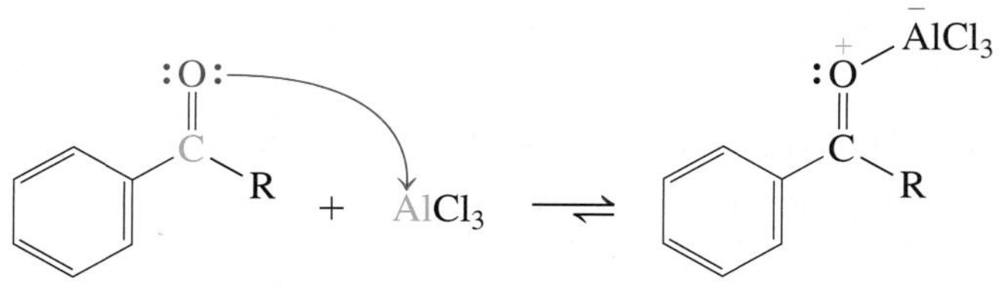

This complexation removes the $AlCl_3$ from the reaction mixture and necessitates the use of at least one full equivalent of the Lewis acid to allow the reaction to go to completion. Acidic aqueous work-up is necessary to liberate the ketone from its aluminum chloride complex, as illustrated by the following examples. Because the work-up step is standard for most organic reactions, we will omit it from future schemes.

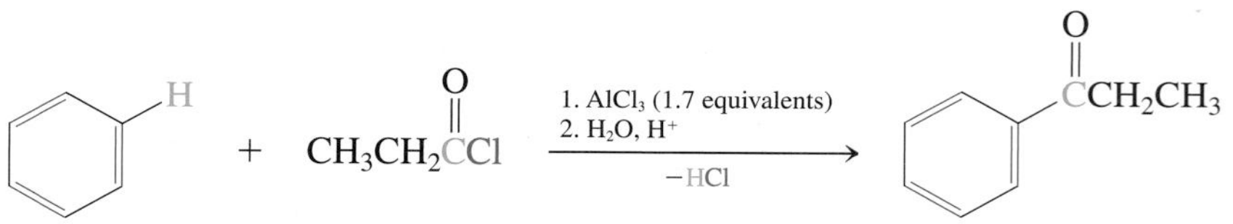

84%
**1-Phenyl-1-propanone
(Propiophenone)**

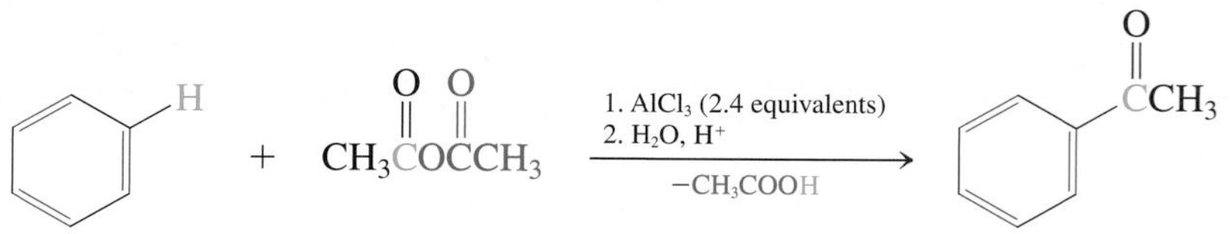

85%
**1-Phenylethanone
(Acetophenone)**

The selectivity of the Friedel-Crafts acylation for single substitution allows for the selective introduction of carbon chains into the benzene nucleus, a task that proved difficult to accomplish by Friedel-Crafts alkylation (Section 15-12). Since we know how to convert the carbonyl function into an alcohol by hydride reduction (Section 8-6) and the hydroxy substituent into a leaving group that can be further reduced by hydride (Section 8-7), we can synthesize the corresponding hydrocarbon. This sequence of acylation–reduction constitutes a roundabout alkylation protocol that occurs selectively. We shall encounter more direct "deoxygenations" of carbonyl groups later (Sections 16-5 and 17-10).

Preparation of Hexylbenzene by Hexanoylation–Reduction of Benzene

Hexylbenzene

Exercise 15-33

The simplest acyl chloride, formyl chloride, H–C(=O)–Cl, is unstable, decomposing to HCl and CO upon attempted preparation. Therefore, direct Friedel-Crafts formylation of benzene is impossible. An alternative process, the Gattermann-Koch reaction, enables the introduction of the formyl group, –CHO, into the benzene ring by treatment with CO under pressure, in the presence of HCl and Lewis acid catalysts. For example, methylbenzene (toluene) can be formylated at the para position in this way in 51% yield. The electrophile in this process was observed directly for the first time in 1997 by treating CO with HF–SbF$_5$ under high pressure: ^{13}C NMR: δ 139.5 ppm; IR: $\tilde{\nu}$ = 2110 cm^{-1}. What is the structure of this species and the mechanism of its reaction with methylbenzene? Explain the spectral data. (**Hints:** Draw the Lewis structure of CO and proceed by considering the species that may arise in the presence of acid. The comparative spectral data for free CO are ^{13}C NMR: δ = 181.3 ppm; IR: $\tilde{\nu}$ = 2143 cm^{-1}.)

In Summary The problems of Friedel-Crafts alkylation (multiple substitution and carbocation rearrangements) are avoided in Friedel-Crafts acylations, in which an acyl halide or carboxylic acid anhydride is the reaction partner, in the presence of a Lewis acid. The intermediate acylium cations undergo electrophilic aromatic substitution to yield the corresponding aromatic ketones.

THE BIG PICTURE

The concept of aromaticity (and antiaromaticity) may seem strange and new to you. However, it is really just an extension of other electronic effects we have encountered earlier, starting with Coulomb's law (Section 1-2), the octet rule (Section 1-3), and the Aufbau principle (Section 1-6). Other examples of orbital overlap and electron delocalization have been shown to be either stabilizing or destabilizing, including the stability ordering of radicals (Section 3-2) and carbocations (Section 7-5) due to hyperconjugation, and the general phenomenon of (de)stabilization by π-electron delocalization (Chapter 14). In this context, aromaticity is simply another important kind of electronic effect in organic chemistry.

We shall explore additional implications of aromaticity in Chapter 16, showing how single substituents on the benzene ring affect further substitutions. As a unit, the benzene ring occurs in organic molecules ranging from polystyrene to aspirin; learning how to incorporate benzene rings into other molecules and changing the substituents on a benzene ring are important aspects of many branches of organic synthesis.

WORKED EXAMPLES: INTEGRATING THE CONCEPTS

15-34. Assigning the Structure of Substituted Benzenes by Spectral Analysis

Compound A, C_8H_{10}, was treated with Br_2 in the presence of $FeBr_3$ to give product B, C_8H_9Br. The spectral data for A and B are given below. Assign structures to A and B.

A: ^1H NMR (CDCl$_3$): $\delta = 2.28$ (s, 6 H), 6.95 (m, 3 H), 7.11 (td, $J = 7.8$, 0.4 Hz, 1 H) ppm.
   ^{13}C NMR (CDCl$_3$): $\delta = 21.3$, 126.1, 128.2, 130.0, 137.7 ppm.
   IR (neat) selected values: $\tilde{v} = 691$, 769, 2921, 2946, 3016 cm^{-1}.
   UV (CH$_3$OH): $\lambda_{max} = 261$ nm.

B: ^1H NMR (CDCl$_3$): $\delta = 2.25$ (s, 3 H), 2.34 (s, 3 H), 6.83 (dd, $J = 7.9$, 2.0 Hz, 1 H), 7.02 (dd, $J = 2.0$, 0.3 Hz, 1 H), 7.36 (dd, $J = 7.9$, 0.3 Hz, 1 H) ppm.
   ^{13}C NMR (CDCl$_3$): $\delta = 20.7$, 22.7, 121.6, 128.1, 131.6, 132.1, 136.9, 137.4 ppm.
   IR (neat) selected values: $\tilde{v} = 2923$, 2961, 3012 cm^{-1}.
   UV (CH$_3$OH): $\lambda_{max} = 265$ nm.

SOLUTION

A cursory glance at the molecular formulas and the spectral data confirms that bromination of a substituted benzene is taking place: One hydrogen in A (C_8H_{10}) is replaced by a bromine atom (C_8H_9Br), and both compounds show aromatic peaks in both the ^1H and ^{13}C NMR spectra. The IR spectra confirm this assignment by the presence of a $C_{aromatic}$–H stretching signal for A and B, and the UV spectra are consistent with a phenyl chromophore. A closer look reveals that both compounds contain two methyl groups attached to the aromatic system at $\delta \approx 2.3$ ppm (Table 10-2). Subtracting $2 \times CH_3$ from C_8H_{10} (A) leaves C_6H_4, a phenyl fragment: Compound A must be a dimethylbenzene isomer. Consequently, B must be a bromo(dimethyl) benzene isomer. The question for both is: Which isomer? To find the answers, it is useful to write down all the possible options. For A, there are three isomers: *ortho-, meta-,* and *para*-dimethylbenzene (xylene; Section 15-1).

The Three Possible Structures for A

Ortho   Meta   Para

A

Is it possible to distinguish between them on the basis of NMR spectroscopy? The answer is yes, because of the varying degrees of symmetry inherent to each ring. Thus, the ortho isomer should exhibit only two types of aromatic hydrogens (two each) and three phenyl carbon signals in the respective NMR spectra. The para isomer is even more symmetrical, containing only one type of benzene hydrogen and two types of phenyl carbons. Both of these predictions are incompatible with the observed data for A. Thus, while some of the aromatic hydrogens are unresolved and appear as a multiplet (3 H), the presence of a single unique proton at $\delta = 7.11$ ppm is consistent only with *meta*-dimethylbenzene. (Which one of the two possible unique hydrogens gives rise to this signal? **Hint:** Look at the coupling pattern of the ^1H NMR signal.) Similarly, this isomer contains four distinct ring carbons, as seen in the ^{13}C NMR spectrum.

Armed with the knowledge of the structure of A, we can now formulate the possible products of its electrophilic aromatic bromination:

The Three Possible Structures for B

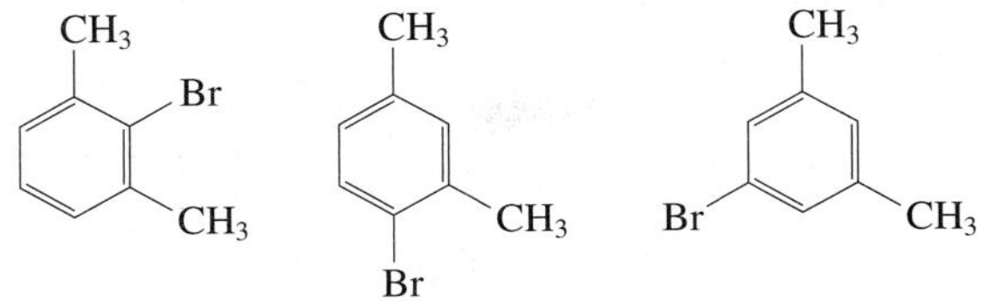

B

Which one is B? Symmetry (or lack thereof) provides the answer. The ^{1}H NMR spectrum of B reveals two distinct methyl groups and three separate ring hydrogen resonances, whereas the ^{13}C NMR spectrum exhibits two methyl carbon peaks, in addition to six aromatic carbon absorptions. The combined data are only compatible with the center structure above, 1-bromo-2,4-dimethylbenzene. The reaction of A to provide B is therefore:

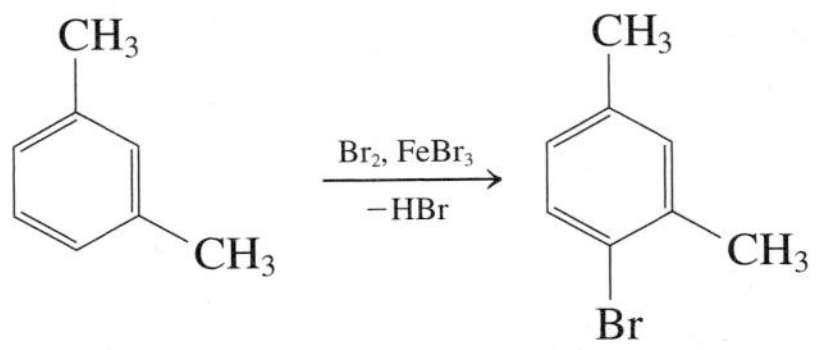

A   **B**

Why does single bromination of A give only isomer B? Read on to Chapter 16! But first try Problem 43 for more practice with spectroscopy.

15-35. Applying Mechanistic Principles to Deduce the Mechanism of a New Reaction

The insecticide DDT (see Real Life 3-2) has been made in ton quantities by treatment of chlorobenzene with 2,2,2-trichloroacetaldehyde in the (required) presence of concentrated H_2SO_4. Formulate a mechanism for this reaction.

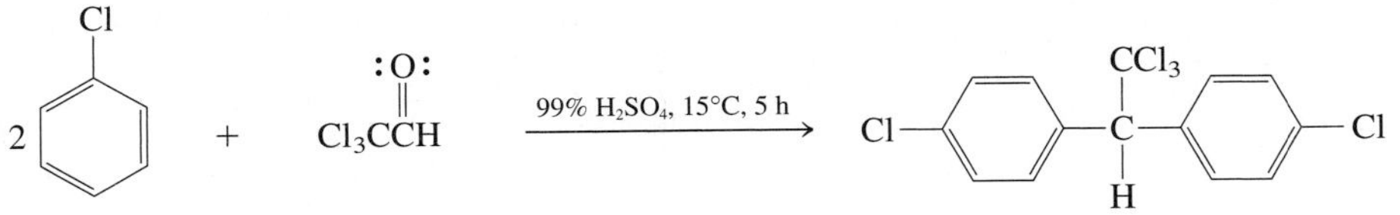

Chlorobenzene   **2,2,2-Trichloro-**   **DDT**
                        acetaldehyde

SOLUTION

Let us first take an inventory of what is given:

1. The product is composed of two subunits derived from chlorobenzene and one subunit derived from the aldehyde.

2. In the process, the starting materials—two chlorobenzenes and one trichloroacetaldehyde, which amount to a combined atomic total of $C_{14}H_{11}Cl_5O$—turn into DDT, with the molecular formula $C_{14}H_9Cl_5$. Conclusion: The elements of H_2O are extruded.

3. Topologically, the transformation constitutes a replacement of an aromatic hydrogen by an alkyl carbon substituent, strongly implicating the occurrence of a Friedel-Crafts alkylation (Section 15-12).

We can now address the details of a possible mechanism. Friedel-Crafts alkylations require positively polarized or cationic carbon electrophiles. In our case, inspection of the product suggests that the carbonyl carbon in the aldehyde is the electrophilic aggressor. This carbon is positively polarized to begin with, because of the presence of the electron-withdrawing oxygen and chlorine substituents. A nonoctet dipolar resonance form (Section 1-5) illustrates this point.

Activation of 2,2,2-Trichloroacetaldehyde as an Electrophile

In the presence of strong acid, protonation of the negatively polarized oxygen of the neutral species generates a positively charged intermediate in which the electrophilic character of the carbonyl carbon is further accentuated in the resonance form depicting a hydroxy carbocation. The stage is now set for the first of two electrophilic aromatic substitution steps (Section 15-11).

First Electrophilic Aromatic Substitution Step

The product of this step is an alcohol, which can be readily converted into the corresponding carbocation by acid (Section 9-2), partly because the ensuing charge is resonance stabilized by the adjacent benzene ring (*benzylic resonance,* Section 22-1, related to *allylic resonance,* Sections 14-1 and 14-3). The cation subsequently enables the second electrophilic aromatic substitution step to provide DDT. Another more advanced mechanism is the subject of Problem 66.

Alcohol Activation and Second Electrophilic Aromatic Substitution Step

New Reactions

1. Hydrogenation of Benzene (Section 15-2)

$\Delta H° = -49.3$ kcal mol^{-1}
Resonance energy: ~ -30 kcal mol^{-1}

Electrophilic Aromatic Substitution

2. Chlorination, Bromination, Nitration, and Sulfonation (Sections 15-9 and 15-10)

$$C_6H_6 \xrightarrow{X_2, FeX_3} C_6H_5X + HX \qquad X = Cl, Br$$

$$C_6H_6 \xrightarrow{HNO_3, H_2SO_4} C_6H_5NO_2 + H_2O$$

$$C_6H_6 \underset{H_2SO_4, H_2O, \Delta}{\overset{SO_3, H_2SO_4}{\rightleftharpoons}} C_6H_5SO_3H \qquad \textbf{Reversible}$$

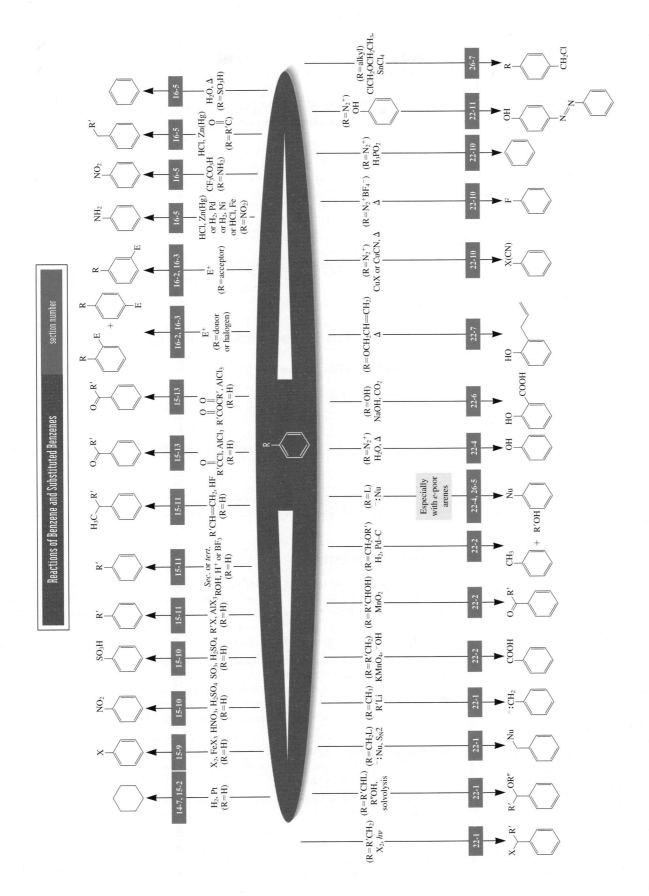

Reactions of Benzene and Substituted Benzenes

3. Benzenesulfonyl Chlorides (Section 15-10)

$$C_6H_5SO_3Na \ + \ PCl_5 \ \longrightarrow \ C_6H_5SO_2Cl \ + \ POCl_3 \ + \ NaCl$$

4. Friedel-Crafts Alkylation (Section 15-11)

$$C_6H_6 \ + \ RX \ \xrightarrow{\text{AlCl}_3} \ C_6H_5R \ + \ HX \ + \ \text{overalkylated product}$$

R⁺ is subject to carbocation rearrangements

Intramolecular

$$+ \ HCl$$

Alcohols and alkenes as substrates

5. Friedel-Crafts Acylation (Section 15-13)

Acyl halides

Requires at least one full equivalent of Lewis acid

Anhydrides

Important Concepts

1. Substituted benzenes are named by adding prefixes or suffixes to the word *benzene*. Disubstituted systems are labeled as 1,2-, 1,3-, and 1,4- or **ortho, meta,** and **para,** depending on the location of the substituents. Many benzene derivatives have common names, sometimes used as bases for naming their substituted analogs. As a substituent, an aromatic system is called **aryl;** the parent aryl substituent, C_6H_5, is called **phenyl;** its homolog $C_6H_5CH_2$ is named **phenylmethyl (benzyl).**

2. Benzene is not a cyclohexatriene but a delocalized cyclic system of six π electrons. It is a **regular hexagon** of six sp^2-hybridized carbons. All six p orbitals overlap equally with their neighbors. Its unusually low heat of hydrogenation indicates a **resonance energy** or **aromaticity** of about 30 kcal mol^{-1} (126 kJ mol^{-1}). The stability imparted by aromatic delocalization is also evident in the transition state of some reactions, such as the Diels-Alder cycloaddition and ozonolysis.

3. The special structure of benzene gives rise to unusual UV, IR, and NMR spectral data. ^{1}H NMR spectroscopy is particularly diagnostic because of the unusual **deshielding** of aromatic hydrogens by an **induced ring current.** Moreover, the substitution pattern is revealed by examination of the o, m, and p coupling constants.

4. The **polycyclic benzenoid hydrocarbons** are composed of linearly or angularly fused benzene rings. The simplest members of this class of compounds are naphthalene, anthracene, and phenanthrene.

5. In these molecules, benzene rings **share** two (or more) carbon atoms, whose π electrons are delocalized over the entire ring system. Thus, naphthalene shows some of the properties characteristic of the aromatic ring in benzene: The electronic spectra reveal extended conjugation, ^{1}H NMR exhibits deshielding ring-current effects, and there is little bond alternation.

6. Benzene is the smallest member of the class of aromatic cyclic polyenes following **Hückel's (4n + 2) rule.** Most of the $4n$ π systems are relatively reactive **anti-** or **nonaromatic** species. Hückel's rule also extends to aromatic charged systems, such as the cyclopentadienyl anion, cycloheptatrienyl cation, and cyclooctatetraene dianion.

7. The most important reaction of benzene is **electrophilic aromatic substitution.** The rate-determining step is addition by the electrophile to give a delocalized hexadienyl cation in which the aromatic character of the original benzene ring has been lost. Fast deprotonation restores the aromaticity of the (now substituted) benzene ring. Exothermic substitution is preferred over endothermic addition. The reaction can lead to halo- and nitrobenzenes, benzenesulfonic acids, and alkylated and acylated derivatives. When necessary, Lewis acid (chlorination, bromination, Friedel-Crafts reaction) or mineral acid (nitration, sulfonation) catalysts are applied. These enhance the electrophilic power of the reagents or generate strong, positively charged electrophiles.

8. **Sulfonation** of benzene is a **reversible** process. The sulfonic acid group is removed by heating with dilute aqueous acid.

9. **Benzenesulfonic acids** are precursors of benzenesulfonyl chlorides. The chlorides react with alcohols to form sulfonic esters containing useful **leaving groups** and with amines to give sulfonamides, some of which are medicinally important.

10. In contrast with other electrophilic substitutions, including Friedel-Crafts acylations, **Friedel-Crafts alkylations** activate the aromatic ring to further electrophilic substitution, leading to product mixtures.

Problems

36. Name each of the following compounds by using the IUPAC system and, if possible, a reasonable common alternative. (**Hint:** The order of functional group precedence is –COOH > –CHO > –OH > –NH$_2$.)

(a)

(b)

(c)

(d)

(e)

(f)

(g)

(h)

(i)

37. Give a proper IUPAC name for each of the following commonly named substances.

(a)
Durene

(b)
Hexylresorcinol

(c)
Eugenol

38. Draw the structure of each of the following compounds. If the name itself is incorrect, give a correct systematic alternative. (**a**) o-Chlorobenzaldehyde; (**b**) 2,4,6-trihydroxybenzene; (**c**) 4-nitro-o-xylene; (**d**) m-isopropylbenzoic acid; (**e**) 4,5-dibromoaniline; (**f**) p-methoxy-m-nitroacetophenone.

39. The complete combustion of benzene is exothermic by approximately -789 kcal mol^{-1}. What would this value be if benzene lacked aromatic stabilization?

40. The ^{1}H NMR spectrum of naphthalene shows two multiplets (Figure 15-16). The upfield absorption ($\delta = 7.49$ ppm) is due to the hydrogens at C2, C3, C6, and C7, and the downfield multiplet ($\delta = 7.86$ ppm) is due to the hydrogens at C1, C4, C5, and C8. Explain why one set of hydrogens is deshielded more than the other.

41. Complete hydrogenation of 1,3,5,7-cyclooctatetraene is exothermic by -101 kcal mol^{-1}. Hydrogenation of cyclooctene proceeds with $\Delta H^\circ = -23$ kcal mol^{-1}. Are these data roughly consistent with the description of cyclooctatetraene presented in the chapter?

42. Which of the following structures qualify as being aromatic, according to Hückel's rule?

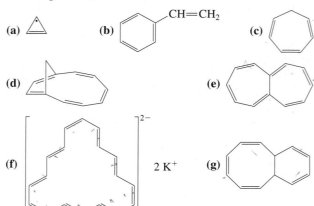

(a)

(b) CH=CH₂

(c)

(d)

(e)

(f) 2 K⁺

(g)

43. Following are spectroscopic and other data for several compounds. Propose a structure for each of them. **(a)** Molecular formula = $C_6H_4Br_2$. ¹H NMR spectrum A. ¹³C NMR: 3 peaks. IR: $\tilde{\nu}$ = 745 (s, broad) cm⁻¹. UV: $\lambda_{max}(\epsilon)$ = 263(150), 270(250), and 278(180) nm. **(b)** Molecular formula = C_7H_7BrO. ¹H NMR spectrum B. ¹³C NMR: 7 peaks. IR: $\tilde{\nu}$ = 680(s) and 765(s) cm⁻¹. **(c)** Molecular formula = $C_9H_{11}Br$. ¹H NMR spectrum C. ¹³C NMR: δ = 20.6 (CH₃), 23.6 (CH₃), 124.2 (C$_{quaternary}$), 129.0 (CH), 136.0 (C$_{quaternary}$), and 137.7 (C$_{quaternary}$) ppm.

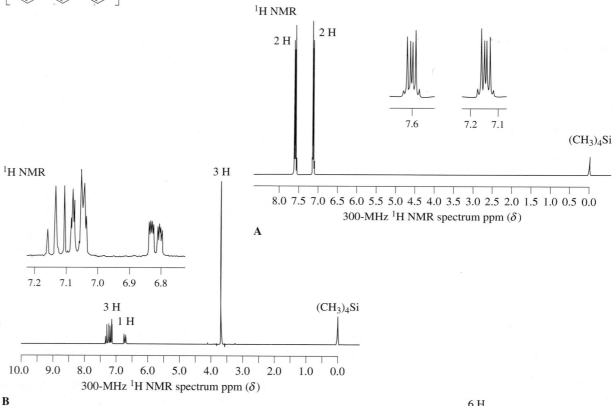

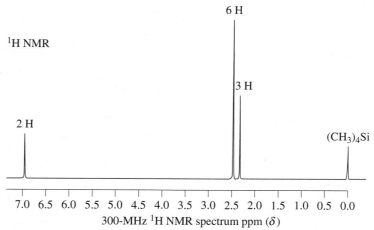

44. Both methylbenzene (toluene) and 1,6-heptadiyne have molecular formulas of C_7H_8 and molecular masses of 92. Which of the two mass spectra shown below corresponds to which compound? Explain your reasoning.

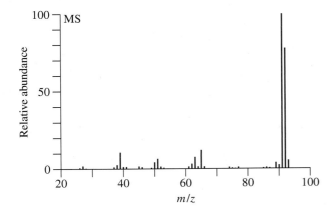

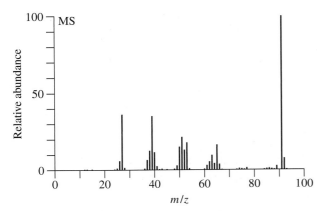

45. (a) Is it possible to distinguish the three isomers of dimethoxybenzene solely on the basis of the number of peaks in their proton-decoupled ^{13}C NMR spectra? Explain. (b) How many different isomers of dimethoxynaphthalene exist? How many peaks should each one exhibit in the proton-decoupled ^{13}C NMR spectrum?

46. The species resulting from the addition of benzene to $HF–SbF_5$ (Exercise 15-24) shows the following ^{13}C NMR absorptions: $\delta = 52.2(CH_2)$, $136.9(CH)$, $178.1(CH)$, and $186.6(CH)$ ppm. The signals at $\delta = 136.9$ and $\delta = 186.6$ are twice the intensity of the other peaks. Assign this spectrum.

47. Reaction review. Without consulting the Reaction Road Map on p. 687, suggest a reagent to convert benzene into each of the following compounds.

(a) C(CH₃)₃ — benzene ring

(b) Cl — benzene ring

(c) cyclohexane

(d) NO₂ — benzene ring

(e) O=C–CH₃ — benzene ring

(f) CH₂CH₃ — benzene ring

(g) SO₃H — benzene ring

(h) Br — benzene ring

48. Give the expected major product of addition of each of the following reagent mixtures to benzene. (**Hint:** Look for analogies to the reactions presented in this chapter.)

(a) $Cl_2 + AlCl_3$

(b) $T_2O + T_2SO_4$ (T = tritium, ^3H)

(c) $(CH_3)_3COH + H_3PO_4$

(d) N_2O_5 (which tends to dissociate into NO_2^+ and NO_3^-)

(e) $(CH_3)_2C{=}CH_2 + H_3PO_4$

(f) $(CH_3)_3CCH_2CH_2Cl + AlCl_3$

(g) $(CH_3)_2\overset{Br}{C}CH_2CH_2\overset{Br}{C}(CH_3)_2 + AlBr_3$

(h) $H_3C{-}\langle\ \rangle{-}COCl + AlCl_3$

49. Write mechanisms for reactions (c) and (f) in Problem 48.

50. Hexadeuteriobenzene, C_6D_6, is a very useful solvent for ^1H NMR spectroscopy because it dissolves a wide variety of organic compounds and, being aromatic, is very stable. Suggest a method for the preparation of C_6D_6.

51. Propose a mechanism for the sulfonation of benzene using chlorosulfuric acid, $ClSO_3H$ (given below).

$$\langle\ \rangle + Cl{-}\overset{O}{\underset{O}{S}}{-}OH \longrightarrow \langle\ \rangle{-}SO_3H + HCl$$

52. Benzene reacts with sulfur dichloride, SCl_2, in the presence of $AlCl_3$ to give diphenyl sulfide, $C_6H_5{-}S{-}C_6H_5$. Propose a mechanism for this process.

53. (a) 3-Phenylpropanoyl chloride, $C_6H_5CH_2CH_2COCl$, reacts with $AlCl_3$ to give a single product with the formula C_9H_8O and an ^1H NMR spectrum with signals at $\delta = 2.53$ (t, $J = 8$ Hz, 2 H), 3.02 (t, $J = 8$ Hz, 2 H), and 7.2–7.7 (m, 4 H) ppm. Propose a structure and a mechanism for the formation of this compound.
(b) The product of the process described in (a) is subjected to the following reaction sequence: (1) $NaBH_4$, CH_3CH_2OH; (2) conc. H_2SO_4, $100\,°C$; (3) H_2, $Pd–C$, CH_3CH_2OH. The resulting molecule exhibits five resonance lines in its ^{13}C NMR spectrum. What is the structure of the substance formed after each of the steps in this sequence?

54. The text states that alkylated benzenes are more susceptible to electrophilic attack than is benzene itself. Draw a graph like that in Figure 15-20 to show how the energy profile of electrophilic substitution of methylbenzene (toluene) would differ quantitatively from that of benzene.

55. Like haloalkanes, haloarenes are readily converted into organometallic reagents, which are sources of nucleophilic carbon.

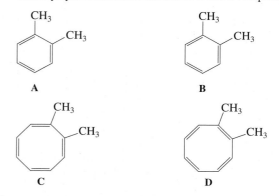

Phenylmagnesium bromide

Mg, (CH₃CH₂)₂O, 25°C

Phenylmagnesium chloride

Mg, THF, 50°C

Grignard reagents

The chemical behavior of these reagents is very similar to that of their alkyl counterparts. Write the main product of each of the following sequences.

(a) C₆H₅Br

1. Li, (CH₃CH₂)₂O
2. CH₃CHO
3. H⁺, H₂O

(b) C₆H₅Cl

1. Mg, THF
2. H₂C—CH₂ (O)
3. H⁺, H₂O

56. Give efficient syntheses of the following compounds, beginning with benzene. **(a)** 1-Phenyl-1-heptanol; **(b)** 2-phenyl-2-butanol; **(c)** 1-phenyloctane. (**Hint:** Use a method from Section 15-14. Why will Friedel-Crafts alkylation not work?)

57. Vanillin, whose structure is shown below and is the subject of the Chapter Opening, is a benzene derivative with several functional groups, each one of which displays its characteristic reactivity. What would you expect to be the products of reaction of vanillin with each of the following reagents?

(a) NaBH₄, CH₃CH₂OH

(b) NaOH, then CH₃I

Vanillin

Vanillin is obtained by extraction from the seed pods of plants in the *Vanilla* genus in a process that dates back at least 500 years: The Mexican Aztecs used it to flavor xocoatl, a chocolate drink. Cortez discovered it in the court of Montezuma and was responsible for introducing it to Europe. Increasing demand for vanillin necessitated development of synthetic procedures, which involve extraction of related compounds from other plant-derived sources. One of the most important of these sources is the wood-derived waste from the manufacture of paper. Treatment with aqueous NaOH, followed by oxidation with pressurized air at 170°C produces sizeable quantities of vanillin. The conversion of eugenol (an extract from cloves) into vanillin is very similar chemically. First, treatment of eugenol with KOH at 150°C in a high-boiling

solvent causes positional isomerization of its side-chain double bond:

Eugenol

KOH, 150°C, 1.5 h

Subsequently, oxidative cleavage, (see Section 12-12) completes the synthesis of vanillin.

(c) Propose a mechanism for the isomerization depicted in the scheme.

58. Because of cyclic delocalization, structures A and B shown here for *o*-dimethylbenzene (*o*-xylene) are simply two resonance forms of the same molecule. Can the same be said for the two dimethylcyclooctatetraene structures C and D? Explain.

A   **B**

C   **D**

59. The energy levels of the 2-propenyl (allyl) and cyclopropenyl π systems (see below) are compared qualitatively in the diagram below. **(a)** Draw the three molecular orbitals of each system, using plus and minus signs and dotted lines to indicate bonding overlap and nodes, as in Figure 15-4. Does either of these systems possess degenerate molecular orbitals? **(b)** How many π electrons would give rise to the maximum stabilization of the cyclopropenyl system, relative to 2-propenyl (allyl)? (Compare Figure 15-5, for benzene.) Draw Lewis structures for both systems with this number of π electrons and any appropriate atomic charges. **(c)** Could the cyclopropenyl system drawn in (b) qualify as being "aromatic"? Explain.

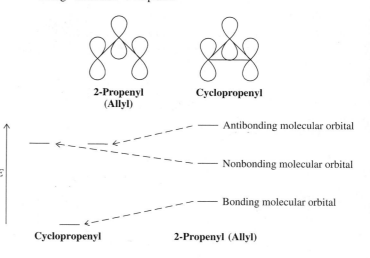

2-Propenyl (Allyl)   **Cyclopropenyl**

E

Antibonding molecular orbital

Nonbonding molecular orbital

Bonding molecular orbital

Cyclopropenyl   **2-Propenyl (Allyl)**

60. 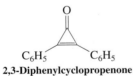 CHALLENGE 2,3-Diphenylcyclopropenone (see structure below) forms an addition product with HBr that exhibits the properties of an ionic substance (i.e., a salt). Suggest a structure for this product and a reason for its existence as a stable entity.

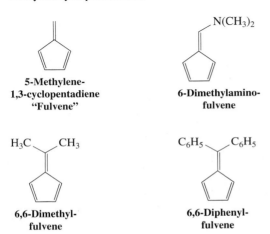

2,3-Diphenylcyclopropenone

61. CHALLENGE Is cyclobutadiene dication ($C_4H_4^{2+}$) aromatic according to Hückel's rule? Sketch its π molecular orbital diagram to illustrate your answer.

62. All the molecules shown below are examples of "fulvenes," or methylenecyclopentadienes.

5-Methylene-1,3-cyclopentadiene "Fulvene"

6-Dimethylamino-fulvene

6,6-Dimethyl-fulvene

6,6-Diphenyl-fulvene

(a) One of these structures is considerably more acidic than the others, with a pK_a around 20. Identify it and its most acidic hydrogen(s), and explain why it is an unusually strong acid for a molecule with only carbon–hydrogen bonds.
(b) None of the 7-methylene-1,3,5-cycloheptatrienes that correspond to the fulvene structures above show any unusual acidity. Explain.

63. A characteristic reaction of fulvenes is nucleophilic addition. To which carbon in a fulvene would you expect nucleophiles to add, and why?

64. CHALLENGE (a) The ^{1}H NMR spectrum of [18]annulene shows two signals, at $\delta = 9.28$ (12 H) and -2.99 (6 H) ppm. The negative chemical shift value refers to a resonance *upfield* (to the *right*) of $(CH_3)_4Si$. Explain this spectrum. (**Hint:** Consult Figure 15-9.) (b) The unusual molecule 1,6-methano[10]annulene (shown below) exhibits two sets of signals in the ^{1}H NMR spectrum at $\delta = 7.10$ (8 H) and -0.50 (2 H) ppm. Is this result a sign of aromatic character?

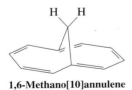

1,6-Methano[10]annulene

65. The ^{1}H NMR spectrum of the most stable isomer of [14]annulene shows two signals, at $\delta = -0.61$ (4 H) and 7.88 (10 H) ppm. Two possible structures for [14]annulene are shown here. How do they differ? Which one corresponds to the NMR spectrum described?

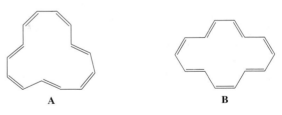

A   **B**

66. Explain the following reaction and the indicated stereochemical result mechanistically.

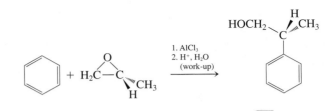

67. Metal-substituted benzenes have a long history of use in medicine. Before antibiotics were discovered, phenylarsenic derivatives were the only treatment for a number of diseases. Phenylmercury compounds continue to be used as fungicides and antimicrobial agents to the present day. On the basis of the general principles explained in this chapter and your knowledge of the characteristics of compounds of Hg^{2+} (see Section 12-7), propose a sensible synthesis of phenylmercury acetate (shown below).

Phenylmercury acetate

Team Problem

68. As a team, discuss the following complementary experimental results as they pertain to the mechanism of electrophilic aromatic substitution.

(a) A solution of HCl and benzene is colorless and does not conduct electricity, whereas a solution of HCl and $AlCl_3$ and benzene is colored and does conduct electricity.
(b) The following are ^{13}C NMR chemical shifts for the species below (see also Exercise 15-24):

Cl and C5: 186.6 ppm
C3: 178.1 ppm
C2 and C4: 136.9 ppm
C6: 52.2 ppm

(c) The relative rates of chlorination of the following compounds are as shown.

Compound	Relative Rate
Benzene	0.0005
Methylbenzene	0.157
1,4-Dimethylbenzene	1.00
1,2-Dimethylbenzene	2.1
1,2,4-Trimethylbenzene	200
1,2,3-Trimethylbenzene	340
1,2,3,4-Tetramethylbenzene	2000
1,2,3,5-Tetramethylbenzene	240,000
Pentamethylbenzene	360,000

(d) When 1,3,5-trimethylbenzene is treated with fluoroethane and one equivalent of BF_3 at $-80\,°C$, an isolable solid salt with a melting point of $-15\,°C$ is produced. Heating the salt results in 1-ethyl-2,4,6-trimethylbenzene.

Preprofessional Problems

69. *o*-Iodoaniline is the common name of which of the following compounds?

(a)

(b)

(c)

(d)

70. The species that is *not* aromatic according to Hückel's rule is

(a)   (b)   (c)   (d)

71. When compound A (shown below) is treated with dilute mineral acid, an isomerization takes place. Which of the following compounds is the new isomer formed?

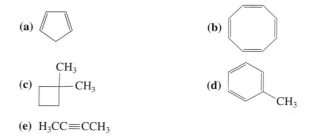

A

(a)   (b)

(c)   (d)

72. Which set of reagents will best carry out the conversion shown?

(a) HBr, peroxides; (b) Br_2, $FeBr_3$; (c) Br_2 in CCl_4; (d) KBr.

73. One of the compounds shown here contains carbon–carbon bonds that are 1.39Å long. Which one?

(a)   (b)

(c)   (d)

(e) $H_3CC\equiv CCH_3$

CHAPTER 16

Electrophilic Attack on Derivatives of Benzene

Substituents Control Regioselectivity

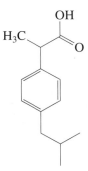

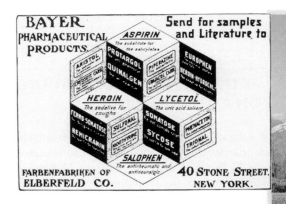

At some time in your life you have probably ingested at least one of the painkillers aspirin, acetaminophen, naproxen, or ibuprofen, perhaps better known under one of their respective brand names, aspirin, Tylenol, Naprosyn, and Advil. Aspirin, acetaminophen, and ibuprofen are ortho- or para-disubstituted benzenes; naproxen is a disubstituted naphthalene. How are such compounds synthesized? The answer is by electrophilic aromatic substitution.

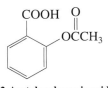

2-Acetyloxybenzoic acid
(Aspirin)

N-(4-Hydroxyphenyl)acetamide
(Acetaminophen)

2-[2-(6-Methoxynaphthyl)]-propanoic acid
(Naproxen)

2-[4-(2-Methylpropyl)-phenyl]propanoic acid
(Ibuprofen)

Aspirin, prepared industrially by selective electrophilic aromatic substitution of phenol, is arguably the blockbuster drug of all times. Its active metabolite, 2-hydroxybenzoic acid (salicylic acid), obtained from the bark of the white willow tree, has been used for four millennia for the treatment of inflammation and to relieve pain or discomfort caused by arthritis, soft-tissue injuries, and fever. Aspirin was discovered by the German company Bayer in the late 19th century and ironically marketed together with another drug, heroin, whose addictive side effects were not recognized then.

Chapter 15 described the use of this transformation in the preparation of monosubstituted benzenes. In this chapter we analyze the effect of such a first substituent on the reactivity and regioselectivity (orientation) of a subsequent electrophilic substitution reaction. Specifically, we shall see that substituents on benzene can be grouped into (1) **activators** (electron donors), which generally direct a second electrophilic attack to the **ortho** and **para positions,** and (2) **deactivators** (electron acceptors), which generally direct electrophiles to the **meta positions.** We will then devise strategies toward the synthesis of polysubstituted arenes, such as the analgesics depicted on the previous page.

Activated ring   **Deactivated ring**

16-1 | ACTIVATION OR DEACTIVATION BY SUBSTITUENTS ON A BENZENE RING

In Section 14-8 we discussed the effect that substituents have on the efficiency of the Diels-Alder reaction: Electron donors on the diene and acceptors on the dienophile are beneficial to the outcome of the cycloaddition. Chapter 15 revealed another manifestation of these effects: Introduction of electron-withdrawing substituents into the benzene ring (e.g., as in nitration) caused further electrophilic aromatic substitution (EAS) to slow down, whereas the incorporation of donors, as in the Friedel-Crafts alkylation, caused substitution to accelerate. What are the factors that contribute to the activating or deactivating nature of substituents in these processes? How do they make a monosubstituted benzene more or less susceptible to further electrophilic attack?

Increasing rate of EAS

The electronic influence of any substituent is determined by an interplay of two effects that, depending on the structure of the substituent, may operate simultaneously: **induction** and **resonance.** *Induction* occurs through the *σ framework,* tapers off rapidly with distance, and is mostly governed by the relative electronegativity of atoms and the resulting polarization of bonds (Tables 1-2 and 8-2). *Resonance* takes place through *π bonds,* is therefore longer range, and is particularly strong in charged systems (Section 1-5, Chapter 14).

Let us look at both of these effects of typical groups introduced by electrophilic aromatic substitution, starting with inductive donors and acceptors. Thus, simple alkyl groups, such as methyl, are donating by virtue of their inductive (Section 11-3) and hyperconjugating σ frame (Sections 7-5 and 11-5). On the other hand, trifluoromethyl (by virtue of its electronegative fluorines) is electron withdrawing. Similarly, directly bound heteroatoms, such as N, O, and the halogens (by virtue of their relative electronegativity), as well as positively polarized atoms, such as those in carbonyl, cyano, nitro, and sulfonyl functions, are inductively electron withdrawing.

Inductive Effects of Some Substituents on the Benzene Ring

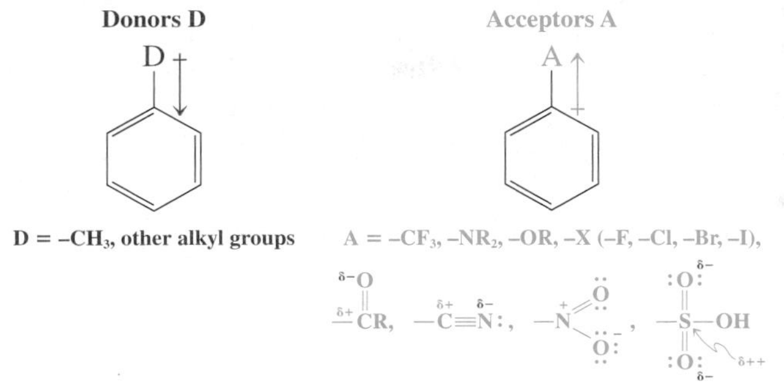

Donors D

Acceptors A

D = –CH₃, other alkyl groups → $D = -CH_3$, other alkyl groups

$A = -CF_3, -NR_2, -OR, -X\ (-F, -Cl, -Br, -I),$

$$\overset{\delta-}{\underset{\delta+}{\text{O}}}\text{CR}, \quad -\overset{\delta+}{C}\equiv\overset{\delta-}{N}:, \quad -\overset{+}{N}\diagdown^{\text{O}}_{\text{O}:^-}, \quad -\overset{:O:^{\delta-}}{\underset{:O:_{\delta-}}{S}}-OH_{\delta++}$$

Now we turn to substituents that resonate with the aromatic π system. Resonance donors bear at least one electron pair capable of delocalization into the benzene ring. Therefore, such groups as –NR₂, –OR, and the halogens belong in that category. You will note that, inductively, these groups are electron withdrawing; in other words, here the two phenomena, induction and resonance, are opposing each other. Which one wins out? The answer depends on the relative electronegativity of the heteroatoms (Table 1-2) and on the ability of their respective p orbitals to overlap with the aromatic π system. For amino and alkoxy groups, resonance overrides induction. For the halogens, induction outcompetes resonance, making them weak electron acceptors.

Resonance Donation to Benzene

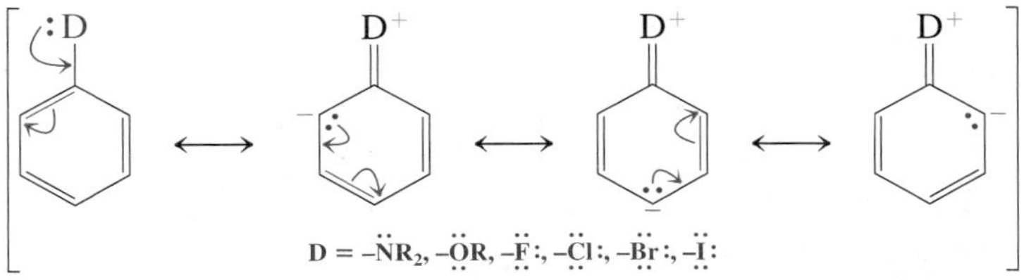

$D = -\ddot{N}R_2, -\ddot{O}R, -\ddot{F}:, -\ddot{C}l:, -\ddot{B}r:, -\ddot{I}:$

Finally, groups bearing a polarized double or triple bond, whose positive (δ^+) end is attached to the benzene nucleus, such as carbonyl, cyano, nitro, and sulfonyl, are electron withdrawing through resonance.

Resonance Acceptance from Benzene

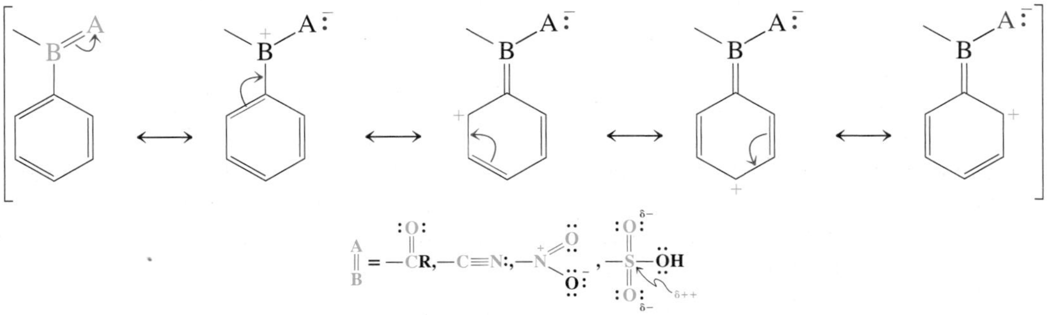

$$\overset{A}{\underset{B}{||}} = -\overset{:O:}{C}R, -C\equiv N:, -\overset{+}{N}\diagdown^{\overset{\delta-}{O}}_{\ddot{O}:}, \quad -\overset{:\overset{\delta-}{O}:}{\underset{:O:_{\delta-}}{S}}-\ddot{O}H_{\delta++}$$

Note that, here, resonance reinforces induction.

Electrostatic potential maps indicate the presence of electron-donating substituents by depicting the benzene ring as shaded relatively red; electron-withdrawing groups make the benzene ring appear shaded relatively blue (green).

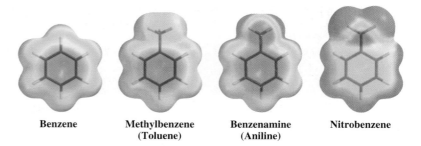

Benzene   Methylbenzene   Benzenamine   Nitrobenzene
(Toluene)   (Aniline)

Exercise 16-1

Explain the ^{1}H NMR spectral assignments in Figures 15-11 and 15-12. (**Hint:** Draw resonance structures involving the substituents on the benzene ring.)

Exercise 16-2

The ^{13}C NMR spectrum of phenol, C_6H_5OH, shows four lines at $\delta = 116.1$ (C2), 120.8 (C4), 130.5 (C3), and 155.6 (C1) ppm. Explain these assignments. (**Hint:** The ^{13}C chemical shift for benzene is $\delta = 128.7$ ppm.)

How do we know whether a substituent functions as a donor or acceptor? In electrophilic aromatic substitution, the answer is simple. Because the attacking species is an electrophile, the more electron rich the arene, the faster the reaction. Conversely, the more electron poor the arene, the slower the reaction. Hence, electron donors activate the ring, whereas electron acceptors deactivate.

Relative Rates of Nitration of C_6H_5X

$$X = NH(C_6H_5) > OH > CH_3 > H > Cl > CO_2CH_2CH_3 > CF_3 > NO_2$$

     10^6 1000   25   1   0.033   0.0037 2.6×10^{-5} 6×10^{-8}

◄ **Increasing rate of nitration**

Exercise 16-3

Specify whether the benzene rings in the compounds below are activated or deactivated.

(a) CH$_2$CH$_3$ / CH$_2$CH$_3$ (b) NO$_2$ / CH$_3$ (c) CO$_2$H / CF$_3$ (d) OCH$_3$ / N(CH$_3$)$_2$

In Summary When considering the effect of substituents on the reactivity of the benzene nucleus, we have to analyze the contributions that occur by induction and resonance. We can group these substituents into two classes: (1) electron donors, which accelerate electrophilic aromatic substitutions relative to benzene, and (2) electron acceptors, which retard them.

16-2 DIRECTING ELECTRON-DONATING EFFECTS OF ALKYL GROUPS

We are now ready to tackle the question of the regioselectivity (orientation) of electrophilic aromatic substitution of substituted benzenes. What controls the position along the benzene ring at which an electrophile will attack? We begin with the electrophilic substitution reactions of alkyl-substituted benzenes, such as methylbenzene (toluene), in which the methyl group is electron donating by induction and hyperconjugation.

Groups that donate electrons by induction and hyperconjugation are activating and direct ortho and para

Electrophilic bromination of methylbenzene (toluene) is considerably faster than the bromination of benzene itself. The reaction is also regioselective: It results mainly in para (60%) and ortho (40%) substitutions, with virtually no meta product.

Electrophilic Bromination of Methylbenzene (Toluene) Gives Ortho and Para Substitution

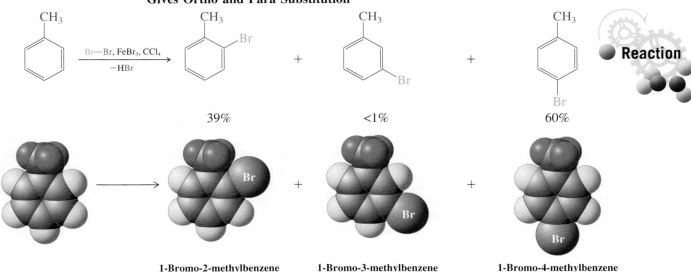

39%   <1%   60%

1-Bromo-2-methylbenzene   1-Bromo-3-methylbenzene   1-Bromo-4-methylbenzene
(*o*-Bromotoluene)   (*m*-Bromotoluene)   (*p*-Bromotoluene)

Is bromination a special case? The answer is no; nitration, sulfonation, and Friedel-Crafts reactions of the alkylbenzene give similar results—mainly ortho and para substitutions (see also Table 16-2 in Section 16-3). Evidently, the nature of the attacking electrophile has little influence on the observed orientation; it is the alkyl group that matters. Because there is virtually no meta product, we say that the activating methyl substituent is **ortho and para directing.**

Can we explain this selectivity by a mechanism? Let us inspect the possible resonance forms of the cations generated after the electrophile, E^+, has attacked the ring in the first, and rate-determining, step.

Ortho, Meta, and Para Attack on Methylbenzene (Toluene)

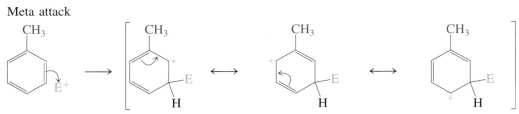

Ortho attack (E^+ = electrophile)

Most significant
resonance contributor

More stable cation

Meta attack

Less stable cation

Methylbenzene (Toluene)

(1,1-Dimethylethyl)-benzene (*tert*-Butylbenzene)

Para attack

Most significant resonance contributor

More stable cation

The alkyl group is electron donating by induction and hyperconjugation (Section 16-1). Electrophilic attack at the ortho and para positions produces an intermediate carbocation in which one of the resonance forms places the positive charge *next* to the alkyl substituent, rendering it tertiary carbocation-like (Section 7-5). Because the alkyl group can donate electron density to stabilize the positive charge, that resonance form is a more important contributor to the resonance hybrid than the others, in which the positive charge is at an unsubstituted carbon. Meta attack, however, produces an intermediate in which *none* of the resonance forms benefits from such direct stabilization. Thus, electrophilic attack on a carbon located ortho or para to the methyl (or another alkyl group) leads to a cationic intermediate that is more stable than the one derived from attack at the meta carbon. The transition state leading to the more stable intermediate is of relatively low energy (Hammond postulate, Section 3-5) and is therefore reached relatively rapidly.

Why are the two favored products, ortho and para, not formed in equal amounts? Frequently, the answer is steric effects. Thus, attack ortho to an existing substituent, especially when it is bulky, by an electrophile (again, especially when bulky) is sterically more encumbered than para attack. Therefore, para products often predominate over their ortho isomers. In the bromination of methylbenzene (toluene), this predominance is small. However, similar halogenation of 1,1-dimethylethyl (*tert*-butyl) benzene results in a much larger para : ortho ratio (~10 : 1).

A particularly impressive "green" example (Real Life 3-1) of this effect is achieved in the Friedel-Crafts acetylation of (2-methylpropyl)benzene with acetic anhydride, in an industrial approach to an ibuprofen intermediate (Chapter Opening; see also Exercise 16-10). Here, a porous zeolite catalyst (see Section 3-3) provides not only the acidic surface sites necessary for the reaction to proceed, but also an environment that enhances para selectivity. This process avoids the use of the corrosive acetyl chloride and AlCl₃ reagents and with it the formation of the toxic HCl by-product of the classical Friedel-Crafts acetylation. Instead, the by-product is acetic acid, itself a valuable commodity.

A Green Acetylation of an Alkylbenzene

$$\xrightarrow{\text{Zeolite, 140°C}}_{-\text{CH}_3\text{COOH}}$$

80% yield, 96% para selectivity

Groups that withdraw electrons inductively are deactivating and meta directing

The strongly electronegative fluorine atoms in (trifluoromethyl)benzene make the trifluoromethyl group inductively electron withdrawing (Section 16-1). In this case, the benzene ring becomes deactivated and reaction with electrophiles is very sluggish. Under stringent conditions, such as heating, substitution does take place—but *only* at the meta positions: The trifluoromethyl group is deactivating and **meta directing.**

Electrophilic Nitration of (Trifluoromethyl)benzene Gives Mainly Meta Substitution

6%	91%	3%
1-Nitro-2-(trifluoromethyl)benzene	**1-Nitro-3-(trifluoromethyl)benzene**	**1-Nitro-4-(trifluoromethyl)benzene**
[*o*-Nitro(trifluoromethyl)benzene]	[*m*-Nitro(trifluoromethyl)benzene]	[*p*-Nitro(trifluoromethyl)benzene]

Once again, the explanation lies in the various resonance forms for the cation produced by ortho, meta, and para attack.

Ortho, Meta, and Para Attack on (Trifluoromethyl)benzene

Ortho attack

Strongly destabilized cation

Mechanism

Meta attack

Less destabilized cation

Para attack

Poor resonance contributor

Strongly destabilized cation

The presence of an inductively electron-withdrawing substituent *de*stabilizes the carbocations resulting from electrophilic attack at *all* positions in the ring. However, ortho and para attack are even less favored than meta attack for the same reasons that they are relatively favored with methylbenzene (toluene): In each case, one of the resonance forms in the intermediate cation places the positive charge next to the substituent. This structure is stabilized by an electron-donating group, but it is *destabilized* by an *electron-withdrawing* substituent—removing electron density from a positively charged center is energetically unfavored. Meta attack avoids this situation. The destabilizing inductive effect is still felt in the meta intermediate, but to a lesser extent. Therefore, the trifluoromethyl group slows substitution and directs the electrophile meta, or, more accurately, *away* from the ortho and para carbons.

Exercise 16-4

Rank these compounds in order of decreasing activity in electrophilic substitution.

(a) [structure: benzene with CF₃ and CH₃ substituents ortho to each other]
(b) [structure: toluene, CH₃ on benzene]
(c) [structure: (trifluoromethyl)benzene, CF₃ on benzene]
(d) [structure: benzene with two CH₃ substituents ortho to each other]

Exercise 16-5

Electrophilic bromination of an equimolar mixture of methylbenzene (toluene) and (trifluoromethyl)benzene with one equivalent of bromine gives only 1-bromo-2-methylbenzene and 1-bromo-4-methylbenzene. Explain.

In Summary Substituents that donate electrons by induction and hyperconjugation activate the benzene ring and direct electrophiles ortho and para; their inductively electron-accepting counterparts deactivate the benzene ring and direct electrophiles to the meta positions.

16-3 DIRECTING EFFECTS OF SUBSTITUENTS IN CONJUGATION WITH THE BENZENE RING

What is the influence of substituents whose electrons are in conjugation with those of the benzene ring? We can answer this question by again comparing the resonance forms of the intermediates resulting from the various modes of electrophilic attack.

Groups that donate electrons by resonance activate and direct ortho and para

Reaction

Benzene rings bearing the groups –NH₂ and –OH are strongly activated. For example, halogenations of benzenamine (aniline) and phenol not only take place in the absence of catalysts but also are difficult to stop at single substitution. The reactions proceed very rapidly and, as in activation by alkyl groups (Section 16-2), furnish exclusively *ortho- and para-substituted products.*

Electrophilic Brominations of Benzenamine (Aniline) and Phenol Give Ortho and Para Substitution

$$\text{Benzenamine (NH}_2) \xrightarrow[\text{−3 HBr}]{3\ \text{Br—Br, H}_2\text{O}} \text{2,4,6-Tribromobenzenamine (NH}_2\text{, 3 Br)}\quad 100\%$$

$$\text{Phenol (OH)} \xrightarrow[\text{−3 HBr}]{3\ \text{Br—Br, H}_2\text{O}} \text{2,4,6-Tribromophenol (OH, 3 Br)}\quad 100\%$$

Benzenamine
(Aniline)

2,4,6-Tribromobenzenamine
(2,4,6-Tribromoaniline)

Phenol

2,4,6-Tribromophenol

Better control of monosubstitution is attained with modified amino and hydroxy substituents, such as in *N*-phenylacetamide (acetanilide) and methoxybenzene (anisole). These groups are ortho and para directing but less strongly activating (Section 16-5).

Electrophilic Nitration of *N*-Phenylacetamide (Acetanilide)

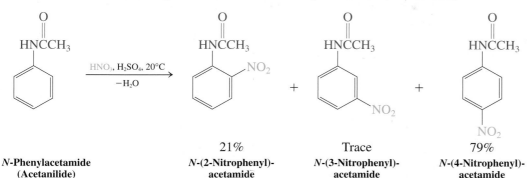

Mechanism

	21%	Trace	79%
N-Phenylacetamide (Acetanilide)	*N*-(2-Nitrophenyl)- acetamide (*o*-Nitroacetanilide)	*N*-(3-Nitrophenyl)- acetamide (*m*-Nitroacetanilide)	*N*-(4-Nitrophenyl)- acetamide (*p*-Nitroacetanilide)

Both the activated nature of these compounds and the observed regioselectivity on electrophilic substitution can be explained by writing resonance forms for the various intermediate cations.

Ortho, Meta, and Para Attack on Benzenamine (Aniline)

Ortho attack

Strongly stabilized cation

Strongly
contributing
all-octet form

Meta attack

Animation

ANIMATED MECHANISM:
Electrophilic aromatic
substitution of benzenamine
(ortho vs meta vs para)

Para attack

Strongly
contributing
all-octet form

Strongly stabilized cation

Because nitrogen is more electronegative than carbon, the amino group in benzenamine (aniline) is inductively electron withdrawing (Section 16-1). However, the lone electron pair on the nitrogen atom may participate in resonance, thereby stabilizing the intermediate cations resulting from ortho and para (but not meta) substitutions. This resonance contribution outweighs the inductive effect. Compared to alkyl groups (Section 16-2), the amino function provides not only an extra resonance contributor to the intermediate cation, but one that features an all-important Lewis octet. The result is a much reduced activation barrier for ortho or para attack. Consequently, benzenamine (aniline) is strongly activated toward electrophilic substitution relative to alkylbenzenes (and benzene), and the reaction is more highly regioselective as well, giving ortho and para substitution products almost exclusively.

Solved Exercise 16-6	**Working with the Concepts: Predicting Regiochemistry in Electrophilic Aromatic Substitutions**

Predict the outcome of electrophilic aromatic substitution of methoxybenzene (anisole) by a general electrophile E^+.

Strategy

In all problems addressing the potential regioselectivity of electrophilic substitution on substituted benzenes, there is a golden rule: When in doubt, write down all intermediates from all possible modes of attack before attempting to formulate a solution.

Solution

• In the case of methoxybenzene, this procedure results in the following cations:

Ortho attack

Meta attack

Para attack

• How do they differ? You can immediately tell that ortho and para attacks give intermediates with four resonance forms, respectively, one of which in each case involves the lone electron pair on the methoxy group.

• In contrast, meta substitution proceeds through a cation described by only three resonance forms, none of which benefits from the participation of the oxygen lone pair of electrons.

• Thus, substitution will occur exclusively ortho and para to the directing methoxy substituent in the starting material.

Exercise 16-7	**Try It Yourself**

In strongly acidic solution, benzenamine (aniline) becomes less reactive to electrophilic attack, and, while ortho, para attack is still dominant, increased meta substitution is observed. Explain. (**Hint:** The nitrogen atom in benzenamine may behave as a base. How would you classify the resulting substituent within the framework of the discussion in Section 16-1?)

REAL LIFE: MATERIALS 16-1 | Explosive Nitroarenes: TNT and Picric Acid

Complete ortho, para nitration of methylbenzene (toluene) or phenol furnishes the corresponding trinitro derivatives, both of which are powerful explosives: TNT (discovered in 1863) and picric acid (1771). Both compounds have long histories as military and industrial explosives.

**2-Methyl-1,3,5-trinitrobenzene
(2,4,6-Trinitrotoluene, TNT)**

**2,4,6-Trinitrophenol
(Picric acid)**

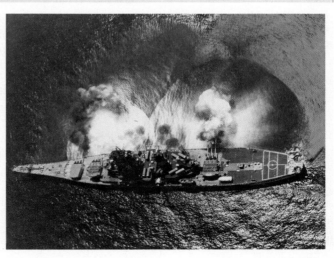

Spherical shock waves generated by the firing of the huge guns of the USS Iowa are clearly visible on the ocean surface.

Explosives are generally high-energy-density compounds capable of extremely rapid decomposition. In contrast to propellants (such as rocket fuel), they do not burn, but detonate under their own power. They frequently generate high heat and a large quantity of gaseous products, producing a (usually destructive) shock wave: TNT has a detonation velocity of 7459 m s^{-1} ($\sim$4.6 mi s^{-1}). An explosion can be initiated by impact (including blasting caps), friction, heat and flame, electrical discharge (including static), and ultraviolet irradiation, depending on the compound. The nitro group features prominently in these materials, because it functions as an oxidizer to the surrounding carbon framework (producing the gases CO and CO_2) and as a precursor to N_2.

TNT is the most widely used military explosive in history. The reasons for its popularity are its low cost and simplicity of preparation, safe handling (low sensitivity to impact and friction), relatively high explosive power (yet good chemical and thermal stability), low volatility and toxicity, compatibility with other explosives, and a low melting point, allowing for melt-casted formulations.

TNT has become such a standard, particularly in military uses, that the destructive power of other explosives, especially in bombs, is often compared to that of an equivalent of TNT. For example, the first atomic bomb—detonated on July 16, 1945, in New Mexico—had the equivalent power of 19,000 tons of TNT. The device exploded over Hiroshima, Japan, which killed more than 140,000 people, had the power of 13,000 tons of TNT. Although these numbers appear huge, comparison with the hydrogen bomb—with the destructive equivalent of 10 million tons of TNT—dwarfs

them. For further calibration, all of the explosions of World War II combined amounted to the equivalent of "only" 2 million tons of TNT.

Picric acid has some commercial applications other than as an explosive—in matches, in the leather industry, in electric batteries, and in colored glass. It is called an acid because of the unusually high acidity of its hydroxy group (pK_a 0.38; Section 22-3), which is increased beyond that of acetic acid (pK_a 4.7) and even hydrogen fluoride (pK_a 3.2; Table 2-2) by the electron-withdrawing effect of the three nitro groups. This property was in part responsible for its replacement by TNT in military uses. For example, in artillery shells, it would corrode the casing and cause leakage, thus creating a hazard.

TNT and picric acid have been replaced gradually by tetryl, RDX, and nitroglycerine (Section 9-11). On the research front, chemists are continuing to explore novel structures. A case in point is octanitrocubane, synthesized in 2000, in which ring strain adds to the brisance of the compound. Its molecular formula, $C_8N_8O_{16}$, indicates the potential to generate 8 CO_2 + 4 N_2 molecules, with an associated 1150-fold volume expansion. A recent example from 2012 is the exotic salt TKX-50, in which two lone carbon atoms are surrounded by 10 nitrogens and the equivalent of 4 H_2O moieties.

**2,4,6-Trinitrophenyl-
N-methylnitramine
(Tetryl)**

**1,3,5-Trinitro-1,3,5-
triazacyclohexane
(RDX)**

Octanitrocubane

TKX-50

Groups that withdraw electrons by resonance deactivate and direct meta

Several groups *deactivate* the benzene ring by resonance (Section 16-1). An example is the carboxy group in benzoic acid, $C_6H_5CO_2H$. Nitration of benzoic acid takes place at only about 1/1000th the rate of benzene nitration and gives predominantly meta substitution. The CO_2H group is deactivating and, as in inductive deactivation (Section 16-2), *meta directing*.

Electrophilic Meta Nitration of Benzoic Acid

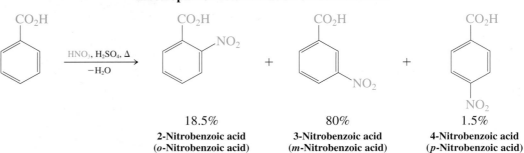

18.5%	80%	1.5%
2-Nitrobenzoic acid (*o*-Nitrobenzoic acid)	**3-Nitrobenzoic acid** (*m*-Nitrobenzoic acid)	**4-Nitrobenzoic acid** (*p*-Nitrobenzoic acid)

Reaction

Animation

ANIMATED MECHANISM:
Electrophilic aromatic substitution of benzoic acid (ortho vs meta vs para)

Let us see how conjugation with the CO_2H function affects the resonance forms of the cations resulting from electrophilic attack on benzoic acid.

Ortho, Meta, and Para Attack on Benzoic Acid

Ortho attack

Poor   **Poor**

Strongly destabilized cation

Meta attack

None is poor
Less destabilized cation

Para attack

Poor   **Poor**

Strongly destabilized cation

Attack at the meta position avoids placing the positive charge next to the electron-withdrawing carboxy group, whereas ortho and para attacks necessitate the formulation of poor resonance contributors. Thus, while the substituent deactivates *all* positions, it does so to a greater extent at the ortho and para positions than at the meta positions. One might say, meta "wins by default."

Exercise 16-8

Electrophilic nitration of nitrobenzene gives almost exclusively 1,3-dinitrobenzene. Formulate the (poor) resonance forms of the intermediate cations resulting from attack by NO_2^+ at the ortho and para positions that explain this result.

There is always an exception: halogen substituents, although deactivating, direct ortho and para

Halogen substituents inductively withdraw electron density (Section 16-1); however, they are donors by resonance. On balance, the inductive effect wins out, rendering haloarenes *deactivated*. Nevertheless, the electrophilic substitution that does take place is mainly at the *ortho and para positions*.

Electrophilic Bromination of Bromobenzene Results in *ortho*- and *para*-Dibromobenzene

13%	2%	85%
1,2-Dibromobenzene (*o*-Dibromobenzene)	**1,3-Dibromobenzene** (*m*-Dibromobenzene)	**1,4-Dibromobenzene** (*p*-Dibromobenzene)

Reaction

The competition between resonance and inductive effects explains this seemingly contradictory reactivity. Again, we must examine the resonance forms for the various possible intermediates.

Mechanism

Ortho, Meta, and Para Attack on a Halobenzene

Ortho attack

More stable cation

Strongly contributing all-octet form

Meta attack

Less stable cation

Para attack

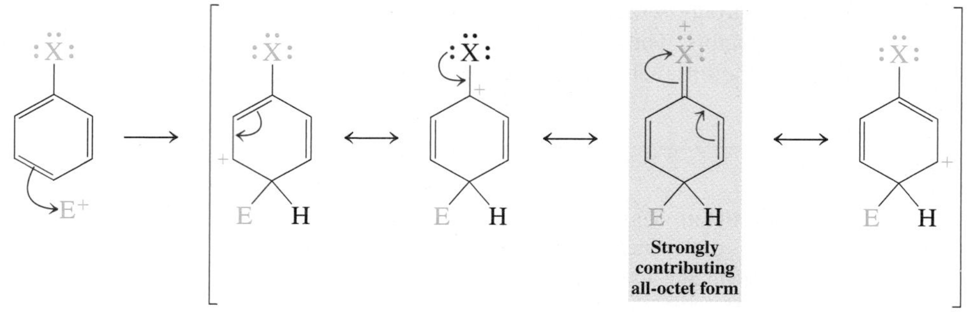

More stable cation

Note that ortho and para attack lead to resonance forms in which the positive charge is placed next to the halogen substituent. Although this might be expected to be unfavorable, because the halogen is inductively electron withdrawing, resonance with the lone electron pairs allows the charge to be delocalized. Therefore, ortho and para substitutions become the preferred modes of reaction. The inductive effect of the halogen is still strong enough to make all three possible cations less stable than the one derived from benzene itself. Therefore, we have the unusual result that halogens are *ortho and para directing,* but *deactivating.*

This section completes the survey of the regioselectivity of electrophilic attack on mono-substituted benzenes, summarized in Table 16-1. Table 16-2 ranks various substituents by their activating power and lists the product distributions obtained on electrophilic nitration of the benzene ring.

Exercise 16-9

Explain why **(a)** $-NO_2$, **(b)** $-\overset{+}{N}R_3$, and **(c)** $-SO_3H$ are meta directing. **(d)** Why should phenyl be activating and ortho and para directing (Table 16-1)? [**Hint:** Draw resonance forms for the appropriate cationic intermediates of electrophilic attack on phenylbenzene (biphenyl).]

Table 16-1	Effects of Substituents in Electrophilic Aromatic Substitution

Ortho and para directors

Moderate and strong activators

$$-\ddot{N}H_2 \sim -\ddot{N}HR \sim -\ddot{N}R_2 > -\ddot{N}HCR \overset{\displaystyle :\overset{..}{O}:}{\overset{\|}{}} > -\overset{..}{\underset{..}{O}}H \sim -\overset{..}{\underset{..}{O}}R$$

⟵ **Increasing activation**

Weak activators   Weak deactivators

Alkyl ≳ phenyl $-\overset{..}{\underset{..}{F}}: \sim -\overset{..}{\underset{..}{C}}l: \sim -\overset{..}{\underset{..}{B}}r: \sim -\overset{..}{\underset{..}{I}}:$

Meta directors

Strong deactivators

$$\underset{\displaystyle}{-\overset{O}{\overset{\|}{C}}OH} \sim -\overset{O}{\overset{\|}{C}}OR \sim -\overset{O}{\overset{\|}{C}}R < -CF_3 < -C\equiv N < -\overset{O}{\underset{O}{\overset{\|}{\underset{\|}{S}}}}OH < -NO_2 < -\overset{+}{N}R_3$$

Increasing deactivation ⟶

Table 16-2 Relative Rates and Orientational Preferences in the Nitration of Some Monosubstituted Benzenes, RC$_6$H$_5$

	R	Relative rate	Percentage of isomer		
			Ortho	Meta	Para
Ortho, para directors	NH(C$_6$H$_5$)	8.4×10^5	71	<0.1	29
	OH	1000	40	<2	58
	CH$_3$	25	58	4	38
	C$_6$H$_5$	8	30	<0.6	70
	H	1			
Special: Ortho, para directors	I	0.18	41	<0.2	59
	Cl	0.033	31	<0.2	69
Meta directors	CO$_2$CH$_2$CH$_3$	0.0037	24	72	4
	CF$_3$	2.6×10^{-5}	6	91	3
	NO$_2$	6×10^{-8}	5	93	2
	$\overset{+}{N}(CH_3)_3$	1.2×10^{-8}	0	89	11

Increasing activation / *Increasing deactivation* / *Increasing rate of nitration*

Exercise 16-10

Compound A is an intermediate in one of the syntheses of ibuprofen (see Chapter Opening). Propose a synthetic route to compound A, starting with (2-methylpropyl)benzene. (**Hint:** You need to introduce the cyano group by nucleophilic substitution.)

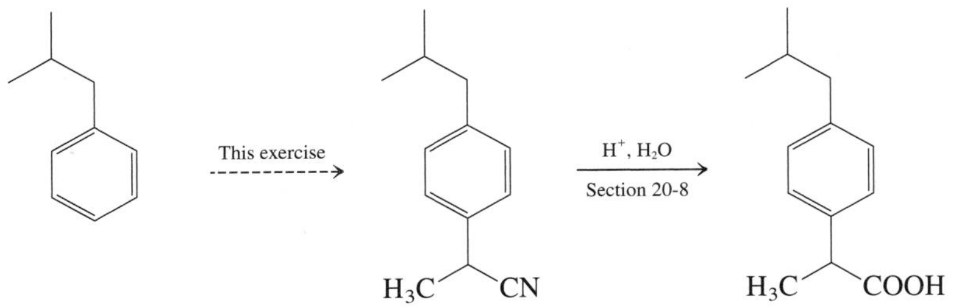

(2-Methylpropyl)-benzene →(This exercise)→ A →(H$^+$, H$_2$O, Section 20-8)→ Ibuprofen

---

In Summary Activating groups, whether operating by induction and hyperconjugation or resonance, direct incoming electrophiles to the ortho and para positions, whereas deactivating groups direct to the meta carbons. This statement is true for all classes of substituents except one—the halogens. They are deactivating by induction, but they stabilize positive charges by resonance, therefore effecting ortho, para substitution.

16-4 | ELECTROPHILIC ATTACK ON DISUBSTITUTED BENZENES

Do the rules developed so far in this chapter predict the reactivity and regioselectivity of still higher substitution? We shall see that they do, provided we take into account the individual effect of each substituent. Let us investigate the reactions of disubstituted benzenes with electrophiles.

The strongest activator wins out

In trying to predict the regioselectivity of electrophilic substitution of disubstituted benzenes, we have to apply the same considerations that were necessary to understand the directing effects of one substituent (Sections 16-1 through 16-3). This may seem rather difficult at first, because the two substituents may be ortho, para, or they may be meta directing, and they can be placed around the ring in three possible ways, 1,2, 1,3, or 1,4. However, the problem becomes much simpler when you remember that ortho, para directors are activators and therefore accelerate attack of the electrophile at the ortho, para positions relative to benzene. In contrast, meta directors are deactivators and exert regiocontrol by retarding ortho, para substitution more than meta substitution. Taking these electronic effects into account, in conjunction with steric considerations, allows us to formulate a set of simple guidelines that enable us to predict the outcome of most electrophilic aromatic substitution reactions.

Guideline 1. The most powerful activator controls the position of attack (indicated by yellow arrows).

(There are two equivalent ortho positions to OH, attack at only one of which is marked.)

Guideline 2. Experimentally, we can rank the directing power of substituents into three groups:

$$NR_2, OR \quad > \quad X, R \quad > \quad \text{meta directors}$$
$$\text{I} \qquad\qquad \text{II} \qquad\qquad \text{III}$$

Members of the higher-ranking groups override the effect of members of lower rank (guideline 1). However, substituents within each group compete to give isomer mixtures (except when such substituents direct to the same position or when symmetry ensures the generation of a single product).

(There are four equivalent ortho positions to the two Br substituents, only one of which is marked.)

(There are four pairwise equivalent, respective ortho, para positions to the two CH₃O substituents, only one set of which is marked.)

(There are four pairwise equivalent, respective meta positions to the two COOH substituents, only two of which are marked.)

Guideline 3. In such cases where product mixtures are predicted on the basis of guidelines 1 and 2, you may typically discount those that constitute ortho attack to bulky groups or between two substituents (dashed black arrows).

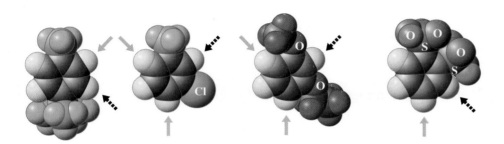

[There are two equivalent
ortho positions to C(CH₃)₃
only one of which is marked.]

(There are four
pairwise equivalent,
respective meta positions
to the two SO₃H
substituents, only one
of which is marked.)

COOH
HO
SO₃H

Sulfosalicylic acid

Did you ever wonder what the doctor does with your urine sample? One test is the gradual addition of sulfosalicylic acid (2-hydroxy-5-sulfobenzoic acid), which precipitates proteins from the solution, as shown in the photo below. High levels of protein may indicate problems with your kidney or in your urinary tract.

Guideline 4. Guidelines 1 through 3 are applicable to more highly substituted benzenes for which the decreasing number of reactive positions reduces the number of possible products.

Solved Exercise 16-11 | **Working with the Concepts: Predicting Regiochemistry in Electrophilic Aromatic Substitutions of Di- and More Highly Substituted Benzenes**

Predict the result of mononitration of 1-bromo-2-methoxybenzene:

OCH₃
Br

Strategy
Following the given guidelines, look at the overall electronic and directing effects of each substituent separately.

Remember WHIP

*W*hat

*H*ow

*I*nformation

*P*roceed

Solution

- Methoxy is a strong activator and directs ortho, para.
- Bromine is a weak deactivator but still directs ortho, para. Clearly, the two substituents conflict in the direction in which they send the incoming electrophile.

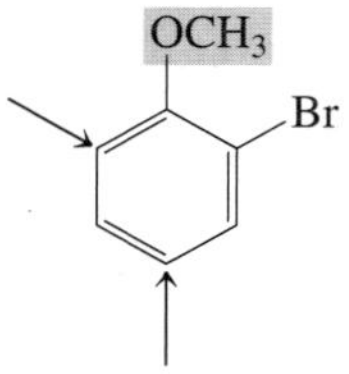

Directing effect of the methoxy group

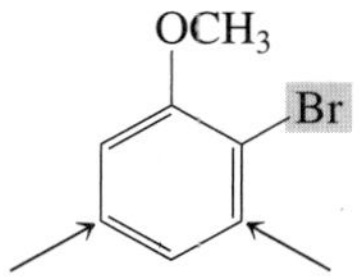

Directing effect of the bromine

- Guidelines 1 and 2 allow for an unambiguous choice between the two with respect to their directing power: Methoxy wins.

Exercise 16-12 | Try It Yourself

Predict the result of mononitration of

(a)

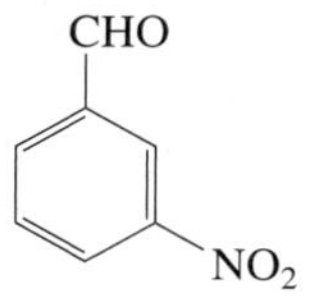

(b)

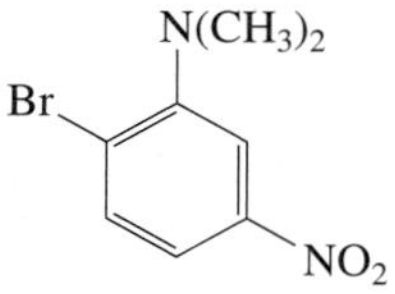

(c)

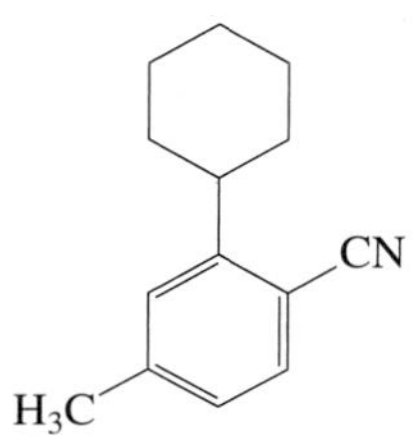

Exercise 16-13

The German chemist Wilhelm Körner (1839–1925) observed in 1874 that each of the three dibromobenzenes A, B, and C furnished a different number of tribromobenzenes upon further bromination, allowing him to assign their respective structures. Try to do the same on the basis of the following results.

(i) A gives two tribromobenzenes in comparable amounts.
(ii) B gives three tribromobenzenes, one of them in minor quantities.
(iii) C gives only one tribromobenzene.

Exercise 16-14

The food preservative BHT (*tert*-butylated hydroxytoluene) has the structure shown here. Suggest a synthesis starting from 4-methylphenol (*p*-cresol).

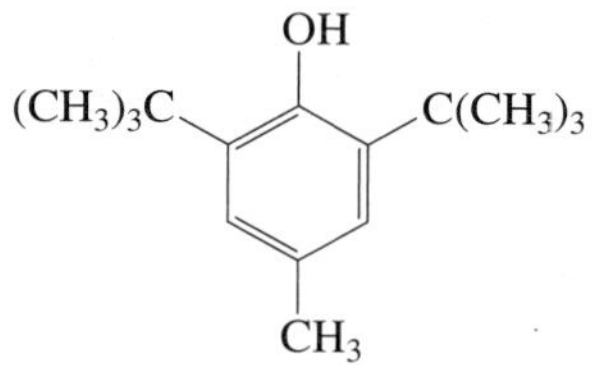

4-Methyl-2,6-bis(1,1-dimethylethyl)phenol
(2,6-Di-*tert*-butyl-4-methylphenol)

In Summary Electrophilic aromatic substitution of multiply substituted benzenes is controlled by the strongest activator and, to a certain extent, by steric effects. The greatest product selectivity appears when there is only one dominant activator or when the substitution pattern minimizes the number of isomers that can be formed.

16-5 KEYS TO SUCCESS: SYNTHETIC STRATEGIES TOWARD SUBSTITUTED BENZENES

The synthesis of substituted benzenes requires planning to ensure that a specific substitution pattern is obtained. How do we approach the problem of targeting a product in which this pattern seems incompatible with the directing sense of the substituents? For example, how can we make a meta-substituted benzenamine (aniline) or an ortho, para–substituted nitrobenzene? To solve problems such as these, we need to know a couple of synthetic "tricks." Among them are the chemical interconversions of ortho, para with meta directors, such as nitro $\rightleftharpoons$ amino or carbonyl $\rightleftharpoons$ methylene, some additional knowledge about the practicality of certain electrophilic substitutions, and the employment of reversible blocking strategies at certain positions with sulfonic acid groups ($-SO_3H$).

We can change the sense of the directing power of substituents

The easiest way to introduce a nitrogen substituent into an arene is by nitration. Yet many desired substituted benzenes have amino functions. Moreover, the nitro group is a meta director, poorly suited for preparing ortho, para–substituted systems. A solution to these problems is provided by the availability of simple reagents that reversibly convert $-NO_2$, a meta director, into $-NH_2$, an ortho, para director. Thus, the nitro group can be reduced to the amino function by either catalytic hydrogenation or exposure to acid in the presence of active metals such as iron or zinc amalgam. The reverse, oxidation of aniline to nitrobenzene, employs trifluoroperacetic acid.

Interconversion of Nitro (Meta Director) with Amino (Ortho, Para Director)

For an example of applying this capability in synthetic strategy, consider the preparation of 3-bromobenzenamine. Direct bromination of benzenamine (aniline) leads to complete ortho and para substitution (Section 16-3) and is therefore useless. However, bromination of nitrobenzene allows preparation of 3-bromonitrobenzene, which can be converted into the required target molecule by reduction. The outcome is a benzene derivative in which two ortho, para directors emerge positioned meta to each other.

3-Bromobenzenamine

Exercise 16-15

Would nitration of bromobenzene be a useful alternative way to begin a synthesis of 3-bromobenzenamine?

Exercise 16-16

Propose a synthesis from benzene of 3-aminobenzenesulfonic acid [metanilic acid, used in the synthesis of azo dyes (Section 22-11), such as "Metanil yellow," and certain sulfa drugs (Section 15-10)].

Exercise 16-17

Use the methods just presented to devise a synthesis of 4-nitrobenzenesulfonic acid from benzene. [**Hint:** Sulfonation proceeds selectively para to activating groups, because the reagent is sterically hindered and the process reversible (Section 15-10).]

Another example of the interconversion of the directing ability of a substituent on a benzene ring is the redox reaction acyl $\rightleftharpoons$ alkyl. Thus, the carbonyl group in acylarenes can be completely reduced by using palladium-catalyzed hydrogenation or treatment with zinc amalgam in concentrated HCl (**Clemmensen* reduction**). Conversely, the methylene group next to the aromatic ring in an alkylarene is susceptible to oxidation to the carbonyl function with the use of CrO_3, H_2SO_4 (Section 22-2).

Interconversion of Acyl (Meta Director) with Alkyl (Ortho, Para Director)

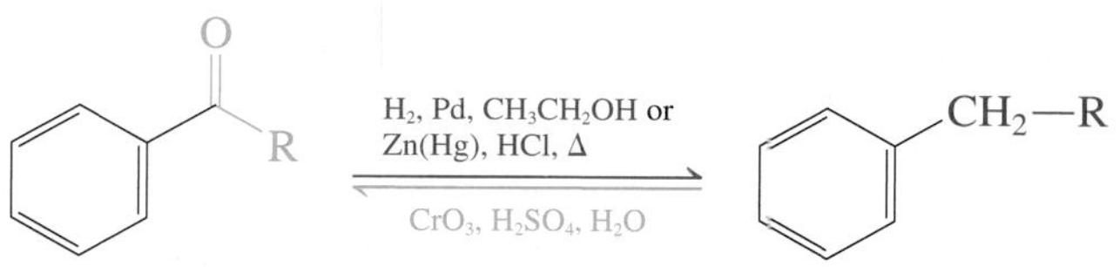

How is this useful? Consider the preparation of 1-chloro-3-ethylbenzene from benzene. Retrosynthetic analysis (Section 8-9) indicates that neither chlorobenzene nor ethylbenzene is suitable as the immediate precursor to product, because each substituent is ortho, para directing. However, we recognize that ethyl can be produced from acetyl, a meta director. Hence, acetylbenzene, readily prepared by Friedel-Crafts acylation, is a perfect relay point, because it can be chlorinated meta. After the carbonyl group has completed its directing job, it is reduced to the desired alkyl group.

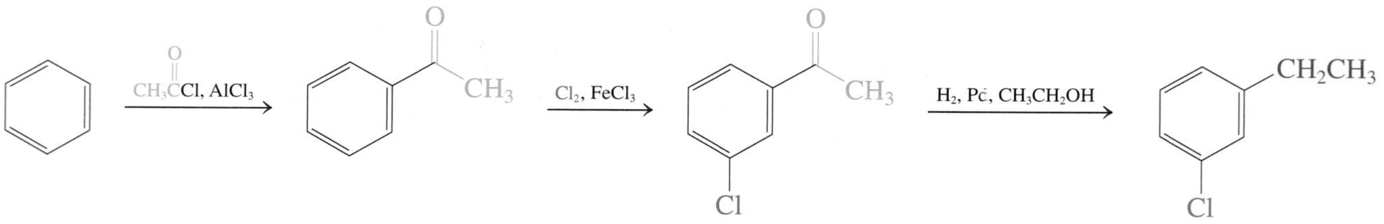

1-Chloro-3-ethylbenzene

Exercise 16-18

Propose a synthesis of 1-chloro-3-propylbenzene, starting from propylbenzene.

The ready reduction of acyl- to alkylarenes also provides a way to synthesize alkylbenzenes without the complication of alkyl group rearrangement and overalkylation. For example, butylbenzene is best synthesized by the sequence of Friedel-Crafts acylation with butanoyl chloride, followed by Clemmensen reduction.

---

*E. C. Clemmensen (1876–1941), president of Clemmensen Chemical Corporation, Newark, New Jersey.

Synthesis of Butylbenzene Without Rearrangement

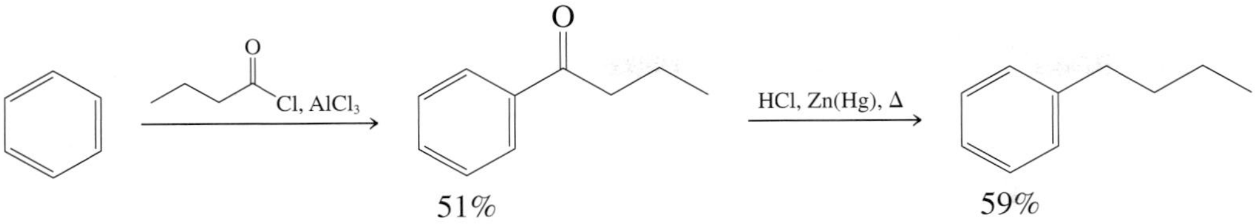

51%   59%

The more direct alternative, Friedel-Crafts butylation of benzene, fails because of the formation of the rearranged product (1-methylpropyl)benzene (*sec*-butylbenzene; Section 15-12) and di- and trialkylation.

Exercise 16-19

Give an efficient synthesis of (2-methylpropyl)benzene (isobutylbenzene, the starting material for the preparation of ibuprofen; see Exercise 16-10), starting from benzene. [**Hint:** What would you expect as the major monosubstitution product of Friedel-Crafts alkylation of benzene with 1-chloro-2-methylpropane (isobutyl chloride)?]

Friedel-Crafts electrophiles do not attack strongly deactivated benzene rings

Let us examine possible syntheses of 1-(3-nitrophenyl)ethanone (*m*-nitroacetophenone). Because both groups are meta directors, two possibilities appear available: nitration of 1-phenylethanone or Friedel-Crafts acetylation of nitrobenzene. However, in practice, only the first route succeeds.

Successful and Unsuccessful Syntheses of 1-(3-Nitrophenyl)ethanone
(*m*-Nitroacetophenone)

86%

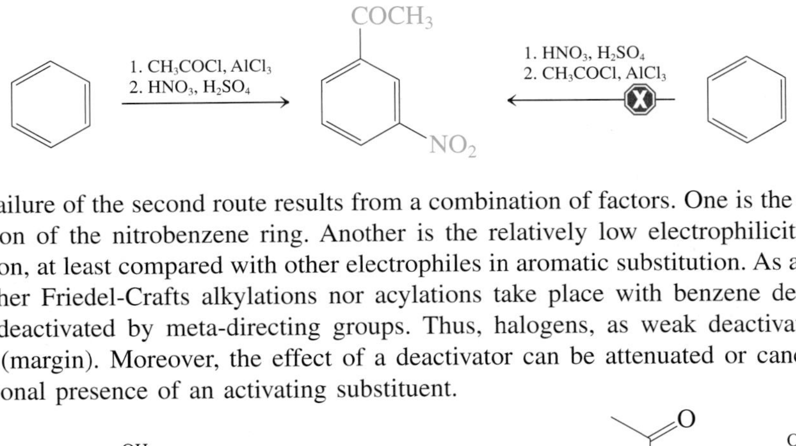

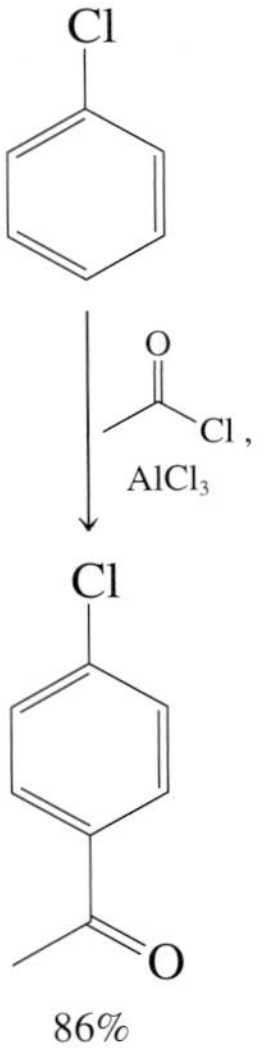

The failure of the second route results from a combination of factors. One is the extreme deactivation of the nitrobenzene ring. Another is the relatively low electrophilicity of the acylium ion, at least compared with other electrophiles in aromatic substitution. As a general rule, neither Friedel-Crafts alkylations nor acylations take place with benzene derivatives strongly deactivated by meta-directing groups. Thus, halogens, as weak deactivators, are tolerated (margin). Moreover, the effect of a deactivator can be attenuated or canceled by the additional presence of an activating substituent.

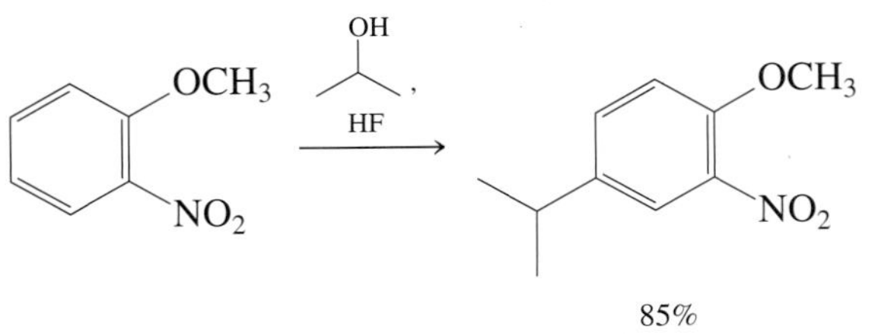

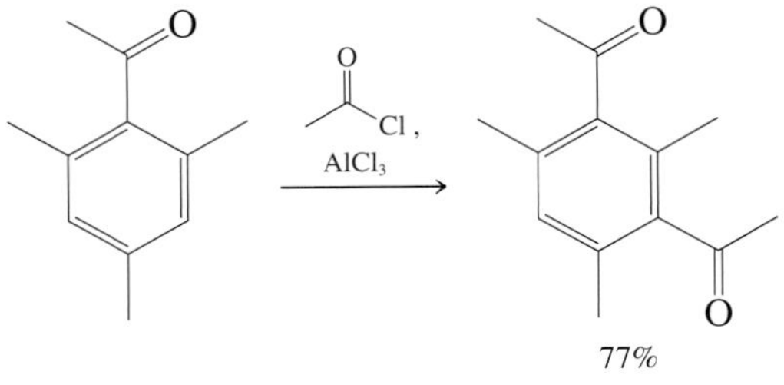

85%

77%

Exercise 16-20

Propose a synthesis of 5-propyl-1,3-benzenediamine, starting from benzene. (**Caution:** Consider carefully the order in which you introduce the groups.)

Reversible sulfonation allows the efficient synthesis of ortho-disubstituted benzenes

A problem of another sort arises in the attempt to prepare an *o*-disubstituted benzene, even when one of the groups is an ortho, para director. Although appreciable amounts of ortho isomers may form in electrophilic substitutions of benzenes containing such groups, the para isomer is the major product in most such cases (Sections 16-2 and 16-3). Suppose you required an efficient synthesis of 1-(1,1-dimethylethyl)-2-nitrobenzene [*o*-(*tert*-butyl)nitrobenzene]. Direct nitration of (1,1-dimethylethyl)benzene (*tert*-butylbenzene) is unsatisfactory.

A Poor Synthesis of 1-(1,1-Dimethylethyl)-2-nitrobenzene
[*o*-(*tert*-Butyl)nitrobenzene]

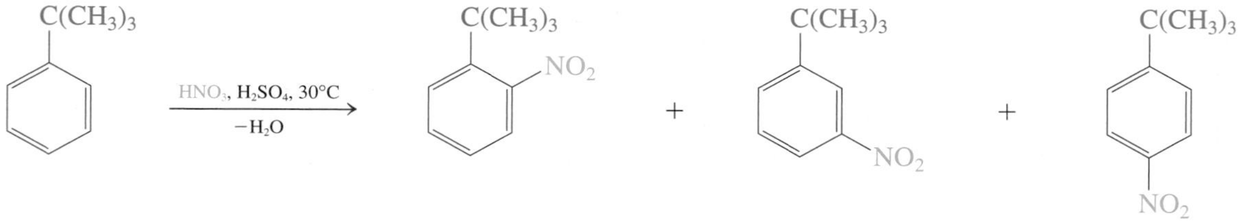

16%	11%	73%
1-(1,1-Dimethylethyl)-2-nitrobenzene	**1-(1,1-Dimethylethyl)-3-nitrobenzene**	**1-(1,1-Dimethylethyl)-4-nitrobenzene**
[*o*-(*tert*-Butyl)nitrobenzene]	**[*m*-(*tert*-Butyl)nitrobenzene]**	**[*p*-(*tert*-Butyl)nitrobenzene]**

A clever solution makes use of reversible sulfonation (Section 15-10) as a blocking procedure. Both the substituent *and* the electrophile are sterically bulky; hence, (1,1-dimethylethyl)benzene is sulfonated almost entirely para, blocking this carbon from further electrophilic attack. Nitration now can occur only ortho to the alkyl group. Heating in aqueous acid removes the blocking group and completes the synthesis.

Reversible Sulfonation as a Blocking Procedure

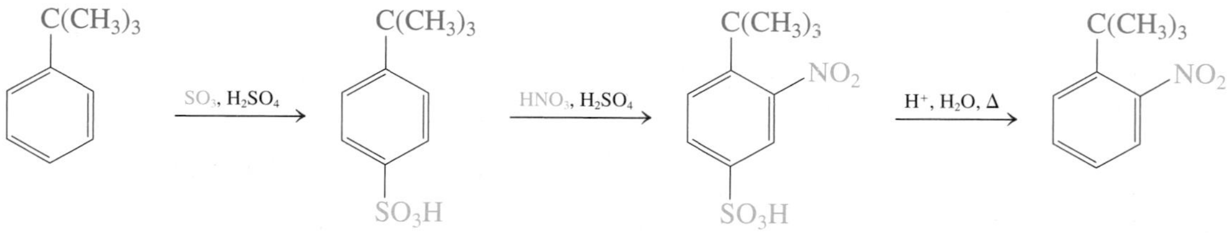

Protecting the Oxygen Atom in Phenol

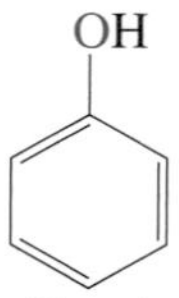

Phenol

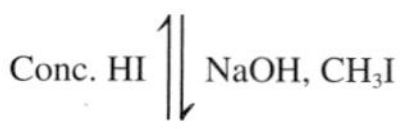

Conc. HI ⇅ NaOH, CH$_3$I

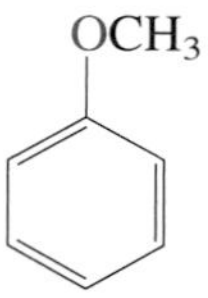

Methoxybenzene

Exercise 16-21

(a) Propofol (below) is a widely used sedative and anesthetic, famously implicated in the accidental death of pop star Michael Jackson in 2009. Suggest a synthetic route to it from benzene.

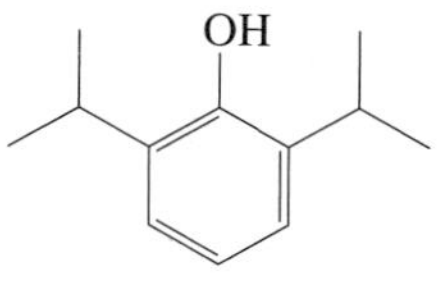

Propofol

(b) Show how you would make 1,3-dibromo-2-nitrobenzene from benzene.

Protection strategies moderate the activating power of amine and hydroxy groups

We noted in Section 16-3 that electrophilic attack on benzenamine (aniline) and phenol are difficult to stop at the stage of monosubstitution because of the highly activating nature of the NH$_2$ and OH groups. In addition, with these Lewis basic functions, complications can arise by direct attack on the heteroatom. To prevent these problems, protecting groups are used: acetyl for benzenamine, as in *N*-phenylacetamide (acetanilide; Sections 20-2 and 20-6), and methyl for phenol, as in methoxybenzene. Deprotection is achieved by basic or acidic hydrolyses, respectively.

In this way, selective halogenation, nitration, and Friedel-Crafts reactions are accomplished. For example, the synthesis of 2-nitrobenzenamine (*o*-nitroaniline) employs this strategy in conjunction with sulfonation to block the para position.

A Synthesis of 2-Nitrobenzenamine (*o*-Nitroaniline)
Through Protected Benzenamine (Aniline)

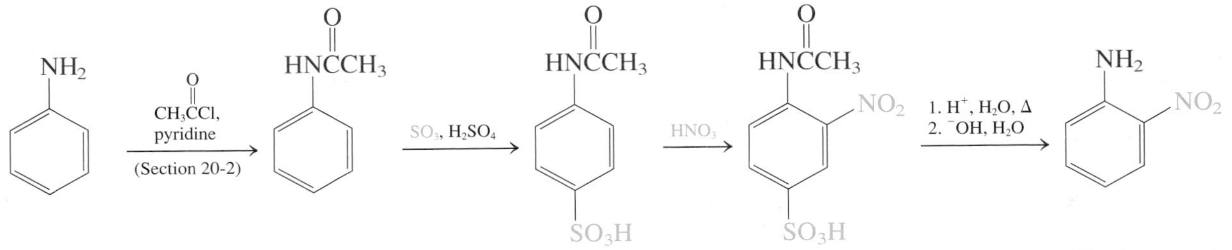

Benzenamine
(Aniline)

N-**Phenylacetamide**
(Acetanilide)

2-Nitrobenzenamine
(*o*-Nitroaniline)

Solved Exercise 16-22 | **Working with the Concepts: Synthetic Strategy in Electrophilic Aromatic Substitution**

What would be a good way of making the benzene derivative A, a synthetic intermediate in the construction of unnatural amino acid oligomers? You may use any monosubstituted benzene derivative as your starting point.

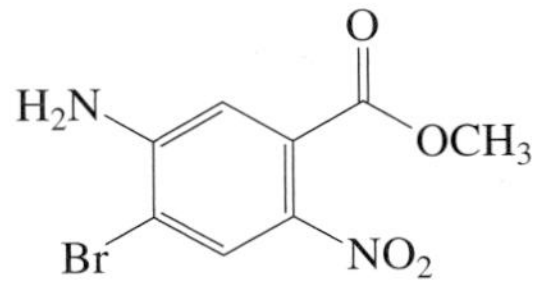

Strategy
At first sight, this target looks formidable. **H**ow do we begin? We could simply select a benzene containing any one of the groups present in A and then see where that compound might lead us by trial and error. Consider, for example, starting with nitrobenzene. The NO₂ group is meta directing, giving us a good way to introduce the bromine. What next? The bromine is an ortho, para director, but can we use that? We do not know yet how to introduce a carboxylic ester function into a benzene ring, so this avenue is closed. We could try to add the second nitrogen by nitration ortho to Br, but para nitration is a likely complication. Regardless, we would be left with two nitro groups in the molecule and no way to reduce one to NH₂ selectively. We would have reached a synthetic dead end, forcing us to start over again, without any indication that our next try would be more successful.

Synthetic Dead Ends

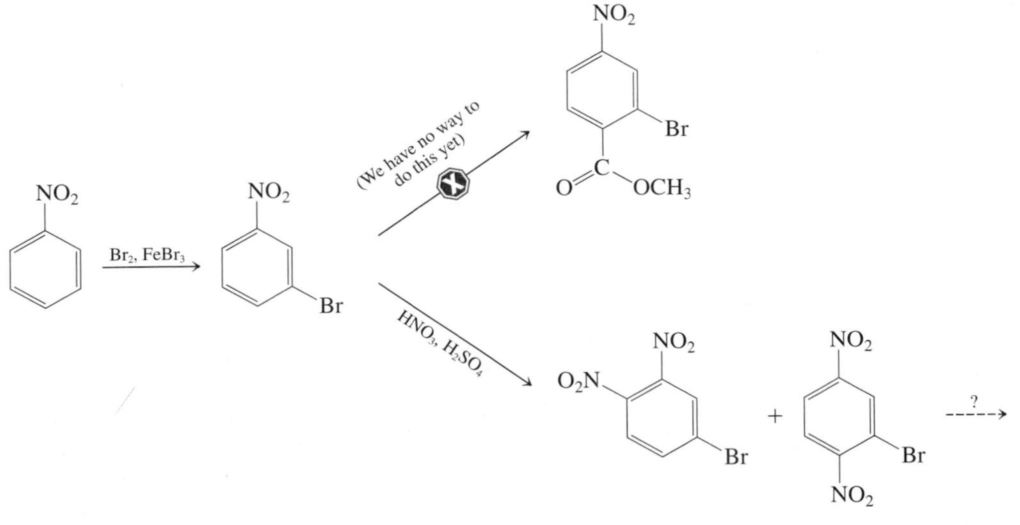

*H*ow do we avoid such pitfalls? We are much better off if we apply the principles of retrosynthesis in conjunction with the *I*nformation we have obtained in this section. The following discussion will offer one solution, but you may come up with more! We can start by evaluating the directing power of the individual substituents, one by one.

Solution

- *P*roceeding clockwise, bromine is an ortho, para director, but a deactivator, and using it as a way to introduce the amino and carbonyl functions does not look promising.
- The amino group is better: Its powerful activating effect could be applied to nitrate para and then brominate ortho.
- Guidelines 1 and 2 allow for an unambiguous choice between the two with respect to their directing power: Amino wins. Early installation of the amino group should be beneficial.
- Next, the ester carbonyl substituent is meta directing and might also be useful, because it could be employed to introduce the meta amino nitrogen in the form of a nitro group.
- Finally, the nitro function is useless (at least as such), because it is meta directing and sufficiently deactivating that substitutions would be very slow at best.
- By this analysis, most fruitful appears to be a retrosynthetic removal of bromine, leading to B. Next we remove nitro, giving C, whose ortho, para-directing amino group is the key to the attachment of those two substituents when we complete the synthesis in the forward direction.
- With C, things have become more transparent, as the amino group may derive from reduction of the nitro group in D, the product of nitration of methyl benzoate, E, our ultimate starting material.

Retrosynthetic Analysis of A

- With this strategy in hand, let us execute the synthesis and see whether some additional practical points have to be considered. We can buy E, and nitration to D is straightforward (Section 16-3).

Execution of the Synthesis of A

- Reduction will give C, the nitration of which was planned to lead to B directly. However, the free amino group in C is sufficiently powerful to cause further nitration of B and therefore has to be moderated by acetamide formation (F). Single nitration is thus ensured, and we anticipate only para substitution to G for steric reasons.
- We deprotect the amine with acid in methanol as solvent to ensure that the ester group stays intact (Section 9-4), rendering B.
- Finally, we expect the final bromination, ortho to the nitrogen function, to take place at the less hindered position, furnishing the desired target.

Exercise 16-23 | Try It Yourself

Apply the strategy discussed above for the synthesis of 2-nitrobenzenamine to a synthesis of 4-acetyl-2-chlorophenol, starting with phenol.

In Summary By careful choice of the sequence in which new groups are introduced, it is possible to devise specific syntheses of multiply substituted benzenes. Such strategies may require changing the directing power of substituents, modifying their activating ability, and reversibly blocking positions on the ring.

16-6 | REACTIVITY OF POLYCYCLIC BENZENOID HYDROCARBONS

In this section, we shall use resonance forms to predict the regioselectivity and reactivity of polycyclic aromatic molecules (Section 15-5), using naphthalene as an example. Some biological implications of the reactivity of these substances will be explored in Section 16-7.

Naphthalene is activated toward electrophilic substitution

The aromatic character of naphthalene is manifest in its reactivity: It undergoes electrophilic substitution rather than addition. For example, treatment with bromine, even in the absence of a catalyst, results in smooth conversion into 1-bromonaphthalene. The mild conditions required for this process reveal that naphthalene is activated with respect to electrophilic aromatic substitution.

75%
1-Bromonaphthalene

Reaction

Other electrophilic substitutions also are readily achieved and, again, are highly selective for reaction at C1. For example,

84%
1-Nitronaphthalene

8%
2-Nitronaphthalene

The highly delocalized nature of the intermediate explains the ease of attack. The cation can be nicely pictured as a hybrid of five resonance forms.

Electrophilic Reactivity of Naphthalene: Attack at C1

However, attack of an electrophile at C2 also produces a cation that may be described by five contributing resonance forms.

Electrophilic Attack on Naphthalene at C2

Why, then, do electrophiles prefer to attack naphthalene at C1 rather than at C2? Closer inspection of the resonance contributors for the two cations reveals an important difference: Attack at C1 allows *two* of the resonance forms of the intermediate to keep an intact benzene ring, with the full benefit of aromatic cyclic delocalization. Attack at C2 allows only *one* such structure, so the resulting carbocation is less stable and the transition state leading to it is less energetically favorable. Because the first step in electrophilic aromatic substitution is rate determining, attack is faster at C1 than at C2.

Electrophiles attack substituted naphthalenes regioselectively

The rules of orientation in electrophilic attack on monosubstituted benzenes extend easily to naphthalenes. *The ring carrying the substituent is the one most affected:* An activating group usually directs the incoming electrophile to the same ring, a deactivating group directs it away. For example, 1-methoxynaphthalene undergoes electrophilic nitration at C2 and C4.

Nitration of 1-Methoxynaphthalene

Deactivating groups in one ring usually direct electrophilic substitutions to the other ring and preferentially in the positions C5 and C8.

30%
1,8-Dinitronaphthalene

60%
1,5-Dinitronaphthalene

Solved Exercise 16-24 | **Working with the Concepts: Predicting Electrophilic Attack on Substituted Naphthalenes**

Predict the position of major electrophilic aromatic nitration in 1-(1-methylethyl)naphthalene.

Strategy
We apply the rules of orientation developed in this section: The more activated (or less deactivated) ring is subject to attack.

Solution
- The substituted ring bears an alkyl group, which is activating and ortho, para directing.
- 1-Methylethyl is moderately bulky, therefore ortho attack is hindered.
- Preferential nitration will occur at C4 to give:

Exercise 16-25 | **Try It Yourself**

Predict the position of electrophilic aromatic nitration in (**a**) 2-nitronaphthalene; (**b**) 5-methoxy-1-nitronaphthalene; (**c**) 1,6-bis(1,1-dimethylethyl)naphthalene. (**Caution:** In problem (c), both rings are activated. **Hint:** Consider steric effects.)

Resonance structures aid in predicting the regioselectivity of larger polycyclic aromatic hydrocarbons

The same principles of resonance, steric considerations, and directing power of substituents apply to larger polycyclic systems, derived from naphthalene by additional benzofusion, such as anthracene and phenanthrene (Section 15-5). For example, the site of preferred electrophilic attack on phenanthrene is C9 (or C10) because the dominant resonance contributor to the resulting cation retains two intact, delocalized benzene rings, whereas all the other forms require disruption of the aromaticity of either one or two of those rings.

Electrophilic Attack on Phenanthrene

Major contributor

Similar considerations pertain to anthracene and the higher polycyclic aromatic hydrocarbons in predicting the relative ease of electrophilic attack at various positions.

Exercise 16-26

Formulate a resonance form of the cation derived by electrophilic attack at C9 of phenanthrene in which the aromaticity of *all* benzene rings is disrupted.

Exercise 16-27

Electrophilic protonation of anthracene exhibits the following relative rates: $k(C9):k(C1):k(C2) \approx$ 11,000 : 7 : 1. Explain. (For the numbering of the anthracene skeleton, see Section 15-5.)

In Summary Naphthalene is activated with respect to electrophilic aromatic substitution; favored attack takes place at C1. Electrophilic attack on a substituted naphthalene takes place on an activated ring and away from a deactivated ring, with regioselectivity in accordance with the general rules developed for electrophilic aromatic substitution of benzene derivatives. Similar considerations apply to the higher polycyclic aromatic hydrocarbons.

16-7 | POLYCYCLIC AROMATIC HYDROCARBONS AND CANCER

Forest fires generate major environmental pollutants, including benzo[*a*]pyrene.

Many polycyclic benzenoid hydrocarbons are carcinogenic. The first observation of human cancer caused by such compounds was made in 1775 by Sir Percival Pott, a surgeon at London's St. Bartholomew's hospital, who recognized that chimney sweeps were prone to scrotal cancer. Since then, a great deal of research has gone into identifying which polycyclic benzenoid hydrocarbons have this physiological property and how their structures correlate with activity. A particularly well studied molecule is benzo[*a*]pyrene, a widely distributed environmental pollutant. It is produced in the combustion of organic matter, such as automobile fuel and oil (for domestic heating and industrial power generation), in incineration of refuse, in forest fires, in burning cigarettes and cigars, and even in roasting meats. The annual release into the atmosphere in the United States alone has been estimated at 3000 tons.

Carcinogenic Benzenoid Hydrocarbons

Benzo[*a*]pyrene   Benz[*a*]anthracene   Dibenz[*a,h*]anthracene

What is the mechanism of carcinogenic action of benzo[*a*]pyrene? An oxidizing enzyme (an oxidase) of the liver converts the hydrocarbon into the oxacyclopropane at C7 and C8. Another enzyme (epoxide hydratase) catalyzes the hydration of the product to the trans diol.

Rapid further oxidation then results in the ultimate carcinogen, a new oxacyclopropane at C9 and C10.

Enzymatic Conversion of Benzo[*a*]pyrene into the Ultimate Carcinogen

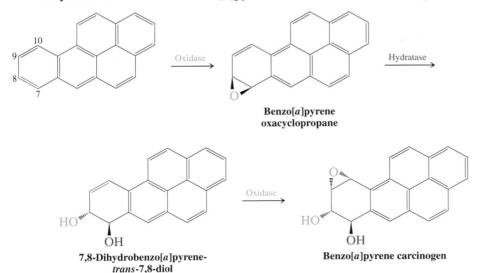

Benzo[*a*]pyrene oxacyclopropane

7,8-Dihydrobenzo[*a*]pyrene-*trans*-7,8-diol

Benzo[*a*]pyrene carcinogen

Polycyclic aromatic hydrocarbons are also prevalent in outer space. This photo shows the spiral galaxy NGC 7793, about 12.7 million light-years away from us, as observed with the Spitzer Infrared Space Telescope. The blue colors embody the $\tilde{\nu} = 2800$ cm^{-1} emission from stars. The green and red colors represent the $\tilde{\nu} = 1700$ and 1250 cm^{-1} emissions from polycyclic aromatic hydrocarbons and possibly dust.

What makes the compound carcinogenic? It is believed that the amine nitrogen in guanine, one of the bases in DNA (see Chapter 26), attacks the oxacyclopropane as a nucleophile. The altered guanine disrupts the DNA double helix, leading to a mismatch in DNA replication.

The Carcinogenic Event

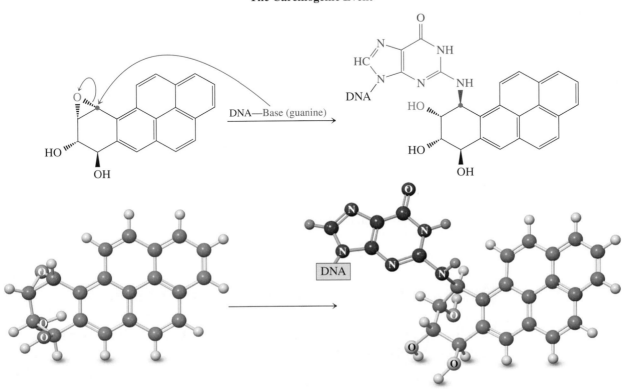

DNA—Base (guanine)

Carcinogenic Alkylating Agents and Sites of Reactivity

BrCH$_2$CH$_2$Br

1,2-Dibromoethane

Oxacyclopropane

ClCH$_2$OCH$_3$

Chloro(methoxy)methane (Chloromethyl methyl ether)

This change can lead to an alteration (mutation) of the genetic code, which may then generate a line of rapidly and indiscriminately proliferating cells typical of cancer. Not all mutations are carcinogenic; in fact, most of them lead to the destruction of only the one affected cell. Exposure to the carcinogen simply increases the likelihood of a carcinogenic event.

Notice that the carcinogen acts as an alkylating agent on DNA. This observation implies that other alkylating agents could also be carcinogenic, and indeed that is found to be the case. The Occupational Safety and Health Administration (OSHA) has published lists of likely carcinogens that include simple alkylating agents such as 1,2-dibromoethane and oxacyclopropane (see Problem 50 of Chapter 1).

The discovery of carcinogenicity in a number of organic compounds necessitated their replacement in synthetic applications. Both 1- and 2-naphthalenamines (naphthylamines) were once widely used in the synthesis of dyes because of the brilliant colors of many of their derivatives (azo dyes; see Section 22-11). These substances were discovered many years ago to be carcinogens, a finding that led to the development both of synthetic routes that avoided their use as intermediates and of new dyes with completely unrelated structures. A more recent example is chloro(methoxy)methane (ClCH$_2$OCH$_3$, chloromethyl methyl ether), once a commonly used reagent for the protection of alcohols by ether formation. The discovery of carcinogenicity in this alkylating agent in the 1970s resulted in the development of several less hazardous reagents.

THE BIG PICTURE

The directing effects of substituents on a benzene ring constitute another case of regioselectivity in organic chemistry. We encountered such selectivity previously in the nucleophilic ring opening of oxacyclopropanes (Section 9-9) and halonium ions (Section 12-6), where it could be controlled by steric or electronic factors (basic or acidic conditions, respectively). We also observed regioselectivity in the nucleophilic trapping of delocalized cations (for example, allylic cations; Section 14-6), where reversibility presented us with the options of kinetic versus thermodynamic regiocontrol. In electrophilic aromatic substitutions, the regiochemical outcome is (usually) kinetic in origin and results mainly from electronic effects. In cases for which several electronically favored sites of attack exist, steric factors often tip the balance. To predict the preferred product of a particular electrophilic aromatic substitution, we need to formulate all the possible cationic intermediates and assess the relative importance of all contributing stabilizing and destabilizing electronic and steric effects.

We have now completed our study of multiple bonds between carbon atoms, including alkenes and alkynes, conjugated dienes, and benzene and other aromatic compounds. Next we examine double bonds between carbon and oxygen atoms, which occur in the important classes of compounds that contain the carbonyl group. We shall see that some of the electronic effects that occur in carbon–carbon double bonds also occur in carbon–oxygen double bonds, but other effects are unique to the polar carbonyl group. As a result, carbonyl chemistry is a rich and very important area of organic chemistry. Understanding reactions involving the carbonyl group is essential to understanding many areas of biochemistry, from the action of pharmaceuticals to the details of molecular genetics. We start in Chapter 17 with a discussion of the carbonyl group in aldehydes and ketones.

WORKED EXAMPLES: INTEGRATING THE CONCEPTS

16-28. Applying Directing-Group Strategies in Electrophilic Aromatic Substitutions

Specifically substituted, functionalized benzenamines (anilines) are important synthetic intermediates in medicinal chemistry and the dye industry. Propose a selective synthesis of 5-chloro-2-methoxy-1,3-benzenediamine, B, from methoxybenzene, A.

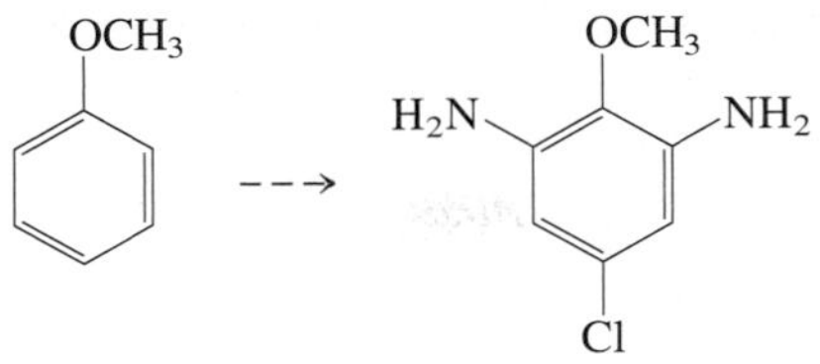

A   B

SOLUTION

What are the feasible bond disconnections that lead to a simpler precursor to B directly–*a*, *b*, or *c*? Our retrosynthetic analysis below (Section 8-9) answers the question: None. Retrosynthetic step *a* proposes a transformation that is impossible to achieve with our present repertory of reactions and, in fact, is very difficult even with special reagents (requiring a source of "CH_3O^+"). Quite apart from this problem, step *a* appears unwise because it cleaves a bond that is given in the starting material. Step *c* is a reverse electrophilic chlorination (Section 15-9), feasible in principle but not in practice because no selectivity for the desired position at C5 can be expected. Thus, while CH_3O directs para (and hence to C5) as wanted, the amino groups activate the other positions ortho, para (and hence to C4 and C6) even more (Section 16-3), ruling out step *c* as a good option, at least as such.

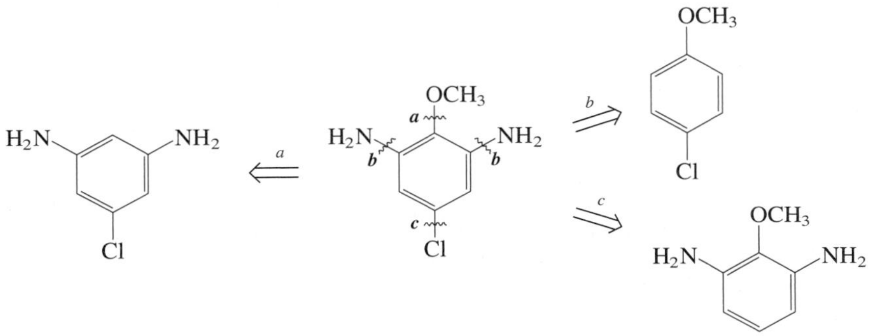

What about disconnection *b*? While direct electrophilic amination of arenes is (like alkoxylation or hydroxylation) not viable, we know that we can achieve it indirectly through a nitration–reduction sequence (Section 16-5). Thus, the problem reduces to a nitration of 1-chloro-4-methoxybenzene. Will it proceed with the desired regioselectivity? Guideline 2 in Section 16-4 answers in the affirmative. This analysis provides 1-chloro-4-methoxybenzene as a new relay point, available, in addition to its ortho isomer, by chlorination of methoxybenzene.

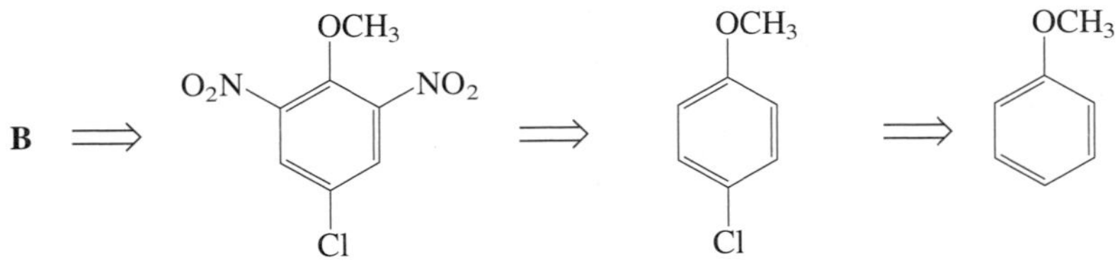

1-Chloro-4-
methoxybenzene

Therefore, a reasonable solution to our synthetic problem is that shown below as a proper synthetic scheme, with reagents, synthetic intermediates, and forward arrows in place.

Solution Synthetic Scheme 1

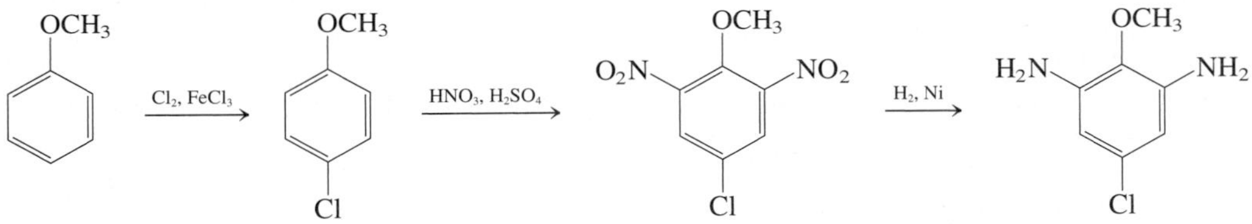

A   (+ ortho
           isomer)   B

As a purist, you might not be satisfied with the lack of regiochemical control of the first step, which not only cuts down on yields but also requires a cumbersome separation. Consideration of a blocking strategy employing a sulfonation may help (Section 16-5). SO_3 is more bulky than chlorine and will furnish exclusively the para substitution product of electrophilic attack on methoxybenzene. Blocking the para position allows selective dinitration ortho to methoxy. After deblocking, the chlorine can be introduced, as shown below.

Solution Synthetic Scheme 2

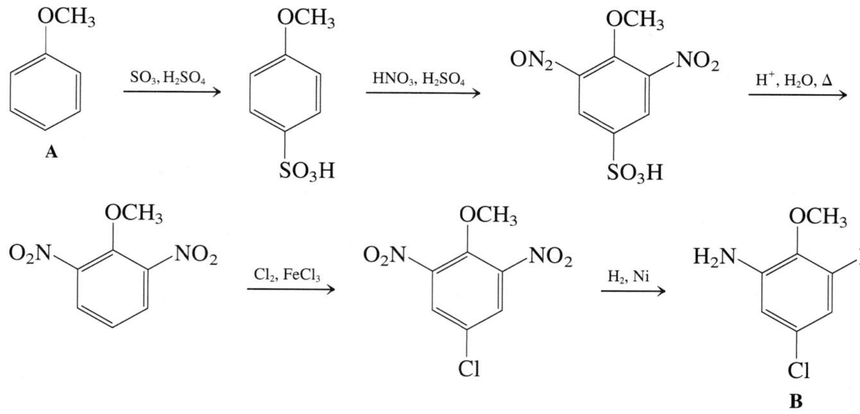

This sequence requires two additional steps and, in practice, overall yields, ease of experimentation, including work-up, disposal cost of waste, and the value and availability of starting materials will determine which route to take, Scheme 1 or Scheme 2.

16-29. Solving a Mechanism Involving a Carbocationic Cascade

Acid-catalyzed cascade reactions are employed by nature and also by synthetic chemists in the construction of complex polycyclic molecules, including steroids (Section 4-7). Provide a mechanism for the following multiple cyclization.

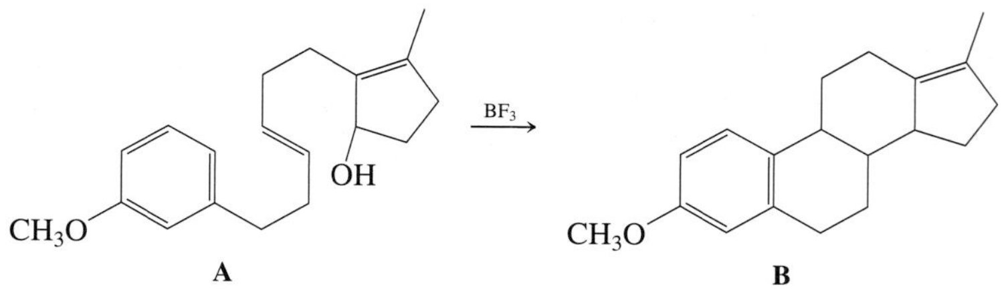

SOLUTION
This is a mechanistic and not a synthetic problem, and therefore we can use only what is given above. Let us look at the general features of this transformation. A starting material with only two rings is converted into a product with four rings by the making of two new C–C bonds, and somewhere along the sequence, the hydroxy function is lost. The reagent BF_3 is a Lewis acid (Section 2-2), "hungry" for an electron pair, and the hydroxy oxygen is an obvious source for such. What are the exact changes in composition during the reaction? This question is answered by assessing the change in the respective molecular formulas: $C_{19}H_{26}O_2$ turns into $C_{19}H_{24}O$. Thus, overall, the process constitutes a dehydration ($-H_2O$).

After this general analysis, we can go about addressing the details of the mechanism. What do we anticipate to happen when a secondary (and allylic) alcohol is treated with a Lewis acid? *Answer:* (Allylic) carbocation (C) formation (Sections 9-2, 9-3, and 14-3):

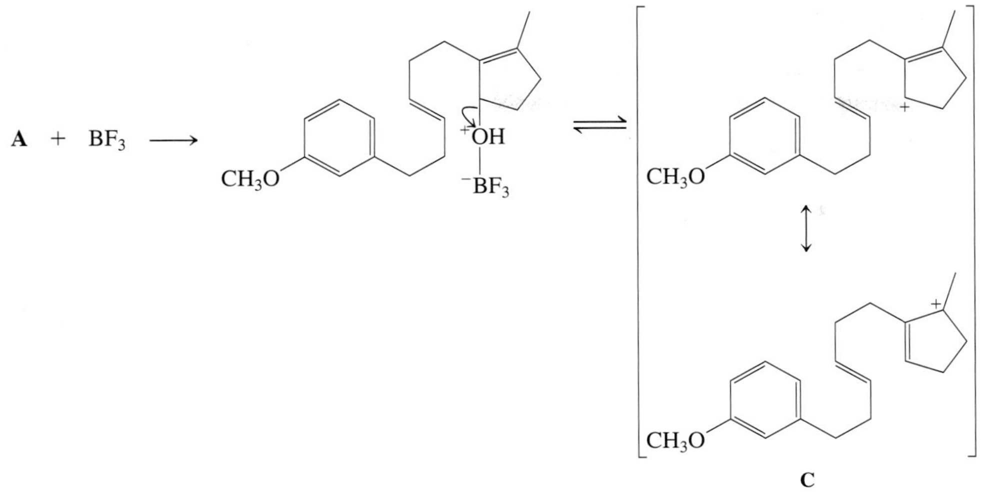

A + BF₃ ⟶

What do we expect from a carbocation in the presence of a nearby double bond? *Answer:* electrophilic addition (Section 12-14) to form a new carbocation (D):

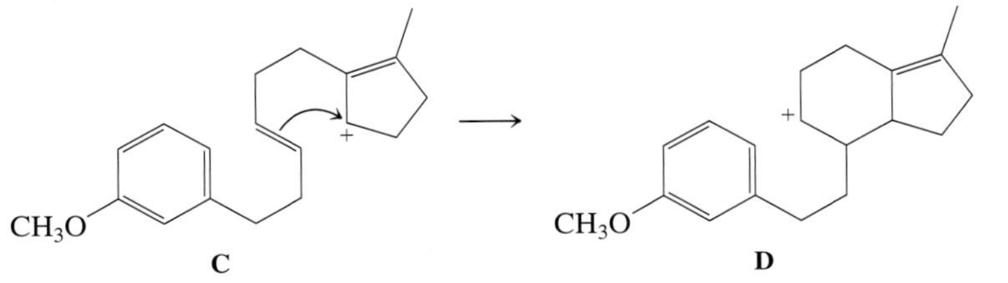

While this answer satisfies the structural requirements for proceeding on to the eventual final product, we can ponder several questions regarding the selectivity of the formation of D. First, C contains two electrophilic sites. Why is only one picked? *Answer:* The less hindered carbon should react faster. Second, conversion to D makes a six-membered ring. Why not attack the other end of the double bond to give a five-membered ring? *Answer:* The observed six-membered ring should be less strained (Section 4-3). And third, when going from C to D, a resonance-stabilized allylic carbocation is converted into an "ordinary" secondary cation. What is the driving force? *Answer:* A new carbon–carbon bond is formed. Having arrived at D, we can recognize the last step as a Friedel-Crafts alkylation of an activated benzene, para to the directing methoxy group.

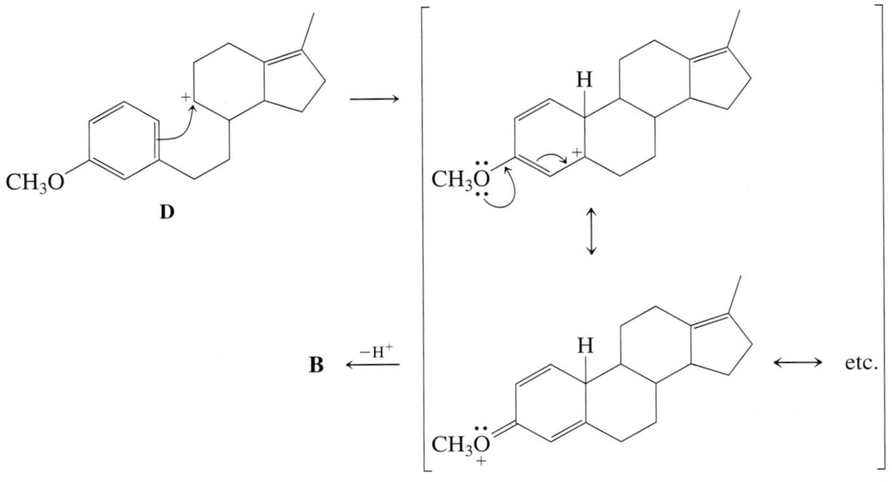

New Reactions

Electrophilic Substitution of Substituted Benzenes

1. Ortho- and Para-Directing Groups (Sections 16-1 through 16-3)

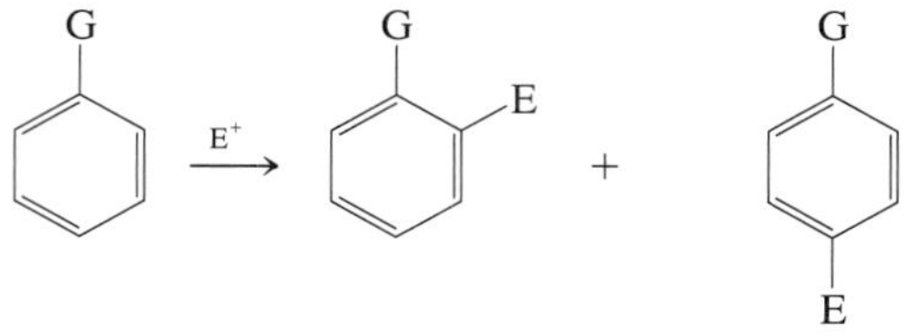
G
E+
G
E
+
G
E

Ortho isomer **Para isomer** **(Usually predominates)**

G = NH_2, OH; strongly activating
= NHCOR, OR; moderately activating
= alkyl, aryl; weakly activating
= halogen; weakly deactivating

2. Meta-Directing Groups (Sections 16-1 through 16-3)

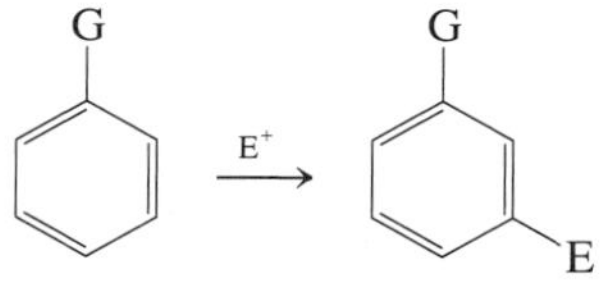
G
E+
G
E

Meta isomer

G = $\overset{+}{N}(CH_3)_3$, NO_2, CF_3, C≡N, SO_3H; very strongly deactivating
= CHO, COR, COOH, COOR, $CONH_2$; strongly deactivating

Synthetic Planning: Switching and Blocking of Directing Power

3. Interconversion of Nitro and Amino Groups (Section 16-5)

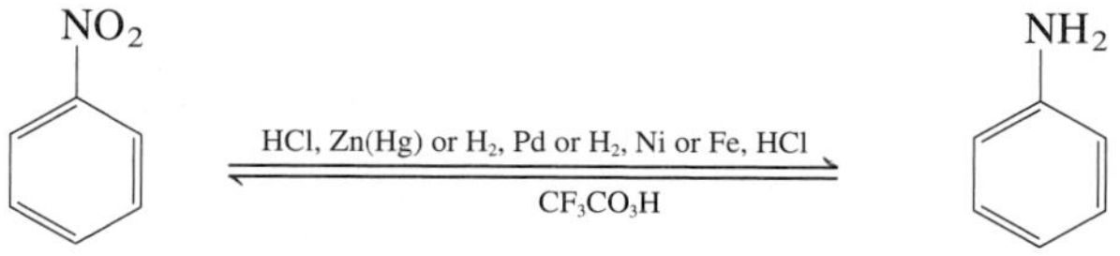
NO2
HCl, Zn(Hg) or H2, Pd or H2, Ni or Fe, HCl
CF3CO3H
NH2

Meta directing **Ortho, para directing**

4. Interconversion of Acyl and Alkyl (Section 16-5)

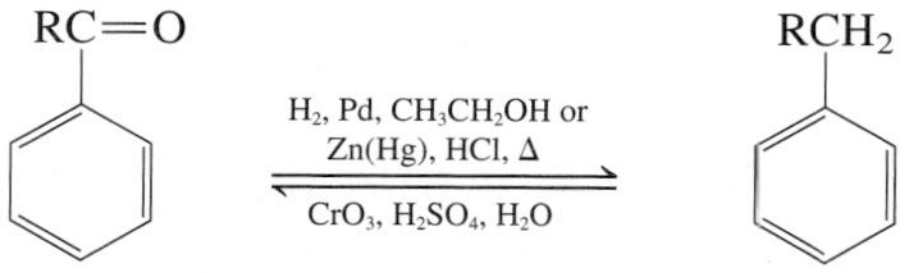
RC=O
H2, Pd, CH3CH2OH or
Zn(Hg), HCl, Δ
CrO3, H2SO4, H2O
RCH2

Meta directing **Ortho, para directing**

5. Blocking by Sulfonation (Section 16-5)

6. Moderation of Strong Activators by Protection (Section 16-5)

7. Electrophilic Aromatic Substitution of Naphthalene (Section 16-6)

Important Concepts

1. Substituents on the benzene ring can be divided into two classes: those that **activate** the ring by **electron donation** and those that **deactivate** it by **electron withdrawal.** The mechanisms of donation and withdrawal are based on **induction** and **hyperconjugation** or **resonance.** These effects may operate simultaneously to either reinforce or oppose each other. Amino and alkoxy substituents are strongly activating, alkyl and phenyl groups weakly so; nitro, trifluoromethyl, sulfonyl, oxo, nitrile, and cationic groups are strongly deactivating, whereas halogens are weakly so.

2. **Activators** direct electrophiles **ortho** and **para; deactivators** direct **meta,** although at a much lower rate. The exceptions are the **halogens,** which deactivate but direct **ortho** and **para.**

3. When there are several substituents, the strongest activator (or weakest deactivator) controls the regioselectivity of attack, the extent of control decreasing in the following order:

$$NR_2, OR > X, R > \text{meta directors}$$

4. Strategies for the **synthesis** of highly substituted benzenes rely on the **directing power** of the substituents, the synthetic ability to **change** the **sense of direction** of these substituents by chemical manipulation, and the use of **blocking** and **protecting** groups.

5. **Naphthalene** undergoes preferred electrophilic substitution at **C1** because of the relative stability of the intermediate carbocation.

6. Electron-donating substituents on one of the naphthalene rings direct electrophiles to the same ring, ortho and para. Electron-withdrawing substituents direct electrophiles away from that ring; substitution is mainly at C5 and C8.

7. The actual carcinogen derived from benzo[*a*]pyrene appears to be an oxacyclopropanediol in which C7 and C8 bear hydroxy groups and C9 and C10 are bridged by oxygen. This molecule alkylates one of the nitrogens of one of the DNA bases, thus causing mutations.

Problems

30. Rank the compounds in each of the following groups in order of decreasing reactivity toward electrophilic substitution. Explain your rankings.

(a) CCl₃   CH₃   CHCl₂   CH₂Cl

(b) OCH₃   O⁻Na⁺   OCCH₃ (O)

(c) CH₂CH₃   CH₂CCl₃   CH₂CF₃   CF₂CH₃

31. The rate of nitration of (chloromethyl)benzene, ⟨benzene⟩—CH₂Cl, is 0.71 relative to the rate of nitration of benzene (=1). The (chloromethyl)nitrobenzene product mixture that results contains 32% ortho, 15.5% meta, and 52.5% para isomers. Explain.

32. Specify whether you expect the benzene rings in the following compounds to be activated or deactivated.

(a) COOH / COOH (para)

(b) NO₂ / NO₂ / F

(c) OH / CH₃

(d) ⟨phenyl⟩—O—⟨phenyl⟩

(e) NH₂ / OH

(f) SO₃H / NO₂

(g) HO / C(CH₃)₃ / CH₃

33. Rank the compounds in each of the following groups in order of decreasing reactivity toward electrophilic aromatic substitution. Explain your answers.

(a) CH₃ / CH₃   COOH / COOH   COOH / CH₃ / CH₃

(b)

34. Halogenation of 1,3-dimethylbenzene (*m*-xylene) takes place 100 times faster than halogenation of either its 1,2- or 1,4-isomers (*o*- or *p*-xylene). Suggest an explanation.

35. Write the structure(s) of the major product(s) that you expect from each of the following electrophilic aromatic substitutions. **(a)** Nitration of methylbenzene (toluene); **(b)** sulfonation of methylbenzene (toluene); **(c)** nitration of 1,1-dimethylethylbenzene (*tert*-butylbenzene); **(d)** sulfonation of 1,1-dimethylethylbenzene (*tert*-butylbenzene).

    In what way does changing the substrate structures from methylbenzene (toluene) in (a) and (b) to 1,1-dimethylethylbenzene (*tert*-butylbenzene) in (c) and (d) affect the expected product distributions?

36. Write the structure(s) of the major product(s) that you expect from each of the following electrophilic aromatic substitutions. **(a)** Sulfonation of methoxybenzene (anisole); **(b)** bromination of nitrobenzene; **(c)** nitration of benzoic acid; **(d)** Friedel-Crafts acetylation of chlorobenzene.

37. Draw appropriate resonance forms to explain the deactivating meta-directing character of the SO₃H group in benzenesulfonic acid.

38. Do you agree with the following statement? "Strongly electron-withdrawing substituents on benzene rings are meta directing because they deactivate the meta positions less than they deactivate the ortho and para positions." Explain your answer.

39. Draw appropriate resonance forms to explain the activating ortho, para-directing character of the phenyl substituent in biphenyl (below).

Biphenyl

40. Give the expected major product(s) of each of the following electrophilic substitution reactions.

(a) ⟨N(CH₃)₂ on benzene⟩ $\xrightarrow{\text{CH}_3\text{CCl, AlCl}_3}$ (O)

(b) ⟨Cl on benzene⟩ $\xrightarrow{\text{Br}_2,\ \text{FeBr}_3}$

(c) ⟨benzene⟩—CCH₂CH₃ (O) $\xrightarrow{\text{HNO}_3,\ \text{H}_2\text{SO}_4}$

(d) 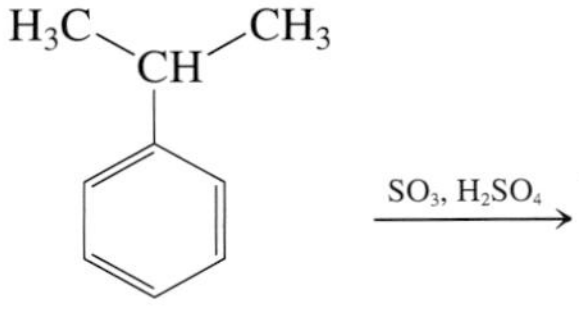 $\xrightarrow{SO_3,\ H_2SO_4}$ **(e)** 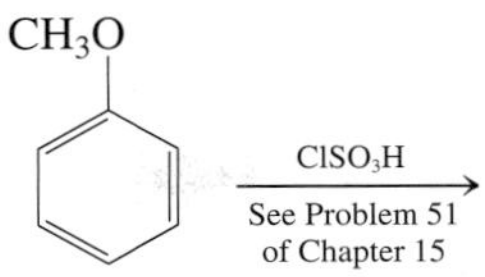 CH_3O— $\xrightarrow[\text{See Problem 51 of Chapter 15}]{ClSO_3H}$

(f) 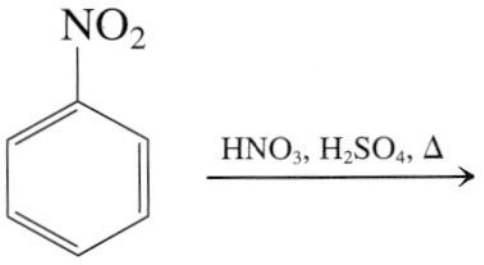 $\xrightarrow{HNO_3,\ H_2SO_4,\ \Delta}$

41. Reaction review. Without consulting the Reaction Road Map on p. 687, suggest a combination of reagent and monosubstituted benzene that would give each of the following compounds. (**Hint:** Refer to Table 16-2 for guidance. **Caution:** Directing effects arise from the group already present on the benzene ring, *not* the group you are trying to attach.)

(a) 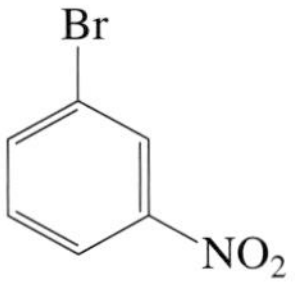 **(b)**

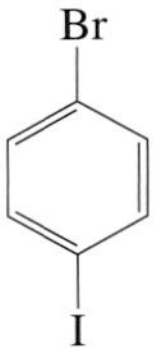

(c) 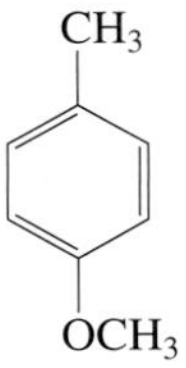 **(d)**

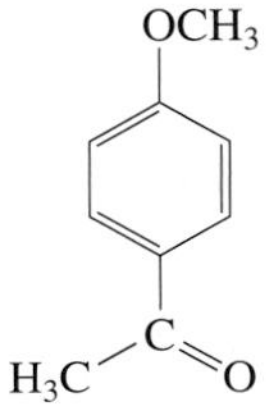

(e) 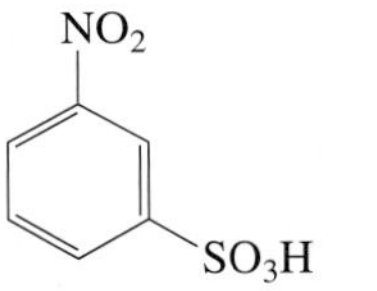 **(f)**

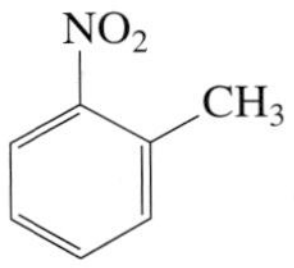

(g) 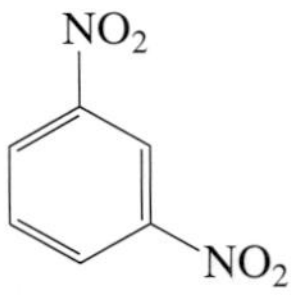 **(h)**

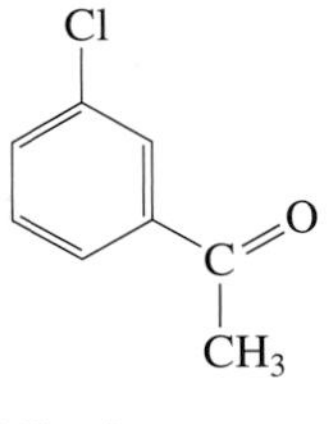

42. Reaction review. The preparation of the following compounds requires more than one step. As in Problem 41, suggest a monosubstituted benzene as starting material and the reagents for all the steps needed to complete their synthesis.

(a) 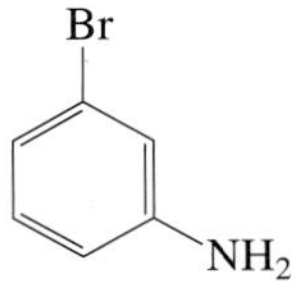 **(b)**

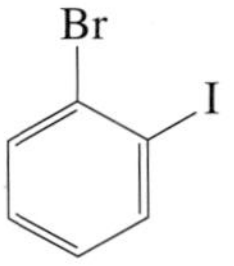

(c) 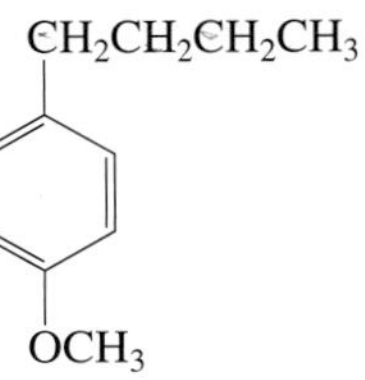 **(d)**

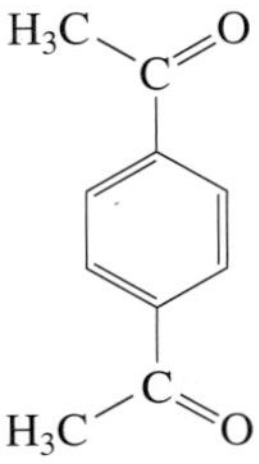

(e)  **(f)**

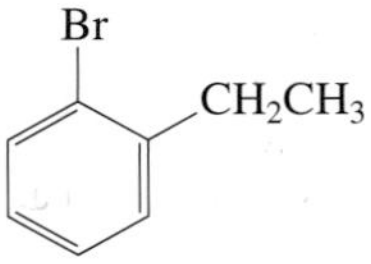

(g)  **(h)**

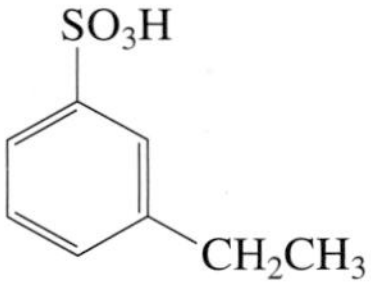

43. Give the expected major product(s) of each of the following reactions.

(a) 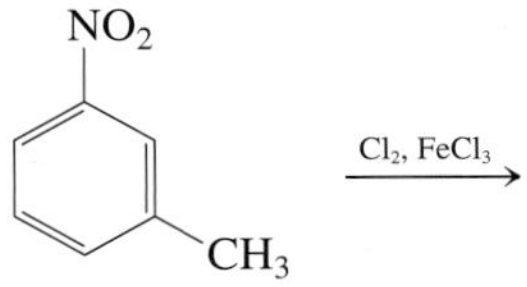 $\xrightarrow{Cl_2,\ FeCl_3}$

(b) 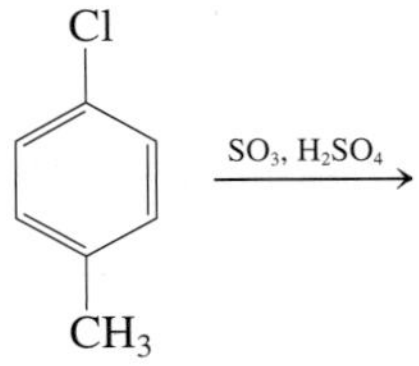 $\xrightarrow{SO_3,\ H_2SO_4}$

(c) 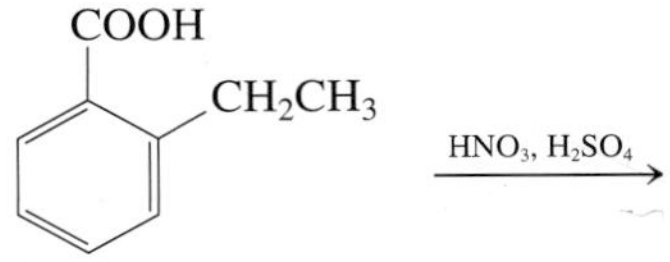 $\xrightarrow{HNO_3,\ H_2SO_4}$

(d) 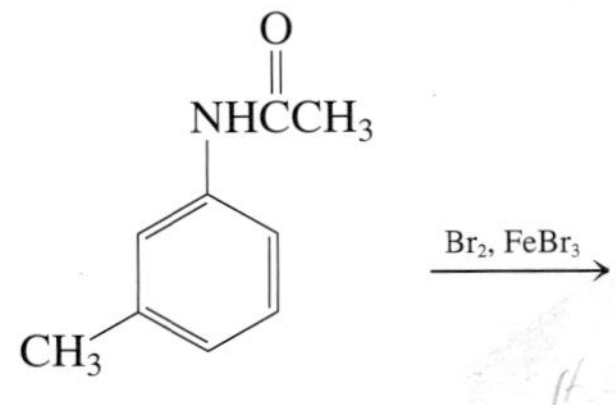 $\xrightarrow{Br_2,\ FeBr_3}$

(e) 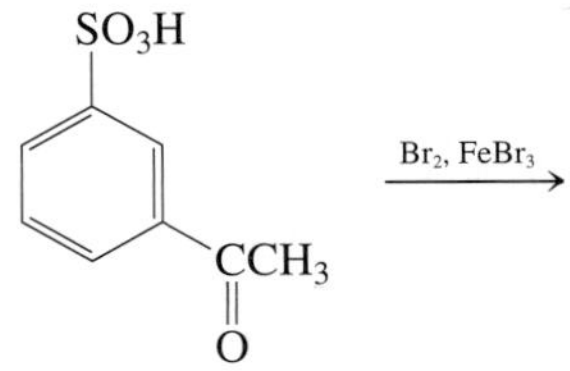 $\xrightarrow{Br_2,\ FeBr_3}$

(f) 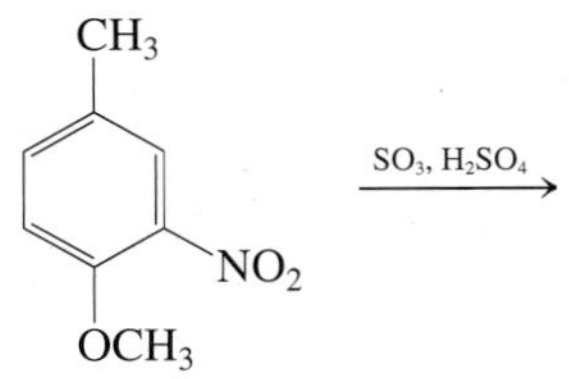 $\xrightarrow{SO_3,\ H_2SO_4}$

(g) 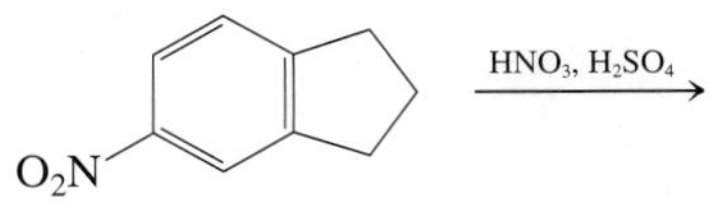 $\xrightarrow{HNO_3,\ H_2SO_4}$

(h) → Cl₂, FeCl₃ → (i) → CH₃Cl, AlCl₃ →

44. **CHALLENGE** (a) When a mixture containing one mole each of the three dimethylbenzenes (*o*-, *m*-, and *p*-xylene) is treated with one mole of chlorine in the presence of a Lewis acid catalyst, one of the three hydrocarbons is monochlorinated in 100% yield, whereas the other two remain completely untouched. Which isomer reacts? Explain the differences in reactivity. (b) The same experiment carried out on a mixture of the following three trimethylbenzenes gives a similar outcome. Answer the questions posed in (a) for this mixture of compounds.

1,2,3-Trimethylbenzene   **1,2,4-Trimethylbenzene**

1,3,5-Trimethylbenzene

45. Propose a reasonable synthesis of each of the following multiply substituted arenes from benzene.

(a)   (b)

(c)   (d)

(e)   (f)

(g)   (h)

46. (4-Methoxyphenyl)methanol (anisyl alcohol) (see below) contributes both to the flavor of licorice and to the fragrance of lavender. Propose a synthesis of this compound from methoxybenzene (anisole). (**Hint:** Consider your range of options for alcohol synthesis. If necessary, refer to Problem 55 of Chapter 15.)

(4-Methoxyphenyl)methanol
(Anisyl alcohol)

47. The science of pain control has evolved dramatically in the past several years. The body deals with pain by releasing anandamide (see Section 20-6). Anandamide binds to the cannabinoid receptor, the same site that recognizes the active ingredients in marijuana. Binding to this site suppresses the perception of pain. The effect is not long lasting, because another enzyme degrades anandamide over time. Thus, recent research has focused on finding therapeutic molecules capable of blocking this anandamide-destroying enzyme. The biphenyl derivative URB597 is an experimental substance that displays some of this desired enzyme-blocking ability, and it has been demonstrated to enhance pain suppression in rats.

Examine the structure of URB597 and address the following question: Would you expect URB597 to be an easy or a difficult target molecule to prepare, beginning with biphenyl (Problem 39)? If you chose easy, illustrate how the reactions of this chapter might be employed in a synthetic approach to this drug. If you chose difficult, explain why.

URB597

48. The NMR and IR spectra for four unknown compounds A through D are presented on the next two pages. Possible empirical formulas for them (not in any particular order) are C_6H_5Br, C_6H_6BrN, and $C_6H_5Br_2N$ (one of these formulas is used twice—two of the unknowns are isomers). Propose a structure and suggest a synthesis of each unknown, starting from benzene.

49. Catalytic hydrogenation of naphthalene over Pd–C results in rapid addition of 2 moles of H_2. Propose a structure for this product.

50. Predict the major mononitration product of each of the following disubstituted naphthalenes. (**a**) 1,3-Dimethylnaphthalene; (**b**) 1-chloro-5-methoxynaphthalene; (**c**) 1,7-dinitronaphthalene; (**d**) 1,6-dichloronaphthalene.

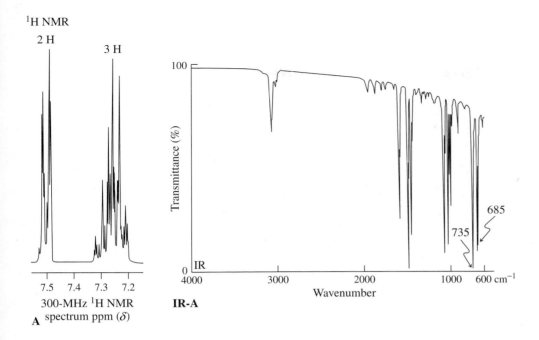

¹H NMR

2 H

3 H

300-MHz ¹H NMR
A spectrum ppm (δ)

IR-A

685

735

Transmittance (%)

4000   3000   2000   1000   600 cm⁻¹

Wavenumber

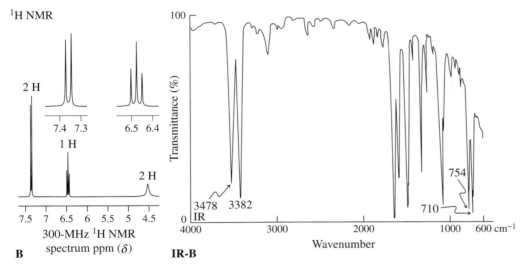

¹H NMR

2 H

7.4 7.3

6.5 6.4

1 H

2 H

7.5   7   6.5   6   5.5   5   4.5

300-MHz ¹H NMR
B spectrum ppm (δ)

IR-B

100

Transmittance (%)

3478   3382

754

710

4000   3000   2000   1000   600 cm⁻¹

Wavenumber

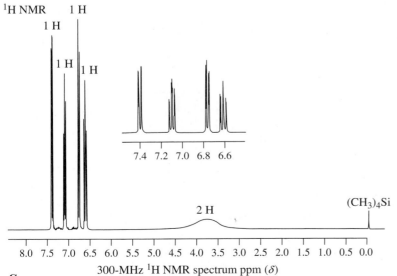

¹H NMR   1 H

1 H

1 H   1 H

7.4 7.2 7.0 6.8 6.6

2 H

(CH₃)₄Si

8.0 7.5 7.0 6.5 6.0 5.5 5.0 4.5 4.0 3.5 3.0 2.5 2.0 1.5 1.0 0.5 0.0

300-MHz ¹H NMR spectrum ppm (δ)

C

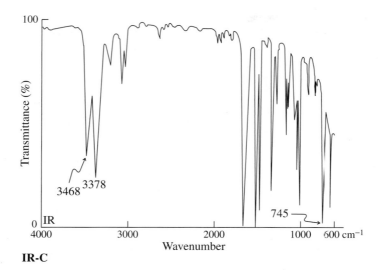

IR-C

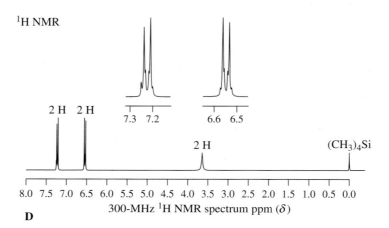

D

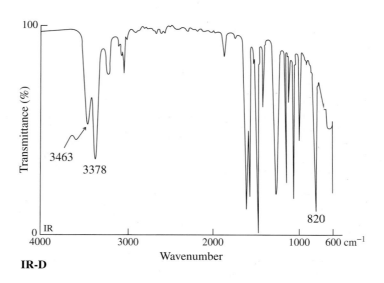

IR-D

51. Write the expected product(s) of each of the following reactions.

(a) [naphthalene] $\xrightarrow{Cl_2,\ CCl_4,\ \Delta}$

(b) [naphthalene with OCH₃] $\xrightarrow{HNO_3}$

(c) [methylnaphthalene, CH₃] $\xrightarrow{Conc.\ H_2SO_4,\ \Delta}$

(d) [naphthalene] $\xrightarrow{CH_3CCl\ (O),\ AlCl_3,\ CS_2}$

(e) [naphthalene with NO₂] $\xrightarrow{Br_2,\ FeBr_3}$

52. **CHALLENGE** Sulfonation of naphthalene at 80°C gives almost entirely 1-naphthalenesulfonic acid, whereas the same reaction at 160°C gives 2-naphthalenesulfonic acid. Propose an explanation. (**Hint:** For the underlying principle, see Section 14-6.)

53. Electrophilic substitution on the benzene ring of benzenethiol (thiophenol, C_6H_5SH) is not possible. Why? What do you think happens when benzenethiol is allowed to react with an electrophile? (**Hint:** Review Section 9-10.)

54. Although methoxy is a strongly activating (and ortho, para-directing) group, the meta positions in methoxybenzene (anisole) are actually slightly *deactivated* toward electrophilic substitution relative to benzene. Explain.

55. Predict the result of mononitration of

(a) [tetrahydronaphthalene]

(b) [tetralone with O]

(c) [difluoro tetrahydronaphthalene, F F / F F]

(d) [diaryl ketone with O and NO₂]

(e) [biphenyl with OCH₃]

56. **CHALLENGE** The *nitroso* group, –NO, as a substituent on a benzene ring acts as an ortho, para-directing group but is deactivating. Use the Lewis structure of the nitroso group and its inductive and resonance interactions with the benzene ring to explain this finding. (**Hint:** Consider possible similarities to another type of substituent that is ortho, para-directing but deactivating.)

57. Typical conditions for nitrosation are illustrated in the following equation. Propose a detailed mechanism for this reaction.

[phenol] $\xrightarrow{NaNO_2,\ HCl,\ H_2O}$ [2-nitrosophenol] + [4-nitrosophenol]

Team Problem

58. Polystyrene (polyethenylbenzene) is a familiar polymer used in the manufacture of foam cups and packing beads. One could, in principle, synthesize polystyrene by cationic polymerization with acid. However, this approach is unsuccessful because of the formation of dimer A.

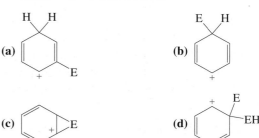

Styrene $\xrightarrow{H_2SO_4,\ H_2O}$ **A**

Split your group into two teams. The first should formulate a mechanism for the cationic polymerization of styrene using acid; the second should do so for the formation of A. Reconvene and compare your notes. At which stage is the normal polymerization sequence diverted to generate A?

Preprofessional Problems

59. Which of the following reactions is an electrophilic aromatic substitution?

(a) $C_6H_{12} \xrightarrow{Se,\ 300°} C_6H_6$

(b) $C_6H_5CH_3 \xrightarrow{Cl_2,\ h\nu} C_6H_5CH_2Cl$

(c) $C_6H_6 + (CH_3)_2CHOH \xrightarrow{BF_3,\ 60°C} C_6H_5CH(CH_3)_2$

(d) $C_6H_5Br \xrightarrow{Mg,\ ether} C_6H_5MgBr$

60. The intermediate cation A in the sequence $C_6H_6 + E^+ \rightarrow A \rightarrow C_6H_5E + H^+$ is best shown as

(a) [cyclohexadienyl cation with H H at top, E and + at bottom]

(b) [cyclohexadienyl cation with E H at top, + at bottom]

(c) [benzene with + and E]

(d) [cyclohexadienyl cation with + and E, EH]

61. The ^{1}H NMR spectrum of an unknown compound shows absorptions at (*no* multiplicities given) $\delta = 0.9$ (6 H), 2.3 (1 H), and 7.3 (5 H) ppm. One of the following five structures satisfies these data. Which one? (**Hint:** The ^{1}H NMR signal for ethane is at $\delta = 0.9$ ppm, that for benzene at $\delta = 7.3$ ppm.)

(**a**) $C_6H_5CH_2CH_2CH_3$

(**b**)

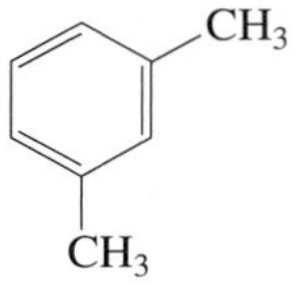

(**c**) $C_6H_5CH(CH_3)_2$

(**d**)

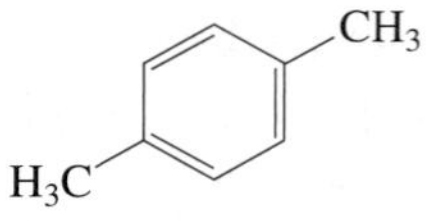

(**e**)

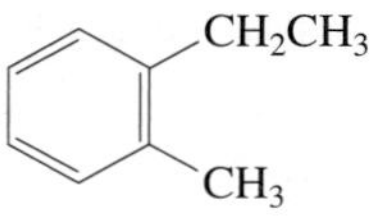

62. A volume of 135 mL of benzene was treated with excess Cl_2 and $AlCl_3$ to yield 50 mL of chlorobenzene. Given the atomic weights of C = 12.0, H = 1.00, and Cl = 35.5, and the densities of benzene, 0.78 g mL^{-1}, and chlorobenzene, 1.10 g mL^{-1}, the percent yield is closest to (**a**) 15; (**b**) 26; (**c**) 35; (**d**) 46; (**e**) 55.

63. Among the following choices, the group that *activates* the benzene ring toward electrophilic aromatic substitution is (**a**) $-NO_2$; (**b**) $-CF_3$; (**c**) $-CO_2H$; (**d**) $-OCH_3$; (**e**) $-Br$.

CHAPTER 17 | Aldehydes and Ketones
The Carbonyl Group

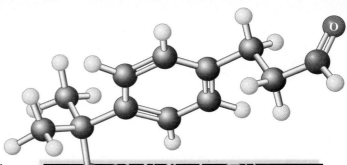

Have you ever sensed an odor that suddenly brought back a long-forgotten memory? If so, you have experienced a phenomenon unique to our sense of smell—it is a primitive sense and the only one for which the related sensory nerves are part of the brain itself. These nerves respond to both the shape and the presence of polar functional groups in volatile molecules. Prominent among the organic compounds with the most potent and varied odors are those possessing the carbon–oxygen double bond, the **carbonyl group.**

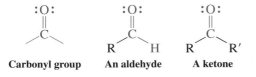

Carbonyl group   **An aldehyde**   **A ketone**

The biochemistry of the reproductive process includes a mechanism by which sperm are selected for fertilization by their ability to "smell" a nearby egg. The aldehyde bourgeonal, shown here, activates this ability, whereas the straight-chain 11-carbon aldehyde, undecanal, shuts it down. The photo shows the process of conception, magnified 3,500 times.

In this chapter and the next, we focus on the two simplest classes of carbonyl compounds: **aldehydes,** in which the carbon atom of the carbonyl group is bound to at least one hydrogen atom, and **ketones,** in which it is bound to two other carbons. These compounds exist throughout nature, contributing to the flavors and aromas of many foods and assisting in the biological functions of many enzymes. Even individual cells, such as the sperm shown in the illustration above, display a sensing ability related to that of smell. The aldehyde bourgeonal mimics as yet unidentified molecules produced by the female reproductive system to attract the most sensitive sperm for fertilization. Understanding these processes may have valuable implications for reproductive medicine. In addition, industry makes considerable use of aldehydes and ketones, both as reagents and as solvents in synthesis. Indeed, the carbonyl group is frequently considered to be the most important functional group in organic chemistry.

After explaining how to name aldehydes and ketones, we look at their structures and physical properties. Like alcohols, the carbonyl group possesses an oxygen atom with two lone pairs, a structural feature that enables it to function as a weak Lewis base. Moreover, the carbon–oxygen double bond is also highly polarized, making the carbonyl carbon quite electrophilic. In the remainder of this chapter, we show how these properties shape the chemistry of this versatile functional group.

17-1 | NAMING THE ALDEHYDES AND KETONES

For the purposes of naming, the carbonyl function is the highest-ranking group we have encountered so far. The aldehyde function takes precedence over that of ketones. For historical reasons, the simpler aldehydes often retain their common names. These names are derived from the common name of the corresponding carboxylic acid, with the ending *-ic* or *-oic acid* replaced by **-aldehyde,** as shown in the examples below.

Formic acid　　**Formaldehyde**　　**Acetic acid**　　**Acetaldehyde**　　***o*-Bromobenzoic acid**　　***o*-Bromobenzaldehyde**

Many ketones also have common names, which consist of the names of the substituent groups followed by the word *ketone.* Dimethyl ketone, the simplest example, is a common solvent best known as **acetone.** Phenyl ketones have common names ending in **-phenone.**

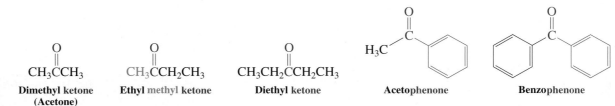

CH_3CCH_3　　$CH_3CCH_2CH_3$　　$CH_3CH_2CCH_2CH_3$

Dimethyl ketone　　**Ethyl methyl ketone**　　**Diethyl ketone**　　**Acetophenone**　　**Benzophenone**
(Acetone)

We will adhere to IUPAC names, which treat aldehydes as derivatives of the alkanes, with the ending *-e* replaced by **-al.** An alkane thus becomes an **alkanal.** Methanal, the systematic name of the simplest aldehyde, is accordingly derived from methane, ethanal from ethane, propanal from propane, and so forth. However, *Chemical Abstracts* retains the common names for the first two, and so shall we. We number the substituent chain beginning with the carbonyl carbon.

Methanal
(Formaldehyde)

HCH　　CH_3CH　　CH_3CH_2CH　　$ClCH_2CH_2CH_2CH$

　　　　Ethanal　　**Propanal**　　**4-Chlorobutanal**　　**4,6-Dimethylheptanal**
　　　　(Acetaldehyde)

Notice that the names parallel those of the 1-alkanols (Section 8-1), except that the position of the aldehyde carbonyl group does not have to be specified; *its carbon is defined as* C1. A dialdehyde is denoted as a **-dial.**

Pentanedial

When the –CHO group is attached to a ring, the compound is called a **carbaldehyde** and the carbon atom bearing the –CHO group is C1. The parent aromatic aldehyde, for instance, is benzenecarbaldehyde, although its common name, benzaldehyde, is so widely used that it is accepted by *Chemical Abstracts.*

Cyclohexanecarbaldehyde　　**Benzenecarbaldehyde**　　**4-Hydroxy-3-methoxy-**
　　　　　　　　　　　　　　　　　　(Benzaldehyde)　　　**benzenecarbaldehyde**
　　　　　　　　　　　　　　　　　　　　　　　　　　　　　　(4-Hydroxy-3-methoxy-
　　　　　　　　　　　　　　　　　　　　　　　　　　　　　　benzaldehyde; vanillin,
　　　　　　　　　　　　　　　　　　　　　　　　　　　　　　see Chapter 15 opening)

Carbon bearing
CHO is C1

Ketones are called **alkanones,** the ending -*e* of the alkane replaced with **-one.** An exception is the smallest ketone, propanone, for which IUPAC has accepted the common name acetone. The carbonyl carbon is assigned the lowest possible number in the chain, regardless of the presence of other substituents or the OH, C=C, or C≡C functional groups. Aromatic ketones are named as aryl-substituted alkanones. Ketones, unlike aldehydes, may also be part of a ring, an arrangement that gives compounds called **cycloalkanones.** Polyketones are labeled **-dione, -trione,** and so on.

2-Pentanone

Numbering starts at end closest to carbonyl

4-Chloro-**6-methyl-3-heptanone**

1-Phenylethanone (Acetophenone)

Carbonyl carbon in ring is C1

2,2-Dimethyl**cyclopentanone**

For purposes of numbering and naming, the carbonyl group of an aldehyde takes precedence over all other groups we have encountered so far. Both aldehydes and ketones take precedence over alcohols, but not over the carboxylic acid function, which we shall encounter in Chapter 19 when we present the entire order of precedence of all the common functional groups.

Heptane-2,6-dione

Aldehydes and Ketones with Other Functional Groups

$$CH_3CCH_2CH=CHCH_2CCH_3$$

7-Hydroxy-**7-methyl-4-octen-2-one**
(Note that the *e* in *-ene* and *-yne* is dropped in *-enone* and *-ynal*)

$HC≡CCH$

Propynal

5-Bromo-**3-**ethynyl**cycloheptanone**

(*R*)-**3-**Formyl**hexanedial**

Systematic nomenclature accepts both **alkanoyl** and **acyl** as names for the fragment RC–. Acyl is more widely used and is the one we will employ (see also Section 15-13). Both the IUPAC and *Chemical Abstracts* retain the common names **formyl** for HC– and **acetyl** for CH_3C–. The term **oxo** denotes the location of a ketone carbonyl group when it is present together with an aldehyde function.

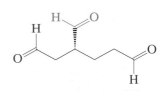

cis-**2-**Acetyl**cyclohexane-carboxylic acid**

$$CH_3CCH_2CH$$

3-Oxo**butanal**

Exercise 17-1

Name or draw the structures of the following compounds.

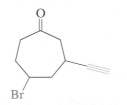

(a)

(b)

(c) 4-Octyn-3-one

(d) 3-Hydroxybutanal   **(e)** 4-Bromocyclohexanone

There are various ways of drawing aldehydes and ketones. As always, condensed formulas or the zigzag (bond-line) notation may be used. In condensed formulas, carbon atoms are followed immediately by the hydrogen atoms to which they are directly attached. Thus, the aldehyde function is abbreviated as –CHO and *never* as –COH, thereby avoiding confusion with the hydroxy group of alcohols. Formaldehyde is written HCHO, acetaldehyde as CH_3CHO, propanal as CH_3CH_2CHO, and so forth.

Various Ways of Writing Aldehyde and Ketone Structures

Butanal:

$$CH_3CH_2CH_2\overset{\overset{\displaystyle O}{\|}}{CH} \qquad CH_3CH_2CH_2CHO$$

Not a hydroxy group

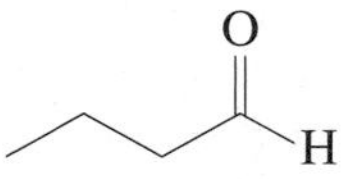

2-Butanone:

$$CH_3CH_2\overset{\overset{\displaystyle O}{\|}}{C}CH_3 \qquad CH_3CH_2COCH_3$$

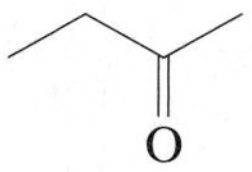

In Summary Aldehydes and ketones are named systematically as alkanals and alkanones. The carbonyl group takes precedence over the hydroxy function and carbon–carbon double and triple bonds in numbering. With these rules, the usual guidelines for numbering the stem and labeling the substituents are followed.

17-2 | STRUCTURE OF THE CARBONYL GROUP

If we think of the carbonyl group as an oxygen analog of the alkene functional group, we can correctly predict its orbital description, the structures of aldehydes and ketones, and some of their physical properties. However, the alkene and carbonyl double bonds do differ considerably in reactivity because of the electronegativity of oxygen and its two lone pairs of electrons.

The carbonyl group contains a short, strong, and very polar bond

Both the carbon and the oxygen of the carbonyl group are sp^2 hybridized. They therefore lie in the same plane as the two additional groups on carbon, with bond angles approximating 120°. Perpendicular to the molecular frame are two p orbitals, one on carbon and one on oxygen, making up the π bond (Figure 17-1).

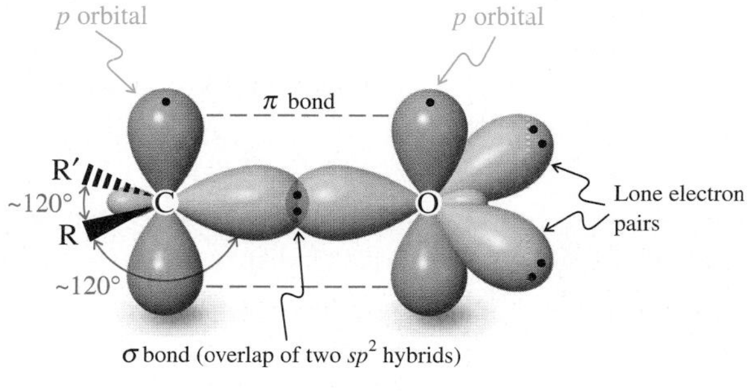

Figure 17-1 Orbital picture of the carbonyl group. The sp^2 hybridization and the orbital arrangement are similar to those of ethene (Figure 11-2). However, both the two lone electron pairs and the electronegativity of the oxygen atom modify the properties of the functional group.

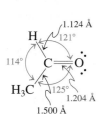

Figure 17-2 Molecular structure of acetaldehyde.

Figure 17-2 shows some of the structural features of acetaldehyde. As expected, the molecule is planar, with a trigonal carbonyl carbon and a short carbon–oxygen bond, indicative of its double-bond character. Not surprisingly, this bond is also rather strong, ranging from 175 to 180 kcal mol^{-1} (732 to 753 kJ mol^{-1}).

Comparison with the electronic structure of an alkene double bond reveals two important differences. First, the oxygen atom bears two lone electron pairs located in two sp^2 hybrid orbitals. Second, oxygen is more electronegative than carbon. This property causes an appreciable polarization of the carbon–oxygen double bond, with a partial positive charge on carbon and an equal amount of negative charge on oxygen. Thus, the carbon is electrophilic, the oxygen nucleophilic and slightly basic. This polarization can be described either by a polar resonance form for the carbonyl moiety or by partial charges. It can also be seen in the electrostatic potential map of formaldehyde in the margin, where the area around the carbon atom appears blue (positive), and that around the oxygen red (negative). As we have seen (Section 16-1), the partial positive charge on the carbonyl carbon renders acyl groups electron withdrawing.

Formaldehyde

Descriptions of a Carbonyl Group

Electrophilic

Nucleophilic and basic

Polarization alters the physical constants of aldehydes and ketones

The polarization of the carbonyl functional group makes the boiling points of aldehydes and ketones higher than those of hydrocarbons of similar size and molecular weight (Table 17-1). Because of their polarity, the smaller carbonyl compounds such as acetaldehyde and acetone are completely miscible with water: An aqueous solution of formaldehyde (Section 17-6) has applications as a disinfectant and a fungicide. As the hydrophobic hydrocarbon part of the molecule increases in size, however, water solubility decreases. Carbonyl compounds with more than six carbons are rather insoluble in water.

Table 17-1	Boiling Points of Aldehydes and Ketones	
Formula	**Name**	**Boiling point (°C)**
$HCHO$	Formaldehyde	−21
CH_3CHO	Acetaldehyde	21
CH_3CH_2CHO	Propanal (propionaldehyde)	49
CH_3COCH_3	Acetone	56
$CH_3CH_2CH_2CHO$	Butanal (butyraldehyde)	76
$CH_3CH_2COCH_3$	Butanone (ethyl methyl ketone)	80
$CH_3CH_2CH_2CH_2CHO$	Pentanal	102
$CH_3COCH_2CH_2CH_3$	2-Pentanone	102
$CH_3CH_2COCH_2CH_3$	3-Pentanone	102

In Summary The carbonyl group in aldehydes and ketones is an oxygen analog of the carbon–carbon double bond. However, the electronegativity of oxygen polarizes the π bond, thereby rendering the acyl substituent electron withdrawing. The arrangement of bonds around the carbon and oxygen is planar, a consequence of sp^2 hybridization.

17-3 | SPECTROSCOPIC PROPERTIES OF ALDEHYDES AND KETONES

What are the spectral characteristics of carbonyl compounds? In ^1H NMR spectroscopy, the formyl hydrogen of the aldehydes is very strongly deshielded, appearing between 9 and 10 ppm, a chemical shift that is unique for this class of compounds. The reason for this effect is twofold. First, the movement of the π electrons, like that in alkenes (Section 11-4), causes a local magnetic field, which strengthens the external field. Second, the charge on

^1H NMR Deshielding in Aldehydes and Ketones

O
||
RCH_2CH

$\delta \sim 2.5$ ~ 9.8 ppm

R′ O
| ||
$RCHCCH_3$

$\delta \sim 2.6$ ~ 2.0 ppm

Figure 17-3 300-MHz ¹H NMR spectrum of propanal. The formyl hydrogen (at $\delta = 9.79$ ppm) is strongly deshielded.

The (green) signal at $\delta = 2.46$ ppm in Figure 17-3 (for the CH₂ hydrogens) illustrates the sequential $N + 1$ rule (Section 10-8). These hydrogens have four neighboring coupled hydrogens in two groups, one (the CH₃) with $J = 7$ Hz and the other (the aldehyde H) with $J = 2$ Hz. The result is $(3 + 1) \times (1 + 1) = 8$ peaks, appearing as a quartet $(J = 7$ Hz) of doublets $(J = 2$ Hz). Note that the two coupling constants appear also in the triplet signals of these neighbors.

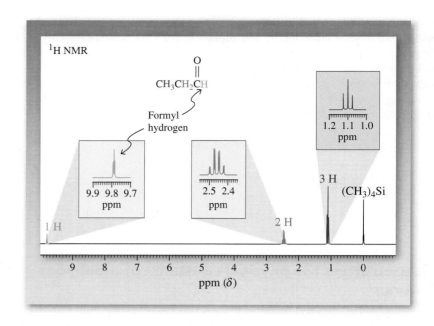

the positively polarized carbon exerts an additional deshielding effect. Figure 17-3 shows the ¹H NMR spectrum of propanal with the formyl hydrogen resonating at $\delta = 9.79$ ppm, split into a triplet $(J = 2$ Hz) because of a small coupling to the neighboring (C2) hydrogens. These hydrogens are also slightly deshielded relative to alkane hydrogens because of the electron-withdrawing character of the carbonyl group. This effect is also seen in the ¹H NMR spectra of ketones: The α-hydrogens normally appear in the region $\delta = 2.0$–2.8 ppm.

Carbon-13 NMR spectra are diagnostic of both aldehydes *and* ketones because of the characteristic chemical shift of the carbonyl carbon. Partly because of the electronegativity of the directly bound oxygen, the carbonyl carbons in aldehydes and ketones appear at even lower field ($\sim$200 ppm) than do the sp^2-hybridized carbon atoms of alkenes (Section 11-4). The carbons next to the carbonyl group also are deshielded relative to those located farther away. The ¹³C NMR spectrum of cyclohexanone is shown in Figure 17-4.

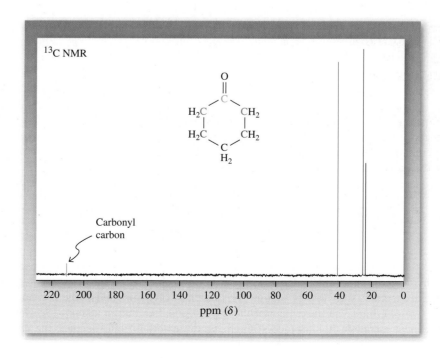

Figure 17-4 ¹³C NMR spectrum of cyclohexanone. The carbonyl carbon at 211.8 ppm is strongly deshielded relative to the other carbons. Because of symmetry, the molecule exhibits only four peaks; the three methylene carbon resonances absorb at increasingly lower field the closer they are to the carbonyl group.

^{13}C NMR Chemical Shifts of Typical Aldehydes and Ketones

$$\underset{\delta\,=\,31.2 \quad 199.6\ \textbf{ppm}}{CH_3-\overset{\displaystyle O}{\overset{\|}{C}}H}$$

$$\underset{\delta\,=\,5.2 \quad 36.7 \quad 201.8\ \textbf{ppm}}{CH_3-CH_2-\overset{\displaystyle O}{\overset{\|}{C}}H}$$

$$\underset{\delta\,=\,30.2 \quad 205.1\ \textbf{ppm}}{CH_3\overset{\displaystyle O}{\overset{\|}{C}}CH_3}$$

$$\underset{\delta\,=\,29.3 \quad 206.6 \quad 45.2 \quad 17.5 \quad 13.5\ \textbf{ppm}}{CH_3\overset{\displaystyle O}{\overset{\|}{C}}-CH_2-CH_2-CH_3}$$

Infrared spectroscopy is a useful way of directly detecting the presence of a carbonyl group. The C=O stretching frequency gives rise to an intense band that typically appears in a relatively narrow range (1690–1750 cm^{-1}; Figure 17-5). The carbonyl absorption for aldehydes appears at about 1735 cm^{-1}; absorptions for acyclic alkanones and cyclohexanone are found at about 1715 cm^{-1}. Conjugation with either alkene or benzene π systems reduces the carbonyl infrared frequency by about 30–40 cm^{-1}; thus 1-phenylethanone (acetophenone) exhibits an IR band at 1680 cm^{-1}. Conversely, the stretching frequency increases for carbonyl groups in rings with fewer than six atoms: Cyclopentanone absorbs at 1745 cm^{-1}, cyclobutanone at 1780 cm^{-1}.

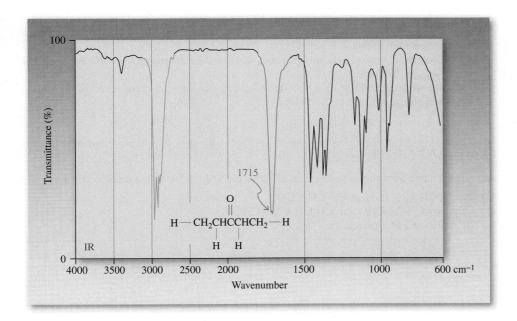

Figure 17-5 IR spectrum of 3-pentanone; $\tilde{\nu}_{C=O\ stretch}$ = 1715 cm^{-1}.

Carbonyl groups also exhibit characteristic electronic (UV) spectra because the nonbonding lone electron pairs on the oxygen atom undergo low-energy $n \rightarrow \pi^*$ transitions (Figure 17-6). For example, acetone shows an $n \rightarrow \pi^*$ band at 280 nm (ϵ = 15) in hexane. The corresponding $\pi \rightarrow \pi^*$ transition appears at about 190 nm (ϵ = 1100). Conjugation with a carbon–carbon double bond shifts absorptions to longer wavelengths. For example, the electronic spectrum of 3-buten-2-one, CH$_2$=CHCOCH$_3$, has peaks at 324 nm (ϵ = 24, $n \rightarrow \pi^*$) and 219 nm (ϵ = 3600, $\pi \rightarrow \pi^*$).

Electronic Transitions of Acetone and 3-Buten-2-one

$$\underset{\textbf{Acetone}}{CH_3\overset{\displaystyle O}{\overset{\|}{C}}CH_3}$$

$$\underset{\textbf{3-Buten-2-one}}{CH_2=CH\overset{\displaystyle O}{\overset{\|}{C}}CH_3}$$

$$\lambda_{max}(\epsilon) = 280(15) \quad n \longrightarrow \pi^*$$
$$190(1100) \quad \pi \longrightarrow \pi^*$$

$$\lambda_{max}(\epsilon) = 324(24) \quad n \longrightarrow \pi^*$$
$$219(3600) \quad \pi \longrightarrow \pi^*$$

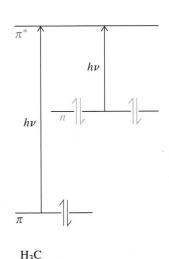

Figure 17-6 The $\pi \rightarrow \pi^*$ and $n \rightarrow \pi^*$ transitions in acetone.

Solved Exercise 17-2 | **Working with the Concepts: Using Spectroscopy Revisited**

How would you apply spectroscopy to differentiate between $CH_3CH_2CH_2CH_2OH$ and $CH_3CH_2CH_2CHO$? Indicate the method and the features in the spectrum that would be most useful.

Strategy

Consider the structural differences in the context of the capabilities of your four spectroscopic methods—NMR, IR, MS, and UV. Judge how each technique contributes useful information to accomplish the task.

Solution

- Structurally, the compounds have very different functional groups: $-CH_2-O-H$ in the primary alcohol and $-CH=O$ in the aldehyde. NMR, IR, and UV spectroscopy each will reflect this structural disparity in a characteristic way. The most obvious distinctions follow.
- In the ^1H NMR spectrum, chemical shift differences will be evident between the $-CH_2-O-H$ of the alcohol at $\delta \approx 3.7$ ppm and the $-CH=O$ of the aldehyde at $\delta \approx 9.7$ ppm (Table 10-2).
- The ^{13}C NMR spectrum will display similarly distinctive peaks for the $-CH_2-O-H$ of the alcohol at $\delta \approx 70$ ppm and for the $-CH=O$ of the aldehyde at $\delta \approx 200$ ppm (Table 10-6).
- IR spectroscopy differentiates functions (Table 11-4). The alcohol O—H will give rise to a strong, broad absorption between 3200 and 3600 cm^{-1} (compare Figure 11-21). In contrast, the aldehyde C=O will give a sharper but equally intense peak around 1735 cm^{-1} (similar to that in the ketone spectrum shown in Figure 17-5).
- Simple alcohols show no significant UV bands whereas the carbonyl group of the aldehyde will exhibit a $\lambda_{max} \approx 280$ nm.
- Finally, the molecules differ in molecular weight by two units. A comparison of their mass spectra will reveal this aspect quite clearly.

Exercise 17-3 | **Try It Yourself**

Answer the same question as in Exercise 17-2 for each of the following pairs of compounds.

(a) $CH_3COCH_2CH_3$ and $CH_3CH_2CH_2CHO$
(b) $CH_3CH=CHCH_2CHO$ and $CH_3CH_2CH=CHCHO$
(c) 2-Pentanone and 3-pentanone

Exercise 17-4

An unknown C_4H_6O exhibited the following spectral data: ^1H NMR: $\delta = 2.03$ (dd, $J = 6.7$, 1.6 Hz, 3 H), 6.06 (ddq, $J = 16.1$, 7.7, 1.6 Hz, 1 H), 6.88 (dq, $J = 16.1$, 6.7 Hz, 1 H), 9.47 (d, $J = 7.7$ Hz, 1 H) ppm; ^{13}C NMR: $\delta = 18.4$, 132.8, 152.1, 191.4 ppm; UV $\lambda_{max}(\epsilon) = 220(15,000)$ and 314(32) nm. Suggest a structure.

Mass spectral fragmentation of aldehydes and ketones provides structural information

The fragmentation patterns of carbonyl compounds are often useful in structural identifications. For example, the mass spectra of the isomeric ketones 2-pentanone, 3-pentanone, and 3-methyl-2-butanone (Figure 17-7) reveal very clean and distinct fragment ions. The predominant decomposition pathway is α cleavage, which severs an alkyl bond to the carbonyl function to give the corresponding resonance-stabilized **acylium cation** and an alkyl radical.

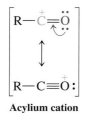

Acylium cation

α Cleavage of Carbonyl Compounds

Acylium cation

$$:\overset{+}{O}\equiv CR' \xleftarrow[-R\,\cdot]{\alpha\ Cleavage} \left[\begin{array}{c} :O: \\ \| \\ R \quad C \quad R' \end{array} \right]^{+\,\cdot} \xrightarrow[-R'\,\cdot]{\alpha\ Cleavage} R\overset{+}{C}\equiv O:$$

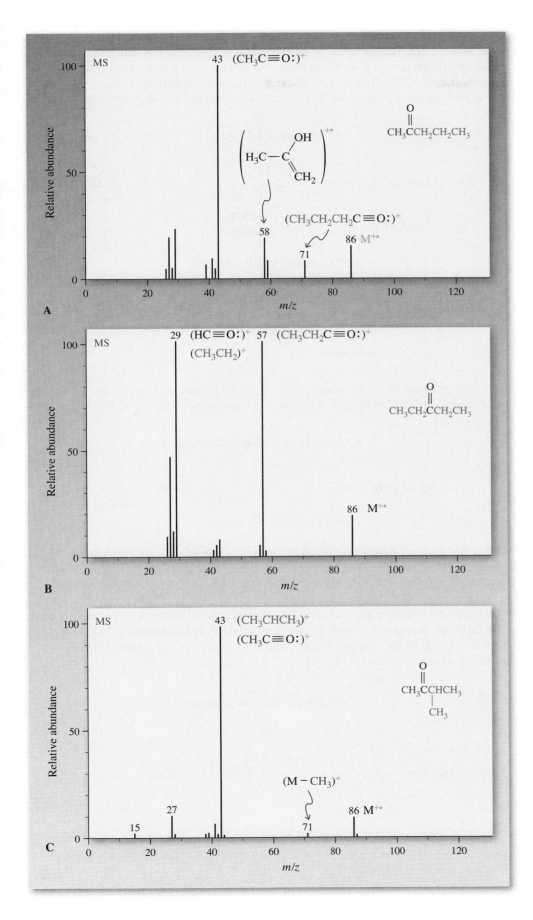

Figure 17-7 Mass spectra of (A) 2-pentanone, showing two peaks for α cleavage and one for McLafferty rearrangement; (B) 3-pentanone, showing only a single α cleavage peak because of symmetry; and (C) 3-methyl-2-butanone, showing two α cleavages.

These fragment ions allow the gross composition of the two alkyl groups in a ketone to be read from the spectrum. In this way, 2-pentanone is readily differentiated from 3-pentanone: α Cleavage of 2-pentanone gives two acylium ions, at $m/z = 43$ and 71, but 3-pentanone gives only one, at $m/z = 57$. (The $m/z = 29$ peak in the mass spectrum of 3-pentanone is due partly to $CH_3CH_2^+$ and partly to $HC\equiv O^+$, which arises by loss of C_2H_4 from $CH_3CH_2C\equiv O^+$.)

α Cleavage in 2-Pentanone

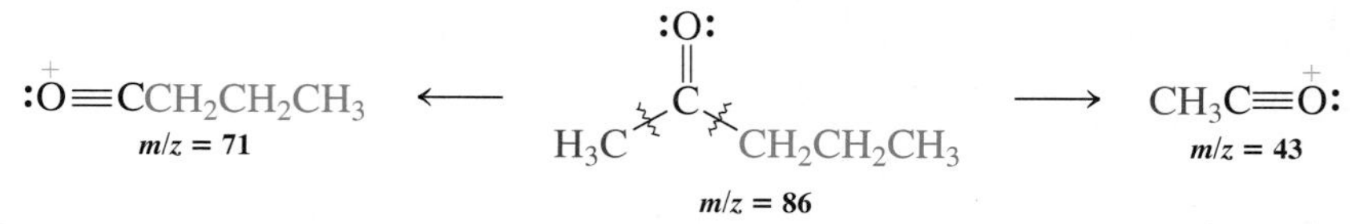

$:O\equiv CCH_2CH_2CH_3$ $\longleftarrow$ ... $\longrightarrow$ $CH_3C\equiv O:$

$m/z = 71$

$m/z = 86$
2-Pentanone

$m/z = 43$

α Cleavage in 3-Pentanone

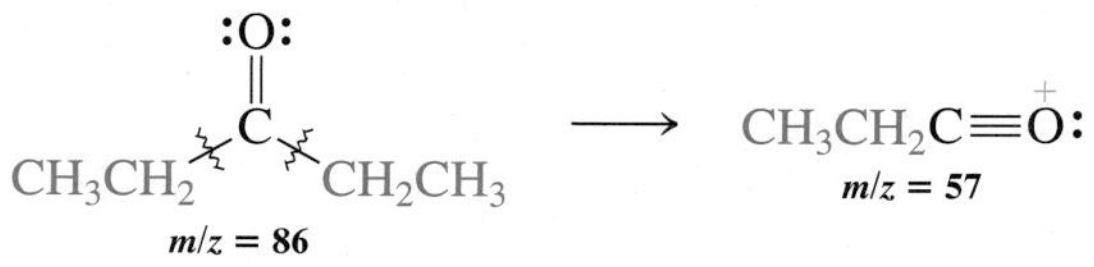

$m/z = 86$
3-Pentanone

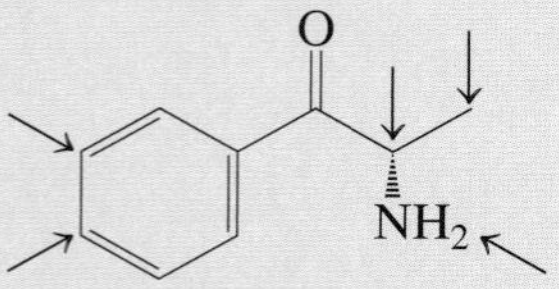

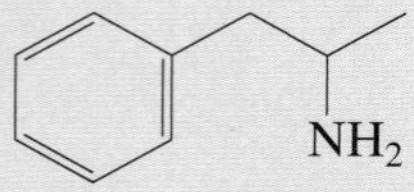

$CH_3CH_2C\equiv O:$ $m/z = 57$

Can 2-pentanone be distinguished from 3-methyl-2-butanone? Not by the observation of α cleavage—in both molecules, the substituent groups are CH_3 and C_3H_7.

α Cleavage in 3-Methyl-2-butanone

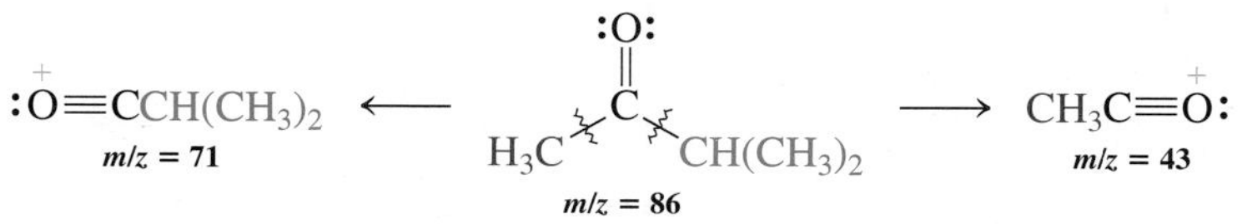

$:O\equiv CCH(CH_3)_2$ $\longleftarrow$ ... $\longrightarrow$ $CH_3C\equiv O:$

$m/z = 71$

$m/z = 86$

$m/z = 43$

However, comparison of the mass spectra of the two compounds (Figure 17-7A and C) reveals an additional prominent peak for 2-pentanone at $m/z = 58$, signifying the loss of a molecular fragment of weight $m/z = 28$. This fragment is absent from the spectra of both other isomers and is characteristic of the presence of hydrogens located γ to the carbonyl group. Compounds with this structural feature and with sufficient flexibility to allow the γ-hydrogen to be close to the carbonyl oxygen decompose by the **McLafferty* rearrangement.** In this reaction, the molecular ion of the starting ketone splits into two pieces (a neutral fragment and a radical cation) in a unimolecular process.

McLafferty Rearrangement

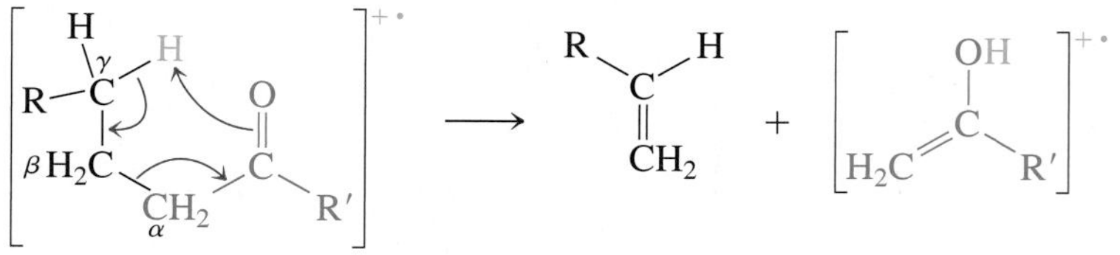

The McLafferty rearrangement yields an alkene and the enol form of a new ketone. In the case of 2-pentanone, ethene and the enol of acetone are produced; the radical cation of

---

*Professor Fred W. McLafferty (b. 1923), Cornell University, Ithaca, New York.

the acetone enol is observed with $m/z = 58$.

$$\left[\underset{\alpha}{CH_3}\overset{O}{\underset{\|}{C}}\underset{\beta}{CH_2}\underset{\gamma}{CH_2}\overset{H}{\underset{|}{CH_2}} \right]^{+\cdot} \longrightarrow \left[H_3C\overset{OH}{\underset{\|}{C}}_{CH_2} \right]^{+\cdot} + CH_2{=}CH_2$$

$m/z = 86$ $m/z = 58$

Neither 3-pentanone nor 3-methyl-2-butanone possesses a γ-hydrogen; therefore, neither is able to undergo McLafferty rearrangement.

Exercise 17-5

How would you tell the difference between **(a)** 3-methyl-2-pentanone and 4-methyl-2-pentanone, and **(b)** 2-ethylcyclohexanone and 3-ethylcyclohexanone, using only mass spectrometry?

In Summary In the NMR spectra of aldehydes and ketones, the formyl hydrogens and the carbonyl carbons show strong deshielding. The carbon–oxygen double bond in ketones gives rise to a strong infrared band at about 1715 cm^{-1}, which is shifted to lower frequency by conjugation and to higher frequency in aldehydes and small rings. The ability of non-bonding electrons to be excited into the π^* molecular orbitals causes the carbonyl group to exhibit characteristic, relatively long-wavelength UV absorptions. Finally, aldehydes and ketones fragment in the mass spectrometer by α cleavage and McLafferty rearrangement.

17-4 | PREPARATION OF ALDEHYDES AND KETONES

Several ways to prepare aldehydes and ketones have already been described in connection with the chemistry of other functional groups (see the Reaction Summary Road Map on pp. 780–781). This section reviews the methods that we have studied, pointing out special features and additional examples. Other routes to aldehydes and ketones will be described in later chapters.

Laboratory syntheses of aldehydes and ketones use four common methods

Table 17-2 summarizes four approaches to synthesizing aldehydes and ketones. First, we have seen (Section 8-6) that oxidation of alcohols by chromium(VI) reagents gives carbonyl compounds. Secondary alcohols give ketones. Primary alcohols give aldehydes, but only in

Table 17-2 Syntheses of Aldehydes and Ketones	
Reaction	**Illustration**
1. Oxidation of alcohols (Section 8-6)	$-CH_2OH \xrightarrow{PCC, CH_2Cl_2} -\overset{O}{\underset{\|}{C}}H$
2. Ozonolysis of alkenes (Section 12-12)	$\underset{/}{\overset{\backslash}{C}}{=}\underset{\backslash}{\overset{/}{C}} \xrightarrow[\text{2. }(CH_3)_2S]{\text{1. }O_3, CH_2Cl_2} \underset{/}{\overset{\backslash}{C}}{=}O + O{=}\underset{\backslash}{\overset{/}{C}}$
3. Hydration of alkynes (Sections 13-7 and 13-8)	$-C{\equiv}C- \xrightarrow{H_2O, H^+, Hg^{2+}} -\overset{O}{\underset{\|}{C}}-CH_2-$
4. Friedel-Crafts acylation (Section 15-13)	$\xrightarrow{RCOCl, AlCl_3}$

A "Green" Oxidation of Alcohols

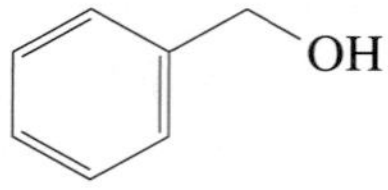

Phenylmethanol (Benzyl alcohol)

Air, 1 atm, Cu catalyst

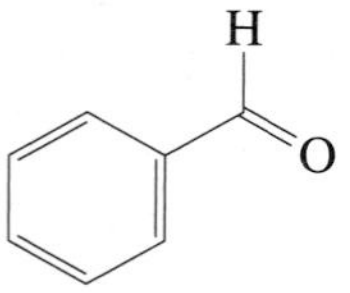

100%

Benzenecarbaldehyde (Benzaldehyde)

Air is the "greenest" oxidizing agent. There is currently much research effort aimed at achieving the oxidation of alcohols under mild aerobic conditions.

the absence of water, to prevent overoxidation to carboxylic acids. Chromium(VI) is selective and does not oxidize alkene and alkyne units.

Selective Alcohol Oxidation

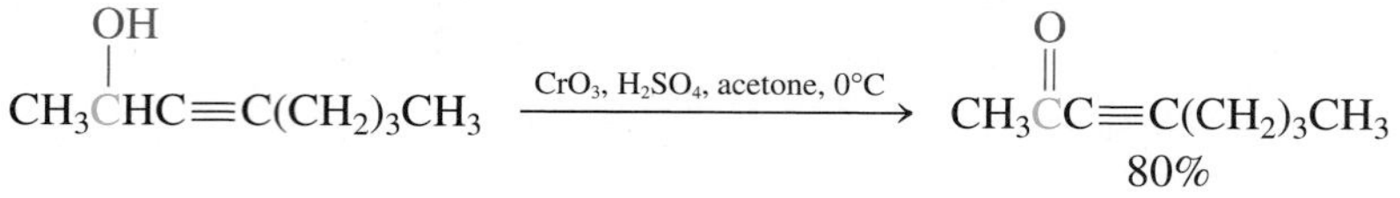

3-Octyn-2-ol   3-Octyn-2-one

Use of PCC (CrO₃ + Pyridine + HCl) to Oxidize a Primary Alcohol to an Aldehyde

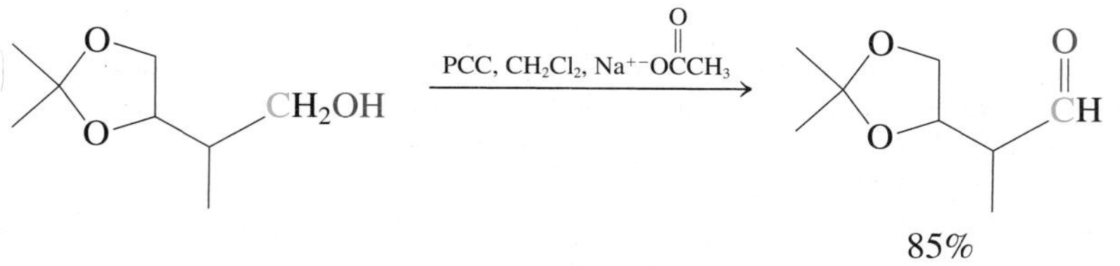

85%

Overoxidation of aldehydes in the presence of water is due to hydration to a 1,1-diol (Section 17-6). Oxidation of this diol leads to the carboxylic acid.

Water Causes the Overoxidation of Primary Alcohols

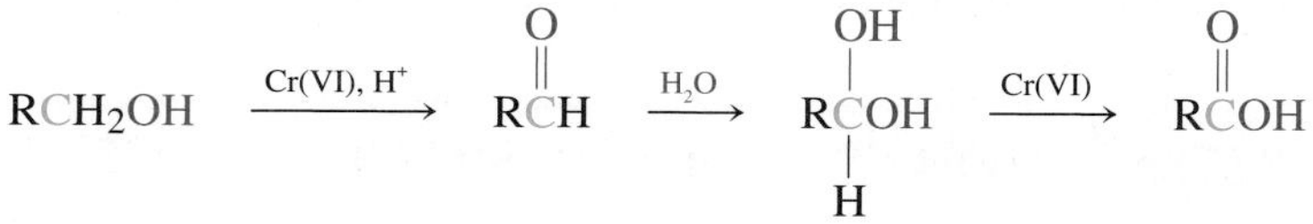

Another mild reagent that specifically oxidizes allylic alcohols (Section 14-3) is manganese dioxide. Ordinary alcohols are not attacked at room temperature, as shown below in the selective oxidation to form a steroid found in the adrenal gland.

Selective Allylic Oxidations with Manganese Dioxide

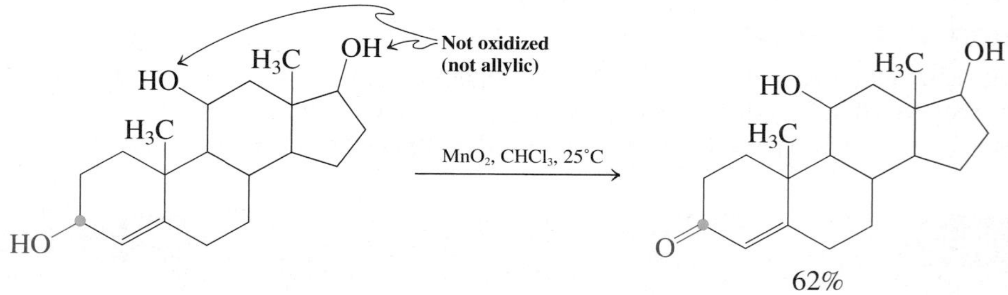

Not oxidized (not allylic)

MnO_2, $CHCl_3$, 25°C

62%

Exercise 17-6

(a) In each of the following retrosynthetic disconnections (⤳), show a reaction that would make the indicated bond.

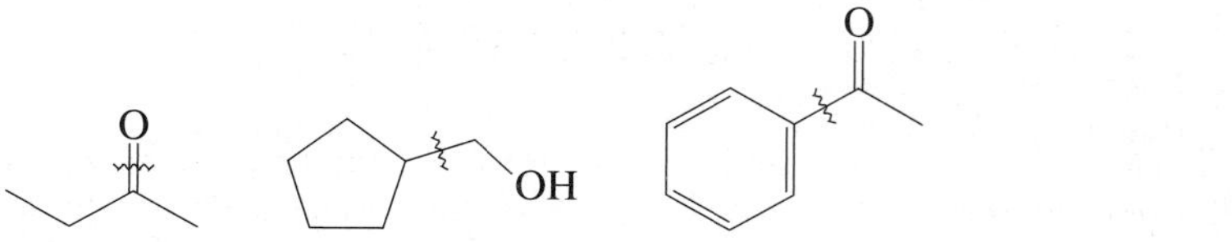

(b) Design a synthesis of 1-cyclohexyl-2-butyn-1-one starting from cyclohexane. You may use any other reagents. Work backwards from the target, along the lines of part (a).

The second method of preparation that we studied is the oxidative cleavage of carbon–carbon double bonds—ozonolysis (Section 12-12). Exposure to ozone followed by treatment with a mild reducing agent, such as zinc metal or dimethyl sulfide, cleaves alkenes to give aldehydes and ketones.

Third, hydration of the carbon–carbon triple bond yields enols that tautomerize to carbonyl compounds (Sections 13-7 and 13-8). In the presence of mercuric ion, addition of water follows Markovnikov's rule to furnish ketones.

Markovnikov Hydration of Alkynes

$$RC \equiv CH \xrightarrow{HOH, \; H^+, \; Hg^{2+}} \left[\begin{array}{c} HO \quad\quad H \\ C=C \\ R \quad\quad H \end{array} \right] \longrightarrow RCCH_3$$

Enol   Ketone

Ozonolysis

$$\xrightarrow[\text{2. Reducing agent}]{\text{1. O}_3}$$

$$CH_3C(CH_2)_4CH$$
85%

Anti-Markovnikov addition is observed in hydroboration–oxidation.

Anti-Markovnikov Hydration of Alkynes

$$RC \equiv CH \xrightarrow{\left(\bigcirc\right)_2 BH} \quad \begin{array}{c} R \quad\quad H \\ C=C \\ H \quad\quad B(\bigcirc)_2 \end{array} \xrightarrow{H_2O_2, \; HO^-} \left[\begin{array}{c} R \quad\quad H \\ C=C \\ H \quad\quad OH \end{array} \right] \longrightarrow RCH_2CH$$

Boron adds to
terminal alkyne carbon

Enol   Aldehyde

Finally, Section 15-13 discussed the synthesis of aryl ketones by Friedel-Crafts acylation, a form of electrophilic aromatic substitution. The following example furnishes an industrially useful perfume additive.

Friedel-Crafts Acylation

$$CH_3O\text{-}\bigcirc \xrightarrow{CH_3COCCH_3, \; AlCl_3, \; CS_2} CH_3O\text{-}\bigcirc\text{-}C\text{-}CH_3$$

Ortho, para-directing
and activating

93%

In Summary Four methods of synthesizing aldehydes and ketones are oxidation of alcohols, oxidative cleavage of alkenes, hydration of alkynes, and Friedel-Crafts acylation. Many other approaches will be considered in later chapters.

17-5 | REACTIVITY OF THE CARBONYL GROUP: MECHANISMS OF ADDITION

How does the carbonyl group's structure (Section 17-2) help us understand the way it functions chemically? We shall see that the carbon–oxygen double bond is prone to additions, just like the π bond in alkenes. Being highly polar, however, it is predisposed toward attack by nucleophiles at carbon and electrophiles at oxygen. This section begins the discussion of the chemistry of the carbonyl group in aldehydes and ketones.

There are three regions of reactivity in aldehydes and ketones

Aldehydes and ketones contain three regions at which most reactions take place: the Lewis basic oxygen, the electrophilic carbonyl carbon, and the adjacent "α"-carbon.

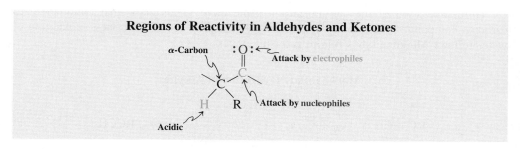

Regions of Reactivity in Aldehydes and Ketones

The remainder of this chapter concerns the first two areas of reactivity, both of which lead to ionic addition to the carbonyl π bond. Other reactions, centered instead on the acidic hydrogen of the α-carbon, will be the subject of Chapter 18.

The carbonyl group undergoes ionic additions

Polar reagents add to the dipolar carbonyl group according to Coulomb's law (Section 1-2) and the fundamental principles of Lewis acid–Lewis base interactions (Section 2-2). Nucleophiles bond to the electrophilic carbonyl carbon and the Lewis basic carbonyl oxygen bonds to electrophiles. Sections 8-6 and 8-8 described several such additions of organometallic and hydride reagents to give alcohols (Table 17-3). These additions fall under the general mechanistic category 4(a), as introduced in Section 2-2.

Reaction

Ionic Additions to the Carbonyl Group

The hydride reagents $NaBH_4$ and $LiAlH_4$ reduce carbonyl groups but not carbon–carbon double bonds. These reagents therefore convert unsaturated aldehydes and ketones into unsaturated alcohols.

1. $LiAlH_4$, $(CH_3CH_2)_2O$, $-10°C$
2. H^+, H_2O

90%

Table 17-3	Additions of Hydride and Organometallic Reagents to Aldehydes and Ketones	
Reaction	**Equation**	
1. Aldehyde + hydride reagent	RCHO $\xrightarrow{NaBH_4,\ CH_3CH_2OH}$ RCH$_2$OH	Primary alcohol
2. Ketone + hydride reagent	R$_2$CO $\xrightarrow{NaBH_4,\ CH_3CH_2OH}$ R$_2$CHOH	Secondary alcohol
3. Formaldehyde + Grignard reagent	H$_2$CO $\xrightarrow{R'MgX,\ (CH_3CH_2)_2O}$ R'CH$_2$OH[a]	Primary alcohol
4. Aldehyde + Grignard reagent	RCHO $\xrightarrow{R'MgX,\ (CH_3CH_2)_2O}$ R'RCHOH[a]	Secondary alcohol
5. Ketone + Grignard reagent	R$_2$CO $\xrightarrow{R'MgX,\ (CH_3CH_2)_2O}$ R'R$_2$COH[a]	Tertiary alcohol

[a] After aqueous acid work-up.

Table 17-4	**Additions of Moderately Basic Nucleophiles to Aldehydes and Ketones**	
Nucleophile	**Intermediate (usually not isolated)**	**Final product (stable)**
1. Water	$\diagdown$C=O $\xrightleftharpoons[\text{Acid or base}]{\text{H}_2\text{O}}$ $\left[\diagdown\text{C}\begin{smallmatrix}\text{OH}\\\text{OH}\end{smallmatrix} \right]$ **Geminal diol (hydrate)**	
2. Alcohol	$\diagdown$C=O $\xrightleftharpoons[\text{Acid or base}]{\text{ROH}}$ $\left[\diagdown\text{C}\begin{smallmatrix}\text{OH}\\\text{OR}\end{smallmatrix} \right]$ **Hemiacetal**	$\xrightleftharpoons[\text{Acid only}]{\text{ROH, }-\text{H}_2\text{O}}$ $\diagdown\text{C}\begin{smallmatrix}\text{OR}\\\text{OR}\end{smallmatrix}$ **Acetal**
3. Ammonia (R = H) or primary amine (R = alkyl or aryl)	$\diagdown$C=O $\xrightleftharpoons[\text{Acid}]{\text{RNH}_2}$ $\left[\diagdown\text{C}\begin{smallmatrix}\text{OH}\\\text{NHR}\end{smallmatrix} \right]$ **Hemiaminal**	$\xrightleftharpoons[\text{Acid}]{-\text{H}_2\text{O}}$ $\diagdown$C=NR **Imine**
4. Secondary amine	$\diagdown$C=O $-$C$\diagdown_\text{H}$ $\xrightleftharpoons[\text{Acid}]{\text{R}_2\text{NH}}$ $\left[\diagdown\text{C}\begin{smallmatrix}\text{OH}\\\text{NR}_2\end{smallmatrix} -\text{C}\diagdown_\text{H} \right]$ **Hemiaminal**	$\xrightleftharpoons[\text{Acid}]{-\text{H}_2\text{O}}$ $\diagdown$C$-$NR$_2$ $-$C$\diagdown$ **Enamine**

Because the nucleophilic reagents illustrated in Table 17-3 are strong bases, their additions are irreversible. This section and Sections 17-6 through 17-9 consider ionic additions of less basic nucleophiles Nu–H, such as water, alcohols, thiols, and amines. These processes are not strongly exothermic but instead establish equilibria that can be pushed in either direction by the appropriate choice of reaction conditions. Table 17-4 summarizes these reactions.

What is the mechanism of the ionic addition of these milder reagents to the carbon–oxygen double bond? Two pathways can be formulated: nucleophilic addition–protonation and electrophilic protonation–addition. The first, which begins with nucleophilic attack, takes place under neutral or, more commonly, basic conditions. As the (frequently anionic) nucleophile approaches the electrophilic carbon, the carbon rehybridizes and the electron pair of the π bond moves over to the oxygen, thereby producing an alkoxide ion. Subsequent protonation, usually from a protic solvent such as water or alcohol, yields the final addition product.

Nucleophilic Addition–Protonation (Basic Conditions)

Mechanism

Note that the new Nu–C bond is made up entirely of the electron pair of the nucleophile. The transformation is reminiscent of an S_N2 reaction. In that process, a leaving group is displaced. Here, an electron pair is moved from a shared position between carbon and oxygen to one solely on the oxygen atom. Additions of strongly basic nucleophiles to carbonyl groups typically follow the nucleophilic addition–protonation pathway.

The second mechanism occurs under acidic conditions and typically features neutral and relatively weak nucleophiles. Here the sequence of events is inverted. First, the Lewis basic oxygen atom undergoes electrophilic attack by a proton, giving the protonated carbonyl compound. This step has an unfavorable equilibrium—the pK_a of the resulting species is around -8. However, it renders the carbonyl carbon strongly electrophilic, thus facilitating the subsequent, very favorable nucleophilic attack that leads to the final addition product.

Mechanism

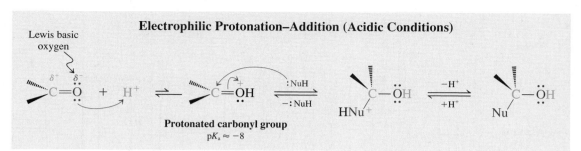

Electrophilic Protonation–Addition (Acidic Conditions)

Protonated carbonyl group
$pK_a \approx -8$

> This mechanism illustrates Le Chatelier's principle: The small amount of protonated carbonyl compound formed initially is depleted continuously by subsequent nucleophilic addition. The initial equilibrium shifts to replenish it, allowing the overall reaction to proceed to product.

By its very nature, the electrophilic protonation–addition mechanism is best suited for reactions of relatively weakly basic nucleophiles.

In Summary There are three regions of reactivity in aldehydes and ketones. The first two are the two atoms of the carbonyl group and are the subject of the remainder of this chapter. The third is the adjacent carbon. The reactivity of the carbonyl group is governed by addition processes. Nucleophilic organometallic and hydride reagents result in irreversible alcohol formation (after protonation). Ionic additions of NuH (Nu = OH, OR, SR, NR_2) are reversible and may begin with nucleophilic attack at the carbonyl carbon, followed by electrophilic trapping of the alkoxide anion so generated. Alternatively, in acidic media, protonation precedes the addition of the nucleophile.

17-6 | ADDITION OF WATER TO FORM HYDRATES

This section and Sections 17-7 and 17-8 introduce the reactions of aldehydes and ketones with water and alcohols. These compounds attack the carbonyl group through the mechanisms just outlined, utilizing either acid or base catalysis.

Water hydrates the carbonyl group

Water is capable of attacking the carbonyl function of aldehydes and ketones. The transformation can be catalyzed by either acid or base and leads to equilibration with the corresponding **geminal diols,** $RC(OH)_2R'$, also called **carbonyl hydrates.**

Reaction

Hydration of the Carbonyl Group

$$\text{C=O} + \text{HOH} \underset{K}{\overset{H^+ \text{ or } HO^-}{\rightleftharpoons}} \text{C-OH}$$

Geminal diol

In the base-catalyzed mechanism, the hydroxide functions as the nucleophile. Water then protonates the intermediate adduct, a hydroxy alkoxide, to give the product diol and to regenerate the catalyst.

Mechanism of Base-Catalyzed Hydration

Mechanism

Hydroxy alkoxide   Geminal diol

In the acid-catalyzed mechanism, the sequence of events is reversed. Here, initial protonation facilitates nucleophilic attack by the weak nucleophile water. Subsequently, the catalytic proton is lost to reenter the catalytic cycle.

Mechanism

Mechanism of Acid-Catalyzed Hydration

$$\overset{\text{\tiny}}{C}=\overset{..}{\underset{..}{O}} \ + \ H^+ \ \rightleftharpoons \ \overset{\text{\tiny}}{C}=\overset{+}{\underset{..}{O}}-H \ \xrightarrow{H_2\overset{..}{O}} \ \underset{\underset{H}{\overset{+}{\underset{|}{HO}:}}}{\overset{|}{\underset{|}{C}}}-\overset{..}{O}H \ \rightleftharpoons \ \underset{\underset{..}{HO}}{\overset{|}{\underset{|}{C}}}-\overset{..}{O}H \ + \ H^+$$

Protonated carbonyl   **Geminal diol**

Hydration is reversible

As these equations indicate, hydrations of aldehydes and ketones are reversible. The equilibrium lies to the left for ketones and to the right for formaldehyde and aldehydes bearing inductively electron-withdrawing substituents. For ordinary aldehydes, the equilibrium constant approaches unity.

How can these trends be explained? Look again at the resonance structures of the carbonyl group, as described in Section 17-2 and repeated here in the margin. In the dipolar resonance form, the carbon possesses carbocation character. Therefore, alkyl groups, which stabilize carbocations (Section 7-5), stabilize carbonyl compounds as well. Conversely, inductively electron-withdrawing substituents, such as CCl_3 and CF_3, increase positive charge at the carbonyl carbon and destabilize carbonyl compounds. The stabilities of the product diols are affected to a lesser extent by substituents. As a net result, therefore, relative to hydration of formaldehyde, the hydration reactions of aldehydes and ketones are progressively more endothermic, whereas hydrations of carbonyl compounds containing electron-withdrawing groups are more exothermic. These thermodynamic effects are paralleled by differences in kinetic reactivity: Carbonyl compounds containing electron-withdrawing groups are the most electrophilic and the most reactive, followed in turn by formaldehyde, other aldehydes, and finally ketones. We shall see that the same trends also govern other addition reactions.

Reminder

Free H^+ does not exist in solution. In the last step of the adjacent scheme, H^+ would be transferred most likely to the oxygen of the water solvent or the starting carbonyl.

$$\left[\overset{\text{\tiny}}{C}=\overset{..}{\underset{..}{O}} \ \longleftrightarrow \ \overset{\text{\tiny}}{C}\overset{+}{-}\overset{..}{\underset{..}{O}}:^- \right]$$

Carbocation-like

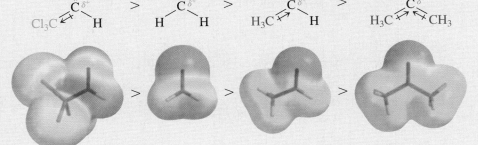

Relative Reactivities of Carbonyl Groups: Aldehydes Are More Reactive than Ketones

Electron-withdrawing CCl_3 group generates additional positive charge at carbonyl carbon

Electron-donating CH_3 groups reduce positive charge at carbonyl carbon

Increasing reactivity

Equilibrium Constants K for the Hydration of Typical Carbonyl Compounds

Cl_3CCH (with O)	$K > 10^4$
HCH (with O)	$K > 10^3$
CH_3CH (with O)	$K \sim 1$
CH_3CCH_3 (with O)	$K < 10^{-2}$

Increasing favorability of addition

Exercise 17-7

(a) Rank in order of increasing favorability of hydration: Cl_3CCH, Cl_3CCCH_3, Cl_3CCCCl_3.
(with respective carbonyl O groups shown)

(b) Treatment of acetone with $H_2{}^{18}O$ results in the formation of labeled acetone, $CH_3\overset{{}^{18}O}{C}CH_3$. Explain.

Despite the fact that hydration is energetically favorable in some cases, it is usually *not* possible to isolate carbonyl hydrates as pure substances: They lose water too easily, regenerating the original carbonyl compound. However, they may participate as intermediates in subsequent chemistry, such as the oxidation of aldehydes to carboxylic acids under aqueous conditions (Sections 8-6 and 17-4).

In Summary The carbonyl group of aldehydes and ketones is hydrated by water. Aldehydes are more reactive than ketones. Electron-withdrawing substituents render the carbonyl group more electrophilic. Hydration is an equilibrium process that may be catalyzed by acids or bases.

17-7 ADDITION OF ALCOHOLS TO FORM HEMIACETALS AND ACETALS

In this section we shall see that alcohols add to carbonyl groups in much the same manner as water does. Both acids and bases catalyze the process. In addition, acids catalyze the further transformation of the initial addition product to furnish acetals by replacement of the hydroxy group with an alkoxy substituent.

Aldehydes and ketones form hemiacetals reversibly

Not surprisingly, alcohols also undergo addition to aldehydes and ketones, by a mechanism virtually identical to that outlined for water. The adducts formed are called **hemiacetals** (*hemi,* Greek, half) because they are intermediates on the way to acetals.

Hemiacetal Formation

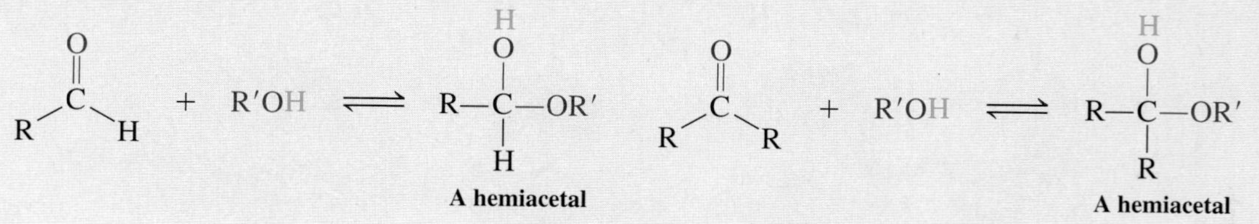

Like hydration, these addition reactions are governed by equilibria that usually favor the starting carbonyl compound. Hemiacetals, like hydrates, are therefore usually not isolable. Exceptions are those formed from reactive carbonyl compounds such as formaldehyde or 2,2,2-trichloroacetaldehyde. Hemiacetals are also isolable from hydroxy aldehydes and ketones when cyclization leads to the formation of relatively strain-free five- and six-membered rings.

Intramolecular Hemiacetal Formation:
Cyclic Hemiacetals Are More Stable than Acyclic Hemiacetals

5-Hydroxypentanal   A cyclic hemiacetal, stable

Glucose

0.003%
Aldehyde form

$\uparrow\downarrow$ H⁺ or HO⁻

> 99%

Cyclic hemiacetal
(Two stereoisomers)

Intramolecular hemiacetal formation is common in sugar chemistry (Chapter 24). For example, glucose, the most common simple sugar in nature, exists as an equilibrium mixture of an acyclic pentahydroxyaldehyde and two stereoisomeric cyclic hemiacetals. The cyclic forms constitute more than 99% of the mixture in aqueous solution.

Acids catalyze acetal formation

In the presence of excess alcohol, the *acid*-catalyzed reaction of aldehydes and ketones proceeds beyond the hemiacetal stage. Under these conditions, the hydroxy function of the initial adduct is replaced by another alkoxy unit derived from the alcohol. The resulting compounds are called **acetals**. (**Ketal** is an older term for an acetal derived from a ketone.)

Acetal Synthesis

$$\underset{\text{RCR}}{\overset{\text{O}}{\parallel}} + 2\,R'OH \overset{H^+}{\rightleftharpoons} R-\underset{\underset{R}{|}}{\overset{\overset{OR'}{|}}{C}}-OR' + H_2O$$

An acetal

Reaction

The net change is the replacement of the carbonyl oxygen by two alkoxy groups and the formation of one equivalent of water.

Let us examine the mechanism of this transformation for an aldehyde. The initial reaction is the ordinary acid-catalyzed addition of the first molecule of alcohol. The resulting hemiacetal can be protonated at the hydroxy group, changing this substituent into water, a good leaving group. Upon loss of water, the resulting carbocation is stabilized by resonance with a lone electron pair on oxygen. A second molecule of alcohol now adds to the electrophilic carbon, initially giving the protonated acetal, which is then deprotonated to the final product.

Mechanism of Acetal Formation

Step 1. Hemiacetal generation: acid-catalyzed addition of first molecule of alcohol

Step 2. Acetal generation: acid-catalyzed S_N1 displacement of water by second molecule of alcohol

Each step is reversible; the entire sequence, starting from the carbonyl compound and ending with the acetal, is an equilibrium process. In the presence of the acid catalyst, the equilibrium may be shifted in either direction: toward acetal by using excess alcohol or removing water; toward starting aldehyde or ketone by adding excess water, a process called **acetal hydrolysis.** *However, in contrast with hydrates and hemiacetals, acetals may be isolated as pure substances by neutralizing the acid catalyst used in their formation.* Reversal of acetal formation cannot occur in the absence of acid. Therefore, acetals may be prepared and put to use synthetically, as described in the next section.

In Summary Alcohols react with aldehydes and ketones to form hemiacetals. This process, like hydration, is reversible and is catalyzed by both acids and bases. Hemiacetals are converted by acid and excess alcohol into acetals. Acetals are stable under neutral or basic conditions but are hydrolyzed by aqueous acid.

17-8 | ACETALS AS PROTECTING GROUPS

In the conversion of an aldehyde or ketone to an acetal, the reactive carbonyl function is transformed into a relatively unreactive ether-like moiety. Because acetal formation is reversible, this process amounts to one of *masking,* or *protecting,* the carbonyl group. Such protection is necessary when selective reactions (for example, with nucleophiles) are required at one functional site of a molecule and the presence of an unprotected carbonyl group elsewhere in the same molecule might interfere. This section describes the execution of such protecting strategies.

Cyclic acetal formation protects carbonyl groups from attack by nucleophiles

Diols, such as 1,2-ethanediol, are particularly effective reagents compared to normal alcohols for forming acetals. Use of diols converts aldehydes and ketones into *cyclic* acetals, which are generally more stable than their acyclic counterparts. The reason lies in part in their relatively more favorable (or, better, less unfavorable) entropy of formation (Section 2-1). Thus, as shown below, two molecules of starting material (carbonyl compound and diol) are converted into two molecules of product (acetal and water) in this process. In contrast, acetal formation with a normal alcohol (Section 17-7) uses up *three* molecules (carbonyl compound and two equivalents of alcohol), resulting in the same number of product molecules.

Acetaldehyde undergoes acid-catalyzed cyclotrimerization to the acetal commonly known as para-ldehyde, a hypnotic sedative that is also used to prevent the growth of mold on leather,

Cyclic Acetalization

$$CH_3\overset{\displaystyle O}{\overset{\|}{C}}H \quad + \quad HOCH_2CH_2OH \quad \xrightarrow[-H_2O]{H^+} \quad \underset{\underset{73\%}{H_3C \quad H}}{\overset{H_2C-CH_2}{\underset{O \quad \quad O}{C}}}$$

$$3\ CH_3CHO \xrightarrow{H^+}$$

Paraldehyde

+ ⟶

A cyclic acetal

Cyclic acetals are readily hydrolyzed in the presence of aqueous acid, but are not attacked by many basic, organometallic, and hydride reagents. These properties make them most useful as **protecting groups** for the carbonyl function in aldehydes and ketones. An example is the alkylation of an alkynyl anion with 3-iodopropanal 1,2-ethanediol acetal.

Use of a Protected Aldehyde in Synthesis

Most electrophilic carbon

$$ICH_2CH_2-CHO \xrightarrow[\text{Protection}]{HOCH_2CH_2OH,\ H^+}$$

3-Iodopropanal

Nucleophilic attack here will fail unless the aldehyde is protected

3-Iodopropanal 1,2-ethanediol acetal

$$\xrightarrow[\substack{-LiI}]{\substack{CH_3(CH_2)_3C\equiv C^-Li^+ \\ \textbf{1-Hexynyllithium} \\ S_N2\ \text{reaction}}}$$

$$CH_3(CH_2)_3C\equiv C-CH_2CH_2 \quad \underset{70\%}{}$$

4-Nonynal 1,2-ethanediol acetal

$$\xrightarrow[\substack{-HOCH_2CH_2OH \\ \text{Deprotection}}]{H^+,\ H_2O} CH_3(CH_2)_3C\equiv C-CH_2CH_2-CHO$$

$$90\%$$

4-Nonynal

When the same alkylation is attempted with unprotected 3-iodopropanal, the alkynyl anion attacks the carbonyl group.

Solved Exercise 17-8 | Working with the Concepts: Using an Acetal Protecting Group in Synthesis

Suggest a convenient way of converting compound A into compound B.

What the problem asks you to do is to design a way to transform one specific compound into another. Your method should be efficient and selective. In other words, it should use a minimum of chemical reactions, consistent with producing the target molecule in good yield and with a minimum of side products.

How to begin? Identify the structural changes that have to be achieved. Caution! Did you notice that the target molecule has one additional carbon? In any synthesis problem, it is advisable to "map" the structural framework of the desired product against that of the starting compound. Try to find structural points of similarity so that you can better define what you need to change, synthetically.

Information needed? The two substances differ only in the functional groups at the right-hand ends: Br has turned into CH_2OH. Therefore, we need to find a synthetic sequence that converts one into the other: $RBr \rightarrow RCH_2OH$. Section 8-8 showed us how to do this, through the reaction of the corresponding Grignard reagent with formaldehyde:

$$RMgBr \xrightarrow[\text{2. } H^+, H_2O]{\text{1. } H_2C=O, \text{ ether}} RCH_2OH$$

The problem with substrate A is that it contains a carbonyl group. This function is incompatible with the generation of a Grignard reagent. How can we prevent its interference? The answer is in this section. In the example immediately preceding this exercise, we note that the acetal group, being similar in function to ethers, is immune to attack by organometallic reagents. Consequently, conversion of the ketone carbonyl of our starting material into an acetal unit should protect it during the formation and use of the Grignard reagent necessary for our synthesis.

Proceed. Treat the initial bromoketone with 1,2-ethanediol in the presence of an acid catalyst to protect the carbonyl group as a cyclic acetal.

• We follow acetal formation with the reactions needed to complete our synthesis: Convert the bromoalkane functional group to the Grignard reagent, react with formaldehyde to add the necessary carbon atom, and protonate the resulting alkoxide with aqueous acid.

• The final step, treatment with aqueous acid, conveniently also hydrolyzes the acetal group and restores the original ketone function. This is the deprotection step referred to in the example in the text above.

Exercise 17-9 | Try It Yourself

Propose a synthetic scheme to convert bromobenzene into .

Work retrosynthetically! What reactions would you use to make the two new C—C bonds to the benzene ring and in what sequence? (**Hint:** See Sections 15-13 and 16-3.)

Thioacetal

Thiols react with the carbonyl group to form thioacetals

Thiols, the sulfur analogs of alcohols (see Section 9-10), react with aldehydes and ketones by a mechanism identical with the one described for alcohols. Instead of a proton catalyst, a Lewis acid, such as BF_3 or $ZnCl_2$, is often used in ether solvent. The reaction produces the sulfur analogs of acetals, **thioacetals,** again, especially well for cyclic systems.

Cyclic Thioacetal Formation from a Ketone

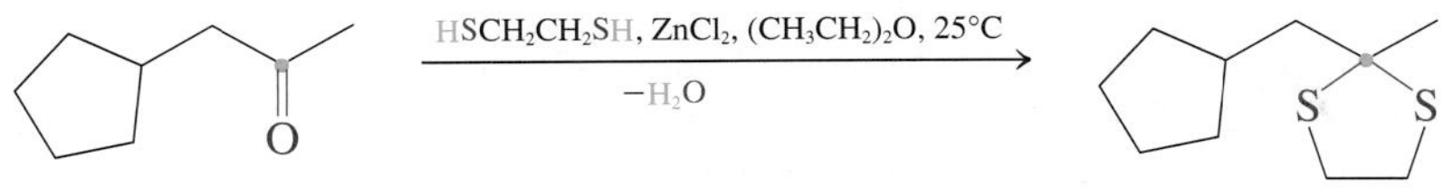

95%
A cyclic thioacetal

These sulfur derivatives are stable in aqueous acid, a medium that hydrolyzes ordinary acetals. The difference in reactivity may be useful in synthesis when it is necessary to differentiate two different carbonyl groups in the same molecule. Hydrolysis of thioacetals is carried out using mercuric chloride in aqueous acetonitrile. The driving force is the formation of insoluble mercuric sulfides.

Thioacetal Hydrolysis

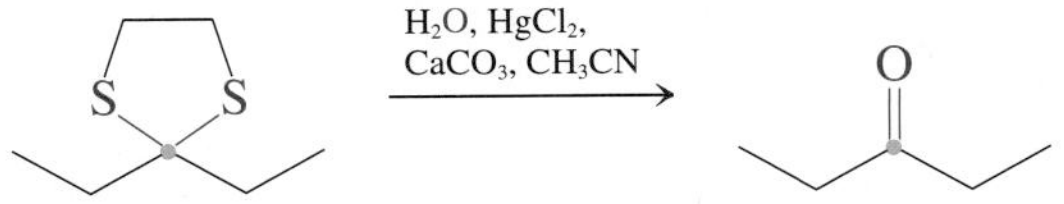

| Solved Exercise 17-10 | Working with the Concepts: Differentiating Carbonyl Groups |

How would you accomplish the following transformation? (**Hint:** See Section 8-8.)

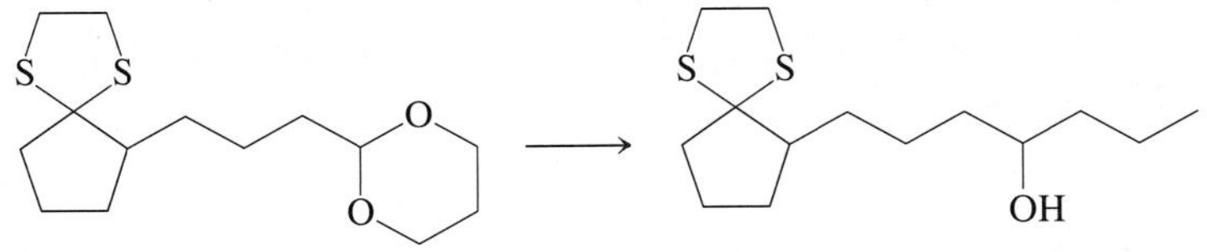

Strategy
Formation of a secondary alcohol may be achieved by addition of a Grignard reagent to an aldehyde. Choose a deprotection method that will release the aldehydic carbonyl function but will leave the thioacetal protected cyclic ketone part untouched.

Solution
- Hydrolyze the acetal with aqueous acid.
- React the resulting aldehyde with the appropriate Grignard reagent to give the desired alcohol.

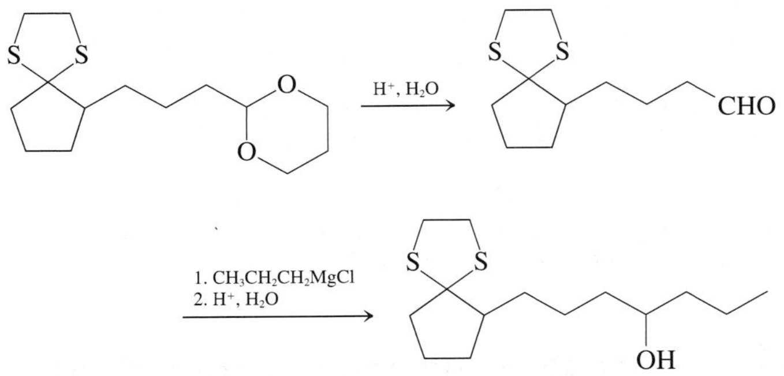

Neither of these steps affects the thioacetal moiety.

Exercise 17-11 | Try It Yourself

Show how the starting material for Exercise 17-10 may be converted into the alcohol in the margin.

Thioacetals are **desulfurized** to the corresponding hydrocarbon by treatment with Raney nickel (Section 12-2). Thioacetal generation followed by desulfurization is used to convert a carbonyl into a methylene group under neutral conditions.

Exercise 17-12

Suggest a possible synthesis of cyclodecane from

(**Caution:** Simple hydrogenation will not do. **Hint:** See Section 12-12.)

In Summary Acetals and thioacetals are useful protecting groups for aldehydes and ketones. Acetals are formed under acidic conditions and are stable toward bases and nucleophiles. They are hydrolyzed by aqueous acid. Thioacetals are usually prepared using Lewis acid catalysts and are stable to both aqueous base and acid. Mercuric salts are required for thioacetal hydrolysis. Thioacetals may also be desulfurized to hydrocarbons with Raney nickel.

17-9 NUCLEOPHILIC ADDITION OF AMMONIA AND ITS DERIVATIVES

Ammonia and the amines may be regarded as nitrogen analogs of water and alcohols. Do they add to aldehydes and ketones? In fact, they do, giving products corresponding to those just studied. We shall see one important difference, however. The products of addition of amines and their derivatives lose water, furnishing either of two new derivatives of the original carbonyl compounds: imines and enamines.

Ammonia   Primary amine

Ammonia and primary amines form imines

Upon exposure to an amine, aldehydes and ketones initially form **hemiaminals,** the nitrogen analogs of hemiacetals. In a subsequent, slower step, hemiaminals of primary amines lose water to form a carbon–nitrogen double bond. This function is called an **imine** (an older name is **Schiff* base**) and is the nitrogen analog of a carbonyl group.

Imine Formation from Amines and Aldehydes or Ketones

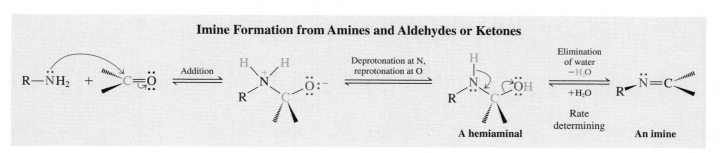

*Professor Hugo Schiff (1834–1915), University of Florence, Italy.

The mechanism of the elimination of water from the hemiaminal is the same as that for the decomposition of a hemiacetal to the carbonyl compound and alcohol. It begins with protonation of the hydroxy group. (Protonation of the more basic nitrogen just leads back to the original carbonyl compound.) Dehydration follows to the intermediate **iminium ion**, which is then deprotonated to furnish the imine.

Mechanism of Hemiaminal Dehydration

Mechanism

Hemiaminal   Iminium ion   Imine

You may have noticed that we have written a first-step direct attack of the amine on the carbonyl function to form the hemiaminal, without the help of a catalyst. This contrasts with the acid- (or base-)catalyzed creation of hemiacetals from alcohols in the preceding section and is a manifestation of the greater nucleophilicity of nitrogen relative to oxygen (Section 6-8). Operationally, many imine preparations are indeed carried out by simply mixing the two components, sometimes in the absence of any solvent. However, most procedures call for catalytic acid (such as HCl) to speed up the rate-determining dehydration step. Under such conditions, it is likely that the hemiaminal arises by initial protonation of the carbonyl oxygen, followed by amine trapping. However, too much acid is detrimental, because amines are basic and will be protonated to ammonium salts (see Table 2-2), reducing the concentration of free amine and hence slowing down the first step (margin). Therefore, overall, the maximum rate acceleration is attained at a pH = 4–5.

$R-\overset{\overset{H}{|}}{N}H_2$ — Retards reaction by decreasing free amine concentration

$R-\ddot{N}H_2$ $\overset{H}{\underset{+\overset{\cdots}{O}=C\overset{R'}{\diagdown R''}}{}}$ — Accelerates reaction by activating the carbonyl function

$\ddot{O}=C\overset{R'}{\diagdown R''}$

Processes such as imine formation from a primary amine and an aldehyde or ketone, in which two molecules are joined with the elimination of water (or other small molecules such as alcohols), are called **condensations.**

Condensation of a Ketone with a Primary Amine

$$R\ddot{N}H_2 \ + \ \ddot{O}{=}C{\overset{R'}{\underset{R''}{\diagup}}} \ \underset{}{\overset{pH \approx 4–5}{\rightleftharpoons}} \ RN{=}C{\overset{R'}{\underset{R''}{\diagup}}} \ + \ H_2\ddot{O}$$

Imines are useful compounds: For example, they are commonly employed in the synthesis of more complex amines (Section 21-6). However, the condensation reaction above is reversible, making imines very susceptible to hydrolysis and often difficult to isolate as pure substances. As a result, when a synthetic sequence requires an imine as an intermediate, the appropriate amine and carbonyl compound are usually mixed in the presence of the reagent that is needed to convert with the imine further. Thus, the imine reacts as soon as it is formed. In this way, the condensation equilibrium is constantly shifted to produce more imine, continuing the reaction until it finally goes to completion.

If care is taken to remove the water that is formed in the condensation process (for example, by continuous distillation from the reaction mixture), imines may be isolated, often in high yield. For example,

$$CH_3\overset{\overset{\text{:O:}}{\|}}{C}CH_3 \ + \ \text{(cyclohexylamine)} \ \xrightarrow{C_6H_6,\ HCl} \ \text{(imine)} \ + \ H_2\ddot{O}$$

95%

REAL LIFE: BIOCHEMISTRY 17-1 | **Imines Mediate the Biochemistry of Amino Acids**

Imines play pivotal roles in the biological transformations of 2-aminocarboxylic acids (amino acids), the building blocks of proteins (Chapter 26). Pyridoxal and pyridoxamine, two derivatives of vitamin B$_6$ (pyridoxine), mediate the interconversion of amino acids and 2-oxocarboxylic acids (ketoacids). Imines are the key intermediates in this transformation, by which amino acids are broken down biologically (a process called catabolism).

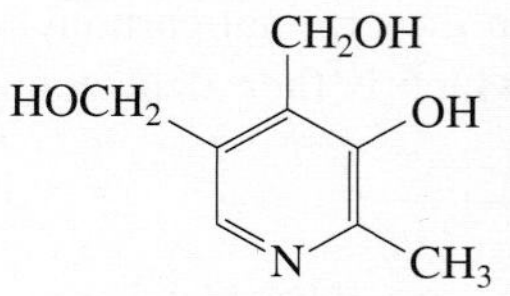

Pyridoxine
(Vitamin B$_6$)

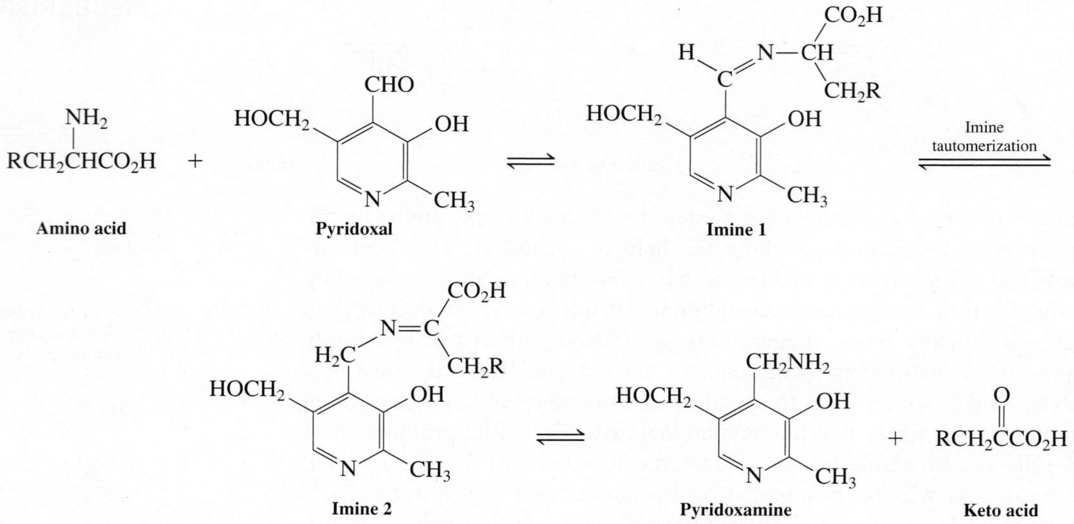

Amino acid

Pyridoxal

Imine 1

Imine
tautomerization

Imine 2

Pyridoxamine

Keto acid

The sequence begins with formation of an imine from the amino acid and the oxidized form of the vitamin, pyridoxal. This imine converts to an isomer via tautomerism: simultaneous proton and double-bond shift (See Sections 13-7 and 13-8). Hydrolysis of the new imine gives pyridoxamine and the ketoacid. Depending on the body's metabolic needs, the pyridoxamine thus formed may proceed to react with other ketoacids to produce required amino acids (the scheme shown above run in reverse), or it may serve to facilitate the ultimate disposal of the nitrogen by excretion.

Exercise 17-13

Reagent A has been used with aldehydes to prepare crystalline imidazolidine derivatives, such as B, for the purpose of their isolation and structural identification. Write a mechanism for the formation of B. [**Hint:** Notice that the product is a nitrogen analog of an acetal. Develop a mechanism analogous to that for acetal formation (Section 17-7), using the two amino groups in the starting diamine in place of two alcohol molecules.]

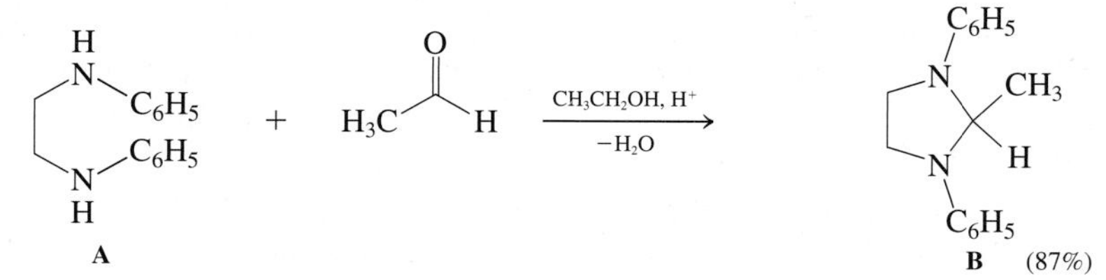

A

N,N′-Diphenyl-1,2-ethanediamine

B   (87%)

2-Methyl-1,3-diphenyl-1,3-diazacyclopentane
(2-Methyl-1,3-diphenylimidazolidine)

(m.p. 102°C)

Special imines and their applications

Certain primary amine derivatives condense with aldehydes and ketones to form products that are less hydrolytically unstable than ordinary imines. Among these are hydroxylamine; hydrazine, H$_2$NNH$_2$ (see also Exercise 3-2), and its derivatives, such as 2,4-dinitrophenylhydrazine; and semicarbazide (all common names). They form so-called **oximes, hydrazones,**

Table 17-5	Imine Derivatives of Aldehydes and Ketones
Reagent	**Product of reaction with** $\overset{\displaystyle\diagdown}{\underset{\displaystyle\diagup}{C}}=O$
H_2NOH **Hydroxylamine**	$\overset{\displaystyle\diagdown}{\underset{\displaystyle\diagup}{C}}=NOH$ **Oxime**
H_2N-NH- ... NO_2 (with O_2N) **2,4-Dinitrophenylhydrazine**	(2,4-Dinitrophenylhydrazone structure) **2,4-Dinitrophenylhydrazone**
$H_2N-NHCNH_2$ (with O double bond) **Semicarbazide**	(Semicarbazone structure) **Semicarbazone**

and **semicarbazones**, respectively (Table 17-5). As shown in the margin for oximes, each of these functions is stabilized by resonance delocalization of a lone pair on the oxygen or nitrogen that is bound to the imine nitrogen atom.

The relative inertness of such modified imine groups in the presence of water is being exploited in the selective delivery of polymer-attached drugs to their respective targets. The polymer is chosen to be hydrophilic, thus ensuring water solubility, and the drug is bound by means of an imine linker. An example is the common antitumor agent adriamycin (Section 24-12), connected to *N*-(2-hydroxypropyl)methacrylamide (HPMA)–methacrylamide (MA) copolymers (for acrylic polymers, see Table 12-3) via a hydrazone unit that is stable at blood pH (7.4). In this way, circulation half-lives are improved, in addition to tumor targeting, because tumor cells are defective in structure and tend to accumulate polymers more than normal tissue. The linker is carefully chosen to hydrolyze at optimum pH = 5, exploiting the finding that cancer cell's environment is more acidic than that of a healthy cell.

Resonance stabilization in oximes

Hydrazone Hydrolysis for Drug Delivery

Another property of oximes, hydrazones, and semicarbazones is that they are highly crystalline and therefore exhibit sharp melting points. These characteristics aid in isolation, purification, and characterization of aldehydes and ketones.

Condensations with secondary amines give enamines

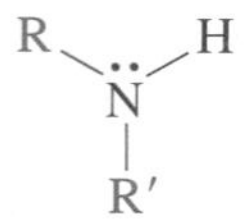

Secondary amine

The condensations of amines described so far are possible only for primary derivatives, because the nitrogen of the amine has to supply both of the hydrogens necessary to form water. Reaction with a secondary amine, such as azacyclopentane (pyrrolidine), therefore takes a different course. After the initial addition, water is eliminated by deprotonation at *carbon* to produce an **enamine.** This functional group incorporates both the *ene* function of an alkene and the *amino* group of an amine.

Enamine Formation

**Azacyclopentane
(Pyrrolidine)**   **Hemiaminal**   **An enamine**

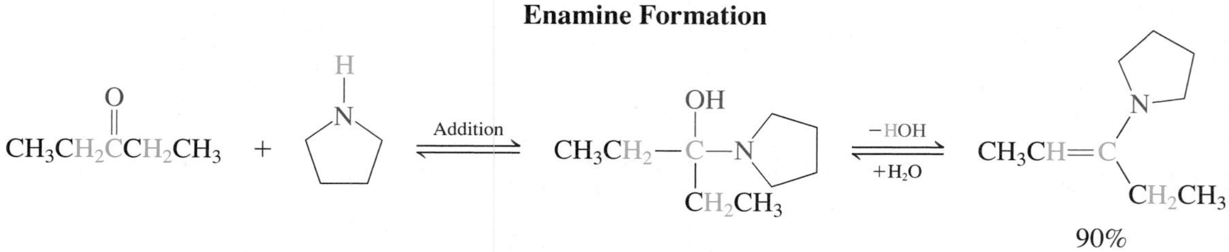

Enamine formation is reversible and hydrolysis occurs readily in the presence of aqueous acid. Enamines are useful substrates in alkylations (Section 18-4).

Solved Exercise 17-14 | **Working with the Concepts: Enamine Formation**

Formulate a detailed mechanism for the acid-catalyzed enamine formation shown above.

Strategy

Begin in a manner identical to the mechanisms of addition of ammonia and primary amines to give imines. As you proceed step by step, look for the point at which the structure of the secondary amine forces the mechanism to diverge from imine formation.

Solution

• Add the nucleophilic amine nitrogen as described earlier in this section. The negatively charged oxygen picks up a proton and the positively charged nitrogen loses one to give the usual hemiaminal intermediate.

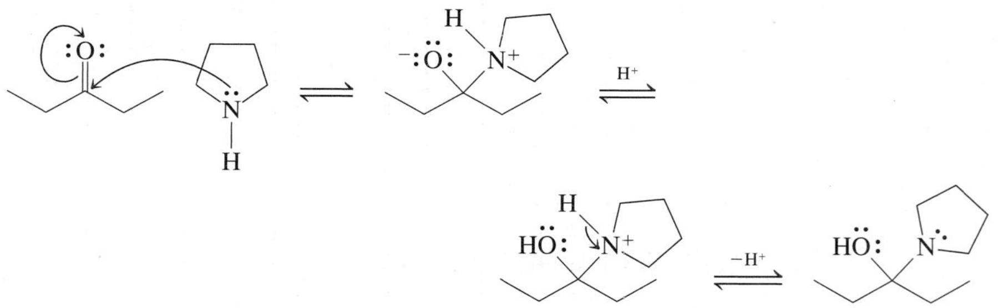

• Using the catalytic acid, protonate the oxygen, converting it into a better leaving group (water). The water departs, leaving an iminium ion, just as in the case of imine formation.
• At this point, however, the nitrogen lacks another hydrogen to lose: Imine formation is impossible. Thus the system must take an alternative path.

- Examine the structure of the actual enamine product: One hydrogen is missing from the carbon adjacent to the one bearing the nitrogen (the α-carbon). This hydrogen can be lost as a proton from the iminium intermediate, giving the final product.

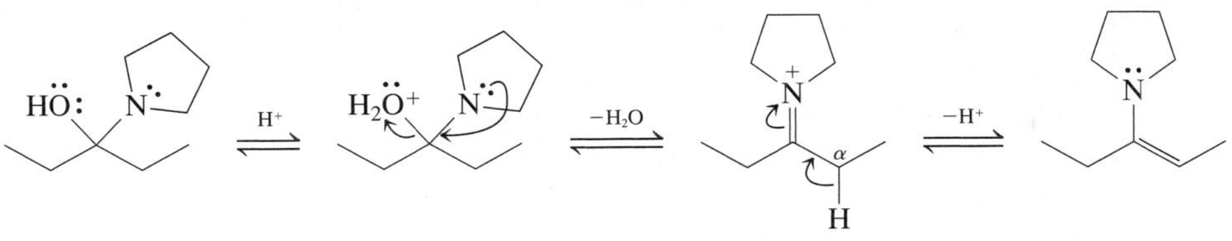

Exercise 17-15 | Try It Yourself

Write the product, in addition to the mechanism of its formation, for the following acid-catalyzed reaction.

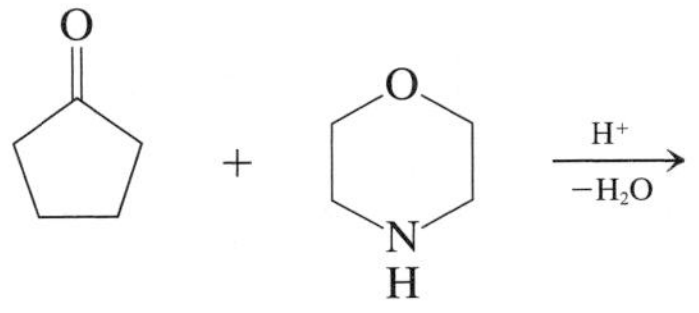

Exercise 17-16 |

Write the products of reactions (a) and (b) and explain the outcome of reaction (c), all occurring under acid-catalyzed conditions.

(a)

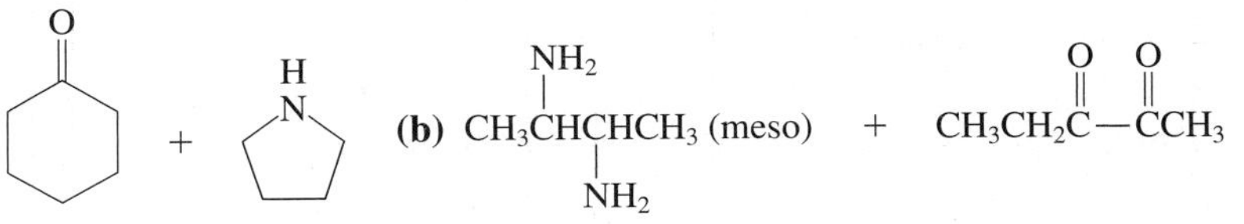

(b) $CH_3CHCHCH_3$ (meso)   + $CH_3CH_2\overset{O}{C}-\overset{O}{C}CH_3$

(c)

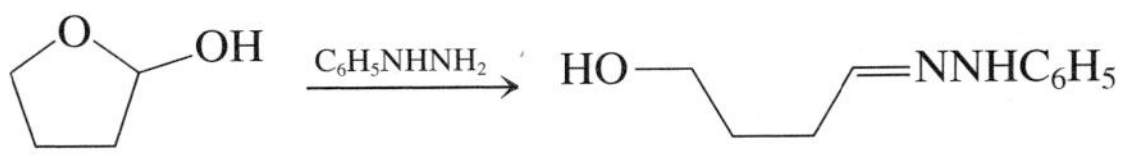

Remember WHIP

*W*hat

*H*ow

*I*nformation

*P*roceed

Recall: "Explain" = think mechanistically

In Summary Primary amines attack aldehydes and ketones to form imines by condensation. Hydroxylamine gives oximes, hydrazines lead to hydrazones, and semicarbazide results in semicarbazones. Secondary amines react with aldehydes and ketones to give enamines.

17-10 | DEOXYGENATION OF THE CARBONYL GROUP

In Section 17-5 we reviewed methods by which carbonyl compounds can be reduced to alcohols. Reduction of the C=O group to CH_2 (**deoxygenation**) also is possible. Two ways in which this may be achieved are Clemmensen reduction (Section 16-5) and thioacetal formation followed by desulfurization (Section 17-8). This section presents a third method for deoxygenation—the Wolff-Kishner reduction.

Strong base converts simple hydrazones into hydrocarbons

Condensation of hydrazine itself with aldehydes and ketones produces simple hydrazones:

Synthesis of a Hydrazone

$$CH_3CCH_3 \;+\; H_2N{-}NH_2 \xrightarrow[{-H_2O}]{CH_3CH_2OH} \text{Acetone hydrazone}$$

Hydrazine

Hydrazones undergo decomposition with evolution of nitrogen when treated with base at elevated temperatures. The product of this reaction, called the **Wolff-Kishner* reduction,** is the corresponding hydrocarbon.

Reaction

Wolff-Kishner Reduction

$$\overset{NH_2}{\underset{RCR'}{N}} \;+\; NaOH \xrightarrow{(HOCH_2CH_2)_2O,\ 180\text{-}200^\circ C} RCH_2R' \;+\; N_2$$

Mechanism

The mechanism of nitrogen elimination includes a sequence of base-mediated hydrogen shifts. The base first removes a proton from the hydrazone to give the corresponding delocalized anion. Reprotonation may occur on nitrogen, regenerating starting material, or on carbon, thereby furnishing an intermediate azo compound (Section 22-11) leading on to the product. The base removes another proton from the azo nitrogen to generate a new anion, which rapidly decomposes irreversibly, with the extrusion of nitrogen gas. The result is a strongly basic alkyl anion (carbanion), which is immediately protonated by the aqueous medium to give the final hydrocarbon product.

Mechanism of Nitrogen Elimination in the Wolff-Kishner Reduction

$$\cdots \;+\; HO{:}^{-} \xrightleftharpoons[{+H_2O}]{{-HOH}} \text{Deprotonation} \quad \longleftrightarrow \quad \xrightleftharpoons{} \text{Reprotonation}$$

Azo intermediate $\;+\; HO{:}^{-} \xrightleftharpoons[{+H_2O}]{{-HOH}}$ **Deprotonation** $\xrightarrow{\text{Irreversible loss of } N_2}$ **Reprotonation** $RCH_2R' \;+\; HO^{-}$

In practice, the Wolff-Kishner reduction is carried out without isolating the intermediate hydrazone. An 85% aqueous solution of hydrazine (hydrazine hydrate) is added to the carbonyl compound in a high-boiling alcohol solvent such as diethylene glycol ($HOCH_2CH_2OCH_2CH_2OH$,

---

*Professor Ludwig Wolff (1857–1919), University of Jena, Germany; Professor N. M. Kishner (1867–1935), University of Moscow.

b.p. 245°C) or triethylene glycol (HOCH$_2$CH$_2$OCH$_2$CH$_2$OCH$_2$CH$_2$OH, b.p. 285°C) containing NaOH or KOH, and the mixture is heated. Aqueous work-up yields the pure hydrocarbon.

1. H$_2$NNH$_2$, H$_2$O, diethylene glycol, NaOH, Δ
2. H$_2$O

69%

Wolff-Kishner reduction complements the Clemmensen and thioacetal desulfurization methods of deoxygenating aldehydes and ketones. Thus, the Clemmensen reduction is unsuitable for compounds containing acid-sensitive groups, and hydrogenation of multiple bonds can accompany desulfurization with hydrogen and Raney nickel. Such functional groups are generally not affected by Wolff-Kishner conditions.

Wolff-Kishner reduction aids in alkylbenzene synthesis

We have already seen that the products of Friedel-Crafts acylations may be converted into alkyl benzenes by using Clemmensen reduction. Wolff-Kishner deoxygenation also is frequently employed for this purpose and is particularly useful for acid-sensitive, base-stable substrates.

Wolff-Kishner Reduction of a Friedel-Crafts Acylation Product

NH$_2$NH$_2$, H$_2$O, KOH, triethylene glycol, Δ

95%

Exercise 17-17

Propose a synthesis of hexylbenzene from hexanoic acid.

In Summary The Wolff-Kishner reduction is the decomposition of a hydrazone by base, the second part of a method of deoxygenating aldehydes and ketones. It complements Clemmensen and thioacetal desulfurization procedures.

17-11 | ADDITION OF HYDROGEN CYANIDE TO GIVE CYANOHYDRINS

Besides alcohols and amines, several other nucleophilic reagents may attack the carbonyl group. Particularly important are carbon nucleophiles, because new carbon–carbon bonds can be made in this way. Section 8-8 explained that organometallic compounds, such as Grignard and alkyllithium reagents, add to aldehydes and ketones to produce alcohols. This section and Section 17-12 deal with the behavior of carbon nucleophiles that are not organometallic reagents—the additions of cyanide ion and of a new class of compounds called ylides.

Hydrogen cyanide adds reversibly to the carbonyl group to form hydroxyalkanenitrile adducts commonly called **cyanohydrins.** The equilibrium may be shifted toward the adduct by the use of liquid HCN as solvent. However, it is dangerous to use such large amounts of HCN,

Really? Burnet moths use the glucose-bound cyanohydrin linamarin as an HCN reservoir for chemical defense. Enzymes catalyze the hydrolysis of the acetal unit to liberate acetone cyanohydrin, which then releases the toxic gas. Females seek out males with high levels of linamarin, which is passed on as a remarkable "nuptial gift" during their mating.

Glucose

Acetone cyanohydrin warhead

Linamarin

which is volatile and highly toxic. Therefore, slow addition of HCl to an excess of the strongly basic NaCN is typically employed to generate HCN in a moderately alkaline mixture.

Cyanohydrin Formation

$$\text{[cyclohexanone]} + \text{Na}^+ \,^-\text{CN} \xrightarrow[-\text{NaCl}]{\text{Conc. HCl}} \text{[1-hydroxycyclohexanecarbonitrile]}$$

60%
1-Hydroxycyclohexanecarbonitrile
(Cyclohexanone cyanohydrin)

Cyanohydrin formation requires the presence of both free cyanide ion and undissociated HCN, a condition satisfied by maintaining a moderately basic pH. Cyanide adds to the carbonyl carbon in the slow, reversible first step of the mechanism. The HCN then protonates the negatively charged oxygen in the fast second step. This protonation is necessary to shift the overall equilibrium toward the product.

Mechanism of Cyanohydrin Formation

$$:N\equiv C:^- \; + \; C=\overset{..}{\underset{..}{O}} \;\; \rightleftharpoons \;\; \underset{NC}{C}-\overset{..}{\underset{..}{O}}:^- \; + \; H-CN \;\; \rightleftharpoons \;\; \underset{NC}{C}-\overset{..}{\underset{..}{O}}H \; + \; ^-:CN$$

Exercise 17-18

Rank the following carbonyl compounds in order of thermodynamic favorability of HCN addition: acetone, formaldehyde, 3,3-dimethyl-2-butanone, acetaldehyde. (**Hint:** See Section 17-6.)

We shall see in subsequent chapters (Sections 19-6 and 26-2) that cyanohydrins are useful intermediates because the nitrile group can be modified by further reaction.

In Summary The carbonyl group in aldehydes and ketones can be attacked by carbon-based nucleophiles. Organometallic reagents give alcohols and cyanide gives cyanohydrins.

17-12 | ADDITION OF PHOSPHORUS YLIDES: THE WITTIG REACTION

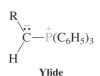

Ylide

Another useful reagent in nucleophilic additions contains a carbanion that is stabilized by an adjacent, positively charged phosphorus group. Such a species is called a **phosphorus ylide,** and its attack on aldehydes and ketones is called the **Wittig* reaction.** The Wittig reaction is a powerful method for the selective synthesis of alkenes from aldehydes and ketones.

Deprotonation of phosphonium salts gives phosphorus ylides

Phosphorus ylides are most conveniently prepared from haloalkanes by a two-step sequence; the first step is the nucleophilic displacement of halide by triphenylphosphine to furnish an alkyltriphenylphosphonium salt.

---

*Professor Georg Wittig (1897–1987), University of Heidelberg, Germany, Nobel Prize 1979 (chemistry).

Phosphonium Salt Synthesis

$$\underset{\substack{\text{Triphenyl-}\\\text{phosphine}}}{H_5C_6-\overset{\displaystyle C_6H_5}{\underset{\displaystyle C_6H_5}{P:}}} \quad + \quad \overset{\displaystyle R}{\underset{\displaystyle H}{\overset{|}{\underset{|}{C}}}}\overset{H}{\underset{}{-X::}} \quad \xrightarrow{\;S_N2\;} \quad \underset{\substack{\text{An alkyltriphenyl-}\\\text{phosphonium}\\\text{halide}}}{\overset{\displaystyle R\quad C_6H_5}{H-\overset{|}{\underset{|}{C}}-\overset{+}{\underset{|}{P}}-C_6H_5}} \quad :X:^-$$

Triphenylphosphine

The positively charged phosphorus atom renders any neighboring proton acidic. In the second step, deprotonation by bases, such as alkoxides, sodium hydride, or butyllithium, gives the ylide. Ylides can be isolated, although they are usually only generated in solutions that are subsequently treated with other reagents.

Ylide Formation

$$\overset{\displaystyle R}{\underset{\displaystyle H}{\overset{|}{\underset{|}{C}}}}\overset{+}{\underset{H}{P(C_6H_5)_3}} \;:X:^- \quad \xrightarrow[\substack{-HCH_2CH_2CH_2CH_3 \\ \text{Deprotonation}}]{\substack{Li-CH_2CH_2CH_2CH_3,\ THF}} \quad \left[\underset{H}{\overset{R}{\overset{|}{C}}}\overset{+}{\underset{}{P(C_6H_5)_3}} \longleftrightarrow \underset{H}{\overset{R}{C}}=P(C_6H_5)_3 \right] \;+\; Li^+\;:X:^-$$

Ylide

Notice that we may formulate a second resonance structure for the ylide by delocalizing the negative charge onto the phosphorus. In this form, the valence shell on phosphorus has been expanded and a carbon–phosphorus double bond is present.

The Wittig reaction forms carbon–carbon double bonds

When an ylide is exposed to an aldehyde or ketone, their reaction ultimately furnishes an alkene by coupling the ylide carbon with that of the carbonyl. The other product of this transformation is triphenylphosphine oxide.

The Wittig Reaction

$$\underset{\substack{\text{Aldehyde}\\\text{or ketone}}}{\overset{}{C=O}} \quad + \quad \underset{\text{Ylide}}{(C_6H_5)_3P=C} \quad \longrightarrow \quad \underset{\text{Alkene}}{C=C} \quad + \quad \underset{\text{Triphenylphosphine oxide}}{(C_6H_5)_3P=O}$$

Reaction

$$\underset{}{CH_3CH_2CH_2\overset{\displaystyle O}{\overset{\|}{C}}H} \quad + \quad CH_3CH_2\overset{\displaystyle CH_3}{\underset{}{\overset{|}{C}}}=P(C_6H_5)_3 \quad \xrightarrow[-(C_6H_5)_3PO]{(CH_3CH_2)_2O,\ 10°C} \quad CH_3CH_2CH_2CH=\overset{\displaystyle CH_3}{\underset{}{\overset{|}{C}}}CH_2CH_3$$

66%
3-Methyl-3-heptene

The Wittig reaction is a valuable addition to our synthetic arsenal because it forms carbon–carbon double bonds. In contrast with eliminations (Sections 11-6 and 11-7), it gives rise to alkenes in which the position of the newly formed double bond is unambiguous. Compare, for example, two syntheses of 2-ethyl-1-butene, one by the Wittig reaction, the other by elimination.

Comparison of Two Syntheses of 2-Ethyl-1-butene

By Wittig reaction

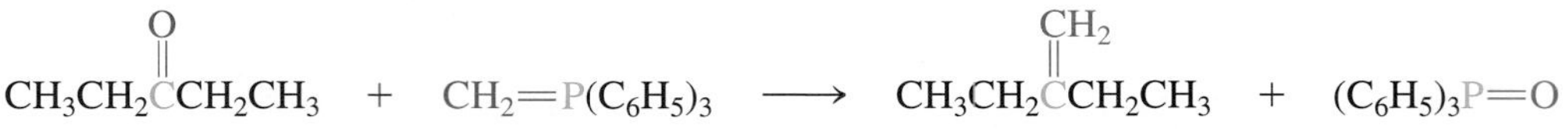

$$CH_3CH_2\overset{\overset{\displaystyle O}{\|}}{C}CH_2CH_3 \ + \ CH_2{=}P(C_6H_5)_3 \longrightarrow CH_3CH_2\overset{\overset{\displaystyle CH_2}{\|}}{C}CH_2CH_3 \ + \ (C_6H_5)_3P{=}O$$

(Only one isomer)

By elimination

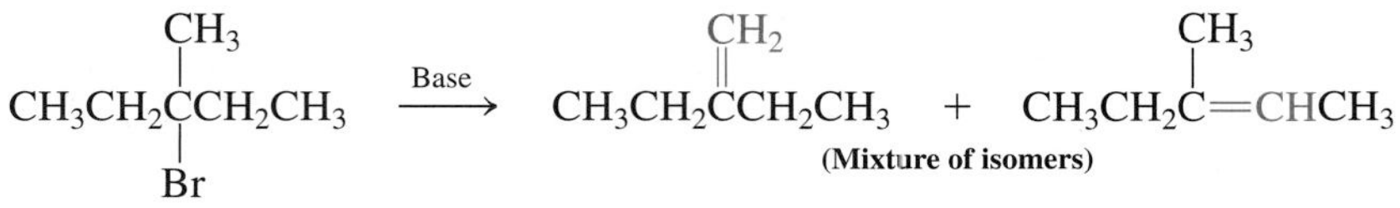

$$CH_3CH_2\overset{\overset{\displaystyle CH_3}{|}}{\underset{\underset{\displaystyle Br}{|}}{C}}CH_2CH_3 \ \xrightarrow{\text{Base}} \ CH_3CH_2\overset{\overset{\displaystyle CH_2}{\|}}{C}CH_2CH_3 \ + \ CH_3CH_2\overset{\overset{\displaystyle CH_3}{|}}{C}{=}CHCH_3$$

(Mixture of isomers)

What is the mechanism of the Wittig reaction? The negatively polarized carbon in the ylide is nucleophilic and can attack the carbonyl group. The result is a **phosphorus betaine,*** a dipolar species of the kind called a *zwitterion* (*Zwitter*, German, hybrid). The betaine is short lived and rapidly forms a neutral **oxaphosphacyclobutane (oxaphosphetane),** characterized by a four-membered ring containing phosphorus and oxygen. This substance then decomposes to the product alkene and triphenylphosphine oxide. The driving force for the last step is the formation of the very strong phosphorus–oxygen double bond.

Mechanism of the Wittig Reaction

Mechanism

Animation

ANIMATED MECHANISM:
The Wittig reaction

A phosphorus betaine

An oxaphosphacyclobutane
(Oxaphosphetane)   Alkene   Triphenylphosphine
oxide

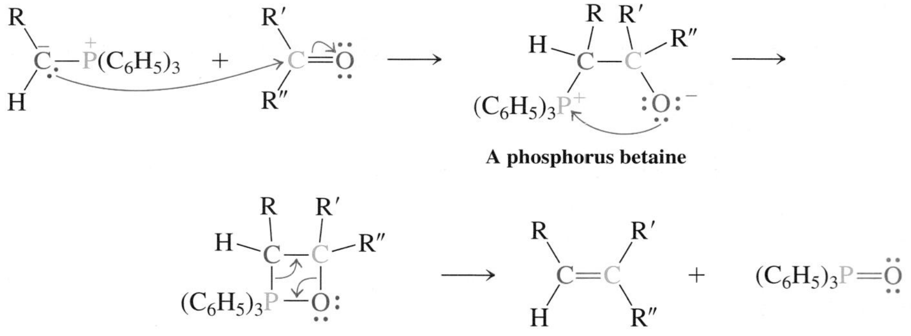

Wittig reaction can be carried out in the presence of ether, ester, halogen, alkene, and alkyne functions. Many display useful stereoselectivity. For example, reactions between non-conjugated ylides and aldehydes typically result in cis (or Z) alkenes with good selectivity.

A Stereoselective Wittig Reaction Giving Mostly cis Product

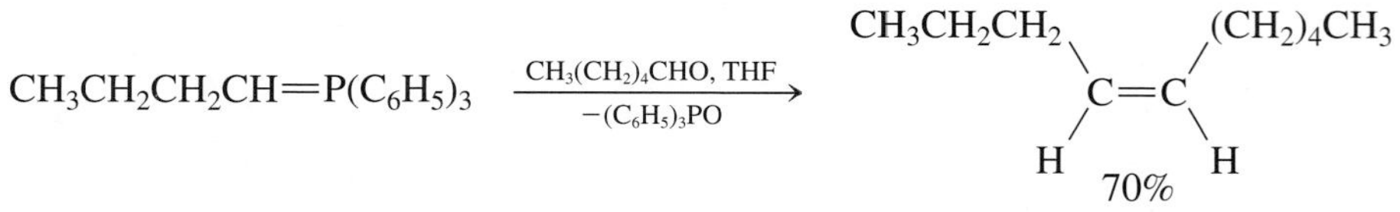

$$CH_3CH_2CH_2CH{=}P(C_6H_5)_3 \xrightarrow[-(C_6H_5)_3PO]{CH_3(CH_2)_4CHO,\ THF}$$

70%

(cis : trans ratio = 6 : 1)

*Betaine is the name of an amino acid, $(CH_3)_3\overset{+}{N}CH_2COO^-$, which is found in beet sugar (*beta*, Latin, beet) and exists as a zwitterion.

In contrast, the presence of conjugation in the ylide frequently results in trans products, as shown in the example below, taken from the commercial synthesis of vitamin A_1 (Section 14-7) by the German chemical company BASF (Badische Anilin and Soda Fabriken).

A Highly trans-Selective Wittig Reaction: BASF Vitamin A_1 Synthesis

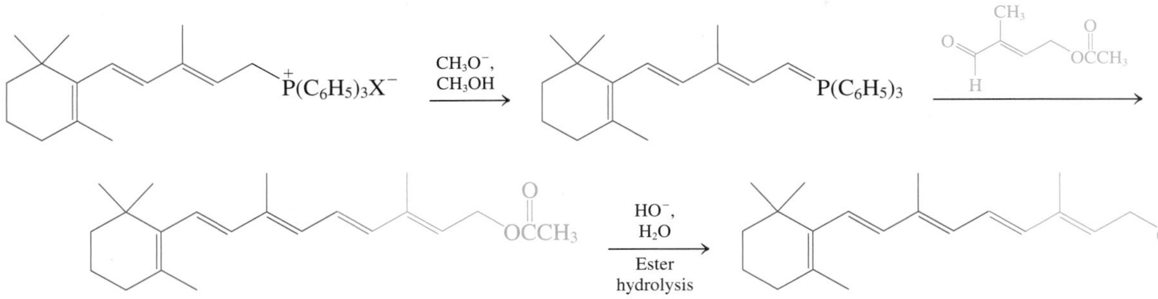

Vitamin A_1

Exercise 17-19

Propose syntheses of 3-methylenecyclohexene from **(a)** 2-cyclohexenone and **(b)** 3-bromocyclohexene, using Wittig reactions.

Exercise 17-20

Propose a synthesis of the following dienone from the indicated starting materials. [**Hint:** Make use of a protecting group (Section 17-8).]

$CH_3\overset{O}{\overset{\|}{C}}CH_2CH_2CH{=}CHCH{=}CH_2$ from $CH_3\overset{O}{\overset{\|}{C}}CH_2CH_2CH_2Br$ and $H\overset{O}{\overset{\|}{C}}CH{=}CH_2$

Reminder

Use retrosynthesis in tackling these problems. It is easier to work backward from the target than forward from the starting material.

Exercise 17-21

Develop concise synthetic routes from starting material to product. You may use any material in addition to the given compound (more than one step will be required).

(a) 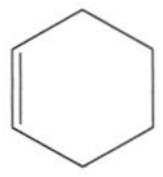 ---> $CH_2{=}CH(CH_2)_4CH{=}CH_2$ (**Hint:** See Section 12-12.)

(b) 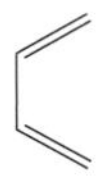 ---> 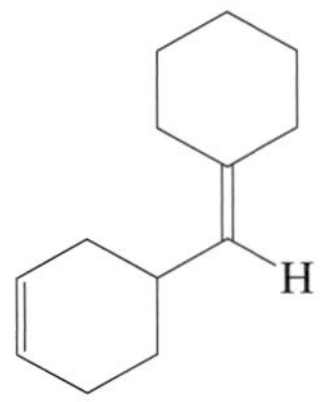

(**Hint:** See Section 14-8.)

In Summary Phosphorus ylides add to aldehydes and ketones to give betaines that decompose by forming carbon–carbon double bonds. The Wittig reaction affords a means of synthesizing alkenes from carbonyl compounds and haloalkanes by way of the corresponding phosphonium salts.

17-13 | OXIDATION BY PEROXYCARBOXYLIC ACIDS: THE BAEYER-VILLIGER OXIDATION

When ketones are treated with peroxycarboxylic acids (Section 12-10), the result is an oxidation of the carbonyl function to an ester, a transformation called **Baeyer-Villiger* oxidation.** The mechanism of this reaction starts by nucleophilic addition of the hydroperoxy end of the peracid to the carbonyl group to generate a reactive peroxide analog of a hemi-acetal. This unstable adduct decomposes through a cyclic transition state in which an alkyl group shifts from the original carbonyl carbon to oxygen to give an ester.

Reaction

Baeyer-Villiger Oxidation

$$CH_3CCH_2CH_3 \xrightarrow{CF_3COOH, CH_2Cl_2} CH_3COCH_2CH_3$$

2-Butanone → Ethyl acetate 72%

Mechanism

R″CR′ + RCOOH ⟶ [cyclic transition state] ⟶ Ester + H—O—C—R

Ketone   Peroxycarboxylic acid

Cyclic ketones are converted into cyclic esters. Attack is at the carbonyl rather than at the carbon–carbon double bond (Section 12-10). Unsymmetric ketones, such as that in the following reaction, can in principle lead to two different esters. Why is only one observed? The answer is that some substituents migrate more easily than others. Experiments have established their relative ease of migration, or **migratory aptitude.**

[reaction scheme: bicyclic ketone → with CH₃COOH, CHCl₃ → lactone, 56%]

The ordering suggests that the migrating carbon possesses carbocationic character in the transition state for rearrangement.

Migratory Aptitudes in the Baeyer-Villiger Reaction

Methyl < primary < phenyl ~ secondary < tertiary

Exercise 17-22

Predict the outcome of the following oxidations with a peroxycarboxylic acid.

(a) [structure]   (b) [structure with CH₃]   (c) [structure]

---

*Professor Johann Friedrich Wilhelm Adolf von Baeyer (1835–1917), University of Munich, Nobel Prize 1905 (chemistry); Victor Villiger (1868–1934), BASF, Ludwigshafen, Germany.

In Summary Ketones can be oxidized with peroxycarboxylic acids to give esters; with unsymmetric ketones, the esters can be formed selectively by migration of only one of the substituents.

17-14 | OXIDATIVE CHEMICAL TESTS FOR ALDEHYDES

Although the advent of NMR and other spectroscopy has made chemical tests for functional groups a rarity, they are still used in special cases in which other analytical tests may fail. Two characteristic simple tests for aldehydes will again turn up in the discussion of sugar chemistry in Chapter 24; they make use of the ready oxidation of aldehydes to carboxylic acids. The first is **Fehling's* test,** in which copper(II) ion (such as in $CuSO_4$) is the oxidant. In a basic medium, the precipitation of red copper(I) oxide indicates the presence of an aldehyde function.

Addition of a molecule containing an aldehyde group to a solution of blue copper(II) sulfate (right) causes the formation of a brick-red precipitate of copper(I) oxide in the Fehling test.

Fehling's Test

$$\underset{RCH}{\overset{O}{\|}} + Cu^{2+} \xrightarrow{\text{NaOH, tartrate (see Real Life 5-3),} \atop H_2O} \underset{\text{Brick-red}}{Cu_2O} + \underset{RCOH}{\overset{O}{\|}}$$

The second is **Tollens's† test,** in which a solution of silver ion (such as in $AgNO_3$) precipitates a silver mirror on exposure to an aldehyde.

Tollens's Test

$$\underset{RCH}{\overset{O}{\|}} + Ag^+ \xrightarrow{NH_3, H_2O} \underset{\text{Mirror}}{Ag} + \underset{RCOH}{\overset{O}{\|}}$$

The Tollens test detects the presence of readily oxidized functional groups in organic molecules, such as aldehydes. Aldehyde addition to a solution of silver(I) in aqueous ammonia (right) causes the rapid deposition of a silver mirror on the glass walls of the tube (left).

The Fehling and Tollens tests are not commonly used in large-scale syntheses. However, the Tollens reaction is employed industrially to produce shiny silver mirrors on glass surfaces, such as the insides of Thermos bottles.

THE BIG PICTURE

We have begun an extended, systematic study of compounds containing the carbonyl group. Nine chapters remain in this book; carbonyl chemistry will be a significant presence in every one of them and will be a main focus of *six* of the nine. Why do these compounds play such a prominent role in organic chemistry? Recall that in Chapter 2 we introduced the properties of electrophilicity and nucleophilicity and discussed how they underlie almost all of the chemical reactivity of organic compounds. In the chapter just completed, we explored the carbonyl group as an *electrophilic* function, illustrating a number of processes that involve the addition of nucleophiles to the carbonyl carbon. In the next chapter, we shall see the other side of the picture of carbonyl chemistry: The carbon atom *adjacent* to the carbonyl group—the α-carbon—is a potential site of *nucleophilic* reactivity. It is this capacity for *both* electrophilic *and* nucleophilic behavior at adjacent sites in the *same molecule* that sets carbonyl compounds apart from almost any other class of organic substances. It is no wonder that the formation of carbon–carbon bonds in nature is in large measure based on carbonyl chemistry.

---

*Professor Hermann C. von Fehling (1812–1885), Polytechnic School of Stuttgart, Germany.
†Professor Bernhard C. G. Tollens (1841–1918), University of Göttingen, Germany.

In the next chapter, we shall find out how to generate nucleophilic reactivity at the α-carbons of aldehydes and ketones and how to use that reactivity in the formation of bonds with a variety of electrophiles, including the carbon atoms of other molecules.

WORKED EXAMPLES: INTEGRATING THE CONCEPTS

17-23. Analyzing an Unexpected Outcome

Oxidation of 4-hydroxybutanal with PCC (pyH$^+$ CrO$_3$Cl$^-$, Section 8-6) does not produce the dialdehyde that one might expect. Instead a cyclic compound called a lactone is formed:

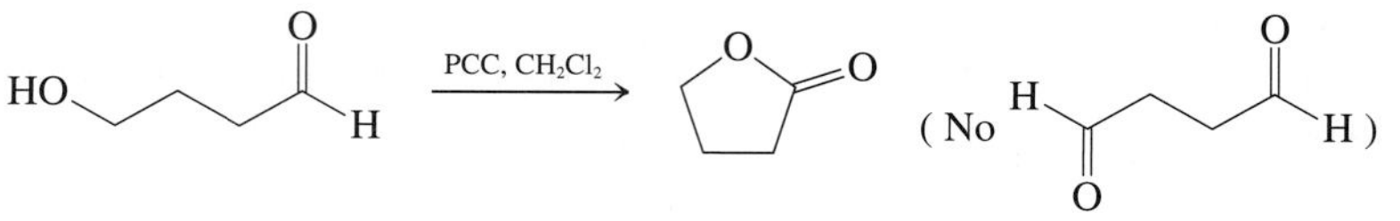

Explain this result.

SOLUTION

Let us first examine the type of compound with which we are dealing. It contains both alcohol and aldehyde functions. Section 17-7 described the reversible addition of alcohols to aldehydes and ketones to give hemiacetals; furthermore, it showed that the equilibrium of the process lies strongly on the product side when it leads to a five- or six-membered ring. For example:

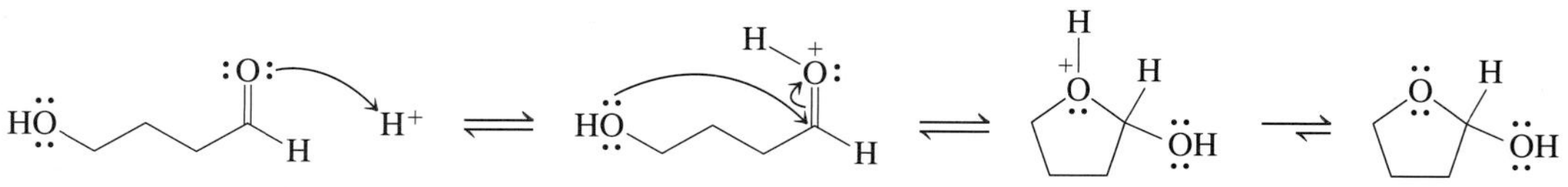

If we compare this hemiacetal with the product of the actual oxidation reaction shown earlier, we see that it is the *hemiacetal* whose oxidation produces the observed product. Notice that, in general, aldehyde hemiacetals possess the same structural feature that is present in ordinary alcohols that can

be oxidized to carbonyl compounds: $-\overset{\overset{\displaystyle H}{\displaystyle |}}{\underset{\displaystyle |}{C}}-O-H$.

17-24. Applying Retrosynthesis Involving Protecting Groups

a. Propose a synthesis of norethynodrel, the major component in one common oral contraceptive preparation (Enovid, compare Real Life 4-3). Use as your starting compound the following nortestosterone derivative (left-hand structure).

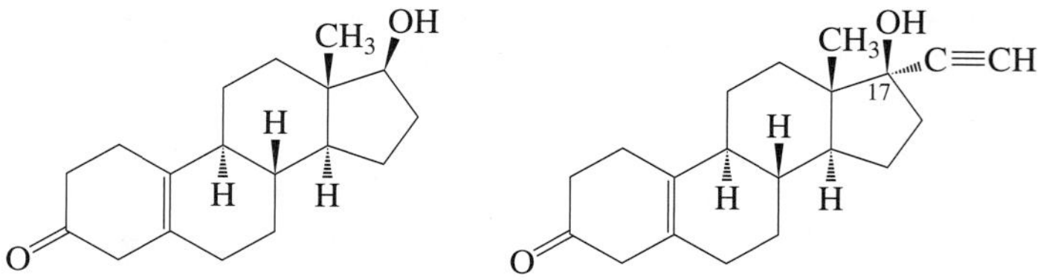

Starting compound

Norethynodrel

SOLUTION

Analyze the problem retrosynthetically (Section 8-9). The target molecule is identical with the starting compound in all but one respect: the additional substituent on the cyclopentane ring (at C17). We know that we cannot add groups directly to secondary alcohols to make tertiary alcohols. However, addition reactions of organometallic reagents to *ketones* afford tertiary alcohols (Section 8-8). The necessary organometallic reagent for our purpose is Li$^+$$^-$C≡CH, an alkynyl anion (Section 13-5). Let us see if we can devise a plan based on this chemistry.

The final step in the synthesis would appear to be addition of LiC≡CH to a C17 ketone, but, as the following reaction shows, there is a major problem: The precursor molecule contains a *second carbonyl group* in the six-membered ring at the lower left (at C3). In such a process, it would be impossible to prevent the alkynyl anion from adding indiscriminately to C3 as well as C17.

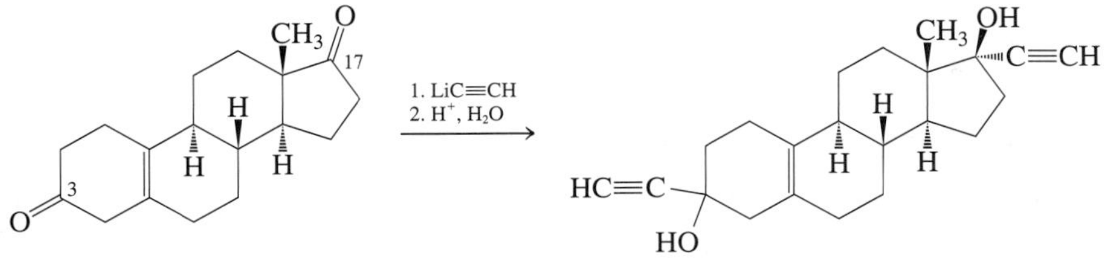

What to do? A similar problem arose in the synthesis of testosterone described in Real Life 9-2. It was solved by the use of protecting groups. We can do something similar here. The underlying conceptual problem is that *we need to avoid having both C3 and C17 exist as carbonyls at the same time in **any** molecule in our synthesis*. Failure to do so will doom a synthetic plan. With this concept in mind, we can imagine a modified ending to our synthesis, in which the carbon–oxygen double bond at C3 is protected in a form unreactive toward organometallic reagents during the previously shown step. After the addition reaction has been completed, the protecting group may be removed. An acetal derived from 1,2-ethanediol should do nicely (Section 17-8):

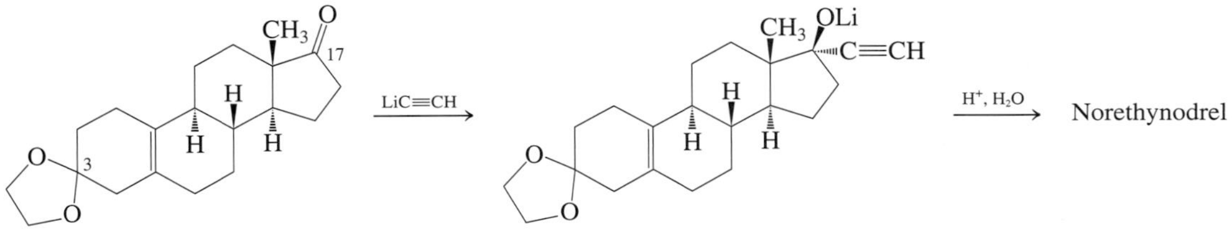

In this scheme, the aqueous acid work-up does double duty: It protonates the alkoxide that results from alkynyllithium addition, and it hydrolyzes the acetal to a ketone.

How does the synthesis begin? The C17 hydroxy group in the original starting material must be oxidized to a carbonyl. However, we cannot perform that oxidation in the presence of the carbonyl function at C3, because we would then violate the aforedescribed principle: We would create a molecule with both carbonyls present at the same time and the need to later react one and not the other with a reagent that cannot discriminate between them. With these considerations in mind, it becomes clear that protection of C3 must come first, before oxidation at C17:

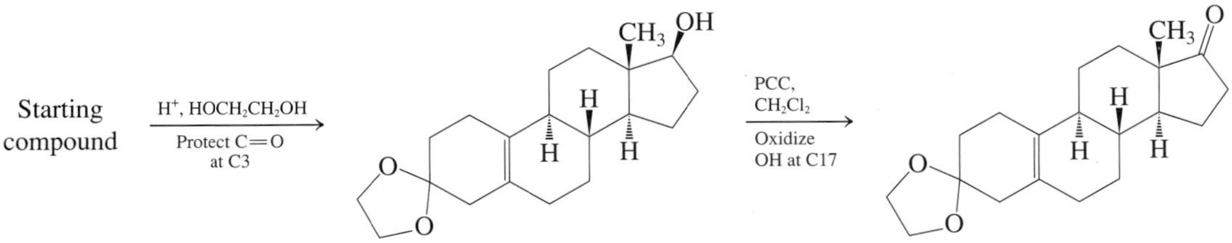

b. Why is pyridinium chlorochromate (PCC) in CH_2Cl_2 a better choice than $K_2Cr_2O_7$ for an oxidant in the second step?

SOLUTION

What could go wrong with $K_2Cr_2O_7$? Consider the conditions under which it is used: typically in aqueous sulfuric acid (Section 8-6). Therefore, this reagent carries the risk of hydrolysis of the acetal group under the acidic conditions. PCC is a neutral, nonaqueous reagent, ideal for substrates containing acid-sensitive functions.

c. Addition of alkynyllithium to the carbonyl at C17 is stereoselective to give the tertiary alcohol shown. Why? Would addition to a carbonyl at C3 also be stereoselective?

SOLUTION

What is the usual origin of stereoselectivity in reactions such as this? Steric hindrance in the immediate vicinity of the reacting center is the most common cause. The methyl substituent at the position adjacent to C17 is axial with respect to the six-membered ring and sterically hinders approach of the alkynyl reagent to the top face of the C17 carbonyl carbon (see Section 4-7). Addition from below is therefore strongly favored. There is no comparable steric hindrance in the vicinity of C3 to give rise to any preference for addition from either side.

New Reactions

Synthesis of Aldehydes and Ketones

1. Oxidation of Alcohols (Section 17-4)

$$
\underset{\substack{| \\ \text{RCHOH} \\ |}}{\overset{R'}{}} \xrightarrow{\text{CrO}_3,\ \text{H}_2\text{SO}_4} \overset{\text{O}}{\underset{}{\text{RCR'}}} \qquad \text{RCH}_2\text{OH} \xrightarrow{\text{PCC, CH}_2\text{Cl}_2} \overset{\text{O}}{\underset{}{\text{RCH}}}
$$

Stable to oxidizing agent

Allylic oxidation

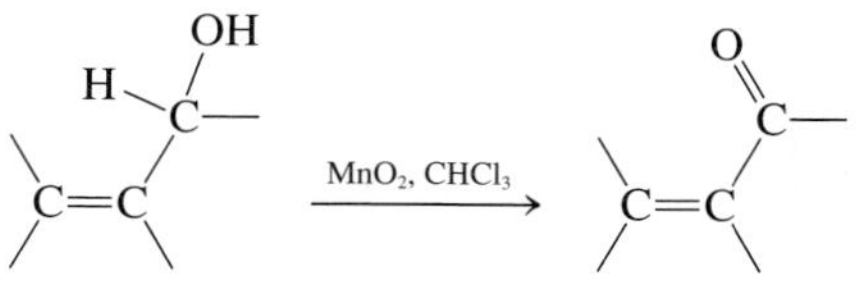

2. Ozonolysis of Alkenes (Section 17-4)

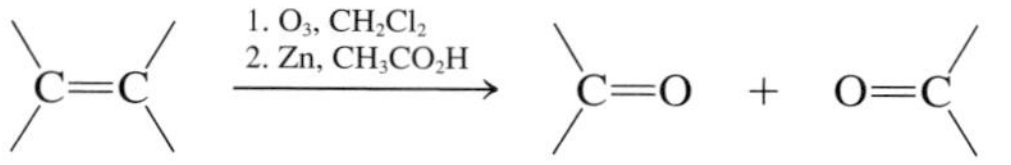

$$
\text{C}=\text{C} \xrightarrow[\text{2. Zn, CH}_3\text{CO}_2\text{H}]{\text{1. O}_3,\ \text{CH}_2\text{Cl}_2} \text{C}=\text{O} \ + \ \text{O}=\text{C}
$$

3. Hydration of Alkynes (Section 17-4)

$$
\text{RC}{\equiv}\text{CH} \xrightarrow{\text{H}_2\text{O, Hg}^{2+},\ \text{H}_2\text{SO}_4} \overset{\text{O}}{\underset{}{\text{RCCH}_3}}
$$

4. Friedel-Crafts Acylation (Section 17-4)

$$
\text{C}_6\text{H}_6 \ + \ \overset{\text{O}}{\underset{}{\text{RCCl}}} \xrightarrow{\text{AlCl}_3} \overset{\text{O}}{\underset{}{\text{C}_6\text{H}_5\text{CR}}} \ + \ \text{HCl}
$$

Reactions of Aldehydes and Ketones

5. Reduction by Hydrides (Section 17-5)

$$
\overset{\text{O}}{\underset{}{\text{RCH}}} \xrightarrow{\text{NaBH}_4,\ \text{CH}_3\text{CH}_2\text{OH}} \text{RCH}_2\text{OH} \qquad \overset{\text{O}}{\underset{}{\text{RCR'}}} \xrightarrow[\text{2. H}^+,\ \text{H}_2\text{O}]{\text{1. LiAlH}_4,\ (\text{CH}_3\text{CH}_2)_2\text{O}} \underset{\underset{\text{H}}{|}}{\overset{\overset{\text{OH}}{|}}{\text{RCR'}}}
$$

Selectivity

$$
\overset{\text{O}}{\underset{}{\text{RCH}{=}\text{CHCR'}}} \xrightarrow[\text{2. H}^+,\ \text{H}_2\text{O}]{\text{1. LiAlH}_4,\ (\text{CH}_3\text{CH}_2)_2\text{O}} \underset{\underset{\text{H}}{|}}{\overset{\overset{\text{OH}}{|}}{\text{RCH}{=}\text{CHCR'}}}
$$

6. Addition of Organometallic Compounds (Section 17-5)

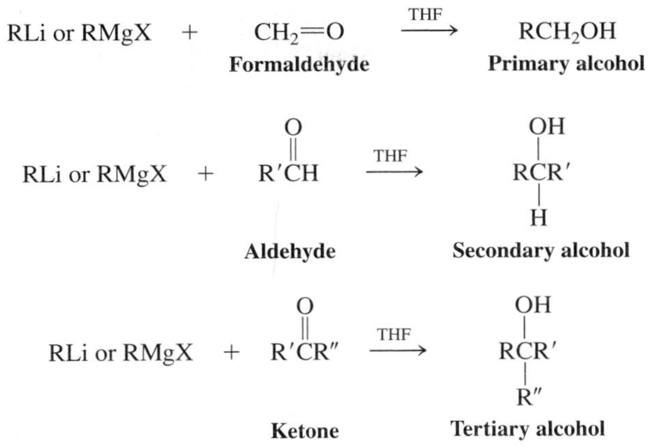

$$\text{RLi or RMgX} \quad + \quad \text{CH}_2\!=\!\text{O} \quad \xrightarrow{\text{THF}} \quad \text{RCH}_2\text{OH}$$

Formaldehyde → Primary alcohol

$$\text{RLi or RMgX} \quad + \quad \underset{\text{Aldehyde}}{\text{R'CH}(=\!\text{O})} \quad \xrightarrow{\text{THF}} \quad \underset{\text{Secondary alcohol}}{\text{RCR'(OH)(H)}}$$

$$\text{RLi or RMgX} \quad + \quad \underset{\text{Ketone}}{\text{R'CR''}(=\!\text{O})} \quad \xrightarrow{\text{THF}} \quad \underset{\text{Tertiary alcohol}}{\text{RCR'(OH)(R'')}}$$

7. Addition of Water and Alcohols—Hemiacetals (Sections 17-6 and 17-7)

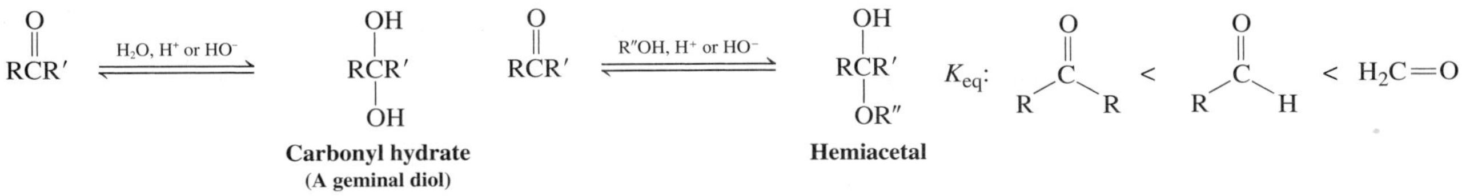

$$\text{RCR'}(=\!\text{O}) \quad \underset{\text{H}_2\text{O, H}^+\text{ or HO}^-}{\rightleftharpoons} \quad \underset{\substack{\text{Carbonyl hydrate} \\ \text{(A geminal diol)}}}{\text{RCR'(OH)(OH)}} \qquad \text{RCR'}(=\!\text{O}) \quad \underset{\text{R''OH, H}^+\text{ or HO}^-}{\rightleftharpoons} \quad \underset{\text{Hemiacetal}}{\text{RCR'(OH)(OR'')}}$$

$$K_{\text{eq}}: \quad \underset{R}{\overset{O}{\text{C}}}\underset{R}{} \quad < \quad \underset{R}{\overset{O}{\text{C}}}\underset{H}{} \quad < \quad \text{H}_2\text{C}\!=\!\text{O}$$

Intramolecular addition

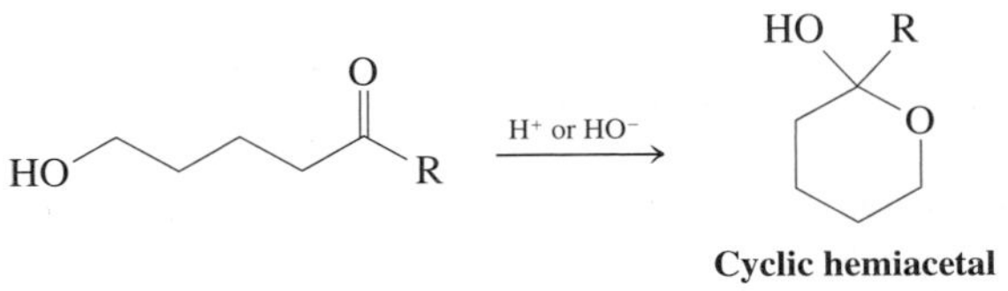

$$\text{HO} \frown \frown \text{CR}(=\!\text{O}) \quad \xrightarrow{\text{H}^+\text{ or HO}^-} \quad \underset{\text{Cyclic hemiacetal}}{\text{(HO)(R) ring O}}$$

8. Acid-Catalyzed Addition of Alcohols—Acetals (Sections 17-7 and 17-8)

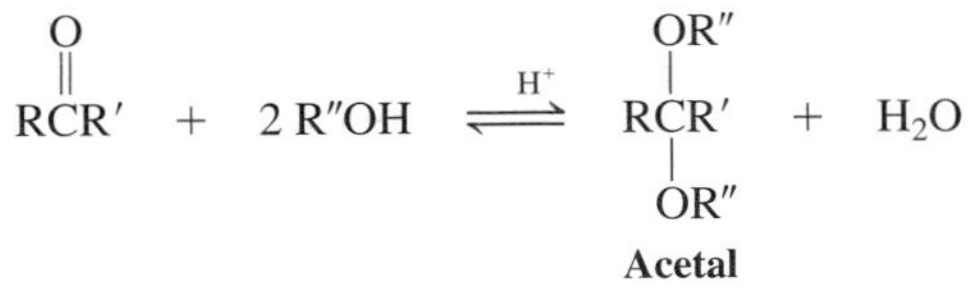

$$\text{RCR'}(=\!\text{O}) \quad + \quad 2\,\text{R''OH} \quad \underset{\text{H}^+}{\rightleftharpoons} \quad \underset{\text{Acetal}}{\text{RCR'(OR'')(OR'')}} \quad + \quad \text{H}_2\text{O}$$

Cyclic acetals

$$\text{RCR'}(=\!\text{O}) \quad + \quad \text{HOCH}_2\text{CH}_2\text{OH} \quad \underset{\text{H}^+}{\rightleftharpoons} \quad \underset{\substack{\text{Ketone, protected as} \\ \text{cyclic acetal} \\ (\text{Stable to base, LiAlH}_4\text{, RMgX})}}{\text{C(R)(R') with O}\frown\frown\text{O ring}} \quad + \quad \text{H}_2\text{O}$$

Ketone

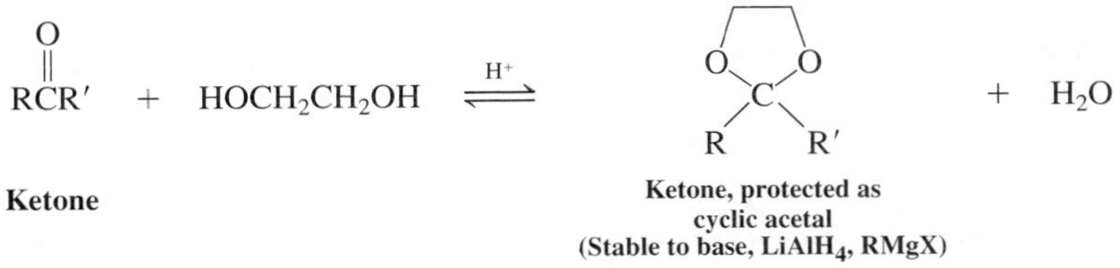

9. Thioacetals (Section 17-8)

Formation

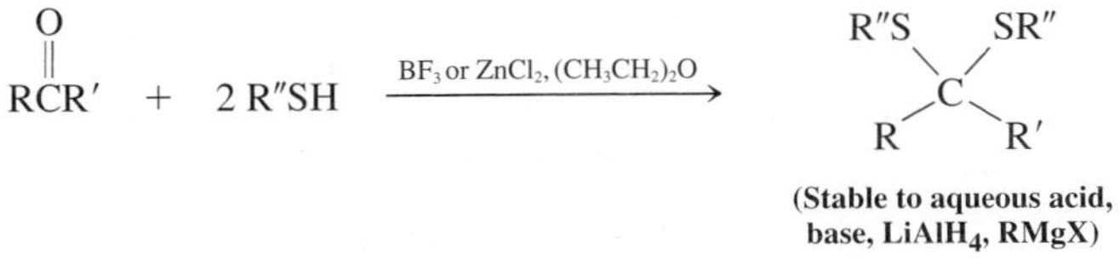

$$\text{RCR'}(=\!\text{O}) \quad + \quad 2\,\text{R''SH} \quad \xrightarrow{\text{BF}_3\text{ or ZnCl}_2,\,(\text{CH}_3\text{CH}_2)_2\text{O}} \quad \underset{\substack{(\text{Stable to aqueous acid,} \\ \text{base, LiAlH}_4\text{, RMgX})}}{\text{C(R)(R')(SR'')(SR'')}}$$

Hydrolysis

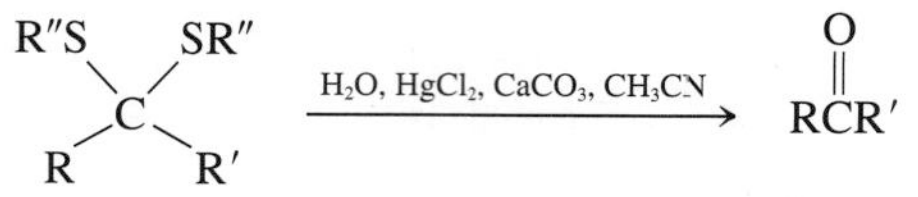

$$\underset{R}{\overset{R''S \quad SR''}{\underset{R'}{\diagdown C \diagup}}} \xrightarrow{\text{H}_2\text{O, HgCl}_2\text{, CaCO}_3\text{, CH}_3\text{CN}} \underset{RCR'}{\overset{O}{\parallel}}$$

10. Raney Nickel Desulfurization (Section 17-8)

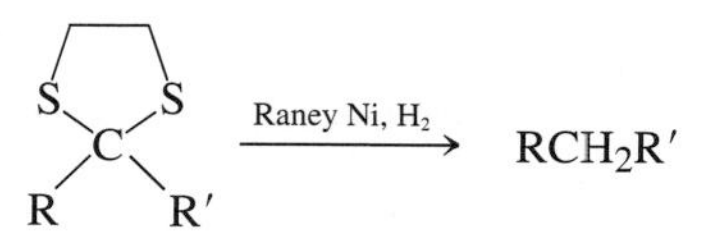

$$\underset{R}{\overset{S \quad S}{\underset{R'}{\diagdown C \diagup}}} \xrightarrow{\text{Raney Ni, H}_2} RCH_2R'$$

11. Addition of Amine Derivatives (Section 17-9)

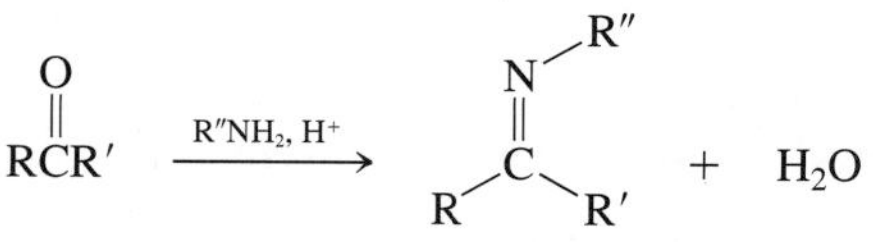

$$\underset{R}{\overset{O}{\underset{}{\parallel}}}{RCR'} \xrightarrow{R''NH_2,\ H^+} \underset{R}{\overset{N-R''}{\underset{R'}{\diagdown C \diagup}}} + H_2O$$

Imine

12. Enamines (Section 17-9)

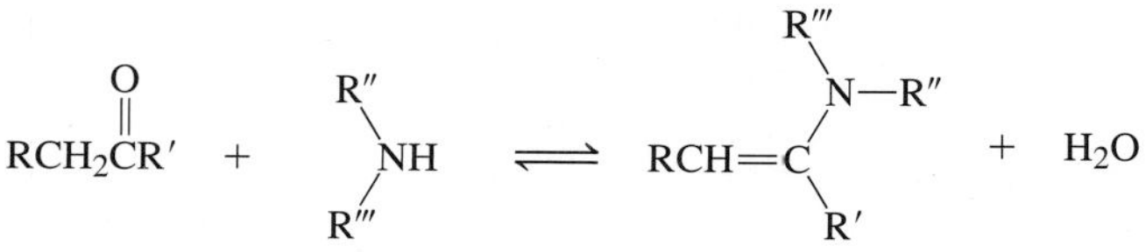

$$\underset{}{\overset{O}{\parallel}}{RCH_2CR'} + \underset{R'''}{\overset{R''}{\diagdown}}NH \rightleftharpoons RCH=\underset{R'}{\overset{\overset{R'''}{\underset{}{\mid}}N-R''}{\underset{}{\diagup C}}} + H_2O$$

Secondary amine   **Enamine**

13. Wolff-Kishner Reduction (Section 17-10)

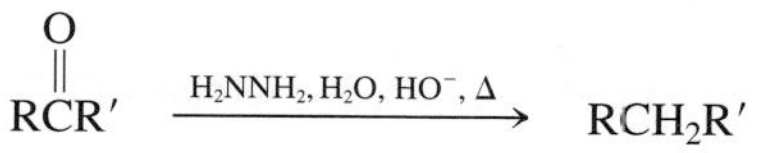

$$\underset{}{\overset{O}{\parallel}}{RCR'} \xrightarrow{H_2NNH_2,\ H_2O,\ HO^-,\ \Delta} RCH_2R'$$

14. Cyanohydrins (Section 17-11)

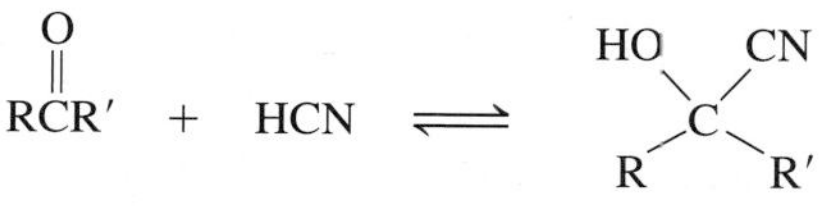

$$\underset{}{\overset{O}{\parallel}}{RCR'} + HCN \rightleftharpoons \underset{R}{\overset{HO \quad CN}{\underset{R'}{\diagdown C \diagup}}}$$

Cyanohydrin

15. Wittig Reaction (Section 17-12)

$$R''CH_2X + P(C_6H_5)_3 \xrightarrow{C_6H_6} R''CH_2\overset{+}{P}(C_6H_5)_3\ X^-$$

Triphenylphosphine   **Phosphonium halide**

Works with primary or secondary haloalkanes

$$R''CH_2\overset{+}{P}(C_6H_5)_3\ X^- \xrightarrow{\text{Base}} R''CH=P(C_6H_5)_3$$

Ylide

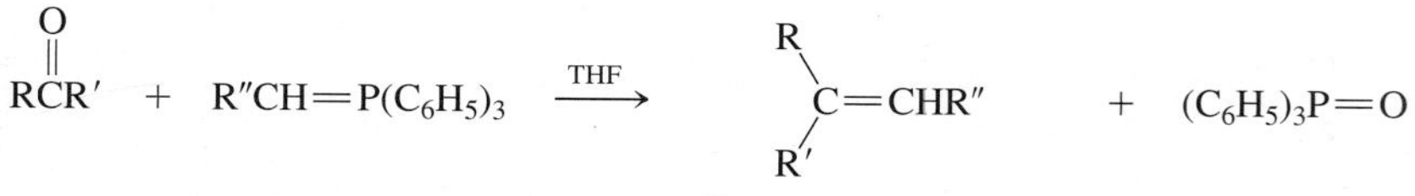

$$\underset{}{\overset{O}{\parallel}}{RCR'} + R''CH=P(C_6H_5)_3 \xrightarrow{\text{THF}} \underset{R'}{\overset{R}{\diagdown}}C=CHR'' + (C_6H_5)_3P=O$$

(Not always stereoselective)

16. **Baeyer-Villiger Oxidation (Section 17-13)**

$$\underset{\text{Ketone}}{\overset{O}{\underset{\|}{RCR'}}} + \underset{}{\overset{O}{\underset{\|}{R''COOH}}} \xrightarrow{CH_2Cl_2} \underset{\text{Ester}}{\overset{O}{\underset{\|}{RCOR'}}} + \overset{O}{\underset{\|}{R''COH}}$$

Migratory aptitudes in Baeyer-Villiger oxidation

Methyl < primary < phenyl ~ secondary < cyclohexyl < tertiary

Important Concepts

1. The **carbonyl group** is the functional group of the **aldehydes (alkanals)** and **ketones (alkanones).** It has precedence over the hydroxy, alkenyl, and alkynyl groups in the naming of molecules.

2. The carbon–oxygen double bond and its two attached nuclei in aldehydes and ketones form a plane. The C=O unit is **polarized,** with a partial negative charge on oxygen and a partial positive charge on carbon.

3. The ^1H NMR **spectra** of aldehydes exhibit a peak at $\delta \approx 9.8$ ppm. In ^{13}C NMR, the carbonyl carbon absorbs at ~200 ppm. Aldehydes and ketones have strong infrared bands in the region 1690–1750 cm^{-1}; this absorption is due to the stretching of the C=O bond. Because of the availability of low-energy $n \rightarrow \pi^*$ transitions, the electronic spectra of aldehydes and ketones have relatively long wavelength bands. This class of compounds displays characteristic mass spectral fragmentations around the carbonyl function.

4. The carbon–oxygen double bond undergoes **ionic additions.** The catalysts for these processes are acids and bases.

5. The reactivity of the carbonyl group increases with increasing **electrophilic character** of the carbonyl carbon. Therefore, aldehydes are more reactive than ketones.

6. Primary amines undergo **condensation** reactions with aldehydes and ketones to imines; secondary amines condense to enamines.

7. The combination of Friedel-Crafts acylation and Wolff-Kishner or Clemmensen reduction allows synthesis of alkylbenzenes free of the limitations of Friedel-Crafts alkylation.

8. The **Wittig reaction** is an important carbon–carbon bond-forming reaction that produces alkenes directly from aldehydes and ketones.

9. The reaction of **peroxycarboxylic acids** with the carbonyl group of ketones produces **esters.**

Problems

25. Draw structures and provide IUPAC names for each of the following compounds.

(a) Methyl ethyl ketone; (b) ethyl isobutyl ketone; (c) methyl *tert*-butyl ketone; (d) diisopropyl ketone; (e) acetophenone; (f) *m*-nitroacetophenone.

26. Name or draw the structure of each of the following compounds.

(a) $(CH_3)_2CHCCH(CH_3)_2$ with O above middle carbon

(b) [structure]

(c) [structure]

(d) [structure with Cl, H, C=C, CH$_2$CHO]

(e) [structure]

(f) [structure]

(g) (Z)-2-Acetyl-2-butenal

(h) *trans*-3-Chlorocyclobutanecarbaldehyde

27. The following spectroscopic data are for two carbonyl compounds with the formula $C_8H_{12}O$. Suggest a structure for each compound. The letter "m" stands for the appearance of this particular part of the spectrum as an uninterpretable multiplet. (a) ^1H NMR: $\delta =$ 1.60 (m, 4 H), 2.15 (s, 3 H), 2.19 (m, 4 H), and 6.78 (t, 1 H) ppm. ^{13}C NMR: $\delta =$ 21.8, 22.2, 23.2, 25.0, 26.2, 139.8, 140.7, and 198.6 ppm. (b) ^1H NMR: $\delta =$ 0.94 (t, 3 H), 1.48 (sex, 2 H), 2.21 (q, 2 H), 5.8–7.1 (m, 4 H), and 9.56 (d, 1 H) ppm. ^{13}C NMR: $\delta =$ 13.6, 21.9, 35.2, 129.0, 135.2, 146.7, 152.5, and 193.2 ppm.

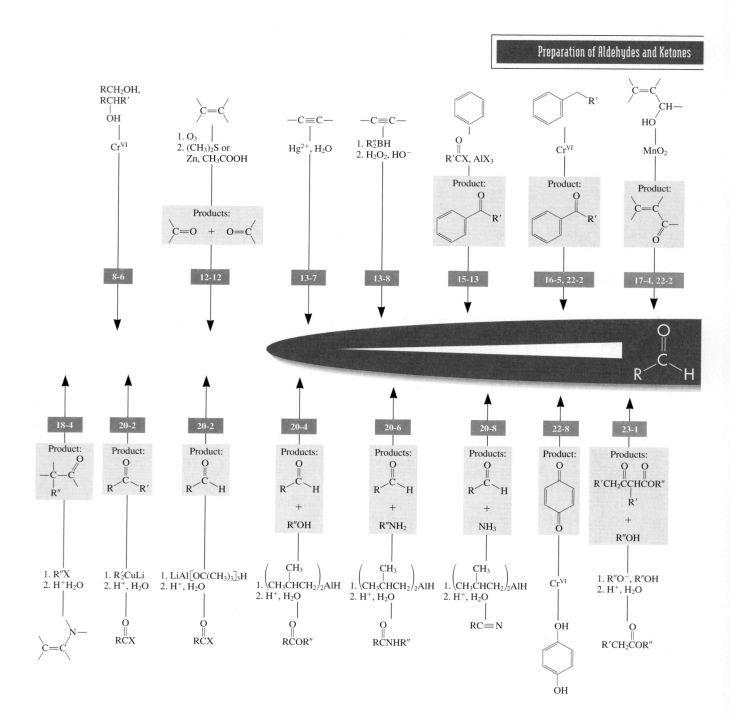

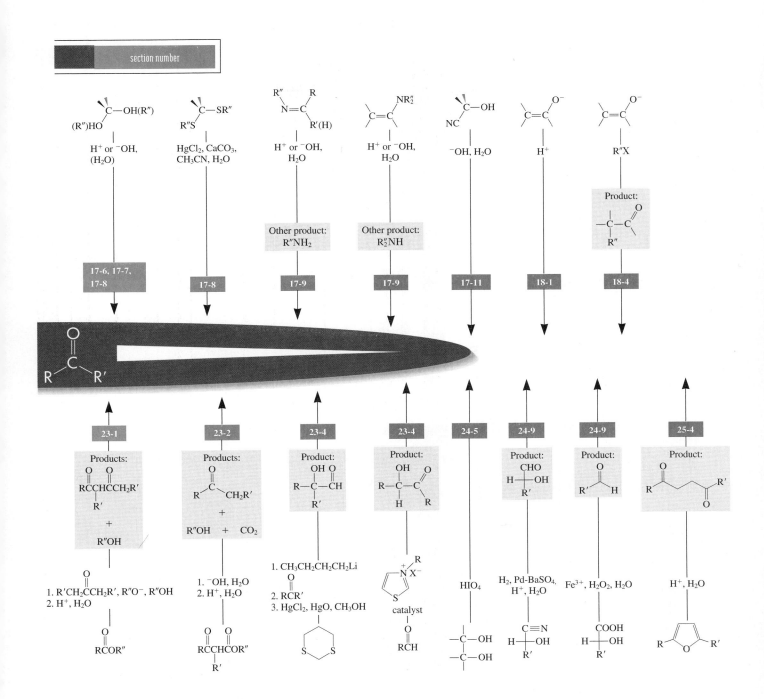

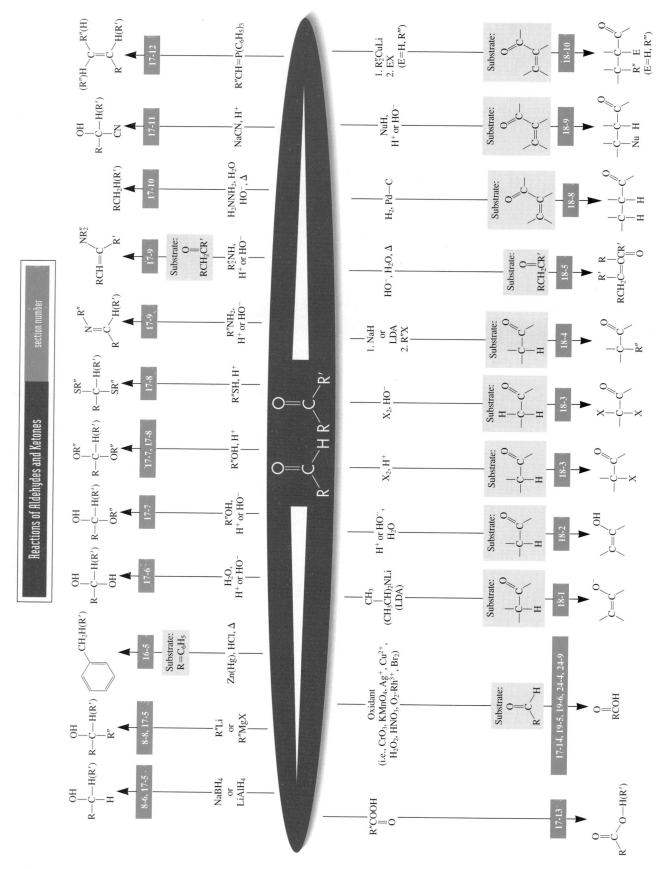

Reactions of Aldehydes and Ketones

section number

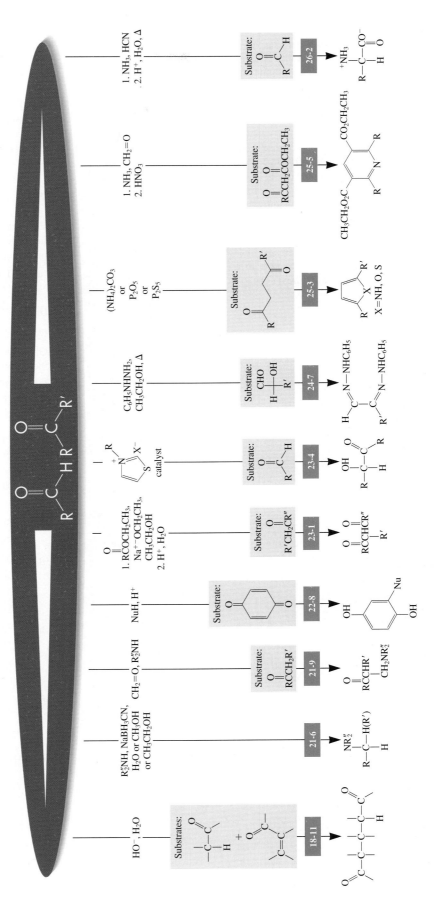

28. The compounds described in Problem 27 have very different ultraviolet spectra. One has $\lambda_{max}(\epsilon) = 232(13,000)$ and $308(1450)$ nm, whereas the other has $\lambda_{max}(\epsilon) = 272(35,000)$ nm and a weaker absorption near 320 nm (this value is hard to determine accurately because of the intensity of the stronger absorption). Match the structures that you determined in Problem 27 to these UV spectral data. Explain the spectra in terms of the structures.

29. Following are spectroscopic and analytical characteristics for an unknown compound. Propose a structure. Empirical formula: $C_8H_{16}O$. ^1H NMR: $\delta = 0.90$ (t, 3 H), 1.0–1.6 (m, 8 H), 2.05 (s, 3 H), and 2.25 (t, 2 H) ppm. IR: 1715 cm^{-1}. UV: $\lambda_{max}(\epsilon) = 280(15)$ nm. MS: $m/z = 128$ (M$^{+\bullet}$); intensity of (M + 1)$^+$ peak is 9% of M$^{+\bullet}$ peak; important fragments are at $m/z = 113$ (M − 15)$^+$ $m/z = 85$ (M − 43)$^+$, $m/z = 71$ (M − 57)$^+$, $m/z = 58$ (M − 70)$^+$ (the second largest peak), and $m/z = 43$ (M − 85)$^+$ (the base peak).

30. Reaction review. Without consulting the Reaction Road Map on pp. 780–781, suggest reagents to convert each of the starting materials below into 3-hexanone.

(a)

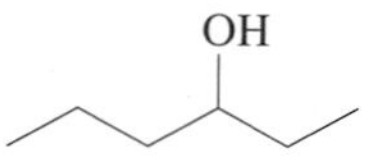

(b)

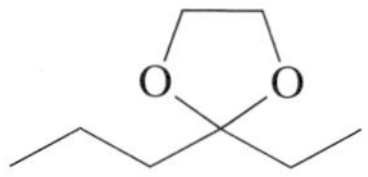

(c)

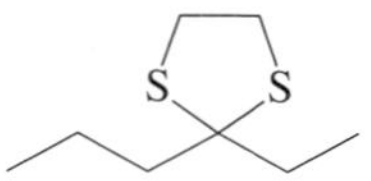

(d)

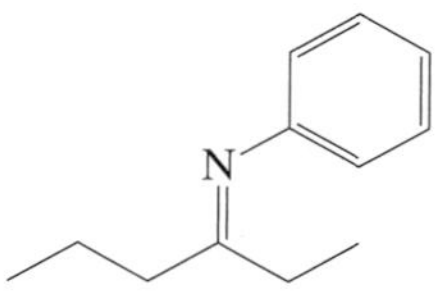

(e)

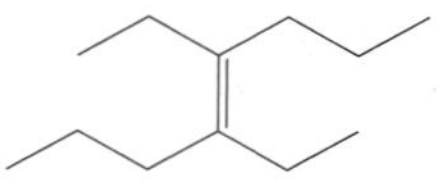

(f)

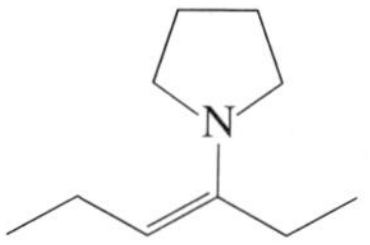

(g)

(h)

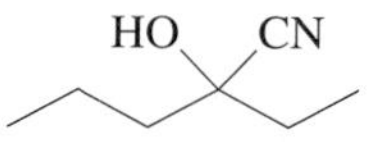

31. Indicate which reagent or combination of reagents is best suited for each of the following reactions.

(a)

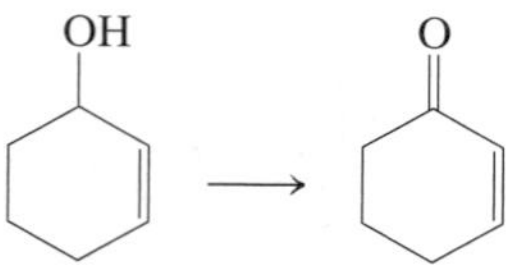

(b)

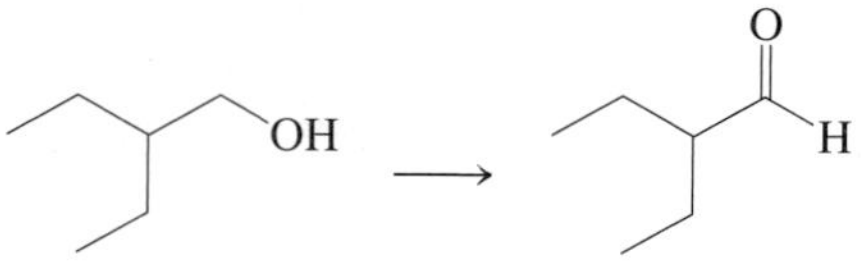

(c) 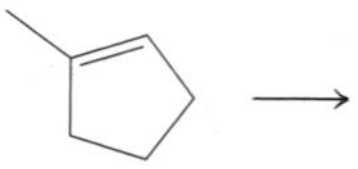 $\longrightarrow$ CH$_3$CCH$_2$CH$_2$CH$_2$CHO

(d)

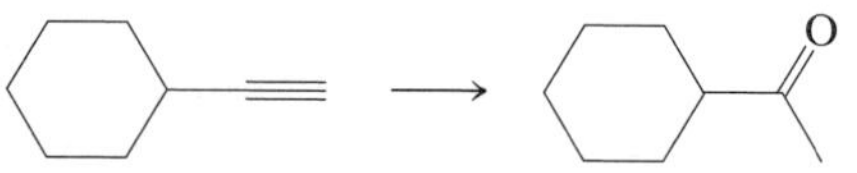

(e)

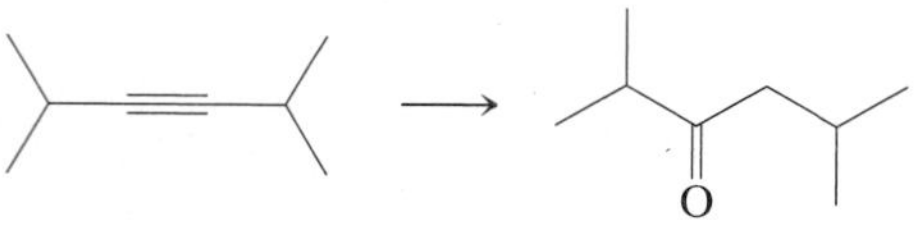

(f)

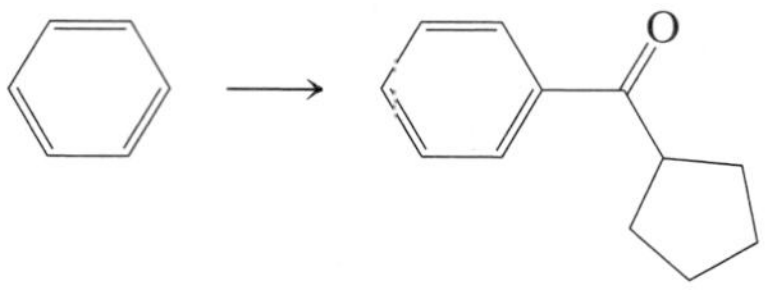

32. Write the expected products of ozonolysis (Section 12-12) of each of the following molecules.

(a) CH$_3$CH$_2$CH$_2$CH=CH$_2$

(b)

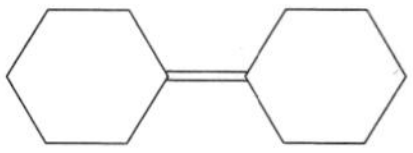

(c)

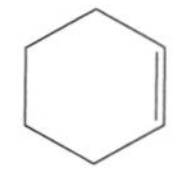

(d)

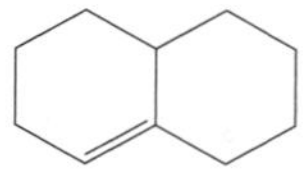

33. For each of the following groups, rank the molecules in decreasing order of reactivity toward addition of a nucleophile to the most electrophilic sp^2-hybridized carbon.

(a) (CH$_3$)$_2$C=O, (CH$_3$)$_2$C=NH, (CH$_3$)$_2$C=$\overset{+}{O}$H

(b) CH$_3$CCH$_3$, CH$_3$CCCH$_3$, CH$_3$CCCCH$_3$

(c) BrCH$_2$COCH$_3$, CH$_3$COCH$_3$, CH$_3$CHO, BrCH$_2$CHO

34. Give the expected products of reaction of butanal with each of the following reagents.

(a) LiAlH$_4$, (CH$_3$CH$_2$)$_2$O, then H$^+$, H$_2$O

(b) CH$_3$CH$_2$MgBr, (CH$_3$CH$_2$)$_2$O, then H$^+$, H$_2$O

(c) HOCH$_2$CH$_2$OH, H$^+$

35. Give the expected products of reaction of 2-pentanone with each of the reagents in Problem 34.

36. Give the expected products of reaction of 4-acetylcyclohexene with each of the reagents in Problem 34.

37. Give the expected product(s) of each of the following reactions.

(a) 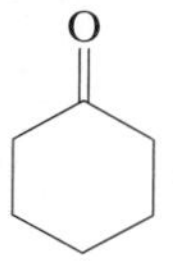 + excess CH$_3$OH $\xrightarrow{\;^-OH\;}$

(b) 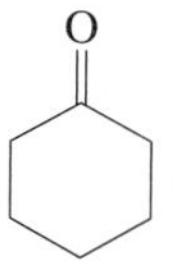 + excess CH$_3$OH $\xrightarrow{\;H^+\;}$

(c) 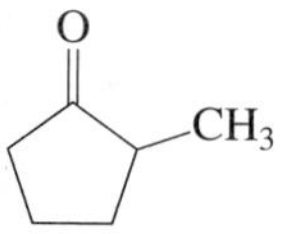 $\xrightarrow{\;H^+\;}$

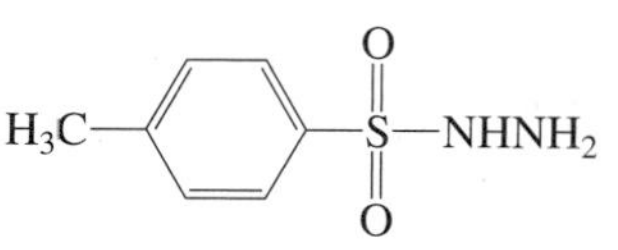

(d) CH$_3$CCH$_3$ + HOCH$_2$CHCH$_2$CH$_2$CH$_3$ $\xrightarrow{\;H^+\;}$

(e) 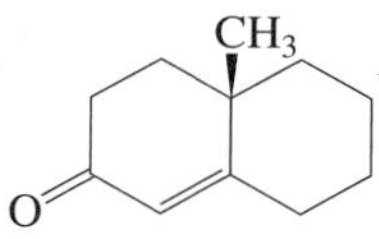 + 2 CH$_3$CH$_2$SH $\xrightarrow{\;BF_3,\ (CH_3CH_2)_2O\;}$

(f) 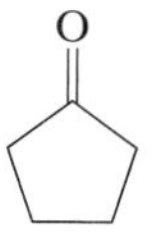 + (CH$_3$CH$_2$)$_2$NH $\longrightarrow$

38. Formulate detailed mechanisms for (a) the formation of the hemiacetal of acetaldehyde and methanol under both acid- and base-catalyzed conditions and (b) the formation of the intramolecular hemiacetal of 5-hydroxypentanal (Section 17-7), again under both acid- and base-catalyzed conditions.

39. Formulate the mechanism of the BF$_3$-catalyzed reaction of CH$_3$SH with butanal (Section 17-8).

40. **CHALLENGE**   Overoxidation of primary alcohols to carboxylic acids is caused by the water present in the usual aqueous acidic Cr(VI) reagents. The water adds to the initial aldehyde product to form a hydrate, which is further oxidized (Section 17-6). In view of these facts, explain the following two observations. **(a)** Water adds to ketones to form hydrates, but no overoxidation follows the conversion of a secondary alcohol into a ketone. **(b)** Successful oxidation of primary alcohols to aldehydes by the water-free PCC reagent requires that the alcohol be added slowly to the Cr(VI) reagent. If, instead, the PCC is added *to the alcohol,* a new side reaction forms an ester. This result is illustrated for 1-butanol.

$$CH_3CH_2CH_2CH_2OH \xrightarrow{\text{PPC, CH}_2\text{Cl}_2} CH_3CH_2CH_2\overset{\overset{\displaystyle O}{\|}}{C}OCH_2CH_2CH_2CH_3$$

(c) Give the products expected from reaction of 3-phenyl-1-propanol and water-free CrO$_3$ (1) when the alcohol is added to the oxidizing agent and (2) when the oxidant is added to the alcohol.

41. Explain the results of the following reactions by means of mechanisms.

(a)

(b)

(c) Explain why hemiacetal formation may be catalyzed by either acid or base, but acetal formation is catalyzed only by acid, not by base.

42. The two isomeric compounds below are naturally occurring insect pheromones. The isomer on the left attracts the male olive fruit fly; the one on the right, the female. **(a)** What kind of isomeric relationship exists between these two structures? **(b)** What functional group is contained in these molecules? **(c)** Both of these compounds are hydrolyzed under aqueous acidic conditions. Draw the products. Are the product molecules from the two starting isomers the same or different?

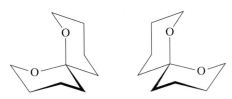

43. Formulate a plausible mechanism for the following reaction. The product is a precursor of mediquox (shown on top of the right

column), an agent used to treat respiratory infections in chickens (no, we are not making this up).

Mediquox

Benzene-1,2-diamine   + $\xrightarrow{\text{CH}_3\text{CH}_2\text{OH}}$

2,3-Dimethylquinoxaline

44. The formation of imines, oximes, hydrazones, and related derivatives from carbonyl compounds is reversible. Write a detailed mechanism for the acid-catalyzed hydrolysis of cyclohexanone semicarbazone to cyclohexanone and semicarbazide.

+   H$_2$O $\xrightarrow{\text{H}^+}$ +   NH$_2$NHCNH$_2$

45. Reaction review II. Without consulting the Reaction Road Map on p. 782, suggest reagents to convert cyclohexanone into each of the compounds below.

(a)   **(b)**   **(c)**

(d)   **(e)**   **(f)**

(g)   **(h)**   **(i)**

(j)   **(k)**   **(l)**

46. Propose reasonable syntheses of each of the following molecules, beginning with the indicated starting material. [**Hint:** Work backward by considering the strategic retrosynthetic C–C disconnections indicated (⌇), and then analyze for the potential need of protecting groups.]

(a) 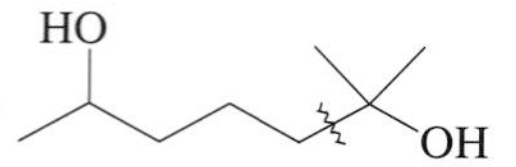 from

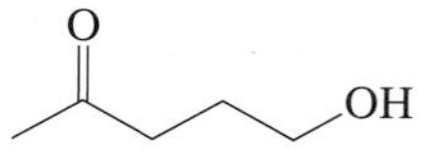

(b) $C_6H_5N{\Large\mbox{$=\!\!\!\!\!\diagup$}}C(CH_2CH_3)_2$ from 3-pentanol

(c) 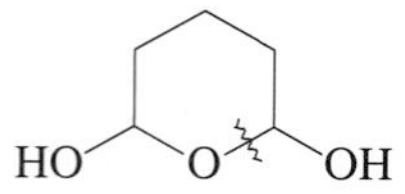 from 1,5-pentanediol

(d) 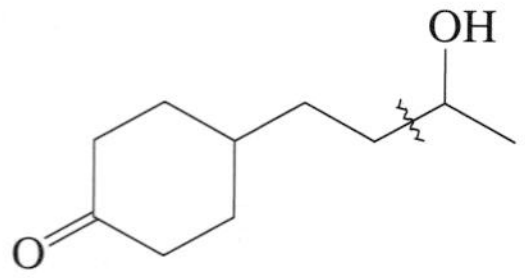 from

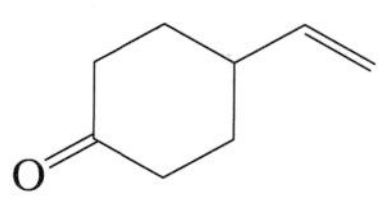

47. The UV absorptions and colors of 2,4-dinitrophenylhydrazone derivatives of aldehydes and ketones depend sensitively on the structure of the carbonyl compound. Suppose that you are asked to identify the contents of three bottles whose labels have fallen off. The labels indicate that one bottle contained butanal, one contained *trans*-2-butenal, and one contained *trans*-3-phenyl-2-propenal. The 2,4-dinitrophenylhydrazones prepared from the contents of the bottles have the following characteristics.

Bottle 1: m.p. 187–188°C; λ_{max} = 377 nm; orange color
Bottle 2: m.p. 121–122°C; λ_{max} = 358 nm; yellow color
Bottle 3: m.p. 252–253°C, λ_{max} = 394 nm; red color

Match up the hydrazones with the aldehydes (*without* first looking up the melting points of these derivatives), and explain your choices. (**Hint:** See Section 14-11.)

48. Indicate the reagent(s) best suited to effect these transformations.

(a)

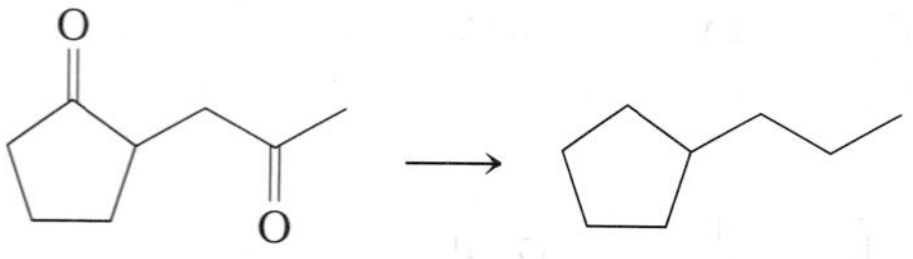

(b) $CH_3CH{=}CHCH_2CH_2\overset{\displaystyle O}{\overset{\|}{C}}H \longrightarrow CH_3CH_2CH_2CH_2CH_2\overset{\displaystyle O}{\overset{\|}{C}}H$

(c) $CH_3CH{=}CHCH_2CH_2\overset{\displaystyle O}{\overset{\|}{C}}H \longrightarrow CH_3CH{=}CHCH_2CH_2CH_2OH$

(d)

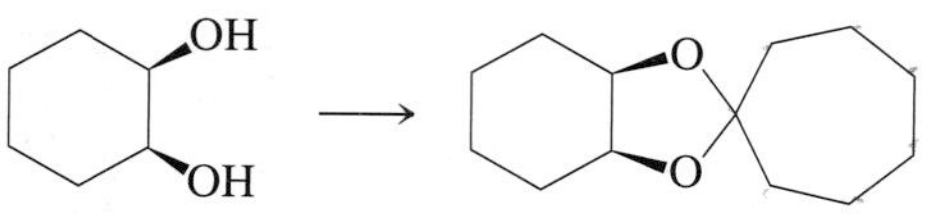

49. The molecule bombykol, whose structure is shown on top of the right column, is a powerful insect pheromone, the sex attractant of the female silk moth (see Section 12-17). It was initially isolated in the amount of 12 mg from extraction of 2 tons of moth pupae. Propose a synthesis from $BrCH_2(CH_2)_9OH$ and $CH_3CH_2CH_2C{\equiv}CCHO$, using a Wittig reaction (which in this system is known to form a trans double bond) as one of the steps of your sequence.

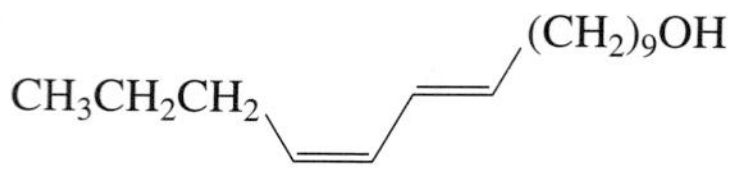

Bombykol

50. For each of the following molecules, propose *two* methods of synthesis from the different precursor molecules indicated.

(a) $CH_3CH{=}CHCH_2CH(CH_3)_2$ from (1) an aldehyde and (2) a different aldehyde

(b) 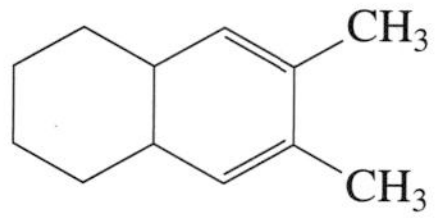 from (1) a dialdehyde and (2) a diketone

51. Three isomeric ketones with the molecular formula $C_7H_{14}O$ are converted into heptane by Clemmensen reduction. Compound A gives a single product upon Baeyer-Villiger oxidation; compound B gives two different products in very different yields; compound C gives two different products in virtually a 1:1 ratio. Identify A, B, and C.

52. Give the product(s) of reaction of hexanal with each of the following reagents.

(a) $HOCH_2CH_2OH$, H^+
(b) $LiAlH_4$, then H^+, H_2O
(c) NH_2OH, H^+
(d) NH_2NH_2, KOH, heat
(e) $(CH_3)_2CHCH_2CH{=}P(C_6H_5)_3$
(f) 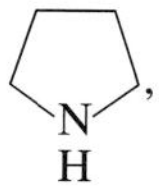 , H^+
(g) Ag^+, NH_3, H_2O
(h) CrO_3, H_2SO_4, H_2O
(i) HCN

53. Give the product(s) of reaction of cycloheptanone with each of the reagents in Problem 52.

54. Formulate in full detail the mechanism for the Wolff-Kishner reduction of 1-phenylethanone (acetophenone) to ethylbenzene (see p. 767).

55. The general equation for the Baeyer-Villiger oxidation (see p. 772) begins with a reaction between a ketone and a peroxycarboxylic acid to form a peroxy analog of a hemiacetal. Formulate a detailed mechanism for this process.

56. (a) Formulate a detailed mechanism for the Baeyer-Villiger oxidation of the ketone shown below (refer to Exercise 17-22).

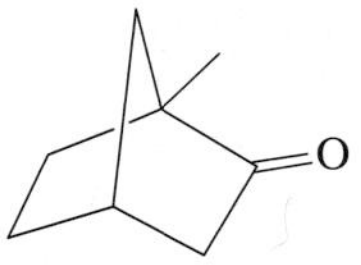

(b) Under Baeyer-Villiger conditions, aldehydes convert to carboxylic acids—for example, benzaldehyde goes to benzoic acid. Explain.

57. Give the two theoretically possible Baeyer-Villiger products from each of the following compounds. Indicate which one is formed preferentially.

(a) 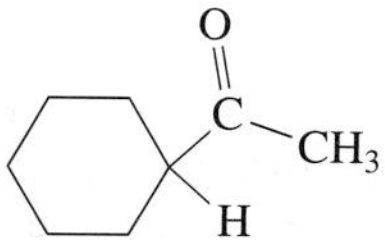 **(b)**

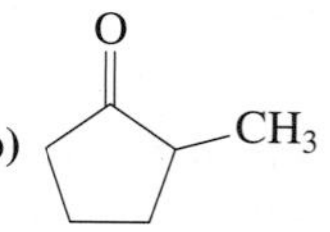

(c)

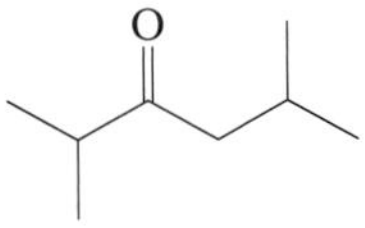

(d)

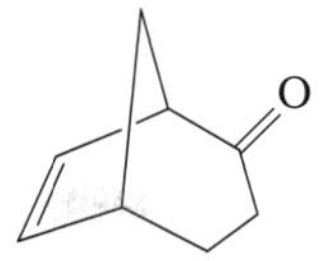

(e) $C_6H_5CCH_3$

58. Propose efficient syntheses of each of the following molecules, beginning with the indicated starting materials.

(a) 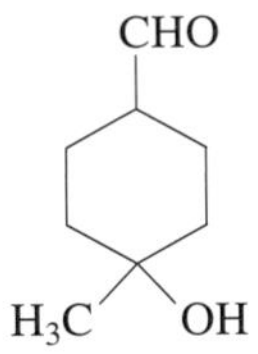 from

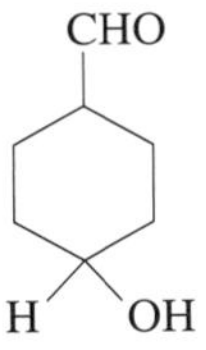

(b) 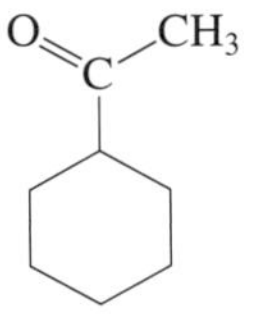 from

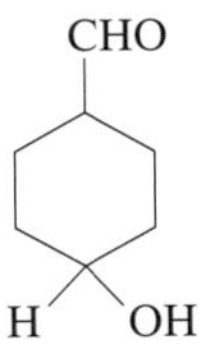

(c) 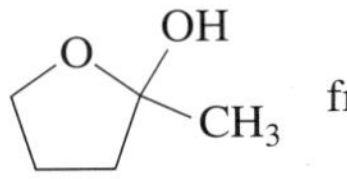 from $ClCH_2CH_2CH_2OH$

59. **CHALLENGE** Explain the fact that, although hemiacetal formation between methanol and cyclohexanone is thermodynamically disfavored, addition of methanol to cyclopropanone goes essentially to completion:

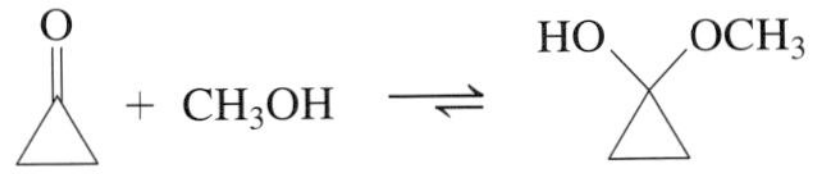

60. The rate of the reaction of NH_2OH with aldehydes and ketones is very sensitive to pH. It is very low in solutions more acidic than pH 2 or more basic than pH 7. It is highest in moderately acidic solution (pH ~ 4). Suggest explanations for these observations.

61. Compound D, formula $C_8H_{14}O$, is converted by $CH_2=P(C_6H_5)_3$ into compound E, C_9H_{16}. Treatment of compound D with $LiAlH_4$ yields *two* isomeric products F and G, both $C_8H_{16}O$, in unequal yield. Heating either F or G with concentrated H_2SO_4 produces H, with the formula C_8H_{14}. Ozonolysis of H produces a ketoaldehyde after Zn–H^+, H_2O treatment. Oxidation of this ketoaldehyde with aqueous Cr(VI) produces the compound shown below.

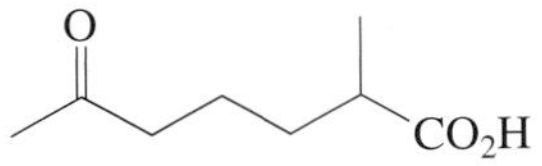

Identify compounds D through H. Pay particular attention to the stereochemistry of D.

62. In 1862, it was discovered that cholesterol (for structure, see Section 4-7) is converted into a new substance named coprostanol by the action of bacteria in the human digestive tract.

Make use of the following information to deduce the structure of coprostanol. Identify the structures of unknowns J through M as well. (i) Coprostanol, upon treatment with Cr(VI) reagents, gives compound J, UV: $\lambda_{max}(\epsilon) = 281(22)$ nm and IR: $\tilde{v} = 1710$ cm^{-1}. (ii) Exposure of cholesterol to H_2 over Pt results in compound K, a stereoisomer of coprostanol. Treatment of K with the Cr(VI) reagent furnishes compound L, which has a UV peak very similar to that of compound J, $\lambda_{max}(\epsilon) = 258(23)$ nm, and turns out to be a stereoisomer of J. (iii) Careful addition of Cr(VI) reagent to cholesterol produces M: UV: $\lambda_{max}(\epsilon) = 286(109)$ nm. Catalytic hydrogenation of M over Pt also gives L.

63. **CHALLENGE** Three reactions that include compound M (see Problem 62) are described here. Answer the questions that follow. **(a)** Treatment of M with catalytic amounts of acid in ethanol solvent causes isomerization to compound N: UV: $\lambda_{max}(\epsilon) = 241(17,500)$ and $310(72)$ nm. Propose a structure for N. **(b)** Hydrogenation of compound N (H_2–Pd, ether solvent) produces compound J (Problem 62). Is this the result that you would have predicted, or is there something unusual about it? **(c)** Wolff-Kishner reduction of compound N (H_2NNH_2, H_2O, HO^-, Δ) leads to 3-cholestene. Propose a mechanism for this transformation.

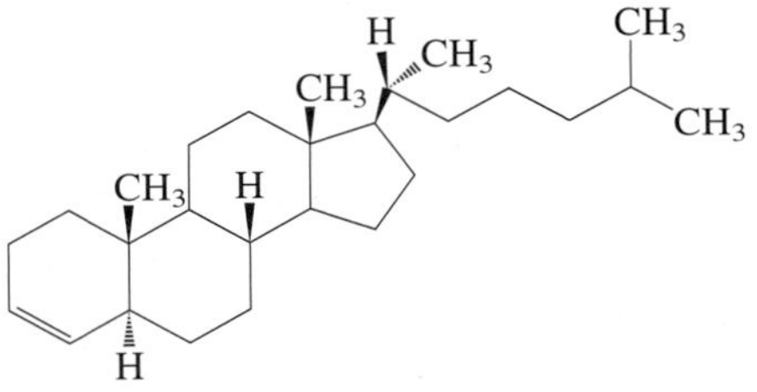

3-Cholestene

Team Problem

64. In acidic methanol, 3-oxobutanal is transformed into a compound with the molecular formula $C_6H_{12}O_3$ (see below).

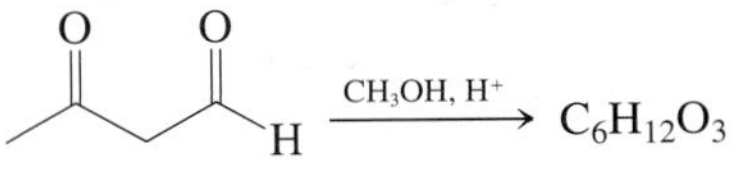

3-Oxobutanal

As a group, analyze the following ^1H NMR and IR spectral data: ^1H NMR (CCl_4): δ = 2.19 (s, 3 H), 2.75 (d, 2 H), 3.38 (s, 6 H), 4.89 (t, 1 H) ppm; IR: 1715 cm^{-1}.

Consider the chemical shifts, the splitting patterns, and the integrations of the signals in the NMR spectrum and discuss possible fragments that could give rise to the observed multiplicities. Use the IR information to assign the functional group that exists in the new molecule. Present an explanation for your structural determination, including reference to the spectral data, and suggest a detailed mechanism for the formation of the new compound.

Preprofessional Problems

65. In the transformation shown here, which of the following compounds is most likely to be compound A (use IUPAC name)? **(a)** 5-Octyn-7-one; **(b)** 5-octyn-2-one; **(c)** 3-octyn-2-one; **(d)** 2-octyn-3-one.

$$\text{3-Octyn-2-ol} \xrightarrow{\text{CrO}_3,\ \text{H}_2\text{SO}_4,\ \text{acetone}} \text{A}$$

66. The reaction

$$\underset{H_3C}{\overset{O}{\underset{}{\|}}}\!\!\!\!C\!\!-\!\!H \rightleftharpoons CH_2{=}\overset{H}{\underset{OH}{C}} \text{ demonstrates}$$

(a) resonance; **(b)** tautomerism; **(c)** conjugation; **(d)** deshielding.

67. Which of the following reagents converts benzenecarbaldehyde (benzaldehyde) into an oxime? **(a)** $H_2NNHC_6H_5$; **(b)** H_2NNH_2; **(c)** O_3; **(d)** H_2NOH; **(e)** $CH_3CH(OH)_2$.

68. Which of the following statements is correct? In the IR spectrum of 3-methyl-2-butanone, the most intense absorption is at **(a)** 3400 cm^{-1}, owing to an OH stretching mode; **(b)** 1700 cm^{-1}, owing to a C$=$O stretching mode; **(c)** 2000 cm^{-1}, owing to a CH stretching mode; **(d)** 1500 cm^{-1}, owing to the rocking of an isopropyl group.

Enols, Enolates, and the Aldol Condensation

α,β-Unsaturated Aldehydes and Ketones

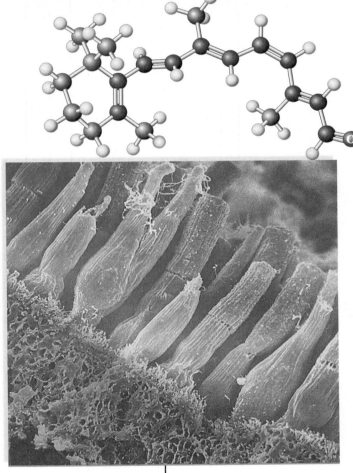

Have another look at the chapter-opening illustration. The very fact that you can see it is made possible by the chemistry alluded to in the caption: Photons impinge on the π system (Section 14-11) of *cis*-retinal bound to a protein to cause cis-trans isomerization. The associated conformational changes result, within picoseconds, in a nerve impulse that is translated by your brain into "vision" (Real Life 18-2). The crucial chemical feature of retinal that makes this process possible is the (in this case extended) conjugative communication between the carbonyl group and the adjacent π system. This chapter will show you that the carbonyl group (much like an ordinary carbon–carbon double bond; Section 14-1) activates adjacent C–H and C=C bonds even in much simpler systems, because of resonance. After you have absorbed the material that follows, you will be able to "look" at this page with quite a different perspective!

In the last chapter, we saw how the structure of the carbonyl group—a multiple bond that is also highly polar—gives rise to a characteristic combination of functional behaviors: addition reactions mediated by electrophilic attack (usually by protons) on the Lewis basic oxygen and attack by nucleophiles on the carbon. We turn now to a third site of reactivity in aldehydes and ketones, the carbon *next to* the carbonyl group, known as the **α-carbon.** The carbonyl group induces enhanced acidity of hydrogens on the α-carbon. Moving these **α-hydrogens** may lead to either of two electron-rich species: unsaturated alcohols called enols or their corresponding conjugate bases, known as enolate ions. Both enols and enolate ions are important nucleophiles, capable of attacking electrophiles such as protons, halogens, haloalkanes, and even other carbonyl compounds.

Photomicrograph of rod and cone cells in the retina. All known image-resolving eyes and, indeed, all known visual systems in nature use a single molecule, *cis*-retinal, for light detection. Absorption of a photon isomerizes the cis double bond to trans. The accompanying massive change in overall molecular geometry is responsible for triggering the nerve impulse that is perceived as vision.

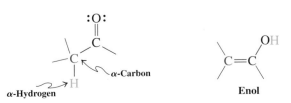

We begin by introducing the chemistry of enolates and enols. Especially important is a reaction between enolate ions and carbonyl compounds, called the aldol condensation. This process is widely used to form carbon–carbon bonds both in the laboratory and in nature. Among the possible products of aldol condensation are α,β-unsaturated aldehydes and ketones, which contain conjugated carbon–carbon and carbon–oxygen π bonds. As expected, electrophilic additions may take place at either π bond. However, more significantly, α,β-unsaturated carbonyl compounds are also subject to nucleophilic attack, a reaction that may involve the *entire* conjugated system.

18-1   ACIDITY OF ALDEHYDES AND KETONES: ENOLATE IONS

The pK_a values of aldehyde and ketone α-hydrogens range from 16 to 21, much lower than the pK_a values of ethene (44) or ethyne (25), but comparable with those of alcohols (15–18). Strong bases can therefore remove an α-hydrogen. The anions that result are known as **enolate ions** or simply **enolates.**

Deprotonation of a Carbonyl Compound at the α-Carbon

α-Carbon

$pK_a \approx$ 16–18 (aldehydes)
19–21 (ketones)

Enolate ion

Why are aldehydes and ketones relatively acidic? We know that acid strength is enhanced by stabilization of the conjugate base (Section 2-3). In the enolate ion, the inductive effect of the positively polarized carbonyl carbon strongly stabilizes the negative charge at the α-position. Aldehydes are stronger acids than ketones because their carbonyl carbon bears a larger partial positive charge (Section 17-6). Further strong stabilization is provided by delocalization of charge onto the electronegative oxygen, as described by the resonance forms just pictured. The effect of delocalization is also reflected in the electrostatic potential map of the acetone enolate shown in the margin (on an attenuated scale), which exhibits negative charge (red) on the α-carbon as well as on the oxygen. An example of quantitative enolate formation is the deprotonation of cyclohexanone (p$K_a \approx$ 19) by lithium diisopropylamide (LDA, pK_a of amine = 36; Section 7-8).

Acetone enolate

Enolate Preparation

$pK_a \approx$ 19   LDA   Cyclohexanone enolate ion $pK_a \approx$ 36

Exercise 18-1

Identify the most acidic hydrogens in each of the following molecules. Give the structure of the enolate ion arising from deprotonation. (**a**) Acetaldehyde; (**b**) propanal; (**c**) acetone; (**d**) 4-heptanone; (**e**) cyclopentanone.

Enolate formation can be regioselective

Deprotonation of an unsymmetrical ketone such as 2-methylcyclopentanone may lead to two isomeric species, a more substituted and a less substituted enolate. The former, featuring the more substituted double bond, is more stable than the latter (Section 11-5).* As in the elimination reaction of a haloalkane via the E2 mechanism (Section 7-7), the choice of base and reaction conditions determines which one is formed. For example, addition of 2-methylcyclopentanone to a cold solution of LDA in THF gives predominantly the less substituted, less stable enolate. The reason is that LDA is a bulky base and prefers to remove a hydrogen from the less-hindered α-carbon, generating the less stable anion, termed the *kinetic enolate*. Under these conditions, namely, the absence of a proton source and at low temperatures, equilibration with the more stable enolate does not occur, and the kinetic enolate can be used as such in practical further transformations.

Deprotonation of 2-Methylcyclopentanone Under Kinetic Conditions

In contrast, at room temperature and in the presence of a proton source such as a slight excess of ketone starting material, equilibration of the kinetic with the more stable *thermodynamic enolate* takes place. It occurs by reversible proton exchange between the unreacted ketone and the respective regioisomeric enolates, furnishing the latter in a ratio that reflects their relative thermodynamic stabilities. In this way, selective reactions, such as alkylations with haloalkanes (Section 18-4), can be executed with the more substituted enolate.

Deprotonation of 2-Methylcyclopentanone Under Equilibrium Conditions

An alternative, more convenient method to access thermodynamic enolates is the use of hydroxide in water or alkoxides in alcohols as bases. These bases are considerably weaker ($pK_a \approx 15$–18; Section 8-3) than LDA and hence deprotonate aldehydes and ketones reversibly in only small equilibrium concentrations. We shall see later in this chapter that certain reagents are able to trap the major enolate to shift this equilibrium to new products.

Enolates are ambident nucleophiles

Each resonance form contributes to the characteristics of the enolate ion and thus to the chemistry of carbonyl compounds. The resonance hybrid possesses partial negative charges on both carbon and oxygen; as a result, it is nucleophilic and may attack electrophiles at either position. A species that can react at two different sites to give two different products is called **ambident** ("two fanged": from *ambi*, Latin, both; *dens*, Latin, tooth). The enolate ion is thus an ambident anion. Its carbon atom is normally the site of reaction, undergoing

Resonance hybrid

---

*Because most of the negative charge is on the oxygen, relative enolate stabilities follow those of alkenes: Alkyl substitution stabilizes the double bond by hyperconjugation.

nucleophilic substitution with S_N2 substrates such as suitable haloalkanes. Because this reaction attaches an alkyl group to the reactive carbon, it is called **alkylation** (more specifically, C-alkylation). As we shall see in Section 18-4, alkylation is a powerful method for carbon–carbon bond formation of ketones. For example, alkylation of cyclohexanone enolate with 3-chloropropene takes place at carbon. Alkylation at oxygen (O-alkylation) is uncommon, although oxygen is typically the site of protonation. The product of protonation is an unsaturated alcohol, called an **alkenol** (or **enol** for short). Enols are unstable and rapidly isomerize back to the original ketones (recall Section 13-7).

Ambient Behavior of Cyclohexanone Enolate Ion

62%

2-(2-Propenyl)cyclohexanone

Cyclohexanone enolate ion

Cyclohexanone enol

Cyclohexanone

Exercise 18-2

(a) Predict the major products of reaction of cyclohexanone enolate made from cyclohexanone and LDA (1:1 ratio at −78°C) with **(i)** iodoethane (reacts by C-alkylation) and **(ii)** chlorotrimethylsilane, $(CH_3)_3Si$–Cl (reacts by O-silylation). **(b)** Do the same for 2-methylcyclopentanone enolate, made, respectively, from 2-methylcyclopentanone and LDA (1:1 ratio at −78°C) and 2-methylcyclopentanone and LDA (1.1:1 ratio at 25°C).

In Summary The hydrogens on the carbon next to the carbonyl group in aldehydes and ketones are acidic, with pK_a values ranging from 16 to 21. Deprotonation leads to the corresponding enolate ions. In the case of unsymmetrical ketones, either the kinetic or the thermodynamic enolate may be obtained. Enolate ions may attack electrophilic reagents at either oxygen or carbon. Protonation at oxygen gives enols.

18-2 KETO–ENOL EQUILIBRIA

We have seen that protonation of an enolate at oxygen leads to an enol. The enol, an unstable isomer of an aldehyde or ketone, rapidly converts into the carbonyl system: It **tautomerizes** (Section 13-7). These isomers are called **enol** and **keto tautomers.** We begin by discussing factors affecting their equilibria, in which the keto form usually predominates. We then describe the mechanism of tautomerism and its chemical consequences.

An enol equilibrates with its keto form in acidic or basic solution

Mechanism

Enol–keto tautomerism proceeds by either acid or base catalysis. Base simply removes the proton from the enol oxygen, reversing the initial protonation. Subsequent (and slower) C-protonation furnishes the thermodynamically more stable keto form.

Base-Catalyzed Enol–Keto Equilibration

Enol form

Enolate ion

Keto form

In the acid-catalyzed process, the enol form is protonated at the double-bonded carbon away from the hydroxy-bearing neighbor. The electrostatic potential map of ethenol (margin) shows that this carbon bears more negative charge (red). Moreover, the resulting cation is resonance-stabilized by the attached hydroxy group, and inspection of the corresponding resonance form reveals it to be simply the protonated carbonyl compound. Deprotonation then gives the product.

Ethenol

Acid-Catalyzed Enol–Keto Equilibration

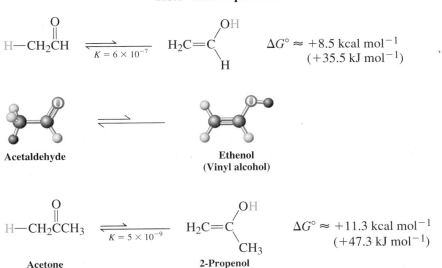

Both the acid- and base-catalyzed enol–keto interconversions occur rapidly in solution whenever there are traces of the required catalysts. Remember that although the keto form (usually) predominates, the enol-to-keto conversion is reversible and the mechanisms by which the keto form equilibrates with its enol counterpart are the exact reverse of the preceding two schemes.

Substituents can shift the keto–enol equilibrium

The equilibrium constants for the conversion of the keto into the enol forms, also called **enolization,** are very small for ordinary aldehydes and ketones, only traces of enol (which is less stable by ca. 8–12 kcal mol^{-1}) being present. However, relative to its keto form, the enol of acetaldehyde is about a hundred times more stable than the enol of acetone, because the less substituted aldehyde carbonyl is less stable than the more substituted ketone carbonyl.

Reaction

Keto–Enol Equilibria

$$H-CH_2CH \underset{K = 6 \times 10^{-7}}{\rightleftharpoons} H_2C=C \qquad \Delta G° \approx +8.5 \text{ kcal mol}^{-1}$$
$$(+35.5 \text{ kJ mol}^{-1})$$

Acetaldehyde	Ethenol (Vinyl alcohol)

$$H-CH_2CCH_3 \underset{K = 5 \times 10^{-9}}{\rightleftharpoons} H_2C=C \qquad \Delta G° \approx +11.3 \text{ kcal mol}^{-1}$$
$$(+47.3 \text{ kJ mol}^{-1})$$

Acetone   **2-Propenol**

Enol formation leads to deuterium exchange and stereoisomerization

What are some consequences of enol formation by tautomerism? One is that treatment of a ketone with traces of acid or base in D_2O solvent leads to the complete exchange of *all* the α-hydrogens.

Enolization Does Not Occur by Direct Proton Shift

The importance of catalysis of the keto–enol tautomerization is brought home by calculations of the energy of the hypothetical transition states for direct intramolecular proton shift between carbon and oxygen. For acetone, it is 53 kcal mol^{-1}! This would require temperatures of ~500°C for the reaction to occur without a catalyst. The calculated transition-state structure reveals unfavorable lengthening of the strong C–H and C=O bonds and bond-angle distortions resulting from the requirement for a four-membered ring atomic array.

Strained ring

Lengthened

Lengthened

Hypothetical transition state structure for the uncatalyzed enolization of acetone

Hydrogen–Deuterium Exchange of Enolizable Hydrogens

α-Hydrogens (exchange in D$_2$O)   Not α-hydrogens (do not exchange)

$$CH_3CCH_2CH_3 \xrightarrow{D_2O, DO^-} CD_3CCD_2CH_3$$

2-Butanone   **1,1,1,3,3-Pentadeuterio-2-butanone**

This reaction can be conveniently followed by ^{1}H NMR, because the signal for these hydrogens slowly disappears as each one is sequentially replaced by deuterium. In this way, the number of α-hydrogens present in a molecule can be readily determined.

Exercise 18-3

Formulate mechanisms for the base- and acid-catalyzed replacement of a single α-hydrogen in acetone by deuterium from D$_2$O.

Exercise 18-4

Write the products (if any) of deuterium incorporation by the treatment of the following compounds with D$_2$O–NaOD.

(a) Cycloheptanone
(b) 2,2-Dimethylpropanal
(c) 3,3-Dimethyl-2-butanone
(d)

Solved Exercise 18-5 | Working with the Concepts: Assigning NMR Signals of a Cyclic Ketone

The ^{1}H NMR spectrum of cyclobutanone consists of a quintet at $\delta = 2.00$ ppm and a triplet at $\delta = 3.13$ ppm. Assign the signals in this spectrum to the appropriate hydrogens in the molecule.

Strategy
Using the information in Section 17-3 regarding chemical shifts of hydrogens in ketones, apply the tools of NMR analysis that you acquired in Chapter 10.

Solution
- The structure of cyclobutanone implies an ^{1}H NMR spectrum displaying two signals: One for the four α-hydrogens (on C2 and C4) and another for the two β-hydrogens (on C3).
- We learned in Section 17-3 that α-hydrogens in carbonyl compounds are more deshielded than β-hydrogens. Furthermore, in accordance with the (N + 1) rule for spin–spin splitting (Section 10-7), the absorption for the α-hydrogens should appear as a triplet as a result of splitting by the two β-hydrogen neighbors; thus we may assign the triplet signal at $\delta = 3.13$ ppm to the four α-hydrogens.
- Conversely, splitting by four neighboring α-hydrogens should cause the signal for the β-hydrogens to appear as a quintet, consistent with the pattern at $\delta = 2.00$ ppm.

Exercise 18-6 | Try It Yourself

What would you expect to observe to change in the ^{1}H NMR spectrum of cyclobutanone upon treatment with D$_2$O–NaOD? (**Hint:** Deuterium atoms do not display signals in an ^{1}H NMR spectrum.)

Another consequence of enol formation is the ease with which stereoisomers at α-carbons interconvert. For example, treatment of *cis*-2,3-disubstituted cyclopentanones with mild base furnishes the corresponding trans isomers. The trans isomers are more stable for steric reasons.

Base-Catalyzed Isomerism of an α-Substituted Ketone

Model Building

>95%

The reaction proceeds through the enolate ion, in which the α-carbon is planar and, hence, no longer a stereocenter. Reprotonation from the side cis to the 3-methyl group results in the trans diastereomer (Sections 4-1 and 5-5).

The process of enolization makes it difficult to maintain optical activity in an aldehyde or ketone whose only stereocenter is an α-carbon. Why? As the (achiral) enol or enolate converts back into the keto form, a racemic mixture of R and S enantiomers is produced. For example, at room temperature, optically active 3-phenyl-2-butanone racemizes with a half-life of minutes in basic ethanol.

Racemization of Optically Active 3-Phenyl-2-butanone

(S)-3-Phenyl-
2-butanone

Enolate ion
(Achiral)

(R)-3-Phenyl-
2-butanone

In the preceding discussion, we have effected acid- or base-catalyzed keto–enol tautomerizations in aqueous or alcoholic solvent. You may have recognized that these are exactly the same conditions that lead to additions of water or alcohol to the carbonyl carbon to form hydrates or hemiacetals, respectively (Sections 17-6 and 17-7). Do these additions not take place here? Indeed they do, in competition with enolization. In both cases we focused only on the mechanisms that explained the transformations under discussion, omitting the competing reaction. The reason we could do so is that *all of these processes are reversible.* Thus, aldehydes or ketones will be in equilibrium both with their hydrates or hemiacetals as well as with the corresponding enols or enolates.

When we described the base-catalyzed stereoisomerization of an α-substituted ketone above, we restricted ourselves to writing only the relevant enolate mechanism. Simultaneous, reversible hemiacetal formation occurs, as described in Section 17-6, but it does not lead to any observable product. In contrast, reaction in the presence of acid drives the hemiacetal equilibrium in the forward direction by thermodynamically favorable acetal generation. We shall continue the practice of formulating mechanisms of new reactions solely to connect starting materials with the observed products. However, you will likely recognize other possible pathways, which, if you follow them, will either constitute reversible, "dead-end" equilibria, or are unlikely thermodynamically because they lead to high-energy species.

Exercise 18-7

Bicyclic ketone A rapidly equilibrates with a stereoisomer upon treatment with base, but ketone B does not. Explain.

A   B

In Summary Aldehydes and ketones are in equilibrium with their enol forms, which are roughly 10 kcal mol^{-1} less stable. Keto–enol equilibration is catalyzed by acid or base. Enolization allows for easy H–D exchange in D_2O and causes isomerization at stereocenters next to the carbonyl group.

18-3 | HALOGENATION OF ALDEHYDES AND KETONES

This section examines a reaction of the carbonyl group that can proceed through the intermediacy of either enols or enolate ions—halogenation. Aldehydes and ketones react with halogens at the α-carbon. In contrast with deuteration, which proceeds to completion with either acid or base, the extent of halogenation depends on whether acid or base catalysis has been used.

In the presence of acid, halogenation usually stops after the first halogen has been introduced, as shown in the following example.

Reaction

Acid-Catalyzed α-Halogenation of Ketones

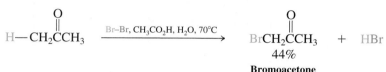

The rate of the acid-catalyzed halogenation is independent of the halogen concentration, an observation suggesting a rate-determining first step involving the carbonyl substrate. This step is enolization. The halogen then rapidly attacks the double bond to give an intermediate oxygen-stabilized halocarbocation. Subsequent deprotonation of this species furnishes the product.

Mechanism of the Acid-Catalyzed Bromination of Acetone

Mechanism

Step 1. Enolization (rate determining)

Step 2. Halogen attack

Step 3. Deprotonation

Why is further halogenation retarded? The answer lies in the requirement for enolization. To repeat halogenation, the halo carbonyl compound must enolize again by the usual acid-catalyzed mechanism. However, the electron-withdrawing power of the halogen reduces the basicity of the carbonyl oxygen and makes protonation, the initial step in enolization, more difficult than in the original carbonyl compound.

Halogenation Slows Down Enolization Because Protonation Is Disfavored

Less basic than unsubstituted ketone

$$BrCH_2CCH_3 \overset{H^+}{\rightleftharpoons} BrCH_2CCH_3$$

Electron withdrawing

Therefore, the singly halogenated product is not attacked by additional halogen until the starting aldehyde or ketone has been used up.

Base-mediated halogenation is entirely different. It proceeds instead by the formation of an enolate ion, which then attacks the halogen. Here the reaction continues until it *completely* halogenates the same α-carbon, leaving unreacted starting material (when insufficient halogen is employed). Why is base-catalyzed halogenation so difficult to stop at the stage of monohalogenation? The electron-withdrawing power of the halogen increases the acidity of the remaining α-hydrogens, accelerating further enolate formation and hence further halogenation.

Mechanism of Halogenation of an Enolate Ion

$$\overset{-:\overset{..}{O}:}{\underset{R}{C}}=CH_2 + Br—Br$$

$$\downarrow$$

$$\overset{:O:}{RCCH_2Br} + Br^-$$

More acidic than unsubstituted ketone

Exercise 18-8

Write the products of the acid- and base-catalyzed bromination of cyclohexanone.

In Summary Halogenation of aldehydes and ketones in acid can proceed selectively to the monohalocarbonyl compounds. In base, *all* α-hydrogens are replaced before another molecule of starting material is attacked.

18-4 | ALKYLATION OF ALDEHYDES AND KETONES

We have seen that enolates may be generated from ketones and aldehydes by treatment with bases, such as LDA (Sections 7-8 and 18-1). The nucleophilic α-carbon of the enolate may participate in S_N2 alkylation reactions with suitable haloalkanes, forming a new carbon–carbon bond in the process. This section presents some features of enolate alkylation and compares it with a related process, alkylation of a species called an *enamine*.

Alkylation of enolates can be difficult to control

In principle, the alkylation of an aldehyde or a ketone enolate is nothing more than a nucleophilic substitution.

Alkylation of an Aldehyde or Ketone

α-Carbon

$$\xrightarrow[\text{(e.g., LDA)}]{\text{Base}} \xrightarrow[S_N2]{R-X}$$

α-Hydrogen   Enolate ion   New carbon–carbon bond

In practice, however, the reaction may be complicated by several factors. The enolate ion is a fairly strong base. Therefore, alkylation is normally feasible using only halomethanes or primary haloalkanes; otherwise, E2 elimination converting the haloalkane into an alkene (Section 7-8) becomes a significant process. Other side reactions can present additional difficulties. Alkylation of aldehydes usually fails because their enolates undergo a highly favorable condensation reaction, which is described in the next section. Even alkylation of ketones may be problematic. You may think, for example, that strict application of conditions for formation of a kinetic enolate (addition of the ketone to LDA at low temperature) might permit clean alkylation at the less substituted α-carbon. However, the product ketone *is a proton source* that may be deprotonated by unreacted starting enolate to give a new enolate ready for a second alkylation (Section 18-1). Repetition of this cycle leads to various regio-isomeric overalkylation products. The reaction of 2-methylcyclohexanone with iodomethane illustrates these complications.

Products of Alkylation of 2-Methylcyclohexanone

Regioisomeric monoalkylation products

Polyalkylation products

Under some circumstances, ketones may be alkylated successfully. In the virtually ideal example below, the ketone possesses only one α-hydrogen and the primary allylic halide is an excellent S_N2 substrate (Section 14-3). The base in this example is sodium hydride, NaH (Section 8-6).

Successful Alkylation of a Ketone

2-Methyl-1-phenyl-1-propanone

2,2,5-Trimethyl-1-phenyl-4-hexen-1-one

Exercise 18-9

The reaction with base of the compound shown in the margin gives three isomeric products $C_8H_{12}O$. What are they? (**Hint:** Try intramolecular alkylations. **Caution:** As stated in Section 18-1, enolate alkylation "normally" proceeds only at carbon. Is the present case "normal"?)

Exercise 18-10

C-alkylation of cyclohexanone enolate with 3-chloropropene (Section 18-1) is much faster than the corresponding reaction with 1-chloropropane. Explain. (**Hint:** See Section 14-3.) What product(s) would you expect from reaction of cyclohexanone enolate with (**a**) 2-bromopropane and (**b**) 2-bromo-2-methylpropane? (**Hint:** See Chapter 7.)

Enamines afford an alternative route for the alkylation of aldehydes and ketones

Section 17-9 showed that the reaction of secondary amines such as azacyclopentane (pyrrolidine) with aldehydes or ketones produces enamines. As the following resonance forms indicate, the nitrogen substituent renders the enamine carbon–carbon double bond electron rich. Furthermore, the dipolar resonance contribution gives rise to significant nucleophilicity at the remote carbon, even though the enamine is neutral. As a result, electrophiles may attack at this position. Let us see how this attack may be used to synthesize alkylated aldehydes and ketones.

Azacyclopentane (Pyrrolidine)

Resonance in Enamines

Carbon is nucleophilic

Ethenamine

Exposure of enamines to haloalkanes results in alkylation at the nucleophilic carbon to produce an iminium salt. Upon aqueous work-up, iminium salts hydrolyze by a mechanism that is the reverse of the one formulated for imine formation in Section 17-9. The results are a new alkylated aldehyde or ketone and the original secondary amine.

Alkylation of an Enamine

Alkylated at α-carbon

$CH_3CH_2CCH_2CH_3$ $CH_3CH_2CCH(CH_3)_2$

3-Pentanone   **Enamine**   **An iminium salt**   **2-Methyl-3-pentanone**

Exercise 18-11

Formulate the mechanism for the final step of the sequence just shown: hydrolysis of the iminium salt.

How does the alkylation of an enamine compare with the alkylation of an enolate? Enamine alkylation is far superior, because it minimizes double or multiple alkylation: Under the conditions of the process, the iminium salt formed after the first alkylation is relatively stable and unable to react with additional haloalkane. Enamines can also be used to prepare alkylated aldehydes, as shown here. (We shall see in the next section that aldehyde enolates undergo a new reaction called the *aldol condensation* and, therefore, cannot be alkylated readily.)

1. $\overset{N}{\underset{H}{}}$
2. CH_3CH_2Br
3. H^+, H_2O

$(CH_3)_2CHCH$ $(CH_3)_2CCH$
                        |
                   CH_3CH_2
                   67%

2-Methylpropanal   **2,2-Dimethylbutanal**

Exercise 18-12

Alkylations of the enolate of ketone A are very difficult to stop before dialkylation occurs, as illustrated here. Show how you would use an enamine to prepare monoalkylated ketone B.

$$\text{A} \xrightarrow[\text{2. BrCH}_2\text{CO}_2\text{C}_2\text{H}_5]{\text{1. LDA, THF}} \text{(94\%)} \qquad \text{B}$$

In Summary Enolates give rise to alkylated derivatives upon exposure to haloalkanes. In these reactions, control of the extent and the position of alkylation, when there is a choice, may be a problem. Enamines derived from aldehydes and ketones undergo alkylation to the corresponding iminium salts, which can hydrolyze to the alkylated carbonyl compounds.

18-5 ATTACK BY ENOLATES ON THE CARBONYL FUNCTION: ALDOL CONDENSATION

We have seen the dual functional capability of carbonyl compounds: electrophilic at the carbonyl carbon, potentially nucleophilic at the adjacent α-carbon. In this section we introduce one of the most frequently employed carbon–carbon bond-forming strategies: attack of an enolate ion at a carbonyl carbon. The product of this process is a β-hydroxy carbonyl compound. Subsequent elimination of water may occur, leading to α,β-unsaturated aldehydes and ketones. The next three sections describe these reactions in detail, and Real Life 18-1 illustrates them in a biological context.

Aldehydes undergo base-catalyzed condensations

Addition of a small amount of dilute aqueous sodium hydroxide to acetaldehyde at low temperature initiates the conversion of the aldehyde into a dimer, 3-hydroxybutanal, with the common name *aldol* (from *ald*ehyde alco*hol*). Upon heating, this hydroxyaldehyde dehydrates to give the final product, the α,β-unsaturated aldehyde *trans*-2-butenal. This reaction is an example of the **aldol condensation.** The aldol condensation is general for aldehydes and, as we shall see, sometimes succeeds with ketones as well. We first describe its mechanism before turning to its uses in synthesis.

Aldol Condensation Between Two Molecules of Acetaldehyde

The aldol condensation highlights the two most important facets of carbonyl group reactivity: enolate formation and attack of nucleophiles at a carbonyl carbon. The base (hydroxide) is not strong enough to convert all of the starting aldehyde into the corresponding enolate ion, but it does bring about an equilibrium between the aldehyde and a small amount of enolate. Because the ion is formed in the presence of a large excess of aldehyde, its nucleophilic α-carbon can attack the carbonyl group of another acetaldehyde molecule. Protonation of the resulting alkoxide furnishes 3-hydroxybutanal.

Mechanism of Aldol Formation

Step 1. Enolate generation

$$HO:^- \quad + \quad H-CH_2-CH \overset{:O:}{\underset{}{\|}} \quad \rightleftharpoons \quad H_2C=C \overset{:\overset{..}{O}:^-}{\underset{H}{\|}} \quad + \quad H\overset{..}{O}H$$

Small equilibrium concentration of enolate

Step 2. Nucleophilic attack

Electrophilic carbonyl carbon

$$CH_3\overset{:O:}{\underset{}{C}}H \; + \; CH_2=C\overset{:\overset{..}{O}:^-}{\underset{H}{}} \; \rightleftharpoons \; CH_3\overset{-:\overset{..}{O}:}{\underset{H}{C}} - CH_2CH\overset{:O:}{\underset{}{\|}}$$

Nucleophilic α-carbon

New carbon–carbon bond

Step 3. Protonation

$$CH_3\overset{:\overset{..}{O}:^-}{\underset{H}{C}} - CH_2CH\overset{:O:}{\underset{}{\|}} + H\overset{..}{-O}H \rightleftharpoons CH_3\overset{:\overset{..}{O}H}{\underset{H}{C}} - CH_2CH\overset{:O:}{\underset{}{\|}} + HO:^-$$

50–60%

3-Hydroxybutanal (Aldol)

Mechanism

Animation

ANIMATED MECHANISM:
Aldol condensation–dehydration

> **Reminder**
>
> We are depicting only the mechanism that describes the formation of observed product. Keep in mind that carbonyl hydration and keto–enol tautomerization are also occurring simultaneously. However, they are reversible dead-end alternatives that do not enter into the scheme (see also Section 18-2).

Note that hydroxide ion functions as a catalyst in this reaction. The last two steps of the sequence drive the initially unfavorable equilibrium toward product, but the overall reaction is not very exothermic. The aldol is formed in 50–60% yield and does not react further when its preparation is carried out at low temperature (5°C).

Exercise 18-13

Give the structure of the hydroxyaldehyde product of aldol condensation at 5°C of each of the following aldehydes: **(a)** propanal; **(b)** butanal; **(c)** 2-phenylacetaldehyde; **(d)** 3-phenylpropanal.

Exercise 18-14

Can benzaldehyde undergo aldol condensation? Why or why not?

At elevated temperature, the aldol is converted into its enolate ion. Elimination of hydroxide ion, normally a poor leaving group, is driven thermodynamically by the formation of the conjugated, relatively stable final product. The net result of this transformation is a hydroxide-catalyzed dehydration of the aldol. We describe the overall process of aldol reaction followed by dehydration as **aldol condensation.** Recall (Section 17-9) that a condensation is a reaction that combines two molecules into one with the elimination of (typically) a molecule of water.

Animation

ANIMATED MECHANISM:
Aldol condensation–dehydration

Mechanism of Aldol Dehydration

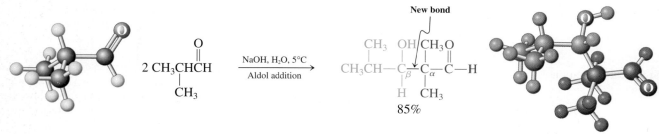

What makes the aldol condensation synthetically useful? It couples two carbonyl compounds, with carbon–carbon bond formation, to furnish a product with a carbonyl group and either an alcohol or an alkene function. Thus, at low temperatures, its outcome is a β-hydroxycarbonyl compound:

2-Methylpropanal   **3-Hydroxy-2,2,4-trimethylpentanal**

At higher temperatures, however, it furnishes an α,β-unsaturated carbonyl compound, which has considerable synthetic utility (Sections 18-9 through 18-11):

Heptanal   **(Z)-2-Pentyl-2-nonenal**

Exercise 18-15

Give the structure of the α,β-unsaturated aldehyde product of aldol condensation of each of the aldehydes in Exercise 18-13.

Ketones can undergo aldol condensation

So far, we have discussed only aldehydes as substrates in the aldol condensation. What about ketones? Treatment of acetone with base does indeed lead to some 4-hydroxy-4-methyl-2-pentanone, but the conversion is poor because of an unfavorable equilibrium with starting material.

Aldol Formation from Acetone

94%

6%

4-Hydroxy-4-methyl-2-pentanone

OH   O
| ||
$CH_3C-CH_2CCH_3$
|
CH_3

↓ NaOH, H_2O, Δ

H_3C
     \
      C=CHCCH_3
     / ||
H_3C O

80%

4-Methyl-3-penten-2-one

+

H_2O (removed)

The lesser driving force of the aldol reaction of ketones is due to the greater stability of a ketone than an aldehyde [about 3 kcal mol^{-1} (12.5 kJ mol^{-1})]. As a result, the aldol addition of ketones is endothermic. To drive the reaction forward, we can extract the product alcohol continuously from the reaction mixture as it is formed. Alternatively, under more vigorous conditions, dehydration and removal of water move the equilibrium toward the α,β-unsaturated ketone (see margin).

Solved Exercise 18-16 | **Working with the Concepts: Practicing Mechanisms of Aldol Reactions**

Formulate the mechanism for the aldol reaction of acetone.

Strategy
As we have recommended previously as a strategy for learning to solve problems, look for a pattern to follow elsewhere in the text or in your notes—in this case, the mechanism for the aldol reaction of acetaldehyde given earlier in this section. Replace the earlier substrate with the new one and follow the identical sequence of steps, *making sure that all your bond-forming and bond-cleavage processes occur at the correct locations with respect to the functional groups in each reacting molecule.*

Solution
- In order, (1) use the base to deprotonate an α-carbon of one acetone molecule to form an enolate ion.
- (2) Add the α-carbon of the negatively charged enolate to the carbonyl carbon of a second molecule of the ketone.
- (3) Protonate the resulting alkoxide oxygen to give the final hydroxyketone product:

Exercise 18-17 | **Try It Yourself**

The aldol reaction above is reversible. Propose a mechanism for the conversion of its product, 4-hydroxy-4-methyl-2-pentanone, back into two molecules of acetone in the presence of $^-$OH.

In Summary Treatment of enolizable aldehydes with catalytic base leads to β-hydroxy aldehydes at low temperature and to α,β-unsaturated aldehydes upon heating. The reaction proceeds by enolate attack on the carbonyl function. Aldol addition to a ketone carbonyl group is energetically unfavorable. To drive the aldol condensation of ketones to product, special conditions have to be used, such as removal of the water or the aldol formed in the reaction.

18-6 | CROSSED ALDOL CONDENSATION

What happens if we try to carry out an aldol condensation between the enolate of one aldehyde and the carbonyl carbon of another? In such a situation, called **crossed aldol condensation,** mixtures ensue, because enolates of both aldehydes are present and may react with the carbonyl groups of either starting compound. For example, a 1:1 mixture of acetaldehyde and propanal gives the four possible aldol addition products in comparable amounts.

Result of Nonselective Crossed Aldol Reaction Between Acetaldehyde and Propanal

(All four reactions occur simultaneously)

1. Propanal enolate adds to acetaldehyde.

$$CH_3CH \quad + \quad CH_3CH=CH \quad \longrightarrow \quad CH_3C-CHCH$$

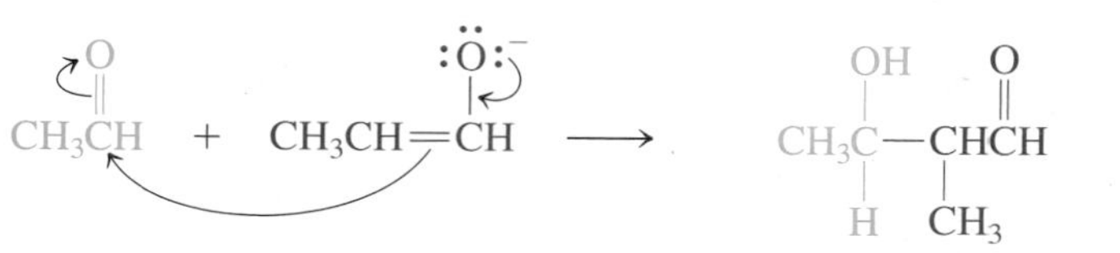

3-Hydroxy-2-methylbutanal

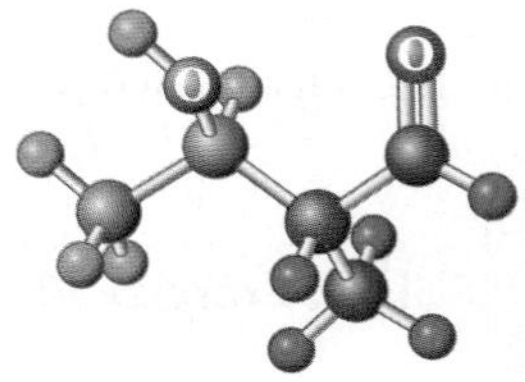

2. Acetaldehyde enolate adds to propanal.

$$CH_3CH_2CH \quad + \quad CH_2=CH \quad \longrightarrow \quad CH_3CH_2C-CH_2CH$$

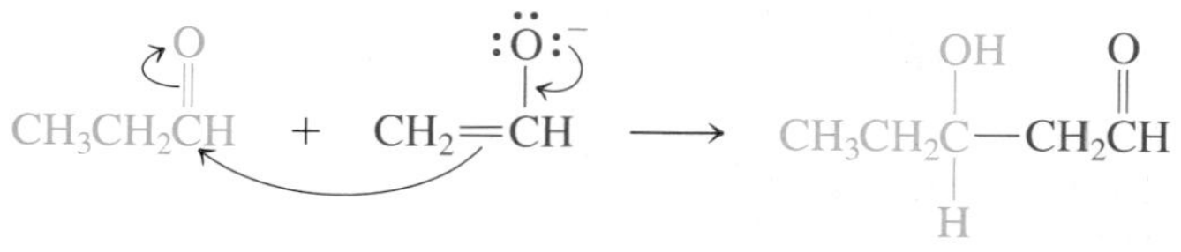

3-Hydroxypentanal

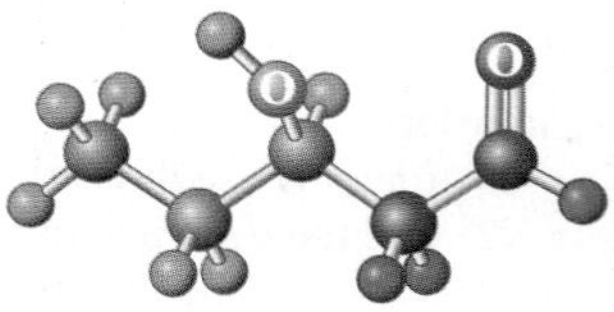

3. Acetaldehyde enolate adds to acetaldehyde.

$$CH_3CH \quad + \quad CH_2=CH \quad + \quad CH_3CH_2CH \quad \longrightarrow \quad CH_3C-CH_2CH$$

Not involved

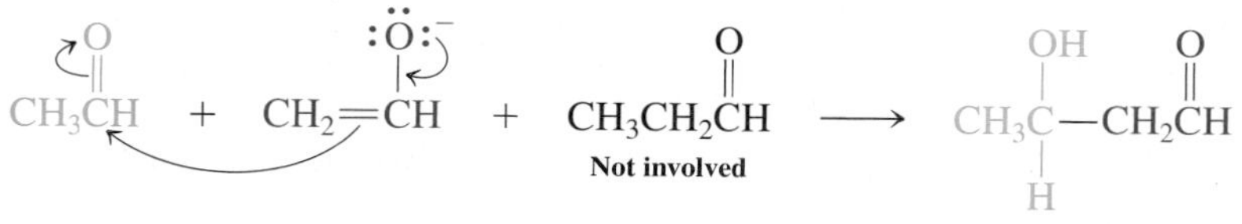

3-Hydroxybutanal

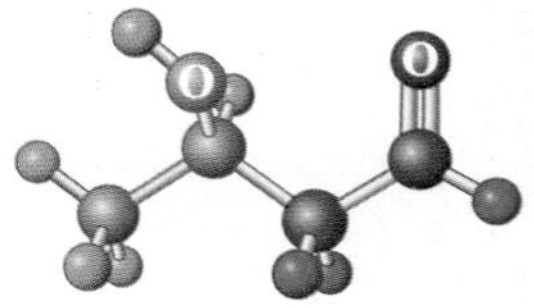

4. Propanal enolate adds to propanal.

$$CH_3CH_2CH \quad + \quad CH_3CH=CH \quad + \quad CH_3CH \quad \longrightarrow \quad CH_3CH_2C-CHCH$$

Not involved

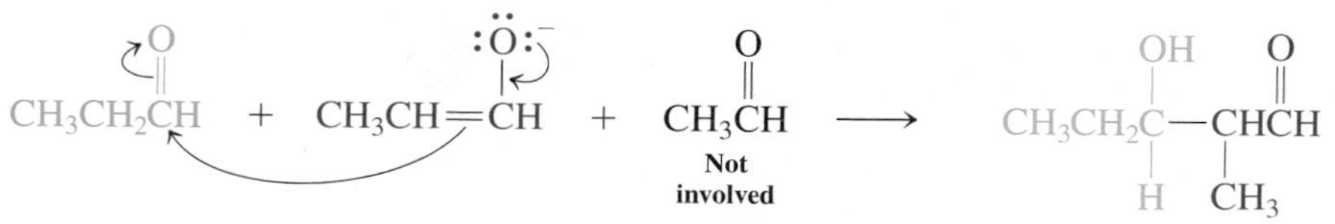

3-Hydroxy-2-methylpentanal

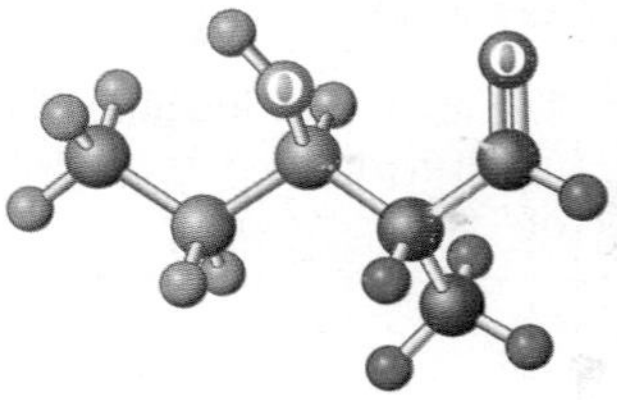

Can we ever efficiently synthesize a single aldol product from the reaction of two different aldehydes? We can, when one of the aldehydes has *no enolizable hydrogens,* because two of the four possible condensation products cannot form. We add the enolizable aldehyde slowly to an excess of the nonenolizable reactant in the presence of base. As soon as the enolate of the addend is generated, it is trapped by the other aldehyde.

REAL LIFE: BIOLOGY AND MEDICINE 18-1 Stereoselective Aldol Reactions in Nature and in the Laboratory: "Organocatalysis"

Glucose ($C_6H_{12}O_6$, Section 24-1), the most common sugar in nature, is absolutely essential to all life on Earth. In many species it is the sole energy source for the major organs, including the entire central nervous system. All species synthesize glucose from smaller molecules. The process is called gluconeogenesis, and it employs a crossed aldol condensation (catalyzed by enzymes appropriately named aldolases) to construct a carbon–carbon bond between two three-carbon precursors.

A primary amino group in the enzyme (supplied by the amino acid lysine, Section 26-1) condenses with the carbonyl group of the monophosphate ester of 1,3-dihydroxyacetone to form an iminium ion. This species isomerizes to an enamine (Section 17-9), whose nucleophilic carbon (red) attacks the aldehyde carbonyl carbon (blue) of 2,3-dihydroxypropanal (glyceraldehyde) 3-phosphate. This nitrogen analog of a crossed aldol condensation proceeds with 100% stereoselectivity: The enzyme causes the two substrate molecules to approach one another in a single, highly ordered geometry. Bond formation occurs exclusively to link the bottom faces of both the (red) enamine carbon and the carbonyl carbon (blue) to each other, resulting in the illustrated stereochemistry in the product. Hydrolysis of the ensuing iminium salt gives the phosphate ester of the six-carbon sugar fructose, which is converted to glucose in a subsequent step.

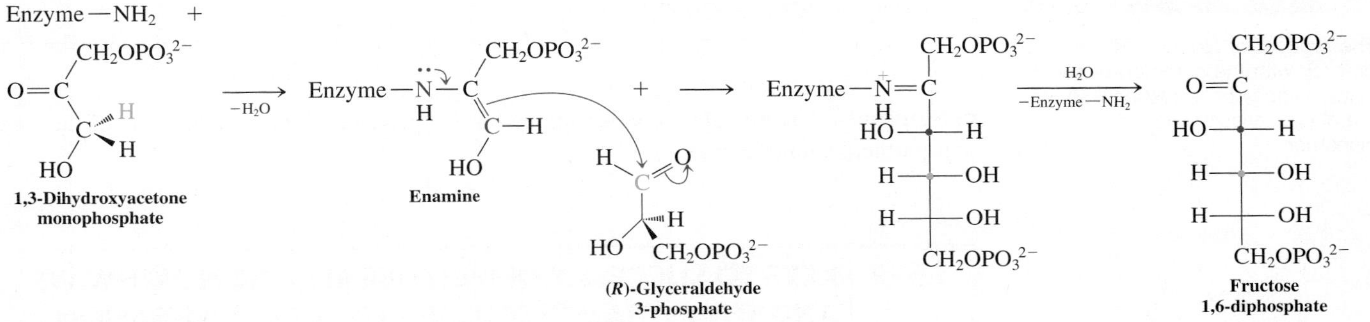

Nature's strategy of transient attachment of a stereodirecting enzyme via an enamine has been exploited by chemists using much simpler molecules, such as the natural amino acid proline (see also Section 26-1). Proline features the secondary amine function necessary for enamine formation, in addition to a carboxy substituent that renders the molecule chiral, the stereocenter exhibiting the *S* configuration. In the example shown below, proline catalyzes the enantioselective aldol addition of acetone to the −CHO group of 2-methylpropanal. As in the enzyme-catalyzed process, the sequence begins with enamine formation, now between acetone and proline. Addition of this enamine to the aldehyde follows. In this step, hydrogen bonding between the carboxy proton and the aldehyde oxygen defines the transition state geometry and leads to the *R* product almost exclusively; hydrolysis regenerates the proline. Notice that proline, as a true catalyst, is not used up in the reaction and is added in only small amounts to effect this transformation. This type of catalysis, termed "organocatalysis," is distinct from the more frequent acid or metal catalysis we have encountered previously (for enantiocatalysis by metals, see Section 12-3 and Real Life 12-2).

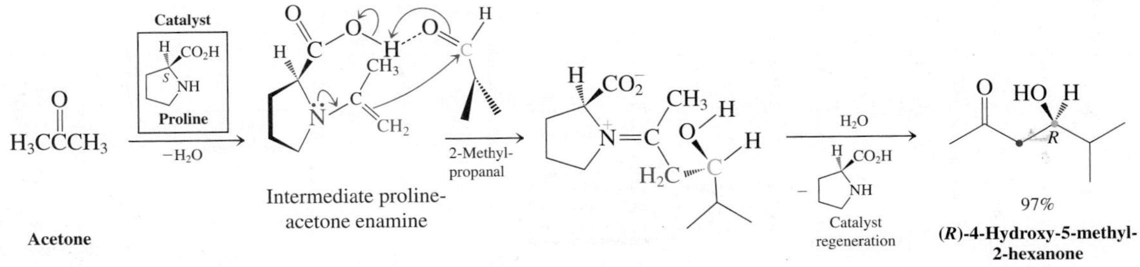

1,3-Dihydroxyacetone monophosphate — **Enamine** — **(*R*)-Glyceraldehyde 3-phosphate** — **Fructose 1,6-diphosphate**

Acetone — Intermediate proline-acetone enamine — **(*R*)-4-Hydroxy-5-methyl-2-hexanone**

A Successful Crossed Aldol Condensation

$$CH_3\underset{\underset{\displaystyle CH_3}{|}}{\overset{\overset{\displaystyle CH_3}{|}}{C}}CHO \quad + \quad CH_3CH_2CHO \xrightarrow{NaOH, H_2O, \Delta} CH_3\underset{\underset{\displaystyle CH_3}{|}}{\overset{\overset{\displaystyle CH_3}{|}}{C}}CH=\underset{\underset{\displaystyle CH_3}{|}}{C}CHO \quad + \quad H_2O$$

Added slowly

2,2-Dimethyl-propanal   **Propanal**   **2,4,4-Trimethyl-2-pentenal**
(No α-hydrogens)

Exercise 18-18

Show the likely products of the following aldol condensations.

(a) [benzaldehyde structure]—CHO   + CH_3CHO (b) 2 [cyclohexanecarbaldehyde structure]—CHO   (reacts with itself)

(c) $CH_2=CHCHO + CH_3CH_2CHO$

The product of Exercise 18-18, part (a), with the common name cinnamaldehyde, is responsible for the characteristic flavor of cinnamon.

In Summary Crossed aldol condensations furnish product mixtures unless one of the reaction partners cannot enolize.

18-7 | KEYS TO SUCCESS: COMPETITIVE REACTION PATHWAYS AND THE INTRAMOLECULAR ALDOL CONDENSATION

It is possible to carry out aldol condensations between two carbonyl groups located *in the same molecule*. Such a reaction is called an **intramolecular aldol condensation** and is important in the synthesis of cyclic compounds, especially five- and six-membered rings.

Intramolecular condensations are favored entropically

Heating a dilute solution of hexanedial with aqueous base results in the formation of the cyclic product 1-cyclopentenecarbaldehyde. Here, one end of the dialdehyde functions as the nucleophilic enolate component (after deprotonation), the other as the electrophilic carbonyl partner. After the initial aldol addition, subsequent dehydration furnishes the product.

$$\underset{\textbf{Hexanedial}}{HC CH_2CH_2CH_2CH_2CH} \xrightarrow[\text{Aldol addition}]{KOH, H_2O, \Delta} \left[\text{cyclic intermediate} \right] \xrightarrow[\text{Dehydration}]{-H_2O} \underset{\substack{\textbf{62\%} \\ \textbf{1-Cyclopentenecarbaldehyde}}}{\text{product}}$$

Site of enolate formation

One molecule transforms into one molecule

New bond

Why does the intramolecular aldol take place in preference to the intermolecular alternative? There are two factors to consider in answering this question. The first is kinetic, that is, relating to the relative speed of the two processes: The intramolecular reaction is faster, because of the kinetic entropic advantage of having both reacting centers in the same molecule (Section 9-6). Using high dilution conditions accentuates this effect by minimizing the rate at which two dialdehyde molecules encounter one another in the reaction mixture. The second is more important and addresses the thermodynamics of the equilibria associated with the two types of reaction, in particular the entropic contribution to the $\Delta G°$ of each ($\Delta G° = \Delta H° - T\Delta S°$; Section 2-1). The intramolecular transformation is more thermodynamically favorable because the associated $\Delta S° \approx 0$: Overall one molecule (the dialdehyde) converts to one molecule (the aldol addition product). In contrast, its intermolecular counterpart features two molecules of dialdehyde transforming into just one, the aldol adduct, with a negative $\Delta S°$ and therefore a less favorable $\Delta G°$ (margin). Since the aldol addition is reversible, the dimeric species will be in equilibrium with the starting dialdehyde, from where it will eventually convert to the cyclic hydroxyaldehyde. Dehydration of the latter to the final product follows.

Intramolecular ketone condensations: Reversibility favors unstrained ring formation

The entropic advantages of intramolecularity are particularly noticeable in intramolecular ketone condensations, transformations that provide a ready source of cyclic and bicyclic α,β-unsaturated ketones. Since ketones can have two enolizable α-carbons, several aldol products may be possible. However, the reversibility of the reaction ensures that usually the least strained ring is generated, typically one that is five or six membered. Thus, reaction of 2,5-hexanedione results in condensation of the terminal methyl (C1 functioning as the enolate carbon) and the C5 carbonyl group to furnish 3-methyl-2-cyclopentenone.

Intramolecular Aldol Condensation of a Dione

2,5-Hexanedione   3-Methyl-2-cyclopentenone

42%

New bond

What happens if the base removes a proton from C3 instead? Does this enolate proceed to react with the C5 carbonyl to give a three-membered ring (margin)? The answer is yes and no: Nucleophilic attack may occur kinetically, but formation of the strained ring renders this reaction pathway prohibitively unfavorable energetically. The equilibrium lies way on the side of starting material, which therefore goes on to the five-membered ring product unperturbed by this alternative option. What about the intermolecular alternative? Again, yes and no. As in the dialdehyde case described above, this trajectory leads to higher-energy species than those attained by cyclization, and the reversibility of the transformation ensures that the intramolecular route prevails.

The foregoing discussion amplifies the importance of reversibility in the chemistry of aldehydes and ketones. Thus, in the aldol condensation, in addition to competing keto–enol tautomerizations and solvent-to-carbonyl additions (Section 18-2), the reaction itself can, in principle, lead to different products. When these alternatives are all reversible, the final outcome is governed by thermodynamics. In accord with Le Chatelier's principle (Section 17-5), as the species that converts to product by the thermodynamically most favored pathway is depleted, the equilibria of all alternative processes will shift to replenish it. Ideally, the result will be only one end molecule; otherwise the method would be of limited use. We shall see that this scenario is common to other transformations involving carbonyl compounds.

Intermolecular Aldol Addition

Two molecules transform into one

Model Building

Unfavorable Strained Ring Formation from 2,5-Hexanedione

Highly endothermic:
Not observed

REAL LIFE: NATURE 18-2 | Absorption of Photons by Unsaturated Aldehydes Enables Vision

trans-**Retinal**   *cis*-**Retinal**

Vitamin A (retinol, see Section 14-7 and the Chapter Opening) is an important nutritional factor in vision. Enzyme-catalyzed oxidation converts it into *trans*-retinal. *Trans*-retinal is present in the light-receptor cells of the human eye, but before it can fulfill its biological function, it has to be isomerized by an enzyme, retinal isomerase, to give *cis*-retinal. *cis*-Retinal behaves like a "loaded spring" by virtue of the steric hindrance associated with the cis double bond (Section 11-5). The molecule fits well into the active site of a protein called opsin (approximate molecular weight 38,000). As shown on the facing page, *cis*-retinal reacts with one of the amine substituents of opsin to form the imine *rhodopsin,* the light-sensitive chemical unit in the eye. The electronic spectrum of rhodopsin, with a λ_{max} at 506 nm ($\epsilon = 40,000$), indicates the presence of a protonated imine group.

When a photon strikes rhodopsin, the *cis*-retinal part isomerizes extremely rapidly, in only picoseconds (10^{-12} s), to the trans isomer. This isomerization induces a geometric change, which appears to severely disrupt the snug fit of the original molecule in the protein cavity. Within nanoseconds (10^{-9} s), a series of new intermediates form from this photoproduct, accompanied by conformational changes in the protein structure, followed by eventual hydrolysis of the ill-fitting retinal unit. This sequence initiates a nerve impulse

perceived by us as light. The *trans*-retinal is then reisomerized to the cis form by retinal isomerase and re-forms rhodopsin, ready for another photon. What is extraordinary about this mechanism is its sensitivity, which allows the eye to register as little as one photon impinging on the retina.

Studies using nerve cells have shown that the geometric change brought about by isomerization of rhodopsin-bound retinal physically opens a "pore" in a special kind of protein called an ion channel, which is located in the cell membrane. Positively charged ions flow through the open pore into the cell, constituting an electrical current. The difference between this process and that of vision is that the geometric change in rhodopsin does not directly open the ion-channel pore in visual system cells. Rather, the dissociation of *trans*-retinal from rhodopsin causes the activation of a substance called G protein, discovered by Gilman* and Rodbell.[†] G protein is bound to the inner surface of the cell membrane and, upon activation, opens the ion channel to initiate the electrical signal of a nerve impulse.

*Professor Alfred G. Gilman (b. 1941), University of Texas Southwestern Medical School, Nobel Prize 1994 (physiology or medicine).
[†]Dr. Martin Rodbell (1925–1998), National Institutes of Health, Nobel Prize 1994 (physiology or medicine).

Exercise 18-19

Predict the outcome of intramolecular aldol condensations of the following compounds.

(a) Cyclodecane-1,5-dione

(b) $C_6H_5\overset{O}{\overset{\|}{C}}(CH_2)_2\overset{O}{\overset{\|}{C}}CH_3$

(c)

(d) 2,7-Octanedione

2-(3-Oxobutyl)cyclohexanone

Exercise 18-20

The intramolecular aldol condensation of 2-(3-oxobutyl)cyclohexanone (margin) can, in principle, lead to four different compounds (ignoring stereochemistry). Draw them and suggest which one would be the most likely to form. (**Hint:** Build models!)

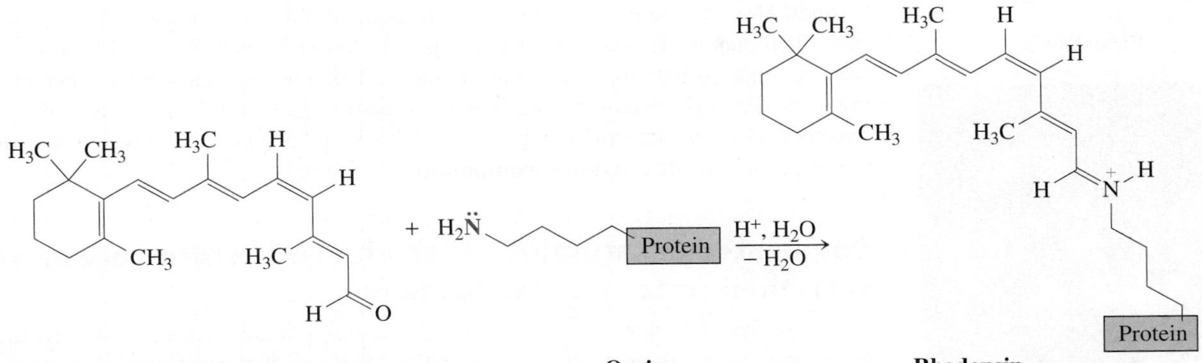

cis-Retinal   **Opsin**   **Rhodopsin**

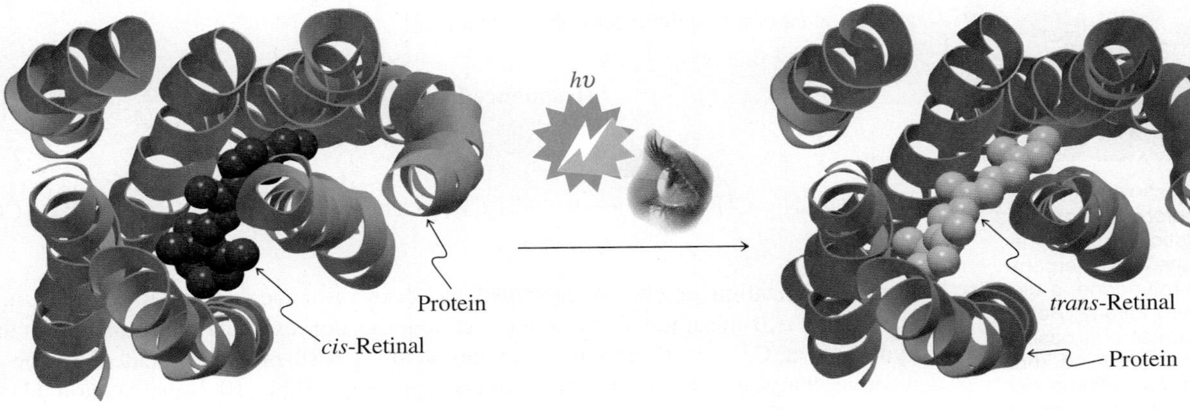

**Rhodopsin before
photon absorption**

**Rhodopsin after
photon absorption**

Exercise 18-21

The following molecules can be dissected by retrosynthetic (single or double) aldol addition or condensation disconnections. Show the starting materials that would make these targets in a forward reaction.

(a) 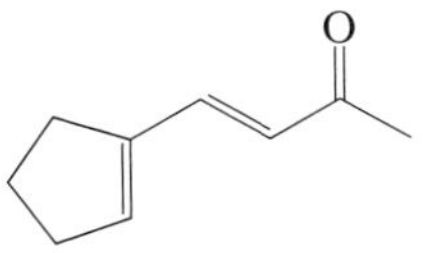   (b) 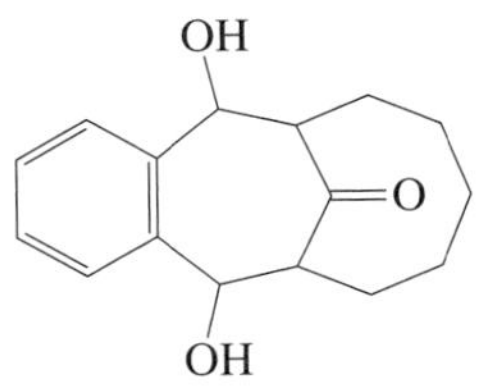   (c)

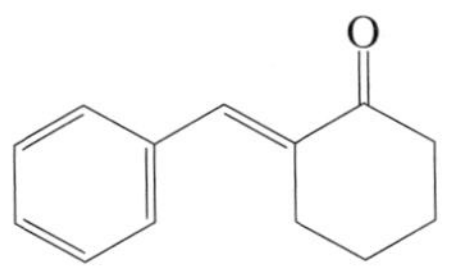

In Summary Intramolecular aldol condensation succeeds with both aldehydes and ketones. It can be highly regioselective and gives the least strained cycloalkenones.

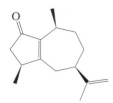

Rotundone

Black pepper (shown in the photo as the fruit before drying) owes its characteristic aroma to the presence of the sesquiterpene (Section 4-7) rotundone, a bicyclic α,β-unsaturated ketone. Rotundone is also present in basil, oregano, and thyme, and imparts a peppery aroma to some red wines. A possible synthesis of rotundone making use of the intramolecular aldol condensation is the subject of Problem 46.

2-Butenal

Reaction

18-8 | PROPERTIES OF α,β-UNSATURATED ALDEHYDES AND KETONES

We have seen that the products of aldol condensation are α,β-unsaturated aldehydes and ketones. How do the properties of α,β-unsaturated aldehydes and ketones compare with the individual characteristics of their two types of double bonds? We shall find that their chemistry in some situations is a simple composite of the two and under other circumstances involves the α,β-unsaturated carbonyl, or **enone**, functional group as a whole. As later chapters will indicate, this complex reactivity is quite typical of molecules with two functional groups, or **difunctional compounds.**

Conjugated unsaturated aldehydes and ketones are more stable than their unconjugated isomers

Like conjugated dienes (Section 14-5), α,β-unsaturated aldehydes and ketones are stabilized by resonance, resulting in the delocalization of the partial positive charge on the carbonyl carbon. As a consequence, and as shown in the electrostatic potential map of 2-butenal in the lower margin, the double bond is relatively electron poor (green, not red; compare Section 12-3), and the β-carbon is quite electrophilic (blue).

Resonance Forms of 2-Butenal

$$\left[CH_3CH=CH-\overset{\overset{\displaystyle :\ddot{O}:}{|}}{CH} \longleftrightarrow CH_3CH\overset{}{=\!\!=}CH\overset{\overset{\displaystyle :\ddot{O}:^-}{|}}{\underset{+}{CH}} \longleftrightarrow CH_3\overset{+}{CH}-CH=\overset{\overset{\displaystyle :\ddot{O}:^-}{|}}{CH} \right]$$

The delocalization of charge described by these resonance forms is reflected in NMR spectra of α,β-unsaturated carbonyl compounds. For example, in *trans*-2-butenal, the hydrogen at C3 (the β-carbon) resonates at δ = 6.88 ppm, deshielded by the partial positive charge relative to a typical alkene hydrogen (Table 10-3 and Section 11-4). The hydrogen at C2, at δ = 6.15 ppm, is also relatively deshielded by the effect of the neighboring carbonyl function (Section 17-3), but much less so. Consistent with these data, the carbon NMR spectrum reveals C3 at δ = 154.3 ppm and C2 at δ = 134.6 ppm. The contribution of the dipolar enolate form suggests that the double bond character of the C=O bond should be attenuated. Indeed, the associated IR stretching frequency in *trans*-2-butenal is at lower wavenumber, $\tilde{\nu} = 1699\ cm^{-1}$, compared to its saturated counterpart butanal, $\tilde{\nu} = 1761\ cm^{-1}$.

Unconjugated β,γ-unsaturated carbonyl compounds rearrange readily to their conjugated isomers. The carbon–carbon double bond is said to "move into conjugation" with the carbonyl group, as the following example shows.

Isomerization of a β,γ-Unsaturated Carbonyl Compound to a Conjugated System

$$\overset{\gamma}{CH_2}\!\!=\!\!\overset{\beta}{CH}CH_2\overset{\overset{\displaystyle O}{\|}}{CH} \xrightarrow{\ H^+\ or\ HO^-,\ H_2O\ } CH_3\overset{\beta}{CH}\!\!=\!\!\overset{\alpha}{CH}\overset{\overset{\displaystyle O}{\|}}{CH}$$

3-Butenal   **2-Butenal: more stable because of conjugation**

The isomerization can be acid or base catalyzed. In the base-catalyzed reaction, the intermediate is the conjugated dienolate ion, which is reprotonated at the carbon terminus.

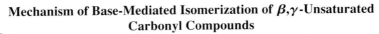

Mechanism of Base-Mediated Isomerization of β,γ-Unsaturated Carbonyl Compounds

Mechanism

CH₂=CH—C—CH + HO:⁻   ⇌   **Deprotonation**

[CH₂=CH—CH—CH ⟷ CH₂=CH—CH=CH ⟷ :CH₂—CH=CH—CH] + H—OH   ⇌   **Reprotonation**

Dienolate ion

Protonation here gives conjugated aldehyde

CH₃—CH=CH—CH + HO:⁻

Exercise 18-22

Propose a mechanism for the acid-catalyzed isomerization of 3-butenal to 2-butenal. (**Hint:** An intermediate is 1,3-butadien-1-ol.)

α,β-Unsaturated aldehydes and ketones undergo the reactions typical of their component functional groups

α,β-Unsaturated aldehydes and ketones undergo many reactions that are perfectly predictable from the known chemistry of the carbon–carbon and carbon–oxygen double bonds. For example, hydrogenation by palladium on carbon gives the saturated carbonyl compound.

H₂, Pd–C, CH₃CO₂CH₂CH₃ (ethyl acetate solvent) →

95%

The carbon–carbon π system also undergoes electrophilic additions. For example, bromination furnishes a dibromocarbonyl compound (compare Section 12-5).

CH₃CH=CHCCH₃   — Br–Br, CCl₄ →   CH₃CHCHCCH₃

60%

3-Penten-2-one   **3,4-Dibromo-2-pentanone**

The carbonyl function can undergo the usual addition reactions (Section 17-5). Thus, nucleophilic addition of hydroxylamine (Section 17-9) results in the expected condensation product (see, however, the next section).

NH₂OH, H⁺ / −H₂O →

4-Phenyl-3-buten-2-one   **Oxime**
                                    (m.p. 115°C)

Exercise 18-23

Propose a synthesis of 3-phenyl-2-methyl-1-propanol starting from propanal.

In Summary α,β-Unsaturated aldehydes and ketones are more stable than their nonconjugated counterparts. Either base or acid catalyzes interconversion of the isomeric systems. Reactions typical of alkenes and carbonyl compounds are also characteristic of α,β-unsaturated aldehydes and ketones.

18-9 | CONJUGATE ADDITIONS TO α,β-UNSATURATED ALDEHYDES AND KETONES

We now show how the conjugated carbonyl group of α,β-unsaturated aldehydes and ketones can enter into reactions that involve the entire functional system. These reactions are 1,4-additions of the type encountered with conjugated dienes, such as 1,3-butadiene (Section 14-6). The reactions proceed by acid-catalyzed, radical, or nucleophilic addition mechanisms, depending on the reagents.

The entire conjugated system takes part in 1,4-additions

Addition reactions in which only one of the π bonds of a conjugated system takes part are classified as 1,2-additions (compare Section 14-6). Examples are the additions of Br_2 to the carbon–carbon double bond and NH_2OH to the carbon–oxygen double bond of α,β-unsaturated aldehydes and ketones (Section 18-8).

1,2-Addition of a Polar Reagent A—B to a Conjugated Enone

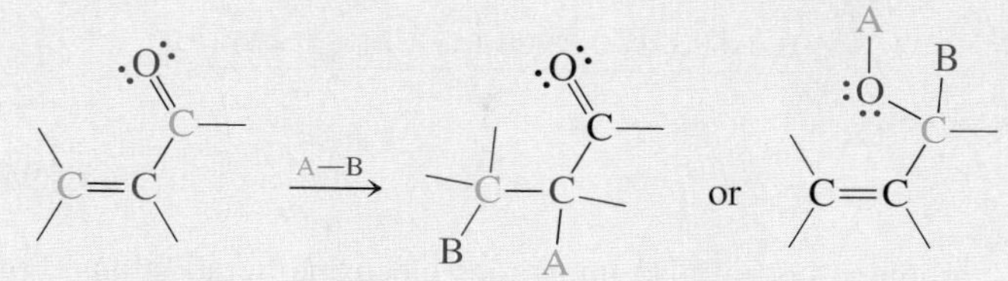

However, several reagents add to the conjugated π system in a 1,4-manner, a result also called **conjugate addition.** In these transformations, the nucleophilic part of a reagent attaches itself to the β-carbon, and the electrophilic part (most commonly, a proton) binds to the carbonyl oxygen. As the resonance forms and the resonance hybrid below illustrate, the β-carbon in an α,β-unsaturated carbonyl compound possesses a partial positive charge and is therefore a second electrophilic site in these molecules, in addition to the carbonyl carbon itself (recall Section 17-2). Addition of nucleophiles to the β-carbon is therefore not a surprising process.

Resonance Forms of α,β-Unsaturated Carbonyl Compounds

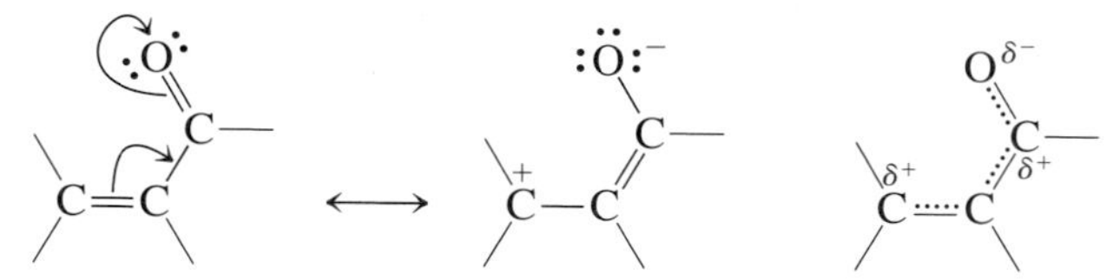

Resonance hybrid:
Both the carbonyl
and β-carbons are
electrophilic.

The initial product of conjugate addition to an α,β-unsaturated carbonyl compound is an enol, which subsequently rapidly tautomerizes to its keto form. Thus the end result *appears* to be that of 1,2-addition of Nu–H to the carbon–carbon double bond.

1,4-Addition of a Polar Reagent A–B to a Conjugated Enone

New bond to carbon

Oxygen and nitrogen nucleophiles undergo conjugate additions

Water, alcohols, amines, and similar nucleophiles undergo 1,4-additions, as the following examples (hydration and amination) show. Although these reactions can be catalyzed by acid or base, the products are usually formed faster and in higher yields with base. These processes are readily reversed at elevated temperatures.

β-Carbon: site of nucleophilic attack

3-Buten-2-one

4-Hydroxy-2-butanone

Reaction

Site of nucleophilic attack

75%

4-Methyl-3-penten-2-one

4-Methyl-4-(methylamino)-2-pentanone

What determines whether 1,2- or 1,4-additions will take place? With the nucleophiles shown, both processes are reversible. Usually the 1,4-products form because they are carbonyl compounds and are more stable than the species arising from 1,2-addition (hydrates, hemiacetals, and hemiaminals, Sections 17-6, 17-7, and 17-9). Exceptions include amine derivatives such as hydroxylamine, semicarbazide, or the hydrazines, for which 1,2-addition eventually leads to an imine product whose precipitation out of solution drives the equilibrium. Once more, the competition between 1,2- and 1,4-addition illustrates Le Chatelier's principle in action, a characteristic of carbonyl chemistry (Sections 18-2 and 18-7).

The mechanism of the base-catalyzed addition to conjugated aldehydes and ketones is direct nucleophilic attack at the β-carbon to give the enolate ion, which is subsequently protonated.

Mechanism of Base-Catalyzed Hydration of α,β-Unsaturated Aldehydes and Ketones

Mechanism

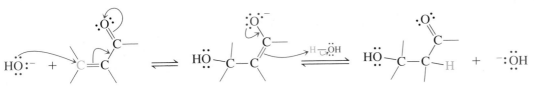

Exercise 18-24

Treatment of 3-chloro-2-cyclohexenone with sodium methoxide in methanol gives 3-methoxy-2-cyclohexenone. Write the mechanism of this reaction. (**Hint:** Start with a conjugate addition.)

The ability to form new bonds with nucleophiles at the β carbon makes α,β-unsaturated aldehydes and ketones extremely useful in synthesis. This ability extends to carbon nucleophiles, permitting the construction of new carbon–carbon bonds. The remainder of the chapter illustrates processes of this type.

Hydrogen cyanide also undergoes conjugate addition

Treatment of a conjugated aldehyde or ketone with cyanide in the presence of acid may result in attack by cyanide at the β-carbon, in contrast with cyanohydrin formation (Section 17-11). This transformation proceeds through a 1,4-addition pathway. The reaction includes protonation of the oxygen, then nucleophilic β-attack, and finally enol–keto tautomerization.

Exercise 18-25

Formulate the mechanism of the acid-catalyzed 1,4-addition of cyanide to 1-phenylpropenone (see margin).

In Summary α,β-Unsaturated aldehydes and ketones are synthetically useful building blocks in organic synthesis because of their ability to undergo 1,4-additions. Hydrogen cyanide addition leads to β-cyano carbonyl compounds; oxygen and nitrogen nucleophiles also can add to the β-carbon.

$$C_6H_5CCH{=}CH_2$$

1-Phenylpropenone

$\Big\downarrow$ KCN, H$^+$

$$C_6H_5CCHCH_2CN$$
$$\text{H}$$

67%

4-Oxo-4-phenyl-butanenitrile

18-10 1,2- AND 1,4-ADDITIONS OF ORGANOMETALLIC REAGENTS

Organometallic reagents may add to the α,β-unsaturated carbonyl function in either 1,2- or 1,4-fashion. Organolithium reagents, for example, react almost exclusively by direct nucleophilic attack at the carbonyl carbon.

Exclusive 1,2-Addition of an Organolithium

$$\underset{\text{H}_3\text{C}}{\overset{\text{H}_3\text{C}}{>}}C{=}CHCCH_3 \quad \xrightarrow[\text{2. H}^+, \text{H}_2\text{O}]{\text{1. CH}_3\text{Li, (CH}_3\text{CH}_2)_2\text{O}} \quad \underset{\text{H}_3\text{C}}{\overset{\text{H}_3\text{C}}{>}}C{=}CHCCH_3$$

4-Methyl-3-penten-2-one   81%

2,4-Dimethyl-3-penten-2-ol

Reactions of Grignard reagents with α,β-unsaturated aldehydes and ketones are not generally very useful, because Grignards may give 1,2-addition, 1,4-addition, or both, depending on the structures of the reacting species and conditions. Fortunately, another type of organometallic reagent, an organocuprate, is very effective in giving rise only to products of conjugate addition. **Organocuprates** have the empirical formula R_2CuLi, and they may be prepared by adding two equivalents of an organolithium reagent to one of copper(I) iodide, CuI.

Example of Preparation of an Organocuprate

$$2\ CH_3Li\ +\ CuI\ \rightarrow\ (CH_3)_2CuLi\ +\ LiI$$

Lithium dimethylcuprate

(An organocuprate reagent)

Organocuprates are highly selective in undergoing 1,4-addition reactions:

Exclusive 1,4-Addition of a Lithium Organocuprate

β-Carbon: site of nucleophilic attack

CH₃(CH₂)₅CH=CCH
|
CH₃

2-Methyl-2-nonenal

1. (CH₃)₂CuLi, THF, −78°C, 4 h
2. H⁺, H₂O

CH₃(CH₂)₅CHCHCH
| |
CH₃ CH₃ O (O on top right)

40%

2,3-Dimethylnonanal

The copper-mediated 1,4-addition reactions are thought to proceed through complex electron-transfer mechanisms. The first isolable intermediate is an enolate ion, which can be trapped by alkylating species, as shown in Section 18-4. Conjugate addition followed by alkylation constitutes a useful sequence for α,β-dialkylation of unsaturated aldehydes and ketones.

α,β-Dialkylation of Unsaturated Carbonyl Compounds

1. R₂CuLi
2. R'X

The following example illustrates this reaction.

1. (CH₃CH₂CH₂CH₂)₂CuLi, THF
2. CH₃I

→   CH₃ ... CH₂CH₂CH₂CH₃   +   CH₃ ... CH₂CH₂CH₂CH₃

84%, 4:1

trans- and *cis-***3-Butyl-2-methylcyclohexanone**

Solved Exercise 18-26	**Working with the Concepts: Conjugate Addition in Synthesis**

Show how you might synthesize [structure] from 3-methyl-2-cyclohexenone.

CH₃

Strategy
The fact that your starting material is an α,β-unsaturated ketone and your product has an additional bond to its β-carbon makes conjugate addition a reasonable place to start. However, for this problem, retrosynthetic analysis (working backward) is especially useful. The presence of an additional ring should suggest an intramolecular aldol condensation as a synthetic possibility. Analyze retrosynthetically first; then proceed in the forward direction.

Solution
• In the target molecule (on the next page, left), consider the C=C double bond of the α,β-unsaturated ketone (arrow) to be the result of an intramolecular aldol condensation between the α-carbon of a cyclohexanone and a side-chain aldehyde function (center structure).

Really?

The root stalks of the tropical plant *Zingiber zerumbet* have been used by Southeast Asian natives as food flavoring in local cuisine, and in herbal remedies for the treatment of inflammation and diarrhea. The juice from the flower heads serves as a shampoo and conditioner ("shampoo ginger"). The medicinal component is zerumbone, an anti-inflammatory and anti-cancer agent. It acts by covalently binding the mercapto (−SH) group of cysteine (Table 26-1) in the polypeptide chain of several regulatory proteins involved in causing inflammation and cancer. Bond formation occurs by conjugate addition to one of the enone units. The triene analog humulene (Problem 72 of Chapter 12), lacking the oxo function, is inactive. Many drugs employ an embedded enone function for attack on nucleophilic centers in biomolecules.

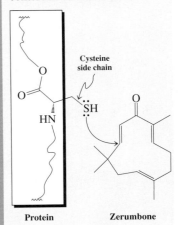

Cysteine side chain

HN   SH

Protein   **Zerumbone**

- Consider the bond connecting this side chain to the β-carbon of the cyclohexanone (center, arrow) as deriving from a conjugate addition to 3-methyl-2-cyclohexenone, the specified starting material:

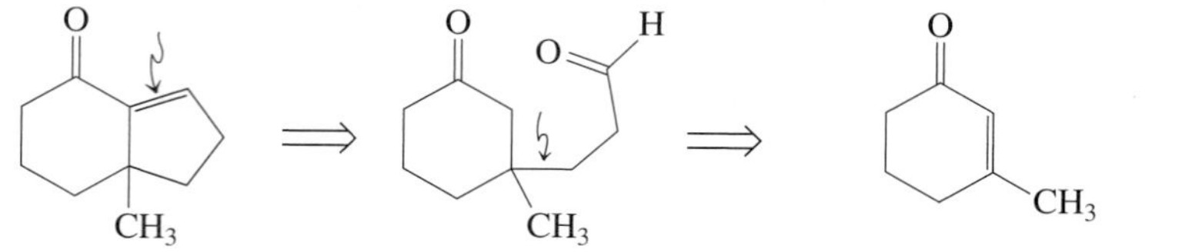

| **Target molecule** | **Ketoaldehyde** | **3-Methyl-2-cyclohexenone** |

- Before translating this retrosynthetic analysis into a synthesis in the forward direction, notice that the side chain, which we intend to use as a nucleophile for 1,4-addition, will have to be introduced as an organocuprate reagent, which in turn has to come from a halide via an organolithium species. However, it also contains a carbonyl function, which will need to be protected, so that the organolithium reagent can be prepared successfully (recall Section 17-8):

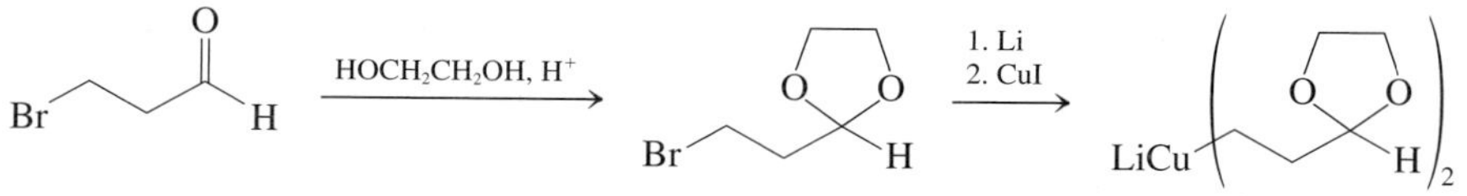

- The synthesis is completed by addition of the cuprate to the cyclohexenone, hydrolysis of the acetal to free the aldehyde group, and intramolecular aldol condensation:

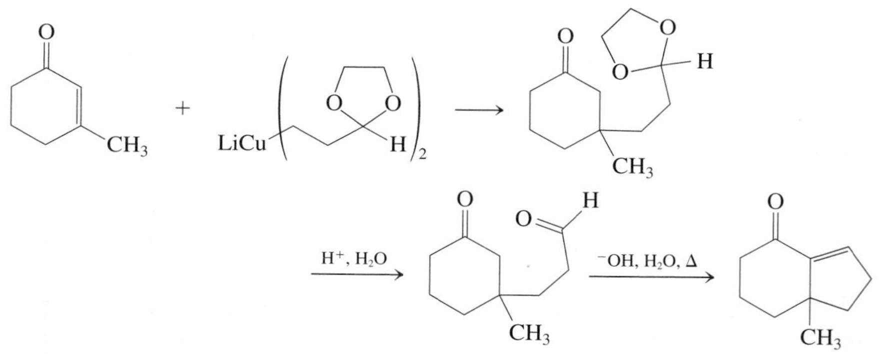

Note: The ketoaldehyde in the final step has more than one α-carbon that can be deprotonated by base. Examine all the aldol possibilities (see Section 18-7). Do you see why the condensation shown is the only one that is observed?

Exercise 18-27 | Try It Yourself

Design a synthesis of [structure] from 3-methyl-2-cyclohexenone. Work backward from the target!

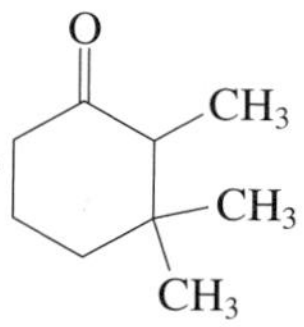

In Summary Organolithium reagents add 1,2 and organocuprate reagents add 1,4 to α,β-unsaturated carbonyl systems. The 1,4-addition process initially gives rise to a β-substituted enolate, which upon exposure to haloalkanes furnishes α,β-dialkylated aldehydes and ketones.

18-11 | CONJUGATE ADDITIONS OF ENOLATE IONS: MICHAEL ADDITION AND ROBINSON ANNULATION

Like other nucleophiles, enolate ions undergo conjugate additions to α,β-unsaturated aldehydes and ketones, in a reaction known as the **Michael* addition.**

Michael Addition

Reaction

64%

In this example, conditions have been chosen to favor formation of the thermodynamic enolate, the one with partial negative charge on the more highly substituted α-carbon (red; recall Section 18-1).

The mechanism of the Michael addition includes nucleophilic attack by the enolate ion on the β-carbon of the unsaturated carbonyl compound (the Michael "acceptor"), followed by protonation of the resulting enolate.

Mechanism of the Michael Addition: 1,4-Addition of an Enolate

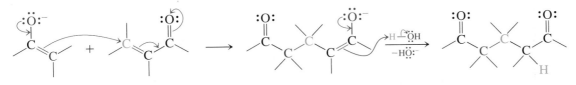

Mechanism

As the mechanism indicates, the reaction works because of the nucleophilic potential of the α-carbon of an enolate and the electrophilic reactivity of the β-carbon of an α,β-unsaturated carbonyl compound.

With some Michael acceptors, such as 3-buten-2-one, the products of the initial addition are capable of a subsequent intramolecular aldol condensation, which creates a new ring.

> Again, we depict only the mechanism that leads to product. Remember, however, that there are several reversible dead ends involving other enolates, hemiacetal formations, and aldol additions.

Michael Addition Followed by Intramolecular Condensation

3-Buten-2-one

54%   86%

*Professor Arthur Michael (1853–1942), Harvard University, Cambridge, MA.

Animation

ANIMATED MECHANISM:
Robinson annulation

The synthetic sequence of a Michael addition followed by an intramolecular aldol condensation is also called a **Robinson* annulation.**

Robinson Annulation

Michael addition

Aldol condensation

$+ \quad H_2O$

The Robinson annulation has found extensive use in the synthesis of polycyclic ring systems, including steroids and other natural products containing six-membered rings.

Solved Exercise 18-28	Working with the Concepts: Using Michael and Robinson Reactions in Synthesis

Propose a synthesis of

C_6H_5

using Michael or Robinson reactions.

Strategy
Use retrosynthetic analysis to identify reasonable starting materials. (**Hint:** *Always* make your first "disconnection" at the C=C double bond in the α,β-unsaturated ketone. Then proceed in a forward direction, following the steps presented in this section.)

Solution
• The C=C double bond in an α,β-unsaturated ketone may be formed from an aldol condensation. Therefore, we begin with its "disconnection" to identify a compound that can be converted to the desired final product. Break the bond, and place a carbonyl group ("=O") at the former β-carbon. Why? In the forward direction, aldol condensation converts the carbonyl group in the starting compound into the β-carbon of the product. The result of this disconnection is the ketoaldehyde below:

Disconnect

CH_3

C_6H_5 C_6H_5 H

In the retrosynthetic analysis, the original β-carbon becomes a carbonyl group

Remember WHIP

*W*hat

*H*ow

*I*nformation

*P*roceed

As in Exercise 18-26, consider and rule out all alternative aldol addition options.

• This compound is an example of a "1,5-dicarbonyl" compound: If you begin counting at either one of the two carbonyl carbons, the other one is the fifth atom away. This type of structure may be prepared by Michael addition of an enolate to an α,β-unsaturated compound. We identify possible starting materials retrosynthetically, by "disconnecting" at either α–β bond between the carbonyl carbons. There are two possible disconnections:

---

*Sir Robert Robinson (1886–1975), Oxford University, England, Nobel Prize 1947 (chemistry).

First Possible Disconnection

O   O
‖   ‖
H₃C   α   β   α   H
              |
             C₆H₅

⇓

O   O
‖   ‖
H₃C   α   CH₃   +   H₂C   β   H
                              |
                             C₆H₅

Second Possible Disconnection

O   O
‖   ‖
H₃C   α   β   α   H
              |
             C₆H₅

⇓

O   O
‖   ‖
H₃C   β   CH₂   +   α   H
                          |
                         C₆H₅

- We have two options from which to choose: Michael addition of the enolate of acetone to 2-phenylpropenal (above, left), and Michael addition of the enolate of 2-phenylacetaldehyde to 3-buten-2-one (above, right). Which is better? Michael additions are normally carried out under basic conditions, and we know (Section 18-5) that aldehydes undergo aldol condensation with base much more readily than do simple ketones. Thus, the second option above risks interference from the aldol reaction between two molecules of 2-phenylacetaldehyde. The first option is better because the enolate of acetone will undergo Michael addition to the aldehyde much more easily than aldol condensation with itself. Thus, we have all the pieces we need to put the synthesis together in the forward direction—it is a Robinson annulation:

O   O
‖   ‖   CH₃CH₂O⁻Na⁺,   O   O
H₃C   CH₃   +   H₂C   H   —————————→   ‖   ‖
                        | CH₃CH₂OH   H₃C   H
                       C₆H₅   Michael |
                                    addition   C₆H₅

———————————————→   O=⟨cyclohexenone⟩—C₆H₅
Intramolecular aldol condensation

Exercise 18-29 | Try It Yourself

Propose retrosynthetic Michael [(a)] or Robinson steps [(b), (c)] to the following compounds. [**Hint:** For (b) and (c), identify and unravel the cyclohexenone ring in a reverse annulation sequence.]

(a)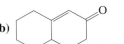

(b) [bicyclic octalone structure]

(c) [polycyclic structure with O]

**Subunit of the antibacterial
platensimycin (margin)**

**Antibacterial Synthesis
by Robinson Annulation**

The power of the Robinson annulation in generating molecular complexity is manifest in the synthesis of the new antibacterial platensimycin [Problem 29, part (c)].

HOOC

HO

HN   OH
         O
O

Platensimycin

The molecule was isolated in 2006 from a strain of *Streptomyces* bacteria and shown to be active against drug-resistant bacterial strains (see Real Life 20-2). It inhibits the growth of cell membranes in the pathogen by a new mechanism that does not occur in humans, and it is therefore essentially nontoxic. A hold-up for advancement to clinical trials is its fast clearance from the body, and current intensive work is devoted to modifying the drug's skeleton to improve its physiological half-life.

In Summary The Michael addition results in the conjugate addition of an enolate ion to give dicarbonyl compounds. The Robinson annulation reaction combines a Michael addition with a subsequent intramolecular aldol condensation to produce new cyclic enones.

THE BIG PICTURE

The aldol condensation and the synthetic sequences that can be built around it provide us with powerful methods for synthesizing complex organic molecules. These reactions provide ways to form carbon–carbon bonds at three sites: the α-, the β-, and the carbonyl carbons. We can construct elaborate acyclic compounds using these processes, and we can also close rings. In the problems that follow, take note of the remarkable variety of compounds that can be formed just using the chemical properties of the carbonyl group and its adjacent atoms.

Continuing our survey of carbonyl chemistry, we turn next to the carboxylic acids, to see how these more highly oxidized systems compare. The principle of enhanced acidity for hydrogens attached to atoms adjacent to a carbonyl carbon is illustrated by the acidity of the O–H hydrogen in these compounds. In addition, the OH portion of the carboxylic acid function is a potential leaving group. The next chapter presents a new method by which substitution reactions can occur, one that is central to the chemistry of all compounds in the carboxylic acid family: the addition–elimination pathway.

WORKED EXAMPLES: INTEGRATING THE CONCEPTS

18-30. Using Retrosynthesis to Design Aldol-Condensation-Based Molecule Construction

Propose as many syntheses as you can of the bicyclic ketone in the margin, using at least one aldol condensation in each. You may use any starting materials you like.

SOLUTION

To approach this problem, it is useful to evaluate the information we have been given. The target compound is a saturated ketone, but aldol condensations give rise to α,β-*unsaturated* carbonyl compounds. Therefore, we can infer that any synthesis we develop must finish with hydrogenation of a C=C double bond in conjugation with the carbonyl group. Based on the target structure, three α,β-unsaturated ketones can be identified as possible precursors.

Having identified these three structures, we can proceed to formulate the compounds that, upon undergoing intramolecular aldol condensation, will produce each one. Retrosynthetically, disconnect each C=C double bond; each disconnected β-carbon derives from a carbonyl group. That gives the following structures, shown in the same left-to-right order as their corresponding aldol enone products above.

All that remains is to complete the syntheses in the forward direction, including necessary reagents. We do so on the next page for the first of these three starting compounds; the other two can be completed in exactly the same way.

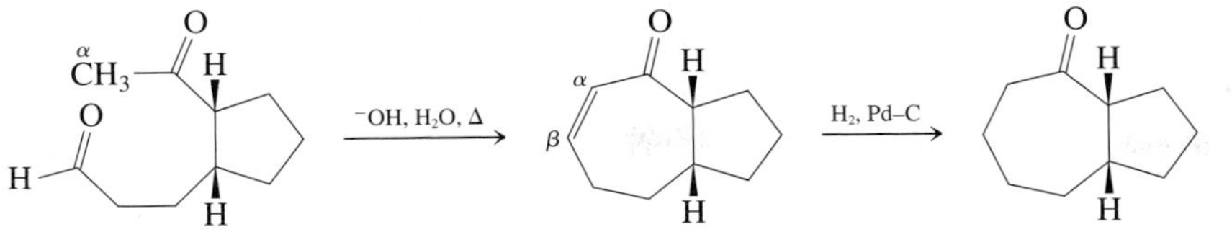

18-31. Building Rings Using Robinson Annulations

The Robinson annulation sequence is a powerful method for the construction of six-membered rings. It is therefore no surprise that it has been widely applied in steroid synthesis. Beginning with the bicyclic ketone shown in the margin, propose a synthesis that makes use of one or more Robinson annulations for the steroid shown here.

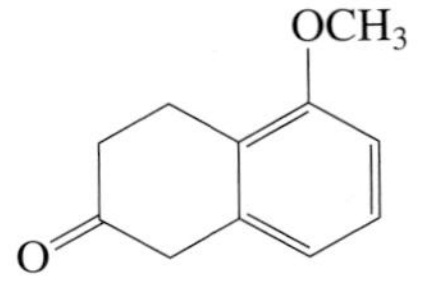

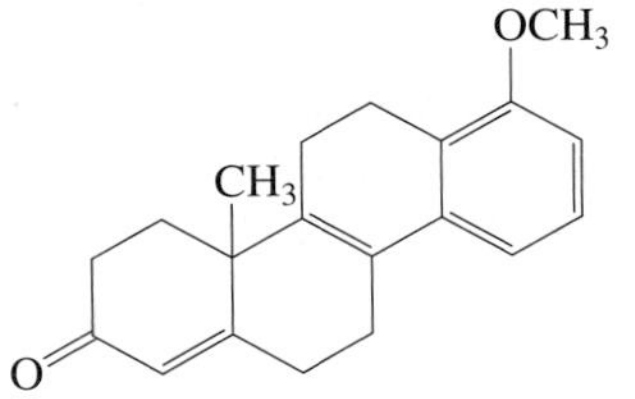

SOLUTION

The Robinson annulation combines Michael addition to an α,β-unsaturated ketone with intramolecular aldol condensation (Section 18-11) to afford a cyclohexenone. Retrosynthetic analysis (Section 8-9) of the target molecule leads to the disconnection of two bonds in ring A: the carbon–carbon double bond, by a retro-aldol condensation, and a single bond by a retro-Michael addition. The Robinson annulation for the construction of ring A is closely related to the example in Section 18-11, which condenses 2-methylcyclohexanone with 3-buten-2-one.

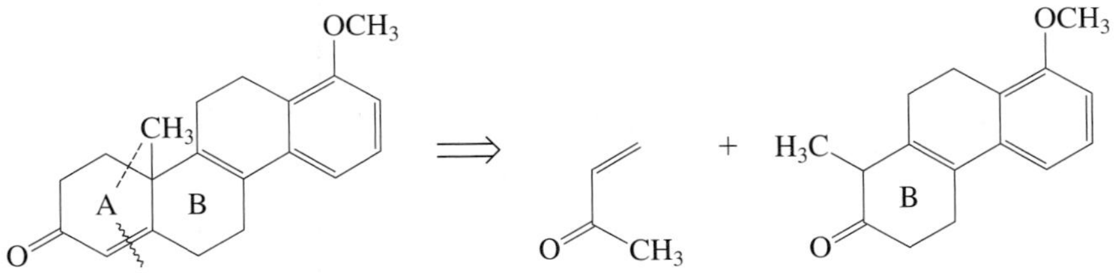

At this stage, we have simplified the target from a tetracyclic to a tricyclic molecule. However, the latter is a β,γ-unsaturated ketone, *not* an obvious product of Robinson annulation for the formation of ring B. Postponing consideration of this problem for the moment, can we formulate a Robinson annulation that transforms the initial ketone into a tricyclic product with the *same molecular skeleton* as that of the product of the last reaction? Yes we can: In fact, we can construct a double-bond isomer:

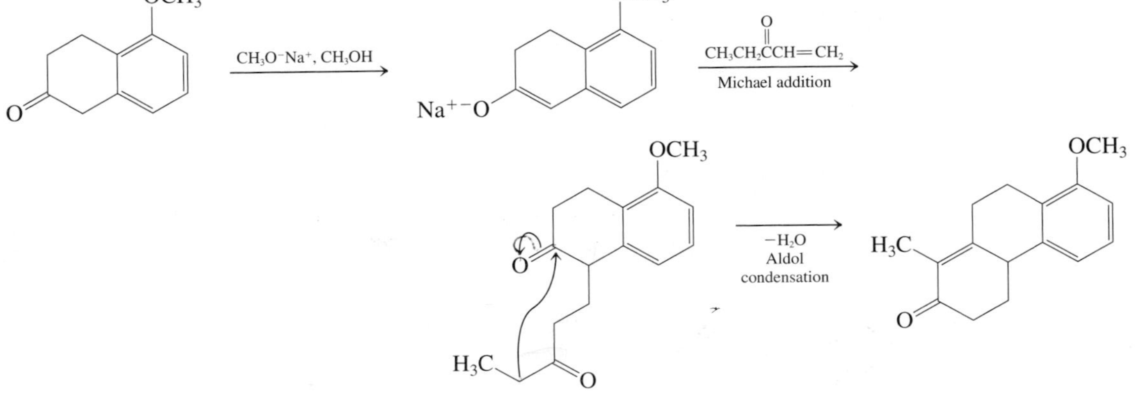

We are quite close to a complete solution. All that remains is to connect the tricyclic ketone that we have just prepared with the necessary second Robinson sequence. To do so, we *think mechanistically*, asking ourselves, "What is the structure of the enolate needed to initiate the annulation?" It is shown at the right in the following reaction, and we note that it is an *allylic* enolate ion (Section 14-4), for which we can write a second resonance contributor. It therefore may be derived by deprotonation of the tricyclic ketone in the preceding reaction at the allylic (γ) position.

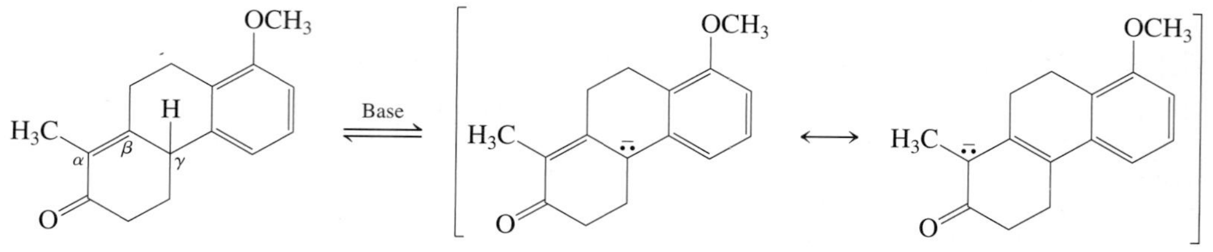

Resonance-stabilized allylic enolate anion

What does this realization mean in practical terms? We can carry out the needed second Robinson annulation starting directly from the preceding α,β-unsaturated ketone; it is not necessary to first prepare the β,γ-unsaturated ketone revealed in our initial retrosynthetic disconnection, because they give the same enolate upon deprotonation. The complete synthesis is shown below.

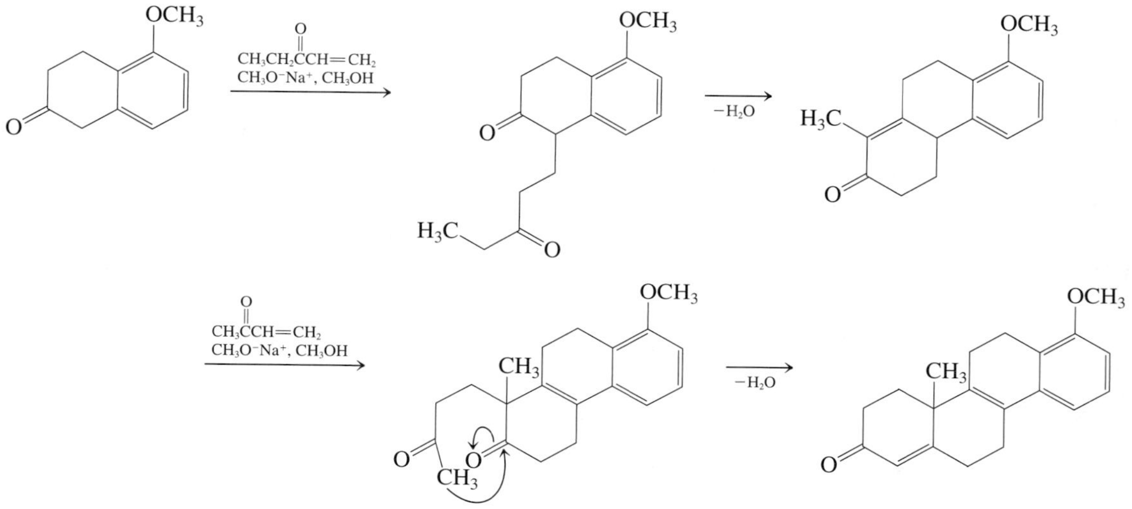

A final note: You may have recognized that the allylic anion discussed above is also benzylic and hence enjoys additional resonance stabilization by conjugation with the benzene ring. Benzylic resonance will be discussed in detail in Section 22-1. For more practice with Robinson annulations, go to Problems 59, 63, and 64.

New Reactions

Synthesis and Reactions of Enolates and Enols

1. **Enolate Ions (Section 18-1)**

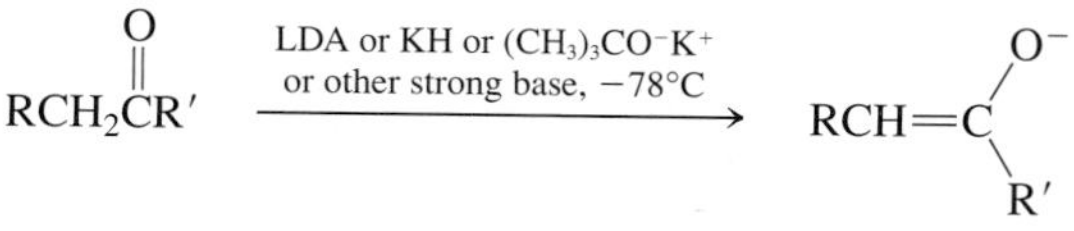

Enolate ion

2. **Keto–Enol Equilibria (Section 18-2)**

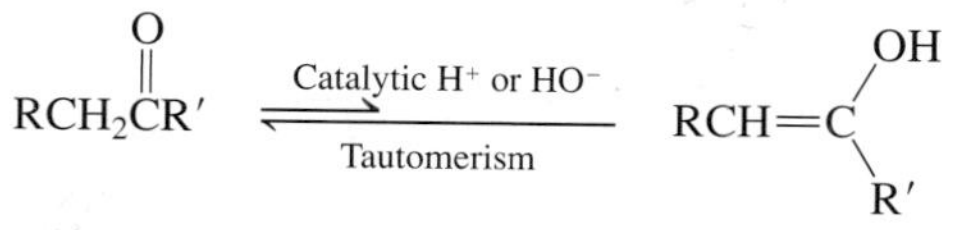

3. Hydrogen–Deuterium Exchange (Section 18-2)

$$RCH_2\overset{O}{\overset{\|}{C}}R' \xrightarrow{\text{D}_2\text{O, DO}^- \text{ or D}^+} RCD_2\overset{O}{\overset{\|}{C}}R'$$

4. Stereoisomerization (Section 18-2)

5. Halogenation (Section 18-3)

$$RCH_2\overset{O}{\overset{\|}{C}}R' \xrightarrow[-\text{HX}]{\text{X}_2,\ \text{H}^+} R\overset{O}{\underset{X}{\overset{\|}{C}}}HCR'$$

6. Enolate Alkylation (Section 18-4)

$$RCH{=}\overset{O^-}{\underset{R'}{C}} \xrightarrow[-\text{X}^-]{\text{R}''\text{X}} R\overset{O}{\underset{R''}{\overset{\|}{C}}}HCR'$$

S$_N$2 reaction: R″X must be methyl or primary halide

7. Enamine Alkylation (Section 18-4)

S$_N$2 reaction: R′X must be methyl, primary, or secondary halide

8. Aldol Condensations (Sections 18-5 through 18-7)

Crossed aldol condensation (one aldehyde not enolizable)

Ketones

Intramolecular aldol condensation

Unstrained rings preferred

Reactions of α,β-Unsaturated Aldehydes and Ketones

9. Hydrogenation (Section 18-8)

$$H_2, Pd, CH_3CH_2OH$$

10. Addition of Halogen (Section 18-8)

$$X_2, CCl_4$$

11. Condensations with Amine Derivatives (Section 18-8)

$$ZNH_2$$

Z = OH, RNH, etc.

Conjugate Additions to α,β-Unsaturated Aldehydes and Ketones

12. Hydrogen Cyanide Addition (Section 18-9)

$$KCN, H^+$$

13. Water, Alcohols, Amines (Section 18-9)

$$H_2O$$

$$ROH$$

$$RNH_2$$

14. Organometallic Reagents (Section 18-10)

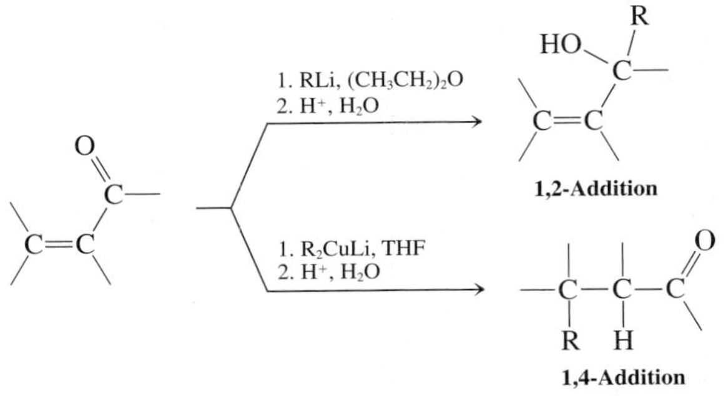

1,2-Addition

1,4-Addition

Additions of RMgX may be 1,2 or 1,4, depending on reagent and substrate structure.

Cuprate additions followed by enolate alkylations

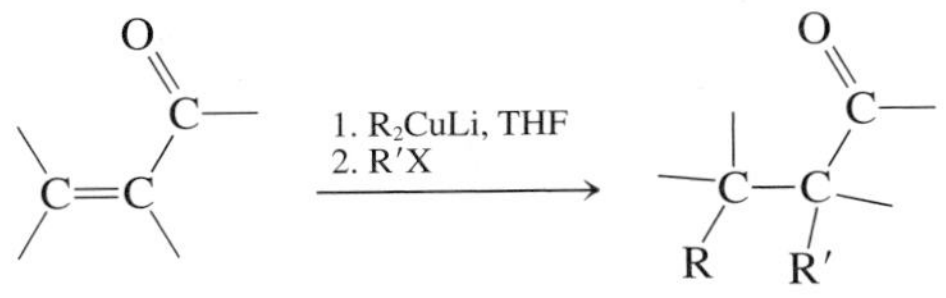

15. Michael Addition (Section 18-11)

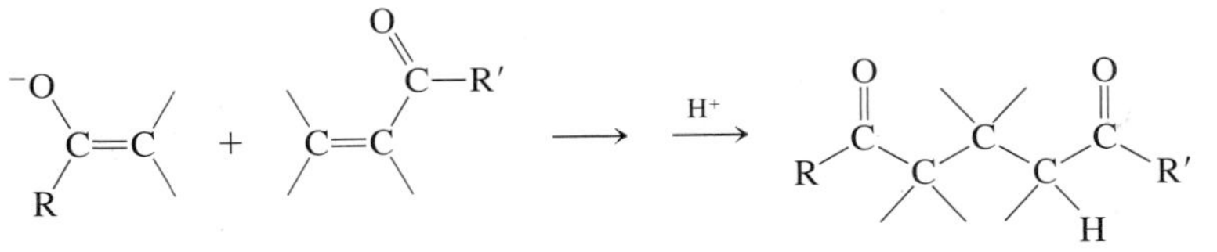

16. Robinson Annulation (Section 18-11)

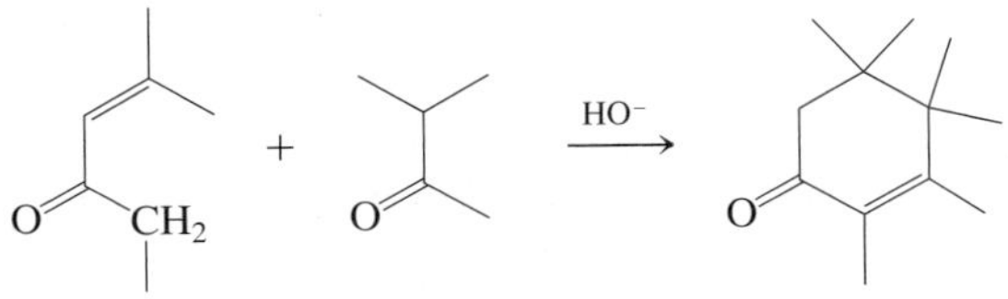

Important Concepts

1. Hydrogens next to the carbonyl group (**α-hydrogens**) are acidic because of the electron-withdrawing nature of the functional group and because the resulting **enolate ion** is resonance stabilized.

2. Electrophilic attack on enolates can occur at both the α-carbon and the oxygen. Haloalkanes usually prefer the α-carbon. Protonation of the oxygen leads to **enols.**

3. **Enamines** are neutral analogs of enolates. Resonance donation of the nitrogen lone pair imparts nucleophilic character on the remote double bond carbon, which can be alkylated to give iminium cations that hydrolyze to aldehydes and ketones on aqueous work-up.

4. Aldehydes and ketones are in equilibrium with their tautomeric enol forms; the **enol–keto conversion** is catalyzed by acid or base. This equilibrium allows for facile α-deuteration and stereochemical equilibration.

5. **α-Halogenation** of carbonyl compounds may be acid or base catalyzed. With acid, the enol is halogenated by attack at the double bond; subsequent renewed enolization is slowed down by the halogen substituent. With base, the enolate is attacked at carbon, and subsequent enolate formation is accelerated by the halogens introduced.

6. Enolates are nucleophilic and reversibly attack the carbonyl carbon of an aldehyde or a ketone in the **aldol condensation.** They also attack the β-carbon of an α,β-unsaturated carbonyl compound in the **Michael addition.**

7. α,β-Unsaturated aldehydes and ketones show the normal chemistry of each individual double bond, but the conjugated system may react as a whole, as revealed by the ability of these compounds to undergo acid- and base-mediated **1,4-additions.** Cuprates add in 1,4-manner, whereas alkyllithiums normally attack the carbonyl function.

Problems

32. Underline the α-carbons and circle the α-hydrogens in each of the following structures.

(a) $CH_3CH_2CCH_2CH_3$ (with O double bond)

(b) $CH_3CCH(CH_3)_2$ (with O double bond)

(c)

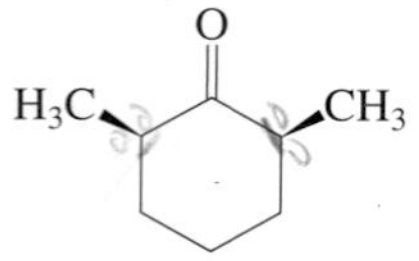

(d)

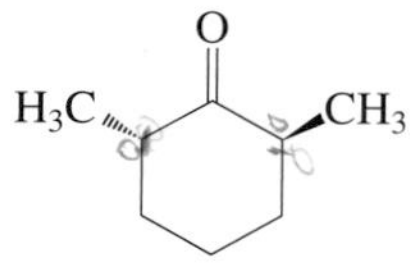

(e)

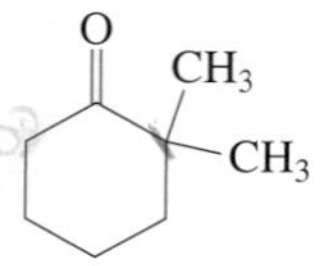

(f)

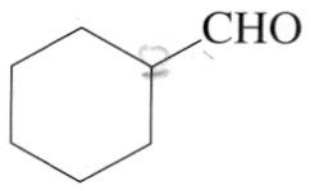

(g) $(CH_3)_3CCH$ (with O double bond)

(h) $(CH_3)_3CCH_2CH$ (with O double bond)

33. Write the structures of every enol and enolate ion that can arise from each of the carbonyl compounds illustrated in Problem 32.

34. What product(s) would form if each carbonyl compound in Problem 32 were treated with (a) alkaline D_2O; (b) 1 equivalent of Br_2 in acetic acid; (c) excess Cl_2 in aqueous base?

35. Describe the experimental conditions that would be best suited for the efficient synthesis of each of the following compounds from the corresponding nonhalogenated ketone.

(a) $C_6H_5CHCCH_3$ (with Br and O)

(b)

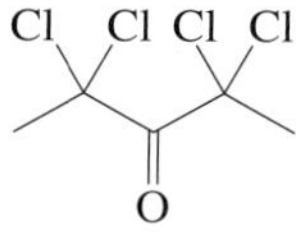

(c)

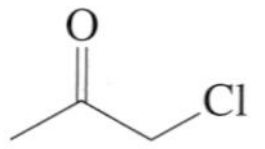

36. Propose a mechanism for the following reaction. (**Hint:** Take note of all of the products that are formed and base your answer on the mechanism for acid-catalyzed bromination of acetone shown in Section 18-3.)

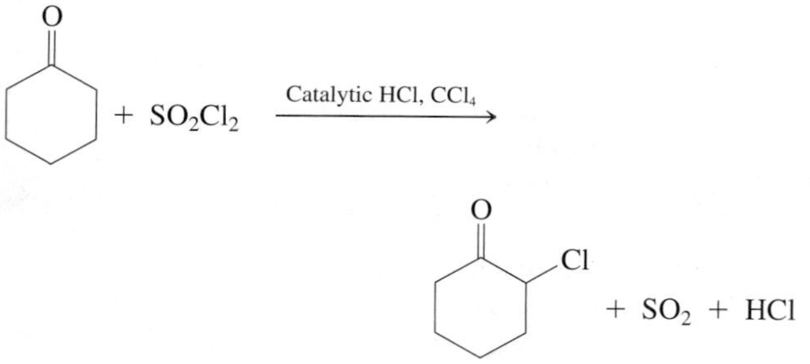

+ SO_2Cl_2 →(Catalytic HCl, CCl_4)

+ SO_2 + HCl

37. Give the product(s) that would be expected on reaction of 3-pentanone with 1 equivalent of LDA, followed by addition of 1 equivalent of

(a) CH_3CH_2Br

(b) $(CH_3)_2CHCl$

(c) 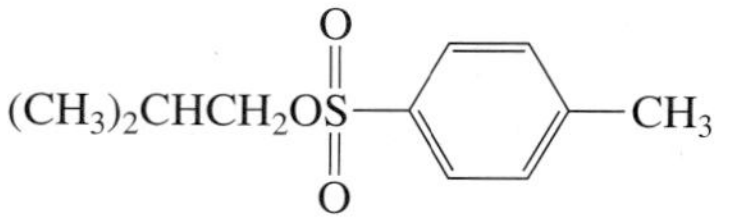 $(CH_3)_2CHCH_2OS$—⟨⟩—CH_3 (sulfonate)

(d) $(CH_3)_3CCl$

38. Give the product(s) of the following reaction sequences.

(a) 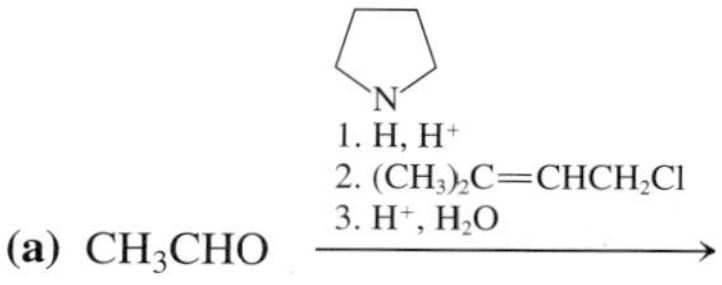 CH_3CHO →
1. H, H$^+$
2. $(CH_3)_2C=CHCH_2Cl$
3. H$^+$, H$_2$O

(b) 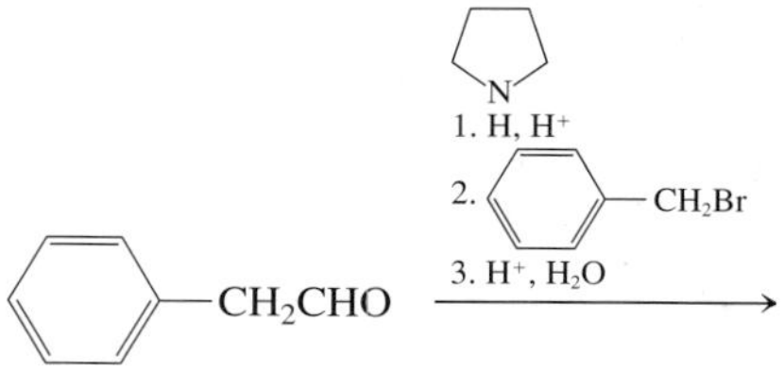 ⟨⟩—CH_2CHO →
1. H, H$^+$
2. ⟨⟩—CH_2Br
3. H$^+$, H$_2$O

39. The problem of double compared with single alkylation of ketones by iodomethane and base is mentioned in Section 18-4. Write a detailed mechanism showing how some double alkylation occurs even when only one equivalent each of the iodide and base is used. Suggest a reason why the enamine alkylation procedure solves this problem.

40. Would the use of an enamine instead of an enolate improve the likelihood of successful alkylation of a ketone by a secondary haloalkane?

41. Formulate a mechanism for the acid-catalyzed hydrolysis of the pyrrolidine enamine of cyclohexanone (shown below).

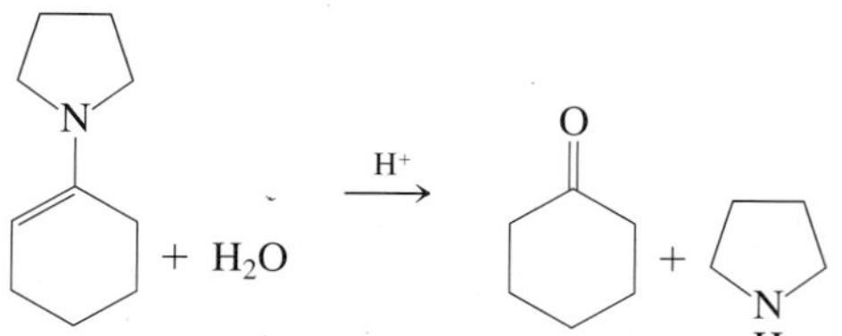

+ H_2O →(H$^+$) + (pyrrolidine, N-H)

42. Write the structures of the aldol condensation products of (a) pentanal; (b) 3-methylbutanal; (c) cyclopentanone.

43. Write the structures of the expected major products of crossed aldol condensation at elevated temperature between excess benzaldehyde and (a) 1-phenylethanone (acetophenone—see Section 17-1 for structure); (b) acetone; (c) 2,2-dimethylcyclopentanone.

44. Formulate a detailed mechanism for the reaction that you wrote in Problem 43, part (c).

45. Give the likely products for each of the following aldol addition reactions.

(a) 2 [benzene]—CH₂CHO $\xrightarrow{\text{NaOH, H}_2\text{O}}$

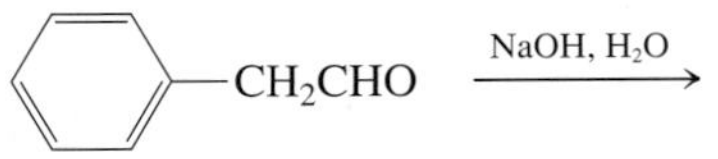

(b) 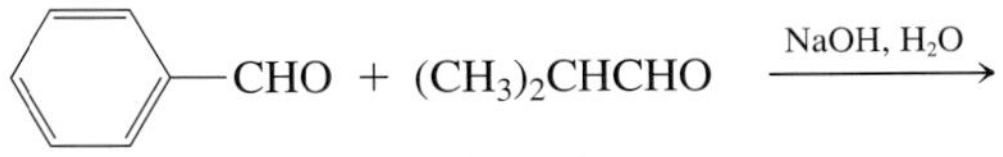 [benzene]—CHO + (CH₃)₂CHCHO $\xrightarrow{\text{NaOH, H}_2\text{O}}$

(c) 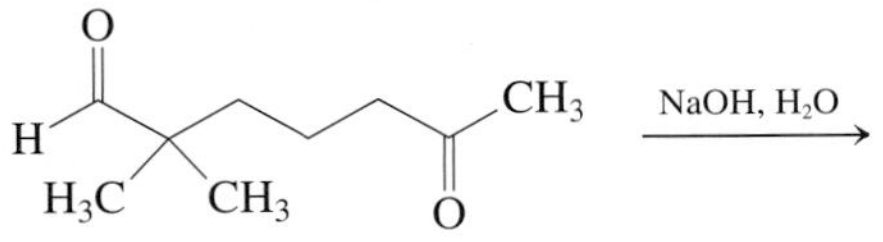 [structure with CHO, H₃C, CH₃, CH₃ and O] $\xrightarrow{\text{NaOH, H}_2\text{O}}$

(d) 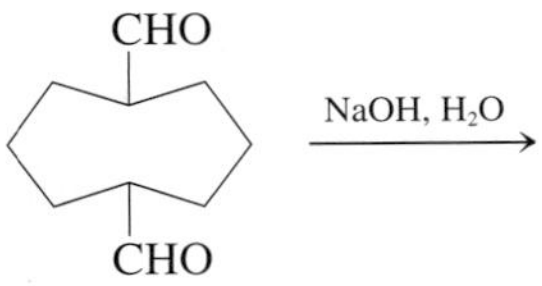 [bicyclic structure with two CHO groups] $\xrightarrow{\text{NaOH, H}_2\text{O}}$

46. Rotundone (below) is the natural product responsible for the peppery aroma in peppers, many herbs, and red wines (p. 810). What cyclic diketone will give rotundone upon intramolecular aldol condensation?

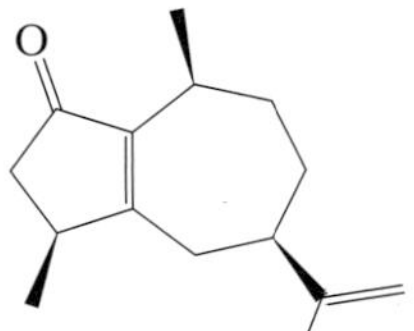

[structure of Rotundone]

Rotundone

47. Write *all* possible products of the base-catalyzed crossed aldol reactions between each pair of reaction partners given below. (**Hint:** Multiple products are possible in every case; be sure to include thermodynamically unfavorable as well as favorable ones.)

(a) Butanal and acetaldehyde

(b) 2,2-Dimethylpropanal and acetophenone

(c) Benzaldehyde and 2-butanone

48. For each of the three crossed aldol reactions described in Problem 47, indicate which, if any, of the multiple possible products should predominate in the reaction mixture and explain why.

49. Aldol condensations may be catalyzed by acids. Suggest a role for H⁺ in the acid-catalyzed version. (**Hint:** Consider what kind of nucleophile might exist in acidic solution, where enolate ions are *unlikely* to be present.)

50. Reaction review. Without consulting the Reaction Road Maps on pp. 782–783, suggest reagents to convert butanal into each of the following compounds.

(a) 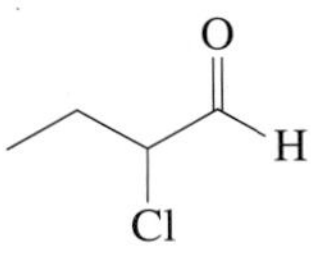 [structure with O, H, Cl]

(b) 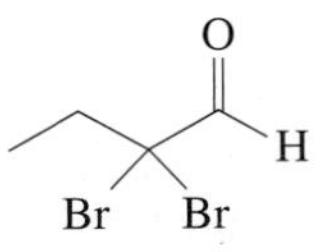 [structure with O, H, Br, Br]

(c) 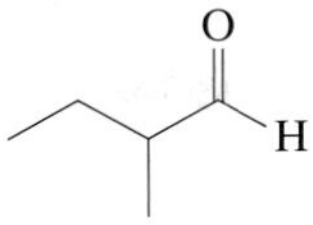 [structure with O, H]

(d) 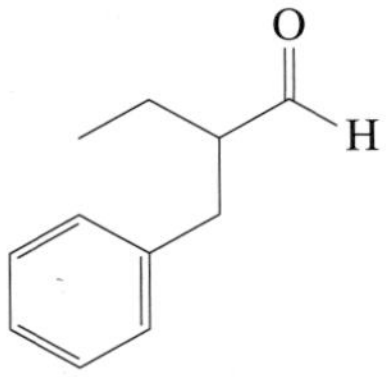 [structure with O, H, benzene]

(e) 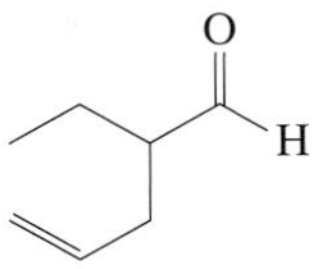 [structure with O, H, alkene]

(f) 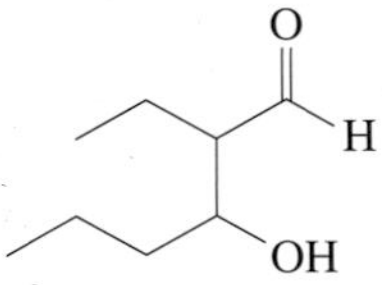 [structure with O, H, OH]

(g) 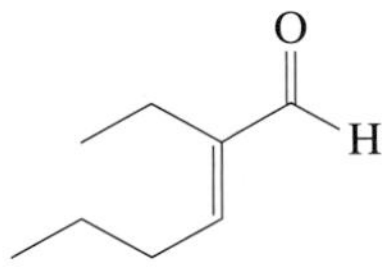 [structure with O, H]

51. Reaction review II. Without consulting the Reaction Road Maps on pp. 782–783, suggest reagents to convert acetophenone into each of the following compounds.

(a) 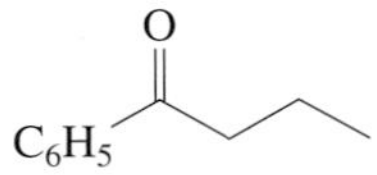 C₆H₅ [structure with O]

(b) 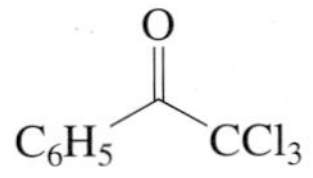 C₆H₅ [structure with O, CCl₃]

(c) 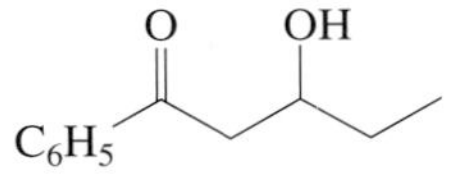 C₆H₅ [structure with O, OH]

(d) 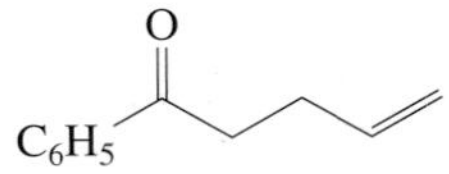 C₆H₅ [structure with O, alkene]

(e) 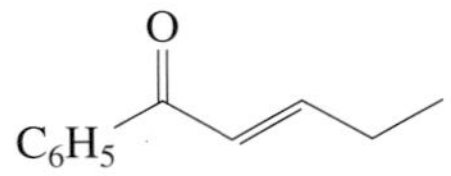 C₆H₅ [structure with O]

(f) 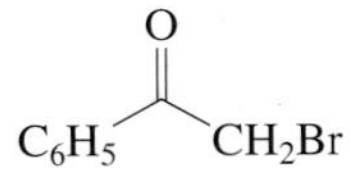 C₆H₅ [structure with O, CH₂Br]

52. Reaction review III. Without consulting the Reaction Road Maps on pp. 782–783, suggest reagents to convert 3-buten-2-one into each of the following compounds.

(a) 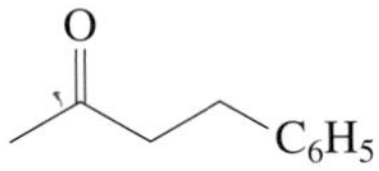 [structure with O, C₆H₅]

(b) 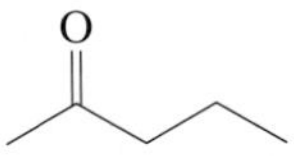 [structure with O]

(c) 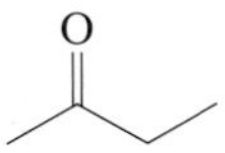 [structure with O]

(d) 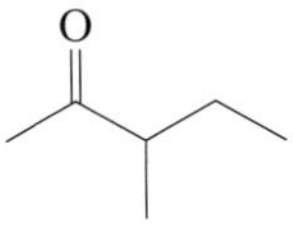 [structure with O]

(e) 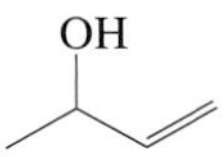 [structure with OH, alkene]

(f) 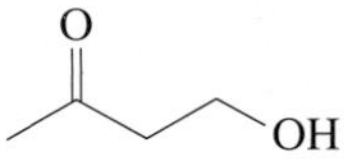 [structure with O, OH]

(g) 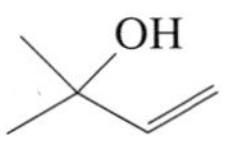 [structure with OH, alkene]

(h) 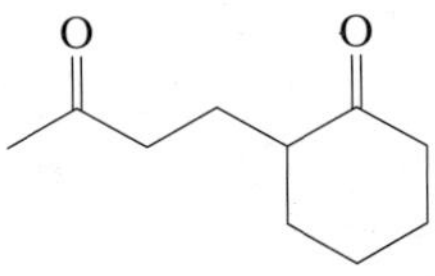 [structure with O, O, cyclohexanone]

53. A number of highly conjugated organic compounds are ingredients of sunscreens. One of the more widely used is 4-methylbenzylidene camphor (4-MBC), whose structure is shown below. This compound is effective in absorbing so-called UV-B radiation (with wavelengths between 280 and 320 nm, responsible for most sunburns). Suggest a simple synthesis of this molecule using a crossed aldol condensation.

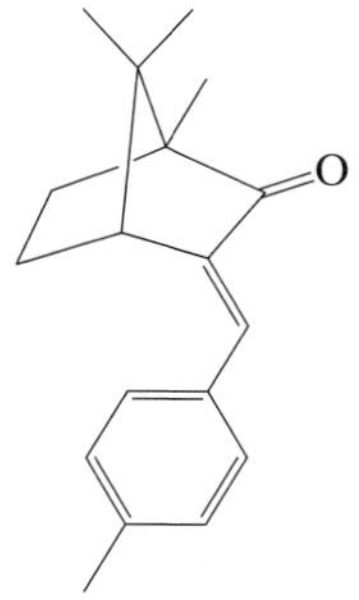

4-MBC

54. CHALLENGE The distillate from sandalwood is one of the oldest and most highly valued fragrances in perfumery. The natural oil is in short supply and, until recently, synthetic substitutes have been difficult to prepare. Polysantol (below) is the most successful of these substitutes.

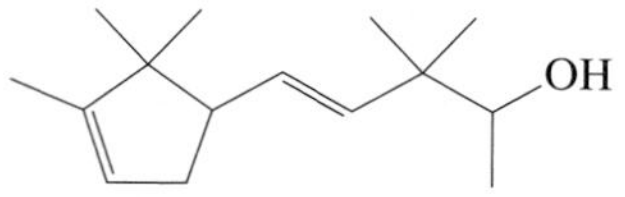

Polysantol

Its synthesis in 1984 involved the aldol condensation shown below.

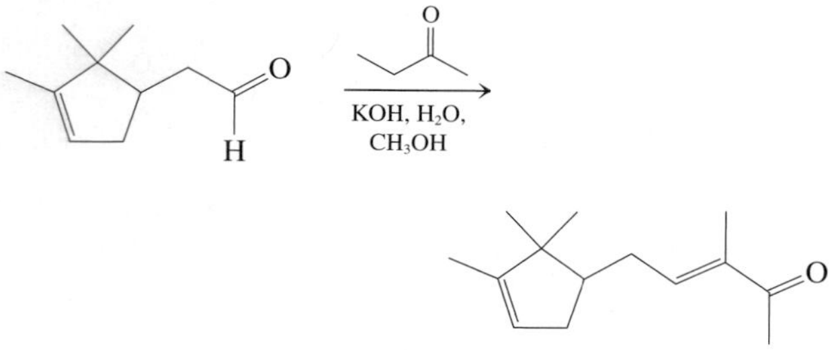

(a) This step, although usable, had a significant drawback that was responsible for its modest yield (60%). Identify this problem in detail.

(b) A procedure that avoids the conventional aldol condensation protocol was published in 2004. A bromoketone is prepared and, upon reaction with Mg metal, gives a magnesium enolate. This enolate then reacts with the aldehyde selectively to give a hydroxyketone that may be dehydrated to the desired product. Discuss how this approach solves the problem outlined in part (a).

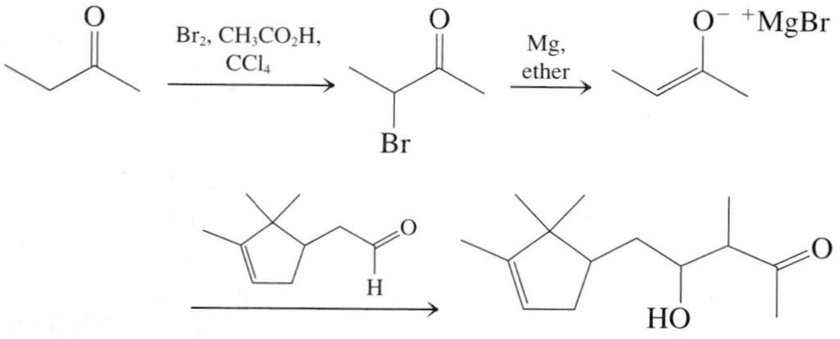

55. Write the expected major product of reaction of each of the carbonyl compounds (i)–(iii) with each of the reagents (a)–(h).

(i) 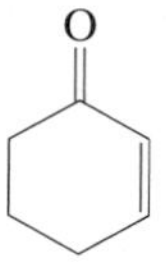

(ii) 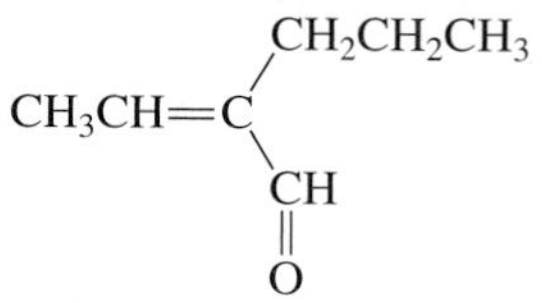

(iii) 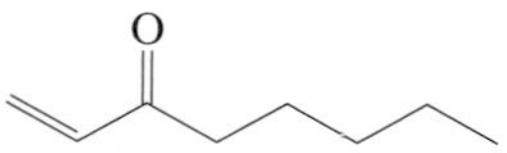

(a) H_2, Pd, CH_3CH_2OH **(b)** $LiAlH_4$, $(CH_3CH_2)_2O$
(c) Cl_2, CCl_4 **(d)** KCN, H^+, H_2O
(e) CH_3Li, $(CH_3CH_2)_2O$ **(f)** $(CH_3CH_2CH_2CH_2)_2CuLi$, THF
(g) $NH_2NH\overset{O}{\overset{\|}{C}}NH_2$, CH_3CH_2OH
(h) $(CH_3CH_2CH_2CH_2)_2CuLi$, followed by treatment with $CH_2{=}CHCH_2Cl$ in THF

56. In each of the following retrosynthetic disconnections (〰), show a reaction or reaction sequence that would make the indicated C—C bonds.

(a) 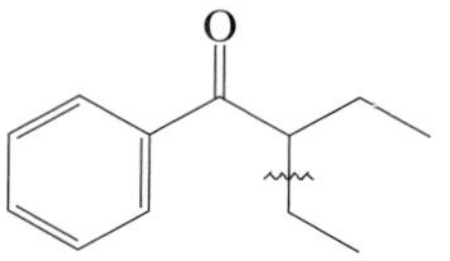

(b) 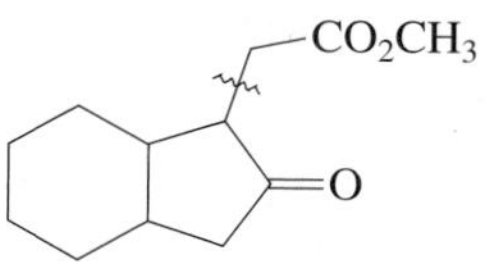

(c) 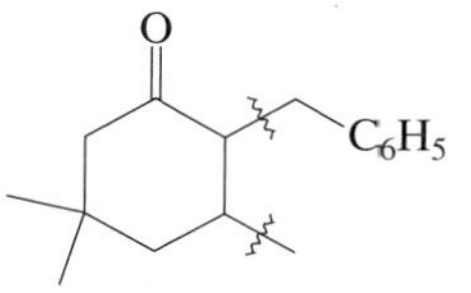

(d) 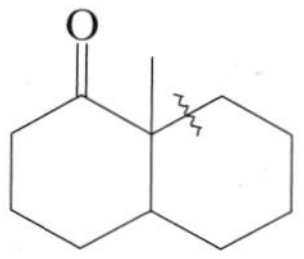

57. Write the products of each of the following reactions after aqueous work-up.

(a) $C_6H_5\overset{O}{\overset{\|}{C}}CH_3 + CH_2{=}CH\overset{O}{\overset{\|}{C}}C_6H_5 \xrightarrow{\text{LDA, THF}}$

(b) 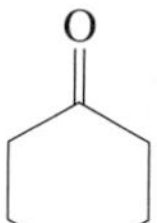 $+ (CH_3)_2C{=}CH\overset{O}{\overset{\|}{C}}H \xrightarrow{\text{NaOH, } H_2O}$

(c) 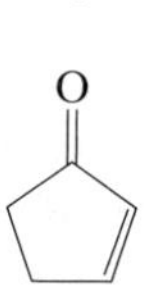 $\xrightarrow[\text{2. } CH_2=CH\overset{O}{\overset{\|}{C}}CH_3]{\text{1. } (CH_2=CH)_2CuLi\text{, THF}}$

(d) 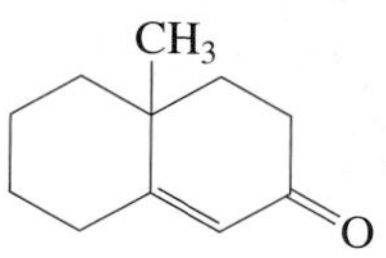

$\xrightarrow[\text{2. } (CH_3)_2C=CH\overset{O}{\overset{\|}{C}}CH_3]{\text{1. } (CH_3)_2CuLi\text{, THF}}$

(e) Write the results that you expect from base treatment of the products of reactions (c) and (d).

58. Write the final products of the following reaction sequences.

(a)

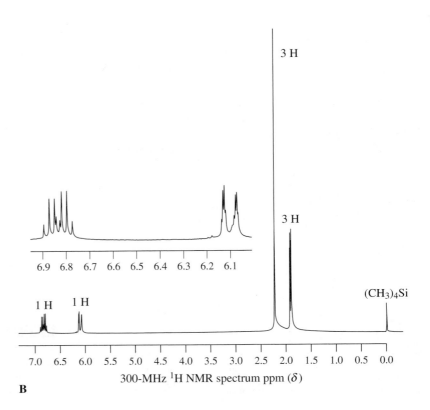

$$\text{O} + \text{CH}_2{=}\text{CHCCH}_3 \xrightarrow{\text{NaOCH}_3, \text{CH}_3\text{OH}, \Delta}$$

(b)

$$\text{H}_3\text{C} \qquad \text{CH}_3 + \text{CH}_2{=}\text{CHCCH}_3 \xrightarrow{\text{KOH, CH}_3\text{OH}, \Delta}$$

(c)

$$\xrightarrow[\substack{2.\ \text{HC}{\equiv}\text{CCCH}_3}]{1.\ \text{LDA, THF}}$$

(d) Write a detailed mechanism for reaction sequence (c). (**Hint:** Treat the 3-butyn-2-one reagent as a Michael acceptor in the first step.)

59. Propose syntheses of the following compounds by using Michael additions followed by aldol condensations (i.e., Robinson annulations). Each of the compounds shown has been instrumental in one or more total syntheses of steroidal hormones. (**Hint:** Identify and unravel the cyclohexenone ring in a reverse annulation sequence.)

(a)

(b)

(c) CH_3O

(d)

60. Would you expect addition of HCl to the double bond of 3-buten-2-one (shown below) to follow Markovnikov's rule? Explain your answer by a mechanistic argument.

$$\overset{\text{O}}{\overset{\|}{\text{CH}_3\text{CCH}{=}\text{CH}_2}}$$
3-Buten-2-one

61. Using the following information, propose structures for each of these compounds. (a) $C_5H_{10}O$, NMR spectrum A, UV: $\lambda_{max}(\epsilon) = 280(18)$ nm; (b) C_5H_8O, NMR spectrum B, UV: $\lambda_{max}(\epsilon) = 220(13{,}200), 310(40)$ nm; (c) C_6H_{12}, NMR spectrum C (next page), UV: $\lambda_{max}(\epsilon) = 189(8{,}000)$ nm; (d) $C_6H_{12}O$, NMR spectrum D (next page), UV: $\lambda_{max}(\epsilon) = 282(25)$ nm.

3 H

3 H

3 H

2.4   1.6   0.9

2 H

2 H

(CH₃)₄Si

3.0 2.5 2.0 1.5 1.0 0.5 0.0

300-MHz ¹H NMR spectrum ppm (δ)

A

3 H

6.9 6.8 6.7 6.6 6.5 6.4 6.3 6.2 6.1

3 H

1 H   1 H

(CH₃)₄Si

7.0 6.5 6.0 5.5 5.0 4.5 4.0 3.5 3.0 2.5 2.0 1.5 1.0 0.5 0.0

300-MHz ¹H NMR spectrum ppm (δ)

B

Next, for each of the following reactions, name an appropriate reagent for the indicated interconversion. (The letters refer to the compounds giving rise to NMR spectra A through D.) **(e)** A → C; **(f)** B → D; **(g)** B → A.

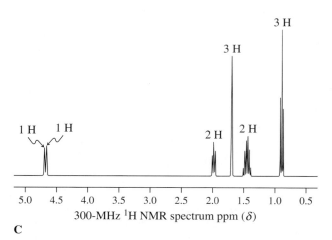

300-MHz ^{1}H NMR spectrum ppm (δ)

C

300-MHz ^{1}H NMR spectrum ppm (δ)

D

62. **CHALLENGE** Treatment of cyclopentane-1,3-dione with iodomethane in the presence of base leads mainly to a mixture of three products.

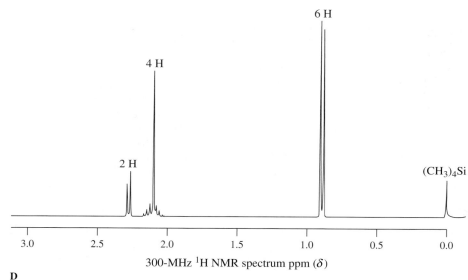

(a) Give a mechanistic description of how these three products are formed.

(b) Reaction of product C with a cuprate reagent results in loss of the methoxy group. For example,

Suggest a mechanism for this reaction, which is another synthetic route to enones substituted at the β-carbon. (**Hint:** See Exercise 18-24.)

63. **CHALLENGE** A somewhat unusual synthesis of cortisone-related steroids includes the two reactions on top of the next page.

(a) Propose mechanisms for these two transformations. Be careful in choosing the initial site of deprotonation in the starting enone. The alkenyl hydrogen, in particular, is *not acidic enough* to be the one initially removed by base in this reaction.

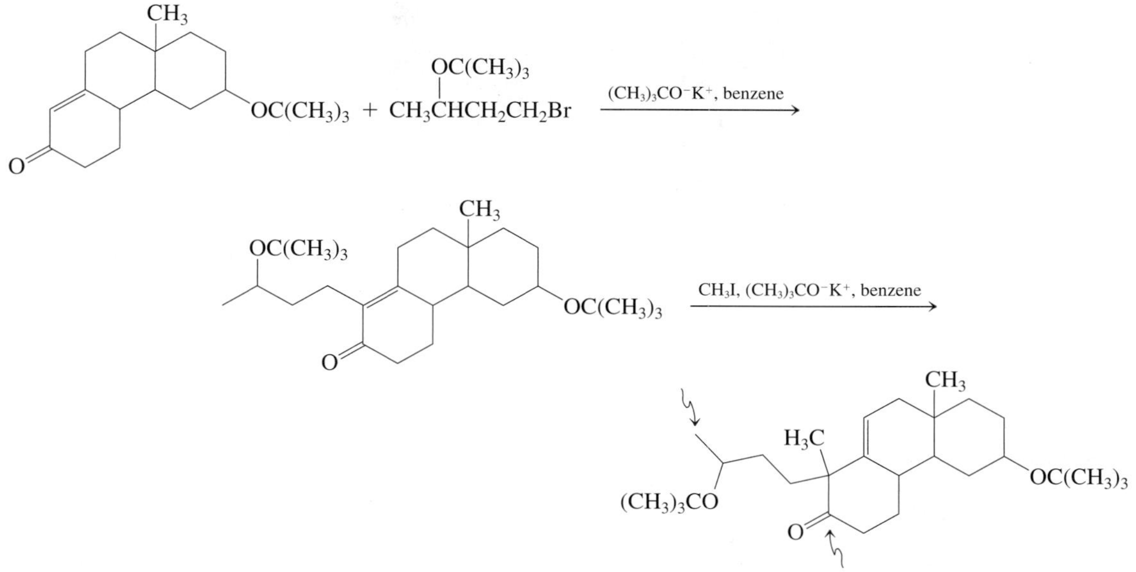

(b) Propose a sequence of reactions that will connect the carbons marked by arrows in the third structure shown to form another six-membered ring.

64. The following steroid synthesis contains modified versions of two key types of reactions presented in this chapter. Identify these reaction types and give detailed mechanisms for each of the transformations shown.

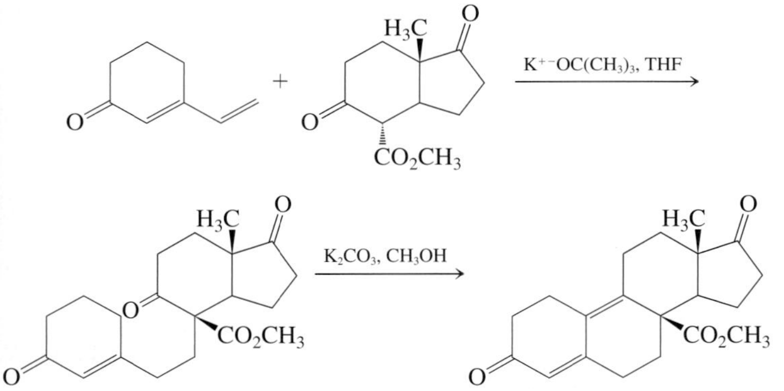

65. Devise reasonable plans for carrying out the following syntheses. Ignore stereochemistry in your strategies.

(a) 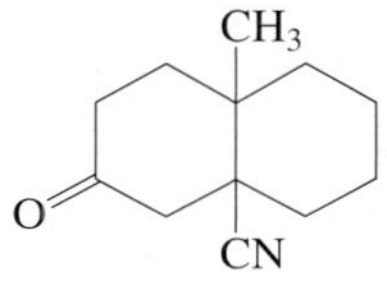 , starting from cyclohexanone

(b) 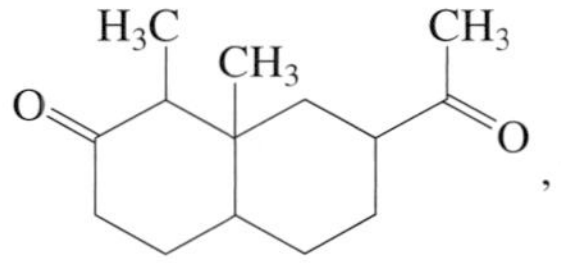 , starting from 2-cyclohexenone

(Hint: Prepare **in your first step.)**

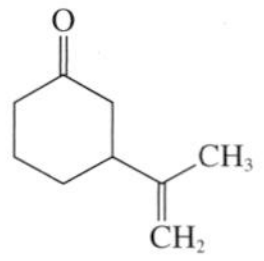

66. Write reagents (a, b, c, d, e) where they have been omitted from the following synthetic sequence. Each letter may correspond to one or more reaction steps. This sequence is the beginning of a synthesis of germanicol, a naturally occurring triterpene. The diol used in the step between (a) and (b) provides selective protection of the more reactive carbonyl group. [**Hint:** See Problem 63 when formulating (b).]

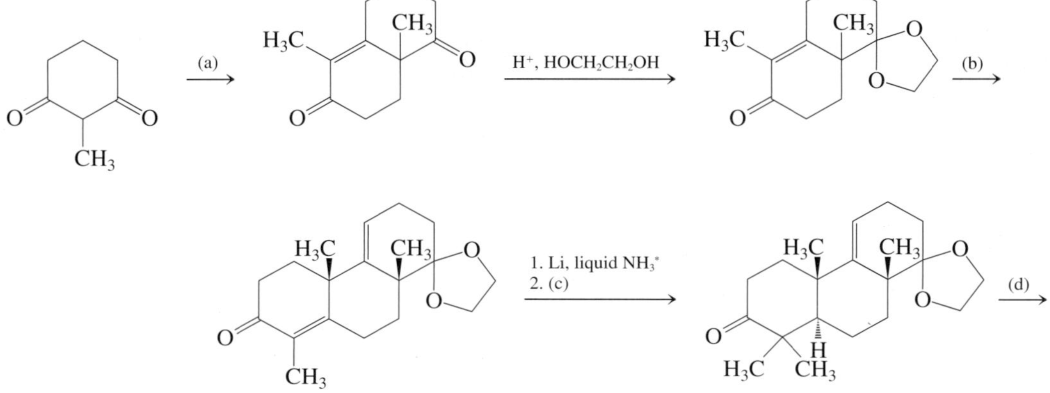

---

*An electron-transfer process (compare alkyne reduction, Section 13-6) that is the equivalent of adding hydride, H:⁻, to the β-carbon. The product is the enolate of the saturated ketone.

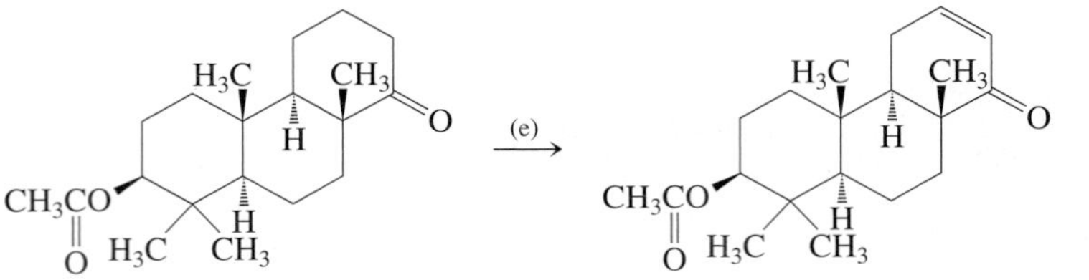

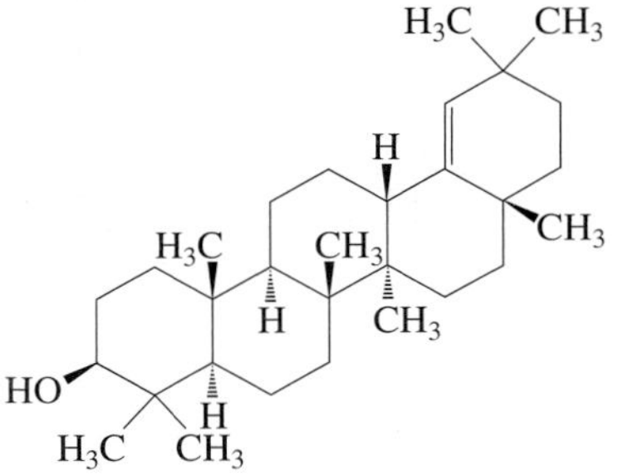

Germanicol

Team Problem

67. When 2-methylcyclopentanone is treated with the bulky base triphenylmethyllithium under the two sets of conditions shown, the two possible enolates are generated in differing ratios. Why is this so?

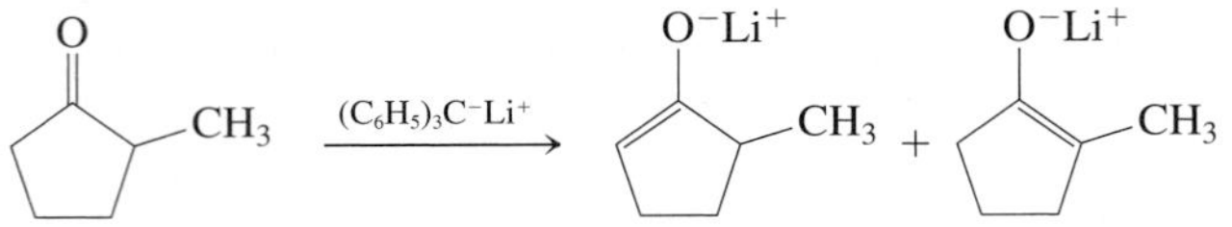

Conditions A: Ketone added to excess base	72%		28%
Conditions B: Excess ketone added to base	6%		94%

To tackle this problem, you have to invoke the principles of kinetic versus thermodynamic control (review Sections 11-6, 14-6, and 18-2); that is, which enolate is formed faster and which one is more stable? Divide your team so that one group considers conditions A and the other conditions B. Use curved arrows to show the flow of electrons leading to each enolate. Then assess whether your set of conditions is subject to enolate equilibration (thermodynamic control) or not (kinetic control). Reconvene to discuss these issues and draw a qualitative potential-energy diagram depicting the progress of deprotonation at the two α sites.

Preprofessional Problems

68. When 3-methyl-1,3-diphenyl-2-butanone is treated with excess D_2O in the presence of catalytic acid, some of its hydrogens are replaced by deuterium. How many? (a) One; (b) two; (c) three; (d) six; (e) eight.

69. How would you best classify the following reaction? (a) Wittig reaction; (b) cyanohydrin formation; (c) conjugate addition; (d) aldol addition.

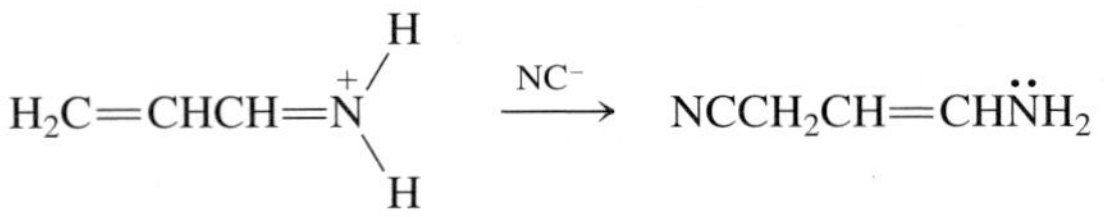

70. The aqueous hydroxide-promoted reaction of the compound shown below with $(CH_3)_3CCHO$ yields exclusively one compound. Which one is it?

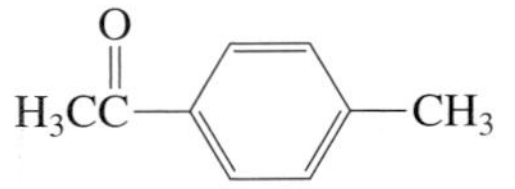

(a)

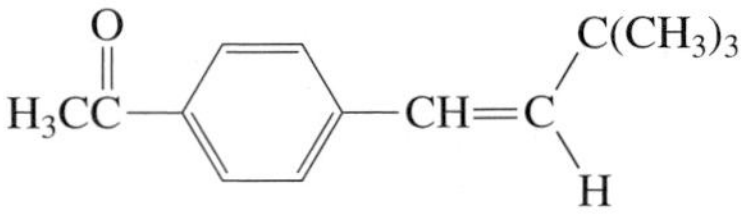

(b)

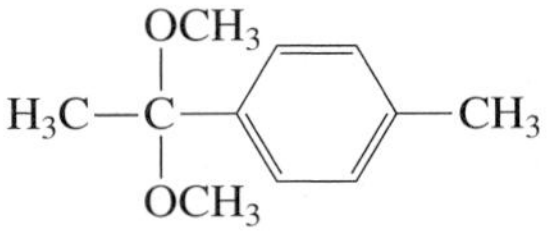

(c) 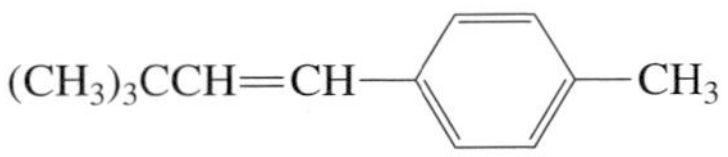 $(CH_3)_3CCH=CH$——CH_3

(d)

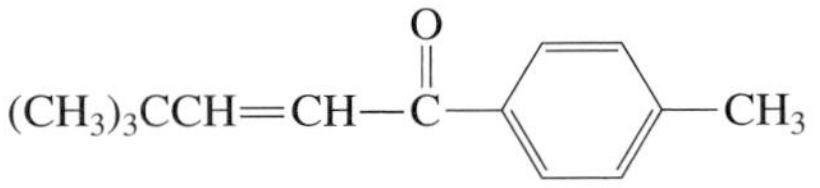

71. The ^1H NMR spectrum of 2,4-pentanedione indicates the presence of an enol tautomer of the dione. What is its most likely structure?

(a)

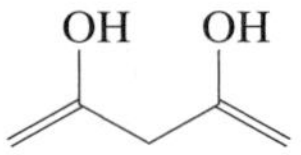

(b)

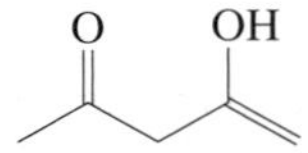

(c)

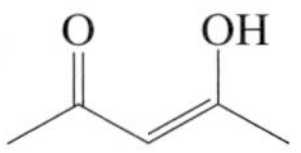

(d)

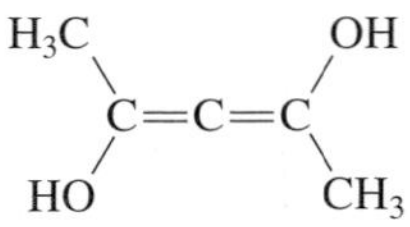

CHAPTER 19

Carboxylic Acids

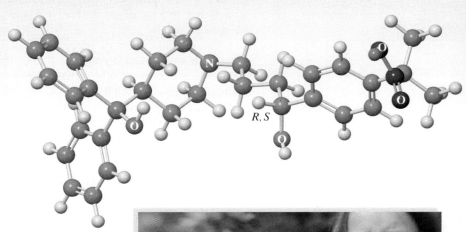

What types of compounds are the end products of biological metabolism? We inhale oxygen and exhale carbon dioxide, one product of the oxidation of organic compounds (Chapter 3). **Carboxylic acids** also are major products of biochemistry occurring under oxidizing conditions and are thus widespread throughout living systems on Earth. You are probably familiar with the bottle of wine that has turned to vinegar, the result of enzymatic oxidation of ethanol to acetic acid. Carboxylic acids are characterized by the presence of the **carboxy group,** a functional group containing a hydroxy unit attached to a carbonyl carbon. This substituent is written COOH or CO_2H; both representations are used in this chapter.

Carboxylic acids not only are widely distributed in nature, they are also important industrial chemicals. For example, besides being the most important building block in the assembly of complex biological molecules, acetic acid is an industrial commodity that is produced in very large quantities.

Much of the functional behavior of carboxylic acids can be anticipated if we view them structurally as hydroxy carbonyl derivatives. Thus, the hydroxy hydrogen is acidic, the oxygens are basic and nucleophilic, and the carbonyl carbon is subject to nucleophilic attack.

The carboxylic acid fexofenadine (Allegra) is a top-selling nonsedating antihistamine drug for the treatment of allergy symptoms. It contains a stereocenter, but both enantiomers are equally effective, hence it is sold as a racemate.

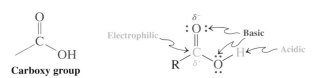

Carboxy group

In this chapter, we first introduce the system of naming carboxylic acids and then list some of their physical and spectroscopic characteristics. We then examine their acidity and basicity, two properties that are strongly influenced by the interaction between the electron-withdrawing carbonyl group and the hydroxy function. Methods for the preparation of the carboxy group are considered next, followed by a survey of its reactivity. Reactions of carboxylic acids will feature a new two-step substitution pathway, *addition–elimination,* for the replacement of the hydroxy group by other nucleophiles, such as halide, alkoxide, and amide. The chemistry of the carboxylic acid derivatives, which result from these transformations, is the subject of Chapter 20.

19-1 | NAMING THE CARBOXYLIC ACIDS

Like other organic compounds, many carboxylic acids have common names that are used frequently in the literature. These names often indicate the natural sources from which the acids were first isolated (Table 19-1). *Chemical Abstracts* retains the common names for the two simplest, **formic** and **acetic** acids. The carboxy function and the functional groups of its derivatives take precedence in naming over any other group discussed so far.

The overall "order of precedence" of the most common functional groups is given below. In any molecule containing more than one functional group, the group highest in precedence is entered as a suffix in the name, and all other substituents as prefixes. For example, consider $HOCH_2CH_2COCH_2CHO$, a molecule containing alcohol, ketone, and aldehyde functions. In its name, the first two groups appear as prefixes (hydroxy- and oxo-, respectively), while the aldehyde group, being of highest precedence, is signified by the suffix -*al:* This compound is 5-hydroxy-3-oxopentanal. Alkenes and alkynes are exceptions. They rank below the amines in order of precedence, but when a carbon–carbon double or triple bond is part of the parent chain or ring of a molecule, we insert -*en(e)*- or -*yn(e)*- just before the suffix for the highest-ranking functional group. The following examples of names for carboxylic acids illustrate these principles.

Order of Precedence of Functional Groups

O ‖ RCOH		O O ‖ ‖ RCOCR		O ‖ RCOR′		O ‖ RCX		O ‖ RCNR′₂		RCN		O ‖ RCH		O ‖ RCR′		ROH		RSH		RNH₂
Carboxylic acid	>	Anhydride	>	Ester	>	Acyl halide	>	Amide	>	Nitrile	>	Aldehyde	>	Ketone	>	Alcohol	>	Thiol	>	Amine

← Increasing precedence in naming

Table 19-1 | Names and Natural Sources of Carboxylic Acids

	Structure	IUPAC name	Common name	Natural source
	HCOOH	Methanoic acid	Formic acid[a]	From the "destructive distillation" of ants (*formica,* Latin, ant)
	CH_3COOH	Ethanoic acid	Acetic acid[a]	Vinegar (*acetum,* Latin, vinegar)
	CH_3CH_2COOH	Propanoic acid	Propionic acid	Dairy products (*pion,* Greek, fat)
	$CH_3CH_2CH_2COOH$	Butanoic acid	Butyric acid	Butter (particularly if rancid) (*butyrum,* Latin, butter)
	$CH_3(CH_2)_3COOH$	Pentanoic acid	Valeric acid	Valerian root
	$CH_3(CH_2)_4COOH$	Hexanoic acid	Caproic acid	Odor of goats (*caper,* Latin, goat)

[a]Used by *Chemical Abstracts.*

The IUPAC system derives the names of the carboxylic acids by replacing the ending -*e* in the name of the alkane by **-oic acid.** The **alkanoic acid** stem is numbered by assigning the number 1 to the carboxy carbon and labeling any substituents along the longest chain incorporating the CO₂H group accordingly.

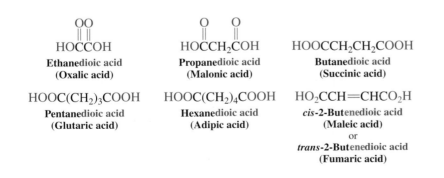

(R)-2-Bromopropanoic acid   **Propenoic acid**   **(2R,3S)-Dimethylpentanoic acid**
(α-Bromopropionic acid)   **(Acrylic acid)**   **(αR,βS-Dimethylvaleric acid)**

To the extent possible, any carbon–carbon double or triple bonds that are present should be included in naming the main chain. Saturated cyclic acids are named as **cycloalkanecarboxylic acids.** Their aromatic counterparts are the **benzoic acids.** In these compounds, the carbon attached to the carboxy functional group is Cl.

5-Butyl-6-heptenoic acid   **4-Methyl-5-oxohexanoic acid**   **1-Bromo-2-chloro-**   **2-Hydroxybenzoic acid**
                                                                cyclopentane-   **(o-Hydroxybenzoic**
                                                                carboxylic acid   **acid, salicylic acid)**

$CH_3\overset{O}{\overset{\|}{C}}CHCH_2CH_2COOH$
$\quad\quad\quad CH_3$

2-Hydroxybenzoic (salicylic) acid has been known for over 2000 years as an analgesic (pain-killing) folk remedy present in willow bark (*salix,* Latin, willow). The acetic acid ester of salicylic acid is better known as aspirin (Real Life 22-2).

Dicarboxylic acids are referred to as **dioic acids.**

$\overset{O\;O}{\overset{\|\;\|}{HOCCOH}}$ $\overset{O\quad O}{\overset{\|\quad\|}{HOCCH_2COH}}$ $HOOCCH_2CH_2COOH$

Ethanedioic acid   **Propanedioic acid**   **Butanedioic acid**
(Oxalic acid)   **(Malonic acid)**   **(Succinic acid)**

$HOOC(CH_2)_3COOH$ $HOOC(CH_2)_4COOH$ $HO_2CCH=CHCO_2H$

Pentanedioic acid   **Hexanedioic acid**   *cis*-**2-Butenedioic acid**
(Glutaric acid)   **(Adipic acid)**   **(Maleic acid)**
                                                          or
                                                          trans-**2-Butenedioic acid**
                                                          (Fumaric acid)

Human beings have known since prehistoric times that chewing on the bark of the white willow relieves pain.

Their common names again derive from their natural origins. For example, butanedioic (succinic) acid was discovered in the distillate from amber (*succinum,* Latin, amber), and *trans*-2-butenedioic (fumaric) acid is found in the plant *Fumaria,* which was burned in ancient times to ward off evil spirits (*fumus,* Latin, smoke).

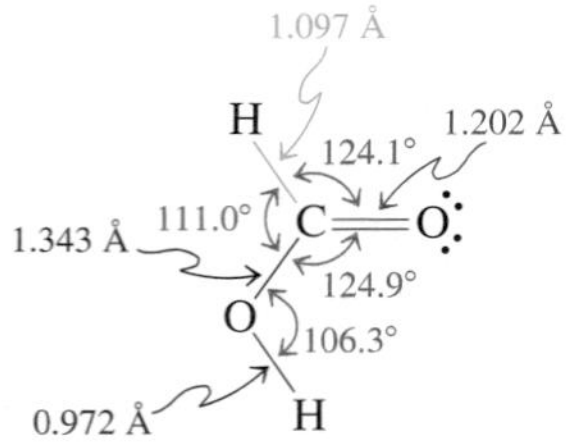

Figure 19-1 The molecular structure of formic acid is planar, with a roughly equilateral trigonal arrangement about the carbonyl carbon.

Allegra

Hydrophilic

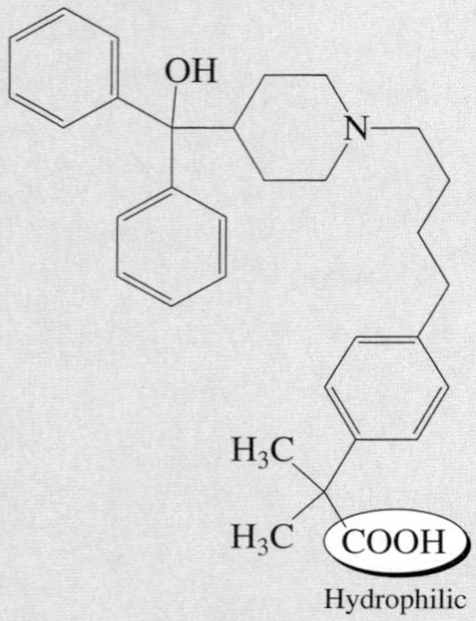

Fexofenadine (Allegra)

Many allergy medications cause drowsiness when ingested, a side effect caused by the crossing of the blood–brain barrier. This barrier is there to prevent large or hydrophilic molecules from entering the cerebral fluid. The blockbuster drug Allegra (see also Chapter Opening) fits this profile, in part because of the presence of the polar carboxy function, which imparts hydrophilic character to the (sizeable) molecule.

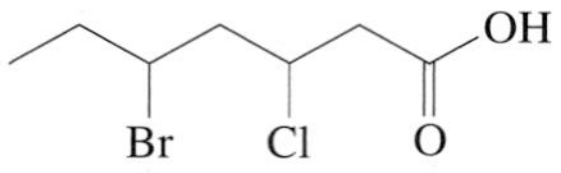

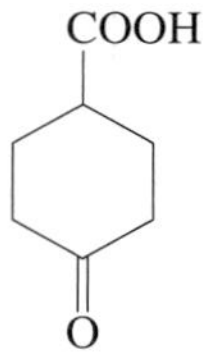

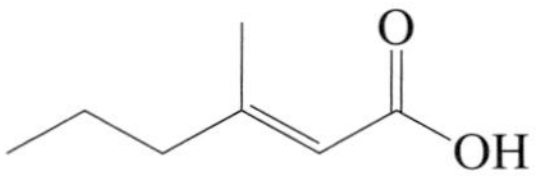

(*E*)-3-Methyl-2-hexenoic acid

Exercise 19-1

Give systematic names or write the structure, as appropriate, of the following compounds.

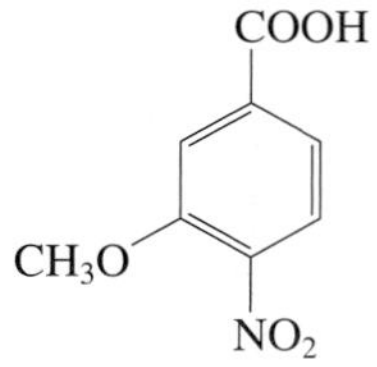

(d) 2,2-Dibromohexanedioic acid
(e) 4-Hydroxypentanoic acid
(f) 4-(1,1-Dimethylethyl)benzoic acid

In Summary The systematic naming of the carboxylic acids is based on the alkanoic acid stem. Cyclic derivatives are called cycloalkanecarboxylic acids, their aromatic counterparts are referred to as benzoic acids, and dicarboxylic systems are labeled alkanedioic acids.

19-2 STRUCTURAL AND PHYSICAL PROPERTIES OF CARBOXYLIC ACIDS

What is the structure of a typical carboxylic acid? What are the characteristic physical properties of carboxylic acids? This section answers these questions, beginning with the structure of formic acid. We shall see that carboxylic acids exist mainly as hydrogen-bonded dimers.

Formic acid is planar

The molecular structure of formic acid is shown in Figure 19-1. It is roughly planar, as expected for a hydroxy-substituted carbonyl compound, with an approximately trigonal carbonyl carbon. (Compare the structure of methanol, Figure 8-1, with that of acetaldehyde, Figure 17-2.) These structural characteristics are found in carboxylic acids in general.

The carboxy group is polar and forms hydrogen-bonded dimers

The carboxy function derives strong polarity from the carbonyl double bond and the hydroxy group, which forms hydrogen bonds to other polarized molecules, such as water, alcohols, and other carboxylic acids. Not surprisingly, therefore, the lower carboxylic acids (up to butanoic acid) are completely soluble in water. As neat liquids and even in fairly dilute solutions (in aprotic solvents), carboxylic acids exist to a large extent as hydrogen-bonded dimers, each O–H$\cdots$O interaction ranging in strength from about 6 to 8 kcal mol^{-1} (25 to 34 kJ mol^{-1}).

Carboxylic Acids Form Dimers Readily

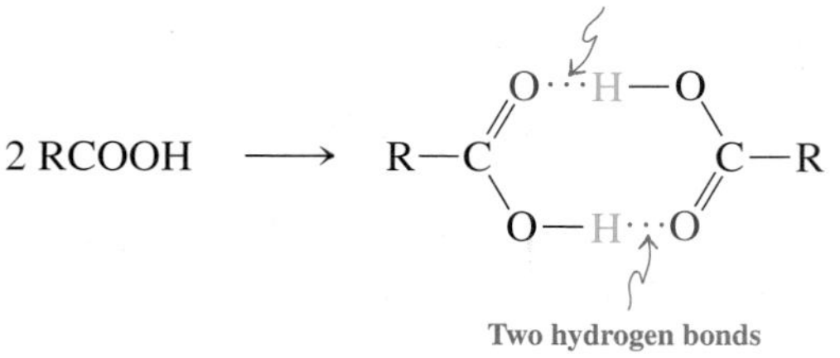

Two hydrogen bonds

Table 19-2 shows that carboxylic acids have relatively high melting and boiling points, because they form hydrogen bonds in the solid as well as the liquid state.

Carboxylic acids, especially those possessing relatively low molecular weights and correspondingly high volatility, exhibit strong odors. For example, the presence of butanoic acid contributes to the characteristic flavor of many cheeses, and (*E*)-3-methyl-2-hexenoic acid is one of the compounds responsible for the smell of human sweat.

Table 19-2	Melting and Boiling Points of Functional Alkane Derivatives with Various Chain Lengths	
Derivative	Melting point (°C)	Boiling point (°C)
CH_4	−182.5	−161.7
CH_3Cl	−97.7	−24.2
CH_3OH	−97.8	65.0
HCHO	−92.0	−21.0
HCOOH	8.4	100.6
CH_3CH_3	−183.3	−88.6
CH_3CH_2Cl	−136.4	12.3
CH_3CH_2OH	−114.7	78.5
CH_3CHO	−121.0	20.8
CH_3COOH	16.7	118.2
$CH_3CH_2CH_3$	−187.7	−42.1
$CH_3CH_2CH_2Cl$	−122.8	46.6
$CH_3CH_2CH_2OH$	−126.5	97.4
CH_3COCH_3	−95.0	56.5
CH_3CH_2CHO	−81.0	48.8
CH_3CH_2COOH	−20.8	141.8

In Summary The carboxy function is planar and contains a polarizable carbonyl group. Carboxylic acids exist as hydrogen-bonded dimers and exhibit unusually high melting and boiling points.

19-3 SPECTROSCOPY AND MASS SPECTROMETRY OF CARBOXYLIC ACIDS

The polarizable double bond and the hydroxy group also shape the spectra of carboxylic acids. In this section, we show how both NMR and IR methods are used to characterize the carboxy group. We also illustrate how carboxylic acids fragment in the mass spectrometer.

The carboxy hydrogen and carbon are deshielded

As in aldehydes and ketones, the hydrogens positioned on the carbon next to the carbonyl group are slightly deshielded in ^1H NMR spectra. The effect diminishes rapidly with increasing distance from the functional group. The hydroxy proton resonates at very low field ($\delta = 10–13$ ppm). As in the NMR spectra of alcohols, its chemical shift varies strongly with concentration, solvent, and temperature, because of the strong ability of the OH group to enter into hydrogen bonding. The ^1H NMR spectrum of pentanoic acid is shown in Figure 19-2.

^1H NMR Chemical Shifts of Alkanoic Acids

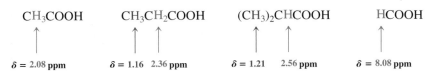

CH_3COOH	CH_3CH_2COOH	$(CH_3)_2CHCOOH$	HCOOH
↑	↑ ↑	↑ ↑	↑
$\delta = 2.08$ ppm	$\delta = 1.16$ 2.36 ppm	$\delta = 1.21$ 2.56 ppm	$\delta = 8.08$ ppm

The ^{13}C NMR chemical shifts of carboxylic acids also are similar to those of the aldehydes and ketones, with moderately deshielded carbons next to the carbonyl group and the typically low-field carbonyl absorptions. However, the amount of deshielding is smaller because the positive polarization of the carboxy carbon is somewhat attenuated by the presence of the extra OH group.

Really? Human bitter taste is mediated by so-called G-protein–coupled receptors, a family of proteins that sense odors, hormones, and neurotransmitters outside the cell to elicit corresponding cellular responses. The simple carboxylic acid pictured below (as the carboxylate anion) has the right three-dimensional shape of its hydrophobic chain and hydrogen-bonding carboxy group to maximize docking to the protein's receptor site, thus blocking this function. It may become useful as an additive to vitamins and other supplements to reduce their often bitter aftertastes.

cis-4-(2,2,3-Trimethylcyclopentyl) butanoate binding to its receptor.

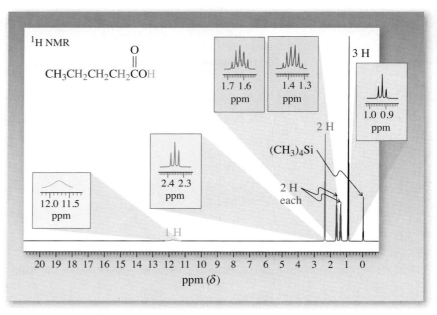

Figure 19-2 300-MHz ^{1}H NMR spectrum of pentanoic acid. The scale has been expanded to 20 ppm to allow the signal for the acid proton at $\delta = 11.75$ ppm to be shown. The methylene hydrogens at C2 absorb at the next lowest field as a triplet ($\delta = 2.35$ ppm, $J = 7$ Hz), followed by a quintet and a sextet, respectively, for the next two sets of methylenes. The methyl group appears as a distorted triplet at highest field ($\delta = 0.92$ ppm, $J = 6$ Hz).

Typical ^{13}C NMR Chemical Shifts of Alkanoic Acids

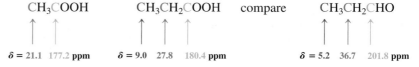

This attenuation is best understood by writing resonance forms. Recall from Section 17-2 that aldehydes and ketones are described by two such representations. The dipolar Lewis structure indicates the polarization of the C=O bond. Even though it is a minor contributor (because of the lack of an octet on carbon), it explains the deshielding of the carbonyl carbon. In carboxylic acids, however, the corresponding dipolar form contributes less to the resonance hybrid: The hydroxy oxygen can donate an electron pair to give a third arrangement in which the carbon and both oxygen atoms have octets. The degree of positive charge on the carbonyl carbon and therefore its deshielding are greatly reduced. The changes in electron density distribution around the carbonyl function when going from acetone to acetic acid are shown in their respective electrostatic potential maps. The strongly positive polarization (blue) of the carbonyl carbon in the ketone is diminished in the acid (green).

Acetone

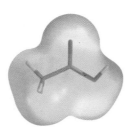

Acetic acid

Resonance in Aldehydes and Ketones

The contribution of the second resonance form, though minor, explains the strong deshielding of the carbonyl and adjacent carbons

Resonance in Carboxylic Acids

The third resonance form explains the attenuated deshielding of the carbonyl carbon relative to that of aldehydes and ketones

Solved Exercise 19-1	Working with the Concepts: Deducing the Structure of Carboxylic Acids from NMR Data

A foul-smelling carboxylic acid with b.p. 164°C gave the following NMR data: ^{1}H NMR: $\delta = 1.00$ (t, $J = 7.4$ Hz, 3 H), 1.65 (sextet, $J = 7.5$ Hz, 2 H), 2.31 (t, $J = 7.4$ Hz, 2 H), and 11.68 (s, 1 H) ppm: ^{13}C NMR: $\delta = 13.4$, 18.5, 36.3, and 179.6 ppm. Assign a structure to it.

Strategy

A good place to start is the ^{13}C NMR spectrum, because it gives you, in addition to confirming the presence of a carboxy carbon, the number of additional unique carbon atoms in the molecule. The ^{1}H NMR spectrum will expand on that information through the number of signals (hence the number of different hydrogens), their chemical shift (hence possibly the presence of other functional groups), integration (hence the relative abundance of each type of hydrogen), and spin–spin splitting (hence the number of equivalent neighbors of the hydrogen giving rise to a particular signal).

Solution

- The ^{13}C NMR spectrum shows a total of four signals, one of which, at $\delta = 179.6$ ppm, is due to the carboxy carbon. This leaves us with three additional peaks, all in the saturated carbon region (Section 10-9). One of them ($\delta = 36.3$ ppm) is slightly deshielded relative to the other two, likely a carbon atom next to the carboxy group.
- Turning to the ^{1}H NMR spectrum, we can identify the carboxy hydrogen at $\delta = 11.68$ ppm, in addition to three additional signals. Again, all of the latter appear in the saturated region (Table 10-2), one of them ($\delta = 2.31$ ppm) slightly deshielded relative to the other two, probably attached to the carbon next to the carboxy group.
- Focusing next on the integrated values for the hydrogen signals, we note that their ratios relative to the single carboxy proton are 3:2:2. The highest-field absorption, at $\delta = 1.00$ ppm, integrates for 3 H and is likely due to the presence of a methyl group. The signal at $\delta = 2.31$ ppm, suspected to be relatively deshielded by the attached carboxy function, integrates for 2 H, pointing to the presence of a –CH$_2$COOH substructure.
- An analysis of the splitting patterns suggests that the methyl group, a triplet, has two equivalent hydrogen neighbors, indicating the presence of **CH$_3$–CH$_2$–**. Moreover, the signal at $\delta = 2.31$ ppm is also a triplet, extending the suggested substructure above to –CH$_2$**CH$_2$**COOH.
- Putting all of this information together leads to the solution: CH$_3$CH$_2$CH$_2$COOH (butanoic or "butyric" acid, with the smell of rancid butter). We can confirm this assignment by looking at the multiplicity of the ^{1}H NMR signal for the CH$_2$ fragment flanked by CH$_3$ on one, and CH$_2$ on the other side, CH$_3$**CH$_2$**CH$_2$COOH. Assuming that the coupling constants in both directions will be very similar, the ($N + 1$) rule predicts a sextet, as is indeed observed.

Exercise 19-3	Try It Yourself

Assign the structure of a carboxylic acid on the basis of the following NMR data: ^{1}H NMR: $\delta = 1.20$ (d, $J = 7.5$ Hz, 6 H), 2.58 (septet, $J = 7.5$ Hz, 1 H), and 12.17 (s, 1 H) ppm; ^{13}C NMR: $\delta = 18.7$, 33.8, and 184.0 ppm. (**Caution:** The ^{13}C NMR spectrum reveals only the number of *unique* carbon atoms in the molecule. **Hint:** The presence of a signal integrating for 6 H in the ^{1}H NMR spectrum is significant. *One* carbon cannot be attached to *six* hydrogens.)

The carboxy group shows two important IR bands

The carboxy group consists of a carbonyl group and an attached hydroxy substituent. Consequently, both characteristic stretching frequencies are seen in the infrared spectrum. The O–H bond gives rise to a very broad band at a lower wavenumber (2500–3300 cm^{-1}) than is observed for alcohols, largely because of strong hydrogen bonding. The IR spectrum of propanoic acid is shown in Figure 19-3.

Figure 19-3 IR spectrum of propanoic acid: $\tilde{\nu}_{\text{O–H stretch}} = 3000 \text{ cm}^{-1}$; $\tilde{\nu}_{\text{C=O stretch}} = 1715 \text{ cm}^{-1}$. The peaks associated with these stretching vibrations are broad because of hydrogen bonding.

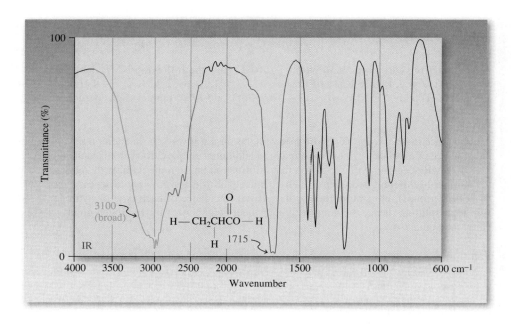

Mass spectra of carboxylic acids show three modes of fragmentation

The mass spectra of carboxylic acids typically show fairly weak molecular ions, because fragmentation occurs readily in several ways. In common with aldehydes and ketones, both α cleavage and McLafferty rearrangement are observed. In addition, cleavage at the C3–C4 bond occurs together with migration of an α-hydrogen to oxygen, because a resonance-stabilized carbocation results:

Mass Spectral Fragmentation of Carboxylic Acids

$m/z = 45$
Protonated CO_2

$m/z = 60$
Acetic acid enol

$m/z = 73$
Protonated propenoic acid

Figure 19-4 shows the mass spectrum of butanoic acid.

Exercise 19-4

Figure 19-5 shows the mass spectrum of pentanoic acid. Identify the peaks resulting from the three cleavage modes illustrated above.

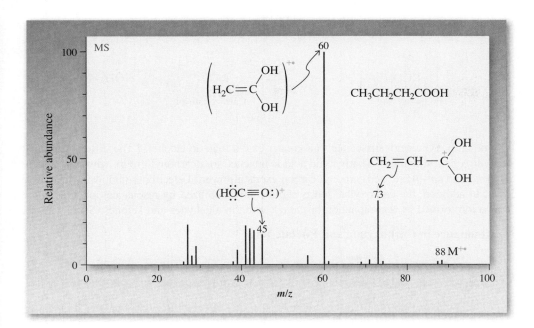

Figure 19-4 Mass spectrum of butanoic acid. The molecular ion and peaks arising from the cleavage modes described in the text are identified.

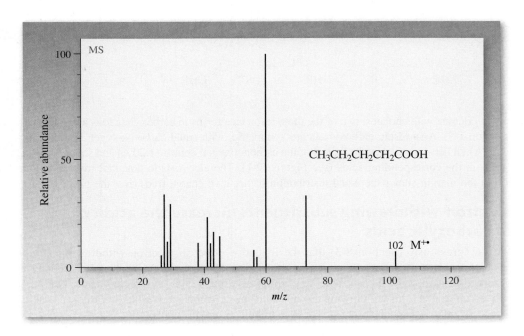

Figure 19-5 Mass spectrum of pentanoic acid. (See Exercise 19-4.) Note the relatively low intensity of the molecular ion peak.

In Summary The NMR signals for carboxylic acids reveal highly deshielded acid protons and carbonyl carbons and moderately deshielded nuclei next to the functional group. The infrared spectrum shows characteristic bands at about 2500–3300 (O–H) and 1710 (C=O) cm^{-1}. The mass spectra of carboxylic acids reflect three characteristic modes of fragmentation.

19-4 | ACIDIC AND BASIC CHARACTER OF CARBOXYLIC ACIDS

Like alcohols (Section 8-3), carboxylic acids exhibit both acidic and basic character: Deprotonation to carboxylate ions is relatively easy, protonation more difficult.

Carboxylic acids are relatively strong acids

Carboxylic acids have much lower pK_a values than alcohols do, even though the relevant hydrogen in each case is that of a hydroxy group.

Carboxylic Acids Dissociate Readily

$$\underset{\substack{K_a \approx 10^{-4}-10^{-5} \\ pK_a \approx 4-5}}{RC\overset{:O:}{\overset{\|}{\ddot{O}}H}} \quad + \quad H_2\ddot{O} \quad \rightleftharpoons \quad \underset{\text{Carboxylate ion}}{RC\overset{:O:}{\overset{\|}{\ddot{O}}}:^-} \quad + \quad H\overset{+}{\ddot{O}}H_2$$

Why do carboxylic acids dissociate to a greater extent than do alcohols? The difference is that the hydroxy substituent of a carboxylic acid is attached to a carbonyl group, whose positively polarized and sp^2-hybridized carbonyl carbon exerts a powerful electron-withdrawing inductive effect. In addition, the carboxylate ion is significantly stabilized by resonance, much as is the enolate ion formed by deprotonation of the α-carbon in aldehydes and ketones (Section 18-1).

Resonance in Carboxylate and Enolate Ions

Carboxylate ion

(B = base)

$$\underset{R}{\overset{:O:}{\underset{:\ddot{O}:}{C}}}\!\!H \; + \; :B^- \;\; \rightleftharpoons \;\; \left[\underset{R}{\overset{:O:}{C}}\overset{}{\underset{\ddot{O}:^-}{}} \;\; \longleftrightarrow \;\; \underset{R}{\overset{:\ddot{O}:^-}{C^+}}\overset{}{\underset{\ddot{O}:^-}{}} \;\; \longleftrightarrow \;\; \underset{R}{\overset{:\ddot{O}:^-}{C}}\overset{}{\underset{\ddot{O}}{}} \right] \; + \; BH \qquad pK_a \approx 4-5$$

Enolate ion

$$\underset{\substack{R' \quad \\ H \quad R}}{\overset{:O:}{C}}\!\!C\!\!H \; + \; :B^- \;\; \rightleftharpoons \;\; \left[\underset{R'}{\overset{:O:}{C}}\overset{}{\underset{CHR}{}} \;\; \longleftrightarrow \;\; \underset{R'}{\overset{:\ddot{O}:^-}{C^+}}\overset{}{\underset{CHR}{}} \;\; \longleftrightarrow \;\; \underset{R'}{\overset{:\ddot{O}:^-}{C}}\overset{}{\underset{CHR}{}} \right] \; + \; BH \qquad pK_a \approx 19-21$$

Acetate ion

In contrast with enolates, two of the three resonance forms in carboxylate ions are equivalent (Section 1-5). As a result, carboxylates are symmetric, with equal carbon–oxygen bond lengths (1.26 Å), in between the lengths typical of the carbon–oxygen double (1.20 Å) and single (1.34 Å) bonds in the corresponding acids (see Figure 19-1). The electrostatic potential map of acetate ion in the margin shows the equal distribution of negative charge (red) over the two oxygens.

Electron-withdrawing substituents increase the acidity of carboxylic acids

We saw previously (Section 8-3) that the inductive effect of electron-withdrawing groups close to the hydroxy function increases the acidity of alcohols. A similar phenomenon is observed in carboxylic acids. Table 19-3 shows the pK_a values of selected acids. Note that two or three electron-withdrawing groups on the α-carbon can result in carboxylic acids whose strength approaches that of some inorganic (mineral) acids.

Exercise 19-5

In the following sets of acids, rank the compounds in order of *decreasing* acidity.

(a) CH_3CH_2COOH $CH_3\overset{\overset{\displaystyle Br}{|}}{C}HCOOH$ CH_3CBr_2COOH

(b) $CH_3\overset{\overset{\displaystyle F}{|}}{C}HCH_2COOH$ $CH_3\overset{\overset{\displaystyle Br}{|}}{C}HCH_2COOH$

(c)

(cyclohexane ring with COOH)   (cyclohexane ring with F and COOH)   (cyclohexane ring with COOH and F)

Table 19-3	pKₐ Values of Various Carboxylic and Other Acids		
Compound	pK_a	**Compound**	pK_a
Alkanoic acids		**Dioic acids**	
HCOOH	3.55	HOOCCOOH	1.27, 4.19
CH_3COOH	4.76	$HOOCCH_2COOH$	2.83, 5.69
$ClCH_2COOH$	2.82	$HOOCCH_2CH_2COOH$	4.20, 5.61
$Cl_2CHCOOH$	1.26	$HOOC(CH_2)_4COOH$	4.35, 5.41
Cl_3CCOOH	0.63		
F_3CCOOH	0.23	**Other acids**	
$CH_3CH_2CH_2COOH$	4.82	H_3PO_4	2.15 (first pK_a)
$CH_3CH_2CHClCOOH$	2.84	HNO_3	−1.4
$CH_3CHClCH_2COOH$	4.06	H_2SO_4	−3.0 (first pK_a)
$ClCH_2CH_2CH_2COOH$	4.52	HCl	−8.0
		H_2O	15.7
Benzoic acids		CH_3OH	15.5
$4\text{-}CH_3C_6H_4COOH$	4.36		
C_6H_5COOH	4.20		
$4\text{-}ClC_6H_4COOH$	3.98		

The dioic acids have two pK_a values, one for each CO_2H group. In ethanedioic (oxalic) and propanedioic (malonic) acids, the first pK_a is lowered by the electron-withdrawing inductive effect of the second carboxy group on the first. In the higher dioic acids, the two pK_a values are close to those of monocarboxylic acids.

The relatively strong acidity of carboxylic acids means that their corresponding **carboxylate salts** are readily made by treatment of the acid with base, such as NaOH, Na_2CO_3, or $NaHCO_3$. These salts are named by specifying the metal and replacing the ending -*ic acid* in the acid name by **-ate.** Thus, $HCOO^-Na^+$ is called sodium formate; $CH_3COO^-Li^+$ is named lithium acetate; and so forth. Carboxylate salts are much more water soluble than the corresponding acids because the polar anionic group is readily solvated.

Carboxylate Salt Formation

CH₃CCH₂CH₂COOH —(NaOH, H₂O)→ CH₃CCH₂CH₂COO⁻Na⁺ + HOH

4,4-Dimethylpentanoic acid
(Slightly water soluble)

Sodium 4,4-dimethylpentanoate
(Highly water-soluble salt)

Carboxylic acids may be protonated on the carbonyl oxygen

The lone electron pairs on both of the oxygen atoms in the carboxy group can, in principle, be protonated, just as alcohols are protonated by strong acids to give alkyloxonium ions (Sections 8-3 and 9-2). It has been found that it is the carbonyl oxygen that is more basic. Why? Protonation at the carbonyl oxygen gives a species whose positive charge is delocalized by resonance. The alternative possibility, resulting from protonation at the hydroxy group, does not benefit from such stabilization (see also Exercise 2-9).

Protonation of a Carboxylic Acid

Positive charge is stabilized by resonance

Resonance stabilization is not possible

Not observed

$K \approx 10^{-6}$

$pK_a \approx -6$

Note that protonation is nevertheless very difficult, as is shown by the very high acidity ($pK_a \approx -6$) of the conjugate acid. We shall see, however, that such protonations are important in many reactions of the carboxylic acids and their derivatives.

Exercise 19-6

The pK_a of protonated acetone is -7.2 and that of protonated acetic acid is -6.1. Explain.

In Summary Carboxylic acids are acidic because the polarized carbonyl carbon is strongly electron withdrawing and deprotonation gives resonance-stabilized anions. Electron-withdrawing groups increase acidity, although this effect wears off rapidly with increasing distance from the carboxy group. Protonation is difficult, but possible, and occurs on the carbonyl oxygen to give a resonance-stabilized cation.

19-5 | CARBOXYLIC ACID SYNTHESIS IN INDUSTRY

Carboxylic acids are useful reagents and synthetic precursors. The two simplest ones are manufactured on a large scale industrially. Formic acid is employed in the tanning process in the manufacture of leather and in the preparation of latex rubber. It is synthesized efficiently by the reaction of powdered sodium hydroxide with carbon monoxide under pressure. This transformation proceeds by nucleophilic addition followed by protonation.

Formic Acid Synthesis

$$NaOH + CO \xrightarrow{150°C,\ 100\ psi} HCOO^-Na^+ \xrightarrow{H^+,\ H_2O} HCOOH$$

There are three important industrial preparations of acetic acid: ethene oxidation through acetaldehyde (Section 12-16); air oxidation of butane; and carbonylation of methanol. The mechanisms of these reactions are complex.

Acetic Acid by Oxidation of Ethene

$$CH_2=CH_2 \xrightarrow[\text{Wacker process}]{O_2,\ H_2O,\ \text{catalytic}\ PdCl_2\ \text{and}\ CuCl_2} CH_3CHO \xrightarrow{O_2,\ \text{catalytic}\ Co^{3+}} CH_3COOH$$

Acetic Acid by Oxidation of Butane

$$CH_3CH_2CH_2CH_3 \xrightarrow{O_2,\ \text{catalytic}\ Co^{3+},\ 15–20\ atm,\ 180°C} CH_3COOH$$

Acetic Acid by Carbonylation of Methanol

$$CH_3OH \xrightarrow[\text{Monsanto process}]{CO,\ \text{catalytic}\ Rh^{3+},\ I_2,\ 30–40\ atm,\ 180°C} CH_3COOH$$

Global demand of acetic acid exceeds 6.5 million tons (6.5×10^9 kg) annually. It is used to manufacture monomers for polymerization, such as methyl 2-methylpropenoate (methyl methacrylate; Table 12-3), as well as pharmaceuticals, dyes, and pesticides. Two dicarboxylic acids in large-scale chemical production are hexanedioic (adipic) acid, used in the manufacture of nylon (see Real Life 21-3), and 1,4-benzenedicarboxylic (terephthalic) acid, whose polymeric esters with diols are fashioned into plastic sheets, films, and bottles for soft drinks.

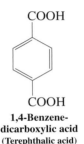

**1,4-Benzene-dicarboxylic acid
(Terephthalic acid)**

19-6 | METHODS FOR INTRODUCING THE CARBOXY FUNCTIONAL GROUP

The oxidation of primary alcohols and aldehydes to carboxylic acids by aqueous Cr(VI) was described in Sections 8-6 and 17-4. This section presents two additional reagents suitable for this purpose. It is also possible to introduce the carboxy function by adding a carbon atom to a haloalkane. This transformation can be achieved in either of two ways: the carbonation of organometallic reagents or the preparation and hydrolysis of nitriles.

Oxidation of primary alcohols and of aldehydes furnishes carboxylic acids

Primary alcohols oxidize to aldehydes, which in turn may readily oxidize further to the corresponding carboxylic acids.

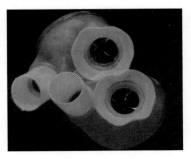

Polyethylene terephthalate is Dacron, used in the Jarvik artificial heart. As a thin film, it is Mylar.

Carboxylic Acids by Oxidation

$$RCH_2OH \xrightarrow{\text{Oxidation}} \overset{\displaystyle O}{\underset{\displaystyle \|}{R}CH} \xrightarrow{\text{Oxidation}} \overset{\displaystyle O}{\underset{\displaystyle \|}{R}COH}$$

In addition to aqueous CrO_3, $KMnO_4$ and nitric acid (HNO_3) are frequently used in this process. Because it is one of the cheapest strong oxidants, nitric acid is often chosen for large-scale and industrial applications.

$$2\ HNO_3\ +\ ClCH_2CH_2\overset{\displaystyle O}{\overset{\displaystyle \|}{C}}H \xrightarrow{25°C} ClCH_2CH_2\overset{\displaystyle O}{\overset{\displaystyle \|}{C}}OH\ +\ 2\ NO_2\ +\ H_2O$$

                         3-Chloropropanal   **3-Chloropropanoic acid**
                                                                  79%

Exercise 19-7

Give the products of nitric acid oxidation of (a) pentanal; (b) 1,6-hexanediol; (c) 4-(hydroxymethyl)-cyclohexanecarbaldehyde.

Organometallic reagents react with carbon dioxide to give carboxylic acids

Organometallic reagents attack carbon dioxide (usually in the solid form known as "dry ice") much as they would attack aldehydes or ketones. The product of this **carbonation** process is a carboxylate salt, which upon protonation by aqueous acid yields the carboxylic acid.

Carbonation of Organometallics

$$\overset{\delta^-}{R}-\overset{\delta^+}{MgX}\ +\ \overset{\delta^-:\ddot{O}:}{\underset{\delta^-:\ddot{O}:}{\overset{\delta^+}{C}}} \xrightarrow{THF} R-\overset{:\ddot{O}:^+MgX}{\underset{\cdot\ddot{O}\cdot}{C}} \xrightarrow[-XMgOH]{H^+,\ HOH} RCOOH$$

$$RLi\ +\ CO_2 \xrightarrow{THF} RCOO^-Li^+ \xrightarrow[-LiOH]{H^+,\ HOH} RCOOH$$

Recall that an organometallic reagent is usually made from the corresponding haloalkane: RX + Mg → RMgX. Hence, carbonation of the organometallic reagent allows the two-step transformation of RX into RCOOH, the carboxylic acid with one more carbon. For example,

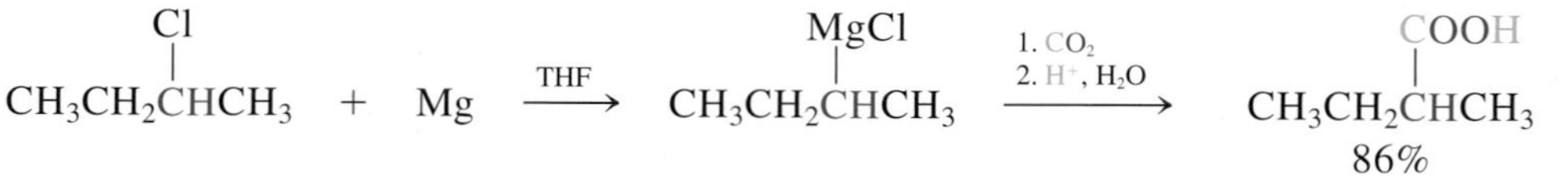

$$
\underset{\substack{\text{Cl}\\|\\\text{CH}_3\text{CH}_2\text{CHCH}_3}}{} + \text{Mg} \xrightarrow{\text{THF}} \underset{\substack{\text{MgCl}\\|\\\text{CH}_3\text{CH}_2\text{CHCH}_3}}{} \xrightarrow[\text{2. H}^+,\text{H}_2\text{O}]{\text{1. CO}_2} \underset{\substack{\text{COOH}\\|\\\text{CH}_3\text{CH}_2\text{CHCH}_3}}{}
$$

2-Chlorobutane

86%
2-Methylbutanoic acid

Nitriles hydrolyze to carboxylic acids

Another method for converting a haloalkane into a carboxylic acid with an additional carbon atom is through the preparation and hydrolysis of a nitrile, RC≡N. Recall (Section 6-2) that cyanide ion, ⁻:C≡N:, is a good nucleophile and may be used to synthesize nitriles through S_N2 reactions. Hydrolysis of the nitrile in hot acid or base furnishes the corresponding carboxylic acid (and ammonia or ammonium ion).

Carboxylic Acids from Haloalkanes Through Nitriles

$$
\text{RX} \xrightarrow[-\text{X}^-]{\text{⁻CN}} \text{RC≡N} \xrightarrow[-\text{NH}_3 \text{ or NH}_4^+]{\substack{\text{1. HO}^-\\\text{2. H}^+,\text{H}_2\text{O}}} \text{RCOOH}
$$

The mechanism of this reaction will be described in detail in Section 20-8.

Carboxylic acid synthesis with the use of nitrile hydrolysis is preferable to Grignard carbonation when the substrate contains other groups that react with organometallic reagents, such as the hydroxy, carbonyl, and nitro functionalities.

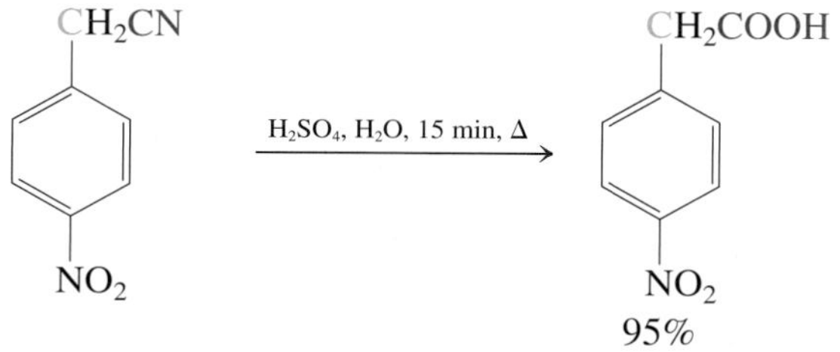

$$
\xrightarrow{\text{H}_2\text{SO}_4,\ \text{H}_2\text{O},\ 15\ \text{min},\ \Delta}
$$

(4-Nitrophenyl)-
ethanenitrile

95%
(4-Nitrophenyl)-
acetic acid

Solved Exercise 19-8 | Working with the Concepts: Carboxylic Acids from Haloalkanes

Suggest a method for accomplishing the conversion shown below. Several steps may be necessary.

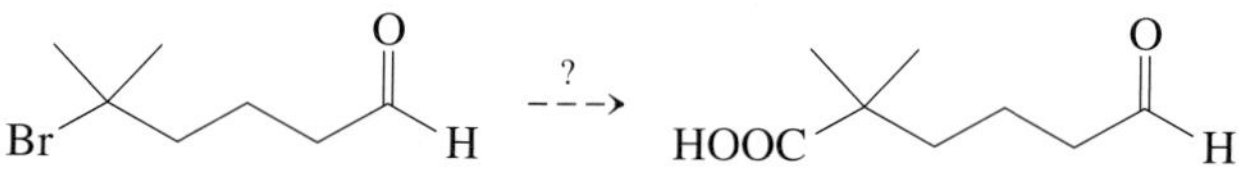

Strategy

Before you start, inspect both structures carefully. Notice that the target molecule has an additional carbon atom that must be introduced. The last two methods in the section above provide possible options, but neither one can be applied directly to the starting material. The carbonyl function will interfere with any attempt to proceed using organometallic reagents and will also react with cyanide ion. Let us look at each of these options.

We know from Section 17-11 that cyanide addition is reversible and can be avoided under basic conditions. Can we therefore displace bromide by cyanide and follow with hydrolysis under basic conditions? Look more closely at the substrate—Br is attached to a *tertiary* carbon. The necessary S_N2 displacement will not occur; only elimination will take place (Section 7-8).

Fortunately, organometallic reagents can be prepared from almost any type of halogenated compound, including tertiary haloalkanes, provided that interference from other functional groups can be avoided. This realization gives us the framework for a solution.

Solution

- Protect the aldehyde group in a way that will allow formation and use of a Grignard reagent: Form the cyclic acetal using 1,2-ethanediol and acid catalysis.
- Following the sequence in this section, prepare the Grignard reagent and add it to CO_2.
- Protonate the resulting carboxylate under acidic, aqueous conditions. The acetal hydrolyzes as well, deprotecting the aldehyde function:

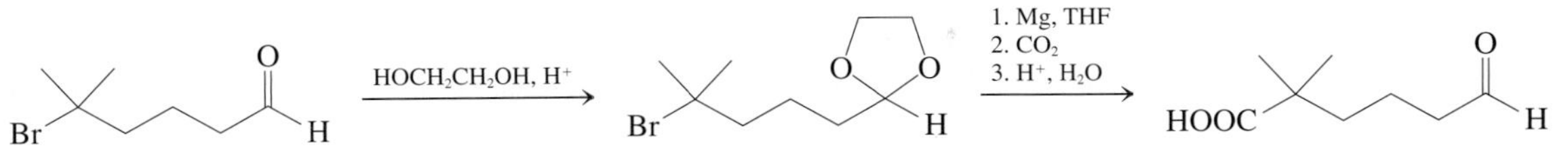

Exercise 19-9 | Try It Yourself

Using chemical equations, show how you would convert each of the following halogenated compounds into a carboxylic acid with one additional carbon atom. If more than one method can be used successfully, show all. If not, give the reason behind your choice of approach. **(a)** 1-Chloropentane; **(b)** iodocyclopentane; **(c)** 4-bromobutanoic acid (**Hint:** See Sections 8-7 and 8-9); **(d)** chloroethene (**Hint:** See Section 13-9); **(e)** bromocyclopropane (**Hint:** See Problem 60 of Chapter 6).

Hydrolysis of the nitrile group in a cyanohydrin, prepared by addition of HCN to an aldehyde or a ketone (Section 17-11), is a general route to 2-hydroxycarboxylic acids, which possess antiseptic properties.

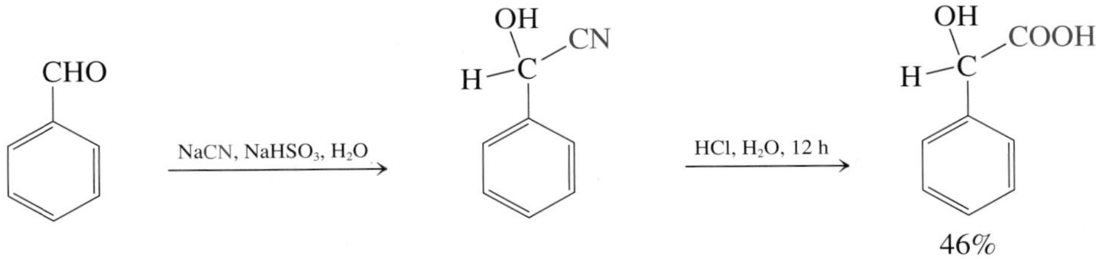

Benzaldehyde **2-Hydroxy-2-phenyl-ethanenitrile (Mandelonitrile)** **2-Hydroxy-2-phenyl-acetic acid (Mandelic acid)**

Exercise 19-10 |

Suggest ways to effect the following conversions (more than one step will be required).

(a) 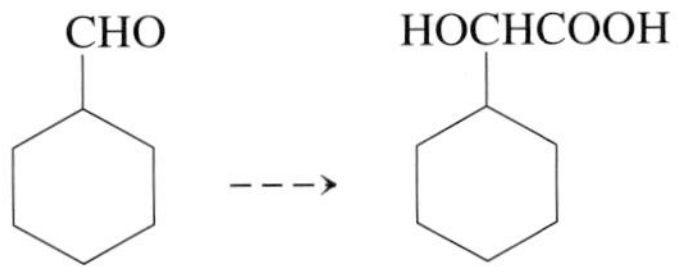

(b) 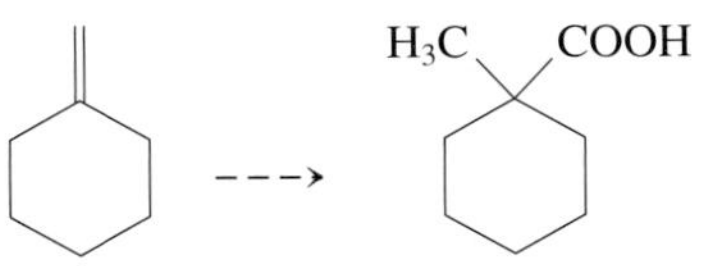

(c) 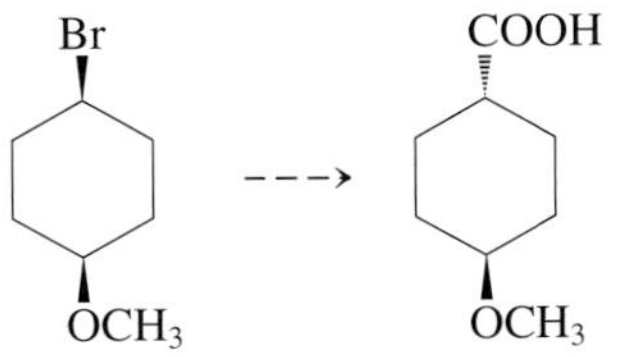

(**Caution:** Pay attention to the stereochemistry!)

In Summary Several reagents oxidize primary alcohols and aldehydes to carboxylic acids. A haloalkane is transformed into the carboxylic acid containing one additional carbon atom either by conversion into an organometallic reagent and carbonation or by displacement of halide by cyanide followed by nitrile hydrolysis.

19-7 | SUBSTITUTION AT THE CARBOXY CARBON: THE ADDITION–ELIMINATION MECHANISM

The carbonyl group in carboxylic acids shows reactivity similar to that in aldehydes and ketones: It is subject to attack by nucleophiles at carbon and electrophiles at oxygen. However, the presence of the carboxy OH group in the structure adds another dimension to the chemical function of carboxylic acids: Just as in alcohols, this OH may be converted into a leaving group (Section 9-2). As a result, after nucleophilic addition to the carbonyl carbon takes place, the leaving group may depart, resulting in a net substitution process and a new carbonyl compound. This section introduces this process and the general mechanisms by which it takes place.

The carbonyl carbon is attacked by nucleophiles

Carboxylic acid derivative

As we have learned from our study of aldehydes and ketones, carbonyl carbons are electrophilic and can be attacked by nucleophiles. This type of reactivity is observed in the carboxylic acids and the **carboxylic acid derivatives,** substances with the general formula RCOL (L stands for leaving group).

Carboxylic Acid Derivatives

$$
\underset{\substack{\textbf{Acyl}\\\textbf{halide}}}{\overset{\overset{\displaystyle O}{\|}}{RCX}} \qquad
\underset{\textbf{Anhydride}}{\overset{\overset{\displaystyle O}{\|}\;\;\overset{\displaystyle O}{\|}}{RCOCR}} \qquad
\underset{\textbf{Ester}}{\overset{\overset{\displaystyle O}{\|}}{RCOR'}} \qquad
\underset{\textbf{Amide}}{\overset{\overset{\displaystyle O}{\|}}{RCNR'_2}}
$$

In contrast to the addition products of aldehydes and ketones (Sections 17-5 through 17-7), the intermediate formed upon attack of a nucleophile on a carboxy carbon can decompose *by eliminating a leaving group.* The result is substitution of the nucleophile for the leaving group via a process called **addition–elimination.**

Substitution at the carboxy carbon occurs by addition–elimination

The carbonyl carbon is trigonal and sp^2 hybridized. As we have seen in the reactions of alkenes and benzenes (Chapters 12 and 15), the substitution mechanisms associated with saturated, tetrahedral, sp^3-hybridized carbon atoms do not typically occur in unsaturated, trigonal planar systems. Their planarity renders backside, S_N2-type displacement geometrically difficult, and sp^2-hybridized carbon atoms form poor carbocations, disfavoring S_N1.

Substitution by the addition–elimination sequence combines two simple mechanistic processes. Addition of a nucleophile to the carbon of a carbonyl group is a reaction we know well. Elimination is just the mechanistic reverse of the addition process. Mechanism type 4(a) from Section 2-2 may be used to illustrate both, as follows.

Forward direction—addition: $\quad X\!:^- \;+\; \overset{\delta^+}{A}\!=\!\overset{\delta^-}{B} \;\longrightarrow\; X\!-\!A\!-\!B\!:^-$

Reverse direction—elimination: $\quad X\!-\!A\!-\!B\!:^- \;\longrightarrow\; X\!:^- \;+\; A\!=\!B$

When the site of attack (atom A) already contains a potential leaving group (Y), then the two steps above can occur one after the other to give net substitution of X for Y:

General Mechanistic Pattern for an Addition–Elimination Sequence

$$
X\!:^- \quad \overset{Y}{\underset{}{A}}\!=\!B \quad \underset{\text{Addition of }X^-}{\rightleftharpoons} \quad X\!-\!\overset{Y}{\underset{}{A}}\!-\!B\!:^- \quad \underset{\text{Elimination of }Y^-}{\rightleftharpoons} \quad \overset{X}{\underset{}{A}}\!=\!B \;+\; Y\!:^-
$$

As this general scheme shows, both steps in an addition–elimination sequence may be reversible. The scheme on the next page shows addition–elimination as applied to a carboxylic acid derivative. The intermediate species formed after addition but before elimination (in contrast to both starting material and product) contains a tetrahedral carbon center. It is therefore called a **tetrahedral intermediate.**

Nucleophilic Substitution by Addition–Elimination

Reaction

Color code reminder:
Nucleophile—red
Electrophile—blue
Leaving group—green

Carboxylic acid
derivative

Tetrahedral
intermediate

Substitution
product

Substitution by addition–elimination is the most important pathway for the formation of carboxylic acid derivatives as well as for their interconversion—that is, RCL → RCL′. The remainder of this section and those that follow will describe how such derivatives are prepared from carboxylic acids, and Chapter 20 will explore their properties and chemistry.

Addition–elimination is catalyzed by both base and acid

Addition–elimination reactions may be carried out under either basic or acidic conditions. We have seen how additions of nucleophiles to aldehydes and ketones (Sections 17-5 through 17-9; Table 17-4) may be catalyzed by either bases or acids. The same is true for additions of nucleophiles to carboxylic acid derivatives. Eliminations from the tetrahedral intermediate are similarly catalyzed: Recall that this process is mechanistically just the reverse of addition; therefore, the same catalytic effects are observed. Let us examine the roles of both base and acid in detail.

All nucleophiles are Lewis bases. However, a nucleophile with a removable proton (Nu–H) may be deprotonated by a strong base (denoted as :B⁻) to give the negatively charged and thus more strongly nucleophilic :Nu⁻, which is the attacking species (Step 1). Addition–elimination occurs (Step 2), and the base may be regenerated (Step 3), making it a catalyst for the overall process. Typical nucleophiles in sequences such as this include water and alcohols, which upon deprotonation give hydroxide and alkoxide ions, respectively.

Mechanism of Base-Catalyzed Addition–Elimination

Step 1. Deprotonation of NuH

$$Nu{-}H \ + \ {^-}{:}B \ \rightleftharpoons \ {^-}{:}Nu \ + \ BH$$

Mechanism

Step 2. Addition–elimination

Step 3. Regeneration of catalyst

$$^-{:}L \ + \ H{-}B \ \rightleftharpoons \ LH \ + \ ^-{:}B$$

(Alternatively, ⁻:L may act as a base in step 1)

In cases where the nucleophile is already strong, base catalysis is unnecessary, and the entire mechanism may be described by Step 2, on the previous page.

Addition–elimination reactions may also benefit from acid catalysis. The acid functions in two ways: First, it protonates the carbonyl oxygen (Step 1), activating the carbonyl group toward nucleophilic attack (Section 17-5). Second, protonation of L (Step 2) makes it a better leaving group (recall Section 6-7 and 9-2).

Mechanism of Acid-Catalyzed Addition–Elimination

Step 1. Protonation

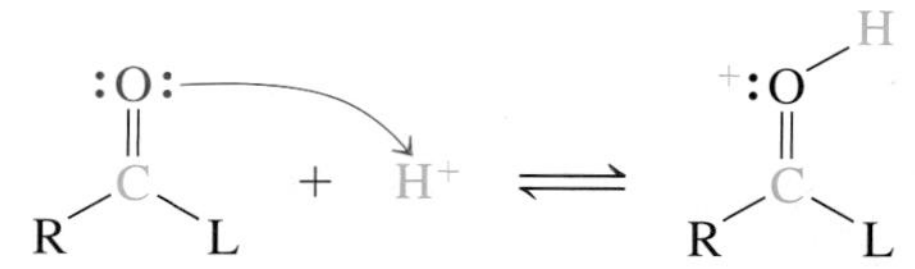

Step 2. Addition–elimination

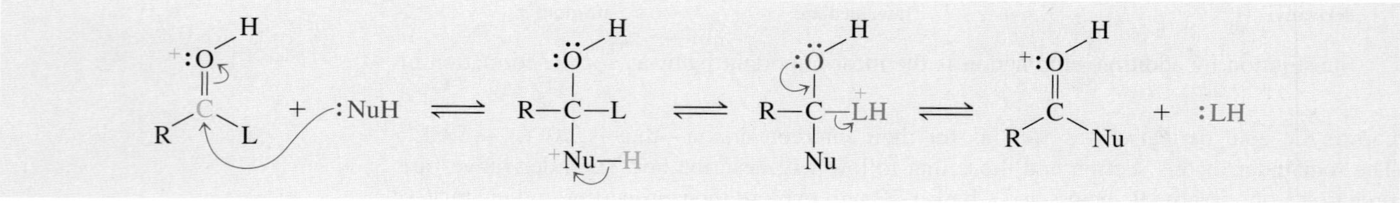

Step 3. Deprotonation; regeneration of catalytic proton

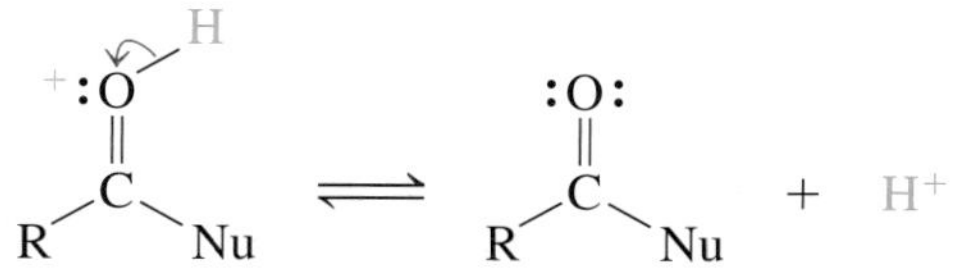

Substitution in carboxylic acids is inhibited by a poor leaving group and the acidic proton

We can apply the general addition–elimination process to the conversion of carboxylic acids into their derivatives. However, we must first overcome two problems. First, we know that the hydroxide ion is a poor leaving group (Section 6-7). Second, the carboxy proton is acidic, and most nucleophiles are bases (Section 6-8). Therefore, the desired nucleophilic attack (path a in the following equation) can encounter interference from an acid-base reaction (path b). Indeed, if a nucleophile is very basic, such as alkoxide, formation of the carboxylate ion will be essentially irreversible, and nucleophilic addition to the carbonyl carbon becomes very difficult.

Competing Reactions of a Carboxylic Acid with a Nucleophile

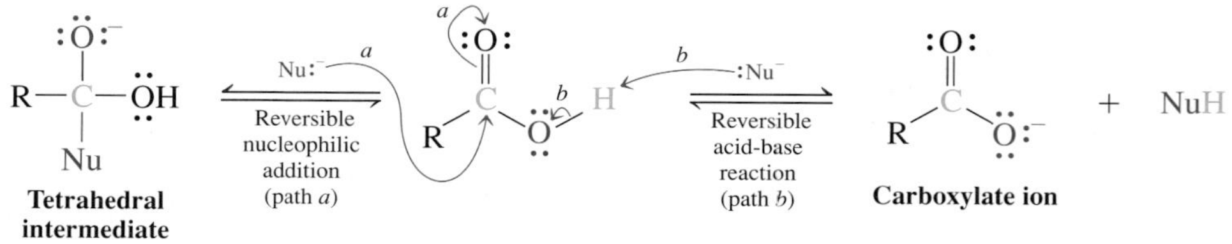

Tetrahedral intermediate   Reversible nucleophilic addition (path a)   Reversible acid-base reaction (path b)   **Carboxylate ion**

Esterification

$$RCOOH \ + \ R'OH$$

$$\xrightarrow{H^+}$$

$$RCOOR' \ + \ H_2O$$

On the other hand, with less basic nucleophiles, especially under acidic conditions, the ready reversibility of carboxylate formation may permit nucleophilic addition to compete and ultimately lead to substitution through the addition–elimination mechanism. A typical example is the **esterification** of a carboxylic acid (Section 9-4), in which an alcohol and a carboxylic acid react to yield an ester and water. The nucleophile, an alcohol, is a weak base, and acid is present to protonate both the carbonyl oxygen, activating it toward nucleophilic addition, and the carboxy OH, converting it into a better leaving group, water.

Acid Catalysis in Esterification

Addition

Starting carboxylic acid → (H⁺, Protonation) → Protonated carboxylic acid → (R'ÖH, Addition) → Protonated tetrahedral intermediate → (−H⁺, Deprotonation) → Tetrahedral intermediate

Elimination

Tetrahedral intermediate → (H⁺, Protonation) → Protonated tetrahedral intermediate → (−H₂Ö, Elimination) → Protonated ester → (−H⁺, Deprotonation) → Final ester product

> **Note:** Every step in this mechanism is reversible, including all protonations and deprotonations. We shall see soon that this fact allows us to drive the overall equilibrium either as written, from the acid to the ester, or in the opposite direction, from the ester back to the acid, by adjusting the reaction conditions.

The sections that follow will examine this and other carboxy substitutions in detail.

Exercise 19-11

Write out the full mechanism for the acid-catalyzed reaction of methanol with acetic acid.

In Summary Nucleophilic attack on the carbonyl group of carboxylic acid derivatives is a key step in substitution by addition–elimination. Either acid or base catalysis may be observed. For carboxylic acids, the process is complicated by the poor leaving group (hydroxide) and competitive deprotonation of the acid by the nucleophile, acting as a base. With less basic nucleophiles, addition can occur.

19-8 CARBOXYLIC ACID DERIVATIVES: ACYL HALIDES AND ANHYDRIDES

With this section, we begin a survey of the preparation of carboxylic acid derivatives. Replacement of the hydroxy group in RCOOH by halide gives rise to **acyl halides;** substitution by alkanoate (RCOO⁻) furnishes **carboxylic anhydrides.** *Both processes first require transformation of the hydroxy functionality into a better leaving group.*

Acyl halides are formed by using inorganic derivatives of carboxylic acids

The conversion of carboxylic acids into acyl halides employs the same reagents used in the synthesis of haloalkanes from alcohols (Section 9-4): $SOCl_2$ and PBr_3. The problem to be solved in both cases is identical—changing a poor leaving group (OH) into a good one.

$$\underset{\text{Acyl halide}}{\overset{\overset{\displaystyle O}{\|}}{RCX}}$$

Acyl Halide Synthesis

$$\underset{\textbf{Butanoic acid}}{CH_3CH_2CH_2\overset{\overset{\displaystyle O}{\|}}{C}OH} \xrightarrow[\substack{-O=S=O \\ -HCl}]{ClSCl, \text{ reflux}} \underset{\substack{\textbf{Butanoyl chloride} \\ 85\%}}{CH_3CH_2CH_2\overset{\overset{\displaystyle O}{\|}}{C}Cl}$$

Reaction

$$3\ \cdot\overset{\overset{\displaystyle O}{\|}}{C}OH \xrightarrow[-H_3PO_3]{PBr_3} 3\ \cdot\overset{\overset{\displaystyle O}{\|}}{C}Br$$

$$90\%$$

(These reactions fail with formic acid, HCOOH, because formyl chloride, HCOCl, and formyl bromide, HCOBr, are unstable. See Exercise 15-33.)

These transformations begin with the conversion of the carboxylic acid into an inorganic derivative in which the substituent on the carbonyl carbon is a good leaving group:

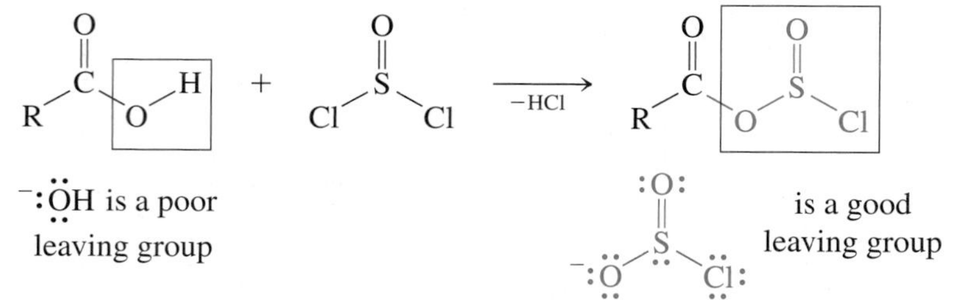

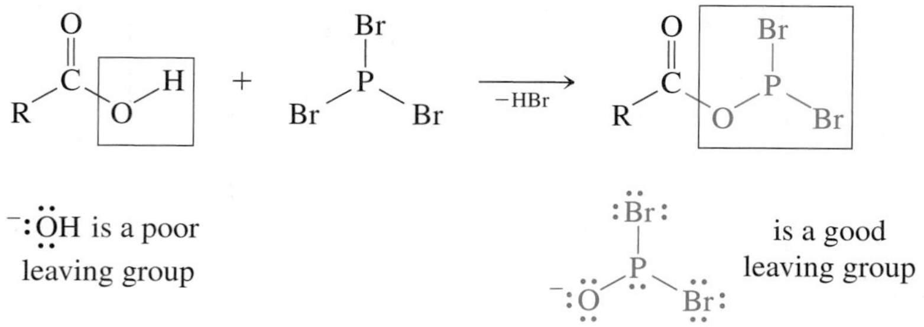

-:ÖH is a poor leaving group

:O: is a good leaving group

-:ÖH is a poor leaving group

:B̈r: is a good leaving group

Solved Exercise 19-12 | **Working with the Concepts: Acyl Chlorosulfite Formation**

Propose a mechanism for the reaction of SOCl$_2$ with a carboxylic acid to give the inorganic derivative (an acyl chlorosulfite) shown above.

Strategy

It is tempting to employ the same steps for this problem as we used in the conversion of an alcohol into the corresponding alkyl chlorosulfite, ROSOCl, upon reaction with SOCl$_2$ (Section 9-4), namely, displacement of chloride from sulfur by the hydroxy group of the alcohol:

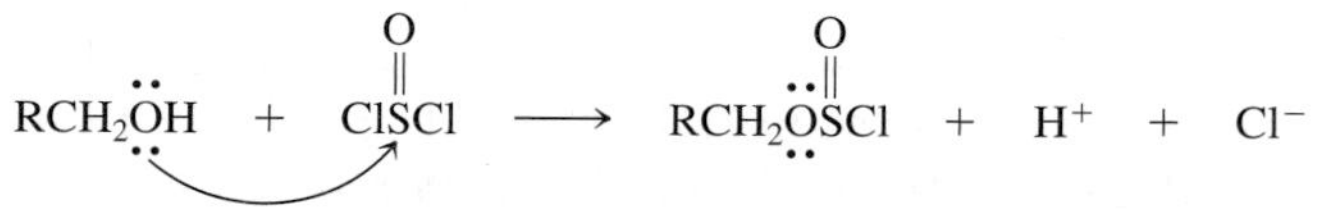

$$RCH_2\ddot{O}H \ + \ Cl\ddot{S}Cl \longrightarrow RCH_2\ddot{O}\overset{..}{S}Cl \ + \ H^+ \ + \ Cl^-$$

However, direct application of this mechanism to a carboxylic acid is not likely to be correct. We saw in Section 19-5 that protonation of a carboxylic acid occurs at the *carbonyl* oxygen atom, *not* on the oxygen of the hydroxy group. The reason is straightforward: Protonation of the carbonyl oxygen gives a resonance-stabilized intermediate, whereas reaction at OH does not. The same can be said of reaction of a carboxylic acid with any electrophile, such as the sulfur in SOCl$_2$ or the phosphorus in PBr$_3$.

Solution

- It is more probable that the transformation in question proceeds as follows:

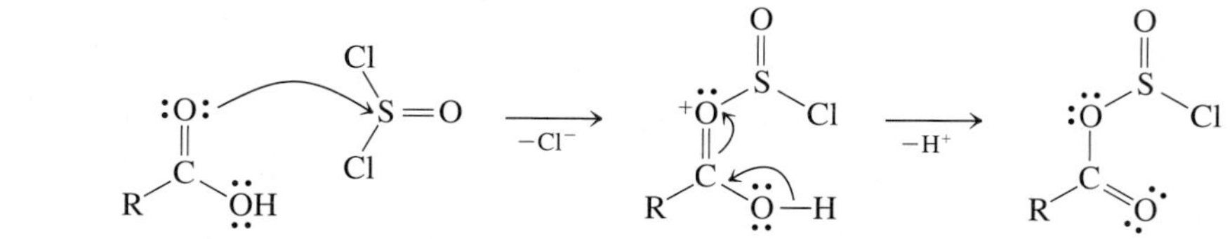

- This reaction pathway converts the carbonyl oxygen, rather than the hydroxy, into the leaving group.
- To our knowledge, this specific mechanism has not been tested experimentally. We present it as an example of a mechanistic hypothesis that is proposed for a set of reacting species, based on experimental observations of other similar species.

Exercise 19-13 | Try It Yourself

Propose a mechanism for the reaction of PBr$_3$ with a carboxylic acid to give the inorganic derivative (an acyl dibromophosphite) shown above Exercise 19-12.

Acid-catalyzed addition–elimination follows.

Mechanism of Acyl Chloride Formation with Thionyl Chloride

Mechanism

Step 1. Addition

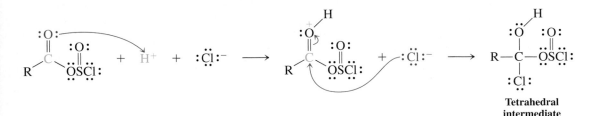

Tetrahedral
intermediate

Step 2. Elimination

The mechanism of acyl bromide formation using phosphorus tribromide (PBr$_3$) is similar (see Section 9-4).

Acids combine with acyl halides to produce anhydrides

The electronegative power of the halogens in acyl halides activates the carbonyl function to attack by other, even weak, nucleophiles (Chapter 20). For example, treatment of acyl halides with carboxylic acids results in **carboxylic anhydrides.**

$$\underset{\text{Carboxylic anhydride}}{\overset{\overset{\displaystyle O \quad O}{\|\quad\|}}{RCOCR}}$$

$$\underset{\text{Butanoic acid}}{CH_3CH_2CH_2\overset{\overset{\displaystyle O}{\|}}{C}OH} \quad + \quad \underset{\text{Butanoyl chloride}}{Cl\overset{\overset{\displaystyle O}{\|}}{C}CH_2CH_2CH_3} \quad \xrightarrow[-HCl]{\Delta,\, 8\, h}$$

$$\underset{\substack{85\% \\ \text{Butanoic anhydride}}}{CH_3CH_2CH_2\overset{\overset{\displaystyle O \quad O}{\|\quad\|}}{C}OCCH_2CH_2CH_3}$$

As the name indicates, carboxylic anhydrides are formally derived from the corresponding acids by loss of water. Although carboxylic acid dehydration is not a general method for anhydride synthesis, cyclic examples may be prepared by heating dicarboxylic acids. A condition for the success of this transformation is that the ring closure lead to a five- or six-membered ring product.

Cyclic Anhydride Formation

$$H_2C \overset{COOH}{\underset{\underset{COOH}{\,}}{|}} \xrightarrow[-H_2O]{300°C} \text{Butanedioic anhydride}$$

95%

Butanedioic acid
(Succinic acid)

Butanedioic anhydride
(Succinic anhydride)

Because the halogen in the acyl halide and the RCO_2 substituent in the anhydride are good leaving groups and because they activate the adjacent carbonyl function, these carboxylic acid derivatives are useful synthetic intermediates for the preparation of other compounds, a topic to be discussed in Sections 20-2 and 20-3.

Exercise 19-14

Suggest two preparations, starting from carboxylic acids or their derivatives, for each of the following compounds.

(a) $CH_3\overset{O}{\overset{\|}{C}}O\overset{O}{\overset{\|}{C}}CH_2CH_3$

(b) $CH_3\overset{H_3C}{\overset{|}{C}}H\overset{O}{\overset{\|}{C}}Cl$

Exercise 19-15

Propose a mechanism for the thermal formation of butanedioic anhydride from the dioic acid.

In Summary The hydroxy group in COOH can be replaced by halogen by using the same reagents used to convert alcohols into haloalkanes—$SOCl_2$ and PBr_3. The resulting acyl halides are sufficiently reactive to be attacked by carboxylic acids to generate carboxylic anhydrides. Cyclic anhydrides may be made from dicarboxylic acids by thermal dehydration.

19-9 | CARBOXYLIC ACID DERIVATIVES: ESTERS

Esters have the general formula $R\overset{O}{\overset{\|}{C}}OR'$. Their widespread occurrence in nature and many practical uses make them perhaps the most important of the carboxylic acid derivatives. This section describes the mineral acid-catalyzed reaction of carboxylic acids with alcohols that forms esters.

Carboxylic acids react with alcohols to form esters

When a carboxylic acid and an alcohol are mixed together, no reaction takes place. However, upon addition of catalytic amounts of a mineral acid, such as sulfuric acid or HCl, the two components combine in an equilibrium process to give an ester and water (Section 9-4). This method of ester formation was first described by the legendary German chemist Emil Fischer and his coworker Arthur Speier in 1895. It is therefore called Fischer-Speier or (more commonly) Fischer esterification.

Reaction

Acid-Catalyzed (Fischer) Esterification

$$R\overset{O}{\overset{\|}{C}}OH \;+\; R'OH \;\underset{}{\overset{H^+}{\rightleftharpoons}}\; R\overset{O}{\overset{\|}{C}}OR' \;+\; H_2O$$

Carboxylic acid   **Alcohol**   **Ester**

Esterification is not very exothermic: $\Delta H°$ is usually close to 0. The associated entropy change is also small; therefore $\Delta G° \approx 0$ and $K \approx 1$. How can the equilibrium be shifted toward the ester product? One way is to use an excess of either of the two starting materials; another is to remove the ester or the water product from the reaction mixture. In practice, esterifications are most often achieved by using the alcohol as a solvent.

$$CH_3\overset{O}{\overset{||}{C}}OH \;+\; CH_3OH \; \underset{-H_2O}{\overset{H_2SO_4,\ \Delta}{\rightleftharpoons}} \; CH_3\overset{O}{\overset{||}{C}}OCH_3$$

Acetic acid   Solvent   Methyl acetate   85%

Exercise 19-16

Show the products of the acid-catalyzed reaction of each of the following pairs of compounds.
(a) Methanol + pentanoic acid; **(b)** formic acid + 1-pentanol; **(c)** cyclohexanol + benzoic acid; **(d)** 2-bromoacetic acid + 3-methyl-2-butanol.

The opposite of esterification is **ester hydrolysis.** This reaction is carried out under the same conditions as esterification, but, to shift the equilibrium, an excess of water is used in a water-miscible solvent.

$$CH_3CH_2CH_2CH_2\overset{CH_3}{\underset{CH_3}{\overset{|}{\underset{|}{C}}}}COOCH_2CH_3 \; \xrightarrow{H_2SO_4,\ HOH,\ acetone,\ \Delta} \; CH_3CH_2CH_2CH_2\overset{CH_3}{\underset{CH_3}{\overset{|}{\underset{|}{C}}}}COOH \;+\; CH_3CH_2OH$$

Ethyl 2,2-dimethylhexanoate   2,2-Dimethylhexanoic acid   85%

Exercise 19-17

Give the products of acid-catalyzed hydrolysis of each of the following esters.

(a) $CH_3(CH_2)_3C\equiv CCH_2CH_2COOCH(CH_3)_2$ **(b)**

(c) $CH_3CH_2\overset{}{\underset{CH_3}{\overset{|}{C}}}HCH_2OOC-\bigcirc$

Esterification proceeds through acid-catalyzed addition–elimination

The presence of the acid catalyst features prominently in the mechanism of ester formation: It causes the carbonyl function to undergo nucleophilic attack by the alcohol (methanol in this example, step 2) and the hydroxy group to leave as water (step 3). We previewed this mechanism earlier (Section 19-7). Here we present it with full details.

Mechanism of Acid-Catalyzed Esterification

Step 1. Protonation of the carboxy group

Reaction

Step 2. Attack by methanol

**Tetrahedral
intermediate**
Relay point:
← Can go back to starting material
Can go forward to product →

Step 3. Elimination of water

Initially, protonation of the carbonyl oxygen gives a delocalized carbocation (step 1). Now the carbonyl carbon is susceptible to nucleophilic attack by methanol. Proton loss from the initial adduct furnishes the tetrahedral intermediate (step 2). This species is a crucial relay point, because it can react in either of *two* ways in the presence of the mineral acid catalyst. First, it can lose methanol by the reverse of steps 1 and 2. This process, beginning with protonation of the methoxy oxygen, leads back to the carboxylic acid. The second possibility, however, is protonation at either hydroxy oxygen, leading to elimination of water and to the ester (step 3). All the steps are reversible; therefore, either addition of excess alcohol or removal of water favors esterification by shifting the equilibria in steps 2 and 3, respectively. Ester hydrolysis proceeds by the reverse of the sequence and is favored by aqueous conditions.

Exercise 19-18

Give the mechanism of esterification of a general carboxylic acid RCOOH with methanol in which the alcohol oxygen is labeled with the ^{18}O isotope ($CH_3{}^{18}OH$). Does the labeled oxygen appear in the ester or in the water product?

Hydroxy acids may undergo intramolecular esterification to lactones

When hydroxy carboxylic acids are treated with catalytic amounts of mineral acid, cyclic esters—or **lactones**—may form. This process is called **intramolecular esterification** and is favorable for formation of five- and six-membered rings. Unlike the intermolecular esterification, the equilibrium is now favored by entropy and therefore particularly facile: Water is expelled from the starting material.

Formation of a Lactone

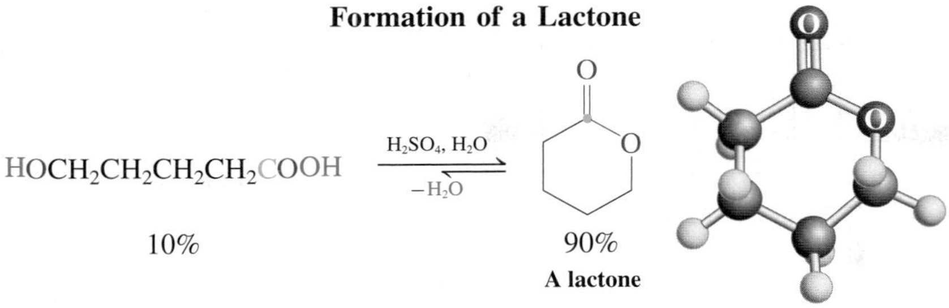
HOCH2CH2CH2CH2COOH
H2SO4, H2O
−H2O
O
O
10%
90%
A lactone

Solved Exercise 19-19 | Working with the Concepts: A Multistep Mechanism

Explain the following result.

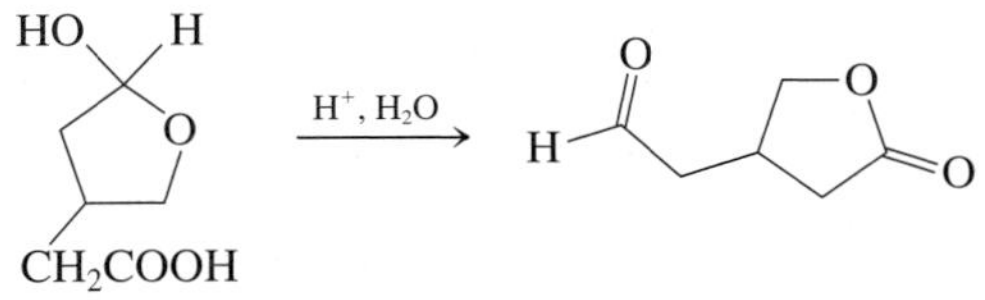
HO
H
O
CH2COOH
H+, H2O
O
H
O
O

Let us employ the WHIP strategy to analyze and solve this problem.

What the question asks for is a *mechanistic* explanation: As always, your clue is the word "explain" in the question.

Strategy

How to start? At first glance the starting material and the product appear to have little in common, structurally. Look critically for elements they share. For starters, they both have the same number of carbon atoms. Looking at their respective molecular formulas, the equation begins with $C_6H_{10}O_4$ and ends with $C_6H_8O_3$: We have lost one (and only one) molecule of H_2O in the process. Turning to the embedded functions, the starting material contains a carboxylic acid group. Does the product contain a carbon atom in the form of a carboxylic acid derivative? Again, yes: It is a lactone, a cyclic ester. Similarly, the product contains an aldehyde carbonyl. Does the starting material contain a carbon atom in the aldehyde oxidation state? Yes: in the form of a five-membered ring hemiacetal. By identifying such features we can establish possible connections between carbon atoms in the starting material and product molecules.

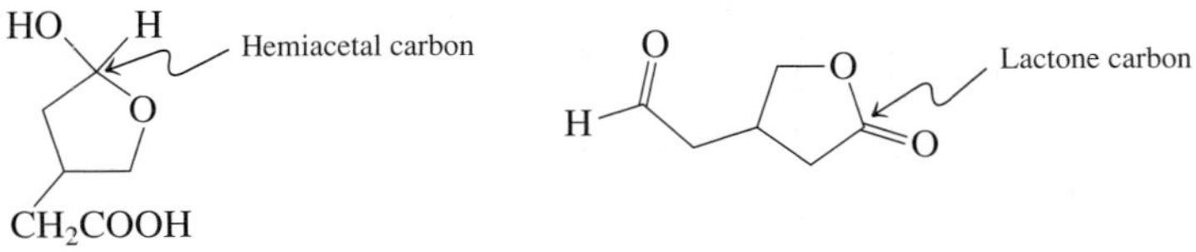
HO
H
Hemiacetal carbon
O
CH2COOH
O
H
O
Lactone carbon
O

Information? Section 17-7 tells us that hemiacetals interconvert with their open-chain hydroxycarbonyl isomers under either acidic or basic conditions. Use this to get started, and then apply information in this section about lactones in order to ***P***roceed.

Solution

Write the mechanism for equilibration of the hemiacetal with its open-chain isomer:

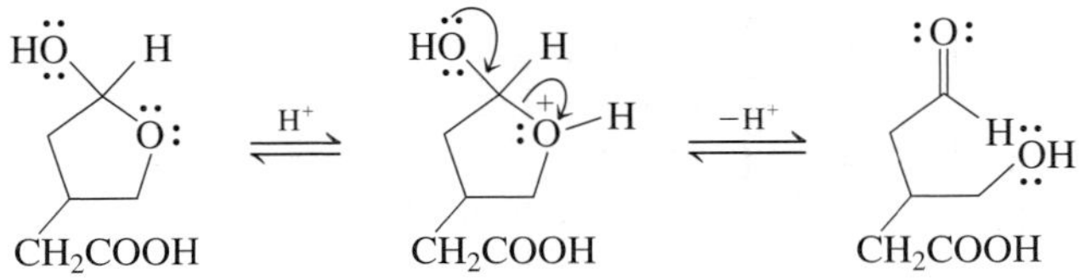
HO
H
O
CH2COOH
H+
HO
H
O
H
CH2COOH
−H+
:O:
H
OH
CH2COOH

The resulting structure contains three functional groups: an aldehyde carbonyl, an alcohol hydroxy, and a carboxylic acid carboxy. Continuing:

- Intramolecular reaction of the hydroxy group with the carboxylic acid function gives the cyclic ester (lactone) product shown above.

• We complete the solution to the exercise by writing out the mechanism, step by step.

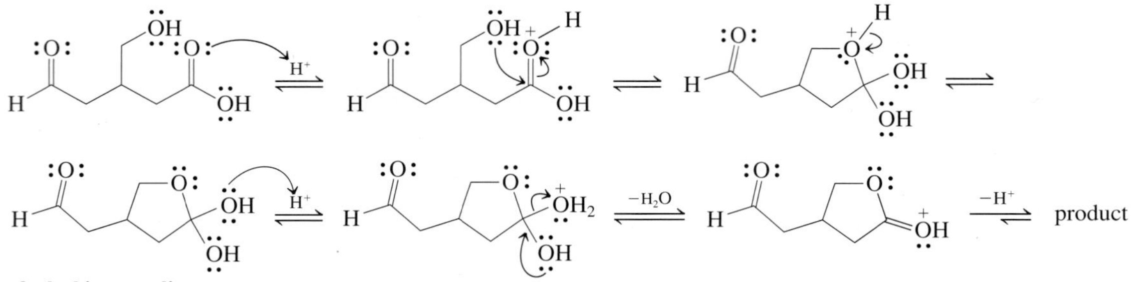

Tetrahedral intermediate

Exercise 19-20 | Try It Yourself

(a) If the starting compound in Exercise 19-19 were labeled with the ^{18}O isotope at its ring oxygen atom, where would the ^{18}O label end up in the product?

(b) Oxandrolone (Anavar) is a widely used synthetic analog of testosterone (see Section 4-7 on anabolic steroids), often prescribed to counteract weight loss due to surgery, injury, or prolonged infection. It works by stimulating the production of protein in the body without the strong androgenic ("virilizing") effect of the male sex hormone. The crucial structural alteration is the substitution of the cyclohexenone A ring by a lactone, made in the last step of its synthesis by the reduction shown below. What is the mechanism of this conversion? (**Hints:** Aqueous $NaBH_4$ has a pH ≈ 10 and consult Section 17-7.)

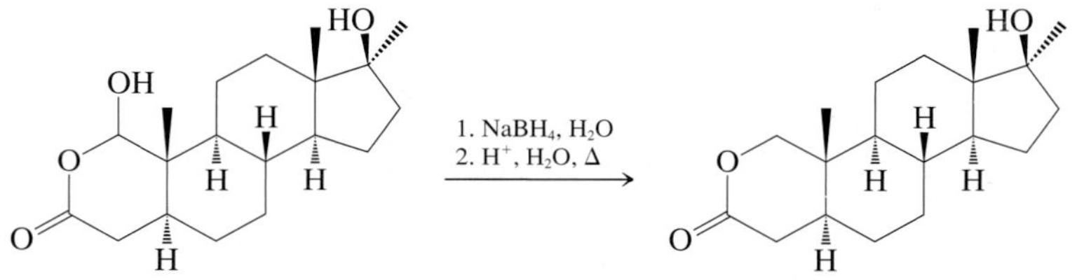

Oxandrolone
(Anavar)

In Summary Carboxylic acids react with alcohols to form esters, as long as a mineral acid catalyst is present. This reaction is only slightly exothermic, and its equilibrium may be shifted in either direction by the choice of reaction conditions. The reverse of ester formation is ester hydrolysis. The mechanism of esterification is acid-catalyzed addition of alcohol to the carbonyl group followed by acid-catalyzed dehydration. Intramolecular ester formation results in lactones, favored only when five- or six-membered rings are produced.

19-10 | CARBOXYLIC ACID DERIVATIVES: AMIDES

Carboxylic amide

As we saw earlier (Section 17-9), amines also are capable of attacking the carbonyl function. When the carbonyl is that of a carboxylic acid, the product is a **carboxylic amide,*** the last major class of carboxylic acid derivatives. The mechanism is again addition–elimination but is complicated by acid-base chemistry.

Amines react with carboxylic acids as bases and as nucleophiles

Nitrogen lies to the left of oxygen in the periodic table. Therefore, amines (Chapter 21) are better bases as well as better nucleophiles than alcohols (Section 6-8). To synthesize carboxylic amides, therefore, we must address the problem discussed in Section 19-7: interference

---

*Do not confuse the names of carboxylic amides with those of the alkali salts of amines, also called amides (e.g., lithium amide, $LiNH_2$).

from competing acid-base reaction. Indeed, exposure of a carboxylic acid to an amine initially forms an ammonium carboxylate salt, in which the negatively charged carboxylate is very resistant to nucleophilic attack.

Ammonium Salts from Carboxylic Acids

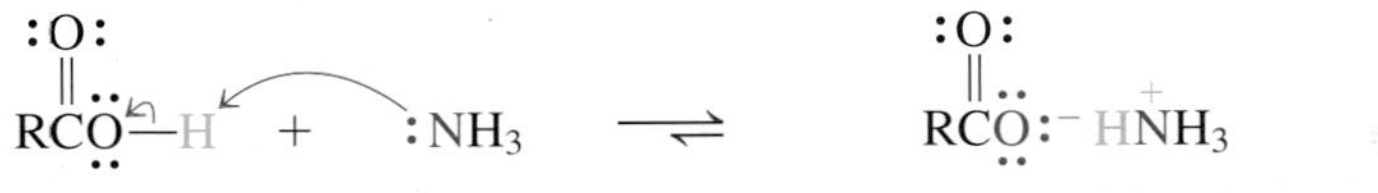

$$\overset{:O:}{\underset{}{\overset{\|}{RCO-H}}} + \quad :NH_3 \quad \rightleftharpoons \quad \overset{:O:}{\underset{}{\overset{\|}{RCO:^-}}} \overset{+}{HNH_3}$$

Ammonia   **An ammonium carboxylate salt**

Notice that salt formation, though very favorable, is nonetheless reversible. Upon heating, a slower but thermodynamically favored reaction between the acid and the amine can take place. The acid and the amine are removed from the equilibrium, and eventually salt formation is completely reversed. In this second mode of reaction, the nitrogen acts as a nucleophile and attacks the carbonyl carbon. Completion of an addition–elimination sequence leads to the amide. Although it is convenient, this method suffers from the high temperatures required to reverse ammonium carboxylate formation. Therefore, better procedures rely on the use of activated carboxylic acid derivatives, such as acyl chlorides (Chapter 20).

Formation of an Amide from an Amine and a Carboxylic Acid

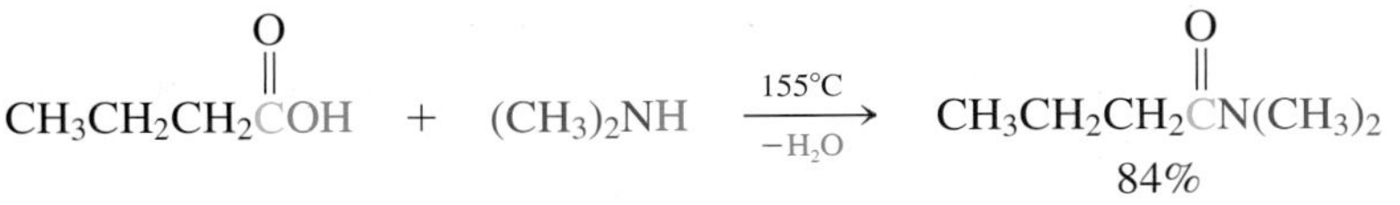

$$\overset{O}{\underset{}{\overset{\|}{CH_3CH_2CH_2COH}}} + \quad (CH_3)_2NH \quad \xrightarrow[-H_2O]{155°C} \quad \overset{O}{\underset{}{\overset{\|}{CH_3CH_2CH_2CN(CH_3)_2}}}$$

84%

N,N-**Dimethylbutan**amide

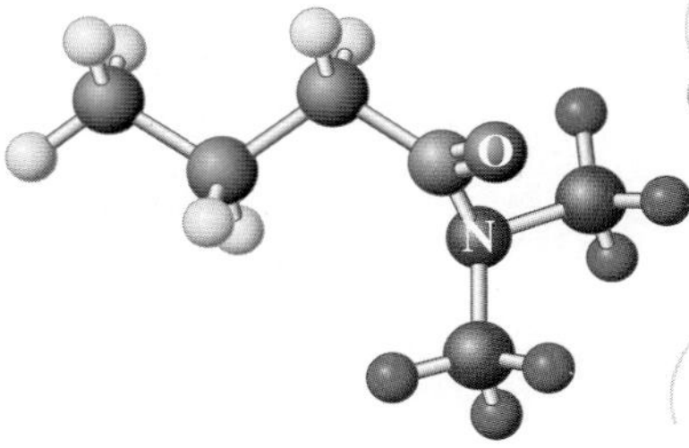

● **Reaction**

⚙ **Mechanism**

Mechanism of Amide Formation

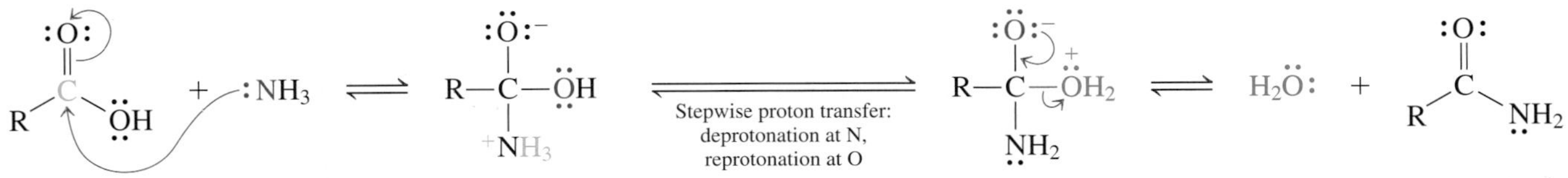

Stepwise proton transfer: deprotonation at N, reprotonation at O

Tetrahedral intermediates   **An amide**

In the mechanism shown above, the elimination step is shown to proceed from a zwitterionic tetrahedral intermediate. However, the prevalence of such a species is likely to be highly pH dependent. Thus, an alternative possibility is to eliminate water from a neutral species, as shown in the margin of the next page. Amide formation is reversible. Thus, treatment of amides with hot aqueous acid or base regenerates the component carboxylic acids and amines.

Dicarboxylic acids react with amines to give imides

Dicarboxylic acids may react twice with the amine nitrogen of ammonia or of a primary amine. This sequence gives rise to **imides,** the nitrogen analogs of cyclic anhydrides (cf. p. 854).

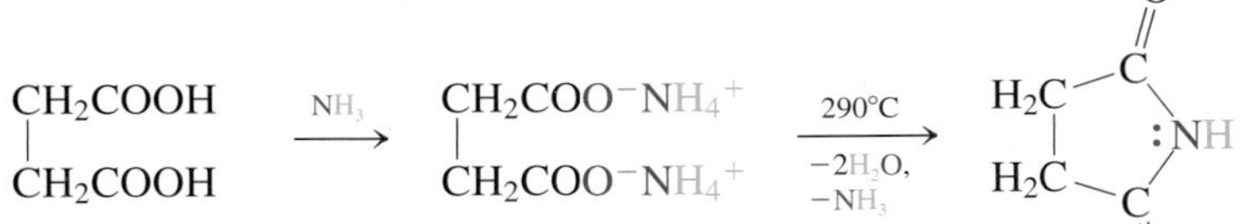

$$\begin{array}{l} CH_2COOH \\ | \\ CH_2COOH \end{array} \xrightarrow{NH_3} \begin{array}{l} CH_2COO^-NH_4{}^+ \\ | \\ CH_2COO^-NH_4{}^+ \end{array} \xrightarrow[-NH_3]{290°C \atop -2H_2O,} $$

Butanedioic acid

$$\begin{array}{c} O \\ \| \\ H_2C \diagdown C \\ | \quad\quad :NH \\ H_2C \diagup C \\ \| \\ O \end{array}$$

83%

Butanimide
(Succinimide)

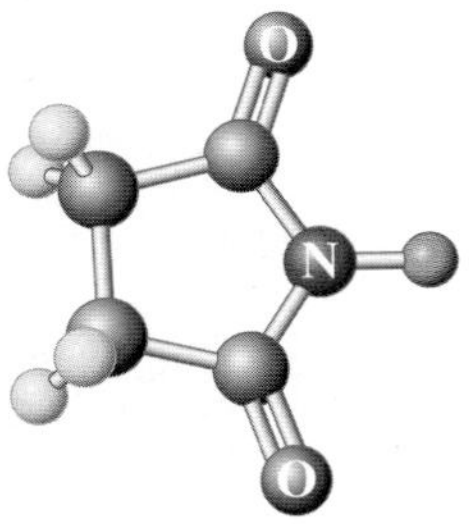

Recall the use of *N*-halobutanimides in halogenations (Section 14-2).

Alternative Mechanism of Amide Formation

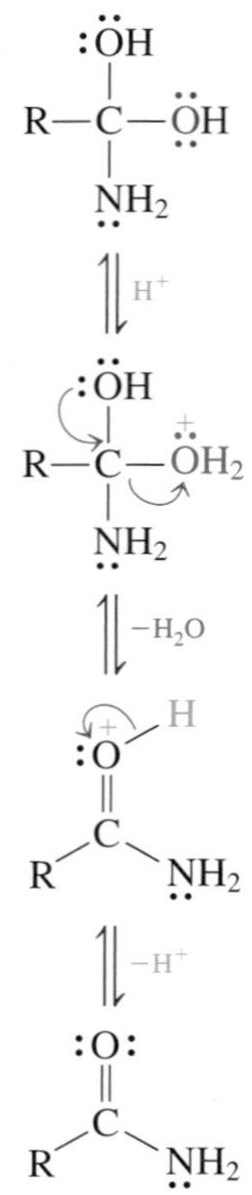

:ÖH
|
R—C—ÖH
|
NH₂

↕ H⁺

:ÖH
|
R—C—ÖH₂
|
NH₂

↕ −H₂O

O⁺—H
‖
C
R   NH₂

↕ −H⁺

:O:
‖
C
R   NH₂

Amino acids cyclize to lactams

In analogy to hydroxycarboxylic acids, some amino acids undergo cyclization to the corresponding cyclic amides, called **lactams** (Section 20-6).

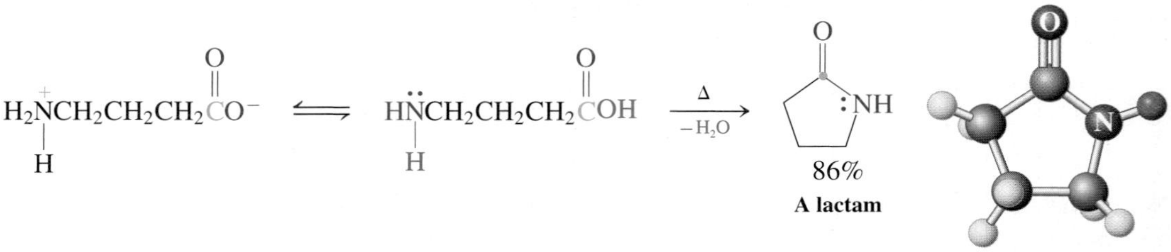

$$H_2\overset{+}{N}CH_2CH_2CH_2\overset{O}{\overset{\|}{C}}O^- \;\rightleftharpoons\; \overset{..}{HN}CH_2CH_2CH_2\overset{O}{\overset{\|}{C}}OH \;\xrightarrow[-H_2O]{\Delta}\;$$

A lactam structure, 86%
A lactam

The penicillin class of antibiotics derives its biological activity from the presence of a lactam function (see Real Life 20-2). Lactams are the nitrogen analogs of lactones (Section 19-9).

Exercise 19-21

Formulate a detailed mechanism for the formation of butanimide from butanedioic acid and ammonia.

In Summary Amines react with carboxylic acids to form amides by an addition–elimination process that begins with nucleophilic attack by the amine on the carboxy carbon. Amide formation is complicated by reversible deprotonation of the carboxylic acid by the basic amine to give an ammonium salt.

19-11 | REDUCTION OF CARBOXYLIC ACIDS BY LITHIUM ALUMINUM HYDRIDE

Lithium aluminum hydride is capable of reducing carboxylic acids all the way to the corresponding primary alcohols, which are obtained upon aqueous acidic work-up.

Reduction of a Carboxylic Acid

$$RCOOH \;\xrightarrow[\text{2. } H^+, H_2O]{\text{1. LiAlH}_4, \text{ THF}}\; RCH_2OH$$

Example:

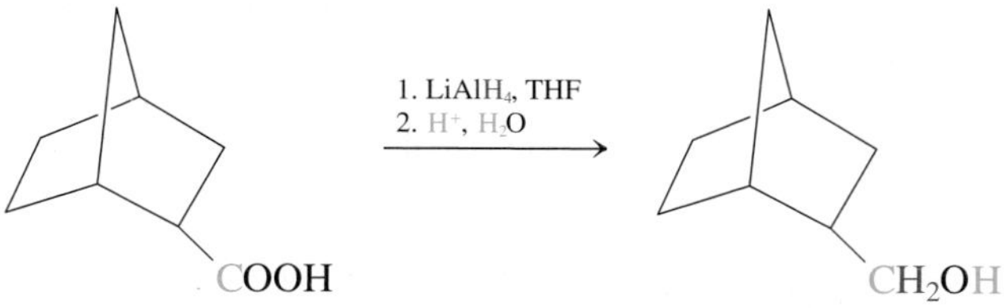

$$\xrightarrow[\text{2. } H^+, H_2O]{\text{1. LiAlH}_4, \text{ THF}}$$

COOH → CH₂OH
65%

Although the exact mechanism of this transformation is not completely understood, it is clear that the hydride reagent first acts as a base, forming the lithium salt of the acid and hydrogen gas. Carboxylate salts are generally resistant to attack by nucleophiles. Yet, despite the negative charge, lithium aluminum hydride is so reactive that it is capable of donating two hydrides to the carbonyl function of the carboxylate, possibly assisted by aluminum (margin; see also Section 8-6). The product of this sequence, a simple alkoxide, gives the alcohol after protonation.

Double Hydride Additions to Carboxylate Ion

Exercise 19-22

Propose synthetic schemes that produce compound B from compound A.

(a) $CH_3CH_2CH_2CN$ $CH_3CH_2CH_2CH_2OH$ **(b)** ▷—CH_2COOH ▷—CH_2CD_2OH
       A   **B**   **A**   **B**

In Summary The nucleophilic reactivity of lithium aluminum hydride is sufficiently great to effect the reduction of carboxylates to primary alcohols.

19-12 | BROMINATION NEXT TO THE CARBOXY GROUP: THE HELL-VOLHARD-ZELINSKY REACTION

Like aldehydes and ketones, alkanoic acids can be monobrominated at the α-carbon by exposure to Br_2. The addition of a trace amount of PBr_3 is necessary to get the reaction started. Because the highly corrosive PBr_3 is difficult to handle, it is often generated in the reaction flask (in situ). This is achieved by the addition of a little elemental (red) phosphorus to the mixture of starting materials; it is converted into PBr_3 instantaneously by the bromine present.

Reaction

A Hell-Volhard-Zelinsky* Reaction

$$CH_3CH_2CH_2CH_2COOH \xrightarrow{\text{Br—Br, trace P}} CH_3CH_2CH_2\overset{\overset{\displaystyle Br}{|}}{C}HCOOH + HBr$$
$$80\%$$
$$\textbf{2-Bromopentanoic acid}$$

The acyl bromide, formed by the reaction of PBr_3 with the carboxylic acid (Section 19-8), is subject to rapid acid-catalyzed enolization. The enol is brominated to give the 2-bromoacyl bromide. This derivative then undergoes an exchange reaction with unreacted acid to furnish the product bromoacid and another molecule of acyl bromide, which reenters the reaction cycle.

Mechanism of the Hell-Volhard-Zelinsky Reaction

Mechanism

Step 1. Acyl bromide formation (see Exercise 19-12 and Section 19-8)

$$3\ RCH_2\overset{\overset{\displaystyle O}{||}}{C}OH + PBr_3 \longrightarrow 3\ RCH_2\overset{\overset{\displaystyle O}{||}}{C}Br + H_3PO_3$$
$$\textbf{Acyl bromide}$$

Step 2. Enolization

$$RCH_2\overset{\overset{\displaystyle O}{||}}{C}Br \underset{}{\overset{H^+}{\rightleftharpoons}} RCH=\overset{\overset{\displaystyle OH}{|}}{\underset{\underset{\displaystyle Br}{|}}{C}}$$
$$\textbf{Enol}$$

*Professor Carl M. Hell (1849–1926), University of Stuttgart, Germany; Professor Jacob Volhard (1834–1910), University of Halle, Germany; Professor Nicolai D. Zelinsky (1861–1953), University of Moscow.

Step 3. Bromination

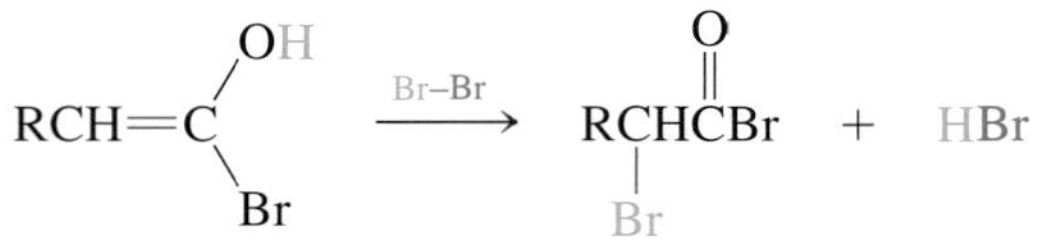

$$RCH{=}C\overset{\displaystyle OH}{\underset{\displaystyle Br}{{\Big|}}} \xrightarrow{\text{Br–Br}} RCH\overset{\displaystyle O}{\underset{\displaystyle Br}{\overset{\|}{C}}}Br \;+\; HBr$$

Step 4. Exchange (see Worked Example 25)

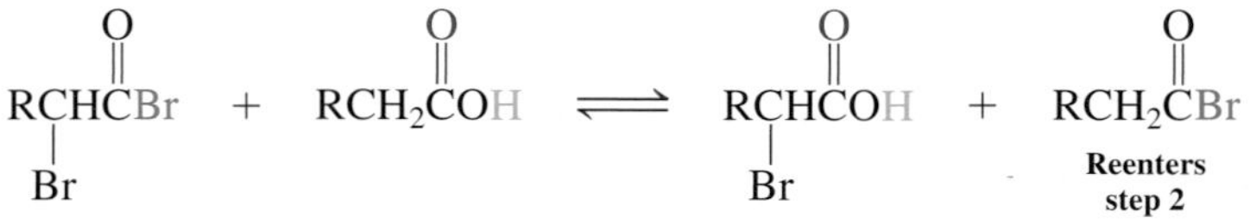

$$RCH\overset{O}{\overset{\|}{C}}Br \;+\; RCH_2\overset{O}{\overset{\|}{C}}OH \;\rightleftharpoons\; RCH\overset{O}{\overset{\|}{C}}OH \;+\; RCH_2\overset{O}{\overset{\|}{C}}Br$$
$$\;\;\;\underset{Br}{\;}\qquad\qquad\qquad\qquad\qquad \underset{Br}{\;}\qquad\qquad \underset{\textbf{Reenters step 2}}{\;}$$

Exercise 19-23

Formulate detailed mechanisms for steps 2 and 3 of the Hell-Volhard-Zelinsky reaction. (**Hints:** Review Sections 18-2 for step 2 and 18-3 for step 3.)

The bromocarboxylic acids formed in the Hell-Volhard-Zelinsky reaction can be converted into other 2-substituted derivatives. For example, treatment with aqueous base gives 2-hydroxy acids, whereas amines yield α-amino acids (Chapter 26). An example of amino acid synthesis is the preparation of racemic 2-aminohexanoic acid (norleucine, a naturally occurring but rare amino acid) shown here.

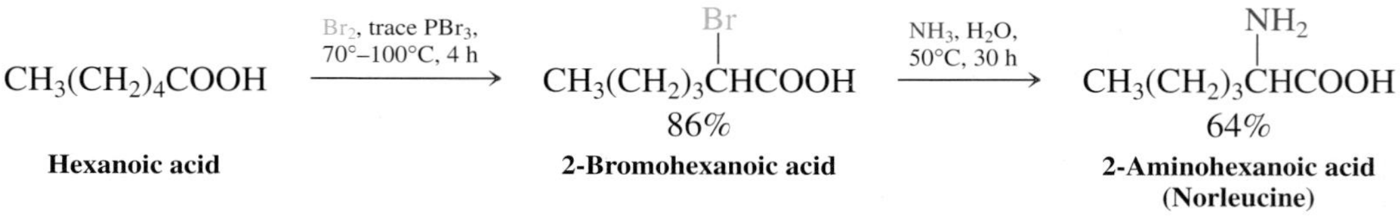

$$CH_3(CH_2)_4COOH \xrightarrow[\text{70°–100°C, 4 h}]{\text{Br}_2,\ \text{trace PBr}_3,} CH_3(CH_2)_3\overset{\text{Br}}{\overset{|}{C}}HCOOH \xrightarrow[\text{50°C, 30 h}]{\text{NH}_3,\ \text{H}_2\text{O},} CH_3(CH_2)_3\overset{\text{NH}_2}{\overset{|}{C}}HCOOH$$

Hexanoic acid   **2-Bromohexanoic acid** 86%   **2-Aminohexanoic acid (Norleucine)** 64%

In Summary With trace amounts of phosphorus (or phosphorus tribromide), carboxylic acids are brominated at C2 (the Hell-Volhard-Zelinsky reaction). The transformation proceeds through 2-bromoacyl bromide intermediates.

19-13 | BIOLOGICAL ACTIVITY OF CARBOXYLIC ACIDS

Considering the variety of reactions that carboxylic acids can undergo, it is no wonder that they are very important, not only as synthetic intermediates in the laboratory, but also in biological systems. This section will provide a glimpse of the enormous structural and functional diversity of natural carboxylic acids. A discussion of amino acids will be deferred to Chapter 26.

As Table 19-1 indicates, even the simplest carboxylic acids are abundant in nature. Formic acid is present not only in ants, where it functions as an alarm pheromone and chemical weapon, but also in plants. For example, one reason why human skin hurts after it touches the stinging nettle is that formic acid is deposited in the wounds.

Acetic acid is formed through the enzymatic oxidation of ethanol produced by fermentation. Vinegar is the term given to the dilute (ca. 4–12%) aqueous solution thus generated in ciders, wines, and malt extracts. Louis Pasteur in 1864 established the involvement of bacteria in the oxidation stage of this ancient process.

Fatty acids are derived from coupling of acetic acid

Acetic acid exhibits diverse biological activities, ranging from a defense pheromone in some ants and scorpions to the primary building block for the biosynthesis of more naturally occurring organic compounds than any other single precursor substance. For example, 3-methyl-3-butenyl pyrophosphate, the crucial precursor in the buildup of the terpenes (Section 14-10), is made by the enzymatic conversion of three molecules of CH_3COOH into an intermediate called mevalonic acid. Further reactions degrade the system to the five-carbon (isoprene) unit of the product.

$$3\ CH_3{-}COOH \xrightarrow[\text{Enzymes}]{} CH_3{-}\underset{\underset{CH_2{-}CH_2OH}{|}}{\overset{\overset{CH_2{-}COOH}{|}}{C}}OH \xrightarrow[\text{Enzymes}]{} CH_3{-}C\underset{CH_2{-}CH_2O{-}P{-}O{-}P{-}OH}{\overset{CH_2}{\diagdown\diagup}}$$

Mevalonic acid   **3-Methyl-3-butenyl pyrophosphate**

A conceptually more straightforward mode of multiple coupling is found in the biosynthesis of **fatty acids.** This class of compounds derives its name from its source, the natural **fats,** which are esters of long-chain carboxylic acids (see Section 20-4). Hydrolysis or **saponification** (so called because the corresponding salts form soaps; *sapo,* Latin, soap—see Real Life 19-1) yields the corresponding fatty acids. The most important of them are between 12 and 22 carbons long and may contain cis carbon–carbon double bonds.

Fatty Acids

$$CH_3(CH_2)_{14}COOH$$

Hexadecanoic acid
(Palmitic acid)

$$\underset{\underset{H}{|}}{\overset{\overset{CH_3(CH_2)_7}{}}{C}}{=}\underset{\underset{H}{|}}{\overset{\overset{(CH_2)_7COOH}{}}{C}}$$

cis-**9-Octadecenoic acid**
(Oleic acid)

In accord with their biosynthetic origin, fatty acids consist mostly of even-numbered carbon chains. A very elegant experiment demonstrated that linear coupling occurred in a highly regular fashion. In it, singly labeled radioactive (^{14}C) acetic acid was fed to several organisms. The resulting fatty acids were labeled only at every other carbon atom.

Reaction

$$CH_3{}^{14}COOH \xrightarrow{\text{Organism}} CH_3{}^{14}CH_2CH_2{}^{14}CH_2CH_2{}^{14}CH_2CH_2{}^{14}CH_2CH_2{}^{14}CH_2CH_2{}^{14}CH_2CH_2{}^{14}CH_2CH_2{}^{14}CH_2CH_2{}^{14}COOH$$

Labeled hexadecanoic (palmitic) acid

The mechanism of chain formation is very complex, but the following scheme provides a general idea of the process. A key player is the mercapto group of an important biological relay compound called coenzyme A (abbreviated HSCoA; Figure 19-6). This function binds acetic acid in the form of a **thiol ester** called acetyl CoA. Carboxylation transforms some

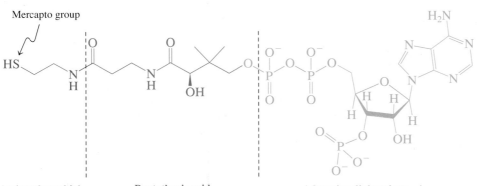

Mercapto group

HS

2-Aminoethanethiol part   **Pantothenic acid part**   **Adenosine diphosphate (ADP) part**

Figure 19-6 Structure of coenzyme A. For this discussion, the important part is the mercapto function. For convenience, the molecule may be abbreviated HSCoA.

REAL LIFE: MATERIALS 19-1 | **Long-Chain Carboxylates and Sulfonates Make Soaps and Detergents**

The sodium and potassium salts of long-chain carboxylic acids have the property of aggregating as spherical clusters called *micelles* in aqueous solution. In such aggregates, the hydrophobic alkyl chains (Section 8-2) of the acids come together in space because of their attraction to one another by London forces (Section 2-7) and their tendency to avoid exposure to polar water. As shown below, the polar, water-solvated carboxylate "head groups" form a spherical shell around the hydrocarbon-like center.

Because these carboxylate salts also create films on aqueous surfaces, they act as soaps. The polar groups stick into the water while the alkyl chains assemble into a hydrophobic

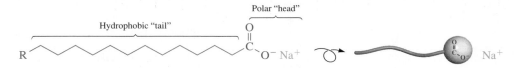

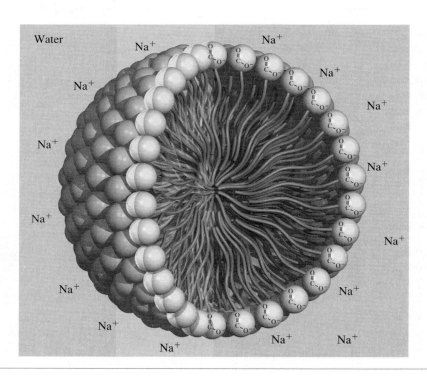

of this thiol ester into malonyl CoA. The two acyl groups are then transferred to two molecules of **acyl carrier protein.** Coupling occurs with loss of CO_2 to furnish a 3-oxobutanoic thiol ester.

Coupling of Acetic Acid Units

Mechanism

Step 1. Thiol ester formation

$$CH_3COOH \;+\; HSCoA \longrightarrow CH_3\overset{O}{\overset{\|}{C}}SCoA \;+\; HOH$$

$$\text{Acetic acid} \qquad \text{Coenzyme A} \qquad \text{Acetyl coenzyme A}$$

Step 2. Carboxylation

$$CH_3\overset{O}{\overset{\|}{C}}SCoA \;+\; CO_2 \xrightarrow{\text{Acetyl CoA carboxylase}} HO\overset{O}{\overset{\|}{C}}CH_2\overset{O}{\overset{\|}{C}}SCoA$$

$$\textbf{Malonyl CoA}$$

False-color scanning electron micrographs of the collar of a pure cotton shirt before washing (left) and after washing (right). Before washing, the threads and subthreads are covered with grime and skin flakes. (Magnified 70 times.)

layer. This construction reduces the surface tension of the water ("surfactant"), permitting it to permeate cloth and other fabrics and giving rise to the foaming typical of soaps. Cleansing is accomplished by dissolving ordinarily water-insoluble materials (oils, fats) in the hydrocarbon interior of the micelles.

One problem with carboxylate soaps has been that they form curd-like precipitates with the ions present in hard water, such as Mg^{2+} and Ca^{2+}. To minimize this drawback, a new generation of detergents was introduced, consisting of branched alkylbenzenesulfonates. Unfortunately, however, these compounds turned out not to be readily biodegradable, because the microorganisms involved in normal sewage treatment are capable of breaking down only straight-chain systems.

Consequently, modern products feature linear alkyl derivatives, such as sodium 4-(5-dodecyl)benzenesulfonate and its positional isomers. A typical formulation contains about 15% of surfactant. The bulk is made up of water softeners, such as Na_2CO_3, which causes Mg^{2+} and Ca^{2+} to precipitate as their carbonates, or sodium triphosphate, which sequesters these ions by tight multiple binding.

Certain steroid *bile acids* such as *cholic acid* (Section 4-7) also have surfactant or detergent-like properties and are found in the bile duct. These substances are released into the upper intestinal tract, where they emulsify water-insoluble fats through the formation of micelles. Hydrolytic enzymes can then digest the dispersed fat molecules.

Sodium 4-(5-dodecyl)-benzenesulfonate

SO_3^-Na+

Magnesium triphosphate

Cholic acid

Step 3. Acetyl and malonyl group transfers

$$CH_3CSCoA \; + \; HS-\boxed{protein} \longrightarrow CH_3CS-\boxed{protein} \; + \; HSCoA$$

Acyl carrier protein

$$HOCCH_2CSCoA \; + \; HS-\boxed{protein} \longrightarrow HOCCH_2CS-\boxed{protein} \; + \; HSCoA$$

Acyl carrier protein

Step 4. Coupling

$$HOCCH_2CS-\boxed{protein} \xrightarrow[-CO_2]{CH_3CS-\boxed{protein}} CH_3CCH_2CS-\boxed{protein}$$

A 3-oxobutanoic thiol ester

More than 90% of the double bonds in naturally occurring unsaturated fatty acids possess the cis configuration, contributing to the reduced melting temperatures of vegetable oils compared with saturated fats (Section 11-3). Exposure of vegetable oil to catalytic hydrogenation conditions produces solid margarine. However, this process does not hydrogenate all of the double bonds: As we saw in Exercise 12-2, a significant fraction of cis double bonds are merely isomerized to trans by the catalyst and remain in the final solid product. For example, the fat in one U.S. brand of synthetic hard (stick) margarine consists of about 18% saturated fatty acids (SFAs) and 23% trans fatty acids (TFAs). The fat in soft (tub) margarines, whose exposure to catalytic hydrogenation conditions is less, has about the same level of SFAs but far less (5–10%) TFAs. For comparison, natural butter is 50–60% SFAs but only 3–5% TFAs.

What, if any, are the health consequences of TFAs in the human diet? It has long been suspected that TFAs are not metabolized by the body in the same way as their cis counterparts; in the 1960s and 1970s, this suspicion was confirmed by studies that indicated that TFAs in foods greatly affect lipid metabolism. Perhaps the most alarming finding was that TFAs accumulate in cell membranes and increase the levels of **low-density lipoproteins** (**LDLs,** popularly if imprecisely known as "bad cholesterol") in the bloodstream while reducing **high-density lipoprotein** levels (**HDLs,** the so-called good cholesterol, see Real Life 4-2).

Studies beginning in the 1990s implicated dietary TFAs in increased risk of breast cancer and heart disease. It is now generally believed that the health effects of TFAs are far more adverse than even those of SFAs. As a result, the U.S. Food and Drug Administration required all food products to list TFA content beginning in 2006. Since that time, numerous communities across the United States, including major cites such as New York and Philadelphia, and the entire state of California, have enacted bans on the use of partially hydrogenated, high-TFA oils for restaurant cooking. Public pressure became so intense that by the end of 2008 most major fast-food chains had switched to low- or no-TFA oils for all purposes. However, a significant practical disadvantage arises in using natural, TFA-free vegetable oils in the preparation of fried foods: These oils

The new French fries: now without trans fatty acids.

contain a high proportion of polyunsaturated fatty acids, such as octadeca-9,12-dienoic (linoleic) acid. While linoleic acid is a relatively healthy dietary component, its doubly allylic CH_2 group at C11 is highly susceptible to radical chemistry such as air oxidation (Section 14-2), which leads to decomposition to rancid by-products and a short shelf life. As a potential solution, the food industry has used both selective plant breeding and genetic engineering to develop crop-derived, inexpensive oils in which most of the linoleic acid is replaced by the comparably healthy oleic acid (octadec-9-enoic) acid, the major component in olive oil. Oleic acid lacks the C12–C13 double bond and is therefore much more resistant to oxidative spoilage. So-called "high oleic" sunflower and safflower oils became commercially available in 2013.

Linoleic acid

Reduction of the ketone function to a methylene group follows. The resulting butanoic thiol ester repeatedly undergoes a similar sequence of reactions elongating the chain, always by two carbons. The eventual product is a long-chain acyl group, which is removed from the protein by hydrolysis.

$$CH_3CH_2-(CH_2CH_2)_{\overline{n}}\,CH_2\overset{\displaystyle O}{\overset{\|}{C}}-S-\boxed{protein} \quad \xrightarrow{H_2O} \quad CH_3CH_2-(CH_2CH_2)_{\overline{n}}\,CH_2\overset{\displaystyle O}{\overset{\|}{C}}-OH \quad + \quad HS-\boxed{protein}$$

Fatty acid

Arachidonic acid is a biologically important unsaturated fatty acid

Naturally occurring unsaturated fatty acids are capable of undergoing further transformations leading to a variety of unusual structures. An example is arachidonic acid, which is the biological precursor to a multitude of important chemicals in the human body, such as prostaglandins (Real Life 11-1), thromboxanes, and prostacyclins. It is released from cell membranes after an injury, by hormonal stimulation, or in the presence of a toxin. The enzyme cyclooxygenase ("COX") then engineers the oxidative cyclization of the tetraene framework through a series of radicals to assemble prostaglandin G2, from which the other derivatives ensue. These metabolites are thought to be the cause of inflammation, swelling, pain, and fever.

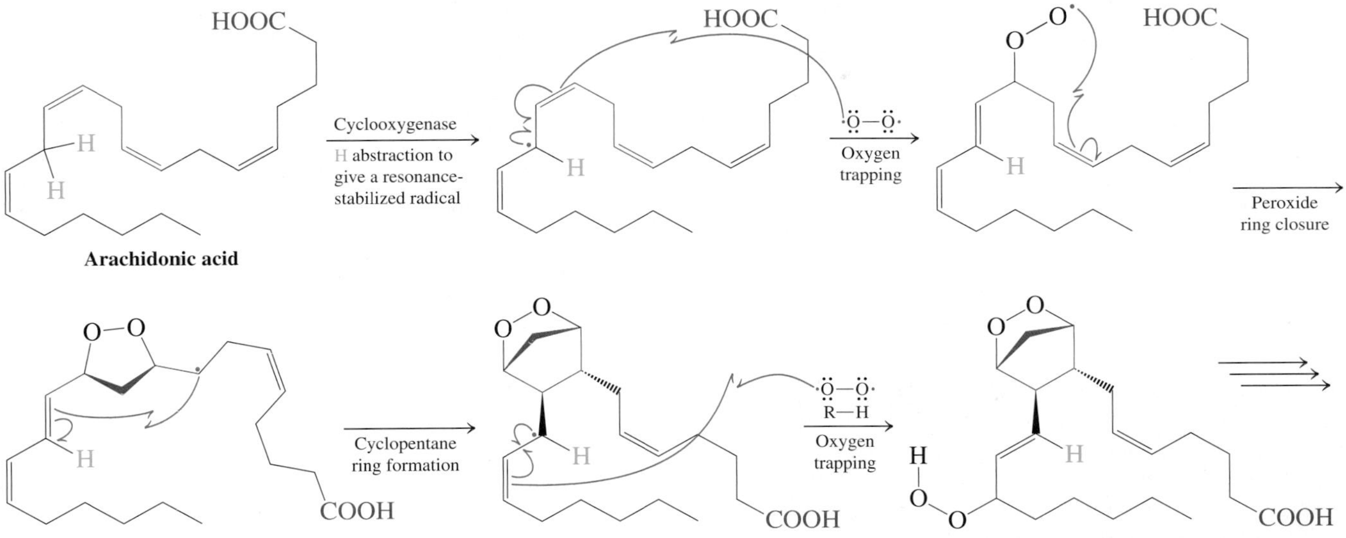

Arachidonic acid

Cyclooxygenase
H abstraction to give a resonance-stabilized radical

Oxygen trapping

Peroxide ring closure

Cyclopentane ring formation

Oxygen trapping

Prostaglandin G₂

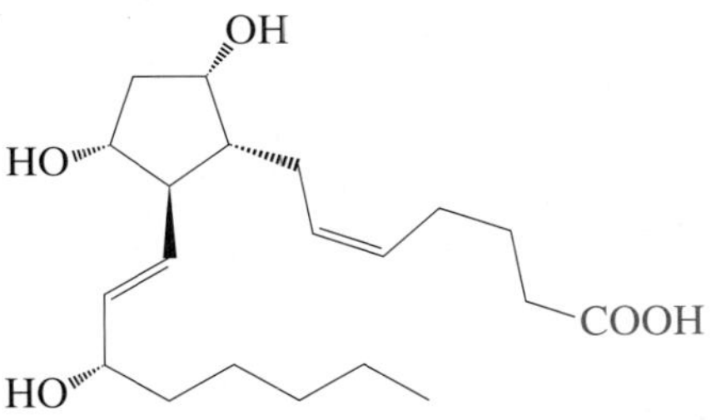

Prostaglandin F₂ₐ
(Induces labor, abortion, menstruation)

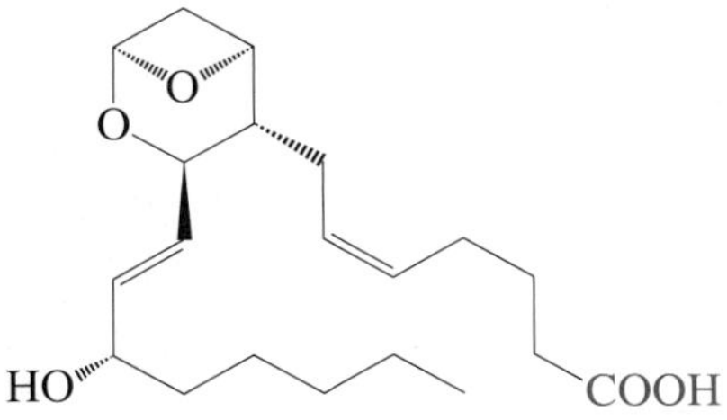

Thromboxane A₂
(Contracts smooth muscles; aggregates blood platelets)

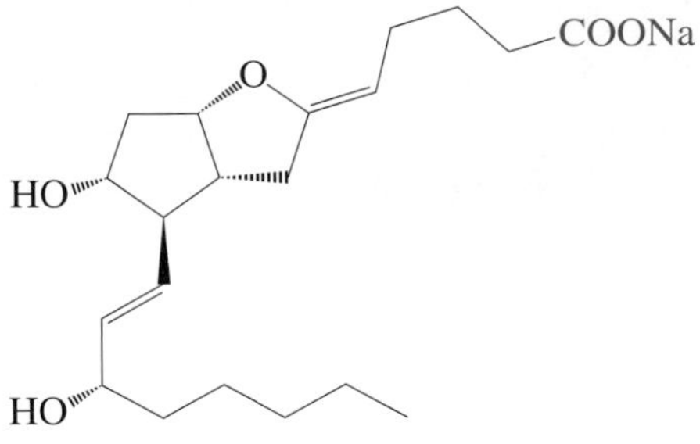

Prostacyclin I₂, sodium salt
(Inhibitor of platelet aggregation; vasodilator, used in heart-bypass operations and in kidney transplant patients)

Not surprisingly, drugs that combat the symptoms caused by the presence of prostaglandins target COX to inhibit its activity, as is the case for aspirin (margin; Real Life 22-2), or the corticosteroids (Section 4-7) which block the biosynthesis of arachidonic acid itself.

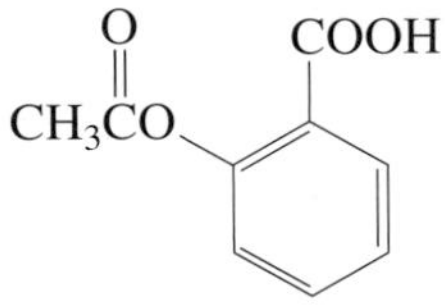

Aspirin

Exercise 19-24

COX effects the oxidation of unsaturated fatty acids other than arachidonic acid. One of the products of the oxidation of eicosapentaenoic acid ("EPA"; a dietary ingredient in oily fish, e.g., salmon and sardine) is shown below. Write a mechanism for its formation.

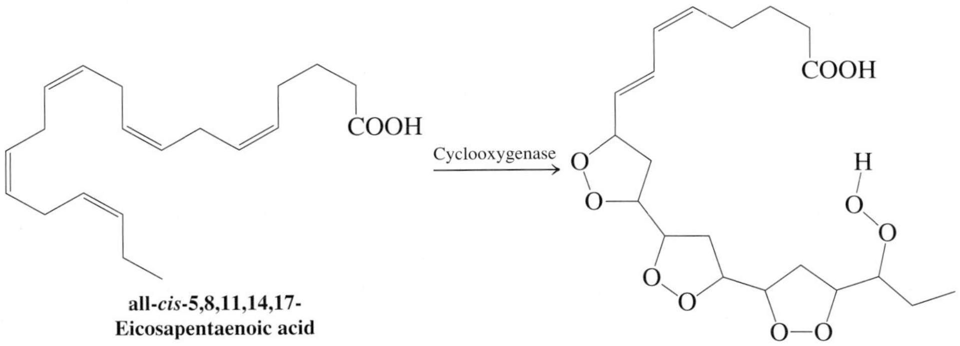

Cyclooxygenase

**all-*cis*-5,8,11,14,17-
Eicosapentaenoic acid**

REAL LIFE: MATERIALS 19-3 | Green Plastics, Fibers, and Energy from Biomass-Derived Hydroxyesters

The first decade of the 21st century saw dramatic developments in the use of naturally occurring but non–petrochemical-derived raw materials for numerous purposes.

Synthetic fibers for fabric manufacture are almost entirely petroleum-derived. Alternative polyesters such as poly(lactic acid) (Ingeo) can be made from vegetable matter—corn initially, but eventually from the starch that may be extracted from otherwise unusable plant residues (stalks, straws, etc.). The preparation of such polyesters is estimated to use two-thirds less fossil fuel than conventional fiber manufacture, and the process emits 80–90% less greenhouse gas.

Ingeo garments made from corn sugar fibers hit the world of fashion.

The persistent problem of conventional plastic waste disposal grows ever more severe as landfills approach capacity with these highly degradation-resistant substances (Section 12-15). Biodegradable plastics provide an option for nonreusable items such as plastic bags, wraps, and bottles. A recently developed and commercialized biobased and biodegradable plastic is poly(β-hydroxybutyrate-*co*-β-hydroxyvalerate) (PHBV), a copolymer of 3-hydroxybutanoic acid and 3-hydroxypentanoic acid. PHBV is a polyester that is produced by bacterial fermentation of mixtures of acetic and propanoic acids. The ratio of the two hydroxy acids controls the properties of the plastic—it is more flexible with more

five-carbon acid, stiffer with more four-carbon acid. PHBV is stable at temperatures up to 140°C, but it is fully degraded to innocuous small organic molecules in 6 months when exposed to microorganisms in soil, compost, or water. One application of PHBV is in controlled drug release: The polymer encapsulates a pharmaceutical, which is released only after the coating has degraded sufficiently by ester hydrolysis to form the two original hydroxyacids, which are natural products of human metabolism and quite harmless.

The promise in this field is yet to be realized: Bioplastic manufacture is not cost competitive with the manufacture of conventional plastics from petrochemicals, and concerns about compostability have also been raised. As a result, as of 2012 some promising materials, such as the polyhydroxyalkanoic acid (PHA)-derived biodegradable plastic Mirel, are no longer being made on a commercial scale, although both production and consumption of these types of materials has seen steady growth worldwide.

Biodegradable plastic bottles would hopefully transform this heap into compost.

$$\left(OCH_2\overset{\displaystyle O}{\overset{\displaystyle \|}{C}} \right)_n \qquad \left(OCHCH_2\overset{\displaystyle O}{\overset{\displaystyle \|}{C}} \right)_n \qquad \left(OCH(CH_2)_x\overset{\displaystyle O}{\overset{\displaystyle \|}{C}} \right)_n$$

PLA (Ingeo)   **PHBV** (R = –CH₃, –CH₂CH₃)   **PHA** (Mirel)

The ergot fungus on rye grains.

Nature also produces complex polycyclic carboxylic acids

Many biologically active natural products that have carboxy groups as substituents of complex polycyclic frames have physiological potential that derives from other sites in the molecule. In these compounds the function of the carboxy group may be to impart water solubility, to allow for salt formation or ion transport, and to enable micellar-type aggregation. Two examples are gibberellic acid, one of a group of plant growth-promoting substances manufactured by fermentation, and lysergic acid, a major product of hydrolyzed extracts of ergot, a fungal parasite that lives on grasses, including rye. Many lysergic acid derivatives possess powerful psychotomimetic activity. In the Middle Ages, thousands who ate rye bread contaminated by ergot experienced the poisonous effects characteristic of these compounds: hallucinations, convulsions, delirium, epilepsy, and death ("St. Anthony's Fire").

The synthetic lysergic acid diethylamide (LSD) is one of the most powerful hallucinogens known. The effective oral dose for humans is only about 0.05 mg.

Gibberellic acid

Lysergic acid

In Summary The numerous naturally occurring carboxylic acids are structurally and functionally diverse. The condensation of multiple acetyl units gives rise to the straight-chain fatty acids. The fatty acids in turn are converted into a wide array of substances exhibiting varied biological and, often, useful medicinal properties.

THE BIG PICTURE

The carboxylic acid functional group combines the carbonyl group of the aldehydes and ketones with the hydroxy group of the alcohols. The carbonyl group enhances the acidity of the hydroxy hydrogen, and the hydroxy group gives rise to the addition–elimination pathway for substitution about the carbonyl carbon. Such substitution reactions result in the formation of the carboxylic acid derivatives known as halides, anhydrides, esters, and amides. These derivatives are important starting points for the synthesis of a variety of other classes of compounds. The next chapter will reveal the remarkable chemical versatility that has made these derivatives mainstays of laboratory chemistry as well as countless naturally occurring biological processes.

After we have completed our presentation of carbonyl chemistry, the only simple compound class left to be discussed will be the amines (Chapter 21). The final five chapters of the text will deal with compounds combining multiple functional groups, with an emphasis on those of synthetic and biochemical importance.

WORKED EXAMPLES: INTEGRATING THE CONCEPTS

19-25. Applying Mechanistic Patterns of the Carboxy Function

Propose a mechanism for the exchange between an acyl bromide and a carboxylic acid, as it occurs in step 4 of the Hell-Volhard-Zelinsky reaction.

$$\underset{\underset{Br}{|}}{RCHCBr} \;+\; RCH_2COH \;\rightleftharpoons\; \underset{\underset{Br}{|}}{RCHCOH} \;+\; RCH_2CBr$$

SOLUTION

First we must consider exactly what changes have occurred and then by which pathways these changes could have taken place. The two starting compounds have exchanged Br and OH substituents on their carbonyl carbons. Given the main chapter theme, addition–elimination pathways for carbonyl substitution reactions, let us look for a logical way to enable that process.

The α-bromoacyl bromide has a strongly electrophilic (δ^+) carbonyl carbon, because of the electron-withdrawing effects of the two bromines. This acyl bromide is therefore a good candidate to be attacked by the most nucleophilic atom in the two compounds, the carbonyl *oxygen* of the carboxylic acid: Recall that the carbonyl oxygen is the most favorable site for attachment of electrophiles, because the species that results is resonance stabilized:

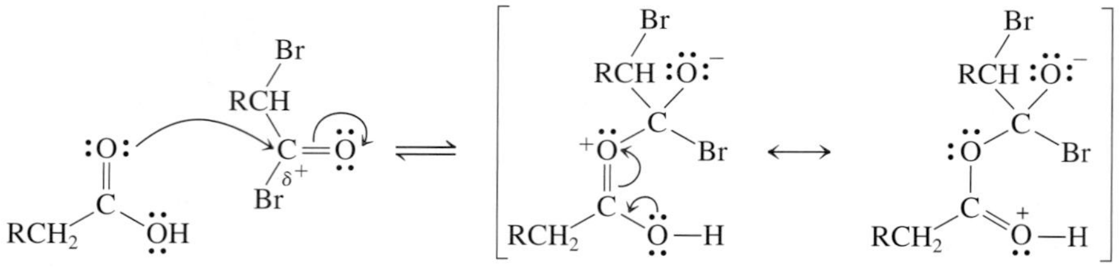

Elimination of bromide ion from this tetrahedral intermediate gives a protonated anhydride. Readdition of bromide to the activated carbonyl group in the anhydride gives a new tetrahedral intermediate. At this stage the exchange sequence can be completed by elimination of the carboxylic acid. This process may be viewed as involving the shift of a proton from one oxygen to another using a six-membered-ring transition state, similar to those we have seen on several previous occasions. This step results directly in the formation of the exchanged product molecules, an α-bromocarboxylic acid and an acyl bromide unsubstituted at the α-carbon:

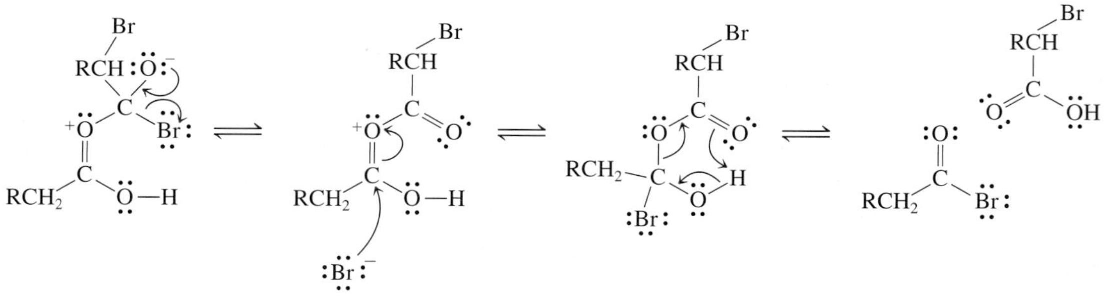

19-26. Interconversion of a Polymeric Polyester with a Large-Ring Lactone

Lactones with rings of more than six members may be synthesized if ring strain and transannular interactions are minimized (see Sections 4-2 through 4-5 and 9-6). One of the most important commercial musk fragrances is a dilactone, ethylene brassylate. Its synthesis, illustrated here, begins with acid-catalyzed reaction of tridecanedioic acid with 1,2-ethanediol, which yields a substance A with the molecular formula $C_{15}H_{28}O_5$. Under the conditions in which it forms, substance A converts into a polymer (lower left structure). Strong heating reverses polymer formation, regenerating substance A. In a slower process, A transforms into the final macrocyclic product (lower right structure), which is distilled from the reaction mixture, thereby shifting the equilibria in its favor.

Commercial Synthesis of a Macrocyclic Musk

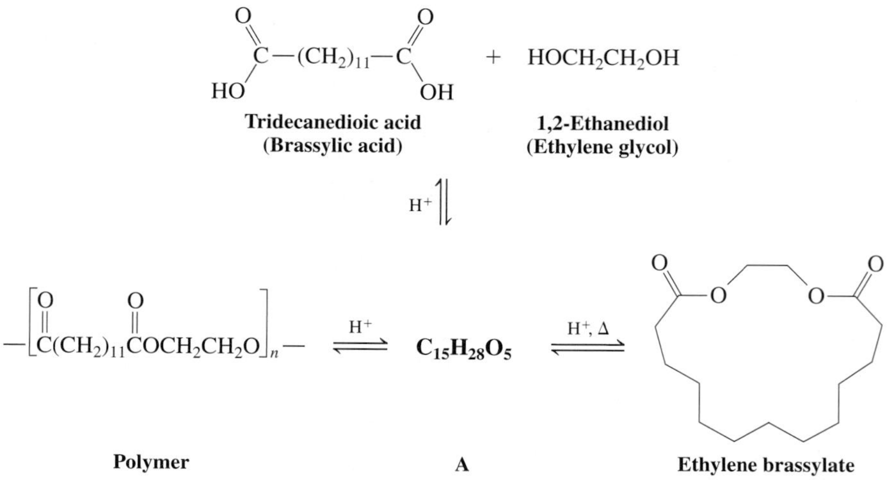

Tridecanedioic acid
(Brassylic acid)

1,2-Ethanediol
(Ethylene glycol)

Polymer

A

Ethylene brassylate

a. Suggest a structure for compound A.

SOLUTION

What information do we possess regarding this compound? We have its molecular formula and the structures both of its immediate precursors and of two of its products. It is possible to arrive at the solution by using any of these clues. For example, the polymeric product is constructed of repeating monomer units, each of which contains an ethylene glycol ester of brassylic acid, just missing the OH from the carboxy group at the left and the H from the hydroxy group at the right. Indeed, the formula of this unit is $C_{15}H_{26}O_4$, or structure A less one molecule of water. If we simply write its structure and then add the missing elements of a molecule of water, we get A, the monoester shown.

**Ethylene glycol monoester of
brassylic acid (A)**

To use another approach, the molecular formulas of the two starting materials add up to $C_{15}H_{30}O_6$, or A + H_2O. Therefore, A is the product of a reaction that combines brassylic acid and ethylene glycol, and *releases* a molecule of water in the process, exactly what is observed in esterification.

b. Propose mechanisms for the reactions that interconvert structure A with both the polymer and the macrocycle.

SOLUTION

Both reactions are examples of esterifications. Polymer formation begins with ester formation between the hydroxy group of one molecule of structure A and the carboxy group of another. We follow the pattern of addition–elimination (Section 19-9), making use of acid catalysis *as specified in the original equation*. Remember: The answers to mechanism problems should make use of only those chemical species actually specified in the reaction. The mechanistic sequence of steps is adapted directly from the text examples: (1) protonation of the carbonyl carbon to be attacked, (2) nucleophilic addition of a hydroxy oxygen to form the tetrahedral intermediate, (3) protonation of one hydroxy group of the intermediate to form a good leaving group, (4) departure of water, and (5) proton loss from oxygen to yield the final product.

Tetrahedral intermediate

The product of this process is a dimer; repetition of such ester formation many times at both ends of this molecule eventually gives rise to the polyester observed in the reaction. The same steps take place in reverse to convert the polymer back into monoester A upon heating.

What about the mechanism for macrocyclic lactone formation? We realize that the process must be intramolecular to form a ring (compare Sections 9-6 and 17-7). We can write exactly the same sequence of steps as we did for esterification, but we use the free hydroxy group at one end of molecule A to attack the carboxy carbon at the other end.

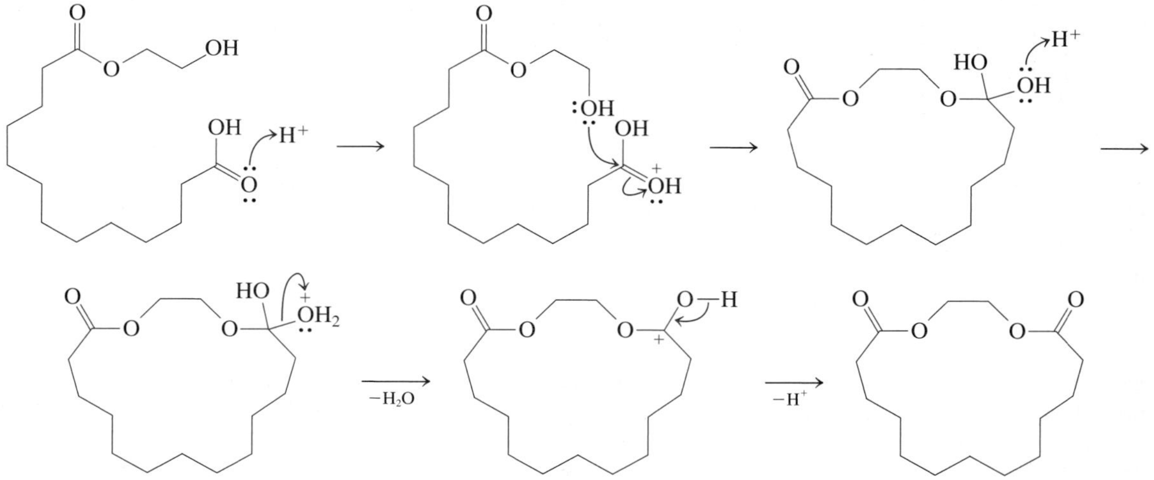

Macrocyclic lactone synthesis is a topic of considerable interest to the pharmaceutical industry because this function constitutes the basic framework of many medicinally valuable compounds. Examples include erythromycin A, a widely used macrolide antibiotic (Real Life 20-2), and tacrolimus (Protopic), a powerful immunosuppressant that shows great promise in controlling the rejection of transplanted organs in human patients.

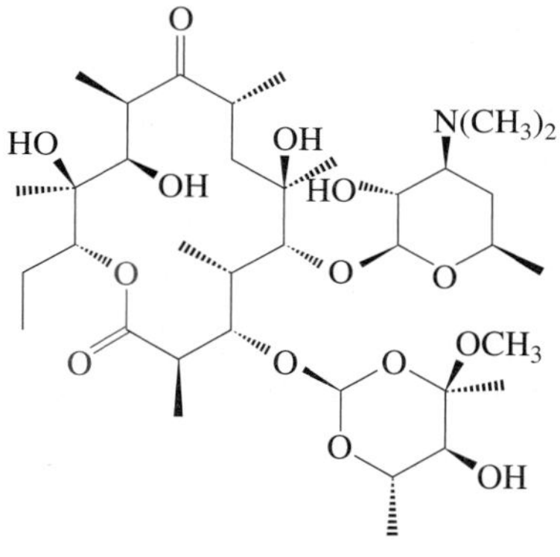

Erythromycin A

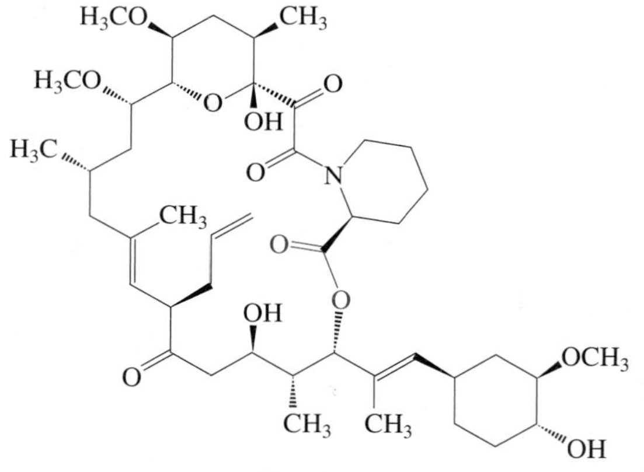

Tacrolimus (Protopic)

New Reactions

1. Acidity of Carboxylic Acids (Section 19-4)

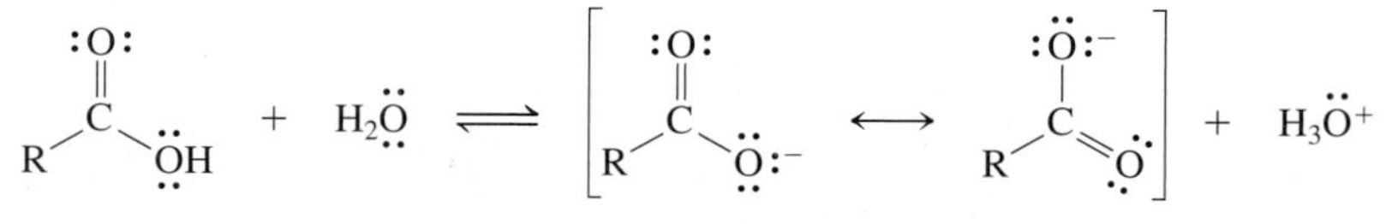

Resonance-stabilized carboxylate ion

$K_a = 10^{-4}–10^{-5}$, p$K_a \approx 4–5$

Salt formation

$$RCOOH \ + \ NaOH \ \longrightarrow \ RCOO^-Na^+ \ + \ H_2O$$

Also with Na₂CO₃, NaHCO₃

2. Basicity of Carboxylic Acids (Section 19-4)

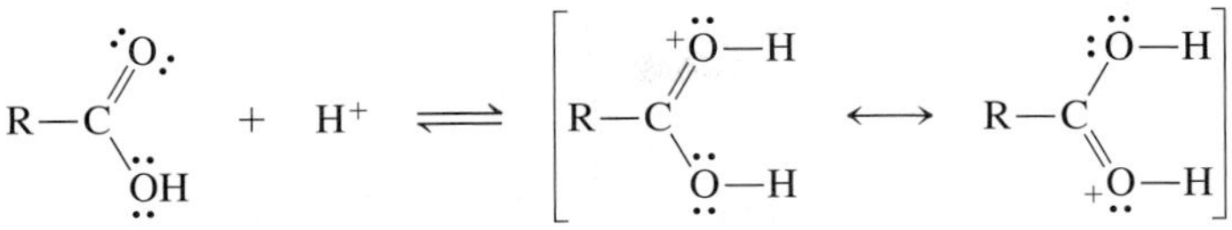

**Resonance-stabilized
protonated carboxylic acid**

Preparation of Carboxylic Acids

3. Oxidation of Primary Alcohols and Aldehydes (Section 19-6)

$$RCH_2OH \ \xrightarrow{\text{Oxidizing agent}} \ RCOOH$$

Oxidizing agents: aqueous CrO₃, KMnO₄, HNO₃

$$RCHO \ \xrightarrow{\text{Oxidizing agent}} \ RCOOH$$

Oxidizing agents: aqueous CrO₃, KMnO₄, Ag⁺, H₂O₂, HNO₃

4. Carbonation of Organometallic Reagents (Section 19-6)

$$RMgX \ + \ CO_2 \ \xrightarrow{\text{THF}} \ RCOO^{-+}MgX \ \xrightarrow{\text{H}^+, \text{H}_2\text{O}} \ RCOOH$$

$$RLi \ + \ CO_2 \ \xrightarrow{\text{THF}} \ RCOO^-Li^+ \ \xrightarrow{\text{H}^+, \text{H}_2\text{O}} \ RCOOH$$

5. Hydrolysis of Nitriles (Section 19-6)

$$RC{\equiv}N \ \xrightarrow{\text{H}_2\text{O}, \Delta, \text{H}^+ \text{or HO}^-} \ RCOOH \ + \ NH_3 \ \text{or} \ NH_4^+$$

Reactions of Carboxylic Acids

6. Nucleophilic Attack at the Carbonyl Group (Section 19-7)

Base-catalyzed addition–elimination

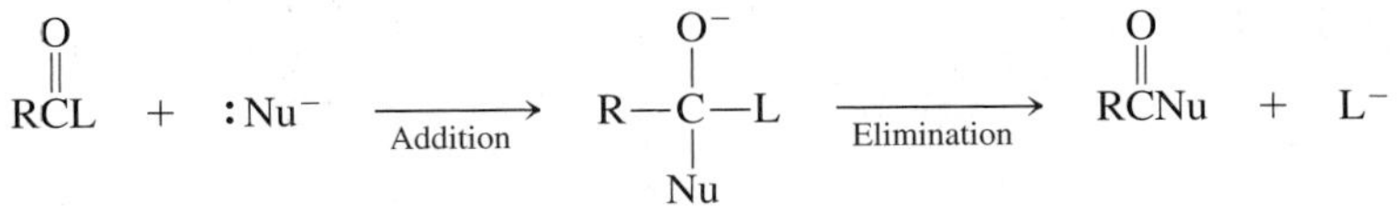

L = leaving group   **Tetrahedral
intermediate**

Acid-catalyzed addition–elimination

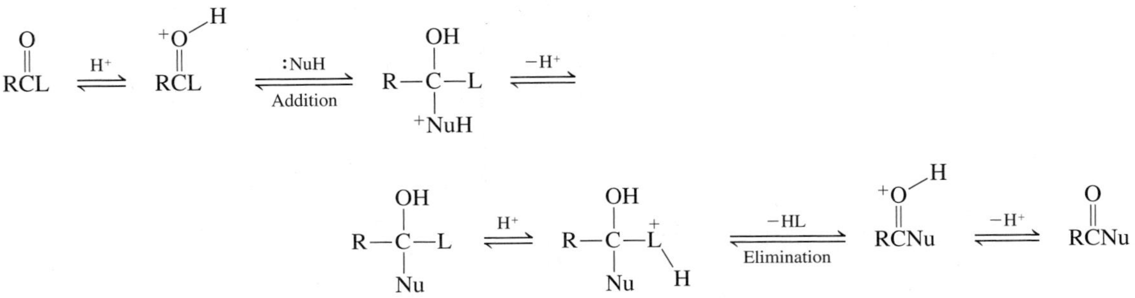

**Tetrahedral
intermediate**

Derivatives of Carboxylic Acids

7. Acyl Halides (Section 19-8)

$$RCOOH \; + \; SOCl_2 \; \longrightarrow \; \underset{\substack{\textbf{Acyl} \\ \textbf{chloride}}}{R\overset{\displaystyle O}{\overset{\|}{C}}Cl} \; + \; SO_2 \; + \; HCl$$

$$3\,RCOOH \; + \; PBr_3 \; \longrightarrow \; \underset{\substack{\textbf{Acyl} \\ \textbf{bromide}}}{3\,R\overset{\displaystyle O}{\overset{\|}{C}}Br} \; + \; H_3PO_3$$

8. Carboxylic Anhydrides (Section 19-8)

$$RCOOH \; + \; R\overset{\displaystyle O}{\overset{\|}{C}}Cl \; \longrightarrow \; \underset{\textbf{Anhydride}}{R\overset{\displaystyle O}{\overset{\|}{C}}O\overset{\displaystyle O}{\overset{\|}{C}}R} \; + \; HCl$$

Cyclic anhydrides

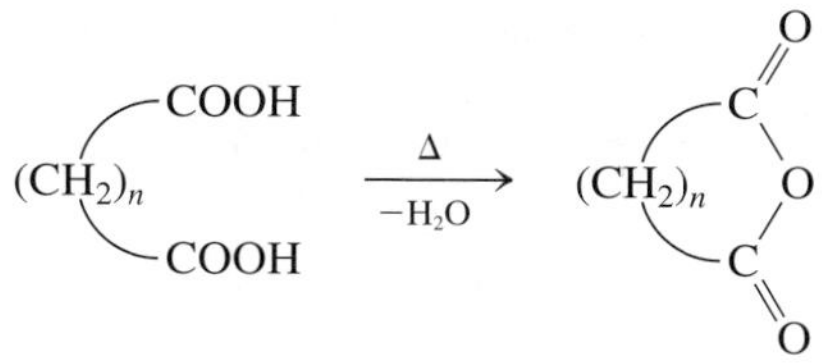

Best for five- or six-membered rings

9. Esters (Section 19-9)

Acid-catalyzed esterification

$$RCO_2H \; + \; R'OH \underset{K\approx 1}{\overset{H^+}{\rightleftharpoons}} \; R\overset{\displaystyle O}{\overset{\|}{C}}OR' \; + \; H_2O$$

Cyclic esters (lactones)

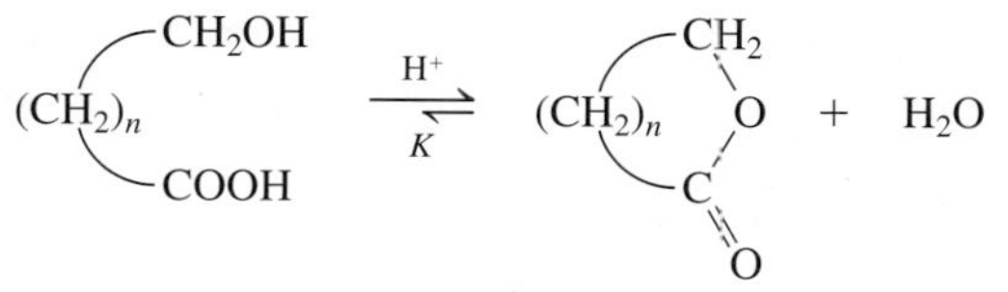

Lactone

$K > 1$ for five- and six-membered rings

10. Carboxylic Amides (Section 19-10)

$$RCOOH \; + \; R'NH_2 \; \longrightarrow \; RCOO^- \; + \; R'NH_3^+ \overset{\Delta}{\longrightarrow} \; R\overset{\displaystyle O}{\overset{\|}{C}}NHR' \; + \; H_2O$$

Imides

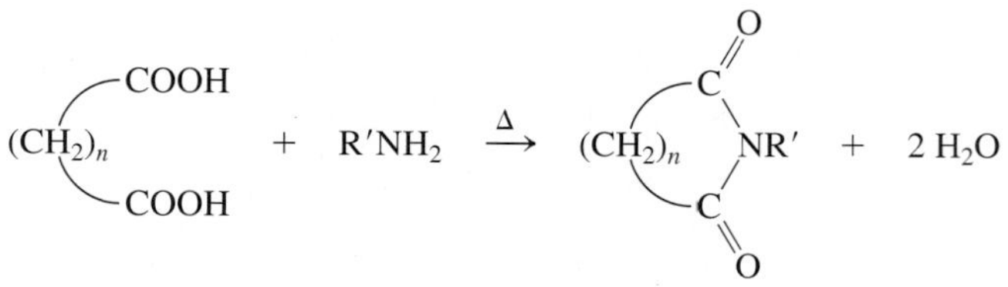

Cyclic amides (lactams)

Lactam

11. Reduction with Lithium Aluminum Hydride (Section 19-11)

$$RCOOH \xrightarrow[\text{2. H}^+, \text{H}_2\text{O}]{\text{1. LiAlH}_4, (\text{CH}_3\text{CH}_2)_2\text{O}} RCH_2OH$$

12. Bromination: Hell-Volhard-Zelinsky Reaction (Section 19-12)

$$RCH_2COOH \xrightarrow{Br_2, \text{ trace P}} R\overset{Br}{\underset{}{C}}HCOOH$$

Important Concepts

1. **Carboxylic acids** are named as **alkanoic acids.** The carbonyl carbon is numbered 1 in the longest chain incorporating the carboxy group. Dicarboxylic acids are called **alkanedioic acids.** Cyclic and aromatic systems are called **cycloalkanecarboxylic** and **benzoic acids,** respectively. In these systems the ring carbon bearing the carboxy group is assigned the number 1.

2. The **carboxy group** is approximately **trigonal planar.** Except in very dilute solution, carboxylic acids form dimers by hydrogen bonding.

3. The carboxylic acid **proton chemical shift** is variable but relatively **high** ($\delta = 10$–13), because of hydrogen bonding. The **carbonyl carbon** is also relatively **deshielded** but not as much as in aldehydes and ketones, because of the resonance contribution of the hydroxy group. The carboxy function shows two important infrared bands, one at about 1710 cm^{-1} for the C=O bond and a very broad band between 2500 and 3300 cm^{-1} for the O–H group. The mass spectrum of carboxylic acids shows facile fragmentation in three ways.

4. The carbonyl group in carboxylic acids undergoes nucleophilic displacement via the **addition–elimination** pathway. Addition of a nucleophile gives an unstable **tetrahedral intermediate** that decomposes by elimination of the hydroxy group to give a **carboxylic acid derivative.**

5. **Lithium aluminum hydride** is a strong enough nucleophile to add to the carbonyl group of carboxylate ions. This process allows the reduction of carboxylic acids to **primary alcohols.**

Problems

27. Name (IUPAC or *Chemical Abstracts* system) or draw the structure of each of the following compounds.

(i) 4-Aminobutanoic acid (also known as GABA, a critical participant in brain biochemistry); (j) *meso*-2,3-dimethylbutanedioic acid; (k) 2-oxopropanoic acid (pyruvic acid); (l) *trans*-2-formylcyclohexanecarboxylic acid; (m) (Z)-3-phenyl-2-butenoic acid; (n) 1,8-naphthalenedicarboxylic acid.

28. Name each of the following compounds according to IUPAC or *Chemical Abstracts.* Pay attention to the order of precedence of the functional groups.

(a) HO–CH₂–CO–CH(Cl)–CH₃

(b) (structure with C=C, CHO, COOH, and two ethyl groups)

(c) (cyclopentene structure with OH and COOH)

(d) (benzene ring with COOH, COCH₃ (acetyl), and NO₂)

29. Rank the group of molecules shown here in decreasing order of boiling point and solubility in water. Explain your answers.

(benzene–COOH, benzene–CH₃, benzene–CHO, benzene–CH₂OH)

30. Rank each of the following groups of organic compounds in order of decreasing acidity.

(a) CH₃CH₂CO₂H, CH₃CCH₂OH, CH₃CH₂CH₂OH

(with O double-bonded to the second carbon in the middle structure)

(b) BrCH₂CO₂H, ClCH₂CO₂H, FCH₂CO₂H

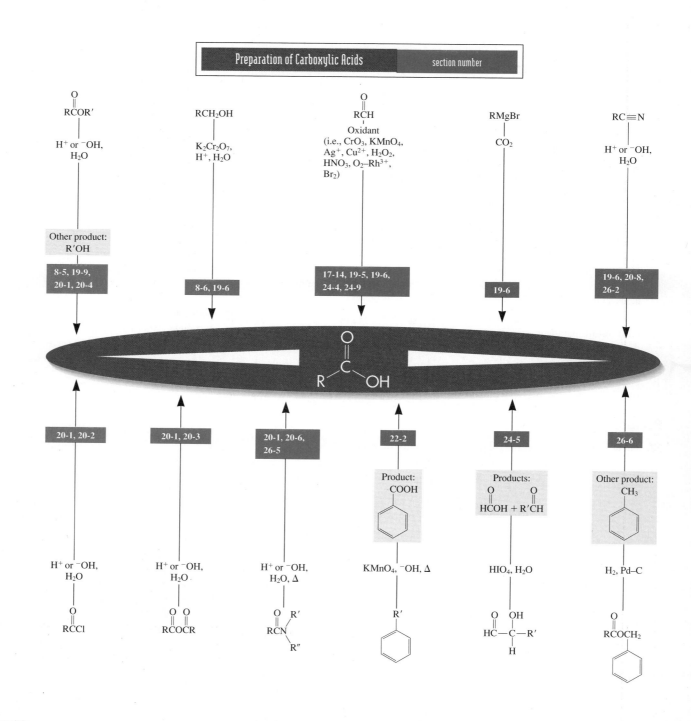

Preparation of Carboxylic Acids   section number

RCOR′ — H⁺ or ⁻OH, H₂O — Other product: R′OH — 8-5, 19-9, 20-1, 20-4

RCH₂OH — K₂Cr₂O₇, H⁺, H₂O — 8-6, 19-6

RCH (O), Oxidant (i.e., CrO₃, KMnO₄, Ag⁺, Cu²⁺, H₂O₂, HNO₃, O₂–Rh³⁺, Br₂) — 17-14, 19-5, 19-6, 24-4, 24-9

RMgBr — CO₂ — 19-6

RC≡N — H⁺ or ⁻OH, H₂O — 19-6, 20-8, 26-2

R–C(=O)–OH

20-1, 20-2 — H⁺ or ⁻OH, H₂O — RCCl

20-1, 20-3 — H⁺ or ⁻OH, H₂O — RCOCR

20-1, 20-6, 26-5 — H⁺ or ⁻OH, H₂O, Δ — RCN(R′)(R″)

22-2 — Product: COOH — KMnO₄, ⁻OH, Δ — R′

24-5 — Products: HCOH + R′CH — HIO₄, H₂O — HC–C(OH)(H)–R′

26-6 — Other product: CH₃ — H₂, Pd–C — RCOCH₂

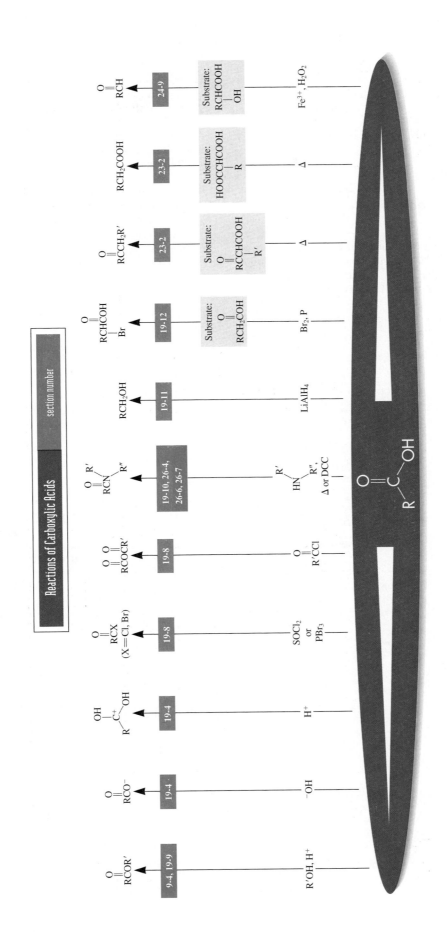

Reactions of Carboxylic Acids

(c) $CH_3CHCH_2CO_2H$, $ClCH_2CH_2CH_2CO_2H$, $CH_3CH_2CHCO_2H$
(with Cl substituents indicated above)

(d) CF_3CO_2H, CBr_3CO_2H, $(CH_3)_3CCO_2H$

(e) [benzene ring]—COOH, O_2N—[benzene ring]—COOH,

O_2N—[benzene ring with NO_2]—COOH, CH_3O—[benzene ring]—COOH

31. Propose a structure for a compound that displays the spectroscopic data that follow. The molecular ion in the mass spectrum appears at $m/z = 116$. IR: $\tilde{\nu} = 1710$ (s) and 3000 (s, broad) cm^{-1}. ^{1}H NMR: $\delta = 0.94$ (t, $J = 7.0$ Hz, 6 H), 1.59 (m, 4 H), 2.36 (quin, $J = 7.0$ Hz, 1 H), and 12.04 (broad s, 1 H) ppm; ^{13}C NMR: $\delta = 11.7, 24.7, 48.7$, and 183.0 ppm.

32. (a) An unknown compound A has the formula $C_7H_{12}O_2$ and infrared spectrum A (below). To which class does this compound belong? (b) Use the other spectra (NMR-B, p. 879, and F, p. 880; IR-D, E, and F, pp. 879–880) and spectroscopic and chemical information in the reaction sequence to determine the structures of compound A and the other unknown substances B through F. References are made to relevant sections of earlier chapters, but do not look them up before you have tried to solve the problem without the extra help. (c) Another unknown compound, G, has the formula $C_8H_{14}O_4$ and the NMR and IR spectra labeled G (p. 880–881). Propose a structure for this molecule. (d) Compound G may be readily synthesized from B. Propose a sequence that accomplishes this efficiently. (e) Propose a completely different sequence from that shown in part (b) for the conversion of C into A. (f) Finally, construct a synthetic scheme that is the reverse of that shown in part (b); namely, the conversion of A into B.

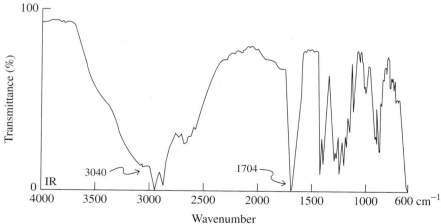

C_6H_{10}
B
^{13}C NMR: $\delta = $ 22.1
24.5
126.2 ppm
^{1}H NMR-B

1. Hg(OCCH$_3$)$_2$, H$_2$O
2. NaBH$_4$
Section 12-7

$C_6H_{12}O$
C
^{13}C NMR: $\delta = $ 24.4
25.9
35.5
69.5 ppm

CrO$_3$, H$_2$SO$_4$,
acetone, 0°C
Section 8-6

$C_6H_{10}O$
D
^{13}C NMR: $\delta = $ 23.8
26.5
40.4
208.5 ppm
IR-D

CH$_2$=P(C$_6$H$_5$)$_3$
Section 17-12

C_7H_{12}
E
IR-E

1. BH$_3$, THF
2. HO$^-$, H$_2$O$_2$
Section 12-8

$C_7H_{14}O$
F
IR-F
^{1}H NMR-F

Na$_2$Cr$_2$O$_7$, H$_2$O, H$_2$SO$_4$
Section 8-6

A

IR-A

(x-axis: Wavenumber, 4000 to 600 cm^{-1}; y-axis: Transmittance (%), 0 to 100; labeled peaks: 3040, 1704)

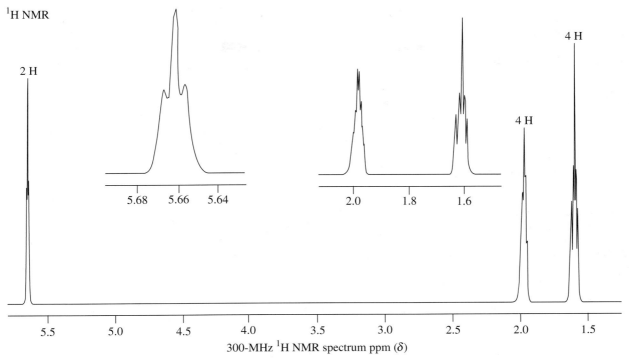

¹H NMR

2 H

5.68 5.66 5.64

2.0 1.8 1.6

4 H

4 H

5.5 5.0 4.5 4.0 3.5 3.0 2.5 2.0 1.5

300-MHz ¹H NMR spectrum ppm (δ)

B

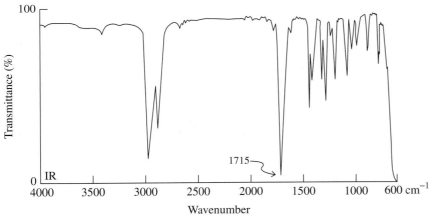

100

Transmittance (%)

1715

0 IR

4000 3500 3000 2500 2000 1500 1000 600 cm⁻¹

Wavenumber

IR-D

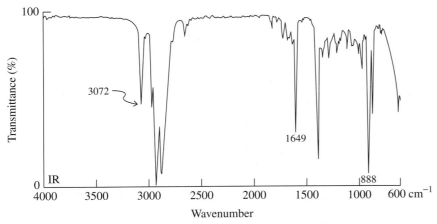

100

Transmittance (%)

3072

1649

888

0 IR

4000 3500 3000 2500 2000 1500 1000 600 cm⁻¹

Wavenumber

IR-E

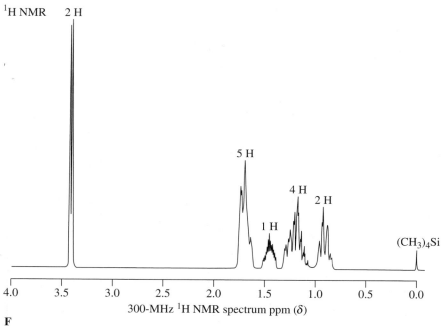

F

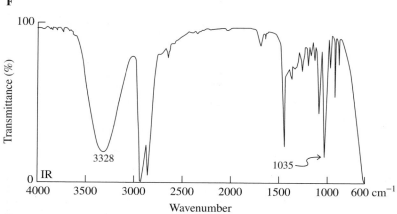

IR-F

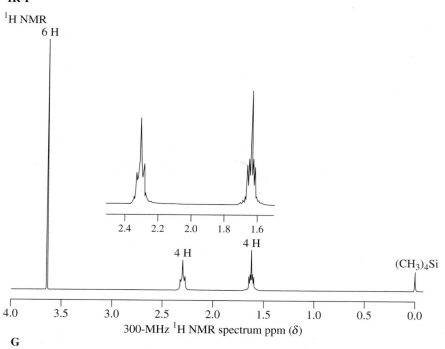

G

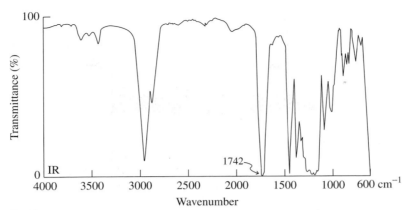

IR-G

33. Give the products of each of the following reactions.

(a) $(CH_3)_2CHCH_2CO_2H + SOCl_2 \longrightarrow$

(b) $(CH_3)_2CHCH_2CO_2H + CH_3COBr \longrightarrow$

(c) [cyclopentane—COOH] $+ CH_3CH_2OH \xrightarrow{H^+}$

(d) CH_3O—[benzene ring]—$COOH + NH_3 \longrightarrow$

(e) Product of (d), heated strongly

(f) Phthalic acid [Problem 27, part (h)], heated strongly

34. When 1,4- and 1,5-dicarboxylic acids, such as butanedioic (succinic) acid (Section 19-8), are treated with $SOCl_2$ or PBr_3 in attempted preparations of the diacyl halides, the corresponding cyclic anhydrides are obtained. Explain mechanistically.

35. Reaction review. Without consulting the Reaction Road Map on p. 876, suggest reagents to convert each of the following starting materials into hexanoic acid: (a) hexanal; (b) methyl hexanoate; (c) 1-bromopentane; (d) 1-hexanol; (e) hexanenitrile.

36. Reaction review II. Without consulting the Reaction Road Map on p. 877, suggest reagents to convert hexanoic acid into each of the following compounds: (a) 1-hexanol; (b) hexanoic anhydride; (c) hexanoyl chloride; (d) 2-bromohexanoic acid; (e) ethyl hexanoate; (f) hexanamide.

37. Fill in suitable reagents to carry out the following transformations.

(a) $(CH_3)_2CHCH_2CHO \longrightarrow (CH_3)_2CHCH_2CO_2H$

(b) [cyclopentane]—CHO $\longrightarrow$ [cyclopentane]—$\overset{OH}{\underset{}{C}}HCO_2H$

(c) [decalin]—Br $\longrightarrow$ [decalin]—CO_2H

(d) [structure with OH]—Cl $\longrightarrow$ [structure with OH and OH, O]

(e) [ethyl 2-methylbutanoate / acid structure] $\longrightarrow$ [anhydride structure]

(f) $(CH_3)_3CCO_2H \longrightarrow (CH_3)_3CCO_2CH(CH_3)_2$

(g) [benzene]—$NHCH_3 \longrightarrow$ [benzene]—$N(CH_3)COCH_3$

38. Propose syntheses of each of the following carboxylic acids that employ at least one reaction that forms a carbon–carbon bond.

(a) $CH_3CH_2CH_2CH_2CH_2CH_2CO_2H$

(b) $CH_3\overset{OH}{\underset{}{C}}HCH_2CO_2H$

(c) $H_3C\overset{CH_3}{\underset{CH_3}{\overset{|}{\underset{|}{C}}}}CO_2H$

39. (a) Write a mechanism for the esterification of propanoic acid with ^{18}O-labeled ethanol. Show clearly the fate of the ^{18}O label. (b) Acid-catalyzed hydrolysis of an unlabeled ester with ^{18}O-labeled water ($H_2^{18}O$) leads to incorporation of some ^{18}O into *both* oxygens of the carboxylic acid product. Explain by a mechanism. (**Hint:** You must use the fact that all steps in the mechanism are reversible.)

40. Give the products of reaction of propanoic acid with each of the following reagents.

(a) $SOCl_2$ (b) PBr_3

(c) CH_3CH_2COBr + pyridine (d) $(CH_3)_2CHOH + HCl$

(e) [benzene]—CH_2NH_2 (f) Product of (e), heated strongly

(g) $LiAlH_4$, then H^+, H_2O (h) Br_2, P

41. Give the product of reaction of cyclopentanecarboxylic acid with each of the reagents in Problem 40.

42. **CHALLENGE** When methyl ketones are treated with a halogen in the presence of base, the three hydrogen atoms on the methyl carbon are replaced to give a CX_3-substituted ketone. This product is not stable under the basic conditions and proceeds to react with hydroxide, ultimately furnishing the carboxylic acid (as its conjugate base) and a molecule of HCX₃, which has the common name *haloform* (i.e., chloroform, bromoform, and iodoform, for X = Cl, Br, and I, respectively). For example:

$$\underset{\text{RCCBr}_3}{\overset{\text{:O:}}{\underset{\|}{}}} + \ ^-\!:\!\ddot{O}H \longrightarrow \underset{\text{RC}\ddot{O}:^-}{\overset{:O:}{\underset{\|}{}}} + \ \text{HCBr}_3$$

Propose a series of mechanistic steps to convert the tribromoketone into the carboxylate. What is the leaving group? Why do you think this species is capable of acting as a leaving group in this process?

43. Interpret the labeled peaks in the mass spectrum of 2-methylhexanoic acid (MS-H).

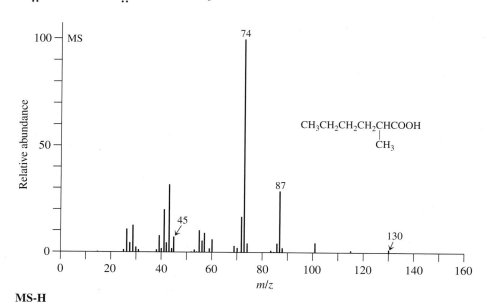

CH₃CH₂CH₂CH₂CHCOOH
  |
  CH₃

MS-H

44. Suggest a preparation of hexanoic acid from pentanoic acid.

45. Give reagents and reaction conditions that would allow efficient conversion of 2-methylbutanoic acid into **(a)** the corresponding acyl chloride; **(b)** the corresponding methyl ester; **(c)** the corresponding ester with 2-butanol; **(d)** the anhydride; **(e)** the *N*-methylamide;

$$\text{(f)} \ \ \underset{\text{CH}_3}{\overset{\text{CH}_3}{\underset{|}{\text{CH}_3\text{CH}_2\text{CHCH}_2\text{OH}}}} \qquad \text{(g)} \ \ \underset{\text{CH}_3}{\overset{\text{Br}}{\underset{|}{\text{CH}_3\text{CH}_2\text{CCO}_2\text{H}}}}$$

46. Treatment of 4-pentenoic acid (below) with Br₂ in the presence of dilute aqueous base yields a nonacidic compound with the formula C₅H₇BrO₂. **(a)** Suggest a structure for this compound and a mechanism for its formation. **(b)** Can you find a second, isomeric product whose formation is also mechanistically reasonable? **(c)** Discuss the issues that contribute to determining which of the two is the major product in this reaction. (**Hint:** Review Section 12-6.)

O
‖
    ⟍⟍⟍C⟍OH

4-Pentenoic acid

47. Show how the Hell-Volhard-Zelinsky reaction might be used in the synthesis of each of the following compounds, beginning in each case with a simple monocarboxylic acid. Write detailed mechanisms for all the reactions in *one* of your syntheses.

(a) CH₃CH₂CHCO₂H
                   |
                   NH₂

(b) C₆H₅—CHCO₂H
                    |
                    CO₂H

(c) (structure: 4-methyl-2-hydroxyhexanoic acid)

(d) HO₂CCH₂SSCH₂CO₂H

(e) (CH₃CH₂)₂NCH₂CO₂H

(f) (C₆H₅)₃P⁺CHCO₂H Br⁻
                            |
                            CH₃

48. The Hell-Volhard-Zelinsky reaction produces only bromocarboxylic acids. However, modified procedures have been developed to convert acyl chlorides into α-chloro- and α-bromoacyl chlorides by reaction with *N*-chloro- and *N*-bromobutanimide (*N*-chloro- and *N*-bromosuccinimide, NCS and NBS: Section 14-2), respectively. Reaction with I₂ gives α-iodoacyl chlorides. (SOCl₂ is used as the solvent to maintain the acyl chloride functional group.) Suggest a mechanism for any one of these processes.

C₆H₅CH₂CH₂COCl

→ NCS, HCl, SOCl₂, 70°C → C₆H₅CH₂CHClCOCl 84%

→ NBS, HBr, SOCl₂, 70°C → C₆H₅CH₂CHBrCOCl 71%

→ I₂, HI, SOCl₂, 85°C → C₆H₅CH₂CHICOCl 75%

49. How would you expect the acidity of acetamide to compare with that of acetic acid? With that of acetone? Which protons in acetamide are the most acidic? Where would you expect acetamide to be protonated by very strong acid?

O
‖
CH₃CNH₂
Acetamide

50. Attempted CrO_3 oxidation of 1,4-butanediol to butanedioic acid results in significant yields of "γ-butyrolactone." Explain mechanistically.

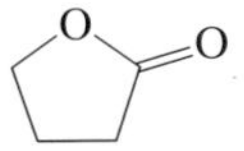

γ-Butyrolactone

51. Following the general mechanistic scheme given in Section 19-7, write detailed mechanisms for each of the following substitution reactions. (Note: These transformations are part of Chapter 20, but try to solve the problem without looking ahead.)

(a)

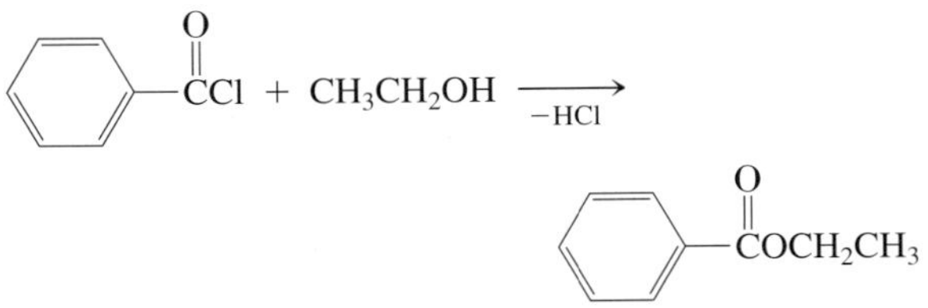

(b) $CH_3\overset{O}{\overset{\|}{C}}NH_2 + H_2O \xrightarrow{H^+} CH_3\overset{O}{\overset{\|}{C}}OH + \overset{+}{N}H_4$

52. Suggest structures for the products of each reaction in the following synthetic sequence.

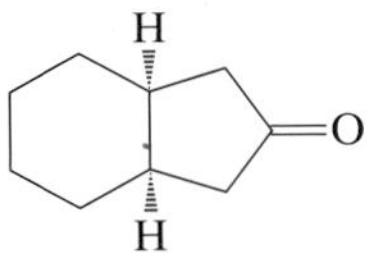

IR: 1745 cm^{-1}

1. 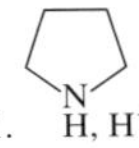, H^+
2. $(CH_3)_2C{=}CHCH_2Br$
3. H^+, H_2O

$\longrightarrow$ $C_{14}H_{22}O$ **H** (**IR: 1675 and 1745 cm^{-1}**) $\xrightarrow{HOCH_2CH_2OH,\ H^+}$

$C_{16}H_{26}O_2$ **I** (**IR: 1670 cm^{-1}**) $\xrightarrow{1.\ O_3,\ CH_2Cl_2;\ 2.\ Zn,\ CH_3COOH;\ 3.\ KMnO_4,\ NaHCO_3}$ $C_{13}H_{20}O_4$ **J** (**IR: 1715 and 3000 (broad) cm^{-1}**) $\xrightarrow{1.\ H^+,\ H_2O;\ 2.\ NaBH_4}$

$C_{11}H_{18}O_3$ **K** (**IR: 1715, 3000 (broad), and 3350 cm^{-1}**) $\xrightarrow{\text{Catalytic } H^+,\ \Delta}$ $C_{11}H_{16}O_2$ **L** (**IR: 1770 cm^{-1}**)

53. S_N2 reactions of simple carboxylate ions with haloalkanes (Section 8-5) in aqueous solution generally do not give good yields of esters. **(a)** Explain why this is so. **(b)** Reaction of 1-iodobutane with sodium acetate gives an excellent yield of ester if carried out in acetic acid (as shown here). Why is acetic acid a better solvent for this process than water?

$$CH_3CH_2CH_2CH_2I + CH_3\overset{O}{\overset{\|}{C}}O^-Na^+ \xrightarrow{CH_3CO_2H,\ 100°C} CH_3CH_2CH_2CH_2O\overset{O}{\overset{\|}{C}}CH_3 + Na^+I^-$$

1-Iodobutane **Sodium acetate** → 95% **Butyl acetate**

(c) The reaction of 1-iodobutane with sodium dodecanoate proceeds surprisingly well in aqueous solution, much better than the reaction with sodium acetate (see the following equation). Explain this observation. (**Hint:** Sodium dodecanoate is a soap and forms micelles in water. See Real Life 19-1.)

$$CH_3CH_2CH_2CH_2I + CH_3(CH_2)_{10}CO_2^-Na^+ \xrightarrow{H_2O} CH_3(CH_2)_{10}\overset{O}{\overset{\|}{C}}OCH_2CH_2CH_2CH_3$$

54. **CHALLENGE** The iridoids are a class of monoterpenes with powerful and varied biological activities. They include insecticides, agents of defense against predatory insects, and animal attractants. The following reaction sequence is a synthesis of neonepetalactone, one of the nepetalactones, which are primary constituents of catnip. Use the information given to deduce the structures that have been left out, including that of neonepetalactone itself.

$C_{10}H_{16}O_2$ **M** (**IR: 890, 1645, 1725 (very strong), and 1705 cm^{-1}**) $\xrightarrow{\text{Base}}$

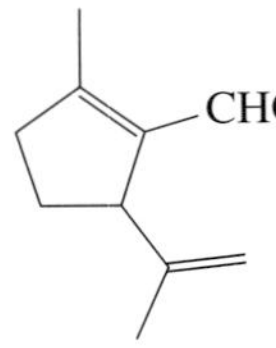

$\xrightarrow{CrO_3,\ H_2SO_4,\ 0°C}$

$C_{10}H_{14}O_2$ **N** (**IR: 890, 1630, 1640, 1720, and 3000 (very broad) cm^{-1}**) $\xrightarrow{CH_3OH,\ H^+}$ $C_{11}H_{16}O_2$ **O** (**IR: 890, 1630, 1640 and 1720 cm^{-1}**) $\xrightarrow{1.\ \text{Dicyclohexylborane, THF};\ 2.\ HO^-,\ H_2O_2}$

$C_{11}H_{18}O_3$ **P** (**IR: 1630, 1720, and 3335 cm^{-1}**) $\xrightarrow{H^+,\ H_2O,\ \Delta}$ **Neonepetalactone** (**IR: 1645 and 1710 cm^{-1}**; **UV: λ_{max} = 241 nm**)

55. **CHALLENGE** Propose *two* possible mechanisms for the following reaction. (**Hint:** Consider the possible sites of protonation in the molecule and the mechanistic consequences of each.) Devise an isotope-labeling experiment that might distinguish your two mechanisms.

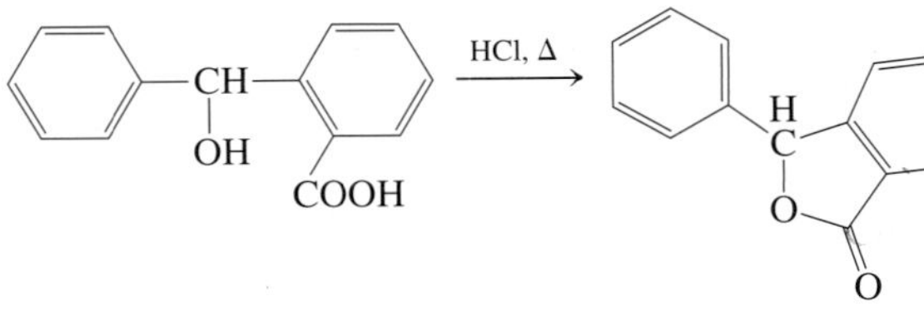

56. Propose a short synthesis of 2-butynoic acid, $CH_3C{\equiv}CCO_2H$, starting from propyne. (**Hint:** Review Sections 13-2 and 13-5.)

57. The benzene rings of many compounds in nature are prepared by a biosynthetic pathway similar to that operating in fatty acid synthesis. Acetyl units are coupled, but the ketone functions are not reduced. The result is a polyketide thiolester, which forms rings by intramolecular aldol condensation.

$$CH_3\overset{O}{\overset{\|}{C}}CH_2\overset{O}{\overset{\|}{C}}CH_2\overset{O}{\overset{\|}{C}}CH_2\overset{O}{\overset{\|}{C}}{-}S{-}\boxed{\text{protein}}$$

Polyketide thiolester

o-Orsellinic acid [for structure, see Problem 27, part (g)] is a derivative of salicylic acid and is prepared biosynthetically from the polyketide thiolester shown. Explain how this transformation might take place. Hydrolysis of the thiolester to give the free carboxylic acid is the last step.

Team Problem

58. Section 19-9 showed you that 4- and 5-hydroxy acids can undergo acid-catalyzed intramolecular esterification to produce the corresponding lactone in good yield. Consider the following two examples of lactonization. Divide the analysis of the reaction sequences among your group. Propose reasonable mechanisms to explain the respective product formation.

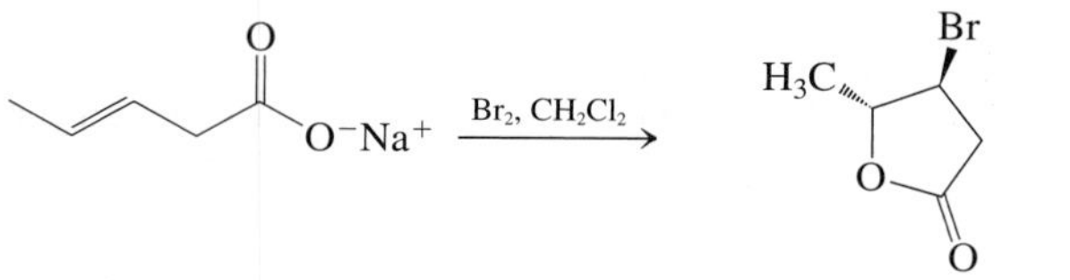

(Note stereochemistry!)

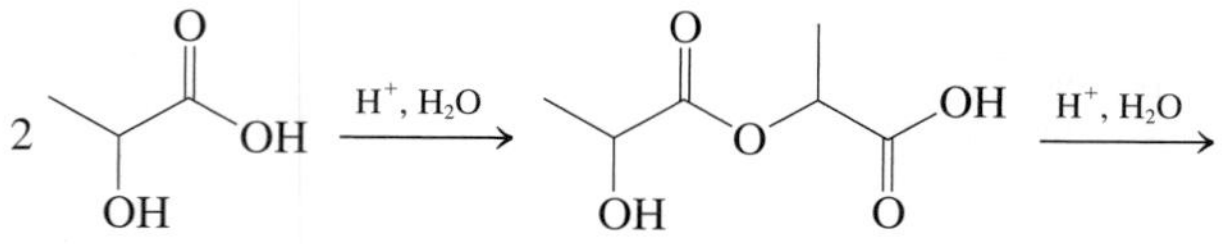

(Two diastereomers)

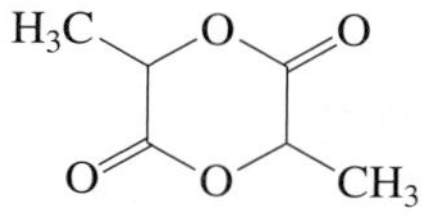

(Cis and trans isomers)

Reconvene to present your mechanisms to each other.

Preprofessional Problems

59. What is the IUPAC name of the compound shown? 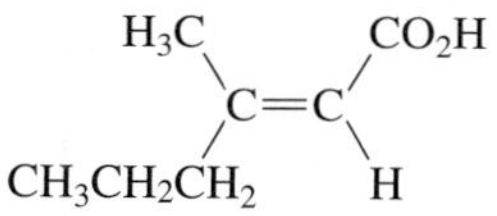
    (a) (E)-3-Methyl-2-hexenoic acid
    (b) (Z)-3-Methyl-2-hexenoic acid
    (c) (E)-3-Methyl-3-hexenoic acid
    (d) (Z)-3-Methyl-3-hexenoic acid

60. Select the acid with the highest K_a (i.e., lowest pK_a).

(a) H_3CCO_2H

(b)

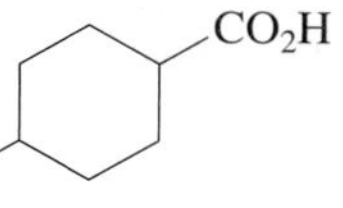

(c)

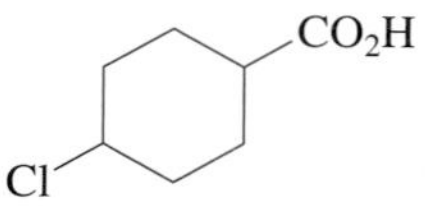

(d) Cl_2CHCO_2H

61. The acid whose structure is shown below can best be prepared via one of these sequences. Which one?

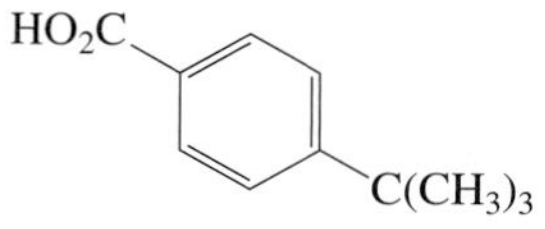

(a) $H_3CBr + Br_3CCO_2H \xrightarrow{K} \xrightarrow{Benzene}$

(b) $(CH_3)_3CI \xrightarrow{Mg, ether} \xrightarrow{CO_2} \xrightarrow{H^+, H_2O \text{ (work-up)}}$

(c)

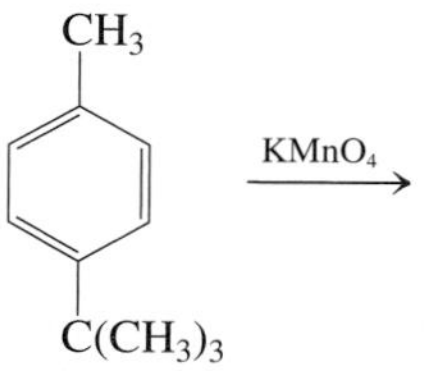

(d)

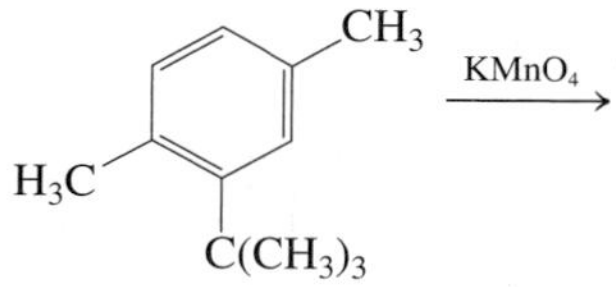

CHAPTER 20

Carboxylic Acid Derivatives

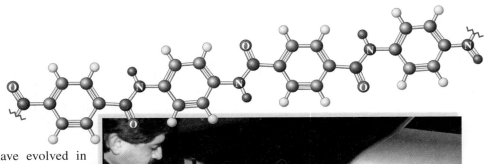

When complex organic systems have evolved in nature, they are frequently made by the combination of carboxy carbonyl groups and the hetero-atoms of alcohols, amines, or thiols. Why? The addition–elimination processes introduced in Chapter 19 provide mechanistic pathways of relatively low activation energy for the interconversion of variously substituted carboxylic acid derivatives, many of which play central roles in biology (Chapter 26). Chemists find these compounds similarly useful, as we shall see in this chapter, which deals with the chemistry of four major carboxylic acid derivatives: halides, anhydrides, esters, and amides. Each has a substituent, L, that can function as a leaving group in substitution reactions. We already know, for instance, that displace-

$$\overset{O}{\underset{\|}{}}$$

ment of the halide in RCX by a carboxy group leads to anhydrides.

$$\overset{\ddot{O}}{\underset{\|}{\underset{R}{\overset{C}{\diagup}}\diagdown L}}$$

$$L = X, \overset{O}{\underset{\|}{O}}CR, OR, NR_2$$

Carboxylic acid derivatives

We begin with a comparison of the structures, properties, and relative reactivities of carboxylic acid derivatives. We then explore the chemistry of each type of compound. Halides and anhydrides are valuable reagents in the synthesis of other carbonyl compounds. Esters and amides are enormously important in nature; for example, the esters include common flavoring agents, waxes, fats, and oils; among the amides we find urea and penicillin. The alkanenitriles, $RC\equiv N$, are also treated here because they have similar reactivity.

The sturdiness of the amide linkage is amply demonstrated by the properties of poly(p-phenylene terephthalamide), which is used in bulletproof vests and body armor under the commercial name of Kevlar. Developed by DuPont chemist Stephanie Kwolek, Kevlar is 16 times as stiff as nylon, the polymer formerly used in such applications. Kevlar's properties derive from the planarity of its benzene rings combined with restricted rotation about the amide linkages (Section 20-6). Thousands of law enforcement officers owe their lives to this remarkable material, which can stop the bullet from a 9-mm handgun traveling at 1200 feet per second.

20-1 RELATIVE REACTIVITIES, STRUCTURES, AND SPECTRA OF CARBOXYLIC ACID DERIVATIVES

Carboxylic acid derivatives undergo substitution reactions with nucleophiles, such as water, organometallic compounds, and hydride reducing agents. These transformations proceed through the familiar (often acid- or base-catalyzed) addition–elimination sequence (Section 19-7).

Addition–Elimination in Carboxylic Acid Derivatives

The relative reactivities of the substrates follow a consistent order: Acyl halides are most reactive, followed by anhydrides, then esters, and finally the amides, which are least reactive.

Relative Reactivities of Carboxylic Acid Derivatives in Nucleophilic Addition–Elimination with Water

The order of reactivity depends directly on the structure of L: how good a leaving group it is and what effect it has on the adjacent carbonyl function. This effect can be understood by looking at the degree to which the various resonance forms A–C contribute to the structure of the carboxylic acid derivative. In this picture, B is a very minor contributor, because it is an electron sextet structure. We have used such forms only to emphasize the polarity of the carbonyl function. This leaves us with A and the dipolar alternative C. A will always be major, because it has no charge separation (Section 1-5). The question is to what extent C helps to stabilize the functional group by delocalization of the lone pairs on the substituent L onto the carbonyl oxygen.

Resonance in Carboxylic Acid Derivatives

The answer to this question lies in the electronegativity of L: Decreasing electronegativity goes with increasing contribution of C. As we saw in Section 6-7, decreasing electronegativity also goes hand in hand with decreasing leaving-group ability, and in turn with decreasing acidity of the conjugate acid.

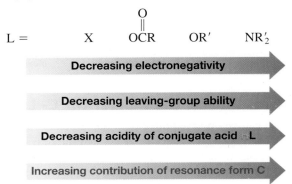

Thus, dipolar resonance form C is most important when L is NR$_2$, because nitrogen is the least electronegative atom in the series. Consequently, amides are the least reactive carboxylic acid derivatives. In esters, the contribution of C is somewhat diminished, because oxygen is more electronegative than nitrogen. Nonetheless, resonance remains strong (see also Section 16-1) and, while esters are more reactive than amides, they are still rather resistant to nucleophilic attack. On the other hand, in anhydrides, the lone pairs on the central oxygen are shared by two carbonyl groups, making them less available for resonance as in C. For that reason, they are more reactive than esters. Finally, acyl halides are the most reactive for two reasons: relative electronegativity (F, Cl; Table 1-2) and poor overlap of the large halogen p orbitals (Cl, Br, I; Section 1-6) with the relatively small $2p$ lobes of the carbonyl carbon. The electrostatic potential maps in the margin depict examples of the two extremes: acetyl chloride and acetamide. In the chloride, the carbonyl carbon is much more positively charged (blue) than in the amide (green). Simultaneously, the extent of electron donation by the L group (Cl versus NH$_2$) is greater in the amide, as indicated by the greater negative charge on the carbonyl oxygen (red).

Acetyl chloride

Acetamide

Solved Exercise 20-1 | Working with the Concepts: Relative Reactivity of Carboxylic Acid Derivatives

Phosgene, phenylmethyl chloroformate, bis(1,1-dimethylethyl) dicarbonate, dimethyl carbonate, and urea are derivatives of carbonic acid H$_2$CO$_3$ (made by dissolving CO$_2$ in water). Rank them in order of decreasing reactivity in nucleophilic addition–eliminations.

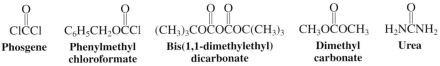

| Phosgene | Phenylmethyl chloroformate | Bis(1,1-dimethylethyl) dicarbonate | Dimethyl carbonate | Urea |

Strategy

All of these compounds possess heteroatom (O, N, or halogen)-containing substituents on *both* sides of the carbonyl carbon, as does carbonic acid, HOCOH. The arguments about relative reactivity of carboxylic acid derivatives made in this section apply here in the same way.

Solution

- To a first approximation, the effects of the substituents on either side of the carbonyl carbon may be considered to be additive. The only difference is that the effect of the second substituent on the carbonyl is somewhat reduced, because the two groups have to compete with each other for resonance with the C=O π bond.
- Thus phosgene, with halogens on both sides, may be considered as a "double acyl halide" and is the most reactive.
- Reactivity decreases in the order in which the compounds are illustrated: The chloroformate is "half halide/half ester"; both carbonyls in the dicarbonate are "half anhydride/half ester"; the carbonate is a "double ester"; and urea is a "double amide" and the least reactive.

Exercise 20-2 | Try It Yourself

Where would you place the compounds below in the reactivity sequence of Exercise 20-1? (**Hint:** Consider steric effects.)

$$(CH_3)_3COCOC(CH_3)_3 \qquad (CH_3)_3COCCl \qquad CH_3OCOCOCH_3$$

Bis(1,1-dimethylethyl) carbonate   **1,1-Dimethylethyl chloroformate**   **Dimethyl dicarbonate**

The greater the resonance, the shorter the C–L bond

The extent of resonance can be observed directly in the structures of carboxylic acid derivatives. In the progression from acyl halides to esters and amides, the C–L bond becomes progressively shorter, owing to increased double-bond character (Table 20-1). The NMR spectra of amides reveal that rotation about this bond has become restricted. For example, *N,N*-dimethylformamide at room temperature exhibits *two* singlets for the two methyl groups, because rotation about the C–N bond is very slow on the NMR time scale. The evidence points to considerable π overlap between the lone pair on nitrogen and the carbonyl carbon, as a result of the increased importance of the dipolar resonance form in amides. The measured barrier to this rotation is about 21 kcal mol^{-1} (88 kJ mol^{-1}).

Slow Rotation in *N,N*-Dimethylformamide

Note that these observations are not consistent with the structure for ammonia shown in Section 1-8, Figure 1-20. Ammonia (and simple amines, as we shall see in Section 21-2) adopt a geometry based on sp^3 hybridization, in order to minimize repulsion between electron pairs. In contrast, the lone pair on nitrogen in an *amide* resides in a *p* orbital, to permit π overlap with the *p* orbital on the carbonyl carbon atom. Consequently, the amide nitrogen atom possesses trigonal planar geometry and sp^2 hybridization: The stabilization due to π overlap in the planar form overrides the normal preference for a tetrahedral arrangement of electron pairs. We shall see in Section 26-4 that the planarity of the nitrogen in amides is the single most important determinator of structure, and therefore function, in peptides and proteins.

Exercise 20-3 |

The methyl group in the ^{1}H NMR spectrum of 1-acetyl 2-phenylhydrazide, shown in the margin, exhibits two singlets at $\delta = 2.02$ and 2.10 ppm at room temperature. Upon heating to 100°C in the NMR probe, the same compound gives rise to only one signal in that region. Explain.

$$\underset{\textbf{1-Acetyl 2-phenylhydrazide}}{CH_3CNHNHC_6H_5}$$

The lack of change in the C—Cl bond length going from the chloroalkane to the acyl chloride is indicative of poorer π overlap between the $2p$ orbital on C and the $3p$ orbital on Cl, compared with the $2p$–$2p$ π overlap for the other acyl derivatives.

Table 20-1	C–L Bond Lengths in $\overset{O}{\underset{\|\|}{RC}}$–L Compared with R–L Single-Bond Distances	
L	Bond length (Å) in R–L	Bond length (Å) in $\overset{O}{\underset{\|\|}{RC}}$–L
Cl	1.78	1.79 (not shorter)
F	1.39	1.35 (shorter by 0.04 Å)
OCH$_3$	1.43	1.36 (shorter by 0.07 Å)
NH$_2$	1.47	1.36 (shorter by 0.11 Å)

Table 20-2	Carbonyl Stretching Frequencies of Carboxylic Acid Derivatives RC–L (O double bond on C)	
L	$\tilde{\nu}_{C=O}\ (cm^{-1})$	
Cl	1790–1815	
$\overset{O}{\underset{\|}{OCR}}$	1740–1790	Two bands are observed, corresponding to asymmetric and symmetric stretching motions
	1800–1850	
OR	1735–1750	
NR'_2	1650–1690	

(Increasing $\tilde{\nu}_{C=O}$ indicated by upward arrow)

Infrared spectroscopy can also be used to probe resonance in carboxylic acid derivatives. The dipolar resonance structure weakens the C=O bond and causes a corresponding decrease in the carbonyl stretching frequency (Table 20-2). The IR data for carboxylic acids reported in Section 19-3 refer to the common dimeric form, in which hydrogen bonding reduces the stretching frequencies of both the O–H and C=O bonds to about 3000 and 1700 cm^{-1}, respectively. A special technique—vapor deposition at very low temperature—allows the IR spectra of carboxylic acid monomers to be measured, for direct comparison with the spectra of carboxylic acid derivatives. Monomeric acetic acid displays $\tilde{\nu}_{C=O}$ at 1780 cm^{-1}, similar to the value for carboxylic anhydrides, higher than that for esters, and lower than that of halides, consistent with the degree of resonance delocalization in carboxylic acids.

In addition to infrared bands due to their carbonyl groups, amides display bands resulting from the stretching of their N–H bonds. In the region of 3100 to 3400 cm^{-1}, amides with two NH bonds show two bands, and those with one NH bond display one (see Section 20-3).

The ^{13}C NMR signals of the carbonyl carbons in carboxylic acid derivatives are less sensitive to polarity differences and fall in a narrow range near 170 ppm.

^{13}C NMR Chemical Shifts of the Carbonyl Carbon in Carboxylic Acid Derivatives

$$CH_3\overset{O}{\overset{\|}{C}}Cl \qquad CH_3\overset{O}{\overset{\|}{C}}O\overset{O}{\overset{\|}{C}}CH_3 \qquad CH_3\overset{O}{\overset{\|}{C}}OH \qquad CH_3\overset{O}{\overset{\|}{C}}OCH_3 \qquad CH_3\overset{O}{\overset{\|}{C}}NH_2$$

$$\delta = 170.3 \qquad\qquad 166.9 \qquad\qquad 177.2 \qquad\qquad 170.7 \qquad\qquad 172.6\ ppm$$

In common with other carbonyl compounds, the mass spectra of carboxylic acid derivatives typically show peaks resulting from both α-cleavage and McLafferty rearrangement.

Exercise 20-4

The mass spectrum of methyl pentanoate (molecular weight = 116) shows fragmentation peaks at $m/z = 85, 74$, and 57. Assign these peaks.

Carboxylic acid derivatives are basic and acidic

The extent of resonance in carboxylic acid derivatives is also seen in their basicity (protonation at the carbonyl oxygen) and acidity (enolate formation). In all cases, protonation requires strong acid, but it gets easier as the electron-donating ability of the L group increases. Protonation is important in acid-catalyzed nucleophilic addition–elimination reactions.

Protonation of a Carboxylic Acid Derivative

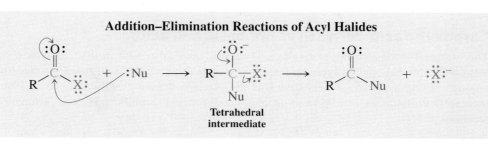

A relatively strong contribution of this resonance form stabilizes the protonated species

Exercise 20-5

The oxygen in acetyl chloride is much less basic than the one in acetamide. Explain, using resonance structures.

For related reasons, the acidity of the hydrogens next to the carbonyl group increases along the series. The acidity of a ketone lies between those of an acyl chloride and an ester.

Acidities of α-Hydrogens in Carboxylic Acid Derivatives in Comparison with Acetone

$$CH_3CN(CH_3)_2 \ < \ CH_3COCH_3 \ < \ CH_3CCH_3 \ < \ CH_3CCl$$

pK_a   ~30   ~25   ~20   ~16

Increasing acidity

Exercise 20-6

Give an example of a reaction from Chapter 19 that takes advantage of the relatively high acidity of the α-hydrogens in acyl halides.

$$CH_3CCl$$

Acetyl chloride

CH₃CHCH₂CBr

3-Methylbutanoyl bromide

Pentanoyl fluoride

In Summary The relative electronegativity and the size of L in RCL controls the extent of resonance of the lone electron pair(s) and the relative reactivity of a carboxylic acid derivative in nucleophilic addition–elimination reactions. This effect manifests itself structurally and spectroscopically, as well as in the relative acidity and basicity of the α-hydrogen and the carbonyl oxygen, respectively.

20-2 CHEMISTRY OF ACYL HALIDES

The **acyl halides**, RCX, are named after the alkan*oic acid* from which they are derived. The halides of cycloalkane*carboxylic acids* are called cycloalkane **carbonyl halides.**

Acyl halides undergo addition–elimination reactions in which nucleophiles displace the halide leaving group. These compounds are so reactive that catalysts are usually not necessary for their conversion.

Addition–Elimination Reactions of Acyl Halides

Tetrahedral intermediate

Cyclohexanecarbonyl fluoride

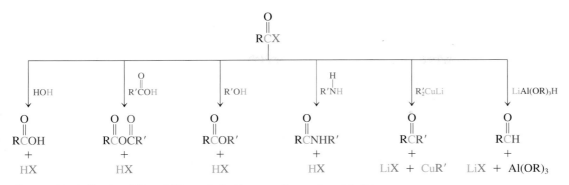

Figure 20-1 Nucleophilic addition–elimination reactions of acyl halides.

Figure 20-1 shows a variety of nucleophilic reagents and the corresponding products. It is because of this wide range of reactivity that acyl halides are useful synthetic intermediates.

Let us consider these transformations one by one (except for anhydride formation, which was covered in Section 19-8). Examples will be restricted to acyl chlorides, which are the most readily accessible, but their transformations can be generalized to a considerable extent to the other acyl halides.

Water hydrolyzes acyl chlorides to carboxylic acids

Acyl chlorides react with water, often violently, to give the corresponding carboxylic acids and hydrogen chloride. This transformation is a simple example of addition–elimination and is representative of most of the reactions of this class of compounds.

Reaction

Acyl Chloride Hydrolysis

$$\underset{\text{Propanoyl chloride}}{CH_3CH_2\overset{O}{\overset{\|}{C}}Cl} + HOH \longrightarrow \underset{\substack{100\% \\ \text{Propanoic acid}}}{CH_3CH_2\overset{O}{\overset{\|}{C}}OH} + HCl$$

Mechanism

Mechanism of Acyl Chloride Hydrolysis

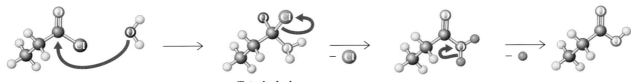

Tetrahedral
intermediate

Alcohols convert acyl chlorides into esters

The analogous reaction of acyl chlorides with alcohols is a highly effective way of producing esters. A base such as an alkali metal hydroxide, pyridine, or a tertiary amine is usually added to neutralize the HCl by-product. Because acyl chlorides are readily made from the corresponding carboxylic acids (Section 19-8), the sequence RCOOH → RCOCl → RCOOR′ is a good method for esterification. By maintaining neutral or basic conditions, this preparation avoids the equilibrium problem of acid-catalyzed ester formation (Fischer esterification, Section 19-9).

Reminder

In the above scheme, free H⁺ does not exist in solution, but is attached to any available electron pair, mostly the oxygen of the solvent water.

**Ester Synthesis from Carboxylic Acids
Through Acyl Chlorides**

$$\underset{\text{RCOH}}{\overset{O}{\|}} \xrightarrow[-\text{HCl}, -\text{SO}_2]{\text{SOCl}_2} \underset{\text{RCCl}}{\overset{O}{\|}} \xrightarrow[-\text{HCl}]{\text{R'OH, base}} \underset{\text{RCOR'}}{\overset{O}{\|}}$$

$$\underset{\substack{\textbf{Acetyl} \\ \textbf{chloride}}}{\overset{O}{\underset{\|}{\text{CH}_3\text{CCl}}}} + \underset{\textbf{1-Propanol}}{\text{HOCH}_2\text{CH}_2\text{CH}_3} \xrightarrow[\text{(See Problem 35)}]{\text{N(CH}_2\text{CH}_3)_3} \underset{\substack{75\% \\ \textbf{Propyl acetate}}}{\overset{O}{\underset{\|}{\text{CH}_3\text{COCH}_2\text{CH}_2\text{CH}_3}}} + \underset{\substack{\textbf{Triethylammonium} \\ \textbf{chloride}}}{\overset{+}{\text{HN}}(\text{CH}_2\text{CH}_3)_3 \; \text{Cl}^-}$$

**1,1-Dimethylethyl
acetate
(*tert*-Butyl acetate)**

Exercise 20-7

You have learned that 2-methyl-2-propanol (*tert*-butyl alcohol) dehydrates in the presence of acid (Section 9-2). Suggest a synthesis of 1,1-dimethylethyl acetate (*tert*-butyl acetate, shown in the margin) from acetic acid. Avoid conditions that might dehydrate the alcohol.

Amines convert acyl chlorides into amides

Secondary and primary amines, as well as ammonia, convert acyl chlorides into amides. As shown in the first example below, aqueous ammonia works quite well for the synthesis of simple amides: NH_3, being a much stronger nucleophile than water, reacts preferentially with the carbonyl derivative. As in ester formation, the HCl formed is neutralized by added base (which can be excess amine).

Reaction

Amides from Acyl Halides

Benzoyl chloride $+ \underset{\textbf{Excess}}{\text{NH}_3} \xrightarrow{\text{H}_2\text{O}}$ **Benzamide** 86% $+ \overset{+}{\text{NH}_4} \text{Cl}^-$

$$\underset{\textbf{Propenoyl chloride}}{\overset{O}{\underset{\|}{\text{CH}_2{=}\text{CHCCl}}}} + 2\,\text{CH}_3\text{NH}_2 \xrightarrow{\substack{\text{Benzene,} \\ 5°\text{C}}} \underset{\substack{68\% \\ \textbf{\textit{N}-Methylpropenamide}}}{\overset{O}{\underset{\|}{\text{CH}_2{=}\text{CHCNHCH}_3}}} + \text{CH}_3\overset{+}{\text{NH}}_3 \, \text{Cl}^-$$

The mechanism of this transformation is, again, addition–elimination, beginning with attack of the nucleophilic amine nitrogen at the carbonyl carbon.

Mechanism

Mechanism of Amide Formation from Acyl Chlorides

**Tetrahedral
intermediate**

Note that in the last step, a proton must be lost from nitrogen to give the amide. Consequently, tertiary amines (which have no hydrogens on nitrogen) cannot form amides. Instead, they convert acyl halides to acyl ammonium salts. In these species, the nitrogen has no lone electron pair to stabilize the carbonyl function by resonance. On the contrary, the positive charge activates the carbonyl carbon to nucleophilic attack. Consequently, acyl ammonium salts show reactivity that is similar to those of acyl halides. A useful consequence of this behavior is that acylations of other functions can be carried out in the presence of tertiary amino substituents.

Acetylation of a *tert*-Amino Alcohol

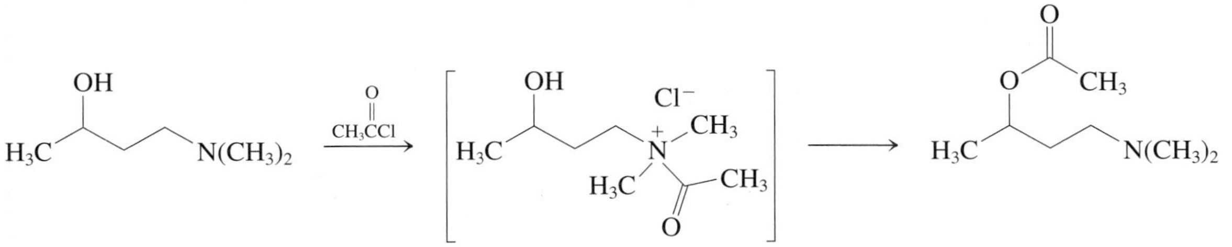

OH
H3C
N(CH3)2
O
CH3CCl
Cl−
CH3
N
+
H3C
CH3
O
O
CH3

Reactive intermediate
acyl ammonium salt

Exercise 20-8

Some amide preparations from acyl halides require a primary or secondary amine that is too expensive to use also as the base to neutralize the hydrogen halide. Suggest a solution to this problem.

Organometallic reagents convert acyl chlorides into ketones

Organometallic reagents (RLi and RMgX) attack the carbonyl group of acyl chlorides to give the corresponding ketones. However, these ketone products are themselves prone to further attack by the relatively unselective organolithium (RLi) and Grignard (RMgX) reagents to give alcohols (see Section 8-8). Ketone formation is best achieved by using diorganocuprates (Section 18-10), which are more selective than RLi or RMgX and do not add to the product ketone.

Formation of a Ketone from an Acyl Halide

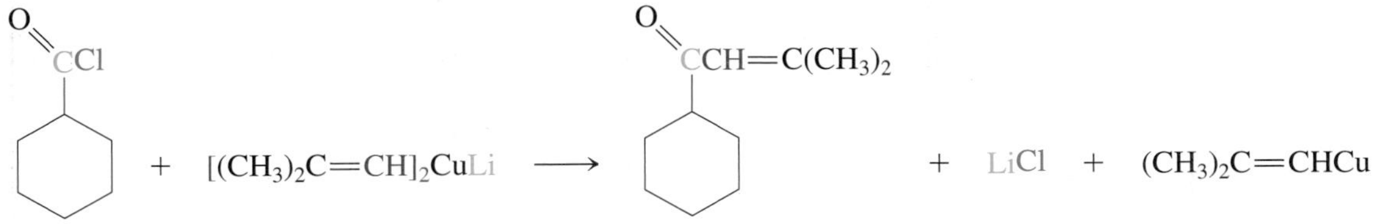

O
CCl
+ [(CH3)2C=CH]2CuLi
O
CCH=C(CH3)2
+ LiCl + (CH3)2C=CHCu

70%

Selective reduction of acyl chlorides results in aldehydes

We can convert an acyl chloride into an aldehyde by hydride reduction. In this transformation, we again face a selectivity problem: Sodium borohydride and lithium aluminum hydride convert aldehydes into alcohols. To prevent such overreduction, we must modify $LiAlH_4$ by letting it react first with three molecules of 2-methyl-2-propanol (*tert*-butyl alcohol; see Section 8-6). This treatment neutralizes three of the reactive hydride atoms, leaving one behind that is nucleophilic enough to attack an acyl chloride but not the resulting aldehyde.

Reductions by Modified Lithium Aluminum Hydride

Preparation of reagent

$$LiAlH_4 + 3\,(CH_3)_3COH \longrightarrow LiAl[OC(CH_3)_3]_3H + 3\,H{-}H$$

Lithium tri(*tert*-butoxy)aluminum hydride

Reduction

$$\underset{\text{RCCl}}{\overset{\text{O}}{\|}} \quad + \quad \text{LiAl[OC(CH}_3)_3]_3\text{H} \quad \xrightarrow{\substack{\text{1. Ether solvent} \\ \text{2. H}^+, \text{H}_2\text{O}}} \quad \underset{\text{RCH}}{\overset{\text{O}}{\|}} \quad + \quad \text{LiCl} \quad + \quad \text{Al[OC(CH}_3)_3]_3$$

Reduction of an Acyl Halide to an Aldehyde

$$\underset{\substack{\text{2-Butenoyl chloride}}}{\text{CH}_3\text{CH}=\text{CH}\overset{\overset{\text{O}}{\|}}{\text{C}}\text{Cl}} \quad \xrightarrow{\substack{\text{1. LiAl[OC(CH}_3)_3]_3\text{H, (CH}_3\text{OCH}_2\text{CH}_2)_2\text{O, } -78°\text{C} \\ \text{2. H}^+, \text{H}_2\text{O}}} \quad \underset{\substack{48\% \\ \text{2-Butenal}}}{\text{CH}_3\text{CH}=\text{CH}\overset{\overset{\text{O}}{\|}}{\text{C}}\text{H}}$$

Exercise 20-9

Prepare the following compounds from butanoyl chloride.

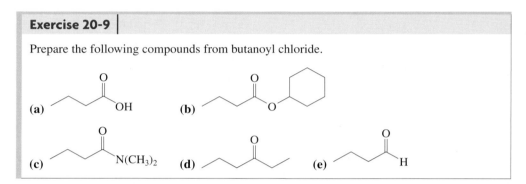

(a)   **(b)**

(c)   **(d)**   **(e)**

In Summary Acyl chlorides are attacked by a variety of nucleophiles, the reactions leading to new carboxylic acid derivatives, ketones, and aldehydes by addition–elimination mechanisms. The reactivity of acyl halides makes them useful synthetic relay points on the way to other carbonyl derivatives.

20-3 | CHEMISTRY OF CARBOXYLIC ANHYDRIDES

Carboxylic anhydrides, $\underset{\text{RCOCR}}{\overset{\text{O O}}{\| \ \|}}$, are named by simply adding the term *anhydride* to the acid name (or names, with regard to mixed anhydrides). This method also applies to cyclic derivatives.

1,2-Benzenedicarboxylic anhydride (Phthalic anhydride)

$$\underset{\substack{\text{Acetic propanoic} \\ \text{anhydride} \\ \text{(A mixed anhydride)}}}{\text{CH}_3\overset{\overset{\text{O}}{\|}}{\text{C}}\text{O}\overset{\overset{\text{O}}{\|}}{\text{C}}\text{CH}_2\text{CH}_3}$$

$$\underset{\text{Acetic anhydride}}{\text{CH}_3\overset{\overset{\text{O}}{\|}}{\text{C}}\text{O}\overset{\overset{\text{O}}{\|}}{\text{C}}\text{CH}_3}$$

Acetic anhydride

2-Butenedioic anhydride (Maleic anhydride)

Pentanedioic anhydride (Glutaric anhydride)

The reactions of carboxylic anhydrides with nucleophiles, although less vigorous, are completely analogous to those of the acyl halides. The leaving group is a carboxylate instead of a halide ion.

Nucleophilic Addition–Elimination of Anhydrides

Typical Reactions of Anhydrides

Acetic anhydride $\xrightarrow[\text{Hydrolysis}]{\text{HOH}}$ Acetic acid

100%

Propanoic anhydride $\xrightarrow[\substack{\text{Alcoholysis}\\\text{(ester formation)}}]{\text{CH}_3\text{OH}}$ Methyl propanoate 83% + Propanoic acid

In every addition–elimination reaction except hydrolysis, the carboxylic acid side product is usually undesired and is removed by work-up with aqueous base. Cyclic anhydrides undergo similar nucleophilic addition–elimination reactions that lead to ring opening.

Nucleophilic Ring Opening of Cyclic Anhydrides

Butanedioic (succinic) anhydride $\xrightarrow[\text{(See Problem 38)}]{\text{CH}_3\text{OH, 100°C}}$ $\text{HOCCH}_2\text{CH}_2\text{COCH}_3$ 96%

Exercise 20-10

Treatment of butanedioic (succinic) anhydride with ammonia at elevated temperatures leads to a compound $C_4H_5NO_2$. What is its structure? (**Hint:** Consult Section 19-10.)

Exercise 20-11

Formulate the mechanism for the reaction of acetic anhydride with methanol in the presence of sulfuric acid.

Although the chemistry of carboxylic anhydrides is very similar to that of acyl halides, anhydrides have some practical advantages. Acyl halides are so reactive that they are difficult to store for extended periods without some hydrolysis occurring due to exposure to atmospheric moisture. As a result, chemists usually prepare acyl halides immediately before they are to be used. Anhydrides, being slightly less reactive toward nucleophiles, are more stable, and several (including all the examples illustrated in this section) are commercially available. Consequently, carboxylic anhydrides are often the preferred reagents for the preparation of many carboxylic acid derivatives.

Anhydride polymers containing embedded drugs have found use as slow-release agents in clinical applications. For example, wafers of the copolymer of decanedioic acid (green) and a dibenzoic acid (blue) containing the anticancer drug carmustine are marketed under the trade name Gliadel. They are implanted in the skull during surgery for brain tumors to facilitate subsequent chemotherapy. The insoluble polymer slowly hydrolyzes to the biodegradable dioic acids, in this way discharging the antitumor agent over periods of time ranging from days to years. Carmustine is a "mustard" (see Exercise 9-25), which cross-links DNA strands by alkylation, thus preventing cell proliferation.

4,4′-(1,3-Propane-dioxy)dibenzoic acid

Decanedioic acid

Mixed polyanhydride

Carmustine

Electrophilic   Electrophilic

In Summary Anhydrides react with nucleophiles in the same way as acyl halides do, except that the leaving group is a carboxylate ion. Cyclic anhydrides furnish dicarboxylic acid derivatives.

20-4 | CHEMISTRY OF ESTERS

As mentioned in Section 19-9, esters, RCOR′, constitute perhaps the most important class of carboxylic derivatives. They are particularly widespread in nature, contributing especially to the pleasant flavors and aromas of many flowers and fruits. Esters undergo typical carbonyl chemistry but with reduced reactivity relative to that of either acyl halides or carboxylic anhydrides. This section begins with a discussion of ester nomenclature. Descriptions follow of the transformations that esters undergo with a variety of nucleophilic reagents.

Esters are alkyl alkanoates

Esters are named as alkyl alkanoates. The ester grouping, –COR, as a substituent is called **alkoxycarbonyl**. A cyclic ester is called a **lactone** (the common name, Section 19-9); the systematic name is **oxa-2-cycloalkanone** (Section 25-1). Its common name is preceded by α-, β-, γ-, δ-, and so forth, depending on ring size.

$$CH_3COCH_3$$

Methyl acetate

$$CH_3CH_2COCH_2CH_3$$

Ethyl propanoate
(Ethyl propionate)

$$CH_3COCH_2CH_2CHCH_3$$

3-Methylbutyl acetate
(Isopentyl acetate)
(A component of banana flavor)

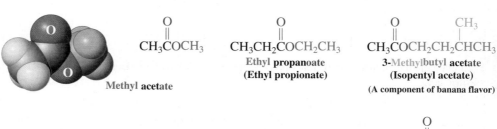

2-Methylpropyl propanoate
(Isobutyl propionate)
(A component of rum flavor)

3-Methylbutyl pentanoate
(Isopentyl valerate)
(A component of apple flavor)

Methyl 2-aminobenzoate
(Methyl anthranilate)
(A component of grape flavor)

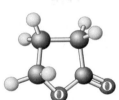

γ-**Valerolactone**
(Systematic: 5-methyloxa-2-cyclopentanone)

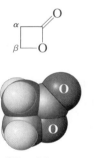

β-**Propiolactone**
(This compound is a carcinogen
and is systematically
called oxa-2-cyclobutanone;
see Section 25-1)

γ-**Butyrolactone**
(Systematic: oxa-2-cyclopentanone)

Lactones: Cyclic esters

(Z)-7-**Dodecenyl acetate**

In addition to their prevalence in plants, esters play biological roles in the animal kingdom. Section 12-17 included several examples of esters that function as insect pheromones. Perhaps the most bizarre of these esters is (Z)-7-dodecenyl acetate, a component of the pheromone mixture in several species of moths. This same compound was recently found to also be the mating pheromone of the elephant (who said nature has no sense of humor?). Section 20-5 will describe a number of more conventional biological functions of esters.

Lactones (cyclic esters) are also widely distributed in nature. The ready availability of γ-valerolactone from renewable plant-derived sources has led to its serious consideration as a potential "green" biofuel. Problem 46, part (d) describes this development in more detail.

This couple is blissfully ignorant of the fact that they are attracted to each other by the same ester pheromone as moths.

Exercise 20-12

Name the following esters.

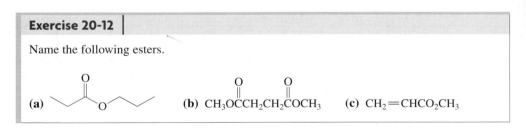

(a)   (b) $CH_3OCCH_2CH_2COCH_3$ (c) $CH_2{=}CHCO_2CH_3$

In industry, lower esters, such as ethyl acetate (b.p. 77°C) and butyl acetate (b.p. 127°C), are used as solvents. For example, butyl butanoate has replaced the ozone-depleting trichloroethane as a cleaning solvent in the manufacture of electronic components such as computer chips. Higher nonvolatile esters are used as softeners (called plasticizers; see Section 12-15) for brittle polymers—in flexible tubing (e.g., Tygon tubing), rubber pipes, and upholstery.

Esters hydrolyze to carboxylic acids; the leaving-group problem

Esters undergo nucleophilic substitution reactions by means of addition–elimination pathways, albeit with reduced reactivity relative to halides and anhydrides. Thus, catalysis by acid or base becomes a frequent necessity. For example, esters are cleaved to carboxylic acids and alcohols in the presence of excess water and strong acid, and the reaction requires heating to proceed at a reasonable rate. The mechanism of this transformation is the reverse of acid-catalyzed esterification (Section 19-9). As in esterification, the acid serves two purposes: It protonates the carbonyl oxygen to make the ester more reactive toward nucleophilic attack, and it protonates the alkoxy oxygen in the tetrahedral intermediate to make it a better leaving group.

In acid-catalyzed esterification, excess alcohol drives the equilibrium toward the ester product. In contrast, the use of excess water pushes the reaction the opposite way, enabling hydrolysis.

Solved Exercise 20-13 | Working with the Concepts: Ester Hydrolysis in Acid

Formulate a mechanism for the acid-catalyzed hydrolysis of γ-butyrolactone.

Strategy
Find the structure of γ-butyrolactone (on the previous page). It is a cyclic ester. Find the general mechanism for addition–elimination under acidic conditions (Section 19-7). Substitute γ-butyrolactone for the general starting material in that mechanism, and proceed.

Solution
• Step by step: Begin with protonation of the carbonyl oxygen,

• Continue with addition–elimination. Water is the nucleophile. Use protonation to turn the *ring oxygen* into a good leaving group. In the tetrahedral intermediate, cleave the bond between the (former) carbonyl carbon and the ring oxygen.

• Deprotonate the carbonyl oxygen. The mechanism is completed. You have your product.

Strong bases also promote ester hydrolysis through an addition–elimination mechanism (Sections 19-7 and 20-1). The base (B) converts the poor nucleophile water into the negatively charged and more highly nucleophilic hydroxide ion.

$$B:^- + H{-}OH \longrightarrow {}^-\!:OH + B{-}H$$

Ester hydrolysis is frequently achieved by using hydroxide itself, in at least stoichiometric amounts, as the base.

Example of Ester Hydrolysis Using Aqueous Base

Reaction

1. KOH, H₂O, CH₃OH, Δ
2. H⁺, H₂O

100%

Methyl 3-methylbutanoate   3-Methylbutanoic acid   +   CH₃OH

Mechanism of Base-Mediated Ester Hydrolysis

Mechanism

Step 1. Addition–elimination

Methoxide: *poor* leaving group

Tetrahedral intermediate

Step 2. Deprotonation

Ester hydrolysis in base (saponification, see Section 19-13) differs from the acid-catalyzed version in several ways. Consider the addition–elimination (step 1) first: The tetrahedral intermediate can revert to starting material by release of a hydroxide ion, or it can go forward toward product through expulsion of methoxide. Either way, the process requires loss of a *poor leaving group*. Both hydroxide and methoxide are strong bases, and *strong bases are poor leaving groups* (Section 6-7). How can these transformations take place at all?

The answer lies in part in the strength of the C=O bond and the associated stability of carbonyl compounds. The tetrahedral intermediate is so much higher in energy that its conversion to a carbonyl compound is typically exothermic even when a poor leaving group is expelled. In addition, the base-mediated process benefits from a very favorable acid-base reaction in step 2: The strong base methoxide, having been released at the addition–elimination stage, rapidly deprotonates the acid to give carboxylate ion. This step is very favorable and drives the entire hydrolysis sequence essentially to completion. Subsequent acidic aqueous work-up gives the carboxylic acid product.

Exercise 20-14 | Try It Yourself

Formulate a mechanism for the base-mediated hydrolysis of γ-butyrolactone.

The latent potential of esters to cleave into their component carboxylic acid and alcohol fragments is exploited by medicinal chemists in the form of ester prodrugs. Prodrugs are inactive drug derivatives that are converted to their active form under physiological conditions. They are employed to increase circulation lifetime, alter solubility, reduce toxicity, improve delivery to the target site, and make the medication more palatable. For example, chloramphenicol (margin) is a widely used antibiotic for the treatment of bacterial infections. However, its administration as such to patients has problems; in particular, intravenous therapy is difficult because of its poor solubility in water. This drawback is circumvented by injecting the butanedioate half ester (R = $COCH_2CH_2COOH$), in which the dangling carboxy group imparts water solubility. Once this compound is in the bloodstream, the active molecule is released by ester hydrolysis. Further, although this trick solves the problem, patients are (understandably) averse to "shots" and generally prefer to take the drug orally. However, chloramphenicol (in aqueous suspension) tastes intensely bitter, rendering ingestion unpleasant, especially for children. This side effect is circumvented with the hexadecanoic ester [R = $CO(CH_2)_{14}CH_3$] prodrug, in which the long, nonpolar fatty acid chain prevents dissolution in the mouth and also interferes with docking to the bitter-taste receptor sites, making the medicine tasteless. Once ingested, an esterase enzyme in the duodenum liberates the actual drug.

Chloramphenicol (R = H)

Transesterification takes place with alcohols

Esters react with alcohols in an acid- or base-catalyzed transformation called **transesterification.** This allows for the direct conversion of one ester into another without proceeding through the free acid. Like esterification, transesterification is a reversible reaction. To shift the equilibrium, a large excess of one of the alcohols is usually employed, sometimes in the form of solvent.

Transesterification

$RCOR'$ + $R''OH$

H$^+$ or $^-OR''$

$RCOR''$ + $R'OH$

Alkoxy groups trade places

Conversion of an Ethyl Ester into a Methyl Ester

$$\underset{\substack{\text{Ethyl}\\\text{octadecanoate}}}{C_{17}H_{35}\overset{\displaystyle O}{\overset{\|}{C}}OCH_2CH_3} + \underset{\substack{\text{Methanol}\\\text{(Solvent)}}}{CH_3OH} \xrightarrow{\;H^+ \text{ or } {}^-OCH_3\;} \underset{\substack{90\%\\\text{Methyl}\\\text{octadecanoate}}}{C_{17}H_{35}\overset{\displaystyle O}{\overset{\|}{C}}OCH_3} + CH_3CH_2OH$$

Lactones are opened to hydroxy esters by transesterification.

Conversion of a Lactone into an Open-Chain Ester

γ-Butyrolactone   3-Bromopropanol   3-Bromopropyl 4-hydroxybutanoate
80%

The mechanisms of transesterification by acid and base are analogous to the mechanisms of the corresponding hydrolyses to the carboxylic acids. Thus, acid-catalyzed transesterification begins with protonation of the carbonyl oxygen, followed by nucleophilic attack of the alcohol on the carbonyl carbon. In contrast, under basic conditions the alcohol is first deprotonated, and the resulting alkoxide ion then adds to the ester carbonyl group.

Exercise 20-15

Formulate mechanisms for the acid- and base-catalyzed transesterifications of γ-butyrolactone by 3-bromo-1-propanol. **Caution:** Under basic conditions, the reaction will actually not proceed as desired. Why? **Hint:** Consult Section 9-6.

Amines convert esters into amides

Amines, which are more nucleophilic than alcohols, readily transform esters into amides. No catalyst is needed, but heating is required.

Amide Formation from Methyl Esters

$$R\overset{\displaystyle O}{\overset{\|}{C}}OCH_3 + R'\overset{\displaystyle H}{\overset{|}{N}}H \xrightarrow{\;\Delta\;} R\overset{\displaystyle O}{\overset{\|}{C}}NHR' + CH_3OH$$

Uncatalyzed Conversion of an Ester into an Amide

$$\underset{\text{Methyl 9-octadecenoate}}{CH_3(CH_2)_7CH=CH(CH_2)_7\overset{\displaystyle O}{\overset{\|}{C}}OCH_3} + \underset{\text{1-Dodecanamine}}{CH_3(CH_2)_{11}\overset{\displaystyle H}{\overset{|}{N}}H} \xrightarrow[\text{(See Problem 43)}]{230°C}$$

$$\underset{\substack{69\%\\ \textit{N}\text{-Dodecyl-9-octadecenamide}}}{CH_3(CH_2)_7CH=CH(CH_2)_7\overset{\displaystyle O}{\overset{\|}{C}}NH(CH_2)_{11}CH_3} + CH_3OH$$

Grignard reagents transform esters into alcohols

Esters can be converted into alcohols by using *two* equivalents of a Grignard reagent. In this way, ordinary esters are transformed into tertiary alcohols, whereas formate esters furnish secondary alcohols.

Reaction

Alcohols from Esters and Grignard Reagents

$$CH_3CH_2\overset{\overset{\displaystyle O}{\|}}{C}OCH_2CH_3 \quad + \quad 2\ CH_3CH_2CH_2MgBr \xrightarrow[-CH_3CH_2OH]{\substack{1.\ (CH_3CH_2)_2O \\ 2.\ H^+,\ H_2O}} CH_3CH_2\overset{\overset{\displaystyle OH}{|}}{\underset{\underset{\displaystyle CH_2CH_2CH_3}{|}}{C}}CH_2CH_2CH_3$$

Two identical alkyl groups, derived from Grignard reagent

69%

Ethyl propanoate   **Propylmagnesium bromide**   **4-Ethyl-4-heptanol**

$$H\overset{\overset{\displaystyle O}{\|}}{C}OCH_3 \quad + \quad 2\ CH_3CH_2CH_2CH_2MgBr \xrightarrow[-CH_3OH]{\substack{1.\ (CH_3CH_2)_2O \\ 2.\ H^+,\ H_2O}} H\overset{\overset{\displaystyle OH}{|}}{\underset{\underset{\displaystyle CH_2CH_2CH_2CH_3}{|}}{C}}CH_2CH_2CH_2CH_3$$

85%

Methyl formate   **Butylmagnesium bromide**   **5-Nonanol**

The reaction begins with addition of the organometallic to the carbonyl function in the usual manner to give the magnesium salt of a hemiacetal (Section 17-7). At room temperature, rapid elimination results in the formation of an intermediate ketone (or aldehyde, from formates). The resulting carbonyl group then immediately adds a second equivalent of Grignard reagent (recall Section 8-8). Subsequent acidic aqueous work-up leads to the observed alcohol.

Mechanism of the Alcohol Synthesis from Esters and Grignard Reagents

Mechanism

$$\overset{\overset{\displaystyle :\!O\!:}{\|}}{\underset{\underset{\displaystyle OCH_3}{}}{R}C} \quad + \quad R'-MgBr \longrightarrow R-\overset{\overset{\displaystyle O-MgBr}{|}}{\underset{\underset{\displaystyle R'}{|}}{C}}OCH_3 \xrightarrow[-CH_3OMgBr]{Fast}$$

Ester   **Tetrahedral intermediate**

$$\overset{\overset{\displaystyle :\!O\!:}{\|}}{\underset{\underset{\displaystyle R'}{}}{R}C} \xrightarrow[Fast]{R'-MgBr} R-\overset{\overset{\displaystyle OMgBr}{|}}{\underset{\underset{\displaystyle R'}{|}}{C}}R' \xrightarrow{H^+,\ H_2O} R-\overset{\overset{\displaystyle OH}{|}}{\underset{\underset{\displaystyle R'}{|}}{C}}R'$$

Ketone   **Alkoxide**   **Alcohol**
(Reaction cannot be   **(Final product)**
stopped at this stage)

Exercise 20-16

Propose a synthesis of triphenylmethanol, $(C_6H_5)_3COH$, beginning with methyl benzoate (shown in the margin) and bromobenzene.

$$C_6H_5\overset{\overset{\displaystyle O}{\|}}{C}OCH_3$$
Methyl benzoate

Esters are reduced by hydride reagents to give alcohols or aldehydes

The reduction of esters to alcohols by $LiAlH_4$ requires 0.5 equivalent of the hydride, because only two hydrogens are needed per ester function. The course of the reaction is similar to that of the double Grignard addition: The first hydride effects addition–elimination to generate the aldehyde, which rapidly picks up a second hydride to furnish the alcohol (after aquous work-up).

Reduction of an Ester to an Alcohol

A milder reducing agent allows the reaction to be stopped at the aldehyde oxidation stage. Such a reagent is bis(2-methylpropyl)aluminum hydride (diisobutylaluminum hydride) when used at low temperatures in toluene.

Reduction of an Ester to an Aldehyde

Esters form enolates that can be alkylated

> The steric bulk of the diisopropylamide anion of LDA ensures that nucleophilic addition to the carbonyl carbon does not compete with deprotonation at the α-carbon.

The acidity of the α-hydrogens in esters is sufficiently high that **ester enolates** are formed by treatment of esters with strong base at low temperatures. Ester enolates react like ketone enolates, undergoing alkylations.

Alkylation of an Ester Enolate

The pK_a of esters is about 25. Consequently, ester enolates exhibit the typical side reactions of strong bases: E2 processes (especially with secondary, tertiary, and β-branched halides) and deprotonations. The most characteristic reaction of ester enolates is the Claisen condensation, in which the enolate attacks the carbonyl carbon of another ester. This process will be considered in Chapter 23.

Exercise 20-17

Give the products of the reaction of ethyl cyclohexanecarboxylate with the following compounds or under the following conditions (and followed by acidic aqueous work-up, if necessary). **(a)** H^+, H_2O; **(b)** HO^-, H_2O; **(c)** CH_3O^-, CH_3OH; **(d)** NH_3, Δ; **(e)** 2 CH_3MgBr; **(f)** $LiAlH_4$; **(g)** 1. LDA, 2. CH_3I.

In Summary Esters are named as alkyl alkanoates. Many of them have pleasant odors and are present in nature. They are less reactive than acyl halides or carboxylic anhydrides and therefore often require the presence of acid or base to transform. With water, esters hydrolyze to the corresponding carboxylic acids or carboxylates; with alcohols, they undergo transesterification and, with amines at elevated temperatures, furnish amides.

Grignard reagents add twice to give tertiary alcohols (or secondary alcohols, from formates). Lithium aluminum hydride reduces esters all the way to alcohols, whereas bis(2-methylpropyl)aluminum (diisobutylaluminum) hydride allows the process to be stopped at the aldehyde stage. With LDA, it is possible to form ester enolates, which can be alkylated by electrophiles.

20-5 | ESTERS IN NATURE: WAXES, FATS, OILS, AND LIPIDS

Esters are essential components in the cells of all living organisms. This section introduces several of the most common types of natural esters with brief descriptions of their biological functions.

Waxes are simple esters, whereas fats and oils are more complex

Esters made up of long-chain carboxylic acids and long-chain alcohols constitute **waxes,** which form hydrophobic (Section 8-2) and insulating coatings on the skin and fur of animals, the feathers of birds, and the fruits and leaves of many plants. Spermaceti and beeswax are liquids or very soft solids at room temperature and are used as lubricants. Sheep's wool provides wool wax, which when purified yields lanolin, a widely used cosmetic base. The leaves of a Brazilian palm are the source of carnauba wax, a tough and water-resistant mixture of several solid esters. Carnauba wax is highly valued for its ability to take and maintain a high gloss and is used as a floor and automobile wax.

$$CH_3(CH_2)_{14}\overset{\overset{\displaystyle O}{\|}}{C}O(CH_2)_{15}CH_3 \qquad CH_3(CH_2)_n\overset{\overset{\displaystyle O}{\|}}{C}O(CH_2)_mCH_3$$

$$n = 24, 26; m = 29, 31$$

Hexadecyl hexadecanoate **Beeswax**
(Cetyl palmitate)
(Wax from the sperm whale)

Triesters of 1,2,3-propanetriol (glycerol) with long-chain carboxylic acids constitute **fats** and **oils** (Sections 11-3 and 19-13). They are also called **triglycerides.** The acids in fats and oils **(fatty acids)** are typically unbranched and contain an even number of carbon atoms; if the fats are unsaturated, the double bonds are usually cis. Fats are biological energy reserves, which are stored in the body's tissues until their metabolism leads ultimately to CO_2 and water. Oils have a similar function in plant seeds. As food components, fats and oils serve as solvents for food flavors and colors and contribute to a sense of "fullness" after eating, because they leave the stomach relatively slowly. Saturated fats containing hexadecanoic (palmitic), tetradecanoic (myristic), and dodecanoic (lauric) acids have been implicated as dietary factors in atherosclerosis (hardening of the arteries; Real Life 4-2). Fortunately for lovers of chocolate, cocoa butter is mainly a low-melting triglyceride that contains two molecules of octadecanoic (stearic) acid—which, despite being saturated, does not contribute to elevated low-density lipoprotein (LDL) levels or atherosclerosis—and one molecule of (Z)-9-octadecenoic (oleic) acid. The latter is also the chief fatty acid component of olive oil, the major source of dietary fat among the Greeks, who have very low rates of heart disease.

Lipids are biomolecules soluble in nonpolar solvents

Extraction of biological material with nonpolar solvents gives a wide assortment of compounds that includes terpenes and steroids (Section 4-7), fats and oils, and a variety of other low-polarity substances collectively called **lipids** (*lipos,* Greek, fat). Lipid fractions include **phospholipids,** important components of cell membranes, which are derived from carboxylic acids and phosphoric acid. In the **phosphoglycerides,** glycerol is esterified with two adjacent fatty acids and a phosphate unit that bears another substituent derived from a

Really? The vicissitude of self-indulgence notwithstanding, consuming chocolate appears to be good for you. Cocoa seems to lower blood pressure, improve cognitive function, and exhibit anticancer and antidiarrheal activity. The only downside is excessive ingestion, because of the calorific content of its coca butter and sugar components. There is a caveat, however, if you have a dog or cat: The presence of the alkaloid stimulant theobromine (an analog of caffeine; see Section 25-8) to the tune of 2–10% makes chocolate toxic to these pets. A 25-g chocolate bar would be enough to cause symptoms of vomiting and diarrhea, and more induces seizures and death.

Theobromine

1,2,3-Propanetriol   **1,2,3-Propanetriol triester**
(Glycerol)   **(Triglyceride)**

low-molecular-weight alcohol such as choline, $[HOCH_2CH_2N(CH_3)_3]^+$ ^-OH. The substance shown here is an example of a **lecithin,** a lipid found in the brain and nervous system.

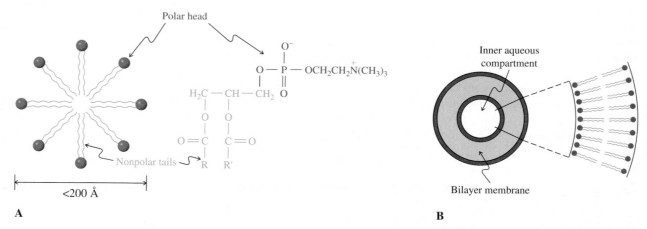

A phosphoglyceride

Hexadecanoic (palmitic) acid unit

cis

$CH_3(CH_2)_7CH{=}CH(CH_2)_7COCH$

cis-9-Octadecenoic
(oleic) acid unit

$CH_2OC(CH_2)_{14}CH_3$

Choline unit

Palmitoyloleoylphosphatidylcholine
(A lecithin)

(Z)-9-Octadecenamide

Compounds such as these play roles in the transmission of nerve impulses and exhibit other biological effects. For example, a lipid containing the amide analog of oleic acid [(Z)-9-octadecenamide] was recently identified as an essential sleep-inducing agent in the brain; indeed, persons with lipid-restricted diets have difficulty experiencing deep sleep.

Because these molecules carry two long hydrophobic fatty acid chains and a polar head group (the phosphate and choline substituent), they are capable of forming micelles in aqueous solution (see Real Life 19-1). In the micelles, the phosphate unit is solubilized by water, and the ester chains are clustered inside the hydrophobic micellar sphere (Figure 20-2A).

Phosphoglycerides can also aggregate in a different way: They may form sheets called a **lipid bilayer** (Figure 20-2B). This capability is significant because, whereas micelles are usually limited in size (<200 Å in diameter), bilayers may be as much as 1 mm (10^7 Å) in length. This property makes them ideal constituents of cell membranes, which act as permeability barriers regulating molecular transport into and out of the cell. Lipid bilayers are relatively stable molecular assemblies. The forces that drive their formation are similar to those at work in micelles: London interactions (Section 2-7) between the hydrophobic alkane chains, and coulombic and solvation forces between the polar head groups and between these polar groups and water.

In Summary Waxes, fats, and oils are naturally occurring, biologically active esters. Lipids are a broad class of biological molecules soluble in nonpolar solvents. They include glycerol-derived triesters, which are components of cell membranes.

Polar head

Nonpolar tails

<200 Å

A

O^-

$O — P — OCH_2CH_2\overset{+}{N}(CH_3)_3$

$H_2C — CH — CH_2$ O

$O=C$ $C=O$

R R'

Inner aqueous compartment

Bilayer membrane

B

Figure 20-2 Phospholipids are substituted esters essential to the structures of cell membranes. These molecules aggregate to form (A) a micelle or (B) a lipid bilayer. The polar head groups and nonpolar tails in phospholipids drive these aggregations. (After *Biochemistry,* 6th ed., by Jeremy M. Berg, John L. Tymoczko, and Lubert Stryer. W. H. Freeman and Company. Copyright © 1975, 1981, 1988, 1995, 2002, 2007, 2012).

REAL LIFE: SUSTAINABILITY 20-1	Moving Away from Petroleum: Green Fuels from Vegetable Oil

The fuel of choice for Rudolph Diesel's internal combustion engine, introduced to the world at the Paris Exposition in 1900, was peanut oil. Coming full circle after over a century, so-called biodiesel fuels have become popular alternatives to the petroleum-derived hydrocarbon mixtures widely used in diesel engines. Rudolph Diesel eventually found that pure vegetable oil was too viscous to be practical because it caused engine fouling in a relatively short time. However, a simple transesterification to generate a mixture of lower-molecular-weight methyl esters of long-chain fatty acids provides a fuel that can be used directly with very little, if any, engine modification. (The glycerol by-product is separated from the ester mixture and used in a variety of commercial applications, such as soap production.)

Soybean oil, french-fry oil, and even used restaurant grease are suitable raw materials for biodiesel—the United States alone discards 3 *billion* gallons of waste cooking oil annually. Biodiesel burns far more cleanly than conventional fuel, emitting no sulfur or volatile organic compounds; it is the only fuel in full compliance with the 1990 U.S. Clean Air Act. It also requires much less energy to produce compared with petroleum diesel, and it is renewable. The price of biodiesel is linked to the costs of the feedstock oils and therefore is currently higher than that of regular fuel. Biodiesel can be blended directly with conventional diesel or just used as is with minimal limitations. Pure biodiesel is more likely to "gel" at low temperatures, so blends are better suited to cold climates. Blends containing 5–20% biodiesel are popular and widely available. As of 2012, over 1600 retail outlets selling biodiesel fuel blends were distributed throughout all of the 50 states in the United States. Sales in 2011 reached nearly 650 million gallons, up from 20 million in 2003, and 500,000 in 1999. Production capacity is more than five times this amount. Expectations are that biodiesel

The 90 million trucks and buses on the roads and highways of the United States consume some 40 billion gallons of diesel fuel annually.

can supplant at least 10% of the U.S. diesel fuel requirement in the short term, thereby helping to provide a "bridge" over the time needed to develop new fuel technologies.

20-6 AMIDES: THE LEAST REACTIVE CARBOXYLIC ACID DERIVATIVES

Among all carboxylic acid derivatives, the amides, $\overset{\displaystyle O}{\overset{\|}{R C N R'_2}}$, are the least susceptible to nucleophilic attack. After a brief introduction to amide naming, this section describes their reactions.

Amides are named alkanamides, cyclic amides are lactams

Amides are called **alkanamides,** the ending -*e* of the alkane stem having been replaced by **-amide.** In common names, the ending -*ic* of the acid name is replaced by the **-amide** suffix. In cyclic systems, the ending -*carboxylic acid* is replaced by **-carboxamide.** Substituents on the nitrogen are indicated by the prefix *N*- or *N,N*-, depending on the number of groups. Accordingly, there are primary, secondary, and tertiary amides.

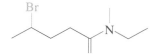

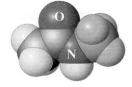

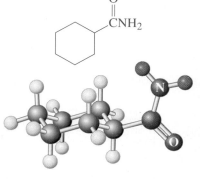

4-Bromo-*N*-ethyl-*N*-methyl**pentanamide**
(A tertiary amide)

Formamide
(A primary amide:
two hydrogens
bound to nitrogen)

N-**Methyl**acet**amide**
(A secondary amide:
one hydrogen
bound to nitrogen)

Cyclohexanecarboxamide
(Another primary amide)

There are several amide derivatives of carbonic acid, H_2CO_3: ureas, carbamic acids, and carbamic esters (urethanes).

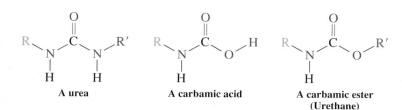

A urea   **A carbamic acid**   **A carbamic ester**
(Urethane)

Cyclic amides are called **lactams** (Section 19-10)—the systematic name is aza-2-cycloalkanones (Section 25-1)—and the rules for naming them follow those used for lactones. The penicillins are annulated β-lactams.

Urethane-based polymers form lightweight, tough, abrasion-resistant, high-performance materials ideal for use in a variety of sporting goods.

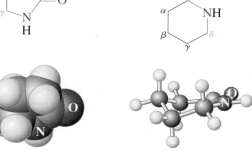

Penicillin
(A β-lactam derivative)

γ-**Butyrolactam**
(Systematic name:
aza-2-cyclopentanone)

δ-**Valerolactam**
(Systematic name:
aza-2-cyclohexanone)

Amides are essential to biochemistry; amide groups link the amino acid subunits that make up the biopolymers called proteins (Chapter 26). Many simpler amides possess varied forms of biological activity. For example, anandamide (after the Sanskrit word *ananda*, bliss), the amide of arachidonic acid (Section 19-13) with 2-aminoethanol, has been found to bind to the same receptor in the brain as does tetrahydrocannabinol (Section 9-11), the active ingredient in marijuana. The release and binding of anandamide is the mechanism by which the body suppresses the perception of pain (Problem 47 of Chapter 16). Anandamide has been isolated from chocolate, suggesting that people who claim to be "addicted to chocolate" may really know what they are talking about.

Anandamide

Amide hydrolysis requires strong heating in concentrated acid or base

The amides are the least reactive of the carboxylic acid derivatives, in part because they are strongly stabilized by delocalization of the nitrogen lone pair (Section 20-1). As a consequence, their nucleophilic addition–eliminations require relatively harsh conditions. For example, hydrolysis to the corresponding carboxylic acid occurs only upon prolonged heating in strong aqueous acid or base by addition–elimination mechanisms. Acid hydrolysis liberates the amine in the form of an ammonium salt.

Animation

ANIMATED MECHANISM:
Amide hydrolysis

Acid Hydrolysis of an Amide

3-Methylpentanamide $\xrightarrow[\text{(See Problem 51)}]{H_2SO_4, H_2O, \Delta, 3\ h}$ **3-Methylpentanoic acid** $+$ $(NH_4)_2SO_4$

Base hydrolysis requires loss of a very poor leaving group, in this case the inorganic amide ion, $^-:\ddot{N}H_2$. The reaction proceeds because (as in ester hydrolysis) the tetrahedral intermediate is high in energy compared to the resonance-stabilized substrate and product molecules. In addition, the equilibrium is driven forward through very rapid protonation of the strongly basic leaving group by the carboxylic acid that is liberated in the elimination step. The overall process therefore furnishes the carboxylate salt, which is protonated during subsequent aqueous work-up to produce the acid.

Reaction

Base Hydrolysis of an Amide

$$CH_3CH_2CH_2\overset{O}{\overset{\|}{C}}\ddot{N}HCH_3 \xrightarrow{H\ddot{O}:^-,\ H_2O,\ \Delta} CH_3CH_2CH_2\overset{O}{\overset{\|}{C}}\ddot{O}:^- + CH_3\ddot{N}H_2 \xrightarrow{H^+,\ H_2O} CH_3CH_2CH_2\overset{O}{\overset{\|}{C}}\ddot{O}H + CH_3\overset{+}{N}H_3$$

N-Methylbutanamide $\qquad\qquad\qquad\qquad\qquad\qquad$ 87%

Butanoic acid

Mechanism of Hydrolysis of Amides by Aqueous Base

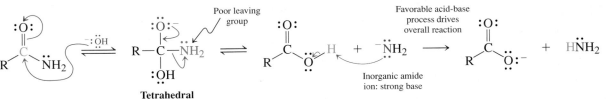

Poor leaving group

Favorable acid-base process drives overall reaction

Tetrahedral intermediate

Inorganic amide ion: strong base

Mechanism

REAL LIFE: MEDICINE 20-2 | Battling the Bugs: Antibiotic Wars

In late summer 1928, the Scottish bacteriologist Sir Alexander Fleming went on vacation. When he returned, the course of human history changed. Fleming had left a culture dish containing the bacterium *Staphylococcus aureus* on a laboratory bench. While he was gone, a cold spell stopped the growth of the bacterium. At the same time, mold spores of *Penicillium notatum* happened to drift up from the floor below, settling into his culture dish. By the time Fleming returned, the weather had warmed, and both microorganisms resumed growing. Intending to clean and sterilize the dish, he fortunately first noticed that *Penicillium* was destroying the bacterial colonies. The substance responsible for this antibiotic effect was isolated in 1939 and named penicillin.

Fleming's original mold produced benzylpenicillin, or penicillin G (R = $C_6H_5CH_2$). Many analogs have been subsequently synthesized, comprising a large class of so-called β-lactam antibiotics, after the strained four-membered lactam ring that characterizes them structurally as well as functionally (see also Chapter 7 Opening). Because ring strain is relieved upon ring opening, β-lactams are unusually reactive compared with ordinary amides. The enzyme transpeptidase catalyzes a crucial reaction in the biosynthesis of a polymer that maintains the structure of bacterial cell walls. A nucleophilic oxygen of the enzyme links with the carboxylic acid function of one amino acid, catalyzing its amide formation with the amine group of another amino acid molecule. Repetition of this process generates the polymer. Penicillin's β-lactam carbonyl group reacts with the crucial enzyme oxygen readily and irreversibly, inactivating the enzyme, stopping cell-wall synthesis, and killing the bacterium.

Some bacteria are resistant to penicillin, because they produce an enzyme, penicillinase, that destroys the β-lactam ring in the antibiotic. Synthesis of analogs afforded a partial solution to this problem. Ultimately, however, it became necessary to turn to antibiotics with completely different modes of action. Erythromycin, produced by a strain of *Streptomyces* bacteria first found in soil samples in the Philippines in 1952, functions in a distinct manner. It is a large ring lactone that interferes with the bacterial *ribosome,* its cell-wall protein synthesis factory. Although erythromycin is unaffected by penicillinase, bacteria resistant to it have developed over the decades since its introduction into the antibiotic arsenal.

An even more complex antibiotic, vancomycin, was discovered in 1956, from fermentation of a bacterium originally found in the soils of the jungles of Borneo. When dangerous strains of *Staphylococcus aureus*

Penicillin in Action

Penicillin

Transpeptidase

enzyme

Inactivated enzyme

possessing resistance to virtually every known antibiotic became serious threats to public health, "Vanco" (for "vanquish") quickly became the "antibiotic of last resort." Its effectiveness arises from an entirely novel chemical interaction: Its shape and structure allow it to form a tight network of hydrogen bonds with the amino acids at the end of the growing cell-wall polymer, preventing them from linking with additional amino acid units. Within a decade vanco-resistant strains of *S. aureus* appeared, in which a small structural change at the end of the polymer disrupts the ability of the antibiotic to bind. However, knowledge of the precise nature of this modification allowed the development in 2006 of a new vancomycin derivative in which the amide carbonyl group shown in red was replaced by a methylene (CH_2). This derivative, while much less active against the normal bacterium, is lethal to the vancomycin-resistant strain.

The battle between scientists and the bacterial world continues: New antibacterials are continually being prepared and tested for potency. Tigecycline, introduced in 2006, is the first of a new class of antibiotics related to the tetracyclines and designed to overcome some of the bugs' resistance strategies. Tigecycline works against some particularly nasty drug-defiant infections of the skin and internal organs, as well as penicillin-resistant *Streptococcus pneumoniae.*

Erythromycin

Vancomycin

**Tigecycline
(Tygacil)**

Another tactic to foil the survival efforts of evolving pathogens was unveiled in 2007. A synthetic polyamide (polypeptide, Section 26-4) was built that causes the immune system to prepare antibodies against resistant strains of *Staphylococcus aureus,* responsible for nearly 100,000 infections and 20,000 deaths each year. Such substances have the potential to give rise to entirely novel vaccination strategies that could save thousands of lives.

Yet another recent strategy has been the development of additives to existing antibiotics, so-called adjuvants that target the biological pathways by which bacteria develop resistance, thus suppressing it.

Meanwhile, microbes continue to adapt and develop modifications to their biochemical machinery to foil antibiotic attack. Whether the "era of antibiotics" that began in the middle of the twentieth century will continue very far into the twenty-first is very much an open question.

The extraordinary expense of developing a new pharmaceutical, estimated at $1.3 billion (a number that includes the cost of failed drugs), coupled with the ever-shortening time for the onset of bacterial resistance, has stimulated the pharmaceutical industry to pursue synthetic approaches to completely novel scaffolds that act through new mechanisms. Such structures are also sought in nature, for example, biologically active compounds produced by microorganisms living in exotic environments, such as the 1200-foot-deep caverns of 4-million-year-old Lechuguilla Cave in New Mexico.

Hunting for exotic bacteria in the Lechuguilla Cave.

Amides can be reduced to amines or aldehydes

In contrast to carboxylic acids and esters, the reaction of amides with lithium aluminum hydride produces amines instead of alcohols.

Reaction

Reduction of an Amide to an Amine

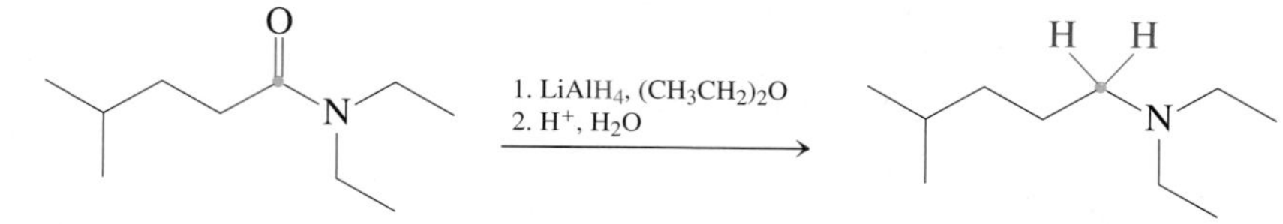

N,N-Diethyl-4-methylpentanamide

85%
N,N-Diethyl-4-methylpentanamine

The mechanism of reduction begins with hydride addition, which gives a tetrahedral intermediate. Elimination of an aluminum alkoxide leads to an **iminium ion.** Addition of a second hydride gives the final amine product.

Mechanism

Mechanism of Amide Reduction by LiAlH₄

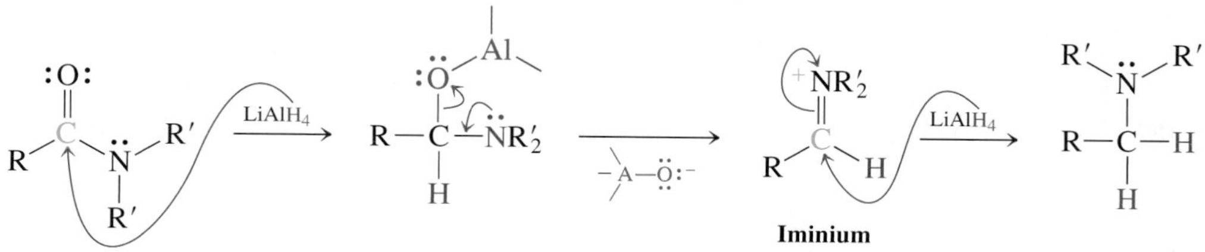

Iminium ion

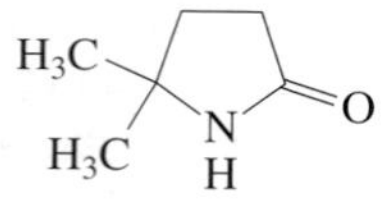

Exercise 20-18

What product would you expect from LiAlH₄ reduction of the compound depicted in the margin?

Reduction of amides by bis(2-methylpropyl)aluminum hydride (diisobutylaluminum hydride) furnishes aldehydes. Recall that esters also are converted into aldehydes by this reagent (Section 20-4).

Reduction of an Amide to an Aldehyde

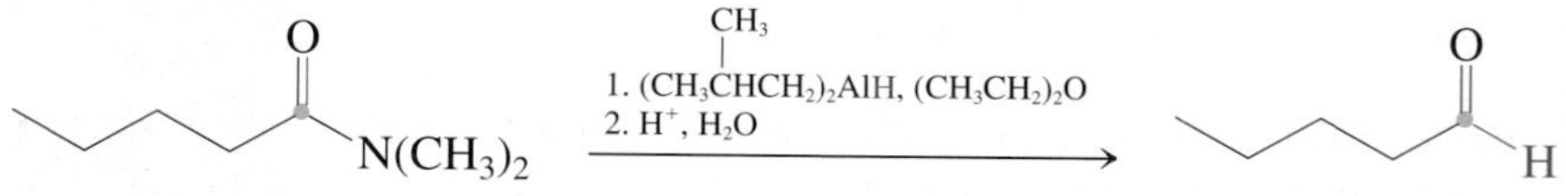

N,N-Dimethylpentanamide

92%
Pentanal

Solved Exercise 20-19 | Working with the Concepts: Reduction of Amides

Treatment of amide A with LiAlH₄, followed by acidic aqueous work-up, gave compound B. Explain.

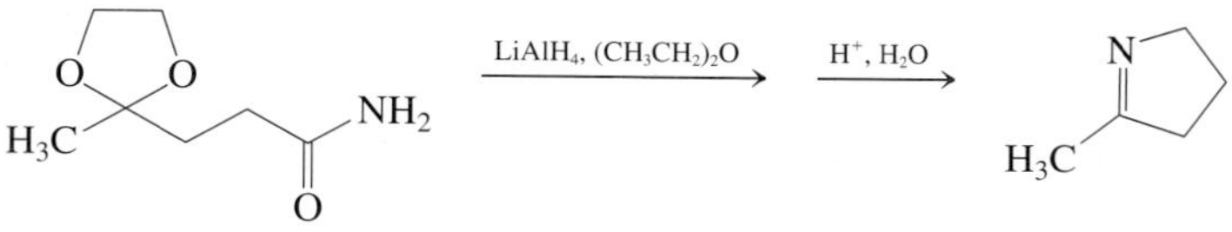

A

B

Remember WHIP
*W*hat
*H*ow
*I*nformation
*P*roceed

Strategy

Consider each step of the sequence individually. Identify the product of the first reaction and consider how it may be able to transform into the final product.

Solution

- What is the result of LiAlH₄ treatment of amide A? As shown in this section, the carboxamide function is reduced to that of a primary amine.
- The acetal group, being immune to strong bases and reducing agents, remains unchanged (Section 17-7). However, it hydrolyzes to give a ketone function upon the aqueous acidic work-up that follows the LiAlH₄ reduction. At this point we have the following aminoketone:

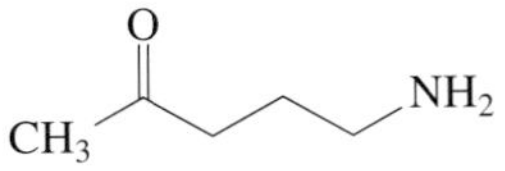

- Comparison of this structure with that of the final product B shows that the amino nitrogen has become bonded to the carbonyl carbon to give a cyclic imine. Acid-catalyzed condensation of the amino with the carbonyl group (Section 17-9) provides a reasonable explanation for this transformation.

Exercise 20-20 | Try It Yourself

Treatment of amide A (Exercise 20-19) with diisobutylaluminum hydride followed by acidic work-up gives an entirely different result from that described in Exercise 20-19: The product is

2-cyclopentenone, [structure]. Explain. (**Hint:** See Section 18-7 and review Problem 49 of Chapter 18.)

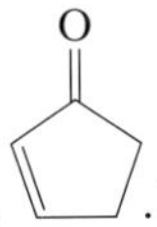

In Summary Carboxylic amides are named as alkanamides or as lactams if they are cyclic. They can be hydrolyzed to carboxylic acids by acid or base and reduced to amines by lithium aluminum hydride. Reduction by bis(2-methylpropyl)aluminum hydride (diisobutyl-aluminum hydride) stops at the aldehyde stage.

20-7 | AMIDATES AND THEIR HALOGENATION: THE HOFMANN REARRANGEMENT

In amides, hydrogens on both the carbon and the nitrogen atoms next to the carbonyl group are acidic. Removal of the NH hydrogen, which has a pK_a of about 22, with base leads to an **amidate ion**. The CH proton is less acidic, with a pK_a of about 30 (Section 20-1); therefore, deprotonation of the α-carbon, leading to an **amide enolate,** is more difficult.

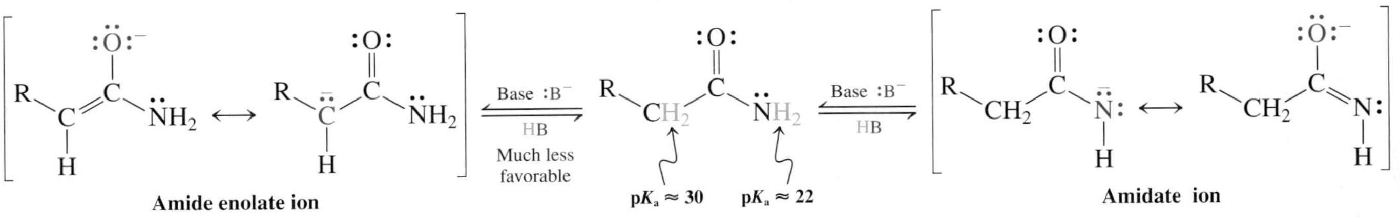

Practically speaking, therefore, a proton may be removed from the α-carbon only with tertiary amides, in which the nitrogen lacks attached hydrogen atoms.

The amidate ion formed by deprotonation of a primary amide is a synthetically useful nucleophile. This section focuses on one of its reactions, the Hofmann rearrangement.

Exercise 20-21

The pK_a of 1,2-benzenedicarboximide (phthalimide, A) is 8.3, considerably lower than the pK_a of benzamide (B). Why?

A

B

In the presence of base, primary amides undergo a special halogenation reaction, the **Hofmann* rearrangement.** In it, the carbonyl group is expelled from the molecule to give a primary amine with one carbon fewer in the chain.

Reaction

Hofmann Rearrangement

$$\underset{RCNH_2}{\overset{O}{\|}} \xrightarrow{X_2,\ NaOH,\ H_2O} RNH_2 \ + \ O{=}C{=}O$$

$$CH_3(CH_2)_6CH_2CONH_2 \xrightarrow{Cl_2,\ NaOH} CH_3(CH_2)_6CH_2NH_2 \ + \ O{=}C{=}O$$

Nonanamide   66%   **Octanamine**

The Hofmann rearrangement begins with deprotonation of nitrogen to form an amidate ion (step 1). Halogenation of the nitrogen follows, much like the α-halogenation of aldehyde and ketone enolates (step 2; see Section 18-3). Subsequently, the second proton on the nitrogen is abstracted by additional base to give an *N*-haloamidate (step 3). This species, containing a weak nitrogen–halogen bond and a good potential leaving group, undergoes loss of halide ion accompanied by migration of the R group from the carbonyl carbon to nitrogen (step 4). The product of this rearrangement is an **isocyanate,** R—N=C=O, a nitrogen analog of carbon dioxide, O=C=O. The *sp*-hybridized carbonyl carbon in the isocyanate is highly electrophilic and is attacked by water to produce an unstable **carbamic acid.** Finally, the carbamic acid decomposes to carbon dioxide and the amine (step 5).

Mechanism of the Hofmann Rearrangement

Mechanism

Animation

ANIMATED MECHANISM:
Hofmann rearrangement

Step 1. Amidate formation

Step 2. Halogenation

---

*This is the Hofmann of the Hofmann rule of E2 reactions (Section 11-6).

Step 3. *N*-Haloamidate formation

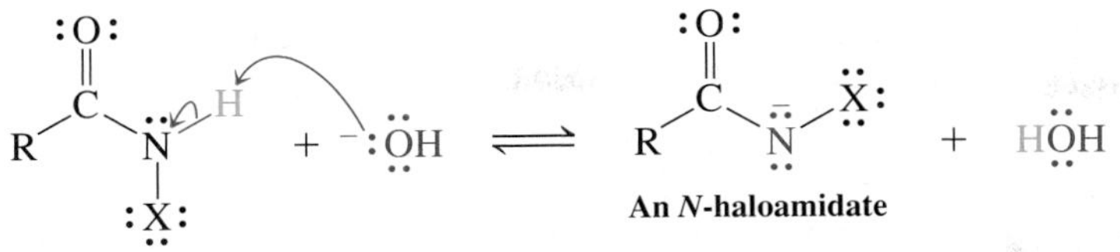

An *N*-haloamidate

Step 4. Rearrangement with halide elimination

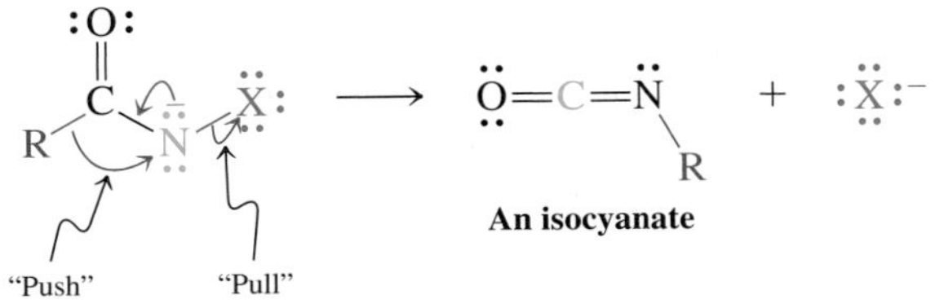

An isocyanate

"Push"   "Pull"

You may wonder why the conditions of the Hofmann rearrangement do not lead to amide hydrolysis. The reason is that amidate formation by deprotonation, followed by halogenation, is much faster then nucleophilic attack by the base on the carbonyl function. Conversely, in the base-mediated hydrolysis of primary and secondary amides (those bearing protons on the nitrogen), equilibration with the amidate is a rapid, nonproductive, and reversible dead-end background reaction.

Step 5. Hydration to the carbamic acid and decomposition

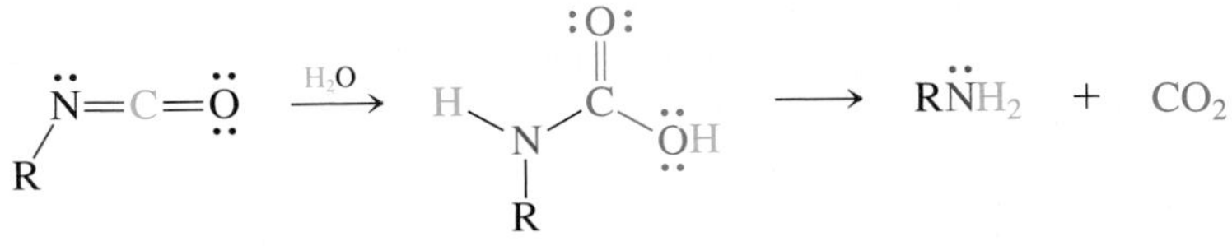

A carbamic acid

In the rearrangement, as the halide ion leaves, the alkyl group literally "slides" over from the carbonyl carbon to the nitrogen, keeping the same "face" toward the nitrogen that was previously bonded to carbon. Therefore, when this alkyl group is chiral, its original stereochemistry is retained during the course of the rearrangement. This kind of electronic "push–pull" process, in which a leaving group (pull) causes a neighboring (alkyl) group to migrate (push), has analogies in the acid-catalyzed rearrangement of 2,2-dimethyl-1-propanol (neopentyl alcohol; Section 9-3), the mechanism of alkylborane oxidation (Section 12-8), and the mechanism of the Baeyer-Villiger oxidation (Section 17-13).

"Push–Pull" Transition States

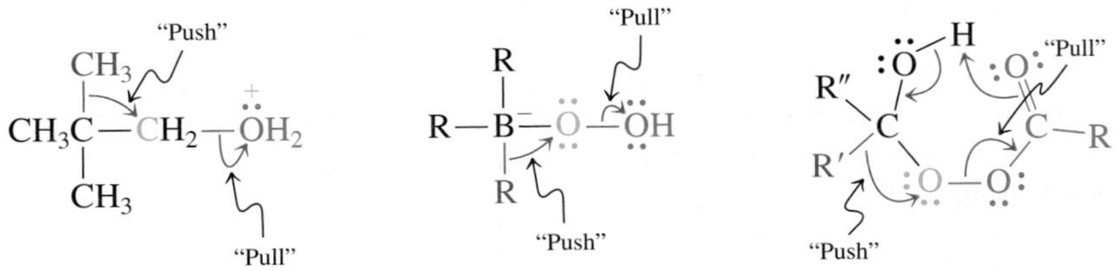

Acid-catalyzed rearrangement of 2,2-dimethyl-1-propanol (neopentyl alcohol)

Alkylborane oxidation

Baeyer-Villiger oxidation

Exercise 20-22

Give the product of the following Hofmann rearrangement process.

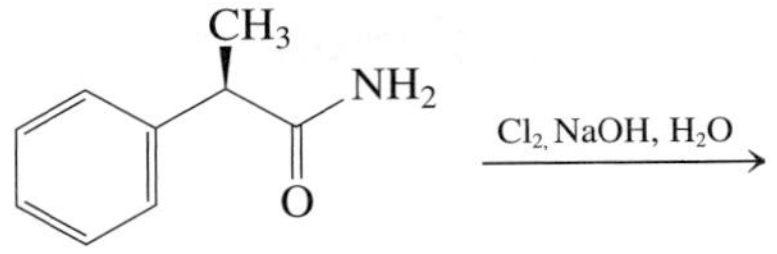

$\xrightarrow{\text{Cl}_2, \text{NaOH, H}_2\text{O}}$

Exercise 20-23

Write a detailed mechanism for the addition of water to an isocyanate under basic conditions and for the decarboxylation of the resulting carbamic acid.

Exercise 20-24

Suggest a sequence by which you could convert ester A into amine B.

COOCH$_3$ NH$_2$

A   --→   B

In Summary Treatment of primary and secondary amides with base leads to deprotonation at nitrogen, giving amidate ions. Bases abstract protons from the α-carbon of tertiary amides. In the Hofmann rearrangement, primary amides react with halogens in base to furnish amines with one fewer carbon. In the course of this process an alkyl shift takes place, which converts an acyl nitrene into an isocyanate.

20-8 ALKANENITRILES: A SPECIAL CLASS OF CARBOXYLIC ACID DERIVATIVES

Nitriles, RC≡N, are considered derivatives of carboxylic acids because the nitrile carbon is in the same oxidation state as the carboxy carbon and because nitriles are readily converted into other carboxylic acid derivatives. This section describes the rules for naming nitriles, the structure and bonding in the nitrile group, and some of its spectral characteristics. Then it compares the chemistry of the nitrile group with that of other carboxylic acid derivatives.

CH$_3$C≡N

**Ethanenitrile
(Acetonitrile)**

CH$_3$CH$_2$C≡N
**Propanenitrile
(Propionitrile)**

CH$_3$
|
CH$_3$CHCH$_2$C≡N
3-Methylbutanenitrile

Benzonitrile

In IUPAC nomenclature, nitriles are named from alkanes

A systematic way of naming this class of compounds is as **alkanenitriles.** In common names, the ending *-ic acid* of the carboxylic acid is usually replaced with **-nitrile.** The chain is numbered like those of carboxylic acids. Similar rules apply to dinitriles derived from dicarboxylic acids. The substituent CN is called **cyano.** Cyanocycloalkanes are called cycloalkane**carbonitriles.** We will retain the common name benzonitrile (from benzoic acid), rather than use the systematic benzenecarbonitrile.

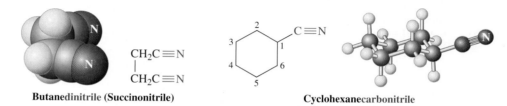

CH$_2$C≡N
|
CH$_2$C≡N

Butanedinitrile (Succinonitrile)   **Cyclohexanecarbonitrile**

The C≡N bond in nitriles resembles the C≡C bond in alkynes

In the nitriles, both atoms in the functional group are *sp* hybridized and a lone electron pair on nitrogen occupies an *sp* hybrid orbital pointing away from the molecule along the C–N

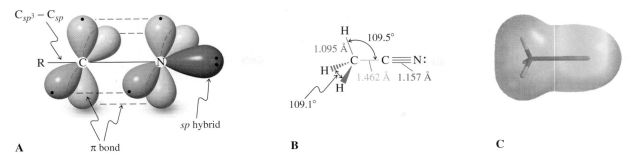

Figure 20-3 (A) Orbital picture of the nitrile group, showing the *sp* hybridization of both atoms in the C≡N function. (B) Molecular structure of ethanenitrile (acetonitrile), which is similar to that of the corresponding alkyne. (C) Electrostatic potential map of ethanenitrile, depicting the positively polarized cyano carbon (blue) and the relatively negatively polarized nitrogen (green) with its lone electron pair (red).

axis. The hybridization and structure of the nitrile functional group very much resemble those of the alkynes (Figure 20-3; see also Figures 13-1 and 13-2).

In the infrared spectrum, the C≡N stretching vibration appears at about 2250 cm^{-1}, in the same range as the C≡C absorption, but much more intense. The ^{1}H NMR spectra of nitriles indicate that protons near the nitrile group are deshielded about as much as those in other carboxylic acid and alkyne derivatives (Table 20-3).

The ^{13}C NMR absorption for the nitrile carbon appears at lower field ($\delta \approx 112-126$ ppm) than that of the alkynes ($\delta \approx 65-85$ ppm), because nitrogen is more electronegative than carbon. Thus, nitriles are polarized, as shown in the electrostatic potential map in Figure 20-3C.

Table 20-3	^{1}H NMR Chemical Shifts of Substituted Methanes CH$_3$X
X	δ_{CH_3} **(ppm)**
—H	0.23
—Cl	3.06
—OH	3.39
O ‖ —CH	2.18
—COOH	2.08
—CONH$_2$	2.02
—C≡N	1.98
—C≡CH	1.80

Exercise 20-25

1,3-Dibromopropane was treated with sodium cyanide in dimethyl sulfoxide-d_6 and the mixture monitored by ^{13}C NMR. After a few minutes, four new intermediate peaks appeared, one of which was located well downfield from the others at $\delta = 117.6$ ppm. Subsequently, another three peaks began growing at $\delta = 119.1, 22.6,$ and 17.6 ppm, at the expense of the signals of starting material and the intermediate. Explain.

Nitriles undergo hydrolysis to carboxylic acids

As mentioned in Section 19-6, nitriles can be hydrolyzed to give the corresponding carboxylic acids. The conditions are usually stringent, requiring concentrated acid or base at high temperatures.

$$N\equiv C(CH_2)_4C\equiv N \xrightarrow{\ H^+,\ H_2O,\ \Delta\ } HOOC(CH_2)_4COOH$$
$$97\%$$

Hexanedinitrile   **Hexanedioic acid**
(Adiponitrile)   **(Adipic acid)**

Reaction

The mechanisms of these reactions proceed through the intermediate amide and include addition–elimination steps.

In the acid-catalyzed process, initial protonation on nitrogen facilitates nucleophilic attack by water. Loss of a proton from oxygen furnishes a neutral intermediate, which is a tautomer of an amide. A second protonation on the nitrogen occurs, followed again by deprotonation of oxygen and formation of the amide. Hydrolysis of the amide proceeds by the usual addition–elimination pathway.

Mechanism of the Acid-Catalyzed Hydrolysis of Nitriles

Mechanism

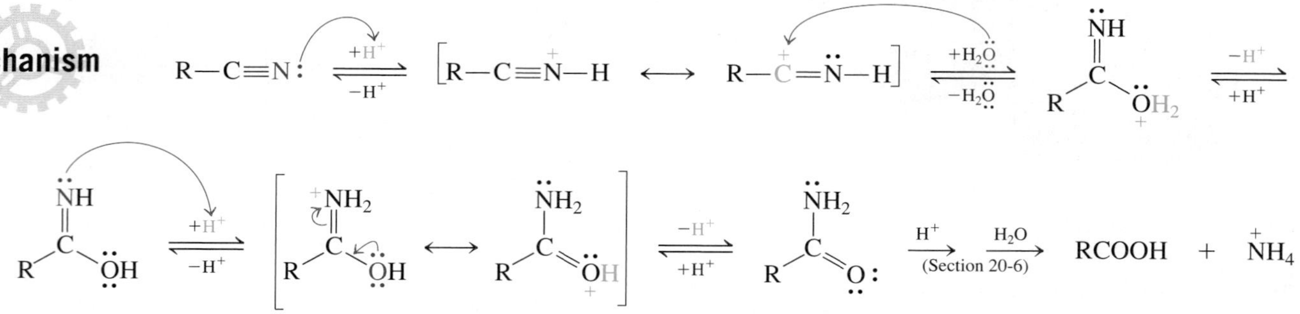

Tautomer of amide

In base-catalyzed nitrile hydrolysis, direct attack of hydroxide gives the anion of the amide tautomer, which protonates on nitrogen. The remaining proton on the oxygen is then removed by base and a second *N*-protonation gives the amide. Hydrolysis is completed as described in Section 20-6.

Mechanism of the Base-Catalyzed Hydrolysis of Nitriles

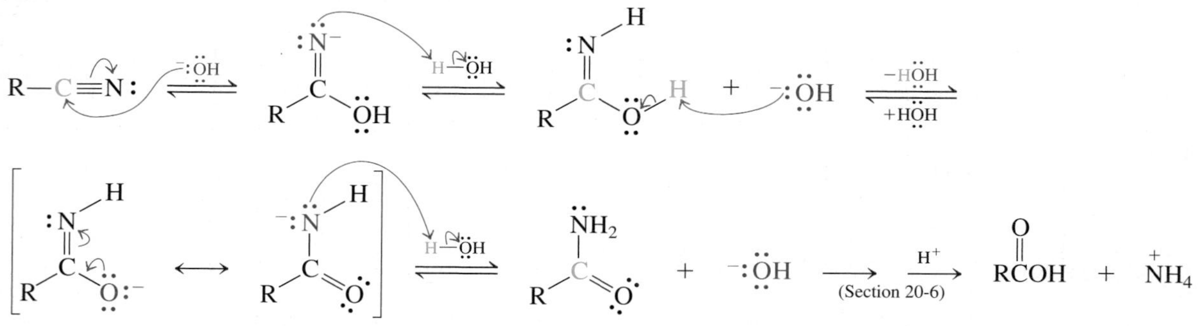

Mechanism

Organometallic reagents attack nitriles to give ketones

Strong nucleophiles, such as organometallic reagents, add to nitriles to give anionic imine salts. Work-up with aqueous acid gives the neutral imine, which is rapidly hydrolyzed to the ketone (Section 17-9).

Ketone Synthesis from Nitriles

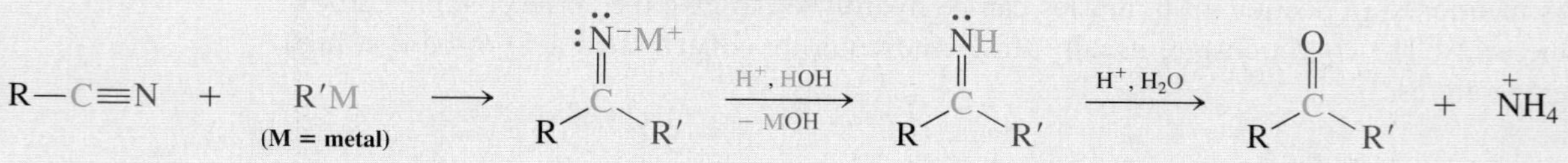

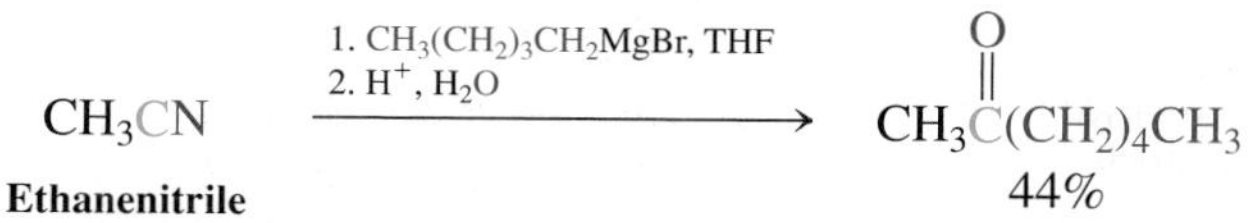

CH₃CN $\xrightarrow[\text{2. H}^+, \text{H}_2\text{O}]{\text{1. CH}_3(\text{CH}_2)_3\text{CH}_2\text{MgBr, THF}}$ CH₃C(CH₂)₄CH₃

Ethanenitrile
(Acetonitrile)

44%
2-Heptanone

Reduction of nitriles by hydride reagents leads to aldehydes and amines

As in its reactions with esters and amides, bis(2-methylpropyl)aluminum (diisobutylaluminum) hydride (DIBAL) adds to a nitrile only once to give an imine derivative. Aqueous hydrolysis then produces an aldehyde.

Aldehyde Synthesis from Nitriles

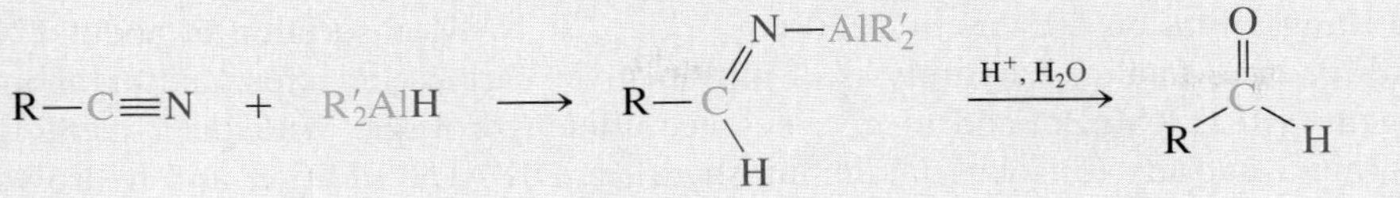

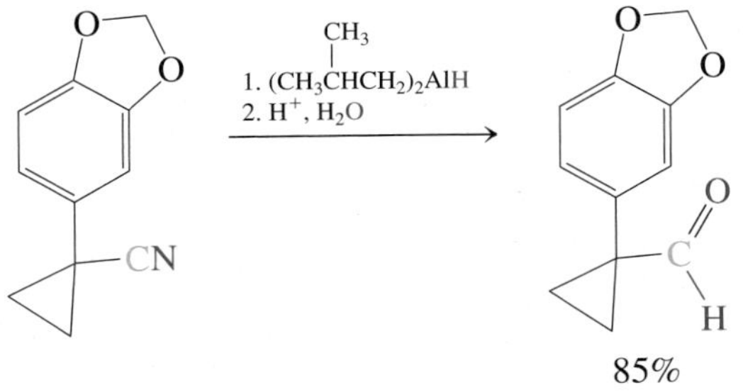

$$R-C\equiv N \;+\; R'_2AlH \;\longrightarrow\; R-\overset{\overset{\displaystyle N-AlR'_2}{\|}}{\underset{\underset{\displaystyle H}{|}}{C}} \;\xrightarrow{H^+, H_2O}\; \underset{R}{\overset{\displaystyle O}{\|}}\overset{}{C}-H$$

1. (CH₃CHCH₂)₂AlH (with CH₃ on central carbon)
2. H⁺, H₂O

85%

Treatment of nitriles with strong hydride reducing agents results in double hydride addition, giving the amine on aqueous work-up. The best reagent for this purpose is lithium aluminum hydride.

$$CH_3CH_2CH_2C\equiv N \;\xrightarrow[\text{2. } H^+, H_2O]{\text{1. LiAlH}_4}\; CH_3CH_2CH_2CH_2NH_2$$

Butanenitrile   85%
                              Butanamine

Exercise 20-26

The reduction of a nitrile by LiAlH₄ to give an amine adds four hydrogen atoms to the C≡N triple bond: two from the reducing agent and two from the water in the aqueous work-up. Formulate a mechanism for this transformation.

Like the triple bond of alkynes (Section 13-6), the nitrile group is hydrogenated by catalytically activated hydrogen. The result is the same as that of reduction by lithium aluminum hydride—amine formation. All four hydrogens are from the hydrogen gas.

$$CH_3CH_2CH_2C\equiv N \;\xrightarrow{H_2,\ PtO_2,\ CH_3CH_2OH,\ CHCl_3}\; CH_3CH_2CH_2CH_2NH_2$$

Butanenitrile   96%
                              Butanamine

Exercise 20-27

Show how you would prepare the following compounds from pentanenitrile.

(a) $CH_3(CH_2)_3COOH$

(b) $CH_3(CH_2)_3\overset{\displaystyle O}{\overset{\|}{C}}(CH_2)_3CH_3$

(c) $CH_3(CH_2)_3\overset{\displaystyle O}{\overset{\|}{C}}H$

(d) $CH_3(CH_2)_3CD_2ND_2$

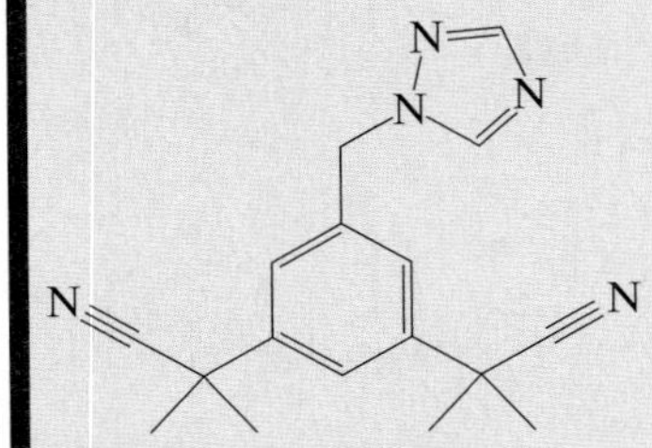

In Summary Nitriles are named as alkanenitriles. Both atoms making up the C–N unit are *sp* hybridized, the nitrogen bearing a lone pair in an *sp*-hybrid orbital. The nitrile stretching vibration appears at 2250 cm^{-1}, the ^{13}C NMR absorption at about 120 ppm. Acid- or base-catalyzed hydrolysis of nitriles gives carboxylic acids, and organometallic reagents (RLi, RMgBr) add to give ketones after hydrolysis. With bis(2-methylpropyl) aluminum hydride (diisobutylaluminum hydride, DIBAL), addition and hydrolysis furnishes aldehydes, whereas LiAlH$_4$ or catalytically activated hydrogen converts the nitrile function into the amine.

THE BIG PICTURE

In the past four chapters we have introduced all of the major compound classes based on the carbonyl functional group. Their similarities and differences are direct reflections of the fact that structure defines function in organic chemistry. The fundamental principles that govern carbonyl chemistry derive from concepts first introduced in Chapters 1 and 2:

- The strength and reactivity of polar bonds (Sections 1-3 and 2-3);
- The effects of resonance delocalization on structure and stability (Section 1-5);
- Direct and indirect applications of acid-base chemistry (Section 2-3).

Carbonyl chemistry illustrates so many basic features of organic chemistry that a careful study of it can serve as an excellent point of departure for developing a more comprehensive grasp of the properties of all of the other compound classes. If at any time in the future you find yourself needing to review your course in organic chemistry for any purpose, carbonyl chemistry will be a good place to start, because it will help you develop the insight to understand much of the rest of the subject matter with considerably greater thoroughness.

In Chapter 21, we turn to the last simple compound class that will be covered in this course, the amines. Again, polar bonds and acid-base chemistry will play important roles.

WORKED EXAMPLES: INTEGRATING THE CONCEPTS

20-28. Extrapolating from Ester Chemistry to Lactone Chemistry

A useful synthesis of certain types of diols includes the reaction of a "bisGrignard" reagent with a lactone:

(a) Formulate a mechanism for this transformation.

SOLUTION

We begin by noticing that a lactone is merely a cyclic ester, and therefore we may refer to the reaction of esters with Grignard reagents described in Section 20-4 as a guide for approaching this question.

Esters react sequentially with two equivalents of Grignard reagent: The first gives rise to an addition–elimination sequence that forms a new carbon–carbon bond, expels an alkoxide leaving group, and produces a ketone. The second Grignard equivalent adds to the carbonyl carbon of this ketone, giving the final product, a tertiary alcohol. The present problem modifies this pattern in two ways: The ester is cyclic; therefore, the first Grignard addition–elimination process opens the ring, and the alkoxide leaving group remains attached by a carbon chain in the intermediate ketone. Instead of two separate molecules of Grignard reagent, we have here a single molecule containing two Grignard functionalities. Therefore, when the second Grignard function adds to the same carbonyl carbon as did the first, it forms a new ring:

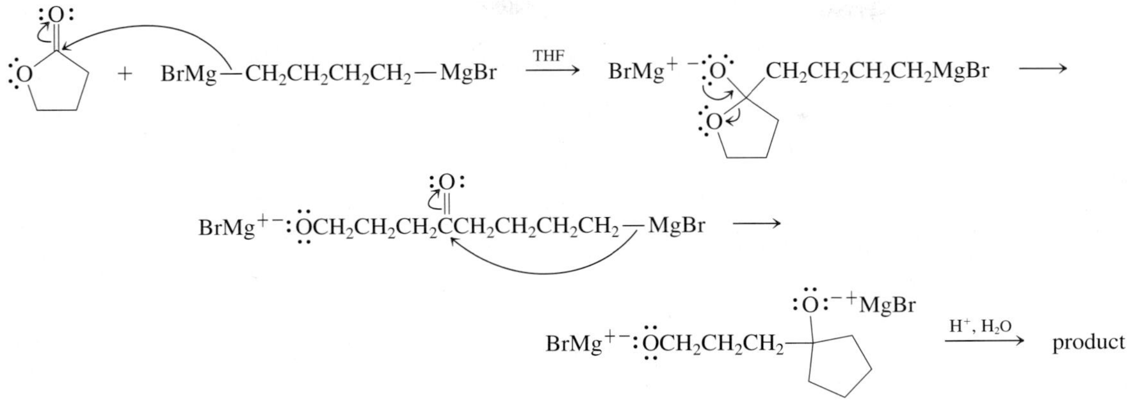

(b) Show how you would apply this general method to the synthesis of diols A and B.

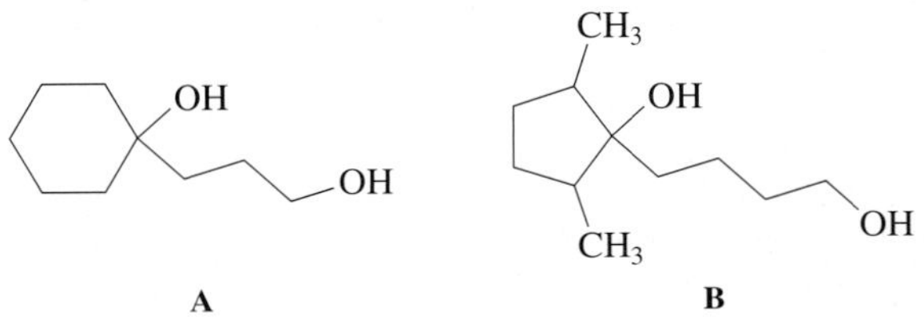

A   B

SOLUTION

Referring to the mechanism in (a), we can generalize the pattern followed by this transformation as shown below.

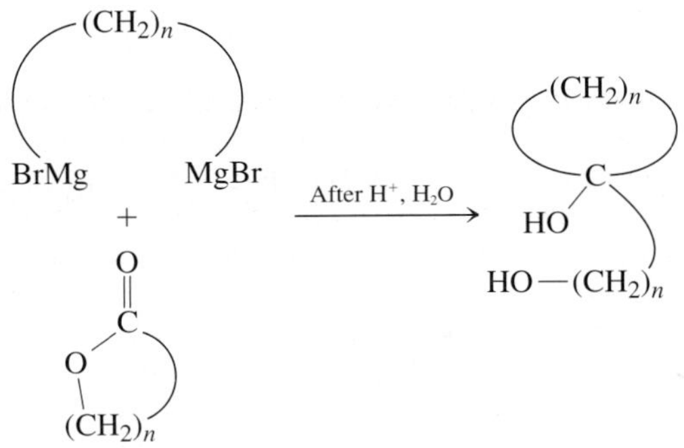

The answers are

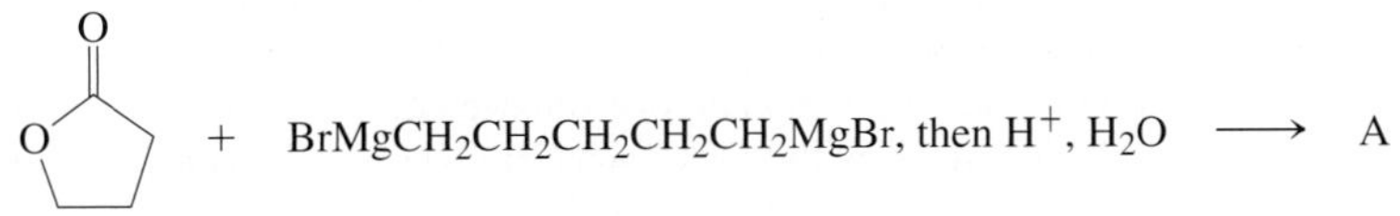

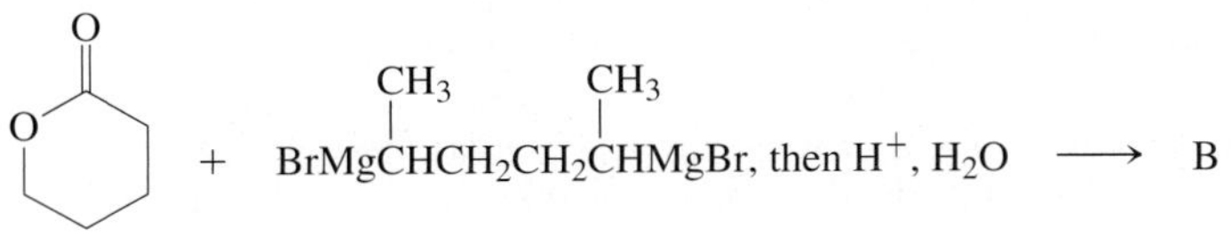

$+$ BrMgCH$_2$CH$_2$CH$_2$CH$_2$CH$_2$MgBr, then H$^+$, H$_2$O $\longrightarrow$ A

$+$ BrMgCHCH$_2$CH$_2$CHMgBr, then H$^+$, H$_2$O $\longrightarrow$ B

20-29. Supplying Reagents for an Extended Synthetic Sequence

Clovene is a sesquiterpene that arises from acid-catalyzed rearrangement of caryophyllene (Problem 74 of Chapter 12—odor of cloves). The following sequence of transformations constitutes a part of a total synthesis of clovene, but the reagents have been omitted. Supply reasonable reagents and reaction conditions for each transformation, some of which require more than one step. In some cases, you may need to refer to earlier chapters, especially Chapters 17 through 19, for information.

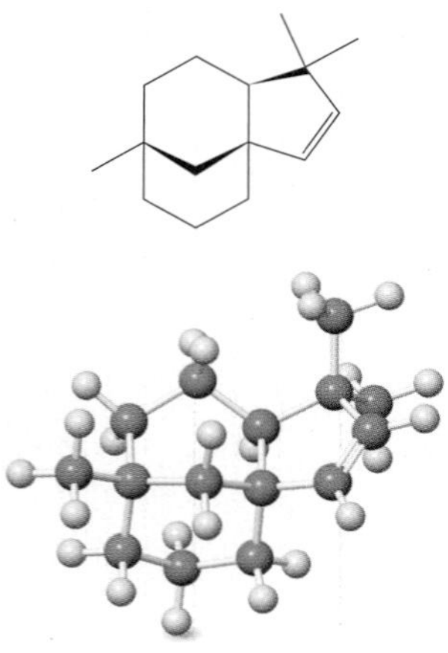

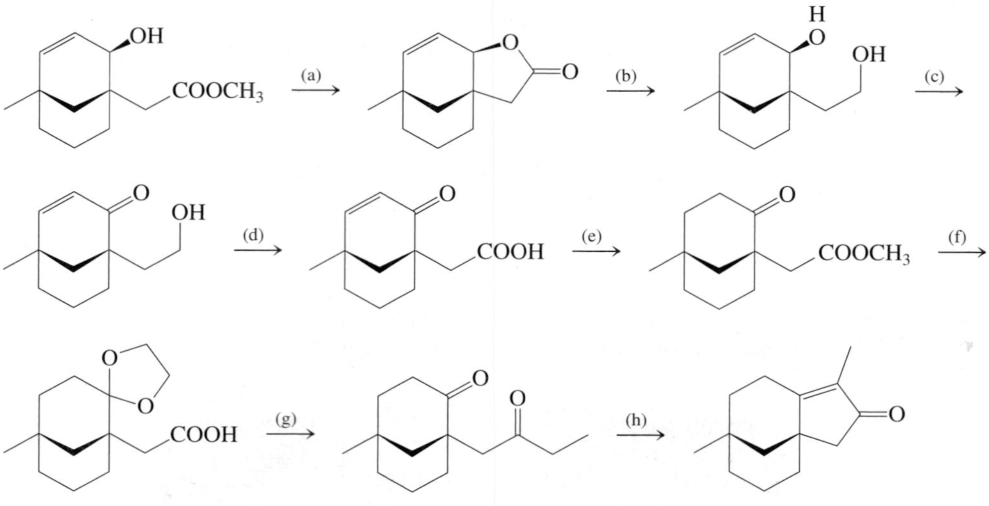

Clovene

SOLUTION

How do we analyze this sort of synthesis problem? We can begin by characterizing each transformation: Identify exactly what has changed. When we have this information, we can refer to the reactions that we know and determine whether more than one step is needed or whether one step is sufficient to carry out the process. Let us proceed in the order shown above.

a. A molecule containing methoxycarbonyl and hydroxy groups converts into a cyclic ester—a lactone. A change from one ester into another would be transesterification (Section 20-4), a reversible process catalyzed equally well by acid or base. How do we shift the equilibrium in the desired direction? Note that methanol is the by-product of the process; it has a low boiling point (65°C, Table 8-1) and may be driven off by heating the reaction mixture. Therefore, (a) is catalytic H^+.

b. The lactone converts into a diol. One could effect this change in two steps: Hydrolysis would cleave the lactone into a hydroxycarboxylic acid (Section 20-4); subsequent reduction by $LiAlH_4$ would furnish the diol (Section 19-11). Much easier is $LiAlH_4$ reduction of the lactone, which would lead to the diol directly. Notice (Section 20-4) that, in the course of these reduction processes alkoxides are the initial products, requiring acidic work-up. Therefore, (b) is 1. $LiAlH_4$, ether (a typical solvent for such reductions) and 2. H^+, H_2O.

c. This step requires some thought. There are two hydroxy groups to start with, and we want to oxidize only one. The usual methods based on, for example, Cr(VI) will not distinguish between them and cannot be used. However, one of the two –OH groups (the one on the ring) is *allylic,* and (Section 17-4) can be oxidized *selectively* to an α,β-unsaturated ketone by MnO_2, leaving the other unchanged. So, (c) is MnO_2, acetone.

d. Now we are back to more straightforward chemistry—oxidation of a primary alcohol to a carboxylic acid. Any Cr(VI) reagent in water will do, such as $K_2Cr_2O_7$ in aqueous acid (Sections 8-6 and 19-6). Thus (d) is $K_2Cr_2O_7$, H_2SO_4, H_2O.

e. Two functional group changes are evident: The enone C–C double bond has been reduced and the carboxy group esterified. We have seen reductions of enones to saturated ketones (Section 18-8); catalytic hydrogenation is the simplest applicable method (a metal–NH_3 reduction is unnecessary, because selectivity is not relevant). Methyl ester formation may be achieved in either of two ways (Sections 19-8, 19-9, and 20-2): via the acyl halide (made with $SOCl_2$) and CH_3OH or by direct acid-catalyzed reaction with CH_3OH. Does the order of steps matter? Neither acids nor esters are affected by catalytic hydrogenation conditions, and esterifications may be carried out in the presence of either enones or ordinary ketones, so the answer is no. For (e), then, 1. H_2, Pd–C, CH_3CH_2OH and 2. CH_3OH, H^+.

f. Here is a curious pair of changes: The carbonyl moiety is protected as a cyclic acetal, and the ester function is hydrolyzed back to the –COOH group, which raises the question of why the latter was esterified in the first place. Most likely, the chemists working on this project tried to form the cyclic acetal on the molecule containing the free carboxy substituent and ran into difficulty. Protection of the carbonyl requires acid catalysis and 1,2-ethanediol (ethylene glycol, Section 17-8). Perhaps complications due to the formation of esters with the –COOH group were observed. In any event, esterification of the latter before acetalization apparently solved the problem. Thus, for (f), 1. $HOCH_2CH_2OH$, H^+ and 2. ^-OH, H_2O (ester hydrolysis *with base,* Section 20-4, to prevent hydrolysis of the acetal unit).

g. The ring carbonyl group has been deprotected, and the carboxy group transformed into a ketone. For the first time in the synthesis, a new carbon–carbon bond has been introduced, between the carboxy carbon and an ethyl group. This process will most likely require the use of an organometallic reagent, from which the ring carbonyl has been protected in sequence (f). From which carboxylic acid derivatives can ketones be prepared? Acyl halides (Section 20-2) and nitriles (Section 20-8). Since you don't know how to convert a carboxylic acid into its nitrile, we choose the following sequence for (g): 1. $SOCl_2$ (gives halide), 2. $(CH_3CH_2)_2CuLi$, ether (converts halo- into ethylcarbonyl), and 3. H^+, H_2O (deprotects ring carbonyl, Section 17-8).

h. Another carbon–carbon bond is made, forming a ring. We trace the pattern of carbon atoms before and after this step to identify which ones were connected (see margin). They are the ring carbonyl carbon and the CH_2 of the ethyl group, α to the side-chain carbonyl. We have here an intramolecular aldol condensation (Section 18-7), readily achieved by aqueous base. So (h), the final step, is ^-OH, H_2O.

Analysis of Step (h) Aldol Condensation

New Reactions

1. Order of Reactivity of Carboxylic Acid Derivatives (Section 20-1)

Esters and amides require acid or base catalysts to react with weak nucleophiles.

2. Basicity of the Carbonyl Oxygen (Section 20-1)

L = leaving group
Basicity increases with increasing contribution of resonance structure C.

3. Enolate Formation (Section 20-1)

Acidity of the neutral derivative generally increases with decreasing contribution of resonance structure C in the anion.

Reactions of Acyl Halides

4. Water (Section 20-2)

5. Carboxylic Acids (Sections 19-8 and 20-2)

6. Alcohols (Section 20-2)

$$\underset{\text{RCX}}{\overset{\overset{\displaystyle O}{\|}}{\text{RCX}}} + R'OH \longrightarrow \underset{\text{Ester}}{\overset{\overset{\displaystyle O}{\|}}{\text{RCOR}'}} + \quad HX$$

(Removed with pyridine, triethylamine, or other base)

7. Amines (Section 20-2)

$$\underset{\text{RCX}}{\overset{\overset{\displaystyle O}{\|}}{\text{RCX}}} + R'NH_2 \longrightarrow \underset{\text{Amide}}{\overset{\overset{\displaystyle O}{\|}}{\text{RCNHR}'}} + \quad HX$$

(Removed with pyridine, triethylamine, excess R′NH₂, or other base)

8. Cuprate Reagents (Section 20-2)

$$\underset{\text{RCX}}{\overset{\overset{\displaystyle O}{\|}}{\text{RCX}}} \xrightarrow[\text{2. H}^+, \text{H}_2\text{O}]{\text{1. R}'_2\text{CuLi, THF}} \underset{\text{Ketone}}{\overset{\overset{\displaystyle O}{\|}}{\text{RCR}'}} + R'Cu + LiX$$

9. Hydrides (Section 20-2)

$$\underset{\text{RCX}}{\overset{\overset{\displaystyle O}{\|}}{\text{RCX}}} \xrightarrow[\text{2. H}^+, \text{H}_2\text{O}]{\text{1. LiAl[OC(CH}_3)_3]_3\text{H, (CH}_3\text{CH}_2)_2\text{O}} \underset{\text{Aldehyde}}{\overset{\overset{\displaystyle O}{\|}}{\text{RCH}}} + LiX + Al[OC(CH_3)_3]_3$$

Reactions of Carboxylic Acid Anhydrides

10. Water (Section 20-3)

$$\underset{\text{RCOCR}}{\overset{\overset{\displaystyle O \quad O}{\| \quad \|}}{\text{RCOCR}}} + H_2O \longrightarrow \underset{\text{Carboxylic acid}}{2\ \overset{\overset{\displaystyle O}{\|}}{\text{RCOH}}}$$

11. Alcohols (Section 20-3)

$$\underset{\text{RCOCR}}{\overset{\overset{\displaystyle O \quad O}{\| \quad \|}}{\text{RCOCR}}} + R'OH \longrightarrow \underset{\text{Ester}}{\overset{\overset{\displaystyle O}{\|}}{\text{RCOR}'}} + \overset{\overset{\displaystyle O}{\|}}{\text{RCOH}}$$

12. Amines (Section 20-3)

$$\underset{\text{RCOCR}}{\overset{\overset{\displaystyle O \quad O}{\| \quad \|}}{\text{RCOCR}}} + R'NH_2 \longrightarrow \underset{\text{Amide}}{\overset{\overset{\displaystyle O}{\|}}{\text{RCNHR}'}} + \overset{\overset{\displaystyle O}{\|}}{\text{RCOH}}$$

Reactions of Esters

13. Water (Ester Hydrolysis) (Sections 19-9 and 20-4)

Acid catalysis

$$\underset{\text{RCOR}'}{\overset{\overset{\displaystyle O}{\|}}{\text{RCOR}'}} + H_2O \xrightarrow{\text{Catalytic H}^+} \underset{\text{Carboxylic acid}}{\overset{\overset{\displaystyle O}{\|}}{\text{RCOH}}} + R'OH$$

Base catalysis

$$\underset{\underset{\textbf{1 equivalent}}{}}{\overset{\overset{O}{\|}}{RCOR'}} + \ ^{-}OH \xrightarrow{H_2O} \underset{\textbf{Carboxylate ion}}{\overset{\overset{O}{\|}}{RCO^-}} + R'OH$$

14. Alcohols (Transesterification) and Amines (Section 20-4)

$$\overset{\overset{O}{\|}}{RCOR'} + R''OH \xrightarrow{H^+ \ or \ ^-OR''} \underset{\textbf{Ester}}{\overset{\overset{O}{\|}}{RCOR''}} + R'OH$$

$$\overset{\overset{O}{\|}}{RCOR'} + R''NH_2 \xrightarrow{\Delta} \underset{\textbf{Amide}}{\overset{\overset{O}{\|}}{RCNHR''}} + R'OH$$

15. Organometallic Reagents (Section 20-4)

$$\overset{\overset{O}{\|}}{RCOR''} \xrightarrow[\text{2. } H^+, H_2O]{\text{1. 2 } R'MgX, (CH_3CH_2)_2O} \underset{\underset{R'}{\overset{|}{\underset{|}{}}}}{\overset{OH}{\underset{|}{R-\overset{|}{\underset{|}{C}}-R'}}} + R''OH$$

Tertiary alcohol

Methyl formate

$$\overset{\overset{O}{\|}}{HCOCH_3} \xrightarrow[\text{2. } H^+, H_2O]{\text{1. 2 } R'MgX, (CH_3CH_2)_2O} \underset{\underset{R'}{\overset{|}{}}}{\overset{OH}{\underset{|}{H-\overset{|}{\underset{|}{C}}-R'}}} + CH_3OH$$

Secondary alcohol

16. Hydrides (Section 20-4)

$$\overset{\overset{O}{\|}}{RCOR'} \xrightarrow[\text{2. } H^+, H_2O]{\text{1. LiAlH}_4, (CH_3CH_2)_2O} \underset{\textbf{Alcohol}}{RCH_2OH}$$

$$\overset{\overset{O}{\|}}{RCOR'} \xrightarrow[\text{2. } H^+, H_2O]{\text{1. } (CH_3\overset{\overset{CH_3}{|}}{CH}CH_2)_2AlH, \text{ toluene, } -60°C} \underset{\textbf{Aldehyde}}{\overset{\overset{O}{\|}}{RCH}}$$

17. Enolates (Section 20-4)

$$\overset{\overset{O}{\|}}{RCH_2COR'} \xrightarrow{\text{LDA, THF}} \left[\overset{\overset{:O:}{\|}}{R\overset{..}{C}H-COR'} \longleftrightarrow \overset{\overset{^-:\overset{..}{O}:}{|}}{RCH=COR'} \right] \xrightarrow{R''X} \underset{}{\overset{\overset{R''}{|}\ \overset{O}{\|}}{RCHCOR'}}$$

Ester enolate ion

Reactions of Amides

18. Water (Section 20-6)

$$\overset{\overset{O}{\|}}{RCNHR'} + H_2O \xrightarrow{H^+, \Delta} \underset{\substack{\textbf{Carboxylic} \\ \textbf{acid}}}{\overset{\overset{O}{\|}}{RCOH}} + R'\overset{+}{N}H_3$$

$$\overset{\overset{O}{\|}}{RCNHR'} + H_2O \xrightarrow{HO^-, \Delta} \overset{\overset{O}{\|}}{RCO^-} + R'NH_2$$

19. Hydrides (Section 20-6)

$$\underset{\text{RCNHR}'}{\overset{\overset{\displaystyle O}{\|}}{}} \xrightarrow[\text{2. H}^+, \text{H}_2\text{O}]{\text{1. LiAlH}_4, \text{(CH}_3\text{CH}_2)_2\text{O}} \underset{\text{Amine}}{\text{RCH}_2\text{NHR}'}$$

$$\underset{\text{RCNHR}'}{\overset{\overset{\displaystyle O}{\|}}{}} \xrightarrow[\text{2. H}^+, \text{H}_2\text{O}]{\overset{\overset{\displaystyle CH_3}{|}}{\text{1. (CH}_3\text{CHCH}_2)_2\text{AlH}, \text{(CH}_3\text{CH}_2)_2\text{O}}} \underset{\text{Aldehyde}}{\overset{\overset{\displaystyle O}{\|}}{\text{RCH}}}$$

20. Enolates and Amidates (Section 20-7)

$$\underset{\underset{\text{p}K_a \approx 30}{\uparrow}}{\overset{\overset{\displaystyle :O:}{\|}}{\text{RCH}_2\text{CNR}'_2}} \xrightarrow{\text{Base}} \underset{\text{Amide enolate ion}}{\text{RCH}=\overset{\overset{\displaystyle \cdot\ddot{O}:^-}{}}{\underset{\underset{\displaystyle NR'_2}{}}{C}}}$$

$$\underset{\underset{\text{p}K_a \approx 22}{\uparrow}}{\overset{\overset{\displaystyle :O:}{\|\cdot\cdot}}{\text{RCH}_2\text{CNHR}'}} \xrightarrow{\text{Base}} \underset{\text{Amidate ion}}{\overset{\overset{\displaystyle :\ddot{O}:^-}{}}{\text{RCH}_2\text{C}=\ddot{N}R'}}$$

21. Hofmann Rearrangement (Section 20-7)

$$\underset{\text{RCNH}_2}{\overset{\overset{\displaystyle O}{\|}}{}} \xrightarrow{\text{Br}_2, \text{NaOH}, \text{H}_2\text{O}, 75°\text{C}} \underset{\text{Amine}}{\text{RNH}_2} + \text{CO}_2$$

Reactions of Nitriles

22. Water (Section 20-8)

$$\text{RC}\equiv\text{N} + \text{H}_2\text{O} \xrightarrow{\text{H}^+ \text{ or HO}^-, \Delta} \underset{\text{Amide}}{\overset{\overset{\displaystyle O}{\|}}{\text{RCNH}_2}} \xrightarrow{\text{H}^+ \text{ or HO}^-, \Delta} \underset{\text{Carboxylic acid}}{\overset{\overset{\displaystyle O}{\|}}{\text{RCOH}}}$$

23. Organometallic Reagents (Section 20-8)

$$\text{RC}\equiv\text{N} \xrightarrow[\text{2. H}^+, \text{H}_2\text{O}]{\text{1. R}'\text{MgX or R}'\text{Li}} \underset{\text{Ketone}}{\overset{\overset{\displaystyle O}{\|}}{\text{RCR}'}}$$

24. Hydrides (Section 20-8)

$$\text{RC}\equiv\text{N} \xrightarrow[\text{2. H}^+, \text{H}_2\text{O}]{\text{1. LiAlH}_4} \underset{\text{Amine}}{\text{RCH}_2\text{NH}_2}$$

$$\text{RC}\equiv\text{N} \xrightarrow[\text{2. H}^+, \text{H}_2\text{O}]{\overset{\overset{\displaystyle CH_3}{|}}{\text{1. (CH}_3\text{CHCH}_2)_2\text{AlH}}} \underset{\text{Aldehyde}}{\overset{\overset{\displaystyle O}{\|}}{\text{RCH}}}$$

25. Catalytic Hydrogenation (Section 20-8)

$$\text{RC}\equiv\text{N} \xrightarrow{\text{H}_2, \text{PtO}_2, \text{CH}_3\text{CH}_2\text{OH}} \underset{\text{Amine}}{\text{RCH}_2\text{NH}_2}$$

Important Concepts

1. The **electrophilic reactivity** of the carbonyl carbon in **carboxylic acid derivatives** is weakened by good electron-donating substituents. This effect, measurable by IR spectroscopy, is responsible not only for the decrease in the reactivity with nucleophiles and acid, but also for the increased basicity along the series: acyl halides–anhydrides–esters–amides. Electron donation by resonance from the nitrogen in amides is so pronounced that there is **hindered rotation** about the amide bond on the NMR time scale.

2. **Carboxylic acid derivatives** are named as **acyl halides, carboxylic anhydrides, alkyl alkanoates, alkanamides,** and **alkanenitriles,** depending on the functional group.

3. **Carbonyl stretching frequencies** in the IR spectra are diagnostic of the carboxylic acid derivatives: Acyl chlorides absorb at 1790–1815 cm^{-1}, anhydrides at 1740–1790 and 1800–1850 cm^{-1}, esters at 1735–1750 cm^{-1}, and amides at 1650–1690 cm^{-1}.

4. Carboxylic acid derivatives generally react with **water** (under acid or base catalysis) to hydrolyze to the corresponding carboxylic acid; they combine with **alcohols** to give esters and with **amines** to furnish amides. With **Grignard** and other **organometallic reagents,** they form ketones; esters may react further to form the corresponding alcohols. Reduction by **hydrides** gives products in various oxidation states: aldehydes, alcohols, or amines.

5. Long-chain esters are the constituents of animal and plant **waxes.** Triesters of glycerol are contained in natural **oils** and **fats.** Their hydrolysis gives **soaps. Triglycerides** containing phosphoric acid ester subunits belong to the class of **phospholipids.** Because they carry a highly polar head group and hydrophobic tails, phospholipids form **micelles** and **lipid bilayers.**

6. **Transesterification** can be used to convert one ester into another.

7. The functional group of **nitriles** is somewhat similar to that of the alkynes. The two component atoms are sp hybridized. The IR stretching vibration appears at about 2250 cm^{-1}. The hydrogens next to the cyano group are deshielded in ^1H NMR. The ^{13}C NMR absorptions for nitrile carbons are at relatively low field ($\delta \sim 112$–126 ppm), a consequence of the electronegativity of nitrogen.

Problems

30. Name (IUPAC system) or draw the structure of each of the following compounds.

(a)

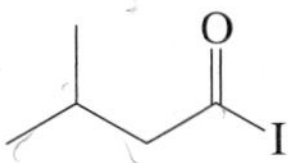

(b)

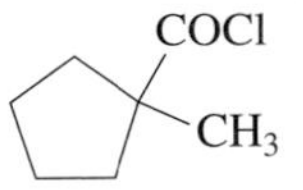

(c) CF_3COCCF_3

(d)

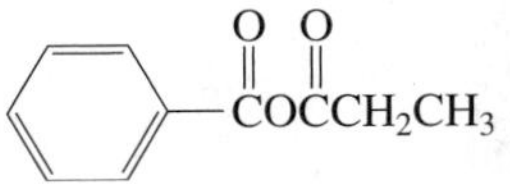

(e) $(CH_3)_3CCOCH_2CH_3$

(f) 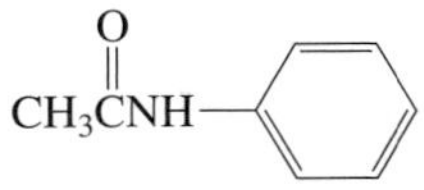 CH_3CNH

(g) Propyl butanoate
(h) Butyl propanoate
(i) 2-Chloroethyl benzoate
(j) N,N-Dimethylbenzamide
(k) 2-Methylhexanenitrile
(l) Cyclopentanecarbonitrile

31. Name each of the following compounds according to IUPAC or *Chemical Abstracts* rules. Pay attention to the order of precedence of the functional groups.

(a)

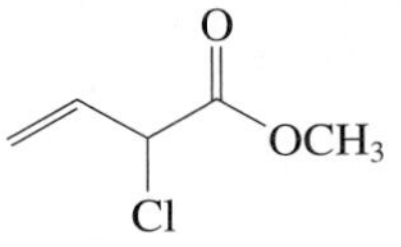

(b)

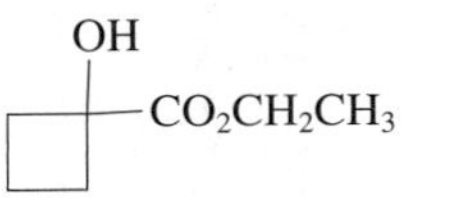

(c)

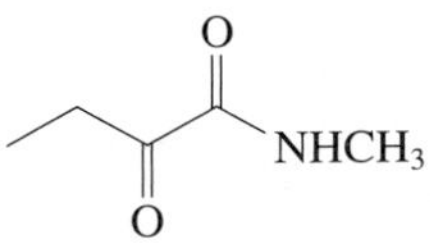

(d)

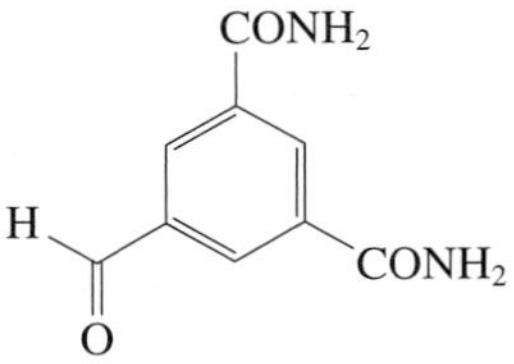

32. **(a)** Use resonance forms to explain in detail the relative order of acidity of carboxylic acid derivatives, as presented in Section 20-1. **(b)** Do the same, but use an argument based on inductive effects.

33. In each of the following pairs of compounds, decide which possesses the indicated property to the greater degree. **(a)** Length of C–X bond: acetyl fluoride or acetyl chloride. **(b)** Acidity of the boldface H: $CH_2(COCH_3)_2$ or $CH_2(COOCH_3)_2$. **(c)** Reactivity toward addition of a nucleophile: (i) an amide or (ii) an imide (as shown below). **(d)** High-energy infrared carbonyl stretching frequency: ethyl acetate or ethenyl acetate.

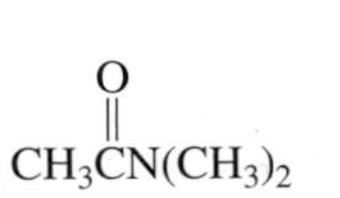

i

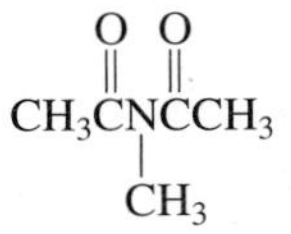

ii

34. Give the product(s) of each of the following reactions.

(a)

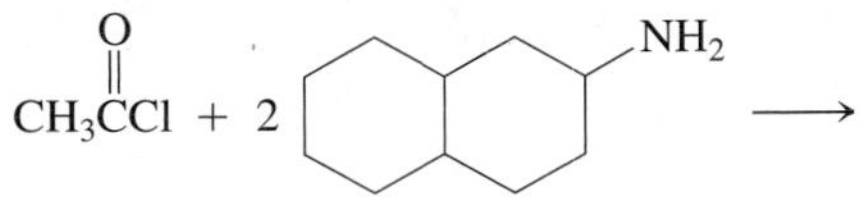

(b)

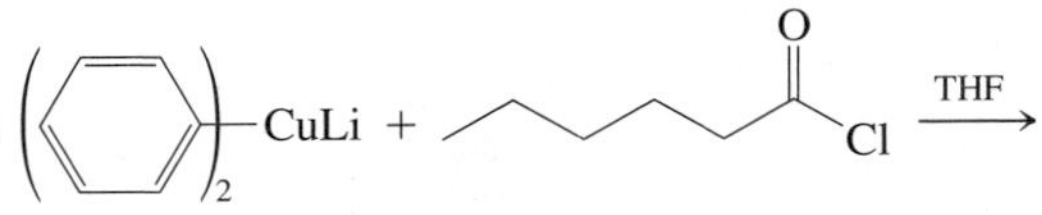

(c)

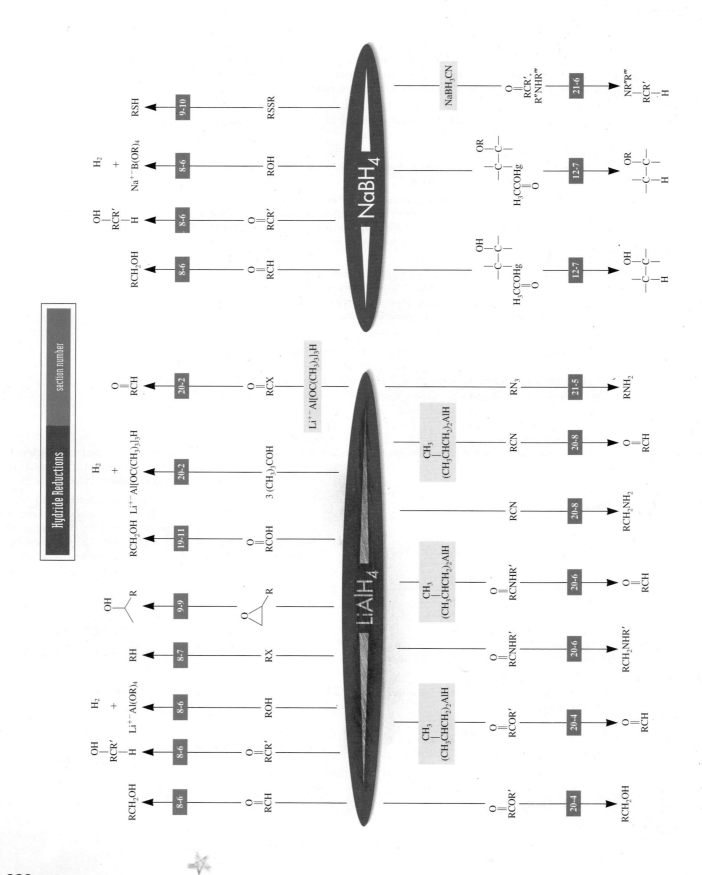

(d)

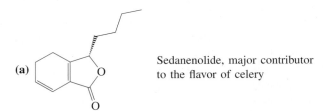

CH$_2$CCl / CH$_2$CCl (anhydride with two C=O) + 2 [benzyl alcohol C$_6$H$_5$CH$_2$OH], pyridine N →

OCH$_3$

H$_3$C

(e) H$_3$C, CCl, C=O $\xrightarrow{\text{LiAl[OC(CH}_3)_3]_3\text{H, THF, }-78°\text{C}}$

35. Formulate a mechanism for the reaction of acetyl chloride with 1-propanol shown on p. 892.

36. Give the product(s) of the reactions of acetic anhydride with each of the following reagents. Assume in all cases that the reagent is present in excess.

(a) (CH$_3$)$_2$CHOH

(b) NH$_3$

(c) [phenyl]—MgBr, THF; then H$^+$, H$_2$O

(d) LiAlH$_4$, (CH$_3$CH$_2$)$_2$O; then H$^+$, H$_2$O

37. Give the product(s) of the reaction of butanedioic (succinic) anhydride with each of the reagents in Problem 36.

38. Formulate a mechanism for the reaction of butanedioic (succinic) anhydride with methanol shown on p. 895.

39. Give the products of reaction of methyl pentanoate with each of the following reagents under the conditions shown.

(a) NaOH, H$_2$O, heat; then H$^+$, H$_2$O
(b) (CH$_3$)$_2$CHCH$_2$CH$_2$OH (excess), H$^+$
(c) (CH$_3$CH$_2$)$_2$NH, heat
(d) CH$_3$MgI (excess), (CH$_3$CH$_2$)$_2$O; then H$^+$, H$_2$O
(e) LiAlH$_4$, (CH$_3$CH$_2$)$_2$O; then H$^+$, H$_2$O
(f) [(CH$_3$)$_2$CHCH$_2$]$_2$AlH, toluene, low temperature; then H$^+$, H$_2$O

40. Give the products of reaction of γ-valerolactone (5-methyloxa-2-cyclopentanone, Section 20-4) with each of the reagents in Problem 39.

41. Draw the structure of each of the following compounds:
(a) β-butyrolactone; **(b)** β-valerolactone; **(c)** δ-valerolactone;
(d) β-propiolactam; **(e)** α-methyl-δ-valerolactam; **(f)** N-methyl-γ-butyrolactam.

42. Formulate a mechanism for the acid-catalyzed transesterification of ethyl 2-methylpropanoate (ethyl isobutyrate) into the corresponding methyl ester. Your mechanism should clearly illustrate the catalytic role of the proton.

43. Formulate a mechanism for the reaction of methyl 9-octadecenoate with 1-dodecanamine shown on p. 900.

44. Reaction review. Suggest reagents to convert each of the following starting materials into the indicated product: **(a)** acetyl chloride into acetic hexanoic anhydride; **(b)** methyl hexanoate

into N-methylhexanamide; **(c)** hexanoyl chloride into hexanal; **(d)** hexanenitrile into hexanoic acid; **(e)** hexanamide into hexanamine; **(f)** hexanamide into pentanamine; **(g)** ethyl hexanoate into 3-ethyl-3-octanol; **(h)** hexanenitrile into 1-phenyl-1-hexanone [C$_6$H$_5$CO(CH$_2$)$_4$CH$_3$].

45. Give the product of each of the following reactions.

(a) *ester* [bicyclic ring with] COOCH$_3$ / COOCH$_3$ $\xrightarrow{\text{1. KOH, H}_2\text{O}}_{\text{2. H}^+, \text{ H}_2\text{O}}$

(b) H$_3$C, O / CH$_3$CH$_2$ [lactone] =O $\xrightarrow{\text{(CH}_3)_2\text{CHNH}_2, \text{ CH}_3\text{OH, }\Delta}$

(c) CH$_3$COCH$_3$ + excess [cyclopentyl]—MgBr $\xrightarrow{\text{1. (CH}_3\text{CH}_2)_2\text{O, 20°C}}_{\text{2. H}^+, \text{ H}_2\text{O}}$

(d) [phenyl]—CH(CH$_3$)COOCH$_2$CH$_3$ $\xrightarrow{\substack{\text{1. LDA, THF, }-78°\text{C} \\ \text{2. CH}_3\text{I, HMPA} \\ \text{3. H}^+, \text{ H}_2\text{O}}}$

(e) [cyclohexyl]—COOCH$_3$ (with CH$_3$) $\xrightarrow{\text{1. (CH}_3\text{CHCH}_2)_2\text{AlH, toluene, }-60°\text{C}}_{\text{2. H}^+, \text{ H}_2\text{O}}$

46. For each of the naturally occurring lactones below, draw the structure of the compound that would result from hydrolysis using aqueous base.

(a) [bicyclic lactone structure] Sedanenolide, major contributor to the flavor of celery

(b) [bicyclic lactone structure with H markings] Nepetalactone, major active component in catnip (**Caution:** Do you notice anything unusual about one of the functional groups after hydrolysis?)

(c) [lactone ring structure] γ-Valerolactone, important in the fragrance industry and a potential biofuel

(d) A novel "green" process converts levulinic acid (4-oxopenta-noic acid, CH$_3$COCH$_2$CH$_2$COOH), which is readily obtained from waste biomass, into γ-valerolactone by treatment with gaseous H$_2$ in the presence of a special hydrogenation catalyst. Speculate on how this reaction may take place.

47. N,N-Diethyl-3-methylbenzamide (N,N-diethyl-m-toluamide), marketed as DEET, is perhaps the most widely used insect repellent in the world and is especially effective in interrupting the transmission of diseases by mosquitoes and ticks. Propose one or more preparations of DEET (see structure below) from 3-methylbenzoic acid and any other appropriate reagents.

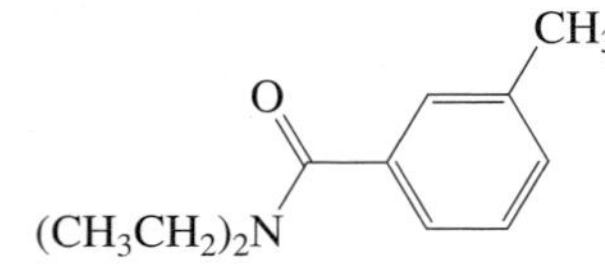

DEET

48. Like many chiral pharmaceuticals, Ritalin, widely used to treat attention deficit disorder, is synthesized and marketed as a racemic mixture. In 2003, a new and potentially practical synthetic route to the active enantiomer (see structure below) was described. The first step is shown below, the reaction of dimethyl sulfate with a six-membered ring lactam. Propose a mechanism for this reaction.

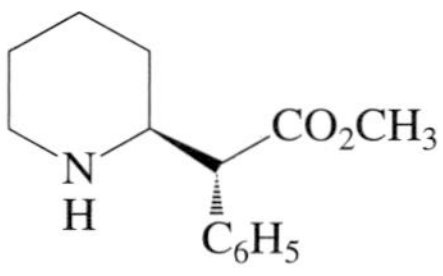

Ritalin
(Active stereoisomer)

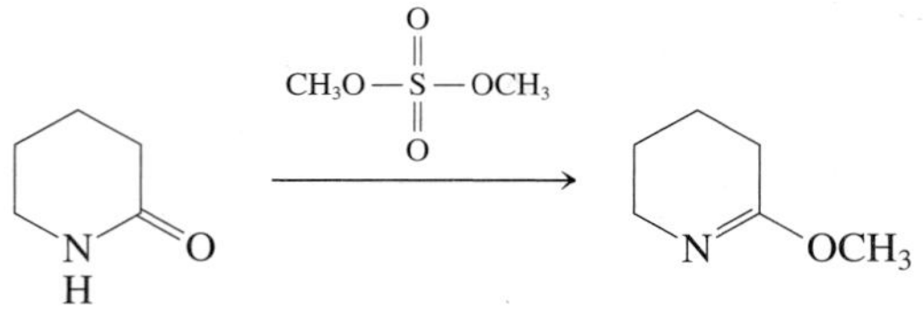

49. Formulate a mechanism for the formation of acetamide, $CH_3\overset{O}{\overset{\|}{C}}NH_2$, from methyl acetate and ammonia.

50. Give the products of the reactions of pentanamide with the reagents given in Problem 39, parts (a, e, f). Repeat for *N,N*-dimethylpentanamide.

51. Formulate a mechanism for the acid-catalyzed hydrolysis of 3-methylpentanamide shown on p. 907. (**Hint:** Use as a model the mechanism for general acid-catalyzed addition–elimination presented in Section 19-7.)

52. What reagents would be necessary to carry out the following transformations? (**a**) Cyclohexanecarbonyl chloride → pentanoyl-cyclohexane; (**b**) 2-butenedioic (maleic) anhydride → (*Z*)-2-butene-1,4-diol; (**c**) 3-methylbutanoyl bromide → 3-methyl-butanal; (**d**) benzamide → 1-phenylmethanamine; (**e**) propaneni-trile → 3-hexanone; (**f**) methyl propanoate → 4-ethyl-4-heptanol.

53. In each of the following retrosynthetic disconnections (⌇), show reactions that would make the indicated C–C bonds.

(**a**) 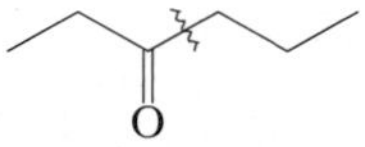
(Two reactions)

(**b**)

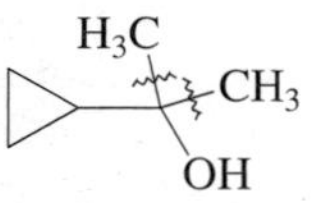

(**c**)

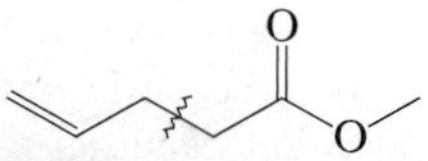

54. Upon treatment with LDA followed by protonation, compounds A and B undergo cis-trans isomerization, but compound C does not. Explain.

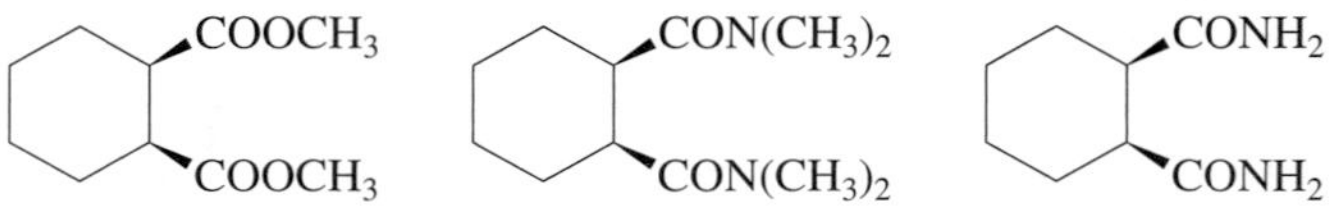

| A | B | C |

55. 2-Aminobenzoic (anthranilic) acid is prepared from 1,2-ben-zenedicarboxylic anhydride (phthalic anhydride) by using the two reactions shown here. Explain these processes mechanistically.

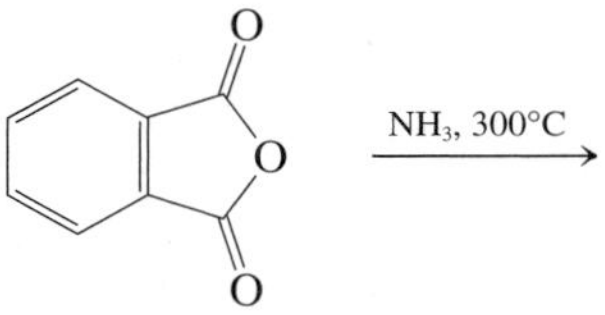

1,2-Benzenedicarboxylic
anhydride
(Phthalic anhydride)

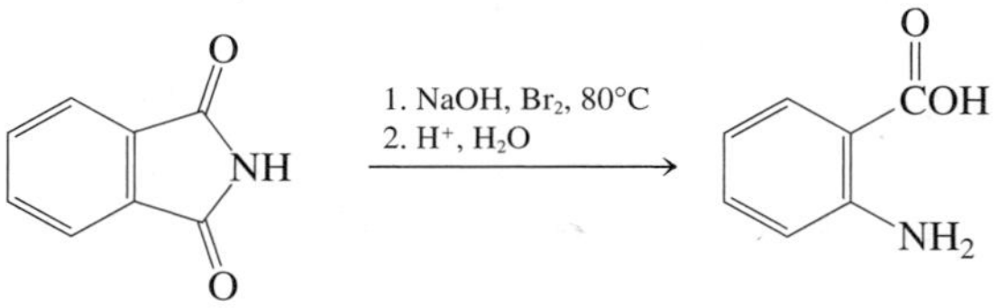

1,2-Benzenedicarboximide   **2-Aminobenzoic acid**
(Phthalimide)   **(Anthranilic acid)**

56. On the basis of the reactions presented in this chapter, write re-action summary charts for esters and amides similar to the chart for acyl halides (Figure 20-1). Compare the number of reactions for each of the compound classes. Is this information consistent with your understanding of the relative reactivity of each of the functional groups?

57. Show how you might synthesize chlorpheniramine (see also p. 700, margin), a powerful antihistamine used in several decon-gestants, from each of carboxylic acids A and B. Use a different carboxylic amide in each synthesis.

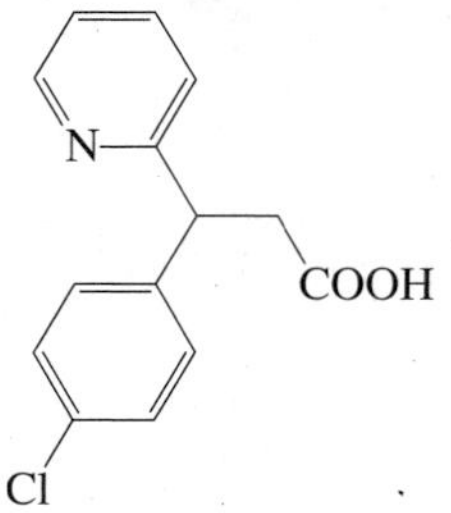

A 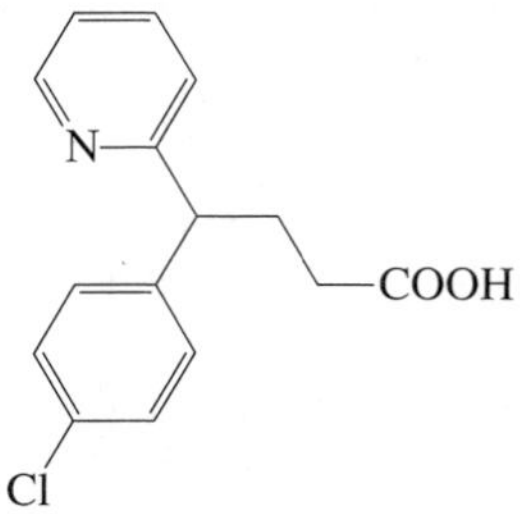   B

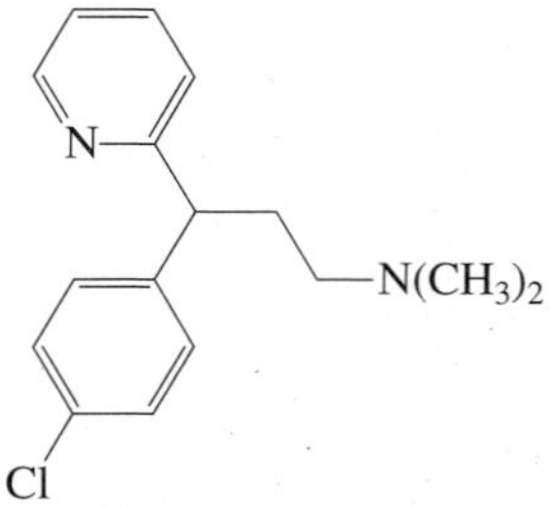

Chlorpheniramine

58. Although esters typically have carbonyl stretching frequencies at about 1740 cm^{-1} in the infrared spectrum, the corresponding value for lactones can vary greatly with ring size. Three examples are shown below. Propose an explanation for the increasing energies of the carbonyl stretching vibrations with decreasing ring size of the lactone.

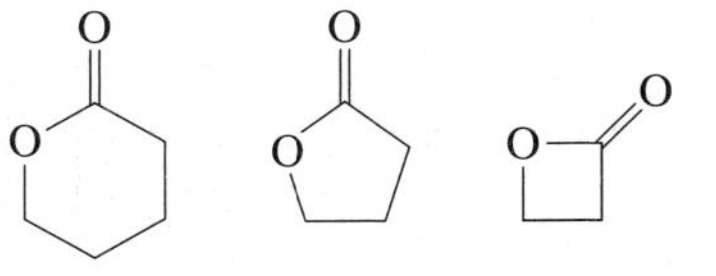

1735 cm^{-1} 1770 cm^{-1} 1840 cm^{-1}

59. Upon completing a synthetic procedure, every chemist is faced with the job of cleaning glassware. Because the compounds present may be dangerous in some way or have unpleasant properties, a little serious chemical thinking is often beneficial before "doing the dishes." Suppose that you have just completed a synthesis of hexanoyl chloride, perhaps to carry out the reaction in Problem 34, part (b); first, however, you must clean the glassware contaminated with this acyl halide. Both hexanoyl chloride and hexanoic acid have terrible odors. **(a)** Would cleansing the glassware with soap and water be a good idea? Explain. **(b)** Suggest a more agreeable alternative, based on the chemistry of acyl halides and the physical properties (particularly the odors) of the various carboxylic acid derivatives.

60. **CHALLENGE**   Show how you would carry out the following transformation in which the ester function at the lower left of the molecule is converted into a hydroxy group but that at the upper right is preserved. (**Hint:** Do not try ester hydrolysis. Look carefully at how the ester groups are linked to the steroid and consider an approach based on transesterification.)

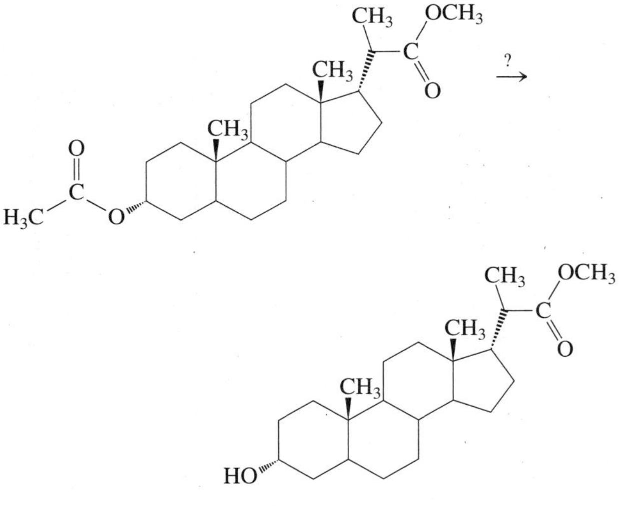

61. The removal of the C17 side chain of certain steroids is a critical element in the synthesis of a number of hormones, such as testosterone, from steroids in the relatively readily available pregnane family.

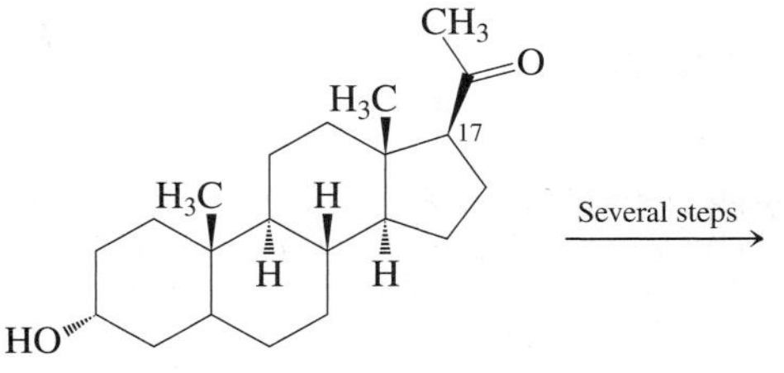

Pregnan-3α-ol-20-one

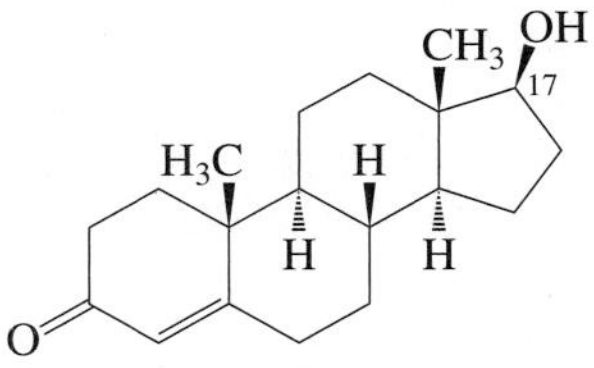

Testosterone

How would you carry out the comparable transformation, shown below, of acetylcyclopentane into cyclopentanol? (**Note:** In this and subsequent synthetic problems, you may need to use reactions from several areas of carbonyl chemistry discussed in Chapters 17–20.)

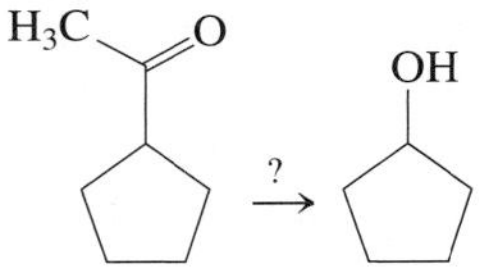

62. Propose a synthetic sequence to convert carboxylic acid A into the naturally occurring sesquiterpene α-curcumene.

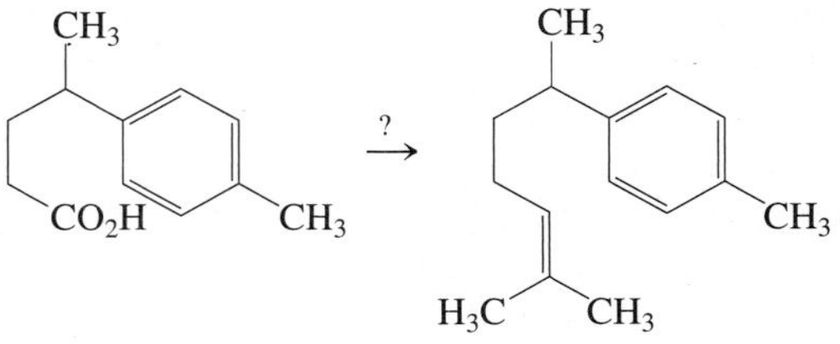

A   α-Curcumene

63. Propose a synthetic scheme for the conversion of lactone A into amine B, a precursor to the naturally occurring monoterpene C.

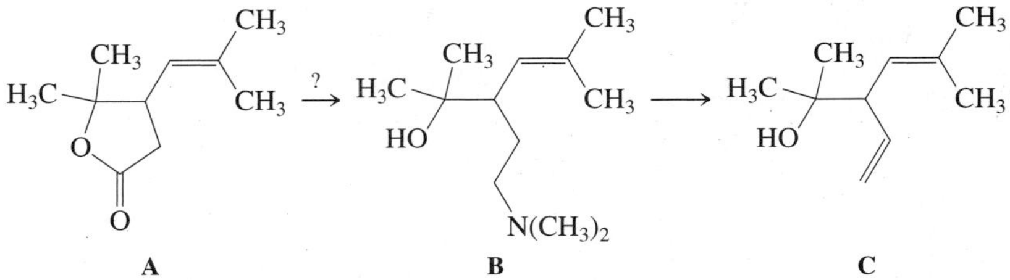

A   B   C

64. **CHALLENGE** Propose a synthesis of β-selinene, a member of a very common family of sesquiterpenes, beginning with the alcohol shown here. Use a nitrile in your synthesis. Inspection of a model may help you choose a way to obtain the desired stereochemistry. (Is the 1-methylethenyl group axial or equatorial?)

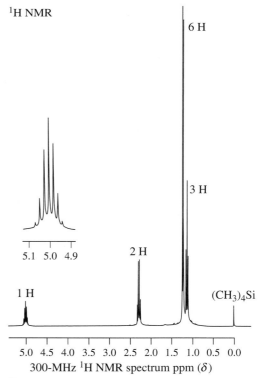

β-**Selinene**

65. Give the structure of the product of the first of the following reactions, and then propose a scheme that will ultimately convert it into the methyl-substituted ketone at the end of the scheme. This example illustrates a common method for the introduction of methyl groups into synthetically prepared steroids. (**Hint:** It will be necessary to protect the carbonyl function.)

$$\xrightarrow{\text{HCN}} \quad C_{11}H_{15}NO \xrightarrow{?}$$

IR: 1715, 2250 cm^{-1}

66. **CHALLENGE** Spectroscopic data for two carboxylic acid derivatives are given in NMR-A and NMR-B. Identify these compounds, which may contain C, H, O, N, Cl, and Br but no other elements. (**a**) ^{1}H NMR: spectrum A (one signal has been amplified to reveal all peaks in the multiplet). IR: 1728 cm^{-1}. High-resolution mass spectrum: m/z for the molecular ion is 116.0837.

See table for important MS fragmentation peaks. (**b**) ^{1}H NMR: spectrum B. IR: 1739 cm^{-1}. High-resolution mass spectrum: The intact molecule gives two peaks with almost equal intensity: $m/z = 179.9786$ and 181.9766. See table for important MS fragmentation peaks.

Mass Spectrum of Unknown A	
m/z	Intensity relative to base peak (%)
116	0.5
101	12
75	26
57	100
43	66
29	34

Mass Spectrum of Unknown B	
m/z	Intensity relative to base peak (%)
182	13
180	13
109	78
107	77
101	3
29	100

^{1}H NMR

3 H

3 H

4.4 4.3 4.2

2 H

1 H

(CH$_3$)$_4$Si

5.0 4.5 4.0 3.5 3.0 2.5 2.0 1.5 1.0 0.5 0.0

300-MHz ^{1}H NMR spectrum ppm (δ)

A

^{1}H NMR

6 H

3 H

5.1 5.0 4.9

2 H

1 H

(CH$_3$)$_4$Si

5.0 4.5 4.0 3.5 3.0 2.5 2.0 1.5 1.0 0.5 0.0

300-MHz ^{1}H NMR spectrum ppm (δ)

B

Team Problem

67. Friedel-Crafts acylations are best carried out with acyl halides, but other carboxylic acid derivatives undergo this process, too, such as carboxylic anhydrides or esters. These reagents may have some drawbacks, however, the subject of this problem.

   Before you start, discuss as a group the mechanisms for forming acylium ions from acyl halides and carboxylic anhydrides in Section 15-13. Then divide your group in two and analyze the outcome of the following two reactions. Use the NMR spectral data given to confirm your product assignments. (**Hint:** D is formed via C.)

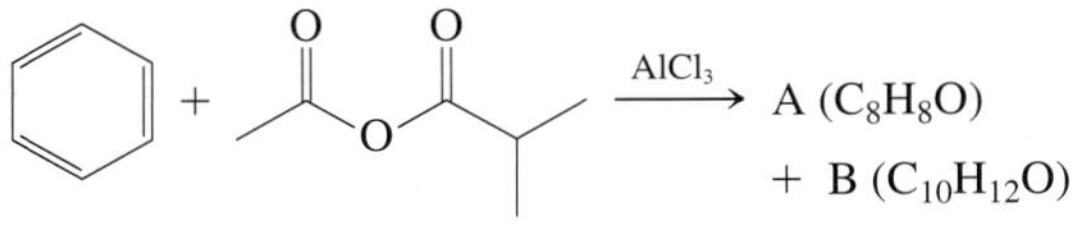

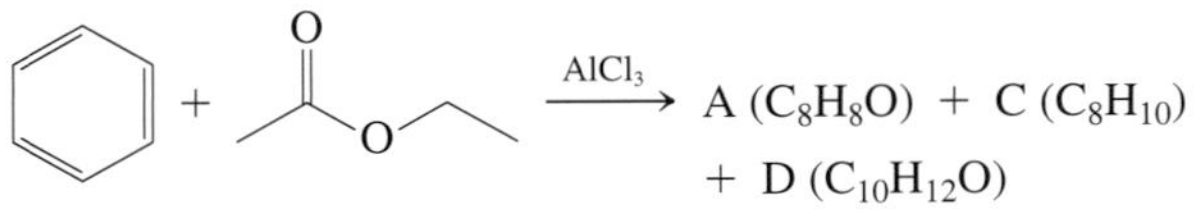

Compound A: ^{1}H NMR: δ 2.60 (s, 3H), 7.40–7.50 (m, 2H), 7.50–7.60 (m, 1H), 7.90–8.00 (m, 2H) ppm.

Compound B: ^{1}H NMR: δ 2.22 (d, 6H), 3.55 (sep, 1H), 7.40–7.50 (m, 2H), 7.50–7.60 (m, 1H), 7.90–8.00 (m, 2H) ppm.

Compound C: ^{1}H NMR: δ 1.20 (t, 3H), 2.64 (q, 2H), 7.10–7.30 (m, 5H) ppm.

Compound D: ^{1}H NMR: δ 1.25 (t, 3H), 2.57 (s, 3H), 2.70 (q, 2H), 7.20 (d, 2H), 7.70 (d, 2H) ppm.

   Reconvene to share your solutions. Then specifically address the nature of the complications ensuing when using the reagents shown. Finally, all together, consider the following reaction sequence. Again, use a mechanistic approach to arrive at the structures of the products.

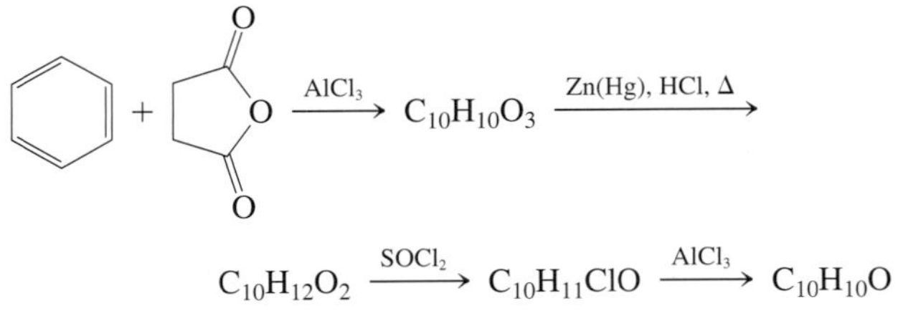

Preprofessional Problems

68. What is the IUPAC name of the compound shown below?
   (a) Isopropyl 2-fluoro-3-methylbutanoate; (b) 2-fluoroisobutanoyl 2-propanoate; (c) 1-methylethyl 2-fluorobutyrate; (d) 2-fluoroisopropyl isopropanoate; (e) 1-methylethyl 2-fluoro-2-methyl-propanoate.

$$(CH_3)_2\overset{\overset{\displaystyle F}{|}}{C}CO_2CH(CH_3)_2$$

69. Saponification of $(CH_3)_2CH\overset{\overset{\displaystyle O}{||}}{C}{}^{18}OCH_2CH_2CH_3$ with aqueous NaOH will give

   (a) $(CH_3)_2CHCO_2{}^-Na^+ + CH_3CH_2CH_2{}^{18}OH$;

   (b) $(CH_3)_2CH\overset{\overset{\displaystyle O}{||}}{C}{}^{18}O^-Na^+ + CH_3CH_2CH_2OH$;

   (c) $(CH_3)_2CHOCH_2CH_2CH_3 + C\equiv{}^{18}O$;

   (d) $(CH_3)_2CHCHO + CH_3CH_2CH_2{}^{18}OH$.

70. The best description for compound A (see below) is (a) an amide; (b) a lactam; (c) an ether; (d) a lactone.

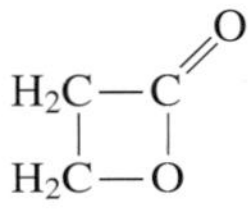

A

71. Which of the three compounds below would be the most reactive toward hydrolysis with aqueous base?

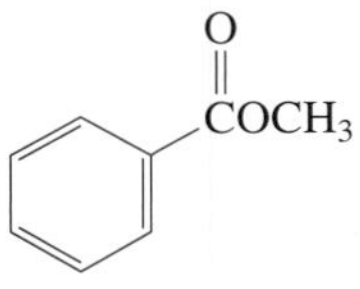

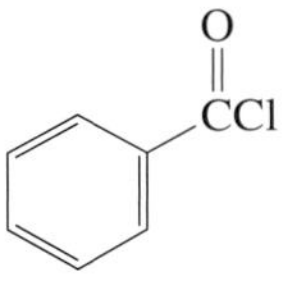

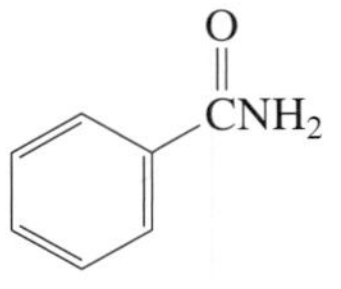

CHAPTER 21

Amines and Their Derivatives

Functional Groups Containing Nitrogen

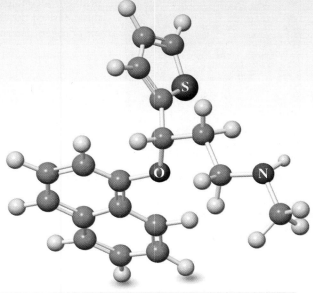

O ur atmosphere consists of about one-fifth oxygen, O_2, and nearly four-fifths nitrogen, N_2. We are fully aware of the importance of the oxygen part: We need it to breathe and nature uses it abundantly in water, alcohols, ethers, and many other organic and inorganic molecules. What about the nitrogen component? Unlike O_2, which is ultimately the reactive ingredient in biological oxidations, N_2 itself is relatively inert. However, in its reduced form of ammonia, NH_3, and its organic derivatives, the amines, it plays as active a role in nature as oxygen. Thus, amines and other nitrogen-bearing compounds are among the most abundant organic molecules. As components of the amino acids, peptides, proteins, and alkaloids, they are essential to biochemistry. Many amines, such as the neurotransmitters, possess powerful physiological activity; related substances have found medicinal uses as decongestants, anesthetics, sedatives, and stimulants (Real Life 21-1). Similar activity is found in cyclic amines in which nitrogen is part of a ring, the nitrogen heterocycles (Chapter 25).

In many respects, the chemistry of the amines is analogous to that of the alcohols and ethers (Chapters 8 and 9). For example, all amines are basic (although primary and secondary amines can also behave as acids), they form hydrogen bonds, and they act as nucleophiles in substitution reactions. However, there are some differences in reactivity, because nitrogen is less electronegative than oxygen. Thus, primary and secondary amines are less acidic and form weaker hydrogen bonds than alcohols, and they are more basic and more nucleophilic. This chapter will show that these properties underlie their physical and chemical characteristics and give us a variety of ways to synthesize amines.

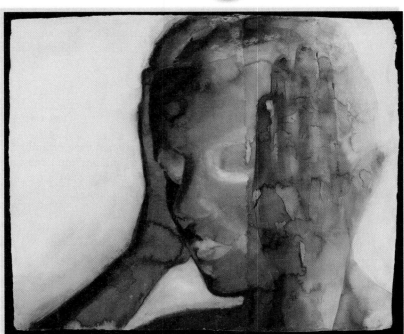

White Noise, by Graham Dean (b. 1951; English painter).

Cymbalta [duloxetine; (+)-(S)-N-methyl-3-(1-naphthalenyloxy)-3-(2-thiophenyl)propan-1-amine] was the top-selling drug in 2010 for the treatment of depression and anxiety (Table 25-1). The presence of the amino function is essential for its activity.

Although higher organisms cannot activate nitrogen by reduction to ammonia, some microorganisms do. Thus, the nodules in the root system of the soybean are the sites of nitrogen reduction by *Rhizobium* bacteria.

21-1 | NAMING THE AMINES

Amines are derivatives of ammonia, in which one, two, or three of the hydrogens have been replaced by alkyl or aryl groups. Therefore, amines are related to ammonia in the same sense as ethers and alcohols are related to water. Note, however, that the designations primary, secondary, and tertiary (see below) are used in a different way. In alcohols, ROH, the nature of the R group defines this designation; in amines, the number of R substituents on nitrogen determines the amine classification.

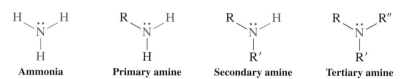

The system for naming amines is confused by the variety of common names in the literature. Probably the best way to name aliphatic amines is that used by *Chemical Abstracts*—that is, as **alkanamines,** in which the name of the alkane stem is modified by replacing the ending -*e* by **-amine.** The position of the functional group is indicated by a prefix designating the carbon atom to which it is attached, as in the alcohols (Section 8-1).

$$CH_3NH_2$$
Methanamine

$$CH_3CHCH_2NH_2 \\ \overset{|}{CH_3}$$
2-Methyl-1-propanamine

(*R*)-*trans*-**3-Penten-2-amine**

Substances with two amine functions are **diamines,** two examples of which are 1,4-butanediamine and 1,5-pentanediamine. Their contribution to the smell of dead fish and rotting flesh leads to their descriptive common names, putrescine and cadaverine, respectively.

$$H_2N\diagdown\diagup\diagdown NH_2$$
1,4-Butanediamine (Putrescine)

$$H_2N\diagdown\diagup\diagdown\diagup NH_2$$
1,5-Pentanediamine (Cadaverine)

The aromatic amines, or anilines, are called **benzenamines** (Section 15-1). For secondary and tertiary amines, the largest alkyl substituent on nitrogen is chosen as the alkanamine stem, and the other groups are named by using the letter *N*-, followed by the name of the additional substituent(s).

The smell of rotting fish is due largely to putrescine.

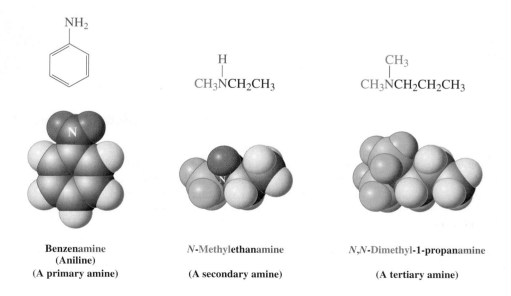

Benzenamine (Aniline) (A primary amine)

N-**Methylethanamine (A secondary amine)**

N,*N*-**Dimethyl-1-propanamine (A tertiary amine)**

An alternative way to name amines treats the functional group, called **amino-,** as a substituent of the alkane stem. This procedure is analogous to naming alcohols as hydroxy-alkanes. It is also used when other functional groups are present in the molecule, because the amine function has the lowest order of precedence of all functional groups discussed in this text (Section 19-1).

CH₃CH₂NH₂
Aminoethane

(CH₃)₂NCH₂CH₂CH₃
N,N-**Dimethylamino**propane

HO ⟍⟍⟍⟍ N(H) ⟍ CH₃CH₂
3-(*N*-Ethylamino)-1-butanol

CH₃NH₂
Methylamine

(CH₃)₃N
Trimethylamine

H₃C⟍N⟍
Benzylcyclohexylmethylamine

Many common names are based on the term **alkylamine** (see margin), as in the naming of alkyl alcohols.

Exercise 21-1

Name each of the following molecules twice, first as an alkanamine, then as an alkyl amine.

(a) CH₃CHCH₂CH₃ with NH₂

(b) N(CH₃)₂ on benzene ring

(c) Br ⟍⟍⟍⟍ with H₃C, H and NH₂

Exercise 21-2

Draw structures for the following compounds (common name in parentheses): **(a)** 2-propynamine (propargylamine); **(b)** (*N*-2-propenyl)phenylmethanamine (*N*-allylbenzylamine); **(c)** *N*,2-dimethyl-2-propanamine (*tert*-butylmethylamine).

In Summary There are several systems for naming amines. *Chemical Abstracts* uses names of the type alkanamine and benzenamine. Alternatives for alkanamine are the terms amino-alkane and alkylamine. The common name for benzenamine is aniline.

21-2 | STRUCTURAL AND PHYSICAL PROPERTIES OF AMINES

Nitrogen often adopts a tetrahedral geometry, similar to carbon, but with one lone electron pair instead of a fourth bond. Thus, amines generally have tetrahedral shapes around the heteroatom. However, this arrangement is not rigid because of a rapid isomerization process called inversion. Nitrogen is not as electronegative as oxygen, so while amines form hydrogen bonds, they are weaker than the hydrogen bonds formed by alcohols. As a result, amines have relatively low boiling points compared to alcohols.

The alkanamine nitrogen is tetrahedral

In contrast to amides (Section 20-1), the nitrogen orbitals in amines are very nearly sp^3 hybridized (see Section 1-8), forming an approximately tetrahedral arrangement. Three vertices of the tetrahedron are occupied by the three substituents, the fourth by the lone electron pair. As we shall see, it is this electron pair that is the source of the basic and nucleophilic properties of the amines. The term **pyramidal** is often used to describe the geometry adopted by the nitrogen and its three substituents. Figure 21-1 depicts the structure of methanamine (methylamine).

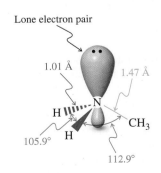

Figure 21-1 Nearly tetrahedral structure of methanamine (methylamine).

REAL LIFE: MEDICINE 21-1 | Physiologically Active Amines and Weight Control

A large number of physiologically active compounds owe their activity to the presence of amino groups. Several simple examples are well-known prescription or illegal drugs, such as adrenaline, Benzedrex, bupropion, amphetamine, mescaline, and Cymbalta (see Chapter Opening). A recurring pattern in many (but not all) of these compounds is the 2-phenylethanamine (β-phenethylamine) unit—that is, a nitrogen connected by a two-carbon chain to a benzene nucleus (highlighted below in green where applicable). This structural feature appears to be crucial for the binding to brain receptor sites responsible for neurotransmitter action at certain nerve terminals. Such amines can affect a wide variety of behavior, from controlling appetite and muscular activity to creating potentially addictive euphoric stimulation.

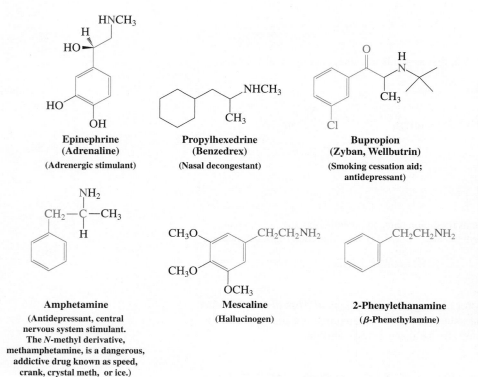

Epinephrine (Adrenaline)
(Adrenergic stimulant)

Propylhexedrine (Benzedrex)
(Nasal decongestant)

Bupropion (Zyban, Wellbutrin)
(Smoking cessation aid; antidepressant)

Amphetamine
(Antidepressant, central nervous system stimulant. The *N*-methyl derivative, methamphetamine, is a dangerous, addictive drug known as speed, crank, crystal meth, or ice.)

Mescaline
(Hallucinogen)

2-Phenylethanamine
(β-Phenethylamine)

Controlling weight by keeping fit: running in place on a treadmill.

The problem of selective targeting of these sites by molecular design has played a major role in the development of drugs that control weight and combat obesity. During the last decade, this effort has taken several directions, many marred by failures due to side effects ranging from depression to adverse heart responses. However, after a 13-year break, a number of promising drugs started to appear on the market in 2012. Thus, whereas many of the classical amphetamine-based "diet pills" relied on their anorectic (appetite-suppressing) effect by acting as stimulants, the new β-phenethylamine derivative lorcaserin (marketed as Belviq) works by increasing satiety. In other words, you feel hungry, but you are more quickly satisfied after you start eating, or you feel full longer. A different strategy is used in the combination of two drugs, phentermine and topiramate (marketed as Qsymia). The former is a stimulant (not surprisingly, considering that it is an isomer of "crystal meth") and appetite suppressant, whose side effects appear to be attenuated by the second, an anticonvulsant and mood stabilizer.

Another approach to controlling weight is to increase metabolic rate (and hence body temperature), a goal most simply and commonly attained by exercising. An arguably less taxing alternative is a thermogenic drug. Caffeine is such a compound, but its enhancing effect on the metabolic rate is short-lived and is followed by a period of actually depressing it.

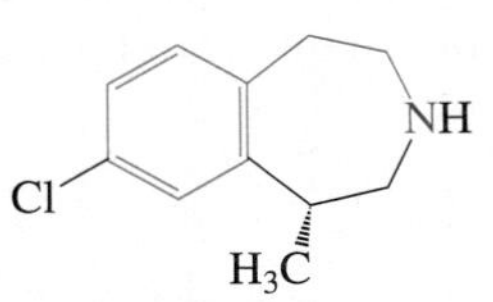

**Lorcaserin
(Belviq)**

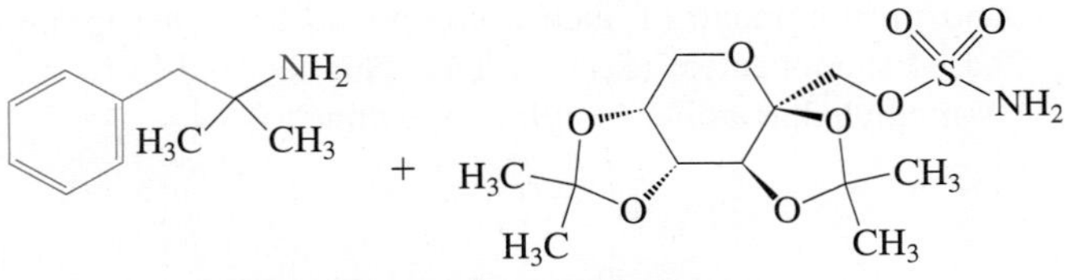

**Phentermine-topimarate (1:6)
(Qsymia)**

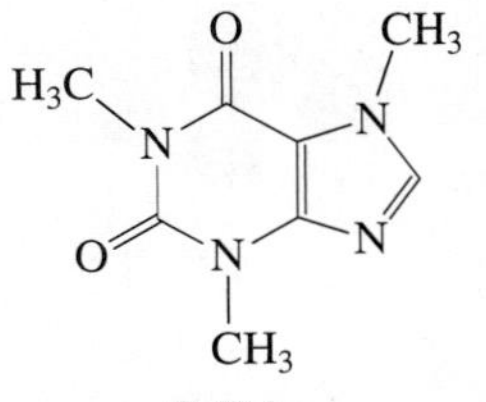

Caffeine

Whereas thermogenic drugs help control body weight by "burning off fat," the exact opposite—namely, inhibition of calorific intake by preventing metabolism—is being attempted with yet another class of drugs. An example is orlistat (approved by the FDA in 1999), a molecule that inhibits enzymes in the intestine called lipases. Lipases catalyze the hydrolysis of triglycerides (triesters of 1,2,3-propanetriol that constitute fats and oils; Section 20-5) to the constituent fatty acids, which are water soluble and hence absorbed by the body. When the enzymes are blocked, triglycerides from the diet are excreted undigested. Initially sold only by prescription, the FDA approved an over-the-counter formulation in 2007 (marketed under the name alli), a first for a weight-loss drug.

Amines are not the only agents that are being investigated for their potential in weight control. For example, the completely nondigestible fat olestra contains no nitrogens, and its development is based on the finding that molecules with more than five long chain ester functions cannot be absorbed through the intestinal wall. Although the structure of olestra looks forbidding, it is actually a simple octaester of common table sugar, sucrose (shown in green; Chapter 24). Olestra is currently used in some foods, especially savory snacks such as potato chips, cheese puffs, and crackers.

You may question the validity of this type of drug research, because it appears to cater to the glutton or the "fashionable" in the search of the perfect body. Not so: Do not forget that obesity is a major health hazard, causing chronic diseases, such as cardiovascular and respiratory problems, hypertension, diabetes, and certain cancers, and that it shortens the life span of those afflicted. It appears to be mostly the result of a metabolic (perhaps in some cases genetic) disposition that cannot otherwise be controlled.

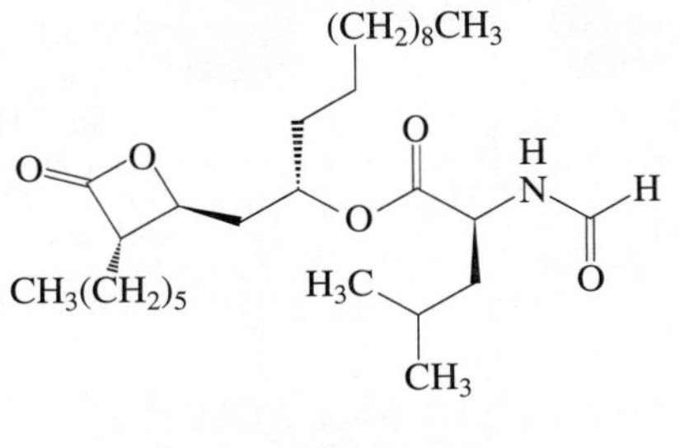

**Orlistat
(Xenical, alli)**

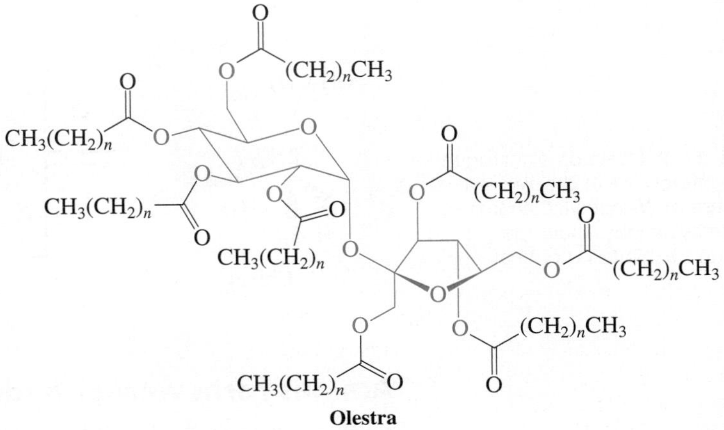

Olestra

Exercise 21-3

The bonds to nitrogen in methanamine (methylamine; see Figure 21-1) are slightly longer than the bonds to oxygen in the structure of methanol (Figure 8-1). Explain. (**Hint:** See Table 1-2.)

Model Building

The tetrahedral geometry around an amine nitrogen suggests that it should be chiral if it bears three different substituents, the lone electron pair serving as the fourth. The image and mirror image of such a compound are not superimposable, by analogy with carbon-based stereocenters (Section 5-1). This fact is illustrated with the simple chiral alkanamine *N*-methylethanamine (ethylmethylamine).

Image and Mirror Image of
N-Methylethanamine (Ethylmethylamine)

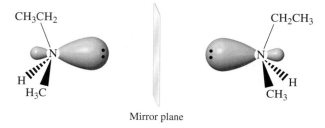

Mirror plane

However, samples of the amine prove *not* to be optically active. Why? Amines are not configurationally stable at nitrogen, because of rapid isomerization by a process called **inversion.** The molecule passes through a transition state incorporating an *sp²*-hybridized nitrogen atom, as illustrated in Figure 21-2. This transition state is similar to that depicted in Figure 6-4 for the inversion observed in S$_N$2 reactions (Section 6-5). The barrier to this motion in ordinary small amines has been measured spectroscopically and found to be between 5 and 7 kcal mol^{-1} (ca. 20–30 kJ mol^{-1}). It is therefore impossible to keep an enantiomerically pure, simple di- or trialkylamine from racemizing at room temperature when the nitrogen atom is the only stereocenter.

Transition State of Inversion

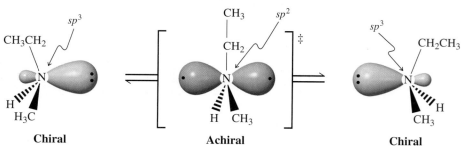

Figure 21-2 Inversion at nitrogen rapidly interconverts the two enantiomers of *N*-methylethanamine (ethylmethylamine). Thus, the compound exhibits no optical activity.

Amines form weaker hydrogen bonds than alcohols do

Because alcohols readily form hydrogen bonds (5–6 kcal mol^{-1}, Section 8-2), they have unusually high boiling points. In principle, so should amines, and indeed Table 21-1 bears out this expectation. However, because nitrogen is less electronegative than oxygen,

Table 21-1	Physical Properties of Amines, Alcohols, and Alkanes					
Compound	Melting point (°C)	Boiling point (°C)	Compound	Melting point (°C)	Boiling point (°C)
CH_4	−182.5	−161.7	$(CH_3)_2NH$	−93	7.4
CH_3NH_2	−93.5	−6.3	$(CH_3)_3N$	−117.2	2.9
CH_3OH	−97.5	65.0			
			$(CH_3CH_2)_2NH$	−48	56.3
CH_3CH_3	−183.3	−88.6	$(CH_3CH_2)_3N$	−114.7	89.3
$CH_3CH_2NH_2$	−81	16.6			
CH_3CH_2OH	−114.1	78.5	$(CH_3CH_2CH_2)_2NH$	−40	110
			$(CH_3CH_2CH_2)_3N$	−94	155
$CH_3CH_2CH_3$	−187.7	−42.1			
$CH_3CH_2CH_2NH_2$	−83	47.8	NH_3	−77.7	−33.4
$CH_3CH_2CH_2OH$	−126.2	97.4	H_2O	0	100

amines form weaker hydrogen bonds* than alcohols do, hence their boiling points are lower and their solubility in water is less. In general, the boiling points of the amines lie between those of the corresponding alkanes and alcohols. The smaller amines are soluble in water and in alcohols because they can form hydrogen bonds to the solvent. If the hydrophobic part of an amine exceeds six carbons, the solubility in water decreases rapidly; the larger amines are essentially insoluble in water.

In Summary Amines adopt an approximately tetrahedral structure in which the lone electron pair occupies one vertex of the tetrahedron. They can, in principle, be chiral at nitrogen but are difficult to maintain in enantiomerically pure form because of fast inversion at the nitrogen. Amines have boiling points higher than those of alkanes of similar size. Their boiling points are lower than those of the analogous alcohols because of weaker hydrogen bonding, and their water solubility is between that of comparable alkanes and alcohols.

21-3 | SPECTROSCOPY OF THE AMINE GROUP

Primary and secondary amines can be recognized by infrared spectroscopy because they exhibit a characteristic broad N–H stretching absorption in the range between 3250 and 3500 cm^{-1}. Primary amines show two strong peaks in this range, whereas secondary amines give rise to only a single line (see also IR spectra of amides, Section 20-1). Primary amines also show a band near 1600 cm^{-1} that is due to a scissoring motion of the NH_2 group (Section 11-8, Figure 11-17, and margin). Tertiary amines do not give rise to such signals, because they do not have a hydrogen that is bound to nitrogen. Figure 21-3 shows the infrared spectrum of cyclohexanamine.

Nuclear magnetic resonance spectroscopy also may be useful for detecting the presence of amino groups. Amine hydrogens often resonate to give broadened peaks, like the OH signal in the NMR spectra of alcohols. Their chemical shift depends mainly on the rate of exchange of protons with water in the solvent and the degree of hydrogen bonding.

Scissoring Motion of the –NH$_2$ Hydrogens

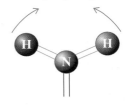

---

*Whereas *all* amines can act as proton acceptors in hydrogen bonding, only primary and secondary amines can function as proton donors, because tertiary amines lack such protons.

Figure 21-3 (A) Infrared spectrum of cyclohexanamine. The molecule exhibits two strong peaks between 3250 and 3500 cm^{-1}, characteristic of the N–H stretching absorptions of the primary amine functional group. The broad band near 1600 cm^{-1} results from scissoring motions of the N–H bonds. (B) *N*-Methylcyclohexanamine shows only one N–H peak at 3300 cm^{-1}. (C) *N,N*-Dimethylcyclohexanamine has no peaks between 3250 and 3500 cm^{-1}.

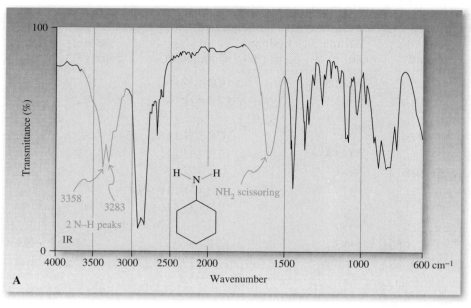

> **Reminder**
>
> In general, the more polar a bond is, the more intense will be its IR stretching absorption (Section 11-8). The N–H bond is less polar than the O–H bond and therefore gives rise to a relatively weaker IR band (compare Figure 21-3 with Figure 11-21).

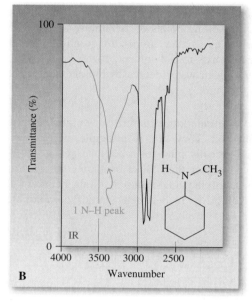

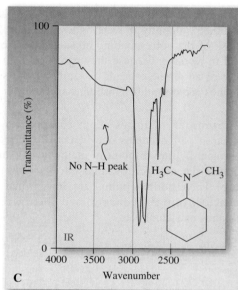

Figure 21-4 shows the ^{1}H NMR spectrum of azacyclohexane (piperidine), a cyclic secondary amine. The amine hydrogen peak appears at $\delta = 1.22$ ppm, and there are two other sets of signals, multiplets at $\delta = 1.34$ and 2.61 ppm. The absorption at lowest field can be assigned to the hydrogens neighboring the nitrogen, which are deshielded by the effect of its electronegativity.

Exercise 21-4

Would you expect the hydrogens next to the heteroatom in an amine, RCH$_2$NH$_2$, to be more or less deshielded than those in an alcohol, RCH$_2$OH? Explain. (**Hint:** See Exercise 21-3.)

The ^{13}C NMR spectra of amines show a similar trend: Carbons bound directly to nitrogen resonate at considerably lower field than do the carbon atoms in alkanes. However, as in ^{1}H NMR (Exercise 21-4), nitrogen is less deshielding than oxygen.

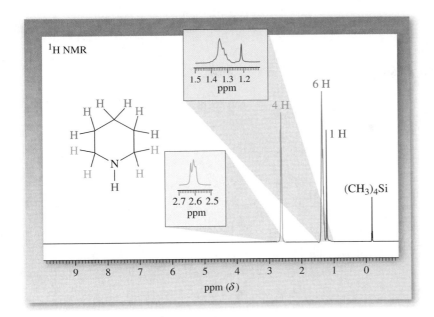

Figure 21-4 300-MHz ^{1}H NMR spectrum of azacyclohexane (piperidine). Like the OH hydrogen signal in alcohols, the amine NH peak may appear almost anywhere in the normal hydrogen chemical-shift range. Here the NH absorption occurs at $\delta = 1.22$ ppm, and the signal is fairly sharp because of the use of dry solvent (CDCl$_3$).

^{13}C Chemical Shifts in Various Amines (ppm)

Mass spectrometry readily establishes the presence of nitrogen in an organic compound. Unlike carbon, which is tetravalent, nitrogen is trivalent. Because of these valence requirements and because nitrogen has an even atomic weight (14), molecules incorporating one nitrogen (or any odd number of nitrogens) have an *odd* molecular weight (review Exercise 11-21). You can readily verify this count by considering that the molecular formula for an alkanamine containing one nitrogen is $C_nH_{2n+3}N$.

The mass spectral fragmentation patterns of amines also aid in structural assignments. For example, the mass spectrum of *N,N*-diethylethanamine (triethylamine) shows the peak for the molecular ion at $m/z = 101$ (Figure 21-5). However, the more prominent base peak is at $m/z = 86$ and is caused by the loss of a methyl group by α-cleavage (Section 11-10). Such fragmentation is favored because it results in a resonance-stabilized **iminium ion.**

Mass-Spectral Fragmentation of *N,N*-Diethylethanamine

$$(CH_3CH_2)_2\overset{+\cdot}{N}CH_2 \!\!\not\!\!-\!\! CH_3 \longrightarrow CH_3\cdot + \left[(CH_3CH_2)_2\overset{+}{N}\!\!=\!\!CH_2 \longleftrightarrow (CH_3CH_2)_2\overset{\cdot\cdot}{N}\!\!-\!\!\overset{+}{C}H_2 \right]$$

$m/z = 101$
N,N-Diethylethanamine
(Triethylamine)

$m/z = 86$
Iminium ion

The rupture of the C–C bond next to nitrogen is frequently so easy that the molecular ion cannot be observed. For example, in the mass spectrum of 1-hexanamine, the molecular ion ($m/z = 101$) is barely visible; the dominating peak corresponds to the methyleneiminium fragment $[CH_2\!\!=\!\!NH_2]^+$ ($m/z = 30$).

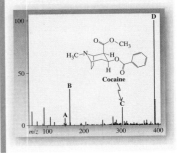

Really? A report that most American paper currency is contaminated with cocaine prompted one of the authors to obtain a crisp $50 bill from a bank (unnamed) and take a mass spectrum. Indeed, there it is (peak C, molecular ion), together with triethanolamine (A; an emulsifier and surfactant in detergents and inks), nicotine (B), and, most prominently, dioctyl phthalate (D; a plasticizer, particularly in PVC plastics). Mind you, the amounts are miniscule, at an average of 16 μg per bill.

Figure 21-5 Mass spectrum of *N*,*N*-diethylethanamine (triethylamine), showing a molecular ion peak at $m/z = 101$. In general, molecules incorporating one nitrogen atom have an odd molecular weight. The base peak is due to loss of a methyl group, resulting in an iminium ion with $m/z = 86$.

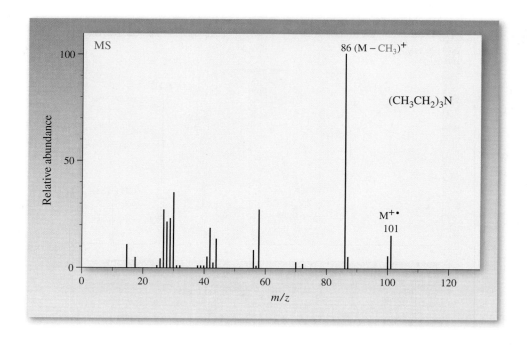

Spectral techniques were used to identify the molecule in Exercise 21-5 in a variety of animals. It was found that the urine of carnivores, such as lions, tigers, and raccoons, contains up to 3000 times as much of this amine as herbivores. Studies on potential prey of carnivores have shown that they have evolved highly sensitive olfactory receptors to detect the chemical and elicit an innate "scurrying" response.

| Solved Exercise 21-5 | Working with the Concepts: Assigning the Structure of an Amine |

A compound with the molecular formula $C_8H_{11}N$ exhibited the following spectral data: MS: m/z (relative intensity) = 121 (M$^+$, 6), 91 (15), 30 (100). ^{1}H NMR: δ = 1.16 (s, 2 H), 2.73 (t, J = 7 Hz, 2 H), 2.93 (t, J = 7 Hz, 2 H), 7.20 (m, 5 H) ppm. ^{13}C NMR: δ 40.2 (CH_2), 43.6 (CH_2), 126.1 (CH), 128.4 (CH), 128.8 (CH), 139.9 (C$_{quaternary}$) ppm. IR (neat): selected $\tilde{\nu}$ = 700, 746, 1603, 2850, 2932, 3026, 3062, 3284, 3366 cm^{-1}.
  What is it?

Strategy
The combined spectral data give us more than enough information to elucidate the structure of this unknown. The following analyzes them in order of appearance in the question, starting with the mass and ^{1}H NMR spectra to reach a tentative assignment. The remaining data are then used to support it. However, in problems like this, you can start with whatever spectral information you deem most relevant, jump from one type of spectroscopy to another, and piece together the solution that way.

Solution
- The mass spectrum shows a molecular ion (m/z = 121), which corresponds to the molecular formula (as it should), $C_8H_{11}N$.
- More instructive are the fragment peaks: the first, at m/z = 91, signifies the loss of a piece with mass = 30, exactly the value of the second fragment ion. In other words, the molecule breaks up very efficiently into two segments of m/z = 91 and 30. How do we dissect $C_8H_{11}N$ to provide such fragments? It should not take you long to realize that m/z = 30 must be due to [CH_2=NH_2]$^+$, as there is no other sensible alternative. (The only other molecule with this mass that can be written is C_2H_6, ethane.) Therefore, m/z = 91 has the composition C_7H_7.
- Turning to the ^{1}H NMR spectrum, we note that the hydrogens are divided into four groups, in the ratio 2:2:2:5. The first group gives rise to a singlet and is therefore due to a pair of hydrogens without immediate hydrogen neighbors or attached to N, as in NH_2 (decoupled by rapid proton exchange). The next two sets of hydrogens appear as triplets and thus must be due to mutually coupled (hence neighboring) CH_2 groups, suggesting the presence of substructure A–CH_2CH_2–B.
- Subtracting the unit C_2H_4 from $C_8H_{11}N$ leaves C_6H_7N for A + B.
- Finally, we observe an aromatic multiplet of 5 H at δ = 7.20 ppm, strongly indicative of a phenyl group, C_6H_5. If this group were A, then B would have to be NH_2, and a possible structure would

be 2-phenylethanamine, $C_6H_5CH_2CH_2NH_2$. Such a solution would certainly be in accord with the mass spectrum, the dominant fragments being methyleneiminium and benzyl cation $[C_6H_5CH_2]^+$. [Parenthetically, benzyl cation rapidly rearranges under mass spectral conditions to its more stable, aromatic annulene isomer, cycloheptatrienyl cation (Section 15-7)].

- Let us see whether the remainder of the spectral data supports our tentative assignment. The ^{13}C NMR spectrum, indeed, exhibits the expected six lines with the appropriate chemical shifts (Table 10-6, Section 21-3).
- The IR spectrum contains bands for the NH_2 group, $\tilde{\nu} = 3366$, 3284, and 1603, very similar to those shown in Figure 21-3(A), in addition to the aromatic and saturated C–H absorptions (Figure 15-7), all fully consistent with the proposed structure, shown in the margin.

NH₂

2-Phenylethanamine

Exercise 21-6 | **Try It Yourself**

What approximate spectral data (IR, NMR, m/z) would you expect for N-ethyl-2,2-dimethylpropanamine, shown in the margin?

N-Ethyl-2,2-dimethyl-propanamine

In Summary The IR stretching absorption of the N–H bond ranges between 3250 and 3500 cm^{-1}; the corresponding ^1H NMR peak is often broad and can be found at variable δ. The electron-withdrawing nitrogen deshields neighboring carbons and hydrogens, although to a lesser extent than the oxygen in alcohols and ethers. The mass spectra of simple alkanamines that contain only one nitrogen atom have odd-numbered molecular ion peaks, because of the trivalent character of nitrogen. Fragmentation occurs in such a way as to produce resonance-stabilized iminium ions.

21-4 | ACIDITY AND BASICITY OF AMINES

Like the alcohols (Section 8-3), amines are both basic and acidic. Because nitrogen is less electronegative than oxygen, the acidity of amines is about 20 orders of magnitude less than that of comparable alcohols. Conversely, the lone pair is much more available for protonation, thereby causing amines to be better bases.

Acidity and Basicity of Amines

Amine acting as an acid: $R\ddot{N}H$ + $^-:B$ $\underset{}{\overset{K_a}{\rightleftharpoons}}$ $R\ddot{\overset{..}{N}}H$ + HB
                       |
                       H
                Acid   **Conjugate base**

Amine acting as a base: $R\ddot{N}H_2$ + HA $\underset{}{\overset{K_b}{\rightleftharpoons}}$ $R\overset{H}{\underset{|}{\overset{|}{N}}}H_2^+$ + $^-:A$
                Base   **Conjugate acid**

Amines are very weak acids

We have seen evidence that amines are much less acidic than alcohols: Amide ions, R_2N^-, are used to deprotonate alcohols (Section 9-1). The equilibrium of this proton transfer lies strongly to the side of the alkoxide ion. The high value of the equilibrium constant, about 10^{20}, is due to the strong basicity of amide ions, which is consistent with the low acidity of amines. The pK_a of ammonia and alkanamines is of the order of 35.

Acidity of Amines

$$RNH + H_2\ddot{O} \underset{}{\overset{K}{\rightleftharpoons}} R\ddot{N}H + H_2\overset{+}{O}H \qquad K_a = \frac{[R\ddot{N}H][H_2\overset{+}{O}H]}{[R\ddot{N}H]} \approx 10^{-35}$$

Amine (Weak acid)	**Amide ion** (Strong base)	$pK_a \approx 35$

The deprotonation of amines requires extremely strong bases, such as alkyllithium reagents. For example, lithium diisopropylamide, the sterically hindered base used in some bimolecular elimination reactions (Section 7-8), is made in the laboratory by treatment of *N*-(1-methylethyl)-2-propanamine (diisopropylamine) with butyllithium.

Preparation of LDA

$$\underset{\substack{\text{*N*-(1-Methylethyl)-} \\ \text{2-propanamine} \\ \text{(Diisopropylamine)}}}{\begin{array}{c} H_3C \quad CH_3 \\ | \quad\quad | \\ CH_3CH\ddot{N}CHCH_3 \\ | \\ H \end{array}} \xrightarrow[-CH_3CH_2CH_2CH_2H]{CH_3CH_2CH_2CH_2Li} \underset{\substack{\text{Lithium} \\ \text{diisopropylamide, LDA}}}{\begin{array}{c} \quad\quad Li^+ \\ H_3C \quad CH_3 \\ | \quad\quad | \\ CH_3CH\overset{-}{N}CHCH_3 \end{array}}$$

An alternative synthesis of amide ions is the treatment of amines with alkali metals. Alkali metals dissolve in amines (albeit relatively slowly) with the evolution of hydrogen and the formation of amine salts (much as they dissolve in water and alcohol, furnishing H_2 and metal hydroxides or alkoxides, Section 9-1). For example, sodium amide can be made in liquid ammonia from sodium metal in the presence of catalytic amounts of Fe^{3+}, which facilitates electron transfer to the amine. In the absence of such a catalyst, sodium simply dissolves in ammonia (labeled "Na, liquid NH_3") to form a strongly reducing solution (Section 13-6).

Amines are moderately basic, ammonium ions are weakly acidic

Amines deprotonate water to a *small* extent to form ammonium and hydroxide ions. Thus, amines are more strongly basic than alcohols but not nearly as basic as alkoxides. Protonation occurs at the site of the free electron pair as pictured in the electrostatic potential map of *N,N*-dimethylmethanamine (trimethylamine) in the margin.

N,N-Dimethylmethanamine
(Trimethylamine)

$CH_3\overset{+}{N}H_3\ Cl^-$
Methylammonium chloride

$$\underset{\substack{\text{Cyclopentylethyl-} \\ \text{methylammonium} \\ \text{iodide}}}{\begin{array}{c} CH_3 \\ | \\ \overset{+}{N}{-}H \quad I^- \\ | \\ CH_2CH_3 \end{array}}$$

Basicity of Amines

$$R\ddot{N}H_2 + H{-}\ddot{O}H \underset{}{\overset{K_b}{\rightleftharpoons}} \underset{\substack{\text{Ammonium} \\ \text{ion}}}{\begin{array}{c} H \\ | \\ R\overset{+}{N}H_2 \end{array}} + \ \ddot{H}\ddot{O}:^-$$

Amine

The resulting ammonium salts can be primary, secondary, or tertiary, depending on the number of substituents on nitrogen.

$R\overset{+}{N}H_3\ Cl^-$	$R_2\overset{+}{N}H_2\ Br^-$	$R_3\overset{+}{N}H\ I^-$
Primary ammonium chloride	**Secondary ammonium bromide**	**Tertiary ammonium iodide**

Ammonium salts are named by attaching the substituent names to the ending *-ammonium* followed by the name of the anion.

It is useful to view the basicity of the amines as a measure of the acidity of their conjugate acids (Section 2-2), the ammonium ions. These species are stronger acids than water ($pK_a = 15.7$; Table 2-2) or alcohols but much weaker than carboxylic acids ($pK_a = 4$–5; Section 19-4).

Acidity of Ammonium Ions

$$RNH_2 + H_2O \rightleftharpoons^{K_a} RNH_2 + H_2OH \qquad K_a = \frac{[RNH_2][H_2OH]}{[RNH_2]} \approx 10^{-10}$$

Conjugate acid   Base

$pK_a \approx 10$

Reminder: The weaker the conjugate acid, the higher its pK_a and the stronger the corresponding base

Any factor, such as substituents or hybridization, that increases the electron density of the amine nitrogen increases the basicity of the amine and therefore the pK_a of the corresponding ammonium salt. Conversely, decreasing the electron density of the amine nitrogen decreases its basicity and the pK_a of the corresponding ammonium salt. For example, alkyl-ammonium salts are slightly less acidic than ammonium ion, NH_4, and therefore the corresponding amines are more basic. The reason is the electron-donating character of alkyl groups (Sections 7-5 and 16-1). However, as shown below, the pK_a values do not increase in a regular way with increasing alkyl substitution. In fact, tertiary amines are typically less basic than secondary systems. Solvation is responsible for this observation. Thus, increasing the number of alkyl groups on the amine nitrogen increases unfavorable steric disruption of the solvent shell (Section 8-3). At the same time, it decreases the number of hydrogens attached to the nitrogen capable of entering into favorable hydrogen bonds (Section 21-2). Both phenomena counteract the inductive donor properties of the alkyl groups in solution. Indeed, in the gas phase, in which there is no solvent, the trend is as expected: (pK_a) $NH_4 < CH_3NH_3 < (CH_3)_2NH_2 \ll (CH_3)_3NH$.

pK_a Values of a Series of Simple Ammonium Ions*

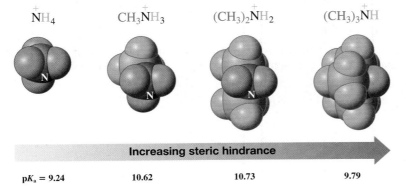

NH_4 CH_3NH_3 $(CH_3)_2NH_2$ $(CH_3)_3NH$

Increasing steric hindrance

$pK_a = 9.24$ 10.62   10.73   9.79

The ready equilibration between free amine and ammonium salt is important for the formulation and activity of amine-based pharmaceuticals. The neutral forms are typically insoluble in water, and therefore the medications are sold in their salt forms, to allow for ready oral or intravenous administration. The pK_a of the amino function also affects the distribution, metabolism, excretion, and binding to receptor sites. Thus, after ingestion, the acidic environment of the stomach (pH ~ 2–4) will keep the drug protonated, therefore minimizing absorption through the nonpolar gastric walls. Once in the intestines (pH ~ 5–7), the acid-base equilibrium

---

*A confusing practice in the literature is to refer to the pK_a value of an ammonium ion as being that of the neutral amine. In the statement "the pK_a of methanamine is 10.62," what is meant is the pK_a of the methylammonium ion. The pK_a of methanamine is actually 35.

is shifted toward the amine, maximizing penetration. This balancing act is continued until the molecule reaches the active site, where it may dock by hydrogen bonding (in the neutral form) or polar interactions (in the salt form). Medicinal chemists often fine-tune lead structures for optimal pK_a profiles. For example, the experimental blood anticoagulant scaffold shown below has been modified extensively to tune its acidity over a pK_a range of 2–11.

R = H, pK_a = 11
R = F, pK_a = 2

Blood anticoagulant

The nitrogen lone pairs in arenamines, carboxamides, and imines are less available for protonation

In contrast to the effect of alkyl groups, the basicity of amines is decreased by electron-withdrawing substituents on the nitrogen. For example, the pK_a of 2-[bis(2-hydroxyethyl)amino]ethanol is only 7.75, a consequence of the presence of three inductively electron-withdrawing oxygens. Similarly, benzenamine (aniline) is considerably less basic (pK_a = 4.63) than its saturated analog cyclohexanamine (pK_a = 10.66) and other primary amines. There are two reasons for this effect. One is the sp^2 hybridization of the aromatic carbon attached to the nitrogen in benzenamine, which renders that carbon relatively electron withdrawing (Sections 11-3 and 13-2), in this way making the nitrogen lone pair less available for protonation. The second reason is the resonance stabilization of the system by delocalization of the electron pair into the aromatic π system (Section 16-1). This resonance is lost upon protonation. A similar example is acetamide, in which the appended acetyl group ties up the nitrogen lone pair by induction (the carbonyl carbon is positively polarized) and resonance (Section 20-1).

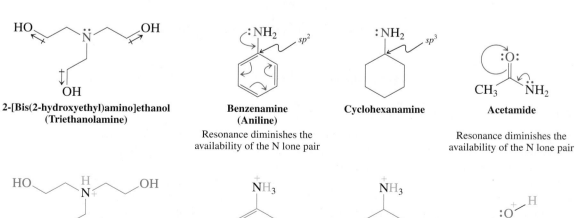

2-[Bis(2-hydroxyethyl)amino]ethanol
(Triethanolamine)

pK_a = 7.75

Benzenamine (Aniline)
Resonance diminishes the availability of the N lone pair

4.63

Cyclohexanamine

10.66

Acetamide
Resonance diminishes the availability of the N lone pair

0.63
Protonation occurs on oxygen

Table 21-2	pK_a Values of Various Amines		
Compound	pK_a of conjugate acid	Compound	pK_a of conjugate acid
$:NH_3$	9.24	:NH₂ (aniline)	4.63
$CH_3\ddot{N}H_2$	10.62		
$(CH_3)_2\ddot{N}H$	10.73		
$(CH_3)_3N:$	9.79	:O: / CH₃—C—NH₂	0.63
$(HOCH_2CH_2)_3N:$	7.75		
:NH₂ (cyclohexyl)	10.66	R₂C=N:R (iminium)	7–9
		$R-C\equiv N:$	< -5

As might be expected, the hybridization at nitrogen itself also drastically affects basicity, in the order $:NH_3 > R_2C=\ddot{N}R' > RC\equiv N:$, a phenomenon that we already encountered in the discussion of the relative acidity of alkanes, alkenes, and alkynes (Section 13-2). Thus, iminium ions (Section 17-9) have pK_a values estimated to be of the order of 7 to 9; *N*-protonated nitriles (Section 20-8) are even more acidic ($pK_a < -5$). Table 21-2 summarizes the pK_a values of the conjugate acids of some representative amines.

Exercise 21-7

Explain the decreasing pK_a values of the following (protonated) amines:

pK_a =   10.7   9.5   8.2

In Summary Amines are poor acids and require alkyllithium reagents or alkali metal treatment to form amide ions. In contrast, they are moderately good bases, leading to the relatively weakly acidic ammonium salts.

21-5 | SYNTHESIS OF AMINES BY ALKYLATION

Amines can be synthesized by alkylating nitrogen atoms. Several such procedures take advantage of an important property of the nitrogen in many compounds: It is *nucleophilic*.

Amines can be derived from other amines

As nucleophiles, amines react with haloalkanes to give ammonium salts (Section 6-2). Unfortunately, this reaction is not clean, because the resulting amine product usually undergoes further alkylation. How does this complication arise?

Consider the alkylation of ammonia with bromomethane. When this transformation is carried out with equimolar quantities of starting materials, the weakly acidic product (methylammonium bromide), as soon as it is formed, (reversibly) donates a proton to the starting, weakly basic ammonia. The small quantities of methanamine generated in this way then compete effectively with the ammonia for the alkylating agent, and this further methylation generates a dimethylammonium salt.

The process does not stop there, either. This salt can donate a proton to either of the other two nitrogen bases present, furnishing *N*-methylmethanamine (dimethylamine). This compound constitutes yet another nucleophile competing for bromomethane; its further reaction leads to *N,N*-dimethylmethanamine (trimethylamine) and, eventually, to tetramethylammonium bromide, a *quaternary* ammonium salt. The final outcome is a mixture of alkylammonium salts and alkanamines.

Methylation of Ammonia

First Alkylation. Two steps give primary amine

Step 1. Nucleophilic Substitution: $H_3N: + CH_3—Br \longrightarrow CH_3\overset{+}{N}H_3 \ Br^-$

Methylammonium bromide

Step 2. Deprotonation: $CH_3\overset{+}{N}H_2 \ Br^- + :NH_3 \rightleftharpoons CH_3\overset{..}{N}H_2 + H\overset{+}{N}H_3 \ Br^-$
$\quad\quad H$

**Methanamine
(Methylamine)**

Subsequent Alkylation. Gives secondary, tertiary, and quaternary amines or ammonium salts, respectively

$CH_3\overset{..}{N}H_2 \xrightarrow[\substack{-\text{Amine} \\ \text{hydrobromide}}]{CH_3Br} (CH_3)_2\overset{..}{N}H \xrightarrow[\substack{-\text{Amine} \\ \text{hydrobromide}}]{CH_3Br} (CH_3)_3N: \xrightarrow{CH_3Br} (CH_3)_4N^+ \ Br^-$

**_N_-Methylmethanamine
(Dimethylamine)** ㅤㅤ **_N,N_-Dimethylmethanamine
(Trimethylamine)** ㅤㅤ **Tetramethylammonium
bromide**

The mixture of products obtained upon treatment of haloalkanes with ammonia or amines is a serious drawback that limits the usefulness of direct alkylation in synthesis. As a result, indirect alkylation methods are frequently applied, particularly in the preparation of primary amines.

Exercise 21-8

Like other amines, benzenamine (aniline) can be benzylated with chloromethylbenzene (benzyl chloride), $C_6H_5CH_2Cl$. In contrast to the reaction with alkanamines, which proceeds at room temperature, this transformation requires heating to between 90 and 95°C. Explain. (**Hint:** See Section 21-4.)

Indirect alkylation leads to primary amines

The synthesis of primary amines requires a nitrogen-containing nucleophile that will undergo reaction *only once* and that can be converted subsequently into the amino group. For example, cyanide ion, ^-CN, turns primary and secondary haloalkanes into nitriles, which are then reduced to the corresponding amines (Section 20-8). This sequence allows the conversion $RX \rightarrow RCH_2NH_2$. Note, however, that this method introduces an additional carbon into the haloalkane framework, because cyanide is alkylated at carbon and not nitrogen.

Conversion of a Haloalkane into the Homologous Amine by Cyanide Displacement–Reduction

$$RX + {}^-CN \longrightarrow RC{\equiv}N + X^-$$

$$RC{\equiv}N \xrightarrow{\text{LiAlH}_4 \text{ or H}_2, \text{ metal catalyst}} RCH_2NH_2$$

$CH_3(CH_2)_8Br + NaCN \xrightarrow[-\text{NaBr}]{\text{DMSO}} CH_3(CH_2)_8CN \xrightarrow{\text{H}_2, \text{Raney Ni}, 100 \text{ atm}} CH_3(CH_2)_8CH_2NH_2$
ㅤㅤㅤㅤㅤㅤㅤㅤㅤㅤㅤㅤㅤㅤㅤㅤㅤㅤ93%ㅤㅤㅤㅤㅤㅤㅤㅤㅤㅤㅤㅤㅤㅤㅤㅤ92%

1-Bromononaneㅤㅤㅤㅤㅤㅤㅤㅤㅤㅤㅤㅤ**Decanenitrile**ㅤㅤㅤㅤㅤㅤㅤㅤㅤㅤㅤㅤ**1-Decanamine**

To transform a haloalkane selectively into the corresponding amine without additional carbons requires a modified nitrogen nucleophile that is unreactive after the first alkylation. Such a nucleophile is the **azide ion, N_3^-**, which reacts with haloalkanes to furnish **alkyl azides.** These azides in turn are reduced by catalytic hydrogenation (Pd–C) or by lithium aluminum hydride to the primary amines.

$$R—N_3$$
Alkyl azide

Azide Displacement–Reduction

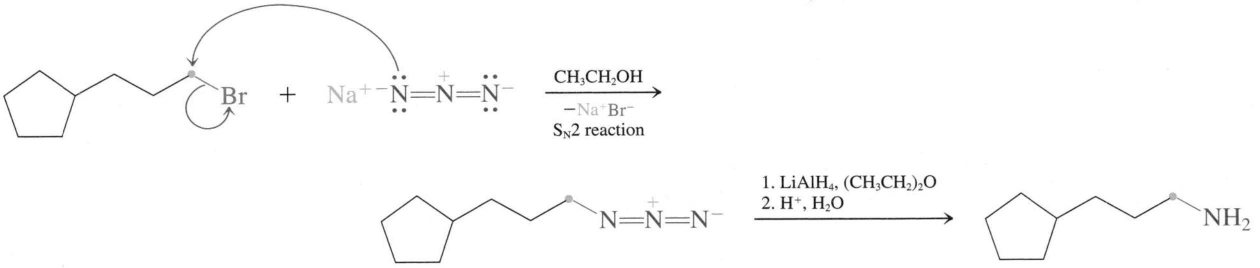

91%
3-Cyclopentylpropyl azide

89%
3-Cyclopentylpropanamine

A nonreductive approach to synthesizing primary amines uses the (commercially available) anion of 1,2-benzenedicarboximide (phthalimide), the imide of 1,2-benzenedicarboxylic (phthalic) acid. This process is also known as the **Gabriel* synthesis.** Because the nitrogen in the imide is adjacent to two carbonyl functions, the acidity of the NH group ($pK_a = 8.3$) is much greater than that of an ordinary amide ($pK_a = 22$, Section 20-7). Deprotonation can therefore be achieved with as mild a base as carbonate ion, the resulting anion being monoalkylated in good yield. The amine can be liberated subsequently by acidic hydrolysis, initially as the ammonium salt. Base treatment of the salt then produces the free amine.

Gabriel Synthesis of a Primary Amine

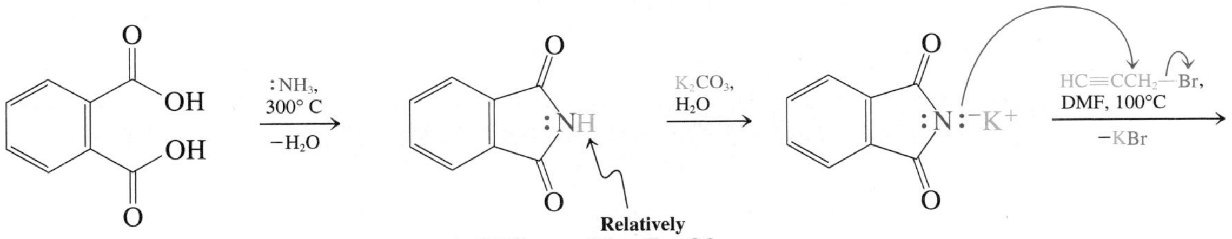

1,2-Benzenedicarboxylic acid
(Phthalic acid)

97%
1,2-Benzenedicarboximide
(Phthalimide)

Relatively acidic: $pK_a = 8.3$

$HC≡CCH_2\ddot{N}H_2$
73%
2-Propynamine
(Propargylamine)

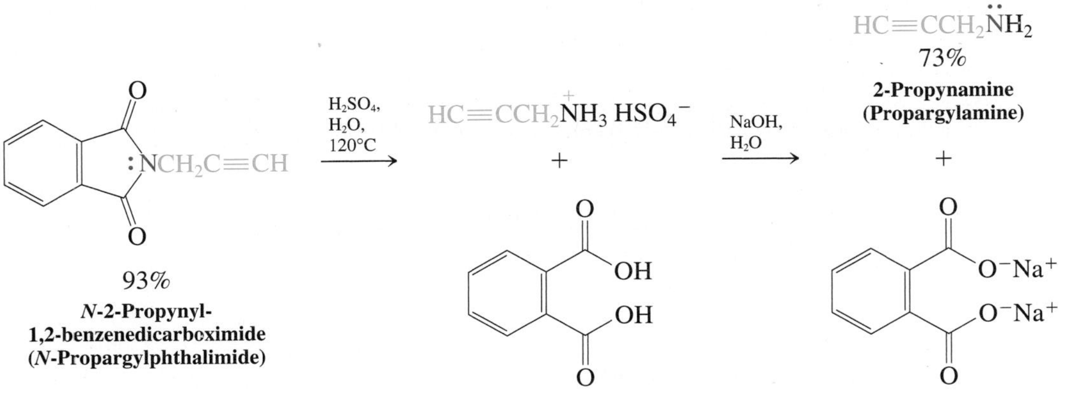

93%
N-2-Propynyl-
1,2-benzenedicarboximide
(*N*-Propargylphthalimide)

Removed in aqueous work-up

---

*Professor Siegmund Gabriel (1851–1924), University of Berlin.

Exercise 21-9

The cleavage of an *N*-alkyl-1,2-benzenedicarboximide (*N*-alkyl phthalimide) is frequently carried out with base or with hydrazine, H_2NNH_2. The respective products of these two treatments are the 1,2-benzenedicarboxylate A or the hydrazide B. Write mechanisms for these two transformations. (**Hint:** Review Section 20-6.)

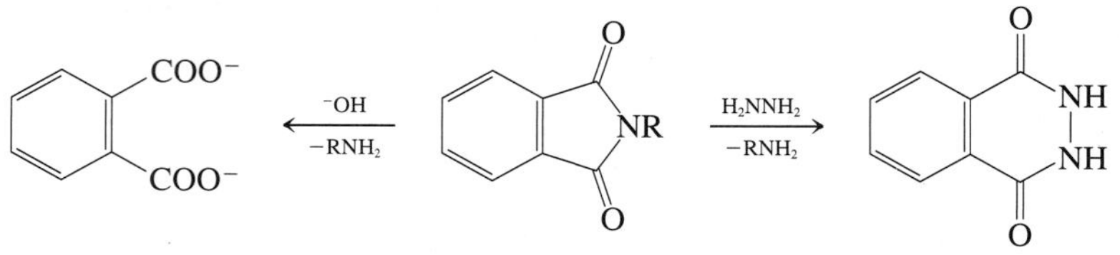

Exercise 21-10

Show how you would apply the Gabriel method to the synthesis of each of the following amines: (**a**) 1-hexanamine; (**b**) 3-methylpentanamine; (**c**) cyclohexanamine; (**d**) $H_2NCH_2CO_2H$, the amino acid glycine. (**Caution:** In the synthesis of glycine, you need to watch out for the carboxylic acid function. Can you see why?) For each of these four syntheses, would the azide displacement–reduction method be equally good, better, or worse?

In Summary Amines can be made from ammonia or other amines by simple alkylation, but this method gives mixtures and poor yields. Primary amines are better made in stepwise fashion, employing nitrile and azide groups, or protected systems, such as 1,2-benzenedicarboxylic imide (phthalimide) in the Gabriel synthesis.

21-6 SYNTHESIS OF AMINES BY REDUCTIVE AMINATION

A more general method of amine synthesis, called **reductive amination** of aldehydes and ketones, allows the construction of primary, secondary, and tertiary amines. In this process, the carbonyl compound is exposed to an amine containing at least one N–H bond (NH_3, primary, secondary amines) and a reducing agent to furnish a new alkylated amine directly (a primary, secondary, or tertiary amine, respectively). The new C–N bond is formed to the carbonyl carbon of the aldehyde or ketone.

General Reductive Amination

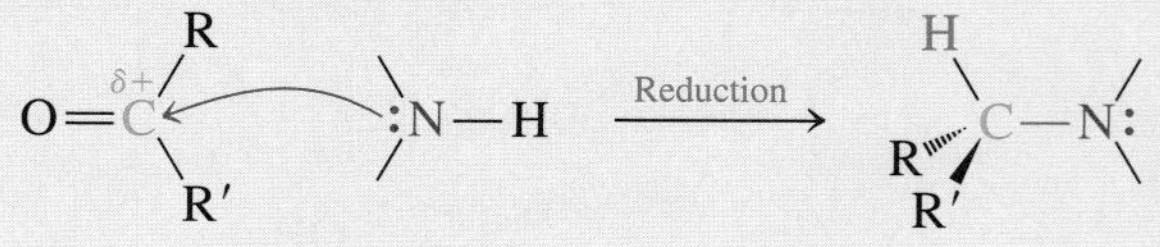

The sequence begins by the initial condensation of the amine with the carbonyl component (Section 17-9) to produce the corresponding imine (for NH_3 and primary amines) or iminium ion (secondary amines). Similar to the carbon–oxygen double bond in aldehydes and ketones, the carbon–nitrogen double bond in these intermediates is then reduced by simultaneous catalytic hydrogenation or by added special hydride reagents.

Reductive Amination of a Ketone with a Primary Amine

$$\underset{R'}{\overset{R}{C}}=O \ + \ H_2NR'' \ \underset{\text{Condensation}}{\overset{}{\rightleftharpoons}} \ \underset{R'}{\overset{R}{C}}=\underset{}{\overset{R''}{N}} \ + \ H_2O$$

$$\underset{R'}{\overset{R}{C}}=\underset{}{\overset{R''}{N}} \ \xrightarrow{\text{Reduction}} \ R'—\underset{H}{\overset{R}{C}}—NHR''$$

Animation

ANIMATED MECHANISM:
Reductive amination

This reaction succeeds because of the selectivity of the reducing agents: either hydrogen gas in the presence of a catalyst (Section 12-2) or sodium cyanoborohydride, $Na^{+-}BH_3CN$. Both react faster with the imine double bond than with the carbonyl group under the conditions employed. With $Na^{+-}BH_3CN$, the conditions are relatively acidic (pH = 2–3), activating the imine double bond by protonation on nitrogen and thus facilitating hydride attack at carbon. As shown in the margin, the relative stability of the modified borohydride reagent at such low pH (at which $NaBH_4$ hydrolyzes; Section 8-6) is due to the presence of the electron-withdrawing cyano group, which renders the hydrogens less basic (hydridic). In a typical procedure, the carbonyl component and the amine are allowed to equilibrate with the imine and water in the presence of the reducing agent. In this way, ammonia furnishes primary amines, and primary amines result in secondary amines.

$$\underset{\delta+}{\overset{N}{\underset{}{\overset{|||}{C}}}} \underset{\uparrow}{\overset{}{}} \text{Electron}$$
withdrawing

$$H \overset{-}{\underset{}{\overset{}{B}}} \underset{H}{\overset{}{\cdots}} H$$

Diminished ability of H to leave as :H⁻, hence reagent is less sensitive to H⁺

Amine Synthesis by Reductive Amination

Phenyl-2-propanone + NH_3 $\underset{+H_2O}{\overset{-H_2O}{\rightleftharpoons}}$ (Not isolated) $\xrightarrow{H_2, Ni, CH_3OH, 90°C, 100 \text{ atm}}$ Amphetamine (Industrial production)

Cyclohexanone + CH_3NH_2 $\underset{+H_2O}{\overset{-H_2O}{\rightleftharpoons}}$ (Not isolated) $\xrightarrow[CH_3CH_2OH, pH = 2-3]{NaBH_3CN,}$ 78% *N*-Methylcyclohexanamine

Similarly, reductive aminations with secondary amines give tertiary amines.

$\xrightarrow{(CH_3)_2NH, NaBH_3CN, CH_3OH}$

89%

Secondary amines cannot form imines with aldehydes and ketones, but are in equilibrium with the corresponding *N,N*-dialkyliminium ions (Section 17-9), which are reduced by addition of H⁻ from cyanoborohydride.

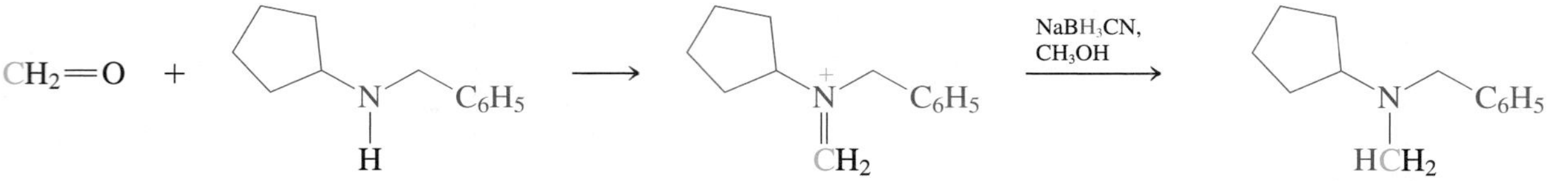

N-(Phenylmethyl)-
cyclopentanamine
(Benzylcyclopentylamine)

Iminium ion

N-Methyl-N-(phenylmethyl)-
cyclopentanamine
(Benzylcyclopentylmethylamine)

Exercise 21-11

Formulate a mechanism for the reductive amination with the secondary amine shown in the preceding example.

Solved Exercise 21-12 | **Working with the Concepts: Applying Reductive Amination**

Explain the following transformation by a mechanism.

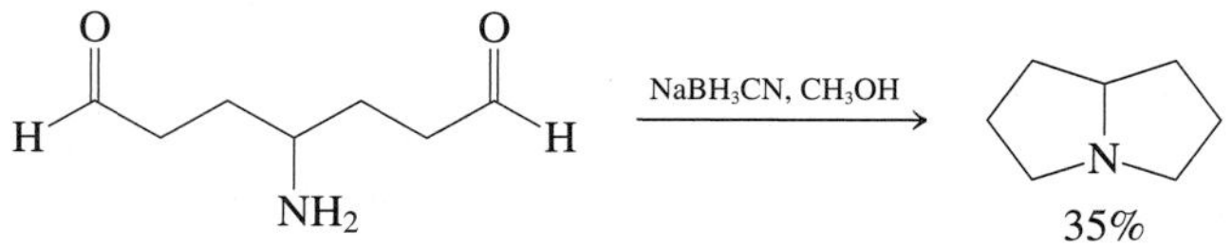

Remember WHIP

*W*hat

*H*ow

*I*nformation

*P*roceed

Strategy
As usual, you need to take an inventory of the functionalities present in the starting material, the reagent, and the product (an amine). The result is that we are dealing with an intramolecular variant of reductive amination. This realization should be the starting point for retrosynthetic analysis.

Solution
• First, write down the generic process retrosynthetically:

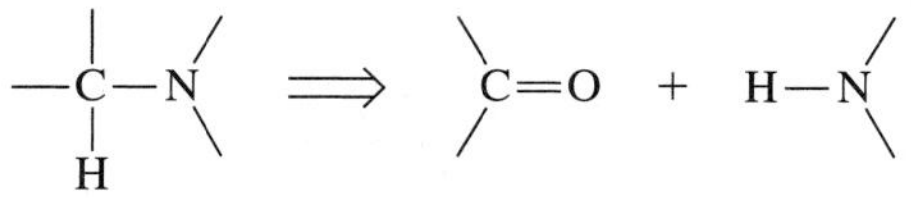

• Thus, the retrosynthetic disconnection from the product is

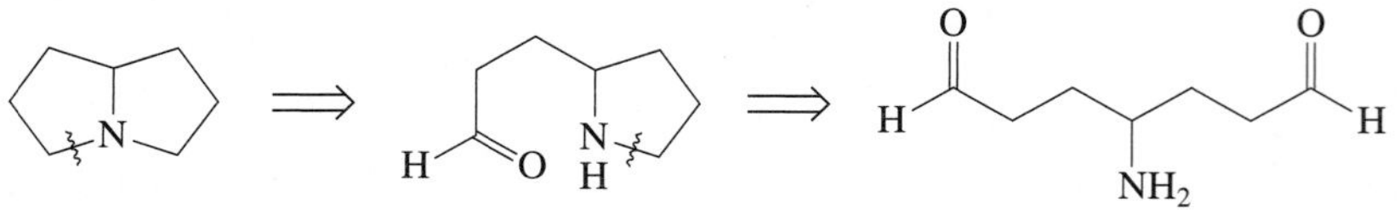

• The mechanism of the forward reaction follows accordingly (not all intermediates are shown):

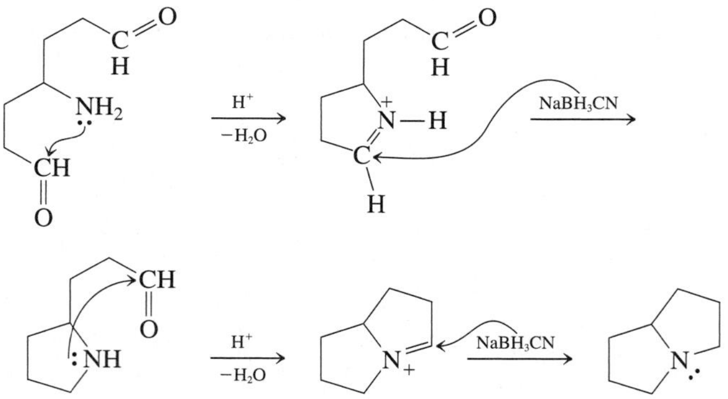

Exercise 21-13 | **Try It Yourself**

Treatment of aldehyde A with the modified sodium cyanoborohydride reagent shown gave the natural product buflavine, a bioactive constituent of the bulbs of the *Boophane flava* plant from southern Africa. Its mass spectral molecular ion occurs at $m/z = 283$. Suggest a structure.

The *Boophane* plant.

In Summary Reductive amination furnishes alkanamines by reductive condensation of amines with aldehydes and ketones.

21-7 | SYNTHESIS OF AMINES FROM CARBOXYLIC AMIDES

Carboxylic amides can be versatile precursors of amines by reduction of the carbonyl unit (Section 20-6). Since amides are in turn readily available by reaction of acyl halides with amines (Section 20-2), the sequence acylation–reduction constitutes a controlled mono-alkylation of amines.

Utility of Amides in Amine Synthesis

$$RCCl + H_2\ddot{N}R' \xrightarrow[-HCl]{Base} RCN\text{HR}' \xrightarrow{LiAlH_4,\ (CH_3CH_2)_2O} RCH_2\ddot{N}HR'$$

Primary amides can also be turned into amines by oxidation with bromine or chlorine in the presence of sodium hydroxide—in other words, by the Hofmann rearrangement (Section 20-7). Recall that in this transformation the carbonyl group is extruded as carbon dioxide, so the resulting amine bears one carbon less than the starting material.

Amines by Hofmann Rearrangement

$$RCNH_2 \xrightarrow{Br_2,\ NaOH,\ H_2O} RNH_2 + O=C=O$$

Exercise 21-14

Show three reactions from any starting materials that would make the C–N bond in *N*-methyl-hexanamine (margin) according to the indicated retrosynthetic disconnection (∿).

In Summary Amides can be reduced to amines by treatment with lithium aluminum hydride. The Hofmann rearrangement converts amides into amines with loss of the carbonyl group.

21-8 | REACTIONS OF QUATERNARY AMMONIUM SALTS: HOFMANN ELIMINATION

Much like the protonation of alcohols, which turns the $-OH$ into the better leaving group $^{\pm}OH_2$ (Section 9-2), protonation of amines might render the resulting ammonium salts subject to nucleophilic attack. In practice, however, amines are not sufficiently good leaving groups (they are more basic than water) to partake in substitution reactions. Nevertheless, they can function as such in the **Hofmann* elimination,** in which a tetraalkylammonium salt is converted to an alkene by a strong base.

Section 21-5 described how complete alkylation of amines leads to the corresponding quaternary alkylammonium salts. These species are unstable in the presence of strong base, because of a bimolecular elimination reaction that furnishes alkenes (Section 7-7). The base attacks the hydrogen in the β-position with respect to the nitrogen, and a trialkylamine departs as a neutral leaving group.

Bimolecular Elimination of Quaternary Ammonium Ions

In the procedure of Hofmann elimination, the amine is first completely methylated with excess iodomethane **(exhaustive methylation)** and then treated with wet silver oxide (a source of HO^-) to produce the ammonium hydroxide. Heating degrades this salt to the alkene. When more than one regioisomer is possible, Hofmann elimination, in contrast to most E2 processes, tends to give less substituted alkenes as the major products. Recall that this result adheres to Hofmann's rule (Section 11-6) and appears to be caused by the bulk of the ammonium group, which directs the base to the less hindered protons in the molecule.

Hofmann Elimination of 1-Butanamine

---

*This is the Hofmann of the Hofmann rule for E2 reactions (Section 11-6) and the Hofmann rearrangement (Section 20-7).

Exercise 21-15

Give the structures of the possible alkene products of the Hofmann elimination of (a) *N*-ethylpropanamine (ethylpropylamine) and (b) 2-butanamine. (**Caution:** In the Hofmann elimination of 2-butanamine, you need to consider several products. **Hint:** Review Section 11-6).

The Hofmann elimination of amines has been used to elucidate the structure of nitrogen-containing natural products, such as alkaloids (Section 25-8). Each sequence of exhaustive methylation and Hofmann elimination cleaves one C–N bond. Repeated cycles allow the heteroatom to be precisely located, particularly if it is part of a ring. In this case, the first carbon–nitrogen bond cleavage opens the ring.

N-Methylazacycloheptane   *N,N*-Dimethyl-5-hexenamine   1,5-Hexadiene

---

Exercise 21-16

Why is exhaustive *methyl*ation and not, say, ethylation used in Hofmann eliminations for structure elucidation? (**Hint:** Look for other possible elimination pathways.)

Exercise 21-17

An unknown amine of the molecular formula $C_7H_{13}N$ has a ^{13}C NMR spectrum containing only three lines of $\delta = 21.0$, 26.8, and 47.8 ppm. Three cycles of Hofmann elimination are required to form 3-ethenyl-1,4-pentadiene (trivinylmethane; margin) and its double-bond isomers (as side products arising from base-catalyzed isomerization). Propose a structure for the unknown.

3-Ethenyl-1,4-pentadiene

In Summary Quaternary methyl ammonium salts, synthesized by amine methylation, undergo bimolecular elimination in the presence of base to give alkenes.

21-9 MANNICH REACTION: ALKYLATION OF ENOLS BY IMINIUM IONS

In the aldol reaction, an enolate ion attacks the carbonyl group of an aldehyde or ketone (Section 18-5) to furnish a β-hydroxycarbonyl product. A process that is quite analogous is the **Mannich* reaction.** Here, however, it is an enol that functions as the nucleophile and an iminium ion, derived by condensation of a second carbonyl component with an amine, as the substrate. The outcome is a β-aminocarbonyl product.

To differentiate the reactivity of the three components of the Mannich reaction, it is usually carried out with (1) a ketone or aldehyde; (2) a relatively more reactive aldehyde (often formaldehyde, $CH_2{=}O$, see Section 17-6); and (3) the amine, all in an alcohol solvent containing HCl. These conditions give the hydrochloride salt of the product. The free amine, called a **Mannich base,** can be obtained upon treatment with base.

---

*Professor Carl U. F. Mannich (1877–1947), University of Berlin.

Reaction

Mannich Reaction

$$\text{Ketone} + CH_2=O + (CH_3)_2NH \xrightarrow{\text{HCl, CH}_3\text{CH}_2\text{OH, }\Delta} \text{Salt of Mannich base}$$

	Ketone	More reactive aldehyde	Amine	Salt of Mannich base

85%

Animation

ANIMATED MECHANISM:
The Mannich reaction

CH_3				

$$CH_3CHCH=O + CH_2=O + CH_3NH_2 \xrightarrow[\text{2. HO}^-,\ \text{H}_2\text{O}]{\text{1. HCl, CH}_3\text{CH}_2\text{OH, }\Delta} $$

2-Methylpropanal

Aldehyde   More reactive aldehyde   Amine

2-Methyl-2-(*N*-methyl-aminomethyl)propanal
Mannich base

70%

> Again, we depict only the mechanism that leads to product. You can formulate multiple alternative reactions steps (protonations; enol, hemiacetal, and heminaminal formations; aldol additions), all of which are reversible dead ends, making the Mannich reaction possible.

The mechanism of this process starts with iminium ion formation between the aldehyde (e.g., formaldehyde) and the amine, on the one hand, and enolization of the ketone, on the other. The electrostatic potential maps in the margin illustrate the electron deficiency of the iminium ion (blue) and the contrasting relative electron abundance (red and yellow) of the enol. One can also recognize that the electron density is greater at the α-carbon (yellow) next to the hydroxy-bearing carbon (green). As soon as the enol is formed, it undergoes nucleophilic attack on the electrophilic iminium carbon, and the resulting species converts to the Mannich salt by proton transfer from the carbonyl oxygen to the amino group.

Mechanism of the Mannich Reaction

Step 1. Iminium ion formation

$$CH_2=O + (CH_3)_2\overset{+}{N}H_2\ Cl^- \longrightarrow CH_2=\overset{+}{N}(CH_3)_2\ Cl^- + H_2O$$

Step 2. Enolization

Mechanism

Iminium ion

Enol

Step 3. Carbon–carbon bond formation

Step 4. Proton transfer

Salt of Mannich base

The following example shows the Mannich reaction in natural product synthesis. In this instance, one ring is formed by condensation of the amino group with one of the two carbonyl functions. Mannich reaction of the resulting iminium salt with the enol form of the other carbonyl function follows. The product has the framework of retronecine (see margin), an alkaloid that is present in many shrubs and is hepatotoxic (causes liver damage) to grazing livestock.

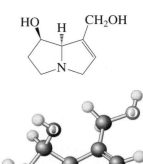

Retronecine

Mannich Reaction in Synthesis

$$\xrightarrow[-H_2O]{H^+} \qquad \xrightarrow{-H^+} \qquad 52\%$$

Solved Exercise 21-18 | **Working with the Concepts: Practicing the Mannich Reaction**

Write the product of the Mannich reaction of ammonia, formaldehyde, and cyclopentanone.

What we need to do is to determine the product of a reaction. Such problems are often relatively straightforward. In the Mannich reaction, however, we have *three* organic components, a complication that requires more care in crafting a solution.

How to begin? To predict the outcome of a Mannich process, it is necessary to follow its mechanism. Several steps are involved, and formulating them in the correct order is essential.

Information needed? The Mannich reaction features the two carbonyl compounds and either ammonia or an amine (see the text section). In particular, we have to identify the more reactive of the carbonyl components, because the process starts by its initial condensation with the amine. The resulting iminium species is then attacked by the enol isomer of the less reactive carbonyl partner.

Proceed. We follow the above steps of the mechanism. Thus:

- The more reactive carbonyl species is formaldehyde.

- Formaldehyde and ammonia condense to $\overset{H}{\underset{H}{C}}{=}\overset{H}{\underset{H}{N^+}}$.

- The enol tautomer of cyclopentanone is .

- Reaction of the two species occurs by attack of the electron-rich enol on the positively polarized iminium ion, and deprotonation on basic work-up finishes the job.

Exercise 21-19 | **Try It Yourself**

Write the products of each of the following Mannich reactions: **(a)** 1-hexanamine + formaldehyde + 2-methylpropanal; **(b)** *N*-methylmethanamine + formaldehyde + acetone; **(c)** cyclohexanamine + formaldehyde + cyclohexanone.

Exercise 21-20

β-Dialkylamino alcohols and their esters are useful local anesthetics. Suggest a synthesis of the anesthetic Tutocaine hydrochloride, beginning with 2-butanone and using any other organic compounds. (**Hint:** Identify the embedded four-carbon unit of 2-butanone in the product and apply retrosynthetic disconnections from it.)

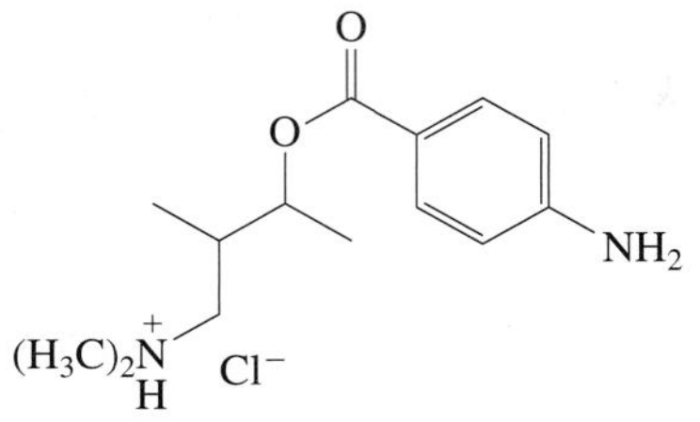

Tutocaine hydrochloride

In Summary Condensation of aldehydes (e.g., formaldehyde) with amines furnishes iminium ions, which are electrophilic and may be attacked by the enols of ketones (or other aldehydes) in the Mannich reaction. The products are β-aminocarbonyl compounds.

21-10 | NITROSATION OF AMINES

Amines react with nitrous acid, through nucleophilic attack on the **nitrosyl cation,** NO^+. The product depends very much on whether the reactant is an alkanamine or a benzenamine (aniline) and on whether it is primary, secondary, or tertiary. This section deals with alkanamines; aromatic amines will be considered in the next chapter.

To generate NO^+, we must first prepare the unstable nitrous acid by the treatment of sodium nitrite with aqueous HCl. In such an acid solution, an equilibrium is established with the nitrosyl cation. (Compare this sequence with the preparation of the nitronium cation from nitric acid; Section 15-10.)

Nitrosyl Cation from Nitrous Acid

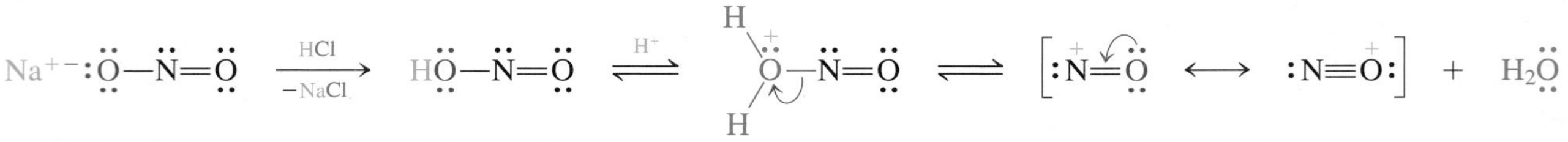

The nitrosyl cation is electrophilic and is attacked by amines to form an *N*-nitrosammonium salt.

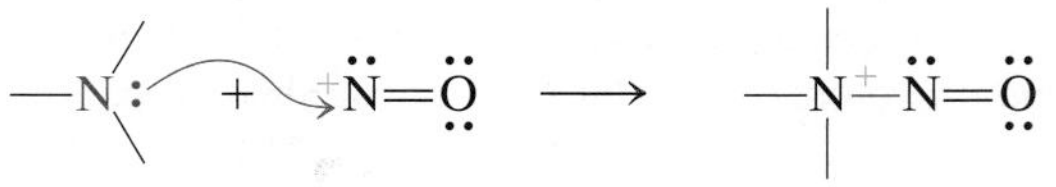

N-Nitrosammonium salt

The course of the reaction now depends on whether the amine nitrogen bears zero, one, or two hydrogens. *Tertiary N*-nitrosammonium salts are stable only at low temperatures and decompose upon heating to give a mixture of compounds. *Secondary N*-nitrosammonium salts are simply deprotonated to furnish the relatively stable **N-nitrosamines** as the major products.

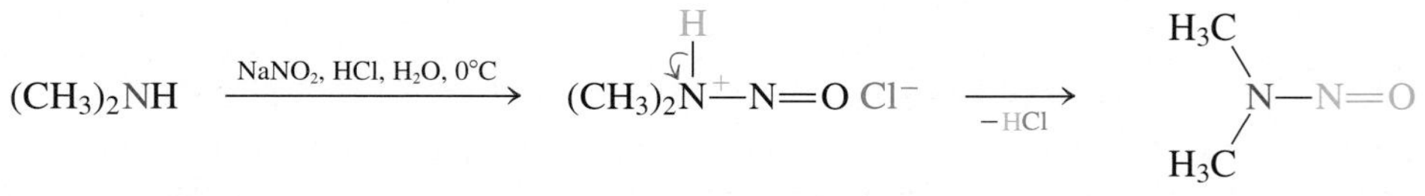

$R_2N-N=O$
N-Nitrosamine

$(CH_3)_2NH \xrightarrow{\text{NaNO}_2,\ \text{HCl},\ \text{H}_2\text{O},\ 0°C} (CH_3)_2\overset{+}{N}-N=O\ Cl^- \xrightarrow{-\text{HCl}}$

H₃C\
 >N—N=O\
H₃C

88–90%
N-Nitrosodimethylamine

REAL LIFE: MEDICINE 21-2 | Sodium Nitrite as a Food Additive, *N*-Nitrosodialkanamines, and Cancer

N-Nitrosodialkanamines are notoriously potent carcinogens in a variety of animals. Although there is no direct evidence, they are suspected of causing cancer in humans as well. Most nitrosamines appear to cause liver cancer, but some of them are very organ specific in their carcinogenic potential (bladder, lungs, esophagus, nasal cavity, etc.). Their mode of carcinogenic action appears to begin with enzymatic oxidation of one of the α-carbons to a hemiaminal (Section 17-9), which allows eventual formation of an unstable monoalkyl-*N*-nitrosamine, as shown here for *N*-nitrosodiethylamine:

Monoalkyl-*N*-nitrosamines are sources of carbocations, as described in this section. Carbocations are powerful electrophiles that are thought to attack one of the bases in DNA to inflict the kind of genetic damage that seems to lead to cancerous cell behavior.

In the early 1970s, farm animals in Norway that had been fed herring meal preserved using relatively large doses of sodium nitrite began to show a high incidence of liver disorders, including cancer. Fish meal contains amines (for example, trimethylamine is primarily responsible for its odor) and it did not take long to realize that these amines reacted with the added nitrite to produce nitrosamines. This finding was disturbing because amines also occur in meats, and sodium nitrite is often added during the meat-curing process. Sodium nitrite inhibits the growth of bacteria that cause botulism, retards the development of rancidity and off-odors during storage, and preserves the flavor of added spices and smoke (if the meat is smoked). In addition, it gives meat an appetizing pink coloration and special flavor. The color has its origin in nitrite-derived nitric oxide, NO, which forms a red complex with the iron in myoglobin. (The oxygen complex of myoglobin gives blood its characteristic color; see Section 26-8).

Does the practice of curing meat lead to nitrosamine contamination? The answer is yes: Nitrosamines have been detected in a variety of cured meats, such as smoked fish, frankfurters (*N*-nitrosodimethylamine), and fried bacon [*N*-nitrosoazacyclopentane (*N*-nitrosopyrrolidine); see Exercise 25-7]. This finding poses an environmental dilemma: Removing sodium nitrite as a food additive would prevent it from becoming a source of nitrosamines, but it would also cause a significant rise in botulism poisoning. Moreover, human exposure to nitrosamines occurs not only through meat; nitrosamines are also present in beer, nonfat dry milk, tobacco products, rubber, some cosmetics, and the gastric juices of the stomach. Their occurrence in the stomach can be called "natural," as it is traced to bacteria in the mouth that reduce nitrate (which is, in turn, prevalent in vegetables such as spinach, beets, radishes, celery, and cabbages) to nitrite, which then reacts with amines present in other foods we ingest.

Because of the potential carcinogenicity of nitrosamines in humans, the level of nitrite in foods is closely regulated (<200 ppm). During the past 20 years, a number of measures, such as the use of additives that inhibit nitrosamine formation in curing (for example, vitamin C; Section 22-9) or changes in production processes (for example, the making of beer), have helped to significantly reduce human exposure to nitrosamines, estimated to be about 0.1 μg per day. It can be argued that we have evolved to tolerate such small quantities, but definitive data are still lacking (see also Real Life 25-4, on natural pesticides).

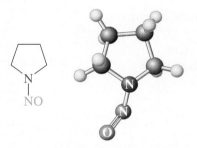

N-Nitrosoazacyclopentane
(*N*-Nitrosopyrrolidine)

Nitrosamines are found in fried bacon.

Similar treatment of *primary* amines initially gives the analogous monoalkyl-*N*-nitrosamines. However, these products are unstable because of the remaining proton on the nitrogen. By a series of hydrogen shifts, they first rearrange to the corresponding diazohydroxides. Then protonation, followed by loss of water, gives highly reactive **diazonium ions,** R–N$_2$⁺. When R is a secondary or a tertiary alkyl group, these ions lose molecular nitrogen, N$_2$, and form the corresponding carbocations, which may rearrange, deprotonate, or undergo nucleophilic trapping (Section 9-3) to yield the observed mixtures of compounds.

Mechanism of Decomposition of Primary *N*-Nitrosamines

Step 1. Rearrangement to a diazohydroxide

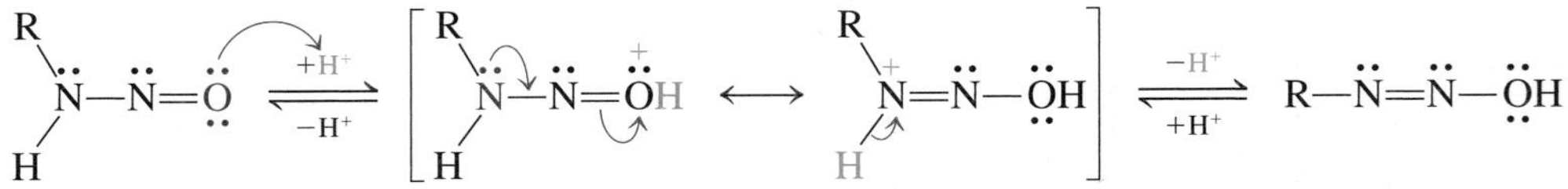

> **Reminder**
>
> In the adjacent scheme, free H⁺ does not exist in solution, but is attached to any available electron pair, mostly the oxygen of the solvent water.

Diazohydroxide

Step 2. Loss of water to give a diazonium ion

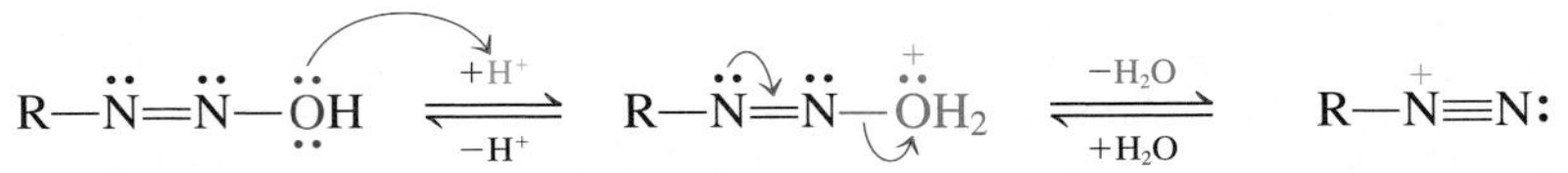

Diazonium cation

Step 3. Nitrogen loss to give a carbocation

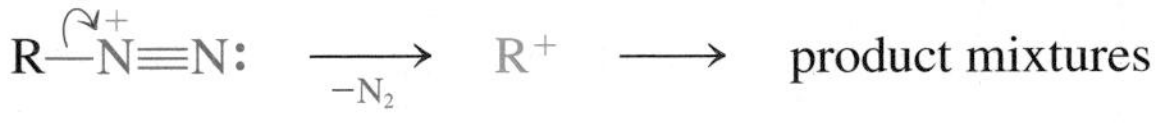

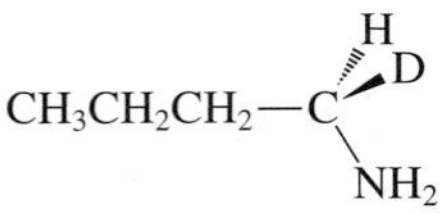

$$R\overset{\frown}{-}\overset{+}{N}\equiv N: \xrightarrow{-N_2} R^+ \longrightarrow \text{product mixtures}$$

Pure *R* enantiomer

NaNO$_2$, HCl, H$_2$O

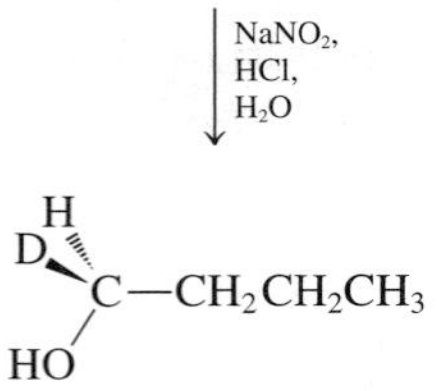

100%
Pure *S* enantiomer

Exercise 21-21

The mechanism just shown pertains to the decomposition of diazonium ions in which the alkyl groups R are secondary or tertiary. The result depicted in the margin was reported in 1991 and addresses the pathway chosen in the case of R = *primary* alkyl. Is the mechanism the same?

The nitrosyl cation also attacks the nitrogen of *N*-methylamides. The products, *N*-methyl-*N*-nitrosamides, are precursors to useful synthetic intermediates.

Nitrosation of an *N*-Methylamide

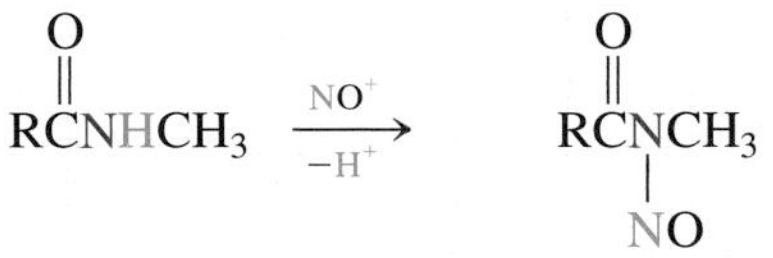

N-Methylamide   *N*-Methyl-*N*-nitrosamide

Solved Exercise 21-22 | Working with the Concepts: Applying Basic Principles: A New Reaction of NO⁺

The ketone shown below undergoes nitrosation to the corresponding oxime (for oximes, see Section 17-9). Explain by a mechanism.

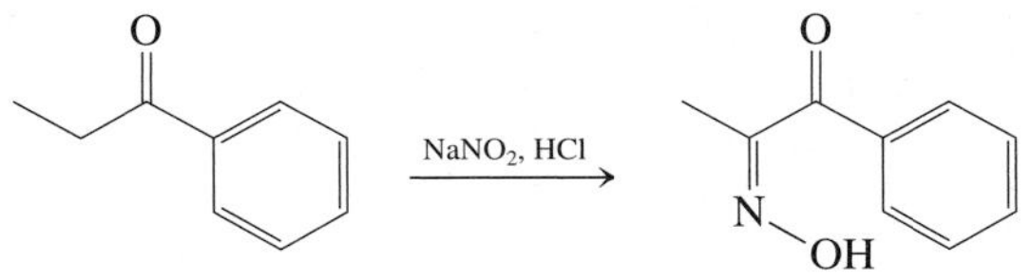

Strategy

This is a new reaction, not discussed in the text. To try to understand it, you need to identify the changes that have occurred in going from starting material to product and to recall the function of the reagent.

Solution

- Topologically, we have attached the NO unit to the α-position of the starting ketone. This becomes clearer if we consider the tautomer of the product:

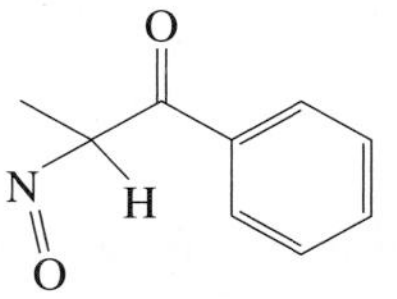

- Turning to the reagent, this section has shown that sodium nitrite in the presence of acid is a source of the electrophilic nitrosyl cation. In our example, it appears that attack has occurred α to the carbonyl function. Do we know anything about how electrophiles might attack ketones in this manner? Or, to put it differently, how do we make the α-carbonyl carbon nucleophilic?

- Recall from Chapter 18 that ketones are in equilibrium with their enol tautomers, an equilibrium that is established rapidly in the presence of catalytic acid. These enols feature electron-rich double bonds that are ready targets for electrophiles, such as iminium ions (Mannich reaction, Section 21-9) or halogens (Section 18-3). Let us formulate a similar attack by NO^+.

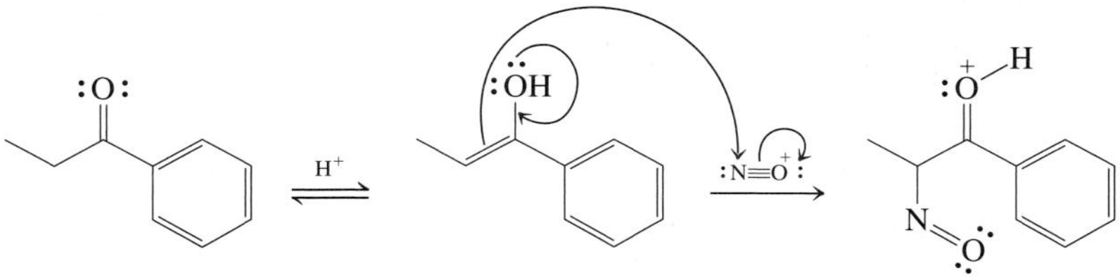

- We have now established the connectivity of the final product. To form it, all we need is to deprotonate the carbonyl function and tautomerize the nitrosyl part to the oxime.

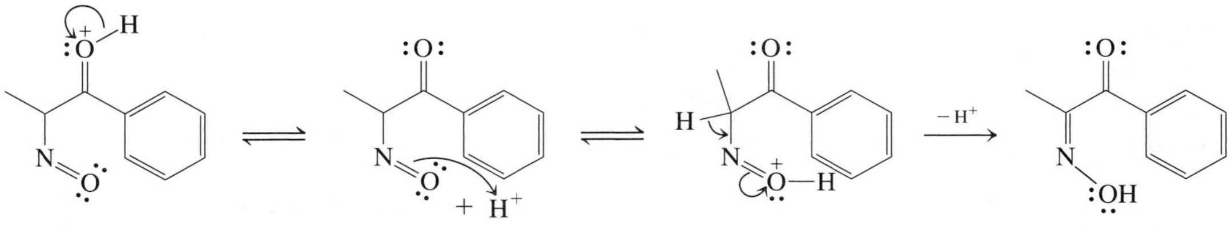

Exercise 21-23 | Try It Yourself

The following reaction has been explored as a possible first step in the preparation of acetaminophen (Chapter 16 Opening). What is its mechanism and what is the mechanistic connection to the process described in Exercise 21-22?

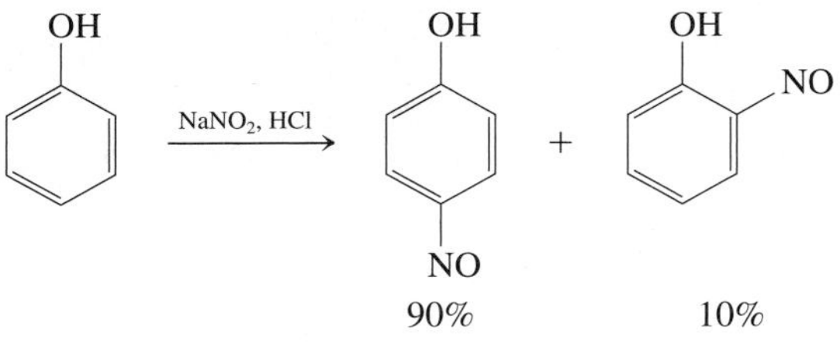

$$\text{NaNO}_2, \text{ HCl}$$

90%   10%

REAL LIFE: MATERIALS 21-3 | Amines in Industry: Nylon, the "Miracle Fiber"

In addition to their significance in medicine (Real Life 21-1), the amines have numerous industrial applications. An example is 1,6-hexanediamine (hexamethylenediamine, HMDA), needed in the manufacture of nylon. The amine is polycondensed with hexanedioic (adipic) acid to produce Nylon 6,6, out of which hosiery, gears, and millions of tons of textile

fiber are made. When it was introduced by du Pont in 1938 as the first synthetic fiber, it was touted to be "as strong as steel, as fine as a spider's web," and dubbed the "miracle fiber," a cheap alternative to natural fabrics, extremely durable, yet readily malleable. It is one of the most-used polymers in the world today.

Polycondensation of Adipic Acid with HMDA

Hexanedioic acid (Adipic acid)

+

1,6-Hexanediamine (Hexamethylenediamine)

Double salt

270°C, 250 psi, −H₂O, Polymerization

Nylon 6,6

The high demand for nylon has stimulated the development of several ingenious cheap syntheses of the monomeric precursors. Thus, hexanedioic (adipic) acid is currently produced from benzene by three different multistep routes, all culminating in the last step in the salt-catalyzed oxidation of cyclohexanol (or cyclohexanol/cyclohexanone mixtures) with nitric acid. A "green" approach was disclosed in 2006 by chemists at Rhodia Chimie, France, in which air is used as the oxidizing agent instead of the toxic and corrosive (and more expensive) HNO₃.

The left leg of movie star Marie Wilson (who can be seen dangling from a crane) advertising Nylon stockings (Hollywood, 1949).

Adipic Acid Production

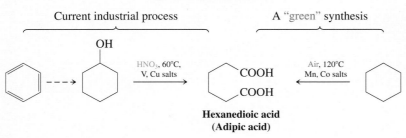

Current industrial process — HNO₃, 60°C, V, Cu salts — **Hexanedioic acid (Adipic acid)** ← Air, 120°C, Mn, Co salts — A "green" synthesis

The hexanediamine component of nylon was originally made by Wallace Carothers* at E. I. du Pont de Nemours from hexanedioic acid, which was turned into hexanedinitrile (adiponitrile) by treatment with ammonia. Finally, catalytic hydrogenation furnished the diamine.

Hexanediamine from Adipic Acid

$$\underset{\substack{\textbf{Hexanedioic acid}\\ \textbf{(Adipic acid)}}}{\overset{\displaystyle O \quad\quad O}{\overset{\displaystyle \| \quad\quad \|}{HOC(CH_2)_4COH}}} \xrightarrow[-4\,H_2O]{NH_3,\,\Delta} \underset{\substack{\textbf{Hexanedinitrile}\\ \textbf{(Adiponitrile)}}}{N\equiv C(CH_2)_4C\equiv N} \xrightarrow{H_2,\,Ni,\,130°C,\,2000\,psi} \underset{\substack{\textbf{1,6-Hexanediamine}\\ \textbf{(Hexamethylene-}\\ \textbf{diamine)}}}{H_2N(CH_2)_6NH_2}$$

Later, a still shorter hexanedinitrile synthesis was discovered that used 1,3-butadiene as a starting material. Chlorination of butadiene furnished a mixture of 1,2- and 1,4-dichlorobutene (Section 14-6). This mixture can be converted directly into only the required 3-hexenedinitrile with sodium cyanide in the presence of cuprous cyanide. Selective hydrogenation of the alkenyl double bond then furnishes the desired product.

Hexanedinitrile (Adiponitrile) from 1,3-Butadiene

$$CH_2{=}CH{-}CH{=}CH_2 \xrightarrow{Cl{-}Cl} ClCH_2CH{=}CHCH_2Cl \;+\; \underset{\substack{\\ \\}}{\overset{\displaystyle Cl \atop \displaystyle |}{ClCH_2CHCH{=}CH_2}} \xrightarrow[-NaCl]{CuCN,\,NaCN}$$

$$\underset{\textbf{3-Hexenedinitrile}}{NCCH_2CH{=}CHCH_2CN} \xrightarrow{H_2,\,catalyst} \underset{\textbf{Hexanedinitrile}}{NC(CH_2)_4CN}$$

In the mid-1960s, Monsanto developed a process that uses a more expensive starting material but takes just one step: the electrolytic hydrodimerization of propenenitrile (acrylonitrile).

Electrolytic Hydrodimerization of Propenenitrile (Acrylonitrile)

$$2\,CH_2{=}CHC\equiv N \;+\; 2\,e \;+\; 2\,H^+ \;\longrightarrow\; N\equiv CCHCH_2{-}CH_2CHC\equiv N$$
$$\underset{\substack{\textbf{Propenenitrile}\\ \textbf{(Acrylonitrile)}}}{\phantom{2\,CH_2{=}CHC\equiv N}} \qquad\qquad\qquad\qquad \underset{}{\overset{}{\;\;\;\;\;\;\;\;\;\;H\;\;\;\;\;\;\;\;\;\;\;\;H}}$$

To counter Monsanto's challenge, du Pont devised yet another synthesis, again starting with 1,3-butadiene, but now eliminating the consumption of chlorine, removing the toxic-waste problems of the disposal of copper salts, and using cheaper hydrogen cyanide rather than sodium cyanide. The synthesis is based on the conceptually simplest approach: direct regioselective anti-Markovnikov addition of two molecules of hydrogen cyanide to butadiene. A transition-metal catalyst, such as iron, cobalt, or nickel, is needed to effect this regiocontrol.

Hydrogen Cyanide Addition to 1,3-Butadiene

$$CH_2{=}CHCH{=}CH_2 \;+\; 2\,HCN \xrightarrow{Catalyst} NCCH_2CHCHCH_2CN$$
$$\qquad\qquad\qquad\qquad\qquad\qquad\qquad\qquad H\;\;H$$

---

*Dr. Wallace H. Carothers (1896–1937), E. I. du Pont de Nemours and Company, Wilmington, Delaware.

Base treatment of *N*-methyl-*N*-nitrosamides gives diazomethane

N-Methyl-*N*-nitrosamides are converted into **diazomethane**, CH_2N_2, upon treatment with aqueous base.

Making Diazomethane

$$CH_2=\overset{+}{N}=\overset{\cdot\cdot}{N}:^- + NH_3 + K_2CO_3 + H_2O$$

N-Methyl-N-nitrosourea   Diazomethane

Diazomethane is used in the synthesis of methyl esters from carboxylic acids. However, it is exceedingly toxic and highly explosive in the gaseous state (b.p. $-24°C$) and in concentrated solutions. It is therefore usually generated in dilute ether solution and immediately allowed to react with the acid. This method is very mild and permits esterification of molecules possessing acid- and base-sensitive functional groups, as shown in the following example.

75%

When it is irradiated or exposed to catalytic amounts of copper, diazomethane evolves nitrogen to generate the reactive carbene methylene, H_2C: Methylene reacts with alkenes by addition to form cyclopropanes stereospecifically (Section 12-9).

Exercise 21-24

Formulate a mechanism for the methyl esterification by diazomethane. (**Hint:** Review the resonance structures for CH_2N_2 in Section 1-5.)

In Summary Nitrous acid attacks amines, thereby causing *N*-nitrosation. Secondary amines give *N*-nitrosamines, which are notorious for their carcinogenicity. *N*-Nitrosamines derived from primary amines decompose through S_N1 or S_N2 processes to a variety of products. *N*-Nitrosation of *N*-methylamides results in the corresponding *N*-nitrosamides, which liberate diazomethane upon treatment with hydroxide. Diazomethane is a reactive substance used in the methylesterification of carboxylic acids and as a source of methylene for the cyclopropanation of alkenes.

THE BIG PICTURE

We have seen the role of the amine function as a base or nucleophile at various points of the text. As early as in Sections 1-3 and 1-8, we learned about its valence electron count and hybridization, using NH_3 as an example. In Section 2-2, we saw its first use as a base or nucleophile, and a resulting practical application in the resolution of ammonium tartrates was featured in Section 5-8. Amine nucleophilicity was an essential part of Chapters 6 and 7, elaborated in its synthetic aspects in Sections 17-9, 18-4, 18-9, 19-10, 20-6, and 20-7.

The present chapter summarizes, reinforces, and extends this material. In particular, it points out the similarities and differences between the amines and the alcohols. Thus, nitrogen is located immediately to the left of oxygen in the periodic table. It therefore has one valence electron less and, correspondingly, lesser positive nuclear charge. Hence, it is trivalent, and not divalent, and less electronegative. However, chemically, it shows many similarities, and the amine–imine functionalities strongly resemble their oxygen counterparts—namely, the alcohol–carbonyl functionalities. In the laboratory, as well as in nature, the two systems complement each other, as described in several Real Life vignettes and problems. Once again, the similarities and differences in how these systems function can be traced back to the similarities and differences in their electronic structures.

We have now completed our survey of the basic organic chemistry compound classes, as defined by simple functional groups. In the remainder of the book, we shall examine compounds with multiple functional groups, whether they be the same or different. We shall see the amino group and amine derivatives again in these chapters, particularly when studying heterocycles, proteins, and nucleic acids. We shall learn that the combination of the amino and carbonyl groups underlies the molecular structure of life itself.

WORKED EXAMPLES: INTEGRATING THE CONCEPTS

21-25. Practicing Retrosynthesis and Synthesis of a Complex Drug

On the basis of the synthetic methods for amines provided in this chapter, carry out retrosynthetic analyses of the antidepressant Prozac, with 4-trifluoromethylphenol and benzene as your starting materials.

Prozac
[*R,S-N*-Methyl-3-(4-trifluoro-
phenoxy)-3-phenyl-1-propanamine]

SOLUTION

As for any synthetic problem, you can envisage many possible solutions. However, the constraints of given starting materials, convergence, and practicality rapidly narrow the number of available options. Thus, it is clear that the 4-trifluoromethylphenoxy group is best introduced by Williamson ether synthesis (Section 9-6) with an appropriate benzylic halide, compound A (Section 22-1), with the use of our first starting compound, 4-trifluoromethylphenol.

Retrosynthetic Step to A

The task therefore reduces to devising routes to A, utilizing the second starting material, benzene, and suitable building blocks. We know that the best way to introduce an alkyl chain into benzene is by Friedel-Crafts acylation (Section 15-13). This reaction would also supply a carbonyl function, as in compound B, which could be readily transformed [by reduction to the alcohol (Section 8-6) and conversion of the hydroxy into a good leaving group, X (Chapter 9)] into compound A (or something similar). On paper, an attractive acylating reagent would be $ClCCH_2CH_2NHCH_3$. However, this reagent carries two functional groups that would react with each other either intermolecularly to produce a polyamide (see Real Life 21-3) or intramolecularly to produce a β-lactam (Section 20-6). This problem could be circumvented by using a leaving group in place of the amino function, the latter to be introduced by one of the methods in Section 21-5.

Retrosynthesis of A

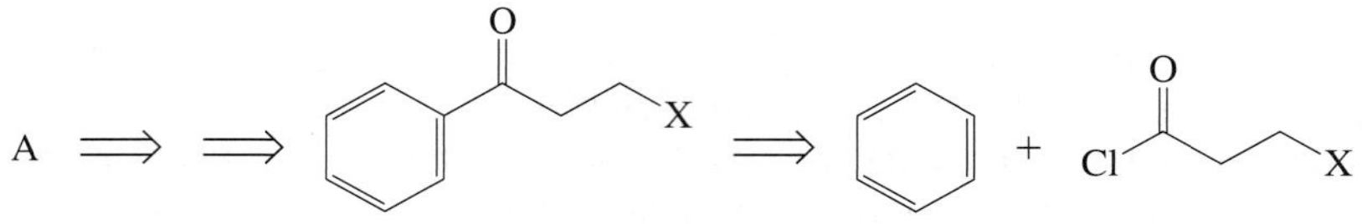

A $\Longrightarrow$ $\Longrightarrow$

B

$\Longrightarrow$

Consideration of cyanide as a building block (Section 21-5) opens up another avenue of retrosynthetic disconnection of compound B (and therefore compound A) through compound C.

Retrosynthesis of B

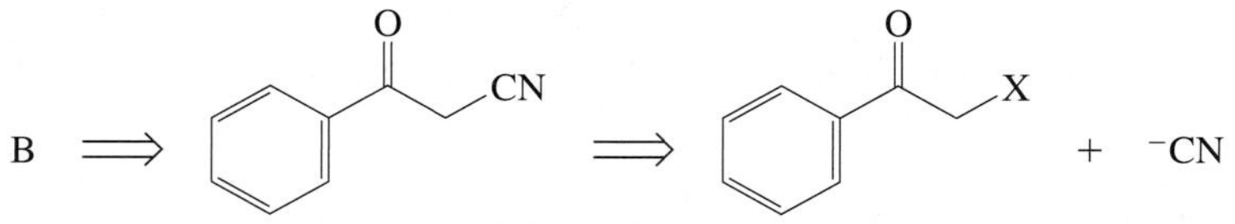

B $\Longrightarrow$

C

$\Longrightarrow$

+ $^-$CN

D

The required compound D could be envisaged to arise from acetylbenzene (made by Friedel-Crafts acetylation of benzene), followed by acid-catalyzed halogenation of the ketone (Section 18-3). Reduction of the nitrile group in compound C could be carried out with concomitant carbonyl conversion to give a primary amine version of compound A, namely, E. *N*-Methylation might be most conveniently accomplished by reductive amination (although, as we shall see in Chapter 22, the benzylic position may be sensitive to the reductive conditions employed).

Retrosynthesis of E

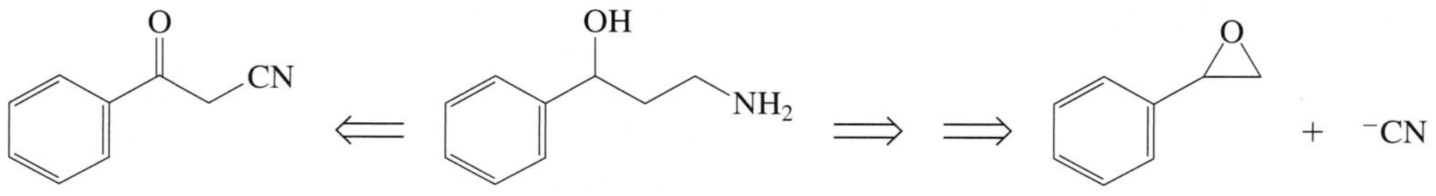

C $\Longleftarrow$ **E** $\Longrightarrow$ $\Longrightarrow$ **Phenyloxacyclo-propane**   + $^-$CN

Intermediate E can be approached by using nucleophilic cyanide in a different manner—namely, by attack at the less hindered position of phenyloxacyclopropane. The latter would arise from phenylethene (styrene) by oxacyclopropanation (Section 12-10), and phenylethene could in turn be readily made from acetylbenzene by reduction–dehydration.

Inspection of the synthetic methods described in Sections 21-5 through 21-8 may give you further ideas about how to tackle the synthesis of Prozac, but they are all variations on the schemes formulated so far. However, consideration of the Mannich reaction (Section 21-9) provides a more fundamental alternative, attractive because of its more highly convergent nature to a derivative of compound B with the intact amino function in place. Indeed, this is the commercial route used by Eli Lilly to the final drug.

Retro-Mannich Synthesis of B (X=CH₃NH)

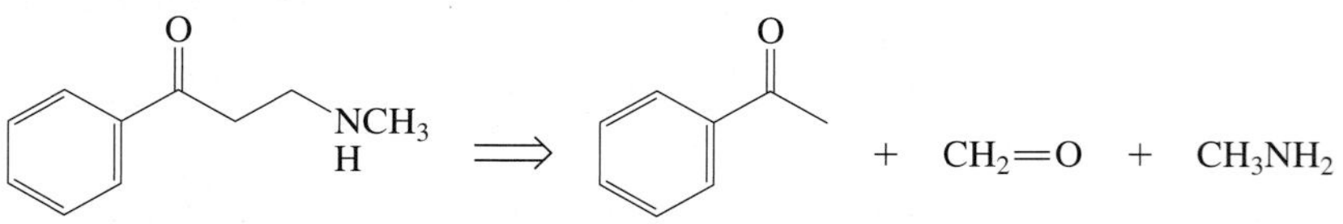

B $\Longrightarrow$ + $CH_2=O$ + CH_3NH_2

21-26. An Enamine Rearrangement Mechanism

Formulate a mechanism for the following reaction.

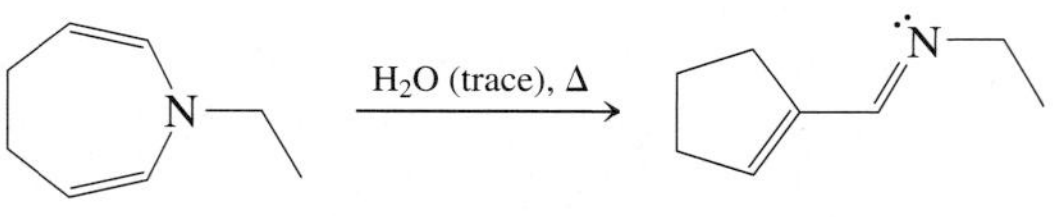

A $\xrightarrow{H_2O \text{ (trace)}, \Delta}$ **B**

SOLUTION

As always, let us first analyze what is given, before delving into a possible solution. We note that $C_8H_{13}N$ on the left turns into the same on the right: We are dealing with an isomerization. There are no reagents, only a catalytic amount of water (a proton source). The topology of a seven-membered ring changes to that of a five-membered one, and the functionality of a (di)enamine (Section 17-9) is transformed into that of an α,β-unsaturated imine (Section 18-8). Our task is to find a way by which water will open the seven-membered ring and eventually close some acyclic intermediate to the five-membered ring product.

What does an enamine do in the presence of water? *Answer:* it hydrolyzes (Section 17-9).

Mechanism of Hydrolysis

Let us formulate the mechanism of this step as applied to our starting material A. Crucial in dealing with enamines is the increased basicity (and nucleophilicity) of the β-carbon, which leads to its preferred protonation (Section 17-9) [and alkylation (Section 18-4)]. Thus, protonation of A gives B, which is attacked by hydroxide to furnish C, thus completing the initial hydration sequence in enamine hydrolysis (Section 17-9). The next step, breaking the C–N bond, has a "quirk" in our system: The nitrogen, usually activated to leave by direct protonation, is part of an enamine unit, which is (again) protonated at carbon to generate D containing an iminium leaving group. Ring opening therefore proceeds through E to F.

Now that we have accomplished ring opening, what is next? Simply looking at the topology of F and the desired B, we can see that we need to form a (double) bond between the β-carbon of the imine and the carbonyl carbon. If the imine were a ketone, we would accomplish this by an intramolecular aldol condensation (Section 18-7). Here, we utilize the imine as a masked ketone, via its enamine form G as the nucleophile. The individual steps are completely analogous.

An "Imine Version" of the Intramolecular Aldol Condensation

Note that the rapid imine F to enamine G conversion obviates an alternative pathway of an intramolecular Mannich reaction (Section 21-9). Formulate this alternative!

New Reactions

1. Acidity of Amines and Amide Formation (Section 21-4)

$$RNH_2 \ + \ H_2O \ \overset{K}{\rightleftharpoons} \ R\overset{-}{N}H \ + \ H_3O^+ \qquad K_a \approx 10^{-35}$$

$$R_2NH \ + \ CH_3CH_2CH_2CH_2Li \ \rightleftharpoons \ R_2N^-Li^+ \ + \ CH_3CH_2CH_2CH_3$$
Lithium
dialkylamide

$$2\ NH_3 \ + \ 2\ Na \ \xrightarrow{\text{Catalytic Fe}^{3+}} \ 2\ NaNH_2 \ + \ H_2$$

2. Basicity of Amines (Section 21-4)

$$RNH_2 \ + \ H_2O \ \rightleftharpoons \ R\overset{+}{N}H_3 \ + \ HO^-$$

$$R\overset{+}{N}H_3 \ + \ H_2O \ \overset{K}{\rightleftharpoons} \ RNH_2 \ + \ H_3O^+ \qquad K_a \approx 10^{-10}$$

Salt formation

$$RNH_2 \ + \ HCl \ \longrightarrow \ R\overset{+}{N}H_3\ Cl^-$$
Alkylammonium
chloride
General for primary, secondary, and tertiary amines

Preparation of Amines

3. Amines by Alkylation (Section 21-5)

$$R\overset{..}{N}H_2 \ + \ R'X \ \longrightarrow \ R\overset{R'}{\underset{|+}{N}}H_2\ X^-$$
General for primary, secondary, and tertiary amines

Drawback: multiple alkylation

$$R\overset{R'}{\underset{|+}{N}}H_2\ X^- \ + \ R'X \ \longrightarrow \ \longrightarrow \ \longrightarrow \ R\overset{+}{N}R'_3\ X^-$$

4. Primary Amines from Nitriles (Section 21-5)

$$RX \ + \ ^-CN \ \xrightarrow[-X^-]{\overset{\text{DMSO}}{S_N2}} \ RCN \ \xrightarrow{\text{LiAlH}_4 \text{ or } H_2, \text{ catalyst}} \ RCH_2NH_2$$
R limited to methyl, primary, and secondary alkyl groups

5. Primary Amines from Azides (Section 21-5)

$$RX \ + \ N_3^- \ \xrightarrow[-X^-]{\overset{\text{CH}_3\text{CH}_2\text{OH}}{S_N2}} \ RN_3 \ \xrightarrow{\overset{1.\ \text{LiAlH}_4,\ (\text{CH}_3\text{CH}_2)_2\text{O}}{2.\ H_2O}} \ RNH_2$$
R limited to methyl, primary, and secondary alkyl groups

6. Primary Amines by Gabriel Synthesis (Section 21-5)

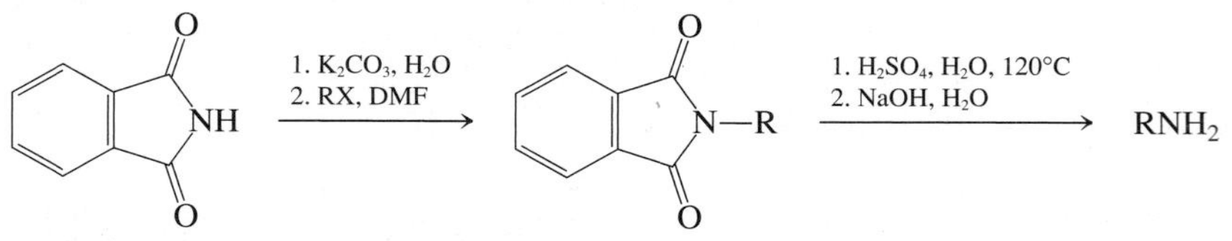

R limited to methyl, primary, and secondary alkyl groups

7. Amines by Reductive Amination (Section 21-6)

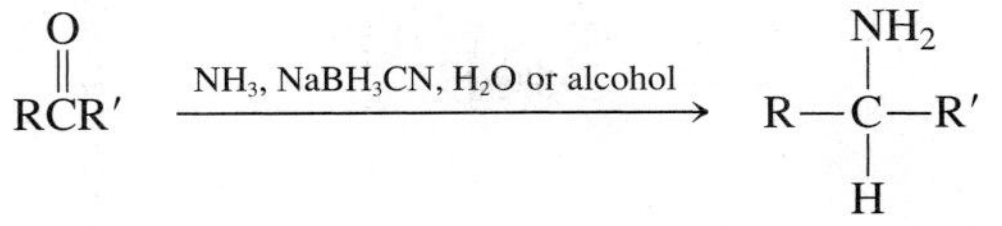

$$\underset{\substack{\| \\ RCR'}}{\overset{O}{}} \xrightarrow{\text{NH}_3, \text{ NaBH}_3\text{CN, H}_2\text{O or alcohol}} \underset{\substack{| \\ H}}{\overset{\text{NH}_2}{R-C-R'}}$$

Reductive methylation with formaldehyde

$$R_2NH \quad + \quad CH_2{=}O \xrightarrow{\text{NaBH}_3\text{CN, CH}_3\text{OH}} R_2NCH_3$$

8. Amines from Carboxylic Amides (Section 21-7)

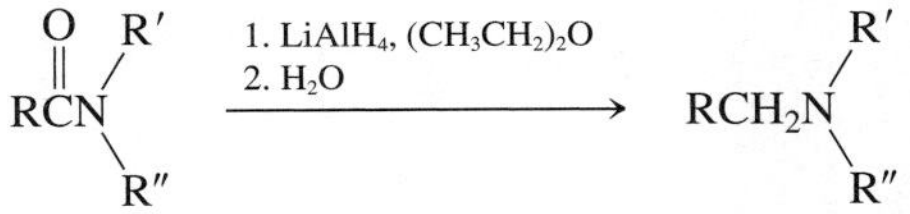

$$\underset{\substack{\| \\ RCN \\ | \\ R''}}{\overset{O \quad R'}{}} \xrightarrow[\text{2. H}_2\text{O}]{\text{1. LiAlH}_4, \text{ (CH}_3\text{CH}_2)_2\text{O}} \underset{\substack{| \\ R''}}{\overset{R'}{RCH_2N}}$$

9. Hofmann Rearrangement (Section 21-7)

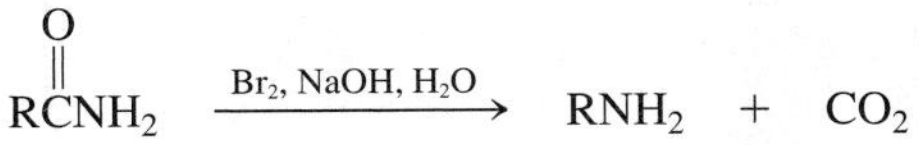

$$\underset{\substack{\| \\ RCNH_2}}{\overset{O}{}} \xrightarrow{\text{Br}_2, \text{ NaOH, H}_2\text{O}} RNH_2 \quad + \quad CO_2$$

Reactions of Amines

10. Hofmann Elimination (Section 21-8)

$$RCH_2CH_2NH_2 \xrightarrow{\text{Excess CH}_3\text{I, K}_2\text{CO}_3} RCH_2CH_2\overset{+}{N}(CH_3)_3 \; I^- \xrightarrow[-\text{AgI}]{\text{Ag}_2\text{O, H}_2\text{O}}$$

$$RCH_2CH_2\overset{+}{N}(CH_3)_3 \; {}^-OH \xrightarrow{\Delta} RCH{=}CH_2 \quad + \quad N(CH_3)_3 \quad + \quad H_2O$$

11. Mannich Reaction (Section 21-9)

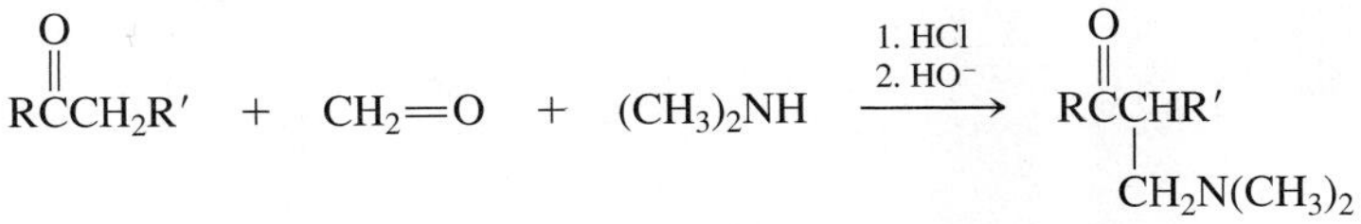

$$\underset{\substack{\| \\ RCCH_2R'}}{\overset{O}{}} + \; CH_2{=}O \; + \; (CH_3)_2NH \xrightarrow[\text{2. HO}^-]{\text{1. HCl}} \underset{\substack{| \\ CH_2N(CH_3)_2}}{\overset{O}{RCCHR'}}$$

12. Nitrosation of Amines (Section 21-10)

Tertiary amines

$$R_3N \xrightarrow{\text{NaNO}_2, \text{ H}^-\text{X}^-} R_3\overset{+}{N}{-}NO \; X^-$$

Tertiary *N*-nitrosammonium salt

Secondary amines

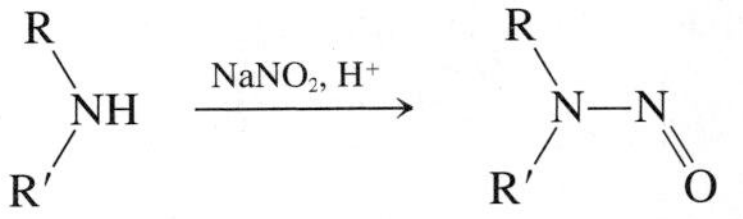

$$\underset{\substack{R' \\}}{\overset{R}{}}{\diagdown}NH \xrightarrow{\text{NaNO}_2, \text{ H}^+} \underset{\substack{R' \\}}{\overset{R}{}}{\diagdown}N{-}N{\diagdown}O$$

***N*-Nitrosamine**

Primary amines

$$RNH_2 \xrightarrow{\text{NaNO}_2, \text{ H}^+} RN{=}NOH \xrightarrow[-\text{H}_2\text{O}]{\text{H}^+} RN_2^+ \xrightarrow{-\text{N}_2} R^+ \longrightarrow \text{mixture of products}$$

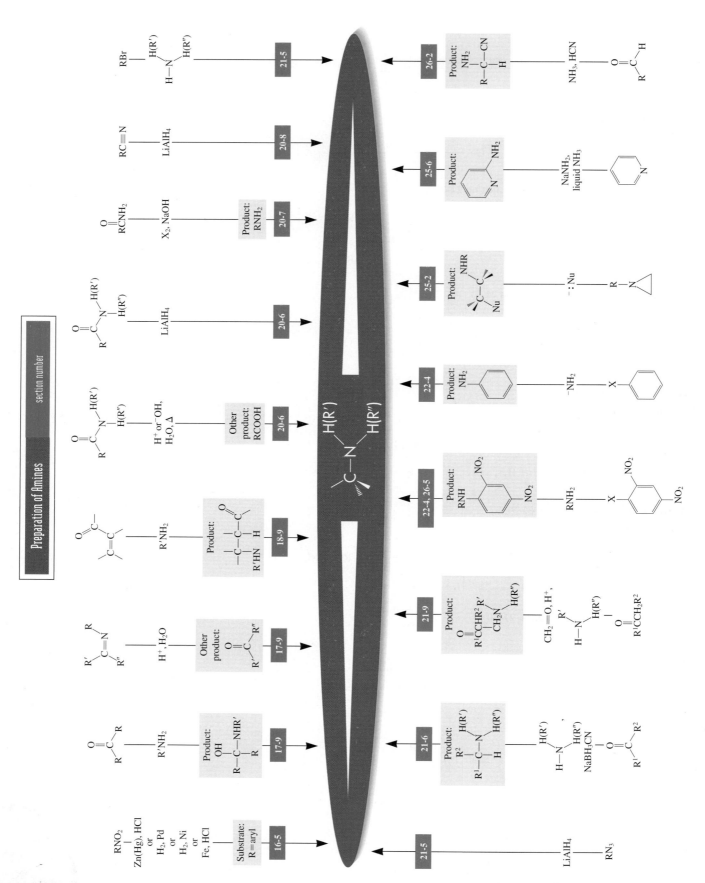

Preparation of Amines

section number

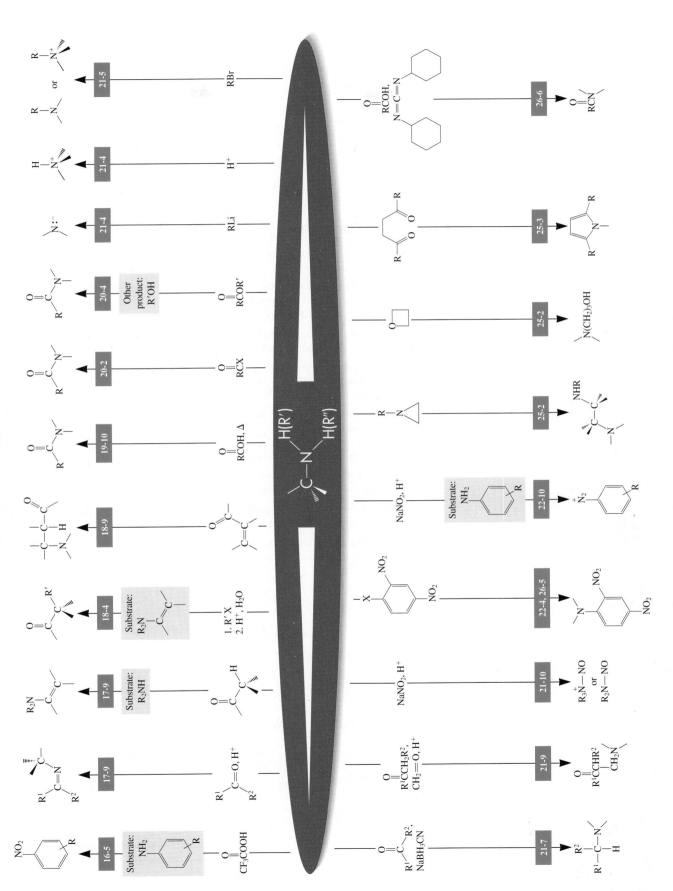

13. Diazomethane (Section 21-10)

$$CH_3NCR \xrightarrow{\text{KOH}} CH_2=\overset{+}{N}=\overset{..}{N}:^-$$

Esterification with diazomethane

$$RCOH \ + \ CH_2N_2 \ \longrightarrow \ RCOCH_3 \ + \ N_2$$

Important Concepts

1. **Amines** can be viewed as derivatives of **ammonia,** just as ethers and alcohols can be regarded as derivatives of water.

2. *Chemical Abstracts* names amines as **alkanamines** (and **benzenamines**), alkyl substituents on the nitrogen being designated as *N*-alkyl. Another system is based on the label aminoalkane. Common names are based on the label alkylamine.

3. The **nitrogen** in amines is sp^3 **hybridized,** the nonbonding electron pair functioning as the equivalent of a substituent. This tetrahedral arrangement **inverts rapidly** through a planar transition state.

4. The **lone electron pair** in amines is less tightly held than in alcohols and ethers, because nitrogen is **less electronegative** than oxygen. The consequences are a diminished capability for hydrogen bonding, higher basicity and nucleophilicity, and lower acidity.

5. **Infrared spectroscopy** helps to differentiate between primary and secondary amines. **Nuclear magnetic resonance** spectroscopy indicates the presence of nitrogen-bound hydrogens; both hydrogen and carbon atoms are **deshielded** in the vicinity of the nitrogen. **Mass spectra** are characterized by **iminium ion** fragments.

6. Indirect methods, such as displacements with azide or cyanide, or reductive amination, are superior to direct alkylation of ammonia for the **synthesis** of amines.

7. The **NR₃** group in a quaternary amine, $R'-\overset{+}{N}R_3$, is a **good leaving group** in E2 reactions; this enables the Hofmann elimination to take place.

8. The **nucleophilic reactivity** of amines manifests itself in reactions with electrophilic carbon, as in haloalkanes, aldehydes, ketones, and carboxylic acids and their derivatives.

Problems

27. Give at least two names for each of the following amines.

(a)

(b)

(c)

(d)

(e) $(CH_3)_3N$

(f) $CH_3CCH_2CH_2N(CH_3)_2$

(g)

(h) $(CH_3CH_2)_2NCH_2CH=CH_2$

28. Give structures that correspond to each of the following names.

(a) *N,N*-Dimethyl-3-cyclohexenamine;

(b) *N*-ethyl-2-phenylethylamine;

(c) 2-aminoethanol;

(d) *m*-chloroaniline.

29. Name each of the following compounds pictured in Real Life 21-1 according to IUPAC or *Chemical Abstracts*. Pay attention to the order of precedence of the functional groups. **(a)** Propylhexedrine; **(b)** amphetamine; **(c)** mescaline; **(d)** epinephrine.

30. As mentioned in Section 21-2, the inversion of nitrogen requires a change of hybridization. **(a)** What is the approximate energy difference between pyramidal nitrogen (sp^3 hybridized) and trigonal planar nitrogen (sp^2 hybridized) in ammonia and simple amines? (**Hint:** Refer to the E_a of inversion.) **(b)** Compare the nitrogen atom in ammonia with the carbon atom in each of the following species: methyl cation, methyl radical, and methyl anion. Contrast the most stable geometries and the hybridizations of each of these species. Using fundamental notions of orbital energies and bond strengths, explain the similarities and differences among them.

31. Use the following NMR- and mass-spectral data to identify the structures of two unknown compounds, A and B.

A: ^{1}H NMR: δ = 0.92 (t, J = 6 Hz, 3 H), 1.32 (broad s, 12 H), 2.28 (broad s, 2 H), and 2.69 (t, J = 7 Hz, 2 H) ppm. Mass spectrum m/z (relative intensity) = 129(0.6) and 30(100).

B: ^{1}H NMR: δ = 1.00 (s, 9 H), 1.17 (s, 6 H), 1.28 (s, 2 H), and 1.42 (s, 2 H) ppm. Mass spectrum m/z (relative intensity) = 129(0.05), 114(3), 72(4), and 58(100).

32. The following spectroscopic data (^{13}C NMR and IR) are for several isomeric amines of the formula $C_6H_{15}N$. Propose a structure for each compound. **(a)** ^{13}C NMR: δ = 23.7 (CH$_3$) and 45.3 (CH) ppm. IR: 3300 cm^{-1}. **(b)** ^{13}C NMR: δ = 12.6 (CH$_3$) and 46.9 (CH$_2$) ppm. IR: no bands in 3250–3500 cm^{-1} range. **(c)** ^{13}C NMR: δ = 12.0 (CH$_3$), 23.9 (CH$_2$), and 52.3 (CH$_2$) ppm. IR: 3280 cm^{-1}. **(d)** ^{13}C NMR: δ = 14.2 (CH$_3$), 23.2 (CH$_2$), 27.1 (CH$_2$), 32.3 (CH$_2$), 34.6 (CH$_2$), and 42.7 (CH$_2$) ppm. IR: 1600 (broad), 3280, and 3365 cm^{-1}. **(e)** ^{13}C NMR: δ = 25.6 (CH$_3$), 38.7 (CH$_3$), and 53.2 (C$_{quaternary}$) ppm. IR: no bands in 3250–3500 cm^{-1} region.

33. The following mass-spectral data are for two of the compounds in Problem 32. Match each mass spectrum with one of them. **(a)** m/z (relative intensity) = 101(8), 86(11), 72(79), 58(10), 44(40), and 30(100). **(b)** m/z (relative intensity) = 101(3), 86(30), 58(14), and 44(100).

34. Is a molecule whose conjugate acid has a high pK_a value a stronger or a weaker base than a molecule whose conjugate acid has a low pK_a value? Explain using a general equilibrium equation.

35. In which direction would you expect each of the following equilibria to lie?

(a) NH_3 + ^-OH $\rightleftharpoons$ NH_2^- + H_2O

(b) CH_3NH_2 + H_2O $\rightleftharpoons$ $CH_3NH_3^+$ + ^-OH

(c) CH_3NH_2 + $(CH_3)_3NH^+$ $\rightleftharpoons$ $CH_3NH_3^+$ + $(CH_3)_3N$

36. How would you expect the following classes of compounds to compare with simple primary amines as bases and acids?

(a) Carboxylic amides; for example, CH_3CONH_2
(b) Imides; for example, $CH_3CONHCOCH_3$
(c) Enamines; for example, CH_2=$CHN(CH_3)_2$

(d) Benzenamines; for example, ⬡—NH_2

37. Several functional groups containing nitrogen are considerably stronger bases than are ordinary amines. One is the amidine group found in DBN and DBU, both of which are widely used as bases in a variety of organic reactions.

Amidine group   **1,5-Diazabicyclo[4.3.0]non-5-ene (DBN)**   **1,8-Diazabicyclo[5.4.0]undec-7-ene (DBU)**

Another unusually strong organic base is guanidine, H_2NCNH_2.
$\overset{NH}{\underset{\|}{}}$
Indicate which nitrogen in each of these bases is the one most

likely to be protonated and explain the enhanced strength of these bases relative to simple amines.

38. Reaction review: Without consulting the Reaction Road Map on p. 970, suggest reagents that convert each of the following starting materials into the indicated product: **(a)** (chloromethyl)benzene (benzyl chloride) into phenylmethanamine (benzylamine); **(b)** benzaldehyde into phenylmethanamine (benzylamine); **(c)** benzaldehyde into *N*-ethylphenylmethanamine (benzylethylamine); **(d)** (chloromethyl)benzene (benzyl chloride) into 2-phenylethanamine (phenethylamine); **(e)** benzaldehyde into *N,N*-dimethylphenylmethanamine (benzyldimethylamine); **(f)** 1-phenylethanone (acetophenone) into 3-amino-1-phenyl-1-propanone; **(g)** benzonitrile into phenylmethanamine (benzylamine); **(h)** 2-phenylacetamide into phenylmethanamine (benzylamine).

39. In the following syntheses of amines, indicate in each case whether the approach will work well, poorly, or not at all. If a method will not work well, explain why.

(a) $CH_3CH_2CH_2CH_2Cl$ $\xrightarrow{\text{1. KCN, CH}_3\text{CH}_2\text{OH} \\ \text{2. LiAlH}_4, (CH_3CH_2)_2O}$ $CH_3CH_2CH_2CH_2NH_2$

(b) $(CH_3)_3CCl$ $\xrightarrow{\text{1. NaN}_3, \text{DMSO} \\ \text{2. LiAlH}_4, (CH_3CH_2)_2O}$ $(CH_3)_3CNH_2$

(c) [cyclohexane-CONH$_2$] $\xrightarrow{\text{Br}_2, \text{NaOH, H}_2\text{O}}$ [cyclohexane-NH$_2$]

(d) [cyclopentane-CH$_2$CH$_2$CH$_2$-Br] $\xrightarrow{\text{CH}_3\text{NH}_2}$ [cyclopentane-CH$_2$CH$_2$CH$_2$-NHCH$_3$]

(e) [decalin-CH$_2$Br, CH$_3$] $\xrightarrow{\text{1. phthalimide N}^-\text{K}^+, \text{DMF} \\ \text{2. H}^+, \text{H}_2\text{O}, \Delta}$ [decalin-CH$_2$NH$_2$, CH$_3$]

(f) [benzene-COCl] $\xrightarrow{\text{1. } \text{benzene-CH}_2\text{NH}_2, \text{HO}^- \\ \text{2. LiAlH}_4, (CH_3CH_2)_2O}$ (benzene-CH$_2$—NH)$_2$

(g) [cyclopentane-CHO] $\xrightarrow{\text{1. } (CH_3)_3CNH_2 \\ \text{2. NaBH}_3\text{CN, CH}_3\text{CH}_2\text{OH}}$ [cyclopentane-CH$_2$NHC(CH$_3$)$_3$]

(h) $H_2NCH_2CH_2CHO$ $\xrightarrow{\text{NaBH}_3\text{CN, CH}_3\text{CH}_2\text{OH}}$ [⬜ NH]

(i)

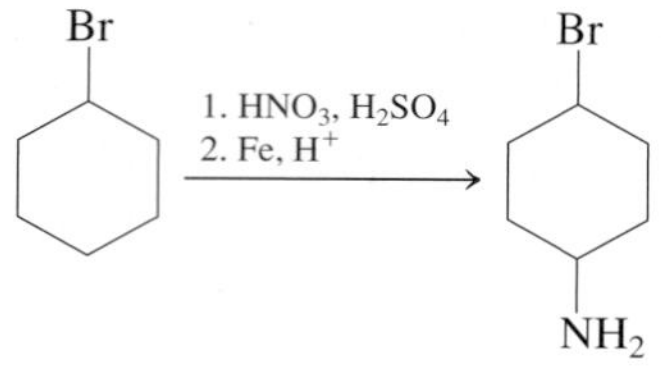

1. HNO₃, H₂SO₄
2. Fe, H⁺

$$\text{1. HNO}_3\text{, H}_2\text{SO}_4 \quad \text{2. Fe, H}^+$$

(j)

1. NaH, THF
2. CH₃I
3. LiAlH₄, (CH₃CH₂)₂O

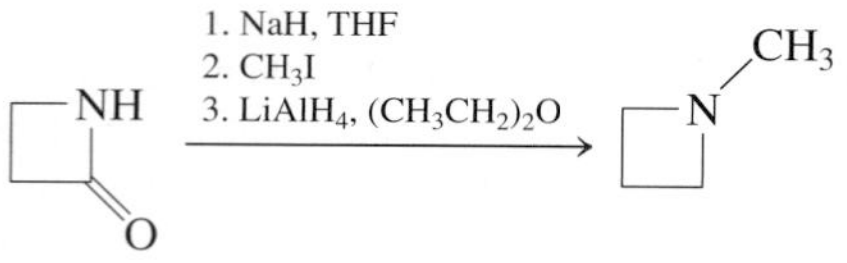

40. For each synthesis in Problem 39 that does not work well, propose an alternative preparation of the final amine, starting either with the same material or with a material of similar structure and functionality.

41. Give the structures of all possible nitrogen-containing organic products that might be expected to form on reaction of chloroethane with ammonia. (**Hint:** Consider multiple alkylations.)

42. Phenylpropanolamine (PPA) has long been an ingredient in many cold remedies and appetite suppressants. In late 2000, the FDA asked manufacturers to remove products containing this compound from the market because of evidence of an increased risk of hemorrhagic stroke. This action has caused a switch to the safer pseudoephedrine as the active component of such medications.

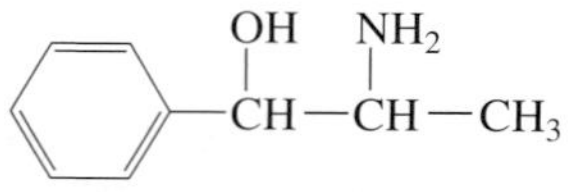

Phenylpropanolamine

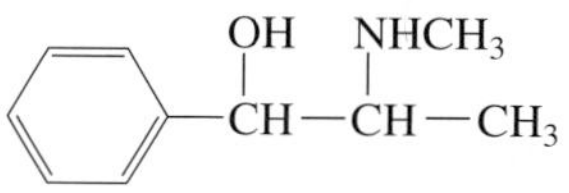

Pseudoephedrine

Suppose that you are the director of a major pharmaceutical laboratory with a huge stock of phenylpropanolamine on hand and the president of the company issues the order, "Pseudoephedrine from now on!" Analyze all your options, and propose the best solution that you can find for the problem.

43. Apetinil, an appetite suppressant (i.e., diet pill; see Real Life 21-1), has the structure shown below. Is it a primary, a secondary, or a tertiary amine? Propose an efficient synthesis of Apetinil from each of the following starting materials. Try to use a variety of methods.

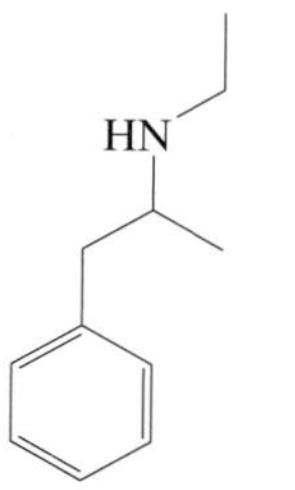

Apetinil

(a) $C_6H_5CH_2COCH_3$

(b) $C_6H_5CH_2\overset{Br}{\underset{|}{C}}HCH_3$

(c) $C_6H_5CH_2\overset{CH_3}{\underset{|}{C}}HCOOH$

44. Suggest the best syntheses that you can for the following amines, beginning each with any organic compounds that do not contain nitrogen. (**a**) Butanamine; (**b**) *N*-methylbutanamine; (**c**) *N,N*-dimethylbutanamine.

45. Give the structures of the possible alkene products of Hofmann elimination of each of the following amines. If a compound can be cycled through multiple eliminations, give the products of each cycle.

(a)

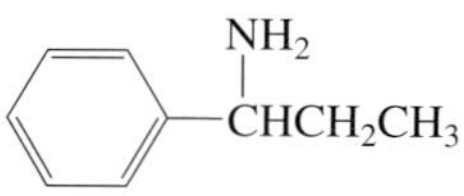

(b)

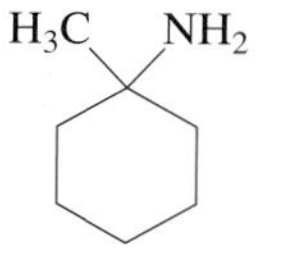

(c) 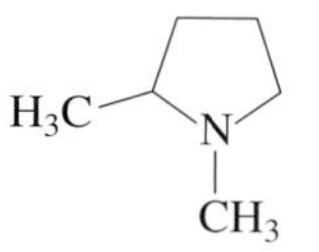 H_3C

(d)

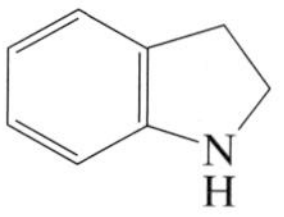

(e)

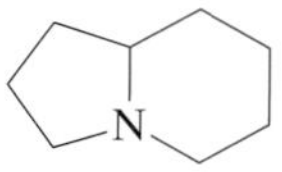

46. What primary amine(s) would give each of the following alkenes or alkene mixtures upon Hofmann elimination? (**a**) 3-Heptene; (**b**) mixture of 2- and 3-heptene; (**c**) 1-heptene; (**d**) mixture of 1- and 2-heptene.

47. Formulate a detailed mechanism for the Mannich reaction between 2-methylpropanal, formaldehyde, and methanamine shown on p. 956.

48. Reaction of the tertiary amine tropinone with (bromomethyl)-benzene (benzyl bromide) gives not one but two quaternary ammonium salts, A and B.

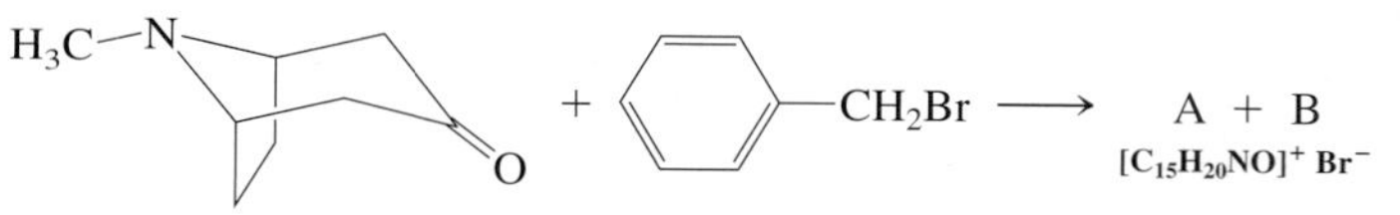

$$H_3C-N \quad + \quad \text{C}_6\text{H}_5-CH_2Br \longrightarrow A + B$$
$$[C_{15}H_{20}NO]^+ \ Br^-$$

Tropinone
$(C_8H_{13}NO)$

Compounds A and B are stereoisomers that are interconverted by base; that is, base treatment of either pure isomer leads to an equilibrium mixture of the two. (**a**) Propose structures for A and B. (**b**) What kind of stereoisomers are A and B? (**c**) Suggest a mechanism for the equilibration of A and B by base. (**Hint:** Think "reversible Hofmann elimination.")

49. Attempted Hofmann elimination of an amine containing a hydroxy group on the β-carbon gives an oxacyclopropane product instead of an alkene.

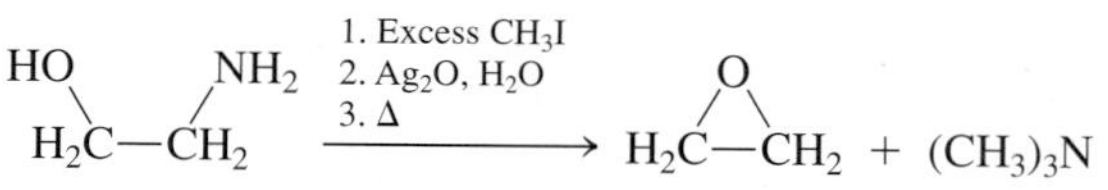

$$HO\underset{H_2C-CH_2}{\overset{NH_2}{}} \quad \xrightarrow[\begin{array}{l}\text{1. Excess CH}_3\text{I}\\\text{2. Ag}_2\text{O, H}_2\text{O}\\\text{3. }\Delta\end{array}]{} \quad \underset{H_2C-CH_2}{\overset{O}{}} + (CH_3)_3N$$

(**a**) Propose a sensible mechanism for this transformation. (**b**) Pseudoephedrine (see Problem 42) and ephedrine are closely related, naturally occurring compounds, as the similar names imply. In fact, they are stereoisomers. From the results of the following reactions, deduce the precise stereochemistries of ephedrine and pseudoephedrine.

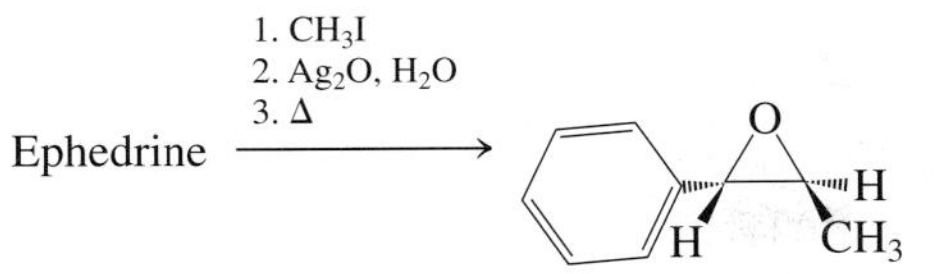

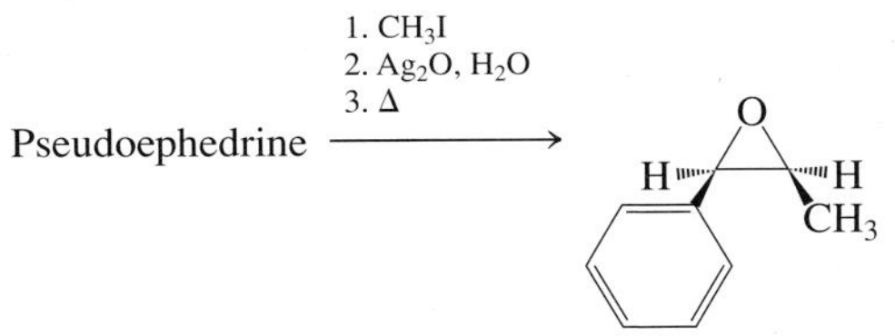

Ephedrine $\xrightarrow[\text{3. }\Delta]{\substack{\text{1. CH}_3\text{I}\\\text{2. Ag}_2\text{O, H}_2\text{O}}}$

Pseudoephedrine $\xrightarrow[\text{3. }\Delta]{\substack{\text{1. CH}_3\text{I}\\\text{2. Ag}_2\text{O, H}_2\text{O}}}$

50. Show how each of the following molecules might be synthesized by Mannich or Mannich-like reactions. (**Hint:** Work backward, identifying the bond made in the Mannich reaction.)

(a)

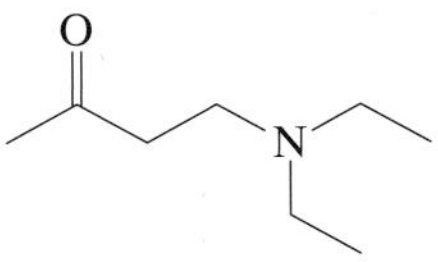

(b)

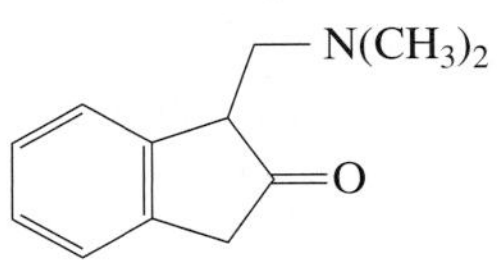

(c) H₃CCHCN
       |
       NH₂

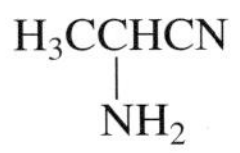

(d)

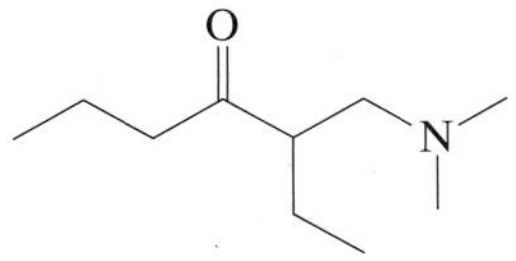

(e)

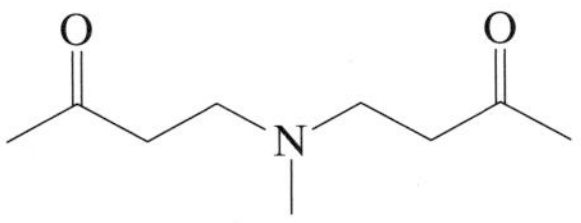

51. Tropinone (Problem 48) was first synthesized by Sir Robert Robinson (famous for the Robinson annulation reaction; Section 18-11), in 1917, by the following reaction. Show a mechanism for this transformation.

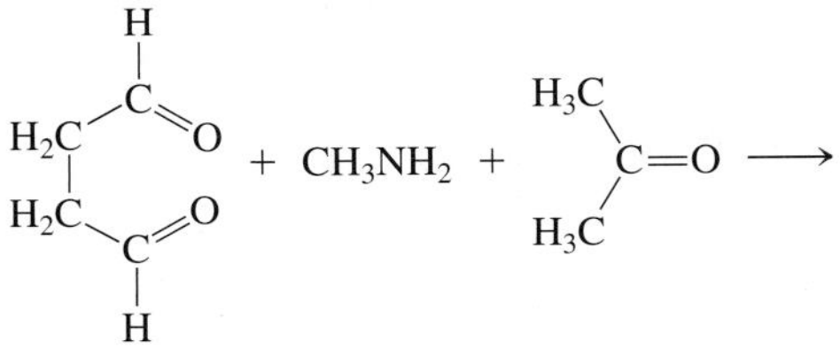

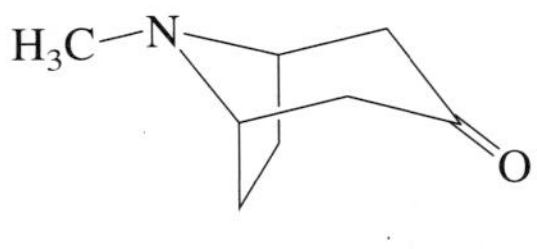

Tropinone is a synthetic precursor to the drug atropine, used topically by ophthalmologists to dilate a patient's pupils. It occurs naturally in the toxic plant belladonna, so named (Italian, beautiful woman) because Mediterranean women allegedly applied its extracts in eyedrops to make them appear more seductive.

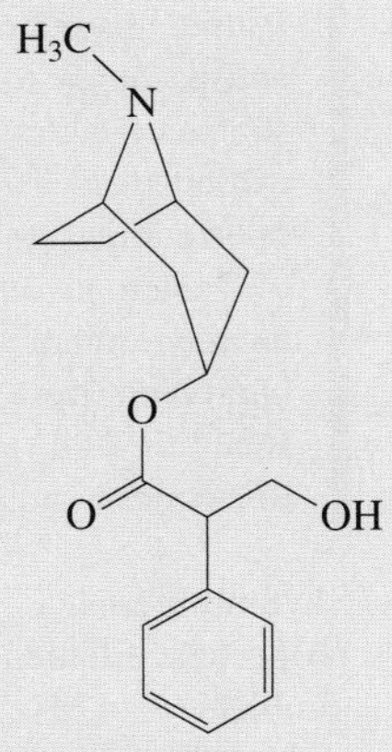

Atropine

52. Illustrate a method for achieving the transformation shown below, using combinations of reactions presented in Sections 21-8 and 21-9.

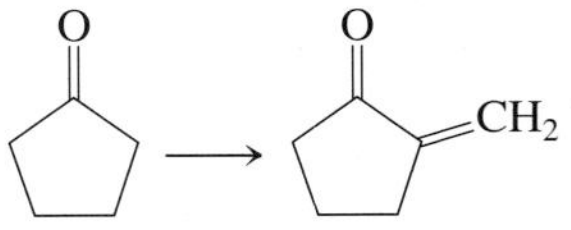

53. Give the expected product(s) of each of the following reactions.

(a) $\xrightarrow{\text{NaNO}_2,\ \text{HCl},\ \text{H}_2\text{O}}$

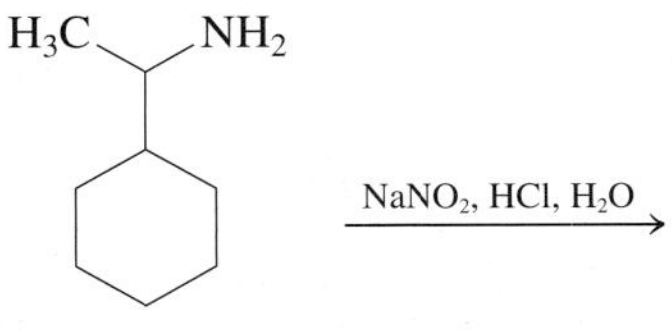

(b) $\xrightarrow{\text{NaNO}_2,\ \text{HCl},\ 0°\text{C}}$

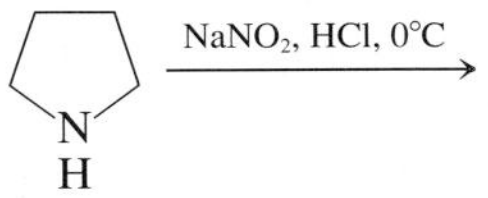

54. Tertiary amines add readily as nucleophiles to carbonyl compounds, but, lacking a hydrogen on nitrogen, they cannot deprotonate to give a stable product. Instead, their addition gives an intermediate that is highly reactive toward other nucleophiles. Thus, tertiary amines are occasionally used as catalysts for the addition of weak nucleophiles to carboxylic acid derivatives.

(a) Fill in the missing structures in the scheme below.

$$(\text{CH}_3)_3\text{N} + \text{CH}_3\text{COCl} \rightleftharpoons \underset{\substack{\textbf{Intermediate A}\\\text{C}_5\text{H}_9\text{ClNO}}}{} \underset{\substack{\textbf{Intermediate B}\\(\text{C}_5\text{H}_9\text{NO})^+}}{\overset{-\text{Cl}^-}{\rightleftharpoons}}$$

(b) Intermediate B readily enters into reactions with weak nucleophiles such as phenol (below). Formulate the mechanism for this process and give the structure of the product.

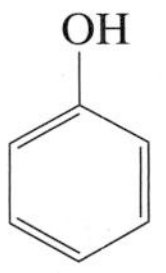

Phenol

55. **CHALLENGE** Tertiary amines undergo reversible conjugate addition to α,β-unsaturated ketones (see Chapter 18). This process is the basis for the Baylis-Hillman reaction, which is catalyzed by tertiary amines, that resembles a crossed aldol reaction. An example is shown below.

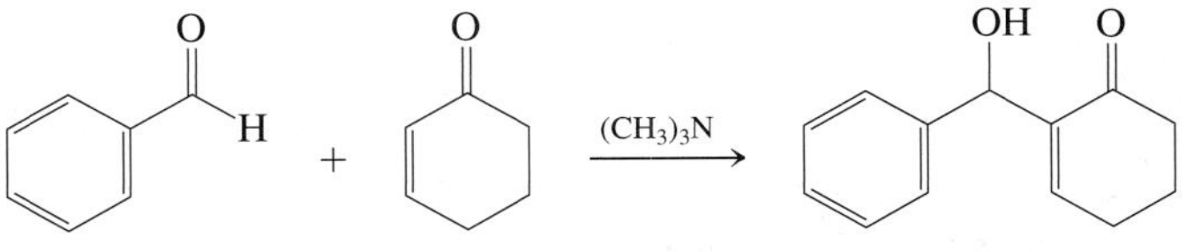

(a) Formulate a mechanism for this process. Begin with conjugate addition of the amine to the enone.

Give the products of each of the following Baylis-Hillman reactions:

(b) CH₃CHO + $\xrightarrow{(\text{CH}_3)_3\text{N}}$

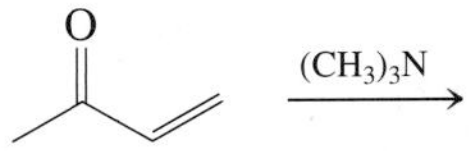

(c) + $\xrightarrow{(\text{CH}_3)_3\text{N}}$

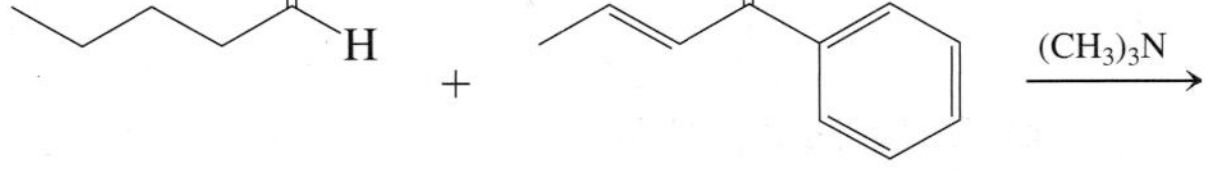

56. **CHALLENGE** Reductive amination of *excess* formaldehyde with a primary amine leads to the formation of a *di*methylated tertiary amine as the product (see the following example). Propose an explanation.

$$(CH_3)_3CCH_2NH_2 \quad + \quad 2\ CH_2=O \xrightarrow{\text{NaBH}_3\text{CN, CH}_3\text{OH}}$$

2,2-Dimethylpropanamine

$$(CH_3)_3CCH_2N(CH_3)_2$$
84%

***N,N,2,2*-Tetramethylpropanamine**

57. Several of the natural amino acids are synthesized from 2-oxocarboxylic acids by an enzyme-catalyzed reaction with a special coenzyme called pyridoxamine. Use electron-pushing arrows to describe each step in the following synthesis of phenylalanine from phenylpyruvic acid.

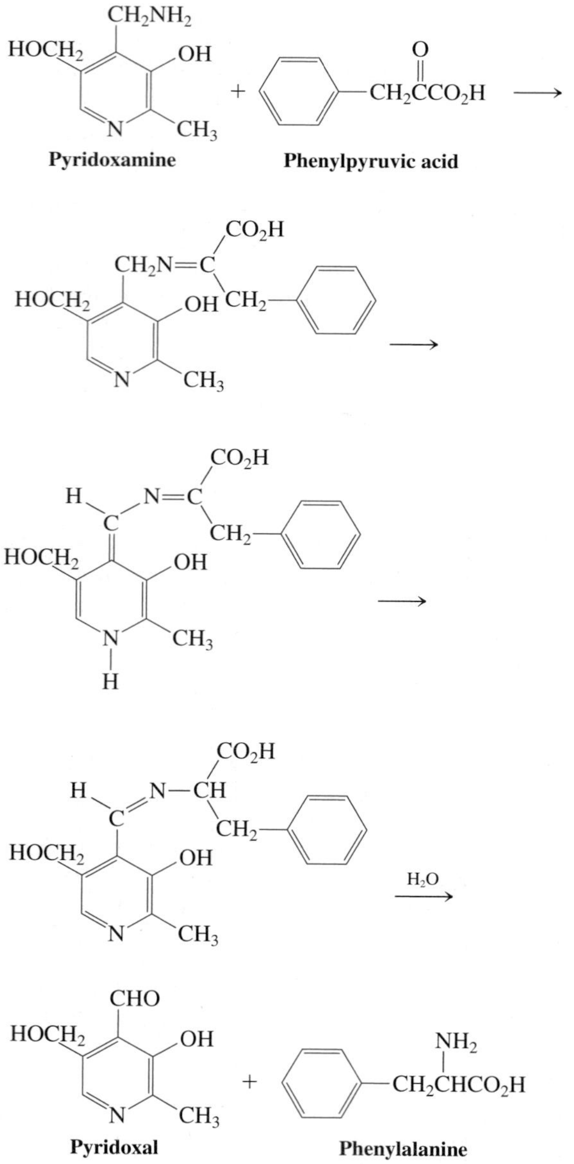

Pyridoxamine   **Phenylpyruvic acid**

Pyridoxal   **Phenylalanine**

58. **CHALLENGE** Using the following information, deduce the structure of coniine, a toxic amine found in poison hemlock, which, deservedly, has a very bad reputation (Section 5-2). IR: 3330 cm^{-1}. ^{1}H NMR: $\delta = 0.91$ (t, $J = 7$ Hz, 3 H), 1.33 (s, 1 H), 1.52 (m, 10 H), 2.70 (t, $J = 6$ Hz, 2 H), and 3.0 (m, 1 H) ppm. MS: m/z (relative intensity) $= 127$ (M$^+$, 43), 84(100), and 56(20).

$$\text{Coniine} \xrightarrow[\substack{\text{1. CH}_3\text{I} \\ \text{2. Ag}_2\text{O, H}_2\text{O} \\ \text{3. }\Delta}]{} \text{mixture of three compounds} \xrightarrow[\substack{\text{1. CH}_3\text{I} \\ \text{2. Ag}_2\text{O, H}_2\text{O} \\ \text{3. }\Delta}]{}$$

$$(CH_3)_3N + \underset{\text{1,4-octadiene and 1,5-octadiene}}{\text{mixture of}}$$

59. Pethidine, the active ingredient in the narcotic analgesic Demerol, was subjected to two successive exhaustive methylation-Hofmann elimination sequences, followed by ozonolysis, with the following results:

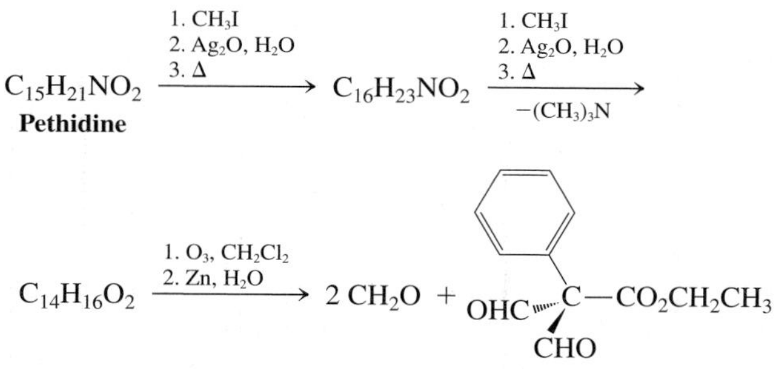

$$C_{15}H_{21}NO_2 \xrightarrow[\substack{\text{1. CH}_3\text{I} \\ \text{2. Ag}_2\text{O, H}_2\text{O} \\ \text{3. }\Delta}]{} C_{16}H_{23}NO_2 \xrightarrow[\substack{\text{1. CH}_3\text{I} \\ \text{2. Ag}_2\text{O, H}_2\text{O} \\ \text{3. }\Delta \\ -(\text{CH}_3)_3\text{N}}]{}$$

Pethidine

$$C_{14}H_{16}O_2 \xrightarrow[\substack{\text{1. O}_3\text{, CH}_2\text{Cl}_2 \\ \text{2. Zn, H}_2\text{O}}]{} 2\ CH_2O + \underset{\text{CHO}}{\text{OHC}\cdots\text{C}-\text{CO}_2\text{CH}_2\text{CH}_3}$$

(a) Propose a structure for pethidine based on this information.
(b) Propose a synthesis of pethidine that begins with ethyl phenylacetate and *cis*-1,4-dibromo-2-butene. (**Hint:** First prepare the dialdehyde ester shown below and then convert it into pethidine.)

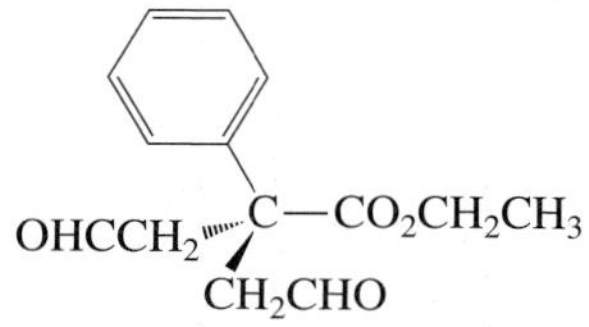

$$\underset{\text{CH}_2\text{CHO}}{\text{OHCCH}_2\cdots\text{C}-\text{CO}_2\text{CH}_2\text{CH}_3}$$

Welcome Side Effects

Pethidine was first investigated in the 1930s as an antispasmodic to relieve cramps (for example, of the stomach). It was noted that mice injected with the drug exhibited an elevated S-shaped tail identical to that observed upon administration of morphine, leading to the development of Demerol as an analgesic. Several successful medicines are the result of such serendipitous "switches," for example, bupropion (from antidepressant to smoking-cessation aid; see Real Life 21-1) and sildenafil (Viagra, from antihypertensive to erectile stimulant; Chapter 25 Opening).

---

60. Skytanthine is a monoterpene alkaloid with the following properties. Elemental analysis: $C_{11}H_{21}N$. ^{1}H NMR: two CH$_3$ doublets ($J = 7$ Hz) at $\delta = 1.20$ and 1.33 ppm; one CH$_3$ singlet at $\delta = 2.32$ ppm; other hydrogens give rise to broad signals at $\delta = 1.3–2.7$ ppm. IR: no bands ≥ 3100 cm^{-1}. Deduce the structures of skytanthine and degradation products A, B, and C (made as shown below) from this information.

Skytanthine $\xrightarrow[\begin{array}{c}\text{1. CH}_3\text{I}\\\text{2. Ag}_2\text{O, H}_2\text{O}\\\text{3. }\Delta\end{array}]{}$ $C_{12}H_{23}N$ **A** $\xrightarrow[\text{2. Zn, H}_2\text{O}]{\text{1. O}_3\text{, CH}_2\text{Cl}_2}$

IR: $\tilde{\nu} = 1646$ cm^{-1}

$CH_2{=}O$ + $C_{11}H_{21}NO$ **B** $\xrightarrow[\text{2. KOH, H}_2\text{O}]{\text{1.}}$

IR: $\tilde{\nu} = 1715$ cm^{-1}

CH_3COOH + $C_9H_{19}NO$ **C** $\xrightarrow{\text{Careful oxidation}}$

IR: $\tilde{\nu} = 3620$ cm^{-1}

IR: $\tilde{\nu} = 1745$ cm^{-1}

61. Many alkaloids are synthesized in nature from a precursor molecule called norlaudanosoline, which in turn appears to be derived from the condensation of amine A with aldehyde B. Formulate a mechanism for this transformation. Note that a carbon–carbon bond is formed in the process. Name a reaction presented in this chapter that features a closely related carbon–carbon bond formation.

A

B

→ alkaloids

Norlaudanosoline

Team Problem

62. Quaternary ammonium salts catalyze reactions between species dissolved in two immiscible phases, a phenomenon called phase-transfer catalysis (see also Real Life 26-2). For example, heating a mixture of 1-chlorooctane dissolved in decane with aqueous sodium cyanide shows no sign of the S_N2 product, nonanenitrile. On the other hand, addition of a small amount of (phenylmethyl)triethylammonium chloride results in a rapid, quantitative reaction.

$CH_3(CH_2)_7Cl$ + $Na^{+-}CN$ $\xrightarrow{}$

1-Chlorooctane   **(Phenylmethyl)-triethylammonium chloride**

$CH_3(CH_2)_7CN$ + Na^+Cl^-
100%
Nonanenitrile

As a team, discuss possible answers to the following questions:
(a) What is the solubility of the catalyst in the two solvents?
(b) Why is the S_N2 reaction so slow without catalyst?
(c) How does the ammonium salt facilitate the reaction?

Preprofessional Problems

63. One of the following four amines is tertiary. Which one?
(a) Propanamine; **(b)** N-methylethanamine; **(c)** N,N-dimethylmethanamine; **(d)** N-methylpropanamine.

64. Identify the best conditions for the transformation shown below.

$$CH_3CH_2\overset{O}{\overset{\|}{C}}NH_2 \longrightarrow CH_3CH_2NH_2 + CO_2$$

(a) H_2, metal catalyst; **(b)** excess CH_3I, K_2CO_3; **(c)** Br_2, NaOH, H_2O; **(d)** $LiAlH_4$, ether; **(e)** CH_2N_2, ether.

65. Rank the basicities of the following three nitrogen-containing compounds (most basic first):

NH_3 CH_3NH_2 $(CH_3)_4N^+ NO_3^-$
A   **B**   **C**

(a) A > B > C; **(b)** B > C > A; **(c)** C > A > B; **(d)** C > B > A; **(e)** B > A > C.

66. Which of the following formulas best represents diazomethane?

(a) $CH_2{=}\overset{+}{N}{=}\overset{..}{\underset{..}{N}}{}^-$

(b) $H{-}\overset{..}{N}{=}C{=}\overset{..}{N}{-}H$

(c) $\overset{-}{\underset{..}{N}}{=}C{=}\overset{+}{N}\overset{\displaystyle H}{\underset{\displaystyle H}{}}$

(d) $:\overset{-}{C}H_2{-}N{\equiv}\overset{+}{N}:$

(e) $CH_2{-}\overset{+}{N}{\equiv}\overset{-}{N}:$

67. Use the following partial IR- and mass-spectral data to identify one of the structures among the selection given. IR: 1690 and 3300 cm^{-1}; MS: $m/z = 73$ (parent ion).

(a) $H\overset{O}{\overset{\|}{C}}N(CH_3)_2$

(b) (structure: epoxide-NHCH$_3$)

(c) $H_2NCH_2C{\equiv}CCH_2NH_2$

(d) $CH_3CH_2\overset{O}{\overset{\|}{C}}NH_2$

(e) (structure: oxetane-NH$_2$)

CHAPTER 22

Chemistry of Benzene Substituents

Alkylbenzenes, Phenols, and Benzenamines

Benzene (Chapter 15) used to be a common laboratory solvent until OSHA (U.S. Occupational Safety and Health Administration) placed it on its list of carcinogens. Chemists now use methylbenzene (toluene) instead, which has very similar solvating power but is *not* carcinogenic. Why not? The reason is the relatively high reactivity of the benzylic hydrogens that renders methylbenzene subject to fast metabolic degradation and extrusion from the body, unlike benzene, which can survive many days embedded in fatty and other tissues. Thus, the benzene ring, though itself quite unreactive because of its aromaticity, appears to activate neighboring bonds or, more generally, affects the chemistry of its substituents. You should not be too surprised by this finding, because it is complementary to the conclusions of Chapter 16. There we saw that substituents affect the behavior of benzene. Here we shall see the reverse.

How does the benzene ring modify the behavior of neighboring reactive centers? This chapter takes a closer look at the effects exerted by the ring on the reactivity of alkyl substituents, as well as of attached hydroxy and amino functions. We shall see that the behavior of these groups (introduced in Chapters 3, 8, and 21) is altered by the occurrence of resonance. After considering the special reactivity of aryl-substituted (benzylic) carbon atoms, we turn our attention to the preparation and reactions of phenols and benzenamines (anilines). These compounds are found widely in nature and are used in synthetic procedures as precursors to substances such as aspirin, dyes, and vitamins.

Ciprofloxacin ("Cipro") is a synthetic tetrasubstituted benzene derivative used widely to treat bacterial infections, especially of the urinary tract. The photo shows a laboratory worker producing tablets of the drug.

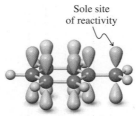

Phenylmethyl (benzyl) system

Sole site of reactivity

Sites of reactivity

2-Propenyl (allyl) system

22-1 | REACTIVITY AT THE PHENYLMETHYL (BENZYL) CARBON: BENZYLIC RESONANCE STABILIZATION

Methylbenzene is readily metabolized because its methyl C–H bonds are relatively weak with respect to homolytic and heterolytic cleavage. When one of these methyl hydrogens has been removed, the resulting **phenylmethyl (benzyl)** group, $C_6H_5CH_2$ (see margin), may be viewed as a benzene ring whose π system overlaps with an extra p orbital located on an attached alkyl carbon. This interaction, generally called **benzylic resonance,** stabilizes adjacent radical, cationic, and anionic centers in much the same way that overlap of a π bond and a third p orbital stabilizes 2-propenyl (allyl) intermediates (Section 14-1). However, unlike allylic systems, which may undergo transformations at either terminus and give product mixtures (in the case of unsymmetrical substrates), benzylic reactivity is regioselective and occurs only at the benzylic carbon. The reason for this selectivity lies in the disruption of aromaticity that goes with attack on the benzene ring.

Benzylic radicals are reactive intermediates in the halogenation of alkylbenzenes

We have seen that benzene will not react with chlorine or bromine unless a Lewis acid is added. The acid catalyzes halogenation of the ring (Section 15-9).

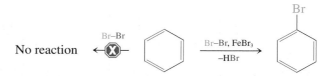

No reaction

In contrast, heat or light allows attack by chlorine or bromine on methylbenzene (toluene) even in the absence of a catalyst. Analysis of the products shows that reaction takes place at the methyl group, *not* at the aromatic ring, and that excess halogen leads to multiple substitution.

Reaction

$$\text{CH}_2\text{H} + \text{Br}_2 \xrightarrow{\Delta} \text{CH}_2\text{Br} + \text{HBr}$$

(Bromomethyl)-benzene

$$\text{CH}_3 \xrightarrow[-\text{HCl}]{\text{Cl}_2,\ h\nu} \text{CH}_2\text{Cl} \xrightarrow[-\text{HCl}]{\text{Cl}_2,\ h\nu} \text{CHCl}_2 \xrightarrow[-\text{HCl}]{\text{Cl}_2,\ h\nu} \text{CCl}_3$$

(Chloromethyl)-benzene   **(Dichloromethyl)-benzene**   **(Trichloromethyl)-benzene**

Each substitution yields one molecule of hydrogen halide as a by-product.

As in the halogenation of alkanes (Sections 3-4 through 3-6) and the allylic halogenation of alkenes (Section 14-2), the mechanism of benzylic halogenation proceeds through radical intermediates. Heat or light induces dissociation of the halogen molecule into atoms. One of them abstracts a benzylic hydrogen, a reaction giving HX and a phenylmethyl (benzyl) radical. This intermediate reacts with another molecule of halogen to give the product, a (halomethyl)benzene, and another halogen atom, which propagates the chain process.

Mechanism of Benzylic Halogenation

Initiation:

$$:\ddot{X}—\ddot{X}: \xrightarrow{\Delta \text{ or } h\nu} 2 :\ddot{X}\cdot$$

Propagation:

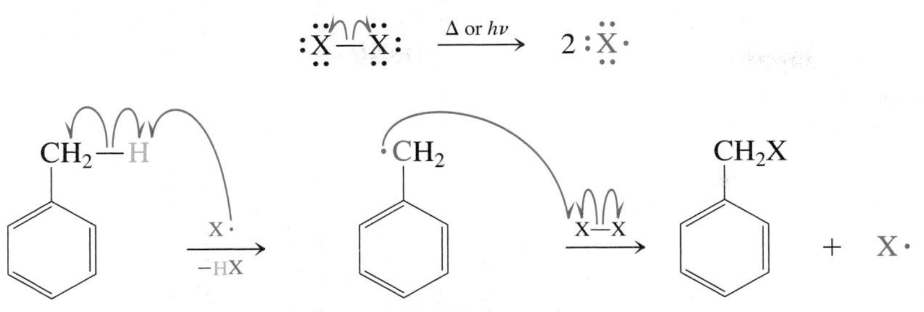

Phenylmethyl (benzyl) radical

Mechanism

Remember: Single-headed ("fishhook") arrows denote the movement of single electrons.

What explains the ease of benzylic halogenation? The answer lies in the stabilization of the phenylmethyl (benzyl) radical by the phenomenon called benzylic resonance (Figure 22-1). As a consequence, the benzylic C–H bond is relatively weak ($DH° = 87$ kcal mol^{-1}, 364 kJ mol^{-1}); its cleavage is relatively favorable and proceeds with a low activation energy.

Inspection of the resonance structures in Figure 22-1 reveals why the halogen attacks only the *benzylic* position and not an aromatic carbon: Reaction at any but the benzylic carbon would destroy the aromatic character of the benzene ring.

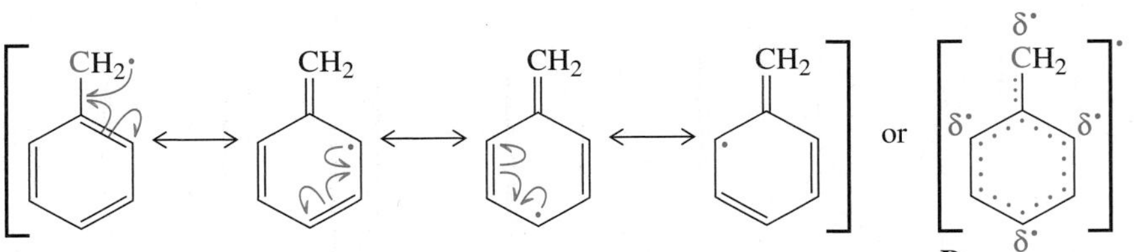

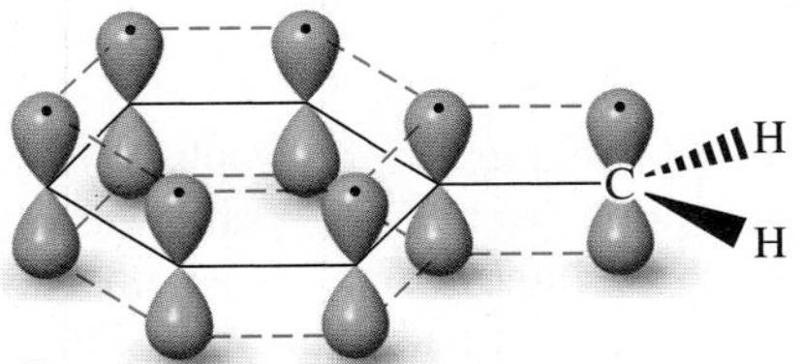

Figure 22-1 The benzene π system of the phenylmethyl (benzyl) radical enters into resonance with the adjacent radical center. The extent of delocalization may be depicted by (A) resonance structures, (B) dotted lines, or (C) orbitals.

Exercise 22-1

For each of the following compounds, draw the structure and indicate where radical halogenation is most likely to occur upon heating in the presence of Br$_2$. Then rank the compounds in approximate descending order of reactivity under bromination conditions. **(a)** Ethylbenzene; **(b)** 1,2-diphenylethane; **(c)** 1,3-diphenylpropane; **(d)** diphenylmethane; **(e)** (1-methylethyl)benzene.

Benzylic cations delocalize the positive charge

Reminiscent of the effects encountered in the corresponding allylic systems (Section 14-3), benzylic resonance can affect strongly the reactivity of benzylic halides and sulfonates in nucleophilic displacements. For example, the 4-methylbenzenesulfonate (tosylate) of 4-methoxyphenylmethanol (4-methoxybenzyl alcohol) reacts with solvent ethanol rapidly via an S$_N$1 mechanism. This reaction is an example of solvolysis, specifically ethanolysis, which we described in Chapter 7.

CH₃O—⟨benzene ring⟩—CH₂OS(=O)(=O)—⟨benzene ring⟩—CH₃ + CH₃CH₂OH $\xrightarrow[\text{Ethanolysis}]{\text{S}_\text{N}1}$

(4-Methoxyphenyl)methyl
4-methylbenzenesulfonate
(A primary benzylic tosylate)

Reaction

CH₃O—⟨benzene ring⟩—CH₂OCH₂CH₃ + HO₃S—⟨benzene ring⟩—CH₃

1-(Ethoxymethyl)-4-methoxybenzene

Mechanism

The reason is the delocalization of the positive charge of the benzylic cation through the benzene ring, allowing for relatively facile dissociation of the starting sulfonate.

Mechanism of Benzylic Unimolecular Nucleophilic Substitution

CH₃Ö—⟨benzene ring⟩—CH₂–L $\xrightarrow{-\text{L}^-}$

[resonance structures of the benzylic cation] $\xrightarrow{\text{CH}_3\text{CH}_2\text{OH}}$ product

Octet form

Benzylic cation

Animation

ANIMATED MECHANISM:
Benzylic nucleophilic
substitution

Several benzylic cations are stable enough to be isolable. For example, the X-ray structure of the 2-phenyl-2-propyl cation (as its SbF₆⁻ salt) was obtained in 1997 and shows the phenyl–C bond (1.41 Å) to be intermediate in length between those of pure single (1.54 Å) and double bonds (1.33 Å), in addition to the expected planar framework and trigonal arrangement of all sp^2 carbons (Figure 22-2), as expected for a delocalized benzylic system.

Exercise 22-2

Which one of the two chlorides will solvolyze more rapidly: (1-chloroethyl)benzene, C₆H₅CHCl, or chloro(diphenyl)methane, (C₆H₅)₂CHCl? Explain your answer.
          |
          CH₃

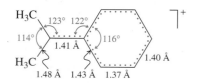

Figure 22-2 Structure of the 2-phenyl-2-propyl cation.

The above S_N1 reaction is facilitated by the presence of the para methoxy substituent, which allows for extra stabilization of the positive charge. In the absence of this substituent, S_N2 processes may dominate. Thus, the parent phenylmethyl (benzyl) halides and sulfonates undergo preferential and unusually rapid S_N2 displacements, even under solvolytic conditions, and particularly in the presence of good nucleophiles. As in allylic S_N2 reactions (Section 14-3), two factors contribute to this acceleration. One is that the benzylic carbon is made relatively more electrophilic by the neighboring sp^2-hybridized phenyl carbon

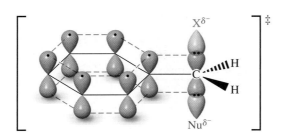

Figure 22-3 The benzene
π system overlaps with the
orbitals of the S_N2 transition state
at a benzylic center. As a result,
the transition state is stabilized,
thereby lowering the activation
barrier toward S_N2 reactions
of (halomethyl)benzenes.

(as opposed to sp^3-hybridized ones; Section 13-2). The second is stabilization of the S_N2 transition state by overlap with the benzene π system (Figure 22-3).

**(Bromomethyl)benzene
(Benzyl bromide)**

**Phenylethanenitrile
(Phenylacetonitrile)**

81%

(~ 100 times faster than S_N2 reactions of primary bromoalkanes)

Animation

ANIMATED MECHANISM:
Benzylic nucleophilic
substitution

Exercise 22-3

Phenylmethanol (benzyl alcohol) is converted into (chloromethyl)benzene in the presence of hydrogen chloride much more rapidly than ethanol is converted into chloroethane. Explain.

Resonance in benzylic anions makes benzylic hydrogens relatively acidic

A negative charge adjacent to a benzene ring, as in the phenylmethyl (benzyl) anion, is stabilized by conjugation in much the same way that the corresponding radical and cation are stabilized. The electrostatic potential maps of the three species (rendered in the margin at an attenuated scale for optimum contrast) show the delocalized positive (blue) and negative (red) charges in the cation and anion, respectively, in addition to the delocalized electron (yellow) in the neutral radical.

**Phenylmethyl (benzyl)
cation**

**Phenylmethyl (benzyl)
radical**

Resonance in Benzylic Anions

p$K_a \approx 41$

+ H⁺

The acidity of methylbenzene (toluene; p$K_a \approx 41$) is therefore considerably greater than that of ethane (p$K_a \approx 50$) and comparable to that of propene (p$K_a \approx 40$), which is deprotonated to produce the resonance-stabilized 2-propenyl (allyl) anion (Section 14-4). Consequently, methylbenzene (toluene) can be deprotonated by butyllithium to generate phenylmethyllithium.

**Phenylmethyl (benzyl)
anion**

Deprotonation of Methylbenzene

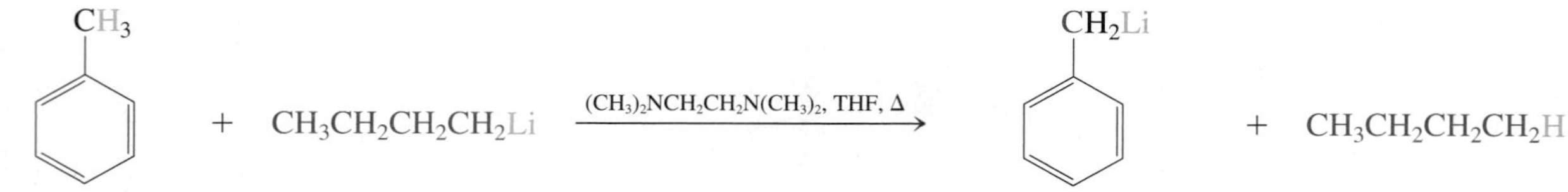

Methylbenzene
(Toluene)

+ $CH_3CH_2CH_2CH_2Li$ $\xrightarrow{(CH_3)_2NCH_2CH_2N(CH_3)_2,\ THF,\ \Delta}$ Phenylmethyllithium
(Benzyllithium)

+ $CH_3CH_2CH_2CH_2H$

Exercise 22-4

Which molecule in each of the following pairs is more reactive with the indicated reagents, and why?

(a) $(C_6H_5)_2CH_2$ or $C_6H_5CH_3$, with $CH_3CH_2CH_2CH_2Li$

(b) 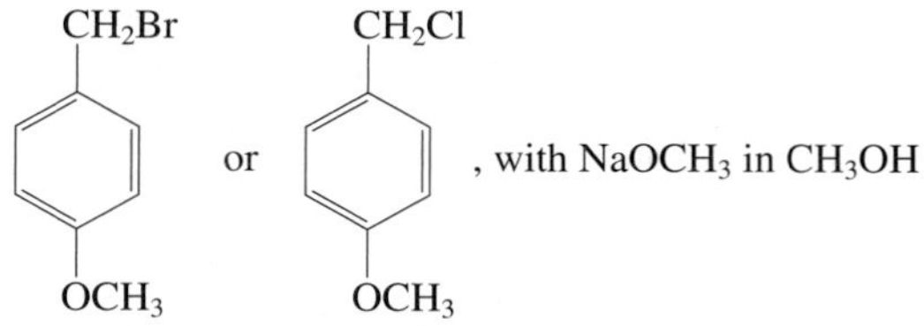   or , with $NaOCH_3$ in CH_3OH

(c) 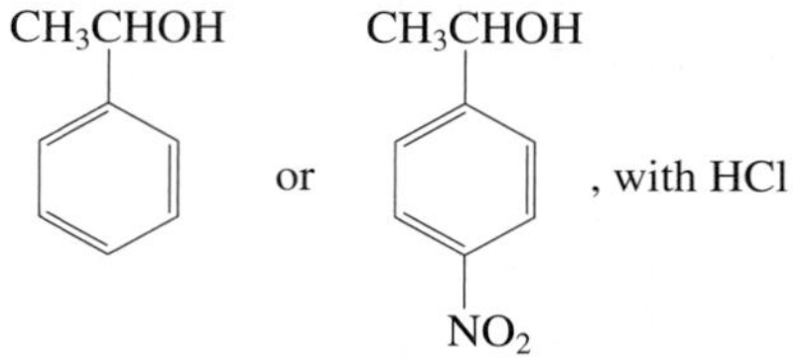   or , with HCl

---

In Summary Benzylic radicals, cations, and anions are stabilized by resonance with the benzene ring. This effect allows for relatively easy radical halogenations, S_N1 and S_N2 reactions, and benzylic anion formation.

22-2 BENZYLIC OXIDATIONS AND REDUCTIONS

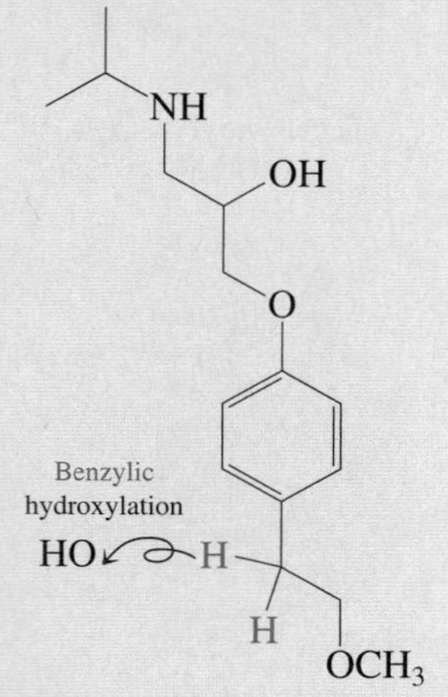

Because it is aromatic, the benzene ring is quite unreactive. While it does undergo electrophilic aromatic substitutions (Chapters 15 and 16), reactions that dismantle the aromatic six-electron circuit, such as oxidations and reductions, are much more difficult to achieve. In contrast, such transformations occur with comparative ease when taking place at *benzylic* positions. This section describes how certain reagents oxidize and reduce alkyl substituents on the benzene ring.

Oxidation of alkyl-substituted benzenes leads to aromatic ketones and acids

Reagents such as hot $KMnO_4$ and $Na_2Cr_2O_7$ may oxidize alkylbenzenes all the way to benzoic acids. Benzylic carbon–carbon bonds are cleaved in this process, which usually requires at least one benzylic C–H bond to be present in the starting material (i.e., tertiary alkylbenzenes are inert).

Complete Benzylic Oxidations of Alkyl Chains

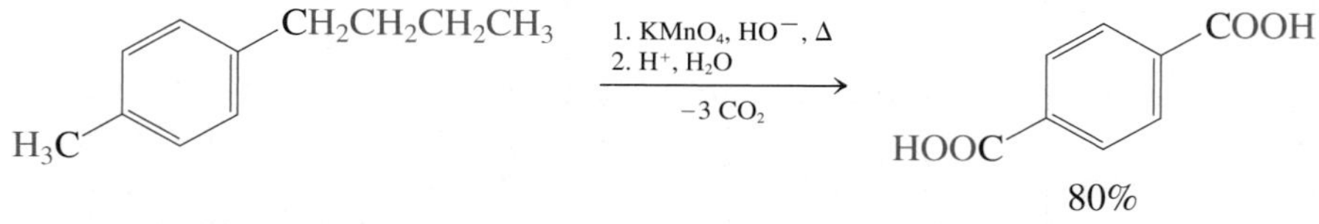

1-Butyl-4-methylbenzene

1. $KMnO_4$, HO^-, Δ
2. H^+, H_2O

$-3\ CO_2$

80%

1,4-Benzenedicarboxylic acid
(Terephthalic acid)

The reaction proceeds through first the benzylic alcohol and then the ketone, at which stage it can be stopped under milder conditions (see margin and Section 16-5).

The special reactivity of the benzylic position is also seen in the mild conditions required for the oxidation of benzylic alcohols to the corresponding carbonyl compounds. For example, manganese dioxide, MnO_2, performs this oxidation selectively in the presence of other (nonbenzylic) hydroxy groups. (Recall that MnO_2 was used in the conversion of allylic alcohols into α,β-unsaturated aldehydes and ketones; see Section 17-4.)

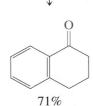

1,2,3,4-Tetra-hydronaphthalene (Tetralin)

CrO_3, CH_3COOH, H_2O, 21°C

Not isolated

71%
1-Oxo-1,2,3,4-tetrahydro-naphthalene (1-Tetralone)

Selective Oxidation of a Benzylic Alcohol with Manganese Dioxide

$\xrightarrow{MnO_2, \text{ acetone, } 25°C, \, 5 \text{ h}}$

94%

Benzylic ethers are cleaved by hydrogenolysis

Exposure of benzylic alcohols, ethers, or esters to hydrogen in the presence of metal catalysts results in rupture of the reactive benzylic carbon–oxygen bond. This transformation is an example of **hydrogenolysis,** cleavage of a σ bond by catalytically activated hydrogen.

Cleavage of Benzylic Ethers by Hydrogenolysis

$\xrightarrow{H_2, \text{ Pd–C, } 25°C}$ + HOR

Exercise 22-5

Write synthetic schemes that would connect the following starting materials with their products.

(a) $- \, - \rightarrow$

(b) $- \, - \rightarrow$

(c) $- \, - \rightarrow$

Hydrogenolysis is not possible for ordinary alcohols, ethers, and esters. Therefore, the phenylmethyl (benzyl) substituent is a valuable protecting group for hydroxy functions. The following scheme shows its use in part of a synthesis of a compound in the eudesmane class of volatile plant oils, which includes substances of importance in both medicine and perfumery.

Phenylmethyl Protection in a Complex Synthesis

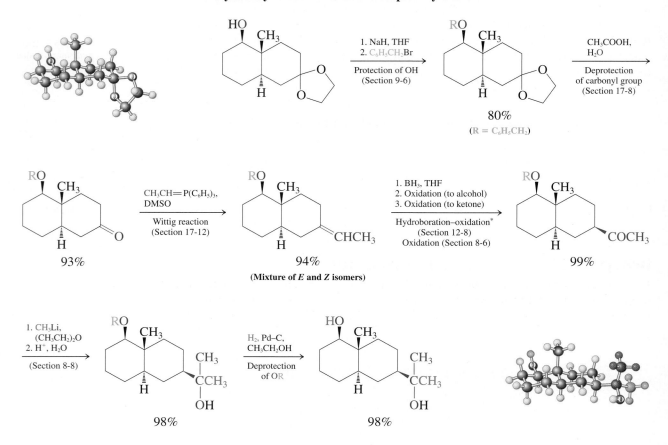

*While the literature reports the use of special oxidizing agents, in principle 2. H_2O_2, ^-OH and 3. CrO_3 would have been satisfactory.

Because the hydrogenolysis of the phenylmethyl (benzyl) ether in the final step occurs under neutral conditions, the tertiary alcohol function survives untouched. A tertiary butyl ether would have been a worse choice as a protecting group, because cleavage of its carbon–oxygen bond would have required acid (Section 9-8), which may cause dehydration (Section 9-2).

In Summary Benzylic oxidations of alkyl groups take place in the presence of permanganate or chromate; benzylic alcohols are converted into the corresponding ketones by manganese dioxide. The benzylic ether function can be cleaved by hydrogenolysis in a transformation that allows the phenylmethyl (benzyl) substituent to be used as a protecting group for the hydroxy function in alcohols.

22-3 NAMES AND PROPERTIES OF PHENOLS

Arenes substituted by hydroxy groups are called **phenols** (Section 15-1). The π system of the benzene ring overlaps with an occupied p orbital on the oxygen atom, a situation resulting in delocalization similar to that found in benzylic anions (Section 22-1). As one result of this extended conjugation, phenols possess an unusual, enolic structure. Recall that enols are usually unstable: They tautomerize easily to the corresponding ketones because of the relatively strong carbonyl bond (Section 18-2). Phenols, however, prefer the enol to the keto form because the aromatic character of the benzene ring is preserved.

Keto and Enol Forms of Acetone and Phenol

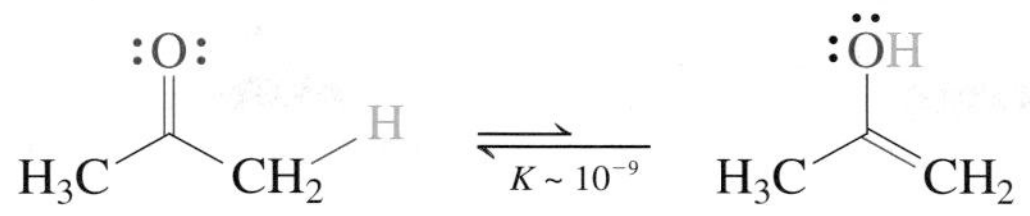

Acetone

2-Propenol

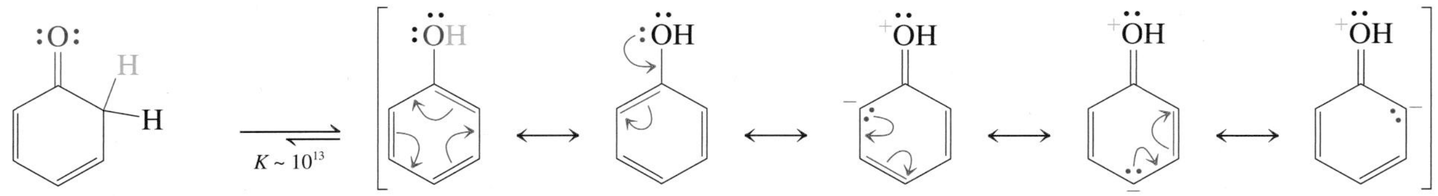

$K \sim 10^{-9}$

2,4-Cyclohexadienone

$K \sim 10^{13}$

Phenol

Exercise 22-6

Many naturally occurring phenols are derived from nonaromatic precursors through rearrangements driven by the aromaticity of the final product. For example, carvone (see Real Life 5-1) undergoes an acid-catalyzed reorganization to carvacrol [2-methyl-5-(1-methylethyl)phenol], a component of the herbs oregano, thyme, and marjoram that contributes to their characteristic odor. Formulate a mechanism.

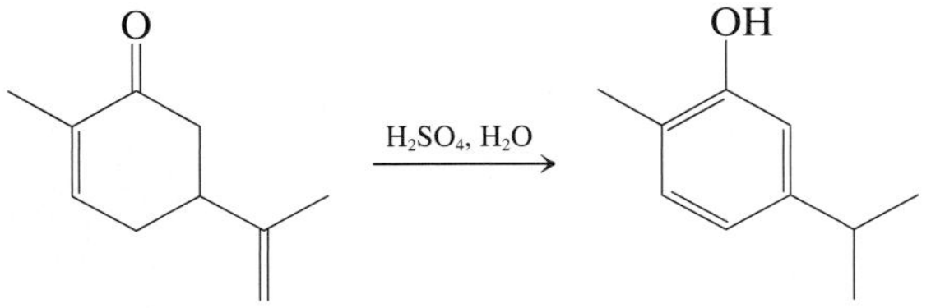

H_2SO_4, H_2O

Carvone

Carvacrol

Phenols and their ethers are ubiquitous in nature; some derivatives have medicinal and herbicidal applications, whereas others are important industrial materials. This section first explains the names of these compounds. It then describes an important difference between phenols and alkanols—phenols are stronger acids because of the neighboring aromatic ring.

Phenols are hydroxyarenes

Phenol itself was formerly known as carbolic acid. It forms colorless needles (m.p. 41°C), has a characteristic odor, and is somewhat soluble in water. Aqueous solutions of it (or its methyl-substituted derivatives) are applied as disinfectants, but its main use is for the preparation of polymers (phenolic resins; Section 22-6). Pure phenol causes severe skin burns and is toxic; deaths have been reported from the ingestion of as little as 1 g. Fatal poisoning may also result from absorption through the skin.

Substituted phenols are named as phenols, benzenediols, or benzenetriols, according to the system described in Section 15-1, although some common names are accepted by IUPAC (Section 22-8). These substances find uses in the photography, dyeing, and tanning industries. The compound bisphenol A (shown in the margin; see also Real Life 22-1) is an important monomer in the synthesis of epoxyresins and polycarbonates, materials that are widely employed in the manufacture of durable plastic materials, food packaging, dental sealants, and coatings inside beverage cans (Real Life 22-1).

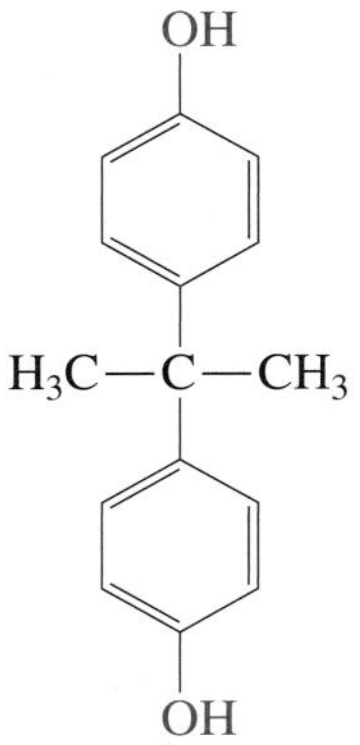

Bisphenol A

Phenols containing the higher-ranking carboxylic acid functionality are called **hydroxy-benzoic acids.** Many have common names. Phenyl ethers are named as **alkoxybenzenes.** As a substituent, C_6H_5O is called **phenoxy.**

4-Methylphenol
(**p-Cresol**)

4-Chloro-3-nitrophenol

3-Hydroxybenz-oic acid
(**m-Hydroxybenzoic acid**)

1,4-Benzenediol
(**Hydroquinone**)

1,2,3-Benzenetriol
(**Pyrogallol**)

Many examples of phenol derivatives, particularly those exhibiting physiological activity, are depicted in this book (e.g., see Real Life 5-4, 9-1, 21-1, 22-1, and 22-2, and Sections 4-7, 9-11, 15-1, 22-9, 24-12, 25-8, and 26-1). You are very likely to have ingested without knowing it the four phenol derivatives shown here.

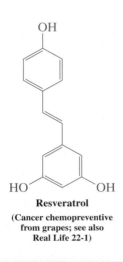

Resveratrol
(**Cancer chemopreventive from grapes; see also Real Life 22-1**)

Capsaicin
(**Active ingredient in hot pepper, as in jalapeño or cayenne pepper; see margin note on the next page**)

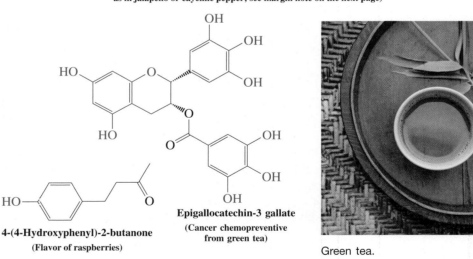

Epigallocatechin-3 gallate
(**Cancer chemopreventive from green tea**)

4-(4-Hydroxyphenyl)-2-butanone
(**Flavor of raspberries**)

A raspberry.

Green tea.

Phenols are unusually acidic

Phenols have pK_a values that range from 8 to 10. Even though they are less acidic than carboxylic acids ($pK_a = 3–5$), they are stronger than alkanols ($pK_a = 16–18$). The reason is resonance: The negative charge in the conjugate base, called the **phenoxide ion,** is stabilized by delocalization into the ring.

Acidity of Phenol

$pK_a \approx 10$

Phenoxide ion

The acidity of phenols is greatly affected by substituents that are capable of resonance. 4-Nitrophenol (*p*-nitrophenol), for example, has a pK_a of 7.15.

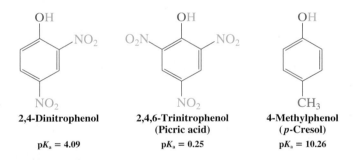

pK$_a$ = 7.15

Negative charge
delocalized into the
nitro group

The 2-isomer has similar acidity (pK_a = 7.22), whereas nitrosubstitution at C3 results in a pK_a of 8.39. Multiple nitration increases the acidity to that of carboxylic or even mineral acids. Electron-donating substituents have the opposite effect, raising the pK_a.

2,4-Dinitrophenol

pK_a = 4.09

**2,4,6-Trinitrophenol
(Picric acid)**

pK_a = 0.25

**4-Methylphenol
(*p*-Cresol)**

pK_a = 10.26

As Section 22-5 will show, the oxygen in phenol and its ethers is also weakly basic, in the case of ethers giving rise to acid-catalyzed cleavage.

Exercise 22-7

(a) Why is 3-nitrophenol (*m*-nitrophenol) less acidic than its 2- and 4-isomers, but more acidic than phenol itself?

(b) Rank in order of increasing acidity: phenol, A; 3,4-dimethylphenol, B; 3-hydroxybenzoic (*m*-hydroxybenzoic) acid, C; 4-(fluoromethyl)phenol [*p*-(fluoromethyl)phenol], D.

In Summary Phenols exist in the enol form because of aromatic stabilization. They are named according to the rules for naming aromatic compounds explained in Section 15-1. Those derivatives bearing carboxy groups on the ring are called hydroxybenzoic acids. Phenols are acidic because the corresponding anions are resonance stabilized.

Some Like it Hot

Chili peppers use capsaicin (see also Problem 60) as a chemical deterrent against herbivores, as well as certain invading fungi. Its burning (to some, painful) sensation is the result of the activation of a protein that is involved in the transmission of pain, specifically that caused by excessive heat, strong acid, or abrasion. Thus, capsaicin "tricks" our neurons into sending a pain signal to the brain, when in fact there is no tissue damage. Real inflammation can ensue, however, due to secondary chemical reactions, not unlike the effect of prostaglandins (Real Life 11-1 and Section 19-13). Remarkably, prolonged exposure to capsaicin desensitizes the pain response by chemical "overload" of our nerves (sort of like a neuroelectrical short), leading to an analgesic effect. The drug Qutenza is a high-potency capsaicin (8%) topical patch for treating pain associated with a shingles infection. The pain of shingles may last for years, and a 30-minute application of the drug is said to bring relief up to 3 months.

REAL LIFE: MEDICINE 22-1 | Two Phenols in the News: Bisphenol A and Resveratrol

This section mentions two phenols as examples of chemicals to which you have potentially frequent exposure: bisphenol A (p. 987) and resveratrol (p. 988). Bisphenol A is the essential ingredient of the familiar polycarbonate plastics used in clear baby bottles, infant formula cans, DVDs, cell phones, eyeglass lenses, auto parts, the linings of food and beverage containers, and reusable plastic bottles. Global demand amounts to 12 billion pounds. Yet there is continuing controversy surrounding this monomer, first made in 1891—a controversy that illustrates the difficulty of interpreting scientific data in terms of human risk assessment.

A polycarbonate plastic bottle in action.

Carbonate unit

Bisphenol A unit
Polycarbonate plastics

   The problem is that bisphenol A behaves as an estrogen mimic in animals. The monomer has been found to leach from the plastic, at increasing rates when heated, such as in a microwave oven. For example, a study carried out in 2003 showed that even very low levels of bisphenol A (about 20 ppb) caused chromosome aberrations in developing mouse eggs. These amounts are in the same range as are encountered in human blood and urine. Because the processes through which human and mouse eggs are prepared for fertilization are very similar, the study is a cause for concern, although it does not prove that humans are at risk. Moreover, in other studies carried out with adult rats, bisphenol A appeared to have no adverse consequences on reproduction and development. A critique of these findings,

dated 2008, pointed out that fetuses and newborns lack the liver enzyme needed to detoxify the chemical, again raising the specter of negative effects on children. As a report by the National Institute of Environmental Health Sciences stresses, the discrepancies in studies of this type may be due to the use of different strains of animals, varying degrees of exposure and differing background levels of estrogenic pollutants, dosing regimens, and housing of the animals (singly versus group). Remember that we are talking about extremely small concentrations of a biologically active compound (parts per billion!) that affect only a certain percentage of animals and to differing degrees. And then there are the big questions of the extent to which animal studies are relevant to humans and whether there is a threshold level of

22-4 | PREPARATION OF PHENOLS: NUCLEOPHILIC AROMATIC SUBSTITUTION

Phenols are synthesized quite differently from the way in which ordinary substituted benzenes are made. Direct *electrophilic* addition of OH to arenes is difficult because of the scarcity of reagents that generate an electrophilic hydroxy group, such as HO^+. Instead, phenols are prepared by formal *nucleophilic* displacement of a leaving group from the arene ring by hydroxide, HO^-, reminiscent of, but mechanistically quite different from, the synthesis of alkanols from haloalkanes. This section considers the ways in which this transformation may be achieved.

Nucleophilic aromatic substitution may follow an addition–elimination pathway

Treatment of 1-chloro-2,4-dinitrobenzene with hydroxide replaces the halogen with the nucleophile, furnishing the corresponding substituted phenol. Other nucleophiles, such exposure that humans can tolerate because of the presence of evolutionary natural detoxification mechanisms. Moreover, finding a replacement for bisphenol A with the same mix of "ideal" properties has its own challenges. For example, reverting to glass for milk and other food containers is clearly inadvisable because of the dangers associated with breakage. Using other (currently inferior) materials as can liners raises the possibility of shorter shelf life and bacterial contamination due to corrosion or cracking. As a result of the concern about prenatal and early childhood exposure, several European countries, Canada, and the United States banned bisphenol A use in baby bottles and cups in 2012.

The issues surrounding potentially harmful chemicals in the environment are equally relevant with respect to their potentially beneficial effects. An example is resveratrol. This compound has been used in traditional medicine to treat conditions of the heart and liver, and recently scientists took an interest in its physiological properties. It is present in various plants and foods, such as eucalyptus, lily, mulberries, peanuts, and, most prominently, in the skin of white and, especially, red grapes, where it is found in concentrations of 50–100 μg/g. The compound is a chemical weapon against invading organisms, such as fungi. Grapes are used for wine making, so resveratrol occurs in red wine, at levels of up to 160 μg per ounce. Studies have suggested that the regular consumption of red wine reduces the incidence of coronary heart disease, a finding described as the "French paradox," namely, the low incidence of heart problems in France despite the relatively high-fat diet. Resveratrol may be the active species, as recent research has indicated its positive cardiovascular effects, including its action as an antioxidant, which inhibits lipid peroxidation (Section 22-9), and as an antiplatelet agent (Real Life 22-2), which prevents atherosclerosis. Indeed, it has been found to hydrogen bond to the same receptor site in cyclooxygenase (COX) as aspirin, which might explain its similar spectrum of activity (see Real Life 22-2 and Section 19-13). Other investigations have shown that the molecule is also an active antitumor agent, involved in retarding the initiation, promotion, and progression of certain cancers, with seemingly minimal toxicity. Perhaps most interesting, it was discovered that resveratrol significantly extended the lifespan of certain species of yeast, worm, the fruit fly, and fish. Along similar lines, it canceled the life-shortening consequences of a high-fat diet in mice. A first glimpse of what might be the origin of these observations was obtained only in 2012, when it was found that the molecule acts through a complicated cascade mechanism on the enzymes responsible for the control of the body's energy balance. The same enzymes are affected by a restricted diet, which has been long known to correlate directly with longevity. As a result of these promising discoveries, resveratrol has been hailed by manufacturers as the "French paradox in a bottle" and pushed for wide-scale consumption. However, experts advise caution. For example, little is known about its metabolism and how it affects the liver, the above results are derived in large part from in vitro and animal experiments, and, like bisphenol A, it has estrogen-like physiological effects, enhancing the growth of breast cancer cells. For the time being, an occasional glass of red wine may be the best course of action, if any!

Resveratrol protects grapes from fungus, such as that shown here.

as alkoxides or ammonia, may be similarly employed, forming alkoxyarenes and arenamines, respectively. Processes such as these, in which a group other than hydrogen is displaced from an aromatic ring, are called **ipso substitutions** (*ipso*, Latin, on itself). The products of these reactions are intermediates in the manufacture of useful dyes.

Nucleophilic Aromatic Ipso Substitution

Ipso position

+ :Nu⁻ ⟶ + Cl⁻

Electron withdrawing

Reaction

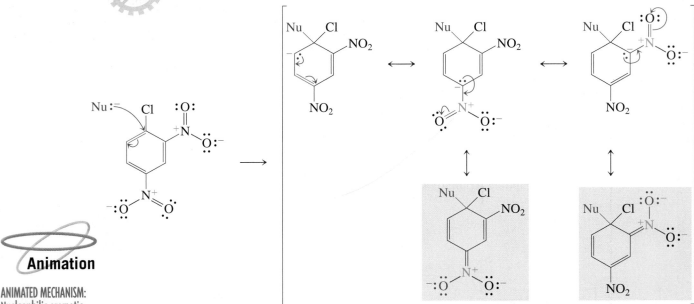

The transformation is called **nucleophilic aromatic substitution.** The key to its success is the presence of one or more strongly electron-withdrawing groups on the benzene ring located ortho or para to the leaving group. Such substituents stabilize an intermediate anion by resonance. In contrast with the S_N2 reaction of haloalkanes, substitution in these reactions takes place by a *two-step mechanism,* an *addition–elimination sequence* similar to the mechanism of substitution of carboxylic acid derivatives (Sections 19-7 and 20-2).

Mechanism of Nucleophilic Aromatic Substitution

Mechanism

Step 1. Addition (facilitated by resonance stabilization)

Animation

ANIMATED MECHANISM:
Nucleophilic aromatic
substitution

The negative charge is strongly stabilized by resonance
involving the ortho- and para-NO₂ groups.

Step 2. Elimination (only one resonance structure is shown)

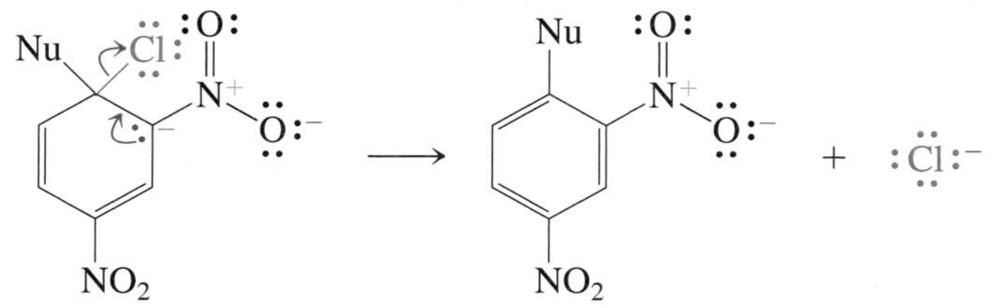

In the first and rate-determining step, ipso attack by the nucleophile produces an anion with a highly delocalized charge, for which several resonance structures may be written, as shown. Note the ability of the negative charge to be delocalized into the electron-withdrawing groups. In contrast, such delocalization is *not* possible in 1-chloro-3,5-dinitrobenzene, in which these groups are located meta; so this compound does *not* undergo ipso substitution under the conditions employed.

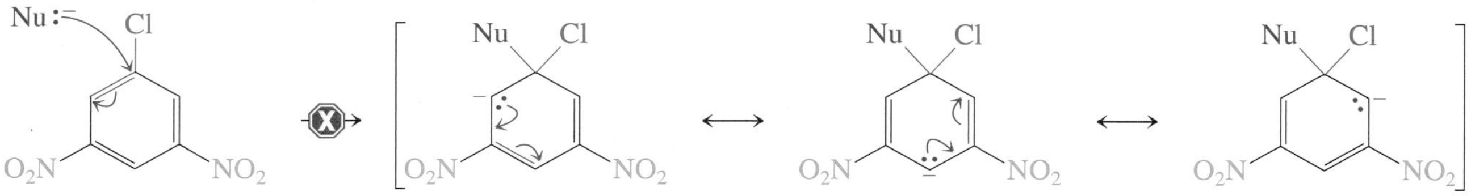

Meta-NO₂ groups do *not* provide resonance stabilization of the negative charge.

In the second step, the leaving group is expelled to regenerate the aromatic ring. The reactivity of haloarenes in nucleophilic substitutions increases with the nucleophilicity of the reagent and the number of electron-withdrawing groups on the ring, particularly if they are in the ortho and para positions.

Exercise 22-8

Write the expected product of reaction of 1-chloro-2,4-dinitrobenzene with NaOCH₃ in boiling CH₃OH.

Solved Exercise 22-9 | **Working with the Concepts: Using Nucleophilic Aromatic Substitution in Synthesis**

Ofloxacin, an antibiotic of the quinolone class (Section 25-7), is used in the treatment of infections of the respiratory and urinary tract, the eyes, the ears, and skin tissue. The quinolone antibiotics are yet another alternative in the ongoing battle against penicillin- (and other drug)-resistant bacteria. The final three steps in the synthesis of ofloxacin are shown below, the last of which is a simple ester hydrolysis (Section 20-4). Formulate mechanisms for the other two transformations from A to B and then to C.

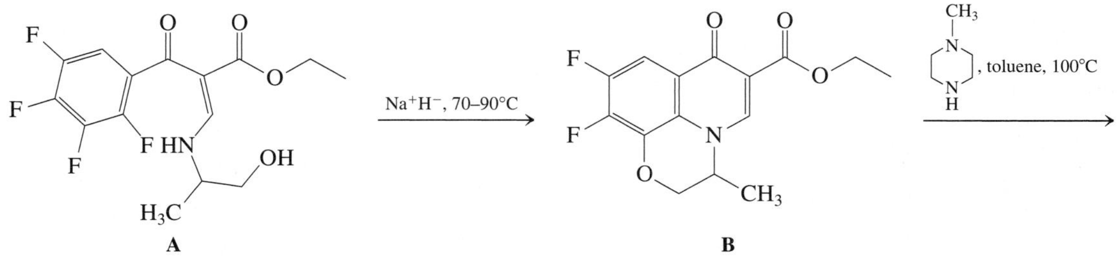

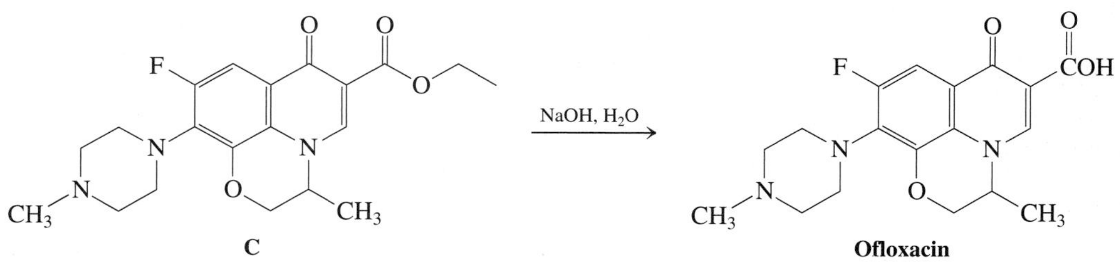

Strategy

Taking an inventory of the topological changes and of the nature of the bonds formed and bonds broken, we note that B and C are the result of two successive intra- and one intermolecular nucleophilic aromatic substitutions, respectively. The first steps are ring closures and therefore especially favored for entropic and enthalpic reasons. (Formation of six-membered rings is a favorable reaction, see Sections 9-6 and 17-7.) The aromatic ring in A should be highly activated with respect to nucleophilic attack, because it bears four inductively electron-withdrawing fluorine substituents, in addition to the resonance-based electron-withdrawing carbonyl group. To propose a mechanism, let's look at the individual steps.

Solution

- In the conversion of A to B, the strong base Na^+H^- is employed. Where are the acidic sites in A? Most obvious is the hydroxy function ($pK_a \sim 15$–16, Section 8-3, Table 8-2), which will certainly be deprotonated.
- What about the amino group? Closer inspection of structure A shows that the nitrogen is connected via a double bond to two carbonyl moieties. Thus, the function is resonance stabilized in much the same way as an ordinary amide (but more so):

Resonance in A

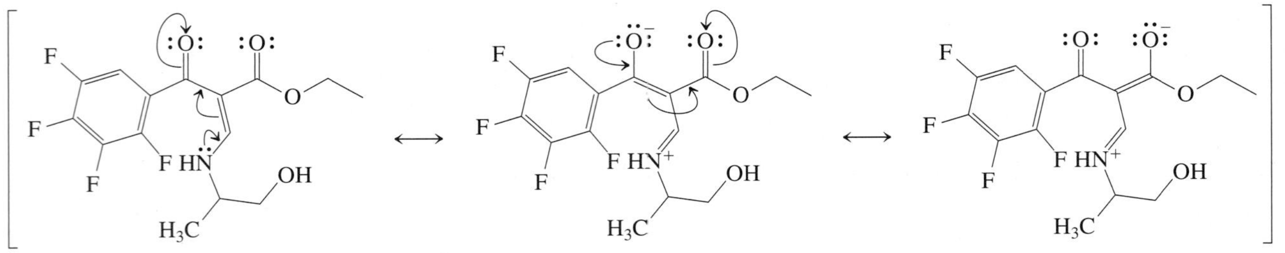

Consequently, it should be relatively acidic, even more acidic than an amide ($pK_a \approx 22$, Section 20-7). It is therefore likely that A is doubly deprotonated before attacking the benzene nucleus. Nucleophilic aromatic substitution is then readily formulated to assemble the first new ring. In the scheme below, only one—the dominating—resonance form of the anion resulting from nucleophilic attack on the arene is shown, that in which the negative charge is delocalized onto the carbonyl oxygen.

First Ring Closure

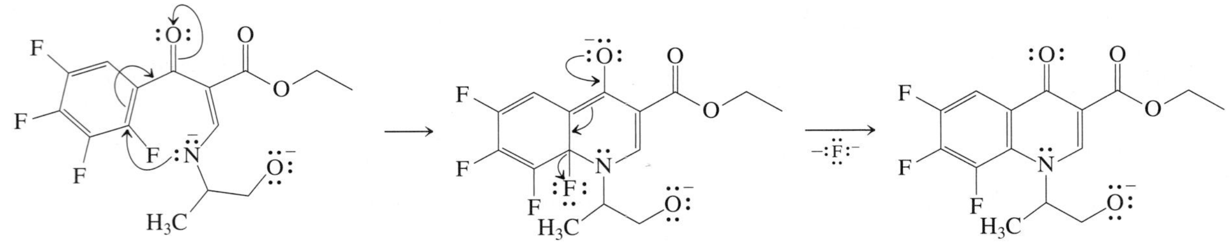

The second nucleophilic substitution by alkoxide is unusual, inasmuch as it has to occur at a position meta to the carbonyl function. Therefore, the intermediate anion is not stabilized by resonance, but solely by inductive effects. The regioselectivity is controlled by strain: The alkoxide cannot "reach" the more favorable carbon para to the carbonyl.

Second Ring Closure to B

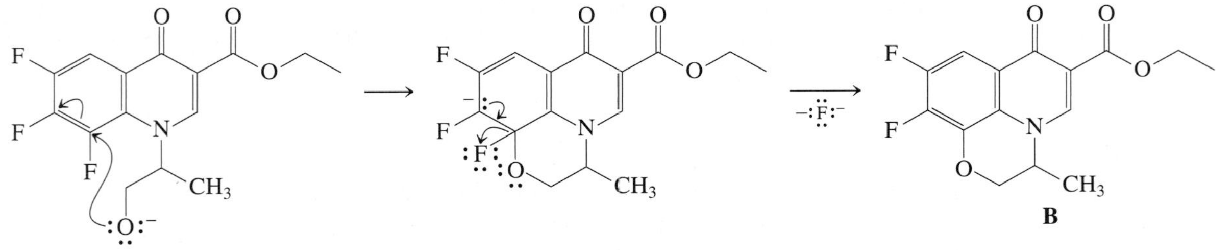

The conversion of B to C features an intermolecular substitution with an amine nucleophile. Of the two carbons bearing the potential fluoride leaving groups, the one para to the carbonyl is picked, because of the resonance stabilization of the resulting intermediate anion. The initial product is formed as the ammonium salt, from which the free base C is liberated by basic work-up.

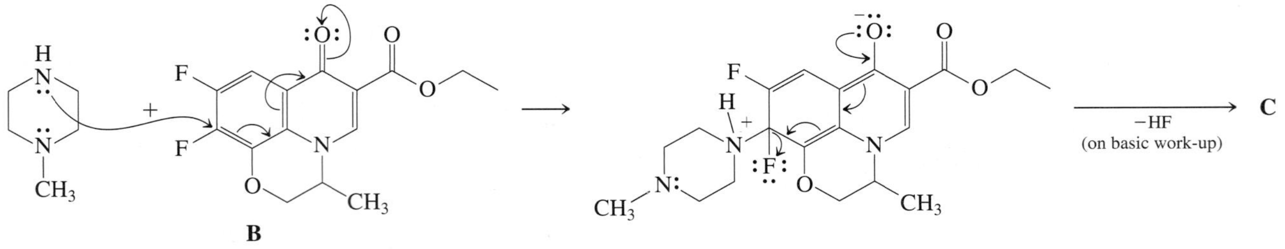

Exercise 22-10 | Try It Yourself

Propose a mechanism for the following conversion. Considering that the first step is rate determining, draw a potential-energy diagram depicting the progress of the reaction. (**Hint:** This is a nucleophilic aromatic substitution.)

Haloarenes undergo substitution through benzyne intermediates

Haloarenes devoid of electron-withdrawing substituents do not undergo simple ipso substitution. Nevertheless, when haloarenes are treated with nucleophiles that are also strong bases, if necessary at highly elevated temperatures, they convert to products in which the halide has been replaced by the nucleophile. For example, if exposed to hot sodium hydroxide followed by neutralizing work-up, chlorobenzene furnishes phenol.

$$\text{Chlorobenzene} \quad \xrightarrow[\text{2. H}^+\text{, H}_2\text{O}]{\text{1. NaOH, H}_2\text{O, 340°C, 150 atm}} \quad \text{Phenol} \quad + \quad \text{NaCl}$$

Reaction

Treatment with potassium amide results in benzenamine (aniline).

$$\text{Chlorobenzene} \quad \xrightarrow[\text{2. H}^+\text{, H}_2\text{O}]{\text{1. KNH}_2\text{, liquid NH}_3} \quad \text{Benzenamine (Aniline)} \quad + \quad \text{KCl}$$

It is tempting to assume that these substitutions follow a mechanism similar to that formulated for nucleophilic aromatic ipso substitution earlier in this section. However, when the last reaction is performed with radioactively labeled chlorobenzene (^{14}C at C1), a very curious result is obtained: Only half of the product is substituted at the labeled carbon; in the other half, the nitrogen is at the *neighboring* position.

^{14}C label

$$\text{Chlorobenzene-1-}^{14}\text{C} \quad \xrightarrow[-\text{KCl}]{\text{KNH}_2\text{, liquid NH}_3} \quad \underset{50\%}{\text{Benzenamine-1-}^{14}\text{C}} \quad + \quad \underset{50\%}{\text{Benzenamine-2-}^{14}\text{C}}$$

Direct substitution does not seem to be the mechanism of these reactions. What, then, is the answer to this puzzle? A clue is the attachment of the incoming nucleophile *only* at the ipso or at the ortho position relative to the leaving group. This observation can be accounted for by an initial base-induced elimination of HX from the benzene ring, a process reminiscent of the dehydrohalogenation of haloalkenes to give alkynes (Section 13-4). In the present case, elimination is not a concerted process, but rather takes place in a sequential manner, deprotonation preceding the departure of the leaving group (step 1 of the mechanism shown). Both stages in step 1 are difficult, with the second being worse than the first. Why is that? With respect to the initial anion formation, recall (Section 11-3) that the acidity of C_{sp2}–H is very low ($pK_a \approx 44$), and the same is true in general for phenyl hydrogens. The presence of the adjacent π system of benzene does not help, because the negative charge in the phenyl anion resides in an sp^2 orbital that is *perpendicular* to the π frame and is therefore incapable of resonance with the double bonds in the six-membered ring. Thus, deprotonation of the haloarene requires a strong base. It takes place ortho to the halogen, because the halogen's inductive electron-withdrawing effect acidifies this position relative to the others.

Although deprotonation is not easy, the second stage of step 1, subsequent elimination of X^-, is even more difficult because of the highly strained structure of the resulting reactive species, called **1,2-dehydrobenzene** or **benzyne.**

Mechanism

Mechanism of Nucleophilic Substitution of Simple Haloarenes

Step 1. Elimination occurs stepwise

$pK_a \sim 44$

sp^2 - Hybrid orbital is perpendicular to aromatic π system

Highly strained triple bond

Phenyl anion
(Intermediate)

Benzyne
(Reactive intermediate; not isolated)

Step 2. Addition occurs to both strained carbons

Animation

ANIMATED MECHANISM:
Nucleophilic aromatic substitution via benzynes

Why is benzyne so strained? Recall that alkynes normally adopt a linear structure, a consequence of the *sp* hybridization of the carbons making up the triple bond (Section 13-2). Because of benzyne's cyclic structure, its triple bond is forced to be bent, rendering it unusually reactive. Thus, benzyne exists only as a reactive intermediate under these conditions, being rapidly attacked by any nucleophile present. For example (step 2), amide ion or even ammonia solvent can add to furnish the product benzenamine (aniline). Because the two ends of the triple bond are equivalent, addition can take place at either carbon, explaining the label distribution in the benzenamine obtained from ^{14}C-labeled chlorobenzene.

Benzyne is too reactive to be isolated and stored in a bottle, but it can be observed spectroscopically under special conditions. Irradiation of benzocyclobutenedione at 77 K

($-196°C$) in frozen argon (m.p. $= -189°C$) produces a species whose IR and UV spectra are assignable to benzyne, formed by loss of two molecules of CO.

Generation of Benzyne, a Reactive Intermediate

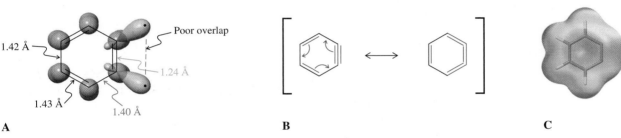

Benzocyclobutene-1,2-dione

$\xrightarrow{hv,\ 77\ K}$ + 2 CO

Although benzyne is usually represented as a cycloalkyne (Figure 22-4A), its triple bond exhibits an IR stretching frequency of 1846 cm^{-1}, intermediate between the values for normal double (cyclohexene, 1652 cm^{-1}) and triple (3-hexyne, 2207 cm^{-1}) bonds. The ^{13}C NMR values for these carbons ($\delta = 182.7$ ppm) are also atypical of pure triple bonds (Section 13-3), indicating a considerable contribution of the cumulated triene (Section 14-5) resonance form (Figure 22-4B). The bond is weakened substantially by poor p orbital overlap in the plane of the ring.

1.42 Å

Poor overlap

1.24 Å

1.43 Å

1.40 Å

A

B

C

Figure 22-4 (A) The orbital picture of benzyne reveals that the six aromatic π electrons are located in orbitals that are perpendicular to the two additional hybrid orbitals making up the distorted triple bond. These hybrid orbitals overlap only poorly; therefore, benzyne is highly reactive. (B) Resonance in benzyne. (C) The electrostatic potential map of benzyne shows electron density (red) in the plane of the six-membered ring at the position of the distorted *sp*-hybridized carbons.

Exercise 22-11

1-Chloro-4-methylbenzene (*p*-chlorotoluene) is not a good starting material for the preparation of 4-methylphenol (*p*-cresol) by direct reaction with hot NaOH, because it forms a mixture of two products. Why does it do so, and what are the two products?

Exercise 22-12

Explain the regioselectivity observed in the following reaction. (**Hint:** Consider the effect of the methoxy group on the selectivity of attack by amide ion on the intermediate benzyne.)

Br

OCH$_3$

$\xrightarrow[\text{$-$KBr}]{\text{KNH}_2,\ \text{liquid NH}_3}$

H$_2$N

OCH$_3$

Major

+

NH$_2$

OCH$_3$

Minor

Phenols are produced from arenediazonium salts

The most general traditional laboratory procedure for making phenols is from arenamines through their **arenediazonium salts,** $ArN_2^+X^-$. Recall that primary alkanamines can be *N*-nitrosated but that the resulting species rearrange to diazonium ions, which are unstable—they lose nitrogen to give carbocations (Section 21-10). In contrast, primary benzenamines (anilines) are attacked by cold nitrous acid, in a reaction called **diazotization,** to give relatively stable, isolable, although still reactive arenediazonium salts. Compared to their alkanediazonium counterparts, these species enjoy resonance stabilization and are prevented from undergoing immediate N_2 loss by the high energy of the resulting **aryl cations** (to be discussed in more detail in Section 22-10).

Diazotization

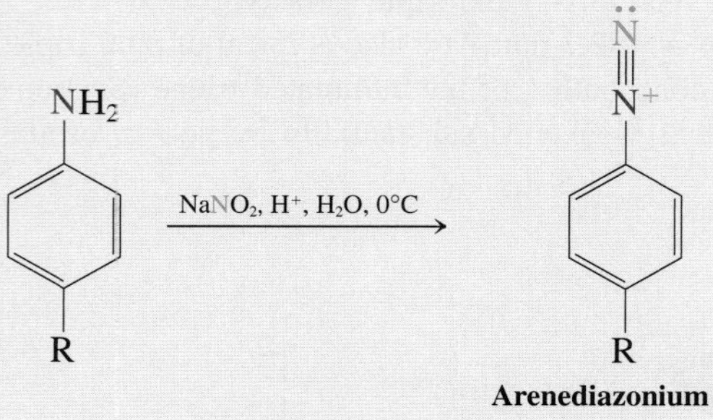

Arenediazonium ion

When arenediazonium ions are gently heated in water, nitrogen is evolved and the resulting aryl cations are trapped extremely rapidly by the solvent to give phenols.

Decomposition of Arenediazonium Salts in Water to Give Phenols

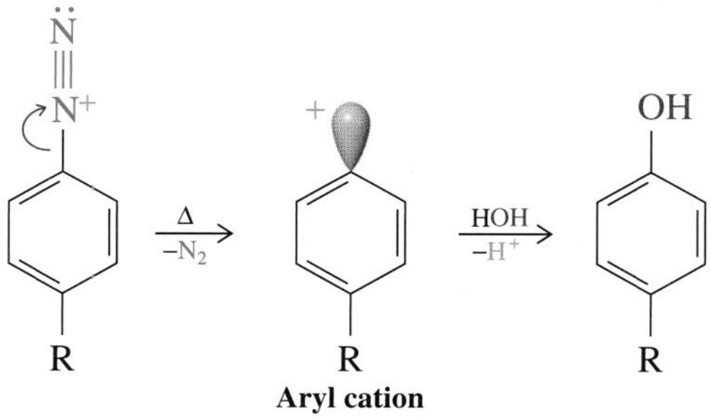

Aryl cation

In these reactions, the "super" leaving group N_2 accomplishes what halides are only able to do when attached to a highly electron-deficient benzene nucleus (nucleophilic aromatic substitution) or under extreme conditions (through benzyne intermediates), namely, replacement by hydroxide. The three mechanisms are completely different. In nucleophilic aromatic substitution, the nucleophile attacks prior to departure of the leaving group. In the benzyne mechanism, the nucleophile acts initially as a base, followed by extrusion of the leaving group and subsequent nucleophilic attack on the strained triple bond. In the arenediazonium-ion decomposition, the leaving group exits first, followed by trapping by water.

The utility of this phenol synthesis is apparent when you recall that arenamines are derived from nitroarenes by reduction, and nitroarenes are made from other arenes by electrophilic aromatic substitution (Chapters 15 and 16). Therefore, retrosynthetically (Section 8-9), we can picture the hydroxy group in any position of a benzene ring that is subject to electrophilic nitration.

Retrosynthetic Connection of Phenols to Arenes

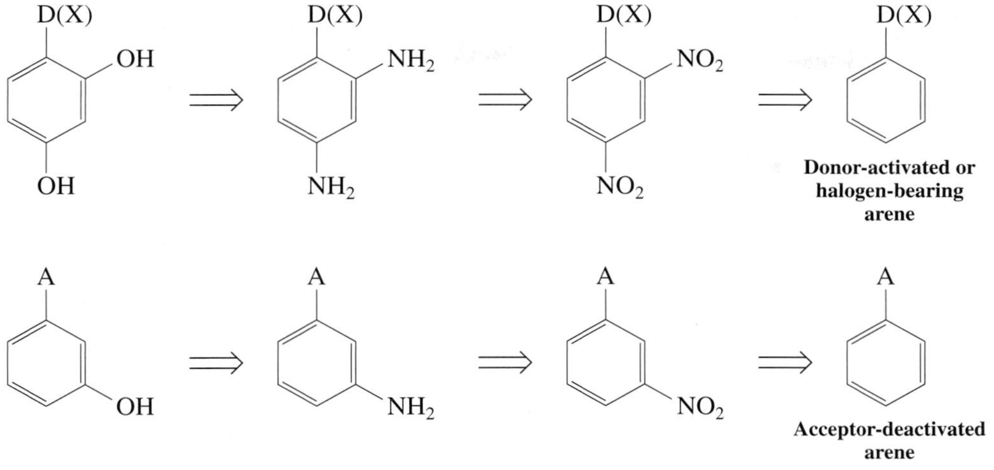

Two examples are depicted below.

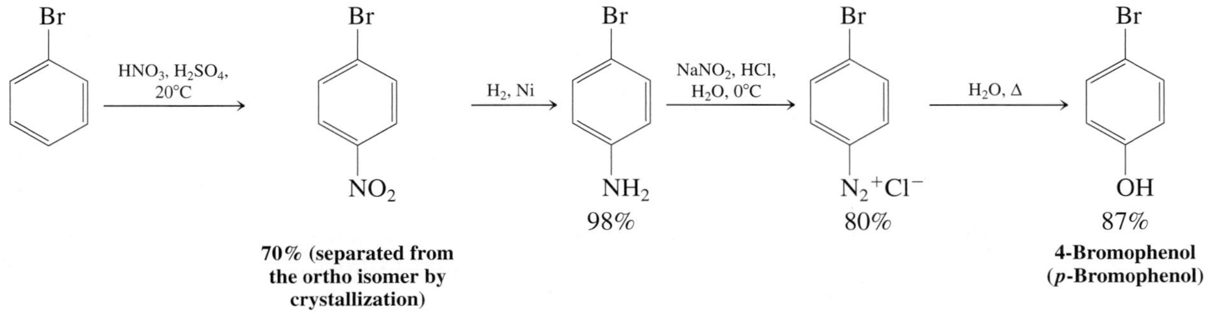

70% (separated from the ortho isomer by crystallization)

4-Bromophenol (p-Bromophenol)

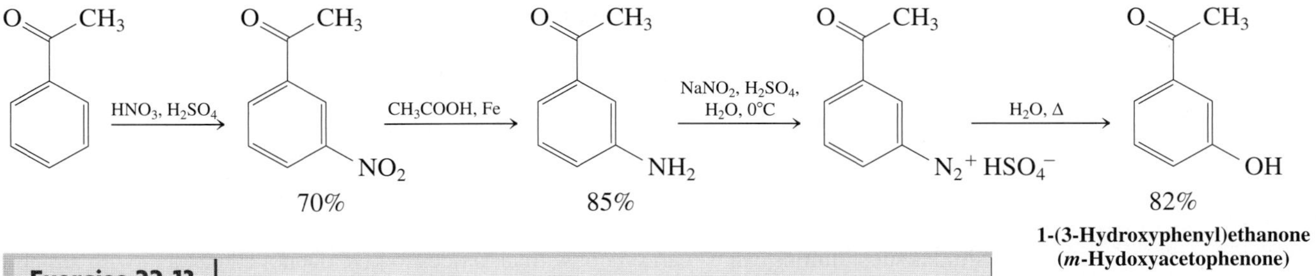

1-(3-Hydroxyphenyl)ethanone (m-Hydoxyacetophenone)

Exercise 22-13

ortho-Benzenediazoniumcarboxylate A (made by diazotization of 2-aminobenzoic acid, Problem 59 of Chapter 20) is explosive. When warmed in solution with *trans,trans*-2,4-hexadiene, it forms compound B. Explain by a mechanism. (**Hint:** Two other products are formed, both of which are gases.)

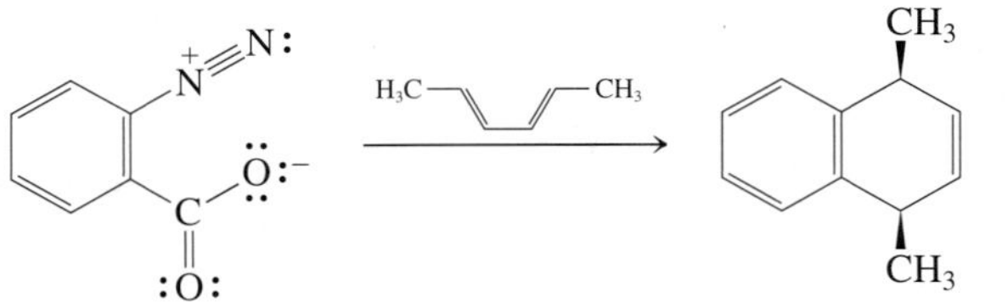

Exercise 22-14

Propose a synthesis of 4-(phenylmethyl)phenol (*p*-benzylphenol) from benzene. (**Caution:** Remember that Friedel-Crafts reactions do not work with deactivated arenes.)

Phenols can be made from haloarenes by Pd catalysis

While we have seen that ordinary halobenzenes are resilient to reaction with hydroxide, they undergo such nucleophilic displacements in the presence of Pd salts and added phosphine ligands PR_3.

Pd-Catalyzed Phenol Synthesis from Haloarenes

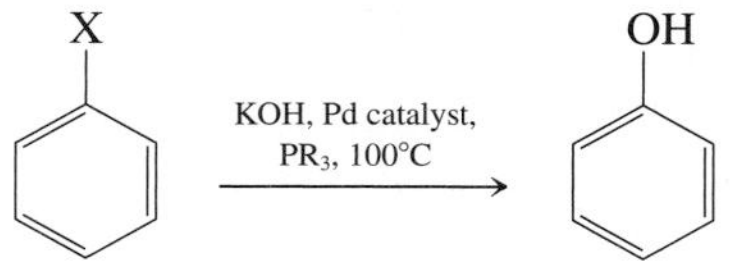

The reaction is general for substituted benzenes, providing a complement to the diazonium method described above.

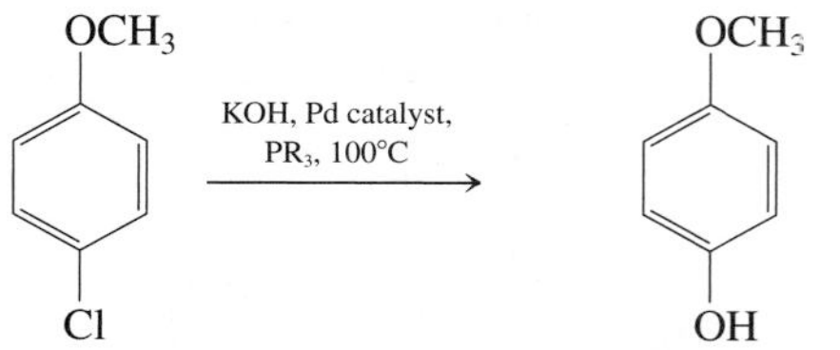

90%

4-Methoxyphenol
(*p*-Methoxyphenol)

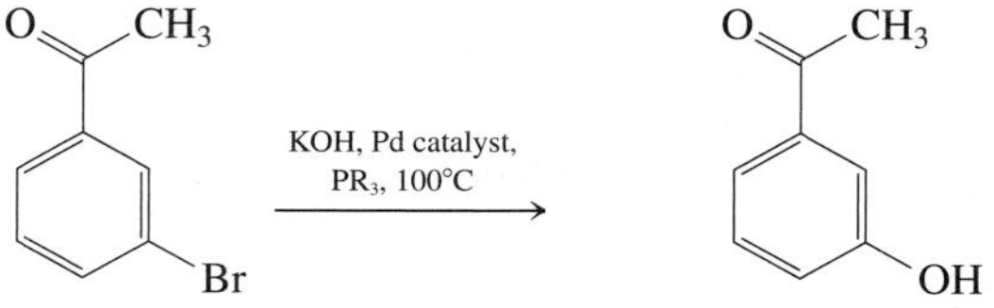

98%

1-(3-Hydroxyphenyl)ethanone
(*m*-Hydroxyacetophenone)

The mechanism is related to that of the Heck and other Pd-catalyzed reaction (Section 13-9; Real Life 13-1). As shown in a simplified manner below, it begins by insertion of the metal into the aryl halide bond, exchange of the halide for hydroxide, and extrusion of the final product with regeneration of the catalyst.

Mechanism of the Pd-Catalyzed Phenol Synthesis from Haloarenes

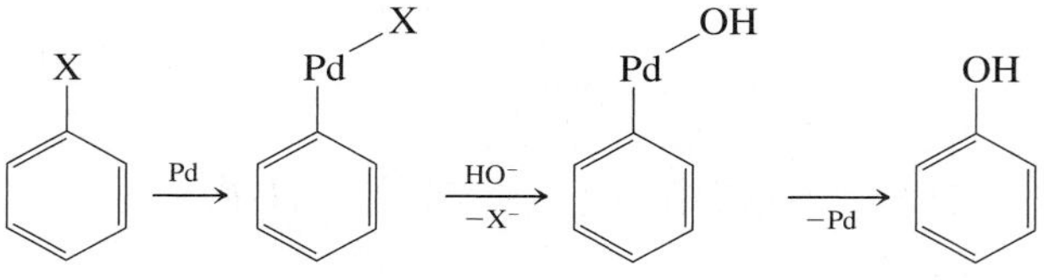

Similar substitutions can be carried out with alkoxides to give phenol ethers, and with amines, including ammonia, to furnish benzenamines.

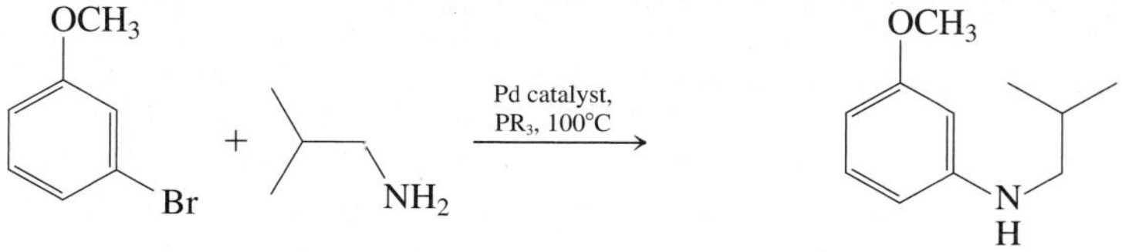

98%

3-Methoxy-*N*-(2-methylpropyl)benzenamine
3-Methoxy-*N*-(2-methylpropyl)aniline

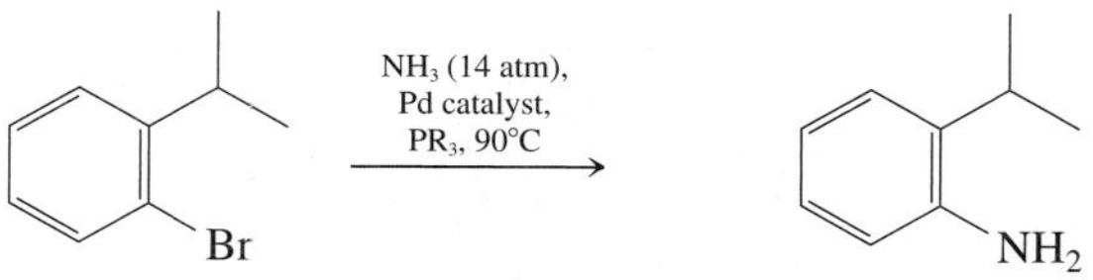

89%

2-(1-Methylethyl)benzenamine
(*o*-Isopropylaniline)

Although the preceding methods are valuable in the preparation of specifically substituted phenols, the parent compound is made industrially by the air oxidation of (1-methylethyl) benzene (isopropylbenzene or cumene; see also Exercise 15-27) to the benzylic hydroperoxide and its subsequent decomposition with acid (margin). The "by-product" acetone is valuable in its own right and makes this process highly cost effective, quite apart from the environmentally benign use of air as an oxidant.

A "Green" Industrial Phenol Synthesis

Exercise 22-15

How would you make the following phenols from the given starting materials? (**Hint:** Consult Chapters 15 and 16.)

In Summary When a benzene ring bears enough strongly electron-withdrawing substituents, nucleophilic addition to give an intermediate anion with delocalized charge becomes feasible, followed by elimination of the leaving group (nucleophilic aromatic ipso substitution). Phenols result when the nucleophile is hydroxide ion, arenamines (anilines) when it is ammonia, and alkoxyarenes when alkoxides are employed. Very strong bases are capable of eliminating HX from haloarenes to form the reactive intermediate benzynes, which are subject to nucleophilic attack to give substitution products. Phenols may also be prepared by decomposition of arenediazonium salts in water and by Pd-catalyzed hydroxylations of haloarenes.

22-5 | ALCOHOL CHEMISTRY OF PHENOLS

The phenol hydroxy group undergoes several of the reactions of alcohols (Chapter 9), such as protonation, Williamson ether synthesis, and esterification.

The oxygen in phenols is only weakly basic

Phenols are not only acidic but also weakly basic. They (and their ethers) can be protonated by strong acids to give the corresponding **phenyloxonium ions.** Thus, as with the alkanols, the hydroxy group imparts amphoteric character (Section 8-3). However, the basicity of phenol is even less than that of the alkanols, because the lone electron pairs on the oxygen are delocalized into the benzene ring (Sections 16-1 and 16-3). The pK_a values for phenyloxonium ions are, therefore, lower than those of alkyloxonium ions.

pK_a Values of Methyl- and Phenyloxonium Ion

Unlike secondary and tertiary alkyloxonium ions derived from alcohols, phenyloxonium derivatives do not dissociate to form phenyl cations, because such ions have too high an energy content (see Section 22-10). The phenyl–oxygen bond in phenols is very difficult to break. However, after protonation of alkoxybenzenes, the bond between the *alkyl* group and oxygen is readily cleaved in the presence of nucleophiles such as Br⁻ or I⁻ (e.g., from HBr or HI) to give phenol and the corresponding haloalkane.

COOH

$\xrightarrow{\text{HBr, }\Delta}$

COOH

OCH₃

OH

90%

$+$ CH₃Br

3-Methoxybenzoic acid
(*m*-Methoxybenzoic acid)

3-Hydroxybenzoic acid
(*m*-Hydroxybenzoic acid)

Exercise 22-16

Why does cleavage of an alkoxybenzene by acid not produce a halobenzene and the alkanol?

Alkoxybenzenes are prepared by Williamson ether synthesis

The Williamson ether synthesis (Section 9-6) permits easy preparation of many alkoxybenzenes. The phenoxide ions obtained by deprotonation of phenols (Section 22-3) are good nucleophiles. They can displace the leaving groups from haloalkanes and alkyl sulfonates.

OH

Cl

$+$ CH₃CH₂CH₂Br

$\xrightarrow[\text{−NaBr, −HOH}]{\text{NaOH, H}_2\text{O}}$

OCH₂CH₂CH₃

Cl

63%

3-Chlorophenol
(*m*-Chlorophenol)

1-Chloro-3-propoxybenzene
(*m*-Chlorophenyl propyl ether)

Esterification leads to phenyl alkanoates

The reaction of a carboxylic acid with a phenol (Section 19-9) to form a phenyl ester is endothermic. Therefore, esterification requires an activated carboxylic acid derivative, such as an acyl halide or a carboxylic anhydride.

OH

CH₃

$+$ CH₃CH₂$\overset{\text{O}}{\overset{\|}{\text{C}}}$Cl

$\xrightarrow[\text{−NaCl, −HOH}]{\text{NaOH, H}_2\text{O}}$

O$\overset{\text{O}}{\overset{\|}{\text{C}}}$CH₂CH₃

CH₃

4-Methylphenol
(*p*-Cresol)

Propanoyl chloride

4-Methylphenyl propanoate
(*p*-Methylphenyl propanoate)

Exercise 22-17

(a) Explain why, in the preparation of acetaminophen (Real Life 22-2), the amide is formed rather than the ester. (**Hint:** Review Section 6-8.)

(b) Salsalate (short for salicyl salicylate), an ester of two salicylic acid molecules (margin), is prescribed to reduce pain and inflammation in rheumatoid arthritis patients as an alternative to naproxen or ibuprofen (Chapter 16 Opening), because it avoids stomach upset. Formulate a synthesis from 2-hydroxybenzoic acid (salicylic acid). (**Hint:** You will need to develop a protecting-group strategy. See Section 9-8 or 22-2.)

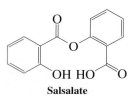

Salsalate

REAL LIFE: MEDICINE 22-2 | **Aspirin: The Miracle Drug**

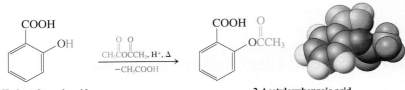

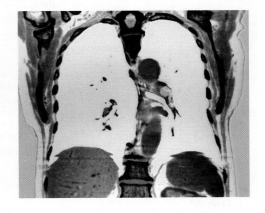

2-Hydroxybenzoic acid
(*o*-Hydroxybenzoic acid,
salicylic acid)

2-Acetyloxybenzoic acid
(*o*-Acetoxybenzoic acid,
acetylsalicylic acid, aspirin)

Aspirin prevents heart attacks by minimizing the formation of blood clots in the coronary arteries. The photograph shows such a blood clot (orange) in the left main pulmonary artery of a patient.

The year 1997 marked the 100th birthday of the synthesis of the acetic ester of 2-hydroxybenzoic acid (salicylic acid); namely, 2-acetyloxybenzoic acid (acetylsalicylic acid), better known as aspirin (see Chapter 16 Opening). Aspirin is the first drug that was clinically tested before it was marketed, in 1899. More than 100 billion tablets per year are taken by people throughout the world to relieve headaches and rheumatoid and other pain, to control fever, and to treat gout and arthritis. Production capacity in the United States alone is 10,000 tons per year.

Salicylic acid (also called spiric acid, hence the name aspirin ["a" for acetyl]) in extracts from the bark of the willow tree or from the meadowsweet plant had been used since ancient times (see Chapter 16 Opening) to treat pain, fever, and swelling. This acid was first isolated in pure form in 1829, then synthesized in the laboratory, and finally produced on a large scale in the nineteenth century and prescribed as an analgesic, antipyretic, and anti-inflammatory drug. Its bitter taste and side effects, such as mouth irritation and gastric bleeding, prompted the search for better derivatives that resulted in the discovery of aspirin.

In the body, aspirin (a relatively reactive phenyl alkanoate, see Problem 59) acts as an acetylating agent of the enzyme cyclooxygenase (COX). It does so by transesterification with the hydroxyl function of the amino acid serine of the polypeptide chain in the active site (see below). Interestingly, while salicylic acid itself is also a COX inhibitor, it works by a different mechanism, namely, suppression of the genetic information that encodes the enzyme, thus attenuating its production.

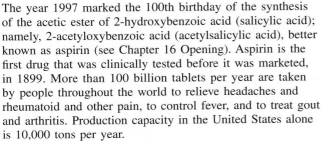

Serine

Aspirin

COX

Disabled COX

Cyclooxygenase mediates the production of prostaglandins (see Real Life 11-1 and Section 19-13), molecules that in turn are inflammatory and pain producing. In addition, one of them, thromboxane A$_2$, aggregates blood platelets, necessary for the clotting of blood when injury occurs. This same process is, however, undesirable inside arteries, causing heart attacks or brain strokes, depending on the location of the clot. Indeed, a large study conducted in the 1980s showed that aspirin lowered the risk of heart attacks in men by almost 50% and reduced the mortality rate during an actual attack by 23%.

Many other potential applications of aspirin are under investigation, such as in the treatment of pregnancy-related complications, viral inflammation in AIDS patients, dementia, Alzheimer's disease, and cancer. Despite its popularity, aspirin can have some serious side effects: It is toxic to the liver, prolongs bleeding, and causes gastric irritation. It is suspected as the cause of Reye's syndrome, a condition that leads to usually fatal brain damage. Because of some of these drawbacks, many other drugs compete with aspirin, particularly in the analgesics market, such as naproxen, ibuprofen, and acetaminophen (see the beginning of Chapter 16). Acetaminophen, better known as Tylenol, is prepared from 4-aminophenol by acetylation.

4-Aminophenol
(*p*-Aminophenol)

N-(4-Hydroxyphenyl)acetamide
[*N*-(*p*-Hydroxyphenyl)acetamide,
acetaminophen, Tylenol]

In Summary The oxygen in phenols and alkoxybenzenes can be protonated even though it is less basic than the oxygen in the alkanols and alkoxyalkanes. Protonated phenols and their derivatives do not ionize to phenyl cations, but the ethers can be cleaved to phenols and haloalkanes by HX. Alkoxybenzenes are made by Williamson ether synthesis, aryl alkanoates by acylation.

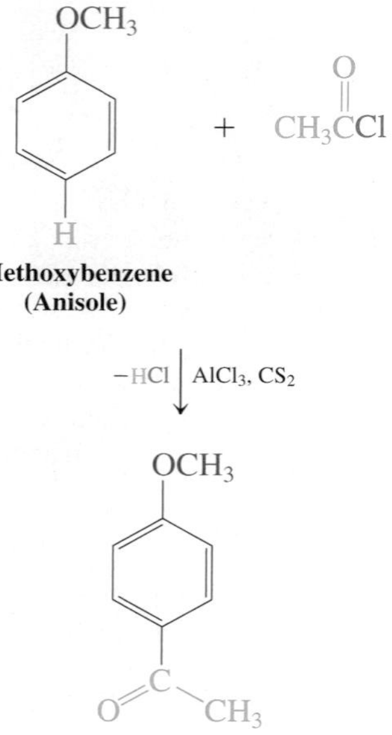

OCH₃

Methoxybenzene
(Anisole)

\+ CH₃CCl

−HCl │ AlCl₃, CS₂

OCH₃

70%
1-(4-Methoxyphenyl)ethanone
(*p*-Methoxyacetophenone)

22-6 | ELECTROPHILIC SUBSTITUTION OF PHENOLS

The aromatic ring in phenols is also a center of reactivity. The interaction between the OH group and the ring strongly activates the ortho and para positions toward electrophilic substitution (Sections 16-1 and 16-3). For example, even dilute nitric acid causes nitration.

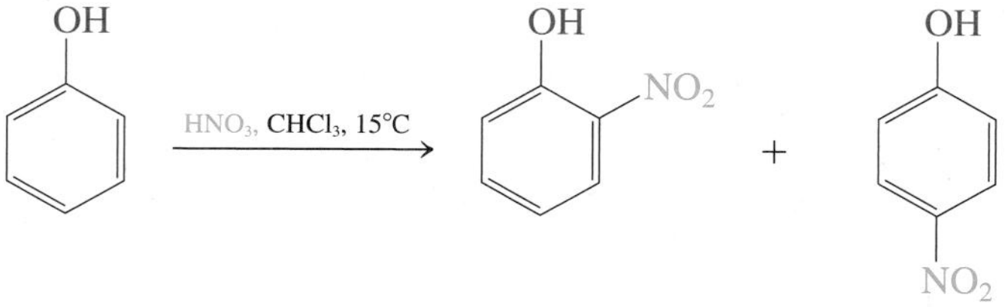

OH → HNO₃, CHCl₃, 15°C →

OH NO₂ 26% **2-Nitrophenol** (*o*-Nitrophenol)

\+

OH NO₂ 61% **4-Nitrophenol** (*p*-Nitrophenol)

Friedel-Crafts acylation of phenols is complicated by ester formation and is better carried out on ether derivatives of phenol (Section 16-5), as shown in the margin.

Phenols are halogenated so readily that a catalyst is not required, and multiple halogenations are frequently observed (Section 16-3). As shown in the following reactions, tribromination occurs in water at 20°C, but the reaction can be controlled to produce the monohalogenation product through the use of a lower temperature and a less polar solvent.

Halogenation of Phenols

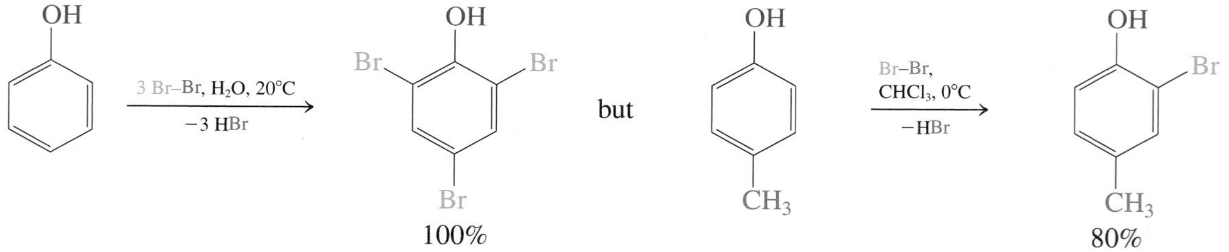

Phenol OH → 3 Br–Br, H₂O, 20°C, −3 HBr →

Br OH Br, Br 100% **2,4,6-Tribromophenol**

but

OH CH₃ **4-Methylphenol** (*p*-Cresol)

→ Br–Br, CHCl₃, 0°C, −HBr →

OH Br CH₃ 80% **2-Bromo-4-methylphenol**

Electrophilic attack at the para position is frequently dominant because of steric effects. However, it is normal to obtain mixtures resulting from both ortho and para substitutions, and their compositions are highly dependent on reagents and reaction conditions.

Exercise 22-18

Friedel-Crafts methylation of methoxybenzene (anisole) with chloromethane in the presence of AlCl₃ gives a 2 : 1 ratio of ortho : para products. Treatment of methoxybenzene with 2-chloro-2-methylpropane (*tert*-butyl chloride) under the same conditions furnishes only 1-methoxy-4-(1,1-dimethylethyl) benzene (*p-tert*-butylanisole). Explain. (**Hint:** Review Section 16-5.)

Solved Exercise 22-19 | **Working with the Concepts: Devising Synthetic Strategies Starting with Substituted Phenols**

Proparacaine is a local anesthetic that is used primarily to numb the eye before minor surgical procedures, such as the removal of foreign objects or stitches. Show how you would make it from 4-hydroxybenzenecarboxylic acid.

4-Hydroxybenzenecarboxylic acid — --→ **Proparacaine**

Strategy

Inspection of the structures of both the starting material and the product reveals three structural modifications: (1) An amino group is introduced into the benzene core, suggesting a nitration-reduction sequence (Section 16-5) that relies on the ortho-directing effect of the hydroxy substituent: (2) the phenol function is etherified, best done by Williamson synthesis (Section 22-5): (3) the carboxy function is esterified with the appropriate amino alcohol (Section 20-2). What is the best sequence in which to execute these manipulations? To answer this question, consider the possible interference of the various functions with the suggested reaction steps.

Solution

• Modification 1 could be carried out with the starting material, resulting in the corresponding amino acid. However, "unmasking" the amino group (from its precursor nitro) early bodes trouble with steps 2 and 3, because amines are better nucleophiles than alcohols. Therefore, attempted etherification of the aminated phenol will lead to amine alkylation (Section 21-5). Similarly, attempts to effect ester formation in the presence of an amine function will give rise to amide generation (Section 19-10). However, early introduction of the nitro group and maintaining it as such until the other functions are protected looks good.

• For modification 2, Williamson ether synthesis in the presence of a carboxy group should be fine, because the carboxylate ion is a poorer nucleophile than phenoxide ion (Section 6-8).

• The preceding protection of the phenolic function is important to the success of modification 3, ester formation with the amino alcohol, especially if we want to activate the carboxy group as an acyl chloride.

• Putting it all together, a possible (and, indeed, the literature) synthesis of proparacaine is as follows:

• Why does the amino alcohol not react with the acyl halide at nitrogen in the last step? After all, amines are more nucleophilic than alcohols. The answer is: It does, but because the amine function is tertiary, it can only form an acyl ammonium salt. This function has reactivity similar to that of an acyl chloride. Thus, attack by the hydroxy group on the acyl ammonium carbonyl carbon wins out thermodynamically to give the ester in the end (Section 20-2).

Exercise 22-20 | **Try It Yourself**

Devise an alternative route to proparacaine from 4-hydroxybenzenecarboxylic acid, using Pd catalysis.

Under basic conditions, phenols can undergo electrophilic substitution, even with very mild electrophiles, through intermediate phenoxide ions. An industrially important application is the reaction with formaldehyde, which leads to *o*- and *p*-hydroxymethylation.

Mechanistically, these processes may be considered enolate condensations, much like the aldol reaction (Section 18-5).

Hydroxymethylation of Phenol

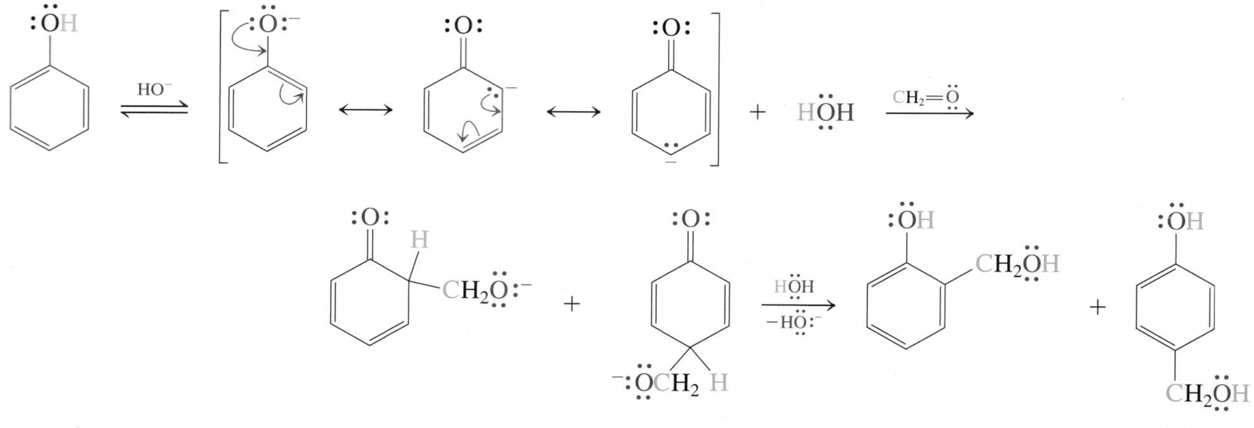

The initial aldol products are unstable: They dehydrate on heating, giving reactive intermediates called **quinomethanes.**

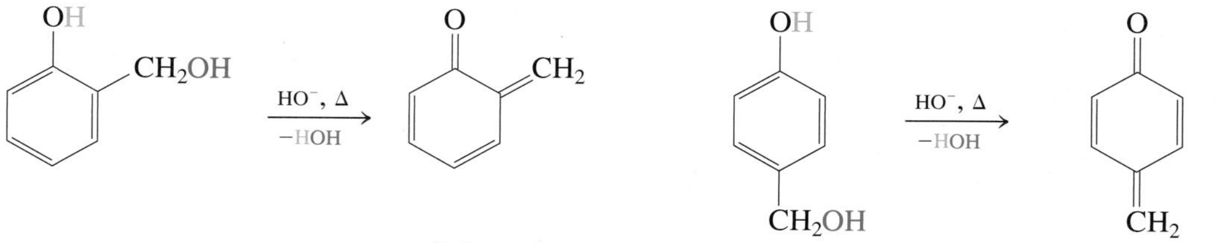

o-Quinomethane

p-Quinomethane

Because quinomethanes are α,β-unsaturated carbonyl compounds, they may undergo Michael additions (Section 18-11) with excess phenoxide ion. The resulting phenols can be hydroxymethylated again and the entire process repeated. Eventually, a complex phenol–formaldehyde copolymer, also called a **phenolic resin** (e.g., Bakelite), is formed. Their major uses are in plywood (45%), insulation (14%), molding compounds (9%), fibrous and granulated wood (9%), and laminates (8%).

Phenolic Resin Synthesis

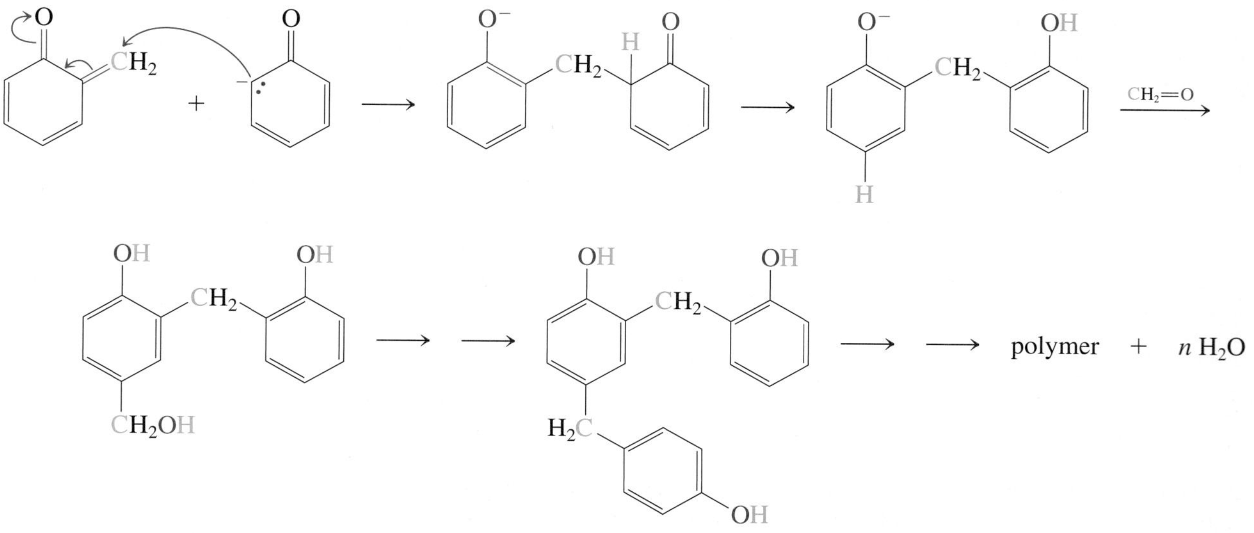

polymer + n H$_2$O

In the **Kolbe*-Schmitt[†] reaction,** phenoxide attacks carbon dioxide to furnish the salt of 2-hydroxybenzoic acid (*o*-hydroxybenzoic acid, salicylic acid, precursor to aspirin; see Real Life 22-2).

Various wood products incorporating Bakelite are used in the construction of houses.

Solved Exercise 22-21 | Working with the Concepts: Recognizing Phenol as an Enol

Formulate a mechanism for the Kolbe-Schmitt reaction.

Strategy

As always, we take an inventory of the components of the reaction: starting materials, other reagents and reaction conditions, and products. Phenol is an electron-rich arene (see this and Section 16-3) and also acidic. Carbon dioxide has an electrophilic carbon that is attacked by nucleophilic carbon atoms, such as in Grignard reagents (Section 19-6). The reaction conditions are strongly basic. Finally, the product looks like that of an electrophilic ortho substitution.

Solution

- Under basic conditions, phenol will exist as phenoxide, which, like an enolate ion (Section 18-1), can be described by two resonance forms (see margin).
- Enolate alkylations occur at carbon (Section 18-10). In analogy, you can formulate a phenoxide attack on the electrophilic carbon of CO_2. Alternatively, you can think of this reaction as an electrophilic attack of CO_2 on a highly activated benzene ring.
- Finally, deprotonation occurs to regenerate the aromatic arene.

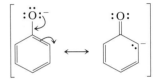

Phenoxide ion

As you will have noticed, the selectivity for ortho attack by CO_2 in this process is exceptional. Although not completely understood, it may involve direction of the electrophile by the Na^+ ion in the vicinity of the phenoxide negative charge.

Exercise 22-22 | Try It Yourself

Phentolamine (as a water-soluble methanesulfonic acid salt) is an antihypertensive that has recently been introduced into dentistry: It cuts in half the time taken to recover from the numbing effect of local anesthetics. The key step in its preparation dates from 1886, the reaction shown below. What is its mechanism? (**Caution:** This is not a nucleophilic aromatic substitution. **Hint:** Think keto–enol tautomerism).

Phentolamine

Hexachlorophene

Exercise 22-23

Hexachlorophene (margin) is a skin germicide formerly used in soaps. It is prepared in one step from 2,4,5-trichlorophenol and formaldehyde in the presence of sulfuric acid. How does this reaction proceed? (**Hint:** Formulate an acid-catalyzed hydroxymethylation for the first step.)

In Summary The benzene ring in phenols is subject to electrophilic aromatic substitution, particularly under basic conditions. Phenoxide ions can be hydroxymethylated and carbonated.

22-7 AN ELECTROCYCLIC REACTION OF THE BENZENE RING: THE CLAISEN REARRANGEMENT

At 200°C, 2-propenyloxybenzene (allyl phenyl ether) undergoes an unusual reaction that leads to the rupture of the allylic ether bond: The starting material rearranges to 2-(2-propenyl)phenol (*o*-allylphenol).

Reaction

$$\xrightarrow{\Delta}$$

2-Propenyloxybenzene
(Allyl phenyl ether)

75%
2-(2-Propenyl)phenol
(*o*-Allylphenol)

This transformation, called the **Claisen* rearrangement,** is another concerted reaction with an aromatic transition state that accommodates the movement of six electrons (Sections 14-8 and 15-3). The initial intermediate is a high-energy isomer, 6-(2-propenyl)-2,4-cyclohexadienone, which enolizes to the final product (Sections 18-2 and 22-3).

Mechanism of the Claisen Rearrangement

Mechanism

6-(2-Propenyl)-
2,4-cyclohexadienone

The Claisen rearrangement is general for other systems. With the nonaromatic 1-ethenyloxy-2-propene (allyl vinyl ether), it stops at the carbonyl stage because there is no driving force for enolization. This is called the **aliphatic Claisen rearrangement.**

*Professor Rainer Ludwig Claisen (1851–1930), University of Berlin, Germany.

Aliphatic Claisen Rearrangement

255°C

50%

1-Ethenyloxy-2-propene
(Allyl vinyl ether)

4-Pentenal

The carbon analog of the Claisen rearrangement is called the **Cope* rearrangement;** it takes place in compounds containing 1,5-diene units.

Cope Rearrangement

178°C

72%

3-Phenyl-1,5-hexadiene

trans-**1-Phenyl-1,5-hexadiene**

Note that all of these rearrangements are related to the electrocyclic reactions that interconvert *cis*-1,3,5-hexatriene with 1,3-cyclohexadiene (see margin and Section 14-9). The only difference is the absence of a double bond connecting the terminal π bonds.

Electrocyclic Reaction of
***cis*-1,3,5-Hexatriene**

Δ

Exercise 22-24

Explain the following transformation by a mechanism. (**Hint:** The Cope rearrangement can be accelerated greatly if it leads to charge delocalization.)

HO

NaOH, H₂O →

OHC

Solved Exercise 22-25 | Working with the Concepts: Applying Claisen and Cope Rearrangements

Citral, B, is a component of lemon grass and as such used in perfumery (lemon and verbena scents). It is also an important intermediate in the BASF synthesis of vitamin A (Section 14-7, Real Life 18-2). The last step in the synthesis of citral requires simply heating the enol ether A. How do you get from A to B?

Δ →

A

B
Citral

---

*Professor Arthur C. Cope (1909–1966), Massachusetts Institute of Technology, Cambridge.

Strategy

As always, we take an inventory of the components of the reaction: starting materials, other reagents and reaction conditions, and products. Here, this is straightforward: There are no reagents, we simply apply heat, and it appears that the reaction is an isomerization. You need to confirm this suspicion, by determining the molecular formulas of A and B. Indeed, it is $C_{10}H_{16}O$ for both. What thermal reactions could you envisage for A?

Solution

- You note that A contains a diene unit connected to an isolated double bond. Hence, in principle, an intramolecular Diels-Alder reaction to C might be feasible (Section 14-8; Exercise 14-24):

Potential Diels-Alder Reaction of A

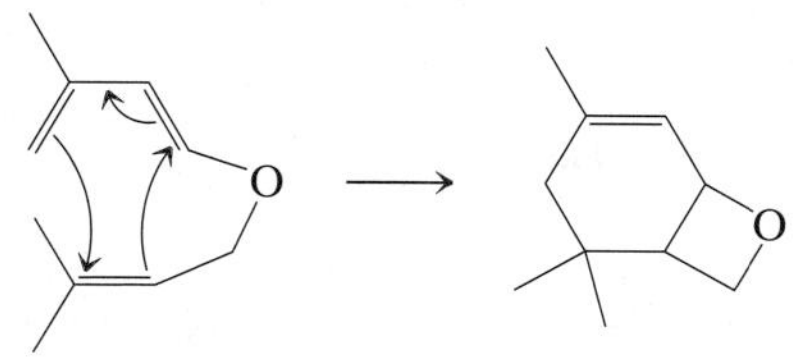

- It is apparent that this pathway is unfavorable for two reasons: (1) Both the diene and the dienophile are electron rich and therefore not good partners for cycloaddition (Section 14-8), and, even more obvious, (2) a strained ring is formed in C.
- An alternative is based on the recognition of a 1,5-hexadiene unit, the prerequisite for a Cope rearrangement. In A, this diene unit contains an oxygen, hence we can write a Claisen rearrangement to see where it leads us.

Claisen Rearrangement of A

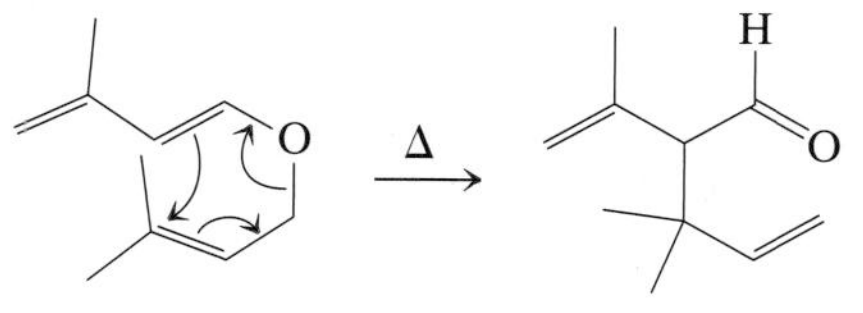

- The product D contains a 1,5-diene substructure, capable of a Cope rearrangement, leading to citral B.

Cope Rearrangement of D

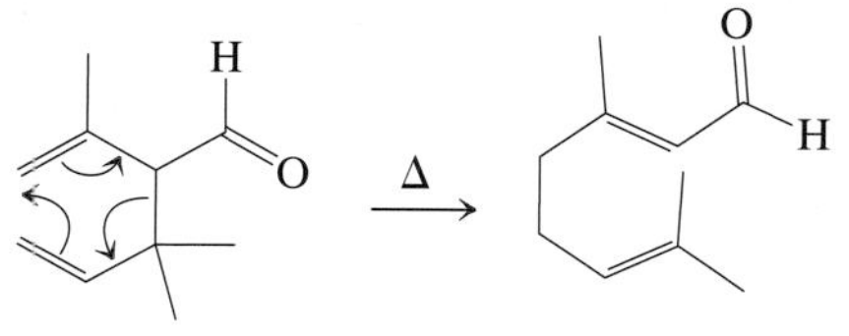

Exercise 22-26 | Try It Yourself

The ether A provides B upon heating to 200°C. Formulate a mechanism. (**Caution:** The terminal alkenyl carbon cannot reach the para position of the benzene ring. **Hint:** Start with the first step of a Claisen rearrangement.)

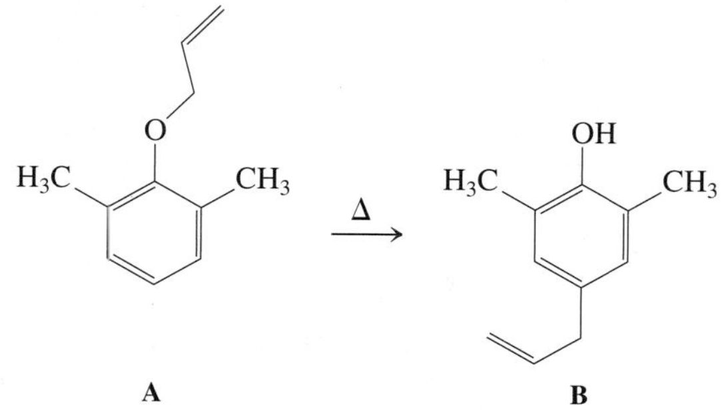

In Summary 2-Propenyloxybenzene rearranges to 2-(2-propenyl)phenol (*o*-allylphenol) by an electrocyclic mechanism that moves six electrons (Claisen rearrangement). Similar concerted reactions are undergone by aliphatic unsaturated ethers (aliphatic Claisen rearrangement) and by hydrocarbons containing 1,5-diene units (Cope rearrangement).

22-8 | OXIDATION OF PHENOLS: BENZOQUINONES

Phenols can be oxidized to carbonyl derivatives by one-electron transfer mechanisms, resulting in a new class of cyclic diketones, called **benzoquinones.**

Benzoquinones and benzenediols are redox couples

The phenols 1,2- and 1,4-benzenediol (for which the respective common names catechol and hydroquinone are retained by IUPAC) are oxidized to the corresponding diketones, *ortho*- and *para*-benzoquinone, by a variety of oxidizing agents, such as sodium dichromate or silver oxide. Yields can be variable when the resulting diones are reactive, as in the case of *o*-benzoquinone, which partly decomposes under the conditions of its formation.

Benzoquinones from Oxidation of Benzenediols

Catechol   *o*-Benzoquinone

Hydroquinone   *p*-Benzoquinone

The redox process that interconverts hydroquinone and *p*-benzoquinone can be visualized as a sequence of proton and electron transfers. Initial deprotonation gives a phenoxide ion, which is transformed into a **phenoxy radical** by one-electron oxidation. Proton dissociation from the remaining OH group furnishes a **semiquinone radical anion,** and a second one-electron oxidation step leads to the benzoquinone. All of the intermediate species in this sequence benefit from considerable resonance stabilization (two forms are shown for the semiquinone). We shall see in Section 22-9 that redox processes similar to those shown here occur widely in nature.

Redox Relation Between *p*-Benzoquinone and Hydroquinone

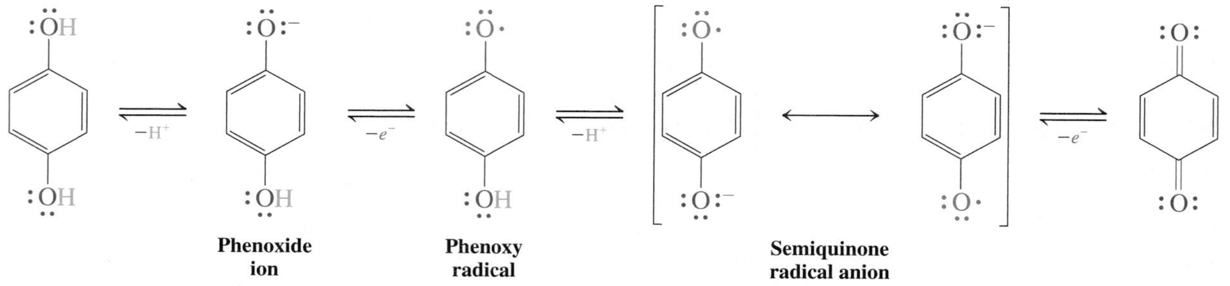

Phenoxide ion   Phenoxy radical   Semiquinone radical anion

Exercise 22-27

Give a minimum of two additional resonance forms each for the phenoxide ion, phenoxy radical, and semiquinone radical anion shown in the preceding scheme.

The enone units in *p*-benzoquinones undergo conjugate and Diels-Alder additions

p-Benzoquinones function as reactive α,β-unsaturated ketones in conjugate additions (see Section 18-9). For example, hydrogen chloride adds to give an intermediate hydroxy dienone that enolizes to the aromatic 2-chloro-1,4-benzenediol.

p-Benzoquinone   6-Chloro-4-hydroxy-2,4-cyclohexadienone   2-Chloro-1,4-benzenediol

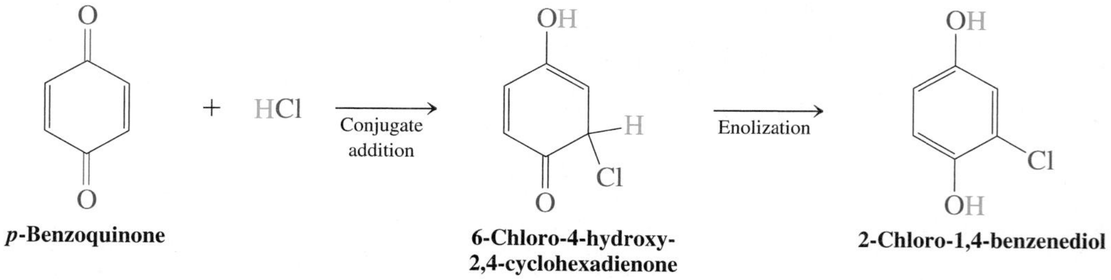

The double bonds also undergo cycloadditions to dienes (Section 14-8). The initial cyclo-adduct to 1,3-butadiene tautomerizes with acid to the aromatic system.

Diels–Alder Reactions of *p*-Benzoquinone

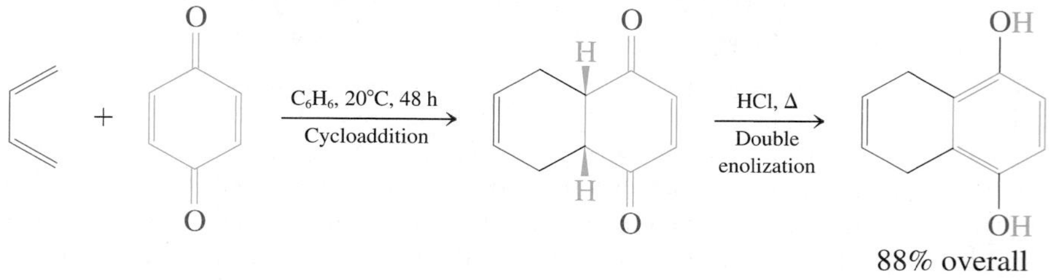

88% overall

REAL LIFE: BIOLOGY 22-3 | Chemical Warfare in Nature: The Bombardier Beetle

The oxidizing power of *p*-benzoquinones is used by some arthropods, such as millipedes, beetles, and termites, as chemical defense agents. Most remarkable among these species is the bombardier beetle. Its name is descriptive of its defense mechanism against predators, usually ants, which involves firing hot corrosive chemicals from glands in its posterior, with amazing accuracy. At the time of an attack (simulated in the laboratory by pinching the beetle with fine-tipped forceps, see photo), two glands located near the end of the beetle's abdomen secrete mainly hydroquinone and hydrogen peroxide, respectively, into a reaction chamber. This chamber contains enzymes that trigger the explosive oxidation of the diol to the quinone and simultaneous decomposition of hydrogen peroxide to oxygen gas and water. This cocktail is audibly expelled at temperatures up to 100°C in the direction of the enemy from the end of the beetle's abdomen, aided for aim by a 270° rotational capability. In some species, firing occurs in pulses of about 500 per second, like a machine gun.

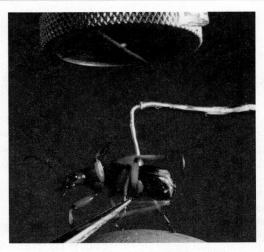

A microphone triggers a flash as a bombardier beetle blasts tweezers with spray.

Exercise 22-28

Explain the following result by a mechanism. (**Hint:** Review Section 18-9.)

$$H_3C \underset{CH_3O}{\overset{O}{\bigcirc}} \overset{CH_3}{\underset{OCH_3}{\bigcirc}} \xrightarrow[\text{CH}_3\text{CH}_2\text{OH}]{\text{CH}_3\text{CH}_2\text{O}^-\text{Na}^+,} H_3C \underset{CH_3CH_2O}{\overset{O}{\bigcirc}} \overset{CH_3}{\underset{OCH_2CH_3}{\bigcirc}}$$

In Summary Phenols are oxidized to the corresponding benzoquinones. The diones enter into reversible redox reactions that yield the corresponding diols. They also undergo conjugate additions and Diels-Alder additions to the double bonds.

22-9 | OXIDATION-REDUCTION PROCESSES IN NATURE

This section describes some chemical processes involving hydroquinones and *p*-benzoquinones that occur in nature. We begin with an introduction to the biochemical reduction of O_2. Oxygen can engage in reactions that cause damage to biomolecules. Natural **antioxidants** inhibit these transformations, as do several synthetic preservatives.

Ubiquinones mediate the biological reduction of oxygen to water

Nature makes use of the benzoquinone–hydroquinone redox couple in reversible oxidation reactions. These processes are part of the complicated cascade by which oxygen is used in biochemical degradations. An important series of compounds used for this purpose are the **ubiquinones** (a name coined to indicate their ubiquitous presence in nature), also collectively called **coenzyme Q (CoQ,** or simply **Q).** The ubiquinones are substituted *p*-benzoquinone derivatives bearing a side chain made up of 2-methylbutadiene units (isoprene; Sections 4-7 and 14-10). An enzyme system that utilizes NADH (Real Life 8-1 and 25-2) converts CoQ into its reduced form (QH_2).

Ubiquinones (n = 6, 8, 10)
(Coenzyme Q)

Enzyme,
reducing agent

Reduced form of coenzyme Q
(Reduced Q, or QH₂)

QH_2 participates in a chain of redox reactions with electron-transporting iron-containing proteins called **cytochromes** (Real Life 8-1). The reduction of Fe^{3+} to Fe^{2+} in cytochrome *b* by QH_2 begins a sequence of electron transfers involving six different proteins. The chain ends with reduction of O_2 to water by addition of four electrons and four protons.

$$O_2 + 4\ H^+ + 4\ e^- \longrightarrow 2\ H_2O$$

Phenol derivatives protect cell membranes from oxidative damage

The biochemical conversion of oxygen into water includes several intermediates, including **superoxide, O_2^{-}**, the product of one-electron reduction, and **hydroxy radical, ·OH**, which arises from cleavage of H_2O_2. Both are highly reactive species capable of initiating reactions that damage organic molecules of biological importance. An example is the phosphoglyceride shown here, a cell-membrane component derived from the unsaturated fatty acid *cis,cis*-octadeca-9,12-dienoic acid (linoleic acid).

Initiation step

Pentadienyl radical

The doubly allylic hydrogens at C11 are readily abstracted by radicals such as ·OH (Section 14-2).

The resonance-stabilized pentadienyl radical combines rapidly with O_2 in the first of two propagation steps. Reaction occurs at either C9 or C13 (shown here), giving either of two peroxy radicals containing conjugated diene units.

Propagation step 1

Peroxy radical

In the second propagation step this species removes a hydrogen atom from C11 of another molecule of phosphoglyceride, or more generally, lipid (Section 20-5), thereby forming a new dienyl radical and a molecule of **lipid hydroperoxide.** The dienyl radical may then reenter propagation step 1. In this way, a large number of lipid molecules may be oxidized following just a single initiation event.

Propagation step 2

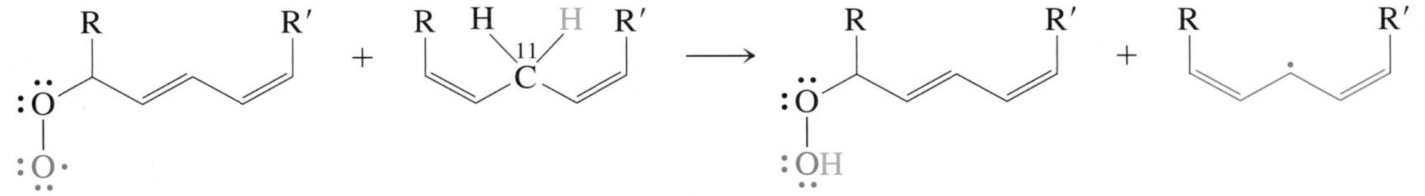

Lipid hydroperoxide

Numerous studies have confirmed that lipid hydroperoxides are toxic, their products of decomposition even more so. For example, loss of ·OH by cleavage of the relatively weak O–O bond gives rise to an alkoxy radical, which may decompose by breaking a neighboring C–C bond (β-scission), forming an unsaturated aldehyde.

β-Scission of a Lipid Alkoxy Radical

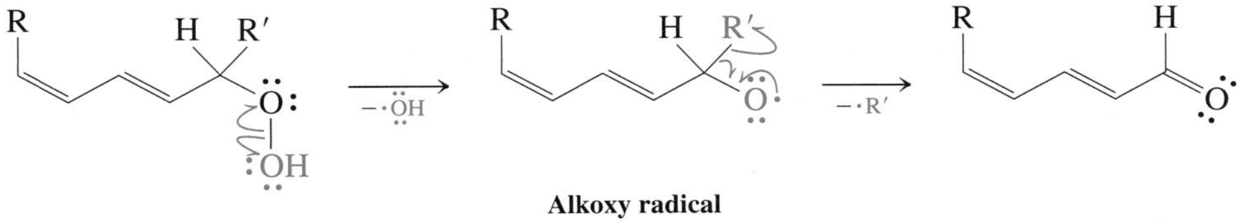

Alkoxy radical

Through related but more complex mechanisms, certain lipid hydroperoxides decompose to give unsaturated hydroxyaldehydes, such as *trans*-4-hydroxy-2-nonenal, as well as the dialdehyde propanedial (malondialdehyde). Molecules of these general types are partly responsible for the smell of rancid fats.

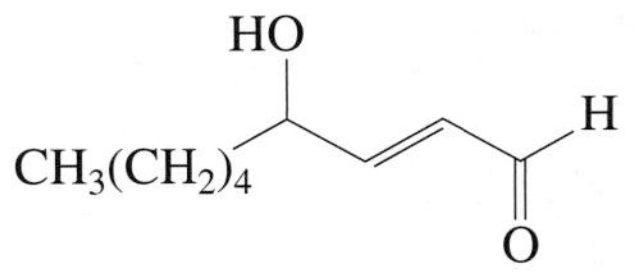

Both propanedial and the α,β-unsaturated aldehydes are extremely toxic, because they are highly reactive toward the proteins that are present in close proximity to the lipids in cell membranes. For example, both dials and enals are capable of reacting with nucleophilic amino and mercapto groups from two different parts of one protein or from two different protein molecules, and these reactions produce cross-linking (Section 14-10). Cross-linking severely inhibits proteins from carrying out their biological functions (Chapter 26).

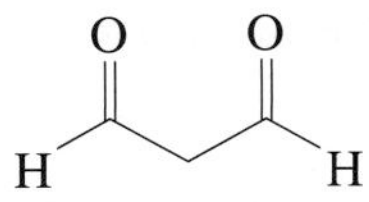

***trans*-4-Hydroxy-2-nonenal**

**Propanedial
(Malondialdehyde)**

Cross-linking of Proteins by Reaction with Unsaturated Aldehydes

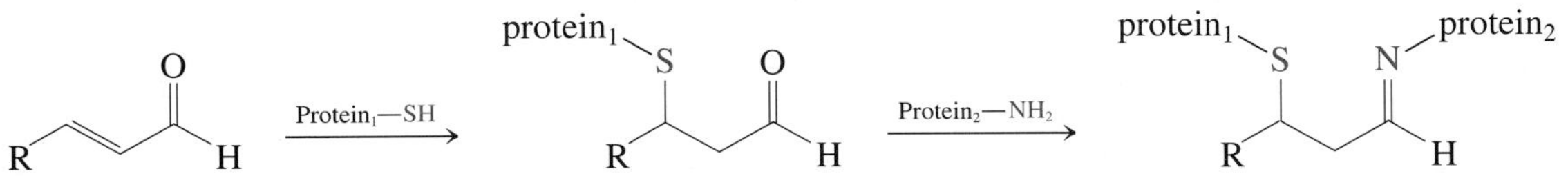

Processes such as these are thought by many to contribute to the development of emphysema, atherosclerosis (the underlying cause of several forms of heart disease and stroke), certain chronic inflammatory and autoimmune diseases, cancer, and, possibly, the process of aging itself.

Does nature provide the means for biological systems to protect themselves from such damage? A variety of naturally occurring antioxidant systems defend lipid molecules inside cell membranes from oxidative destruction. The most important is **vitamin E,** a collection

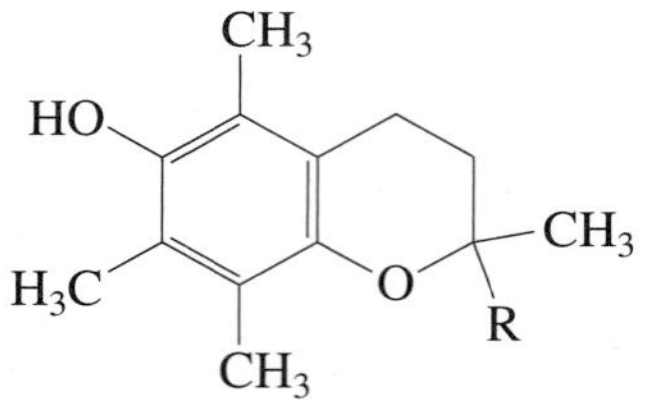

Vitamin E
(α-Tocopherol)

R = branched C$_{16}$H$_{33}$ chain

of eight compounds with very similar structures, commonly represented by one of them, α-tocopherol (margin). They all possess a long hydrocarbon chain (see Problem 46 of Chapter 2), a feature that makes them lipid soluble. Their reducing qualities stem from the presence of the hydroquinone-like aromatic ring (Section 22-8). The corresponding phenoxide ion is an excellent electron donor. The protective qualities of vitamin E rest on its ability to break the propagation chain of lipid oxidation by the reduction of radical species.

Reactions of Vitamin E with Lipid Hydroperoxy and Alkoxy Radicals

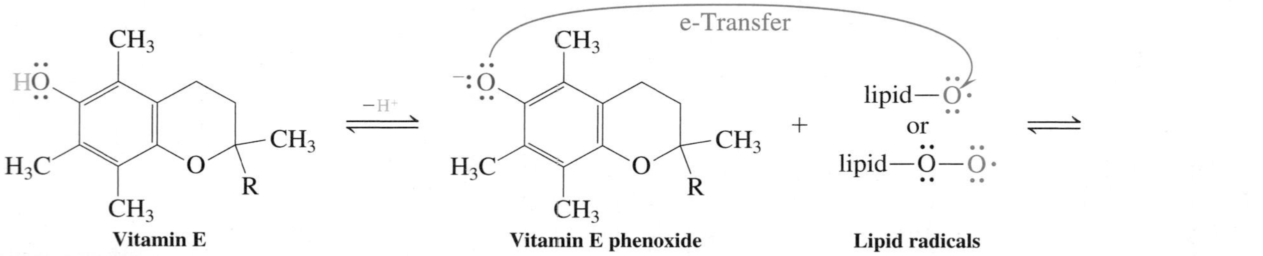

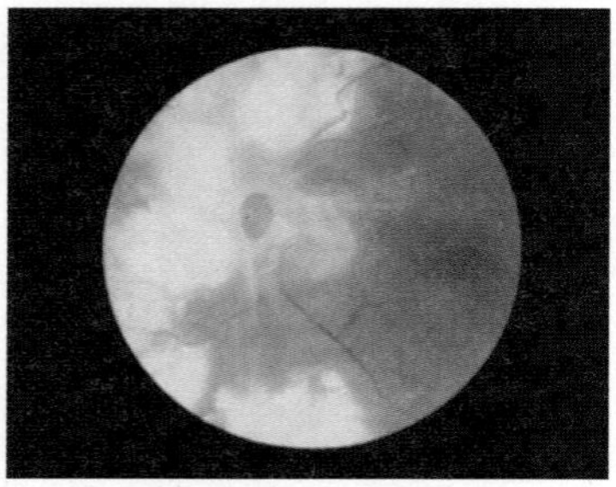

Vitamin E   **Vitamin E phenoxide**   **Lipid radicals**

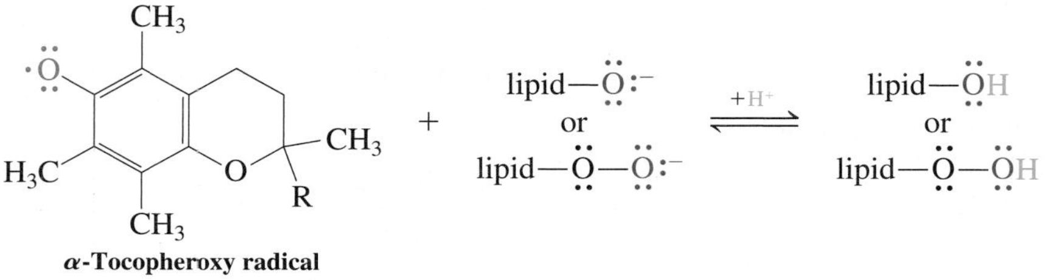

α-Tocopheroxy radical

Lipid peroxidation has been implicated in retinal diseases, and clinicians prescribe antioxidants to help in the treatment of diabetic retinopathy. This condition is characterized by a compromised blood supply that causes the development of fatty exudates and fibrous tissue on the retina (yellow areas in photo of the eye).

In this process, lipid radicals are reduced and protonated. Vitamin E is oxidized to an α-tocopheroxy radical, which is relatively unreactive because of extensive delocalization and the steric hindrance of the methyl substituents. Vitamin E is regenerated at the membrane surface by reaction with water-soluble reducing agents such as **vitamin C.**

Regeneration of Vitamin E by Vitamin C

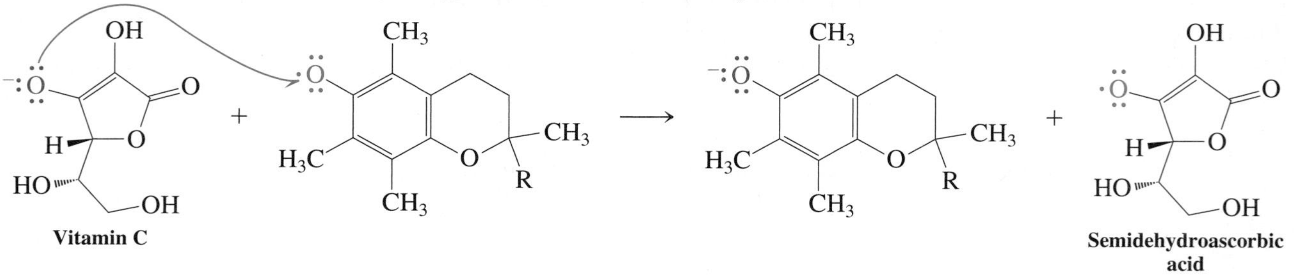

Vitamin C   **Semidehydroascorbic acid**

The product of vitamin C oxidation eventually decomposes to lower-molecular-weight water-soluble compounds, which are excreted by the body.

Exercise 22-29

Vitamin C is an effective antioxidant because its oxidation product semidehydroascorbic acid is stabilized by resonance. Give other resonance forms for this species.

Glutathione, an intracellular antioxidant, and why too much acetaminophen is toxic

Another way in which cells protect themselves from oxidative damage is through the production of **glutathione,** a peptide that incorporates a mercapto group, –SH (see Section 9-10), arising from the presence of the amino acid cysteine (Table 26-1 and Section 26-4). This function serves as a scavenger of radicals and other oxidants, such as hydrogen peroxide, H_2O_2. The molecule is converted into the corresponding disulfide in this process (Section 9-10), but is regenerated by enzyme-mediated reduction.

The mercapto group can also undergo conjugate additions to α,β-unsaturated carbonyl compounds (see Section 18-9 and Really? on p. 815, including benzoquinones and related compounds (Section 22-8). Such a process is part of the metabolism of the popular analgesic acetaminophen (Tylenol) in the liver. The enzyme cytochrome P-450 (Real Life 8-1) oxidizes acetaminophen to the reactive imine NAPQI, which in turn is trapped through nucleophilic attack by glutathione. The resulting addition product is then further metabolized to a water-soluble derivative, to be excreted in the urine. At therapeutic doses of the drug, the lost glutathione is readily replaced by biosynthesis. However, at higher doses, especially in cases of accidental or intentional overdosage, acetaminophen can cause potentially fatal liver failure due largely to oxidative stress caused by the dearth of glutathione. Such poisoning accounts for about one-half of all cases of acute liver failure in the United States today. Ironically, a study in 2011 pinpointed NAPQI as the active agent in the mechanism of pain suppression by acetaminophen: It suppresses proteins involved in the sensing of pain.

Synthetic analogs of vitamin E are preservatives

Synthetic phenol derivatives are widely used as antioxidants and preservatives in the food industry. Perhaps two of the most familiar are 2-(1,1-dimethylethyl)-4-methoxyphenol (butylated hydroxyanisole, or **BHA**) and 2,6-bis(1,1-dimethylethyl)-4-methylphenol (butylated hydroxytoluene, or **BHT;** see Exercise 16-14). For example, addition of BHA to butter increases its storage life from months to years. Both BHA and BHT function like vitamin E, reducing oxygen radicals and interrupting the propagation of oxidation processes.

In Summary Oxygen-derived radicals are capable of initiating radical chain reactions in lipids, thereby leading to toxic decomposition products. Vitamin E is a naturally occurring phenol derivative that functions as an antioxidant to inhibit these processes within membrane lipids. Vitamin C and glutathione are biological reducing agents located in the intra- and extracellular aqueous environments. High concentrations of benzoquinones can bring about cell death by consumption of glutathione; vitamin C can protect the cell by reduction of the benzoquinone. Synthetic food preservatives are structurally designed to mimic the antioxidant behavior of vitamin E.

22-10 | ARENEDIAZONIUM SALTS

As mentioned in Section 22-4, *N*-nitrosation of primary benzenamines (anilines) furnishes arenediazonium salts, which can be used in the synthesis of phenols. Arenediazonium salts are stabilized by resonance of the π electrons in the diazo function with those of the aromatic ring. They are converted into haloarenes, arenecarbonitriles, and other aromatic derivatives through replacement of nitrogen by the appropriate nucleophile.

Arenediazonium salts are stabilized by resonance

The reason for the stability of arenediazonium salts, relative to their alkane counterparts, is resonance and the high energy of the aryl cations formed by loss of nitrogen. One of the electron pairs making up the aromatic π system can be delocalized into the functional group, which results in charge-separated resonance structures containing a double bond between the benzene ring and the attached nitrogen.

Resonance in the Benzenediazonium Cation

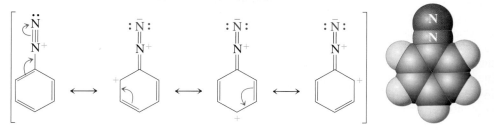

At elevated temperatures (>50°C), nitrogen extrusion does take place, however, to form the very reactive phenyl cation. When this is done in aqueous solution, phenols are produced (Section 22-4).

Why is the phenyl cation so reactive? After all, it is a carbocation that is part of a benzene ring. Should it not be resonance stabilized, like the phenylmethyl (benzyl) cation? The answer is no, as may be seen in the molecular-orbital picture of the phenyl cation (Figure 22-5). The empty orbital associated with the positive charge is one of the sp^2 hybrids

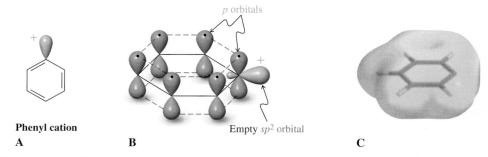

p orbitals

Phenyl cation

A   **B**   Empty sp^2 orbital   **C**

Figure 22-5 (A) Line structure of the phenyl cation. (B) Orbital picture of the phenyl cation. The alignment of its empty sp^2 orbital is perpendicular to the six-π-electron framework of the aromatic ring. As a result, the positive charge is not stabilized by resonance. (C) The electrostatic potential map of the phenyl cation, shown at an attenuated scale for better contrast, pinpoints much of the positive charge (blue edge on the right) as located in the plane of the six-membered ring.

aligned in *perpendicular* fashion to the π framework that normally produces aromatic resonance stabilization. Hence, this orbital cannot overlap with the π bonds, and the positive charge cannot be delocalized. Moreover, the cationic carbon would prefer *sp* hybridization, an arrangement precluded by the rigid frame of the benzene ring. We used a similar argument to explain the difficulty in deprotonating benzene to the corresponding phenyl anion (Section 22-4).

Arenediazonium salts can be converted into other substituted benzenes

When arenediazonium salts are decomposed in the presence of nucleophiles other than water, the corresponding substituted benzenes are formed. For example, diazotization of arenamines (anilines) in the presence of hydrogen iodide results in the corresponding iodoarenes.

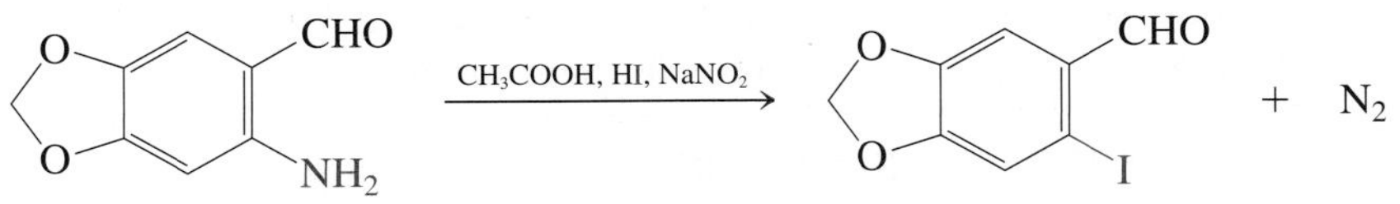

Attempts to obtain other haloarenes in this way are frequently complicated by side reactions. One solution to this problem is the **Sandmeyer* reaction,** which makes use of the fact that the exchange of the nitrogen substituent for halogen is considerably facilitated by the presence of cuprous [Cu(I)] salts. The detailed mechanism of this process is complex, and radicals are participants. Addition of cuprous cyanide, CuCN, to the diazonium salt in the presence of excess potassium cyanide gives aromatic nitriles.

Sandmeyer Reactions

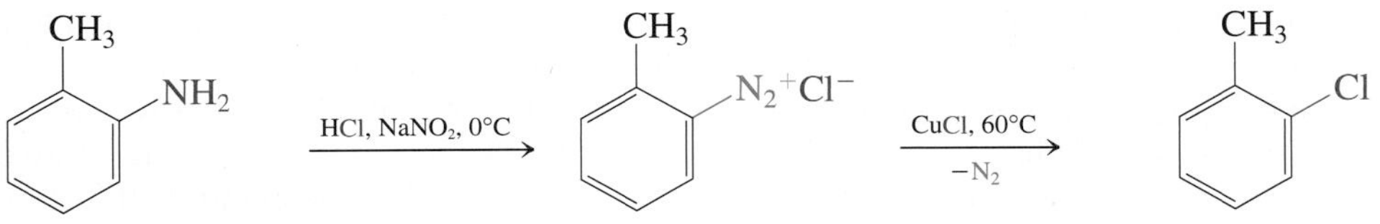

2-Methylbenzenamine
(*o*-Methylaniline)

1-Chloro-2-methylbenzene
(*o*-Chlorotoluene)

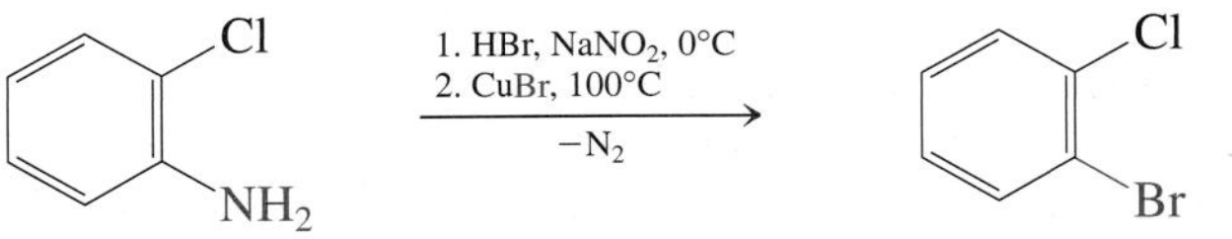

79% overall

2-Chlorobenzenamine
(*o*-Chloroaniline)

73%

1-Bromo-2-chlorobenzene
(*o*-Bromochlorobenzene)

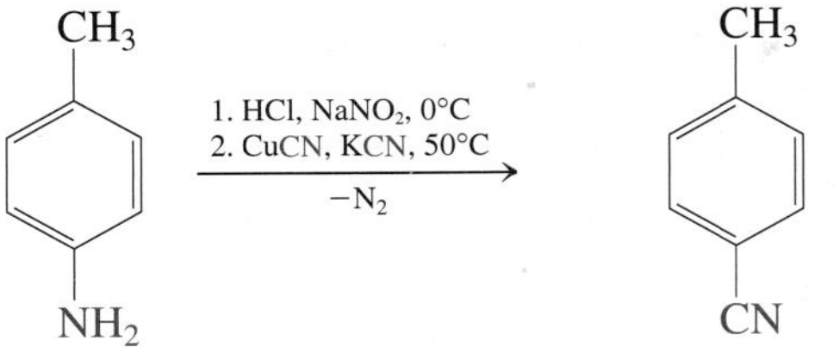

70%

4-Methylbenzonitrile
(*p*-Tolunitrile)

---

*Dr. Traugott Sandmeyer (1854–1922), Geigy Company, Basel, Switzerland.

Exercise 22-30

Propose syntheses of the following compounds, starting from benzene.

(a) [structure: benzene ring with ethyl group and I in meta positions]

(b) [structure: benzene ring with CN and CN in meta positions]

(c) [structure: benzene ring with OH and SO₃H in meta positions]

Although it is written as H_3PO_2, hypophosphorous acid is actually a mixture of two tautomers:

[structures: two phosphorus tautomers]

Major   **Minor**

Hypophosphorous acid

It reduces diazonium formally by hydride transfer to give the arene and H_3PO_3 (see Section 9-4).

The diazonium group can be removed by reducing agents. The sequence diazotization–reduction is a way to replace the amino group in arenamines (anilines) with hydrogen. The reducing agent employed is aqueous hypophosphorous acid, H_3PO_2. This method is especially useful in syntheses in which an amino group is used as a removable directing substituent in electrophilic aromatic substitution (Section 16-5).

Reductive Removal of a Diazonium Group

$$\text{(structure with CH}_3\text{, Br, NH}_2) \xrightarrow{\text{NaNO}_2,\ \text{H}^+,\ \text{H}_2\text{O}} \text{(structure with CH}_3\text{, Br, N}_2^+) \xrightarrow[\text{$-$N}_2,\ -\text{H}_3\text{PO}_3]{\text{H}_3\text{PO}_2,\ \text{H}_2\text{O},\ 25°\text{C}} \text{(structure with CH}_3\text{, Br, H)}$$

85%

1-Bromo-3-methylbenzene
(*m*-Bromotoluene)

Another application of diazotization in synthetic strategy is illustrated in the synthesis of 1,3-dibromobenzene (*m*-dibromobenzene). Direct electrophilic bromination of benzene is not feasible for this purpose; after the first bromine has been introduced, the second will attack ortho or para. What is required is a meta-directing substituent that can be transformed eventually into bromine. The nitro group is such a substituent. Double nitration of benzene furnishes 1,3-dinitrobenzene (*m*-dinitrobenzene). Reduction (Section 16-5) leads to the benzenediamine, which is then converted into the dihalo derivative.

Synthesis of 1,3-Dibromobenzene by Using a Diazotization Strategy

$$\text{benzene} \xrightarrow{\text{HNO}_3,\ \text{H}_2\text{SO}_4,\ \Delta} \text{(m-dinitrobenzene)} \xrightarrow{\text{H}_2,\ \text{Pd}} \text{(m-diaminobenzene)} \xrightarrow[\text{2. CuBr, 100°C}]{\text{1. NaNO}_2,\ \text{H}^+,\ \text{H}_2\text{O}} \text{(m-dibromobenzene)}$$

Exercise 22-31

Propose a synthesis of 1,3,5-tribromobenzene from benzene.

In Summary Arenediazonium salts, which are more stable than alkanediazonium salts because of resonance, are starting materials not only for phenols, but also for haloarenes, arenecarbonitriles, and reduced aromatics by displacement of nitrogen gas. The intermediates in some of these reactions may be aryl cations, highly reactive because of the absence of any electronically stabilizing features, but other, more complicated mechanisms may be followed. The ability to transform arenediazonium salts in this way gives considerable scope to the regioselective construction of substituted benzenes.

22-11 | ELECTROPHILIC SUBSTITUTION WITH ARENEDIAZONIUM SALTS: DIAZO COUPLING

Being positively charged, arenediazonium ions are electrophilic. Although they are not very reactive in this capacity, they can accomplish electrophilic aromatic substitution when the substrate is an activated arene, such as phenol or benzenamine (aniline). This reaction, called **diazo coupling,** leads to highly colored compounds called **azo dyes.** For example, reaction of *N,N*-dimethylbenzenamine (*N,N*-dimethylaniline) with benzenediazonium chloride gives the brilliant orange dye Butter Yellow. This compound was once used as a food coloring agent but has been declared a suspect carcinogen by the Food and Drug Administration. Like many azo dyes, it is employed as a pH indicator, yellow above pH = 4.0, red below pH = 2.9. The reason for the color change is the protonation of one of the diazo nitrogens at low pH to generate a resonance-stabilized cation. A convenient way to quantify the color of dyes is by UV-visible spectroscopy (Section 14-11). Thus, the two colors of Butter Yellow are manifested in absorptions at λ_{max} = 420 nm (yellow) and 520 nm (red), respectively.

Dyes are important additives in the textile industry. Azo dyes, while still in widespread use, are becoming less attractive for this purpose because some have been found to degrade to carcinogenic benzenamines.

Diazo Coupling

$$\text{(diazo coupling reaction scheme)}$$

4-Dimethylaminoazobenzene
(*p*-Dimethylaminoazobenzene, Butter Yellow)

Yellow   Red

Dyes used in the clothing industry usually contain sulfonic acid groups that impart water solubility and allow the dye molecule to attach itself ionically to charged sites on the polymer framework of the textile.

Industrial Dyes

$(CH_3)_2N$ — ⟨N=N⟩ — $SO_3^-Na^+$

Methyl Orange

pH ≤ 3.1, red; λ_{max} = 520 nm
pH ≥ 4.4, yellow; λ_{max} = 450 nm

$Na^{+-}O_3S$ .... $SO_3^-Na^+$

NH_2 .... H_2N

Congo Red

pH ≤ 3.0, blue-violet; λ_{max} = 590 nm
pH ≥ 5.2, red; λ_{max} = 497 nm

| **William Perkin's Synthetic Dyes and the Beginning of Medicinal Chemistry**

In 1851 William Perkin* was a precocious 13-year-old pupil at City of London School. Perkin's teacher recognized the boy's interest and aptitude in science and encouraged him to attend lectures presented by legendary scientist Michael Faraday.[†] Two years later Perkin was hired as a lab assistant by the brilliant young chemist August von Hofmann,[‡] a

founding member of London's Royal College of Chemistry. The rest, as they say, is history.

Hofmann and Perkin began a study of aromatic amines derived from coal tar. The latter was readily available as the by-product of the extraction from coal of the gas that fueled London's streetlights. Amines were of interest because of the

Quinine

2-Naphthalenamine

Mauvein A (R = H)
(major component of mauve)
Mauvein B (R = CH$_3$)

need for quinine (Section 28-5), the only known treatment for malaria, which was endemic in England at the time. The Quechua tribes of the Peruvian Andes had used the bark of the chinchona tree, a rich source of quinine, for this purpose, and Europeans had been importing this resource for over 200 years.

Lacking any knowledge of molecular structure, Perkin and Hofmann compared the composition of quinine, $C_{20}H_{24}N_2O_2$, with that of 2-naphthalenamine, $C_{10}H_9N$, and embarked on a series of attempts to oxidatively dimerize it. As the structures of these compounds show, these efforts were doomed from the start: There is very little structural resemblance between them. However, Perkin made the startling discovery that the products of this and other arenamine oxidations contained brightly colored substances that could be used as dyes in numerous applications. One such dye, a delicate shade of purple called mauve, took the fashion world by storm and made Perkin a very rich man.

---

*Sir William Henry Perkin, 1838–1907, Royal College of Chemistry, London, England.

[†]This is the same Faraday as the discoverer of benzene (Chapter 15 opening).

[‡]This is the same Hofmann as that of the Hofmann rule (Section 11-6) and the Hofmann rearrangement (Section 20-7).

Jessica Alba at the Metropolitan Museum of Art's Costume Institute Gala: λ_{max} = 550 nm.

Exercise 22-32

Write the products of diazo coupling of benzenediazonium chloride with each of the following molecules. **(a)** Methoxybenzene; **(b)** 1-chloro-3-methoxybenzene; **(c)** 1-(dimethylamino)-4-(1,1-dimethylethyl)benzene. (**Hint:** Diazo couplings are quite sensitive to steric effects.)

In Summary Arenediazonium cations attack activated benzene rings by diazo coupling, a process that furnishes azobenzenes, which are often highly colored.

More significant in the long run was the realization by several prominent late-19th-century biologists that Perkin's as well as other dyes could be used to stain biological tissue, particularly bacterial cells, making them clearly visible under a microscope. For example, degradation of mauvein B furnished safranin, to this day used to differentiate bacterial species into Gram-positive and Gram-negative (named after the Danish bacteriologist Hans Christian Gram, who discovered it in 1882). It works because the dye attaches itself only to the walls of the former, giving a purple-blue color, while Gram-negative bacteria elicit a pink-red color (see photo).

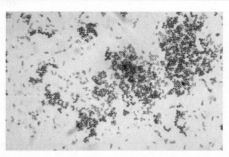

A Gram stain of mixed *Staphylococcus aureus* (Gram-positive; purple-blue) and *Escherichia coli* (Gram-negative; pink-red).

Safranin   **Prontosil**   **Sulfanilamide**

The finding that dyes stained bacteria encouraged researchers to find out whether these compounds also had clinical efficacy. In 1932, Gerhard Domagk,* at the pharmaceutical company Bayer (which also produced aspirin, see Chapter 16 Opening and Real Life 22-2) tested the wool and leather dye 4-[(2,4-diaminophenyl)azo] benzenesulfonamide against *Streptococcus*, responsible for the (then) often lethal scarlet fever, pneumonia, and strep throat. The results were phenomenal, and the first antibacterial drug, prontosil, was introduced to the world. It was soon found that prontosil is actually a prodrug (see p. 899), rapidly degraded in the body to sulfanilamide—marking the advent of the "miracle" sulfa drugs (Sections 9-11 and 15-10). Although they are now used only in special applications, the discovery of these medicines propelled the modern development of organic molecules as medicinal agents.

---

*Professor Gerhard J. P. Domagk, 1895–1964, University of Münster, Germany, Nobel Prize 1939 (medicine).

Like other azo dyes, the drug prontosil is brightly colored—in this case, bright red. (Photo courtesy of Chemical Heritage Foundation.)

THE BIG PICTURE

This chapter completes our understanding of the interplay between the benzene ring and its alkyl, hydroxy, and amino or modified amino substituents. Just as much as the electronic character of such substituents may activate or deactivate the ring with respect to substitution (Chapter 16; Sections 22-4 and 22-6) and the way in which substitutents are positioned controls the location of attack by electrophiles and nucleophiles, the ring imparts special reactivity on attached nuclei because of its resonating capacity. In other words, and to repeat the underlying theme of the text, the structure of the substituted aromatic compound determines its function. In many respects, this behavior is a simple extension of the chemistry of delocalized systems (Chapter 14) with the added feature of aromaticity (Chapter 15).

In the next chapter, we shall continue to examine the effect that two functional groups in the same molecule have on each other, now turning to carbonyl groups. In Chapter 24, we shall study molecules containing the carbonyl and several hydroxy groups and their biological relevance. Then, in Chapters 25 and 26, we conclude with biologically important compounds containing an array of functionality.

WORKED EXAMPLES: INTEGRATING THE CONCEPTS

22-33. Practicing Retrosynthesis and Synthesis of a Tetrasubstituted Arene

5-Amino-2,4-dihydroxybenzoic acid A is a potential intermediate in the preparation of natural products of medicinal value (Section 22-3). Propose syntheses, starting from methylbenzene (toluene).

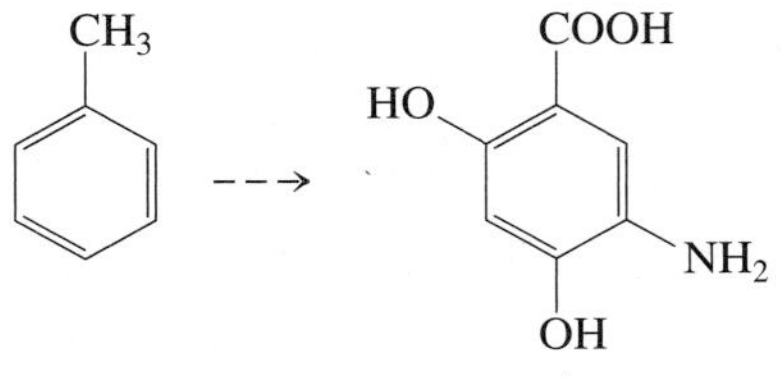

SOLUTION

This problem builds on the expertise that you gained (Chapter 16) in controlling the substitution patterns of target benzenes, but now with a greatly expanded range of reactions. The key is, again, recognition of the directing power of substituents—ortho, para or meta (Section 16-2)—and their interconversion (Section 16-5).

Retrosynthetic analysis of compound A reveals one carbon-based substituent, the carboxy group, which can be readily envisaged to be derivable from the methyl group in the starting material (by oxidation, Section 22-2). In the starting material, the carbon-based substituent is ortho, para directing, suggesting its use (retrosynthesis 1) in the introduction of the two hydroxy functions (as in compound B, through nitration–reduction–diazotization–hydrolysis; Sections 22-4 and 22-10). In compound A, it is meta directing and potentially utilizable (retrosynthesis 2) for the amination at C3 (as in compound C, through nitration–reduction).

Retrosynthesis 1

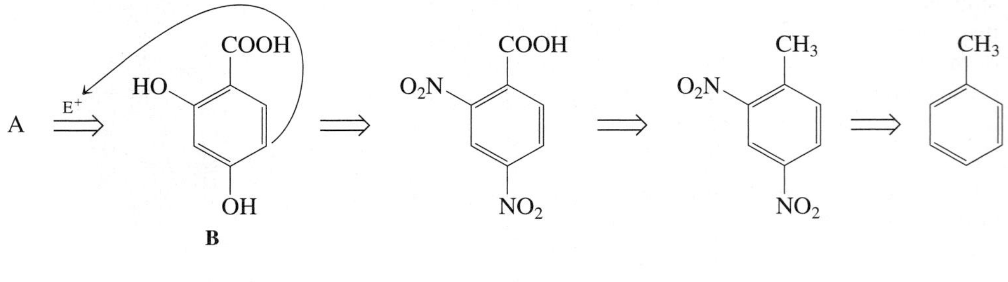

Retrosynthesis 2

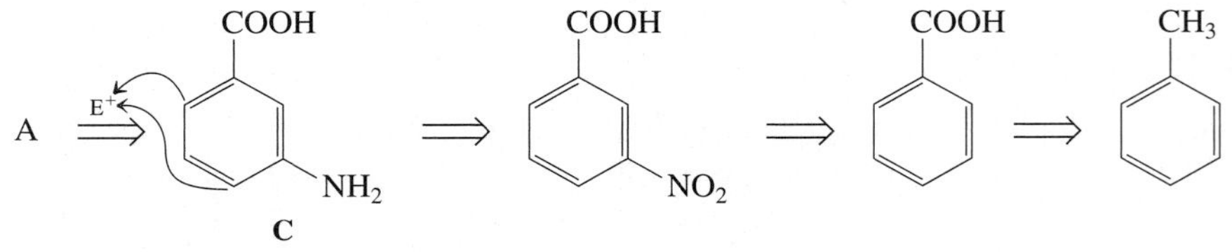

The question is, Are compounds B and C effective precursors of compound A? The answer is yes. Nitration of compound B should take place at the desired position (C3 in the product), ortho and para, respectively, to the two hydroxy substituents, thus placing the nitrogen at its position in compound A.

Electrophilic attack between the OH groups would be expected to be sterically hindered (Section 16-5). Conversely, the amino group in compound C, especially when protected as an amide, should direct electrophilic substitution to the less hindered ortho carbon and the para carbon, again yielding the desired pattern. The actual proposed synthetic schemes would be then as follows.

Synthesis 1

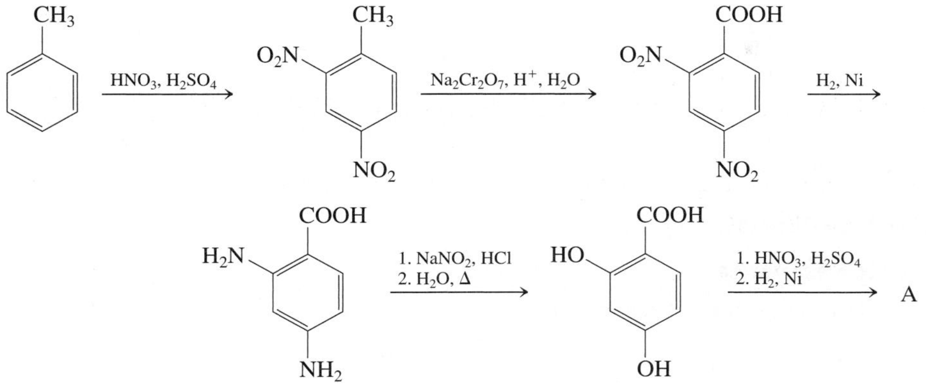

Synthesis 2

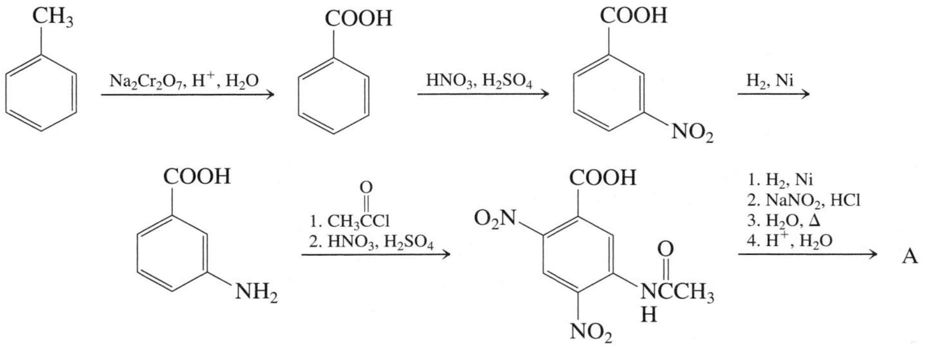

22-34. A Rearrangement Mechanism Toward a Phenol

A key reaction in the early development of the "pill" (Section 4-7; Real Life 4-3) was the "dienone-phenol" rearrangement shown below. Write a mechanism.

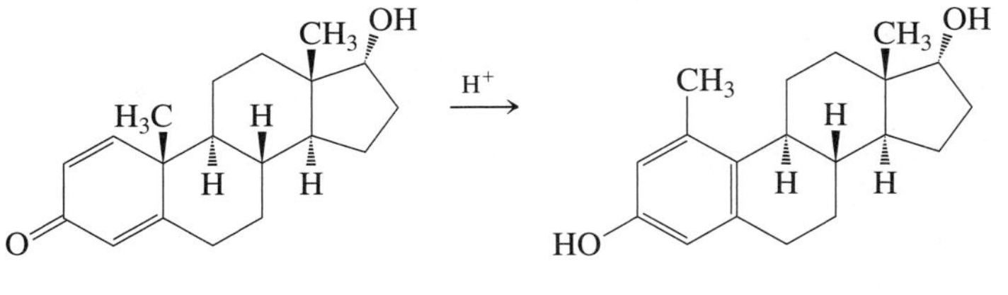

SOLUTION

This is a mechanistic problem involving an acid-catalyzed rearrangement featuring a migrating alkyl (methyl) group. Sound familiar? Review Section 9-3 on carbocation rearrangements! How do we obtain a suitable carbocation from our starting material? *Answer:* Protonation of the carbonyl group, which furnishes a resonance-stabilized hydroxypentadienyl cation (Section 14-6, 14-7, and Exercise 18-22).

Protonation of Dienone Function

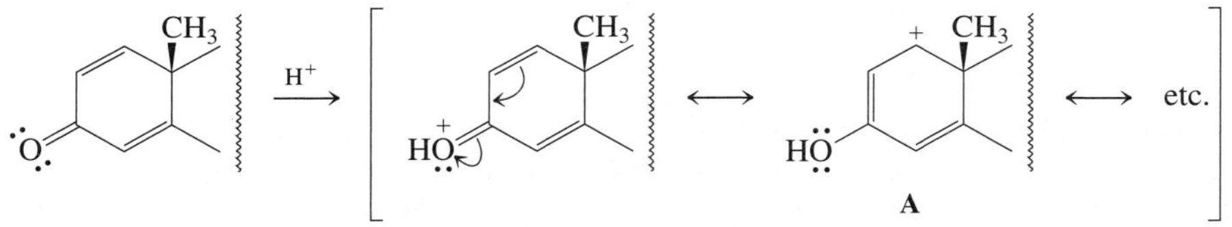

Two of the resonance structures (write all of them) place a positive charge next to the methyl group. Only form A is "productive," inasmuch as it leads to product by aromatization, which is the major driving force for the whole process.

Methyl Shift and Phenol Formation

New Reactions

Benzylic Resonance

1. Radical Halogenation (Section 22-1)

Requires heat, light, or a radical initiator

2. Solvolysis (Section 22-1)

3. S$_N$2 Reactions of (Halomethyl)benzenes (Section 22-1)

Through delocalized transition state

4. Benzylic Deprotonation (Section 22-1)

p$K_a \approx$ 41

Phenylmethyllithium (Benzyllithium)

Oxidation and Reduction Reactions on Aromatic Side Chains

5. Oxidation (Section 22-2)

Benzylic alcohols

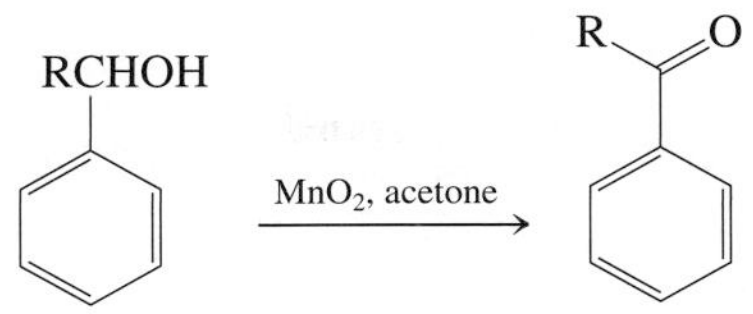
RCHOH
MnO_2, acetone
R
O

6. Reduction by Hydrogenolysis (Section 22-2)

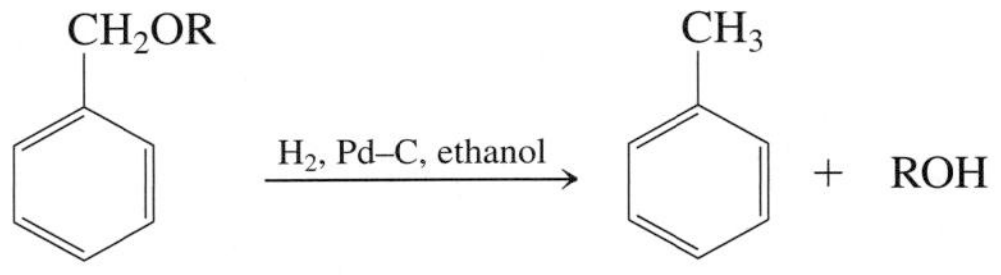
CH_2OR
H_2, Pd–C, ethanol
CH_3
+ ROH

$C_6H_5CH_2$ is a protecting group for ROH.

Phenols and Ipso Substitution

7. Acidity (Section 22-3)

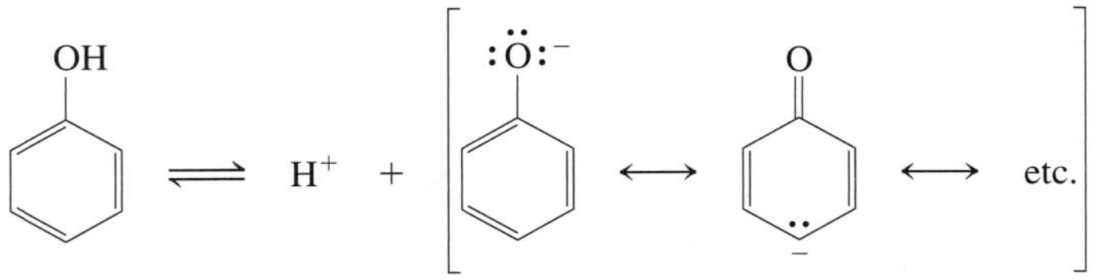
OH
H^+ +
$:\ddot{O}:^-$
O
etc.

p$K_a \approx 10$
Much stronger acid than simple alkanols.

Phenoxide ion

8. Nucleophilic Aromatic Substitution (Section 22-4)

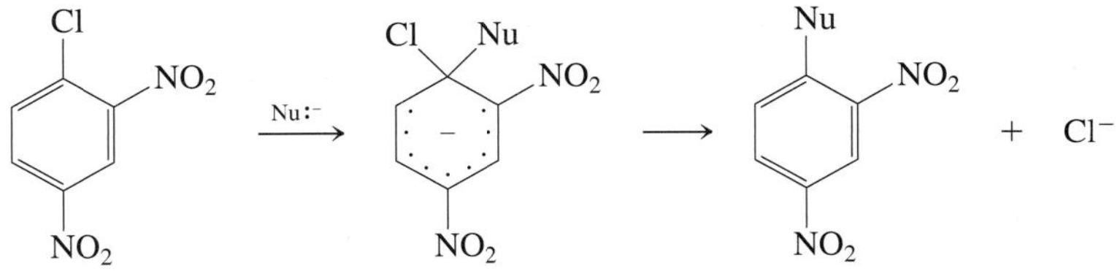
Cl
NO_2
$Nu:^-$
Cl
Nu
NO_2
Nu
NO_2
+ Cl^-
NO_2
NO_2
NO_2

Nucleophile attacks at ipso position.

9. Aromatic Substitution Through Benzyne Intermediates (Section 22-4)

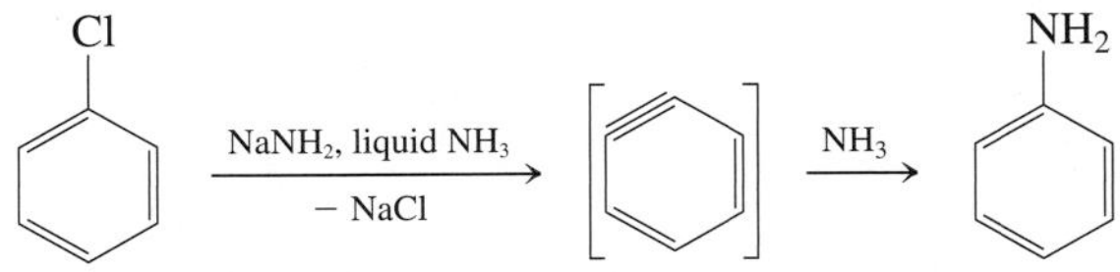
Cl
$NaNH_2$, liquid NH_3
− NaCl
NH_3
NH_2

Nucleophile attacks at both ipso and ortho positions.

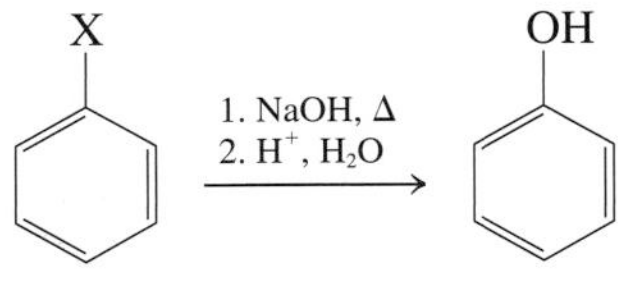
X
1. NaOH, Δ
2. H^+, H_2O
OH

10. Arenediazonium Salt Hydrolysis (Section 22-4)

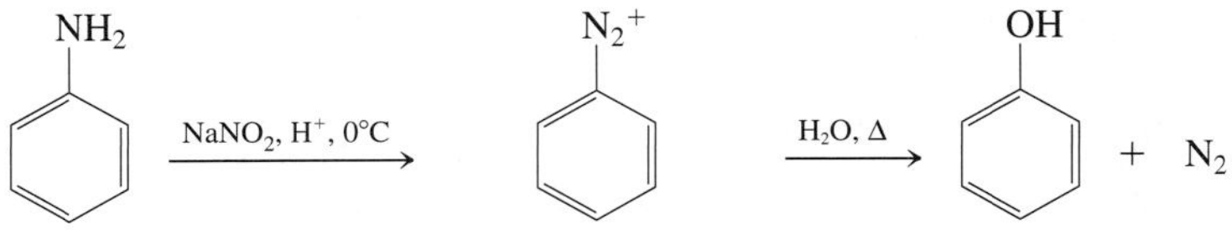
NH_2
$NaNO_2$, H^+, 0°C
$N_2{}^+$
H_2O, Δ
OH
+ N_2

Benzenediazonium cation

11. Substitution with Pd Catalysis

$$\text{(Aryl–X)} \xrightarrow[\text{PR}_3,\ 100°\text{C}]{\text{KOH, Pd catalyst,}} \text{(Phenol)}$$

Reactions of Phenols and Alkoxybenzenes

12. Ether Cleavage (Section 22-5)

$$\text{(Ar–OR)} \xrightarrow{\text{HBr, }\Delta} \text{(Ar–OH)} + \text{RBr}$$

Aryl C–O bond is not cleaved.

13. Ether Formation (Section 22-5)

$$\text{(Ar–OH)} + \text{RX} \xrightarrow{\text{NaOH, H}_2\text{O}} \text{(Ar–OR)}$$

Alkoxybenzene
Williamson method (Section 9-6)

14. Esterification (Section 22-5)

$$\text{(Ar–OH)} + \overset{\text{O}}{\underset{\|}{\text{RCCl}}} \xrightarrow{\text{Base}} \text{Phenyl alkanoate}$$

Phenyl alkanoate

15. Electrophilic Aromatic Substitution (Section 22-6)

$$\text{(OCH}_3\text{)} + \text{E}^+ \longrightarrow \text{(ortho-E)} + \text{(para-E)} + \text{H}^+$$

16. Phenolic Resins (Section 22-6)

$$\text{(Phenol)} + \text{CH}_2{=}\text{O} \xrightarrow{\text{NaOH}} \text{(o-CH}_2\text{OH phenol)} \xrightarrow[-\text{H}_2\text{O}]{\text{HO}^-}$$

$$\xrightarrow[\text{, NaOH}]{} \longrightarrow \longrightarrow \text{polymer}$$

17. Kolbe Reaction (Section 22-6)

18. Claisen Rearrangement (Section 22-7)

Aromatic Claisen rearrangement

Aliphatic Claisen rearrangement

19. Cope Rearrangement (Section 22-7)

20. Oxidation (Section 22-8)

2,5-Cyclohexadiene-1,4-dione
(*p*-Benzoquinone)

21. Conjugate Additions to 2,5-Cyclohexadiene-1,4-diones (*p*-Benzoquinones) (Section 22-8)

22. Diels-Alder Cycloadditions to 2,5-Cyclohexadiene-1,4-diones (*p*-Benzoquinones) (Section 22-8)

23. Lipid Peroxidation (Section 22-9)

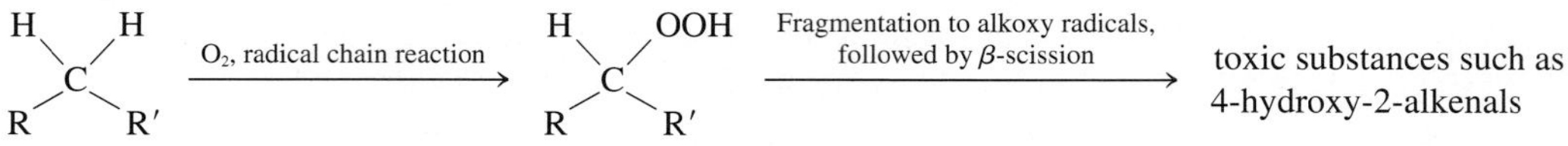

24. Inhibition by Antioxidants (Section 22-9)

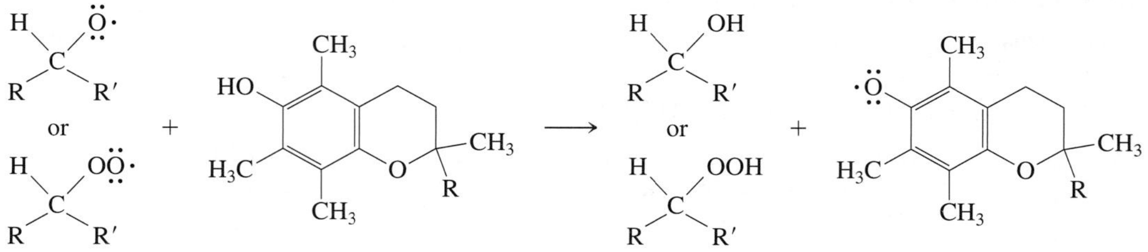

Vitamin E
(or BHA or BHT)

25. Vitamin C as an Antioxidant (Section 22-9)

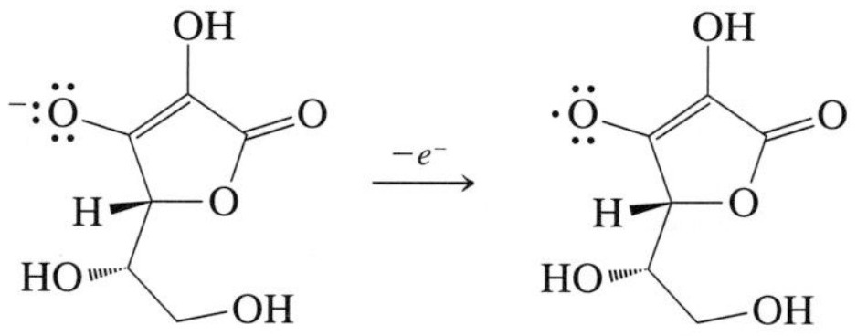

$-e^-$

Semidehydroascorbic
acid

Arenediazonium Salts

26. Sandmeyer Reactions (Section 22-10)

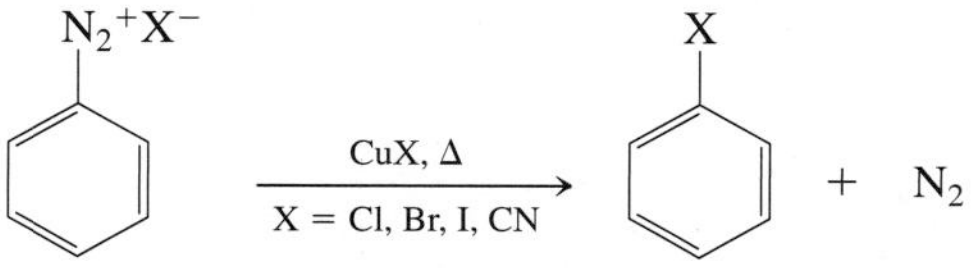

$$\xrightarrow[\text{X = Cl, Br, I, CN}]{\text{CuX, }\Delta} \quad + \quad N_2$$

27. Reduction (Section 22-10)

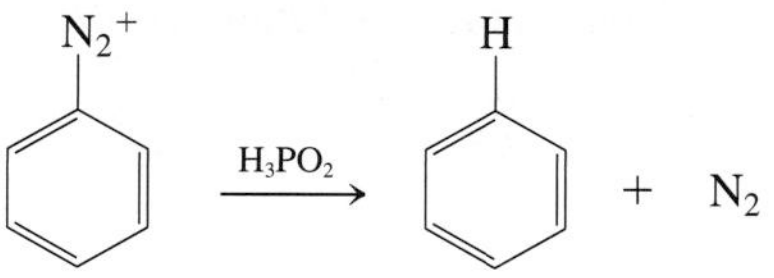

$$\xrightarrow{H_3PO_2} \quad + \quad N_2$$

28. Diazo Coupling (Section 22-11)

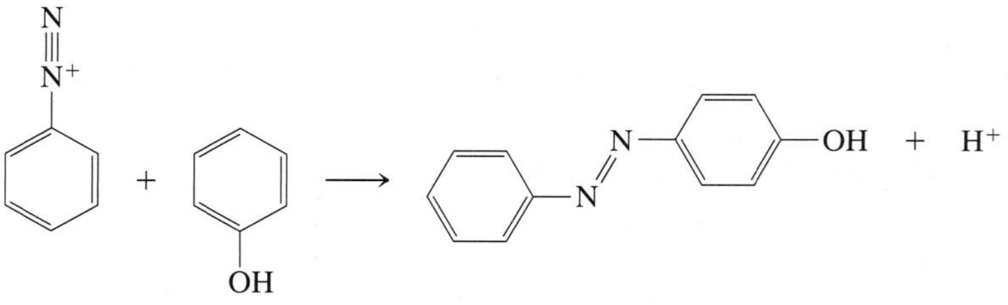

Azo compound

Occurs only with strongly activated rings

Important Concepts

1. Phenylmethyl and other **benzylic radicals, cations,** and **anions** are reactive intermediates stabilized by **resonance** of the resulting centers with a benzene π system.

2. **Nucleophilic aromatic ipso substitution** accelerates with the nucleophilicity of the attacking species and with the number of electron-withdrawing groups on the ring, particularly if they are located ortho or para to the point of attack.

3. **Benzyne** is destabilized by the strain of the two distorted carbons forming the triple bond.

4. **Phenols** are aromatic enols, undergoing reactions typical of the hydroxy group and the aromatic ring.

5. **Benzoquinones** and benzenediols function as redox couples in the laboratory and in nature.

6. Vitamin E and the highly substituted phenol derivatives BHA and BHT function as inhibitors of the **radical-chain oxidation of lipids.** Vitamin C also is an antioxidant, capable of regenerating vitamin E at the surface of cell membranes.

7. **Arenediazonium ions** furnish reactive **aryl cations** whose positive charge cannot be delocalized into the aromatic ring.

8. The amino group can be used to direct electrophilic aromatic substitution, after which it is replaceable by diazotization and substitution, including reduction.

Problems

35. Give the expected major product(s) of each of the following reactions.

(a) 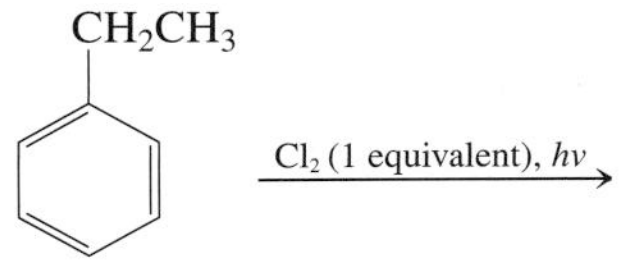 CH_2CH_3 (on benzene ring) $\xrightarrow{Cl_2 \text{ (1 equivalent)}, \, h\nu}$

(b) 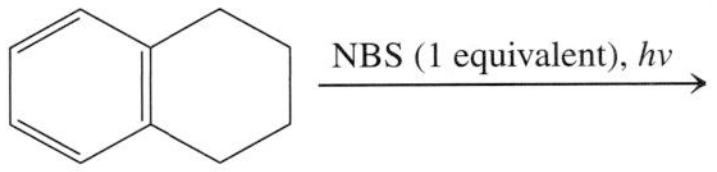   (tetralin structure) $\xrightarrow{NBS \text{ (1 equivalent)}, \, h\nu}$

36. Formulate a mechanism for the reaction described in Problem 35, part (b).

37. Propose syntheses of each of the following compounds, beginning in each case with ethylbenzene. **(a)** (1-Chloroethyl)-benzene; **(b)** 2-phenylpropanoic acid; **(c)** 2-phenylethanol; **(d)** 2-phenyloxacyclopropane.

38. Predict the order of relative stability of the three benzylic cations derived from chloromethylbenzene (benzyl chloride), 1-(chloromethyl)-4-methoxybenzene (4-methoxybenzyl chloride), and 1-(chloromethyl)-4-nitrobenzene (4-nitrobenzyl chloride). Rationalize your answer with the help of resonance structures.

39. By drawing appropriate resonance structures, illustrate why halogen atom attachment at the para position of the phenylmethyl (benzyl) radical is unfavored compared with attachment at the benzylic position.

40. Triphenylmethyl radical, $(C_6H_5)_3C\cdot$, is stable at room temperature in dilute solution in an inert solvent, and salts of triphenylmethyl cation, $(C_6H_5)_3C^+$, can be isolated as stable crystalline solids. Propose explanations for the unusual stabilities of these species.

41. Give the expected products of the following reactions or reaction sequences.

(a)  $BrCH_2CH_2CH_2$—(benzene ring)—CH_2Br $\xrightarrow{H_2O, \, \Delta}$

(b) 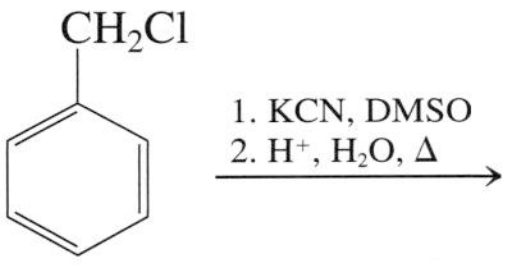 CH_2Cl (on benzene ring) $\xrightarrow[\text{2. } H^+, H_2O, \Delta]{\text{1. KCN, DMSO}}$

(c) 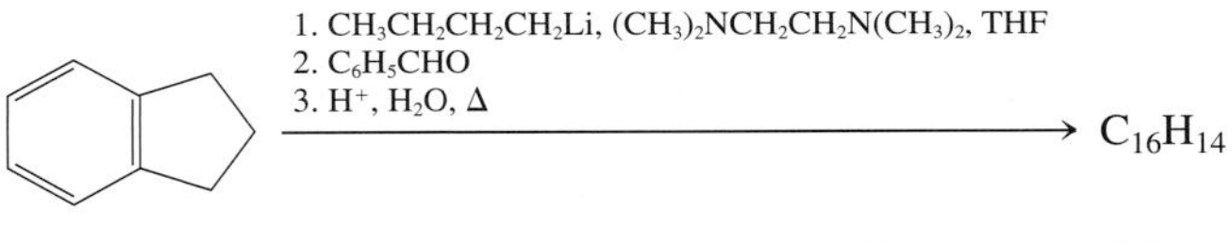   (indane structure) $\xrightarrow[\begin{array}{l}\text{2. } C_6H_5CHO \\ \text{3. } H^+, H_2O, \Delta\end{array}]{\text{1. } CH_3CH_2CH_2CH_2Li, (CH_3)_2NCH_2CH_2N(CH_3)_2, THF}$ $C_{16}H_{14}$

42. The hydrocarbon with the common name fluorene is acidic enough ($pK_a \approx 23$) to be a useful indicator in deprotonation reactions of compounds of greater acidity. Indicate the most acidic hydrogen(s) in fluorene. Draw resonance structures to explain the relative stability of its conjugate base.

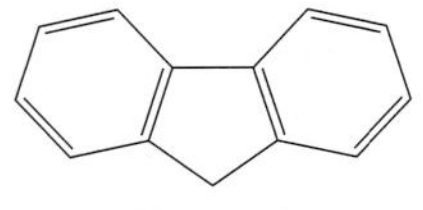

(fluorene structure)

Fluorene

43. Outline a straightforward, practical, and efficient synthesis of each of the following compounds. Start with benzene or methylbenzene. Assume that the para isomer (but *not* the ortho isomer) may be separated efficiently from any mixtures of ortho and para substitution products. Work backward from the target.

(a) 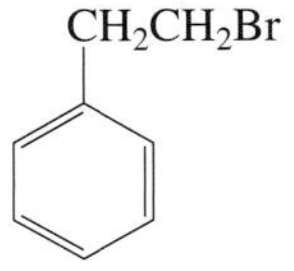 CH_2CH_2Br (on benzene ring)

(b) 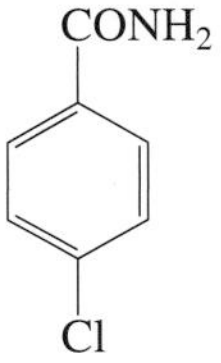 $CONH_2$ / Cl (para-substituted benzene)

(c) 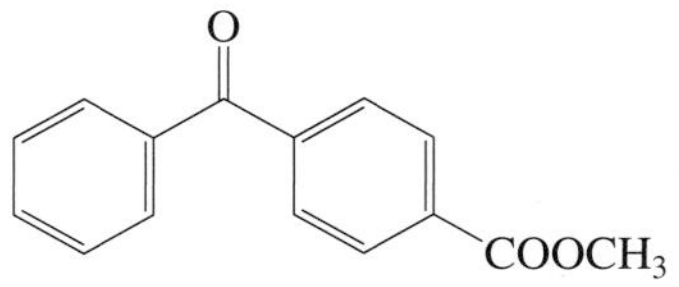 (diaryl ketone with $COOCH_3$ substituent)

(d) 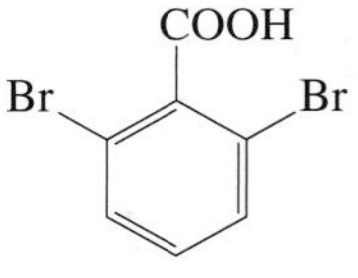 $COOH$ with Br at both ortho positions

44. Rank the following compounds in descending order of reactivity toward hydroxide ion.

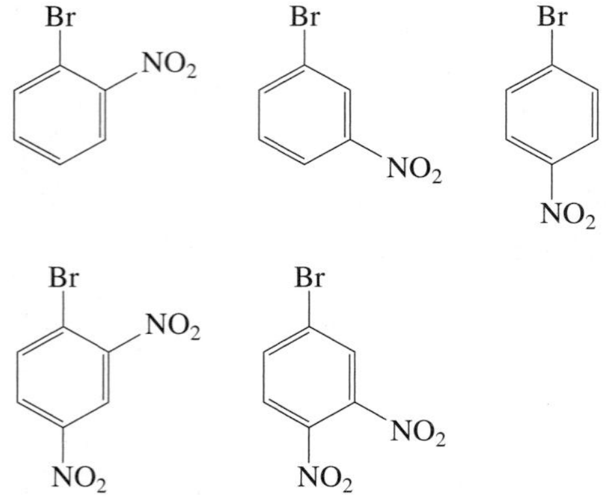

45. Predict the main product(s) of the following reactions. In each case, describe the mechanism(s) in operation.

(a)

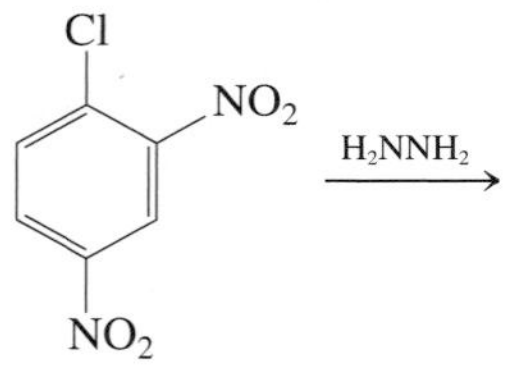

(b)

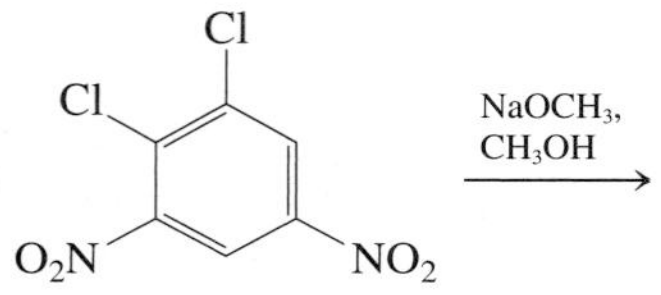

(c)

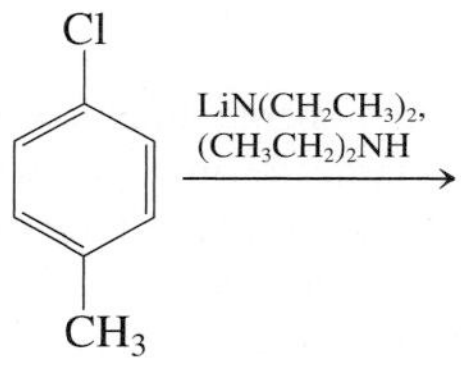

46. Starting with benzenamine, propose a synthesis of aklomide, an agent used to treat certain exotic fungal and protozoal infections in veterinary medicine. Several intermediates are shown to give you the general route. Fill in the blanks that remain; each requires as many as three sequential reactions. (**Hint:** Review the oxidation of amino- to nitroarenes in Section 16-5.)

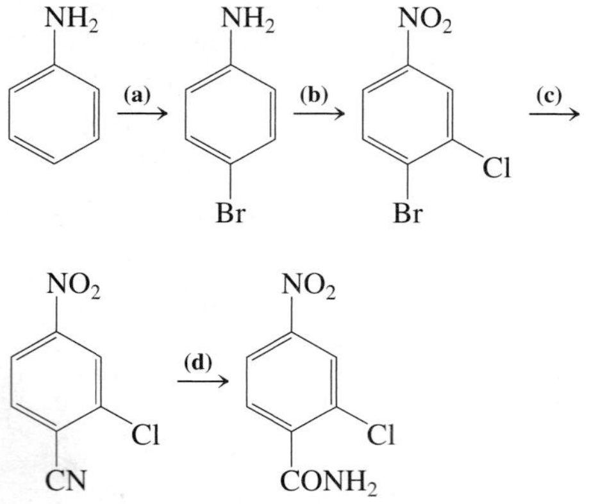

Aklomide

47. Explain the mechanism of the following synthetic transformation. (**Hint:** Two equivalents of butyllithium are used.)

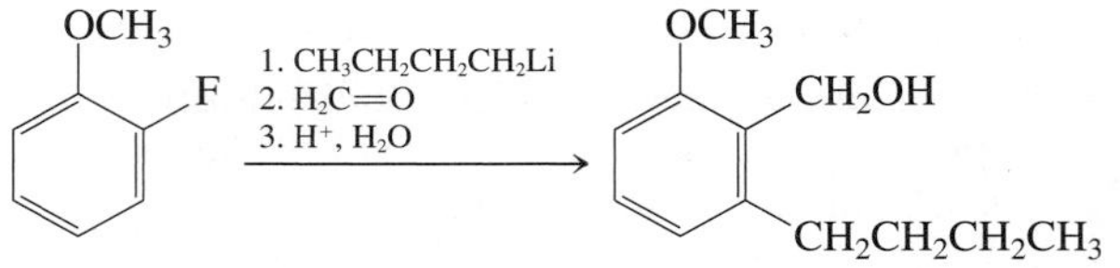

48. In nucleophilic aromatic substitution reactions that proceed by the addition–elimination mechanism, fluorine is the most easily replaced halogen in spite of the fact that F⁻ is by far the worst leaving group among halide ions. For example, 1-fluoro-2,4-dinitrobenzene reacts much more rapidly with amines than does the corresponding chloro compound. Suggest an explanation. (**Hint:** Consider the effect of the identity of the halogen on the rate-determining step.)

49. Based on the mechanism presented for the Pd-catalyzed reaction of a halobenzene with hydroxide ion, write out a reasonable mechanism for the Pd-catalyzed reaction of 1-bromo-3-methoxybenzene with 2-methyl-1-propanamine shown in Section 22-4.

50. Give the likely products of each of the following reactions. Each one is carried out in the presence of a Pd catalyst, a phosphine, and heat.

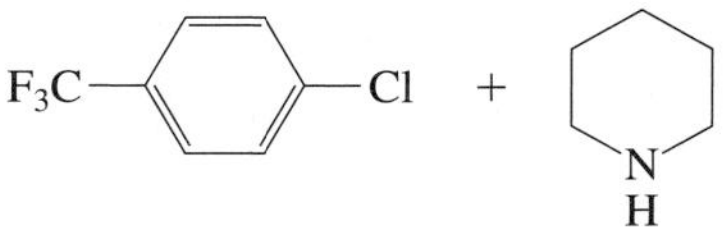

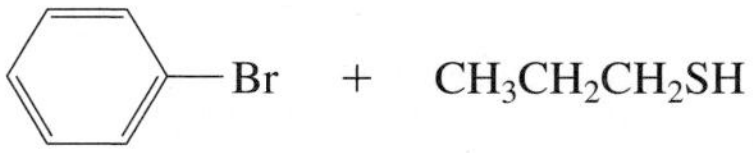

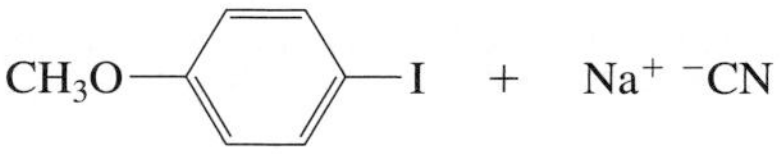

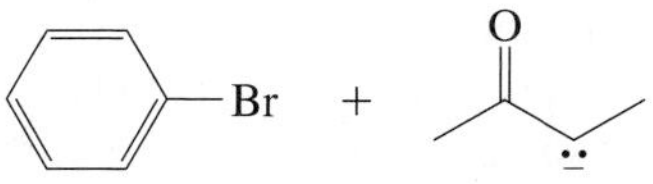

51. A very efficient short synthesis of resveratrol (Real Life 22-1) was reported in 2006. Fill in reasonable reagents for steps (a) through (d). Refer to the referenced text sections as needed.

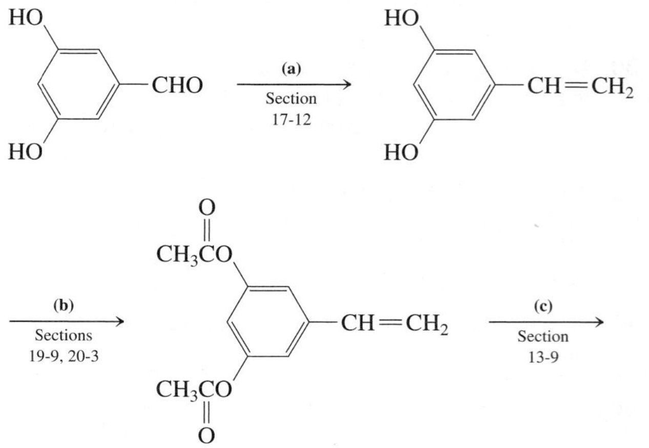

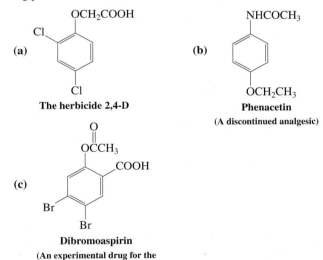

$$\xrightarrow[\text{Section} \atop \text{20-4}]{\textbf{(d)}} \quad \text{resveratrol}$$

52. The reaction sequence shown below illustrates the synthesis of 2,4,5-trichlorophenoxyacetic acid (2,4,5-T), a powerful herbicide. A 1:1 mixture of the butyl esters of 2,4,5-T and its dichlorinated analog 2,4-D was used between 1965 and 1970 as a defoliant during the Vietnam War under the code name Agent Orange. Propose mechanisms for the reactions in the synthesis of this substance, whose effects on the health of those exposed to it remain topics of considerable controversy.

$$\xrightarrow[\text{$-$NaCl}]{\text{1. NaOH, 150°C} \atop \text{2. H}^+,\text{ H}_2\text{O}}$$

2,4,5-Trichlorophenol
(2,4,5-TCP)

$$\xrightarrow[\text{$-$NaCl}]{\text{ClCH}_2\text{COOH,} \atop \text{NaOH, H}_2\text{O, Δ}}$$

85%
2,4,5-Trichlorophenoxyacetic acid
(2,4,5-T)

53. Give the expected major product(s) of each of the following reactions and reaction sequences.

(a)
$$\xrightarrow{\text{1. KMnO}_4, {}^-\text{OH}, \Delta \atop \text{2. H}^+, \text{H}_2\text{O}}$$

(b) CH$_2$OH / CH$_2$CCH$_3$ (O)
$$\xrightarrow{\text{1. MnO}_2, \text{ acetone} \atop \text{2. KOH, H}_2\text{O}, \Delta}$$

(c) CH$_3$
$$\xrightarrow{\text{1. (CH}_3)_2\text{CHCl, AlCl}_3 \atop \text{2. HNO}_3, \text{H}_2\text{SO}_4 \atop \text{3. KMnO}_4, \text{NaOH}, \Delta \atop \text{4. H}^+, \text{H}_2\text{O}}$$

54. Rank the following compounds in order of descending acidity.

(a) CH$_3$OH   **(b)** CH$_3$COOH

(c) OH with SO$_3$H (para)   **(d)** OH with OCH$_3$ (para)

55. Design a synthesis of each of the following phenols, starting with either benzene or any monosubstituted benzene derivative.

(a) OH with CH$_3$ (para)

(b) OH with Br, Br (2,6)

(c) The three benzenediols

(d) OH with Cl, NO$_2$, NO$_2$

56. Starting with benzene, propose syntheses of each of the following phenol derivatives.

(a) OCH$_2$COOH with Cl, Cl
The herbicide 2,4-D

(b) NHCOCH$_3$ with OCH$_2$CH$_3$
Phenacetin
(A discontinued analgesic)

(c) OCCH$_3$ (O), COOH, Br, Br
Dibromoaspirin
(An experimental drug for the treatment of sickle-cell anemia)

57. Name each of the following compounds.

(a) OH with Cl, Br

(b) OH with CH$_2$OH

(c) HO, OH, SO$_3$H

(d) OH with O—phenyl

(e) quinone with SCH$_3$

(e) OH with CF$_3$ (para)

(f) OH (phenol)

58. Give the expected product(s) of each of the following reaction sequences.

(a)

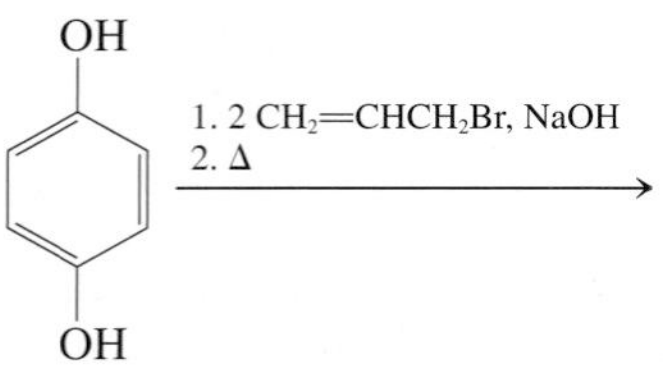

1. 2 CH₂=CHCH₂Br, NaOH
2. Δ

(b)

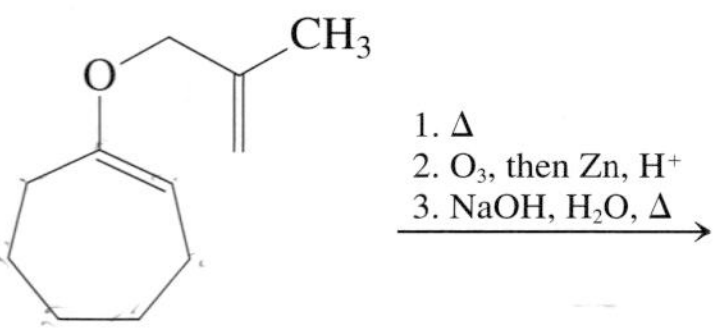

1. Δ
2. O₃, then Zn, H⁺
3. NaOH, H₂O, Δ

(c)

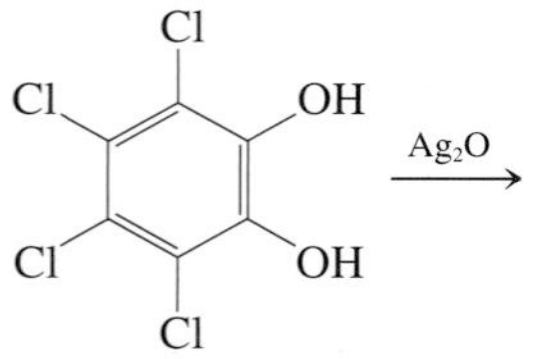

Ag₂O

(d)

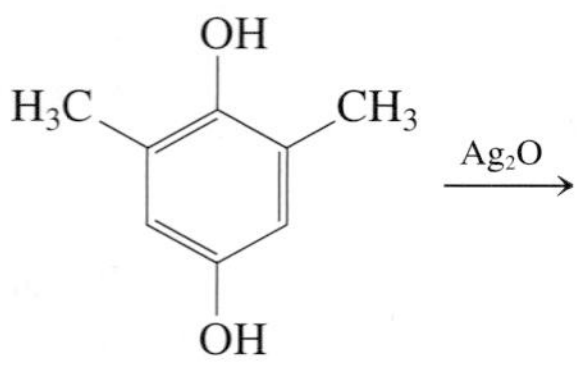

Ag₂O

(e)

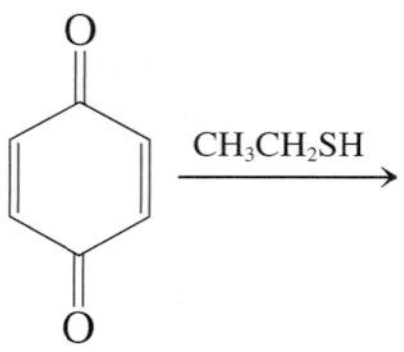

CH₃CH₂SH (two possibilities)

(f)

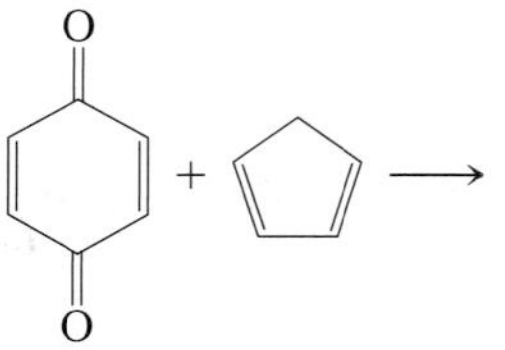

59. As a children's medicine, acetaminophen (Tylenol) has a major marketing advantage over aspirin: Liquid Tylenol preparations (essentially, acetaminophen dissolved in flavored water) are stable, whereas comparable aspirin solutions are not. Indeed, phenyl alkanoates undergo hydrolysis (and transesterification) considerably more rapidly than alkyl alkanoates, reactivity that underlies the mechanism of action of aspirin (Real Life 22-2). Explain.

60. Chili peppers contain significant quantities of vitamins A, C, and E, as well as folic acid and potassium. They also contain small quantities of capsaicin, the fiery essence responsible for the "hot" in hot peppers (see p. 989). The pure substance is in fact quite dangerous: Chemists handling capsaicin must work in special filtered-air enclosures and wear full body protection. One milligram placed on the skin will cause a severe burn. Although capsaicin has no odor or flavor of its own, its pungency—in the form of stimulation of the nerves in the mucous membranes of the mouth—can be detected even when the substance is diluted

to one part in 17 million of water. The hottest peppers exhibit about 1/20th of this level of pungency.

The structure of capsaicin is shown on p. 988. Some of the data that were used in its elucidation are presented below. Interpret as much of this information as you can.

MS: m/z = 122, 137 (base peak), 195 (tricky!), 305.

IR: $\tilde{\nu}$ = 972, 1660, 3016, 3445, 3541 cm⁻¹.

¹H NMR: δ = 0.93 (6H, d, J = 8 Hz), 1.35 (2H, quin, J = 7 Hz), 1.62 (2H, quin, J = 7 Hz), 1.97 (2H, q, J = 7 Hz), 2.18 (2H, t, J = 7 Hz), 2.20 (1H, m), 3.85 (3H, s), 4.33 (2H, d, J = 6 Hz), 5.33 (2H, m), 5.82 (2H, broad s), 6.73 (1H, d, J = 9 Hz), 6.78 (1H, s), 6.85 (1H, d, J = 9 Hz) ppm.

61. **CHALLENGE** Biochemical oxidation of aromatic rings is catalyzed by a group of liver enzymes called aryl hydroxylases. Part of this chemical process is the conversion of toxic aromatic hydrocarbons such as benzene into water-soluble phenols, which can be easily excreted. However, the primary purpose of the enzyme is to enable the synthesis of biologically useful compounds, such as the amino acid tyrosine from its relative phenylalanine (below).

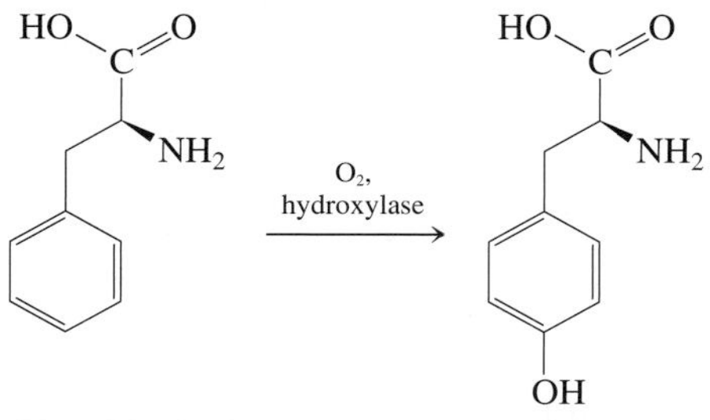

O₂, hydroxylase

Phenylalanine → **Tyrosine**

(a) Extrapolating from your knowledge of benzene chemistry, which of the following three possibilities seems most reasonable: The oxygen is introduced by electrophilic attack on the ring; the oxygen is introduced by free-radical attack on the ring; or the oxygen is introduced by nucleophilic attack on the ring? **(b)** It is widely suspected that oxacyclopropanes play a role in arene hydroxylation. Part of the evidence is the following observation: When the site to be hydroxylated is labeled with deuterium, a substantial proportion of the product still contains deuterium atoms, which have apparently migrated to the position ortho to the site of hydroxylation.

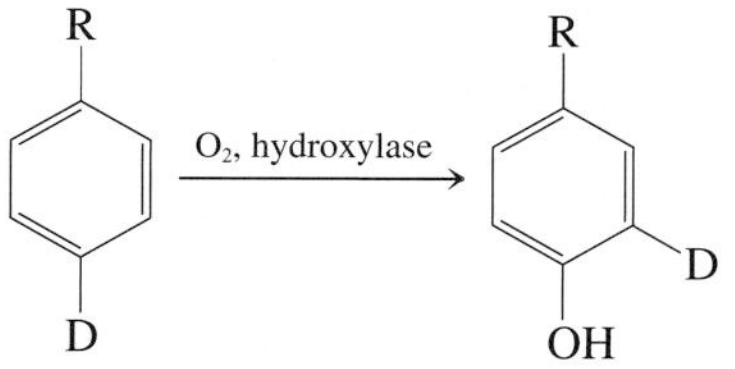

O₂, hydroxylase

through

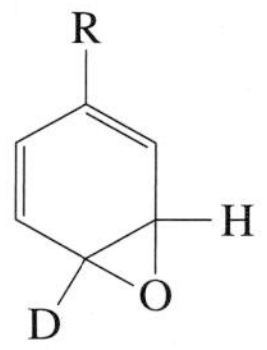

as an intermediate

Suggest a plausible mechanism for the formation of the oxacyclopropane intermediate and its conversion into the observed product. (**Hint:** Hydroxylase converts O_2 into hydrogen peroxide, HO–OH.) Assume the availability of catalytic amounts of acids and bases, as necessary.

Note: In victims of the genetically transmitted disorder called phenylketonuria (PKU), the hydroxylase enzyme system described here does not function properly. Instead, phenylalanine in the brain is converted into 2-phenyl-2-oxopropanoic (phenyl-pyruvic) acid, the reverse of the process shown in Problem 57 of Chapter 21. The buildup of this compound in the brain can lead to severe retardation; thus people with PKU (which can be diagnosed at birth) must be restricted to diets low in phenylalanine.

62. A common application of the Cope rearrangement is in ring-enlargement sequences. Fill in the reagents and products missing from the following scheme, which illustrates the construction of a 10-membered ring.

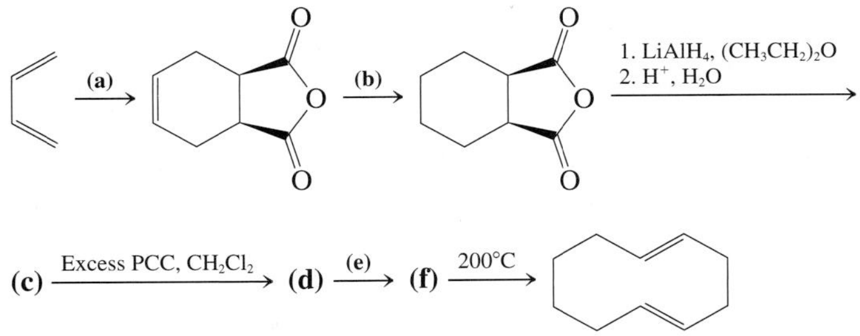

63. As mentioned in Section 22-10, the Sandmeyer reactions, which entail the substitution of the nitrogen in arenediazonium salts by Cl, Br, or CN, require copper(I) ion as a catalyst and proceed by complex radical mechanisms. Why do such substitutions not follow an S_N2 pathway? Why is the S_N1 mechanism, which operates in the corresponding displacements with OH and I, not effective?.

64. Formulate a detailed mechanism for the diazotization of benzenamine (aniline) in the presence of HCl and $NaNO_2$. Then suggest a plausible mechanism for its subsequent conversion into iodobenzene by treatment with aqueous iodide ion (e.g., from K^+I^-). Bear in mind your answer to Problem 63.

65. Show how you would convert 3-methylbenzenamine into each of the following compounds: (**a**) methylbenzene; (**b**) 1-bromo-3-methylbenzene; (**c**) 3-methylphenol; (**d**) 3-methylbenzonitrile; (**e**) *N*-ethyl-3-methylbenzenamine.

66. Devise a synthesis of each of the following substituted benzene derivatives, starting from benzene.

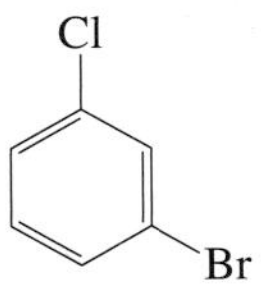

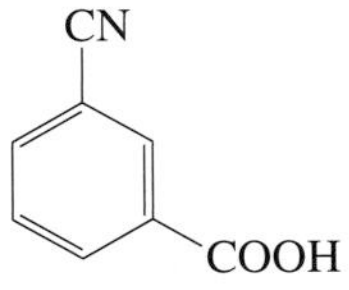

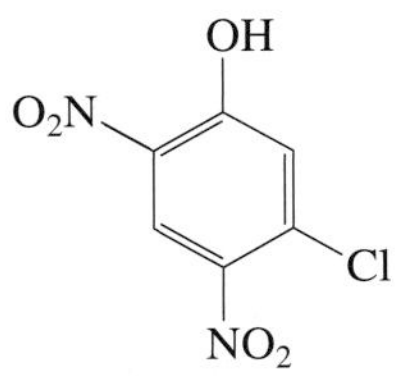

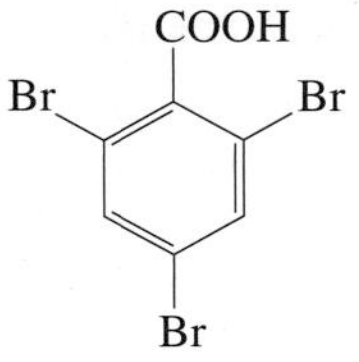

67. Write the most reasonable structure of the product of each of the following reaction sequences.

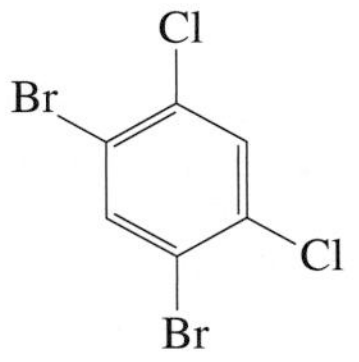

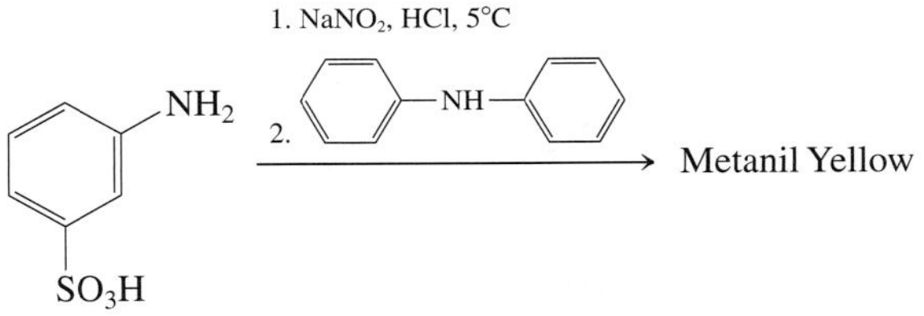

For the following reaction, assume that electrophilic substitution occurs preferentially on the most activated ring (Section 16-6).

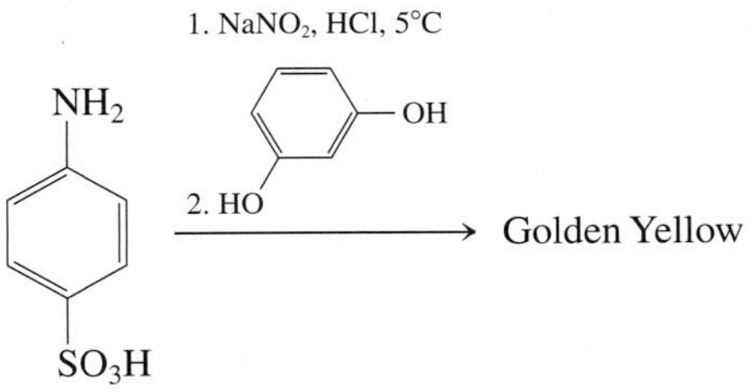

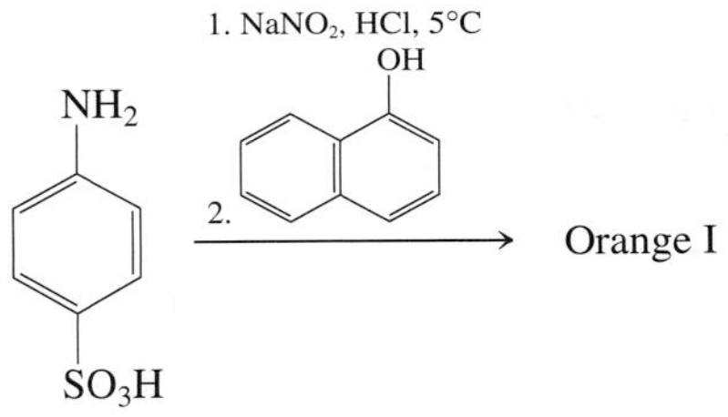

68. Show the reagents that would be necessary for the synthesis by diazo coupling of each of the following three compounds. For structures, see Section 22-11.

(**a**) Methyl Orange   (**b**) Congo Red

(**c**) Prontosil, H_2N-

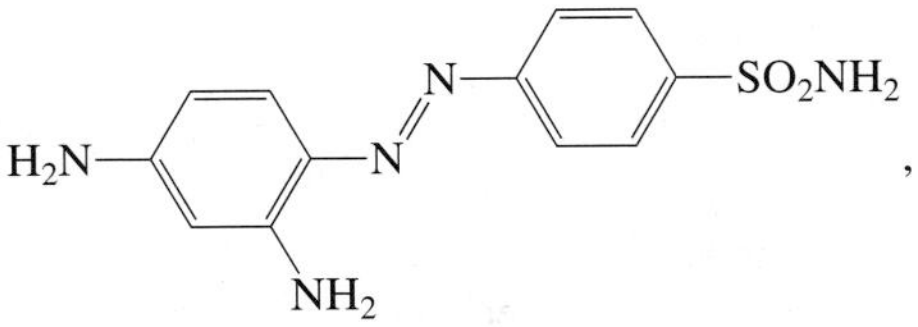

the first commercial antibiotic (Real Life 22-4).

69. **CHALLENGE** (**a**) Give the key reaction that illustrates the inhibition of fat oxidation by the preservative BHT. (**b**) The extent to which fat is oxidized in the body can be determined by measuring the amount of *pentane* exhaled in the breath. Increasing the amount of vitamin E in the diet decreases the amount of pentane exhaled. Examine the processes described in Section 22-9 and identify one that could produce pentane. You will have to do some extrapolating from the specific reactions shown in the section.

70. **CHALLENGE** The urushiols are the irritants in poison ivy and poison oak that give you rashes and make you itch upon exposure. Use the following information to determine the structures of urushiols I ($C_{21}H_{36}O_2$) and II ($C_{21}H_{34}O_2$), the two major members of this family of unpleasant compounds.

$$\text{Urushiol II} \xrightarrow{\text{H}_2,\ \text{Pd–C, CH}_3\text{CH}_2\text{OH}} \text{urushiol I}$$

$$\text{Urushiol II} \xrightarrow[\text{NaOH}]{\text{Excess CH}_3\text{I,}} \underset{\textbf{Dimethylurushiol II}}{C_{23}H_{38}O_2} \xrightarrow[\text{2. Zn, H}_2\text{O}]{\text{1. O}_3,\ \text{CH}_2\text{Cl}_2}$$

$$CH_3CH_2CH_2CH_2CH_2CH_2CHO \quad + \quad \underset{\textbf{Aldehyde A}}{C_{16}H_{24}O_3}$$

Synthesis of Aldehyde A

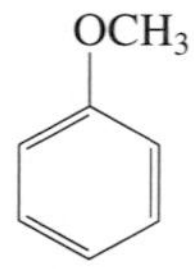

OCH₃ (structure)

$$\xrightarrow[\text{2. HNO}_3,\ \text{H}_2\text{SO}_4]{\text{1. SO}_3,\ \text{H}_2\text{SO}_4} \underset{\textbf{B}}{C_7H_7NSO_6} \xrightarrow{\text{H}^+,\ \text{H}_2\text{O},\ \Delta}$$

$$\underset{\textbf{C}}{C_7H_7NO_3} \xrightarrow[\begin{array}{l}\text{2. NaNO}_2,\ \text{H}^+,\text{H}_2\text{O}\\\text{3. H}_2\text{O},\ \Delta\end{array}]{\text{1. H}_2,\ \text{Pd, CH}_3\text{CH}_2\text{OH}}$$

$$\underset{\textbf{D}}{C_7H_8O_2} \xrightarrow[\begin{array}{l}\text{2. NaOH, CH}_3\text{I}\\\text{3. H}^+,\ \text{H}_2\text{O}\end{array}]{\text{1. CO}_2,\ \text{pressure, KHCO}_3,\ \text{H}_2\text{O}}$$

$$\underset{\textbf{E}}{C_9H_{10}O_4} \xrightarrow[\begin{array}{l}\text{2. H}^+,\ \text{H}_2\text{O}\\\text{3. MnO}_2,\ \text{acetone}\end{array}]{\text{1. LiAlH}_4,\ (\text{CH}_3\text{CH}_2)_2\text{O}}$$

$$\underset{\textbf{F}}{C_9H_{10}O_3} \xrightarrow[\begin{array}{l}\text{2. Excess H}_2,\ \text{Pd–C, CH}_3\text{CH}_2\text{OH}\\\text{3. PCC, CH}_2\text{Cl}_2\end{array}]{\text{1. C}_6\text{H}_5\text{CH}_2\text{O(CH}_2)_6\text{CH}=\text{P(C}_6\text{H}_5)_3} \text{aldehyde A}$$

71. Is the site of reaction in the biosynthesis of norepinephrine from dopamine (see Problem 65 of Chapter 5) consistent with the principles outlined in this chapter? Would it be easier or more difficult to duplicate this transformation nonenzymatically? Explain.

Team Problem

72. As a team, consider the following schemes that outline steps toward the total synthesis of taxodone D, a potential anticancer agent isolated from the seeds of the bald cypress tree. For the first scheme, divide your team into two groups, one to discuss the best option for A to effect the initial reduction step, the second to assign a structure to B, using the partial spectral data provided.

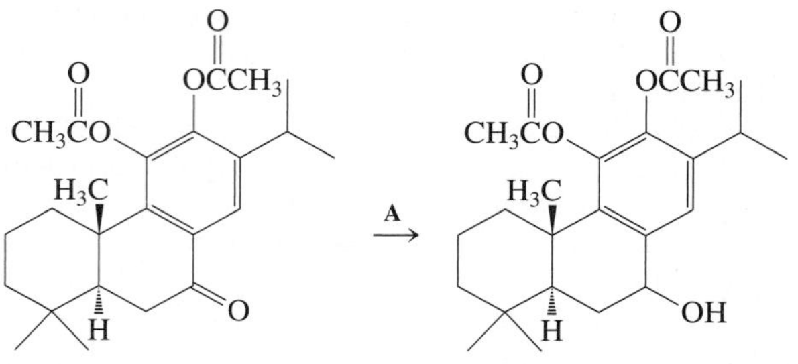

$$\xrightarrow{\text{A}}$$

$$\xrightarrow{\text{H}^+,\ \text{toluene, }\Delta} \textbf{B}$$

^{1}H NMR: $\delta = 5.99$ (dd, 1 H), 6.50 (d, 1 H) ppm.
IR: $\tilde{\nu} = 1720$ cm^{-1}.
MS: $m/z = 384$ (M$^+$).

Reconvene to discuss both parts of the first scheme. Then, as a group, analyze the remainder of the synthesis shown in the second scheme below. Use the spectroscopic data to help you determine the structures of C and taxodone, D.

$$\textbf{B} \longrightarrow \textbf{C}$$

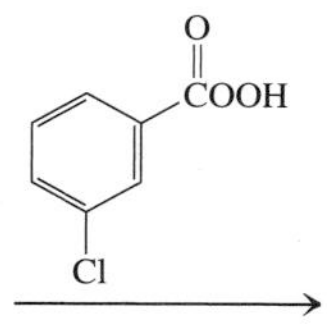

^{1}H NMR: $\delta = 3.51$ (dd, 1 H), 3.85 (d, 1 H) ppm.
MS: $m/z = 400$ (M$^+$).

$$\xrightarrow{^-\text{OH, H}_2\text{O}} \textbf{D}$$

Taxodone
^{1}H NMR: $\delta = 6.55$ (d, 1 H), 6.81 (s, 1 H) ppm, no other alkenyl or aromatic signals.
IR: $\tilde{\nu} = 1628, 3500, 3610$ cm^{-1}.
UV-Vis: λ_{max} (ε) = 316(20,000) nm.
MS: $m/z = 316$ (M$^+$).

Propose a mechanism for the formation of D from C. (**Hint:** After ester hydrolysis, one of the phenoxide oxygens can donate its electron pair through the benzene ring to effect a reaction at the para position. The product contains a carbonyl group in its enol form. For the interpretation of some of the spectral data, see Section 17-3.)

Preprofessional Problems

73. After chlorobenzene has been boiled in water for 2 h, which of the following organic compounds will be present in greatest concentration?

(a) C_6H_5OH

(b)

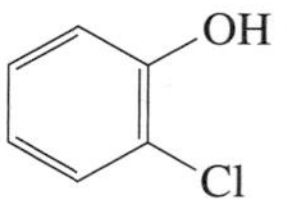

(c)

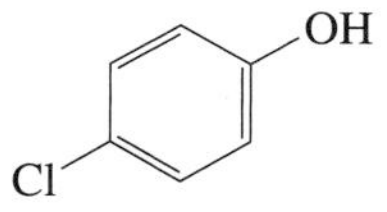

(d) C_6H_5Cl

(e)

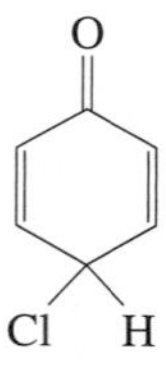

74. What are the products of the following reaction?

$C_6H_5OCH_3 \xrightarrow{HI, \Delta} ?$

(a) $C_6H_5I + CH_3OH$ (b) $C_6H_5OH + CH_3I$

(c) $C_6H_5I + CH_3I$ (d)

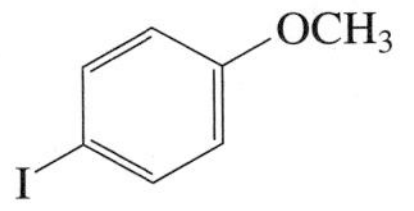

[structure: 4-iodoanisole with OCH₃] $+ H_2$

75. The transformation of 4-methylbenzenediazonium bromide to toluene is best carried out by using:

(a) H^+, H_2O (b) H_3PO_2, H_2O

(c) H_2O, ^-OH (d) Zn, NaOH

76. What is the principal product after the slurry obtained on treating benzenamine (aniline) with potassium nitrite and HCl at 0°C has been added to 4-ethylphenol?

(a) 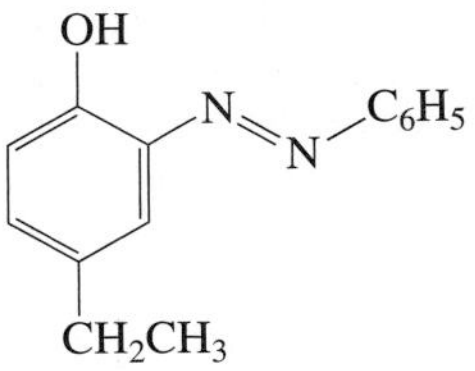 [structure: phenol with N=N–C₆H₅ and CH₂CH₃]

(b) 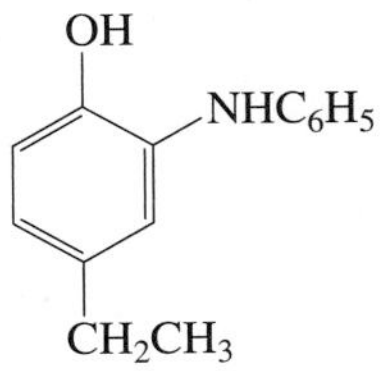 [structure: phenol with NHC₆H₅ and CH₂CH₃]

(c) 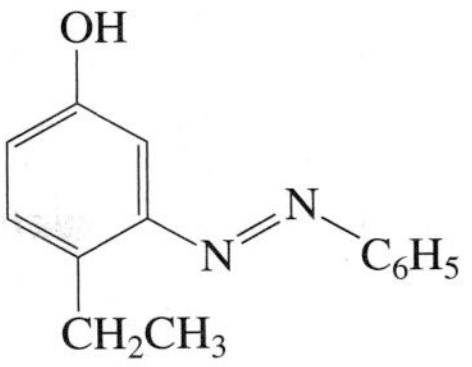 [structure: phenol with N=N–C₆H₅ and CH₂CH₃]

(d) 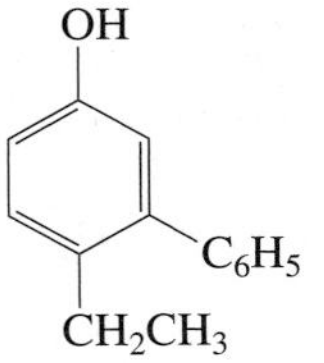 [structure: phenol with C₆H₅ and CH₂CH₃]

77. Examination of the ^1H NMR spectra of the following three isomeric nitrophenols reveals that one of them displays a hydroxy (phenolic) proton at substantially lower field than do the other two. Which one?

(a) 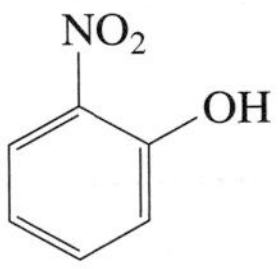 [structure: 2-nitrophenol, NO₂ and OH]

(b) 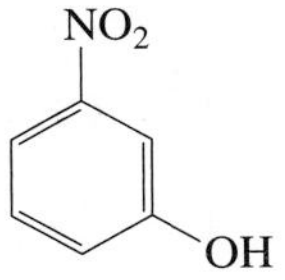 [structure: 3-nitrophenol, NO₂ and OH]

(c)  [structure: 4-nitrophenol, NO₂ and OH]

CHAPTER 23

Ester Enolates and the Claisen Condensation

Synthesis of β-Dicarbonyl Compounds; Acyl Anion Equivalents

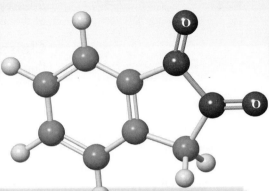

I n Chapter 18, during our survey of carbonyl chemistry, we learned that many techniques developed by the synthetic organic chemist are in fact based on natural processes for the construction of carbon–carbon bonds in biological systems. The aldol condensation (Sections 18-5 through 18-7) is one such process, a powerful method for converting aldehydes and ketones into β-hydroxycarbonyl compounds. In this chapter we shall examine first the related **Claisen condensation,** in which attack of an ester enolate on a carbonyl group generates a new carbon–carbon bond. We have already

seen its use in the biosynthesis of long-chain carboxylic acids (Section 19-13). The products of Claisen condensations are 1,3-dicarbonyl compounds, more commonly known as β-dicarbonyl compounds, which are important for their versatility in synthesis.

Examples of β-Dicarbonyl Compounds

$$CH_3CCH_2CCH_3$$

2,4-Pentanedione (Acetylacetone)

(A β-diketone)

$$CH_3CCH_2COCH_3$$

Methyl 3-oxobutanoate (Methyl acetoacetate)

(A β-ketoester)

$$HOCCH_2COH$$

Propanedioic acid (Malonic acid)

(A β-dicarboxylic acid)

Another natural carbon–carbon bond-forming process that has been adapted for use by synthetic chemists involves thiamine, or vitamin B_1, shown on p. 1058. Thiamine plays an essential role in several biochemical processes, including the biosynthesis of sugars, as we shall see in Chapter 24. Real Life 23-2 describes how thiamine also mediates sugar metabolism by converting pyruvic acid, a product of sugar metabolism, into acetyl CoA (Section 19-13). The relevant carbon–carbon bond-forming process makes use of a new kind of nucleophile derived from aldehydes and ketones, an **acyl anion equivalent.**

The diketone 1,2-indanedione was developed by Professor Madeleine Joullié at the University of Pennsylvania for the detection of latent fingerprints on porous surfaces such as paper. The reactive carbonyl groups undergo condensation with traces of amino acids deposited by human touch, and the adducts exhibit fluorescence on exposure to light (as shown in the photo on the left, courtesy of Professor Joullié). This advance in forensic science was spectacularly successful in pinpointing the perpetrator in the assassination of the Israeli Minister of Tourism in 2001, who was shot to death in the Jerusalem Hyatt Hotel. Initial investigations led to a room, in which a newspaper produced clear fingermarks (see photo on the right, courtesy of the division of identification and forensic science of the Israel Police, with the help of Professor Joseph Almog, The Hebrew University of Jerusalem), which were successfully compared by the AFIS (Automated Fingerprint Identification System) to point out the assassin.

These species illustrate how a normally electrophilic carbonyl carbon atom can be temporarily made nucleophilic and used to form a new carbon–carbon bond with the carbonyl carbon of another aldehyde or ketone molecule, giving α-hydroxyketones as the ultimate products.

23-1 | β-DICARBONYL COMPOUNDS: CLAISEN CONDENSATIONS

Ester enolates undergo addition–elimination reactions with ester functions, furnishing β-ketoesters. These transformations, known as **Claisen* condensations,** are the ester analogs of the aldol reaction (Section 18-5).

Claisen condensations form β-dicarbonyl compounds

Ethyl acetate reacts with a stoichiometric amount of sodium ethoxide to give ethyl 3-oxobutanoate (ethyl acetoacetate).

Reaction

Claisen Condensation of Ethyl Acetate

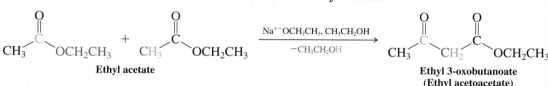

Unlike the aldol condensation (Section 18-5), in which two molecules are joined with elimination of water, the Claisen condensation proceeds by expelling a molecule of alcohol. It is also not catalytic: As we shall see, its mechanism requires the presence of a little over a full equivalent of base in order for the starting materials to proceed to product.

Keys to success: The Claisen condensation works because hydrogens flanked by two carbonyl groups are acidic

The mechanism of the Claisen condensation begins with deprotonation of the ester by ethoxide to form the ester enolate ion (step 1). This step is unfavorable because of the large difference in acidity between the α-hydrogens of the ester ($pK_a \approx 25$) and ethanol ($pK_a = 15.9$). The small equilibrium concentration of enolate thus formed adds to the carbonyl group of another ester molecule (step 2); loss of ethoxide completes an addition–elimination sequence, giving ethyl 3-oxobutanoate (commonly described as a 3- or β-ketoester) (step 3).

Mechanism

Animation

ANIMATED MECHANISM:
The Claisen condensation

Mechanism of the Claisen Condensation

Step 1. Ester enolate formation

Step 2. Nucleophilic addition

---

*This is the Claisen of the Claisen rearrangement (Section 22-7).

Step 3. Elimination

At this stage, however, the sequence is endothermic. Because each step is reversible, the overall equilibrium lies strongly on the side of the initial starting materials. It is the hydrogens at the center of the 3-ketoester that are the key to driving these equilibria forward. These hydrogens are unusually acidic, with a $pK_a \approx 11$, less than that of ethanol. The reason is the presence of the two adjacent, inductively electron-withdrawing carbonyl groups and the extensive delocalization of the negative charge of the anion resulting upon deprotonation. Thus, every molecule of 3-ketoester that forms in step 3 is acidic enough to react immediately with the coproduct ethoxide in step 4.

Step 4. Deprotonation of ketoester drives equilibrium

Le Chatelier's principle again: Step 4 depletes the ketoester from the reaction mixture, and steps 1 through 3 shift to replenish it.

Because this step depletes the mixture of the ethoxide necessary for step 1, the need for (a little more than) a stoichiometric equivalent of ethoxide base now becomes clear: It is to assure that enough base will be present to deprotonate all of the intermediate 3-ketoester, in this way pushing the overall equilibrium forward to the deprotonated ketoester.

With the equilibrium driven forward to the deprotonated condensation product, all that remains is to restore a proton to its central carbon by work-up with aqueous acid (step 5), thus completing the process.

Step 5. Protonation upon acidic aqueous work-up

You will have noted that the conditions of the Claisen condensation, ester plus alkoxide in alcohol, are identical to those used in base-catalyzed transesterifications (Section 20-4). Why, then, did we not see condensation products in such ester exchanges? The answer is that transesterification is much faster than the Claisen process (see step 1 above). Indeed, in order to avoid product mixtures, the latter always employs alkoxides that are identical to that in the ester function, in this way relegating transesterification to a nonproductive background transformation.

Exercise 23-1

Give the products of Claisen condensation of (a) ethyl propanoate; (b) ethyl 3-methylbutanoate; (c) ethyl pentanoate. For each, the base is sodium ethoxide, the solvent ethanol.

The phenomenon of the unusual acidity of the hydrogens in ethyl 3-oxobutanoate described above is general for all β-dicarbonyl compounds and extends to other groups that are electron withdrawing by induction and resonance. Table 23-1 lists pK_a values of the central hydrogens of several such systems. In the adjacent margin are illustrative NMR data, showcasing their (and their attached carbons') electron deficient nature through relative deshielding. The corresponding enolates and related carbanions are relatively nonbasic and, as we shall see in subsequent sections, are useful nucleophiles in organic synthesis.

Retro-Claisen condensation: β-Ketoesters lacking central hydrogens are cleaved by alkoxide

Attempted Claisen condensation using an ester with only one α-hydrogen fails. As the reaction below shows, the product would be a 2,2-disubstituted 3-ketoester. Lacking a central hydrogen that may be removed by base to shift the overall equilibrium to product (as in step 4 above), the trial is doomed by the unfavorable thermodynamics of steps 1–3. It leads to a reversible "dead end," and no Claisen condensation product is observed.

Failure of a Claisen Condensation—A Mechanistic "Dead End"

Ethyl 2-methylpropanoate

Ethyl 2,2,4-trimethyl-3-oxopentanoate
(Not observed)

Although 2,2-disubstituted β-ketoesters cannot be formed directly using the Claisen condensation, we shall see in Section 23-2 that they can be synthesized in another way. What happens if such a ketoester is treated with alkoxide base? *Complete reversal* of the Claisen condensation ensues, giving two molecules of simple ester through a mechanism that is the exact reverse of the forward reaction. Having no central hydrogen for the base to remove, addition of alkoxide to the ketone carbonyl group occurs instead. The resulting tetrahedral

NMR Chemical Shifts of β-Dicarbonyls Compared to Their Monocarbonyl Analogs

^{13}C and ^1H NMR δ values (ppm)

Table 23-1	pK_a Values for β-Dicarbonyl and Related Compounds	
Name	**Structure**	**pK_a**
2,4-Pentanedione (Acetylacetone)	$CH_3CCH_2CCH_3$ (with two C=O)	9
Methyl 2-cyanoacetate	$NCCH_2COCH_3$ (with C=O)	9
Ethyl 3-oxobutanoate (Ethyl acetoacetate)	$CH_3CCH_2COCH_2CH_3$ (with two C=O)	11
Propanedinitrile (Malonodinitrile)	$NCCH_2CN$	13
Diethyl propanedioate (Diethyl malonate)	$CH_3CH_2OCCH_2COCH_2CH_3$ (with two C=O)	13

More acidic ↑

intermediate fragments to release the enolate ion of the original ester. This process, called **retro-Claisen condensation,** confirms the thermodynamic basis for the failure of the Claisen condensation "dead end" shown above.

Reversal of a Claisen Condensation (Retro-Claisen Condensation)

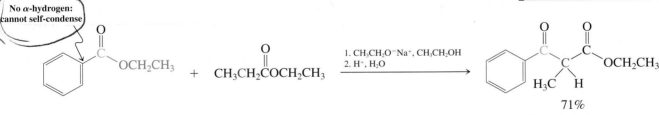

Exercise 23-2

Explain the following observation.

$$2\ CH_3C(\!=\!O)C(CH_3)_2\text{-COOCH}_3 \xrightarrow[\text{2. H}^+,\ H_2O]{\text{1. CH}_3O^-Na^+,\ CH_3OH} CH_3CCH_2COOCH_3 + 2\ (CH_3)_2CHCOOCH_3$$

Claisen condensations can have two different esters as reactants

Mixed Claisen condensations start with two different esters. Like crossed aldol condensations (Section 18-6), they are typically unselective and furnish product mixtures. However, a selective mixed condensation is possible when one of the reacting partners has no α-hydrogens, as in ethyl benzoate.

Note: To minimize self-condensation of ethyl propanoate, the benzoate ester is used in excess.

A Selective Mixed Claisen Condensation

No α-hydrogen: cannot self-condense

$$\text{Ethyl benzoate} + CH_3CH_2COCH_2CH_3 \xrightarrow[\text{2. H}^+,\ H_2O]{\text{1. CH}_3CH_2O^-Na^+,\ CH_3CH_2OH} \text{product}$$

71%
Ethyl 2-methyl-3-oxo-3-phenylpropanoate

Exercise 23-3

Give all the Claisen condensation products that would result from treatment of a mixture of ethyl acetate and ethyl propanoate with sodium ethoxide in ethanol.

Exercise 23-4

Is the mixed Claisen condensation between ethyl formate and ethyl acetate likely to afford one major product? Explain and give the structure(s) of the product(s) you expect.

Intramolecular Claisen condensations result in cyclic compounds

The intramolecular version of the Claisen reaction, called the **Dieckmann*** condensation, produces cyclic 3-ketoesters. As expected (Section 9-6), it works best for the formation of five- and six-membered rings.

Model Building

$$CH_3CH_2OCCH_2CH_2CH_2CH_2CH_2COCH_2CH_3$$

1. $CH_3CH_2O^-Na^+$, CH_3CH_2OH
2. H^+, H_2O

60%

Diethyl heptanedioate   **Ethyl 2-oxocyclohexanecarboxylate**

Solved Exercise 23-5 | Working with the Concepts: Predicting a Successful Claisen Condensation

Two cyclic products are possible from the Dieckmann (intramolecular Claisen) condensation shown below, but only one of them actually forms. Describe and explain briefly the outcome of this reaction.

1. $CH_3CH_2O^-Na^+$, CH_3CH_2OH
2. H^+, H_2O

Strategy

To draw the possible products, use the mechanistic pattern described throughout this section.

Solution

• Form a new carbon–carbon bond between the carbon atom α to one of the ester carbonyl groups and the carbonyl carbon atom of the other. The products of the two possible options are shown below:

No α-hydrogen

1. $CH_3CH_2O^-Na^+$, CH_3CH_2OH
2. H^+, H_2O

1. $CH_3CH_2O^-Na^+$, CH_3CH_2OH
2. H^+, H_2O

• Notice that the product of the first condensation lacks an acidic α-hydrogen between its two carbonyl groups. This molecule is not isolated under Claisen condensation conditions, because the equilibrium associated with its formation cannot be driven to completion by deprotonation (step 4 of the mechanism presented earlier in this section; see also "Failure of a Claisen Condensation"). Only the product of the second condensation process is obtained.

---

*Professor Walter Dieckmann (1869–1925), University of Munich, Germany.

REAL LIFE: NATURE 23-1 | Claisen Condensations Assemble Biological Molecules

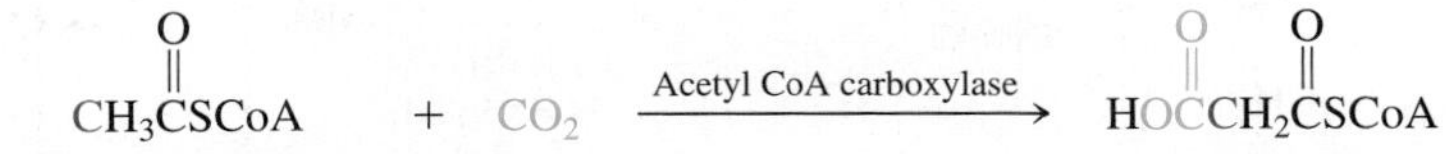

The coupling processes that build fatty acid chains from thioesters of coenzyme A (Section 19-13) are forms of Claisen condensations. The carboxylation of acetyl CoA into malonyl CoA (shown above) is a variant in which the carbon of CO_2 rather than that of an ester carbonyl group is the site of nucleophilic attack.

The methylene group in the carboxylated species is much more reactive than the methyl group in acetyl thioesters and participates in a wide variety of Claisen-like condensations. Although these processes require enzyme catalysis, they may be written in simplified form, shown below.

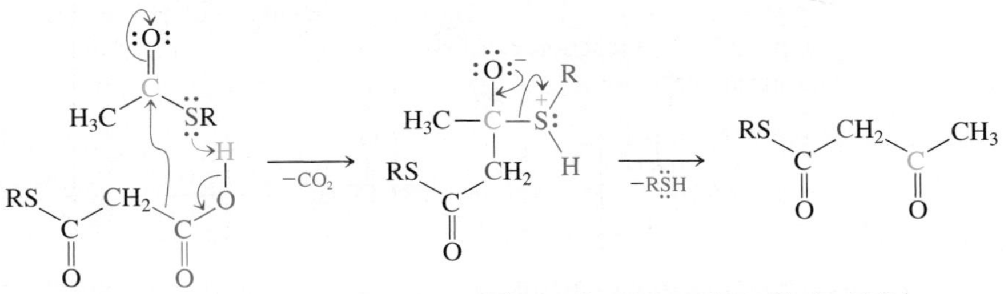

(RSH = acyl carrier protein; see Section 19-13.)

The product of the reaction above, the acetoacetyl derivative of the acyl carrier protein, is the starting point for the biosynthesis of other compounds besides fatty acids. Steroids derive from a sequence of enzyme-catalyzed Claisen-like condensations that produce the branched five-carbon skeleton of 2-methyl-1,3-butadiene (isoprene), the building block of

the terpenes (Section 4-7). In this variation, the enolate of acetyl CoA adds to the ketocarbonyl of acetoacetyl CoA, giving β-hydroxy-β-methylglutaryl CoA (HMG CoA). A sequence of enzyme-catalyzed reduction, decarboxylation, dehydration, and phosphorylation finally gives 3-methyl-3-butenyl (isoprenyl) pyrophosphate.

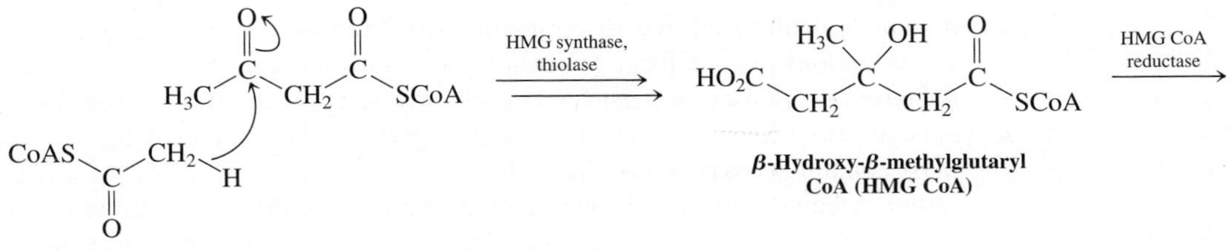

β-Hydroxy-β-methylglutaryl CoA (HMG CoA)

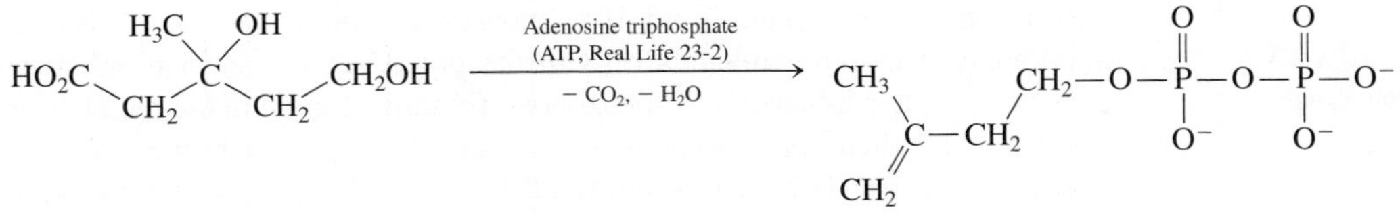

Mevalonic acid

Isoprenyl pyrophosphate

In animals, six molecules of isoprenyl pyrophosphate are used to construct the hydrocarbon squalene, which contains a 24-carbon chain and six methyl substituents. Squalene, in

turn, undergoes a complex sequence of cyclization, rearrangement, and bond cleavage to give steroids, such as cholesterol (Sections 4–7 and 14–10).

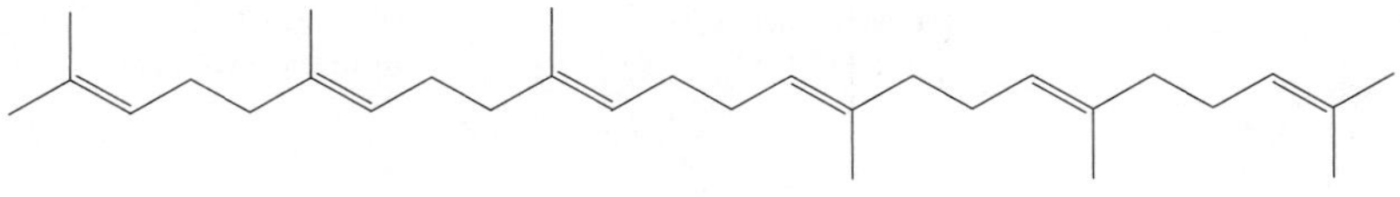

Squalene

Exercise 23-6 | Try it Yourself

Formulate a mechanism for the following reaction.

Diethyl 1,2-benzenedicarboxylate
(Diethyl phthalate)

$+ \quad CH_3CO_2CH_2CH_3$

1. CH₃CH₂O⁻Na⁺,
   CH₃CH₂OH
2. H⁺, H₂O

⟶

$CO_2CH_2CH_3$

60–80%

Ketones undergo mixed Claisen reactions

Ketones can participate in the Claisen condensation. Because they are more acidic than esters, they are deprotonated before the ester has a chance to undergo self-condensation. The products (after acidic work-up) may be β-diketones, β-ketoaldehydes, or other β-dicarbonyl compounds. The reaction can be carried out with a variety of ketones and esters both inter- and intramolecularly.

$$CH_3COCH_2CH_3 \quad + \quad CH_3CCH_3 \xrightarrow[\text{2. H}^+,\text{ H}_2O]{\text{1. NaH, (CH}_3\text{CH}_2)_2\text{O}} CH_3CCH_2CCH_3$$

85%

Exercise 23-7 |

1,3-Cyclohexanedione (margin) can be prepared by an intramolecular mixed Claisen condensation between the ketone carbonyl and ester functions of a single molecule. What is the structure of this substrate molecule?

1,3-Cyclohexanedione

Retrosynthetic analysis clarifies the synthetic utility of the Claisen condensation

Having seen a variety of types of Claisen condensations, we can now ask how this process may be logically analyzed for synthetic use. Three facts are available to help us: (1) Claisen condensations always form 1,3-dicarbonyl compounds; (2) one of the reaction partners in a Claisen condensation must be an ester, whose alkoxide group is lost in the course of the condensation; and (3) the other reaction partner (the source of the nucleophilic enolate) must contain at least two acidic hydrogens on an α-carbon. In addition, if a mixed condensation is being considered, one reaction partner should be incapable of self-condensation (e.g., it should lack α-hydrogens). If we are given the structure of a target molecule and wish to determine whether (and, if so, how) it can be made by a Claisen condensation, we must analyze it retrosynthetically with the preceding points in mind. For example, let us consider whether 2-benzoylcyclohexanone (margin) can be made by a Claisen condensation.

It is a 1,3-dicarbonyl compound, meeting the first requirement. What bond forms in a Claisen condensation? By examining all the examples in this section, we find that the new bond in the product always connects one of the carbonyl groups of the 1,3-dicarbonyl moiety to the carbon atom *between* them. Our target molecule contains two such bonds, which we label *a* and *b*. As we continue our analysis by disconnecting each of these strategic bonds in turn, we must employ the second point: The carbonyl group at which the new carbon–carbon bond forms starts out as part of an ester function. Thus, working backward, *we must imagine reattaching an alkoxy group to this carbonyl carbon.* Thus:

2-Benzoylcyclohexanone

Disconnect here
Reattach —OR here

Disconnect here
Reattach —OR here

$\overset{a}{\Longleftarrow}$

$\overset{b}{\Longrightarrow}$

$+ \quad$ RO

Disconnection of bond *a* reveals a ketoester, which undergoes intramolecular Claisen condensation in the forward direction, whereas disconnection of bond *b* gives cyclohexanone and a benzoic ester. Both condensations are quite feasible; however, the second is preferable because it constructs the target from two smaller pieces:

| **Solved Exercise 23-8** | **Working with the Concepts: Claisen Retrosynthetic Analysis** |

Show a synthesis of using a Claisen or a Dieckmann condensation.

Strategy
Begin by identifying the strategic bonds—the ones that may be made in a Claisen-type condensation. Once you have found them (there will be two), disconnect each one in turn and attach an alkoxy group to the separated carbonyl carbon. This procedure will give you the two possible starting points for the Claisen-based synthesis.

Solution
• The strategic bonds are those connecting the two carbonyl groups to the shared atom between them (arrows). Label them *a* and *b*.
• (1) Break each of these bonds, and (2) attach an ethoxy group to the carbonyl fragment (use an ethoxy group because the target molecule is an ethoxy ester). You should come up with something like this, which constitutes your retrosynthetic analysis:

The simplest β-dicarbonyl compound, propanedial (malondialdehyde, MDA), is a natural by-product of both oxidative and photochemical degradation of biological molecules. It is also highly toxic (Section 22-9): MDA is damaging to the eye, and failure of natural processes to clear it from the retina is the root cause of age-related macular degeneration, a leading cause of blindness in the elderly.

**Propanedial
(Malondialdehyde)**

National Eye Institute/National Institutes of Health-Age-related Macular degeneration

- At this stage you have identified the two possible starting points. To complete the exercise, write the condensation reactions in the forward direction:

Synthesis a:

CH₃CH₂OC— (α) —CCO₂CH₂CH₃ → 1. CH₃CH₂ONa, CH₃CH₂OH 2. H⁺, H₂O → product with CCO₂CH₂CH₃

Synthesis b:

cyclohexanone + CH₃CH₂OCCO₂CH₂CH₃ → 1. CH₃CH₂ONa, CH₃CH₂OH 2. H⁺, H₂O → product with CCO₂CH₂CH₃

Synthesis a is a Dieckmann condensation featuring attack of the α-carbon of the ketone enolate on the ester carbonyl at the other end of the molecule. Synthesis b is a mixed Claisen condensation in which cyclohexanone enolate attacks one of the two ester groups of diethyl ethanedioate (diethyl oxalate).

Exercise 23-9 | **Try It Yourself**

Suggest syntheses of the following molecules by Claisen or Dieckmann condensations.

(a) CH₃CCH₂CH (with two O above)

(b) eight-membered ring ketone with CO₂CH₂CH₃

(c) H₃C—C(=O)— cyclopentanone

In Summary Claisen condensations are endothermic and therefore would not take place without a stoichiometric amount of base strong enough to deprotonate the resulting 3-ketoester. Mixed Claisen condensations between two esters are nonselective, unless they are intramolecular (Dieckmann condensation) or one of the components is devoid of α-hydrogens. Ketones also participate in selective mixed Claisen reactions because they are more acidic than esters.

23-2 | β-DICARBONYL COMPOUNDS AS SYNTHETIC INTERMEDIATES

Having seen how to prepare β-dicarbonyl compounds, let us explore their synthetic utility. This section will show that the corresponding anions are readily alkylated and that 3-ketoesters are hydrolyzed to the corresponding acids, which can be decarboxylated to give ketones or new carboxylic acids. These transformations open up versatile synthetic routes to other functionalized molecules.

β-Dicarbonyl anions are nucleophilic

The unusually high acidity of the methylene hydrogens of β-ketocarbonyl compounds can be used to synthetic advantage. Their unusually low pK_a values (ca. 9–13, see Table 23-1) allow alkoxide bases to remove a proton from this methylene group essentially quantitatively, giving an enolate ion that may be alkylated to produce substituted derivatives. For example, treatment of ethyl 3-oxobutanoate (ethyl acetoacetate) with $NaOCH_2CH_3$ effects complete deprotonation to the enolate, which reacts via the S_N2 mechanism with iodomethane to afford the methylated derivative shown. The remaining acidic hydrogen at this position can be abstracted with the somewhat stronger base KOC(CH₃)₃. The resulting anion may undergo another S$_N$2 alkylation, shown here with benzyl bromide, to yield the doubly substituted product.

β-Ketoester Alkylations

O   O
‖   ‖
CH₃C—CH—COCH₂CH₃
        |
        H

1. Na⁺⁻OCH₂CH₃,
   CH₃CH₂OH
2. CH₃I
———————————→
−CH₃CH₂OH
−NaI

O   O
‖   ‖
CH₃C—CH—COCH₂CH₃
        |
   New CH₃
   bond

1. K⁺⁻OC(CH₃)₃,
   (CH₃)₃COH
      CH₂Br
2. ⟨benzyl⟩
———————————→
−(CH₃)₃COH
−KBr

O   CH₃ O
‖ | ‖
CH₃C—C—COCH₂CH₃
     |
 New CH₂
 bond
  |
 ⟨phenyl⟩

Ethyl 3-oxobutanoate
(Ethyl acetoacetate)

65%
Ethyl 2-methyl-
3-oxobutanoate

77%
Ethyl 2-methyl-
2-(phenylmethyl)-
3-oxobutanoate

These alkylations provide synthetic access to 2,2-disubstituted β-ketoesters, which, as we saw in Section 23-1, cannot be prepared directly by means of the Claisen condensation.

Other β-dicarbonyl compounds undergo similar reactions.

O   O
‖   ‖
CH₃CH₂OC—CH—COCH₂CH₃
        |
        H

1. Na⁺⁻OCH₂CH₃, CH₃CH₂OH
         CH₃
2. CH₃CH₂CHBr
———————————→
−CH₃CH₂OH
−NaBr

O   O
‖   ‖
CH₃CH₂OC—CH—COCH₂CH₃
        |
   CH₃CH₂CH
        |
        CH₃
       84%

Diethyl propanedioate
(Diethyl malonate)

Diethyl 2-(1-methylpropyl)propanedioate

Exercise 23-10

Give a synthesis of 2,2-dimethyl-1,3-cyclohexanedione from methyl 5-oxohexanoate.

3-Ketoacids readily undergo decarboxylation

Hydrolysis of 3-ketoesters furnishes 3-ketoacids, which in turn readily undergo decarboxylation under mild conditions. The products—ketones and carboxylic acids—contain the alkyl groups introduced in prior alkylation steps.

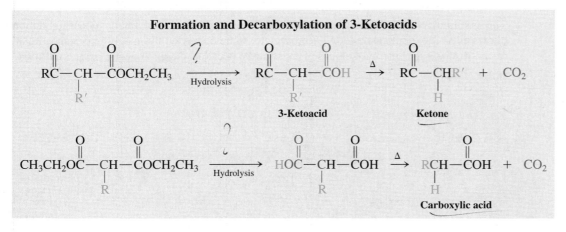

Formation and Decarboxylation of 3-Ketoacids

Reaction

$$\underset{\substack{(CH_2)_3CH_3}}{CH_3\overset{O}{\overset{\|}{C}}-CH-\overset{O}{\overset{\|}{C}}OCH_2CH_3} \xrightarrow[\substack{-CH_3CH_2OH \\ -CO_2}]{\substack{1.\ NaOH,\ H_2O \\ 2.\ H_2SO_4,\ H_2O,\ 100°C}} \underset{\substack{H \\ 60\%}}{CH_3\overset{O}{\overset{\|}{C}}-CH(CH_2)_3CH_3}$$

Ethyl 2-butyl-3-oxobutanoate   **2-Heptanone**

$$\underset{\substack{CH_3CH_2CH \\ | \\ CH_3}}{CH_3CH_2O\overset{O}{\overset{\|}{C}}-CH-\overset{O}{\overset{\|}{C}}OCH_2CH_3} \xrightarrow[\substack{-CH_3CH_2OH \\ -CO_2}]{H_2SO_4,\ H_2O,\ \Delta} \underset{\substack{H \\ 65\%}}{CH_3CH_2\overset{CH_3}{\overset{|}{C}HCH}-COOH}$$

Diethyl 2-(1-methylpropyl)propanedioate   **3-Methylpentanoic acid**

Decarboxylation, or loss of CO_2, is not a typical reaction of carboxylic acids under ordinary conditions. However, β-ketoacids are unusually prone to decarboxylation for two reasons. First, the Lewis basic oxygen of the 3-keto function is ideally positioned to bond with the carboxy hydrogen by means of a cyclic six-atom transition state. Second, this transition state has aromatic character (Section 15-3), because three electron pairs shift around the cyclic six-atom array. The species formed in decarboxylation are CO_2 and an enol, which tautomerizes rapidly to the final ketone product.

Mechanism of Decarboxylation of 3-Ketoacids

Mechanism

$$\underset{\substack{H\quad H}}{H_3C-\overset{:O:}{\overset{\|}{C}}\cdots\overset{\overset{H}{\cdots}}{C}-\overset{\ddot{O}:}{\overset{\|}{C}}-\ddot{O}:} \longrightarrow \underset{CH_2}{H_3C-C} + \overset{:O:}{\underset{:O:}{\overset{\|}{C}}} \longrightarrow \underset{\substack{CH_2 \\ H}}{H_3C-\overset{O}{\overset{\|}{C}}} + CO_2$$

Note that *only* carboxylic acids with a second carbonyl group in the 3- or β-position are structurally capable of reacting in this way. Carboxylic acids lacking a β-carbonyl function *do not* decarboxylate, regardless of the presence of C=O groups elsewhere in the molecule.

Loss of CO_2 occurs readily only from the neutral carboxylic acid. If the ester is hydrolyzed with base, acid must be added to protonate the resulting carboxylate salt in order for decarboxylation to occur. Decarboxylation of substituted propanedioic (malonic) acids follows the same mechanism.

Exercise 23-11

Formulate a detailed mechanism for the decarboxylation of $CH_3CH(COOH)_2$ (methylmalonic acid).

The acetoacetic ester synthesis leads to methyl ketones

The combination of alkylation followed by ester hydrolysis and finally decarboxylation allows ethyl 3-oxobutanoate (ethyl acetoacetate) to be converted ultimately into 3-substituted or 3,3-disubstituted methyl ketones. This strategy is called the **acetoacetic ester synthesis.**

Acetoacetic Ester Synthesis

$$CH_3\overset{O}{\overset{\|}{C}}-CH_2-\overset{O}{\overset{\|}{C}}OCH_2CH_3 \dashrightarrow \underset{\substack{R'}}{CH_3\overset{O}{\overset{\|}{C}}-\overset{R}{\overset{|}{C}}-\overset{O}{\overset{\|}{C}}OCH_2CH_3} \dashrightarrow \underset{R'}{CH_3\overset{O}{\overset{\|}{C}}-CH{\overset{R}{\diagup}}_{\diagdown R'}}$$

3,3-Disubstituted methyl ketone

Methyl ketones with either one or two substituent groups on C3 can be synthesized by using the acetoacetic sequence.

Syntheses of Substituted Methyl Ketones

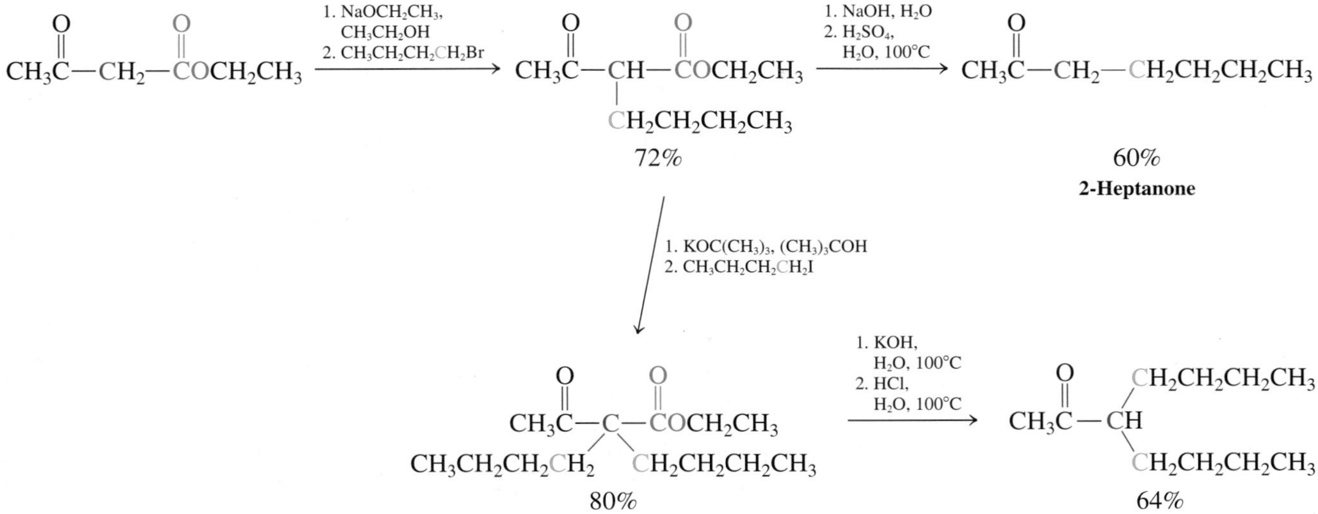

Exercise 23-12

Propose syntheses of the following ketones, beginning with ethyl 3-oxobutanoate (ethyl acetoacetate). **(a)** 2-Hexanone; **(b)** 2-octanone; **(c)** 3-ethyl-2-pentanone; **(d)** 4-phenyl-2-butanone.

The malonic ester synthesis furnishes carboxylic acids

Diethyl propanedioate (malonic ester) is the starting material for preparing 2-alkylated and 2,2-dialkylated acetic acids, a method called the **malonic ester synthesis.**

Malonic Ester Synthesis

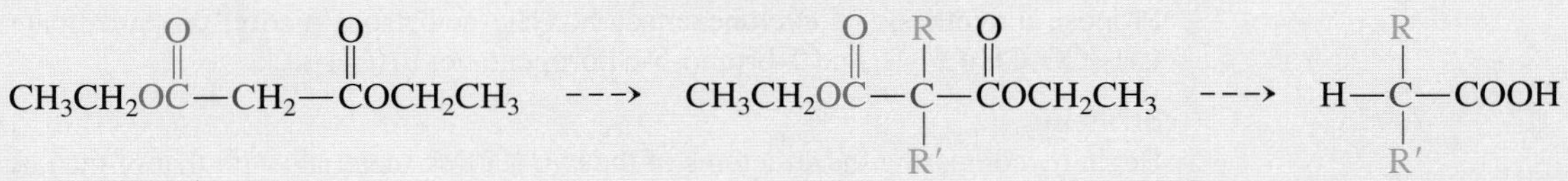

Like the acetoacetic ester route to ketones, the malonic ester synthesis can lead to carboxylic acids with either one or two substituents at C2.

Synthesis of a 2,2-Dialkylated Acetic Acid

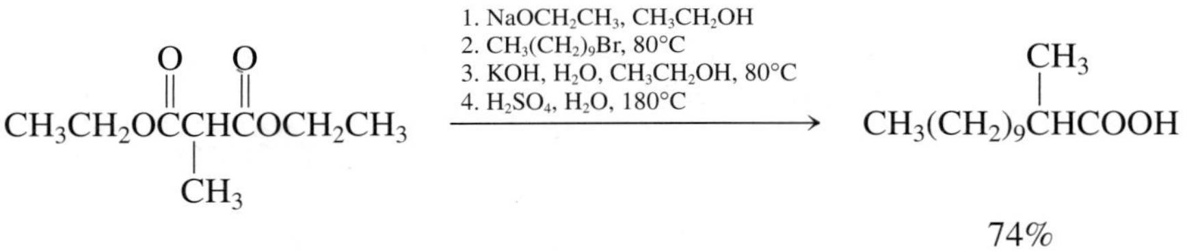

Diethyl 2-methylpropanedioate
(Diethyl methylmalonate)

74%
2-Methyldodecanoic acid

Exercise 23-13

(a) Give the structure of the product formed after each of the first three steps in the preceding synthesis of 2-methyldodecanoic acid. (b) How would you make the starting material, diethyl 2-methylpropanedioate?

The rules and limitations governing S_N2 reactions apply to the alkylation steps. Thus, tertiary haloalkanes exposed to β-dicarbonyl anions give mainly elimination products. However, the anions can successfully attack acyl halides, α-bromoesters, α-bromoketones, and oxacyclopropanes.

Exercise 23-14

The first-mentioned compound in each of the following parts is treated with the subsequent series of reagents. Give the final products. (**Note:** The choice of the conditions for the last step(s), either direct acid-catalyzed hydrolysis–decarboxylation or stoichiometric base hydrolysis–acidification–decarboxylation, are arbitrary. The conditions given below are those that gave the best yields experimentally.)

(a) $CH_3CH_2O_2C(CH_2)_5CO_2CH_2CH_3$: $NaOCH_2CH_3$; $CH_3(CH_2)_3I$; $NaOH$; and H^+, H_2O, Δ

(b) $CH_3CH_2O_2CCH_2CO_2CH_2CH_3$: $NaOCH_2CH_3$; CH_3I; KOH; and H^+, H_2O, Δ

(c) $CH_3\overset{\displaystyle O}{\overset{\|}{C}}CHCO_2CH_3$: NaH, C_6H_6; $C_6H_5\overset{\displaystyle O}{\overset{\|}{C}}Cl$; and H^+, H_2O, Δ
        $|$
      CH_3

(d) $CH_3\overset{\displaystyle O}{\overset{\|}{C}}CH_2CO_2CH_2CH_3$: $NaOCH_2CH_3$; $BrCH_2CO_2CH_2CH_3$; $NaOH$; and H^+, H_2O, Δ

(e) $CH_3CH_2CH(CO_2CH_2CH_3)_2$: $NaOCH_2CH_3$; $BrCH_2CO_2CH_2CH_3$; and H^+, H_2O, Δ

(f) $CH_3\overset{\displaystyle O}{\overset{\|}{C}}CH_2CO_2CH_2CH_3$: $NaOCH_2CH_3$; $BrCH_2\overset{\displaystyle O}{\overset{\|}{C}}CH_3$; and H^+, H_2O, Δ

Solved Exercise 23-15 | Working with the Concepts: Syntheses Using β-Dicarbonyl Compounds

Propose a synthesis of cyclohexanecarboxylic acid from diethyl propanedioate (malonate), $CH_2(CO_2CH_2CH_3)_2$, and 1-bromo-5-chloropentane, $Br(CH_2)_5Cl$.

Strategy

Begin by comparing the structures of the two starting materials with that of the target molecule. Identify the strategic bonds in the product—the ones that may be made in a β-dicarbonyl-based carboxylic acid synthesis. Once you have done so, find the carbon atoms *in the starting materials* that give rise to these bonds.

Solution

• The strategic bonds are those at the α-carbon of the target carboxylic acid (arrows). Line up the product structure with those of the starting molecules in a way that clarifies how they are to be linked. You should come up with something like this, which constitutes your retrosynthetic analysis:

| Cyclohexane-carboxylic acid | 1-Bromo-5-chloropentane | Diethyl propanedioate (Diethyl malonate) |

- What makes this particular synthesis challenging to analyze? To construct a *cyclic* product, we must connect the malonate *α*-carbon to *two* functionalized atoms of a *single* substrate molecule, in this case the 1-bromo-5-chloropentane. The five carbon atoms of this compound combine with the *α*-carbon of the malonate to furnish the desired six-membered ring.
- The actual synthetic sequence can be depicted as follows. Deprotonation of the malonate *α*-carbon followed by S$_N$2 displacement at the C–Br bond (better leaving group; recall Section 6-7) of the substrate gives a 5-chloropentyl malonate derivative. Subsequent treatment with another equivalent of base removes the remaining *α*-hydrogen, and the resulting enolate ion displaces chloride intramolecularly to form the ring. Hydrolysis and decarboxylation complete the synthesis:

$$H_2C(CO_2CH_2CH_3)_2 \xrightarrow[\text{2. Br(CH}_2)_5\text{Cl}]{\text{1. NaOCH}_2\text{CH}_3,\ \text{CH}_2\text{CH}_3\text{OH}} Cl(CH_2)_5\!-\!CH(CO_2CH_2CH_3)_2$$

$$\xrightarrow[\text{CH}_3\text{CH}_2\text{OH}]{\text{NaOCH}_2\text{CH}_3,}
\begin{array}{c}\text{CO}_2\text{CH}_2\text{CH}_3 \\ \text{CO}_2\text{CH}_2\text{CH}_3\end{array}
\xrightarrow[\text{2. H}_2\text{SO}_4,\ \text{H}_2\text{O},\ \Delta]{\text{1. NaOH, H}_2\text{O}}
\text{—CO}_2\text{H}$$

Exercise 23-16 | Try It Yourself

How would you modify the synthesis in Exercise 23-15 in order to use it to prepare heptanedioic acid, $HOOC(CH_2)_5COOH$? (**Hint:** The same starting materials may be used.)

In Summary *β*-Dicarbonyl compounds such as ethyl 3-oxobutanoate (acetoacetate) and diethyl propanedioate (malonate) are versatile synthetic building blocks for elaborating more complex molecules. Their unusual acidity makes it easy to form the corresponding anions, which can be used in nucleophilic displacement reactions with a wide variety of substrates. Their hydrolysis produces 3-ketoacids that are unstable and undergo decarboxylation on heating.

23-3 | *β*-DICARBONYL ANION CHEMISTRY: MICHAEL ADDITIONS

Reaction of the stabilized anions derived from *β*-dicarbonyl compounds and related analogs (Table 23-1) with *α,β*-unsaturated carbonyl compounds leads to 1,4-additions. This transformation, an example of **Michael addition** (Section 18-11), is base catalyzed and works with *α,β*-unsaturated ketones, aldehydes, nitriles, and carboxylic acid derivatives, all of which are termed **Michael acceptors.**

Michael Addition

$$(CH_3CH_2O_2C)_2CH_2 \ + \ CH_2\!=\!CHCCH_3 \xrightarrow[\text{CH}_3\text{CH}_2\text{OH, }-10\text{ to }25°\text{C}]{\text{Catalytic CH}_3\text{CH}_2\text{O}^-\text{Na}^+,} (CH_3CH_2O_2C)_2CH\!-\!CH_2CH_2CCH_3$$

71%

Diethyl propanedioate
(Diethyl malonate)

3-Buten-2-one
(Methyl vinyl ketone)
(Michael acceptor)

Diethyl 2-(3-oxobutyl)propanedioate

Why do stabilized anions undergo 1,4- rather than 1,2-addition to Michael acceptors? 1,2-Addition occurs, but is reversible with relatively stable anionic nucleophiles, because it leads to a relatively high-energy alkoxide. Conjugate addition is favored thermodynamically because it produces a resonance-stabilized enolate ion.

Exercise 23-17

Formulate a detailed mechanism for the Michael addition process just depicted. Why is the base required in only catalytic amounts?

Exercise 23-18

Give the products of the following Michael additions [base in square brackets].

(a) $CH_3CH_2CH(CO_2CH_2CH_3)_2$ + $CH_2=CHCH$ (with C=O above) $[Na^{+-}OCH_2CH_3]$

(b) 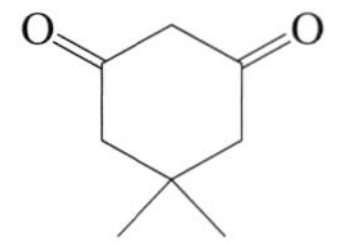 + $CH_2=CHC\equiv N$ $[Na^{+-}OCH_3]$

(c) 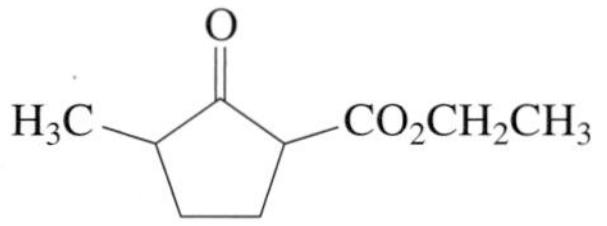 + $CH_3CH=CHCO_2CH_2CH_3$ $[K^{+-}OCH_2CH_3]$

Exercise 23-19

Explain the following observation. (**Hint:** Consider proton transfer in the first Michael adduct.)

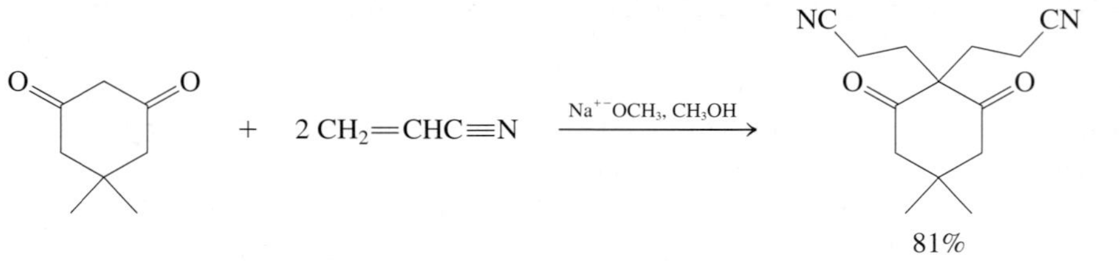

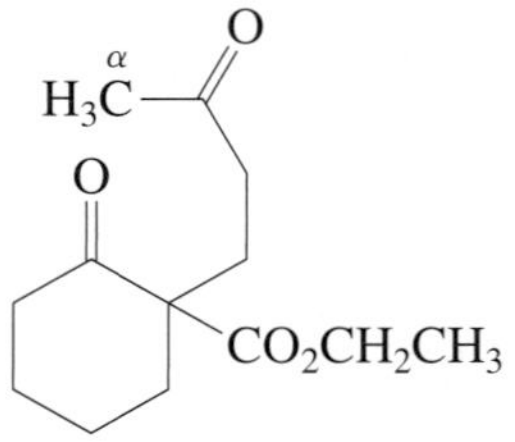

+ $2\ CH_2=CHC\equiv N$ $\xrightarrow{Na^{+-}OCH_3,\ CH_3OH}$

81%

The following reaction is a useful application of Michael addition of anions of β-ketoesters to α,β-unsaturated ketones. The addition gives initially the diketone shown in the margin. Deprotonation of the α-methyl group in the side chain gives an enolate that is perfectly positioned to react with the carbonyl carbon of the cyclohexanone ring. This intramolecular aldol condensation forms a second six-membered ring. Recall (Section 18-11) that the synthesis of six-membered rings by Michael addition followed by aldol condensation is called **Robinson annulation.**

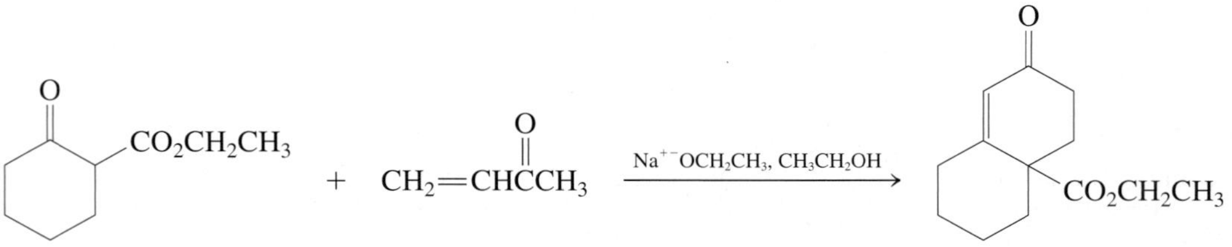

+ $CH_2=CHCCH_3$ (with C=O above) $\xrightarrow{Na^{+-}OCH_2CH_3,\ CH_3CH_2OH}$

Solved Exercise 23-20 | Working with the Concepts: Mechanism of Robinson Annulation

Formulate a detailed mechanism for the preceding transformation.

Strategy
You need to show every step of the process in sequence, starting with the Michael addition and following with the intramolecular aldol condensation. Review the details of these reactions in Section 18-11.

Solution
• Following the cues in the text, we begin with a step-by-step description of the Michael addition: deprotonation of the α-carbon between the ketone and ester carbonyl groups in the cyclic β-ketoester, addition to the β-carbon of the α,β-unsaturated ketone, and protonation of the resulting enolate by a molecule of alcohol.

• The intramolecular aldol reaction follows: Ethoxide deprotonates the α-methyl group, giving the corresponding enolate, which adds to the ring carbonyl carbon. The resulting alkoxide is protonated by alcohol. Ethoxide then initiates loss of a molecule of water by removal of an α-hydrogen and expulsion of hydroxide from the β-carbon, completing the aldol condensation and giving the final product.

Exercise 23-21 | Try It Yourself

Illustrate a Robinson annulation that ensues upon treatment of a mixture of and

3-buten-2-one with sodium ethoxide in ethanol. Draw the intermediate of the initial Michael addition.

In Summary β-Dicarbonyl anions, like ordinary enolate anions, undergo Michael additions to α,β-unsaturated carbonyl compounds. Addition of a β-ketoester to an enone gives a diketone, which can generate six-membered rings by intramolecular aldol condensation (Robinson annulation).

Acyl anion

23-4 | ACYL ANION EQUIVALENTS: PREPARATION OF α-HYDROXYKETONES

Throughout our study of carbonyl compounds, we have been used to thinking of the carbonyl carbon atom as being electrophilic and its α-carbon, in the form of an enol or an enolate, as being nucleophilic. These tendencies govern the rich chemistry of the functional groups built around the C=O unit. However, although the array of transformations available to carbonyl compounds is large, it is not without limits. For example, we have no way of making a direct connection between two carbonyl carbon atoms, because both are electrophilic: Neither can serve as a nucleophilic electron source to attack the other. One may imagine a hypothetical carbonyl-derived nucleophilic species, such as an **acyl anion** (margin), which might add to an aldehyde or ketone to give an α-hydroxyketone, as follows.

A Plausible (but Unfeasible) Synthesis of α-Hydroxyketones

$$:O: \qquad :\ddot{O}: \qquad :O: :\ddot{O}:^{-} \qquad :O: :\ddot{O}H$$
$$\overset{\parallel}{\underset{}{RC:^-}} \xrightarrow[\text{HCR}]{} \overset{\parallel \quad \mid}{RC-CHR} \xrightarrow[-HO]{HOH} \overset{\parallel \quad \mid}{RC-CHR}$$

Acyl anion
(cannot be made)

The ability to generate such an anion would expand the versatility of carbonyl chemistry immensely, making a wide variety of 1,2-difunctionalized systems available, in analogy to the 1,3-difunctional products that we can prepare by using aldol and Claisen condensations. Unfortunately, acyl anions are high-energy species and cannot be generated readily for synthetic applications. Consequently, chemists have explored the construction of other chemical species containing negatively charged carbon atoms that can undergo addition reactions *and later be transformed into carbonyl groups*. These special nucleophiles, called **masked acyl anions** or **acyl anion equivalents,** are the subject of this section.

Exercise 23-22

Why are acyl anions not formed by reaction of a base with an aldehyde? (**Hint:** See Sections 17-5 and 18-1.)

Cyclic thioacetals are masked acyl anion precursors

Cyclic thioacetals are formed by the reaction of dithiols with aldehydes and ketones (Section 17-8). The hydrogens on the carbon positioned between the two sulfur atoms in thioacetals are acidic enough ($pK_a \simeq 31$) to be removed by suitably strong bases, such as alkyllithiums. The negative charge in the conjugate base is stabilized inductively by the highly polarizable sulfur atoms.

Deprotonation of 1,3-Dithiacyclohexane, a Cyclic Thioacetal, by Butyllithium

1,3-Dithiacyclohexane

The anion of 1,3-dithiacyclohexane, as well as the anions of its substituted derivatives, are nucleophilic and add to aldehydes and ketones, furnishing alcohols with an adjacent thioacetal group. The example below shows a sequence that begins with the formation of a substituted 1,3-dithiacyclohexane (also known as a 1,3-dithiane) from an aldehyde. Deprotonation gives the masked acyl anion, which adds to the carbonyl group of 2-cyclohexenone to give an alcohol. Finally, hydrolysis of the thioacetal function in this substance returns the original carbonyl group, now as part of an α-hydroxyketone product.

[Reaction scheme: CH₃CHO (Electrophilic) reacts with HS⌢SH, CHCl₃, HCl, −H₂O (Section 17-8) to give 2-Methyl-1,3-dithiacyclohexane (91%) with H₃C and H substituents. This reacts with CH₃CH₂CH₂CH₂Li, THF to give the Masked acyl anion (Nucleophilic, with CH₃ and Li⁺). This reacts with 1. 2-cyclohexenone, 2. H⁺, H₂O to give an alcohol intermediate with S–S, OH, H₃C (New bond) (70%), which with H₂O, HgCl₂, CaCO₃, CH₃CN gives 1-Acetyl-2-cyclohexen-1-ol (H₃C–C(=O), OH) (93%).]

In this synthesis, the electrophilic carbonyl carbon of the starting aldehyde is transformed into a *nucleophilic* atom, the negatively charged C2 of a 1,3-dithiacyclohexane anion. After this anion is added to the ketone, hydrolysis of the thioacetal function regenerates the original electrophilic carbonyl group. The sequence therefore employs the *reversal of the polarization* of this carbon atom to form the carbon–carbon bond. Reagents exhibiting reverse polarization greatly increase the strategies available to chemists in planning syntheses. We have in fact seen this strategy before: Conversion of a haloalkane into an organometallic (e.g., Grignard) reagent (Section 8-7) reverses the polarity of the functionalized carbon from electrophilic ($^{\delta+}$C–X$^{\delta-}$) to nucleophilic ($^{\delta-}$C–M$^{\delta+}$).

While this section describes the application of dithiacyclohexane anions as masked acyl anions to the preparation of α-hydroxyketones only, it is readily apparent that their alkylation with other electrophilic reagents allows for a general ketone synthesis (see Worked Example 23-26).

Exercise 23-23

Formulate a synthesis of 2-hydroxy-2,4-dimethyl-3-pentanone, beginning with simple aldehydes and ketones and using a 1,3-dithiacyclohexane anion.

Thiazolium salts catalyze aldehyde coupling

Masked acyl anions feature as reactive intermediates in the dimerization of aldehydes to α-hydroxyketones catalyzed by **thiazolium salts.** Thiazole is a heteroaromatic compound (Section 25-4) containing sulfur and nitrogen. The salts are derived from thiazoles by alkylation at nitrogen and contain a resonance-stabilized positive charge (margin).

The thiazolium ion has an unusual feature: a relatively acidic proton located between the two heteroatoms (at C2). The corresponding carbanion is stabilized inductively by the adjacent positive charge that is distributed over both heteroatoms by resonance. The charge-neutral carbene form (Section 12-9) is a minor contributor to the resonance hybrid.

Thiazole

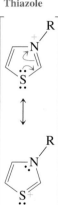

Thiazolium salt

REAL LIFE: NATURE 23-2 | Thiamine: A Natural, Metabolically Active Thiazolium Salt

Thiamine (A = H)

Thiamine pyrophosphate (TPP) A = $-\overset{\displaystyle O}{\underset{\displaystyle OH}{P}}-O-\overset{\displaystyle O}{\underset{\displaystyle OH}{P}}-OH$

The catalytic activity of thiazolium salts in aldehyde dimerization has an analogy in nature: the action of thiamine, or vitamin B_1. Thiamine, in the form of its pyrophosphate, is a coenzyme* for several biochemical transformations, including the breakdown of the sugar glucose (Chapter 24).

In living systems, the **tricarboxylic acid (TCA) cycle,** also known as the **citric acid** or **Krebs[†] cycle,** generates two-thirds of the energy derived from food in higher organisms. The TCA cycle combines acetyl CoA

2-Oxobutanedioic (Oxaloacetic) acid

Citric acid

(Section 19-13) with 2-oxobutanedioic (oxaloacetic) acid to form citric acid. In each turn of the cycle, two carbons of citric acid are oxidized to CO_2, regenerating oxaloacetic acid and giving rise in a coupled process to a molecule of

---

*A molecule required by an enzyme for its biological function (see also Sections 19-13, 22-9, and Real Life 23-1).

[†]Sir Hans Adolf Krebs, 1900–1981, Oxford University, Nobel Prize 1953 (physiology or medicine.)

adenosine triphosphate (ATP), the main energy source in cells. (For the structure of adenosine, see Section 26-9.) Acetyl CoA is the only molecule capable of entering the TCA cycle; therefore, for any food-derived compound to serve as an energy source, it must first be converted into acetyl CoA.

The metabolism of glucose leads to pyruvic (2-oxopropanoic) acid. The role of thiamine is to catalyze the conversion of pyruvic acid into acetyl CoA by first facilitating loss of a molecule of CO_2 and then activating the remaining acetyl fragment as an acyl anion equivalent.

The first transformation begins with addition of the thiamine conjugate base to the α-keto carbon of pyruvate. The adduct thus formed decarboxylates readily, generating CO_2 and a resonance-stabilized zwitterion. This product is an acetyl anion equivalent, which participates in nucleophilic attack on a sulfur atom of a substance called lipoamide. The result is a tetrahedral intermediate from which the thiamine catalyst, having done its job, is eliminated. Finally, in a transesterification-like process (Section 20-4), the thiol group of coenzyme A replaces the dihydrolipoamide (DHLPA) function at the acetyl carbonyl carbon, furnishing acetyl CoA.

Under nonoxidative (anaerobic, or "oxygen debt") conditions, such as those found in muscle tissue undergoing extreme exertion, an alternative process drives the formation of energy-rich ATP: Pyruvic acid is *reduced* enzymatically

Pyruvate Adduct Formation

Deprotonated thiamine pyrophosphate (TPP)

Pyruvate

Addition compound

Decarboxylation

$$
\left[\begin{array}{c} \text{structure} \end{array}\right] \rightleftharpoons \left[\begin{array}{c} \text{structures} \end{array}\right] + CO_2
$$

Acetyl anion equivalent

Acetyl Transfer and Formation of Acetyl CoA

$$ H^+ \ + \ \text{(lipoamide)} \ + \ \text{(structure)} \ \rightleftharpoons \ \text{(structure)} \ \underset{-TPP}{\rightleftharpoons} $$

Lipoamide
(R″ = amide linkage to enzyme)

Tetrahedral intermediate

$$ \text{(S-Acetyl dihydrolipoamide)} \quad \xrightarrow[-\text{DHLPA}]{\text{ACoSH}} \quad CH_3\underset{O}{\overset{O}{C}}SCoA $$

S-Acetyl dihydrolipoamide

to (S)-(+)-2-hydroxypropanoic (lactic) acid. Excessive buildup of lactic acid in muscle tissue causes fatigue and cramps. Removal of the lactic acid from the muscle occurs both by slow diffusion into the bloodstream and by enzyme-catalyzed conversion back into pyruvic acid, once the condition of oxygen debt has been relieved. This need to alleviate oxygen debt is the reason why you breathe hard during and after physical exercise.

The combined effects of oxygen debt and lactic acid buildup in muscles: the aftermath of the the women's 3000m Steeplechase at the London 2012 Olympic Games.

$$
CH_3\overset{O}{\overset{\|}{C}}COOH \quad \underset{}{\overset{\text{Lactic acid dehydrogenase}}{\rightleftharpoons}} \quad \overset{HO}{\underset{H_3C}{\overset{H}{\underset{}{\text{C}}}}}{-}COOH
$$

(S)-(+)-2-Hydroxy-propanoic (lactic) acid

Thiazolium Salts Are Acidic

In the presence of thiazolium salts, aldehydes undergo conversion into α-hydroxy-ketones. An example of this process is the conversion of two molecules of butanal into 5-hydroxy-4-octanone. The catalyst is *N*-dodecylthiazolium bromide, which contains a long-chain alkyl substituent to improve its solubility in organic solvents.

Aldehyde Coupling

Reaction

2 CH₃CH₂CH₂CHO → CH₃CH₂CH₂C(O)−CH(OH)CH₂CH₂CH₃

Butanal

N-Dodecylthiazolium bromide, NaOH, H₂O

76%
5-Hydroxy-4-octanone

The mechanism of this reaction begins with reversible addition of C2 in the deprotonated thiazolium salt to the carbonyl function of an aldehyde.

Mechanism of Thiazolium Salt Catalysis in Aldehyde Coupling

Step 1. Deprotonation of thiazolium salt

Step 2. Nucleophilic attack by catalyst

Step 3. Masked acyl anion formation

Acyl anion equivalent

Step 4. Nucleophilic attack on second aldehyde

Step 5. Liberation of α-hydroxyketone

The product alcohol of step 2 contains a thiazolium unit as a substituent. This group is electron withdrawing and increases the acidity of the adjacent proton. Deprotonation leads to an unusually stable masked acyl anion. Nucleophilic attack by this anion on another molecule of aldehyde, followed by loss of the thiazolium substituent, liberates the α-hydroxyketone. The thiazolium moiety thus released may initiate another catalytic cycle. This process represents another example of organocatalysis—catalysis utilizing exclusively organic species (Real Life 18-1).

Comparison of the thiazolium method for synthesis of α-hydroxyketones with the use of dithiacyclohexane anions is instructive. Thiazolium salts have the advantage in that they are needed in only catalytic amounts. However, their use is limited to the synthesis of

$$\underset{\substack{\text{O} \\ \|}}{\text{R}-\text{C}}-\underset{\substack{\text{OH} \\ |}}{\text{CH}}-\text{R}$$

molecules R–C–CH–R in which the two R groups are identical. The dithiacyclohexane method is more versatile and can be used to prepare a much wider variety of substituted α-hydroxyketones.

Exercise 23-24

Which of the following compounds can be prepared by using thiazolium ion catalysts, and which are accessible only from 1,3-dithiacyclohexane anions? Formulate syntheses of at least two of these substances, one by each route.

In Summary α-Hydroxyketones are available from addition of masked acyl anions to aldehydes and ketones. The conversion of aldehydes into the anions of the corresponding 1,3-dithiacyclohexanes (1,3-dithianes) illustrates the method of reverse polarization. The electrophilic carbon changes into a nucleophilic center, thereby allowing addition to an aldehyde or ketone carbonyl group. Thiazolium ions catalyze the dimerization of aldehydes, again through the transformation of the carbonyl carbon into a nucleophilic atom.

THE BIG PICTURE

Returning to the chemistry of the C=O functional group, we have seen further evidence of the central role that carbonyl compounds play in synthesis, both in the laboratory and in nature. While there are a variety of reactions introduced in this chapter that appear to be new, they have close analogs in processes we have seen before. The most prominent of these is the Claisen condensation, which is the ester counterpart to the aldol condensation of aldehydes and ketones and similarly gives a product with multiple functional groups. Notice how this topic again brings us back to the concepts of acid-base chemistry, especially with regard to the deprotonation of α-carbons whose hydrogens are rendered acidic by either one or (in the case of β-dicarbonyl compounds) two neighboring C=O functions. Decarboxylation is an elimination reminiscent of other reactions with aromatic transition states, including several described in Chapter 15, as well as the rearrangements just discussed in Section 22-7. Finally, the two types of acyl anion equivalents illustrate very different ways to modify a carbonyl carbon such that it can support a negative charge and thus behave as a nucleophile. This is a more sophisticated form of functional group modification, but conceptually it relies on nothing more complicated than an application of the principles discussed at the beginning of Chapter 2: To make a hydrogen acidic, stabilize the negative charge associated with its conjugate base. The innovation in this chemistry is the use of *reversible* modification of the C=O group to accomplish this task, such that the carbonyl function can be restored after its modified, nucleophilic "alter ego" has done its job.

Chapter 24 begins the final phase of our introductory course in organic chemistry, wherein we have a closer look at some of the organic compounds of nature. Fittingly, we start with the carbohydrates, compounds combining the familiar features of alcohols and carbonyl compounds.

WORKED EXAMPLES: INTEGRATING THE CONCEPTS

23-25. Step-by-Step Analysis and Application of the Mixed Claisen Condensation

a. Consider the process shown below. Although it is a mixed Claisen condensation, it forms a single product with the molecular formula $C_{11}H_{14}O_6$ in 80% yield. Propose a structure for the product and rationalize the pathway by which it forms. The first component, diethyl ethanedioate (diethyl oxalate), is used in excess.

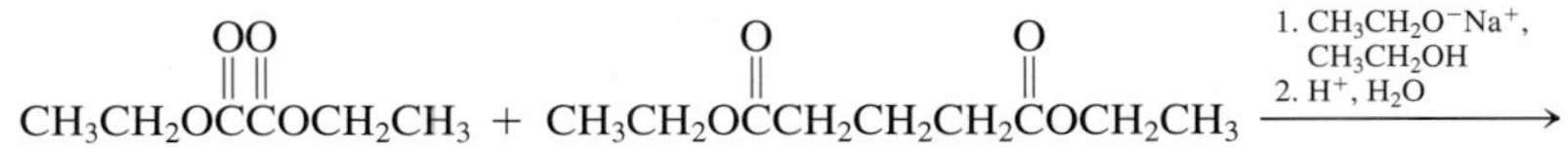

SOLUTION

We begin by noticing that diethyl oxalate lacks α-hydrogens and therefore cannot undergo Claisen condensation on its own. Next we see that the formula of the product equals the sum of the formulas of the two starting materials ($C_{15}H_{26}O_8$) minus *two* molecules of ethanol ($C_4H_{12}O_2$)—a change that would be expected for the product of a process involving two Claisen condensations. Because the oxalate is present in excess, it is logical to assume that the first reaction is a Claisen condensation between one of its ester carbonyl groups and an enolate ion derived from the pentanedioate ester:

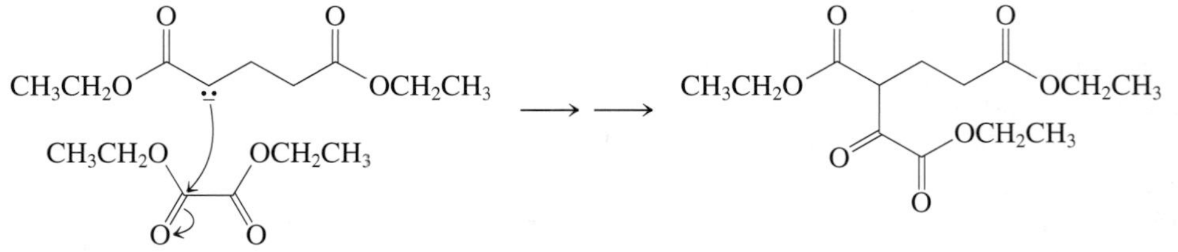

The formula of this intermediate product is $C_{13}H_{20}O_7$—one molecule of ethanol has been eliminated so far. A second condensation must still take place, but it cannot involve addition to another molecule, because the product we are aiming for does not contain enough atoms in its formula. The only option is a *second intramolecular* Claisen condensation, between the remaining ester group of the oxalate-derived portion of this intermediate and the other α-carbon of the pentanedioate-derived moiety:

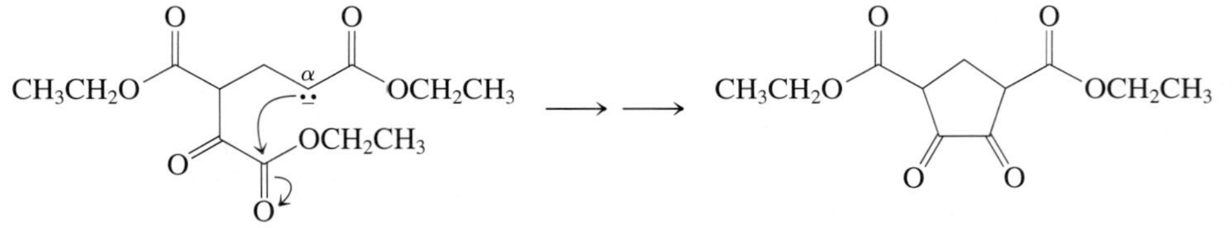

b. Propose a synthesis of the following compound:

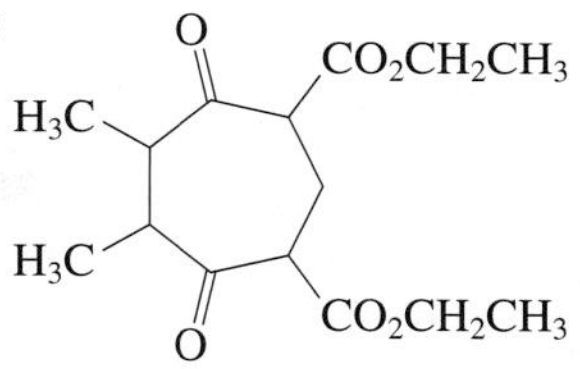

SOLUTION

Using the approach illustrated in (a), we can analyze this target retrosynthetically in the context of ring-formation by double Claisen condensation:

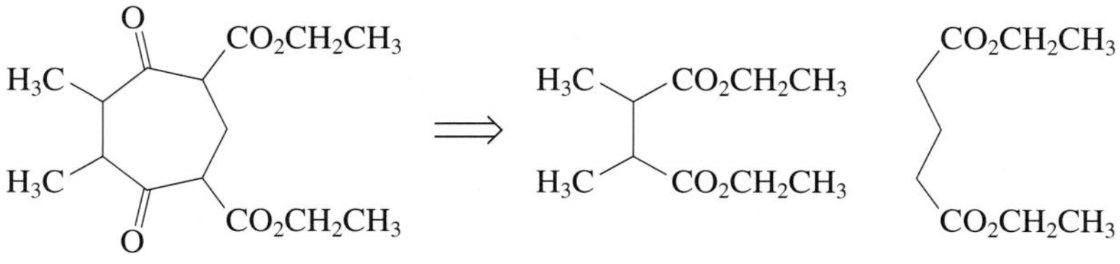

What makes this mixed Claisen condensation feasible? Although both diesters possess α-hydrogens and can therefore give rise to enolate ions upon treatment with base, the first (dimethyl substituted) ester will not undergo successful condensation with a second identical molecule, because the product will lack the necessary additional α-hydrogen that needs to be removed to drive the equilibrium forward. Thus, this diester may be used in excess and will participate only as the carbonyl partner in reaction with enolate ions derived from the pentanedioate. Reaction of the ester mixture with excess ethoxide in ethanol, followed by treatment with aqueous acid, gives rise to the desired product.

23-26. Retrosyntheses of Unsymmetrical Ketones

Propose syntheses of the two ketones in (a) and (b), using the methodology introduced in this chapter. Ground rules for these problems: You may use any organic compounds *provided that they contain no more than six carbons each*. Any inorganic reagents are allowable.

a. 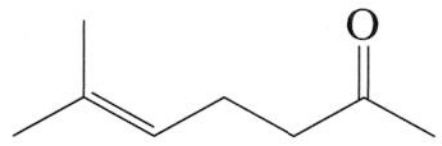 (an intermediate in the synthesis of the widely used perfume essence linalool—see Problem 52 of Chapter 13)

SOLUTION

Where do we start? We note that the target molecule is a 2-alkanone—a *methyl* ketone. As a result, we *may* be able to utilize the acetoacetic ester chemistry introduced in Section 23-2, which is specific in the preparation of methyl ketones. Recall that the acetoacetic ester synthesis affords ketones of the general formulas RCH_2COCH_3 and $RR'CHCOCH_3$, depending on whether we alkylate the starting ester once or twice. The desired product fits the first of these general formulas, so let us examine a retrosynthetic analysis along these lines. The strategic bond (Section 8-9) is the one between the R group and the C3 α-carbon of the ketone:

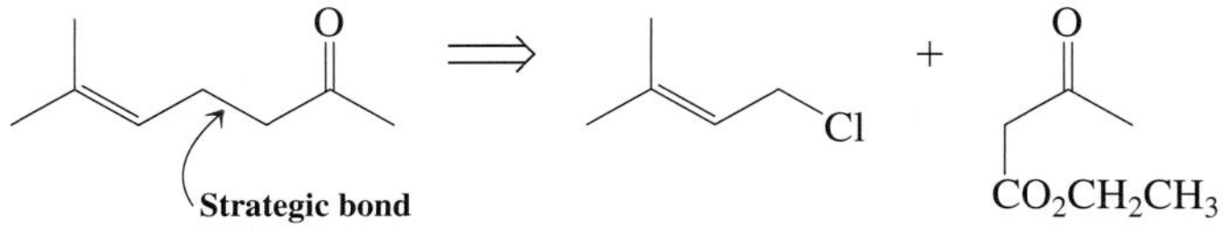

With a general strategy in hand, let us look at its details: A β-ketoester alkylation is necessary and we know (Section 18-4) that alkylations of enolates in general follow the S_N2 mechanism. Is the proposed substrate suitable for this process? We note that the leaving group, Cl, is located on a carbon atom that is both primary and allylic; hence this substance should be an excellent S_N2 substrate (Sections 6-9, 7-9, and 14-3). The final synthesis can therefore be laid out in a manner virtually identical with that of 2-heptanone on page 1051, replacing 1-bromobutane with the allylic chloride:

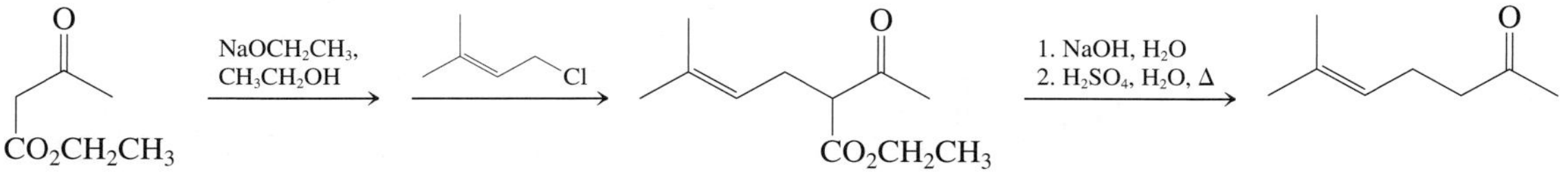

b. 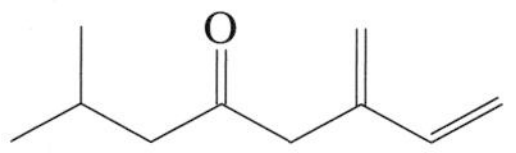 (an intermediate in the synthesis of the sex attractant of the bark beetle *Ips confusus*)

SOLUTION

We note immediately that this target molecule is not a methyl ketone; thus the acetoacetic ester synthesis is not conveniently applicable. We have made simple ketones before by adding organometallic reagents to aldehydes to give secondary alcohols, followed by oxidation (Section 8-9). By this route, the strategic bonds are those to the corresponding alcohol carbon, giving us two potential retrosynthetic disconnections, as shown here.

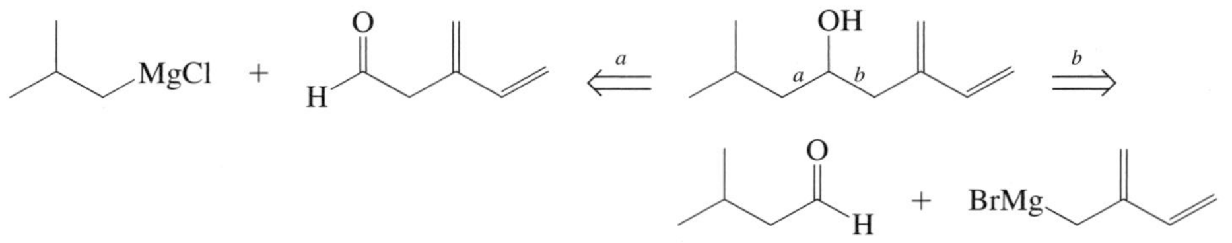

While the preceding provides perfectly good solutions, the problem requires us to employ methodology from the current chapter. Having ruled out acetoacetic ester chemistry, we have an alternative in the use of acyl anion equivalents, for which we saw examples applied to the synthesis of α-hydroxyketones (Section 23-4). Can simple aldehydes and ketones be prepared with these equivalents as well? The answer is yes, by simple alkylation with RX, followed by hydrolysis. Thus,

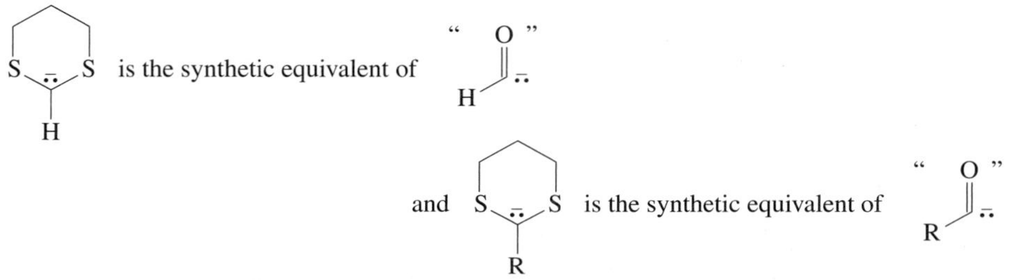

Turning to the problem at hand, we analyze retrosynthetically, identifying the two possible strategic bonds in the disubstituted 1,3-dithiacyclohexane corresponding to our target. The resulting monosubstituted dithianes are readily derived from the corresponding aldehydes by thioacetalization (Section 17-8):

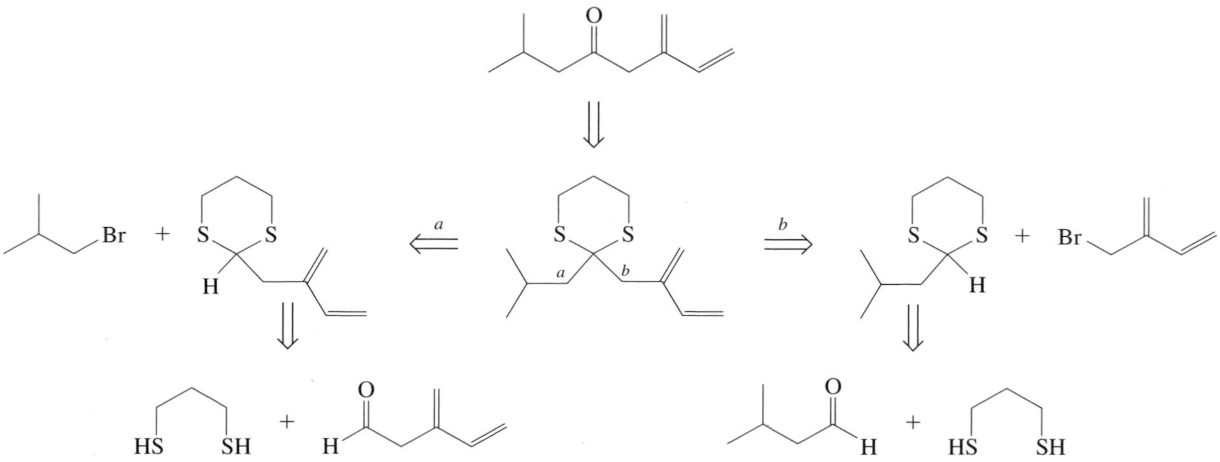

We have now to decide which strategic bond construction, a or b, is better before designing our final answer. Two considerations are important in this decision: (1) ease of bond formation and (2) size and structural and functional complexity of starting materials. Bond a forms in an S_N2 displacement of a branched, primary haloalkane—not the best situation, especially with strongly basic nucleophiles (Section 7-9), and dithiacyclohexanes have pK_a values in the 30s. Bond b employs a primary, allylic substrate—a far better alternative. Approach b is also preferable on the basis of our second consideration:

two even-sized (five-carbon-unit) pieces with separated functional groups. Our synthetic scheme for the solution is therefore as shown.

New Reactions

Synthesis of β-Dicarbonyl Compounds

1. Claisen Condensation (Section 23-1)

Endothermic reaction; equilibrium driven to anion of product by excess base.

2. Dieckmann Condensation (Section 23-1)

3. β-Diketone Synthesis (Section 23-1)

Intramolecular

3-Ketoesters as Synthetic Building Blocks

4. Enolate Alkylation (Section 23-2)

5. 3-Ketoacid Decarboxylation (Section 23-2)

6. Acetoacetic Ester Synthesis of Methyl Ketones (Section 23-2)

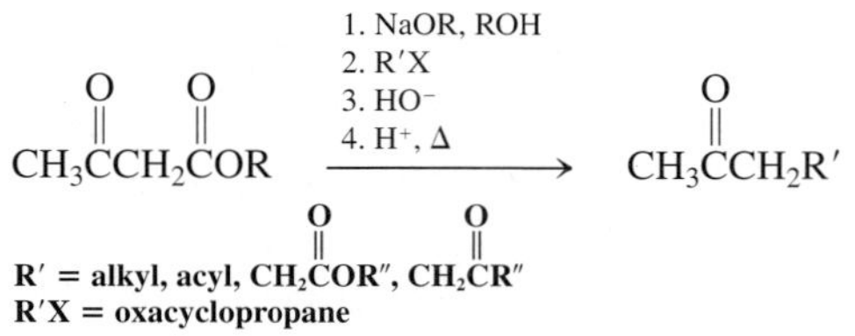

$$\underset{\text{CH}_3\text{CCH}_2\text{COR}}{\overset{\text{O} \quad \text{O}}{||\quad||}} \xrightarrow[\substack{\text{3. HO}^- \\ \text{4. H}^+, \Delta}]{\substack{\text{1. NaOR, ROH} \\ \text{2. R}'\text{X}}} \underset{\text{CH}_3\text{CCH}_2\text{R}'}{\overset{\text{O}}{||}}$$

$$\text{R}' = \text{alkyl, acyl, } \overset{\overset{\text{O}}{||}}{\text{CH}_2\text{COR}''}, \overset{\overset{\text{O}}{||}}{\text{CH}_2\text{CR}''}$$
R′X = oxacyclopropane

7. Malonic Ester Synthesis of Carboxylic Acids (Section 23-2)

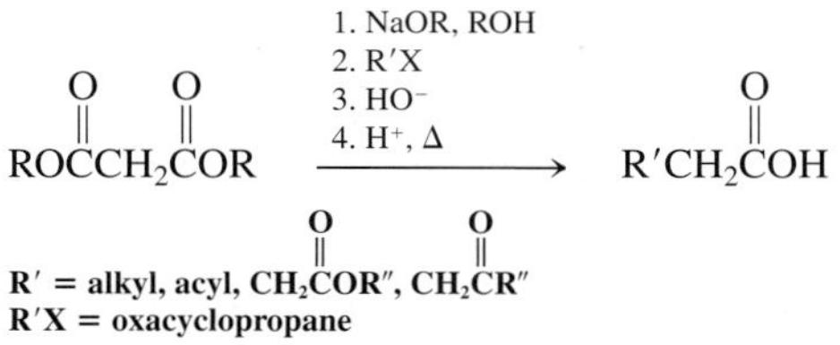

$$\underset{\text{ROCCH}_2\text{COR}}{\overset{\text{O} \quad \text{O}}{||\quad||}} \xrightarrow[\substack{\text{3. HO}^- \\ \text{4. H}^+, \Delta}]{\substack{\text{1. NaOR, ROH} \\ \text{2. R}'\text{X}}} \underset{\text{R}'\text{CH}_2\text{COH}}{\overset{\text{O}}{||}}$$

$$\text{R}' = \text{alkyl, acyl, } \overset{\overset{\text{O}}{||}}{\text{CH}_2\text{COR}''}, \overset{\overset{\text{O}}{||}}{\text{CH}_2\text{CR}''}$$
R′X = oxacyclopropane

8. Michael Addition (Section 23-3)

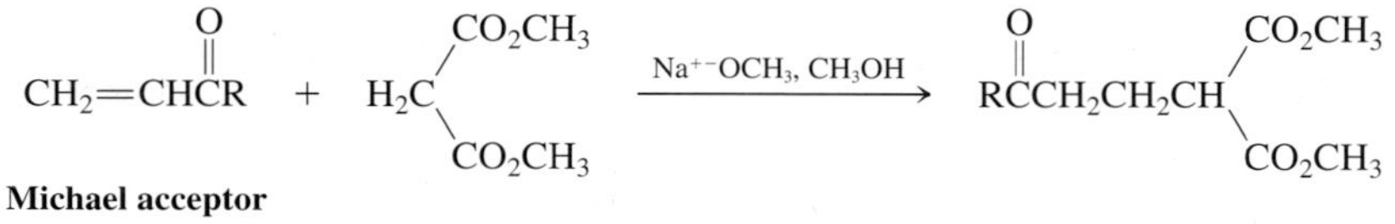

Michael acceptor

Acyl Anion Equivalents

9. 1,3-Dithiacyclohexane (1,3-Dithiane) Anions as Acyl Anion Equivalents (Section 23-4)

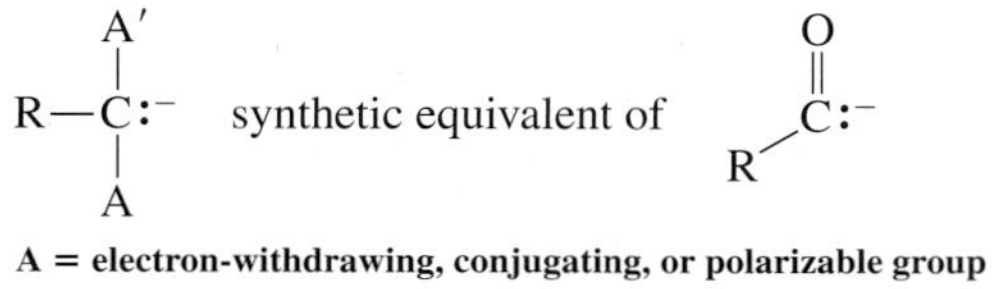

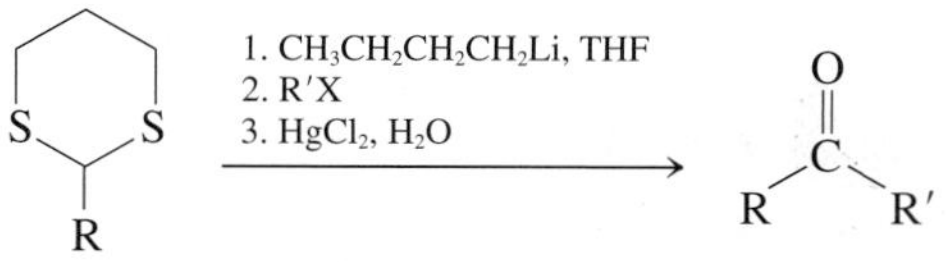

A = electron-withdrawing, conjugating, or polarizable group

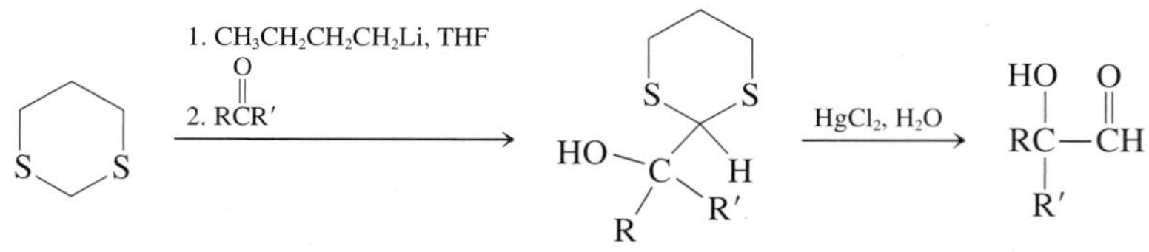

10. Thiazolium Salts in Aldehyde Coupling (Section 23-4)

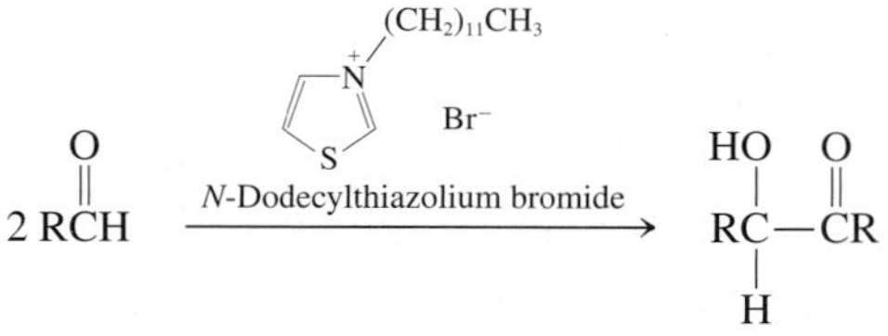

N-Dodecylthiazolium bromide

Important Concepts

1. The **Claisen condensation** is driven by the stoichiometric generation of a stable **β-dicarbonyl anion** in the presence of excess base.

2. **β-Dicarbonyl compounds** contain acidic hydrogens at the carbon between the two carbonyl groups because of the inductive electron-withdrawing effect of the two neighboring carbonyl functions and because the anions resulting from deprotonation are resonance stabilized.

3. Although **mixed Claisen condensations** between esters are usually not selective, they can be so with certain substrates (nonenolizable esters, intramolecular versions, ketones).

4. 3-Ketoacids are unstable; they **decarboxylate** in a concerted process through an aromatic transition state. This property, in conjunction with the nucleophilic reactivity of 3-ketoester anions, allows the synthesis of substituted ketones and acids.

5. Because **acyl anions** are not directly available by deprotonation of aldehydes, they have to be made as masked reactive intermediates or stoichiometric reagents by transformations of functional groups.

Problems

27. Arrange the following compounds in order of increasing acidity. Estimate pK_a values for each.

(a)

(b) CH_3CO_2H

(c) CH_3OH

(d)

(e) CH_3CHO

(f)

(g) $CH_3O_2CCH_2CO_2CH_3$

(h) $CH_3O_2CCO_2CH_3$

28. Give the expected results of the reaction of each of the following molecules (or combinations of molecules) with excess $NaOCH_2CH_3$ in CH_3CH_2OH, followed by aqueous acidic work-up.

(a) $CH_3CH_2CH_2COOCH_2CH_3$

(b) $C_6H_5\overset{\displaystyle CH_3}{\underset{\displaystyle |}{CH}}CH_2COOCH_2CH_3$

(c) $C_6H_5CH_2\overset{\displaystyle CH_3}{\underset{\displaystyle |}{CH}}COOCH_2CH_3$

(d) $CH_3CH_2O\overset{\displaystyle O}{\overset{\|}{C}}(CH_2)_4\overset{\displaystyle O}{\overset{\|}{C}}OCH_2CH_3$

(e) $CH_3CH_2O\overset{\displaystyle O}{\overset{\|}{C}}\overset{}{\underset{\displaystyle \underset{|}{CH_3}}{CH}}(CH_2)_4\overset{\displaystyle O}{\overset{\|}{C}}OCH_2CH_3$

(f) $C_6H_5CH_2CO_2CH_2CH_3 + HCO_2CH_2CH_3$

(g) $C_6H_5CO_2CH_2CH_3 + CH_3CH_2CH_2CO_2CH_2CH_3$

(h)

(i)

29. The following mixed Claisen condensation works best when one of the starting materials is present in large excess. Which of the two starting materials should be present in excess? Why? What side reaction will compete if the reagents are used in comparable amounts?

$$CH_3CH_2\overset{O}{\overset{\|}{C}}OCH_3 + (CH_3)_2CH\overset{O}{\overset{\|}{C}}OCH_3 \xrightarrow{\text{NaOCH}_3,\ \text{CH}_3\text{OH}}$$

$$(CH_3)_2CH\overset{O}{\overset{\|}{C}}\underset{\displaystyle \underset{|}{CH_3}}{CH}\overset{O}{\overset{\|}{C}}OCH_3$$

30. Suggest a synthesis of each of the following β-dicarbonyl compounds by Claisen or Dieckmann condensations.

(a)

(b) $C_6H_5\overset{O}{\overset{\|}{C}}\underset{\displaystyle \underset{|}{C_6H_5}}{CH}\overset{O}{\overset{\|}{C}}OCH_2CH_3$

(c)

(d) $H\overset{O}{\overset{\|}{C}}CH_2\overset{O}{\overset{\|}{C}}OCH_2\overset{O}{\overset{\|}{C}}CH_3$

(e) $C_6H_5\overset{O}{\overset{\|}{C}}CH_2\overset{O}{\overset{\|}{C}}C_6H_5$

(f) $CH_3CH_2OCCH_2COCH_2CH_3$ (with two C=O groups)

(g) 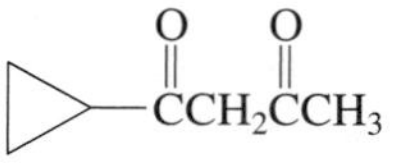 (cyclopropyl)$-CCH_2CCH_3$ (with two C=O groups)

31. Do you think that propanedial, shown below, can be easily prepared by a simple Claisen condensation? Why or why not?

$HCCH_2CH$ (with two C=O groups)

Propanedial

32. Devise a preparation of each of the following ketones by using the acetoacetic ester synthesis.

(a) 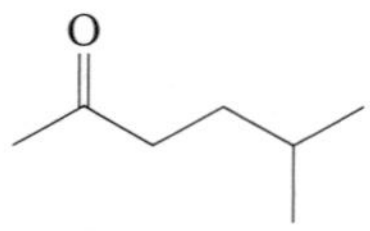 (structure of a ketone)

(b) 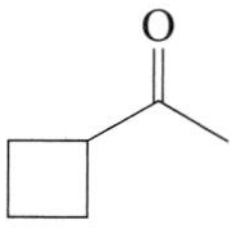 (cyclobutyl methyl ketone structure)

(c) 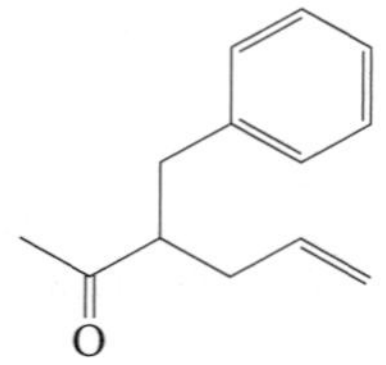 (structure with benzyl group, ketone, and allyl group)

(d) 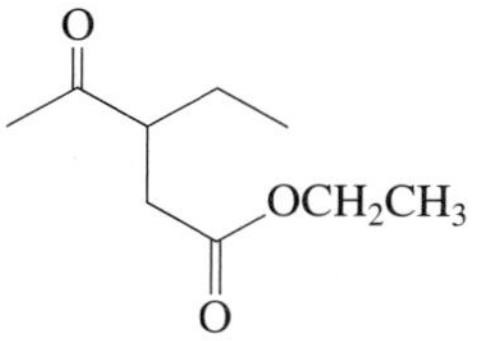 (structure with ketone, ethyl group, and OCH_2CH_3 ester)

33. Devise a synthesis for each of the following four compounds by using the malonic ester synthesis.

(a) 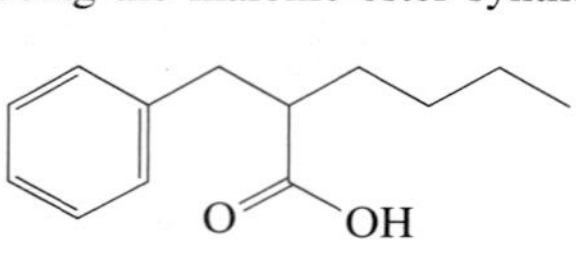 (benzyl-substituted carboxylic acid, $O\,\,OH$)

(b) 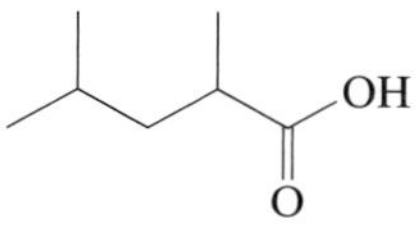 (branched carboxylic acid with OH, O)

(c) 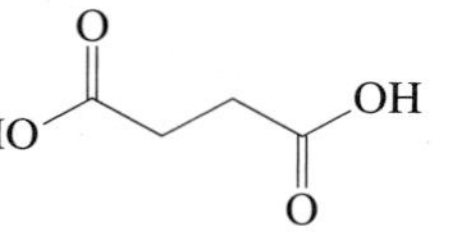 HO—(dicarboxylic acid structure)—OH

(d) 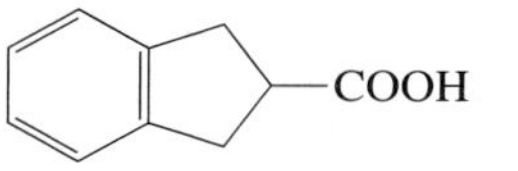 (indane structure with $-COOH$)

34. Use the methods described in Section 23-3, with other reactions if necessary, to synthesize each of the following compounds. In each case, your starting materials should include one aldehyde or ketone and one β-dicarbonyl compound.

(a) 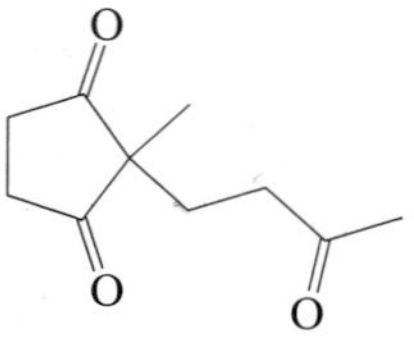 (cyclopentanedione structure with side chain ketone)

(b) 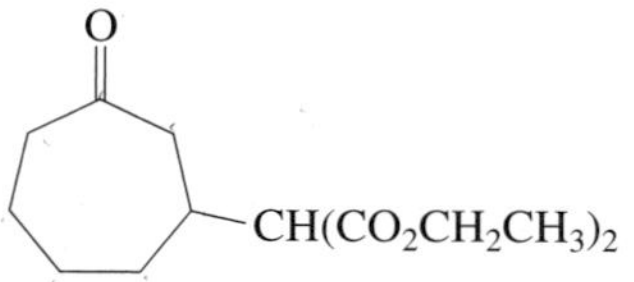 (cycloheptanone structure with $-CH(CO_2CH_2CH_3)_2$)

(c) 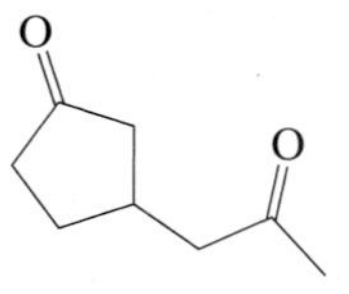 (cyclopentanone with side chain ketone)

(**Hint:** A decarboxylation is necessary.)

35. **CHALLENGE** Carbonic acid, H_2CO_3, $\left(HO-\overset{\displaystyle O}{\underset{}{C}}-OH\right)$, is commonly assumed to be an unstable compound that easily decomposes into a molecule of water and a molecule of carbon dioxide:

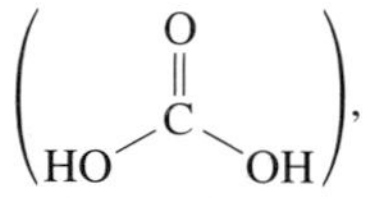

$H_2CO_3 \rightarrow H_2O + CO_2 \uparrow$. Indeed, the evidence of our own experience in opening a container of any carbonated beverage supports this apparently obvious notion. However, in 2000 it was discovered that this assumption is not quite correct: Carbonic acid is actually a perfectly stable, isolatable compound *in the complete absence of water*. Its decomposition, which is a decarboxylation reaction, is strongly catalyzed by water. It is exceedingly difficult to completely exclude moisture without the use of special techniques, which explains why carbonic acid has been such an elusive species to obtain in pure form.

Based on the discussion of the mechanism of decarboxylation of 3-ketocarboxylic acids in Section 23-2, suggest a role for a molecule of water to play in catalyzing the decarboxylation of carbonic acid. (**Hint:** Try to arrange one water and one carbonic acid molecule to give a six-membered ring stabilized by hydrogen bonds, and then see if a cyclic aromatic transition state for decarboxylation exists.)

> **How strong is carbonic acid?** The pK_a value usually quoted, about 6.4, is misleading because it refers to the acidity of the *equilibrium mixture* of aqueous CO_2 and H_2CO_3, which contains mostly molecular CO_2. The true acidity of carbonic acid is much greater, as confirmed by a study in 2009. It places the pK_a of H_2CO_3 at around 3.5, close to that of formic acid (3.6).

36. Based on your answer to Problem 35, predict whether or not water should catalyze the decarboxylation of each of the following compounds. For each case in which your answer is yes, draw the transition state and the final products.

(a) 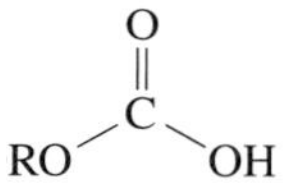 $RO-\overset{\displaystyle O}{\underset{}{C}}-OH$ (alkyl carbonate: a monoester of carbonic acid)

(b) 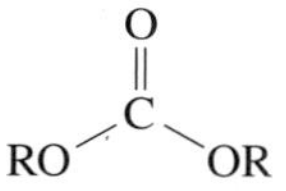 $RO-\overset{\displaystyle O}{\underset{}{C}}-OR$ (dialkyl carbonate: a diester of carbonic acid)

(c) 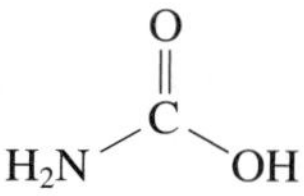 $H_2N-\overset{\displaystyle O}{\underset{}{C}}-OH$ (carbamic acid)

(d) 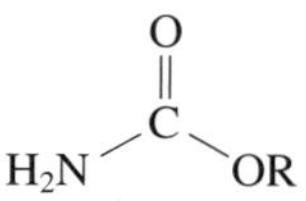 $H_2N-\overset{\displaystyle O}{\underset{}{C}}-OR$ (a carbamate ester)

37. Write out, in full detail, the mechanism of the Michael addition of malonic ester to 3-buten-2-one in the presence of ethoxide ion. Be sure to indicate all steps that are reversible. Does the overall reaction appear to be exo- or endothermic? Explain why only a catalytic amount of base is necessary.

38. Give the likely products of each of the following reactions. All are carried out in the presence of a Pd catalyst, a ligand for the metal such as a phosphine, and heat. (**Hint:** Refer to Section 22-4 for related Pd-catalyzed reactions of halobenzenes with a nucleophile.)

(a) 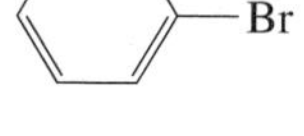 (benzene)$-Br$ $+$ $^-:CH(CO_2CH_2CH_3)_2$

(b) 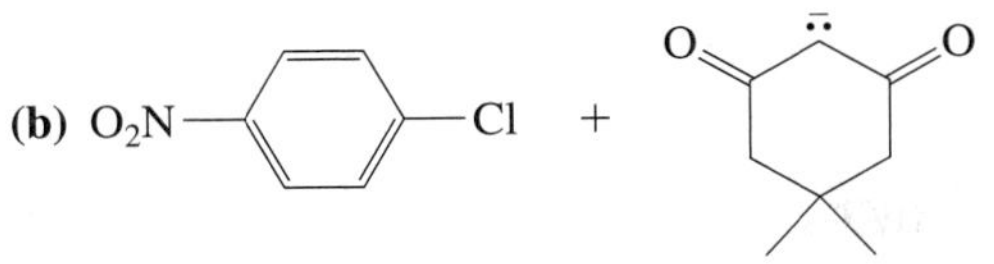 O_2N—⟨benzene ring⟩—Cl   +

39. Based on the mechanism presented for the Pd-catalyzed reaction of a halobenzene with hydroxide ion (Section 22-4), write out a reasonable mechanism for the Pd-catalyzed reaction of Problem 38, part (a).

40. Using the methods described in this chapter, design a multistep synthesis of each of the following molecules, making use of the indicated building blocks as the sources of all the carbon atoms in your final product.

(a) 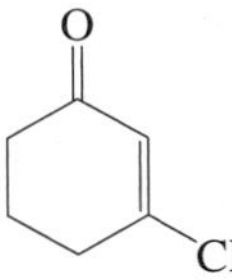 , from $CH_3CO_2CH_2CH_3$ and $CH_3CCH=CH_2$

(b) 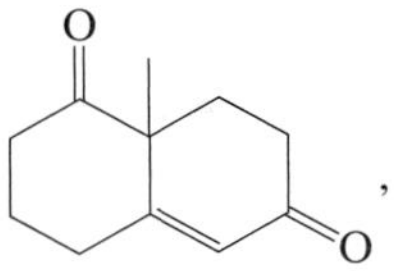 , from CH_3I, $CH_2(CO_2CH_2CH_3)_2$ and

$CH_3CCH=CH_2$ (**Hint:** First make 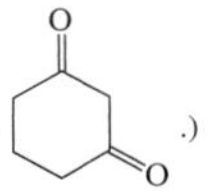 .)

(c) 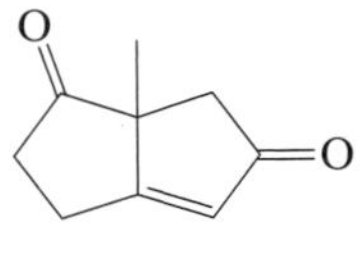 =O , from CH_3I, $CH_2(CO_2CH_2CH_3)_2$ and

$BrCH_2CCH_3$ (**Hint:** First make 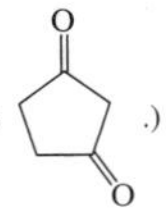 .)

41. Give the products of reaction of the following aldehydes with catalytic *N*-dodecylthiazolium bromide. **(a)** $(CH_3)_2CHCHO$; **(b)** C_6H_5CHO; **(c)** cyclohexanecarbaldehyde; **(d)** $C_6H_5CH_2CHO$.

42. Give the products of the following reactions.

(a) $C_6H_5CHO + HS(CH_2)_3SH \xrightarrow{BF_3}$

(b) Product of (a) + $CH_3CH_2CH_2CH_2Li \xrightarrow{THF}$

What are the results of reaction of the substance formed in (b) with each aldehyde in Problem 41, followed by hydrolysis in the presence of $HgCl_2$?

43. (a) On the basis of the following data, identify unknowns A, found in fresh cream prior to churning, and B, possessor of the characteristic yellow color and buttery odor of butter.
   A: MS: m/z (relative abundance) $= 88(M^+, weak)$, 45(100), and 43(80).
   ^1H NMR: $\delta = 1.36$ (d, $J = 7$ Hz, 3 H), 2.18 (s, 3 H), 3.73 (broad s, 1 H), 4.22 (q, $J = 7$ Hz, 1 H) ppm.
   IR: $\tilde{v} = 1718$ and 3430 cm^{-1}.
   B: MS: m/z (relative abundance) $= 86(17)$ and 43(100).
   ^1H NMR: $\delta = 2.29$ (s) ppm.
   IR: $\tilde{v} = 1708$ cm^{-1}.

(b) What kind of reaction is the conversion of compound A into compound B? Does it make sense that this should take place in the churning of cream to make butter? Explain. **(c)** Outline laboratory syntheses of A and B, starting with molecules containing only two carbons. **(d)** The UV spectrum of A has a λ_{max} at 271 nm, whereas that of B has a λ_{max} at 290 nm. [Tailing of the latter absorption into the visible region of the spectrum (Section 14-11) is responsible for the yellow color of B.] Explain the difference in λ_{max}.

44. Write chemical equations to illustrate all primary reaction steps that can occur between a base such as ethoxide ion and a carbonyl compound such as acetaldehyde. Explain why the carbonyl carbon is not deprotonated to any appreciable extent in this system.

45. The nootkatones are bicyclic ketones that contribute to the flavor and aroma of grapefruits [in addition to 2-(4-methyl-3-cyclohexenyl)-2-propanethiol, Section 9-11]. Nootkatone is also an environmentally friendly insecticide repellent against ticks, mosquitos, and termites. Fill in the missing reagents in the partial synthetic sequence for the preparation of isonootkatone shown below. More than one synthetic step may be needed for each transformation.

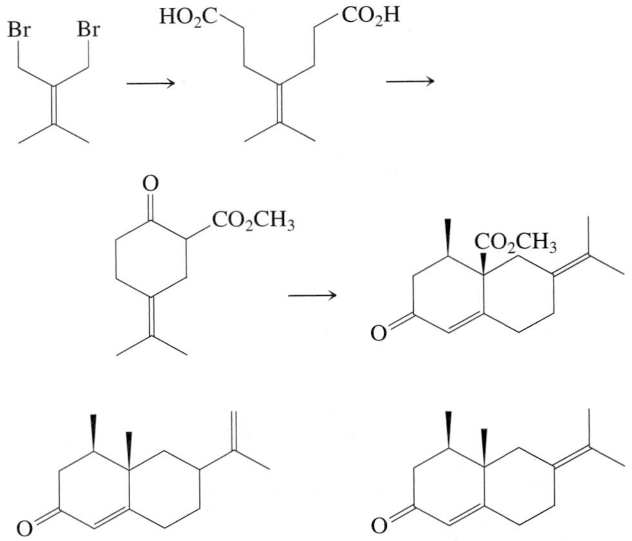

Nootkatone   **Isonootkatone**

46. β-Dicarbonyl compounds condense with aldehydes and ketones that do not undergo self-aldol reaction. The products are α,β-unsaturated dicarbonyl compounds, and the process goes by the colorful name of Knoevenagel condensation. **(a)** An example of a Knoevenagel condensation is given below. Propose a mechanism.

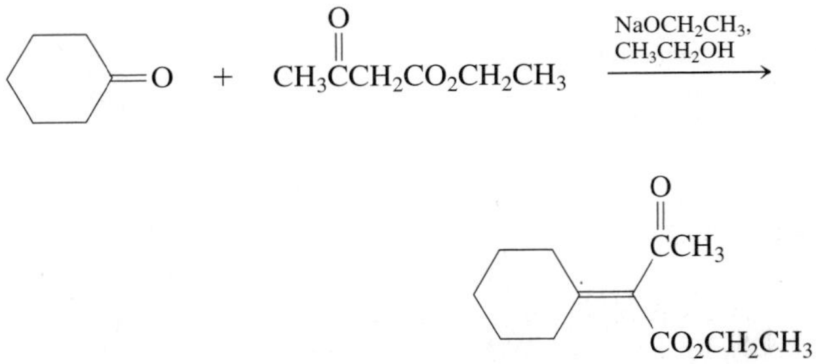

(b) Give the product of the Knoevanagel condensation shown below.

$$\text{C}_6\text{H}_5\text{—CHO} \;+\; \text{CH}_2(\text{CO}_2\text{CH}_2\text{CH}_3)_2 \xrightarrow{\text{NaOCH}_2\text{CH}_3,\;\text{CH}_3\text{CH}_2\text{OH}}$$

(c) The diester shown below is the starting material for the dibromide used in the synthesis of isonootkatone (Problem 45). Suggest a preparation of this diester using the Knoevanagel condensation. Propose a sequence to convert it into the dibromide of Problem 45.

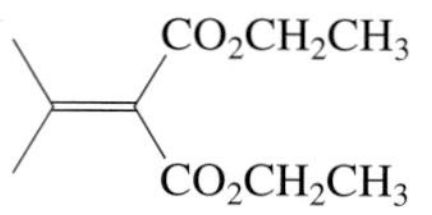

47. The following ketones cannot be synthesized by the acetoacetic ester method (why?), but they can be prepared by a modified version of it. The modification includes the preparation (by Claisen condensation) and use of an appropriate 3-ketoester,

$$\overset{\text{O}\quad\;\;\text{O}}{\underset{}{\text{RCCH}_2\text{COCH}_2\text{CH}_3}},$$ containing an R group that appears in the final product. Synthesize each of the following ketones. For each, show the structure and synthesis of the necessary 3-ketoester as well.

(a)

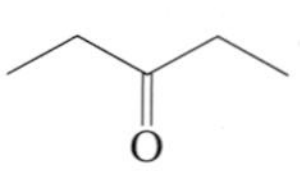

(b)

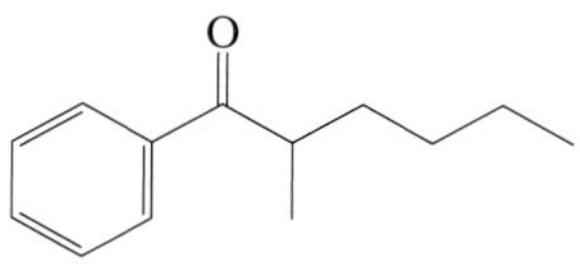

(c)

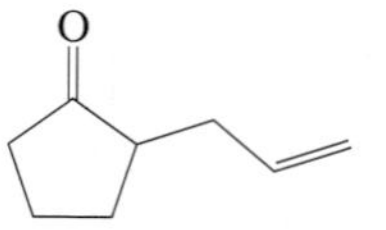

(**Hint:** Use a Dieckmann condensation.)

(d)

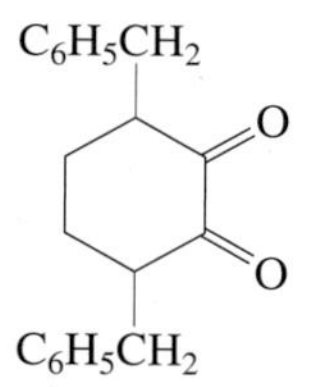

(**Hint:** Use a double Claisen condensation.)

48. Some of the most important building blocks for synthesis are very simple molecules. Although cyclopentanone and cyclohexanone are readily available commercially, an understanding of how they can be made from simpler molecules is instructive. The following are possible retrosynthetic analyses (Section 8-9) for both of these ketones. Using them as a guide, write out a synthesis of each ketone from the indicated starting materials.

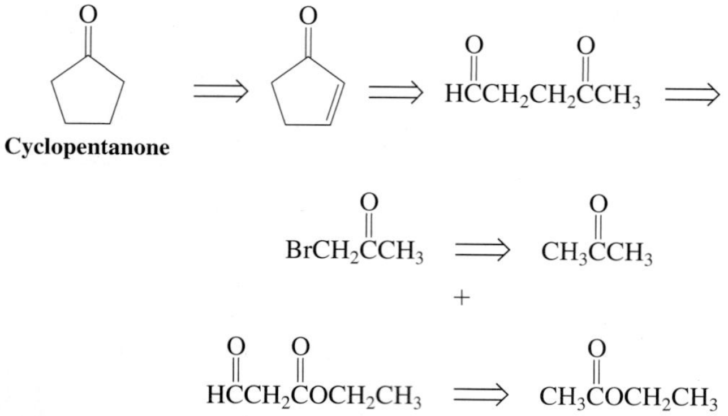

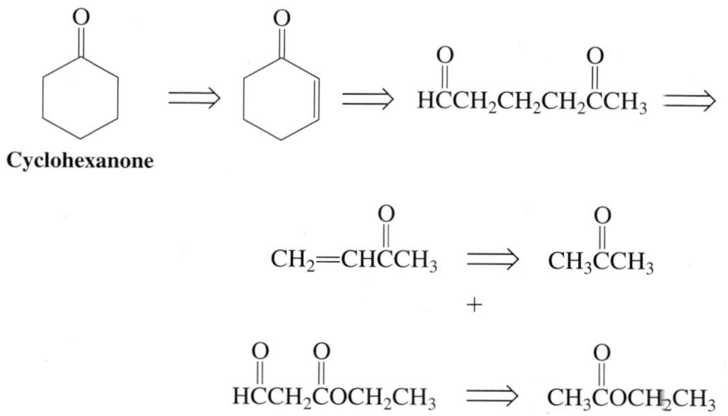

Cyclopentanone

$$\text{BrCH}_2\overset{\text{O}}{\underset{}{\text{CCH}_3}} \;\Longrightarrow\; \text{CH}_3\overset{\text{O}}{\underset{}{\text{CCH}_3}}$$

$$+$$

$$\overset{\text{O}\quad\;\text{O}}{\underset{}{\text{HCCH}_2\text{COCH}_2\text{CH}_3}} \;\Longrightarrow\; \text{CH}_3\overset{\text{O}}{\underset{}{\text{COCH}_2\text{CH}_3}}$$

Cyclohexanone

$$\overset{\text{O}\qquad\qquad\qquad\;\;\text{O}}{\underset{}{\text{HCCH}_2\text{CH}_2\text{CH}_2\text{CCH}_3}} \;\Longrightarrow\;$$

$$\text{CH}_2{=}\text{CHCCH}_3 \;\Longrightarrow\; \text{CH}_3\overset{\text{O}}{\underset{}{\text{CCH}_3}}$$

$$+$$

$$\overset{\text{O}\quad\;\;\text{O}}{\underset{}{\text{HCCH}_2\text{COCH}_2\text{CH}_3}} \;\Longrightarrow\; \text{CH}_3\overset{\text{O}}{\underset{}{\text{COCH}_2\text{CH}_3}}$$

49. A short construction of the steroid skeleton (part of a total synthesis of the hormone estrone) is shown here. Formulate mechanisms for each of the steps. (**Hint:** A process similar to that taking place in the second step is presented in Problem 48 of Chapter 18.)

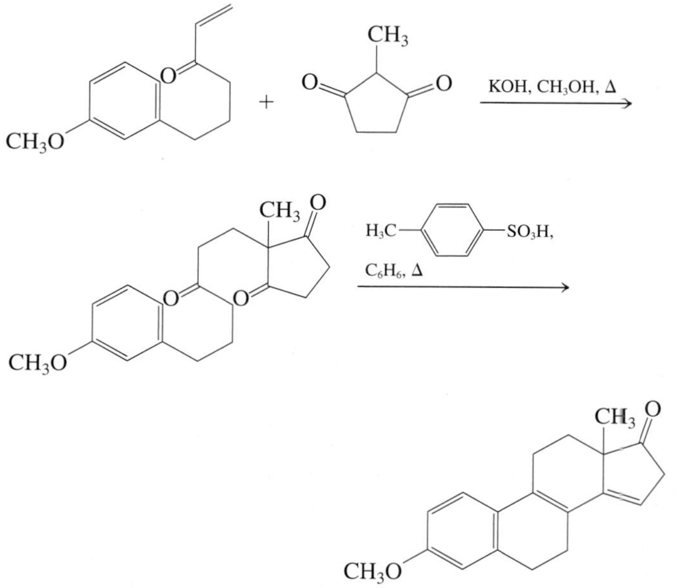

50. **CHALLENGE** Using methods described in Section 23-4 (i.e., reverse polarization), propose a simple synthesis of each of the following molecules.

(a)

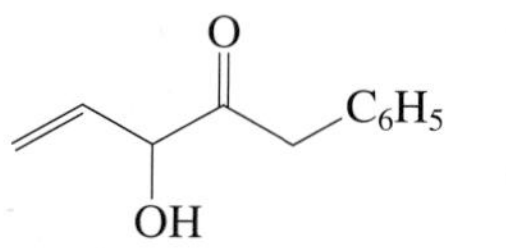

(b)

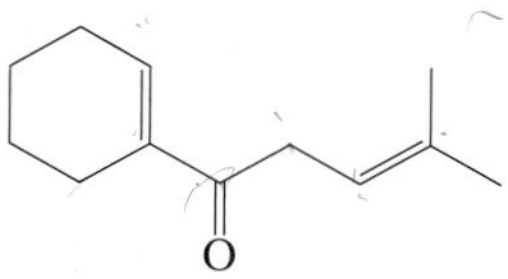

(c)

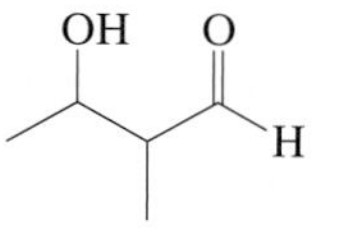

51. **CHALLENGE** Propose a synthesis of ketone C, which was central in attempts to synthesize several antitumor agents. Start with aldehyde A, lactone B, and anything else you need.

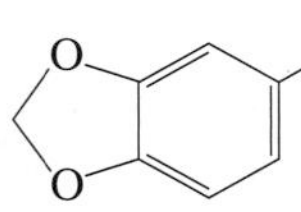

A 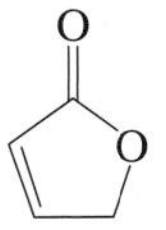   B 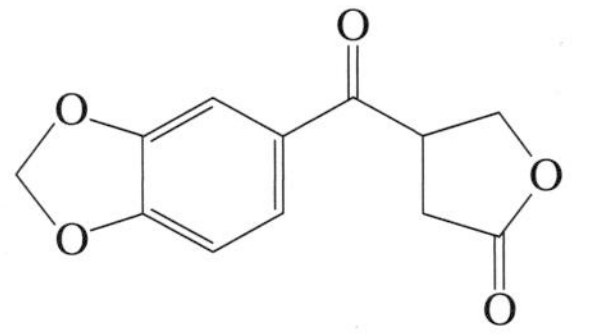   C

Team Problem

52. Split your team in two, each group to analyze one of the following reaction sequences by a mechanism (^{13}C = carbon-13 isotope).

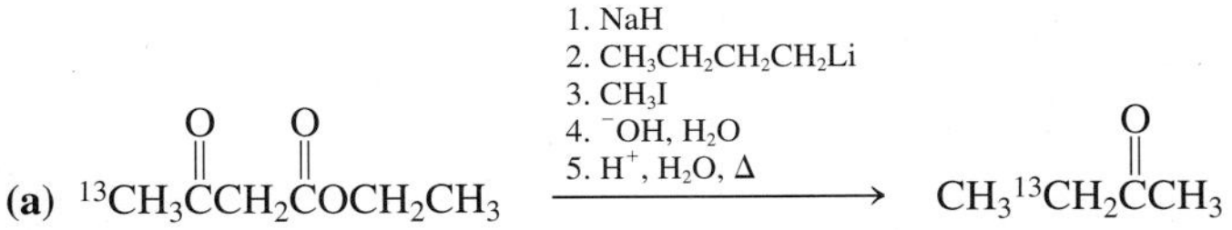

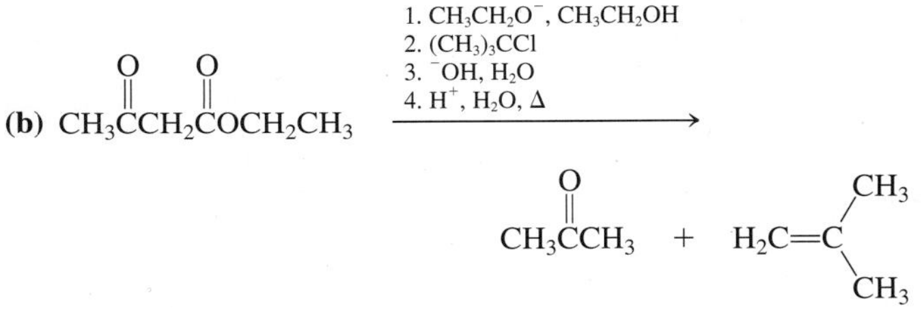

(a) $^{13}CH_3CCH_2COCH_2CH_3$
1. NaH
2. $CH_3CH_2CH_2CH_2Li$
3. CH_3I
4. $^-OH, H_2O$
5. H^+, H_2O, Δ
$\longrightarrow$ $CH_3{}^{13}CH_2CCH_3$

(b) $CH_3CCH_2COCH_2CH_3$
1. $CH_3CH_2O^-, CH_3CH_2OH$
2. $(CH_3)_3CCl$
3. $^-OH, H_2O$
4. H^+, H_2O, Δ
$\longrightarrow$

CH_3CCH_3 + $H_2C=C(CH_3)CH_3$

Reconvene and discuss your results. Specifically, address the position of the ^{13}C label in the product of (a) and the failure to obtain alkylation in (b).

As a complete team, also discuss the mechanism of the following transformation. (**Hint:** For the first step, a minimum of three equivalents of KNH_2 is required.)

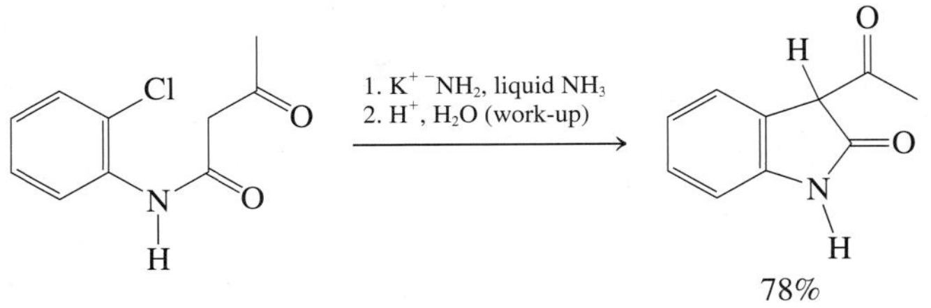

1. $K^+ {}^-NH_2$, liquid NH_3
2. H^+, H_2O (work-up)

78%

Preprofessional Problems

53. Two of the following four compounds are more acidic than CH_3OH (i.e., two of these have K_a *greater* than methanol). Which ones?

$CH_3CH_2OCH_2CH_3$ (cyclic O ether) CH_3CCH_2CHO CF_3CH_2OH
A   B   C   D

(a) A and B; **(b)** B and C; **(c)** C and D; **(d)** D and A; **(e)** D and B.

54. The reaction of ethyl butanoate with sodium ethoxide in CH_3CH_2OH gives

(a) 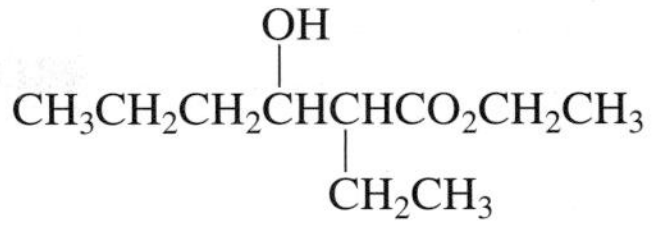 $CH_3CH_2CH_2\overset{OH}{\underset{CH_2CH_3}{CHCHCO_2CH_2CH_3}}$

(b) 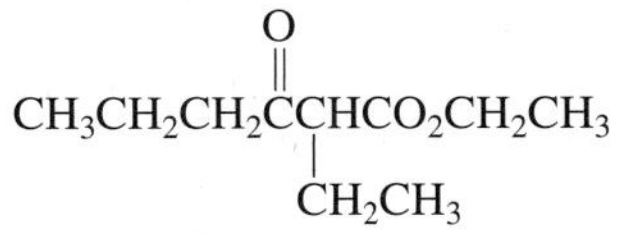 $CH_3CH_2CH_2\overset{O}{C}\underset{CH_2CH_3}{CHCO_2CH_2CH_3}$

(c) $CH_3CH_2\overset{O}{C}CH_2CH_2CO_2CH_2CH_3$

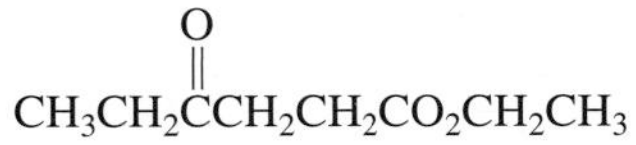

(d) 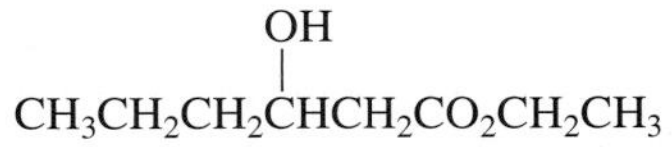 $CH_3CH_2CH_2\overset{OH}{CH}CH_2CO_2CH_2CH_3$

55. When acid A (below) is heated to 230°C, CO_2 and H_2O are evolved and a new compound is formed. Which one?

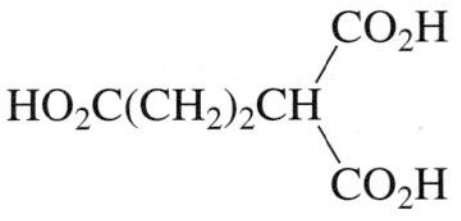

$HO_2C(CH_2)_2\overset{CO_2H}{\underset{CO_2H}{CH}}$

A

(a) 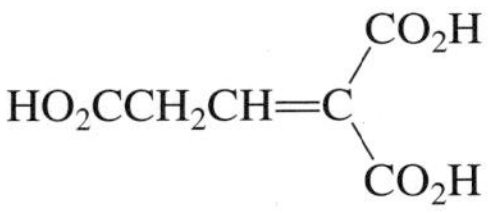 $HO_2CCH_2CH=C\overset{CO_2H}{\underset{CO_2H}{}}$ **(b)** $HO_2CCH_2CH_2CH_2CH_3$

(c) 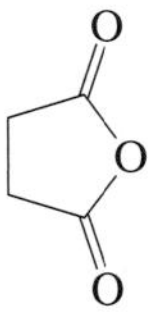   **(d)** $CH_3CH_2CH(CO_2H)_2$ **(e)**

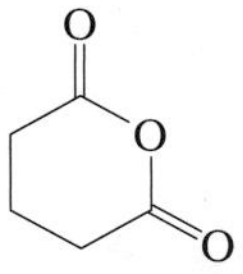

56. A compound with m.p. = −22°C has a parent peak in its mass spectrum at m/z = 113. The ^1H NMR spectrum shows absorptions at δ = 1.2 (t, 3 H), 3.5 (s, 2 H), and 4.2 (q, 2 H) ppm. The IR spectrum exhibits significant bands at $\tilde{v}$ = 1750, 2250, and 3000 cm^{-1}. What is its structure?

(a) 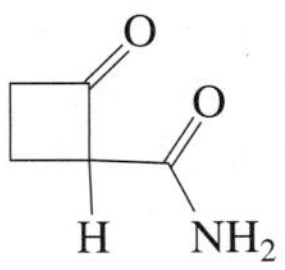   **(b)**

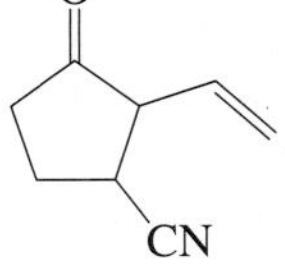

(c) 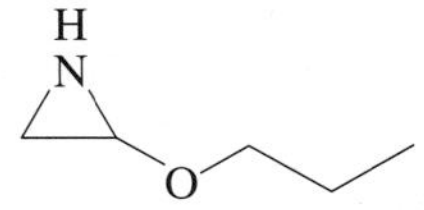   **(d)**

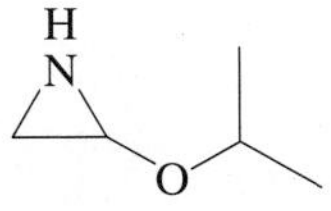

(e)

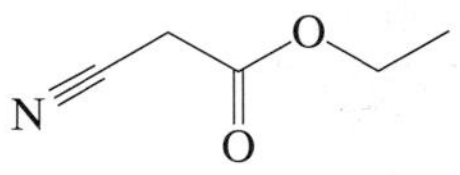

CHAPTER 24 | Carbohydrates
Polyfunctional Compounds in Nature

Take a piece of bread and place it in your mouth. After a few minutes it will begin to taste distinctly sweet, as if you had added sugar to it. Indeed, in a way, this is what happened. The acid and enzymes in your saliva have cleaved the starch in the bread into its component units: glucose molecules. You all know glucose as dextrose or grape sugar. The polymer, starch, and its monomer, glucose, are two examples of carbohydrates.

Carbohydrates are most familiar to us as major contributors to our daily diets, in the form of sugars, fibers, and starches, such as bread, rice, and potatoes. In this capacity, they function as chemical energy-storage systems, being metabolized to water, carbon dioxide, and heat or other energy. Members of this class of compounds give structure to plants, flowers, vegetables, and trees. They also serve as building units of fats (Sections 19-13 and 20-5) and nucleic acids (Section 26-9). Carbohydrates are considered to be **polyfunctional,** because they possess multiple functional groups. Glucose, $C_6(H_2O)_6$, and many related simple members of this compound class form the building blocks of the complex carbohydrates and have the empirical formulas $C_n(H_2O)_n$, essentially hydrated carbon. As a result, they are highly water soluble.

We shall first consider the structure and naming of the simplest carbohydrates—the sugars. We shall then turn our attention to their chemistry, which is governed by the presence of carbonyl and hydroxy functions along carbon chains of various lengths. Sections 24-1 through 24-3 will address aspects of their properties and chemical behavior. Beginning with Section 24-4, we shall discuss reactions of carbohydrates that are useful preparatively, both for elucidating their structure as well as converting them into other substances. We have already seen an example of the biosynthesis of carbohydrates (Real Life 18-1). Finally, we shall describe a sampling of more complex types of carbohydrates found in nature.

In everyday life, when we say the word "sugar," we usually refer to sucrose, the most widely occurring disaccharide in nature. Obtained from sugar beet and sugar cane (illustrated here), sucrose is prepared commercially in pure form in greater quantities than any other chemical substance. In addition, the fibrous "waste" material of the sugar cane plant, composed of more complex carbohydrates, is now widely exploited as a "green" biofuel.

24-1 | NAMES AND STRUCTURES OF CARBOHYDRATES

The simplest carbohydrates are the sugars, or **saccharides.** As chain length increases, the increasing number of stereocenters gives rise to a multitude of diastereomers. Fortunately for chemists, nature deals mainly with only one of the possible series of enantiomers. Sugars are polyhydroxycarbonyl compounds and many form stable cyclic hemiacetals, which affords additional structural and chemical variety.

CHO
|
H—C—OH
|
CH₂OH

2,3-Dihydroxypropanal
(Glyceraldehyde)
(An aldotriose)

CH₂OH
|
C=O
|
CH₂OH

1,3-Dihydroxyacetone
(A ketotriose)

Sugars are classified as aldoses and ketoses

Carbohydrate is the general name for the monomeric (monosaccharides), dimeric (disaccharides), trimeric (trisaccharides), oligomeric (oligosaccharides), and polymeric (polysaccharides) forms of sugar (*saccharum,* Latin, sugar). A **monosaccharide,** or **simple sugar,** is an aldehyde or ketone containing at least two additional hydroxy groups. Thus, the two simplest members of this class of compounds are 2,3-dihydroxypropanal (glyceraldehyde) and 1,3-dihydroxyacetone (margin). **Complex sugars** (Section 24-11) are those formed by the linkage of simple sugars through ether bridges.

Aldehydic sugars are classified as **aldoses;** those with a ketone function are called **ketoses.** On the basis of their chain length, we call sugars **trioses** (three carbons), **tetroses** (four carbons), **pentoses** (five carbons), **hexoses** (six carbons), and so on. Therefore, 2,3-dihydroxypropanal (glyceraldehyde) is an aldotriose, whereas 1,3-dihydroxyacetone is a ketotriose.

Glucose, also known as dextrose, blood sugar, or grape sugar (*glykys,* Greek, sweet), is a pentahydroxyhexanal and hence belongs to the class of aldohexoses. It occurs naturally in many fruits and plants and in concentrations ranging from 0.08 to 0.1% in human blood. A corresponding isomeric ketohexose is **fructose,** the sweetest natural sugar (some synthetic sugars are sweeter), which is also present in many fruits (*fructus,* Latin, fruit) and in honey. Another important natural sugar is the aldopentose **ribose,** a building block of the ribonucleic acids (Section 26-9).

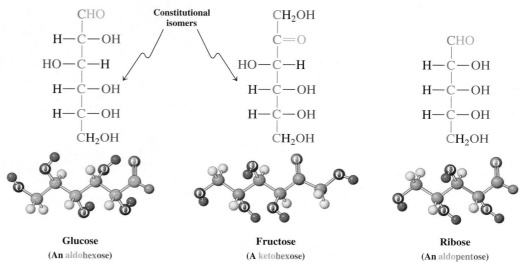

CHO
|
H—C—OH
|
HO—C—H
|
H—C—OH
|
H—C—OH
|
CH₂OH

Glucose
(An aldohexose)

Constitutional isomers

CH₂OH
|
C=O
|
HO—C—H
|
H—C—OH
|
H—C—OH
|
CH₂OH

Fructose
(A ketohexose)

CHO
|
H—C—OH
|
H—C—OH
|
H—C—OH
|
CH₂OH

Ribose
(An aldopentose)

Exercise 24-1

To which classes of sugars do the following monosaccharides belong?

(a)
CHO
|
HCOH
|
HCOH
|
CH₂OH
Erythrose

(b)
CHO
|
HOCH
|
HOCH
|
HCOH
|
CH₂OH
Lyxose

(c)
CH₂OH
|
C=O
|
HOCH
|
HCOH
|
CH₂OH
Xylulose

A **disaccharide** is derived from two monosaccharides by the formation of an ether (usually, acetal) bridge (Sections 17-7 and 24-11). Hydrolysis regenerates the monosaccharides. Ether formation between a mono- and a disaccharide results in a trisaccharide, and repetition of this process eventually produces a natural polymer (polysaccharide). Polysaccharides constitute the framework of cellulose and starch (Section 24-12).

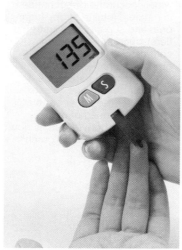

Diabetics need to monitor the glucose levels in their blood carefully, as with the simple "glucometer" shown.

Most sugars are chiral and optically active

With the exception of 1,3-dihydroxyacetone, all the sugars mentioned so far contain at least one stereocenter. The simplest chiral sugar is 2,3-dihydroxypropanal (glyceraldehyde), with one asymmetric carbon. Its dextrorotatory form is found to be *R* and the levorotatory enantiomer *S*, as shown in the Fischer projections of the molecule. Recall that, by convention, the horizontal lines in Fischer projections represent bonds to atoms *above the plane of the page* (Section 5-4).

Fischer Projections of the Two Enantiomers of 2,3-Dihydroxypropanal (Glyceraldehyde)

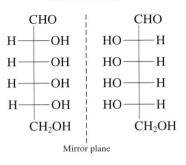

Model Building

Enantiomers

Even though *R* and *S* nomenclature is perfectly satisfactory for naming sugars, an older system is still in general use. It was developed before the absolute configuration of sugars was established, and it relates all sugars to 2,3-dihydroxypropanal (glyceraldehyde). Instead of *R* and *S*, it uses the prefixes D for the (+) enantiomer of glyceraldehyde and L for the (−) enantiomer (Real Life 5-2). Those monosaccharides whose *highest-numbered stereocenter* (i.e., the one farthest from the aldehyde or keto group) has the same absolute configuration as that of D-(+)-2,3-dihydroxypropanal [D-(+)-glyceraldehyde] are then labeled D; those with the opposite configuration at that stereocenter are named L. Two diastereomers that differ in the stereochemistry of *only one stereocenter* (the same in each one) are also called **epimers** (margin).

Designation of a D and an L Sugar

Epimers

The D,L nomenclature divides the sugars into two groups. As the number of stereocenters increases, so does the number of stereoisomers. For example, the aldotetrose 2,3,4-trihydroxybutanal has two stereocenters and hence may exist as four stereoisomers: two diastereomers, each as a pair of enantiomers. The next higher homolog, 2,3,4,5-tetrahydroxypentanal, has three stereocenters, and therefore eight stereoisomers are possible: four diastereomeric pairs of enantiomers. Similarly, 16 stereoisomers (as eight enantiomeric pairs) may be formulated for the corresponding pentahydroxyhexanal.

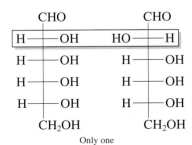

Like many natural products, these diastereomers have common names that are often used, mainly because the complexity of these molecules leads to long systematic names. This chapter will therefore deviate from our usual procedure of labeling molecules systematically. The isomer of 2,3,4-trihydroxybutanal with 2*R*,3*R* configuration is called erythrose; its diastereomer, threose. Note that each of these isomers has two enantiomers, one belonging to the family of the D sugars, its mirror image to the L sugars. The sign of the optical rotation is not correlated with the D and L label (just as in the *R,S* notation: (−) Does not necessarily correspond to *S*, and (+) does not necessarily correspond to *R*; see Section 5-3). For example, D-glyceraldehyde is dextrorotatory, but D-erythrose is levorotatory.

Model Building

Stereoisomeric 2,3,4-Trihydroxybutanals:
Erythrose (2 Enantiomers) and Threose (2 Enantiomers)

Diastereomers

Enantiomers   Enantiomers

```
      CHO              OHC              CHO              OHC
   H—R—OH          HO—S—H           HO—S—H          H—R—OH
   H—R—OH          HO—S—H           H—R—OH          HO—S—H
     CH2OH            HOCH2            CH2OH            HOCH2
     2R,3R            2S,3S            2S,3R            2R,3S
```

D-(−)-Erythrose   L-(+)-Erythrose   D-(−)-Threose   L-(+)-Threose

Mirror   Mirror
plane   plane

As mentioned previously, an aldopentose has three stereocenters and hence $2^3 = 8$ stereoisomers. There are $2^4 = 16$ such isomers in the group of aldohexoses. Why then use the D,L nomenclature even though it designates the absolute configuration of only one stereocenter? Probably because *almost all naturally occurring sugars have the D configuration.* Evidently, somewhere in the structural evolution of the sugar molecules, nature "chose" only one configuration for one end of the chain. The amino acids are another example of such selectivity (Chapter 26).

Figure 24-1 shows Fischer projections of the series of D-aldoses up to the aldohexoses. To prevent confusion, chemists have adopted a standard way to draw these projections: The carbon chain extends vertically and the aldehyde terminus is placed at the top. In this convention, the hydroxy group at the highest-numbered stereocenter (at the bottom) points to the right in all D sugars. Figure 24-2 shows the analogous series of ketoses.

Exercise 24-2

Give a systematic name for **(a)** D-(−)-ribose and **(b)** D-(+)-glucose. Remember to assign the *R* and *S* configuration at each stereocenter.

```
  HO  H  H  OH
HOCH2        CHO
     H  OH
     A
```

Exercise 24-3

Redraw the hashed-wedged line structure of sugar A (shown in the margin) as a Fischer projection and find its common name in Figure 24-1.

In Summary The simplest carbohydrates are sugars, which are polyhydroxy aldehydes (aldoses) and ketones (ketoses). They are classified as D when the highest-numbered stereocenter is *R*, L when it is *S*. Sugars related to each other by inversion at one stereocenter are called epimers. Most of the naturally occurring sugars belong to the D family.

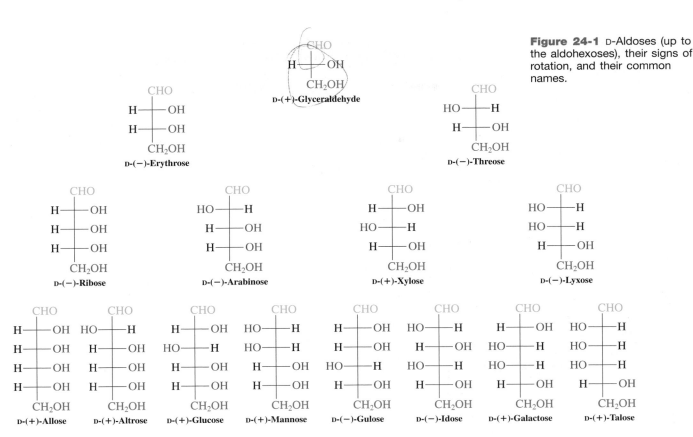

Figure 24-1 D-Aldoses (up to the aldohexoses), their signs of rotation, and their common names.

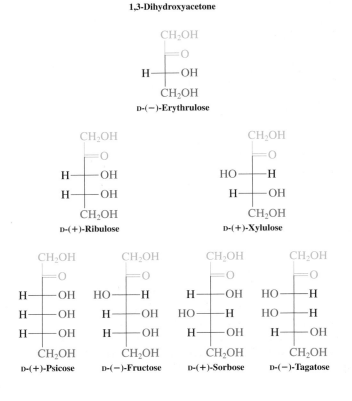

Figure 24-2 D-Ketoses (up to the ketohexoses), their signs of rotation, and their common names.

24-2 | CONFORMATIONS AND CYCLIC FORMS OF SUGARS

Sugars are molecules with multiple functional groups and multiple stereocenters. This structural complexity gives rise to a variety of chemical properties. To enable chemists to focus on the portion of a sugar molecule involved in any given chemical process, several ways of depicting sugars have been developed. We have seen the Fischer representation in Section 24-1; this section shows how to interconvert Fischer and hashed-wedged representations. In addition, it introduces the cyclic isomers that exist in solutions of simple sugars.

Fischer projections depict all-eclipsed conformations

Recall (Section 5-4) that the Fischer projection represents the molecule in an *all-eclipsed arrangement*. It can be translated into an all-eclipsed hashed-wedged line picture.

Model Building

Fischer Projection and Hashed-Wedged Line Structures
for D-(+)-Glucose

A molecular model can help you see that the all-eclipsed form actually possesses a roughly circular shape, as illustrated by the "scrolled" rendition above (center picture). In the subsequent all-eclipsed hashed-wedged line structure, notice that the groups on the *right* of the carbon chain in the original Fischer projection now project *upward* (wedged bonds). From this conformer, we can reach the all-staggered form by 180° rotations of C3 and C5.

Sugars form intramolecular hemiacetals

Sugars are hydroxycarbonyl compounds that should be capable of intramolecular hemiacetal formation (see Section 17-7). Indeed, glucose and the other hexoses, as well as the pentoses, exist as an equilibrium mixture with their cyclic hemiacetal isomers, in which the hemiacetals strongly predominate. In principle, any one of the five hydroxy groups could add to the carbonyl group of the aldehyde. However, three- and four-membered rings are too strained, and five- and six-membered rings are the products, the latter usually dominating. The six-membered ring structure of a monosaccharide is called a **pyranose,** a name derived from *pyran,* a six-membered cyclic ether (see Sections 9-6 and 25-1). Sugars in the five-membered ring form are called **furanoses,** from *furan* (Section 25-3).

Five- and Six-Membered Cyclic Hemiacetal Formation

New stereocenter

Furanose ring   **Furan**

New stereocenter

Pyranose ring   **Pyran**

To depict a D-series sugar correctly in its cyclic form, draw the hashed-wedged line representation of the all-eclipsed structure on the previous page and flip it upside down, as shown below. Rotation of C5 places its hydroxy group in position to form a six-membered cyclic hemiacetal by addition to the C1 aldehyde carbon. Similarly, a five-membered ring can be made by rotation of C4 to place its OH group in position to bond to C1. The resulting drawings depict the ring in a somewhat artificial way: It is shown as if it were planar and from a perspective in which the bottom-most ring bonds (between C2 and C3 in the structures below) are interpreted as being *in front* of the plane of the page. As we shall see shortly, a slightly modified version of this type of drawing is quite commonly used to depict the cyclic forms of sugars.

Model Building

Animation

ANIMATED MECHANISM: Cyclic hemiacetal formation by glucose

Cyclic Hemiacetal Formation by Glucose

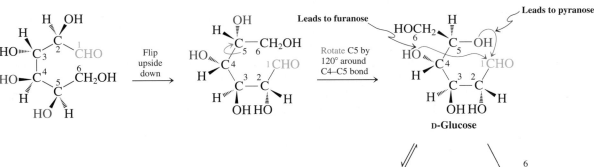

Leads to furanose   **Leads to pyranose**

D-Glucose

D-Glucofuranose
(Less stable, 0.4%)

New stereocenter

D-Glucopyranose
(More stable, 99.6%)

New stereocenter

(Groups on the *right* in the original Fischer projection [circled] *point downward* in the cyclic hemiacetal except at C5, which has been rotated)

Exercise 24-4

Draw the Fischer projection of L-(−)-glucose and illustrate its transformation into the corresponding six-membered cyclic hemiacetal.

In contrast with glucose, which exists primarily as the pyranose, fructose forms both fructopyranose and fructofuranose in a rapidly equilibrating 68:32 mixture.

Model Building

Really? High-fructose corn syrup (HFCS), was introduced in the 1980s as a cheaper alternative to table sugar (sucrose; Section 24-11) and is used in many packaged foods and beverages. The label "high-fructose" is commonly misinterpreted. HFCS is produced from corn to yield a product that is almost entirely glucose. Because glucose is less sweet than sucrose, corn syrup is treated with enzymes that isomerize the glucose to fructose (which is sweeter than sucrose), resulting in HFCS 42 (42% fructose), used in some beverages, processed foods, cereals, and baked goods. Further enrichment gives the ingredient in soft drinks, HFCS 55 (55% fructose), which is about as sweet as sugar. Compare this to sucrose, which is a disaccharide composed of fructose and glucose held together by a mutual acetal linkage. This labile bond is hydrolyzed in the intestinal tract to its component sugars, equivalent to HFCS 50. In short, the two sweeteners are in essence the same, at least chemically.

Cyclic Hemiacetal Formation by Fructose

D-Fructose

D-Fructofuranose
(Furanose = five-membered
cyclic sugar)
32%

D-Fructopyranose
(Pyranose = six-membered
cyclic sugar)
68%

New stereocenter (anomeric carbon)

Note that, upon cyclization, the carbonyl carbon turns into a new stereocenter. As a consequence, hemiacetal formation leads to *two* new compounds, two diastereomers (epimers) differing in the configuration of the hemiacetal group. If that configuration is *S* in a D-series sugar, that diastereomer is labeled α; when it is *R* in a D sugar, the isomer is called β. Hence, for example, D-glucose may form α- or β-D-glucopyranose or -furanose. Because this type of diastereomer formation is unique to sugars, such isomers have been given a separate name: **anomers.** The new stereocenter is called the **anomeric carbon.**

Solved Exercise 24-5 | **Working with the Concepts: Relationships Between Sugar Stereoisomers**

The anomers α- and β-glucopyranose should form in equal amounts because they are enantiomers. True or false? Explain your answer.

Strategy
Be sure that you understand the meanings of enantiomer and diastereomer, and how physical properties of molecules relate to those stereoisomeric relationships, before you answer this question about anomers.

Solution
False! Anomers differ in configuration at *only* the anomeric carbon (C1). The configurations at the remaining stereocenters (C2 through C5) are the same in α-glucopyranose as they are in β-glucopyranose. Anomers are therefore diastereomers—not enantiomers—and should not form in equal amounts. Enantiomers differ from each other in configuration at *every* stereocenter.

Exercise 24-6 | **Try It Yourself**

How is α-L-glucopyranose related to α-D-glucopyranose? What is the configuration of the anomeric carbon in α-L-glucopyranose, R or S? Answer the same question for β-L-glucopyranose and β-D-glucopyranose.

Fischer, Haworth, and chair cyclohexane projections help depict cyclic sugars

How can we represent the stereochemistry of the cyclic forms of sugars? One approach uses Fischer projections. We simply draw elongated lines to indicate the bonds formed upon cyclization, preserving the basic "grid" of the original formula. Be careful not to confuse the three new apices of the ensuing rectangle with carbon atoms.

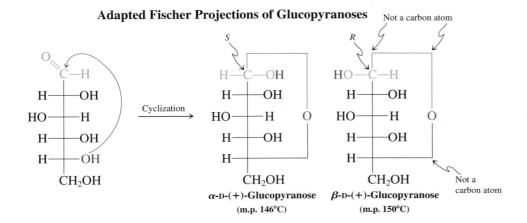

Adapted Fischer Projections of Glucopyranoses   Not a carbon atom

α-D-(+)-Glucopyranose (m.p. 146°C) β-D-(+)-Glucopyranose (m.p. 150°C)   Not a carbon atom

> In the Fischer projection of the α form of a D sugar, the anomeric OH points toward the *right*. In the Fischer projection of the β form, the anomeric OH is on the *left*.

Haworth* projections more accurately represent the real three-dimensional structure of the sugar molecule. The cyclic ether is written in line notation as a pentagon or a hexagon, the anomeric carbon (in a D sugar) placed on the right, and the ether oxygen put on top. The substituents located above or below the ring are attached to vertical lines. In relating the Haworth projection to a three-dimensional structure, we interpret the ring bond at the bottom (between C2 and C3) to be *in front* of the plane of the paper, and the ring bonds containing the oxygen are understood to be in back.

Haworth Projections

α-D-(−)-Erythrofuranose α-D-(+)-Glucopyranose β-D-(+)-Glucopyranose

> Groups on the *right* in the Fischer projection point *downward* in the Haworth formula.

In a Haworth projection, the α anomer has the OH group at the anomeric carbon pointing down, whereas the β anomer has it pointing up.

---

*Sir W. Norman Haworth (1883–1950), University of Birmingham, England, Nobel Prize 1937 (chemistry).

Exercise 24-7

Draw Haworth projections of (**a**) α-D-fructofuranose; (**b**) β-D-glucofuranose; and (**c**) β-D-arabinopyranose.

Haworth projections are used extensively in the literature, but here, to make use of our knowledge of conformation (Sections 4-3 and 4-4), the cyclic forms of sugars will be presented mostly as envelope (for furanoses) or chair (for pyranoses) conformations (Figures 4-4 and 4-6). As in Haworth notation, the ether oxygen will be placed usually top right and the anomeric carbon at the right vertex of the envelope or chair.

Conformational Pictures of Glucofuranose and -pyranose

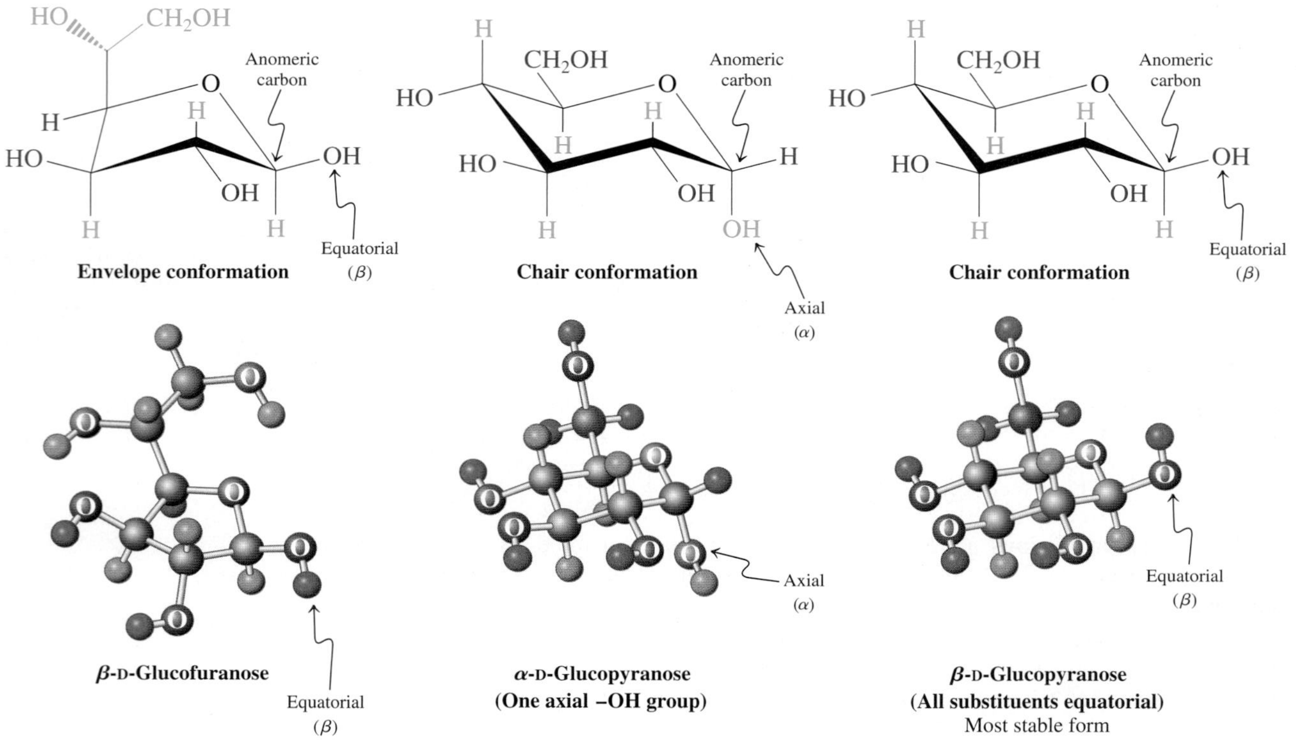

Envelope conformation

Anomeric carbon

Equatorial (β)

Chair conformation

Anomeric carbon

Axial (α)

Chair conformation

Anomeric carbon

Equatorial (β)

β-D-**Glucofuranose**

Equatorial (β)

α-D-**Glucopyranose**
(One axial –OH group)

Axial (α)

β-D-**Glucopyranose**
(All substituents equatorial)
Most stable form

Equatorial (β)

Although there are exceptions, most aldohexoses adopt the chair conformation that places the bulky hydroxymethyl group at the C5 terminus in the equatorial position. For glucose, this preference means that, in the α form, four of the five substituents can be equatorial, and one is forced to lie axial; in the β form, *all* substituents can be equatorial. This situation is unique for glucose; the other seven D aldohexoses (see Figure 24-1) contain one or more axial substituents.

Exercise 24-8

Using the values in Table 4-3, estimate the difference in free energy between the all-equatorial conformer of β-D-glucopyranose and that obtained by ring flip (assume that $\Delta G^\circ_{CH_2OH} = \Delta G^\circ_{CH_3} = 1.7$ kcal mol^{-1} and that the ring oxygen mimics a CH$_2$ group).

In Summary Hexoses and pentoses can take the form of five- or six-membered cyclic hemiacetals. These structures rapidly interconvert through the open-chain polyhydroxy-aldehyde or ketone, with the equilibrium usually favoring the six-membered (pyranose) ring.

24-3 | ANOMERS OF SIMPLE SUGARS: MUTAROTATION OF GLUCOSE

Glucose precipitates from concentrated solutions at room temperature to give crystals that melt at 146°C. Structural analysis by X-ray diffraction reveals that these crystals contain only the α-D-(+)-glucopyranose anomer (Figure 24-3). When crystalline α-D-(+)-glucopyranose is dissolved in water and its optical rotation measured immediately, a value $[\alpha]_D^{25°C} = +112$ is obtained. Curiously, this value decreases with time until it reaches a constant +52.7. The process that gives rise to this effect is the interconversion of the α and β anomers.

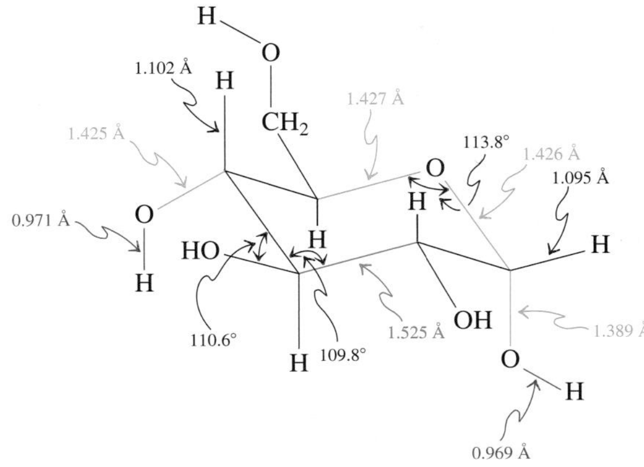

Figure 24-3 Structure of α-D-(+)-glucopyranose, with selected bond lengths and angles.

In solution, the α-pyranose rapidly establishes an equilibrium (in a reaction that is catalyzed by acid and base; see Section 17-7) with a small amount of the open-chain aldehyde, which in turn undergoes reversible ring closure to the β anomer.

Interconversion of Open-Chain and Pyranose Forms of D-Glucose

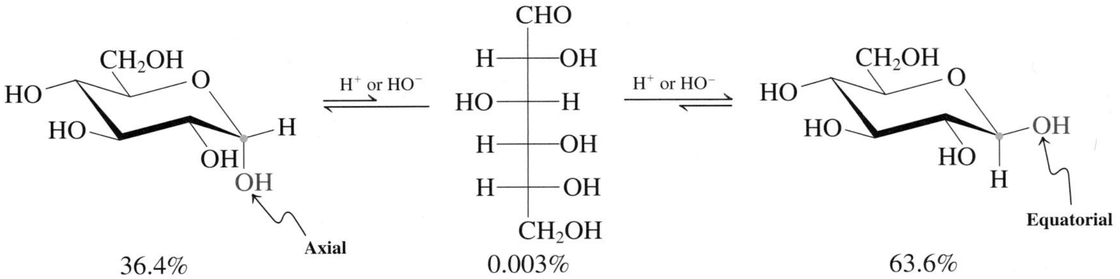

36.4%	0.003%	63.6%
α-D-(+)-Glucopyranose	**Aldehyde form**	**β-D-(+)-Glucopyranose**
($[\alpha]_D^{25°C} = +112$)		($[\alpha]_D^{25°C} = +18.7$)

The β form has a considerably lower specific rotation (+18.7) than its anomer; therefore, the observed α value in solution decreases. Similarly, a solution of the pure β anomer (m.p. 150°C, obtainable by crystallizing glucose from acetic acid) gradually increases its specific rotation from +18.7 to +52.7. At this point, a final equilibrium has been reached, with 36.4% of the α anomer and 63.6% of the β anomer. The change in optical rotation observed when a sugar equilibrates with its anomer is called **mutarotation** (*mutare*, Latin, to change). Interconversion of α and β anomers is a general property of sugars. This includes all monosaccharides capable of existing as cyclic hemiacetals.

Exercise 24-9

An alternative mechanism for mutarotation bypasses the aldehyde intermediate and proceeds through oxonium ions. Formulate it.

NMR Spectra of Glucose

The composition of glucose in water is readily measured by ^{1}H NMR spectroscopy. The five hydrogens at C2–C6 bearing only one attached oxygen appear as overlapping multiplets at $\delta = 3$–4 ppm and are difficult to interpret. However, the unique anomeric hydrogens with two appended oxygens show up at lower field (see, for example, Table 10-3), for the α anomer at $\delta = 5.13$ (d, $J = 3$ Hz) ppm, and for the β anomer at $\delta = 4.55$ (d, $J = 8$ Hz) ppm, in the ratio of ~1:2.

Solved Exercise 24-10	Working with the Concepts: Calculating Specific Rotations of Diastereomeric Sugar Mixtures

Calculate the equilibrium ratio of α- and β-glucopyranose (which has been given in the text) from the specific rotations of the pure anomers and the observed specific rotation at mutarotational equilibrium.

Strategy

The specific rotation of an equilibrium mixture is the average of the rotations of the contributing isomers—however, *weighted* by the respective mole fractions in the mixture. All the raw data you need are given in the problem statement.

Solution

• Let's designate the mole fraction of the α form as x_α and the mole fraction of its β isomer as x_β. Their respective specific rotations are given as $+112$ (α) and $+18.7$ (β), whereas the equilibrium mixture exhibits a value of $+52.7$. Thus, we have

$$+52.7 = (+112)(x_\alpha) + (+18.7)(x_\beta)$$

• By the definition of mole fraction, $x_\alpha + x_\beta = 1$, so we can substitute one for the other in the above equation. Solving gives $x_\alpha = 0.364$ and $x_\beta = 0.636$. Therefore the equilibrium ratio is $x_\alpha/x_\beta = (0.636)/(0.364) = 1.75$.

Exercise 24-11	Try It Yourself

The pure α and β forms of D-galactose exhibit $[\alpha]_D$ values of $+150.7$ and $+52.8$, respectively. The equilibrium mixture after mutarotation in water shows a specific rotation of $+80.2$. Calculate the composition of the equilibrium mixture.

Exercise 24-12	

By using Table 4-3, estimate the difference in energy between α- and β-glucopyranose at room temperature (25°C). Then calculate it by using the equilibrium percentage.

In Summary The hemiacetal carbon (anomeric carbon) can have two configurations: α or β. In solution, the α and β forms of the sugars are in equilibrium with each other. The equilibration can be followed by starting with a pure anomer and observing the changes in specific rotation, a phenomenon also called mutarotation.

24-4	POLYFUNCTIONAL CHEMISTRY OF SUGARS: OXIDATION TO CARBOXYLIC ACIDS

Simple sugars exist as isomers: the open-chain structure and the α and β anomers of various cyclic forms. Because all of these isomers equilibrate rapidly, the relative rates of their individual reactions with various reagents determine the product distribution of a particular transformation. We can therefore divide the reactions of sugars into two groups, those of the linear form and those of the cyclic forms, *because the two structures contain different functional groups*. Although the two forms may sometimes react competitively, we see in this section that reactions of aldoses with oxidizing agents take place at the aldehyde moiety of the open-chain form, not the hemiacetal function of the cyclic isomers.

Fehling's and Tollens's tests detect reducing sugars

Because they are polyfunctional compounds, the open-chain monosaccharides undergo the reactions typical of each of their functional groups. For example, aldoses contain the oxidizable formyl group and therefore respond to the standard oxidation tests such as exposure to Fehling's or Tollens's solutions (Section 17-14). The α-hydroxy substituent in ketoses is similarly oxidized.

Results of Fehling's and Tollens's Tests on Aldoses and Ketoses

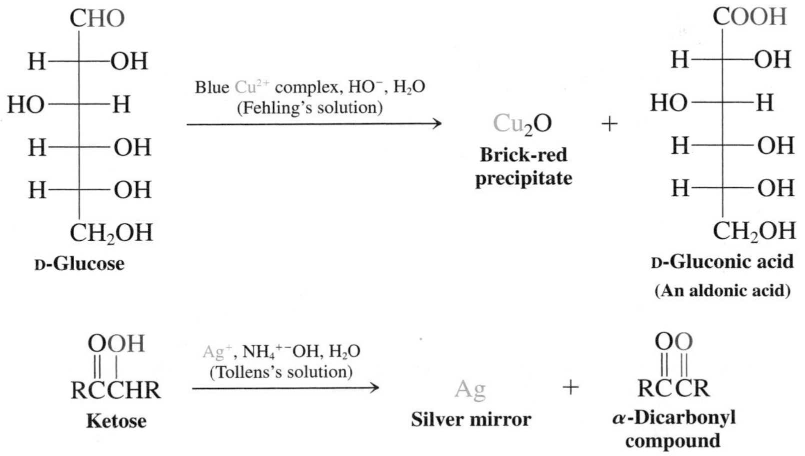

In these reactions, the aldoses are transformed into **aldonic acids,** ketoses into α-dicarbonyl compounds. Sugars that respond positively to these tests are called **reducing sugars.** All ordinary monosaccharides are reducing sugars.

Oxidation of aldoses can give mono- or dicarboxylic acids

Aldonic acids are made on a preparative scale by oxidation of aldoses with bromine in buffered aqueous solution (pH $= 5-6$). For example, D-mannose yields D-mannonic acid in this way. Upon subsequent evaporation of solvent from the aqueous solution of the aldonic acid, the γ-lactone (Section 20-4) forms spontaneously.

Aldonic Acid Preparation and Subsequent Dehydration to Give an Aldonolactone

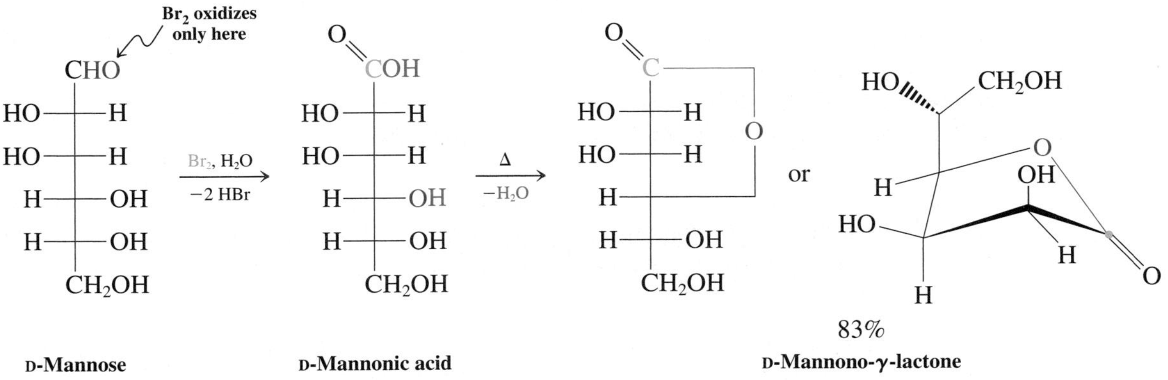

Exercise 24-13

What would the product be if the oxidation of D-mannose took place at the hemiacetal hydroxy group of the cyclic pyranose form, rather than at the carbonyl group of the open-chain isomer, as shown above?

More vigorous oxidation of an aldose leads to attack at the primary hydroxy function as well as at the formyl group. The resulting dicarboxylic acid is called an **aldaric,** or

saccharic, acid. This oxidation can be achieved with warm dilute aqueous nitric acid (see Section 19-6). For example, D-mannose is converted into D-mannaric acid under these conditions.

Preparation of an Aldaric Acid

HNO₃ oxidizes only here

CHO
HO——H
HO——H
H——OH
H——OH
CH₂OH

$\xrightarrow{\text{HNO}_3, \text{H}_2\text{O}, 60°\text{C}}$

COOH
HO——H
HO——H
H——OH
H——OH
COOH
44%

D-Mannose   D-Mannaric acid

The preceding syntheses of aldonic and aldaric acids directly from the corresponding aldoses are notable in that they occur in the presence of unprotected secondary hydroxy substituents. This selectivity arises from the intrinsically higher reactivity of the aldehyde function in oxidations (via the corresponding hydrates) and the lesser steric hindrance of primary versus secondary alcohols. To achieve selective conversion of the internal hydroxy groups (leaving the remaining hydroxycarbonyl frame intact) requires more complex protecting-group strategies. You will get a glimpse of such approaches in Section 24-8.

Exercise 24-14

The two sugars D-allose and D-glucose (Figure 24-1) differ in configuration only at C3. If you did not know which was which and you had samples of both, a polarimeter, and nitric acid at your disposal, how could you distinguish the two? (**Hint:** Write the products of oxidation.)

In Summary The chemistry of the sugars is largely that expected for carbonyl compounds containing several hydroxy substituents. Oxidation (by Br₂) of the formyl group of aldoses gives aldonic acids; more vigorous oxidation (by HNO₃) converts sugars into aldaric acids.

24-5 | OXIDATIVE CLEAVAGE OF SUGARS

The methods for oxidation of sugars discussed so far leave the basic skeleton intact. A reagent that leads to C–C bond rupture is periodic acid, HIO₄. This compound oxidatively degrades vicinal diols to give carbonyl compounds.

Oxidative Cleavage of Vicinal Diols with Periodic Acid

$\xrightarrow{\text{HIO}_4, \text{H}_2\text{O}}$

77%

cis-1,2-Cyclohexanediol   Hexanedial

The mechanism of this transformation proceeds through a cyclic **periodate ester,** which decomposes to give two carbonyl groups.

Mechanism of Periodic Acid Cleavage of Vicinal Diols

Cyclic periodate ester

Mechanism

Because most sugars contain several pairs of vicinal diols, oxidation with HIO_4 can give complex mixtures. Sufficient oxidizing agent causes complete degradation of the chain to one-carbon compounds, a technique that has been applied in the structural elucidation of sugars. For example, treatment of glucose with five equivalents of HIO_4 results in the formation of five equivalents of formic acid and one of formaldehyde. Similar degradation of the isomeric fructose consumes an equal amount of oxidizing agent, but the products are three equivalents of the acid, two of the aldehyde, and one of carbon dioxide.

Periodic Acid Degradation of Sugars

It is found that (1) the breaking of each C–C bond in the sugar consumes one molecule of HIO_4, (2) each aldehyde and secondary alcohol unit furnishes an equivalent of formic acid, and (3) the primary hydroxy function gives formaldehyde. The carbonyl group in ketoses gives CO_2. The number of equivalents of HIO_4 consumed reveals the size of the sugar molecule, and the ratios of products are important clues to the number and arrangement of hydroxy and carbonyl functions. Notice in particular that, after degradation, each carbon fragment retains the same number of attached hydrogen atoms as were present in the original sugar.

Exercise 24-15

Write the expected products (and their ratios), if any, of the treatment of the following compounds with HIO_4. **(a)** 1,2-Ethanediol (ethylene glycol); **(b)** 1,2-propanediol; **(c)** 1,2,3-propanetriol; **(d)** 1,3-propanediol; **(e)** 2,4-dihydroxy-3,3-dimethylcyclobutanone; **(f)** D-threose.

Exercise 24-16

Would degradation with HIO_4 permit the following sugars to be distinguished? Explain. (For structures, see Figures 24-1 and 24-2.) **(a)** D-Arabinose and D-glucose; **(b)** D-erythrose and D-erythrulose; **(c)** D-glucose and D-mannose.

In Summary Oxidative cleavage with periodic acid degrades the sugar backbone to formic acid, formaldehyde, and CO_2. The ratio of these products depends on the structure of the sugar.

24-6 | REDUCTION OF MONOSACCHARIDES TO ALDITOLS

Aldoses and ketoses are reduced by the same types of reducing agents that convert aldehydes and ketones into alcohols. The resulting polyhydroxy compounds are called **alditols.** For example, D-glucose gives D-glucitol (older name, D-sorbitol) when treated with sodium borohydride.

The hydride reducing agent traps the small amount of the open-chain form of the sugar, in this way shifting the equilibrium from the unreactive cyclic hemiacetal to the product.

Preparation of an Alditol

Red seaweed contains relatively large amounts of D-glucitol.

Many alditols are found in nature. D-Glucitol is present in red seaweed in concentrations as high as 14%, as well as in many berries (but not in grapes), in cherries, in plums, in pears, and in apples. It is prepared commercially from D-glucose by high-pressure hydrogenation or by electrochemical reduction. Glucitol is widely used as a sweetener in products such as mints, cough drops, mouthwashes, and chewing gum, where it is often identified by its alternative name, sorbitol. Glucitol is similar in caloric content to glucose. However, the types of bacteria present in the mouth that cause dental caries are less capable of metabolizing glucitol than glucose.

Exercise 24-17

(a) Reduction of D-ribose with $NaBH_4$ gives a product without optical activity. Explain. **(b)** Similar reduction of D-fructose gives two optically active products. Explain.

In Summary Reduction of the carbonyl function in aldoses and ketoses (by $NaBH_4$) furnishes alditols.

24-7 | CARBONYL CONDENSATIONS WITH AMINE DERIVATIVES

As might be expected, the carbonyl function in aldoses and ketoses undergoes condensation reactions with amine derivatives (Section 17-9). For example, treatment of D-mannose with phenylhydrazine gives the corresponding **hydrazone,** D-mannose phenylhydrazone. Surprisingly, the reaction does not stop at this stage but can be induced to continue with additional phenylhydrazine (two extra equivalents). The final product is a double phenylhydrazone, also called an **osazone** (here, phenylosazone). In addition, one equivalent each of benzenamine (aniline), ammonia, and water is generated.

Phenylhydrazone and Phenylosazone Formation

D-Mannose → D-Mannose phenylhydrazone (75%) → A phenylosazone (95%)

The mechanism of osazone synthesis is complex and will not be discussed in detail here. Formally, it constitutes an oxidation at C2 by one equivalent of phenylhydrazine, which in turn is reduced by N–N bond rupture to the component amines. Once they are formed, the osazones do not continue to react with excess phenylhydrazine but are stable under the conditions of the reaction.

Historically, the discovery of osazone formation marked a significant advance in the practical aspects of sugar chemistry. Sugars are well known for their reluctance to crystallize from syrups. Their osazones, however, readily form yellow crystals with sharp melting points, thus simplifying the isolation and characterization of many sugars, particularly if they have been formed as mixtures or are impure.

Exercise 24-18

Compare the structures of the phenylosazones of D-glucose, D-mannose, and D-fructose. In what way are they related?

In Summary One equivalent of phenylhydrazine converts a sugar into the corresponding phenylhydrazone. Additional hydrazine reagent causes oxidation of the center adjacent to the hydrazone function to furnish the osazone.

24-8 | ESTER AND ETHER FORMATION: GLYCOSIDES

Because of their multiple hydroxy groups, sugars can be converted into alcohol derivatives. This section explores the formation of simple esters and ethers of monosaccharides and also addresses selective reactions at the anomeric hydroxy group in their cyclic isomers.

Sugars can be esterified and methylated

Esters can be prepared from monosaccharides by standard techniques (Sections 19-9, 20-2, and 20-3). Excess reagent will completely convert all hydroxy groups, including the hemiacetal function. For example, acetic anhydride transforms β-D-glucopyranose into the pentaacetate.

Complete Esterification of Glucose

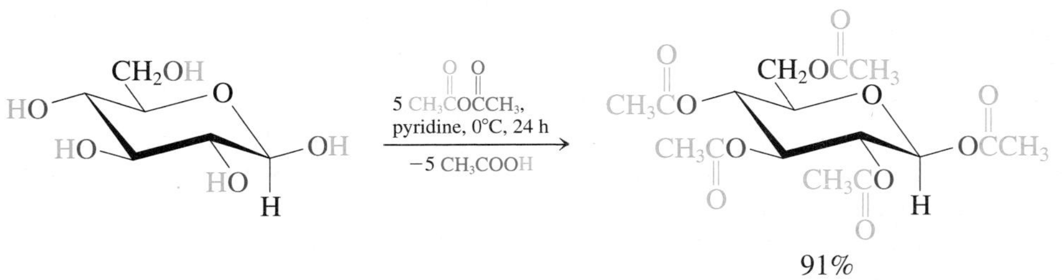

β-D-Glucopyranose   91%
                    β-D-Glucopyranose pentaacetate

Williamson ether synthesis (Section 9-6) allows complete methylation.

Complete Methylation of a Pyranose

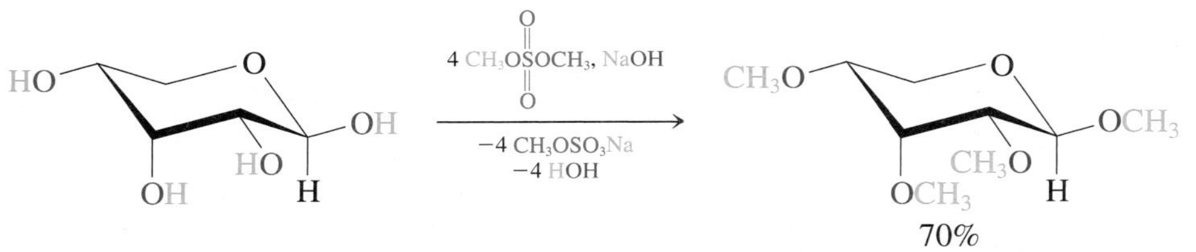

β-D-Ribopyranose   70%
                    β-D-Ribopyranose tetramethyl ether

Notice that the hemiacetal function at C1 is converted into an acetal group. The acetal function can be selectively hydrolyzed back to the hemiacetal (see Section 17-7).

Selective Hydrolysis of a Sugar Acetal

Denotes a mixture
of stereoisomers
at this carbon

$$\text{8\% HCl, HOH, } \Delta$$

D-Ribopyranose trimethyl ether
(Mixture of α and β forms)

67%

$+ \quad CH_3OH$

Animation

ANIMATED MECHANISM:
Methyl glycoside formation

It is also possible to convert the hemiacetal unit of a sugar selectively into the acetal. For example, treatment of D-glucose with acidic methanol leads to the formation of the two methyl acetals. Sugar acetals are called **glycosides.** Thus, glucose forms **glucosides.**

Selective Preparation of a Glycoside (Sugar Acetal)

$$\begin{array}{c} CH_3OH, \\ 0.25\% \text{ HCl}, \\ H_2O \\ \hline -HOH \end{array}$$

α- or β-D-Glucopyranose   Reactive hemiacetal function

Methyl α-D-glucopyranoside
(m.p. 166°C, $[\alpha]_D^{25°C} = +158$)

$+$

Methyl β-D-glucopyranoside
(m.p. 105°C, $[\alpha]_D^{25°C} = -33$)

Because glycosides contain a blocked anomeric carbon atom, they do not show mutarotation in the absence of acid, they test negatively to Fehling's and Tollens's reagents (they are *non*reducing sugars), and they are unreactive toward reagents that attack carbonyl groups. Such protection can be useful in synthesis and in structural analysis (see Exercise 24-19). An example (shown below) is the selective oxidation of the terminal hydroxy function of glucose when protected at C1 as the methyl pyranoside (see also Problem 62). After deprotection, this process results in glucuronic acid, and you can readily imagine the extension of the strategy to other "uronic" acids. Glucuronic acid plays an important role in the metabolism of drugs and other foreign materials in the body, and the presence of the reactive hemiacetal unit and the water-solubilizing carboxy group are crucial for its function. A glycosyl transferase enzyme catalyzes displacement of the anomeric –OH by the targeted material, for example, the –OH group in acetaminophen (Section 22-9, Real Life 22-2). Such derivatization renders a drug highly water soluble to allow for rapid excretion through urination.

$$\xrightarrow[\text{Oxidation}]{K_2Cr_2O_7, \, H^+, \, H_2O}$$

$$\xrightarrow[\text{Deprotection}]{H^+, \, H_2O}$$

Methyl D-glucopyranoside

Glucuronic acid pyranoside

$$\xrightarrow[\text{transferase}]{\text{Glycosyl-}}$$

Glucuronic acid

Acetaminophen

Acetaminophen glucuronide

Exercise 24-19

The same mixture of glucosides is formed in the methylation of D-glucose with acidic methanol, regardless of whether you start with the α or β form. Why?

Exercise 24-20

Draw the structure of methyl α-D-arabinofuranoside.

Exercise 24-21

Methyl α-D-glucopyranoside consumes two equivalents of HIO_4 to give one equivalent each of formic acid and dialdehyde A (shown in the margin). An unknown aldopentose methyl furanose reacted with one equivalent of HIO_4 to give dialdehyde A, but no formic acid. Suggest a structure for the unknown. Is there more than one solution to this problem?

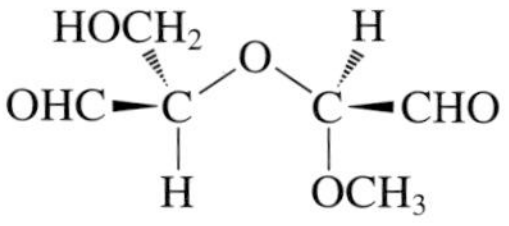

A

Neighboring hydroxy groups in sugars can be linked as cyclic ethers

The presence of neighboring pairs of hydroxy groups in the sugars allows for the formation of cyclic ether derivatives. For example, it is possible to synthesize five- or six-membered cyclic sugar acetals from the vicinal (and also from some β-diol) units by treating them with carbonyl compounds (Section 17-8).

Cyclic Acetal Formation from Vicinal Diols

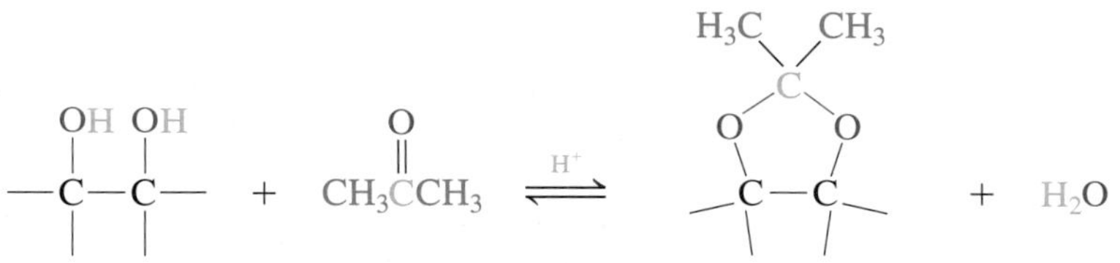

Acetone acetal

Such processes work best when the two OH groups are positioned cis to allow a relatively unstrained five- or six-membered ring to form. Cyclic acetal and ether formation is often employed to protect selected alcohol functions. The remaining hydroxy moieties can then be converted into leaving groups, transformed by elimination, or oxidized to carbonyl compounds. For example, an important step in the commercial synthesis of vitamin C (Section 22-9 and margin) is the selective oxidation of the C1 hydroxy group of L-sorbose to the corresponding carboxylic acid (2-keto-L-gulonic acid; see also Problem 63 of Chapter 23). To accomplish this task, all other alcohol functions have to be protected in the form of two adjacent acetal units.

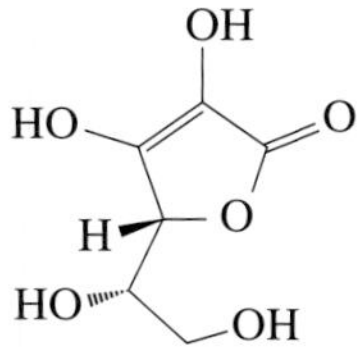

Vitamin C

Acetalization of L-Sorbose

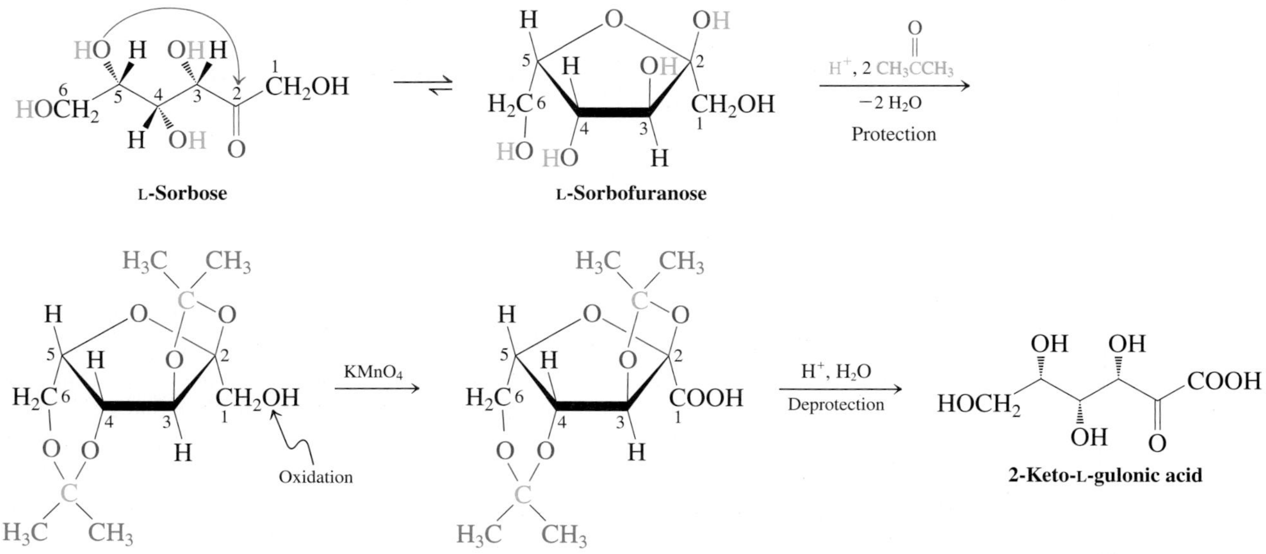

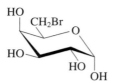

Exercise 24-22

Suggest a synthesis of the compound shown in the margin from D-galactose. (**Caution:** You will need protecting groups. **Hint:** Consider a strategy that uses cyclic acetals for protection.)

In Summary The various hydroxy groups of sugars can be esterified or converted into ethers. The hemiacetal unit can be selectively protected as the acetal, also called a glycoside. Finally, the various diol units in the sugar backbone can be linked as cyclic acetals, depending on steric requirements.

24-9 | STEP-BY-STEP BUILDUP AND DEGRADATION OF SUGARS

Larger sugars can be made from smaller ones and vice versa, by chain lengthening and chain shortening. These transformations can also be used to structurally correlate various sugars, a procedure applied by Fischer to prove the relative configuration of all the stereocenters in the aldoses shown in Figure 24-1.

Cyanohydrin formation and reduction lengthens the chain

To lengthen the chain of a sugar, an aldose is first treated with HCN to give the corresponding cyanohydrin (Section 17-11). Because this transformation forms a new stereocenter, two diastereomers (epimers) appear. Separation of the diastereomers and partial reduction of the nitrile group by catalytic hydrogenation in aqueous acid then gives the aldehyde groups of the chain-extended sugars.

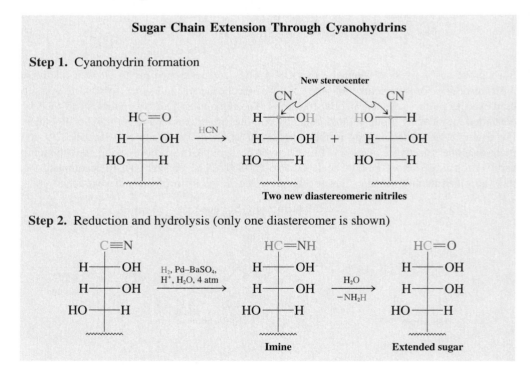

Sugar Chain Extension Through Cyanohydrins

Step 1. Cyanohydrin formation

Step 2. Reduction and hydrolysis (only one diastereomer is shown)

In this hydrogenation, a modified palladium catalyst (similar to the Lindlar catalyst, Section 13-6) allows selective reduction of the nitrile to the imine, which hydrolyzes under the reaction conditions. The special catalyst is necessary to prevent the hydrogenation from proceeding all the way to the amine (Section 20-8). The preceding chain-lengthening sequence is an improved and shortened version of what is known as the **Kiliani-Fischer* synthesis** of

---

*Professor Heinrich Kiliani (1855–1945), University of Freiburg, Germany; Professor Emil Fischer (see Section 5-4).

chain-extended sugars. In the late 1800s, Kiliani demonstrated that addition of cyanide to an aldose gives the cyanohydrin, which, upon hydrolysis (Sections 19-6 and 20-8), is converted into the chain-extended aldonic acid. Fischer subsequently succeeded in converting this aldonic acid into an aldose by lactone formation (Sections 19-9 and 24-4) followed by reduction (Section 20-4; for Fischer's reduction method, see Problem 62 at the end of this chapter).

Solved Exercise 24-23 | Working with the Concepts: Sugar Synthesis

What are the products of chain extension of D-erythrose?

Strategy
As the text above indicates, Kiliani-Fischer chain extension adds a new formyl group to C1 of an aldose, converting the original formyl function into that of a secondary alcohol. Find D-erythrose in Figure 24-1 and identify the sugar(s) that should result from operation of this process.

Solution
• D-Erythrose is a four-carbon aldose. Addition of a formyl substituent to C1 leaves C2, C3, and C4 unchanged, while changing the original carbonyl group to that of a secondary alcohol. The emerging new stereocenter may form with either the *R* or *S* configuration.
• Figure 24-1 shows that both D-ribose and D-arabinose have five carbons, and in both of them the "bottom" three carbons (C3, C4, and C5) are identical in structure and stereochemistry to the corresponding three carbon atoms (C2, C3, and C4) in D-erythrose. D-Ribose and D-arabinose differ only at C2, the new stereocenter. They are the products of D-erythrose chain extension.

Exercise 24-24 | Try It Yourself

What are the chain-extension products of D-arabinose?

Ruff degradation shortens the chain

Whereas the preceding approach synthesizes higher sugars, complementary strategies degrade higher sugars to lower sugars one carbon at a time. One of these strategies is the **Ruff* degradation.** This procedure removes the carbonyl group of an aldose and converts the neighboring carbon into the aldehyde function of the new sugar.

The Ruff degradation is an oxidative decarboxylation. The sugar is first oxidized to the aldonic acid by aqueous bromine. Exposure to hydrogen peroxide in the presence of ferric ion then leads to the loss of the carboxy group and oxidation of the new terminus to the aldehyde function of the lower aldose. During this process, the stereochemistry of the hydroxy carbon next to the carbonyl is lost. Consequently, two epimeric aldoses at that center will give the same new sugar.

Ruff Degradation of Sugars

$$HC{=}O \qquad \xrightarrow[\text{Br}_2,\ \text{H}_2\text{O}]{} \qquad COOH \qquad \xrightarrow[-CO_2]{Fe^{3+},\ H_2O_2,\ H_2O} \qquad HC{=}O$$

Stereochemistry lost

Reaction

The oxidative decarboxylation takes place by means of two one-electron oxidations. The first gives a carboxy radical, which is unstable and rapidly loses CO_2. The resulting hydroxy-substituted radical is oxidized again to give the aldehyde (see next page).

---

*Professor Otto Ruff (1871–1939), University of Danzig, Germany.

REAL LIFE: NATURE 24-1 | Biological Sugar Synthesis

In nature, carbohydrates are produced primarily by a reaction sequence called *photosynthesis* (see also Section 9-11). In this process, sunlight impinging on the chlorophyll of green plants is absorbed. The photochemical energy thus obtained is used to convert carbon dioxide and water into oxygen and the polyfunctional structure of carbohydrates.

The detailed mechanism of this transformation is complicated and takes many steps, the first of which is the absorption of one quantum of light by the extended π system (Section 14-11) of chlorophyll (*chloros*, Greek, green, and *phillon*, Greek, leaf).

Photosynthesis of Glucose in Green Plants

$$6\ CO_2 + 6\ H_2O \underset{\substack{\text{Released} \\ \text{metabolic} \\ \text{energy} \\ (670\ \text{kcal mol}^{-1})}}{\overset{\substack{\text{Sunlight,} \\ \text{chlorophyll}}}{\rightleftharpoons}} \underset{\text{Glucose}}{C_6(H_2O)_6} + 6\ O_2$$

Chlorophyll a

Real Life 18-1 described an aldol-based biosynthesis of carbohydrates, and Real Life 23-2 introduced thiamine pyrophosphate (TPP) in its role in the metabolism of glucose to produce biochemical energy. Here we illustrate another biochemical sequence that uses thiamine catalysis to interconvert ketoses.

In ketose interconversion, the enzyme transketolase employs a molecule of thiamine to catalyze the transfer of a (hydroxymethyl)carbonyl unit from xylulose to erythrose, producing fructose in the process.

Mechanistically, the deprotonated thiazolium ion first attacks the carbonyl group of the donor sugar (xylulose) to form an addition product, in a way completely analogous to addition to aldehydes (Section 23-4). Because the donor sugar contains a hydroxy group next to the reaction site, this initial product can decompose by the reverse of the addition process to an aldehyde (glyceraldehyde) and a new thiamine intermediate. This new intermediate attacks another aldehyde (erythrose) to produce a new addition product. The catalyst then dissociates as TPP, releasing the new sugar molecule (fructose).

The cycle of photosynthesis, carbohydrate interconversion, and carbohydrate metabolism is a beautiful example of how nature reuses its resources. First, CO_2 and H_2O are consumed to convert solar energy into chemical energy and O_2. When an organism needs some of the stored energy, it is generated by conversion of carbohydrate into CO_2 and H_2O, using up roughly the same amount of O_2 originally liberated.

Mechanism of Oxidative Decarboxylation

Mechanism

Ruff degradation gives low yields because of the sensitivity of the products to the reaction conditions. Nevertheless, the procedure is useful in structural elucidations (Exercise 24-25). Fischer originally carried out such studies to establish the relative configurations of the monosaccharides (the **Fischer proof**). The next section will describe some of the logic behind his approach to the problem.

Transketolase-Catalyzed Biosynthesis of Fructose from Erythrose Using Thiamine Pyrophosphate

Sugar Activation   **Removal of Old Aldehyde**

Introduction of New Aldehyde

Exercise 24-25

Ruff degradation of two D pentoses, A and B, gave two new sugars, C and D. Oxidation of C with HNO₃ gave *meso*-2,3-dihydroxybutanedioic (tartaric) acid, that of D resulted in an optically active acid. Oxidation of either A or B with HNO₃ furnished an optically active aldaric acid. What are compounds A, B, C, and D?

In Summary Sugars can be made from other sugars by step-by-step one-carbon chain lengthening (cyanohydrin formation and reduction) or shortening (Ruff degradation).

24-10 | RELATIVE CONFIGURATIONS OF THE ALDOSES: AN EXERCISE IN STRUCTURE DETERMINATION

Imagine that we have been presented with 14 jars, each filled with one of the aldoses in Figure 24-1, each labeled with a name, but *no structural formula*. How would we establish the structure of each substance *without using spectroscopy*?

```
        CHO
   H ——R—— OH
   H ———— OH
        CH₂OH
   Structure 1
  (D-Erythrose)
```

↓ HNO₃

```
        COOH
   H ———— OH
   H ———— OH
        COOH
```
meso-Tartaric acid
(Not optically active)

```
        CHO
  HO ——S—— H
   H ———— OH
        CH₂OH
   Structure 2
  (D-Threose)
```

↓ HNO₃

```
        COOH
  HO ———— H
   H ———— OH
        COOH
```
(−)-Tartaric acid
(Optically active)

This was the challenge faced by Fischer over a century ago. In an extraordinary display of scientific logic, Fischer solved this problem by interpreting the results of a combination of carefully thought out synthetic manipulations, designed to convert the aldoses he had available to him into one another and into related substances. Fischer made one assumption: That the dextrorotatory enantiomer of 2,3-dihydroxypropanal (glyceraldehyde) had the D (and not the L) configuration. This assumption was proved correct only after a special kind of X-ray analysis was developed in 1950, long after Fischer's time. Fischer's was a lucky guess; otherwise, all the structures of the D sugars in Figure 24-1 would have had to be changed into their mirror images. Fischer's goal was to establish the *relative* configurations of all the stereoisomers—that is, to associate each unique sugar with a unique *relative* sequence of stereocenters.

The structures of the aldotetroses and aldopentoses can be determined from the optical activity of the corresponding aldaric acids

Beginning with the structure of D-glyceraldehyde, we shall now set out to prove unambiguously the structures of the higher D aldoses. We shall use Fischer's logic, although not his procedures, because most of these aldoses were unavailable in his time. (Consult Figure 24-1 as required while following these arguments.) We begin with chain extension of D-glyceraldehyde, obtaining a mixture of D-erythrose and D-threose. Oxidation of D-erythrose with nitric acid gives *optically inactive meso*-tartaric acid, which tells us that D-erythrose must have structure 1, with both –OH groups on the *same side* in the Fischer representation. In contrast, oxidation of D-threose forms an *optically active* acid. Thus, D-threose must have the opposite stereochemistry at C2 (structure 2). Both must have the same (*R*) configuration at C3, the carbon derived from the C2 stereocenter of our starting material, D-glyceraldehyde. The difference is at C2: D-Erythrose is 2*R*; D-threose, 2*S*.

Now that we know the structures of D-erythrose and D-threose, we continue as follows. Chain extension of D-erythrose (Exercise 24-23) gives a mixture of two pentoses: D-ribose and D-arabinose. Their configurations at C3 and C4 must be the same as those of C2 and C3 in D-erythrose; they differ at C2, the new stereocenter created in the lengthening procedure. Nitric acid oxidation of D-ribose gives an optically inactive (meso) acid; thus D-ribose has structure 3. D-Arabinose oxidizes to an optically active acid; it must have structure 4.

```
      COOH              CHO          D-Erythrose          CHO               COOH
  H ——— OH         H ——— OH                          HO ——— H          HO ——— H
  H ——— OH    ←    H ——— OH         chain extension    H ——— OH    →    H ——— OH
  H ——— OH   HNO₃  H ——— OH                           H ——— OH   HNO₃    H ——— OH
      COOH              CH₂OH                              CH₂OH             COOH
      Meso         Structure 3                         Structure 4      Optically active
(Not optically active)  (D-Ribose)                    (D-Arabinose)
```

Proceeding in the same way from D-threose, chain extension gives a mixture of D-xylose and D-lyxose. The oxidation product of D-xylose is meso; therefore D-xylose has structure 5. The optically active oxidation product of D-lyxose confirms its structure as 6.

```
      COOH              CHO          D-Threose            CHO               COOH
  H ——— OH         H ——— OH                          HO ——— H          HO ——— H
 HO ——— H    ←    HO ——— H          chain extension   HO ——— H    →    HO ——— H
  H ——— OH   HNO₃  H ——— OH                           H ——— OH   HNO₃    H ——— OH
      COOH              CH₂OH                              CH₂OH             COOH
      Meso         Structure 5                         Structure 6      Optically active
(Not optically active)  (D-Xylose)                     (D-Lyxose)
```

Symmetry properties also define the structures of the aldohexoses

We now know the structures of the four aldopentoses and can extend the chain of each of them. This process gives us four pairs of aldohexoses, each pair distinguished from the others by the unique sequence of stereocenters at C3, C4, and C5. The members of each pair are epimers and differ only in their configuration at C2.

The structural assignment for the four sugars obtained from D-ribose and D-lyxose, respectively, is again accomplished by oxidation to the corresponding aldaric acids. Both D-allose and D-galactose give optically inactive oxidation products, in contrast with D-altrose and D-talose, which give optically active dicarboxylic acids.

D-Allose   D-Galactose   D-Altrose   D-Talose
(From D-ribose)   (From D-lyxose)   (From D-ribose)   (From D-lyxose)
(Both give meso dicarboxylic acids)   (Both give optically active dicarboxylic acids)

The structural assignments of the four remaining hexoses (with structures 7–10) cannot be based on the approach taken so far, because all give optically active diacids upon oxidation. We join the actual strategy devised by Fischer to solve a problem that had become substantially more complicated. He found that chain extension of D-arabinose gave a mixture of D-glucose and D-mannose, one of which must have structure 7 and the other structure 8. Displaying synthetic creativity decades ahead of his time, he developed a procedure to *exchange* the functionalities at C1 and C6 of a hexose—converting C1 to a primary alcohol and C6 to an aldehyde. Carrying out this procedure starting with both D-glucose and D-mannose, he found that the result of C1/C6 exchange in glucose was a *new sugar*, which did not exist in nature and was given the name gulose. On the other hand, exchange of C1 and C6 in mannose returned mannose, unchanged. Examination of structures 7 and 8 reveals why this is so and gives us a glimpse of Fischer's genius. If we perform a C1/C6 exchange in 8 and then rotate the resulting Fischer projection by 180°, we realize that the initial and final structures are identical: The exchange procedure gives the *original sugar* back. Thus 8 is identified as the structure of D-mannose, and 7, in which C1/C6 exchange gives a new structure, must correspond to D-glucose.

7   New sugar   8   Same as 8
(D-Glucose)   (Gulose)   (D-Mannose)

Exercise 24-26

Was the gulose that Fischer synthesized from D-glucose a D-sugar or an L-sugar? (**Caution:** In a very uncharacteristic oversight, Fischer himself got the wrong answer at first, and that confused *everybody* for *years*.)

CHO
H——OH
H——OH
HO——H
H——OH
CH$_2$OH
9

CHO
HO——H
H——OH
HO——H
H——OH
CH$_2$OH
10

> **Exercise 24-27**
>
> Fischer proved that his structural reasoning was correct by comparing the nitric acid oxidation of D-glucose to that of his synthetic gulose. What do these two reactions produce?

With the synthesis of gulose, one of the two remaining hexoses (the products of chain lengthening of D-xylose) was identified as structure 9. By process of elimination, the other, D-idose, could be assigned structure 10, thus completing the exercise.

> **Exercise 24-28**
>
> In the preceding discussion, we assigned the structures of D-ribose and D-arabinose by virtue of the fact that, upon oxidation, the first gives a meso diacid, the second an optically active diastereomer. Could you arrive at the same result by ^{13}C NMR spectroscopy?

In Summary   Step-by-step one-carbon chain lengthening or shortening, in conjunction with the symmetry properties of the various aldaric acids, allows the stereochemical assignments of the aldoses.

24-11 | COMPLEX SUGARS IN NATURE: DISACCHARIDES

A substantial fraction of the natural sugars occur in dimeric, trimeric, higher oligomeric (between 2 and 10 sugar units), and polymeric forms. The sugar most familiar to us, so-called table sugar, is a dimer.

Sucrose is a disaccharide derived from glucose and fructose

Sucrose, ordinary table sugar (see Chapter Opening), is one of the few natural chemicals consumed in unmodified form (water and NaCl are examples of others). Its average yearly consumption in the United States is about 77 pounds per person. Sugar is isolated from sugar cane and sugar beets, in which it is particularly abundant (about 14–20% by weight), although it is present in many plants in smaller concentrations. World production is about 168 million tons a year.

Sucrose has not been considered in this chapter so far, because it is not a simple monosaccharide; rather, it is a disaccharide composed of two units, glucose and fructose. The structure of sucrose can be deduced from its chemical behavior. Acidic hydrolysis splits it into glucose and fructose. It is a nonreducing sugar. It does not form an osazone. It does not undergo mutarotation. These findings suggest that the component monosaccharide units are linked by an acetal bridge (Section 17-7) connecting the two anomeric carbons; in this way, the two cyclic hemiacetal functions block each other. X-ray structural analysis confirms this hypothesis: Sucrose is a disaccharide in which the α-D-glucopyranose form of glucose is attached to β-D-fructofuranose in this way.

Sucrose is a pure chemical compound and forms large crystals from saturated aqueous solutions.

Both anomeric carbons in sucrose are parts of **acetal** functional groups.

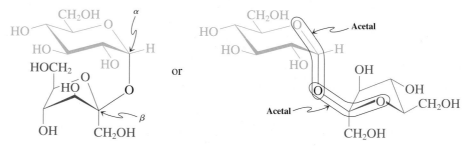

Sucrose, a β-D-fructofuranosyl-α-D-glucopyranoside

Sucrose has a specific rotation of +66.5. Treatment with aqueous acid decreases the rotation until it reaches a value of −20. The same effect is observed with the enzyme invertase. The phenomenon, known as the **inversion of sucrose,** is related to mutarotation of monosaccharides. It includes three separate reactions: hydrolysis of the disaccharide to the component monosaccharides α-D-glucopyranose and β-D-fructofuranose; mutarotation of α-D-glucopyranose to the equilibrium mixture with the β form; and mutarotation of β-D-fructofuranose to the slightly more stable β-D-fructopyranose. Because the value for the specific rotation of fructose (-92) is more negative than the value for glucose ($+52.7$) is positive, the resulting mixture, sometimes called **invert sugar,** has a net negative rotation, *inverted* from that of the original sucrose solution.

Inversion of Sucrose

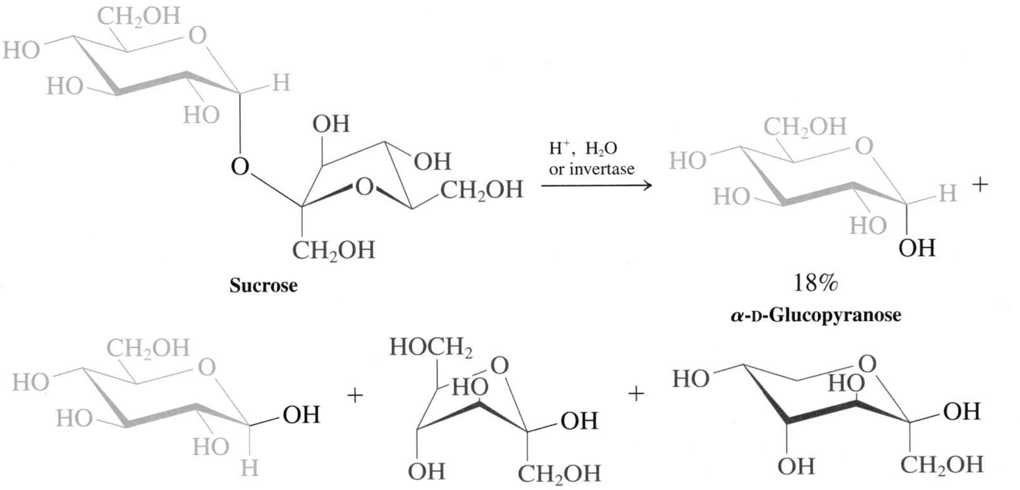

Sucrose

18% **α-D-Glucopyranose**

32% **β-D-Glucopyranose**

16% **β-D-Fructofuranose**

34% **β-D-Fructopyranose**

The noncaloric, edible fat substitute olestra (Real Life 21-1) consists of a mixture of sucroses esterified with seven or eight fatty acids obtained from vegetable oils, such as hexadecanoic (palmitic) acid. The fatty acids shield the sucrose core of olestra so effectively that it is completely immune to attack by digestive enzymes and passes through the digestive tract unchanged.

Caramel!

Sucrose is unstable at temperatures $\geq 160°C$ (320°F) and thermolyzes to the brown candy that we appreciate as caramel. Caramel contains a myriad of decomposition products formed by simultaneous cleavage of sucrose to glucose and fructose, dehydrations, fragmentations (such as retroaldol additions), isomerizations through enols (apply the last three reactions to glucose and see what you get), and polymerizations. Two of the odorants that have been identified in the mixture are shown below.

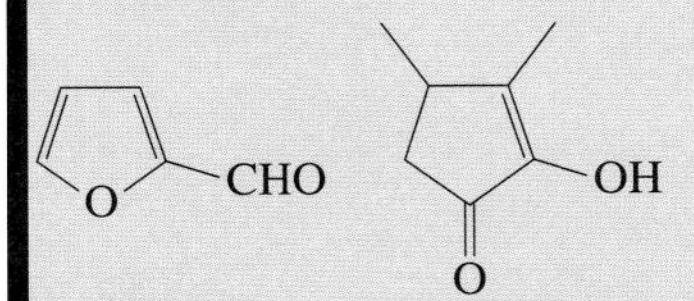

Furan-2-carbaldehyde (Almond flavor) — **2-Hydroxy-3,4-dimethyl-2-cyclopentenone (Sweet, burnt aroma)**

Solved Exercise 24-29 | Working with the Concepts: Practicing Sugar Chemistry

Write the products (if any) of the reaction of sucrose with excess $(CH_3)_2SO_4$, NaOH.

Strategy

The chemistry of disaccharides and other more complex carbohydrates is largely the same as that of monosaccharides. Identify the functional groups present and consider processes comparable to those described earlier for simple sugars.

Solution

- Dimethyl sulfate is a reagent that is commonly used to convert alcohols into methyl ethers via the Williamson ether synthesis (Sections 9-6 and 22-8). (Iodomethane would be equally effective but is more costly.)
- As illustrated by the example in Section 24-8, dimethyl sulfate methylates all free –OH groups in a molecule, converting them into $-OCH_3$ groups by S_N2 processes. Carrying out this reaction with sucrose gives as the product

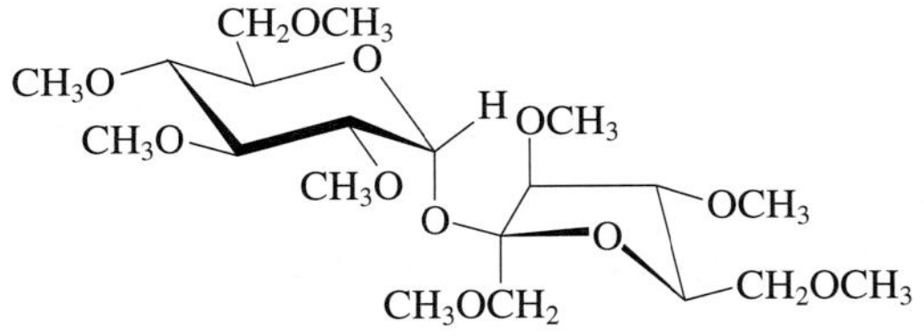

For the follow-up Exercise 24-30, see p. 1101.

REAL LIFE: FOOD CHEMISTRY 24-2 | Manipulating Our Sweet Tooth

Sweetness is one of our five basic tastes; it is complemented by sour (acidic, as in lemon), bitter (as in coffee), salty, and umami (savory, as in meats). A great variety of structures are perceived as sweet, ranging from amino acids to salts. Sugars rank prominently, probably because we have evolved to perceive foods with high nutritional content as pleasurable to the palate. Indeed, most simple saccharides taste sweet to varying degrees. The mechanism by which we detect sweetness has been elucidated only during the past decade. It features so-called G-protein–coupled receptors* (see p. 837, margin) that occur on the outside of the gustatory cells decorating the surface of the tongue. Two such proteins act in concert to encapsulate a sugar such as sucrose like a Venus flytrap (as shown in the illustration). The associated molecular motion is transmitted through the cell membrane to another protein, which, upon activation, starts a series of processes that lead ultimately to the release of a neurotransmitter and a signal to the brain. Other sweet substances can interact with this complex assembly in different ways, providing a rationale for the diversity of their structures.

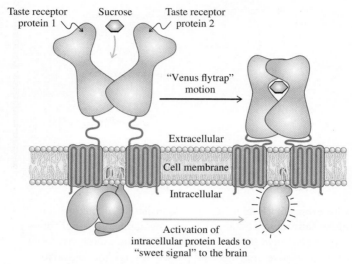

The sweetness receptor. (Edited after *Lehninger Principles of Biochemistry,* 6th ed., by David L. Nelson and Michael M. Cox. W. H. Freeman and Company. Copyright © 2000, 2005, 2008, 2013.)

Obesity is a global epidemic (see Real Life 21-1), and a great deal of effort has gone into developing less calorific synthetic and natural sugar substitutes. Well-known representatives of the former are saccharin and aspartame. These compounds exhibit 160–170 times the sweetness of sucrose, but they do not reproduce the exact sensation or texture of table sugar and also taste slightly differently (the "sweetness profile"), generally described as "bitter" or "metallic." As a result, confections containing these sweeteners must be supplemented with other materials to give them perceived characteristics acceptable to the consumer. For example, the formulation of low-calorie chocolate is an unusual challenge, because much of chocolate's appealing "melt-in-your-mouth" character comes from its fat content. To simulate fat, mixtures of alditols such as xylitol and maltitol (together with synthetic glucopyranose polymers) are added. Apart from imparting a rich creamy texture to the product, such alditols are almost as sweet as sucrose but have only two-thirds the calories. An added benefit is improved dental health, because the bacteria that cause tooth decay do not ferment alditols as well as normal sugar.

*Professor Brian K. Kobilka (b. 1955), Stanford University, California, and Professor Robert J. Lefkowitz (b. 1943), Duke University, North Carolina, Nobel Prizes 2012 (chemistry).

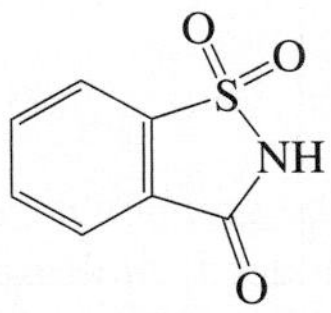

Saccharin

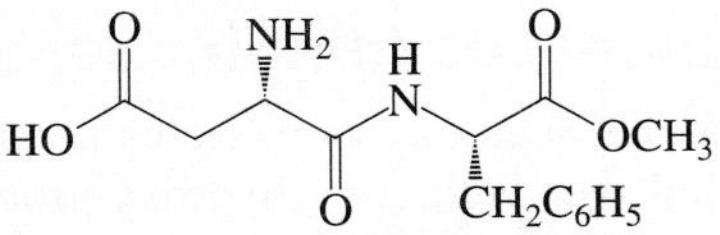

Aspartame

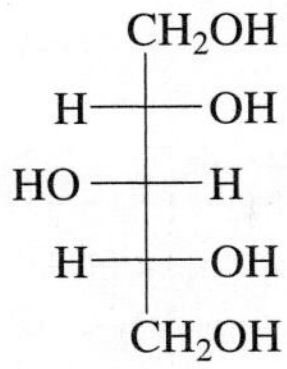

Xylitol

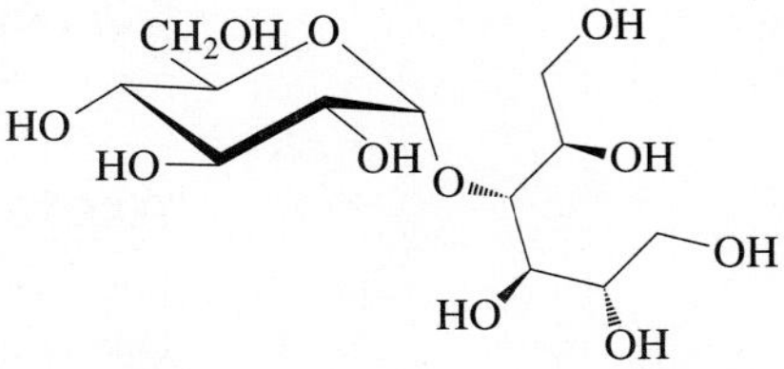

Maltitol
(Glucopyranosylglucitol)

An increasingly competitive market rival to aspartame is sucralose (sold in the United States as Splenda). A trichlorinated derivative of sucrose, sucralose is 600 times sweeter. It has the benefit of sufficient heat stability to be usable in cooking and baking, and it is available in a variety of forms appropriate to its diverse potential uses. It is absorbed to a limited extent by the digestive tract, but there is no evidence of any ill effects in humans. Sucralose has been approved for safe consumption in dozens of countries, including the United States and Canada, and can be found in more than 4500 food and beverage products.

A similar competition, but from natural sources, arrived with the isolation of two remarkable hybrid terpene-carbohydrates from the leaves of the Paraguayan herb *Stevia rebaudiana*, stevioside and its close relative rebaudioside A (commercialized as Truvia and PureVia). They are 3–400 times as sweet as sucrose, and their structures render them relatively resistant to digestive processes and therefore virtually noncaloric. Rebaudioside A was designated as "GRAS" ("generally regarded as safe") by the U.S. FDA in 2008. Stevioside is widely used in Asia and was approved for use in the European Union in 2011.

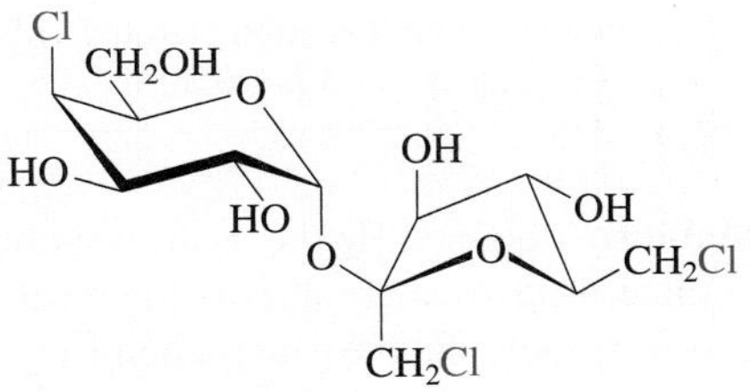

Sucralose

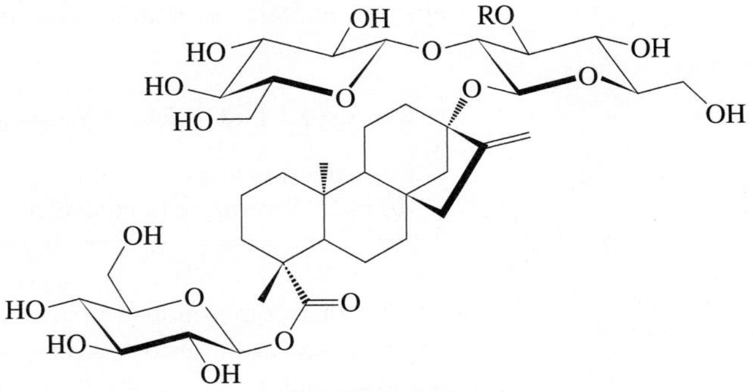

Stevioside (R = H)
Rebaudioside A (R = glucopyranosyl)

The newest frontier in sweetener research is the area of "sweet enhancers." These substances, while tasteless themselves, cause sweetness receptors to respond more strongly to sweet substances. For example, the simple experimental molecule dubbed SE-2 enhances the sweetness of sucralose up to eightfold. It is thought that SE-2 and related structures bind to the Venus flytrap complex at a remote site from the sweetener itself, causing the "trap" to stay shut for longer periods of time. Sucrose and sucralose enhancers are beginning to be commercialized around the world, adding another dimension to how to tweak our taste buds.

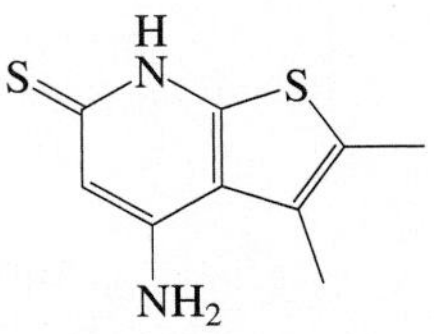

SE-2

Write the products (if any) of the reaction of sucrose with (**a**) 1. H⁺, H₂O, 2. NaBH₄; and (**b**) NH₂OH.

Acetals link the components of complex sugars

Sucrose contains an acetal linkage between the anomeric carbons of the component sugars. One could imagine other acetal linkages with other hydroxy groups. Indeed, **maltose** (malt sugar), which is obtained in 80% yield by enzymatic (amylase) degradation of starch (to be discussed in the next section), is a dimer of glucose in which the hemiacetal oxygen of one glucose molecule (in the α anomeric form) is bound to C4 of the second.

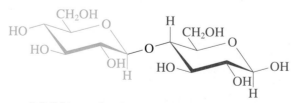

β-Maltose, an *α*-D-glucopyranosyl-*β*-D-glucopyranose

In this arrangement, one glucose retains its unprotected hemiacetal unit with its distinctive chemistry. For example, maltose is a reducing sugar; it forms osazones, and it undergoes mutarotation. Maltose is hydrolyzed to two molecules of glucose by aqueous acid or by the enzyme maltase. It is about one-third as sweet as sucrose.

Draw the structure of the initial product of β-maltose when it is subjected to (**a**) Br₂ oxidation; (**b**) phenylhydrazine (3 equivalents); (**c**) conditions that effect mutarotation.

Another common disaccharide is **cellobiose,** obtained by the hydrolysis of cellulose (to be discussed in the next section). Its chemical properties are almost identical with those of maltose, and so is its structure, which differs only in the stereochemistry at the acetal linkage—β instead of α.

β-Cellobiose, a *β*-D-glucopyranosyl-*β*-D-glucopyranose

Aqueous acid cleaves cellobiose into two glucose molecules just as efficiently as it hydrolyzes maltose. However, enzymatic hydrolysis requires a different enzyme, β-glucosidase, which specifically attacks only the β-acetal bridge. In contrast, maltase is specific for α-acetal units of the type found in maltose.

After sucrose, the most abundant natural disaccharide is **lactose** (milk sugar). It is found in human and most animal milk (about 5% solution), constituting more than one-third of the solid residue remaining on evaporation of all volatiles. Its structure is made up of galactose and glucose units, connected in the form of a β-D-galactopyranosyl-α-D-glucopyranose. Crystallization from water furnishes only the α anomer.

Lactose contains an unprotected hemiacetal unit and is therefore a reducing sugar that undergoes mutarotation. The first step of its degradation in the body is hydrolysis of the glycosidic bond by the enzyme lactase to give galactose and glucose. Lack of this enzyme causes lactose intolerance, a condition that is associated with abdominal cramps and diarrhea after the consumption of milk and other dairy products.

Got lactose?

Crystalline α-lactose, a β-D-galactopyranosyl-α-D-glucopyranose

In Summary Sucrose is a dimer derived from linking α-D-glucopyranose with β-D-fructofuranose at the anomeric centers. It shows inversion of its optical rotation on hydrolysis to its mutarotating component sugars. The disaccharide maltose is a glucose dimer in which the components are linked by a carbon–oxygen bond between an α anomeric carbon of a glucose molecule and C4 of the second. Cellobiose is almost identical with maltose but has α β configuration at the acetal carbon. Lactose has a β-D-galactose linked to glucose in the same manner as in cellobiose.

24-12 | POLYSACCHARIDES AND OTHER SUGARS IN NATURE

Polysaccharides are the polymers of monosaccharides. Their possible structural diversity exceeds that of alkene polymers (Sections 12-14 and 12-15), particularly in variations of chain length and branching. Nature, however, has been remarkably conservative in its construction of such macromolecules. The three most abundant natural polysaccharides—cellulose, starch, and glycogen—are derived from the same monomer: glucose.

Cellulose and starch are glucose polymers

Cellulose is a poly-β-glucopyranoside linked at C4, containing about 3000 monomeric units and having a molecular weight of about 500,000. It is largely linear.

Cellulose

Individual strands of cellulose tend to align with one another and are connected by multiple hydrogen bonds. The development of so many hydrogen bonds is responsible for the highly rigid structure of cellulose and its effective use as the cell-wall material in organisms. Thus, cellulose is abundant in trees and other plants. Cotton fiber is almost pure cellulose, as is filter paper. Wood and straw contain about 50% of the polysaccharide.

Like cellulose, **starch** is a polyglucose, but its subunits are connected by α-acetal linkages. It functions as a food reserve in plants and (like cellulose) is readily cleaved by aqueous acid into glucose. Major sources of starch are corn, potatoes, wheat, and rice. Hot water swells granular starch and allows the separation of the two major components: **amylose** (~20%) and **amylopectin** (~80%). Both are soluble in hot water, but amylose is less soluble in cold water. Amylose contains a few hundred glucose units per molecule

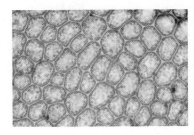

Cell walls (like those depicted in a leaf of moss) rely on cellulose for rigidity.

(molecular weight, 150,000–600,000). Its structure is different from that of cellulose, even though both polymers are unbranched. The difference in the stereochemistry at the anomeric carbons leads to the strong tendency of amylose to form a helical polymer arrangement (not the straight chain shown in the formula). Note that the disaccharide units in amylose are the same as those in maltose.

Amylose

In contrast with amylose, amylopectin is branched, mainly at C6, about once every 20 to 25 glucose units. Its molecular weight runs into the millions.

Amylopectin

Glycogen is a source of energy

Another polysaccharide similar to amylopectin but with greater branching (1 per 10 glucose units) and of much larger size (as much as 100 million molecular weight) is **glycogen.** This compound is of considerable biological importance because it is one of the major energy-storage polysaccharides in humans and animals and because it provides an immediate source of glucose between meals and during (strenuous) physical activity. It is accumulated in the liver and in rested skeletal muscle in relatively large amounts. The manner in which cells make use of this energy storage is a fascinating story in biochemistry.

A special enzyme, glycogen phosphorylase, first breaks glycogen down to give a derivative of glucose, α-D-glucopyranosyl 1-phosphate. This transformation takes place at one of the nonreducing terminal sugar groups of the glycogen molecule and proceeds step by step—one glucose molecule at a time. Because glycogen is so highly branched, there are many such end groups at which the enzyme can "nibble" away, making sure that, at a time of high-energy requirements, a sufficient amount of glucose becomes available quickly.

Usain Bolt degrading glycogen.

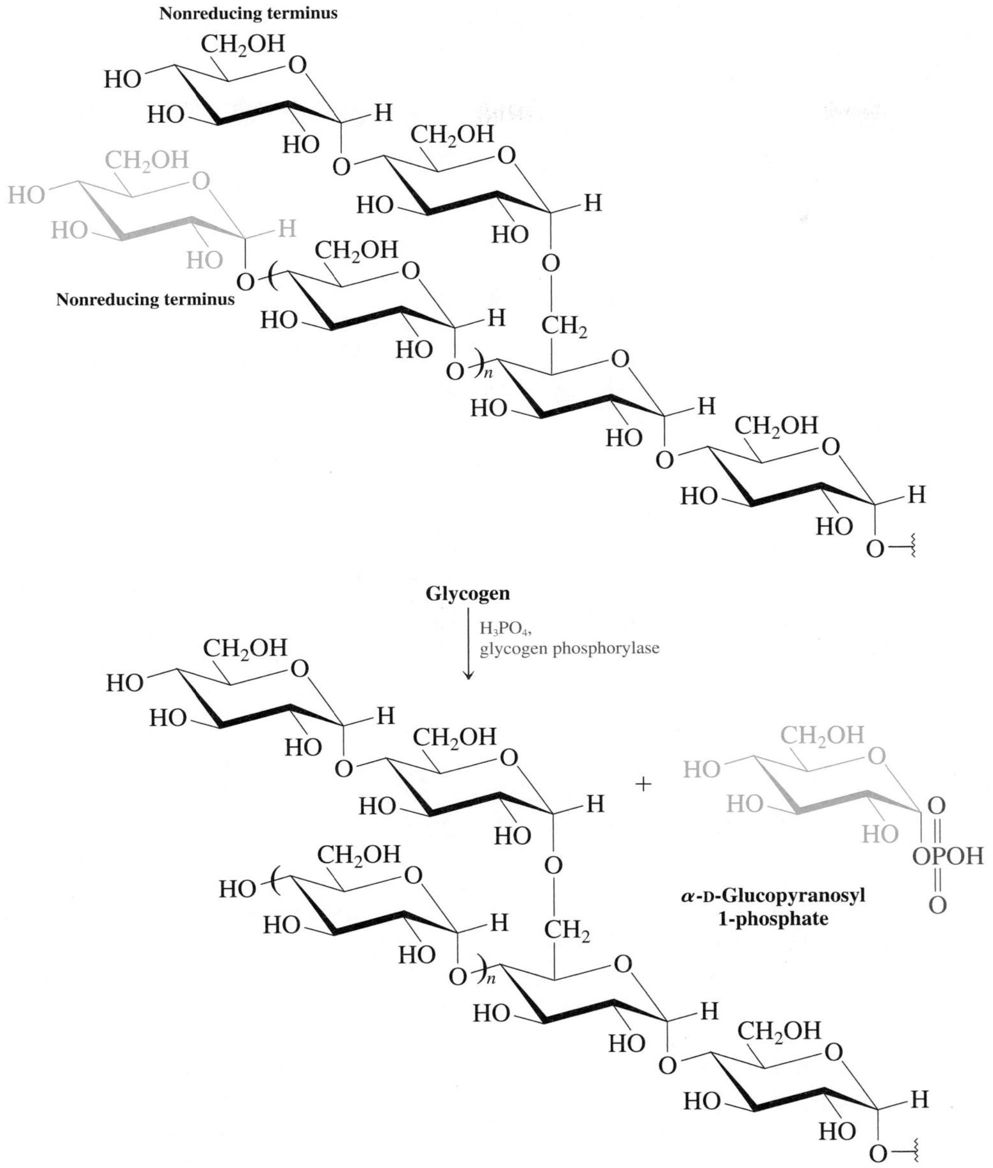

Glycogen

H₃PO₄,
glycogen phosphorylase

α-D-**Glucopyranosyl
1-phosphate**

Phosphorylase cannot break α-1,6-glycosidic bonds. As soon as it gets close to such a branching point (in fact, as soon as it reaches a terminal residue four units away from that point), it stops (Figure 24-4). At this stage, a different enzyme comes into play, transferase, which can shift blocks of three terminal glucosyl residues from one branch to another. One glucose substituent remains at the branching point. Now a third enzyme is required to remove this last obstacle to obtaining a new straight chain. This enzyme is specific for the kind of bond at which cleavage is needed; it is α-1,6-glucosidase, also known as the debranching enzyme. When this enzyme has completed its task, phosphorylase can continue degrading the glucose chain until it reaches another branch, and so forth.

Cell-surface carbohydrates mediate cell-recognition processes

For multicellular organisms to function, each of the great variety of cells must be able to recognize and interact specifically with other cells and with various chemical species. These specific interactions are described as **cell-recognition** processes. Cell recognition usually involves noncovalent binding, commonly by means of hydrogen bonds, with molecules on the exterior surface of the cell. After binding takes place, movement of extracellular material into the interior of the cell may occur.

Figure 24-4 Steps in the degradation of a glycogen side chain. Initially, phosphorylase removes glucose units 1 through 5 and 15 through 17 step by step. The enzyme is now four sugar units away from a branching point (10). Transferase moves units 6 through 8 in one block and attaches them to unit 14. A third enzyme, α-1,6-glucosidase, debranches the system at glucose unit 10, by removing glucose 9. A straight chain has been formed and phosphorylase can continue its degradation job.

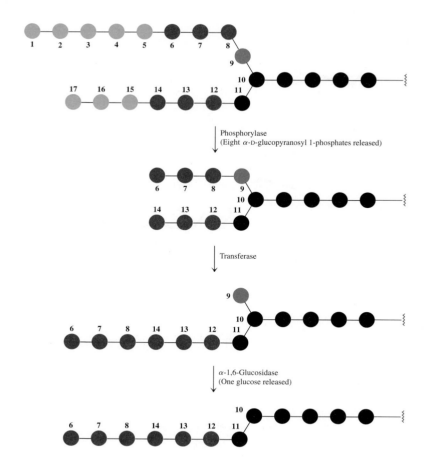

Phosphorylase
(Eight α-D-glucopyranosyl 1-phosphates released)

Transferase

α-1,6-Glucosidase
(One glucose released)

Cell recognition is central to virtually every cell function. Examples of recognition in cell–cell interactions include the response of the immune system to invasion by "foreign" cells, infection of cells by bacteria or viruses, and fertilization of ova by sperm. Certain bacteria, such as cholera and pertussis ("whooping cough"), do not invade cells directly, but the toxins they produce bind to cell surfaces and are subsequently "escorted" inside.

These recognition processes are based on carbohydrates that are present on cell surfaces. The carbohydrates are linked either to lipids (**glycolipids;** Section 20-5) or to proteins (**glycoproteins;** Chapter 26) that are embedded in the cell wall (Figure 24-5). Glucosyl cerebroside is the simplest of the glycolipids and illustrates key features of this class of compounds: a polar carbohydrate "head" that resides on the cell surface, attached to two virtually nonpolar "tails" that embed in the cell membrane. Differences in carbohydrate composition at the "head" give rise to the different functions that these species display. For example, cholera toxin binds specifically to a glycolipid called GM1 pentasaccharide. This substance is found naturally at the surface of the cells of the intestine that are responsible for water absorption. The result is disruption of this function and severe diarrhea.

Glucosyl cerebroside

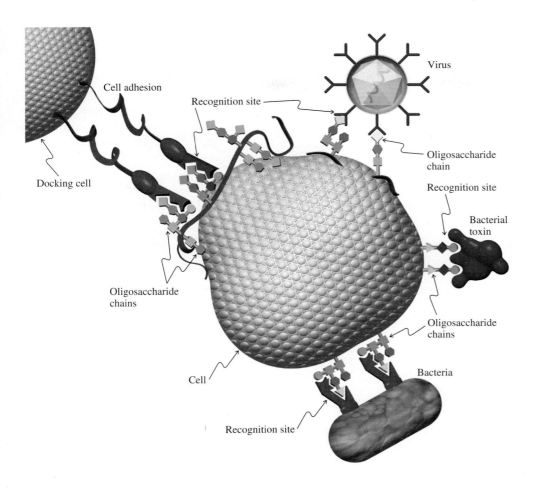

Figure 24-5 Schematic drawing illustrating carbohydrate recognition at the cell surface, enabling cell–cell binding, in addition to adhesion of other species, such as microbes (bacteria, viruses, and bacterial toxins). The sugar chains can be linked to proteins (red ribbons).

The human blood groups—O, A, B, and AB—are each characterized by different oligosaccharides on the red blood cell surface. When an incompatible blood type is introduced by transfusion, the body's immune system produces antibodies that bind to the oligosaccharides on the surfaces of the foreign red cells, causing them to clump together. The result is widespread clogging of blood vessels and death. Type O cells are "coated" with a hexasaccharide whose structure is common to all four blood groups; therefore, type O cells are not recognized as "foreign" by any recipient, regardless of blood type. In type A and type B blood cells, two different additional saccharides are linked to the type O hexasaccharide. It is this "seventh" saccharide that triggers the immune response. Therefore, persons of types A and B cannot donate to each other, and neither can donate to someone of type O. By contrast, type AB cells contain both the type A and type B saccharides, and therefore individuals with type AB blood can receive transfusions from any donor without risk of immune system reaction.

Oddly enough, the biological purpose (if any) of the blood cell surface oligosaccharides is a mystery. Some scientists speculate that the presence of different blood groups (and the associated antibodies) may play a role in inhibiting viral infection: Viruses transmitted between persons with different blood groups might have a greater likelihood of being recognized as "foreign" and destroyed by the recipient's immune system. Nevertheless, people with a rare genetic abnormality that prevents their blood cells from forming *any* surface oligosaccharides are still capable of living normal, healthy lives.

Modified sugars may contain nitrogen

Many of the naturally occurring sugars have a modified structure or are attached to some other organic molecule. In one large class of sugars, at least one of the hydroxy groups has replaced by an amine function. They are called **glycosylamines** when

An Amino Deoxysugar Drug

The galactose mimic migalastat is a drug in clinical trials for the treatment of the rare Fabry disease. The ailment originates from a genetic mutation that prevents the proper folding of the enzyme α-galactosidase A. This enzyme catalyzes the hydrolysis of lipid galactose glycosides, and its absence causes lipid to accumulate in blood vessels and other tissues, impairing function. Migalastat acts by chaperoning the evolution of the correct enzyme conformation.

Migalastat

Galactose

REAL LIFE: MEDICINE 24-3 | Sialic Acid, "Bird Flu," and Rational Drug Design

In 1918 an influenza strain of unprecedented virulence jumped from its normal avian hosts to humans. By the time the outbreak subsided a year later, this "flu" killed as many as 100 million people—the numbers were so staggering that only estimates are possible—nearly 5% of the world's population. The flu virus is a simple entity, its biochemistry governed by a half-dozen or so segments of RNA (Chapter 26) that control the synthesis of a handful of proteins. Two proteins in particular, hemagglutinin (HA) and neuraminidase (NA), determine the virus's ability to bind to and infect cells and to reproduce by release of new virus particles from infected cells, respectively.

Influenza viruses mutate rapidly, giving rise to strains that generate modified versions of HA and NA. The flu responsible for the 1918–1919 pandemic is labeled H1N1. In all, 16 forms of HA and 9 of NA have been identified. The H5N1 strain is of most current concern: It infects vast populations of birds worldwide, especially in Southeast Asia. At present, H5N1 rarely jumps to humans, but, like H1N1, it is highly lethal when it does. The fear that a mutant variety of H5N1 will gain the ability of human-to-human transmission has sparked considerable research into vaccines to prevent infection, as well as antiviral drugs that inhibit the function of viral proteins such as NA once infection has occurred. The vaccine route is made more difficult by the tendency of the virus to mutate in almost every reproductive cycle. The antiviral strategy relies on the known biochemistry of a carbohydrate that occurs naturally in the blood and tissues of all vertebrate animals: **sialic acid.**

Sialic acid describes not one but a group of compounds, *N*- or *O*-substituted derivatives of neuraminic acid, a monosaccharide with a nine-carbon backbone. These compounds have essential biological functions in the blood, in the brain (especially for learning and memory), and elsewhere. The term "sialic acid" is often used to refer to the most common member of the family, *N*-acetylneuraminic acid (Neu5Ac). Sialic acid also protects organisms from invading pathogens;

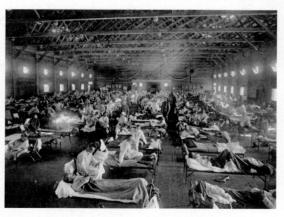

Emergency hospital ward set up during the 1918 Spanish flu pandemic at the US Army's Camp Funston in Kansas, USA.

paradoxically, however, its presence on cell surfaces facilitates the binding of HA of the influenza virus, which leads to infection.

In the 1970s, studies identified a sialic acid derivative, DANA, that bound to NA and inhibited viral reproduction. An X-ray crystal structure study of the DANA–NA complex showed that the hydroxy group at C4 of the bound derivative was located close to a negatively charged carboxylate ion that was essential for NA activity. Carboxylic acids have a pK_a of about 4 (Section 19-4); therefore, at the nearly neutral pH of the body's fluids, an acid will be in the form of its conjugate base. In an effort to make this binding stronger, a group of Australian chemists converted the C4 hydroxy into an amino group, which would be mostly protonated at pH $\approx$ 7 (Section 21-4). The amine analog was found to bind to and inhibit NA 100 times more strongly. In a final modification, this substituent was converted to a guanidino function,

the nitrogen is attached to the anomeric carbon and **amino deoxy sugars** when it is located elsewhere.

β-D-Glucopyranosylamine
(A glycosylamine)

2-Amino-2-deoxy-D-glucopyranose
(An amino deoxy sugar)

When a sugar is attached by its anomeric carbon to the hydroxy group of another complex residue, it is called a **glycosyl group.** The remainder of the molecule (or the product after removal of the sugar by hydrolysis) is called the **aglycon.** An example is adriamycin, a member of the anthracycline family of antibiotics. Adriamycin and its deoxy analog daunomycin have been remarkably effective in the treatment of a wide variety of human cancers. They now constitute a cornerstone of combination cancer chemotherapy (see also p. 763). The aglycon part of these systems is a linear tetracyclic framework incorporating an

$$\overset{\text{NH}}{\underset{\|}{}}$$
—NH—C—NH₂, which is a far stronger base than a simple amine, because the positive charge of its conjugate acid is highly delocalized by resonance (Section 21-6). This compound became the drug zanamivir, now marketed as Relenza by the GlaxoSmithKline pharmaceutical company. Relenza is quite likely to be the front-line antiviral compound in the worldwide effort to prepare for a bird flu outbreak.

Relenza is the result of a process now commonly followed by pharmaceutical companies called **rational drug design.** Sophisticated software packages model both the three-dimensional shape and the surface charge distribution of target molecules, such as NA. These models are in turn employed to "design" drug "candidates" that have complementary shapes and charge distributions, so that they will "dock" with the target in a way that inhibits its function. This strategy enables chemists to identify optimal structural features of a potential drug candidate *before* actually going into the lab to synthesize the molecule. The graphic below displays schematically the process that led from sialic acid to the drug Relenza, based on the experimental and computational evaluations of the attraction of each drug-candidate structure to the carboxylate ion and other groups in the critical region of the neuraminidase (NA) target molecule.

Rational Design of the Neuraminidase (NA) Inhibitor Zanamivir (Relenza) $(R = CH_3\overset{O}{\overset{\|}{C}})$

Sialic acid (Neu5Ac) ⟶ DANA ⟶ 4-Amino-DANA ⟶ Zanamivir (Relenza)

| Weak inhibitor (hydrogen bonding) | Slightly better inhibitor (hydrogen bonding) | 100× better inhibitor (ionic attraction) | Final pharmaceutical (strong ionic attraction) |

anthraquinone moiety (derived from anthracene; see Section 15-5). The amino sugar is called daunosamine.

Adriamycin (R = OH)
Daunomycin (R = H)

Streptomycin

An unusual group of antibiotics, the aminoglycoside antibiotics, is based on oligosaccharide structures almost exclusively. Of particular therapeutic importance is streptomycin (an antituberculosis agent), isolated in 1944 from cultures of the mold *Streptomyces griseus*. The molecule consists of three subunits: the furanose streptose, the glucose derivative 2-deoxy-2-methylamino-L-glucose (an example of the rare L form), and streptidine, which is actually a hexasubstituted cyclohexane.

In Summary The polysaccharides cellulose, starch, and glycogen are polyglucosides. Cellulose consists of repeating dimeric cellobiose units. Starch may be regarded as a polymaltose derivative. Its occasional branching poses a challenge to enzymatic degradation, as does glycogen. Metabolism of these polymers first gives monomeric glucose, which is then further degraded. Finally, many sugars exist in nature in modified form or as simple appendages to other structures. Examples include cell-surface carbohydrates and amino sugars. The aminoglycoside antibiotics consist entirely of saccharide molecules, modified and unmodified.

THE BIG PICTURE

The carbohydrates are a broad and diverse class of compounds, responsible for countless functions in biological chemistry. Nonetheless, their chemical behavior *and their biological activity* arise directly from their structures as polyfunctional molecules with well-defined shapes, which in turn result from their stereochemical characteristics. Their study provides an excellent opportunity to review the properties of the carbonyl and hydroxy functional groups, and to explore how such simple structural components can give rise to a rich array of molecular properties.

We continue in Chapter 25 with a look at heterocyclic compounds, also powerful contributors to biological function, but with structures built around rings. These systems contain oxygen, sulfur, or nitrogen atoms in addition to carbon. In Chapter 26, we will return to systems in which carbohydrates play an important role when we discuss the nucleic acids.

WORKED EXAMPLE: INTEGRATING THE CONCEPTS

24-32. An Extended Exercise in Sugar Structure Determination

Rutinose is a sugar that is part of several bioflavonoids, compounds found in many plants that have significant therapeutic value in maintaining cardiovascular health in general and the strength of the walls of blood vessels in particular. Rutin is one rutinose-containing bioflavonoid found in buckwheat and eucalyptus. Hesperidin is another, derived from the peels of lemons and oranges. Each contains rutinose bound to a tricyclic aglycon (Section 24-12).

The peels of these citrus fruits are rich in bioflavonoids.

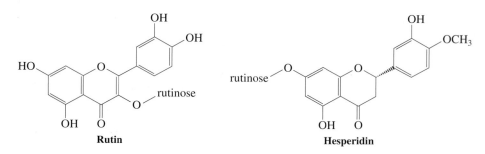

Rutin   **Hesperidin**

Use the following information to deduce a structure for the sugar rutinose.

a. Rutinose is a reducing sugar that, upon acid hydrolysis, gives one equivalent each of D-glucose and a sugar A with the formula $C_6H_{12}O_5$. Sugar A reacts with four equivalents of HIO_4 to give four equivalents of formic acid and one equivalent of acetaldehyde. What can we conclude regarding sugar A at this stage?

SOLUTION

What does the result of HIO_4 degradation tell us? Each equivalent of HIO_4 cleaves a bond between two oxygen-bearing carbon atoms. As the examples in Section 24-5 illustrate, formic acid can arise from either a terminal formyl group or an internal secondary hydroxy group. Acetaldehyde is an unusual degradation product. Its formation implies a terminal methyl substituent, attached to a secondary hydroxy carbon. A logical reconstruction of sugar A from this information may look like this:

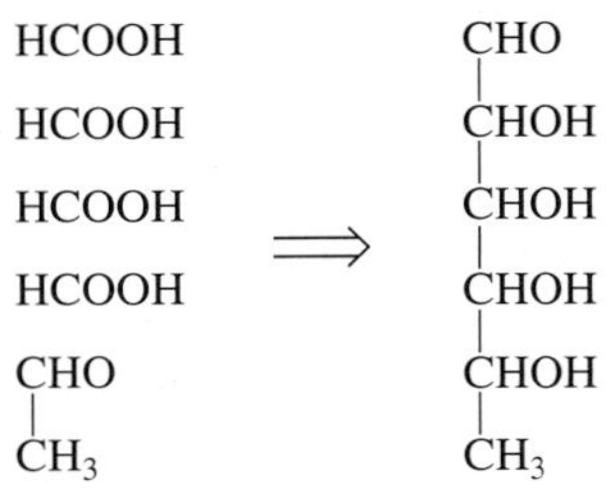

As a check, we note that our structure does indeed have the molecular formula $C_6H_{12}O_5$.

b. Sugar A can be synthesized from L-(−)-mannose as shown in the margin. Step 3 (noted by an asterisk) is a special reaction that converts the terminal primary alcohol into a carboxylic acid group. What does this result reveal about the stereocenters in sugar A?

L-(−)-Mannose

1. HSCH₂CH₂SH, ZnCl₂
2. Raney Ni (Section 17-8)
3. O₂, Pt*
4. Δ (−H₂O) (Section 24-4)
5. [(CH₃)₂CHCH₂]₂AlH (Section 20-4)

Sugar A

SOLUTION

We follow the reaction sequence step by step to obtain the full open-chain structure of sugar A:

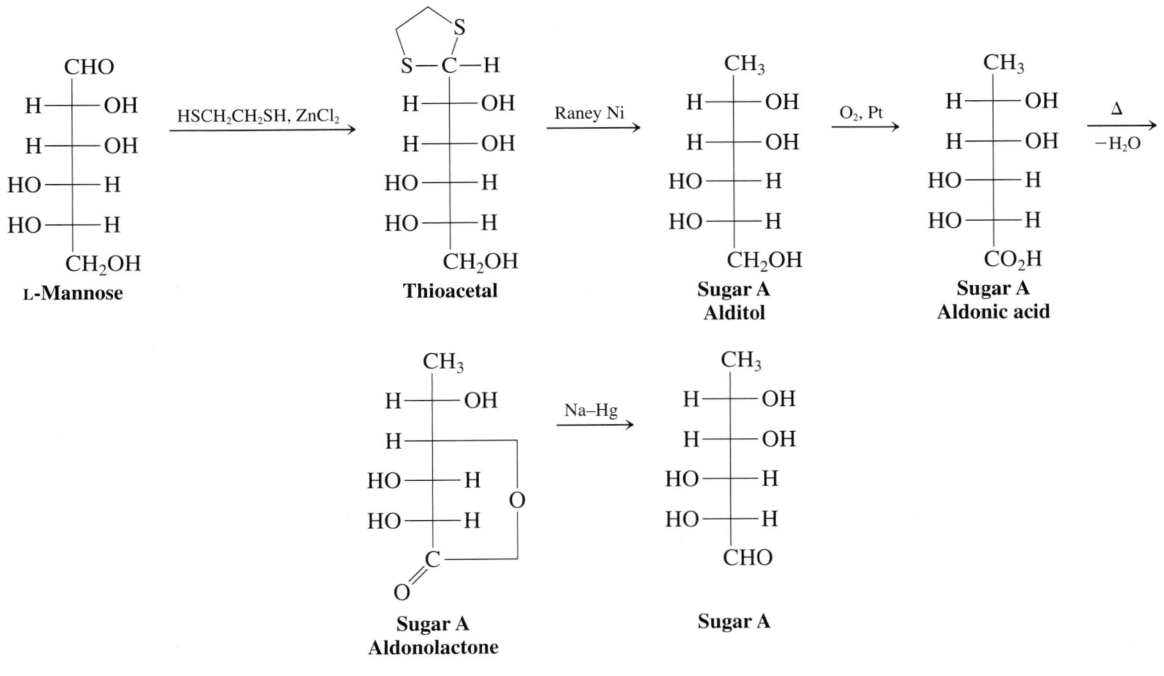

Sugar A has the name 6-deoxy-L-mannose.

c. Complete methylation of rutinose with the use of excess dimethyl sulfate (Section 24-8) gives a heptamethylated derivative. On subsequent mild acid hydrolysis, one equivalent of 2,3,4-tri-*O*-methyl-D-glucose and one equivalent of the 2,3,4-tri-*O*-methyl derivative of sugar A are obtained. What possible structure(s) of rutinose is consistent with these data?

SOLUTION

Dimethyl sulfate treatment converts all free –OH groups into –OCH₃ moieties (Section 24-8). Therefore, we may conclude that rutinose possesses seven hydroxy substituents, and (at least) one must be part of a cyclic hemiacetal function. Recall that rutinose is a reducing sugar (Section 24-4). We may also conclude that *both* monosaccharide components of rutinose are in cyclic forms. Why?

The open-chain forms of glucose and sugar A contain a total of *nine* –OH groups. For rutinose to possess only seven, two of the original nine must be built into a cyclic glycosidic acetal function linking the sugars, similar to the linkages in the disaccharides maltose, cellobiose, and lactose (Section 24-11).

Acid hydrolysis of the heptamethylated rutinose gives two *trimethylated* monosaccharides. Acid cleaves the glycoside linkage between the two sugars (Section 24-11), but where did the seventh methyl group go? It must have been attached to a hemiacetal oxygen, in the form of a methyl glycoside—an acetal—which we know is cleaved by acid much more easily than an ordinary methyl ether (Section 24-8). The six methyl groups (at carbons 2, 3, and 4 in the two resulting monosaccharides) must be bound to oxygen atoms that are *not* part of either the glycosidic linkage between the sugars or their rings. Only the oxygen atoms left unmethylated by dimethyl sulfate treatment can constitute these parts of rutinose. Thus, we may conclude that the oxygens at C5 in both methylated monosaccharide products are contained in pyranose (six-membered) rings, because five-membered rings would have contained the oxygens at C4 instead. We are almost finished. There remain three hydroxy groups that are candidates for the linkage between the sugars: the hemiacetal –OH at C1 of sugar A and either the C1 hemiacetal hydroxy or the C6 primary –OH of the glucose. To summarize:

6-Deoxy-L-mannopyranose
Sugar A

D-Glucopyranose

In fact, we already have the answer: Rutinose is a *reducing sugar* and must therefore possess *at least one hemiacetal function.* The only option therefore is to assign the latter to the glucose moiety and the glycoside bond to C6–OH and C1 of sugar A. We conclude by redrawing the structures in more descriptive Haworth projections and chair conformations (the stereochemistry at C1 of sugar A is not defined) by following the procedures in Section 24-2. In the following three drawings, the glucopyranose moieties are below the sugar A plane. Because A is an L sugar, care is necessary to preserve the proper absolute stereochemistry. One method is to follow the same procedure as for D sugars: Place the ether linkage in back and rotate the Fischer projection 90° clockwise, such that groups originally on the right in the Fischer projection now point downward and those on the left point upward. Indicators D and L refer to the configuration at C5. Whereas the substituent (C6) at this position points upward in a D sugar, with C5 inverted, this group must now point downward. The two structures I and II below result. An equally correct alternative is to rotate the Fischer projection of an L sugar the opposite way—counterclockwise. In this manner structure III at the right is obtained. Notice that the C6 methyl group now points *upward,* the anomeric carbon (C1) is to the *left,* and the structure of sugar A is presented in a view rotated by 180° about an axis perpendicular to the page, relative to the drawings at the right. Build models!

Rutinose

I

II

III

New Reactions

1. Cyclic Hemiacetal Formation in Sugars (Section 24-2)

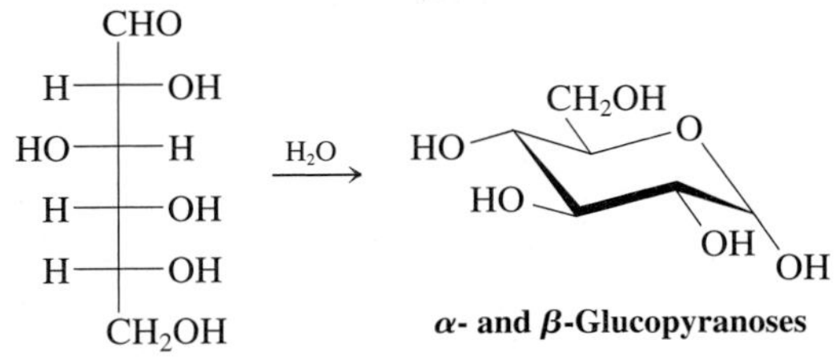

α- and β-Glucopyranoses

2. Mutarotation (Section 24-3)

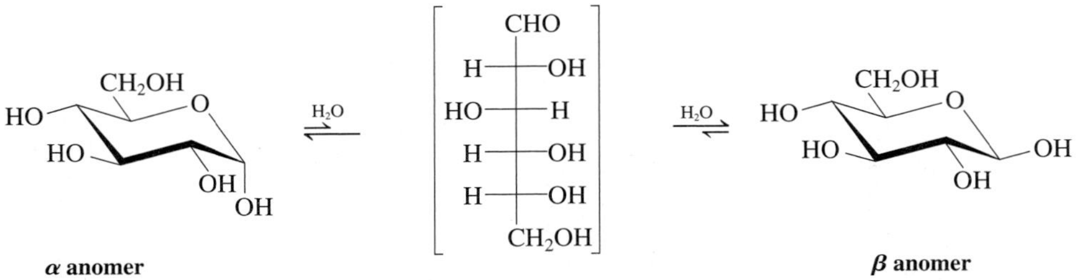

α anomer
$([\alpha]_D^{25°C} = +112)$

(Equilibrium $[\alpha]_D^{25°C} = +52.7$)

β anomer
$([\alpha]_D^{25°C} = +18.7)$

3. Oxidation (Section 24-4, most nonessential H and OH substituents have been omitted)

Tests for reducing sugars

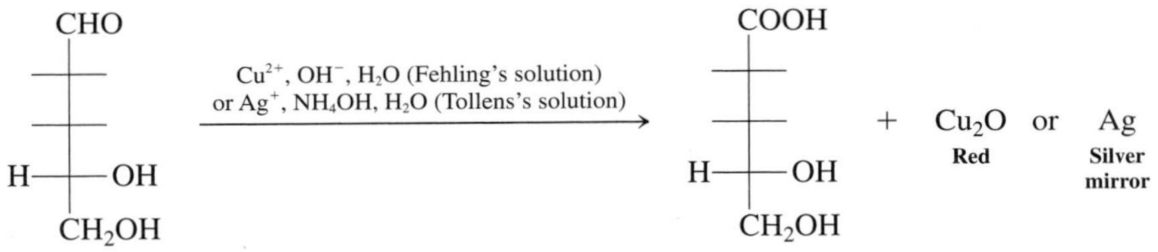

Cu^{2+}, OH^-, H_2O (Fehling's solution)
or Ag^+, NH_4OH, H_2O (Tollens's solution)

+ Cu_2O or   Ag
        Red   Silver
                     mirror

Aldonic acid synthesis

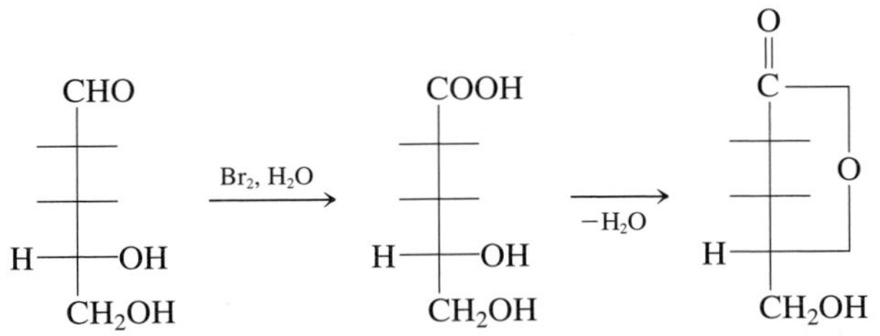

Br_2, H_2O

$-H_2O$

Aldonic acid   γ-**Lactone**

Aldaric acid synthesis

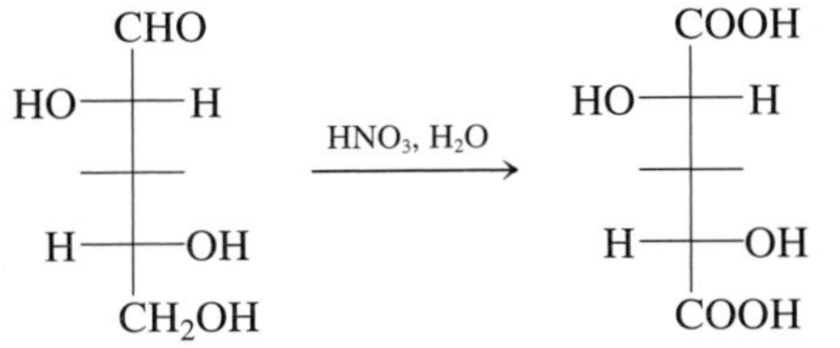

HNO_3, H_2O

Aldaric acid

4. Sugar Degradation (Section 24-5)

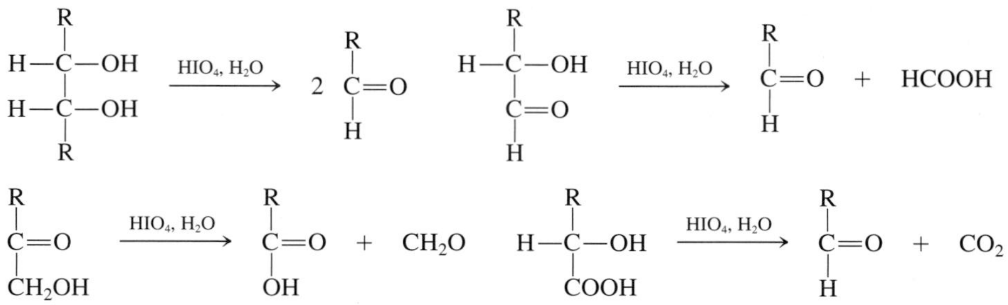

$$\begin{array}{c} R \\ | \\ H-C-OH \\ | \\ H-C-OH \\ | \\ R \end{array} \xrightarrow{\text{HIO}_4,\ \text{H}_2\text{O}} 2 \begin{array}{c} R \\ | \\ C=O \\ | \\ H \end{array}$$

$$\begin{array}{c} R \\ | \\ H-C-OH \\ | \\ C=O \\ | \\ H \end{array} \xrightarrow{\text{HIO}_4,\ \text{H}_2\text{O}} \begin{array}{c} R \\ | \\ C=O \\ | \\ H \end{array} + \text{HCOOH}$$

$$\begin{array}{c} R \\ | \\ C=O \\ | \\ CH_2OH \end{array} \xrightarrow{\text{HIO}_4,\ \text{H}_2\text{O}} \begin{array}{c} R \\ | \\ C=O \\ | \\ OH \end{array} + CH_2O$$

$$\begin{array}{c} R \\ | \\ H-C-OH \\ | \\ COOH \end{array} \xrightarrow{\text{HIO}_4,\ \text{H}_2\text{O}} \begin{array}{c} R \\ | \\ C=O \\ | \\ H \end{array} + CO_2$$

5. Reduction (Section 24-6)

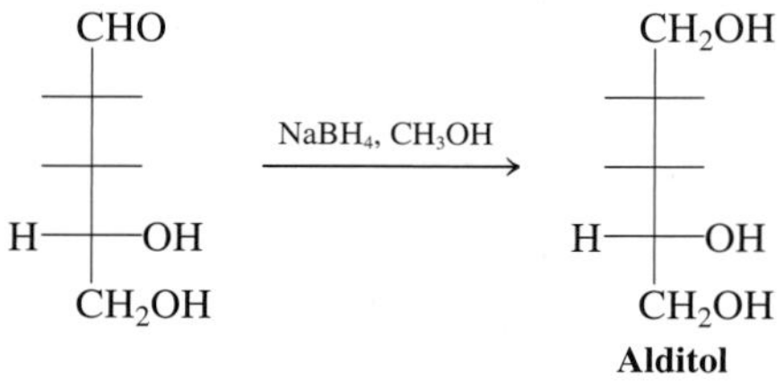

$$\begin{array}{c} CHO \\ | \\ H-\!\!\!-OH \\ | \\ CH_2OH \end{array} \xrightarrow{\text{NaBH}_4,\ \text{CH}_3\text{OH}} \begin{array}{c} CH_2OH \\ | \\ H-\!\!\!-OH \\ | \\ CH_2OH \end{array}$$

Alditol

6. Hydrazones and Osazones (Section 24-7)

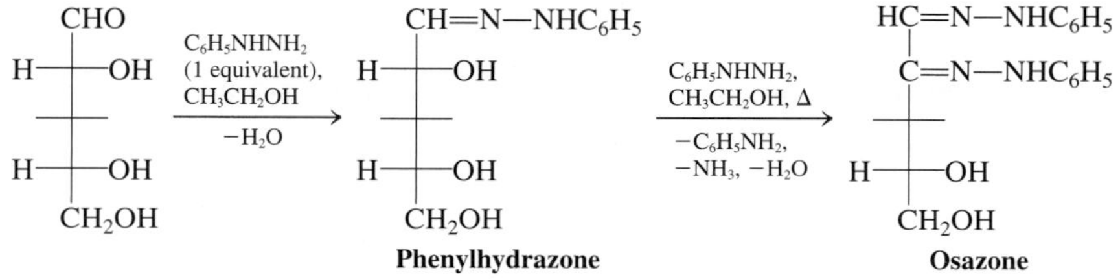

$$\begin{array}{c} CHO \\ | \\ H-\!\!\!-OH \\ | \\ H-\!\!\!-OH \\ | \\ CH_2OH \end{array} \xrightarrow[\substack{\text{CH}_3\text{CH}_2\text{OH} \\ -\text{H}_2\text{O}}]{\substack{\text{C}_6\text{H}_5\text{NHNH}_2 \\ \text{(1 equivalent),}}} \begin{array}{c} CH=N-NHC_6H_5 \\ | \\ H-\!\!\!-OH \\ | \\ H-\!\!\!-OH \\ | \\ CH_2OH \end{array} \xrightarrow[\substack{\text{CH}_3\text{CH}_2\text{OH},\ \Delta \\ -\text{C}_6\text{H}_5\text{NH}_2, \\ -\text{NH}_3,\ -\text{H}_2\text{O}}]{\text{C}_6\text{H}_5\text{NHNH}_2,} \begin{array}{c} HC=N-NHC_6H_5 \\ | \\ C=N-NHC_6H_5 \\ | \\ H-\!\!\!-OH \\ | \\ CH_2OH \end{array}$$

Phenylhydrazone   **Osazone**

7. Esters (Section 24-8)

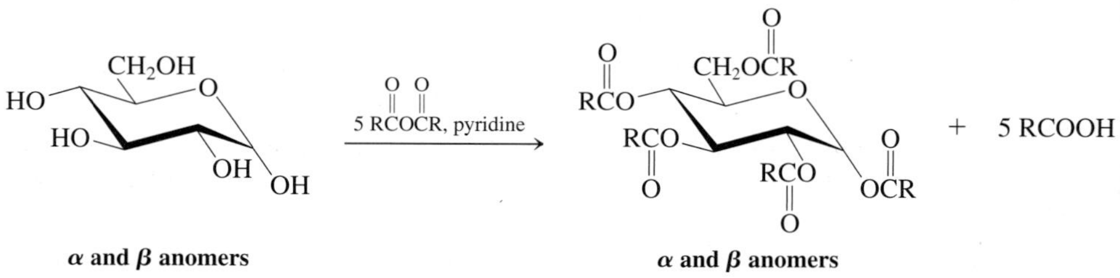

α and β anomers $\xrightarrow[\text{5 RCOCR, pyridine}]{}$ α and β anomers   +   5 RCOOH

8. Glycosides (Section 24-8)

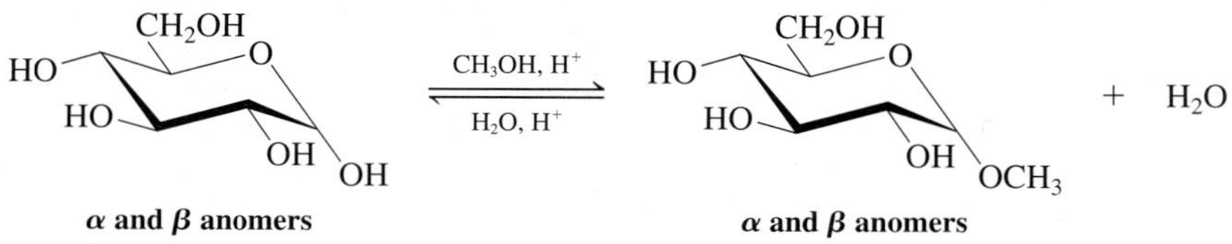

α and β anomers $\underset{\text{H}_2\text{O, H}^+}{\overset{\text{CH}_3\text{OH, H}^+}{\rightleftharpoons}}$ α and β anomers   + H_2O

9. Ethers (Section 24-8)

α and β anomers $\xrightarrow[-\text{Na}_2\text{SO}_4]{\text{5 (CH}_3)_2\text{SO}_4,\ \text{Na}^{+-}\text{OH}}$ α and β anomers

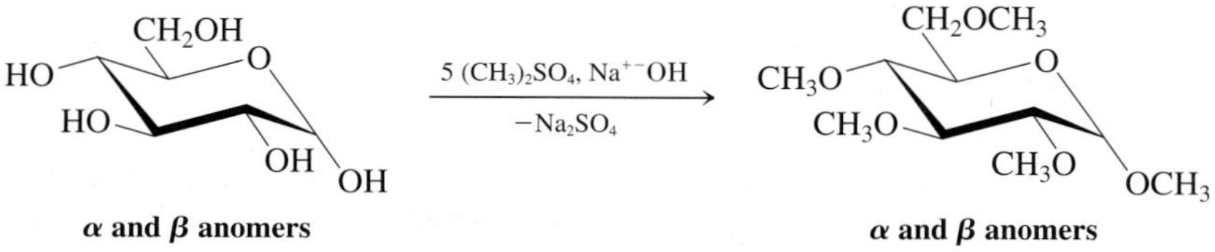

10. Cyclic Acetals (Section 24-8)

11. Chain Extension Through Cyanohydrins (Section 24-9)

12. Ruff Degradation (Section 24-9)

Important Concepts

1. **Carbohydrates** are naturally occurring **polyhydroxycarbonyl** compounds that can exist as monomers, dimers, oligomers, and polymers.

2. **Monosaccharides** are called **aldoses** if they are aldehydes and **ketoses** if they are ketones. The chain length is indicated by the prefix tri-, tetr-, pent-, hex-, and so forth.

3. Most natural carbohydrates belong to the **D family;** that is, the stereocenter farthest from the carbonyl group has the same configuration as that in (R)-$(+)$-2,3-dihydroxypropanal [D-$(+)$-glyceraldehyde].

4. The keto forms of carbohydrates exist in equilibrium with the corresponding five-membered **(furanoses)** or six-membered **(pyranoses)** cyclic hemiacetals. The new stereocenter formed by cyclization is called the **anomeric carbon,** and the two **anomers** are designated α and β.

5. **Haworth projections** of D sugars depict the cyclic ether in line notation as a pentagon or a hexagon, the anomeric carbon placed on the right and the ether oxygen at the top. The substituents located above or below the ring are attached to vertical lines. The ring bond at the bottom (between C2 and C3) is understood to be in front of the plane of the paper, and the ring bonds containing the oxygen are understood to be in back. The α anomer has the OH group at the anomeric carbon pointing downward, whereas the β anomer has it pointing upward.

6. Equilibration between anomers in solution gives rise to changes in the measured optical rotation called **mutarotation.**

7. The reactions of the saccharides are characteristic of carbonyl, alcohol, and hemiacetal groups. They include oxidation of the aldehyde to the carboxy function of **aldonic acids,** double oxidation to **aldaric acids,** oxidative cleavage of vicinal diol units, reduction to **alditols,** condensations, esterifications, and acetal formations.

8. Sugars containing hemiacetal functions are called **reducing sugars,** because they readily reduce Tollens's and Fehling's solutions. Sugars in which the anomeric carbon is acetalized are nonreducing.

9. The synthesis of higher sugars is based on **chain lengthening,** the new carbon being introduced by cyanide ion. The synthesis of lower sugars relies on **Ruff chain shortening,** a terminal carbon being expelled as CO_2.

10. The **Fischer proof** uses the techniques of chain lengthening and shortening together with the symmetry properties of aldaric acids to determine the structures of the alsoses.

11. **Di-** and **higher saccharides** are formed by the combination of hemiacetal and alcohol hydroxy groups of two sugars to give an acetal linkage.

12. The change in optical rotation observed in acidic aqueous solutions of sucrose, called the **inversion of sucrose,** is due to the equilibration of the starting sugar with the various cyclic and anomeric forms of its component monomers.

13. Many sugars contain modified backbones. Amino groups may have replaced hydroxy groups, there may be substituents of various complexity **(aglycons),** the backbone carbon atoms of a sugar may lack oxygens, and (rarely) the sugar may adopt the L configuration.

Problems

33. The designations D and L as applied to sugars refer to the configuration of the highest-numbered stereocenter. If the configuration of the highest numbered stereocenter of D-ribose (Figure 24-1) is switched from D to L, is the product L-ribose? If not, what is the product? How is it related to D-ribose (i.e., what kind of isomers are they)?

34. To which classes of sugars do the following monosaccharides belong? Which are D and which are L?

(a) HOCH₂ (+)-Apiose

(b) HO (−)-Rhamnose

(c) (+)-Mannoheptulose

35. Draw open-chain (Fischer-projection) structures for L-(+)-ribose and L-(−)-glucose (see Exercise 24-2). What are their systematic names?

36. Identify the following sugars, which are represented by unconventionally drawn Fischer projections. **(Hint:** It will be necessary to convert these projections into more conventional representations *without* inverting any of the stereocenters.)

(a) HOCH₂—CHO

(b) HOCH₂—H

(c) HO—

(d) HOCH₂—OH

(e) HOCH₂—H

37. Redraw each of the following sugars in open-chain form as a Fischer projection, and find its common name.

(a) OHC—CH₂OH

(b) H—OH—CH₂OH

(c)

(d) HO—CH₂OH

38. For each of the following sugars, draw all reasonable cyclic structures, using either Haworth or conformational formulas; indicate which structures are pyranoses and which are furanoses; and label α and β anomers.

(a) (−)-Threose; (b) (−)-allose; (c) (−)-ribulose; (d) (+)-sorbose; (e) (+)-mannoheptulose (Problem 34).

39. Are any of the sugars in Problem 38 incapable of mutarotation? Explain your answer.

40. Draw the most stable pyranose conformation of each of the following sugars. (a) α-D-Arabinose; (b) β-D-galactose; (c) β-D-mannose; (d) α-D-idose.

41. Ketoses show positive Fehling's and Tollens's tests not only by oxidation to α-dicarbonyl compounds, but through a second process: Ketoses isomerize to aldoses in the presence of base. The aldose then undergoes oxidation by the Fehling's or Tollens's solution. Using any ketose in Figure 24-2, propose a base-catalyzed mechanistic pathway to the corresponding aldose. [**Hint:** Review Section 18-2.]

42. What are the products (and their ratios) of periodic acid cleavage of each of the following substances: **(a)** 1,3-dihydroxyacetone; **(b)** rhamnose (Problem 34); **(c)** glucitol.

43. Write the expected products of the reaction of each of the following sugars with (i) Br_2, H_2O; (ii) HNO_3, H_2O, 60°C; (iii) $NaBH_4$, CH_3OH; and (iv) excess $C_6H_5NHNH_2$, CH_3CH_2OH, Δ. Find the common names of all the products. **(a)** D-(−)-Threose; **(b)** D-(+)-xylose; **(c)** D-(+)-galactose.

44. Draw the Fischer projection of an aldohexose that will give the same osazone as **(a)** D-(−)-idose and **(b)** L-(−)-altrose.

45. **(a)** Which of the aldopentoses (Figure 24-1) would give optically active alditols upon reduction with $NaBH_4$? **(b)** Using D-fructose, illustrate the results of $NaBH_4$ reduction of a ketose. Is the situation more complicated than reduction of an aldose? Explain.

46. Which of the following glucoses and glucose derivatives are capable of undergoing mutarotation? **(a)** α-D-Glucopyranose; **(b)** methyl α-D-glucopyranoside; **(c)** methyl α-2,3,4,6-tetra-O-methyl-D-glucopyranoside (i.e., the tetramethyl ether at carbon 2, 3, 4, and 6); **(d)** α-2,3,4,6-tetra-O-methyl-D-glucopyranose; **(e)** α-D-glucopyranose 1,2-acetone acetal.

47. **(a)** Explain why the oxygen at C1 of an aldopyranose can be methylated so much more easily than the other oxygens in the molecule. **(b)** Explain why the methyl ether unit at C1 of a fully methylated aldopyranose can be hydrolyzed so much more easily than the other methyl ether functions in the molecule. **(c)** Write the expected product(s) of the following reaction.

$$D\text{-Fructose} \xrightarrow{CH_3OH,\ 0.25\%\ HCl,\ H_2O}$$

48. Of the four aldopentoses, two form diacetals readily when treated with excess acidic acetone, but the other two form only monoacetals. Explain.

49. D-Sedoheptulose is a sugar intermediate in a metabolic cycle (the pentose oxidation cycle) that ultimately converts glucose into 2,3-dihydroxypropanal (glyceraldehyde) plus three equivalents of CO_2. Determine the structure of D-sedoheptulose from the following information.

$$D\text{-Sedoheptulose} \xrightarrow{6\ HIO_4} 4\ HCOH + 2\ HCH + CO_2$$

$$D\text{-Sedoheptulose} \xrightarrow{C_6H_5NHNH_2} \text{an osazone identical with that formed by another sugar, aldoheptose A}$$

$$\text{Aldoheptose A} \xrightarrow{\text{Ruff degradation}} \text{aldohexose B}$$

$$\text{Aldohexose B} \xrightarrow{HNO_3,\ H_2O,\ \Delta} \text{an optically active product}$$

$$\text{Aldohexose B} \xrightarrow{\text{Ruff degradation}} D\text{-ribose}$$

50. Is it possible for two different diastereomeric aldoses to give the same product upon Kiliani-Fischer chain elongation? Why or why not?

51. In each of the following groups of three D aldoses, identify the two that give the same product upon Ruff degradation. **(a)** Galactose, gulose, talose; **(b)** glucose, gulose, idose; **(c)** allose, altrose, mannose.

52. Illustrate the results of chain elongation of D-talose through a cyanohydrin. How many products are formed? Draw them. After treatment with warm HNO_3, are the resulting dicarboxylic acids optically active or inactive?

53. **CHALLENGE** **(a)** Write a detailed mechanism for the isomerization of β-D-fructofuranose (from the hydrolysis of sucrose) into an equilibrium mixture of the β-pyranose and β-furanose forms. **(b)** Although fructose usually appears as a furanose when it is part of a polysaccharide, in the pure crystalline form, fructose adopts a β-pyranose structure. Draw β-D-fructopyranose in its most stable conformation. In water at 20°C, the equilibrium mixture contains about 68% β-D-pyranose and 32% β-D-furanose. **(c)** What is the free-energy difference between the pyranose and furanose forms at this temperature? **(d)** Pure β-D-fructopyranose has $[\alpha]_D^{20°C} = -132$. The equilibrium pyranose–furanose mixture has $[\alpha]_D^{20°C} = -92$. Calculate $[\alpha]_D^{20°C}$ for pure β-D-fructofuranose.

54. Classify each of the following sugars and sugar derivatives as either reducing or nonreducing. **(a)** D-Glyceraldehyde; **(b)** D-arabinose; **(c)** β-D-arabinopyranose 3,4-acetone acetal; **(d)** β-D-arabinopyranose acetone diacetal; **(e)** D-ribulose; **(f)** D-galactose; **(g)** methyl β-D-galactopyranoside; **(h)** β-D-galacturonic acid (shown below); **(i)** β-cellobiose; **(j)** α-lactose.

β-D-Galacturonic acid

55. Is α-lactose capable of mutarotation? Write an equation to illustrate your answer.

56. Trehalose, sophorose, and turanose are disaccharides. Trehalose is found in the cocoons of some insects, sophorose turns up in a few bean varieties, and turanose is an ingredient in low-grade honey made by bees with indigestion from a diet of pine tree sap. Identify among the following structures those that correspond to trehalose, sophorose, and turanose on the basis of the following information: (i) Turanose and sophorose are reducing sugars. Trehalose is nonreducing. (ii) Upon hydrolysis, sophorose and trehalose give two molecules each of aldoses. Turanose gives one molecule of an aldose and one molecule of a ketose. (iii) The two aldoses that constitute sophorose are anomers of each other.

(a)

(b)

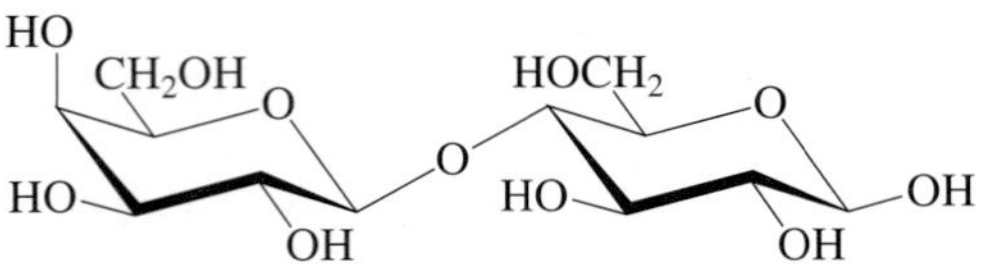

(c)

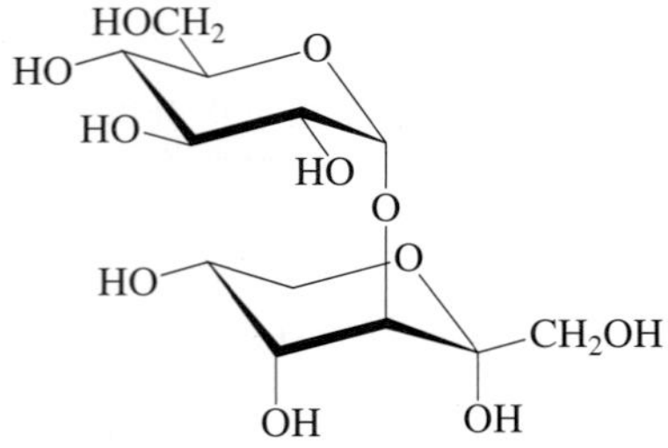

(d)

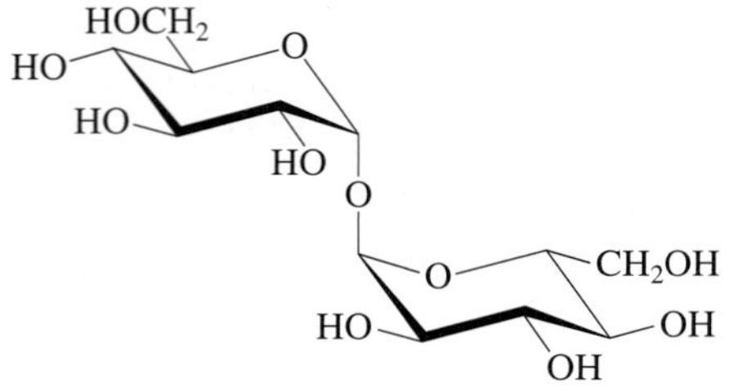

57. Identify and, using the information in Sections 24-1 and 24-11, provide names for the carbohydrates that are contained in the structure of stevioside (Real Life 24-2).

58. The disaccharide at the terminus of the oligosaccharide attached to the surface of type B red blood cells is an α-D-galactopyranosyl-β-D-galactopyranose. The linkage is an acetal from C1 of the first galactose to the C3 hydroxy group of the second. In other words, the full name is 3-(α-D-galactopyranos-1-yl)-β-D-galactopyranose. Draw the structure of this disaccharide, using conformationally accurate chairs for the six-membered rings.

59. Glucose reacts with ammonia in the presence of a trace of acid to give predominantly β-D-glucopyranosylamine (Section 24-12). Propose a reasonable mechanism for this transformation. Why is only the hydroxy group at C1 replaced?

60. **(a)** A mixture of (R)-2,3-dihydroxypropanal (D-glyceraldehyde) and 1,3-dihydroxyacetone that is treated with aqueous NaOH rapidly yields a mixture of three sugars: D-fructose, D-sorbose, and racemic dendroketose (below; only one enantiomer is shown). Explain this result by means of a detailed mechanism. **(b)** The same product mixture is also obtained if either the aldehyde or the ketone *alone* is treated with base. Explain. [**Hint:** Closely examine the intermediates in your answer to part (a).]

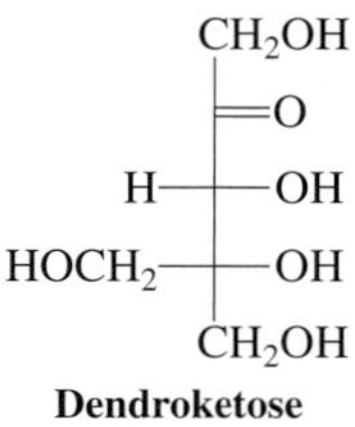

Dendroketose

61. Write or draw the missing reagents and structures (a) through (g). What is the common name of (g)?

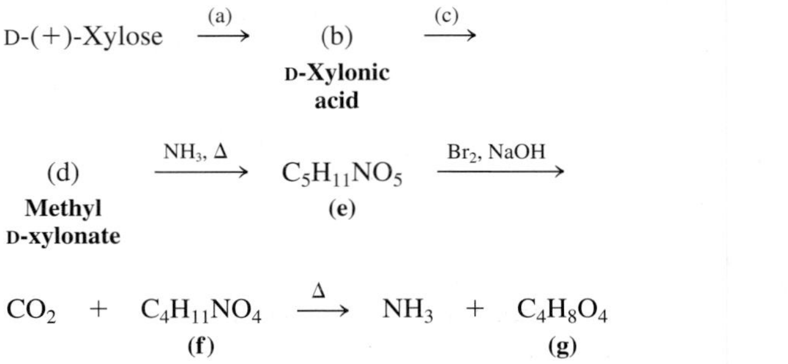

D-(+)-Xylose $\xrightarrow{\text{(a)}}$ **(b)** $\xrightarrow{\text{(c)}}$
D-Xylonic acid

(d) $\xrightarrow{\text{NH}_3,\ \Delta}$ $C_5H_{11}NO_5$ $\xrightarrow{\text{Br}_2,\ \text{NaOH}}$
Methyl D-xylonate **(e)**

CO_2 + $C_4H_{11}NO_4$ $\xrightarrow{\Delta}$ NH_3 + $C_4H_8O_4$
(f) **(g)**

The preceding sequence (called the *Weerman degradation*) achieves the same end as what procedure described in this chapter?

62. **CHALLENGE** Fischer's solution to the problem of sugar structures was actually much more difficult to achieve experimentally than Section 24-10 implies. For one thing, the only sugars that he could obtain readily from natural sources were glucose, mannose, and arabinose. (Erythrose and threose were, in fact, not then available at all, either naturally or synthetically.) His ingenious solution required a way to exchange the functionalities at C1 and C6 of glucose and mannose in order to make the critical distinction described at the end of the section. (Of course, had gulose existed in nature, all this effort would have been unnecessary, but Fischer wasn't so lucky.) Fischer's plan led to unexpected difficulties, because at a key stage he got a troublesome mixture of products. Nowadays we solve the problem in the manner shown below. Fill in the missing reagents and structures (a) through (g). Use Fischer projections for all structures. Follow the instructions and hints in parentheses.

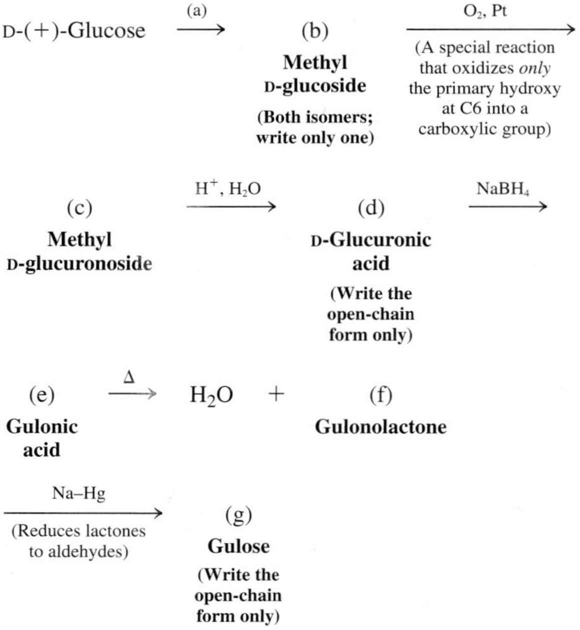

D-(+)-Glucose $\xrightarrow{\text{(a)}}$ **(b)** $\xrightarrow{\text{O}_2,\ \text{Pt}}$
Methyl D-glucoside
(Both isomers; write only one)
(A special reaction that oxidizes *only* the primary hydroxy at C6 into a carboxylic group)

(c) $\xrightarrow{\text{H}^+,\ \text{H}_2\text{O}}$ **(d)** $\xrightarrow{\text{NaBH}_4}$
Methyl D-glucuronoside **D-Glucuronic acid**
(Write the open-chain form only)

(e) $\xrightarrow{\Delta}$ H_2O + **(f)**
Gulonic acid **Gulonolactone**

$\xrightarrow[\text{(Reduces lactones to aldehydes)}]{\text{Na–Hg}}$ **(g)**
Gulose
(Write the open-chain form only)

63. **CHALLENGE** Vitamin C (ascorbic acid, Section 22-9) is present almost universally in the plant and animal kingdoms. (According to Linus Pauling, mountain goats biosynthesize from 12 to 14 g of it per day.) Animals produce it from D-glucose in the liver by the four-step sequence D-glucose → D-glucuronic acid (see Problem 62) → D-glucuronic acid γ-lactone → l-gulonic acid γ-lactone → vitamin C.

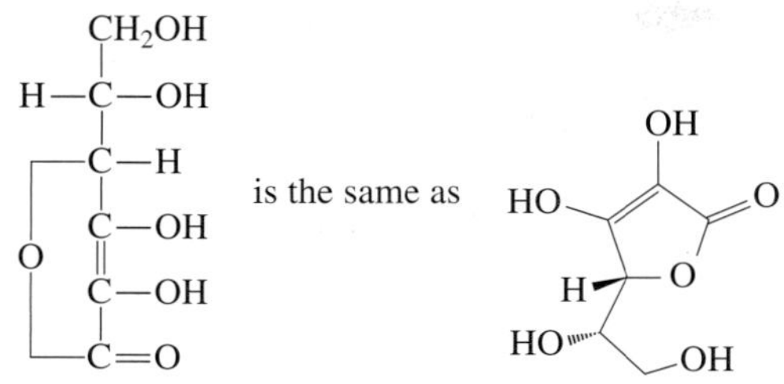

Vitamin C

The enzyme that catalyzes the last reaction, L-gulonolactone oxidase, is absent from humans, some monkeys, guinea pigs, and birds, presumably because of a defective gene resulting from a mutation that may have occurred some 60 million years ago. As a result, we have to get our vitamin C from food or make it in the laboratory. In fact, the ascorbic acid in almost all vitamin supplements is synthetic. An outline of one of the major commercial syntheses follows. Draw the missing reagents and products (a) through (f). (**Hint:** See Section 24-8, sugar acetals)

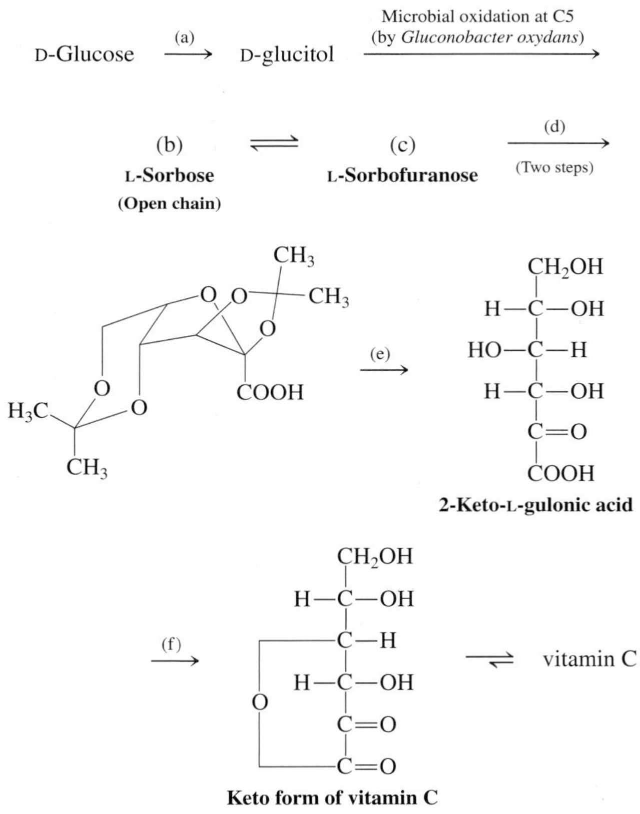

Team Problem

64. This problem is meant to encourage you to think as a team about how you might establish the structure of a simple disaccharide, with some additional information at your disposal. Consider D-lactose (Section 24-11) and assume you do not know its structure. You are given the knowledge that it is a disaccharide, linked in a β manner to the anomeric carbon of only one of the sugars, and you are given the structures of all of the aldohexoses (Figure 24-1), as well as all of their possible methyl ethers. Deal with the following questions as a team or by dividing the work, before joint discussion, as appropriate.

(a) Mild acid hydrolyzes your "unknown" to D-galactose and D-glucose. How much information can you derive from that result?

(b) Propose an experiment that tells you that the two sugars are not connected through their respective anomeric centers.

(c) Propose an experiment that tells you which one of the two sugars contains an acetal group used to bind the other. (**Hint:** The functional group chemistry of the monosaccharides described in the chapter can be applied to higher sugars as well. Specifically, consider Section 24-4 here.)

(d) Making use of the knowledge of the structure of all possible methyl ethers of the component monosaccharides, design experiments that will tell you which (nonanomeric) hydroxy group is responsible for the disaccharide linkage.

(e) Similarly, can you use this approach to distinguish between a furanose and pyranose structure for the component of the disaccharide that can mutarotate?

Preprofessional Problems

65. Most natural sugars have a stereocenter that is identical to that in (R)-2,3-dihydroxypropanal, shown as a Fischer projection below. What is the (very popular) common name for this compound? **(a)** D-(+)-Glyceraldehyde; **(b)** D-(−)-glyceraldehyde; **(c)** L-(+)-glyceraldehyde; **(d)** L-(−)-glyceraldehyde

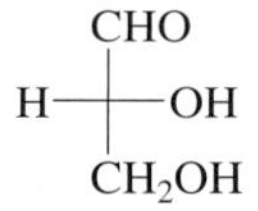

66. What kind of sugar is the compound shown below? **(a)** An aldopentose; **(b)** a ketopentose; **(c)** an aldohexose; **(d)** a ketohexose

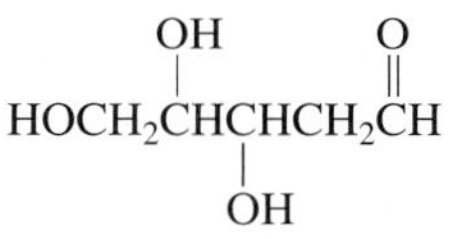

$$HOCH_2CHCHCH_2CH$$

(with OH, OH, and O substituents)

67. Which one of the following statements is *true* about the oxacyclohexane conformer of the sugar β-D-(+)-glucopyranose?

(a) One OH group is axial, but all remaining substituents are equatorial. **(b)** The CH_2OH group is axial, but all remaining groups are equatorial. **(c)** All groups are axial. **(d)** All groups are equatorial.

68. The methyl glycoside of mannose is made by treating the sugar with

(a) $AlBr_3$, CH_3Br; **(b)** dilute aqueous CH_3OH; **(c)** CH_3OCH_3 and $LiAlH_4$; **(d)** CH_3OH, HCl; **(e)** oxacyclopropane, $AlCl_3$

69. One of the statements below is correct about the sugar shown. Which one?

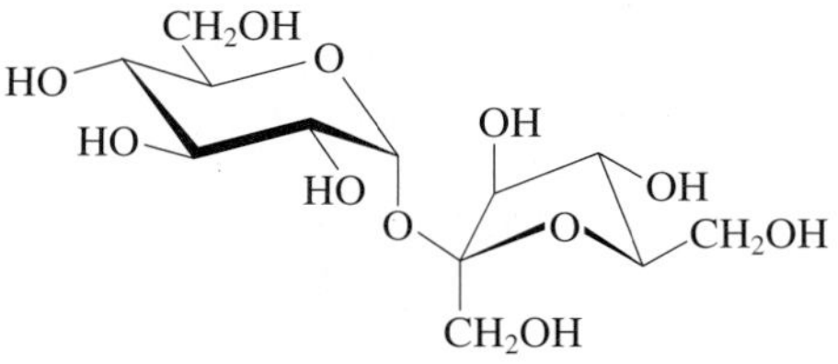

(a) It is a nonreducing sugar. **(b)** It forms an osazone. **(c)** It exists in two anomeric forms. **(d)** It undergoes mutarotation.

CHAPTER 25 | Heterocycles

Heteroatoms in Cyclic Organic Compounds

How many of you can do without that morning cup of coffee? The stimulating ingredient in coffee is the heterocycle caffeine, 135 mg of it in a normal 8-oz serving. Of course, there is also Roman espresso, which has 100 mg of caffeine in less than an ounce of lio'

L ook at the list of the United States's top ten prescription drugs (Table 25-1). What do most of their structures have in common? Apart from the presence of heteroatoms, such as oxygen, nitrogen, and sulfur, they contain at least two rings. Moreover, these rings are made up not solely of carbon, called **carbocycles** (Chapter 4), but also of heteroatoms and are thus called **heterocycles.**

This chapter describes the naming, syntheses, and reactions of some saturated and aromatic heterocyclic compounds in order of increasing ring size, starting with the heterocyclopropanes. Some of this chemistry is a simple extension of transformations presented earlier for carbocycles. However, the heteroatom often causes heterocyclic compounds to exhibit special chemical behavior.

Most physiologically active compounds owe their biological properties to the presence of heteroatoms, mainly in the form of heterocycles. A majority of the known natural products are heterocyclic. It is therefore not surprising that more than half the published chemical literature deals with such compounds—their synthesis, isolation, and interconversions. Indeed, we have already encountered many examples—the cyclic ethers (Section 9-6), acetals (Sections 17-8, 23-4, and 24-8), carboxylic acid derivatives (Chapters 19 and 20), and amines (Chapter 21). The bases in DNA, whose sequence stores hereditary information, are heterocycles (Section 26-9); so are many vitamins, such as B_1 (thiamine, Real Life 23-2), B_2 (riboflavin, Real Life 25-3), B_6 (pyridoxine), the spectacularly complex B_{12}, and vitamins C and E (Section 22-9). The structures of vitamins B_6 and B_{12}, as well as additional examples of heterocyclic systems and their varied uses, are depicted here.

Table 25-1 Top Ten U.S. Prescription Drugs (Ranked by 2011 Sales) by Generic and, in Parentheses, Selected Brand Names[a]

1. Atorvastatin (Lipitor)

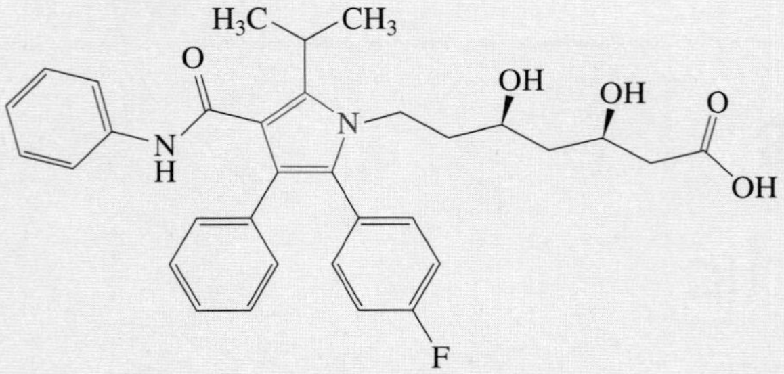

Cholesterol reducer

2. Clopidogrel (Plavix)

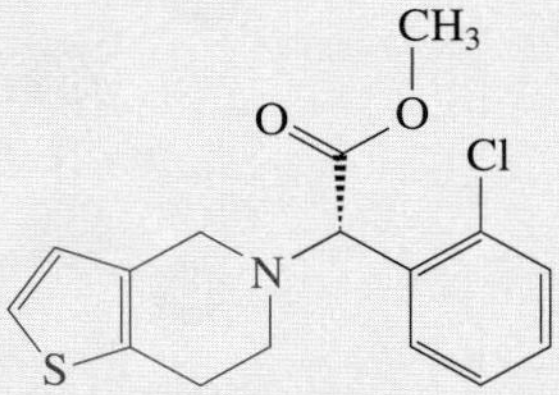

Antiplatelet agent

3. Esomeprazole (Nexium)

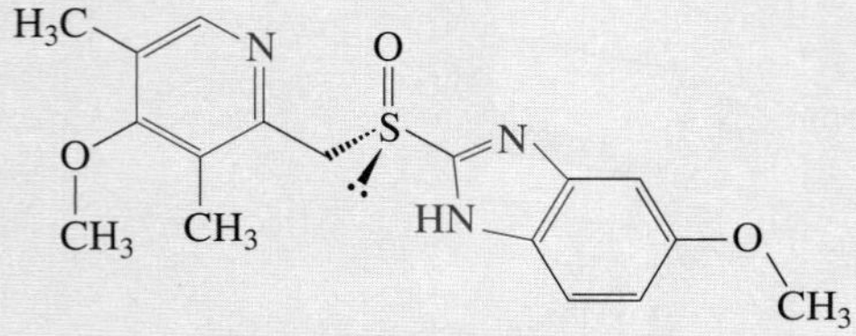

Stomach acid reducer

4. Aripiprazole (Abilify)

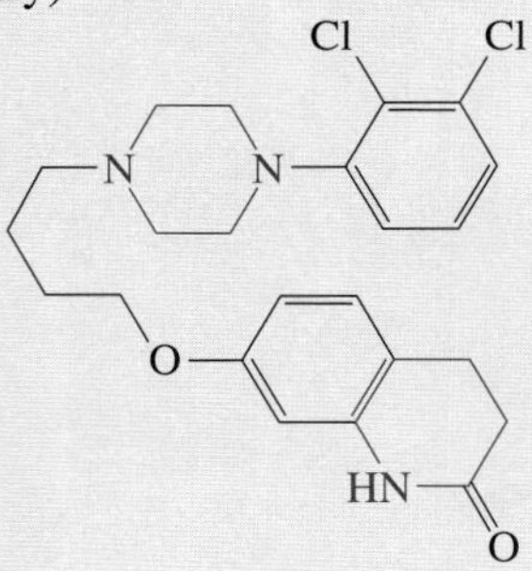

Antipsychotic

5. Fluticasone and salmeterol mixture (Advair Diskus)

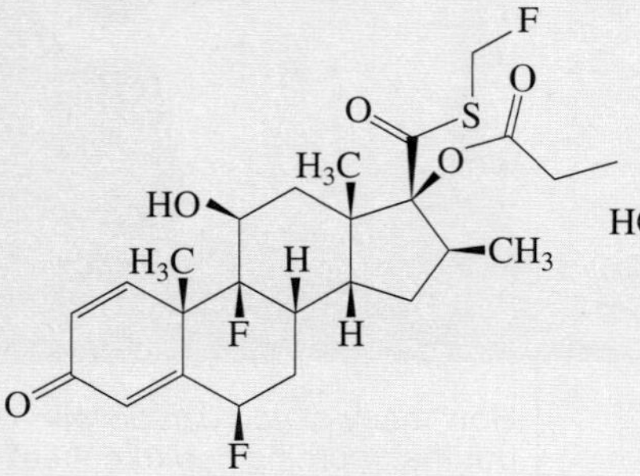

Fluticasone

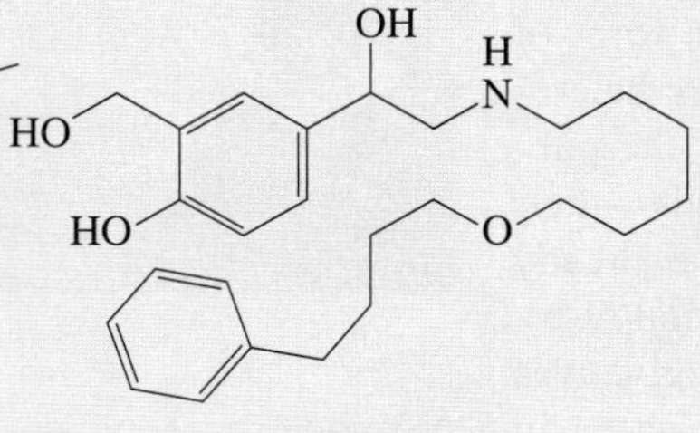

Salmeterol (Racemate)

Antiasthmatic

6. Quetiapine (Seroquel)

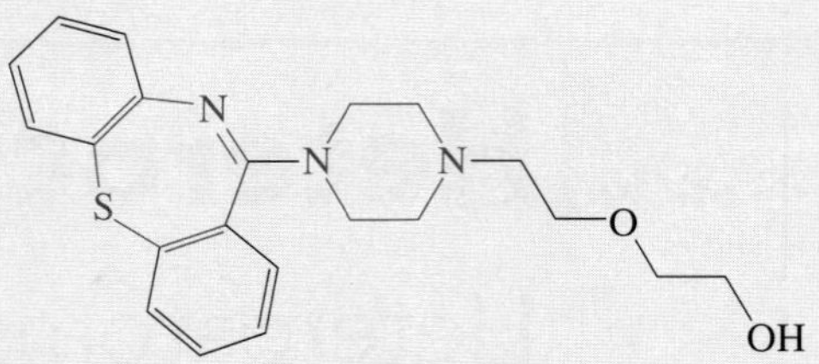

Antipsychotic

7. Montelukast (Singulair)

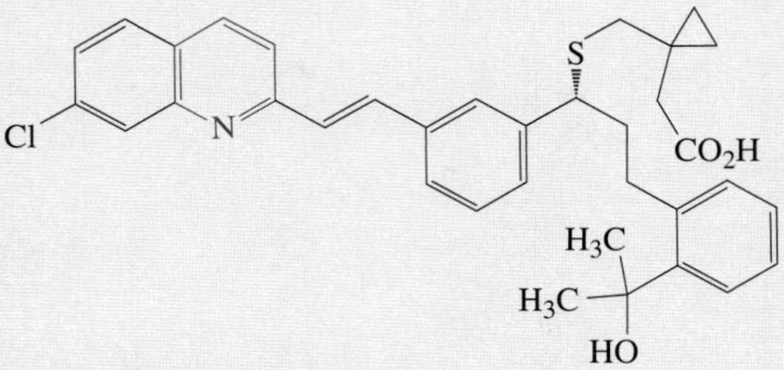

Antiasthmatic

8. Rosuvastatin (Crestor)

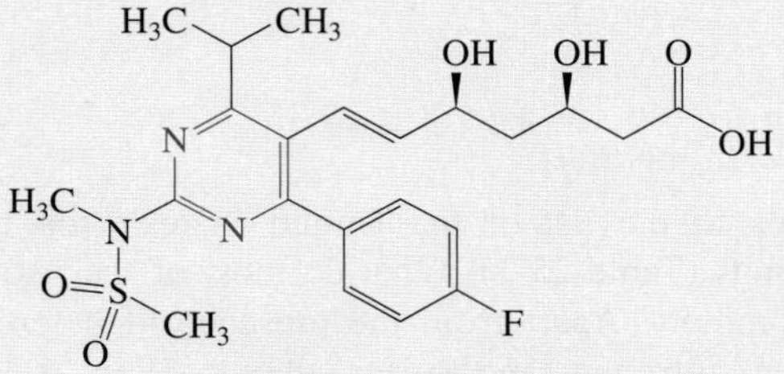

Cholesterol reducer

9. Duloxetine (Cymbalta)

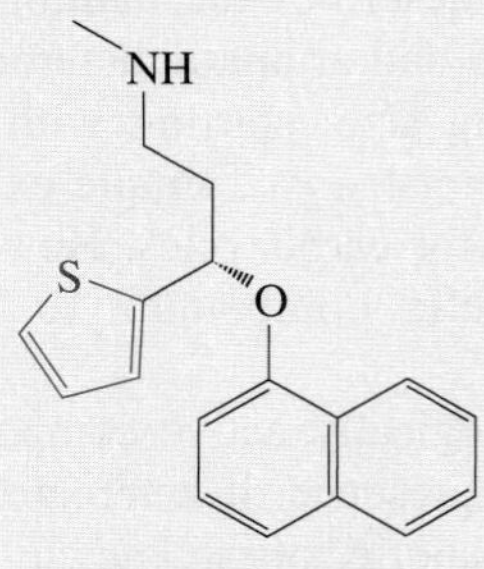

Antidepressant

10. Adalimumab (Humira)

Monoclonal antibody

Antirheumatic

[a]Total U.S. sales of pharmaceuticals reached \$340 billion in 2011, with Lipitor leading at \$7.7 billion.

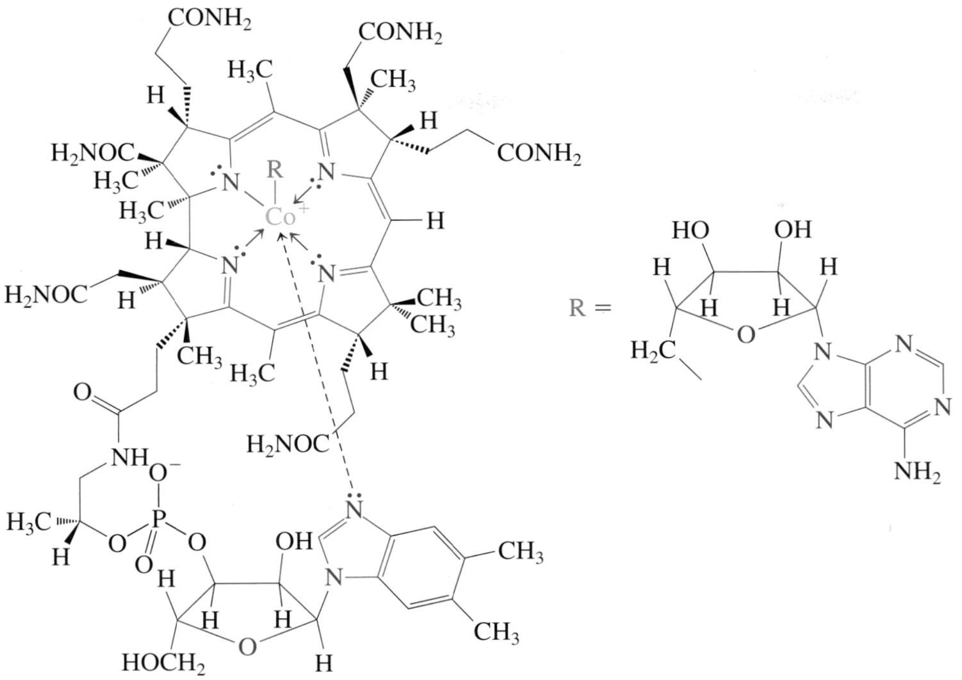

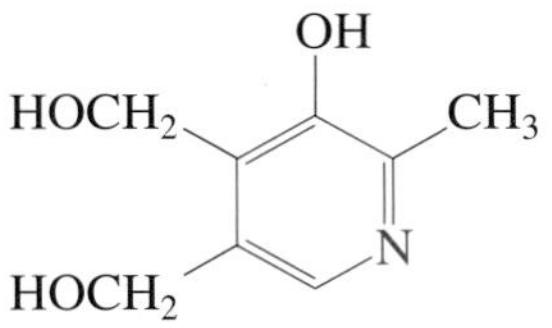

Pyridoxine, vitamin B$_6$

(Enzyme cofactor vitamin with multiple functions)

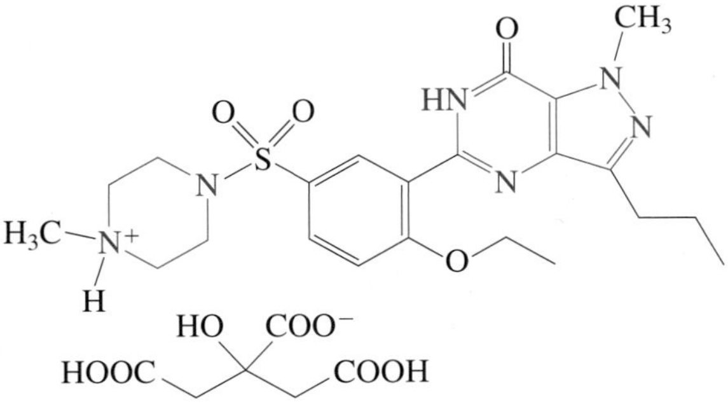

R =

Vitamin B$_{12}$
(Cobalamin)

(Catalyzes biological rearrangements and methylations)

Sildenafil citrate
(Viagra)

(Treats erectile dysfunction;
see also Worked Example 25-29)

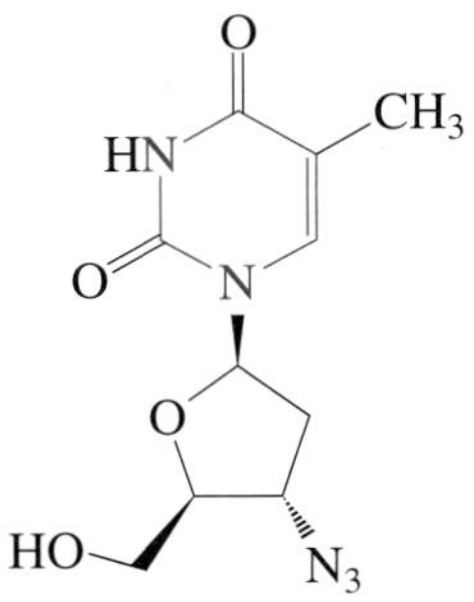

Zidovudine
(AZT)

(Antiviral AIDS drug,
see Real Life 26-3)

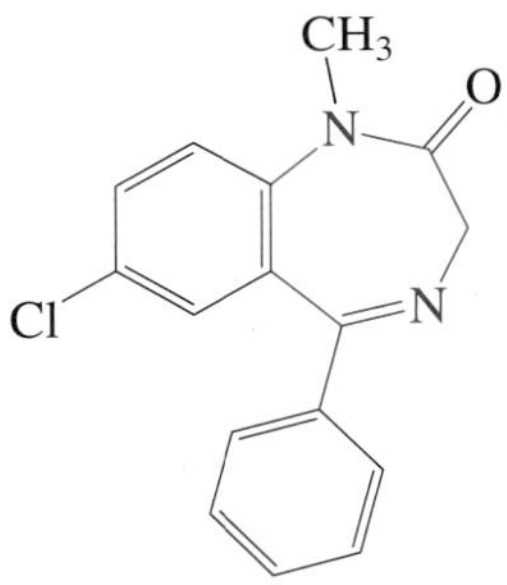

Diazepam
(Valium)

(Tranquilizer)

25-1 | NAMING THE HETEROCYCLES

Like all other classes of compounds we have encountered, the heterocycles contain many members bearing common names. Moreover, there are several competing systems for naming heterocycles, which is sometimes confusing. We shall adhere to the simplest system. We regard saturated heterocycles as derivatives of the related carbocycles and use a prefix to denote the presence and identity of the heteroatom: **aza-** for nitrogen, **oxa-** for oxygen, **thia-** for sulfur, **phospha-** for phosphorus, and so forth. Other widely used names will be given in parentheses. The location of substituents is indicated by numbering the ring atoms, starting with the heteroatom. In the following examples, space-filling renditions of the most representative ring sizes, from three to six membered, give you a perspective of their actual shape.

How Drugs Are Named

Drug naming follows its own (not so) systematic rules. In 1961, the American Medical Association, in conjunction with other groups, founded the U.S. Adopted Names (USAN) Council to formulate generic names. This organization coined a list of standardized syllables (the "stem") that is related to the structure, function, or target of the drug. Stems are usually endings to the name, but they may occur as prefixes or in the middle. For example, -vastatin (as in atorvastatin, see Table 25-1) refers to a cholesterol-lowering medication; -grel (as in clopidogrel) to a platelet aggregation inhibitor; -tiapine (as in quetiapine) to an antipsychotic; estr- to an estrogen; -vudine (as in zidovudine on the left) to an antiviral; and so on. There are over 600 such stems in existence, and more are being added. The final name includes additional syllables that are at the company's or discoverers' discretion, such as "es" in esomeprazole indicating the S enantiomer, and "zido" in zidovudine as short form of azido.

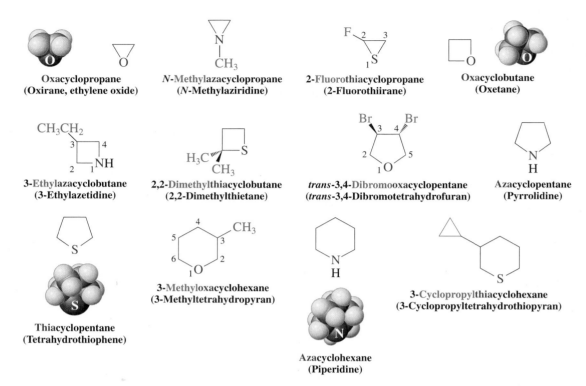

Oxacyclopropane
(Oxirane, ethylene oxide)

N-**Methylazacyclopropane**
(*N*-**Methylaziridine**)

2-Fluorothiacyclopropane
(**2-Fluorothiirane**)

Oxacyclobutane
(**Oxetane**)

3-Ethylazacyclobutane
(**3-Ethylazetidine**)

2,2-Dimethylthiacyclobutane
(**2,2-Dimethylthietane**)

trans-**3,4-Dibromooxacyclopentane**
(*trans*-**3,4-Dibromotetrahydrofuran**)

Azacyclopentane
(**Pyrrolidine**)

Thiacyclopentane
(**Tetrahydrothiophene**)

3-Methyloxacyclohexane
(**3-Methyltetrahydropyran**)

Azacyclohexane
(**Piperidine**)

3-Cyclopropylthiacyclohexane
(**3-Cyclopropyltetrahydrothiopyran**)

The common names of unsaturated heterocycles are so firmly entrenched in the literature that we shall use them here.

Pyrrole   **Furan**   **Thiophene**   **Pyridine**   **Quinoline**   **Indole**   **Adenine**
(See Section 26-9)

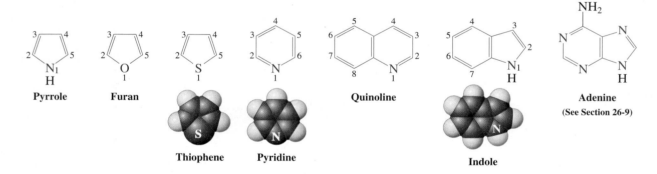

Exercise 25-1

Name or draw the following compounds. **(a)** *trans*-2,4-Dimethyloxacyclopentane (*trans*-2,4-dimethyltetrahydrofuran); **(b)** *N*-ethylazacyclopropane;

(c) O_2N —⟨pyridine⟩— NO_2

(d) ⟨4-bromoindole structure with Br⟩

25-2 | NONAROMATIC HETEROCYCLES

As illustrated by the chemistry of the oxacyclopropanes (Section 9-9), ring strain allows the three- and four-membered heterocycles to undergo nucleophilic ring opening readily. In contrast, the larger, unstrained systems are relatively inert to attack.

Ring strain makes heterocyclopropanes and heterocyclobutanes reactive

Heterocyclopropanes are relatively reactive because ring strain is released by nucleophilic ring opening. Under basic conditions, this process gives rise to inversion at the less substituted center (Section 9-9).

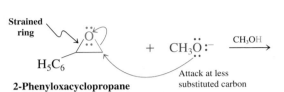

2-Phenyloxacyclopropane   →   2-Methoxy-1-phenylethanol
85%

Attack at less substituted carbon

N-Ethyl-(2*S*,3*S*)-*trans*-2,3-dimethylazacyclopropane

70% CH₃CH₂NH₂, H₂O, 120°C, 16 days

Attack with inversion

meso-*N*,*N'*-Diethyl-2,3-butanediamine
55%

Model Building

Heterocyclopropane Drug War Heads

The antitumor agent azinomycin A acts by cross-linking double-stranded DNA through nucleophilic heterocyclopropoane ring openings by DNA bases.

Azinomycin A

Exercise 25-2

Explain the following result by a mechanism. (**Caution:** This is not an oxidation. **Hints:** Calculate the molecular formulas of starting material and product. Try a ring opening catalyzed by the Lewis acid and consider the options available to the resulting intermediate. See Section 9-3.)

MgBr₂, (CH₃CH₂)₂O

100%

Exercise 25-3

2-(Chloromethyl)oxacyclopropane reacts with hydrogen sulfide ion (HS⁻) to give thiacyclobutan-3-ol. Explain by a mechanism.

Solved Exercise 25-4 | Working with the Concepts: Practicing Mechanisms That Involve Heterocycles

Isomeric cylindricines A and B, isolated in 1993, are the two main alkaloids (Section 25-8) present in extracts from the Australian marine plant *Clavelina cylindrica*. The two compounds equilibrate to a 3:2 mixture. Formulate a mechanism for this process. (**Hint:** Check Exercise 9-25.)

Cylindricine A   ⇌   Cylindricine B

REAL LIFE: MEDICINE 25-1 | Smoking, Nicotine, Cancer, and Medicinal Chemistry

Nicotine is an azacyclopentane derivative that is responsible for the addictive character of cigarette smoking. Like other drugs of abuse, its presence activates the brain's pleasure centers by stimulating the release of the neurotransmitter dopamine, while its absence activates the body's stress systems ("withdrawal"). The causal link between smoking and cancer is well established, and there are over 40 known carcinogens among the thousands of compounds contained in cigarette smoke, among them benz[*a*]pyrene (Section 16-7).

Nicotine appears to play a dual contributory role, because its metabolites are outright carcinogens and because the parent system itself, while not causing cancer, is a tumor promoter.

The metabolic pathway has as the initial step the *N*-nitrosation of the azacyclopentane (pyrrolidine) nitrogen. Oxidation and ring opening (compare Real Life 21-2) then take place, giving a mixture of two *N*-nitrosodialkanamines (*N*-nitrosamines), each of which is a known powerful carcinogen.

Nicotine

(Section 21-10) → NO⁺ → Ring opening and oxidation (Real Life 21-2) →

4-(*N*-Methyl-*N*-nitrosamino)-1-(3-pyridyl)-1-butanone

+

4-(*N*-Methyl-*N*-nitrosamino)-4-(3-pyridyl)butanal

Upon protonation of the oxygen in the nitroso group, these substances become reactive alkylating agents, capable of transferring methyl groups to nucleophilic sites in biological molecules such as DNA, as shown below. The diazohydroxide that remains decomposes through a diazonium ion to a carbocation, which may inflict additional molecular damage (Section 21-10).

$$\text{:N} \rightleftharpoons \text{H}^+ \quad \left[\cdots \right] \quad \xrightarrow{\text{Nu : (in DNA)}} \quad + \quad \overset{+}{\text{Nu}}{-}\text{CH}_3$$

The ability of nicotine to act as a tumor promoter has surfaced in several studies. For example, certain human lung cancer cells show 50% acceleration in their rate of division if nicotine is added to the culture medium. Nicotine also inhibits the "suicide" of malignant cells, one of the main repair and clean-up mechanisms of the body in case of cell damage. Thus, nicotine actually helps cells with genetic damage to

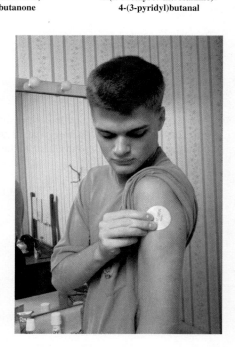

Nicotine patches are used to wean individuals from their addiction to cigarette smoking.

Strategy

Following the hint, Exercise 9-25 points out that a 2-chloroethylsulfide can undergo intramolecular S_N2 reaction to an intermediate sulfonium ion, which is subject to opening by external nucleophilic attack. Recognizing that the "action" part of cylindricine A is a 2-chloroethylamine substructure, we can apply the same principle here.

Solution

- Intramolecular attack by the amino nitrogen on the chlorine-bearing carbon produces reactive intermediate ammonium ion A (next page).
- The activated azacyclopropane ring may be attacked at either of its two carbons by the chloride ion liberated in the first step.
- Ring opening via the primary carbon regenerates starting material (*a*).
- However, attack on the neighboring position (*b*) unravels A to cylindricine B.
- Build a model!

survive and, presumably, proliferate. This finding has caused concern, because nicotine is the crucial ingredient in a number of smoking-cessation therapies that feature gums, patches, lozenges, and inhalators.

There are about 1.3 billion smokers worldwide, a staggering number, to which the United States contributes 60 million. Among them, the majority wants to and has tried to quit. Because tobacco-related illnesses are the second leading cause of death (after heart disease), there is a tremendous need for new therapies. A new concept in this vein is the

drug varenicline (Chantix), which came on the market in 2006. It is a nicotine receptor "partial agonist," that is, this molecule targets the same receptor sites of the central nervous system as nicotine, but with only partial efficacy. As such, it both reduces the craving for "a smoke" and decreases the pleasure associated with it, but not completely, thus minimizing the chances of a relapse.

The synthesis of varenicline is a showcase of reactions that you have encountered in this book, as practiced by medicinal chemists.

Synthesis of Varenicline

Varenicline

Cylindricine A   A   Cylindricine B

Exercise 25-5 | **Try It Yourself**

Treatment of thiacyclobutane with chlorine in $CHCl_3$ at $-70°C$ gives $ClCH_2CH_2CH_2SCl$ in 30% yield. Suggest a mechanism for this transformation. [**Hint:** The sulfur in sulfides is nucleophilic (Section 9-10).]

The reactivity of the four-membered heterocycloalkanes bears out expectations based on ring strain: They undergo ring opening, as do their three-membered cyclic counterparts, but more stringent reaction conditions are usually required. The reaction of oxacyclobutane with CH_3NH_2 is typical.

$$\square\!\!\overset{..}{\underset{..}{O}} + CH_3\overset{..}{N}H_2 \xrightarrow{150°C} H\overset{..}{O}\diagdown\!\!\diagup\!\!\diagdown\overset{..}{N}HCH_3$$

45%

***N*-Methyl-3-amino-1-propanol**

The β-lactam antibiotic penicillin functions through related ring-opening processes (see Real Life 20-2).

Exercise 25-6

2-Methyloxacyclobutane reacts with HCl to give two products. Write their structures.

Heterocyclopentanes and heterocyclohexanes are relatively unreactive

The unstrained heterocyclopentanes and heterocyclohexanes are relatively inert. Recall that oxacyclopentane (tetrahydrofuran, THF) is used as a solvent. However, the heteroatoms in aza- and thiacycloalkanes allow for characteristic transformations (see Sections 9-10, 17-8, 18-4, and Chapter 21). In general, ring opening occurs by conversion of the heteroatom into a good leaving group.

Exercise 25-7

Treatment of azacyclopentane (pyrrolidine) with sodium nitrite in acetic acid gives a liquid, b.p. 99–100°C (15 mm Hg), that has the composition $C_4H_8N_2O$. Propose a structure for this compound. (**Hint:** Review Section 21-10.)

In Summary The reactivity of heterocyclopropanes and heterocyclobutanes results from the release of strain by ring opening. The five- and six-membered heterocycloalkanes are less reactive than their smaller-ring counterparts.

Aromatic Heterocyclopentadienes

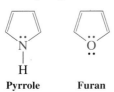

Pyrrole   **Furan**

Thiophene

| 25-3 | STRUCTURES AND PROPERTIES OF AROMATIC HETEROCYCLOPENTADIENES |

Pyrrole, furan, and **thiophene** are 1-hetero-2,4-cyclopentadienes. Each contains a butadiene unit bridged by an sp^2-hybridized heteroatom bearing lone electron pairs. These systems contain delocalized π electrons in an aromatic six-electron framework. This section considers the structures and methods of preparation of these compounds.

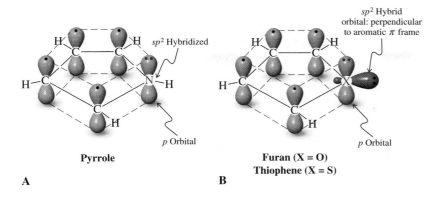

sp² Hybrid orbital: perpendicular to aromatic π frame

sp² Hybridized

p Orbital

p Orbital

Pyrrole

Furan (X = O)
Thiophene (X = S)

A   B

Figure 25-1 Orbital pictures of (A) pyrrole and (B) furan (X = O), and thiophene (X = S). The heteroatom in each is sp² hybridized and bears one delocalized lone electron pair.

Pyrrole, furan, and thiophene contain delocalized lone electron pairs

The electronic structure of the three heterocycles pyrrole, furan, and thiophene is similar to that of the aromatic cyclopentadienyl anion (margin and Section 15-7). The cyclopentadienyl anion may be viewed as a butadiene bridged by a negatively charged carbon whose electron pair is delocalized over the other four carbons. The heterocyclic analogs contain a neutral atom in that place, again bearing lone electron pairs. One of these pairs is similarly delocalized, furnishing the two electrons needed to satisfy the $4n + 2$ rule (Section 15-7). To maximize overlap, the heteroatoms are hybridized sp^2 (Figure 25-1), the delocalized electron pair being assigned to the remaining p orbital. In pyrrole, the sp^2-hybridized nitrogen bears a hydrogen substituent in the plane of the molecule. For furan and thiophene, the second lone electron pair is placed into one of the sp^2 hybrid orbitals, again in the plane and therefore with no opportunity to achieve overlap. This arrangement is much like that in the phenyl anion (Section 22-4). As a consequence of delocalization, pyrrole, furan, and thiophene exhibit properties that are typical of aromatic compounds, such as unusual stability, deshielded protons in the ¹H NMR spectra due to the presence of ring currents, and the ability to undergo electrophilic aromatic substitution (Section 25-4).

The delocalization of the lone pair in the 1-hetero-2,4-cyclopentadienes can be described by charge-separated resonance forms, as shown for pyrrole.

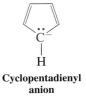

Cyclopentadienyl anion

Resonance Forms of Pyrrole

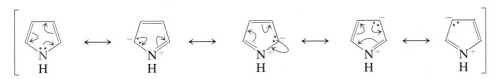

Model Building

Notice that there are four dipolar forms in which a positive charge is placed on the heteroatom and a negative charge successively on each of the carbons. This picture suggests that the heteroatom should be relatively electron poor and the carbons relatively electron rich. The electrostatic potential maps below confirm this expectation. Thus (on the same scale), the nitrogen in pyrrole is less electron rich (orange) than that in its saturated counterpart azacyclopentane (red), whereas the diene portion in pyrrole is more electron rich (red) than that in 1,3-cyclopentadiene (yellow).

Pyrrole   **Azacyclopentane**   **1,3-Cyclopentadiene**

Borole Is Antiaromatic

In contrast to furan, pyrrole, and thiophene, the heteroatom in borole does not contain an electron pair that can participate in cyclic delocalization. The p orbital in the sp^2-hybridized boron atom (see Figure 1-17) is empty and the molecule is isoelectronic with the antiaromatic cyclopentadienyl cation (Section 15-7). Indeed, the parent compound borole is not known.

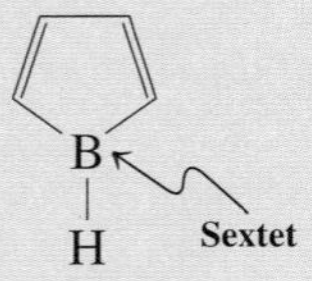

Borole
(4π Electrons: antiaromatic)

As this description suggests, the extent of aromaticity in these systems depends on the relative ability of the heteroatom to donate its lone electron pair, in turn described by its respective electronegativity (Table 1-2). Consequently, the aromatic character increases along the series furan < pyrrole < thiophene, a trend that is reflected in relative reactivity and stability.

Aromaticity in Heterocyclopentadienes

Increasing aromaticity

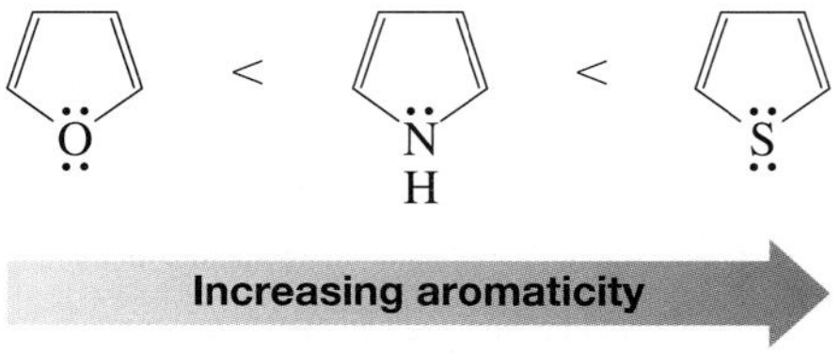

Exercise 25-8

Azacyclopentane and pyrrole are both polar molecules. However, the dipole vectors in the two molecules point in opposite directions. What is the sense of direction of this vector in each structure? Explain your answer.

Pyrroles, furans, and thiophenes are prepared from γ-dicarbonyl compounds

Syntheses of the heterocyclopentadienes use a variety of cyclization strategies. A general approach is the **Paal-Knorr* synthesis** (for pyrroles) and its variations (for the other heterocycles). The target molecule is made from an enolizable γ-dicarbonyl compound that is treated with an amine derivative (for pyrroles) or P_2O_5 (for furans) or P_2S_5 (for thiophenes).

Cyclization of a γ-Dicarbonyl Compound to a 1-Hetero-2,4-cyclopentadiene

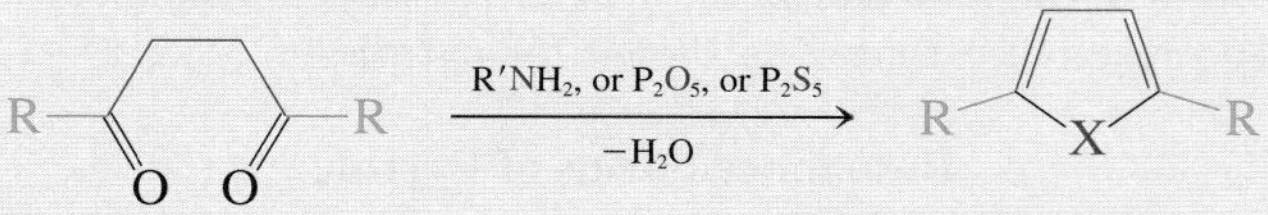

$$R'NH_2, \text{ or } P_2O_5, \text{ or } P_2S_5$$
$$-H_2O$$

X = NR′, O, S

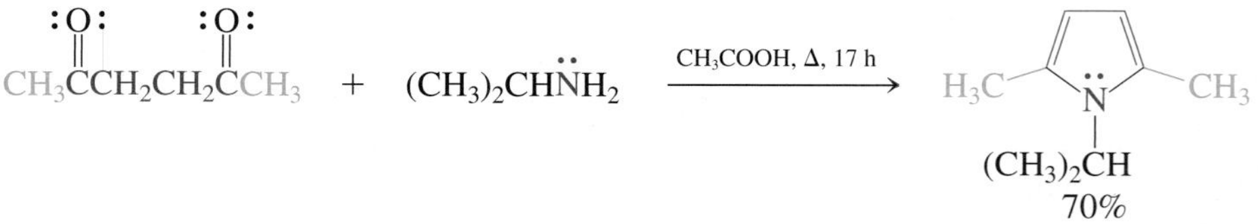

$$CH_3COOH, \Delta, 17\text{ h}$$

$(CH_3)_2CH$
70%

N-(1-Methylethyl)-2,5-dimethylpyrrole

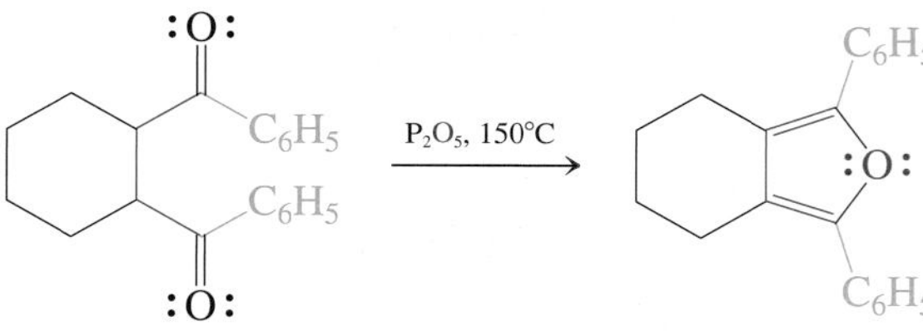

$$P_2O_5, 150°C$$

62%

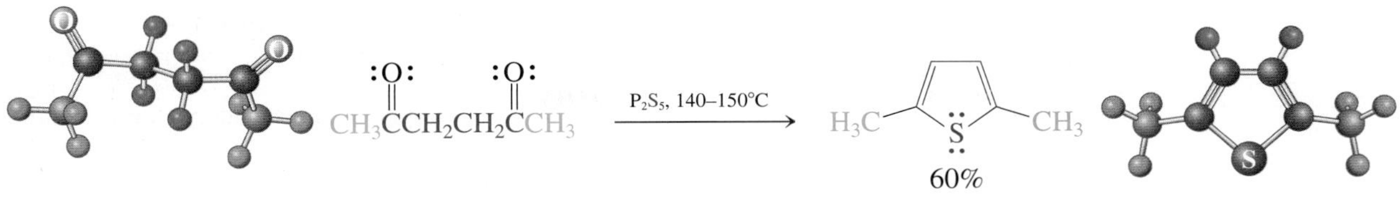
:O: :O:
$CH_3CCH_2CH_2CCH_3$
P_2S_5, 140–150°C
H_3C
S
CH_3
60%
S

2,5-Dimethylthiophene

Exercise 25-9

Formulate a possible mechanism for the acid-catalyzed dehydration of 2,5-hexanedione to 2,5-dimethylfuran. (**Hint:** The crucial ring closure is accomplished by the oxygen of one carbonyl group attacking the carbon of the second. See also Section 18-2.)

Exercise 25-10

4-Methylpyrrole-2-carboxylic acid (compound B) is the trail pheromone of the ant species *Atta texana*. A third of a milligram has been estimated to be sufficient to mark a path around Earth, and each ant carries only 3.3 ng (10^{-9} g). Propose a synthesis starting from 3-methylcyclobutene-1-carboxylic acid (compound A). (**Hint:** What dione is the retrosynthetic precursor to compound B, and how can you make it from compound A?)

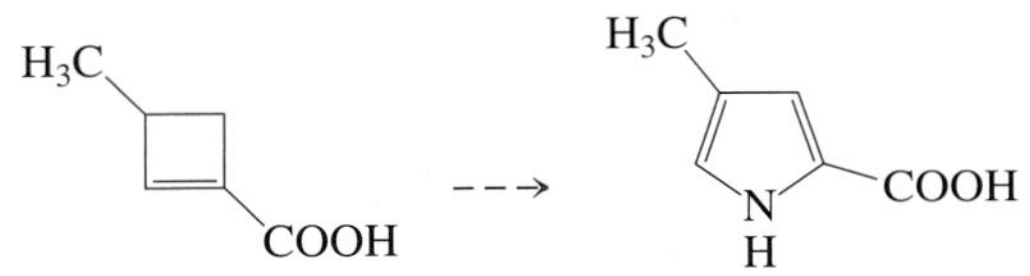
H_3C
COOH
H_3C
N
H
COOH

A **B**

Exercise 25-11

The following equation is an example of another synthesis of pyrroles. Write a mechanism for this transformation. (**Hint:** Refer to Section 17-9.)

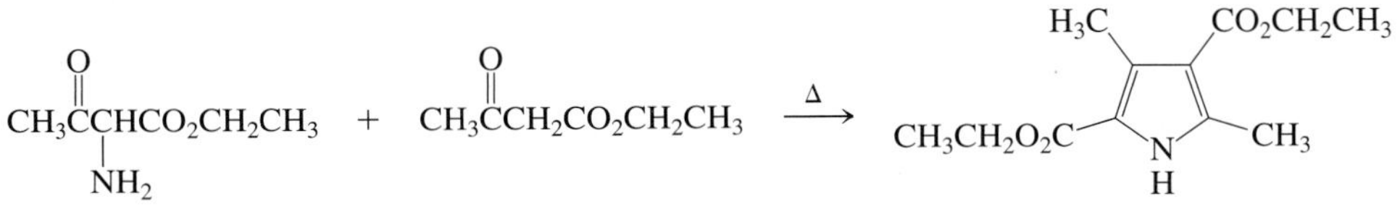
O
$CH_3CCHCO_2CH_2CH_3$
NH_2
+
O
$CH_3CCH_2CO_2CH_2CH_3$
Δ
H_3C
$CO_2CH_2CH_3$
$CH_3CH_2O_2C$
N
H
CH_3

Ethyl 2-amino-3-oxobutanoate **Ethyl 3-oxobutanoate** **Diethyl 3,5-dimethylpyrrole-2,4-dicarboxylate**

In Summary Pyrrole, furan, and thiophene contain delocalized aromatic π systems analogous to that of the cyclopentadienyl anion. A general method for the preparation of 1-hetero-2,4-cyclopentadienes is based on the cyclization of enolizable γ-dicarbonyl compounds.

25-4 REACTIONS OF THE AROMATIC HETEROCYCLOPENTADIENES

The reactivity of pyrrole, furan, and thiophene and their derivatives is governed largely by their aromaticity and is based on the chemistry of benzene. This section describes some of their reactions, particularly electrophilic aromatic substitution, and introduces indole, a benzofused analog of pyrrole.

Pyrroles, furans, and thiophenes undergo electrophilic aromatic substitution

As expected for aromatic systems, the 1-hetero-2,4-cyclopentadienes undergo electrophilic substitution. There are two sites of possible attack, at C2 and at C3. Which one should be more reactive? An answer can be found by the same procedure used to predict the regioselectivity of electrophilic aromatic substitution of substituted benzenes (Chapter 16): enumeration of all the possible resonance forms for the two modes of reaction.

Consequences of Electrophilic Attack at C2 and C3 in the Aromatic Heterocyclopentadienes

Mechanism

Attack at C2

Three resonance forms

Attack at C3

Two resonance forms

Both modes benefit from the presence of the resonance-contributing heteroatom, but attack at C2 leads to an intermediate with an additional resonance form, thus indicating this position to be the preferred site of substitution. Indeed, such selectivity is generally observed. However, because C3 also is activated to electrophilic attack, mixtures of products can form, depending on conditions, substrates, and electrophiles.

Electrophilic Aromatic Substitution of Pyrrole, Furan, and Thiophene

Reaction

50%
2-Nitropyrrole

13%
3-Nitropyrrole

64%
2-Chlorofuran

64%
2-Acetyl-5-methylthiophene

The relative reactivity of benzene and the three heterocycles in electrophilic substitutions is the result of contributions from the aromaticity of the respective rings and the stabilization of the intermediate cations. It increases in the order benzene ≪ thiophene < furan < pyrrole.

Solved Exercise 25-12 | **Working with the Concepts: Predicting the Position of Electrophilic Aromatic Substitution of Substituted Thiophene**

The monobromination of thiophene-3-carboxylic acid gives only one product. What is its structure, and why is it the only product formed?

Strategy
The two preferred sites of reaction should be at the carbon atoms next to sulfur, in our unsymmetrical thiophene C2 and C5. To see which one of these is more active, we need to compare the respective resonance forms of the intermediate cation resulting from electrophilic attack at either position.

Solution
• Attack at C2

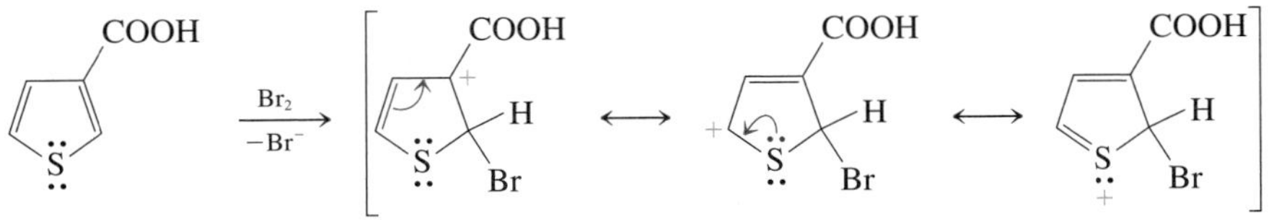

• Attack at C5

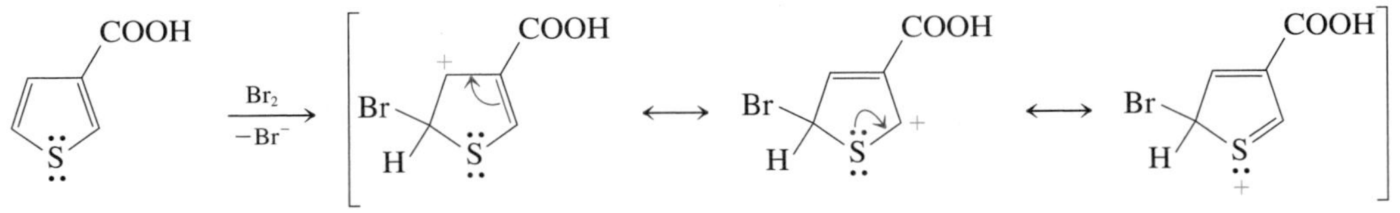

• Result: Attack on C5 avoids placing the positive charge on C3, as in A, bearing the electron-withdrawing carboxy function. Therefore, the only product is 5-bromo-3-thiophenecarboxylic acid.

Exercise 25-13 | **Try It Yourself**

The monobromination of 3-methylfuran gives only one product. What is its structure, and why is it the only product formed?

Pyrrole is extremely nonbasic compared to ordinary amines (Section 21-4), because the lone electron pair on nitrogen is tied up by conjugation. Very strong acid is required to effect protonation, and it takes place on C2, not on the nitrogen.

The Protonation of Pyrrole Occurs on Carbon

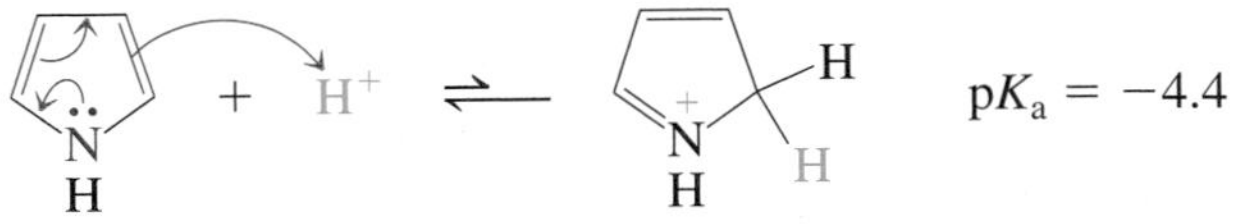

$pK_a = -4.4$

Pyrroles are not only very nonbasic, they are, in fact, relatively acidic. Thus, while azacyclopentane has a $pK_a = 35$ (normal for an amine, Section 21-4), the corresponding value for pyrrole is 16.5! The reasons for this increase in acidity are the change in hybridization from sp^3 to sp^2 (see Section 11-3) and the delocalization of the negative charge (as in cyclopentadienyl anion, Section 15-7).

Pyrrole Is Relatively Acidic

sp^3

$pK_a = 35$
Azacyclopentane

sp^2

$pK_a = 16.5$

Five resonance forms

1-Hetero-2,4-cyclopentadienes can undergo ring opening and cycloaddition reactions

Furans can be hydrolyzed under mild conditions to γ-dicarbonyl compounds. The reaction may be viewed as the reverse of the Paal-Knorr-type synthesis of furans. Pyrrole polymerizes under these reaction conditions, whereas thiophene is stable.

Hydrolysis of a Furan to a γ-Dicarbonyl Compound

H_3C —(O)— CH_3 $\xrightarrow{CH_3COOH, H_2SO_4, H_2O, \Delta}$ $CH_3CCH_2CH_2CCH_3$

90%

2,5-Hexanedione

Raney nickel desulfurization (Section 17-8) of thiophene derivatives results in sulfur-free acyclic saturated compounds.

$\xrightarrow{\text{Raney Ni, } (CH_3CH_2)_2O, \Delta}{-NiS}$

50%

Being least aromatic (Section 25-3), the π system of furan (but not of pyrrole or thiophene) possesses sufficient diene character to undergo Diels-Alder cycloadditions (Section 14-8).

$+$ $\xrightarrow{(CH_3CH_2)_2O, 25°C}$

95%

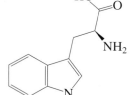

Tryptophan

Indole is a benzopyrrole

Indole is the most important *benzannulated* (fused-ring) derivative of the 1-hetero-2,4-cyclopentadienes. It forms part of many natural products, including the amino acid tryptophan (Section 26-1).

Indole is related to pyrrole in the same way that naphthalene is related to benzene. Its electronic makeup is indicated by the various possible resonance forms that can be formulated for the molecule. Although those resonance forms that disturb the cyclic six-π-electron system of the fused benzene ring are less important, they indicate the electron-donating effect of the heteroatom.

Resonance in Indole

Exercise 25-15

Predict the preferred site of electrophilic aromatic substitution in indole. Explain your choice.

Exercise 25-16

Irradiation of compound A in ethoxyethane (diethyl ether) at $-100°C$ generates the enol form, B, of acetylbenzene and a new compound, C, which isomerizes to indole upon warming to room temperature.

The ^{1}H NMR spectrum of compound C shows signals at $\delta = 3.79$ (d, 2 H) and 8.40 (t, 1 H) ppm, in addition to four aromatic absorptions. Indole has peaks at $\delta = 6.34$ (d, 1 H), 6.54 (broad d, 1 H), and 7.00 (broad s, 1 H) ppm. What is compound C? (**Hint:** This reaction proceeds by a mechanism similar to the mass spectral McLafferty rearrangement, Section 17-3.)

In Summary The donation of the lone electron pair on the heteroatom to the diene unit in pyrrole, furan, and thiophene makes the carbon atoms in these systems electron rich and therefore more susceptible to electrophilic aromatic substitution than the carbons in benzene. Electrophilic attack is frequently favored at C2, but substitution at C3 is also observed, depending on conditions, substrates, and electrophiles. Some rings can be opened by hydrolysis or by desulfurization (for thiophenes). The diene unit in furan is reactive enough to undergo Diels-Alder cycloadditions. Indole is a benzopyrrole containing a delocalized π system.

25-5 STRUCTURE AND PREPARATION OF PYRIDINE: AN AZABENZENE

Pyridine can be regarded as a benzene derivative—an **azabenzene**—in which an sp^2-hybridized nitrogen atom replaces a CH unit. The pyridine ring is therefore aromatic, but its electronic structure is perturbed by the presence of the electronegative nitrogen atom. This section describes the structure, spectroscopy, and preparation of this simple azabenzene.

Pyridine is aromatic

Pyridine contains an sp^2-hybridized nitrogen atom like that in an imine (Section 17-9). In contrast to pyrrole, only one electron in the p orbital completes the aromatic π-electron arrangement of the aromatic ring; as in the phenyl anion, the lone electron pair is located

Indole-Based Neurotransmitters

Similar to the 2-phenylethanamine fragment (Real Life 21-1), the 2-(3-indolyl)ethanamine (tryptamine) nucleus is embedded in the structure of a number of neurotransmitters, including psychedelic (such as LSD, Section 25-8) and other drugs.

Tryptamine (R = H)
Serotonin (R = OH)
(Regulates intestinal movements, mood, appetite, and sleep)

Melatonin
(Regulates circadian rhythm, the organism's internal 24-hour clock)

Sumatriptan
(Migraine drug)

sp^2 **Pyridine**

Figure 25-2 (A) Orbital picture of pyridine. The lone electron pair on nitrogen is in an sp^2-hybridized orbital and is *not* part of the aromatic π system. (B) The electrostatic potential map of pyridine reveals the location of the lone electron pair on nitrogen (red) in the molecular plane and the electron-withdrawing effect of the electronegative nitrogen on the aromatic π system (green; compare to the electrostatic potential map of pyrrole in Section 25-3).

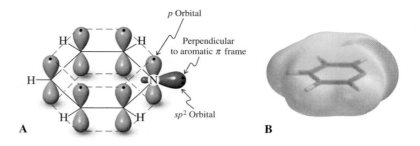

in one of the sp^2 hybrid atomic orbitals in the molecular plane (Figure 25-2). Therefore, in pyridine, the nitrogen does not donate excess electron density to the remainder of the molecule. Quite the contrary: Because nitrogen is more electronegative than carbon (Table 1-2), it withdraws electron density from the ring, both inductively and by resonance.

Resonance in Pyridine

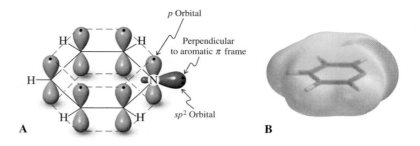

Exercise 25-17

Azacyclohexane (piperidine) is polarized because of the presence of the relatively electronegative nitrogen. The polarization in pyridine is twice as large. Explain.

Aromatic delocalization in pyridine is evident in the ^{1}H NMR spectrum, which reveals the presence of a ring current. The electron-withdrawing capability of the nitrogen is manifest in larger chemical shifts (more deshielding) at C2 and C4, as expected from the resonance picture.

^{1}H NMR Chemical Shifts (ppm) in Pyridine and Benzene

Pyridine Is a Weak Base

Pyridinium ion
$pK_a = 5.29$

Because the lone pair on nitrogen is not tied up by conjugation (as it is in pyrrole, Exercise 25-14), pyridine is a weak base. (It is used as such in numerous organic transformations.) Compared with alkanamines (pK_a of ammonium salts $\approx$ 10; Section 21-4), the pyridinium ion has a low pK_a, because the nitrogen is sp^2 and not sp^3 hybridized (see Section 11-3 for the effect of hybridization on acidity).

Pyridine is the simplest azabenzene. Some of its higher aza analogs are shown here. They behave like pyridine but show the increasing effect of aza substitution—in particular, increasing electron deficiency. Minute quantities of several 1,4-diazabenzene (pyrazine) derivatives are responsible for the characteristic odors of many vegetables. One drop of 2-methoxy-3-(1-methylethyl)-1,4-diazabenzene (2-isopropyl-3-methoxypyrazine) in a large swimming pool would be more than adequate to give the entire pool the odor of raw potatoes.

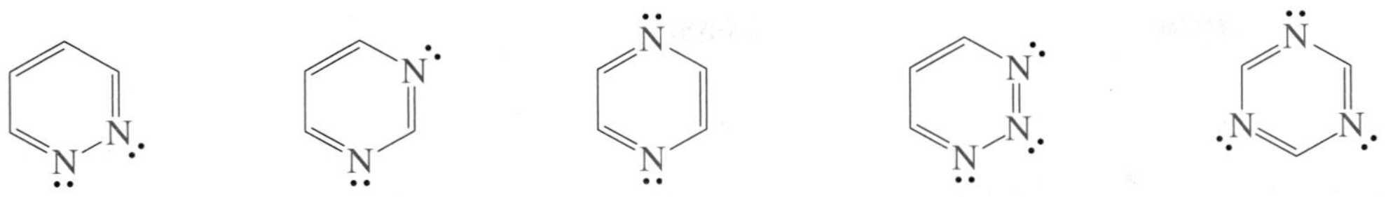

1,2-Diazabenzene
(Pyridazine)

1,3-Diazabenzene
(Pyrimidine)

1,4-Diazabenzene
(Pyrazine)

1,2,3-Triazabenzene
(1,2,3-Triazine)

1,3,5-Triazabenzene
(1,3,5-Triazine)

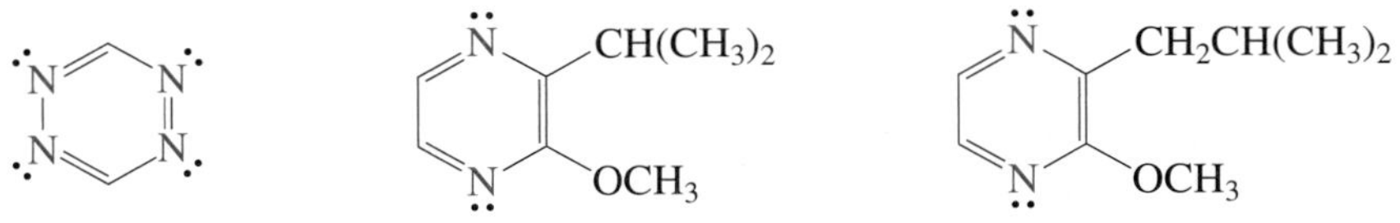

1,2,4,5-Tetraazabenzene
(1,2,4,5-Tetrazine)

2-Methoxy-3-(1-methylethyl)-
1,4-diazabenzene
(Potatoes)

2-Methoxy-3-(2-methylpropyl)-
1,4-diazabenzene
(Green peppers)

Pyridines are made by condensation reactions

Pyridine and simple alkylpyridines are obtained from coal tar. Many of the more highly substituted pyridines are in turn made by both electrophilic and nucleophilic substitution of the simpler derivatives.

Pyridines can be made by condensation reactions of acyclic starting materials such as carbonyl compounds with ammonia. The most general of these methods is the **Hantzsch* pyridine synthesis.** In this reaction, two molecules of a β-dicarbonyl compound, an aldehyde, and ammonia combine in several steps (Worked Example 25-28) to give a substituted dihydropyridine, which is readily oxidized by nitric acid to the aromatic system. When the β-dicarbonyl compound is a 3-ketoester, the resulting product is a 3,5-pyridinedicarboxylic ester. Hydrolysis followed by pyrolysis of the calcium salt of the acid causes decarboxylation.

> **Really?**
>
> The ultimate azaanalog of benzene is hexaazabenzene, an allotrope of nitrogen gas. It has been calculated to be aromatic by the ring-current criterion (Section 15-4), but exceedingly unstable: The activation energy for retrocyclization to three molecules of N_2 is estimated to be only ~2 kcal mol^{-1}, releasing about 200 kcal mol^{-1} during the process. The primary reasons are the relatively weak N—N bonds (48 kcal mol^{-1}), destabilized by repulsion of the six lone electron pairs, and the strength of the N≡N bond (225 kcal mol^{-1}). 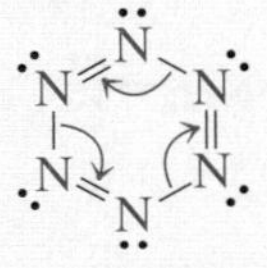
>
> **Hexaazabenzene**
> **(Hexazine)**
>
> 3 :N≡N:

Hantzsch Synthesis of 2,6-Dimethylpyridine

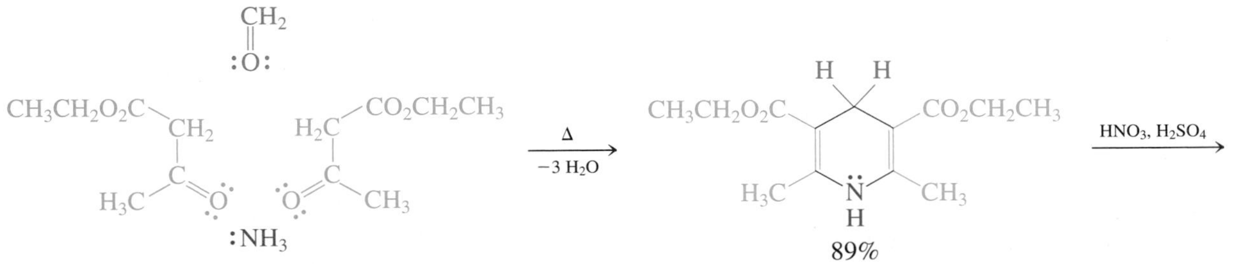

89%
Diethyl 1,4-dihydro-2,6-dimethyl-
3,5-pyridinedicarboxylate

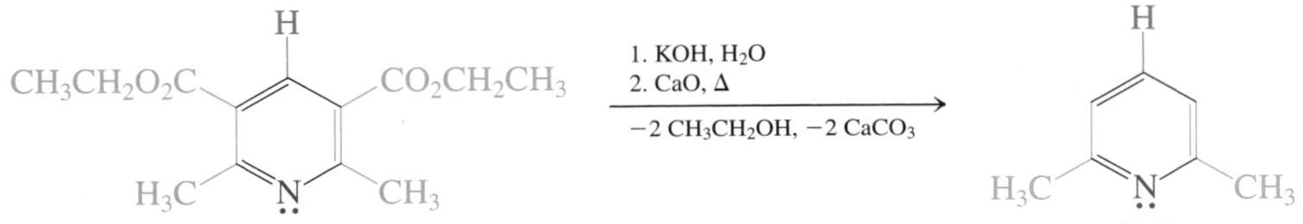

65%
Diethyl 2,6-dimethyl-3,5-
pyridinedicarboxylate

65%
2,6-Dimethylpyridine

---

*Professor Arthur R. Hantzsch (1857–1935), University of Leipzig, Germany.

The first step of the Hantzsch pyridine synthesis is an example of a four-component reaction: Four molecules combine in a specific fashion to form a single product. This is remarkable considering the number of possible condensation pathways that are available to the starting materials. The reason for its success is the reversibility of channels that do not lead to the observed dihydropyridine and the careful tuning of reactivity of the participating molecules: ammonia (or ammonium acetate); a relatively reactive aldehyde; and a β-dicarbonyl constituent that is used twice. (For a stepwise tour through the mechanism, see Worked Example 25-28). Multicomponent reactions of this type, in which water is the only by-product, are by their very nature atom economical and hence "green" (Real Life 3-1). Even more environmentally friendly is the use of water as a solvent and, in the aromatization step to the substituted pyridine, simply oxygen in the presence of activated carbon (a form of highly porous carbon derived from charcoal).

A "Super Green" Hantzsch Pyridine Synthesis

Exercise 25-18

What starting materials would you use in the Hantzsch synthesis of the following pyridines?

Solved Exercise 25-19 | Working with the Concepts: Practicing Mechanisms of Pyridine Syntheses

The Hantzsch synthesis of pyridines features 1,4-dihydropyridines in the first step. A variant of the method uses hydroxylamine (Table 17-5), which can be regarded as an oxidized version of ammonia. With this reagent, pyridines are formed directly from 1,5-dicarbonyl compounds, in turn readily made by Michael additions of enolates to α,β-unsaturated aldehydes and ketones (Section 18-11). Formulate the mechanisms of the two steps in the pyridine synthesis shown.

Strategy

Looking at the first step of the reaction, you should recognize a Michael addition of cyclohexanone to an unsaturated ketone. (You can review this reaction in Section 18-11). The transformation in the second step looks similar to the Paal-Knorr synthesis of pyrroles

(Exercise 25-9), except that the result is the construction of a six-membered (not a five-membered) ring. The mechanism you are looking for should be based on the mechanisms for these two reactions.

Solution

- In the first step, the product is formed by C–C bond construction between the carbon next to the carbonyl in cyclohexanone and the terminal position of the enone; in other words, cyclohexanone is alkylated at the α-position. Remember that ketones are alkylated through their enolates (Section 18-4) and, in this case, alkylation occurs by attack of cyclohexanone enolate on the positively polarized enone β-carbon, as sketched below.

- For the second step, recall that primary amines react with ketones reversibly to furnish imines by loss of water (Section 17-9) and that, in fact, hydroxylamine makes oximes (Table 17-5). These condensations proceed through the intermediacy of hemiaminals by nucleophilic attack of the amine nitrogen on the carbonyl carbon.

Hemiaminal   **Oxime**

- Imines (like carbonyl compounds that form enols) are in equilibrium with their tautomeric forms, enamines (Section 17-9). If we write this form for our oxime, we arrive at an aminocarbonyl compound that is poised to undergo fast intramolecular formation of another hemiaminal. Dehydration produces a new cyclic enamine, which, upon inspection, is nothing but a hydrated pyridine. Finally, the driving force of aromaticity facilitates the rapid loss of water and generation of the pyridine product.

Oxime   **Enamine**   **Intramolecular hemiaminal**

Exercise 25-20 | Try It Yourself

Sketch a plausible mechanism by which the following condensation reaction occurs. [**Hints:** Start with an aldol condensation of the aldehyde with the enamine. (Review enamine alkylation, Section 18-4). Then use the resulting product as a Michael acceptor of the second enamine.]

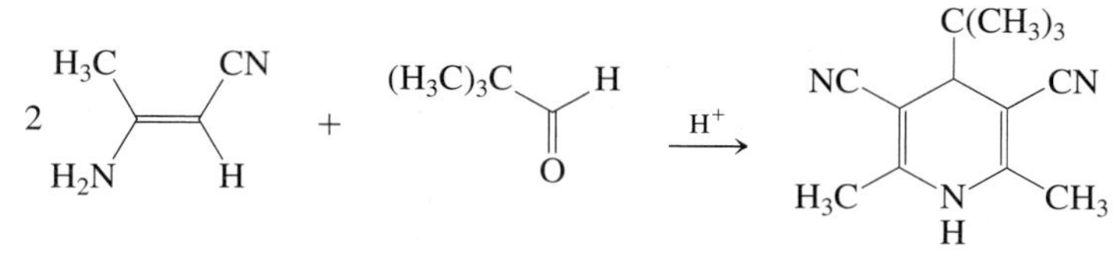

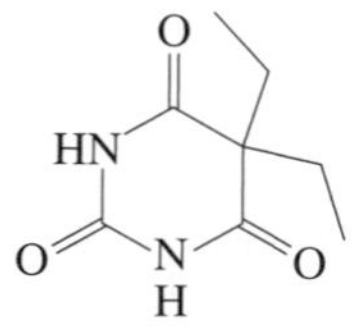

Exercise 25-21 |

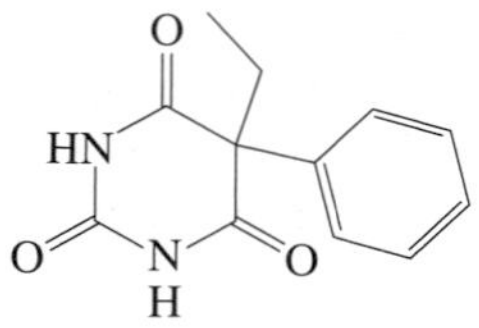

1,3-Diazabenzene-2,4,6-triol prefers the triketo tautomeric form (Section 22-3) by about 29 kcal mol^{-1}. It is commonly known as barbituric acid (pK_a = 7.4) and constitutes the basic frame of a group of sedatives and sleep inducers called barbiturates, of which veronal and phenobarbital are two examples (margin).

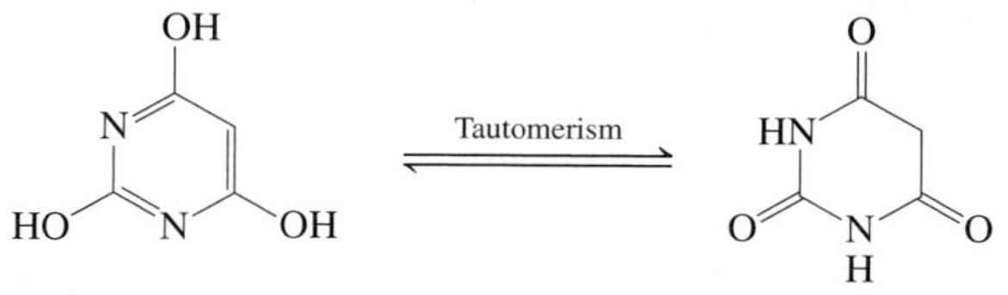

Barbituric acid

Propose a synthesis of veronal from diethyl propanedioate (malonic ester; Section 23-2) and urea (Section 20-6). (**Hint:** In the presence of a base, such as an alkoxide, amides are in equilibrium with the corresponding amidates; Section 20-7.)

Veronal

Phenobarbital

In Summary Pyridines are aromatic but electron poor. The lone pair on nitrogen makes the heterocycle weakly basic. Pyridines are prepared by condensation of a β-dicarbonyl compound with ammonia and an aldehyde.

25-6 | REACTIONS OF PYRIDINE

The reactivity of pyridine derives from its dual nature as both an aromatic molecule and a cyclic imine. Both electrophilic and nucleophilic substitution processes may occur, leading to a variety of substituted derivatives.

Pyridine undergoes electrophilic aromatic substitution only under extreme conditions

Because the pyridine ring is electron poor, the system undergoes electrophilic aromatic substitution only with great difficulty, several orders of magnitude more slowly than benzene, and at C3 (see Section 15-8).

Electrophilic Aromatic Substitution of Pyridine

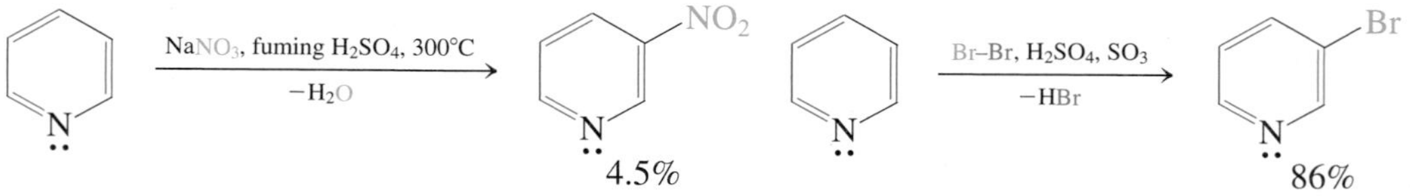

Exercise 25-22

Explain why electrophilic aromatic substitution of pyridine occurs at C3.

Activating substituents allow for milder conditions or improved yields.

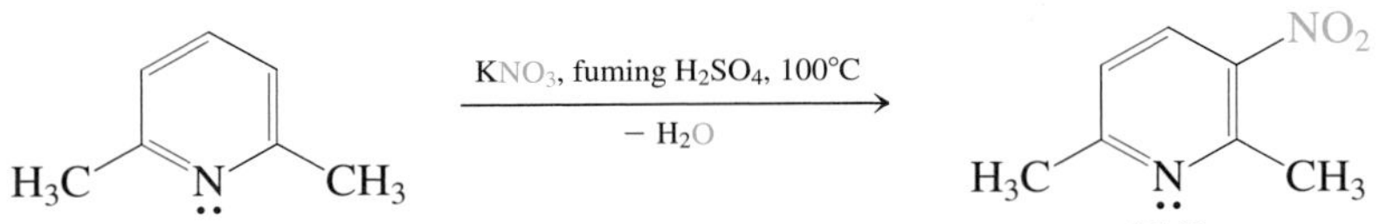
KNO3, fuming H2SO4, 100°C
− H2O
H3C
N
CH3
NO2

81%

2,6-Dimethylpyridine **2,6-Dimethyl-3-nitropyridine**

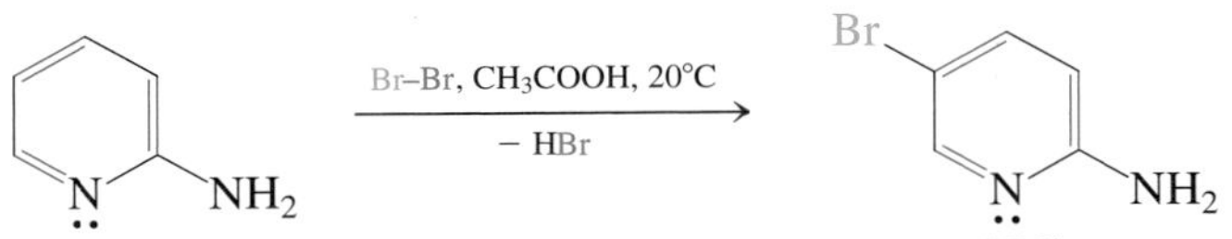
Br–Br, CH3COOH, 20°C
− HBr
N
NH2
Br

90%

2-Aminopyridine **2-Amino-5-bromopyridine**

Pyridine undergoes relatively easy nucleophilic substitution

Because the pyridine ring is relatively electron deficient, it undergoes nucleophilic substitution much more readily than does benzene (Section 22-4). Attack at C2 and C4 is preferred because it leads to intermediates in which the negative charge is on the nitrogen. An example of nucleophilic substitution of pyridine is the **Chichibabin* reaction,** in which the heterocycle is converted into 2-aminopyridine by treatment with sodium amide in liquid ammonia.

Chichibabin Reaction

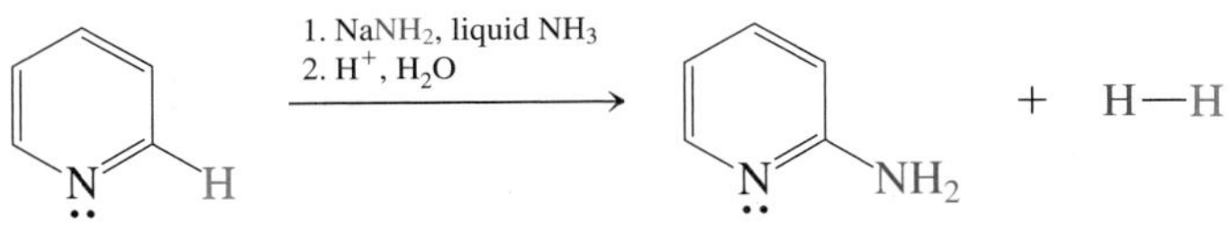
1. NaNH2, liquid NH3
2. H+, H2O
N
H
NH2
+ H—H

70%

2-Aminopyridine

This reaction proceeds by the addition–elimination mechanism. The first step is attack by $^{-}\!:\ddot{N}H_2$ at C2, a process that resembles 1,2-addition to an imine function. Expulsion of a hydride ion, $H\!:^{-}$, from C2 is followed by deprotonation of the amine nitrogen to give H_2 and a resonance-stabilized 2-pyridineamide ion. Protonation by aqueous work-up furnishes the final product. Note the contrast with *electrophilic* substitutions, which include *proton* loss, not expulsion of hydride as a leaving group.

Transformations related to the Chichibabin reaction take place when pyridines are treated with Grignard or organolithium reagents.

Mechanism of the Chichibabin Reaction

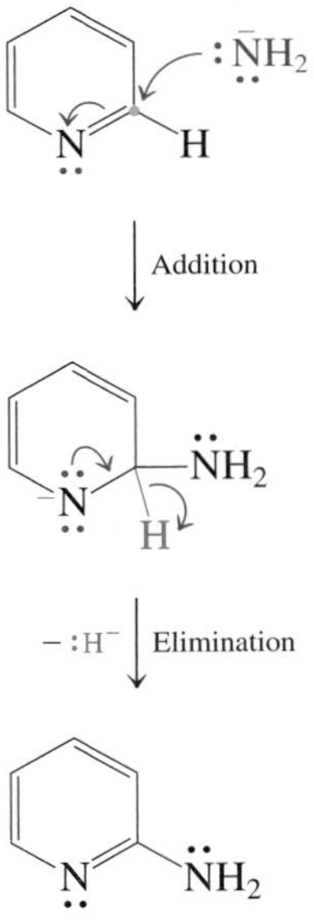
:NH2
N
H
Addition
NH2
H
−:H−
Elimination
NH2

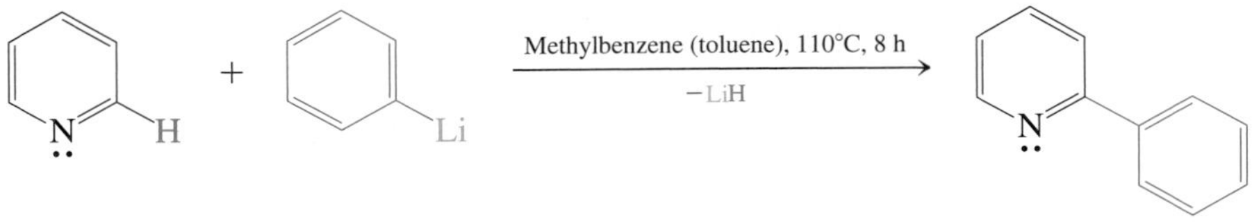
N
H
+
Li
Methylbenzene (toluene), 110°C, 8 h
−LiH
N

49%

2-Phenylpyridine

*Professor Alexei E. Chichibabin (1871–1945), University of Moscow, Russia.

Nicotinamide adenine dinucleotide

A complex pyridinium derivative, *nicotinamide adenine dinucleotide* (NAD$^+$) is an important biological oxidizing agent. The structure consists of a pyridine ring [derived from 3-pyridinecarboxylic (nicotinic) acid], two ribose molecules (Section 24-1) linked by a pyrophosphate bridge, and the heterocycle adenine (Section 26-9).

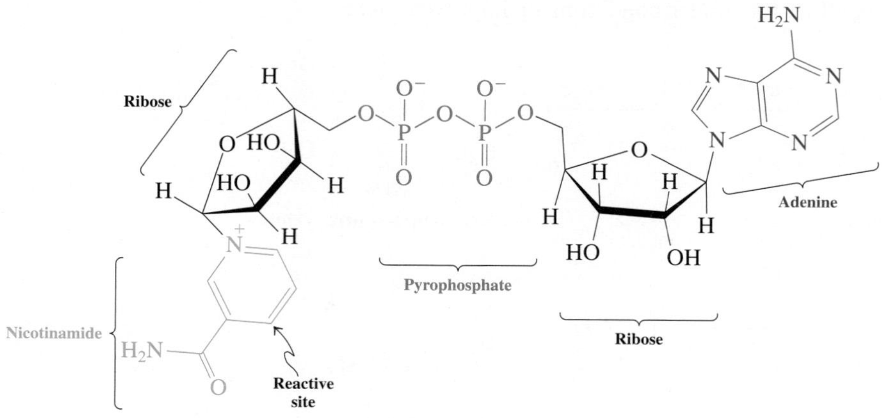

Most organisms derive their energy from the oxidation (removal of electrons) of fuel molecules, such as glucose or fatty acids; the ultimate oxidant (electron acceptor) is oxygen, which gives water. Such biological oxidations proceed through a cascade of electron-transfer reactions requiring the intermediacy of special redox reagents. NAD$^+$ is one such molecule. In the oxidation of a substrate, the pyridinium ring in NAD$^+$ undergoes a two-electron reduction with simultaneous protonation.

NAD$^+$ is the electron acceptor in many enzymatic oxidations of alcohols to aldehydes (including the conversion of vitamin A into retinal, Real Life 18-2; see also Real Life 8-1). This reaction can be seen as a transfer of hydride from C1 of the alcohol to C4 of the pyridinium nucleus with simultaneous deprotonation to give the aldehyde and the

Reduction of NAD$^+$

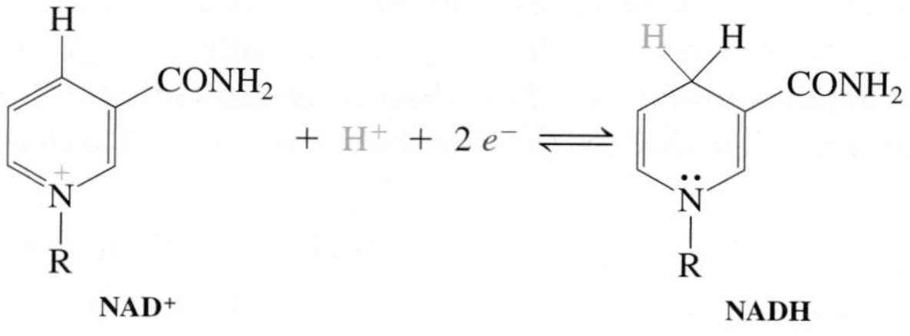

dihydropyridine, NADH. With other enzymes, the reverse is achieved, namely, the reduction of aldehydes and ketones to alcohols with NADH (see Problems 58 and 59 of Chapter 8, and Section 22-9).

The "action" part of NADH constitutes a simple dihydropyridine of the type readily accessible in the first step of the Hantzsch pyridine synthesis (Section 25-5). Therefore chemists have probed whether such compounds can be used as metal-free alternatives to hydride reagents, such as

In most nucleophilic substitutions of pyridines, halides are leaving groups, the 2- and 4-halopyridines being particularly reactive.

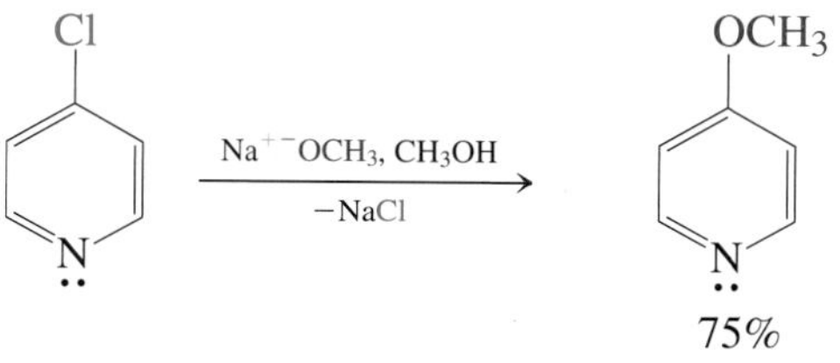

75%
4-Methoxypyridine

LiAlH$_4$, in carbonyl reductions, or as surrogates for catalytically activated (e.g., Pd or Pt) hydrogen in hydrogenations. Indeed, such so-called *biomimetic* reactions (because they mimic the principles applied by nature) have been achieved with "Hantzsch esters" to reduce 2-oxoesters to the corresponding alcohols. Similarly, conjugate hydride addition to α,β-unsaturated aldehydes, followed by protonation provides the saturated aldehydes.

Hantzsch Esters in Reductions

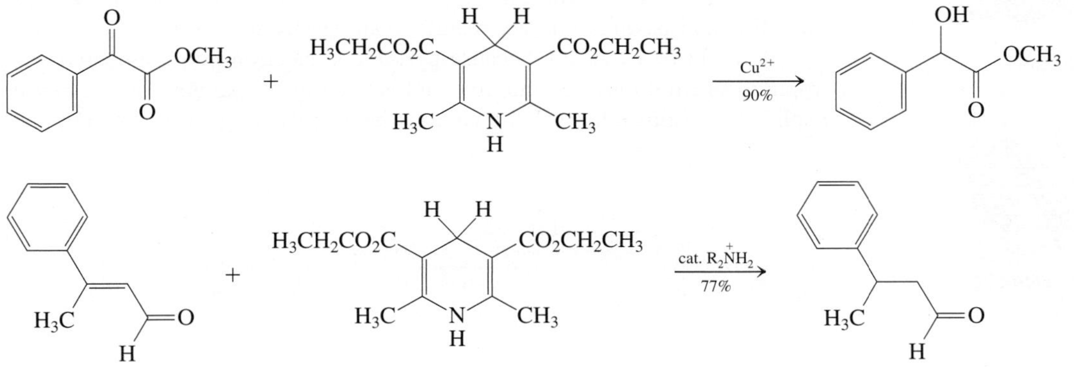

The first process is not quite metal free, as it requires catalytic amounts of Lewis acidic Cu^{2+} to activate the carbonyl function. This has the advantage that in the presence of chiral ligands (see Section 12-2 and Real Life 5-4 and 12-2) to the Cu^{2+}, the hydride adds enantioselectively from only one side of the molecule to generate only one enantiomer of the alcohol, much as nature does it. The second process constitutes another example of organocatalysis (Real Life 18-1), and similar activation takes place using catalytic amounts of an ammonium salt, which converts intermittently the aldehyde function to the corresponding iminium ion, in which the positive charge serves to render the β-carbon a better hydride acceptor. Traditionally, such activations are performed with H$^+$ or Lewis acids, and hydrogenation of alkenes normally uses heterogeneous metal catalysts (Section 12-2).

The NAD$^+$–NADH Redox Couple

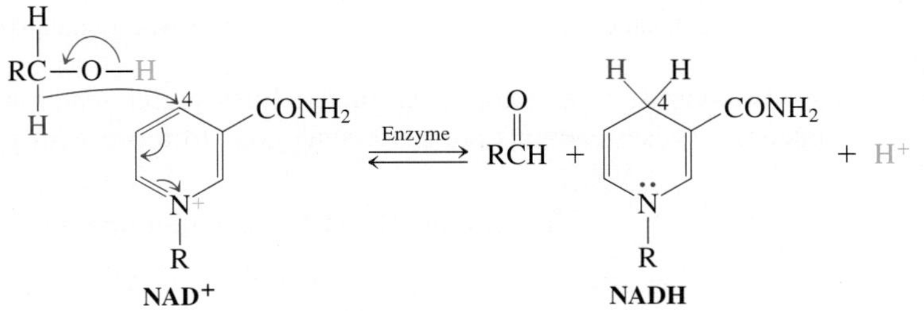

Exercise 25-23

Propose a mechanism for the reaction of 4-chloropyridine with methoxide. [**Hint:** Think of the pyridine ring as containing an α,β-unsaturated imine function (see Sections 17-9 and 18-9).]

Exercise 25-24

The relative rates of the reactions of 2-, 3-, and 4-chloropyridine with sodium methoxide in methanol are 3000 : 1 : 81,000. Explain.

In Summary Pyridine undergoes slow electrophilic aromatic substitution preferentially at C3. Nucleophilic substitution reactions occur more readily to expel hydride or another leaving group from either C2 or C4.

25-7 | QUINOLINE AND ISOQUINOLINE: THE BENZOPYRIDINES

Quinoline

Isoquinoline

We can imagine the fusion of a benzene ring to pyridine in either of two ways, giving us **quinoline** and **isoquinoline** (1- and 2-azanaphthalene, according to our systematic nomenclature). Both are liquids with high boiling points. Many of their derivatives are found in nature or have been synthesized in the search for physiological activity. Like pyridine, quinoline and isoquinoline are readily available from coal tar.

As might be expected, because pyridine is electron poor compared with benzene, electrophilic substitutions on quinoline and isoquinoline take place at the *benzene* ring. As with naphthalene, substitution at the carbons next to the ring fusion predominates.

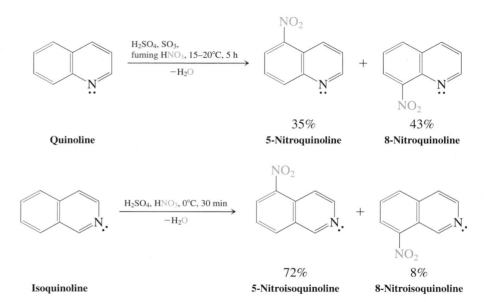

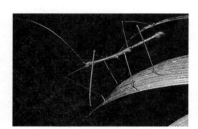

The Peruvian fire stick secretes quinoline from glands behind the head as a chemical defense against predators, such as frogs, cockroaches, spiders, and ants.

In contrast to electrophiles, nucleophiles prefer reaction at the electron-poor *pyridine* nucleus. These reactions are quite analogous to those with pyridine.

Chichibabin Reaction of Quinoline and Isoquinoline

REAL LIFE: BIOLOGY 25-3 | Folic Acid, Vitamin D, Cholesterol, and the Color of Your Skin

Folic acid is essential to human life. Its structure incorporates a 1,3,5,8-tetraazanaphthalene (pteridine) ring system, together with 4-aminobenzoic acid (Section 15-4) and (S)-aminopentanedioic (glutamic) acid (Section 26-1). Folic acid is necessary for the proper development of the nervous system in the very early stages of pregnancy. One of its functions is the transfer of one-carbon fragments between

biomolecules (see Problems 69 and 70 of Chapter 9). A deficiency of this substance, which must be obtained from the diet, is associated with crippling and often fatal birth defects such as spina bifida ("open spine") and anencephaly (a failure of the brain to develop normally). The U.S. Public Health Service recommends that all women of child-bearing age take 400 μg (0.4 mg) of folic acid daily.

1,3,5,8-Tetraaza-
naphthalene part

4-Aminobenzoic
acid part

(S)-2-Aminopentanedioic
(glutamic) acid part

Folic acid

Vitamin D$_2$

Vitamin D is also an essential nutrient for good health. It supports the growth of healthy bones in children, and its lack leads to a deforming condition known as **rickets.** Throughout our lives, vitamin D plays a crucial role in maintaining proper levels of calcium and phosphorus in the body. Unlike folic acid, your body can manufacture vitamin D. The starting material is cholesterol (Real Life 4-2), and the other indispensable ingredient is sunlight—specifically, the ultraviolet B radiation (the range from 280 to 315 nm) that falls on the Earth's surface when the sun is high in the sky. From these facts derives a truly remarkable story.

The shades of human skin range from nearly jet black in some tropical populations to the palest of skin tones in redheads from northern Europe (the United Kingdom in particular). How and why is this the case?

The extended aromatic π system in the tetraazanaphthalene ring of folic acid absorbs ultraviolet light strongly (Section 15-5) and undergoes subsequent structural changes; in short, sunlight, which is necessary for the production of vitamin D, destroys the body's stores of folic acid. The variations in human skin color around Earth represent the response of human evolution to the varying amounts of UV light to which humans are exposed. At the evolutionary dawn of humanity, as the amount of protective hair covering the body diminished, individuals who possessed greater amounts of skin-darkening pigments had a higher likelihood of giving birth to healthy babies. Their slightly darker skin provided some protection for the folic acid in their bodies, while still allowing the synthesis of adequate amounts of vitamin D.

What about lighter-skinned populations? Assuming the accuracy of the "out of Africa" hypothesis, humans moving northward to latitudes with less direct sunlight would have faced vitamin D deficiencies as a function of the darkness of

their skin. In this situation, individuals with slightly lighter skin would have had the advantage, their increased vitamin D production increasing the odds of reaching reproductive age. Over the generations of northward human migration, the lightening of the skin would have reached the present balance that allows both the preservation of necessary amounts of folic acid and the synthesis of adequate quantities of vitamin D.

How do the darkest-skinned individuals synthesize enough vitamin D, when their skin permits only very little UV light to penetrate? An adaptation developed in this population, which favored individuals who produced higher levels of cholesterol, the precursor of vitamin D, in their blood. We see the consequences of this modification in the higher incidence of cholesterol-related heart disease and stroke in this group.

The diversity of the interplay between folic acid and vitamin D production: The French soccer team celebrates after winning the Under-20 World Cup Final on July 13, 2013 in Istanbul, Turkey.

Exercise 25-25

Quinoline and isoquinoline react with organometallic reagents exactly as pyridine does (Section 25-6). Give the products of their reaction with 2-propenylmagnesium bromide (allylmagnesium bromide).

The following structures are representative of higher aza analogs of naphthalene.

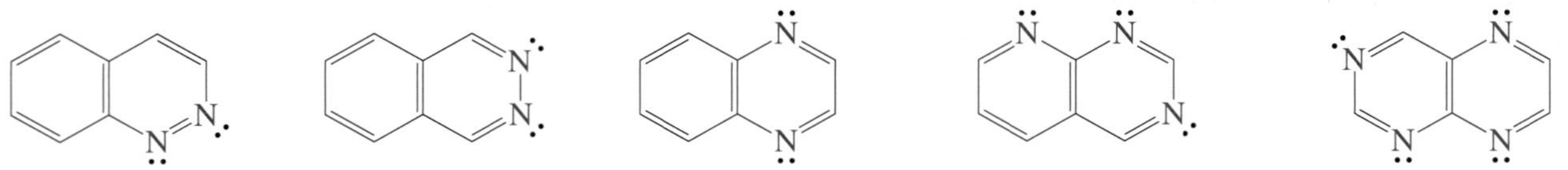

1,2-Diazanaphthalene
(**Cinnoline**)

2,3-Diazanaphthalene
(**Phthalazine**)

1,4-Diazanaphthalene
(**Quinoxaline**)

1,3,8-Triazanaphthalene
(**Pyrido[2,3-d]pyrimidine**)

1,3,5,8-Tetraazanaphthalene
(**Pteridine**)

Solved Exercise 25-26 | Working with the Concepts: Recognizing Retrosynthetic Disconnections in the Azanaphthalenes

As described in Section 25-5, some pyridine syntheses are based on a variety of condensation reactions, typically aldol (Section 18-5) and imine condensations (Section 17-9). Bearing this in mind, how would you dissect the pyridine ring of quinoline A retrosynthetically to two appropriate starting materials for its construction?

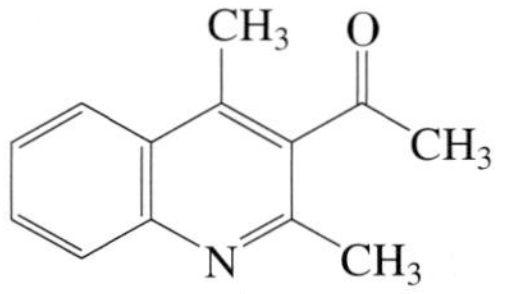

Strategy
The most obvious disconnection is the imine linkage, which could arise from the corresponding amine and carbonyl component. The latter reveals an α,β-unsaturated ketone that might be assembled by an aldol condensation.

Solution
• First, we disconnect the imine double bond (step 1). This produces a benzenamine on one side and a carbonyl group on the other.
• Second, we apply a retroaldol step to the remaining double bond (step 2) to generate 2-acetylbenzeneamine and 2,4-pentanedione.
• Are these two materials suitable starting materials for the synthesis of A? The answer is yes: This reaction is an example of the so-called Friedländer synthesis of quinolines, and it proceeds with either acid or base catalysis. While one can envisage several competing condensations of the starting materials, they are reversible and there is considerable driving force toward the aromatic cyclization product. Moreover, the aminoketone component contains the two functions in conjugation, thus preventing its self-condensation (write the dipolar ammonium enolate resonance form).

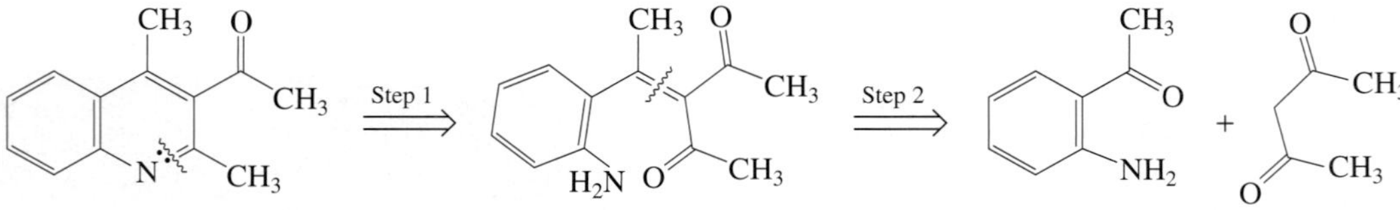

Exercise 25-27 | Try It Yourself

Sketch plausible retrosynthetic disconnections for the following molecules. [**Hints:** For part b, see Section 17-10; for part c, see Problem 43 of Chapter 17.]

(a) 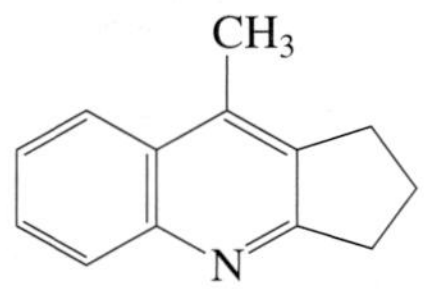   (b) 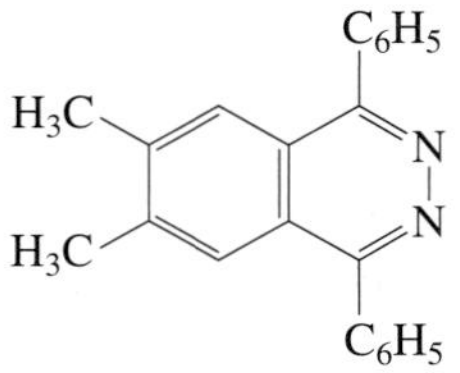   (c)

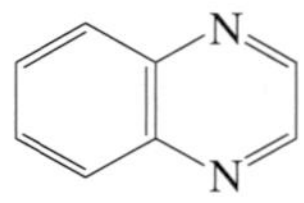

In Summary The azanaphthalenes quinoline and isoquinoline may be regarded as benzo-pyridines. Electrophiles attack the benzene ring, nucleophiles attack the pyridine ring.

25-8 | ALKALOIDS: PHYSIOLOGICALLY POTENT NITROGEN HETEROCYCLES IN NATURE

The **alkaloids** are bitter-tasting, natural nitrogen-containing compounds found particularly in plants. The name is derived from their characteristic basic properties (alkali-like), which are induced by the lone electron pair of nitrogen.

As with acyclic amines (Chapter 21), the (Lewis) basic nature of the alkaloids, in conjunction with their particular three-dimensional architecture, gives rise to often potent physiological activity. We have already noted some examples of this behavior in the narcotics morphine and heroin (Section 9-11), the psychoactive lysergic acid and LSD (Section 19-13), and the antibiotic penicillins (Real Life 20-2).

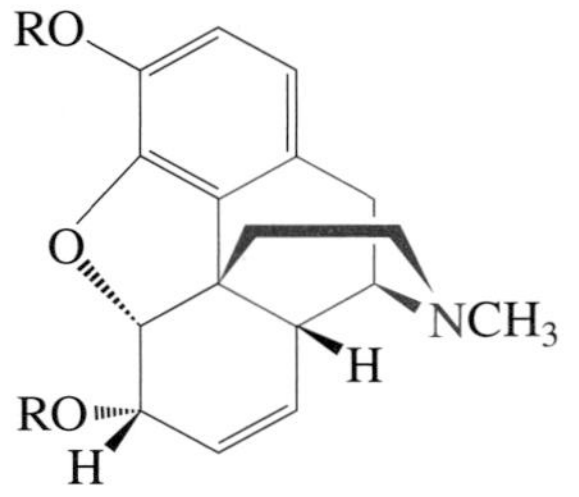

Morphine (R = H)

O
||
Heroin (R = CH₃C)

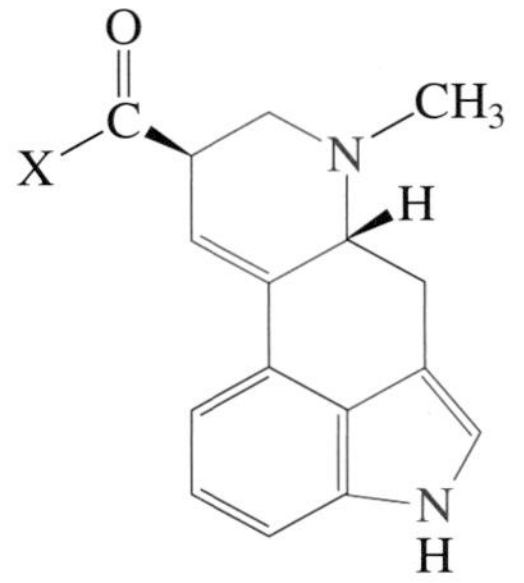

Lysergic acid (X = OH)
Lysergic acid diethylamide, LSD
[X = (CH₃CH₂)₂N]

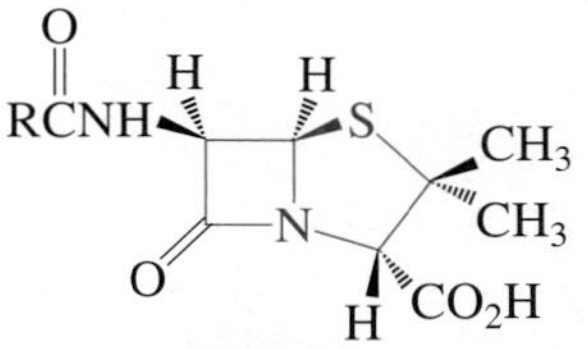

Penicillin

Medicinal chemists strive to strip down the structures of complex natural drugs to identify the minimal requirements for activity: the pharmacophore. For morphine this approach has yielded hundreds of simpler analogs with a varying pharmacological spectrum.

The Pharmacophore of Morphine

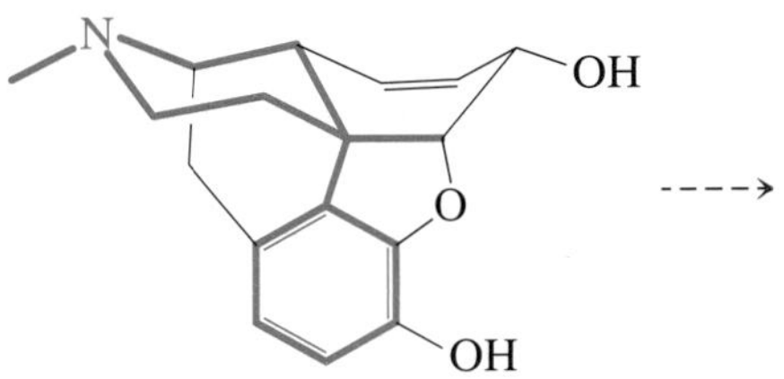

Morphine
(Isolated from the opium poppy)

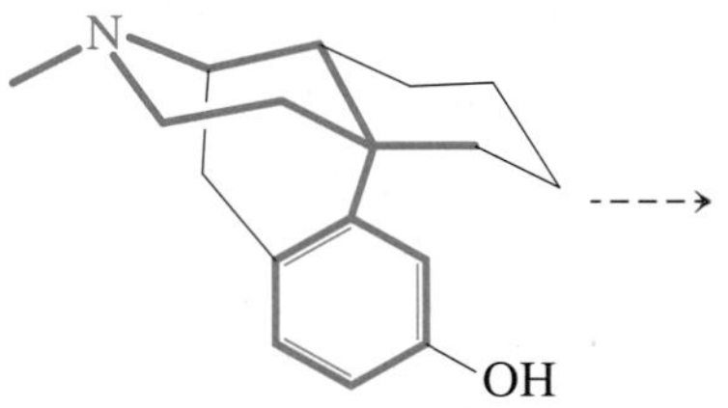

Levorphanol
(Levo-Dromoran)

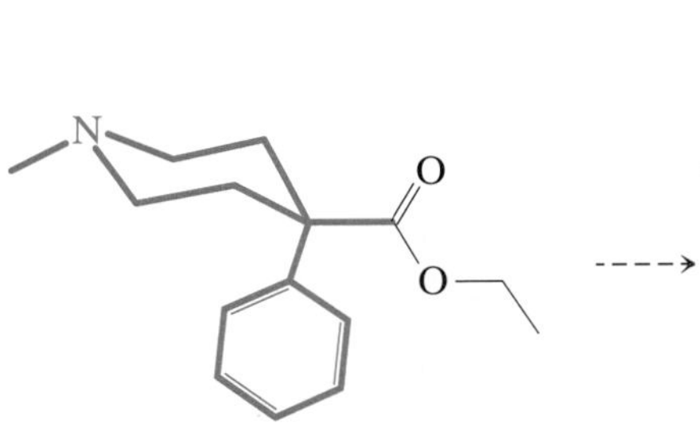

Pethidine
(Demerol)

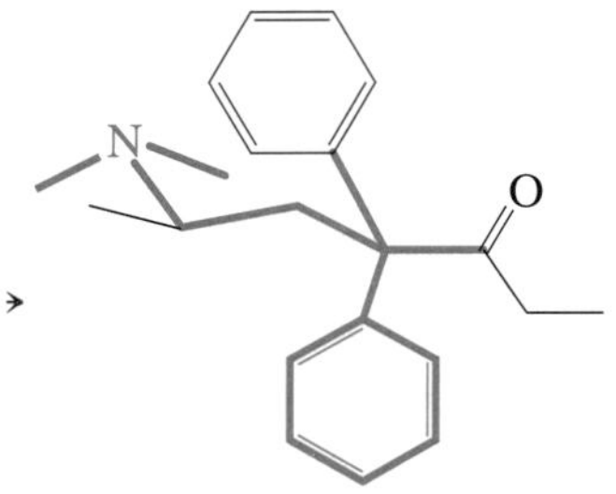

Methadone

REAL LIFE: NATURE 25-4 | Nature Is Not Always Green: Natural Pesticides

Many people believe that everything synthetic is somehow suspect and "bad," and that all of nature's chemicals are benign. As pointed out by Ames* and others, this is a misconception. While we have seen that, indeed, many manufactured chemicals have problems with toxicity and adverse effects on the environment, nature's chemicals are not any different from synthetic ones. Nature has its own highly productive laboratory, which puts out compounds by the millions, many of which are highly toxic, such as quite a few of the alkaloids found in plants. Consequently, there are numerous (sometimes lethal) cases of poisoning (especially of children) due to the accidental ingestion of plant material, the eating of green potatoes (exposed to sunlight, which increases their toxin level), the drinking of herbal teas, the consumption of "poison" mushrooms, and so forth. Abraham Lincoln's mother died from drinking milk from a cow that had grazed on the toxic snakeroot plant. The issue strikes at the heart of the

argument that "organic" food, especially fruits and vegetables, is more healthy than conventional produce. Proponents advocate that organic products are better because they are grown without added synthetic pesticides, and critics claim that this very fact makes them prone to be contaminated with higher levels of bacteria and natural toxins.

Green potatoes are toxic because of the presence of the alkaloid solanine.

Solanine

Lycopsamine
(Liver toxin in Comfrey tea)

Tremetone
(Toxin in snakeroot plant)

In addition to the problem of immediate potential toxicity of plant chemicals, there is a growing body of evidence that points to adverse interactions between foods and pharmaceuticals. For example, bergamottin in grapefruit juice is responsible for the "grapefruit juice effect" on the bioavailability of several prescription drugs. The compound inhibits cytochrome P-450 enzymes responsible for drug metabolism

(Real Life 8-1), preventing or slowing the removal of certain substances. This can cause the effective concentration of a drug to be greatly increased, reaching dangerous—even lethal—levels. Another example is the herbal antidepressant hypericum (St. John's wort), which can cause abortions and also interferes with the action of the birth control pill (Real Life 4-3).

Bergamottin

Hypericin
(Active component of hypericum)

*Professor Bruce N. Ames (b. 1928), University of California at Berkeley.

What is the purpose of these compounds in plant life? Plants cannot run away from predators and invading organisms, such as fungi, insects, animals, and humans, and they have no organs with which to defend themselves. Instead, they have developed an array of chemical weapons, "natural pesticides," with which to mount an effective defense strategy. Tens of thousands of these chemicals are now known. They are either already present in the existing plant or are generated in a primitive "immune response" to external damage, such as by caterpillars or herbivorous insects. For example, in the tomato plant, a small polypeptide (Section 26-4) containing 18 amino acids, systemin, is the chemical alarm signal for external attack. The molecule travels rapidly through the plant, initiating a cascade of reactions that produce chemical poisons. The effect is either to fend off attackers completely or to slow them down sufficiently so that other predators will consume them. Interestingly, one of these compounds is salicylic acid, the core of aspirin (Real Life 22-2), which prevents the point of damage (much like a wound) from being infected. Plants in distress have evolved to use chemicals as alarm pheromones (Section 12-17), activating the chemical weapons complex of (as yet) undamaged neighbors by air- or waterborne molecular signals. They may also develop resistance (immunity) by chemical pathways.

Americans consume about 1.5 g of natural pesticides per person per day, in the form of vegetables, fruit, tea, coffee, and so forth—10,000 times more than their intake of synthetic pesticide residues. The concentration of these natural compounds ranges in the parts per million (ppm), orders of magnitude above the levels at which water pollutants (e.g., chlorinated hydrocarbons) and other synthetic pollutants (e.g., detergents, Real Life 19-1) are usually measured (parts per billion, or ppb). Few of these plant toxins have been tested for carcinogenicity but, of those tested (in rodents), roughly half are carcinogenic, the same proportion as that of synthetic chemicals. Many have proven toxicity. Examples of some (potentially) toxic pesticides in common foods are given below.

Why, then, have we all not been exterminated by these poisons? One reason is that the level of our exposure to any one of these natural pesticides is very small. More important, we, like plants, have evolved to defend ourselves against this barrage of chemical projectiles. Thus, for starters, our first line of defense, the surface layers of the mouth, esophagus, stomach, intestine, skin, and lungs, is discarded once every few days as "cannon fodder." In addition, we have multiple detoxifying mechanisms, rendering ingested poisons nontoxic; we excrete a lot of material before it does any harm; our DNA has many ways of repairing damage; and finally, our ability to smell and taste "repugnant" substances (such as the "bitter" alkaloids, rotten food, milk that is "off," eggs that smell of "sulfur") serves as an advance warning signal. In the final analysis, we each must judge what we put into our bodies, but the age-old wisdoms still hold: Avoid anything in excess, and maintain variety in your diet.

Natural Plant Pesticides and Their Occurrence (in ppm)

Caffeic acid
(Carcinogen)

Apple, carrot, celery, grapes, lettuce, potato (50–200); basil, dill, sage, thyme, and other herbs (>1000); coffee (roasted beans, 1800)

Allyl isothiocyanate
(Carcinogen)

Cabbage (35–590); cauliflower (12–66); Brussels sprouts (110–1560); brown mustard (16,000–72,000); horseradish (4500)

(R)-Limonene
(Carcinogen)

Orange juice (31); black pepper (8000)

Serotonin
(Neurotransmitter, vasoconstrictor)

Carrot (10–20)

Carotatoxin
(Neurotoxin)

Parsley (11,000–112,000); celery (1,300–46,000)

Psoralen
(Carcinogen)

Banana (15,000)

Nicotine (see also Real Life 25-1), present in dried tobacco leaves in 2–8% concentration, is the stimulating ingredient in cigarettes and other tobacco products.

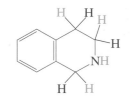

Nicotine   **Caffeine (R = CH₃)**   **Cocaine**
Theobromine (R = H)

Even more stimulating than nicotine are caffeine and theobromine, present in coffee and tea or cocoa (chocolate), respectively. Perhaps the most dangerous stimulant is cocaine, extracted from the leaves of the coca shrub, which is cultivated mainly in South America for the purpose of illegal drug trafficking. Cocaine is shipped and sold in the form of the water-soluble hydrochloride salt ("street cocaine"), which may be ingested through the nasal passages by "snorting" or orally and intravenously. The actual alkaloid is known as "freebase" or "crack" and is inhaled by smoking. There are severe physical and psychological side effects of the drug, such as brain seizures, respiratory collapse, heart attack, paranoia, and depression. The compound has some good uses, nevertheless. For example, it functions as a very effective topical anesthetic in eye operations.

Quinine, isolated from cinchona bark (as much as 8% concentration), is the oldest known effective antimalarial agent. A malaria attack consists of a chill accompanied or followed by a fever, which terminates in a sweating stage. Such attacks may recur regularly. The name "malaria" is derived from the Italian *malo,* bad, and *aria,* air, referring to the old belief that the disease was caused by noxious effluent gases from marshland. The actual culprit is a protozoan parasite (*Plasmodium* species) transmitted by the bite of an infected female mosquito of the genus *Anopheles* (see Real Life 3-2). It is estimated that from 300 to 500 million people are affected by this disease, which kills over 2 million each year, more than half of them children.

Strychnine is a powerful poison (the lethal dose in animals is about 5–8 mg kg⁻¹), the lethal ingredient of many a detective novel.

An *Anopheles stephensi* mosquito lunching on a human host through its pointed feeding tube. Engorgement has proceeded to the point that a droplet of blood emanates from the abdomen.

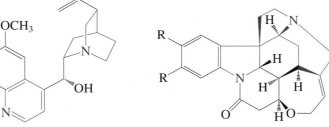

Quinine   **Strychnine (R=H)**   **1,2,3,4-**Tetrahydro**isoquinoline**
Brucine (R=CH₃O)

The isoquinoline and 1,2,3,4-tetrahydroisoquinoline nuclei are abundant among the alkaloids, and their derivatives are physiologically active, for example, as hallucinogens, central nervous system agents (depressants and stimulants), and hypotensives. Note that the pharmacophoric 2-phenylethanamine unit (see Real Life 21-1) is part of these nuclei and is also present in most of the other alkaloids considered in this section. (Find it in morphine, lysergic acid, quinine—there is a quirk here—and strychnine.)

In Summary The alkaloids are natural nitrogen-containing compounds, many of which are physiologically active.

THE BIG PICTURE

The discussion of saturated and aromatic heterocycles pulls together several concepts and applications in this book—in particular, the ring opening of strained heterocycles; the extension of the principles of aromaticity to heteroaromatic systems, including electrophilic and nucleophilic aromatic substitution; the use of condensation methodology in the construction of heteroaromatic systems; and a glimpse at their structural diversity in nature and as synthetic drugs. The field of heterocyclic chemistry is vast, and this chapter has only highlighted a few of its aspects. We shall see in the next and final chapter that heterocycles are integral parts of the nucleic acids DNA and RNA.

WORKED EXAMPLES: INTEGRATING THE CONCEPTS

25-28. Applying Condensation Reactions in Heterocyclic Syntheses

As we have seen for the heterocyclopentadienes (Section 25-3) and pyridine (Section 25-5), heteroaromatic compounds can almost invariably be made by condensation reactions of carbonyl substrates with appropriate heterofunctions.

a. Write a plausible mechanism for the first step of the Hantzsch synthesis of 2,6-dimethylpyridine (Section 25-5), shown again here.

SOLUTION

By following the fate of the four components in the starting mixture, you can see that the ammonia has reacted at the two keto carbonyl carbons, presumably by imine and then enamine formation (Section 17-9), whereas the formaldehyde component has made bonds to the acidic methylene of the 3-oxobutanoates (Section 23-2), presumably initially by an aldol-like condensation process (Section 18-5). Let us formulate these steps one at a time.

Step 1. Aldol-like condensation of formaldehyde with ethyl 3-oxobutanoate

Step 2. Enamine formation of ammonia with ethyl 3-oxobutanoate

We note that step 1 furnishes a Michael acceptor, whereas step 2 results in an enamine. The latter can react with the former, in analogy to the reaction of enolates, by Michael addition, in this case to a neutral oxoenamine.

Step 3. Michael addition of the enamine

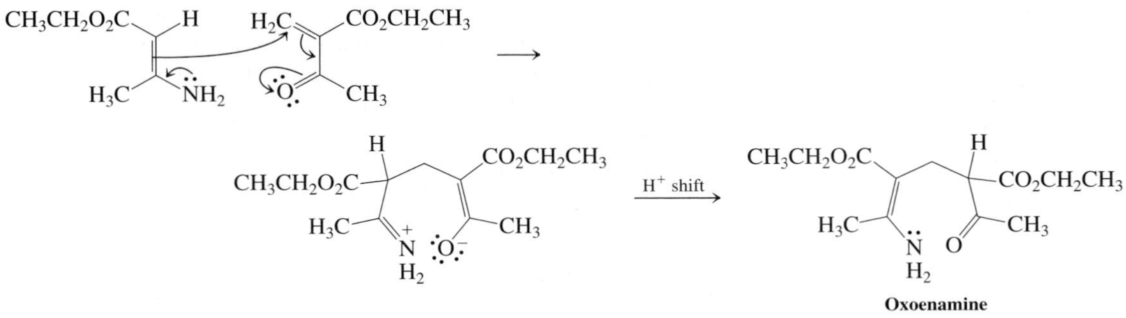

This species is now perfectly set up to undergo an intramolecular imine condensation to the 3,4-dihydropyridine, which tautomerizes (Sections 13-7 and 18-2) to the more stable 1,4-dihydro product.

Step 4. Intramolecular imine formation and tautomerization

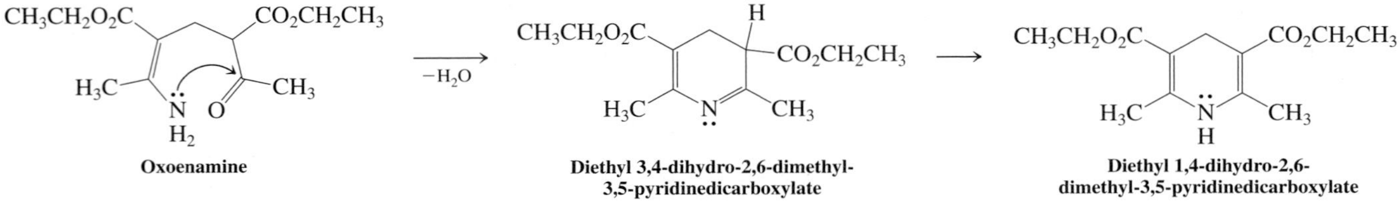

Oxoenamine

Diethyl 3,4-dihydro-2,6-dimethyl-3,5-pyridinedicarboxylate

Diethyl 1,4-dihydro-2,6-dimethyl-3,5-pyridinedicarboxylate

b. On the basis of the discussion in part a and Exercises 25-11 and 25-14, suggest some simple retro-synthetic (retrocondensation) reactions to indole, quinoline, and 1,4-diazanaphthalene (quinoxaline) from ortho-disubstituted benzenes.

SOLUTION

You can view indole as a benzofused enamine. The enamine part is retrosynthetically connected to the corresponding enolizable carbonyl and amino function (Section 17-9).

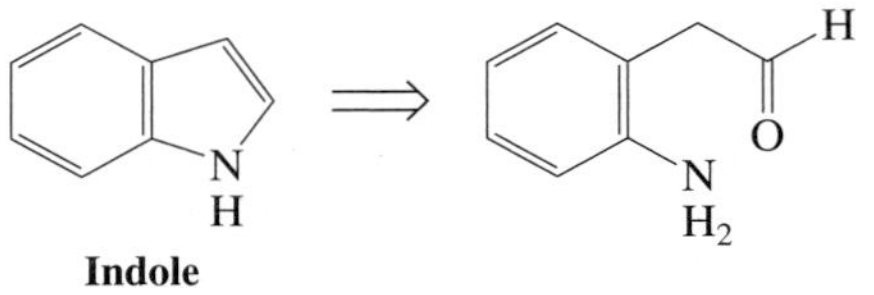

Indole

Quinoline can be viewed as a benzofused α,β-unsaturated imine. The imine nucleus is retrosynthetically opened to the corresponding amine and α,β-unsaturated carbonyl fragments (Section 17-9), which can be constructed by an aldol condensation (Section 18-5) employing acetaldehyde.

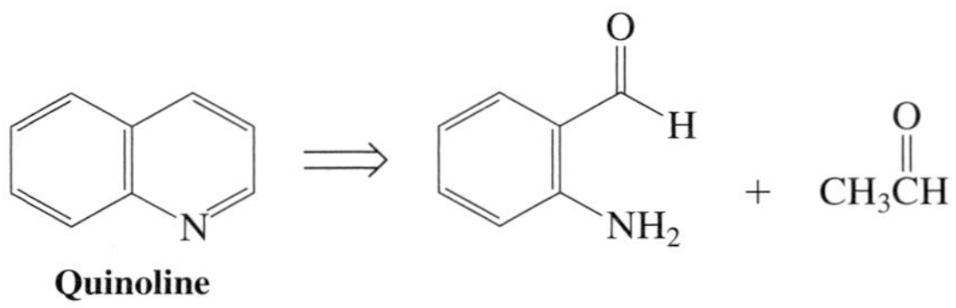

Quinoline

1,4-Diazanaphthalene can be dissected by retrosynthetic hydrolysis of the two imine functions (Section 17-9) to 1,2-benzenediamine and ethanedione (glyoxal).

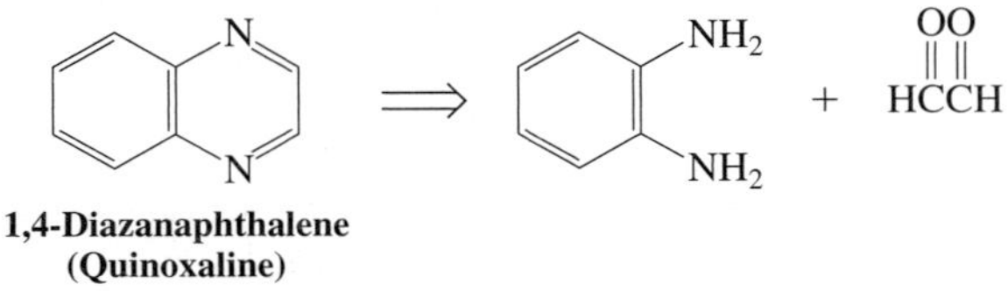

**1,4-Diazanaphthalene
(Quinoxaline)**

25-29. Understanding the Synthesis of Viagra

Viagra (sildenafil citrate) was introduced in 1998 as an effective way to treat male erectile dysfunction (MED). It was discovered accidentally during clinical trials of the compound as a treatment for coronary heart disease and works by enhancing the production of nitric oxide (NO) in erectile tissue, ultimately leading to vasodilation (see Real Life 26-1). The route by which Viagra was made originally in the research laboratory is shown below.

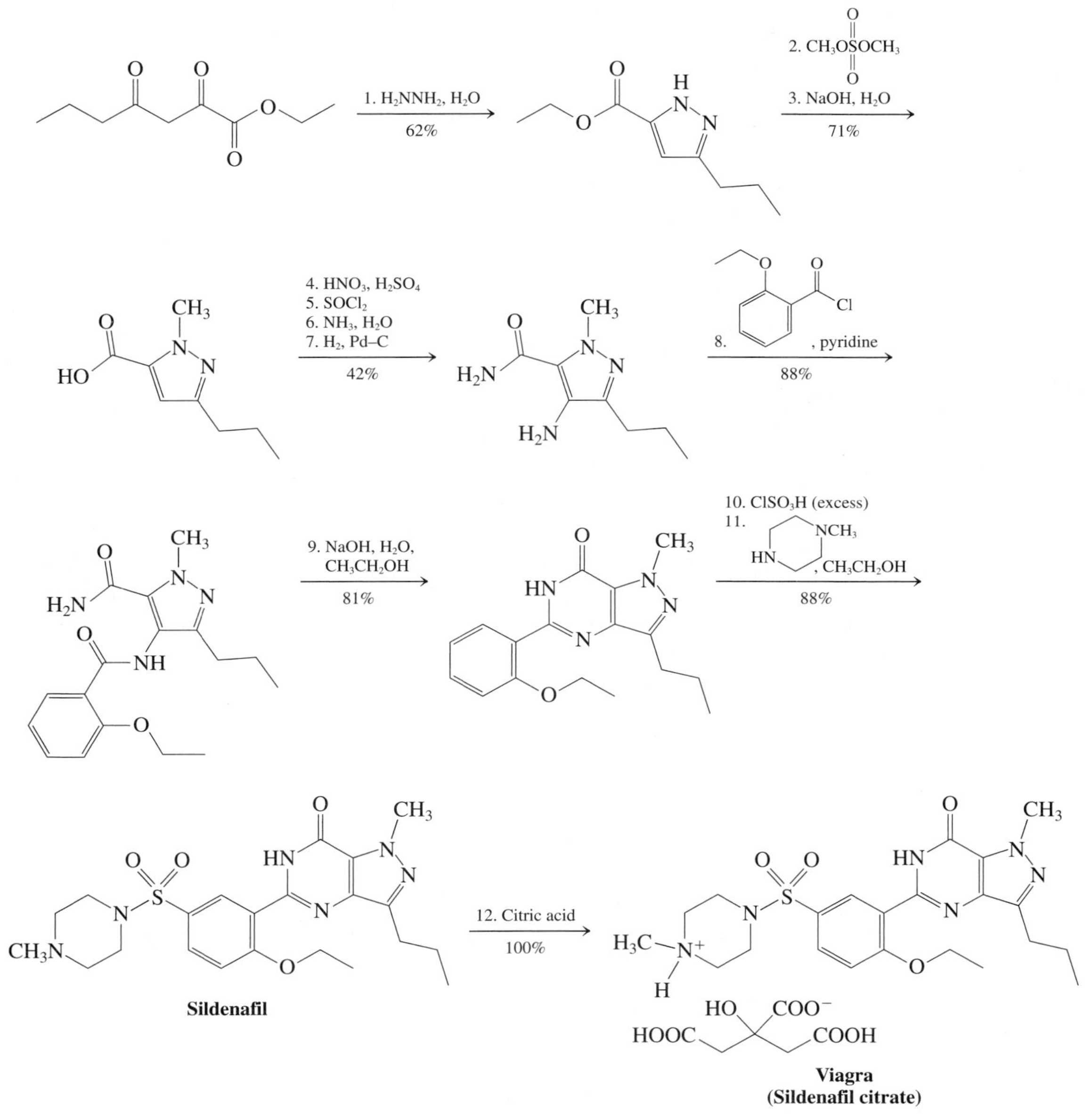

Sildenafil

**Viagra
(Sildenafil citrate)**

With your current knowledge of synthetic chemistry, you should be able to understand each of the 12 steps in this synthesis, even though some of the functional units are new to you. Identify the essential features of each step and rationalize its outcome.

SOLUTION

This problem is typical of what the practitioner of organic synthesis might encounter while reading the literature. He or she may not be versed in the particular class of substances described, or the specific reagents employed, but will nevertheless be able to follow the narrative by extrapolating from basic principles.

Step 1. We can see topologically that the N_2 unit in the reagent hydrazine, H_2NNH_2, has been added across the β-dicarbonyl function (Chapter 23) in some way. Comparison of the respective molecular formulas reveals that two molecules of water are lost on going from starting materials to product, suggestive of a double condensation. Substituted hydrazines undergo just that with ketones (Section 17-10; see also Exercise 25-27, part b), normally only once; here we do it twice. Application to a 1,3-dione produces a diazacyclopentadiene, which aromatizes by deprotonation (to an aromatic anion, see Section 15-7), followed by reprotonation. The resulting 2-azapyrrole (Section 25-4) is called pyrazole, and this is one way of making it.

General Pyrazole Synthesis

Step 2. Dimethylsulfate, $(CH_3O)_2SO_2$, is a methylating agent (Section 6-7) and serves to alkylate the pyrazole nitrogen.

Step 3. This step is a simple base-mediated ester hydrolysis (Section 20-4).

Step 4. The product of the sequence 4–7 contains an additional amino substituent on the pyrazole ring. The requisite nitrogen is introduced by way of an electrophilic nitration (Sections 15-10 and 25-4).

Steps 5 and 6. This sequence converts the carboxylic acid to the carboxylic amide via the acid chloride (Sections 19-8 and 20-2).

Step 7. We now convert the nitro substituent introduced in step 4 into the amino function by catalytic hydrogenation (Section 16-5).

Step 8. Next, we introduce the 2-ethoxybenzoyl fragment by amide formation (Section 20-2).

Step 9. This condensation is unusual, as it involves deactivated amino and carbonyl groups. It is facilitated by its intramolecular nature, much like that of cyclic imide formation (Section 19-10), and the dipolar (Section 20-1) aromatic resonance stabilization of the product, shown in a general way below.

Step 10. This step is a variant of aromatic sulfonation (Section 15-10) and simultaneous sulfonyl chloride generation, using chlorosulfonic acid, $ClSO_3H$, the acid chloride of sulfuric acid.

Step 11. We arrive at sildenafil, the active ingredient of Viagra, by sulfonamide production (Section 15-10).

Step 12. To allow for water solubility, sildenafil is administered in the protonated form of the ammonium salt of citric acid, the drug Viagra itself.

New Reactions

1. Reactions of Heterocyclopropanes (Section 25-2)

2. Ring Opening of Heterocyclobutanes (Section 25-2)

Less reactive than heterocyclopropanes

3. Paal-Knorr Synthesis of 1-Hetero-2,4-cyclopentadienes (Section 25-3)

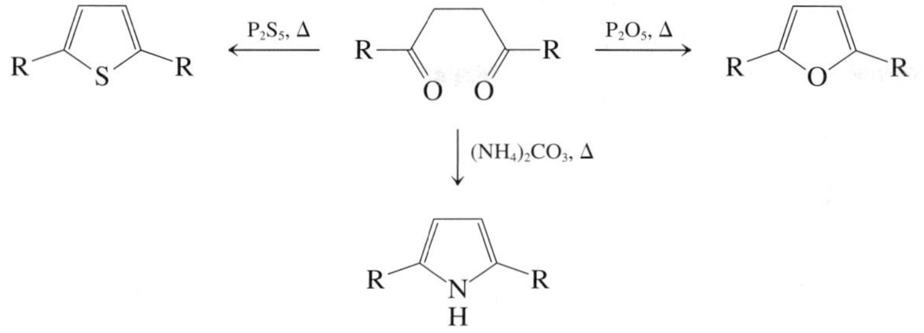

4. Reactions of 1-Hetero-2,4-cyclopentadienes (Section 25-4)

Electrophilic substitution

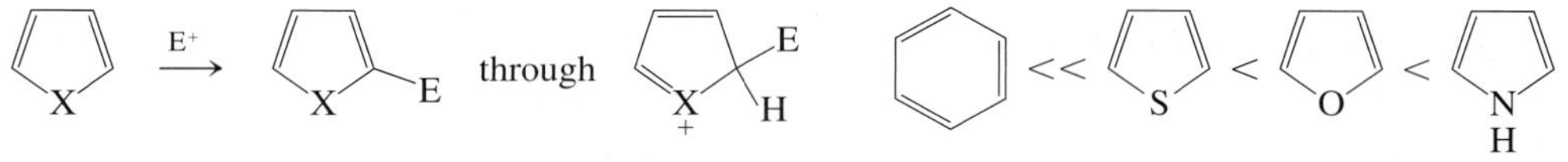

Main product   **Relative reactivity**

Ring opening

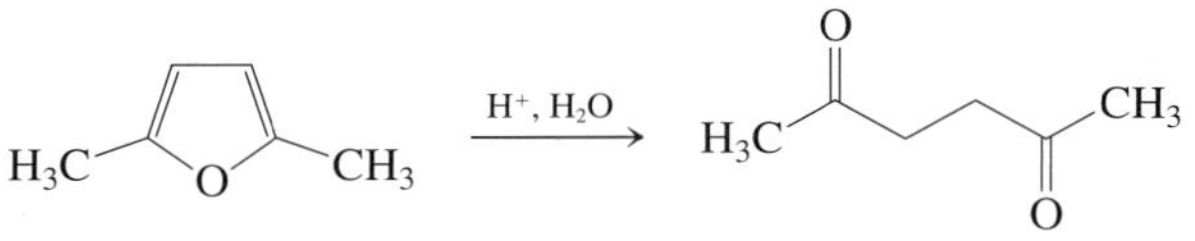

Cycloaddition

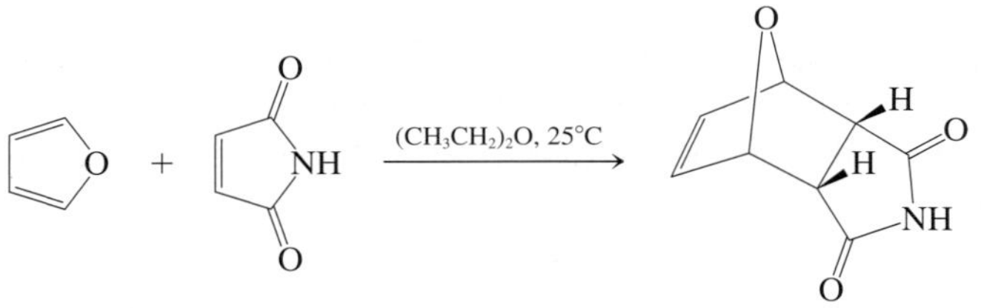

5. Hantzsch Synthesis of Pyridines (Section 25-5)

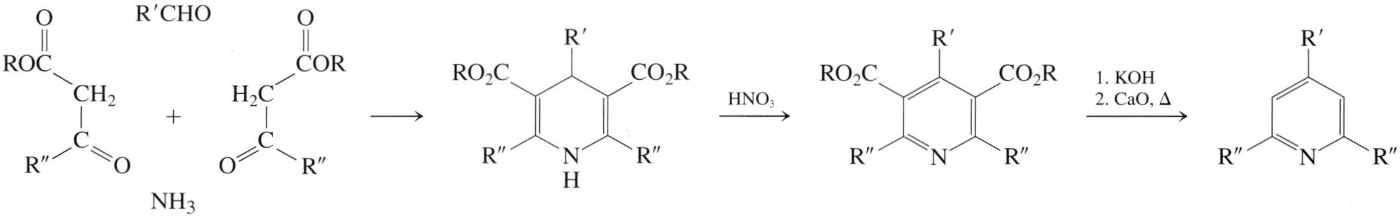

6. Reactions of Pyridine (Sections 25-5 and 25-6)

Protonation (Section 25-5)

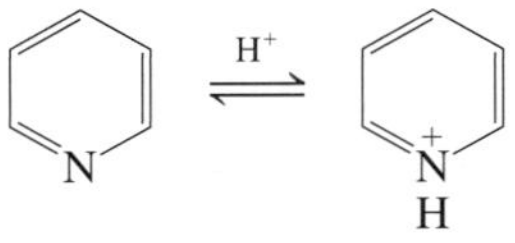

Pyridinium ion
(pK_a = 5.29)

Electrophilic substitution (Section 25-6)

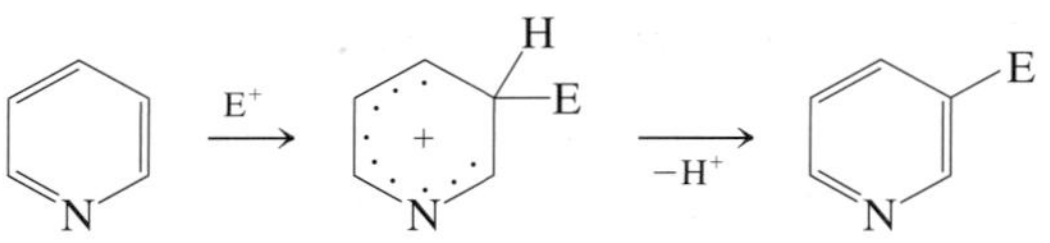

Ring is deactivated relative to benzene.

Nucleophilic substitution (Section 25-6)

$$\text{pyridine} \xrightarrow[\text{Chichibabin reaction}]{\text{NaNH}_2,\ \text{liquid NH}_3} \text{intermediate} \xrightarrow{-\text{H}^-} \xrightarrow{\text{H}^+,\text{H}_2\text{O}} \text{2-aminopyridine}$$

$$\text{pyridine} \xrightarrow[\text{Organometallic reagent}]{\text{RLi}} \text{2-R-pyridine}$$

$$\text{Halopyridine} (X = Br, Cl) \xrightarrow{:\text{Nu}^-} \text{4-Nu-pyridine} + X^-$$

Halopyridine
(X = Br, Cl)

7. Reactions of Quinoline and Isoquinoline (Section 25-7)

Electrophilic substitution   Nucleophilic substitution

Important Concepts

1. The **heterocycloalkanes** can be named using cycloalkane nomenclature. The prefix aza- for nitrogen, oxa- for oxygen, thia- for sulfur, and so forth, indicates the heteroatom. Other systematic and common names abound in the literature, particularly for the aromatic heterocycles.

2. The **strained three-** and **four-membered heterocycloalkanes** undergo **ring opening** with nucleophiles easily.

3. The **1-hetero-2,4-cyclopentadienes** are **aromatic** and have an arrangement of six π electrons, similar to that in the cyclopentadienyl anion. The heteroatom is sp^2 hybridized, the p orbital contributing two electrons to the π system. As a consequence, the diene unit is electron rich and reactive in electrophilic aromatic substitutions.

4. Replacement of one (or more) of the CH units in benzene by an sp^2-hybridized nitrogen gives rise to **pyridine** (and other azabenzenes). The p orbital on the heteroatom contributes one electron to the π system; the lone electron pair is located in an sp^2 hybrid atomic orbital in the molecular plane. Azabenzenes are **electron poor,** because the electronegative nitrogen withdraws electron density from the ring by induction and by resonance. Electrophilic aromatic substitution of azabenzenes is sluggish. Conversely, nucleophilic aromatic substitution occurs readily; this is shown by the Chichibabin reaction, substitutions by organometallic reagents next to the nitrogen, and the displacement of halide ion from halopyridines by nucleophiles.

5. The azanaphthalenes (benzopyridines) **quinoline** and **isoquinoline** contain an electron-poor pyridine ring, susceptible to nucleophilic attack, and an electron-rich benzene ring that enters into electrophilic aromatic substitution reactions, usually at the positions closest to the heterocyclic unit.

Problems

30. Name or draw the following compounds. (a) *cis*-2,3-Diphenyloxacyclopropane; (b) 3-azacyclobutanone; (c) 1,3-oxathiacyclopentane; (d) 2-butanoyl-1,3-dithiacyclohexane;

(g)

(h)

31. Identify by name (either IUPAC or common) as many of the heterocyclic rings contained in the structures shown in Table 25-1 as you can.

32. Give the expected product of each of the following reaction sequences.

(a)

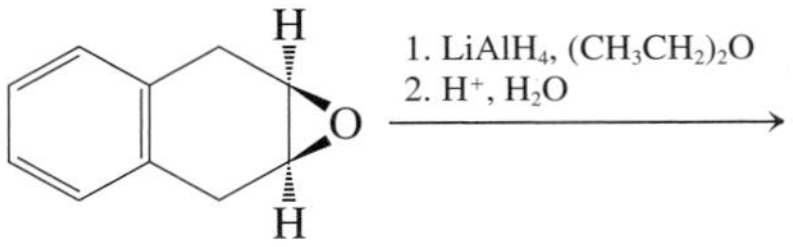

1. LiAlH₄, (CH₃CH₂)₂O
2. H⁺, H₂O

(b)

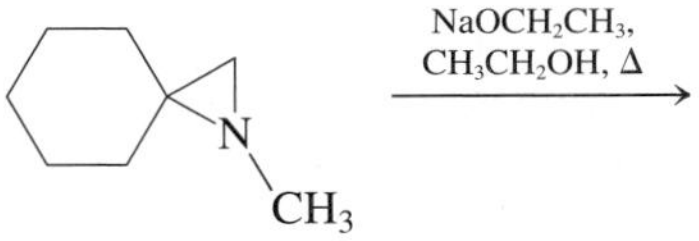

NaOCH₂CH₃,
CH₃CH₂OH, Δ

(c)

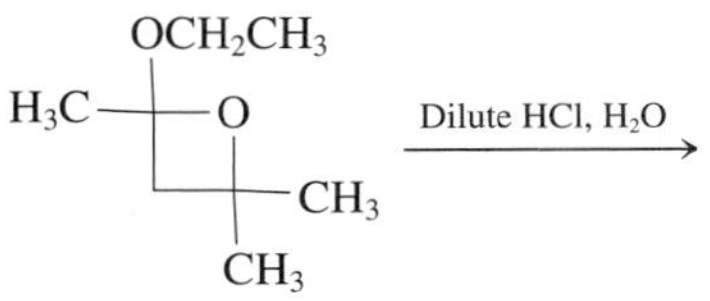

Dilute HCl, H₂O

33. The penicillins are a class of antibiotics containing two heterocyclic rings that interfere with the construction of cell walls by bacteria (Real Life 20-2). The interference results from reaction of the penicillin with an amino group of a protein that closes gaps that develop during construction of the cell wall. The insides of the cell leak out, and the organism dies. **(a)** Suggest a reasonable product for the reactions of penicillin G with the amino group of a protein (protein–NH₂). (**Hint:** First identify the most reactive electrophilic site in penicillin.)

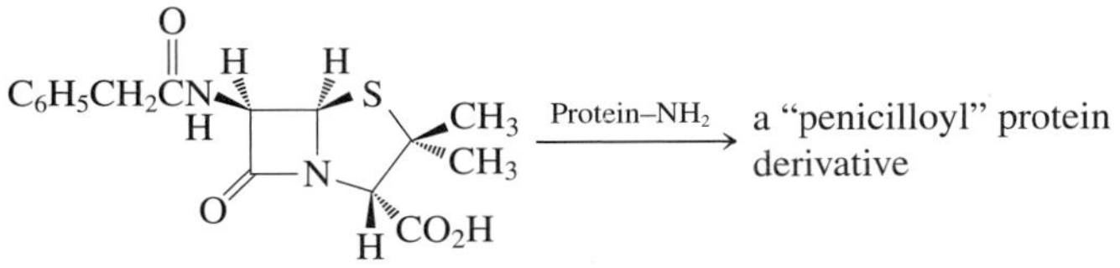

Protein–NH₂ a "penicilloyl" protein derivative

Penicillin G

(b) Penicillin-resistant bacteria secrete an enzyme (penicillinase) that catalyzes hydrolysis of the antibiotic faster than the antibiotic can attack the cell-wall proteins. Propose a structure for the product of this hydrolysis and suggest a reason why hydrolysis destroys the antibiotic properties of penicillin.

Penicillin G →[H₂O penicillinase] penicilloic acid

(**Hydrolysis product; no
antibiotic activity**)

34. Propose reasonable mechanisms for the following transformations.

(a)

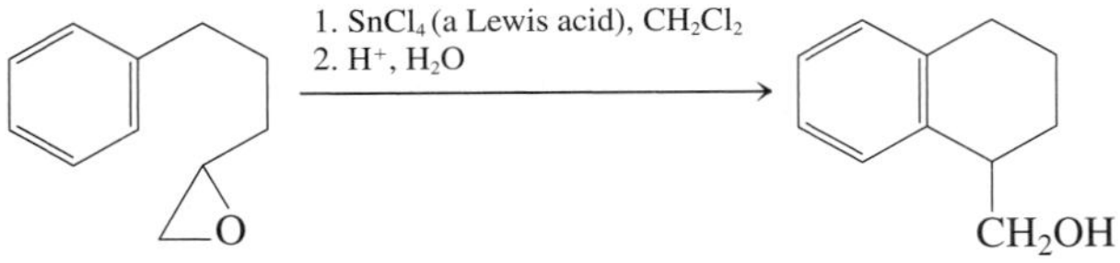

1. SnCl₄ (a Lewis acid), CH₂Cl₂
2. H⁺, H₂O

(b)

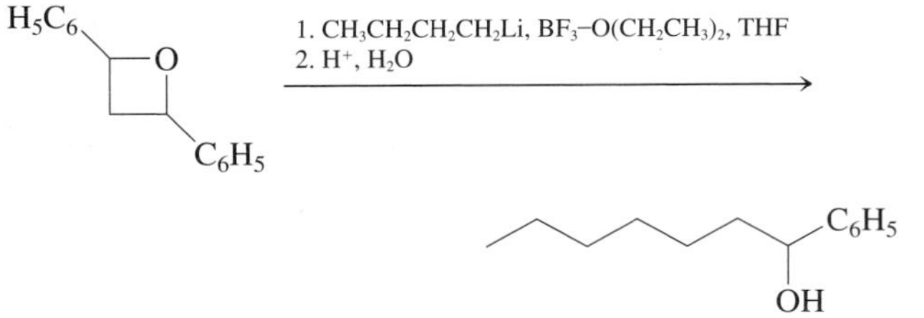

1. CH₃CH₂CH₂CH₂Li, BF₃–O(CH₂CH₃)₂, THF
2. H⁺, H₂O

(c)

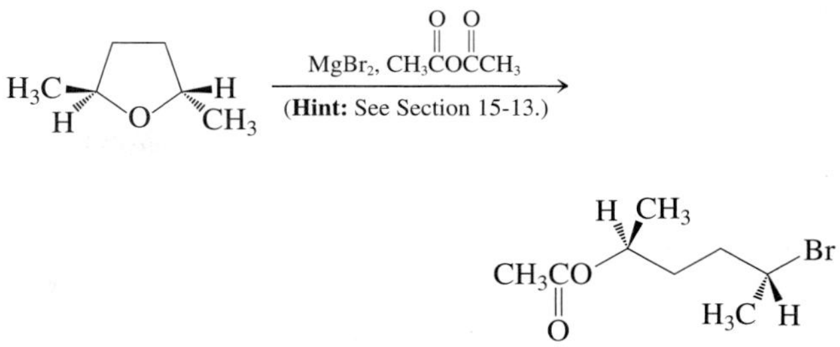

MgBr₂, CH₃COCCH₃

(**Hint:** See Section 15-13.)

35. Rank the following compounds in increasing order of basicity: water, hydroxide, pyridine, pyrrole, ammonia.

36. The heterocyclopentadienes below contain more than one heteroatom. For each molecule, identify the orbitals occupied by all lone electron pairs on the heteroatoms and determine whether the molecule qualifies as aromatic. Are any of these heterocycles more basic than pyrrole?

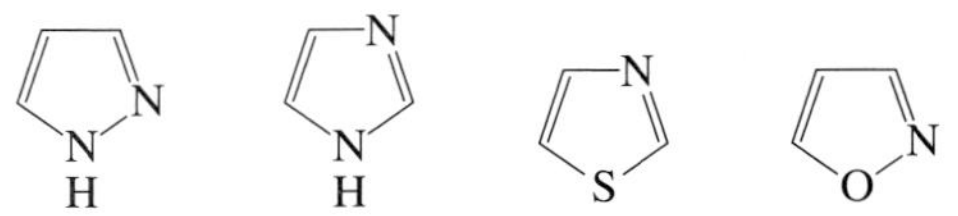

Pyrazole   Imidazole   Thiazole   Isoxazole

37. Give the product of each of the following reactions.

(a)

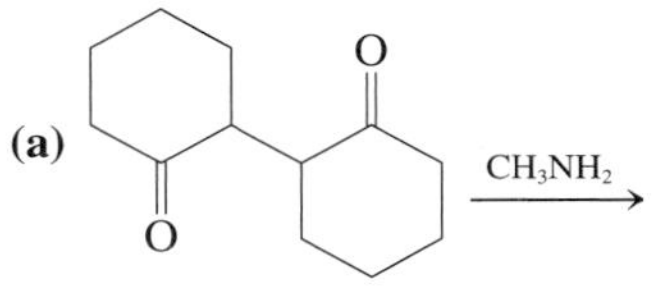

CH₃NH₂

(b)

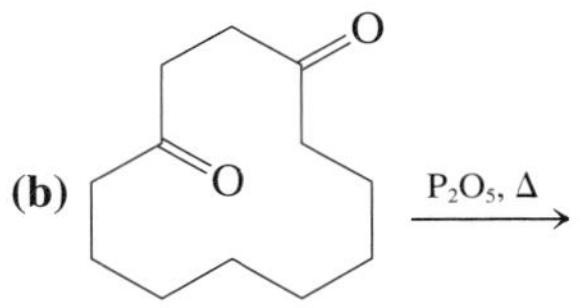

P₂O₅, Δ

38. 1-Hetero-2,4-cyclopentadienes can be prepared by condensation of an α-dicarbonyl compound and certain heteroatom-containing diesters. Propose a mechanism for the following pyrrole synthesis.

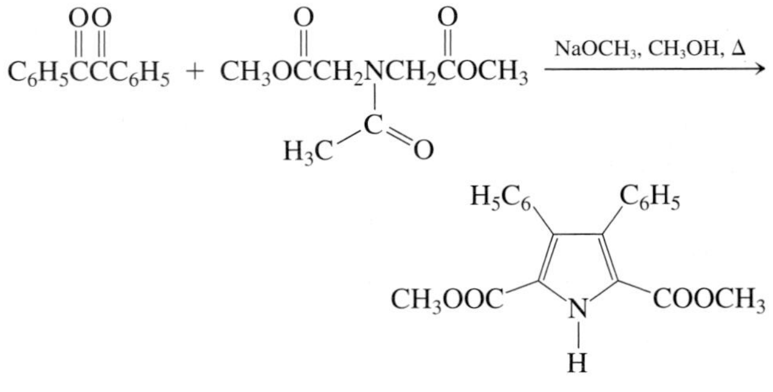

C₆H₅CCC₆H₅ + CH₃OCCH₂NCH₂COCH₃ →[NaOCH₃, CH₃OH, Δ]

How would you use a similar approach to synthesize 2,5-thiophenedicarboxylic acid?

39. Give the expected major product(s) of each of the following reactions. Explain how you chose the position of substitution in each case.

(a) 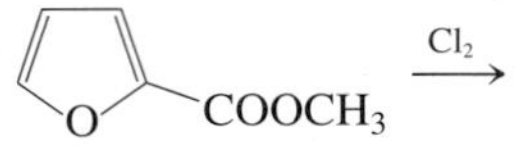 $\xrightarrow{Cl_2}$

(b) 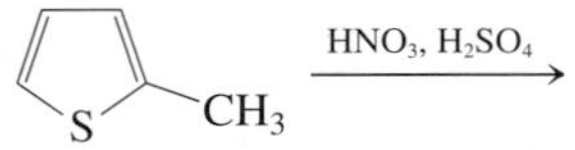 $\xrightarrow{HNO_3, H_2SO_4}$

(c) 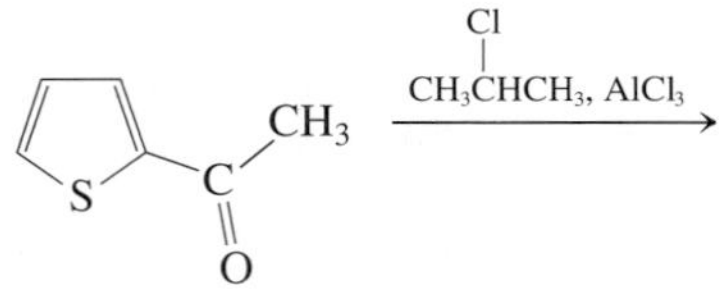 $\xrightarrow{\underset{CH_3CHCH_3, AlCl_3}{\overset{Cl}{|}}}$

(d) 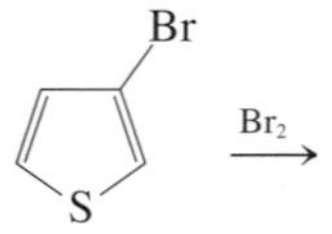 $\xrightarrow{Br_2}$

(e) 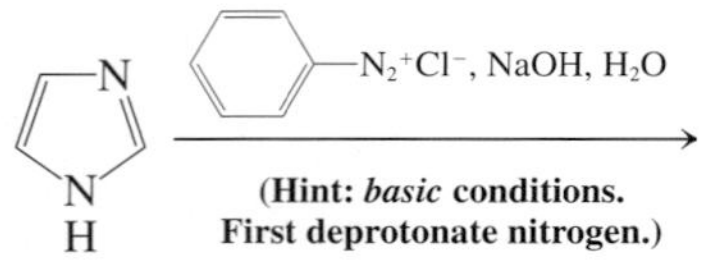 $\xrightarrow{-N_2^+Cl^-, NaOH, H_2O}$

(**Hint:** *basic* conditions. First deprotonate nitrogen.)

40. Give the products expected of each of the following reactions.

(a) 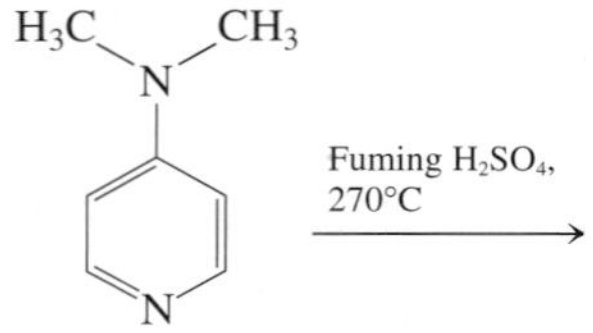 $\xrightarrow[270°C]{\text{Fuming } H_2SO_4,}$

(b) 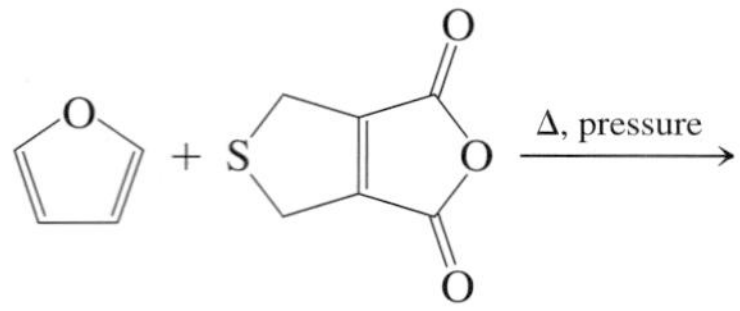 $\xrightarrow{\Delta, \text{ pressure}}$

(c) $\xrightarrow{KSH, CH_3OH, \Delta}$

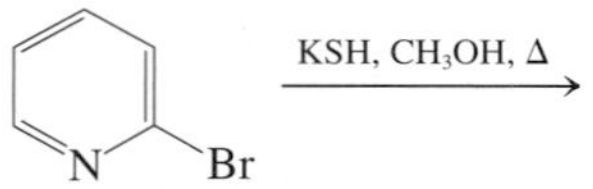

(d) 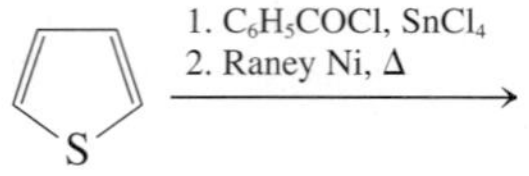 $\xrightarrow[\text{2. Raney Ni, } \Delta]{\text{1. } C_6H_5COCl, SnCl_4}$

(e) 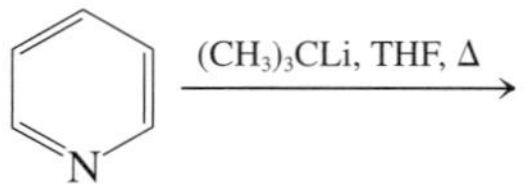 $\xrightarrow{(CH_3)_3CLi, THF, \Delta}$

41. Propose a synthesis of each of the following substituted heterocycles, using synthetic sequences presented in this chapter.

(a)

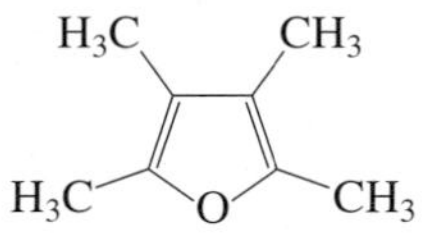

(b)

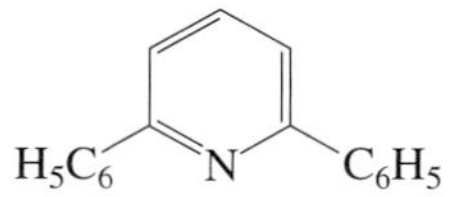

(c)

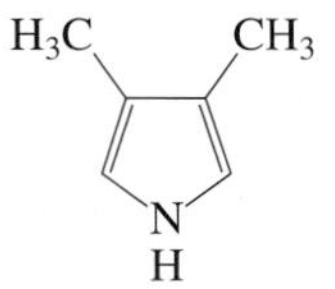

(d)

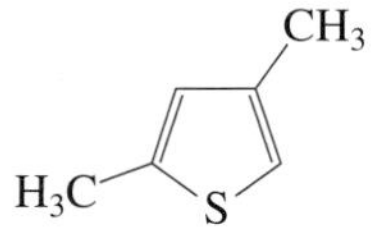

42. The structures of caffeine, the principal stimulant in coffee, and theobromine, its close relative in chocolate, are given in Section 25-8. Propose an efficient synthetic method to convert theobromine into caffeine.

43. Melamine, or 2,4,6-triamino-1,3,5-triazabenzene, is a heterocyclic compound that has been implicated in the illnesses and deaths due to kidney failure of both house pets and humans who ingested melamine-contaminated food. As its formula shows, melamine has a very high nitrogen content. Proteins (Chapter 26) are the main natural source of nitrogen in foods, and nitrogen analysis is commonly used to determine protein content in foods. The illegal addition of melamine to packaged foods increases their nitrogen content and makes them appear richer in protein; in reality, they are deadly. (a) A typical protein in food contains about 15% nitrogen. What is the % N in melamine? Does this result explain the motivation behind the melamine "doping" of packaged foods? (b) Melamine is synthesized by addition of ammonia to cyanuric chloride (2,4,6-trichloro-1,3,5-triazabenzene, see equation below). What kind of reaction is this? Formulate a mechanism and explain why cyanuric chloride displays this type of chemistry.

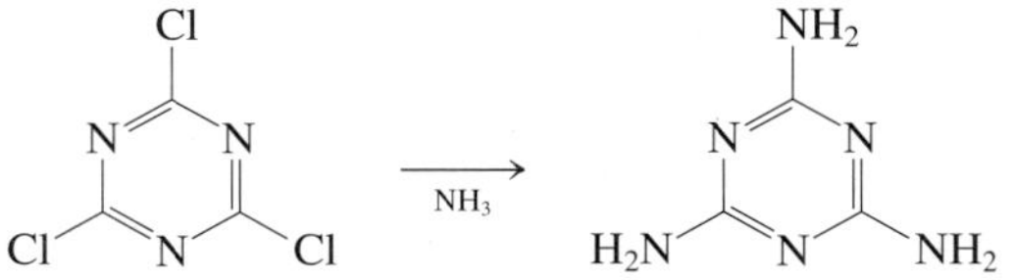

Cyanuric chloride   **Melamine**

44. Chelidonic acid, a 4-oxacyclohexanone (common name, γ-pyrone), is found in a number of plants and is synthesized from acetone and diethyl ethanedioate. Formulate a mechanism for this transformation.

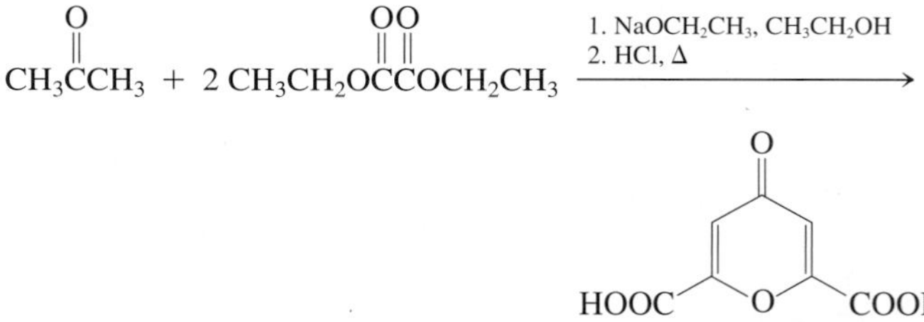

Chelidonic acid

45. Porphyrins are polyheterocyclic constituents of hemoglobin and myoglobin, the molecules that transport molecular oxygen in living systems (Section 26-8), of the cytochromes, which also play central roles in biological redox processes (Section 22-9), and of chlorophyll (Real Life 24-1), which mediates photosynthesis in all green plants.

Porphyrins are the products of a remarkable reaction between pyrrole and an aldehyde in the presence of acid:

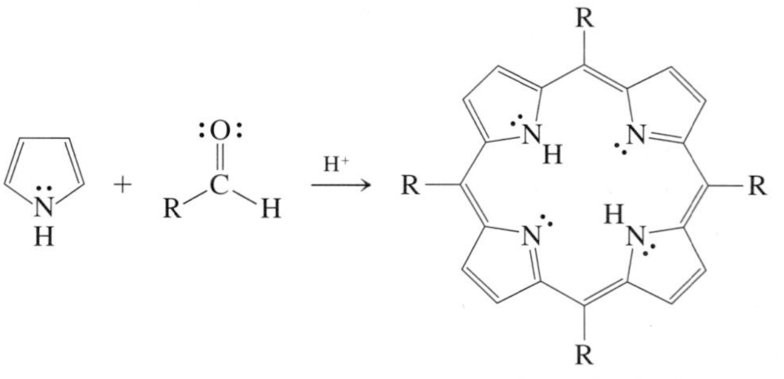

A porphyrin

This reaction is complicated and has many steps. The simpler condensation of one molecule of benzaldehyde and two of pyrrole to give the product shown below, a dipyrrylmethane, is illustrative of the first stage in porphyrin formation. Propose a step-by-step mechanism for this process.

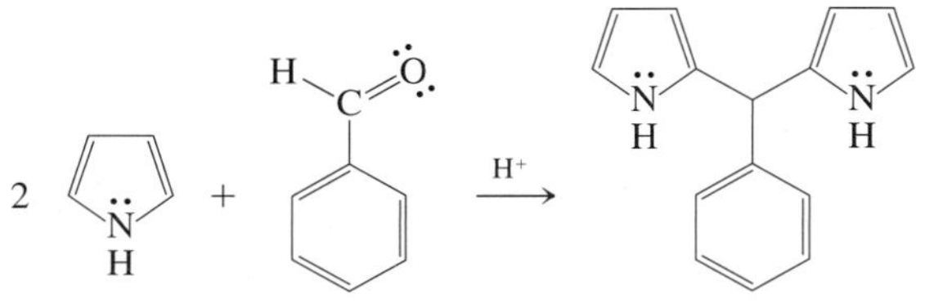

A dipyrrylmethane

46. Isoxazoles (Problem 36) have taken on increased significance as synthetic targets, because they are found in the structures of some recently discovered naturally occurring molecules that show promise as antibiotics (see Real Life 20-2). Isoxazoles may be prepared by reaction of alkynes with compounds containing the unusual nitrile oxide functional group:

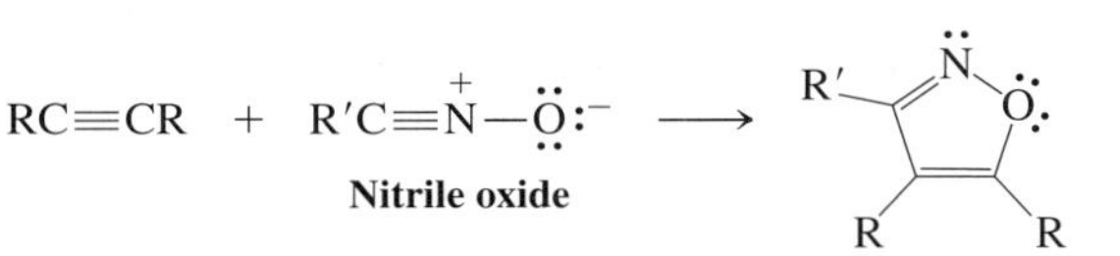

Nitrile oxide

Isoxazole

Suggest a mechanism for this process.

47. Reserpine is a naturally occurring indole alkaloid with powerful tranquilizing and antihypertensive activity. Many such compounds possess a characteristic structural feature: one nitrogen atom at a ring fusion separated by two carbons from another nitrogen atom (see also p. 1135).

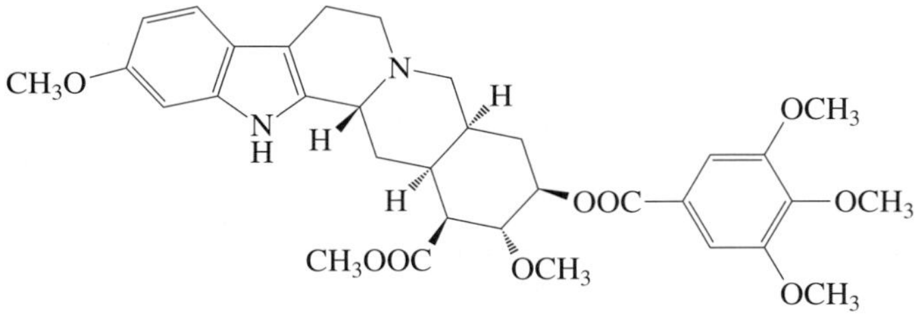

Reserpine

A series of compounds with modified versions of this structural feature have been synthesized and also shown to have antihypertensive activity, as well as antifibrillatory properties. One such synthesis is shown here. Name or draw the missing reagents and products (a) through (c).

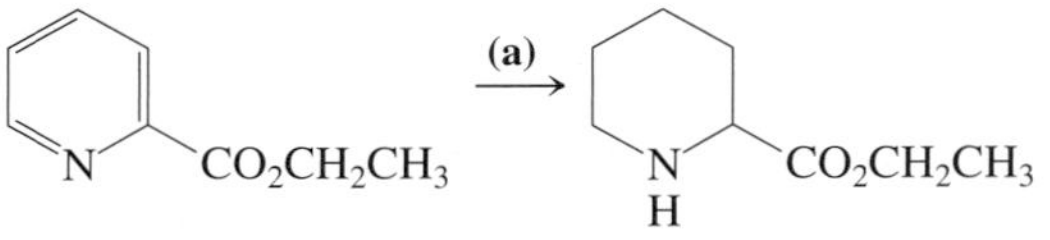

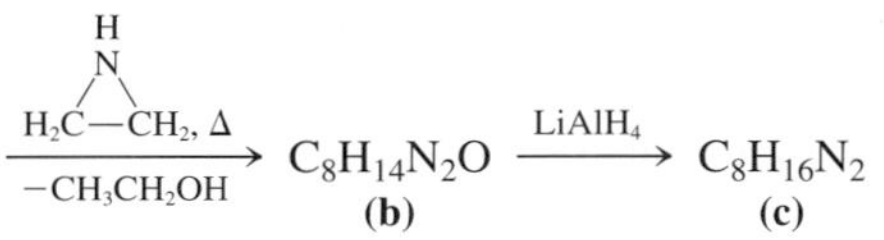

48. Starting with benzenamine (aniline) and pyridine, propose a synthesis of the antimicrobial sulfa drug sulfapyridine.

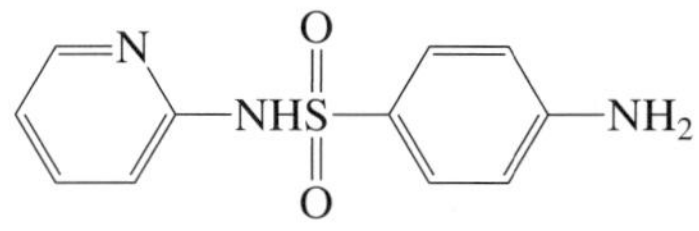

Sulfapyridine

49. Derivatives of benzimidazole possess biological activity somewhat like that of indoles and purines (of which adenine, Section 25-1, is an example). Benzimidazoles are commonly prepared from benzene-1,2-diamine. Devise a short synthesis of 2-methylbenzimidazole from benzene-1,2-diamine.

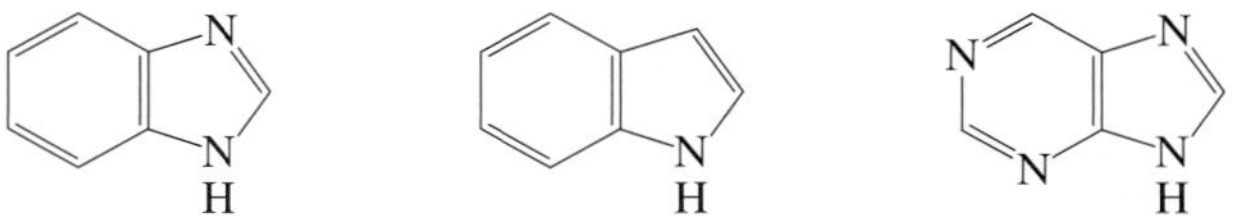

Benzimidazole   **Indole**   **Purine**

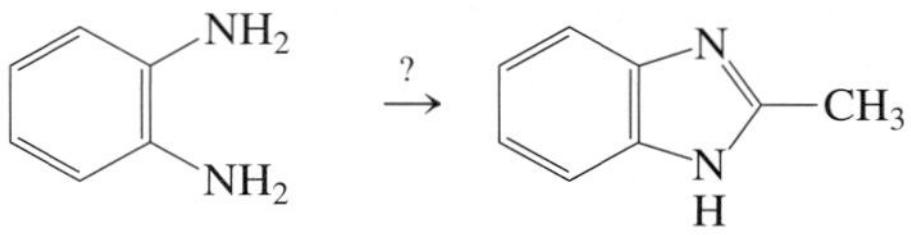

Benzene-1,2-diamine   **2-Methylbenzimidazole**

50. The Darzens condensation is one of the older methods (1904) for the synthesis of three-membered heterocycles. It is most commonly the reaction of a 2-halo ester with a carbonyl derivative in the presence of base. The following examples of the Darzens condensation show how it is applied to the synthesis of oxacyclopropane and azacyclopropane rings. Suggest a reasonable mechanism for each of these reactions.

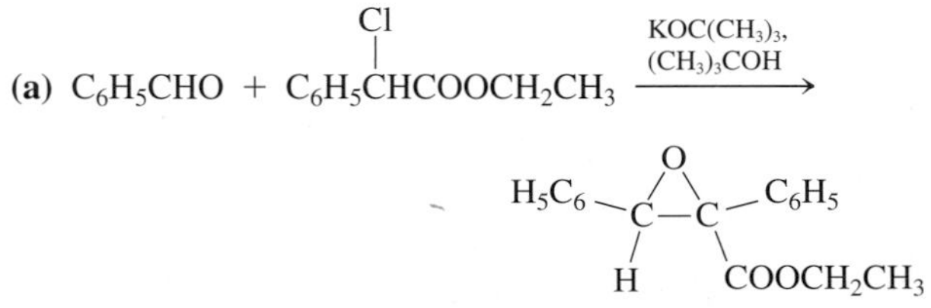

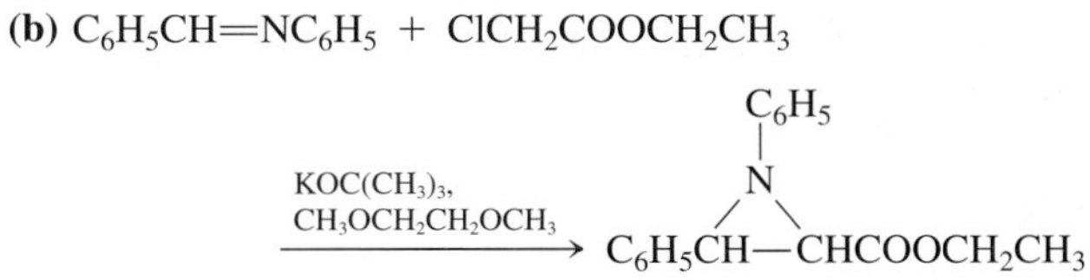

51. (a) The compound shown below, with the common name 1,3-dibromo-5,5-dimethylhydantoin, is useful as a source of electrophilic bromine (Br$^+$) for addition reactions. Give a more systematic name for this heterocyclic compound. (b) An even more remarkable heterocyclic compound (ii) is prepared by the following reaction sequence. Using the given information, deduce structures for compounds i and ii, and name the latter.

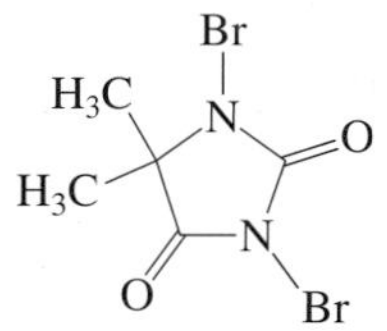

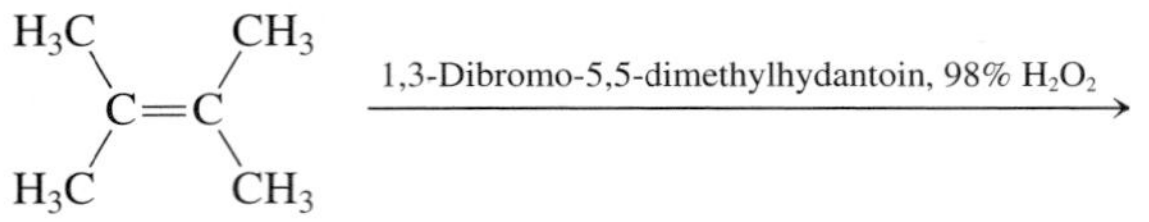

Heterocycle **ii** is a yellow, crystalline, sweet-smelling compound that decomposes upon gentle heating into two molecules of acetone, one of which is formed directly in its $n \rightarrow \pi^*$ excited state (Sections 14-11 and 17-3). This electronically excited product is chemiluminescent.

Heterocycles similar to compound **ii** are responsible for the chemiluminescence produced by a number of species [e.g., fireflies (see Real Life 9-1) and several deep-sea fish]; they also serve as the energy sources in commercial chemiluminescent products.

52. Azacyclohexanes (piperidines) can be synthesized by reaction of ammonia with cross-conjugated dienones: ketones conjugated on both sides with double bonds. Propose a mechanism for the following synthesis of 2,2,6,6-tetramethylaza-4-cyclohexanone.

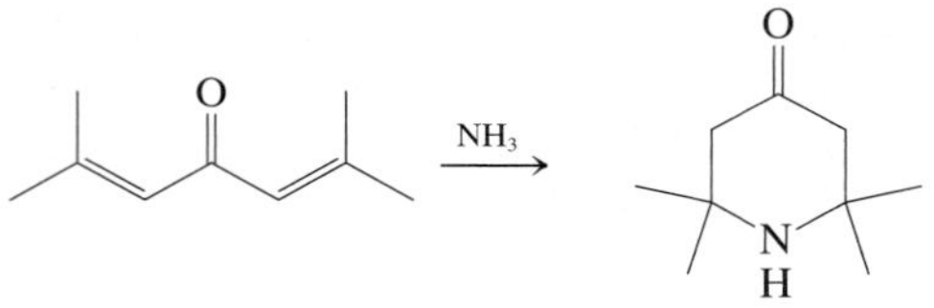

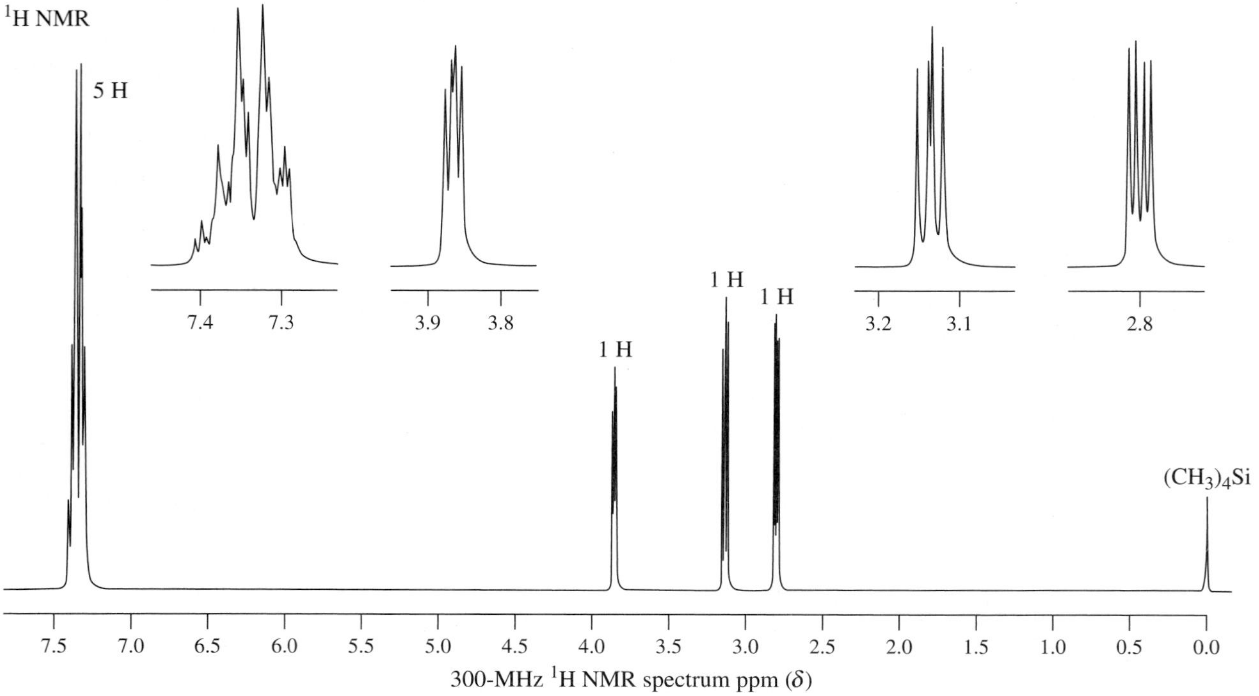

53. Quinolines (Section 25-7) are heterocycles that are widely used in medicinal chemistry because of the diversity of biological activity—including anticancer utility—that their derivatives display. A short synthesis of 3-acyldihydroquinolines (which may be converted into 3-acylquinolines by mild oxidation) is shown below. Propose a mechanism for this process. (**Hint:** Review Section 18-10.)

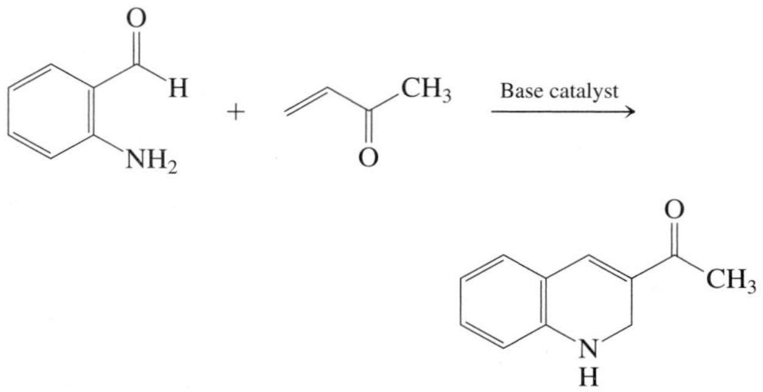

54. CHALLENGE At which position(s) do you expect 3-acetylquinoline (below) to undergo nitration in the presence of a mixture of sulfuric and fuming nitric acids? Will this reaction be faster or slower than nitration of quinoline itself?

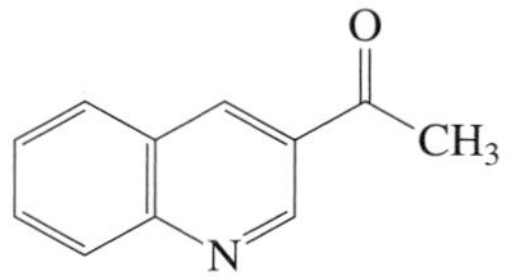

3-Acetylquinoline

55. Compound A, C_8H_8O, exhibits ^1H NMR spectrum A. Upon treatment with concentrated aqueous HCl, it is converted almost instantaneously into a compound that exhibits spectrum B (p. 1161). What is compound A, and what is the product of its treatment with aqueous acid?

300-MHz ^1H NMR spectrum ppm (δ)

A

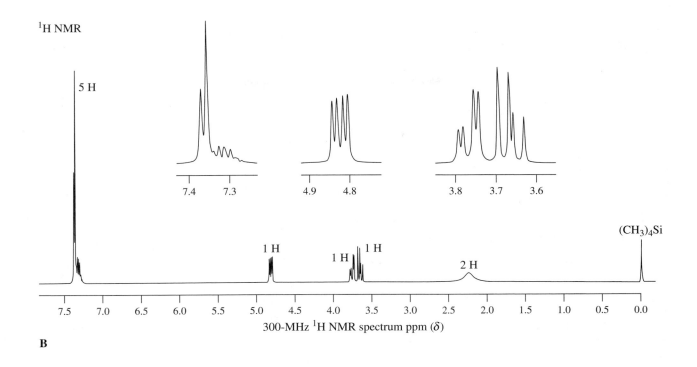

¹H NMR

5 H

7.4   7.3   4.9   4.8   3.8   3.7   3.6

(CH₃)₄Si

1 H   1 H   1 H   2 H

7.5 7.0 6.5 6.0 5.5 5.0 4.5 4.0 3.5 3.0 2.5 2.0 1.5 1.0 0.5 0.0

300-MHz ¹H NMR spectrum ppm (δ)

B

56. Heterocycle C, C_5H_6O, exhibits ¹H NMR spectrum C and is converted by H_2 and Raney nickel into compound D, $C_5H_{10}O$, with spectrum D (p. 1162). Identify compounds C and D. (Note: The coupling constants of the compounds in this problem and the next one are rather small; they are therefore not nearly as useful in structure elucidation as those around a benzene ring.)

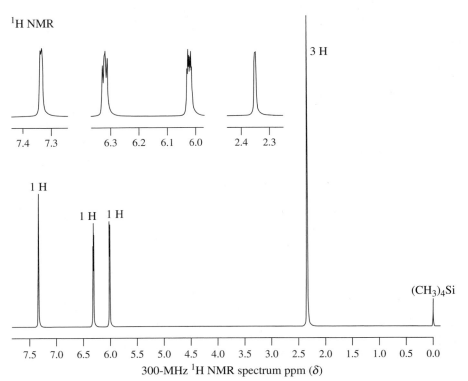

¹H NMR

7.4   7.3   6.3 6.2 6.1 6.0   2.4   2.3

3 H

1 H

1 H   1 H

(CH₃)₄Si

7.5 7.0 6.5 6.0 5.5 5.0 4.5 4.0 3.5 3.0 2.5 2.0 1.5 1.0 0.5 0.0

300-MHz ¹H NMR spectrum ppm (δ)

C

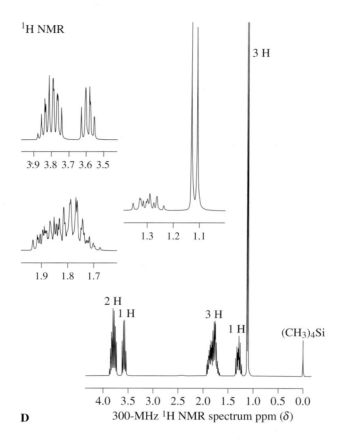

¹H NMR

3 H

(CH₃)₄Si

3.9 3.8 3.7 3.6 3.5

1.3 1.2 1.1

1.9 1.8 1.7

2 H

1 H

3 H

1 H

4.0 3.5 3.0 2.5 2.0 1.5 1.0 0.5 0.0

D

300-MHz ¹H NMR spectrum ppm (δ)

57. [CHALLENGE] The commercial synthesis of a useful heterocyclic derivative requires treatment of a mixture of aldopentoses (derived from corncobs, straw, etc.) with hot acid under dehydrating conditions. The product, E, has ¹H NMR spectrum E, shows a strong IR band at 1670 cm⁻¹, and is formed in nearly quantitative yield. Identify compound E and formulate a mechanism for its formation.

$$\text{Aldopentoses} \xrightarrow{\text{H}^+,\ \Delta} \text{C}_5\text{H}_4\text{O}_2$$

E

Compound E is a valuable synthetic starting material. The following sequence converts it into furethonium, which is useful in the treatment of glaucoma. What is the structure of furethonium?

$$\text{E} \xrightarrow[\text{2. Excess CH}_3\text{I, (CH}_3\text{CH}_2)_2\text{O}]{\text{1. NH}_3,\ \text{NaBH}_3\text{CN}} \text{furethonium}$$

¹H NMR

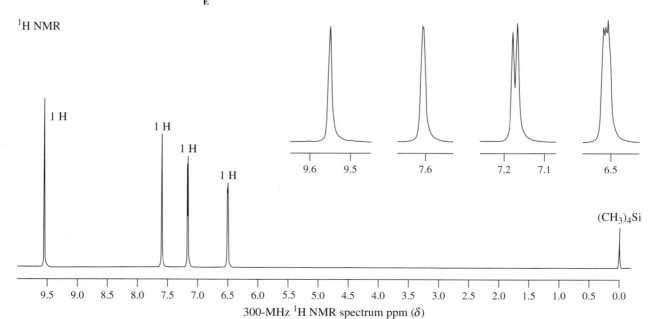

1 H

1 H

1 H

1 H

9.6 9.5

7.6

7.2 7.1

6.5

(CH₃)₄Si

9.5 9.0 8.5 8.0 7.5 7.0 6.5 6.0 5.5 5.0 4.5 4.0 3.5 3.0 2.5 2.0 1.5 1.0 0.5 0.0

300-MHz ¹H NMR spectrum ppm (δ)

E

58. [CHALLENGE]   Treatment of a 3-acylindole with LiAlH₄ in $(CH_3CH_2)_2O$ reduces the carbonyl all the way to a CH_2 group. Explain by a plausible mechanism. (**Hint:** Direct S_N2 displacement of alkoxide by hydride is *not* plausible.)

59. The sequence below is a rapid synthesis of one of the heterocycles in this chapter. Draw the structure of the product, which has ¹H NMR spectrum F.

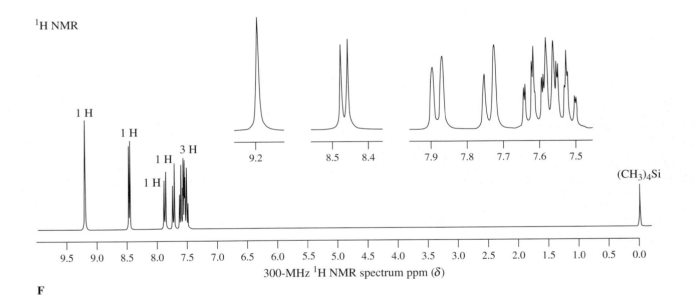

Team Problem

60. This problem introduces two literature syntheses of indole derivatives, and you are asked to come up with plausible mechanisms for them. Divide your team in two, each group concentrating on one of the methods.

Fischer Indole Synthesis of 2-Phenylindole

2-Phenylindole

In this procedure, a hydrazone of an enolizable aldehyde or ketone is heated in strong acid, causing ring closure with simultaneous expulsion of ammonia to furnish the indole nucleus. [**Hints:** The mechanism of the reaction proceeds in three stages: (1) an imine–enamine tautomerization (recall

Section 17-9); (2) an electrocyclic reaction (a "diaza-Cope" rearrangement; recall Section 22-7); (3) another imine–enamine (in this case, benzenamine) tautomerization; (4) ring closure to the heterocycle; and (5) elimination of NH_3.]

Reissert Indole Synthesis of Ethyl Indole-2-carboxylate

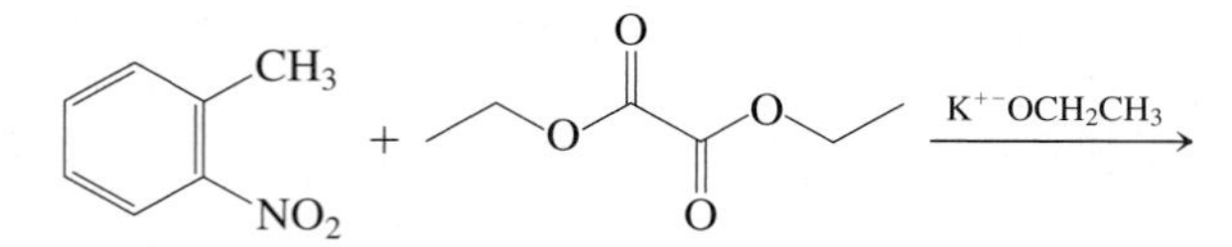

2-Methylnitrobenzene
(o-Nitrotoluene)

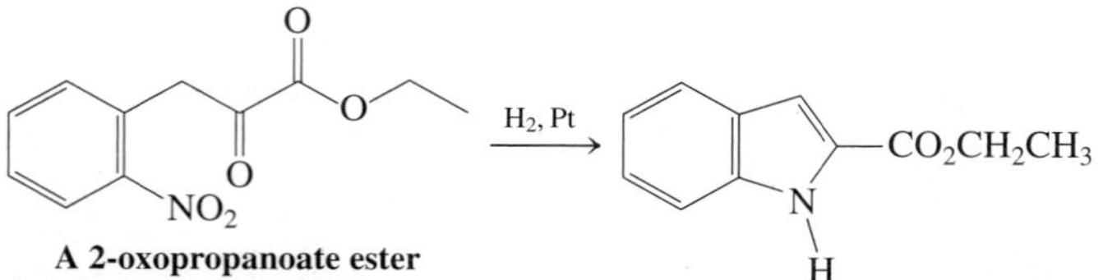

A 2-oxopropanoate ester
(A pyruvate ester)

Ethyl indole-2-carboxylate

In this sequence, a 2-methylnitrobenzene (o-nitrotoluene) is first converted into an ethyl 2-oxopropanoate (pyruvate, Real Life 23-2) ester, which, on reduction, is transformed into the target indole. [**Hints:** (1) The nitro group is essential for the success of the first step. Why? Does this step remind you of another reaction? Which one? (2) Which functional group is the target of the reduction step (recall Section 16-5)? (3) The ring closure to the heterocycle requires a condensation reaction.]

Preprofessional Problems

61. The proton decoupled ^{13}C NMR spectrum of pyridine will display how many peaks? (a) One; (b) two; (c) three; (d) four; (e) five.

62. Pyrrole is a much weaker base than azacyclopentane (pyrrolidine) for which of the following reasons? (a) The nitrogen in pyrrole is more electropositive than that in pyrrolidine; (b) pyrrole is a Lewis acid; (c) pyrrole has four electrons; (d) pyrrolidine can give up the proton on the nitrogen atom more readily than can pyrrole; (e) pyrrole is aromatic.

Pyrrole   **Azacyclopentane**
(Pyrrolidine)

63. Which of the following compounds would you expect to be the major organic product of the two-step sequence shown here?

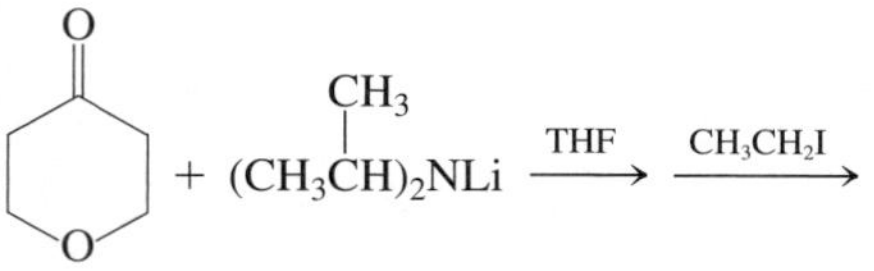

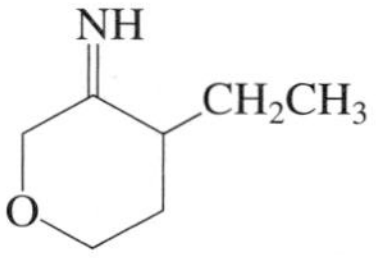

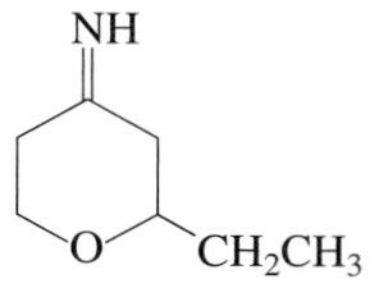

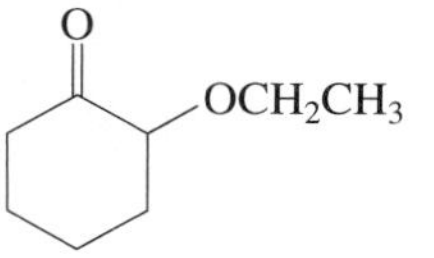

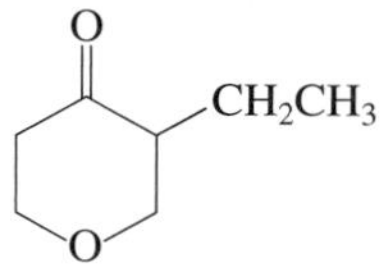

64. This reaction yields one main organic product. Which of the following compounds is it?

2-Phenylthiophene $\xrightarrow{SnCl_4, CH_3CCl}$

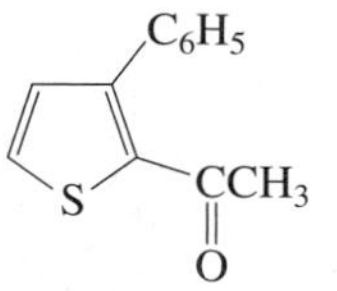

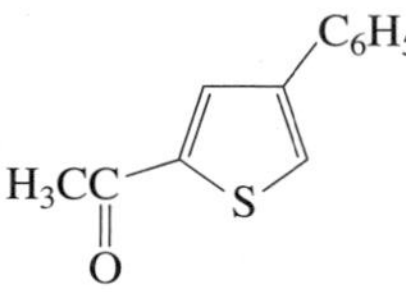

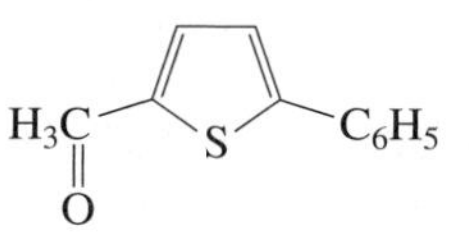

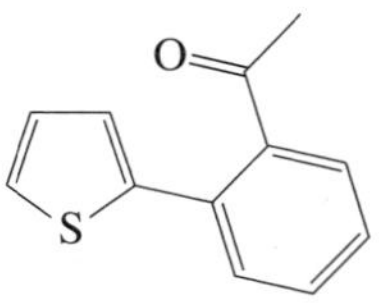

CHAPTER 26

Amino Acids, Peptides, Proteins, and Nucleic Acids

Nitrogen-Containing Polymers in Nature

On p. 1 of this text we defined organic chemistry as the chemistry of carbon-containing compounds. We then went on to point out that organic molecules constitute the chemical bricks of life. Indeed, a historical definition of organic chemistry restricts it to living organisms. What is life and how do we, as organic chemists, approach its study? A functional definition of life refers to it as a condition of matter manifested by growth, metabolism, reproduction, and evolution. The underlying basic processes are chemical, and researchers hope to decipher their complexity by investigating specific reactions or reaction sequences. The "whole," however, is much more complex than these individual pieces, because they interact by multiple feedback loops in a manner that makes it impossible to show that each effect has a simple cause. This final chapter will provide a glimpse of that complexity by taking you from amino acids to their polymers, the polypeptides—in particular, the large natural polypeptides called **proteins**—and to their biological origin, **DNA.**

Proteins have an astounding diversity of functions in living systems. As **enzymes,** they catalyze transformations ranging in complexity from the simple hydration of carbon dioxide to the replication of entire chromosomes—great coiled strands of DNA, the genetic material in living cells. Enzymes can accelerate certain reactions many millionfold.

We have already encountered the protein rhodopsin, the photoreceptor that generates and transmits nerve impulses in retinal cells (Real Life 18-2). Other proteins serve for transport and storage. Thus, hemoglobin carries oxygen; iron is transported in the blood by transferrin and stored in the liver by ferritin. Proteins play a crucial role in coordinated motion, such as muscle contraction. They give mechanical support to skin and bone; they are the antibodies responsible for our immune protection; and they control growth and differentiation—that is, which part of the information stored in DNA is to be used at any given time.

We begin with the structure and preparation of the 20 most common amino acids, the building blocks of proteins. We then show how amino acids are linked by peptide bonds in the three-dimensional structure of hemoglobin and other polypeptides. Some proteins contain thousands of amino acids, but we shall see how to determine the sequence of amino acids in many polypeptides and synthesize these molecules in the laboratory. Finally, we consider how other polymers, the nucleic acids DNA and RNA, direct the synthesis of proteins in nature.

Pregabalin [Lyrica; (S)-3-(aminomethyl)-5-methylhexanoic acid] is an unnatural amino acid that is effective in the treatment of chronic pain. The molecule is shown in its zwitterionic ammonium carboxylate form.

α-Amino acid

26-1 | STRUCTURE AND PROPERTIES OF AMINO ACIDS

Amino acids are carboxylic acids that bear an amine group. The most common of these in nature are the **2-amino acids,** or **α-amino acids,** which have the general formula $RCH(NH_2)COOH$; that is, the amino function is located at C2, the α-carbon. The R group can be alkyl or aryl, and it can contain hydroxy, amino, mercapto, sulfide, carboxy, guanidino, or imidazolyl groups. Because of the presence of both amino and carboxy functions, amino acids are both acidic and basic.

The stereocenter of common 2-amino acids has the S configuration

More than 500 amino acids exist in nature, but the proteins in all species, from bacteria to humans, consist mainly of only 20. Adult humans can synthesize all but eight, and two only in insufficient quantities. This group is often called the **essential amino acids** because they must be included in our diet. Although amino acids can be named in a systematic manner, they rarely are; so we shall use their common names. Table 26-1 lists the 20 most common amino acids, along with their structures, their pK_a values, and the three- and (the newer) one-letter codes that abbreviate their names. We shall see later how to use these codes to describe peptides conveniently.

Table 26-1 Natural (2S)-Amino Acids							
R	Name	Three-letter code	One-letter code	pK_a of α-COOH	pK_a of α-$^+NH_3$	pK_a of acidic function in R	Isoelectric point, pI
H	Glycine	Gly	G	2.3	9.6	—	6.0
Alkyl group							
CH_3	Alanine	Ala	A	2.3	9.7	—	6.0
$CH(CH_3)_2$	Valine[a]	Val	V	2.3	9.6	—	6.0
$CH_2CH(CH_3)_2$	Leucine[a]	Leu	L	2.4	9.6	—	6.0
$CHCH_2CH_3$ (S) $\vert$ CH_3	Isoleucine[a]	Ile	I	2.4	9.6	—	6.0
H_2C-⟨benzene⟩	Phenylalanine[a]	Phe	F	1.8	9.1	—	5.5
$COOH^b$ HN—H ⟨ring⟩ CH_2	Proline	Pro	P	2.0	10.6	—	6.3
Hydroxy containing							
CH_2OH	Serine	Ser	S	2.2	9.2	—	5.7
$CHOH$ (R) $\vert$ CH_3	Threonine[a]	Thr	T	2.1	9.1	—	5.6
H_2C-⟨benzene⟩$-OH$	Tyrosine	Tyr	Y	2.2	9.1	10.1	5.7
Amino containing							
$CH_2\overset{O}{\overset{\|}{C}}NH_2$	Asparagine	Asn	N	2.0	8.8	—	5.4

Continued

Table 26-1   Natural (2S)-Amino Acids (continued)

R	Name	Three-letter code	One-letter code	pKa of α-COOH	pKa of α-+NH₃	pKa of acidic function in R	Isoelectric point, pI
Amino containing (*continued*)							
CH₂CH₂C(=O)NH₂	Glutamine	Gln	Q	2.2	9.1	—	5.7
(CH₂)₄NH₂	Lysine[a]	Lys	K	2.2	9.0	10.5[c]	9.7
(CH₂)₃NHC(=NH)NH₂	Arginine[a]	Arg	R	2.2	9.0	12.5[c]	10.8
(indole-CH₂ structure)	Tryptophan[a]	Trp	W	2.8	9.4	—	5.9
(imidazole-CH₂ structure)	Histidine[a]	His	H	1.8	9.2	6.1[c]	7.6
Mercapto or sulfide containing							
CH₂SH	Cysteine[d]	Cys	C	2.0	10.3	8.2	5.1
CH₂CH₂SCH₃	Methionine[a]	Met	M	2.3	9.2	—	5.7
Carboxy containing							
CH₂COOH	Aspartic acid	Asp	D	1.9	9.6	3.7	2.8
CH₂CH₂COOH	Glutamic acid	Glu	E	2.2	9.7	4.3	3.2

[a]Essential amino acids. [b]Entire structure. [c]pKa of conjugate acid. [d]The stereocenter is *R* because the CH₂SH substituent has higher priority than the COOH group.

Model Building

Amino acids may be depicted by either hashed-wedged line structures or by Fischer projections.

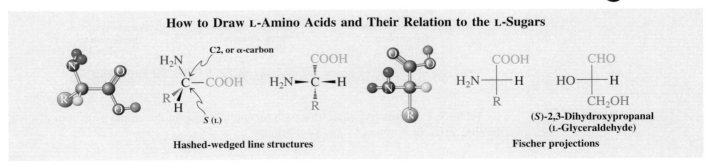

How to Draw L-Amino Acids and Their Relation to the L-Sugars

Hashed-wedged line structures

Fischer projections

(*S*)-2,3-Dihydroxypropanal (L-Glyceraldehyde)

In all but glycine, the simplest of the amino acids, C2 is a stereocenter and usually adopts the *S* configuration. Other stereocenters located in the substituent R may have either *R* (as in threonine) or *S* configuration (as in isoleucine).

As in the names of the sugars (Section 24-1), an older amino acid nomenclature uses the prefixes D and L, which relate all the L-amino acids to (*S*)-2,3-dihydroxypropanal (L-glyceraldehyde). As emphasized in the discussion of the natural D sugars, a molecule belonging to the L family is not necessarily levorotatory. For example, both valine ($[\alpha]_D^{25°C} = +13.9$) and isoleucine ($[\alpha]_D^{25°C} = +11.9$) are dextrorotatory.

Proton transfer

Zwitterion

Glycine zwitterion

Amino acids are acidic and basic: zwitterions

Because of their two functional groups, the amino acids are both acidic and basic; that is, they are **amphoteric** (Section 8-3). The carboxylic acid group protonates the amine function, thus forming a **zwitterion.** This ammonium carboxylate form is favored because an ammonium ion is much less acidic ($pK_a \approx 10\text{–}11$) than a carboxylic acid ($pK_a \approx 2\text{–}5$). The highly polar zwitterionic structure allows amino acids to form particularly strong crystal lattices. Most of them therefore are fairly insoluble in organic solvents, and they decompose rather than melt when heated. The electrostatic potential map of glycine in the margin depicts its highly dipolar nature, originating from the juxtaposition of the electron-rich (red) carboxylate function with the electron-poor (blue) ammonium group.

The structure of an amino acid in aqueous solution depends on the pH. Consider, for example, the simplest member of the series, glycine. The major form in neutral solution is the zwitterion. However, in strong acid (pH < 1), glycine exists predominantly as the cationic ammonium carboxylic acid, whereas strongly basic solutions (pH > 13) contain mainly the deprotonated 2-aminocarboxylate ion. These forms interconvert by acid-base equilibria (Section 2-2).

$$\overset{+}{H_2N}CH_2COOH \underset{H^+}{\overset{HO^-}{\rightleftharpoons}} \overset{+}{H_2N}CH_2COO^- \underset{H^+}{\overset{HO^-}{\rightleftharpoons}} H_2NCH_2COO^-$$

| An ammonium ion predominates at pH < 1 | A neutral zwitterion predominates at pH ≈ 6 | A carboxylate ion predominates at pH > 13 |

Increasing pH

Table 26-1 records pK_a values for each functional group of the amino acids. For glycine, the first value (2.3) refers to the equilibrium

$$\overset{+}{H_3N}CH_2COOH + H_2O \rightleftharpoons \overset{+}{H_3N}CH_2COO^- + H_2\overset{+}{O}H$$
$$pK_a = 2.3$$

$$K_1 = \frac{[\overset{+}{H_3N}CH_2COO^-][H_2\overset{+}{O}H]}{[\overset{+}{H_3N}CH_2COOH]} = 10^{-2.3}$$

Note that this pK_a is more than two units less than that of an ordinary carboxylic acid (pK_a $CH_3COOH = 4.74$), an observation that is true for all the other α-aminocarboxy groups in

Table 26-1. This difference is a consequence of the electron-withdrawing effect of the protonated amino group. The second pK_a value (9.6) describes the second deprotonation step:

$$\overset{+}{H_3}NCH_2COO^- + H_2O \rightleftharpoons H_2NCH_2COO^- + H_2\overset{+}{O}H$$
$$pK_a = 9.6$$

$$K_2 = \frac{[H_2NCH_2COO^-][H_2\overset{+}{O}H]}{[\overset{+}{H_3}NCH_2COO^-]} = 10^{-9.6}$$

At the isoelectric point, the net charge is zero

The pH at which the extent of protonation equals that of deprotonation is called the **isoelectric pH** or the **isoelectric point (p*I*;** Table 26-1). At this pH, the amount of positive charge balances that of negative charge and the concentration of the charge-neutralized zwitterionic form is at its greatest. For an amino acid without any additional acidic or basic groups, such as glycine, the value of its p*I* is the average of its two pK_a values.

$$pI = \frac{pK_{a\text{-COOH}} + pK_{a\text{-}\overset{+}{N}H_2H}}{2} = \text{(for glycine)} \frac{2.3 + 9.6}{2} = 6.0$$

pH at which concentration of zwitterion is at its maximum

When the side chain of the acid bears an additional acidic or basic function, the p*I* is either decreased or increased, respectively, as one would expect. Table 26-1 shows seven amino acids in which this is the case. Specifically, for the four amino acids with an acidic side chain, the p*I* is the average of its two lowest pK_a values. Conversely, for the three amino acids with a basic side chain, the p*I* is the average of its two highest pK_a values.

Assignment of pK_a Values in Selected Amino Acids

Aspartic acid **Tyrosine** **Lysine** **Arginine**

Why is this so? Picture these amino acids in their fully protonated form and then raise the pH (in other words, add base) to observe what happens to their net charge. For example, the amino dicarboxylic acid, aspartic acid, will be positively charged at low pH due to the presence of the ammonium substituent. To reach a charge-neutral form, on average one of the carboxy functions has to be deprotonated. This will happen at a pH midpoint between the two respective pK_a values (1.9 and 3.7), at p*I* = 2.8. The similar glutamic acid has the corresponding value at 3.2. At physiological pH, both of the carboxy functions are deprotonated, and the molecules exist as the zwitterionic anions aspartate and glutamate. (Monosodium glutamate, MSG, is used as a flavor enhancer in various foods.) Tyrosine, which bears the relatively nonacidic neutral phenol substituent (at low pH), has a p*I* = 5.7, which is midway between the pK_a's of the other two more acidic groups.

REAL LIFE: MEDICINE 26-1 | Arginine and Nitric Oxide in Biochemistry and Medicine

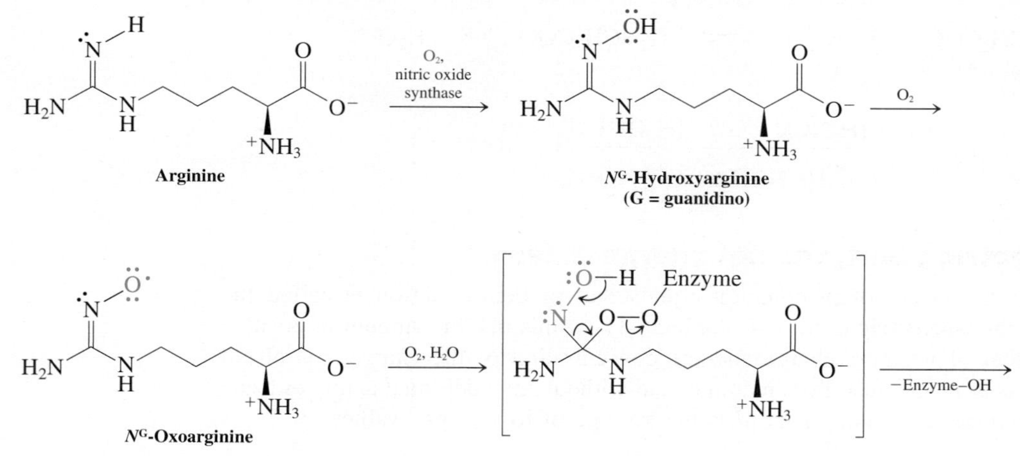

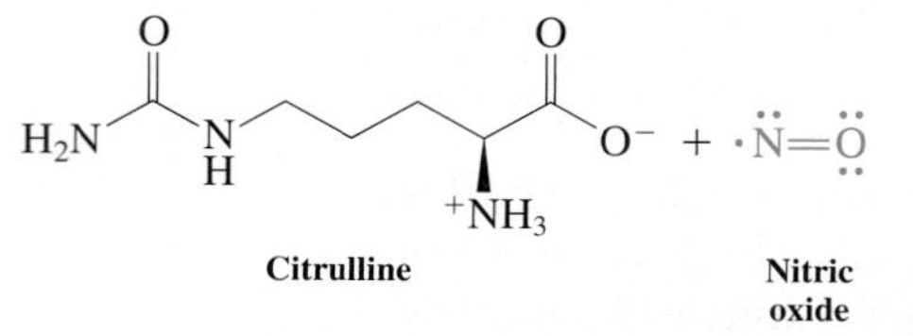

Arginine

N^G-**Hydroxyarginine**
(G = guanidino)

N^G-**Oxoarginine**

Citrulline   **Nitric oxide**

Nitric oxide is synthesized by the enzyme-catalyzed oxidation of arginine, as shown above. The enzyme involved, nitric oxide synthase, may well be essential for cell survival. Thus, a recent study shows that fruit flies, *Drosophila*, genetically modified to be incapable of making this enzyme, die as embryos.

Nitric oxide is released by cells on the inner walls of blood vessels and causes adjacent muscle fibers to relax. This 1987 discovery explains the effectiveness of nitroglycerin and other organic nitrates as treatments for angina and heart attacks, a nearly century-old mystery: These substances are metabolically converted into NO, which dilates the blood vessels (see also Worked Example 25-29).

Paradoxically, considering the essential benefits of NO, its overproduction leads to septic shock. Uncontrolled release may also be the cause of brain damage after a stroke and disorders such as Alzheimer's and Huntington's diseases. The occurrence of high levels of nitrite ion, NO_2^- (the product of NO oxidation), in the joints of people with rheumatoid arthritis indicates overproduction of NO as a response to inflammation. Similar associations have been established for schizophrenia, urinary disorders, and multiple sclerosis. The rapidly evolving story of NO is an example of how little we still know about the functions of the body, but also how rapidly a discovery leads to the evolution of an entire field.

In the late 1980s and early 1990s, scientists, among them the 1998 Nobelists (medicine) Furchgott, Ignarro, and Murad,* made a series of startling discoveries. The simple but highly reactive and exceedingly toxic molecule nitric oxide, $:N=\overset{\cdot\cdot}{O}:$, is synthesized in a wide variety of cells in mammals, including humans.

In the body, nitric oxide performs several critical biological functions as diverse as control of blood pressure, inhibition of platelet aggregation, cell differentiation, neurotransmission, and penile erection. It also plays a major role in the activity of the immune system. For example, macrophages (cells associated with the body's immune system) destroy bacteria and tumor cells by exposing them to nitric oxide.

*Professor Robert F. Furchgott (1916–2009), State University of New York, Brooklyn; Professor Louis J. Ignarro (b. 1941), University of California at Los Angeles; Professor Ferid Murad (b. 1936), University of Texas, Dallas.

In contrast to the above examples, lysine bears an additional basic amino group that is protonated in a strongly acidic medium to furnish a dication. When the pH of the solution is raised, deprotonation of the carboxy group occurs first, followed by respective proton loss from the nitrogen at C2 and the remote ammonium function. The isoelectric point is located halfway between the last two pK_a values, at $pI = 9.7$. Arginine carries a substituent new to us: the relatively basic guanidino group, $-NHCNH_2$, derived from the molecule guanidine (margin). The pK_a of its conjugate acid is 12.5, almost three units greater than that of the ammonium ion (Section 21-4). The pI of arginine lies midway between that of the guanidinium and ammonium groups, at 10.8.

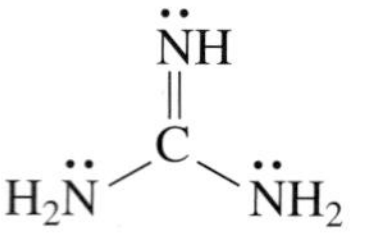

Guanidine

Exercise 26-4

Guanidine is found in turnip juice, mushrooms, corn germ, rice hulls, mussels, and earthworms. Its basicity is due to the formation of a highly resonance-stabilized conjugate acid. Draw its resonance forms. (**Hint:** Review Section 20-1.)

Histidine (margin) contains another new subsituent, the basic **imidazole** ring (see Problem 6 of Chapter 25). In this aromatic heterocycle, one of the nitrogen atoms is hybridized as in pyridine and the other is hybridized as in pyrrole.

Exercise 26-5

Draw an orbital picture of imidazole. (**Hint:** Use Figure 25-1 as a model.)

The imidazole ring is relatively basic because the protonated species is stabilized by resonance.

Resonance in Protonated Imidazole

$pK_a = 7.0$

This resonance stabilization is related to that in amides (Sections 20-1 and 26-4). Imidazole is significantly protonated at physiological pH ($pI = 7.6$). It can therefore function as a proton acceptor and donor at the active site of a variety of enzymes (see, e.g., chymotrypsin, Section 26-4).

The amino acid cysteine bears a relatively nucleophilic and acidic mercapto substituent ($pK_a = 8.2$, $pI = 5.1$). In addition, thiols can be oxidized to disulfides under mild conditions (Section 9-10). In nature, various enzymes are capable of oxidatively coupling and reductively decoupling the mercapto groups in the cysteines of proteins and peptides, thereby reversibly linking peptide strands (Section 9-10). We have highlighted previously the importance of this amino acid in biological function (see Really? on p. 815 and Section 22-9).

In Summary There are 20 elementary L-amino acids, all of which have common names. Unless there are additional acid-base functions in the side chain, their acid-base behavior is governed by two pK_a values, the lower one describing the deprotonation of the carboxy group. At the isoelectric point, the number of amino acid molecules with net zero charge is maximized. Some amino acids contain additional acidic or basic functions, such as hydroxy, amino, guanidino, imidazolyl, mercapto, and carboxy.

26-2 | SYNTHESIS OF AMINO ACIDS: A COMBINATION OF AMINE AND CARBOXYLIC ACID CHEMISTRY

Chapter 21 treated the chemistry of amines, Chapters 19 and 20 that of carboxylic acids and their derivatives. We use both in preparing 2-amino acids.

Hell-Volhard-Zelinsky bromination followed by amination converts carboxylic acids to 2-amino acids

What would be the quickest way of introducing a 2-amino substituent into a carboxylic acid? Section 19-12 pointed out that simple 2-functionalization of an acid is possible by

Histidine

Imidazole

Cysteine

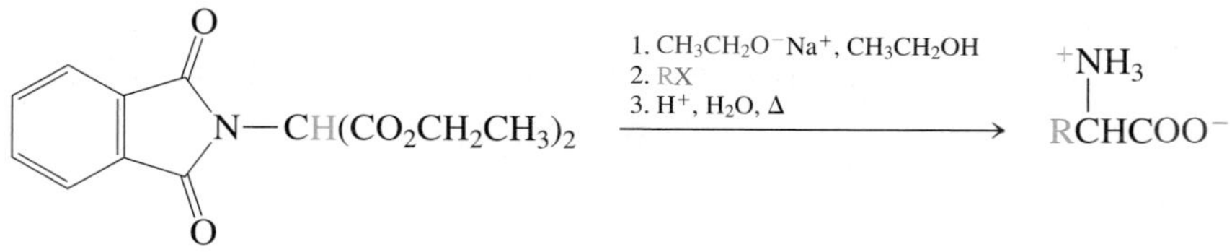

Penicillamine in Chelation Therapy

The dimethyl-(S)-cysteine derivative penicillamine (cuprimine), a metabolite of penicillin, is administered to patients suffering from a rare genetic disorder known as Wilson disease. In these individuals, normal copper removal is impaired, and the accumulation of the metal leads to neurological and liver problems. The drug acts by coordination (see Section 9-5) involving the lone pairs on S and N, in a treatment called chelation therapy (*khēlē*, Greek, claw). The ensuing strong complex sequesters the metal and is then excreted in the urine. A number of similar "chelators" are used in the treatment of acute mercury, iron, arsenic, lead, uranium, plutonium, and other toxic-metal poisoning.

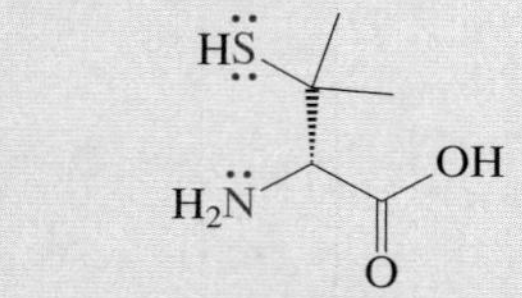

Penicillamine

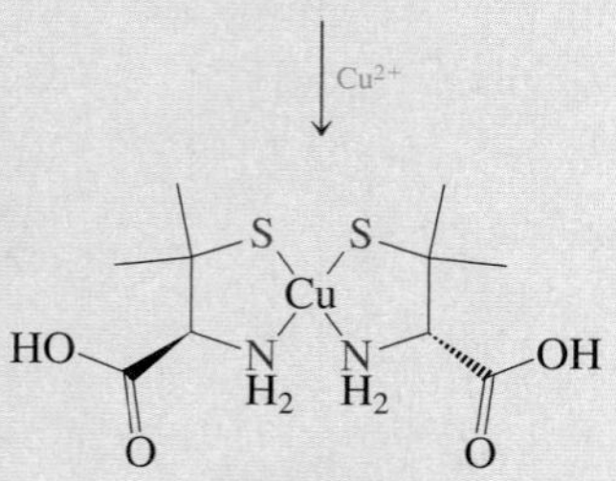

Copper chelate

the Hell-Volhard-Zelinsky bromination. Furthermore, the bromine in the product can be displaced by nucleophiles, such as ammonia. In these two steps, propanoic acid can be converted into racemic alanine.

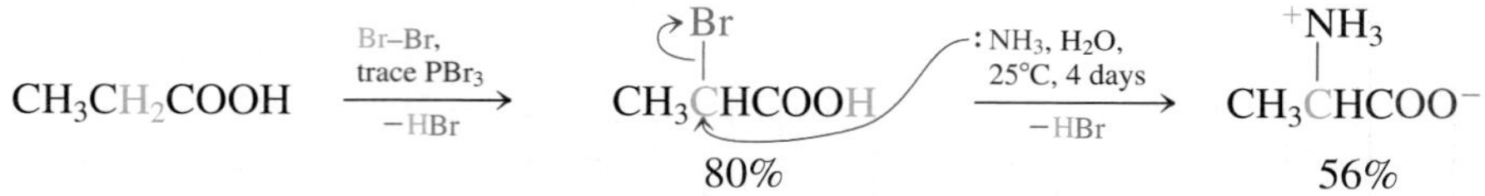

$$CH_3CH_2COOH \xrightarrow[-HBr]{\substack{Br-Br, \\ trace\ PBr_3}} CH_3CHCOOH \xrightarrow[-HBr]{\substack{:NH_3,\ H_2O, \\ 25°C,\ 4\ days}} CH_3CHCOO^-$$

Propanoic acid   2-Bromopropanoic acid 80%   Alanine 56%

Unfortunately, this approach suffers frequently from relatively low yields. A better synthesis utilizes Gabriel's procedure for the preparation of primary amines (Section 21-5).

The Gabriel synthesis can be adapted to produce amino acids

Recall that *N*-alkylation of 1,2-benzenedicarboxylic imide (phthalimide) anion followed by acid hydrolysis furnishes amines (Section 21-5). To prepare an amino acid instead, we can use diethyl 2-bromopropanedioate (diethyl 2-bromomalonate) in the first step of the reaction sequence. This alkylating agent is readily available from the bromination of diethyl propanedioate (malonate). Now the alkylation product can be hydrolyzed and decarboxylated (Section 23-2). Hydrolysis of the imide group then furnishes an amino acid.

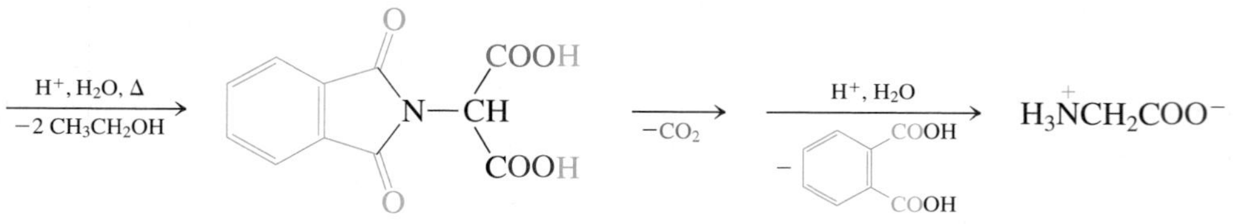

Gabriel Synthesis of Glycine

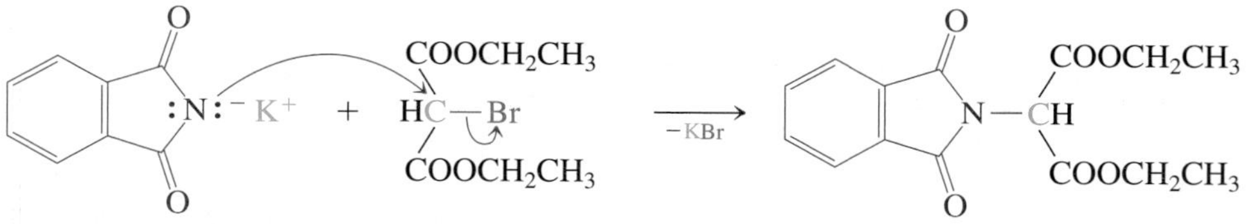

One of the advantages of this approach is the versatility of the initially formed 2-substituted propanedioate. This product can itself be alkylated, thus allowing for the preparation of a variety of substituted amino acids.

Exercise 26-5

Propose Gabriel syntheses of methionine, aspartic acid, and glutamic acid.

| **Solved Exercise 26-7** | **Working with the Concepts: Practicing Variants of the Gabriel Amino Acid Synthesis** |

A variation of the Gabriel synthesis uses diethyl *N*-acetyl-2-aminopropanedioate (acetamidomalonic ester), A. Propose a synthesis of racemic serine from this reagent.

$$
\text{A} \quad \dashrightarrow \quad \underset{\text{Serine}}{\text{HOCH}_2\overset{+\text{NH}_3}{\underset{|}{\text{CH}}}\text{COO}^-}
$$

Strategy

We are starting with a substituted malonic ester. You should be asking yourself the following questions: What synthesis should I be looking at? What substituent needs to be introduced? How do I achieve this substitution?

Solution

- We apply the principles of the malonic ester synthesis of substituted carboxylic acids (Section 23-2).
- The substituent to be introduced is hydroxymethyl.
- Hydroxymethylation is achieved by crossed aldol addition reaction (Section 18-6) of the malonic ester enolate with formaldehyde.
- Decarboxylation (Section 23-2) and amide hydrolysis (Section 20-6) is achieved with aqueous acid.

$$
\text{A} \xrightarrow{\text{CH}_2=\text{O, NaOH}} \quad \xrightarrow{\text{H}^+, \text{H}_2\text{O}, \Delta} \quad
$$

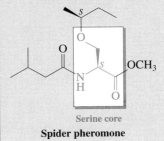
Exercise 26-8 | Try It Yourself

In the search for lead compounds for the treatment of certain neurodegenerative disorders, medicinal chemists used diethyl *N*-acetyl-2-aminopropanedioate (acetamidomalonic ester, reagent A in Exercise 26-7) for the synthesis of the heterocyclic amino acid B from starting compound A. Suggest how they might have done so. [**Caution:** You need to introduce a leaving group into A. **Hint:** The heterocycle is aromatic (Section 25-3) and therefore attached carbons are reactive (Section 22-1).]

A → B

Amino acids are prepared from aldehydes by the Strecker synthesis

The crucial step in the **Strecker*** synthesis is a variation of the cyanohydrin formation from aldehydes and hydrogen cyanide (Section 17-11).

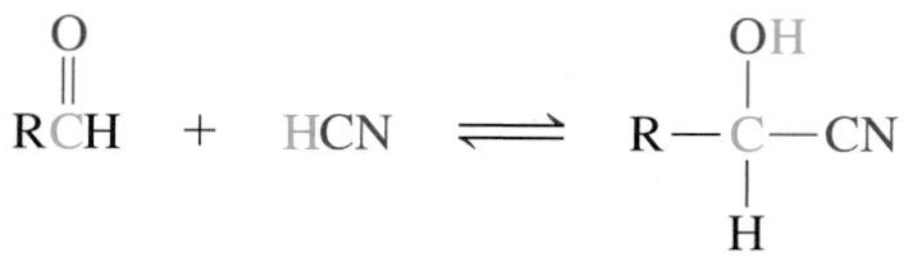

$$RCH \overset{O}{\underset{}{||}} + HCN \rightleftharpoons R-\underset{\underset{H}{|}}{\overset{\overset{OH}{|}}{C}}-CN$$

Cyanohydrin

When the same reaction is carried out in the presence of ammonia or ammonium cyanide, $H_4N^+CN^-$, it is the intermediate imine that undergoes addition of hydrogen cyanide, to furnish the corresponding 2-amino nitriles. Subsequent acidic or basic hydrolysis (Section 20-8) results in the desired amino acids.

Strecker Synthesis of Alanine

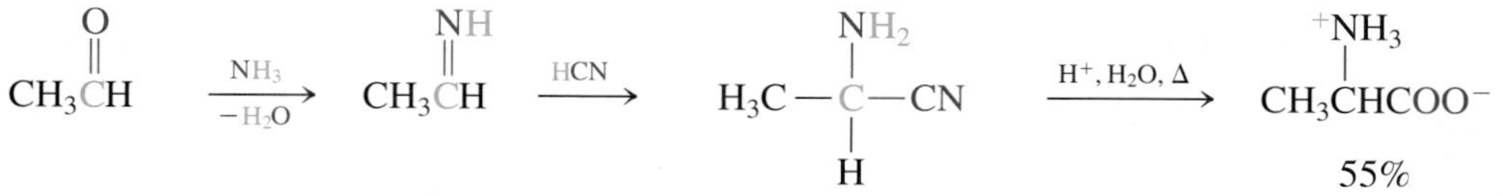

| Acetaldehyde | Imine intermediate | 2-Aminopropanenitrile | Alanine 55% |

Animation

ANIMATED MECHANISM:
The Strecker synthesis

Exercise 26-9

Propose Strecker syntheses of glycine (from formaldehyde) and methionine (from 2-propenal). (**Hint:** Review Section 18-9.)

In Summary Racemic amino acids are made by the amination of 2-bromocarboxylic acids, applications of the Gabriel synthesis of amines, and the Strecker synthesis. The last method proceeds through an imine variation of the preparation of cyanohydrins, followed by hydrolysis.

26-3 | SYNTHESIS OF ENANTIOMERICALLY PURE AMINO ACIDS

All the methods of the preceding section produce amino acids in racemic form. However, we noted that most of the amino acids in natural polypeptides have the *S* configuration. Thus, many synthetic procedures—in particular, peptide and protein syntheses—require enantiomerically pure compounds. To meet this requirement, either the racemic amino acids must be resolved (Section 5-8) or a single enantiomer must be prepared by enantioselective reactions.

A conceptually straightforward approach to the preparation of pure enantiomers of amino acids would be resolution of their diastereomeric salts. Typically, the amine group is first protected as an amide and the resulting product is then treated with an optically active amine, such as the inexpensive alkaloid brucine (Section 25-8 and margin). The two diastereomers formed can be separated by fractional crystallization. Unfortunately, in practice, this method can be tedious and can suffer from poor yields.

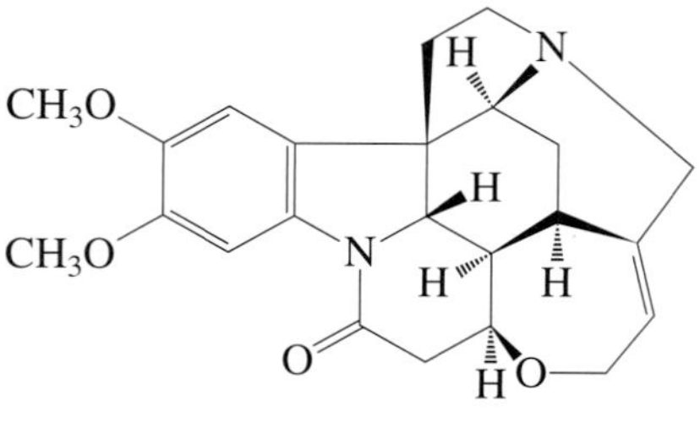

Brucine

---

*Professor Adolf Strecker (1822–1871), University of Würzburg, Germany.

Resolution of Racemic Valine

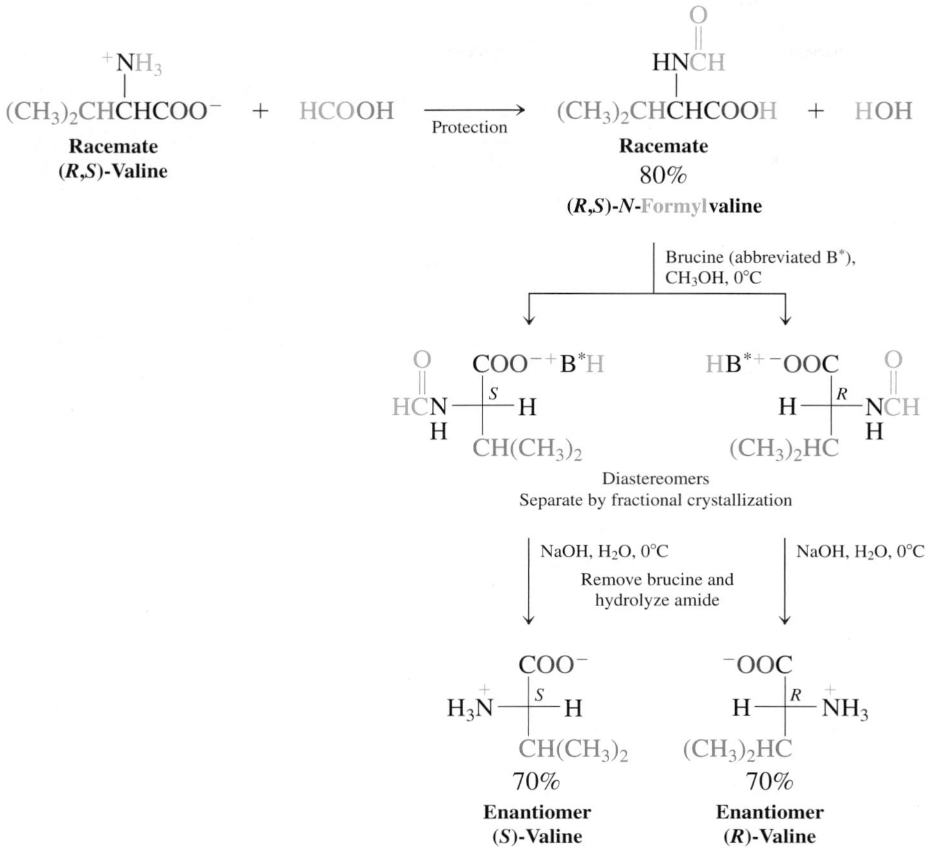

$$(CH_3)_2CHCHCOO^- \overset{+NH_3}{|} \quad + \quad HCOOH \quad \xrightarrow{\text{Protection}} \quad (CH_3)_2CHCHCOOH \overset{HNCH}{\underset{\parallel}{|}}{\overset{O}{}} \quad + \quad HOH$$

Racemate
(R,S)-Valine

Racemate
80%
(R,S)-N-Formyl valine

Brucine (abbreviated B*),
CH₃OH, 0°C

Diastereomers
Separate by fractional crystallization

NaOH, H₂O, 0°C
Remove brucine and
hydrolyze amide

NaOH, H₂O, 0°C

Enantiomer **70%**
(S)-Valine

Enantiomer **70%**
(R)-Valine

In an alternative approach, the stereocenter at C2 is formed enantioselectively, such as in enantioselective hydrogenations of α,β-unsaturated amino acids (Section 12-2). Nature makes use of this strategy in the biosynthesis of amino acids. Thus, the enzyme glutamate dehydrogenase converts the carbonyl group in 2-oxopentanedioic acid into the amine substituent in (S)-glutamic acid by a biological reductive amination (for chemical reductive aminations, see Section 21-6). The reducing agent is NADH (Real Life 25-2).

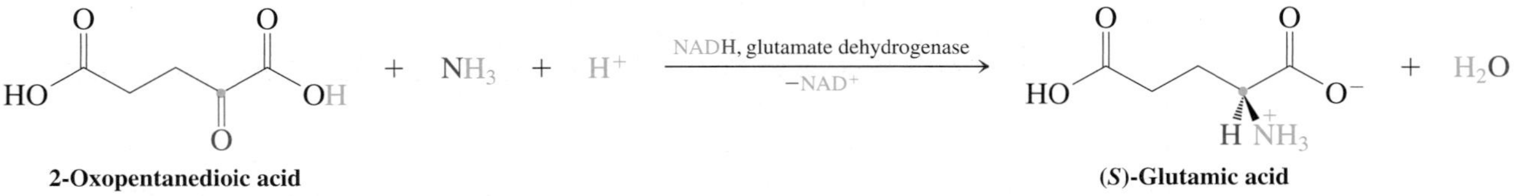

2-Oxopentanedioic acid

(S)-Glutamic acid

(S)-Glutamic acid is the biosynthetic precursor of glutamine, proline, and arginine. Moreover, it functions to aminate other 2-oxo acids, with the help of another type of enzyme, a transaminase, making additional amino acids available.

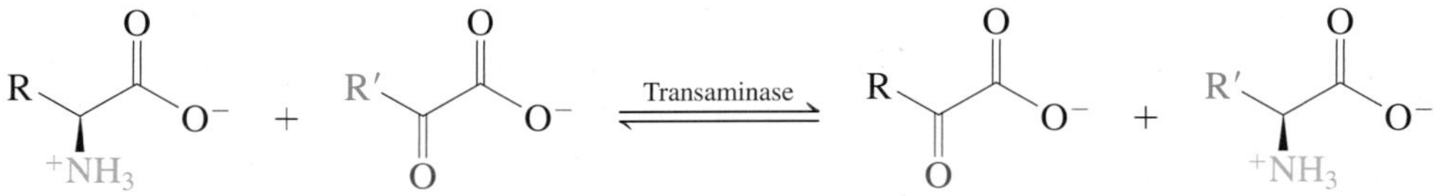

In Summary Optically pure amino acids can be obtained by resolution of a racemic mixture or by enantioselective formation of the C2 stereocenter.

| **Enantioselective Synthesis of Optically Pure Amino Acids: Phase-Transfer Catalysis**

A better alternative to the preparation of pure enantiomers by resolution of α-amino acid racemates is their direct generation from achiral precursors. For this purpose, we can use enantiopure chiral catalysts in the synthetic step that engenders chirality at C2. An example is the catalytic asymmetric ("enantioselective") hydrogenation of α-enamino acids (Real Life 5-4 and 12-2; Section 12-2). An alternative is the alkylation of suitable derivatives of glycine in the presence of optically active ammonium salts. For example, the imine-ester derivative of glycine shown on the facing page was benzylated to give 66% enantiomeric excess (Section 5-2) of the *R* product, using a biphasic aqueous CH_2Cl_2 system and 0.1 equivalent of an added ammonium salt of cinchonine. Cinchonine is an alkaloid obtained cheaply from the South American cinchona tree. Crystallization and hydrolysis of the product's imine and ester functions gave optically pure phenylalanine.

The first step of this process is carried out in a rapidly stirred mixture of dichloromethane, which contains the organic starting materials, and basic water. The ammonium salt, composed of hydrophobic substituents around the positively charged nitrogen, is soluble in both phases and shuttles back and forth between them, carrying with it either chloride or hydroxide as counterion. Hydroxide in the organic phase is responsible for α-deprotonation of the protected glycine. The enolate so formed is not free, but is

A cinchona tree in the Amazon rainforest.

ion paired with the chiral ammonium ion. As a result, alkylation by S_N2 occurs preferentially from one side (by diastereomeric transition states; Section 5-7) to generate an excess of one enantiomer of protected alanine. In other words, the handedness of the optically pure catalyst ensures preferential formation of one enantiomer (here *R*), purified subsequently by crystallization.

Quaternary ammonium salts, such as tetrabutylammonium, have been used more generally in such biphasic

---

26-4 | **PEPTIDES AND PROTEINS: AMINO ACID OLIGOMERS AND POLYMERS**

Amino acids are very versatile biologically because they can be polymerized. In this section, we describe the structure and properties of such **polypeptide** chains. Long polypeptide chains are called proteins (somewhat arbitrarily defined as >50 amino acids) and are one of the major constituents of biological structures. Proteins serve an enormous variety of biological functions; these functions are often facilitated by the twists and folds of the component chains.

Amino acids form peptide bonds

2-Amino acids are the monomer units in polypeptides. The polymer forms by repeated reaction of the carboxylic acid function of one amino acid with the amine group in another to make a chain of amides (Section 20-6). The amide linkage joining amino acids is also called a **peptide bond.**

$$2n \ HN-\underset{\underset{H}{|}}{\overset{\overset{R}{|}}{C}}-COH \longrightarrow -(NH-\underset{\underset{H}{|}}{\overset{\overset{R}{|}}{C}}-\overset{O}{\overset{||}{C}} \overset{\text{Peptide bond}}{\diagup} NH-\underset{\underset{H}{|}}{\overset{\overset{R}{|}}{C}}-\overset{O}{\overset{||}{C}})_n- \ + \ 2n \ H_2O$$

2-Amino acid
(α-Amino acid)

Polyamino acid
(Polypeptide)

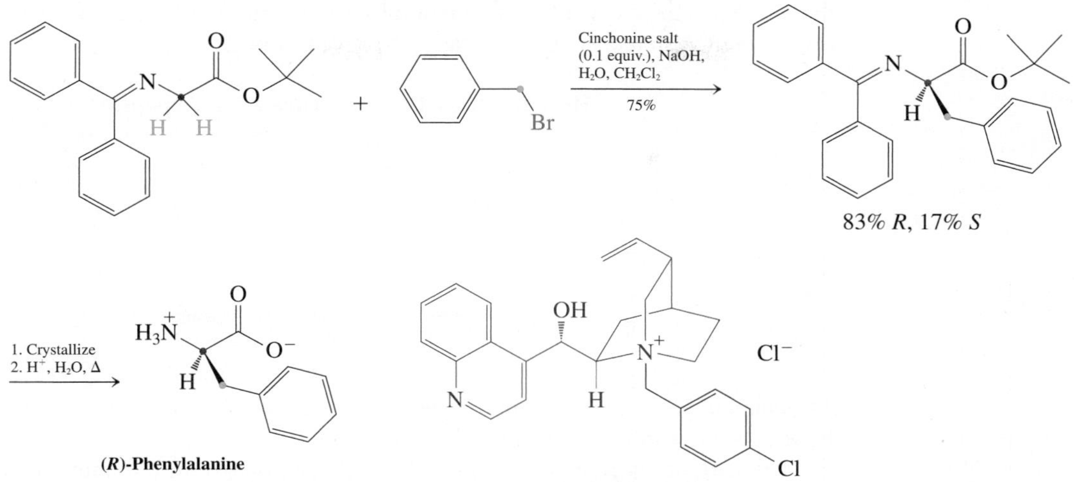

83% *R*, 17% *S*

(*R*)-Phenylalanine

N-4-Chlorobenzylcinchonium chloride
(Cinchonine marked red)

systems to enable normally insoluble counterions to enter the organic phase and react with organic substrates. The observed reactivity is called **phase-transfer catalysis,** of which the above is an example. Other applications are the S_N2 reaction (Problem 62 of Chapter 21), MnO_4^- oxidations (Section 12-11), dichlorocarbene additions (Section 12-9), and hydride reductions (Section 8-6). The employment of minimal quantities of organic solvents in conjunction with water, the simplicity of the reaction procedure, the catalytic nature of the process, and its selectivity makes phase-transfer catalysis a "greener" alternative to many homogeneous transformations.

The oligomers formed by linking amino acids in this way are called **peptides.** For example, two amino acids give rise to a **dipeptide,** three to a **tripeptide,** and so forth. The individual amino acid units forming the peptide are referred to as **residues.** In some proteins, two or more polypeptide chains are linked by disulfide bridges (Sections 9-10 and 26-1).

Of great importance for the structure of polypeptides is the fact that the peptide bond is fairly rigid at room temperature and planar, a result of the conjugation of the amide nitrogen lone electron pair with the carbonyl group (Section 20-1). The N–H hydrogen is almost always located trans to the carbonyl oxygen, and rotation about the C–N bond is slow because the C–N bond has partial double-bond character. The result is a relatively short bond (1.32 Å), between the length of a pure C–N single bond (1.47 Å, Figure 21-1) and that of a C–N double bond (1.27 Å). On the other hand, the bonds adjacent to the amide function enjoy free rotation. Thus, polypeptides are relatively rigid but nevertheless sufficiently mobile to adopt a variety of conformations. Hence, they may fold in many different ways. Most biological activity is due to such folded arrangements; straight chains are usually inactive.

Planarity Induced by Resonance in the Peptide Bond

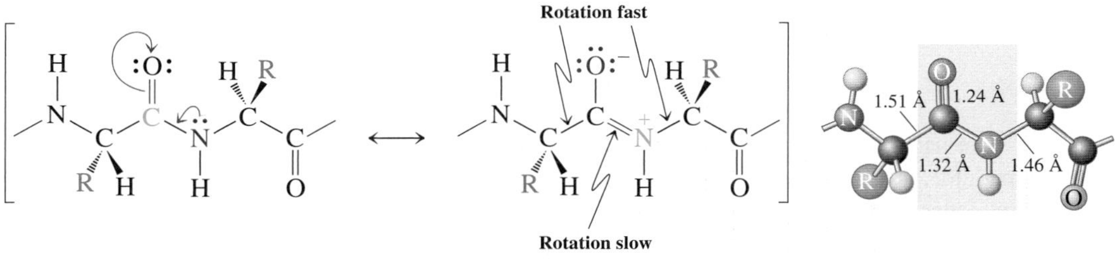

Polypeptides are characterized by their sequence of amino acid residues

In drawing a polypeptide chain, the **amino end,** or **N-terminal amino acid,** is placed at the left. The **carboxy end,** or **C-terminal amino acid,** appears at the right. The configuration at the C2 stereocenters is usually presumed to be *S*.

How to Draw the Structure of a Tripeptide

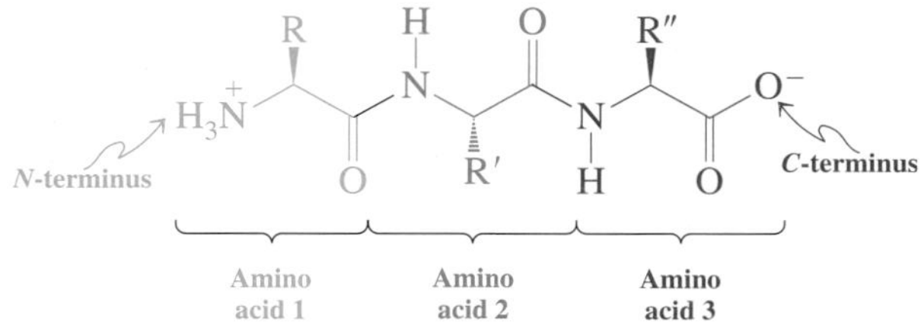

The chain incorporating the amide (peptide) bonds is called the **main chain,** the substituents R, R′, and so forth, are the **side chains.**

The naming of peptides is straightforward. Starting from the amino end, the names of the individual residues are simply connected in sequence, each regarded as a substituent to the next amino acid, ending with the *C*-terminal residue. Because this procedure rapidly becomes cumbersome, the three-letter abbreviations listed in Table 26-1 are used for larger peptides.

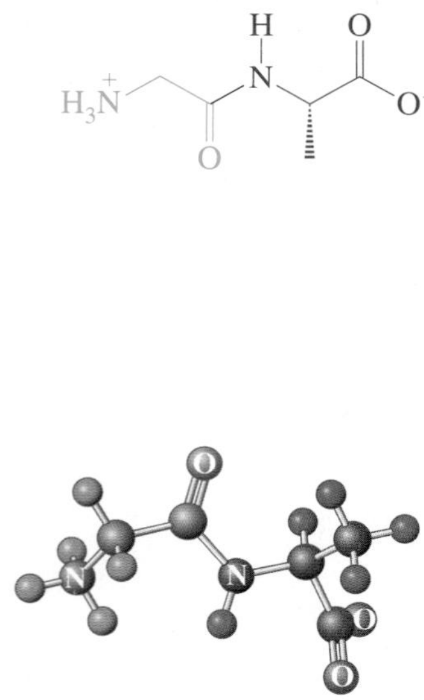

Glycylalanine
Gly-Ala

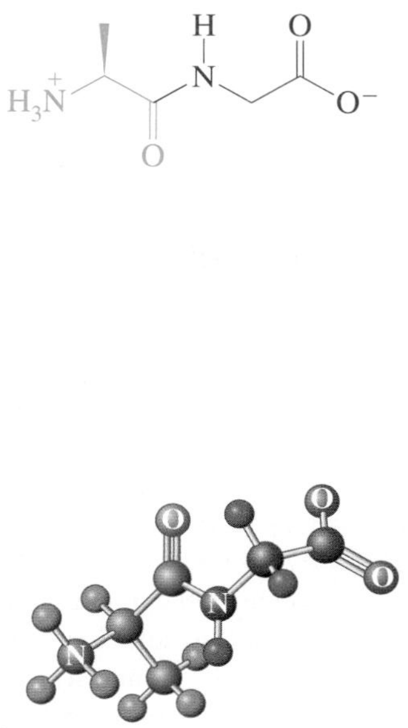

Alanylglycine
Ala-Gly

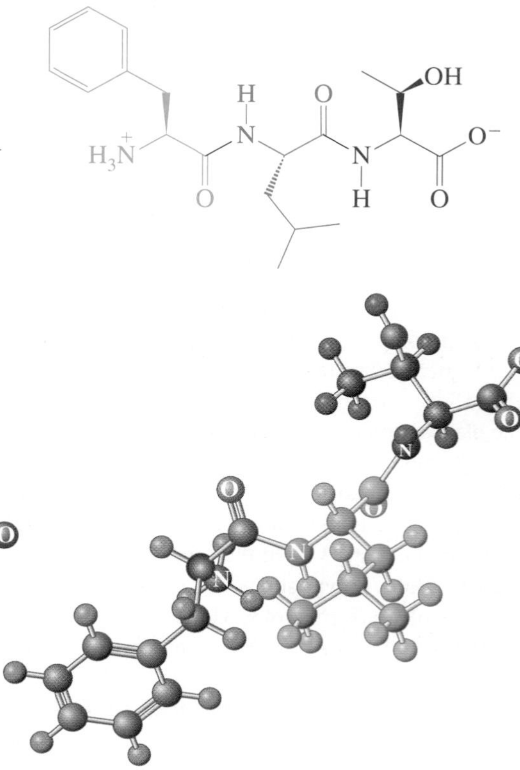

Phenylalanylleucylthreonine
Phe-Leu-Thr

Let us look at some examples of peptides and their structural variety. A dipeptide ester, aspartame, is a low-calorie artificial sweetener (NutraSweet; see Worked Example 26-28). In the three-letter notation, the ester end is denoted by OCH₃. Glutathione (Section 22-9), a tripeptide, is found in all living cells, and in particularly high concentrations in the lens of the eye. It is unusual in that its glutamic acid residue is linked at the γ-carboxy group (denoted γ-Glu) to the rest of the peptide. It functions as a biological reducing agent by being readily oxidized enzymatically at the cysteine mercapto unit to the disulfide-bridged dimer.

Aspartylphenylalanine methyl ester
Asp-Phe-OCH₃
(Aspartame)

γ-Glutamylcysteinylglycine
γ-Glu-Cys-Gly
(Glutathione)

Model Building

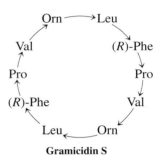

Gramicidin S is a cyclic peptide topical antibiotic constructed out of two identical pentapeptides that have been joined head to tail. It contains phenylalanine in the *R* configuration and a rare amino acid, ornithine [Orn, a lower homolog (one less CH₂ group) of lysine]. In the short notation in which gramicidin S is shown in the margin, the sense in which the amino acids are linked (amino-to-carboxy direction) is indicated by arrows.

Orn ⟶ Leu
Val ↗   ↘ (*R*)-Phe
Pro   Pro
(*R*)-Phe   Val
Leu ⟵ Orn

Gramicidin S

> Gramicidin S acts by disrupting cellular membranes, which causes cell death. The cell membrane separates the interior of cells from the outside surroundings and is permeable only to selected ions and molecules.

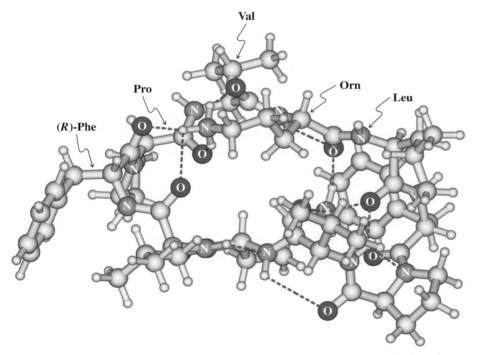

Gramicidin S. (Computer-generated model, courtesy Professor Evan R. Williams and Dr. Richard L. Wong, University of California at Berkeley. The red dashed lines represent hydrogen bonds.)

Insulin illustrates the three-dimensional structure adopted by a complex sequence of amino acids (Figure 26-1). This protein hormone is an important drug in the treatment of diabetes because of its ability to regulate glucose metabolism. Insulin contains 51 amino acid residues incorporated into two chains, denoted A and B. The chains are connected by two disulfide bridges, and there is an additional disulfide linkage connecting the cysteine residues at positions 6 and 11 of the A chain, causing it to loop. Both chains fold up in a way that minimizes steric interference and maximizes electrostatic, London, and hydrogen-bonding attractions. These forces give rise to a fairly condensed three-dimensional structure (Figure 26-2).

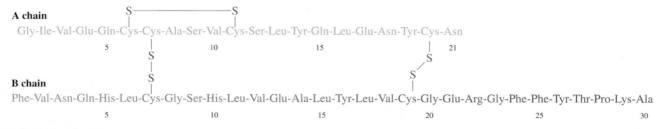

A chain

S————————S
| |
Gly-Ile-Val-Glu-Gln-Cys-Cys-Ala-Ser-Val-Cys-Ser-Leu-Tyr-Gln-Leu-Glu-Asn-Tyr-Cys-Asn

5   10   15   21

S   S
| |
S   S

B chain

Phe-Val-Asn-Gln-His-Leu-Cys-Gly-Ser-His-Leu-Val-Glu-Ala-Leu-Tyr-Leu-Val-Cys-Gly-Glu-Arg-Gly-Phe-Phe-Tyr-Thr-Pro-Lys-Ala

5   10   15   20   25   30

Figure 26-1 Bovine (cattle) insulin is made up of two amino acid chains, linked by disulfide bridges. The amino (*N*-terminal) end is at the left in both chains.

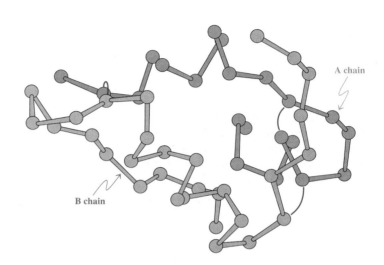

Figure 26-2 Three-dimensional structure of insulin. Residues in chain A are blue, those in B green. The disulfide bridges are indicated in red. (After *Biochemistry,* 6th ed., by Jeremy M. Berg, John L. Tymoczko, and Lubert Stryer. W.H. Freeman and Company. Copyright © 1975, 1981, 1988, 1995, 2002, 2007.)

Because synthetic methods gave only low yields, insulin used to be isolated from the pancreas of slaughtered cows and pigs, purified, and sold as such. In the 1980s, the development of genetic engineering methods (Section 26-11) allowed the cloning of the human gene that codes for insulin. The gene was placed into bacteria modified to produce the drug continuously. In this way, enough material is generated to treat millions of diabetics throughout the world.

Exercise 26-10

Vasopressin, also known as antidiuretic hormone, controls the excretion of water from the body. Draw its complete structure. (**Caution:** There is a disulfide bridge between the two cysteine residues.)

S————————S
| |
Cys-Tyr-Phe-Gln-Asn-Cys-Pro-Arg-Gly-NH$_2$

Vasopressin

A Hydrogen Bond Between Two Polypeptide Strands

Hydrogen bond

Proteins fold into pleated sheets and helices: secondary and tertiary structure

Insulin and other polypeptide chains adopt well-defined three-dimensional structures. Whereas the sequence of amino acids in the chain defines the **primary structure,** the folding pattern of the chain induced by the spatial arrangement of close-lying amino acid residues gives rise to the **secondary structure** of the polypeptide. The secondary structure results mainly from the rigidity of the amide bond and from hydrogen (and other noncovalent) bonding (margin) along the chain(s). Two important arrangements are the pleated sheet, or β configuration, and the α helix.

In the **pleated sheet** (also called β sheet; Figure 26-3), two chains line up with the amino groups of one peptide opposite the carbonyl groups of a second, thereby allowing hydrogen

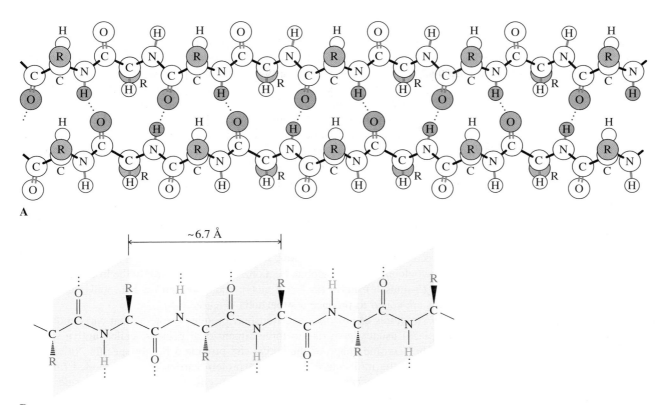

A

B

Figure 26-3 (A) The pleated sheet, or β configuration, which is held in place by hydrogen bonds (dotted lines) between two polypeptide strands. (After "Proteins," by Paul Doty, *Scientific American,* September 1957. Copyright © 1957, Scientific American, Inc.) (B) The peptide bonds define the individual pleats (shaded in yellow); the positions of the side chains, R, are alternately above and below the planes of the sheets. The dotted lines indicate hydrogen bonds to a neighboring chain or to water.

bonds to form. Such bonds can also develop within a single chain if it loops back on itself. Multiple hydrogen bonding of this type can impart considerable rigidity to a system. The planes of adjacent amide linkages form a specific angle, a geometry that produces the observed pleated-sheet structure, in which the R groups protrude above and below at each kink.

The **α helix,** as shown in Figure 26-4, allows for intramolecular hydrogen bonding between nearby amino acids in the chain: The carbonyl oxygen of each amino acid is

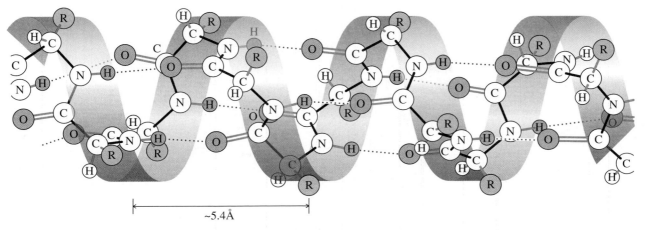

Figure 26-4 The α helix, in which the polymer chain is arranged as a right-handed spiral held rigidly in shape by intramolecular hydrogen bonds (red dotted lines). (After "Proteins," by Paul Doty, *Scientific American,* September 1957. Copyright © 1957, Scientific American, Inc.)

interacting with the amide hydrogen four residues ahead. There are 3.6 amino acids per turn of the helix, two equivalent points in neighboring turns being about 5.4 Å apart. The C=O and N—H bonds point in opposite directions and are roughly aligned with the helical axis. On the other hand, the (hydrophobic) R groups point away from the helix.

Not all polypeptides adopt idealized structures such as these. If too much charge of the same kind builds up along the chain, charge repulsion will enforce a more random orientation. In addition, the rigid proline, because its amino nitrogen is part of the substituent ring and has no N—H available for hydrogen bonding, can cause a kink or bend in an α helix.

Further folding, coiling, and aggregation of polypeptides is induced by distant residues in the chain end and give rise to their **tertiary structure.** A variety of forces, all arising from the R group, come into play to stabilize such molecules, including disulfide bridges, hydrogen bonds, London forces, and electrostatic attraction and repulsion. There are also **micellar effects** (Real Life 19-1 and Section 20-5): The polymer adopts a structure that maximizes exposure of polar groups to the aqueous environment while minimizing exposure of hydrophobic groups (e.g., alkyl and phenyl), the "hydrophobic effect" (Section 8-2, Real Life 19-1). Pronounced folding is observed in the **globular proteins,** many of which perform chemical transport and catalysis (e.g., myoglobin and hemoglobin, Section 26-8, Figure 26-8). In the **fibrous proteins,** such as myosin (in muscle), fibrin (in blood clots), and α-keratin (in hair, nails, and wool), several α helices are coiled to produce a **superhelix** (Figure 26-5).

The tertiary structures of enzymes and transport proteins (proteins that carry molecules from place to place) usually give rise to three-dimensional pockets, called **active sites** or **binding sites.** The size and shape of the active site provide a highly specific "fit" for the **substrate** or **ligand,** the molecule on which the protein carries out its intended function. The inner surface of the pocket typically contains a specific arrangement of the side chains of polar amino acids that attracts functional groups in the substrate by hydrogen bonding or ionic interactions. In enzymes, the active site aligns functional groups and additional molecules in a way that promotes their reactions with the substrate.

An example is the active site of chymotrypsin, a mammalian digestive enzyme responsible for the degradation of proteins in food. Chymotrypsin accomplishes the hydrolysis of peptide bonds at body temperature and at physiological pH. Recall that ordinary amide hydrolysis requires much more drastic conditions (Section 20-6). Moreover, the enzyme also recognizes specific peptide linkages that are targeted for selective cleavage, such as the carboxy end of phenylalanine residues (see Section 26-5, Table 26-2). How does it do that?

A simplified picture of the active part of this large molecule (of dimensions $51 \times 40 \times 40$ Å) in this process is shown in the following scheme.

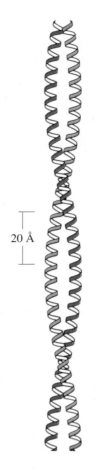

Figure 26-5 Idealized picture of a superhelix, a coiled coil.

20 Å

Peptide Hydrolysis in the Active Site of Chymotrypsin

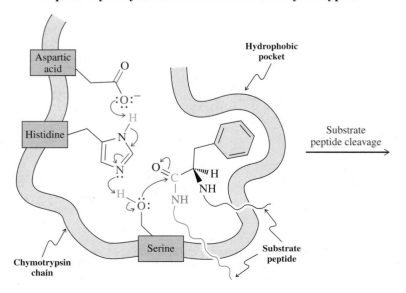

Bound ester hydrolysis

+ ∿∿∿∿NH₂

Note: In this scheme, the red arrows depict only the respective addition steps of the two addition–elimination sequences.

The enzyme has four important, close-lying parts, all of which work together to facilitate the hydrolysis reaction: a hydrophobic pocket, and the residues of aspartic acid, histidine, and serine. The hydrophobic pocket (see Section 8-2) helps to bind the polypeptide to be "digested" by attraction of the hydrophobic phenyl substituent of one of its component phenylalanine residues. With the phenyl group held in this pocket, the three amino acid residues cooperate in a proton-transfer relay sequence to effect nucleophilic addition–elimination (Sections 19-7 and 20-6) of the serine hydroxy group to the carbonyl function of Phe, releasing the amine part of the cleaved polypeptide. The remainder of the substrate is held by an ester linkage to the enzyme, positioned to undergo ester hydrolysis (Section 20-4) by a water molecule. This reaction is aided by a tandem proton-transfer sequence similar to that used for hydrolyzing the peptide bond. With the link to the enzyme broken, the carboxy segment of the original substrate is now free to leave the intact active site of chymotrypsin, making room for another polypeptide.

Exercise 26-11

The scheme just shown omits the respective elimination steps of the two nucleophilic addition–elimination reactions. Show how the enzyme aids in accelerating them as well. [**Hint:** Draw the result of the "electron pushing" depicted in the first (or second) picture of the scheme and think about how a reversed electron and proton flow might help.]

Misfolded Proteins and "Mad Cow" Disease

The correct tertiary structure of proteins is essential to their function (see also Section 13-11, and p. 1107, margin). Misfolding, while frequently simply eliminating activity, can sometimes have deleterious consequences. An example is prions, a class of misfolded proteins that propagate by inducing other proteins to convert to their so-called prion form, triggering a chain reaction that produces a polymer that causes tissue damage and cell death. Prions are responsible for the fatal "mad cow" disease in cattle and the related Creutzfeldt-Jakob disease in humans, conditions characterized by degeneration of the brain and spinal cord.

Denaturation, or breakdown of tertiary structure, usually causes precipitation of the protein and destroys its catalytic activity. Denaturation is caused by exposure to excessive heat or extreme pH values. Think, for example, of what happens to clear egg white when it is poured into a hot frying pan or to milk when it is added to lemon tea.

Some molecules, such as hemoglobin (Section 26-8), also adopt a **quaternary structure** (see Figure 26-9), in which two or more polypeptide chains, each with its own tertiary structure, combine to form a larger assembly. A simplified picture of the progression from primary to quaternary structures is given on the next page.

The Progression from Primary to Secondary, Tertiary, and Quaternary Polypeptide Structures

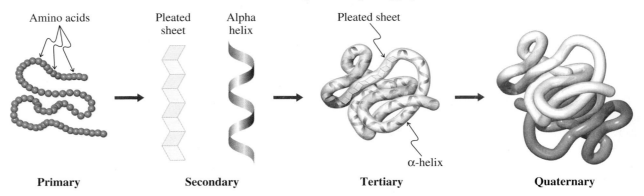

Amino acids

Pleated sheet   Alpha helix

Pleated sheet

α-helix

Primary   **Secondary**   **Tertiary**   **Quaternary**

The three-dimensional constellation of polypeptides is a direct consequence of primary structure: In other words, the amino acid sequence specifies in which way the chain will coil, aggregate, and otherwise interact with internal and external molecular units. Therefore, knowledge of this sequence is of paramount importance in the understanding of protein structure and function. How to obtain this knowledge is the subject of the next section.

In Summary Polypeptides are polymers of amino acids linked by amide bonds. Their amino acid sequences can be described in a shorthand notation using the three- or one-letter abbreviations compiled in Table 26-1. The amino end group is placed at the left, the carboxy end at the right. Polypeptides can be cyclic and can also be linked by disulfide and hydrogen bonds. The sequence of amino acids is the primary structure of a polypeptide, folding gives rise to its secondary structure, further folding and coiling produce its tertiary structure, and aggregation of several polypeptides results in the quaternary structure.

26-5 | DETERMINATION OF PRIMARY STRUCTURE: AMINO ACID SEQUENCING

Biological function in polypeptides and proteins requires a specific three-dimensional shape and arrangement of functional groups, which, in turn, necessitate a definite amino acid sequence. One "monkey wrench" residue in an otherwise normal protein can completely alter its behavior. For example, sickle-cell anemia, a potentially lethal condition, is the result of changing a *single* amino acid in hemoglobin (Section 26-8). The determination of the primary structure of a protein, called amino acid or polypeptide **sequencing,** can help us to understand the protein's mechanism of action.

In the late 1950s and early 1960s, amino acid sequences were discovered to be predetermined by DNA, the molecule containing our hereditary information (Section 26-9). Thus, through a knowledge of protein primary structure, we can learn how genetic material expresses itself. Functionally similar proteins in related species should, and do, have similar primary structures. The closer their sequences of amino acids, the more closely the species are related. Polypeptide sequencing therefore strikes at the heart of the question of the evolution of life itself. This section shows how chemical means, together with analytical techniques, allow us to obtain this information.

First, purify the polypeptide

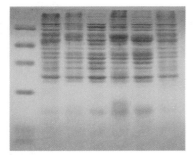

Electrophoresis of proteins on polyacrylamide. Each column of blue bands illustrates the separation of a mixture of proteins into their individual components.

The problem of polypeptide purification is an enormous one, and attempts at its solution consume many days in the laboratory. Several techniques can separate polypeptides on the basis of size, solubility in a particular solvent, charge, or ability to bind to a support. Although detailed discussions are beyond the scope of this book, we shall briefly describe some of the more widely used methods.

In **dialysis,** the polypeptide is separated from smaller fragments by filtration through a semipermeable membrane. A second method, **gel-filtration chromatography,** uses a carbohydrate polymer in the form of a column of beads as a support. Smaller molecules diffuse more easily into the beads, spending a longer time on the column than large ones do; thus, they emerge from the column later than the large molecules. In **ion-exchange chromatography,** a charged support separates molecules according to the amount of charge that they carry. Another method based on electric charge is **electrophoresis.** A spot of the mixture to be separated is placed on a plate covered with a thin layer of chromatographic material (such as polyacrylamide) that is attached to two electrodes. When the voltage is turned on, positively charged species (e.g., polypeptides rich in protonated amine groups) migrate toward the cathode, negatively charged species (carboxy-rich peptides) toward the anode. The separating power of this technique is extraordinary. More than a thousand different proteins from one species of bacterium have been resolved in a single experiment.

Finally, **affinity chromatography** exploits the tendency of polypeptides to bind very specifically to certain supports by hydrogen bonds and other attractive forces. Peptides of differing sizes and shapes have differing retention times in a column containing such a support.

Second, determine which amino acids are present

When the polypeptide strand has been purified, the next step in structural analysis is to establish its composition. To determine which amino acids and how much of each is present in the polypeptide, the entire chain is degraded by amide hydrolysis (6 N HCl, 110°C, 24 h) to give a mixture of the free amino acids. The mixture is then separated and its composition recorded by an automated **amino acid analyzer.**

This instrument consists of a column bearing a negatively charged support, usually containing carboxylate or sulfonate ions. The amino acids pass through the column in slightly acidic solution. They are protonated to a greater or lesser degree, depending on their structure, and therefore are more or less retained on the column. This differential retention separates the amino acids, and they come off the column in a specific order, beginning with the most acidic and ending with the most basic. At the end of the column is a reservoir containing a special indicator. Each amino acid produces a violet color (see Worked Example 26-29) whose intensity is proportional to the amount of that acid present and is recorded in a chromatogram (Figure 26-6). The area under each peak is a measure of the relative amount of a specific amino acid in the mixture.

The amino acid analyzer can readily establish the composition of a polypeptide. For example, the chromatogram of hydrolyzed glutathione (Section 26-4) gives three equal-sized peaks, corresponding to Glu, Gly, and Cys.

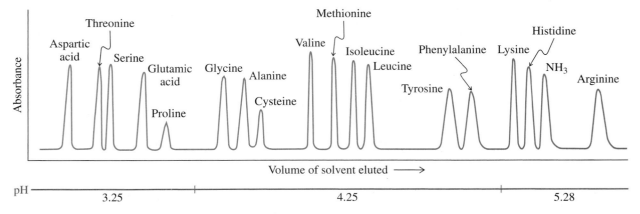

Figure 26-6 A chromatogram showing the presence of various amino acids separated by an amino acid analyzer using a polysulfonated ion-exchange resin. The more acidic products (e.g., aspartic acid) are generally eluted first. Ammonia is included for comparison.

Exercise 26-12

Hydrolysis of the A chain in insulin (Figure 26-1) gives a mixture of amino acids. What are they and what is their relative abundance in the mixture?

Sequence the peptide from the amino (*N*-terminal) end

Once we know the gross makeup of a polypeptide, we can determine the order in which the individual amino acids are bound to one another—the amino acid sequence.

Several different methods can reveal the identity of the residue at the amino end. Most exploit the uniqueness of the free amino substituent, which may enter into specific chemical reactions that serve to "tag" the *N*-terminal amino acid. One such procedure is the **Edman* degradation,** and the reagent used is phenyl isothiocyanate, $C_6H_5N{=}C{=}S$ (a sulfur analog of an isocyanate; Section 20-7).

Recall (Section 20-7) that isocyanates are very reactive with respect to nucleophilic attack, and the same is true of their sulfur analogs. In the Edman degradation, the terminal amino group adds to the isothiocyanate reagent to give a thiourea derivative (refer to Section 20-6 for the urea function). Mild acid causes extrusion of the tagged amino acid as a phenylthiohydantoin, leaving the remainder of the polypeptide unchanged (see also Exercise 26–13). The phenylthiohydantoins of all amino acids are well known, so the *N*-terminal end of the original polypeptide can be readily identified. The new chain, carrying a new terminal amino acid, is now ready for another Edman degradation to tag the next residue, and so forth. The entire procedure has been automated to allow the routine identification of polypeptides containing 50 or more amino acids.

Beyond that number, the buildup of impurities becomes a serious impediment. The reason for this drawback is that each degradation round, though proceeding in high yield, is not completely quantitative, thus leaving small quantities of incompletely reacted peptide admixed with the new one. You can readily envisage that, with each step, this problem increases until the mixtures become intractable.

Edman Degradation of the A Chain of Insulin

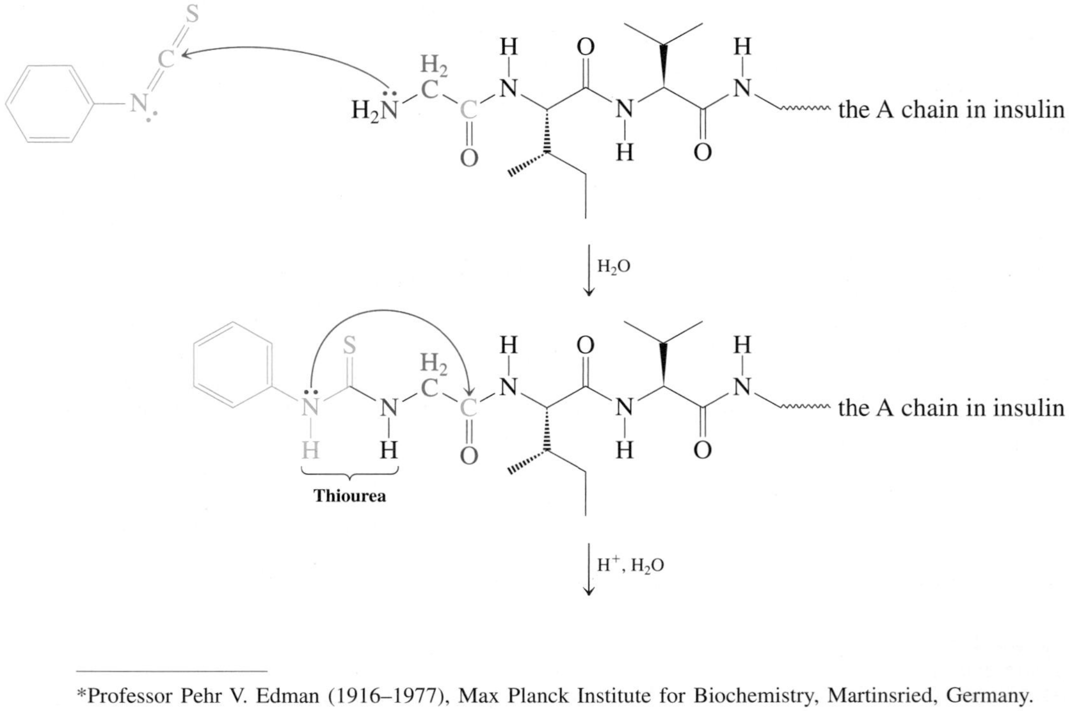

---

*Professor Pehr V. Edman (1916–1977), Max Planck Institute for Biochemistry, Martinsried, Germany.

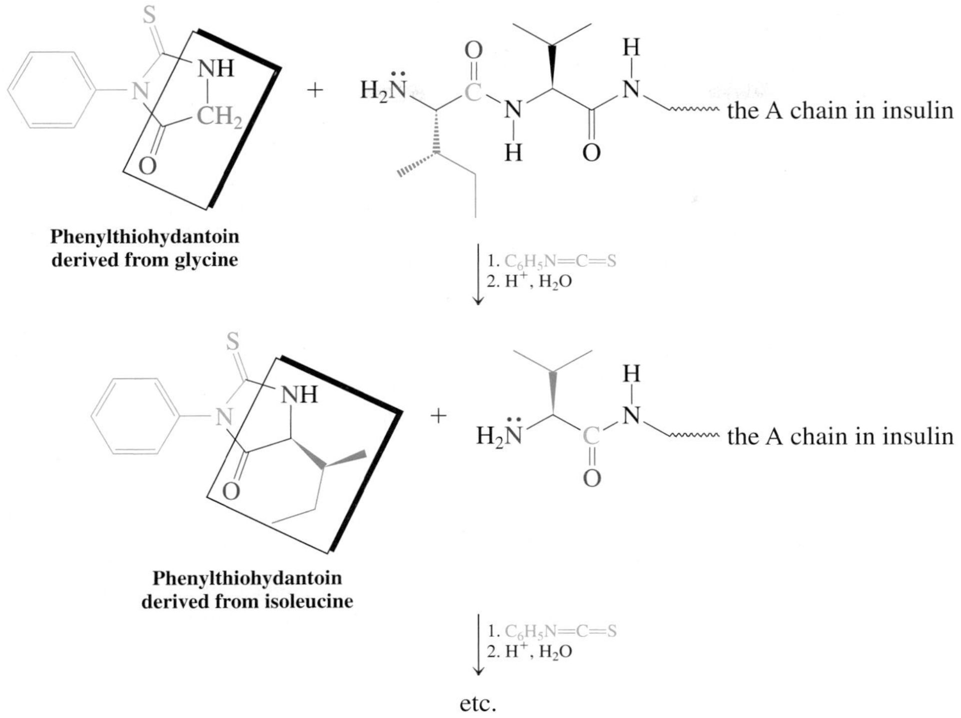

**Phenylthiohydantoin
derived from glycine**

+ H₂N̈ ... the A chain in insulin

1. C₆H₅N=C=S
2. H⁺, H₂O

**Phenylthiohydantoin
derived from isoleucine**

+ H₂N̈ ... the A chain in insulin

1. C₆H₅N=C=S
2. H⁺, H₂O

etc.

Exercise 26-13

The pathway of phenylthiohydantoin formation is not quite as straightforward as indicated by the red arrow in step 2 of the preceding scheme, but rather proceeds through the intermediacy of an isomeric thiazolinone (margin), which then rearranges under acidic conditions to the more stable phenylthiohydantoin. Formulate a mechanism for the formation of this compound (R = H) from the acid treatment of the phenylthiourea of glycine amide, $C_6H_5NHC(=S)NHCH_2C(=O)NH_2$. [**Hint:** Sulfur is more nucleophilic than nitrogen (Section 6-8).]

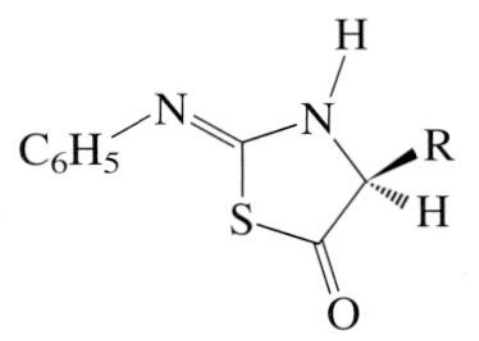

Exercise 26-14

Polypeptides can be cleaved into their component amino acid fragments by treatment with dry hydrazine. This method reveals the identity of the carboxy end. Explain.

The chopping up of longer chains is achieved with enzymes

The Edman procedure allows for the ready sequencing of only relatively short polypeptides. For longer ones (e.g., those with more than 50 residues), it is necessary to cleave the larger chains into shorter fragments in a selective and predictable manner. These cleavage methods rely mostly on hydrolytic enzymes. For example, trypsin, a digestive enzyme of intestinal liquids, cleaves polypeptides only at the carboxy end of arginine and lysine.

Selective Hydrolysis of the B Chain of Insulin by Trypsin

Phe-Val-Asn-Gln-His-Leu-Cys-Gly-Ser-His-Leu-Val-Glu-Ala-Leu-Tyr-Leu-Val-Cys-Gly-Glu-Arg-Gly-Phe-Phe-Tyr-Thr-Pro-Lys-Ala

     5   10   15   20   25   30

Trypsin, H₂O

Phe-Val-Asn-Gln-His-Leu-Cys-Gly-Ser-His-Leu-Val-Glu-Ala-Leu-Tyr-Leu-Val-Cys-Gly-Glu-Arg + Gly-Phe-Phe-Tyr-Thr-Pro-Lys + Ala

Peptide Cleavage to Overlapping Sequences

Table 26-2	Specificity of Hydrolytic Enzymes in Polypeptide Cleavage
Enzyme	**Site of cleavage**
Trypsin	Lys, Arg, carboxy end
Clostripain	Arg, carboxy end
Chymotrypsin	Phe, Trp, Tyr, carboxy end
Pepsin	Asp, Glu, Leu, Phe, Trp, Tyr, carboxy end
Thermolysin	Leu, Ile, Val, amino end

A more selective enzyme is clostripain, which cleaves only at the carboxy end of arginine. In contrast, chymotrypsin, which, like trypsin, is found in mammalian intestines, is less selective and cleaves at the carboxy end of phenylalanine (see Section 26-4), tryptophan, and tyrosine. Other enzymes have similar selectivity (Table 26-2). In this way, a longer polypeptide is broken down into several shorter ones, which may then be sequenced by the Edman procedure.

After a first enzymatic cleavage, sequences of segments of the polypeptide under investigation are determined, but the order in which they are linked is not. For this purpose, selective hydrolysis is carried out a second time, by using a different enzyme that provides pieces in which the connectivities broken under the first conditions are left intact, so-called *overlap peptides.* The solution is then found by literally "piecing" together the available information like a puzzle (margin).

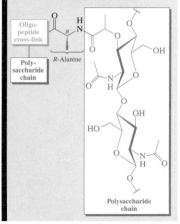
Solved Exercise 26-15	Working with the Concepts: Determining Amino Acid Sequences

A polypeptide containing 21 amino acids was hydrolyzed by thermolysin. The products of this treatment were Gly, Ile, Val-Cys-Ser, Leu-Tyr-Gln, Val-Glu-Gln-Cys-Cys-Ala-Ser, and Leu-Glu-Asn-Tyr-Cys-Asn. When the same polypeptide was hydrolyzed by chymotrypsin, the products were Cys-Asn, Gln-Leu-Glu-Asn-Tyr, and Gly-Ile-Val-Glu-Gln-Cys-Cys-Ala-Ser-Val-Cys-Ser-Leu-Tyr. Give the amino acid sequence of this molecule.

Strategy
Piecing together puzzles of this type is best done visually. First lay out all the individual products, best in order of decreasing length. Starting with the largest oligopeptide resulting from the hydrolysis of the starting material by one enzyme, look for matching partial sequences of pieces obtained in the hydrolysis by the second enzyme. One of the remaining oligopeptides should provide an overlap sequence with a second by lining it up against one of the termini of the starting large fragment, thus allowing the gradual lengthening of it, until all degradation products have been reassembled to the original polypeptide.

Solution
• Product tabulation:

Thermolysin Hydrolysis	**Chymotrypsin Hydrolysis**
Val-Glu-Gln-Cys-Cys-Ala-Ser (a)	Gly-Ile-Val-Glu-Gln-Cys-Cys-Ala-Ser-Val-Cys-Ser-Leu-Tyr (g)
Leu-Glu-Asn-Tyr-Cys-Asn (b)	Gln-Leu-Glu-Asn-Tyr (h)
Leu-Tyr-Gln (c)	Cys-Asn (i)
Val-Cys-Ser (d)	
Gly (e)	
Ile (f)	

• The largest fragment is (g), which arises from the chymotrypsin hydrolysis, and we can readily line up several of the pieces of the thermolysin treatment against it:

Gly-Ile-Val-Glu-Gln-Cys-Cys-Ala-Ser-Val-Cys-Ser-Leu-Tyr- (g)
(e) (f)   (a)   (d)   Leu-Tyr-Gln (c)

- This process establishes overlap peptide (c) as the link between (g) and (h), which furnishes the larger substructure (g) + (h):

$$\underbrace{\text{Gly-Ile-Val-Glu-Gln-Cys-Cys-Ala-Ser-Val-Cys-Ser-Leu-Tyr-}}_{(g)}\underbrace{\text{Gln-Leu-Glu-Asn-Tyr-}}_{(h)}$$

$$\underset{(i)}{\text{Leu-Glu-Asn-Tyr-}\underline{\text{Cys-Asn}}\ \text{(b)}}$$

- Inspection of the right terminus of (g) + (h) points to (b) as the overlap peptide to the remainder of the chain and (i). The polypeptide sequence is therefore

  Gly-Ile-Val-Glu-Gln-Cys-Cys-Ala-Ser-Val-Cys-Ser-Leu-Tyr-Gln-Leu-Glu-Asn-Tyr-Cys-Asn

- If you check Figure 26-1, you will recognize the A chain of insulin.

Exercise 26-16 | Try It Yourself

What are the polypeptide products obtained in the hydrolysis of the A chain of insulin by pepsin?

Protein sequencing is made possible by recombinant DNA technology

Despite the success of the techniques for polypeptide sequencing described so far, allowing for the structure elucidation of hundreds of proteins, their application to large systems (i.e., those containing more than 1000 residues) is an expensive, laborious, and time-consuming business. Progress in this field was significantly impeded until the advent of recombinant DNA technology (Section 26-11). As we shall see (Section 26-10, Table 26-3), the sequences of the four bases in DNA—adenine, thymine, guanine, and cytosine—are directly correlated to the amino acid sequences of the proteins encoded by genes or the corresponding messenger RNA. Modern developments have led to the rapid automated analysis of DNA, and the knowledge gained can be immediately translated into primary protein structure. In this way, tens of thousands of proteins have been sequenced in the past few years.

In Summary The structure of polypeptides is established by various degradation schemes. First, the polymer is purified; then the kind and relative abundance of the component amino acids are determined by complete hydrolysis and amino acid analysis. The *N*-terminal residues can be identified by Edman degradation. Repeated Edman degradation gives the sequence of shorter polypeptides, which are made from longer polypeptides by specific enzymatic hydrolysis. Finally, recombinant DNA technology has made primary structure analysis of larger proteins relatively easy.

26-6 | SYNTHESIS OF POLYPEPTIDES: A CHALLENGE IN THE APPLICATION OF PROTECTING GROUPS

In a sense, the topic of peptide synthesis is a trivial one: Only one type of bond, the amide linkage, has to be made. The formation of this linkage was described in Section 19-10. Why discuss it further? This section shows that, in fact, achieving selectivity poses great problems, for which specific solutions have to be found.

Consider even as simple a target as the dipeptide glycylalanine. Just heating glycine and alanine to make the peptide bond by dehydration would result in a complex mixture of di-, tri-, and higher peptides with random sequences. Because the two starting materials can form bonds either to their own kind or to each other, there is no way to prevent random oligomerization.

An Attempt at the Synthesis of Glycylalanine by Thermal Dehydration

$$\text{Gly} + \text{Ala} \xrightarrow[-\text{H}_2\text{O}]{\Delta} \underset{\substack{\text{Desired} \\ \text{product}}}{\text{Gly-Gly}} + \text{Ala-Gly} + \text{Gly-Ala} + \text{Ala-Ala} + \text{Gly-Gly-Ala} + \text{Ala-Gly-Ala etc.}$$

Selective peptide synthesis requires protecting groups

To form peptide bonds selectively, the functional groups of the amino acids have to be protected. There are several amino- and carboxy-protecting strategies available, two of which are described here (see also Problem 55).

The amino end is frequently blocked by a **phenylmethoxycarbonyl group** (abbreviated **carbobenzoxy** or **Cbz**), introduced by reaction of an amino acid with phenylmethyl chloroformate (benzyl chloroformate).

Protection of the Amino Group in Glycine

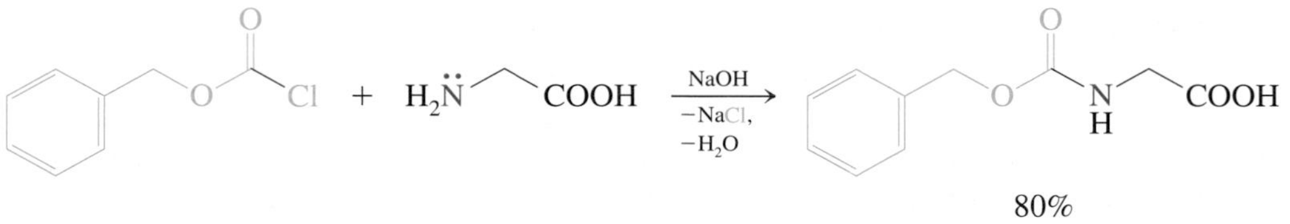

Phenylmethyl chloroformate
(Benzyl chloroformate)

Glycine

80%

Phenylmethoxycarbonylglycine
(Carbobenzoxyglycine, Cbz-Gly)

The amino function is deprotected by hydrogenolysis (Section 22-2), which initially furnishes the carbamic acid as a reactive intermediate (Section 20-7). Decarboxylation occurs instantly to restore the free amine.

Deprotection of the Amino Group in Glycine

Carbamic acid function

95%

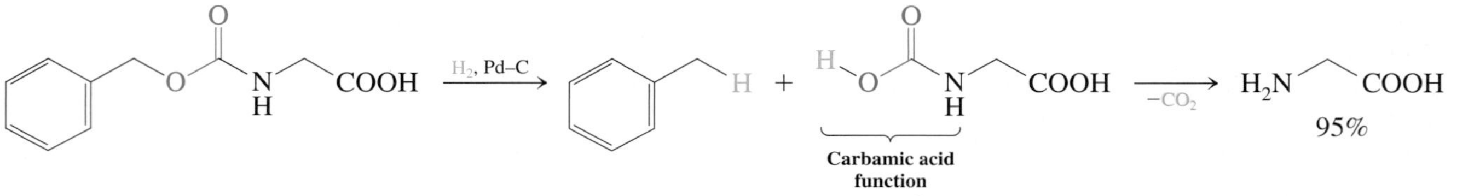

Another amino-protecting group is **1,1-dimethylethoxycarbonyl (*tert*-butoxycarbonyl, Boc),** introduced by reaction with bis(1,1-dimethylethyl) dicarbonate (di-*tert*-butyl dicarbonate). Similar to carbamic acids, the leaving group, $(CH_3)_3COCOOH$, is an unstable monoester of carbonic acid (see Problem 36, part (a), of Chapter 22). It decomposes spontaneously to CO_2 and $(CH_3)_3COH$.

Protection of the Amino Group in Amino Acids as the Boc Derivative

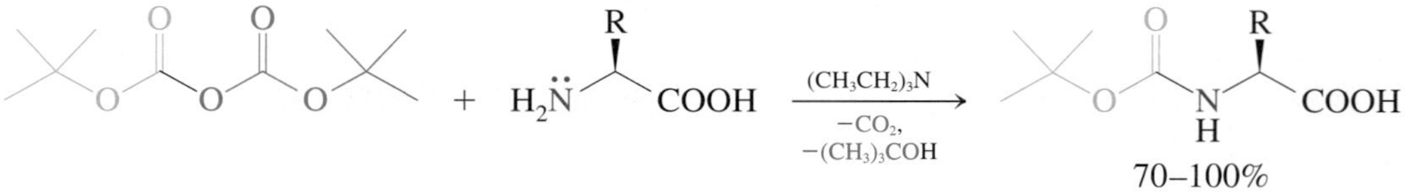

Bis(1,1-dimethylethyl) dicarbonate
(Di-*tert*-butyl dicarbonate)

70–100%

1,1-Dimethylethoxy-carbonylamino acid
(*tert*-Butoxycarbonylamino acid, Boc-amino acid)

Deprotection in this case is achieved by treatment with acid under conditions mild enough to leave peptide bonds untouched.

Deprotection of Boc-Amino Acids

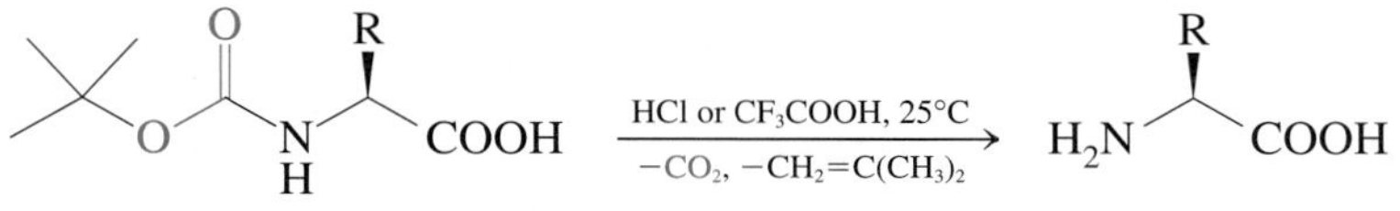

Exercise 26-17

The mechanism of the deprotection of Boc-amino acids is different from that of the normal ester hydrolysis (Section 20-4): It proceeds through the intermediate 1,1-dimethylethyl (*tert*-butyl) cation. Formulate this mechanism.

The carboxy terminus of an amino acid is protected by the formation of a simple ester, such as methyl or ethyl. Deprotection results from treatment with base. Phenylmethyl (benzyl) esters can be cleaved by hydrogenolysis under neutral conditions (Section 22-2).

Peptide bonds are formed by using carboxy activation

With the ability to protect either end of the amino acid, we can synthesize peptides selectively by coupling an amino-protected unit with a carboxy-protected one. Because the protecting groups are sensitive to acid and base, the peptide bond must be formed under the mildest possible conditions. Special carboxy-activating reagents are used.

Perhaps the most general of these reagents is **dicyclohexylcarbodiimide (DCC).** The electrophilic reactivity of this molecule is similar to that of an isocyanate (Section 20-7); it is ultimately hydrated to N,N'-dicyclohexylurea.

Peptide Bond Formation with Dicyclohexylcarbodiimide (DCC)

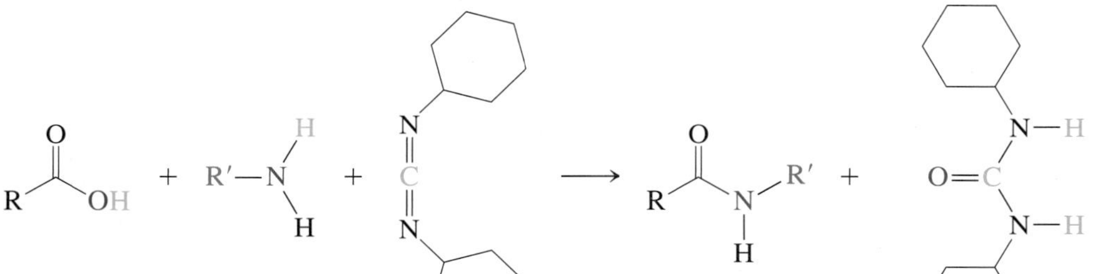

Dicyclohexyl-
carbodiimide (DCC)

N,N'-Dicyclohexylurea

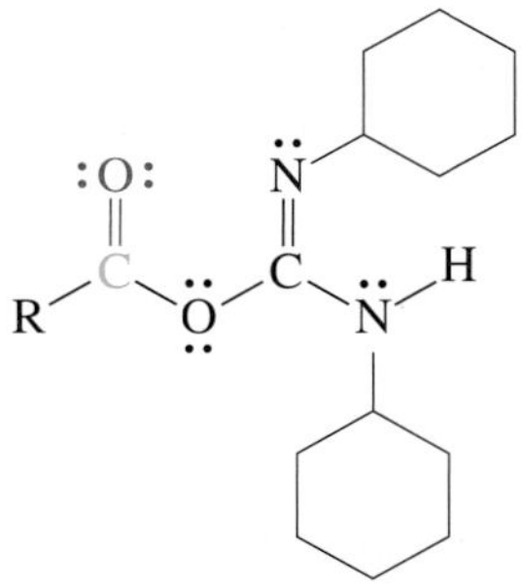

An *O*-acyl isourea

The role of DCC is to activate the carbonyl group of the acid to nucleophilic attack by the amine. This activation arises from the formation of an ***O*-acyl isourea,** in which the carbonyl group possesses reactivity similar to that in an anhydride (Section 20-3).

Armed with this knowledge, let us solve the problem of the synthesis of glycylalanine. We add amino-protected glycine to an alanyl ester in the presence of DCC. The resulting product is then deprotected to give the desired dipeptide.

Preparation of Gly-Ala

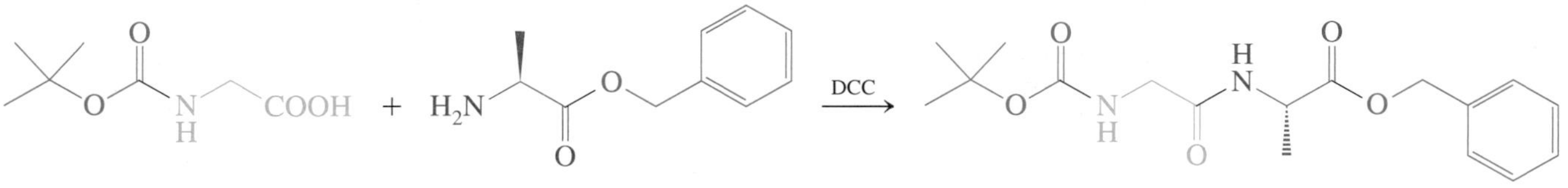

Boc-Gly   Ala-OCH₂C₆H₅   Boc-Gly-Ala-OCH₂C₆H₅

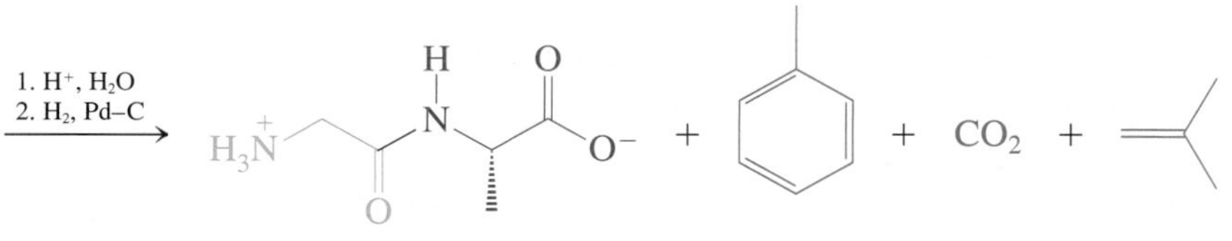

Gly-Ala

For the preparation of a higher peptide, deprotection of only one end is required, followed by renewed coupling, and so forth.

Solved Exercise 26-18 | Working with the Concepts: Writing Mechanisms of DCC Coupling Reactions

The mechanism of peptide coupling with DCC proceeds through the *O*-acyl isourea shown in this section. What happens next depends on reaction conditions: In some cases amide formation occurs by direct attack of amine; in other cases a carboxylic anhydride is an intermediate, which is then attacked by the amine to generate the amide product. Formulate the two competing pathways.

Strategy
Since the *O*-acyl isourea is an analog of an anhydride, you should review the mechanisms of nucleophilic attack on this functional group (Section 20-3). We can then extrapolate to the *O*-acyl isourea, using either the amine or the carboxylic acid as the respective nucleophiles.

Solution
- The *O*-acyl isourea is activated by protonation at the imine nitrogen. This site is unusually basic because of resonance stabilization in the protonated species, not unlike the situation encountered in the case of the guanidino group (Section 26-1). The resulting protonated urea substituent is an excellent leaving group. We can then write an addition–elimination sequence for the attack of the amine on the carbonyl carbon to eventually furnish the amide.

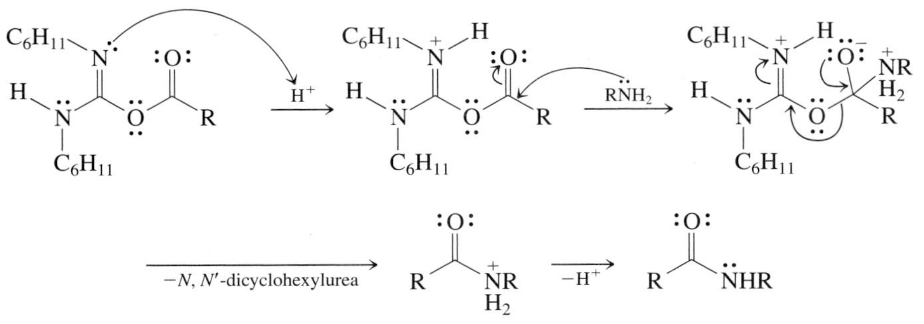

- The same protonated *O*-acyl isourea as above can also be attacked by carboxylate ion. Addition–elimination results in the carboxylic anhydride.

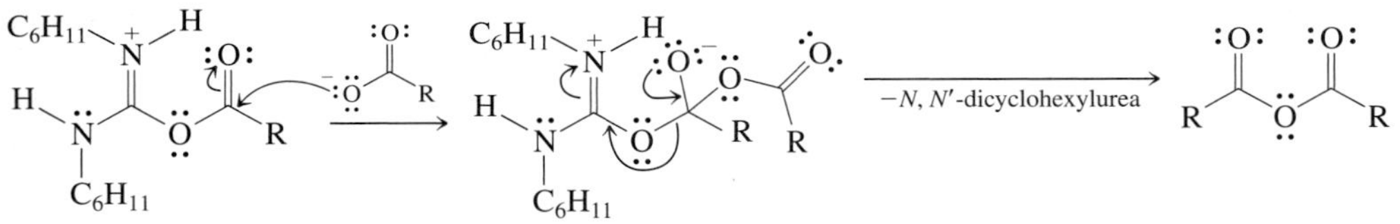

- The carboxylic anhydride is attacked by the amine through the usual addition–elimination mechanism to produce the amide (Section 20-3).

Exercise 26-19 | Try It Yourself

DCC is also used for the synthesis of esters from carboxylic acids and alcohols. Write a mechanism.

Exercise 26-20 |

Propose a synthesis of Leu-Ala-Val from the component amino acids.

In Summary Polypeptides are made by coupling an amino-protected amino acid with another in which the carboxy end is protected. Typical protecting groups are readily cleaved esters and related functions. Coupling proceeds under mild conditions with dicyclohexylcarbodiimide as a dehydrating agent.

26-7 | MERRIFIELD SOLID-PHASE PEPTIDE SYNTHESIS

Polypeptide synthesis has been automated. This ingenious method, known as the **Merrifield* solid-phase peptide synthesis,** uses a solid support of polystyrene to anchor a peptide chain.

Polystyrene is a polymer (Section 12-15) whose subunits are derived from ethenylbenzene (styrene). Although beads of polystyrene are insoluble and rigid when dry, they swell considerably in certain organic solvents, such as dichloromethane. The swollen material allows reagents to move in and out of the polymer matrix easily. Thus, its phenyl groups may be functionalized by electrophilic aromatic substitution. For peptide synthesis, a form of Friedel-Crafts alkylation is used to chloromethylate a few percent of the phenyl rings in the polymer.

Electrophilic Chloromethylation of Polystyrene

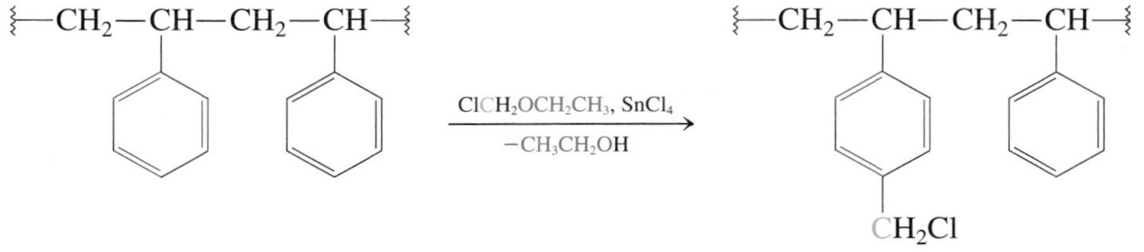

Polystyrene   **Functionalized polystyrene**

Exercise 26-21

Formulate a plausible mechanism for the chloromethylation of the benzene rings in polystyrene. (**Hint:** Review Section 15-11.)

A dipeptide synthesis on chloromethylated polystyrene proceeds as follows.

Solid-Phase Synthesis of a Dipeptide

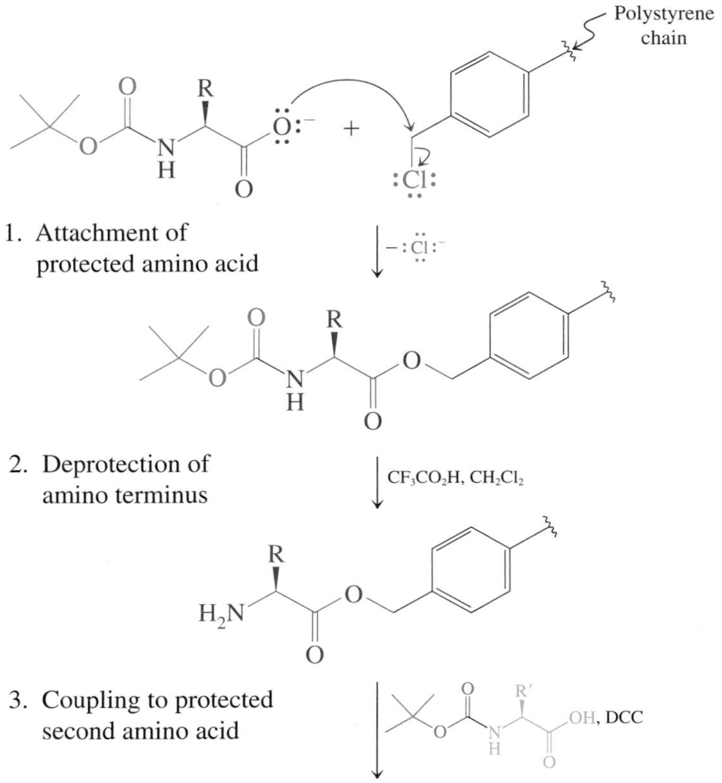

1. Attachment of protected amino acid

2. Deprotection of amino terminus

3. Coupling to protected second amino acid

Animation

ANIMATED MECHANISM:
Merrified synthesis of peptides

---

*Professor Robert B. Merrifield (1921–2006), Rockefeller University, New York, Nobel Prize 1984 (chemistry).

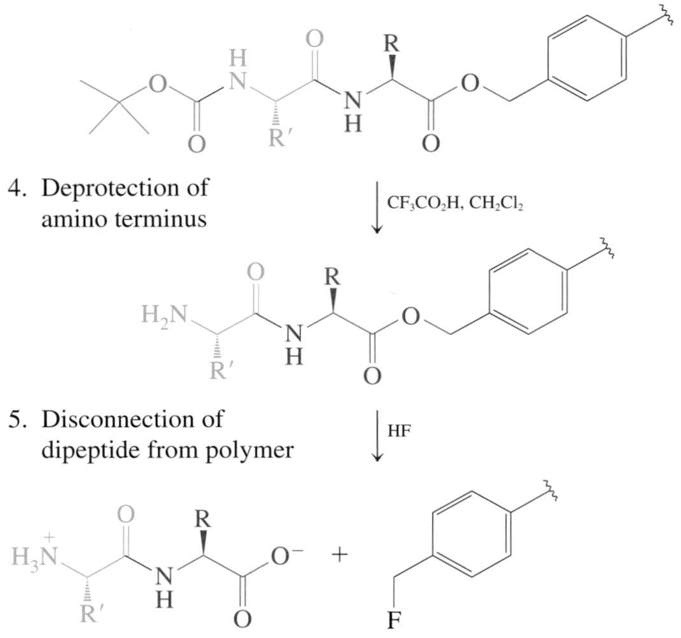

4. Deprotection of amino terminus

CF_3CO_2H, CH_2Cl_2

5. Disconnection of dipeptide from polymer

HF

First an amino-protected amino acid is anchored on the polystyrene by nucleophilic substitution of the benzylic chloride by carboxylate. Deprotection is then followed by coupling with a second amino-protected amino acid. Renewed deprotection and final removal of the dipeptide by treatment with hydrogen fluoride complete the sequence.

The great advantage of solid-phase synthesis is the ease with which products can be isolated. Because all the intermediates are immobilized on the polymer, the products can be purified by simple filtration and washing.

Obviously, it is not necessary to stop at the dipeptide stage. Repetition of the deprotection–coupling sequence leads to larger and larger peptides. Merrifield designed a machine that would carry out the required series of manipulations automatically, each cycle requiring only a few hours. In this way, the first total synthesis of the protein insulin was accomplished. More than 5000 separate operations were required to assemble the 51 amino acids in the two separate chains; thanks to the automated procedure, this took only several days.

Automated protein synthesis has opened up exciting possibilities. First, it is used to confirm the structure of polypeptides that have been analyzed by chain degradation and sequencing. Second, it can be used to construct synthetic proteins that might be more active and more specific than natural ones. Such proteins could be invaluable in the treatment of disease or in understanding biological function and activity.

In Summary Solid-phase synthesis is an automated procedure in which a carboxy-anchored peptide chain is built up from amino-protected monomers by cycles of coupling and deprotection.

26-8 POLYPEPTIDES IN NATURE: OXYGEN TRANSPORT BY THE PROTEINS MYOGLOBIN AND HEMOGLOBIN

Two natural polypeptides function as oxygen carriers in vertebrates: the proteins myoglobin and hemoglobin. Myoglobin is active in the muscle, where it stores oxygen and releases it when needed. Hemoglobin is contained in red blood cells and facilitates oxygen transport. Without its presence, blood would be able to absorb only a fraction (about 2%) of the oxygen needed by the body.

How is the oxygen bound in these proteins? The secret of the oxygen-carrying ability of myoglobin and hemoglobin is a special nonpolypeptide unit, called a **heme group**, attached to the protein. Heme is a cyclic organic molecule (called a **porphyrin**) made of

The Aroma of Fried Steak

Broiling meat, baking bread, and roasting coffee all initiate chemistry between sugars and amino acids called the Maillard reaction. This reaction, with its characteristic browning and release of aromas, is distinct from caramelization (p. 1099, margin), which involves only the decomposition of sucrose. The Maillard reaction begins with imine formation between the amino group of the peptide and the carbonyl group of the sugar, followed by dehydrations, fragmentations, isomerization through enols, and polymerizations. The resulting products contain nitrogen or sulfur (derived from cysteine). Some of the familiar odors are caused by the compounds shown below.

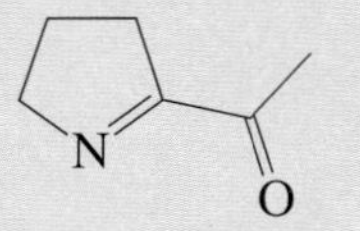

1-(1-Azacyclopent-1-en-2-yl)-1-ethanone (Bread, roasted meat and fish flavor)

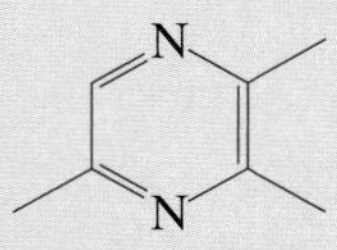

2,3,5-Trimethyl-1,4-diazabenzene (Roasted meat and fish, soy flavor)

Porphine

Heme

Figure 26-7 Porphine is the simplest porphyrin. Note that the system forms an aromatic ring of 18 delocalized π electrons (indicated in red). A biologically important porphyrin is the heme group, responsible for binding oxygen. Two of the bonds to iron are dative (coordinate covalent), indicated by arrows.

four linked, substituted pyrrole units surrounding an iron atom (Figure 26-7). The complex is red, giving blood its characteristic color.

The iron in the heme is attached to four nitrogens but can accommodate two additional groups above and below the plane of the porphyrin ring. In myoglobin, one of these groups is the imidazole ring of a histidine unit attached to one of the α-helical segments of the protein (Figure 26-8A). The other is most important for the protein's function—bound oxygen. Close to the oxygen-binding site is a second imidazole of a histidine unit, which protects this side of the heme by steric hindrance. For example, carbon monoxide, which also binds to the iron in the heme group, and thus blocks oxygen transport, is prevented from binding as strongly as it normally would because of the presence of the second imidazole group. Consequently, CO poisoning can be reversed by administering oxygen to a person who has been exposed to the gas. The two imidazole substituents in the neighborhood of the iron atom in the heme group are brought into close proximity by the unique folding pattern of the protein. The rest of the polypeptide chain serves as a mantle, shielding and protecting the active site from unwanted intruders and controlling the kinetics of its action (Figure 26-8B and C).

Myoglobin and hemoglobin offer excellent examples of the four structural levels in proteins. The primary structure of myoglobin consists of 153 amino acid residues of known sequence. Myoglobin has eight α-helical segments that constitute its secondary structure, the longest having 23 residues. The tertiary structure has the bends that give myoglobin its three-dimensional shape.

Fresh meat rapidly assumes a dark, sometimes grayish appearance. However, when it is treated with carbon monoxide, it retains its bright red color, a feature that appeals to consumers. The reason is the formation of a stable Fe–CO complex of the heme in hemoglobin.

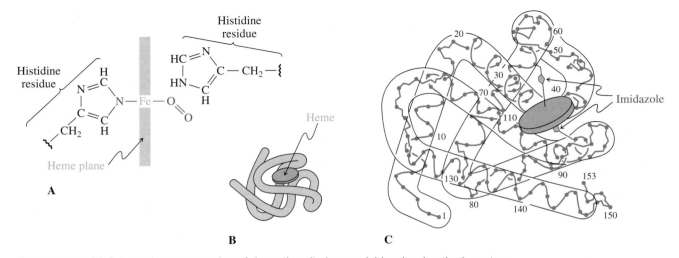

Figure 26-8 (A) Schematic representation of the active site in myoglobin, showing the iron atom in the heme plane bound to a molecule of oxygen and to the imidazole nitrogen atom of one histidine residue. (B) Schematic representation of the tertiary structures of myoglobin and its heme. (C) Secondary and tertiary structure of myoglobin. (After "The Hemoglobin Molecule," by M. F. Perutz, *Scientific American,* November 1964. Copyright © 1964, Scientific American, Inc.)

Figure 26-9 The quaternary structure of hemoglobin. Each α and β chain has its own heme group. (After R. E. Dickerson and I. Geis, 1969, *The Structure and Action of Proteins,* Benjamin-Cummings, p. 56. Copyright © 1969 by Irving Geis.)

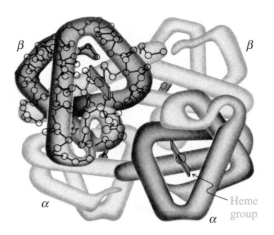

Hemoglobin contains four protein chains: two α chains of 141 residues each, and two β chains of 146 residues each. Each chain has its own heme group and a tertiary structure similar to that of myoglobin. There are many contacts between the chains; in particular, α_1 is closely attached to β_1, as is α_2 to β_2. These interactions give hemoglobin its quaternary structure (Figure 26-9).

The folding of the hemoglobin and myoglobin of several living species is strikingly similar even though the amino acid sequences differ. This finding implies that this specific tertiary structure is an optimal configuration around the heme group. The folding allows the heme to absorb oxygen as it is introduced through the lung, hang on to it as long as necessary for transport, and release it when required.

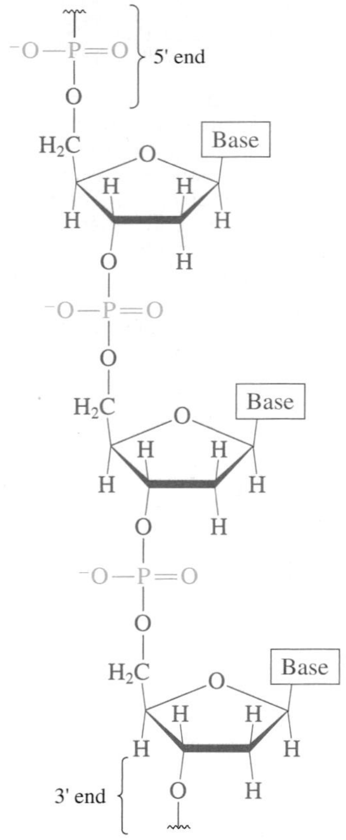

26-9 | BIOSYNTHESIS OF PROTEINS: NUCLEIC ACIDS

How does nature assemble proteins? The answer to this question is based on one of the most exciting discoveries in science, the nature and workings of the genetic code. All hereditary information is embedded in the **deoxyribonucleic acids (DNA).** The expression of this information in the synthesis of all proteins, including the many enzymes necessary for cell function, is carried out by the **ribonucleic acids (RNA).** After the carbohydrates and polypeptides, the nucleic acids are the third major type of biological polymer. This section describes their structure and function.

Four heterocycles define the structure of nucleic acids

Considering the structural diversity of natural products, the structures of DNA and RNA are simple. All their components, called **nucleotides,** are polyfunctional, and it is one of the wonders of nature that evolution has eliminated all but a few specific combinations. Nucleic acids are polymers in which phosphate units link sugars, which bear various heterocyclic nitrogen **bases** (Figure 26-10).

In DNA, the sugar units are 2-deoxyriboses, and only four bases are present: **cytosine (C), thymine (T), adenine (A),** and **guanine (G).** The sugar characteristic of RNA is ribose, and again there are four bases, but the nucleic acid incorporates **uracil (U)** instead of thymine.

Figure 26-10 Part of a DNA chain. The base is a nitrogen heterocycle. The sugar is 2-deoxyribose.

Nucleic Acid Sugars and Bases

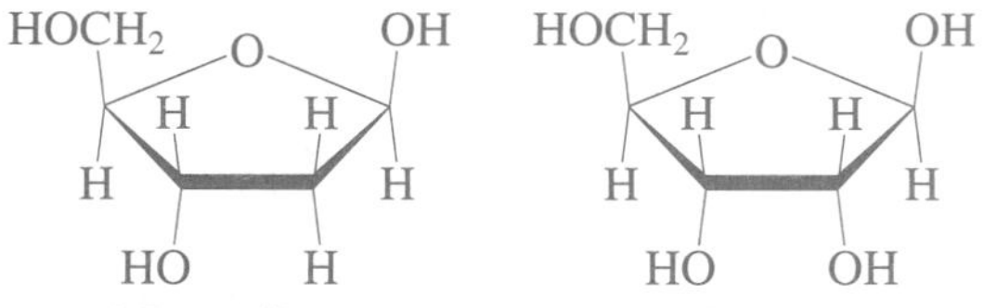

2-Deoxyribose   Ribose

Cytosine (C)   **Thymine (T)**   **Adenine (A)**   **Guanine (G)**   **Uracil (U)**

Exercise 26-22

Even though the preceding structures do not (except for adenine) indicate so, cytosine, thymine, guanine, and uracil are aromatic, albeit somewhat less so than the corresponding azapyridines. Explain. (**Hint:** Recall the discussions about amide resonance in Sections 20-1 and 26-4 and Worked Example 25-29.)

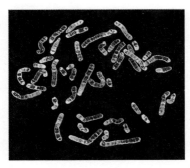

Human male chromosomes, the structures in the cell that carry the genetic information encoded in their component DNA.

We construct a nucleotide from three components. First, we replace the hydroxy group at C1 in the sugar with one of the base nitrogens. This combination is called a **nucleoside.** Second, a phosphate substituent is introduced at C5. In this way, we obtain the four nucleotides of both DNA and RNA. The positions on the sugars in nucleosides and nucleotides are designated $1'$, $2'$, and so forth, to distinguish them from the carbon atoms in the nitrogen heterocycles.

Adenosine
(A nucleoside)

Thymidine
(A nucleoside)

Nucleotides of DNA

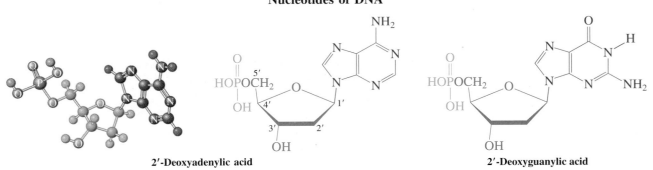

2′-Deoxyadenylic acid

2′-Deoxyguanylic acid

2′-Deoxycytidylic acid

Thymidylic acid

Nucleotides of RNA

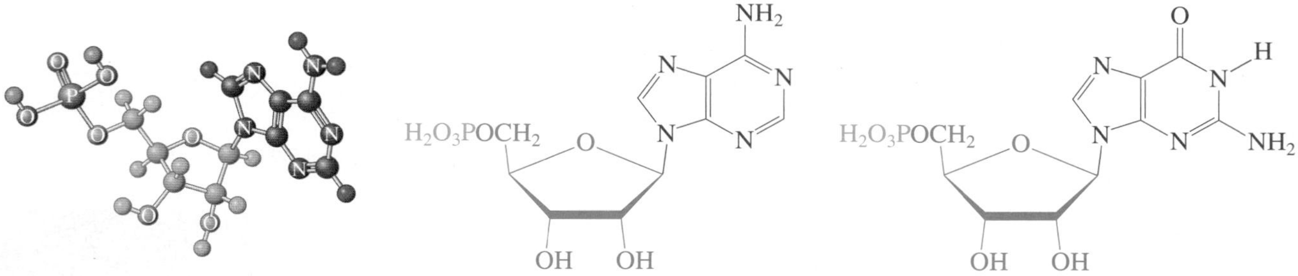

Adenylic acid

Guanylic acid

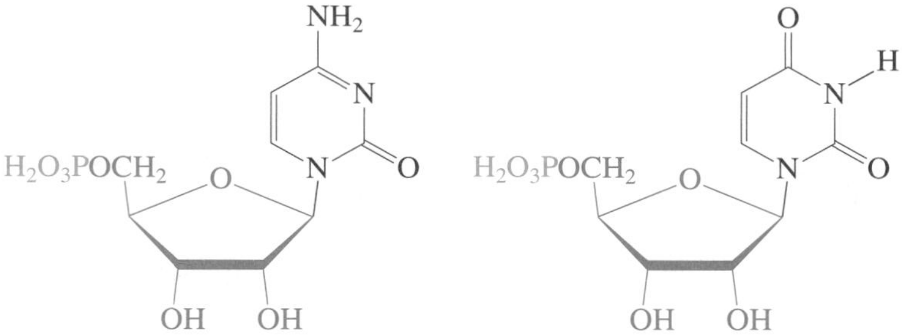

Cytidylic acid

Uridylic acid

The polymeric chain shown in Figure 26-10 is then readily derived by repeatedly forming a phosphate ester bridge from C5′ (called the **5′ end**) of the sugar unit of one nucleotide to C3′ (the **3′ end**) of another (Figure 26-10).

In this polymer, the bases adopt the same role as that of the 2-substituent in the amino acids of a polypeptide: Their sequence varies from one nucleic acid to another and determines the fundamental biological properties of the system. Viewed from the perspective of storing information, polypeptides do so by using an amide polymer backbone along which differing side chains are attached, like the letters in a word. In the case of polypeptides there are 20 such letters, for the 20 natural amino acids (Table 26-1). The same goal is achieved in nucleic acids through a sugar-phosphate polymeric array that bears a sequence of amine bases, except that now we have only four letters: C, T, A, G for DNA and C, U, A, G for RNA.

Information Storage in Polypeptides and Nucleic Acids

A polypeptide

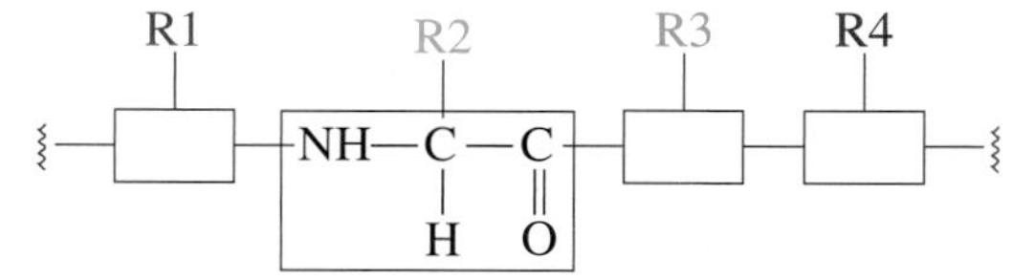

A nucleic acid

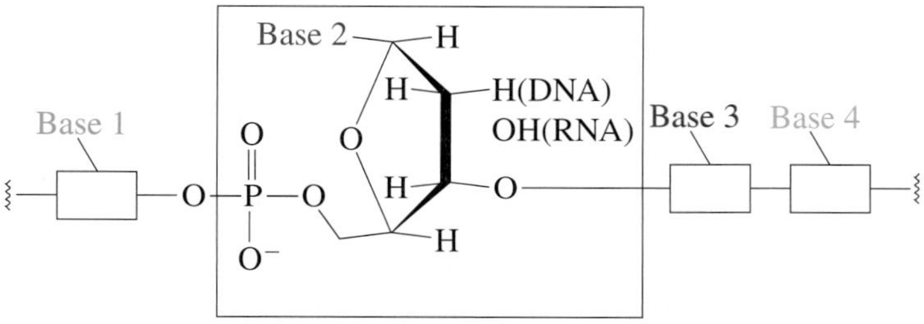

REAL LIFE: MEDICINE 26-3 | Synthetic Nucleic Acid Bases and Nucleosides in Medicine

5-Fluorouracil
(Fluracil)

9-[(2-Hydroxyethoxy)methyl]guanine
(Acyclovir)

3'-Azido-3'-deoxythymidine
(Zidovudine, or AZT)

1-β-D-Ribofuranosyl-1,2,4,-triazole-3-carboxamide
(Ribavirin)

The central role played by nucleic acid replication in biology has been exploited in medicine. Many hundreds of synthetically modified bases and nucleosides have been prepared and their effects on nucleic acid synthesis investigated. Some of them in clinical use include 5-fluorouracil (fluracil), an anticancer agent, 9-[(2-hydroxyethoxy)methyl]guanine (acyclovir), which is active against two strains of herpes simplex virus, 3'-azido-3'-deoxythymidine (zidovudine, or AZT), the first approved drug (in 1987) that combats the AIDS virus, and 1-β-D-ribofuranosyl-1,2,4-triazole-3-carboxamide (ribavirin), used to treat hepatitis C and respiratory viral infections.

Substances such as these may interfere with nucleic acid replication by masquerading as legitimate nucleic acid building blocks. The enzymes associated with this process are fooled into incorporating the drug molecule, and the synthesis of the biological polymer cannot continue. For example, the base thymine is derived biosynthetically by

methylation of uracil at C5 catalyzed by the enzyme thymidylate synthetase. The activity of 5-fluorouracil is derived from its ability to act as an impostor of uracil in the mechanism of this reaction. The first step in this mechanism is the conjugate addition (Section 18-9) of the mercapto group of a cysteine unit in the enzyme's backbone to deoxyuridylic acid. The ensuing enolate then undergoes a Michael addition (Section 18-11) to the iminium form of the biological methylating agent 5,10-methylenetetrahydrofolate (Problems 69 and 70 of Chapter 9; Real Life 25-3). This adduct subsequently falls apart by a sequence that starts with deprotonation at the original enolate carbon (red H), hydride transfer, and elimination of the mercapto group to generate the products deoxythymidylic acid and dihydrofolic acid. With fluorine in place of the hydrogen, this deprotonation is foiled, and the synthesis of the thymidine nucleotide necessary for DNA replication sabotaged, resulting in (cancer) cell death.

Melamine Toxicity and Multiple Hydrogen Bonding

The illicit adulteration of pet food milk imported from China with melamine (to make it appear nitrogen = protein rich) in 2007 resulted in the numerous deaths of dogs, cats, and, especially tragic, also infants. The cause of death was renal failure. It turned out that, while melanine itself is relatively nontoxic, it forms three strong hydrogen bonds to a by-product of its synthesis, cyanuric acid (see Problem 43 of Chapter 25). The complex accumulates in the kidneys and precipitates as obstructing large crystals.

2,4,6-Triamino-1,3,5-triazabenzene (Melamine)   1,3,5-Triazacyclo-hexane-2,4,6-trione (Cyanuric acid)

Figure 26-11 Hydrogen bonding between the base pairs adenine–thymine and guanine–cytosine. The components of each pair are always present in equal amounts. The electrostatic potential maps of the respective pairs adenine–thymine and guanine–cytosine show how regions of opposite charges (red and blue) come to lie face to face.

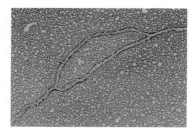

Transmission electron microscope picture of replicating DNA. Unwinding of the two complementary strands generates a "bubble" that enlarges to form a Y-shaped molecule termed a replication fork.

Nucleic acids form a double helix

Nucleic acids, especially DNA, can form extraordinarily long chains (as long as several centimeters) with molecular weights of as much as 150 billion. Like proteins, they adopt secondary and tertiary structures. In 1953, Watson and Crick* made their ingenious proposal that DNA is a double helix composed of two strands with complementary base sequences. A crucial piece of information was that, in the DNA of various species, the ratio of adenine to thymine, like that of guanine to cytosine, is always one to one. Watson and Crick concluded that two chains are held together by hydrogen bonding in such a way that adenine and guanine, respectively, in one chain always face thymine and cytosine in the other, and vice versa (Figure 26-11). Thus, if a piece of DNA in one strand has the sequence -A-G-C-T-A-C-G-A-T-C-, this entire segment is hydrogen bonded to a complementary strand running in the opposite direction, -T-C-G-A-T-G-C-T-A-G-, as shown.

$$\text{——A—G—C—T—A—C—G—A—T—C——}$$
$$\text{——T—C—G—A—T—G—C—T—A—G——}$$

Because of other structural constraints, the arrangement that maximizes hydrogen bonding and minimizes steric repulsion is the double helix (Figure 26-12). The cumulative hydrogen-bonding energies involved are substantial considering that just single base pairing, as in Figure 26-11, is favorable by 5.8 kcal mol^{-1} (24 kJ mol^{-1}) for G–C and 4.3 kcal mol^{-1} (18 kJ mol^{-1}) for A–T.

Adenine–thymine   Guanine–cytosine

DNA replicates by unwinding and assembling new complementary strands

There is no restriction on the variety of sequences of the bases in the nucleic acids. Watson and Crick proposed that the specific base sequence of a particular DNA contained all genetic information necessary for the duplication of a cell and, indeed, the growth and development of the organism as a whole. Moreover, the double-helical structure suggested a way in which DNA might **replicate**—make exact copies of itself—and so pass on the genetic code. In this mechanism, each of the two strands of DNA functions as a template. The double helix partly unwinds, and enzymes called DNA polymerases then begin to assemble the new DNA by coupling nucleotides to one another in a sequence complementary to that in the template, always juxtaposing C to G and A to T (Figure 26-13). Eventually, two complete double helices are produced from the original. This process is at work throughout the entire human genetic material, or **genome**—some 2.9 billion base pairs—with an error frequency of less than 1 in 10 billion base pairs.

*Professor James D. Watson (b. 1928), Harvard University, Cambridge, Massachusetts, Nobel Prize 1962 (medicine); Professor Sir Francis H. C. Crick (1916–2004), Cambridge University, England, Nobel Prize 1962 (medicine).

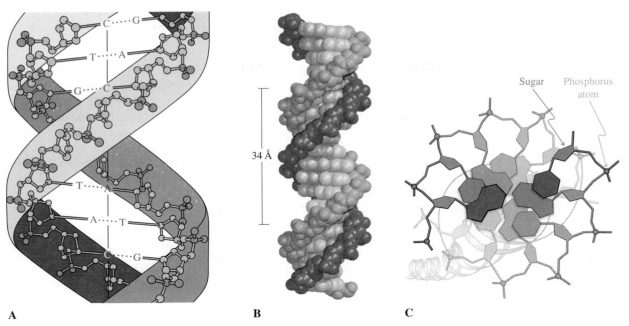

A   B   C

Figure 26-12 (A) The two nucleic acid strands of a DNA double helix are held together by hydrogen bonding between the complementary sets of bases. Note: The two chains run in opposite directions and all the bases are on the inside of the double helix. The diameter of the helix is 20 Å; base-base separation across the strands is ~3.4 Å; the helical turn repeats every 10 residues, or 34 Å. (B) Space-filling model of DNA double helix (green and red strands). In this picture, the bases are shown in lighter colors than the sugar-phosphate backbone. (C) The DNA double helix, in a view down the axis of the molecule. The color scheme in (A) and (C) matches that in Figure 26-10. (After *Biochemistry*, 6th ed., by Jeremy M. Berg, John L. Tymoczko, and Lubert Stryer, W. H. Freeman and Company. Copyright © 1975, 1981, 1988, 1995, 2002, 2007.)

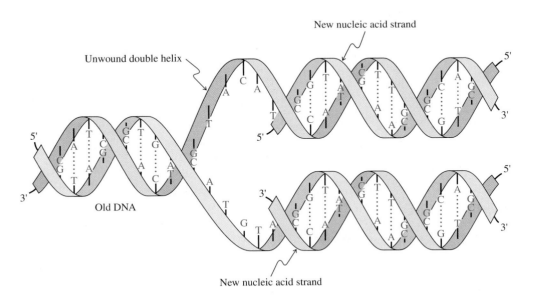

Figure 26-13 Simplified model of DNA replication. The double helix initially unwinds to two single strands, each of which is used as a template for reconstruction of the complementary nucleic acid sequence.

In Summary The nucleic acids DNA and RNA are polymers containing monomeric units called nucleotides. There are four nucleotides for each, varying only in the structure of the base: cytosine (C), thymine (T), adenine (A), and guanine (G) for DNA; cytosine, uracil (U), adenine, and guanine for RNA. The two nucleic acids differ also in the identity of the sugar unit: deoxyribose for DNA, ribose for RNA. DNA replication and RNA synthesis from DNA is facilitated by the complementary character of the base pairs A–T, G–C, and A–U. The double helix partly unwinds and functions as a template for replication.

26-10	PROTEIN SYNTHESIS THROUGH RNA

The mechanism of duplicating the entire nucleotide sequence in DNA replication is used by nature and by chemists to obtain partial copies of the genetic code for various purposes. In nature, the most important application is the assembly of RNA, called **transcription,** which transcribes the parts of the DNA that contain the information (the **genes**) necessary to synthesize proteins in the cell. The process by which this transcribed information is decoded and used to construct proteins is called **translation.** The three key players in protein synthesis are the "DNA transcript" **messenger RNA (mRNA),** the "delivery unit" for the specific amino acids to be connected by peptide bonds, **transfer RNA (tRNA),** and the catalyst that enables amide bond formation—the **ribosome.**

Protein synthesis starts with mRNA, a transcript of a piece of a single strand of partly unwound DNA (Figure 26-14). Its chain is much shorter than that of DNA, and it does not stay bound to the DNA but breaks away as its synthesis is finished. The mRNA is the template responsible for the correct sequencing of the amino acid units in proteins. How does mRNA do that? Each sequence of three bases, called a **codon,** specifies a particular amino acid (Table 26-3). Simple permutation of this three-base code with a total of four bases gives $4^3 = 64$ possible distinct sequences. That number is more than enough, because only 20 different amino acids are needed for protein synthesis. This might seem like overkill, but consider that the next lower alternative—namely, a two-base code—would give only $4^2 = 16$ combinations, too few for the number of different amino acids found in natural proteins.

Table 26-3	Three-Base Code for the Common Amino Acids Used in Protein Synthesis				
Amino acid	**Base sequence**	**Amino acid**	**Base sequence**	**Amino acid**	**Base sequence**
Ala (A)	GCA	His (H)	CAC	Ser (S)	AGC
	GCC		CAU		AGU
	GCG				UCA
	GCU	Ile (I)	AUA		UCG
			AUC		UCC
Arg (R)	AGA		AUU		UCU
	AGG				
	CGA	Leu (L)	CUA	Thr (T)	ACA
	CGC		CUC		ACC
	CGG		CUG		ACG
	CGU		CUU		ACU
			UUA		
Asn (N)	AAC		UUG	Trp (W)	UGG
	AAU				
		Lys (K)	AAA	Tyr (Y)	UAC
Asp (D)	GAC		AAG		UAU
	GAU				
		Met (M)	AUG	Val (V)	GUA
Cys (C)	UGC				GUG
	UGU	Phe (F)	UUU		GUC
			UUC		GUU
Gln (Q)	CAA				
	CAG	Pro (P)	CCA	Chain initiation	AUG
			CCC		
Glu (E)	GAA		CCG	Chain termination	UGA
	GAG		CCU		UAA
					UAG
Gly (G)	GGA				
	GGC				
	GGG				
	GGU				

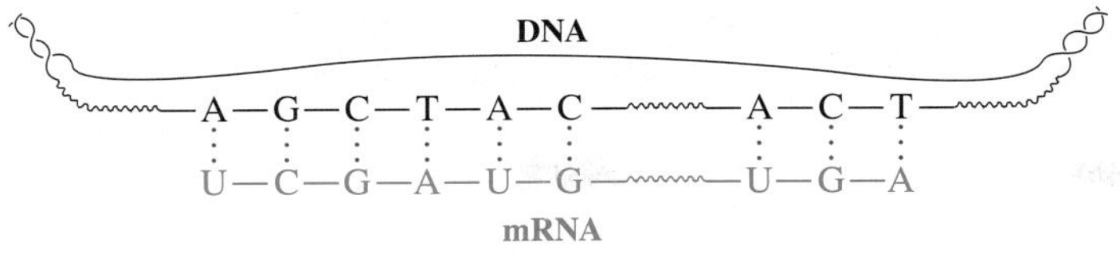

DNA

—A—G—C—T—A—C——————A—C—T——————
‖ ‖ ‖ ‖ ‖ ‖   ‖ ‖ ‖
U—C—G—A—U—G——————U—G—A

mRNA

Figure 26-14 Simplified picture of messenger RNA synthesis from a single strand of (partly unwound) DNA.

Codons do not overlap; in other words, the three bases specifying one amino acid are not part of another preceding or succeeding codon. Moreover, the "reading" of the base sequence is consecutive; each codon immediately follows the next, uninterrupted by genetic "commas" or "hyphens." Nature also makes full use of all 64 codons by allowing for several of them to describe the same amino acid (Table 26-3). Only tryptophan and methionine are characterized by single three-base codes. Some codons act as signals to initiate or terminate production of a polypeptide chain. Note that the initiator codon (AUG) is also the codon for methionine. Thus, if the codon AUG appears *after* a chain has been initiated, methionine will be produced. The complete base sequence of the DNA in a cell defines its **genetic code.**

Mutations in the base sequence of DNA can be caused by physical (radiation) or chemical (carcinogens; see, e.g., Section 16-7) interference. Mutations can either replace one base with another or can add or delete one base or more. Here lies some of the potential value of redundant codons. If, for example, the RNA sequence CCG (proline) were changed as a result of a DNA mutation to the RNA sequence CCC, proline would still be correctly synthesized.

Exercise 26-23

What is the sequence of an mRNA molecule produced from a DNA template strand with the following composition?

5'-ATTGCTCAGCTA-3'

Exercise 26-24

(a) What amino acid sequence is encoded in the following mRNA base sequence (starting at the left end)?

A-A-G-U-A-U-G-C-A-U-C-A-U-G-C-U-U-A-A-G-C

(b) Identify the mutation that would have to occur for Trp to be present in the resulting peptide.

With a copy of the requisite codons in hand, proteins are then synthesized along the mRNA template with the help of a set of other important nucleic acids, the tRNAs. These molecules are relatively small, containing from 70 to about 90 nucleotides. Each tRNA is specifically designed to carry one of the 20 amino acids for delivery to the mRNA in the course of protein buildup. The amino acid sequence encoded in mRNA is read codon by codon by a complementary three-base sequence on tRNA, called an **anticodon.** In other words, the individual tRNAs, each carrying its specific amino acid, line up along the mRNA strand in the correct order. At this stage, catalytic ribosomes (very large enzymes) containing their own RNA facilitate peptide-bond formation (Figure 26-15). As the polypeptide chain grows longer, it begins to develop its characteristic secondary and tertiary structure (α helix, pleated sheets, etc.), helped by enzymes that form the necessary disulfide bridges.

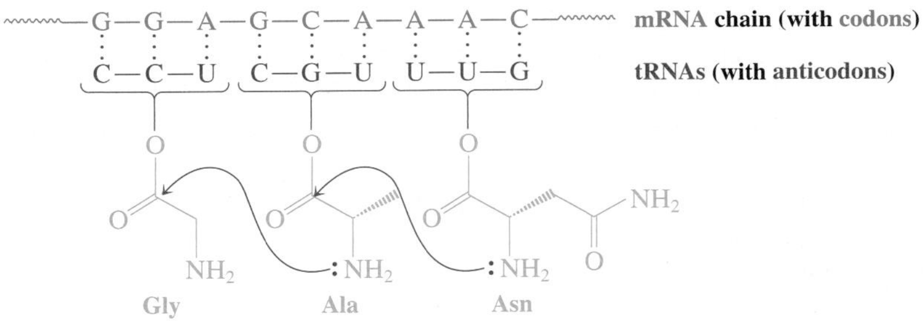

The sheep Dolly, the world's first clone of an adult mammal cell, died in 2003 at the age of 7.

—G—G—A—G—C—A—A—A—C—— mRNA **chain (with codons)**
  ‖ ‖ ‖   ‖ ‖ ‖   ‖ ‖ ‖
  C—C—U   C—G—U   U—U—G   tRNAs **(with anticodons)**

Gly   Ala   Asn

Figure 26-15 Representation of the biosynthesis of the tripeptide Gly-Ala-Asn. The tRNAs carrying their specific amino acids line up their anticodons along the codons of mRNA, before ribosomal enzymes accomplish amide linkage.

All of this happens with remarkable speed. It is estimated that a protein made up of approximately 150 amino acid residues can be biosynthesized in less than 1 minute, and that without protecting groups (Section 26-6)! Clearly, nature still has the edge over the synthetic organic chemist, at least in this domain.

In Summary RNA is responsible for protein biosynthesis; each three-base sequence, or codon, specifies a particular amino acid. Codons do not overlap, and more than one codon can specify the same amino acid.

26-11   DNA SEQUENCING AND SYNTHESIS: CORNERSTONES OF GENE TECHNOLOGY

Molecular biology is undergoing a revolution because of our capability to decipher, reproduce, and alter the genetic code of organisms. Individual genes or other DNA sequences from a genome can be copied *(cloned),* often on a large scale. Genes of higher organisms can be expressed (i.e., they can be made to start protein synthesis) in lower organisms, or they can be modified to produce "unnatural" proteins. Modified genes have been reintroduced successfully into the organism of their origin, causing changes in the physiology and biochemistry of the "starting" host. Much of this development is due to biochemical advances such as the discovery of enzymes that can selectively cut, join, or replicate DNA and RNA. For example, *restriction enzymes* cut long molecules into small defined fragments that can be joined by *DNA ligases* to other DNAs *(recombinant DNA technology). Polymerases* catalyze DNA replication, and selected DNA sequences are made on a large scale by the *polymerase chain reaction* (described later in this section). The list goes on and on, and its contents are beyond the scope of this text.

The essential foundation for these discoveries has been the knowledge of the primary structure of nucleic acids and the development of methods for their synthesis.

Rapid DNA sequencing has deciphered the human genome

The sequencing of DNA can be accomplished by using both chemical (introduced by Gilbert* with his student Maxam) and enzymatic methods, especially the Sanger[†] dideoxynucleotide protocol. In both approaches, and in analogy to protein analysis (Section 26-5), the unwieldy DNA chain is first cleaved at specific points by enzymes called **restriction endonucleases.**

There are more than 200 such enzymes, providing access to a multitude of overlapping sequences. In the Maxam–Gilbert procedure, described here in a simplified manner, the pieces of DNA obtained are then tagged at their 5′-phosphate ends (the starting end) with a radioactive phosphorus (^{32}P) label for analytical detection in the next step (oligonucleotides are colorless and cannot be identified visually). This step consists of subjecting multiple copies of the pure, labeled fragments in four separate experiments to chemicals that cleave only next to specific bases (i.e., only at A, G, T, and C, respectively). The conditions of cleavage are controlled so that cleavage occurs essentially only once on each copy, but, overall, at all possible places at which the particular base occurs. The products of the individual experiments are mixtures of oligonucleotides whose ends are known: Half of them contain the radioactive label at the beginning end, while the other half do not have any label and are ignored. The position of this end nucleotide in the original DNA is deduced from the length of the radioactive fragment, a measurement accomplished by electrophoresis (Section 26-5), which separates mixtures of oligonucleotides by size. The patterns of products that are made and identified in this way reveal the base sequence.

For a simple analogy, picture a measuring tape that has been marked randomly by spots of four different colorless labels. You can make multiple copies of this tape, and you have

---

*Professor Walter Gilbert (b. 1932), Harvard University, Nobel Prize 1980 (chemistry).
[†]Professor Frederick Sanger (b. 1918), Cambridge University, England, Nobel Prizes 1958 (chemistry; for the structure of insulin) and 1980 (chemistry; for nucleotide sequencing).

four special automatic scissors that cut the tape at only one specific label. You divide your collection of tapes into four batches and let each type of scissors do its work. The result from each experiment will be a collection of tapes of different lengths indicating where on the original a specific label occurs.

In the Sanger method, the piece of DNA to be sequenced, the *template strand,* is subjected to an enzyme, DNA polymerase, that replicates it many times (as the complementary strand, Section 26-9), starting from one end (the 3′ end). Replication requires addition of the four ingredient nucleotides (abbreviated as N below) A, G, T, and C [in the form of their reactive triphosphates (TP)], respectively, for chain buildup. The trick is to add very small amounts of a modified nucleotide to the mixture, which causes termination of chain growth at the very spot at which the normal nucleotide would have been incorporated. The modification is simple: Instead of the usual 2′-deoxyribose (d), we use 2′,3′-dideoxyribose (dd). Moreover, to visualize the end result, the base attached to the dideoxyribose is labeled with a fluorescent dye molecule that can be detected by exposure to light of the appropriate wavelength.

The Nucleotide Ingredients in the Sanger DNA Sequencing Method

"Normal" 2′-deoxyribonucleoside triphosphate (dNTP)

Fluorescent-dye-labeled 2′,3′-dideoxyribonucleoside triphosphate (ddNTP)

The stepwise procedure is as follows (Figure 26-16). In the first step, multiple copies of the template strand are made by a procedure that also adds a short known nucleotide sequence to its 3′-end. This auxiliary is necessary to bind to a complementary section of the template, making a short double strand that serves as a primer for DNA polymerase; in other words, it tells the enzyme where to start replicating (namely, where the known double strand ends and the unknown single strand starts). We then divide our supply of template strands to be sequenced into four portions and carry out polymerase reactions on each of them separately. In the first, we use dATP, dTTP, dGTP, and dCTP, in addition to 1% of ddATP, and let the reaction run its course. The polymerase will begin to synthesize a complementary strand to the template; that is, when it encounters a T, it will add an A; when it encounters a C, it will add a G, and vice versa. The key is that 1% of the time, when it finds a T, it will add a defunct ddA, lacking the OH group necessary for further chain lengthening. Replication will stop and the incomplete strand will fall off the template.

Note that, since 99% of the time T will be matched by dA and therefore replication will continue uninterrupted, all Ts along the sequence have essentially the same chance of being "hit" by ddA. In this way, all Ts are tagged and a collection of oligonucleotides of different lengths is assembled, all of which end with ddA. In step 2, this mixture is then separated by electrophoresis and the position of the bands visualized by fluorescent light emission of the dye that is part of ddA. The position of the bands gives the length of the fragment strand and therefore the positions of the nucleotide A in the replicate strand and T in the template strand.

Having established the location of all the Ts in the template strand, the same experiment is carried out three additional times, with ddT, ddG, and ddC, respectively, as the chain-stopping additives, thus giving away the position of the other three nucleotides in the sequence.

The Sanger method has been automated such that DNAs with thousands of base pairs can be sequenced reliably in a day or less. The keys to automation have been the attachment of dyes of differing color to the ddNTP building blocks, for example, red for T, green for A, yellow for G, and blue for C; the use of laser detection; parallel analysis; and the ever-increasing power of computers. The sequencing machine produces chromatographic print-outs providing the sequence under investigation on simple sheets of paper (Figure 26-17). The high demand for low-cost sequencing generated by the initiation of various whole-genome projects in the 1990s has given rise to a number of high-throughput technologies.

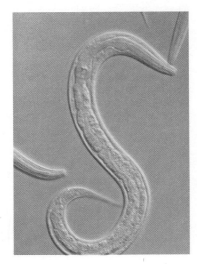

The ordinary soil-dwelling worm *Caenorhabditis elegans* has about as many genes as we do. Surprised?

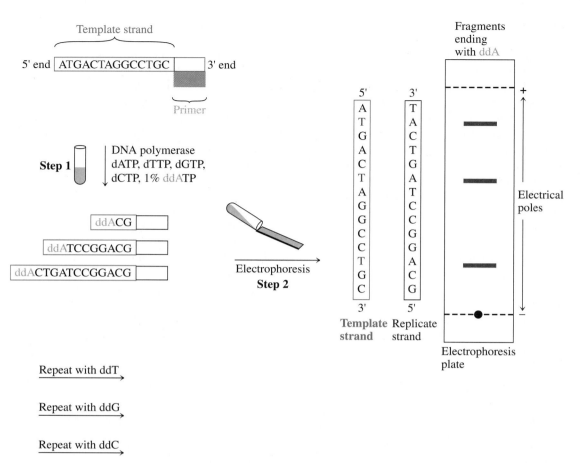

Figure 26-16 Sanger dideoxynucleotide sequencing of a sample oligonucleotide, called the template strand. The attached known primer sequence tells the enzyme DNA polymerase where to start replication. Replication stops when a 2′,3′-dideoxyribonucleotide (in the experiment shown, it is ddA) is added to the growing chain. Electrophoresis of the resulting mixture reveals the length of the fragments with the corresponding end nucleotide and, therefore, the position of this nucleotide in the replicate strand and its complement in the template strand.

The genetic code of the soybean was determined in 2008. It has about 66,000 genes, more than twice that of the human genome. Surprised again?

Here, the DNA is cleaved randomly and the individual pieces (up to 1000 bases long) are examined in parallel. Thus, millions of sequences can be determined simultaneously and the final connectivity established by computerized alignments of overlapping fragments. As a result, human DNA can now be analyzed 1–2 days, for about $1000, and bacterial genomes in a matter of hours, for <$100. The cost of analysis per million base pairs has plummeted to a few cents. These techniques have become so convenient that DNA sequencing, rather than peptide sequencing, is now the method of choice for structural elucidation of the encoded proteins (see Sections 26-5 and 26-10). Moreover, they herald the advent of a time in which disease may be tackled starting at its genome "source," by the ready detection of

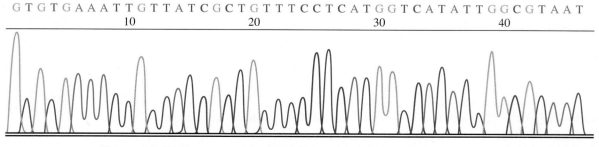

GTGTGAAATTGTTATCGCTGTTTCCTCATGGTCATATTGGCGTAAT

Figure 26-17 The color of the peaks in a printout from an automated DNA sequencer tells us the sequence of an oligonucleotide.

a mutation in a gene, or the pinpointing of abnormal base sequences affecting the on–off switches that ensure the proper functioning of our DNA.

The first genome of a live organism to be sequenced (in 1995) was that of the bacterium *Haemophilus influenzae Rd.* (which causes ear infections and meningitis), with 1.8 million base pairs (Mb). After this milestone, larger genomes were tackled, including those of yeast (12 Mb; 1996), the worm *Caenorhabditis elegans* (97 mB; 1998), and the fruit fly (160 Mb; 1999), as a preamble for the "real thing," our own blueprint of hereditary information.

In 1990, the Human Genome Project was begun—an international effort aimed at establishing the sequence of the 2.9 billion bases in human DNA and hence identifying the associated genes. The project was originally expected to take about 15 years, but thanks to the above advances in nucleotide sequencing, it was finished in April 2003. Only about 1.5% of the sequence codes for proteins (Section 26-10), which are responsible for what we are and how we function. Most of our genome (about 95%) consists of long chains of often-repeating segments that do not code for amino acids and are active in regulating gene expression, affecting cellular machinery in the reading of nearby genes and protein synthesis, and functioning as the "glue" of chromosomal structure. A big surprise is that we have only about 25,000 genes, many fewer than the expected 100,000, and a rather sobering number, considering that it is in the ballpark of that of the worm *Caenorhabditis elegans* (19,500 genes) or of the first two other mammalian genomes to be unraveled, those of the mouse (2002) and rat (2004), both with about 29,000 genes. It now appears that the quality of genes is as important as their quantity: Human complexity may arise from the fact that our genes can produce not only more, but also more sophisticated proteins, capable of fulfilling more than one function. Moreover, complexity is also dependent on the interplay of proteins, evidently more varied in humans. To shed light on how and where one human differs from the other in these respects, the 1000 Genomes Project was launched in 2008 and completed in October 2012 with the announcement of the sequencing of 1092 genomes. The analysis of the resulting data will help us understand diversity, disease, and the origin of our "humanity."

The deciphering of the human genome has sparked the development of several new research areas attempting to grapple with the enormous quantity of data becoming available on a daily basis. For example, *functional genomics* aims to find out what genes do, especially as it pertains to human health. *Comparative genomics* is establishing similarities in the genomes of a variety of species, from pigs to Japanese puffer fish.

Equally or arguably even more important is *proteomics,* the identification and functional study of the abundant new proteins that are being discovered as a consequence of the knowledge of the encoding gene. Compared to the genome, the corresponding proteome may be much more complex because proteins exist in a multitude of substitutionally "decorated" forms, such as phosphorylated, glycosylated, sulfated, and other variants. Some estimates place their number at 1 million. The Protein Structure Initiative is a U.S. national program devoted to determining the three-dimensional shapes of these molecules. Started in the year 2000 and greatly aided by robotic technology, it had determined over 5000 structures of proteins derived from plants, mice, yeast, and bacteria by 2012.

Finally, *bioinformatics* is the application of computer technology to the discovery of more deep-seated information in biological data. To use language as an analogy: The genome has provided words; now is the time to decipher sentences and meanings. Among the tasks that will occupy the attention of scientists for decades are the determination of the number of genes, their exact locations, and their regulation and function. Unraveling these aspects of genetics may help us understand an organism's susceptibility to disease; the details of gene expression and protein synthesis; the cooperation between proteins to elicit complex function; the evolutionary relationship among the species on our planet; the relationship of single point mutations to disease; and the details of multiple gene expression for the development of complex traits.

Diversity? These individuals differ by about one in every 1500 DNA bases.

Really? The human body is inhabited by ~10,000 (mainly) bacterial species, the "microbiome," residing in the gastrointestinal tract, our skin, mouth, eyes, and other areas. Bacterial cells outnumber human cells by at least 10 to 1, but because they are smaller, their combined mass is only about 1 kg. The gut alone contains about 100 trillion bacteria. The total gene pool exceeds the human genome by a factor of 100. Thus, from a cellular perspective, we are walking microbiomes. The relationship is of mutual benefit, as the microorganisms are essential in the metabolism of carbohydrates, the development of the immune system, protection against alien microbes, gut function, production of vitamins, and organization of fat storage. Perturbations of the microbiome are being recognized as being associated with disease, but little is known about its composition, distribution, and genetic makeup. The Human Microbiome Project was initiated in 2008 to explore this uncharted territory.

Like sequencing, DNA synthesis is automated

You can order your custom-made oligonucleotide today and receive it tomorrow. The reason for this efficiency is the use of automated nucleotide coupling by so-called DNA synthesizers that operate on the same principle as that of the Merrifield polypeptide synthesis

(Section 26-7): solid-phase attachment of the growing chain and the employment of protected nucleotide building blocks. Protection of the bases cytosine, adenine, and guanine is required at their amino functions (absent in thymine, which needs no protection), in the form of their amides.

Protected (Except for Thymine) DNA Bases

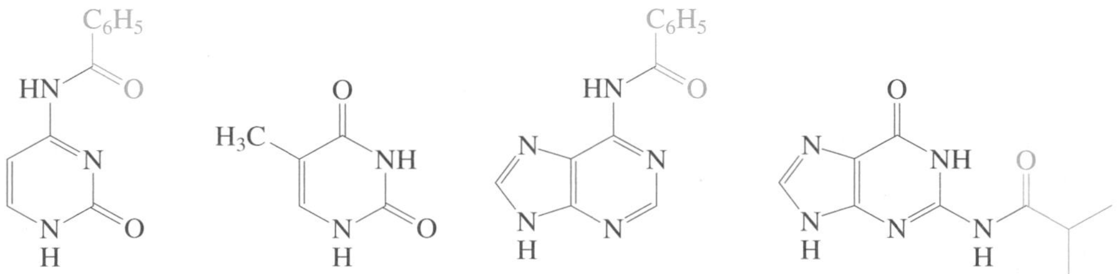

Cytosine (C)   Thymine (T)   Adenine (A)   Guanine (G)

The sugar moiety is blocked at C5′ as a dimethoxytrityl [di(4-methoxyphenyl)phenylmethyl, DMT] ether, readily cleaved by mild acid through an S_N1 mechanism (Section 22-1 and Problem 40 of Chapter 22), much like 1,1-dimethylethyl (*tert*-butyl) ethers (Section 9-8, Real Life 9-2). To anchor the first protected nucleoside on the solid support, the C3′-OH is attached to an activated linker, a diester. Unlike the Merrifield medium of polystyrene, the solid used in oligonucleotide synthesis is surface-functionalized silica (SiO_2) bearing an amino substituent as a "hook." Coupling to the anchor nucleoside is by amide formation.

Anchoring the Protected Nucleoside on SiO$_2$

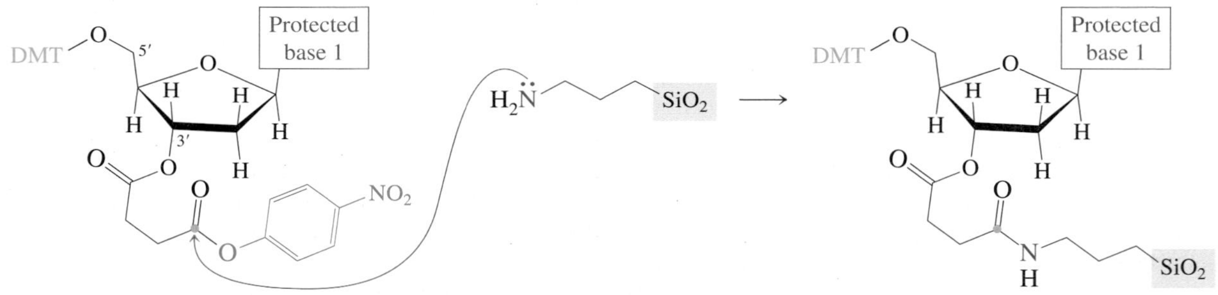

Silica-anchored first
nucleoside

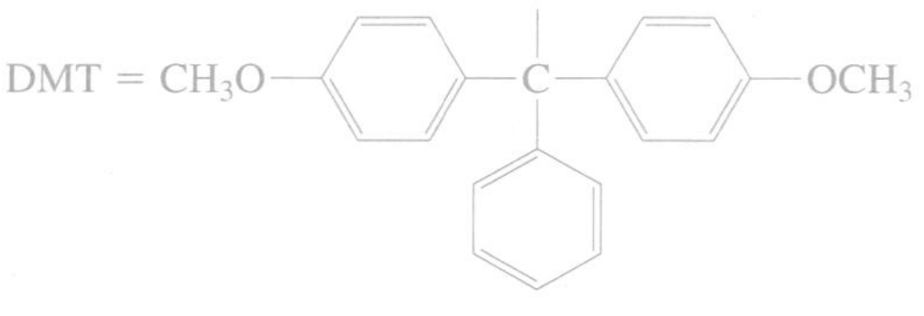

Dimethoxytrityl
[Di(4-methoxyphenyl)phenylmethyl]

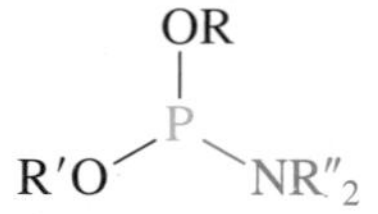

A phosphoramidite

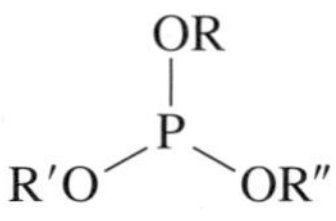

A phosphite

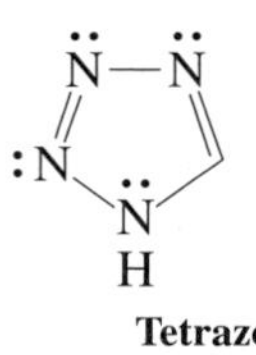

Tetrazole

With the first nucleoside in place, we are ready to attach to it the second. For this purpose, the point of attachment, the 5′-OH, is deprotected with acid. Subsequent addition of a 3′-OH activated nucleoside effects coupling. The activating group is an unusual phosphoramidite [containing P(III)], which, as we shall see shortly, also serves as a masked phosphate [P(V)] for the final dinucleotide and is subject to nucleophilic substitution, not unlike PBr_3 (recall Sections 9-4 and 19-8). The displacement reaction is catalyzed and furnishes a phosphite derivative; the catalyst is the, again unusual, aromatic heterocycle tetrazole, a tetrazacyclopentadiene related to pyrrole (Section 25-3) and imidazole (Section 26-1). Finally, the phosphorus is oxidized with iodine to the phosphate oxidation state.

Dinucleotide Synthesis: Deprotection, Coupling, and Oxidation

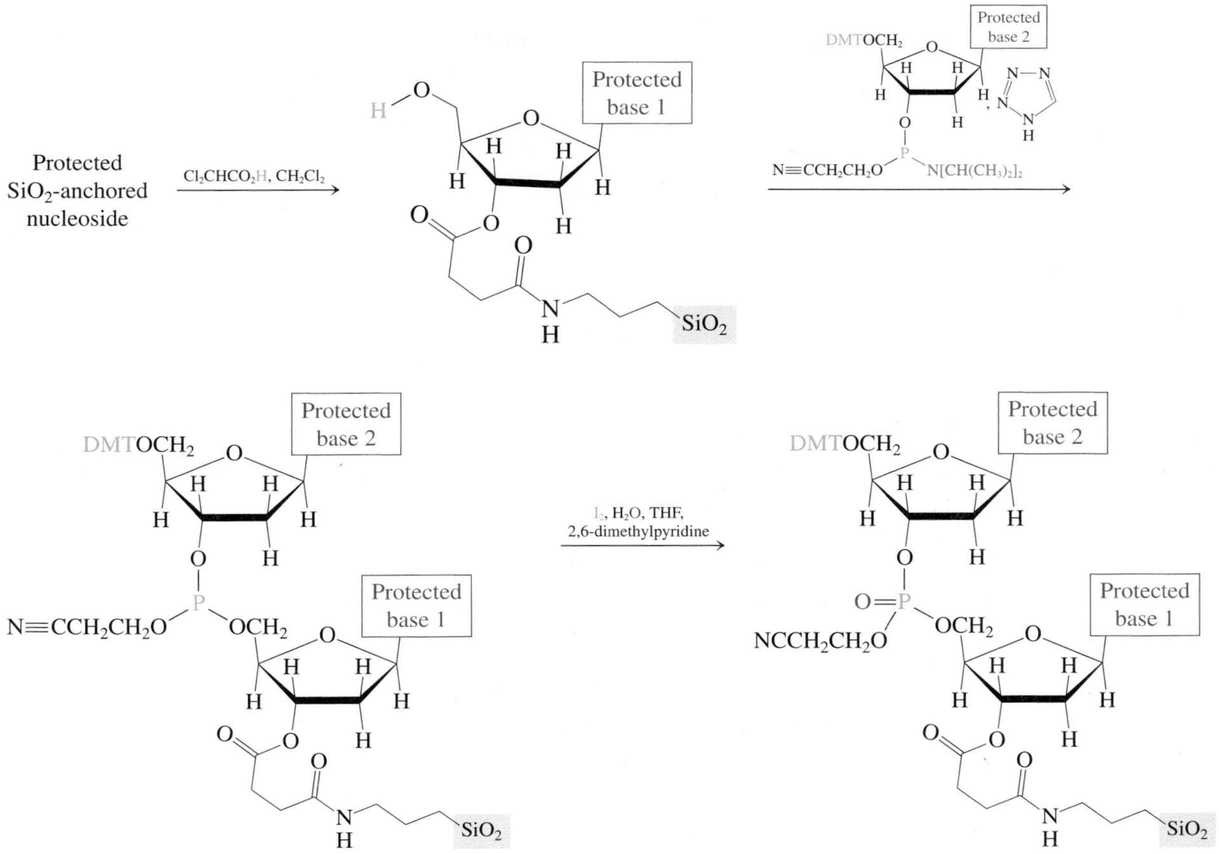

The sequence (1) DMT hydrolysis, (2) coupling, and (3) oxidation can be repeated multiple times in the synthesizer machine until the desired oligonucleotide is assembled in its protected and immobilized form. The final task is to remove the product from the silica and deprotect the DMT-bearing terminal sugar, all the bases, and the phosphate group, without cleaving any of the other bonds. Remarkably, this task can be done in just two steps, first treatment with acid and then with aqueous ammonia, as shown here for the dinucleotide made in the preceding scheme.

Liberation of the Solid-Phase-Supported and Protected Dinucleotide

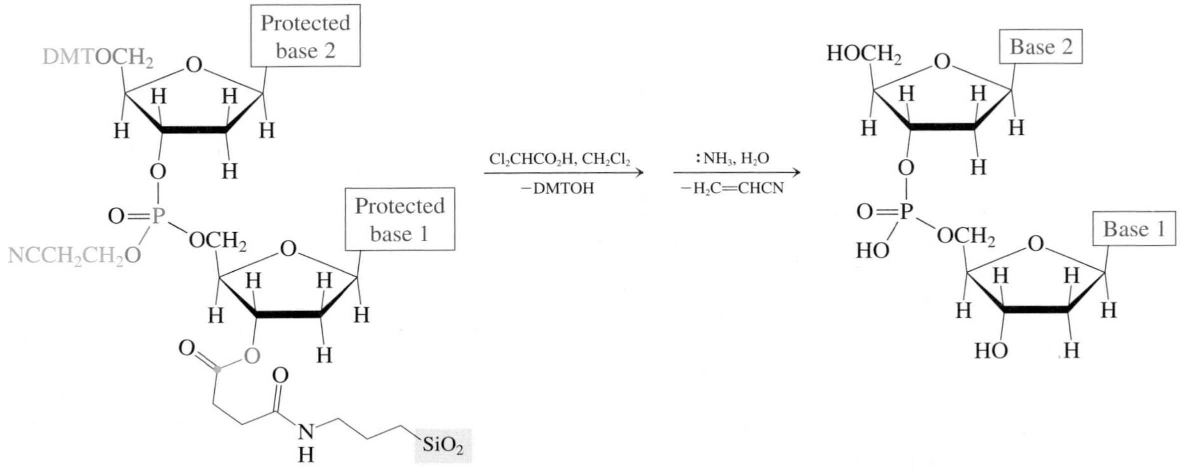

Remember WHIP

*W*hat

*H*ow

*I*nformation

*P*roceed

Solved Exercise 26-25 | **Working with the Concepts: Exploring Mechanistic Features of Solid-Phase Dinucleotide Synthesis**

As shown in this section, oligonucleotide synthesis features as the key step a tetrazole-catalyzed coupling of an alcohol (the first nucleotide) with a phosphoramidite (the second nucleotide). On first sight (focusing on all the nitrogen lone pairs), the function of tetrazole appears to be that of a base (to deprotonate the alcohol), but this heterocycle is actually quite acidic (pK_a = 4.8), about as much as acetic acid (Section 19-4). Moreover, kinetic experiments indicate a rate-determining step involving both the tetrazole and phosphoramidite. Explain these observations and suggest a plausible mechanism for the coupling process. (**Hint:** Review pyrrole in Sections 25-3 and 25-4.)

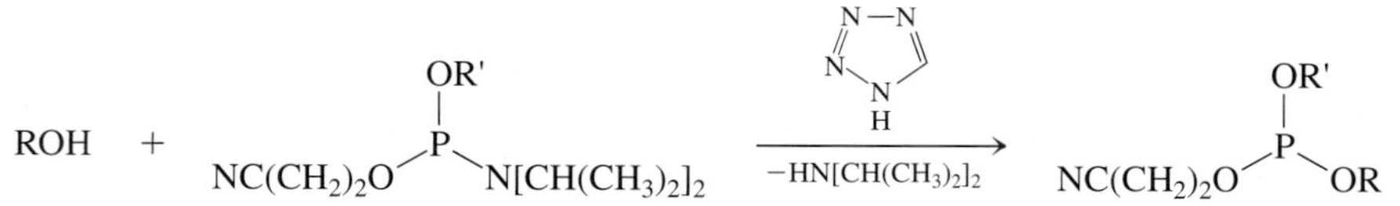

Strategy

We follow the hint and try to extrapolate the chemistry of pyrrole to that of tetrazole, in particular its acidity and, considering the kinetic evidence that points to attack of phosphoramidite on the heterocycle, its most nucleophilic site.

Solution

- Why is tetrazole so acidic? This molecule is obviously no ordinary amine, which would exhibit a pK_a of about 35 (Section 21-4). A clue is provided by the observation of similar relatively high acidity in pyrrole, with a pK_a of 16.5 (Section 25-4). The explanation was delocalization of the charge in the anion and the sp^2 hybridization of the substituent carbons. The same reasoning holds here, accentuated by the presence of the additional three electron-withdrawing nitrogen atoms on the ring.

- Turning to the mechanism of the coupling reaction and considering the acidity of tetrazole, it is reasonable to assume that the most basic amine nitrogen on the phosphoramidite is protonated rapidly (reversibly), changing the substituent into an excellent leaving group (the first step in the scheme below).

- Formulation of this protonated species suggests the nature of the rate-determining second step: nucleophilic displacement of the protonated amine from the phosphorus.

- Which nitrogen of tetrazole will attack? Recalling the electrophilic aromatic substitution chemistry of pyrrole (Section 25-4), we note that optimal resonance stabilization of the resulting cation is attained by involving the position at the terminus of the diene portion, in the present case N2.

- Rapid proton loss (step 3) then furnishes a new intermediate, in which the amino substituent on the original phosphorus reagent has been replaced by tetrazolyl.

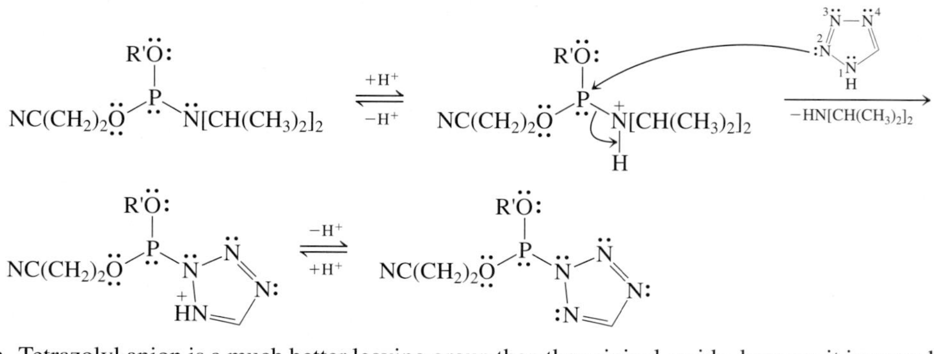

- Tetrazolyl anion is a much better leaving group than the original amide, because it is a much weaker base. Thus, the actual coupling with the alcohol takes place through this intermediate by nucleophilic displacement, as shown next.

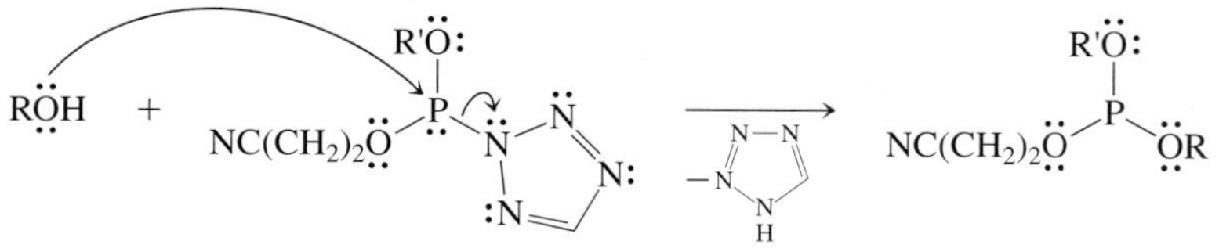

Exercise 26-26 | Try It Yourself

Formulate mechanisms for all the hydrolysis reactions that effect deprotection of the solid-phase supported dinucleotide in the scheme at the bottom of p. 1209. [**Caution:** The cyanoethyl protecting group is not removed by nucleophilic hydrolysis of the phosphate ester, because such a process would cleave all P–O bonds indiscriminately. Can you imagine another way? **Hint:** Like other carboxylic acid derivatives (Section 20-1), the α hydrogen in nitriles is acidic.]

Exercise 26-27 |

In the dinucleotide synthesis on page 1209, the attachment of the first nucleoside to silica employs a 4-nitrophenyl ester as a leaving group. Why would the use of this substituent be advantageous?

The polymerase chain reaction (PCR) makes multiple copies of DNA

DNA cloning—the preparation of large numbers of identical DNA molecules for the study of their sequence, expression, and regulation—revolutionized molecular biology in the mid-1970s. Cloning needs living cells to amplify inserted DNA. Therefore, the discovery of a procedure by Mullis* in 1984 that reproduces DNA segments in vitro by the millions, without the necessity of living cells, marked a stunning advance: the **polymerase chain reaction (PCR).**

The key to this reaction is the ability of some DNA-copying enzymes to remain stable at temperatures as high as 95°C. The original enzyme used in PCR is Taq DNA polymerase, found in a bacterium, *Thermus aquaticus,* in a hot spring in Yellowstone National Park (now other improved enzymes have become available). As described earlier in the Sanger dideoxyribose method of DNA sequencing, polymerases need a supply of the four nucleotides and a short primer to start their job. In nature, a primase enzyme takes care of the priming and the entire DNA sequence behind the primer is reproduced. In PCR, the primer consists of a short (20 bases) oligonucleotide strand that is complementary to a short piece of the DNA to be copied.

In practice, the PCR is carried out as shown schematically in Figure 26-18 (p. 1214). The reaction flask is charged with the (double-stranded) DNA to be copied, the four nucleotides, the primer, and Taq polymerase. In the first step, the mixture is heated to 90–95°C to induce the DNA to separate into two strands. Cooling to 54°C allows the primers to attach themselves to the individual DNA molecules. Raising the temperature to 72°C provides optimum (in fact, "native") conditions for Taq polymerase to add nucleotides to the primer along the attached DNA strand until the end, resulting in a complementary copy of the template. All of this takes only a few minutes and can be repeated in automated temperature-regulated vessels multiple times. Because the products of each cycle are again separated into their component strands to function as additional templates in the subsequent cycle, the amount of DNA produced with time increases exponentially: After n cycles, the quantity of DNA is 2^{n-1}. For example, after 20 cycles (less than 3 h), it can be about 1 million; after 32 cycles, 1 billion.

There are many practical applications of this technique. In medical diagnostics, bacteria and viruses (including HIV) are readily detected through the use of specific primers. Early cancer detection is made possible by identifying mutations in growth-control genes. Inherited disorders can be pinpointed, starting from unborn babies to dead bodies. For example, blood stains on the clothing of Abraham Lincoln have been analyzed to show that he had a genetic disorder called Marfan syndrome, which affects the structure of the connective tissue. In forensics and legal medicine, PCR has been used to identify the origin of blood, saliva, and other biological clues left by the perpetrator of a crime. Paternity and genealogy can be proved fairly conclusively (Real Life 26-4). In molecular evolution, ancient DNA

---

*Dr. Kary B. Mullis (b. 1944), La Jolla, California, Nobel Prize 1993 (chemistry).

REAL LIFE: FORENSICS 26-4 | DNA Fingerprinting

The human genome is unique and unalterable for each individual and is the same for every cell in his or her body. Therefore, in principle, it should be usable to establish identity unambiguously, much like conventional fingerprinting. However, sequencing 3 billion base pairs per person is still too time consuming and costly to be practicable. Instead, *DNA fingerprinting* applies abbreviated and much faster techniques.

An older but instructive approach, called Southern blotting (after its inventor, E. M. Southern), is probably best known from its application in crime laboratories and in paternity cases in the late 1980s and early 1990s. It is based on the finding that those pieces of DNA that seemingly have no coding function contain repeating sequences (from 20 to 100 base pairs) called *variable-number tandem repeats* or VNTR. The key is that the number of repeats, from 1 to 30 in a row, at certain places (also called *loci*) differs among individuals. The size of the corresponding DNA section will differ accordingly. The DNA fingerprint then is the visualization of the number of VNTR at a number of loci.

In practice, the sample of DNA (for example, from the blood at the scene of a crime or from the suspected father of a child) was cut into smaller pieces by restriction enzymes and the smaller fragments sorted by size using electrophoresis through a slab of agarose gel. To save the electrophoresis pattern on a more stable surface, it was literally "blotted" onto a nylon membrane (hence the "Southern blot"). To locate a specific VNTR sequence on one of the DNA fragments, a DNA sequence complementary to that of a VNTR locus was generated, bearing a radioactive or chemiluminescent label. This probe was then allowed to bind to complementary DNA sequences on the membrane and the pattern visualized on a photographic film (see photo). Standard forensic analysis used between 3 and 5 loci, and analysis of the combined data placed the odds of finding two individuals with the same observed pattern at 1 to about 10–100 million, unless they were biologically related.

A similar procedure can be followed to establish genealogy, even over several generations. Thus, your VNTR could be derived from the DNA of either or both of your parents,

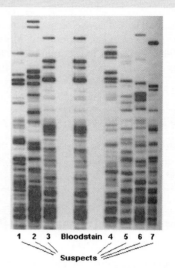

1 2 3 Bloodstain 4 5 6 7

Suspects

A bloodstain on a murder victim is suspected to belong to the perpetrator. Which one of the suspects is it? (Image credited to Cellmark Diagnostics, Abingdon, England.)

but you can't have any VNTR that are not present in your mother or father. A hypothetical case is shown in the simplified Southern blot patterns on the facing page. You can see that the parents' sequences (and no others) show up in part in their joint biological offspring (red and blue). However, the son from another marriage exhibits none of the current father's traits (no blue), but those of his biological father (green). Even more strikingly, the adopted daughter has a completely incongruent pattern (orange). In paternity disputes, the identity of the father can be ascertained with a probability of up to 100,000 to 1.

Currently, forensic DNA analysis uses the polymerase chain reaction to amplify 13 specific DNA loci. The VNTR at these loci are only four base pairs, called *short tandem repeats* (STR). The amplification primers are chemically

can be multiplied or even reconstructed by amplification of isolated fragments. In this way, the agent of the bubonic plague was successfully identified to be the bacterium *Yersinia pestis,* isolated from the teeth of victims buried in French mass graves dating from 1590 to 1722.

In 2010 the complete genome of a Neanderthal was announced, and comparison with that of modern humans proved that interbreeding occurred between the two species 60,000 years ago: We carry a few percent of Neanderthal genes. Similarly, DNA from a single finger bone found in a Siberian cave in 2008 led to the discovery of a previously unknown species of humans that lived 40,000 years ago, the Denisovans. Remnants of their DNA are also embedded in ours, presenting a rapidly developing picture of human ancestry. Turning to extinct animals, in that same year, part of the DNA of the woolly mammoth was determined. In 2011, Japanese scientists announced plans to clone the species within six years. What will be next?

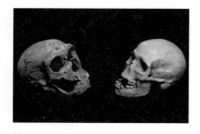

Homo neanderthalensis and *homo sapiens:* Which is which?

VNTR Patterns of a Family

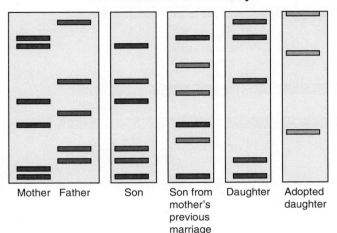

Mother Father   Son   Son from   Daughter Adopted
                          mother's   daughter
                          previous
                          marriage

purported daughters revealed an mtDNA sequence consistent with the purported parents, confirming their family relationship. Moreover, the sequences relating the three daughters to the Tsarina were also found in Prince Philip, the Duke of Edinburgh, who is a distant relative of the late Tsarina through a maternal line. The Tsar's remains were similarly analyzed by comparison to the DNA fingerprints of two living relatives of the Tsar. The fate of the youngest, fourth daughter and the hemophiliac son (and Crown Prince) of the Tsar remained unknown until 2008, when their burned and shattered skeletons were found in a forest near the grave of the other members of the family. Despite the poor condition of the bone fragments recovered, sufficient amounts of DNA could be extracted to confirm their identity.

labeled with differently colored fluorescent dyes so that the PCR products can be subjected to electrophoresis into the path of a camera, through a capillary or a thin slab of polyacrylamide gel, and detected by their fluorescence. Elapsed electrophoresis time to detection enables measurement of the molecular weight of the PCR products, from which the number of STR at any of the loci can be determined. The odds of obtaining a "false positive" with this method can be as low as 1 in 100 billion.

A case of an application of DNA fingerprinting that had historical impact is the identification of the remains of Tsar Nicholas II of Russia and some of his family members. It had been believed that the Tsar's entire family, along with the Royal Physician and three servants, had been executed by the Bolshevik revolutionaries during the night of July 16, 1918, and their remains hastily buried in a makeshift grave. This story could not be confirmed until the grave was found in 1991 and nine bodies exhumed. Mitochondrial DNA (mtDNA) sequencing using bone samples enabled identification of the Tsar, his wife (the Tsarina), their three daughters, and the other four skeletons belonging to the non-family members. Each of the three

Tsar Nicholas II of Russia and his family were executed in 1918 during the Russian revolution, as was verified in 1991 and 2008 by DNA fingerprinting.

THE BIG PICTURE

This concluding chapter drives home the message we have expressed in many places: Life is based on organic molecules. From a chemical standpoint, life at the level of amino acids and nucleic acids is astonishingly simple. It is the structures and properties of the polymeric arrays that add the more complex dimensions and functions typical of living organisms. To use an analogy, think of a modern desktop computer. At the level of individual chips, the interactions are well defined and understood, and can be analyzed in terms of on and off pulses of current. However, when we add some controlling software, guiding the microcircuits to act in unison, we achieve the complex problem-solving ability we desire from a computer. Similarly, at the molecular level of biology, all molecules, no matter how complicated, react according to the same chemical principles we have studied in this book. However, when the molecules and cells act in unison, the unpredictable behavior of life occurs.

Figure 26-18 Polymerase chain reaction (PCR). A cycle consists of three steps: strand separation, attachment of primers, and extension of primers by DNA synthesis. The reactions are carried out in a closed vessel. The cycle is driven by changes in temperature. Sequences on one strand of the original DNA are denoted by *abcde* and those on the complementary strand by *a'b'c'd'e'*. The primers are shown in blue and the new DNA extending from the primers in the respective complementary color to the old strands (red or green). (After *Biochemistry*, 6th ed., by Jeremy M. Berg, John L. Tymoczko, Lubert Stryer, W. H. Freeman and Company. Copyright © 1975, 1981, 1988, 1995, 2002, 2007.)

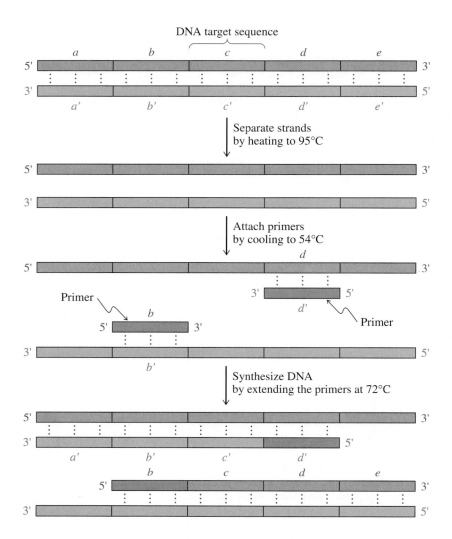

For those of you who will go on to specialize in the life sciences, as biochemists, molecular cell biologists, doctors, and so on, this chapter will have whetted your appetite for more and will also have laid a molecular foundation for your future studies. For others, it will have provided a glimpse of the molecular basis for all life-related biochemical processes, including protein synthesis, function, and biological origin in the nucleic acids. In today's society, as we grapple with issues involving genetic engineering, development of drugs to fight disease, environmental dangers, energy resources—an entire range of issues involving science and technology—it is useful to understand the chemical principles that underlie these developments and make them possible for good and for ill.

WORKED EXAMPLE: INTEGRATING THE CONCEPTS

26-28. Aspartame as a Synthetic Target

Aspartame (NutraSweet), Asp-Phe-OCH$_3$, appears to be a simple target of synthesis. That this is not so becomes apparent when you try to devise routes to it, starting simply from aspartic acid (Asp) and phenylalanine (Phe). Analyze the problem and formulate various approaches to the synthesis of this compound.

Aspartame

SOLUTION

Simple retrosynthetic analysis dissects the molecule into its two amino acid components, Asp and Phe-OCH$_3$ (made by methyl esterification of Phe), which might be envisaged to be coupled with DCC (Section 26-6). There is a problem, however: Asp has an additional β-carboxy group, bound to interfere. We are therefore facing the task of preparing a selectively carboxy-protected Asp. We can think about accomplishing it by two strategies: (A) the direct protection of the Asp nucleus or, if strategy A is not successful, (B) a total synthesis of an appropriately derivatized Asp from simpler starting materials that avoids Asp altogether. The following strategies focus on practical solutions to our problem found in the literature, although you may come up with alternatives that might be even better.

A. Selective Protection of the Asp Nucleus

Are the two carboxy groups chemically sufficiently different to warrant the investigation of selective monoprotection of one of the carboxy groups in Asp? The answer is yes, perhaps. Thus, we have learned that an α-amino group causes an increase in the acidity of a carboxylic acid by about 2 pK units (Section 26-1). For Asp, pK_a (α-COOH) = 1.9, whereas pK_a (β-COOH) = 3.7 (Table 26-1). One might therefore be tempted to try a selective ammonium salt formation with the α-carboxylate function at carefully controlled pH, followed by thermolysis (Section 19-10). The problem is that, under the thermal conditions of the relatively slow amide formation, fast proton exchange takes place. It is better to hope for differing reactivity of the two carbonyl groups in addition–elimination reactions of an appropriate derivative. The electron-withdrawing effect of the α-amino group should manifest itself here in increased electrophilicity of the adjacent carbonyl carbon (Sections 17-6, 19-4, and 20-1), rendering the corresponding ester function more susceptible to basic hydrolysis. Indeed, this approach has been realized successfully with the N-Cbz-protected (phenylmethyl) diester of Asp (prepared by standard procedures, see Section 26-6), as shown below.

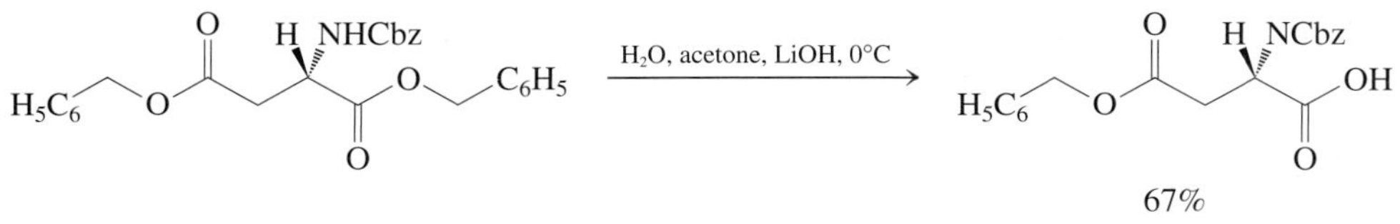

Cbz-Asp-(OCH$_2$C$_6$H$_5$)$_2$

H$_2$O, acetone, LiOH, 0°C

67%

The product was coupled with Phe-OCH$_3$ and DCC, and then deprotected by catalytic hydrogenolysis (Section 26-6) to give aspartame.

A simplified synthesis relies on the potential to protect difunctional compounds as cyclic derivatives. For example, 1,2-diols are "masked" as cyclic acetals (Section 24-8), hydroxy acids as lactones (Section 19-9), amino acids as lactams (Section 19-10), and dicarboxylic acids as anhydrides (Section 19-8). The last two possibilities merit consideration as applied to Asp. However, direct lactam formation can be quickly ruled out because of the complications of ring strain (although β-lactams have been used in the preparation of aspartame). This problem is absent with respect to dehydration to the five-membered ring anhydride. Because anhydrides are activated carboxylic acid derivatives (Section 20-3), the Asp anhydride can be coupled directly with Phe-OCH$_3$ without the help of added DCC. Nucleophilic attack of the amino end of Phe-OCH$_3$ occurs preferentially at the desired position, albeit not completely so; 19% of the product derives from peptide-bond formation at the β-carboxy group of Asp.

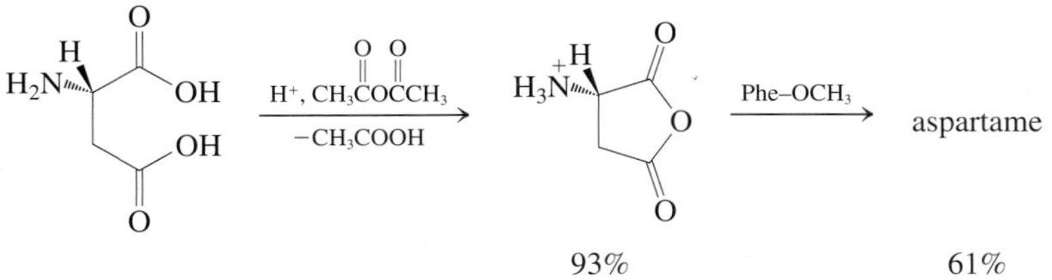

H$^+$, CH$_3$COCCH$_3$

−CH$_3$COOH

Phe–OCH$_3$ aspartame

93%   61%

Asp anhydride

B. Total Syntheses of Carboxy-Differentiated Asp Derivatives

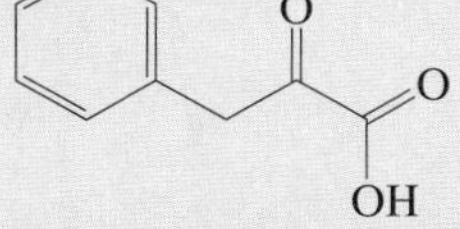

The alternative to approach A is to start from scratch and build up the Asp framework by using the methods of Section 26-2 in such a way as to provide as a final product a selectively monoprotected carboxy derivative. There are many possible strategies for the solution of this problem, but, as you contemplate them, you will discover that finding an appropriate masked β-carboxy group that is not unmasked during the manipulations of the α-amino acid part is not easy. In the following Gabriel synthesis, recourse was made to the relatively innocuous 2-propenyl substituent, which is eventually elaborated by oxidative cleavage (Sections 12-12 and 24-5) to the free β-carboxy moiety.

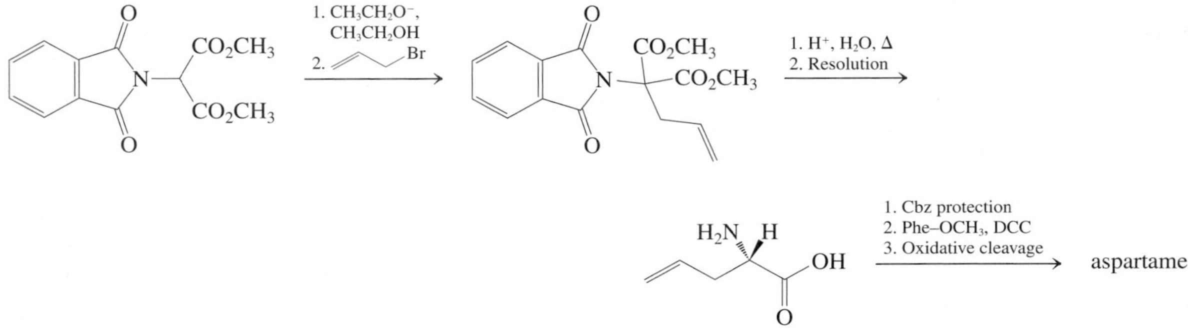

26-29. Writing out a Mechanistic Problem Involving Amino Acids

The amino acid analyzer (Section 26-5) "detects" the elution of any amino acid by the presence of an indicator that turns a deep violet color. This indicator is ninhydrin A, which reacts with amino acids to the purple compound B ("Ruhemann's purple"), generating as by-products an aldehyde and CO_2.

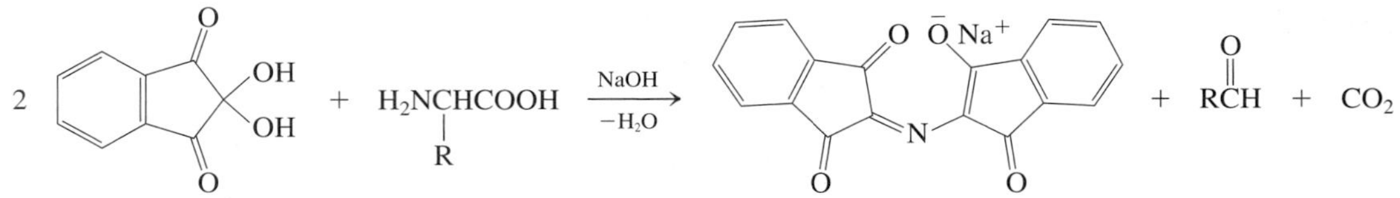

A	Amino acid	B
Ninhydrin		Ruhemann's purple

Write a reasonable mechanism for this process.

SOLUTION

Let's take an inventory: (1) Ninhydrin is a carbonyl hydrate (Section 17-6) of the corresponding trione, whose central carbonyl group is activated by its two neighbors. (2) Two molecules of the indicator disassemble one molecule of the amino acid into B (containing the amino nitrogen), an aldehyde (accounting for the RCH part), and CO_2 (derived from the carboxy function). (3) Decarboxylations of carboxylic acids occur readily in the presence of 3-oxo (or similar) functions (Section 23-2). (4) The highly delocalized (hence colored; Section 14-11) B is the enolate (Section 18-1) of an imine (Section 17-9).

The recognition of the last point allows us to simplify the problem somewhat, as B must be derived by a condensation of the dehydrated A with the corresponding amine C.

Simplifying B

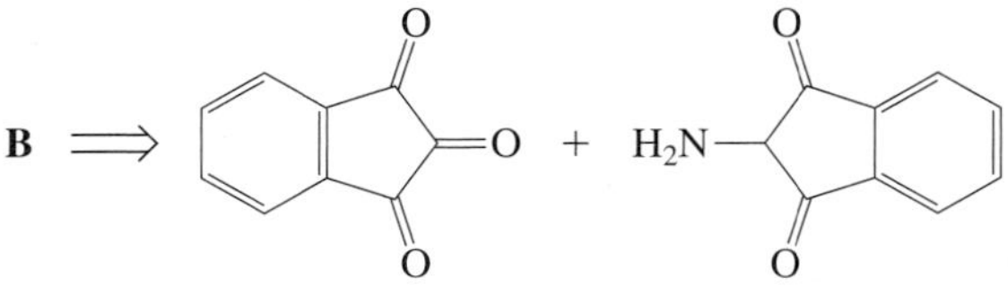

Coupling this realization with point 1 suggests that the first step of the mechanism is a base-catalyzed dehydration (Section 17-6) to the trione form of ninhydrin, which condenses with the amino acid to the corresponding imine D.

Steps 1 and 2

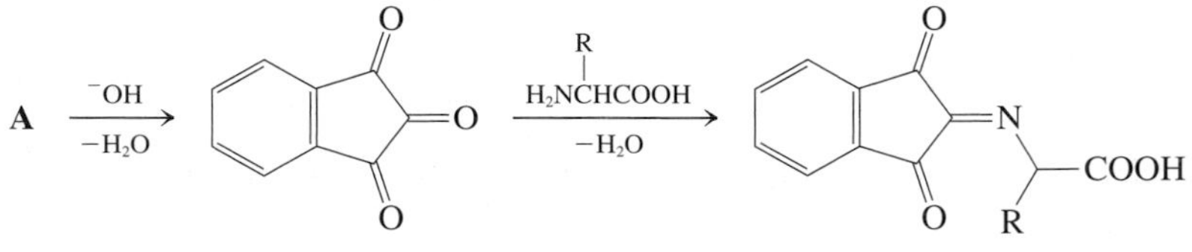

Point 3 encourages us to look for a 3-oxocarboxylic acid-type intermediate that would spontaneously lose CO_2 via an aromatic transition state (Section 23-2). Indeed, D fits the bill: It is an imine derivative of such a species, electronically quite equivalent, and smoothly rendering E.

Step 3

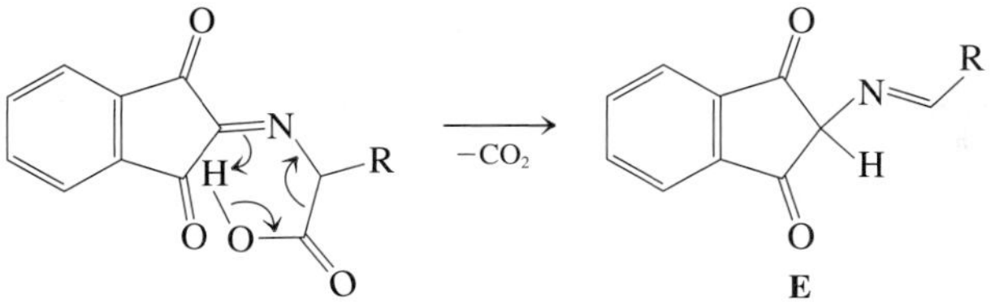

E

Now we can see the step that provides a connection to C and also the aldehyde product: simple hydrolysis of E.

Step 4

$$\text{E} \xrightarrow{H_2O,\ HO^-} \text{C} + \text{RCH}(\text{=O})$$

As already discussed, C then condenses with ninhydrin under the basic conditions to generate the salt B.

New Reactions

1. Acidity of Amino Acids (Section 26-1)

$$H_3\overset{+}{N}CH(R)COOH \qquad H_3\overset{+}{N}CH(R)COO^-$$

$$pK_a \approx 2\text{–}3 \qquad pK_a \approx 9\text{–}10$$

$$\text{Isoelectric point } pI = \frac{pK_{COOH} + pK_{\overset{+}{N}H_3}}{2}$$

2. Strongly Basic Guanidino Group in Arginine (Section 26-1)

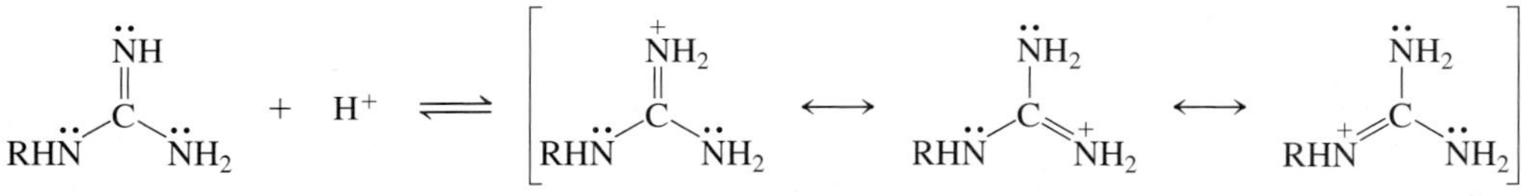

$pK_a \approx 13$

3. Basicity of Imidazole in Histidine (Section 26-1)

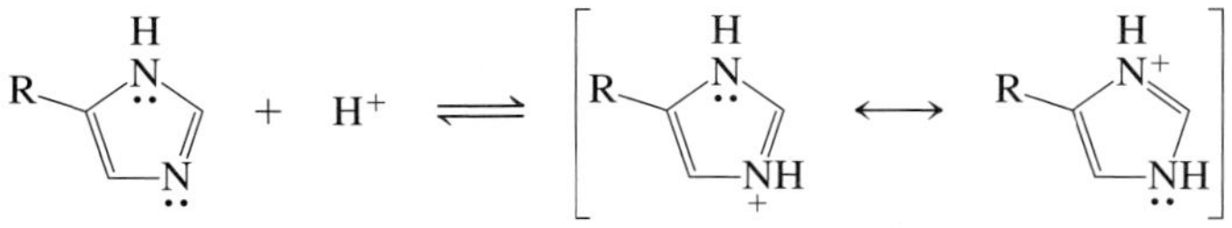

$pK_a = 7.0$

Preparation of Amino Acids

4. Hell-Volhard-Zelinsky Bromination Followed by Amination (Section 26-2)

$$RCH_2COOH \xrightarrow[\text{2. } NH_3,\ H_2O]{\text{1. } Br_2,\text{ trace } PBr_3} RCH(\overset{+}{N}H_3)COO^-$$

5. Gabriel Synthesis (Section 26-2)

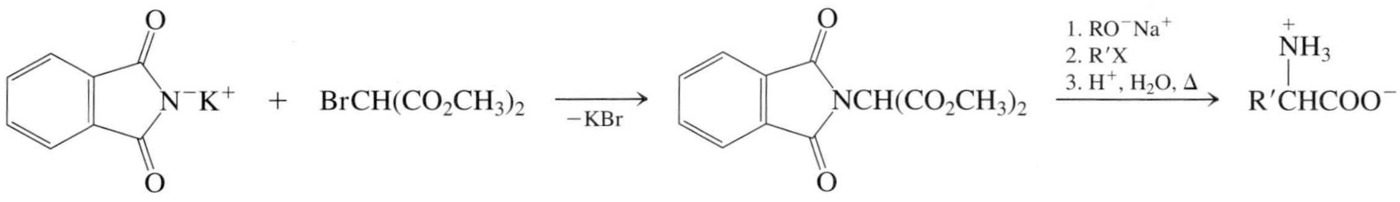

6. Strecker Synthesis (Section 26-2)

$$\underset{\underset{RCH}{||}}{O} \xrightarrow{\text{HCN, NH}_3} \underset{\underset{RCHCN}{|}}{NH_2} \xrightarrow{\text{H}^+, \text{H}_2\text{O}, \Delta} \underset{\underset{RCHCOO^-}{|}}{\overset{+}{NH_3}}$$

Polypeptide Sequencing

7. Hydrolysis (Section 26-5)

$$\text{Peptide} \xrightarrow{\text{6 N HCl, 110°C, 24 h}} \text{amino acids}$$

8. Edman Degradation (Section 26-5)

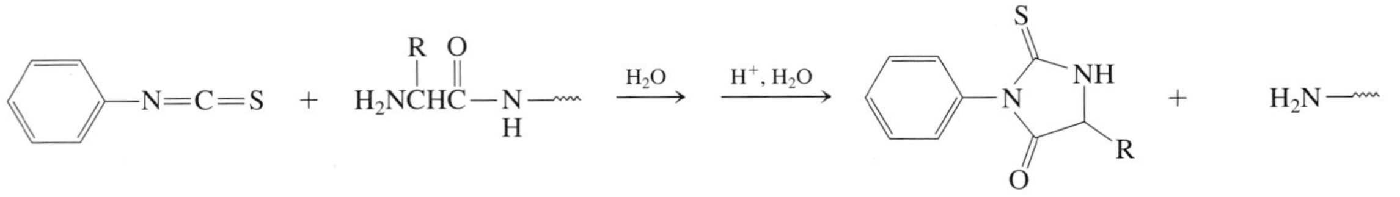

Phenylthiohydantoin   Lower polypeptide

Preparation of Polypeptides

9. Protecting Groups (Section 26-6)

Phenylmethyl
chloroformate
(Benzyl
chloroformate)

Cbz-protected
amino acid

Bis(1,1-Dimethylethyl)
dicarbonate
(Di-*tert*-butyl dicarbonate)

Boc-protected
amino acid

10. Peptide-Bond Formation with Dicyclohexylcarbodiimide (DCC; Section 26-6)

$$\text{Cbz-Gly} \quad + \quad \text{Ala-OCH}_2\text{C}_6\text{H}_5 \quad + \quad \underset{\textbf{DCC}}{\text{C}_6\text{H}_{11}\text{N}=\text{C}=\text{NC}_6\text{H}_{11}} \longrightarrow$$

$$\text{Cbz-Gly-Ala-OCH}_2\text{C}_6\text{H}_5 \quad + \quad \text{C}_6\text{H}_{11}\text{NH}\overset{\overset{O}{||}}{\text{C}}\text{NHC}_6\text{H}_{11}$$

11. Merrifield Solid-Phase Synthesis (Section 26-7)

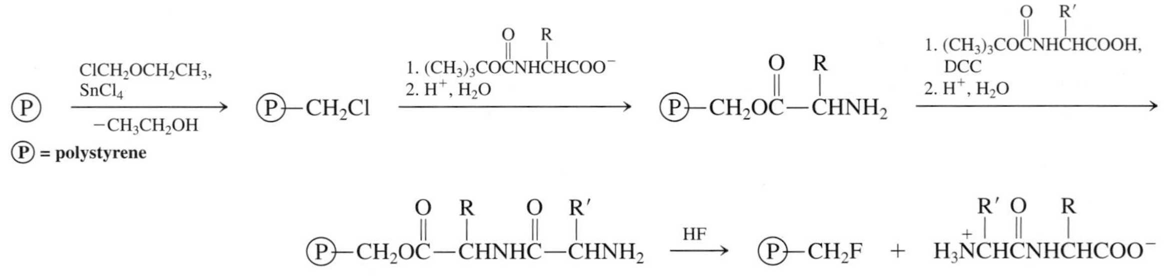

Ⓟ = polystyrene

Important Concepts

1. **Polypeptides** are poly(amino acids) linked by **amide bonds.** Most natural polypeptides are made from only 19 different L-amino acids and glycine, all of which have common names and three- and one-letter abbreviations.

2. Amino acids are **amphoteric;** they can be protonated and deprotonated.

3. **Enantiomerically pure** amino acids can be made by classical fractional crystallization of diaste-reomeric derivatives or by enantioselective reactions of appropriate achiral precursors.

4. The **structures** of **polypeptides** are varied; they can be linear, cyclic, disulfide bridged, pleated sheet, α-helical or superhelical, or disordered, depending on size, composition, hydrogen bonding, and electrostatic and London forces.

5. **Amino** and **nucleic acids** are **separated** mainly by virtue of their size- or charge-based differences in ability to bind to solid supports.

6. **Polypeptide sequencing** entails a combination of selective chain cleavage and amino acid analysis of the resulting shorter polypeptide fragments.

7. **Polypeptide synthesis** requires end-protected amino acids that are coupled by dicyclohexyl-carbodiimide. The product can be selectively deprotected at either end to allow for further extension of the chain. The use of **solid supports,** as in the Merrifield synthesis, can be automated.

8. The proteins myoglobin and hemoglobin are polypeptides in which the amino acid chain envelops the active site, **heme.** The heme contains an iron atom that reversibly binds oxygen, allowing for oxygen uptake, transport, and delivery.

9. The **nucleic acids** are biological polymers made of phosphate-linked base-bearing sugars. Only four different bases and one sugar are used for DNA and RNA, respectively. Because the base pairs adenine–thymine (uracil for RNA) and guanine–cytosine are held by particularly favorable **hydrogen bonding,** a nucleic acid can adopt a dimeric helical structure containing **complementary base sequences.** In DNA, this arrangement unwinds and functions as a template during **DNA replication** and **RNA synthesis.** In **protein synthesis,** each amino acid is specified by a set of three consecutive RNA bases, called a **codon.** Thus, the base sequence **(genetic code)** in a strand of RNA translates into a specific amino acid sequence in a protein.

10. **DNA sequencing** relies on restriction enzymes, radioactive labeling, and specific chemical cleavage reactions to small fragments, analyzed by electrophoresis.

11. **DNA synthesis** employs silica as a support on which the growing oligonucleotide sequence is built up with the help of base, alcohol, and phosphite-phosphate **protecting groups.**

12. The **polymerase chain reaction (PCR)** makes multiple copies of DNA.

Problems

30. Draw stereochemically correct structural formulas for isoleucine and threonine (Table 26-1). What is a systematic name for threonine?

31. The abbreviation *allo* means *diastereomer* in amino acid terms. Draw allo-L-isoleucine, and give it a systematic name.

32. Draw the structure that each of the following amino acids would have in aqueous solution at the indicated pH values. **(a)** Alanine at pH = 1, 7, and 12; **(b)** serine at pH = 1, 7, and 12; **(c)** lysine at pH = 1, 7, 9.5, and 12; **(d)** histidine at pH = 1, 5, 7, and 12; **(e)** cysteine at pH = 1, 7, 9, and 12; **(f)** aspartic acid at pH = 1, 3, 7, and 12; **(g)** arginine at pH = 1, 7, 12, and 14; **(h)** tyrosine at pH = 1, 7, 9.5, and 12.

33. Group the amino acids in Problem 32 according to whether they are **(a)** positively charged, **(b)** neutral, or **(c)** negatively charged at pH = 7.

34. Show how the pI values for the amino acids in Problem 32 (see Table 26-1) are derived. For each amino acid that possesses more than two pK_a values, give the reason for your choice when calculating pI.

35. Show how Hell-Volhard-Zelinsky bromination followed by amination can be used to synthesize each of the following amino acids in racemic form: **(a)** Gly; **(b)** Phe; **(c)** Ala.

36. Show how the use of a Strecker synthesis can give each of the following amino acids in racemic form: **(a)** Gly; **(b)** Ile; **(c)** Ala.

37. What amino acid would result from carrying out the following synthetic sequence?

38. Using either one of the methods in Section 26-2 or a route of your own devising, propose a reasonable synthesis of each of the following amino acids in racemic form. **(a)** Val; **(b)** Leu; **(c)** Pro; **(d)** Thr; **(e)** Lys.

39. **(a)** Illustrate the Strecker synthesis of phenylalanine. Is the product chiral? Does it exhibit optical activity? **(b)** It has been found that replacement of NH_3 by an optically active amine in the Strecker synthesis of phenylalanine leads to an excess of one enantiomer of the product. Assign the *R* or *S* configuration to each stereocenter in the following structures, and explain why the use of a chiral amine causes preferential formation of one stereoisomer of the final product.

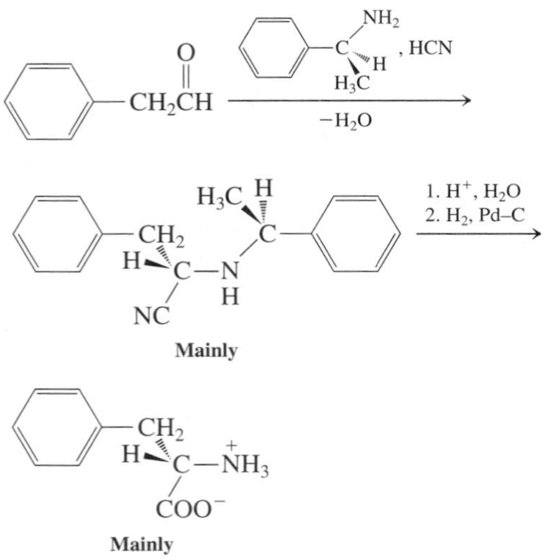

40. The antibacterial agent in garlic, allicin (Real Life 9-4, Problem 73 of Chapter 9), is synthesized from the unusual amino acid alliin by the action of the enzyme allinase. Because allinase is an extracellular enzyme, this process takes place only when garlic cells are crushed. Propose a reasonable synthesis for the amino acid alliin from the starting material in Problem 37. (**Hint:** Begin by designing a synthesis of an amino acid from Table 26-1 that is structurally related to alliin. To deal with the sulfur function, review Section 9-10.)

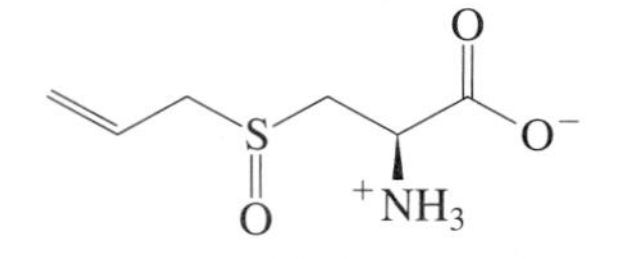

Alliin

41. Devise a procedure for separating a mixture of the four stereoisomers of isoleucine into its four components: (+)-isoleucine, (−)-isoleucine, (+)-alloisoleucine, and (−)-alloisoleucine (Problem 31). (Note: Alloisoleucine is much more soluble in 80% ethanol at all temperatures than is isoleucine.)

42. Identify each of the following structures as a dipeptide, tripeptide, and so forth, and point out all the peptide bonds.

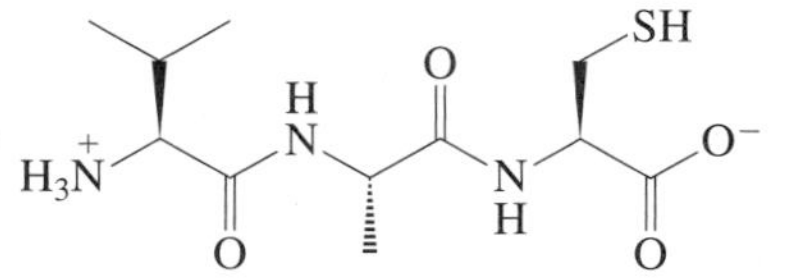

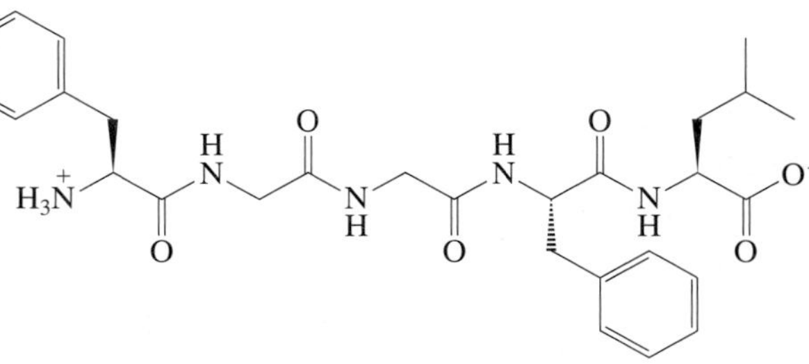

43. Using the three-letter abbreviations for amino acids, write the peptide structures in Problem 42 in short notation.

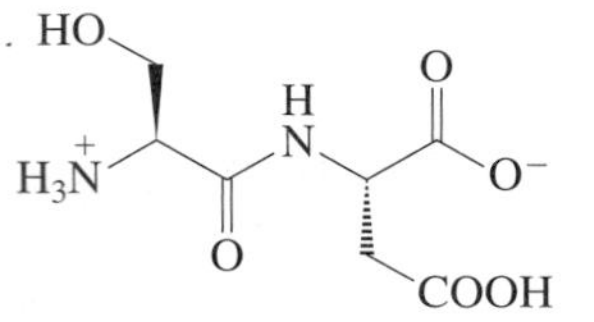

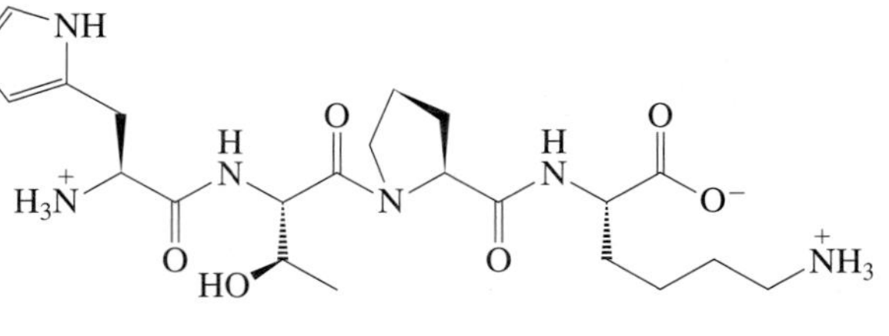

44. Indicate which of the amino acids in Problem 32 and the peptides in Problem 42 would migrate in an electrophoresis apparatus at pH = 7 **(a)** toward the anode or **(b)** toward the cathode.

45. Silk is composed of β sheets whose polypeptide chains consist of the repeating sequence Gly-Ser-Gly-Ala-Gly-Ala. What characteristics of amino acid side chains appear to favor the β-sheet configuration? Do the illustrations of β-sheet structures (Figure 26-3) suggest an explanation for this preference?

46. Identify as many stretches of α helix as you can in the structure of myoglobin (Figure 26-8C). Prolines are located in myoglobin at positions 37, 88, 100, and 120. How does each of these prolines affect the tertiary structure of the molecule?

47. Of the 153 amino acids in myoglobin, 78 contain polar side chains (i.e., Arg, Asn, Asp, Gln, Glu, His, Lys, Ser, Thr, Trp, and Tyr). When myoglobin adopts its natural folded conformation, 76 of these 78 polar side chains (all but those of two histidines) project outward from its surface. Meanwhile, in addition to the two histidines, the interior of myoglobin contains only Gly, Val, Leu, Ala, Ile, Phe, Pro, and Met. Explain.

48. Explain the following three observations. **(a)** Silk, like most polypeptides with sheet structures, is water insoluble. **(b)** Globular proteins such as myoglobin generally dissolve readily in water. **(c)** Disruption of the tertiary structure of a globular protein (denaturation) leads to precipitation from aqueous solution.

49. In your own words, outline the procedure that might have been followed by the researchers who determined which amino acids were present in vasopressin (Exercise 26-10).

50. Write the products of a single Edman degradation of the peptides in Problem 42.

51. What would be the outcome of reaction of gramicidin S with phenyl isothiocyanate (an attempted Edman degradation)? (**Hint:** With which functional group does this substance react?)

52. The polypeptide bradykinin is a tissue hormone that can function as a potent pain-producing agent. By means of a single treatment with the Edman reagent, the *N*-terminal amino acid in bradykinin is identified as Arg. Incomplete acid hydrolysis of the intact polypeptide causes random cleavage of many bradykinin molecules into an assortment of peptide fragments that includes Arg-Pro-Pro-Gly, Phe-Arg, Ser-Pro-Phe, and Gly-Phe-Ser.

Complete hydrolysis followed by amino acid analysis indicates a ratio of 3 Pro, 2 Phe, 2 Arg, and one each of Gly and Ser. Deduce the amino acid sequence of bradykinin.

53. The amino acid sequence of met-enkephalin, a brain peptide with powerful opiate-like biological activity, is Tyr-Gly-Gly-Phe-Met. What would be the products of step-by-step Edman degradation of met-enkephalin?

The peptide shown in Problem 42, part (d) is leu-enkephalin, a relative of met-enkephalin with similar properties. How would the results of Edman degradation of leu-enkephalin differ from those of met-enkephalin?

54. Secreted by the pituitary gland, corticotropin is a hormone that stimulates the adrenal cortex. Determine its primary structure from the following information. (i) Hydrolysis by chymotrypsin produces six peptides: Arg-Trp, Ser-Tyr, Pro-Leu-Glu-Phe, Ser-Met-Glu-His-Phe, Pro-Asp-Ala-Gly-Glu-Asp-Gln-Ser-Ala-Glu-Ala-Phe, and Gly-Lys-Pro-Val-Gly-Lys-Lys-Arg-Arg-Pro-Val-Lys-Val-Tyr. (ii) Hydrolysis by trypsin produces free lysine, free arginine, and the following five peptides: Trp-Gly-Lys, Pro-Val-Lys, Pro-Val-Gly-Lys, Ser-Tyr-Ser-Met-Glu-His-Phe-Arg, and Val-Tyr-Pro-Asp-Ala-Gly-Glu-Asp-Gln-Ser-Ala-Glu-Ala-Phe-Pro-Leu-Glu-Phe.

55. Propose a synthesis of leu-enkephalin [see Problem 42, part (d)] from the component amino acids.

56. The following molecule is thyrotropin-releasing hormone (TRH). It is secreted by the hypothalamus, causing the release of thyrotropin from the pituitary gland, which, in turn, stimulates the thyroid gland. The thyroid produces hormones, such as thyroxine, that control metabolism in general.

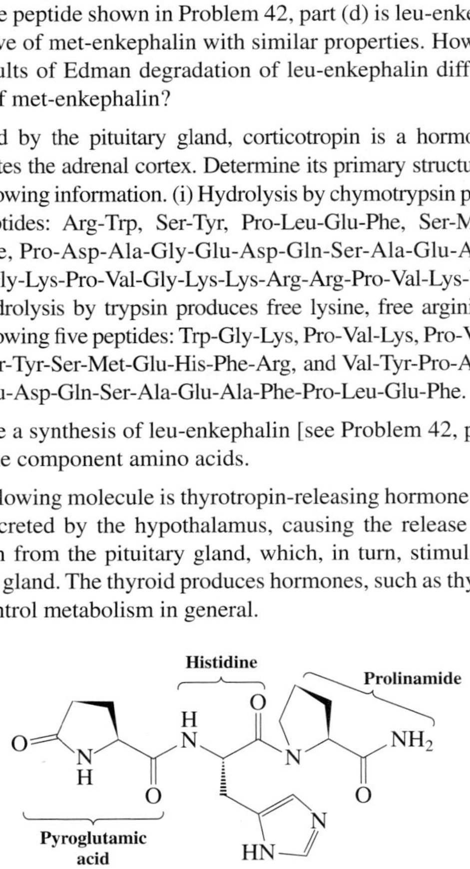

The initial isolation of TRH required the processing of 4 tons of hypothalamic tissue, from which 1 mg of the hormone was obtained. Needless to say, it is a bit more convenient to synthesize TRH in the laboratory than to extract it from natural sources. Devise a synthesis of TRH from Glu, His, and Pro. Note that pyroglutamic acid is the lactam of Glu and may be readily obtained by heating Glu to 140°C.

57. (a) The structures illustrated for the four DNA bases (Section 26-9) represent only the most stable tautomers. Draw one or more alternative tautomers for each of these heterocycles (review tautomerism, Sections 13-7 and 18-2). (b) In certain cases, the presence of a small amount of one of these less stable tautomers can lead to an error in DNA replication or mRNA synthesis due to faulty base pairing. One example is the imine tautomer of adenine, which pairs with cytosine instead of thymine. Draw a possible structure for this hydrogen-bonded base pair (see Figure 26-11). (c) Using Table 26-3, derive a possible nucleic acid sequence for an mRNA that would code for the five amino acids in met-enkephalin (see Problem 53). If the mispairing described in part (b) were at the first possible position in the synthesis of this mRNA sequence, what would be the consequence in the amino acid sequence of the peptide? (Ignore the initiation codon.)

58. Factor VIII is one of the proteins participating in the formation of blood clots. A defect in the gene whose DNA sequence codes for Factor VIII is responsible for classic hemophilia. Factor VIII contains 2332 amino acids. How many nucleotides are needed to code for its synthesis?

59. In addition to the conventional 20 amino acids in Tables 26-1 and 26-3, two others, selenocysteine (Sec) and pyrrolysine (Pyl), are incorporated into proteins using the nucleic acid-based cell machinery described in Section 26-10. The three-base codes for Sec and Pyl are UGA and UAG, respectively. These codes normally serve to terminate protein synthesis. However, if they are preceded by certain specific base sequences, they instead cause incorporation of these unusual amino acids and continued growth of the peptide chain.

Pyl is rare, occurring only in some ancient bacteria. Sec, containing the side chain CH_2SeH, is widespread; in fact, it occurs in at least two dozen human proteins that rely on the reactivity of the essential trace element selenium for their function. The pK_a for the SeH group in Sec is 5.2. (For comparison, the pK_a for the SH in Cys is 8.2.) (a) Draw the structure of Sec at pH 7; compare it with that of Cys at the same pH. (b) Determine pI for Sec. (c) Given the relationship between sulfur and selenium in the periodic table, and the pK_a differences between SH and SeH, how do you expect the chemical reactivities of Sec and Cys to compare?

60. **CHALLENGE** Hdroxyproline (Hyp), like many other amino acids that are not "officially" classified as essential, is nonetheless a very necessary biological substance. It constitutes about 14% of the amino acid content of the protein collagen. Collagen is the main constituent of skin and connective tissue. It is also present, together with inorganic substances, in nails, bones, and teeth. (a) The systematic name for hydroxyproline is (2S,4R)-4-hydroxyazacyclopentane-2-carboxylic acid. Draw a stereochemically correct structural formula for this amino acid. (b) Hyp is synthesized in the body in peptide-bound form from peptide-bound proline and O_2, in an enzyme-catalyzed process that requires vitamin C. In the absence of vitamin C, only a defective, Hyp-deficient collagen can be produced. Vitamin C deficiency causes scurvy, a condition characterized by bleeding of the skin and swollen, bleeding gums.

In the following reaction sequence, an efficient laboratory synthesis of hydroxyproline, fill in the necessary reagents (i) and (ii), and formulate detailed mechanisms for the steps marked with an asterisk.

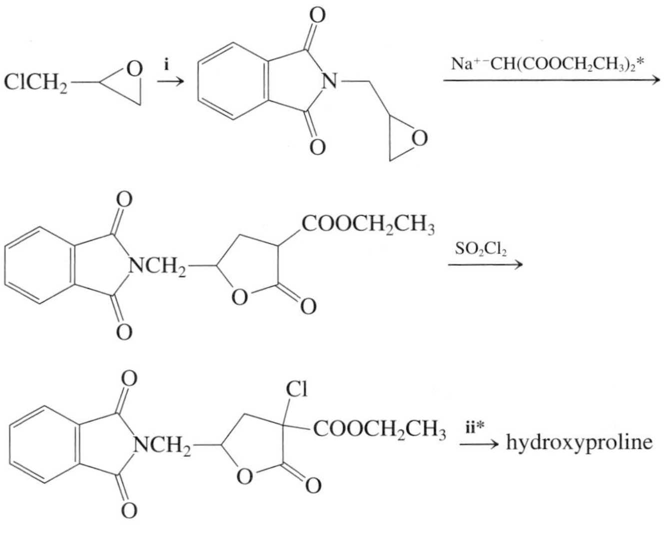

(c) Gelatin, which is partly hydrolyzed collagen, is rich in hydroxyproline and, as a result, is often touted as a remedy for split or brittle nails. Like most proteins, however, gelatin is almost completely broken down into individual amino acids in the stomach and small intestine before absorption. Is the free hydroxyproline thus introduced into the bloodstream of any use to the body in the synthesis of collagen? (**Hint:** Does Table 26-3 list a three-base code for hydroxyproline?)

61. **CHALLENGE**   The biosynthesis of oligosaccharides (Chapter 24) employs the chemical constituents of proteins and nucleic acids as well as carbohydrates. In the example shown, a disaccharide linkage is created between a molecule of galactose and a molecule of N-acetylgalactosamine. The galactose (the "donor" sugar) is carried into the process as a uridine diphospate ester, and the "acceptor" galactosamine is held in place by a glycoside linkage to the hydroxyl group of a serine residue in a protein. The enzyme galactosyl transferase specifically forms a disaccharide linkage between C1 of the donor and C3 of the acceptor:

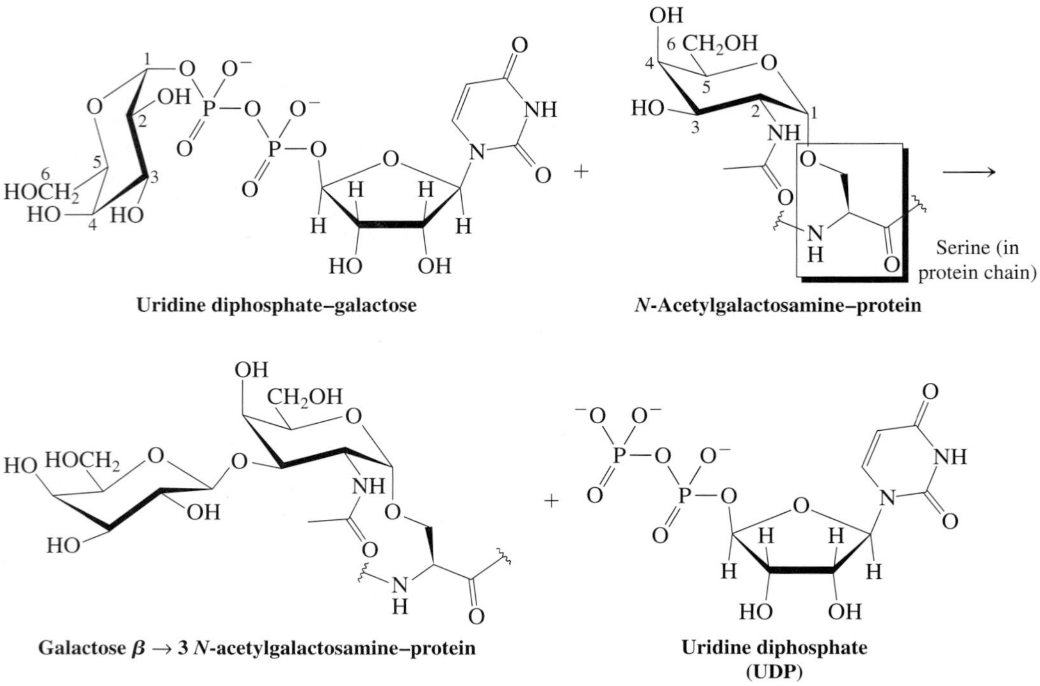

Uridine diphosphate–galactose

N-Acetylgalactosamine–protein

Galactose $\beta \rightarrow 3$ *N*-acetylgalactosamine–protein

Uridine diphosphate
(UDP)

What type of basic mechanistic process does this reaction resemble? Discuss the roles played by the various participants of the reaction.

62. **CHALLENGE**   Sickle-cell anemia is an often fatal genetic condition caused by a single error in the DNA gene that codes for the β chain of hemoglobin. The correct nucleic acid sequence (read from the mRNA template) begins with AUGGUGCACCUGA-CUCCUGA**G**GAGAAG . . . , and so forth. **(a)** Translate this into the corresponding amino acid sequence of the protein. **(b)** The mutation that gives rise to the sickle-cell condition is replacement of the boldface A in the preceding sequence by U. What is the consequence of this error in the corresponding amino acid sequence? **(c)** This amino acid substitution alters the properties of the hemoglobin molecule—in particular, its polarity and its shape. Suggest reasons for both these effects. (Refer to Table 26-1 for amino acid structures and to Figure 26-8C for the structure of myoglobin, which is similar to that of hemoglobin. Note the location of the amino acid substitution in the tertiary structure of the protein.)

Team Problem

63. Amino acids can be used as enantiomerically pure starting materials in organic synthesis. Scheme I depicts the first steps in the synthesis of a reagent employed in the preparation of enantiomerically pure β-amino acids, such as that occurring in the side chain of taxol (Section 4-7). Scheme II features an ester of the same amino acid for the preparation of an unusual heterocyclic dipeptide used in the study of polypeptide conformations.

Scheme I: Synthesis of an enantiomerically pure reagent

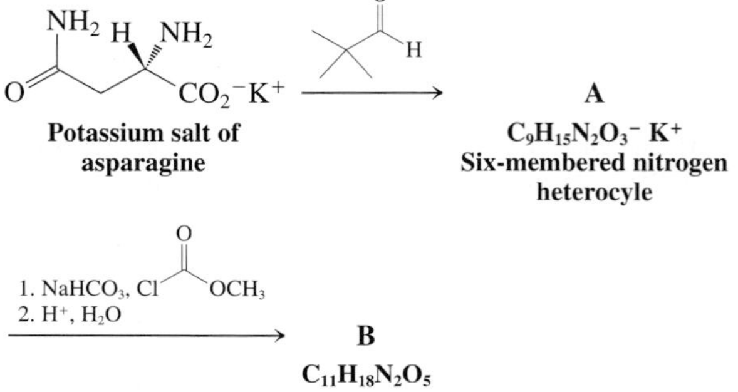

**Potassium salt of
asparagine**

A
$C_9H_{15}N_2O_3^-$ K$^+$
**Six-membered nitrogen
heterocyle**

1. NaHCO$_3$, Cl—C(O)—OCH$_3$
2. H$^+$, H$_2$O

B
$C_{11}H_{18}N_2O_5$

Consider the following questions:

1. There are two diastereomers of A formed in a 90:10 ratio. The major isomer is the one that can attain the most stable chair conformation. Are the two substituents on the ring cis or trans to each other? Label their positions as equatorial or axial.

2. Which nitrogen is nucleophilic and yields the carbamic ester (Section 20-6) B?

Scheme II: Synthesis of an unusual heterocyclic dipeptide

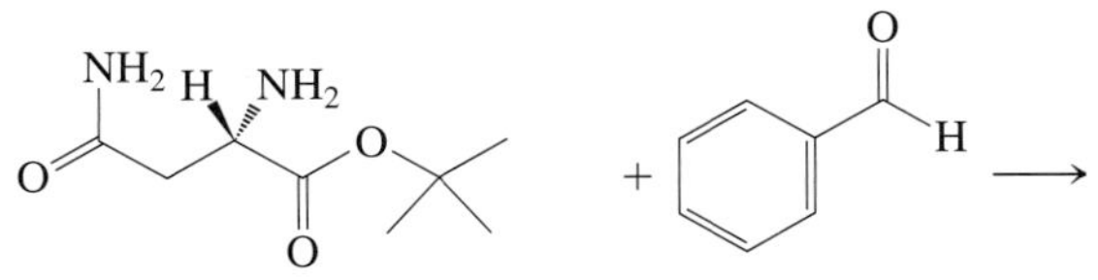

1,1-Dimethylethyl (*tert*-butyl) ester of
asparagine

$$\mathbf{C} \xrightarrow[\text{acid chloride}]{\text{Fmoc–amino}} \mathbf{D}$$

$C_{15}H_{20}N_2O_3$ **Heterocyclic**
(Acyclic compound)   **dipeptide**

The Fmoc protecting group (highlighted in the box) is used instead of Cbz or Boc, with which you are familiar, because the amino *acid chloride* is necessary to make the new amide bond. Neither the Cbz nor the Boc group is stable under these conditions.

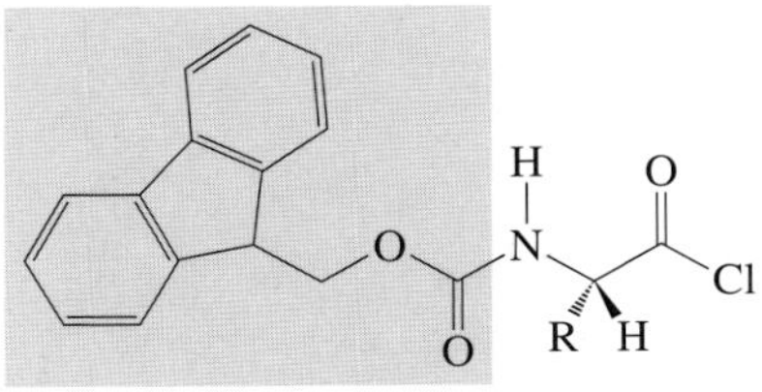

Fluorenylmethyloxycarbonyl(Fmoc)–
amino acid chloride

Consider the following questions:

1. What functional group is generated in C?
2. Where is the peptide bond in D? Circle it.

Reconvene to discuss the answers to the questions posed in each scheme and the structures you proposed for A–D.

Preprofessional Problems

64. Structure A (shown below) is that of a naturally occurring α-amino acid. Select its name from the following list. **(a)** Glycine; **(b)** alanine; **(c)** tyrosine; **(d)** cysteine.

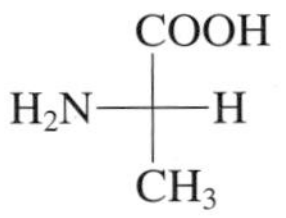

65. The primary structure of a protein refers to: **(a)** cross-links with disulfide bonds; **(b)** presence of an α helix; **(c)** the α-amino acid sequence in the polypeptide chain; **(d)** the orientation of the side chains in three-dimensional space.

66. Which one of the following five structures is a zwitterion?

(a) $^-O_2CCH_2\overset{O}{\overset{\|}{C}}NH_2$ **(b)** $^-O_2CCH_2CH_2CO_2^-$

(c) $H_3\overset{+}{N}CH_2CO_2^-$ **(d)** $CH_3(CH_2)_{16}CO_2^-K^+$

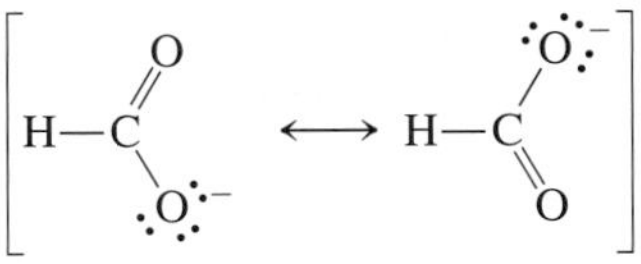

67. When an α-amino acid is dissolved in water and the pH of the solution adjusted to 12, which of the following species is predominant?

(a) 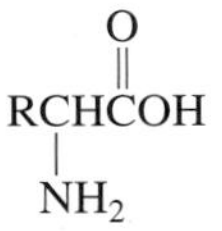 $\underset{NH_2}{R\overset{O}{\overset{\|}{C}}HCOH}$ **(b)** $\underset{^+NH_3}{R\overset{O}{\overset{\|}{C}}HCOH}$

(c) $\underset{^+NH_3}{R\overset{O}{\overset{\|}{C}}HCO^-}$ **(d)** 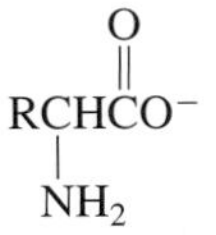 $\underset{NH_2}{R\overset{O}{\overset{\|}{C}}HCO^-}$

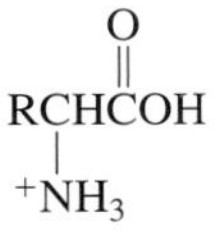

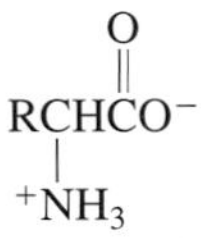

68. How many stereocenters are present in the small, naturally occurring peptide glycylalanylalanine? **(a)** Zero; **(b)** one; **(c)** two; **(d)** three.

Answers to Exercises

Chapter 1

1-1

(a)

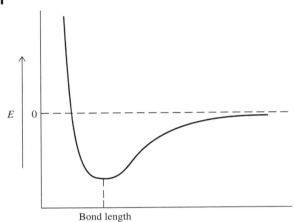

(b) Self-explanatory.

1-2

Li^+ :Br:$^-$ $[Na]_2^+$:O:$^{2-}$ Be^{2+}[:F:]$_2^-$

Al^{3+}[:Cl:]$_3^-$ Mg^{2+} :S:$^{2-}$

1-3

:F:F: :F:C:F: with :F: top and :F: bottom   H:C:H with :Cl: top and :Cl: bottom   H:P:H with H bottom

:Br:I: $^-$:O:H   H:N:H with $^-$ charge $^-$:C:H with H bottom (and H)

1-4

H→O←H   SC→O   S→O   I→Br   H→C←H (with H top ↓ and H bottom ↑)

Cl←C→Cl (with H top ↓ and Cl bottom ↑)   H→C→Cl (with H top ↓ and Cl bottom ↑)   H→C←H (with H top ↓ and Cl bottom ↑)

1-5

You can view NH_3 as being isoelectronic with H_3C^-, H_2O with H_2C^{2-}. Electron repulsion by the free electron pairs causes the bonding electrons to "bend away," giving rise to the respective pyramidal and bent structures.

1-6

H:I: H:C:C:C:H (with H H H top and H H H bottom)   H:C:O:H (with H top and H bottom)   H:S:S:H

O::Si::O   O::O   S::C::S

1-7

Worked out in chapter.

1-8

S::O   :F:O:F: :F:B:N:H (with :F:H top and :F:H bottom)$^+$ H:C:O:H (with H top and H H bottom)$^+$

:Cl: C::O :Cl: $^-$:C:::N: $^-$:C:::C:$^-$

1-9

(a) The geometry should be close to trigonal (counting the lone electron pair), with equal N–O bond lengths and 1/2 of a negative charge on each oxygen atom.

$$\left[:\overset{..}{O} \quad \overset{..}{O}:^- \overset{\displaystyle N}{} \quad \longleftrightarrow \quad {}^-:\overset{..}{O} \quad \overset{..}{O}: \overset{\displaystyle N}{} \right]$$

(b) From Exercise 1-8: S=O; bond length 1.48 Å.

For SO_2:

$$\left[:\overset{-}{\underset{..}{O}}:\overset{+}{S}::\overset{..}{O} \quad \longleftrightarrow \quad \overset{..}{O}::\overset{+}{S}::\overset{-}{\underset{..}{O}} \quad \longleftrightarrow \quad \overset{..}{O}::\overset{+}{S}:\overset{-}{\underset{..}{O}}: \quad \longleftrightarrow \quad :\overset{-}{\underset{..}{O}}:\overset{2+}{S}:\overset{-}{\underset{..}{O}}: \right] $$
A

bond length of 1.43 Å fits structure A best, as one would expect all other forms to exhibit a value >1.48 Å.

1-10

Worked out in chapter.

1-11

(a) $\left[{}^-:C\equiv\overset{+}{N}-\overset{..}{O}:^- \quad \longleftrightarrow \quad {}^{2-}\overset{..}{C}=\overset{+}{N}=\overset{..}{O} \right]$

The left-hand structure is preferred because the charges are more evenly distributed and a negative charge resides on the relatively more electronegative oxygen.

(b) $\left[{}^-\overset{..}{N}=\overset{..}{O} \quad \longleftrightarrow \quad \overset{..}{N}-\overset{..}{O}:^- \right]$

The left-hand structure is preferred because the right-hand structure has no octet on nitrogen.

1-12

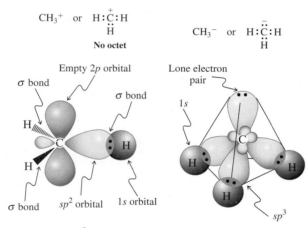

1-13

Draw according to the following electronic configurations:
$S\ (1s)^2(2s)^2(2p)^6(3s)^2(3p)^4$; $P\ (1s)^2(2s)^2(2p)^6(3s)^2(3p)^3$

1-14

Worked out in chapter.

1-15

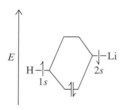

1-16

CH_3^+ or $H:\overset{+}{\underset{\cdot\cdot}{C}}:H$
No octet

CH_3^- or $H:\overset{\cdot\cdot}{\underset{}{C}}:H$

Trigonal, sp^2 hybridized, electron deficient like BH_3

Tetrahedral, sp^3 hybridized, closed shell

1-17

Worked out in chapter.

1-18

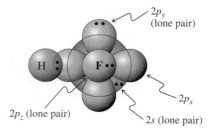

Hydrogen fluoride

1-19

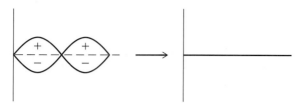

Butane   **Isobutane**

1-20

Self-explanatory. Note that the molecules are flexible and can adopt a variety of arrangements in space.

1-21

$CH_3CH_2CH_2CH_3$ $CH_3\overset{\overset{CH_3}{|}}{C}HCH_3$

1-22

(a)

(b)

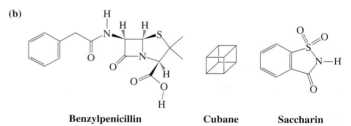

Benzylpenicillin   **Cubane**   **Saccharin**

Chapter 2

2-1

$\Delta G° = \Delta H° - T\Delta S°$
$= 22.4\text{ kcal mol}^{-1} - (298\text{ K} \times 33.3\text{ cal K}^{-1}\text{ mol}^{-1})$
$= 12.5\text{ kcal mol}^{-1}$

The reaction is unfavorable at 25°C. At higher temperatures, $\Delta G°$ is less positive, eventually becoming negative. The crossover point, $\Delta G° = 0$, is reached at 400°C, when $\Delta H° = T\Delta S°$.

2-2

$\Delta G° = \Delta H° - T\Delta S°$
$= -15.5\text{ kcal mol}^{-1}$
$\quad - (298\text{ K} \times -31.3\text{ cal K}^{-1}\text{ mol}^{-1})$
$= -6.17\text{ kcal mol}^{-1}$

The entropy is negative because two molecules are converted into one in this reaction.

2-3

Worked out in chapter.

2-4

We are given the stoichiometry of the reaction in the chemical equation: The two starting materials react in a 1:1 ratio. Consumption of one-half of the CH_3Cl reduces its concentration from 0.2 to 0.1 M. At this point an equal amount of NaOH must have been consumed. Therefore its concentration drops from 1.0 to 0.9 M, 90% of the initial value. The new rate is thus found by the following equation, in which [CH$_3$Cl] and [NaOH] refer to the initial concentrations of these materials.

$$\text{New rate} = k(0.5[\text{CH}_3\text{Cl}]) \, (0.9[\text{NaOH}]) = (0.45) \, k[\text{CH}_3\text{Cl}][\text{NaOH}]$$
$$= 0.45 \text{ (initial rate)}$$
$$= 0.45 \, (1 \times 10^{-4})$$
$$= 4.5 \times 10^{-5} \text{ mol L}^{-1}\text{s}^{-1}$$

It is equally valid (although not necessary in this particular case) to calculate the value of the rate constant from the given information and use it to arrive at the final answer:

$$k = (1 \times 10^{-4} \text{ mol L}^{-1}\text{s}^{-1}) \, / \, (0.2 \text{ M}) \, (1.0 \text{ M}) = 5 \times 10^{-4} \text{ L mol}^{-1}\text{s}^{-1}$$

$$\text{New rate} = k(0.5[\text{CH}_3\text{Cl}]) \, (0.9[\text{NaOH}]) = (5 \times 10^{-4}) \, (0.1) \, (0.9)$$
$$= 4.5 \times 10^{-5} \text{ mol L}^{-1}\text{s}^{-1}$$

2-5

(a) $+6.17$ kcal mol^{-1}

(b) $\Delta G° = 15.5 - (0.773 \times 31.3)$
$$= -8.69 \text{ kcal mol}^{-1}$$

Hence, the dissociation equilibrium lies on the side of ethene and HCl at this high temperature, where the entropy factor overrides the $\Delta H°$ term.

2-6

$k = 10^{14}e^{-58.4/1.53} = 3.03 \times 10^{-3} \text{ s}^{-1}$

2-7

(a) H$_3$C—Ö: $\quad+\quad$ $^+$H $\quad\longrightarrow\quad$ H$_3$C—Ö—H $\quad$ (Compare to Example 2)

(b) H$^+$ + CH$_3$CH=CHCH$_3$ $\quad\longrightarrow\quad$ CH$_3$CH—$\overset{+}{\text{C}}$HCH$_3$

with H above the first carbon

$\quad\quad\quad\quad\quad\quad\quad\quad\quad\quad$ (Compare to Example 4b)

(c) (CH$_3$)$_2$N: + H—Cl: $\quad\longrightarrow\quad$ (CH$_3$)$_2$N—H + :Cl:$^-$

$\quad\quad\quad\quad\quad\quad\quad\quad\quad\quad$ (Compare to Example 3)

(d) CH$_3$—Ö:$^-$ + H$_2$C=Ö $\quad\longrightarrow\quad$ CH$_3$—Ö—CH$_2$—Ö:$^-$

$\quad\quad\quad\quad\quad\quad\quad\quad\quad\quad$ (Compare to Example 4a)

2-8

(a) HSO$_3^-$ (b) ClO$_3^-$ (c) HS$^-$ (d) (CH$_3$)$_2$O (e) SO$_4^{2-}$

2-9

(a) (CH$_3$)$_2$NH (b) HS$^-$ (c) $^+$NH$_4$ (d) (CH$_3$)$_2$C=$\overset{+}{\text{Ö}}$H
(e) CF$_3$CH$_2$OH

2-10

Phosphorous acid is stronger. It has the smaller pK_a value, which corresponds to a larger acid dissociation constant, K_a. K_a(HNO$_2$) = $10^{-3.3}$; K_a(H$_3$PO$_3$) = $10^{-1.3}$.

2-11

Protonation of the oxygen at the end of the double bond gives a structure with three contributing resonance forms:

Protonation of the oxygen atom of the OH group gives a structure with only two contributing forms, the second of which is poor because of the two adjacent positively charged atoms:

Protonation at the double-bonded oxygen is preferred.

2-12

Worked out in chapter.

2-13

(a) pK_a = $-\log K_a$
K_a (acetic acid) = $10^{-4.7}$; K_a (benzoic acid) = $10^{-4.2}$. Benzoic acid is a stronger acid by a factor of $10^{0.5}$ = 3.2.

(b) Relatively electron withdrawing

Stronger acid

2-14

Worked out in chapter.

2-15

CH$_3$CH$_2$—Ï: + :NH$_3$ $\quad\longrightarrow\quad$ CH$_3$CH$_2$—$\overset{\overset{\text{H}}{|+}}{\underset{|}{\text{N}}}$H + :Ï:$^-$

2-16

(a)

(b) Higher homologs:

Lower homologs:

2-17

Isohexane $\quad\quad$ **Neopentane**

2-18

2-19

Self-explanatory.

2-20

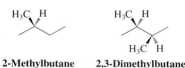

2-21

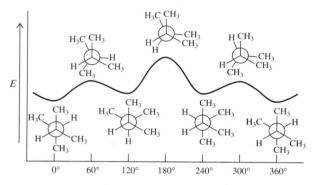

2-Methylbutane   **2,3-Dimethylbutane**

2-22

Worked out in chapter.

2-23

In this example (graphed below), the energy difference between the two staggered conformers turns out to be quite small.

Chapter 3
3-1

Worked out in chapter.

3-2

On the basis of orbital size and energy match, and therefore covalent overlap, the C–C, N–N, and O–O bonds should be quite similar. Furthermore, each bond is between two identical atoms, so all three will be nonpolar and lack Coulombic attraction. However, each N in the N–N bond contains one lone pair, and each O in hydrogen peroxide bears two. Lone pair–lone pair repulsion will weaken the N–N bond in hydrazine relative to the C–C bond in ethane, and the effect will be even more severe in hydrogen peroxide.

3-3

First: $CH_3 \dashv C(CH_3)_3$ $DH° = 87$ kcal mol^{-1}
Second: $CH_3 \dashv CH_3$ $DH° = 90$ kcal mol^{-1}

3-4

Worked out in chapter.

3-5

$CH_3Cl + Cl_2 \xrightarrow{hv} CH_2Cl_2 + HCl$

Initiation $:\!\ddot{C}l\!-\!\ddot{C}l\!: \xrightarrow{hv} 2\ :\!\ddot{C}l\cdot$

Propagation 1 $ClCH_2\!-\!H + \cdot\ddot{C}l\!: \longrightarrow ClCH_2\cdot + H\!-\!\ddot{C}l\!:$

Propagation 2 $ClCH_2\cdot + :\!\ddot{C}l\!-\!\ddot{C}l\!: \longrightarrow ClCH_2\!-\!\ddot{C}l\!: + \cdot\ddot{C}l\!:$

3-6

The answer is found by inspection of Figure 3-7 and Table 3-5. The chlorine atom is much more reactive than the bromine atom, as indicated by the ΔH values by the hydrogen abstraction step: $+2$ kcal mol^{-1} versus $+18$ kcal mol^{-1}. Since the transition states of atom abstraction reflect the relative stabilities of the products, it is evident that it is much easier to get to $(CH_3 + HCl)$ than to $(CH_3 + HBr)$.

3-7

$CH_3CH_2CH_2CH_3 + Cl_2 \xrightarrow{hv}$

$CH_3CH_2CH_2CH_2Cl + CH_3CH_2\overset{\displaystyle Cl}{\overset{|}{C}}HCH_3 + HCl$

The ratio of primary to secondary product is calculated by multiplying the number of respective hydrogens in the starting material by their relative reactivity:

$(6 \times 1):(4 \times 4) = 6:16 = 3:8$

In other words, 2-chlorobutane : 1-chlorobutane $= 8:3$.

3-8

Worked out in chapter.

3-9

The starting compound has four distinct groups of hydrogens:

Four possible monochlorination products are therefore possible, corresponding to replacement of a hydrogen atom in each of these four groups by chlorine:

A   **B**

C   **D**

The table summarizes the determination of the relative amounts of each product that would be expected to form.

Position (group)	Number of hydrogens	Relative reactivity	Relative yield	Percentage yield
A	3	1	3	10%
B	6	1	6	20%
C	4	4	16	53%
D	1	5	5	17%

3-10

Worked out in chapter.

3-11

2,3-Dimethylbutane, $(CH_3)_2CH—CH(CH_3)_2$, contains 12 identical primary and 2 identical tertiary hydrogen atoms, a 6:1 ratio. However, the tertiary:primary selectivity for monochlorination is only 5:1. Therefore, monochlorination will give roughly equal amounts of the two possible monochlorination products, $(ClCH_2)(CH_3)CH—CH(CH_3)_2$ and $(CH_3)_2CCl—CH(CH_3)_2$, and will not be useful synthetically. In contrast, the tertiary:primary selectivity for monobromination is about 1800:1. Monobromination will produce $(CH_3)_2CBr—CH(CH_3)_2$ with high selectivity, (1800/6) or 300:1, and will be an excellent synthetic procedure.

3-12

In this isomerization, a secondary hydrogen and a terminal methyl group in butane switch positions:

Hence,

$\Delta H° =$ (sum of the strengths of the bonds broken)
      $-$ (sum of the strengths of the bonds made)
$= (98.5 + 89) - (88 + 101)$
$= -1.5$ kcal mol^{-1}

Chapter 4

4-1

Aspects of ring strain and conformational analysis are discussed in Sections 4-2 through 4-5.

Note that the cycloalkanes are much less flexible than the straight-chain alkanes and thus have less conformational freedom. Cyclopropane must be flat and all hydrogens eclipsed. The higher cycloalkanes have increasing flexibility, more hydrogens attaining staggered positions and the carbon atoms of the ring eventually being able to adopt *anti* conformations.

4-2

Worked out in chapter.

4-3

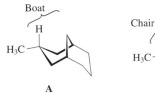

trans-1-Bromo-2-methylcyclohexane   *cis*-1-Bromo-3-methylcyclohexane   *trans*-1-Bromo-3-methylcyclohexane

cis-1-Bromo-4-methylcyclohexane   *trans*-1-Bromo-4-methylcyclohexane

4-4

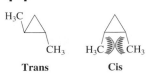

Trans   **Cis**

The cis isomer suffers from steric hindrance and has a larger heat of combustion (by about 1 kcal mol^{-1}).

4-5

Worked out in chapter.

4-6

This problem is a variant of Exercise 4-5, except that x is now given and the $\Delta H°$ of the reaction is the unknown. Since the product cyclohexane is strain free, the heat of the reaction of A with H_2 will equal the strain of the shared bond. With the central bond in 2,3-dimethylbutane, $DH° = 85.5$ kcal mol^{-1} (Table 3-2) as a reference, the strength of this bond will be $85.5 - 50.7 = 34.8$ kcal mol^{-1}. The formation of cyclohexane from A will therefore be exothermic by $\Delta H° = (104 + 34.8) - 197 = -58.2$ kcal mol^{-1}.

4-7

(a) One carbon bridge

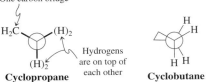

Hydrogens are on top of each other

Cyclopropane   **Cyclobutane**

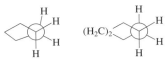

Cyclopentane   **Cyclohexane**

The respective C—H torsional angles are roughly 0° for cyclopropane; 37° for cyclobutene; variable (depending on the bond under inspection) for cyclopentane, ranging from ~10–40°; and 57° for cyclohexane.

(b)

4-8

$\log K = -1.7/1.36 = -1.25$
$K = 10^{-1.25} = 0.056$. Compare $K = 5/95 = 0.053$

4-9

Worked out in chapter.

4-10

Boat   Chair

A   **B**

4-11

(a) $\Delta G° =$ energy difference between an axial methyl and axial ethyl group: $1.75 - 1.70 =$ about 0.05 kcal mol^{-1}, that is, very small.
(b) Same as (a).
(c) $1.75 + 1.70 = 3.45$ kcal mol^{-1}

4-12

(a)

Both axial–equatorial

(b)

Diequatorial   **Diaxial**

(c)

Diequatorial   **Diaxial**

(d)

Both axial–equatorial

4-13

Worked out in chapter.

4-14

In the molecular model for *trans*-1,2-dimethylcyclohexane, perform the equatorial-to-axial ring flip and compare the respective methyl environments with those in methylcyclohexane. Clearly, in the diaxial form, the two methyls are completely out of each other's way, each encountering the same two 1,3-diaxial hydrogens across the respective faces of the molecule as in the monomethyl derivative. Hence, additivity applies. In the diequatorial structure, the two substituents appear to be very close. To visualize this, recall Figure 4-12, focus on the ring bond connecting the two methyl-bearing carbons, and ignore the remainder: We are looking at a *gauche* butane interaction (Section 2-9, Figures 2-12 and 2-13), destabilizing this conformer by ~0.9 kcal mol^{-1}. Hence, the $\Delta G°$ for ring flip is less than 3.4 kcal mol^{-1}.

} **Gauche**

0.9 kcal mol^{-1}

4-15

trans-Decalin is fairly rigid. Full chair – chair conformational "flipping" is not possible. In contrast, the axial and equatorial positions in the cis isomer can be interchanged by conformational isomerization of both rings. The barrier to this exchange is small ($E_a = 14$ kcal mol^{-1}). Because one of

the appended bonds is always axial, the cis isomer is less stable than the trans isomer by 2 kcal mol^{-1} (as measured by combustion experiments).

Ring flip in *cis*-decalin

4-16

All equatorial

4-17

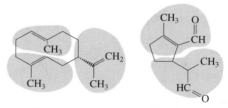

Sesquiterpene   **Monoterpene**

4-18

Chrysanthemic acid: $\mathrm{C{=}C}$, —COOH, —COOR

Grandisol: $\mathrm{C{=}C}$, —OH

Menthol: —OH

Camphor: $\mathrm{C{=}O}$

β-Cadinene: $\mathrm{C{=}C}$

Taxol: —OH, —O—, aromatic benzene rings, $\mathrm{C{=}O}$, —COOR, —CONHR

Chapter 5
5-1

Cyclopropyl-cyclopentane   **Cyclobutyl-cyclobutane**

Both hydrocarbons have the same molecular formula: C_8H_{14}. Therefore, they are (constitutional) isomers.

5-2

There are several boat and twist-boat forms of methylcyclohexane, some of which are shown:

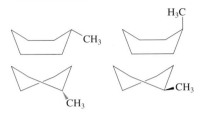

5-3

All are chiral. Note, however, that 2-methylbutadiene (isoprene) itself is achiral. Number of stereocenters: chrysanthemic acid, 2; grandisol, 2; menthol, 3; camphor, 2; β-cadinene, 3; taxol, 11; epiandrosterone, 7; cholesterol, 8; cholic acid, 11; cortisone, 6; testosterone, 6; estradiol, 5; progesterone, 6; norethindrone, 6; ethynylestradiol, 5; RU-486, 5.

5-4

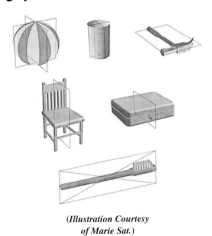

(Illustration Courtesy of Marie Sat.)

5-5

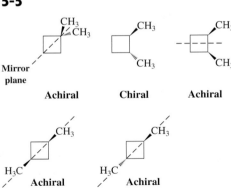

5-6

$$[\alpha] = \frac{6.65}{1 \times 0.1} = 66.5$$

The enantiomer of natural sucrose has $[\alpha] = -66.5$.

5-7

Worked out in chapter.

5-8

Optical purity (%)	Ratio (+/−)	$[\alpha]_{obs.}$
75	87.5/12.5	+17.3°
50	75/25	+11.6°
25	62.5/37.5	+5.8°

5-9

(a) —CH₂Br > —CCl₃ > —CH₂CH₃ > —CH₃

(b) [cyclohexyl] > —CH(CH₃)CH₃ > —CH₂CH(CH₃)CH₃

(c) —C(CH₃)₃ > —CH(CH₃)CH₂CH₃ > —CH₂CH(CH₃)CH₃ > —CH₂CH₂CH₂CH₃

(d) —CH(Br)CH₃ > —CH(Cl)CH₃ > —CH₂CH₂Br > —CH₂CH₃

5-10

Worked out in chapter.

5-11

(+)-2-Bromobutane: *S*
(+)-2-Aminopropanoic acid: *S*
(−)-2-Hydroxypropanoic acid: *R*

5-12

S   *R*   *S*

5-13

5-14

120°

5-15

Worked out in chapter.

5-16

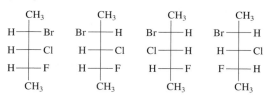

5-17

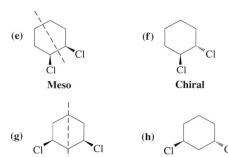

Placing the lowest-priority substituent d at the top of a Fischer projection means that it is located behind the plane of the page, the place required for the correct assignment of absolute configuration by visual inspection.

5-18

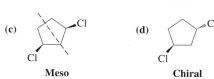

They are diastereomers.

5-19

1: (2S,3S)-2-Fluoro-3-methylpentane.
2: (2R,3S)-2-Fluoro-3-methylpentane.
3: (2R,3R)-2-Fluoro-3-methylpentane.
4: (2S,3S)-2-Fluoro-3-methylpentane.
1 and 2 are diastereomers; 1 and 3 are enantiomers; 1 and 4 are identical; 2 and 3 and 2 and 4 are diastereomers; 3 and 4 are enantiomers. With the inclusion of the mirror image of 2, there are four stereoisomers.

5-20

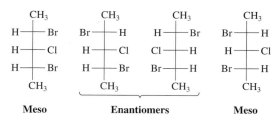

With the inclusion of the four mirror images, there are four enantiomeric pairs of diastereomers.

5-21

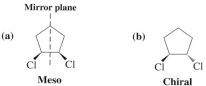

5-22

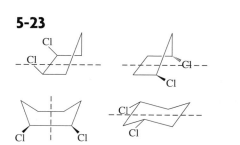

5-23

5-24

Almost any halogenation at C2 gives a racemate; the exception is bromination, which results in achiral 2,2-dibromobutane. In addition, bromination at C3 gives the two 2,3-dibromobutane diastereomers, one of which, 2R,3S, is meso.

5-25
Worked out in chapter.

5-26

Attack at C1:

CH₂Br
H——Br
H——H
H——H
CH₃

(R)-1,2-Dibromopentane
Chiral, optically active

Attack at C2:

CH₃
Br——Br
H——H
H——H
CH₃

2,2-Dibromopentane
Achiral

Attack at C3:

CH₃   CH₃
H——Br   H——Br
H——Br   Br——H
H——H   H——H
CH₃   CH₃

(2S,3R)-2,3-   **(2S,3S)-2,3-**
Dibromopentane   **Dibromopentane**
Chiral, optically   **Chiral, optically**
active   **active**

Diastereomers, formed
in unequal amounts

Attack at C4:

CH₃   CH₃
H——Br   H——Br
H——H   H——H
H——Br   Br——H
CH₃   CH₃

(2S,4R)-2,4-   **(2S,4S)-2,4-**
Dibromopentane   **Dibromopentane**
Achiral, meso,   **Chiral,**
optically inactive   **optically active**

Diastereomers, formed
in unequal amounts

Attack at C5:

CH₃
H——Br
H——H
H——H
CH₂Br

(S)-1,4-Dibromopentane
Chiral, optically active

5-27

Bromocyclohexane has a mirror plane and is therefore achiral. Chlorination at C2 should give the diastereomers *cis*- and *trans*-1-bromo-2-chlorocyclohexane in unequal ratios. In practice, the less hindered trans product dominates. Since we are starting with achiral material, the products are racemic. Indeed,

attack from the left (a) gives the set of enantiomers shown, whereas the equally likely attack from the right (b) gives the mirror images (not shown).

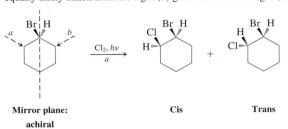

Mirror plane: Cis   Trans
achiral

Chapter 6

6-1

(a) $CH_3CH_2CH_2CH_2\ddot{\underset{..}{I}}:$

(b) $CH_3CH_2CH_2CH_2\ddot{\underset{..}{O}}CH_2CH_3$

(c) $CH_3CH_2CH_2CH_2\overset{..}{N}=\overset{+}{N}=\overset{..}{\underset{..}{N}}^-$

(d) $\left[CH_3CH_2CH_2CH_2\underset{\underset{CH_3}{|}}{\overset{\overset{CH_3}{|}}{As}}CH_3 \right]^+ \quad :\ddot{\underset{..}{Br}}:^-$

(e) $\left[CH_3CH_2CH_2CH_2\underset{\underset{CH_3}{|}}{\overset{..}{Se}}CH_3 \right]^+ \quad :\ddot{\underset{..}{Br}}:^-$

6-2
Worked out in chapter.

6-3

Delete C—N bond

$H_3C\overset{\overset{CH_3}{|}}{\underset{\underset{CH_3}{|}}{-\overset{+}{N}}}CH_3 \ \ I^- \ \Rightarrow \ H_3C\overset{\overset{CH_3}{|}}{\underset{\underset{CH_3}{|}}{-N}}: \ + \ H_3C-I$

Connect I to C of CH₃

6-4

(a) H⁺ ⁻:ÖH ⟶ H₂Ö

(b) :F̈:⁻ BF₃ ⟶ ⁻BF₄

(c) H₃N: H—C̈l: ⟶ ⁺NH₄ :C̈l:⁻

(d) Na⁺ ⁻:ÖCH₃ H—S̈—H ⟶ CH₃ÖH Na⁺ ⁻:S̈H

(e) (CH₃)₂Ö̈⁺—H H₂Ö: ⟶ (CH₃)₂Ö̈ H₃O:⁺

(f) H₂O: H—ÖH ⟶ H₃O:⁺ ⁻:ÖH

6-5

In mechanisms 1 and 3, oxygen is the nucleophile and carbon is the electrophile. In mechanism 4, the carbon–carbon double bond is the nucleophile and the proton is the electrophile. In mechanism 2, the dissociation, no external nucleophiles or electrophiles are pictured, but the carbon atom is electrophilic at the start and becomes more so after the chloride leaves.

6-6

(a) $(CH_3)_3N:$ $CH_3\overset{\frown}{\underset{..}{I}}:$

(b) Only the first approach is shown.

$H_3C—\underset{..}{\overset{..}{S}}:^-$ $\overset{\frown}{\underset{..}{Br}}:$ and $—\underset{..}{\overset{..}{S}}:^-$ $H_3C\overset{\frown}{\underset{..}{Br}}:$

6-7

Only two examples are shown.

4. $:N≡C:^-$ $\overset{\frown}{\underset{..}{I}}:$ 7. $(CH_3)_3P:$ $CH_3\overset{\frown}{\underset{..}{Br}}:$

6-8

(a) $—\overset{|}{C^+}$ + $:\underset{..}{\overset{..}{Cl}}:^-$ ⟶ $—\overset{|}{\underset{|}{C}}—\underset{..}{\overset{..}{Cl}}:$

(b) $H\underset{..}{\overset{..}{O}}:^-$ + $\overset{H}{\underset{|}{C}}—\overset{+}{C}—$ ⟶ $H_2\underset{..}{\overset{..}{O}}$ + $C=C$

6-9

Worked out in chapter.

6-10

(a) 9×10^{-10} mol L^{-1} s^{-1}
(b) 1.2×10^{-9} mol L^{-1} s^{-1}
(c) 2.7×10^{-9} mol L^{-1} s^{-1}

6-11

Frontside displacement

$\begin{matrix} H_3C \\ \quad \\ H \end{matrix} \overset{:\underset{..}{\overset{..}{I}}:^-}{\underset{CH_2CH_3}{C}}—\underset{..}{\overset{..}{Br}}:$ ⟶ $\begin{matrix} H_3C \\ \quad \\ H \end{matrix} \overset{}{\underset{CH_2CH_3}{C}}—\underset{..}{I}:$ + $:\underset{..}{\overset{..}{Br}}:^-$

Backside displacement

$:\underset{..}{\overset{..}{I}}:^-$ $\overset{H_3C}{\underset{H}{\underset{CH_2CH_3}{C}}}—\underset{..}{\overset{..}{Br}}:$ ⟶ $:\underset{..}{\overset{..}{I}}—C\overset{CH_3}{\underset{CH_3CH_2}{\underset{H}{}}}$ + $:\underset{..}{\overset{..}{Br}}:^-$

6-12

(a) $Cl—\!\!\begin{matrix} CH_2CH_3 \\ \quad \\ H \\ CH_2CH_2CH_2CH_3 \end{matrix}$ + $Na^+ {}^-SH$ ⟶

$H—\!\!\begin{matrix} CH_2CH_3 \\ \quad \\ SH \\ CH_2CH_2CH_2CH_3 \end{matrix}$ + $Na^+ Cl^-$

(b) $\overset{H \quad Br}{\diagdown \quad}$ + $:N(CH_3)_3$ ⟶

$(CH_3)_3\overset{+}{N}\ H$ + Br^-

(c) $\overset{H\ \ I}{\diagup}\cdots\overset{}{\underset{H\ \ CH_3}{}}$ + $K^+ {}^-SeCH_3$ ⟶

$\overset{CH_3Se\ \ H}{}\cdots\overset{}{\underset{H\ \ CH_3}{}}$ + $K^+ I^-$

6-13

$\begin{matrix} CH_3 \\ H\!—\!Br \\ H\!—\!H \\ H\!—\!Br \\ CH_3 \end{matrix}$ $\xrightarrow[-Br^-]{^-CN}$ $\begin{matrix} CH_3 \\ NC\!—\!H \\ H\!—\!H \\ NC\!—\!H \\ CH_3 \end{matrix}$

Meso   **Meso**

$\begin{matrix} CH_3 \\ \bigcirc \\ I \end{matrix}$ $\xrightarrow[-I^-]{^-CN}$ $\begin{matrix} CH_3 \\ \bigcirc \\ CN \end{matrix}$

Trans   **Cis**

6-14

$\overset{CH_2CH_3}{H\!\!—\!\!\overset{S}{\underset{(CH_2)_4CH_3}{}}\!\!—\!\!I}$ + $CH_3\overset{O}{\overset{||}{C}}O^-$ ⟶ $CH_3\overset{O}{\overset{||}{C}}O\overset{CH_2CH_3}{\underset{(CH_2)_4CH_3}{\overset{R}{—}}}H$ + I^-

$\downarrow Br^-$

$Br\!\!—\!\!\overset{CH_2CH_3}{\underset{(CH_2)_4CH_3}{\overset{R}{}}}\!\!—\!\!H$ + $CH_3\overset{O}{\overset{||}{C}}O^-$ ⟶ $H\overset{CH_3CH_2}{\underset{(CH_2)_4CH_3}{\overset{S}{—}}}O\overset{O}{\overset{||}{C}}CH_3$

6-15

Worked out in chapter.

6-16

The structure of S-alanine follows from the priority rules presented in Chapter 5: $–NH_3^+ > –COO^- > –CH_3 > –H$, giving as the answer

$\overset{H_3C}{\underset{H_3N^+}{\underset{H}{C}}}—COO^-$

Because S_N2 displacement inverts the stereochemistry at the site of reaction, the stereoisomer of 2-bromopropanoic acid necessary for the synthesis of S-alanine is

$\overset{H_3C}{\underset{H}{\underset{Br^{'''}}{C}}}—COOH$

the R enantiomer.

6-17

$\begin{matrix} CH_3 \\ Br\!—\!H \\ H\!—\!H \\ H\!—\!Cl \\ CH_3 \end{matrix}$ + ^-CN $\xrightarrow[\text{Excess}]{\text{Ethanol}}$ $\begin{matrix} CH_3 \\ H\!—\!CN \\ H\!—\!H \\ NC\!—\!H \\ CH_3 \end{matrix}$ + Br^- + Cl^-

2R,4R   **2S,4S**

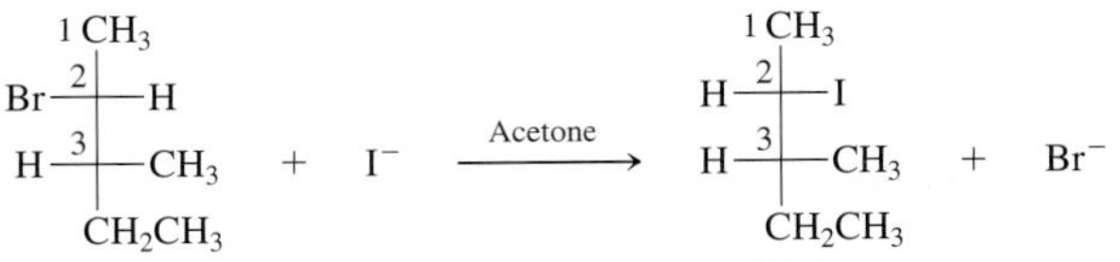

2R,3R → **2S,3R**

All four components are diastereomers of their counterparts on p. 226.

6-18
I⁻ is a better leaving group than Cl⁻. Hence the product is $Cl(CH_2)_6SeCH_3$.

6-19
The relative acidities of the acids are listed first, then the relative basicities of their conjugate bases. In each case, the weaker of the two bases (the last compound listed) is the better leaving group.
(a) $H_2Se > H_2S$, $HS^- > HSe^-$
(b) $H_2S > PH_3$, $PH_2^- > HS^-$
(c) $HClO_3 > HClO_2$, $ClO_2^- > ClO_3^-$
(d) $HBr > H_2Se$, $HSe^- > Br^-$
(e) $H_3O^+ > {}^+NH_4$, $NH_3 > H_2O$

6-20
(a) $^-OH > {}^-SH$ (b) $^-PH_2 > {}^-SH$

(c) $^-SeH > I^-$ (d) $HO\overset{O}{\overset{\|}{S}}O^- > HO\overset{O}{\underset{\|}{\overset{\|}{S}}}O^-$

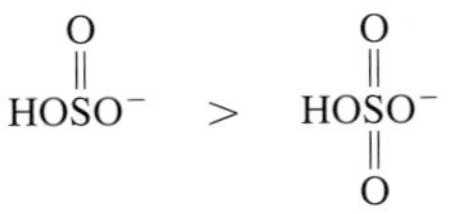

The relative acidities of the respective conjugate acids follow the inverse order.

6-21
(a) $HS^- > H_2S$ (b) $CH_3S^- > CH_3SH$
(c) $CH_3NH^- > CH_3NH_2$ (d) $HSe^- > H_2Se$

6-22
(a) $CH_3S^- > Cl^-$ (b) $P(CH_3)_3 > S(CH_3)_2$
(c) $CH_3CH_2Se^- > Br^-$ (d) $H_2O > HF$

6-23
(a) $CH_3SeH > CH_3SH$ (b) $(CH_3)_2PH > (CH_3)_2NH$

6-24
(a) CH_3S^- (b) $(CH_3)_2NH$

6-25
Worked out in chapter.

6-26
The substrate contains a nucleophilic nitrogen atom that is well positioned to react in an intramolecular manner with a carbon bearing a good leaving group at the other terminus. Following the model reaction 6 in Table 6-3, we can write the following mechanism:

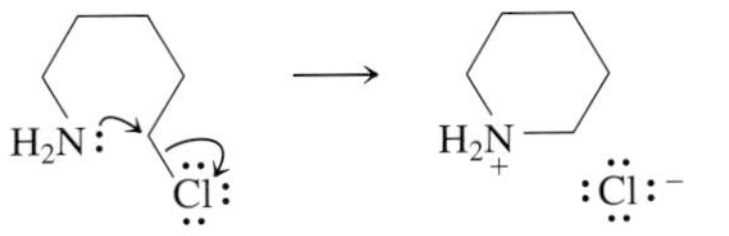

The product is a salt, which is only poorly soluble in the weakly polar ether solvent and precipitates as a white solid.

6-27

The more reactive substrates are (a) [structure] and
(b) $CH_3CH_2CH_2Br$.

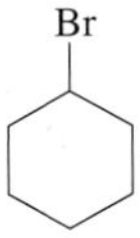

6-28

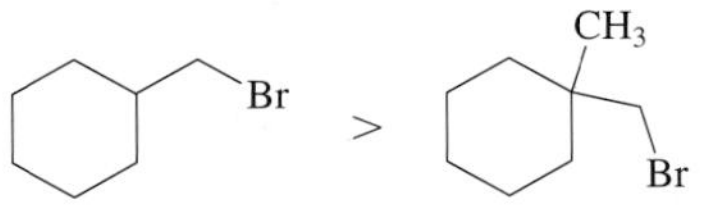

Chapter 7

7-1
Compound A is a 2,2-dialkyl-1-halopropane (neopentyl halide) derivative. The carbon bearing the potential leaving group is primary but very hindered and therefore very unreactive with respect to any substitution reactions. Compound B is a 1,1-dialkyl-1-haloethane (*tert*-alkyl halide) derivative and undergoes solvolysis.

7-2
Bonds broken [R = $(CH_3)_3C$]: 71 (R–Br) + 119 (H–OH) = 190 kcal mol⁻¹
Bonds made: 96 (R–OH) + 87 (H–Br) = 183 kcal mol⁻¹
$$\Delta H° = +7 \text{ kcal mol}^{-1}$$

By this calculation, the reaction should be endothermic. It still proceeds because of the excess water employed and the favorable solvation energies of the products.

7-3

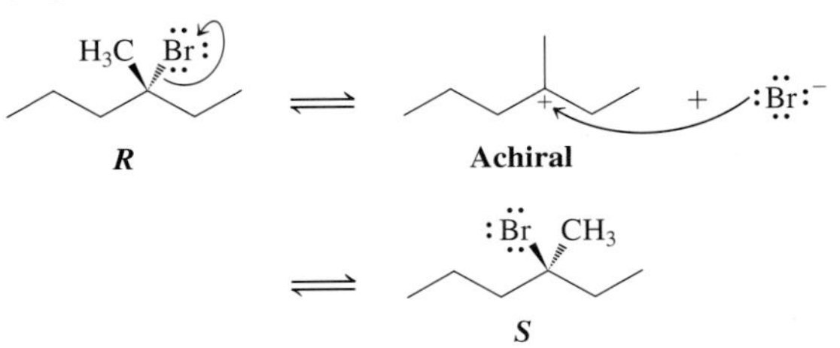

The molecule dissociates to the achiral tertiary carbocation. Recombination gives a 1:1 mixture of R and S product.

7-4
Worked out in chapter.

7-5

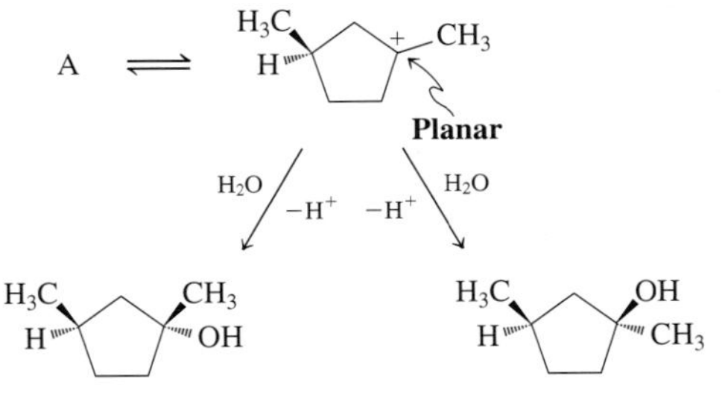

7-6
Worked out in chapter.

7-7
Concentrated aqueous ammonia contains three potential nucleophiles: water, ammonia, and hydroxide ion. The **caution** indicates that the concentration of hydroxide is very low, suggesting that the major S_N1 product is not likely to result from reaction of the intermediate carbocation with hydroxide. Both water and ammonia are present in sizable concentrations, however, and ammonia is the superior nucleophile (Table 6-7). Thus the major product should be the amine, $(CH_3)_3CNH_2$.

7-8
Worked out in chapter.

7-9

It is true that the reaction described will yield *some* (R)-2-butanamine. Ammonia is a good nucleophile, so both S_N1 and S_N2 mechanisms may occur to give both the *R* and *S* enantiomers of the product. It would be incorrect, however, to describe this as a "synthetically useful" preparation of the *R* enantiomer. The outcome will be a mixture of enantiomers with the *S* predominating, and they will require a somewhat laborious separation (Section 5-8). A more efficient method, one that makes better use of the enantiomeric purity of the starting chloro compound, would employ exclusively stereospecific reactions. We know that the S_N2 process is stereospecific and proceeds with inversion. If we choose to react the starting material with a good S_N2 nucleophile such as I^-, *which can also function as a good leaving group,* we may cleanly obtain a product with an *S* stereocenter:

$$R\text{-}CH_3CH_2CHClCH_3 + NaI \longrightarrow S\text{-}CH_3CH_2CHICH_3 + NaCl$$

Carried out in acetone with stoichiometric NaI, this process is driven to completion by the precipitation of NaCl, which is insoluble in that solvent (Section 6-8).

Having prepared $S\text{-}CH_3CH_2CHICH_3$ with high enantiomeric purity, one can subsequently perform a second, clean S_N2 inversion using ammonia as the nucleophile in, preferably, a polar aprotic solvent such as an ether to arrive at the desired product.

7-10

$$(CH_3)_3CBr \rightleftharpoons (CH_3)_3C^+ + Br^-$$

$(CH_3)_3COCH_2CH_3$ $(CH_3)_3COH$ $H_2C=C(CH_3)_2$

+ H^+ + H^+ + H^+

7-11

S_N2 E2

7-12

$CH_2=CH_2$; no E2 possible; $CH_2=C(CH_3)_2$; no E2 possible.

7-13

I^- is a better leaving group, thus allowing for selective elimination of HI by E2.

7-14

Worked out in chapter.

7-15

All chlorines are equatorial, lacking *anti* hydrogens.

7-16

(a) $N(CH_3)_3$, stronger base, worse nucleophile
(b) $(CH_3CH)_2N^-$, more hindered base (with CH3 on carbon)
(c) Cl^-, stronger base, worse nucleophile (in protic solvent)

7-17

Thermodynamically, eliminations are usually favored by entropy, and the entropy term in $\Delta G° = \Delta H° - T\Delta S°$ is temperature dependent. Kinetically, eliminations have higher activation energies than do substitutions, so their rates rise more rapidly with increasing temperature (see Problem 49 of Chapter 2).

7-18

(a) $CH_3CH_2CH_2CN$
(b) $CH_3CH_2CH_2OCH_3$
(c) $CH_3CH=CH_2$

7-19

(a) $(CH_3)_2CHCH_2I$ (b) $(CH_3)_2C=CH_2$

7-20

(a) $(CH_3)_2CHOCH_2CH_3$
(b) $(CH_3)_2CHSCH_3$
(c) $CH_3CH=CH_2$

7-21

(a)   (b)

7-22

(a) The second reaction will give more E2 product, because a stronger base is present.
(b) The first reaction will give E2 product, mainly because of the presence of a strong, hindered base, which is absent in the second reaction.

Chapter 8

8-1

(a)   (b)   (c) $(CH_3)_3CCH_2OH$

8-2

(a) 4-Methyl-2-pentanol
(b) *cis*-4-Ethylcyclohexanol
(c) 3-Bromo-2-chloro-1-butanol

8-3

Worked out in chapter.

8-4

All the bases whose conjugate acids have pK_a values $\geqslant 15.5$—i.e., $CH_3CH_2CH_2CH_2Li$, LDA, and KH.

8-5

8-6

In condensed phase, $(CH_3)_3COH$ is a weaker acid than CH_3OH. The equilibrium lies to the right.

8-7

(a) $NaOH$, H_2O
(b) 1. CH_3CO_2Na,   2. $NaOH$, H_2O
(c) H_2O

8-8

(a)

(b) $CH_3CH_2CHCH_2CH_3$ (with OH)   (c)

8-9

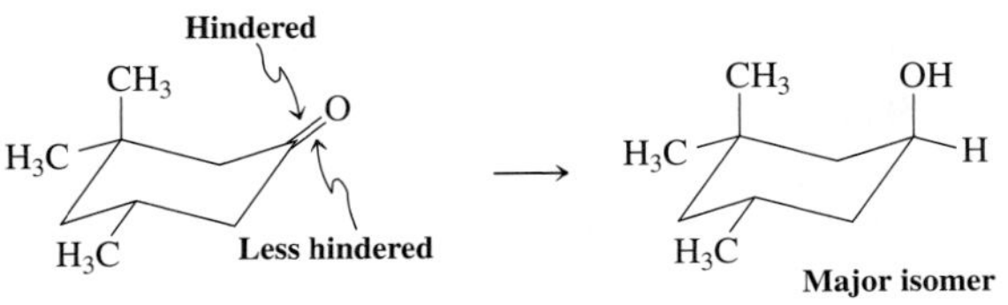

Hindered
CH3
O
H3C
H3C
Less hindered
CH3
OH
H3C
H
H3C
Major isomer

8-10

(a) $CH_3(CH_2)_8CHO + NaBH_4$ **(b)**

$+ \quad NaBH_4$

(c)

$+ \quad NaBH_4$ **(d)**

$+ \quad NaBH_4$

8-11

(a)

OH
Na2Cr2O7,
H2SO4, H2O
O
NaBH4
OH
OH
+
CH3
CH3
CH3
CH3
cis
No stereochemistry
cis
+
trans

The lesson is that oxidation of an alcohol will remove any stereochemistry associated with the hydroxy-bearing carbon. Reduction may generate two stereoisomers, unless there is strong stereoselectivity (see Exercise 8-9).

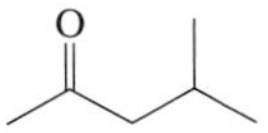

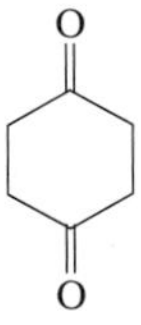

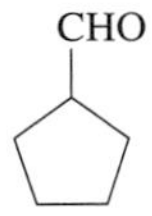

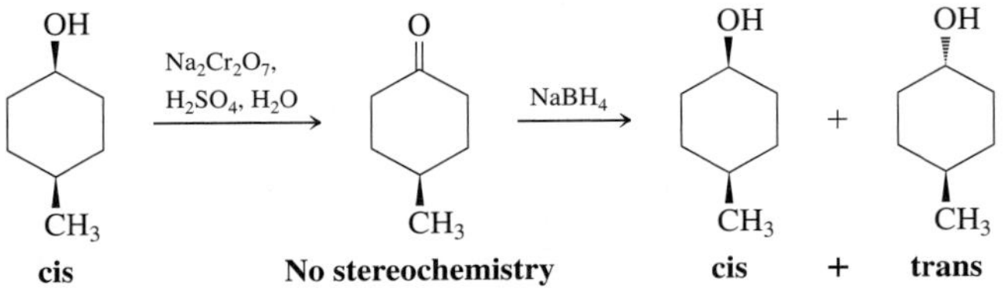

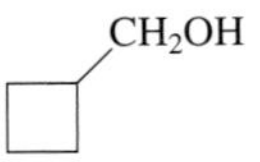

(b)

HO H HO H
Meso
HO H H OH
+ Mirror image

(c)

O
O
H
H
Achiral

8-12

(a) $CH_3CH_2CH(OH)CH(CH_3)_2 \quad + \quad Na_2Cr_2O_7$

(b)

$+ \quad PCC$

(c)

CH3CH2
OH
CH3

$+ \quad Na_2Cr_2O_7$

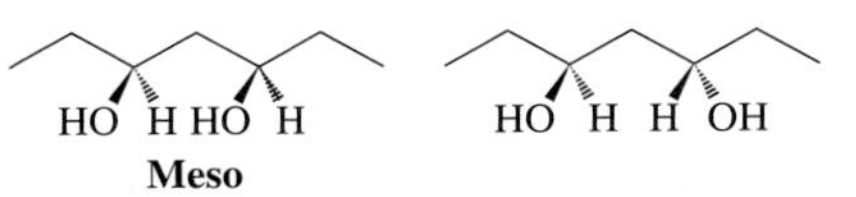

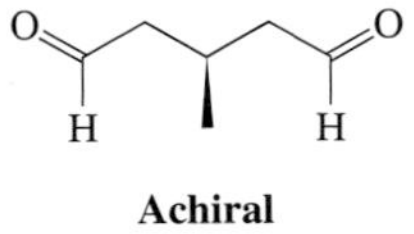

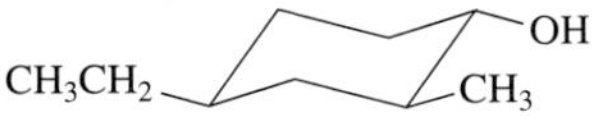

8-13

Worked out in chapter.

8-14

$$CD_3OH \xrightarrow[2.\ D_2O]{1.\ CH_3Li} CD_3OD$$

8-15

$$(CH_3)_2CHBr \xrightarrow{Mg} (CH_3)_2CHMgBr \xrightarrow{CH_2{=}O} (CH_3)_2CHCH_2OH$$

8-16

(a) $CH_3CH_2CH_2CH_2Li + CH_2{=}O$

(b) $CH_3CH_2CH_2MgBr + CH_3CH_2CH_2CHO$

(c) $(CH_3)_3CLi \quad +$

(d) $CH_3CH_2CH_2MgBr \quad + \quad CH_3\overset{O}{\overset{\|}{C}}CH_2CH_3$

8-17

Worked out in chapter.

8-18

(a) Product:

$ClCH_2CH_2CH_2C(OCH_2CH_3)(CH_3)_2$ By S_N1

(b) Product:

$(CH_3)_2C(OH)CH_2CH_2CHO$

The second hydroxy function is tertiary.

8-19

The desired alcohol is tertiary and is therefore readily made from 4-ethylnonane by 1. Br_2, $h\nu$; 2. hydrolysis (S_N1). However, the starting hydrocarbon is itself complex and would require an elaborate synthesis. Thus, the retrosynthetic analysis by C–O disconnection is poor.

8-20

Worked out in chapter.

8-21

$$CH_4 \xrightarrow{Br_2,\ h\nu} CH_3Br \xrightarrow{Mg} CH_3MgBr$$

CH_3Br → (1. NaOH; 2. PCC) → $H_2C{=}O$

CH_3MgBr → (1. $H_2C{=}O$; 2. PCC) → CH_3CHO

$$CH_3CHO \xrightarrow[2.\ Na_2Cr_2O_7]{1.\ CH_3MgBr} CH_3\overset{O}{\overset{\|}{C}}CH_3 \xrightarrow{CH_3MgBr} (CH_3)_3COH$$

Chapter 9

9-1

$$\underset{pK_a\,=\,15.5}{CH_3OH} + HO^- \overset{K}{\rightleftharpoons} CH_3O^- + \underset{pK_a\,=\,15.7}{H_2O}$$

$$K = \frac{[CH_3O^-]\,[H_2O]}{[CH_3OH]\,[HO^-]}$$

The pK_a values for CH_3OH and H_2O are essentially the same, hence we can simplify $K = 1$. Remember, however, that this K refers to *equimolar concentrations of starting materials.* Since CH_3OH is the solvent, its

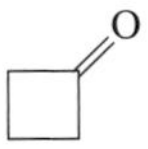

concentration is its molarity: $1000/32 \cong 31$, a 3100-fold excess over the starting concentration of HO^-. Hence the equilibrium will be shifted strongly to the right, and essentially all of the added HO^- will be converted to CH_3O^-.

9-2

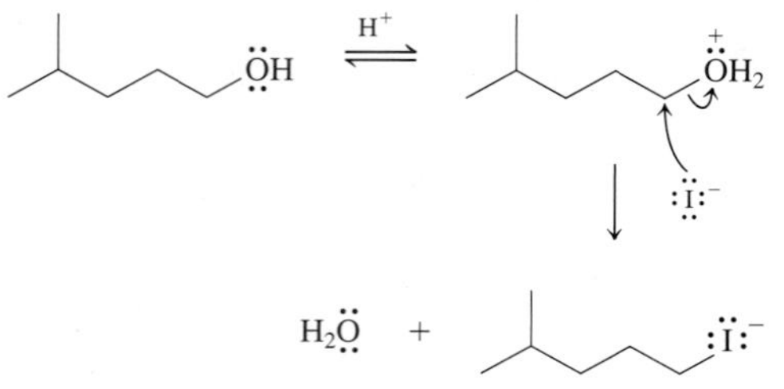

9-3

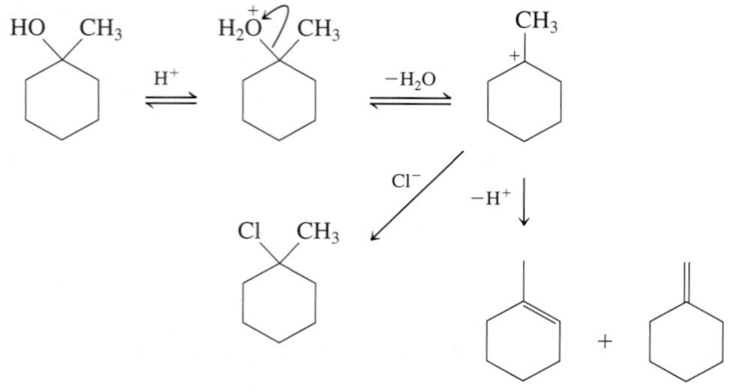

The tertiary carbocation is either trapped by the nucleophile (Cl^-) or undergoes E1. (HSO_4^- is a poor nucleophile.)

9-4

Worked out in chapter.

9-5

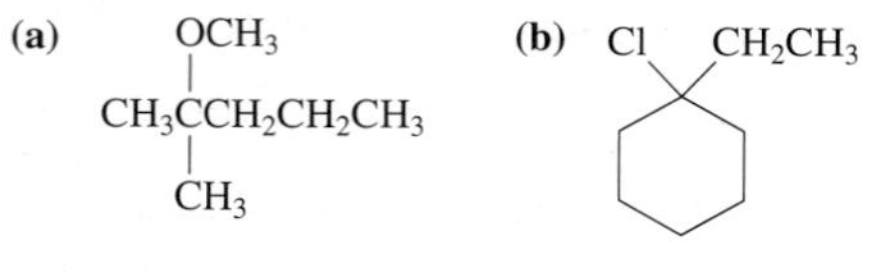

(a)

(b)

9-6

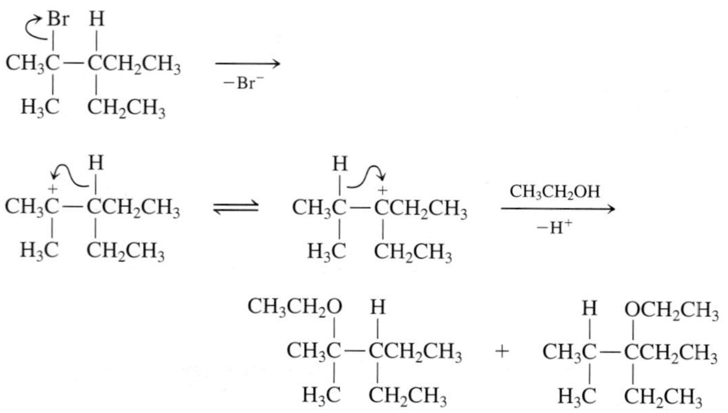

Similarly,

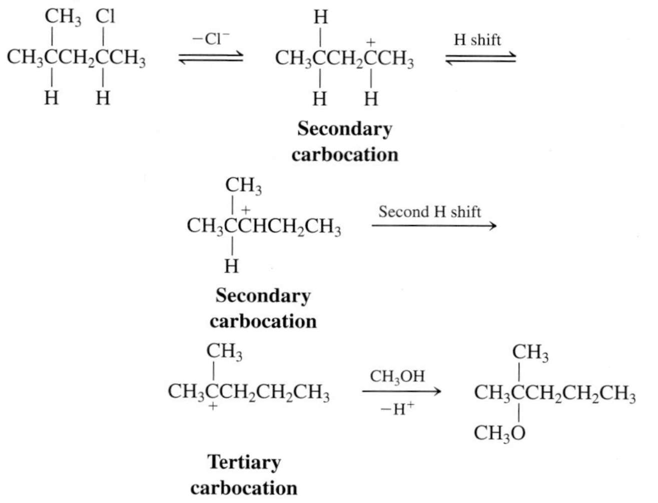

9-7

(a)

(b)

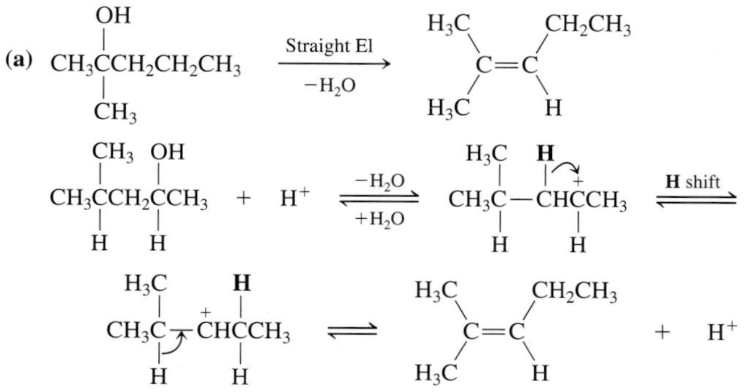

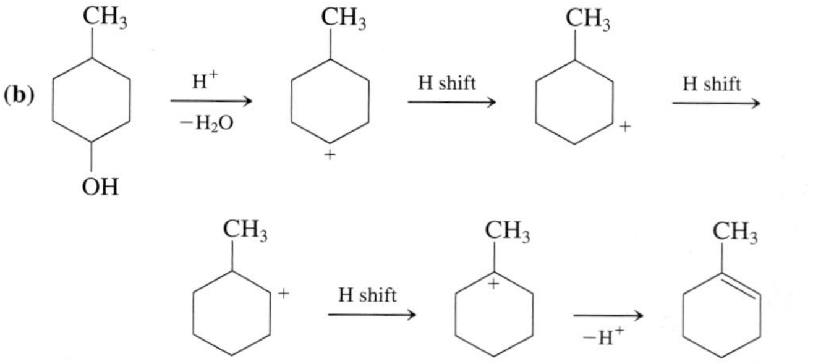

9-8

Worked out in chapter.

9-9

$(CH_3)_3CCH=CH_2$, $CH_2=C(CH_3)CH(CH_3)_2$, and $(CH_3)_2C=C(CH_3)_2$

9-10

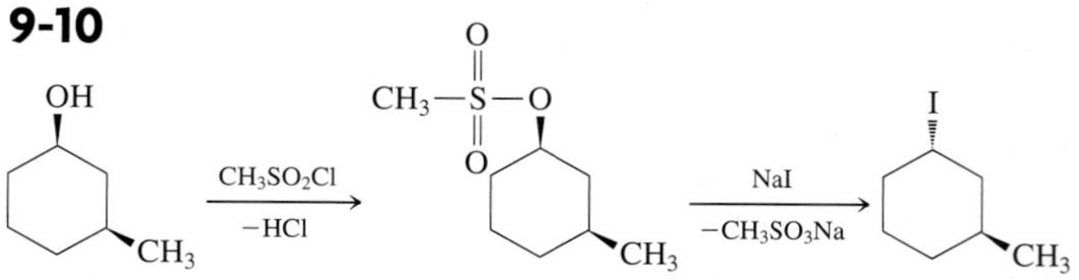

9-11

(a) 1. CH_3SO_2Cl, 2. NaI (b) HCl (c) PBr_3

9-12

(a) 1. $CH_3CH_2CH_2CH_2O^-Na^+ + CH_3CH_2I$,
2. $CH_3CH_2O^-Na^+ + CH_3CH_2CH_2CH_2I$

(b) Best is $+$ CH_3I

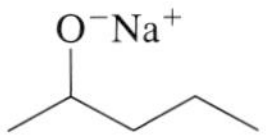

The alternative, $CH_3O^-Na^+$ $+$, suffers from competing E2.

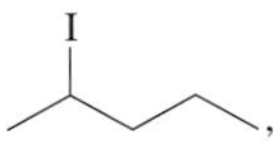

(c) Best is $+$ $CH_3CH_2CH_2Br$

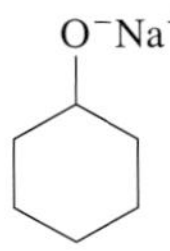

The alternative, $+$ $CH_3CH_2CH_2O^-Na^+$,

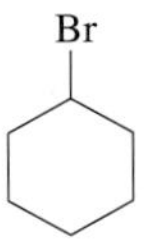

suffers from competing E2.

(d) 1. 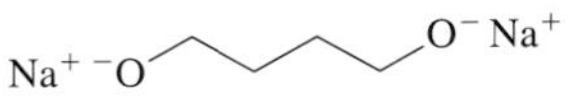 $Na^+ {}^-O$———$O^- Na^+$ $+$ $CH_3CH_2OSO_2CH_3$

2. 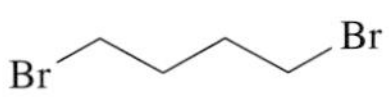 Br———Br $+$ $2\ CH_3CH_2O^-\ Na^+$

The problem with HO———Br as a starting material is that

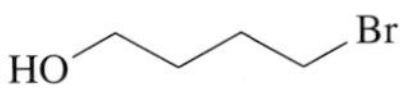

it will cyclize to  .

9-13

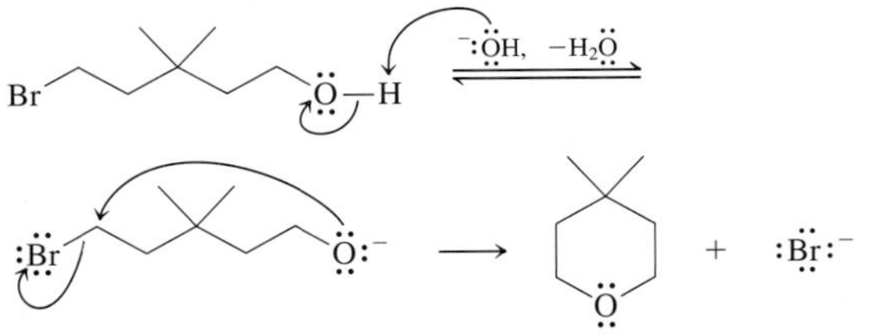

9-14

Worked out in chapter.

9-15

The *tert*-butyl group "locks" the conformations of the cyclohexane rings (Table 4-3). For A, this means a trans-diaxial arrangement of the alkoxide function and the leaving bromide, allowing for a relatively unconstrained S_N2 reaction. For B, the alkoxide function and the leaving bromide are forced into a trans-diequatorial arrangement, making the S_N2 reaction much more difficult.

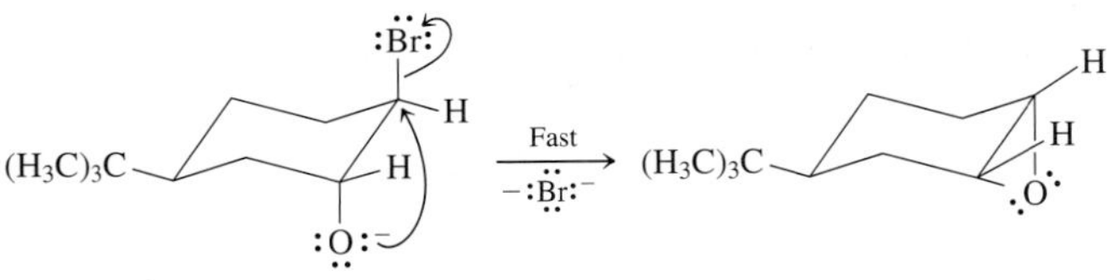

A: Good alignment for backside displacement

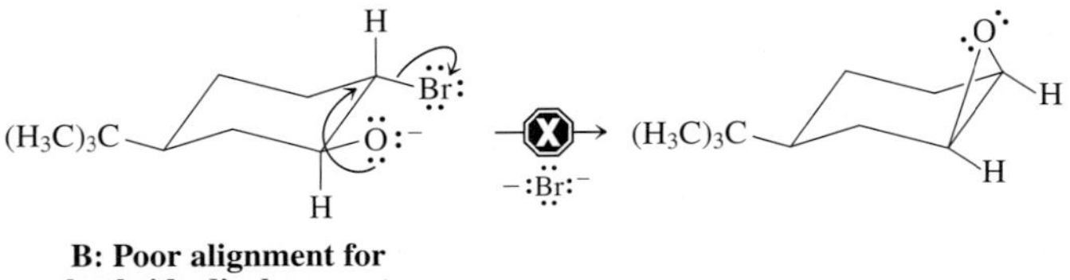

B: Poor alignment for backside displacement

9-16

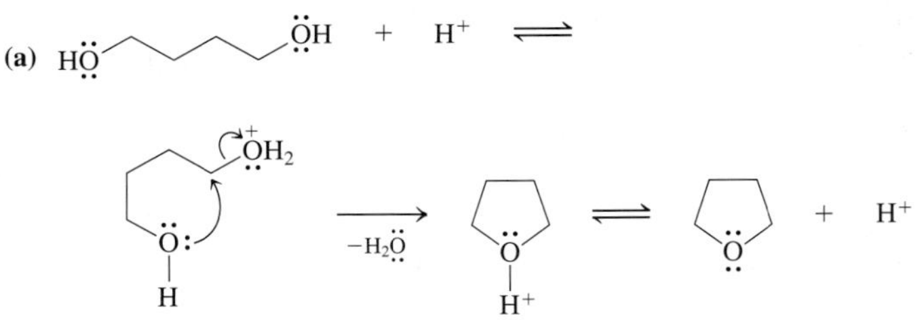

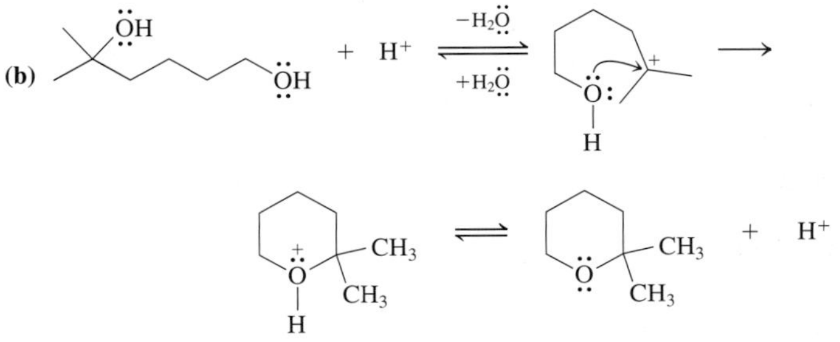

(a) HO———$ÖH$ $+$ H^+ $\rightleftharpoons$

(b)

9-17

(a) This ether is best synthesized by solvolysis:

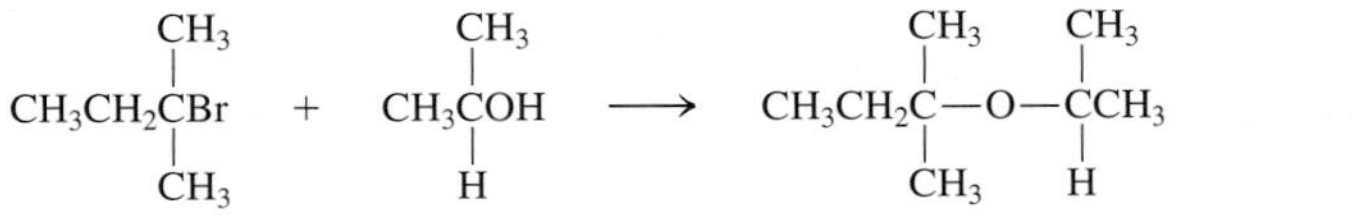

2-Methyl-2-(1-methylethoxy)butane

The alternative, an S_N2 reaction, would give elimination:

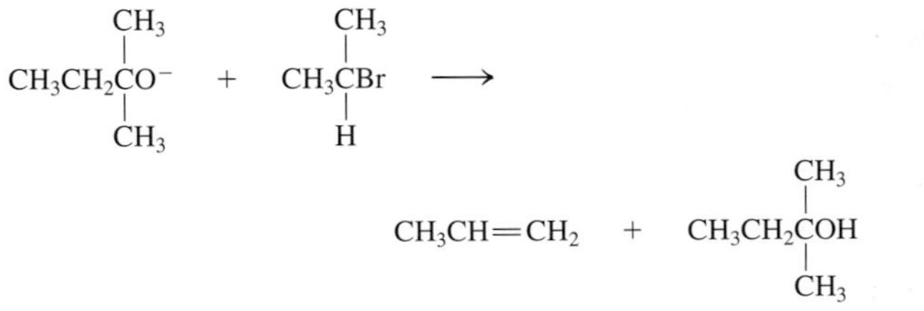

(b) This target is best prepared by an S_N2 reaction with a halomethane, because such an alkylating agent cannot undergo elimination. The alternative would be nucleophilic substitution of a 1-halo-2,2-dimethylpropane, a reaction that is normally too slow.

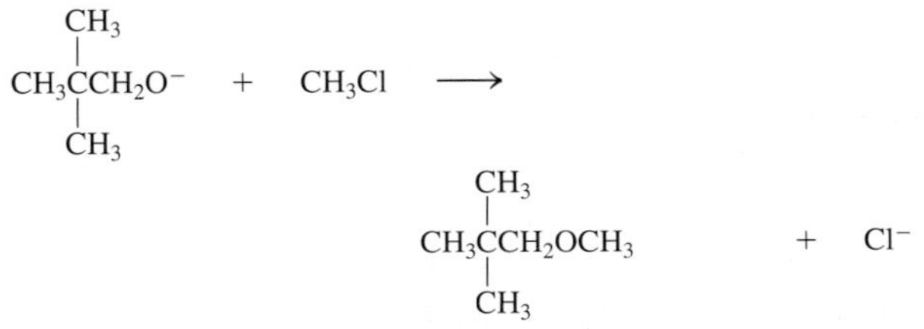

1-Methoxy-2,2-dimethylpropane

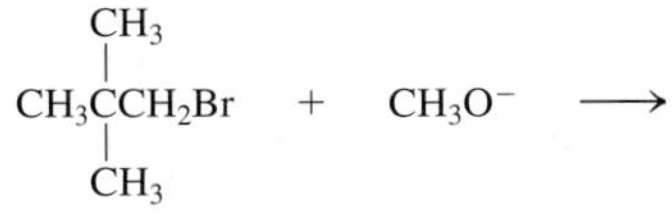

$\longrightarrow$ slow reaction, impractical

9-18

$$CH_3OCH_3 + 2\ HI \xrightarrow{\Delta} 2\ CH_3I + H_2O$$

Mechanism

$$CH_3\ddot{O}CH_3 + H\ddot{I}: \rightleftharpoons CH_3\overset{H}{\overset{|+}{O}}CH_3 + :\ddot{I}:^-$$

$$:\ddot{I}:^- \quad CH_3-\overset{H}{\overset{|+}{O}}-CH_3 \longrightarrow CH_3\ddot{I}: + H\ddot{O}CH_3$$

$$CH_3\ddot{O}H + H\ddot{I}: \rightleftharpoons CH_3\overset{H}{\overset{|+}{O}}H + :\ddot{I}:^-$$

$$:\ddot{I}:^- \quad CH_3-\overset{H}{\overset{|+}{O}}-H \longrightarrow CH_3\ddot{I}: + H_2\ddot{O}$$

9-19

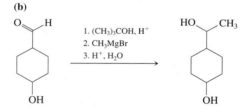

$$H_2O + I\diagdown\diagup\diagdown\diagup\diagdown I$$

9-20

(a)

1. $(CH_3)_3COH$, H^+
2. Mg
3. D_2O
4. H^+, H_2O

$$BrCH_2CH_2CH_2OH \longrightarrow DCH_2CH_2CH_2OH$$

(b)

1. $(CH_3)_3COH$, H^+
2. CH_3MgBr
3. H^+, H_2O

(reaction of 4-hydroxycyclohexanecarbaldehyde giving 1-(4-hydroxycyclohexyl)ethanol)

9-21

Worked out in chapter.

9-22

$$\overset{O}{\underset{H_3C\ \ R\ \ R\ CH_3}{\triangle}} \xrightarrow{LiAlH_4} \overset{HO}{\underset{H_3C}{H^{\cdots}R}}$$

Attack at either end of the oxacyclopropane will give the same product. Note that *cis*-dimethylcyclopropane is not suitable as a precursor, because it is achiral and therefore gives racemic alcohol.

9-23

$$(CH_3)_3CLi + \overset{O}{\triangle} \longrightarrow \text{(neopentyl-type alcohol) OH}$$

9-24

(a) $(CH_3)_3COH$ (b) $CH_3CH_2CH_2CH_2C(CH_3)_2OH$
(c) $CH_3SCH_2C(CH_3)_2OH$ (d) $HOCH_2C(CH_3)_2OCH_2CH_3$
(e) $HOCH_2C(CH_3)_2Br$

9-25

(a)

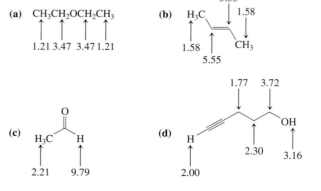

(b) Intramolecular sulfonium salt formation

$$ClCH_2CH_2\ddot{S}CH_2 \overset{CH_2-Cl}{\curvearrowright} \longrightarrow ClCH_2CH_2-\overset{+}{\underset{..}{S}}\overset{CH_2}{\underset{CH_2}{\diagdown}} \quad Cl^-$$

Nucleophiles attack by ring opening

$$ClCH_2CH_2\overset{+}{\underset{..}{S}}\overset{CH_2}{\underset{CH_2}{\diagdown}} :Nu \longrightarrow ClCH_2CH_2\ddot{S}CH_2CH_2Nu^+$$

Chapter 10

10-1

There are quite a number of isomers, e.g., several butanols, pentanols, hexanols, and heptanols. Examples include

$$\underset{\substack{H_3C\quad CH_3 \\ | \quad\quad | \\ CH_3C-COH \\ | \quad\quad | \\ H_3C\quad CH_3}}{\text{}}$$

2,3,3-Trimethyl-2-butanol

$$CH_3\underset{\underset{CH_3}{|}}{CH}\underset{\underset{CH_3}{|}}{CH}CH_2CH_2OH$$

3,4-Dimethyl-1-pentanol

$$CH_3(CH_2)_4\underset{\underset{\ }{|}}{\overset{\overset{CH_3}{|}}{C}}HOH$$

2-Heptanol

10-2

$$DH^\circ_{Cl_2} = 58\ \text{kcal mol}^{-1} = \Delta E$$
$$\Delta E = 28{,}600/\lambda$$
$$\lambda = 28{,}600/58 = 493\ \text{nm, in the ultraviolet-visible range}$$

10-3

$\delta = 80/90 = 0.89$ ppm; $\delta = 162/90 = 1.80$ ppm; $\delta = 293/90 = 3.26$ ppm; identical with the δ values measured at 300 MHz.

10-4

The methyl group resonates at higher field; the methylene hydrogens are relatively deshielded because of the cumulative electron-withdrawing effect of the two heteroatoms.

10-5

Indicated below are the actually measured chemical shifts δ (ppm).

(a) $CH_3CH_2OCH_2CH_3$
     1.21 3.47 3.47 1.21

(b) H_3C ... 5.55 ... 1.58 ... CH_3 ... 1.58 ... 5.55

(c) H_3C $\overset{O}{\underset{}{\diagup}}$ H ... 2.21 ... 9.79

(d) H (alkyne) ... 2.00 ... 1.77 ... 3.72 ... OH ... 2.30 ... 3.16

10-6

Worked out in chapter.

10-7

(a) 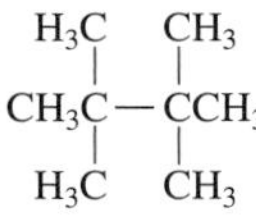
$$H_3C \quad CH_3$$
$$CH_3C—CCH_3 \quad \text{One peak}$$
$$H_3C \quad CH_3$$

(b) 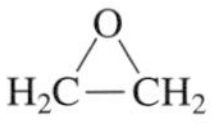
$$H_2C—CH_2 \quad \text{One peak}$$
(with O bridging)

10-8

The answer to the **caution** is "No." The substituent divides the originally all-equivalent hydrogens into seven new sets! First (as shown below), C1, C2, C3, and C4 are now different, and, second, all the hydrogens located on the same side of bromine are different from those located on the other side. In practice, this gives rise to a pretty messy spectrum, because the chemical shifts of all the hydrogens, except that at C1, are quite similar. See Real Life 10-3 for the actual ^{1}H NMR spectrum of bromocyclohexane.

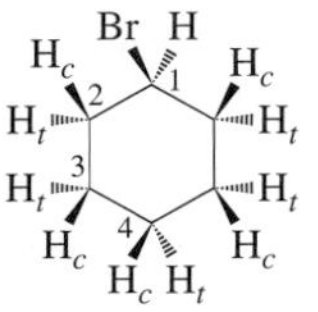

10-9

Worked out in chapter.

10-10

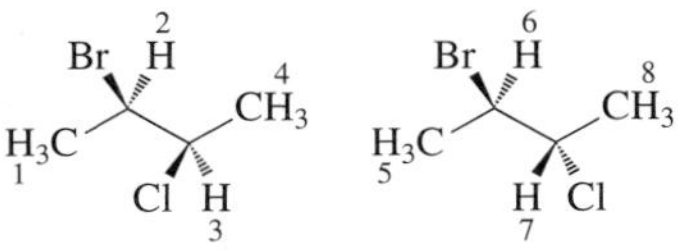

There is no symmetry, and both molecules will give rise to their distinct sets of four different signals, respectively.

10-11

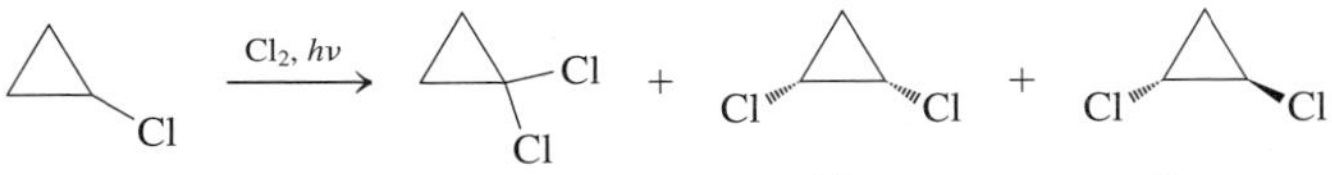

1,1-Dichlorocyclopropane shows only one signal for the four equivalent hydrogens. *Cis*-1,2-Dichlorocyclopropane exhibits three in the ratio 2:1:1. The lowest-field absorption is due to the two equivalent hydrogens next to the chlorine atoms at C1 and C2. The two hydrogens at C3 are not equivalent: One lies cis to the chlorine atoms, the other trans. In contrast, the trans isomer reveals only two signals (integration ratio, 1:1). Now the hydrogens at C3 *are* equivalent, as shown by a 180° rotational symmetry operation:

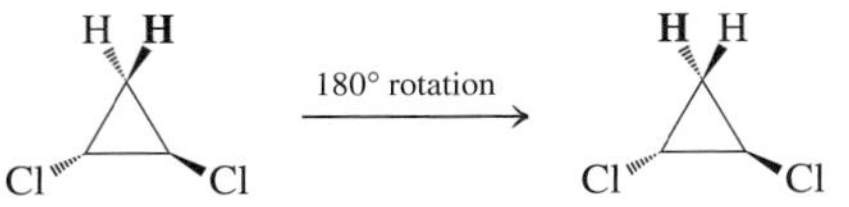

10-12

The following δ values were recorded in CCl$_4$ solution. You could not have predicted them exactly, but how close did you get?
(a) δ = 3.38 (q, *J* = 7.1 Hz, 4 H) and 1.12 (t, *J* = 7.1 Hz, 6 H) ppm
(b) δ = 3.53 (t, *J* = 6.2 Hz, 4 H) and 2.34 (quin, *J* = 6.2 Hz, 2 H) ppm
(c) δ = 3.19 (s, 1 H), 1.48 (q, *J* = 6.7 Hz, 2 H), 1.14
(s, 6 H), and 0.90 (t, *J* = 6.7 Hz, 3 H) ppm
(d) δ = 5.58 (t, *J* = 7 Hz, 1 H) and 3.71 (d, *J* = 7 Hz, 2 H) ppm

10-13

Worked out in chapter.

10-14

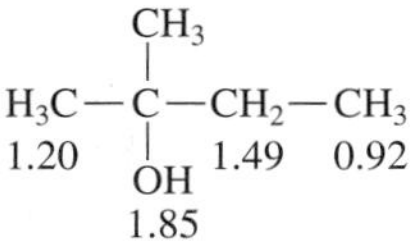

The ^{1}H NMR δ values (ppm) are given next to the corresponding hydrogens.

10-15

Worked out in chapter.

10-16

(a) Quintet (quin); triplet of triplets (tt); (b) quintet (quin); doublet of quartets (dq, or quartet of doublets, qd, which is the same); (c) triplet (t); doublet of doublets (dd); (d) sextet (sex); doublet of doublet of quartets (ddq).

10-17

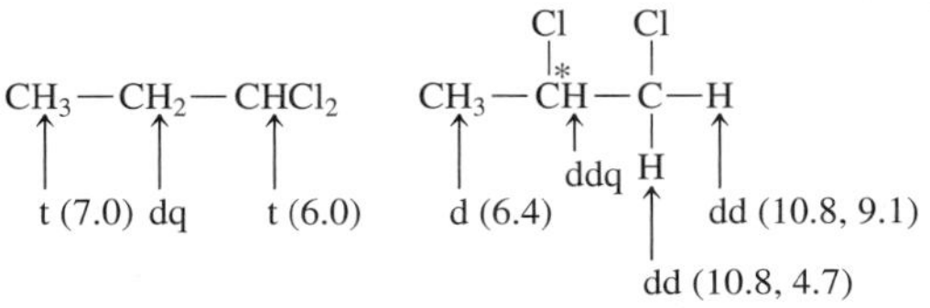

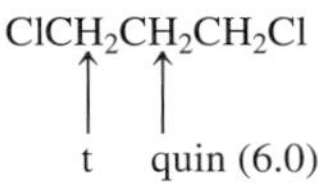

The hydrogens at Cl are not equivalent, because of the presence of the adjacent stereocenter (see Real Life 10-3)

The experimental *J* values (Hz) are given in parentheses.

10-18

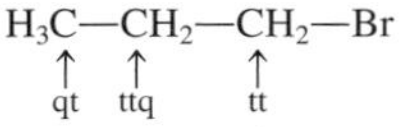

10-19

(a) 3; (b) 3; (c) 7; (d) 2;
(e) At 20°C (fast ring flip): 3. At −60°C (slow ring flip): 6.

10-20

Worked out in chapter.

10-21

Na$^+$ $^-$BH$_4$ reduction of ribose gives ribitol, which is a pentol with a plane of symmetry and therefore only three ^{13}C peaks. On the other hand, reduction of arabinose gives the unsymmetrical compound arabitol, with five peaks. We shall learn more about sugars in Chapter 24.

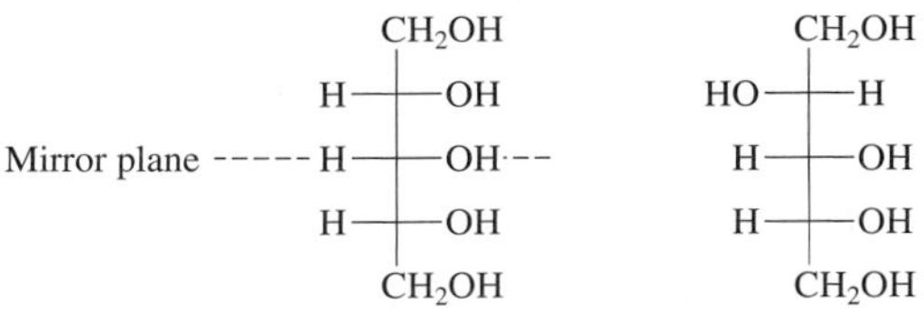

Ribitol
Meso: three ^{13}C signals

Arabitol
Chiral: five ^{13}C signals

10-22

For compound A, three lines, one of them at relatively high field (CH$_3$); DEPT would confirm CH$_3$ and two CH units. For compound B, three lines, no CH$_3$ absorption; DEPT would confirm the absence of CH$_3$ and the presence of two CH$_2$ units and one CH unit.

Chapter 11

11-1

(a) 2,3-Dimethyl-2-heptene   (b)   3-Bromocyclopentene

11-2

(a) *cis*-1,2-Dichloroethene
(b) *trans*-3-Heptene
(c) *cis*-1-Bromo-4-methyl-1-pentene

11-3

(a) (*E*)-1,2-Dideuterio-1-propene
(b) (*Z*)-2-Fluoro-3-methoxy-2-pentene
(c) (*E*)-2-Chloro-2-pentene

11-4

(a)   (b)   OH

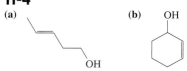

11-5

(a)

(b) (1-Methylethenyl)cyclopentene [(1-methylvinyl)cyclopentene]

11-6

$$CH_2{=}CHLi \ + \ CH_3\overset{O}{\overset{\|}{C}}CH_3 \longrightarrow CH_2{=}CH\underset{CH_3}{\overset{OH}{\underset{|}{\overset{|}{C}}}}CH_3$$

The reaction of ethenyllithium (vinyllithium) with carbonyl compounds is like that of other alkyllithium organometallics.

11-7

The induced local magnetic field strengthens H_0 in the region occupied by the methyl hydrogens.

11-8

Worked out in chapter.

11-9

Only the signal at $\delta = 7.23$ ppm shows two larger J values, 14.4 and 6.8 Hz, for trans and cis vicinal coupling, respectively. This signal must therefore correspond to the lone hydrogen of the CH group. By process of elimination we thus have the assignments shown below (δ values in ppm).

2.10 (s, 3 H) $\longrightarrow$ CH$_3$$\overset{O}{\overset{\|}{C}}$O   H $\longleftarrow$ 4.73 (dd, J = 14.4, 1.6 Hz, 1 H)

C=C

H   H $\longleftarrow$ 4.52 (dd, J = 6.8, 1.6 Hz, 1 H)

7.23 (dd, J = 14.4, 6.8 Hz, 1 H)

11-10

1-Hexene < *cis*-3-hexene < *trans*-4-octene < 2,3-dimethyl-2-butene.

11-11

If you can make a model of alkene A (without breaking your plastic sticks), you will notice its extremely strained nature, much of which is released on hydrogenation. You can estimate the excess strain in A (relative to B) by subtracting the $\Delta H°$ of the hydrogenation of a "normal" tetrasubstituted double bond (≈ -27 kcal mol^{-1}) from the $\Delta H°$ of the A-to-B transformation: 38 kcal mol^{-1}.

11-12

Worked out in chapter.

11-13

(a) We expect the less hindered base ethoxide to give predominantly the Saytzev rule product, while *tert*-butoxide should give the Hofmann rule product:

Preferred site of attack by ethoxide $\longrightarrow$ More likely to be $\longleftarrow$ attacked by *tert*-butoxide

$$H_3C-\underset{CH_3}{\overset{H}{\underset{|}{\overset{|}{C}}}}-\underset{Br}{\overset{CH_3}{\underset{|}{\overset{|}{C}}}}-CH_3 \longrightarrow$$

H$_3$C   CH$_3$ H$_3$C   CH$_2$
      C=C   +   CH—C
H$_3$C   CH$_3$ H$_3$C   CH$_3$

A   **B**
Saytzeff product   **Hofmann product**

(b) The base is even more bulky than *tert*-butoxide and thus gives a larger proportion of Hofmann product B.

11-14

H   H,,CH$_3$
   C—C   Base $\longrightarrow$
H$_3$C''' *R*   *R* Br
      D   or
      S   *S*

H$_3$C,,,   H   H,,,   H
      C=C   +   C=C
      D   CH$_3$ H$_3$C   CH$_3$
       E   **Cis (Z)**

H   H,,CH$_3$
   C—C   Base $\longrightarrow$
D''' *S*   *R* Br
H$_3$C   or
      R   *S*

D,,,   H   H$_3$C,,,   H
      C=C   +   C=C
H$_3$C   CH$_3$ H   CH$_3$
       Z   **Trans (E)**

Note that, in the first case, a pair of isomers is formed with the configuration *opposite* that generated in the second. The *E* and *Z* isomers of 2-deuterio-2-butene are isotopically pure in each case; none of the protic 2-butene with the same configuration is generated. The protio-2-butenes are also pure, devoid of any deuterium.

11-15

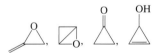

$$\underset{\substack{| \quad | \quad |\\ H \ H \ H}}{\overset{\substack{H_3C \ H \ OH\\ | \quad | \quad |}}{CH_3C-C-CCH_3}} \xrightarrow[-H_2O]{H^+}$$

$$\underset{\substack{| \quad | \quad |\\ H \ H \ H}}{\overset{\substack{H_3C \ H \quad CH_3\\ | \quad | \quad \overset{+}{}}}{CH_3C-C-C}} \longrightarrow \underset{\substack{| \quad |\\ H \quad H}}{\overset{\substack{CH_3 \quad H\\ | \quad |}}{H_3CC \overset{+}{\underset{}{C}} CCH_3}}$$

$$-H^+ \downarrow \qquad\qquad -H^+ \downarrow$$

$(CH_3)_2CHCH=CHCH_3$

$$\underset{H_3C}{\overset{H_3C}{>}}C=C\underset{CH_2CH_3}{\overset{H}{<}}$$

11-16

(a)

$$CH_3CH_2CH_2\ddot{O}H \underset{-H^+}{\overset{+H^+}{\rightleftharpoons}}$$

$$\underset{H}{\overset{}{CH_3CHCH_2\overset{+}{\ddot{O}}H_2}} \xrightarrow[-H_2SO_4]{+HOSO_3^-} CH_3CH=CH_2 + H_2\ddot{O}$$

(b) $CH_3CH_2CH_2OCH_2CH_2CH_3 \underset{}{\overset{H^+}{\rightleftharpoons}} CH_3CH=CH_2 + CH_3CH_2CH_2OH$
in analogy to (a). The propanol may then be dehydrated as in (a).

11-17

Alkene A: $\underset{H}{\overset{H_3C}{>}}C=C\underset{CH_3}{\overset{H}{<}}$

B: $CH_3CH_2CH=CH_2$

C: $CH_2=C\underset{CH_3}{\overset{CH_3}{<}}$

11-18

(a) CH_3OCH_3, CH_3CH_2OH, $\overset{O}{\triangle}$, $O—O$, $H\overset{O}{C}OH$

(b) $H_2C=O$

(c) (structures), $CH_2=CHCH$,

$HC\equiv CCH_2OH$, $CH_3C\equiv COH$, $HC\equiv COCH_3$,

(structures)

11-19

(a) $C_7H_{12}O$ **(b)** C_6H_{14}

11-20

CH_2Br_2: $m/z = 176, 174, 172$; intensity ratio $1:2:1$

11-21

For most elements in organic compounds, such as C, H, O, S, P, and the halogens, the mass (of the most abundant isotopes) and valence are either both even or both odd, so even molecular weights always result. Nitrogen is a major exception: the atomic weight is 14, but the valence is 3. This phenomenon has led to the **nitrogen rule** in mass spectrometry, as expressed in this exercise.

11-22

Mass spectrum of 3-methyl-3-heptanol

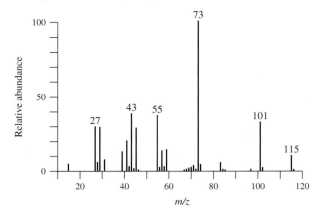

The major primary fragments are due to cleavage of the bonds α to the hydroxy group. Why? Consider their strength and the electronic structure of the resulting radical cations. (Draw resonance forms.) Do these cations fragment by loss of water?

11-23

$$H_3C-CH=CH\underset{H}{\overset{\overset{+}{CH_3}}{-C}}CH_2CH_3$$

The observed peaks are due to the molecular ion and two fragmentations (as indicated in the drawing) leading to the respective resonance stabilized allylic cations: $m/z = 98\ M^+$, $83\ (M-CH_3)^+$, $69\ (M-CH_2CH_3)^+$.

11-24

(a) $H_{sat} = 12$; degree of unsaturation $= 1$
(b) $H_{sat} = 20$; degree of unsaturation $= 4$
(c) $H_{sat} = 17$; degree of unsaturation $= 5$
(d) $H_{sat} = 19$; degree of unsaturation $= 2$
(e) $H_{sat} = 8$; degree of unsaturation $= 0$

11-25

Worked out in chapter.

11-26

Based on the information in Table 11-4 (no absorption from 1620 to 1680 cm^{-1} for a carbon–carbon double bond and none between 2100 and 2260 cm^{-1} for a triple bond), this particular compound lacks π bonds. The degrees of unsaturation (still two) leave us no alternative: It must contain two rings. There are only two options:

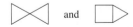 and

Both molecules are known, but they are quite reactive (recall Exercise 4-5).

Chapter 12

12-1

Estimating the strengths of the bonds broken and the bonds made

$$CH_2{=}CH_2 \;+\; HO{-}OH \;\longrightarrow\; \begin{array}{c} HO \quad OH \\ H{-}C{-}C{-}H \\ H \quad H \end{array}$$

65   49 $2 \times (\sim 94)$ kcal mol^{-1}

gives $\Delta H° = -74$ kcal mol^{-1}. Even though it is very exothermic, this reaction requires a catalyst.

12-2

Worked out in chapter.

12-3

Following the mechanism shown for Exercise 12-2, we envision complexation of the substrate to the catalyst surface and transfer of one hydrogen to an alkene carbon:

Single-hydrogen transfer intermediate

The problem states that 2-methyl-2-butene forms—a molecule with a double bond between C2 and C3. This can arise from transfer of the hydrogen at C3 to the catalyst surface. A reasonable scenario is shown here.

12-4

$$\begin{array}{c} CH_3CH_2 \quad\quad CH_3 \\ \underset{\underset{CH_3}{|}}{C}{-}\underset{\underset{CH_2}{\,}}{C} \\ H \,{}^{S} \end{array} \xrightarrow{H_2,\ catalyst}$$

$$\begin{array}{c} CH_3CH_2 \\ \underset{\underset{CH_3}{|}}{C}{-}CH(CH_3)_2 \\ H\,{}^{S} \end{array}$$

Not a stereocenter

12-5

Using H$^+$:

Using H—I:

12-6

(a) Both enantiomers
Br

(b) Br ... + ... Br
Both enantiomers

(c) $(CH_3)_2CCH_2CH_3$
Br

(d) Both cis and trans

12-7

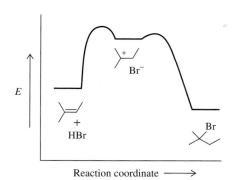

E

Reaction coordinate $\longrightarrow$

12-8

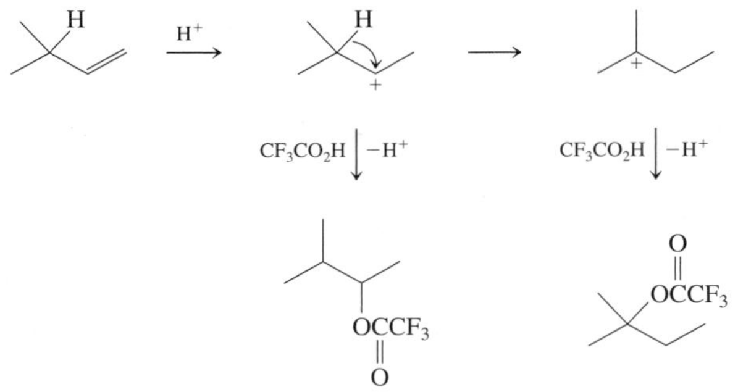

12-9

Protonation to the 1,1-dimethylethyl (*tert*-butyl) cation is reversible. With D^+, fast exchange of all hydrogens for deuterium will take place.

$$CH_2=C(CH_3)_2 \underset{-D^+}{\overset{+D^+}{\rightleftharpoons}} \overset{+}{D}CH_2C(CH_3)_2 \underset{+H^+}{\overset{-H^+}{\rightleftharpoons}}$$

$$DCH=C(CH_3)_2 \underset{-D^+}{\overset{+D^+}{\rightleftharpoons}} D_2\overset{+}{C}HC(CH_3)_2 \underset{+H^+}{\overset{-H^+}{\rightleftharpoons}}$$

$$D_2C=C(CH_3)_2 \underset{-D^+}{\overset{+D^+}{\rightleftharpoons}} D_3\overset{+}{C}C(CH_3)_2 \underset{+H^+}{\overset{-H^+}{\rightleftharpoons}}$$

$$\begin{array}{c} D_3C \\ C=CH_2 \\ H_3C \end{array} \underset{-D^+}{\overset{+D^+}{\rightleftharpoons}} \quad \text{and so on} \quad \text{-->}$$

$$(CD_3)_3C^+ \underset{-D_2O}{\overset{D_2O}{\rightleftharpoons}} (CD_3)_3COD + D^+$$

12-10

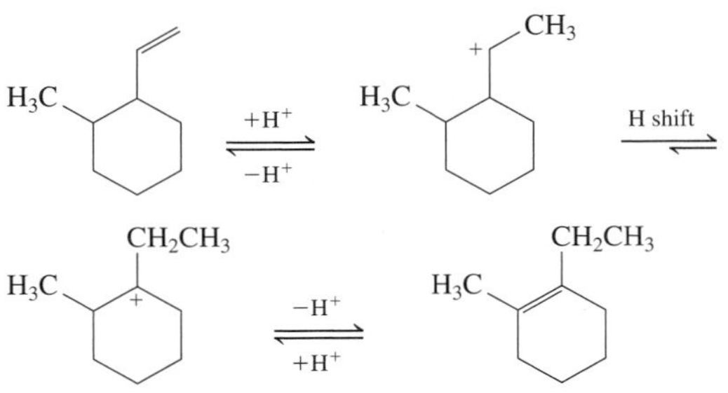

Tetrasubstituted alkene, most stable

12-11

$$CH_2=CH_2 + F-F \longrightarrow \begin{array}{cc} F & F \\ | & | \\ CH_2-CH_2 \end{array}$$
$$\quad 65 \qquad\quad 38 \qquad\qquad 2 \times (\approx 111) \text{ kcal mol}^{-1}$$
$$\qquad\qquad\qquad\qquad\qquad \Delta H^\circ = -119 \text{ kcal mol}^{-1}$$

$$CH_2=CH_2 + I-I \longrightarrow \begin{array}{cc} I & I \\ | & | \\ CH_2-CH_2 \end{array}$$
$$\quad 65 \qquad\quad 36 \qquad\qquad 2 \times (\approx 56) \text{ kcal mol}^{-1}$$
$$\qquad\qquad\qquad\qquad\qquad \Delta H^\circ = -11 \text{ kcal mol}^{-1}$$

12-12

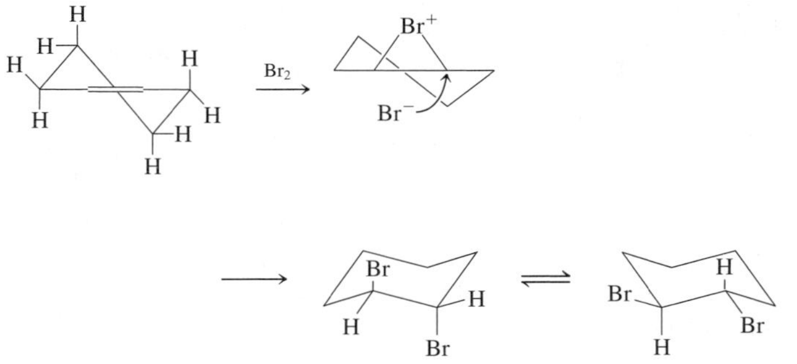

(1S,2S)-*trans*-1,2-Dibromocyclohexane

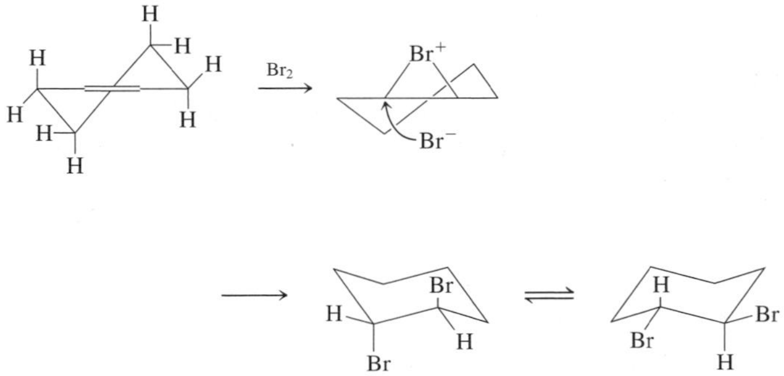

(1R,2R)-*trans*-1,2-Dibromocyclohexane

Anti addition to either conformation gives the trans-diaxial conformer initially.

12-13

(a) Only one diastereomer is formed (as a racemate):

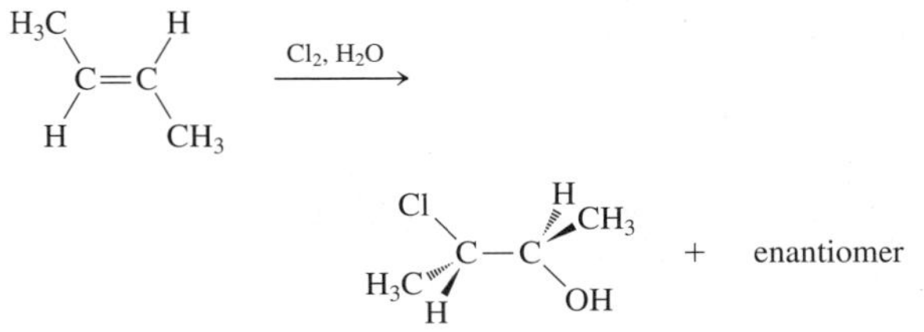

(b) Two isomers are formed, but only one diastereomer of each (as racemates):

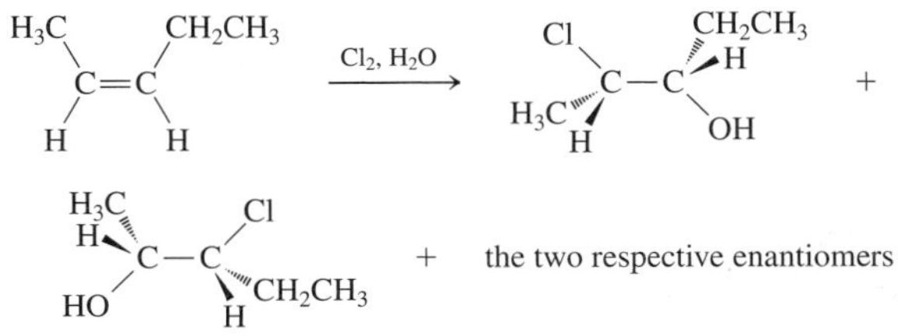

12-14

(a) CH_3CHCH_2Cl (both enantiomers)

(b) 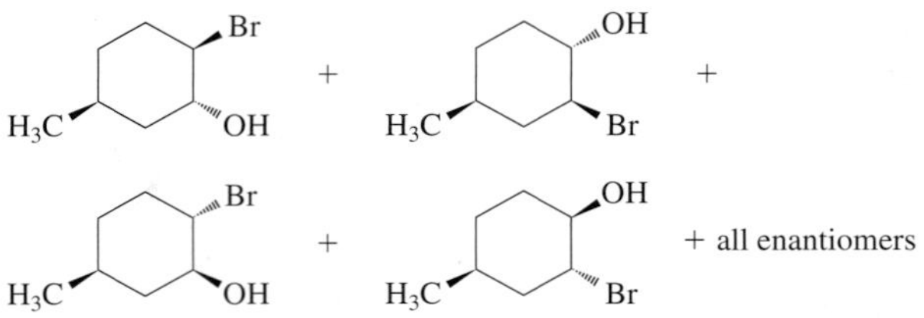 + +

+ + all enantiomers

12-15

Worked out in chapter.

12-16

The first step in the mechanism is attack by Br_2 on the alkene double bond. It can occur from the same ("top" face in the drawing below) or the opposite face ("bottom") of the ring that contains the methyl group. Although it is relatively remote, the methyl group exerts some steric hindrance. Thus, while both isomeric bromonium ions will be formed, the trans isomer will be the major one. In accord with convention (Section 5-7), all steps are illustrated for only one of the two enantiomers of the (racemic) starting compound.

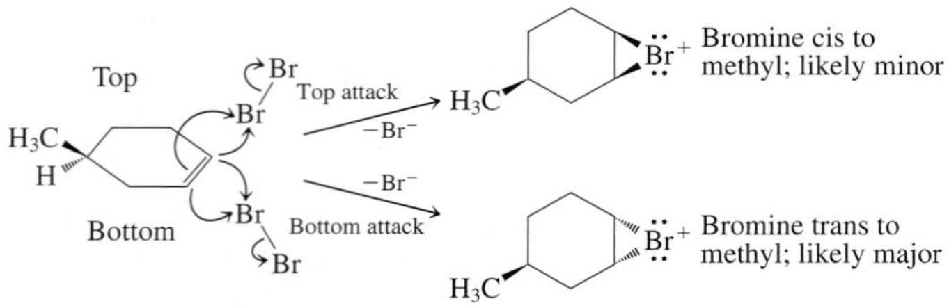

Anti-attack by the oxygen of a molecule of water follows. The two carbons of the bromonium ion rings are similar but not identical: One is closer to the methyl group than the other. Therefore, both positions will be reactive, but not equally so, producing two regioisomers. After proton loss, the ultimate result is four isomeric products (each formed as a racemate):

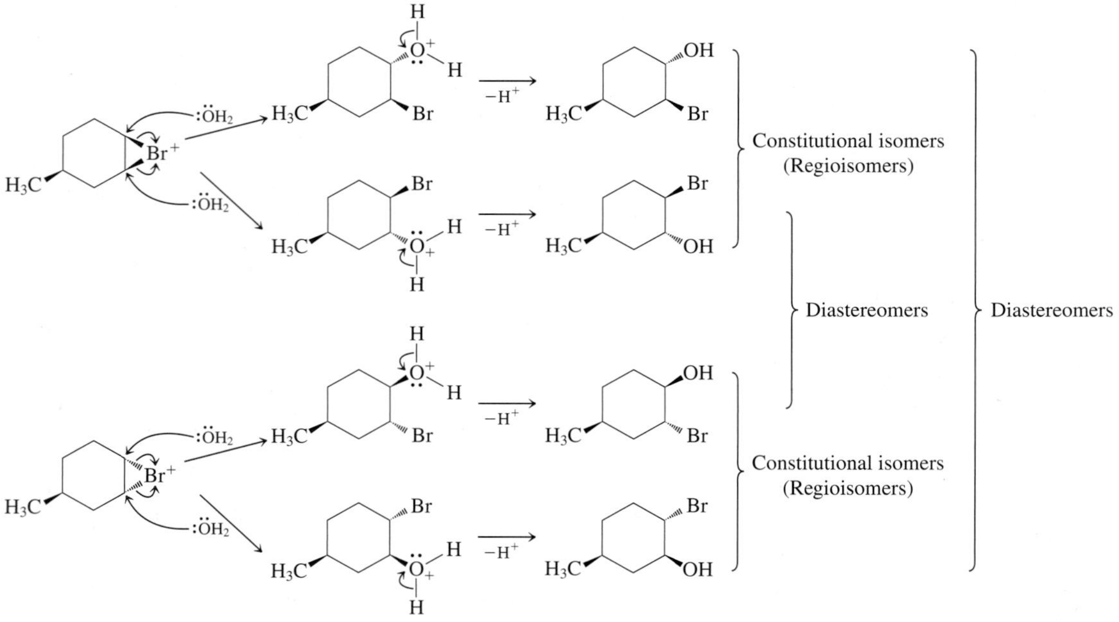

12-17

+ enantiomer

cis-2-Pentene

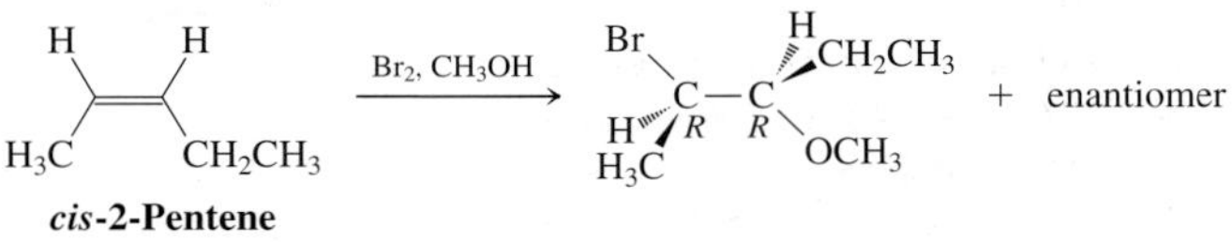

Opening of the bromonium ion can also give (3R,2R)- and (3S,2S)-3-bromo-2-methoxypentane.

12-18

Worked out in chapter.

12-19

For the acetate group, $H_3C-\overset{\overset{\displaystyle O}{\|}}{C}-O-$, the common abbreviation AcO— is used below. Dissociation of mercuric acetate gives the electrophile, which attacks the double bond to result in the mercurinium ion.

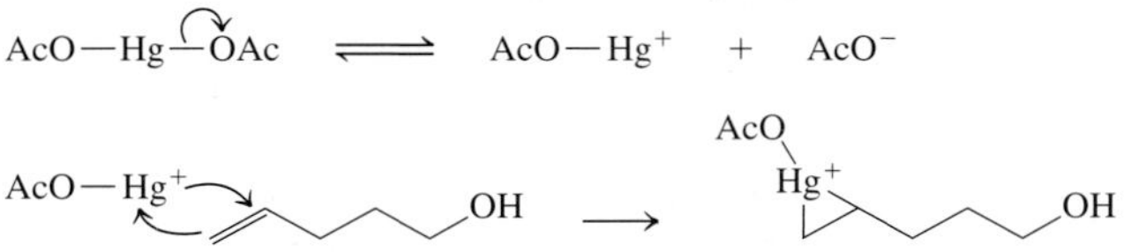

Intramolecular nucleophilic attack by the remote hydroxy oxygen follows. The regiochemical issue arises at this point. Either a five-membered ring (pathway *a*) or a six-membered ring (pathway *b*) can form:

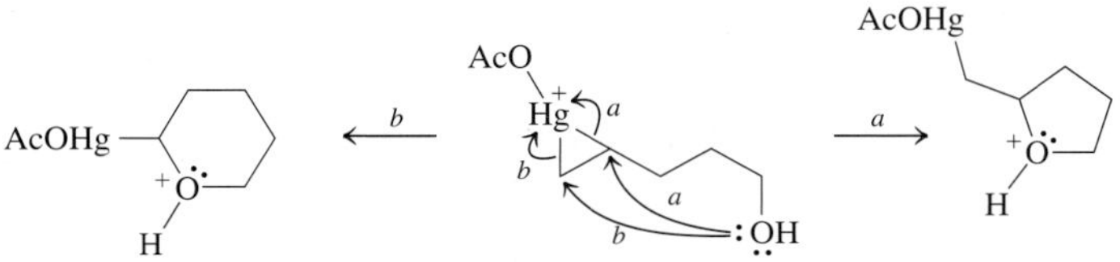

In general, we know that six-membered rings are more stable. However, we also know that five-membered rings form more quickly (Section 9-6). In addition, formation of the five-membered ring in this situation follows Markovnikov's rule: The oxygen attacks the more substituted carbon of the

Since we are starting with racemic material, all intermediates and products are racemic, that is, they are generated as an equimolar mixture of enantiomers.

mercurinium ion, where greater partial positive charge resides. This pathway gives the observed product:

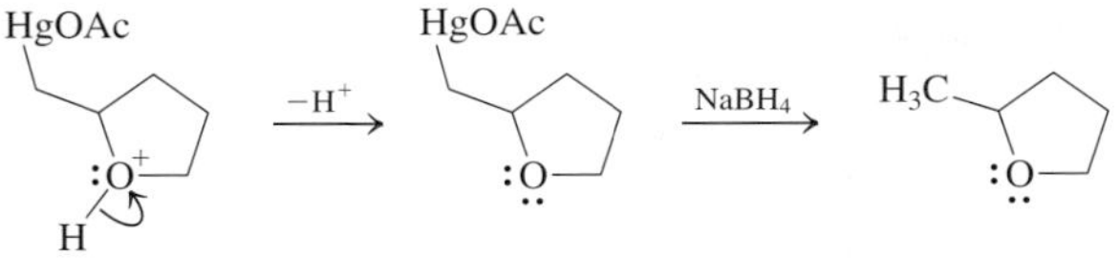

Both starting material and product have the formula $C_5H_{10}O$; they are isomers.

12-20

(a) $CH_3CH_2CH_2OH$

(b) 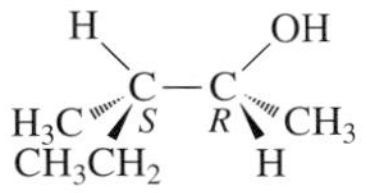 + enantiomer

12-21

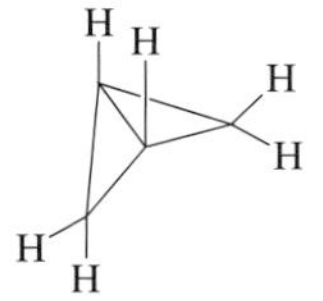

Bicyclo[1.1.0]butane

12-22

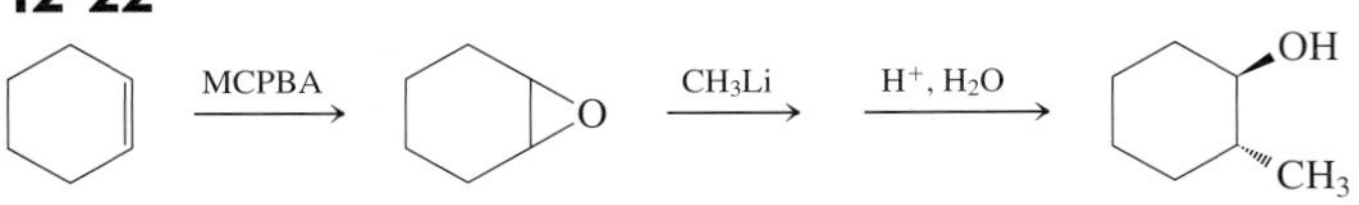

12-23

(a) 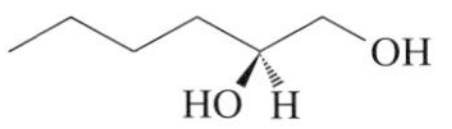 + enantiomer

(b) 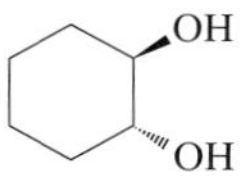 + enantiomer

70%

(c) 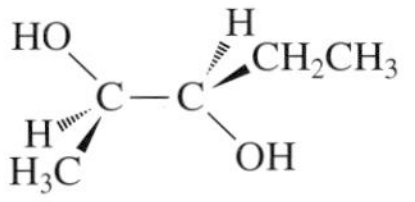 + enantiomer

(d) 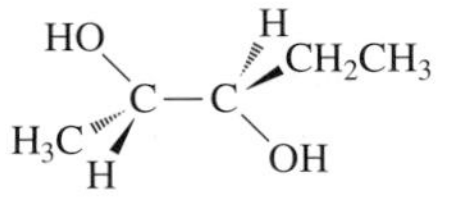 + enantiomer

12-24

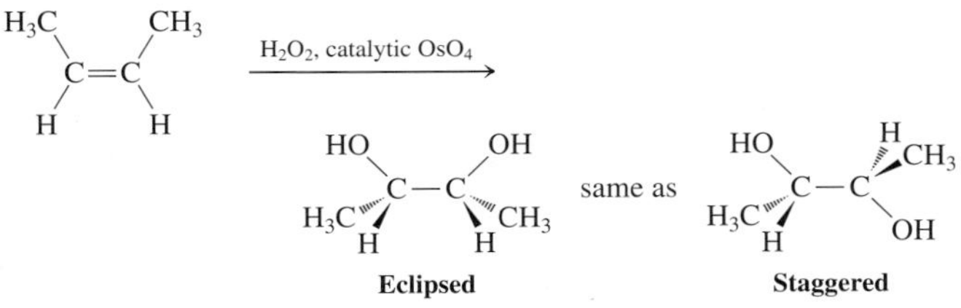

H_2O_2, catalytic OsO_4

Eclipsed   same as   Staggered

Meso

H_2O_2, catalytic OsO_4

Eclipsed   same as   Staggered

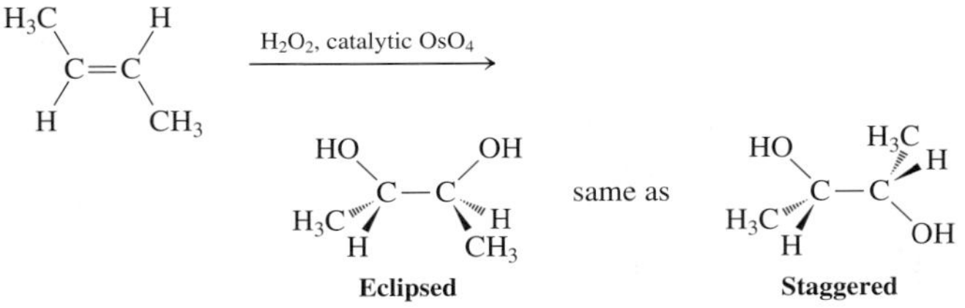

(R,R), (S,S)

12-25

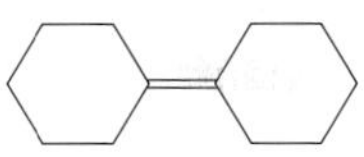

$C_{12}H_{20}$

12-26

(a) 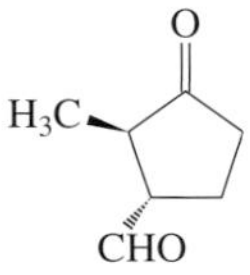 + $H_2C=O$ (b)  + $H_2C=O$

(c)

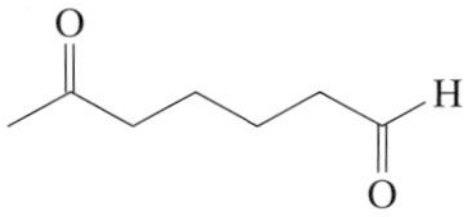

12-27

Worked out in chapter.

12-28

The product, drawn out, is [structure]. Carbons atoms 1 and 6 are each double bonded to an oxygen atom. Therefore, prior to ozonolysis they must have been double bonded to each other. We (1) remove the oxygens and (2) rewrite the chain to allow carbons 1 and 6 to be in position (3) to be connected with a double bond, giving the starting material, the solution to the problem:

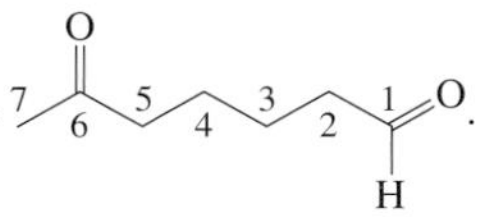

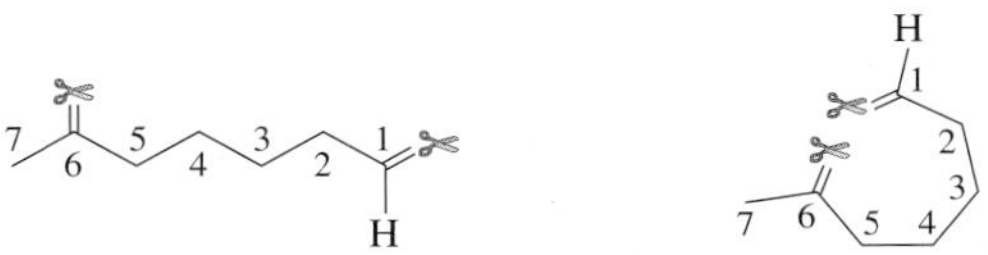

(1) Oxygen atoms removed from carbons atoms 1 and 6

(2) Chain rewritten to bring carbons 1 and 6 into proximity

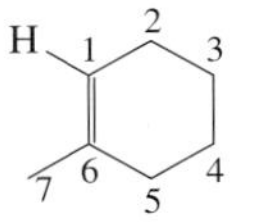

(3) Carbons 1 and 6 joined by a double bond

Hint: Keep numbering the carbon atoms as you go from one structure to the next, to avoid losing or gaining a carbon atom.

12-29

Initiation

$$(C_6H_5)_2PH \xrightarrow{h\nu} (C_6H_5)_2P\cdot + H\cdot$$

Chain carrier

Propagation

$$CH_3(CH_2)_5CH=CH_2 + (C_6H_5)_2P\cdot \longrightarrow$$

$$CH_3(CH_2)_5\dot{C}HCH_2P(C_6H_5)_2$$

More stable radical

$$CH_3(CH_2)_5\dot{C}HCH_2P(C_6H_5)_2 + (C_6H_5)_2PH \longrightarrow$$

$$CH_3(CH_2)_5CH_2CH_2P(C_6H_5)_2 + (C_6H_5)_2P\cdot$$

Product

12-30

This is an irregular copolymer with both monomers incorporated in random numbers but regioselectively along the chain. Write a mechanism for its formation.

$$\left[(-CH_2\overset{\overset{\displaystyle Cl}{|}}{\underset{\underset{\displaystyle Cl}{|}}{C}})_m (-CH_2\overset{\overset{\displaystyle H}{|}}{\underset{\underset{\displaystyle Cl}{|}}{C}})_n - \right]$$

Chapter 13

13-1

(a)

1-Hexyne   2-Hexyne

3-Hexyne   4-Methyl-1-pentyne

(R)-3-Methyl-1-pentyne   (S)-3-Methyl-1-pentyne

4-Methyl-2-pentyne   3,3-Dimethyl-1-butyne

(b) (R)-3-Methyl-1-penten-4-yne

(c)

3-Butyn-1-ol   (S)-3-Butyn-2-ol   (R)-3-Butyn-2-ol

2-Butyn-1-ol   1-Butyn-1-ol
**(This compound is highly unstable
and does not exist in solution)**

13-2

From the data in Section 11-5, we can calculate the heat of hydrogenation of the first π bond in the butynes.

$CH_3CH_2C\equiv CH + H_2 \longrightarrow CH_3CH_2CH=CH_2$
$\Delta H° = -(69.9 - 30.3) = -39.6$ kcal mol^{-1}

$CH_3C\equiv CCH_3 \quad + \quad H_2 \quad \longrightarrow \quad$
(product: cis-2-butene)

$\Delta H° = -(65.1 - 28.6) = -36.5$ kcal mol^{-1}

In both cases, more heat is released than expected for a simple C–C double bond.

13-3

Worked out in chapter.

13-4

Only those bases whose conjugate acids have a pK_a higher than that of ethyne (pK_a = 25) will deprotonate it: (CH$_3$)$_3$COH has a pK_a ≈ 18, so (CH$_3$)$_3$CO$^-$ is too weak; but [(CH$_3$)$_2$CH]$_2$NH has a pK_a ≈ 40, and therefore LDA is a suitable base.

13-5

Worked out in chapter.

13-6

$H_3C-C\equiv C-CH_2-CH_3$
s   q   t

13-7

The starting materials in each case can be

(a)   (b) $\diagup\hspace{-1mm}(CH_2)_5CH_3$

(c)

13-8

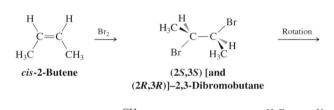

cis-2-Butene   (2S,3S) [and
                              (2R,3R)]–2,3-Dibromobutane

(Z)-2-Bromo-2-butene

trans-2-Butene   (E)-2-Bromo-2-butene

13-9

(a) $CH_3(CH_2)_3C\equiv CH$

1. CH$_3$CH$_2$MgBr
2. H$_2$C=O
3. PCC, CH$_2$Cl$_2$
4. CH$_3$(CH$_2$)$_3$C≡CMgBr

$$CH_3(CH_2)_3C\equiv C\overset{\overset{\displaystyle OH}{|}}{\underset{\underset{\displaystyle H}{|}}{C}}C\equiv C(CH_2)_3CH_3$$

(b) $HC\equiv CLi \xrightarrow{CH_3CH_2CH_2Br}$

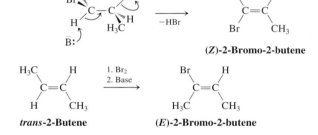

13-10

$\equiv\!\!\!-Li \quad + \quad CH_3CHO$

13-11

(structure: a cyclopropane ring bearing a CH₃ group and a chain with a ketone C=O and an ethyl ester —CO—O—CH₂CH₃, with a cis-propenyl chain)

13-12

$CH_3CH_2C{\equiv}CH$
1. $CH_3CH_2CH_2CH_2Li$
2. (epoxide, O on three-membered ring)
3. H_2, Lindlar cat.
→ HO— (cis-alkenol chain)

13-13

Worked out in chapter.

13-14

In contrast to the diyne in Exercise 13-13, 2,7-undecadiyne has only internal triple bonds.

(structure with two internal triple bonds) Na, NaNH₂, liquid NH₃ → (trans,trans-diene chain)

13-15

$CH_3C{\equiv}CCH_3 \xrightarrow{H^+} CH_3CH{=}\overset{+}{C}CH_3 \xleftarrow{:\overset{..}{\underset{..}{Br}}:^-}$

$CH_3CH{=}\overset{Br}{\underset{CH_3}{C}} \xrightarrow{H^+}$

$\left[CH_3CH_2\overset{+}{C}\overset{\overset{..}{\underset{..}{Br}}:}{\underset{CH_3}{}} \longleftrightarrow CH_3CH_2C\overset{\overset{+}{\underset{..}{Br}}:}{\underset{CH_3}{}} \right] \xrightarrow{Br^-} CH_3CH_2\overset{Br}{\underset{Br}{C}}CH_3$

13-16

$CH_3CH_2C{\equiv}CH \xrightarrow{Cl_2}$ (C=C with Cl, H, CH₃CH₂, Cl substituents) $\xrightarrow{Cl_2} CH_3CH_2\overset{Cl\ Cl}{\underset{Cl\ Cl}{C{-}CH}}$

13-17

(cyclohexane ring bearing OH and a C(OH)=CH₂ group)

13-18

(a) CH_3CHO (b) $CH_3\overset{O}{\overset{\|}{C}}CH_3$ (c) $CH_3CH_2\overset{O}{\overset{\|}{C}}CH_3$

(d) $CH_3CH_2\overset{O}{\overset{\|}{C}}CH_3$

(e) (2-methyl-3-hexanone structure) + (2-methyl-3-hexanone structure)

13-19

Worked out in chapter.

13-20

As in the previous exercise, first construct the necessary carbon–carbon bond by deprotonation of the terminal alkynyl hydrogen, followed by S_N2 reaction, and then modify the triple bond to the desired new functional group.

(structure)—H $\xrightarrow[\text{liquid NH}_3]{\text{LiNH}_2,}$ (structure)—Li $\xrightarrow[\text{2. H}_2\text{O}]{\text{1. CH}_3\text{CH}_2\text{Br}}$

(internal alkyne structure) $\xrightarrow[\text{liquid NH}_3]{\text{Na,}}$ (trans-alkene structure)

13-21

2 (cyclohexene) + BH_3

13-22

(a) CH_3CHO (b) CH_3CH_2CHO
(c) $CH_3CH_2CH_2CHO$

13-23

$(CH_3)_3CC{\equiv}CH \xrightarrow[\text{2. H}_2\text{O}_2,\ HO^-]{\text{1. Dicyclohexylborane}} (CH_3)_3CCH_2\overset{O}{\overset{\|}{C}}H$

13-24

(cis-1-bromo-1-alkene) $\xrightarrow{Pd}$ (cis-alkenyl–Pd–Br) $\xrightarrow{\text{(methyl acrylate } CH_3O_2C{-}CH{=}CH_2)}$

(structure with CH₃O, O, Pd—Br inserting into acrylate)

(structure with CH₃O, O, H, Pd—Br) → (final diene ester product CH₃O, O)

Chapter 14

14-1

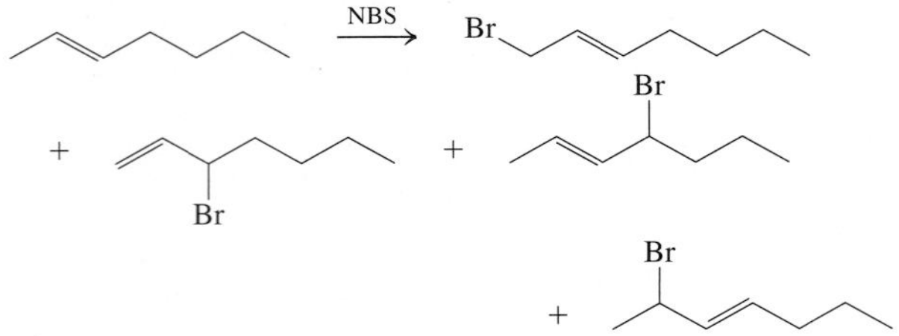

14-2

(a)   (b) 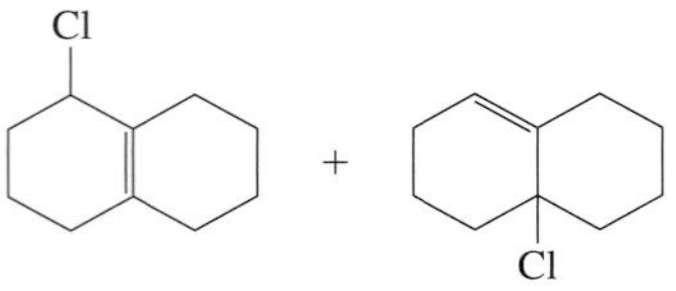   (c)

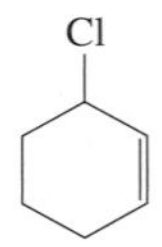

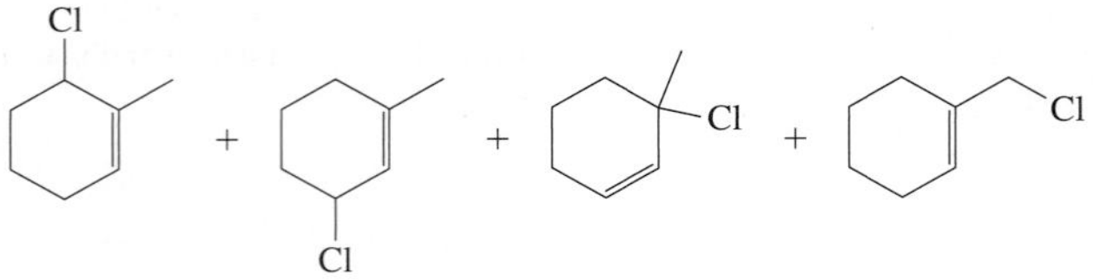

14-3

The intermediate allylic cation is achiral.

14-4

Worked out in chapter.

14-5

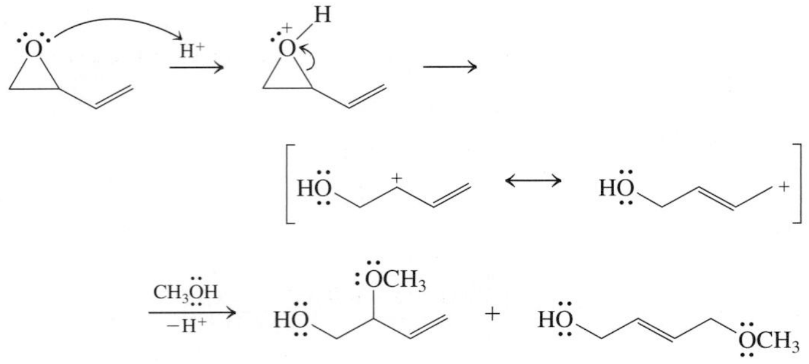

14-6

Upon ionization, chloride ion does not immediately diffuse away from the intermediate allylic cation. Reattachment to give either starting material or its allylic isomer occurs—recall the reversibility of S_N1 reactions (Exercise 7-3). However, chloride continues to dissociate, allowing acetate, a good nucleophile *but a poor leaving group* (Table 6-4), to win out ultimately.

14-7

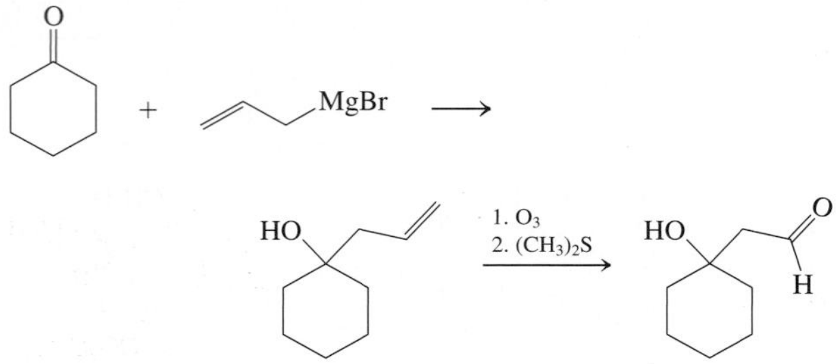

14-8

(a) 5-Bromo-1,3-cycloheptadiene
(b) (E)-2,3-Dimethyl-1,3-pentadiene

(c)   (d)

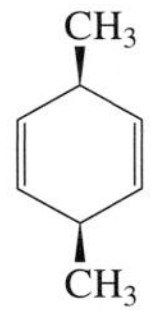

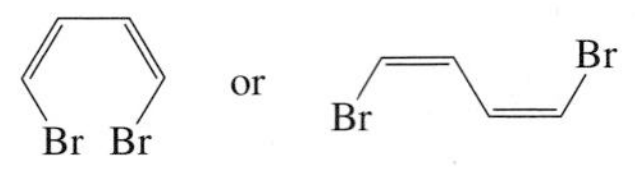

14-9

An internal trans double bond is more stable than a terminal double bond by about 2.7 kcal mol^{-1} (see Figure 11-12). This difference plus the expected stabilization energy of 3.5 kcal mol^{-1} add up to 6.2 kcal mol^{-1}, pretty close to the observed value.

14-10

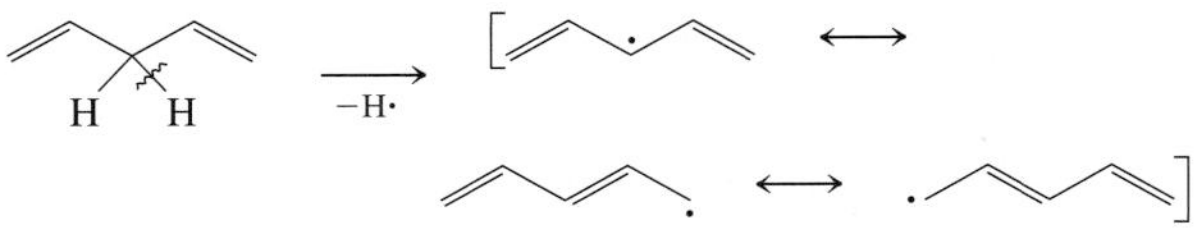

The product is the delocalized pentadienyl radical.

14-11

(a) $HOCH_2CHCHCH_2OH$ $\xrightarrow{PBr_3}$

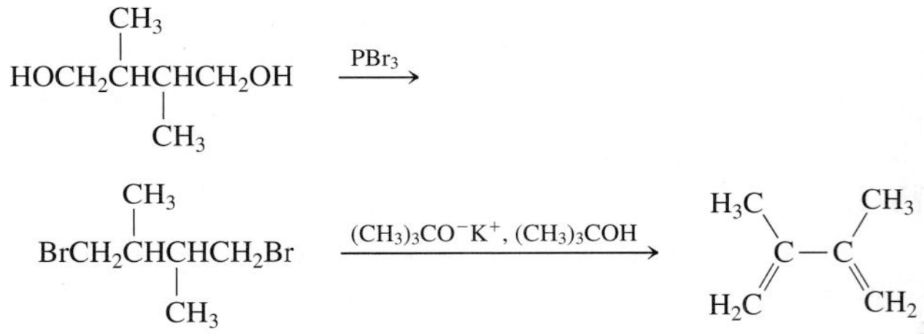

(b)

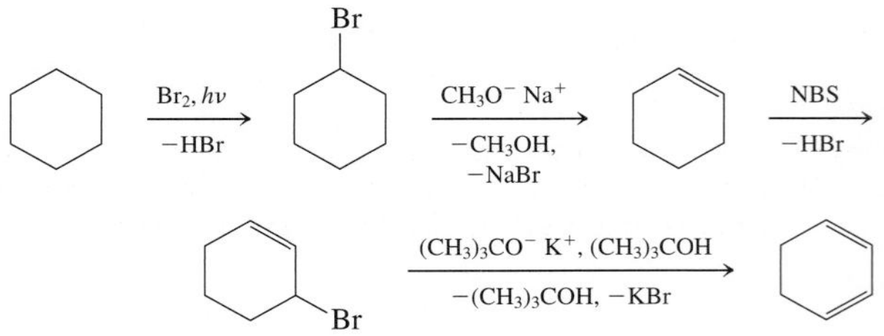

14-12

(a)   Same product for both modes of addition

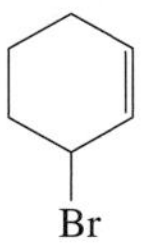

(b)   Both cis and trans

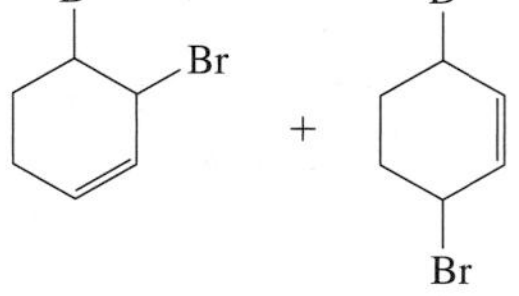

Addition of HX to unsubstituted cycloalka-1,3-dienes in either 1,2- or 1,4-manner gives the same product because of symmetry.

14-13

Worked out in chapter.

14-14

A is a 1,2-adduct and the kinetic product.
B is a 1,4-adduct and the thermodynamic product.

14-15

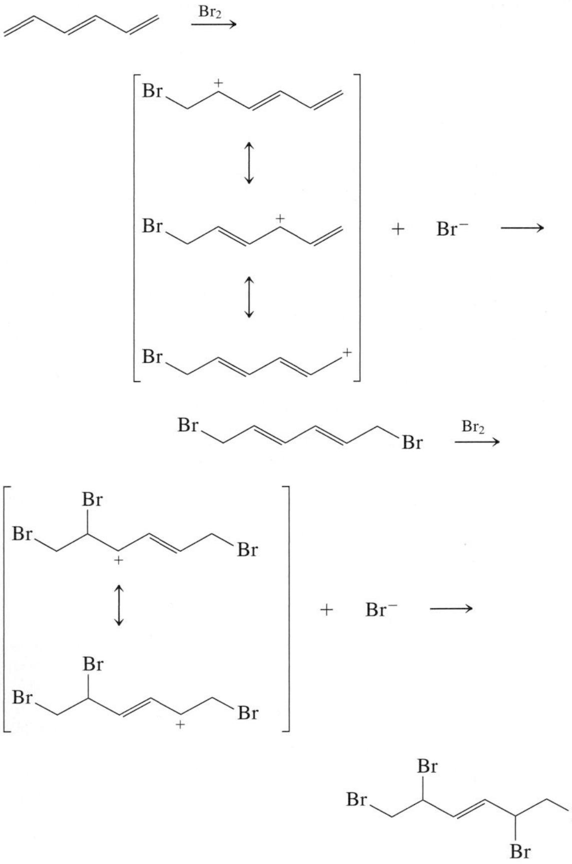

14-16

(a), (b) Electron rich, because alkyl groups are electron donors.
(c), (d) Electron poor, because the carbonyl group is electron withdrawing by induction and resonance and the fluoroalkyl group is so by induction only.

14-17

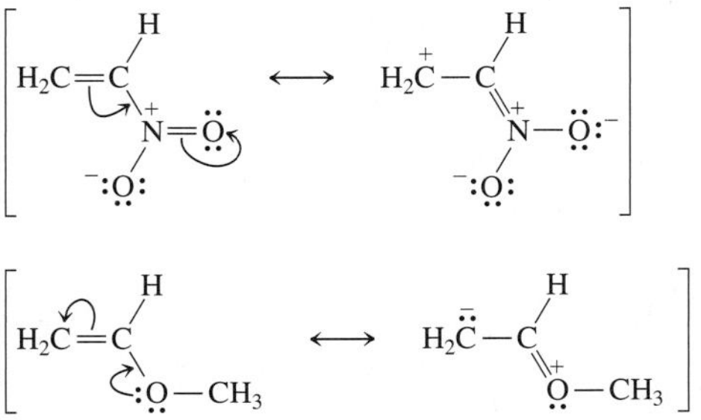

14-18

(a)

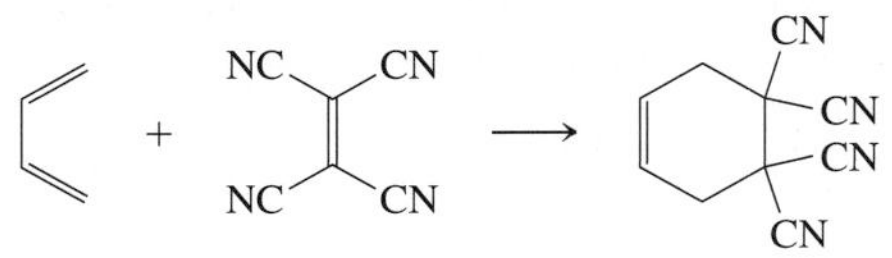

(b)

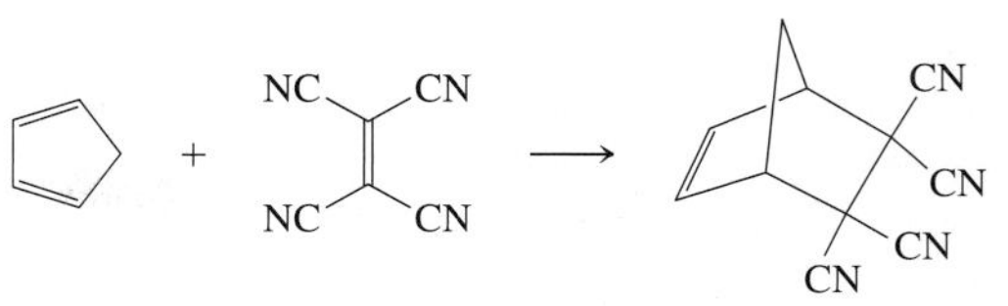

(Make a model
of this product)

(c)

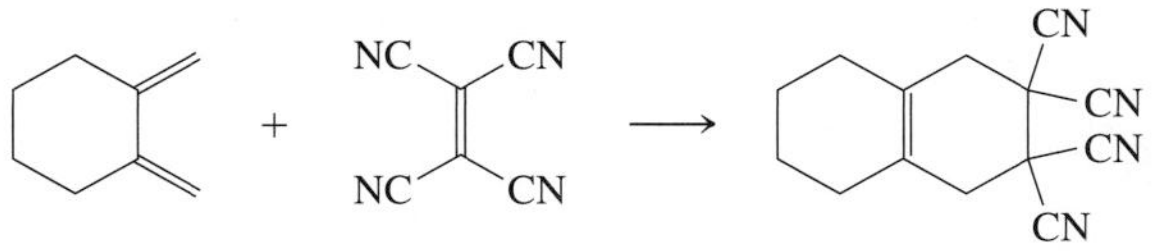

14-19

Worked out in chapter.

14-20

(a)

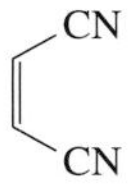

(b)

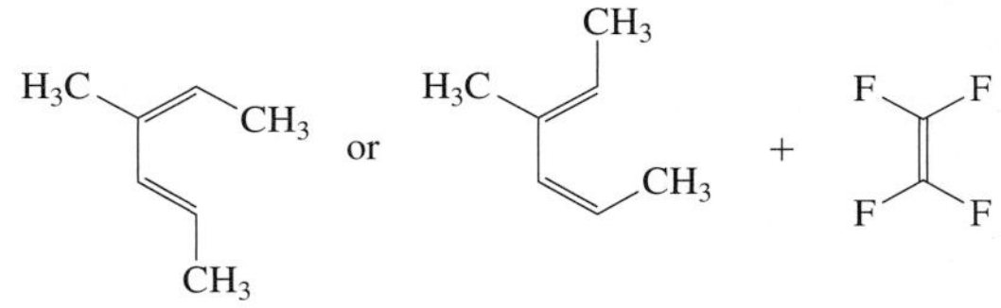

14-21

The cis,trans isomer cannot readily reach the s-cis conformation because of steric hindrance.

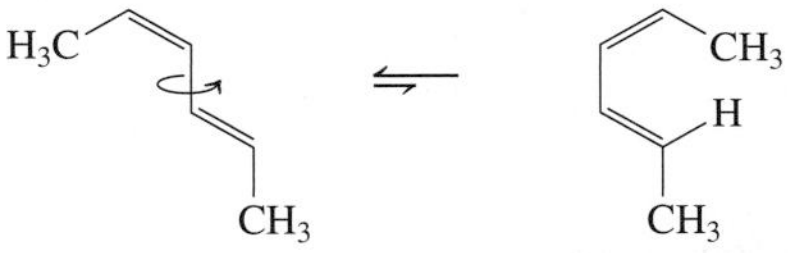

Sterically hindered

14-22

Worked out in chapter.

14-23

(a)

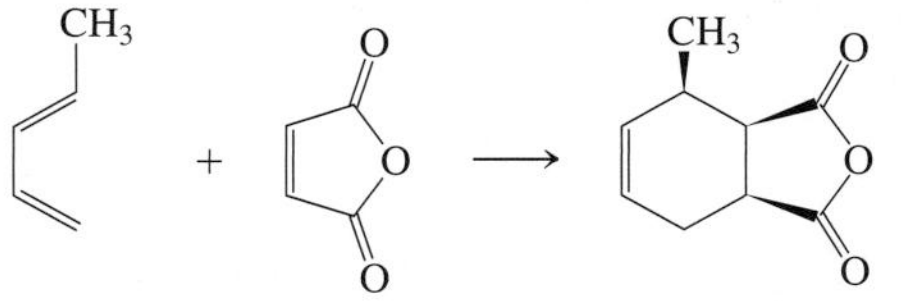

(b)

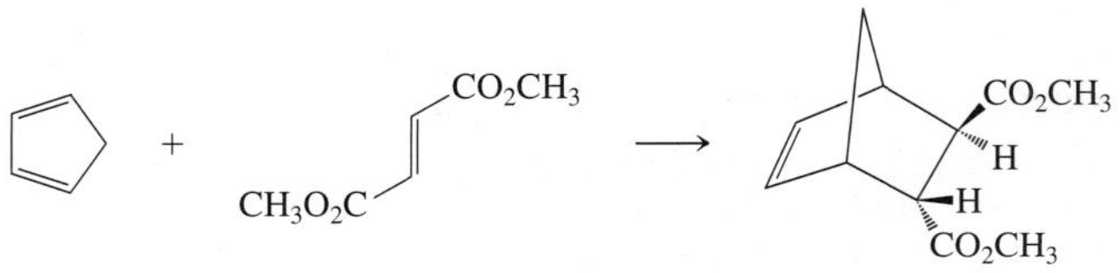

14-24

The first product is the result of exo addition; the second product, the outcome of endo addition.

14-25

(a)　(b)

14-26

A $\xrightarrow{hν}$ B

Two con-
rotatory ring
closures

A $\xrightarrow{\Delta}$ C

Two disro-
tatory ring
closures

14-27

Worked out in chapter.

14-28

Conrotatory. Make a model.

14-29

(b), (c), (a), (f), (d), (e)

Chapter 15

15-1

(a) 1-Chloro-4-nitrobenzene (*p*-chloronitrobenzene)
(b) 1-Deuterio-2-methylbenzene (*o*-deuteriotoluene)
(c) 2,4-Dinitrophenol

15-2

(a) CH_3 / $CHCH_2CH_2CH_3$

(b) $CH=CH_2$ / NO_2

(c) CH_3 / O_2N　NO_2 / NO_2

15-3

(a) 1,3-Dichlorobenzene (*m*-dichlorobenzene)
(b) 2-Fluorobenzenamine (*o*-fluoroaniline)
(c) 1-Bromo-4-fluorobenzene (*p*-bromofluorobenzene)

15-4

1,2-Dichlorobenzene

and

1,2,4-Trichlorobenzene

and

15-5

Compound B has lost its cyclic arrangement of six π electrons and therefore its aromaticity. Thus, ring opening is endothermic.

15-6

The unsymmetrically substituted 1,2,4-trimethylbenzene exhibits the maximum number, nine, of ^{13}C NMR lines. Symmetry reduces this count to six in 1,2,3- and three in 1,3,5-trimethylbenzene.

15-7

Worked out in chapter.

15-8

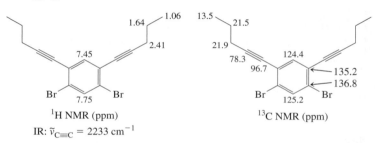

^{1}H NMR (ppm)　　　　　^{13}C NMR (ppm)

IR: $\tilde{v}_{C\equiv C} = 2233$ cm^{-1}

15-9

(a)　H_3C / CH_3

(b)　NO_2 / Br

(c)　C_6H_5 / C_6H_5

(d) 9-Bromophenanthrene
(e) 5-Nitro-2-naphthalenesulfonic acid

15-10

HO　　OH

15-11

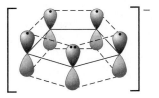

The maximum number of aromatic benzene Kekulé rings is two, in three of the resonance forms (the first, third, and fourth).

15-12

This is an unusual Diels-Alder reaction in which one molecule acts as a diene, the other as a dienophile. Make models.

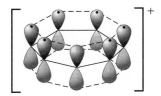

Note the surprising result of applying the general stereochemical scheme on p. 607 here (replace the i's in starting material and product by a bond).

15-13

No. Cyclooctatetraene has localized double bonds. Double-bond shift results in geometrical isomerization and not in a resonance form, as shown for 1,2-dimethylcyclooctatetraene.

15-14

15-15

They are all nonplanar because of bond angle (e.g., a flat all-cis-[10]annulene requires C_{sp^2} bond angles of 144°, a considerable distortion from the normal value), eclipsing, and transannular strain (e.g., the two inside hydrogens of trans,cis,trans,cis,cis-[10]annulene occupy the same region in space).

15-16

(a), (c), (d) Aromatic; (b), (e) antiaromatic

15-17

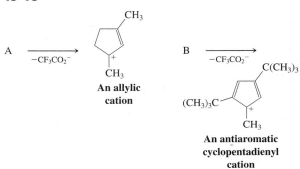

15-18

A → −CF₃CO₂⁻ → An allylic cation

B → −CF₃CO₂⁻ → An antiaromatic cyclopentadienyl cation

15-19

(a), (b) Aromatic; (c) antiaromatic

15-20

Worked out in chapter.

15-21

The dianion is an aromatic system of 10 π electrons, but pentalene has $4n$ π electrons.

15-22

The difference is that the dihydrogenation of benzene involves the energetic cost of losing its aromaticity, ~30 kcal mol⁻¹. This value more than offsets the heat of hydrogenation contribution, making the reaction endothermic. You can also arrive at this conclusion by inspection of Figure 15-3: The heat of hydrogenation of benzene to 1,3-cyclohexadiene is endothermic by 54.9 − 49.3 = 5.6 kcal mol⁻¹.

15-23

Molecular weight = 84

15-24

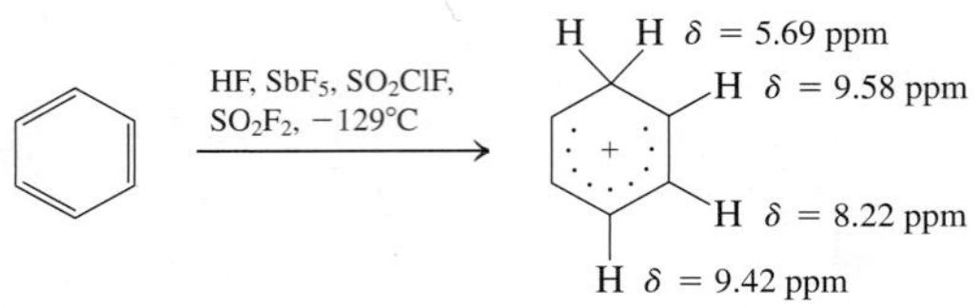

H   H $\delta = 5.69$ ppm
H $\delta = 9.58$ ppm
H $\delta = 8.22$ ppm
H $\delta = 9.42$ ppm

HF, SbF$_5$, SO$_2$ClF, SO$_2$F$_2$, $-129°$C

The NMR assignments correspond to the amount of charge expected at the various hexadienyl cation carbons on the basis of resonance.

15-25

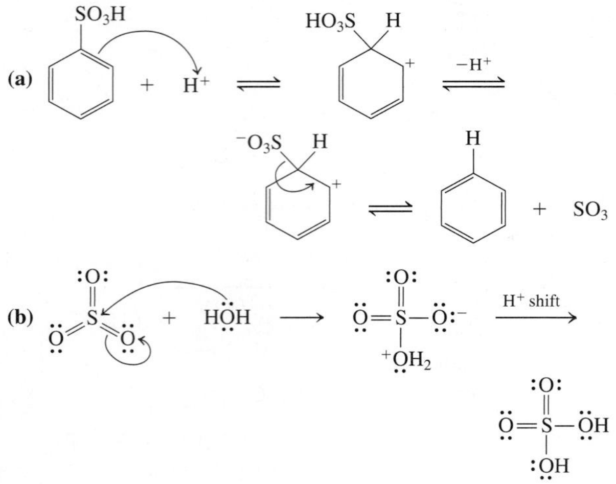

(a)

(b)

15-26

$(CH_3)_3CCl + AlCl_3 \longrightarrow (CH_3)_3C^+ + AlCl_4^-$

**1,1-Dimethylethyl
(*tert*-butyl) cation**

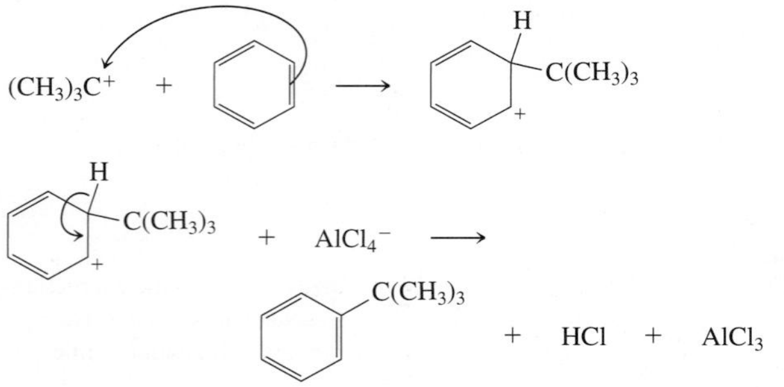

$+ HCl + AlCl_3$

15-27

$CH_3CH{=}CH_2 + H^+ \xrightarrow{\text{Markovnikov addition}}$

$+ H^+$

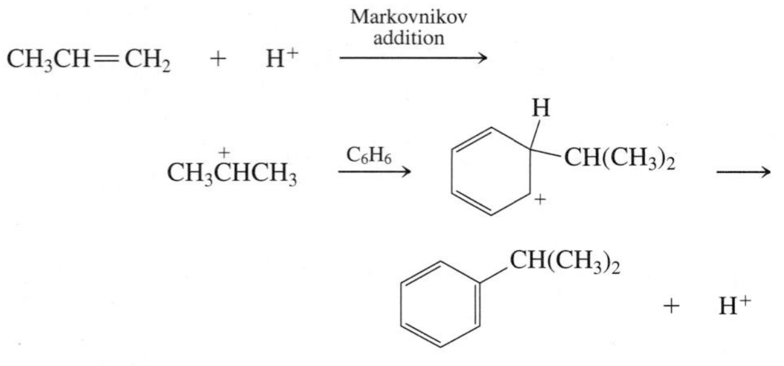

15-28

Worked out in chapter.

15-29

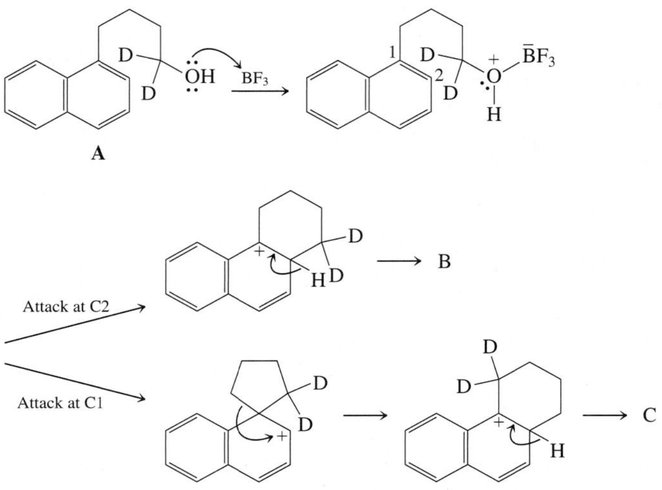

A

Attack at C2 → B

Attack at C1 → C

15-30

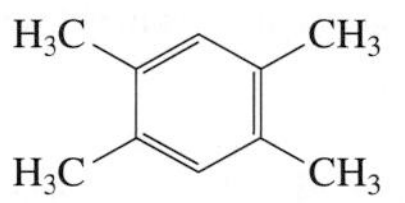

1,2,4,5-Tetramethylbenzene (Durene)

15-31

Worked out in chapter.

15-32

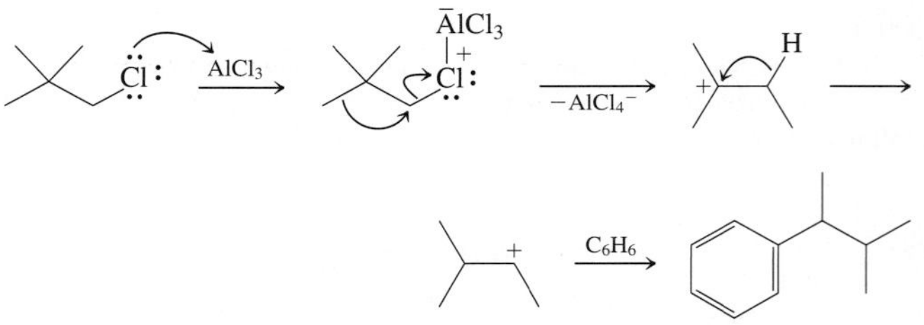

C_6H_6

15-33

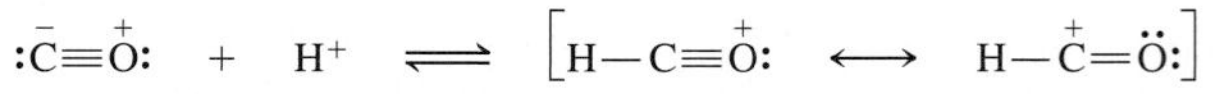

$:\overset{-}{C}{\equiv}\overset{+}{O}: + H^+ \rightleftharpoons [H{-}C{\equiv}\overset{+}{O}: \longleftrightarrow H{-}\overset{+}{C}{=}\overset{..}{O}:]$

Formyl cation

$+ H^+$

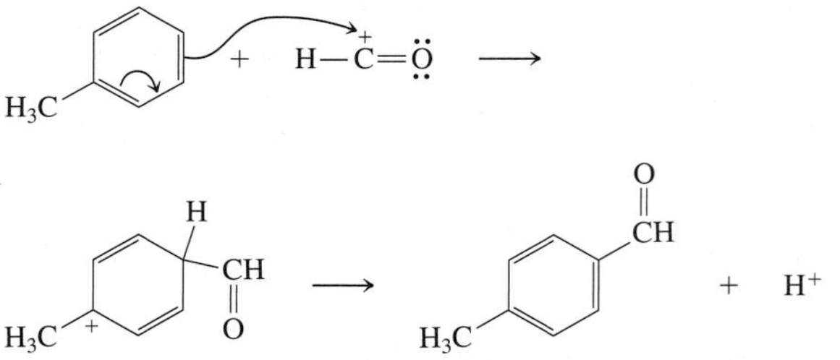

The spectral data for the formyl cation indicate the dominant contribution of $[H{-}C{\equiv}O:^+]$ in its resonance-structural description. This species can be viewed as a positively charged oxygen analog of ethyne ($\delta_{^{13}C} = 71.9$ ppm.; $\tilde{\nu}_{C{\equiv}C} = 1974$ cm^{-1}), the oxygen and the charge causing the observed relatively deshielded carbon resonance, in addition to the strengthening of the triple bond and the associated relatively high wavenumber band in the IR spectrum.

Chapter 16

16-1

4-(N,N-Dimethylamino)benzaldehyde: Formulation of dipolar resonance forms involving the electron-withdrawing carbonyl group reveals the presence of partial positive charges at its ortho (but not meta) carbons, reflected in the relative deshielding of the associated two (green) hydrogens. Conversely, formulation of dipolar resonance forms involving the electron-donating amino substituent indicates the presence of partial negative charges at its ortho (but not meta) carbons, causing relative shielding of the corresponding (red) hydrogens.

1-Methoxy-2,4-dinitrobenzene: By a similar argument, the relatively deshielded hydrogens (blue and green) are those located ortho and para to the electron-withdrawing nitro functions, the third hydrogen (red) is placed ortho to the electron-donating methoxy substituent. The "extra" deshielding of the hydrogen (blue) at C3 may be ascribed to its relative proximity to the two nitro groups, maximizing their inductive effect on their common neighbor.

16-2

Relative to benzene, C1 is deshielded by the strong inductive withdrawing effect of the oxygen, C2 and C4 are shielded because resonance of the benzene π system with the oxygen lone electron pairs places partial negative charges at these positions, and C3 is essentially unaffected.

16-3

(a), (d) Activated; (b), (c) deactivated

16-4

(d) > (b) > (a) > (c)

16-5

Methylbenzene (toluene) is activated and will consume all of the electrophile before the latter has a chance to attack the deactivated ring of (trifluoromethyl)benzene.

16-6

Worked out in chapter.

16-7

Benzenamine (aniline) is basic and therefore becomes increasingly protonated at decreasing pH. The resulting decreasing concentration of aniline in the solution causes the rate of ortho,para substitution to decrease. The anilinium ion resulting from protonation no longer has a lone electron pair available for resonance with the ring and is fully positively charged. As such, the ammonium substituent is an inductive deactivator and meta director, causing the relative proportion of meta products to increase.

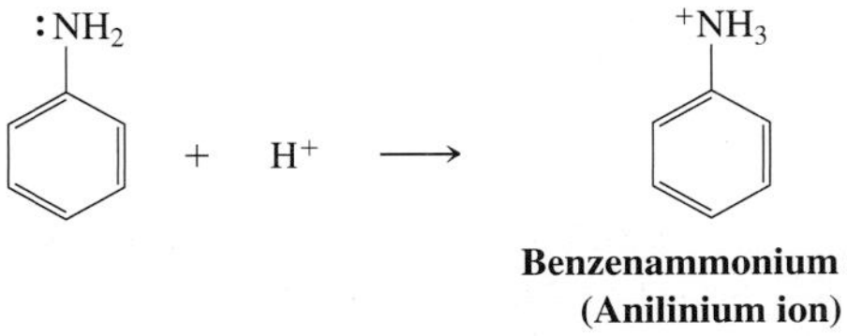

Benzenammonium ion (Anilinium ion)

$pK_a = 4.60$

Note, however, that the amino group is such a powerful activator (see also Table 16-2) that ortho,para substitution still dominates even in strong acid, in which the concentration of free amine is very low.

16-8

Ortho attack:

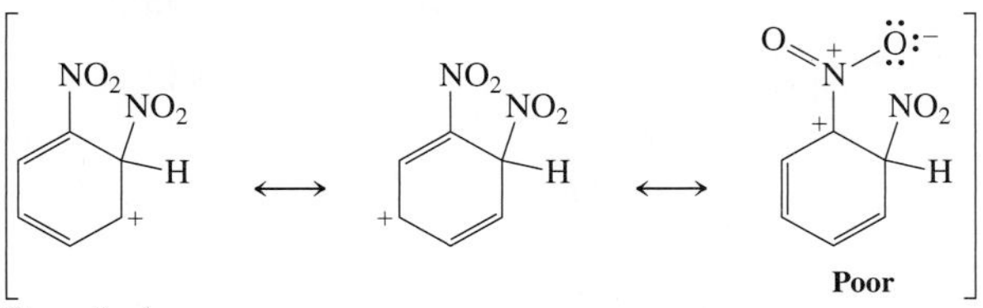

Poor

Para attack:

Poor

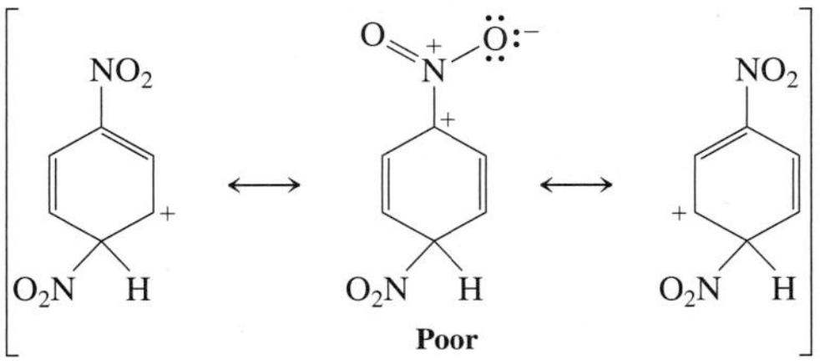

16-9

(a), (c) Electron withdrawing by resonance and induction.
(b) Electron withdrawing by induction.
(d) The phenyl substituent acts as a resonance donor.

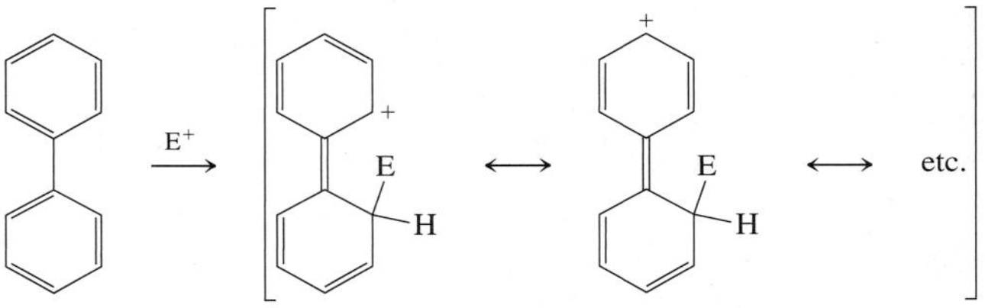

Total of six resonance structures

16-10

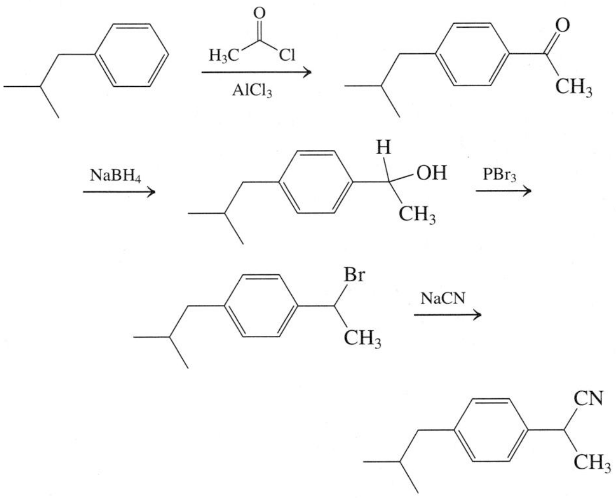

A

16-11

Worked out in chapter.

16-12

(a)

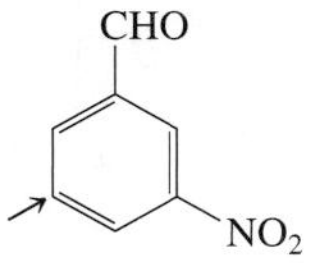

(b) 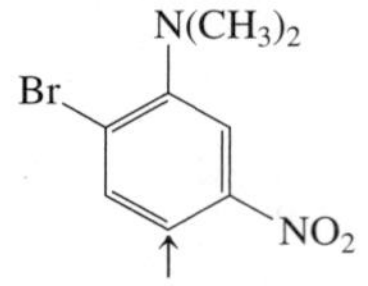
N(CH3)2
Br
NO2

(c) 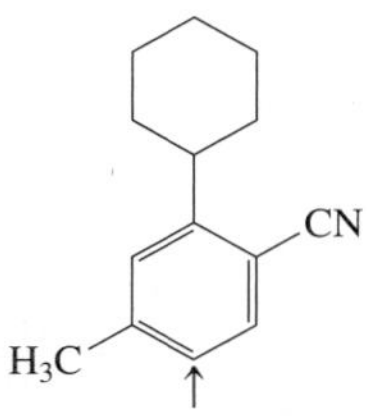
CN
H3C

16-13

(i) 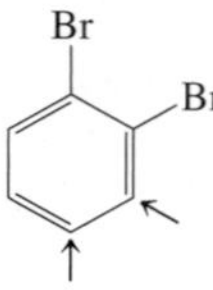
Br
Br

A

(ii) 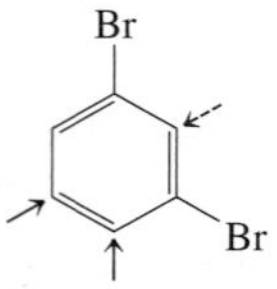
Br
Br

B

(iii) 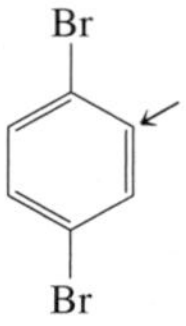
Br
Br

C

16-14

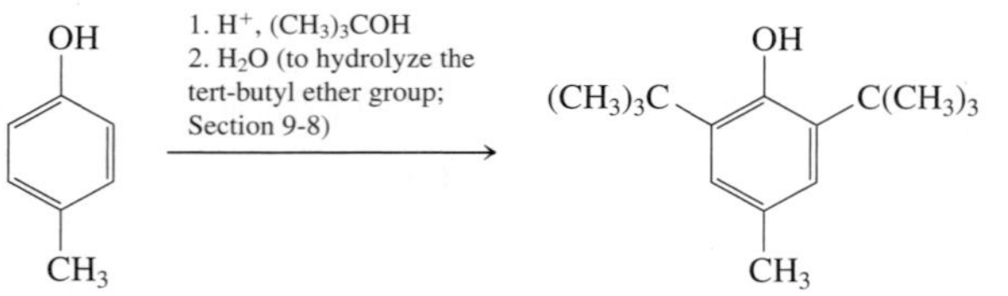
OH
CH3
1. H+, (CH3)3COH
2. H2O (to hydrolyze the tert-butyl ether group; Section 9-8)
OH
(CH3)3C
C(CH3)3
CH3

16-15

No, because the nitrogen is introduced ortho and para to bromine only.

16-16

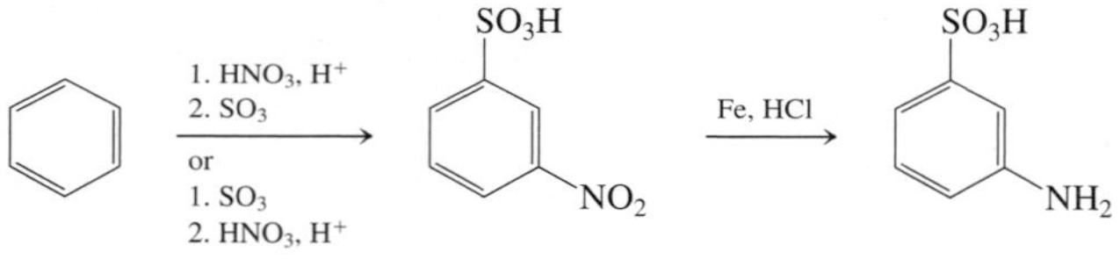
1. HNO3, H+
2. SO3
or
1. SO3
2. HNO3, H+
SO3H
NO2
Fe, HCl
SO3H
NH2

16-17

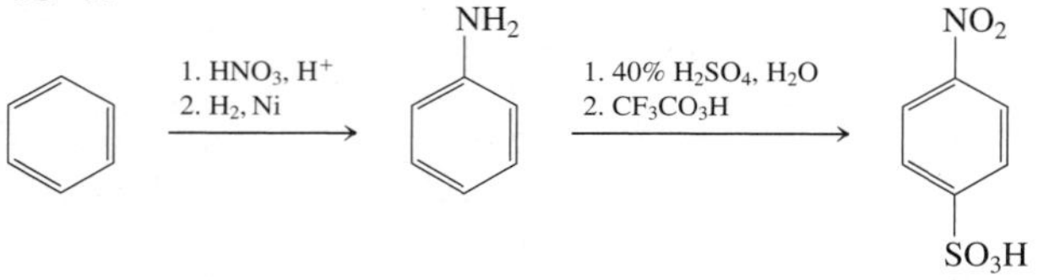
1. HNO3, H+
2. H2, Ni
NH2
1. 40% H2SO4, H2O
2. CF3CO3H
NO2
SO3H

16-18

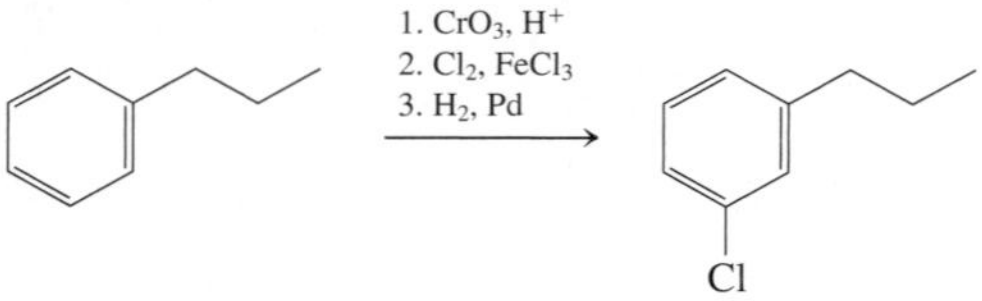
1. CrO3, H+
2. Cl2, FeCl3
3. H2, Pd
Cl

16-19

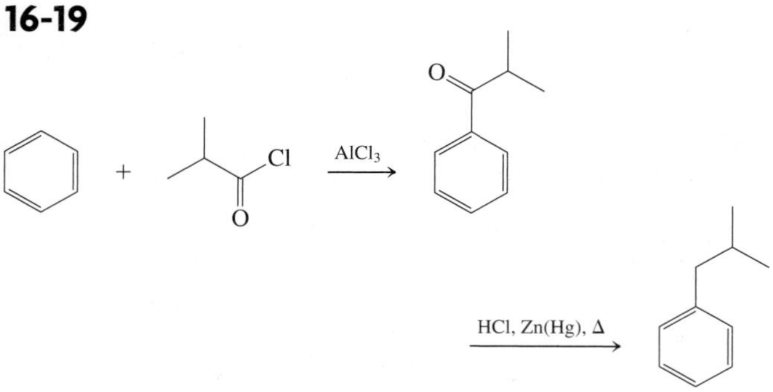
+
Cl
O
AlCl3
O
HCl, Zn(Hg), Δ

Direct Friedel-Crafts alkylation of benzene with 1-chloro-2-methylpropane gives (1,1-dimethylethyl)benzene (*tert*-butylbenzene) by rearrangement of the carbon electrophile (see Section 15-12).

16-20

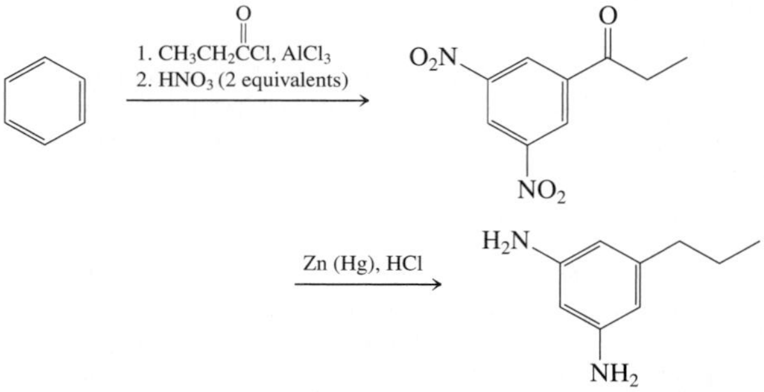
O
1. CH3CH2CCl, AlCl3
2. HNO3 (2 equivalents)
O2N
O
NO2
Zn (Hg), HCl
H2N
NH2

16-21

(a) 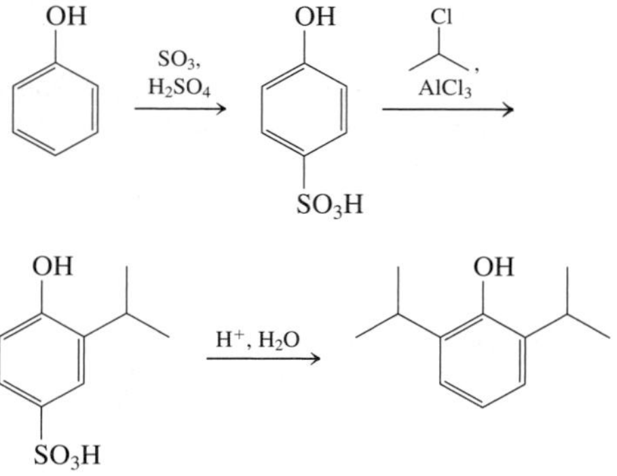
OH
SO3,
H2SO4
OH
SO3H
Cl
AlCl3
OH
SO3H
H+, H2O
OH

(b) 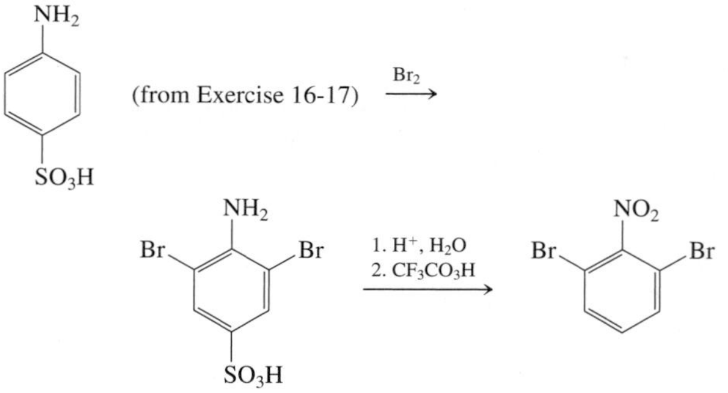
NH2
SO3H
(from Exercise 16-17)
Br2
NH2
Br
Br
SO3H
1. H+, H2O
2. CF3CO3H
NO2
Br
Br

16-22

Worked out in chapter.

16-23

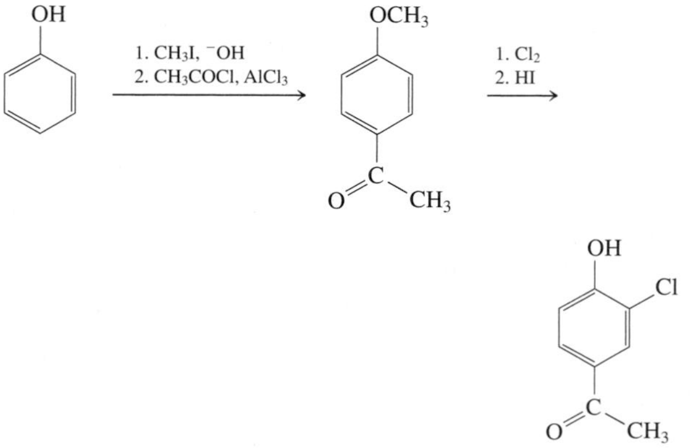
OH
1. CH3I, −OH
2. CH3COCl, AlCl3
OCH3
C
O
CH3
1. Cl2
2. HI
OH
Cl
C
O
CH3

16-24
Worked out in chapter.

16-25
(a) At C5 and C8; (b) at C6 and C8; (c) at C4.

16-26

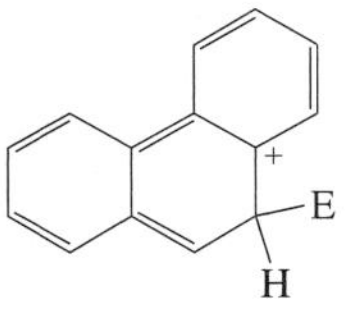

16-27
The answer is derived by inspection of the respective sets of resonance forms of the cations generated by protonation at the given positions; only the crucial ones are depicted here, although we recommend that you write all possible forms as an additional exercise.

At C9:

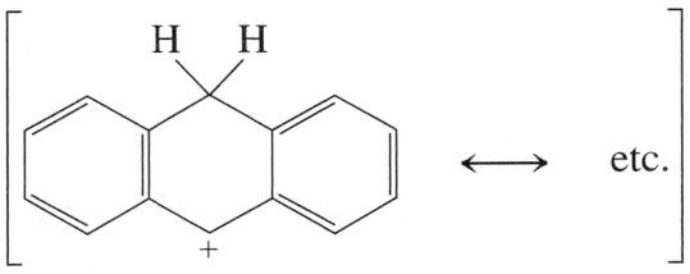

The essence of the analysis is, then, in each case, to count up the number of resonance contributors that contain intact benzene rings. You will recognize the important difference between protonation at C9 (which generates two separate benzene nuclei) and that at C1 and C2, which furnishes naphthalene fragments ("less aromatic" than two benzene rings, see Section 15-5). The $[C9-H]^+$ structure shown here already has four benzenoid resonance forms, even without moving the positive charge into one of the adjacent benzene rings. Moving the positive charge still leaves one benzene ring untouched (allowing for the formulation of its two contributing resonance forms every time the charge is moved around the other benzene ring). Attack at C1 and C2 has considerably fewer such benzenoid substructures. The preferred protonation at C1 over C2 has the same explanation as that given in this section for the same preference observed for naphthalene (i.e., regard anthracene as a benzo[b]naphthalene).

At C1:

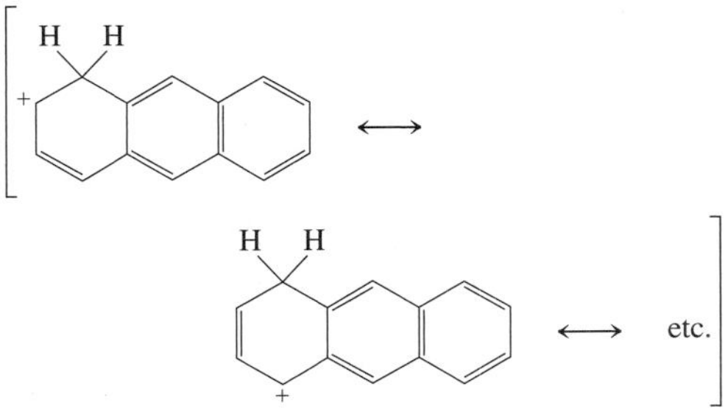

At C2:

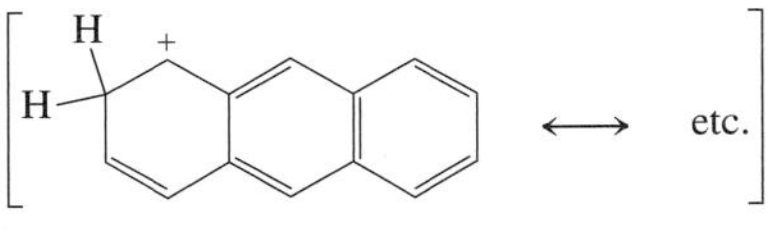

Chapter 17

17-1
(a) 2-Cyclohexenone
(b) (E)-4-Methyl-4-hexenal

(c)   (d) 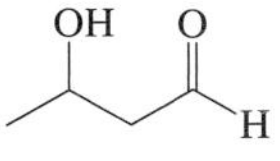   (e)

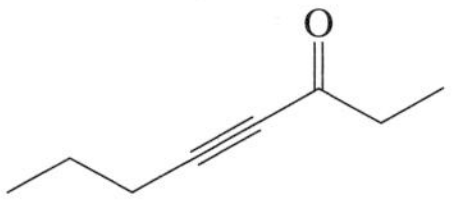

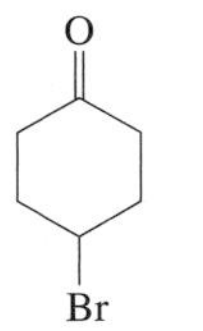

17-2
Worked out in chapter.

17-3
(a) ^{1}H NMR: absence versus presence of aldehyde resonance; three signals versus four signals; multiplicity differences are dramatic—i.e., the ketone exhibits a singlet (CH_3), triplet (CH_3), and quartet (CH_2), whereas the aldehyde shows a triplet (CH_3), sextet (CH_2 of C3), doublet of triplets (α-CH_2), and triplet (CHO).

(b) UV: $\lambda_{max} \approx 280$ (unconjugated carbonyl) versus 325 nm (conjugated carbonyl). ^{1}H NMR: most drastic differences would be the spin–spin splitting patterns, e.g., CH_3 doublet versus triplet, CHO triplet versus doublet, etc.

(c) The lack of symmetry in 2-pentanone versus 3-pentanone is evident in ^{1}H NMR: s (CH_3), t (CH_3), sex (CH_2), t (CH_2) versus t (2 CH_3s), q (2 CH_2s), and in ^{13}C NMR: five lines versus three lines.

17-4

^{1}H NMR: $J = 6.7$ Hz $J = 7.7$ Hz

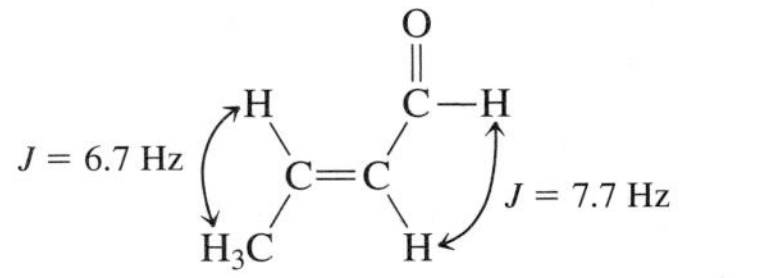

$J_{trans} = 16.1$ Hz
$J(CH_3–H2) = 1.6$ Hz (allylic coupling; Table 11-2)

^{13}C NMR: $CH_3 — CH{=}CH — CH$
   18.4   152.1   132.8   191.4

UV: Absorptions are typical for a conjugated enone.

17-5
(a) Both show the same α cleavage patterns but different McLafferty rearrangements.

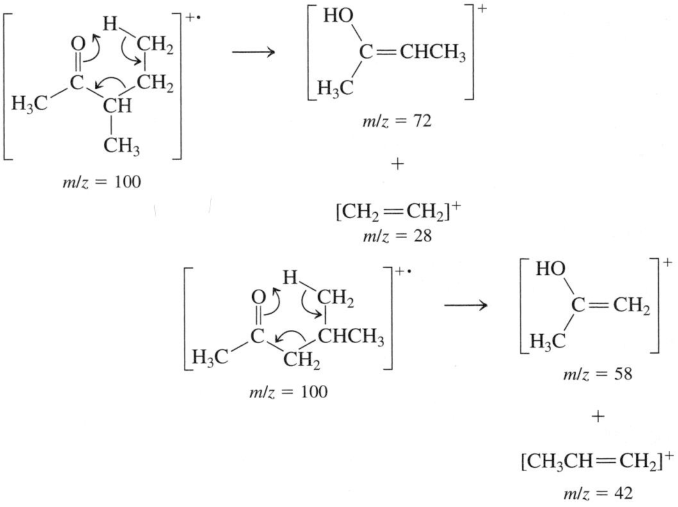

(b) Both show the same α cleavage patterns, but only 2-ethylcyclohexanone has an accessible γ-hydrogen for the McLafferty rearrangement.

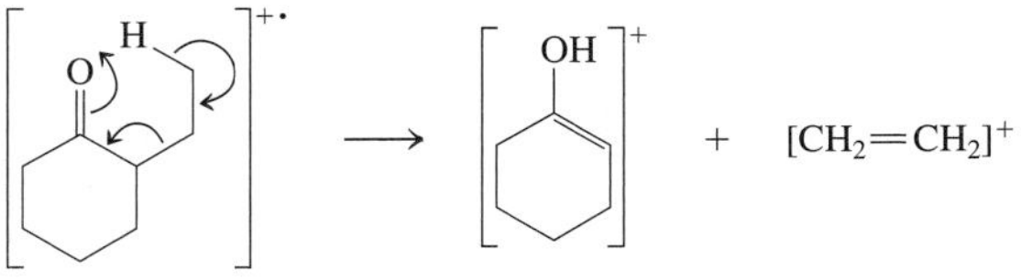

17-6

(a)

$$\text{HC}\equiv\text{CH} \xrightarrow[\text{Hg}^{2+}]{\text{H}_2\text{O, H}^+,} \text{CH}_3\text{CCH}_2\text{CH}_3 \xleftarrow{\begin{array}{c}1.\ \text{O}_3\\2.\ (\text{CH}_3)_2\text{S}\end{array}} \text{CH}_2=\text{C(CH}_3)\text{CH}_2\text{CH}_3$$

$$\text{C}_5\text{H}_9\text{MgBr} \xrightarrow{\text{H}_2\text{C}=\text{O}} \text{C}_5\text{H}_9\text{CH}_2\text{OH}$$

$$\text{C}_6\text{H}_6 \xrightarrow[\text{AlCl}_3]{\text{CH}_3\text{COCl},} \text{C}_6\text{H}_5\text{COCH}_3$$

(b)

$$\text{cyclohexane} \xrightarrow[\text{2. Mg}]{1.\ \text{Br}_2,\ h\nu} \text{cyclohexyl-MgBr} \xrightarrow{\text{CH}_3\text{C}\equiv\text{CCH}} $$

$$\text{HOCC}\equiv\text{CCH}_3 \text{ (cyclohexyl)} \xrightarrow{\text{MnO}_2} \text{OCC}\equiv\text{CCH}_3 \text{ (cyclohexyl)}$$

1-Cyclohexyl-2-butyn-1-one

17-7

(a) $\text{Cl}_3\text{CCCH}_3 \ < \ \text{Cl}_3\text{CCH} \ < \ \text{Cl}_3\text{CCCCl}_3$

(b)
$$\text{CH}_3\text{CCH}_3 + \text{H}_2{}^{18}\text{O} \rightleftharpoons \rightleftharpoons \text{CH}_3\overset{\text{OH}}{\underset{{}^{18}\text{OH}}{\text{C}}}\text{CH}_3 \rightleftharpoons$$

$$\rightleftharpoons \text{CH}_3\overset{{}^{18}\text{O}}{\text{C}}\text{CH}_3 + \text{H}_2\text{O}$$

17-8

Worked out in chapter.

17-9

$$\text{Br-C}_6\text{H}_5 \xrightarrow[\substack{\text{Friedel-Crafts}\\\text{acylation}\\(\text{Section 15-13})}]{\text{CH}_3\text{COCl, AlCl}_3} \text{Br-C}_6\text{H}_4\text{-COCH}_3 \xrightarrow[\text{Protection}]{\text{HOCH}_2\text{CH}_2\text{OH, H}^+}$$

$$\text{Br-C}_6\text{H}_4\text{-C(CH}_3)(\text{OCH}_2\text{CH}_2\text{O}) \xrightarrow[\substack{\text{Grignard}\\\text{reagent}\\\text{formation}}]{\text{Mg, (CH}_3\text{CH}_2)_2\text{O}} \text{BrMg-C}_6\text{H}_4\text{-C(CH}_3)(\text{OCH}_2\text{CH}_2\text{O})$$

$$\xrightarrow[\substack{\text{Grignard}\\\text{reaction;}\\\text{deprotection}}]{\begin{array}{c}1.\ \text{H}_2\text{C}=\text{O}\\2.\ \text{H}^+,\ \text{H}_2\text{O}\end{array}} \text{HOCH}_2\text{-C}_6\text{H}_4\text{-COCH}_3$$

17-10

Worked out in chapter.

17-11

$$\xrightarrow{\text{H}_2\text{O, HgCl}_2, \text{CaCO}_3}$$

$$\xrightarrow{\text{NaBH}_4, \text{CH}_3\text{CH}_2\text{OH}}$$

17-12

$$\xrightarrow{\begin{array}{c}1.\ \text{O}_3\\2.\ \text{CH}_3\text{SCH}_3\end{array}}$$

$$\xrightarrow{\begin{array}{c}1.\ \text{HSCH}_2\text{CH}_2\text{SH}\\2.\ \text{Raney Ni}\end{array}}$$

17-13

The mechanism of imidazolidine formation is similar to that formulated for cyclic acetal synthesis.

$$\text{H}_3\text{C}-\text{CHO} + \text{C}_6\text{H}_5\text{NHCH}_2\text{CH}_2\text{NHC}_6\text{H}_5 \longrightarrow$$

$$\text{H}_3\text{C}-\overset{\text{OH}}{\underset{\underset{\text{C}_6\text{H}_5}{\overset{|}{\text{N}}}}{\underset{|}{\text{C}}}}\text{H}-\text{NCH}_2\text{CH}_2\text{NHC}_6\text{H}_5 \xrightarrow[-\text{H}_2\text{O}]{\text{H}^+}$$

$$\overset{\text{H}_3\text{C}}{\underset{\text{H}}{\text{C}}}=\overset{+}{\underset{\text{C}_6\text{H}_5}{\text{N}}}\text{CH}_2\text{CH}_2\ddot{\text{N}}\text{HC}_6\text{H}_5 \xrightarrow{-\text{H}^+} \text{imidazolidine (C}_6\text{H}_5, \text{CH}_3, \text{H, C}_6\text{H}_5)$$

17-14

Worked out in chapter.

17-15

The amine is called morpholine; it is secondary. Enamine formation will be the result.

17-16

(a)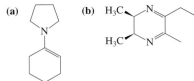

(b) H₃C, CH₃, ethyl, N ring structure (2,3-dimethyl-5-ethyl-6-methylpyrazine derivative)

(c) [mechanism sequence showing furanose oxygen with OH, protonation by H⁺, loss of H₂O, oxocarbenium ion]

C₆H₅N̈HN̈H₂ → [intermediate with ⁺NHNHC₆H₅] → −H⁺ → [O—NNHC₆H₅]

→ H⁺ → [ring O⁺ with NHNHC₆H₅] ⇌ HÖ—⁺—NNHC₆H₅

→ −H⁺ → HÖ— —=NNHC₆H₅

17-17

CH₃(CH₂)₄COOH $\xrightarrow[\text{2. C}_6\text{H}_6, \text{AlCl}_3]{\text{1. SOCl}_2}$

[phenyl ketone: C₆H₅—C(=O)—(CH₂)₄CH₃]

$\xrightarrow{\text{H}_2\text{NNH}_2, \text{KOH}, \Delta}$

[phenylhexane chain]

17-18

Formaldehyde > acetaldehyde > acetone > 3,3-dimethyl-2-butanone

17-19

(a) [cyclohex-2-enone] + CH₂=P(C₆H₅)₃

(b) [3-bromocyclohexene] treated successively with 1. P(C₆H₅)₃,
2. CH₃CH₂CH₂CH₂Li, 3. H₂C=O   or
1. H₂O,   2. MnO₂,   3. CH₂=P(C₆H₅)₃.

17-20

[ketone with Br chain]
$\xrightarrow{\substack{\text{1. HO—CH}_2\text{CH}_2\text{—OH, H}^+ \\ \text{2. P(C}_6\text{H}_5)_3 \\ \text{3. H}^+, \text{H}_2\text{O}}}$

[dioxolane with ⁺P(C₆H₅)₃Br⁻ chain]
$\xrightarrow{\substack{\text{1. CH}_3\text{Li} \\ \text{2. H}_2\text{C}=\text{CHCHO} \\ \text{3. H}^+, \text{H}_2\text{O}}}$

CH₃CCH₂CH₂CH=CHCH=CH₂

17-21

(a) [cyclohexene] $\xrightarrow[\text{2. (CH}_3)_2\text{S}]{\text{1. O}_3}$ HC(CH₂)₄CH (with two C=O) $\xrightarrow{\text{CH}_2=\text{P(C}_6\text{H}_5)_3}$

CH₂=CH(CH₂)₄CH=CH₂

(b) [butadiene] + [acrolein CHO] → [cyclohexenyl aldehyde CH] $\xrightarrow{\text{P(C}_6\text{H}_5)_3}$

[bis-cyclohexenyl methylene structure with H]

17-22

(a) CH₂=CHCH₂CH₂OCCH₃ (with O)

(b) [bicyclic lactone structure with O and O]

(c) (CH₃)₃COCCH₂CH₃ (with O)

Chapter 18

18-1

(a) CH₂=C (with O⁻ and H)

(b) CH₃CH=C (with O⁻ and H)

(c) CH₂=C (with O⁻ and CH₃)

(d) CH₃CH₂CH=C (with O⁻ and CH₂CH₂CH₃)

(e) [cyclopentenyl O⁻]

18-2

(a)

(i) [2-ethylcyclohexanone]

(ii) [cyclohexenyl OSi(CH₃)₃]

(b)
Kinetic conditions

(i) [2-ethyl-5-methylcyclopentanone]

(ii) [cyclopentenyl OSi(CH₃)₃ with methyl]

Thermodynamic conditions

(i) [2-ethyl-2-methylcyclopentanone]

(ii) [cyclopentenyl OSi(CH₃)₃ with methyl]

18-3

Base catalysis

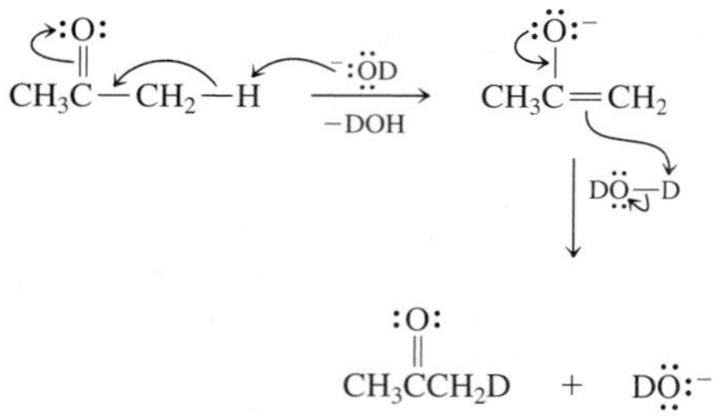

$$CH_3CCH_2D \quad + \quad D\ddot{O}:^-$$

Acid catalysis

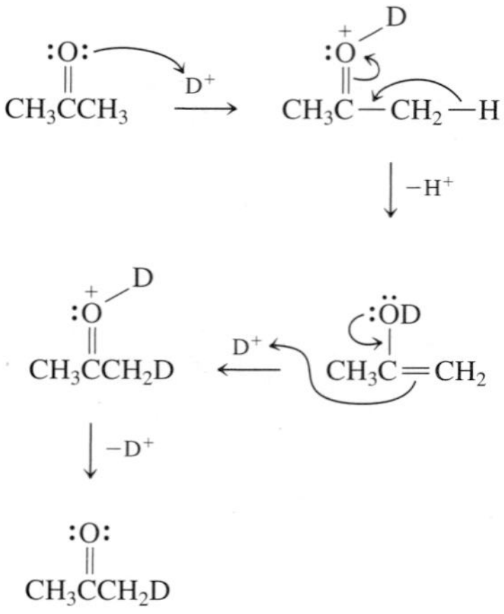

$$\overset{:O:}{\underset{\|}{CH_3CCH_2D}}$$

18-4

(a)

(b) $(CH_3)_3C\overset{O}{\underset{\|}{C}}CH$

No enolizable hydrogen

(c) $(CH_3)_3C\overset{O}{\underset{\|}{C}}CCD_3$

(d)

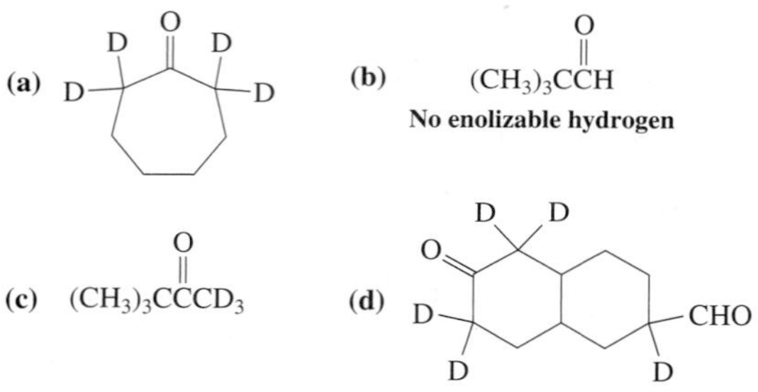

18-5

Worked out in chapter.

18-6

NaOD in D_2O, a solution containing the strongly basic ^-OD, causes removal of α-hydrogens to give enolate ions. These react with D_2O, eventually replacing all of the original α-hydrogen atoms with deuterium. Thus, in the NMR spectrum, the signal for the α-hydrogens will disappear and only the signal for the β-hydrogens at $\delta = 2.00$ ppm will remain as a somewhat broadened singlet.

18-7

Ketone A undergoes cis-to-trans isomerization by inversion (R to S) at the tertiary α-position. In ketone B, this position is not enolizable, because it is quaternary.

18-8

Acid-catalyzed:

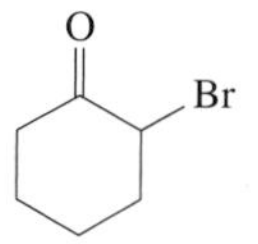

Base-catalyzed:

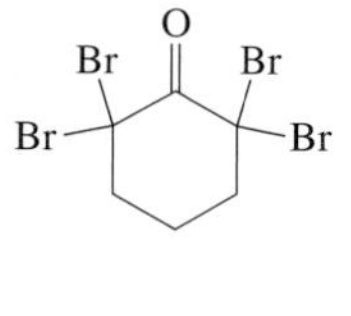

18-9

The two possible C-alkylations are retarded, because they lead to strained products. As a result, O-alkylation, which constructs a relatively unstrained six-membered ring, becomes competitive.

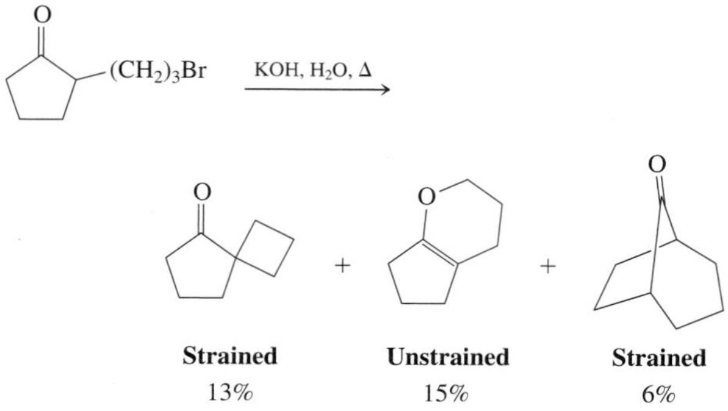

Strained	**Unstrained**	**Strained**
13%	15%	6%

18-10

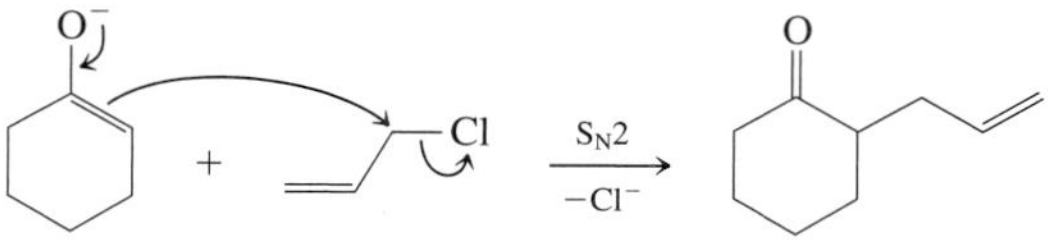

The transition state is stabilized by overlap of the inverting C_{sp^3} orbital with the π system.

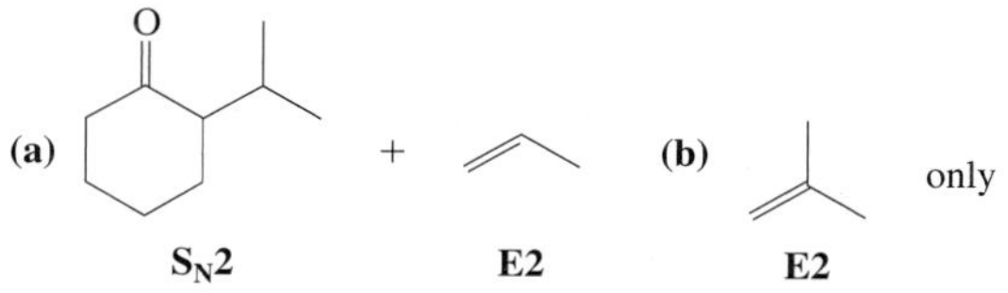

(a) + (b) only

S_N2 E2   E2

18-11

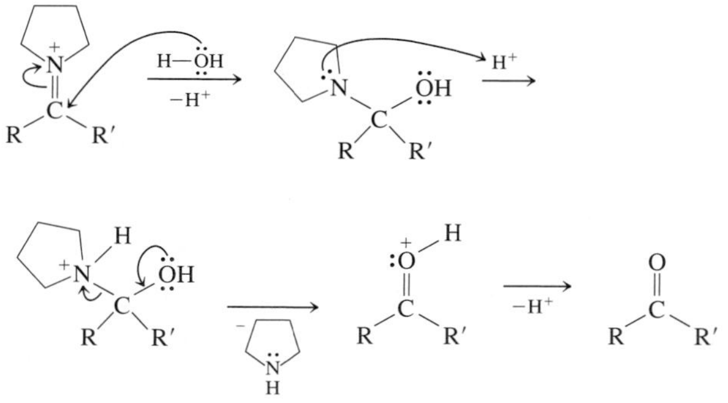

18-12

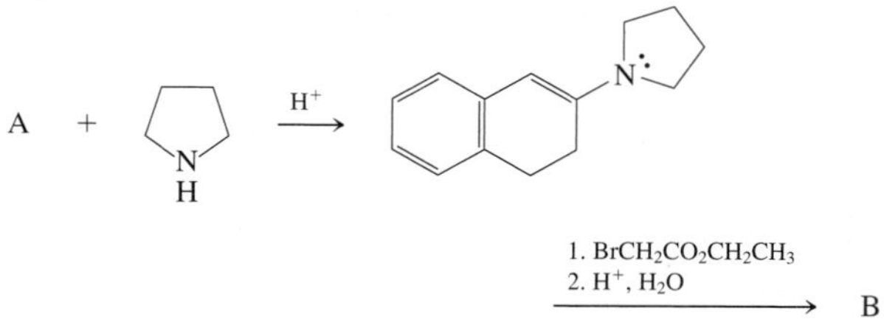

A + $\xrightarrow{H^+}$

$\xrightarrow[\text{2. } H^+, H_2O]{\text{1. } BrCH_2CO_2CH_2CH_3}$ B

18-13

(a) $CH_3CH_2\overset{OH}{\underset{H}{\overset{|}{C}}}-\overset{}{\underset{CH_3}{\overset{|}{C}}}HCHO$ (b) $CH_3CH_2CH_2\overset{OH}{\underset{H}{\overset{|}{C}}}-\overset{}{\underset{CH_2CH_3}{\overset{|}{C}}}HCHO$

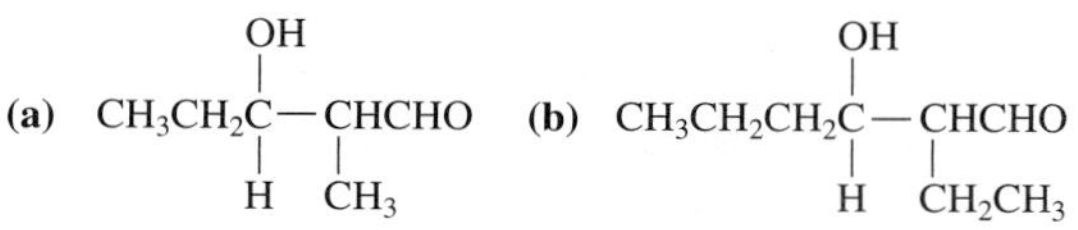

$$\underset{H}{\overset{OH}{\text{(c)} \ C_6H_5CH_2\overset{|}{C}-\underset{\underset{C_6H_5}{|}}{C}HCHO}} \qquad \underset{H}{\overset{OH}{\text{(d)} \ C_6H_5CH_2CH_2\overset{|}{C}-\underset{\underset{CH_2C_6H_5}{|}}{C}HCHO}}$$

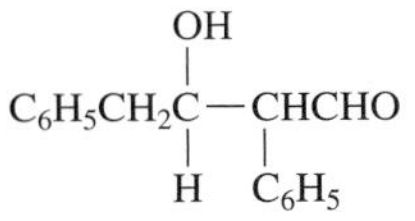

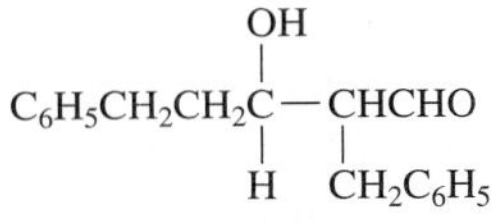

18-14

It cannot with itself, because it does not contain any enolizable hydrogens. It can, however, undergo crossed aldol condensations (Section 18-6) with enolizable carbonyl compounds.

18-15

(a) 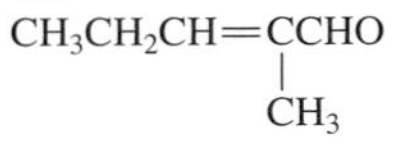 $CH_3CH_2CH{=}\underset{\underset{CH_3}{|}}{C}CHO$

(b) $CH_3CH_2CH_2CH{=}\underset{\underset{CH_2CH_3}{|}}{C}CHO$

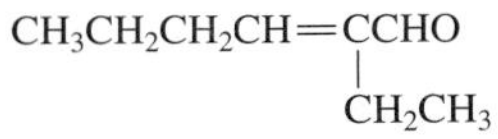

(c) $C_6H_5CH_2CH{=}\underset{\underset{C_6H_5}{|}}{C}CHO$

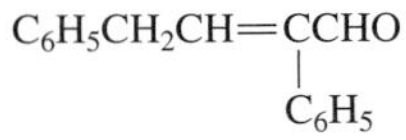

(d) 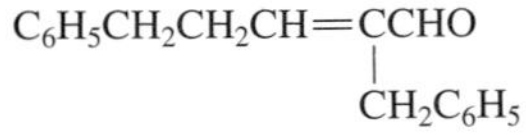 $C_6H_5CH_2CH_2CH{=}\underset{\underset{CH_2C_6H_5}{|}}{C}CHO$

18-16

Worked out in chapter.

18-17

The reverse, "retro-aldol" process follows the same pathway, each step run in reverse:

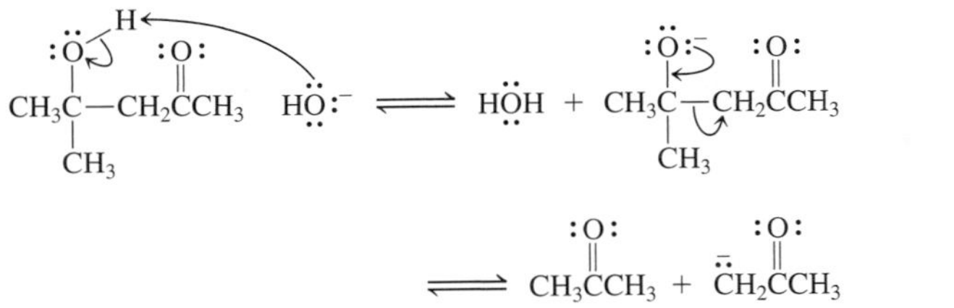

18-18

(a)

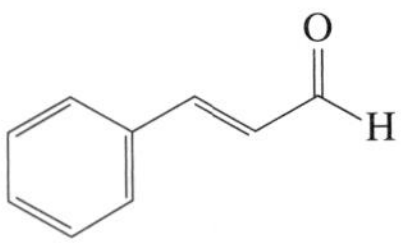

(b)

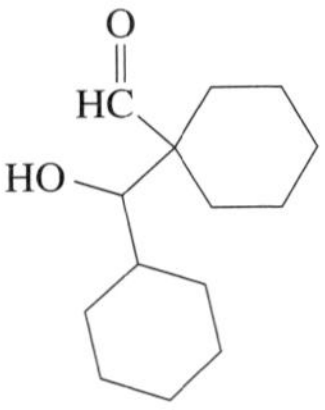

(c) 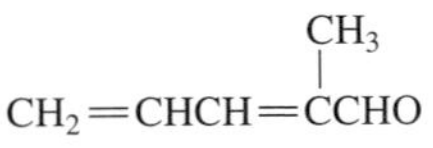 $CH_2{=}CHCH{=}\underset{\overset{|}{CH_3}}{C}CHO$

18-19

(a) 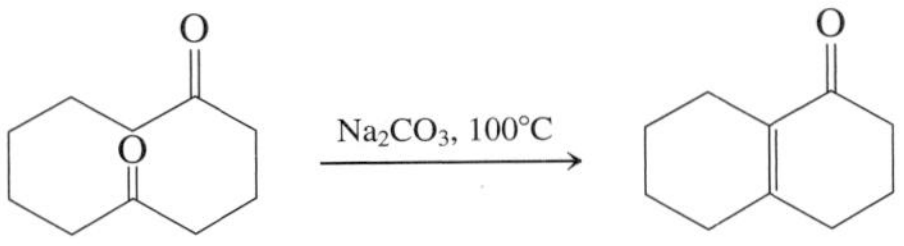 $\xrightarrow{Na_2CO_3, \ 100°C}$

(b)

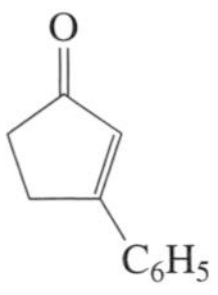

(c)

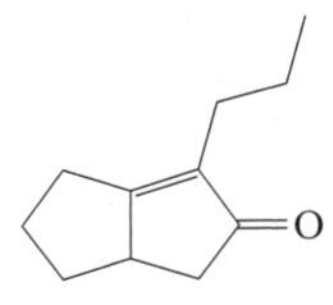

(d)

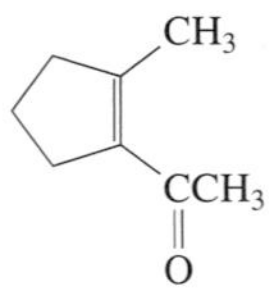

18-20

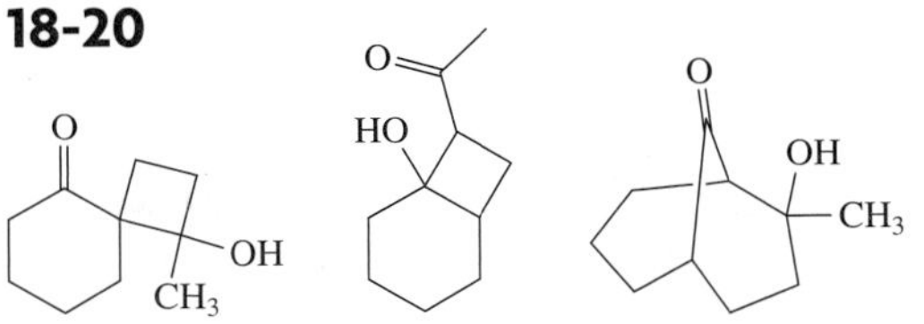

These three compounds are not formed, because of strain. Dehydration is, in addition, prohibited, again because of strain (or, in the first structure, because there is no proton available). The fourth possibility is most facile.

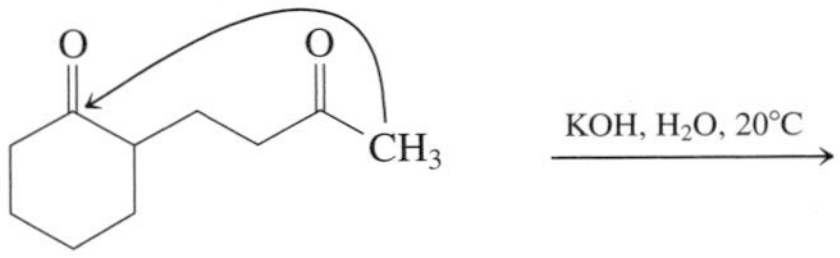

$\xrightarrow{KOH, \ H_2O, \ 20°C}$

2-(3-Oxobutyl)cyclohexanone

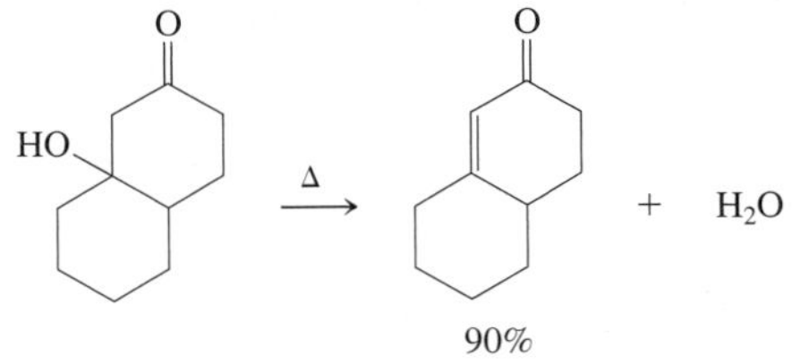

$\xrightarrow{\Delta}$ $+$ H_2O

90%

18-21

(a) 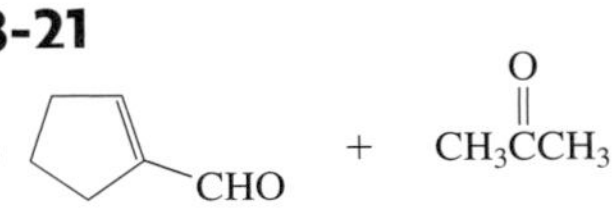 $+$ $CH_3\overset{\overset{O}{\|}}{C}CH_3$

(b)

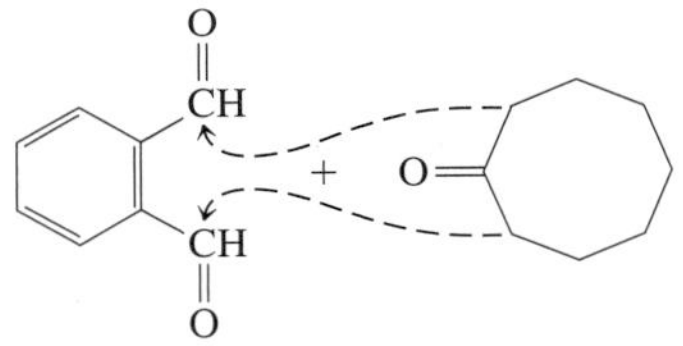

(c) 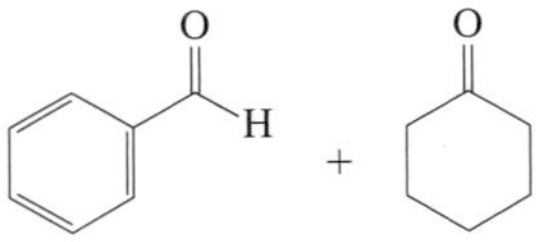 $+$

18-22

Mechanism of acid-mediated isomerization of β,γ-unsaturated carbonyl compounds:

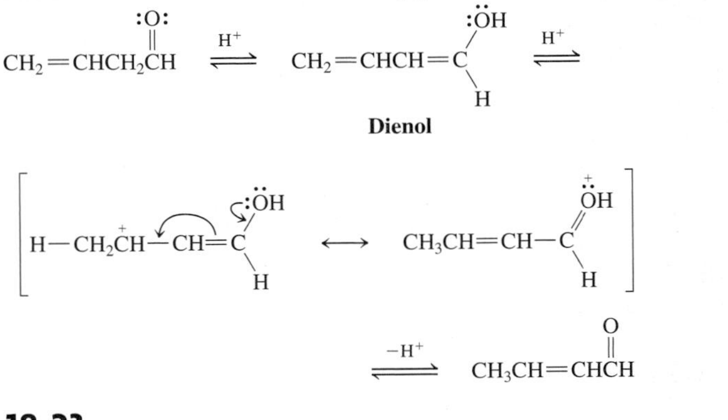

Dienol

$\xrightarrow{-H^+}$ $CH_3CH{=}CH\overset{\overset{O}{\|}}{C}H$

18-23

CH_3CH_2CHO $\xrightarrow{\text{—CHO, NaOH, H}_2\text{O, }\Delta}$

$\xrightarrow[\text{2. LiAlH}_4]{\text{1. H}_2, \text{Pd—C}}$

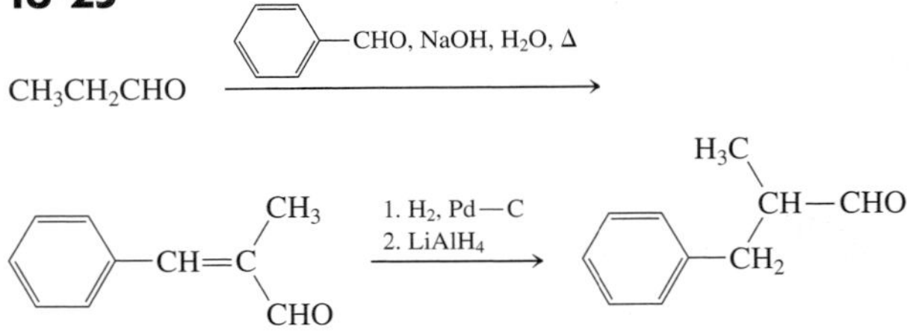

18-24

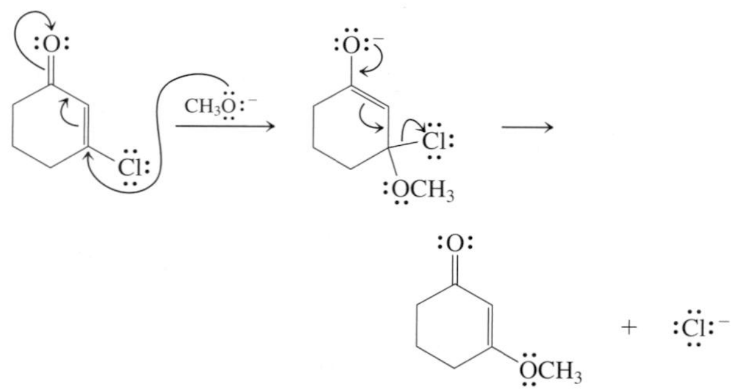

18-25

1. Protonation

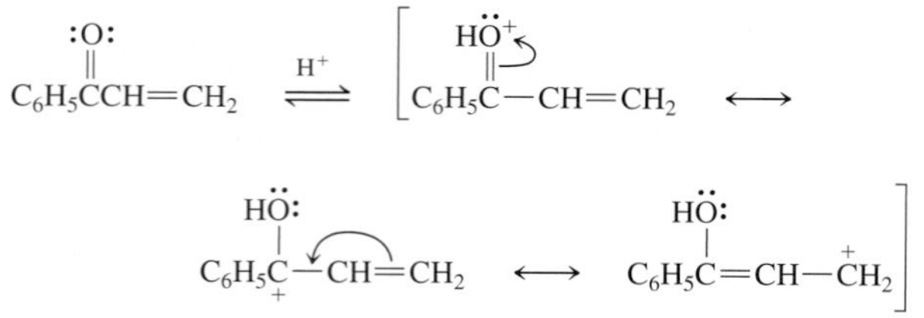

2. Cyanide attack

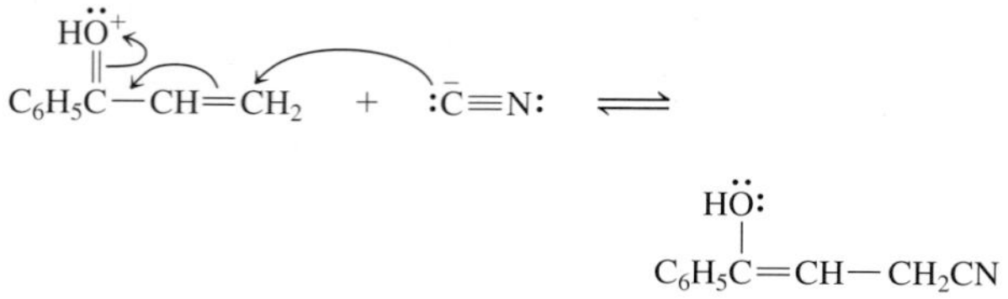

3. Enol–keto tautomerization

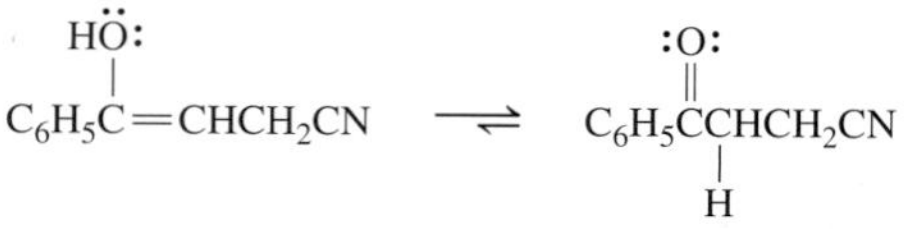

18-26

Worked out in chapter.

18-27

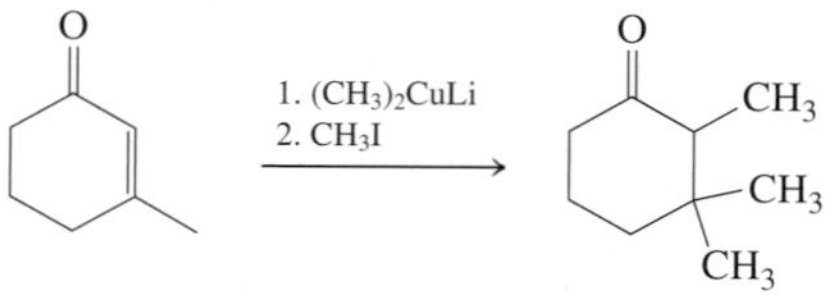

18-28

Worked out in chapter.

18-29

(a)

(b)

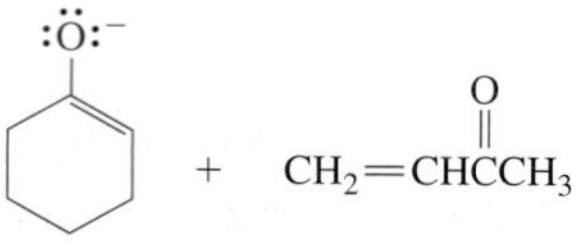

(c)

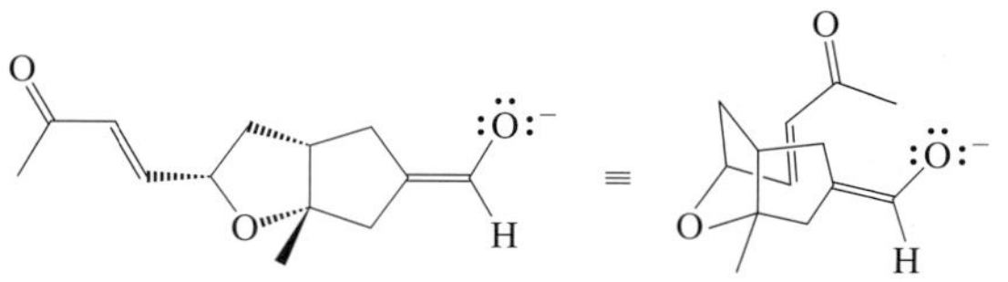

Chapter 19

19-1

(a) 5-Bromo-3-chloroheptanoic acid
(b) 4-Oxocyclohexanecarboxylic acid
(c) 3-Methoxy-4-nitrobenzoic acid

(d)

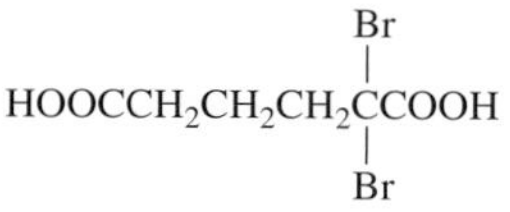

(e) 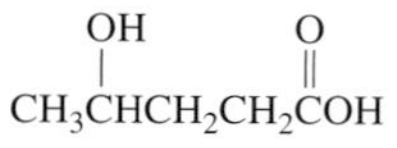 **(f)**

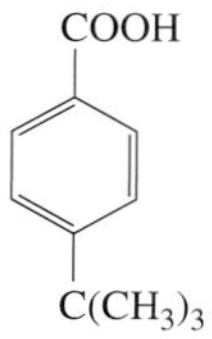

19-2

Worked out in chapter.

19-3

2-Methylpropanoic acid. Spectral assignments (ppm):

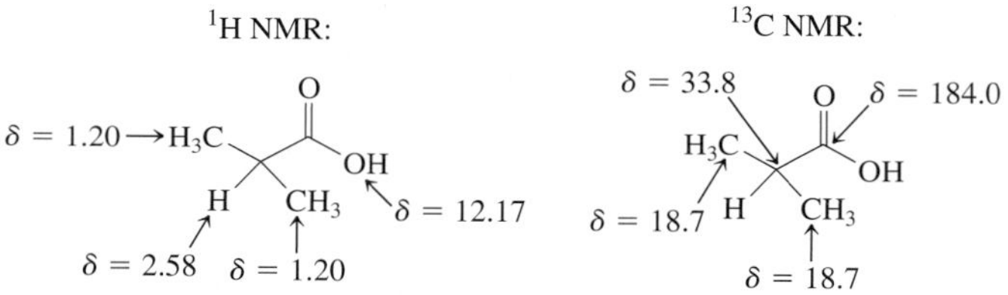

19-4

Applying the example of pentanoic acid to the three generic cleavage modes presented immediately preceding the exercise means replacing R by CH_3. As you can see, this change is inconsequential with respect to the identity of the fragment ions formed: $m/z = 45, 60,$ and 73, all visible in Figure 19-4.

19-5

(a) $CH_3CBr_2COOH > CH_3CHBrCOOH > CH_3CH_2COOH$

(b)

(c)

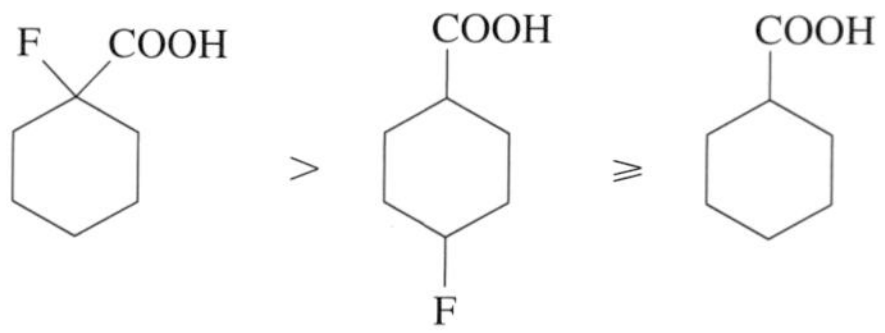

19-6

Protonated acetone has fewer resonance forms.

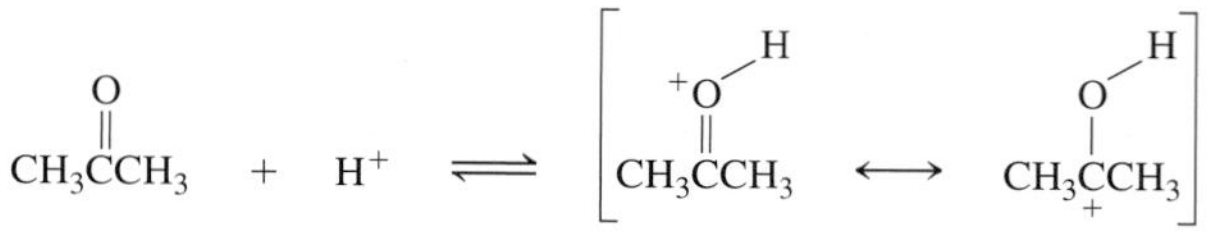

19-6 (continued, reaction scheme)

$$CH_3\overset{\displaystyle O}{\overset{\|}{C}}OH \;+\; H^+ \;\rightleftharpoons$$

$$\left[\; CH_3\overset{\displaystyle \overset{+}{O}H}{\underset{}{C}}\!-\!OH \;\longleftrightarrow\; CH_3\overset{\displaystyle OH}{\underset{}{\overset{}{C}}}\!-\!OH \;\longleftrightarrow\; CH_3\overset{\displaystyle OH}{\underset{+}{C}}\!=\!\overset{+}{O}H \;\right]$$

19-7

(a) $CH_3(CH_2)_3COOH$ **(b)** $HOOC(CH_2)_4COOH$

(c)

COOH
[cyclohexane ring]
COOH

19-8

Worked out in chapter.

19-9

(a) [CH₂=CH–CH₂–CH₂–CH₂–Cl chain] Cl   1. Mg,   2. CO_2,   3. H^+, H_2O
                                            or 1. ^-CN,   2. H^+, H_2O

(b) [cyclopentane ring with I] I   Same conditions as in (a)

(c) Br[—CH₂CH₂CH₂—]OH   1. ^-CN,   2. H^+, H_2O
                  O

The alternative (1. Mg,   2. CO_2,   3. H^+, H_2O) fails because Grignard formation is not possible in the presence of a carboxy group.

(d) [CH₂=CH–CH₂–Cl] Cl   1. Mg,   2. CO_2,   3. H^+, H_2O

The alternative (1. ^-CN,   2. H^+, H_2O) fails because S_N2 reactions are not possible on sp^2-hybridized carbons.

(e) [cyclopropane]—Br   1. Mg,   2. CO_2,   3. H^+, H_2O

The alternative (1. ^-CN,   2. H^+, H_2O) fails because S_N2 reactions of strained halocycloalkanes are very slow.

19-10

(a) 1. HCN,   2. H^+, H_2O

(b) [methylenecyclohexane] $\xrightarrow{HBr}$ [1-bromo-1-methylcyclohexane, H_3C, Br] $\xrightarrow[\text{3. } H^+, H_2O]{\text{1. Mg, 2. } CO_2}$ [1-methylcyclohexanecarboxylic acid, H_3C, COOH]

(c) [cis-4-bromo-methoxycyclohexane, Br / OCH₃] $\xrightarrow[\substack{-Br^- \\ S_N2}]{^-CN}$ [CN / OCH₃ trans] $\xrightarrow[\text{2. } H^+, H_2O]{\text{1. HO}^-, H_2O}$ [COOH / OCH₃ trans]

19-11

As in the scheme immediately preceding this exercise ($R = R' = CH_3$).

19-12

Worked out in chapter.

19-13

[reaction scheme with PBr_3, showing attack of carboxylic acid oxygen on phosphorus]

$$\xrightarrow{-\,:\!Br\!:^-}\qquad \xrightarrow{-H^+}$$

19-14

(a) 1. $CH_3\overset{\displaystyle O}{\overset{\|}{C}}Cl \;+\; Na^{+\,-}O\overset{\displaystyle O}{\overset{\|}{C}}CH_2CH_3$

2. $CH_3\overset{\displaystyle O}{\overset{\|}{C}}O^-Na^+ \;+\; Cl\overset{\displaystyle O}{\overset{\|}{C}}CH_2CH_3$

(b) $CH_3\overset{\displaystyle H_3C\;\;O}{\underset{}{\overset{\;\;\;\;\|}{C}H}}OH \;+\; SOCl_2$

19-15

The reaction is self-catalyzed by acid.

[mechanism scheme showing formation of succinic anhydride]

$$+\; H^+ \;\longrightarrow\; \xrightarrow{-H^+}\; \xrightarrow{+H^+}$$

$$\xrightarrow{-H_2O}\;\longrightarrow\; +\; H^+$$

19-16

(a) [butanoic acid methyl ester] OCH₃, O

(b) [pentyl formate] H—C(=O)—O—pentyl

(c) [cyclohexyl benzoate] benzene ring—C(=O)—O—cyclohexyl

(d) Br—CH₂—C(=O)—O—[3-methyl-2-butyl]

19-17

(a) $CH_3(CH_2)_3C\!\equiv\!CCH_2CH_2CO_2H + (CH_3)_2CHOH$ (or $CH_3CH=CH_2$)

(b) $CH_3\overset{\displaystyle O}{\overset{\|}{C}}(CH_2)_5OH \;+\; CH_3COOH$

(c) [cyclopentane]—$CO_2H \;+\; CH_3CH_2\underset{\underset{\displaystyle CH_3}{|}}{C}HCH_2OH$

19-18

Label appears in the ester.

$$R\overset{\displaystyle O}{\overset{\|}{C}}OH \;+\; H^{18}OCH_3 \;\underset{}{\overset{H^+}{\rightleftharpoons}}\; R\overset{\displaystyle O}{\overset{\|}{C}}{}^{18}OCH_3 \;+\; H_2O$$

19-19
Worked out in chapter.

19-20
(a) Follow that oxygen through each step of the mechanism. It becomes the oxygen of the hydroxy group in the intermediate alcohol and then the ring oxygen in the final product.

(b) Using an abbreviated steroid stencil:

$$H^+, H_2O \xrightarrow[\text{Lactonization as in Exercise 19-19}]{} \text{Oxandrolone}$$

19-21

$$+\ 2\ NH_3 \rightleftharpoons \quad \xrightarrow{\Delta}$$

$$+\ :NH_3 \rightleftharpoons \quad \xrightarrow{-H_2O}$$

$$\xrightarrow[+NH_3]{-NH_3} \rightleftharpoons$$

$$\rightleftharpoons \quad \rightleftharpoons \quad +\ H_2O$$

19-22
(a) 1. H^+, H_2O, 2. $LiAlH_4$, 3. H^+, H_2O
(b) 1. $LiAlD_4$, 2. H^+, H_2O

19-23

2. $RCH_2CBr + H^+ \rightleftharpoons RCH_2CBr$

$$RCH-C \overset{OH}{\underset{Br}{}} \rightleftharpoons RCH=C \overset{OH}{\underset{Br}{}} + H^+$$

3. $RCH=C$ ⟶ $RCHCBr + Br^-$ ⟶

$$RCHCBr + H^+ + Br^-$$

19-24

Chapter 20
20-1
Worked out in chapter.

20-2

$$C_6H_5CH_2OCCl > (CH_3)_3COCCl > CH_3OCOCOCH_3 >$$

$$(CH_3)_3COCOCOC(CH_3)_3 > CH_3OCOCH_3 > (CH_3)_3COCOC(CH_3)_3$$

20-3
At room temperature, rotation around the amide bond is slow on the NMR time scale and two distinct conformers can be observed.

$$H_3C-C(=O)-N(H)-NHC_6H_5 \rightleftharpoons H_3C-C(=O)-N(H)-NHC_6H_5$$

Heating makes the equilibration so fast that the NMR technique can no longer distinguish between the two species.

20-4

$m/z = 85$, $[CH_3CH_2CH_2CH_2C{=}O]^+$
(α cleavage)

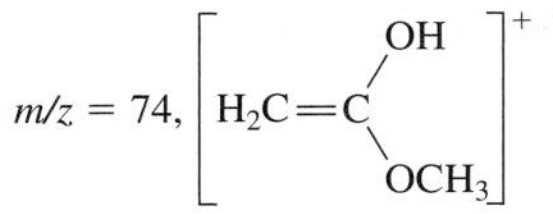

$m/z = 74$, $\left[H_2C{=}C \begin{smallmatrix} OH \\ \\ OCH_3 \end{smallmatrix} \right]^+$

(McLafferty rearrangement)

$m/z = 57$, $[CH_3CH_2CH_2CH_2]^+$
(α cleavage)

20-5

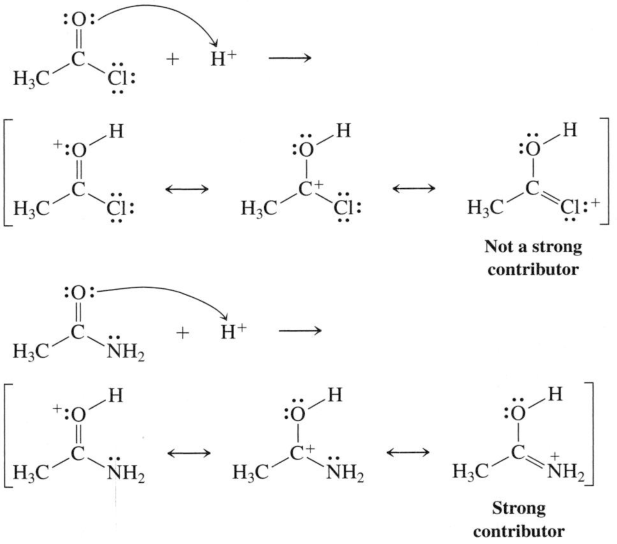

Not a strong contributor

Strong contributor

20-6

Section 19-12, step 2 of the Hell-Volhard-Zelinsky reaction.

20-7

20-8

Use *N,N*-diethylethanamine (triethylamine) to generate the acyl triethylammonium salt; then add the expensive amine.

20-9

Use the following reagents.

(a) H_2O (b) 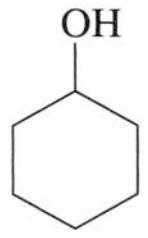 (c) $(CH_3)_2NH$

(d) $(CH_3CH_2)_2CuLi$ (e) $LiAl[OC(CH_3)_3]_3H$

20-10

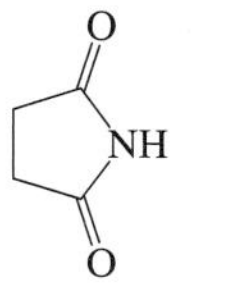

Butanimide (succinimide)

(See Section 19-10)

20-11

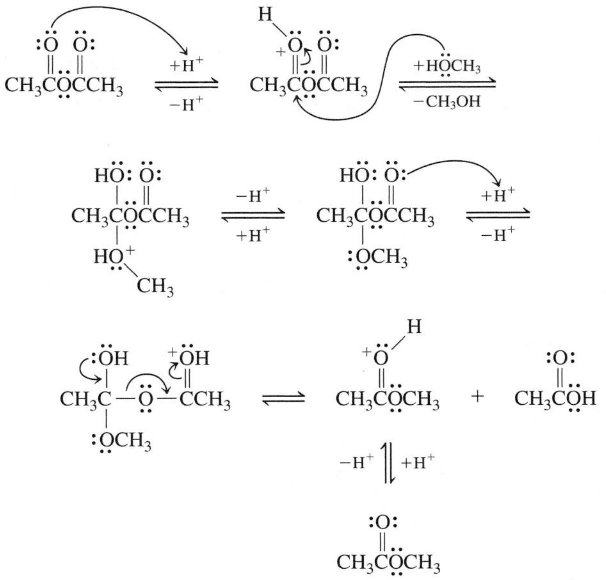

20-12

(a) Propyl propanoate
(b) Dimethyl butanedioate
(c) Methyl 2-propenoate (methyl acrylate)

20-13

Worked out in chapter.

20-14

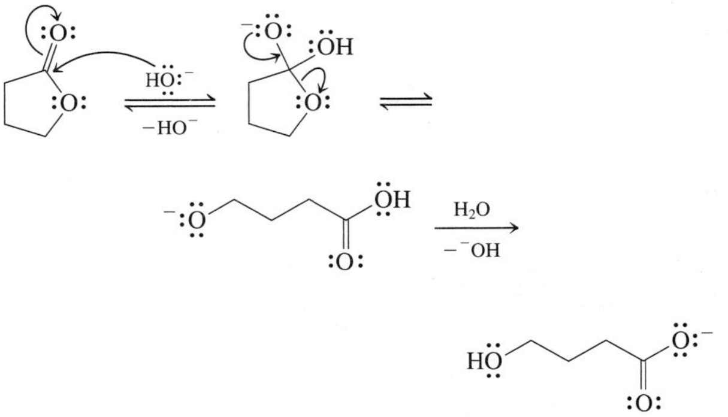

20-15

Acid catalysis: As in Exercise 20-13, but use $BrCH_2CH_2CH_2OH$ instead of H_2O as the nucleophile in the second step.

Base catalysis: In principle as in Exercise 20-14, but use $BrCH_2CH_2CH_2O^-$ instead of HO^- as the nucleophile in the first step. However, in practice, the bromoalkoxide will undergo intramolecular Williamson ether formation to oxacyclobutane (see Section 9-6).

20-16

$C_6H_5Br \xrightarrow{Mg} C_6H_5MgBr$

$C_6H_5COCH_3 + 2\,C_6H_5MgBr \xrightarrow{H^+, H_2O} (C_6H_5)_3COH$

20-17

(a) CO_2H on cyclohexane

(b) CO_2^- on cyclohexane

(c) CO_2CH_3 on cyclohexane

(d) $\overset{O}{\overset{\|}{C}}NH_2$ on cyclohexane

(e) CH_3CHOH on cyclohexane (with CH_3)

(f) CH_2OH on cyclohexane

(g) $H_3C \quad CO_2CH_2CH_3$ on cyclohexane

20-18

H_3C, H_3C on a pyrrolidine ring with N—H

20-19

Worked out in chapter.

20-20

Diisobutylaluminum hydride converts the amide into an aldehyde. Acidic hydrolysis releases the carbonyl group from the acetal. The product is an oxoalkanal capable of intramolecular aldol condensation under the acidic conditions to give the observed product:

H_3C—acetal—NH_2 —Diisobutylaluminum hydride→ H_3C—acetal—H —H^+, H_2O→

H_3C—(O)—CH$_2$...—H ⇌ [enol intermediate] → cyclopentanone derivative with $:O^-$

—Proton transfer→ cyclopentanone with OH —$-H_2O$→ cyclopentenone

20-21

In A, the negative charge can be delocalized over two carbonyl groups; not possible for B.

phthalimide N—H —$:\ddot{O}H$, $-H_2O$→

[three resonance structures of phthalimide anion]

20-22

CH_3, phenyl, NH_2 (chiral center, 1-phenylethylamine)

20-23

$R\ddot{N}=C=\ddot{O}$ + $H_2\ddot{O}:$ ⟶

$\left[R\ddot{N}=C\overset{\ddot{O}:^-}{\underset{\overset{+}{O}H_2}{}} \longleftrightarrow R\ddot{N}-C\overset{\ddot{O}}{\underset{\overset{+}{O}H_2}{}} \right]$ —Double H^+ shift→

$R-\overset{H}{\underset{H}{N^+}}-C\overset{\ddot{O}:}{\underset{\ddot{O}:^-}{}}$ ⟶ $R\ddot{N}H_2$ + $\ddot{O}=C=\ddot{O}$

20-24

$COOCH_3$ (on branched chain) —H^+, H_2O, Δ→ $COOH$ (on branched chain) —1. $SOCl_2$ 2. NH_3→

or NH_3, H_2O, Δ ↓

$\overset{O}{\overset{\|}{C}}NH_2$ (on branched chain) —Br_2, HO^-→ NH_2 (on branched chain)

20-25

$BrCH_2CH_2CH_2Br$ —^-CN, $-Br^-$→

$BrCH_2CH_2CH_2C{\equiv}N$ —^-CN, $-Br^-$→ $NCCH_2CH_2CH_2CN$

$\delta = 117.6$ and 3 other signals — 119.1 22.6 17.6 ppm

20-26

The exact details of this reduction are not known. A possible mechanism is

$R-C{\equiv}N$ —$LiAlH_4$→ $\overset{^-AlH_3Li^+}{\underset{}{N}}$=$C$ (R, H) —$LiAlH_4$→

$Li^+H_3Al^-$—$\overset{^-AlH_3Li^+}{N}$—$R-\overset{}{C}-H$ (with H) —H^+, H_2O→ RCH_2NH_2

20-27

(a) 1. H_2O, HO^-, 2. H^+, H_2O
(b) 1. $CH_3CH_2CH_2CH_2MgBr$, 2. H^+, H_2O
(c) 1. $(CH_3\overset{CH_3}{CH}CH_2)_2AlH$, 2. H^+, H_2O
(d) D_2, Pt

Chapter 21
21-1
(a) 2-Butanamine, *sec*-butylamine
(b) *N,N*-Dimethylbenzenamine, *N,N*-dimethylaniline
(c) (*R*)-6-Bromo-2-hexanamine, (*R*)-(5-bromo-1-methyl)pentylamine

21-2

$$CH_2NHCH_2CH=CH_2$$

(a) $HC\equiv CCH_2NH_2$ (b) 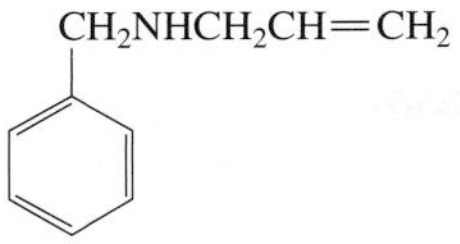   (c) $(CH_3)_3CNHCH_3$

21-3

The lesser electronegativity of nitrogen, compared with oxygen, allows for slightly more diffuse orbitals and hence longer bonds to other nuclei.

21-4

Less, because, again, nitrogen is less electronegative than oxygen. See Tables 10-2 and 10-3 for the effect of the electronegativity of substituent atoms on chemical shifts.

21-5

Worked out in chapter.

21-6

IR: Secondary amine, hence a weak band at ≈ 3400 cm^{-1}.

^{1}H NMR: s for the 1,1-dimethylethyl (*tert*-butyl) group at high field; s for the attached methylene group at $\delta \approx 2.7$; q for the second methylene unit close to the first; t for the unique methyl group at high field, closest to the 1,1-dimethylethyl (*tert*-butyl) signal.

^{13}C NMR: Five signals, two at low field, $\delta \approx 45–50$ ppm.

MS: $m/z = 115$ (M$^+$), 100 [(CH$_3$)$_3$CCH$_2$NH=CH$_2$]$^+$, and 58 (CH$_2$=NHCH$_2$CH$_3$)$^+$. In this case, two different iminium ions can be formed by fragmentation.

21-7

The answer lies in the hybridization at C2 of all compounds, which changes from sp^3 to sp^2 to sp. Remember that the inductive electron withdrawing power of carbon increases in that sequence (Section 13-2). Therefore, along the series of amines in this problem, the electron density at nitrogen decreases, hence its basicity, or, conversely, the pK_a value of the conjugate acid.

21-8

As discussed in Section 21-4 (and Sections 16-1 and 16-3), the nitrogen lone electron pair in benzenamine is tied up by resonance with the benzene ring. Therefore, the nitrogen is less nucleophilic than the one in an alkanamine.

21-9

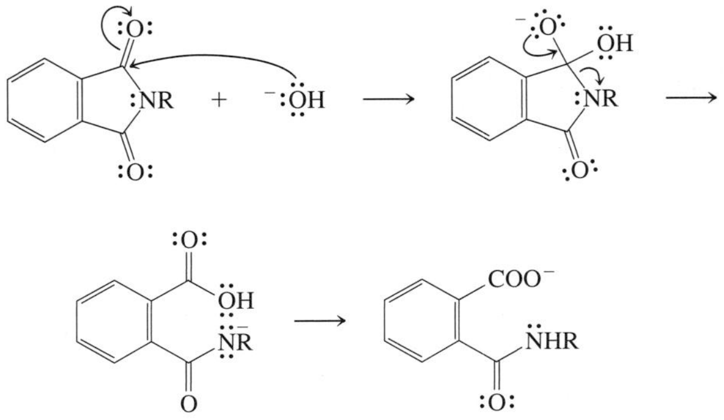

Continue as in the normal amide hydrolysis by base (Section 20-6).

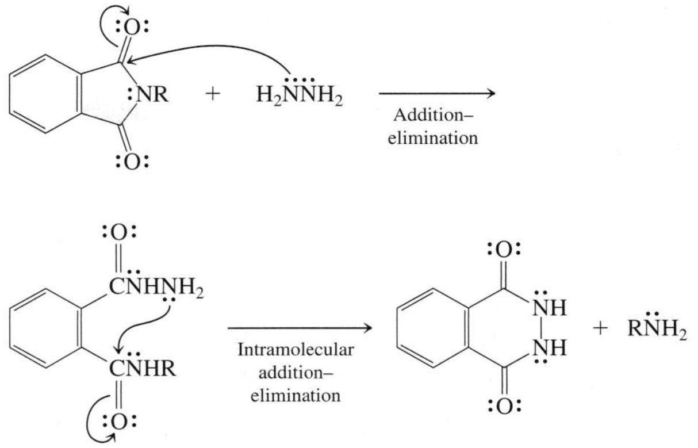

21-10

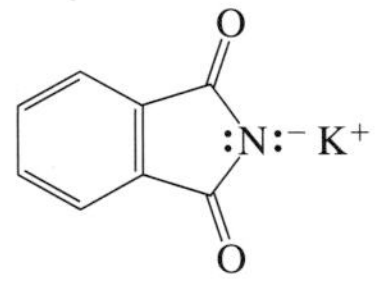

A

(a) 1. A + CH$_3$(CH$_2$)$_5$Br;   2. H$^+$, H$_2$O;   3. HO$^-$, H$_2$O

(b) 1. A   + 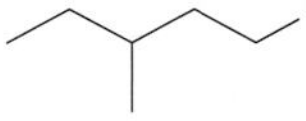   Br;   2. H$^+$, H$_2$O   3. HO$^-$, H$_2$O

(c) 1. A   + 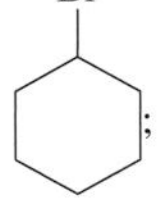   ;   2. H$^+$, H$_2$O;   3. HO$^-$, H$_2$O

(d) 1. A + BrCH$_2$CO$_2$CH$_2$CH$_3$;   2. H$^+$, H$_2$O

Protection of the carboxy group is necessary to prevent the acidic proton from reacting with A (see also Section 26-2). The azide method should work well for (a)–(c). For (d), the reduction step requires catalytic hydrogenation because LiAlH$_4$ would also attack the ester function.

21-11

In abbreviated form:

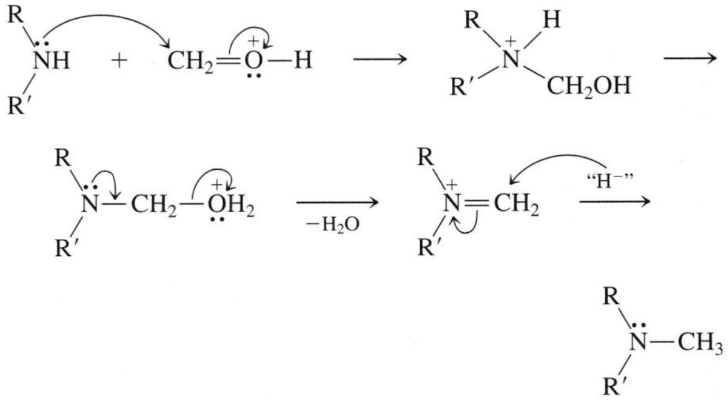

21-12

Worked out in chapter.

21-13

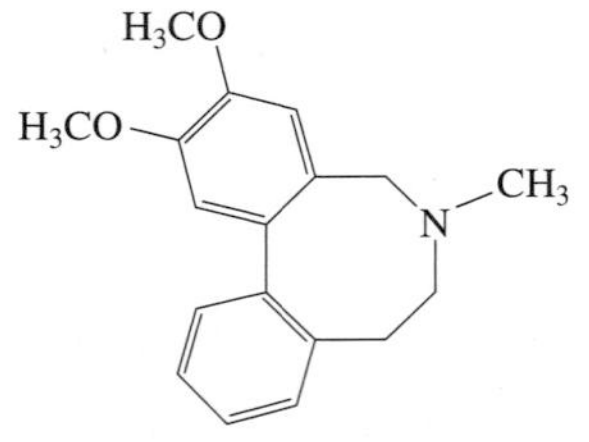

Buflavine

21-14

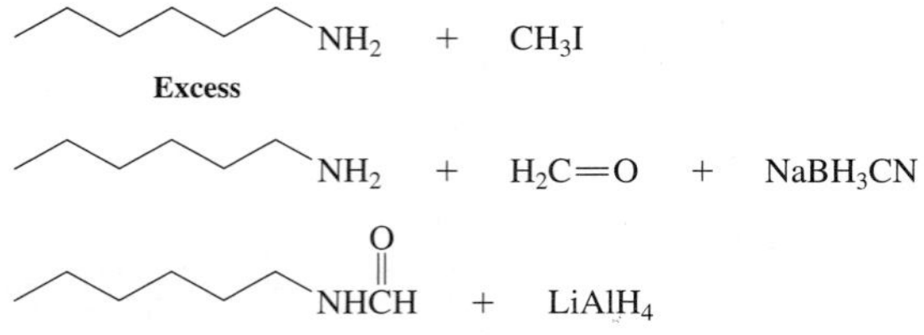

21-15

(a) $CH_3CH{=}CH_2$ and $CH_2{=}CH_2$
(b) $CH_3CH_2CH{=}CH_2$ and $CH_3CH{=}CHCH_3$ (cis and trans). The terminal alkene predominates. This reaction is kinetically controlled according to the Hofmann rule (Section 11-1). Thus, the base prefers attack at the less bulky end of the sterically encumbered quaternary ammonium ion.

21-16

The ethyl group can be extruded as ethene by Hofmann elimination [see Exercise 21-15, part (a)], thus giving product mixtures. Generally, any alkyl substituent with hydrogens located β to the quaternary nitrogen is capable of such elimination, an option absent for methyl.

21-17

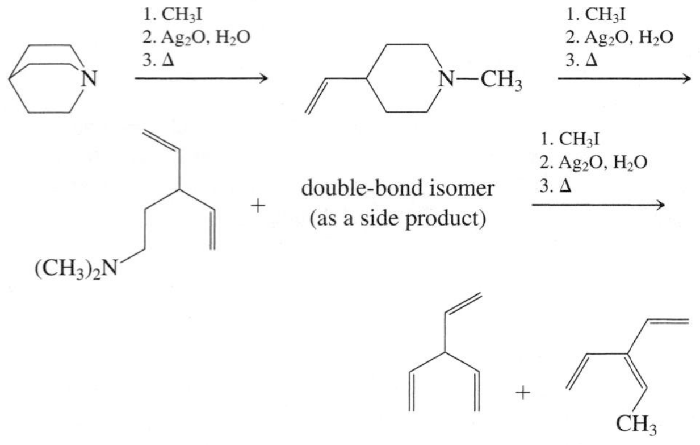

21-18

Worked out in chapter.

21-19

(a) $CH_3\overset{CH_3}{\underset{CH_2NH(CH_2)_5CH_3}{C}}CHO$ **(b)**

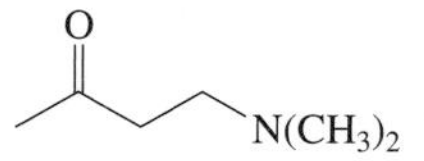

(c)

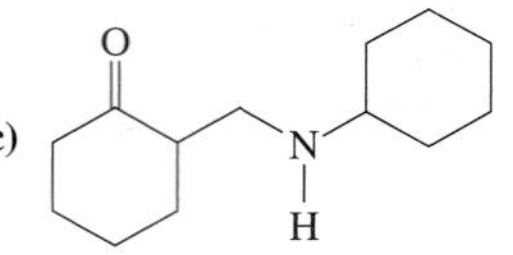

21-20

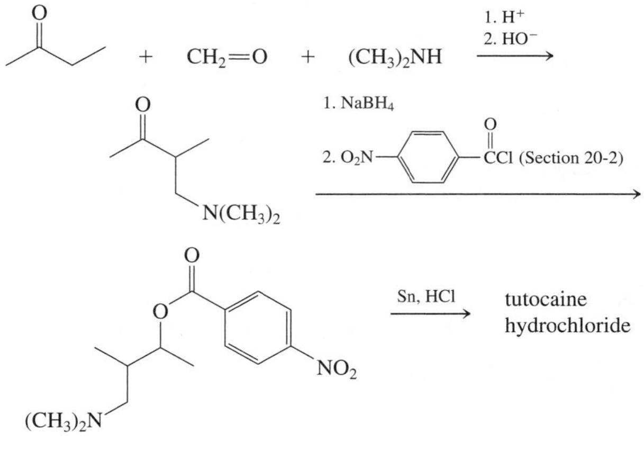

21-21

Nucleophilic displacement of N_2 in RN_2^+ by water occurs by an S_N2 mechanism with inversion.

21-22

Worked out in chapter.

21-23

The reaction is a straightforward electrophilic aromatic substitution (the first stage of which, for ortho attack, is sketched in the structure below), similar to nitration (Section 15-10), except that the attacking electrophile is the nitrosyl, :N≡O:$^+$ (and not nitronium, NO_2^+) cation. The relationship to Problem 21-22 becomes evident when you recognize phenol as an enol (hence the name; Section 22-3).

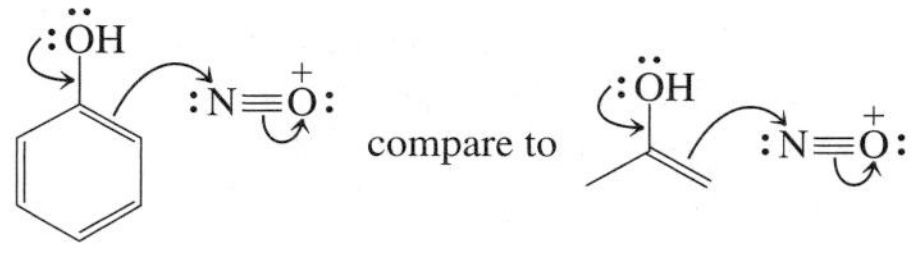

21-24

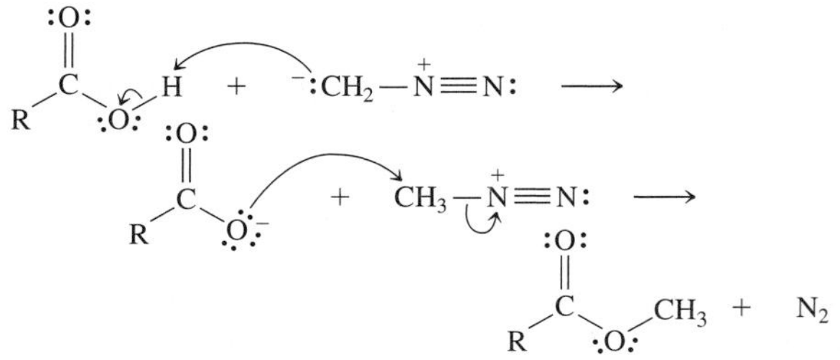

Chapter 22

22-1

(a)

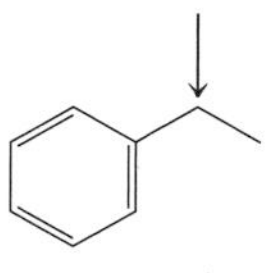

(b)

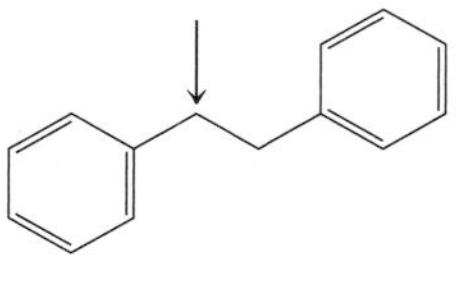

(c)

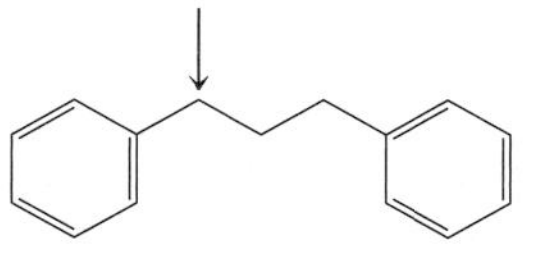

(d) 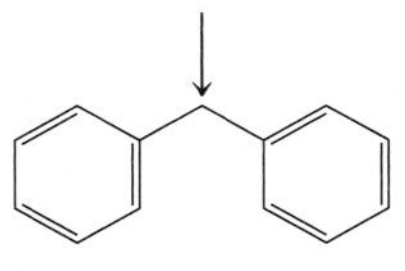 **(e)**

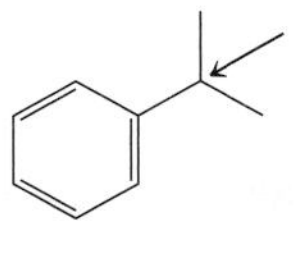

Order of reactivity: (d) > (e) > (a), (b), (c)

22-2

$(C_6H_5)_2CHCl$ solvolyzes faster because the additional phenyl group causes extra resonance stabilization of the intermediate carbocation.

22-3

Both react through an S_N2 mechanism that includes chloride attack on the protonated hydroxy group. The conversion of phenylmethanol is accelerated relative to that of ethanol because of a delocalized transition state.

22-4

(a) $(C_6H_5)_2CH_2$, because the corresponding anion is better resonance stabilized.
(b) $4\text{-}CH_3OC_6H_4CH_2Br$, because it contains a better leaving group.
(c) $C_6H_5CH(CH_3)OH$, because the corresponding benzylic cation is not destabilized by the extra nitro group (draw resonance structures).

22-5

(a) 1. $KMnO_4$, 2. $LiAlH_4$, 3. H_2, Pd–C
(b) 1. $KMnO_4$, 2. H^+, H_2O, 3. Δ ($-2\ H_2O$)
(c) 1. H_2, Pd–C, 2. $SOCl_2$

22-6

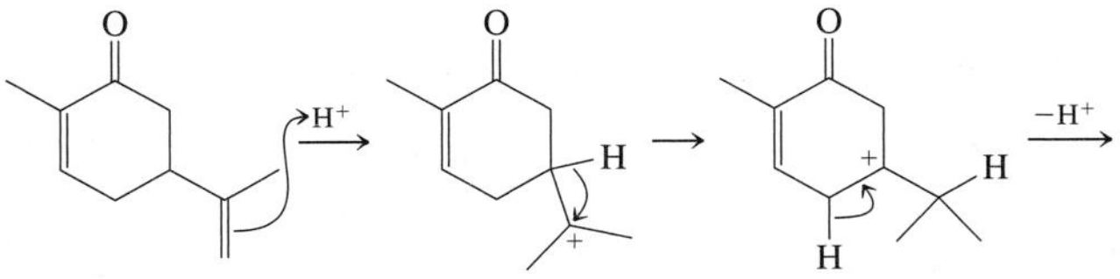

Carvone

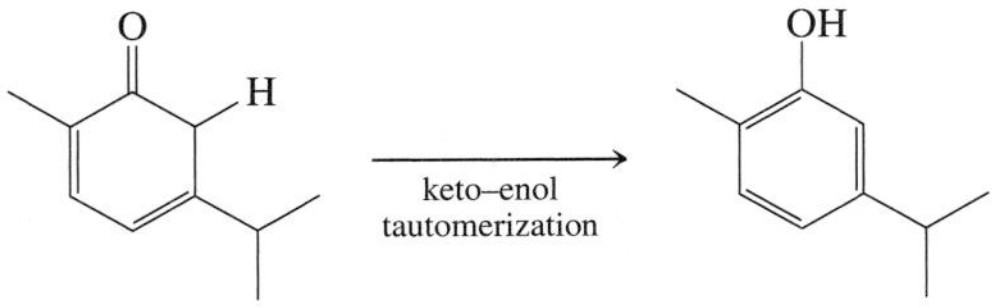

Carvacrol

22-7

(a) The nitro group is inductively electron withdrawing in all positions but can stabilize the negative charge by resonance only when at C2 or C4.
(b) B, A, D, C

22-8

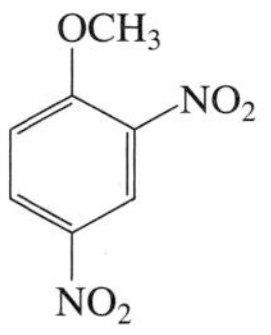

22-9

Worked out in chapter.

22-10

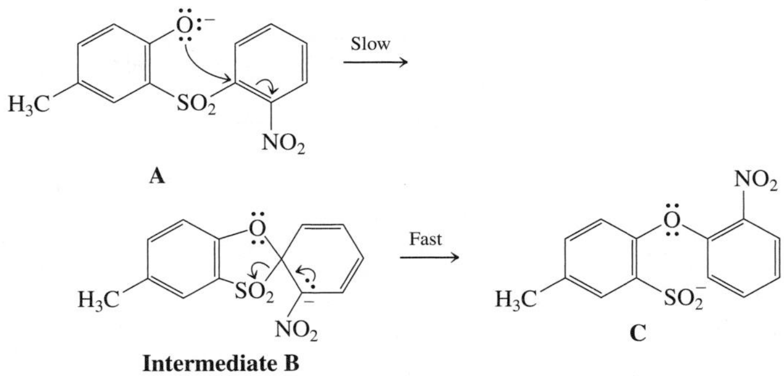

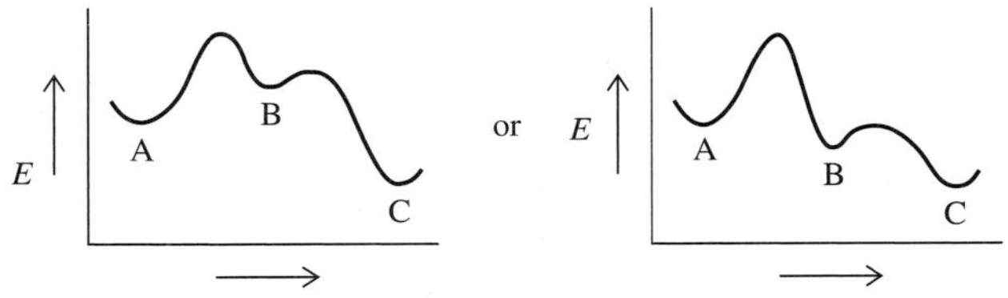

A

Intermediate B

C

22-11

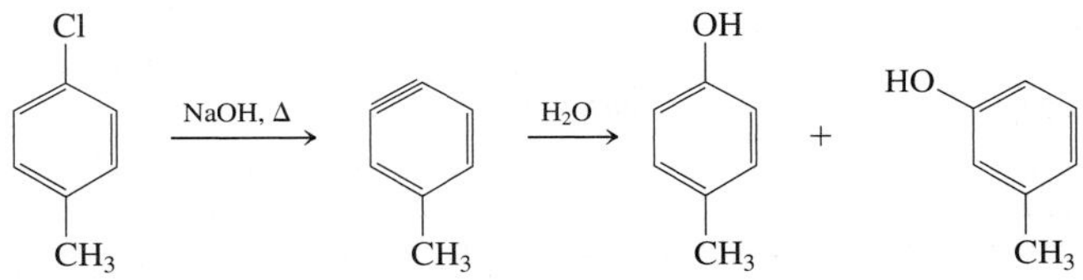

22-12

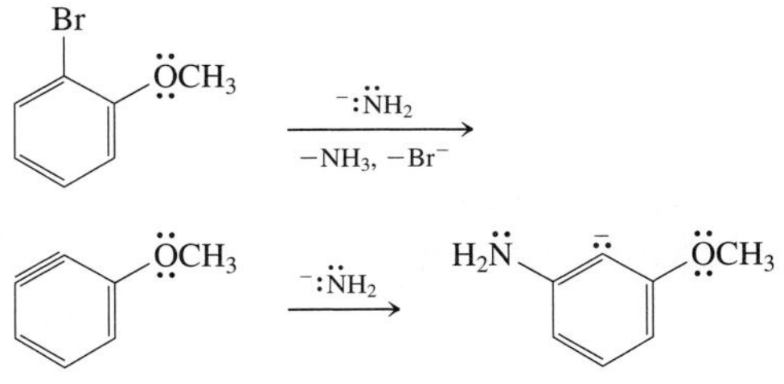

A

The product A of amide addition to the intermediate benzyne is stabilized by the inductively electron-withdrawing effect of the methoxy oxygen; therefore, it is formed regioselectively. Protonation gives the major final product. Note that there is no possibility for delocalization of the negative charge in A because the reactive electron pair is located in an sp^2 orbital that lies perpendicular to the aromatic π system.

22-13

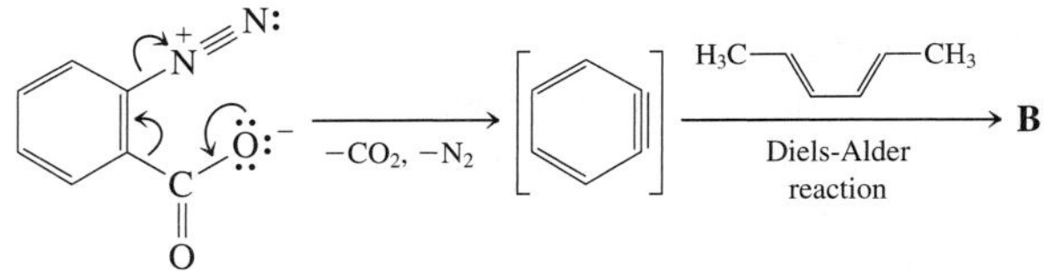

A   **Benzyne**

22-14

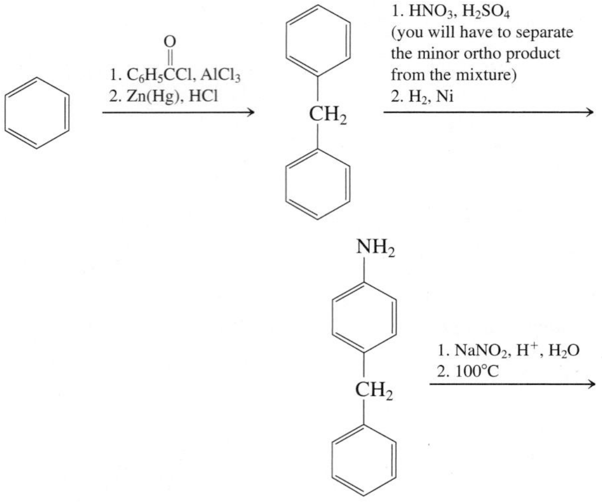

1. C₆H₅CCl, AlCl₃
2. Zn(Hg), HCl

1. HNO₃, H₂SO₄
(you will have to separate the minor ortho product from the mixture)
2. H₂, Ni

1. NaNO₂, H⁺, H₂O
2. 100°C

22-15

(a) 1. Br₂, 2. Pd catalyst, KOH
(b) 1. HNO₃, H₂SO₄, 2. Br₂, 3. Pd catalyst, KOH
(c) 1. Br₂, 2. Pd catalyst, KOH

22-16

Such a process would require nucleophilic attack by halide ion on a benzene ring, a transformation that is not competitive.

22-17

(a) Amines are more nucleophilic than alcohols; this rule also holds for benzenamines (anilines) relative to phenols.

(b)

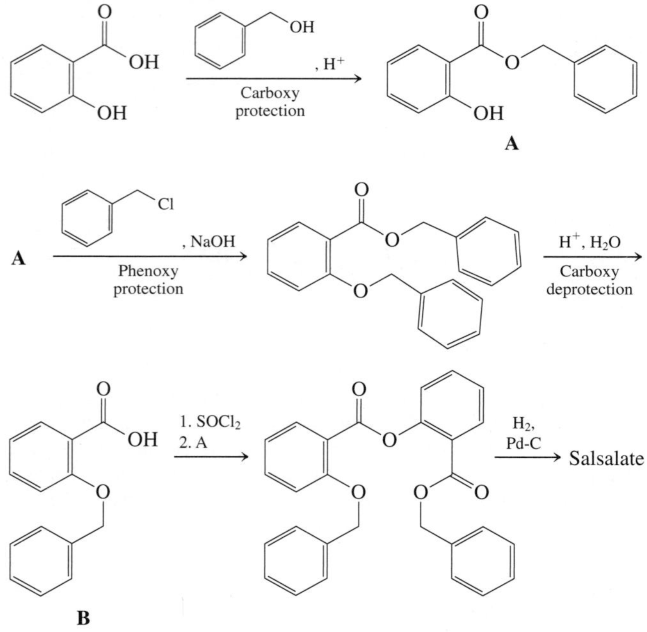

, H⁺
Carboxy protection

A

A → , NaOH
Phenoxy protection

H⁺, H₂O
Carboxy deprotection

1. SOCl₂
2. A

H₂, Pd-C → Salsalate

B

22-18

The 1,1-dimethylethyl (*tert*-butyl) group is considerably larger than methyl, attacking preferentially at C4.

22-19

Worked out in chapter.

22-20

1. Br₂,   2. CH₃CH₂CH₂Cl, NaOH,   3. SOCl₂, HOCH₂CH₂N(CH₂CH₃)₂,
4. Pd catalyst, NH₃

22-21

Worked out in chapter.

22-22

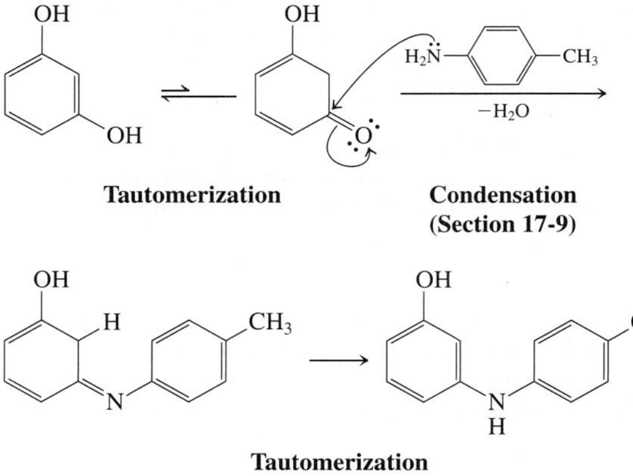

−H₂O

Tautomerization

Condensation (Section 17-9)

Tautomerization

22-23

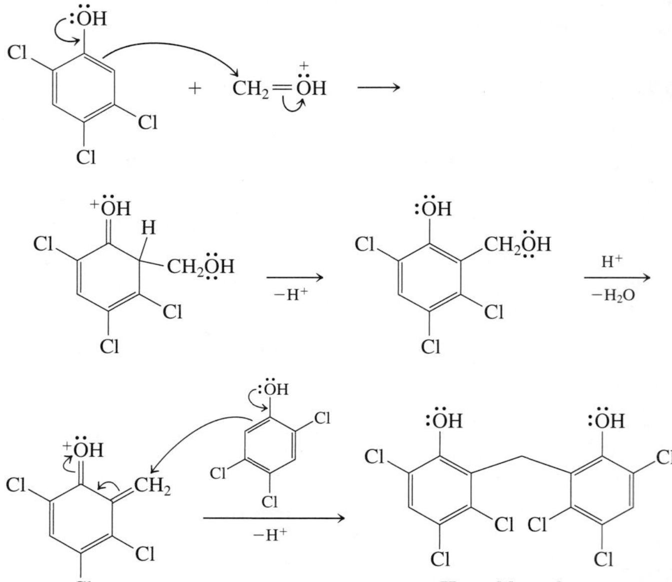

+ CH₂=ÖH →

−H⁺

H⁺ −H₂O

−H⁺

Hexachlorophene

22-24

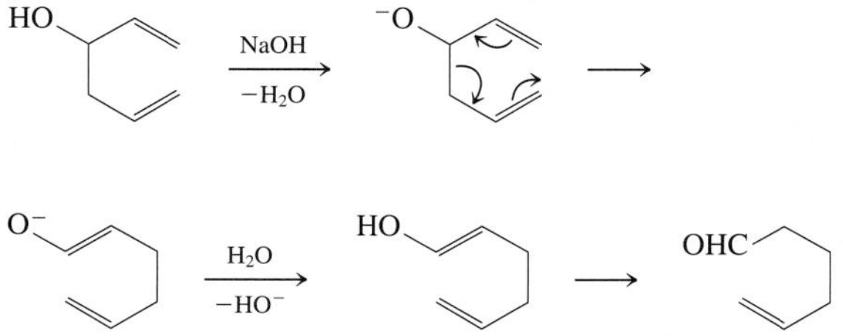

NaOH −H₂O

H₂O −HO⁻

The Cope rearrangement is especially fast in this case because the negative charge in the initial enolate ion is delocalized.

22-25

Worked out in chapter.

22-26

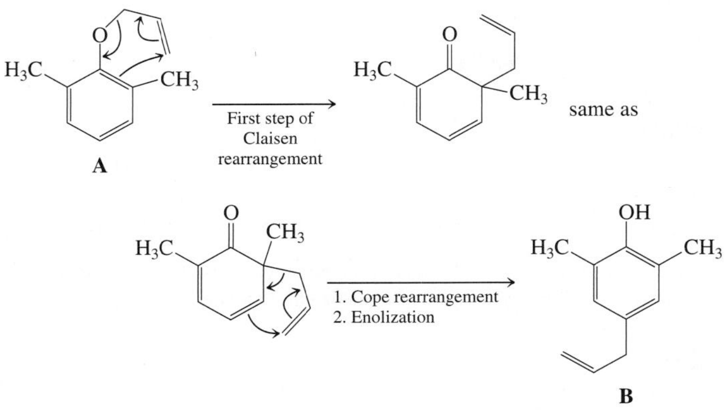

H$_3$C— ... CH$_3$
A

First step of Claisen rearrangement →

H$_3$C— ... CH$_3$ same as

H$_3$C— ... CH$_3$

1. Cope rearrangement
2. Enolization →

H$_3$C— OH —CH$_3$

B

H$_3$C— CH$_3$ CH$_3$O OCH$_2$CH$_3$

Repeat →

H$_3$C— CH$_3$ CH$_3$CH$_2$O OCH$_2$CH$_3$

22-29

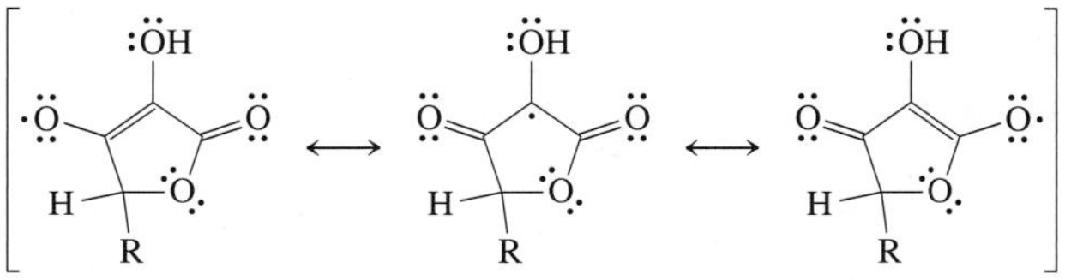

[OH ... O ... ↔ O ... O ... ↔ O ... O]
H R, H R, H R

22-27

[:O:$^-$ ↔ :O: ↔ :O: ↔ :O:]
:OH :OH :OH :OH

[:Ö· ↔ :O: ↔ :O: ↔ :O:]
:OH :OH :OH :OH

[:Ö· ↔ :O: ↔ :O: ↔ :O: ↔
:O:$^-$:O:$^-$:O:$^-$:O:$^-$

:Ö· ↔ :Ö·· ↔ :Ö:$^-$ ↔ etc.]
:O: :O: :O:

22-30

(a) 1. CH$_3$CCl, AlCl$_3$, 2. HNO$_3$, H$_2$SO$_4$, 3. H$_2$, Pd, 4. NaNO$_2$, HI

(b) 1. HNO$_3$ (2 equivalents), H$_2$SO$_4$, 2. H$_2$, Ni, 3. NaNO$_2$, HCl, 4. CuCN, KCN

(c) 1. HNO$_3$, H$_2$SO$_4$, 2. SO$_3$, H$_2$SO$_4$, 3. H$_2$, Ni, 4. NaNO$_2$, HCl, 5. H$_2$O, Δ

22-31

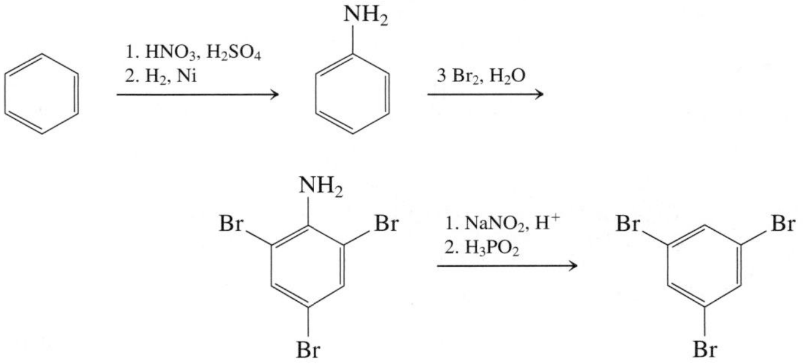

1. HNO$_3$, H$_2$SO$_4$
2. H$_2$, Ni → (NH$_2$)

3 Br$_2$, H$_2$O →

Br (NH$_2$) Br / Br

1. NaNO$_2$, H$^+$
2. H$_3$PO$_2$ →

Br / Br / Br

22-32

(a) OCH$_3$ ... N=N ... (phenyl)

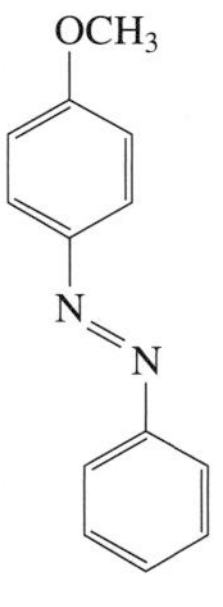

(b) OCH$_3$ / Cl ... N=N ... + Cl / OCH$_3$ ... N=N

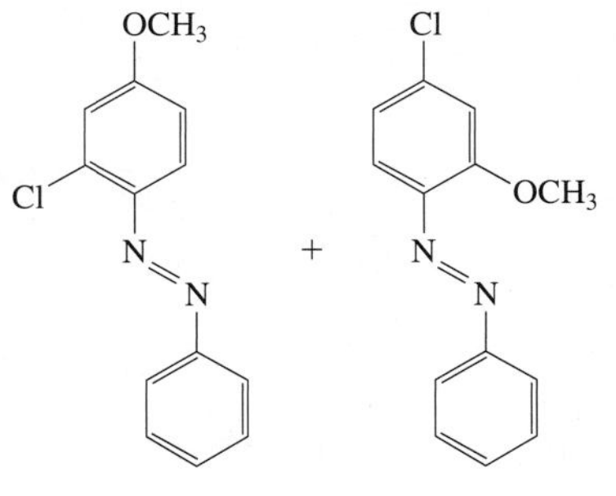

(c) (CH$_3$)$_3$C— N(CH$_3$)$_2$ / N=N

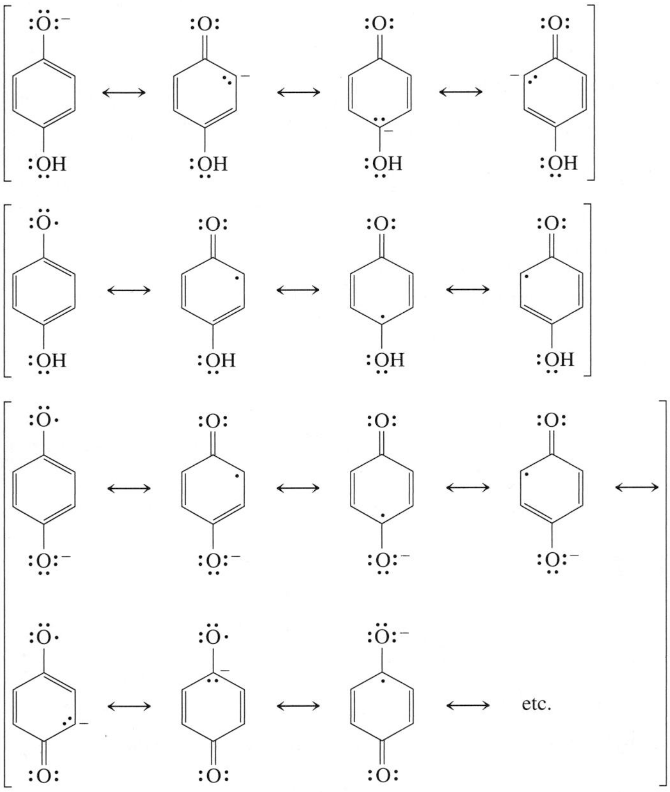

22-28

This exchange goes through two conjugate addition–elimination cycles.

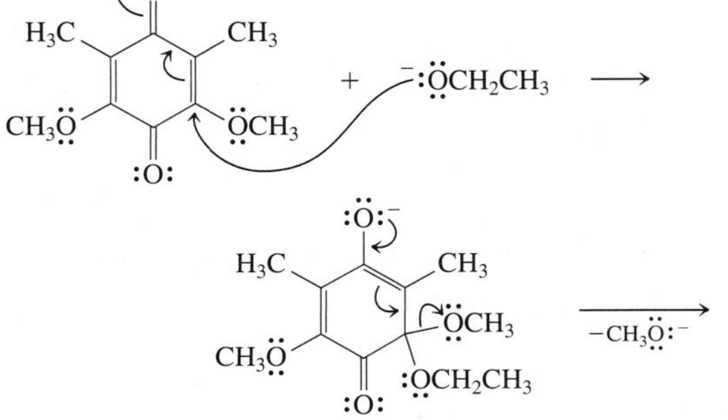

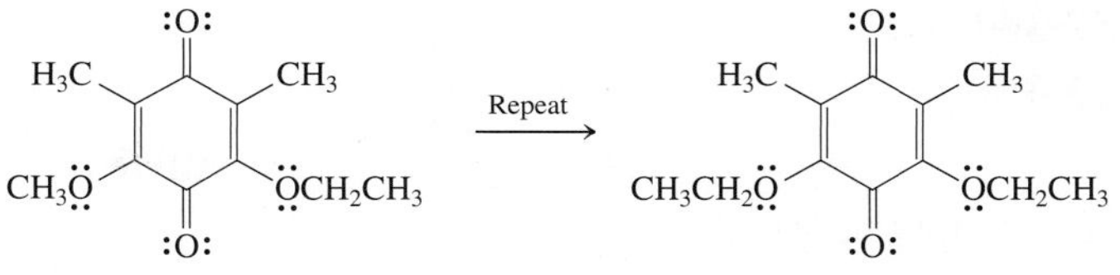

H$_3$C— :O: —CH$_3$ CH$_3$Ö OCH$_3$:O: + $^-$:ÖCH$_2$CH$_3$ →

H$_3$C— :O:$^-$ —CH$_3$ CH$_3$Ö ÖCH$_3$ / :OCH$_2$CH$_3$:O:

−CH$_3$Ö:$^-$ →

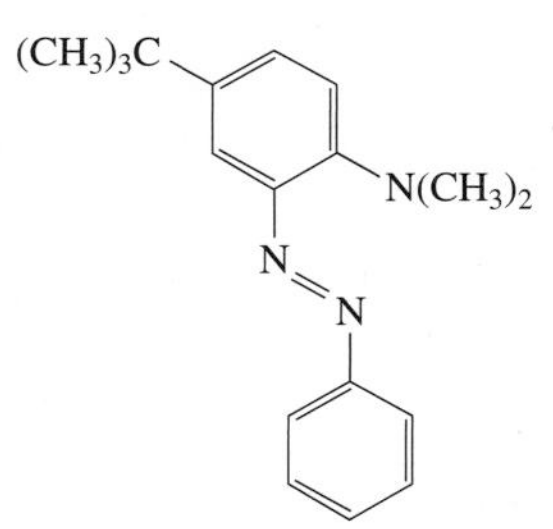

Chapter 23
23-1

(a) $CH_3CH_2CCHCOCH_2CH_3$
        with two O (carbonyls) and CH_3 substituent

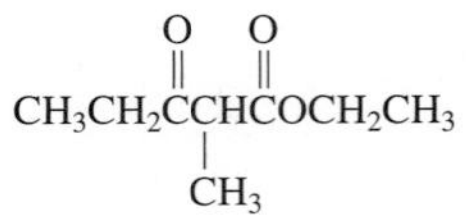

(b) $CH_3CHCH_2CCHCOCH_2CH_3$
        with CH_3, two O, and $(CH_3)_2CH$ substituent

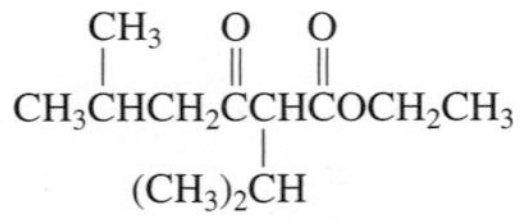

(c) $CH_3(CH_2)_3CCHCOCH_2CH_3$
        with two O and $(CH_2)_2CH_3$ substituent

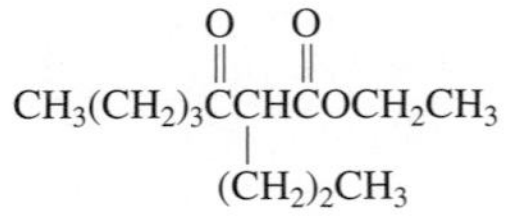

23-2
Retro-Claisen condensation

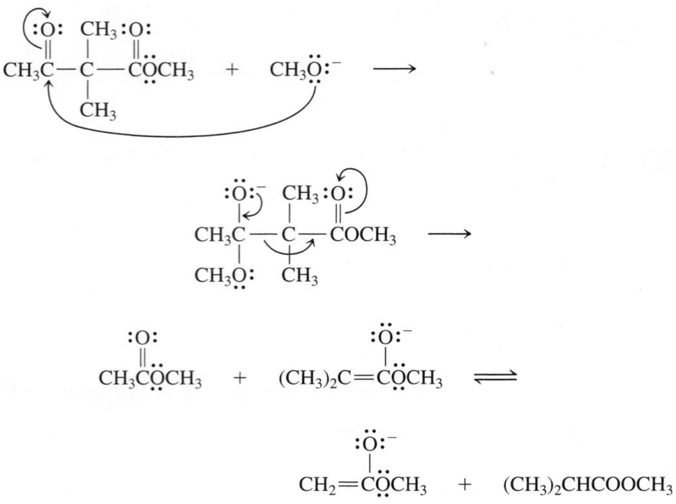

$CH_3C-C-COCH_3$ (with $:O:$, CH_3, $:O:$) $+$ $CH_3\ddot{O}:^-$ $\longrightarrow$

$CH_3C-C-COCH_3$ (with $:\ddot{O}:^-$, $CH_3\ddot{O}:$, CH_3) $\longrightarrow$

$CH_3\overset{:O:}{C}CH_3$ $+$ $(CH_3)_2C=\overset{:\ddot{O}:^-}{C}OCH_3$ $\rightleftharpoons$

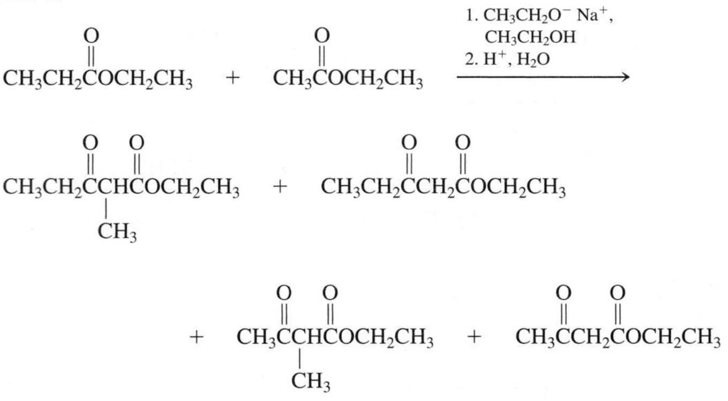

$CH_2=\overset{:\ddot{O}:^-}{C}OCH_3$ $+$ $(CH_3)_2CHCOOCH_3$

Forward Claisen condensation

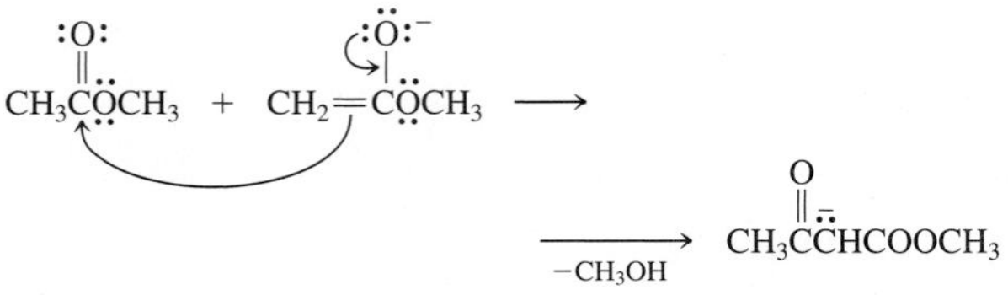

$CH_3\overset{:O:}{C}OCH_3$ $+$ $CH_2=\overset{:\ddot{O}:^-}{C}OCH_3$ $\longrightarrow$

$\xrightarrow{-CH_3OH}$ $CH_3\overset{O}{C}CHCOOCH_3$

23-3

$CH_3CH_2COCH_2CH_3$ $+$ $CH_3COCH_2CH_3$ $\xrightarrow{\begin{array}{l}1.\ CH_3CH_2O^-\ Na^+,\\ \ \ CH_3CH_2OH\\ 2.\ H^+, H_2O\end{array}}$

$CH_3CH_2CCHCOCH_2CH_3$ (with two O, CH_3) $+$ $CH_3CH_2CCH_2COCH_2CH_3$ (with two O)

$+$ $CH_3CCHCOCH_2CH_3$ (with two O, CH_3) $+$ $CH_3CCH_2COCH_2CH_3$ (with two O)

23-4

$HCOCH_2CH_3$ $+$ $CH_3COCH_2CH_3$ $\xrightarrow{\begin{array}{l}1.\ CH_3CH_2O^-Na^+,\\ \ \ CH_3CH_2OH\\ 2.\ H^+, H_2O\end{array}}$

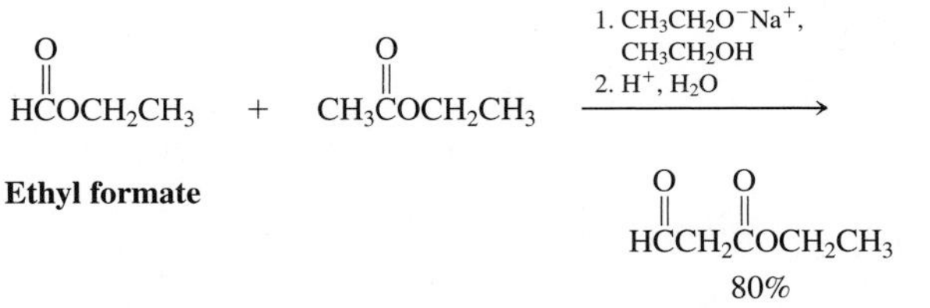

Ethyl formate

$HCCH_2COCH_2CH_3$ (with two O)
80%
Ethyl 3-oxopropanoate

Ethyl formate is not enolizable and the carbonyl group is more electrophilic than that of the "methyl substituted" analog ethyl acetate.

23-5
Worked out in chapter.

23-6
This mechanism is abbreviated.

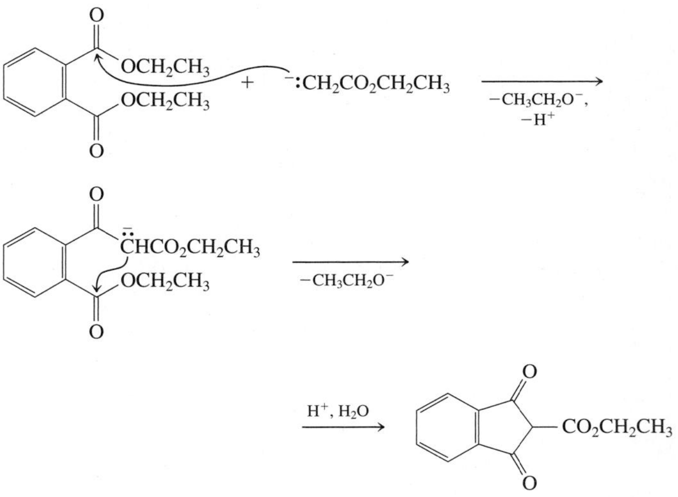

(benzene ring with two $C=O$ groups and OCH_2CH_3 groups) $+$ $^-:CH_2CO_2CH_2CH_3$ $\xrightarrow{\begin{array}{l}-CH_3CH_2O^-,\\ -H^+\end{array}}$

(benzene ring with $\ddot{C}HCO_2CH_2CH_3$, OCH_2CH_3, $C=O$) $\xrightarrow{-CH_3CH_2O^-}$

$\xrightarrow{H^+, H_2O}$ (indanedione ring with $CO_2CH_2CH_3$)

23-7

(methyl 5-oxohexanoate structure with CH_3, O, CO_2CH_3) $\xrightarrow{\begin{array}{l}1.\ CH_3O^-Na^+, CH_3OH\\ 2.\ H^+, H_2O\end{array}}$ (1,3-cyclohexanedione structure)

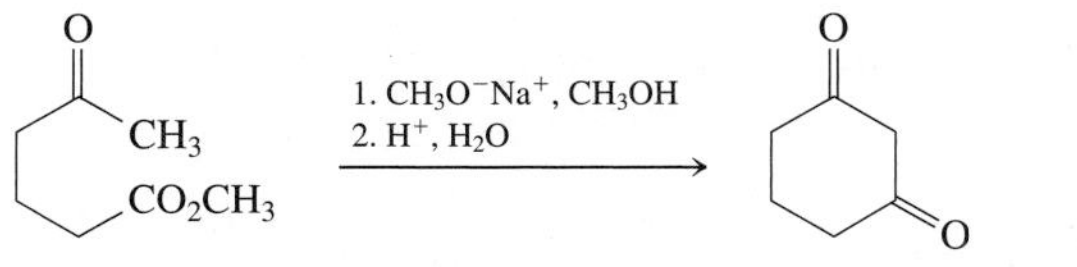

Methyl 5-oxohexanoate   **1,3-Cyclohexanedione**

23-8
Worked out in chapter.

23-9

(a) CH_3CCH_3 (with O) $+$ $HCO_2CH_2CH_3$

    1. $CH_3CH_2O^-$,   2. H^+, H_2O

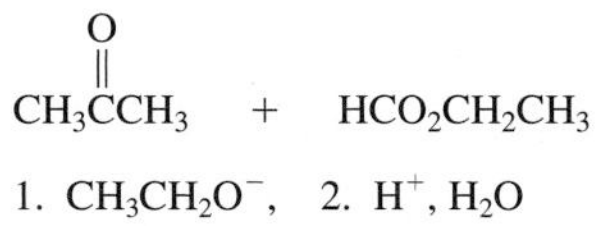

(b) (cyclooctanone ring with O) $+$ $CH_3CH_2OCOCH_2CH_3$ (with O)

    1. NaH,   2. H^+, H_2O

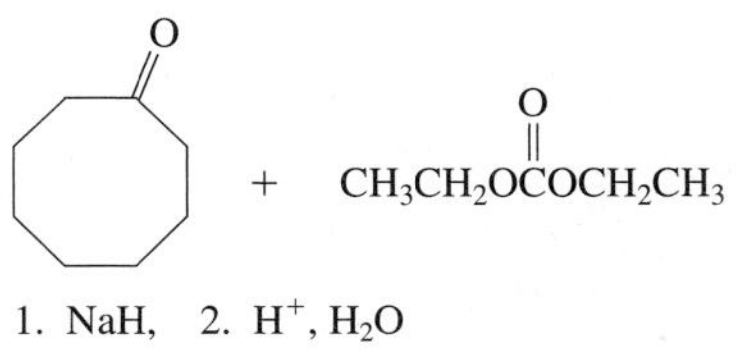

(c) H₃C–C(=O)–CH₂CH₂CH₂–CO₂CH₂CH₃

$H_3C\overset{O}{\overset{\|}{C}}CH_2CH_2CH_2CO_2CH_2CH_3$

1. $CH_3CH_2O^-$, 2. H^+, H_2O

23-10

$CH_3\overset{O}{\overset{\|}{C}}CH_2CH_2CH_2CO_2CH_3$ $\xrightarrow{\text{Exercise 23-7}}$

100%

$\xrightarrow{\text{2 NaOCH}_3,\ \text{CH}_3\text{I},\ \text{CH}_3\text{OH}}$

80%

2,2-Dimethyl-1,3-cyclohexanedione

23-11

23-12

(a) 1. NaOCH₂CH₃, 2. CH₃CH₂CH₂Br, 3. NaOH,
4. H⁺, H₂O, Δ; **(b)** 1. NaOCH₂CH₃, 2. CH₃(CH₂)₄Br,
3. NaOH, 4. H⁺, H₂O, Δ; **(c)** 1. 2 NaOCH₂CH₃,
2. 2 CH₃CH₂Br, 3. NaOH, 4. H⁺, H₂O, Δ; **(d)** 1. NaOCH₂CH₃,
2. C₆H₅CH₂Cl, 3. NaOH, 4. H⁺, H₂O, Δ

23-13

(a) 1. CH₃CH₂OOC–C̈H–COOCH₂CH₃,
             |
             CH₃

2. CH₃CH₂OOC–C(–(CH₂)₉CH₃)–COOCH₂CH₃,
             |
             CH₃

3. K⁺⁻OOC–C(–(CH₂)₉CH₃)–COO⁻K⁺
             |
             CH₃

(b) CH₃CH₂OOCCH₂COOCH₂CH₃ $\xrightarrow[\text{2. CH}_3\text{Br}]{\text{1. NaOCH}_2\text{CH}_3}$

CH₃CH₂OOCCHCOOCH₂CH₃
           |
           CH₃

23-14

(a)

2-Butylcyclohexanone

(b) CH₃CH₂CO₂H
Propanoic acid

(c) $CH_3\overset{O}{\overset{\|}{C}}\overset{O}{\overset{\|}{C}}HC_6H_5$ with CH₃ substituent

2-Methyl-1-phenyl-1,3-butanedione

(d) This sequence is general for 2-halo esters.

$CH_3\overset{O}{\overset{\|}{C}}CH_2\overset{O}{\overset{\|}{C}}OCH_2CH_3$ $\xrightarrow[\text{2. BrCH}_2\text{CO}_2\text{CH}_2\text{CH}_3]{\text{1. CH}_3\text{CH}_2\text{O}^-\text{Na}^+}$

$CH_3\overset{O}{\overset{\|}{C}}CH\overset{O}{\overset{\|}{C}}OCH_2CH_3$
       |
       CH₂COCH₂CH₃
          |
          O

$\xrightarrow[\text{2. H}^+,\ \text{H}_2\text{O},\ \Delta]{\text{1. NaOH}}$ $CH_3\overset{O}{\overset{\|}{C}}CH_2CH_2\overset{O}{\overset{\|}{C}}OH$

4-Oxopentanoic acid

Note that only the carboxy group attached to the α-carbonyl carbon can undergo decarboxylation.

(e) $HO\overset{O}{\overset{\|}{C}}CHCH_2\overset{O}{\overset{\|}{C}}OH$
              |
              CH₃CH₂

2-Ethylbutanedioic acid

Excessive heating may dehydrate this product to the anhydride (Section 19-8).

(f) $CH_3\overset{O}{\overset{\|}{C}}CH_2CH_2\overset{O}{\overset{\|}{C}}CH_3$

2,5-Hexanedione

23-15

Worked out in chapter.

23-16

$CH_2(CO_2CH_2CH_3)_2$ $\xrightarrow[\text{CH}_3\text{CH}_2\text{OH}]{\text{CH}_3\text{CH}_2\text{O}^-\ \text{Na}^+,}$ $Na^{+\ -}{:}CH(CO_2CH_2CH_3)_2$

$\xrightarrow{\text{2 Na}^{+\ -}{:}\text{CH(CO}_2\text{CH}_2\text{CH}_3)_2}$

$\xrightarrow[\substack{-2\ \text{CH}_3\text{CH}_2\text{OH},\\ -2\ \text{CO}_2}]{\text{H}^+,\ \text{H}_2\text{O}}$

Heptanedioic acid

23-17

$^-:CH(CO_2CH_2CH_3)_2$ + $CH_2=CH-CCH_3$ (with O double bond) ⟶

$(CH_3CH_2O_2C)_2CHCH_2CH=CCH_3$ (with O^-) $\xrightarrow{CH_2(CO_2CH_2CH_3)_2}$

$(CH_3CH_2O_2C)_2CHCH_2CH_2CCH_3$ (with O double bond) + $^-:CH(CO_2CH_2CH_3)_2$

The enolate of the product regenerates the enolate of the starting malonic ester.

23-18

(a) $(CH_3CH_2O_2C)_2CCH_2CH_2CH$ (with CH_3CH_2 and O), 40%

(b) , 56%

(c) H_3C , 66%

23-19

+ $CH_2=CHCN$ ⟶

NC— (chain) — **Acidic** H

⟶

$CH_2=CHCN$ / Second Michael addition ⟶

23-20

Worked out in chapter.

23-21

Michael addition ⟶

Aldol condensation ⟶

23-22

Nucleophiles add to the carbonyl group; bases deprotonate next to it.

23-23

1. $CH_3CH_2CH_2CH_2Li$
2. $(CH_3)_2CHBr$

1. $CH_3CH_2CH_2CH_2Li$
2. CH_3CCH_3 (with O)

H $CH(CH_3)_2$

$(CH_3)_2C$ $CH(CH_3)_2$ (with OH) $\xrightarrow{Hg^{2+},\ H_2O}$ (product with OH and O)

23-24

(a) CH_3CH_2CHO + thiazolium ion catalyst or 1,3-dithiacyclohexane and 1. $CH_3CH_2CH_2CH_2Li$, 2. CH_3CH_2Br, 3. $CH_3CH_2CH_2CH_2Li$, 4. CH_3CH_2CHO, 5. Hg^{2+}, H_2O

(b) 1,3-Dithiacyclohexane and 1. $CH_3CH_2CH_2CH_2Li$, 2. $CH_3(CH_2)_3Br$, 3. $CH_3CH_2CH_2CH_2Li$, 4. CH_3CH_2CHO, 5. Hg^{2+}, H_2O

(c) 1,3-Dithiacyclohexane and 1. $CH_3CH_2CH_2CH_2Li$, 2. $CH_3CH_2CHCH_3$ (with Br), 3. $CH_3CH_2CH_2CH_2Li$, 4. $CH_3CCH_2CH_3$ (with O), 5. Hg^{2+}, H_2O

(d) C_6H_5CHO + thiazolium ion catalyst

(e) $(CH_3)_2CHCHO$ + thiazolium ion catalyst or 1,3-dithiacyclohexane and 1. $CH_3CH_2CH_2CH_2Li$, 2. $(CH_3)_2CHBr$, 3. $CH_3CH_2CH_2CH_2Li$, 4. $(CH_3)_2CHCHO$, 5. Hg^{2+}, H_2O

Chapter 24

24-1

(a) Aldotetrose
(b) Aldopentose
(c) Ketopentose

24-2

(a) (2R,3R,4R)-2,3,4,5-Tetrahydroxypentanal
(b) (2R,3S,4R,5R)-2,3,4,5,6-Pentahydroxyhexanal

24-3

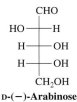

D-(−)-**Arabinose**

24-4

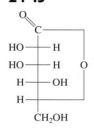

L-(−)-Glucose

24-5

Worked out in chapter.

24-6

α-L-Glucopyranose and α-D-glucopyranose are enantiomers. As for the configurations, the text above Exercise 24-5 provides the answer (although of course you may—and probably should—draw the structures for yourself to confirm): The α diastereomer in a D-series aldohexose is S at the anomeric (hemiacetal) carbon. Everything else follows. If α-D-glucopyranose has the S configuration at that carbon, then its enantiomer α-L-glucopyranose is correspondingly R. In the β anomers the stereochemistry at that position is inverted. Therefore, β-L-glucopyranose has the S configuration and β-D-glucopyranose the R at their anomeric carbons, respectively.

24-7

(a)

(b)

(c)

24-8

Four axial OH groups, $4 \times 0.94 = 3.76$ kcal mol^{-1}; one axial CH$_2$OH, 1.70 kcal mol^{-1}; $\Delta G = 5.46$ kcal mol^{-1}. The concentration of this conformer in solution is therefore negligible by this estimate.

24-9

Only the anomeric carbon and its vicinity are shown.

Planar

24-10

Worked out in chapter.

24-11

Following the procedure outlined in Exercise 24-10 results in $x_\alpha = 0.28$ and $x_\beta = 0.72$.

24-12

$\Delta G^\circ_{\text{estimated}} = -0.94$ kcal mol^{-1} (one axial OH); $\Delta G^\circ = -RT \ln K = -1.36 \log 63.6/36.4 = -0.33$ kcal mol^{-1}. The difference between the two values is due to the fact that the six-membered ring is a cyclic ether (not a cyclohexane).

24-13

24-14

Oxidation of D-glucose should give an optically active aldaric acid, whereas that of D-allose leads to loss of optical activity. This result is a consequence of turning the two end groups along the sugar chain into the same substituent.

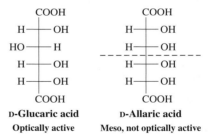

D-Glucaric acid   **D-Allaric acid**

Optically active   **Meso, not optically active**

This operation may cause important changes in the symmetry of the molecule. Thus, D-allaric acid has a mirror plane. It is therefore a meso compound and not optically active. (This also means that D-allaric acid is identical with L-allaric acid.) On the other hand, D-glucaric acid is still optically active.

Other simple aldoses that turn into meso-aldaric acids are D-erythrose, D-ribose, D-xylose, and D-galactose (see Figure 24-1).

24-15

(a) 2 H$_2$C=O   (b) CH$_3$CH=O + H$_2$C=O
(c) 2 H$_2$C=O + HCOOH   (d) No reaction
(e) OHCC(CH$_3$)$_2$CHO + CO$_2$
(f) 3 HCOOH + H$_2$C=O

24-16

(a) D-Arabinose $\longrightarrow$ 4 HCO$_2$H + CH$_2$O
     D-Glucose $\longrightarrow$ 5 HCO$_2$H + CH$_2$O
(b) D-Erythrose $\longrightarrow$ 3 HCO$_2$H + CH$_2$O
     D-Erythrulose $\longrightarrow$ HCO$_2$H + 2 CH$_2$O + CO$_2$
(c) D-Glucose or D-mannose $\longrightarrow$ 5 HCO$_2$H + CH$_2$O

24-17

(a) Ribitol is a meso compound.
(b) Two diastereomers, D-mannitol (major) and D-glucitol

24-18

All of them are the same.

24-19

The mechanism of acetal formation proceeds through the same intermediate cation in both cases.

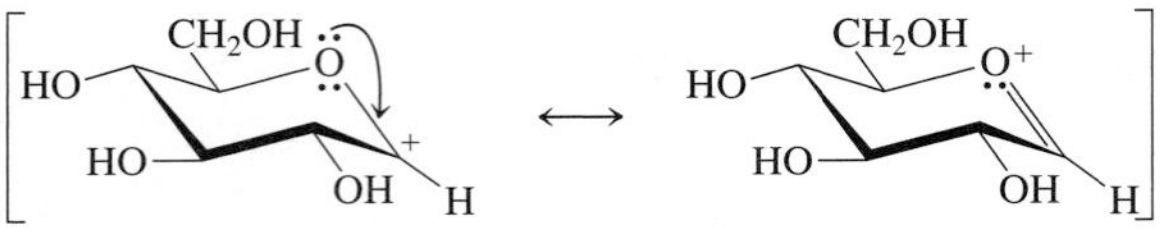

24-20

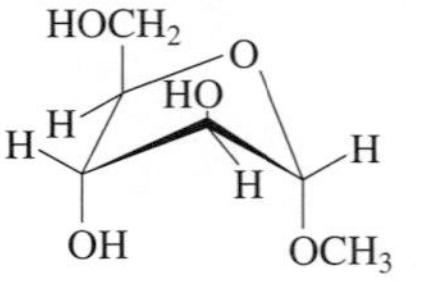

24-21

Same structure as that in Exercise 24-20 or its diastereomers with respect to C2 and C3.

24-22

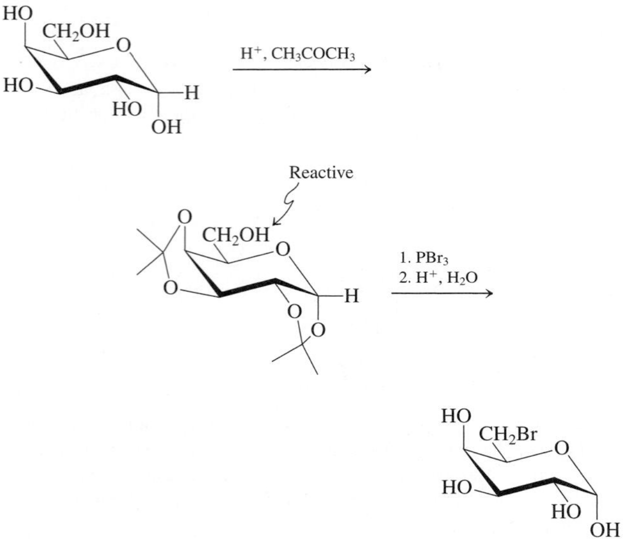

24-23

Worked out in chapter.

24-24

D-Glucose and D-mannose

24-25

A, D-arabinose; B, D-lyxose; C, D-erythrose; D, D-threose.

24-26

The answer is obtained by rotating the Fischer projection shown by 180° and comparing it with that of D-gulose in Figure 24-1: It is L-gulose.

24-27

Both oxidations produce the same molecule: glucaric acid.

24-28

^{13}C NMR would show only three lines for ribaric acid, but five for arabinaric acid.

24-29

Worked out in chapter.

24-30

(a) As shown earlier in this section, aqueous acid (step 1) converts sucrose into the invert sugar mixture of four major monosaccharide isomers, two each of D-glucose and of D-fructose. Sodium borohydride reduces the aldehyde or ketone function of the *open-chain* form of a sugar to a hydroxy group (Section 24-6). Reduction of the carbonyl group of D-glucose gives D-glucitol; reduction of the C2 keto group of D-fructose generates a new stereocenter at C2, giving the two epimeric polyalcohols D-glucitol (again) and D-mannitol.

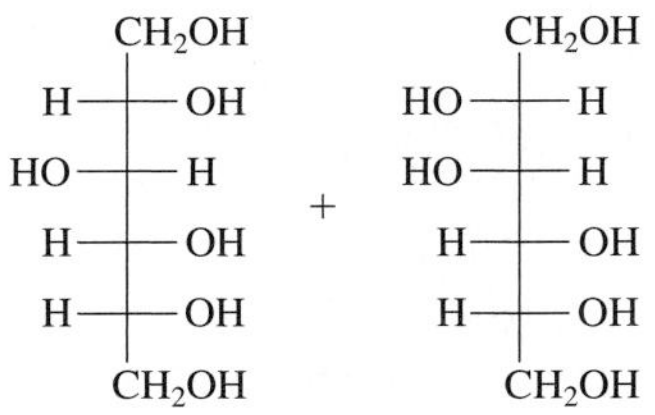

(b) Hydroxylamine reacts with the carbonyl groups of the open chain forms of aldoses and ketoses to give oximes (Sections 17-9 and 24-7). Sucrose, however, contains acetal but *not* hemiacetal functions and does not equilibrate with any structure having a free carbonyl group. As a result, sucrose does not react with hydroxylamine.

24-31

(a)

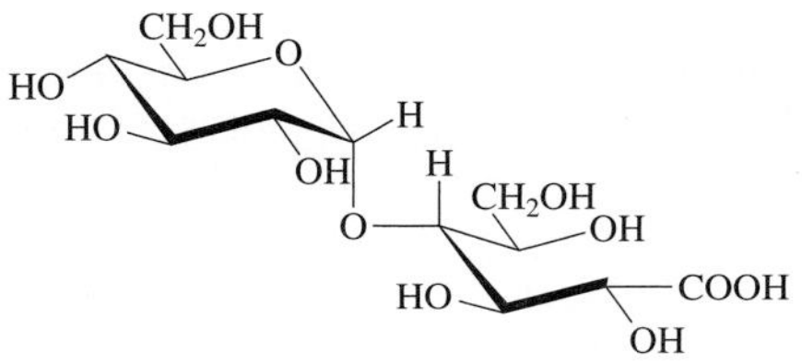

(b)

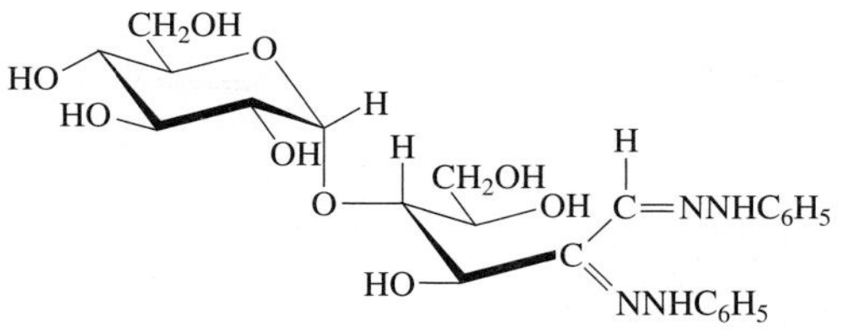

(c)

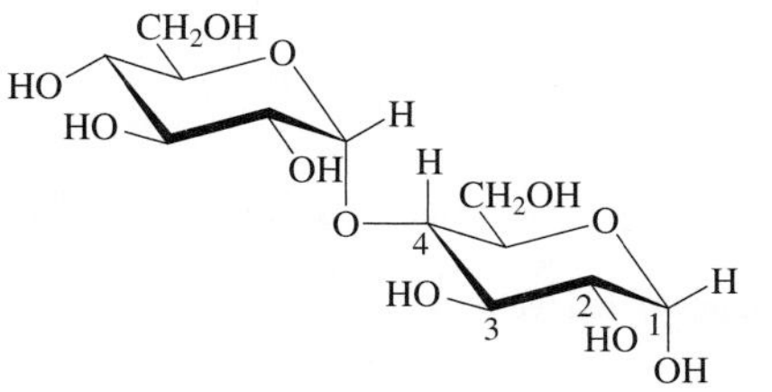

α-Maltose

Chapter 25
25-1

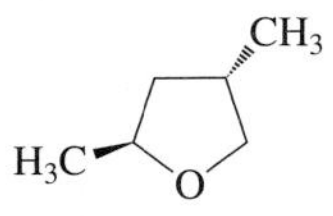

(c) 2,6-Dinitropyridine   (d) 4-Bromoindole

25-2

(The reaction scheme showing the epoxide + MgBr₂ rearrangement to ketone is depicted.)

25-3

(Reaction scheme: epoxide + ⁻:SH → with H⁺ transfer from S to O → thietane product + Cl⁻)

25-4

Worked out in chapter.

25-5

(Reaction scheme: thietane + Cl–Cl → sulfonium + Cl⁻ → Cl–(CH₂)₃–SCl)

25-6

(Reaction scheme: 2-methyloxetane + HCl → protonated oxetane + Cl⁻ → ring-opened products)

25-7

(Structure: N-nitrosopyrrolidine)

25-8

(Two pyrrolidine resonance structures)

Nitrogen is more electronegative than carbon   **Because of resonance, the molecule is now polarized in the opposite direction**

25-9

(Reaction mechanism scheme showing acid-catalyzed cyclization of 2,5-hexanedione to 2,5-dimethylfuran)

25-10

(Reaction scheme: 3-methylcyclobutene carboxylic acid with 1. O₃ 2. H₂, Pd → **A** (aldehyde-acid) → NH₃, Δ → B)

25-11

(Reaction scheme showing β-Keto amine + β-Keto ester → −H₂O → pyrrole product)

β-Keto amine   **β-Keto ester**

25-12

Worked out in chapter.

25-13

(Reaction mechanism: 3-methylfuran + Br₂, −Br⁻ → resonance-stabilized cation intermediates → −H⁺ → 2-bromo-3-methylfuran)

25-14

Protonation at the α-carbon generates a cation described by three resonance forms. Protonation of the nitrogen produces an ammonium ion devoid of resonance stabilization.

25-15

Only attack at C3 produces the iminium resonance form without disrupting the benzene ring

25-16

25-17

Because of the electronegativity of nitrogen, the dipole vector in both compounds points toward the heteroatom. The dipole moment of pyridine is larger than that of azacyclohexane (piperidine), because it is enhanced by the dipolar resonance forms in pyridine. In addition, the nitrogen is sp^2 hybridized. (See Sections 11-3 and 13-2 for the effects of hybridization on electron-withdrawing power.)

25-18

(a) $CH_3CCH_2CO_2CH_2CH_3$, NH_3, , CH_3CCH_2CN

(b) CH_3CCH_2CN, NH_3, $(CH_3)_3CCHO$

(c) $CH_3CH_2CCH_2CO_2CH_2CH_3$, NH_3, CH_3CHO

25-19

Worked out in chapter.

25-20

25-21

$$CH_3CH_2O_2CCH_2CO_2CH_2CH_3 + CH_3CH_2I \xrightarrow[CH_3CH_2OH]{Na^{+-}OCH_2CH_3,}$$

25-22

C3 is the least deactivated position in the ring. Attack at C2 or C4 generates intermediate cations with resonance forms that place the positive charge on the electronegative nitrogen.

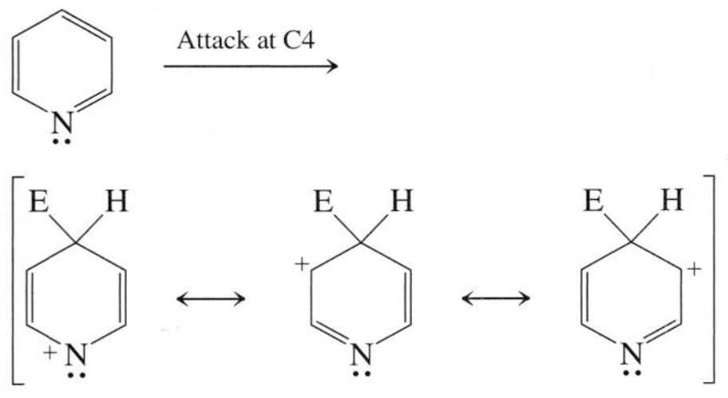

Attack at C4

25-23

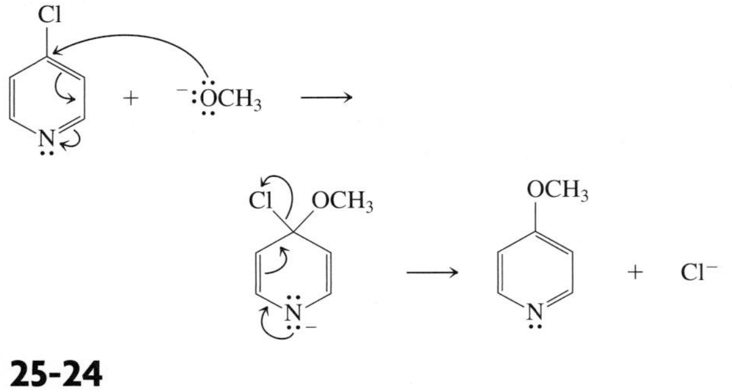

2-Chloropyridine + $^-$:OCH$_3$ →

25-24

Attack at C2 and C4 produces the more highly resonance stabilized anions (only the most important resonance forms are shown). The relatively slightly diminished reactivity of C2 versus C4 is ascribed to steric disruption of solvation of the partial negative charge on the adjacent N.

2-Chloropyridine →

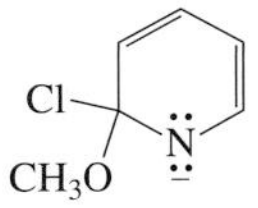

3-Chloropyridine →

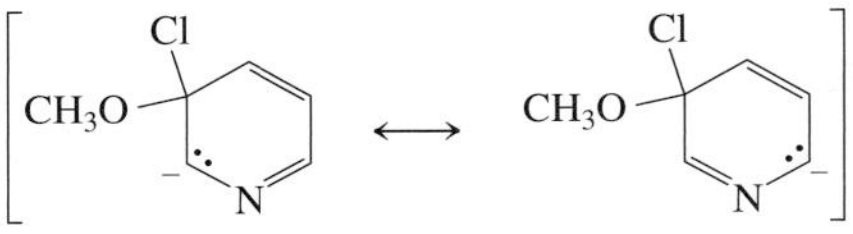

4-Chloropyridine →

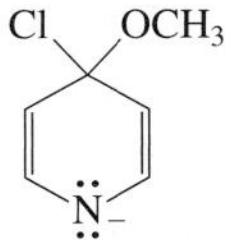

25-25

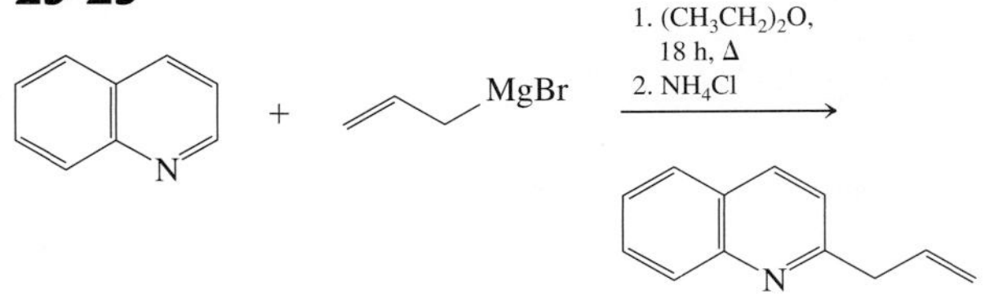

+ allyl MgBr →
1. (CH$_3$CH$_2$)$_2$O, 18 h, Δ
2. NH$_4$Cl

56%
2-(2-Propenyl)quinoline

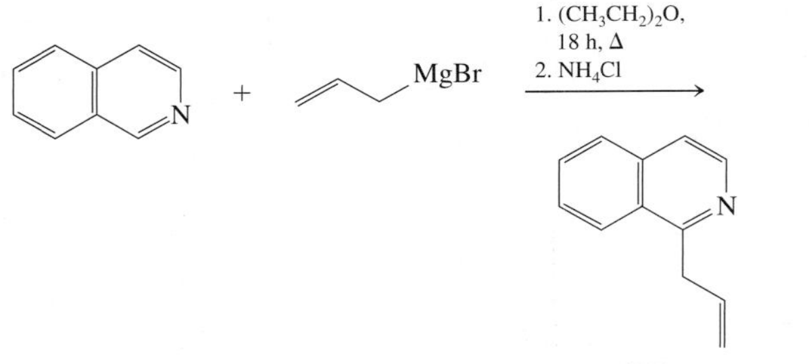

+ allyl MgBr →
1. (CH$_3$CH$_2$)$_2$O, 18 h, Δ
2. NH$_4$Cl

57%
1-(2-Propenyl)isoquinoline

25-26
Worked out in chapter.

25-27

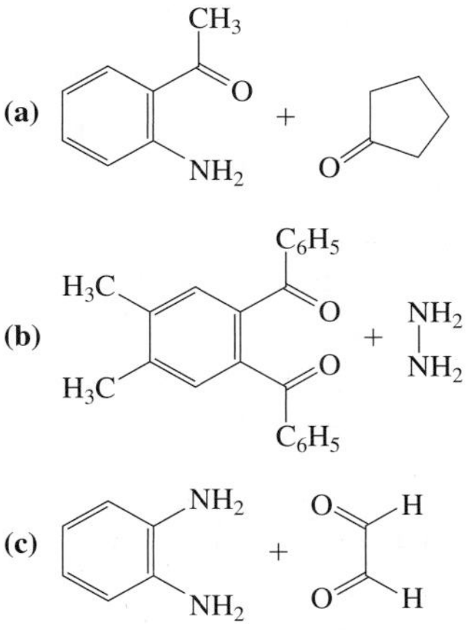

(a)

(b)

(c)

Chapter 26
26-1
(2*S*)-Aminopropanoic acid; (2*S*)-amino-3-methylbutanoic acid; (2*S*)-amino-4-methylpentanoic acid; (2*S*)-amino-3-methylpentanoic acid; (2*S*)-amino-3-phenylpropanoic acid; (2*S*)-amino-3-hydroxypropanoic acid; (2*S*)-amino-3-(4-hydroxyphenyl)propanoic acid; (2*S*),6-diaminohexanoic acid; (2*R*)-amino-3-mercaptopropanoic acid; (2*S*)-amino-4-(methylthio)butanoic acid; (2*S*)-aminobutanedioic acid; (2*S*)-aminopentanedioic acid

26-2

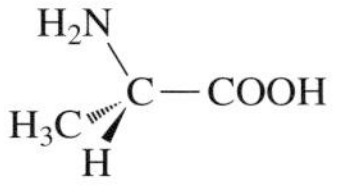

(*S*)-Alanine 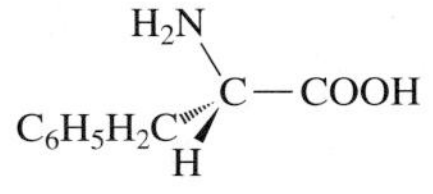   (*S*)-Phenylalanine

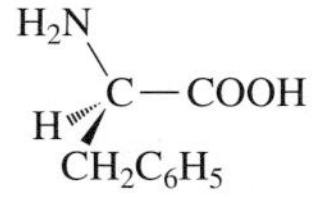

(*R*)-Phenylalanine 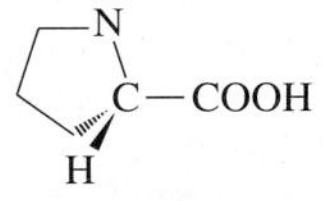   (*S*)-Proline

26-3
All natural amino acids shown in Table 26-1 contain a primary amino group, except proline, where it is secondary. In their respective carboxylates, this group exists as –NH$_2$, displaying *two* (symmetric and asymmetric stretching) $\tilde{\nu}_{NH_2}$ bands at ≈3400 cm^{-1}, whereas proline carboxylate shows only one.

26-4

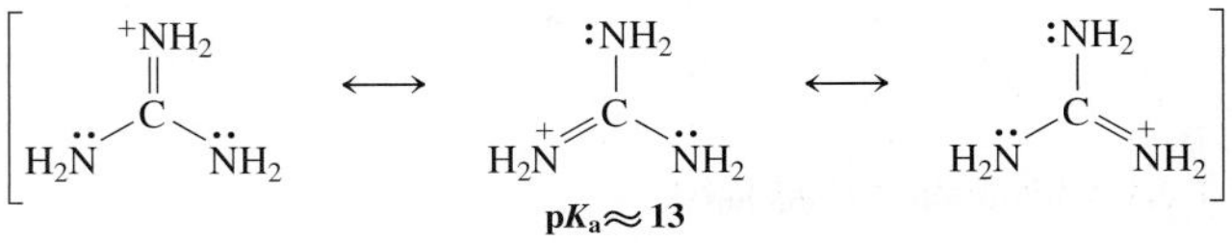

pK_a ≈ 13

26-5

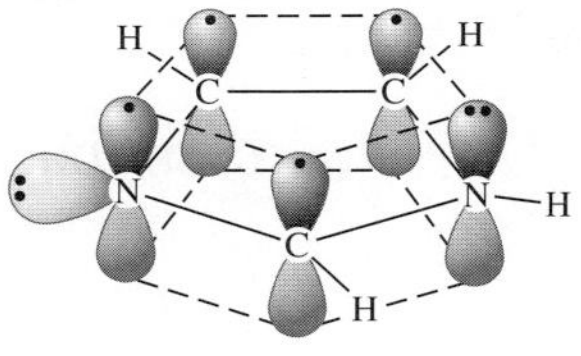

26-6

The yields given are those found in the literature.

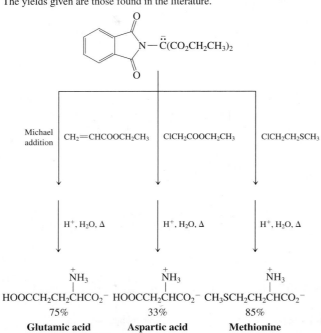

Michael addition: CH$_2$=CHCOOCH$_2$CH$_3$ → H$^+$, H$_2$O, Δ → $\overset{+}{NH_3}$ HOOCCH$_2$CH$_2$CHCO$_2^-$ **75%** **Glutamic acid**

ClCH$_2$COOCH$_2$CH$_3$ → H$^+$, H$_2$O, Δ → $\overset{+}{NH_3}$ HOOCCH$_2$CHCO$_2^-$ **33%** **Aspartic acid**

ClCH$_2$CH$_2$SCH$_3$ → H$^+$, H$_2$O, Δ → $\overset{+}{NH_3}$ CH$_3$SCH$_2$CH$_2$CHCO$_2^-$ **85%** **Methionine**

CH$_3$SH + CH$_2$=CHCH(=O) →(Michael addition)→ CH$_3$SCH$_2$CH$_2$CH(=O) **84%** **3-(Methylthio)propanal**

1. Na^{+-}CN, (NH$_4$)$_2$CO$_3$
2. NaOH
→ $\overset{+}{NH_3}$ CH$_3$SCH$_2$CH$_2$CHCOO$^-$ **58%** **Methionine**

26-7

Worked out in chapter.

26-8

H$_3$CO, OCH$_3$ (A) → Br$_2$ → H$_3$CO, OCH$_3$, Br

With acetamido diester:
$\overset{O}{\underset{H}{CH_3C-N}}$(CO$_2CH_2CH_3$)$_2$ → H$_3$CO, OCH$_3$ with N-acetyl CO$_2$CH$_2$CH$_3$ group

→ H$^+$, H$_2$O, Hydrolysis of ester and ether functions, and decarboxylation → HO, OH, H$_2$N, COOH **B**

26-9

These syntheses are found in the literature.

CH$_2$=O → NH$_4$$^{+-}CN, H_2SO_4$ → H$_2$NCH$_2$CN **2-Aminoethanenitrile** → BaO, H$_2$O, Δ → H$_3$$\overset{+}{N}CH_2COO^-$ **42%** **Glycine**

26-10

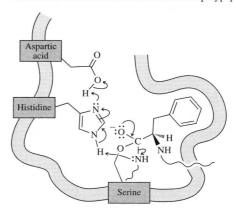

26-11

Elimination of the amino end of the cleaved polypeptide:

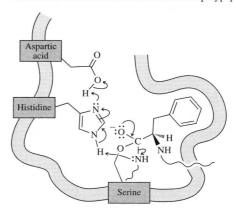

Release of the carboxy end of the cleaved polypeptide:

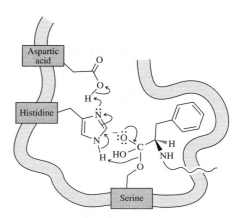

26-12

Hydrolysis of the A chain in insulin produces one equivalent each of Gly, Ile, and Ala, two each of Val, Glu, Gln, Ser, Leu, Tyr, and Asn, and four of Cys.

26-13

$C_6H_5NHCNHCH_2CNH_2$ $\xrightleftharpoons{H^+}$ ... $\xrightleftharpoons{-H^+}$ ...

... $\xrightleftharpoons[\text{Section 17-7}]{H^+}$ ... $+$ $^+NH_4$

26-14

All peptide bonds are cleaved by "transamidation" to give all component amino acids as their corresponding hydrazides. Only the carboxy end retains its free carboxy function.

26-15

Worked out in chapter.

26-16

The applicable sites of cleavage are at the carboxy ends of
1. Glu: Gly-Ile-Val-Glu, Gln-Cys-Cys-Ala-Ser-Val-Cys-Ser-Leu-Tyr-Gln-Leu-Glu, Asn-Tyr-Cys-Asn;
2. Leu: Gly-Ile-Val-Glu-Gln-Cys-Cys-Ala-Ser-Val-Cys-Ser-Leu, Tyr-Gln-Leu, Glu-Asn-Tyr-Cys-Asn;
3. Tyr: Gly-Ile-Val-Glu-Gln-Cys-Cys-Ala-Ser-Val-Cys-Ser-Leu-Tyr, Gln-Leu-Glu-Asn-Tyr, Cys-Asn

26-17

$(CH_3)_3COCNHCHCOOH$ $\xrightleftharpoons{H^+}$ ...

$\xrightleftharpoons{}$ $(CH_3)_3C^+$ $+$ $O=CNHCHCOOH$ $\xrightarrow{-H^+}$

$H_3\overset{+}{N}CHCOO^-$ $+$ CO_2 $+$ $H_2C=C(CH_3)_2$

26-18

Worked out in chapter.

26-19

... $\xrightarrow{H^+}$ ... $\xrightarrow{ROH}$

... $\xrightarrow{-N,N'\text{-dicyclohexylurea}}$ $R\overset{+}{C}\underset{H}{OR}$ $\xrightarrow{-H^+}$ $RCOR$

26-20

1. Ala + $(CH_3)_3COCOCOC(CH_3)_3$ $\longrightarrow$ Boc-Ala + CO_2 + $(CH_3)_3COH$

2. Val + CH_3OH $\xrightarrow{H^+}$ Val-OCH$_3$ + H_2O

3. Boc-Ala + Val-OCH$_3$ $\xrightarrow{DCC}$ Boc-Ala-Val-OCH$_3$

4. Boc-Ala-Val-OCH$_3$ $\xrightarrow{H^+}$ Ala-Val-OCH$_3$ + CO_2 + $CH_2=C(CH_3)_2$

5. Leu + $(CH_3)_3COCOCOC(CH_3)_3$ $\longrightarrow$ Boc-Leu + CO_2 + $(CH_3)_3COH$

6. Boc-Leu + Ala-Val-OCH$_3$ $\xrightarrow{DCC}$ Boc-Leu-Ala-Val-OCH$_3$

7. Boc-Leu-Ala-Val-OCH$_3$ $\xrightarrow[\text{2. HO}^-,\ H_2O]{\text{1. H}^+,\ H_2O}$ Leu-Ala-Val

26-21

Polystyrene + $ClCH_2\overset{SnCl_3}{-}O-CH_2CH_3$ $\longrightarrow$

... CH_2Cl + $CH_3CH_2\ddot{O}SnCl_3$

$\xrightarrow{-H^+}$ ... CH_2Cl

26-22

The various dipolar amide resonance forms provide cyclic aromatic electron sextets.

Cytosine **Thymine** **Guanine** **Uracil**

26-23

5′-UAGCUGAGCAAU-3′

26-24

(a) Lys-Tyr-Ala-Ser-Cys-Leu-Ser
(b) Mutation of C to G in UGC (Cys) to become UGG (Trp).

26-25

Worked out in chapter.

26-26

1. DMT–OR deprotection (R = sugar of nucleotide): hydrolysis by S_N1

$$DMT–OR \xrightarrow[-ROH]{H^+} DMT^+ \xrightarrow{^+NH_4{}^-OH} DMT–OH$$

2. Nucleic acid base deprotection (RNH$_2$ = base, R′CO$_2$H = carboxylic acid): amide hydrolysis

3. Phosphate deprotection (R and R′ = sugars of nucleotides): E_2

4. Disconnection from the solid support (ROH = sugar of nucleotide): ester hydrolysis

26-27

The para nitro group is electron withdrawing by resonance, thus activating the carbonyl carbon with respect to nucleophilic attack, relative to a simple benzoate:

Photograph Credits

Index

Definitions page references are bold, table page references are followed by *t*, figure page references are italicized, Real Life page references are underlined, and footnote page references are followed by *n*.

1 Structure and Bonding

Ionic bonds: Formed by transfer of electrons from one atom to another (1-3)

Covalent bonds: Formed by electron sharing between two atoms; **polar covalent bonds:** Between atoms of differing electronegativity (1-3)

Lewis structures: Electron dot descriptions of molecules; **resonance forms:** Multiple Lewis structures for a molecule; **octet rule:** Preferred valence electron total for atoms (2 for H) (1-4, 5)

VSEPR: Governs molecular shape and geometry around atoms (1-8)

Hybrid orbitals: Explain geometry; sp^3: Tetrahedral (109°, CH_4); sp^2: Trigonal planar (120°, BH_3 and $H_2C=CH_2$); sp: Linear (180°, BeH_2 and $HC\equiv CH$) (1-8)

σ bonds: From end-to-end orbital overlap; **π bonds:** From side-by-side orbital overlap (1-8)

Constitutional isomers: Different connectivity (1-9)

2 Structure and Reactivity; Acids and Bases

Thermodynamics: Govern equilibria; $\Delta G° = -RT \ln K = -1.36 \log K$ (at 25°C) (2-1)

Gibbs free energy: $\Delta G° = \Delta H° - T\Delta S°$; **enthalpy:** $\Delta H°$; **exothermic** if < 0; **endothermic** if > 0; **entropy:** $\Delta S°$; measures energy dispersal (disorder) (2-1)

Kinetics: Govern rates; **first order:** Rate $= k[A]$; **second order:** Rate $= k[A][B]$ (2-1)

Reaction coordinate: Graph of energy vs. structural change; **transition state:** Maximum-energy point on reaction coordinate; **activation energy:** $E_a = E$ (transition state) $- E$ (starting point) (2-1, 7, 8)

Brønsted acids: Proton donors; **Brønsted bases:** Proton acceptors (2-3)

Strong acids HA: Have weak conjugate bases A^-; **acid dissociation constant K_a:** $pK_a = -\log K_a$; **HA acid strength:** ↑ for A larger, more electronegative, resonance-delocalized negative charge; ↓ pK_a = stronger acid (2-3)

Lewis acids/electrophiles: Electron pair acceptors; **Lewis bases/nucleophiles:** Electron pair donors (2-2, 3)

2-3-4 Alkanes and Cycloalkanes

IUPAC Nomenclature: Longest continuous chain; **numbering:** Lowest for first substituent (2-6)

Alkanes: C_nH_{2n+2} hydrocarbons; straight-chain or branched; acyclic; nearly non-polar; weak intermolecular **London forces** (2-7)

Conformations: "Free" rotation about single bonds; **staggered** 2.9 kcal mol^{-1} more stable (lower energy) than **eclipsed** (2-8, 9)

Cycloalkanes: C_nH_{2n} hydrocarbons; **n = 3, 4:** Cyclopropane, cyclobutane (strained bond angles); **n = 6:** Cyclohexane (most stable, chair conformation); **axial** (less stable) & **equatorial** (more stable) substituent positions (4-2, 3)

Stereoisomers: Same atom connectivity, different three-dimensional arrangement; **cis:** Two substituents on same ring face; **trans:** Opposite faces (4-1)

Reactions: Few (nonpolar, lack functional groups); combustion, halogenation (3-3 to 11)

Halogenation: Radical chain mechanism with initiation, propagation, termination steps (3-4)

$$-\overset{|}{\underset{|}{C}}-H + X_2 \xrightarrow{h\nu \text{ or } \Delta} -\overset{|}{\underset{|}{C}}-X + HX$$
$$X = Cl, Br$$
$$(DH°_{C-H} + DH°_{X-X}) - (DH°_{C-X} + DH°_{H-X}) = \Delta H°$$

Reactivity: $F_2 > Cl_2 > Br_2$ (I_2 none); **selectivity:** $Br_2 > Cl_2 > F_2$ (3-8)

Reactivity: Tert > sec > prim > methane C—H; follows radical stability due to **hyperconjugation** (3-7)

Bond-dissociation energy: $DH°$; bond dissociation gives radicals or free atoms (3-1)

5 Stereoisomers

Chiral: Handed; not superimposable on mirror image; **enantiomers:** Mirror-image stereoisomers; **stereocenter:** Center of chirality in a molecule, such as a carbon with 4 different groups attached (5-1); **nomenclature:** R/S system (5-3)

Optical activity: Rotation of the plane of polarized light by an enantiomer; **racemate (racemic mixture):** Optically inactive 1:1 mixture of two enantiomers (5-2)

Diastereomers: Non-mirror image stereoisomers (5-5)

Meso compound: Achiral molecule with multiple stereocenters (5-6)

Resolution: Separation of enantiomers (5-8)

6-7 Haloalkanes

Functional group: $^{\delta+}C-X^{\delta-}$; electrophilic C, leaving group X^- (6-1, 2)

Reactions: Nucleophilic substitution (6-2 to 11, 7-1 to 5, 7-8, 9), elimination (7-6 to 9)

$$H-\overset{|}{\underset{|}{C}}-\overset{|}{\underset{|}{C}}-Nu \xleftarrow[-X^-]{Subst} Nu^- + H-\overset{|}{\underset{|}{C}}-\overset{|}{\underset{|}{C}}-X \xrightarrow[-HX]{Elim} \overset{\diagup}{\underset{\diagdown}{C}}=\overset{\diagup}{\underset{\diagdown}{C}}$$

S_N2: R = Me > prim > sec, backside displacement;
S_N1: **R** = tert > sec, racemization; **E2:** Nu^- strong base;
E1: Side rxn to S_N1, rates of S_N1, E1 both follow carbocation stabilization by hyperconjugation

8-9 Alcohols and Ethers

Nomenclature: Alkan*ol*—longest chain containing OH; **numbering:** Lowest for OH (8-1)

Functional group: $^{\delta+}C^{\delta-}O-H^{\delta+}$; Lewis basic O; acidic H ($pK_a = 16$–18, like H_2O), hydrogen bonding (8-2, 3)

Preparation: Hydride/Grignard addn to C=O; **prim:** RCHO + $LiAlH_4$ or H_2C=O + RMgX; **sec:** RR′C=O + $LiAlH_4$ or RCHO + R′MgX; **tert:** RR′C=O + R″MgX (8-6, 8)

$$H-\overset{|}{\underset{|}{C}}-OH \xleftarrow[\text{2. H}^+, \text{H}_2\text{O}]{\text{1. LiAlH}_4} \overset{|}{\underset{|}{C}}=O \xrightarrow[\text{2. H}^+, \text{H}_2\text{O}]{\text{1. RMgX}} R-\overset{|}{\underset{|}{C}}-OH$$

Oxidation: Cr(VI) reagents; RCH_2OH (prim) + PCC → RCHO (aldehyde); prim + $Na_2Cr_2O_7$ → RCO_2H (carboxylic acid); RR′CHOH (sec) + $Na_2Cr_2O_7$ → RR′C=O (ketone) (8-6)

Substitution: Prim, sec + $SOCl_2$, PBr_3, P/I_2 → RX; tert + HX → RX (9-2 to 4)

Dehydration: Conc H_2SO_4 + prim (180°C), sec (100°C), tert (50°C) → alkene; carbocations rearrange (9-2, 3, 7)

Ether synthesis: Williamson, RO^- + R′X (R′ = Me, prim) → ROR′ (9-6)

Ether cleavage: ROR′ + HX (X = Br, I) → RX + R′X (9-8)

10-11 Spectroscopy

High-resolution mass spectrometry: Gives molecular formula (11-9, 10)

Degree of unsaturation: Gives number of rings + number of π bonds; degree of unsaturation = $(H_{sat} - H_{actual})/2$, where $H_{sat} = 2n_C + 2 - n_X + n_N$ (11-11)

Infrared spectroscopy: Gives bonds and functional groups (11-8)

Wavenumber	3650–3200	3150–3000	3000–2840	2260–2100	1760–1690	1680–1620	< 1500
Bond	O—H (s, br)	=C—H (m)	—C—H (s)	C≡C (w)	C=O (s)	C=C (m)	Fingerprint
	N—H (m, br)			C≡N (m)			
	≡C—H (s)						

w = weak; m = medium; s = strong; br = broad

Nuclear magnetic resonance: Gives hydrogen and carbon signals; **chemical shift:** Structural environment; **integration:** Number of H for each signal; **splitting:** Number of H neighbors ($N + 1$ rule) (10-3 to 9)

Chem shift	9.9–9.5	9.5–6.0	5.8–4.6
H type	$\overset{O}{\underset{}{\parallel}}$ —C—H	H (aromatic)	C=C—H
4.0–3.0	**2.6–1.6**	**1.7–0.8**	**variable**
H C—O,Br,Cl	H—C—C,O	alkyl C—H	O—H N—H

11-12 Alkenes

Nomenclature: Longest chain containing C=C (OH takes precedence); **stereochemistry:** Cis/trans or E/Z systems (11-1)

Functional group: C=C, π bond electron pair nucleophilic, electrophiles add (11-2, 12-3 to 13)

Stability: ↑ with increased substitution (R_2C=CR_2 most stable; H_2C=CH_2 least stable); trans disubstituted > cis disubstituted

Preparation: Haloalkane + strong base (bulky for prim RX), E2, anti stereospecific, most stable alkene favored (**Saytzev rule**) except with bulky base (**Hofmann rule**) (7-7, 11-6)

Preparation: Alcohol + conc H_2SO_4, product mixtures (**Saytzev rule**) (9-2, 11-7)

Hydrogenation: H_2 with catalytic Pd or Pt, syn addition → alkane (12-2)

Electrophilic addition mechanism: Electrophile adds to less substituted alkene carbon, nucleophile to more substituted alkene carbon (12-3)

$$CH_3CH=CH_2 \quad E-Nu \longrightarrow$$
more highly substituted

$$CH_3\overset{+}{C}H-CH_2E \quad {}^-:Nu \longrightarrow CH_3CH-\underset{Nu}{CH_2E}$$

Hydrohalogenation: Markovnikov regioselectivity, except HBr + peroxides (ROOR) (12-3, 13)

Hydration: Markovnikov with aq acid or oxymercuration; **anti-Markovnikov** with borane (12-4, 7, 8)

Halogenation: Anti stereochemistry of addition via cyclic halonium ion (12-5)

Dihydroxylation: Anti using peroxycarboxylic acid; syn using OsO_4 (12-10, 11)

Ozonolysis: Cleavage using ozone followed by reduction (12-12)

13 Alkynes

Functional group: $C{\equiv}C$, two π bonds; ${\equiv}C{-}H$ bond acidic ($pK_a = 25$) (13-2)

Preparation: Alkene + halogen → 1,2-dihaloalkane (12-5), then double elimination ($NaNH_2$) → alkyne (13-4)

Alkynyl anions: $RC{\equiv}CH + NaNH_2 \rightarrow RC{\equiv}C{:}^-$, then $R'X$ (R' = Me, prim) → $RC{\equiv}CR'$ (13-5)

Reduction: H_2, Pt → alkane; H_2, Lindlar's → cis alkene; Na, NH_3 → trans alkene; (13-6)

Addition: HX, X_2 add twice (13-7)

Hydration: Hg^{2+}, H_2O (Markovnikov) or R_2BH, then ^-OH, H_2O_2 (anti-Markovnikov) → enol → ketone or aldehyde via tautomerism (13-7, 8; 18-2)

14 Dienes

1,2- and 1,4-addition to 1,3-dienes: Via delocalized allylic intermediate; kinetic product forms fastest; thermodynamic is more stable (14-6)

Diels-Alder reaction: Concerted, stereospecific cycloaddition (14-8)

15-16-22 Benzene and Aromaticity

Nomenclature: Special common names, ortho/o (1,2), meta/m (1,3), para/p (1,4) for disubstituted (15-1)

Aromaticity: Special stability, properties, and reactions, based on Hückel's rule, $4n + 2\pi$ electrons in a circle of p orbitals (15-2 to 7)

Electrophilic aromatic substitution mechanism:
Electrophile adds to position favored by groups previously present (directing effects, below), followed by loss of H^+ (15-8)

Five electrophilic aromatic substitutions below, clockwise from bottom right: Sulfonation, nitration, halogenation, Friedel-Crafts (F-C) alkylation, Friedel-Crafts acylation (15-9 to 13)

Directing effects: Ortho/para-directing $G_{o,p}$ include alkyl, aryl, halogen, —OR, —NR$_2$; meta-directing G_m include —SO$_3$H, —NO$_2$, >C=O, —CF$_3$, —NR$_4^+$ (16-2, 3)

Activation/deactivation: All $G_{o,p}$ except halogen are activating; halogen and all G_m are deactivating (16-1 to 3)

Alkylbenzenes: F-C alkylation (may rearrange); F-C acylation, then reduction (16-5)

Benzylic reactivity: Halogenation (22-1); oxidation to benzoic acid (22-2)

Benzenamines: Preparation (16-5); diazotization (22-4, 10, 11)

Phenols: Acidic ($pK_a = 10$; more acidic with deactivating substituents); phenoxide + R'X (R' = Me, prim) → ether (22-4, 5)

17-18 Aldehydes and Ketones

Nomenclature: Alkan*al* and alkan*one*—longest chain containing C=O; common: Formaldehyde (methanal), acetaldehyde (ethanal), acetone (propanone) (17-1)

Preparation: ROH oxidation (8-6); C=C + O$_3$ (12-12); C≡C hydration (13-7, 8): F-C acylation (15-13)

Functional group: $^{\delta+}$C=O$^{\delta-}$; electrophilic C, Nu$^-$ adds (17-2)

Addition: H$_2$O → hydrate, ROH → hemiacetal (both unstable); ROH, H$^+$ → acetal (17-6 to 8)

Addition: RNH$_2$ → $\diagdown$C=NR; R$_2$NH → enamine (17-9)

Deoxygenation: C=O → CH$_2$, Clemmensen (16-5); dithioacetal + Raney Ni (17-8); hydrazone + base, Δ (17-10)

Wittig: C=O + CH$_2$=PPh$_3$ → C=C (17-12)

Aldol: 2 Aldehydes, aldehyde + ketone, intramolecular → α,β-unsaturated aldehyde/ketone (18-5 to 7)

1,2- vs 1,4-addition: RLi → alcohol (1,2); H$_2$O, ROH, RNH$_2$, R$_2$CuLi (cuprate) → 3-substituted carbonyl product (1,4); **Michael addition:** Enolate → 1,5-dicarbonyl (1,4); **Robinson annulation:** Intramolecular aldol → 6-membered ring (18-8 to 11)

19-20-23 Carboxylic Acids and their Derivatives

Nomenclature: Alkan*oic acid*—longest chain ending with CO_2H; common: Formic (methanoic), acetic (ethanoic), benzoic (benzenecarboxylic) (19-1)

Preparation: ROH oxidation; $RMgX + CO_2$; RCN hydrolysis (8-6, 19-6)

Functional group: Highly polar, H-bonding (19-2); Lewis basic $\rightarrow {}^{\delta-}O{=}C{-}O{-}H^{\delta+} \leftarrow$ acidic ($pK_a = 4$) (19-4); Nu adds to C=O, OH may leave (addition–elimination) (19-7)

Reduction: $LiAlH_4 \rightarrow RCH_2OH$ (19-11)

Derivatives: RCO_2H (acid) $+ SOCl_2 \rightarrow RCOCl$ (halide); $RCO_2H + R'OH, H^+ \rightarrow RCO_2R'$ (ester); $RCO_2H + R'NH_2, \Delta \rightarrow RCONHR'$ (amide) (19-8 to 10)

Reactivity order: Toward Nu^- addition/elimination of leaving group (**bold**); halides (and anhydrides)—no catalyst needed; esters, amides—H^+ or ^-OH catalyst required (20-1)

Hydrolysis: All $+$ excess $H_2O \rightarrow RCO_2H$ (acid) (20-2 to 6)

Esterification: Halide, ester (transesterification) $+$ excess $R''OH \rightarrow RCO_2R''$ (20-2 to 4)

Amide formation: Halide, ester $+$ excess $R'NH_2 \rightarrow RCONHR'$ (20-2 to 4)

Hydride reduction to aldehyde: Halide $+ LiAl(O^tBu)_3$; ester, amide $+$ DIBAL (20-2 to 6)

Hydride reduction to prim alcohol: Halide, ester $+ LiAlH_4$ (20-2 to 4)

Hydride reduction to amine: $RCONHR' + LiAlH_4 \rightarrow RCH_2NHR'$ (20-6)

Cuprate addition: Ketones from halides (20-2)

Grignard addition: Tert alcohols from esters (20-4)

Claisen condensation: 2 Esters $\rightarrow \beta$-ketoester; 1,7-diester $\rightarrow$ cyclic β-ketoester (intramolecular: **Dieckmann condensation**) (23-1)

Acetoacetic Ester Synthesis of Methyl Ketones: Alkylate, hydrolyze, decarboxylate (23-2)

Malonic Ester Synthesis of Carboxylic Acids: Same (23-2)

21 Amines

Nomenclature: Alkan*amine,* follow rules for alcohols, use *N*- for additional groups on nitrogen; common, alkyl amine (21-1)

Functional group: Nonplanar (pyramidal), sp^3 N, inverts easily; Lewis basic, nucleophilic N; N—H $\leftarrow$ weakly δ^+ ($pK_a = 35$) (21-2, 4)

Basicity: Amine $+ H^+ \rightarrow$ ammonium ion ($pK_a = 10$) (21-4)

Preparation: $RX + N_3^- \rightarrow RN_3$ (azide), then $LiAlH_4 \rightarrow RNH_2$ (primary amine) (21-5)

One less carbon: $RCONH_2$ (amide) $+ Cl_2$, NaOH $\rightarrow RNH_2$ (**Hofmann**) (20-7, 21-7)

One more carbon: $RX + {}^-CN \rightarrow RCN$ (nitrile), then $LiAlH_4 \rightarrow RCH_2NH_2$ (20-8)

Reductive amination: $\diagdown C{=}O + NH_3, NaBH_3CN \rightarrow$

$\diagdown CHNH_2$ (prim amine); $\diagdown C{=}O + RNH_2, NaBH_3CN \rightarrow$

$\diagdown CHNHR$ (sec amine); $\diagdown C{=}O + RR'NH, NaBH_3CN \rightarrow$

$\diagdown CHNRR'$ (tert amine) (21-6)

24 Carbohydrates

Conventional representations: For D-(+)-glucose (24-1, 2)

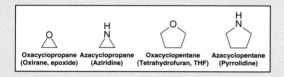

25 Heterocycles

Nomenclature: *Hetero*cycloalkane; common names for many (25-1)

Oxacyclopropane (Oxirane, epoxide) Azacyclopropane (Aziridine) Oxacyclopentane (Tetrahydrofuran, THF) Azacyclopentane (Pyrrolidine)

Heteroaromatics: Common names (25-1)

Furan Pyrrole Thiophene Pyridine

26 Amino Acids, Peptides, Proteins, and Nucleic Acids

α-Amino acids: Most have L (*S*) stereocenter, exist as zwitterions, common names (26-1)

H_3N^+ —|— H with COO⁻ above, R below   same as   R —|— COO⁻ with H_3N^+, H

Acid/base properties: At low pH, H^+ converts COO^- to CO_2H; at high pH, ^-OH converts NH_3^+ to NH_2 (26-1)

Polypeptides: Contain amino acids linked by amide bonds; amide N trigonal planar, sp^2 (26-4)

Peptide linkage
Bold atoms coplanar

Proteins: Polypeptides with polar amino acid side chains in an **active site** that catalyze biochemical reactions (26-8)

Nucleic acids: Polymers made of phosphate-linked sugars bearing the heterocyclic bases adenine (A), guanine (G), cytosine (C), and thymine (T) in DNA or uracil (U) in RNA; contain the **genetic code** for protein biosynthesis (26-9)